EARTH SCIENCE

The Earth, The Atmosphere, and Space

EARTH SCIENCE

The Earth, The Atmosphere, and Space

Stephen Marshak

UNIVERSITY OF ILLINOIS AT URBANA-CHAMPAIGN

Robert Rauber

UNIVERSITY OF ILLINOIS AT URBANA-CHAMPAIGN

w. w. norton & company

new york   london

W. W. Norton & Company has been independent since its founding in 1923, when William Warder Norton and Mary D. Herter Norton first published lectures delivered at the People's Institute, the adult education division of New York City's Cooper Union. The firm soon expanded its program beyond the Institute, publishing books by celebrated academics from America and abroad. By midcentury, the two major pillars of Norton's publishing program—trade books and college texts—were firmly established. In the 1950s, the Norton family transferred control of the company to its employees, and today—with a staff of four hundred and a comparable number of trade, college, and professional titles published each year—W. W. Norton & Company stands as the largest and oldest publishing house owned wholly by its employees.

Editors: Eric Svendsen and Jake Schindel

Senior project editor: Thom Foley

Developmental editor: Sunny Hwang

Assistant editor: Rachel Goodman

Production manager: Benjamin Reynolds

Managing editor, college: Marian Johnson

Managing editor, college digital media: Kim Yi

Media editor: Robert Bellinger

Associate media editor: Cailin Barrett-Bressack

Media project editor: Marcus van Harpen

Media editorial assistant: Liz Vogt

Marketing manager: Katie Sweeney

Design director: Rubina Yeh

Designer (interior and cover): Anne DeMarinis

Photo editor: Stephanie Romeo

Photo researchers: Jane Miller and Elyse Rider

Permissions manager: Megan Schindel

Composition: MPS Limited

Illustrations: Stan Maddock

Manufacturing: LSC Communications—Kendallville, IN

Permission to use copyrighted material is included at the back of this book.

Library of Congress Cataloging-in-Publication Data

Names: Marshak, Stephen, 1955- | Rauber, Robert M.

Title: Earth science : the Earth, the atmosphere, and space / Stephen
  Marshak, University of Illinois at Urbana-Champaign, Robert Rauber,
  University of Illinois at Urbana-Champaign.

Description: First edition. | New York : W.W. Norton & Company, [2017] |
  Includes bibliographical references and index.

Identifiers: LCCN 2016051308 | ISBN 9780393928136 (pbk.)

Subjects: LCSH: Earth sciences—Textbooks. | Astronomy--Textbooks.

Classification: LCC QE28 .M34148 | DDC 550—dc23 LC record available at https://lccn.loc.gov/2016051308

W. W. Norton & Company, Inc., 500 Fifth Avenue, New York, NY 10110

wwnorton.com

W. W. Norton & Company Ltd., 15 Carlisle Street, London W1D 3BS

234567890

DEDICATION

To Kathy, Emma, David, and Michelle

(My bedrock, through uplift and subsidence.)

—STEVE MARSHAK

To Ruta, Carolyn, Josh, Molly, Stacy, and Fabian

(My sunshine, no matter how big the storms.)

—BOB RAUBER

BRIEF CONTENTS

PART V
OUR SOLAR SYSTEM . . . AND BEYOND

CONTENTS

Prelude

Chapter 1

Chapter 2

THE WAY THE EARTH WORKS
Plate Tectonics 50

Chapter 3

INTRODUCING MINERALS AND
THE NATURE OF ROCK 88

Chapter 4

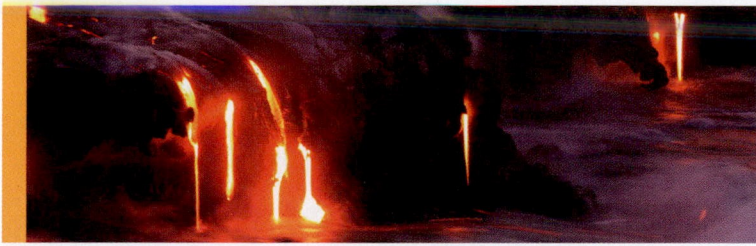

UP FROM THE INFERNO
Volcanism and Igneous Rocks 114

Chapter 5

A SURFACE VENEER
Sediments and Sedimentary Rocks 154

Chapter 8

A VIOLENT PULSE
Earthquakes 246

Chapter 9

DEEP TIME
How Old Is Old? 282

Chapter 10

A BIOGRAPHY OF THE EARTH 312

Chapter 11

RICHES IN ROCK
Energy and Mineral Resources 344

PART II
Ever-Changing Landscapes (continued)

PART III
Restless Seas

Chapter 14

EXTREME REALMS
Desert and Glacial Landscapes 462

Chapter 15

OCEAN WATERS
The Blue of the Blue Marble 510

Chapter 20

CLIMATE AND CLIMATE CHANGE 696

Chapter 21

INTRODUCING ASTRONOMY
Looking Beyond the Earth 738

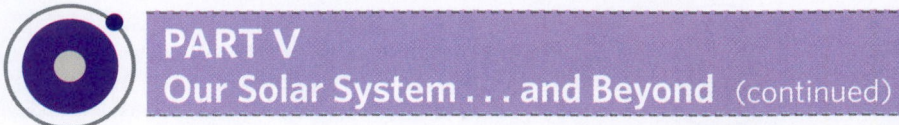
Chapter 22

OUR NEIGHBORHOOD IN SPACE
The Solar System 776

Chapter 23

THE SUN, THE STARS, AND DEEP SPACE 814

PREFACE

Narrative Themes

Why do earthquakes, volcanoes, floods, and landslides happen? What causes mountains to rise? Does climate change through time? Where's the record of evolution? When did the Earth form and by what process? How can we find valuable metals and where do we drill to find oil? What drives violent storms? Is there an edge to the Universe? Could there be life on other planets or moons? Why are there oceans? The study of Earth Science addresses these important questions and many more.

Earth Science provides an introduction to the study of our planet and its context in space. It addresses the essence of four different disciplines: geology, oceanography, atmospheric science (including meteorology), and astronomy. As such, this book will help you understand what's beneath your feet, what's over the horizon, what is up in the sky, and what is far beyond the highest clouds.

Because of the diversity of topics covered in *Earth Science*, it's important to keep in mind a set of narrative themes while reading this book. These themes, listed below, serve as the building blocks from which you can construct your personal understanding of our amazing planet and of the Universe that it flies through.

1. Our planet's land, water, atmosphere, and living inhabitants are dynamically interconnected, and materials constantly cycle among various living and nonliving reservoirs on, above, and within the planet. Researchers refer to this complex network of interconnecting features and phenomena as the *Earth System*.

2. The Earth is a planet, formed like many other planets from dust and gas. But, in contrast to other planets, the Earth is a dynamic place on which new geologic features continue to form and old ones continue to be destroyed. A combination of special conditions, such as the distance from the Sun, chemical composition, the appearance of liquid oceans, and the presence of life, makes our planet truly unique in our Solar System. Quite literally, there's no place like home! While there may be exoplanets (planets orbiting other stars) that resemble the Earth, we'll very likely never interact with them.

3. The Universe, Solar System, and the Earth are all very old. Researchers suggest that the Universe began in the Big Bang about 13.8 billion years ago, that our Solar System originated around 4.57 billion years ago, and that the birth of the Earth took place 4.54 billion years ago. During this deep time, our planet's surface, subsurface, and atmosphere have changed, and life has evolved.

4. Unlike all other planets or moons in the Solar System, the Earth's outer shell, the lithosphere, consists of about 20 plates. The *theory of plate tectonics* states that these plates slowly move relative to one another so that the map of our planet continuously changes. Plate interactions cause earthquakes and volcanoes, build mountains, shift oceans, provide gases that make up the atmosphere, and affect the distribution of life on Earth.

5. Internal processes (driven by the Earth's internal heat—and manifested by plate motions) and external processes (driven by heat from the Sun) interact at the Earth's surface to produce complex landscapes.

6. Knowledge of the Earth, its oceans, and its atmosphere can help society to understand such dangerous natural hazards as earthquakes, tsunamis, volcanoes, landslides, storms, and floods. In some cases, this knowledge can help to reduce the injury and devastation that these hazards can cause.

7. Energy and mineral resources come from the Earth and are formed by geologic phenomena. Geologic studies can help locate these resources and mitigate the consequences of their use.

8. Physical features of the Earth are linked to life processes, and vice versa. Therefore, the history of life links intimately to the history of the physical Earth. The careful study of rocks can provide a rich record of this history.

9. The water in the oceans constantly moves and moderates the climate by transporting heat around the world. And where the ocean interacts with the land, fascinating coastal landscapes develop.

10. Our planet's atmosphere has changed dramatically over the course of the Earth's history. If you went back a billion years in time, the atmosphere would be unbreathable and you would suffocate. In modern times, this precious blanket of air has been significantly affected by the activities of society.

11. The atmosphere constantly moves, sometimes smoothly, sometimes violently. A variety of dangerous storms can develop in association with atmospheric movements, and these cannot be predicted more than a few days in advance.

12. Climates on the Earth vary with latitude, elevation, and distance from the sea, and the position of continents with respect to climate belts has changed over the course of geologic time. On a shorter time scale, both climate and sea level can change, sometimes very significantly, so that regions that were once dry can become shallow seas, and places that were once warm can later become covered in ice. Human society, during the past couple of centuries, has become a major agent of change on our planet.

13. We have much to learn by looking upward and examining the Universe beyond the Earth. Modern instruments and spacecraft have answered many questions, but with each answer, more questions arise about our Solar System, stars, galaxies, and the array of bizarre and beautiful objects that lie at inconceivable distances from us.

14. Science comes from observation, and people make scientific discoveries based on those observations. Earth Science utilizes ideas from physics, chemistry, and biology, so the study of Earth Science provides an excellent means to improve science literacy.

These narrative themes serve as the *take-home message* of this book, a message that we hope students will remember long after they finish their introductory Earth Science course. In effect, the themes provide a mental framework on which students can organize and connect ideas to develop a modern, coherent image of our planet in its context.

Pedagogical Approach

Students learn best from textbooks when they can actively engage with a combination of narrative text and narrative art. Some students respond more to words, which help them to organize information, provide answers to questions, fill in the essential steps that link ideas together, and develop a personal context for understanding information. Other students respond more to images that illustrate processes. Still others respond to question-and-answer-based active learning, an approach by which students "practice" their knowledge in real time. *Earth Science* provides all three of these learning tools. The text has been crafted to be engaging and to carry students forward along a narrative arc, the art has been configured to tell a story (see "Narrative Art and *What an Earth Scientist Sees*"), the chapters are laid out to help students internalize key principles (see "Organization"), and the online activities have been designed both to engage students and to provide active feedback (see "New *Smartwork5 Online Tutorial and Assessment System* for *Earth Science*"). In-text features have also been thoughtfully developed to guide students to a richer understanding of and fuller engagement with unfamiliar concepts (see "Pedagogical Features Designed to Help Students Master the Concepts").

A note about units: Scientists the world over use the metric system for describing distances, weights, temperatures, and other physical features. But many of the students using this book grew up using the English system of units. To help students visualize and appreciate units, and to see the relationship between metric and English units, we provide both. The conversions we provide are sometimes exact and sometimes approximate, to reflect the precission of a measurement provided. For example, if we describe a distance as more or less 1,000 kilometers, we'll provide the conversion as 600 miles (instead of 621.37 miles), for it would be misleading to specify one unit as an approximation and the other is a precise one.

Organization

The topics covered in this book have been arranged so that students can build their knowledge of Earth Science on a foundation of overarching principles. This book contains five parts.

Part I begins by considering how the Earth formed, and how it's structured, overall, from surface to center. With this basic background, students can delve into plate tectonics, the grand unifying theory of Earth Science. Plate tectonics appears early in the book, so that students can gain an appreciation for the "action" underlying geology, and they can then use plate tectonics theory as a foundation from which they can interpret and link the ideas presented in subsequent chapters. Knowledge of plate tectonics, for example, helps students understand the suite of chapters on minerals, rocks, and the rock cycle. Knowledge of plate tectonics and rocks together, in turn, provides a basis for understanding volcanoes, earthquakes, and mountains. And then, with this background, students can see how the map of the Earth, and its inhabitants, have changed throughout the vast expanse of geologic time. We dedicate a unique chapter to energy and mineral resources, an especially relevant topic given present-day global interest in sustainability.

Part II hones in on the landscapes of the Earth's surface, the visible part of our planet. We'll understand how landscapes result from a never-ending battle between uplift (ultimately driven by plate tectonics) and erosion (ultimately driven by energy from the Sun). We begin by understanding surface movements in general, most dramatically manifested by landslides. Then, we consider our planet's freshwater, the landscapes associated with water movement, and the challenges presented by floods and the loss of water supplies. We conclude this part by looking at the dramatic landscapes developed in realms of extreme climates.

Part III takes us offshore, into the ocean. We begin by examining the water of this realm, focusing on its characteristics, its movements, and the life within it. We conclude by considering the interface between water and the solid Earth, both beneath the oceans and along its shores.

Part IV of the book introduces students to the Earth's atmosphere. We begin by describing the character and evolution of the atmosphere and the overall composition of air. We then develop an understanding of how and why air moves, and how this movement leads to weather systems and storms. The book provides a particular focus on the devastating drama of thunderstorms, tornados, hurricanes, and mid-latitude cyclones, because of the impact these can have on society. This part then culminates with an examination of the scientific and social elements of climate change.

Part V, this book's final part, goes beyond the confines of the Earth and heads into outer space. We'll see how new discoveries provide answers to age-old questions about where our planet came from, and what the distant future holds. Students will learn how our knowledge of space has advanced over the centuries, what the other objects of our Solar System look like, and how the Sun and the stars produce energy. This part ends by taking a voyage to the farthest reaches of the Universe, which we learn about by analyzing light that began its journey to Earth billions of years before our planet had even formed.

Although we organized the chapters in *Earth Science* in a particular way that provides an overall direction to our narrative, we have ensured that this book is flexible, and that chapters can be assigned in a different sequence. Therefore, each chapter is self-contained, and we reiterate relevant material where necessary. This flexibility allows instructors to choose their own strategies for teaching Earth Science.

Special Features of This Text

Narrative Art and *What an Earth Scientist Sees*

To help students visualize topics, this book is lavishly illustrated. The figures are designed to provide a realistic context for interpreting features and phenomena without overwhelming students with extraneous detail. In this edition, many drawings and photographs have been integrated into *narrative art* that has been laid out, labeled, and annotated to tell a story—the figures are drawn to teach! Subcaptions are positioned adjacent to the relevant parts of a figure, labels point out key features, and balloons provide important detail. Subparts have been arranged to convey time progression, where relevant. The color schemes in the drawings have been tied to those of relevant photos, so that students can easily visualize the relationships between drawings and photos. In some examples, photographs are accompanied by annotated sketches labeled *What an Earth Scientist Sees*, which help students to be certain that they actually see the specific features that the photo was intended to show. The in-text art also serves as the foundation for the robust suite of videos, animations, and simulations that give students a fuller and more dynamic means for visualizing complicated processes that happen over long periods of time (see "Narrative Art Videos, Animations, and Simulations").

Earth Science at a Glance

In addition to individual figures, each chapter contains at least one dramatic *Earth Science at a Glance* illustration. These illustrations either expand on a particular topic or provide a synopsis of many topics in a beautiful, artistic rendering. Renowned British artist Gary Hincks handpainted the majority of these. Others were developed by the book's lead artist, Stan Maddock, and his colleagues. These illustrations provide a way for students to visualize key concepts . . . at a glance.

How Can I Explain? Features

Unique **How Can I Explain** features in every chapter provide simple, fun, hands-on projects that help students to better visualize and master key concepts. These exercises are particularly useful for future educators, as the projects can easily translate into effective lesson plans for grade-school classes.

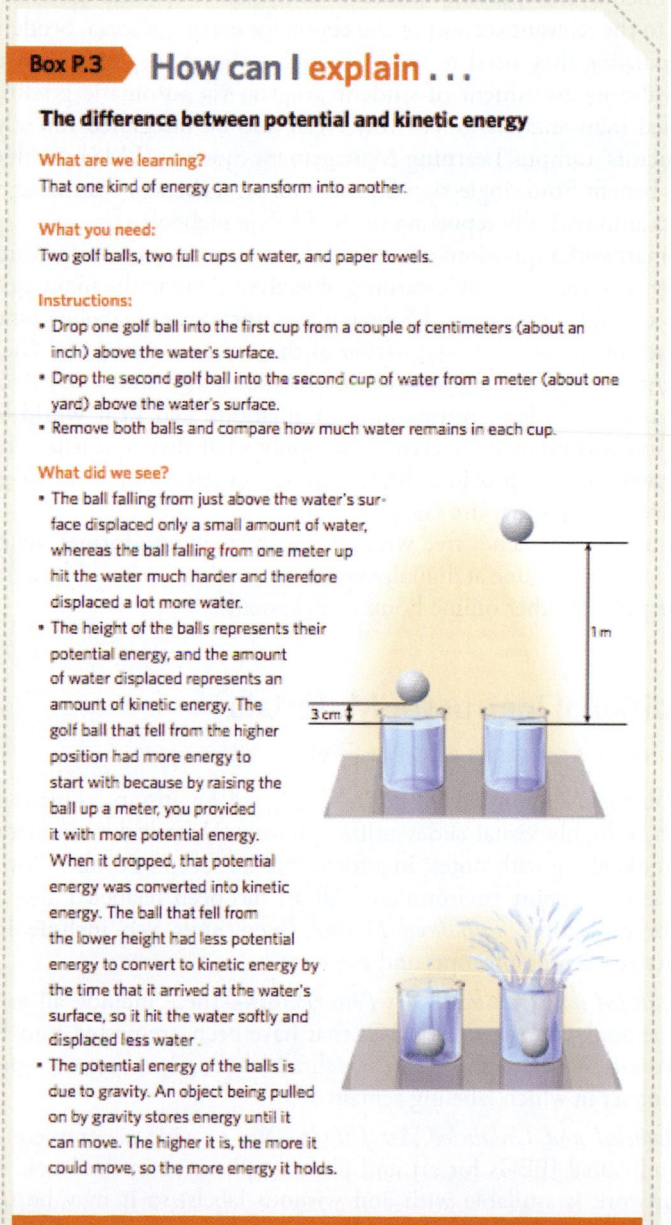

Box P.3   How can I **explain . . .**

The difference between potential and kinetic energy

What are we learning?
That one kind of energy can transform into another.

What you need:
Two golf balls, two full cups of water, and paper towels.

Instructions:
- Drop one golf ball into the first cup from a couple of centimeters (about an inch) above the water's surface.
- Drop the second golf ball into the second cup of water from a meter (about one yard) above the water's surface.
- Remove both balls and compare how much water remains in each cup.

What did we see?
- The ball falling from just above the water's surface displaced only a small amount of water, whereas the ball falling from one meter up hit the water much harder and therefore displaced a lot more water.
- The height of the balls represents their potential energy, and the amount of water displaced represents an amount of kinetic energy. The golf ball that fell from the higher position had more energy to start with because by raising the ball up a meter, you provided it with more potential energy. When it dropped, that potential energy was converted into kinetic energy. The ball that fell from the lower height had less potential energy to convert to kinetic energy by the time that it arrived at the water's surface, so it hit the water softly and displaced less water.
- The potential energy of the balls is due to gravity. An object being pulled on by gravity stores energy until it can move. The higher it is, the more it could move, so the more energy it holds.

1m

3 cm

Pedagogical Features Designed to Help Students Master the Concepts

Each chapter begins with a series of ***learning objectives*** that frame the major concepts of the chapter for the students. Every chapter section ends with a ***Take-Home Message***, a brief summary that helps students identify and remember the highlight of the section before moving on to the next. These also contain a quick question, challenging students to engage. ***See for Yourself*** features guide students on virtual field trips, via *Google Earth™*, to locations around the globe or in the sky where they can apply their newly acquired knowledge to the interpretation of real-world geologic features. ***How Can I Explain*** features, described above, drive student interaction with the course material by providing simple hands-on projects that illustrate important concepts. ***Did You Ever Wonder*** panels prompt students to connect new information to their existing knowledge base by asking Earth Science-related questions that they have probably already thought about. ***Consider This*** boxes help students connect the course concepts to real-life applications, or to understand the concepts more thoroughly. ***Science Toolboxes*** give students brief, accessible introductions or reviews of basic science terms and concepts that can help them better grasp the chapter

Take-home message . . .

An ordinary thunderstorm has a vertical, non-rotating updraft and tends to dissipate fairly quickly. Squall-line thunderstorms form in a row along a strong front or along storm-generated cold pools. The most violent thunderstorms, supercells, form where there is vertical wind shear in addition to strong instability. A supercell has a tilted rotating updraft, a particularly broad anvil, and an overshooting top.

Quick Question -
Why do supercell thunderstorms tend to survive for a relatively long time?

material. And the two-page ***Chapter Review*** spreads are useful for easy reference and as study guides for students. All the review questions are tagged to the learning objectives from the beginning of the chapter in order to make them more useful as assignable questions for the instructor. Every review section also includes visual and applied questions designed not only to test basic knowledge but also to stimulate critical thinking.

Up-to-Date Coverage of Current Topics

Earth Science reflects the latest research and discoveries in the discipline, to help students understand the events and discoveries that have been featured in news headlines and recognize the relevance of these events to their own lives.

Media and Assessment Resources
Narrative Art Videos, Animations, and Simulations

Accompanying *Earth Science* at no cost to students or instructors is a rich collection of over sixty new animations and videos that illustrate

Earth Science processes and course concepts. Animations developed by Stephen Marshak and Alex Glass (Duke University) utilize a consistent style, applying a 3-D perspective to help students grasp active earth science phenomena. Some of these animations are simulations that allow students to control aspects and variables of an Earth Science

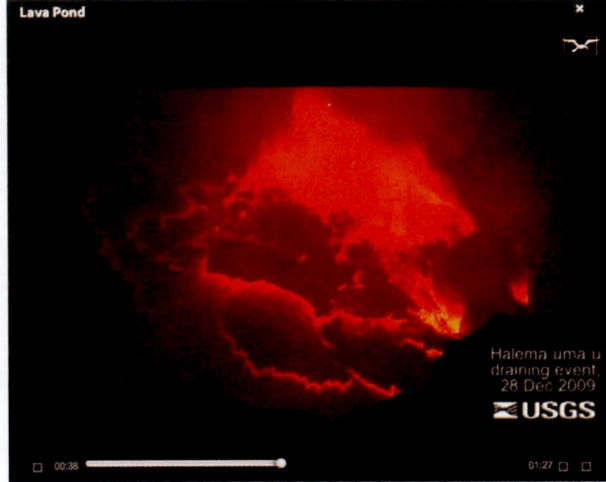

process. In the **Narrative Art Videos**, Stephen Marshak enhances explanations of core concepts in the text by describing the processes displayed in animated versions of the book's figures. This full suite of visual resources is available to students at digital.wwnorton.com/earthscience, through the Marshak YouTube channel, and in the LMS Coursepacks. The suite is additionally available to instructors at the book's Instructor Resources page and on an easy-to-use USB flash drive.

Real-World Videos

In addition to the suite of original videos, animations, and simulations that we described above, instructors and students will have access to over a hundred real-world videos, showing earth science—and earth scientists—in action. These videos have been carefully selected by Tobin Hindle of Florida Atlantic University. Teaching notes and classroom discussion questions to accompany the videos are available in the Interactive Instructors Guide, and selected videos are available in the Smartwork5 tutorial and assessment system, with accompanying questions and feedback. Over eighty real world videos are available at digital.wwnorton.com/earthscience, in the LMS Coursepacks, in the online Interactive Instructors Guide, and on the instructor USB flash drive.

New Smartwork5 Online Tutorial and Assessment System for *Earth Science*

The Smartwork5 online tutorial and assessment system features visual assignments that provide students with answer-specific feedback and links to the relevant section of the ebook for every question. Students get the coaching they need to work through assignments, while instructors get real-time assessment of student progress via automatic grading and detailed item analysis. Smartwork5 can also be integrated directly into instructors' campus Learning Management Systems (LMS), so that students benefit from single-sign on and instructors benefit from assignment results automatically reporting to the LMS gradebook.

Smartwork5 questions are written by and for earth science educators, in service of the textbook's learning objectives. Among the many question types available in Smartwork5 are ranking, sorting, and labeling tasks that take advantage of the visual nature of the course material and challenge students to think about real-world situations. Questions based on the Narrative Art Videos, Animations, Simulations, and Real World Videos help students prepare for class, or to apply what they've learned. Finally, Smartwork5 also provides basic reading quizzes and *Geotour*-guided inquiry activities that use *Google Earth*™.

Smartwork5 comes free with all new texts in any format, or can be purchased standalone at digital.wwnorton.com/earthscience for a fraction of the price of other online homework systems.

Additional Instructor Materials
Lecture PowerPoints and Art Files

- *Enhanced Lecture PowerPoints*—Designed for instant classroom use, these highly visual slides utilize photographs and line art from the book, along with notes, in a form that has been optimized for use in the PowerPoint environment. All art has been relabeled and resized for projection. *Enhanced Lecture PowerPoints* also include in-class active learning prompts and exercises.

- *Labeled and Unlabeled Art PowerPoints*—These include all art from the book formatted as JPEGs that have been pre-pasted into PowerPoints. We offer one set in which all labeling has been stripped and one set in which labeling remains.

- *Labeled and Unlabeled Art JPEGs*—We provide a complete file of individual JPEGs for art and photographs used in the book. Again, artwork is available with and without labels, so it may be used in presentations, quizzes, exams, etc.

- *Semesterly PowerPoint Update Service*—W. W. Norton & Company offers a semesterly update service that provides new, current-event-based PowerPoint slides, with instructor support, tying events in the news to core concepts from the text.

Instructor's Manual

The instructor's manual, prepared by Heather Cook of California State University San Marcos, Geoff Cook of University of California, San Diego, and Julia Domenech-Eckberg, is designed to help instructors prepare lectures, homework, and exams. Each chapter contains:

- Learning objectives
- Complete answers to end-of-chapter Review and On Further Thought questions
- Descriptions of all Animations and Simulations, plus suggested classroom uses and discussion questions
- Teaching notes and discussion questions for the Real World Videos
- Activity ideas to implement active learning in class
- A correlation between Marshak/Rauber's *Earth Science* and Tarbuck and Lutgen's *Earth Science 14th Edition* for those looking to make the switch.

Interactive Instructor's Guide (IIG)

Searchable by chapter, phrase, topic, or learning objective, the Interactive Instructor's Guide instantly provides multiple ideas for teaching: video clips, powerpoints, animations, and other class activities and exercises. This repository of lecture and teaching materials functions both as a course prep tool and as a means of tracking the latest ideas in teaching the Earth Science course.

Test Bank

The Test Bank, authored by Heather Cook, Geoff Cook, and Julia Domenech-Eckberg, has been written to assess the text's learning objectives using carefully vetted and well-rounded questions. Every item in the Test Bank has been reviewed to ensure scientific accuracy. Each chapter features 50 multiple-choice questions, 10 short-answer or essay questions that test student's critical thinking and knowledge-application skills, and several art-based questions using modified images from the text. Finally, each question is tagged by text section, learning objective, difficulty level, and Bloom's taxonomy level.

LMS Coursepacks

Available at no cost to professors or students, Norton Coursepacks bring high-quality Norton digital media into a new or existing course in your campus Learning Management System. For each chapter, the Norton Coursepack offers:

- Reading Quizzes by Marianne Caldwell of Hillsborough Community College
- Links to Videos, Animations, and Simulations
- Links to the ebook
- Vocabulary Flashcards
- Link to download the See for Yourself and GeoTours kmz files
- GeoTours activities
- Test Bank questions in your campus LMS format

Norton Coursepacks are available for Blackboard, Canvas, Moodle, and Desire2Learn.

Geotours Workbook, Second Edition

Created by Scott and Beth Wilkerson of DePauw University, and Stephen Marshak, *Geotours Workbook* contains active-learning exercises arranged by topic that take students on virtual field trips in *Google Earth™* to see outstanding examples of Earth science at locations around the world. Each Geotour is accompanied by a worksheet that includees instructions and multiple choice questions. The Workbook accompanies a custom-made *Google Earth™* kmz file created by Scott and Beth Wilkerson, available for free download by all instructors and students using *Earth Science* at digital.wwnorton.com/earth science. The *Geotours Workbook* can be packaged for free with *Earth Science* and includes complete user instructions and advanced instruction. Request a sample copy from your local Norton representative to preview each worksheet.

See for Yourself *Google Earth™* Sample Site File.

Earth Science users who simply want access to sample field sites for classroom presentations or distribution to students can download the sites from digital.wwnorton.com/earthscience.

ebook-digital.wwnorton.com/earthscience

Compatible with all computers and mobile devices, the Norton ebook reader provides intuitive highlighting, note taking, and bookmarking functionalities. The Norton ebook for *Earth Science* includes dynamic features that engage students, such as our animations, Narrative Art Videos, and links to *Google Earth™* See for Yourself sites. To help focus student reading, instructors can share notes with their class, including images and video. Reports on student and class-wide time-on-task allow instructors to monitor student reading and engagement. The Norton ebook reader can also be integrated into your campus learning management system. When integration is enabled, students can click on a link to the ebook from their campus LMS and be redirected. The *Earth Science* ebook is available for purchase at **digital.wwnorton.com/earthscience**

See for yourself—
Using *Google Earth™*

Visiting Field Sites Identified in the Text

There's no better way to appreciate geology than to see it firsthand in the field. The challenge is that the great variety of geologic features that we discuss in this book can't be visited from any one locality. So even if your class can take geology field trips during the semester, you'll at most see just a few geologic settings. Fortunately, *Google Earth™* makes it possible for you to fly to spectacular geologic field sites anywhere in the world in a matter of seconds—you can take a virtual field trip electronically. Using related options, you can also look at stars in the sky.

In each chapter in this book, a *See for Yourself* provides sites that you can explore on your own computer (Mac or PC) using *Google Earth™* software, or on your Apple/Android smartphone or tablet with the appropriate *Google Earth™* app.

To get started, follow these three simple steps:

1 Check to see whether *Google Earth™* is installed on your personal computer, smartphone, or tablet. If not, download the software from **earth.google.com** or the app from the Apple or Android app store.

2 Each *See for Yourself* site provides a thumbnail photo and brief description of the site (highlighting what you will see), as well as the latitude and longitude of the site.

3 Open *Google Earth™*, and enter the coordinates of the site in the search window. As an example, let's find Mt. Fuji, a beautiful volcano in Japan. We specify the coordinates in the book as follows:

Latitude   35°21′41.78″N,   Longitude   138°43′50.74″E

Type these coordinates into the search window as:

35 21 41.78N, 138 43 50.74E

Note that the degree (°), minute (′), and second (″) symbols are optional and can be left as blank spaces.

When you click Enter or Return, your device will bring you to the viewpoint right above Mt. Fuji, illustrated by the thumbnail on the left. Note that you can use the tools built into *Google Earth™* to vary the elevation, tilt, orientation, and position of your viewpoint. The thumbnail on the right shows the view you'll see of the same location if you tilt your viewing direction and look north.

View looking down.

View looking north.

Need More Help?

Please visit https://digital.wwnorton.com/earthscience to find a video showing you how to download and install *Google Earth™*, more detailed instructions on how to find the *See for Yourself* sites, additional sites not listed in this book, links to *Google Earth™* videos describing basic functions, and links to any hardware and software requirements. Also, notes addressing important *Google Earth™* updates will be available at this site.

We also offer a separate book—the *Geotours Workbook, Second Edition* (ISBN 978-1-324-00096-9)—that identifies even more interesting geologic sites to visit, provides active-learning exercises linked to the sites, and explains how you can create your own virtual field trips.

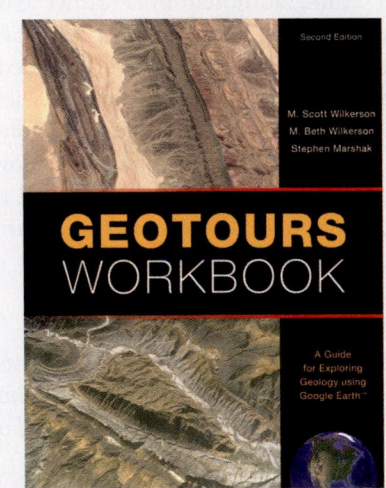

Acknowledgments

Many people helped the authors bring this book from the concept stage to the shelf, and kept the the momentum of this sometimes overwhelming project moving forward.

First and foremost, we wish to thank Kathy Marshak, who served as the book's in-home production editor. She helped coordinate much material from the lead author's previous books; edited and proofed text and figures; monitored the flow of manuscript and proofs on behalf of the authors; helped coordinate the art program; queried stylistic choices; and served as an invaluable extra set of eyes at every stage. This book would not have happened without Kathy.

The staff of W. W. Norton & Company have, as always, been wonderful to work with, and we thank them for their insightful input and enduring patience during the development of this book. It has been a privilege to work with an employee-owned company whose editorial staff remains willing to work directly with authors. In particular, we would like to thank Thom Foley, the senior project editor for the book, who did a Herculean job of overseeing an unexpectedly complicated process of managing the manuscript, chapter proofs, art, and photos, all while remaining incredibly calm. Thom invested untold hours in sorting out composition and design issues, and in making sure every word and figure ended up where it should be. This book greatly benefited from the input of three supervising editors. The idea for the book nucleated in discussion with the late Jack Repcheck, a mentor who we will always remember. Eric Svendsen energetically pushed the book through the starting gate, injecting many ideas that contributed to its style and organization. Eric's experience and skill guided the book in new directions and connected the project to new trends in science pedagogy and book design. Jake Schindel took over the reins of the book in mid-course, and brought the project over the finish line. Jake's ability to solve problems and keep the project moving forward is more than impressive and we greatly appreciate the enthusiasm and skill that he brought with him when he jumped into the deep end. We wish to thank Rob Bellinger, Cailin Barrett-Bressack, and Liz Vogt for their innovative approach to ancillary development and for overseeing the development of the online educational supplements. We are also very grateful to all of the instructors who have helped to develop accompanying materials; Sunny Hwang, for helpful contributions to the writing and developmental organization of the book; Norma Sims Roche, who completed a spectacular job of copyediting, patiently sorting through and resolving edits upon edits upon edits upon edits; Stephanie Romeo for her expert and thoughtful management of the photo program; Jane Miller and Elyse Rieder, the photo researchers, for seeking out many of the new images; Ben Reynolds for coordinating the back-and-forth between the publisher and various suppliers; Debra Morton Hoyt for overseeing the book's design; Rachel Goodman, assistant editor, for dealing with expected and unexpected details throughout the book's development; Megan Schindel for expertly tracking down permissions; and Katie Sweeney, who has ably assumed the mantle of marketing manager for the book.

Production of the illustrations has involved many people over many years. Stan Maddock has been the primary artist for Norton's geoscience books for two decades, and it's been a delight to work with him and the entire art team. It has also been great fun to interact with Gary Hincks, who painted many of the two-page spreads, in part using his own designs and geologic insights. Versions of several of these paintings originally appeared in *Earth Story* (BBC Worldwide, 1998) and were based on illustrations conceived with Simon Lamb and Felicity Maxwell. Some of the chapter quotes were found in *Language of the Earth*, compiled by F. T. Rhodes and R. O. Stone (Pergamon, 1981).

This book and its parent, *Earth: Portrait of a Planet*, have benefited greatly from input by expert reviewers for specific chapters, by general reviewers of the entire book, and by other reviewers who have provided helpful feedback for this volume. These reviewers include:

Erica Barrow, *Ivy Tech Community College*
Celso Batalha, *Evergreen Valley College*
Karin Block, *City College of New York*
Susan Bratcher, *Suny Buffalo*
Brett Burkett, *Collin College*
Karen Busen, *Tallahassee Community College*
Marianne Caldwell, *Hillsborough Community College*
Jennifer J. Charles-Tollerup, *Rowan College at Gloucester County*
Laura Chartier, *University of Alaska, Anchorage*
Robert L. Dennison, *Heartland Community College*
Todd Feeley, *Montana State University*
Kerry Workman Ford, *California State University, Fresno*
Nicholas Frankovits, *University of Akron*
Kenneth Galli, *Boston College*
Bryan Gibbs, *Richland College*
Alexander Glass, *Duke University*
Alessandro Grippo, *Santa Monica College*
Donald Hellstern, *Brookhaven College*
Tobin Hindle, *Florida Atlantic University*
Ashanti Johnson, *University of Texas, Arlington*
Leslie Kanat, *Johnson State College*
Zoran Kilibarda, *Indiana University Northwest*
Michael Kozuch, *California State University, East Bay*
Michael Kruge, *Montclair State University*
Kody Kuehnl, *Franklin University*
Brett C. Latta, *Franklin University*
Kristine Larsen, *Central Connecticut State University*
Rita Leafgren, *University of Northern Colorado*
Maureen Lemke, *Texas State University*
Judy McIlrath, *University of South Florida*
Jamie Mitchem, *University of North Georgia, Gainesville*
Jacob Napieralski, *University of Michigan, Dearborn*
Joseph Osborn, *Century College*
Mark Peebles, *St. Petersburg College*
Julia Sankey, *California State University, Stanislaus*
Steven H. Schimmrich, *SUNY Ulster County Community College*
Marcia Schulmeister, *Emporia State University*
Jennifer K. Sheppard, *Moraine Valley Community College*
Andrew Smith, *Vincennes University*
Steven Stemle, *Palm Beach State College*
Keith Sverdrup, *University of Wisconsin, Milwaukee*
Anthony J. Vega, *Clarion University*
Adil M. Wadia, *The University of Akron*
Shizuko Watanabe, *Eastfield College*
Terry R. West, *Purdue University*
James Wysong, *Hillsborough Community College*

We also wish to thank the instructors who joined us for focus groups regarding this text and the course it serves; their input helped shape the form and content of this text. These instructors include:

Johnsely S. Cyrus, *North Carolina Agricultural and Technical State University*

Frank DeCourten, *Sierra College*

David Douglass, *Pasadena City College*

Todd Feeley, *Montana State University*

Garry Hayes, *Modesto Junior College*

Richard W. Hurst, *California Lutheran University*

Melissa Lobegeier, *Middle Tennessee State University*

John McDaris, *Carleton College*

Rosemary Millham, *SUNY New Paltz*

Kent Murray, *University of Michigan, Dearborn*

Jeffrey Myers, *Western Oregon University*

Caroline Pew, *University of Washington*

Nancy Price, *Portland State University*

Eric Jonathan Pyle, *James Madison University*

René Shroat-Lewis, *University of Arkansas at Little Rock*

Carol Anne Stein, *University of Illinois, Chicago*

Rebecca Teed, *Wright State University*

Leanne Sue Teruya, *San Jose State University*

Suzanne Traub-Metlay, *Front Range Community College*

ABOUT THE AUTHORS

Steve Marshak is a professor of geology at the University of Illinois at Urbana-Champaign, where he also serves as the director of the School of Earth, Society, & Environment. He holds an A.B. from Cornell University, an M.S. from the University of Arizona, and a Ph.D. from Columbia University. Steve's research interests in structural geology and tectonics have led to his participation in field projects on many continents. He has won several campus teaching awards and received the Neil Miner Award from the National Association of Geoscience Teachers for "exceptional contributions to the stimulation of interest in the Earth Sciences." He is a Fellow of the Geological Society of America. In addition to research papers and *Earth Science*, Steve has authored *Earth: Portrait of a Planet,* and *Essentials of Geology*, and has co-authored *Laboratory Manual for Introductory Geology, Earth Structure: An Introduction to Structural Geology and Tectonics*, and *Basic Methods of Structural Geology*.

Bob Rauber is a professor of atmospheric sciences at the University of Illinois at Urbana-Champaign, where he also serves as the department head. He holds a B.S. in Physics and a B.A. in English from the Pennsylvania State University, as well as M.S. and Ph.D. degrees in Atmospheric Science from Colorado State University. In addition to teaching and writing, Bob oversees a research program that focuses on the development and behavior of storms. To carry out this work, Bob flies into hazardous weather in specially equipped airplanes—during these flights, he's so intent on recording data that he usually doesn't notice the bouncing and lightning. Bob has won several campus teaching awards, is a Fellow of the American Meteorological Society (AMS), and serves as Publication Commissioner for the AMS. In addition to research papers and *Earth Science*, his textbook writing credits include lead authorship of *Severe and Hazardous Weather: An Introduction to High Impact Meteorology*.

EARTH SCIENCE

The Earth, The Atmosphere, and Space

PRELUDE

Welcome to Earth Science!

By the end of the Prelude you should be able to . . .

A. describe the variety of subjects that an Earth Science course encompasses.

B. evaluate whether a news story pertains to an aspect of Earth Science, and how.

C. understand the concept of the Earth System and describe its key components.

D. explain why studying our Universe can help us understand our home planet and to address practical issues.

E. analyze an example of a scientific investigation.

F. recall some of the key themes of Earth Science.

P.1 Introduction

Our C-130 Hercules transport plane rose from the frozen surface of the Ross Sea, along the coast of Antarctica, and turned south. We were heading to a field site about 250 km away where, with luck, we'd be able to spend the next month studying cliff exposures of some very unusual rocks (Fig. P.1). The plane climbed past the smoking summit of Mt. Erebus, Earth's southernmost volcano, and for the next hour, flew along the Transantarctic Mountains, a long range of rugged ridges that divides the continent into two parts, East Antarctica and West Antarctica. Over millions of years, snow that accumulated in Antarctica's cold climate built into vast *glaciers*, sheets and rivers of solid ice that last all year and slowly flow across the ground. The glacier of East Antarctica covers most of the continent and locally attains a thickness of over 3 km (2 miles). Between the Transantarctic Mountains and the far side of East Antarctica, over 2,500 km (1,500 miles) away, the surface of this glacier forms a high plain of blinding white called the Polar Plateau.

While marveling at this stark panorama—so different from the forests, grasslands, farms, and cities of more populated regions on Earth—we heard the engines slow and felt the Hercules begin to descend. As the plane approached a glacier's surface, just below the cliff that we hoped to study, the pilot lowered the landing gear, a huge tricycle of Teflon-coated skis. Shouting above the engine noise, a crew member reminded us of the emergency procedure: "If you hear three short blasts of the siren, hold on for dear life!"

Seconds later, the skis slammed into small frozen snowdrifts that formed wave-like ripples on the surface of

A plane dropping geologists on a snowfield

Figure P.1 Geologic fieldwork in Antarctica unlocks the mysteries of an icebound continent.

The Transantarctic Mountains separate East Antarctica from West Antarctica

Weddell Sea Ice Shelf

to Africa

Antarctic Peninsula

to South America

West Antarctica

South Pole

East Antarctica

Ross Sea Ice Shelf

Transantarctic Mountains

Mt. Erebus

to Australia

0   500   mi
0   500   1,000 km

The smoking summit of Mt. Erebus

Sledding to a field site with crates of supplies

the glacier. *Wham, wham, wham, wham!* It felt as though a fairy-tale giant was shaking the plane. Then, as fast as it began, the shaking stopped, and we were airborne again, looking for a softer landing surface. We finally touched down at a location where soft snow blankets the hard glacial ice of the Polar Plateau. The ramp at the tail of the plane was lowered and, bundled against the frigid gale generated by the plane's still-roaring propellers, we jumped out and unloaded food, fuel, stoves, tents, sledges, and snowmobiles. The instant the cargo was out, the ramp rose, the Hercules trundled off, and though it struggled to attain takeoff speed in the thick snow, it finally rose skyward.

When the glint of the plane's metal skin had passed beyond the horizon, the silence of Antarctica hit us—no dogs barked, no leaves rustled, and no traffic rumbled in this land of black rock, white ice, and blue sky. It would take us almost two days, even with the aid of powerful snowmobiles, to haul sledges of food and equipment to our field site, where we would spend the next month collecting rock samples and mapping their distribution. Why go to so much effort and expense to study an exposure of rocks? The Scottish poet Walter Scott (1771–1832) asked the same question, and provided a colorful answer: "Some rin uphill and down dale, knapping the chucky stanes to pieces wi' hammers, like sae mony road-makers run daft—they say it is to see how the warld was made!"

Indeed, to see how the world was made. Field expeditions like the one we've just described, along with analyses carried out with instruments in laboratories and calculations run on computers, have led to an explosion of discoveries about the land, sea, and atmosphere of Earth. When we add observations made with telescopes and space probes of the myriad objects that populate the **Universe** (all of space, and everything within it), we can begin to understand not only the wonder of our home planet, but also its context in the broadest sense **(Fig. P.2)**. Read on, and you can share in these discoveries and develop a personal image of the land beneath your feet, the landscapes between you and the horizon, the sea that surrounds the land, the atmosphere that surrounds Earth, and the objects that sparkle in the night sky above.

P.2 What's in an Earth Science Course?

Defining Earth Science and Its Components

For most of humanity's existence, speculation about the natural world lay in the realm of philosophy, and people attributed natural features in their surroundings to supernatural phenomena. But, beginning a few thousand years ago, and accelerating in the past few hundred years, study of the natural world and its surroundings became the focus of **science**, the systematic analysis of natural phenomena based on observation, experiment, and calculation. Science has evolved into several distinct disciplines, most of which may already be familiar to you. For example, if someone asks, "What is *chemistry*?" you might respond that it's the study of chemicals and reactions between them. Similarly, if someone asks, "What is *physics*?" you might respond that it concerns the study of matter and energy, and if someone asks, "What is *biology*?" you might respond that it's the study of living organisms.

If someone asks, "What is Earth Science?" you might be stumped. That's because **Earth Science** combines many disciplines **(Fig. P.3)**. It includes **geology**, the study of our planet with a focus on the materials that compose it, the phenomena that change it, and its long-term history. It includes **oceanography**, the study of the water and life in the oceans, as well as the way in which ocean water moves and interacts with land and air. And it includes **atmospheric science**, the study of the air layer that surrounds the Earth. Atmospheric science, in turn, includes both meteorology and climate science. **Meteorology** focuses on the *weather*, meaning the condition of the atmosphere at a given location and time, as well as the movement of air and its consequences. **Climate science** focuses on a region's overall annual pattern of weather averaged over many years—its *climate*—and how that pattern changes over time. High school or college courses entitled Earth Science also commonly introduce the basics of **astronomy**, or *space science*, the study of planets, stars, and other objects of the Universe. Learning about astronomy provides a framework of knowledge on which to build an understanding of the formation and history of our own

Figure P.2 A spectacular view from NASA's Hubble Space Telescope of a distant region of space.

(a) A geologist studies a rock face.

(b) Oceanographers sample the seas.

(c) Atmospheric scientists monitor and analyze the atmosphere.

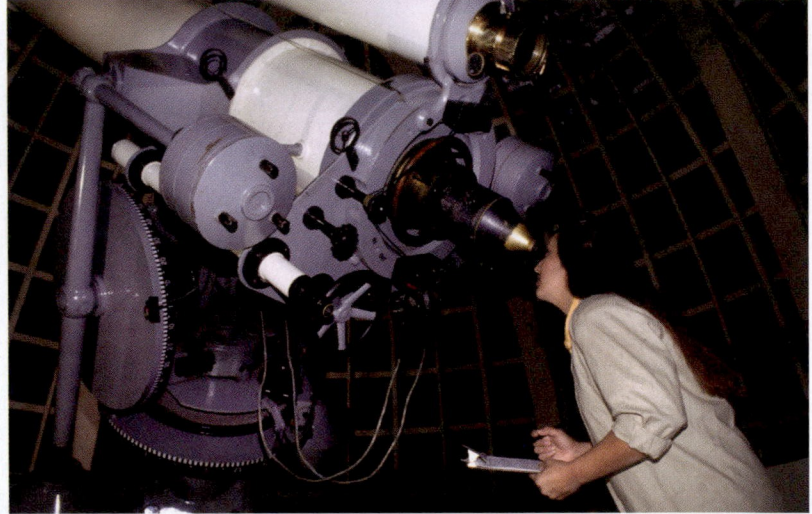

(d) Astronomers explore the cosmos.

planet. In sum, we see that an Earth Science course covers the nature, origin, and evolution of all of our natural surroundings. It's a broad subject, indeed!

How Scientists Work

The information discussed in this book comes from the work of scientists. Popular media often characterize scientists as awkward loners with poor taste in clothing. Who are they, really? In an Earth Science course, you'll see that **scientists** are people who spend their careers in search of ideas to explain the way our Universe operates. They carry out their search in many different ways. At any given time, at locations all around the world, *field scientists* scale cliffs **(Fig. P.4a)**, fly into storms **(Fig. P.4b)**, plow through stormy seas, or stargaze from mountaintops. Meanwhile, *laboratory scientists* peer

down microscopes, adjust electronic equipment, or mix test tubes of chemicals **(Fig. P.4c)**; and *computational scientists* program computers or devise equations to provide *models*—simulations—of natural phenomena. In Earth Science, both computer and physical models can help us visualize processes that take place too slowly or too quickly to see in real time, or can characterize objects that are too small or too large to measure directly **(Fig. P.5)**.

As you study Earth Science, you'll have an opportunity to see how scientists conduct **research**, the process of seeking to understand natural phenomena through observation, experiment, and calculation. You'll see that research does not involve reliance on subjective guesses, but rather anchors in the development of a consistent set of testable concepts, following the basic tenets of the

(a) Cliff exposures in the desert of Utah provide a record of the Earth's past.

(b) To learn about storms, researchers fly into one and take measurements.

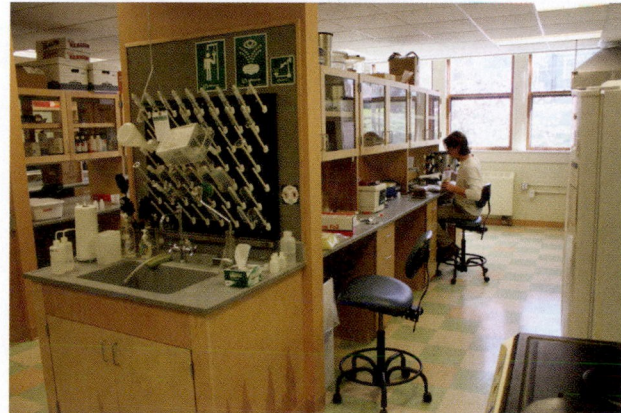

(c) Laboratories provide an opportunity to carry out controlled experiments.

protect our environment. To drive home the point that science is a human endeavor, this book highlights where, when, and how ideas originated so that you can answer the question, "How do we know that?"

Who are the scientists that make the discoveries we'll be focusing on in this book? When a headline begins with "Scientists say . . ." and then continues with "an earthquake shook Japan today," or "the supply of oil may be running out," the scientists under discussion are *geologists*. If the

Figure P.5 Use of models in Earth Science.

(a) A stream of water flowing over sand can simulate the evolution of a river.

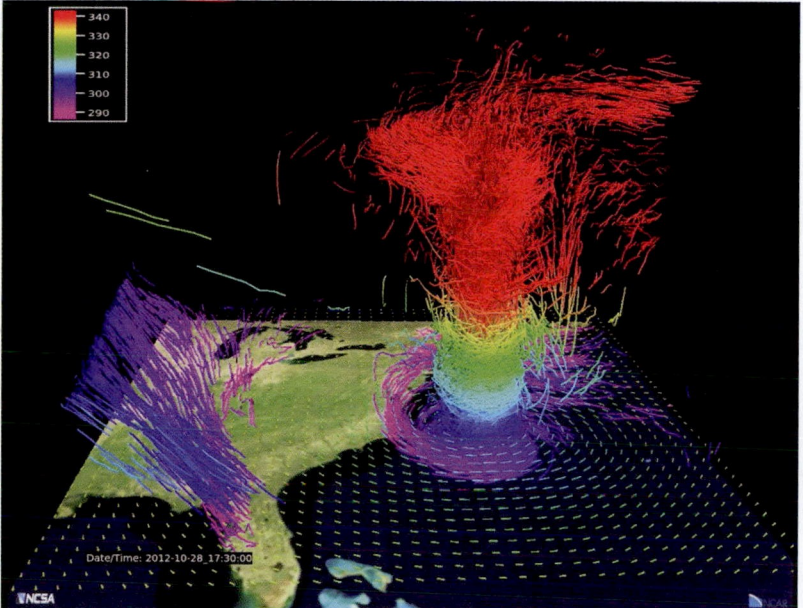

(b) A computer model simulating air flow and air potential-temperature variation during Hurricane Sandy of 2012. Lines indicate air movement. Colors indicate potential temperature, the temperature that a volume of air would have if it had the same pressure as air does at sea level (purple = cool; red = warm).

scientific method **(Box P.1)**. Some scientists focus their research on defining new principles that can profoundly change our understanding of nature—think of Charles Darwin, Marie Curie, or Albert Einstein—whereas others focus on practical matters, using the results of their research to improve our lives, find our resources, and

Box P.1 ▶ Consider this . . .

The scientific method

Sometime during the past 200 million years, a large block of rock or metal zoomed in from outer space and slammed into our planet at a site in what is now the midwestern United States, a landscape that today hosts flat cornfields. The impact of this *meteorite* blasted debris skyward and carved a deep, bowl-shaped depression called a *crater*. The impact also shattered rock beneath the crater, and it caused layers of rock that had been buried deeply below the ground surface to spring upward and tilt on end. In the millions of years that followed, flowing water and blowing wind carried away the debris and wore down the crater until the depression had disappeared entirely. But this process of *erosion* did not carve deeply enough to remove the fractures and tilted rock layers that the impact produced. Some 15,000 years ago, sand, gravel, and mud carried by a vast glacier, much like the one that covers Antarctica today,

buried the area and completely hid evidence of the impact from view **(Fig. BxP.1)**. So much history beneath a cornfield! How do we know this? It took years of scientific research!

Scientists commonly guide their research by using the **scientific method.** Let's consider the idealized components of the scientific method and see how researchers applied them to come up with and verify the meteorite-impact story.

• *Recognizing the problem:* Any scientific project, like any detective story, generally begins by identifying a problem, the scientific word for "mystery." The cornfield mystery came to light when workers drilling a water well discovered that limestone, a rock commonly made of shell fragments, lies just below a thin blanket of 15,000-year-old sediment left behind by glaciers. In surrounding regions, the rock directly beneath the glacial sediment consists of sandstone, a rock made of sand grains. In fact, outside of the mysterious cornfield, the limestone layer lies underneath the sandstone layer. Because limestone can be used

Figure BxP.1 An ancient meteorite impact has excavated a crater and permanently changed rock beneath the surface.

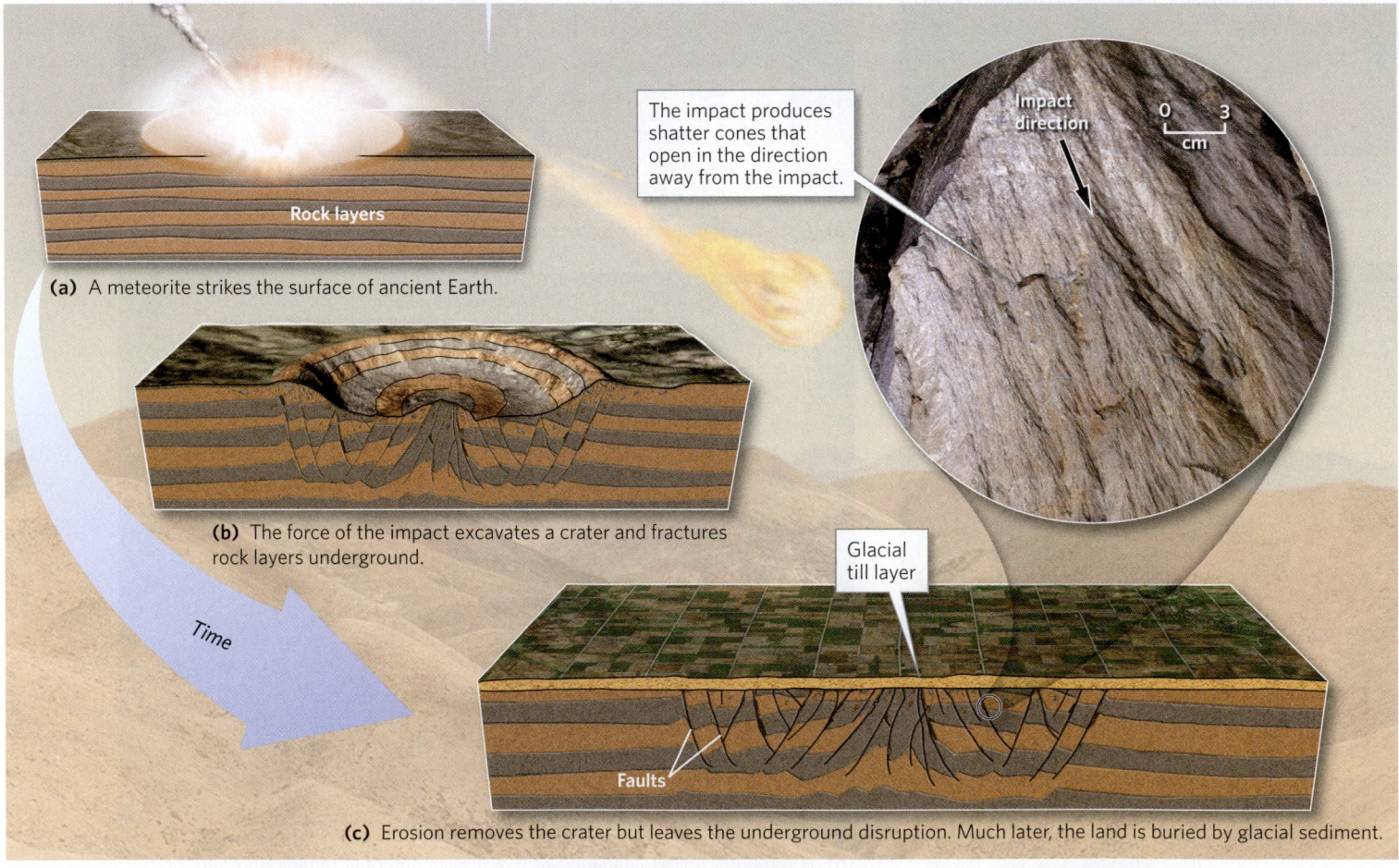

(a) A meteorite strikes the surface of ancient Earth.

Rock layers

The impact produces shatter cones that open in the direction away from the impact.

Impact direction

0   3
cm

(b) The force of the impact excavates a crater and fractures rock layers underground.

Time

Glacial till layer

Faults

(c) Erosion removes the crater but leaves the underground disruption. Much later, the land is buried by glacial sediment.

to make cement and to produce the agricultural lime used in improving soil, it's valuable, so workers bulldozed off the glacial sediment and dug a quarry to excavate the limestone. They were surprised to find that rock layers exposed in the quarry were tilted steeply and had been shattered pervasively by large cracks, for in surrounding regions, the rock layers are horizontal, like the layers in a birthday cake, and contain relatively few cracks. What phenomenon brought limestone up close to the Earth's surface, tilted the layering in the rocks, and shattered the rocks? Curious geologists, like crime-scene investigators, journeyed to the quarry to find out.

- *Collecting data:* To better characterize a problem, and to help solve a problem, scientists collect **data**, sets of observations, measurements, or calculations. Data serve as clues. To collect data on the quarry problem, geologists went into the quarry, measured the orientation of the rock layers, and *documented* (made a written or photographic record of) the fractures that broke up the rocks. Note that in this example, data comes from analyzing natural features formed long ago. In other cases, data might come from experiments or computer models. Observations, experiments, and calculations should be *repeatable*, in the sense that another researcher following the same procedure should obtain the same result.

- *Proposing hypotheses:* A scientific **hypothesis** is merely a possible natural explanation that can explain a set of data. Scientists may propose hypotheses before, during, or after their initial data collection. In this example, the geologists came up with two alternative hypotheses to explain the features in the quarry: (1) the features formed during an ancient volcanic explosion; and (2) the features are a consequence of an ancient meteorite impact.

- *Testing hypotheses:* Because a hypothesis is just an idea that can be either right or wrong, scientists must put a hypothesis through a series of tests to determine if it might be correct, or if it can't be correct. Often, these tests involve making *predictions* of what scientists will see if they make more observations or if they conduct an experiment or calculation. If none of the original hypotheses passes the tests, the problem remains unsolved, and researchers try to come up with new hypotheses. If a hypothesis does pass a test, it could be right.

The geologists at the quarry compared their field observations with previously published observations made at known sites of volcanic explosions and meteorite impacts, and they conducted experiments designed to simulate such events. They learned that if the geologic features visible in the quarry were the result of volcanism, the quarry should contain certain distinctive types of rocks. But no such rocks were found. If, however, the features were the result of an impact, the rocks should contain *shatter cones*, uniquely shaped cracks. Shatter

cones can be overlooked, so the geologists returned to the quarry specifically to search for them—and found them in abundance. The impact hypothesis passed this test, and remained a possibility!

Scientists don't always see the entire path to solving a problem when they start working on it. In fact, at first they might not recognize the problem to be solved. So they don't always follow the components of the scientific method in a specific sequence. Furthermore, serendipity often plays a role in research, in that scientists may stumble onto a new problem or solution without planning on it. And while ideally, scientists try to confirm results by repeating observations, experiments, or calculations, in Earth Science this isn't always possible because the phenomena under study may have happened in the very distant past, or in locations that are very far away. In fact, some questions just can't be answered fully using available resources and methods, so we have to accept that some interpretations remain uncertain.

In common English, the word *theory* often substitutes for the word *hypothesis*. For example, you may see a sentence like, "The scientist's proposal is only a theory," or "The author of the article proposes many theories," with the implication that a theory is just an idea that's as likely to be wrong as it is to be right. In scientific discussion, however, a **theory** is a scientific idea supported by an abundance of evidence. In other words, it's a robust idea that has passed many tests and has, so far, failed none. Scientists have much more confidence in the validity of a theory than they do in the validity of a hypothesis. In the example we described above, continued study in the midwestern quarry eventually yielded so much evidence favoring the impact hypothesis that the hypothesis came to be viewed as a theory. Scientists continue to test theories over a long time. Commonly, the tests involve making predictions based on the theory. Successful theories, those that are supported by many observations and lead to many successful predictions, eventually become part of a discipline's foundation. However, some theories may eventually fail a test, and if so, scientists discard them and search for better ideas.

In some cases, scientists have been able to devise concise statements that completely describe a specific relationship or phenomenon. Such a statement, called a **scientific law**, applies without exception for a defined range of conditions. Newton's law of gravitation serves as an example: it's a simple mathematical expression that always defines how fast an object accelerates (speeds up) when pulled on by another mass. Note that scientific laws do not, in themselves, explain a phenomenon, and in this regard, they differ from theories. For example, the "law of gravity" does not explain how *gravity*, the force that causes masses to attract each other, operates, but the "theory of evolution" does provide an explanation of why species evolve over time.

headline continues with "severe storms are approaching," or "the length of the growing season will change," the scientists referred to are *atmospheric scientists*. If the headline continues, "a shift in the pattern of ocean currents was detected," or a "seawater has become more acidic in recent years," the scientists involved are *oceanographers*. And if the sentence continues, "a new planet has been found in orbit around a distant star," or "a solar storm might disrupt communications tomorrow," the scientists making news are *astronomers*. In this book, for simplicity, we'll often refer to geologists, atmospheric scientists, and oceanographers together as "Earth scientists" because their interests focus on our home planet in particular.

Take-home message . . .

Earth Science encompasses geology, oceanography, atmospheric science, and astronomy. Field, laboratory, and computational scientists all perform research in these disciplines to try to understand our surroundings, from the ground below to the sky above, and the stars beyond.

Quick Question -
Why is it harder to define Earth Science than it is to define, say, chemistry?

P.3 Narrative Themes of This Book

To understand a novel, you'll need to learn names of the characters, and you'll need to pay attention to how the characters interact and relate to one another. Similarly, when you read a textbook, you'll need to learn some new words, or *terminology*, in order to be able to discuss ideas efficiently. You'll also need to understand how ideas relate to one another in order to appreciate the story that the textbook has been written to tell. In this book, we hope to tell the story of the Earth and its setting in space. This story incorporates several *narrative themes*, concepts that appear and reappear and serve to tie ideas together. These themes, which we introduce next, provide this book's overall take-home message. As you will see, understanding these themes requires an awareness of several elementary concepts of other scientific disciplines, which you may have learned in earlier courses, but which we'll briefly reintroduce as a refresher, when needed.

Matter and Energy Behave in Predictable Ways

The Earth, and the Universe as a whole, consist of matter and energy **(Box P.2)**. Throughout this book, we discuss the nature of matter and energy and the many ways that they interact. These interactions happen in predictable,

natural ways. For example, when you push a boulder off the edge of a cliff, gravity pulls it down the cliff and it tumbles to the valley floor below—during this process, potential energy is converted to kinetic energy **(Box P.3)**. When the Sun shines on a dark-colored car, the car's surface become very hot; when you fill a balloon with helium gas, the balloon rises because it is buoyant; and when you place liquid water in a freezer, it goes through a change in state and turns into solid ice. Because such phenomena happen in the same way under the same circumstances every time, scientists can predict—to some extent—how natural processes operate.

We say "to some extent" in the previous paragraph because *unpredictability* is a part of nature. For example, while we can predict that a boulder pushed over a cliff will tumble downslope, we can't predict exactly where it will land, for it may bounce or bang into other boulders on its way down. Similarly, when we make a measurement repeatedly, we may find that the values we obtain vary slightly, because there's a randomness to natural phenomena. For this reason, scientists try to characterize the *degree of uncertainty*, the likelihood of variation, when describing a measurement or prediction.

The Earth System Contains Many Interacting Realms

Look around you, and you'll instantly realize that our home planet is a complicated place. It includes a variety of *realms*, defined as follows:

- *Geosphere:* The part of the Earth that starts at the solid surface and goes down to the center makes up the **geosphere**. Geologists divide the geosphere into concentric layers, nested one within the other like the layers of an onion. The outermost layer, the layer we live on, is the *crust*. It overlies the *mantle*, which in turn surrounds the *core*.

- *Hydrosphere:* The **hydrosphere** consists of the liquid or solid water of the Earth. Water takes the form of *surface water* (oceans, lakes, streams, rivers, swamps, and glaciers) and **groundwater**, the water that fills open holes and cracks underground in the upper several kilometers of the crust.

- *Cryosphere:* Sometimes, Earth scientists refer to the frozen component of the hydrosphere separately as the **cryosphere**. (The prefix *cryo–* comes from the Greek word for ice-like.) The cryosphere includes regions where glaciers cover the land surface, where the ground remains frozen all year, and where sea ice covers the ocean.

- *Atmosphere:* The **atmosphere** is a layer of gas that extends from the Earth's surface upward. We refer to the mixture of gases in the atmosphere as *air*.

Science toolbox . . .

Matter and energy

Think about your surroundings as you sit reading this book. If you're holding a paper copy of the book, you can heft its weight and you can see its words and pictures. The book has weight because it contains matter, and you can see its pages because it's being exposed to light, a form of energy. By **matter**, we mean material substance, the stuff that occupies space. The amount of matter in an object is its **mass**, so an object with greater mass contains more matter. (**Weight** refers to the force applied by a mass due to a gravitational pull.) If you were to look at a piece of matter that is greatly magnified, you would find that it consists of tiny particles called *atoms*, which may bond or stick together into *molecules*.

Scientists recognize different **states of matter**, meaning different ways in which matter behaves. In a **solid**, atoms or molecules remain fixed in position, so a solid retains its shape, regardless of changes in the size of its container **(Fig. BxP.2a)**. In a **liquid**, atoms or molecules can move relative to one another, but remain in contact, so a liquid can flow and conform to the shape of its container **(Fig. BxP.2b)**. In a **gas**, atoms and molecules move about freely, so not only can a gas flow, but it can also expand or contract when the size of its container changes. In other words, a gas will spread out and completely fill whatever container it occupies **(Fig. BxP.2c)**. We'll discuss a fourth state of matter, *plasma*, later in the book.

From your everyday experience, you know that not all objects of the same size weigh the same amount—it's harder to lift a block of rock than a block of Styrofoam. That's because the objects have different densities. **Density** refers to the amount of matter within a given volume or, put another way, the mass per unit volume. A volume of space that contains hardly any matter, meaning that its density approaches zero, is a **vacuum**. The word comes from the Latin *vacuus*, which means vacant. On Earth, scientists produce vacuums by pumping out the gas from a chamber. The space between stars is a profound vacuum, containing less matter than the best vacuums produced in a laboratory.

Energy can make matter move or change its character. For example, energy can cause a car to roll along the highway, transform a liquid into a gas, or produce glowing images on your television screen. Physicists distinguish between several forms of energy, including *kinetic energy*, the energy of motion; *potential energy*, the energy stored in a material that can be released later; *radiant energy*, the energy carried from one location to another in the form of *radiation* (such as light); and *thermal energy*, or *heat*, the energy that can cause a material to warm up.

Figure BxP.2 Three states of matter.

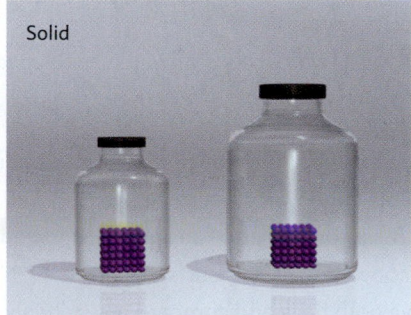

(a) A solid retains its shape and density.

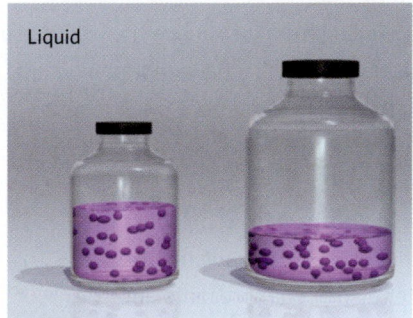

(b) A liquid conforms to its container, but retains its density.

(c) A gas fills its container, so its density can change.

• *Biosphere:* All living organisms on Earth, together with the portion of the Earth in which living organisms exist, constitute the **biosphere**. The biosphere encompasses the land surface, the sea, the uppermost several kilometers of the geosphere, and the lower several kilometers of the atmosphere. Life, as we know it, survives only under conditions in which water can remain in liquid form, so the bottom of the biosphere lies at depths where groundwater approaches boiling, and the top of the biosphere lies at elevations where water stays perpetually frozen.

How can I explain . . .

The difference between potential and kinetic energy

What are we learning?

That one kind of energy can transform into another.

What you need:

Two golf balls, two full cups of water, and paper towels.

Instructions:

- Drop one golf ball into the first cup from a couple of centimeters (about an inch) above the water's surface.
- Drop the second golf ball into the second cup of water from a meter (about one yard) above the water's surface.
- Remove both balls and compare how much water remains in each cup.

What did we see?

- The ball falling from just above the water's surface displaced only a small amount of water, whereas the ball falling from one meter up hit the water much harder and therefore displaced a lot more water.

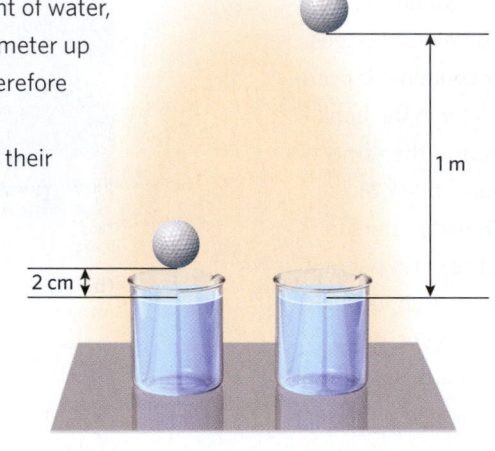

- The height of the balls represents their potential energy, and the amount of water displaced represents an amount of kinetic energy. The golf ball that fell from the higher position had more energy to start with because by raising the ball up a meter, you provided it with more potential energy. When it dropped, that potential energy was converted into kinetic energy. The ball that fell from the lower height had less potential energy to convert to kinetic energy by the time that it arrived at the water's surface, so it hit the water softly and displaced less water.

- The potential energy of the balls is due to gravity. An object being pulled on by gravity stores energy until it can move. The higher it is, the more it could move, so the more energy it holds.

Earth scientists refer to all the realms that we've just described, along with the beautifully complex ways in which the realms interact with one another, as the **Earth System** (**Earth Science at a Glance**, pp. 16–17). During a **cycle** in the Earth System, materials pass among different components of a given realm, or among different realms, over time. For example, during the *hydrologic cycle*, ocean water evaporates to become water vapor (a gas) in the air. This water vapor, in turn, condenses to form clouds, and then falls as rain or snow on the land or the sea. Once on land, the water may flow in streams, become part of a glacier, sink underground to become groundwater, or become incorporated into a living organism. Water in each of these *reservoirs* eventually moves into others and may ultimately end up back in the ocean.

The Universe and the Earth Are Very Old

Scientific research indicates that the physical Universe as we know it originated about 13.8 billion years ago. Our planet is much younger—its birth took place about 4.54 billion years ago. Nevertheless, the Earth is vastly older than recorded human history, which extends back only about 6,000 years. Scientists use the term **cosmologic time** (from the word *cosmos*, a synonym for Universe) for time intervals related to events in the history of the Universe and the term **geologic time** for time intervals related to the history of the Earth. The immense span of cosmologic time means that there's more than enough time for stars to form, mature, and die. Similarly, the immense span of geologic time means that there's more than enough time for mountain ranges to rise and then be removed by **erosion** (the grinding away or removal of material by flowing water, ice, or air), and for species of living organisms to appear and then go extinct. Because numbers used in discussing geologic or cosmologic time can be huge, researchers use the following abbreviations:

1 Ka = 1 thousand years ago
(Ka stands for *kilo-annum*)

1 Ma = 1 million years ago
(Ma stands for *mega-annum*)

1 Ga = 1 billion years ago
(Ga stands for *giga-annum*)

To discuss geologic time, geologists use the **geologic time scale**, which divides the Earth's history into *eons* and *eras* (**Fig. P.6**). We'll see in Chapter 9 that eras are separated into smaller subdivisions called *periods*. Some periods, such as the Jurassic, have familiar names in popular culture.

Plate Tectonics Provides the Foundation of Geology

The crust, along with the uppermost portion of the mantle, forms a relatively rigid shell called the **lithosphere**. This shell lies above the warmer, softer part of the mantle. According to the **theory of plate tectonics** (commonly called simply *plate tectonics*), the lithosphere is not intact, but rather consists of roughly 20 discrete pieces, called *plates*, which move relative to one another **(Fig. P.7)**. Plate interactions yield earthquakes, volcanoes, and mountain ranges.

Because of plate motions, the map of the Earth's surface constantly changes as continents move relative to one another and as ocean basins between them open and close. At certain times in the past, continents accumulated into supercontinents, which later broke up. At 200 Ma, for example, the Atlantic Ocean didn't exist, and a dinosaur could have walked from the location of what is now New York to the location of what is now Paris without getting its feet wet. Plate motions take place slowly—at rates between 1 and 15 cm (0.5 to 6 inches) per year—so they're almost imperceptible during the course of a human lifetime. Nevertheless, there's plenty of time for accumulated displacements to become large. In fact, the Atlantic Ocean grew to its present width in 180 million years, about the last 4% of Earth's history.

The Earth Is a Planet, and Each Planet Is Unique

The Earth is one of four planets, the ones closest to the Sun, which are known as the *terrestrial planets* and have solid, rocky surfaces. The outer four planets, known as the *Jovian planets*, consist of gas and various kinds of ice (such as frozen ammonia, methane, and water), and do not have rocky surfaces. No two planets look the same. Of the terrestrial planets, only the Earth has liquid water in abundance on its surface, an oxygen-rich atmosphere, and moving plates, and only the Earth harbors complex life. In recent years, astronomers have discovered more than 3,500 *exoplanets*, planets associated with stars other than our own. Some of these planets may be Earth-like in size, but whether any host moving plates, liquid water, or life remains unknown.

Various Sources of Energy Affect the Earth and Other Planets

The Earth is a dynamic place. Interactions among moving plates generate earthquakes, volcanoes, and mountains, while circulation of water and air erodes our planet's surface and redistributes materials (see Earth Science at a Glance, pp. 16–17). What sources of energy drive all this geologic activity? Plate motion, with all its consequences, happens because the Earth remains hot

Figure P.6 Time scales in Earth Science.

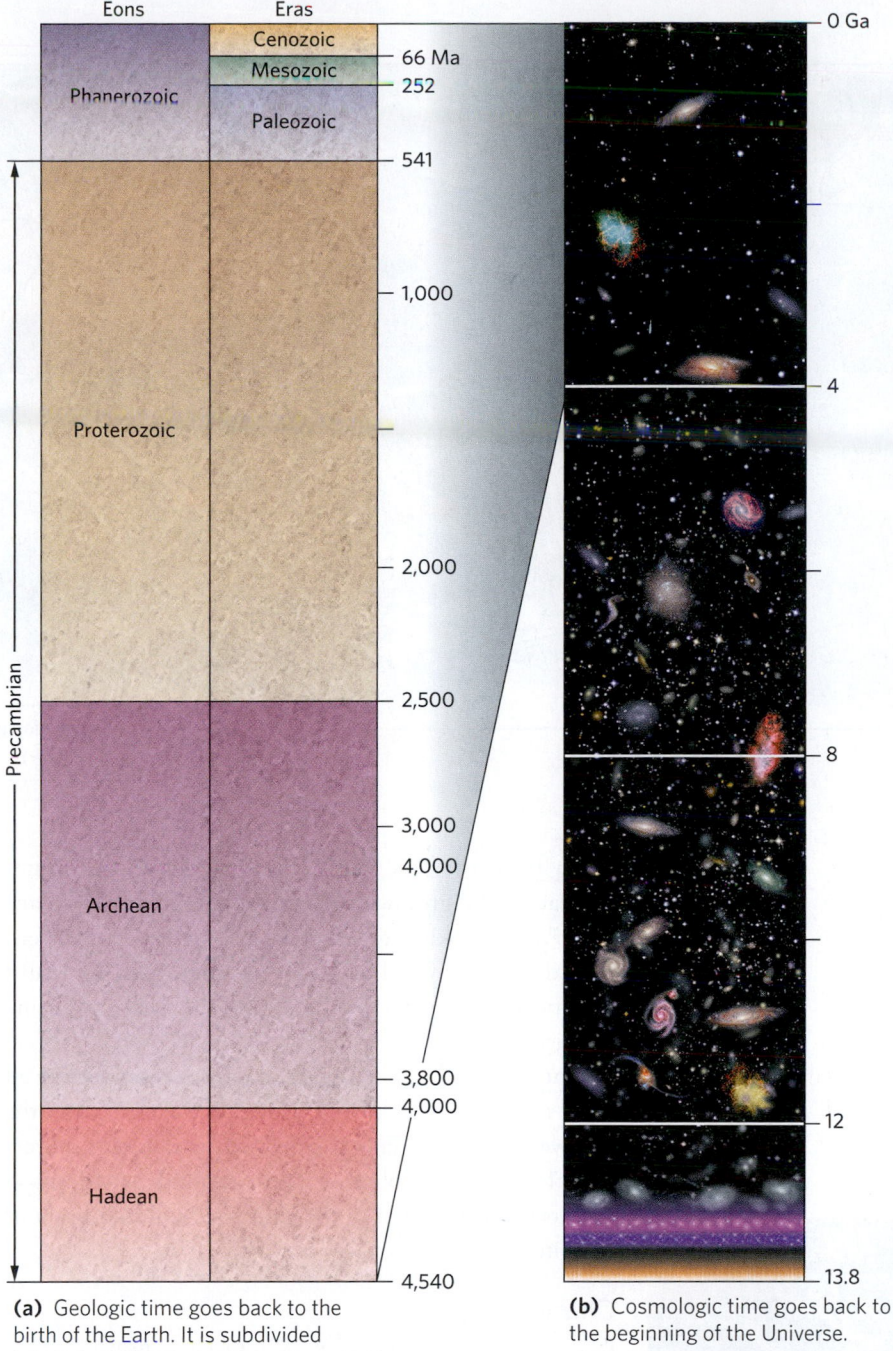

(a) Geologic time goes back to the birth of the Earth. It is subdivided into eons and eras.

(b) Cosmologic time goes back to the beginning of the Universe.

and relatively soft inside due to its *internal energy*. Some of this energy remains from the squeezing together, or *compression*, of materials during our planet's formation, and some comes from radioactive atoms. *Gravity*, the pull that one mass exerts on another, holds the Earth together, keeps it in orbit around the Sun, contributes to tides, causes rivers to flow, and makes rocks tumble downslope. *External energy*, which travels to the

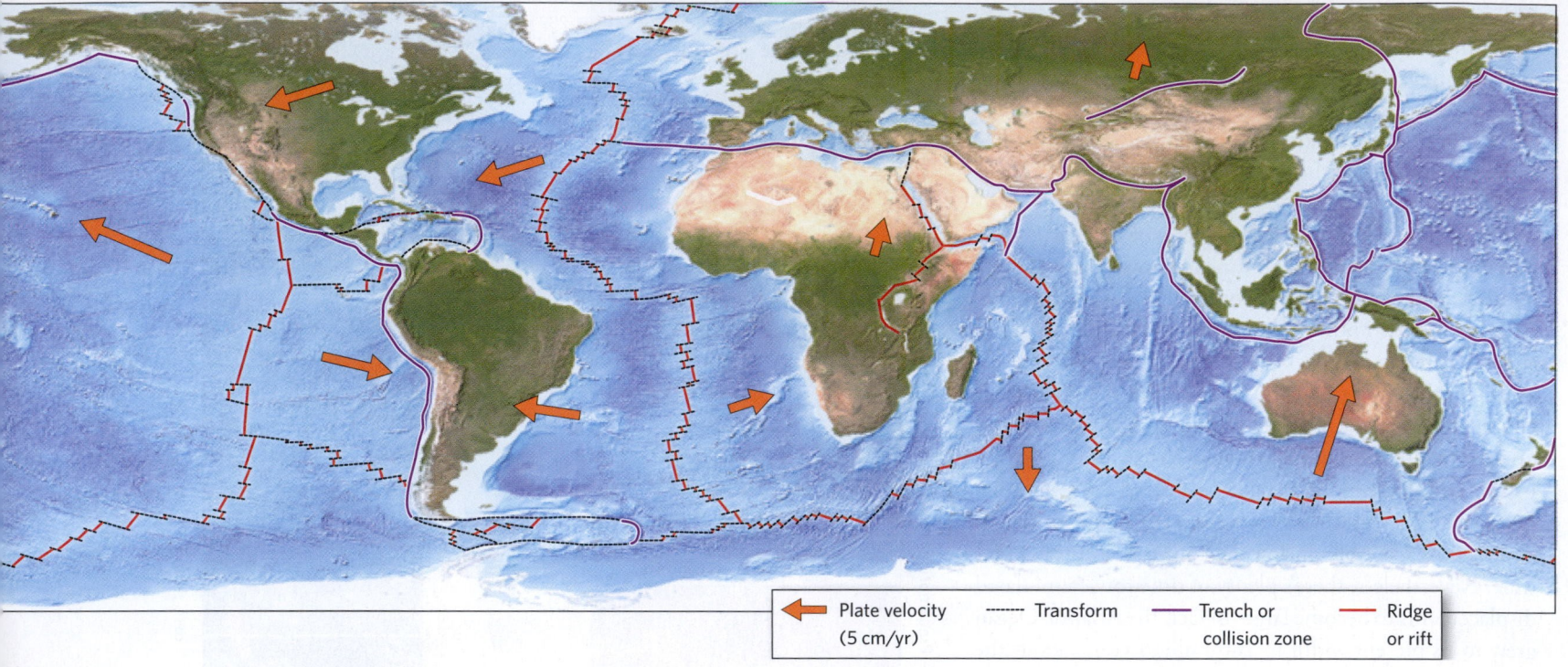

| → Plate velocity (5 cm/yr) | ----- Transform | — Trench or collision zone | — Ridge or rift |

Earth from the Sun in the form of light, contributes to heating the ground, the atmosphere, and the oceans **(Fig. P.8)**. Working together with gravity, external energy drives the wind and the currents, which in turn transport *thermal energy* (heat) and materials from one location to another.

Internal energy, gravity, and external energy affect all other objects of the Universe as well as the Earth. But the relative amounts of energy provided by different sources vary for different objects. We can understand many aspects of the behavior of the Universe by thinking about the interplay among energy sources.

Change Happens

In space, stars and planets form, objects collide, stars explode, and the Universe expands. On Earth, mountains rise, continents move, landscapes undergo erosion, the climate warms and cools, sea level rises and falls, and life evolves. Clearly, change happens everywhere—but rates of change vary immensely. For example, some stars have remained virtually unchanged since the Universe began, whereas others burned furiously for just a few million years and then blew up. On Earth, it takes hundreds of millions of years for continents to merge and drift apart, but a landslide may remove a mountainside in a matter of minutes, and an earthquake may last only seconds. If you

look at a satellite image of cloud patterns in the atmosphere, you can see that clouds circulate about the Earth over the course of days **(Fig. P.9a)**, yet hidden within some of those clouds may be a tornado, which can destroy everything in its path in less than a minute **(Fig. P.9b)**. And if you study the ocean, you'll see that some currents carry water across entire ocean basins over the course of months, but others swirl in nearly circular eddies over the course of days. Furthermore, 600 waves may arrive at a beach every hour **(Fig. P.9c)**.

Witness the aftermath of a large flood or hurricane, and it's clear that the might of natural phenomena can change human-made structures **(Fig. P.10a)**. But watch a bulldozer clear a swath of forest, hear dynamite blast apart a mountain, try to sail through rafts of garbage floating in the ocean, walk in a human-made canyon of skyscrapers, or measure the increasing carbon dioxide concentration in the atmosphere, and it's clear that people can play a key role in changing the Earth **(Fig. P.10b)**. When discussing change on the Earth, therefore, we distinguish between *natural change*, meaning change due to geologic, meteorological, or astronomical events, and *anthropogenic change*, meaning change that results from human activity. For all but the last 0.000001% of geologic time, all change that took place in the Earth System was natural. In the past few centuries, however,

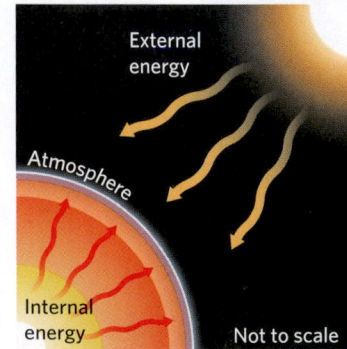

Figure P.8 The difference between internal and external energy in the Earth System.

our species has developed the ability to modify the Earth System in ways, and at rates, that do not happen naturally. The last few centuries of Earth history, because of human activities, differ so much from previous times that some scientists refer to them as a separate period, the *Anthropocene*.

Take-home message . . .

Many narrative themes tie together concepts in Earth Science. For example, matter and energy behave in predictable ways, the Earth is a complex system with interacting realms, the Universe and the Earth are so old that there's plenty of time for significant change to take place, plate tectonics causes geologic phenomena on Earth, and similar energy sources affect all objects in the Universe.

Quick Question --------------------------------

What examples emphasize that humans can now change the Earth at significant rates?

P.4 Why Study Earth Science?

Hundreds of thousands of people pursue careers as professionals in various aspects of geology, oceanography, atmospheric science, and astronomy. These careers are anchored in corporations, universities, government agencies, observatories, consulting firms, and nongovernmental organizations. Nevertheless, most students reading this book will not have such careers as a goal. So why should future teachers, construction workers, executives, volunteers, lawyers, office personnel, politicians, or doctors study Earth Science? This question has many answers.

Figure P.9 The atmosphere and oceans of the Earth System are in constant motion at all time and distance scales.

(a) A satellite view shows swirling regions of clouds.

(b) In a tornado, air swirls up a funnel at immense speed.

(c) Ocean waves crash on the rocky shore of a Hawaiian island.

Figure P.10 Natural change and anthropogenic change.

(a) The natural power of a hurricane can level a beachside community.

(b) The human-generated power of a chain saw and a bulldozer can level a forest.

The Earth System

Thunderhead

Lightning

Rain and snow

External energy

Mountain uplift

Continental glacier

City

Sun

Rocky coastline

Ocean

Desert

Valley

Arid mountains

Mining

Field pattern

Lakes

Deciduous forest

Forested mountains

Beach

Shark

Tropical rainforest

Coral reef

Internal energy

When you stand on the surface of the Earth, you can see the wondrous ways in which components of the Earth System interact. The geosphere consists of the solid part of our planet. You see it wherever you see exposed rock, sediment, or soil. Most of it lies underground, in the internal layers of our planet. The hydrosphere consists of all liquid water at or near the surface of the Earth. It fills oceans, lakes, underground pores, and occurs as gas in the atmosphere. The cryosphere consists of frozen water, mostly in glaciers. The biosphere consists of living organisms, from bacteria to whales. The atmosphere is the envelope of gas that encircles the planet. Flow in the air and sea transfer heat and water around the planet.

Internal energy rising from the interior and external energy coming to the Earth from the Sun keep the Earth System dynamic, so that materials cycle from component to component over time. Human society is having a growing impact on the Earth System, by extracting resources, building and farming on its surface, and emitting waste.

Cirrus clouds

Jet stream

Moon

Aurora

Ice and snow

Wind system

Coniferous forest

Evaporation

Volcanic islands

Industrial pollution

Cold surface current

Surface waters

Delta

Swamps

Warm surface current

Twilight zone

Abyssal zone

Whale

Seafloor

Bacteria and plankton

Giant squid

Deep-sea current

Black smokers

(a) Knowledge of Earth Science may help homeowners avoid building or buying homes in risky settings.

(b) The collapse of this building during an earthquake might have been prevented if engineers had taken into account the earthquake hazard of the region and designed the building to withstand shaking.

Earth Science Addresses Practical Issues

Civilization exists by geological consent, subject to change without notice.

—WILL DURANT (AMERICAN HISTORIAN, 1885–1981)

An Earth Science course may be among the most practical courses you can take. Let's consider a few examples of issues you may encounter that might be affected by what you learn in an Earth Science course.

- *Where you live:* Is the location of an apartment that you'd like to rent safe from flooding, landslides, or earthquakes **(Fig. P.11)**? Earth Science can help you to analyze risk due to natural hazards.

- *What you drink:* Does your community pump drinking water from wells? Earth Science can help you evaluate whether or not a new landfill project outside of town might contaminate your town's water supply.

- *What your weather will be:* Are you worried that thunderstorms may disrupt an upcoming activity? Earth Science can help you to use a radar display on your smartphone to determine if such weather phenomena might be approaching **(Fig. P.12)**.

Figure P.12 Storms can arrive suddenly, and can be a dangerous natural hazard.

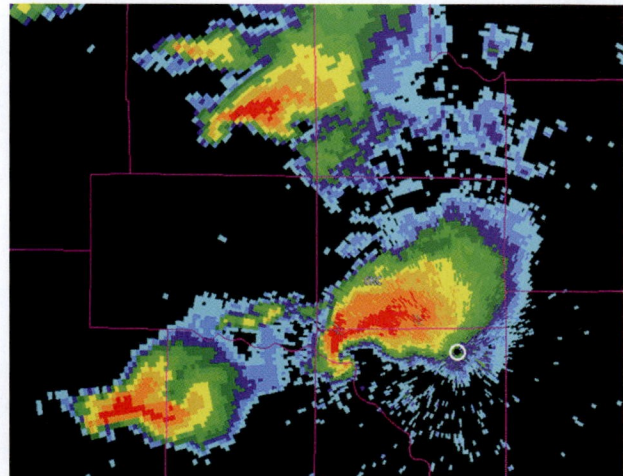

(a) A thunderstorm approaches, bringing with it wind, rain, and lightning.

(b) Radar images, as we'll see in Chapter 19, show the distribution of rain in a thunderstorm, because the radar beam reflects off of rain drops. Red indicates places where rain is more intense, blue indicates places where rain is less intense. Black areas are rain free. The width of the area shown is about 200 km (125 miles).

- *When to worry about natural hazards:* Do you get nervous when flying in an airplane? Earth Science can help you understand why planes shake when passing through certain types of clouds. Has someone announced that an earthquake, meteorite, or hurricane will strike the day after tomorrow? Earth Science can help you evaluate the reliability of such a prediction.

- *How to interpret issues with societal consequences:* Have you read news articles about climate change? Earth Science introduces the methods that researchers use to study the issue and describes the history of climate change over geologic time. This knowledge can help you understand predictions of climate change in the future.

We could continue adding to this list for pages. Hopefully, by the end of your course, you'll have insights into many Earth Science issues that could affect your livelihood, your family, your country, and your planet in the future. Because the Earth is our only home, all citizens of the 21st century have a stake in its future, and the study of Earth Science can help people make informed choices.

Earth Science Helps You Understand Our Context in the Universe

This book can give you a perspective on how your home planet, you, and all your surroundings came to be. You'll learn not only about the Earth, but also about the Sun, the planets of our Solar System, the Milky Way galaxy, and the hundreds of billions of other galaxies in the visible Universe **(Fig. P.13)**. You will come to appreciate the immense size of the Universe, and you will realize that, although you feel motionless, you, and everyone you know, are on an incredible journey through space.

Earth Science Addresses Resources and the Environment

We take it for granted that we can construct buildings, roads, and dams out of concrete; that we can go to a store and buy metal cooking pots or paper clips; that we can run a laptop computer for hours on a built-in battery; and that we can generate the power to light a city from materials pumped out of the ground. Do you ever wonder where all this "stuff" comes from? Where do we obtain the lime in concrete, the clay in bricks, the copper of pipes and wires, or the gasoline that powers a car? All of these resources come from the Earth **(Fig. P.14a)**. In this book, we'll discuss why certain resources occur where they do, how we can find them, and how we transform them into usable products. You may be in for some surprises—did you know, for example, that much of the concrete we use today comes from the shells of sea creatures that lived hundreds of millions of years ago?

Because Earth Science discusses the origin and supplies of resources, it provides a basis for understanding **sustainability**, the question of whether we can continue to maintain or improve our standard of living without running out of resources. You'll learn why some resources are *renewable*, in that natural processes replace them in years or decades, while others are *nonrenewable*, meaning that the rate at which we consume them exceeds the rate at which natural processes produce them. In addition, you'll see why many activities of society impact *environmental quality* (the breathability of air, the drinkability of water, and the ability of a region to support life) and may change the character of the Earth's future climate **(Fig. P.14b)**.

Figure P.13 Earth Science may help you understand the context of your home planet in space. This view of the Earth, taken by an astronaut orbiting the Moon, emphasizes that we live on an island in space.

Figure P.14 The activities of human society can impact environmental quality and the climate.

(a) At a coal mine, miners excavate energy.

(b) Gas emitted from coal-burning power plants in China.

Earth Science Helps You Appreciate Your Surroundings

When you finish reading this book, we hope your view of your surroundings is forever colored by scientific curiosity. When you drive down a highway, roadcuts through rocks will no longer be merely faceless cliffs, but rather will be open books telling a story of Earth's long history **(Fig. P.15)**. When you stand on the beach, you won't just enjoy a pretty view, but may ponder the dynamic and ever-changing interface between land and sea. When you gaze at clouds drifting by on a windy day, you won't just see puffs of white, but will think about what they're telling you about a change in the weather. And when you look at the night sky, you'll know that the stars aren't just points of light, but rather are suns lighting their own solar systems, and that these very distant solar systems just might host planets like our home, Earth.

Now, it's onward to Earth Science. We'll begin the book at the beginning, with the formation of the Universe and the elements of which it is made. With this background, we will then approach our own planet, and describe its basic architecture, before moving on to the grand unifying theory of geology, plate tectonics. Part I of the book covers topics related to the land, the materials below it, and the features formed on its surface. Subsequent parts address the oceans, the atmosphere, and finally, space. Enjoy your tour!

Take-home message . . .

The study of Earth Science can help you understand many practical issues you will face in life, such as whether your home is subject to natural hazards. It can also help you to appreciate your surroundings and our place in the Universe, and to understand the context of sustainability and environmental issues.

Quick Question --------------------------------
What factors may influence the long-term safety of a home site?

Figure P.15 Geologic exploration provides not only beautiful views, but a look into the Earth's past. If you were to hike by the rock exposure shown on the left, it might look like "just a cliff of tan rock." To a geologist, however, this cliff reveals many events in Earth's history: sand was transformed into sandstone layers (beds), and the beds were bent, due to plate motion, into curves called folds.

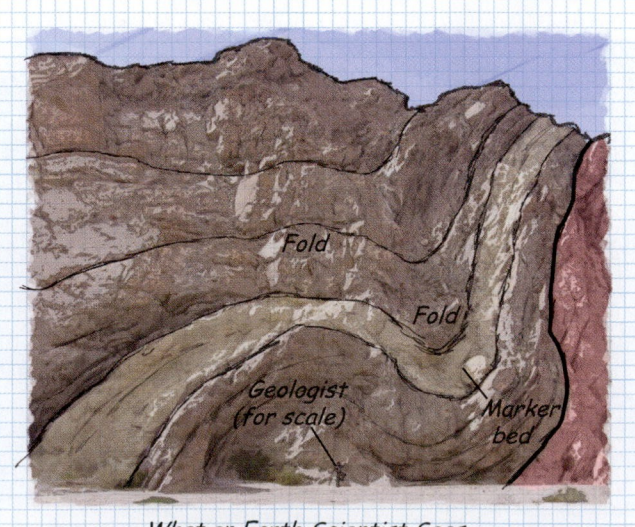

Fold

Fold

Geologist (for scale)

Marker bed

What an Earth Scientist Sees

PRELUDE REVIEW

- Earth Science is a composite of many disciplines, including geology, oceanography, and atmospheric science. An Earth Science course also typically includes astronomy.

- There are many kinds of scientists—some work in the field, some in the lab, and some in the office. They try to interpret the natural world on the basis of observations, measurements, and calculations, generally by following the scientific method.

- The Universe—everything that exists, both on Earth and in all of space—contains matter and energy, which behave predictably, according to physical laws.

- Our home planet can be envisioned as a complex system, the Earth System, which includes the geosphere, hydrosphere, cryosphere, atmosphere, and biosphere.

- Scientific measurements indicate that the birth of the Universe happened 13.8 billion years ago, while the Earth formed 4.54 billion years ago. Geologic time, the time since Earth's formation, has been divided into named increments.

- The outer shell of the Earth, the lithosphere, is divided into about 20 plates that move relative to one another. This theory of plate tectonics explains earthquakes, volcanoes, and mountain building.

- Planets are affected both by energy coming from their interior and by energy coming from the Sun. On Earth, internal energy drives plate tectonics, while external energy drives the flow of oceans and the air.

- The Earth is a dynamic planet on which change is the norm. Because geologic time is so long, the Earth has been able to change in major ways since its birth.

- Earth Science helps you understand features and processes that affect the environment and addresses many practical issues, such as where resources come from and how seriously to consider predictions of natural disasters.

- Most of all, Earth Science will help you appreciate and understand your surroundings, from the ground beneath your feet to the stars in the sky above.

Key Terms

astronomy (p. 5)
atmosphere (p. 10)
atmospheric science (p. 5)
biosphere (p. 11)
climate science (p. 5)
cosmologic time (p. 12)
cryosphere (p. 10)
cycle (p. 12)
data (p. 9)
density (p. 11)
Earth Science (p. 5)

Earth System (p. 12)
energy (p. 11)
erosion (p. 12)
gas (p. 11)
geologic time (p. 12)
geologic time scale (p. 12)
geology (p. 5)
geosphere (p. 10)
groundwater (p. 10)
hydrosphere (p. 10)
hypothesis (p. 9)

liquid (p. 11)
lithosphere (p. 13)
mass (p. 11)
matter (p. 11)
meteorology (p. 5)
oceanography (p. 5)
research (p. 6)
science (p. 5)
scientific law (p. 9)
scientific method (p. 8)
scientist (p. 6)

solid (p. 11)
states of matter (p. 11)
sustainability (p. 19)
theory (p. 9)
theory of plate tectonics (p. 13)
Universe (p. 5)
vacuum (p. 11)
weight (p. 11)

Review Questions

The letters following each Review Question refer to the corresponding Learning Objective from the Chapter Opener.

1. What are the various subjects that an Earth Science course might cover? **(A)**

2. Describe the different components of the Earth System. Name the three principal layers of the geosphere. **(C)**

3. What are the various kinds of energy that play a role in driving the activity of the Earth System and causing aspects of the system to change over time? What is a "cycle" in the Earth System? **(C)**

4. Describe some of the ways you might use your understanding of Earth Science to address problems that you might face in your community. **(D)**

5. Explain the difference between cosmologic and geologic time. Imagine that a continent moves at 5 cm per year. How long would it take to move 2,000 km? What percentage of geologic time does this represent? **(F)**

6. Name some of the practical applications of Earth Science. How can Earth Science help you understand sustainability and environmental issues? **(D)**

7. This mine truck carries 100 tons of coal. Where does this resource, and others like it, come from? **(F)**

PART I

Our Dynamic Planet

In Part I, we begin our introduction to geology, the study of the Earth. Chapter 1 discusses how the Earth, and the elements that comprise it, formed. We then turn our attention to the Theory of Plate Tectonics—the grand unifying idea that provides a foundation for understanding how and why geologic processes operate. In Chapters 3–6, we introduce the materials that make up the solid Earth–minerals (the building blocks of rock), then igneous, sedimentary, and metamorphic rocks. This background allows us to discuss the Earth's dynamic behavior—mountain building and earthquakes for example (Chapters 7–8). Chapters 9–10 introduce the Earth's history and the methods that geologists use to study it, and we then conclude Part I by applying our knowledge of the Earth's materials, processes, and history, to the more practical side of geology—the search for energy and mineral resources.

The Matterhorn, a peak in the Alps of Switzerland and France, consists of rock that once lay kilometers below the surface. Some of the rocks are from continental crust, some from the ocean floor. Our modern understanding of how these rocks formed and evolved, and how and when they moved to their present position in this mountain range, has its foundation in the theory of plate tectonics.

23

FROM THE BIG BANG
TO THE BLUE MARBLE

By the end of the chapter you should be able to . . .

A. provide the scientific explanation for the origin of the Universe and the elements in it.

B. describe how, according to the nebular theory, the Earth formed.

C. interpret features that a space probe would detect when approaching the Earth from space.

D. classify the great diversity of materials that the Earth contains.

E. create a model of the basic internal layers of the Earth.

1.1 Introduction

Sometime in the distant past, perhaps more than fifty thousand generations ago, our ancestors developed the capacity for conscious thought. This amazing ability led them to pose broad questions about everything that physically exists both in and on the Earth and throughout space beyond—about the **Universe**. Why is there land and sea? What shaped the mountains and set the courses of rivers? What are the objects that cross the sky at night? In early cultures, such musings spawned legends in which heroes with supernatural powers sculpted the landscape or pushed stars across the heavens. In recent centuries, scientific research based on observations, experiments, and calculations has led to new ideas that explain characteristics of the Earth and the Universe, and how they have changed over time.

In this chapter, we set the stage for discussing broad questions about our home planet and its context by focusing on three goals. First, we introduce aspects of *scientific cosmology* (study of the overall structure, formation, and evolution of the Universe and all objects within it), with a focus on understanding the nature and origin of the materials that make up the Earth. Second, we characterize the overall architecture of the Earth System by taking an imaginary journey aboard a space probe that rockets from interplanetary space to the Earth and then goes into orbit. And third, we explore the Earth's interior, from its surface to its center. We revisit these topics in greater detail in later chapters.

Research by astronomers, physicists, and geologists suggests that our Solar System formed when gravity pulled gas, dust, and ice into a disk. The central ball became so large and dense that it turned into a nuclear furnace, our Sun. Farther out, matter accumulated into planetesimals, which then collided. The largest planetesimals cleared their orbits of debris and eventually became planets.

Box 1.1

Science toolbox . . .

What is a force?

It's hard to talk about the behavior of the physical Universe without using the concept of a force. Simplistically, we can think of a **force** as a push or pull acting on an object. Physicists define force by an equation, $F = ma$, or in words, force equals mass times acceleration. An *acceleration* is a change in the velocity (speed or direction) of an object's movement, so the equation means that application of force can cause an object to move at a different speed and/or in a different direction than it was once moving. The larger the force, the greater the amount of change it can cause.

Physicists distinguish between contact forces and field forces. You apply a **contact force** when you push or pull on one object with another. To picture a contact force, imagine a bat striking a ball—the impact causes the ball to head off in a different direction. A **field force** emanates from an object and can act at a distance from an object, even in a vacuum. **Gravity**, for example, is a field force in that it is an invisible attraction between two objects and can act across a distance. Gravity keeps the Earth in orbit around the Sun, even though no visible rope or chain connects the Earth to the Sun. **Electromagnetism** is a field force produced by magnetic objects or by electric currents. Unlike gravity, electromagnetic force can be either attractive, pulling objects together, or repulsive, pushing objects apart. A magnet, for example, can attract or repel another magnet at a distance.

A *force field* is the region affected by an invisible push or pull created by a force. The Earth lies within the gravitational field of the Sun, and you are standing in the gravitational field of the Earth. Magnets also produce force fields, as you'll see when you move a magnet close to a metal paper clip: when the magnet gets close enough, the clip moves toward the magnet.

Figure Bx1.1 Examples of field forces in every day life.

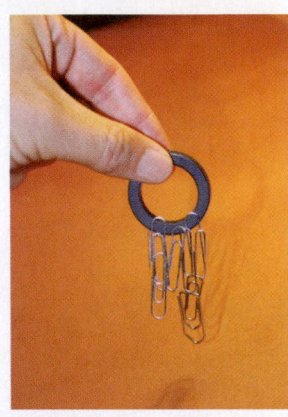

(a) Gravity, a field force, pulls a person down a zip line.

(b) A magnet produces a field force sufficient to hold onto these clips.

Figure 1.1 The Milky Way, as seen in the dark night sky of Washington state.

1.2 A Basic Image of the Universe

Think about the spectacle of a clear night sky and imagine how incredibly mysterious it must have seemed to human observers in the days before telescopes and spacecraft. At times, the Moon dominates the night sky, and its white glow hides dimmer objects. But on a moonless night, away from the glare of a city, you can see countless sparkling points and spots of light, and you might even see a band of faint, glowing haze arcing from horizon to horizon **(Fig. 1.1)**. What are these **celestial objects**—these objects and features visible in the night sky—and how are they arranged?

Introducing Stars, Galaxies, and Nebulae

Prior to the Renaissance, most people assumed that the Earth lay at the center of the Universe and that all celestial objects moved around the Earth. This view changed when new methods for studying space became available. Scientific observations demonstrated that our **Sun**, rather than being a chariot of fire driven across the sky by a god, is an incredibly immense, amazingly hot ball of gas that emits

(a) The Sun, like all stars, radiates immense amounts of energy.

(b) A view of the Andromeda Galaxy.

Figure 1.2 Stars and galaxies.

vast amounts of energy into the space around it **(Fig. 1.2a)**. What about the tiny points of light that twinkle in the night sky? Each point of light whose position relative to the others remains fixed night after night is called a **star**. A star is a sun, like our Sun, but much farther away. Stars cluster in immense groups, called **galaxies**. Our Sun, along with over 300 billion other stars, belongs to the **Milky Way Galaxy**. The individual stars that we can see lie within the nearer parts of our galaxy. The rest of its stars, all together, constitute the glowing band that spans a dark night sky (see Fig. 1.1). Viewed from a great distance, the Milky Way would look like a giant, slowly swirling, flattened spiral similar in shape to its neighbor, the Andromeda Galaxy **(Fig. 1.2b)**. Our Sun, near the outer edge of a curving arm of the Milky Way, slowly revolves around the center of the galaxy, making a full circuit every 250 million years. Astronomers estimate that the Universe contains hundreds of billions of galaxies.

Scientific observations also demonstrate that the Earth is a planet, one of eight associated with our Sun. In the night sky, planets stand out because, over the course of many nights, we can see them move relative to the backdrop of stars. A **planet**, by modern definition, is a large spherical mass that follows a path, or **orbit**, around a star and has incorporated all the matter that lies in or near its orbit. All the planets that orbit our Sun lie in roughly the same plane.

Astronomers refer to the four planets lying nearer to the Sun as the *terrestrial planets* because they resemble the Earth. These planets are Mercury, Venus, the Earth itself, and Mars. All are relatively small (the Earth is the largest) and have rocky surfaces. The four planets that lie farther from the Sun are relatively large and consist of gas and ice (in this context, ice is a solid form of materials, such

as water and methane, that can exist as gas at the Earth's surface). These planets—Jupiter, Saturn, Uranus, and Neptune—are known as the *Jovian planets* after Jupiter, the largest of them. All the planets, except Mercury and Venus, have moons. A **moon** is a sizable solid object that orbits a planet. The Earth has one (we call it simply the Moon), whereas Jupiter and Saturn each have more than sixty. Moons range in size from larger than the planet Mercury to less than 10 km across. The Sun, the planets, and the moons, as well as various other smaller objects grasped by the gravitational force of the Sun **(Box 1.1)**, together constitute the **Solar System (Fig. 1.3)**.

When we see the Sun, or any star, we are seeing light that has traveled through space. (As we will see later, visible light is one of many types of *electromagnetic radiation*; other types include radio waves, X-rays, and ultraviolet light.) The distances between stars and between galaxies are so immense that, to avoid using colossal numbers, astronomers specify these distances in terms of light-years. One **light-year** represents the distance that electromagnetic radiation, such as light, travels in one year. Because the **speed of light** is so incredibly fast—300,000 km (186,000 miles) per second—one light-year represents a huge distance, roughly 9.46 trillion kilometers (5.88 trillion miles). The nearest star to our Sun lies 4.2 light-years away, and Andromeda lies over 2.5 million light-years away. This means that when we look at Andromeda with a telescope, we are seeing light that began its journey toward Earth 2.5 Ma, before *Homo sapiens*, our own species, even existed. In effect, we are seeing back in time.

What lies between the stars? Most of space consists of a **vacuum**, a volume that contains virtually no matter. To get a sense of the profundity of this vacuum, keep

Science toolbox . . .

The components of matter

We've already introduced the concept of matter because most students have a general sense of what this terms means from earlier science courses. But in order to talk about the formation of the Universe and the stars and planets within it, it's necessary to clarify the specific definitions of elements, compounds, and the particles that it is made of. Here's a quick review.

Elements, Atoms, and Nuclear Bonds

An **element** is piece of matter that cannot be subdivided into other components with different properties. Copper, for example, fits the definition because we cannot subdivide copper into different materials. In contrast, salt is not an element because chemists can separate salt into other materials (sodium and chlorine), neither of which looks or behaves anything like salt. If you keep subdividing an element into smaller and smaller pieces with the same properties, you end up with an **atom**, the smallest piece of an element that has the properties of that element. We refer to different elements using abbreviations called *chemical symbols*: for example, H = hydrogen; He = helium; C = carbon; O = oxygen; and Fe = iron.

It wasn't until the early 20th century that researchers realized that atoms themselves can be subdivided into *subatomic particles*: **protons** that have a positive charge, **neutrons** that have a neutral charge, and **electrons** that have a negative charge **(Fig. Bx1.2a)**. (Simplistically, the term *charge* refers to the electrical behavior of a particle. Think of a positive charge as the + end of a battery and a negative charge as the – end of a battery.) Opposite charges attract, meaning they pull toward each other, whereas like charges repel, meaning they push away from each other. Neutrons weigh slightly more than protons, and both are much larger than electrons. Specifically, a proton is about 1,836 times the size of an electron.

Protons and neutrons stick together to form a dense ball, the **nucleus**, at the center of an atom (see Fig. Bx1.2a). Electrons swirl around the nucleus in an *electron cloud*—the outer edge of the electron cloud is what defines the surface of an atom. The nucleus of an atom is very tiny compared with the complete volume within the surface of the electron cloud, so atoms actually consist mostly of empty space.

The attractive force that holds particles together in a nucleus is called a *nuclear bond*. Nuclear bonds act only over extremely small distances, so in order for **nuclear fusion**—the merging of two nuclei—to take place, the nuclei must be stripped of their electron shells and must collide at high velocity **(Fig. Bx1.2b)**. When this happens, the nuclei get close enough for nuclear bonds to overcome electrical repulsion.

Figure Bx1.2 Atoms and nuclear bonds.

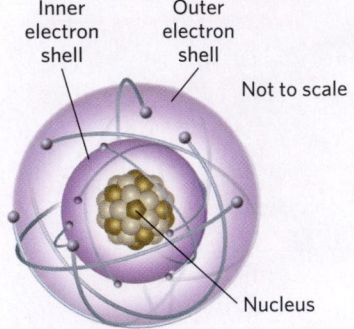

(a) An image of an atom with a nucleus orbited by electrons.

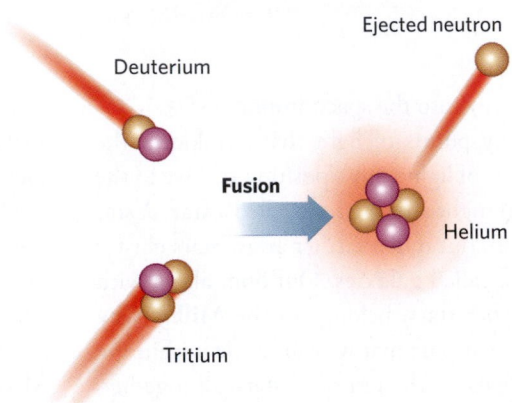

(b) Fusion happens when colliding particles combine to form a new, larger nucleus. (Deuterium and tritium are versions of hydrogen nuclei.)

Modern studies have shown that atoms are so small that a single gram (0.035 ounces) of hydrogen contains about 602,000,000,000,000,000,000,000 atoms. To picture how tiny atoms are, keep in mind that 5 to 10 trillion atoms fit within the area of the period at the end of this sentence.

Atomic Number and Atomic Weight

An atom of one element differs from an atom of another element because it contains a different number of protons. Scientists refer to the number of protons in an element's nucleus as its *atomic number*. Hydrogen has an atomic number of 1, helium has an atomic number of 2, carbon has an atomic number of 6, and iron has an atomic number of 26. Uranium, with an atomic number of 92, is the largest naturally occurring atom. When nuclear fusion takes place, two nuclei with small atomic numbers bond together to form a new nucleus with a larger atomic number; the resulting atom is a different element.

The number of protons plus the number of neutrons approximately represents the *atomic weight* (atomic mass) of an atom. The sum is not exactly the same as the mass because neutrons are slightly more massive than protons and because electrons also have mass. (In detail, atomic weight and atomic mass have slightly different meanings.) All atoms, except for the most common form of hydrogen, contain neutrons. Therefore, the atomic weight of an atom exceeds the atomic number. Small atoms generally have the same number of neutrons as protons, whereas large atoms contain more neutrons than protons.

Molecules, Compounds, and Chemical Bonds

Atoms can attach to other atoms by an attractive force called a *chemical bond*. Scientists use the word **molecule** for a particle composed of two or more atoms bonded together. Chemical bonds, unlike nuclear bonds, involve the interaction of electron clouds, not nuclei. Therefore, the formation of a chemical bond does not change the identity of the elements involved. Some molecules contain atoms of only one element, whereas others contain atoms of two or more different elements. Chemists refer to a substance in which molecules consist of two or more elements as a *compound*.

We can specify the composition of a molecule by providing its *chemical formula*, a "recipe" that uses abbreviations to represent the elements in a compound and their relative proportions. For example, hydrogen gas consists of H_2 molecules; oxygen gas of O_2 molecules; water of H_2O molecules; and methane of CH_4 molecules. Methane's chemical formula indicates that it contains one carbon atom for every four hydrogen atoms. Note that a molecule is the smallest piece of a compound that has the characteristics of that compound.

in mind that a cubic meter of air at sea level on Earth contains 300,000,000,000,000,000,000,000 atoms or molecules (Box. 1.2). By pumping air out of a closed container in a laboratory, scientists can produce a partial vacuum, but the best one ever made still contains about 100,000,000 atoms or molecules per cubic meter. In contrast, the empty space between stars contains only about 3 atoms or molecules per cubic meter.

Although most of the region between stars is a profound vacuum, enough matter has gathered in some places to form visible clouds of gas, ice, and dust (tiny solid particles), called **nebulae**. Nebulae come in many shapes and sizes (Fig. 1.4). The density of matter in a nebula may be 1,000 to 10,000 times that of matter in the vacuum of space, but it's still vastly less than that of the air you breathe.

The Expanding Universe

In the 1920s, astronomers such as Edwin Hubble braved many a frosty night beneath the open dome of a mountaintop observatory in order to chart distant galaxies too faint to be seen with the naked eye, but visible through

Figure 1.3 The relative positions of planets in the Solar System. Note that the orbits all lie in the same plane, which has been highlighted.

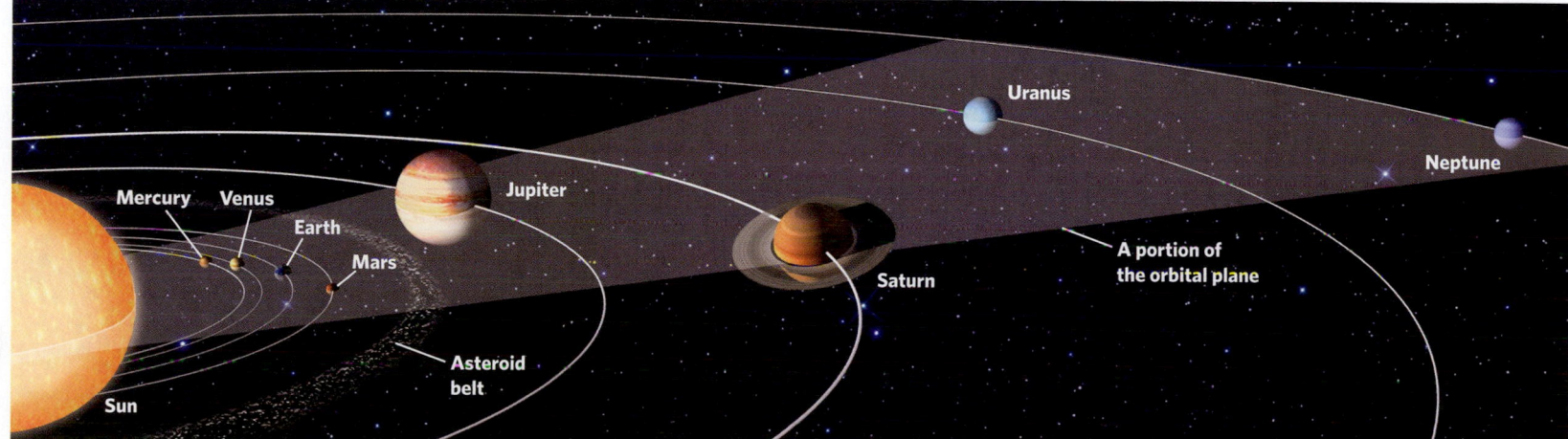

Figure 1.4 The Orion Nebula. The Hubble Space Telescope captured this image of the Orion Nebula, a cloud of gas and dust 24 light-years across, in which new stars are forming.

large telescopes. At first, they hoped only to document the locations and shapes of the galaxies. But eventually, these astronomers learned how to determine whether the galaxies were moving toward or away from the Earth, and to calculate how fast they were moving (see Chapter 23). Hubble pondered the results of these measurements and realized that all distant galaxies—regardless of their direction from the Earth—are moving away from us at great velocity. He deduced that this phenomenon could happen only if the Universe were expanding. Up until Hubble's eureka moment, astronomers assumed that the size of the Universe remained unchanged. To picture the expansion of the Universe, imagine a ball of dough that you've just mixed to make raisin bread. The ball is compact, and the raisins within it lie close together. When you put the dough in a warm oven, the dough starts to expand **(Fig. 1.5)**. During this process, each raisin moves away from all of its neighbors.

Hubble's hypothesis eventually passed so many tests that it gained the status of a theory (see Box P.1), now known as the **expanding Universe theory**. It led astronomers to realize that the Earth is merely one planet in one of over 300 billion star systems in our galaxy, and that more than a trillion galaxies are zooming away from one another through the unfathomable distances of space.

Take-home message . . .

The Earth is one of eight planets orbiting the Sun in the Solar System, which lies at the edge of the Milky Way Galaxy. There are more than a trillion galaxies in the Universe. Despite the immense numbers of objects it contains, the Universe is made up mostly of empty space, a vacuum. All distant galaxies are hurtling away from the Earth at high velocity, so the Universe is expanding.

Quick Question -
"When you look at the night sky, you are seeing the past." Why?

Figure 1.5 A raisin-bread analogy for the expanding Universe. As the dough expands, each raisin moves farther away from the others.

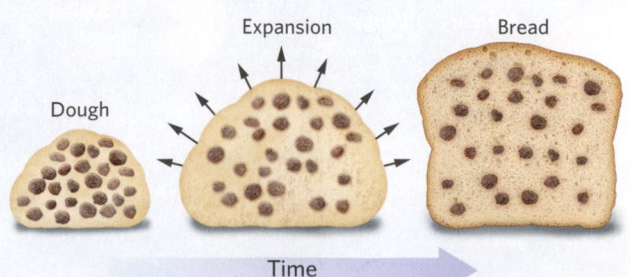

Dough Expansion Bread

Time

1.3 Formation of the Universe and Its Elements

The Big Bang

Hubble's expanding Universe theory triggers the following question: Did the expansion of the Universe begin at a specific time in the past? If so, the instant that expansion began would mark the beginning of our physical Universe. Astronomers conclude that expansion did indeed begin at a specific instant, which they picture as a cataclysmic blast called the **Big Bang (Fig. 1.6).** According to this idea, which has been supported by so much evidence that it has become a theory, all the matter and energy that we can see today—everything that now constitutes our Universe—initially existed within an infinitesimally small point. The point exploded in the Big Bang at about 13.8 Ga. Of course, no human was present during the Big Bang, so no one actually saw it happen. But by combining clever calculations with careful observations, researchers have developed a model of how the Universe evolved once the explosion had happened.

During the first instant after the Big Bang, the Universe was so small, so dense, and so hot that it consisted entirely of energy. Within a few seconds, however, the smallest atoms, hydrogen atoms, began to form. Only a few minutes later, when the Universe's temperature had fallen below 1 billion °C (1.8 billion °F) and its diameter had grown to over 50 million km (35 million miles), most of the hydrogen atoms that exist today had formed. For the next several minutes, some of the hydrogen atoms collided and underwent **nuclear fusion** (binding together of atomic nuclei) to form helium atoms as well as tiny amounts of other atoms with small atomic numbers. By the time the Universe was a mere 20 minutes old, virtually all of the new atomic nuclei that would ever appear had formed. This process, known as **Big Bang nucleosynthesis**, yielded mostly hydrogen (74% by mass) and helium (24% by mass). As time progressed, chemical bonds linked hydrogen atoms together into H_2 molecules (see Box 1.2).

Birth of the First Stars: The Nebular Theory

By the time the Universe reached its 200 millionth birthday, most of its H_2 and He gas had collected into immense, slowly swirling nebulae **(Fig. 1.7).** Eventually, at many places in the young Universe, a patch of nebula cooled sufficiently that gravity could cause the gas in the patch to collapse inward **(Box 1.3).** When this happened, atoms and molecules moved closer together, and the collapsing patch became progressively denser. Furthermore, as the gas compacted into a smaller region, its initial

Figure 1.6 An artist's rendering of Universe expansion from the Big Bang through the present and on into the future.

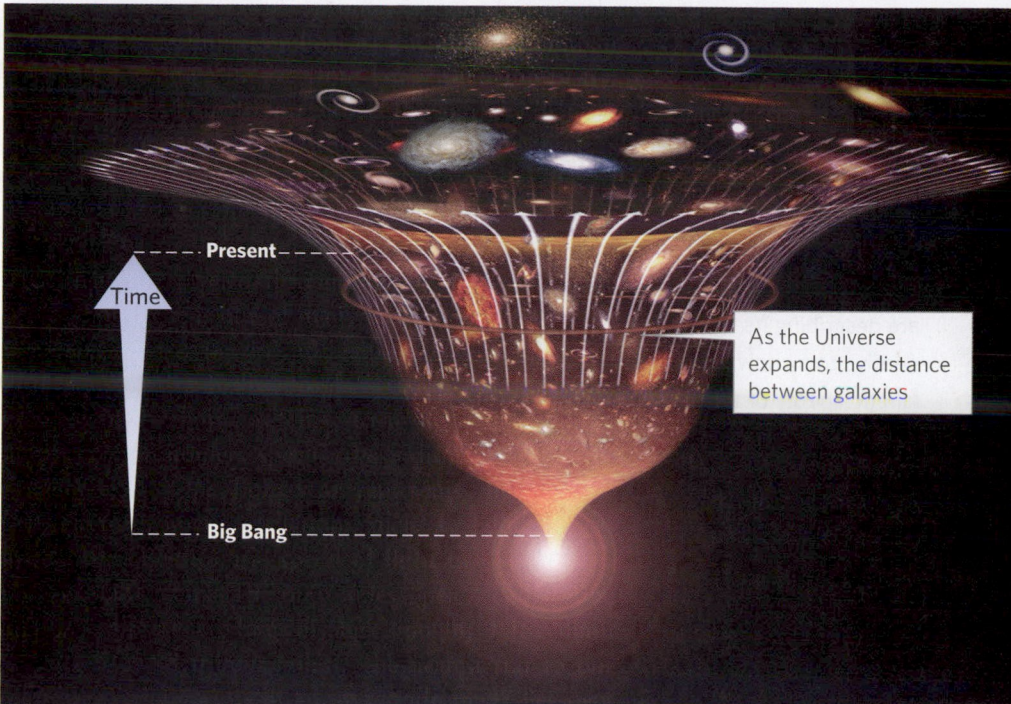

Present

Time

As the Universe expands, the distance between galaxies

Big Bang

Figure 1.7 An early nebula might have looked like this, when the first star formed. It's mostly swirling gas.

Box 1.3 ▶ **Science toolbox . . .**

Heat and temperature

The atoms and molecules that make up a solid object do not stay rigidly fixed in place; instead, they vibrate. Atoms in a liquid or gas not only vibrate, but can also move about with respect to one another. All this movement represents **thermal energy**, defined as the total kinetic energy of all the substance's atoms in a given volume.

When we say that one object is hotter or colder than another, we are describing its temperature. **Temperature** is a measure of warmth relative to a known reference value, and it represents the average kinetic energy of atoms in the material. To specify temperature—that is, to indicate relative hotness or coldness—scientists use the **centigrade scale** (also called the *Celsius scale*), which is divided into *degrees* (°C). Scientists have calibrated the centigrade scale so that the boiling point of water at sea level is 100°C and its freezing point is 0°C at sea level. An increment of 1°C, therefore, equals 1/100 of the temperature difference between the boiling and freezing points of water. The English system uses the **Fahrenheit scale** for temperature. On this scale, water boils at 212°F and freezes at 32°F at sea level. Note that a degree on

the Fahrenheit scale is smaller than a degree on the centigrade scale—the former is 5/9 of the latter.

Scientists commonly use the *Kelvin scale,* named for Lord Kelvin (1824–1907), a British physicist. On the Kelvin scale, the coldest possible temperature, called *absolute zero,* is designated as 0 K. (Note that when discussing measurements on the Kelvin scale, we do not use the degree sign.) A material simply can't get colder than absolute zero, meaning that you cannot extract any thermal energy from a substance at 0 K—in other words, the motion of its atoms and molecules has effectively ceased. The increment of temperature on the Kelvin scale has the same magnitude as in the centigrade scale. 0 K equals –273.15°C, so on the Kelvin scale, the freezing point of water is 273.15 K, and the boiling point of water is 373.15 K.

We use the term **heat** in reference to thermal energy transferred from one object or region to another. Heat always flows from a location of higher temperature to a region of lower temperature. For example, a freezer doesn't add cold to water to make it freeze, but rather removes heat. When any material warms, its atoms or molecules move faster, and when a material undergoes compression without losing thermal energy, it heats up.

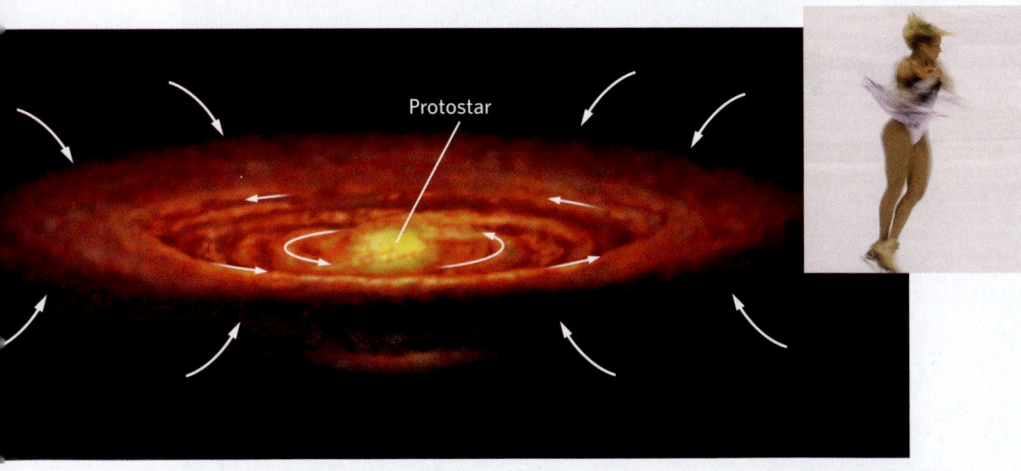

Figure 1.8 Artist's rendition of a star forming as matter falls into a rotating accretionary disk. As the mass grows, the orbital velocity increases, as occurs when a spinning skater pulls in her arms.

swirling motion transformed into rotational motion around an axis, and as more and more gas moved inward, the rotation rate increased. (This increase is an example of a phenomenon that physicists refer to as the *conservation of angular momentum*—the same thing happens when a spinning ice skater pulls her arms inward.) Because of its increase in rotation rate, a collapsing patch of nebula evolved a disk shape with a bulbous center **(Fig. 1.8)**. Astronomers refer to such a disk as an *accretionary disk*

because matter *accretes,* or attaches, to the disk over time. (Chapter 23 describes this process in more detail.)

What happened next? When enough matter had attached to the accretionary disk, gravity caused the central ball of gas to collapse even more, so that the gas within it underwent extreme compression and became very dense. Gas heats up when compressed, so the temperature within the central ball of gas increased dramatically, until eventually the ball became hot enough to glow, and at that point, it became a **protostar**.

A protostar continues to grow by pulling in successively more mass until its central part reaches a temperature of about 10,000,000°C. At such extremely high temperatures, hydrogen atoms move so fast that they slam together very forcefully, and nuclear fusion binds them together to form helium nuclei. Because fusion generates prodigious amounts of energy, when fusion began in the first protostar, the body ignited and became a fearsome nuclear furnace—a true star. Star formation first took place perhaps 800 million years after the Big Bang. The concept that a star forms from the collapse of a nebula is now known as the **nebular theory**.

Star formation happened in galaxies throughout the young Universe, producing vast numbers of stars. A

(a) A photo of solar (stellar) wind streaming into space.

(b) The Crab Nebula, a gas cloud ejected by a supernova whose light reached the Earth in 1054 C.E.

Figure 1.9 Element factories in space.

large proportion of these *first-generation stars* were very massive, some with 100 or more times the mass of our Sun. Astronomers have determined that, because larger stars contain more mass than smaller ones, gravitational pull within larger stars causes more compression and higher temperatures. At these higher temperatures, more fusion takes place because atoms are moving faster, so hydrogen gets consumed by fusion more rapidly. Therefore, while a medium-sized star, like our Sun, can survive for billions of years, a huge star may survive only millions of years before its nuclear furnace runs out of hydrogen. Gas, when heated, expands, so fusion in a star causes an outward push or pressure. When fusion slows, the outward pressure produced by heat generation in the star's interior decreases, so gravity rapidly pulls much of the star's mass inward. This triggers an immense explosion, which ejects huge quantities of matter back into space. Astronomers refer to such an exploding star as a **supernova**, from the Latin *nova*, meaning new, because to observers on Earth, the explosion appears as a new, bright, but short-lived point of light. Supernova explosions peppered the young Universe.

Formation of the Elements: We Are All Made of Stardust!

The Big Bang produced all of the matter that exists in the Universe, but it did not produce all of the elements that exist today. Big Bang nucleosynthesis yielded only hydrogen, helium, and trace amounts of other elements with atomic numbers of less than 5, whereas the Universe today contains 92 naturally occurring elements. Where do the elements with larger atomic numbers come from? In other words, when and how did elements such as carbon (atomic number 6), iron (atomic number 26), and uranium (atomic number 92) form? Elements up to

and including iron formed, and continue to form, inside stars by a process called **stellar nucleosynthesis**, during which smaller nuclei fuse together to form larger ones. In other words, we can consider stars to be "element factories." Production of larger atoms (carbon through iron) typically requires the superhot conditions that develop during later stages in a star's life, and formation of the largest atoms, from iron through uranium, occurs mostly during supernova explosions.

What makes the atoms formed in stars or in supernovae available to be incorporated into planets? Some escape from a star during the star's lifetime simply by moving fast enough to overcome the star's gravitational pull. These atoms make up a stream of charged particles, called *stellar wind*, that flows from a star. We refer to the stream from our own Sun as the **solar wind** (Fig. 1.9a). Atoms that remain within a star through its lifetime may be expelled or blasted into space when the star dies, such as occurs during a supernova explosion (Fig. 1.9b). Once ejected into space, these atoms mix with existing nebulae that contain hydrogen and helium left over from Big Bang nucleosynthesis. In these nebulae, atoms combine into various molecules, which in turn bond together to form countless tiny specks of ice and dust. As we've seen, **ice** in this context refers to solids made of materials that tend to exist as gas or liquid at the pressures and temperatures of the Earth's surface, such as water, methane, and ammonia. Chemists call such materials *volatile materials*, for they evaporate relatively easily. **Dust** refers to particles made of rocky or metallic solids, known to chemists as *refractory materials* because they melt only at high temperatures.

A second generation of stars eventually formed out of these new, compositionally more diverse nebulae. Second-generation stars lived and died and contributed elements

This truth within thy mind rehearse, That in a boundless Universe Is boundless better, boundless worse.

—ALFRED, LORD TENNYSON (BRITISH POET, 1809–1892)

to nebulae from which third-generation stars formed, and so on. Because of stellar nucleosynthesis and supernova nucleosynthesis, each generation of stars contains an increasing proportion of heavier elements. Different stars live for different amounts of time because the lifetime of a star depends on its mass, so at any given moment, the Universe contains many generations of stars. Our own Sun may be a third-, fourth-, or fifth-generation star. Think of it—since the mix of elements we find on Earth, including within our own bodies, comes from the hearts of extinct stars, we are all made of stardust!

Did you ever wonder . . .

where the atoms that make up your body first formed?

Take-home message . . .

Research indicates that the Universe originated at the moment of the Big Bang, about 13.8 billion years ago. Big Bang nucleosynthesis produced mostly hydrogen and helium. According to the nebular theory, nebulae composed of these elements collapsed to produce the first stars, which generated energy by nuclear fusion. These stars later exploded as supernovae. Nucleosynthesis in stars and supernovae produced larger atoms, so atoms in your body came from the interiors of stars.

Quick Question -
Could our Sun be a first-generation star? Explain.

Figure 1.10 Formation of solid bodies in the Solar System.

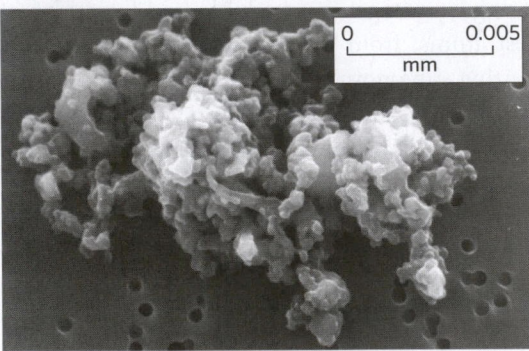

(a) At first, solids collected into dust-sized aggregates.

(b) Eventually, the dust collected into grainy masses similar to those in this meteorite fragment.

1.4 Formation of the Earth

Applying the Nebular Theory to Planet Formation

Around 4.57 billion years ago, the central ball of an accretionary disk near the edge of the Milky Way Galaxy collapsed and ignited to become our Sun. What happened to the material in the flattened outer part of the disk, the material that did not get drawn into the Sun by gravity? This part of the disk, the **protoplanetary disk**, became the source of the planets in our Solar System.

When the Sun first became a nuclear furnace, it emitted a strong solar wind, which evaporated the ice specks in the inner part of the protoplanetary disk and blew the volatile materials into the outer part of the disk, where some froze into ice again. As a result, the inner part of the disk ended up containing mostly dust. Eventually, particles in the protoplanetary disk began to clump and bind together due to gravity, growing from soot-sized specks into boulder-sized blocks, and eventually into **planetesimals**, defined as solid bodies with diameters greater than 1 km (**Fig. 1.10**). The gravitational pull of a large planetesimal was enough to attract objects that were in its path or that crossed its orbit. As a result, several planetesimals grew large enough to become **protoplanets**, bodies approaching the size of today's planets. Some of these protoplanets collided with each other and broke apart. But

(c) The protoplanetary disk consisted of gas, ice, and dust.

eight protoplanets succeeded in gathering up virtually all the debris within their own orbits to become full-fledged planets (**Earth Science at a Glance,** pp. 36–37).

Calculations suggest that planetary growth took place quickly and had finished by the time the Solar System was only a small percentage of its present age. All the planets lie in roughly the same orbit because they formed within the same protoplanetary disk. Those planets in the inner orbits, where the protoplanetary disk consisted mostly of flecks of solid dust, became the terrestrial planets, consisting almost entirely of rock and metal. In the outer part of the Solar System, where large volumes of gas and ice had accumulated, the giant Jovian planets, which consist mostly of gas and ice, grew.

Not all of the solid materials that had been in the protoplanetary disk are within planets. For example, large numbers of planetesimals and planetesimal fragments lie in the *asteroid belt* between the orbits of Mars and Jupiter. These objects, called **asteroids**, consist of rock and metal and serve as a source of meteorites. Beyond the orbit of Neptune are millions of icy fragments, planetesimals, and even planet-sized spheres known as *dwarf planets*. *Comets*, long-tailed glowing objects that occasionally arc across the sky, come from these distant reaches of the Solar System.

Internal Differentiation of the Earth

As a planetesimal grows, it gets progressively warmer inside. Its internal temperature increases for three reasons. First, as the mass of a planetesimal becomes greater, gravity squeezes its interior together more tightly, and as we have seen, compression of a material causes it to warm up. Second, during the early history of the Solar System, planetesimals frequently collided, and at each impact, kinetic energy transformed into thermal energy. And third, planetesimals contain radioactive atoms, which produce heat when they decay. (**Radioactive decay** refers to the process by which an unstable atomic nucleus spontaneously breaks apart or emits one or more subatomic particles. Energy is released when nuclear bonds break during the decay process.)

Smaller planetesimals remained homogeneous because they did not become very hot inside. In contrast, the interiors of large planetesimals became hot enough that they began to melt internally, and droplets of molten *iron alloy* (a blend of iron and other elements) formed. These droplets were denser than the rocky material around them, so they sank downward, eventually accumulating at the center of the planetesimal (**Box 1.4**). As time passed, enough metal accumulated to produce a metallic ball—a *core*—at the center of the planetesimal. The remaining materials formed a rocky shell, the *mantle*, that surrounded the core. By this process of **differentiation**, large planetesimals, including the one that eventually became the Earth,

---

| Box 1.4 | **How can I explain . . .** |

Differentiation of the Earth's interior

What are we learning?
Why the Earth, and other planets, separated into distinct internal layers.

What you need:
- An empty 16-fluid-ounce (0.5-liter) glass bottle with a screw cap.
- Vegetable oil (about 1½ cup; 0.350 liter), to represent the mantle.
- Red vinegar (about ¼ cup; 0.60 liter), to represent the core.

Instructions:
- Pour the oil and the vinegar into the bottle. The amounts suggested represent a very approximate proportion, by volume, of mantle to core. (In the Earth, the core accounts for 15% of total volume, the mantle 84%, and the crust 1%.)
- Place the bottle in a refrigerator for at least an hour. Remove the bottle and shake it hard to homogenize the contents. While the bottle remains cold, the oil and vinegar remain thoroughly mixed, like a homogeneous planetesimal.
- Heat the bottle to a warm temperature (about 30°C, or 86°F). If it's a warm, sunny day, just put it outside; otherwise, place it in a pan of hot water.
- Wait, and you will see the oil and water separate, with the denser material (vinegar) sinking to the bottom and the less dense material (oil) rising to the top. The liquid in the bottle is no longer homogeneous.

What did we see?
- The difference between homogeneous and nonhomogeneous materials.
- How heat can change the *viscosity* (resistance to flow) of a material.
- How warming a material decreases its viscosity so that materials of different density can separate into layers of different density.
- An image of Earth differentiation.

Homogeneous oil and vinegar

Oil has separated from the vinegar

Forming the Earth-Moon System

2. Gravity pulls gas and dust inward to form an accretionary disk. Eventually a glowing ball—the proto-Sun—forms at the center of the disk.

1. Forming the Solar System, according to the nebular theory: A nebula forms from hydrogen and helium left over from the Big Bang, as well as from heavier elements that were produced by fusion reactions in stars or during explosions of stars.

6. Gravity reshapes the proto-Earth into a sphere. The interior of the Earth differentiates into a core and mantle.

5. Forming the planets from planetesimals: Planetesimals grow by continuous collisions. Gradually, an irregularly shaped proto-Earth develops. The interior heats up and becomes soft.

7. Soon after the Earth forms, a protoplanet collides with it, blasting debris that forms a ring around the Earth.

8. The Moon forms from the ring of debris.

3. "Dust" (particles of refractory materials) concentrates in the inner rings, while "ice" (particles of volatile materials) concentrates in the outer rings. Eventually, the dense ball of gas at the center of the disk becomes hot enough for fusion reactions to begin. When it ignites, it becomes the Sun.

4. Dust and ice particles collide and stick together, forming planetesimals.

9. Eventually, the atmosphere develops from volcanic gases. When the Earth becomes cool enough, moisture condenses and rains to produce the oceans. Some gases may be added by passing comets.

37

Figure 1.11 Differentiation of the Earth's interior.

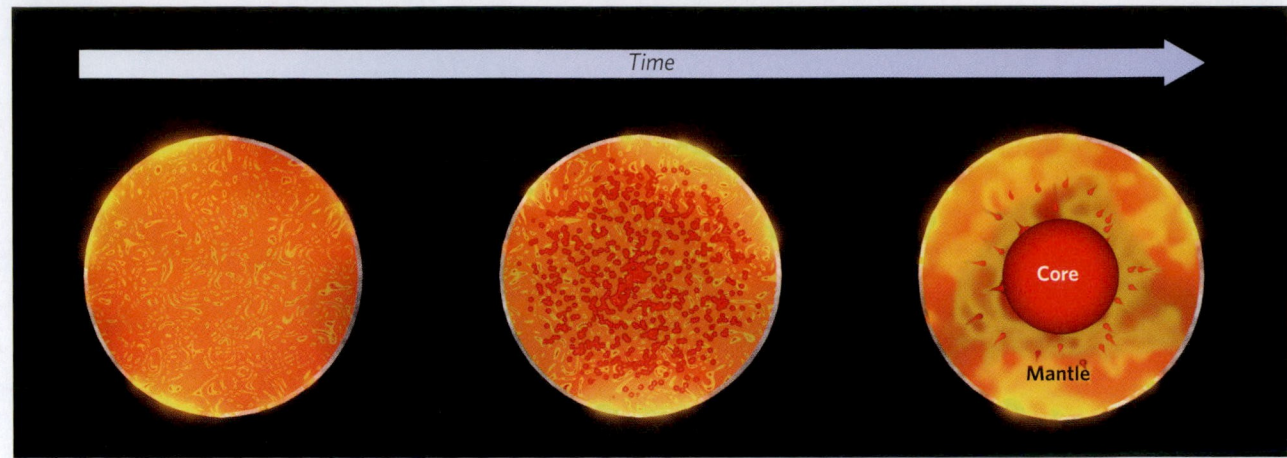

(a) Early on, the Earth was fairly homogeneous inside.

(b) When the temperature got hot enough, iron began to melt.

(c) The iron accumulated at the center of the planet to form a metallic core.

developed internal layering early in their history **(Fig. 1.11)**. This process was probably complete by 4.54 Ga, and geologists consider that date to be the birthday of the Earth. On the Earth's birthday, the Universe was already over 9 billion years old.

Making the Earth Round

Small planetesimals had irregular shapes, as do asteroids and small moons today. Planets and large moons, on the other hand, are spherical. Why? A small planetesimal is cool and rigid, so the weak pull of gravity cannot cause its material to flow. But once a planetesimal has a diameter of about 400 to 800 km (250 to 500 miles), its interior becomes warm and soft enough, so the force of gravity can make it flow. When this happens, bulges sink and dimples rise, until the planetesimal evens out into a sphere whose mass is evenly distributed around its center, so that the force of gravity is nearly the same at all points on its surface (see Earth Science at a Glance, pp. 36–37).

Take-home message . . .

According to the nebular theory, the planets of the Solar System formed from material in the outer part of the accretionary disk that formed the Sun. Gravity brought materials together into planetesimals, then protoplanets, and finally planets. In the outer part of the Solar System, large amounts of gas and ice collected around rocky and metallic cores to form the Jovian planets. Once large enough, inner planetesimals differentiated into a metallic core and a rocky mantle, and gravity rounded them into spheres.

Quick Question -
Why did the Jovian planets form in the outer part of the protoplanetary disk?

1.5 The Blue Marble: Introducing the Earth

Imagine that you're on board a space probe flying from a distant planet toward the Earth. You're reaching the final phase of your voyage, with the Earth in sight. As you approach the planet, you'll first detect how it influences the space around it. Then, when you enter its orbit, you'll be able to recognize several distinct components, or realms, of the Earth. Finally, with measurements made from space, you'll even be able to define the basic internal structure of the planet. Let's look at your discoveries.

Welcome to the Neighborhood!

From the orbit of Mars, the Earth looks like a large star with a bluish tint, but at a distance of about 30 million km (20 million miles), you can finally see the Earth as a sphere with a distinctive, beautiful bluish glow—in fact, it looks like a blue glass marble. Finally, at a distance of perhaps 1 million km (600 thousand miles), three times the distance to the Moon, you can begin to see the complexities of the Earth's surface and atmosphere **(Fig. 1.12)**.

Flying inward from the orbit of the Moon, your instruments discover that the Earth produces a magnetic field, like a signpost saying "Welcome to Earth!" A **magnetic field** is the region affected by the invisible force (push or pull) emanating from a magnet (see Box 1.1). Earth's magnetic field, like the familiar magnetic field around a bar magnet, is a *dipole*, meaning that it has a north pole and a south pole. We can portray the magnetic field by drawing curving lines around the magnet, running from the north pole to the south pole. These lines represent the directions along which magnetic materials (such as iron filings or compass needles) would align when placed in the

field **(Fig. 1.13a, b)**. The solar wind, which contains charged particles (protons and electrons), warps the Earth's magnetic field into a huge teardrop pointing away from the Sun. Fortunately, the magnetic field deflects most (but not all) of the solar wind from the Earth. In this way, the magnetic field acts like a shield against the charged particles of the solar wind, which can harm living organisms. The region inside this shield is called the *magnetosphere* **(Fig. 1.13c)**.

Though it protects the Earth from most of the solar wind, the magnetic field does not stop your spacecraft, and you continue to speed toward the planet. At distances of between 10,500 and 3,000 km (6,500 and 18,000 miles) out from the Earth, you encounter the *Van Allen radiation belts,* named for the physicist who first recognized them in 1958. These belts form where the Earth's magnetic field becomes strong enough to trap both particularly energetic solar wind particles and *cosmic rays* (nuclei of atoms shot into space from supernova explosions) that were moving so fast that they could penetrate the weaker, outer part of

Figure 1.12 The Earth, as seen from space.

(a) During the day, the Earth reflects sunlight and clouds. Oceans, land, and ice are all visible.

(b) At night, the lights of human civilization stand out. (This image is a composite of cloudless photos.)

Figure 1.13 The Earth's magnetic field.

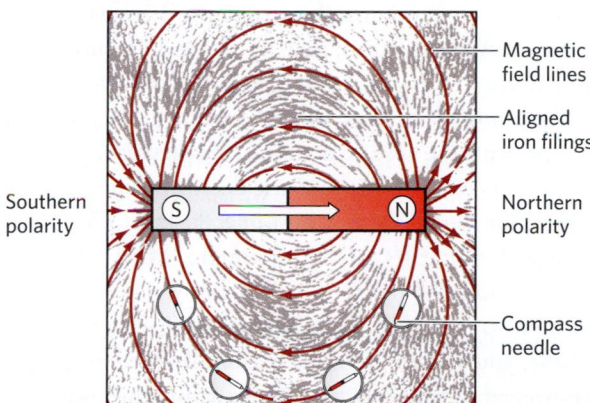

Magnetic field lines

Aligned iron filings

Southern polarity

Northern polarity

Compass needle

(a) Magnetic field lines produced by a bar magnet point into the magnet's south pole and out from the magnet's north pole.

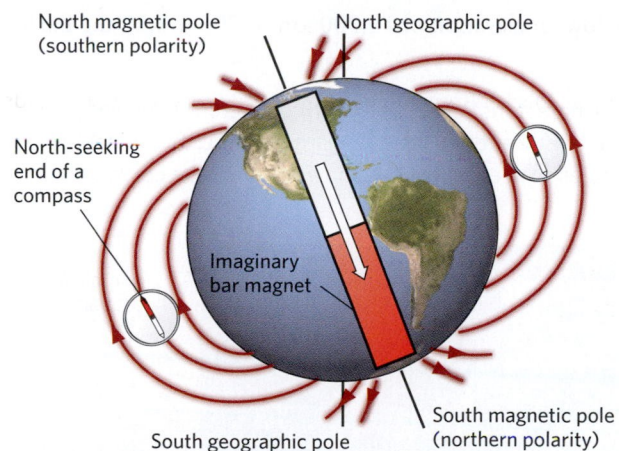

North magnetic pole (southern polarity)

North geographic pole

North-seeking end of a compass

Imaginary bar magnet

South geographic pole

South magnetic pole (northern polarity)

(b) We can represent the Earth's magnetic field as an imaginary bar magnet inside the planet.

Did you ever wonder . . .

why compasses work?

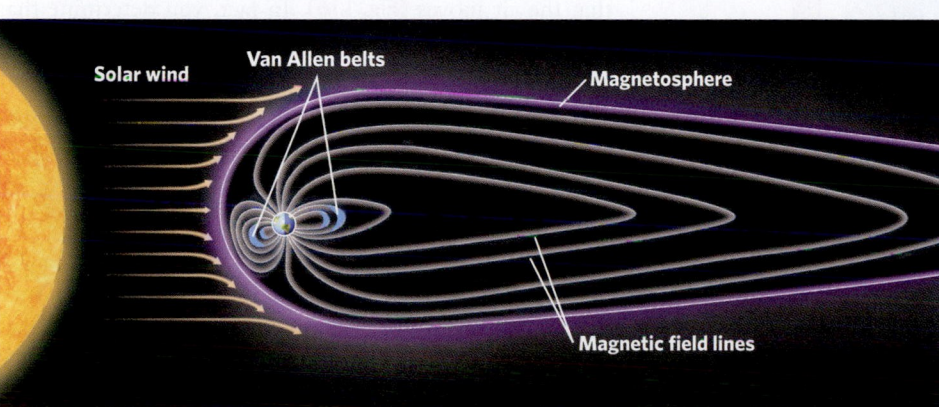

Solar wind

Van Allen belts

Magnetosphere

Magnetic field lines

(c) Earth behaves like a magnetic dipole, but its magnetic field lines are distorted by the solar wind. The Van Allen radiation belts trap charged particles.

(d) A view of aurorae as seen from space.

the magnetic field. Some of these particles do, however, make it past the Van Allen belts. Many of them flow along magnetic field lines to the Earth's north and south magnetic poles, where they cause gas atoms in the upper atmosphere to glow, producing spectacular *aurorae* **(Fig. 1.13d)**.

Introducing the Atmosphere

At a distance of about 15,000 km (9,300 miles), you can still see the whole Earth as a sphere, but as you descend below an elevation of 10,000 km (6,200 miles), the Earth

Figure 1.15 Characteristics of Earth's atmosphere. Molecules pack together more tightly at the base of the atmosphere, so atmospheric pressure changes with elevation (as shown by the blue curve).

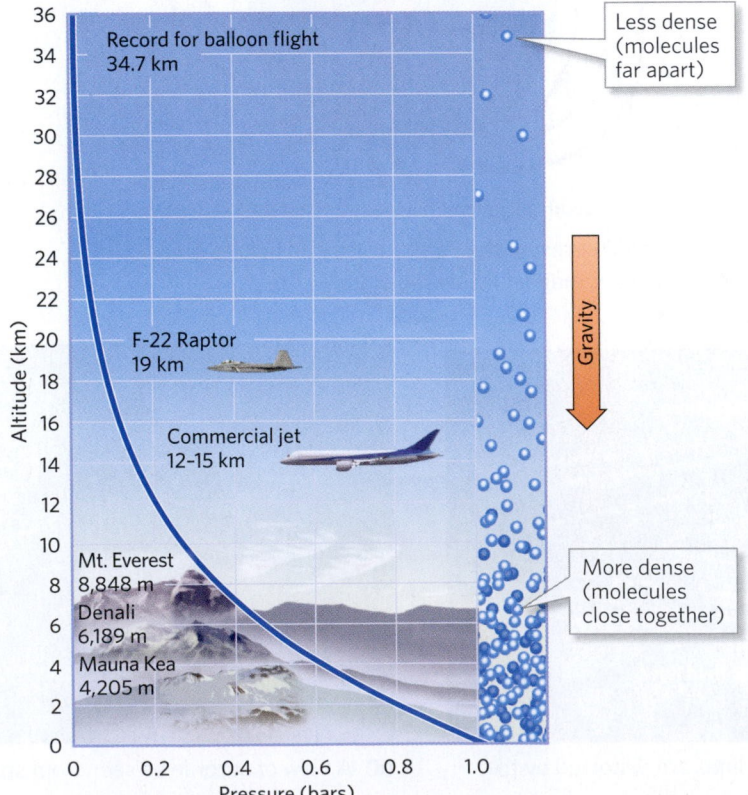

more than fills your field of view. At this elevation, your instruments detect that the concentration of gas outside the spacecraft has increased slightly. This change means that you've reached the uppermost limit of the **atmosphere**, the layer of gas that surrounds the Earth and is held there by gravity **(Fig. 1.14)**. From the outer edge of the atmosphere on down, the density of atmospheric gas progressively increases because the weight of the gas higher in the atmosphere pushes down on the gas below and squeezes its molecules closer together. As a result, 99.9% of the gas molecules in the Earth's atmosphere lie below an elevation of 50 km (30 miles). To avoid entering the denser part of the atmosphere, where your space probe would slow down and heat up, you aim into orbit around the Earth at an elevation of about 400 km (250 miles), the position of the International Space Station. Here, the air remains so thin that it barely affects your space probe, and you can study the atmosphere and map the surface below in detail.

First, you analyze variations in the composition of **air**, the mixture of gases making up the atmosphere, and find that it consists of 78% nitrogen molecules (N_2), 21% oxygen molecules (O_2), and 1% *trace gases*, including argon (Ar), carbon dioxide (CO_2), and methane (CH_4). Air also contains varying amounts of water vapor. Then, you measure **air pressure**, the push that air exerts on its surroundings. **Pressure** in any material (air, water, or rock) can be described in units of *force per unit area*. One such unit, a **bar**, represents the approximate average pressure of the atmosphere at sea level. Air pressure decreases with increasing elevation, so at an elevation of 8.85 km (5.5 miles), on the peak of Earth's highest mountain, Mt. Everest, air pressure is only 0.3 bar **(Fig. 1.15)**. Climbers need oxygen tanks to help them breathe at that elevation.

Next, you analyze the distribution of water vapor and the movement of air. You can see that within the lower part of the atmosphere, below an elevation of about 12 km (7.5 miles), the air contains *clouds*, wisps or billows composed of tiny water droplets or ice crystals. The shape and distribution of clouds change constantly, providing a visual clue that the air moves **(Fig. 1.16)**. In fact, you determine that the *wind*, the movement of air parallel to the Earth's surface, generally has speeds between 1 and 50 km per hour (0.6 and 30 mph) near the ground, but at higher altitudes, it may increase to over 300 km per hour (180 mph).

The Hydrosphere and Cryosphere—Only on Earth!

Having made a quick analysis of the atmosphere, you turn your attention to the Earth's surface and map its characteristics. Immediately, you realize that the Earth differs from all the other terrestrial planets in the Solar System in that water covers about 70% of its surface. Of this water, about 97% is salty, and the remainder is fresh. Of the

Figure 1.16 The Earth as seen from orbit. Clouds tower above the ocean, which reflects sunlight.

liquid *freshwater*, only a small part is in lakes, ponds, rivers, and streams; the rest resides underground, as **groundwater**, in tiny holes and cracks within rock and sediment.

Your mapping emphasizes that not all of Earth's water occurs in liquid form. In polar regions and at high elevations, water freezes into ice **(Fig. 1.17)**. When ice builds up on the land surface in a thick enough layer that it lasts all year and starts to flow due to gravity, it becomes a **glacier**. In very cold regions, near-surface groundwater also freezes, producing *permafrost* (permanently frozen ground), and the sea surface freezes to form a thin layer of *sea ice*. As we noted in the Prelude, the Earth's ice makes up the **cryosphere**, and its liquid and frozen water together constitute the **hydrosphere**.

Materials of the Geosphere

If you sent down probes to sample and analyze the composition of the solid Earth, the **geosphere**, you would detect all 92 naturally occurring elements, but you would find that only four of these—iron, oxygen, silicon, and

Ice

Land

Ocean

25 km

Figure 1.17 Not all of the hydrosphere is liquid water. The liquid part of the hydrosphere (in the form of ocean water) and the frozen part (glaciers of the cryosphere) come in contact along the coast of Alaska.

Figure 1.18 The proportions of elements making up the solid mass of the Earth. Note that iron and oxygen account for most of the mass.

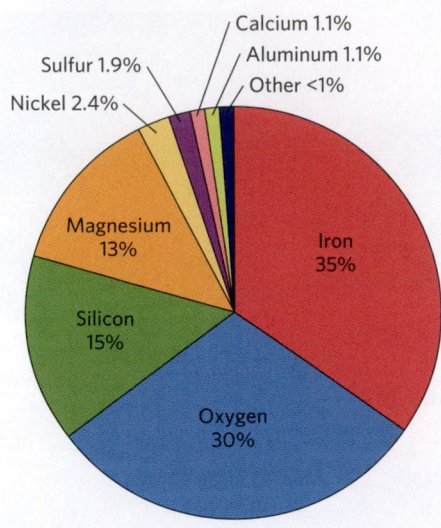

- Calcium 1.1%
- Aluminum 1.1%
- Other <1%
- Sulfur 1.9%
- Nickel 2.4%
- Magnesium 13%
- Iron 35%
- Silicon 15%
- Oxygen 30%

magnesium—make up 90% of the Earth's solid mass **(Fig. 1.18)**. You would also learn that the geosphere contains a great variety of different materials. Let's take a quick look at some basic categories of these materials **(Fig. 1.19)**—all will be discussed in more detail later in the book.

- *Melts:* A **melt**, or *molten material*, forms when a solid becomes hot enough to transform into liquid.

- *Minerals:* A **mineral** is a solid, naturally occurring substance in which atoms are arranged in an orderly pattern. Further, a mineral has a definable chemical composition, in that we can write a chemical formula defining the proportions of elements that it contains. A sample of a mineral that has smooth, naturally grown faces is a *crystal*, whereas an irregularly shaped sample, or a fragment broken off a crystal or group of crystals, is a *grain*.

- *Glasses:* A solid in which atoms are not arranged in an orderly pattern is a **glass**.

- *Sediments and soils:* A **sediment** is an accumulation of loose mineral grains that are not stuck together. Sediment can react with air, water, and life at the surface of the Earth to evolve into **soil**.

- *Metals:* A solid composed of metallic atoms (such as iron, aluminum, copper, or tin) is a **metal**. Metals can be pounded into sheets. In a metal, electrons flow freely, so metals conduct electricity. An **alloy** is a blend containing more than one type of metal.

- *Rocks:* A **rock** is a solid aggregate (collection) composed of mineral crystals or grains, or is a mass of natural glass. Geologists recognize three basic kinds of rock:

  1. **Igneous rock** is formed by the solidification of molten rock, melt that forms inside the Earth. Igneous rock can solidify underground or at the surface of the Earth. Molten rock underground is called *magma*, whereas molten rock that spills out at the Earth's surface is *lava*.

  2. **Sedimentary rock** is formed by the cementing together of solid grains, or by the precipitation of minerals out of water solutions, at or near the Earth's surface.

  3. **Metamorphic rock** is formed when existing rock undergoes changes, primarily due to an increase in temperature, pressure, or both, beneath the Earth's surface.

The Surface of the Geosphere

As you orbit the Earth in your space probe, your instruments detect variations in the elevation of the geosphere, both beneath the ocean and where is exposed on land. From these data, you can construct a map showing **bathymetry,** variation in the depth of the seafloor, and a

Figure 1.19 Materials of the geosphere.

Melt

Mineral

Glass

Sediment

Metal

Rock

Figure 1.20 Topography and bathymetry. This map shows the Earth's topography (variation in land elevation) and bathymetry (variation in seafloor depth); it also shows where glaciers cover the land (white areas).

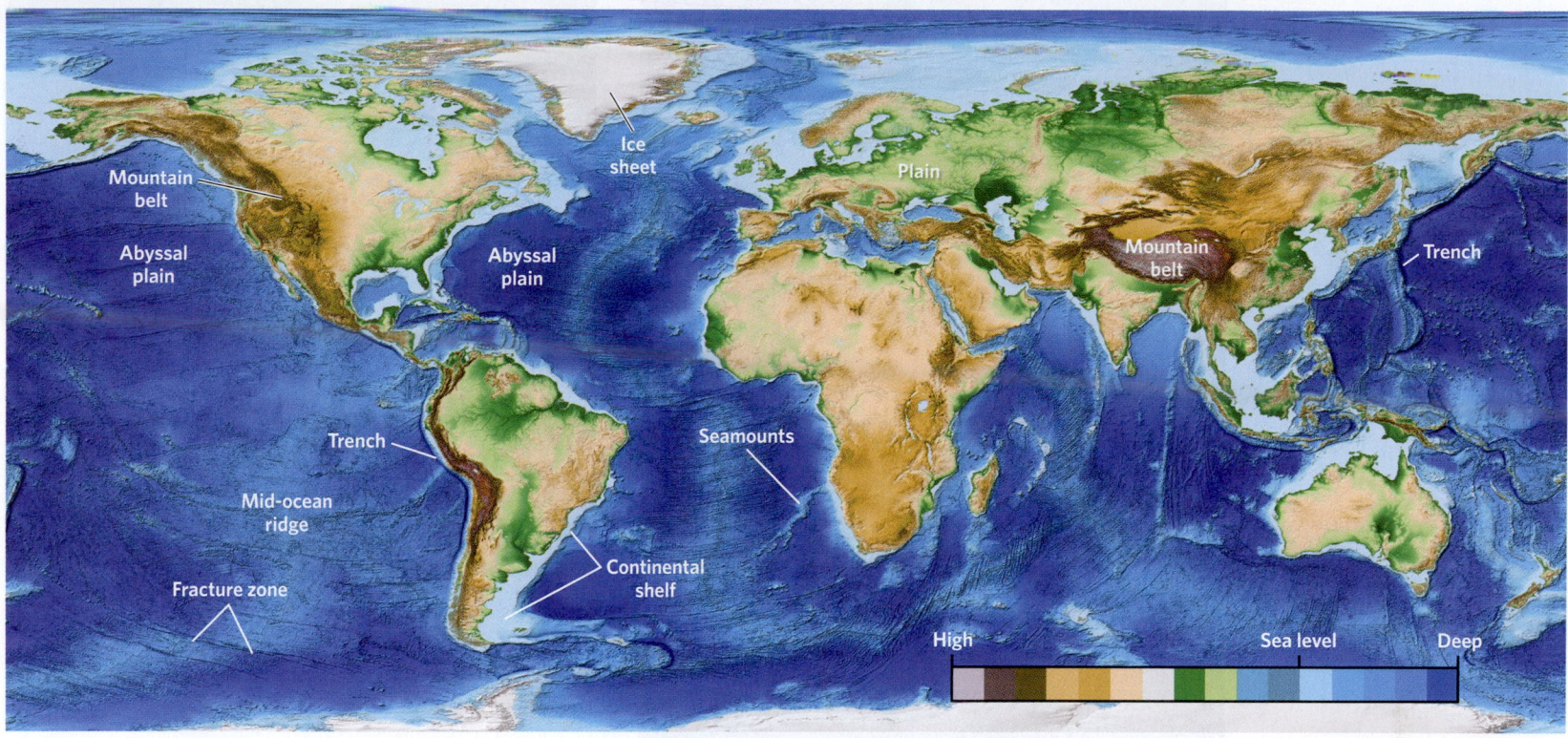

map showing **topography,** variation in the elevation of the land **(Fig. 1.20)**.

The bathymetric map reveals that most of the seafloor consists of flat *abyssal plains*, at a depth of 4 to 5 km (2 to 3 miles), but that the seafloor also hosts submarine mountain ranges, called *mid-ocean ridges*, that rise a couple of kilometers (1 mile) above the abyssal plains; deep troughs, called *trenches*, that descend to depths of 7 to 11 km (4 to 7 miles); linear *fracture zones* where the seafloor has been broken by numerous cracks; and, along the edges of some continents, broad *continental shelves*, where water depths are less than 0.5 km (0.3 miles).

Most land on Earth lies within large blocks called *continents*. The remainder occurs in much smaller pieces, called *islands*. A topographic map reveals plains, mountain ranges (the highest of which rises to a height of 8.85 km; 5.5 miles), plateaus (flat areas of elevated land), volcanic peaks (some of which are erupting molten rock), valleys, and many other landscapes. Looking at the land surface, you see that some is covered by soil, some is exposed rock, and some has a veneer of sediment. Volcanoes occur in some places, but not everywhere.

Together, the bathymetry and topography of the Earth make it look very different from the other terrestrial planets **(Fig. 1.21)**. Most notably, the surfaces of the other planets remain pockmarked by craters, while very few

craters appear on Earth. This difference leads you to realize that processes must be taking place that can uplift the land, **erode** (grind or bevel away) uplifted areas, and bury features over time. In other words, unlike the other terrestrial planets, the Earth remains a dynamic place, where natural phenomena actively modify its surface.

The Biosphere

Of course, even a cursory examination reveals that, unlike all other planets in the Solar System, the Earth has a **biosphere**, defined as the realm of living organisms. It extends from a few kilometers below the surface to a few kilometers above. In recent years, researchers have started referring to the spatial realm of the biosphere as the **critical zone** to emphasize that the resources and conditions that sustain life exist within this zone. From your orbiting space probe, you would see plants and animals, and you would detect chemicals that can be produced only by living organisms.

How much living material exists in the biosphere? Researchers estimate that the total mass of Earth's *live* organisms, not counting microbes, comes to about 570 billion metric tons (1 metric ton = 1,000 kg, or 2,204 pounds). Microbes may account for as much biomass as visible organisms, or even more, so the Earth's total biomass probably exceeds 1 trillion metric tons. This

See for yourself

Meteor Crater

Latitude: 35°1′37.18″ N
Longitude: 111°1′20.17″ W

If you look at the Moon through a telescope, you see thousands of craters, the bowl-shaped depressions that mark the locations of meteorite impacts. On Earth, most craters have been eroded away, but a few remain. Zoom to an elevation of 5.7 km (19,000 feet) and look straight down. You are seeing Meteor Crater, Arizona, which was formed by the impact of a 50-m-wide meteorite less than 50,000 years ago.

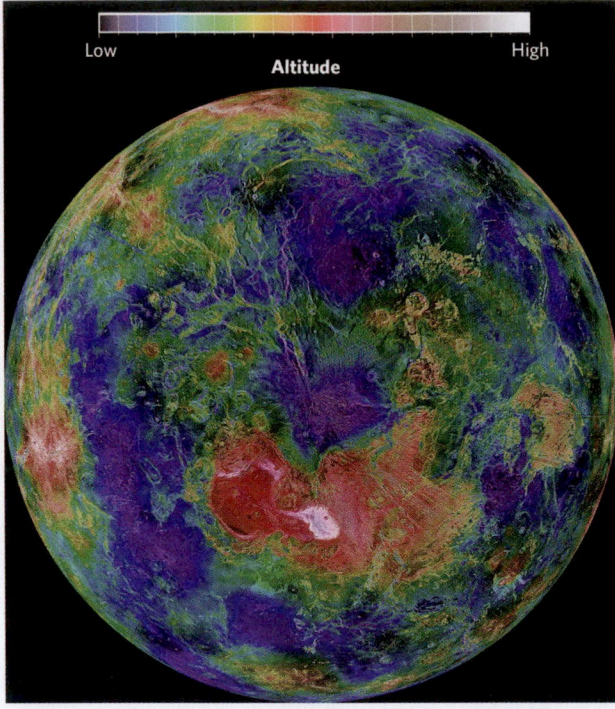

(a) Venus

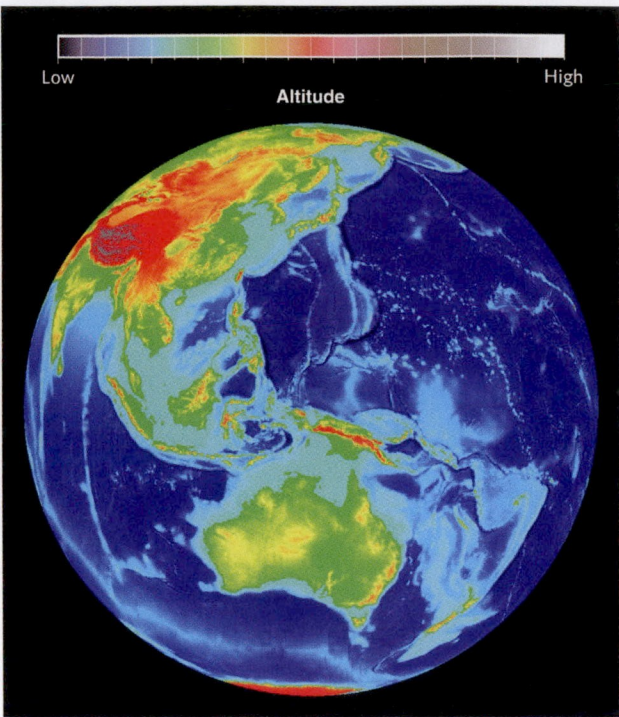

(b) Earth

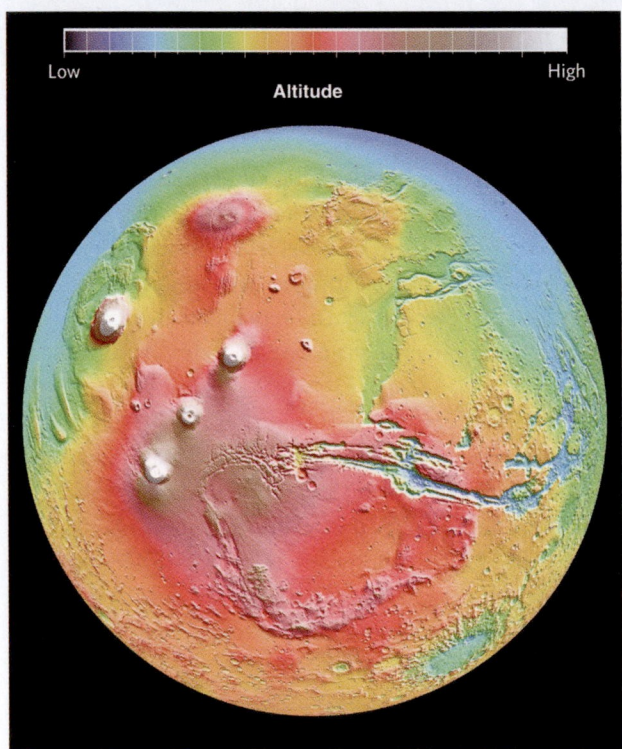

(c) Mars

quantity roughly equals the weight of 10 volcanoes the size of Mt. Fuji!

You would also notice the changes that human society has made to the planet. For example, you would see farm fields, cities and towns, highways, canals, dams, and mines, and you would detect a variety of anthropogenic materials such as concrete, asphalt, brick, lumber, tile,

and sheet metal. A view of the Earth at night, showing the lights of cities, emphasizes the reach of civilization (see Fig. 1.12b).

Take-home message . . .

A space probe approaching the Earth would learn that it has many characteristics that distinguish it from other planets. It has a magnetic field and an atmosphere composed principally of nitrogen and oxygen. The elevation of its surface varies, both on land and on the seafloor. Overall, the Earth consists of several realms: the atmosphere, hydrosphere, cryosphere, geosphere, and biosphere.

Quick Question -
What are the various materials that make up the geosphere?

1.6 A First Glance at the Earth's Interior

Hints as to What's Inside the Earth

Can you discover anything about the Earth's interior—the material between the Earth's surface and its center—without drilling a hole through it? Fortunately, the answer is yes, because the distance from the surface to the center—6,371 km (3,959 miles)—is comparable to the distance from the east coast of the United States to the west coast of Europe.

Box 1.5

Consider this . . .

Temperatures and pressures inside the Earth

When the Earth first formed, it was very hot due to the compression of matter accompanying planet formation and the energy provided by the impacts of meteorites. In fact, the planet was probably so hot that its surface and much of its interior was molten. Over time, as heat from the Earth radiated into space, the planet cooled and solidified. However, the decay of radioactive elements in rock has prevented the Earth from cooling entirely, so it still remains hot inside. For example, while Earth's average ground-surface temperature is about 14°C (57°F), the temperature at a depth of 40 km (25 miles) ranges between 600°C and 1,200°C (1,100°F to 2,200°F), and the temperature at the center of the Earth ranges between 5,000°C and 6,000°C (9,000°F to 11,000°F). Geologists refer to the rate at which temperature increases with increasing depth in the Earth as the **geothermal gradient (Fig Bx 1.5)**.

Pressure also increases with depth due to the weight of overlying material. Because the materials of the geosphere are denser than the gas of the atmosphere, pressure increases more rapidly with depth in the geosphere than in the atmosphere. Geologists estimate that at a depth of 40 km, the pressure reaches 10,000 bars, and at the center of the Earth, it reaches about 3,500,000 bars.

Figure Bx1.5 The geotherm indicates how temperature changes with depth.

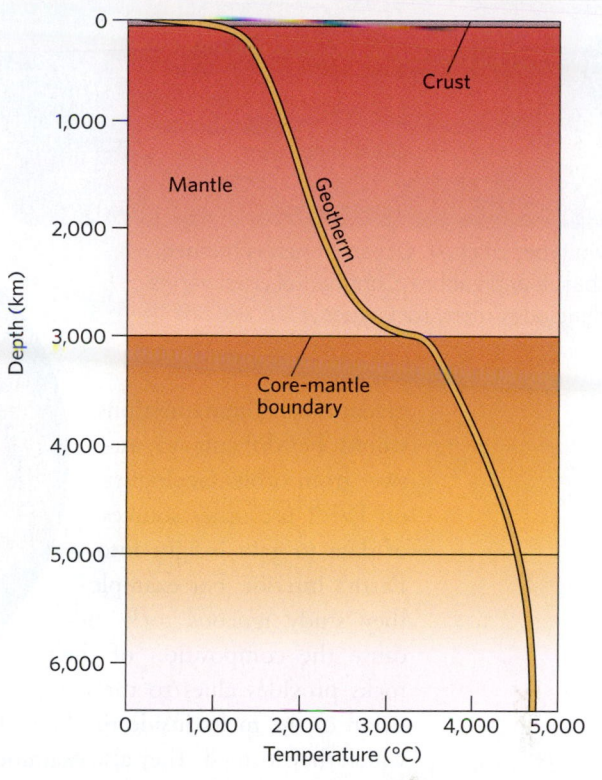

First, by measuring the mass and volume of the whole Earth, you can calculate its average density, and by sampling rocks exposed at the surface of the planet, you can measure the density of those rocks. When you compare these two measurements, you'll find that the average density of the whole Earth is about twice the density of the common rocks found at its surface. Material inside the planet, therefore, must be much denser than the material at its surface. Second, by measuring the overall shape of the Earth, you'll find that the planet is nearly a sphere, even though it spins rapidly on its axis. This observation means that the Earth's interior must be mostly solid and that the densest material must be concentrated near its center. If the Earth were entirely molten beneath its surface, or if its mass were distributed evenly throughout, spinning would cause it to flatten into a disk.

Eventually, putting all these deductions together, you could come up with a model in which the Earth resembles a hard-boiled egg, with three principal layers **(Fig. 1.22)**: (1) a thin, not-so-dense *crust* (the eggshell) composed mostly of relatively low-density rock; (2) a solid *mantle* in the middle (the white) composed of relatively high-density rock; and (3) a very dense *core* (the yolk) consisting of metal alloy.

Refining the Picture of Earth's Interior

Clearly, many questions would remain after your initial survey. Exactly what are the materials inside the Earth, how thick are the layers, and are the boundaries between layers sharp or

Figure 1.22 An early image of Earth's internal layers.

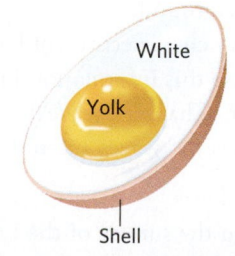

(a) The hard-boiled egg analogy for the Earth's interior.

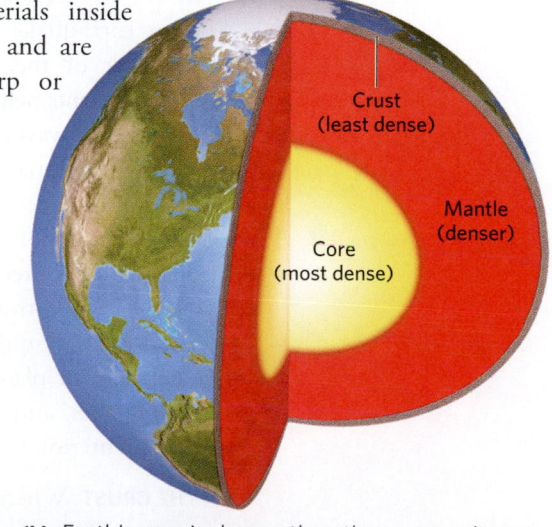

(b) Earth's core is denser than the surrounding regions.

Figure 1.23 A modern view of Earth's interior layers.

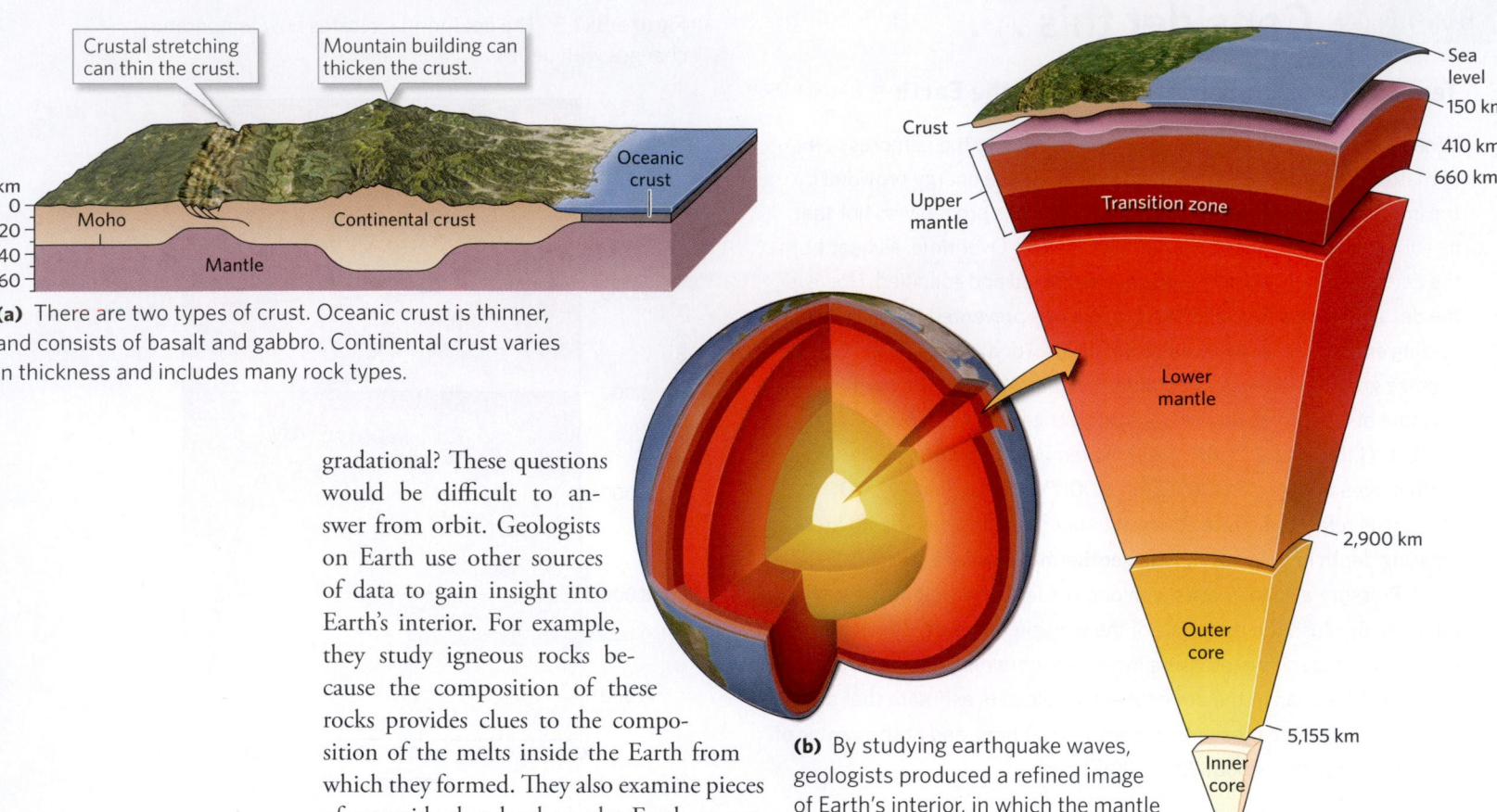

(a) There are two types of crust. Oceanic crust is thinner, and consists of basalt and gabbro. Continental crust varies in thickness and includes many rock types.

(b) By studying earthquake waves, geologists produced a refined image of Earth's interior, in which the mantle and core are subdivided.

gradational? These questions would be difficult to answer from orbit. Geologists on Earth use other sources of data to gain insight into Earth's interior. For example, they study igneous rocks because the composition of these rocks provides clues to the composition of the melts inside the Earth from which they formed. They also examine pieces of asteroids that land on the Earth as meteorites because asteroids are fragments of planetesimals. Finally, they conduct experiments and develop models in order to calculate how temperature and pressure vary inside the Earth **(Box 1.5)** and to determine the nature of materials that can exist under such conditions.

Geologists have been able to refine the image of the Earth's interior by measuring the speed and path of *seismic waves*, vibrations that pass through the Earth from earthquakes. Study of seismic waves reveals the character of the Earth's interior much as ultrasound measurements help doctors study the insides of a patient, for seismic waves travel at different velocities through different materials, and they bend or reflect when they reach boundaries between different materials. Therefore, the behavior of seismic waves defines the depths of boundaries between layers underground.

Here, we introduce some key characteristics of Earth's interior layers so that we can use this information in our discussion of plate tectonics in Chapter 2. We'll revisit earthquakes, and the use of seismic waves to study the Earth's interior, in Chapter 8.

THE CRUST When you stand on the surface of the Earth, you are on top of its outermost layer, the **crust (Fig. 1.23a)**. The base of the crust, a boundary called the **Moho,** lies

at different depths in different locations. *Oceanic crust,* the crust beneath the seafloor, has a thickness that ranges between 7 and 10 km (4 and 6 miles). *Continental crust,* the crust beneath most land, has a thickness that ranges between 25 and 70 km (15 and 45 miles). Compared with the radius of the Earth, the crust is so thin that if the Earth were the size of a balloon, the crust would be about the thickness of the balloon's skin. What is the crust made of? The crust is not simply cooled mantle, like the skin on chocolate pudding, but rather consists of a variety of rocks.

The top of the oceanic crust is a blanket of sediment, generally less than 1 km (600m) thick. This sediment consists of tiny fragments of dust and plankton shells that have settled like snow out of the sea. Beneath this blanket, oceanic crust consists of two kinds of dark gray, relatively dense igneous rock: **basalt,** which contains tiny crystals, and *gabbro*, with coarser crystals. (We'll describe these rocks, and others mentioned in this section, in more detail in Chapter 3.) Most of the continental crust, in contrast, consists of different kinds of igneous and metamorphic rocks. In many locations, a blanket of sedimentary rock covers the igneous and metamorphic rocks. To simplify our discussion for now, we can say that the continental crust has an average composition similar to that of **granite,** a light-colored igneous rock with a density less than that of basalt or gabbro.

THE MANTLE The mantle of the Earth forms a 2,885-km (1,792-mile)-thick layer beneath the crust. In terms of volume, it is the largest part of the Earth **(Fig. 1.23b).** The mantle consists entirely of a very dark, very dense igneous rock called **peridotite.** This means that peridotite, though rare at the Earth's surface, is actually the most abundant rock on our planet!

Studies of seismic waves suggest that the mantle contains two sublayers—the *upper mantle* above a depth of 660 km (410 miles), and the *lower mantle* below that depth. Geologists refer to the bottom portion of the upper mantle, the interval between 410 km (255 miles) and 660 km deep, as the *transition zone,* for within this zone, several changes take place in the character of the minerals making up mantle peridotite. Almost the entire mantle is solid, but most of it is hot enough to be able to flow very slowly (at a rate of less than 15 cm (6 inches) per year), a process known as *plastic flow.*

THE CORE Early calculations suggested that the core, the central ball deep inside the Earth, had the same density as gold, so for many years people held the fanciful hope that vast riches lay at the heart of our planet. Alas, geologists eventually concluded that the core consists of a far less glamorous material—*iron alloy* (a mix of iron with 4% nickel and up to 10% oxygen, silicon, or sulfur).

Studies of seismic waves led geologists to divide the core into two parts, the *outer core* that is between 2,900 km (1,802 miles) and 5,155 km (3,203 miles) deep and the *inner core* that is from 5,155 km deep down to the Earth's center at 6,371 km (3,959 miles) (see Fig. 1.23b). The iron alloy of the outer core is molten, so it can flow fairly rapidly. This flow generates the Earth's magnetic field. The inner core, in contrast, is solid.

Hopefully, our whirlwind tour of the Earth has left you with many questions. Why are continents high and ocean basins low? What processes led to the formation of the distinctive topographic features of land, such as mountain belts, and the distinctive bathymetric features of the ocean basins, such as ridges and trenches? Why do volcanoes occur where they do? How are all these features related to processes happening inside the Earth? We'll try to answer some of these questions in the next chapter.

Take-home message . . .

The distinct layers of the Earth's interior, from outside to inside, are the crust, mantle, and core. The crust beneath the oceans differs from crust beneath the continents. The mantle, the largest part of the Earth, consists of very dense, almost entirely solid rock, but it is so hot that it flows plastically. The core consists of iron alloy. Flow in the liquid outer core generates the magnetic field. The inner core is solid.

Quick Question ------------------------------
What sources of data reveal the character of the Earth's interior?

Another View Flying into New York City, an airline passenger can see many key components of the Earth System—the atmosphere, the hydrosphere, the geosphere, and the biosphere. The unique conditions of these components made the explosion of human cities possible.

CHAPTER REVIEW

Chapter Summary

- The Earth is one of eight planets orbiting our Sun in the Solar System. The four terrestrial planets are relatively small, have rocky surfaces. The Jovian planets consist mostly of gas and ice, and are relatively large.

- Modern cosmologic studies place our Solar System on an outer arm of the spiral Milky Way Galaxy. The Universe contains more than a trillion galaxies. Distances between stars are so large that we measure them in light-years.

- Galaxies zoom away from one another as the Universe has been expanding since the Big Bang, at about 13.8 Ga. Big Bang nucleosynthesis produced small atoms.

- Portions of nebulae collapse inward to produce accretionary disks, the centers of which become dense balls called protostars. When a protostar becomes hot enough inside for nuclear fusion to begin, it becomes a star.

- Stellar nucleosynthesis produces larger atoms, and the explosion of a supernova produces the largest atoms.

- Planets formed in the outer portions of the solar accretionary disk. Dust or ice collected into clumps, which merged to form planetesimals. Some of these planetesimals grew into protoplanets, then true planets.

- The solar wind blew volatile materials to the outer part of the protoplanetary disk, and the Jovian planets formed from those materials. The terrestrial planets formed nearer the Sun.

- When a terrestrial planetesimal became large enough for compression to make its interior very hot, it underwent differentiation into a metallic core and a rocky mantle. The heating and softening of the interior also allowed the planetesimal to become spherical.

- A probe approaching the Earth from space would detect its magnetic field extending into space around the planet. The magnetic field protects the Earth from solar wind and cosmic rays.

- An envelope of air, a mixture consisting mostly of nitrogen and oxygen gas, surrounds the Earth. 99.9% of air lies within 50 km (30 miles) of the surface. Atmospheric pressure increases downward.

- The surface of the Earth consists of land and sea. Parts of the land are covered by surface water and have groundwater beneath.

- A great variety of different materials—including minerals, rocks, sediments, metals, glass, and melts—make up the geosphere.

- The elevation of the geosphere varies across its surface. Variation in the land elevation is topography, and variation in seafloor depth is bathymetry.

- The interior of the Earth consists of a thin crust, a thick mantle, and an inner core.

- Pressure and temperature both increase with depth in the Earth. The rate of temperature increase is the geothermal gradient.

- The crust beneath the oceans differs in thickness and composition from the crust beneath the continents. Rocks of the crust are less dense than those of the mantle. The core consists of iron alloy.

- The mantle can be subdivided into upper and lower mantles, and the core can be subdivided into a liquid outer core and solid inner core. Flow in the outer core produces a magnetic field.

Key Terms

air (p. 40)
air pressure (p. 40)
alloy (p. 42)
asteroid (p. 35)
atmosphere (p. 40)
atom (p. 28)
basalt (p. 46)
bar (p. 40)
bathymetry (p. 42)
Big Bang (p. 31)
Big Bang nucleosynthesis (p. 31)
biosphere (p. 43)
celestial object (p. 26)
centigrade scale (p. 32)
contact force (p. 26)
critical zone (p. 43)

crust (p. 46)
cryosphere (p. 41)
differentiation (p. 35)
dust (p. 33)
electromagnetism (p. 26)
electron (p. 28)
element (p. 28)
erode (p. 43)
expanding Universe theory (p. 30)
Fahrenheit scale (p. 32)
field force (p. 26)
force (p. 26)
galaxy (p. 27)
geosphere (p. 41)
geothermal gradient (p. 45)
glacier (p. 41)

glass (p. 42)
granite (p. 47)
gravity (p. 26)
groundwater (p. 41)
heat (p. 32)
hydrosphere (p. 41)
ice (p. 33)
igneous rock (p. 42)
light-year (p. 27)
magnetic field (p. 38)
melt (p. 42)
metal (p. 42)
metamorphic rock (p. 42)
Milky Way Galaxy (p. 27)
mineral (p. 42)
moho (p. 46)

molecule (p. 29)
moon (p. 27)
nebula (p. 29)
nebular theory (p. 32)
neutron (p. 28)
nuclear fusion (pp. 28, 31)
nucleus (p. 28)
orbit (p. 27)
peridotite (p. 47)
planet (p. 27)
planetesimal (p. 34)
pressure (p. 40)
proton (p. 28)
protoplanet (p. 34)
protoplanetary disk (p. 34)
protostar (p. 32)

radioactive decay (p. 35)
rock (p. 42)
sediment (p. 42)
sedimentary rock (p. 42)
soil (p. 42)

Solar System (p. 27)
solar wind (p. 33)
speed of light (p, 27)
star (p. 27)
stellar nucleosynthesis (p. 33)

Sun (p. 26)
supernova (p. 33)
temperature (p. 32)
thermal energy (p. 32)
topography (p. 43)

Universe (p. 25)
vacuum (p. 27)

Review Questions

The letters following each Review Question refer to the corresponding Learning Objective from the Chapter Opener.

1. How many planets does our Solar System contain, and what is the position of the Solar System in the Milky Way Galaxy? **(A)**

2. What is the difference between a nebula and a vacuum? **(A)**

3. Explain the expanding Universe theory and its relationship to the Big Bang theory. According to the theory, when did the Universe form? **(A)**

4. Distinguish between Big Bang nucleosynthesis and stellar nucleosynthesis. Why is it fair to say that we are all made of stardust? **(A)**

5. The image shows a nebula formed by a supernova explosion. How many elements does it contain? **(B)**

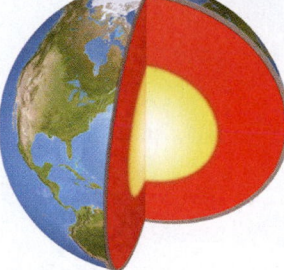

6. Describe the steps in the formation of the Solar System according to the nebular theory. **(B)**

7. Why isn't the Earth homogeneous, and why is it round? **(B)**

8. What is the Earth's magnetic field? How does the magnetic field interact with the solar wind? **(C)**

9. What is the Earth's atmosphere composed of? Why would you die of suffocation if you were to jump from a jet? **(C)**

10. What is the proportion of land area to sea area on Earth? Is the seafloor completely flat? How about the land surface? **(C)**

11. Describe the major categories of materials constituting the geosphere. **(D)**

12. Distinguish among the various realms of the Earth System. **(C)**

13. How do temperature and pressure change with increasing depth in the Earth? **(E)**

14. Label the principal layers of the Earth on the figure. **(E)**

15. What sources provide geologists with information about the character of the Earth's interior? **(E)**

16. What is the Moho? Describe basic differences between continental crust and oceanic crust. **(E)**

17. What is the mantle composed of? Is the mantle rigid and unmoving? **(E)**

18. What is the core composed of? How do the inner and outer cores differ? Which produces the magnetic field? **(E)**

On Further Thought

19. Could a planet like the Earth have formed in association with a first-generation star? **(B)**

20. Why are the Jovian planets, which contain abundant gas and ice, farther from the Sun than the terrestrial planets? **(B)**

21. At highway speeds (100 km per hour), how long would it take for you to drive a distance equal to the thickness of continental crust? Of the entire mantle? From the base of the mantle to the core? **(E)**

Online Resources

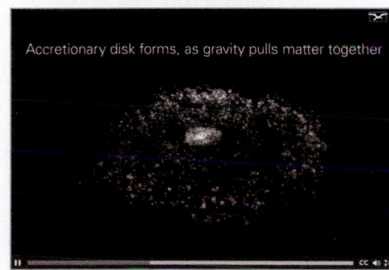

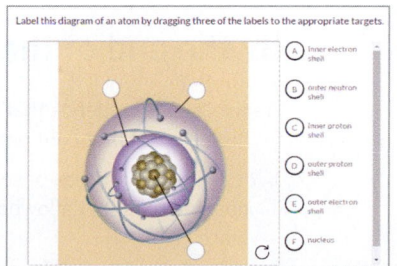

Videos

This chapter features videos showing how our Solar System and the Earth formed, including coverage of nebular theory, proto-Earth, the formation of the Moon, and the beginning of the atmosphere.

Smartwork5

This chapter features visual labeling exercises on topics such as atomic structures and our Solar System.

2 THE WAY THE EARTH WORKS

Plate Tectonics

By the end of the chapter you should be able to . . .

A. discuss the evidence that Alfred Wegener used to justify his proposal that continents drift.

B. describe the process of sea-floor spreading and the observations that allowed geologists to confirm it takes place.

C. contrast the lithosphere with the asthenosphere, identify major plates of the lithosphere, and explain how the boundaries between plates can be recognized.

D. sketch the three types of plate boundaries, and describe the nature of motion that occurs across them.

E. relate types of geologic activity to types of plate boundaries, and explain how new plate boundaries can form and existing ones can cease activity.

F. outline the major ideas now included in the modern theory of plate tectonics, and reinterpret Wegener's observations in the context of his theory.

G. describe how measurements of paleomagnetism have helped to prove plate tectonics happens.

H. explain the methods scientists use to describe and measure the velocity of plate motion.

2.1 Introduction

In September 1930, a German meteorologist, Alfred Wegener, along with several colleagues, sledged across the ice sheet of Greenland to resupply weather observers stranded at a remote camp **(Fig. 2.1a)**. After dropping off crates of food, Wegener and one companion decided to head back immediately. Sadly, they were never seen again. At the time of his death, Wegener was well known, not only to researchers studying climate, but also to geologists, because some 15 years earlier he had published a book in which he challenged geologists' long-held assumption that the location of a continent remains fixed for all of geologic time. Wegener proposed instead that in the past, continents fit together like pieces of a giant jigsaw puzzle in one vast supercontinent, and that this

supercontinent—which he named **Pangaea** (pronounced Pan-JEE-ah; Greek for all land)—later fragmented into separate continents. Over time, these continents moved apart into their present positions. He named this process **continental drift** (Fig. 2.1b).

Initially, hardly any geologists accepted Wegener's hypothesis, and Wegener died without knowing that the idea would later become the foundation of a scientific revolution. Today, geologists take for granted that the map of the Earth constantly changes—continents do indeed slowly waltz around its surface, and they have combined, and then broken apart, more than once over geologic time.

The revolution that led to our modern image of a dynamic Earth began in 1960, when an American professor, Harry Hess (1906–1969), suggested that new ocean

The great cleft in the land represents the trace of the Mid-Atlantic Ridge, where it cuts through Iceland. The theory of plate tectonics explains why land on the near side moves away from land on the far side at a rate of 2 cm (1 inch) per year.

Figure 2.1 Alfred Wegener and his model of continental drift.

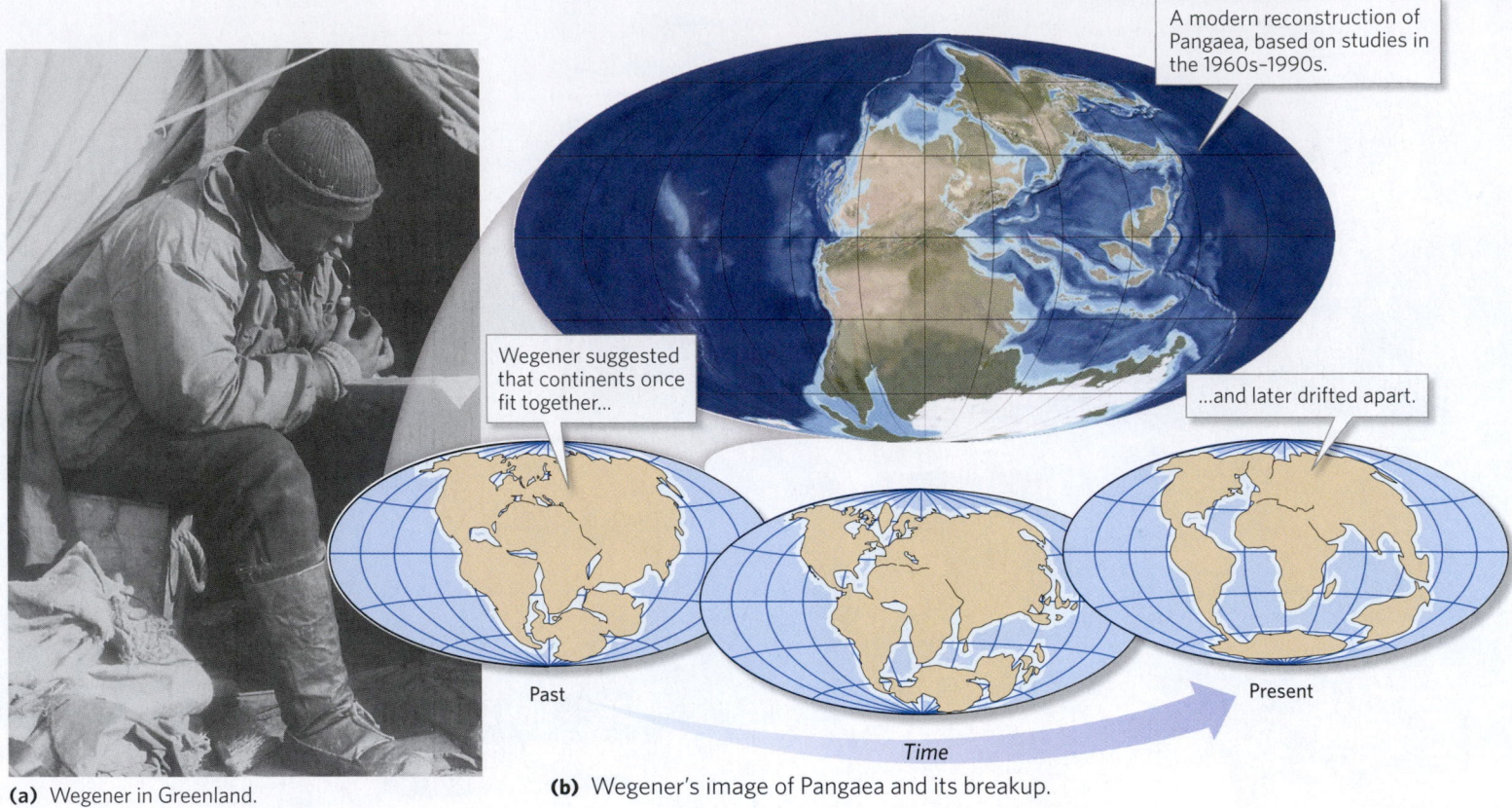

A modern reconstruction of Pangaea, based on studies in the 1960s–1990s.

Wegener suggested that continents once fit together...

...and later drifted apart.

Past

Present

Time

(a) Wegener in Greenland.

(b) Wegener's image of Pangaea and its breakup.

See for **yourself**

The Fit of Continents

Latitude: 33°31′25.77″ N
Longitude: 39°26′23.32″ W

Zoom to an elevation of 14,800 km (9,200 miles) and look straight down.

Note how northwest Africa could fit snugly along eastern North America. Wegener used this fit as evidence for Pangaea.

floor forms between two continents as they move apart, a process now known as *seafloor spreading*. Hess also suggested that continents can move toward each other when the old ocean floor between them sinks back down into the Earth's interior, a process now known as *subduction*. Many geologists began to explore the implications of Hess's suggestions, and by 1968, they had developed a fairly complete model describing how seafloor spreading and subduction operate. In this model, the Earth's *lithosphere*—its outer, relatively rigid shell—consists of several pieces, or *plates*, that slowly move relative to one another. Because geologists have confirmed this model by many observations, it has gained the status of a theory, called the *theory of plate tectonics*, or simply *plate tectonics*. The name comes from the Greek word *tekton*, which means builder, because plate movements "build" regional geologic features.

We begin this chapter by introducing the observations that led Wegener to propose his continental-drift hypothesis. Then we discuss subsequent discoveries that led to plate tectonics theory. Next, we describe the nature of lithosphere plates and the boundaries between them, the way that measurements of magnetism prove plate tectonics theory, and the way geologists describe plate motions. You'll see that plate tectonics provides a basis for understanding much of geology.

2.2 Continental Drift

Wegener's Evidence

Why did Wegener propose that Pangaea once existed and later separated into continents that then drifted apart? To find an answer, let's look at Wegener's key observations.

FIT OF THE CONTINENTS. Almost as soon as maps depicting the Atlantic Ocean became available, scholars noted that Africa and Europe looked like they would fit snugly against the Americas if the Atlantic Ocean didn't exist. Wegener took this concept further by showing that all the continents could fit together, with remarkably few overlaps or gaps, to form Pangaea **(Fig. 2.2)**. He argued that that this fit was too good to be coincidence.

DISTRIBUTION OF CLIMATE BELTS IN THE PAST. Wegener knew that different climate belts occur at different latitudes. For example, polar climates lie at high latitudes, hot and dry climates at subtropical latitudes, and steamy wet climates at tropical latitudes. Wegener speculated that if a continent drifted from one latitude to another, the climate at a location on the continent would change over time, and that a succession of sedimentary-rock beds (layers) preserved at the location would record this

Figure 2.2 The "Bullard fit" of the continents. In 1965, Edward Bullard used a computer to fit the continents together and demonstrate how minor the gaps and overlaps are, although the match still isn't perfect.

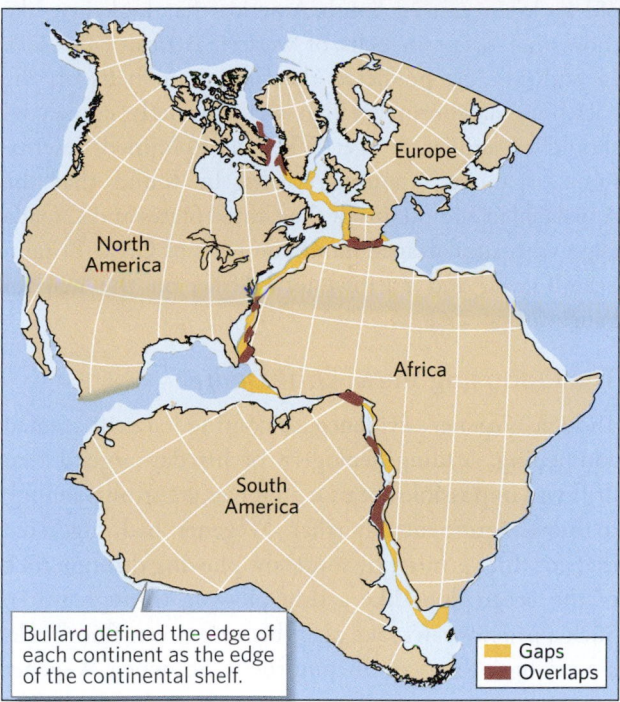

Bullard defined the edge of each continent as the edge of the continental shelf.

Gaps
Overlaps

Figure 2.3 Evidence for a supercontinent.

(a) Glacial striations of late Paleozoic age on the surface of bedrock along the southern coast of Australia.

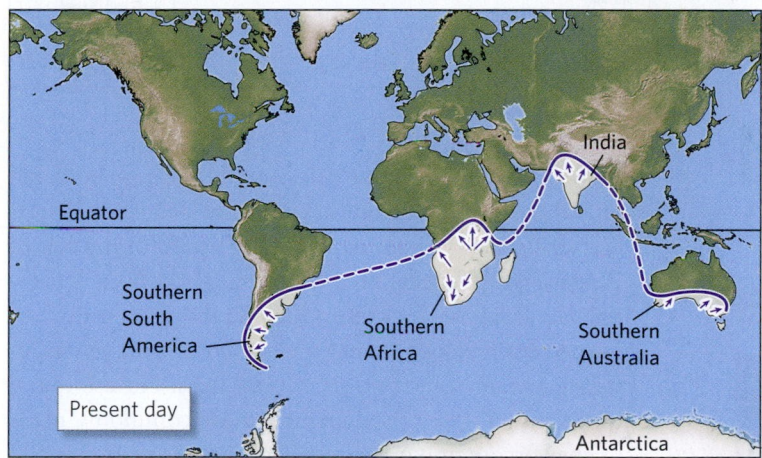

(b) A map showing the distribution of late Paleozoic glacial deposits and the orientation of associated striations.

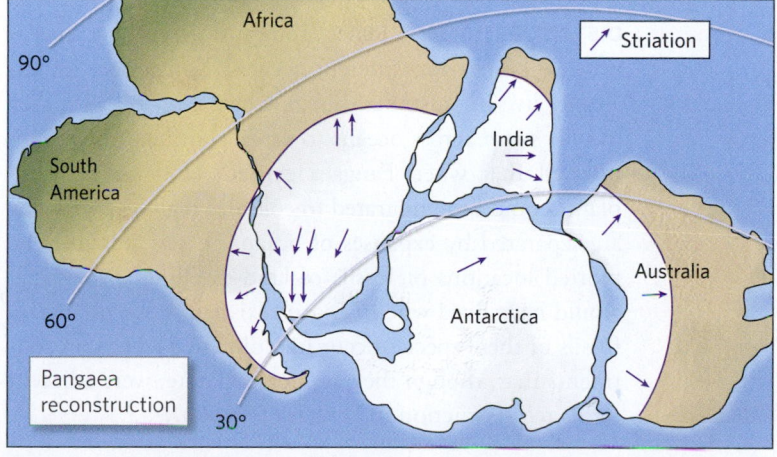

(c) In Wegener's reconstruction of Pangaea, the glaciated areas connect to outline a region of late Paleozoic southern polar ice caps.

change, for sedimentary rocks contain clues to the environment in which they formed (see Chapter 5).

To confirm his idea, Wegener first looked for evidence defining the distribution of Paleozoic-age ice sheets, or *continental glaciers*, for such glaciers tend to develop at high (polar) latitudes. His search paid off. Wegener found *glacial striations*, scratches carved by glaciers, in South America, Africa, India, and Australia, land areas that all now lie in nonpolar latitudes **(Fig. 2.3a)**. Further, the striations indicated that the Paleozoic ice moved from what is now ocean onto the land. Glaciers can't move this way—instead, they must flow from the land to the sea. Wegener also found Paleozoic sedimentary beds composed of a distinctive type of sediment left by glaciers. He realized that Paleozoic glacial features could not have formed at the latitudes where they now occur, and suggested that their present locations make sense if the continents were once united in Pangaea, with the southern part of Pangaea lying beneath a polar ice sheet **(Fig. 2.3b, c)**.

If the southern part of Pangaea straddled the South Pole in the late Paleozoic, then at that time, southern North America, southern Europe, and northwestern Africa would have straddled the equator and would have hosted tropical climates in which swamps and reefs would

Figure 2.4 Climate belts, as indicated by distinct rock types, make sense on a map of Pangaea.

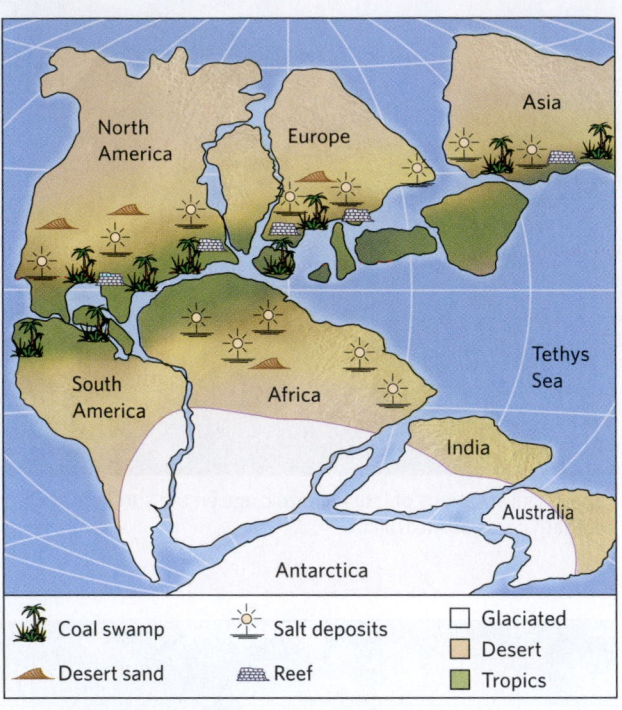

Coal swamp

Desert sand

Salt deposits

Reef

☐ Glaciated

Desert

Tropics

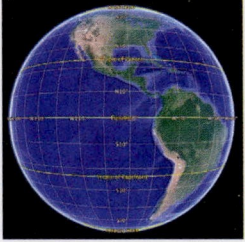

See for yourself

Climate Belts

Latitude: 0°10′24.19″ S
Longitude: 93°7′7.22″ W

Zoom to 15,000 km (9,200 miles), add the grid, and look straight down. Click "grid."

North America is now at temperate to arctic latitudes. Wegener noted that in the late Paleozoic it was tropical, as if near the equator.

grow. The tropical climate belts would have been bordered on either side by subtropical climate belts hosting deserts and shallow seas of particularly salty water. In the belt of Pangaea that Wegener predicted would have been equatorial, late Paleozoic rock layers include abundant coal (rock formed from woody plant remains) containing *fossils* (remains preserved in rock) of tropical trees. He also found limestone (a type of sedimentary rock commonly made of shells) containing the fossils of organisms that prefer warm water. In the portions of Pangaea that Wegener predicted would be subtropical, late Paleozoic sedimentary beds include deposits formed in desert dunes or very salty water **(Fig. 2.4)**.

DISTRIBUTION OF FOSSILS. Today, different continents provide homes for different species. Kangaroos, for example, live in Australia, but nowhere else, because they cannot swim across oceans to other continents. Wegener inferred that when Pangaea existed, land animals and plants could have migrated to colonize regions that today are separated by expanses of ocean. To test this idea, he plotted locations of fossils of land-dwelling species that would have lived when Pangaea existed. As he predicted, fossils of these species occur in sedimentary beds on continents that, though they are now separate, were adjacent in his reconstruction of Pangaea **(Fig. 2.5)**.

MATCHING GEOLOGIC UNITS AND MOUNTAIN BELTS. Wegener speculated that distinct rock types should be traceable from the coast of one continent to the coast of the adjacent continent on his reconstruction of Pangaea. He was correct again. Specifically, he found belts of very old rocks in eastern South America that look just like those that occur in belts of western Africa **(Fig. 2.6a)**. In addition, he predicted that mountain belts on the coastlines of continents that were formerly connected should be similar in age. When he examined descriptions of ancient mountain belts, he found that the Appalachian Mountains of the United States and Canada align with similar-aged mountain belts of Great Britain, Scandinavia, and northwestern Africa on the map of Pangaea **(Fig. 2.6b, c)**.

The Opposing View: Drift Denial

Though Wegener's evidence for continental drift seemed compelling, leading geologists of his day argued that drift was impossible because no forces are strong enough to move continents. Further, Wegener had suggested that continents move by somehow plowing through rock of the ocean floor like a ship plowing through water. Such a process can't take place because ocean-floor rocks are stronger than continental rocks. So, when Wegener perished in a Greenland blizzard, most geologists remained unconvinced of his proposals and preferred to think that continents do not move relative to one another. It would take three more decades before this *fixist view* would finally die, a victim of new observations made possible by research technologies not available in Wegener's day. Geologists came to prefer a *mobilist view*, that movement of continents is happening, one of many consequences of a broader process: plate tectonics. The first step in this revolution came with the recognition of seafloor spreading.

Take-home message . . .

Alfred Wegener proposed his hypothesis of continental drift based on his observations of the shape of coastlines, on the record of past climates preserved in sedimentary rocks, on the distribution of fossils, and on the matching of rocks and mountains across oceans. He argued that the continents formerly made up one supercontinent, Pangaea, that later broke apart. The hypothesis was not widely accepted until decades after Wegener's death.

Quick Question -
Why were geologists of his day opposed to Wegener's hypothesis?

Figure 2.5 Fossil evidence for seafloor spreading.

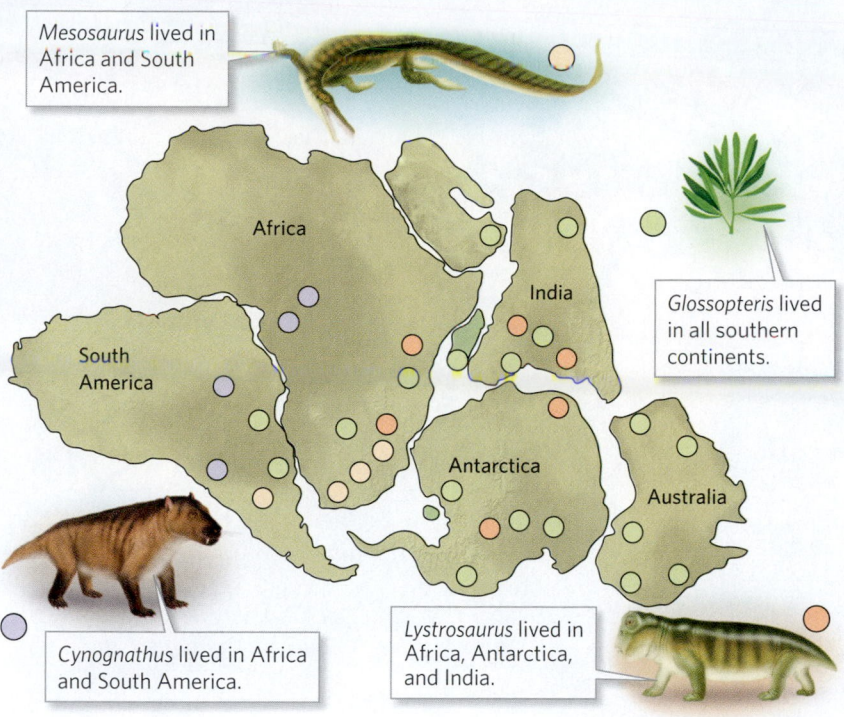

Mesosaurus lived in Africa and South America.

Africa

India

Glossopteris lived in all southern continents.

South America

Antarctica

Australia

Cynognathus lived in Africa and South America.

Lystrosaurus lived in Africa, Antarctica, and India.

(a) Fossil locations show that Mesozoic land-dwelling organisms occurred on more than one continent.

(b) Fossil leaves of *Glossopteris*, a land plant, from rock in Australia. The presence of such fossils provided supporting evidence for the proposal that the continents were attached in the past.

Figure 2.6 Further evidence of drift: rocks on different sides of the ocean match.

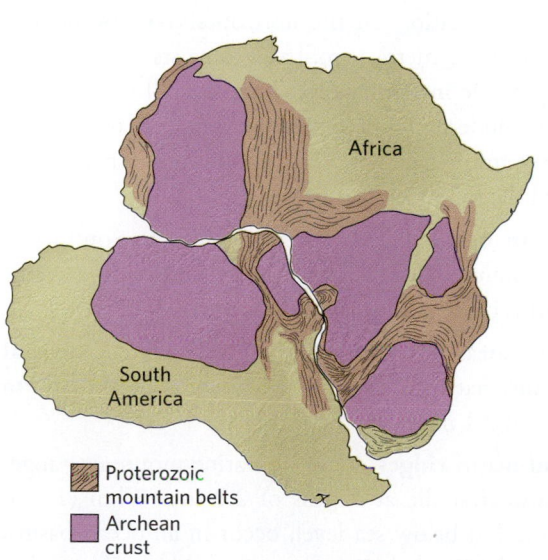

Africa

South America

▨ Proterozoic mountain belts

▨ Archean crust

(a) Distinctive belts of rock in South America align with similar ones in Africa without the Atlantic Ocean. Proterozoic rocks are ~2.5 to 0.5 Ga. Archean rocks are >2.5 Ga.

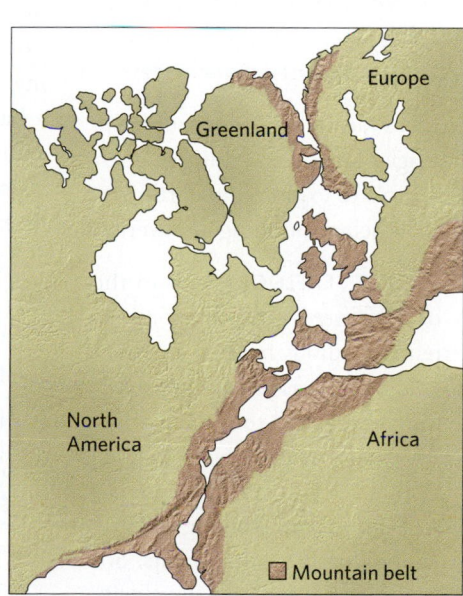

Europe

Greenland

North America

Africa

▨ Mountain belt

(b) If the Atlantic Ocean didn't exist, Paleozoic mountain belts on both coasts would be adjacent.

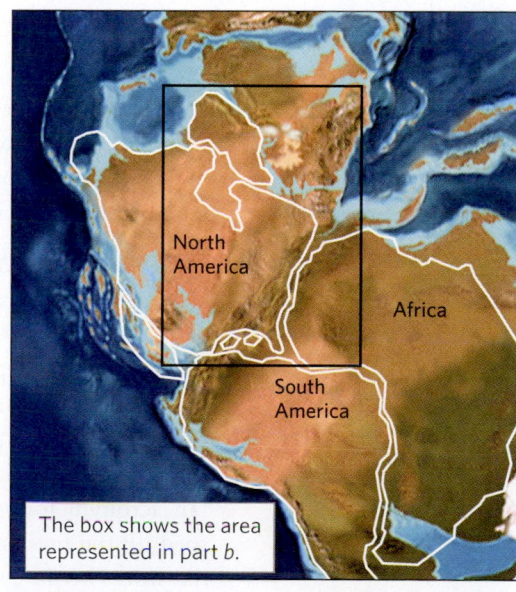

North America

Africa

South America

The box shows the area represented in part *b*.

(c) A modern reconstruction showing the positions of mountain belts in Pangaea. Modern continents are outlined in white.

Figure 2.7 Bathymetry of the whole ocean.

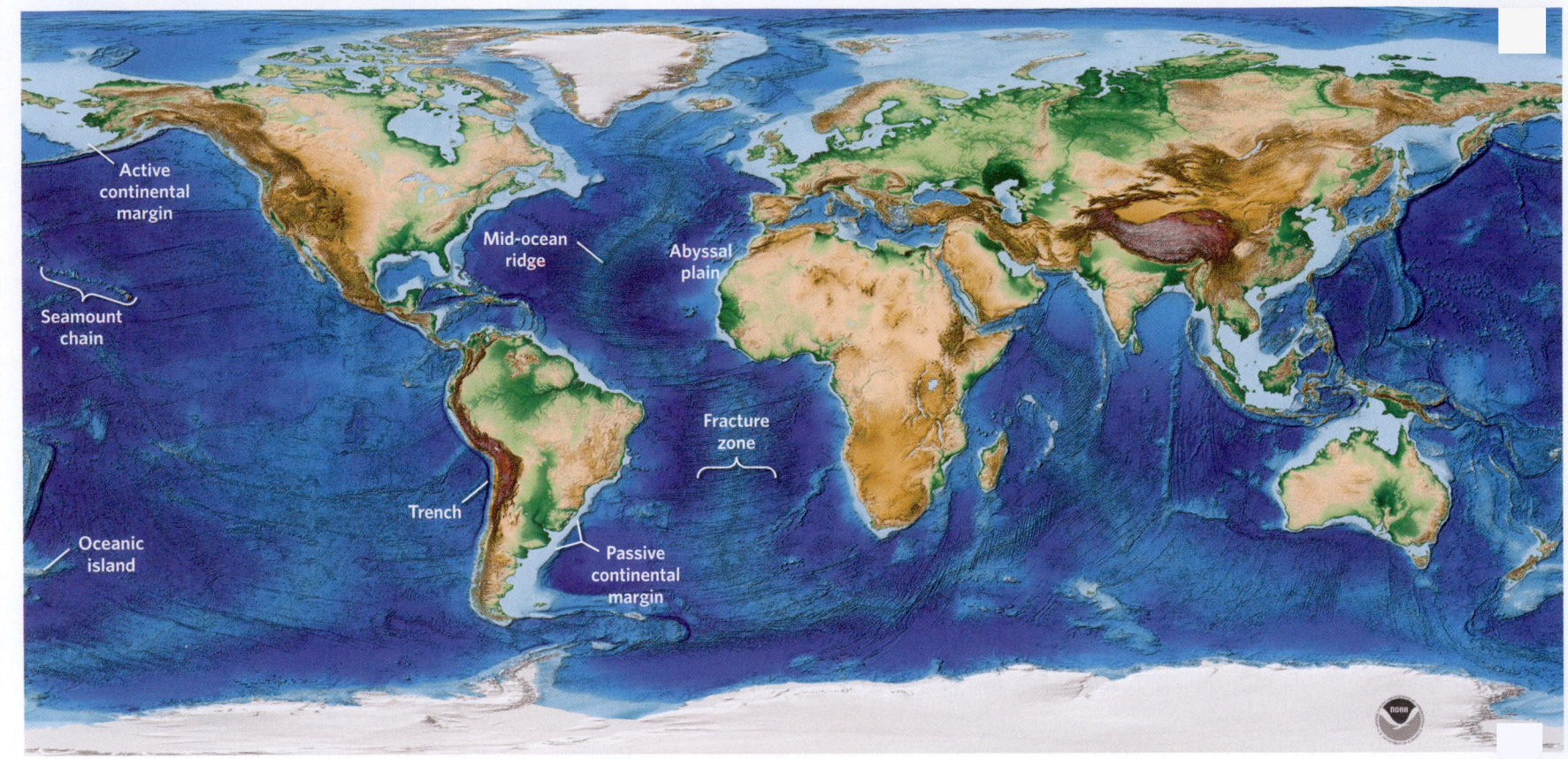

Active
continental
margin

Mid-ocean
ridge

Abyssal
plain

Seamount
chain

Fracture
zone

Trench

Oceanic
island

Passive
continental
margin

2.3 The Discovery of Seafloor Spreading

New Geologic Observations (1930–1960)

From the time of Wegener's death in 1930 until the 1960s, geology entered an age of discovery, during which researchers developed new tools to explore the Earth System. Here, we review some of their astounding discoveries.

BATHYMETRIC FEATURES OF THE SEAFLOOR. Prior to the 20th century, the study of *bathymetry*—depth variations of the seafloor—was extremely tedious because the only way to measure depth at a location was to make a *sounding* by lowering a heavy weight attached to a cable down to the seafloor. In the deeper parts of the ocean, a single measurement could take hours. The invention of *sonar* (an acronym for *so*und *na*vigation *a*nd *r*anging) greatly sped up the process of making depth measurements. Sonar works because sound waves travel at a known velocity in water. Therefore, the interval between the time when a transmitter on a ship emits a sound pulse aimed at the seafloor and the time when the echo returns to a receiver on the ship after bouncing off the seafloor represents

the seafloor depth. With sonar, it's possible to obtain a continuous record of the depth of the seafloor while a ship cruises, and from this record, to produce a *bathymetric profile*, a diagram that plots depth on the vertical axis against location on the horizontal axis. By obtaining many bathymetric profiles, geologists can produce a **bathymetric map** that reveals the overall shape of the seafloor. Modern technology allows satellites to map bathymetry from space **(Fig. 2.7)**. Bathymetric maps have revealed the following features:

- **Ocean basins** are distinctly lower than continents. The boundary between an ocean basin and a continent is called a **continental margin**.

- Large areas of ocean basins consist of broad **abyssal plains**, flat regions that lie at depths of 4 to 5 km (2.5 to 3.1 miles) below sea level.

- **Mid-ocean ridges**, long submarine mountain ranges whose crests lie at depths of 2.0 to 2.5 km (1.2 to 1.5 miles) below sea level, occur in all ocean basins. The adjective *mid-ocean* can be a bit confusing because it gives the impression that these features all run down the center of ocean basins. This isn't always the case—some ridges lie closer to one side of an ocean basin than they do to the other. Mid-ocean ridges

Figure 2.8 Faulting and earthquakes.

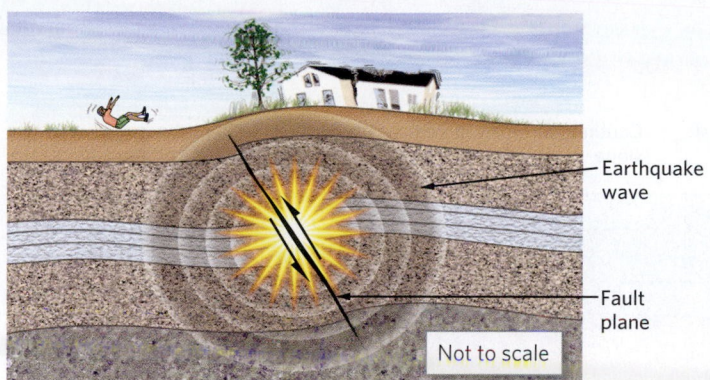

(a) When the rock inside the Earth suddenly breaks and slips, forming a fracture called a fault, it generates shock waves that pass through the Earth and shake the surface.

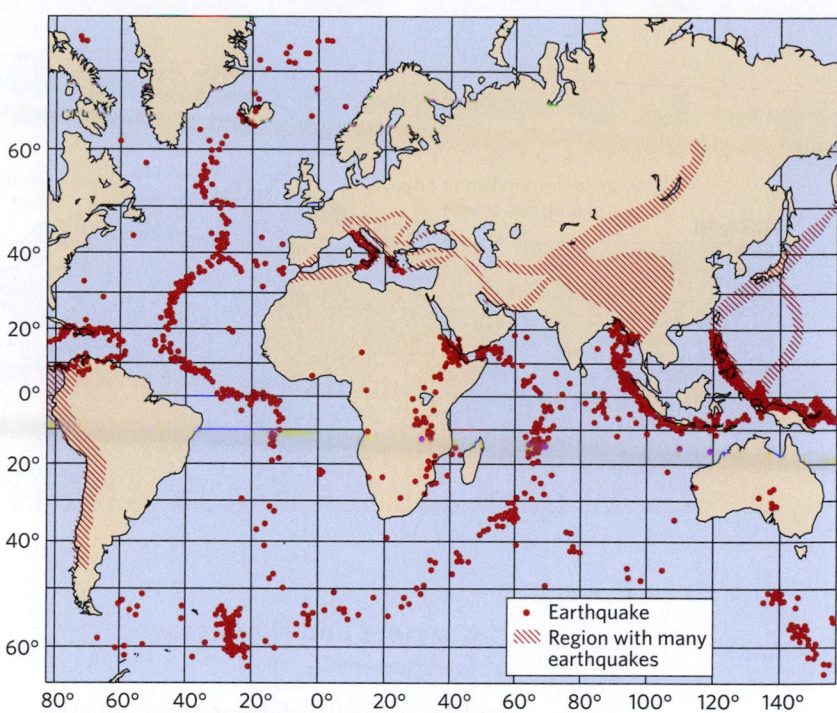

(b) A 1953 map showing the distribution of earthquake locations in the ocean basins. Note that earthquakes occur in belts.

tend to be symmetrical, in that the bathymetry on one side of a *ridge axis* (the line following the crest) is a mirror image of the bathymetry on the other side. In addition, they tend to be segmented, in that along its length, a mid-ocean ridge consists of segments that do not align end to end.

- Narrow bands of broken-up oceanic crust, containing vertical fractures bounded by steep cliffs, cross mid-ocean ridges and trend roughly at right angles to the ridge axis. These features are known as **fracture zones**. The segments of a mid-ocean ridge terminate at fracture zones.

- Elongate troughs, known as **deep-sea trenches**, in which the ocean floor reaches depths of 7 to 11 km (4 to 7 miles), occur in or along the edges of ocean basins. Some trenches border continents, but some do not. Further, not all continental edges border trenches.

- At hundreds of locations in the world's oceans, volcanic eruptions have produced peaks of igneous rock that built up above the surrounding seafloor. If the top of a peak rises above sea level, it forms an **oceanic island**, whereas if its top lies below sea level, it's a **seamount**. Oceanic islands and seamounts typically occur in chains.

CHARACTER OF OCEANIC CRUST. In addition to obtaining information about the shape of the seafloor surface, geologists dredged the seafloor, drilled into the seafloor, and used various instruments to characterize subsurface layers within the oceanic crust. This work led to the realization that the upper layer of the oceanic crust beneath abyssal plains consists of a relatively thin sediment that settled down from overlying seawater. Almost no sediment occurs at the ridge axis. The sediment layer thickens

progressively away from a mid-ocean ridge. Beneath the sediment layer, oceanic crust consists of *basalt*, a dense gray igneous rock.

Researchers also measured the temperature in holes drilled into the seafloor. From these measurements, they calculated **heat flow**, the rate at which heat rises from the Earth's interior. By comparing results from many locations, researchers learned that heat flow beneath mid-ocean ridges is greater than that beneath abyssal plains.

DISTRIBUTION OF EARTHQUAKES. A **fault** is a fracture in the Earth's crust on which *slip* (sliding) takes place. Most **earthquakes**—episodes of ground shaking—are due to fault slip **(Fig. 2.8a)** because the sudden movement generates vibrations that travel through the Earth in the form of waves, and when these waves reach the ground surface, they cause the shaking we feel as an earthquake.

When researchers learned how to find the location of an earthquake (see Chapter 8), it became possible to produce maps that show the global distribution of earthquakes. Even early versions of such maps demonstrated that earthquakes do not occur randomly, but rather cluster in distinct belts, called **seismic belts**, from the Greek word *seismos*, which means earthquake **(Fig. 2.8b)**. Seismic belts lie along trenches, along mid-ocean ridge axes, along parts of fracture zones, and on other faults.

See for yourself

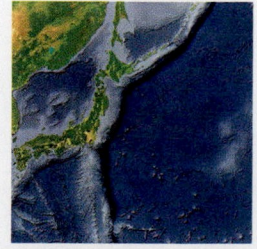

Trenches

Latitude: 35°37′59.86″ N
Longitude: 145°36′12.78″ E

Fly to an elevation of 5,500 km (3,500 miles) and look straight down. Zoom in and use the elevation tool to measure trench depth. The dark bands that you see are the trenches of the western Pacific Ocean, near Japan.

Figure 2.9 Harry Hess's basic concept of seafloor spreading (1962). Hess implied, incorrectly, that only the crust moved. We will see that this sketch is an oversimplification and contains errors. But it does introduce, for the first time, the concept that ocean basins grow.

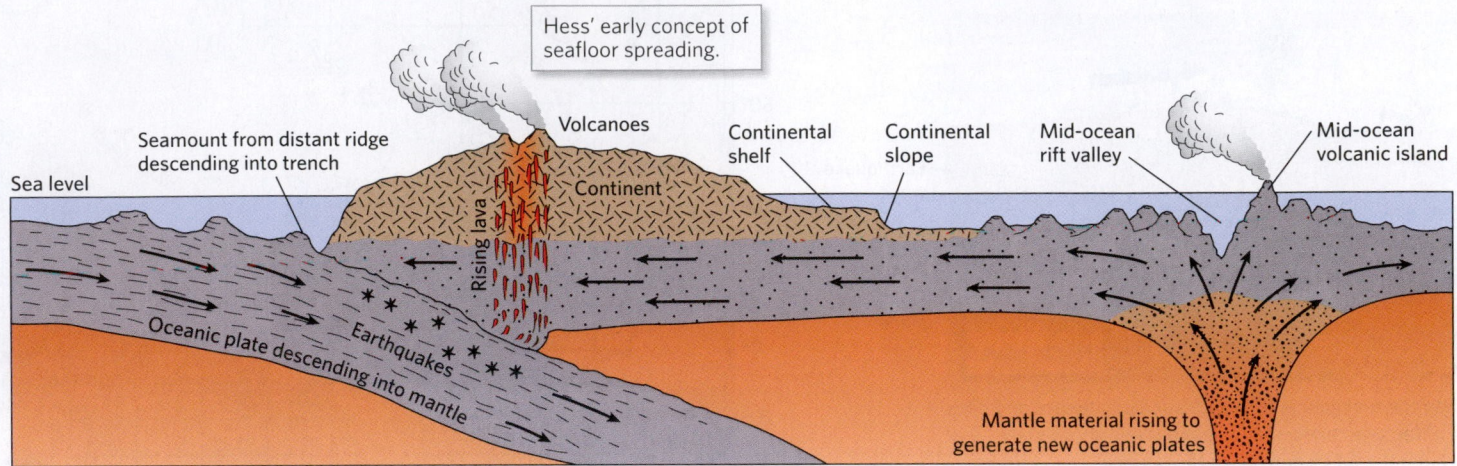

Hess' early concept of seafloor spreading.

Seamount from distant ridge descending into trench

Volcanoes

Continental shelf

Continental slope

Mid-ocean rift valley

Mid-ocean volcanic island

Sea level

Continent

Rising lava

Earthquakes

Oceanic plate descending into mantle

Mantle material rising to generate new oceanic plates

Harry Hess and His "Essay in Geopoetry"

Initially, geologists did not see relationships among the observations described above. That changed when Harry Hess realized that the observations were clues that could ultimately help answer the question of how continents move. Hess began by thinking about the implications of seafloor-sediment thickness. He thought that if the ocean basins were as old as the Earth itself, they should be covered by thick sediment. The thinness of the sediment found on the ocean floor indicated that the ocean floor couldn't be as old as the Earth. Further, the progressive increase in thickness of the sediment away from mid-ocean ridges suggested that ridges are younger than abyssal plains. Hess concluded, therefore, that new ocean floor must be forming at the ridges, so that an ocean basin can grow wider with time. But how? The association of earthquakes with mid-ocean ridges provided an answer to Hess—seafloor must be cracking and splitting apart at the ridges. The high heat flow along mid-ocean ridge axes further suggested that molten rock was rising beneath the zone where this splitting took place.

Hess put the clues together and proposed that ocean basins form by a process that came to be known as **seafloor spreading**. According to Hess's proposal, seafloor stretched apart along the axis of a mid-ocean ridge, and new oceanic crust formed from magma that rose and filled the space. Once formed, new seafloor moved away from the ridge, and as a consequence, an ocean basin could grow wider over time. Hess realized that if new seafloor forms at the ridge axis, then somewhere else, old seafloor must be consumed—otherwise, the Earth would have to be expanding significantly, and no evidence indicated that

it had. So he proposed that old seafloor sank back into the mantle at deep-sea trenches, and that this movement generated the seismic belts observed along trenches **(Fig. 2.9)**.

Hess, along with other geologists, recognized that seafloor spreading provided the long-sought explanation of how continental drift occurs, an explanation that had eluded Wegener. Simply put, continents move apart due to seafloor spreading at mid-ocean ridges, and continents move toward each other as the seafloor between them is consumed at trenches. The hypothesis seemed so elegant that Hess referred to it as "an essay in geopoetry."

Take-home message . . .

Observations of seafloor bathymetry, sediment cover, heat flow, and seismicity led to the proposal that ocean basins open by seafloor spreading, during which new seafloor forms at mid-ocean ridges, and they close by subduction, during which old seafloor sinks back into the mantle at trenches.

Quick Question -
What is the relationship between seismic belts and bathymetric features of the ocean basins?

2.4 Modern Plate Tectonics Theory

The proposal of seafloor spreading set off a revolution in geologic thought. Over the course of the next decade, most geologists came to realize that their long-held interpretations of the Earth, based on the fixist premise that the positions of continents did not change and that the ocean basins that we see today date back to the birth of the

Figure 2.10 The nature and behavior of the lithosphere.

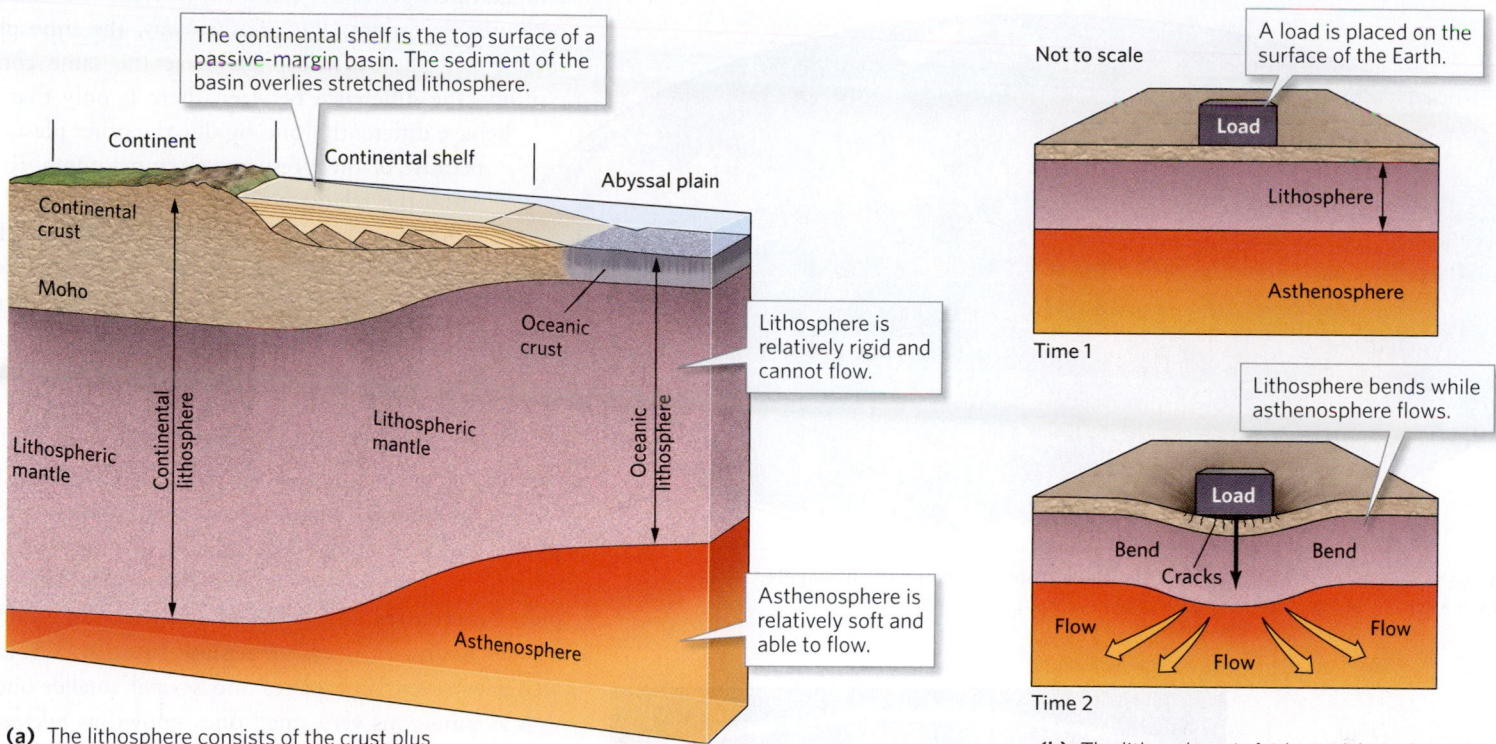

The continental shelf is the top surface of a passive-margin basin. The sediment of the basin overlies stretched lithosphere.

Continent

Continental shelf

Abyssal plain

Continental crust

Moho

Oceanic crust

Lithospheric mantle

Continental lithosphere

Lithospheric mantle

Oceanic lithosphere

Asthenosphere

Lithosphere is relatively rigid and cannot flow.

Asthenosphere is relatively soft and able to flow.

(a) The lithosphere consists of the crust plus the uppermost mantle. It is thicker beneath continents than beneath oceans.

Not to scale

A load is placed on the surface of the Earth.

Load

Lithosphere

Asthenosphere

Time 1

Lithosphere bends while asthenosphere flows.

Load

Bend   Bend

Cracks

Flow   Flow

Flow

Time 2

(b) The lithosphere is fairly rigid, but when a heavy load, such as a glacier or volcano, builds on its surface, the surface bends down. This can happen because the underlying plastic asthenosphere can flow out of the way.

Earth, were simply wrong! If seafloor spreading and subduction take place, and as a consequence, the continents move, then the outer layer of the Earth, the shell that we live on, must be quite mobile indeed.

By 1968, researchers had converged on a new model of the Earth. In this model, the Earth has a rigid outer shell. (A *rigid material*, in this context, is one that bends and breaks when subjected to a force, but does not flow.) This shell is not completely intact, but rather consists of separate pieces, or *plates*, that move relative to one another. After this model passed many tests, it came to be known as the *theory of plate tectonics*, or simply **plate tectonics**. Geologists realized that the theory could explain not only all of Wegener's observations, but also many other important geologic phenomena, such as earthquakes, volcanoes, and mountain building. In effect, plate tectonics became the *grand unifying theory* of geology. In this section, we describe key aspects of the theory. We begin by clarifying what we mean by a plate.

The Concept of Lithosphere and Asthenosphere

As we noted in Chapter 1, geologists back in the 19th century concluded that the interior of the Earth consists of distinct layers (crust, mantle, and core). At first, geologists

thought that the rigid shell of the Earth consisted only of the crust—in fact, that's how Hess portrayed plates in his sketch depicting seafloor spreading. But this image is not correct. The rigid outer shell of the Earth includes not only the crust, but also the top part of the upper mantle directly beneath the crust. Geologists refer to this combination—crust plus the uppermost part of the upper mantle—as the **lithosphere**. The crust, as we've seen, is about 7 to 10 km (4 to 6 miles) thick beneath the oceans and about 25 to 70 km (15 to 45 miles) thick beneath continents. The lithosphere, in contrast, is about 100 km (60 miles) thick beneath the oceans and 150 km (90 miles) thick beneath the continents **(Box 2.1)**. Geologists refer to the part of the mantle within the lithosphere as the **lithospheric mantle (Fig. 2.10a)**. The portion of the mantle that lies beneath the lithosphere is called the **asthenosphere**. In contrast to the lithosphere, the mantle of the asthenosphere displays *plastic behavior*. (A *plastic material*, in this context, is a material can flow slowly without breaking.) The asthenosphere flows plastically at rates of only 1 to 15 cm (0.5 to 6 inches) per year.

The distinctions between crust and mantle and between lithosphere and asthenosphere can be confusing. Crust differs from mantle by its composition, meaning that mantle rock contains a different mix of chemicals

Figure 2.11 Lithosphere plates and continental margins.

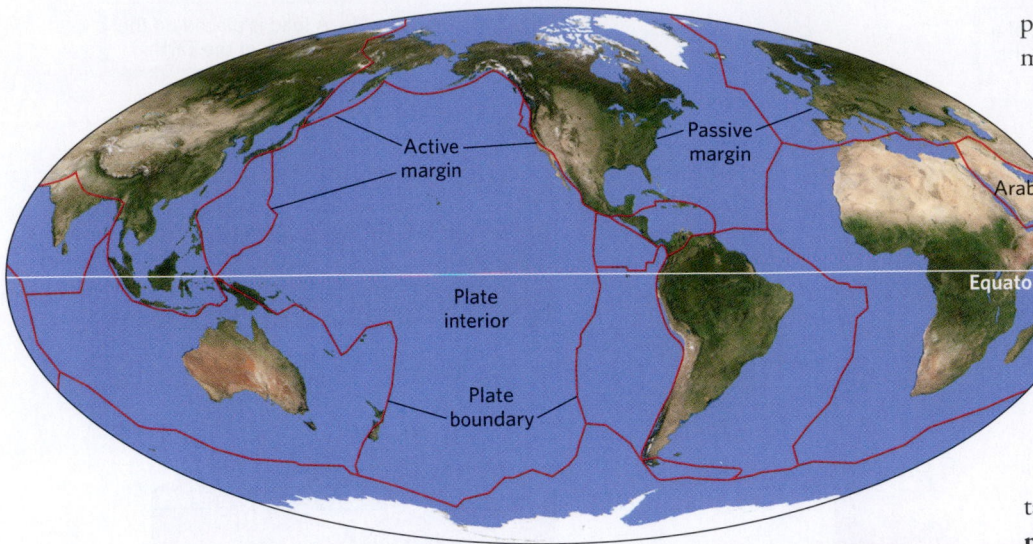

(a) An image of the globe, on which some of the major plate boundaries are highlighted. Active continental margins lie along plate boundaries; passive margins do not.

(b) A bathymetric image of the continental shelf along the east coast of North America.

than does crustal rock, whereas lithosphere differs from asthenosphere by its response to a force (lithosphere behaves rigidly, whereas asthenosphere behaves plastically). If an imaginary heavy load were placed on the surface of the lithosphere, the lithosphere would bend down and crack, while the underlying asthenosphere would flow out of the way **(Fig. 2.10b)**. Note that the Moho is the base of the crust, not the base of the lithosphere (see Chapter 1).

Why is the lithospheric mantle rigid while the asthenosphere is not? This contrast in behavior exists because the Earth gets hotter with increasing depth, and as rock gets hotter, it gets softer. At the depth of the lithosphere's

base, mantle rock (peridotite) reaches a temperature of 1,280°C (2,340°F). At this temperature, and at hotter temperatures, peridotite can flow, whereas at cooler temperatures, it cannot. Put another way, the lithospheric mantle and the asthenosphere have the same composition. The difference between them is only that they behave differently (one rigidly, the other plastically) because of their respective temperatures. To picture the relationship between temperature and rigidity, imagine a simple experiment: Place one wax candle in a freezer and one in a warm oven. After a while, take them out and try to bend them. The cold candle resists bending and breaks because it's rigid, whereas the warm one changes shape without breaking because it's plastic. Heat makes rock, like wax, softer.

As we've mentioned, the Earth's lithosphere contains a number of major breaks that separate it into **plates**, which geologists also refer to as *lithosphere plates* **(Fig. 2.11a)**. The dividing lines between plates are called **plate boundaries**, and a portion of a plate that is not near a plate boundary is called a **plate interior**. Geologists recognize seven large plates and several smaller ones, as well as numerous very small ones known as *microplates*. Some plates have familiar-sounding names, such as the North American Plate and the African Plate, while some have less familiar names, such as the Cocos and Juan de Fuca Plates.

Note that some plates consist entirely of oceanic lithosphere, whereas others consist of both oceanic and continental lithosphere. In addition, some continental margins are plate boundaries but some are not, so geologists distinguish between **active margins** of continents, which coincide with plate boundaries, and **passive margins**, which do not. Typically, a thick wedge of sediment accumulates along a passive margin. This wedge includes sand and mud carried to the sea by rivers as well as the shells of marine organisms. The surface of such a wedge is a **continental shelf**, a broad region of shallow sea between the continent and the abyssal plain **(Box 2.1 and Fig. 2.11b)**.

Identifying Plate Boundaries

How do we recognize the location of a plate boundary? All plate boundaries are identified by the presence of *active faults*, meaning faults where slip has occurred recently and will probably occur in the future. Slip on faults at plate boundaries permits one plate to move relative to its neighbor, and this slip generates earthquakes **(Fig. 2.12a)**. Seismic belts, therefore, define the positions of plate boundaries **(Fig. 2.12b)**. Plate interiors, the regions away from the plate boundaries, remain relatively earthquake-free because the plates are rigid.

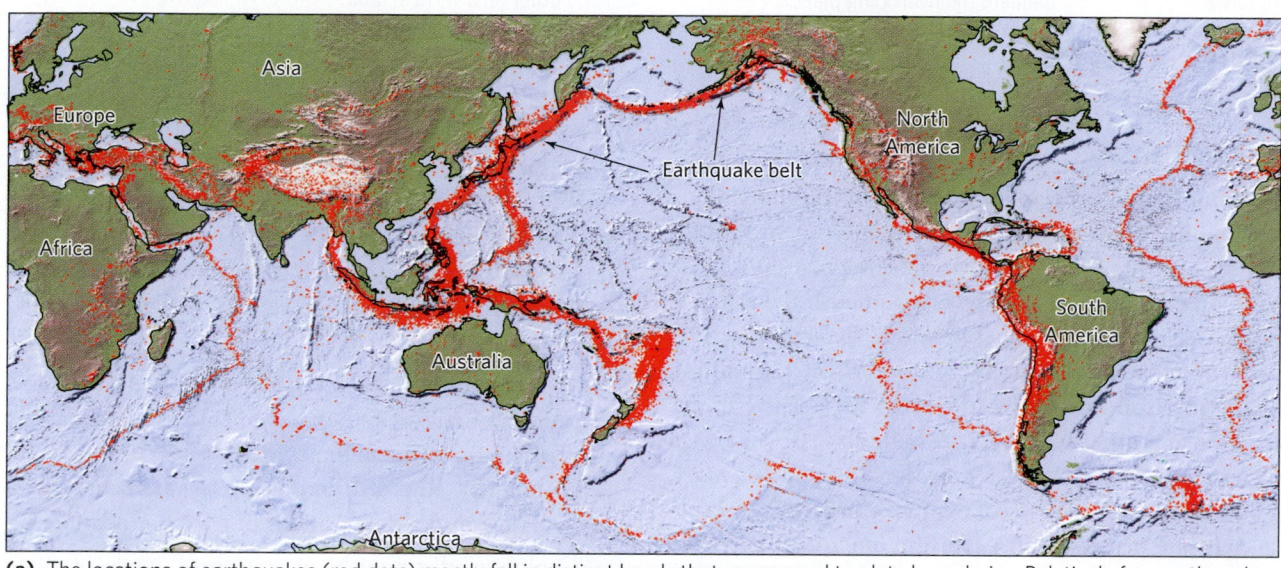

Figure 2.12 The locations of plate boundaries and the distribution of earthquakes.

(a) The locations of earthquakes (red dots) mostly fall in distinct bands that correspond to plate boundaries. Relatively few earthquakes occur in the stabler plate interiors.

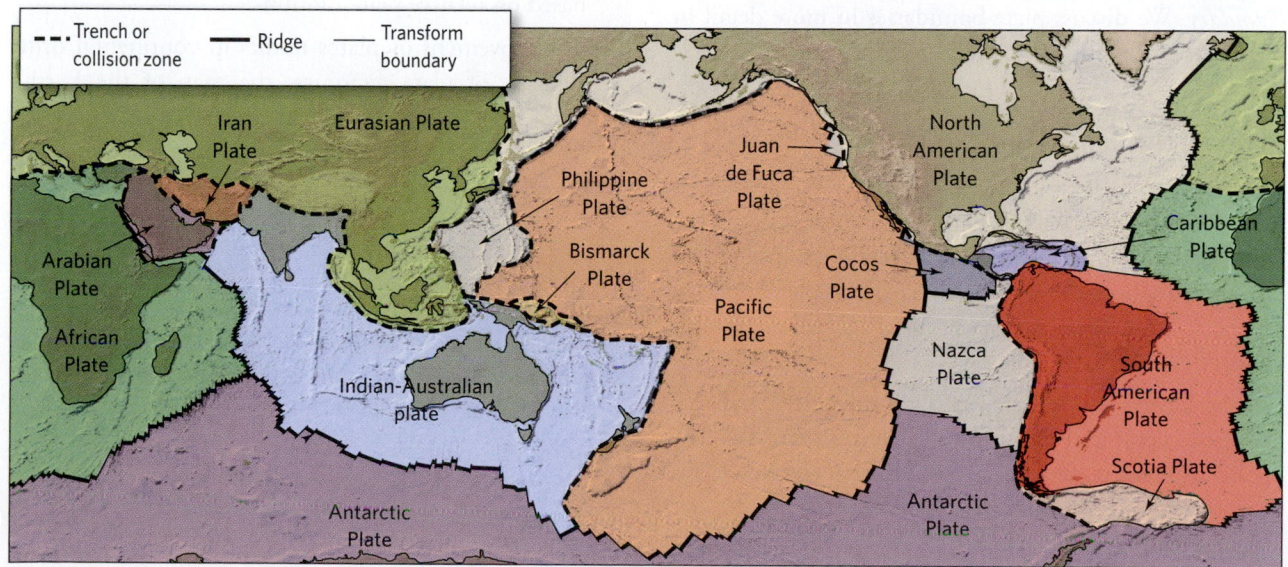

(b) A map of major plates shows that some consist entirely of oceanic lithosphere, whereas others consist of both continental and oceanic lithosphere.

Figure 2.13 The three types of plate boundaries.

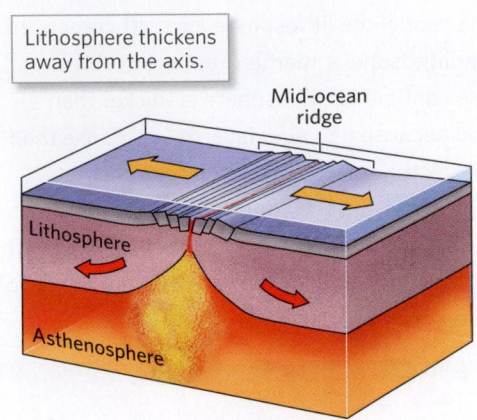

Lithosphere thickens away from the axis.

Mid-ocean ridge

Lithosphere

Asthenosphere

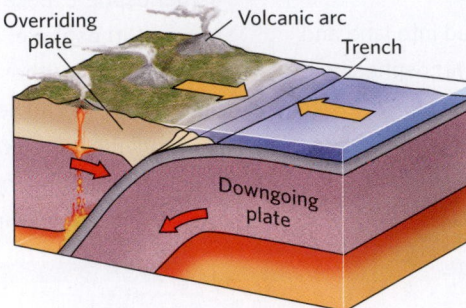

The process of consuming a plate is called subduction.

Overriding plate

Volcanic arc

Trench

Downgoing plate

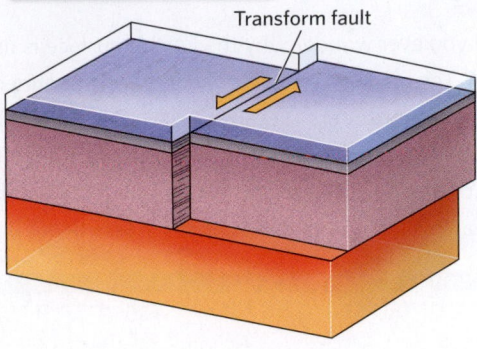

No new plate forms, and no old plate is consumed.

Transform fault

(a) At a divergent boundary, two plates move away from the axis of a mid-ocean ridge, where new oceanic lithosphere forms.

(b) At a convergent boundary, two plates move toward each other; the downgoing plate sinks beneath the overriding plate.

(c) At a transform boundary, two plates slide past each other on a vertical fault.

We've stated that plates move relative to one another. Such movement can take place because the underlying plastic asthenosphere can't hold a plate in a fixed position. To picture this movement, imagine placing a block of wood on a layer of molasses—if you push the wood, it moves relative to the molasses below. Geologists classify plate boundaries into three types based on the motion of the plate on one side of the boundary relative to the plate on the other **(Fig. 2.13)**. A boundary at which two plates move apart from each other is a *divergent boundary*, a boundary at which two plates move toward each other so that one plate sinks beneath the other is a *convergent boundary*, and a boundary at which one plate slips horizontally along the edge of another plate is a *transform boundary*. We discuss plate boundaries in more detail in the next section of this chapter.

Plate Tectonics Theory Today: A Synopsis

It took years of work by many researchers around the world to refine the theory of plate tectonics. In light of our introduction to lithosphere, asthenosphere, plates, and plate boundaries, we can summarize the key components of the theory as follows:

- The Earth's lithosphere, its rigid outer shell, consists of the crust and underlying uppermost mantle. The lithosphere is divided into plates that move relative to one another.
- Plates are on the order of 100 to 150 km (60 to 90 miles) thick and range from hundreds to thousands

of kilometers across. Some consist of only oceanic lithosphere, whereas others consist of both oceanic and continental lithosphere.

- Plates sit on the underlying asthenosphere, the portion of the mantle that is warm enough to be plastic enough to flow.
- As plates move, their interactions at plate boundaries cause earthquakes and other geologic phenomena, but plate interiors remain intact (**Earth Science at a Glance**, pp. 64–65). The distribution of seismic belts, therefore, determines the positions of plate boundaries.
- Geologists distinguish among three types of plate boundaries (convergent, divergent, and transform) based on relative plate motions.
- The movement of plates results in continental drift. Because of plate tectonics, the map of the Earth's surface constantly changes.

Take-home message . . .

Earth's lithosphere—its rigid outer shell—is divided into more than a dozen plates that move relative to one another. As a plate moves, its interior remains mostly unchanged, but slip occurs along the plate boundary, causing earthquakes. This overall concept is the theory of plate tectonics.

Quick Question -
What does the lithosphere consist of? How does it differ from the asthenosphere?

Figure 2.14 Bathymetry of a mid-ocean ridge.

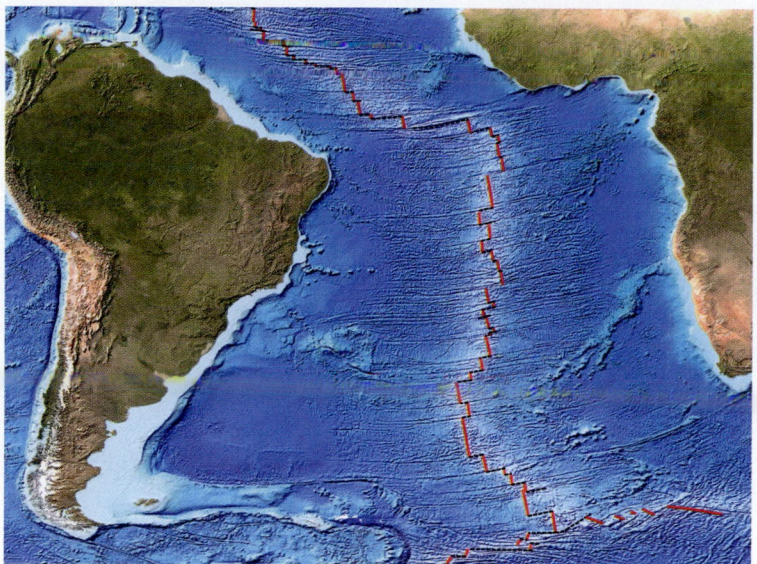

What an Earth Scientist Sees

(a) A bathymetric map of the Mid-Atlantic Ridge in the South Atlantic Ocean. The lighter shades of blue are shallower water depths. Red lines are the axis of the ridge. The sketch labels key features.

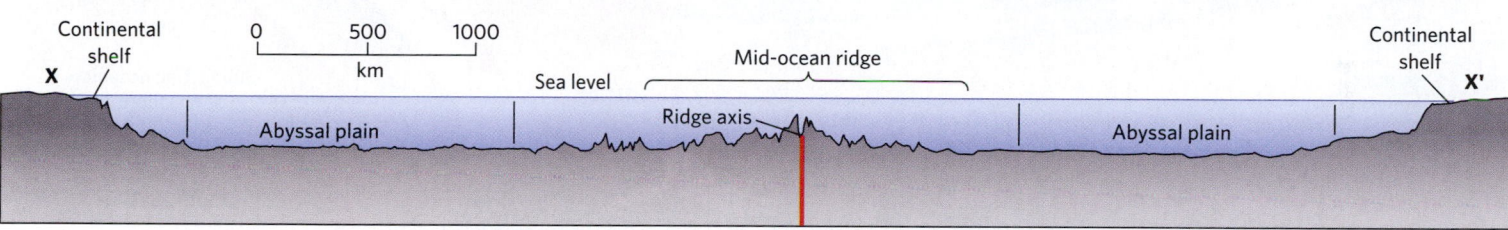

(b) A bathymetric profile (vertical slice) across the Mid-Atlantic Ridge, from continent to continent. Note that the ridge is higher than the abyssal plains.

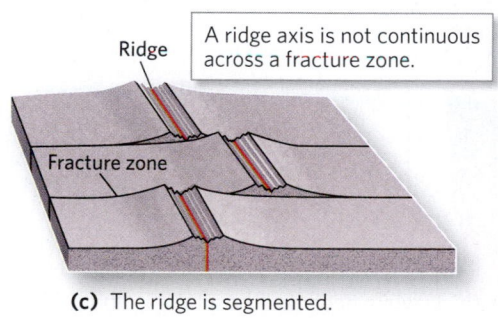

(c) The ridge is segmented.

This feature, which rose about 2 km (1.2 miles) above the abyssal plains, came to be known as the Mid-Atlantic Ridge (**Fig. 2.14a, b**). This ridge extends from the latitude of far northern Greenland to the latitude of far southern South America. Significantly, the ridge is not a continuous line. Rather, along its length, it consists of distinct segments whose ends terminate at a fracture zone. Further, the segments do not line up, so the end of one segment may be offset along the fracture zone by a few kilometers to several hundred kilometers (**Fig. 2.14c**).

Bathymetric studies over the next several decades found that all the major oceans contain such ranges, which came to be known in general as *mid-ocean ridges* or mid-ocean rises, although not all occupy the center line of the basin. Until the theory of plate tectonics came along, no one knew why mid-ocean ridges existed. Plate tectonics theory, however, explains their geologic origin. A mid-ocean ridge delineates a **divergent boundary**, also known as a spreading boundary. At this type of plate boundary, two oceanic plates move apart by the process of seafloor spreading. Note that during this process, no open space

2.5 Geologic Features of Plate Boundaries

Divergent-Plate Boundaries

In 1872, HMS *Challenger* set sail from England to begin the world's first research cruise. At the time, no one knew how deep the ocean was at more than a few localities, so every now and then the ship's crew took a sounding. The results hinted that a submerged mountain range ran from north to south down the center of the Atlantic Ocean basin.

See for yourself

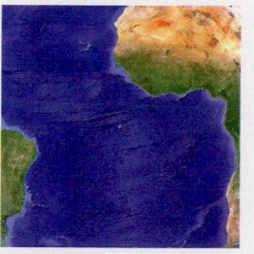

The Mid-Atlantic Ridge

Latitude: 2°34'26.28" S
Longitude: 15°37'59.86" W

Fly to an elevation of 11,300 km (6,800 miles) and look straight down to see a view of the Mid-Atlantic Ridge in the equatorial Atlantic Ocean. Zoom in to see a fracture zone.

The Theory of Plate Tectonics

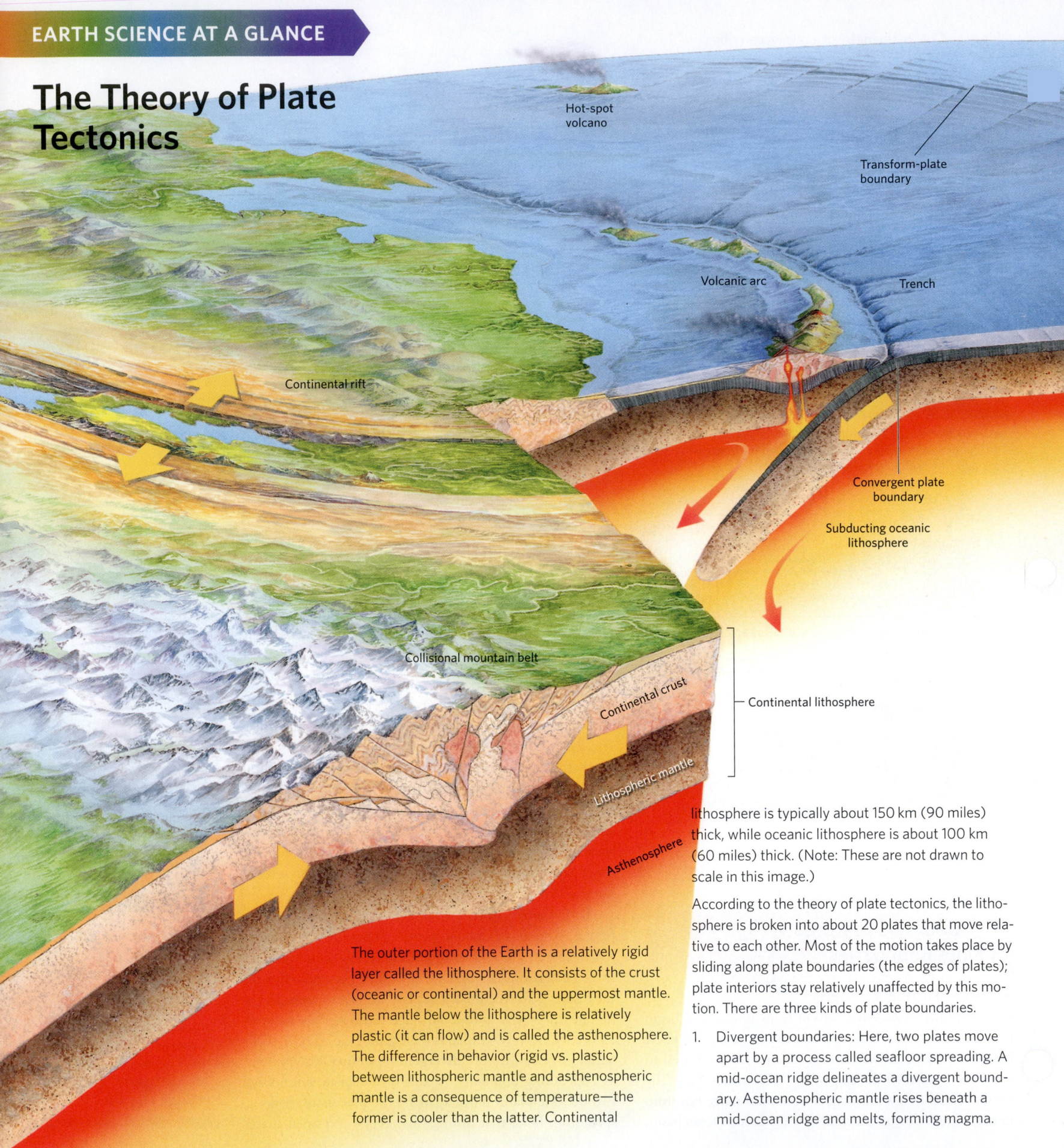

Hot-spot volcano

Transform-plate boundary

Volcanic arc

Trench

Continental rift

Convergent plate boundary

Subducting oceanic lithosphere

Collisional mountain belt

Continental crust

Continental lithosphere

Lithospheric mantle

Asthenosphere

The outer portion of the Earth is a relatively rigid layer called the lithosphere. It consists of the crust (oceanic or continental) and the uppermost mantle. The mantle below the lithosphere is relatively plastic (it can flow) and is called the asthenosphere. The difference in behavior (rigid vs. plastic) between lithospheric mantle and asthenospheric mantle is a consequence of temperature—the former is cooler than the latter. Continental

lithosphere is typically about 150 km (90 miles) thick, while oceanic lithosphere is about 100 km (60 miles) thick. (Note: These are not drawn to scale in this image.)

According to the theory of plate tectonics, the lithosphere is broken into about 20 plates that move relative to each other. Most of the motion takes place by sliding along plate boundaries (the edges of plates); plate interiors stay relatively unaffected by this motion. There are three kinds of plate boundaries.

1. Divergent boundaries: Here, two plates move apart by a process called seafloor spreading. A mid-ocean ridge delineates a divergent boundary. Asthenospheric mantle rises beneath a mid-ocean ridge and melts, forming magma.

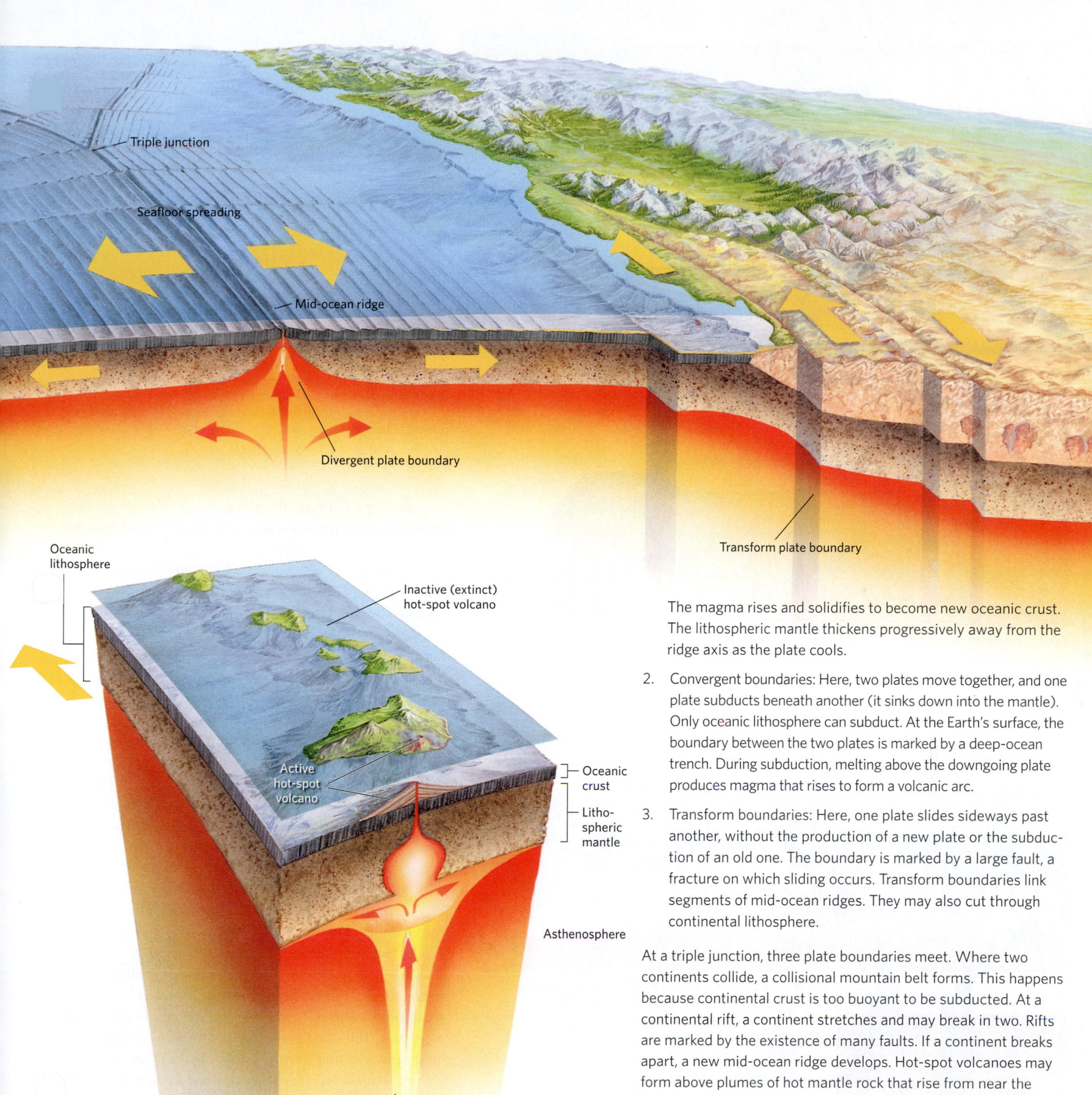

Triple junction

Seafloor spreading

Mid-ocean ridge

Divergent plate boundary

Transform plate boundary

Oceanic lithosphere

Inactive (extinct) hot-spot volcano

Active hot-spot volcano

Oceanic crust

Lithospheric mantle

Asthenosphere

Mantle plume

The magma rises and solidifies to become new oceanic crust. The lithospheric mantle thickens progressively away from the ridge axis as the plate cools.

2. Convergent boundaries: Here, two plates move together, and one plate subducts beneath another (it sinks down into the mantle). Only oceanic lithosphere can subduct. At the Earth's surface, the boundary between the two plates is marked by a deep-ocean trench. During subduction, melting above the downgoing plate produces magma that rises to form a volcanic arc.

3. Transform boundaries: Here, one plate slides sideways past another, without the production of a new plate or the subduction of an old one. The boundary is marked by a large fault, a fracture on which sliding occurs. Transform boundaries link segments of mid-ocean ridges. They may also cut through continental lithosphere.

At a triple junction, three plate boundaries meet. Where two continents collide, a collisional mountain belt forms. This happens because continental crust is too buoyant to be subducted. At a continental rift, a continent stretches and may break in two. Rifts are marked by the existence of many faults. If a continent breaks apart, a new mid-ocean ridge develops. Hot-spot volcanoes may form above plumes of hot mantle rock that rise from near the core-mantle boundary. As a plate drifts over a hot spot, it leaves a chain of extinct volcanoes.

Figure 2.15 The process of seafloor spreading at a mid-ocean ridge.

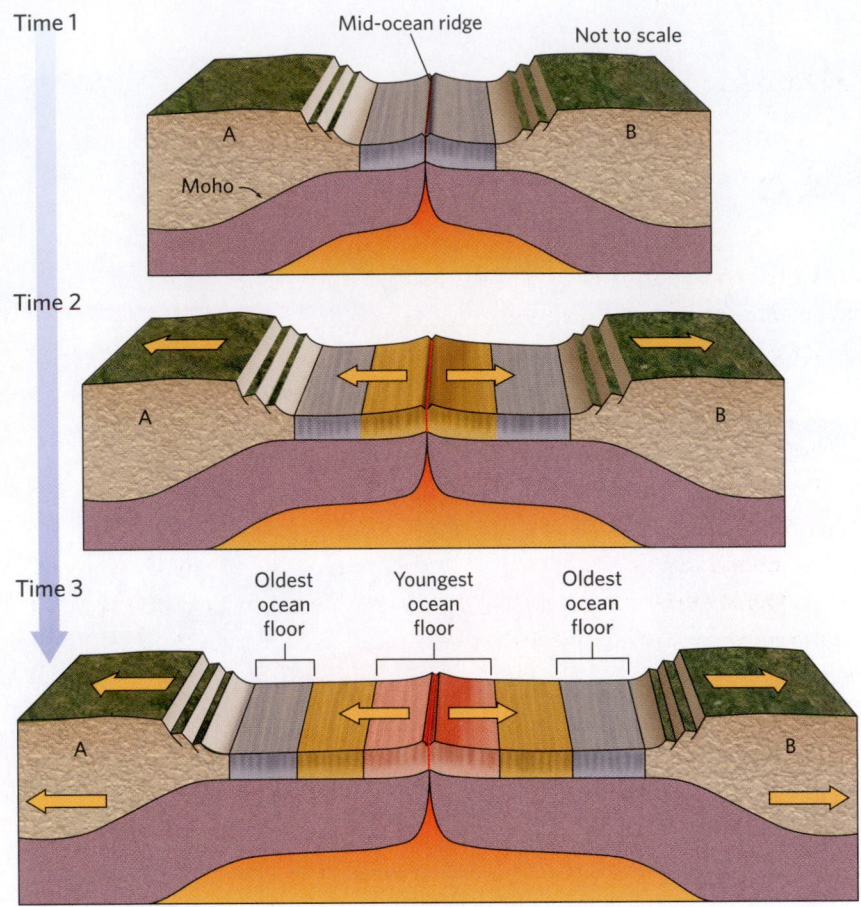

(a) During seafloor spreading, the ocean floor gets wider and continents on either side move apart. New oceanic crust forms at the ridge axis.

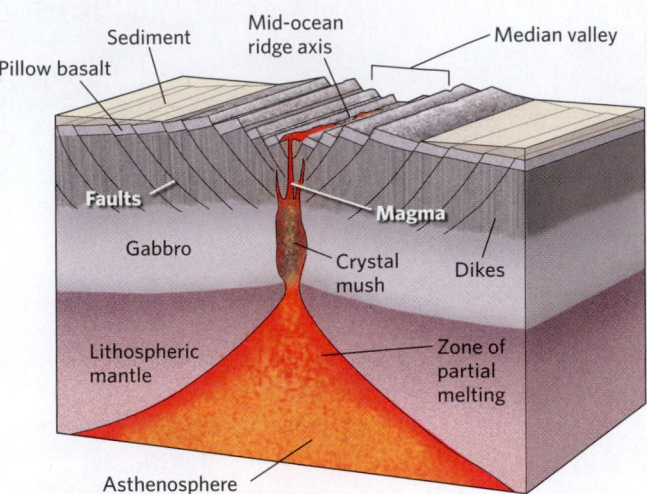

(b) Beneath a mid-ocean ridge is a magma chamber. Gabbro forms on the side of the magma chamber. Basalt dikes protrude upward. Pillow lava forms on the seafloor surface.

Figure 2.16 Changes accompanying the aging of lithosphere.

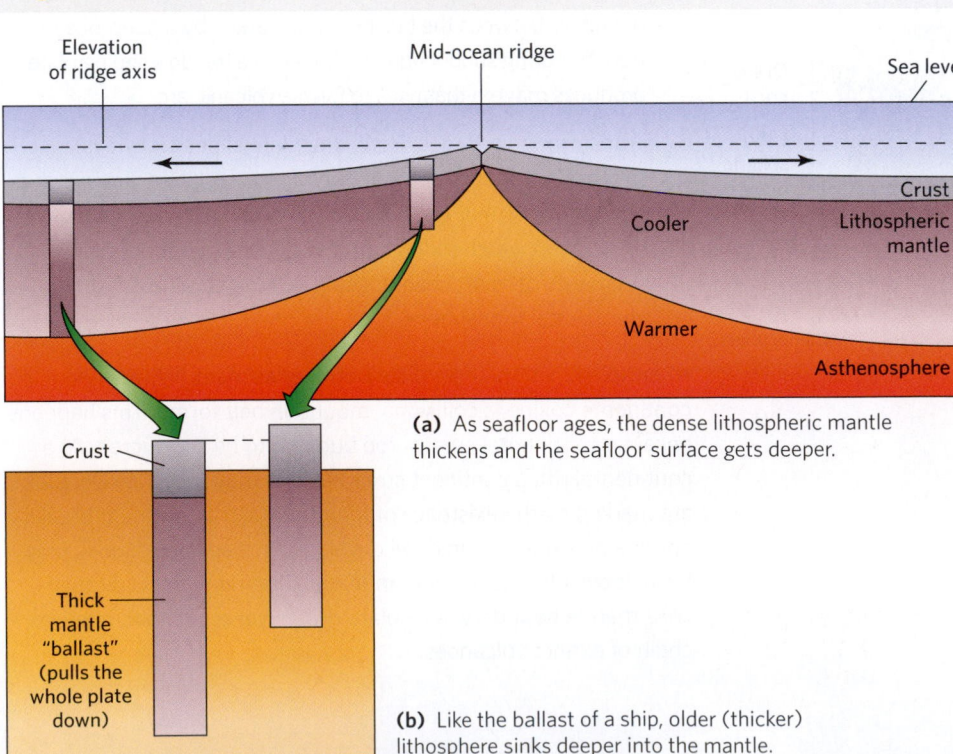

(a) As seafloor ages, the dense lithospheric mantle thickens and the seafloor surface gets deeper.

(b) Like the ballast of a ship, older (thicker) lithosphere sinks deeper into the mantle.

ever develops between diverging plates. Rather, as the plates move apart, new oceanic lithosphere forms between them **(Fig. 2.15a)**.

It is along the *ridge axis*, a line following the crest of the ridge, that new seafloor forms. Typically, a narrow valley, bordered by faults, defines the ridge axis **(Fig. 2.15b)**. Overall, the seafloor slopes gently away from the ridge, reaching the abyssal-plain depth (4.5 km; 2.8 miles) at up to 800 km (500 miles) from the ridge axis. The Mid-Atlantic Ridge is roughly symmetrical, in that its eastern half looks like a mirror image of its western half.

What happens at the ridge during seafloor spreading? As the process takes place, hot asthenosphere rises beneath the ridge. As this asthenosphere rises, it begins to melt (for reasons we'll explain in Chapter 4). The resulting molten rock, or *magma*, rises and accumulates in a magma chamber below the ridge axis (see Fig. 2.15b); a *magma chamber* can be pictured as a volume filled with a mush of crystals mixed with red-hot melt. Some of this mush solidifies into gabbro along the sides of the chamber, while some rises still higher, along vertical cracks, and solidifies into wall-like sheets called *dikes*, that are made of basalt. And some makes it all the way to the surface of the seafloor and seeps out as lava from small submarine volcanoes that follow cracks along the ridge axis. This lava cools when it comes in contact with cold seawater and forms glassy, melon-shaped blobs known as *pillows*, which accumulate into mounds of *pillow basalt*. The entire layer of igneous rock formed at the ridge axis, from the base of the gabbro to the top of the pillow basalt, represents new oceanic crust.

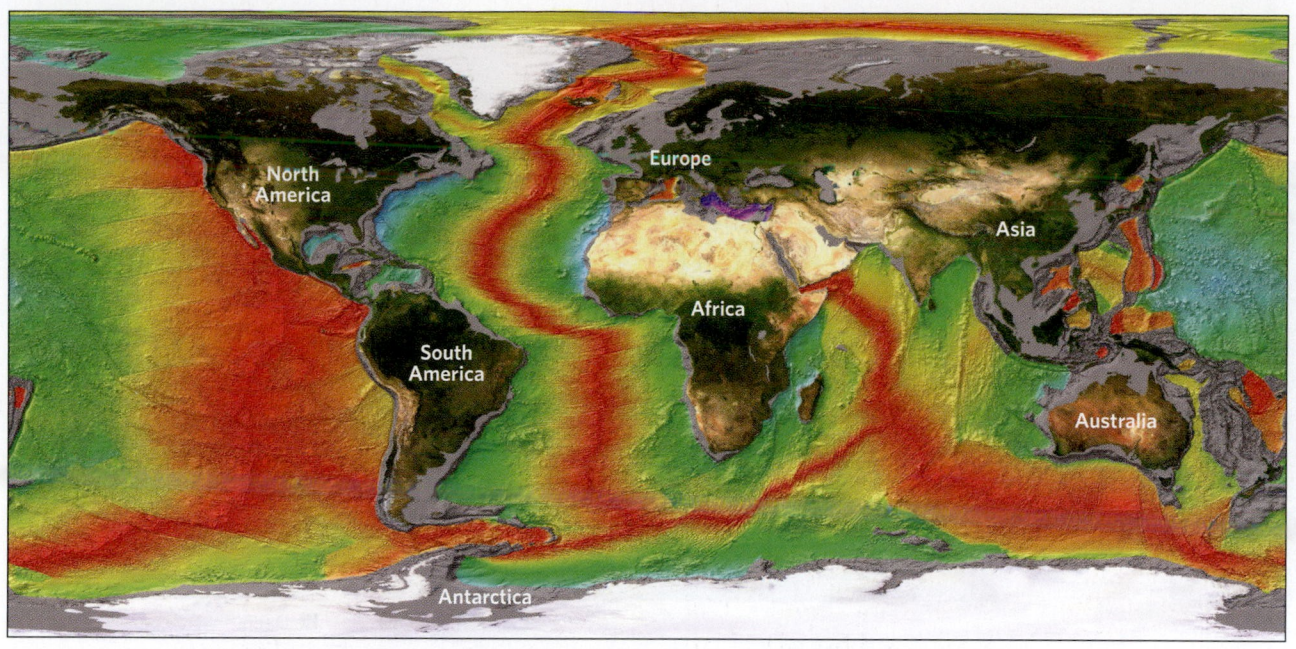

Figure 2.17 Age of the seafloor.

(a) Note that the seafloor grows older with increasing distance from a mid-ocean ridge axis.

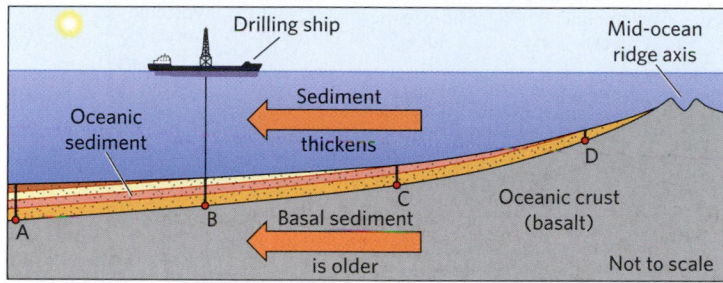

(b) Drilling into the sediment layer of the ocean floor confirms that the basal sediment in contact with basalt gets older the farther away it is from the ridge.

As it forms, oceanic crust moves away from the ridge axis, and as this happens, more asthenosphere rises, more magma forms and rises, and still more crust forms. In other words, like a vast, continuously moving conveyor belt, oceanic crust forms at the ridge and then moves away from the ridge. The stretching force, or *tension*, applied to newly formed solid crust as spreading takes place breaks up this new crust, resulting in the formation of faults bordering the ridge axis. Slip on these faults causes divergent-boundary earthquakes.

At the ridge axis, a lithosphere plate consists only of new crust. As this crust moves away from the ridge axis, no additional igneous rock is added to it, so its thickness stays the same. Therefore, the crust, along with the mantle directly beneath, progressively cools as it ages. Loss of heat from the mantle causes the boundary between cooler (<1,280°C; 2,340°F), rigid mantle and warmer (>1,280°C), plastic mantle to become deeper. Because this boundary defines the base of the lithosphere, the oceanic lithosphere thickens as it ages **(Fig. 2.16)**. Further, because rock becomes denser as it cools, the lithosphere, overall, gets denser as it ages. So, like a cargo ship whose keel sinks deeper into the water, and whose deck moves downward as workers load it with heavy cargo, the lithosphere moves deeper into the asthenosphere, and the seafloor becomes lower, as the lithosphere ages. As a result, abyssal plains, which are underlain by old lithosphere, are deeper than mid-ocean ridges, which are underlain by young lithosphere.

Because all seafloor forms at mid-ocean ridges, the youngest seafloor borders the ridge axis, and the oldest seafloor lies farthest from the ridge **(Fig. 2.17a)**. For this reason, the layer of deep-sea sediment thickens away from the ridge axis. Drilling into the seafloor confirms that the basal (lowest) sediment layer becomes progressively older with increasing distance from the ridge. This observation serves as a proof of seafloor spreading **(Fig. 2.17b)**.

Convergent-Plate Boundaries

During its epic voyage of discovery, HMS *Challenger* made soundings in the western Pacific, and it found a location where ocean depth exceeded 8 km (5 miles). Seventy-six years later, HMS *Challenger II* revisited the location and surveyed it with sonar, finding

Figure 2.18 The bathymetry of a convergent boundary.

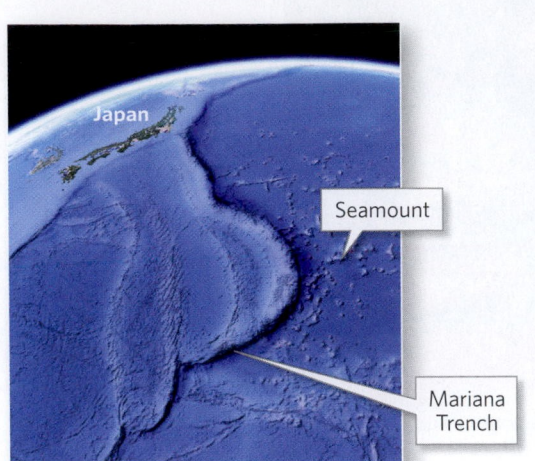

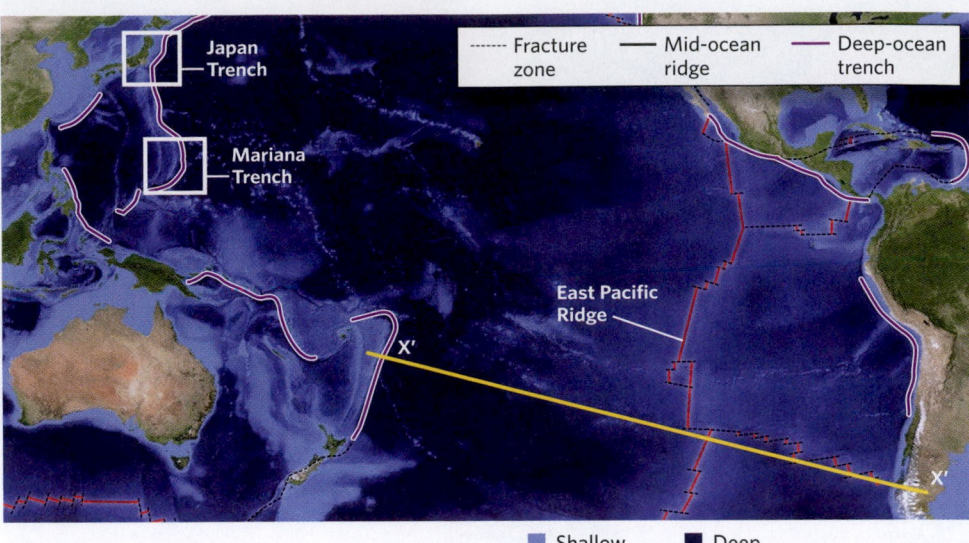

(a) A bathymetric map of the western Pacific, showing the trace of the Mariana Trench. Dark blue is very deep water.

(b) Trenches occur on both sides of the Pacific Ocean. The yellow line shows the location of the bathymetric profile.

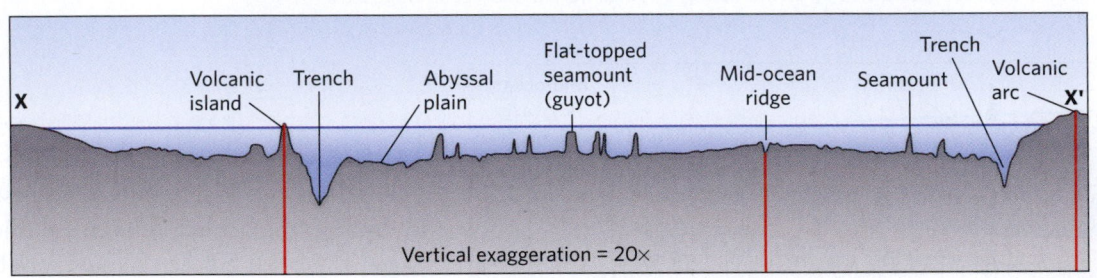

(c) A bathymetric profile showing the contrast between the depth of a trench and the depth of an abyssal plain. This profile goes from the Tonga Trench, east of Australia, to the coast of South America. Red lines are the axis of the ridge. The sketch labels key features.

that the region of deep water delineated a 2,500-km (1,500-miles)-long trough, now known as the Mariana Trench, whose deepest point was almost 11 km (7 miles) below sea level—almost three times the average depth of the seafloor **(Fig. 2.18a)**! Similar deep-sea trenches define most of the boundary of the Pacific Plate and occur along portions of the margins of several other plates as well, though none are as deep as the Mariana Trench **(Fig. 2.18b, c)**. **Volcanic arcs**, chains of volcanoes that typically trace out a curving arc in map view, border one side of a trench, and massive earthquakes happen with unnerving frequency beneath trenches and volcanic arcs.

As was the case with mid-ocean ridges, the origin and geologic meaning of deep-sea trenches remained a mystery before the advent of plate tectonics. But once the theory had been proposed, geologists realized that trenches mark locations where the lithosphere of an oceanic plate bends down and slides under the edge of another plate. As a consequence, plates on either side of

a trench move toward each other. Due to this relative motion, geologists refer to the plate boundary associated with a trench as a **convergent boundary** **(Fig. 2.19a)**. Keep in mind that, at a convergent boundary, plates don't butt into each other like angry rams. Rather, the *downgoing plate* sinks into the asthenosphere beneath the *overriding plate*. Geologists refer to the sinking process as **subduction**, so a convergent boundary can also be called a **subduction zone**.

Subduction occurs for a simple reason: lithospheric mantle, once it has cooled and thickened for at least 10 million years, is denser than the underlying asthenosphere, so it can sink down through the asthenosphere. Subduction starts when one part of a plate pushes over another part along a fault. Once the edge of the downgoing plate has entered the asthenosphere, it begins to sink like an anchor **(Fig. 2.19b)**. To make room for the sinking lithosphere, soft asthenosphere flows out of its way, just as water flows out of the way of a sinking anchor.

Figure 2.19 During the process of subduction, oceanic lithosphere slips back into the deeper mantle.

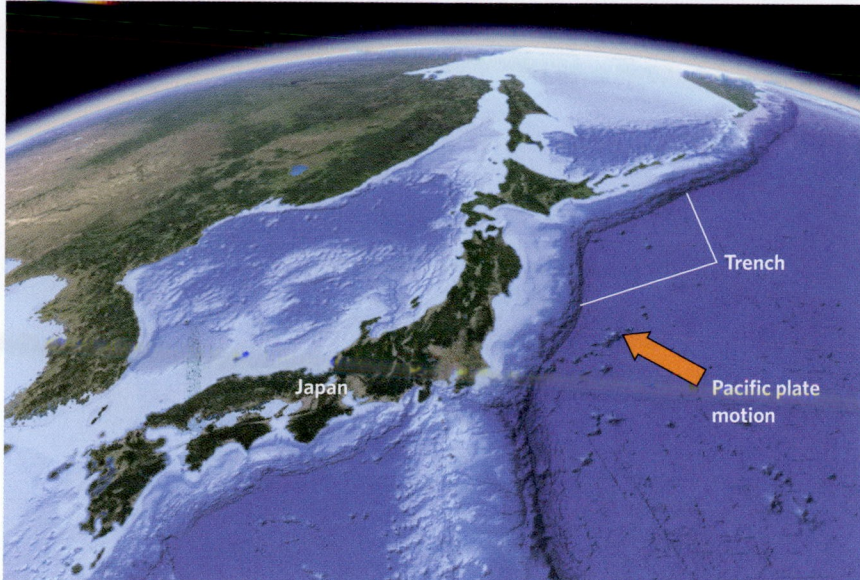

(a) A trench delineates the place where the Pacific plate is being subducted beneath Japan.

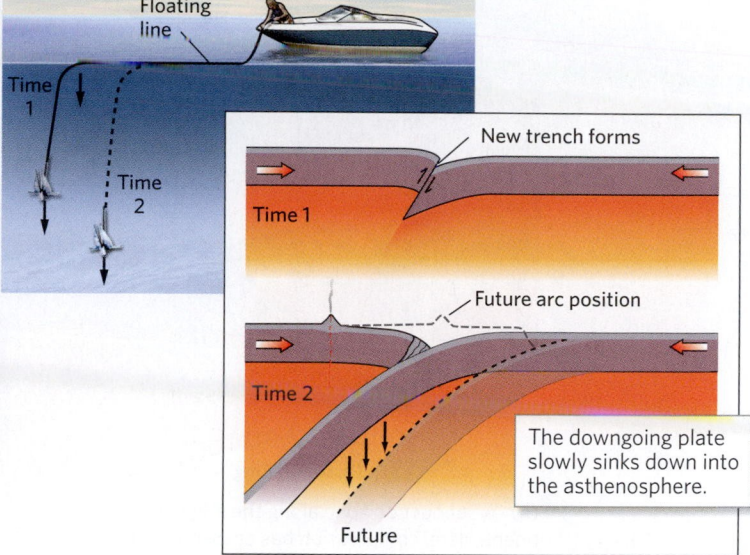

(b) Sinking of the downgoing plate resembles the sinking of an anchor attached to a rope.

Asthenosphere can flow only very slowly, however, so subduction cannot proceed any faster than about 15 cm (6 inches) per year.

All ocean floor eventually undergoes subduction, so the oldest ocean floor on Earth is only about 200 million years old, much less than the age of the Earth. Because the circumference (and therefore volume) of the Earth remains nearly constant over time, the amount of subduction that takes place worldwide must equal the overall amount of seafloor spreading.

Let's look more closely at the geologic activity that takes place at a convergent boundary. As subduction takes place, seafloor sediment, as well as sand and mud that wash into the trench from nearby land, gets scraped up and incorporated into a wedge-shaped mass, known as an **accretionary prism**, along the margin of the overriding plate **(Fig. 2.20a, b)**. In effect, an accretionary prism resembles a pile of sand that builds in front of a bulldozer. A volcanic arc develops on the edge of the overriding plate, on the opposite side of the accretionary prism from the trench. The supply of magma for this arc forms just above the surface of the downgoing plate when the plate reaches a depth of about 150 km (90 miles) below the Earth's surface (see Chapter 4). Note that subduction takes place beneath the entire length of an arc at any given time, so all the volcanoes in the arc are *active volcanoes*, meaning that they all are either erupting, recently erupted, or are likely to erupt in the future.

Where an oceanic plate subducts beneath a continent, a **continental arc** develops along the edge of

the continent—the Andes arc along the coast of South America and the Cascade arc of Oregon and Washington serve as examples. At places where an oceanic plate subducts beneath another oceanic plate, volcanoes build up on the seafloor of the overriding plate and grow into a volcanic **island arc**—the Mariana arc and the Aleutian arc serve as examples **(Fig. 2.20c)**. In a few locations, a volcanic arc grows on a sliver of continental crust that separated from the main continent due to the growth of a small ocean basin, called a *marginal sea* **(Fig. 2.20d)**—the Japan arc is an example.

The boundary between the downgoing plate and the base of the overriding plate at a convergent boundary contains many faults, and slip on these faults generates earthquakes. Earthquakes also happen at greater depths in downgoing plates. In fact, geologists detect earthquakes in downgoing plates to a depth of 660 km (410 miles) **(Fig. 2.21)**. Downgoing plates eventually sink below a depth of 660 km, but they do so without generating earthquakes. The lower mantle may be a "graveyard" for subducted plates.

Transform-Plate Boundaries

We noted earlier that mid-ocean ridges are segmented and that the segments are linked by fracture zones **(Fig. 2.22a)**. Originally, researchers incorrectly assumed that the entire length of each fracture zone was an active fault. But when information about the distribution of earthquake epicenters along mid-ocean ridges became available, it became clear that slip occurs only on the segment of a fracture

Figure 2.20 The nature of a convergent boundary varies with location.

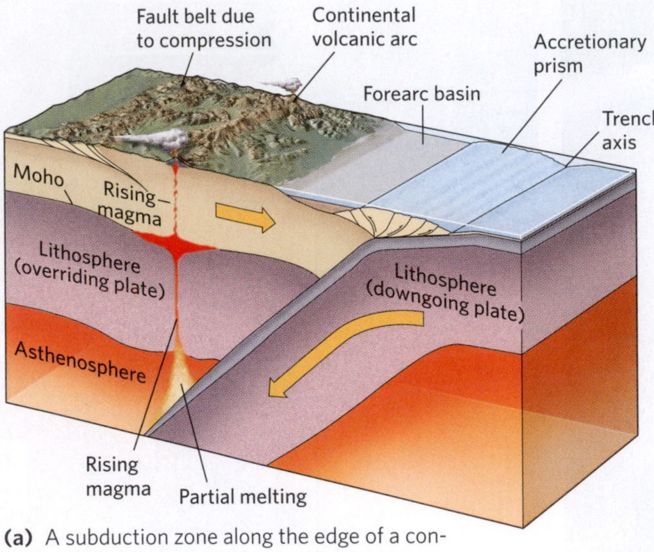

Fault belt due to compression

Continental volcanic arc

Forearc basin

Accretionary prism

Trench axis

Moho

Rising magma

Lithosphere (overriding plate)

Lithosphere (downgoing plate)

Asthenosphere

Rising magma

Partial melting

(a) A subduction zone along the edge of a continent. Here, compression has caused faulting behind the continental arc.

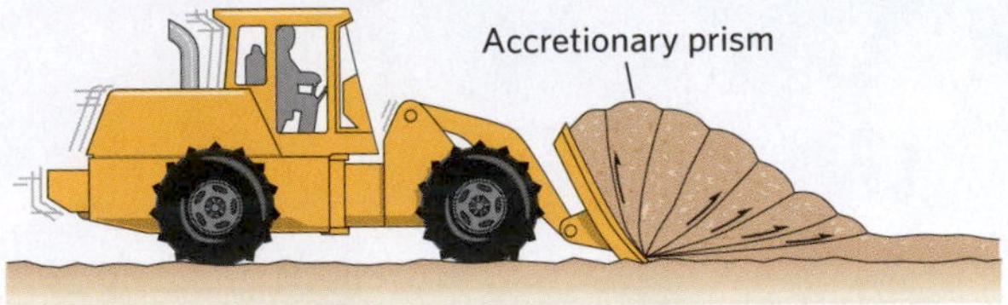

Accretionary prism

(b) The overriding plate acts like a bulldozer, scraping sediment off the downgoing plate to build an accretionary prism.

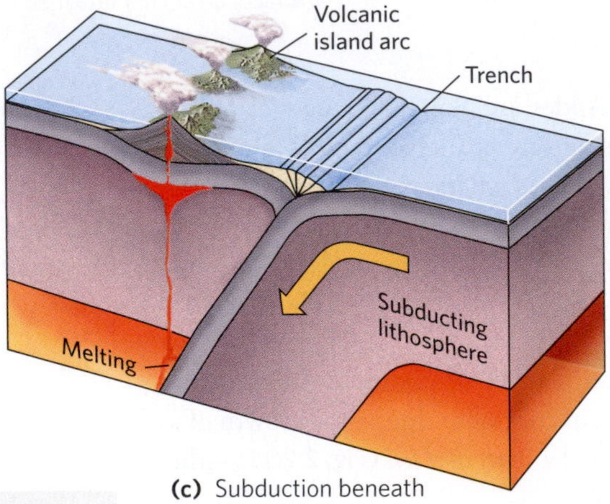

Volcanic island arc

Trench

Subducting lithosphere

Melting

(c) Subduction beneath oceanic lithosphere produces an island arc.

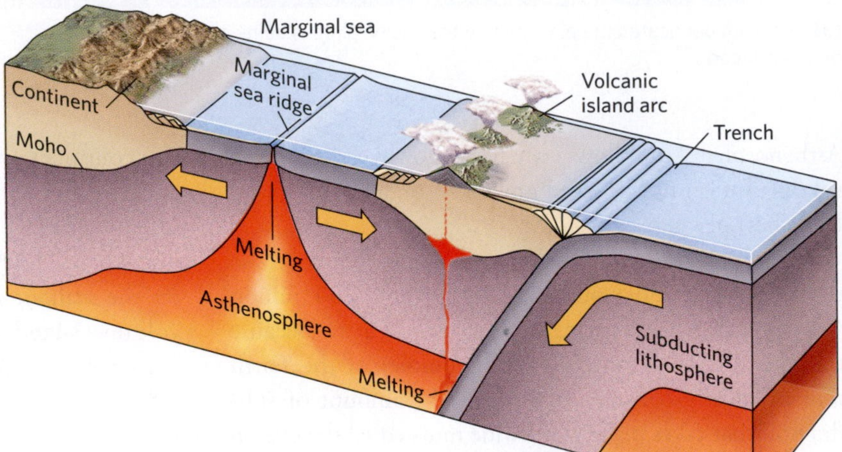

Marginal sea

Marginal sea ridge

Volcanic island arc

Continent

Moho

Trench

Melting

Asthenosphere

Subducting lithosphere

Melting

(d) Marginal seas form by seafloor spreading behind some volcanic arcs.

See for yourself

Himalayan Collision

Latitude: 27°59'17.95" N
Longitude: 86°55'30.71" E

Zoom to an elevation of 7000 km (4300 miles) and look straight down.

You're seeing the Himalayas. Zoom closer, and you'll be at the peak of Mt. Everest.

zone that lies between the ends of two ridge segments. The portions of fracture zones that extend beyond the ends of ridge segments, into the abyssal plain, do not generate earthquakes (Fig. 2.22b).

The distribution of slip along fracture zones remained a mystery until a Canadian geologist, J. Tuzo Wilson (1908–1993), began to think about fracture zones in the context of seafloor spreading. To understand Wilson's interpretation, imagine a map showing two north-south-trending mid-ocean ridge segments linked by an east-west-trending fracture zone (Fig. 2.22c). Plate A is moving east, and Plate B is moving west, relative to the ridge axis. The northern ridge segment intersects the fracture zone at Point X, and the southern one intersects it at Point Y. Along the portion of the fracture zone between X and Y, Plate A moves to the right relative to Plate B, so this portion must be an active fault

along which earthquakes take place. In other words, this segment of the fracture zone is a plate boundary. To the east of Point X, and to the west of Point Y, however, seafloor north and south of the fracture zone moves in the same direction and at the same rate. These portions of the fracture zone, therefore, are not active faults. In fact, when you cross these segments, you stay on the same plate.

Wilson introduced the term **transform fault** (or *transform*) for the actively slipping segment of a fracture zone between two ridge segments, and he emphasized that transform faults are a third type of plate boundary, a **transform boundary**. At such a location, one plate slips sideways relative to the other, but no new plate forms, and no old plate subducts. Transform boundaries are, therefore, vertical faults on which the slip direction parallels the Earth's surface.

Figure 2.21 Subducted lithosphere sinks into the mantle like a denser fluid sinking into a less dense fluid. The subducted plate may eventually separate into pieces as it passes through the layers of the mantle. Earthquakes occur in the subducted plate down to a depth of about 660 km (410 miles).

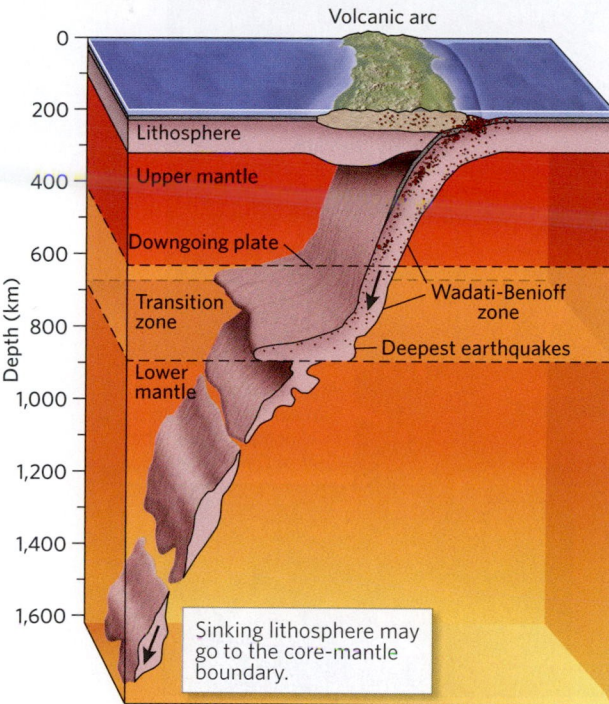

Sinking lithosphere may go to the core-mantle boundary.

Figure 2.22 Oceanic transform faulting

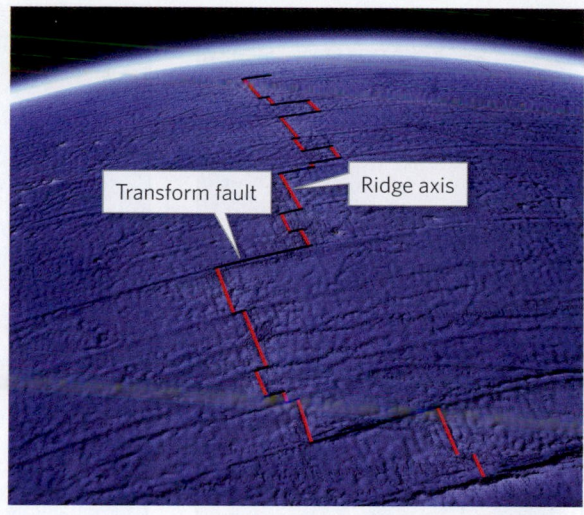

(a) Transform faults along the Mid-Atlantic Ridge. Notice that the transform faults terminate at ridge segments. Fracture zones continue beyond the ridge segments.

Wilson and others soon realized that not all transforms link ridge segments, and that not all transforms are submarine. For example, the Alpine fault cuts across New Zealand and links two trenches. The San Andreas fault of California serves as a transform boundary between the North American Plate and the Pacific Plate and links a ridge segment to a trench. The portion of California that lies to the west of the fault moves with the Pacific Plate, whereas the portion to the east of the fault moves with the North American Plate **(Fig. 2.23)**.

(b) Note that only the segment of the fracture zone between the two ridge segments is an active transform fault.

Take-home message . . .

Geologists distinguish among three types of plate boundaries based on relative plate movement: divergent, convergent, and transform. Seafloor spreading occurs at divergent boundaries, marked by mid-ocean ridges. At a convergent boundary, an oceanic plate sinks beneath the edge of another plate. At transform boundaries, one plate slips sideways along the edge of another.

Quick Question --------------------------------
Why do earthquakes occur at all three types of plate boundaries?

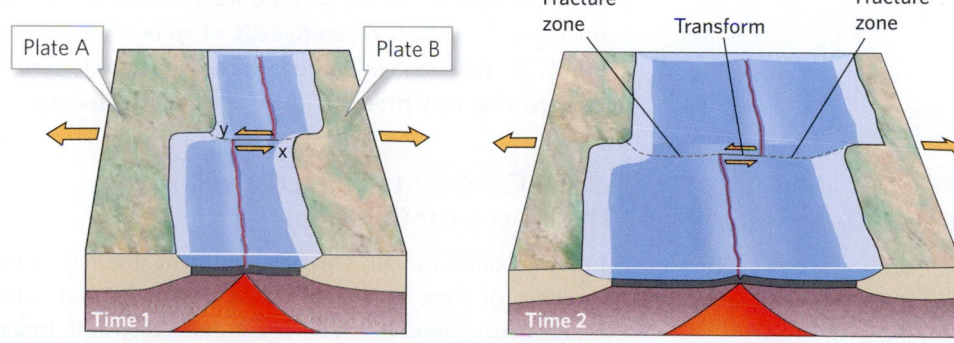

(c) Wilson's interpretation of a transform fault.

Figure 2.23 The San Andreas fault is a continental transform boundary.

(a) In southern California, the San Andreas fault cuts a dry landscape. The fault trace is in the narrow valley. The land has been pushed up slightly along the fault.

(b) The San Andreas fault is a transform boundary between the North American and Pacific Plates. The Pacific Plate is moving northwest relative to North America.

2.6 The Birth and Death of Plate Boundaries

The configuration of plates and plate boundaries visible on our planet today has not existed for all of geologic history, and it will not stay the same in the future. Because of plate motion, new oceanic plates form, only to be consumed by subduction later on, and continents merge, then later separate. In this section, we first examine how a convergent boundary ceases to exist when two continents collide. Then we look at how continents split apart, or *rift*, a process that may produce a new divergent boundary.

When Continents Converge: The Process of Collision

India was once a small, separate continent that lay far to the south of Asia. Over time, however, subduction consumed ocean floor between India and Asia, and India moved northward. Around 40 to 50 million years ago, all the ocean floor between the two continents had been subducted, and India itself pushed into Asia. This type of event, leading to the merging of two landmasses after subduction of the intervening ocean floor, is called **collision**, and if the event involves two continents, geologists refer to it as *continental collision*.

Collision happens when two relatively buoyant pieces of crust converge at a plate boundary. Buoyant crust, such as continental crust or an island arc, cannot be subducted **(Fig. 2.24a, b)**. When collision happens, the convergent boundary that once lay between the buoyant pieces ceases to exist. For example, when India collided with Asia, the oceanic lithosphere that was attached to the northern coast of India broke off and sank down into the mantle, so subduction along the southern coast of Asia stopped. In addition, the rock and sediment that once lay along the margins of the two landmasses, or had been scraped off the seafloor between them, underwent squeezing as if in a giant vise, resulting in the huge (approximately 8-km-, or 5-miles-high) Himalayan mountain belt. The continued push of India into Asia causes continued growth of the Himalayas and also contributes to the uplift of the

Figure 2.24 Continental collision.

Time 1: Before

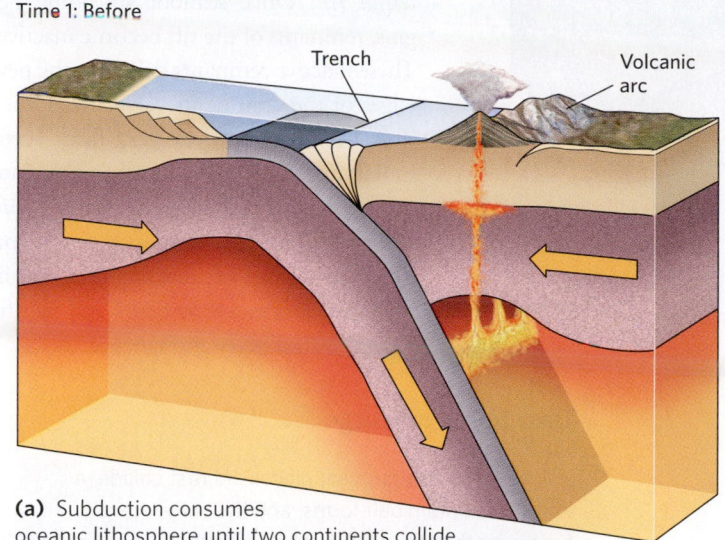

(a) Subduction consumes oceanic lithosphere until two continents collide.

Time 2: After

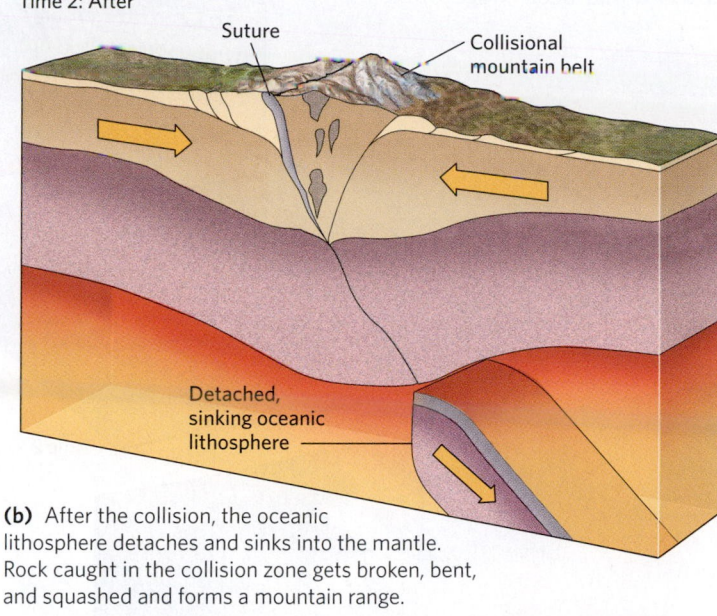

(b) After the collision, the oceanic lithosphere detaches and sinks into the mantle. Rock caught in the collision zone gets broken, bent, and squashed and forms a mountain range.

(c) An oblique satellite view showing the collision of India with Asia to produce the Himalayas.

Tibetan Plateau. As continents collide, not only does the surface of the Earth go up, but the base of the crust goes down, so the continental crust becomes thicker.

Collision has yielded some of the most spectacular mountain ranges on the planet today; examples include the Himalayas and the Alps **(Fig. 2.24c)**. It also produced major mountain ranges in the past, ranges that have since been eroded away. For example, the Appalachian Mountains in the eastern United States initially grew as a consequence of three collisions. After the most recent collision, between Africa and North America around 280 million years ago, North America became part of Pangaea.

Stretching a Continent: The Process of Rifting

Continents, once formed, may eventually break apart—as Wegener realized when he proposed that Pangaea broke into several smaller continents. The process of stretching and breaking a continent apart is called **rifting (Fig. 2.25)**. This process tends to be confined to a distinct linear belt called a **rift**. The manner in which rock responds to rifting varies with depth. Near the surface of the continent, stretching leads to the development of many faults. Blocks of crust tilt and slip downward on these faults so that low areas, called *rift basins*, develop. The basins fill with eroded sediment from the rift's margin as well as from the tilted blocks. Deeper down, where rock is warmer and softer, stretching takes place by plastic flow (see Fig. 2.25). As continental lithosphere thins during rifting, hot asthenosphere rises beneath the rift, so no open space develops. Some of this asthenosphere melts, producing magma that erupts as lava from volcanoes in the rift (see Chapter 4).

Rifting is taking place today at several localities on the Earth. For example, along the *East African Rift*, easternmost Africa is separating from central Africa. A fault-bounded valley, containing elongate lakes in places, delineates the axis of this rift **(Fig. 2.26a)**. The *Basin and Range Province* of the western United States is a rift containing many elongate rocky ridges (the edges of tilted crustal blocks) that are separated from each other by narrow, sediment-filled basins **(Fig. 2.26b)**. Rifting in the Basin and Range Province has stretched the crust by a few hundred kilometers in an east-west direction.

Figure 2.25 During the process of rifting, continental lithosphere stretches and thins. If rifting succeeds, a new mid-ocean ridge forms.

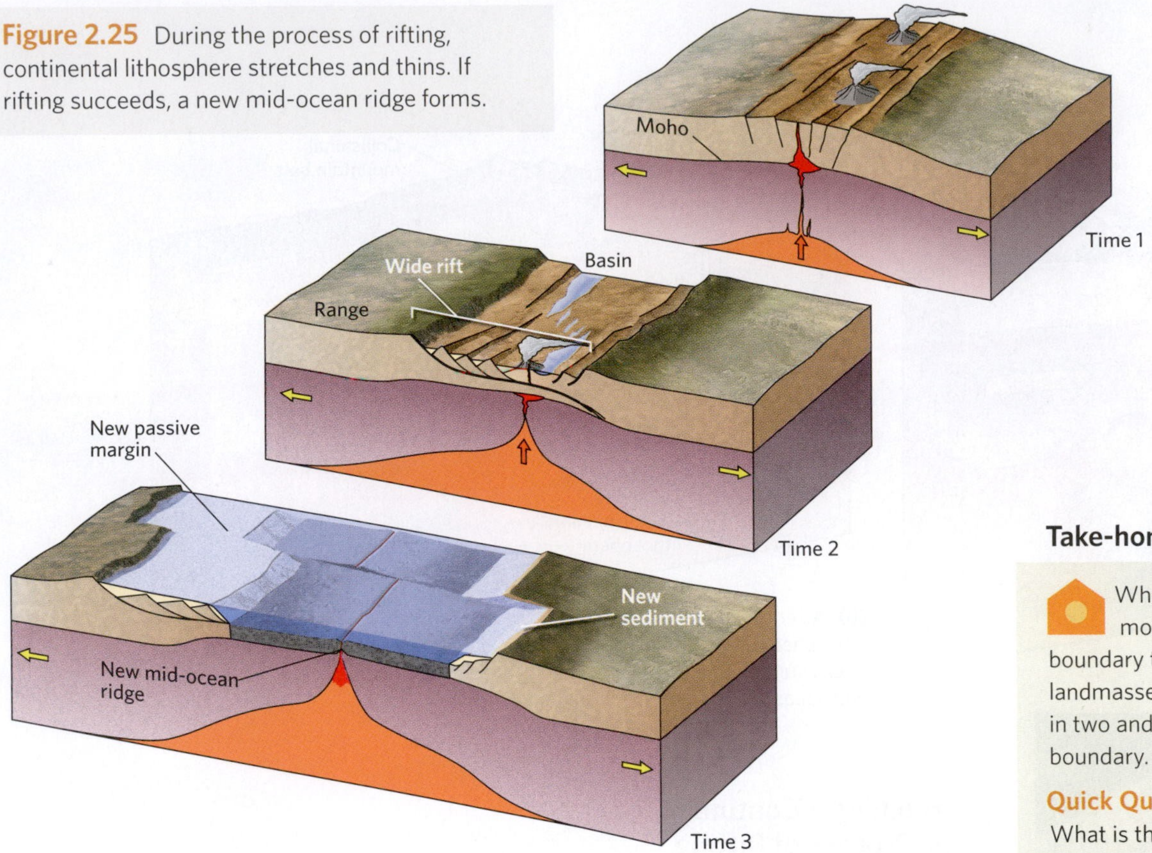

Moho

Time 1

Wide rift

Basin

Range

New passive margin

Time 2

New sediment

New mid-ocean ridge

Time 3

A rift that continues stretching until the continent breaks apart and a new mid-ocean ridge develops is called a *successful rift*. Once seafloor spreading begins, remnants of the rift become inactive. These inactive remnants delineate the new edges of the continents. Over time, they sink, become buried by very thick accumulations of sediment, and evolve into passive margins. Along an *unsuccessful rift*, rifting ceases before it splits a continent. The trace of the rift remains as a scar in the crust, delineated by faults and by rift-related sedimentary and igneous rocks.

Take-home message . . .

When two buoyant pieces of crust collide, a mountain belt forms, and the convergent boundary that once existed between the two colliding landmasses ceases to exist. Rifting can split a continent in two and can lead to the formation of a new divergent boundary.

Quick Question

What is the difference between a successful and an unsuccessful rift?

Figure 2.26 Examples of present-day crustal rifting.

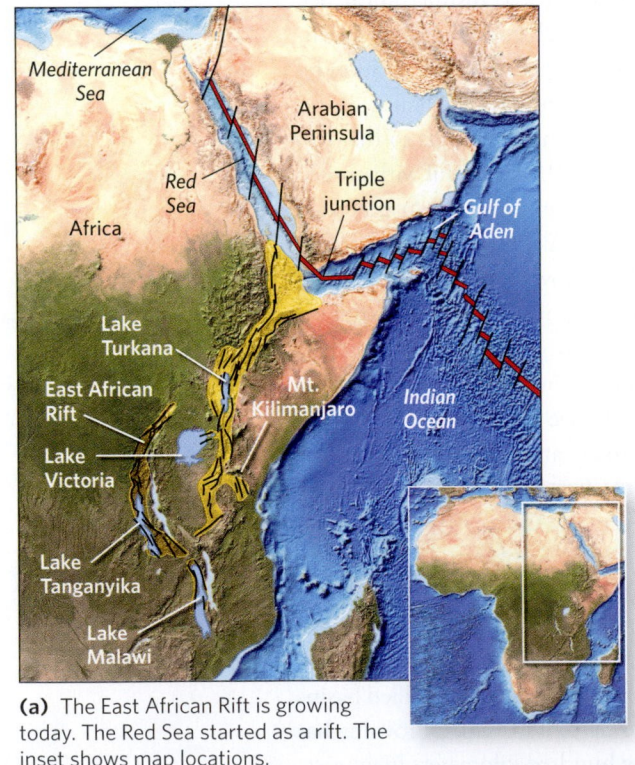

Mediterranean Sea
Arabian Peninsula
Red Sea
Triple junction
Gulf of Aden
Africa
Lake Turkana
East African Rift
Mt. Kilimanjaro
Indian Ocean
Lake Victoria
Lake Tanganyika
Lake Malawi

(a) The East African Rift is growing today. The Red Sea started as a rift. The inset shows map locations.

Snake River Plain
Reno
Salt Lake City
Basin and Range
Sierra Nevada
Colorado Plateau
San Andreas fault
Rio Grande Rift
N
250 km

(b) The Basin and Range Province is a rift. Faulting bounds the narrow north-south-trending mountains, separated by basins. The arrows indicate the direction of stretching.

2.7 Special Locations in the Plate Mosaic

Triple Junctions

At several localities on Earth, three plate boundaries intersect. Each of these intersections, a **triple junction**, stands out on a global map of plate boundaries. For example, one occurs at the point where three ridges intersect in the western Indian Ocean **(Fig. 2.27a)**. Another occurs along the coast of California, where the San Andreas fault, a transform fault, intersects the Cascade subduction zone and an oceanic transform fault **(Fig. 2.27b)**.

Hot Spots

On a global basis, most *subaerial volcanoes* (volcanoes that protrude into the atmosphere) occur along volcanic arcs bordering subduction zones or within rifts. Similarly, most *submarine volcanoes* (volcanoes hidden by the sea) occur along mid-ocean ridges. The volcanoes of volcanic arcs and mid-ocean ridges are *plate-boundary volcanoes* in that they form as a direct consequence of interactions at a plate boundary. Not all volcanoes are plate-boundary volcanoes. Geologists have identified about a hundred locations where volcanoes do not form directly as a consequence of plate interactions. These locations are called **hot spots**.

Most hot spots occur in the interior region of a plate, away from plate boundaries. Examples include Hawaii **(Fig. 2.28a)**, a location within the Pacific Plate thousands of kilometers from the nearest plate boundary, and Yellowstone National Park, which occurs in the interior of the North American Plate. A few hot spots occur along mid-ocean ridges, but the volcanoes formed at these hot spots differ from normal ridge volcanoes in that they produce much more lava than normal seafloor spreading produces. Iceland serves as an example of such a hot spot. So much lava erupts in Iceland, in fact, that it has built a plateau that rises 2 km (1.2 miles) above the rest of the Mid-Atlantic Ridge.

Why do hot spots exist? Researchers still debate this question. Most favor a model in which hot spots lie at the tops of **mantle plumes**, narrow columns of particularly hot rock that flow upward through the mantle **(Fig. 2.28b)**. When the hot peridotite in a mantle plume reaches the base of the lithosphere, it begins to melt, producing magma that seeps up through the lithosphere and erupts as a *hot-spot volcano* (see Chapter 4).

J. Tuzo Wilson noticed that most hot-spot volcanoes in plate interiors lie at the end of a chain of *extinct volcanoes*, volcanoes that will never erupt again. In the Hawaiian Islands, for example, active volcanoes

Figure 2.27 Triple junctions.

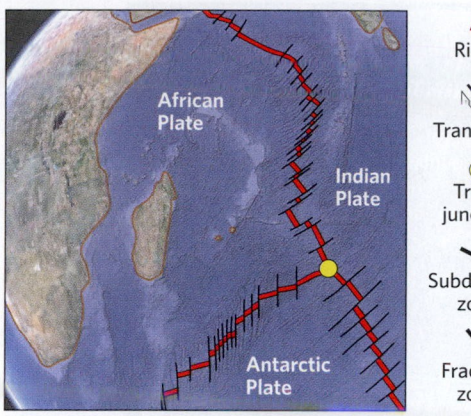

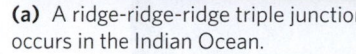
Ridge
Transform
Triple junction
Subduction zone
Fracture zone

(a) A ridge-ridge-ridge triple junction occurs in the Indian Ocean.

(b) A trench-transform-transform triple junction occurs at the northern end of the San Andreas fault.

erupt at the southeastern end of a chain of extinct volcanic islands and seamounts (see Fig. 2.28a). (This configuration differs from that of an island arc, whose component volcanoes are all active.) To explain the configuration of the Hawaiian-Emperor seamount chain, Wilson proposed that the position of a hot spot stays fixed, more or less, relative to the moving plate above it. As the plate moves, the volcano forming above the hot spot is eventually carried off the hot spot and becomes extinct **(Fig. 2.28c)**. A new volcano then starts to grow over the hot spot, and this volcano remains active for a while, until it, too, gets carried off the hot spot and goes extinct **(Box 2.2)**. The process repeats for as long at the hot spot survives, which can be tens of millions of years. Extinct volcanic islands slowly sink below sea level and become seamounts. The resulting chain of inactive volcanic islands and seamounts is a **hot-spot track (Fig. 2.29)**.

Take-home message . . .

Some features of the Earth occur at specific points. For example, a triple junction is the point where three plate boundaries join. A hot spot is a point where volcanism is not the direct result of plate-boundary processes. Hot spots may form due to melting at the top of a mantle plume. As a plate moves over a plume, a hot-spot track develops.

Quick Question -
How can you distinguish a volcanic arc from a hot-spot track?

Figure 2.28 Hot spots.

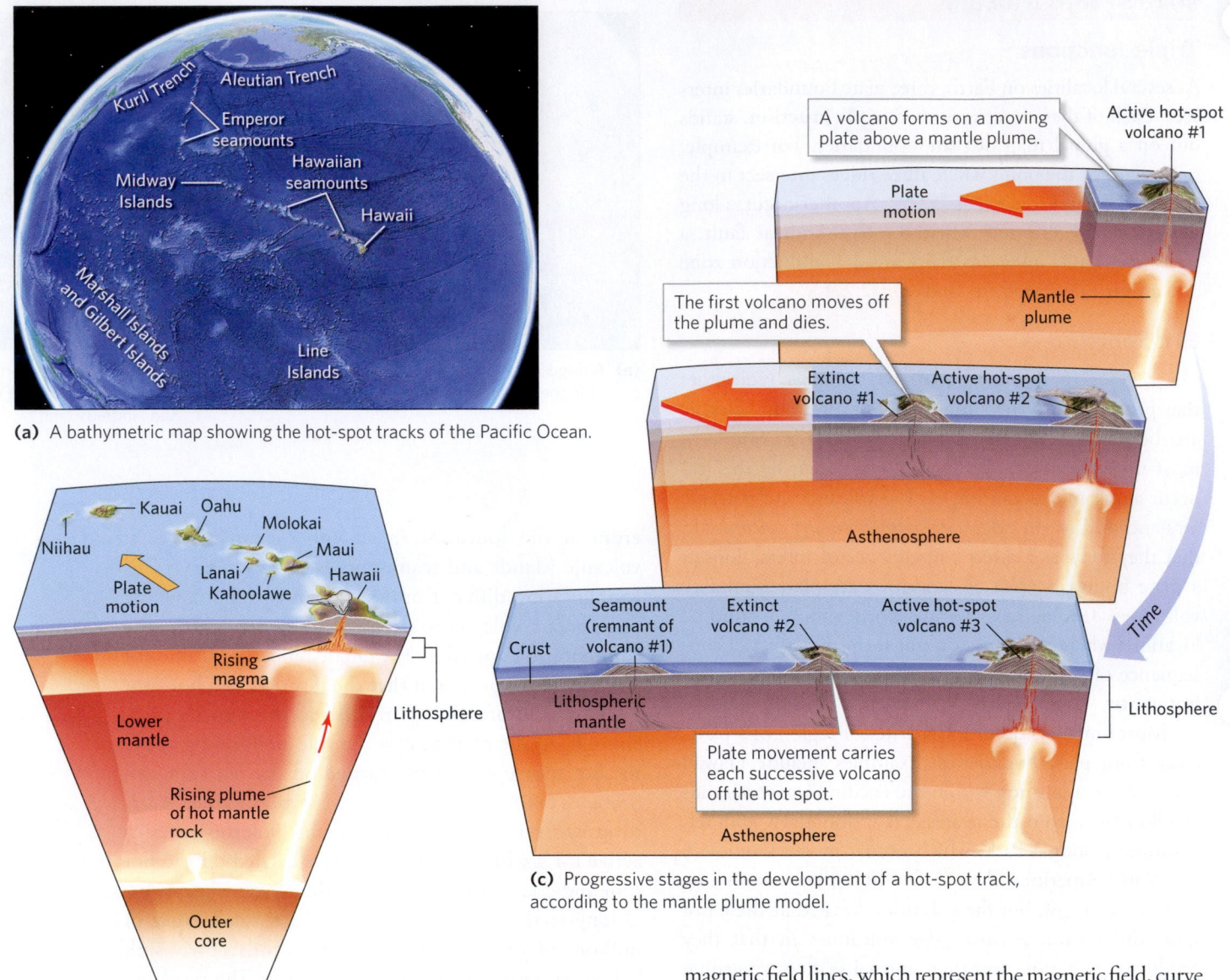

(a) A bathymetric map showing the hot-spot tracks of the Pacific Ocean.

(b) According to the mantle-plume model, Hawaii lies above a rising column of hot mantle.

(c) Progressive stages in the development of a hot-spot track, according to the mantle plume model.

2.8 Paleomagnetism: A Proof of Plate Tectonics

For a hypothesis to become a theory, it must be subjected to many tests. Two key tests of plate tectonics came from the study of paleomagnetism. Here, we introduce paleomagnetism and then show how paleomagnetic tests helped prove plate tectonics.

Introducing Paleomagnetism

As we saw in Chapter 1, the flow of liquid iron alloy in the Earth's outer core generates a magnetic field. Invisible magnetic field lines, which represent the magnetic field, curve through space around the Earth. Like the field produced by a bar magnet, Earth's magnetic field is dipolar, meaning that it has a north pole and a south pole. We can define these **magnetic poles** as the points on the Earth's surface where magnetic field lines point straight down (Fig. 2.30a). Geologists, by convention, refer to the pole near the north end of the Earth as the *north magnetic pole* and to the pole at the south end as the *south magnetic pole*. We can represent the **dipole** of Earth's magnetic field with an arrow that passes through the center of the Earth and points from the north magnetic pole to the south magnetic pole.

At any given time, the magnetic poles of the Earth do not exactly overlie the **geographic poles**, the points where the rotational axis of the Earth intersects its surface. For example, during recorded history, the north magnetic pole has followed a circuitous route to its present location in the Arctic Ocean off the northern coast

Box 2.2 ▶ How can I explain . . .

The evolution of hot-spot tracks

What are we learning?
How a hot-spot track forms in the interior of a plate.

What you need:
- A disposable aluminum-foil baking pan.
- Cornstarch and water.
- A short candle and some matches.
- Two wood blocks that are slightly taller than the candle.

Instructions:
- Mix the water and cornstarch to make a paste.
- Cover the base of the pan with a 2-cm (1-inch)-thick layer of paste.
- Configure the blocks so that they border the candle and can support the pan.
- Light the candle.

- Slide the pan slowly over the candle. Bubbles form in the paste above the heat and remain as mounds (volcanoes) in the paste. Number the volcanoes from oldest to youngest.

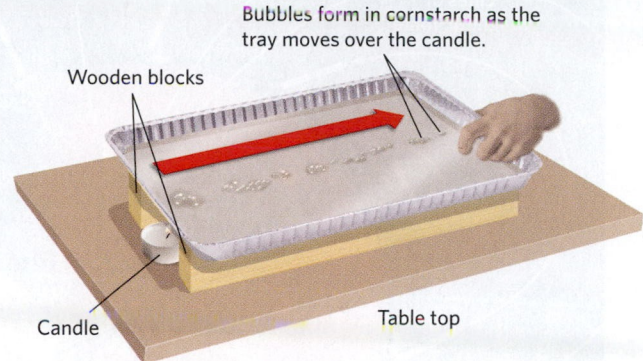

What did we see?
As the paste layer (plate) moves over the fixed heat source, bubbles (representing volcanoes) form. But no bubble erupts for long, because as soon as it gets carried off the heat source, it becomes inactive (extinct).

Figure 2.29 Locations of hot-spot volcanoes and hot-spot tracks. The most recent volcano (dot) is at one end of each track (red line).

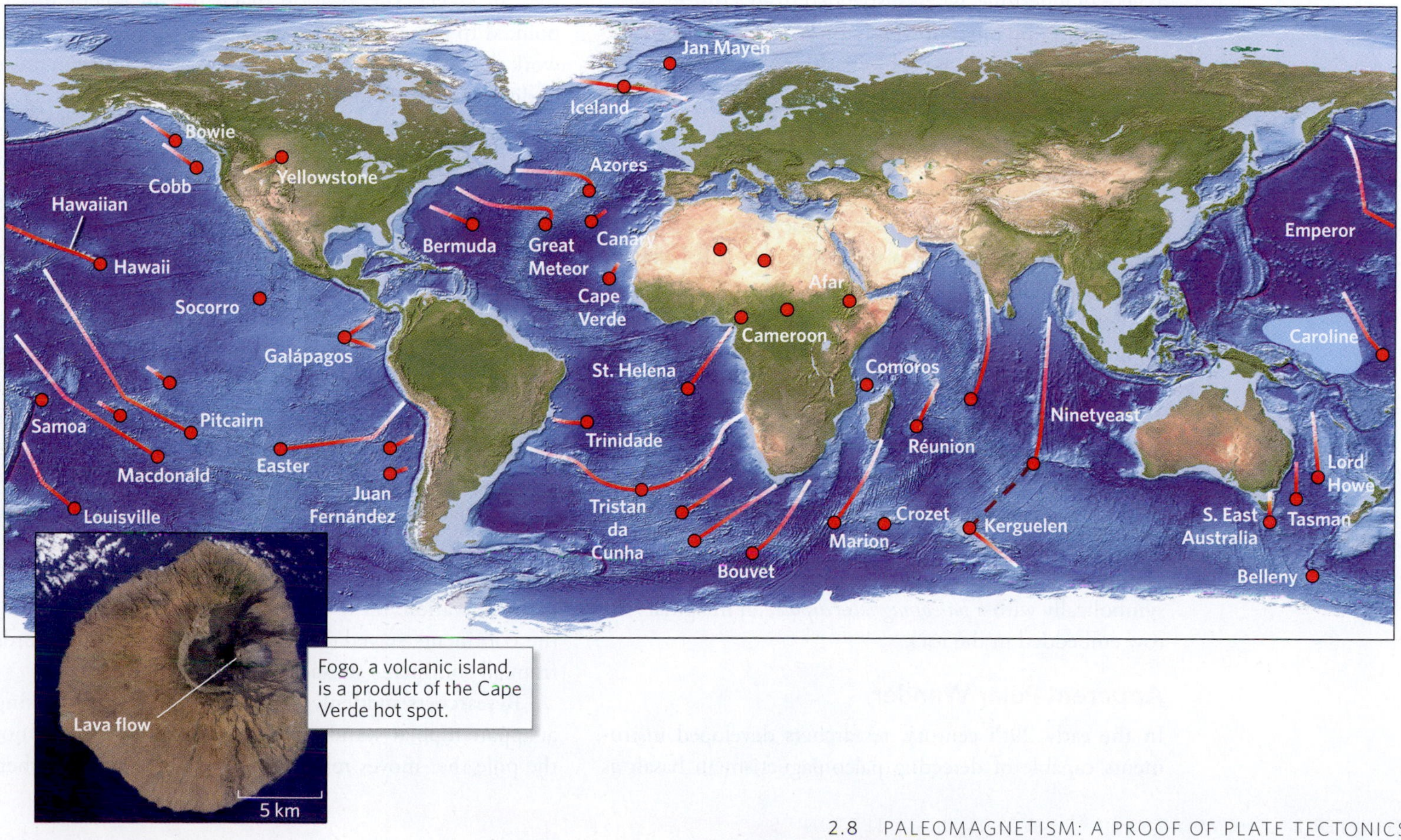

Fogo, a volcanic island, is a product of the Cape Verde hot spot.

Lava flow

5 km

Figure 2.30 Features of the Earth's magnetic field.

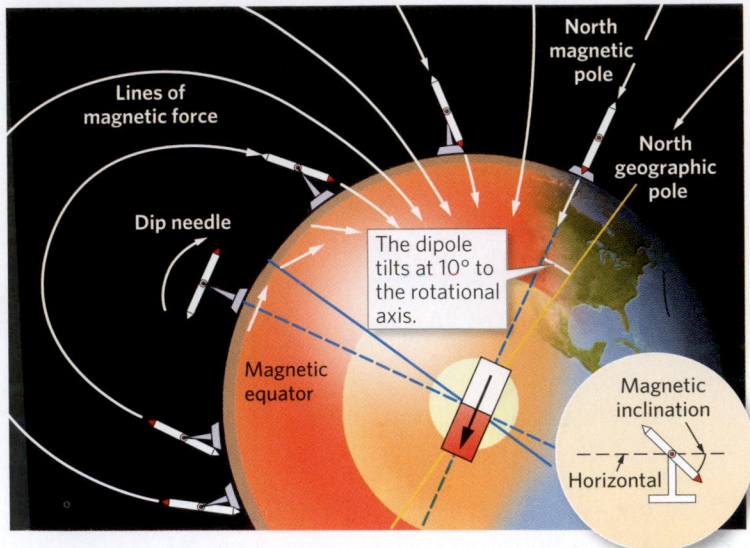

(a) The Earth's magnetic axis is not parallel to its axis of rotation.

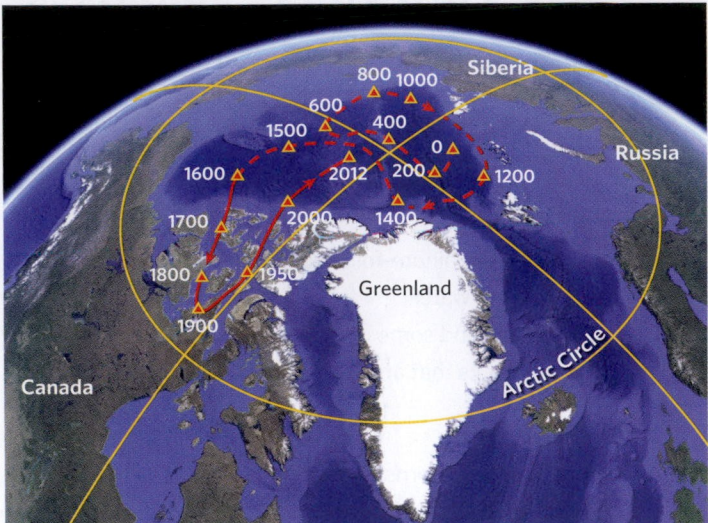

(b) A simplified map showing the changing position of the north magnetic pole over the past 2,000 years. Before about 1600, its position was not as well known, so the path is dashed.

of Canada **(Fig. 2.30b)**. But magnetic poles don't seem to stray far from the geographic poles, and we can assume that, when averaged over several thousand years, the position of a magnetic pole does correspond with that of the geographic pole.

Over 1,500 years ago, Chinese sailors discovered that a piece of lodestone, when suspended from a thread, swivels until it points in a northerly direction, so it can serve as a compass. Lodestone exhibits this behavior because it consists of *magnetite*, an iron oxide mineral that is naturally magnetic. Any magnetic material, if it can move freely, will line up with magnetic field lines when placed in a magnetic field. Some rock types contain enough tiny specks of magnetite or other magnetic minerals that the rock behaves, overall, like a weak magnet. This behavior allows the rock to preserve a record of the orientation of the Earth's magnetic field, at the time the rock forms, for millions or even billions of years. This record of past magnetism is called **paleomagnetism**.

How does paleomagnetism develop? Simplistically, during the rock's formation, magnetic mineral specks in the rock align with the Earth's magnetic field. Once the rock-forming process is finished, the specks can't move, even if the orientation of the Earth's magnetic field, or of the rock, changes. We can represent paleomagnetism symbolically with a *paleomagnetic dipole*, an imaginary arrow embedded in the rock.

Apparent Polar Wander

In the early 20th century, researchers developed instruments capable of detecting paleomagnetism in basalt as well as in certain types of sedimentary rock. From this work, they made a surprising discovery: in rocks that formed long ago, the paleomagnetic dipole does not point to the present-day magnetic poles. At first, they interpreted this discovery to mean that at the time the rock formed, the Earth's magnetic poles were in different locations than they are today. The location that a paleomagnetic dipole pointed to came to be known as a **paleopole**. Additional work showed that paleopole positions, as recorded in rocks of different ages from the same region, appeared to change over time. In fact, when plotted on a map, paleopole positions for successively younger rocks from a region trace out a curving line on the surface of the Earth. This line is called an **apparent polar-wander path (Fig. 2.31a)**.

To interpret apparent polar-wander paths, researchers first took the fixist view and assumed that the position of the continent from which a sequence of rocks had been collected had stayed the same through geologic time. If this assumption was correct, then the apparent polar-wander path would represent how the position of the Earth's magnetic pole migrated over time **(Fig. 2.31b)**. But the researchers were in for a surprise! When they obtained apparent polar-wander paths from different continents, they found that each was different from the others **(Fig. 2.31c)**. The hypothesis that continents are fixed cannot explain this observation, for if the magnetic pole moved while all the continents stayed fixed, then paleomagnetic records from all continents should produce the same paths.

Researchers suddenly realized that they were looking at apparent polar-wander paths in the wrong way. It's not the pole that moves relative to fixed continents—rather, it's the continents that move relative to a more or less fixed magnetic pole! Moreover, since each continent has a unique apparent polar-wander path, continents not only move relative to the magnetic pole (Fig. 2.31d), but must also move relative to each other. This discovery proved that Wegener was right all along—continents do move! Therefore, the discovery served as a test of plate tectonics.

Magnetic Reversals and Marine Magnetic Anomalies

Will the north-seeking end of your compass always point north? No. Study of paleomagnetism has revealed that at various times in the past, the Earth's magnetic field suddenly flipped. Sometimes the Earth has **normal polarity**, as it does today, with the north magnetic pole near the north geographic pole, and sometimes it has **reversed polarity**, with the north magnetic pole near the south geographic pole (Fig. 2.32). Changes in the Earth's magnetic field from normal to reversed polarity, or from reversed back to normal polarity, are **magnetic reversals**. They may be due to changes in the circulation pattern of liquid iron alloy in the outer core. Note that during a reversal, only the magnetic field polarity changes—the Earth itself does not turn upside down.

Not long before the discovery of magnetic reversals, geologists developed a new tool, called **isotopic dating** (or *radiometric dating*), that can provide the age of a rock in years (see Chapter 9). By recording the paleomagnetic polarity in successions of progressively older layers of basalt whose ages had been determined by isotopic dating, researchers established a **magnetic-reversal chronology**, a chart showing when reversals had happened and, therefore, the durations of *polarity intervals* between

Figure 2.31 Apparent polar-wander paths and their interpretation.

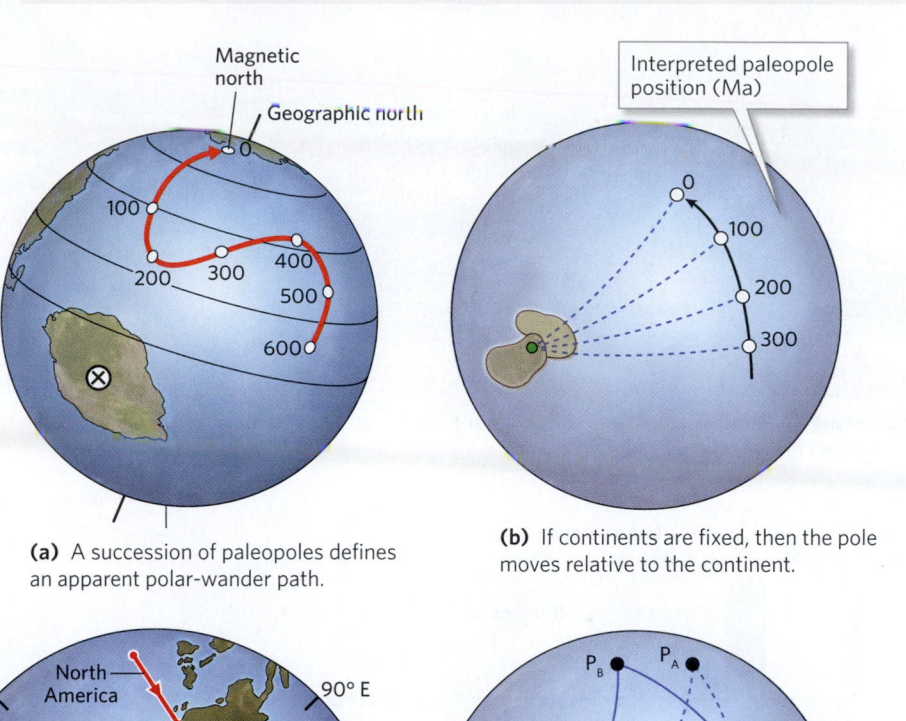

(a) A succession of paleopoles defines an apparent polar-wander path.

(b) If continents are fixed, then the pole moves relative to the continent.

(c) The apparent polar-wander path of North America is not the same as that of Europe or Africa.

(d) If the pole is fixed, then the continents must drift relative to the pole (and to one another).

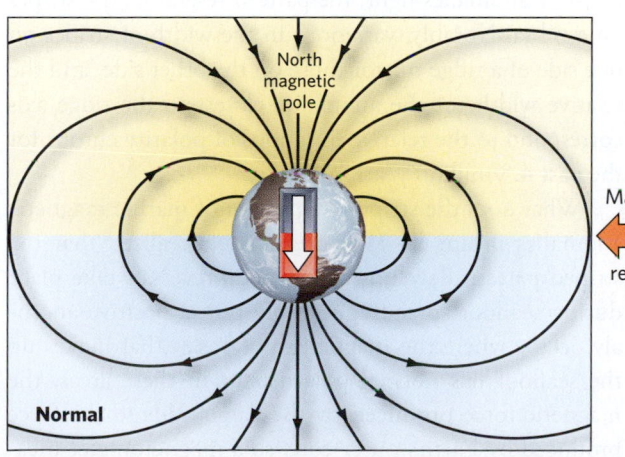

(a) During normal polarity, the dipole points to the south.

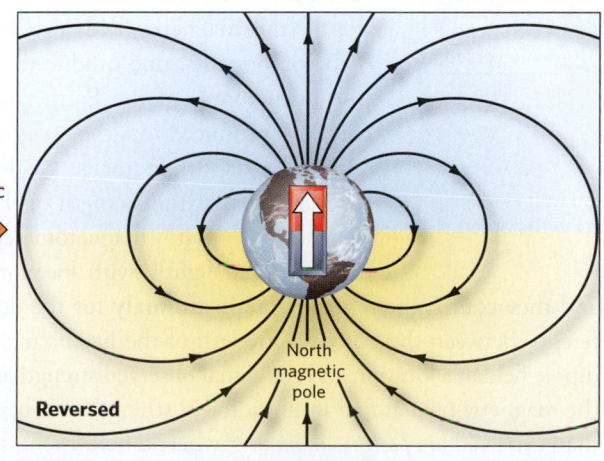

(b) During reversed polarity, the dipole points to the north.

Figure 2.32 The polarity of the Earth's magnetic field reverses every now and then.

Figure 2.33 Magnetic reversals and their chronology.

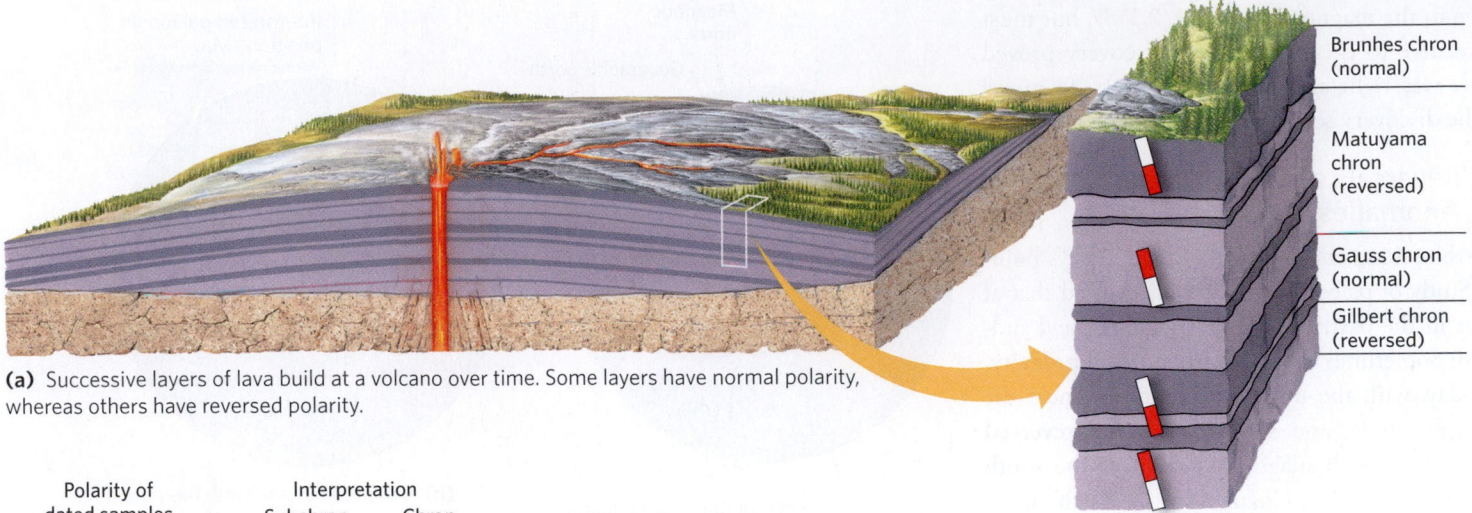

(a) Successive layers of lava build at a volcano over time. Some layers have normal polarity, whereas others have reversed polarity.

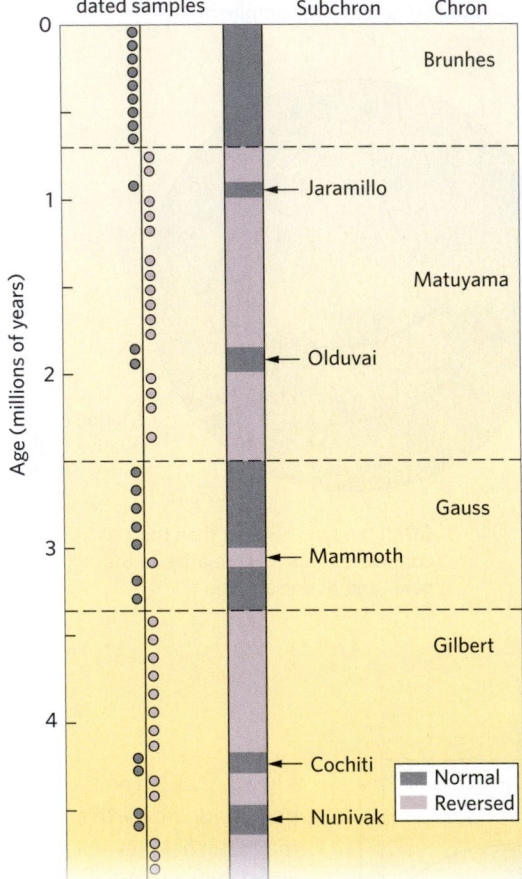

(b) Observations of such layers led to the production of a reversal chronology with named polarity intervals, or chrons.

reversals (Fig. 2.33a). A major polarity interval is called a **polarity chron**. Some chrons include a few short-duration intervals, known as *polarity subchrons* (Fig. 2.33b). Because of limitations on the precision of isotopic dating, the first magnetic-reversal chronology chart represented only the most recent 4.5 million years of Earth history.

At about the same time that some geologists were studying the timing of magnetic reversals, other geologists were measuring regional variations in the strength of Earth's magnetic field by using an instrument called a *magnetometer*. At any given location on the surface of the Earth, the magnetic field includes two components: one produced by the main dipole of the Earth, and another produced by the magnetism of near-surface rocks. Geologists found that the strength of the magnetism that magnetometers detect varies slightly with location, and they coined the term **magnetic anomaly** for the difference between the expected strength of the Earth's main dipole field at a location and the actual observed strength of the magnetic field at that location. Field strengths stronger than expected are *positive anomalies*, whereas field strengths weaker than expected are *negative anomalies*.

On continents, the pattern of magnetic anomalies reflects the composition of the rock underlying the ground at different locations. Rocks that contain more iron, for example, tend to be more magnetic, so their presence produces a positive magnetic anomaly. Since magnetic rock bodies can have somewhat irregular shapes, the pattern of magnetic anomalies on a continent tends to be rather complicated. When geologists first mapped magnetic anomalies on the ocean floor, they expected to see the same sort of complicated pattern that they saw on the continents. But they were in for another surprise! When they towed magnetometers back and forth across the ocean and mapped variations in measured magnetic field strength above the seafloor (Fig. 2.34a), they found that **marine magnetic anomalies** define a distinct pattern of alternating bands aligned parallel to mid-ocean ridge axes (Fig. 2.34b–d). If we color positive anomalies dark and negative anomalies light, the pattern resembles the stripes on a zebra. Notably, variations in the width of stripes on one side of a ridge mirror those on the other side, and the relative widths of the anomalies closest to the ridge axis correspond to the relative durations of polarity chrons for the past 4.5 million years.

What does the stripe-like pattern of marine magnetic anomalies mean? In 1963, researchers realized that the striped pattern develops as magnetic reversals take place during seafloor spreading. Simply put, a positive anomaly occurs where the magnetism of basalt that makes up the seafloor has normal polarity, for in these areas, the magnetic force produced by the basalt adds to the force produced by Earth's outer core, so a magnetometer measures a stronger signal than expected. A negative anomaly

Figure 2.34 Marine magnetic anomalies.

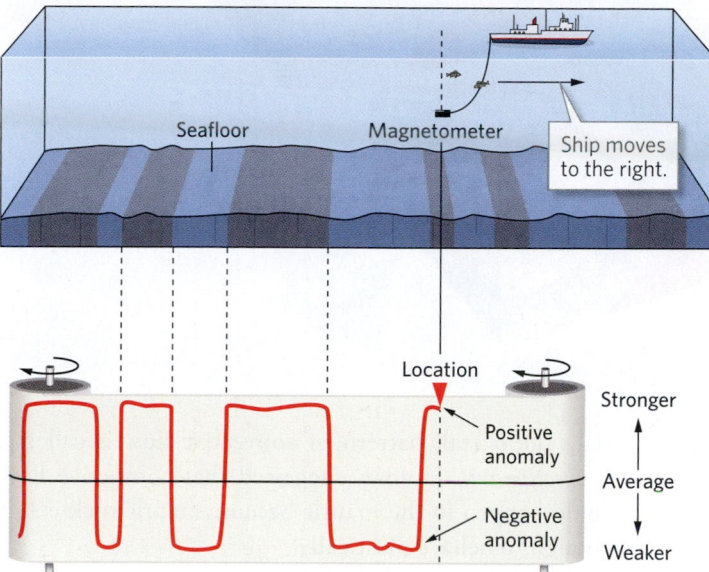

(a) A ship towing a magnetometer detects changes in the strength of the magnetic field on the seafloor.

(b) On a paper record, intervals of stronger magnetism (positive anomalies) alternate with intervals of weaker magnetism (negative anomalies).

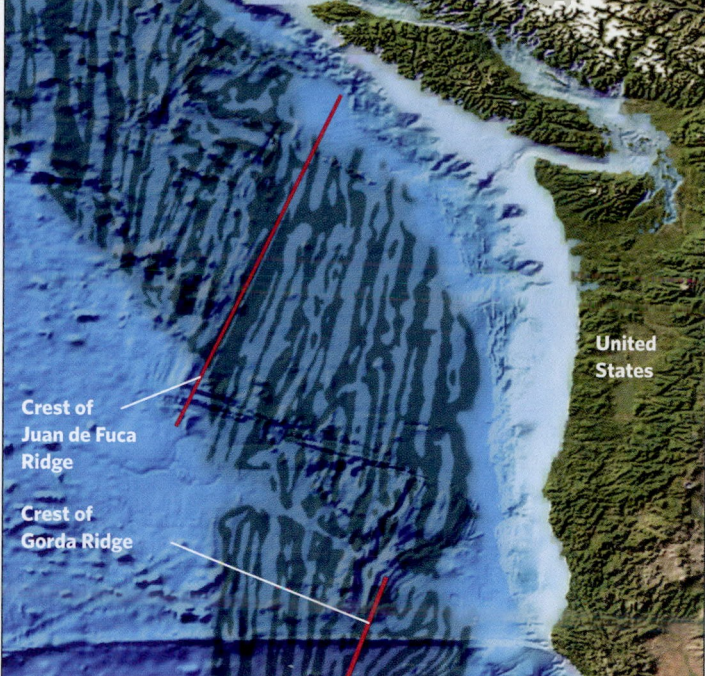

(c) A map showing areas of positive anomalies (dark) and negative anomalies (light) off the west coast of North America.

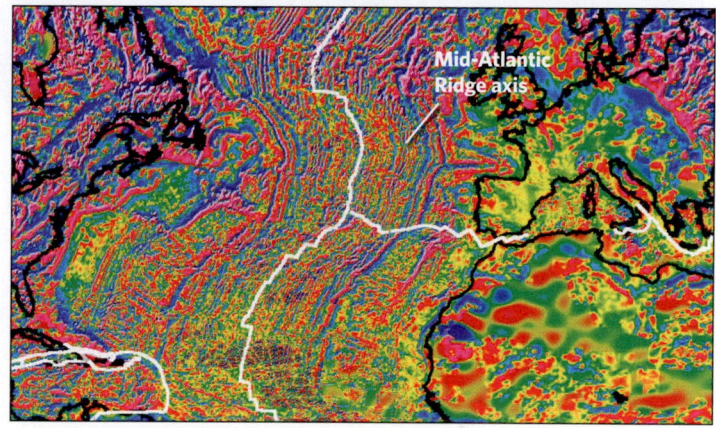

(d) A magnetic anomaly map of the North Atlantic, and adjacent continents. The colors indicate field strength (red is strong, blue is weak).

occurs over regions of the seafloor where the basalt has reversed polarity. In these areas, the magnetic force produced by the basalt subtracts from the force produced by the Earth's outer core, so the magnetometer measures a weaker signal **(Fig 2.35a)**. Therefore, seafloor yielding positive anomalies formed at times when the Earth had normal polarity, whereas seafloor yielding negative anomalies formed when the Earth had reversed polarity.

As seafloor spreading takes place, bands of seafloor with different polarities form and then move away from the ridge axis **(Fig. 2.35b)**. Since seafloor spreading along a given segment of ridge takes place at a fairly constant rate, the relative widths of the stripes correspond to the duration of polarity chrons **(Fig. 2.35c)**. The discovery of the pattern of stripes and the confirmation of the relationship between the widths of the stripes and the durations of chrons serve as proof that seafloor spreading takes place, and therefore also serve as proof of plate tectonics. By assuming constant rates of plate motion, geologists can estimate the age of seafloor and, therefore, the times of magnetic reversals recorded by seafloor older than 4.5 Ma. These estimates indicate that the oldest seafloor currently on Earth formed about 200 Ma.

Take-home message . . .

Rocks record the position of the Earth's magnetic poles at the time the rock formed. The study of paleomagnetism indicates that the continents move relative to the Earth's magnetic poles. Each continent displays a different apparent polar-wander path, so the continents also move relative to one another. The polarity of the Earth's magnetic field reverses every now and then. Ocean floor formed at mid-ocean ridges records these reversals as marine magnetic anomalies, confirming the existence of seafloor spreading.

Quick Question -
What determines the widths of marine magnetic anomalies?

Figure 2.35 The progressive development of marine magnetic anomalies.

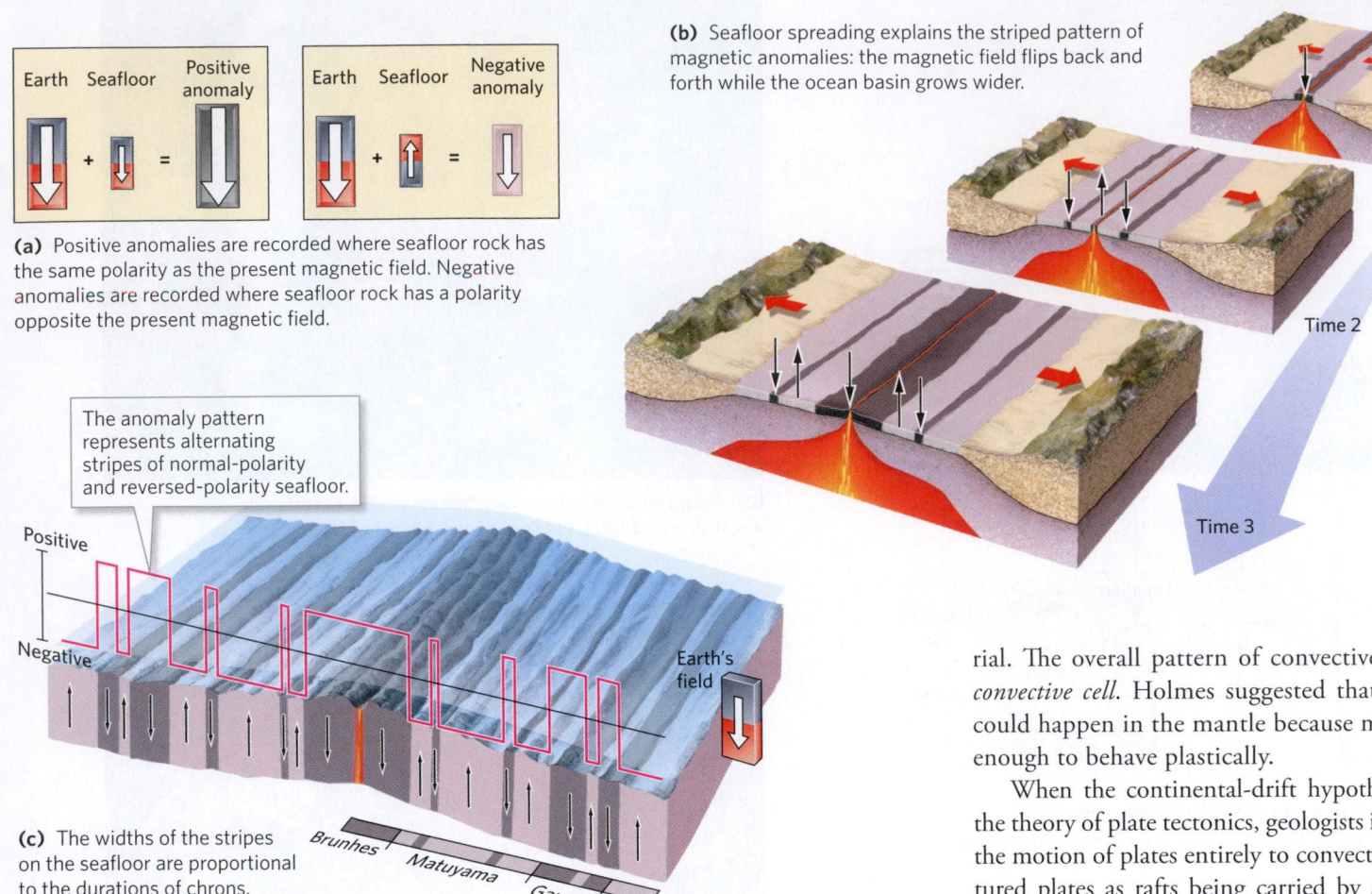

(a) Positive anomalies are recorded where seafloor rock has the same polarity as the present magnetic field. Negative anomalies are recorded where seafloor rock has a polarity opposite the present magnetic field.

(b) Seafloor spreading explains the striped pattern of magnetic anomalies: the magnetic field flips back and forth while the ocean basin grows wider.

Time 1

Time 2

Time 3

The anomaly pattern represents alternating stripes of normal-polarity and reversed-polarity seafloor.

Positive

Negative

Earth's field

(c) The widths of the stripes on the seafloor are proportional to the durations of chrons.

Brunhes   Matuyama   Gauss   Gilbert

2.9 The Velocity of Plate Motions

What Drives the Plates?

Because Alfred Wegener had not been able to answer the question of what makes continents drift, most geologists of his era didn't believe that continents could drift. But the minority who did agree with Wegener continued to speculate about how the Earth could be mobile. In 1928, a British researcher, Arthur Holmes, suggested that continental drift may be due to convective flow in the mantle. **Convective flow**, in general, refers to a process of circulation and heat transfer that happens when a deeper layer of material warms up and, therefore, expands and becomes less dense than the overlying layer of cooler material. As a result, the deeper material becomes buoyant, and if the material is weak enough to flow, it rises. When this happens, cooler material sinks to take the place of the rising warm mate-

> The least movement is of importance to all nature.
>
> —BLAISE PASCAL (FRENCH MATHEMATICIAN, 1880–1930)

rial. The overall pattern of convective flow is called a *convective cell*. Holmes suggested that convective flow could happen in the mantle because mantle rock is hot enough to behave plastically.

When the continental-drift hypothesis evolved into the theory of plate tectonics, geologists initially attributed the motion of plates entirely to convection, and they pictured plates as rafts being carried by simple convective cells in the asthenosphere. As a result, early drawings depicting plate motion implied that warm rock rose at the mid-ocean ridge and cool rock sank at subduction zones **(Fig. 2.36a)**. But it's not that simple **(Fig. 2.36b)**! Eventually, geologists realized that it's impossible to draw a global system of convective cells that can explain the complexity of real plate-boundary configurations. So, while convective flow in the mantle does happen, and probably does influence the long-term movement of plates, the details of plate motion are controlled by forces applied along plate boundaries. Here, we describe two of these forces: ridge-push force and slab-pull force.

Ridge-push force develops because the lithosphere beneath mid-ocean ridges sits higher than the lithosphere beneath adjacent abyssal plains. The elevated lithosphere spreads sideways due to gravity, much as a mound of honey spreads sideways and moves horizontally across a table. This motion drives plates in a direction pointing away from the ridge axis **(Fig. 2.36c)**. **Slab-pull force** develops because oceanic lithospheric mantle is cooler, and therefore denser, than the warmer asthenosphere below. So, once a plate starts to subduct, it sinks, like an anchor into water, though much more slowly.

Figure 2.36 Forces driving plate motions.

Old idea

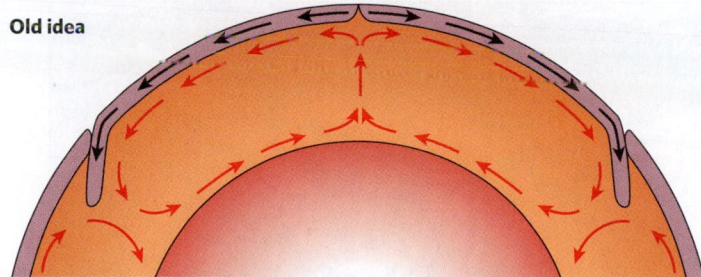

(a) The old, incorrect image of simple convective cells in the asthenosphere. In this model, the plates were viewed as passive rafts carried along by convective flow in the mantle.

Modern image

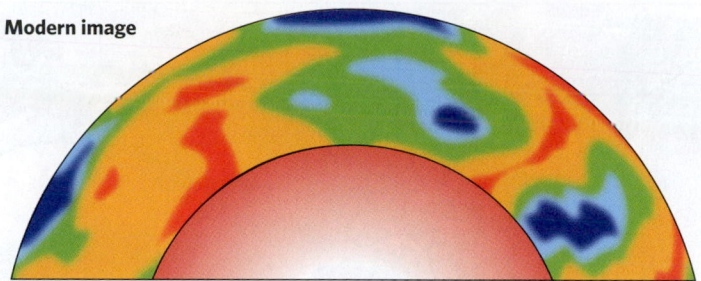

(b) Modern studies show that convective flow in the mantle is complicated. Warm (redder) areas rise, and cooler (blue) areas sink.

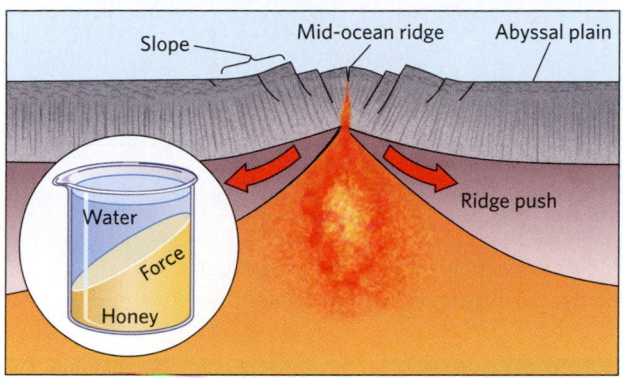

(c) Ridge push develops because the region of a rift is elevated. Like a wedge of honey with a sloping surface, the mass of the ridge pushes sideways.

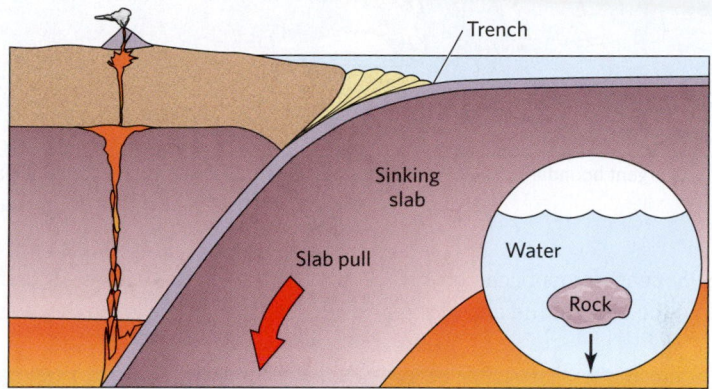

(d) Slab pull develops because lithosphere is denser than the underlying asthenosphere and sinks like a stone in water (though much more slowly).

The subducted section slowly pulls the rest of the plate behind it (Fig. 2.36d). Geologists still debate the relative roles of the different forces acting on plates. Most likely, the motions that we see reflect a complex combination of forces. A growing consensus, however, suggests that slab pull is the dominant force applied to plates.

Relative versus Absolute Plate Velocities

How fast do plates move? The answer depends on your *reference frame*. To illustrate this concept, imagine two cars speeding in the same direction down the highway. From the viewpoint of a tree along the side of the road, Car A zips by at 100 km per hour, while Car B moves at 80 km per hour. But relative to Car B, Car A moves at only 20 km per hour. How you describe the velocity of a car depends on whether the reference frame is another moving car or a fixed tree. Similarly, geologists use two different reference frames for describing plate velocity. If we describe the movement of Plate A relative to Plate B, then we are talking about **relative plate velocity**. But if we describe the movement of both plates relative to a

fixed reference point that is not on one of the plates, then we are speaking of **absolute plate velocity (Fig. 2.37)**. Note that to completely specify velocity, you need to indicate both the rate of movement and the direction of movement. Physicists define rate (or speed) by the equation: rate = distance ÷ time.

One way to determine the rate of relative plate velocities is to measure the distance of marine magnetic anomalies of a known age from a mid-ocean ridge. If the spreading rate at the ridge is constant, then the rate of a point on the plate relative to the ridge axis can be calculated from the equation for rate. The rate of the plate on one side of the ridge relative to the plate on the other is twice this value. The direction of motion is perpendicular to the ridge axis.

To measure absolute plate velocities, at least approximately, we can measure the movement of a plate relative to a hot spot. Specifically, if we assume that the position of a hot spot does not change much for a long time, then a hot-spot track on a plate moving over a mantle plume provides a record of the plate's rate and direction of movement. For example, if we assume that the Hawaiian hot spot's position

Figure 2.37 Plate velocities worldwide. The black arrows show relative plate velocities and the red arrows show absolute plate velocities.

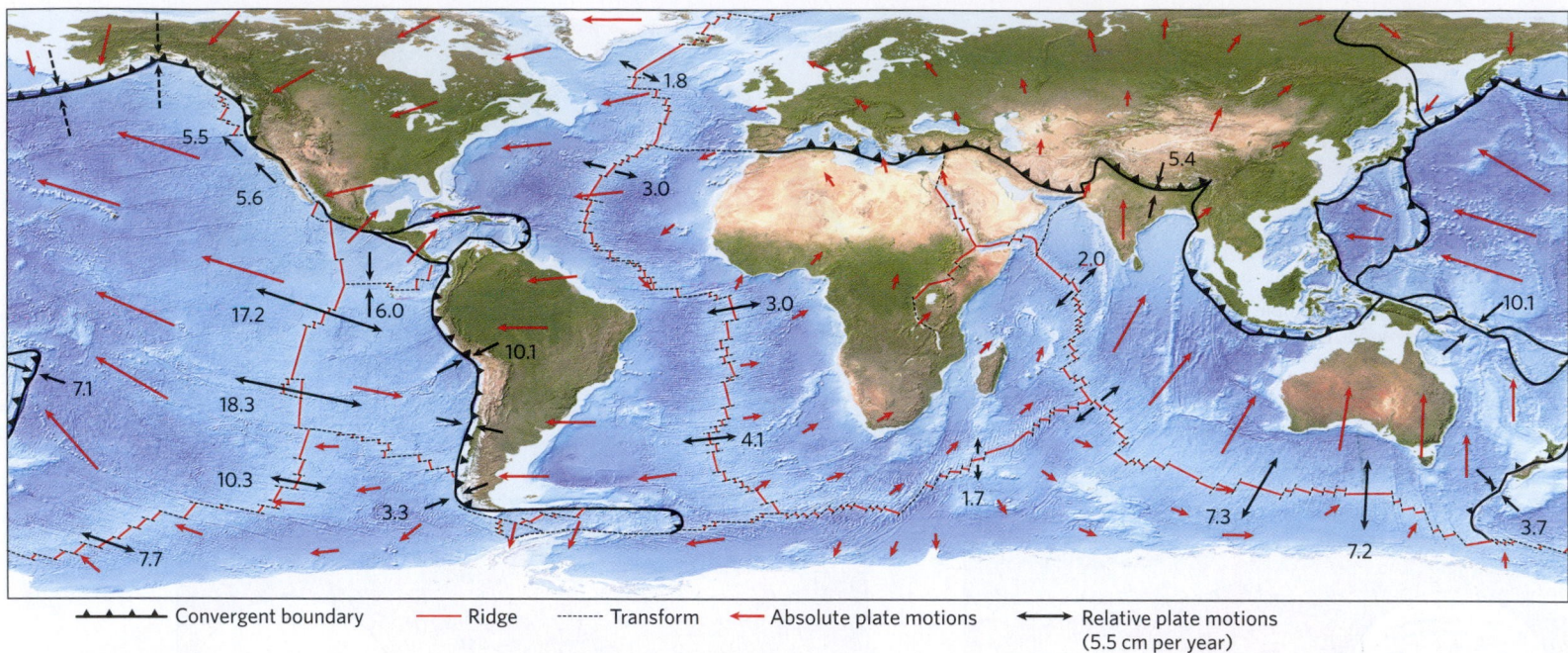

| △△△ Convergent boundary | ─── Ridge | ------- Transform | ← Absolute plate motions | ← → Relative plate motions (5.5 cm per year) |

Figure 2.38 GPS measurements can now track modern-day plate motions accurately. The length of each arrow on this map indicates the velocity at which a point on the crust at the end of the arrow is moving. Turkey is moving westward, and the Greek islands are moving southwestward.

Figure 2.39 Due to plate tectonics, the map of the Earth's surface slowly changes. Here we see the assembly, and later the breakup, of Pangaea over the past 400 million years.

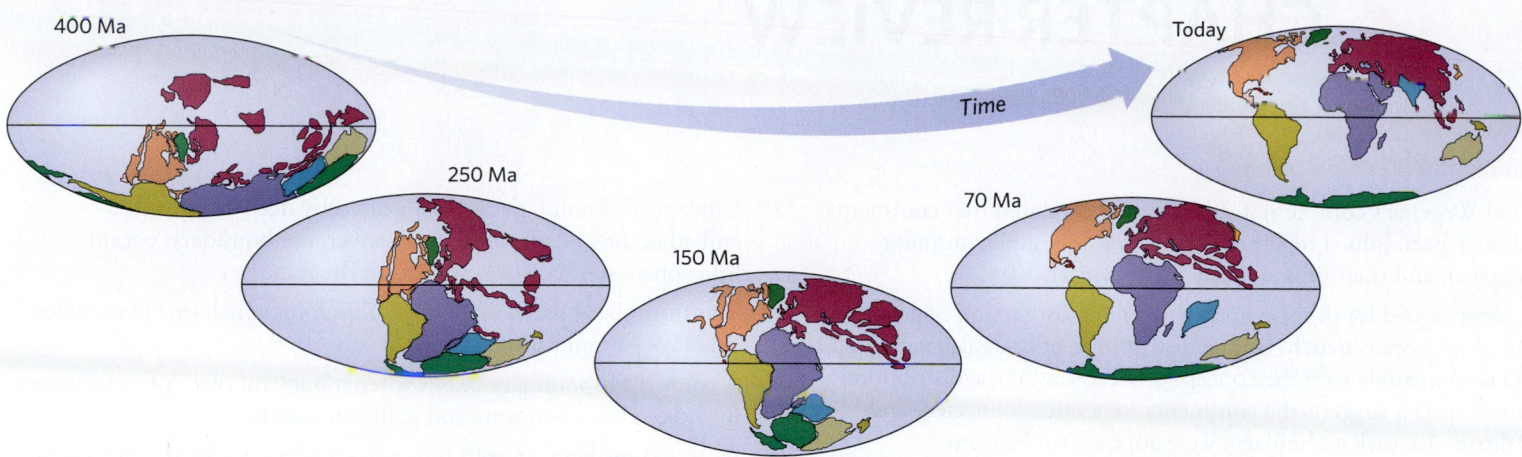

is fixed, then the orientation of the Hawaiian-Emperor seamount chain indicates the absolute direction of Pacific Plate motion, and by dating the rocks of each volcano in the chain, we can calculate the rate of plate movement. Note that the Hawaiian chain runs northwest, whereas the Emperor chain curves north-northwest (see Fig. 2.28a). Isotopic dates of volcanic rocks from the bend indicate that they formed about 43 million years ago, suggesting that the direction in which the Pacific Plate was moving changed at that time.

GPS: Observing Plate Motions in Real Time

Using the methods described above, geologists determined that relative plate motions on Earth today occur at rates of about 1 to 15 cm (0.4 to 6 inches) per year, about the rate at which your fingernails grow. These rates, though small, can yield large displacements during the immensity of geologic time. For example, in 1 million years, a plate moving at 1 cm per year travels 10 km (6 miles).

Can we detect very slow plate motions by direct observation? Until the 1990s, the answer was no. But now, by using the **global positioning system** (**GPS**)—the same technology that drivers use to find their destinations—geologists can detect displacements as small as a few millimeters per year **(Fig. 2.38)**. With such resolution, it's actually possible to see movement of plates over the course of a year! Our ability to observe plate motions directly serves as the ultimate proof of plate tectonics.

Paleogeography

Knowledge of the positions of marine magnetic anomalies allowed geologists to reconstruct the opening of modern oceans back to about 200 Ma. Defining the relative positions and latitudes of continents earlier in

Earth history, however, requires the use of other data sources. For example, paleomagnetic studies can indicate the latitudes of continents at times in the past, and they can indicate whether two continents were moving as one or were moving separately (demonstrating that they were apart). Study of rock units and fossils also helps determine when continents were merged and when they were separate.

By taking such data into account, geologists have refined the image of continental drift as Wegener had imagined it, and they can describe the complex ways in which the map of our planet's surface has changed over time **(Fig. 2.39)**. This work shows that Pangaea was not the only supercontinent during Earth history. In fact, continents merged to form supercontinents, which later broke up, at least four times during the past few billion years. Plate interactions have produced a multitude of different rock types, as we will see in the next few chapters, as well as many mountain belts and other geologic features, which we will discuss later in the book.

Take-home message . . .

Plate motion takes place because plates are acted on by ridge push, slab pull, and convective flow in the asthenosphere. This motion happens at rates of 1 to 15 cm per year. Relative plate velocity refers to the rate at which a plate moves relative to its neighbor, whereas absolute plate velocity refers to the rate at which a plate moves relative to a fixed reference point in the mantle. GPS measurements can now detect relative plate motions directly. Paleomagnetic, fossil, and rock unit data have allowed us to track the movement of continents over Earth's history.

Quick Question -
Was Pangaea the only supercontinent in history?

2 CHAPTER REVIEW

Chapter Summary

- Alfred Wegener's continental-drift hypothesis stated that continents had once been joined together to form a single supercontinent (Pangaea), and then subsequently drifted apart.

- Wegener argued for drift by noting that (1) coastlines on opposite sides of the oceans match; (2) the distribution of late Paleozoic climate belts is compatible with the concept of Pangaea; (3) the distribution of fossil species suggests the continents were once connected; and (4) distinctive rock assemblages were adjacent on Pangaea.

- Most geologists did not initially accept Wegener's ideas because he couldn't explain how continents move.

- Research in the mid-20th century led to the proposal of seafloor spreading—the idea that new ocean floor forms at mid-ocean ridges. Old ocean floor must sink back into the mantle at trenches.

- The lithosphere, Earth's rigid outer layer, consists of several plates that move relative to one another over the underlying, relatively soft asthenosphere. This concept forms the basis of plate tectonics theory.

- Plate interactions occur along plate boundaries; the interiors of plates remain relatively rigid and stable. Earthquakes, therefore, delineate the position of plate boundaries.

- There are three types of plate boundaries—divergent, convergent, and transform—distinguished by the movement of the plate on one side of the boundary relative to the plate on the other side.

- Divergent boundaries, where seafloor spreading takes place, are marked by mid-ocean ridges. As a consequence of seafloor spreading, continents on either side of the ocean drift apart.

- Convergent boundaries are delineated by deep-sea trenches and adjacent volcanic arcs. At a convergent boundary, oceanic lithosphere subducts and sinks into the mantle.

- Transform boundaries are large faults along which one plate slides sideways past another.

- A convergent boundary ceases when a buoyant piece of crust moves into the subduction zone and collision occurs.

- A large continent can split into two smaller ones by the process of rifting. During rifting, continental lithosphere stretches and thins.

- Three plate boundaries intersect at a triple junction.

- Hot spots are places where volcanism not related to plate-boundary processes takes place. As a plate moves over the hot spot, the volcano moves off the hot spot and becomes extinct. Hot spots may form over mantle plumes, columns of rising asthenosphere.

- By measuring paleomagnetism in successively older rocks, geologists can define an apparent polar-wander path.

- Apparent polar-wander paths are different for different continents because continents move relative to one another, while the Earth's magnetic poles remain roughly fixed.

- Marine magnetic anomalies on the seafloor develop because magnetic reversals happen while seafloor spreading is taking place.

- Plate motion is probably the result of several forces, including ridge push, slab pull, and convective flow in the asthenosphere. Plates move at rates of about 1 to 15 cm (0.4 to 6 inches) per year.

- Modern GPS methods can measure plate motions in real time.

Key Terms

absolute plate velocity (p. 83)
abyssal plain (p. 56)
accretionary prism (p. 69)
active margin (p. 60)
apparent polar-wander path (p. 78)
asthenosphere (p. 59)
bathymetric map (p. 56)
collision (p. 72)
continental arc (p. 69)
continental drift (p. 51)
continental margin (p. 56)
continental shelf (p. 60)
convective flow (p. 82)
convergent boundary (p. 68)
deep-sea trench (p. 57)
dipole (p. 76)

divergent boundary (p. 63)
earthquake (p. 57)
fault (p. 57)
fracture zone (p. 57)
geographic pole (p. 76)
global positioning system (p. 85)
heat flow (p. 57)
hot spot (p. 75)
hot-spot track (p. 75)
island arc (p. 69)
isotopic dating (p. 79)
lithosphere (p. 59)
lithospheric mantle (p. 59)
magnetic anomaly (p. 80)
magnetic pole (p. 76)
magnetic reversal (p. 79)

magnetic-reversal chronology (p. 79)
mantle plume (p. 75)
marine magnetic anomaly (p. 80)
mid-ocean ridge (p. 56)
normal polarity (p. 79)
ocean basin (p. 56)
oceanic island (p. 57)
paleomagnetism (p. 78)
paleopole (p. 78)
Pangaea (p. 51)
passive margin (p. 60)
plate (p. 60)
plate boundary (p. 60)
plate interior (p. 60)
plate tectonics (p. 59)

polarity chron (p. 80)
relative plate velocity (p. 83)
reversed polarity (p. 79)
ridge-push force (p. 82)
rift (p. 73)
rifting (p. 73)
seafloor spreading (p. 58)
seamount (p. 57)
seismic belt (p. 57)
slab-pull force (p. 82)
subduction (p. 68)
subduction zone (p. 68)
transform boundary (p. 70)
transform fault (p. 70)
triple junction (p. 75)
volcanic arc (p. 68)

Review Questions

The letters following each Review Question refer to the corresponding Learning Objective from the Chapter Opener.

1. What was Wegener's continental-drift hypothesis? What was his evidence? How did he construct this map? **(A)**

2. Describe discoveries made from the 1930s to the 1960s about the bathymetry of the seafloor. **(B)**

3. Do earthquakes occur randomly, or are they associated with bathymetric and topographic features? Does heat flow vary randomly, or are variations associated with bathymetric features? **(B)**

4. What is the hypothesis of seafloor spreading as defined by Harry Hess? How did Hess explain how seafloor spreading could take place without an increase in Earth's circumference? **(B)**

5. How did drilling into the seafloor help prove seafloor spreading? **(B)**

6. What are the characteristics of a lithosphere plate? Is it composed of crust alone? What characteristic defines the boundary between lithosphere and asthenosphere? **(C)**

7. How do active and passive continental margins differ? **(C)**

8. How does oceanic lithosphere differ from continental lithosphere, and how does this explain the existence of ocean basins? **(C)**

9. How do we identify a plate boundary? What plates appear on this map? Describe the three types of plate boundaries. For each, be sure to indicate the nature of relative plate motion, and name a bathymetric or topographic feature associated with that type of plate boundary. **(D)**

10. How does oceanic crust form along a mid-ocean ridge? How does oceanic lithospheric mantle form? **(C)**

11. Identify the major geologic features of a convergent boundary on this drawing. **(E)**

12. Is new plate material formed or consumed at a transform boundary? Are all transform boundaries submarine? **(D)**

13. Describe the process of continental collision and give examples of where this process has occurred. **(D)**

14. Describe the characteristics of a rift and give examples of where rifting takes place today. **(E)**

15. What is a marine magnetic anomaly? How is it detected? **(G)**

16. How is a hot-spot track produced, and how can hot-spot tracks be used to determine the past absolute motion of a plate? **(H)**

17. What is paleomagnetism? How did the discovery of apparent polar-wander paths serve as a proof that continents move? **(G)**

18. How did the observed pattern of marine magnetic anomalies form, and how did its existence help prove plate tectonics? **(G)**

19. Discuss the major forces that move lithosphere plates. **(F)**

20. Explain the difference between relative plate velocity and absolute plate velocity. Can we measure plate velocity directly? **(H)**

On Further Thought

21. Explain the bend in the Hawaiian-Emperor seamount chain in terms of the absolute motion of the Pacific Plate. **(F)**

22. How does the existence of the Appalachian Mountains indicate that there was an ocean separating North America from Africa and Europe prior to the formation of Pangaea? **(E)**

Online Resources

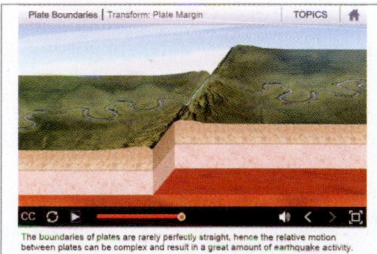

Animations
This chapter features animations of the three types of plate boundaries and their effects.

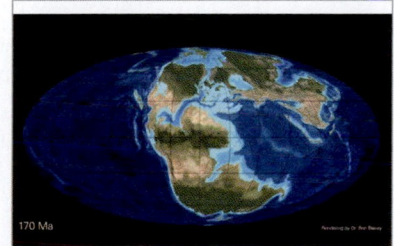

Videos
This chapter also features a video on the breakup of Pangaea.

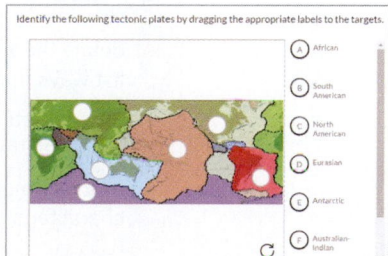

Smartwork5
This chapter includes visual questions on continental drift, collision, and tectonic plates.

3 INTRODUCING MINERALS AND THE NATURE OF ROCK

A museum specimen of red quartz crystals, colored by trace amounts of iron oxide.

3.1 Introduction

The track got narrower and bumpier as the caravan of four-wheel-drive vehicles moved along uphill. The vehicles carried a group of geology students—their destination, an abandoned mine containing beautiful mineral specimens **(Fig. 3.1)**. Finally, the group reached the mine—but it wasn't abandoned! A one-armed watchman glared at them from behind an old refrigerator door on which he'd painted the words, "Is there life after death? Pass this sign and find out!"

After a few moments of awkward silence, one of the students spoke up. "We're just poking around looking for minerals," she said. The watchman replied, "I know . . . I've been watching." Undeterred, the student asked, "Any chance we can take a look in the mine?" There was

another long silence. Then, the man whispered, "You forgot the magic word!" In unison, the students looked up and said, "Please?" With a grin, the watchman said, "What the heck, I could use the company. You can look but don't collect," and led them into the mine.

With headlamps on and eyes wide open, the students explored. Some walls were decorated with bright blue and green streaks, and others with bronze-like crystals. The students spent a wonderful hour marveling at the great variety of shapes and colors their lamps illuminated. The trip had not been wasted!

The minerals in the old mine are but one example of the solid "stuff" that makes up our planet. In fact, if you look at any natural landscape, you'll see a great variety of solid "Earth materials," most of which have familiar names: minerals, rock, soil, glass, gravel, mud, sand,

Figure 3.1 Searching for minerals.

(a) These geology students are jouncing over a dirt track looking for an old mine.

(b) Colorful minerals found on the wall of a mine. These are copper-bearing minerals.

ash, ore. These materials form in many different ways, and over time they react with one another, as well as with water, air, and living organisms. In order to understand the natural phenomena that operate in the Earth System, we must first develop a foundation to understand the natural solid material that makes up our planet. In this chapter, we begin that task. We start by introducing minerals. We explain what they are and how they form, discuss how to distinguish among different types, and introduce the basis for classifying them. After introducing minerals, we turn our attention to rocks. We explain what rock is, introduce the three basic classes of rocks, and learn how geologists study them. This chapter sets the stage for the next three chapters, in which we discuss rocks in greater detail.

I died as a mineral and became a plant. I died as plant and rose to animal. I died as animal and I was man.

—JALAL-UDDIN RUMI (PERSIAN MYSTIC AND POET, 1207–1273)

3.2 What Is a Mineral?

Defining Minerals

Why start our study of the solid Earth by studying minerals? Without exaggeration, we can say that minerals are the building blocks of this planet. Also, minerals are important from a practical standpoint because, as we've alluded to already, they serve as the raw materials for manufacturing many of the items—from windows to wires, and from bricks to concrete—used to sustain modern society.

We've used the word *mineral* in this chapter without formally defining the term because it's a common English word. But in everyday conversation, the meaning of the word is vague. In fact, people refer to a variety of different materials as minerals—you may talk about minerals in a

Figure 3.2 Examples of minerals, both dramatic and common.

(a) A royal crown containing a variety of valuable jewels.

(b) A cluster of clear to white quartz crystals from Brazil. Some mineral specimens are as beautiful as sculptures.

Figure 3.3 What is a crystalline solid?

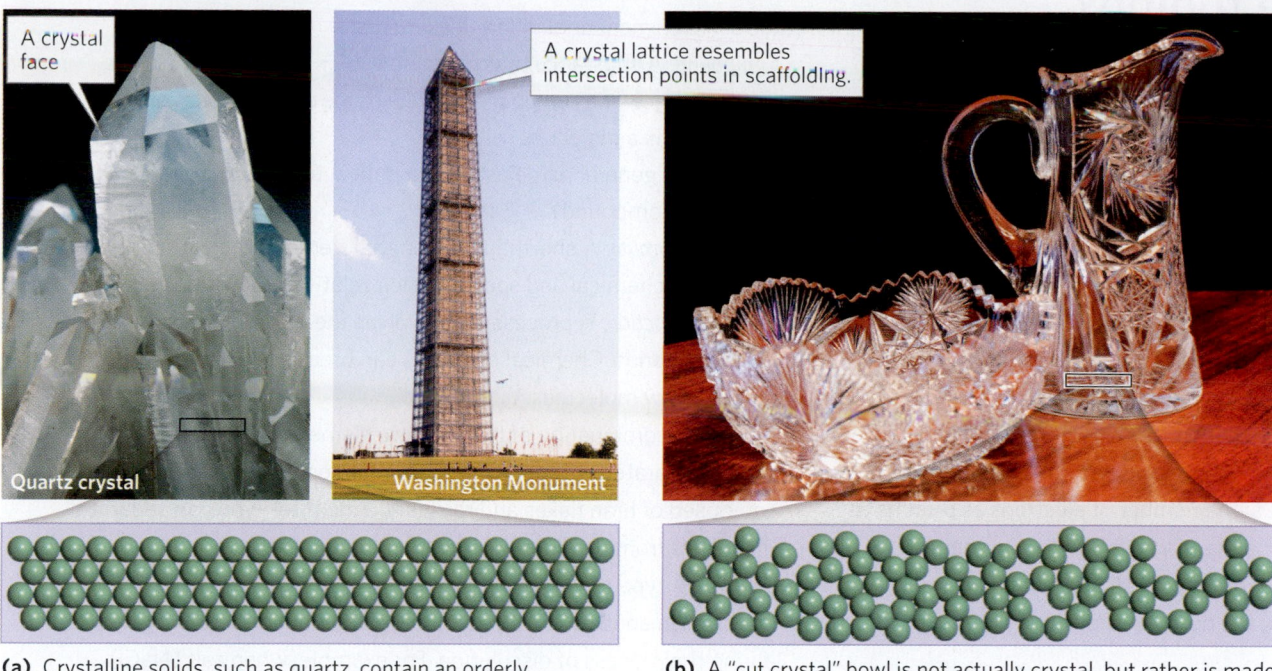

A crystal face

A crystal lattice resembles intersection points in scaffolding.

Quartz crystal

Washington Monument

(a) Crystalline solids, such as quartz, contain an orderly arrangement of atoms. The geometry of the arrangement defines the crystal lattice.

(b) A "cut crystal" bowl is not actually crystal, but rather is made of glass. The atoms within do not have an orderly arrangement.

Halite Diamond Staurolite Quartz

Garnet Stibnite Calcite Kyanite

(c) Crystals come in a variety of shapes, including cubes, prisms, blades, and pyramids. Some terminate at a point and some terminate with flat surfaces.

rock, or minerals in food, or minerals as simply anything that cannot be considered an animal or plant. To a geologist, however, the term has a more restricted definition. A **mineral** is a naturally occurring, homogeneous, crystalline solid that has a definable chemical composition and, in most cases, is inorganic **(Fig. 3.2)**. Let's pull apart this mouthful of a definition and examine each of its components in detail:

- *Naturally occurring:* Real minerals are produced in nature, not in factories. We need to emphasize this point because in recent decades, chemists have synthesized materials that have characteristics identical to those of real minerals. Technically, these new materials can be referred to as *synthetic minerals*.

- *Homogeneous:* A piece of a mineral has the same composition and structure throughout.

- *Solid:* A solid is a state of matter that can maintain its shape indefinitely, so it does not conform to the shape of its container. Liquids, such as oil or water, and gases, such as air, are not minerals.

- *Crystalline solid:* A solid in which atoms are not distributed randomly, but rather remain fixed in a specific, orderly pattern, is a **crystalline solid (Fig. 3.3a)**.

- *Definable chemical composition:* This component of the definition emphasizes that a given mineral contains specific elements in specific proportions. As a result, we can write a chemical formula for a mineral **(Box 3.1)**. Some minerals contain only one element, but most are compounds of two or more elements. For example, *diamond* and *graphite* both have the formula C because they consist entirely of carbon. *Quartz* has the formula SiO_2 (meaning that it contains the elements silicon and oxygen in the proportion of one silicon atom for every two oxygen atoms), and *calcite* has the formula $CaCO_3$ (meaning that it consists of calcium ions bonded to carbonate ions). Quartz and calcite have fairly simple formulas. In comparison, some other minerals have rather complicated formulas. For example, biotite's formula is $K(Mg,Fe)_3(AlSi_3O_{10})(OH)_2$.

- *Inorganic:* To understand this aspect of the definition, we must first distinguish between organic and inorganic chemicals. **Organic chemicals** are carbon-containing compounds that either occur in living organisms or resemble compounds in living organisms. In many organic chemicals, the carbon atoms are arranged in chains or rings and are bonded to hydrogen, nitrogen,

Box 3.1

Science toolbox . . .

Basic chemistry terminology

To describe minerals, we need to use terminology developed by chemists. We list common terms here for ready reference (see also Box 1.2).

- *Element:* A pure substance that cannot be separated into other elements.
- *Atom:* The smallest piece of an element that retains the characteristics of that element. An atom consists of a nucleus surrounded by orbiting electrons. The nucleus of an atom contains both protons and neutrons. (There's one exception to the previous statement—hydrogen's nucleus contains only a proton.) Electrons have a negative charge, protons have a positive charge, and neutrons have a neutral charge. An atom that has the same number of electrons as protons is *neutral,* meaning that it does not have an overall electric charge.
- *Atomic number:* The number of protons in an atom of an element.
- *Atomic weight:* Approximately the number of protons plus neutrons in an atom of an element.
- *Ion:* An atom that is not neutral. An *anion* has excess negative charge because it has more electrons than protons, whereas a *cation* has an excess positive charge because it has more protons than electrons. We indicate charge with a superscript. For example, Cl^- has one excess electron, and Fe^{2+} has lost two electrons.
- *Chemical bond:* An attractive force that holds two or more atoms together. There are several types. *Covalent bonds* form when atoms share electrons. *Ionic bonds* form when a cation and anion get close together and attract each other because opposite charges attract. In materials with *metallic bonds,* some of the electrons can move freely.

- *Molecule:* Two or more atoms bonded together. The atoms may be of the same element or of different elements.
- *Compound:* A substance that contains two or more elements. The smallest piece of a compound that retains the characteristics of that compound is a molecule.
- *Chemical:* A general name used for a pure substance (either an element or a compound).
- *Chemical formula:* A shorthand recipe that itemizes the various elements in a chemical and specifies their relative proportions.
- *Chemical reaction:* A process that involves the breaking or forming of chemical bonds. Chemical reactions can break molecules apart or produce new molecules.
- *Mixture:* A combination of two or more elements or compounds that can be separated without a chemical reaction. For example, a cereal composed of bran flakes and raisins is a mixture—you can separate the raisins from the flakes without destroying either.
- *Solution:* A type of material in which one chemical (the *solute*) dissolves in another (the *solvent*). A solute may separate into ions during the process of dissolution. For example, when salt (NaCl) dissolves in water, it separates into sodium (Na^+) and chloride (Cl^-) ions. In a solution, atoms or molecules of the solvent surround atoms or molecules of the solute.
- *Precipitate:* (noun) A solid formed when ions in liquid solution join together to form solid grains by separation and settling from a solution. Salt remaining when seawater evaporates is a precipitate.
- *Precipitate:* (verb) The process of forming a solid that separates out of a solution or a gas. When saltwater evaporates, salt crystals precipitate from the water.

---

and/or oxygen atoms. Sugar ($C_{12}H_{22}O_{11}$), for example, is an organic chemical. Almost all minerals are inorganic—sugar, for example, is not a mineral. But we have to add the qualifier "almost all" because *mineralogists* (geologists who specialize in studying minerals) consider about 50 organic substances formed when crystals grow naturally in organic debris to be minerals.

With the formal definition of a mineral in mind, we can make an important distinction between a mineral and a glass. A **glass** is an inorganic solid that does not have a crystal structure. In other words, the atoms, ions, or molecules in a mineral lock into a regular arrangement, like soldiers standing in formation, whereas those in a glass are arranged in a semi-chaotic way, like people standing around at a party **(Fig. 3.3b)**.

If you ever need to figure out whether a substance is a mineral or not, just check it against the criteria of the definition. Is motor oil a mineral? No—it consists of organic chemicals and is a liquid. Is table salt a mineral? Yes—it's a solid crystalline compound with the formula NaCl, and it's formed by a geologic process (in this case, precipitation from seawater).

Beauty in Patterns: Crystals and Their Structure

We've mentioned already that minerals, by definition, are crystalline. As a result, they can occur in the form of a crystal. We've already used the word *crystal* without providing a formal definition because it's a common English word. But as was the case for the word *mineral,* geologists

Did you ever wonder . . .

whether there's a difference between a crystal of quartz and "crystal glassware"?

use a more limited definition: specifically, a **crystal** is a single, continuous (that is, uninterrupted) piece of a solid inside which atoms, molecules, or ions are fixed in an orderly arrangement. Note that geologists use another word, grain, in a more general sense. A **grain** can mean a piece of a material that either grew into its present shape, and thus could also be called a crystal, or one that broke off of a larger piece. The larger piece could be a single crystal or a mass composed of many crystals, or even of glass.

A crystal may be bounded by flat surfaces, called **crystal faces**, that intersect at sharp edges and that developed naturally as the crystal grew. Or a crystal may have an irregularly shaped or rounded surface. Crystals with flat faces are called *euhedral crystals*, and those without these faces are called *anhedral crystals*. Each face of a euhedral crystal makes a specific angle relative to its neighbor. Euhedral crystals may come in many fascinating shapes—cubes, trapezoids, pyramids, octahedrons, hexagonal columns, blades, needles, columns, and obelisks—that look like they belong in the pages of a geometry book **(Fig. 3.3c)**. Note that the crystal faces of a euhedral crystal grew to their present shape. They are not formed by breaking, cutting, or grinding. Therefore, if you purchase "crystal glassware" for your dining room, you're not buying mineral crystals, for two reasons: first, glassware consists of glass, so it does not have a crystal structure, and second, the flat faces of the glassware were produced by artisans using a grinding wheel, not by growth.

The orderly arrangement of atoms inside a crystal—its **crystal structure**—provides one of nature's most spectacular examples of a *pattern*, the repetition of a feature in a geometric arrangement. Just as the pattern on wallpaper is defined by the regular spacing of a motif, the pattern in a crystal is defined by the regular spacing of atoms **(Fig. 3.4a, b)**. You can picture atoms in minerals as tiny balls held in place by chemical bonds. The architecture of the arrangement, simplistically, resembles the grid of scaffolding used for a construction project (see Fig. 3.3a). A geometric grid defining the relative positions of atoms in a crystal is called a *crystal lattice*.

As an example of a mineral crystal, let's consider the internal structure of *halite* (NaCl), the salt you use to season food or melt ice on the sidewalk. Within a crystal of halite, chloride ions (Cl^-) alternate with sodium ions (Na^+) in a lattice that looks like an array of lines that intersect one another at a 90° angle **(Fig. 3.4c)**. This internal configuration causes salt crystals to have the shape of a cube, with crystal faces that intersect at 90° angles. Note that the pattern of atoms or ions in a mineral displays **symmetry**, meaning that the shape of one part of a mineral is a mirror image of the shape of another part **(Fig. 3.4d)**.

Figure 3.4 The internal structure of minerals.

(a) The repetition of a flower motif on wallpaper.

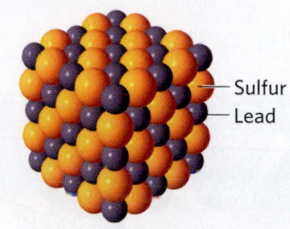

(b) The repetition of alternating sulfur and lead atoms in the mineral galena (PbS).

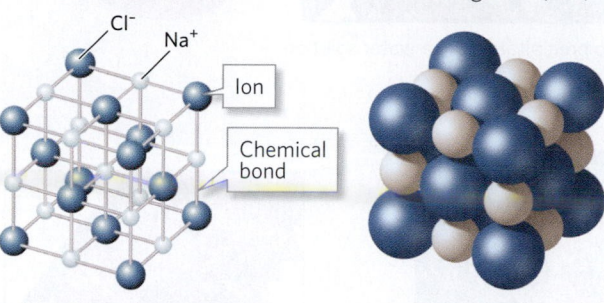

(c) Models of halite (NaCl): The model on the left portrays ions as balls and chemical bonds as sticks. The model on the right gives a sense of how ions fit together in a crystal.

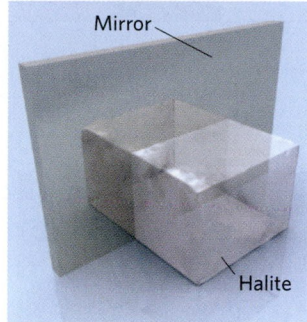
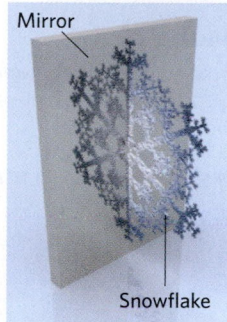

(d) Minerals display symmetry. One half of a halite crystal or a snowflake (an ice crystal) is a mirror image of the other.

How Do Minerals Form?

Where do minerals come from? Geologists recognize five different mineral-forming processes. Which process happens at a given place and time depends on the geologic setting of the mineral-forming environment (**Earth Science at a Glance**, p. 95):

- *Solidification of a melt:* The process by which a liquid transforms into a solid is called *solidification* or *freezing*. During solidification, atoms or ions lock into lattice positions at the surface of a crystal. The growth of ice in water that you've placed in a freezer is an example of this process.

- *Precipitation from a liquid solution:* During precipitation, ions dissolved in liquid bond together, forming a solid crystal that separates from the liquid. Salt crystals, for example, develop when saltwater in a pan evaporates and leaves behind a whitish crust.

Figure 3.5 The growth of crystals.

(a) New crystals nucleate and begin to precipitate from a water solution. As time progresses, they grow into the open space.

Ions attach to the crystal face.

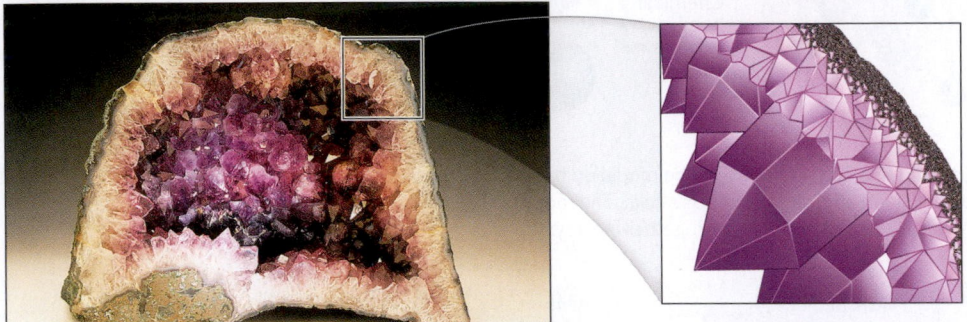

(b) A geode from Brazil consists of purple quartz crystals (amethyst) that grew from the wall into the center. The enlargement sketch indicates that the crystals are euhedral.

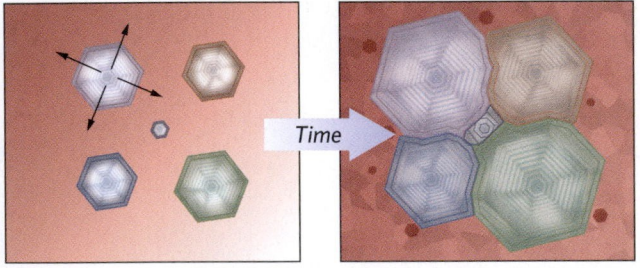
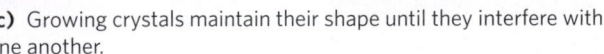

(c) Growing crystals maintain their shape until they interfere with one another.

(d) A crystal growing in a confined space will be anhedral.

- *Diffusion in solids:* **Diffusion** involves the migration of atoms or molecules through a material. For example, if you drip a drop of ink into water, the ink molecules diffuse or disperse through the water, and if you spray perfume into the air, the perfume molecules diffuse through the air. Formation of certain minerals involves diffusion of atoms through a solid, a process that happens much more slowly than does diffusion through air or water. Diffusing atoms can move to a new location within a solid and attach to other atoms, or can rearrange within a solid. Both processes allow a new crystal of a different mineral to grow. For example, garnet—a purple to maroon mineral sometimes used in jewelry—grows by diffusion in solid rock when the rock has warmed to a high temperature.

- *Biomineralization:* Some minerals form at the interface between the physical and biological components of the Earth System. This process, called *biomineralization*, occurs when metabolism in living organisms causes atoms to precipitate either within or on the organisms' cells or immediately adjacent to their cells. For example, clams extract calcium (Ca^+) and carbonate (CO_3^{-2}) ions from water to produce shells of calcite.

- *Precipitation from a gas:* We usually think of precipitation as taking place in liquid. But around volcanic vents, where volcanic gases and steam enter the atmosphere and cool very quickly, minerals precipitate directly from the gas to form deposits around the vent. Bright yellow deposits of sulfur crystals grow by this mechanism. Snowflakes also form by this process.

How do crystals grow **(Box 3.2)**? The process starts with the chance appearance of a *seed*, meaning an extremely tiny crystal. Once the seed exists, other atoms in the surrounding material attach themselves to the faces of the seed. As the crystal grows, its faces move outward, so that at any given time, the youngest part of the crystal lies at the crystal face **(Fig. 3.5a)**. Crystal faces maintain the same orientation relative to each other as they grow—for example, the angle between two adjacent faces of a quartz crystal is 120°. A crystal's shape (needle-like, sheet-like, blade-like) depends both on the geometry of the crystal lattice and on whether the crystal can grow faster in one direction than the others.

If a mineral grows without being inhibited by its surroundings, it becomes a euhedral crystal. Minerals protruding from the walls of a **geode**, a mineral-lined cavity in rock, are commonly euhedral crystals, for example **(Fig. 3.5b)**. If, however, the surroundings inhibit the growth of a mineral, an anhedral crystal develops **(Fig. 3.5c, d)**.

Take-home message . . .

Minerals are homogeneous, naturally occurring solids with a crystal structure and a definable chemical formula; most are inorganic. They grow in a variety of ways, including solidification from a melt and precipitation from a solution. Because of their crystal structure, minerals can develop into crystals whose faces have a specific orientation relative to each other.

Quick Question -
Why isn't naturally occurring petroleum considered to be a mineral?

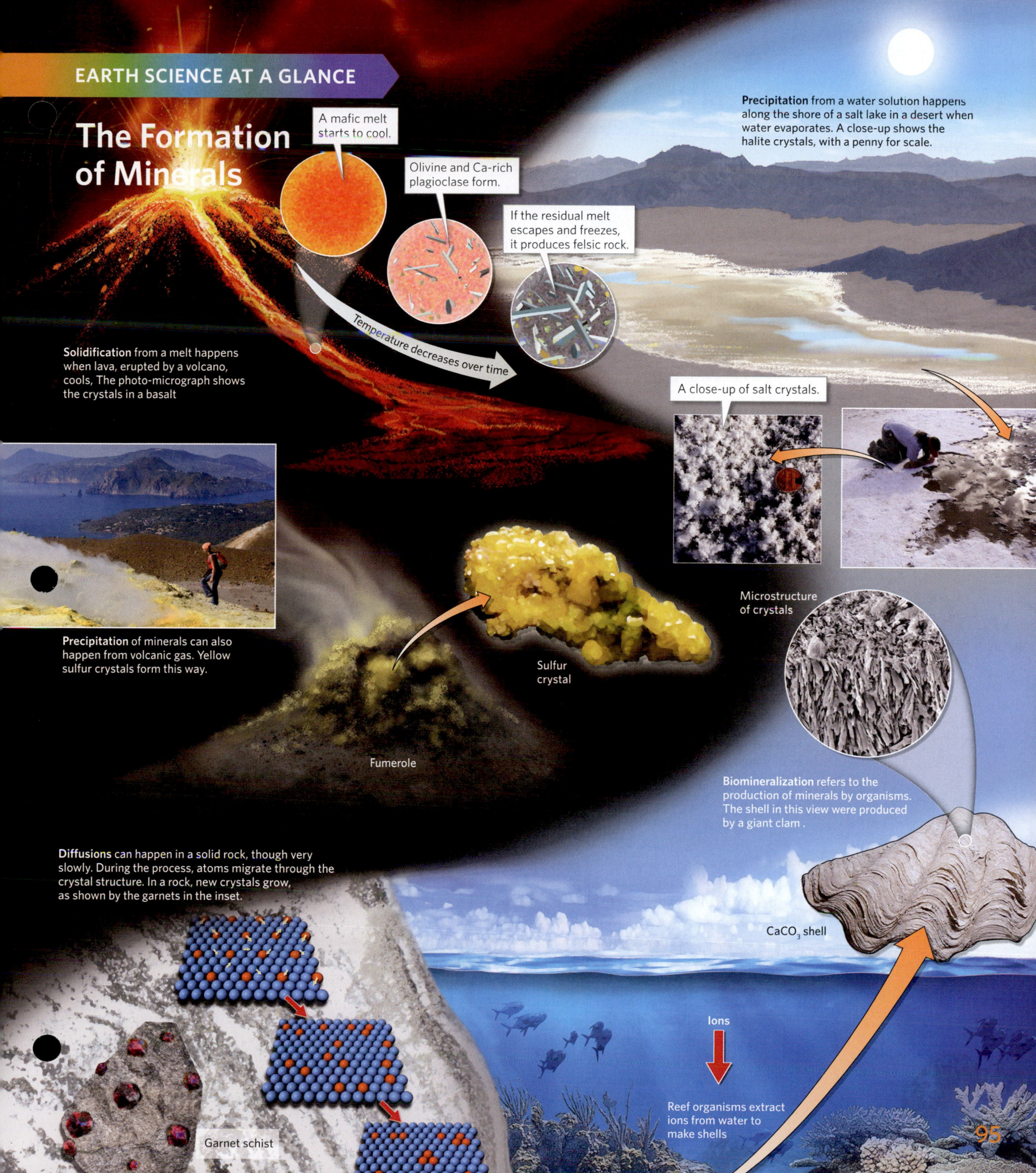

The Formation of Minerals

A mafic melt starts to cool.

Olivine and Ca-rich plagioclase form.

If the residual melt escapes and freezes, it produces felsic rock.

Temperature decreases over time

Solidification from a melt happens when lava, erupted by a volcano, cools, The photo-micrograph shows the crystals in a basalt

Precipitation from a water solution happens along the shore of a salt lake in a desert when water evaporates. A close-up shows the halite crystals, with a penny for scale.

A close-up of salt crystals.

Precipitation of minerals can also happen from volcanic gas. Yellow sulfur crystals form this way.

Sulfur crystal

Fumerole

Microstructure of crystals

Biomineralization refers to the production of minerals by organisms. The shell in this view were produced by a giant clam .

Diffusions can happen in a solid rock, though very slowly. During the process, atoms migrate through the crystal structure. In a rock, new crystals grow, as shown by the garnets in the inset.

$CaCO_3$ shell

Garnet schist

Ions

Reef organisms extract ions from water to make shells

95

How can I explain . . .

Crystal growth

What are we learning?

The concept that crystals grow. This example, crystals growing in a solution, exemplifies one of many environments in which crystals grow.

What you need:

- Salt and water.
- A 16-fluid-ounce (about 0.5 liter) glass jar.
- A 2-mm-diameter glass rod.

Instructions:

- Fill the jar with room-temperature water and dissolve as much salt in the water as it can hold.
- Place the rod in the saltwater.
- Place the jar, uncovered, in a refrigerator, and let the water evaporate over a period of days. Salt crystals will form on the base of the jar and on the glass rod.
- Remove the rod to examine the crystals with a magnifying glass.

Repeat the experiment, only this time place the jar in a warm water bath on a hot plate. Let the water in the jar evaporate quickly.

What did we see?

As the water evaporates, the solution becomes saturated, so that it has no more capacity to hold dissolved salt. The Na$^+$ and Cl$^-$ ions combine to form salt molecules, and crystals grow as those molecules precipitate on the glass. Under magnification, the cubic shape of the crystals may be visible. Environmental conditions, such as the rate of evaporation, affect crystal size.

Stir in salt | Water evaporates | Crystal grows

3.3 How Can You Tell One Mineral from Another?

Basics of Mineral Identification

Mineralogists have identified and named about 4,000 different minerals so far, and they discover new ones every year. Minerals can be practical, beautiful, or even unhealthy (Box 3.3). Minerals differ from one another because they have different chemical compositions, different crystal structures, or both. Amateurs and professionals alike get a kick out of recognizing distinctive minerals. They might hover around a display case in a museum and name specimens without bothering to look at the labels. How do they do it? The trick lies in learning to recognize the basic physical properties—the visual and material characteristics—that distinguish one mineral from another. You can characterize some physical properties, such as shape and color, from a distance. Others, such as hardness and magnetization, can be determined only by handling the specimen or by performing an *identification test* on it. Identification tests include scratching the mineral against another object, placing it near a magnet, lifting it, tasting it, or placing a drop of acid on it. Let's examine some of the physical properties used in mineral identification.

COLOR. Color results from the way a mineral interacts with light. The color you see when looking at a mineral represents the colors of the light spectrum that the mineral doesn't absorb. Some minerals always have the same color, but others come in a range of colors (Fig. 3.6a). Color variations in a mineral may be due to impurities in the crystal lattice. The presence of iron atoms in quartz, for example, makes the quartz red or purple.

STREAK. The **streak** of a mineral refers to the color of a powder produced by scraping the mineral against an unglazed ceramic plate (Fig. 3.6b). Streaks for a given mineral tend to be less variable than the color of whole specimens, so they provide a more reliable clue to a mineral's identity. Calcite, for example, always yields a white streak, even though pieces of calcite may be white, pink, or clear.

LUSTER. Luster refers to the way a mineral's surface reflects light. You can describe luster by comparing the appearance of the mineral with the appearance of a familiar substance. For example, minerals that look like metal have a *metallic luster*, whereas those that do not have a *nonmetallic luster* (Fig. 3.6c, d). Various self-explanatory adjectives used for different types of

Did you ever wonder . . .

How collectors can name a mineral by looking at it?

Figure 3.6 Physical characteristics of minerals.

(a) Color is diagnostic of some minerals, but not all. For example, quartz can come in many colors.

(b) To obtain the streak of a mineral, rub it against an unglazed porcelain plate. The streak consists of mineral powder.

(c) Pyrite has a metallic luster because it gleams like metal.

(d) Feldspar has a nonmetallic luster.

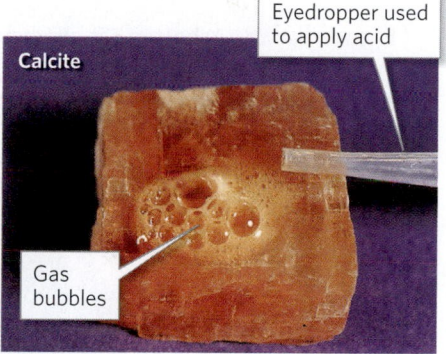

(f) Calcite reacts with hydrochloric acid to produce carbon dioxide gas.

(e) Crystal habit refers to the shape or character of a crystal. These giant gypsum crystals, from a cave in Mexico, are columnar.

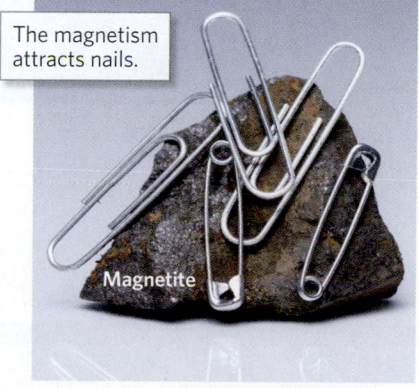

(g) Magnetite is magnetic.

Box 3.3

Consider this . . .

Unhealthy minerals

Most minerals are harmless, so you can ingest them routinely. For example, we use halite to salt food, calcite as an antacid, and certain types of clay to thicken milkshakes. But some minerals pose hazards to human health due to their composition, crystal habit, or grain size. Let's consider a couple of examples.

Some minerals contain chemicals that can poison you. For example, a version of pyrite (FeS_2), called arsenopyrite (FeAsS), contains arsenic, a deadly poison. Arsenopyrite can reside unchanged in rock buried deep in the Earth for millions of years. But if the mineral is exposed to oxygen-bearing groundwater or to the air, it undergoes a chemical reaction with oxygen to become an acid that dissolves in water. Arsenic-bearing groundwater can cause arsenic poisoning.

Concerns about asbestos have been featured in the news for decades. That's because asbestos was once used widely to manufacture insulation, brakes, roof shingles, and floor tiles. *Asbestos* refers to a group of several silicate minerals that grow in clumps of very fine, flexible needles or fibers **(Fig. Bx3.3a, b)**. Minerals in this group have a very high melting temperature and are very strong. That's why builders installed asbestos floor tiles and shingles in the early and mid-20th century. Unfortunately, some (not all) kinds of asbestos, if inhaled, become embedded in human lungs and can cause cancer. For this reason, asbestos has been banned for most applications, and most remodeling or demolition projects for pre-1980s buildings require contractors to follow strict rules for asbestos abatement—they must enclose the areas where asbestos is being removed and must train workers to wear protective clothing and minimize production of asbestos dust **(Fig. Bx3.3c)**.

Quartz and feldspar, and many other silicate minerals, are safe under most circumstances, but can pose a danger if pulverized. Specifically, the dust of pulverized *silicate* minerals can, if inhaled, become embedded in lungs and cause silicosis. Because the minerals are very stable, neither metabolic processes in lungs nor coughing can remove the dust. The irritation caused by the dust triggers the growth of fibrous masses that eventually block access to air and make breathing difficult. Everyone breathes a little dust now and then with no ill effects. But prolonged exposure to dust can be dangerous.

Figure Bx3.3 Asbestos has characteristics that can make it both useful and hazardous.

(a) Asbestos is a fibrous mineral.

(b) At high magnification, asbestos fibers look like tiny needles.

(c) Workers removing asbestos must wear protective gear, and the work area must be sealed off.

nonmetallic luster include silky, glassy, satiny, resinous, pearly, or earthy.

HARDNESS. Hardness, a measure of the relative ability of a mineral to resist scratching, represents the resistance of bonds in the mineral to being broken. Hard minerals can scratch soft minerals, but soft minerals cannot scratch hard ones. In the early 1800s, a mineralogist named Friedrich Mohs listed some minerals in sequence of relative hardness and assigned numbers (1 to 10) to those minerals. For example, a mineral with a hardness of 5 can scratch all minerals with a hardness of less than 5. We can add familiar objects (fingernails, copper coins, window glass, and steel knives) to the list to provide context for judging the hardness of unfamiliar minerals (Table 3.1). For example, if glass has a hardness of about 5.5, then any mineral that can scratch glass has a Mohs hardness higher than 5.5, and any mineral the glass can scratch has a hardness of less than 5.5. We now refer to this list of reference minerals and objects, ordered by relative hardness, as the **Mohs hardness scale**.

SPECIFIC GRAVITY. Specific gravity represents the density of a mineral, as specified by the ratio between the weight of a volume of the mineral and the weight of an equal volume of water. For example, 1 cm³ of quartz has a weight of 2.65 g, whereas 1 cm³ of water has a weight of 1.00 g, so the specific gravity of quartz is 2.65. In practice, you can develop a feel for specific gravity by hefting mineral specimens in your hand. A piece of galena (lead ore, with the formula PbS) "feels" heavier than a similar-sized piece of quartz.

CRYSTAL HABIT. The **crystal habit** of a mineral refers to the shape of a single euhedral crystal, or to the character of an aggregate of many well-formed crystals that grew together as a group **(Fig. 3.6e)**. Crystal habit depends on the internal arrangement of atoms in the crystal. A description of crystal habit generally includes adjectives that define the relative dimensions and the geometric shape of the crystal. For example, crystals that have roughly the same length in all directions are called equant or blocky; those that are much longer in one dimension than in others are columnar or, in the extreme, needle-like. Similarly, specimens shaped like sheets of paper are called platy, and those shaped like knives are called bladed.

SPECIAL PROPERTIES. Some minerals have distinctive properties that readily distinguish them from other minerals. For example, calcite ($CaCO_3$) fizzes in dilute hydrochloric acid (HCl) to produce carbon dioxide (CO_2) gas **(Fig. 3.6f)**, graphite makes a gray mark on paper, magnetite attracts iron objects **(Fig. 3.6g)**, halite tastes salty,

Table 3.1   Mohs hardness scale

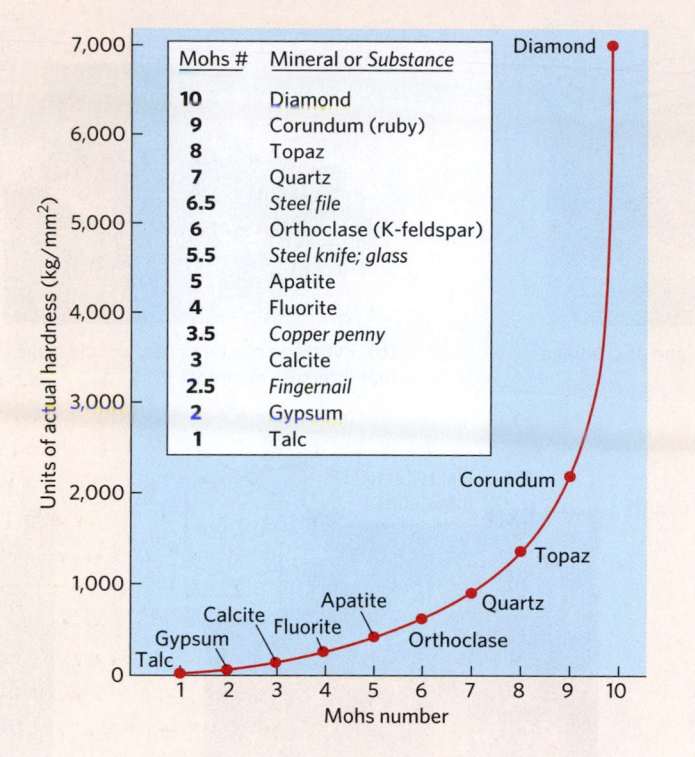

Mohs #	Mineral or *Substance*
10	Diamond
9	Corundum (ruby)
8	Topaz
7	Quartz
6.5	*Steel file*
6	Orthoclase (K-feldspar)
5.5	*Steel knife; glass*
5	Apatite
4	Fluorite
3.5	*Copper penny*
3	Calcite
2.5	*Fingernail*
2	Gypsum
1	Talc

Note: Mohs's numbers are relative—in reality, diamond is 3.5 times harder than corundum, as the graph shows.

and plagioclase has striations (thin parallel corrugations or stripes) on its surface.

FRACTURE AND CLEAVAGE. Different minerals fracture (break) in different ways, depending on the internal arrangement of their atoms. If a mineral breaks to form distinct planar surfaces that have a specific orientation relative to the crystal structure, then we say that the mineral has **cleavage** and refer to each surface as a *cleavage plane*.

Cleavage planes form in the directions where the bonds holding atoms together in the crystal are the weakest **(Fig. 3.7a–e)**. Some minerals have one direction of cleavage. For example, mica has very weak bonds in one direction but strong bonds in the other two directions, so it easily splits into parallel sheets; the surface of each sheet is a cleavage plane. Other minerals have two or three directions of cleavage that intersect at a specific angle. In halite, for example, three directions of cleavage intersect at right angles, so halite crystals break into little cubes. Cleavage planes are not crystal faces, though sometimes the two types of surfaces can be mistaken for each other. By examining a specimen carefully, you can

Figure 3.7 The nature of mineral cleavage and fracture.

(a) Mica has one strong plane of cleavage and splits into sheets.

(b) Pyroxene has two planes of cleavage that intersect at almost 90°.

(c) Amphibole has two planes of cleavage that intersect at 60°.

(d) Halite has three mutually perpendicular planes of cleavage.

(e) Calcite has three planes of cleavage, one of which is inclined.

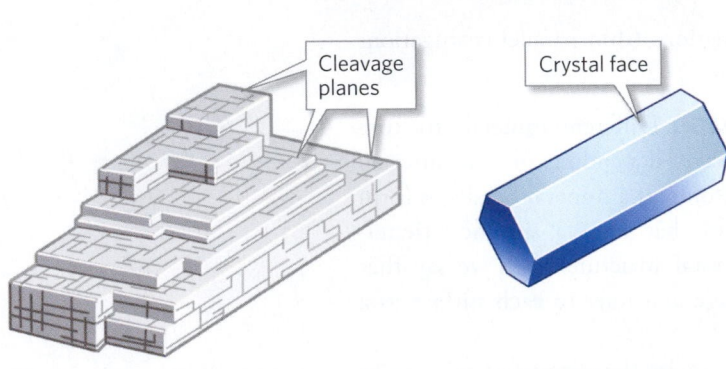

(f) How do you distinguish between cleavage planes and crystal faces? Cleavage planes can be repeated, whereas a crystal face is a single surface.

(g) Minerals without cleavage can develop irregular or conchoidal fractures.

generally distinguish between them: cleavage planes of the same orientation are repeated throughout the specimen, while crystal faces are not **(Fig. 3.7f)**. Minerals that have no cleavage at all (because bonding is equally strong in all directions) break either by forming irregular fractures or by forming **conchoidal fractures (Fig. 3.7g)**, which produce a smoothly curving, clamshell-shaped surface.

Classifying Minerals by Their Chemical Composition

THE PRINCIPAL CHEMICAL CLASSES. Figuring out a way to *classify* minerals—to group them in a meaningful way that emphasizes common features—wasn't easy because, at first glance, there are so many characteristics that could conceivably provide a basis for classification. If you were given a tray of 20 random mineral specimens, you might try to classify them based on obvious physical characteristics such as color—but you would soon find that such a basis isn't terribly useful, for many minerals with the same color are otherwise very different from one another.

A Swedish chemist, Jöns Berzelius (1779–1848), proposed that minerals can be grouped by similarities in their chemical composition. Mineralogists have found this approach helpful, and they now distinguish several principal *chemical classes* of minerals that differ from one another in terms of the anion (negative ion) or anionic group (negatively charged molecule) that the mineral contains (Table 3.2). Different minerals within a class differ from one another in terms of the cations (positively charged ions) that they contain or in terms of their crystal structures, or both.

THE SILICATE MINERALS. Of the various mineral classes, **silicate minerals** account for the largest proportion of the Earth. In fact, over 95% of rocks in the continental crust consist of silicate minerals, and the rocks that make up oceanic crust and the Earth's mantle consist almost entirely of silicate minerals. Internally, silicate minerals contain various arrangements of a fundamental building block called the **silicon-oxygen tetrahedron** (SiO_4^{4-}). These blocks, sometimes informally called *silica tetrahedra*, each consist of a silicon atom surrounded by four oxygen atoms, which produce a pyramid-like shape with four triangular faces (Fig. 3.8a). In a silicate mineral, tetrahedra can link together by sharing oxygen atoms. Mineralogists distinguish among different groups of silicate minerals based on how the tetrahedra link together, which in turn determines the ratio of silicon to oxygen in the mineral (Fig. 3.8b). In olivine, the tetrahedra do not share any oxygen atoms and are held together only by the attraction of cations, whereas in pyroxene, tetrahedra link to form a single chain; in amphibole, they link to form double chains; in mica, they link to form two-dimensional sheets; and in quartz or feldspar, they link to form a three-dimensional jungle-gym-like network.

Take-home message . . .

We can distinguish among different minerals based on their physical properties such as luster, color, cleavage, hardness, and specific gravity. The 4,000 or so minerals can be grouped into a relatively small number of classes on the basis of their chemical composition. Silicate minerals are the most abundant minerals on Earth.

Quick Question -
What is the difference between a cleavage plane and a crystal face in a mineral?

Table 3.2 Some of the major mineral classes

Class Name	Anion or Anionic Group	Examples
Silicates	SiO_4^{4-}	Quartz (SiO_2); garnet group ($Mg_3Al_2Si_3O_{12}$)*
Oxides	O^{2-}	Hematite (Fe_2O_3); magnetite (Fe_3O_4)
Sulfides	S^{2-}	Galena (PbS); pyrite (FeS_2)
Sulfates	SO_4^{2-}	Gypsum ($CaSO_4 \cdot 2H_2O$)
Halides	F^-, Cl^-, Br^-	Halite (NaCl); fluorite (CaF_2)
Carbonates	CO_3^{2-}	Calcite ($CaCO_3$); dolomite ($CaMg[CO_3]_2$)
Native metals	—	Composed entirely of metal atoms

*The formula provided is only one of many for the variety of minerals in the garnet group.

3.4 Something Special: Gems

Mystery and romance follow famous **gems**, mineral specimens that have been cut and polished and are particularly beautiful or valuable. Consider the gem now known as the Hope Diamond, recognized by name the world over. No one knows who first dug it out of the ground (Box 3.4; p. 104). What we do know is that in the 1600s, a French trader obtained a large (112.5 carats, where 1 carat = 200 mg/0.007 ounce), rare blue diamond in India and carried it back to France. King Louis XIV bought the diamond and had it fashioned into a gem of 68 carats. This gem vanished in 1762 during a burglary. Perhaps it was lost forever—perhaps not. In 1830, a 44.5-carat blue diamond mysteriously appeared for sale. Henry Hope, a British banker, purchased the stone, which then became known as the Hope Diamond (Fig. 3.9a). It changed hands several times until 1958, when a New York jeweler, Harry Winston, donated it to the Smithsonian Institution in Washington, DC, where it now sits behind bulletproof glass.

Some gems originate as unique minerals. Others are merely pretty and rare versions of more common minerals. For example, ruby is a special version of the mineral corundum, and emerald is a special version of the mineral beryl (Fig. 3.9b). The beauty of a gem lies in its color and, in the case of transparent gems, its "fire"—the way that light bends and reflects within the gem.

A *gemstone*, a mineral specimen that will be a gem when cut and polished, can form in many ways, like any other mineral. Some solidify from a melt, some form by diffusion, some precipitate from a water solution, and some are a consequence of the chemical interaction of rock with water near the Earth's surface. Some gemstones form only in association with an unusual igneous rock,

See for yourself

Kimberley Diamond Mine

Latitude: 28°44'17.06" S
Longitude: 24°46'30.77" E

Zoom to an elevation of 13 km (~8 miles) and look straight down.

The field of view shows the town of Kimberley, South Africa, and its inactive diamond mine. The mine looks like a circular pit. You can also see the tailings pile of excavated rock debris.

Did you ever wonder . . .

how gems in jewelry get all those shiny faces?

Figure 3.8 The structure of silicate minerals.

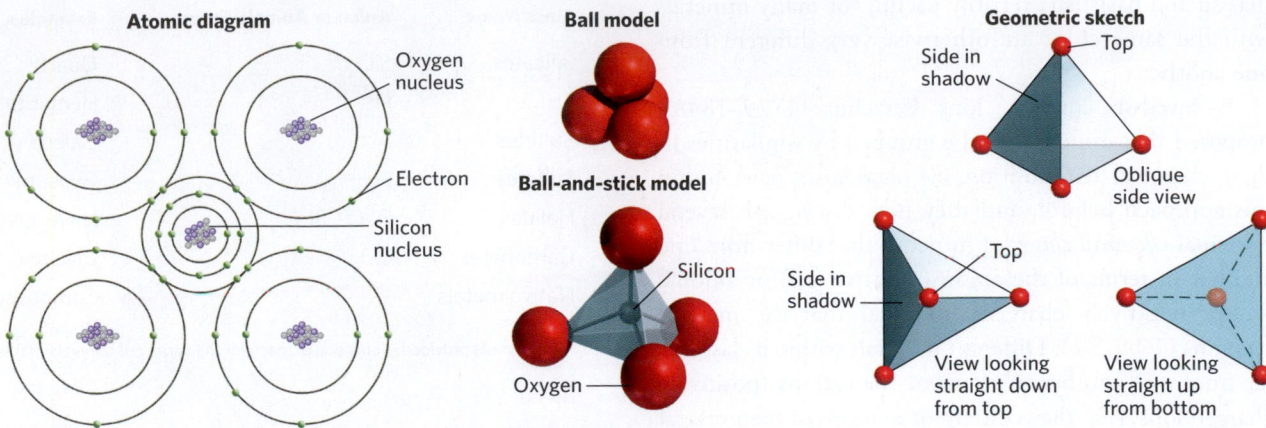

Atomic diagram

Oxygen nucleus
Electron
Silicon nucleus

Ball model

Ball-and-stick model

Silicon
Oxygen

Geometric sketch

Top
Side in shadow
Oblique side view
Top
Side in shadow
View looking straight down from top
View looking straight up from bottom

(a) The fundamental building block of a silicate mineral is the silicon-oxygen tetrahedron. Oxygen atoms occupy the corners of the tetrahedron, and a silicon atom lies at the center. Geologists portray the tetrahedron in a number of different ways.

Isolated tetrahedra
(e.g., olivine, garnet)

Tetrahedron facing down
Tetrahedron facing up

Single chain
(e.g., pyroxene)

Double chain
(e.g., amphibole)

Two-dimensional sheet
(e.g., mica)

Three-dimensional framework
(e.g., quartz, feldspar)

(Oxygens, not shown)

(b) The classes of silicate minerals differ from one another by the way in which the silicon-oxygen tetrahedra are linked. Where the tetrahedra link, they share an oxygen atom. Oxygen atoms are shown in red. Positive silicon ions (not shown) occupy spaces between tetrahedra.

See for yourself

Ekati Diamond Mine, Canada

Latitude: 64°43'15.44" N
Longitude: 110°36'56.27" W

Zoom to an elevation of 100 km (~62 miles) and look straight down.

The Ekati Diamond Mine is in a remote, largely uninhabited region of the Northwest Territories of Canada. Prospectors found diamond pipes here in the early 1990s after a 20-year search. The mine opened in 1998 and within 10 years had produced more than 40 million carats (8,000 kg) of diamonds. Zoom down to 10 km (~6 miles) to see details of the mining operation.

called pegmatite, that appears to form by precipitation from very hot, steamy melts.

As we've noted, gems used in jewelry are "cut" stones, meaning that they have many smooth, shiny faces that form sharp angles with their neighbors. The smooth faces, technically known as **facets**, are not natural cleavage planes, nor are they crystal faces. Rather, they are produced when gem cutters carefully grind and polish specimens by using a faceting machine **(Fig. 3.10a)**. To produce a facet, the gem cutter sticks a gemstone to the end of a *doping stick* and then holds it against a spinning grinding plate called a *lap* until the facet is complete. Then, the gem cutter rotates the gemstone by a specific angle and grinds away another face. Grinding facets on a gem is a lot of work—a typical engagement-ring diamond has 57 individual facets **(Fig. 3.10b, c)**!

Figure 3.9 Gems.

(a) Some gems, like the Hope Diamond, are unique minerals.

Non-gem-quality beryl

(b) Others, like emeralds, are simply clear and well-colored versions of common minerals.

Figure 3.10 The process of creating facets on gems.

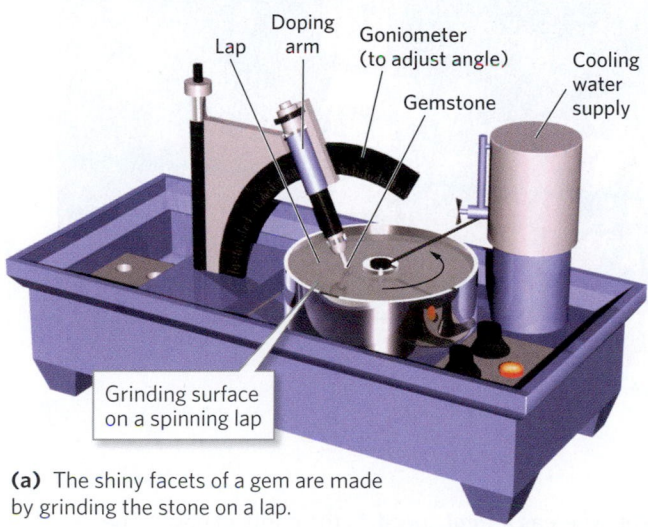

Lap
Doping arm
Goniometer (to adjust angle)
Gemstone
Cooling water supply

Grinding surface on a spinning lap

(a) The shiny facets of a gem are made by grinding the stone on a lap.

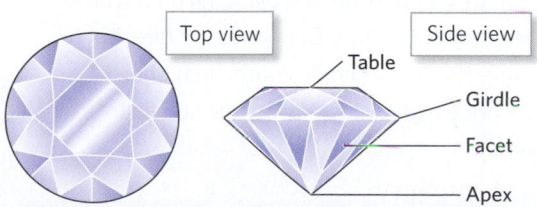

Top view
Side view
Table
Girdle
Facet
Apex

(c) There are many different "cuts" for a gem. Here we see the top and side views of a brilliant-cut diamond.

Sapphires and rubies are varieties of the common mineral, corundum (Al_2O_3). Rubies are the red variety.

(b) Corundum (Al_2O_3) comes in many colors. Gem-quality versions, including ruby and sapphire, can be cut into many shapes.

Box 3.4

Consider this . . .

Where do diamonds come from?

Diamond crystals can grow from carbon—an element concentrated by geologic processes at the surface of the Earth. But such growth takes place only at temperatures and pressures found at depths greater than 150 km (90 miles) below the Earth's surface. (Engineers can duplicate these conditions in the laboratory—corporations manufacture several tons of synthetic diamonds a year for use as abrasives.) How does carbon get from the Earth's surface down into the mantle, where it transforms into diamond? The theory of plate tectonics provides answers. Geologists suggest that subduction carries small amounts of carbonate minerals and organic debris down into the mantle at convergent plate boundaries. As they subduct, the carbon atoms first form crystals of graphite, but when the graphite reaches depths of about 150 km, the intense pressure causes the graphite structure to collapse and produce the more compact and denser diamond structure. Some of the resulting diamonds become trapped in the mantle beneath continents.

What geologic processes bring diamonds up from the mantle back to the top of the crust? Geologists speculate that the transfer takes place when rifting thins the continental crust and causes some of the underlying mantle to melt. Certain types of magma carry diamonds up into the upper crust, where the magma solidifies to form a special kind of igneous rock called *kimberlite* (named for Kimberley, South Africa, where it was first found).

In places where diamonds occur in solid kimberlite, they can be obtained by digging up the kimberlite and crushing it to separate out the diamonds **(Fig. Bx3.4a, b)**. But nature can also break diamonds free from the Earth. In places where kimberlite has been exposed at the ground surface, the rock chemically reacts with water and air (a process called *weathering*, discussed in Chapter 5). Over time, these reactions cause most of the minerals in the kimberlite to disintegrate, producing sandy debris that washes away in rivers. Diamonds are so strong, however, that they remain as solid grains in river gravel, so many diamonds have been obtained simply by separating them from deposits of river gravel **(Fig. Bx3.4c)**.

Gem-quality diamonds come in a range of sizes. Jewelers measure gemstone size in carats, where one carat equals 200 mg (0.2 g), and one ounce equals 142 carats. (Note that a *carat* measures

Figure Bx3.4 Where diamonds come from.

(a) A diamond mine pit.

(b) A diamond embedded in solid kimberlite.

(c) Sifting for diamonds in a streambed.

gemstone weight, whereas a *karat* specifies the purity of gold.) The largest known diamond was discovered in South Africa in 1905. This specimen, called the Cullinan Diamond, weighed 3,106 carats (621 g; 22 ounces.) before being cut. By comparison, the diamond on a typical engagement ring weighs less than one carat. Diamonds are rare, but not as rare as their price suggests. A worldwide consortium of diamond producers keeps the price high by stockpiling the stones to prevent flooding of the market.

Figure 3.11 The two different textures of rocks.

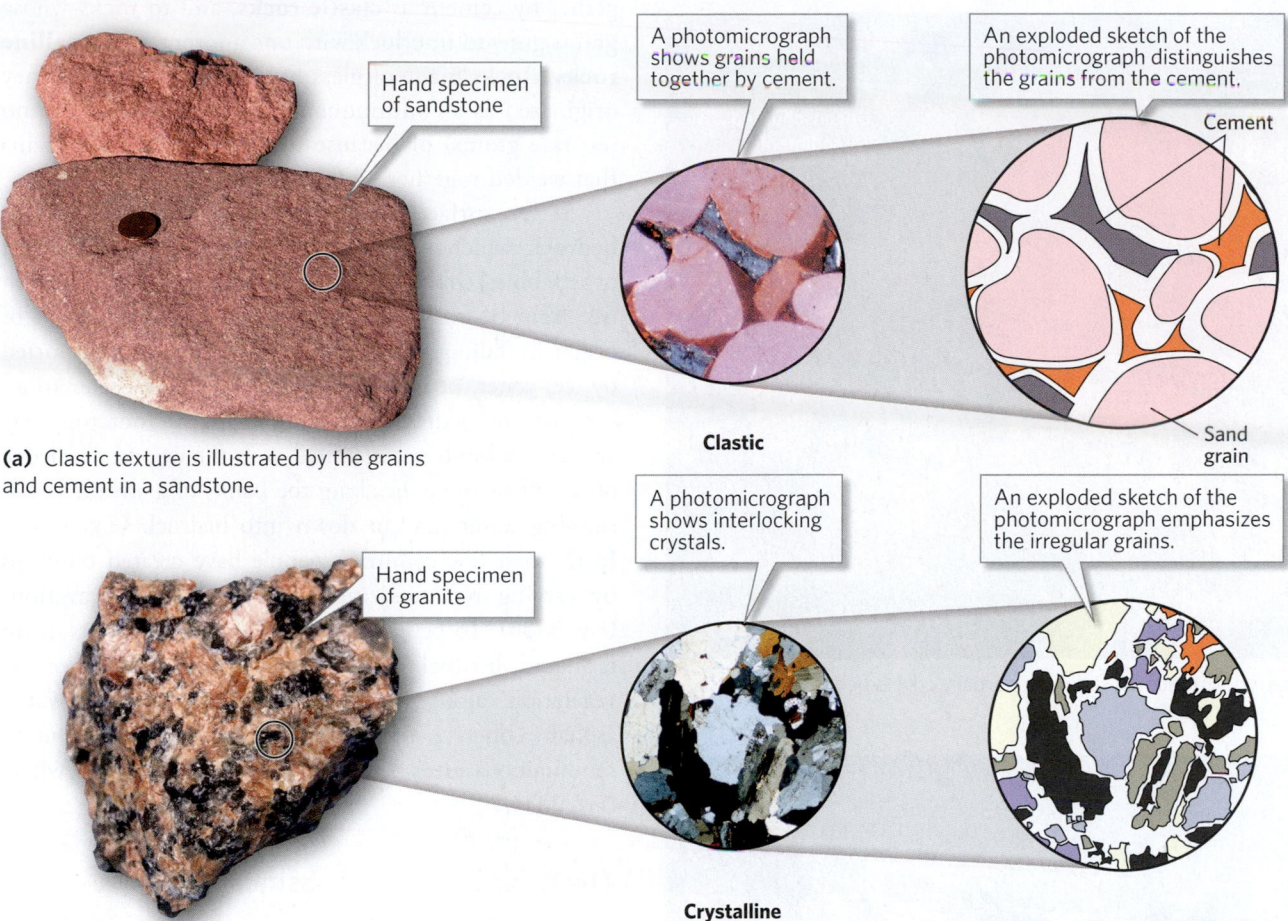

Hand specimen of sandstone

A photomicrograph shows grains held together by cement.

An exploded sketch of the photomicrograph distinguishes the grains from the cement.

Cement

Clastic

Sand grain

(a) Clastic texture is illustrated by the grains and cement in a sandstone.

Hand specimen of granite

A photomicrograph shows interlocking crystals.

An exploded sketch of the photomicrograph emphasizes the irregular grains.

Crystalline

(b) Crystalline texture is illustrated by the interlocking crystals in a granite.

Take-home message . . .

Gemstones are particularly rare and beautiful minerals. The gems or jewels found in jewelry have been faceted using a lap. The facets are not natural crystal faces or cleavage surfaces. The fire of a jewel comes from the way it reflects light internally.

Quick Question --------------------------------
What's the difference between a facet on a gem and a crystal face?

3.5 Introducing Rocks

What Is a Rock?

The thousands of workers who faced death daily as they blasted and chiseled their way through the Sierra Nevada mountain range of California to make a path for the first transcontinental railroad simply thought of rock as a hard, heavy, solid mass that's difficult to dig into or to move. Geologists use a more precise definition: a **rock** is a coherent, naturally occurring solid consisting of an aggregate of mineral grains or, less commonly, a mass of glass. To clarify this definition, let's look at its components:

- *Coherent:* A rock holds together as a solid mass. It's not a pile of loose grains.

- *Naturally occurring:* Real rocks form only during geologic processes. Manufactured materials, such as concrete and brick, are not rocks, although they contain materials similar to those in rocks.

- *An aggregate of mineral grains or a mass of glass:* The vast majority of rocks consist of an aggregate, or collection, of many mineral grains that are attached to one another. Some rocks contain only one kind of mineral, whereas others contain several different kinds. A few types of rock, formed at volcanoes, consist of glass (see Chapter 4).

How does a rock hold together? Most rocks are a coherent mass either because their grains are bonded to one another by natural **cement**, mineral crystals that precipitate from water in the space between grains **(Fig. 3.11a)**, or because their grains grow together and

Did you ever wonder . . .
----
if the brick used in building your house is a kind of rock?

Figure 3.12 What is bedrock, and what isn't?

(a) In the desert landscape of Arizona, unconsolidated sediment partially covers bedrock of reddish sedimentary rocks.

(b) The loose boulders on the ground at the base of this cliff in Utah are not bedrock, but the top of the cliff behind them is.

interlock with one another like pieces in a jigsaw puzzle **(Fig. 3.11b)**. We refer to rocks made of grains held together by cement as **clastic rocks**, and to rocks whose grains grew to interlock with one another as **crystalline rocks**. Rocks made of glass hold together because they originated as a continuous mass (that is, they have no separate grains) or because they consist of glass grains that welded together while still hot.

At the surface of the Earth, rock occurs either as **bedrock**, which is still attached to the Earth's crust below, or as pebbles, cobbles, blocks, or boulders that have broken free from bedrock and have moved from their point of origin by falling down a slope or by being transported by ice, water, or wind **(Fig. 3.12)**. Geologists refer to an exposure of bedrock as an **outcrop**. An outcrop may appear as a knob out in a field or in the woods, as a ledge on a cliff or ridge, or along the banks of a stream where running water has cut down into bedrock **(Fig. 3.13a)**. In the past few centuries, people have created outcrops by carving roadcuts, rail cuts, and other excavations **(Fig. 3.13b)**. To people who live in cities or forests or on farmland, bedrock outcrops may be unfamiliar because vegetation, along with sand, mud, gravel, soil, water, asphalt, concrete, or buildings may cover them. But in mountainous areas, outcrops may be nearly everywhere **(Fig. 3.13c)**.

The Basics of Rock Classification

Beginning in the 18th century, geologists struggled to develop a sensible way to classify rocks. Eventually, it became clear that a *genetic scheme*—a scheme that focuses on the origin (genesis) of rocks—provides the best approach for classifying rocks, and this is the approach that we continue to use today. In this scheme, geologists recognize three basic rock groups:

- **Igneous rocks** form by the freezing (solidification) of molten rock, or *melt* **(Fig. 3.14a)**. The process of solidification can take place underground or at the Earth's surface.

- **Sedimentary rocks** form either by the cementing together of grains broken off pre-existing rocks or by the precipitation of mineral crystals out of water solutions at or near the Earth's surface **(Fig. 3.14b)**.

- **Metamorphic rocks** form when pre-existing rocks undergo changes in response to a modification of their environment, without first melting. For example, metamorphism takes place in response to the rise in temperature and pressure that happens when rock that was near the Earth's surface ends up deep beneath a mountain belt. Metamorphism produces new minerals and new arrangements of crystals **(Fig. 3.14c)**.

Figure 3.13 Common types of outcrops.

(a) A stream cut near Catskill, New York, exposes rock that would otherwise be hidden by trees.

(b) To keep the slope (grade) of this highway gentle, engineers cut an artificial canyon through rock west of Denver.

(c) Natural cliff exposures on a mountain face near Salt Lake City, Utah. The cliff slopes are too steep for trees to take root.

Figure 3.14 The three basic rock groups.

Igneous

(a) Lava (molten rock) freezes to form igneous rock. Here, the molten tip of a brand-new flow still glows red.

Sedimentary

(b) Sand, formed from grains eroded from the cliffs above, collects on the beach. If buried and turned to rock, it becomes layers of sandstone, like those making up the cliffs.

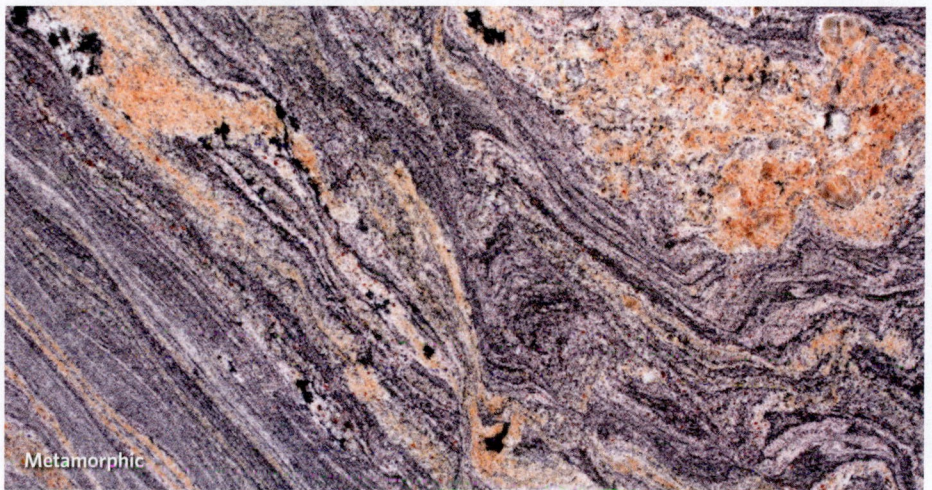

Metamorphic

(c) Metamorphic rock forms when pre-existing rocks endure changes in environmental conditions that cause new minerals and textures to form.

Figure 3.15 Describing grains in rock.

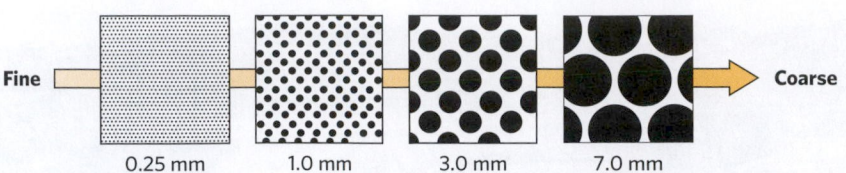

0.25 mm   1.0 mm   3.0 mm   7.0 mm

(a) Geologists define grain size by using this comparison chart.

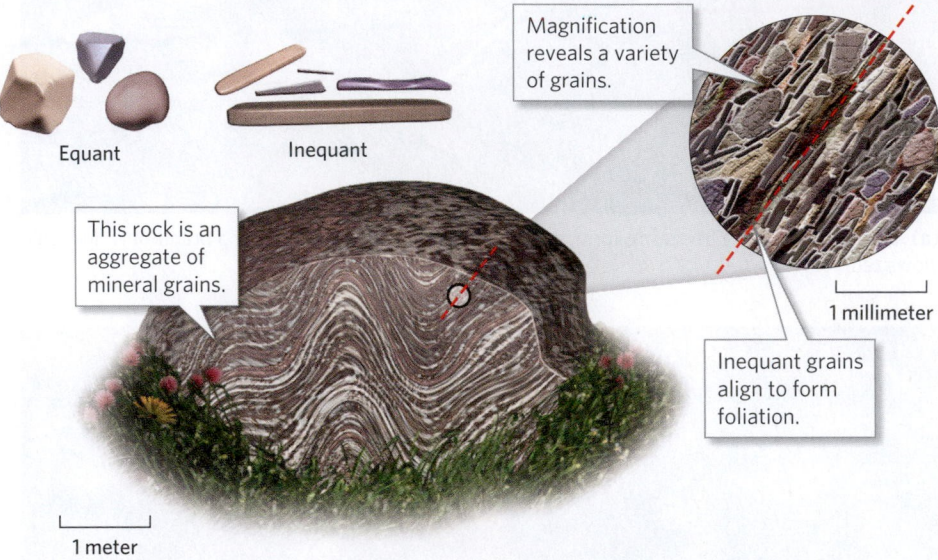

Magnification reveals a variety of grains.

Equant   Inequant

This rock is an aggregate of mineral grains.

Inequant grains align to form foliation.

1 millimeter

1 meter

(b) Grains in rock come in a variety of shapes. Some are equant, whereas some are inequant. In this example of metamorphic rock, inequant grains align to define metamorphic foliation.

Fundamental Characteristics of Rocks

Each of the three basic rock groups contains many different individual rock types, each of which has a name. Rock names come from various sources—some reflect the dominant component making up the rock, some indicate a region where the rock was first discovered, some derive from a root word of Latin origin, and some come from local traditional names. All told, there are hundreds of different rock names, but in this book we'll be introducing only about 30 of the more common ones. Specific rock types can be distinguished from one another by a number of physical characteristics, including the following:

- *Grain size and shape:* The dimensions of individual grains in a rock may be measured in millimeters or centimeters. Some grains are so small that they can't be seen without a microscope, whereas others are as big as a fist or larger **(Fig. 3.15a)**. Some grains are **equant**, meaning that they have similar dimensions in all directions, whereas some are **inequant**, meaning that their dimensions are not the same in all directions **(Fig. 3.15b)**. In some rocks, all the grains have the same size, whereas other rocks contain a variety of different-sized grains.

- *Composition:* The term *rock composition* refers to the proportions of different chemicals making up the minerals or glass in a rock.

- *Texture:* The term **texture** refers to both the arrangement of grains in a rock (the way grains connect to one another) and the degree to which inequant grains align parallel to one another.

Figure 3.16 Layering in rock.

Horizontal bedding

Younger beds

Older beds

Tilted bedding

(a) Bedding in a sedimentary rock, here defined by alternating layers of coarser and finer grains, as exposed on a cliff along an Oregon beach. In this example, older beds were tilted before younger ones were deposited.

Foliation plane

(b) Foliation in this outcrop of metamorphic rock near Mecca, California, is defined by alternating light and dark layers. The color of each layer depends on the minerals it contains.

Figure 3.17 Basic tools for studying rocks in the field.

(a) A field geologist examining a hand specimen (on a cold day).

(b) A hand specimen.

(c) A hand lens.

- *Layering:* Some rock bodies appear to contain distinct layering, defined either by bands of different compositions or textures or by the alignment of inequant grains. The layering in sedimentary rocks is called *bedding,* whereas the layering in metamorphic rocks is called *metamorphic foliation* **(Fig. 3.16)**.

Studying Rocks

To begin a study of a rock, it's best to examine it in an outcrop. If the outcrop is big enough, you'll be able to see layering, and you can study relationships between the rock you're interested in and other rocks around it **(Fig. 3.17a)**. To gain insight into the composition and texture of the rock, you may need to break off a **hand specimen**, a fist-sized piece you can look at more closely with a *hand lens* (magnifying glass) **(Fig. 3.17b, c)**. Observations made with a hand lens will enable you to identify individual mineral grains and see connections between grains.

Commonly, the magnification offered by a hand lens can't provide the level of detail necessary to complete your study. To address this challenge, you can take a hand specimen back to the lab and prepare a **thin section**, a very thin slice (about 0.03 mm thick, the thickness of a human hair) glued to a glass slide **(Fig. 3.18a–c)**. By placing the thin section beneath the lens of a *petrographic microscope,* a special microscope used to study rocks, you can see grains at high magnification. A petrographic microscope differs from an ordinary microscope in that it illuminates the thin section with transmitted *polarized light.* This means that the illuminating light consists of light waves aligned in the same direction, and that the light passes up through the thin section from underneath. When viewed with this type of lighting, each type of mineral displays a unique suite of colors **(Fig. 3.18d)**. The specific color the observer sees depends on the specific mineral making up the grain, the orientation of the crystal lattice relative to the light

Figure 3.18 Studying rocks in thin section.

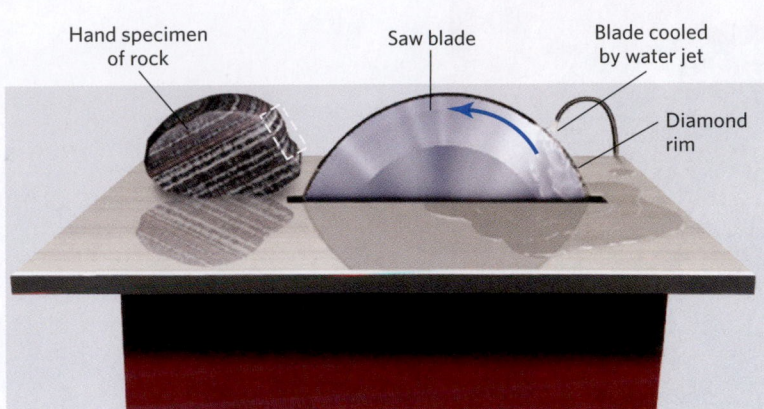

(a) Using a special saw, a geologist cuts a thin chip of a rock specimen.

Hand specimen of rock

Saw blade

Blade cooled by water jet

Diamond rim

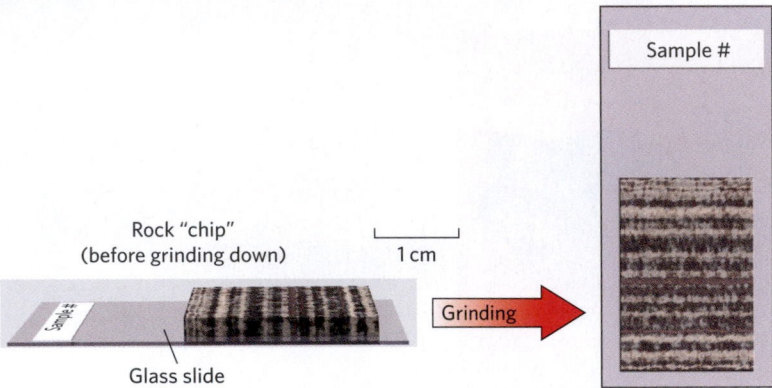

Sample #

Rock "chip" (before grinding down)

1 cm

Glass slide

Grinding

Sample #

(b) The geologist glues the chip to a glass slide and grinds it down until it is so thin that light can pass through it.

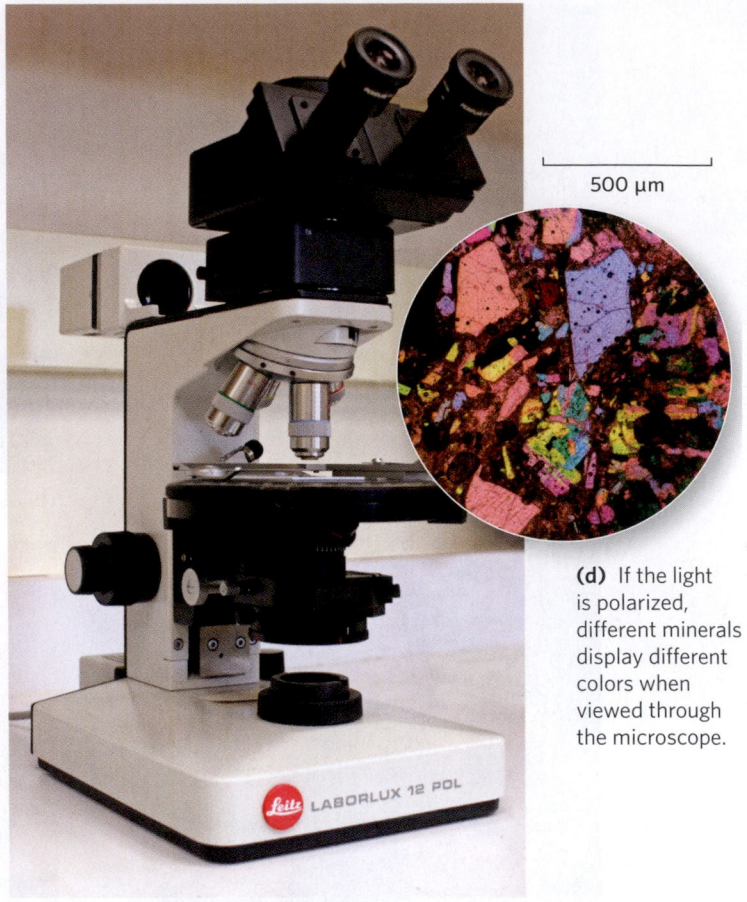

500 µm

(d) If the light is polarized, different minerals display different colors when viewed through the microscope.

(c) With a petrographic microscope, it's possible to view thin sections with light that shines through the sample from below.

beam, and the thickness of the grain. The brilliant colors and strange shapes in a thin section rival the beauty of stained glass. As a record of your observations, you can obtain a **photomicrograph**, a photograph taken through the lens of the petrographic microscope.

Since the 1950s, new high-tech electronic instruments have enabled geologists to examine rocks on an even finer scale than is possible with a petrographic microscope. Modern research laboratories typically house electron microprobes, which focus a beam of electrons on a small part of a grain and reveal the chemical composition of the mineral **(Fig. 3.19)**; mass spectrometers, which analyze the proportions of atoms with different atomic weights; and X-ray diffractometers, which help identify minerals by detecting how an X-ray beam interacts with crystal structures.

From here, we go from the general to the specific. We will see how centuries of research, first in the field with hammer and hand lens, and later in the lab with microscope and spectrometer, have taught geologists to read the record of Earth's processes and history preserved in rock.

Take-home message . . .

Geologists classify rocks into three groups—igneous, sedimentary, and metamorphic—based on the way the rock forms. Each group includes many different kinds of rocks, distinguished from one another by physical characteristics. You can study rocks both at outcrops in the field and by taking samples back to the lab for analysis with petrographic microscopes and high-tech instruments.

Quick Question -
Considering the definition of a rock, why isn't sand a rock? Why isn't concrete?

Figure 3.19 High-tech equipment for analyzing rocks.

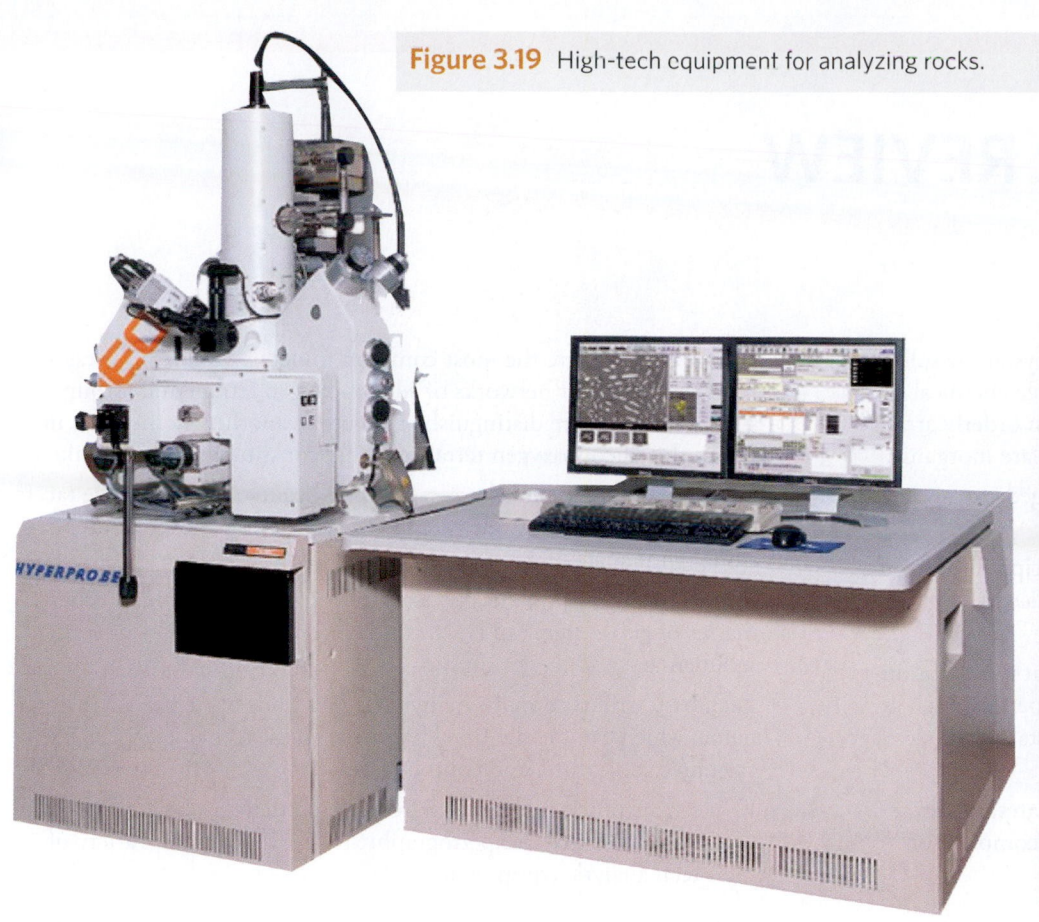

(a) An electron microprobe uses a beam of electrons to analyze the chemical composition of minerals.

(b) An X-ray diffractometer can define the crystal structure of minerals in rocks.

Another View Stone buildings of Machu Picchu, Peru, have survived for centuries, whereas the cliffs behind have stood for millions of years. Both speak to the durability of rock.

3 CHAPTER REVIEW

Chapter Summary

- Minerals are naturally occurring, homogeneous crystalline solids formed by geologic processes. They have a definable chemical composition and a crystal structure characterized by an orderly arrangement of atoms, ions, or molecules. Most minerals are inorganic.

- In the crystal lattice of a mineral, atoms are arranged in a specific pattern.

- Minerals can form by solidification of a melt, precipitation from a water solution, diffusion through a solid, metabolism of organisms, or precipitation from a gas.

- Mineralogists have identified about 4,000 different types of minerals. They can be identified by their physical properties, such as color, streak, luster, hardness, specific gravity, crystal habit, cleavage, magnetism, and reactivity with acid.

- Minerals can be grouped into classes (such as silicates, oxides, sulfides, and carbonates) based on their chemical composition.

- Silicate minerals are the most common minerals on Earth. These minerals consist of networks of silicon-oxygen tetrahedra. Groups of silicate minerals are distinguished from one another by the ways in which the silicon-oxygen tetrahedra that constitute them are linked.

- Gemstones are minerals known for their beauty and rarity. The facets on cut gems used in jewelry are made by grinding and polishing the stones with a faceting machine.

- Rocks are coherent, naturally occurring aggregates of minerals or masses of glass. They can be classified into three groups—igneous, sedimentary, and metamorphic—based on how they formed.

- Each basic group contains many different kinds of rocks, each with a name, which can be identified based on physical characteristics such as grain size, texture, and composition.

- Rocks are exposed in outcrops, from which they can be collected for detailed studies that use petrographic microscopes and a variety of high-tech analytic equipment.

Key Terms

bedrock (p. 106)
cement (p. 105)
clastic rock (p. 106)
cleavage (p. 99)
conchoidal fracture (p. 100)
crystal (p. 93)
crystal face (p. 93)
crystal habit (p. 99)
crystalline rock (p. 106)
crystalline solid (p. 91)

crystal structure (p. 93)
diffusion (p. 94)
equant (p. 108)
facet (p. 102)
gem (p. 101)
geode (p. 94)
glass (p. 92)
grain (p. 93)
hand specimen (p. 109)
hardness (p. 99)

igneous rock (p. 106)
inequant (p. 108)
luster (p. 96)
metamorphic rock (p. 106)
mineral (p. 91)
Mohs hardness scale (p. 99)
organic chemicals (p. 91)
outcrop (p. 106)
photomicrograph (p. 110)
rock (p. 105)

sedimentary rock (p. 106)
silicate mineral (p. 101)
silicon-oxygen tetrahedron (p. 101)
specific gravity (p. 99)
streak (p. 96)
symmetry (p. 93)
texture (p. 108)
thin section (p. 109)

Review Questions

The letters following each Review Question refer to the corresponding Learning Objective from the Chapter Opener.

1. What is a mineral, as geologists understand the term? How is this definition different from the everyday usage of the word? **(A)**

2. Why is glass not a mineral? Salt is a mineral, but the plastic making up an inexpensive pen is not. Why not? **(A)**

3. Describe several ways that mineral crystals can form. **(A)**

4. Why do some minerals occur as euhedral crystals, whereas others occur as anhedral grains? **(A)**

5. How can you determine the hardness of a mineral? What is the Mohs hardness scale? **(C)**

6. List and define the principal physical properties used to identify a mineral. Which minerals react with acid to produce CO_2 (see figure)? **(C)**

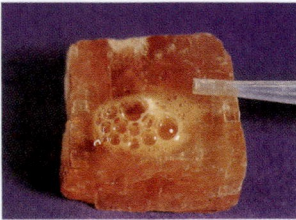

7. What is the key characteristic that geologists use to separate minerals into classes? **(B)**

8. How do you distinguish cleavage surfaces from crystal faces on a mineral? Identify the planes on the accompanying figure. How does each type of surface form? **(C)**

9. What is a silicon-oxygen tetrahedron? On what basis do mineralogists organize silicate minerals into distinct groups? **(B)**

10. Why are some minerals considered gemstones? How do jewelers make the facets on a gem? **(C)**

11. What holds the grains together in a rock? Identify the textures shown in the accompanying figures. **(E)**

12. On what basis can you distinguish rock from loose sediment? **(D)**

13. On what basis do geologists classify rocks into three basic groups? What are the groups? **(F)**

14. What is an outcrop? Provide examples of several different types of outcrops. **(E)**

15. What are the various tools that modern geologists can use to study rocks in the field and lab? **(D)**

On Further Thought

16. Imagine that you find two milky-white crystals, each about 2 cm across. One consists of plagioclase and the other of quartz. How can you determine which is which? **(C)**

17. Could you use crushed calcite to grind facets on a diamond? Why or why not? **(C)**

18. You are an architect who has been hired to build a decorative plaza of rock. Heavy traffic will be driving over the plaza frequently, so the rock needs to be very durable. Which of the following would make the strongest and most durable surface for the plaza—limestone or granite? (Hint: Think about the minerals that make up these rocks. Limestone contains calcite and granite contains quartz and feldspar.) **(C)**

Online Resources

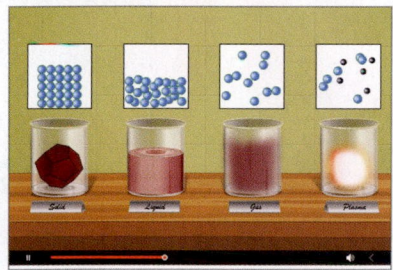

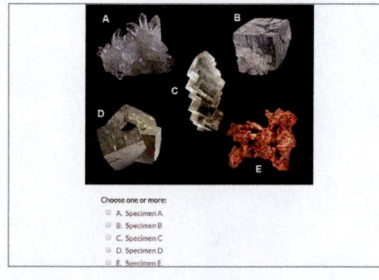

Animations
This chapter features animations on assessing various mineral types and distinguishing rock groups.

Smartwork5
This chapter includes visual matching and labeling exercises designed to help students better understand crystal structure, how crystals grow, and the physical characteristics of minerals.

4 UP FROM THE INFERNO
Volcanism and Igneous Rocks

By the end of the chapter you should be able to . . .

A. explain why melting happens inside the Earth.

B. illustrate how molten rock rises and, eventually, cools to produce igneous rock.

C. differentiate between different compositions of magma.

D. explain the difference between extrusive and intrusive igneous rock.

E. discuss why there are so many different types of igneous rocks and how we can identify and interpret them.

F. distinguish among various ways that volcanoes erupt, and explain why some produce rivers of lava while others explode to produce ash.

G. interpret the various hazards that develop during or after a volcanic eruption.

H. evaluate how we can, to some extent, predict eruptions and mitigate volcanic hazards.

Lava from an active volcano on Hawaii spills into the sea at night. Contact with the red-hot lava instantly turns the water to steam. Cooling of the lava forms new igneous rock.

4.1 Introduction

In 79 C.E., a Roman town called Pompeii sprawled at the foot of Mt. Vesuvius, in what is now southern Italy. Pompeiians thought that Vesuvius, which at the time towered 3 km (10,000 feet) above the nearby sea, was just another scenic mountain. They were wrong—very wrong! Mt. Vesuvius is a **volcano**. Geologists use this term for (1) a *vent* or opening from which **melt** (molten rock), fragments of solidified melt, and gas emerge from underground; and (2) a hill or mountain built from the materials that came out of a vent.

For several weeks, earthquakes jolted Pompeii with unnerving frequency, and Vesuvius grumbled like distant thunder. But people shopping in the town's markets paid little heed until 1:00 P.M. on August 24, when Vesuvius suddenly roared, and a dark, mottled cloud churned out of its summit. This was no normal cloud. Instead of just water mist, it contained hot gas, tiny glass flakes, and marble-sized rock fragments. As lightning sparked in its crown, the cloud spread over Pompeii, turning day into night. Sulfurous fumes filled the air, the glass flakes sifted down like dust, and the rock fragments fell like hail. Panic ensued as inhabitants rushed to escape. Sadly, it was too late. Vesuvius suddenly exploded, blasting even more hot gas and debris skyward. Most of this material swirled high into the atmosphere, but some rushed downslope in a scalding avalanche that swept over and buried Pompeii minutes later (Fig. 4.1a).

A blanket of volcanic debris protected the ruins of the town so well that 18 centuries later, when archaeologists began excavations there, they exposed an amazingly

Figure 4.1 The eruption of Vesuvius buried Pompeii in 79 C.E.

(a) A painting depicting the drama of the explosion.

(b) Today, the relict of Mt. Vesuvius overlooks the excavated ruins of Pompeii.

Glowing waves rise and flow, burning all life on their way, and freeze into black, crusty rock which... builds the land, thereby adding another day to the geologic past.... I became a geologist forever, by seeing with my own eyes—the Earth is alive!

—HANS CLOOS (GEOLOGIST, 1886–1951), ON SEEING AN ERUPTION OF MT. VESUVIUS

complete record of daily life in the Roman Empire (Fig. 4.1b). Occasionally, they also found open spaces within the debris, left when bodies that had been surrounded by ash turned to dust. By filling the spaces with plaster, archaeologists produced casts of Pompeii's inhabitants, twisted in agony or huddled in despair, at the moment of death (Fig. 4.1c).

On that fateful day in 79 C.E., Pompeii experienced a violent **volcanic eruption**, defined as an event during which molten rock flows or sprays out of a vent, and/or solid debris blasts out of a vent. The ancient Romans thought that eruptions happened when Vulcan, the god of fire, fueled his subterranean forges to manufacture weapons. No one believes the Roman myth anymore, but the god's name has been immortalized as the root of the English word *volcano*. Geologic studies demonstrate that volcanic eruptions are a manifestation of **igneous activity**, the name geologists use for the overall process during which rock deep underground melts, producing molten rock that rises up into the crust, and in some cases, makes it all the way to the Earth's surface. Geologists refer to molten rock underground as **magma** and to molten rock erupted at the Earth's surface at a volcano as **lava**. Any rock whose formation involves freezing of a melt is an **igneous rock**.

In this chapter, we examine both the process and the products of igneous activity. We begin by explaining why rocks melt inside the Earth to produce magma in the first place, and we show why such melting takes place only at certain locations. We also examine a variety of different types of igneous rocks and show what the characteristics of these rocks can tell us about the environment in which

(c) Plaster cast of an unfortunate victim who was buried beneath volcanic ash and debris.

they solidified. Next, we focus on volcanic eruptions themselves, the manifestation of igneous activity that we can see directly. We conclude the chapter by discussing how human society can be affected by volcanic eruptions and how we can predict and prepare for them.

Figure 4.2 Decompression melting.

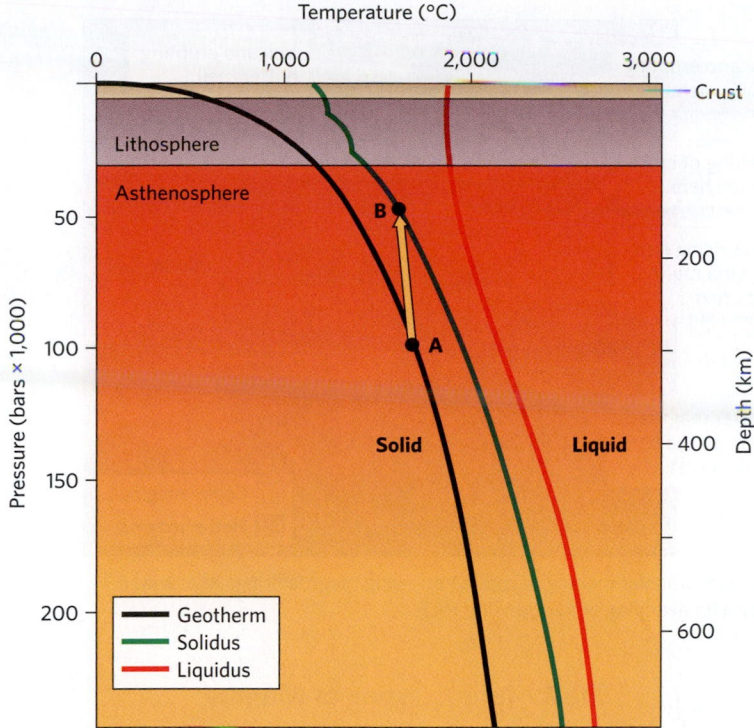

(a) Decompression melting takes place when the pressure acting on hot rock decreases. When rock rises from point A to point B, the pressure decreases a lot, but the rock cools only a little, so the rock begins to melt. The *solidus* line indicates conditions at which melt begins to form, and the *liquidus* line indicates conditions at which rock has completely melted.

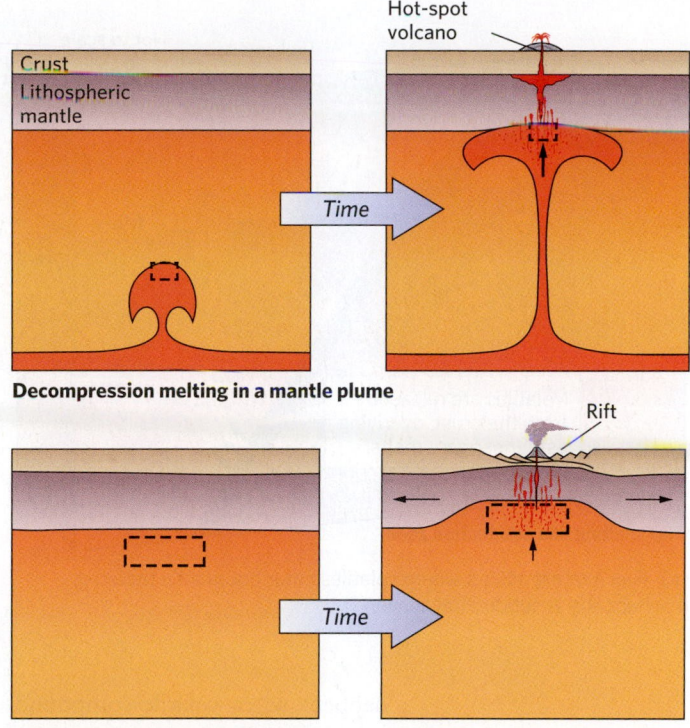

Decompression melting in a mantle plume

Decompression melting beneath a rift

Decompression melting beneath a mid-ocean ridge

(b) The conditions leading to decompression melting occur in several different geologic environments. In each case, a volume of hot asthenosphere (outlined by dashed lines) rises to a shallower depth, and magma (red dots) forms.

4.2 Melting and Formation of Magma

Causes of Melting Inside the Earth

Popular media give the false impression that the mantle beneath the Earth's crust is completely molten and that igneous activity happens when a conduit opens up between an "underground magma sea" and the Earth's surface. This image is wrong! The crust and mantle both consist mostly of solid rock. You can see why by examining a graph that shows how the **geotherm**, the line representing how temperature changes with depth in the Earth, compares with the *solidus* (or melting curve), the line representing the pressure and temperature conditions at which melting starts **(Fig. 4.2a)**. The geotherm generally lies entirely within the solid field, even though rock at greater depth is very hot. Why? Both temperature and pressure increase with increasing depth in the Earth. Though an increase in temperature causes atoms to vibrate faster and try to break free of crystals, a process that would form a liquid, an increase in pressure, simplistically, holds the atoms together and prevents them from breaking free, keeping the crystals solid.

Though the crust and mantle are mostly solid, large amounts of magma do form. Why? Melting of pre-existing rock to produce magma takes place at special locations in response to local changes in temperature, pressure, or chemical composition. Let's look at each of these causes individually.

- *Melting due to decompression:* At lower pressures, atoms are held together less tightly, so bonds between them can break more easily (see Fig. 4.2a). Therefore, a decrease in pressure in very hot rock can trigger melting. Such **decompression melting** happens when hot rock from deep in the mantle rises to shallower depths without undergoing cooling **(Fig. 4.2b)**.

- *Melting due to the addition of volatiles:* If you live in a cold climate, you may have thrown salt on a sidewalk to melt ice, for adding salt to water decreases the melting temperature of the water. A similar process

Did you ever wonder . . .

where the molten rock erupting from a volcano comes from?

Figure 4.3 Flux melting and heat-transfer melting.

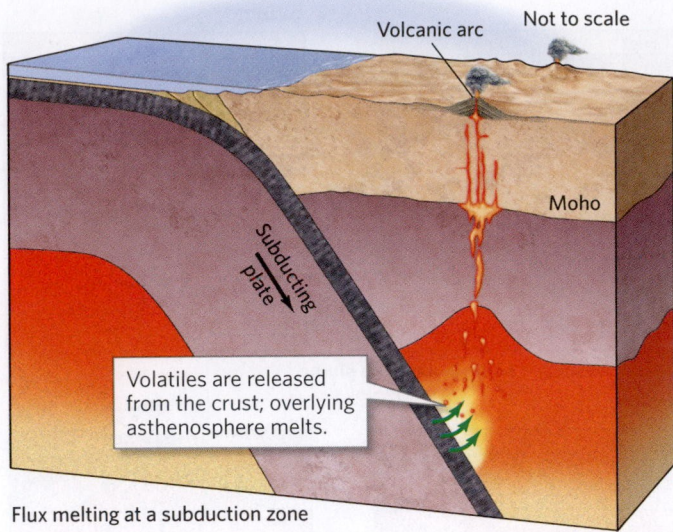

(a) Flux melting occurs where volatiles enter hot mantle material; this happens at subduction zones.

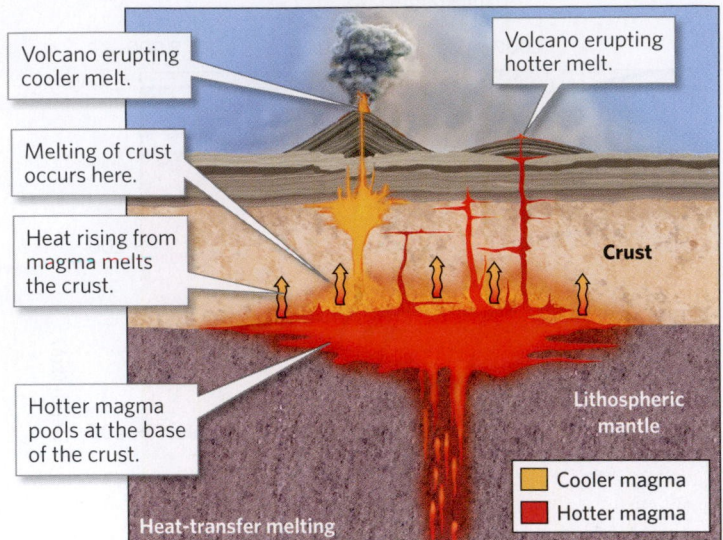

(b) Heat-transfer melting occurs when rising magma brings heat with it and melts overlying or surrounding rock.

happens when volatile compounds (ones that easily exist in gaseous forms) such as water (H_2O) and carbon dioxide (CO_2) seep into solid, very hot, rock. Geologists refer to melting due to addition of volatile compounds as **flux melting** (Fig. 4.3a).

- *Melting due to heat transfer:* Not all rock types start to melt at the same temperature. Specifically, mantle rocks begin to melt at higher temperatures (about 1,200°C, or 2,200°F) than do rocks of the continental crust (about 700°C, or 1,300°F). Therefore, magma rising from the mantle is much hotter than the melting temperature of continental crust. The heat that magma brings with it may increase the temperature of the surrounding crust sufficiently to cause it to melt (Fig. 4.3b). To picture the process, imagine injecting hot fudge into a ball of ice cream—the fudge raises the temperature in the ice cream and causes it to melt. We call such melting **heat-transfer melting**, because it results from the transfer of heat from a hotter material to a cooler one.

Under the temperature and pressure conditions that occur in our planet, only a small percentage of the rock at a given location can melt before the magma migrates away. Geologists emphasize this fact by saying that magma tends to be the product of **partial melting**. Put another way, the temperatures in the Earth at a given depth do not reach the *liquidus*, the temperature at which rock completely melts (see Fig. 4.2a). Molten rock in the Earth tends to represent 2% to 20% partial melting. We refer to the rock that melted to produce a magma as the *magma source*.

Did you ever wonder . . .
what's in the "smoke" that rises above a volcano?

The Different Types of Magma

Magmas are liquids composed of many chemicals (O, Si, Al, Fe, Mg, K, Na, and Ca) that tend to bond and form oxide compounds, such as silica (SiO_2), iron oxide (FeO or Fe_2O_3), and magnesium oxide (MgO). Not all magma has the same composition, meaning that not all magma contains the same proportions of different oxides. Geologists distinguish four major types of magma based on the proportion of silica relative to iron oxide and magnesium oxide that the melt contains. **Mafic magma** has a relatively high proportion of iron oxide and magnesium oxide relative to silica—the letters *ma–* stand for magnesium, and the letters *–fic* stand for iron (from the Latin *ferric*). **Ultramafic magma** contains even more iron and magnesium oxide than does mafic magma, so we add the prefix *ultra–*. **Felsic magma** derives its name from the words *feldspar* and *silica*. It contains a relatively high proportion of silica, so geologists sometimes use the term *silicic* in place of felsic. **Intermediate magma** gets its name simply because this type has a composition between that of felsic and mafic magma. In sum, magma compositions, listed in order, from lowest to highest silica content, are ultramafic, mafic, intermediate, and felsic (Table 4.1).

Significantly, partial melting yields magma that is more felsic than the magma source because a higher proportion of chemicals needed to form felsic minerals migrate into the magma. For this reason, partial melting of ultramafic rock in the mantle produces mafic magma, and partial melting of intermediate rock in continental crust produces felsic magma. The specific composition of a magma (or lava) found at a given location depends on many factors

(Box 4.1). Notably, the temperature at which a magma can remain liquid depends on its composition, so different types of magma have different temperatures (see Table 4.1).

Melting Also Yields Volcanic Gases

The dark cloud rising above a volcano may look like the smoke of a house fire, but it differs greatly in composition (Fig. 4.4). Wood smoke consists of soot (tiny carbon particles), various organic chemicals, H_2O, and CO_2. In contrast, volcanic emissions contain volcanic ash (glass flakes), very tiny (<0.001 mm) particles of liquid or solid called **aerosols**, and a variety of **volcanic gases**, gases that originate underground and rise from a volcanic vent. Volcanic gases include H_2O, CO_2, sulfur dioxide (SO_2), and hydrogen sulfide (H_2S).

Where do volcanic gases come from? The atoms in them start out in volatile compounds bonded to minerals in the magma source. When the magma source undergoes partial melting, volatiles separate from solid minerals. Under the high pressure that exists at depth, these compounds remain dissolved in the magma, just as carbon dioxide remains dissolved in an unopened can of beer. As magma approaches the Earth's surface, the pressure imposed on it by the weight of overlying rock and magma decreases, and at a depth of about 5 km (3 miles) below the surface, volatiles come out of solution and form bubbles, just like the bubbles that form in an open bottle of beer. In some cases, so many bubbles form that erupting lava becomes frothy, like the foam of a beer that's been poured too quickly. What happens to those bubbles? Some remain trapped in magma or lava even after the melt freezes. The resulting holes in the rock are called **vesicles**. Bubbles that reach the surface of lava enter the atmosphere.

Take-home message . . .

Crust and mantle remain solid except in special locations where pre-existing solid rock, the magma source, undergoes melting. Melting can be triggered by a decrease in pressure, addition of volatiles, or injection of hot magma from deeper in the Earth. Geologists classify magma based on the proportion of silica that it contains.

Quick Question -----------------------------
Where does volcanic gas come from?

4.3 Formation of Igneous Rock

What Causes Magma to Rise?

Once formed, magma moves upward, away from the magma source, and rises toward the Earth's surface. Why? First, magma is less dense than solid rock, so it's buoyant

Table 4.1 The four major types of magma

Name	% Silica	Temperature
Felsic	66–76	700°C
Intermediate	52–66	900°C
Mafic	45–52	1,100°C
Ultramafic	38–45	1,300°C

relative to its surroundings. Buoyancy lifts magma upward through denser rock just as buoyancy lifts less dense Styrofoam upward through denser water. Second, magma rises because the weight of the overlying rock produces pressure at depth that literally squeezes the magma upward, just as the weight of your foot squeezes mud up between your toes when you step in a puddle.

Intrusive versus Extrusive Igneous Rock

Most people have seen photos or movies of volcanic eruptions. Such dramatic images highlight floods of lava and avalanches of debris. New igneous rock forms when the lava flowing from a volcano solidifies after coming in contact with air or water, or when debris blasted from a volcano accumulates and either welds together or becomes cemented together. Geologists refer to such igneous rock, formed from materials erupted by a volcano, as **extrusive igneous rock** (Fig. 4.5a, b).

Not all igneous rock, however, is extrusive. In fact, a vastly greater proportion of the Earth's igneous rock forms

Figure 4.4 A cloud of gas, mixed with ash, rising above a volcano in the Aleutian Islands, Alaska.

H_2O, CO_2, and SO_2

Box 4.1 ▶ Consider this . . .

Why does magma composition vary?

Magma varies significantly in composition from place to place, as reflected by the rock that eventually solidifies from it. Such differences are important because they tell us about the origin and evolution of the magma. Let's look at some of the factors that control magma composition.

- *Composition of the magma source:* The chemicals available to go into a melt depend on the chemicals available in the magma source. Not all magmas are derived from the same rock type, so not all have the same composition.

- *Degree of partial melting:* The percentage of a magma source that melts to form magma affects the proportions of chemicals in the magma. The greater the percentage of partial melting, the more mafic the magma.

- *Chemical interaction with surroundings:* Magma may incorporate chemicals dissolved from the solid rock through which it rises, or from blocks of rock that fall into the magma. This process is called **assimilation**.

- *Fractional crystallization:* Magma contains many different chemical compounds, so when it freezes to become igneous rock, many different minerals can form. In the 1920s, an American geologist named Norman Bowen discovered that not all of these minerals form at the same time. Specifically, when a mafic magma starts to freeze, crystals of mafic minerals such as olivine and pyroxene grow first **(Fig. Bx4.1a)**, and their formation removes iron and magnesium from the melt. As cooling continues, amphiboles form, and micas, quartz, and potassium feldspar are the final minerals to appear. While all these minerals are forming, plagioclase crystals also grow. As the melt cools, however, the composition of the plagioclase crystals changes—early-formed plagioclase contains more calcium (Ca), while later-formed plagioclase contains more sodium (Na). This sequence of crystallization is known as **Bowen's reaction series (Fig. Bx4.1b)**, and the process of crystal formation and associated removal of chemicals from the melt is called **fractional crystallization**. Note that because fractional crystallization progressively extracts iron and magnesium from the magma, the magma that remains as fractional crystallization takes place becomes progressively more felsic.

Figure Bx4.1 Bowen's reaction series is the sequence of crystallization in cooling magma.

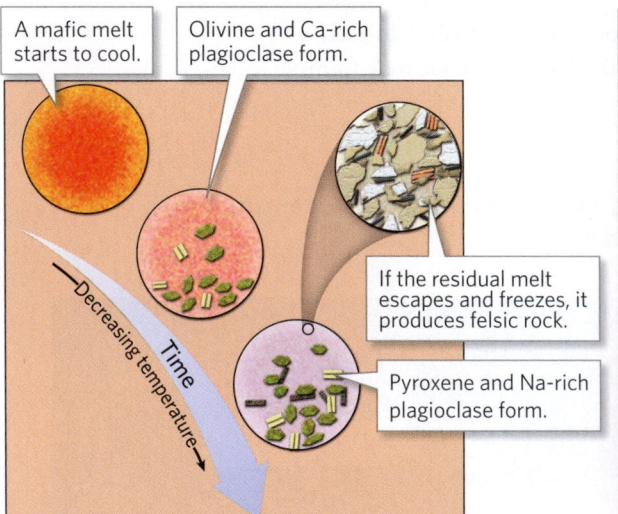

(a) With decreasing temperature, fractional crystallization begins, and the composition of the remaining magma becomes increasingly felsic.

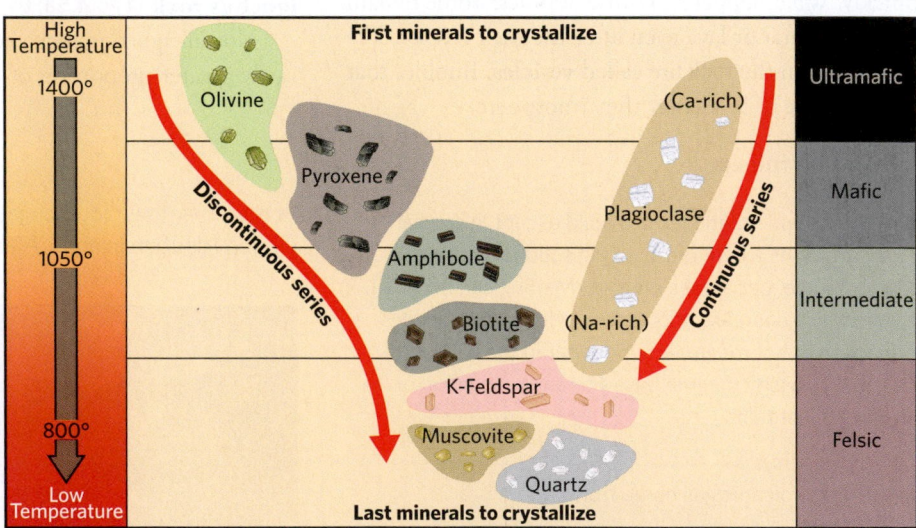

(b) This chart displays the discontinuous and continuous reaction series. Rocks formed from minerals at the top of the series are mafic, whereas rocks formed from the bottom of the series are felsic. Here, the term "discontinuous" indicates that a succession of different minerals forms, whereas the term "continuous" means the same mineral (plagioclase) with different compositions (Ca vs. Na) forms.

Figure 4.5 Formation of extrusive and intrusive igneous rock.

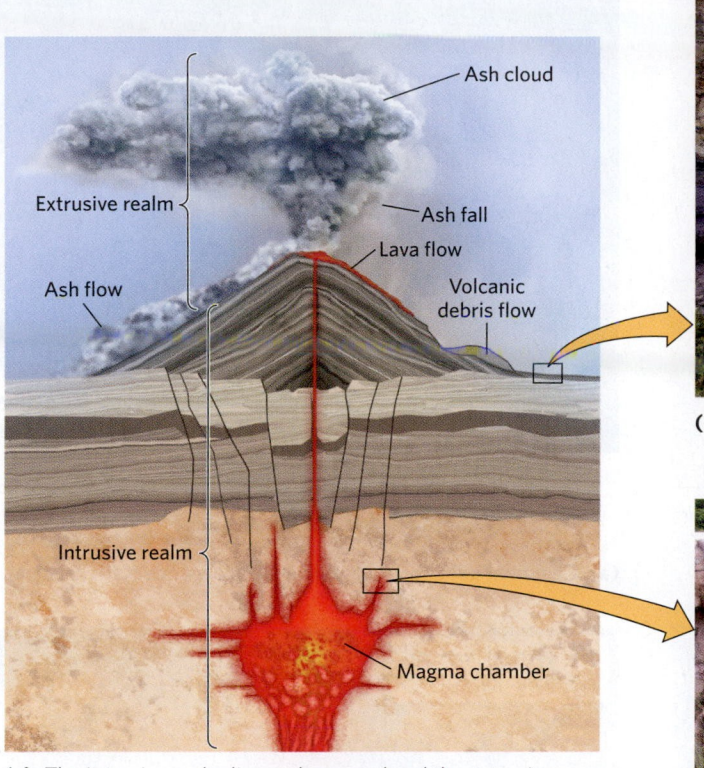

(a) The intrusive realm lies underground and the extrusive realm lies aboveground. Lava flows, as well as various types of ash eruptions, all produce extrusive rocks.

(b) Extrusive rocks include lava flows and pyroclastic layers.

(c) An intrusion of basalt (dark rock) cuts across a mass of granite (light rock).

by solidification of magma underground, out of view **(Fig. 4.5c)**. Rock produced by the freezing of magma underground, after it has *intruded* (pushed its way into) cooler overlying rock, is **intrusive igneous rock**. Geologists refer to the pre-existing rock surrounding an intrusion as **wall rock**.

Transforming Melt into Rock

How long it takes for a melt to cool depends on how fast heat transfers from the melt into its surroundings. The rate of heat transfer, in turn, depends on three factors: the temperature of the environment in which cooling takes place, the shape and size of the molten mass, and the ability of the surroundings to extract heat.

The temperature of a melt's cooling environment depends on where the cooling takes place. For example, an intrusion that cools deep in the crust is surrounded by hot wall rock, for the temperature of the crust increases with depth. In such settings, not only does the wall rock insulate the intrusion, for rock is an excellent insulator, but the temperature contrast between the magma and the wall rock is small, so heat moves out of the melt very slowly. In contrast, lava flowing on the surface of the Earth cools quickly, both because it's not well insulated

and because the temperature contrast between the lava and its surroundings is very large.

At a given location, cooling rate also depends on the shape and size of a body of melt **(Fig. 4.6a)**. Because the body loses heat to its surroundings only at its surface, bodies with a large surface area per unit volume cool faster. Therefore, a pancake-shaped body cools faster than a spherical body of the same volume, a shoebox-sized intrusion cools more quickly than a building-sized intrusion, and small blobs of lava sprayed into the air cool faster than a thick lava flow **(Fig. 4.6b)**. Finally, the cooling rate of an intrusion depends on whether it comes in contact with water, which can absorb and carry away a lot of heat. If groundwater circulates through an intrusion, it removes heat much as coolant removes heat from a car's engine, so a wet igneous intrusion cools faster than a dry one. Similarly, lava extruded underwater cools more quickly then lava extruded under air.

Figure 4.6 Factors affecting the cooling rate of intrusions.

Figure 4.7 Columnar jointing.

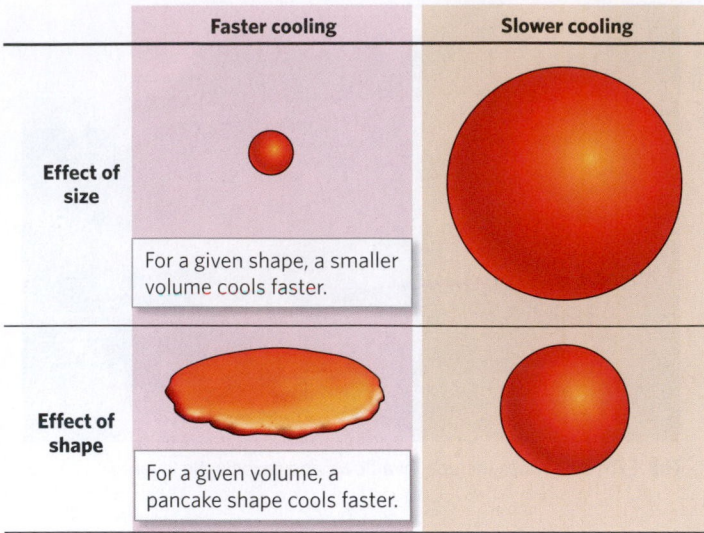

	Faster cooling	Slower cooling
Effect of size	For a given shape, a smaller volume cools faster.	
Effect of shape	For a given volume, a pancake shape cools faster.	

(a) The shape and size of an intrusion affect its surface area per unit volume, which in turn affects its rate of cooling.

Sky — Rubbly top of flow — Columnar-jointed interior — Rubbly base of flow — Older flow

(a) Columnar jointing in a lava flow in Yellowstone National Park.

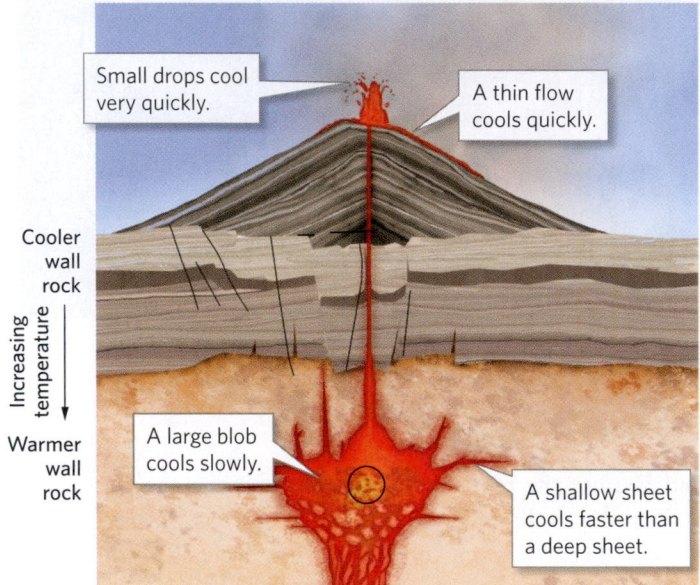

Small drops cool very quickly.

A thin flow cools quickly.

Cooler wall rock

Increasing temperature

Warmer wall rock

A large blob cools slowly.

A shallow sheet cools faster than a deep sheet.

(b) The cooling rate of molten rock depends on the size and shape of a lava or magma body. For intrusions, it also depends on depth.

The vertical lines are columnar joints.

(b) Huge columnar joints in Devils Tower, Wyoming.

Development of Columnar Jointing

Cooling of lava or magma eventually produces solid rock. But right after solidification, the new rock remains very hot. As a hot rock cools, it shrinks slightly, and this shrinking causes the rock to crack. In fine-grained igneous rocks, cracking may cause the rock to break into roughly hexagonal columns, producing a distinctive pattern called **columnar jointing** (Fig. 4.7a, b). Examples of columnar jointing can be so visually striking that they become tourist destinations.

Take-home message . . .

Magma rises because it is buoyant and because pressure from the overlying rocks squeezes it upward. When molten rock enters a cooler environment, it freezes. Intrusive igneous rock solidifies underground, whereas extrusive igneous rock forms from lava or debris. The rate of cooling depends on the environment and on the size and shape of the melt body.

Quick Question -
Which cools faster, a large blob of magma intruded at depth or a thin flow of lava extruded at the surface? Why?

Figure 4.8 Igneous sills and dikes: examples of tabular intrusions.

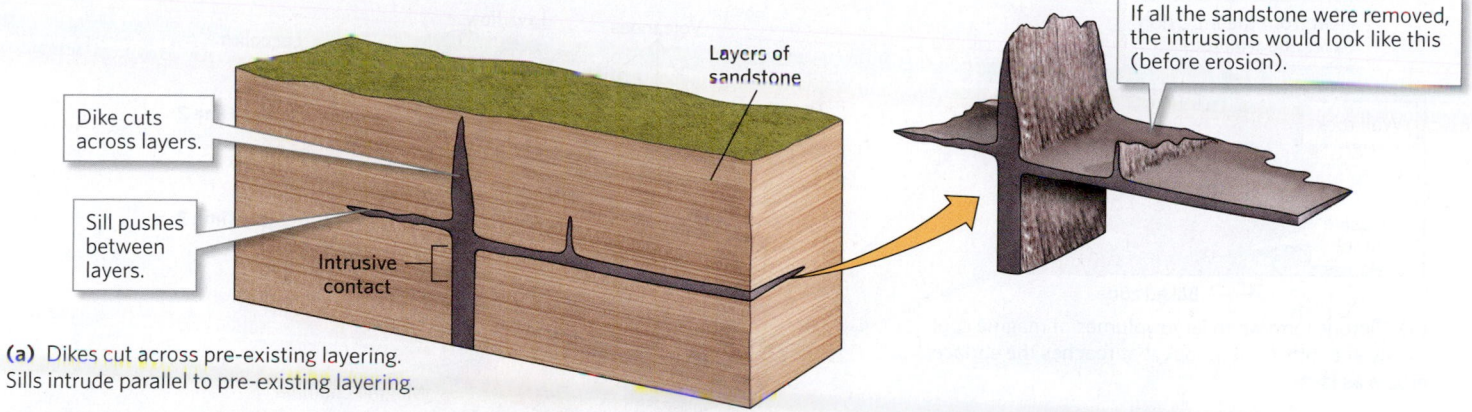

If all the sandstone were removed, the intrusions would look like this (before erosion).

Dike cuts across layers.

Sill pushes between layers.

Layers of sandstone

Intrusive contact

(a) Dikes cut across pre-existing layering. Sills intrude parallel to pre-existing layering.

(b) An example of an igneous dike cutting across sedimentary rocks in Montreal, Canada.

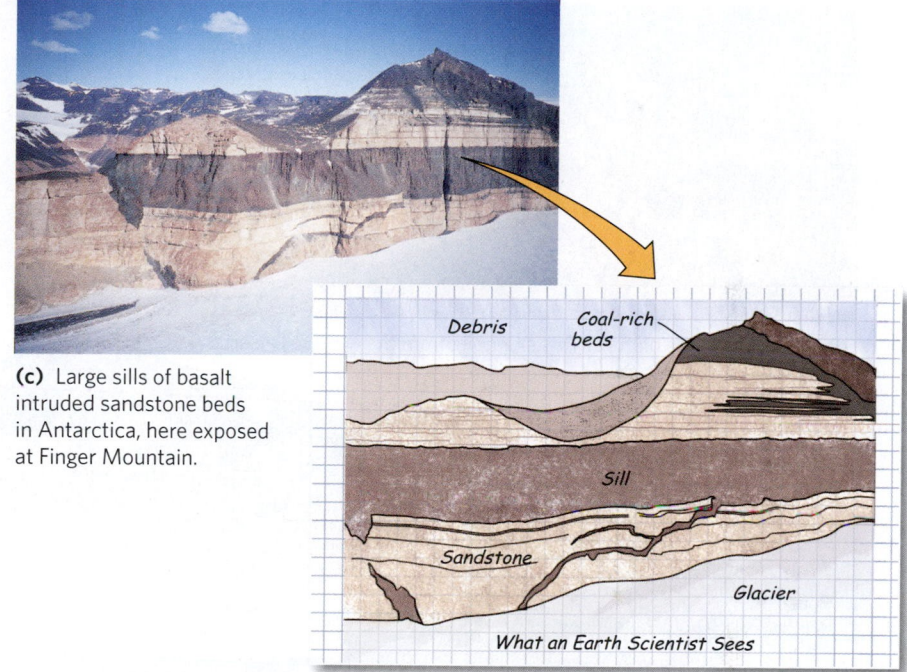

(c) Large sills of basalt intruded sandstone beds in Antarctica, here exposed at Finger Mountain.

Debris   Coal-rich beds

Sill

Sandstone

Glacier

What an Earth Scientist Sees

4.4 The Products of Igneous Activity

Igneous activity plays a major role in the Earth System because it transfers material from deep in the Earth up to shallower levels, or onto the surface. In this section, we examine some of the products of igneous activity more closely.

Igneous Intrusions

Intrusions form underground, so we can't see them as they form. But over the course of geologic time, erosion may expose rock that had formed deeper in the crust. So, in many locations, geologists can actually walk across ancient igneous intrusions and study their characteristics. Based on such research, geologists distinguish among different types of intrusions on the basis of shape.

Tabular intrusions, or *sheet intrusions*, have relatively planar surfaces and roughly uniform thickness. They generally range in length from meters to tens of kilometers, and in thickness from centimeters to tens of meters, but in some places, they are much larger. Geologists distinguish between two types of tabular intrusions, based on the orientation of the intrusion relative to surrounding rock. A **dike** cuts across pre-existing layering (bedding or foliation) of wall rock, whereas a **sill** intrudes parallel to pre-existing layering (**Fig. 4.8a–c**). In places where wall rock does not have layering, geologists refer to steep or vertical tabular intrusions as dikes and to near-horizontal bodies as sills. In some locations, an intrusion builds into a blister-like shape, called a **laccolith**, which is flat on the bottom and round on the top.

Plutons are irregular or blob-shaped intrusions that range in size from tens of meters across to several kilometers across (**Fig. 4.9a–c**). In some cases, plutons may form from the solidification of magma that filled a

Figure 4.9 Plutons and batholiths.

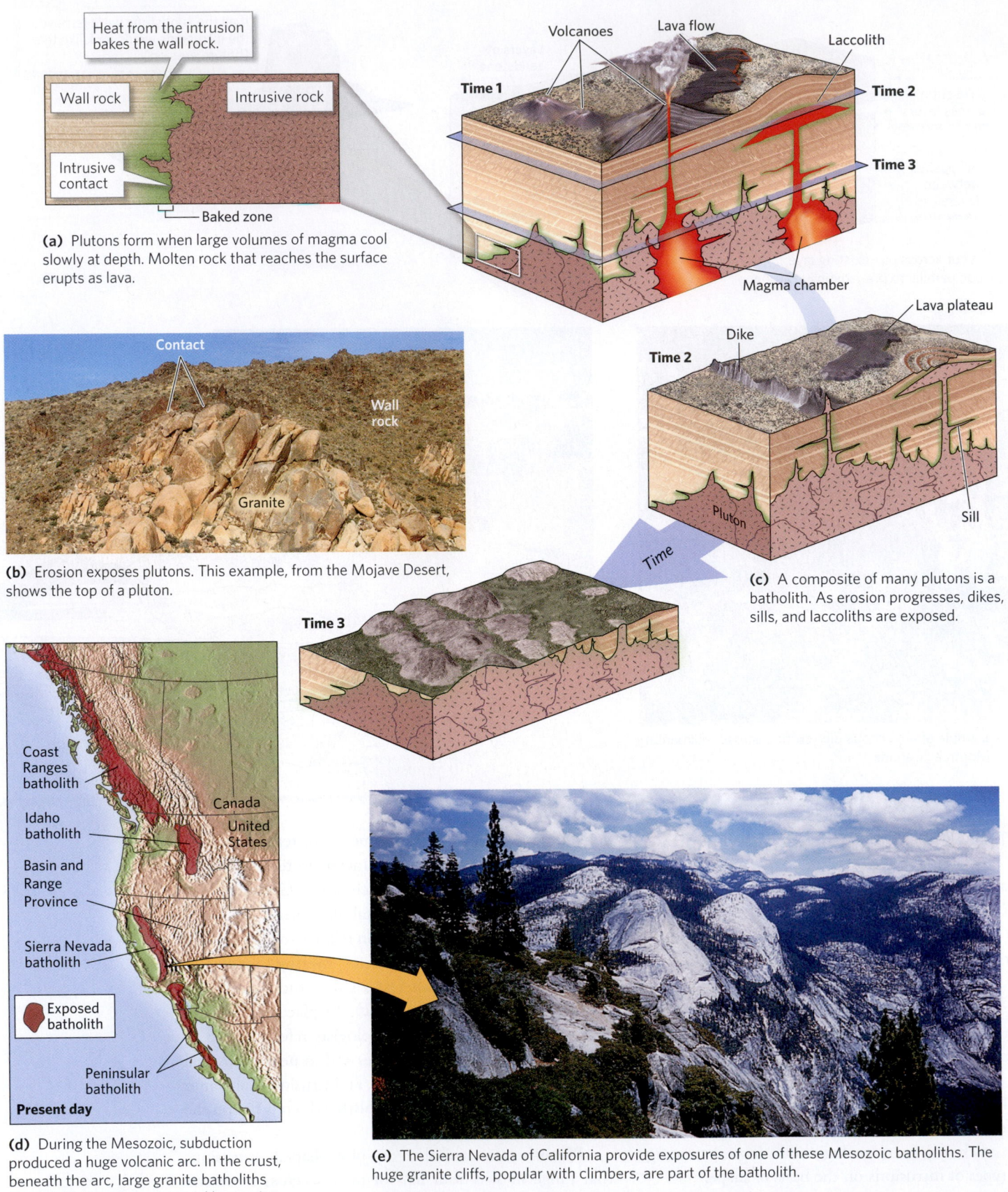

Heat from the intrusion bakes the wall rock.

Wall rock

Intrusive rock

Intrusive contact

Baked zone

(a) Plutons form when large volumes of magma cool slowly at depth. Molten rock that reaches the surface erupts as lava.

Volcanoes

Lava flow

Laccolith

Time 1

Time 2

Time 3

Magma chamber

Contact

Wall rock

Granite

(b) Erosion exposes plutons. This example, from the Mojave Desert, shows the top of a pluton.

Lava plateau

Dike

Time 2

Pluton

Sill

Time

Time 3

(c) A composite of many plutons is a batholith. As erosion progresses, dikes, sills, and laccoliths are exposed.

Coast Ranges batholith

Idaho batholith

Basin and Range Province

Sierra Nevada batholith

Canada

United States

Exposed batholith

Peninsular batholith

Present day

(d) During the Mesozoic, subduction produced a huge volcanic arc. In the crust, beneath the arc, large granite batholiths formed. They are now exposed by erosion.

(e) The Sierra Nevada of California provide exposures of one of these Mesozoic batholiths. The huge granite cliffs, popular with climbers, are part of the batholith.

How can I explain . . .

The meaning of viscosity

What are we learning?

The viscosity of a magma or lava affects how it flows and how easily gas can escape from the magma.

What you need:

- A squeeze bottle of ketchup and a jar of smooth peanut butter.
- Large plastic cutting board; stopwatch; scoop or spoon; cup; plastic straws; microwave oven.

Instructions:

- Place the cutting board on a counter with one end propped up so that the surface of the board tilts at an angle of 45° into a sink or tray.
- Place a scoop of one of the substances near the upslope end of the cutting board. See how far it flows down the cutting board in a given time, and estimate its ending thickness. Repeat with the second substance. Which is more viscous?
- Based on your reading, which substance serves as a model of felsic magma and which of mafic magma?
- Pour a cup full of ketchup, and smooth out the peanut butter in the jar. Insert a straw deep into each and blow hard. In which substance do bubbles form and rise more easily? What does this mean in terms of how gas escapes from a substance or gets trapped and builds up pressure?

- Heat the materials briefly in a microwave oven. Repeat the experiment on the tilted cutting board, using the warmed products. How does temperature affect viscosity?

What did we see?

This exercise illustrates the concept of using physical models to represent the behavior of real systems. In detail, it illustrates the contrast between high viscosity and low viscosity and the effect of temperature on viscosity.

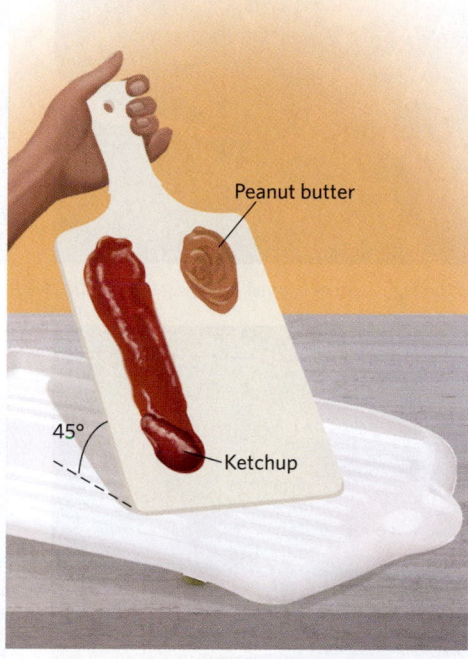

Peanut butter

45°

Ketchup

Bubbles in ketchup

Bubbles don't form easily in peanut butter

---

magma chamber, a space containing a high proportion of magma. Alternatively, plutons may be built by successive intrusions of many adjacent sills. When numerous plutons form within a region, the result can be a vast igneous body hundreds of kilometers long and tens to a hundred kilometers wide; such an immense compound intrusion is known as a called **batholith** (Fig. 4.9d, e). Plutons that intruded between 145 and 80 million years ago form the Sierra Nevada batholith of California—the spectacular cliffs of Yosemite National Park expose parts of this batholith.

Lava Flows

Geologists use the term **lava flow** both for molten lava moving over the Earth's surface and for the layer of solid igneous rock formed when the lava freezes. Some lava flows ooze very slowly, like a very sticky paste, while others spill down the *flank* (side) of a volcano in a fast-moving stream. These contrasts reflect differences in the **viscosity**, or resistance to flow, of lavas (Box 4.2). A higher-viscosity lava is stickier and does not flow as easily as a lower-viscosity lava. (As an analogy, higher-viscosity molasses flows more slowly than lower-viscosity water.) Differences

Figure 4.10 A lava dome surrounded by a ring of volcanic debris.

Figure 4.11 Mafic lava flows.

in lava viscosity depend, in turn, on the silica content and the temperature of molten rock. The proportion of silica affects viscosity because silica tends to form long, chain-like molecules that tangle with one another and slow the flow. Therefore, the greater the silica concentration, the more viscous the lava. Temperature affects viscosity because heat causes chemical bonds to break more easily. Therefore, a hotter lava of a given composition is less viscous than a cooler lava of the same composition.

With these concepts in mind, you will not be surprised to learn that a cool felsic lava tends to build into

(a) A satellite image of the big island of Hawaii, showing long mafic lava flows.

(b) A lava flow coming from Mt. Etna, Sicily.

(c) A mafic lava flow covers a highway on Hawaii.

(d) In a lava tube, molten lava flows under a crust of solid rock.

(e) A drained lava tube exposed by a road cut on Hawaii.

"Ropes" of pahoehoe

0.5 m

(a) Pahoehoe developing on a new lava flow in Hawaii.

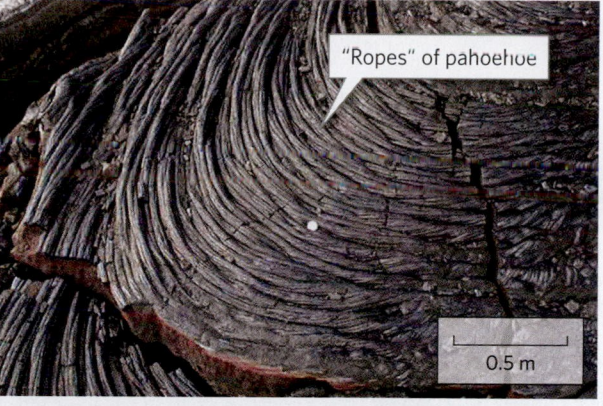

"Ropes" of pahoehoe

0.5 m

(b) The surface of a pahoehoe flow after it has cooled and solidified. Note the coin for scale.

Figure 4.12 Surface textures of mafic lava flows.

(c) The rubbly surface of an a'a' flow, Sunset Crater, Arizona.

Older pillows

(d) Pillow basalt develops when lava erupts underwater. Later uplift may expose pillows above sea level, as in this Oregon outcrop.

a bulbous mound, called a **lava dome**, that does not extend far from the vent **(Fig. 4.10)**. Rock on the surface of a lava dome solidifies as lava continues to be injected within it. In contrast, hot mafic lava can flow far from the vent before it solidifies to produce long, relatively thin flows **(Fig. 4.11a, b)**. Notably, on steep slopes near the summit of a volcano, hot mafic lava flows reach speeds of over 30 km (18 miles) per hour. After mafic lava has started to cool and has become more viscous, it slows to walking pace or slower. Cooling eventually causes the surface of a flow to crust over **(Fig. 4.11c)**. The new solid crust insulates the interior of the flow, allowing the interior to remain liquid. As cooling progresses, lava moves only through **lava tubes**, tunnel-like conduits in the interior of a larger lava flow **(Fig. 4.11d)**. Lava tubes permit lava to travel many kilometers from the vent before it freezes. In some cases, lava tubes eventually drain and become empty tunnels **(Fig. 4.11e)**.

The surface texture of a mafic lava flow depends on whether or not the flow develops a solid surface before it stops moving. Flows with soft, pasty surfaces wrinkle into smooth, glassy, rope-like ridges. Geologists refer to such flows by their Polynesian name, **pahoehoe** (pronounced pa-hoy-hoy) **(Fig. 4.12a, b)**. If the surface layer of the lava

solidifies while the flow is still moving, the layer breaks into a jumble of jagged fragments, creating a rubbly flow also known by its Polynesian name, **a'a'** (pronounced ah-ah) **(Fig. 4.12c)**.

Mafic flows that erupt underwater look different from those that erupt on land because the lava cools much more quickly when in contact with water. Moments after extrusion, submarine mafic lava forms a glass-encrusted blob, or *pillow*. The rind of the pillow stops the flow's advance until the pressure of the lava squeezing into the pillow breaks the rind, and then a new blob of lava squirts out, which itself freezes into a pillow. Geologists refer to layers composed of these blobs as **pillow basalt** **(Fig. 4.12d)**.

Pyroclastic Deposits

In 1943, as Dionisio Pulido prepared to sow his field west of Mexico City, an earthquake jolted the ground, and the surface of the field bulged upward by a few meters and cracked. Volcanic ash, along with larger fragments, filled the air, and Dionisio fled. By the following morning, the field lay buried beneath a 40-m (130-foot)-high mound of gray-black debris. Dionisio had witnessed the birth of a new volcano, named Paricutín. Within a year, debris erupted from Paricutín had built into a steep-sided cone 330 m (1,100 feet)

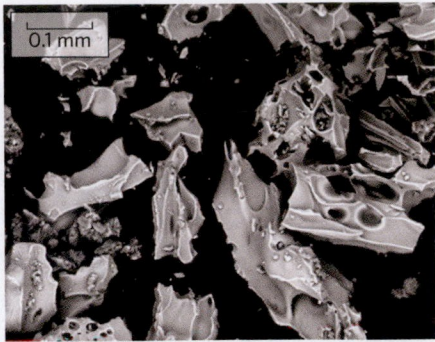

(b) Electron photomicrograph of ash.

(c) Pumice lapilli.

(a) Pyroclastic debris billowing from the 2008 eruption of Chaiten in Chile.

(d) Accretionary lapilli.

high, and by the time the volcano ceased erupting, nine years later, its deposits covered 25 km² (10 square miles).

Geologists formally refer to all kinds of debris produced by an eruption as either **pyroclastic debris** or **tephra** (Fig. 4.13a). Pyroclastic debris includes solid chunks formed when small blobs of lava freeze in midair or shortly after falling on the ground, as well as fragments formed when already solid igneous rock from a volcano breaks up and gets ejected during an eruption. Different names indicate different sizes of pyroclastic debris: **ash** consists of tiny flakes or slivers (<2 mm, or 0.1 inch diameter) (Fig. 4.13b); **lapilli** are marble- to golf-ball-sized (between 2 and 64 mm, or 0.1 and 2.5 inches) pieces; and **blocks** consist of large chunks (>64 mm). Formation of pyroclastic debris involves a few different processes.

Felsic lava typically contains a high concentration of tiny gas bubbles. When such frothy lava freezes, it produces a Styrofoam-like type of igneous rock called *pumice*. The force of an eruption can blast apart pumice and other rock that previously formed in and around the vent to produce ash, lapilli, and blocks (Fig. 4.13c). Ash that erupts into rain may clump together into little balls to form another type of lapilli, known as *accretionary lapilli* (Fig. 4.13d).

During eruptions of gas-rich mafic lava, the bursting of rising bubbles causes the melt to spurt skyward in a **lava fountain**, which may rise a few hundred meters above the vent (Fig. 4.14a). (A similar phenomenon happens when you pour soda into a glass and your face gets wet from spray as you take a drink.) Some of the lava in the fountain separates into clots that freeze as they fall, or immediately after they fall, to form a type of lapilli informally known as *cinders* (Fig. 4.14b). When a fountain ejects a melon-sized chunk that is still hot and soft, the chunk becomes streamlined as it falls and freezes to become a **volcanic bomb** (Fig. 4.14c). If a chunk sent skyward was firm before being ejected, it lands as an angular block (Fig. 4.14d).

Newly deposited layers of pyroclastic debris tend to be fairly weak, so they can slip downslope in landslides. If ash mixes with water coming from rain or melting ice

(a) Fountains of mafic lava may erupt from volcanoes.

Figure 4.14 Pyroclastic debris from mafic eruptions.

5 cm

(b) Small clots freeze into tephra ("cinders").

(c) Clots of lava that freeze in the air form bombs.

(d) Fragments of rock blasted off the volcano form angular blocks.

and snow, a muddy slurry called a **lahar**, resembling very wet concrete, may form and flow downslope **(Fig. 4.15a, b)**. Lahars can travel down stream valleys for tens of kilometers. Long after an eruption, streams and rivers flowing down the flanks of volcanoes pick up and transport pyroclastic debris, redepositing it as sediment farther downslope. Geologists use the general term **volcaniclastic deposits** for all fragmental material emitted by a volcano. The term refers not only to pyroclastic debris, but also to deposits of landslides and lahars and to water-transported sediments made of volcanic materials.

Figure 4.15 Examples of lahars.

(a) A lahar fills a riverbed in New Zealand after an eruption in 2007.

(b) The deposits of a lahar, 20 years after the 1980 eruption of Mt. St. Helens, include logs ripped from hillslopes.

Take-home message . . .

🏠 Igneous rocks form both underground and at the Earth's surface. Intrusive igneous rocks form when magma solidifies underground in dikes, sills, or plutons. Extrusive igneous rocks include lava flows and pyroclastic debris (also known as tephra). Ash mixed with water forms a lahar.

Quick Question ---------------------------

What is the difference between a dike, a sill, and a pluton?

Did you ever wonder . . .

why building stones come in different colors?

4.5 Classifying Igneous Rocks

Because melts can have a variety of compositions, and because they can freeze to form igneous rocks in many different environments above and below the surface of the Earth, a wide spectrum of different igneous rock types occur on the Earth. Geologists organize these rocks into classes on the basis of two characteristics: texture and composition. A description of texture indicates whether an igneous rock is *crystalline* (meaning that it consists of interlocking crystals), *fragmental* (meaning that it consists of pieces that are either welded or cemented to one another), or *glassy* (meaning that it consists mostly of glass) **(Fig. 4.16a–c)**. A description of composition

characterizes the relative proportions of different chemicals making up the rock. Studying an igneous rock's texture tells us about the environment in which it formed, and studying its composition tells us about the magma source and the way in which the magma evolved.

Crystalline Igneous Rocks

A **crystalline igneous rock** consists of mineral crystals that grew in a melt and fit together like pieces in a jigsaw puzzle. Geologists distinguish between fine-grained (*aphanitic*) igneous rocks, meaning rocks in which the crystals are too small to be identified without a microscope, and coarse-grained (*phaneritic*) igneous rocks, in which the minerals making up individual grains can be distinguished with the naked eye. Some igneous rocks have a uniform texture, in that they contain mostly grains of about the same size. But some contain large grains surrounded by tiny grains. In such *porphyritic* igneous rocks, the larger grains are called *phenocrysts*, and the surrounding finer grains constitute *groundmass*. The grain size of an igneous rock reflects the cooling rate: finer-grained rocks cooled quickly, whereas coarser-grained rocks cooled slowly. A porphyritic rock develops when a magma starts cooling slowly, so that early-forming crystals become large, and then cools quickly, so that the remaining melt forms a fine-grained rock.

We distinguish different compositional classes of crystalline igneous rocks based on the relative proportion of silica to iron and magnesium oxide in the rock, so we can use the same names for describing igneous rock composition as we do for magma composition. In order from least to most silica, the compositional classes are ultramafic, mafic, intermediate, and felsic.

There are many different names for igneous rocks. The name used for a given rock

Crystalline   **Fragmental**   **Glassy**

Figure 4.16 Igneous rock textures as viewed through a microscope. The field of view is about 3 mm. The black area in the glassy rock is glass.

(a) Granite is crystalline.

(b) Tuff is fragmental.

(c) Obsidian is glassy.

sample depends on both the chemical composition and the grain size of the sample **(Fig. 4.17a, b)**. For example, a fine-grained felsic rock is a *rhyolite*, while a coarse-grained one is a *granite*; a fine-grained intermediate rock is an *andesite*, while a coarse-grained one is a *diorite*; a fine-grained mafic rock is a *basalt*, while a coarse-grained one is a *gabbro*; and a fine-grained ultramafic rock is a *komatiite*, while a coarse-grained one is a *peridotite*. Note that both members of each pair have the same chemical composition, but they cooled at different rates. The color and density of an igneous rock provide a clue to its composition. Samples of felsic rock (rhyolite or granite), for example, tend to be light tan, light gray to pink, or maroon and have a lower density, whereas samples of mafic rock (basalt or gabbro) tend to be black or dark gray overall and have a higher density.

Glassy Igneous Rocks

Glassy igneous rock forms when lava cools so quickly that atoms or molecules do not attain the orderly arrangement that they have in minerals, but rather freeze in place with a disordered configuration. Some glassy rocks consist of a solid mass of glass through and through, whereas others contain tiny crystals surrounded by a groundmass of glass. Glassy rocks formed from gas-rich lava include abundant vesicles.

Geologists distinguish among several different kinds of glassy rocks, but we'll describe only a few of the more common ones here. **Obsidian** is a mass of solid, felsic glass (see Fig. 4.16c). It tends to be black or brown and to break conchoidally, meaning that fracture surfaces tend to be clamshell shaped. Because of its breaking pattern, pre-industrial people fashioned arrowheads, scrapers, and knife blades from obsidian. **Pumice** is a light-colored, felsic glassy rock that contains an abundance of tiny vesicles. As we noted earlier, it forms by the cooling of frothy lava that resembles the head of foam on a glass of beer. In some specimens of pumice, vesicles make up most of the rock's volume, so solid glass occurs only in the thin walls separating vesicles. Lapilli made of such pumice can be light enough to float on water **(Fig. 4.18a)**. **Scoria** is a mafic volcanic rock that contains abundant vesicles. Typically, the vesicles in scoria are bigger than those in pumice, and the glass walls between vesicles are thicker. Scoria tends to be dark gray in color **(Fig. 4.18b)**.

Fragmental Igneous Rocks

Fragmental igneous rock forms either when hot fragments of glass weld together while still extremely hot, or when cool fragments accumulate and later undergo compaction and cementation. Geologists use the term **pyroclastic rock** for fragmental igneous rocks formed exclusively from igneous debris that accumulated directly from an eruption—examples include **tuff (Fig. 4.19a)**,

Figure 4.17 Classification of crystalline igneous rocks.

(a) Types of crystalline igneous rocks, arranged by grain size and composition.

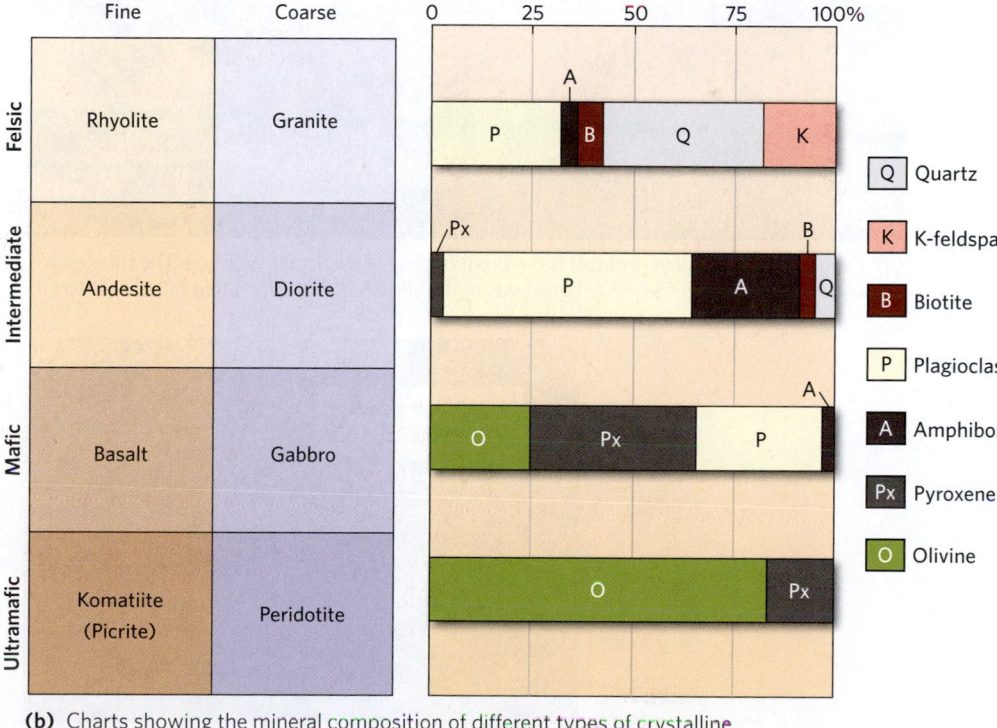

(b) Charts showing the mineral composition of different types of crystalline igneous rocks. The right chart shows typical percentages of the minerals in the rock types on the left.

Figure 4.18 Igneous rocks containing vesicles.

(a) Pumice is so light that paper can hold it up.

(b) Scoria looks like a dark sponge, though it is very hard.

Figure 4.19 Fragmental igneous rocks.

(a) Each of these thick layers of tuff is the product of a huge volcanic eruption. The rubble-strewn slopes that cover the lower two-thirds of the cliff consist of debris that fell from the cliff.

A golf-ball-sized fragment of pumice

(b) A close-up of the tuff; it consists of ash as well as pumice lapilli.

a fine-grained rock composed mostly of volcanic ash (Fig. 4.19b), and **volcanic breccia**, a coarser rock composed of volcanic blocks. Rocks formed from lahar deposits, landslide deposits, or stream-transported volcanic sediments are known as *volcaniclastic rocks.*

Take-home message . . .

Igneous rocks form from material that solidified from a melt. They can have three different textures: crystalline, glassy, or fragmental. Grain size of crystalline rocks reflects cooling rate. The name used for a given sample of crystalline rock depends on its grain size and composition. Fragmental igneous rocks form either directly from pyroclastic debris or from debris that was later transported.

Quick Question -
What rock forms from felsic magma cooled in a large pluton at depth?

4.6 The Nature of Volcanoes

Nearly all of the islands that dot the azure waters of the Lesser Antilles, along the eastern edge of the Caribbean Sea, were built from the products of eruptions at volcanoes. Eruptions do not occur continuously at each volcano. An individual eruption may take place over a period of days, weeks, or even years. Between eruptions, the islands appear tranquil, so people have built communities along their shores. In fact, when Mt. Pelée, one of the Lesser Antilles volcanoes, began to emit clouds of lapilli, pumice fragments, and ash in late April of 1902, 28,000 people lived in the port of St. Pierre, 10 km (6 miles) from Pelée's summit.

In the first week of May, felsic magma oozed up the throat of Mt. Pelée and built into a lava dome that obstructed the vent, so pressure began to build inside the volcano. Suddenly, on May 8, the dome cracked and released the pressure. In the same way that champagne bursts out of a bottle when the cork pops, a cloud of ash burst out of Mt. Pelée and mushroomed skyward. Some of the ash rose to stratospheric heights, but some rushed down the volcano's flank in a scalding avalanche, known as a **pyroclastic flow**, much like the one that destroyed Pompeii. When this turbulent mix of glowing ash and superheated air slammed into St. Pierre, it toppled and burned buildings and incinerated or asphyxiated the inhabitants. Only two people survived—one was a prisoner protected by the stout walls of his underground cell.

Volcanoes are the most visible manifestations of igneous activity. Some can explode suddenly, like Mt. Pelée

Figure 4.20 Examples of effusive eruptions.

(a) An effusive eruption on Hawaii. Lava fountains out of a small crater, and collects in a fast-moving flow.

(b) An example of a lava lake. Note that a thin layer of dark new rock has formed over the molten lava.

and Mt. Vesuvius, but others don't. Now that we have introduced igneous materials and their origins, let's focus our attention on volcanoes. In this section, we examine the different ways in which volcanoes erupt.

Will It Flow, or Will It Blow?

The character of a volcanic eruption—whether it emits flows of lava or blasts out billows of ash and debris—is known as a volcano's **eruptive style**. Geologists have traditionally assigned names to different eruptive styles based on well-known examples (**Earth Science at a Glance**, pp. 134–135). Here, we distinguish between two main styles of eruptions.

During an **effusive eruption**, low-viscosity lava fountains or spills out of the vent **(Fig. 4.20a)**. Once extruded, the lava may pool within the crater that surrounds

the vent, producing a *lava lake* **(Fig. 4.20b)**. Typically, cooling causes the surface of a lava lake to crust over with dark rock, which cracks to reveal still-molten rock beneath. When lava from an effusive eruption reaches a slope, it starts to stream down the flanks of the volcano as a lava flow. Some lava flows may extend for kilometers to tens of kilometers. Only mafic lava has low enough viscosity to erupt effusively, so effusive eruptions tend to produce basalt. Impressive examples of effusive eruptions happen frequently on the island of Hawaii. In localities where the mafic lava contains a lot of gas, lava fountains out of the vent.

As the name emphasizes, an **explosive eruption** involves an energetic blast that forcefully ejects material from a volcano. Such eruptions produce clouds and avalanches of pyroclastic debris **(Fig. 4.21a, b)**. Explosive

Figure 4.21 Examples of explosive eruptions.

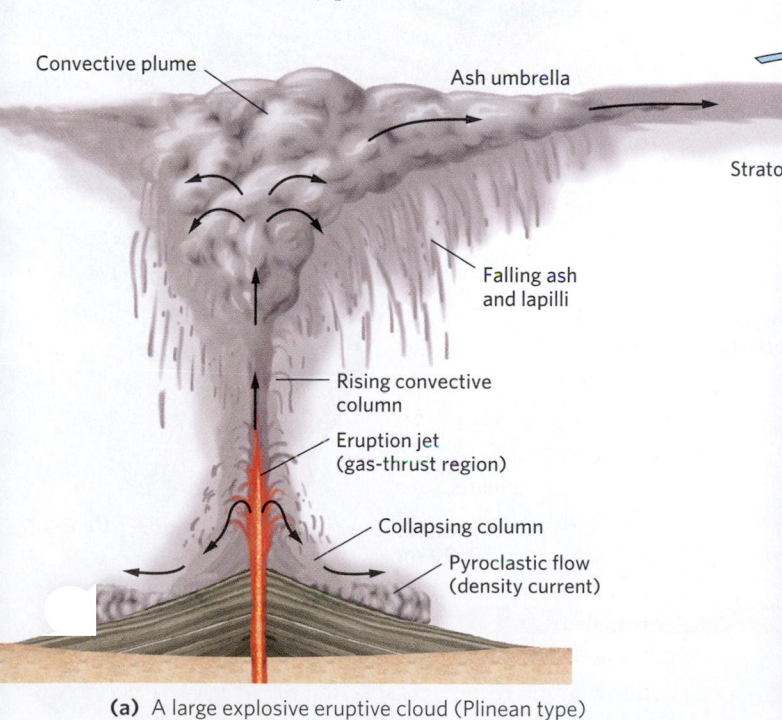

Convective plume

Ash umbrella

Wind

Stratospheric haze

Falling ash and lapilli

Rising convective column

Eruption jet (gas-thrust region)

Collapsing column

Pyroclastic flow (density current)

(a) A large explosive eruptive cloud (Plinean type) contains several components.

(b) The eruptive cloud of an eruption on Mt. Etna, Italy.

Ash cloud

Pyroclastic flow

Volcanoes

Beneath a volcano, magma rises to fill a pervasively cracked region of crust and forms a magma chamber. Some of the magma erupts at a surface vent. Once molten rock has erupted at the surface, it becomes lava. Some lava spills down the side of the volcano in lava flows. Some fountains out of a vent to form scoria fragments that pile up in a cone around the vent. Eruptions may eject larger chunks as blocks or bombs. The nature of eruptions depends on

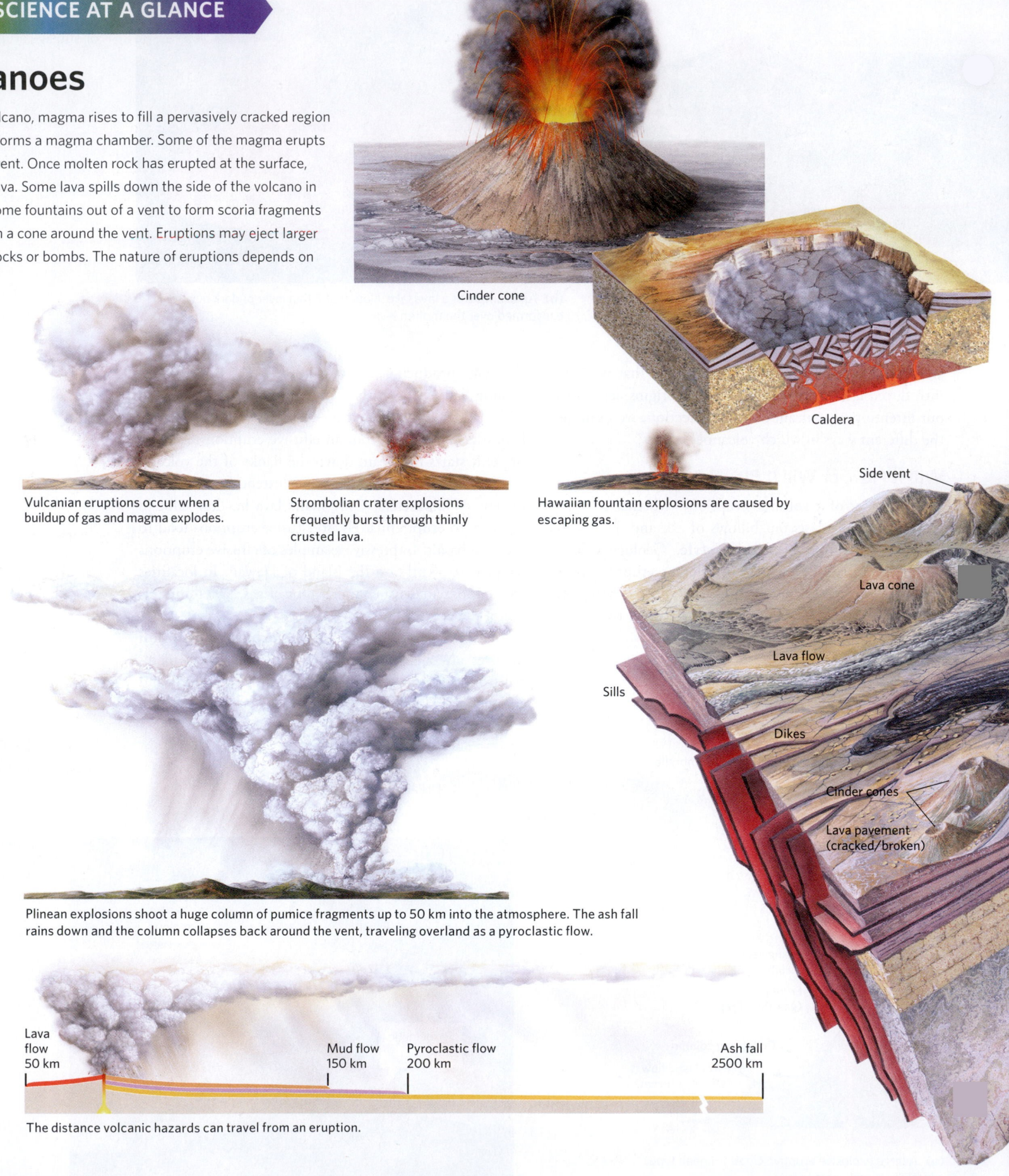

Cinder cone

Caldera

Vulcanian eruptions occur when a buildup of gas and magma explodes.

Strombolian crater explosions frequently burst through thinly crusted lava.

Hawaiian fountain explosions are caused by escaping gas.

Side vent

Lava cone

Lava flow

Sills

Dikes

Cinder cones

Lava pavement (cracked/broken)

Plinean explosions shoot a huge column of pumice fragments up to 50 km into the atmosphere. The ash fall rains down and the column collapses back around the vent, traveling overland as a pyroclastic flow.

Lava flow 50 km

Mud flow 150 km

Pyroclastic flow 200 km

Ash fall 2500 km

The distance volcanic hazards can travel from an eruption.

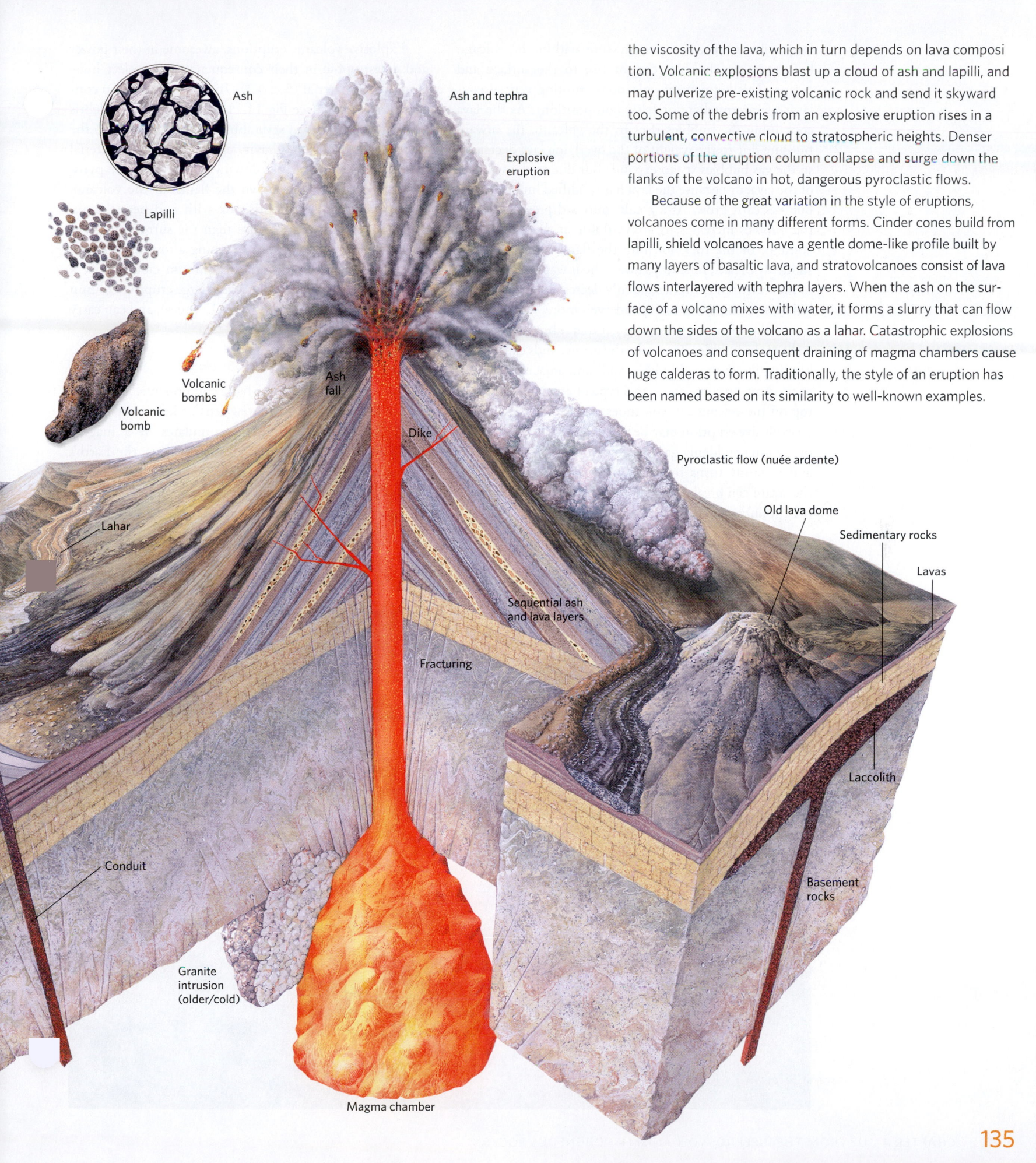

the viscosity of the lava, which in turn depends on lava composition. Volcanic explosions blast up a cloud of ash and lapilli, and may pulverize pre-existing volcanic rock and send it skyward too. Some of the debris from an explosive eruption rises in a turbulent, convective cloud to stratospheric heights. Denser portions of the eruption column collapse and surge down the flanks of the volcano in hot, dangerous pyroclastic flows.

Because of the great variation in the style of eruptions, volcanoes come in many different forms. Cinder cones build from lapilli, shield volcanoes have a gentle dome-like profile built by many layers of basaltic lava, and stratovolcanoes consist of lava flows interlayered with tephra layers. When the ash on the surface of a volcano mixes with water, it forms a slurry that can flow down the sides of the volcano as a lahar. Catastrophic explosions of volcanoes and consequent draining of magma chambers cause huge calderas to form. Traditionally, the style of an eruption has been named based on its similarity to well-known examples.

Ash

Lapilli

Volcanic bombs

Volcanic bomb

Ash and tephra

Explosive eruption

Ash fall

Dike

Lahar

Pyroclastic flow (nuée ardente)

Old lava dome

Sedimentary rocks

Lavas

Sequential ash and lava layers

Fracturing

Laccolith

Conduit

Basement rocks

Granite intrusion (older/cold)

Magma chamber

eruptions occur when gas pressure within the volcano builds up and bubbles cannot rise to the surface and escape. This happens at volcanoes erupting very viscous lava (of intermediate or felsic composition). As the lava containing the bubbles rises in the volcano, the inward pressure due to the weight of the overlying lava decreases, so the gas bubbles try to expand. But due to the strength of the lava, or because the lava has solidified into glass, the bubbles can't grow. As a result, outward pressure within the bubbles increases. If a lava dome at the crest of the volcano, or the rock forming the flank of the volcano, suddenly breaks, or slumps away, the inward pressure due to the weight of the lava suddenly decreases further, so the bubbles suddenly try to expand even more, and the pressure within them becomes so great that it breaks the glassy walls of the bubbles. Instantly, the gas expands explosively, blasting the shattered bubble walls and adjacent rock out of the vent. In some cases, this type of eruption blows the top off the volcano. In volcanoes erupting mafic magma, an explosive eruption may be triggered when groundwater or seawater suddenly gains access to the magma chamber and instantly turns to steam. The sudden expansion of the steam can break up the rock forming the volcano and blast it outward.

Did you ever wonder . . .

whether all volcanoes look the same?

Explosive volcanic eruptions, awesome in their power and catastrophic in their consequences, may eject huge amounts of material **(Box 4.3)**. The eruptive column consists of two parts (see Fig 14.21a, b). At the base, debris blasts upward as far as several hundred meters due to the force of the explosion. Gravity causes part of the material in the column to collapse downward, forming the pyroclastic flows that surge down the flanks of the volcano. Finer components of debris mix with and heat the air. This hot mixture is less dense than the surrounding air, so it rises upward buoyantly, forming a towering *convective plume* that resembles the mushroom cloud above a nuclear explosion. The top of a very large eruptive column can reach heights of up to 27 km (17 miles), and can carry dust into the stratosphere **(Fig. 4.22)**.

The Architecture of Volcanoes

All volcanoes share the same basic components. Anywhere from a few kilometers to a few tens of kilometers below the Earth's surface, magma accumulates in a magma chamber. Magma rises through a conduit to the Earth's surface and erupts from a vent. The material that builds up around the conduit constitutes a *volcanic edifice* (a hill or mountain of extrusive igneous rock).

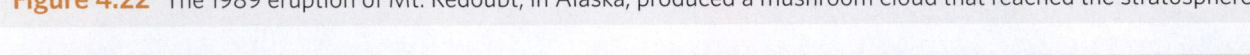

Figure 4.22 The 1989 eruption of Mt. Redoubt, in Alaska, produced a mushroom cloud that reached the stratosphere.

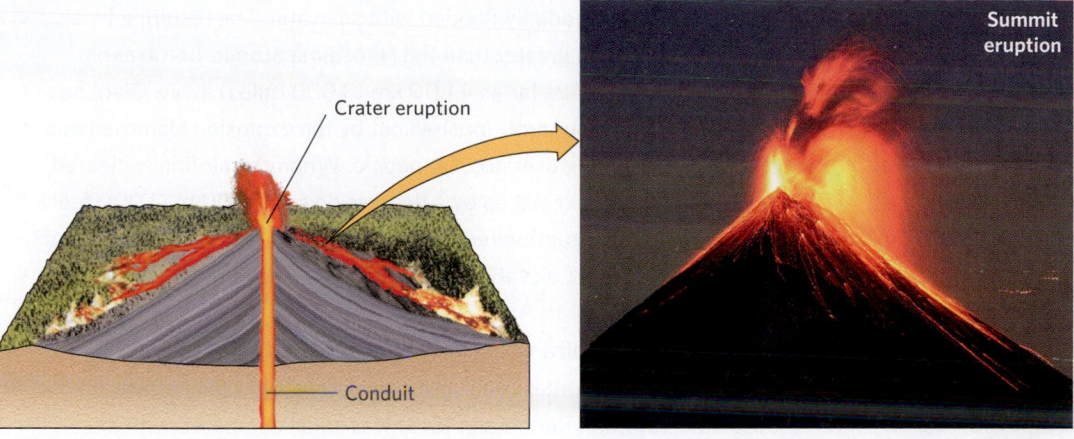

(a) In a crater eruption, lava spouts from a chimney-shaped conduit.

In some volcanoes, the conduit through which lava and gas reach the Earth's surface has a chimney-like shape, and is topped by a circular depression, or **crater**, that resembles a bowl surrounding the vent **(Fig. 4.23a)**. Craters, which may be up to 500 m (1,600 feet) across and 200 m (650 feet) deep, develop in two ways: some form during eruptions when pyroclastic debris builds up around the vent, whereas others form after an eruption when the volcano's summit collapses into the drained conduit. Not all eruptions come from a single crater, however. During **fissure eruptions**, curtains of lava spew from an elongate crack, or *fissure* **(Fig. 4.23b)**. Though eruptions commonly take place from a *summit vent* at the peak of a volcano, not all do—some eruptions come from *flank vents* on the side slope of a volcano.

The overall shapes of volcanoes vary as well. Geologists distinguish among three different volcano shapes. **Shield volcanoes**, so named because they resemble a

(b) In a fissure eruption, lava comes out in a curtain along the length of a crack.

soldier's shield lying on the ground, are broad, gentle domes. They form from layer upon layer of low-viscosity mafic lava produced by effusive eruptions **(Fig. 4.24a)**. **Cinder cones** consist of symmetrical, cone-shaped piles of lapilli formed from clots fountaining out of a vent during an effusive eruption. Typically, cinder cones have deep, pit-like craters at their summits **(Fig. 4.24b)**.

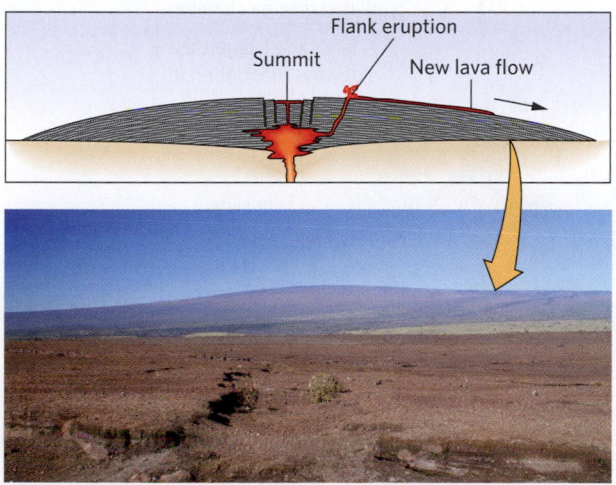

(a) A shield volcano, made from successive layers of low-viscosity basaltic lava, has very gentle slopes.

Figure 4.24
Volcano shapes.

(b) A cinder cone in Arizona. A lava flow covers the land surface in the distance.

Consider this . . .

Volcanic explosions to remember

Mt. St. Helens, a snow-crested volcano in the Cascade Range of the northwestern United States, had not erupted since 1857. However, geologic evidence suggested that the mountain had a violent past, punctuated by many explosive eruptions. On March 20, 1980, an earthquake announced that the volcano was awakening once again. A week later, the summit began emitting gas and pyroclastic debris. Geologists who set up monitoring stations to observe the volcano noted that its north side was beginning to bulge markedly, suggesting that the volcano was filling with magma and that the magma was making the volcano expand like a balloon. Their concern that an eruption was imminent led local authorities to evacuate people from the area.

The climactic eruption came suddenly. At 8:32 A.M. on May 18, David Johnston, a geologist monitoring the volcano from a station 10 km away, shouted over his two-way radio, "Vancouver, Vancouver, this is it!" An earthquake had triggered a huge landslide that caused 3 km³ (0.7 miles³) of the volcano's weakened north side to slide away. The landslide released pressure on the magma in the volcano, causing a sudden and violent expansion of gases that blasted through the side of the volcano **(Fig. Bx4.3a)**. Rock, steam, and ash screamed north at the speed of sound and flattened a forest and everything in it over an area of 600 km² (140 miles²) **(Fig. Bx4.3b)**. Tragically, Johnston, along with 60 others, vanished forever.

Seconds after the sideways blast, a vertical column carried over 500 million tons of ash (about 1 km³) up to the stratosphere, where strong high-altitude winds transported it rapidly around the globe. In towns near the volcano, a blizzard of ash buried fields. Water-saturated ash formed lahars that flooded river valleys, carrying away everything in their path. When the eruption was finally over, the peak of Mt. St. Helens had disappeared—the summit now lay 440 m (1,450 feet) lower, and the once snow-covered mountain was a gray mound with a large gouge in one side.

An even greater explosion happened in 1883. Krakatau (also known as Krakatoa), a volcano in the sea between Java and Sumatra, where the Indian Ocean floor subducts beneath Southeast Asia, had grown to become a 9-km (6-mile)-long island rising 800 m (2,600 feet) above the sea. On May 20, the island began to erupt with a series of large explosions, yielding ash that settled as far as 500 km (300 miles) away. Smaller explosions continued through June and July, and steam and ash rose from the island, forming a huge black cloud that rained ash into the surrounding straits. Krakatau's demise came at

10 A.M. on August 27, perhaps when the volcano cracked and the magma chamber suddenly flooded with seawater. The resulting blast, five thousand times greater than the Hiroshima atomic-bomb explosion, could be heard as far as 4,800 km (3,000 miles) away. Giant sea waves—a type of **tsunami**—pushed out by the explosion slammed into coastal towns, killing over 36,000 people. When the air finally cleared, Krakatau was gone, replaced by a depression some 300 m (1,000 feet) deep. All told, the eruption redistributed 20 km³ of rock. The ash that it sent to stratospheric elevations caused spectacular sunsets during the next several years.

Both of the volcanic explosions we've just described pale in comparison to the largest ones in Earth history. Even the largest observed eruption (Tambora in 1815) was small compared with an explosion that took place over 600,000 years ago in what is now Yellowstone National Park, Wyoming **(Fig. Bx4.3c)**. Geologists use the term **supervolcano** for a volcano that yields an explosive eruption that ejects more than 1,000 km³ (240 miles³) of debris. At least ten of these events have been identified from the occurrence of very large calderas or of very thick and widespread tuffs. The most recent occurred about 26,500 years ago.

Figure Bx4.3 Examples of explosive eruptions.

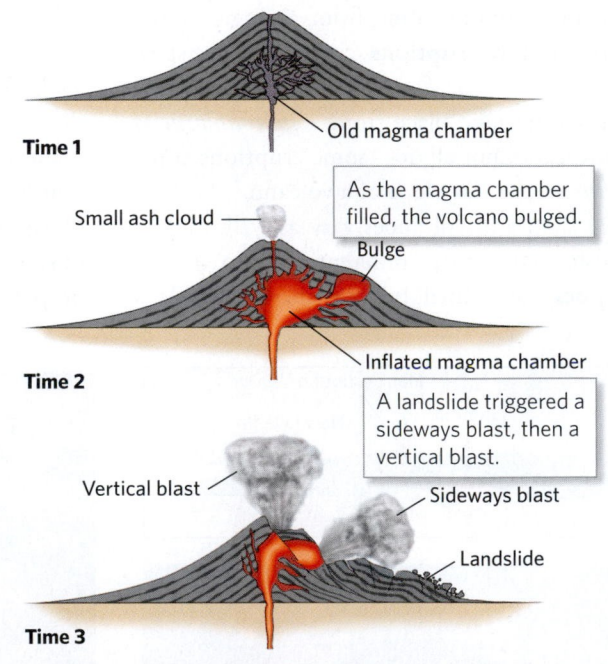

Time 1 — Old magma chamber

As the magma chamber filled, the volcano bulged.

Small ash cloud — Bulge

Time 2 — Inflated magma chamber

A landslide triggered a sideways blast, then a vertical blast.

Vertical blast — Sideways blast

Landslide

Time 3

(a) Stages during the eruption of Mt. St. Helens, 1980.

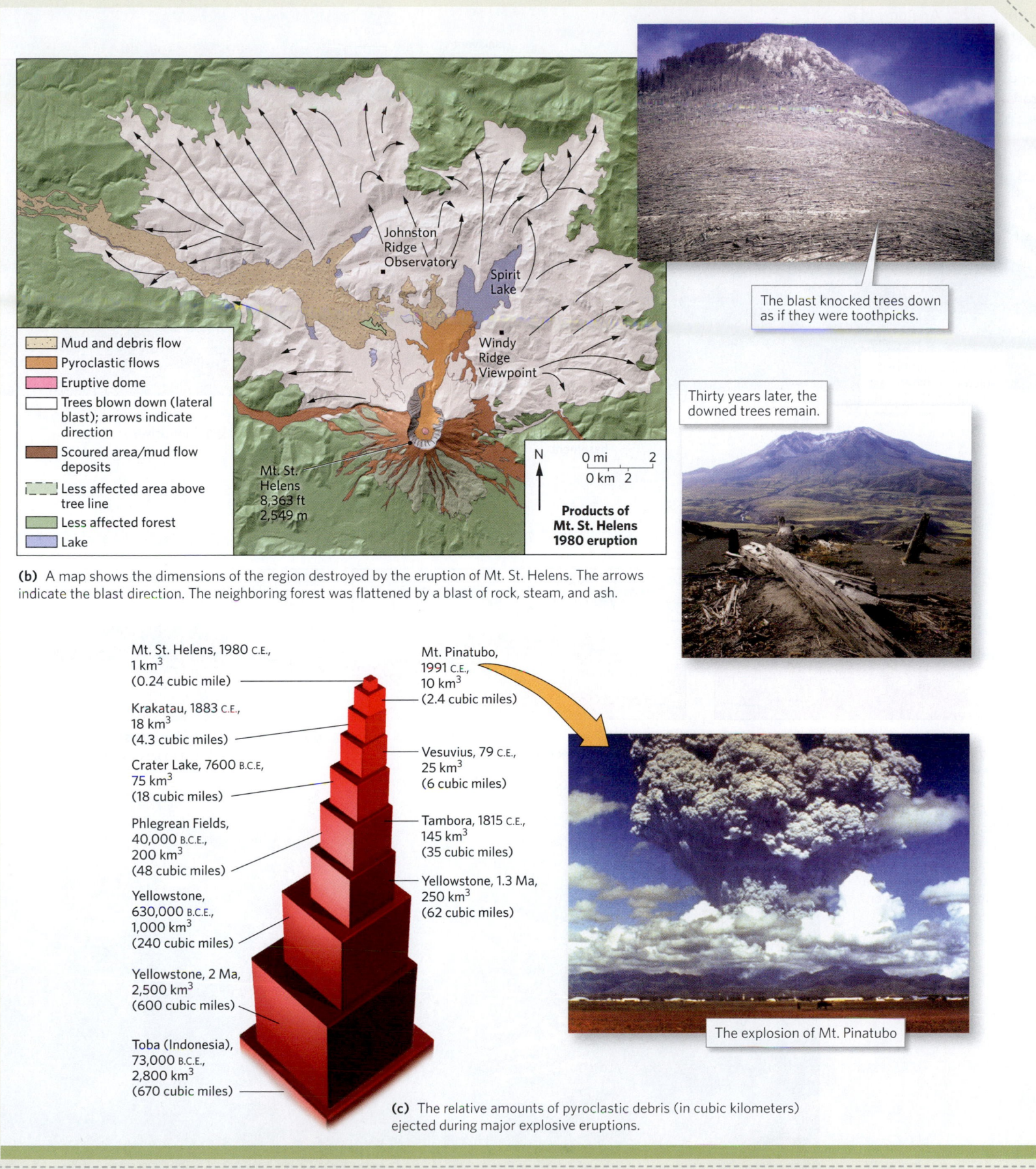

The blast knocked trees down as if they were toothpicks.

Thirty years later, the downed trees remain.

(b) A map shows the dimensions of the region destroyed by the eruption of Mt. St. Helens. The arrows indicate the blast direction. The neighboring forest was flattened by a blast of rock, steam, and ash.

Mud and debris flow
Pyroclastic flows
Eruptive dome
Trees blown down (lateral blast); arrows indicate direction
Scoured area/mud flow deposits
Less affected area above tree line
Less affected forest
Lake

Johnston Ridge Observatory

Spirit Lake

Windy Ridge Viewpoint

Mt. St. Helens 8,363 ft 2,549 m

N

0 mi 2
0 km 2

Products of Mt. St. Helens 1980 eruption

Mt. St. Helens, 1980 C.E., 1 km^3 (0.24 cubic mile)

Krakatau, 1883 C.E., 18 km^3 (4.3 cubic miles)

Crater Lake, 7600 B.C.E, 75 km^3 (18 cubic miles)

Phlegrean Fields, 40,000 B.C.E., 200 km^3 (48 cubic miles)

Yellowstone, 630,000 B.C.E., 1,000 km^3 (240 cubic miles)

Yellowstone, 2 Ma, 2,500 km^3 (600 cubic miles)

Toba (Indonesia), 73,000 B.C.E., 2,800 km^3 (670 cubic miles)

Mt. Pinatubo, 1991 C.E., 10 km^3 (2.4 cubic miles)

Vesuvius, 79 C.E., 25 km^3 (6 cubic miles)

Tambora, 1815 C.E., 145 km^3 (35 cubic miles)

Yellowstone, 1.3 Ma, 250 km^3 (62 cubic miles)

The explosion of Mt. Pinatubo

(c) The relative amounts of pyroclastic debris (in cubic kilometers) ejected during major explosive eruptions.

Figure 4.25 Anatomy of a stratovolcano. A stratovolcano consists of layers of tephra and lava. Landslides occasionally occur and transport material downslope.

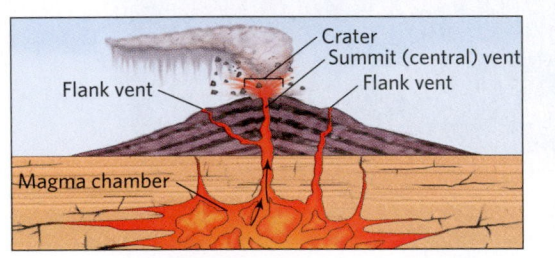

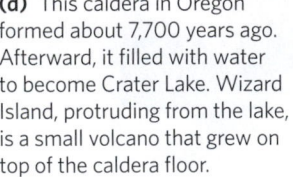

Mt. Fuji, a composite volcano in Japan, last erupted in 1707.

Crater

Flank vent

Vent

Recent landslide

Alluvial apron

▦	Older volcano	▦ Lava flows	▦ Alluvium	\	Faults
▦	Pre-volcanic basement	▦ Tephra	▦ Landslides	▮	Intrusives

Figure 4.26 Formation of a volcanic caldera.

Time

 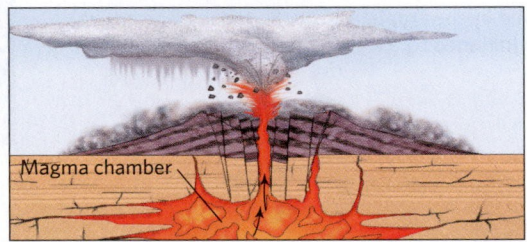 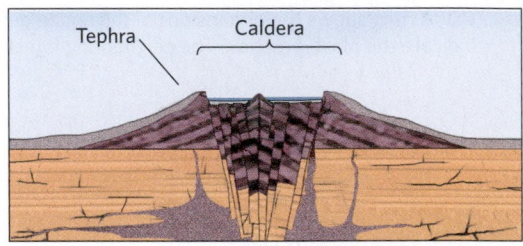

Crater
Summit (central) vent
Flank vent
Flank vent
Magma chamber

Magma chamber

Tephra
Caldera

(a) As an eruption begins, the magma chamber inflates with magma. There may be a central vent and one or more flank vents.

(b) During the eruption, the magma chamber drains, and the central portion of the volcano collapses downward.

(c) The collapsed area becomes a caldera. Later, a new volcano may begin to grow within the caldera.

(d) This caldera in Oregon formed about 7,700 years ago. Afterward, it filled with water to become Crater Lake. Wizard Island, protruding from the lake, is a small volcano that grew on top of the caldera floor.

Stratovolcanoes, also known as *composite volcanoes*, tend to be large (up to a few kilometers high and 15 km, or 9 miles across), cone-shaped mountains made from alternating layers of lava and pyroclastic debris erupted during alternating effusive and explosive eruptions (Fig. 4.25). The prefix *strato*– emphasizes the layered character of the igneous rock that makes up the volcano. The shape of a stratovolcano, exemplified by Japan's Mt. Fuji, serves as the classic image most people have of a volcano. In most examples, the upper, steeper part of the cone consists of erupted layers of lava, tephra, and tuff that collected in place, whereas the lower, less steep part consists of landslide or lahar deposits. Large explosions occasionally blast off the summit or side of the volcano. When this happens, a new volcanic edifice may build on the remnants of its predecessor.

After major eruptions, especially ones that explosively blast out large volumes of debris, the summit area of a volcano may collapse into the large, drained magma chamber below, producing a **caldera**. Calderas are circular to elliptical depressions, up to tens of kilometers across and several hundred meters deep, that generally have a fairly flat floor covered by lava or pyroclastic debris (Fig. 4.26).

Take-home message . . .

At a volcano, lava rises from a magma chamber and erupts from a chimney-like conduit or a crack-like fissure. Eruptive style varies—during effusive eruptions, low-viscosity magma erupts, whereas during explosive eruptions, large volumes of pyroclastic debris blast skyward. The shapes of volcanoes vary from gently sloping shields to cone-like stratovolcanoes. Collapse following a major explosion can produce a caldera.

Quick Question --------------------------------
What phenomena can trigger a volcanic explosion?

4.7 Where Does Igneous Activity Occur?

Before the theory of plate tectonics was proposed, geologists didn't really understand what caused igneous activity on Earth. They knew that many (though not all) volcanoes occurred around the edge of the Pacific, forming a pattern they referred to as the "Ring of Fire." We now realize that most igneous activity occurs along divergent and convergent plate boundaries, but some occurs at hot spots and in rifts (Fig. 4.27). In this section, we relate the character of igneous activity to different geologic settings.

Mid-Ocean Ridges

During seafloor spreading at a mid-ocean ridge, underlying hot asthenosphere rises from deeper down (see Fig. 4.2b). When the peridotite of the asthenosphere reaches a relatively shallow depth, it undergoes decompression melting and produces mafic magma, which rises. Some of this magma accumulates in a magma chamber at a depth between 7 and 4 km (4 and 2.5 miles). Slow cooling along the walls of the chamber produces gabbro. Some magma rises still farther, filling vertical cracks and solidifying to form basalt dikes, and some reaches the seafloor and erupts to form pillow basalt.

We don't generally see the volcanic activity of mid-ocean ridges because most of it occurs beneath a 2-km (1.2 mile)-deep blanket of seawater. Where this submarine volcanism has been mapped by explorers in submersibles, it appears to occur along fissures parallel to the ridge axis (Fig. 4.28a, b). An individual fissure, which remains active for tens to hundreds of years before becoming inactive, yields elongate mounds of pillow basalt. New oceanic crust along ridges is riddled with cracks, into which seawater percolates. This water, warmed by magma, dissolves minerals in the crust, before reentering the sea at *hydrothermal vents*, cracks that emit hot-water solutions from below the seafloor. When these solutions come in contact with cold seawater, they cool, and the dissolved minerals they contain precipitate to form a dark cloud of finely suspended grains. As a consequence, hydrothermal vents are known as **black smokers** (Fig. 4.28c).

Convergent Boundaries

Most *subaerial volcanoes* (those that protrude into the air) occur along convergent boundaries. The magma that feeds these volcanoes forms because volatile compounds seep from the subducting plate into the overlying asthenosphere and trigger flux melting in the asthenosphere (see Fig. 4.3a). These volatiles come from the oceanic crust, which incorporates water that circulates through the crust at a mid-ocean ridge. When the crust undergoes subduction at a convergent boundary, the hitchhiking volatiles sink with the plate into the asthenosphere. At a depth of about 150 km (90 miles), the subducted plate gets hot enough for the bonds holding the volatiles to the crustal minerals to break, and the volatiles then seep into the overlying asthenosphere.

Once magma forms in the asthenosphere above the subducted plate, it rises and eventually seeps through the lithosphere of the overriding plate. Some of it rises all the way to the surface and erupts as lava and ash in a chain of volcanoes called a **volcanic arc**. If the overriding plate consists of oceanic lithosphere, the volcanoes are islands, and geologists refer to the chain as an *island arc*

Did you ever wonder . . .
----
whether we can see all the volcanic eruptions taking place on Earth today?

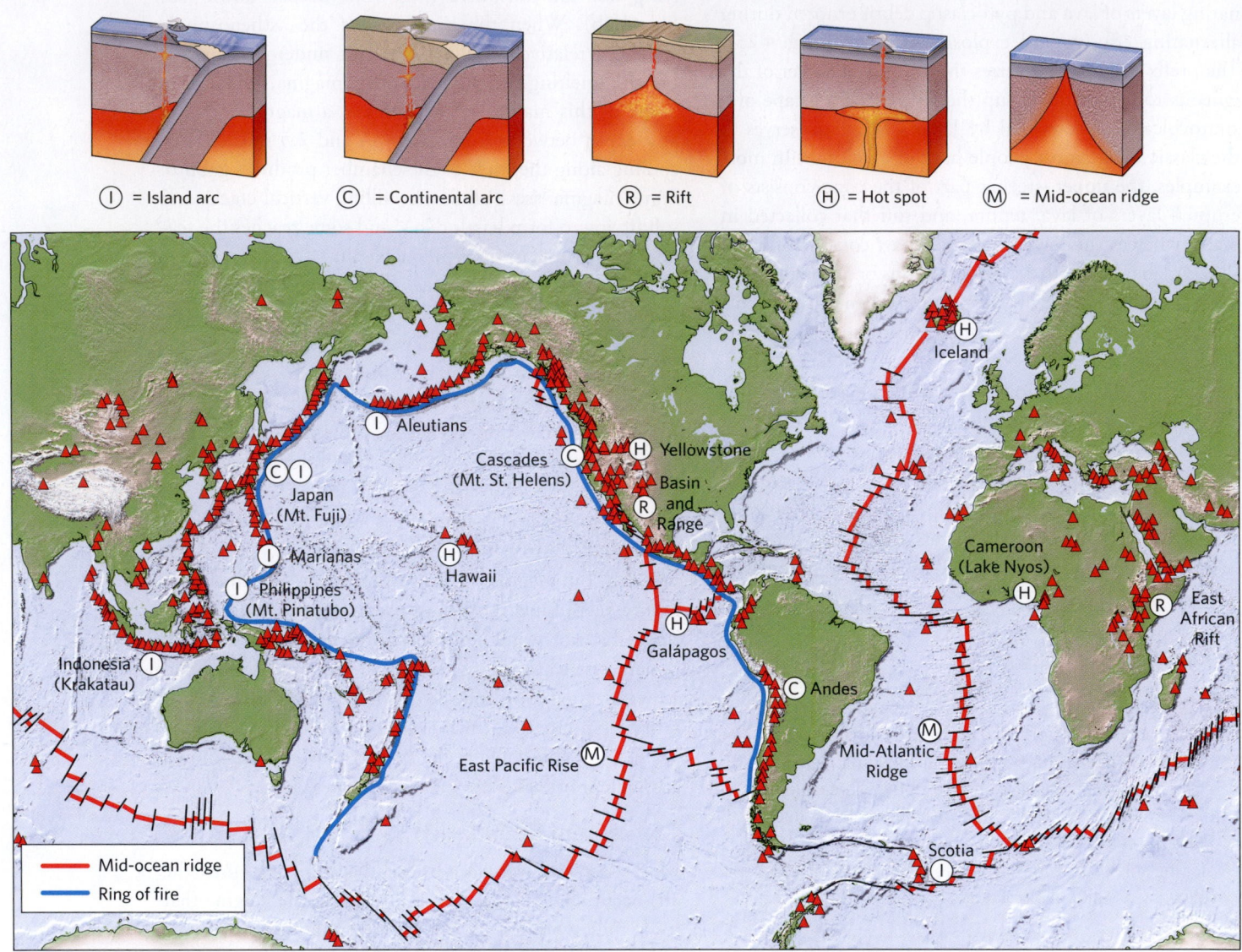

Figure 4.27 The geologic settings of volcanoes. The map shows the distribution of volcanoes around the world. The diagrams above the map show the five basic types of geologic settings in which volcanoes form, in the context of plate tectonics theory.

(I) = Island arc   (C) = Continental arc   (R) = Rift   (H) = Hot spot   (M) = Mid-ocean ridge

(Fig. 4.29a), whereas if the overriding plate consists of continental lithosphere, the chain is a *continental arc* **(Fig. 4.29b)**.

Many different kinds of magma form at volcanic arcs. Mafic basaltic lava erupts when magma rises directly from the mantle. Magma, however, may evolve and react with the crust as it moves upward, and in continental arcs, it may trigger heat-transfer melting of the crust itself. Therefore, continental arcs commonly erupt andesite and, on occasion, rhyolite. Depending on the magma composition, eruptions may be effusive or explosive, so the products typically build stratovolcanoes. Large volumes of granite and diorite intrude to form plutons in the crust beneath continental arcs. In some cases, large batholiths eventually develop there.

Continental Rifts

During rifting, continental lithosphere stretches horizontally and thins vertically (see Fig 4.2b). The thinning of lithosphere causes decompression melting of the underlying asthenosphere and the production of mafic magma. Some of this magma rises straight to the surface and erupts as basalt **(Fig 4.30a)**. But some magma undergoes fractional crystallization during its rise or causes heat-transfer melting within the crust, both of which yield felsic magma. As a consequence, rifts host both basaltic fissure eruptions, in which curtains of lava fountain up or linear chains of cinder cones develop, and explosive rhyolitic volcanoes, which produce vast sheets of tuff. In a few locations, large stratovolcanoes, such as Mt. Kilimanjaro, grow in rifts **(Fig 4.30b)**.

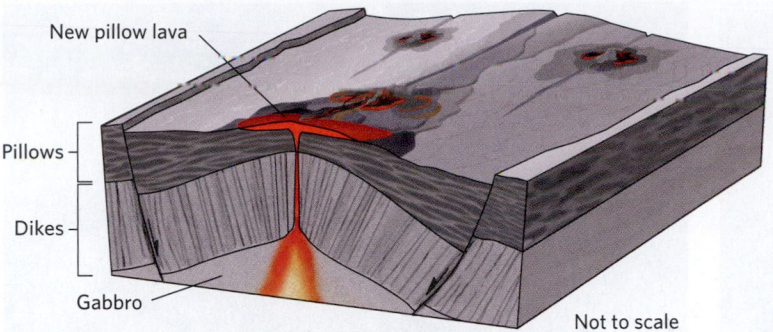

(a) Mounds of pillow basalt erupt along fissures.

(b) Pillow basalt on the seafloor along the Juan de Fuca Ridge.

Hot Spots

According to the mantle plume hypothesis, hot spots form where a column of very hot asthenosphere rises from deep in the mantle up to the base of the lithosphere. The rock at the top, or *head*, of the plume undergoes decompression melting and produces large volumes of mafic magma. When a hot-spot volcano begins to form on oceanic lithosphere, basaltic magma erupts underwater and yields a mound of pillow lava. With time, the volcano grows up above the sea surface and becomes an island. When the volcano emerges from the sea, the lava no longer cools as quickly, so it flows as a thin sheet over a great distance. Successive effusive eruptions produce thousands of thin basalt flows that pile up to build a shield volcano **(Fig. 4.31a)**. As the volcano grows, portions of it succumb to the pull of gravity and slip seaward, creating large submarine slumps and landslides of volcanic debris. Each of the Hawaiian Islands formed in this way **(Fig. 4.31b)**. Since Hawaii is a hot-spot track, only

(c) An example of an active black smoker along the Mid-Atlantic Ridge.

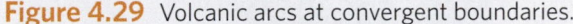
Figure 4.29 Volcanic arcs at convergent boundaries.

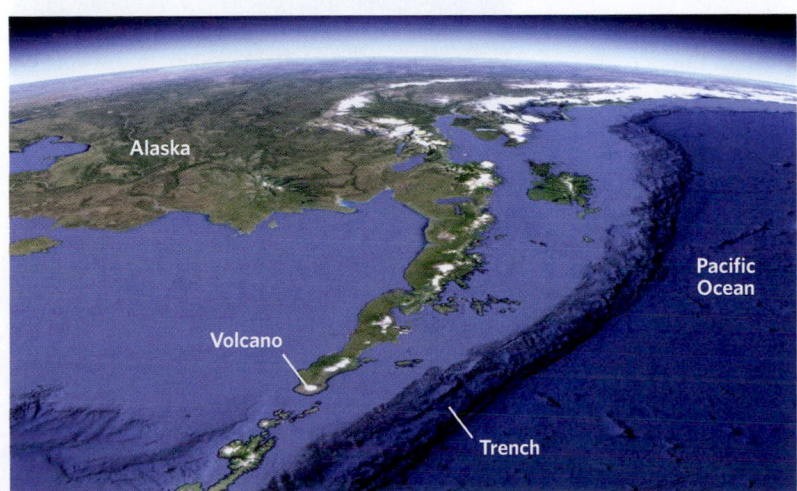

(a) The Aleutian island arc of Alaska, viewed looking northeast. The white peaks are volcanoes. This chain formed at a convergent boundary.

(b) The Andean continental arc lies near the western coast of South America.

Figure 4.30 Igneous activity in continental rifts.

(a) These recent cinder cones have been cut by faults associated with rifting.

(b) Mt. Kilimanjaro, the highest mountain in Africa (5.9 km, or 3.7 miles high), is a stratovolcano that has formed in a rift zone.

the Big Island still erupts—the other islands ceased being active long ago and have been gradually eroding away and collapsing into the sea (see Chapter 2).

Not all oceanic hot-spot volcanoes occur in the middle of plates, however. Iceland, for example, formed over a hot spot under the axis of the Mid-Atlantic Ridge. Because of this hot spot, far more magma erupts in Iceland than in other places along the ridge, so eruptions in Iceland have built a broad plateau of basalt **(Fig. 4.32a)**. Iceland straddles a divergent boundary, so plate motion actively stretches the island apart. Indeed, the central part of the island is a narrow rift, within which the youngest volcanic rocks of the island can be found **(Fig. 4.32b)**. This rift is the trace of the Mid-Atlantic Ridge (see Chapter 2 opening photo).

Furthermore, not all hot spots occur beneath oceanic lithosphere. The best-known continental hot spot underlies Yellowstone National Park. The park lies at the northeastern end of a track of volcanism marked by a string of calderas **(Fig. 4.33a–c)**. Unlike the Hawaiian hot spot, the Yellowstone hot spot has erupted both basaltic and rhyolitic lava and pyroclastic debris because basaltic

magma rising from the asthenosphere causes heat-transfer melting in the continental crust, a process that yields felsic magma. Huge supervolcanic explosions have occurred in association with its eruptions of felsic magma (see Box 4.3). The most recent of these, which occurred about 640,000 years ago, produced immense pyroclastic flows, which accumulated as yellow, tan, and orange layers of tuff tens of meters thick, and a huge caldera almost 100 km (60 miles) across.

(b) The bathymetry of Kauai, in Hawaii, shows the underwater part of the island.

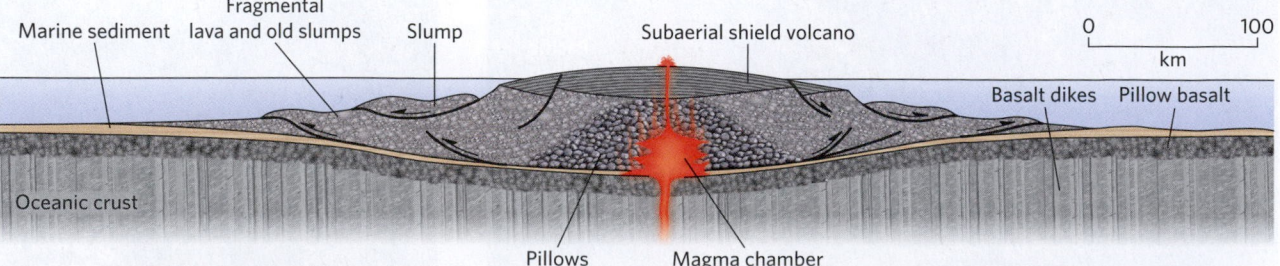

Figure 4.31 The structure of an oceanic hot-spot volcano is complicated.

(a) Initially, eruption produces pillow basalts. When the volcano emerges above sea level, it becomes a shield volcano. The margins of the island frequently undergo slumping, and the weight of the volcano pushes down the surface of the lithosphere.

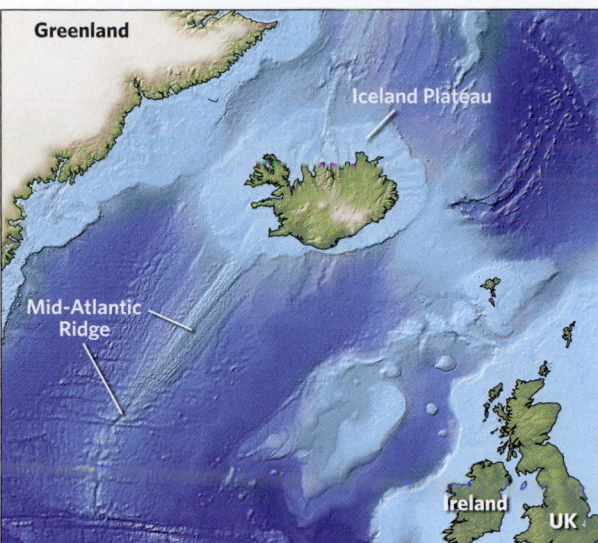

(a) A bathymetric map shows that Iceland sits atop a huge oceanic plateau straddling the Mid-Atlantic Ridge. Light blue is shallower water; dark blue is deeper.

(b) A geologic map of Iceland shows that the youngest volcanoes occur in the central rift, which is effectively the on-land portion of the Mid-Atlantic Ridge.

Recent sediment	
<0.7 Ma volcanics	
0.7–3.1 Ma volcanics	
>3.1 Ma volcanics	

Volcanoes
Fissures
Glaciers

Figure 4.32 Iceland lies over a hot spot on the Mid-Atlantic Ridge.

Large Igneous Provinces

In several locations around the world, immense volumes of low-viscosity mafic lava have erupted over a relatively short time and spread out in vast flows, some of which extend over 500 km (300 miles) from the vent. Geologists refer to the rock formed from these flows as **flood basalt (Fig. 4.34a)**. Over time, many successive eruptions of flood basalt can build up a broad *basalt plateau*. The total volume of rock in such a plateau may be so great (over 175,000 km³, 42,000 cubic miles), that geologists also refer to the region as a **large igneous province** or **LIP (Fig. 4.34b)**.

Figure 4.33 Volcanic activity at the Yellowstone hot spot.

A resurgent dome is a bulge formed when a magma chamber inflates.

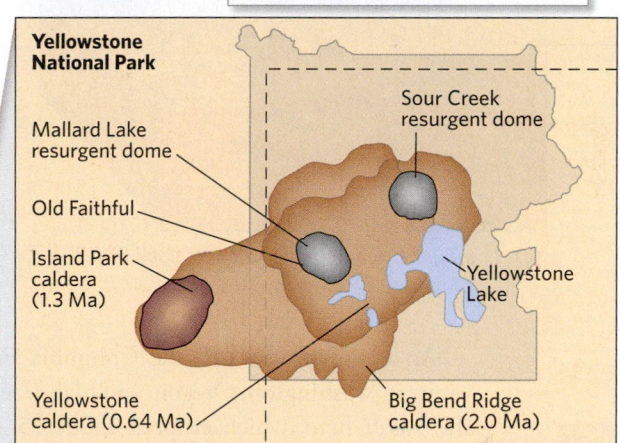

(b) Yellowstone overlies a huge caldera. Immense eruptions occurred here at least three times during the past 2 million years.

(c) Felsic tuffs form the colorful walls of Yellowstone Canyon.

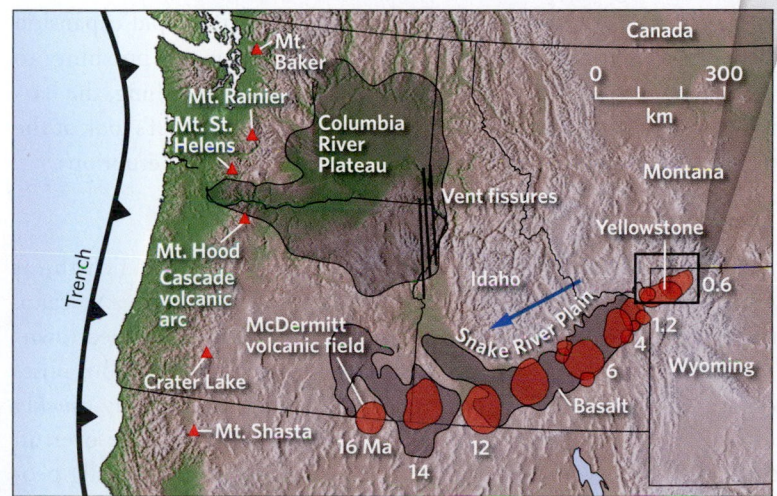

(a) Yellowstone National Park lies at the end of a continental hot-spot track. Progressively older calderas follow the Snake River Plain to the west. The blue arrow indicates plate motion.

Figure 4.34 Large igneous provinces (LIPs).

(a) Flood basalts form the layers exposed in Palouse Canyon, Washington.

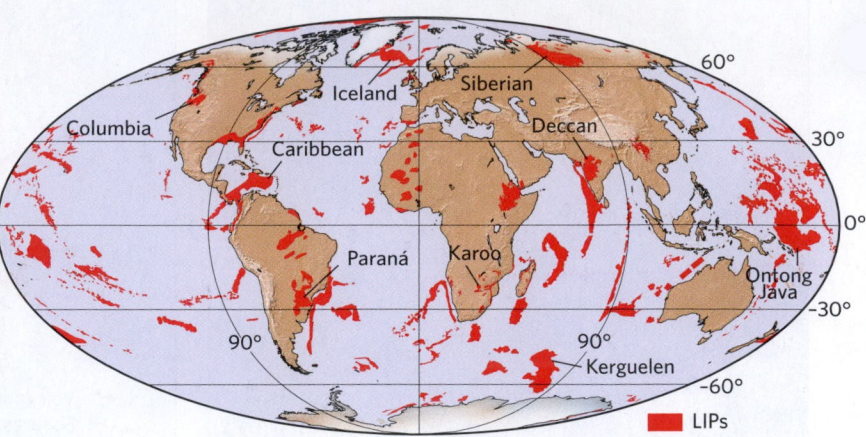

(b) A map showing the distribution of LIPs on Earth. The red areas are or once were underlain by immense volumes of basalt; not all of this basalt is exposed.

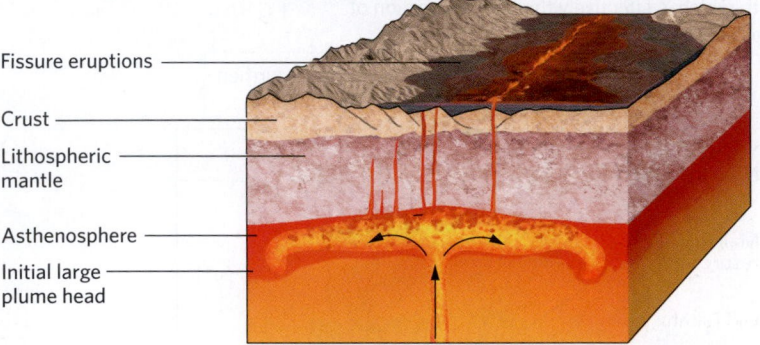

(c) The plume model for formation of flood basalts.

Fissure eruptions
Crust
Lithospheric mantle
Asthenosphere
Initial large plume head

Take-home message . . .

Plate tectonics theory helps explain why igneous activity occurs where it does. Flux melting produces melts at convergent boundaries. Melting of the mantle at hot spots, mid-ocean ridges, and rifts is probably due to decompression. Injection of hot mantle-derived magma into continental crust causes heat-transfer melting of the crust.

Quick Question ----------------------------
What are large igneous provinces?

An example of a LIP, the Columbia River Plateau, occurs in Washington, Oregon, and Idaho (see Fig. 4.33a). The basalt here, which erupted about 15 million years ago, reaches a thickness of 3.5 km (2 miles). Geologists have identified about 300 individual flows in the Columbia River Plateau. Even larger LIPs occur in eastern Siberia (an occurrence known as the Siberian Traps), India (the Deccan Plateau), Brazil (the Paraná region), and South Africa (the Karoo Plateau). Several immense submarine basalt plateaus have also been identified on the ocean floor. The largest of these oceanic plateaus, the Ontong-Java Plateau in the western Pacific, has an area of 2,000,000 km² (770,000 square miles).

What causes LIP flood-basalt eruptions? A popular hypothesis suggests that flood basalts form when the head of a large mantle plume rises beneath a region that is undergoing rifting (Fig. 4.34c). As the plume reaches the base of the lithosphere, a huge volume of mantle undergoes decompression, and because rock in the plume is particularly hot, more partial melting takes place than is usual. When conduits form and provide access for the melt to reach the surface, immense fissure eruptions take place.

4.8 Beware: Volcanoes Are Hazards!

Volcanoes are natural hazards with the potential to cause great destruction, as illustrated by the calamities at Mt. Pelée and Mt. Vesuvius. Because of the rapid expansion of cities, far more people live in dangerous proximity to volcanoes today than ever before, so if anything, the hazard posed by volcanoes has gotten worse. Let's look at the different kinds of threats posed by volcanic eruptions.

Hazards Due to Lava and Ash

When you think of an eruption, perhaps the first threat that comes to mind is the lava that flows from a volcano. Indeed, on many occasions, lava has overwhelmed towns (Fig. 4.35a–d). Basaltic lava from effusive eruptions poses the greatest threat because it can flow relatively quickly and spread over a broad area. Lava flows have overrun roads, housing developments, and vehicles. Usually people have time to get out of the way of such flows, but not necessarily with their possessions. Sometimes, buildings burst into flames from the intense heat of a nearby flow.

Figure 4.35 Volcanic hazards: lava and ash.

Lava Flows

(a) A lava flow reaches a house in Hawaii and sets it on fire.

(b) Lava from Mt. Etna threatens a town and olive grove in Sicily.

(c) Residents rescue household goods after a lava flow filled the streets of Goma, along the East African Rift.

(d) This empty school bus was engulfed by lava in Hawaii.

Pyroclastic Debris

(e) A pyroclastic flow rushes down the slope of the Soufriere Hills Volcano, on the island of Monserrate.

(f) A blizzard of ash fell from the cloud erupted by Mt. Pinatubo in the Philippines.

(g) Lapilli falls from an eruption in Iceland.

(h) A lahar submerges farmland in Colombia.

Figure 4.36 Effects of the 1991 Mt. Pinatubo eruption on climate.

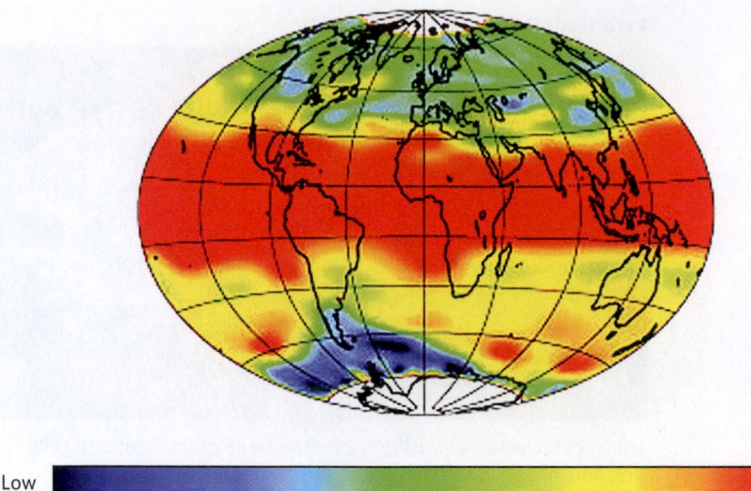

Low ▬▬▬▬▬▬▬▬▬▬▬▬ High

(a) Aerosols from the eruption encircled the globe. The colors indicate concentrations of aerosols in the atmosphere two months after the eruption.

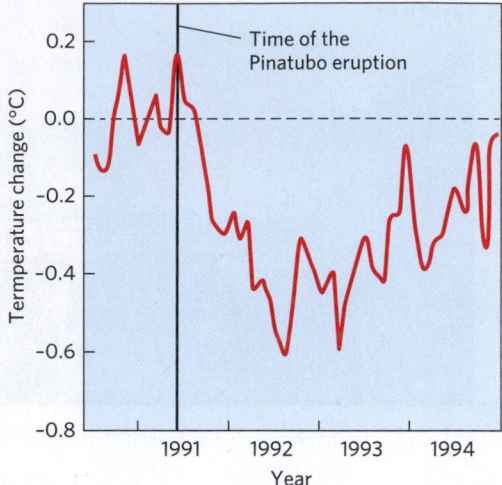

(b) Global air temperature decreased notably relative to average temperature (dashed line) for several years after the eruption.

Did you ever wonder . . .

whether volcanic eruptions affect air travel?

As we've noted, some volcanoes explode catastrophically (see Box 4.3). Explosions can flatten forests, bury the landscape with debris, generate tsunamis in the sea, trigger large landslides, and produce lahars. During an explosive eruption, large quantities of pyroclastic debris erupt into the air **(Fig. 4.35e–g)**. Close to the volcano, lapilli tumble from the sky, smashing through buildings or crushing them beneath a blanket up to several meters thick. Winds can carry fine ash over a broad region. In the Philippines, for example, a typhoon spread ash from the 1991 eruption of Mt. Pinatubo over a 4,000-km² (1,500 square miles) area. When this ash sifts down from the sky, it buries crops, poisons soil, and insidiously infiltrates machinery, causing moving parts to wear out. A pyroclastic flow, racing down the flank of a volcano, can be so hot and poisonous that it means instant death to anyone caught in its path. In addition, because it moves so fast, the force of its impact can flatten buildings and forests.

If ash from an eruption mixes with water, it can start to flow down river valleys in a viscous slurry called a lahar, as noted earlier **(Fig. 4.35h)**. Because lahars are denser than water, they pack more force than clear water flowing at the same velocity and can therefore carry away everything in their path, including bridges, boulders, and huge trees. Perhaps the most destructive lahar of recent times accompanied the eruption of a snow-crested volcano in Colombia in 1985. The lahar surged down the mountainside into a valley, burying the sleeping town of Armero, 60 km (40 miles) away, with a 5-m (16 foot)-thick layer of mud and entombing the town's 25,000 citizens.

The cloud of ash and gas that rises to stratospheric heights above a volcano is hazardous to airplanes. The gritty ash and acidic aerosols in the cloud scratch windows and damage the fuselage, and ash sucked into a jet engine melts, yielding a glassy coating that restricts air flow. As a consequence, temperature sensors indicate that the engine is overheating, making the engine shut down. This happened to a British Airways 747 that flew through the ash cloud over a volcano on Java in 1982. All four engines failed, and for 13 minutes, the plane glided earthward, dropping from 11.5 km (37,000 feet), while the pilots tried repeatedly to restart the engines. Finally, at 3.7 km (12,000 feet), the engines had cooled sufficiently and suddenly roared back to life. The plane headed for an emergency landing in Jakarta where, without functioning instruments, the pilot brought the 263 passengers and crew back to the ground safely. Because of safety concerns following the 2010 ash-rich eruption of Eyjafjallajökull in Iceland, all airspace in Europe was closed for days, stranding millions of passengers and disrupting the global economy.

Volcanoes and Climate

In 1783, Benjamin Franklin was serving as the American ambassador to France. The summer of that year seemed to be unusually cool and hazy. Franklin, who was an accomplished scientist as well as a statesman, couldn't resist seeking an explanation for this phenomenon, and learned that in June of 1783, a huge volcanic eruption had taken place in Iceland. He wondered if the "smoke" from the eruption had prevented sunlight from reaching the Earth, thus causing the cooler temperatures. Franklin reported this idea at a scientific meeting, and by doing so, may have been the first scientist to suggest a link between volcanic eruptions and climate.

Franklin's idea seemed to be confirmed in 1815, when Mt. Tambora in Indonesia exploded, making the sky so hazy that temperatures dipped by several degrees. It remained so cold in the northern hemisphere that 1816 became known as "the year without a summer." Studies of atmospheric conditions following the 1991 eruption of Mt. Pinatubo in the Philippines provide clear documentation of the short-term effects that an eruption can have on global atmospheric temperature (Fig. 4.36a, b).

How does volcanic activity cool the climate? When a large explosive eruption takes place, fine ash as well as aerosols (including microscopic droplets of sulfuric acid) enter the stratosphere. It takes only about two weeks for ash and aerosols to encircle the planet. Because this material occupies the layer of the atmosphere that lies above the weather, it does not get washed away by rainfall and stays suspended for as long as a few years. The haze it produces reflects light from the Sun back to space during the day, keeping it from reaching the Earth's surface, so the surface doesn't warm as much as it would if the eruption hadn't occurred.

Take-home message . . .

Volcanoes can be dangerous! Lava flows, pyroclastic flows, ash clouds, explosions, lahars, landslides, and tsunamis produced during eruptions can destroy cities and farmland. Ash in the air can be a hazard for air travel. Ash and aerosols thrown high into the atmosphere reflect sunlight and can cause a temporary global drop in temperature.

Quick Question -
Why can a lahar do so much more damage than a similarly sized flood of clear water?

4.9 Protecting Ourselves from Vulcan's Wrath

The Minoan culture thrived in the eastern Mediterranean beginning in 3650 B.C.E. It vanished during the century following explosive eruptions of the nearby Santorini volcano in 1645 B.C.E. All that remains of Santorini today is a huge caldera whose rim projects above sea level as the Greek island of Thera. Archeologists speculate that ash clouds, tsunamis, and earthquakes generated by Santorini may have disrupted the daily lives of the seafaring Minoans so much that survivors eventually deserted their elaborately decorated palaces and moved elsewhere. Clearly, volcanic eruptions are a natural hazard of extreme danger. Can anything be done to protect

Figure 4.37 The shape of a volcano changes as it erodes over time.

Time

An active volcano is a smooth cone.

Erosion carves gullies into the volcano.

Eventually, only hills of volcanic rock remain.

lives and property from this danger? The answer is yes. In this section, we examine evidence that geologists use to determine whether a volcano has the potential to erupt.

Active or Extinct?

The first step in determining whether a volcano may be a hazard is to determine whether it has the potential to erupt. Geologists refer to volcanoes that erupted during the last 10,000 years and might erupt in the future as **active volcanoes**. In contrast, **extinct volcanoes** are ones that have shut off entirely and can never erupt again because the geologic conditions leading to eruption no longer exist. Geologists use the term **dormant volcano** for an active volcano that is not currently erupting. Extinct volcanoes are not a worry—active volcanoes are.

To decide whether a volcano is active or extinct, it's necessary to determine when past eruptions took place. If a volcano erupted during historic time, it's probably still active. In regions where the historic record doesn't extend back more than a couple of centuries, the ages of eruptions can determined by using isotopic dating methods to obtain numerical ages for minerals in pyroclastic debris and lava or for wood buried by pyroclastic debris. The surface landscape of the volcano may also provide clues to its activity (Fig. 4.37). A volcano whose surface exposes lava or pyroclastic debris that has not been forested probably erupted recently, whereas a volcano whose surface has become forested and incised by streams has not erupted for a while. A volcano that has completely eroded away, so that underlying intrusions crop out, probably has become extinct.

Did you ever wonder . . .
whether geologists can predict volcanic eruptions?

Predicting Eruptions

From data on timing of the past several eruptions at a given active volcano, geologists can calculate the **recurrence interval** of activity at a volcano, meaning the average time between eruptions. Knowledge of the recurrence interval provides a rough estimate of the frequency of eruptions. If, for example, the recurrence interval is 50 years, then a volcano will likely erupt a few times over the course of a couple of centuries, but if the recurrence interval is 5,000 years, the likelihood of an eruption taking place in your lifetime is slim, though not impossible. Because the recurrence interval merely gives the average time between eruptions, it is not possible to predict the exact timing of an eruption years in the future. However, short-term (weeks to months) predictions of impending volcanic activity may be feasible. Some volcanoes send out distinct warning signals of impending eruptions that geologists can monitor:

- *Earthquake activity:* When magma flows into a volcano, rocks surrounding the magma chamber break and slip, and small explosions may take place in the magma chamber. All this movement causes earthquakes, so in the days or weeks preceding an eruption, the region beneath a volcano becomes seismically active.

- *Changes in heat flow:* The intrusion of hot magma into a volcano increases local *heat flow*, the amount of heat passing through rock. In some cases, the increase in heat flow melts snow or ice on the volcano.

- *Increases in gas and steam emission:* Even though magma remains below the surface, gases bubbling out of the magma, or steam formed by the heating of groundwater by the magma, percolate upward through cracks in the Earth and rise from the volcanic vent **(Fig. 4.38a)**. So an increase in the volume of gas emission, or the formation of new hot springs, indicates that magma has entered the ground below.

- *Changes in shape:* As magma fills the magma chamber inside the volcano, it pushes outward and can cause the surface of the volcano to bulge **(Fig. 4.38b)**.

Mitigating Volcanic Hazards

Let's say we determine that a given active volcano has the potential to erupt in the near future. There's no way to stop an eruption, so what can we do to prevent the loss

Figure 4.38 Monitoring the activity of volcanoes.

(a) Monitoring volcanic activity is a risky business, for the gases can be deadly.

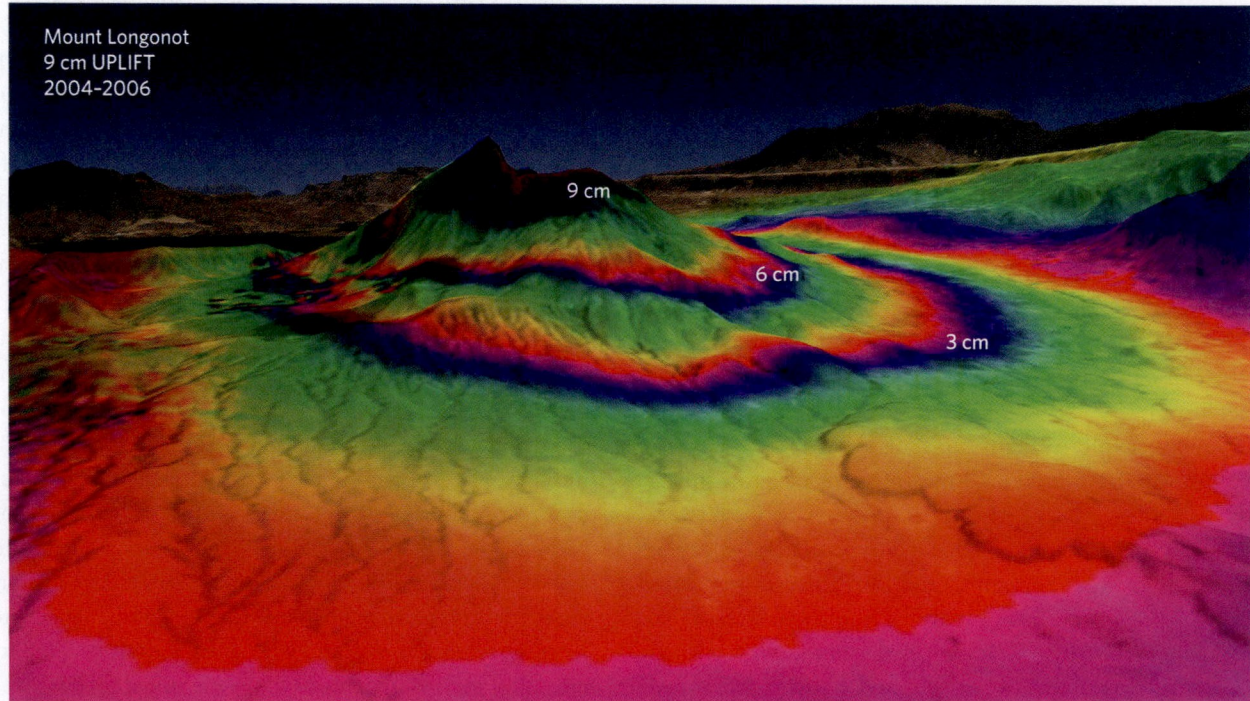

Mount Longonot
9 cm UPLIFT
2004–2006

9 cm

6 cm

3 cm

(b) Monitoring ground movements that may reflect intrusion of magma below. Each color-spectrum band represents 3 cm (1 inch) of uplift. The measurements were made by satellite radar measurements.

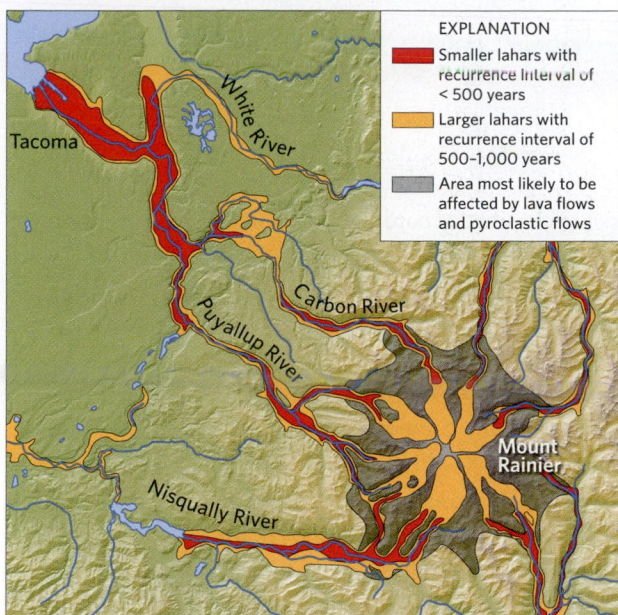

EXPLANATION

🟥 Smaller lahars with recurrence interval of < 500 years

🟧 Larger lahars with recurrence interval of 500–1,000 years

⬛ Area most likely to be affected by lava flows and pyroclastic flows

of Mt. Etna, an active volcano on the island of Sicily, a basaltic lava flow approached the town of Catania. Fifty townspeople boldly hacked through the solidified side of the flow to create an opening through which lava could exit. They hoped to cut off the supply of lava feeding the end of the flow that was approaching their homes. Their strategy worked, but unfortunately, the diverted flow began to move toward the neighboring town of Paterno. Five hundred Paternoans chased away the Catanians. The hole was closed, and the flow headed back toward Catania, eventually overrunning part of the town.

More recently, people have used high explosives to blast breaches in the flanks of lava flows and have employed bulldozers to build dams and channels to divert lava. Major efforts to divert flows from a 1983 eruption of Mt. Etna, and from another in 1992, were successful (Fig. 4.40a). Inhabitants of Iceland used a particularly creative approach in 1973 to stop a flow before it overran a town: they sprayed cold seawater onto the flow to freeze it in its tracks (Fig. 4.40b).

of life and property? As an important first step, geologists compile a volcanic **hazard assessment map** (Fig. 4.39), which delineates areas that lie in the path of potential lava flows, lahars, debris flows, or pyroclastic flows. If an eruption is imminent, people within these danger zones should be evacuated. Unfortunately, because of the uncertainty of predicting eruptions, the decision about whether or not to evacuate can be difficult.

In a very few cases, people have tried to divert or stop a lava flow. For example, during a 1669 eruption

Take-home message . . .

🏠 Geologists distinguish among active and extinct volcanoes, and they can provide near-term predictions of eruptions, allowing people to take precautions. Hazard maps help determine who should be evacuated. Rarely, it's possible to divert lava flows.

Quick Question -
What evidence suggests that a volcano may be about to erupt?

Figure 4.40 Efforts to divert lava flows away from inhabited locations.

(a) Workers spray a lava flow to solidify it, and use a bulldozer to build an embankment to divert it, on the flanks of Mt. Etna.

(b) Firefighters pumped 6 million m³ (1.5 billion gallons) of seawater on a lava flow in Iceland in an effort to freeze it and stop it.

4 CHAPTER REVIEW

Chapter Summary

- Volcanoes are vents from which molten rock (lava) erupts at the Earth's surface or are built of erupted material.

- Magma is liquid rock (melt) under the Earth's surface, whereas lava is melt that has erupted from a volcano at the Earth's surface.

- Magma forms when hot rock in the Earth melts. Melting occurs due to decompression, addition of volatiles, or heat transfer.

- Only a small proportion of the magma source melts to form magma. Magmas tend to contain more silica than their source rock.

- Magma composition reflects the composition of the magma source and the way the magma evolves.

- Magma rises because of its buoyancy and because of pressure caused by the weight of overlying rock.

- Intrusive igneous rocks form when magma intrudes pre-existing rock below the Earth's surface. Extrusive igneous rocks form from lava or pyroclastic debris that erupts from a volcano.

- The rate at which magma cools depends on the depth, size, and shape of the intrusion, and groundwater circulation. The cooling rate influences the texture of igneous rock.

- Tabular intrusions include dikes and sills. Blob-shaped intrusions are called plutons. Huge intrusions are batholiths.

- Lava may solidify to form flows, or it may explode into the air to form ash, lapilli, and other pyroclastic debris.

- The viscosity of lava depends largely on its composition. Viscous felsic lava flows tend to pile into domes at a volcano's vent, whereas less viscous mafic lava can flow great distances.

- Igneous rocks are classified according to texture and composition. Some igneous rocks consist of glass or of pyroclastic debris.

- Some volcanic eruptions are effusive (dominated by lava flows or fountains), whereas some are explosive.

- A volcano's shape depends on its eruptive style. Shield volcanoes are broad, gentle domes formed by effusive eruptions. Cinder cones are symmetrical hills of lapilli. Stratovolcanoes can become large and consist of alternating layers of pyroclastic debris and lava.

- Eruptions may occur in a crater at a volcano's summit or from fissures on its flanks. Collapse of a volcano produces a huge, bowl-shaped depression called a caldera.

- Plate tectonics theory can explain where igneous activity happens. Magma forms at convergent boundaries due to flux melting, and at divergent boundaries, rifts, and hot spots due to decompression melting. In continental crust, heat-transfer melting can take place.

- Volcanic eruptions pose many hazards: lava flows overrun roads and towns, ash falls blanket the landscape, pyroclastic flows incinerate everything in their path, and landslides and lahars bury the land surface.

- We can distinguish between active and extinct volcanoes based on their eruption history. Imminent eruptions can be predicted by earthquake activity, changes in heat flow, changes in shape of the volcano, and the emission of gas and steam.

Key Terms

a'a' (p. 127)
active volcano (p. 149)
aerosol (p. 119)
ash (p. 128)
assimilation (p. 120)
batholith (p. 125)
black smoker (p. 141)
block (p. 128)
Bowen's reaction series (p. 120)
caldera (p. 141)
cinder cone (p. 137)
columnar jointing (p. 122)
crater (p. 137)
crystalline igneous rock (p. 130)
decompression melting (p. 117)
dike (p. 123)

dormant volcano (p. 149)
effusive eruption (p. 133)
eruptive style (p. 133)
explosive eruption (p. 133)
extinct volcano (p. 149)
extrusive igneous rock (p. 119)
felsic magma (p. 118)
fissure eruption (p. 137)
flood basalt (p. 145)
flux melting (p. 118)
fractional crystallization (p. 120)
fragmental igneous rock (p. 131)
geotherm (p. 117)
glassy igneous rock (p. 131)
hazard assessment map (p. 151)
heat-transfer melting (p. 118)

igneous activity (p. 116)
igneous rock (p. 116)
intermediate magma (p. 118)
intrusive igneous rock (p. 121)
laccolith (p. 123)
lahar (p. 129)
lapilli (p. 128)
large igneous province (LIP) (p. 145)
lava (p. 116)
lava dome (p. 127)
lava flow (p. 125)
lava fountain (p. 128)
lava tube (p. 127)
mafic magma (p. 118)
magma (p. 116)

magma chamber (p. 125)
melt (p. 115)
obsidian (p. 131)
pahoehoe (p. 127)
partial melting (p. 118)
pillow basalt (p. 127)
pluton (p. 123)
pumice (p. 131)
pyroclastic debris (p. 128)
pyroclastic flow (p. 132)
pyroclastic rock (p. 131)
recurrence interval (p. 150)
scoria (p. 131)
shield volcano (p. 137)
sill (p. 123)
stratovolcano (p. 141)

Review Questions

The letters following each Review Question refer to the corresponding Learning Objective from the Chapter Opener.

1. Describe the three processes that are responsible for the formation of magma. **(A)**

2. Why are there so many different compositions of magma? Does partial melting produce magma with the same composition as the magma source from which it was derived? **(C)**

3. Why does magma rise from depth to the surface of the Earth? **(B)**

4. Explain the process of fractional crystallization. **(C)**

5. What factors control the viscosity of a melt, and how does viscosity affect the behavior of magma or lava? **(F)**

6. What factors control the cooling rate of a magma? **(D)**

7. What is the difference between a sill and a dike, and how do both differ from a pluton or batholith? Identify the intrusions shown. **(E)**

8. How does grain size reflect the cooling rate of a magma? **(E)**

9. Why do magmas form in association with subduction? **(A)**

10. Why does melting take place beneath the axis of a mid-ocean ridge? **(A)**

11. What process in the mantle may be responsible for causing hot-spot volcanoes to form? **(B)**

12. Describe how magmas are produced at continental rifts. Why can you find both basalt and rhyolite in such settings? **(B)**

13. What is a large igneous province (LIP), and how might it form? **(E)**

14. Describe three different kinds of material that can erupt from a volcano. Identify them on the figure. **(F)**

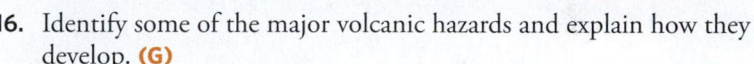

15. Describe the differences among shield volcanoes, stratovolcanoes, and cinder cones. How do the composition of their lavas and other factors explain these differences? **(F)**

16. Identify some of the major volcanic hazards and explain how they develop. **(G)**

17. To what extent can geologists predict volcanic eruptions, and what observations provide the basis for a prediction? **(H)**

18. Explain how steps can be taken to protect people from the effects of eruptions. **(H)**

On Further Thought

19. The Cascade Range of the northwestern United States is only about 800 km (500 miles) long. The volcanic chain of the Andes is several thousand kilometers long. Look at a map showing the Earth's plate boundaries and explain why the Andes volcanic chain is so much longer than the Cascade volcanic chain. **(B)**

20. Do people living near the volcanoes of Hawaii face the same kind of volcanic hazards as do people living near Mt. Rainier in the northwestern United States? **(G)**

Online Resources

Animations
This chapter features animations covering the various types of volcanoes and lava flows, lava composition, plate boundaries, and the formation and activity of hot spots.

Videos
This chapter features a video on the topic of partial melting.

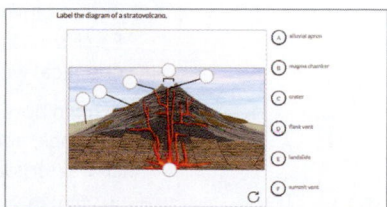

Smartwork5
Features include visual and labeling exercises on the processes involved in magma formation, the creation of various igneous rocks, and volcanic activity.

5 A SURFACE VENEER
Sediments and Sedimentary Rocks

By the end of the chapter you should be able to . . .

A. understand how weathering produces sediment at the Earth's surface, and distinguish between physical and chemical weathering.

B. explain how soil forms, how it differs from sediment and rock, and how environmental factors control its characteristics.

C. describe how sediment transforms into sedimentary rock, list the key types of sedimentary rocks, and explain the basis for classifying sedimentary rock.

D. relate examples of sedimentary structures to the environments in which they form.

E. apply your knowledge of how depositional environment controls the composition, textures, and layering of sedimentary rock to interpret an outcrop or description of sedimentary rock.

F. explain the concept of a sedimentary basin, and explain why transgression and regression occur.

5.1 Introduction

What lies beneath the floor of the Mediterranean Sea? No one really knew until the 1970s, when researchers drilled a 1-km (0.6-mile)-deep hole into the sea's floor **(Fig. 5.1a)**. They expected to find layers of clay, sand, and gravel carried into the sea by rivers, as well as layers of plankton shells that settled out of seawater when the plankton died. The researchers did find these layers, but they also found thick layers of halite and gypsum, salts that crystallize when seawater dries up **(Fig. 5.1b)**. The presence of these salts puzzled the researchers because to form them, the Mediterranean would have had to evaporate almost completely multiple times. Could that really have happened?

The answer proved to be yes. When the African Plate, which was moving north relative to the Eurasian Plate,

started to collide with the Eurasian Plate several million years ago, land blocked off the Straits of Gibraltar, the only opening between the Atlantic Ocean and the Mediterranean. Since rivers supply only 10% of the water needed to keep the Mediterranean full, cutting off inflow from the Atlantic indeed caused the Mediterranean to dry up until it became a 2-km (1.2-mile)-deep *basin* (a low area) whose surface was a salt-encrusted desert **(Fig. 5.1c)**. This basin remained dry until sea level in the Atlantic rose sufficiently for water to spill over the land bridge between Africa and Europe, causing the Straits of Gibraltar to become a gigantic waterfall that refilled the Mediterranean. The process of filling and drying up, in the Mediterranean, repeated several times, for sea level rises and falls over time.

Geologists refer to materials such as those found beneath the floor of the Mediterranean—clay, gravel,

This highway in Utah descends through layer upon layer of sedimentary strata, formed from sediment that was originally deposited as mud, sand, and gravel on the surface of the Earth. After being buried, the sediment turned into sedimentary rock.

Figure 5.1 Evidence that the Mediterranean Sea was once dry.

The *Glomar Challenger*

(a) The *Glomar Challenger* drilling ship drilled a 1-km-deep hole into the sea's floor.

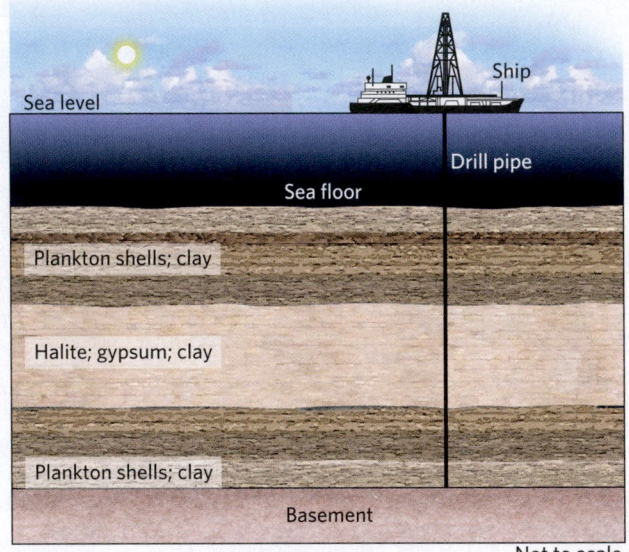

Sea level

Ship

Drill pipe

Sea floor

Plankton shells; clay

Halite; gypsum; clay

Plankton shells; clay

Basement

Not to scale

(b) A cross section showing the sediment beneath the floor of the Mediterranean Sea. The vertical line represents the drill hole.

Salt basin

Europe

Atlantic Ocean

Mediterranean basin

Future Strait of Gibraltar

(c) The Mediterranean basin at 6 Ma. The deepest parts of the basin were saline lakes; the rest of the basin was a desert.

Table 5.1 Names for clasts of different diameters

Clast Name	Diameter
Boulder	More than 256 millimeters (mm)
Cobble	Between 64 mm and 256 mm
Pebble	Between 2 mm and 64 mm
Sand	Between $\frac{1}{16}$ mm and 2 mm
Silt	Between $\frac{1}{256}$ mm and $\frac{1}{16}$ mm
Clay	Less than $\frac{1}{256}$ mm

256 mm = 10 inches; 64 mm = 2.5 inches; 2 mm = 0.0787 inches; 1/16 mm = 0.0025 inches; 1/256 mm = 0.00015 inches.

sand, shell accumulations, and salt—as sediment. More precisely, **sediment** consists of one or more of the following: loose fragments of rocks or minerals; shells and shell fragments; or mineral crystals that precipitate directly from water. A loose fragment of any of these materials is called a **clast** (from the Greek *klastos*, broken). Clasts come in a variety of sizes, ranging from too small to see, even with a microscope, to car-sized, or even larger. Each clast size category has a name **(Table 5.1)**. The term **grain** can be used as a synonym for *clast*, as well as for a single crystal of a mineral.

The researchers who drilled into the Mediterranean seafloor also noted that the character of seafloor sediments changes progressively with depth. In the upper few hundred meters, sediment is *unconsolidated*, meaning that it easily separates into grains. Deeper down, sediment is *consolidated*, meaning that its grains are packed together. Deeper still, minerals precipitated from water fill gaps between the grains and hold the grains together. As a consequence, the sediment has transformed into solid *sedimentary rock*.

Like pages from a history book, sediments and sedimentary rocks record the story of Earth's past. To help you understand how to read this record, we begin this chapter by discussing the production of sediment by *weathering*, a process that breaks down pre-existing rock to form clasts and dissolved ions. Next, we discuss soil, the life-hosting layer that develops due to the interaction of air, rain, and living organisms with sediment and rock. Then we turn our attention to the formation of sedimentary rock and show how the study of sedimentary rock can help us to characterize environmental conditions in the past. Finally, we answer the question of why thick accumulations of sediment occur by introducing sedimentary basins, regions of the crust that slowly sink over time.

Figure 5.2 Evidence of weathering.

(a) A statue that shows the effects of weathering.

(b) This outcrop shows the contrast between weathered and fresh granite. Note the hammer for scale. The white dashed line is the boundary between weathered and fresh.

5.2 Weathering and the Formation of Sediment

Have you ever seen an old marble statue that has been standing outside for a long time? Look closely and you'll notice that the details the sculptor worked so hard to create have been blunted and that cracks have formed in the stone (Fig. 5.2a). The statue is undergoing **weathering**, the combination of processes that gradually break up or chemically change and weaken rock exposed to air and water at or just below the Earth's surface. We refer to the cracking and breaking up of rock as **physical weathering** (or *mechanical weathering*) and to the chemical reaction of rock with air and water as **chemical weathering**. We'll discuss these two aspects of weathering in turn to see how weathering ultimately causes *fresh rock*, meaning unweathered rock, to disintegrate and turn into new sediment (Fig. 5.2b). Weathering makes rock susceptible to **erosion**, the process by which solid material is ground away by glaciers, rivers, wind, or waves.

Prying Rock Apart: The Process of Physical Weathering

The process of physical weathering begins with the formation of natural cracks, or **joints**, in rock. Joints develop for many reasons. For example, when erosion removes *overburden* (overlying rock), rock that was once many kilometers below the surface of the Earth rises to a shallower depth. In the process, the rock undergoes slight expansion and cooling because pressures and temperatures progressively decrease toward the surface. The resulting change in shape can cause the rock to crack. (Chapter 7 describes additional reasons for joint formation.)

Almost all rock outcrops contain joints, which are generally spaced from centimeters to meters apart (Fig. 5.3a, b). In sedimentary rocks, some joints may be planar and perpendicular to bedding, so that the rock breaks into rectangular blocks. In bodies of homogeneous crystalline rock, such as granite, *exfoliation joints* split the rock into onion-like sheets oriented roughly parallel to the ground surface. Joints may also be irregular surfaces that

Figure 5.3 Jointing breaks rocks into blocks or sheets that can separate from an outcrop.

100-m-high sandstone cliff, Ireland.

(a) Vertical joints break beds of sedimentary rock into blocks that fall to the base of this 100-m-high cliff.

Joints in granite, California.

(b) Exfoliation joints on this granite hillslope break the rock into onion-like sheets.

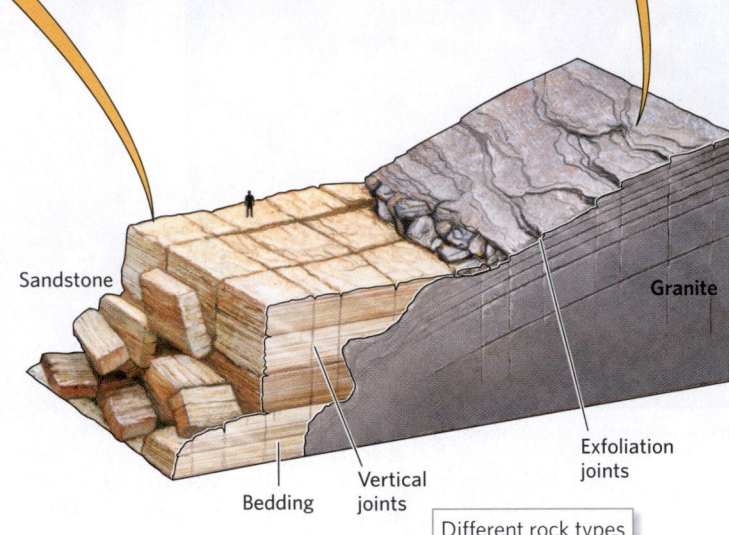

Sandstone

Granite

Bedding

Vertical joints

Exfoliation joints

Different rock types display different types of joints.

Figure 5.4 Wedging is one type of physical weathering.

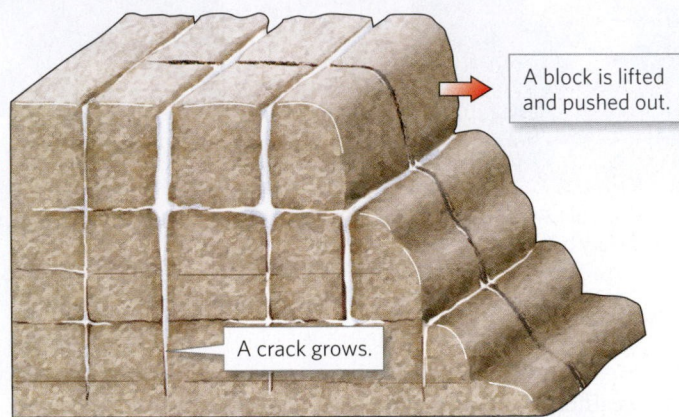

A block is lifted and pushed out.

A crack grows.

(a) When the water that fills cracks freezes, it expands and wedges the cracks open.

Tree growing in a joint.

(b) Root wedging pushes open a joint, slowly separating a block from the cliff.

Eventually, the blocks tumble to the base of the cliff.

(c) Salt wedging led to the disintegration of these gravestones in Whitby, England.

break rocks into jagged pieces. A number of phenomena contribute to the widening and lengthening of joints. For example, when the water trapped in a joint freezes and expands, it forces the joint open, a process called **frost wedging** (Fig. 5.4a). A tree root growing in a joint can also push it open, a process known as **root wedging** (Fig. 5.4b). Development of joints eventually transforms formerly intact bedrock into separate blocks. These blocks separate from the bedrock and tumble downslope, commonly breaking into smaller blocks as they fall.

Physical weathering can also act at the grain scale. For example, in dry climates or along the seashore, salty water seeps into rock. As the water evaporates, the salt precipitates as crystals in the tiny open spaces, or *pores*, within the rock. Growth of salt crystals causes **salt wedging**, by pushing apart the surrounding grains. This process weakens the rock so that when it's exposed to wind and rain, it breaks up into separate grains (Fig. 5.4c).

Chemical Weathering

During chemical weathering, molecules in air and water react with minerals in rock by breaking chemical bonds. Examples of such reactions include:

- *Dissolution:* During **dissolution**, minerals in a water solution break down into ions. This reaction primarily affects relatively soluble salts and carbonate minerals (Fig. 5.5a, b), but even quartz grains eventually dissolve slightly.

- *Hydrolysis:* During **hydrolysis**, reactions with water break down existing minerals and produce new minerals. For example, hydrolysis transforms feldspar and many other silicate minerals into clay.

- *Oxidation:* Reactions of iron-bearing minerals with atmospheric oxygen or with oxygen dissolved in water cause **oxidation**, or "rusting," of rock, yielding a reddish-brown mixture of weak iron oxide and iron hydroxide minerals.

- *Hydration:* Hydration happens when water molecules form bonds within mineral crystals. Bonds between water molecules and other compounds tend to be weaker than other types of chemical bonds, so hydration weakens minerals.

Recent research emphasizes that organisms play a major role in the chemical weathering process in certain environments. For example, the roots of plants, fungi, and lichens secrete organic acids that help dissolve minerals in rocks. And some bacteria and archaea literally eat minerals for lunch—they pluck molecules from minerals and use the energy from chemical bonds to support their metabolism.

Physical and Chemical Weathering Working Together

So far, we've looked at the processes of chemical and physical weathering separately. In the natural world, however, physical and chemical weathering generally work together. Physical weathering speeds up chemical weathering because, by breaking up rock, it provides more surface area that can come in contact with air or water (Fig. 5.6a). In turn, chemical weathering speeds up physical weathering, either by dissolving cements that hold rock grains together or by transforming hard minerals such as feldspar

Figure 5.5 Dissolution is a form of chemical weathering.

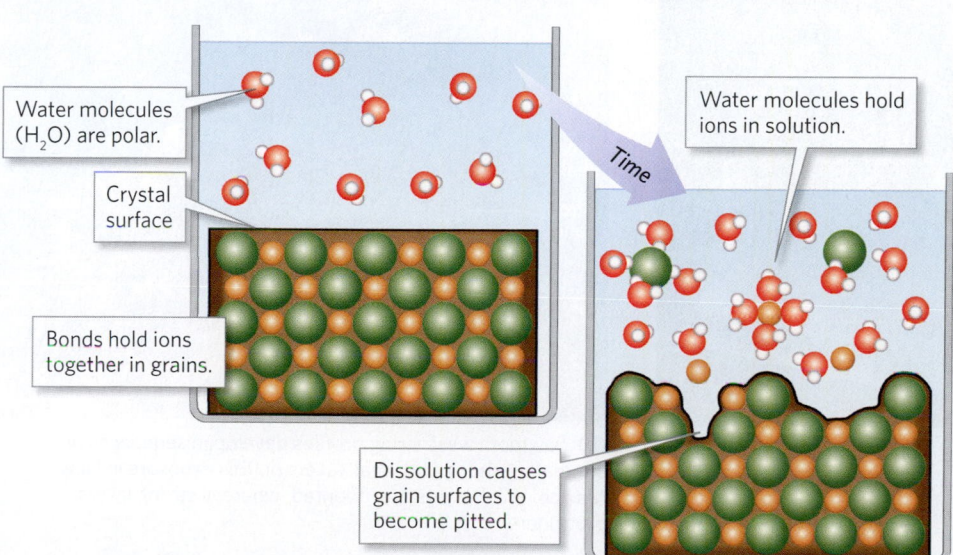

Water molecules (H₂O) are polar.

Crystal surface

Bonds hold ions together in grains.

Time

Water molecules hold ions in solution.

Dissolution causes grain surfaces to become pitted.

(a) Dissolution occurs when water molecules pluck ions from grain surfaces.

Water seeping into joints in limestone produced troughs.

(b) Dissolution along joints intersecting the surface of limestone bedrock in Ireland produced these troughs.

Figure 5.6 Physical and chemical weathering processes work together.

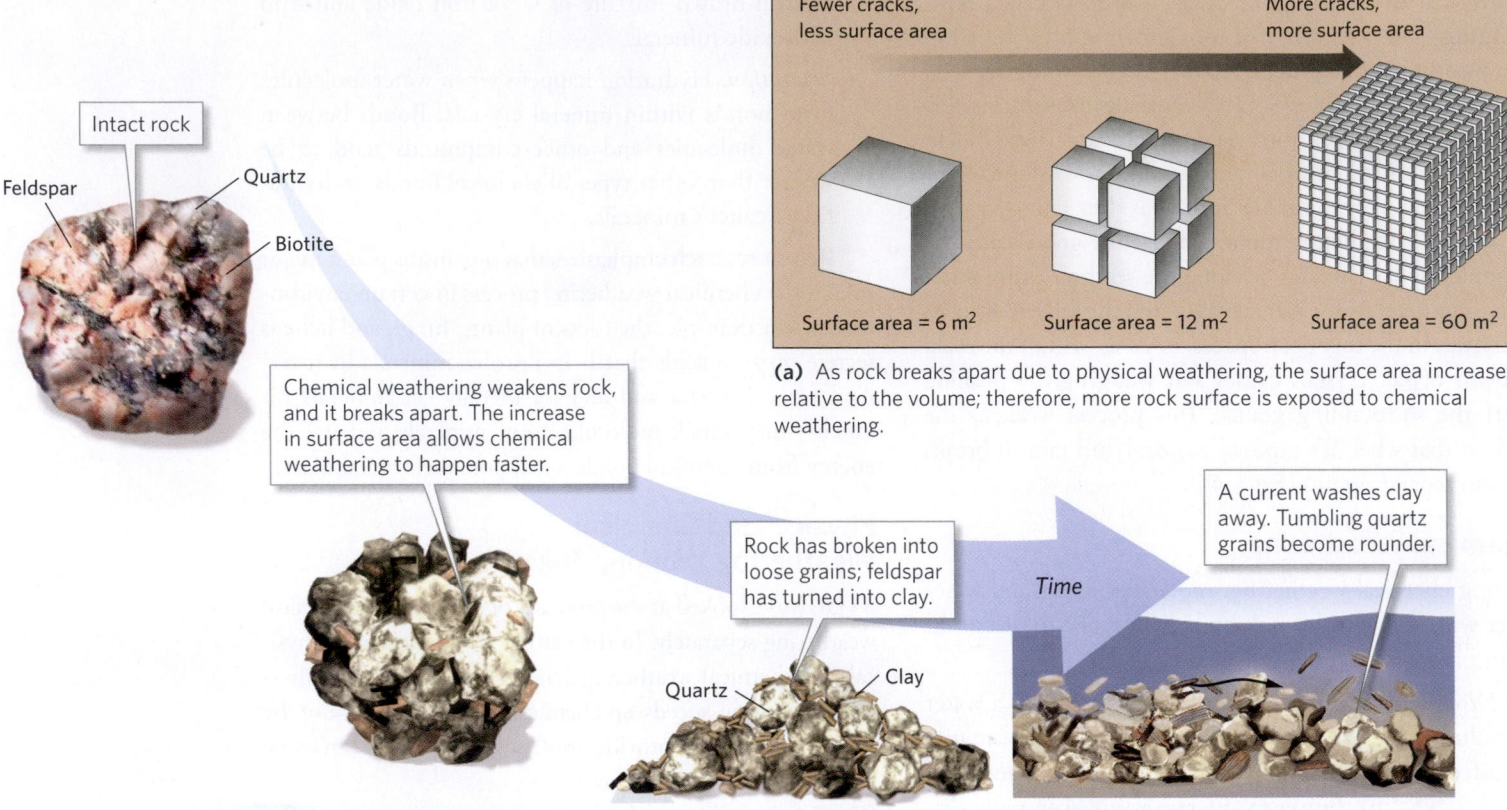

Intact rock

Feldspar

Quartz

Biotite

Fewer cracks,
less surface area

More cracks,
more surface area

Surface area = 6 m²   Surface area = 12 m²   Surface area = 60 m²

(a) As rock breaks apart due to physical weathering, the surface area increases relative to the volume; therefore, more rock surface is exposed to chemical weathering.

Chemical weathering weakens rock, and it breaks apart. The increase in surface area allows chemical weathering to happen faster.

Rock has broken into loose grains; feldspar has turned into clay.

Time

A current washes clay away. Tumbling quartz grains become rounder.

Quartz   Clay

(b) Chemical weathering weakens rock, so it breaks apart. As this happens, the surface area increases, so chemical weathering happens still faster. Eventually, the rock completely disaggregates to form sediment. Weathering of granite produces quartz sand and clay.

Figure 5.7 Differential weathering.

Weak shale

Strong sandstone

(a) Inscriptions on a granite headstone (left) last for centuries, but those on a marble headstone (right) may weather away in decades. These gravestones are in the same cemetery and are about the same age.

(b) Sawtooth weathering profiles develop in sequences of alternating strong and weak layers on this exposure in New Mexico. Weak layers are indented, whereas strong layers protrude.

Figure 5.8 Weathering may result in rounded forms.

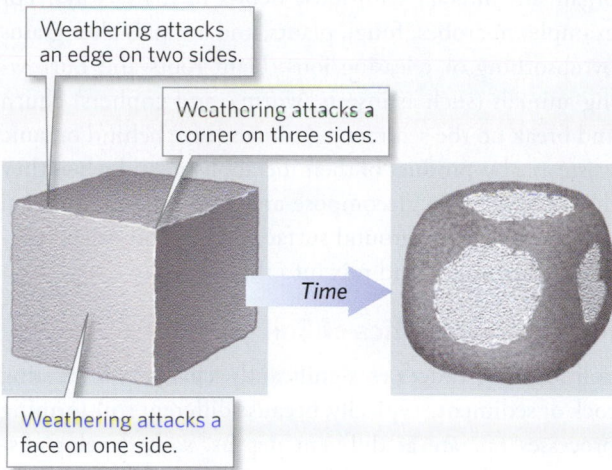

Weathering attacks an edge on two sides.

Weathering attacks a corner on three sides.

Time

Weathering attacks a face on one side.

(a) Weathering attacks a block of rock more vigorously at the edges than at the flat faces, and most vigorously at the corners. Thus, homogeneous rocks tend to weather into rounded blocks.

(b) Weathering along cracks in this granite in the Mojave Desert led to the formation of rounded blocks.

into soft minerals such as clay. Both processes weaken rock, allowing it to disintegrate more easily **(Fig. 5.6b)**.

Not all rock types have the same composition or contain the same number of joints, so various rock types weather at different rates, a phenomenon known as *differential weathering*. We say that rocks that weather slowly are *resistant* to weathering, whereas rocks that weather quickly are *nonresistant*. You can easily see the consequences of differential weathering in a graveyard. Inscriptions on granite headstones remain sharp and clear for centuries, whereas those on marble headstones become blunted within decades **(Fig. 5.7a)**. That's because granite contains lots of quartz and other durable minerals, whereas marble consists of calcite, which dissolves relatively easily in acidic rainwater. As a result of differential weathering, cliffs composed of alternating resistant and nonresistant rock layers take on a stair-step or sawtooth shape **(Fig. 5.7b)**. Furthermore, weathering attacks a flat rock face from only one direction, an edge from two directions, and a corner from three directions. Therefore, weathering happens faster at edges and even faster at corners, so over time, edges of blocks become blunt and corners become rounded **(Fig. 5.8a, b)**.

Take-home message . . .

During physical weathering, rock breaks into pieces. The process commonly involves the formation of joints. During chemical weathering, air and water chemically react with minerals in rock, causing them to dissolve, rust, or transform into softer minerals. Working together, chemical and physical weathering cause rock to disintegrate and transform into sediment.

Quick Question -
Why do the corners of blocks tend to become blunted over time?

5.3 Soil and its Formation

If you've ever had the chance to dig in a garden, field, or forest floor, you've seen firsthand that the soil in which flowers, crops, and trees grow looks and feels different from rock, beach sand, or potter's clay. **Soil** consists of sediment that has been modified over time by physical and chemical interaction with rainwater, air, organisms, and decaying organic matter. Let's look at how nature produces soil, why not all soils look the same, and why soil is a resource under threat.

How Does Soil Form?

Soil formation, which takes place at or just below the Earth's surface, represents an interaction among several components (lithosphere, hydrosphere, atmosphere, and biosphere) of the Earth System. Soil can form either from weathered rock or from a pre-existing sediment deposit.

Three key processes contribute to soil formation. First, chemical and physical weathering break up pre-existing material (rock or sediment), forming loose grains. Second, rainwater falls on the accumulation of grains, sinks in, and slowly percolates downward. As water moves down, it dissolves ions and picks up clay flakes and carries them along **(Box 5.1)**. The region from which downward transport removes ions and clay is the **zone of leaching**. Farther underground, new mineral crystals precipitate from the downward-percolating water, and the clay that was leached from above gets left behind. The

Figure 5.9 Formation of soil.

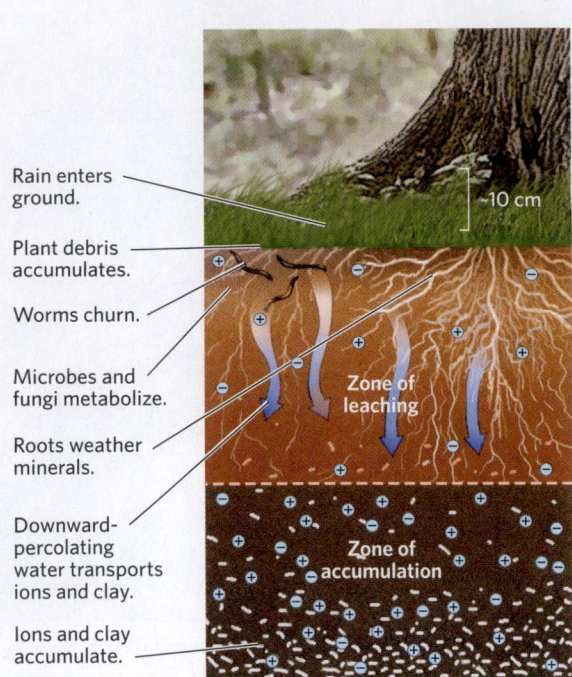

Rain enters ground.

Plant debris accumulates.

Worms churn.

Microbes and fungi metabolize.

Roots weather minerals.

Downward-percolating water transports ions and clay.

Ions and clay accumulate.

~10 cm

Zone of leaching

Zone of accumulation

(a) Soil formation involves many processes, including water percolation and leaching as well as interaction with organisms and organic matter.

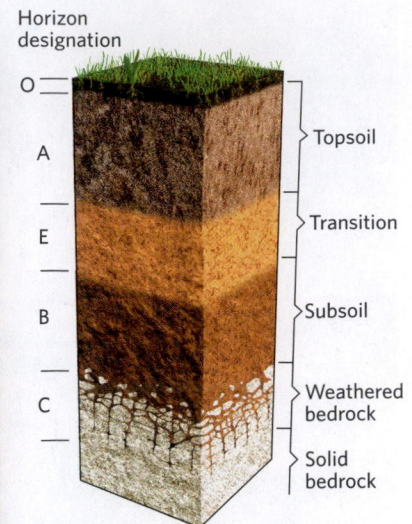

Horizon designation

O
A
E
B
C

Topsoil
Transition
Subsoil
Weathered bedrock
Solid bedrock

(b) Soils develop distinct horizons. This example is from a temperate forest.

region in which new minerals grow and transported clay collects is the **zone of accumulation** (Fig. 5.9a). Third, organisms interact with loose debris in many ways. For example, microbes, fungi, plants, and animals alter grains by absorbing or releasing ions. Plant roots and burrowing animals (such as insects, worms, and gophers) churn and break up the sediment. Animals leave behind organic waste as a by-product of their metabolism, and when they die, their remains decompose and mix into the mineral debris. And at the ground surface, leaves and other vegetation decompose and mix into the debris.

Key Characteristics of Soil

Soil-forming processes significantly change pre-existing rock or sediment. Typically, because different soil-forming processes operate at different depths, soils develop distinct zones, known as **soil horizons**, arranged in a vertical sequence, collectively called a **soil profile**.

Let's look at an example of a soil profile formed in a temperate forest (Fig. 5.9b). The highest horizon, the O-horizon (the prefix stands for *organic*), consists mostly of *humus* (decayed plant debris). Below the O-horizon, we find the A-horizon, in which humus has decayed further and has mixed with mineral grains (clay, silt, and sand). In some locations, you may find an E-horizon

Figure 5.10 Factors that control the character of soil.

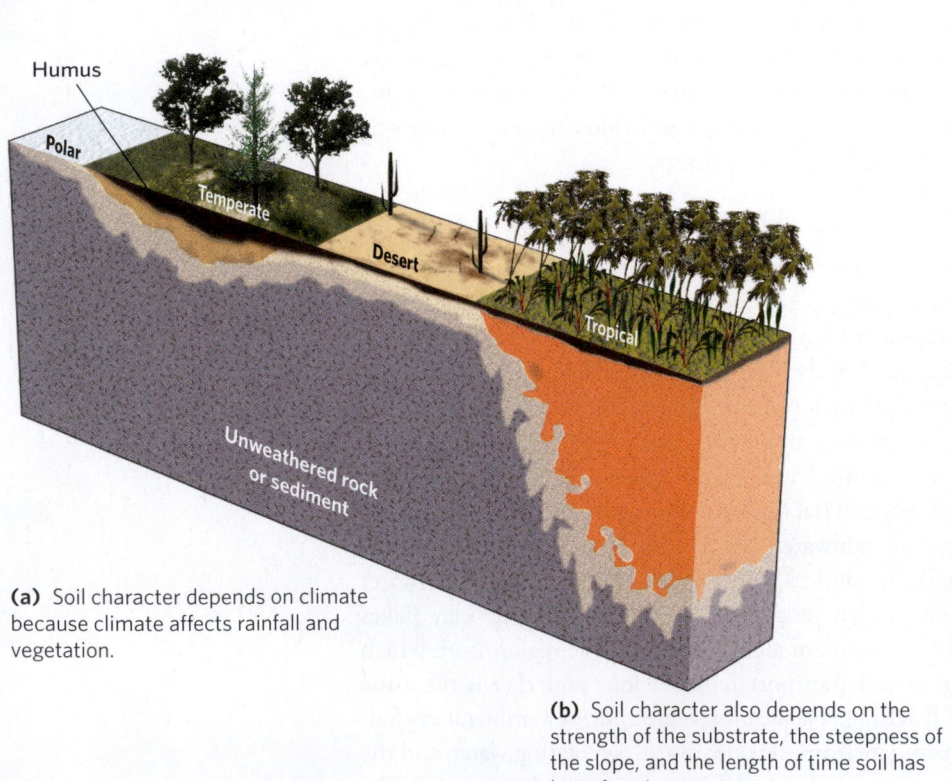

Humus
Polar
Temperate
Desert
Tropical
Unweathered rock or sediment

(a) Soil character depends on climate because climate affects rainfall and vegetation.

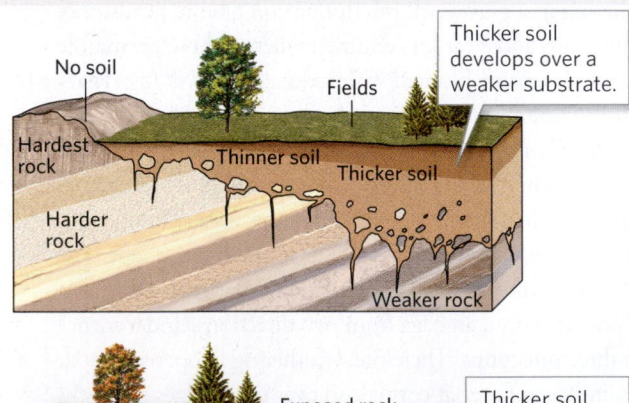

No soil
Hardest rock
Harder rock
Thinner soil
Fields
Thicker soil
Weaker rock

Thicker soil develops over a weaker substrate.

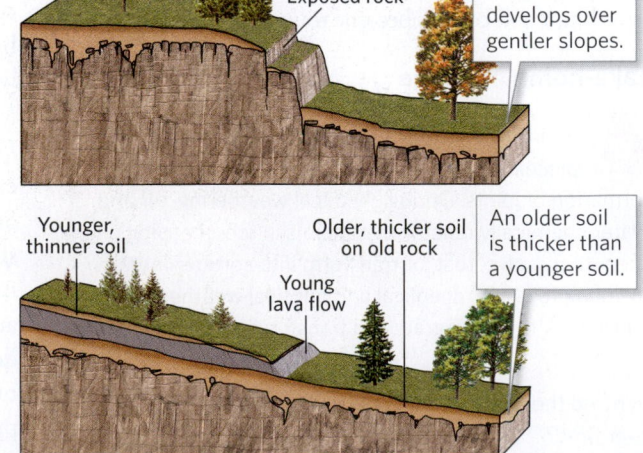

Exposed rock

Thicker soil develops over gentler slopes.

Younger, thinner soil
Older, thicker soil on old rock
Young lava flow

An older soil is thicker than a younger soil.

(b) Soil character also depends on the strength of the substrate, the steepness of the slope, and the length of time soil has been forming.

Box 5.1

How can I explain . . .

Soil formation

What are we learning?
How leaching and addition of organic matter changes sediment into soil.

What you need:
Pint-sized (16 fl oz) disposable thin plastic cup; mixing bowl; sand; crushed antacid tablets; green-colored clay powder (available online); fine gravel; cheesecloth; coarsely ground coffee; tablespoon; plate; tape; rubber band; plastic sandwich bag; tray; utility scissors.

Instructions:
- Mix 20 tablespoons (10 fl oz equivalent) of sand with 6 tablespoons (3 fl oz) of clay, 2 tablespoons (1 fl oz) of crushed antacid tablets, and 2 tablespoons (1 fl oz) of fine gravel.
- Cut off the bottom of the cup and cover it with cheesecloth, held in place by tape. Place the base of the cup on a plate and pour in the sediment mixture. Stretch the rubber band around the cheesecloth to ensure that the cheesecloth can hold the weight of the sediment.
- Hold the cup over a sink and slowly pour 2 to 3 cups of water onto the top of the sediment. Have a friend stir coffee (a total of 2 tablespoons) into the top inch (2 cm) of the sediment as you pour the water. As you pour, water should drain out the bottom. Pour slowly enough so that the sediment does not become saturated.
- Press the plastic bag against the top of the cup and lay the cup on its side on the tray.

- Cut the cup lengthwise with the scissors and spread it open to see the material within. Describe how the material varies from top to bottom. How is it different from the original mixture?

What did we see?
Soil formation involves water percolation down through sediment and addition of organic matter. Water leaches soluble materials (antacid powder) and moves fine clay. Insoluble and coarser material (sand and gravel) remain behind. Organic matter mixes in at the surface. You should see a change in the character of the sediment from top to bottom, and you may see horizons. You could check for changes in the concentration of antacid by dropping dilute acid onto the sediment and seeing if it fizzes as much on top as on the bottom.

Plastic cup
Coffee
Sand with some green clay
Sand
Cheesecloth held in place by rubber band

Plastic cup
Some coffee has moved down
Coffee-stained sand
Clay has migrated down

---

at the base of the A-horizon. In the E-horizon, leaching has taken place, but not much organic matter has mixed in. The O-, A-, and E-horizons together represent the zone of leaching and can form **topsoil**, the part of the soil that farmers till when planting crops. Beneath the A-horizon (or, in some cases, the E-horizon) lies the B-horizon, or **subsoil**, representing the zone of accumulation. Here, transported clay and very fine organic particles collect and new minerals such as iron oxide grow. Finally, at the base of a soil profile, we find the C-horizon, which consists of weathered rock or sediment that has not yet undergone leaching or accumulation. The C-horizon grades downward into unweathered bedrock or unweathered sediment.

Not All Soils Are the Same

As farmers, foresters, and ranchers know well, the soil in one locality may differ greatly from the soil in another in terms of composition, thickness, and texture. Such diversity exists because the makeup of a soil depends on several factors, including the following **(Fig. 5.10a, b)**:

- *Climate:* Rainfall and temperature affect the rate of chemical weathering, the amount of leaching, and the abundance of organisms in soil. Therefore, climate affects the rate of soil development, the degree to which soluble minerals are redistributed during soil formation, and the amount of organic matter contained in a soil.

Figure 5.11 Examples of soil classification.

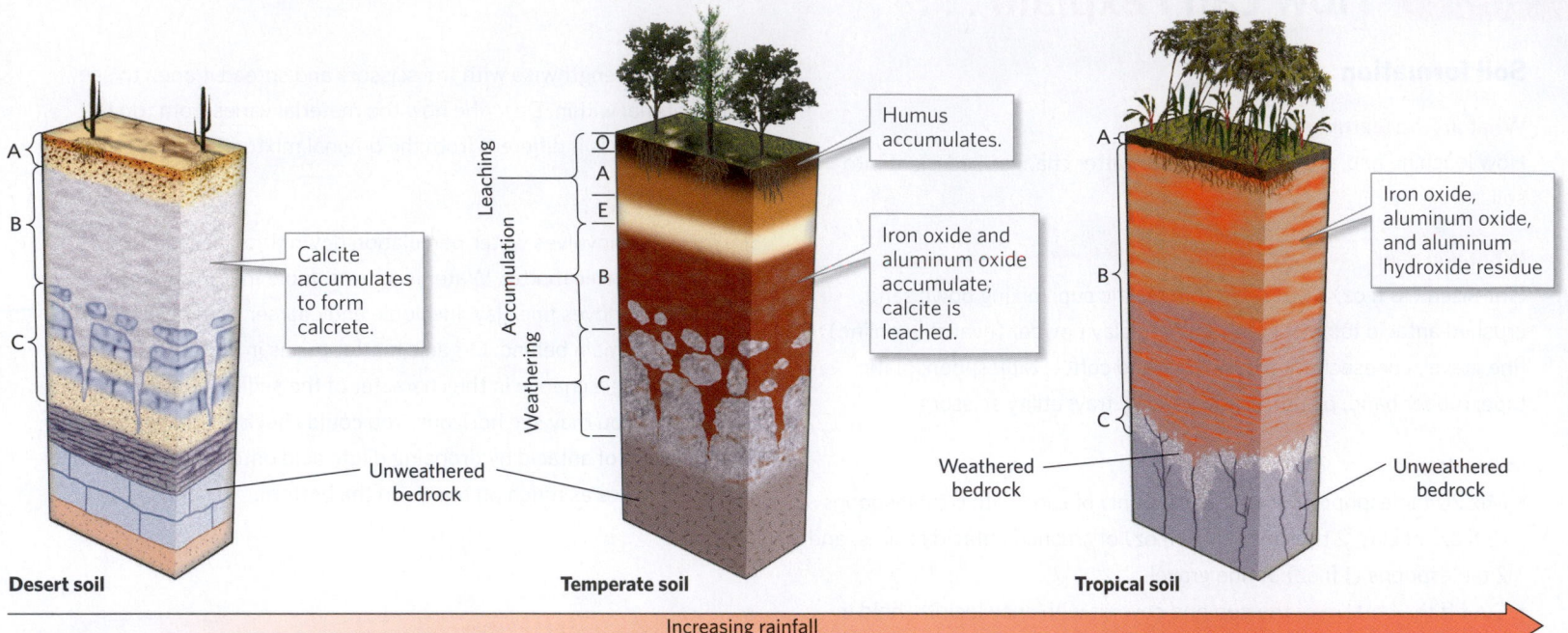

(a) Aridisol forms in deserts. Rainfall is so low that no O-horizon forms, and soluble minerals accumulate in the B-horizon.

(b) Alfisol forms in temperate climates. An O-horizon forms, and relatively insoluble materials accumulate in the B-horizon.

(c) Oxisol forms in tropical climates, where percolating rainwater leaches all soluble minerals, leaving only iron- and aluminum-rich residues.

- *Substrate composition:* Some soils form on basalt, some on granite, and some on recently deposited quartz sand. These substrates contain different minerals and, therefore, different proportions of chemical compounds. Soils formed on different substrates end up with different chemical compositions.

- *Slope steepness:* A thick soil can accumulate on flat land, but on a steep slope, debris washes or slumps away before it can develop into soil. So, all other factors being equal, soil thickness increases as slope angle decreases.

- *Time:* Soil takes time to form, so a young soil tends to be thinner and less developed than an old soil. A thick soil takes millennia to form.

- *Vegetation type:* Different kinds of plants extract or add different nutrients and quantities of organic matter and have different kinds of root systems, all of which affect soil. Plants with large, deep root systems help prevent soil from washing away.

With the effects of these factors in mind, let's try to understand the character of a few types of soils. Researchers have developed classification schemes for soils, using specific names for different types.

- *Desert soil:* In deserts, where hardly any rain falls and little or no vegetation grows, an *aridisol* forms (Fig. 5.11a). Such a soil has no O-horizon because no organic matter accumulates, and it has no A-horizon because hardly any leaching takes place. In fact, soluble minerals such as calcite, which would be completely leached out of soil where rainfall is heavy, tend to accumulate in the B-horizon of an aridisol. This calcite can cement grains into a rock-like mass called a *caliche*.

- *Temperate soil:* In temperate environments, where moderate amounts of rain fall, an *alfisol* forms (Fig. 5.11b). Because a large quantity of organic matter can accumulate in a temperate environment, an alfisol has a significant O-horizon. And because a relatively large amount of water percolates through an alfisol, the B-horizon accumulates only relatively insoluble minerals.

- *Tropical soil:* In a tropical climate, where heavy rains fall, a distinct brick-red soil called an *oxisol* (known traditionally as *laterite*) develops (Fig. 5.11c). The large volume of water percolating down through an oxisol leaches out most soluble minerals. As a result, the A-horizon consists mostly of an insoluble residue of iron and aluminum oxide, which give oxisol a brick-red color (Fig. 5.12).

Destroying Soils

Soil is one of our planet's most valuable resources. Unfortunately, destruction of native plants and their root systems (by grazing animals, lack of rainfall, and

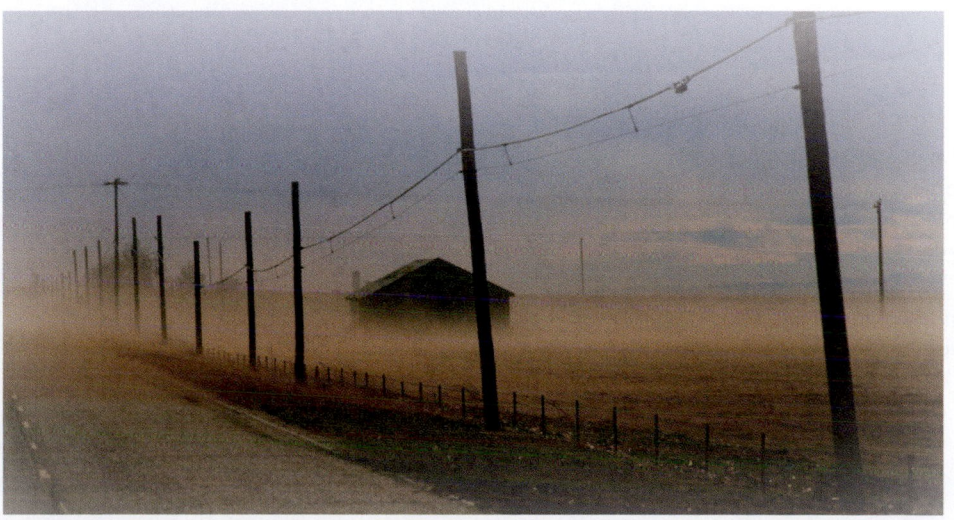

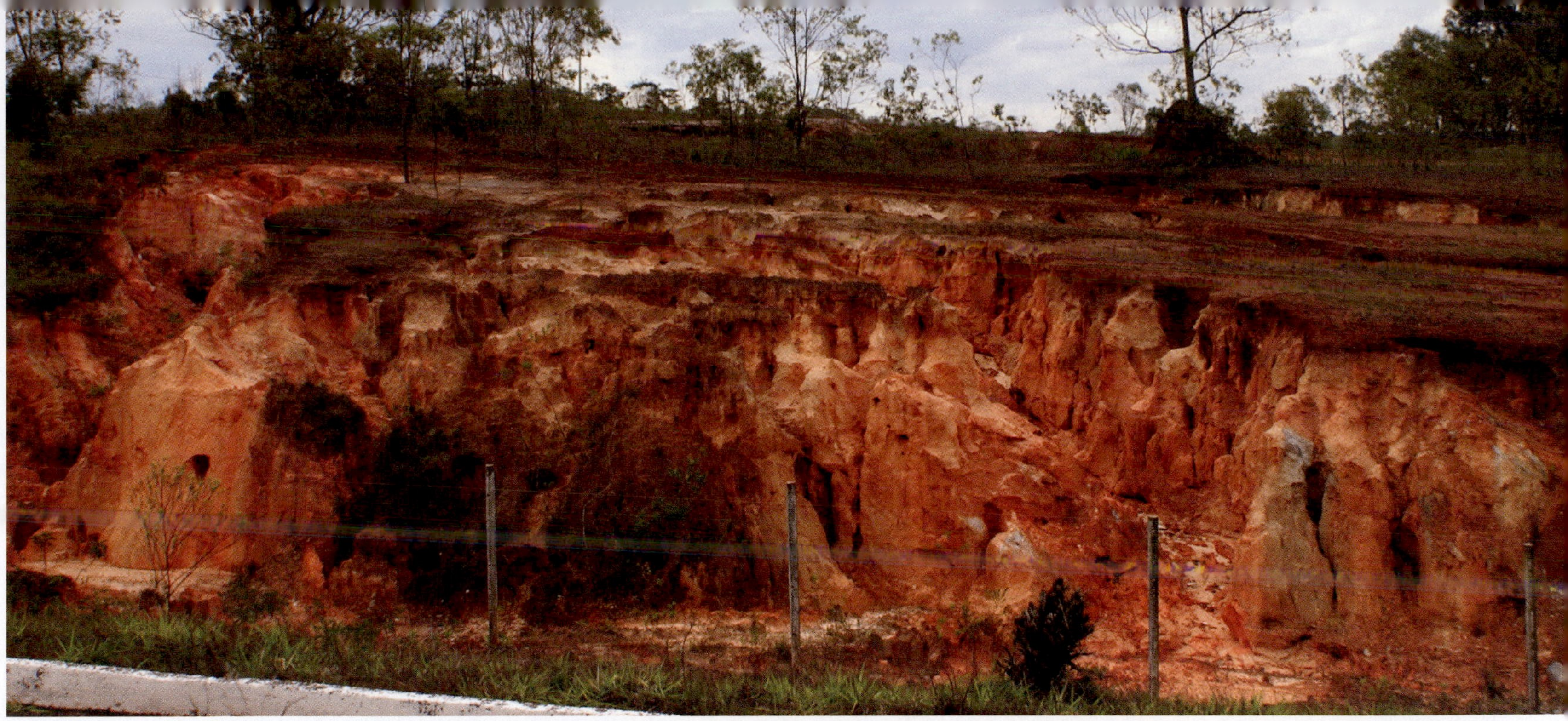

Figure 5.12 An exposure of laterite in Brazil. Note the hard crust that has formed, in places, at the land surface.

plows) leaves soils exposed at the ground surface. As a result, heavy rains or strong winds can carry soil away, a process called **soil erosion** (Fig. 5.13). Because human activities, such as farming, ranching, and clear cutting, increase the rate of soil erosion, modifying these activities could conserve soil. For example, planting grass or winter crops in a field, instead of leaving it barren, decreases soil-erosion rates.

The consequences of rainforest destruction have had particularly profound effects on soil. In an established rainforest, lush growth provides sufficient organic debris so that trees can grow. Logging the forest, or burning off vegetation to provide fields for agriculture, leads to the disappearance of humus, which decays quickly in tropical climates. The exposed laterite contains relatively few nutrients, which crop plants consume quickly. As a result, the land soon may become unsuitable for agriculture or for regrowth of the rainforest.

Take-home message . . .

Soil forms when downward-percolating rainwater and living organisms interact with rock or sediment just below the Earth's surface. Percolating water leaches ions and clay from the topsoil and deposits them in the subsoil, producing distinct soil horizons. Organic matter mixes into the top horizons. Soil character depends not only on the material from which a soil forms, but also on climate, time, and slope. Human activity can accelerate erosion of soil.

Quick Question
How does a desert soil differ from a rainforest soil?

5.4 Introducing Sedimentary Rock

Look closely at the colorful cliffs forming the walls of the Grand Canyon, and you'll notice distinct layers that vary in texture and composition (Fig. 5.14). These layers consist of sedimentary rock formed by the burial and hardening of sediment deposited between 541 and 200 million years ago. Formally defined, **sedimentary rock** is rock that forms at or near the surface of the Earth in one of the following ways: (1) by the compacting and

Figure 5.13 When fields are barren, wind can erode the soil and produce dust storms.

Figure 5.14 Layers of sedimentary rock form the cliffs of the Grand Canyon, Arizona, as seen from an airplane.

cementing together of loose fragments produced by the weathering of pre-existing rock; (2) by the growth of shell masses or by the cementing together of shells and shell fragments; (3) by the accumulation and alteration of organic matter left after the death of plants or plankton; or (4) by the precipitation of minerals directly from water solutions.

A layer of sedimentary rock is called a **bed**. A succession of beds records tales of ancient events and ancient environments. For example, features preserved in the sedimentary rocks exposed in the Grand Canyon tell us that beaches, reefs, lakes, mud flats, river floodplains, and desert dunes all existed at this location in the geologic past. Sedimentary rocks occur only in the upper part of the crust and effectively form a *cover* that buries the underlying *basement* of igneous and/or metamorphic rock **(Fig. 5.15)**. Geologists divide sedimentary rocks into four major classes based on how they formed: clastic, biochemical, organic, and chemical. Let's look at how each of these classes formed.

Take-home message . . .

Sedimentary rock forms at or near the Earth's surface by a variety of processes. Sediment accumulates in layers called beds. Geologists distinguish different classes of sedimentary rocks based on how the rocks formed.

Quick Question -----------------------------
Can living organisms contribute to sedimentary rock formation?

5.5 Making New Rocks from the Debris of Others

How Do Clastic Sedimentary Rocks Form?

Nine hundred years ago, a thriving community of Native Americans inhabited the high plateau of Mesa Verde, Colorado. In hollows beneath a huge overhanging ledge, they built stone-block buildings that survive to this day **(Fig. 5.16)**. Clearly, the blocks are solid and durable—they are, after all, rock. But if you were to rub your thumb along one, it would feel gritty, and small rounded grains of quartz would break free and roll under your thumb. The blocks consist of *sandstone*, a rock consisting of sand grains cemented together. Geologists refer to all rocks formed from once-separate grains that have been packed together and then cemented to one another as **clastic sedimentary rocks**, and to the individual grains or fragments that form them as *clasts*. Formation of clastic sedimentary rock involves the following steps **(Fig. 5.17a–c)**:

- *Weathering:* Clasts form by disintegration of bedrock due to weathering.

- *Downslope movement:* Gravity pulls clasts from higher to lower elevations. They may tumble individually or as part of a landslide.

Figure 5.15 Standing at the edge of the inner gorge in the Grand Canyon, you can see that the sedimentary rocks form a "cover" layer that has buried a basement of old metamorphic rocks.

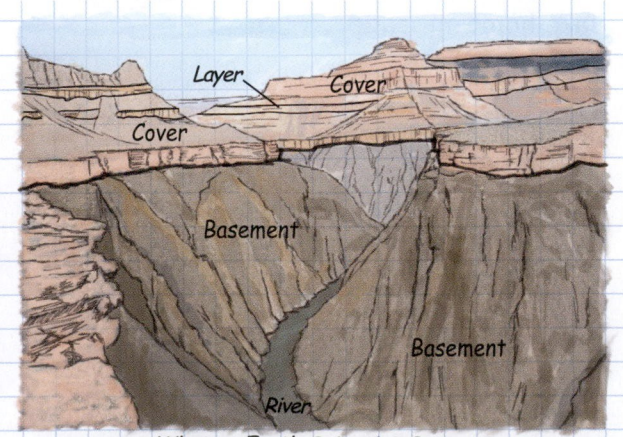

What an Earth Scientist Sees

Figure 5.16 Between 600 and 1300 c.e., Pueblo people inhabited these dwellings made of sandstone blocks beneath a protective ledge of sandstone bedrock. The inset shows the grainy character of the rock.

Sandstone layer

- *Erosion:* Moving air, water, or ice (a *transporting medium*) can carve into bedrock or debris accumulations and pick up clasts. This process is called *erosion*.

- *Transportation:* Clasts can be carried far from their origin by a transporting medium **(Fig. 5.18)**. The ability of the medium to carry sediment depends on the medium's viscosity and its velocity. Ice is solid, so it can transport clasts of any size, regardless of how slowly the ice moves. In contrast, water and air are fluids whose ability to carry clasts depends on their speed of flow and viscosity—fast-moving, muddy water transports large clasts, whereas slow-moving, clear water only transports small ones.

- *Deposition:* The accumulation of clasts to form layers of sediment at one location is the process of **deposition**. It happens when a transporting medium loses its ability to carry clasts. For example, when flowing water slows, it drops the clasts it's carrying.

- *Lithification:* **Lithification** refers to the transformation of unconsolidated sediment (loose clasts) into solid rock. The process involves a few stages. First, sediment undergoes *burial*, meaning that new sediment layers accumulate above it. The weight of the overlying sediment eventually squeezes out most water or air that had been trapped in pores between clasts. Because of such **compaction**, the clasts fit together more tightly. Finally, minerals precipitate from groundwater passing through the remaining pores. Like glue, these minerals bind the clasts together to make an overall solid rock. This binding process is called **cementation**.

Describing and Classifying Clastic Sedimentary Rocks

Say that you pick up a clastic sedimentary rock and want to describe it so clearly that, from your words alone, another person can picture the rock. Geologists find the following characteristics most useful for classifying clastic sedimentary rocks:

- *Clast size:* The fragments or grains that make up rock range in diameter from microscopic to very large. Table 5.1 indicates the names assigned to specific clast sizes.

- *Clast composition:* Grains in sedimentary rock may consist of rock fragments (lithic clasts) containing many mineral grains, or they may consist of individual mineral grains. Some sedimentary rocks contain mostly clasts of one composition, whereas some contain clasts of many different compositions.

- *Angularity and sphericity: Angularity* refers to the degree to which clasts have angular, as opposed to rounded, corners and edges, and *sphericity* refers to the degree to which the shape of a clast resembles a sphere.

- *Sorting:* Geologists use the term **sorting** in reference to the range of grain sizes in a sediment. Well-sorted sediment contains clasts that are all the same size, whereas poorly sorted sediment contains clasts of many different sizes.

In every grain of sand, there is the story of the Earth.

—RACHEL CARSON (AMERICAN NATURALIST, 1907–1964)

Figure 5.17 Steps in clastic sedimentary rock formation.

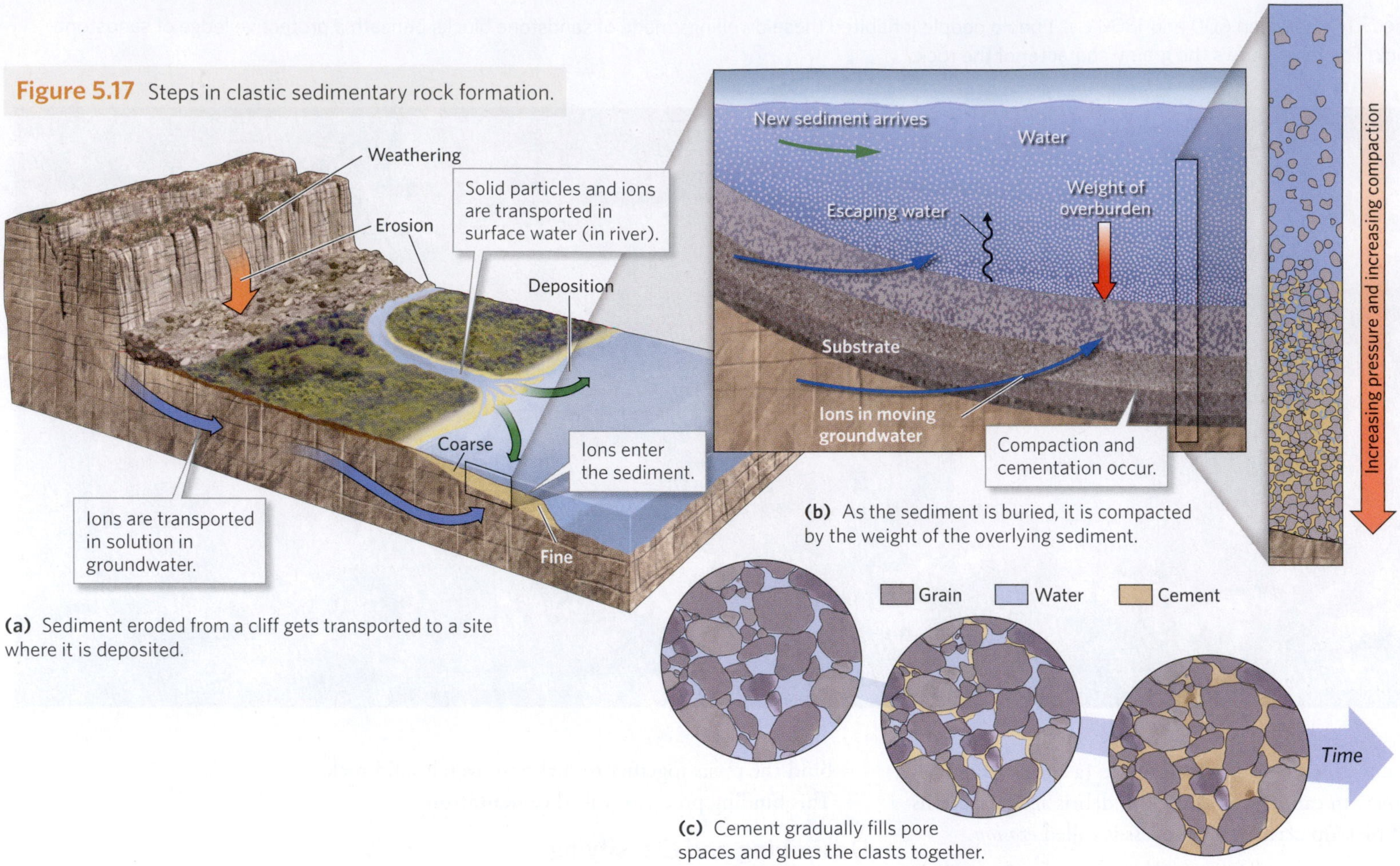

Weathering

Erosion

Solid particles and ions are transported in surface water (in river).

Deposition

Coarse

Ions enter the sediment.

Ions are transported in solution in groundwater.

Fine

New sediment arrives

Water

Escaping water

Weight of overburden

Substrate

Ions in moving groundwater

Compaction and cementation occur.

Increasing pressure and increasing compaction

(a) Sediment eroded from a cliff gets transported to a site where it is deposited.

(b) As the sediment is buried, it is compacted by the weight of the overlying sediment.

| Grain | Water | Cement |

Time

(c) Cement gradually fills pore spaces and glues the clasts together.

- *Character of cement:* Not all clastic sedimentary rocks contain the same kind of cement. In some, the cement consists predominantly of quartz, whereas in others, it consists predominantly of calcite.

With the above characteristics in mind, we can distinguish among several common types of clastic sedimentary rocks (Table 5.2).

The character of the sediment that accumulates at a location depends not only on the composition of the sediment's source, but also on the character of the transporting medium and on the distance that clasts move. For example, grains carried by a faster and more viscous fluid are coarser than those carried by a slower and less viscous fluid. The farther a sediment has been transported (by a current or waves), the better sorted it becomes, the less angular its clasts become, and the greater the chance that its less resistant clasts have weathered and disintegrated (Fig. 5.19a, b). Geologists use the term **maturity** to indicate the degree to which a sediment has changed from the place it originated to the place it was deposited. More mature sediments tend to occur farther from the sediment source and contain better-sorted sediment and fewer grains of nonresistant minerals (Fig. 5.19c).

Did you ever wonder . . .

where beach sand comes from?

Figure 5.18 The production and transportation of sediment along a creek in California. Landslides dump debris onto the valley floor. When the stream floods, water carries the clasts downstream.

Figure 5.19 Grain characteristics change with transportation.

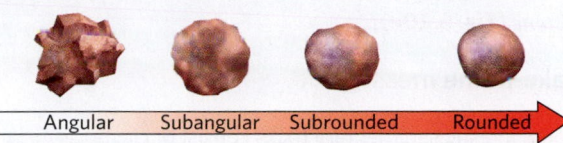

Angular   Subangular   Subrounded   Rounded

(a) Transportation decreases angularity.

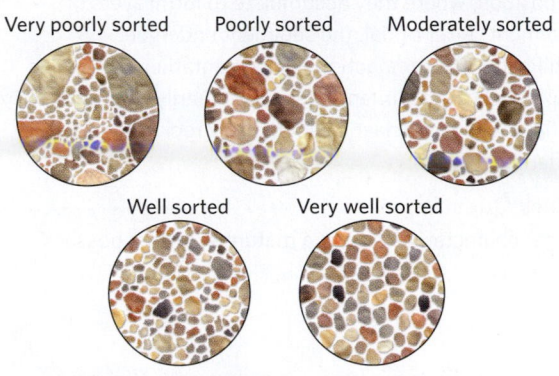

Very poorly sorted   Poorly sorted   Moderately sorted

Well sorted   Very well sorted

(b) Transportation sorts grains by size.

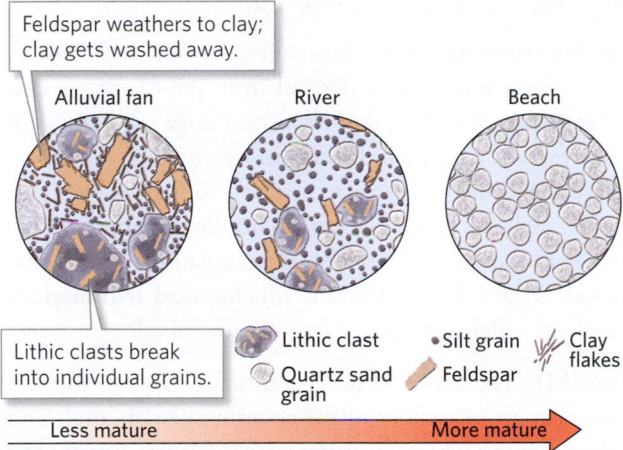

Feldspar weathers to clay; clay gets washed away.

Alluvial fan   River   Beach

Lithic clasts break into individual grains.

○ Lithic clast   • Silt grain   ☆ Clay flakes
○ Quartz sand grain   ▬ Feldspar

Less mature   More mature

(c) A less mature sediment consists of fragments of the original rock and contains both resistant and nonresistant minerals. A mature sediment contains only well-sorted resistant minerals.

In order to understand how different kinds of sediment can be deposited, let's follow the fate of rock fragments as they gradually move from a cliff face in the mountains to the seashore via a river. You'll see that different kinds of sediment develop along the route. Each kind, if buried and lithified, would yield a different type of sedimentary rock.

To start, imagine that blocks of granite tumble off a cliff and break up on the way down. If the pile of angular fragments that accumulates at the base of the cliff were to be cemented together, the resulting rock would be *breccia*

(Fig. 5.20a). Now imagine that a landslide moves the fragments into a turbulent river, which carries them away. In the water, the clasts collide, break into smaller pieces, and become rounded pebbles and cobbles. Where the current slows downstream, these clasts are deposited as gravel. Burial and lithification of the clasts would produce *conglomerate* (Fig. 5.20b). Over time, the clasts undergo chemical weathering and break apart into individual mineral grains, eventually producing a mixture of quartz, feldspar, and clay. Clay is so fine that flowing water easily picks it up and carries it away, leaving sand composed of quartz and some feldspar grains—if this sediment were buried and lithified, it would become *arkose* (Fig. 5.20c). The feldspar grains continue to weather into clay, and the sand grains wear down into smaller pieces, so that gradually, during successive events that wash the sediment downstream, the sediment carried by the river includes clay as well as silt and sand composed of durable quartz. Sand accumulates in the river channel, and if lithified, it becomes sandstone, whereas silt and clay collect on *floodplains*, the broad, flat areas bordering a river, and if lithified, they become *siltstone* and *shale* or *mudstone*.

Sediment carried all the way to the sea settles out when the flowing water enters the sea and slows. This sediment builds a *delta*, a wedge of sediment at the mouth of the river. Over time, waves and current may redistribute the sediment. Some of the silt settles in quiet lagoons or mud flats along the shore, or farther offshore in quieter water. The silt, if lithified, becomes siltstone, and the clay, if lithified, becomes shale or mudstone (Fig. 5.20d). Some of the clay may be carried far out into the ocean, where it accumulates along with plankton shells on the deep seafloor. Occasionally, underwater avalanches carry chaotic mixtures of coarser grains, along with clay, down submarine slopes. This sediment, if lithified, becomes *wacke*. Along the coast, ocean waves wash away finer grains and

Table 5.2   Classification of clastic sedimentary rocks

Clast Size	Clast Character	Rock Name
Very coarse	Rounded pebbles and cobbles	Conglomerate
	Angular clasts	Breccia
	Large clasts in a clay matrix*	Diamictite
Medium	Sand-sized grains	Sandstone
	Quartz grains only	Quartz sandstone
	Quartz and feldspar grains	Arkose
	Sand grains in a clay matrix*	Wacke
Fine	Silt-sized clasts	Siltstone
Very fine	Clay and/or very fine silt	Shale or mudstone

*We use the word *matrix* for the finer-grained material surrounding larger grains.

Figure 5.20 Different kinds of sediments lithify into different kinds of sedimentary rocks.

Sediment ———— Lithification ————▶ Sedimentary rock

(a) Lithification of an accumulation of angular clasts yields breccia.

(b) Layers of river gravel lithify into conglomerate.

Alluvial fan

(c) Sediment deposited close to its source, like the sediment in this alluvial fan, can be rich in feldspar. Lithification of this sediment yields arkose.

Sandstone

Shale

(d) Layers of mud, like these layers exposed beneath marsh grass, lithify to form shale. Here, thin beds of shale are interbedded with sandstone.

(e) Layers of beach or dune sand lithify into sandstone.

leave behind beaches of well-sorted sand. Winds may pick up some of this sand and deposit it in desert dunes. Any of these sand deposits, if buried and lithified, become *sandstone* **(Fig. 5.20e)**.

Take-home message . . .

Clastic sedimentary rocks consist of clasts weathered and eroded from pre-existing rocks. Wind, water, or ice transports clasts to a site of deposition, where they accumulate to form layers of sediment. After burial, the sediment undergoes lithification by compaction and cementation. Geologists classify clastic sedimentary rocks primarily by grain size. Different types of clastic sedimentary rocks form in different settings.

Quick Question -
What characteristics does a mature sediment possess?

5.6 When Life Builds Rock

Biochemical Sedimentary Rocks

So far, we've discussed the making of new sedimentary rocks from solid grains derived from pre-existing rocks. Weathering, however, doesn't just produce solid grains. It also yields ions, which dissolve in water. Some organisms have the ability to extract dissolved ions from water to make shells. When these organisms die, the mineral material in their shells survives. Accumulations of this material, when lithified, become **biochemical sedimentary rock**, of which geologists recognize several different types:

BIOCHEMICAL LIMESTONE. A snorkeler gliding above a reef sees an incredibly diverse community of corals and algae, around which creatures such as clams, oysters, and snails live, and above which plankton float **(Fig. 5.21a)**. These organisms share an important characteristic: they make shells of calcium carbonate ($CaCO_3$), either as calcite or as aragonite. When the organisms die, the shells remain. Rocks formed predominantly from calcium carbonate shells are the biochemical version of **limestone**. Because the principal compound making up limestone is $CaCO_3$, geologists refer to limestone as a type of **carbonate rock**.

Some limestone forms from mounds of shell-secreting organisms that grew in place, some from debris composed of coarse shell fragments, some from extremely fine shell fragments, and some from plankton shells that settled out of water. These different types of carbonate sediments yield different types of limestone, such as *fossiliferous limestone*, containing abundant visible fossil shells or shell

Figure 5.21 The formation of limestone, a type of carbonate rock.

(a) Corals and other organisms living on this reef make their shells out of calcium carbonate.

(b) Broken shells were cemented together to form this 415-Ma fossiliferous limestone.

(c) Ancient limestone tends to occur in gray, blocky layers, like those exposed in this roadcut in New York State.

fragments **(Fig. 5.21b)**; *micrite*, consisting of very fine carbonate mud; and *chalk*, consisting of calcite plankton shells. Experts use alternative, more complex classification schemes for these rocks.

In outcrops, limestone typically looks like a massive light-gray to dark-bluish-gray rock that breaks into chunky blocks, rather than a pile of shell fragments **(Fig. 5.21c)**. That's because underground processes change the texture of the rock over time. For example, water passing through the rock dissolves some grains, causes new ones to grow, and precipitates cement.

BIOCHEMICAL CHERT. Along the California coast, just northwest of San Francisco, roadcuts expose a reddish, almost porcelain-like rock occurring in 3- to 15-cm (1- to 6-inch)-thick layers **(Fig. 5.22a)**. Hit it with a hammer, and the rock breaks to form smooth, clam-shell-shaped (conchoidal) fractures. Geologists call this rock *biochemical chert*—it's made from quartz grains that are too small to be seen without extreme magnification. The chert of coastal California formed from the shells of silica-secreting plankton that accumulated on the seafloor.

Organic Sedimentary Rocks

We've seen how shells become biochemical sedimentary rock. What becomes of the cellulose, fats, carbohydrates, proteins, or other organic compounds of organisms? Commonly, organic debris either gets eaten by other organisms or decomposes at the Earth's surface. But in some environments, organic debris accumulates with other sediment and undergoes burial before completely decomposing. When lithified, such organic-rich sediment becomes **organic sedimentary rock**. Examples include **coal**, a black rock composed primarily of carbon from woody plant remains **(Fig. 5.22b)**, and *oil shale*, a dark gray to black rock formed from compacted clay mixed with the organic remains of plankton and algae.

Figure 5.22 Examples of biochemical and organic sedimentary rocks.

(a) This bedded chert developed from deep-sea sediment made up of the shells of silica-secreting plankton.

Sandstone and shale

Coal layer

(b) Coal is deposited in layers (beds), just like other kinds of sedimentary rocks.

Take-home message . . .

Living organisms can be involved in forming sedimentary rocks. Biochemical sedimentary rocks form from the shells of organisms. Organic sedimentary rocks include coal, made from the remains of woody material, and oil shale, made from a mixture of clay and plankton or algal remains.

Quick Question -
What is the difference between limestone and chert?

5.7 Solids from Solutions: Chemical Sedimentary Rocks

The colorful terraces that grow around the vents of hot springs, the layers of salt that underlie the floor of the Mediterranean Sea, the smooth, sharp point of an ancient arrowhead—all these materials have something in common. They consist of rock formed by the direct precipitation of minerals from water solutions. We call such rocks **chemical sedimentary rocks**. They commonly have a crystalline texture. Geologists distinguish among chemical sedimentary rock types on the basis of their composition.

EVAPORITES (SALT DEPOSITS). In 1965, Craig Breedlove became the first person to drive a vehicle on land at a speed of 600 mph (966 km/h). Such high-speed trials must take place on extremely long, flat surfaces. Not many places provide such conditions—but the Bonneville Salt Flats of Utah do. They formed when an ancient salt lake evaporated: under the heat of the Sun, water molecules drifted up into the atmosphere, but the salt that had been dissolved in the water stayed behind.

Salt precipitation occurs when saltwater becomes *supersaturated*, meaning that it has exceeded its capacity to contain dissolved ions. In a supersaturated solution, ions bond together to form solid grains that either settle out of the water or grow on the floor of the water body. This process takes place in desert lakes and along the edges of restricted seas where evaporation removes water from a water body faster than new water enters **(Fig. 5.23a, b)**. Because salt deposits, which include both halite and gypsum, form as a consequence of evaporation, geologists refer to them as **evaporites**.

TRAVERTINE (CHEMICAL LIMESTONE). *Travertine* consists of crystalline calcite formed by precipitation from groundwater that has seeped out of the ground. For example, the emergence of groundwater from hot springs leads to travertine precipitation because when the water degasses (loses some dissolved CO_2), evaporates, and cools, it becomes supersaturated. Travertine produced at hot springs builds terraces and mounds **(Fig. 5.24a)**. Travertine also forms where water seeps out of the walls or ceilings of caves, producing *speleothems* **(Fig. 5.24b)**. In some environments, microbes assist the precipitation process.

DOLOSTONE. A rock that contains a large proportion of dolomite, a carbonate mineral that contains magnesium and calcium, with the chemical formula $CaMg(CO_3)_2$, is a **dolostone**. Dolostone forms when limestone reacts with magnesium-bearing groundwater. This change can take place beneath lagoons along a shore soon after the limestone has formed, or a long time later, after the limestone has been buried deeply. (Many geologists prefer to use the term *dolomite* for both the rock and the mineral.)

Figure 5.23 The formation of evaporite deposits.

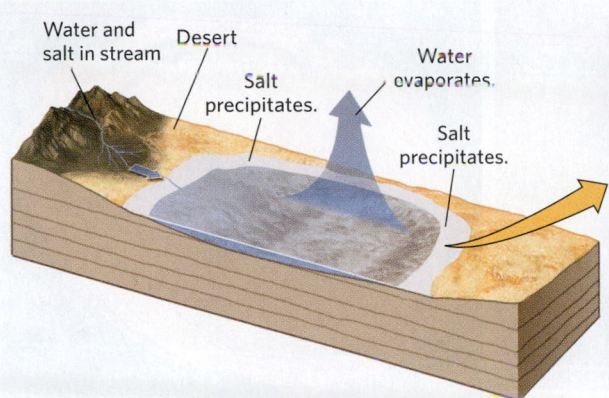

(a) When saltwater evaporates, various mineral salts, including halite and gypsum, precipitate.

(b) Under the desert sun, a salt lake on the floor of Death Valley has dried up, leaving a white crust of new salt crystals.

REPLACEMENT CHERT. Like obsidian, chert can be used by artisans for making sharp-edged tools because it fractures conchoidally. As we've seen, some chert forms from the shells of silica-secreting plankton. Chert can also form when silica chemically precipitates from groundwater and replaces other minerals within a solid rock. Geologists refer to this rock as *replacement chert*. Some replacement chert grows into layers of nodules (lumpy blobs) in limestone **(Fig. 5.25a)**; some grows within logs buried by silica-rich volcanic ash, yielding *petrified wood* **(Fig. 5.25b)**; and some grows in open spaces to form colorful *agate* **(Fig. 5.25c)**. Replacement chert can come in many colors—black, red, brown, and green—depending on the chemical impurities it contains.

Take-home message . . .

Chemical sedimentary rocks form by precipitation from a water solution. Evaporites contain halite and gypsum precipitated from salt lakes or restricted seas. Travertine consists of carbonate precipitated from groundwater seeping out at springs or in caves. Dolostone forms when magnesium-rich groundwater reacts with limestone.

Quick Question -

How is replacement chert different from biochemical chert?

Did you ever wonder . . .

how the flint used for arrowheads first formed?

Figure 5.24 Examples of travertine (chemical limestone) deposits.

(a) Travertine accumulates in terraces at Mammoth Hot Springs in Yellowstone National Park, Wyoming.

(b) Travertine speleothems form as calcite-rich water drips from the ceiling of Timpanogos Cave in Utah.

Figure 5.25 Replacement chert.

(a) Replacement chert forms a layer of black nodules in this gray limestone in New York.

Layer of black chert

Tilted limestone beds

(b) This 14-cm (5.5-inch)-diameter log of petrified wood from Wyoming formed about 50 Ma. The texture of the bark is still visible.

(c) A thin slice of Brazilian agate, lit from behind, shows distinct growth rings.

5.8 Interpreting Sedimentary Structures

Sedimentary rock isn't just a homogeneous, featureless mass. Rather, it contains layering as well fascinating textures and shapes that formed during or immediately after deposition. Geologists refer to such features as **sedimentary structures**.

Bedding and Stratification

If you look at an outcrop of sedimentary rock, you'll see a set of parallel planes that separate the rock into distinct layers. A single layer of sedimentary rock is called a **bed**, and the boundary between two beds is a *bedding plane*. Geologists commonly use the term **strata** (from the Latin *stratum*, meaning pavement) for a succession of several beds, and **bedding** for the occurrence of beds **(Fig. 5.26a)**.

Why does bedding form? To answer this question, remember how sediment accumulates. The sediment deposited at any given time reflects the composition of the sediment and the conditions of deposition (such as water depth and current velocity). Therefore, changes in the sediment source or the conditions of deposition can change the type of sediment deposited. Such changes are recorded in sedimentary rock by a change in composition or grain size, or both, and this change can delineate bedding. As an example, picture the sediment layers deposited by a river **(Fig. 5.26b)**. On a normal day, the river flows slowly and transports only silt, so layers of silt collect along the river. After a heavy rainstorm, the river flows faster and transports pebbles, so a layer of gravel accumulates over the silt layer. Later, when the river slows again, another layer of silt buries the gravel. This succession of sediments,

when lithified, becomes alternating beds of siltstone and conglomerate **(Box 5.2)**. Sediments settle due to gravity, so beds tend to be nearly horizontal when deposited.

Current-Related Structures

In quiet water or calm air, sediment settles like snow, and when it does, the sediment layer that forms has a smooth surface and a homogeneous texture. In contrast, sediment layers deposited by flowing water or air tend to develop distinctive sedimentary structures both on the bed surface and within the bed. Such structures provide clues to the setting in which the sediment was deposited, distinguish the top of the bed from the bottom, and characterize the direction in which currents flowed at the time of deposition. Let's consider a few examples.

RIPPLE MARKS, DUNES, AND CROSS BEDS. The top surface of a sediment layer deposited beneath flowing air or water typically displays wave-like ridges and troughs **(Fig. 5.27a)**. These **ripple marks** tend to be relatively small (less than 10 cm, or 4 inches, high) and align perpendicular to the current flow. Burial of ripple marks by another layer of sediment preserves them, so you can find them on bedding planes of ancient rocks **(Fig. 5.27b)**. **Dunes** are relatively large ridges built of sediment transported by a current **(Fig. 5.28a)**. On the bed of a stream, dunes may be tens of centimeters high, and in windy deserts, they may be tens of meters or even a hundred meters high **(Fig. 5.28b)**.

If you slice into a ripple mark or dune, you'll find distinct *laminations* (thin layers within a thicker layer) known as **cross beds**, inclined at an angle to the top of the sediment layer. To see how cross beds develop, imagine a current of air or water moving uniformly in one direction **(Fig. 5.28c)**. The current picks up sand from the upstream or windward face of the ripple or dune and deposits it on

Figure 5.26 Bedding in sedimentary rocks.

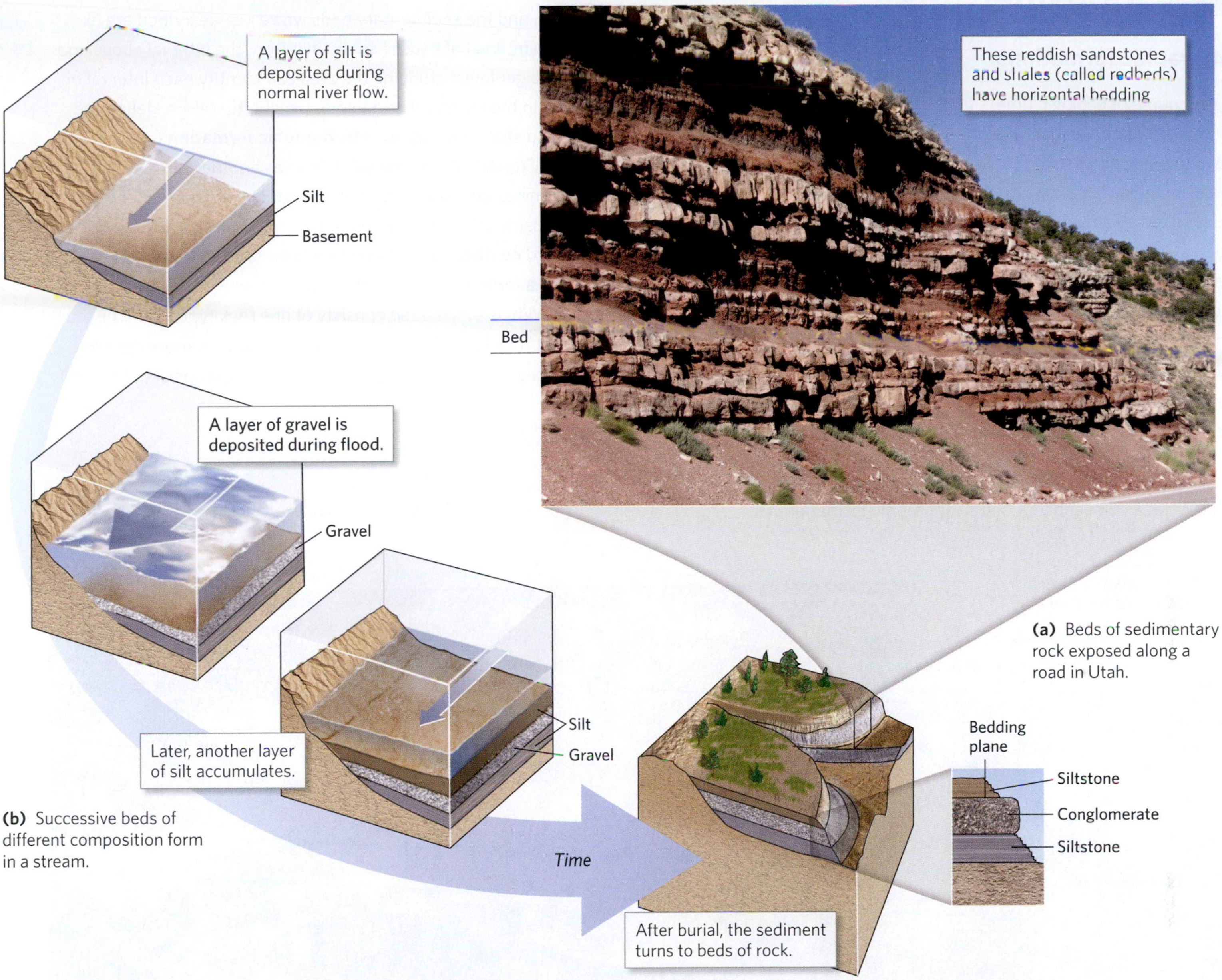

A layer of silt is deposited during normal river flow.

Silt
Basement

A layer of gravel is deposited during flood.

Gravel

Later, another layer of silt accumulates.

Silt
Gravel

(b) Successive beds of different composition form in a stream.

Time

After burial, the sediment turns to beds of rock.

These reddish sandstones and shales (called redbeds) have horizontal bedding

Bed

(a) Beds of sedimentary rock exposed along a road in Utah.

Bedding plane

Siltstone
Conglomerate
Siltstone

Figure 5.27 Ripples and ripple marks.

(a) Modern ripples, exposed at low tide, on a sandy beach on Cape Cod, Massachusetts.

(b) Ripples on a tilted, 145-Ma sandstone bed.

Box 5.2 ▶ Consider this . . .

Stratigraphic formations

Over the course of geologic time, significant long-term changes in conditions of deposition can take place. As a result, a given succession of beds may differ markedly from the strata above or below. For example, imagine that a region contains alternating beds of silt and mud deposited by a river. Because of a long-term rise in sea level, beds of beach sand accumulate on top of the river deposits. When sea level rises still higher, layers of deeper-water marine shells bury the sand. Beds of one or more particular rock types may take millions of years to accumulate. If the deposits are eventually buried deep enough to lithify, they will transform into a succession of siltstone and shale, overlain by sandstone, in turn overlain by limestone. Each component of this succession represents deposition during a given interval of geologic time.

Fast-forward a few hundred million years. Erosion has removed overburden, and the sedimentary beds we've just described are exposed. Each interval of beds is so distinct from the interval above and below that a geologist studying the area can identify each interval on many cliffs in the vicinity. The geologist might, therefore, define the siltstone and shale beds as one **stratigraphic formation** (or simply a *formation*) **(Fig. Bx5.2a, b)**, the sandstone as another formation, and the limestone as another formation. Formations are commonly named after the locality where they were first found and studied. Geologists use the word **contact** for the boundary between two formations or between a formation and another rock unit (such as an igneous intrusion). When a formation consists of one rock type, its name may include that rock type. The name of a formation with more than one rock type may include the word *formation*. A *group* consists of several formations.

Figure Bx5.2 The concept of a stratigraphic formation.

The surface between two units is called a contact.

(a) Examples of stratigraphic formations exposed on the wall of the Grand Canyon.

(b) A panoramic view of the Grand Canyon, showing how contacts (yellow lines) can be traced along the cliff faces.

the downstream or leeward face. Sand builds up until the downstream slope becomes too steep and gravity causes the sand to tumble down the leeward face, or *slip face*. As a dune or ripple builds downstream, more sand buries each successive slip face, preserving the previous one as a cross bed. When new layers of dunes or ripples build over existing ones, a succession of cross-bedded layers, separated from each other by main bedding planes, accumulates **(Fig. 5.28d)**. Cross-bed orientation serves as a clue to the direction in which the current flowed during deposition, and cross-bed shape (concave up) distinguishes the top of a bed from the bottom.

Figure 5.28 The formation of dunes and cross beds.

(a) A small dune developing in Death Valley. Small ripples have formed on the top of the dune.

(b) Large sand dunes in a windstorm.

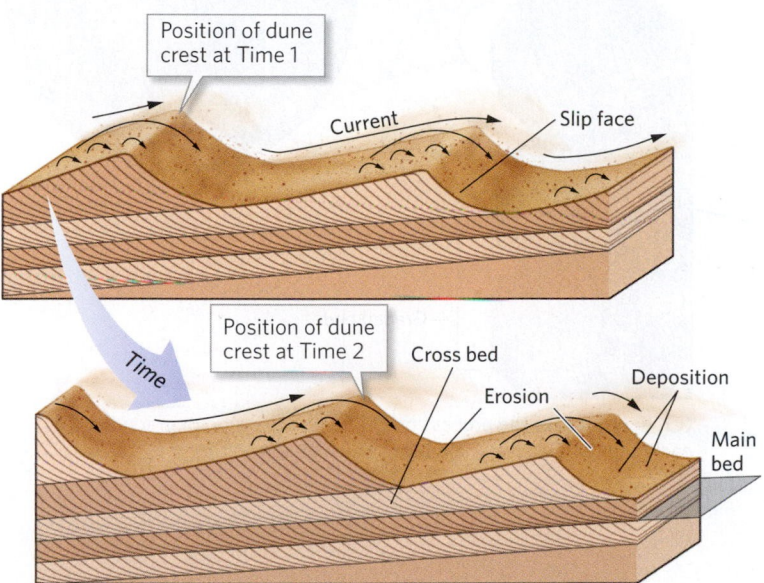

(c) Cross beds form when a current carries sand up the windward side of a ripple or dune, where it builds up until it falls down the slip face.

What an Earth Scientist Sees

(d) This sandstone exposed in a cliff face in Zion Canyon, Utah, contains large cross beds formed 200 Ma.

GRADED BEDDING. When unconsolidated sediment slides down an underwater slope, it mixes with water to produce a murky, turbulent cloud. This cloud of suspended sediment is denser than clear water, so it flows downslope as an underwater avalanche known as a **turbidity current** (**Fig. 5.29a–c**). At the base of the slope, the turbidity current slows, so the sediment that it carries settles out. Larger grains sink faster than finer grains, so the coarsest sediment accumulates first, and progressively finer grains accumulate on top, with the finest sediment (clay) settling out last. This process forms a **graded bed**, a layer of sediment in which grain size varies from coarse at the bottom to fine at the top. A succession of deposits from turbidity currents is called a *turbidite*.

Sediments Exposed to the Air

In some environments, mud (clay mixed with water) becomes exposed to air after it has been deposited. If it rains before the mud dries completely, small impact craters form where raindrops strike the mud surface. If the sediment dries out, mud layers shrink and break into hexagonal plates that warp up at their edges. Gaps, known as **mudcracks**, develop between the plates (**Fig. 5.30a, b**). Burial of the mud layer and its

Figure 5.29 The development of graded beds from turbidity currents.

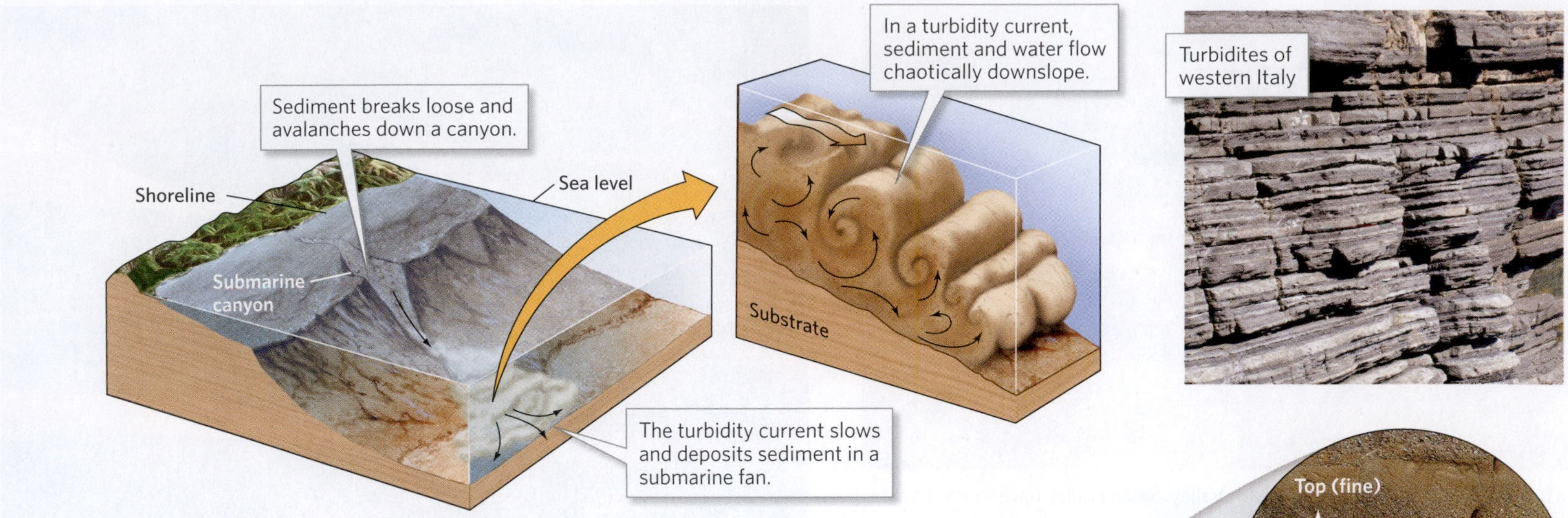

Sediment breaks loose and avalanches down a canyon.

Shoreline

Sea level

Submarine canyon

In a turbidity current, sediment and water flow chaotically downslope.

Substrate

Turbidites of western Italy

The turbidity current slows and deposits sediment in a submarine fan.

(a) An earthquake or storm can trigger a turbidity current (underwater avalanche). These currents commonly flow down submarine canyons.

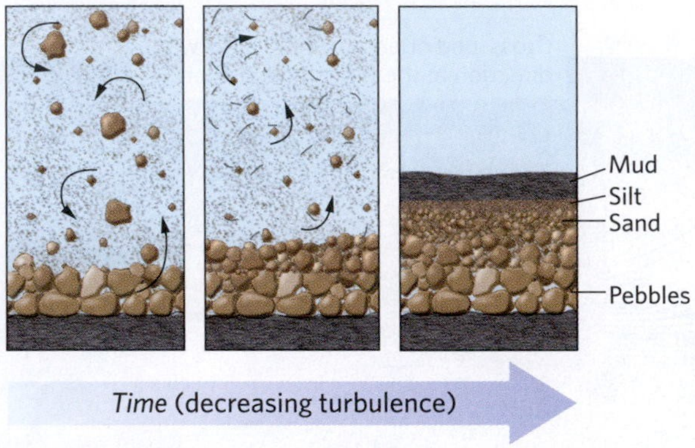

Mud
Silt
Sand

Pebbles

Time (decreasing turbulence)

(b) As the turbidity current slows, the coarsest grains settle out first, and the finest grains settle out last.

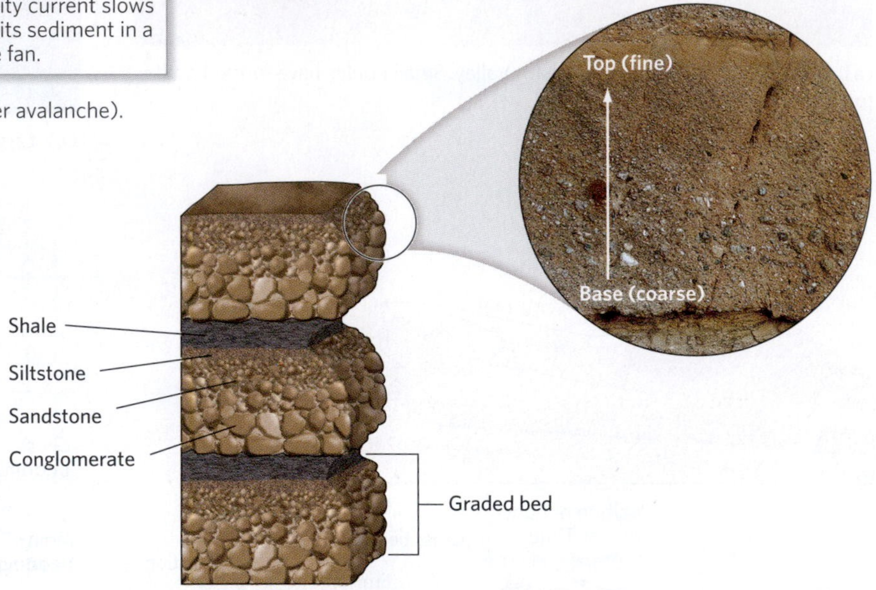

Top (fine)

Base (coarse)

Shale

Siltstone

Sandstone

Conglomerate

Graded bed

(c) Repetition of the process produces a succession of graded beds; each bed is coarser at the bottom and finer at the top.

Figure 5.30 Formation of mudcracks.

(a) Mudcracks in red mud at Bryce Canyon, Utah. As the mud dries, it contracts, and the cracks form.

(b) Mudcracks preserved in a 410-Ma bed exposed on the base of a cliff in New York.

subsequent lithification can preserve raindrop impressions and mudcracks.

Take-home message . . .

🏠 Sedimentary structures include bedding, ripple marks and dunes, cross bedding, mudcracks, and graded beds, all formed during or soon after deposition of sediment. Some of these features form on the surfaces of beds, some within beds.

Quick Question -
What role does a current play in developing sedimentary structures?

5.9 Recognizing Depositional Environments

By looking at strata, can we characterize the **depositional environment**, meaning the conditions in which the sediments that formed the strata accumulated? Yes. To identify depositional environments, geologists, like crime-scene investigators, look for clues: detectives seek fingerprints and bloodstains to identify a culprit, while geologists examine grain size, composition, sorting, sedimentary structures, and **fossils** (the relicts of organisms buried with the sediment) to identify a depositional environment. In this section, we journey through several depositional environments (**Earth Science at a Glance**, pp. 182–183), each of which leaves a distinct fingerprint in strata.

When Sediments Accumulate on Land

We begin our exploration of depositional environments by considering those that develop inland from the seashore, so that they are not affected by ocean tides and waves. Such nonmarine settings, or *terrestrial depositional environments*, include dry land, lakes, streams, and swamps. Commonly, oxygen in surface water or groundwater reacts with iron in terrestrial sediments to produce rust-like iron oxide minerals, which give the sediments an overall reddish hue. Geologists informally refer to strata with this hue as **redbeds** (see Fig. 5.26a).

GLACIAL DEPOSITS. If we travel high in the mountains, or to polar regions, where it's so cold that more snow collects in the winter than melts away, we find glaciers growing and slowly flowing. The solid ice of a glacier can carry along all the sediment—from clay-sized to boulder-sized—that falls on the glacier from adjacent cliffs or that gets eroded from the ground at its base or sides. When the ice melts away, the sediment that was in it or on it accumulates as

glacial till (**Fig. 5.31a**). Since ice does not sort sediment by size, rocks formed from till typically contain larger clasts randomly distributed through mudstone or siltstone. Such rock is called **diamictite.**

MOUNTAIN STREAM DEPOSITS. Traveling beyond the end of the glacier, we enter a realm where turbulent streams rush down valleys. This fast-moving water carries large clasts during floods, and the clasts grind together and become rounded. Between floods, when water flow slows, the cobbles and boulders settle out, while the stream carries sand and mud farther downstream (**Fig. 5.31b**). When lithified, stream gravel becomes conglomerate.

ALLUVIAL-FAN DEPOSITS. When we arrive at a mountain front, where fast-moving streams empty onto a plain, the water slows, causing sediments to accumulate. In arid regions, this accumulation produces a wedge-shaped pile called an *alluvial fan*, consisting of sand and gravel (**Fig. 5.31c**). The sand of an alluvial fan may contain feldspar grains that have not yet weathered into clay, so alluvial-fan sediments, when lithified, become arkose and conglomerate.

SAND-DUNE DEPOSITS. In very dry climates, few plants grow, so the ground surface lies exposed to wind. The moving air builds dunes composed of well-sorted, well-rounded sand. If buried, these dunes become thick, cross-bedded sandstone (**Fig. 5.31d**).

RIVER DEPOSITS. In climates where rivers flow, we find several distinctive depositional environments. Rivers transport gravel, sand, silt, and mud. Coarser sediments tumble along the riverbed and collect in cross-bedded, rippled layers, while finer sediments drift along in the water. These fine sediments settle out along the banks of the river or on the floodplain, where they dry out between floods. River sediments lithify to form sandstone, siltstone, and shale (or mudstone). Typically, the coarser sediments collect in the river channel, surrounded by layers of fine-grained floodplain deposits. In cross section, channel deposits have a lens-like shape (**Fig. 5.31e**). Shales in these deposits may contain mudcracks. Geologists commonly refer to river deposits as *fluvial sediments* (from the Latin *fluvius*, meaning river).

LAKE DEPOSITS. The quiet offshore water of lakes can't transport coarse sediment, so only fine clay reaches the middle of the lake, where it settles to form mud on the lake bed. Lake sediments, therefore, become finely laminated shale (**Fig. 5.31f**). A small delta forms where a stream enters a lake, for the moving water of the stream slows as it enters standing water of the lake. In a lake delta, gravel typically collects near the shore in horizontal beds, sand on sloping beds offshore, and silt on horizontal beds where the delta merges with the deep lake bed (**Fig. 5.32**).

Figure 5.31 Examples of terrestrial depositional environments.

(a) Glacial till at the toe of a glacier in France.

(b) Coarse mountain stream deposits in Colorado.

(c) An alluvial fan in Death Valley, California.

(d) Sand dunes in Brazil.

(e) A 310-Ma river channel exposed in a quarry in Indiana.

Edge of photo

Younger floodplain deposits

Channel fill

Older floodplain deposits

(Talus)

What an Earth Scientist Sees

(f) Laminated mud that accumulated on a lake bed.

When Sediments Accumulate in the Sea

A variety of depositional environments occur along coasts, in shallow water offshore, and beneath the deep sea. All of these settings are types of *marine depositional environments*. The type of sediment deposited in a specific marine setting depends on the water depth, wave energy, and water temperature, and on whether or not clasts of quartz and clay are present.

MARINE DELTA DEPOSITS. Where a river empties into the sea, a large *marine delta* may accumulate. Such deltas are more complex than the small lake deltas that we mentioned earlier, for marine deltas host a variety of depositional environments, including swamps, channels, floodplains, and submarine slopes. Sea-level changes may cause the positions of these environments to change over time. The deposits of a marine delta, therefore, produce a great variety of sedimentary rock types (Fig. 5.33a).

COASTAL CLASTIC DEPOSITS. Ocean currents and waves transport sand along the coastline. The sand washes back and forth in the surf, so it becomes well sorted and well rounded (Fig. 5.33b). Where the coast has a very gentle slope, broad tidal flats develop, on which ripple-marked beds of silt accumulate, and in lagoons protected from

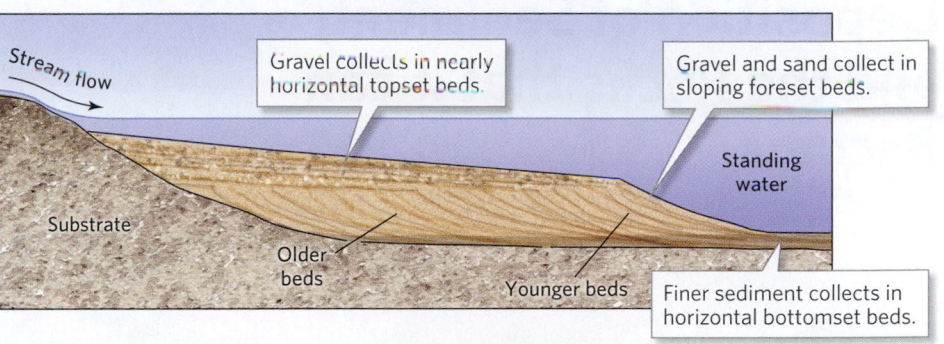

Figure 5.32 A cross section of a small delta that formed where a stream enters a lake. Different types of sediments accumulate in different parts of the delta.

Stream flow

Gravel collects in nearly horizontal topset beds.

Gravel and sand collect in sloping foreset beds.

Standing water

Substrate

Older beds

Younger beds

Finer sediment collects in horizontal bottomset beds.

Figure 5.33 Examples of shallow-marine depositional environments.

Not to scale

River-mouth sand and silt
Marsh (organic-rich mud)
River bank
River channel
Organic-rich mud
Shoreline
Submarine mudflows
Sea
Delta face
Fluvial channel sand and silt
Shallow-marine mud and silt
Silt, interbedded with mudflows and turbidites
Turbidite
Deeper-water mud and silt

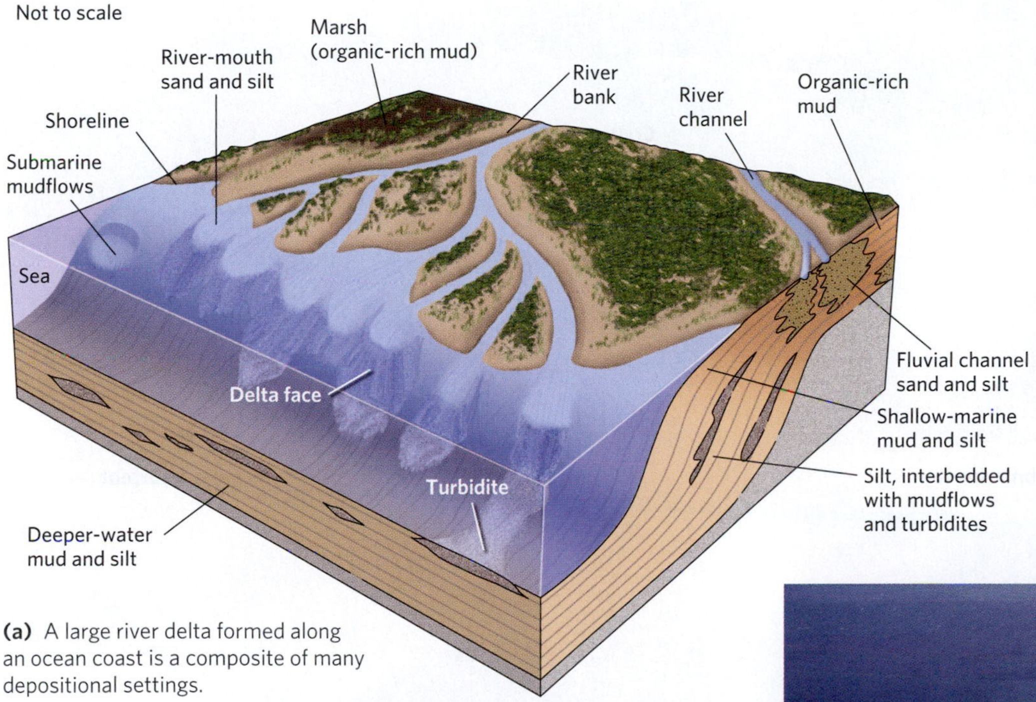

(a) A large river delta formed along an ocean coast is a composite of many depositional settings.

(b) Waves along the coast sort beach sand.

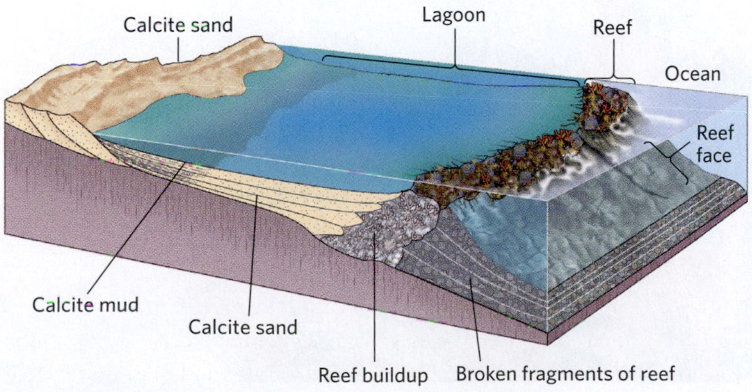

Calcite sand
Lagoon
Reef
Ocean
Reef face
Calcite mud
Calcite sand
Reef buildup
Broken fragments of reef

(c) Carbonate reefs form along shorelines where the water is clear and warm.

Reef

(d) A reef surrounds a tropical island in the Pacific.

Deposition of Sediment, and the Rocks that it can Become

Glacial environment

Estuary

Beach

Bar

Continental shelf

Coastal erosion

Turbidity current

Submarine fan

Deep-sea current

Redbeds

Bedding

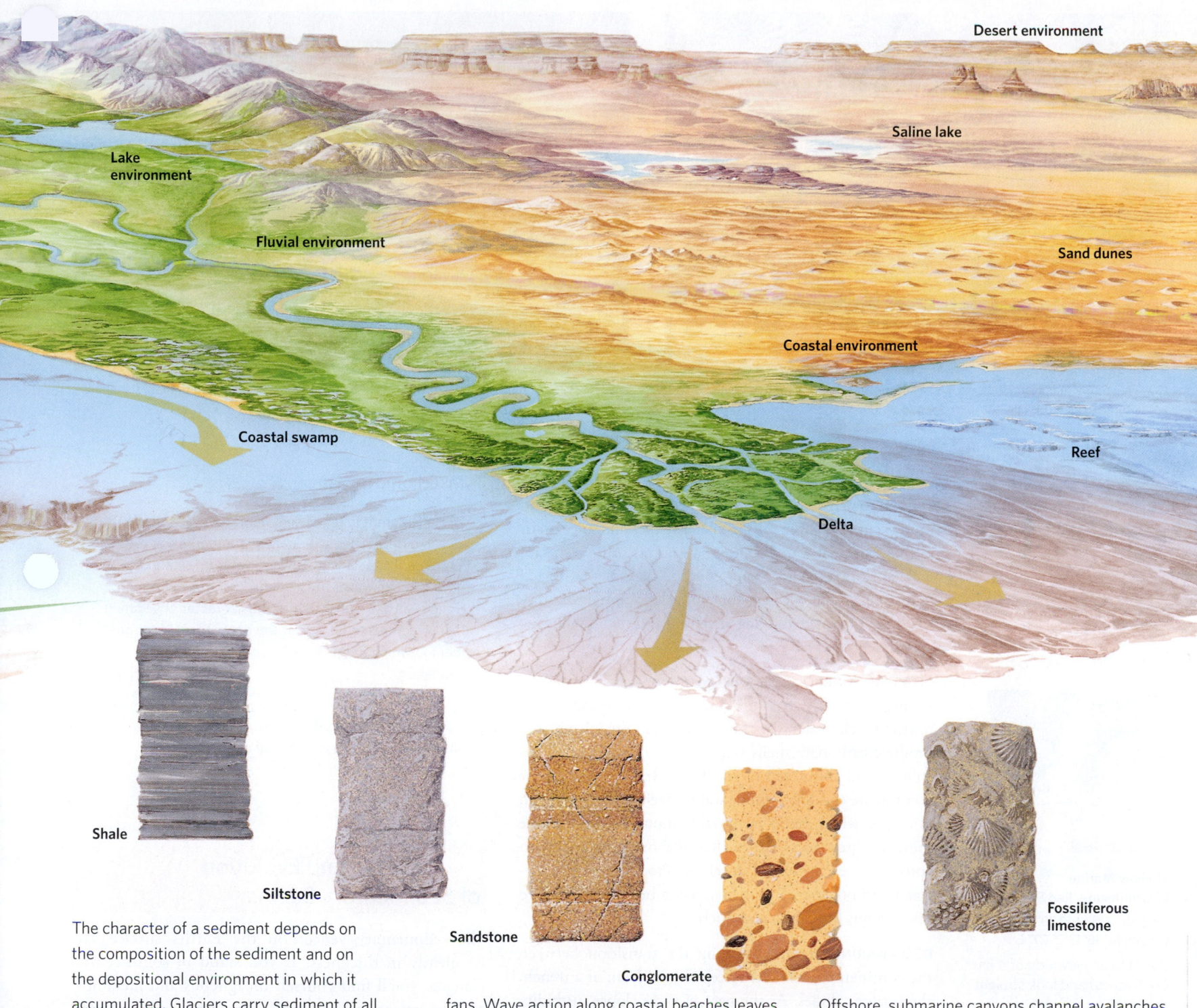

Desert environment

Saline lake

Lake environment

Fluvial environment

Sand dunes

Coastal environment

Coastal swamp

Reef

Delta

Shale

Siltstone

Sandstone

Conglomerate

Fossiliferous limestone

The character of a sediment depends on the composition of the sediment and on the depositional environment in which it accumulated. Glaciers carry sediment of all sizes, so they leave deposits of poorly sorted till. Streams deposit coarser sands and gravels in their channels and finer ones on floodplains. In the quiet water of lakes, fine-grained mud accumulates. In desert environments, sand builds into dunes, evaporates precipitate in saline lakes, and gravel and sand gather in alluvial fans. Wave action along coastal beaches leaves behind well-sorted sand. In swampy areas, large volumes of plant matter accumulate. Where a river flows into the sea, its water slows and deposits a large delta. In warmer coastal marine environments, carbonate sediments, from the shells of organisms, are deposited. Locally, corals and other organisms build carbonate reefs. Offshore, submarine canyons channel avalanches of sediment, or turbidity currents, out to the deep seafloor. Far from shore, pelagic marine sediment, commonly from the shells of plankton, slowly settles out. Burial and lithification of these various types of sediments turn them into examples of the different classes of sedimentary rocks— clastic, chemical, biochemical, and organic.

Figure 5.34 Deep-marine carbonate sediments.

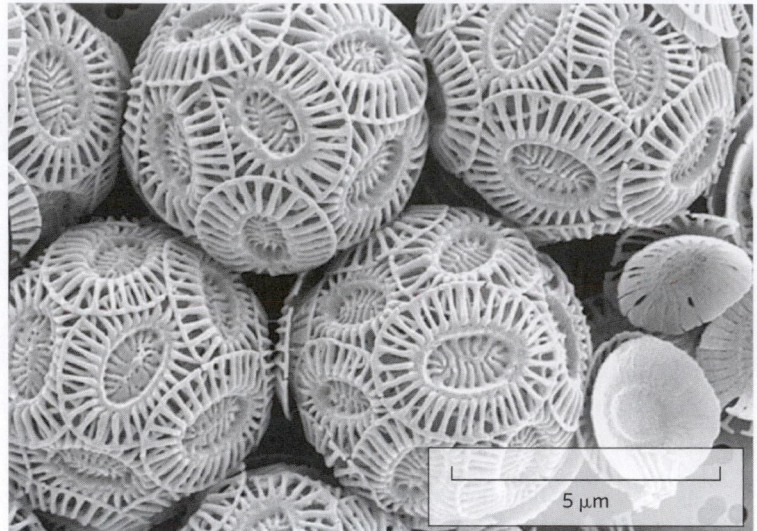

(a) These tiny plankton shells consist of carbonate minerals. The shells are so small that they could pass through the eye of a needle.

(b) The white chalk cliffs of southeastern England consist of carbonate plankton shells deposited about 80 Ma.

waves, mud settles out. Offshore, where the water is deep and wave energy does not much affect the seafloor, silt and mud accumulate. Organisms may live in abundance on or in this sediment, but because the water is still, its surface is free of ripple marks. So, if you find beds of well-sorted sandstone, along with siltstone and shale containing marine fossils, you are looking at coastal and shallow-marine clastic deposits.

COASTAL CARBONATE DEPOSITS. In shallow-marine settings relatively free of sand and clay, warm, clear, nutrient-rich water can host abundant organisms that produce carbonate shells **(Fig. 5.33c, d)**. Nearby beaches collect sand composed of small shell fragments, lagoons accumulate carbonate mud, and on reefs, organisms such as corals, which build immovable mounds of carbonate minerals, grow in place. Farther offshore from a reef, an apron of reef fragments develops. Products of shallow-water carbonate environments, when buried, lithify into fossiliferous limestone and micrite.

DEEP-MARINE DEPOSITS. Along the transition between the continental shelf and a deep abyssal plain or a trench, the seafloor slopes. At the base of these slopes, turbidity currents deposit graded beds. On abyssal plains, far from land, only fine clay and plankton provide a source of sediment. The clay eventually settles on the deep seafloor, forming deposits that become finely laminated mudstones. The plankton shells settle and lithify to form chert or chalk **(Fig. 5.34a, b)**. Deposits of mudstone, chalk, or bedded chert, therefore, indicate a deep-marine depositional environment.

See for yourself

Shallow Marine Environments, Red Sea, Egypt

Latitude: 22°38′17.70″ N
Longitude: 36°13′21.27″ E

Zoom to an elevation of 4 km (~2.5 miles) and look straight down.

The desert sands of the Sahara abut the blue waters of the Red Sea. What types of sedimentary rocks would form if the sediments visible in this view were to be buried and preserved?

Take-home message . . .

Different types of sedimentary rocks form in different depositional environments. For example, strata deposited along a river differ from strata deposited by ocean waves, by glaciers, or in the deep sea. Geologists use clues such as the character of sedimentary structures to figure out the depositional environment in which the sediments under study were deposited.

Quick Question ------------------------------
What characteristics distinguish the deposits of an alluvial fan from the deposits of a marine delta?

5.10 Origin and Evolution of Sedimentary Basins

The sedimentary veneer on the Earth's surface varies greatly in thickness. If you stand in south-central Canada, you'll find yourself on igneous and metamorphic basement that's over a billion years old, and sedimentary rocks will be nowhere in sight. But if you stand along the southern coast of Texas, you'll have to drill through 15 km (10 miles) of sedimentary beds before reaching igneous and metamorphic basement. Thick accumulations of sediment form only in certain regions where the surface of the Earth sinks, or undergoes **subsidence**, forming a depression in which sediment

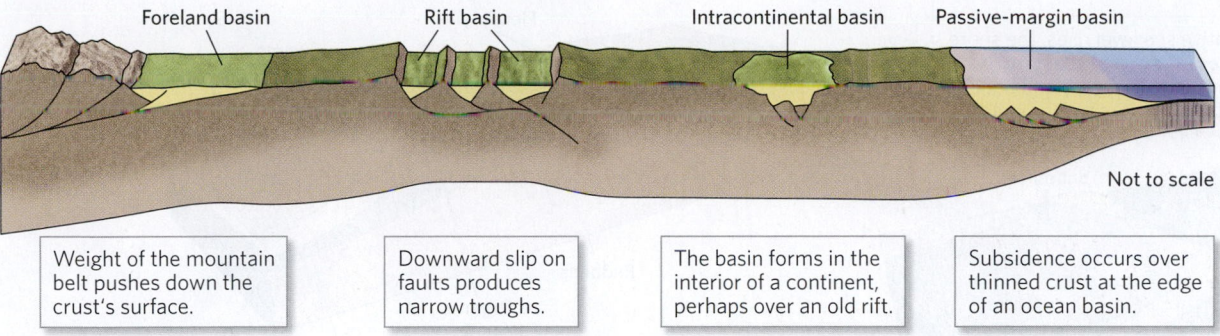

Figure 5.35 The geologic settings of sedimentary basins.

Foreland basin   Rift basin   Intracontinental basin   Passive-margin basin

Not to scale

Weight of the mountain belt pushes down the crust's surface.

Downward slip on faults produces narrow troughs.

The basin forms in the interior of a continent, perhaps over an old rift.

Subsidence occurs over thinned crust at the edge of an ocean basin.

collects. Geologists refer to a sediment-filled depression as a **sedimentary basin**. In what geologic settings do sedimentary basins form, and why does the character of the sediments deposited at a location in them change over time? We have to consider plate tectonics theory and sea-level change to find the answers.

Why Do Sedimentary Basins Form?

Geologists distinguish among different kinds of sedimentary basins in the context of plate tectonics theory **(Fig. 5.35)**:

- *Rift basins:* In continental rifts, regions where the lithosphere has stretched horizontally and, therefore, thinned vertically, blocks of crust slide downward along faults, producing low troughs bordered by narrow mountain ridges (see Fig. 2.25). These troughs, which fill with sediment, are called rift basins.

- *Passive-margin basins:* Passive continental margins, as we saw in Chapter 2, are continental margins that are not plate boundaries. They form after continental rifting has finished and a new mid-ocean ridge has formed. The continental crust that had stretched and heated during the rifting event slowly cools and undergoes subsidence, creating a *passive-margin basin*. Such basins range from 150 to 400 km (90 to 250 miles) wide and fill with various types of marine sediment. The top surface of a passive-margin basin is a continental shelf.

- *Intracontinental basins:* Bowl-shaped depressions in the interiors of continents are called *intracontinental basins*. They probably formed over places where continents underwent unsuccessful rifting during the latter part of the Precambrian, for when rifting ceased, the rift and surrounding continent cooled and subsided.

- *Foreland basins:* Along the edges of mountain belts that formed during continental collisions, and sometimes during the interaction of plates at a convergent boundary, large slices of rock push up faults and onto the surface of the continent. The weight of these slices pushes down on the surface of the lithosphere, producing a wedge-shaped depression adjacent to the mountain range that fills with sediment eroded from the range. Fluvial and deltaic strata accumulate in foreland basins.

Transgression and Regression

When relative sea level rises, the shoreline migrates inland, a process called **transgression**, and when relative sea level falls, the shoreline migrates seaward, a process called **regression** (Fig. 5.36). We can see the consequences of transgression and regression because they affect the depositional environment at a location. The process of transgression and regression can lead to the formation of a broad blanket of sediment, not all of which was deposited at the same time. At times in Earth history, sea level rose so far that large areas of continental interiors were submerged. Later, sea level dropped and these regions became dry land again. Therefore, we see the record of transgression and regression in the strata of continental interiors.

Changes Underground: Diagenesis

Earlier in this chapter, we discussed the process of lithification, by which sediment hardens into rock. Lithification represents one aspect of a broader process that geologists call **diagenesis**. This term includes not only all the physical, chemical, and biological processes that transform sediment into sedimentary rock, but also the processes that alter the characteristics of a sedimentary rock after the rock has formed. In sedimentary basins, strata may become very deeply buried. As a result,

Figure 5.36 The effects of transgression and regression during deposition of a sedimentary sequence.

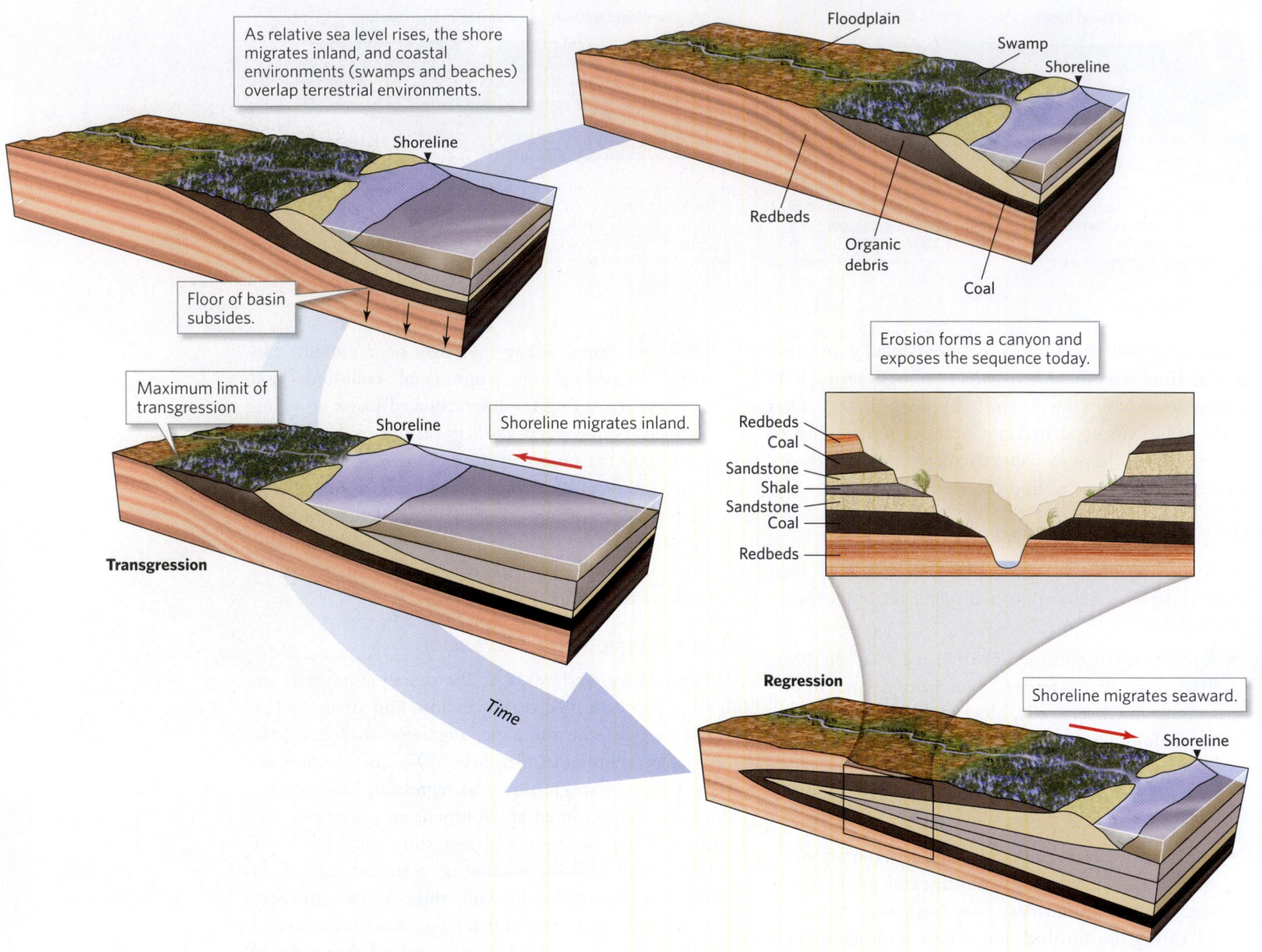

As relative sea level rises, the shore migrates inland, and coastal environments (swamps and beaches) overlap terrestrial environments.

Shoreline

Floodplain

Swamp

Shoreline

Redbeds

Organic debris

Coal

Floor of basin subsides.

Maximum limit of transgression

Shoreline

Shoreline migrates inland.

Erosion forms a canyon and exposes the sequence today.

Redbeds
Coal
Sandstone
Shale
Sandstone
Coal
Redbeds

Transgression

Regression

Time

Shoreline migrates seaward.

Shoreline

the rocks endure high pressures and temperatures and come in contact with warm groundwater. Diagenesis under such conditions can cause chemical reactions that can produce new minerals in the rock and can also cause existing cements to dissolve and new ones to precipitate.

When geologists use the term *diagenesis*, they are referring to processes that take place in sedimentary rocks and do not produce the textures and minerals found in metamorphic rocks. The gradual transition from diagenesis to metamorphism happens at temperatures between 200°C and 300°C (390°F and 570°F). In the next chapter, we enter the realm of true metamorphism.

Take-home message . . .

Sedimentary basins form where the surface of the Earth sinks to form a depression in which thick deposits of sediment accumulate. Basins form in rifts, along passive margins, and along the fronts of mountain ranges. The type of sediment deposited at a location changes as sea level rises or falls. Once lithified, strata may undergo diagenesis.

Quick Question -
Explain why marine strata underlie large regions now in the middle of continents.

Another View These cliffs in New Mexico formed from huge sand dunes that were deposited at 180 Ma.

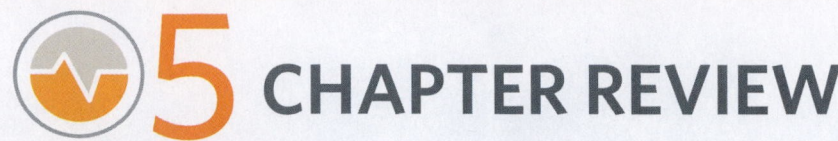

5 CHAPTER REVIEW

- Rocks at or near the Earth's surface undergo physical and chemical weathering and, as a result, eventually disintegrate into loose clasts. The process also produces ions in solution.

- Soil forms when downward-percolating rainwater and living organisms interact with rock and debris. Leaching takes place nearer the surface and accumulation farther down.

- The character of soil depends on climate, on time, source composition, and other factors. Soil develops distinct horizons. Environments affect the nature of these horizons.

- Soil is an essential resource, but is undergoing destruction by erosion in some locations.

- Sediment consists of fragments of pre-existing rocks, minerals precipitated from water, or shells and fragments of shells.

- Geologists recognize four major types of sedimentary rocks.

- Clastic sedimentary rocks form when sediment, produced by weathering and erosion, undergoes transportation deposition, burial, and finally, lithification. Such rocks are classified by grain size.

- Biochemical sedimentary rocks develop from the shells of organisms, and organic sedimentary rocks consist of plant debris or of altered plankton remains. Chemical sedimentary rocks precipitate directly from water. Organic sedimentary rocks are formed from materials produced by living organisms.

- Different sedimentary rocks have different compositions. Limestone consists of calcite, chert of silica, coal of carbon, shale of clay, sandstone of quartz, and evaporites of salt.

- Sedimentary structures formed during deposition include bedding, cross bedding, graded bedding, ripple marks, dunes, and mudcracks. These structures serve as clues to depositional environments.

- Glaciers, streams, alluvial fans, deserts, rivers, lakes, deltas, beaches, shallow seas, and deep seas each accumulate a different, distinctive assemblage of sedimentary strata.

- Thick layers of sedimentary rocks accumulate in sedimentary basins.

- Transgressions occur when sea level rises and the coastline migrates inland. Regressions occur when sea level falls and the coastline migrates seaward.

Key Terms

bed (pp. 166, 174)
bedding (p. 174)
biochemical sedimentary rock (p. 170)
carbonate rock (p. 170)
cementation (p. 167)
chemical sedimentary rock (p. 172)
chemical weathering (p. 157)
clast (p. 156)
clastic sedimentary rock (p. 166)
coal (p. 171)
compaction (p. 167)
contact (p. 176)
cross bed (p. 174)

deposition (p. 167)
depositional environment (p. 179)
diagenesis (p. 185)
diamictite (p. 179)
dissolution (p. 159)
dolostone (p. 172)
dune (p. 174)
erosion (p. 157)
evaporite (p. 172)
fossil (p. 179)
frost wedging (p. 159)
graded bed (p. 177)
grain (p. 156)
hydrolysis (p. 159)
joint (p. 157)

limestone (p. 170)
lithification (p. 167)
maturity (p. 168)
mudcrack (p. 177)
organic sedimentary rock (p. 171)
oxidation (p. 159)
physical weathering (p. 157)
redbed (p. 179)
regression (p. 185)
ripple mark (p. 174)
root wedging (p. 159)
salt wedging (p. 159)
sediment (p. 156)
sedimentary basin (p. 185)
sedimentary rock (p. 165)

sedimentary structure (p. 174)
soil (p. 161)
soil erosion (p. 165)
soil horizons (p. 162)
soil profile (p. 162)
sorting (p. 167)
strata (p. 174)
stratigraphic formation (p. 176)
subsidence (p. 184)
subsoil (p. 163)
topsoil (p. 163)
transgression (p. 185)
turbidity current (p. 177)
weathering (p. 157)
zone of accumulation (p. 162)
zone of leaching (p. 161)

The letters following each Review Question refer to the corresponding Learning Objective from the Chapter Opener.

1. Explain the differences between physical and chemical weathering. What are the features shown in the illustration? **(A)**

2. What are the various reactions that can be involved in chemical weathering? **(A)**

3. Explain the process of soil formation. How does soil differ from sediment? **(B)**

4. What are soil horizons, and what factors determine the character of soil? **(B)**

5. How does a soil formed in a tropical environment differ from one formed in an arid climate? **(B)**

6. How have soil erosion rates been affected by human activity? **(B)**

7. Describe how a clastic sedimentary rock forms from its unweathered parent rock. **(C)**

8. Explain how biochemical sedimentary rocks form. **(C)**

9. How do grain size, sorting, and angularity change as sediments move downstream? **(C)**

10. Describe the two different kinds of chert. How are they similar? How are they different? **(C)**

11. Do all chemical sedimentary rocks have the same composition? What conditions yield evaporites? **(C)**

12. How does dolostone differ from limestone, and how does dolostone form? What kinds of rock form in the environment shown by the illustration? **(D)**

13. What are cross beds, and how do they form? How can you read the current direction from cross beds? **(D)**

14. Describe how a turbidity current forms and moves. How does it produce graded bedding? **(D)**

15. Compare the sediments of an alluvial fan with those of a deep-marine deposit. What kinds of deposits can form along a coast? What conditions lead to the deposition of sediment that will eventually turn into limestone? **(E)**

16. Why don't thick layers of sediment accumulate everywhere? What types of geologic settings lead to the formation of sedimentary basins? **(F)**

17. Exploration of Mars by robotic vehicles suggests that layers of sedimentary rock cover portions of the planet's surface. Some researchers claim that the layers contain cross bedding and relicts of gypsum crystals. At face value, what do these features suggest about depositional environments on Mars in the past? **(E)**

18. The Gulf Coast of the United States hosts a passive-margin basin. Drilling reveals that the base of the sedimentary succession in this basin consists of redbeds. These are overlain by a thick layer of evaporite, over which lie beds of sandstone and shale containing fossils of marine organisms. Be a sedimentary detective and describe the environments in which the sediments in this basin were deposited. Has the environment of deposition at this location changed over time? **(E)**

19. Examine the Bahamas with *Google Earth*™ at latitude 23°58′40.98″ N, longitude 77°30′20.37″ W. Note that broad expanses of very shallow water surround the islands, that white-sand beaches occur along the coasts of the islands, and that small reefs occur offshore. What does the sand consist of, and what rock will it become if it eventually becomes buried and lithified? **(E)**

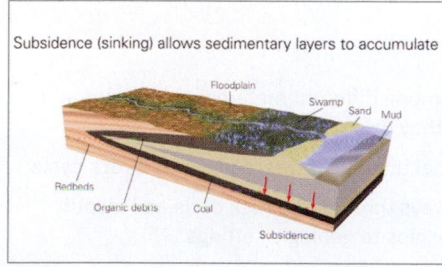

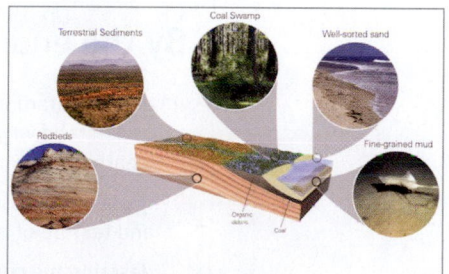

Videos
This chapter features videos on the formation of sedimentary rocks and soil, transgression and regression, and how these processes appear in the stratigraphic record.

Smartwork5
This chapter features questions on recognizing types of sedimentary rocks, structures, and environments.

6 A PROCESS OF CHANGE
Metamorphism and the Rock Cycle

By the end of the chapter you should be able to . . .

A. define metamorphism and metamorphic rock, and understand why metamorphism is different from weathering or melting.

B. explain processes that change the mineral assemblage and texture of rock during metamorphism.

C. describe the conditions and geologic phenomena in the Earth that cause metamorphism.

D. recognize a metamorphic rock and explain how geologists classify metamorphic rocks.

E. distinguish among different metamorphic grades and explain why they exist.

F. design a model that illustrates stages in the rock cycle.

G. sketch pathways through the rock cycle and identify their relationships to geologic settings.

H. discuss the sources of energy that drive the rock cycle on Earth.

6.1 Introduction

After a caterpillar grows to full size, it attaches to a branch and becomes encased in a rigid chrysalis. Inside, the organic material of what was the caterpillar reorganizes and a butterfly with broad, colorful wings grows **(Fig. 6.1a)**. Biologists refer to this amazing change as *metamorphosis*, from the Greek words *meta*, meaning change, and *morphe*, meaning form. Under appropriate geologic conditions, rocks can undergo changes every bit as dramatic as those that take place during the transformation of a caterpillar into a butterfly. Geologists use the term **metamorphism** for this process of change in rocks and the term *metamorphic rock* for the product of this change. More precisely, a **metamorphic rock** is a rock that forms when a pre-existing rock, or **protolith**, undergoes a solid-state change in response to the modification of its environment **(Fig. 6.1b)**.

Let's consider the components of this definition more closely. By *solid state*, we mean that a metamorphic rock does not form from molten rock (magma or lava) for, by definition, rocks that solidify from melt are igneous. By *change*, we mean that metamorphism produces new minerals that did not exist in the protolith and/or produces a new *texture* (an arrangement or alignment of mineral grains) that differs from that of the protolith. And by *modification of its environment*, we mean that metamorphism takes place when the temperature, pressure, or fluid composition affecting the rock becomes different from what it had been before.

This outcrop, on a hill in Scotland, reveals metamorphic rock that formed when sedimentary rock was subjected to high temperatures and pressures after being buried very deeply, during mountain building. In response to a change in environment, a new group of metamorphic minerals grew.

Figure 6.1 Dramatic transformations.

Before

Before

After

After

(a) A caterpillar undergoes metamorphosis to become a butterfly.

(b) A shale undergoes metamorphism to become a garnet schist.

In the previous two chapters, we discussed the two other principal rock types (igneous and sedimentary). Because the Earth remains geologically active, the processes that form igneous, sedimentary, or metamorphic rocks are always happening somewhere. Over time, as a result, atoms in one kind of rock may eventually end up in another kind of rock. We refer to the transfer of materials among the three basic rock types due to geologic processes as the *rock cycle*.

This chapter begins by explaining the causes of metamorphism and the basis for classifying metamorphic rocks. We then focus on the geologic settings in which these rocks form in the context of plate tectonics theory. We conclude this chapter—and our introduction to Earth materials overall—by examining the rock cycle.

Nothing in the world lasts, save eternal change.

—HONORAT DE BUEIL (FRENCH POET, 1589–1650)

6.2 Causes and Consequences of Metamorphism

What Is a Metamorphic Rock?

If someone were to put a hand specimen of rock on a table in front of you, how would you know that it is metamorphic? First, metamorphic rocks may possess **metamorphic minerals**, new minerals that grow in place within solid rock during metamorphism. In fact, metamorphism may produce a *metamorphic mineral assemblage*, meaning a group of minerals that were not in the protolith. Second, metamorphic rocks can display **metamorphic textures**, distinctive arrangements and orientations of mineral grains not found in other rock types. For example, metamorphism may produce *metamorphic foliation*, a distinct type of layering. When metamorphic minerals or textures develop, a metamorphic rock can become as different from its protolith as a butterfly is from a caterpillar.

Processes Involved in Metamorphism

The process of forming metamorphic minerals and textures takes place very slowly (over thousands to millions of years) and involves several processes, which sometimes occur alone and sometimes together. Examples of metamorphic processes include the following:

- *Recrystallization:* The process of recrystallization changes the shape and size of certain mineral grains in a rock without changing the identity of the minerals (Fig. 6.2a). For example, recrystallization may transform a sandstone, composed of tiny, rounded quartz grains held together by quartz cement, into quartzite, a metamorphic rock with larger interlocking quartz crystals.

- *Phase change:* If a type of mineral in a rock transforms into another mineral that has the same chemical composition, but a different crystal structure, we say that a **phase change** has taken place. For example, the transformation of graphite into diamond represents a phase change, for while both minerals consist of pure carbon, they have different crystal lattices. A phase change due to compression (squeezing) involves the placement of atoms in a more compact arrangement.

- *Metamorphic reactions:* When subjected to elevated pressures and temperatures during metamorphism, clay in a shale may be replaced by a mixture of mica and garnet. Such a change, involving the growth of new minerals that were not present in the protolith, exemplifies a *metamorphic reaction* (Fig. 6.2b). During a metamorphic reaction, atoms separate from pre-existing minerals and re-bond to form new minerals. Typically, new metamorphic minerals grow to interlock with one another, so metamorphic rocks have a crystalline texture.

- *Pressure solution:* In water, mineral grains tend to dissolve more rapidly at points where the grains are being pressed together more tightly. Because of this phenomenon, known as *pressure solution*, grains may change shape (Fig. 6.2c). Because different minerals dissolve at different rates, pressure solution can also alter the overall composition of a rock.

- *Plastic deformation:* At elevated temperatures and pressures, mineral grains can change shape very slowly without breaking (Fig. 6.2d). This process, called *plastic deformation*, can produce new textures.

Figure 6.2 Metamorphic processes as seen through a microscope.

Protolith ———————→ Metamorphic rock

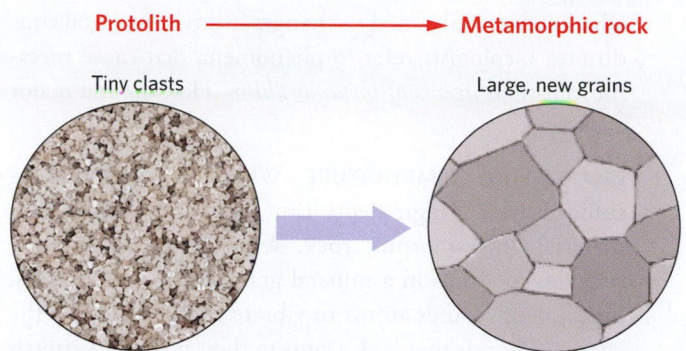

Tiny clasts / Large, new grains

(a) Mineral grains may recrystallize to form new, interlocking grains of the same mineral. Typically, the new grains are larger.

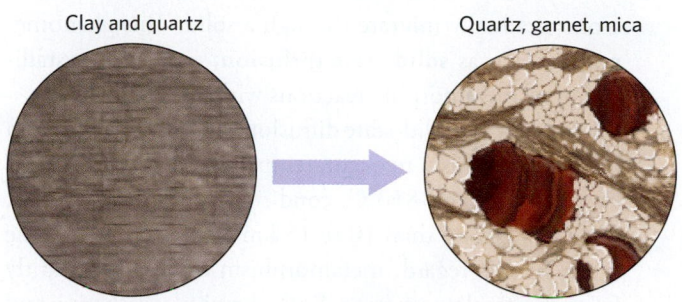

Clay and quartz / Quartz, garnet, mica

(b) Metamorphic reactions may change the original mineral assemblage into a new, metamorphic mineral assemblage.

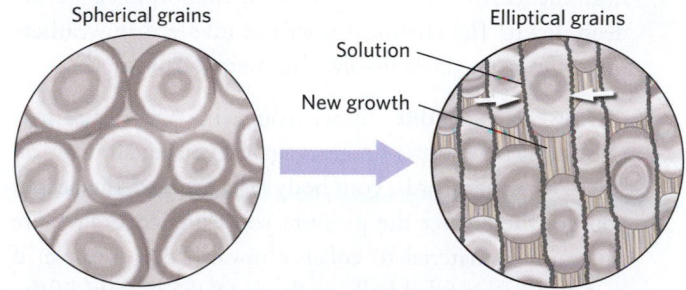

Spherical grains / Elliptical grains
Solution
New growth

(c) Pressure solution dissolves grains on the sides being pressed together more tightly. White arrows indicate the squeezing direction.

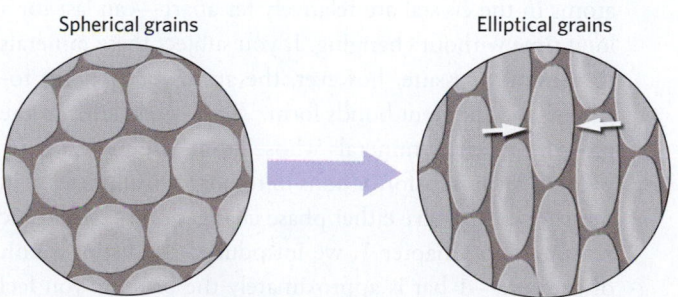

Spherical grains / Elliptical grains

(d) Plastic deformation changes the shape of grains, without breaking them. White arrows indicate the compression direction.

**Did you
ever wonder . . .**

how temperature and
pressure change with
depth?

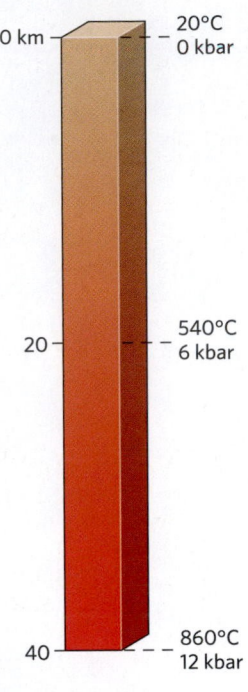

Conditions That Cause Metamorphism

Caterpillars undergo metamorphosis because of hormonal changes in their bodies. Rocks undergo metamorphism when they are subjected to changes in environmental conditions. Geologists refer to phenomena that cause metamorphism as *agents of metamorphism*. Here are the major agents:

CHANGE IN TEMPERATURE. When a rock heats sufficiently, its ingredients can transform into a new material—metamorphic rock. Why? Think about what happens to atoms in a mineral grain as the grain warms. Heat energy causes atoms to vibrate more rapidly, so the chemical bonds that lock atoms to their neighbors stretch and bend. If the bonds stretch or bend too far and break, atoms detach from their original neighbors, move slightly, and form new bonds with other atoms. As a consequence, atoms gradually migrate through a solid. This phenomenon, known as **solid-state diffusion**, enables recrystallization or metamorphic reactions within solid rock.

In order for solid-state diffusion to happen fast enough for metamorphism to occur, temperatures must rise to between 250°C and 850°C, conditions that generally exist at depths greater than 10 to 15 km (6 to 10 miles) in the crust. In this regard, metamorphism differs significantly from chemical weathering. Both chemical weathering and metamorphism can change the minerals in a rock, but chemical weathering occurs under low temperatures at or near the Earth's surface, whereas metamorphism occurs deep down. The chemical reactions involved in weathering differ from those involved in metamorphism.

CHANGE IN PRESSURE. When you swim underwater in a swimming pool, water squeezes against you equally from all sides—in other words, your body feels *pressure*. The deeper you go, the greater the pressure you experience. Pressure can cause a material to collapse inward. For example, if you pull an air-filled balloon down to the bottom of the pool, the balloon becomes smaller. Pressure can have the same effect on minerals. Near the Earth's surface, minerals with relatively open crystal structures—meaning that the atoms in the crystal are relatively far apart—can last for a long time without changing. If you subject these minerals to extreme pressure, however, the atoms push closer together, and different bonds form. This process leads to the growth of denser minerals whose atoms are more tightly packed. Such transformations in response to an increase in pressure can involve either phase changes or metamorphic reactions. In Chapter 1, we introduced the *bar* as a unit of pressure—1 bar is approximately the pressure you feel standing in air at sea level. Pressures at which metamorphism takes place are so large that geologists specify them in *kilobars* (1 kbar = 1,000 bars).

CHANGE IN BOTH TEMPERATURE AND PRESSURE. So far, we've looked separately at changes caused by pressure and by temperature. In the Earth, however, pressure and temperature both increase with increasing depth. So, when geologic processes cause a rock that had been at the land surface to end up deep below the surface, the rock experiences changes in both pressure and temperature. For example, if a continental collision shoves the margin of one continent under the margin of another, a shale that was once at the surface may end up at a depth of 20 km (12 miles). As a consequence, not only does the temperature affecting the shale increase from 20°C (70°F) to 550°C (1,000°F), but the pressure affecting the shale increases from 1 bar to 5 kbar.

APPLICATION OF STRESS. Put a ball of dough on the floor, lay a book on it, and step on the book. You'll see the ball flatten into a pancake, oriented parallel to the floor, because the downward push you apply with your foot exceeds the push provided by air in other directions. You have subjected the dough to **compression** in a vertical direction **(Fig. 6.3a)**. If you had compressed the ball of dough between a wall and a vertical book, horizontal compression would have flattened the ball into a pancake shape parallel to the wall, for the direction of flattening depends on the direction of compression **(Fig. 6.3b)**.

Compression is an example of **stress**, formally defined as the application of force over an area **(Box 6.1)**. (*Pressure* is a special condition of stress that exists when the amount of compression applied to a body is the same in all directions.) Not all stress in the Earth is compression. For example, if you grab each side of the dough ball with your hands and pull, you're applying **tension** to the ball, so it stretches and becomes longer. And if you place the dough on a table, set your hand on top of it, and move your hand parallel to the table, you're applying *shear* to the ball. **Shear** happens when one part of a material moves sideways relative to another part **(Fig. 6.3c)**.

At the pressures and temperatures that exist below a depth of 10 to 15 km in the crust, slowly subjecting a rock to stress can cause the rock's shape to change without breaking it, somewhat like the ball of dough you've just pictured. This shape change may involve pressure solution or plastic deformation. As a rock's overall shape changes, its internal texture may also change. For example, *platy* (pancake-shaped) grains may become parallel to one another, and *elongate* (cigar-shaped) grains may align in the same direction. (Platy and elongate grains are examples of *inequant grains*, meaning that the length of the grain is not the same in all directions; inequant grains differ from *equant grains* in that the latter have roughly the same

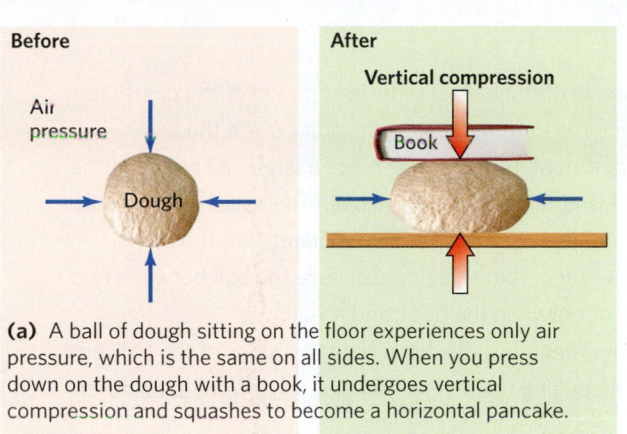

Before

Air pressure

Dough

After

Vertical compression

Book

(a) A ball of dough sitting on the floor experiences only air pressure, which is the same on all sides. When you press down on the dough with a book, it undergoes vertical compression and squashes to become a horizontal pancake.

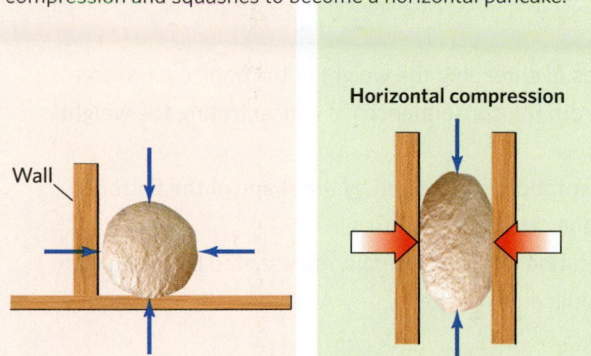

Wall

Horizontal compression

(b) If the book compresses the dough ball horizontally, it produces a vertical pancake.

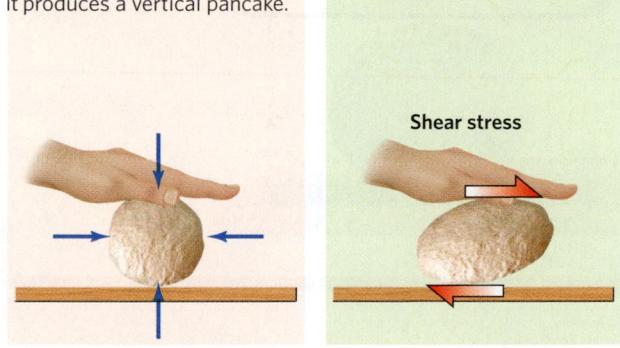

Shear stress

(c) Shear stress acts parallel to a surface. Here, shear smears out the dough ball parallel to the floor.

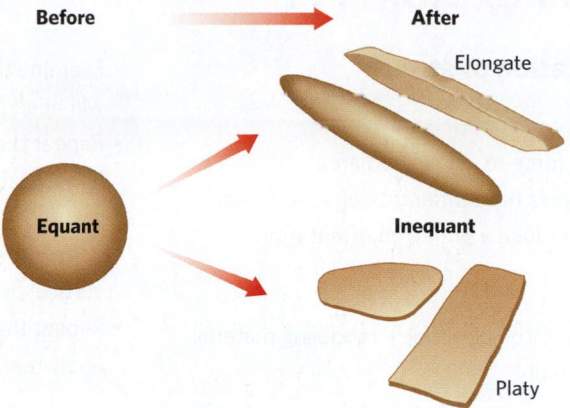

Before

After

Elongate

Equant

Inequant

Platy

(d) Stress can transform equant grains into inequant grains. Inequant grains can be elongate (cigar-shaped) or platy (pancake-shaped).

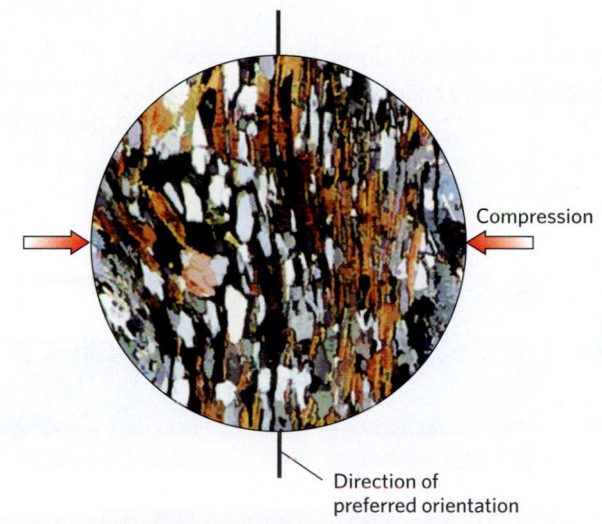

Compression

Direction of preferred orientation

(e) In metamorphic rock, inequant grains may be aligned to form a preferred orientation, as seen here through a microscope. The inequant grains are perpendicular to the compression direction.

dimensions in all directions, as shown in **Fig. 6.3d**.) When inequant grains in a rock become aligned, we say that they have attained a **preferred orientation (Fig. 6.3e)**. So, stress during metamorphism can produce a preferred orientation of inequant mineral grains. As we'll see shortly, development of a preferred orientation yields metamorphic foliation.

INTERACTION WITH HYDROTHERMAL FLUIDS. Sometimes metamorphism takes place in the presence of very hot water solutions, known as **hydrothermal fluids**. The

presence of hydrothermal fluids speeds up metamorphic reactions because the atoms involved in such reactions can diffuse (migrate) faster through the fluid than through a solid mineral grain. In addition, hydrothermal fluids passing through a rock may pick up some dissolved ions and drop others off, just as a bus picks up and drops off passengers. So interaction with hydrothermal fluids can change the overall chemical composition of a rock during metamorphism. Geologists refer to compositional changes that take place during metamorphism as **metasomatism**.

Box 6.1 ▶ How can I explain . . .

Changes due to the application of stress

What are we learning?
- That stress is the application of force to a specified area.
- That different orientations of stress have different consequences.
- That application of stress can produce a preferred orientation.

What you need:
- A baseball-sized mass of *Play-Doh*™ or *plasticine*™ modeling material.
- A heavy book; six coins; a large nail.

Instructions:
- Mold the modeling material into a spherical ball.
- Insert the coins, each perpendicular to the ball's surface, at different locations along a circumference on the surface of the ball, as shown in the figure.
- Place the ball on a table; place the book horizontally on the surface of the ball; press down, if the book isn't heavy enough.

- Examine the shape of the ball, and the orientation of the coins. You will need to slice the flattened ball open, adjacent to the coins.
- Repeat the experiment, but this time place a large nail between the horizontal book and the surface of the ball. The nail should be vertical. The weight of the book pushes the nail deep into the ball.
- Repeat the experiment, but this time squeeze the ball between two vertical books, or between the book and a wall.
- Repeat the experiment, but this time, hold two sides of the ball, and apply tension by pulling the sides of the ball away from each other.

What did we see?
- The consequences of spreading the weight of the book over a large area is different from the consequences of concentrating the weight of the book on a nail.
- The preferred orientation, manifested by the shape of the flattened ball, depends on the orientation of stress.
- A preferred orientation of platy minerals, represented by dimes, can develop due to application of stress.

Dough ball   Coin

Take-home message . . .

🏠 A metamorphic rock contains mineral assemblages and/or textures that are different from those found in its protolith. Many processes, such as recrystallization, metamorphic reactions, phase changes, pressure solution, and plastic deformation can take place during metamorphism. These processes happen in response to changes in temperature and pressure, application of stress, or interaction with hydrothermal fluids.

Quick Question -
Can the overall composition of rock change during metamorphism?

6.3 Types of Metamorphic Rocks

Now that we've introduced the processes involved in metamorphism, and the conditions that cause metamorphism, let's examine the various types of rocks formed by metamorphism. Coming up with a way to classify and name the great variety of metamorphic rocks on Earth has not been easy. After decades of debate, geologists agreed to divide metamorphic rocks into two fundamental classes—foliated and nonfoliated—based on whether or not the rock contains metamorphic foliation. We distinguish among different types of foliated rocks based on the character of the foliation, and we distinguish among different types of nonfoliated rocks based on their composition.

Foliated Metamorphic Rocks

To understand foliated metamorphic rocks, we first need to consider the nature of foliation in more detail. The word *foliation* comes from the Latin *folium*, meaning leaf. Geologists use the term **metamorphic foliation** in reference to parallel surfaces or layers that form during the process of metamorphism. Foliation can give metamorphic rock a striped or streaked appearance in an outcrop, and in some cases, it makes a rock susceptible to splitting into thin sheets. Not all foliation looks the same. Its character depends on grain size and on whether the foliation is primarily due to the alignment of platy minerals or to the development of alternating layers of different metamorphic minerals. There are several common types of foliated metamorphic rocks:

- *Slate:* The finest-grained foliated metamorphic rock, **slate**, forms by metamorphism of shale or mudstone (rocks composed predominantly of clay), under relatively low pressures and temperatures, while the rock is undergoing compression. Geologists refer to the foliation in slate as *slaty cleavage*. It forms when clay flakes, which are very tiny, platy grains, become aligned parallel to one another and perpendicular to the direction of compression. (This alignment happens partly because pre-existing flakes rotate into planes of foliation, and partly because new flakes grow within planes of foliation.) Horizontal compression of a sequence of shale beds under appropriate metamorphic conditions, for example, produces vertical slaty cleavage **(Fig. 6.4a)**. The presence of slaty cleavage allows slate to split into thin sheets, so the rock can be used to make excellent roofing shingles **(Fig. 6.4b)**.

Did you ever wonder . . .

why slate makes such nice roofing shingles?

Figure 6.4 Slate is a foliated metamorphic rock that forms at relatively low temperatures and pressures.

(a) Slaty cleavage forms in response to compression. In this example, layers of shale also bend to form folds as slaty cleavage develops. The cleavage tends to be oriented parallel to the axial plane, an imaginary surface that, simplistically, divides the fold in half.

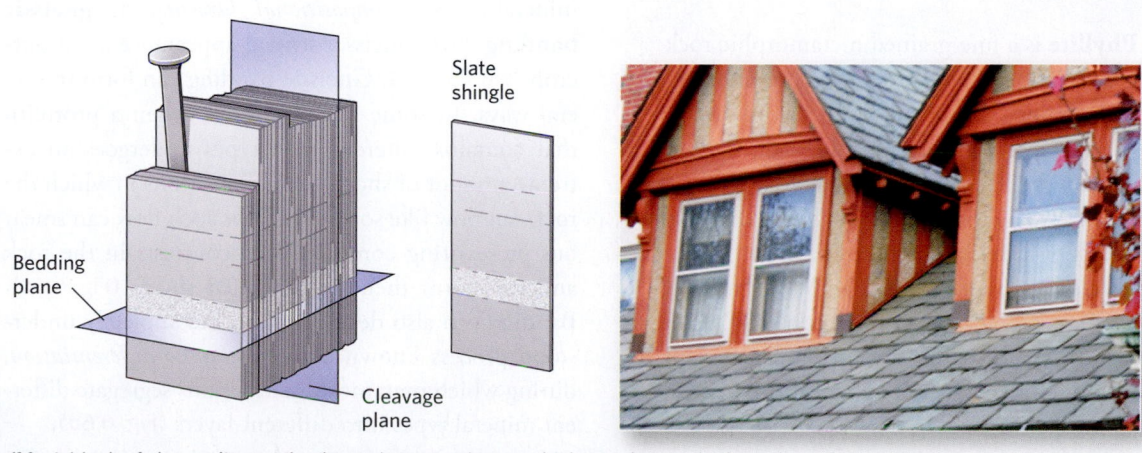

(b) A block of slate splits easily along cleavage planes, which may be at a high angle to the bedding planes. Workers split slate to produce roof shingles that, when overlapped, make a watertight surface (inset).

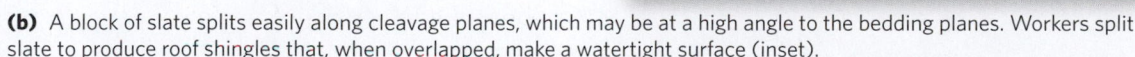

Figure 6.5 Examples of foliated metamorphic rocks.

(a) During formation of phyllite, clay recrystallizes to form tiny mica flakes that reflect light, giving the rock a sheen.

(b) Schist contains coarse mica flakes along with other metamorphic minerals.

(c) In metaconglomerate, pebbles and cobbles flatten into a pancake shape without cracking. Note the coin for scale.

- *Phyllite:* **Phyllite** is a fine-grained metamorphic rock containing a foliation caused by the preferred orientation of fine-grained white mica. The word comes from the Greek word *phyllon*, meaning leaf, as does the word *phyllo*, the flaky dough in Greek pastry. The name *mica* refers to a group of minerals that occur in very thin sheets or flakes because they have one strong direction of cleavage (see Chapter 3). White mica is translucent, so its presence gives phyllite a silky luster **(Fig. 6.5a)**. Phyllite forms by the metamorphism of slate at a temperature high enough to cause clay to recrystallize into white mica.

- *Schist:* **Schist** is a medium- to coarse-grained metamorphic rock that possesses *schistosity*, a type of foliation defined by the preferred orientation of large crystals of mica (generally muscovite or biotite) **(Fig. 6.5b)**. Schist forms at a higher temperature than phyllite, for at higher temperatures, mica can grow into larger crystals that lie in the plane of foliation.

- *Metaconglomerate:* Under the metamorphic conditions that produce slate, phyllite, or schist, a protolith of conglomerate becomes **metaconglomerate**. During the formation of this rock, pressure solution and plastic deformation flatten pebbles and cobbles into pancake-like shapes or stretch them into cigar-like shapes. These flattened clasts align to produce the foliation in metaconglomerate **(Fig. 6.5c)**.

- *Gneiss:* **Gneiss** is a metamorphic rock containing alternating layers of dark-colored and light-colored minerals. This *compositional layering*, or **gneissic banding**, gives gneiss a striped appearance at an outcrop **(Fig. 6.6a, b)**. Gneissic banding can form in several ways. In some cases, it forms when a protolith that contains different rock types undergoes an extreme amount of shear under conditions in which the rock can flow like soft plastic, for such flow can smear out pre-existing compositional contrasts in the rock and transform them into aligned sheets **(Fig. 6.6c)**. Banding can also develop by an incompletely understood process known as *metamorphic differentiation*, during which metamorphic reactions segregate different mineral types into different layers **(Fig. 6.6d)**.

Figure 6.6 Gneiss.

(a) In this outcrop of gneiss in Brazil, some of the layers are only centimeters across.

(b) An outcrop of 2.7-Ga gneiss in Ontario, Canada, displays distinct foliation that was later bent.

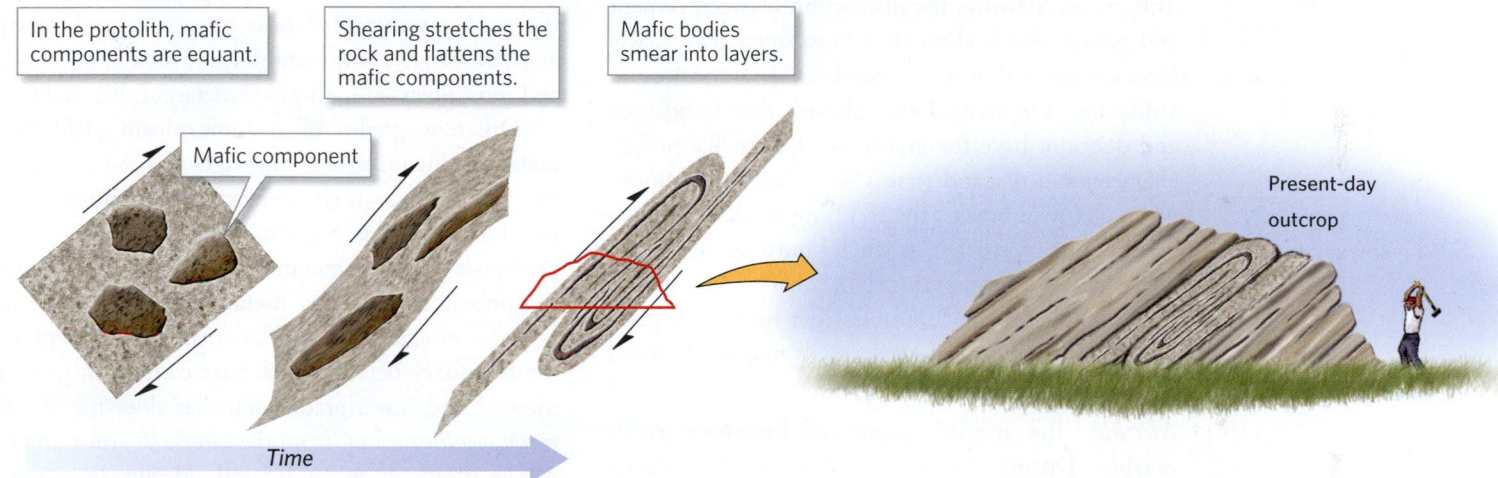

In the protolith, mafic components are equant.

Shearing stretches the rock and flattens the mafic components.

Mafic bodies smear into layers.

Mafic component

Present-day outcrop

Time

(c) Formation of gneiss, in some cases, involves extreme shear. Original contrasting rock types are smeared into parallel layers.

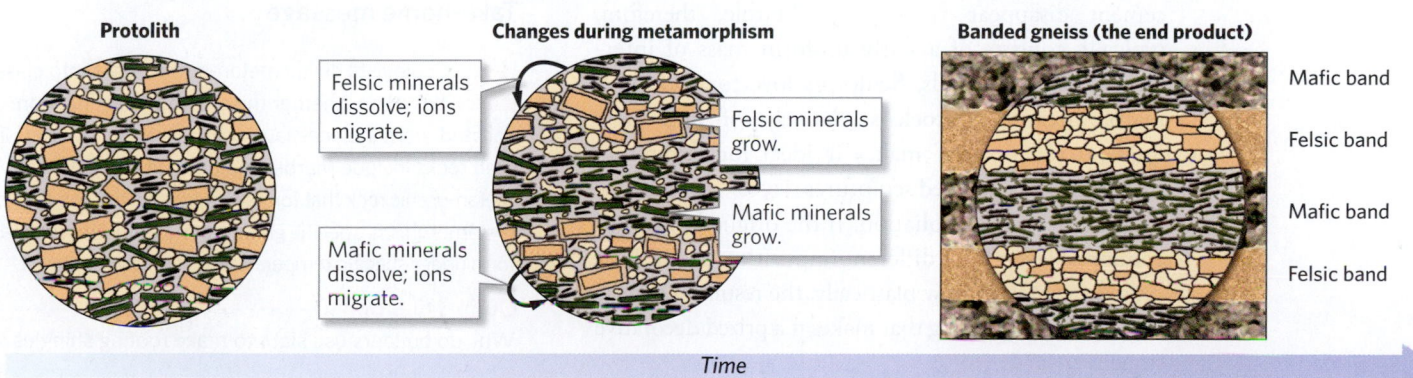

Protolith

Changes during metamorphism

Banded gneiss (the end product)

Felsic minerals dissolve; ions migrate.

Felsic minerals grow.

Mafic minerals grow.

Mafic minerals dissolve; ions migrate.

Mafic band

Felsic band

Mafic band

Felsic band

Time

(d) Gneiss may also form by metamorphic differentiation, during which metamorphic reactions cause felsic and mafic mineral crystals to grow in distinct, separate layers.

Nonfoliated Metamorphic Rocks

In a nonfoliated metamorphic rock, metamorphic minerals are either equant or, if inequant because of crystal shape, have a random orientation. Examples include the following:

- *Hornfels:* **Hornfels** is a fine-grained nonfoliated rock that contains a variety of metamorphic minerals. The specific mineral assemblage in a hornfels depends on the composition of the protolith and on the temperature and pressure of metamorphism. Some of the minerals in hornfels, such as feldspar, grow as inequant crystals with a random orientation. Hornfels forms when a rock undergoes heating without being subjected to compression, shear, or tension.

- *Quartzite:* **Quartzite** is a metamorphic rock consisting entirely of quartz. It forms by the metamorphism of pure quartz sandstone when, during metamorphism, sand grains and cements recrystallize and are replaced by new, larger, interlocking crystals of quartz. This process destroys the distinction between cement and grains, and it eliminates most open pore space. How can you tell the difference between quartzite and sandstone? Quartzite looks glassier than sandstone and does not have the grainy, sandpaper-like surface characteristic of sandstone **(Fig. 6.7a)**. In addition, when quartzite breaks, the resulting cracks cut across grain boundaries, whereas when sandstone breaks, its cracks curve around grains. Note that not all quartzite lacks foliation; in examples that have undergone significant shear, the quartz crystals may be flattened and aligned.

- *Marble:* The metamorphism of limestone yields **marble**. During the formation of marble, calcite grains in the protolith recrystallize, so fossil shells, pore space, and the distinction between grains and cement disappear **(Fig. 6.7b)**. Marble, therefore, typically consists of a fairly uniform mass of interlocking calcite crystals. Sculptors love to work with marble because the rock is relatively soft and has a uniform texture that makes it ideal for fashioning smooth, highly detailed sculptures **(Fig. 6.7c)**. Notably, not all marble lacks foliation. If the original protolith contained layers with different impurities, and if shear caused the rock to flow plastically, the resulting marble develops color banding that makes it a prized decorative stone **(Fig. 6.7d)**.

Metamorphic Grade

Not all metamorphism takes place under the same physical conditions. For example, rock at a depth of 40 km (24 miles) beneath a mountain range undergoes more intense metamorphism than rock at a depth of 20 km (12 miles). Geologists use the term **metamorphic grade** in an informal way to indicate the intensity of metamorphism, meaning the amount or degree of metamorphic change. A more formal specification of the intensity of metamorphism uses the concept of a *metamorphic facies* **(Box 6.2)**.

The metamorphic grade that develops at a locality depends primarily on the temperature reached in rock at that locality during metamorphism, for temperature plays the dominant role in determining the extent of recrystallization and the nature of metamorphic reactions that take place. *Low-grade* metamorphic rocks form at temperatures of 250°C to 400°C (480°F to 750°F), *intermediate-grade* metamorphic rocks form at 400°C to 600°C (750°F to 1,100°F), and *high-grade* metamorphic rocks form at 600°C to 850°C (1,100°F to 1,560°F) **(Fig. 6.8a)**. Geologists consider slate and phyllite as examples of low-grade metamorphic rocks, most schist and some gneiss as intermediate-grade metamorphic rocks, and some schist and most gneiss as high-grade metamorphic rocks.

Different grades of metamorphism yield different metamorphic mineral assemblages **(Fig. 6.8b)**. On a map, we can outline areas, known as **metamorphic zones**, in which rock has a given grade **(Fig. 6.8c)**. Typically, geologists identify metamorphic zones by looking for the presence of specific metamorphic minerals, known as *index minerals*. A line drawn on a map representing the boundary between rock that contains a given index mineral and lower-grade rock that does not is called a *metamorphic isograd* (from the Greek *iso*, meaning equal). All locations along an isograd, ideally, have the same metamorphic grade.

Take-home message . . .

Geologists divide metamorphic rocks into classes based on whether the rock contains foliation. Foliated rocks include slate, schist, and gneiss. Nonfoliated rocks include marble and quartzite. The type of metamorphic rock that forms depends on the conditions of metamorphism. Specific grades of metamorphic minerals form under specific temperature and pressure ranges.

Quick Question -
Why do builders use slate to make roofing shingles?

Figure 6.7 Examples of nonfoliated metamorphic rocks.

Quartz sandstone—the protolith of quartzite.

(a) In this sandstone from Kuwait (left), the sand grains stand out. In contrast, this nonfoliated maroon quartzite (right) looks glassy.

In this unmetamorphosed limestone, fossils and bedding are visible.

Italian marble quarry sliced into a mountain

(b) In this unmetamorphosed limestone from New York (left), you can see fossils and shell fragments. Such grains are not visible in the white marble (right) exposed in an Italian quarry.

(c) Sculptors like to work with nonfoliated white marble.

(d) This marble floor has color banding inherited from original bedding.

Figure 6.8 Intensity of metamorphism is indicated by metamorphic grade.

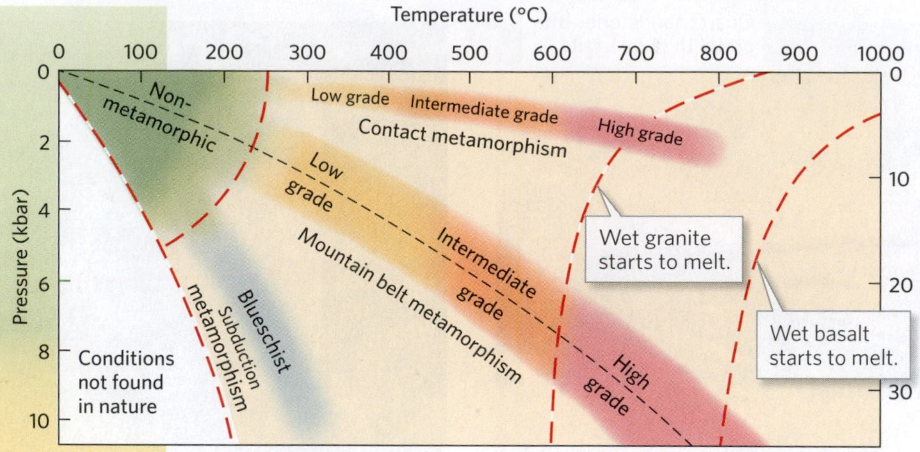

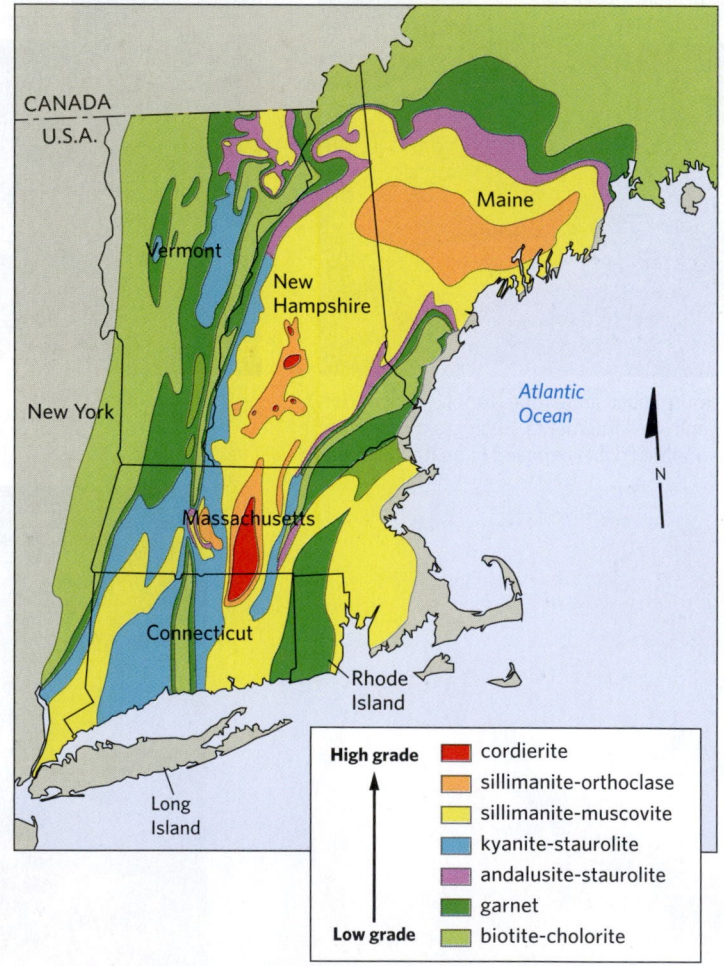

(a) This graph depicts the approximate temperatures and pressures of metamorphic grades. "Wet" means that the rock contains water. Wet rock melts at a lower temperature than does dry (water-free) rock.

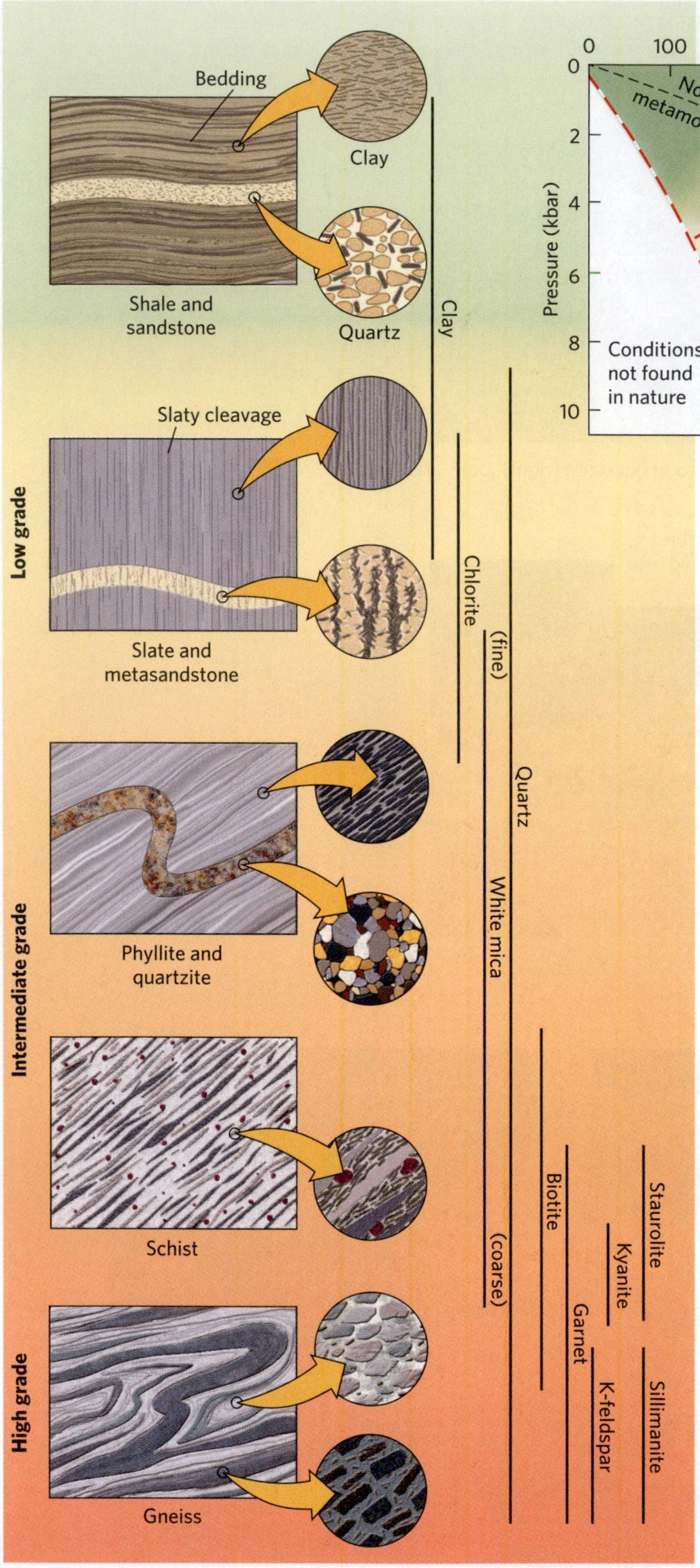

(b) Here we see the consequences of the progressive metamorphism of shale and sandstone, from low grade to high grade, during mountain building. Lines on the right indicate the range of grades in which certain diagnostic metamorphic minerals grow.

(c) Approximate metamorphic zones and isograds in New England, in the northeastern United States. The minerals listed are index minerals.

Box 6.2

Consider this . . .

Metamorphic facies

In the early years of the 20th century, geologists working in Scandinavia—where erosion by glaciers has left beautiful, nearly unweathered exposures of metamorphic rocks—came to realize that metamorphic rocks, in general, do not consist of a hodgepodge of minerals formed at different times and in different places. Rather, they consist of distinct sets of minerals that grew in association with one another at a certain pressure and temperature. It seemed that such mineral assemblages more or less represent a condition of *chemical equilibrium*, meaning that the chemicals making up the rock had organized into a group of mineral grains that were—to anthropomorphize a bit—comfortable together and with their surroundings and did not feel the need to change further. The geologists also determined that the specific mineral assemblage present in a rock depends both on the pressure and temperature conditions and on the composition of the protolith.

This discovery led the geologists to propose the concept of metamorphic facies. A **metamorphic facies** is a set of metamorphic mineral assemblages typical of a certain range of pressure and temperature. Each specific assemblage in a facies reflects the original protolith composition. According to this definition, a given metamorphic facies includes several different kinds of rocks that differ from one another in terms of chemical composition and, therefore, mineral content. But all the rocks of a given facies formed under roughly the same temperature and pressure conditions. Geologists recognize several facies, including zeolite, hornfels, greenschist, amphibolite, blueschist, eclogite, and granulite. The names of the different facies are based on a distinctive feature or mineral found in some of the rocks of the facies.

We can represent the approximate conditions under which metamorphic facies form by using a graph **(Fig. Bx6.2)**. The horizontal axis represents temperature, and the vertical axis represents pressure. Each area on the graph labeled with a facies name represents the approximate range of temperatures and pressures in which mineral

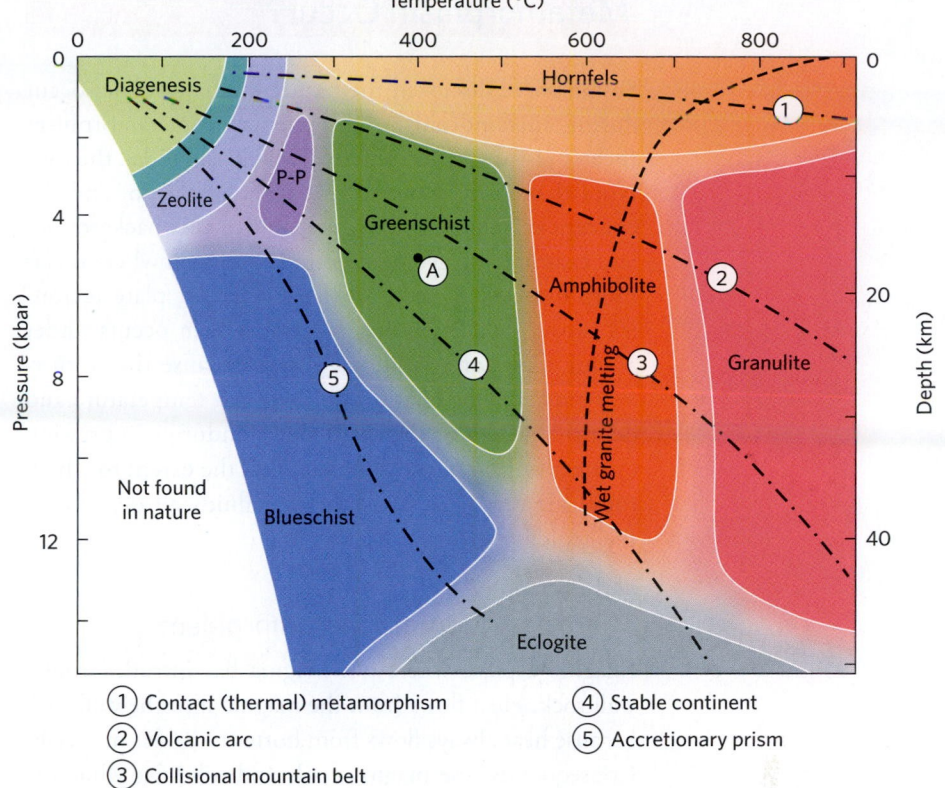

① Contact (thermal) metamorphism
② Volcanic arc
③ Collisional mountain belt
④ Stable continent
⑤ Accretionary prism

Figure Bx6.2 The common metamorphic facies. The boundaries between the facies are depicted as wide bands because they are gradational and approximate. Note that some amphibolite-facies rocks and all granulite-facies rocks form at pressure-temperature conditions under which granite containing water will melt. Thus, such metamorphic rocks develop only if the protolith is dry. One of the facies depicted on the graph is not mentioned in the text: specifically, the P-P (prehnite-pumpellyite) facies, named for two metamorphic minerals. The numbers refer to different geothermal gradients.

assemblages characteristic of that particular facies form. For example, rock subjected to the pressure and temperature at Point A (4.5 kbar and 400°C) develops a mineral assemblage characteristic of the greenschist facies. As the graph implies, pressures and temperatures at boundaries between facies can't be determined precisely, and transitions between facies are gradual.

We can portray the *geothermal gradients* (the change in temperature with depth) of different crustal regions using the same graph. Beneath mountain belts, for example, the geothermal gradient passes through the zeolite, greenschist, amphibolite, and granulite facies. In contrast, in accretionary prisms that form during subduction, temperature increases slowly with increasing depth, so blueschist assemblages form.

6.4 Where Does Metamorphism Occur?

So far, we've discussed the nature of changes that occur during metamorphism, the agents of metamorphism (heat, pressure, stress, and hydrothermal fluids), the rock types that form as a result of metamorphism, and the concept of metamorphic grade. With this background, let's examine the geologic settings on Earth where metamorphism takes place, in the context of plate tectonics theory. You'll see that metamorphism occurs under a wide range of conditions. That's because the *geothermal gradient* (the relationship between temperature and depth), the extent to which rocks endure compression or shear during metamorphism, and the extent to which rocks interact with hydrothermal fluids all depend on the geologic setting.

Thermal or Contact Metamorphism

Imagine a place where hot magma has intruded cooler wall rock. Heat flows from the magma into the wall rock because heat always flows from hotter to colder materials. Consequently, the magma cools and solidifies while the wall rock heats up. As this happens, hydrothermal fluids may be circulating through both the intrusion and the wall rock. Heat and hydrothermal circulation cause the wall rock to undergo metamorphism. The highest-grade metamorphic rock forms immediately adjacent to the intrusion, where the temperatures are highest, and progressively lower-grade rocks form farther away. The band of metamorphic rock that forms around an igneous intrusion is called a **metamorphic aureole**—the word *aureole* comes from the Latin *aureola*, meaning crown or halo **(Fig. 6.9a)**. The width of a metamorphic aureole depends on the amount of heat the intrusion releases, which in turn depends on the size and shape of the intrusion, as well as on the amount of hydrothermal circulation that takes place. For example, a large pluton provides more heat, and therefore produces a wider aureole, than does a thin sill. Similarly, an intrusion into wet rock may produce a wider aureole than does an intrusion into dry rock because hydrothermal fluids transport heat and cause metasomatism.

The local metamorphism caused by heat from an igneous intrusion can be called either **thermal metamorphism**, to emphasize that it develops in response to heat without a change in pressure and without compression or shear, or **contact metamorphism**, to emphasize that it develops adjacent to the contact between an intrusion and its wall rock. Because this type of metamorphism

takes place without application of compression or shear, preferred orientation doesn't develop. Therefore, metamorphic aureoles typically contain hornfels, a nonfoliated metamorphic rock. As explained by the theory of plate tectonics, magma intrudes the crust at convergent boundaries, in rifts, and during certain stages of mountain building, so these are the places where you can find examples of contact metamorphism.

Dynamic Metamorphism

As we've discussed, faults are fractures on which one piece of crust slides, or shears, past another. Near the Earth's surface (in the upper 10 to 15 km of the crust), this movement fractures rock, breaking it into angular fragments or even crushing it to a powder. But because of the geothermal gradient, rock at greater depth is so warm that it undergoes plastic deformation when shear along a fault takes place. During shearing under metamorphic conditions, the minerals in the rock recrystallize, so the rock develops a new texture. We call this process **dynamic metamorphism** because it occurs as a consequence of shear (movement) alone under metamorphic conditions, without requiring a change in temperature or pressure. Dynamically metamorphosed rock typically develops foliation that is oriented nearly parallel to the fault **(Fig. 6.9b)**.

Dynamothermal or Regional Metamorphism

Along convergent boundaries or during continental collision, large slices of continental crust slip along faults and move up and over other portions of the crust. As a consequence, a protolith that was once near the Earth's surface along the margin of a continent can end up at great depth beneath a mountain range **(Fig. 6.9c)**. In this environment, three processes can affect the protolith: (1) it heats up, because temperature increases with depth and because igneous activity happens nearby; (2) it endures greater pressure, because the weight of the overburden increases with depth; and (3) it undergoes compression and shear. As a result, a protolith that ends up at depth beneath a mountain belt transforms into foliated metamorphic rock. The type of foliated rock that forms depends on the grade of metamorphism: slate forms at shallower depths, whereas schist and gneiss form at greater depths.

Since the metamorphism we've just described involves not only heat but also compression and shear, geologists refer to it as **dynamothermal metamorphism**. Typically, such metamorphism affects a large region—often the length and breadth of a whole mountain belt—so it is also

Figure 6.9 Mechanisms of metamorphism.

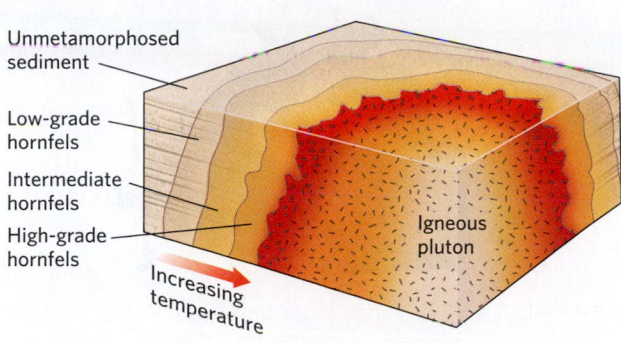

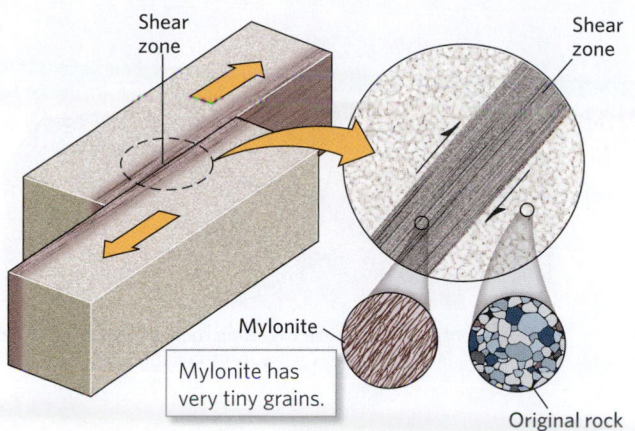

(a) Heat from a large pluton can produce a metamorphic aureole in which hornfels develops. Metamorphic grade decreases progressively away from the pluton.

(b) Shearing of a rock under metamorphic conditions causes original crystals to divide into tiny crystals without breaking to form a type of schist called mylonite.

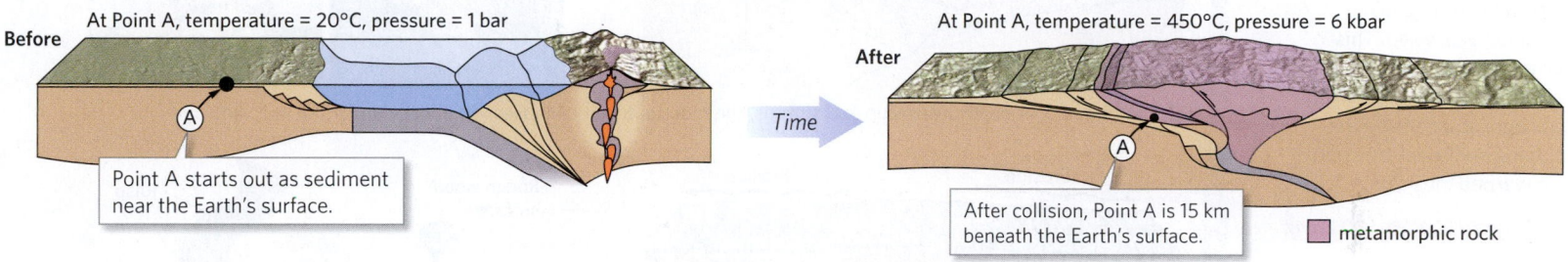

(c) Dynamothermal metamorphism occurs during the development of mountain belts. The process is also called regional metamorphism.

known as **regional metamorphism**. Erosion eventually removes the mountains, exposing the belt of metamorphic rock that once lay at depth.

Metamorphism in Subduction Zones

Blueschist is a relatively rare rock that contains an unusual blue-colored version of the mineral amphibole. Laboratory experiments indicate that this mineral forms only under conditions of very high pressure and relatively low temperature. Such conditions cannot develop in continental crust because, at the high pressures needed to produce blue amphibole, the temperature in continental crust is also high (see Box 6.2). So, to figure out where blueschist forms, we must ask where on Earth high pressure can develop at a relatively low temperature.

Plate tectonics theory provides the answer to this puzzle. Researchers have found that blueschist occurs only in the accretionary prisms that form at subduction zones **(Fig. 6.9d)**. Such prisms grow to be over 20 km (12 miles) thick, so rock at the base of the prism endures high pressure (due to the weight of the overburden). But because subducted oceanic crust beneath the prism is cool, temperatures in the prism remain relatively low.

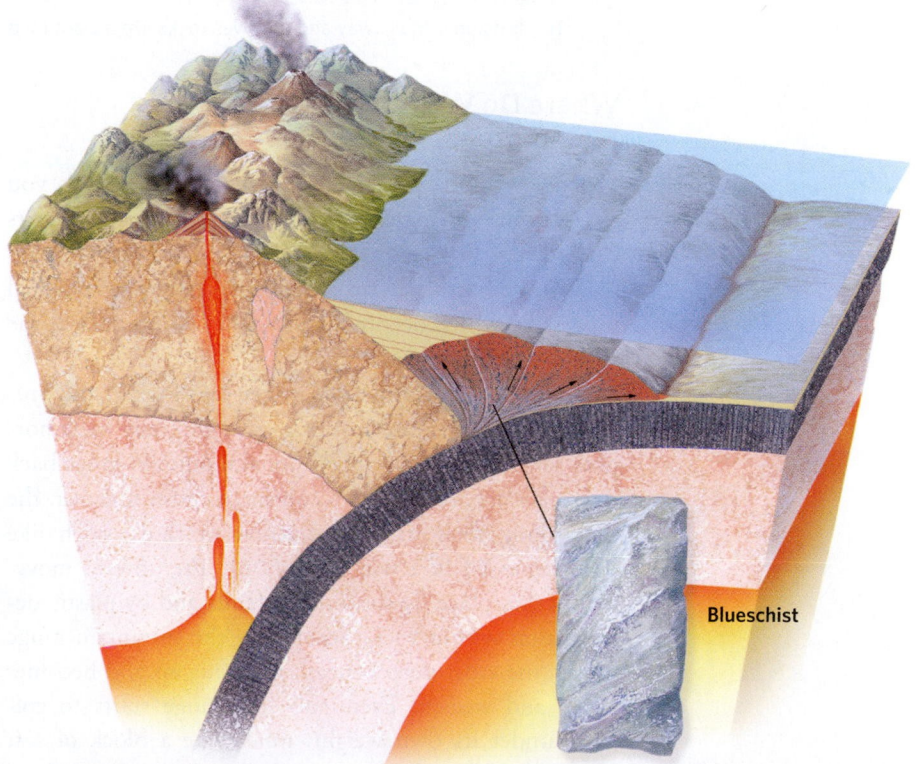

(d) Blueschist forms at the base of an accretionary prism at a convergent boundary.

Wind River Mountains, Wyoming

Latitude: 43°6'7.22" N
Longitude: 109°21'36.45" W

Zoom to 20 km (~12.5 miles) and look obliquely.

Faulting uplifted the Precambrian basement of western Wyoming 40 to 80 million years ago. This rock underwent high-grade metamorphism over 2 billion years ago. You can see that overlying sedimentary strata have warped into a huge fold.

Figure 6.10 Processes that exhume metamorphic rock. Note how the red dot (representing metamorphic rock formed at the base of a mountain range) moves progressively closer to the surface over time.

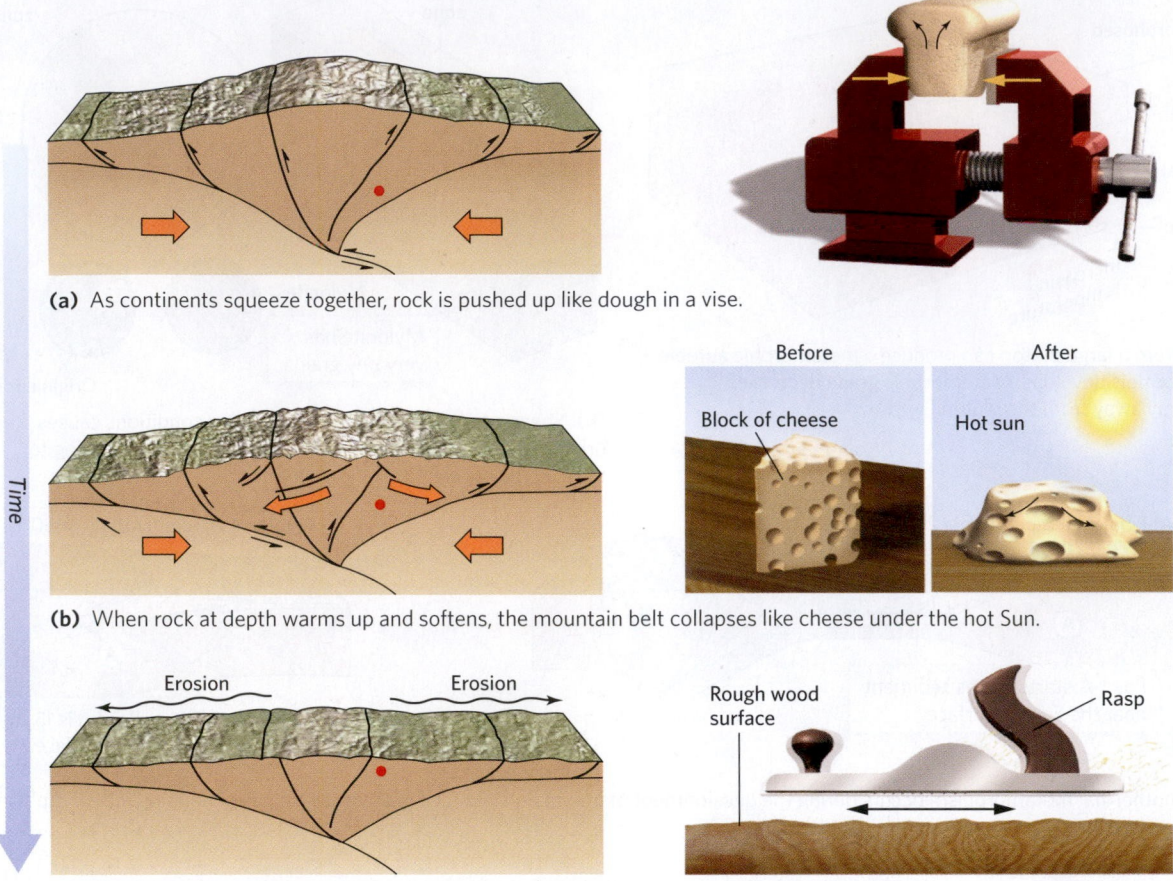

(a) As continents squeeze together, rock is pushed up like dough in a vise.

(b) When rock at depth warms up and softens, the mountain belt collapses like cheese under the hot Sun.

(c) Erosion grinds away and removes rocks like a giant rasp.

Where Do We Find Metamorphic Rocks?

When you stand on an outcrop of metamorphic rock, you are standing on material that once lay many kilometers beneath the surface of the Earth. How did that rock return to the Earth's surface? Geologists refer to the overall process by which deeply buried rocks end up back at the surface as **exhumation**.

To see how exhumation works, let's look at the processes that contribute to bringing high-grade metamorphic rocks from below a collisional mountain range back to the surface. First, as two continents push together, the rock along their margins squeezes upward, much like dough pressed in a vise **(Fig. 6.10a)**. The upward movement takes place by shear along faults and by plastic deformation within rock. Second, as the mountain range grows, the crust deep beneath it warms up and becomes softer and weaker. Eventually, the range starts to collapse under its own weight, much like a block of soft cheese placed in the hot sun **(Fig. 6.10b)**. As a result of this collapse, the upper crust spreads out sideways, a process that involves both faulting and plastic deformation.

Horizontal stretching of the upper part of the crust causes it to become thinner in the vertical direction, and as the upper part of the crust becomes thinner, the deeper crust ends up closer to the surface. Third, erosion takes place at the surface—landslides, river flow, glacial flow, and wind together act like a giant rasp that removes rock at the surface and exposes rock that was once below it **(Fig. 6.10c)**.

Keeping in mind the processes that form metamorphic rock and cause exhumation, let's ask the question: Where are metamorphic rocks presently exposed? You can start your quest to find metamorphic rock outcrops by hiking into a mountain range. As we've seen, the processes involved in mountain building produce metamorphic rocks, so the towering cliffs in the interior of a mountain range typically consist of schist, gneiss, quartzite, and marble **(Fig. 6.11a)**. You can find even more metamorphic rocks by walking across a **shield**, a broad expanse of Precambrian continental crust that includes rocks last metamorphosed during the many mountain-building events that took place over a billion years ago **(Fig. 6.11b, c)**. Effectively, a shield is a portion of the old and relatively stable part of a continent where sedimentary cover has been stripped from the basement.

Figure 6.11 Exposures of metamorphic rock.

(a) The Gunnison River has carved a deep canyon into the Precambrian rock in Colorado.

(b) A photograph from an airplane window of the flat landscape of the eastern Canadian Shield.

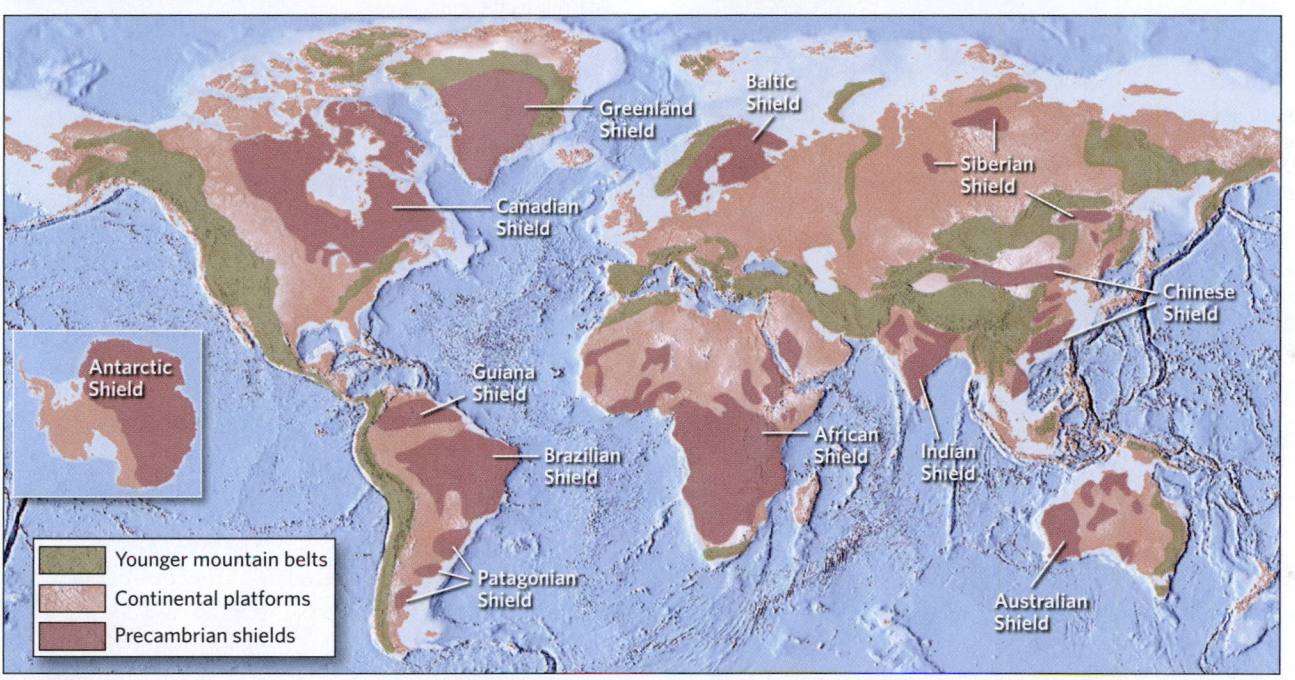

Younger mountain belts
Continental platforms
Precambrian shields

(c) A map showing the distribution of shields, areas where broad expanses of Precambrian crust, including Precambrian metamorphic rocks, crop out.

See for **yourself**

Canadian Shield, East of Hudson Bay

Latitude: 61°09′50.15″ N
Longitude: 76°42′43.88″ W

Zoom to 225 km (140 miles) and look down.

The Canadian shield at this locality is covered by sparse vegetation. Differential erosion of the gneiss makes the foliation stand out. The east-west band of foliation delineates a zone of dynamic metamorphism associated with a fault zone.

Take-home message . . .

Thermal (contact) metamorphism develops around igneous intrusions due to heat from the intrusion. Dynamothermal (regional) metamorphism develops beneath mountain ranges where rock undergoes compression and shear at high temperatures and pressures. Erosion and uplift may eventually expose metamorphic rock in mountain ranges or shields.

Quick Question -
How does metamorphism at the base of an accretionary prism differ from metamorphism at the base of a collisional mountain range?

6.5 The Rock Cycle

Even though rock seems to be strong and durable, in the time frame of Earth history, it doesn't last forever. Due to a great variety of geologic processes, atoms making up minerals of one rock type may eventually be rearranged to become other minerals, or may move elsewhere (**Earth Science at a Glance**, pp. 208–209). As a result, the material that starts out in one rock type may end up in another rock type at the same, or even at a different, location. Later, the same atoms may rearrange or move again to form a third rock type, and so on. Geologists refer to the progressive transformations that result in the passage of atoms through

Did you ever wonder . . .
whether rocks, once formed, last forever?

Rock-Forming Environments and the Rock Cycle

Rocks form in many different environments. Igneous rocks develop where melt rises from depth and cools. Intrusive igneous rocks form where magma cools underground; extrusive igneous rocks form where lava and ash erupt at the surface.

Weathering and erosion break up existing rock and produce sediment. Different kinds of sediments develop in different places, reflecting both the composition of the source and the setting in which the sediment accumulates. When this sediment eventually gets buried and undergoes lithification, new sedimentary rocks form.

Under certain conditions, pre-existing rocks can undergo change in the solid state—metamorphism—which produces metamorphic rocks. Contact metamorphism is due to heat released by an intrusion of magma. Regional metamorphism occurs where tectonic processes cause rocks from the surface to be buried very deeply.

Drainage networks collect surface water that can transport sediment to the ocean.

Sand dunes form from grains carried by the wind.

In a desert environment, rock weathers and fragments. Debris falls in landslides.

Flash floods carry sediment out of canyons to form an alluvial fan.

Volcanic eruptions emit lava and ash, which form new igneous rock at Earth's surface.

Sedimentary rocks make a cover on the surface of continents.

The crust and lithospheric mantle stretch and thin in a rift.

Magma rises from the mantle. Heat from this magma causes contact metamorphism.

Deep levels of continents consist of ancient metamorphic and igneous rocks. This is the basement of the continents.

Continental margins slowly sink and are buried by new sediment.

Partial melting occurs in the asthenosphere to produce new magma.

km
0
10
20
30
40
50
60
70
80
90
100

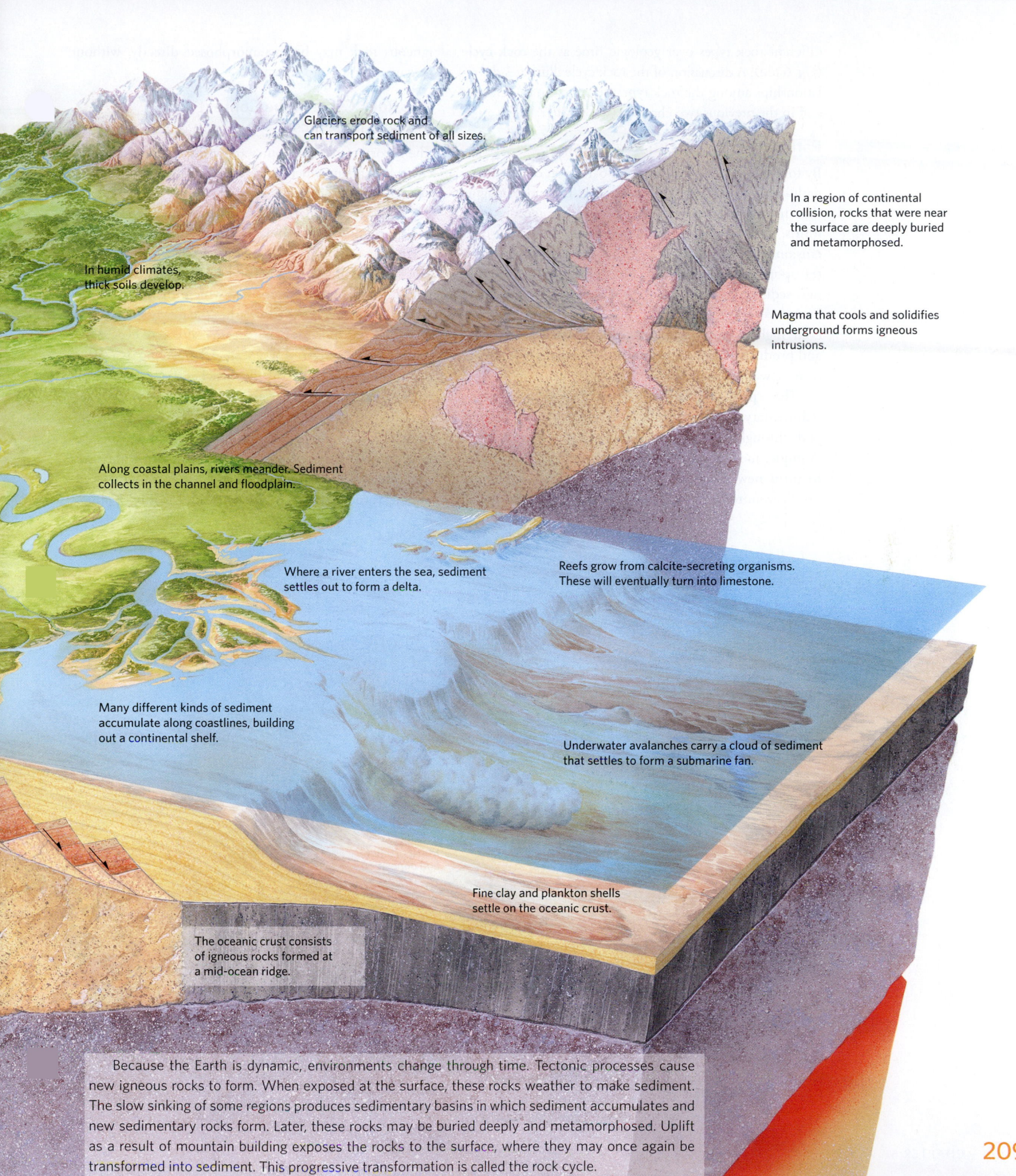

Glaciers erode rock and can transport sediment of all sizes.

In a region of continental collision, rocks that were near the surface are deeply buried and metamorphosed.

In humid climates, thick soils develop.

Magma that cools and solidifies underground forms igneous intrusions.

Along coastal plains, rivers meander. Sediment collects in the channel and floodplain.

Where a river enters the sea, sediment settles out to form a delta.

Reefs grow from calcite-secreting organisms. These will eventually turn into limestone.

Many different kinds of sediment accumulate along coastlines, building out a continental shelf.

Underwater avalanches carry a cloud of sediment that settles to form a submarine fan.

Fine clay and plankton shells settle on the oceanic crust.

The oceanic crust consists of igneous rocks formed at a mid-ocean ridge.

Because the Earth is dynamic, environments change through time. Tectonic processes cause new igneous rocks to form. When exposed at the surface, these rocks weather to make sediment. The slow sinking of some regions produces sedimentary basins in which sediment accumulates and new sedimentary rocks form. Later, these rocks may be buried deeply and metamorphosed. Uplift as a result of mountain building exposes the rocks to the surface, where they may once again be transformed into sediment. This progressive transformation is called the rock cycle.

209

different rock types over geologic time as the **rock cycle** (Fig. 6.12a). A discussion of the rock cycle illustrates the relationships among the rock types described in this chapter and in the previous two chapters (Fig. 6.12b).

Pathways through the Rock Cycle

By following the arrows in Figure 6.12, you can see many pathways through the rock cycle. For example, imagine that magma intrudes the crust and new igneous rock forms. If that igneous rock undergoes weathering and erosion, it becomes sediment that eventually undergoes transport, deposition, burial, and lithification to produce new sedimentary rock. If that sedimentary rock becomes buried very deeply, it turns into metamorphic rock. At even higher temperatures, the rock may partially melt and produce magma that rises and eventually solidifies to form new igneous rock.

The path we've just described (igneous → sedimentary → metamorphic → igneous) is not the only path through the rock cycle. There can be shortcuts. For example, metamorphic rock may be uplifted and eroded to form new sediment that later undergoes burial and lithification to form new sedimentary rock. Likewise, an igneous rock may be metamorphosed directly, without first weathering into sediment.

The Rock Cycle in the Context of Plate Tectonics

The theory of plate tectonics allows us to understand the environments in which stages of the rock cycle take place. Let's consider an example.

Suppose that magma formed beneath a continental hot-spot volcano reaches the surface and erupts to form basalt lava (Fig. 6.13a). Interaction with wind, rain, and vegetation gradually weathers the basalt, physically breaking it into smaller fragments, which then chemically weather to form clay. Water washes the newly formed clay away and transports it downstream—if you've ever seen a muddy-looking river, you've seen clay traveling to a site of deposition. Eventually, the river reaches the sea, where the water slows down and the clay settles out.

Let's imagine that the clay accumulates along the coast of Continent X. Gradually, through time, the clay layer undergoes burial, so the clay flakes within the layer pack together and become cemented to form a new sedimentary rock, shale. The shale resides 6 km (3 miles)

Figure 6.12 The rock cycle.

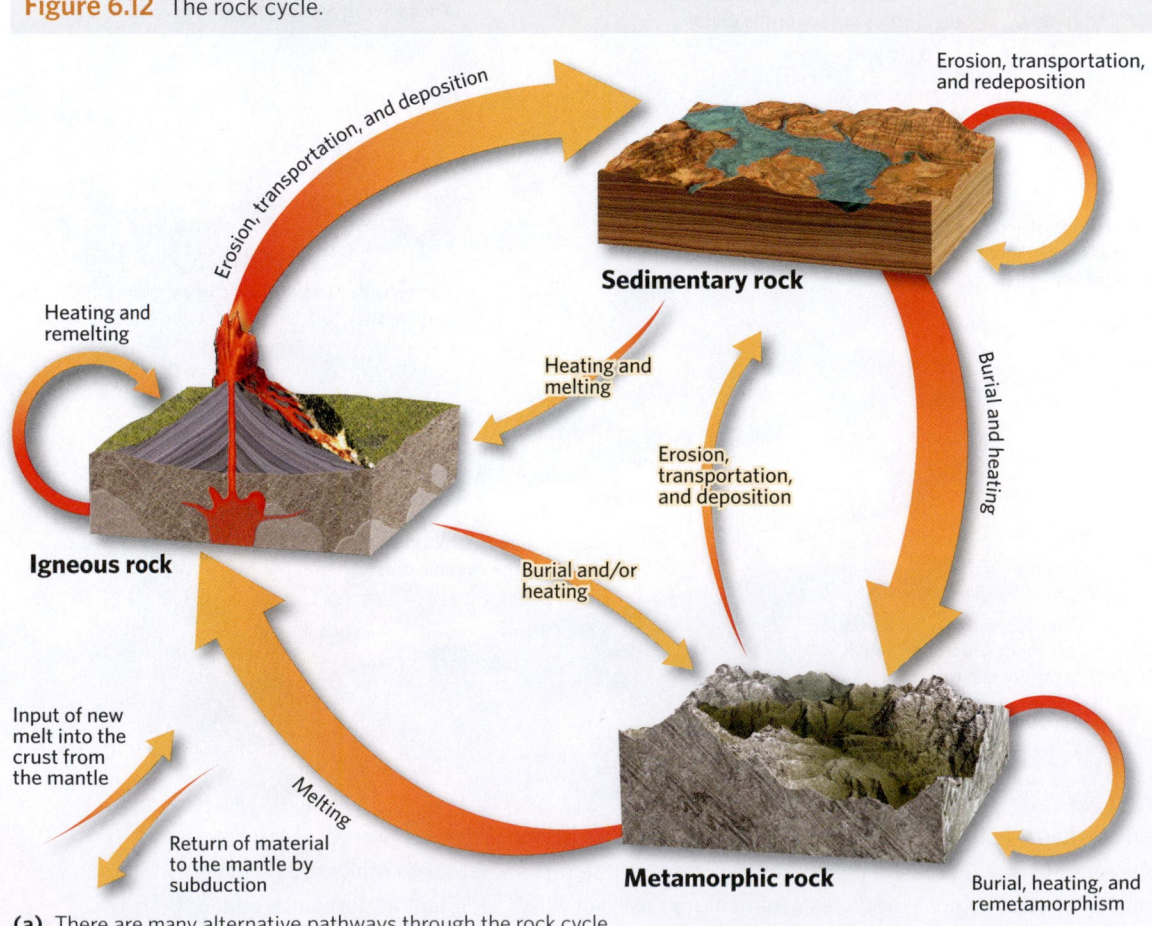

(a) There are many alternative pathways through the rock cycle.

below the continental shelf in the passive-margin basin (see Chapter 5) for millions of years, until Continent Y collides with Continent X. The margin of Continent Y pushes over the shale at the margin of Continent X. As a result, the shale becomes deeply buried and undergoes compression and shear. As mountains grow, the shale that was once 6 km below the surface may end up at a depth of 25 km (15 miles). As it becomes progressively more deeply buried, the shale metamorphoses into slate, then phyllite, and then schist (Fig. 6.13b).

But the story is not over. Once mountain building stops, exhumation returns some of the schist to the ground surface. This schist erodes to form sediment, which a river carries to a sedimentary basin, where it undergoes deposition and burial, eventually lithifying to form a new sedimentary rock. Some of the schist, however, remains below the surface (Fig. 6.13c). Eventually, continental rifting takes place at the site of the former mountain range, and the crust containing the schist begins to split apart. Rifting causes decompression melting in the underlying mantle. The rising magma brings heat into the crust. Magma that intrudes at shallow crustal levels causes the schist in adjacent wall rock to undergo contact metamorphism. Near the base of the crust, the magma from the mantle may cause heat-transfer melting of the schist (see Chapter 4), a process that generates a new felsic magma. This felsic magma then rises to the surface of the crust, erupts, and freezes to form rhyolite, a new igneous rock (Fig. 6.13d). In terms of the rock cycle, we've gone full circle, for the atoms that were once part of one igneous rock (a basalt) are now incorporated in another igneous rock (a rhyolite). Note that even though the rock cycle primarily recycles rocks within the crust, some material can enter the crustal rock cycle when melting within the mantle produces magma that rises into the crust, and some material can leave the rock cycle when subduction carries it back into the mantle.

Not all atoms pass through the rock cycle at the same rate, and for that reason, we find rocks of many different ages at the surface of the Earth. Some rocks remain in one form for less than a few million years, while others have stayed unchanged for most of Earth history. For example, a rock in the Appalachian Mountains has passed through stages of the rock cycle many times during the past billion years because the eastern margin of North America has been subjected to multiple episodes of basin formation, mountain building, and rifting. In contrast, 3-Ga igneous rocks found in the shield of central Canada have not yet passed the first stage of the rock cycle.

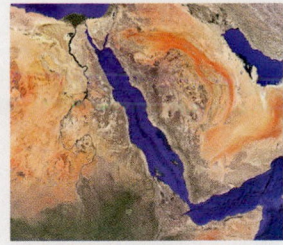

Igneous rock forming, Hawaii

Sedimentary strata, Utah

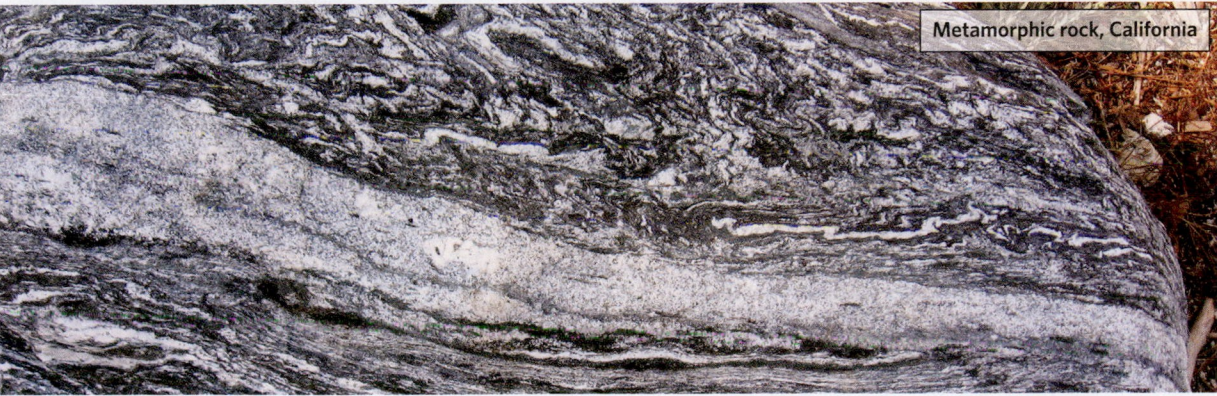

Metamorphic rock, California

(b) Examples of the three classes of rock.

Figure 6.13 An example of the rock cycle in the context of plate tectonics.

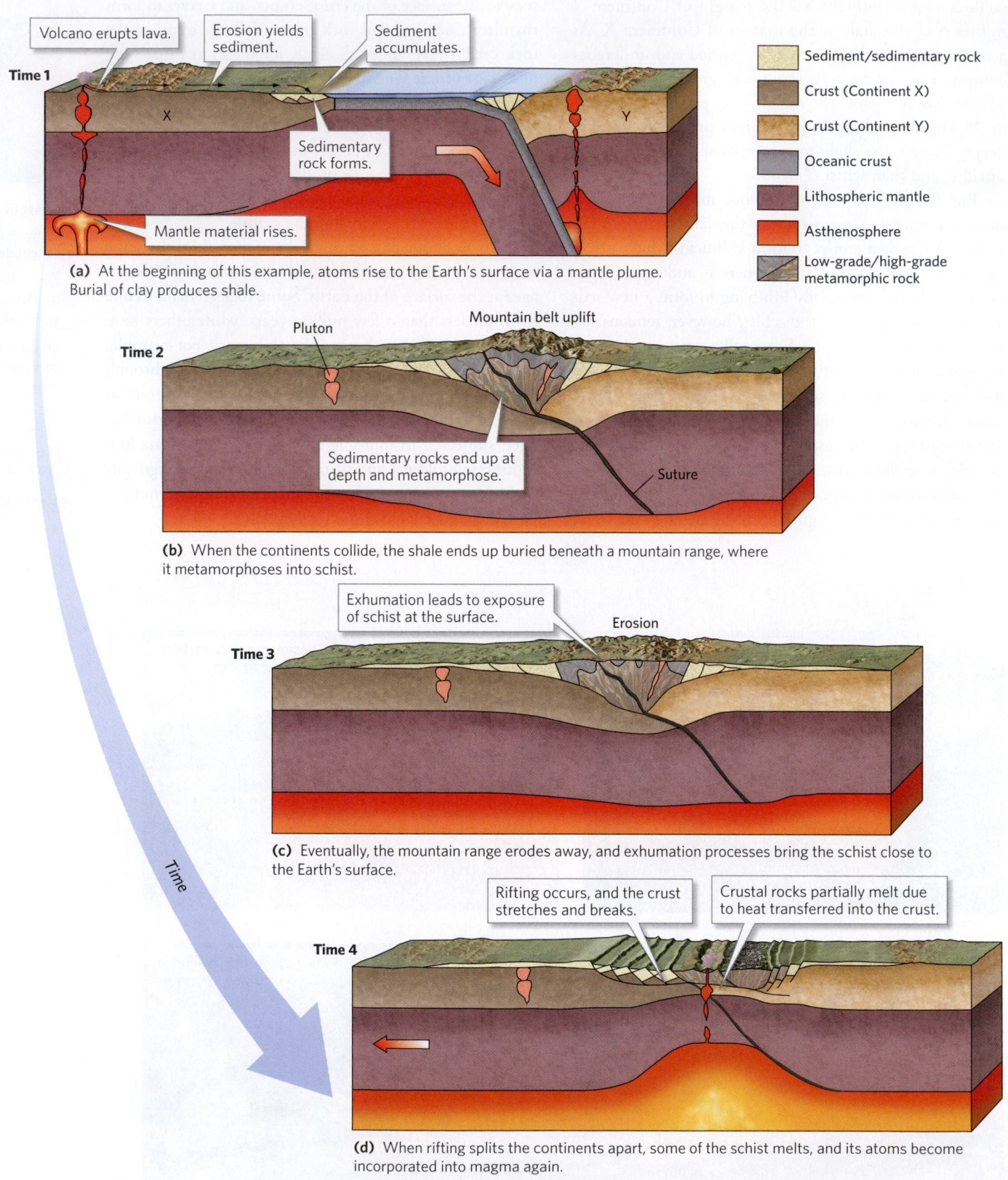

Sediment/sedimentary rock

Crust (Continent X)

Crust (Continent Y)

Oceanic crust

Lithospheric mantle

Asthenosphere

Low-grade/high-grade metamorphic rock

Volcano erupts lava.

Erosion yields sediment.

Sediment accumulates.

Time 1

Sedimentary rock forms.

Mantle material rises.

(a) At the beginning of this example, atoms rise to the Earth's surface via a mantle plume. Burial of clay produces shale.

Pluton

Mountain belt uplift

Time 2

Sedimentary rocks end up at depth and metamorphose.

Suture

(b) When the continents collide, the shale ends up buried beneath a mountain range, where it metamorphoses into schist.

Exhumation leads to exposure of schist at the surface.

Erosion

Time 3

(c) Eventually, the mountain range erodes away, and exhumation processes bring the schist close to the Earth's surface.

Rifting occurs, and the crust stretches and breaks.

Crustal rocks partially melt due to heat transferred into the crust.

Time 4

Time

(d) When rifting splits the continents apart, some of the schist melts, and its atoms become incorporated into magma again.

What Energy Drives the Rock Cycle?

The rock cycle occurs because the Earth is a dynamic, ever-changing planet with many rock-forming environments. External energy (solar radiation), internal energy (Earth's internal heat), and gravitational energy all play a role in driving the rock cycle by keeping the mantle, crust, atmosphere, and oceans in motion. Specifically, our planet's internal heat and gravitational field work together to drive plate movements and mantle plumes. Plate interactions cause the uplift of mountain ranges, a process that leads to weathering, erosion, and sediment production, as well as to metamorphism and igneous activity. Circulation of the atmosphere—driven by a combination of solar radiation and gravity—produces rain, snow, and wind. The rain feeds streams, and the snow accumulates to form glaciers, which, along with wind and wind-driven waves, act as agents of erosion that produce sediment. In the Earth System, life also plays a key role by contributing to weathering and mineral precipitation.

Take-home message . . .

Rocks don't last forever. Melting, weathering, burial, and metamorphism can change rock so that the atoms in one rock type end up in another, and later, in another still. This progression, the rock cycle, is driven by energy coming both from inside the Earth and from the Sun.

Quick Question -
How do steps in the rock cycle relate to plate interactions?

Another View This rock outcrop in Death Valley consists of marble. The banding was inherited from original bedding, but it was sheared during metamorphism. The outcrop was sculpted and polished by a stream.

~15 cm

6 CHAPTER REVIEW

Chapter Summary

- Metamorphism refers to changes in a rock that result in the formation of a new mineral assemblage or a new texture due to changes in temperature or pressure, the application of stress, and interaction with hydrothermal fluids.

- Metamorphism involves recrystallization, phase changes, metamorphic reactions, pressure solution, or plastic deformation. If hydrothermal fluids bring in or remove ions, we say that metasomatism has occurred.

- A preferred orientation develops where compression and shear cause inequant grains in a rock to align parallel to one another.

- Geologists separate metamorphic rocks into two classes, foliated rocks and nonfoliated rocks. Foliated rocks include slate, phyllite, schist, and gneiss. Nonfoliated rocks include hornfels, quartzite, and marble.

- Metamorphic rocks formed under relatively low temperatures are low-grade, whereas those formed under high temperatures are high-grade. Intermediate-grade rocks develop between these two extremes. Different mineral assemblages form at different grades.

- Geologists identify different grades of rock by looking for index minerals. Isograds indicate the location at which index minerals first appear. A metamorphic zone is the region between two isograds and represents an area of rock with a given grade.

- A metamorphic facies is a group of metamorphic mineral assemblages that develop under a specified range of temperature and pressure conditions.

- Thermal (contact) metamorphism occurs in an aureole surrounding an igneous intrusion. Dynamic metamorphism occurs along faults. Dynamothermal (regional) metamorphism results when rocks undergo heating and shearing during mountain building.

- We find belts of metamorphic rocks in mountain ranges. Blueschist forms in accretionary prisms. Shields expose broad areas of Precambrian metamorphic rocks.

- Over geologic time, atoms in one rock type may end up in different rock types. This transfer from one rock type to another is the rock cycle. Not all material follows the same path through the rock cycle.

- The rock cycle operates because the Earth System is dynamic. Solar energy, Earth's internal energy, gravity, and life all play roles in driving the rock cycle.

Key Terms

compression (p. 194)
contact metamorphism (p. 204)
dynamic metamorphism (p. 204)
dynamothermal metamorphism (p. 204)
exhumation (p. 206)
gneiss (p. 198)
gneissic banding (p. 198)
hornfels (p. 200)
hydrothermal fluid (p. 195)

marble (p. 200)
metaconglomerate (p. 198)
metamorphic aureole (p. 204)
metamorphic facies (p. 203)
metamorphic foliation (p. 197)
metamorphic grade (p. 200)
metamorphic mineral (p. 192)
metamorphic rock (p. 191)
metamorphic texture (p. 192)
metamorphic zone (p. 200)

metamorphism (p. 191)
metasomatism (p. 195)
phase change (p. 193)
phyllite (p. 198)
preferred orientation (p. 195)
protolith (p. 191)
quartzite (p. 200)
regional metamorphism (p. 205)
rock cycle (p. 210)
schist (p. 198)

shear (p. 194)
shield (p. 206)
slate (p. 197)
solid-state diffusion (p. 194)
stress (p. 194)
tension (p. 194)
thermal metamorphism (p. 204)

Review Questions

The letters following each Review Question refer to the corresponding Learning Objective from the Chapter Opener.

1. How are metamorphic rocks different from igneous and sedimentary rocks? **(A)**

2. What phenomena can cause metamorphism? **(C)**

3. How can the assemblage of minerals in a metamorphic rock differ from that of its protolith? **(B)**

4. What is metamorphic foliation, and how does it form? Draw the orientation of foliation on the thin section. **(D)**

5. How does slate differ from phyllite? How does phyllite differ from schist? How does schist differ from gneiss? **(D)**

6. Why is hornfels nonfoliated? **(D)**

7. How does marble differ from quartzite? **(D)**

8. What is meant by metamorphic grade, and how can it be determined? How does the concept of metamorphic grade differ from the concept of metamorphic facies? Identify the axes of the graph, and identify areas representing conditions in which low-, intermediate-, and high-grade rocks form. **(E)**

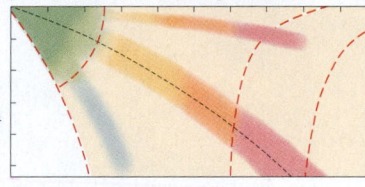

9. Describe the geologic settings where thermal, dynamic, and dynamothermal metamorphism take place. Label the aureole of highest-grade rock on the diagram. **(C)**

10. How does plate tectonics theory explain the formation of blueschist? **(C)**

11. Where would you go if you wanted to find exposed metamorphic rocks, and how did such rocks return to the surface of the Earth after undergoing metamorphism at depth in the crust? **(D)**

12. Once formed, does a rock necessarily last for all of Earth's history? Explain your answer. **(F)**

13. Describe a pathway through the rock cycle. **(G)**

14. Have all continental rocks passed through the rock cycle the same number of times? Explain your answer. **(G)**

On Further Thought

15. Do you think that you would be likely to find a broad region (hundreds of kilometers across) in which outcrops consist of high-grade hornfels? Why or why not? (Hint: Think about the causes of metamorphism and the conditions under which hornfels forms.) **(C)**

16. Could you find a layer of metamorphic rock between the layers of sedimentary rock in a sedimentary basin? Why or why not? **(A)**

17. The geothermal gradient of Mars is 8°C/km (i.e., temperature increases by 8°C per kilometer of depth), and the crust of Mars is 30 km thick. Do high-grade metamorphic rocks form in this crust? **(E)**

18. Does basalt of the oceanic crust pass through stages of the rock cycle as it moves from a mid-ocean ridge to a trench? **(G)**

19. Does the rock cycle happen on the Moon? **(H)**

Online Resources

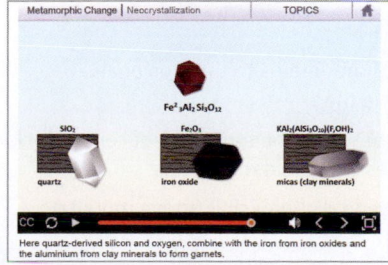

Animations

This chapter features animations on the six major types of metamorphic change, focusing on each process close up.

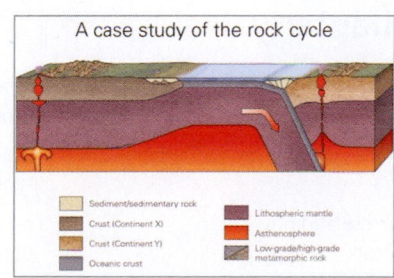

Video

This chapter features a video that explains the rock cycle, and includes a case study.

Smartworks5

This chapter features visual exercises on metamorphism and its effects.

7 CRAGS, CRACKS, AND CRUMPLES
Mountain Building and Geologic Structures

By the end of the chapter **you should be able to . . .**

A. describe how rocks can deform (crack, bend, or flow) in response to stress.

B. identify basic geologic structures (joints, faults, folds, foliation) and sketch them.

C. explain why mountain belts grow and how their formation relates to plate tectonics.

D. relate the shapes of mountains to the erosional processes that carve them.

E. draw connections between the formation of certain rock types and processes that build mountains.

F. recognize a craton and distinguish it from a mountain belt.

Folds involving sedimentary strata exposed along the coast of southern Japan. The layers were originally horizontal, but processes associated with subduction of the Pacific Ocean floor caused them to be bent.

7.1 Introduction

Geographers call the peak of Mt. Everest "the top of the world," for this mountain, which lies in the Himalayas of south-central Asia, rises higher than any other on Earth. The ice on Mt. Everest's summit glistens at an elevation 8.85 km (29,029 feet) above sea level, almost the cruising height of modern jets. In 1953, Sir Edmund Hillary, from New Zealand, and Tenzing Norgay, a Nepalese guide, became the first to reach the summit. Since then, thousands of other people have also succeeded, but over 250 have died trying. Mountains appeal to nonclimbers as well, for almost everyone loves a vista of snow-crested peaks. The stark cliffs, clear air, meadows, forests, streams, and glaciers of mountainous landscapes provide a refuge from the mundane.

With the exception of large volcanoes formed over hot spots, mountains do not occur in isolation, but rather in ranges of elevated land called **mountain belts**, or **orogens** (from the Greek words *oros*, meaning mountain, and *genesis*, meaning birth). To build such belts, thousands of cubic kilometers of rock must be pushed skyward against the pull of gravity. Clearly, the existence of mountains serves as one of the most obvious indications of Earth's dynamic activity! A map of the present-day Earth reveals about a dozen major mountain belts and numerous smaller ones **(Fig. 7.1)**. In addition, many places that are now lowlands were mountains in the distant geologic past. So **orogeny**, the process of mountain-belt formation, has happened many times and in many places over Earth's history.

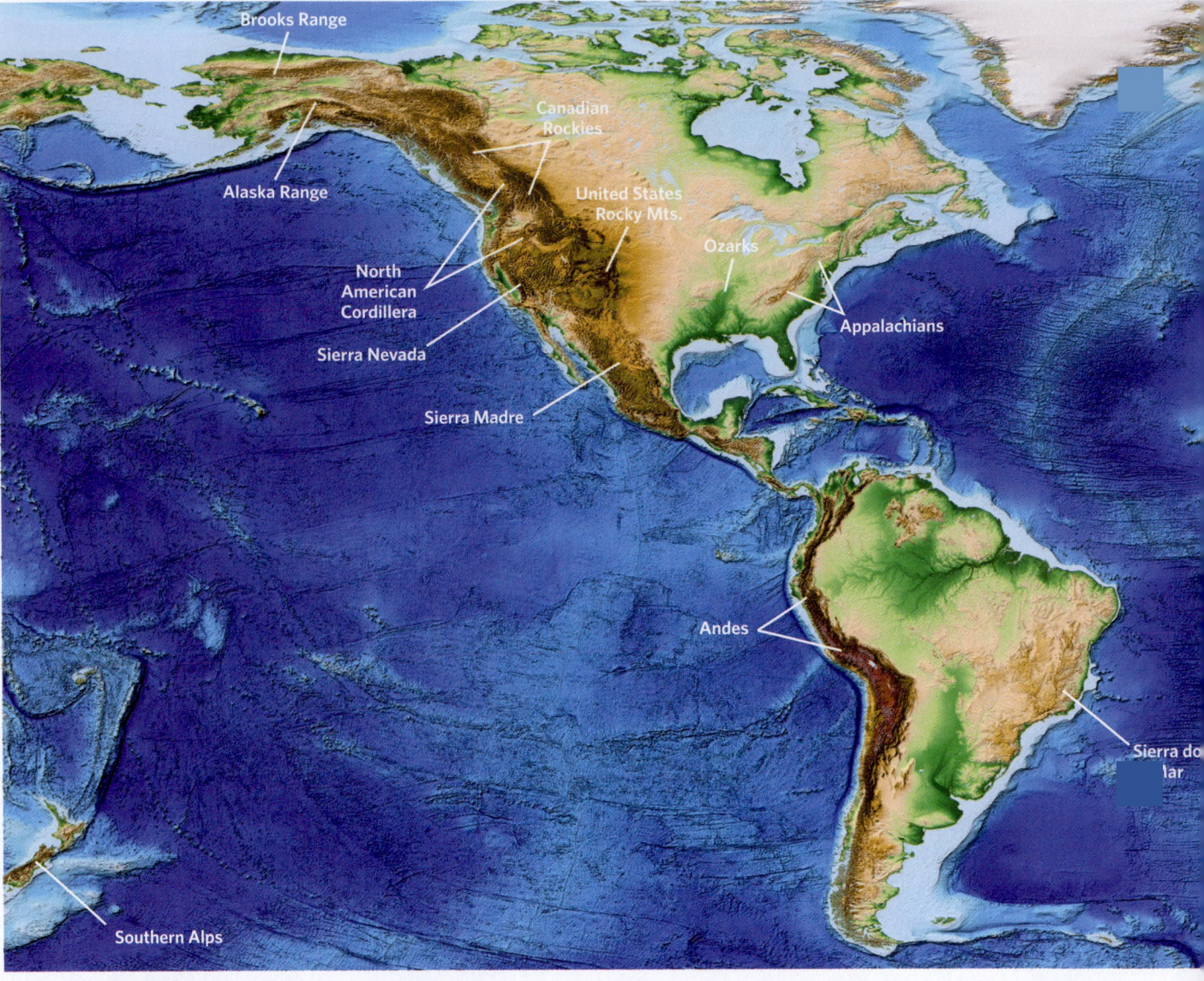

Brooks Range

Canadian Rockies

Alaska Range

United States Rocky Mts.

Ozarks

North American Cordillera

Appalachians

Sierra Nevada

Sierra Madre

Andes

Sierra do lar

Southern Alps

Orogeny not only leads to **uplift**, the vertical rise of the land surface and the rock beneath, but also causes rocks to undergo **deformation**, a process during which rocks bend, break, or flow. **Geologic structures**, the features produced by deformation, include *joints* (natural cracks), *faults* (fractures on which one body of rock slides past another), *folds* (bends, curves, or wrinkles of rock layers), and *foliation* (the layering in rock that develops when metamorphism happens during deformation).

A given mountain-building event may last for tens of millions of years. As soon as land starts to rise into a mountain belt, however, *erosion* begins to grind it away, producing sediment and sculpting spectacular, jagged landscapes. Erosion is the force that carves the cliffs and valleys and sharp peaks of mountains—without it, a mountain belt might be just a region of elevated land. Erosion and uplift battle as a mountain belt forms. When uplift ceases, erosion can eventually shave a mountain range down to near sea level. But even after the high peaks are gone, a low-lying

belt of fractured, contorted, and metamorphosed rock remains. These scars serve as a permanent monument to what was once a region of high peaks.

In this chapter, we first learn about the processes that deform rocks and how to describe and interpret the geologic structures that result from their deformation. Next, we turn our attention to the question of why mountain belts form, and find that answers come from plate tectonics theory. Finally, we consider the phenomena that sculpt regions of uplift to form the terrains that we can see when visiting mountainous landscapes.

7.2 Rock Deformation

What Is Deformation?

To get a visual sense of what geologists mean by *deformation*, let's contrast rock that has not been affected by mountain building with rock that has been affected.

Our undeformed example comes from a roadcut in the interior plains of North America, and our deformed example comes from a cliff in the Alps, a mountain belt of Europe.

The roadcut exposes nearly horizontal beds of sandstone, shale, and limestone cut by a few joints (Fig. 7.2a). These beds have the same orientation that they had when first deposited. The sand grains in the sandstone of this outcrop are nearly spherical (the same shape as when they were deposited), and the clay flakes in the shale lie roughly parallel to the bedding plane. When we say that this outcrop exposes *undeformed rock*, we mean that it contains no geologic structures, other than a few joints (visible as thin vertical cracks cutting across the bedding).

The rocks of the Alpine cliff look very different (Fig. 7.2b). Here, we find layers of quartzite, slate, and marble (the metamorphic equivalents of sandstone, shale, and limestone) that have been contorted into folds, so

that the rock layers, instead of being horizontal, trace out wave-like shapes. On microscopic examination, we may find that the grains in the quartzite are not spherical, but have been flattened into elliptical shapes. And in the slate, the clay flakes no longer align with the bedding plane, but rather align in a new orientation that defines slaty cleavage, a type of foliation (see Chapter 6). Finally, we see that the quartzite and slate layers end abruptly at a sloping fault, below which the outcrop consists of folded marble. Slip on the fault brought the slate and quartzite from another location and juxtaposed it against the marble. The folds, faults, foliation, and flattened grains in the Alpine cliff are all manifestations of deformation, so we say that this outcrop exposes *deformed rock*. The character of the geologic structures visible in the cliff emphasizes that during deformation, rocks can undergo one or more of the following changes (Fig. 7.3): (1) a change in location, or *displacement*; (2) a change in orientation, or *rotation*; and (3) a change in shape, or *distortion*.

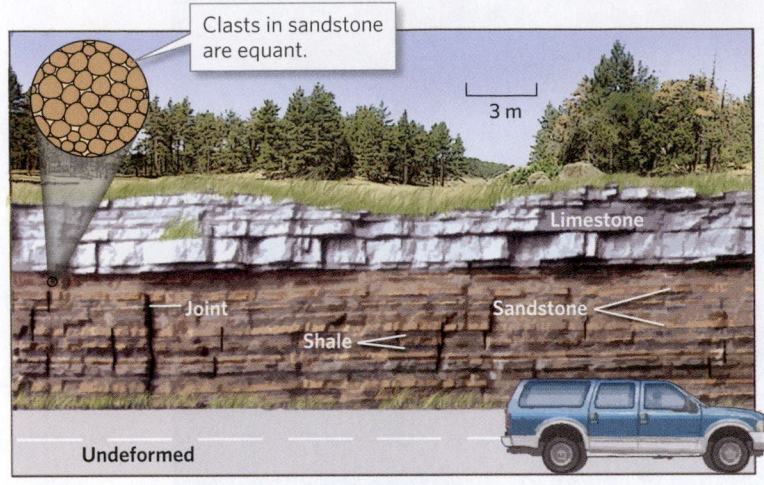

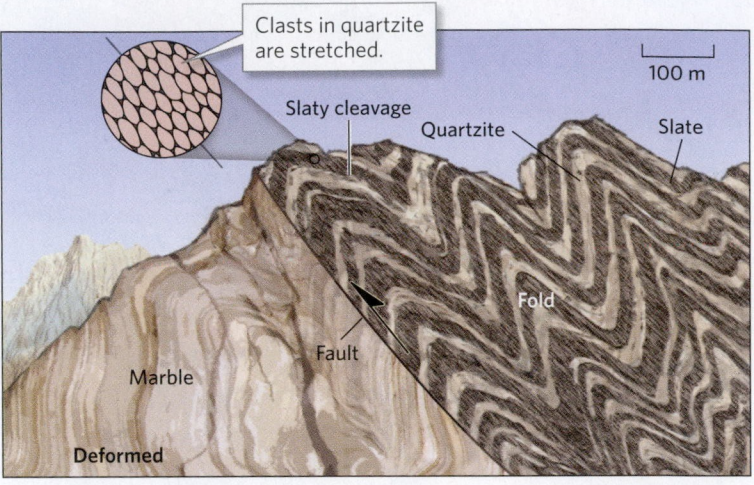

(a) These flat-lying beds along a highway in the interior plains of North America are essentially undeformed. A few joints, formed when overlying rock eroded away, are visible.

(b) In this Alpine cliff, deformation has caused rock layers to undergo folding and faulting. In addition, foliation (in this case, slaty cleavage) has developed, and clasts have been stretched.

Figure 7.3 The components of deformation include displacement, rotation, and distortion.

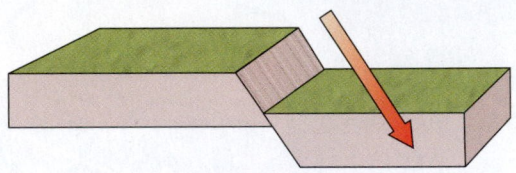

(a) Displacement occurs when a block of rock moves from one location to another.

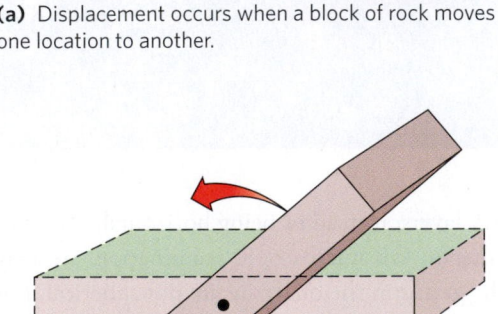

(b) Rotation occurs when a body of rock undergoes tilting.

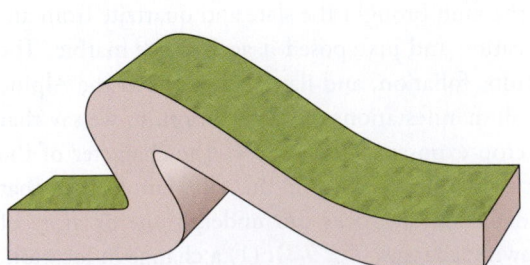

(c) Distortion occurs when a body of rock changes shape. The development of a fold represents one type of distortion.

Geologists refer to the amount of distortion that a body of rock or a region of crust develops during deformation as its **strain**. We can distinguish among different kinds of strain according to the nature of the overall shape change. Specifically, if a layer becomes longer, it has undergone *stretching*, or extension, whereas if the layer becomes shorter, it has undergone *shortening* (Fig. 7.4a–c). And if a change in shape involves the movement of one part of a rock body sideways past another, it has undergone *shear strain* (Fig. 7.4d, e).

Brittle versus Plastic Deformation

Drop a porcelain plate on a hard floor, and it shatters into pieces. You've just observed **brittle deformation**, the process by which a material cracks or fractures to form pieces that no longer hold together (Fig. 7.5a, b). The formation of joints and faults serves as an example of brittle deformation in rocks. Now, squeeze a ball of soft dough between a book and a tabletop, or bend a stick of chewing gum. You've just observed **plastic deformation**, the process during which objects change shape without visibly breaking (Fig. 7.5c, d). The formation of folds serves as a

Figure 7.4 Strain is a measure of the distortion, or change in shape, that takes place in rock during deformation.

(a) An unstrained cube and an unstrained fossil brachiopod shell.

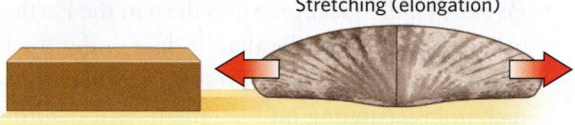

(b) Horizontal stretching changes the cube into a horizontal brick and elongates the shell.

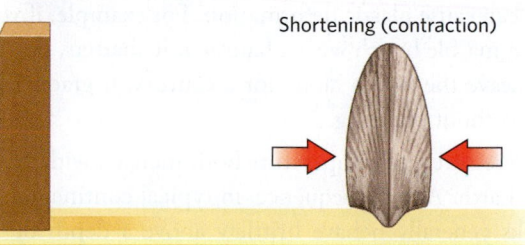

(c) Horizontal shortening changes the cube into a vertical brick and makes the shell narrower.

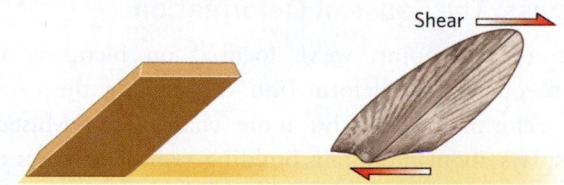

(d) Shear strain transforms the cube into a parallelogram, and changes angular relationships in the shell.

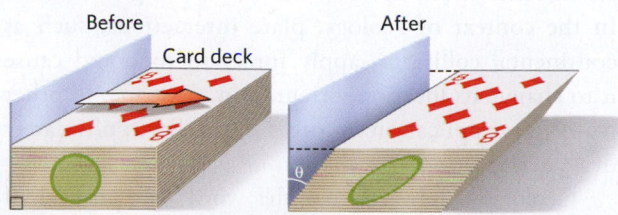

(e) You can produce shear strain by moving a deck of cards so that each card slides a little with respect to the one below.

Figure 7.5 Brittle versus plastic deformation.

(a) Brittle deformation occurs when you drop a plate and it shatters.

(b) The cracks in these quartzite beds in Utah are due to brittle deformation.

(c) Plastic deformation occurs when you squash a ball of dough.

(d) The quartzite cobbles in this conglomerate were flattened plastically.

Figure 7.6 Types of stress.

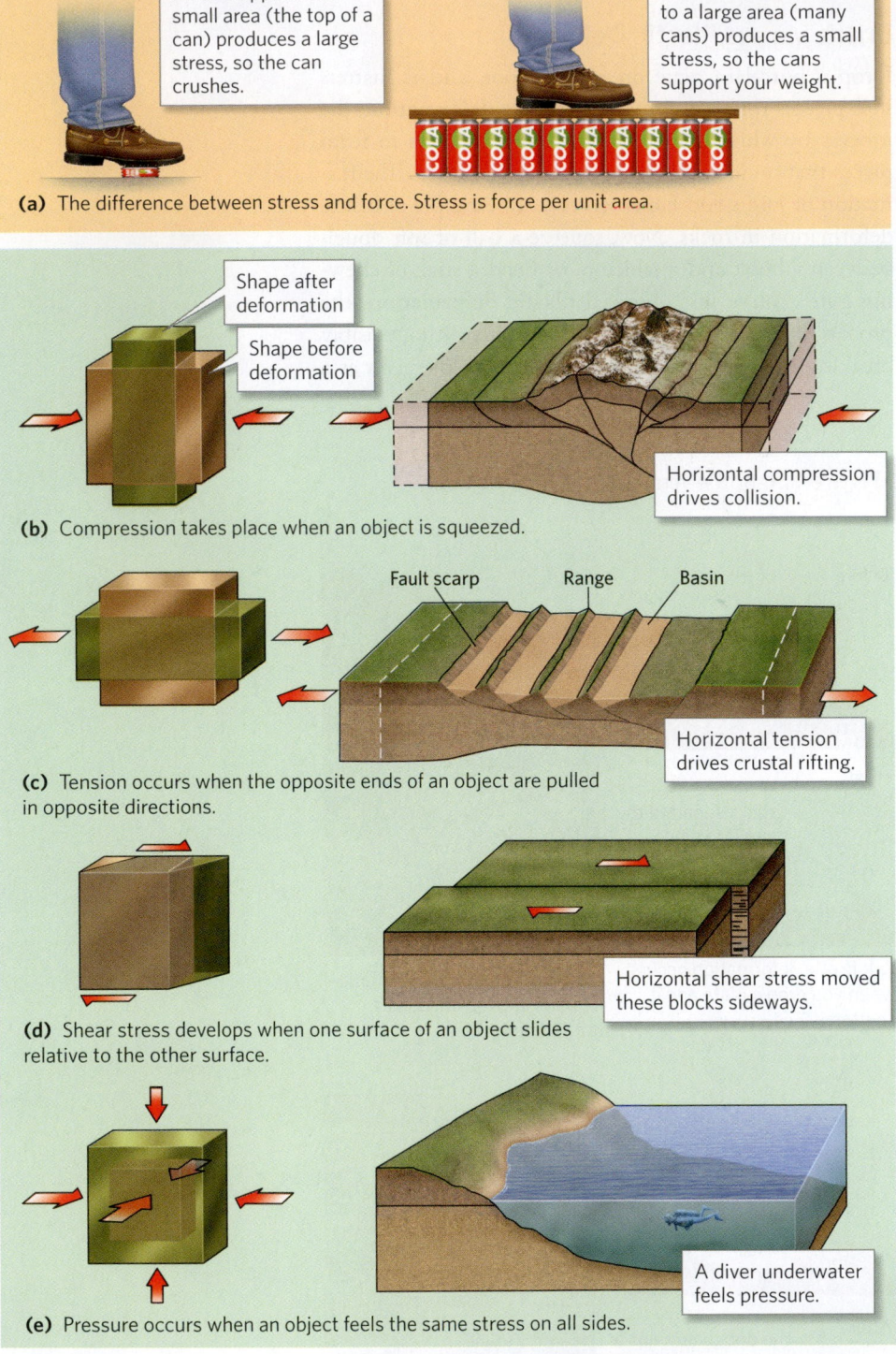

(a) A force applied to a small area (the top of a can) produces a large stress, so the can crushes.

The same force applied to a large area (many cans) produces a small stress, so the cans support your weight.

The difference between stress and force. Stress is force per unit area.

Shape after deformation

Shape before deformation

Horizontal compression drives collision.

(b) Compression takes place when an object is squeezed.

Fault scarp   Range   Basin

Horizontal tension drives crustal rifting.

(c) Tension occurs when the opposite ends of an object are pulled in opposite directions.

Horizontal shear stress moved these blocks sideways.

(d) Shear stress develops when one surface of an object slides relative to the other surface.

A diver underwater feels pressure.

(e) Pressure occurs when an object feels the same stress on all sides.

geologic example of plastic deformation. Everyday examples of deformation, such as breaking plates or squeezing dough, take place fast enough for you to see. In contrast, geologic deformation in mountain ranges generally takes place so slowly (at rates of less than a couple of millimeters to a few centimeters a year) that you can't see it in the course of a human lifetime.

What actually happens within rock during deformation? Recall that the atoms that make up mineral grains are connected to each other by chemical bonds. During brittle deformation, many bonds break and stay broken, leading to the formation of a permanent crack across which material no longer connects. During plastic deformation, in contrast, some bonds break, but new ones quickly form. In this way, the atoms within a rock are rearranged, and the rock gradually changes shape without breaking into pieces.

What determines whether rock deforms brittlely or plastically? The deformation behavior of a rock depends on several factors:

- *Temperature:* Heat makes materials softer, so warmer rocks tend to deform plastically, whereas colder rocks tend to deform brittlely. To see the effect of temperature for yourself, compare the behavior of a candle warmed in an oven with that of one cooled in a freezer when you bend it with your hands—the former easily changes shape, but the latter snaps in two.

- *Pressure:* Under great pressures deep in the Earth, rock behaves more plastically than it does under the lower pressures near the surface. Pressure effectively prevents rock from separating into fragments as it deforms.

- *Deformation rate:* A sudden change in shape can cause brittle deformation, whereas a slow change in shape can cause plastic deformation. For example, if you hit a marble bench with a hammer, it shatters, but if you leave the bench alone for a century, it gradually sags without breaking.

Pressure and temperature both increase with depth in the Earth. As a consequence, in typical continental crust, rocks generally behave brittlely above a depth of 10 to 15 km (6 to 9 miles) and plastically below that depth.

Stress: The Cause of Deformation

Up to this point, we've focused on picturing the consequences of deformation. Describing the causes of deformation is a bit more challenging. Museum displays about mountain building typically bypass the issue by saying something like, "The mountains were produced by forces deep within the Earth." What does this mean? Isaac Newton stated that a **force** can cause an object to speed up, slow down, or change direction. In the context of geology, plate interactions, such as continental collisions, apply forces to rock and cause it to change its location, orientation, or shape. In other words, the application of forces in the Earth ultimately drive deformation.

Geologists, however, use the word *stress* instead of force when talking about generation of deformation, where we define **stress** as force per unit area. We must distinguish between stress and force because the actual consequences of applying a force depend not only on the amount of force applied, but also on the area over which the force acts. For example, the force caused by your body weight applied to the top of a can crushes the can, but the same force spread over the tops of 100 cans does not (Fig. 7.6a). How does the concept of stress apply to geology? During mountain building, the force of one plate interacting with another is spread across the contact between the two plates, so the deformation resulting at any specific location actually depends on the type of stress developed at that location, not on the overall force.

Different types of stress develop in the Earth's crust in different geologic settings (Fig. 7.6b–e). Squeezing a rock generates **compression**, pulling on a rock yields **tension**, and shoving one part of a rock sideways with respect to another produces **shear**. The familiar word **pressure** actually refers to a special stress condition that happens when the stress acting on all sides of an object is the same.

Note that *stress* and *strain*, in the context of geology, have very different meanings, even though, in everyday English, the terms may seem interchangeable. In geology, *stress* refers to compression, tension, or shear, whereas *strain* is a measure of the shape changes resulting from deformation. Put simply, stress causes strain. Specifically, compression causes shortening, tension generates stretching, and shear produces shear strain. Pressure can cause an object to become smaller but will not cause it to change shape. With this knowledge of stress and strain, we can now look at the nature and origin of various classes of geologic structures.

Take-home message . . .

Because of stress (compression, tension, shear) that develops in the Earth, rocks may change their shape, position, or orientation. All of these changes are manifestations of deformation. Strain is a measure of the change in shape. During brittle deformation, rocks break, whereas during plastic deformation, rocks bend and distort without breaking.

Quick Question -
When you sit on a chair, does the weight of your body apply vertical compression or vertical tension?

7.3 Geologic Structures Formed by Brittle Deformation

Joints: Natural Cracks in Rocks

If you examine just about any outcrop of rock, you'll find cracks, meaning surfaces across which rock broke so that the pieces on either side of the break no longer connect.

Geologists refer to a natural crack in rock as a **joint** if there has been no movement of the rock on either side of the crack (Fig. 7.7a, b). Because rock is not connected across a joint, joints are planes of weakness, so their presence can affect the stability of cliffs (Fig. 7.7c).

Joints develop in response to tension in brittle rock. In other words, a rock cracks and a joint forms because the rock has been pulled slightly apart. The tension that causes joints may exist for a variety of geologic reasons. For example, some joints develop when the rock of a region cools and shrinks—the same process happens when you heat up a glass plate and then plunge it into cold water. Other joints develop when rock formerly at depth undergoes a decrease in pressure as overlying rock erodes away, and therefore changes shape slightly. (Note that joints can form in response to cooling or a decrease in pressure even in regions where no mountain building has occurred—that's why we see the traces of joints in the undeformed outcrop depicted in Figure 7.2a.) Finally, some joints form when rock layers bend and stretch during mountain building.

Faults: Surfaces of Slip

After the San Francisco earthquake of 1906, geologists found a rupture that had torn through the land surface near the city. Where this rupture crossed orchards, it offset rows of trees, and where it crossed a fence, it broke the fence in two—the western side of the fence moved northward by about 2 m (6 feet) relative to the eastern side. The rupture represents the trace of the San Andreas fault. A **fault**, as we have seen, is a fracture surface on which sliding occurs. **Slip**, the sliding movement along a fault, can generate earthquakes. Geologists refer to the amount of movement that takes place across a fault as the fault's **displacement**.

Faults have formed throughout Earth history. *Active faults* are those on which sliding has been occurring in recent geologic time, whereas *inactive faults* ceased to slip long ago. Some faults, such as the San Andreas, displace the ground surface when they move. Others slip only at depth, and remain invisible at the surface unless they are later exposed by erosion.

Fault Classification

Not all faults result in the same kind of crustal deformation. Specifically, some accommodate shortening, some accommodate stretching, and some accommodate horizontal shear. It's important for geologists to distinguish among different kinds of faults in order to interpret their geologic significance. Fault classification focuses on two characteristics of faults: (1) the *dip,* or slope, of the fault plane (Box 7.1) and (2) the shear sense across the fault. By *shear sense*, we mean the direction in which material on

Figure 7.7 Examples of joints.

(a) Prominent vertical joints cut these red sandstone beds in Arches National Park, Utah.

(b) Vertical joints in shale on a cliff face near Ithaca, New York.

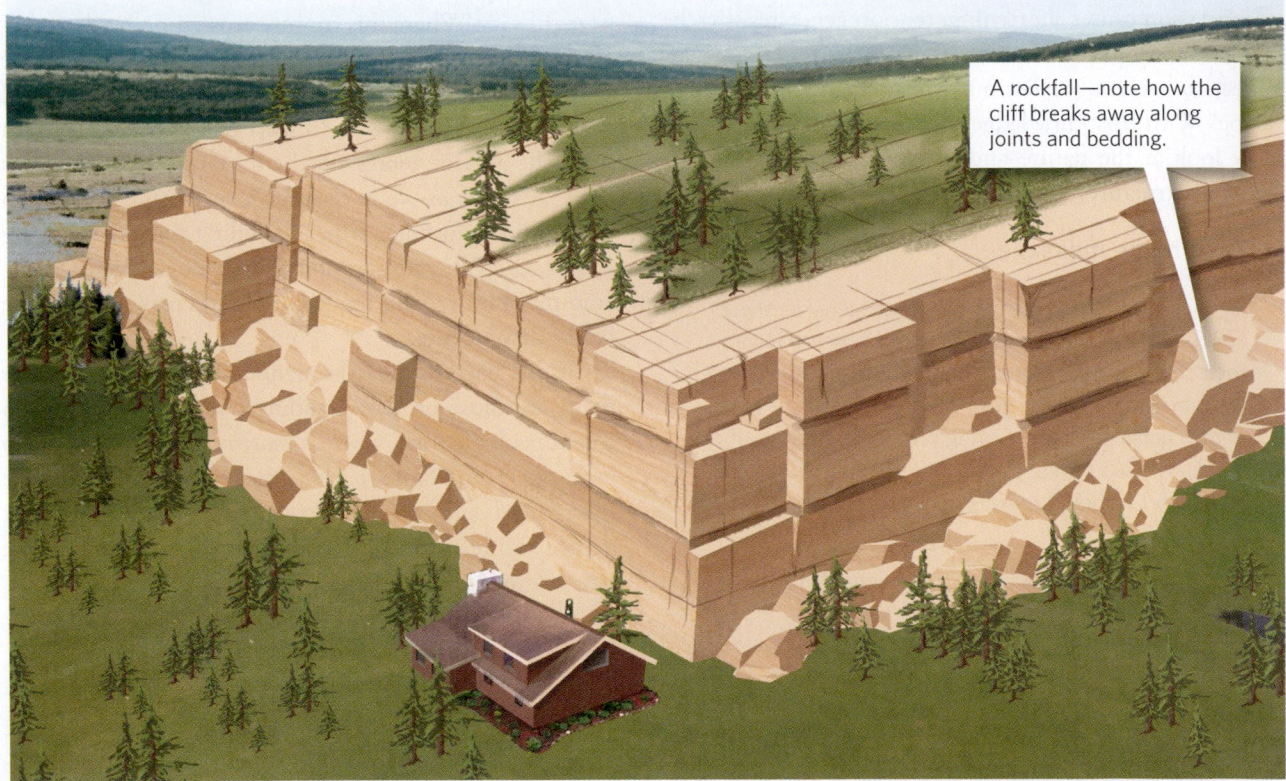

A rockfall—note how the cliff breaks away along joints and bedding.

(c) The orientation of joints controls the stability of cliff faces. In this example, there are two sets of joints. These joints, together with horizontal beds, cause the outcrop to separate into rectangular blocks.

Box 7.1 ▶ Consider **this . . .**

Describing the orientation of geologic structures

When discussing geologic structures, it's important to be able to communicate information about their orientation. For example, does a fault exposed in an outcrop at the edge of town continue beneath the nuclear power plant 3 km to the north, or does it go beneath the hospital 2 km to the east? If we know the fault's orientation, we may be able to answer such questions. To describe the orientation of a geologic structure, geologists picture the structure as a simple geometric shape, then specify the angles that the shape makes with respect to a horizontal plane (a flat surface parallel to sea level), a vertical plane (a flat surface perpendicular to sea level), and the north direction (a line of longitude).

Let's start by considering planar structures such as faults, beds, joints, and foliation. We call these *planar structures* because they are two-dimensional surfaces, so a portion of them can be represented as a geometric plane. A planar structure's orientation can be specified by its strike and its dip. The **strike** is the angle between an imaginary horizontal line, the strike line, on the planar structure and an imaginary horizontal line pointing to true north **(Fig. Bx7.1a, b)**. The **dip** is the angle of the planar structure's slope, or more precisely, the angle between a horizontal plane and the *dip line* (an imaginary line parallel to the steepest slope on the structure), as measured in a vertical plane perpendicular to the strike. A horizontal planar

structure has a dip of 0°, whereas a vertical one has a dip of 90°. We can measure the strike with a special type of compass **(Fig. Bx7.1c)**, and the dip with a clinometer, a type of protractor that uses a bubble to indicate horizontal, and we can represent strike and dip on a map using the symbol shown in Figure Bx7.1b.

We can envision a *linear structure* as a geometric line rather than a plane. Examples of linear structures include scratches or grooves on a rock surface. Geologists specify the orientation of a linear structure by giving its plunge and bearing **(Fig. Bx7.1d)**. The **plunge** is the angle between a linear structure and an imaginary horizontal line in the vertical plane that contains the structure. A horizontal linear structure has a plunge of 0°, and a vertical one has a plunge of 90°. The **bearing** is the angle between due north and the projection of the linear structure on a horizontal plane. A projection is the shadow of a line on a horizontal plane if lit from above.

Figure Bx7.1 Specifying the orientation of planar and linear structures.

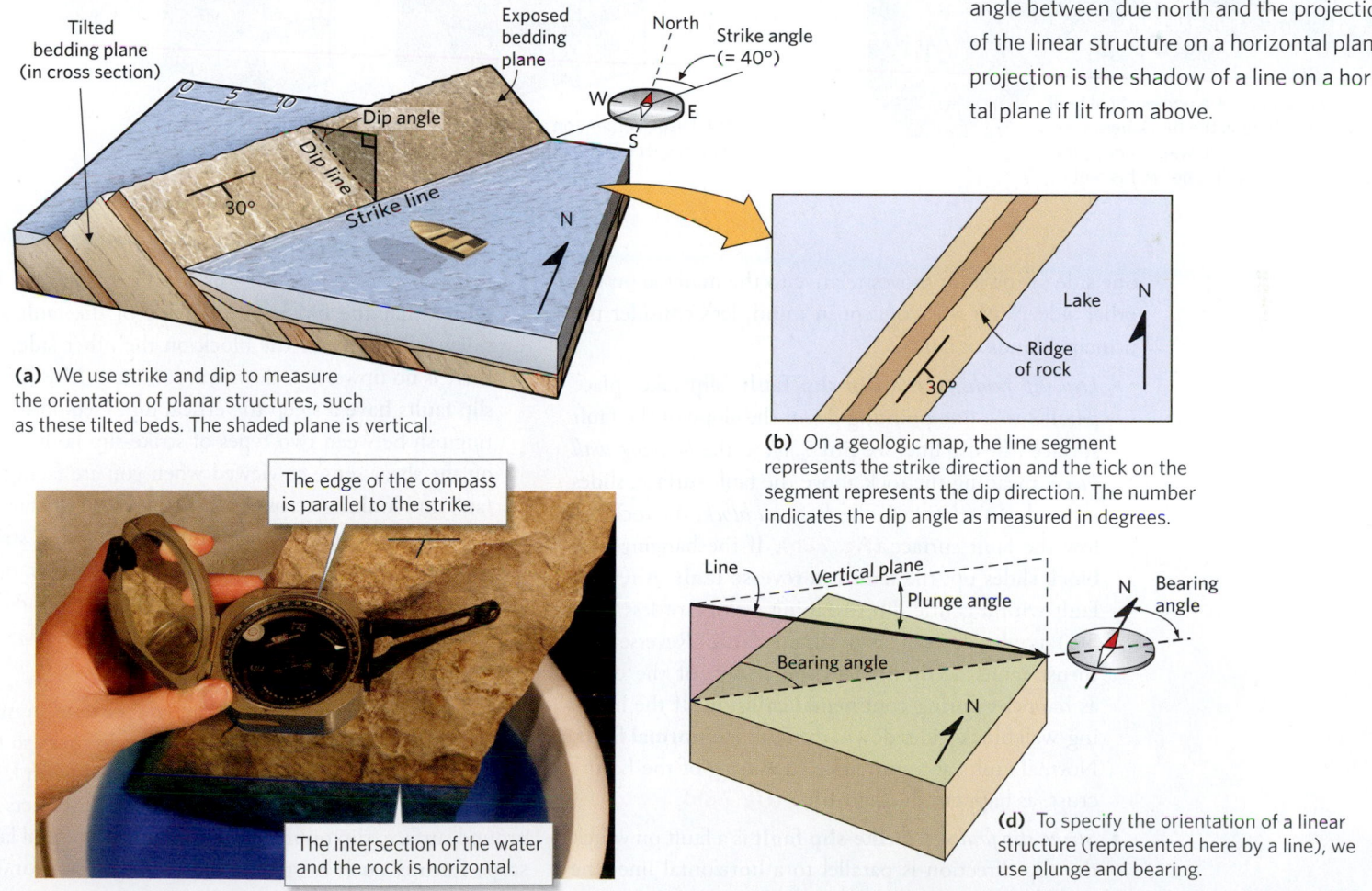

(a) We use strike and dip to measure the orientation of planar structures, such as these tilted beds. The shaded plane is vertical.

(b) On a geologic map, the line segment represents the strike direction and the tick on the segment represents the dip direction. The number indicates the dip angle as measured in degrees.

(c) Geologists use a special compass to measure strike and dip.

The edge of the compass is parallel to the strike.

The intersection of the water and the rock is horizontal.

(d) To specify the orientation of a linear structure (represented here by a line), we use plunge and bearing.

Figure 7.8 The principal categories of faults.

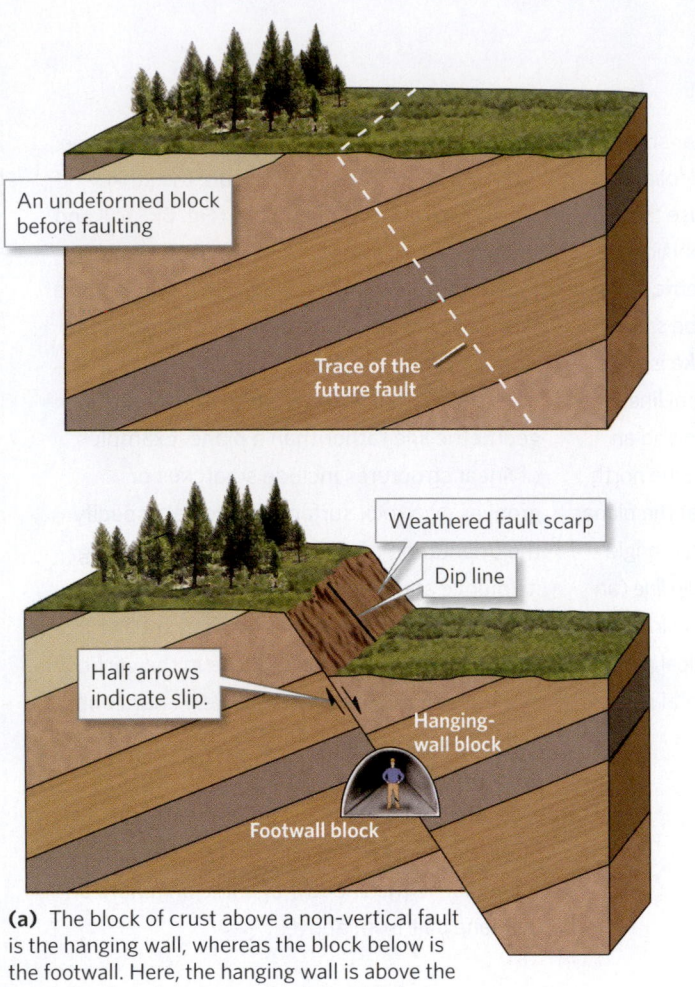

(a) The block of crust above a non-vertical fault is the hanging wall, whereas the block below is the footwall. Here, the hanging wall is above the man's head, and the footwall is below his feet.

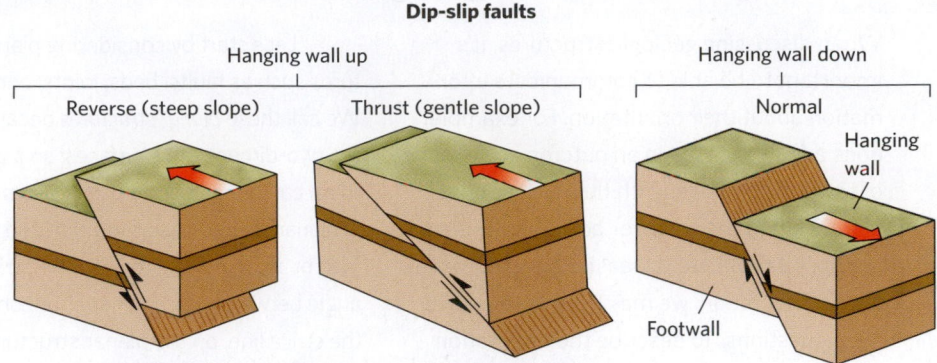

(b) Displacement on a dip-slip fault is parallel to the fault's slope. Reverse faults and thrust faults cause crustal shortening, whereas normal faults cause crustal stretching (red arrows).

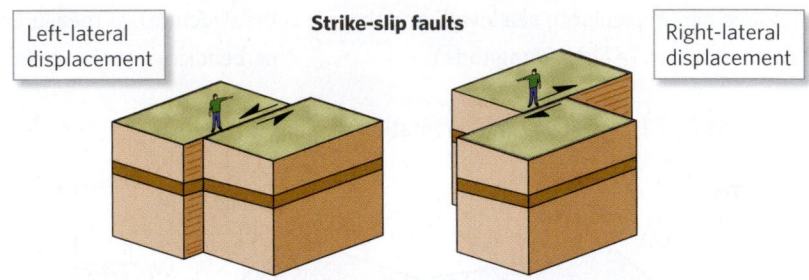

(c) Displacement on a strike-slip fault moves one block horizontally with respect to the other. There is no up-and-down motion.

one side of the fault moves relative to the material on the other side. With this concept in mind, let's consider the principal kinds of faults.

- *Dip-slip faults:* On a **dip-slip fault**, slip takes place parallel to a line pointing down the slope of the fault surface (the dip line; see Box 7.1), so the *hanging-wall block*, meaning the rock above the fault surface, slides up or down relative to the *footwall block*, the rock below the fault surface (**Fig. 7.8a**). If the hanging-wall block slides up, the fault is a **reverse fault**. A reverse fault with a gentle dip (meaning a slope of less than 30°) is also known as a **thrust fault**. Reverse and thrust faults accommodate shortening of the crust, as happens during continental collision. If the hanging-wall block slides down, the fault is a **normal fault**. Normal faults accommodate stretching of the Earth's crust, as happens during rifting (**Fig. 7.8b**).

- *Strike-slip faults:* A **strike-slip fault** is a fault on which the slip direction is parallel to a horizontal line (the

strike line; see Box 7.1) on the fault surface. This means that the block on one side of the fault slips sideways relative to the block on the other side, and there is no upward or downward motion. Most strike-slip faults have a steep to vertical dip. Geologists distinguish between two types of strike-slip faults based on the shear sense as viewed when you are facing the fault and looking across it. If the block on the far side has slipped to your left, the fault is a *left-lateral* strike-slip fault, and if the block has slipped to your right, the fault is a *right-lateral* strike-slip fault (**Fig. 7.8c**).

Recognizing Faults

How do you recognize a fault when you see one? In many cases, faulting visibly shifts distinct layers in rocks so that layers on one side of a fault are not continuous with layers on the other side (**Fig. 7.9a, b**). If a fault intersects the ground surface, slip on the fault can displace natural landscape features such as stream valleys or glacial moraines

Figure 7.9 Recognizing fault displacement in the field.

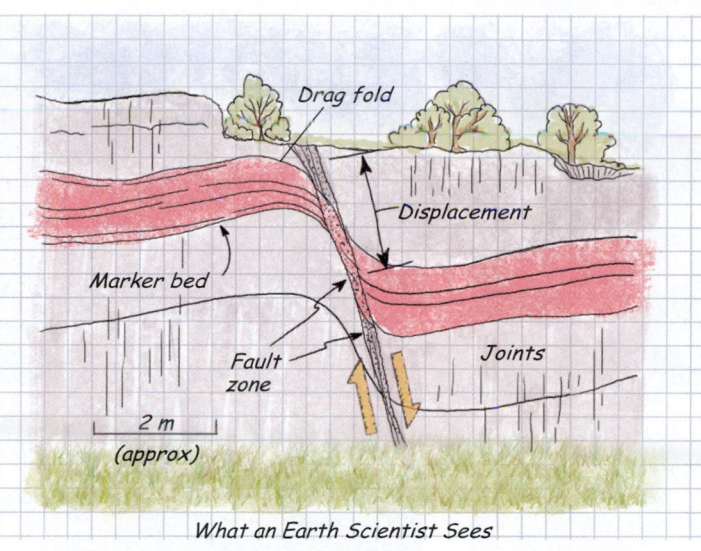

What an Earth Scientist Sees

(a) A steep normal fault has displaced a distinctive redbed (a marker layer). Note that the faulting has formed a 0.5-m-wide fault zone of broken rock. Folds have developed adjacent to the fault.

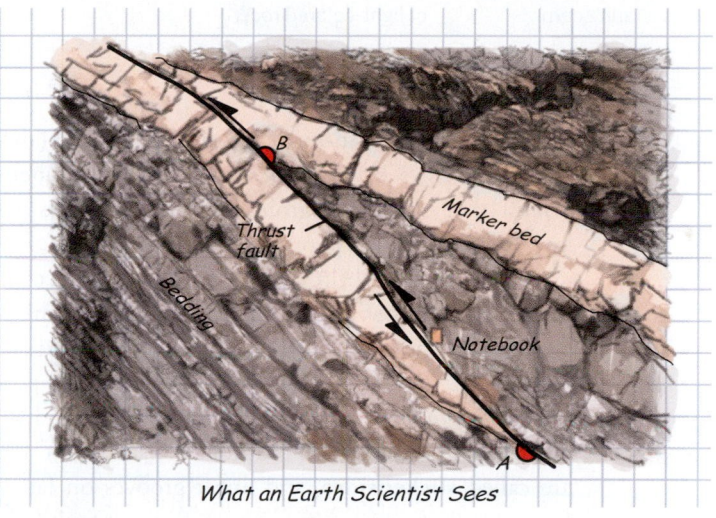

What an Earth Scientist Sees

(b) Slip on a thrust fault has caused one part of the light-colored marker bed to be shoved over another part, as emphasized in the drawing. Note that the beds are tilted to the right. The distance between the red dots is the displacement.

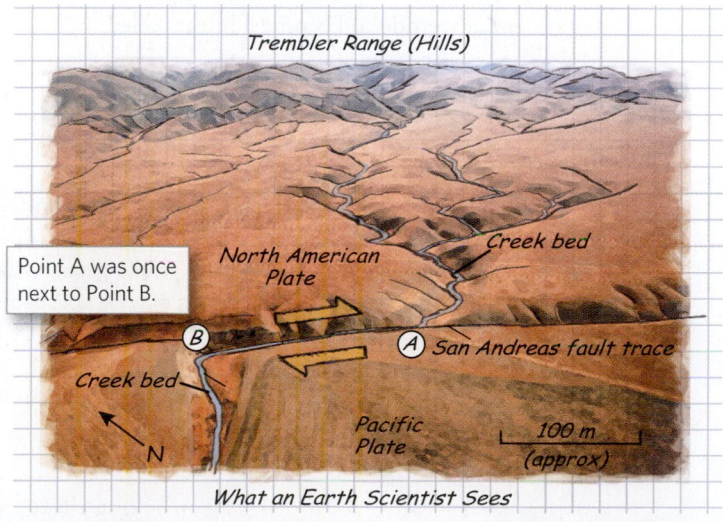

What an Earth Scientist Sees

(c) An aerial photograph of a portion of the San Andreas fault. As emphasized by the sketch, the fault offsets a stream channel in a right-lateral shear sense.

Figure 7.10 Features of exposed fault surfaces.

A scarp due to slip on a normal fault in Nevada.

(a) This normal fault has displaced the ground surface. The exposed step is a fault scarp.

(b) This fault breccia consists of irregular fragments of light-colored rock.

Striation orientation

(c) Slip lineations or fault striations on the surface of a strike-slip fault may look like grooves or scratches.

(Fig. 7.9c) or human-made features such as highways, fences, or rows of trees in orchards. Displacement on a fault that offsets the ground surface produces a step, known as a **fault scarp**, on the ground surface **(Fig. 7.10a)**.

Faulting under brittle conditions may crush or break rock in a band or zone bordering the fault. If this shattered rock consists of visible angular fragments, it's called *fault breccia* **(Fig. 7.10b)**, but if it consists of a fine powder, it's called *fault gouge*. Some fault surfaces are polished and grooved by movement on the fault. Polished fault surfaces are called **slickensides**, and linear grooves on fault surfaces are *slip lineations* **(Fig. 7.10c)**. Because faults tend to break up and weaken rock, the *fault trace*, the line of intersection between the fault and the ground surface, may preferentially erode to become a linear valley.

Take-home message . . .

Geologic structures formed by brittle deformation include joints (natural cracks) and faults (fractures on which sliding occurs). Geologists distinguish among different kinds of faults based on the relative displacement across the fault. Faulting can break up rock, and fault surfaces themselves may be polished and grooved due to slip.

Quick Question -
How can you distinguish between a joint and a fault in the field?

7.4 Folds and Foliation

Basic Fold Shapes

Imagine a carpet lying flat on the floor. Push on one end of the carpet, and it will wrinkle or contort into a series of wave-like curves. Stresses developed during mountain building can cause bedding, foliation, or other features in rock to warp or bend in a similar manner. The result—a curve in the shape of a rock layer—is called a **fold** (see Fig. 7.2b).

Not all folds look the same: some look like arches, some look like troughs, and some have other shapes. To describe these shapes, we must first label the parts of a fold **(Fig. 7.11a)**. The **limbs** are the sides of the fold that have the least curvature, and the **hinge** refers to a line along which the fold has the greatest curvature. The *axial plane*, an imaginary plane that contains the hinges of successive layers, effectively divides the fold into two halves. With these terms in hand, we distinguish among the following types of folds:

• *Anticlines, synclines, and monoclines:* Folds that have an arch-like shape in which the limbs dip away from the hinge are called **anticlines** (see Fig. 7.11a), whereas folds with a trough-like shape in which the limbs dip toward the hinge are called **synclines** **(Fig. 7.11b)**. A **monocline** has the shape of a carpet draped over a stair step **(Fig. 7.11c)**.

Figure 7.11 Geometric characteristics of folds.

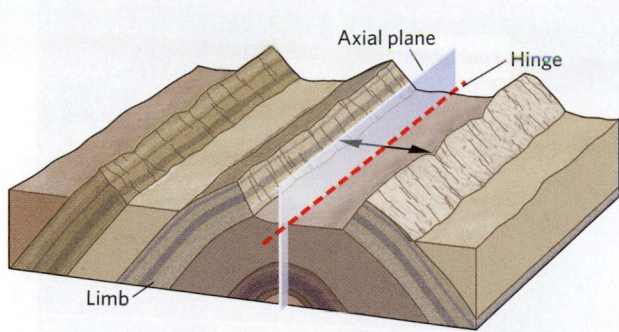

(a) An anticline looks like an arch. The limbs dip away from the hinge, as indicated by the outward-pointing arrows. Note that because of erosion, the oldest strata are exposed along the hinge.

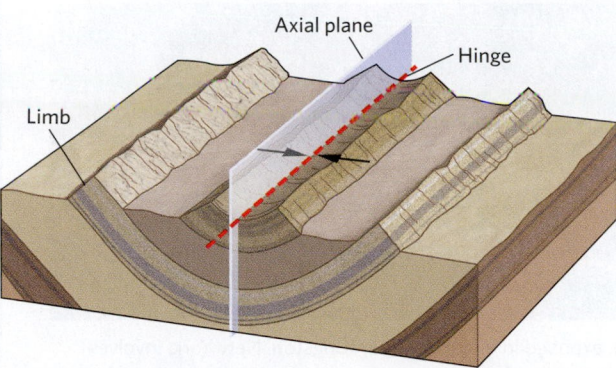

(b) A syncline looks like a trough. The limbs dip toward the hinge, as indicated by the inward-pointing arrows. Note that because of erosion, the youngest strata are exposed along the hinge.

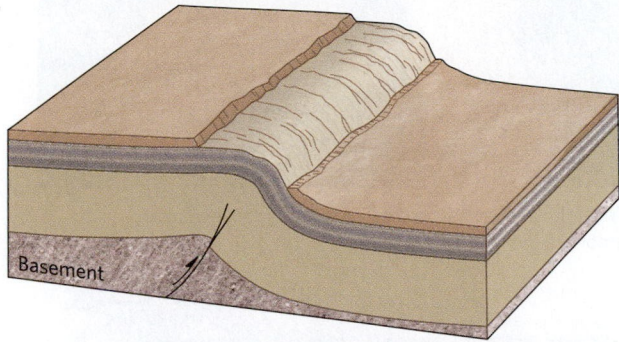

(c) A monocline looks like a stair step and is commonly draped over a fault block.

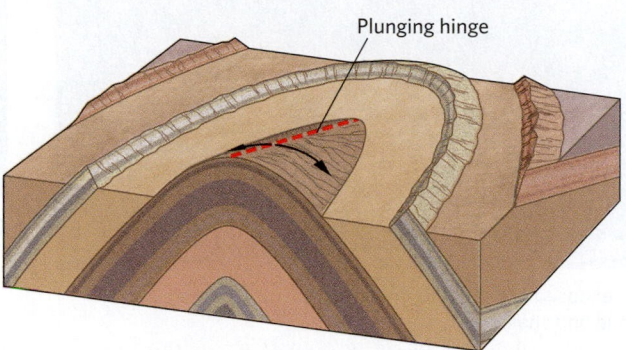

(d) A plunging anticline has a tilted hinge.

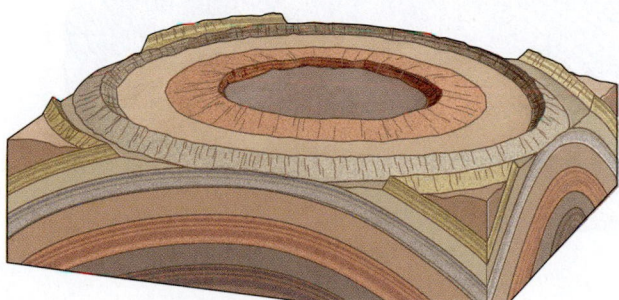

(e) A dome has the shape of an overturned bowl.

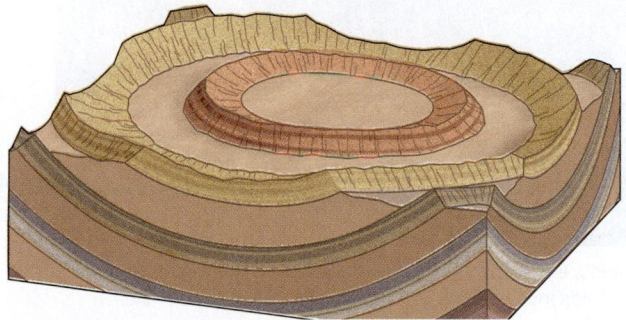

(f) A basin has the shape of an upright bowl.

- *Nonplunging and plunging folds:* If the hinge is horizontal, the fold is called a *nonplunging fold*, but if the hinge is tilted, the fold is called a *plunging fold* (Fig. 7.11d).

- *Domes and basins:* A fold with the shape of an overturned bowl is called a **dome**, whereas a fold shaped like an upright bowl is called a **basin** (Fig. 7.11e, f). Domes and basins both form circular patterns that look like bull's-eyes on a map if the land surface cuts across the structure. If domes and basins involve sedimentary beds, the oldest beds will crop out in the center of a dome, whereas the youngest beds will crop out in the center of a basin.

Using these terms, you can identify the various folds shown in Figure 7.12a–e.

Formation of Folds

Why do folds form? Some rock layers wrinkle up, or *buckle*, in response to end-on compression. In such cases, a "train" of anticlines and synclines may form, with their hinges perpendicular to the direction of compression

Figure 7.12 Examples of folds on outcrops and in the landscape.

(a) This anticline, exposed in a roadcut near Kingston, New York, involves beds of Paleozoic limestone.

(b) This syncline, exposed in a roadcut in Maryland, involves beds of Paleozoic sandstone and shale.

(d) This train of folds, exposed in sea cliffs in eastern Ireland, includes anticlines and synclines. The folds involve beds of Paleozoic sandstone and shale.

Fold hinge

Fold limb

(c) This fold, exposed along the coast of Brazil, occurs in Precambrian gneiss.

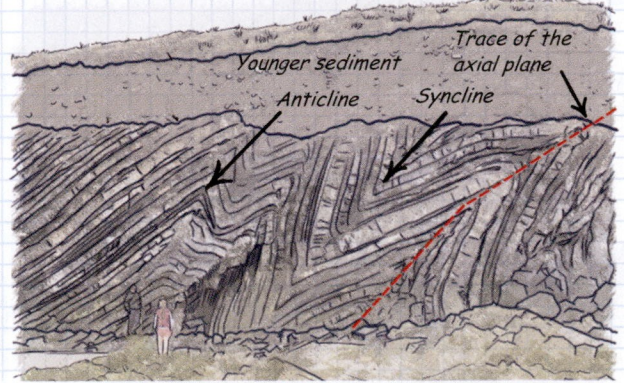

Younger sediment

Trace of the axial plane

Anticline   Syncline

What an Earth Scientist Sees

Aerial view

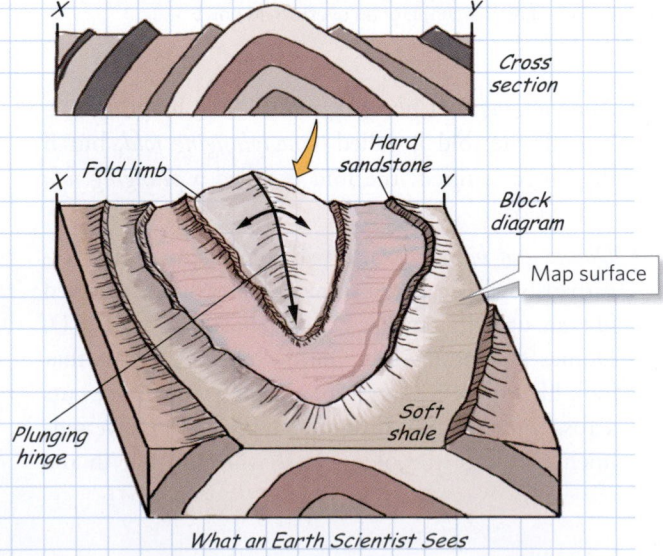

X   Y

Cross section

Hard sandstone

Fold limb

X   Y

Block diagram

Map surface

Plunging hinge

Soft shale

What an Earth Scientist Sees

(e) The plunging anticline of Sheep Mountain, Wyoming, is easy to see because of the lack of vegetation. Resistant rock layers (sandstone) stand out as ridges, whereas weaker rock layers (shale) erode away. A block diagram shows how the surface exposures relate to underground structures.

Figure 7.12 continued

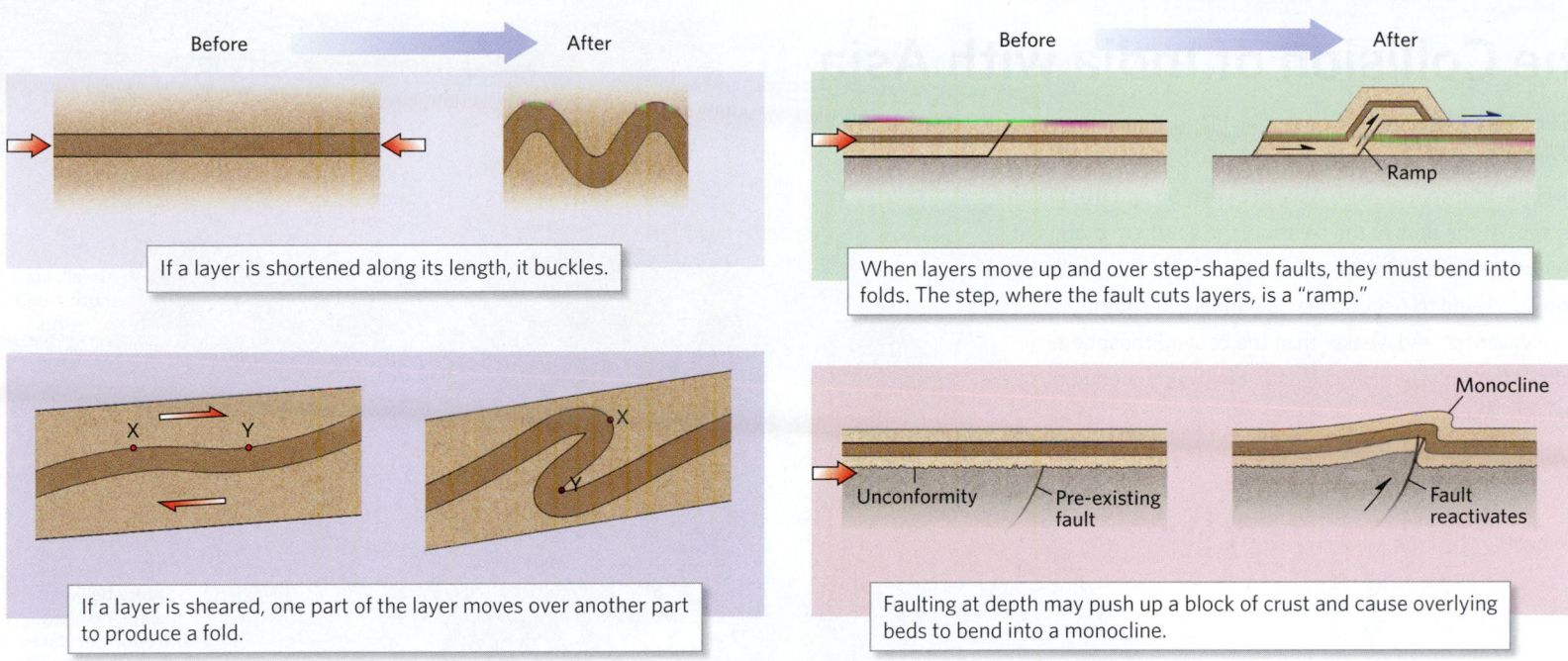

Before · After

If a layer is shortened along its length, it buckles.

When layers move up and over step-shaped faults, they must bend into folds. The step, where the fault cuts layers, is a "ramp." Ramp

If a layer is sheared, one part of the layer moves over another part to produce a fold.

Faulting at depth may push up a block of crust and cause overlying beds to bend into a monocline. Unconformity · Pre-existing fault · Monocline · Fault reactivates

(f) Folding can be caused by many different processes, illustrated here in cross section.

Figure 7.13 Development of slaty cleavage may accompany folding as compress on shortens beds. Slaty cleavage tends to be parallel to a fold's axial plane.

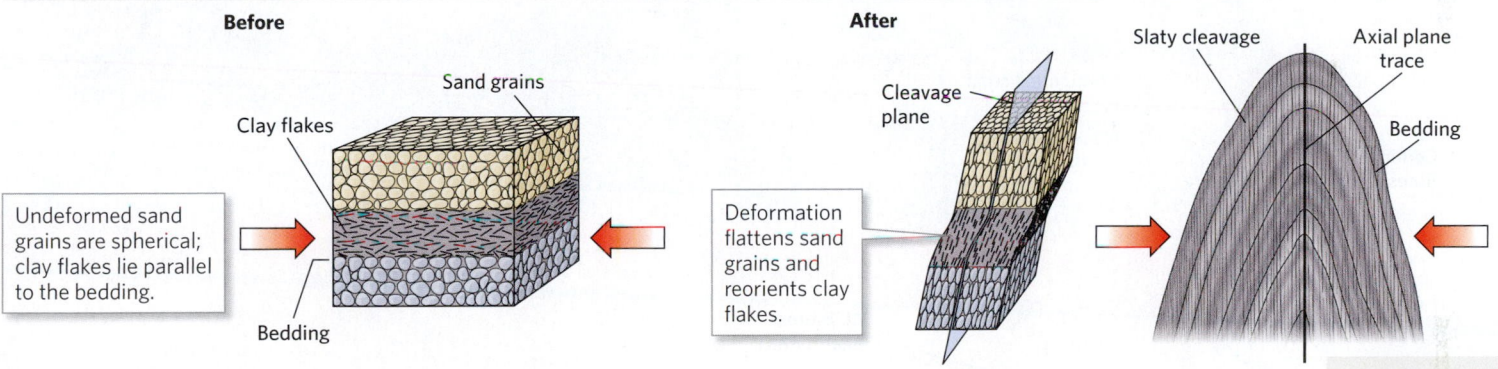

Before · After

Sand grains · Clay flakes

Undeformed sand grains are spherical; clay flakes lie parallel to the bedding. · Bedding

Cleavage plane · Deformation flattens sand grains and reorients clay flakes.

Slaty cleavage · Axial plane trace · Bedding

(Fig. 7.12f). Other folds form where shear gradually shifts one part of a rock body relative to another—layers caught up in the zone of shear undergo folding as they are effectively dragged along. Still other folds develop where rock layers move up and over step-like bends in a thrust fault, and must bend to conform to the fault's shape. Finally, some folds form when slip on a fault causes a block of basement, overlain by beds of sedimentary rock, to move upward relative to a neighboring block; when this happens, the beds bend.

Foliation Produced by Deformation

As we discussed in Chapter 6, *foliation*—parallel surfaces or layering such as slaty cleavage, schistosity, and gneissic banding—can develop during metamorphism. During the formation of foliation, stress can cause elongate or platy grains to align parallel to one another **(Fig. 7.13)**.

Take-home message . . .

Folds, such as anticlines and synclines, are bends or curves in rock layers that form in response to stress. Stress can also cause the shape and orientation of grains to change, yielding foliation.

Quick Question -----------------------------
What processes cause rock layers to undergo folding?

7.5 Causes of Mountain Building

Before plate tectonics theory became established, geologists were just plain confused about how mountains formed. In the context of the new theory, however, the many processes driving mountain building became clear: mountains form primarily in response to convergent-boundary deformation, continental collision, and

See for **yourself**

Folds, Central Australia

Latitude: 24°18′44.08″ S
Longitude: 132°10′34.02″ E

Zoom to 30 km (~18.5 miles) and look down.

Alternating beds of resistant and nonresistant sedimentary strata have been warped into plunging folds.

The Collision of India with Asia

The Himalaya Mountains and other highlands of southern Asia are a consequence of the collision of India, a block of old, cold continental lithosphere, with Asia, between 55 and 40 million years ago. At the time of the collision, the southern margin of Asia consisted of several smaller crustal blocks that had relatively recently accreted to Asia, so the lithosphere of southern Asia was relatively warmer and weaker than the cooler lithosphere of India. After India initially collided, it has continued to push slowly into Asia.

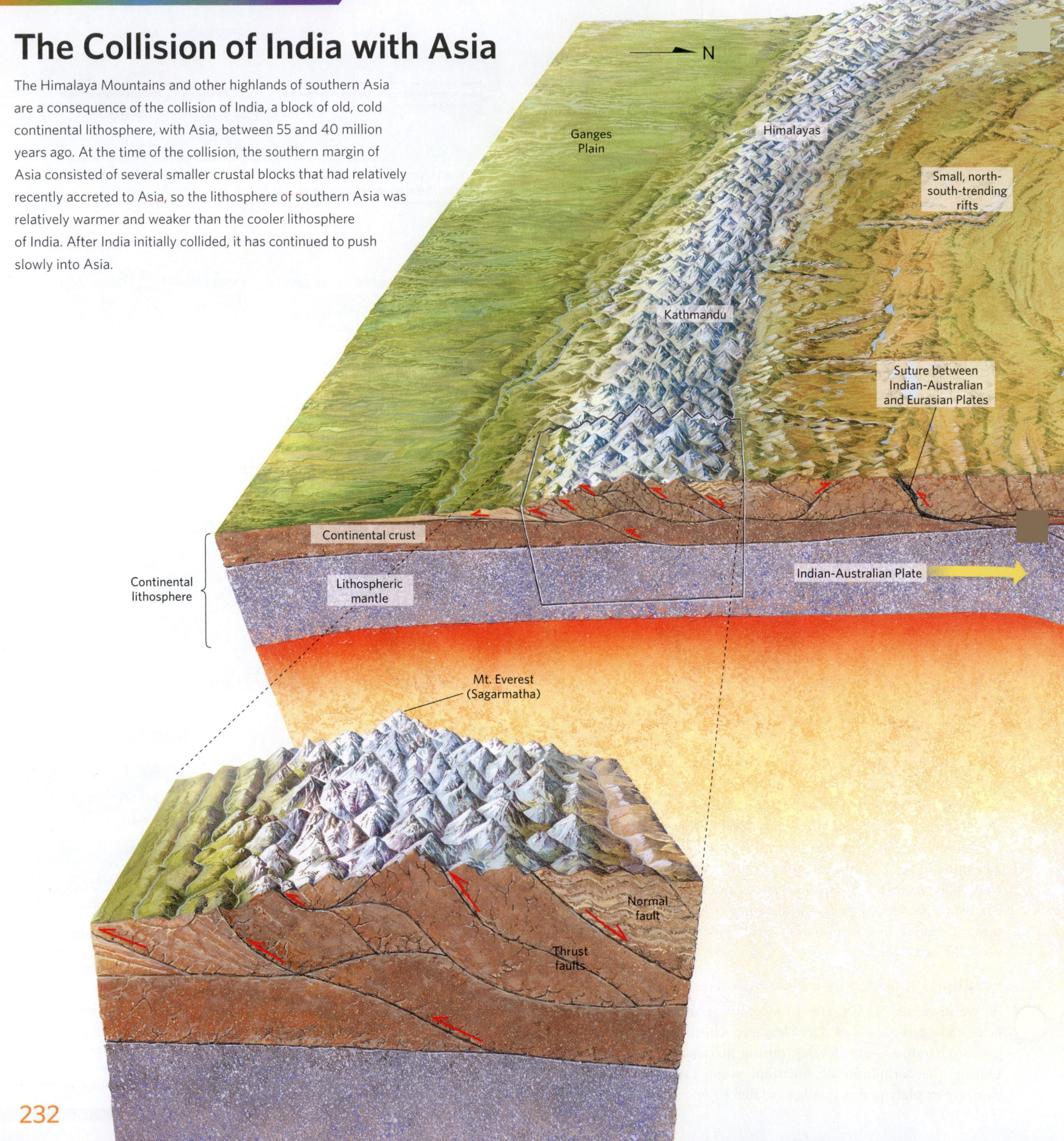

N

Ganges Plain

Himalayas

Small, north-south-trending rifts

Kathmandu

Suture between Indian-Australian and Eurasian Plates

Continental crust

Continental lithosphere

Lithospheric mantle

Indian-Australian Plate

Mt. Everest (Sagarmatha)

Normal fault

Thrust faults

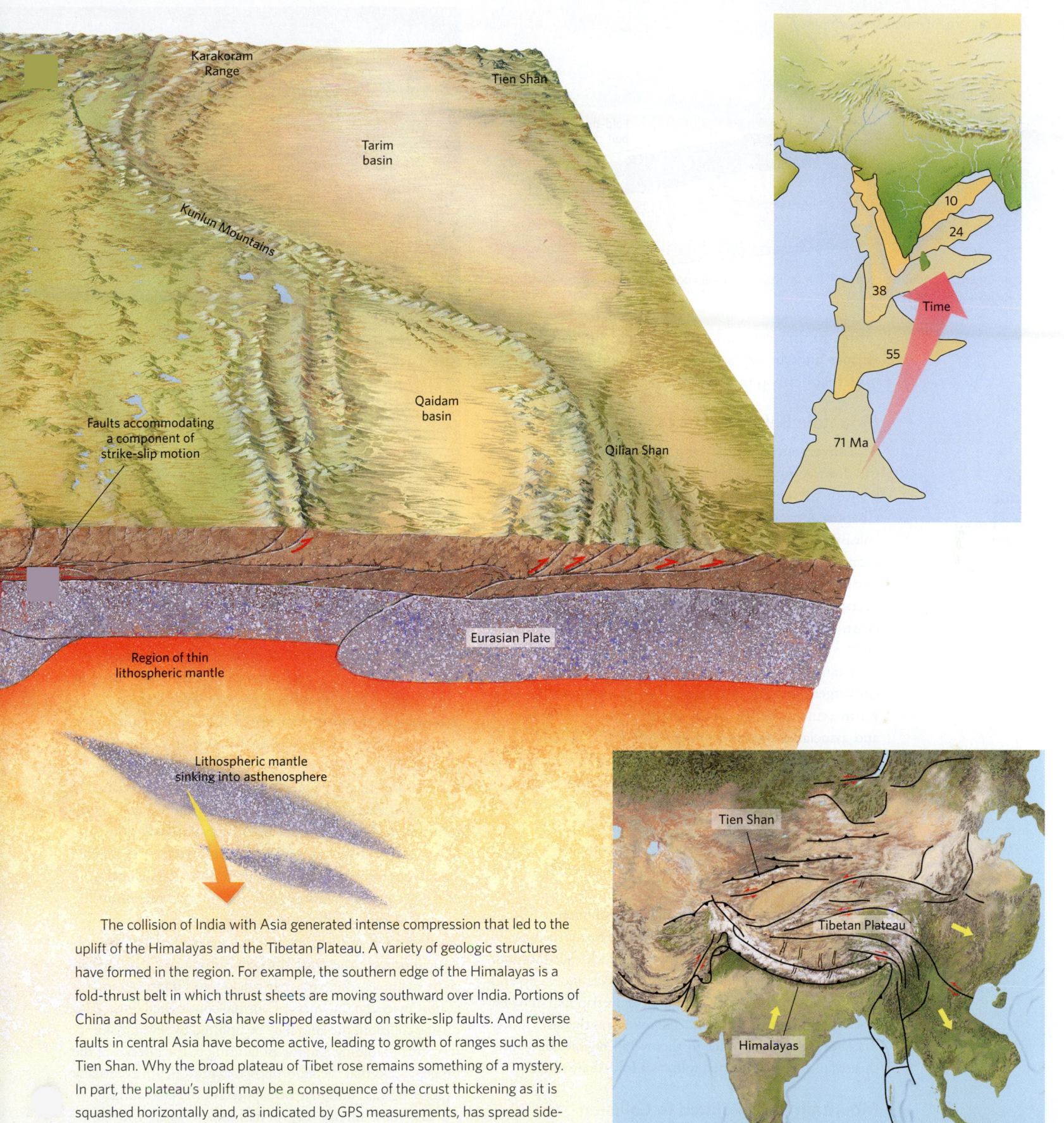

Karakoram
Range

Tien Shan

Tarim
basin

Kunlun Mountains

Faults accommodating
a component of
strike-slip motion

Qaidam
basin

Qilian Shan

Eurasian Plate

Region of thin
lithospheric mantle

Lithospheric mantle
sinking into asthenosphere

10

24

38

Time

55

71 Ma

Tien Shan

Tibetan Plateau

Himalayas

The collision of India with Asia generated intense compression that led to the uplift of the Himalayas and the Tibetan Plateau. A variety of geologic structures have formed in the region. For example, the southern edge of the Himalayas is a fold-thrust belt in which thrust sheets are moving southward over India. Portions of China and Southeast Asia have slipped eastward on strike-slip faults. And reverse faults in central Asia have become active, leading to growth of ranges such as the Tien Shan. Why the broad plateau of Tibet rose remains something of a mystery. In part, the plateau's uplift may be a consequence of the crust thickening as it is squashed horizontally and, as indicated by GPS measurements, has spread sideways to the east. Uplift may also be due to the heating of the region, when slabs of the underlying lithospheric mantle dropped off and sank; hot asthenosphere rose to fill the space. Heating of the overlying, remaining lithosphere caused it to rise.

Figure 7.14 Convergent-boundary orogeny.

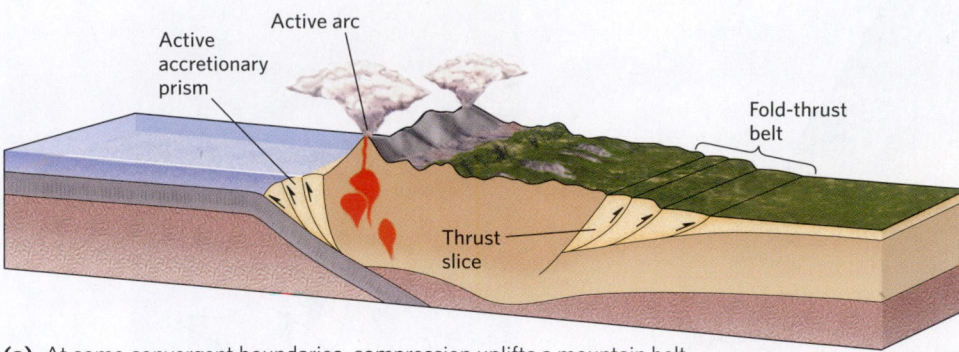

Active
accretionary
prism

Active arc

Fold-thrust
belt

Thrust
slice

(a) At some convergent boundaries, compression uplifts a mountain belt.

(b) These mountains in Chile are part of the Andes, which formed along a convergent boundary.

rifting. In this section, we look at three different geologic settings and the types of mountains and geologic structures that develop in each one.

Mountain Belts Related to Subduction

At convergent boundaries, the oceanic lithosphere of the downgoing plate subducts and sinks into the mantle beneath the overriding plate. This process, as we saw in Chapter 2, triggers melting in the mantle and growth of a volcanic arc along the edge of the overriding plate. In many locations where the overriding plate consists of continental crust, the interaction between the downgoing plate and the overriding plate produces compression in the continental crust. This compression, in turn, causes the crust to shorten and produces a **convergent-boundary orogen** (Fig. 7.14a).

Many types of geologic structures form within convergent-boundary orogens. For example, in the very warm crust at depth in the orogen, plastic deformation and associated metamorphism cause folds, foliation, and reverse faults to form. This deformation also moves rocks from deeper levels up toward the Earth's surface, producing a belt of uplifted land that undergoes erosion by rivers and glaciers to produce sharp peaks. The Andes serve as an example of convergent-boundary deformation, for they have formed due to the subduction of Pacific Ocean floor beneath South America (Fig. 7.14b).

Notably, along the continental side of such a belt, at shallower depths, numerous thrust faults form, each carrying a wide, and relatively thin, sheet of rock called a *thrust slice*. As Figure 7.14a shows, each thrust slice moves up and over the one in front of it, so the slices overlap like shingles. Rocks within thrust slices undergo folding as the slices move. Therefore, geologists refer to the overall region containing thrust faults and associated folds as a **fold-thrust belt**.

Mountain Belts Related to Collision

Once the oceanic lithosphere between two blocks of relatively buoyant continental crust subducts completely, the blocks, which were once separated by an ocean, collide with each other and produce a **collisional orogen** (Fig. 7.15a, b). The boundary between what had been separate blocks is called a **suture**. Large thrust faults form during a collisional orogeny, as the edge of one block slips up and over the margin of the other. As a result, rocks in the footwall block may end up tens of kilometers below the land surface, where they undergo folding and metamorphism. Typically, these rocks develop foliation. The most intense deformation and metamorphism in a collisional mountain belt occurs in the internal or central zone of the orogen. On both sides of the internal metamorphic zone, fold-thrust belts tend to develop.

Because of the deformation that takes place within a collisional orogen, the crust thickens substantially. In fact, in some examples, the crust may thicken to 70 km (43 miles), almost twice the thickness of normal continental crust. Also during collision, the land may be uplifted by several kilometers. Due to folding and thrust faulting in the interior of the orogen, rocks that metamorphosed at great depth eventually squeeze upward and may be exposed at the land surface.

The most intense collisions happen when two continents come together. For example, continental collision yielded the Himalayas and the Alps during the Cenozoic

Figure 7.15 Collisional orogeny.

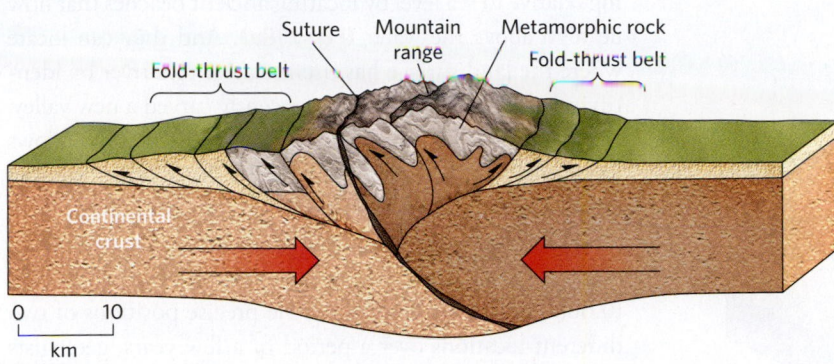

(a) During continental collision, continents squeeze together and deform. Thrust faulting brings metamorphic rock up from beneath the mountain belt to shallower levels.

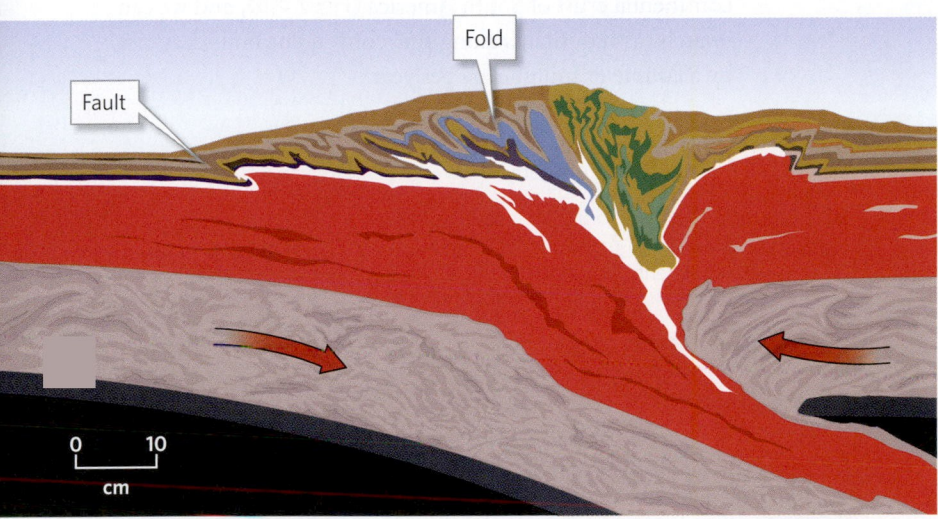

(b) Geologists can simulate collisional orogeny in the laboratory using layers of colored sand. Dragging the left side of such a model under the right side produces structures and uplift, as shown in this sketch.

(c) This view of the Himalayas from space makes clear that these mountains were uplifted when India collided with and pushed into Asia.

(Fig. 7.15c; **Earth Science at a Glance**, pp. 232–233). But collisions also occur between island arcs, between small continental blocks, between island arcs and continents, and between oceanic plateaus and continents. Along some convergent boundaries, numerous collisions over geologic time suture a series of blocks to the edge of the overriding plate. Geologists refer to the suturing of smaller blocks to a larger one as **accretion**. An incoming buoyant crustal block is called an *exotic terrane* when it lies offshore and an *accreted terrane* once it has been sutured to the overriding plate (Fig. 7.16a). The process of accretion can add a substantial width of new crust to the edge of a continent (Fig. 7.16b).

Mountain Belts Related to Continental Rifting

A *continental rift*, as we saw in Chapter 2, is a place where a continent stretches. This movement causes normal faults to develop in the upper crust (Fig. 7.17a). Movement on these faults causes blocks of crust to drop down, and in the process, strata in the blocks undergo tilting. Commonly, a nearly horizontal fault, or *detachment*, underlies all of the blocks. The low areas between the tilted blocks fill with sediment eroded from the blocks. As a result, rifts typically contain sets of narrow, elongate mountain ranges, composed of tilted blocks, separated by deep, sediment-filled rift basins. Because these ranges form due to slip on faults, they are sometimes called *fault-block mountain ranges*. Rifting produced the Basin and Range Province of Utah, Nevada, and Arizona (Fig. 7.17b). The Basin and Range Province lies within the North American Cordillera, one of North America's mountain belts (Box 7.2).

Measuring Mountain Building in Progress

The rumblings of earthquakes and the eruptions of volcanoes in some ranges attest to present-day, continuing movements that geologists can measure

Figure 7.16 The accretion of exotic terranes can widen a mountain belt.

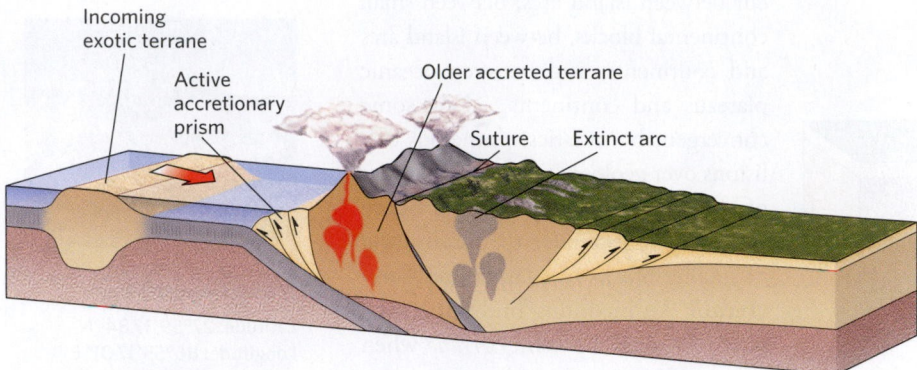

(a) An example of exotic terranes attaching to a continent as a consequence of subduction.

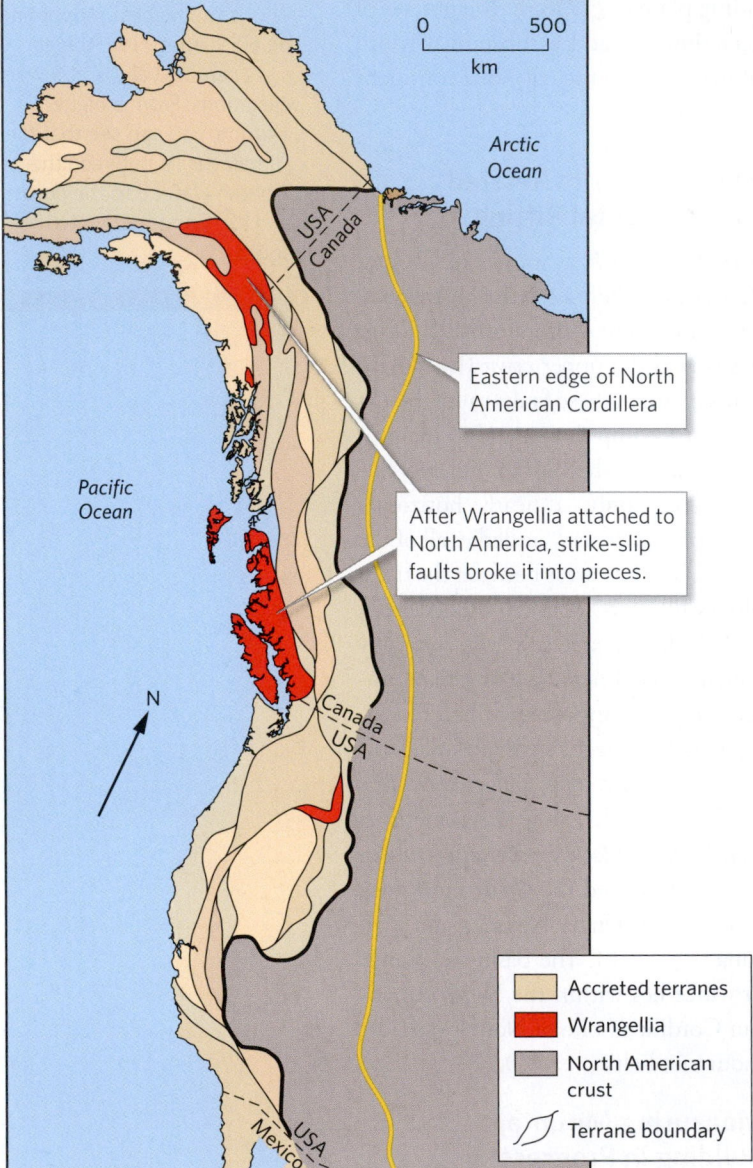

(b) The western portion of North America consists of accreted terranes that attached to the continent during the Mesozoic. A distinct terrane called Wrangellia (highlighted in red) was sliced into pieces that were displaced by strike-slip faults.

through field studies and satellite technology. For example, researchers can determine where coastal areas have been rising relative to sea level by locating ancient beaches that now lie high above the water **(Fig. 7.18a)**. And they can locate where the land surface has risen relative to a river by identifying places where a river has recently carved a new valley. In addition, GPS (the global positioning system) allows direct measurement of uplift and horizontal-shortening rates. While standard hand-held GPS devices provide locations with accuracies of only about ±2 m (6 feet), research-quality GPS devices can specify locations to within ±2 mm (0.008 inches). By comparing the precise positions of two different locations over a period of a few years, geologists can detect crustal motion. In effect, we can "see" the Andes shorten horizontally at a rate of a couple of centimeters per year as the subducting oceanic lithosphere compresses the continental crust of South America **(Fig. 7.18b)**, and we can "watch" as mountains along this convergent boundary rise by a couple of millimeters per year.

Figure 7.17 Rift-related orogeny.

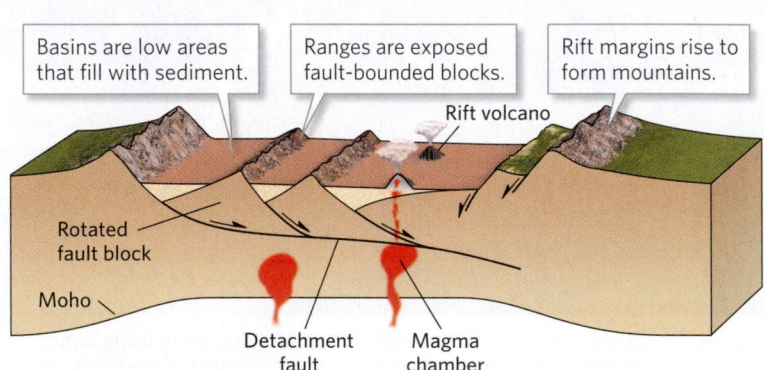

(a) Rifting leads to the development of numerous narrow mountain ranges separated by rift basins. Rift-margin mountains may also form.

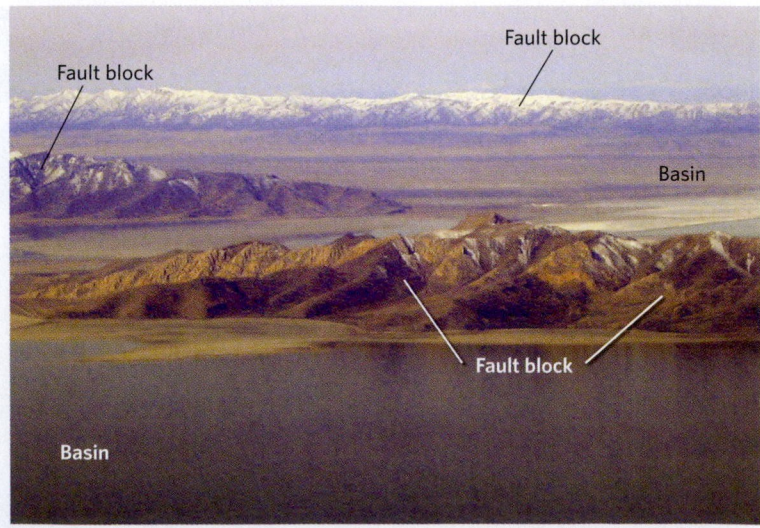

(b) The Basin and Range Province of the western United States formed during Cenozoic rifting.

Box 7.2

Consider this . . .

Forming the mountain belts of North America

If you ever travel coast-to-coast across the United States (for instance, from Washington, DC to San Francisco) you will have the opportunity to discover firsthand that North America hosts two major mountain belts—the *Appalachian Mountains* in the east, and the *North American Cordillera* in the west, separated by the Great Plains. Why and when did these mountain belts form? Geologists have determined that both resulted from multiple mountain-building events, which we describe below. Each mountain-building event took tens of millions of years, so the movements during a human lifespan would be imperceptible to our eyes, but movement of just a centimeter a year yields 10 km (6 miles) of movement in 1,000,000 years.

Between 600 and 470 Ma, the east coast of North America was a passive-margin sedimentary basin, a quiet environment in which over 15 km (9 miles) of sediment accumulated. This quiet period ended about 470 Ma, when an offshore volcanic arc collided with the east coast. The collision took perhaps 20 million years, and by the end of it, the strata of the passive-margin basin had been folded and metamorphosed. Between 400 and 370 Ma, small blocks of continental crust collided with and were sutured to eastern North America, causing more deformation. The final, most dramatic event happened at the end of the Paleozoic (330 to 300 Ma), when Africa collided with North America. This collision squashed the eastern edge of North America so much that it formed a 300-km (200-mile)-wide fold-thrust belt. After the collision, North America and Africa were sutured together in Pangaea **(Fig. Bx7.2a)**. The high peaks formed during the Alleghanian orogeny eventually eroded away, so today, we can see exposed rocks that were once 5 to 15 km (3 to 9 miles) below the mountain peaks.

The region that is now the North American Cordillera was a passive margin until nearly the end of the Paleozoic. Beginning at that time, the region endured mountain-building events associated with convergent boundaries. During some of these events, numerous blocks of crust—mostly island arcs and oceanic plateaus—accreted to North America, so that the continent grew westward by up to 200 km (120 miles). Starting about 165 Ma, a convergent boundary formed along the west coast, along which Pacific Ocean floor subducted beneath North America. The granites of the Sierra Nevada are a remnant of the volcanic arc formed at that time. To the east, affecting an area that is now Nevada and Utah, as well as western Wyoming and Montana, a broad fold-thrust belt developed **(Fig. Bx7.2b)**. Between 80 and 40 Ma, deformation began to affect large areas of Wyoming, Colorado, and New Mexico, and Precambrian metamorphic rock was pushed up over Paleozoic and Mesozoic sedimentary rock on large reverse faults. This event produced the present-day Rocky Mountains. During the past 25 million years, a large swath of the North American Cordillera in the United States has undergone rifting to form the Basin and Range Province.

Figure Bx7.2 Formation of North America's mountain belts.

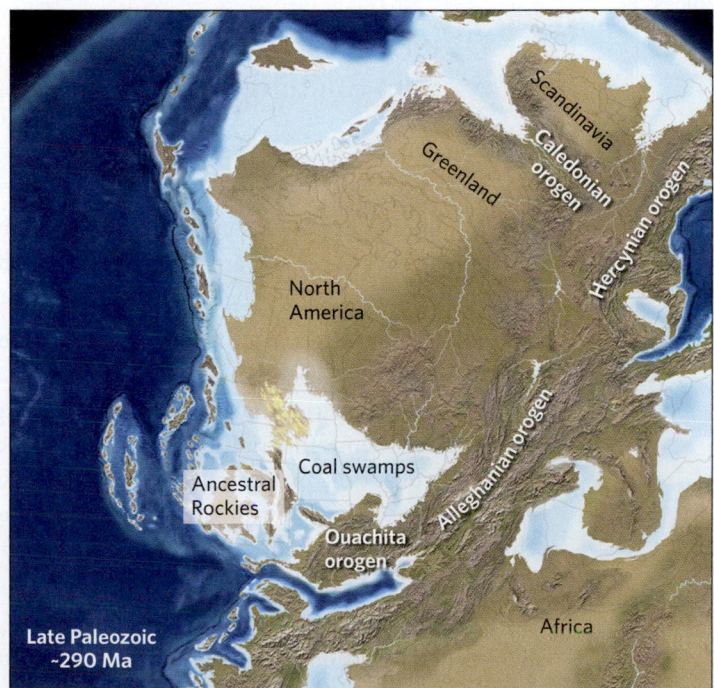

(a) Near the end of the Paleozoic (about 290 Ma), North America had collided with Africa, and the Appalachians were a high mountain range.

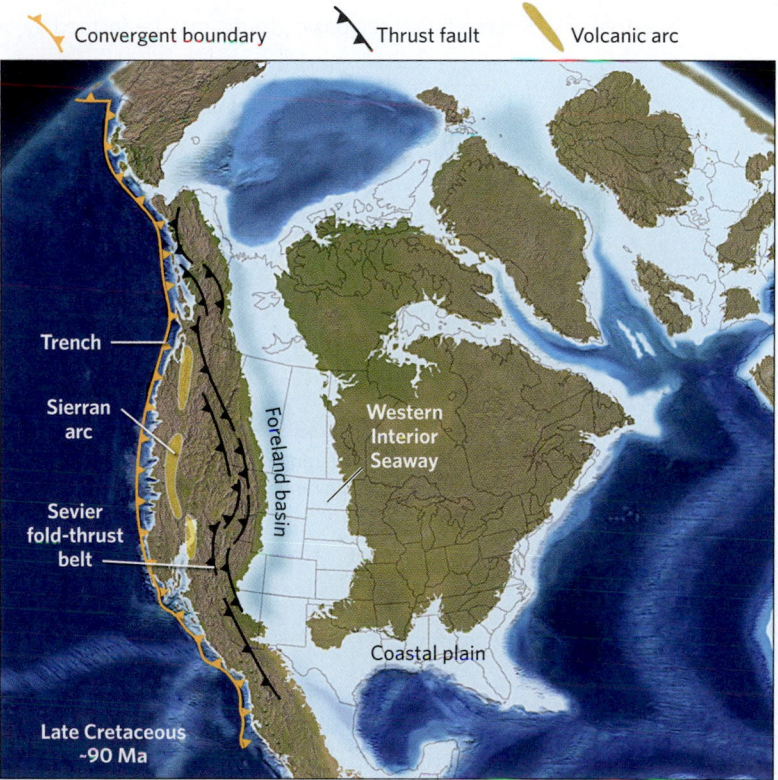

(b) Near the end of the Mesozoic (about 90 Ma), western North America was an Andes-like convergent boundary, and a fold-thrust belt had formed there. A shallow sea traversed the continent.

Figure 7.18 Evidence of present-day mountain building.

Uplift due to 1703 earthquake

Uplift due to 1923 earthquake

Present sea level

(a) Wave erosion cut these terraces along a beach. During earthquakes, the terraces are uplifted above sea level.

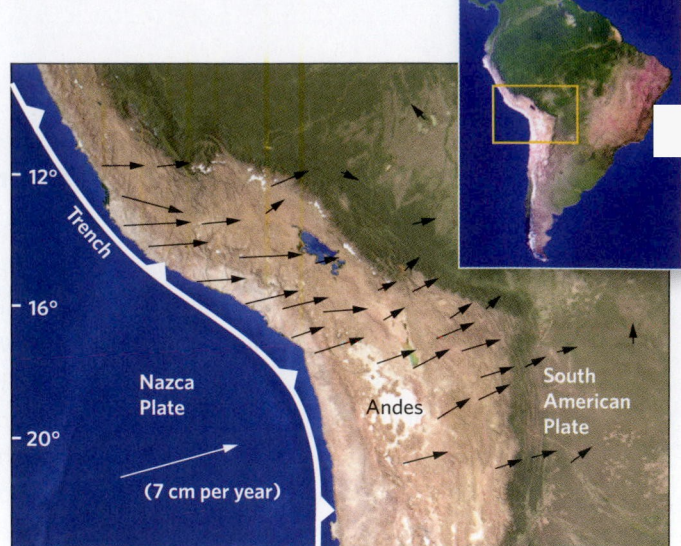

(b) GPS measurements of shortening in the Andes. The arrows indicate the velocity of locations in the Andes relative to the interior of South America. The white arrow indicates relative plate motion.

Take-home message . . .

Mountain belts form in association with convergent boundaries, continental collisions, and rifting. During convergent-boundary and collisional orogeny, continental crust thickens, large thrust faults and folds form, and regional metamorphism takes place. Rifting yields fault-block mountains separated by rift basins.

Quick Question ------------------------------
Could fold-thrust belts develop in rifts?

Figure 7.19 Rocks of all three basic groups may be formed by orogeny.

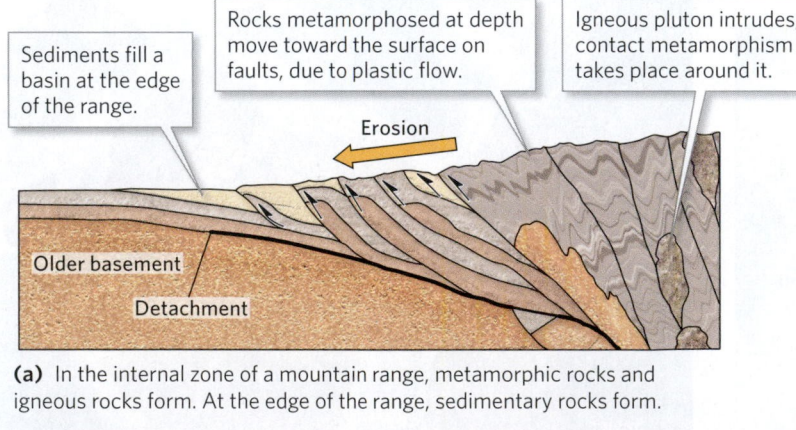

Sediments fill a basin at the edge of the range.

Rocks metamorphosed at depth move toward the surface on faults, due to plastic flow.

Igneous pluton intrudes; contact metamorphism takes place around it.

Erosion

Older basement

Detachment

(a) In the internal zone of a mountain range, metamorphic rocks and igneous rocks form. At the edge of the range, sedimentary rocks form.

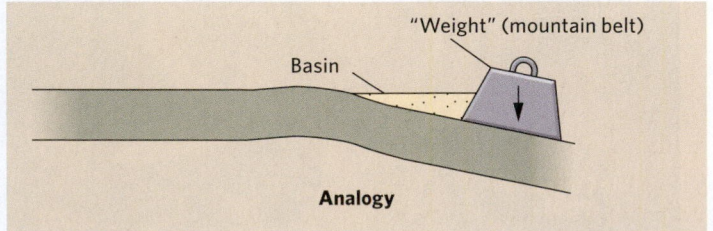

"Weight" (mountain belt)

Basin

Analogy

(b) A sedimentary basin develops because the mountain range acts as a weight that pushes down the surface of the lithosphere.

7.6 Other Consequences of Mountain Building

Formation of Rocks In and Near Mountains

Mountain building produces conditions that can form a great variety of rock types. In fact, rocks of all three basic groups can form in mountain belts (Fig. 7.19a):

- *Igneous:* Melting takes place in the mantle and lower crust in regions where mountains are forming. For example, flux melting takes place at convergent boundaries. The magma rises to form plutons in continental crust of the overriding plate. Decompression melting takes place beneath rifts, producing magma that rises into the rifted crust. When the magma in orogens freezes, it becomes igneous rock.

- *Sedimentary:* Weathering and erosion in mountain belts generate vast quantities of sediment, which tumbles down slopes. Glaciers or streams transport sediment to low areas, where it accumulates in alluvial

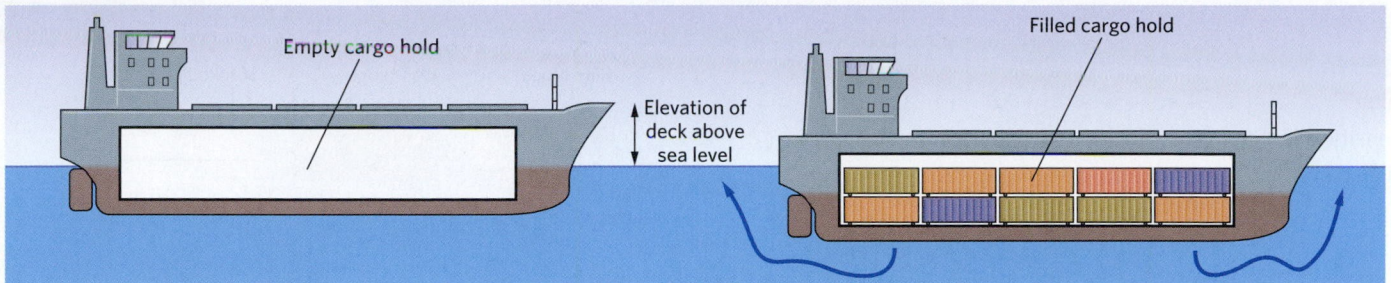

Figure 7.20 A ship analogy for the concept of isostasy. The elevation of a ship's deck above sea level changes when a load is added to the ship. Water flows out of the way as the ship moves downward.

fans or deltas. In some locations, the weight of the mountain belt itself pushes down the surface of the nearby lithosphere, thereby producing a deep sedimentary basin (Fig. 7.19b).

- *Metamorphic:* Contact metamorphism occurs near plutons intruded during mountain building. Regional metamorphism occurs where mountain building thrusts one part of the crust over another. When this happens, rock that was once near the surface ends up at great depths, where it endures high temperature and pressure. Because this metamorphism occurs in rock undergoing shearing and compression, the resulting metamorphic rocks contain foliation.

Processes That Cause Uplift and Produce Mountainous Topography

Leonardo da Vinci, the great Renaissance artist and scientist, enjoyed walking in the mountains, where he sketched ledges and examined the rocks he found there. To his surprise, he discovered marine shells (fossils) in limestone beds cropping out a kilometer above sea level. He puzzled over this observation and finally concluded that the limestone had been lifted from below sea level up to its present elevation. Modern geologists agree with Leonardo, and refer to vertical movement of the Earth's surface from a lower to a higher elevation as *uplift*. What processes can cause the surface of the Earth to rise? The list is long because, as we have seen, mountain building happens in numerous geologic settings. But to understand how uplift processes work, we must begin by introducing the concept of *isostasy*.

Imagine a ship anchored in a port. Its deck lies at a specific elevation above the sea surface. If we make the ship heavier by adding cargo, or shorter by removing a deck, the position of its top surface becomes lower. In contrast, if we make the ship lighter by removing cargo, or taller by adding a deck, the position of its top surface becomes higher. Vertical movements of the lithosphere are somewhat similar, for the lithosphere, which consists of relatively rigid crust and lithospheric mantle, rests on

the softer asthenosphere below. The elevation of the surface of the lithosphere (the land surface, on continents), like the elevation of the surface of a ship, depends on a balance between forces pulling the lithosphere down and forces pushing it up (Fig. 7.20). This balance is known as **isostasy**. Put another way, isostasy exists where the elevation of the Earth's surface reflects the level at which the lithosphere naturally "floats." Anything that changes the lithosphere's thickness or density affects the elevation of the lithosphere's surface (Box 7.3). So, to answer the question of why mountain belts can rise, we must identify geologic processes that can change the thickness or density of layers in the lithosphere. Let's consider some ways in which such changes can take place.

CRUSTAL SHORTENING AND THICKENING. As two blocks of continental crust collide, the associated compression can cause the crust in the collision zone to shorten horizontally and thicken vertically. As a result, the surface of the crust goes up, and the base of the crust goes down; the base of the lithosphere goes down as well (Fig. 7.21a). For example, in the Himalayas, the surface of the crust has risen to an elevation of about 8 km (5 miles), and the base of the crust now lies at a depth of over 60 km (37 miles) (Fig. 7.21b). This downward protrusion of crust is called a *crustal root*, and the downward protrusion of the lithosphere is called the *lithospheric mantle root*.

REMOVAL OF LITHOSPHERIC MANTLE. The weight of the lithospheric mantle (composed of denser rock than the crust) effectively pulls the lithosphere down, just as heavy cargo makes a ship settle deeper into the water. Therefore, removing some or all of the lithospheric mantle from the bottom of a plate causes the surface of the remaining, thinner lithosphere (with the crust on top) to rise in order to maintain isostasy, even if the crust thickness remains unchanged (Fig. 7.22). Such removal—a process that geologists call *delamination*—resembles removal of cargo or ballast from the hold of a ship: as the weight of the cargo decreases, the deck of the ship rises.

Innumerable peaks, black and sharp, rose grandly into the dark blue sky, their bases set in solid white, their sides streaked and splashed with snow, like ocean rocks with foam.... [Mountains] are nature's poems carved on tables of stone.

—JOHN MUIR (AMERICAN NATURALIST, *1838–1914*)

Isostasy

What are we learning?

How a change in density and a change in thickness can be accommodated to maintain isostasy.

What you need:

- A tub of water.
- Wooden blocks, some of dense hardwood, like oak, and some of less dense wood, like pine. Begin with two pine blocks and one oak block, all of the same thickness.

Instructions:

- Place a pine block in the water, and let it come to equilibrium, meaning that isostasy has been achieved. Use a ruler to measure the distance between the water surface and the top of the block. Record your result.
- Add a second pine block on top of the first. (You may have to prop the blocks against the corner of the tub to keep them from tipping.) With the second block in place, measure the distance between the water surface and the top of the top block. You will see that this distance has increased.
- Now, place a block of oak in the water. Using the ruler, determine whether the distance from the top of this block to the water surface is greater or less than the distance you measured for the single pine block.
- By lining up a row of floating blocks of different thickness but the same density, you can see how the high elevations of a collisional mountain belt come to overlie the thickest crust.

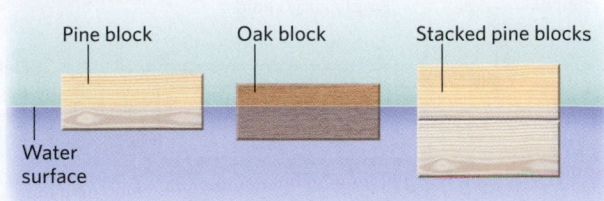

Pine block   Oak block   Stacked pine blocks

Water surface

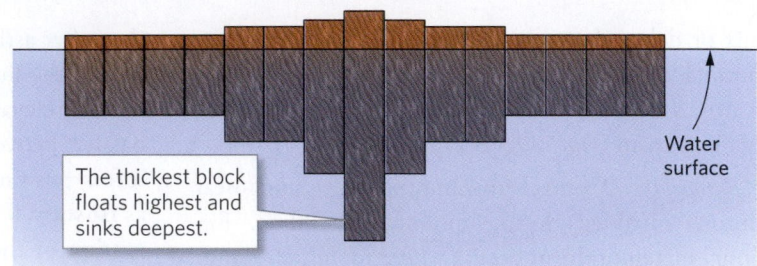

The thickest block floats highest and sinks deepest.

Water surface

What did we see?

The distance of a block's top surface from the water surface depends on both the thickness and the density of the block. Increase the block's thickness (by adding a second block), and the surface is higher. Increase the block's density (by replacing a block with one of denser wood), and the surface is lower. If you were able to measure thicknesses, volumes, and densities, you would see that a wood block sinks until the mass of the block below the water is equal to the mass of the water displaced. This relationship bears the name *Archimedes' principle*, recognizing its discoverer, the ancient Greek scientist Archimedes.

THINNING AND HEATING OF THE LITHOSPHERE. In rifts, the lithosphere undergoes stretching and thinning. As a result, less dense asthenosphere rises beneath the rift, and the remaining lithosphere heats up. Heating this thinned lithosphere causes it to expand and become even less dense. These two processes cause the overall region to rise.

What Goes Up Must Come Down

When the land surface rises significantly for any reason, gravity and the Sun's energy begin to drive erosion through the action of wind and precipitation. As a slope steepens, for example, landslides cause rock and debris to tumble down the slope. When winds blow clouds over the mountains, rain provides water that collects in streams whose flow carries away debris and sculpts valleys and canyons. If temperatures remain cold enough during the year, glaciers grow and slowly flow, carving peaks and deepening valleys. The net effect of all these processes is to grind away elevated areas and produce the jagged landscapes that we associate with mountainous terrain (Fig. 7.23). It's important to keep in mind that uplift and erosion happen simultaneously in active mountain belts, so for the elevation of a range to increase over time, the range must uplift faster than it erodes. If the uplift rate becomes less than the erosion rate, the elevation of the range decreases.

The highest point on Earth, the peak of Mt. Everest, lies 8.85 km (5.5 miles) above sea level. Can our planet's

Figure 7.21 The concept of isostasy as applied to collisional mountain ranges.

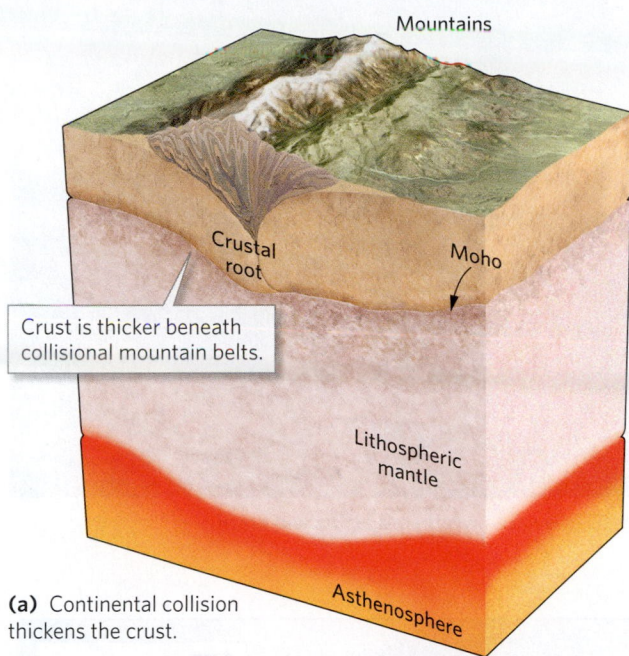

(a) Continental collision thickens the crust.

(b) The Himalayas constitute the world's highest mountain range. The crust beneath them is almost twice the average thickness.

mountain ranges get significantly higher? Probably not, for at the high temperatures that occur at depths of 15 to 32 km (9 to 20 miles), quartz-rich crustal rocks become so weak that it is relatively easy for them to flow slowly. When this flow begins, the overlying mountains begin to collapse under their own weight. They spread sideways like soft cheese that has been left out in the summer Sun. Geologists call this process *orogenic collapse*.

Take-home message . . .

Mountain building is typically accompanied by igneous activity and metamorphism, as well as by deposition of sediment. Beneath some mountain belts, the continental crust has thickened, so the surface of the crust "floats" higher. Once uplift has occurred, erosion sculpts rugged topography. Because deep crust is heated and weakened, mountain belts may eventually collapse under their own weight.

Quick Question ------------------------------
How can thinning of the lithosphere cause uplift?

7.7 Basins and Domes in Cratons

So far in this chapter, we've discussed deformation caused by dramatic mountain-building events. A **craton** consists of continental lithosphere that has not been affected by mountain-building events for at least the last 1 billion years. Cratons, therefore, tend to have relatively subdued topography. The interior of North America, the region between the North American Cordillera on the west and the Appalachians on the east, is a craton **(Fig. 7.24)**. Cratons form when crust has been able to cool substantially and, therefore, has become relatively strong and

Figure 7.22 Uplift due to delamination may happen after collision.

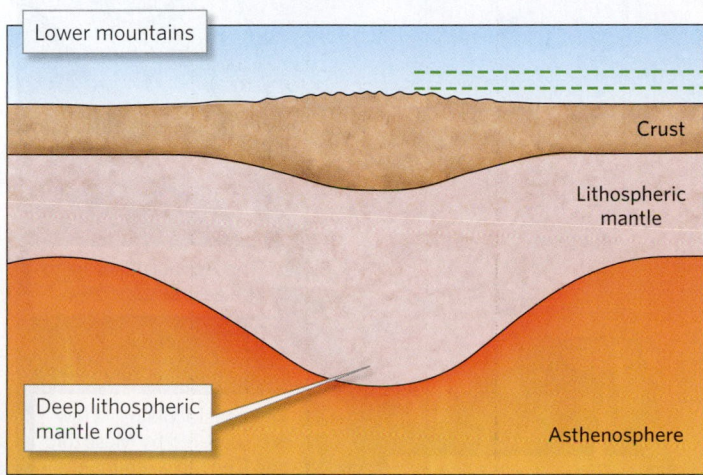

(a) Collision thickens the crust and lithospheric mantle. The lithospheric mantle root acts like ballast, holding down the surface of the crust.

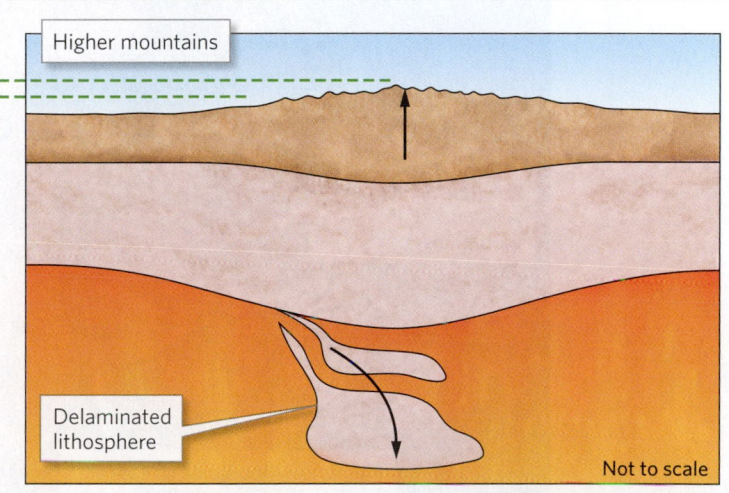

(b) If part of the lithospheric mantle root detaches and sinks, the surface of the lithosphere may rise, like a balloon dropping ballast.

Figure 7.23 As soon as land rises, water and ice begin eroding it.

(a) Glaciers carved these rugged peaks in Switzerland.

(b) Streams cut these valleys into weathered bedrock in Brazil.

Figure 7.24 North America's craton consists of a shield, where Precambrian rock is exposed, and a cratonic platform, where Paleozoic sedimentary rock covers the Precambrian rock.

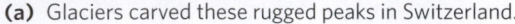

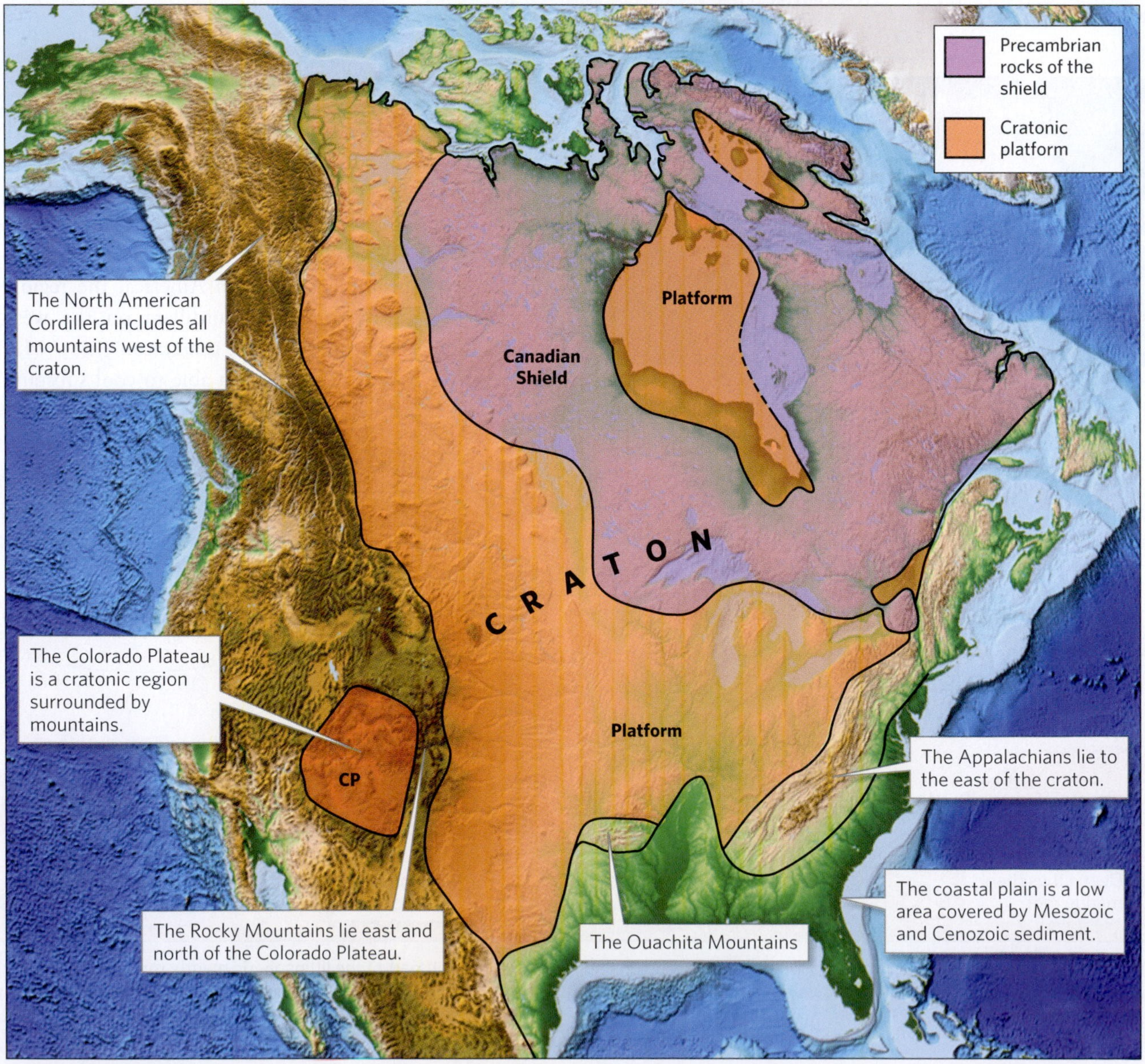

Precambrian rocks of the shield

Cratonic platform

The North American Cordillera includes all mountains west of the craton.

The Colorado Plateau is a cratonic region surrounded by mountains.

Platform

Canadian Shield

CRATON

Platform

CP

The Appalachians lie to the east of the craton.

The Rocky Mountains lie east and north of the Colorado Plateau.

The Ouachita Mountains

The coastal plain is a low area covered by Mesozoic and Cenozoic sediment.

Figure 7.25 Domes and basins of the North American cratonic platform.

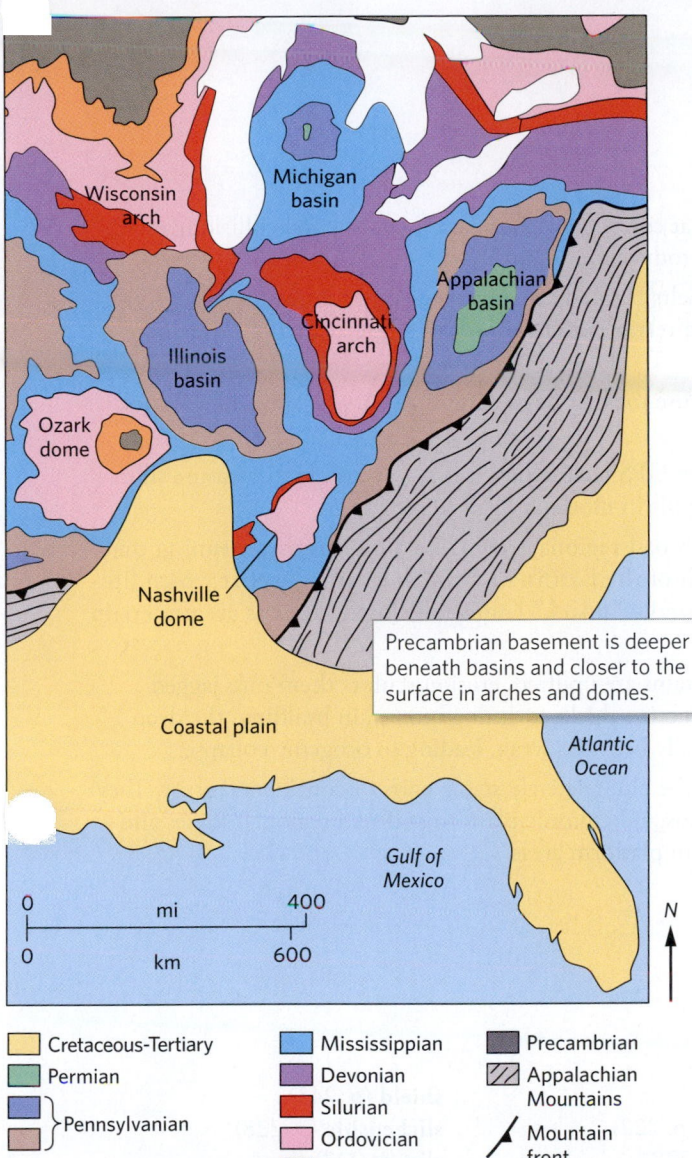

Precambrian basement is deeper beneath basins and closer to the surface in arches and domes.

Legend:
- Cretaceous-Tertiary
- Permian
- Pennsylvanian
- Mississippian
- Devonian
- Silurian
- Ordovician
- Cambrian
- Precambrian
- Appalachian Mountains
- Mountain front

(a) A geologic map of the mid-continent platform region, showing the locations of basins and domes.

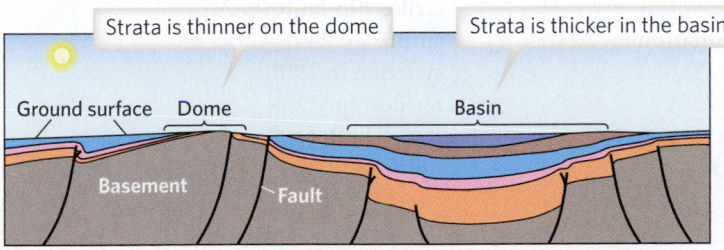

Strata is thinner on the dome — Strata is thicker in the basin

(b) A cross section showing how strata thin toward the crest of a dome and thicken toward the center of a basin. The cross section is vertically exaggerated. This cross section symbolically represents the Ozark dome and the Illinois basin.

stable. Geologists divide cratons into **shields**, in which Precambrian metamorphic and igneous rocks are exposed at the ground surface, and **cratonic platforms**, where a relatively thin layer of younger Phanerozoic sediment covers the Precambrian rocks (see Fig. 7.24).

In the shield areas of cratons, we find widespread exposures of intensively deformed metamorphic rocks with abundant examples of folds and foliation. That's because the crust making up the cratons was deformed during multiple mountain-building events in the distant past. These mountain-building events are so old that erosion has worn away the original mountains, in the process exhuming deep crustal rocks.

In cratonic platforms, the sedimentary beds typically define regional domes and basins—broad areas (over a few hundred kilometers across) that gradually sank or uplifted, respectively, over geologic time **(Fig. 7.25)**. For example, in Missouri, strata arch across a broad uplift, the Ozark dome, whose diameter is 320 km (200 miles). The thickness of strata decreases toward the top of the dome because less sediment accumulated within the dome than in adjacent basins. When erosion removes the top of the dome, contacts between stratigraphic formations display a bull's-eye shape, and the oldest exposed rocks lie at the center of the dome at the ground surface. In the Illinois basin, sedimentary strata warp downward into a huge bowl that is also about 300 km across. The strata get thicker toward the basin's center, indicating that the floor of the basin was sinking, providing space, at the same time sediment was accumulating. The basin, like the dome discussed above, displays a bull's-eye shape, but here the youngest strata are exposed in the center.

Take-home message . . .

Cratons are portions of continents that consist of old (Precambrian) and relatively stable crust. Parts of cratons may be covered by Phanerozoic sedimentary strata. A basin occurs where the strata are warped down into a bowl shape, and a dome occurs where the strata are warped up.

Quick Question -
Imagine that coal occurs in a particular stratigraphic formation. Will mines to reach the coal be deeper or shallower in the center of a basin?

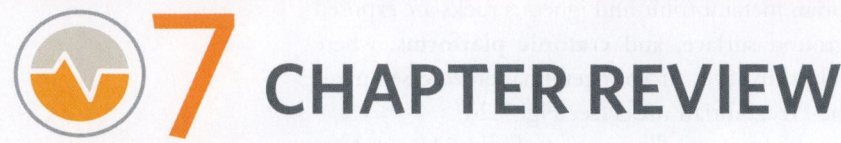

7 CHAPTER REVIEW

The letters following each Review Question refer to the corresponding Learning Objective from the Chapter Opener.

1. What changes do rocks undergo during formation of a mountain belt such as the Alps? **(C)**

2. Contrast brittle and plastic deformation. **(A)**

3. What factors determine whether a rock will behave in brittle or plastic fashion? **(A)**

4. How are stress and strain different? **(A)**

5. How is a fault different from a joint? **(B)**

6. Compare normal, reverse, and strike-slip faults. Which type of fault does the diagram show? **(B)**

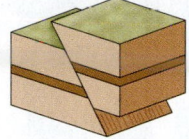

7. How can you recognize faults in the field? **(B)**

8. Describe the differences among an anticline, a syncline, and a monocline. Which type of fold does the figure show? **(B)**

9. Discuss the relationship between foliation and deformation. **(A)**

10. Describe the principle of isostasy. **(D)**

11. Discuss the processes by which mountain belts form at convergent boundaries, during continental collisions, and in continental rifts. **(C)**

12. How are the structures affecting the sedimentary strata of a craton different from those of a mountain belt? Which type of structure does the diagram show? **(F)**

13. Why do sedimentary basins form along the margins of orogens? Where might you find metamorphic and igneous rocks in an orogeny? **(E)**

14. Can we measure mountain building in progress? If so, how? **(C)**

On Further Thought

15. Imagine that a geologist sees two outcrops of resistant sandstone, as depicted in the cross-section sketch. The region between the outcrops is covered by soil. A distinctive bed of cross-bedded sandstone occurs in both outcrops, so the geologist has correlated the western outcrop (on the left) with the eastern outcrop. The curving lines in the bed indicate the shape of the cross beds. Keeping in mind how cross beds form (see Chapter 5), sketch the connection of the cross-bedded layer from one outcrop to the other, before erosion. What geologic structure have you drawn? **(E)**

16. The Pyrenees, an east-west-trending mountain range along the border between Spain and France, uplifted during the Cenozoic. On both sides of the range, fold-thrust belts have developed. In the interior of the range, metamorphic rocks are exposed. What kind of stress caused the range to form? What direction did Spain move, relative to France, to cause this stress? **(C)**

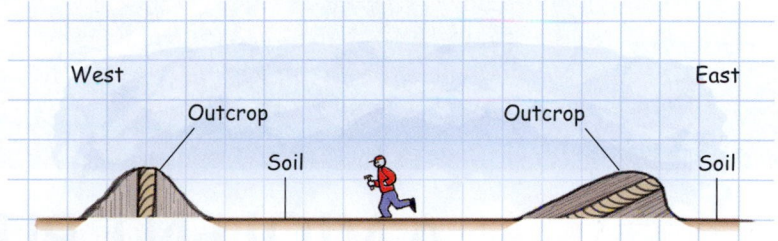

Online Resources

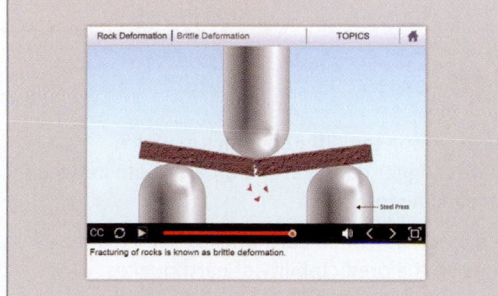

Animations
This chapter features animations on types of rock deformation and faults.

Videos
This chapter features a video on continental collision and the formation of mountains.

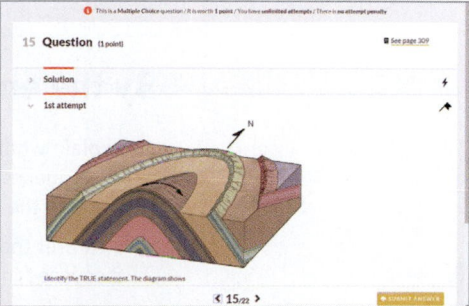

Smartwork5
This chapter features identification exercises on types of faults, folds, and deformations.

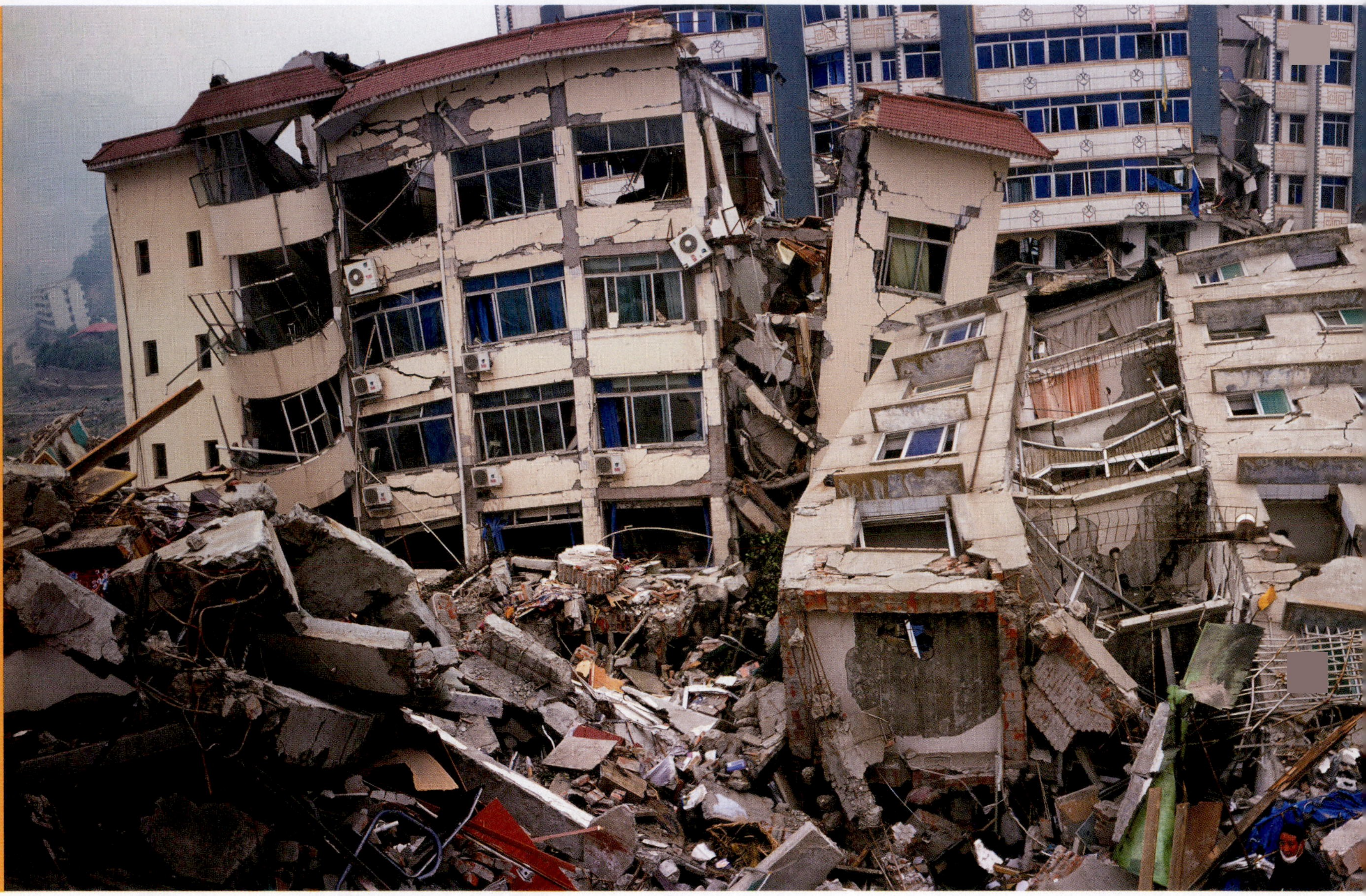

8   A VIOLENT PULSE

Earthquakes

8.1 Introduction

It was midafternoon on March 11, 2011, and in many seaside towns along the eastern coast of Honshu, the northern island of Japan, fishing fleets unloaded their catch, factories churned out goods, shoppers browsed the stores, and office workers tapped at computers. No one realized that their surroundings would soon change forever. Honshu lies near a convergent boundary where the Pacific Plate slips beneath the edge of Japan and sinks back into the mantle. Averaged over time, this movement takes place at about 8 cm (3 inches) per year. But the movement doesn't happen smoothly. Rather, for a while, rocks adjacent to the boundary quietly and subtly bend and warp. Then, suddenly, just as a wooden stick snaps after you've bent it too far, a measurable amount of movement takes place in a matter of seconds to minutes **(Fig. 8.1)**. On March 11, at 2:46 P.M., the "snap" started at a point located about 130 km (80 miles) east of Japan's coast and about 24 km (15 miles) below the Earth's surface. When it happened, Japan lurched eastward by a few meters and a portion of seafloor rose vertically by several centimeters. The stage had been set for a disaster.

The instant that movement took place, vibrations began to pass through the surrounding rock, just like the vibrations that pass along the two pieces of the bent stick when you snap it between your hands. The vibrations traveled at an average speed of 11,000 km (7,000 miles) per hour—10 times the speed of sound in air—and transported energy from the site of the "snap" outward. When they reached the surface of the Earth, they caused an episode of ground shaking—an **earthquake**.

This apartment building broke apart and tipped over during a major earthquake (magnitude 8.0) that struck Beichuan, Sichuan, China in 2008. Violent earthquakes, which are inevitable on our dynamic planet, can have catastrophic consequences.

247

Figure 8.1 What happens during an earthquake?

Stick

(a) Before deformation, the rock layers in this example are not bent.

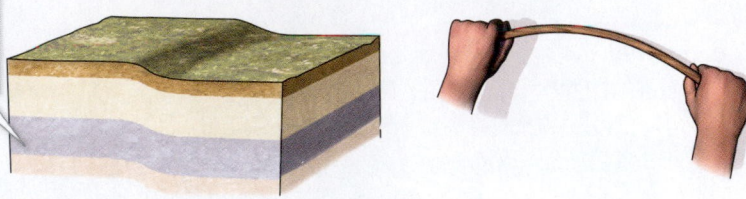

The layer of rock bends elastically (exaggerated here).

(b) As deformation occurs, rock bends elastically, like a stick that you arch between your hands. The drawing exaggerates the amount of bending.

Due to elastic rebound, the rock layers return to their initial shape.

(c) Eventually, the rock breaks, and sliding suddenly occurs on a fault. This break generates vibrations like those you feel when you break a stick.

So much energy was released by the March 11 event that the resulting earthquake—named the Tōhoku earthquake, after the eastern province of Honshu—had historic consequences. In the area affected by the earthquake, shaking caused people to lose their balance and fall. Bottles and plates flew off shelves and crashed to the floor, bookshelves tipped over, and furniture danced around rooms. Buildings twisted and swayed, and in some cases their ceilings and facades collapsed in a shower of debris **(Fig. 8.2a)**. Outside, dust rose from the ground and landslides tumbled down hillslopes. In addition, gas pipelines broke, sending flammable vapors into the air. Some of the gas ignited in billows of flame.

Meanwhile, the sudden displacement of the seafloor pushed up the surface of the ocean itself. This movement produced a *tsunami*, a distinctive, very broad wave. When this wave approached the shore, it grew to a height of over 10 m (33 feet). It overtopped seawalls and battered the landscape as far as several kilometers inland **(Fig. 8.2b, c)**.

Earthquakes are a fact of life on our dynamic planet. Fortunately, most cause no damage or casualties, either because they are too small or because they take place in unpopulated areas. But a few hundred earthquakes per year rattle the ground sufficiently to crack or topple buildings and injure their occupants, and every 5 to 20 years, on average, a great earthquake, such as the Tōhoku event, becomes a horrific calamity.

What geologic phenomena trigger earthquakes? Why do earthquakes take place where they do? How do they

Figure 8.2 The Tōhoku earthquake and tsunami, Japan, 2011.

The tsunami filled harbors and spilled over seawalls.

Some buildings collapsed due to ground shaking.

cause damage? Can we predict when earthquakes will happen or even prevent them from happening? What can they tell us about the interior of the Earth? *Seismologists* (from the Greek word *seismos*, meaning shock or earthquake), researchers who study earthquakes, have addressed many of these questions. In this chapter, we present some of their answers.

8.2 What Causes Earthquakes?

Ancient cultures offered a variety of explanations for **seismicity** (earthquake activity), most of which involved the restless gyrations of mythical subterranean animals. Modern research has replaced these supernatural explanations with the view that, while volcanic eruptions and nuclear explosions can trigger some earthquakes, most take place when one body of rock suddenly moves past another on a **fault**, a fracture surface on which sliding, or *slip*, takes place **(Fig. 8.3)**. The energy released by an earthquake travels through the Earth as **seismic waves** or *earthquake waves*.

How Does Faulting Generate Seismic Waves?

As we discussed in Chapter 7, slip on a fault displaces the rock on one side of the fault relative to the rock on the other side. You can find faults in many locations—but don't panic! Very few faults act as a source of earthquakes at any particular time, so geologists distinguish between *active faults*, which have moved relatively recently or might move in the near future, and *inactive faults*, which moved in the distant geologic past and probably won't move again in the near future.

Faulting produces seismic waves in two ways: (1) when previously intact rock suddenly breaks, forming a new fault on which slip takes place; and (2) when a pre-existing fault suddenly slips again. To picture the first of these processes, imagine that you grip each side of a brick-shaped block of rock with a clamp. Now, apply a slight upward push on one of the clamps and a slight downward push on the other. This action generates stress in the rock (see Chapter 7), and the rock bends slightly, but doesn't break **(Fig. 8.4a)**. If you stop pushing, and therefore decrease the stress, the rock "relaxes" and returns to its original shape. Geologists refer to a change in shape that can be reversed by removal of stress as **elastic deformation**—the same phenomenon happens when you bend a stick and then let it go. Now repeat the experiment, pushing one side of the block up and the other side down even more, so that the stress becomes greater. This time, small cracks start to develop in the rock. With continued pushing, the cracks begin to connect to one another until, suddenly, a fracture cuts across the entire block of rock **(Fig. 8.4b)**. The instant that happens, the block breaks in two, and rock on one side of the fracture slides past rock on the other side. Because sliding takes place, the fracture becomes a fault.

Figure 8.3 An example of a fault surface in Arizona that became exposed because the rock above the fault surface has eroded away. Slip lineations formed on this fault surface during movement.

An aerial view shows the tsunami advancing across the shore.

Coast

Slip lineations

Fault surface

Fault trace

What an Earth Scientist Sees

Figure 8.4 A model representing the development of a new fault.

Time

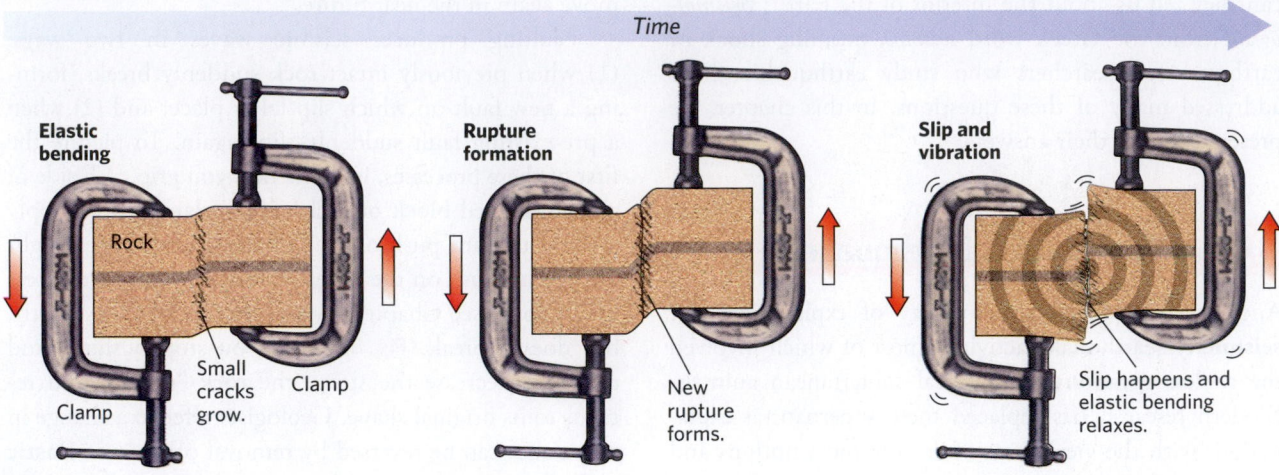

Elastic bending

Rock

Small cracks grow.

Clamp

Clamp

Rupture formation

New rupture forms.

Slip and vibration

Slip happens and elastic bending relaxes.

(a) Imagine a block of rock gripped by two clamps. Move one clamp up, and the rock starts to bend. Small cracks develop along the bend.

(b) Eventually, the cracks link. When this happens, a rupture cuts completely through the rock.

(c) The instant that the rupture forms, the rock breaks into two pieces that slide past each other. The energy that is released generates vibrations (seismic waves).

If you look closely at Figure 8.4b, you'll notice that the portion of the rock in which bending takes place, and elastic deformation accumulates, is wider than the fault itself. When the new fault suddenly forms and slips, the bent rock on either side of the fault straightens out, or *rebounds*, so the accumulated elastic deformation relaxes and the stress decreases (Fig. 8.4c). When relaxation takes place, however, the rock doesn't move smoothly back to its original shape; rather, it "twangs" back and forth, like a bouncing spring. It's this back-and-forth movement of rock on either side of the fault that generates seismic waves. You've seen the same action when you bend a stick until it snaps. Geologists refer to the idea that a fractured rock vibrates when a fault slips, and that this vibration produces seismic waves, as the **elastic-rebound theory**. Note that rebound does not return the rock to its original position, for as we've seen, faulting causes displacement. Furthermore, once a fault forms, it doesn't continue to slip forever because *friction*—the resistance to sliding caused by bumps and irregularities on a surface—eventually slows and stops the movement.

Once a fault has formed, it serves as a scar in the Earth's crust that remains weaker than the surrounding, intact crust. So when stress builds up in the crust, slip can take place on a pre-existing fault before the stress becomes large enough to generate a new fault. For this reason, many earthquakes happen when sudden slip takes place on a pre-existing fault, the second way of producing seismic waves. Notably, a pre-existing fault won't slide again until the stress overcomes the resistance of friction. Therefore, as the magnitude of stress increases, rock on either side of the fault undergoes elastic deformation, so when

the pre-existing fault finally slips, elastic rebound takes place. Seismologists refer to the cycle of stress buildup, then sliding and stress release, as **stick-slip behavior**.

The main rupturing event along a fault produces the **mainshock**. Commonly, a number of smaller earthquakes, called **foreshocks**, precede the mainshock. They may be due to the development and growth of cracks in the zone that will become the principal surface of rupturing. In the days to months following a large earthquake, the region affected by that earthquake endures a series of **aftershocks**. The largest aftershock tends to be 10 times smaller than the mainshock, and most are much smaller than that. Aftershocks happen because slip during the mainshock doesn't relax all the elastic deformation in rock adjacent to the fault—in fact, the movement during the mainshock may cause local increases in elastic deformation. The occurrence of aftershocks relaxes these remaining patches of instability.

The Area and Amount of Slip during an Earthquake

How large is the area of a fault surface that slips during an earthquake? The answer depends on the size of the earthquake: generally, the larger the earthquake, the larger the slipped area and the greater the displacement. For example, the major earthquake that hit San Francisco, California, in 1906 ruptured a segment of the San Andreas fault that was 430 km (270 miles) long, as measured parallel to the Earth's surface, by 15 km (9 miles) deep, as measured perpendicular to the Earth's surface, so the area that slipped was about 6,500 km² (2,500 square miles). During the 2011 Tōhoku earthquake, an area 300 km (180 miles) long by 100 km

Figure 8.5 Examples of displacement on faults in California.

What an Earth Scientist Sees

(a) A wooden fence built across the San Andreas fault was offset during the 1906 San Francisco earthquake. The displacement indicates strike-slip movement.

These ruptures formed where the fault broke the ground surface.

Dirt road

(b) This aerial photograph shows a dirt road offset by slip on a fault in the Mojave Desert in 1999.

(60 miles) wide slipped. During a small earthquake, an area less than a square kilometer may slip.

The amount of displacement varies with location along a fault: it tends to be greatest near where the slip begins and to die out progressively toward the edge of the slipped area, beyond which the displacement is zero. The maximum observed displacement associated with very large earthquakes can be meters to tens of meters **(Fig. 8.5)**. For example, up to 7 m (23 feet) of slip occurred during the 1906 San Francisco earthquake, and up to 30 m (100 feet) of slip occurred during the 2011 Tōhoku earthquake. Smaller earthquakes, such as the one that struck Northridge, California, in 1994, may result in only about 0.5 m (1.6 feet) of slip. Nevertheless, this earthquake toppled homes, ruptured pipelines, and caused casualties. The smallest earthquakes that people can feel result from displacements measured in millimeters to centimeters.

Defining the Location of an Earthquake

The location where the generation of seismic waves begins is called the **focus**, or *hypocenter*, of an earthquake **(Fig. 8.6a)**. An earthquake whose focus lies at a depth of less than 70 km (43 miles) is a *shallow-focus earthquake*, one that occurs at a depth of between 70 km and 300 km (185 miles) is an *intermediate-focus earthquake*, and one whose focus lies between 300 and 660 km (410 miles) is a *deep-focus earthquake*. Because earthquake foci do not lie on the Earth's surface, we can't plot their positions directly on a map. What you are actually looking at on a

map when you see a dot representing the position of an earthquake is the **epicenter** of the earthquake, the point on the surface of the Earth that lies vertically above the focus **(Fig. 8.6b)**.

Take-home message . . .

Most earthquakes happen when stress causes either sudden formation of a new fault or slip on a pre-existing fault. When the slip takes place, elastically deformed rock on either side of the fault rebounds and generates seismic waves. The area of a fault that slips, and the amount of displacement that takes place on a fault, are greater for larger earthquakes. The focus of an earthquake is the point within the Earth where slip begins, and the epicenter is the point on the Earth's surface that is vertically above the focus.

Quick Question -
How does stick-slip behavior operate?

8.3 Seismic Waves and Their Measurement

The Different Types of Seismic Waves

We've mentioned that the energy produced by slip on a fault moves through the Earth in the form of seismic waves. Seismologists distinguish between two general categories of seismic waves based on where the waves move: **body waves** pass through the interior of the Earth,

Figure 8.6 Earthquake foci and epicenters.

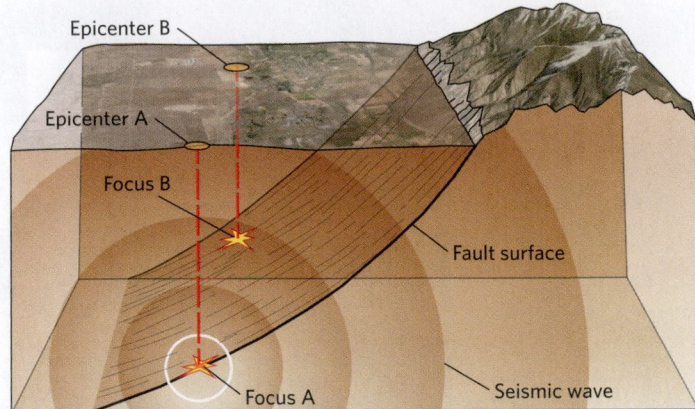

(a) The focus is the point on a fault where slip begins and from which seismic waves begin radiating. The epicenter is the point on the Earth's surface directly above the focus.

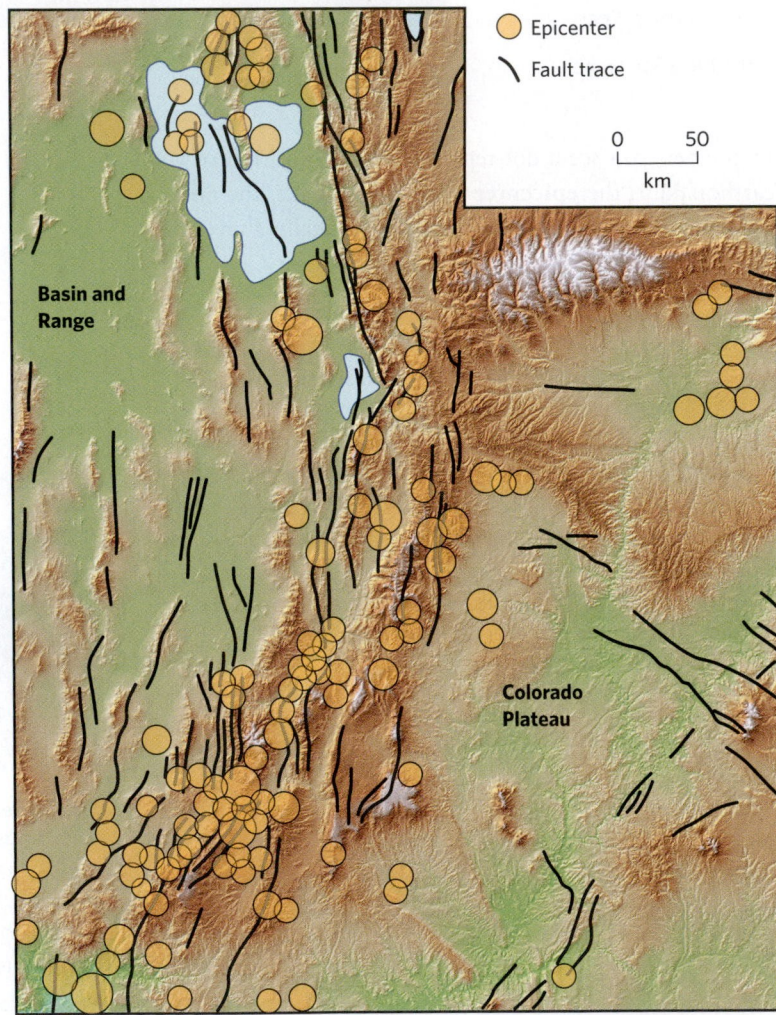

(b) A map of Utah, showing the distribution of earthquake epicenters recorded over several years. The distribution indicates that seismic activity happens mostly in a distinct belt following the boundary between the Basin and Range Province and the Colorado Plateau.

whereas **surface waves** travel along the Earth's surface. As a wave passes through a rock, the chemical bonds holding atoms together in the rock's minerals elastically stretch and bend.

Body waves cause rock to vibrate in two different ways. Seismologists refer to waves that cause back-and-forth vibrations *parallel* to the direction in which the wave itself moves as **compressional waves (Fig. 8.7a)**. As a compressional wave passes through a material, it first contracts (squeezes together), then dilates (expands). To see this kind of motion in action, push on the end of a spring to compress the coils together, then watch as the pulse of contraction moves along the length of the spring. Body waves that cause up-and-down vibration perpendicular to the direction of wave motion are known as **shear waves (Fig. 8.7b)**. To see shear-wave motion, jerk the end of a rope up and down repeatedly and watch how the up-and-down motion travels along the rope. Surface waves also have two different forms: some cause the ground surface to move up and down in rolling undulations, whereas others cause the ground surface to shimmy back and forth sideways, like a snake **(Fig. 8.7c)**. Seismologists assign names to the different kinds of waves we've just described:

- *P-waves* (P, for primary) are compressional body waves.
- *S-waves* (S, for secondary) are shear body waves.
- *R-waves* (R, for Rayleigh) are surface waves that cause the ground to undulate up and down.
- *L-waves* (L, for Love) are surface waves that cause the ground to shimmy back and forth.

The different types of seismic waves travel at different velocities. P-waves move fastest, which is why they are called primary waves. S-waves travel at about 60% of the speed of a P-wave, which is why they are called secondary waves. Surface waves are slower than body waves, so both R-waves and L-waves arrive at a location on Earth's surface substantially after the body waves have arrived there.

Using Seismometers to Record Earthquakes

If you're standing at the epicenter of a significant earthquake, you'll certainly feel it. The same earthquakes may be imperceptible to a person farther away. That's because seismic waves weaken as they travel, for Earth materials act like shock absorbers, dampening vibrations as they pass. (You've seen a similar loss in energy if you've thrown a pebble in a pond—the waves gets smaller as they move farther away from the point of impact.) In order to detect and study earthquakes of all sizes, even far from the epicenter, seismologists use a **seismometer**, an instrument that can measure the ground motion produced by an earthquake, even if that motion is very tiny.

Figure 8.7 Different types of seismic waves.

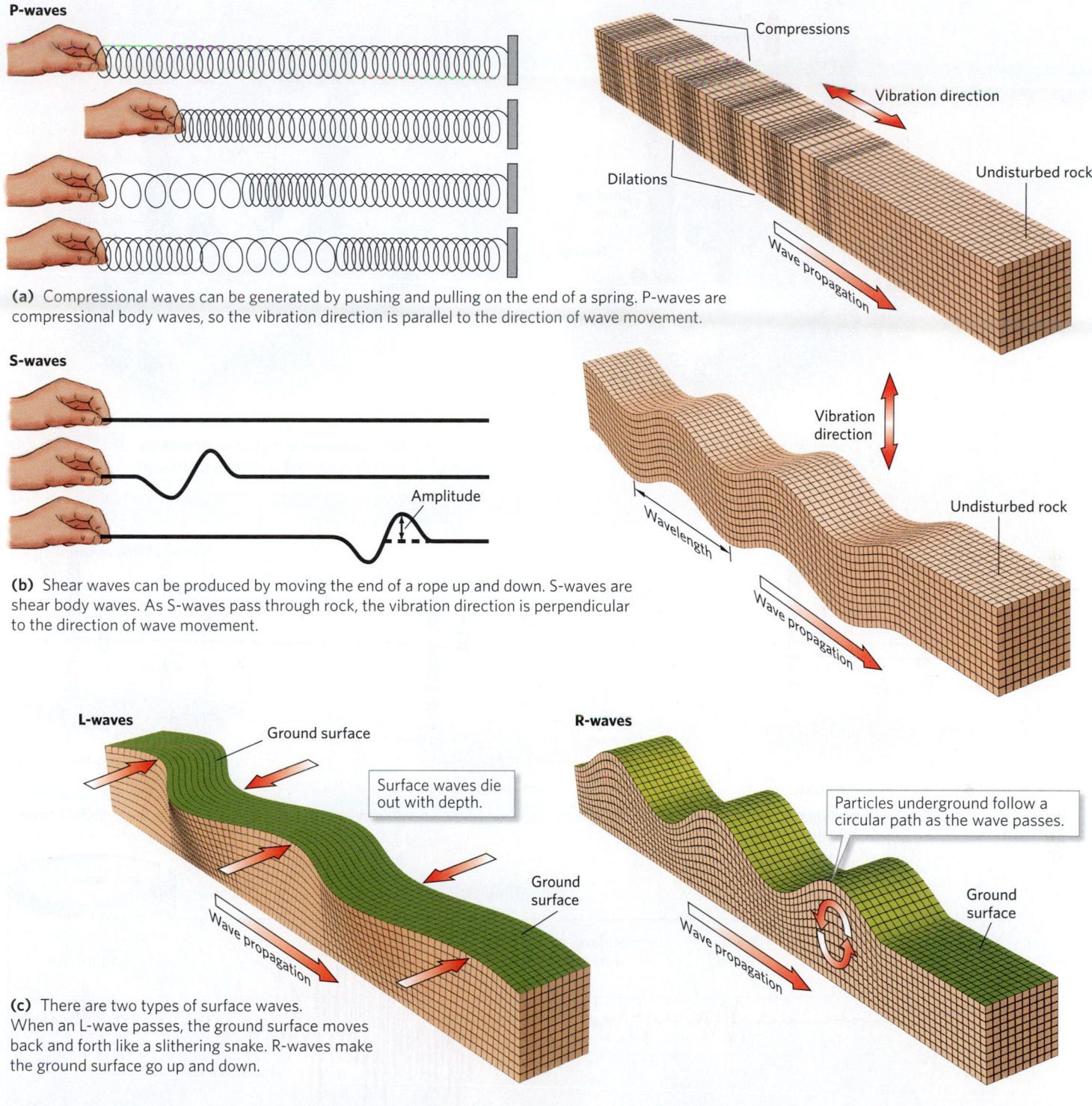

P-waves

(a) Compressional waves can be generated by pushing and pulling on the end of a spring. P-waves are compressional body waves, so the vibration direction is parallel to the direction of wave movement.

Compressions

Vibration direction

Dilations

Undisturbed rock

Wave propagation

S-waves

Amplitude

(b) Shear waves can be produced by moving the end of a rope up and down. S-waves are shear body waves. As S-waves pass through rock, the vibration direction is perpendicular to the direction of wave movement.

Vibration direction

Wavelength

Undisturbed rock

Wave propagation

L-waves

Ground surface

Surface waves die out with depth.

Ground surface

Wave propagation

R-waves

Particles underground follow a circular path as the wave passes.

Ground surface

Wave propagation

(c) There are two types of surface waves. When an L-wave passes, the ground surface moves back and forth like a slithering snake. R-waves make the ground surface go up and down.

Seismometers can be configured in two ways: a vertical-motion seismometer detects up-and-down ground motion **(Fig. 8.8a)**, whereas a horizontal-motion seismometer detects back-and-forth ground motion **(Fig. 8.8b)**. The heart of a traditional mechanical seismometer consists of a heavy weight suspended from a spring. The spring, in turn, hangs from a sturdy frame that has been bolted to the ground. A pen extends from the weight and touches a revolving paper-covered cylinder that is also attached to the frame. Before an earthquake, when the ground is steady, the pen traces out a straight reference line on the paper as the cylinder turns, but when a seismic wave arrives and causes the ground surface to move, the seismometer frame—along with the paper-covered cylinder—moves with it **(Fig. 8.8c)**. Because of its *inertia* (the tendency of an object at rest

Figure 8.8 The basic operation of a seismometer.

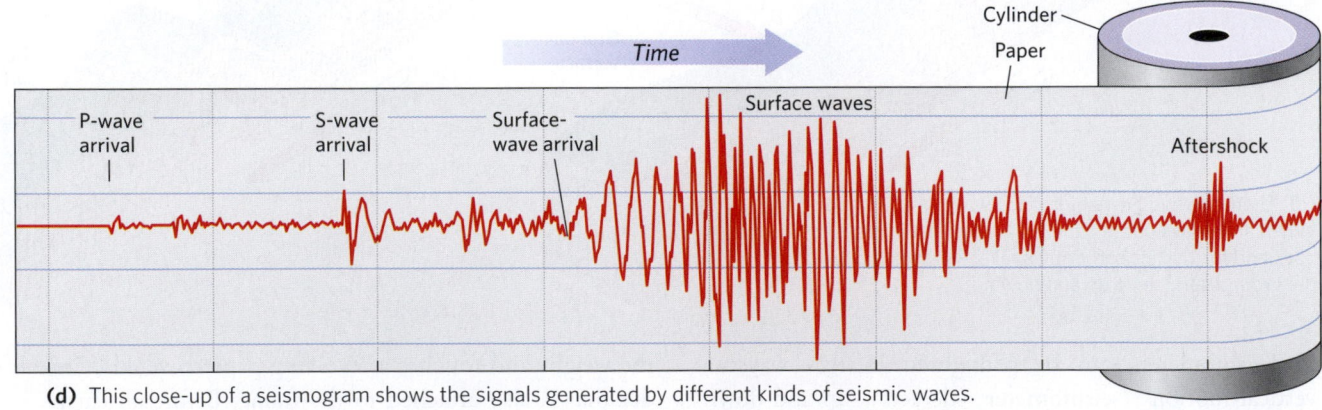

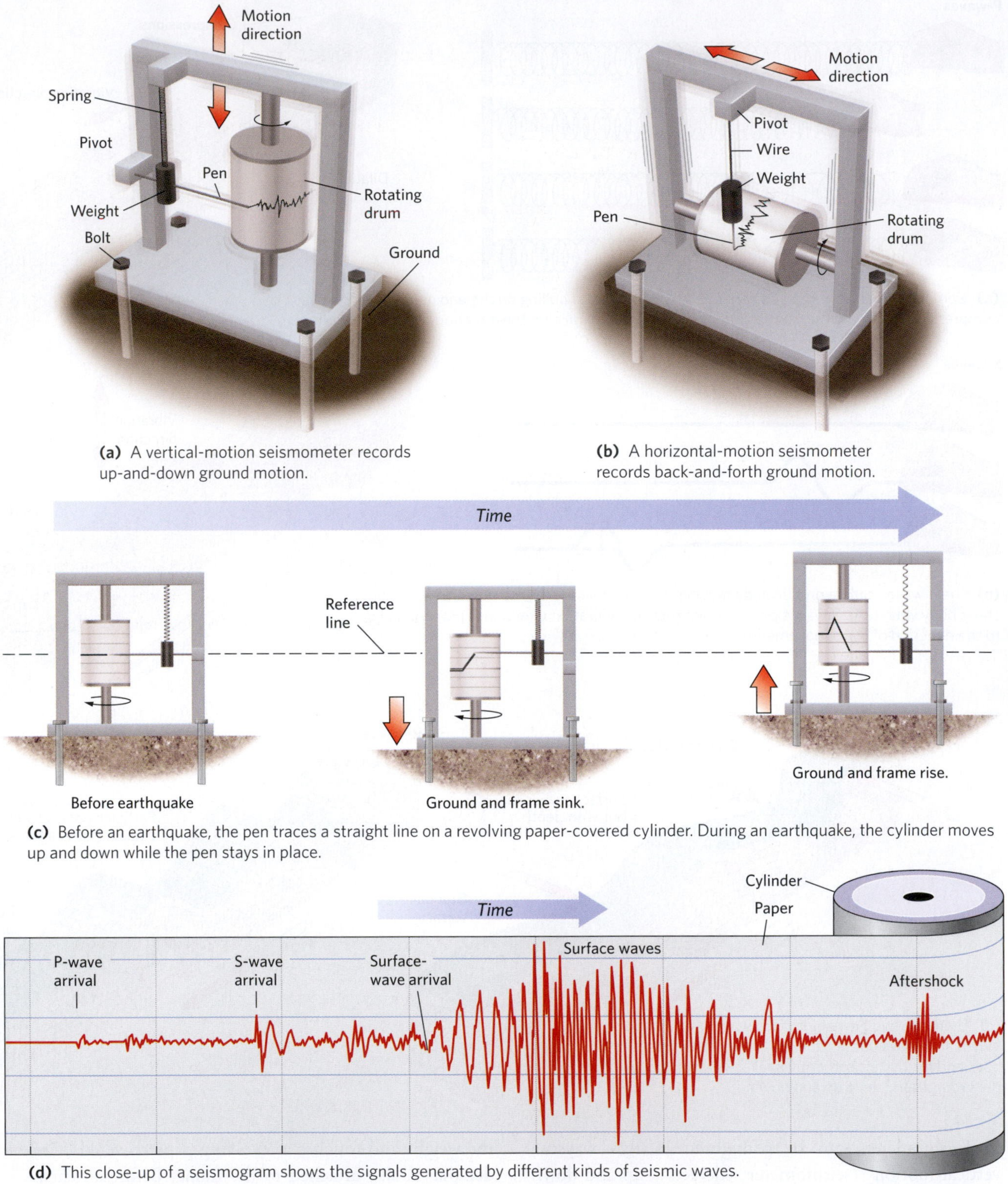

Motion direction

Spring
Pivot
Pen
Weight
Bolt
Rotating drum
Ground

(a) A vertical-motion seismometer records up-and-down ground motion.

Motion direction

Pivot
Wire
Weight
Pen
Rotating drum

(b) A horizontal-motion seismometer records back-and-forth ground motion.

Time

Reference line

Before earthquake

Ground and frame sink.

Ground and frame rise.

(c) Before an earthquake, the pen traces a straight line on a revolving paper-covered cylinder. During an earthquake, the cylinder moves up and down while the pen stays in place.

Time

Cylinder
Paper

P-wave arrival

S-wave arrival

Surface-wave arrival

Surface waves

Aftershock

(d) This close-up of a seismogram shows the signals generated by different kinds of seismic waves.

to remain at rest), however, the weight, with its attached pen, remains fixed. As the revolving cylinder moves with respect to the fixed pen, the pen traces a line on the paper that represents the ground motion. If the cylinder were not revolving, the pen would go back and forth, or up and down, in place. But because the paper moves under the pen, the pen traces out a line that looks somewhat like a wave **(Box 8.1)**. Modern electronic seismometers work on the same principle, but the weight is a magnet that moves relative to a wire coil, thereby producing an electrical signal that can be recorded digitally. Such seismometers can record ground movements as small as a millionth of a millimeter (only 10 times the diameter of an atom).

An earthquake record produced by a seismometer is called a **seismogram** **(Fig. 8.8d)**. At first glance, a typical seismogram looks like a messy squiggle of lines, but to a seismologist it contains a wealth of information. The horizontal axis on a seismogram represents time, and the vertical axis represents the **amplitude** (technically, one-half the wave height) of seismic waves. We refer to the instant at which a seismic wave appears at a seismometer station as the **arrival time** of the wave. The first squiggles on the record represent P-waves because P-waves travel the fastest and arrive first. Next come the S-waves, and finally the surface waves (R-waves and L-waves). Typically, the surface waves have the largest amplitude and arrive over a relatively long interval of time.

Finding the Epicenter

How do we find the location of an earthquake's epicenter? We begin by calculating the difference between the arrival time of P-waves and that of S-waves recorded at a seismometer **(Fig. 8.9a)**. This time delay depends on distance between the seismometer and the epicenter because the waves travel at different velocities **(Fig. 8.9b)**. To picture this, imagine two cars traveling in the same direction, with one traveling at 60 km per hour (37 mph) and the other at 50 km per hour (31 mph). At a distance of 60 km from the start line, the faster car is 10 km (6 miles) ahead, but at a distance of 120 km (75 miles) from the start line, the faster car is 20 km (12 miles) ahead. The time delay recorded at one seismometer, however, tells us only the distance between the epicenter and the seismometer—it does not tell us the direction from the seismometer to the epicenter. To determine the map location of the epicenter, therefore, we must calculate the distance from the epicenter to three different seismometer stations. We can then draw a circle around each station on a map such that the radius of the circle is the distance between that station and the epicenter at the scale of the map. The epicenter lies at the intersection of the three circles, for this is the only point that has the appropriate measured distance from all three stations **(Fig. 8.9c)**.

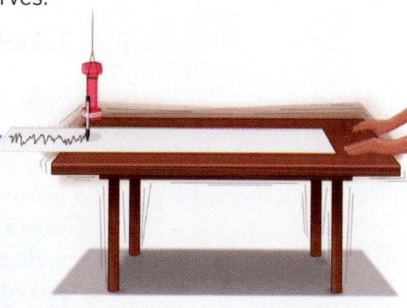

Figure 8.9 Locating an earthquake's epicenter.

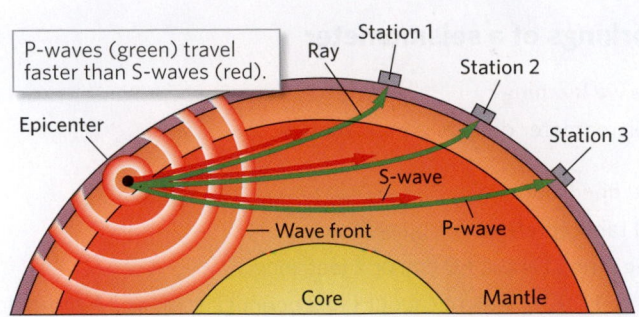

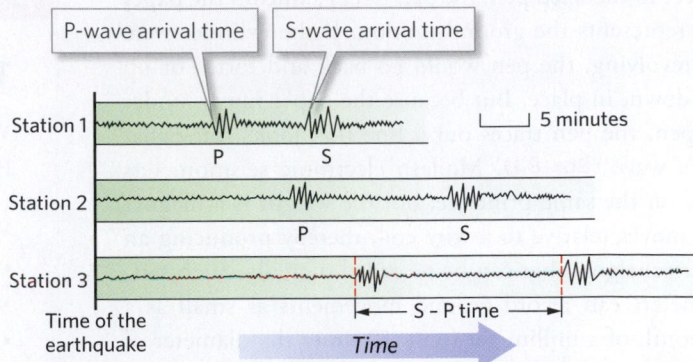

(a) Seismic waves of different types travel at different velocities. The greater the distance between the epicenter and the seismometer station, the longer it takes for seismic waves to arrive there, and the greater the delay between the P-wave and S-wave arrival times. This difference is called the S – P time.

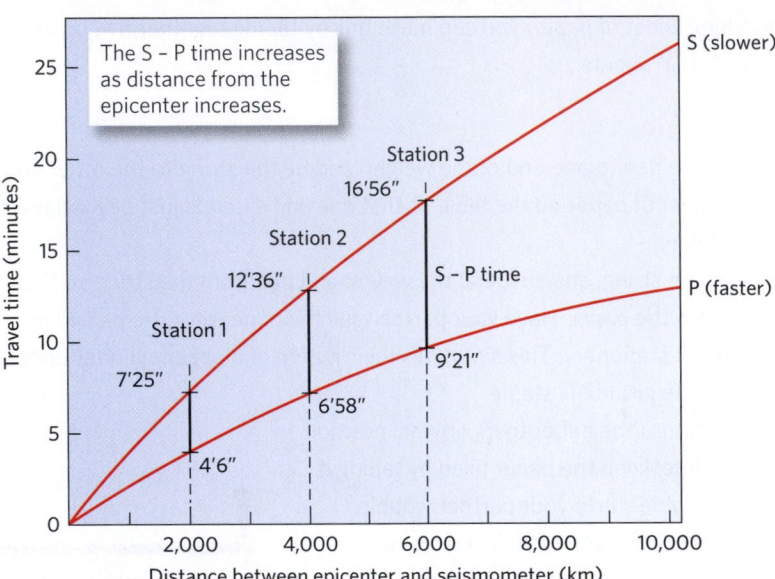

(b) We can represent the different arrival times of P-waves and S-waves on a graph of travel-time curves. The red lines are called travel-time curves. The S-P time at a given distance from the epicenter is represented by the vertical distances between the S-wave travel-time curve and the P-wave travel-time curve.

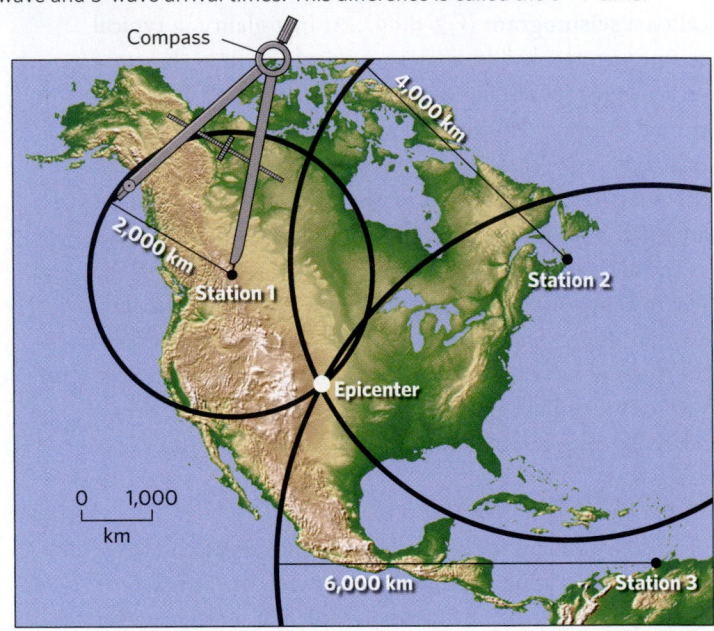

(c) If an earthquake epicenter lies 2,000 km from Station 1, we draw a circle with a radius of 2,000 km around that station at the scale of the map. We repeat for the other two stations. The intersection of the three circles is the epicenter.

Take-home message . . .

Earthquake energy travels as seismic waves. Body waves travel through the interior of the Earth, whereas surface waves travel along the ground surface. Seismologists distinguish between two kinds of body waves (P-waves and S-waves) and between two kinds of surface waves (L-waves and R-waves). Seismometers record ground shaking.

Quick Question -
How can you determine the location of an earthquake's epicenter?

8.4 Defining the Size of Earthquakes

During a typical earthquake, the ground shakes when body waves coming up through the solid Earth reach the surface, as well as when surface waves moving along the ground surface pass by. At a given location, some earthquakes are "large," in that they shake the ground violently, whereas others are "small," in that they can barely be felt. Seismologists have developed two different scales—the intensity scale and the magnitude scale—to define earthquake size in a uniform way so that we can compare different earthquakes.

Table 8.1 Modified Mercalli intensity scale

MMI	Destructiveness (Perceptions of the Extent of Shaking and Damage)
I	Detected only by seismic instruments; causes no damage.
II	Felt by a few stationary people, especially in upper floors of buildings; suspended objects, such as lamps, may swing.
III	Felt indoors; standing automobiles sway on their suspensions; it seems as though a heavy truck is passing.
IV	Shaking awakens some sleepers; dishes and windows rattle.
V	Most people awaken; some dishes and windows break, unstable objects tip over; trees and poles sway.
VI	Shaking frightens some people; plaster walls crack, heavy furniture moves slightly, and a few chimneys crack, but overall little damage occurs.
VII	Most people are frightened; plaster cracks, windows break, some chimneys topple, and unstable furniture overturns; poorly built buildings sustain considerable damage.
VIII	Many chimneys and factory smokestacks topple; heavy furniture overturns; substantial buildings sustain damage, and poorly built buildings suffer severe damage.
IX	Frame buildings separate from their foundations; most buildings sustain damage, and some buildings collapse; the ground cracks, underground pipes break, and rails bend; some landslides occur.
X	Most masonry structures are destroyed; the ground cracks in places; landslides occur; bridges collapse; facades on buildings collapse; railways and roads suffer severe damage.
XI	Few masonry buildings remain; many bridges collapse; broad fissures form in the ground; most pipelines break; severe liquefaction of sediment occurs; some dams collapse.
XII	Earthquake waves cause visible undulations of the ground surface; objects fly up off the ground; there is complete destruction of buildings and bridges of all types.

Modified Mercalli Intensity Scale

Earthquake **intensity** refers to the degree of ground shaking at a locality. In 1902, an Italian scientist named Giuseppe Mercalli devised a scale for defining earthquake intensity by assessing both the damage that the earthquake caused and people's perception of the shaking. A version of this scale, called the **Modified Mercalli Intensity (MMI) scale**, continues to be used today (Table 8.1). We represent earthquake intensities on this scale by Roman numerals. Because earthquake waves lose energy as they travel, the intensity value tends to be greatest near the epicenter and to decrease progressively away from the epicenter. Seismologists can draw a map to show how the intensity of an earthquake varies over a region (Fig. 8.10).

Earthquake Magnitude Scales

When you hear a report of an earthquake disaster in the news, you will probably encounter a phrase like, "An earthquake with a magnitude of 7.2 struck the city yesterday." What does this phrase mean? Earthquake **magnitude** is a number that represents the amount of energy released by an earthquake, as determined from a measurement of the amplitude of ground shaking as recorded by a seismometer. To calculate a magnitude, seismologists first measure the height of the largest spike on a seismogram to obtain

Figure 8.10 Modified Mercalli Intensity (MMI) contours for the 1886 Charleston, South Carolina, earthquake. Note that ground shaking reached an MMI of X at the epicenter, but in New York City, ground shaking reached an MMI of only II to III.

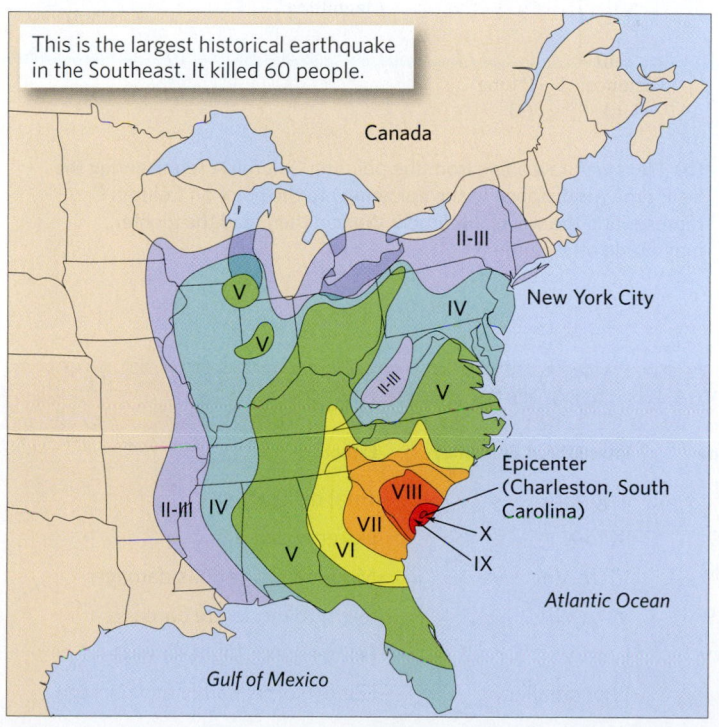

This is the largest historical earthquake in the Southeast. It killed 60 people.

Figure 8.11 Using the Richter magnitude scale.

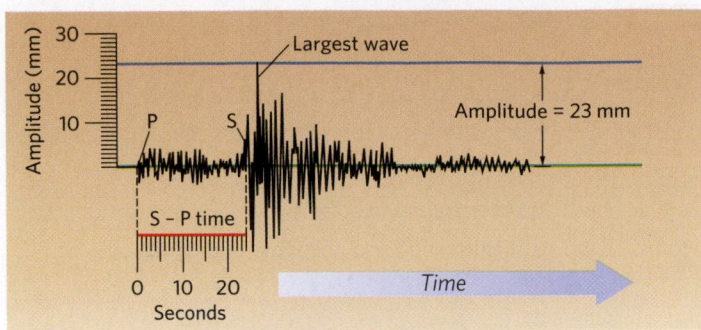

(a) To calculate the Richter magnitude from a seismogram, we first measure the amplitude of the largest wave. We then calculate the difference between the arrival times of S-waves and P-waves (S – P time) to determine the distance to the epicenter (see Figure 8.9).

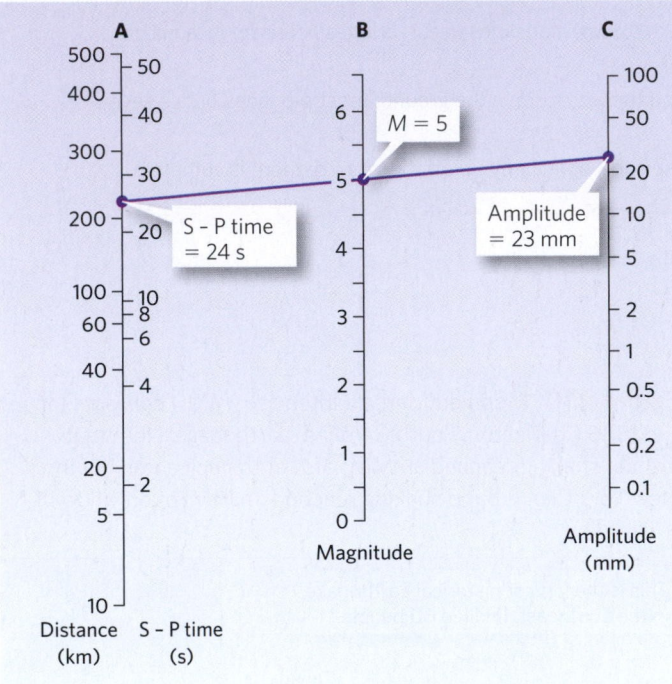

(b) Next we draw a line from the point on Column A representing the S – P time (or distance to the epicenter) to the point on Column C representing the wave amplitude. We can then read the Richter magnitude off Column B.

Table 8.2 Adjectives for describing earthquakes

Adjective	Magnitude	Intensity at Epicenter	Effects
Great	>8.0	X to XII	Total destruction
Major	7.0 to 7.9	IX to X	Extreme damage
Strong	6.0 to 6.9	VII to VIII	Moderate to serious damage
Moderate	5.0 to 5.9	VI to VII	Slight to moderate damage
Light	4.0 to 4.9	IV to V	Felt by most; slight damage
Minor	<3.9	III or smaller	Felt by some; hardly any damage

the maximum amplitude of the ground motion. Then, after they have determined the distance between the epicenter and the seismometer, they adjust the measurement to be equivalent to the maximum amplitude that would be recorded by a seismometer positioned a certain distance—the *reference distance*—from the epicenter. Because of this adjustment, seismologists obtain the same magnitude for a particular earthquake from a measurement at any seismometer. In other words, a given earthquake has only one magnitude number, so, unlike intensity, magnitude does not depend on distance from the epicenter. We specify magnitude (M) using Arabic numerals.

In 1935, an American seismologist, Charles Richter, developed a scale for defining earthquake magnitude. This scale, now known as the **Richter scale**, is logarithmic, meaning that an increase of one unit of magnitude represents a tenfold increase in the maximum amplitude of ground motion. So, a magnitude 8 earthquake results in ground motion that is 10 times greater than that of a magnitude 7 earthquake and 1,000 times greater than that of a magnitude 5 earthquake. Richter used 100 km (62 miles) as the reference distance. Since there's not necessarily a seismometer at exactly 100 km from an epicenter, he developed a simple chart to adjust for the distance of a seismometer station from the epicenter **(Fig. 8.11)**.

These days, seismologists use a somewhat different, more accurate scale, called the **moment magnitude scale**, to represent an earthquake's size. To calculate an earthquake's moment magnitude (abbreviated M_w), seismologists take into account the amplitudes of several different seismic waves, the dimensions of the slipped area on the fault, and the displacement that occurred. The moment magnitude scale, like the Richter scale, is logarithmic.

To make discussion of earthquakes easier, seismologists use familiar adjectives to describe an earthquake's magnitude **(Table 8.2)**. Earthquakes that most people can feel have a magnitude greater than M_w 4, and earthquakes that can cause moderate damage have a magnitude greater than M_w 5. The catastrophic 2011 Tōhoku earthquake registered as M_w 9.0, and the largest recorded earthquake in history, the 1960 Chilean quake, registered as M_w 9.5. News reporters sometimes incorrectly state that the moment magnitude scale "goes from 1 to 10." In fact, microearthquakes (M_w −1 or M_w −2) can be detected by seismometers positioned close to the epicenter, and there is no defined upper limit to the magnitude scale. That said, seismologists estimate that M_w 9.5 is about as big as an earthquake can get, given the known dimensions of faults on Earth.

Energy Release by Earthquakes

To give a sense of the amount of energy released by an earthquake, seismologists compare earthquakes to other energy-releasing events. For example, according to some estimates, an M_w 5.3 earthquake releases about as much energy as the Hiroshima atomic bomb, and an M_w 9.0 earthquake releases significantly more energy than the largest hydrogen bomb ever detonated. Notably, although an increase of one unit of magnitude represents a tenfold increase in the maximum amplitude of ground motion, it represents a 32-fold increase in energy release. Therefore, an M_w 8 earthquake releases about 1 million times more energy than an M_w 4 earthquake **(Fig. 8.12)**. Fortunately, very large earthquakes occur much less frequently than small earthquakes.

Take-home message . . .

We can specify earthquake size by intensity (a measure based on perception of damage caused and shaking felt) or by magnitude (a representation of energy released by the earthquake, based on a measurement of ground motion). An increase of one magnitude unit represents a tenfold increase in the amplitude of shaking and a 32-fold increase in energy. A great earthquake releases vastly more energy than a giant hydrogen bomb.

Quick Question -
Can you specify the size of an earthquake by giving just one intensity number? How about a single magnitude number?

8.5 Where and Why Do Earthquakes Occur?

Earthquakes do not take place everywhere on the globe. By plotting the distribution of earthquake epicenters on a map, seismologists have found that most, but not all, earthquakes occur in **seismic belts (Fig. 8.13)**. Most seismic belts correspond to plate boundaries, and earthquakes within these belts are known as *plate-boundary earthquakes*. Those earthquakes that occur away from plate boundaries are called *intraplate earthquakes*.

Earthquakes at Plate Boundaries

The relative motion of plates causes slip on faults. Plates move relative to their neighbors at rates of 1 to 15 cm (0.4 to 6 inches) per year. So, over a period of decades to centuries, large stresses build along plate boundaries, and these stresses cause sudden slip on faults. As we saw in Chapter 7, geologists distinguish among different types of faults **(Fig. 8.14)**. The fault type that occurs at a given plate boundary depends on the relative motion across that boundary.

DIVERGENT-BOUNDARY SEISMICITY. At a divergent boundary (mid-ocean ridge), two plates form and move apart. Divergent boundaries are broken into ridge

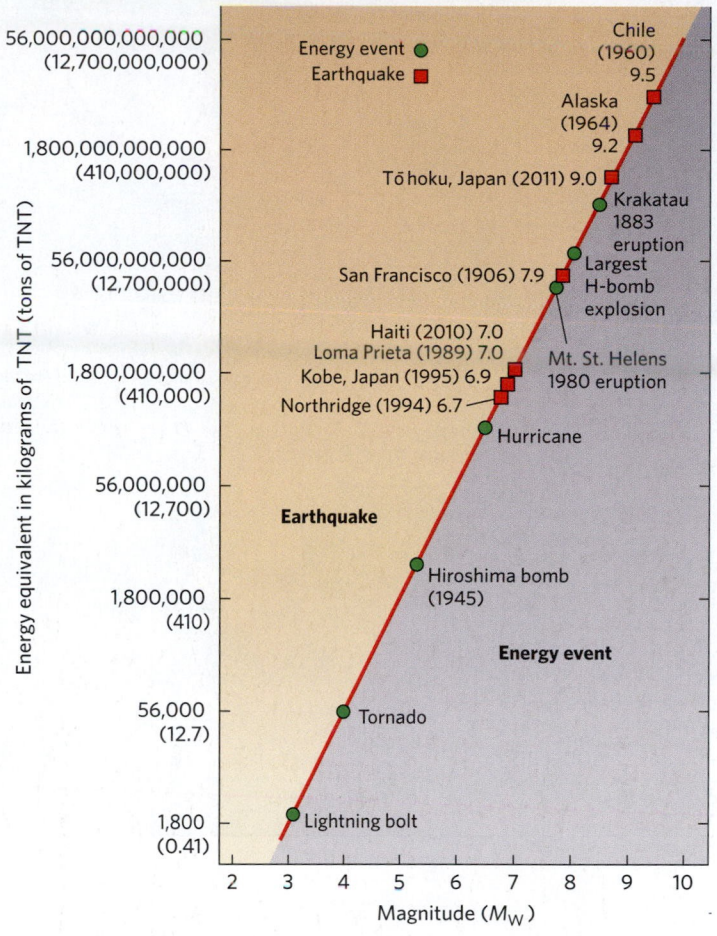

Figure 8.12 The energy released by earthquakes increases dramatically with magnitude. Great earthquakes release vastly more energy than the largest nuclear bombs.

segments linked by transform boundaries. Therefore, two kinds of faults develop at divergent boundaries: along the ridge segments, stretching generates normal faults, whereas along the transform faults, strike-slip displacement occurs **(Fig. 8.15)**. Earthquakes on these faults have shallow foci.

TRANSFORM-BOUNDARY SEISMICITY ON CONTINENTS. At transform boundaries, where one plate slides past another without producing or consuming lithosphere, strike-slip displacement takes place on transform faults. (In other words, transform faults are a type of strike-slip fault.) The majority of transform faults in the world link segments of mid-ocean ridges, as we've seen. But a few, such as the San Andreas fault of California, the Alpine fault of New Zealand, and the Anatolian faults in Turkey, cut through continental crust. Large earthquakes on continental transform faults can be very destructive, for they have shallow foci. This means that the seismic waves they produce can still carry a large amount of energy when they reach the ground surface. Further, because they're on land, they may lie near population centers.

Did you ever wonder . . .
whether an earthquake will happen near where you live?

Figure 8.13 A map of epicenters for a period of several decades emphasizes that most earthquakes occur in distinct seismic belts along plate boundaries. The color of an epicenter dot represents the depth to the focus of the earthquake.

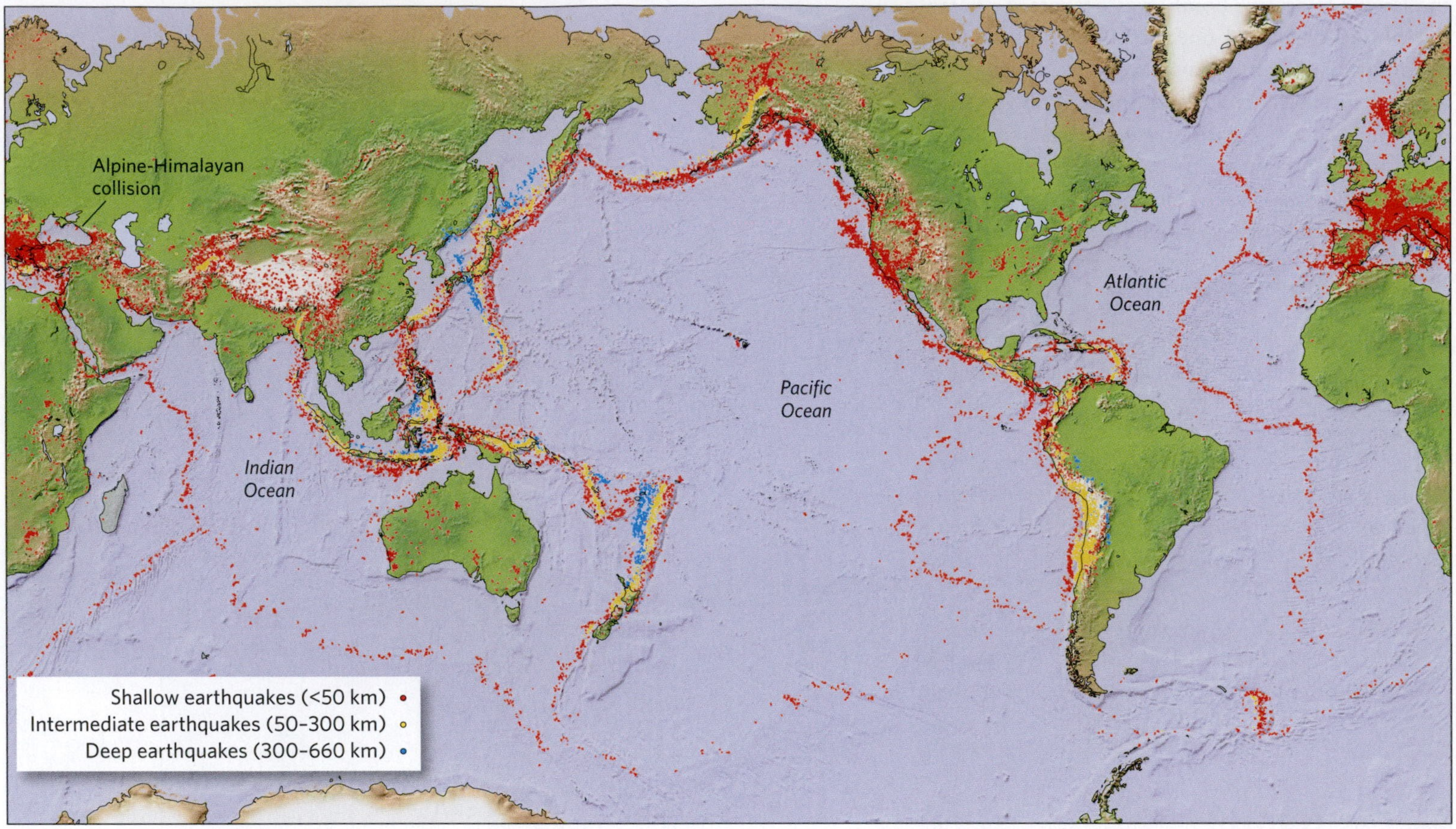

Alpine-Himalayan collision

Atlantic Ocean

Pacific Ocean

Indian Ocean

Shallow earthquakes (<50 km) •
Intermediate earthquakes (50–300 km) •
Deep earthquakes (300–660 km) •

Figure 8.14 The basic types of faults. Fault types are distinguished from one another by the direction of slip relative to the fault surface (see Chapter 7).

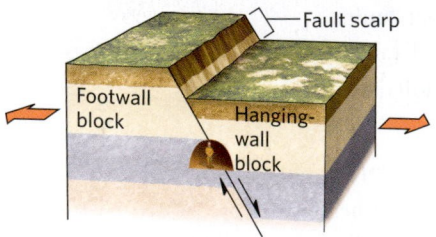

Fault scarp

Footwall block

Hanging-wall block

(a) Normal faults form during extension of the crust. The hanging wall moves down.

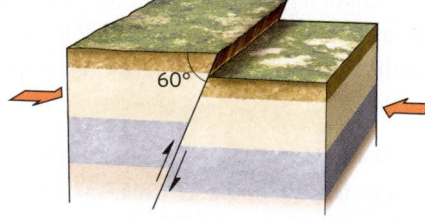

60°

(b) Reverse faults form during shortening of the crust. The hanging wall moves up, and the fault is steep.

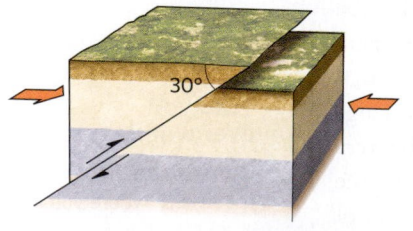

30°

(c) Thrust faults also form during shortening. The fault's slope is gentle (less than 30°).

Strike-slip faults tend to be vertical.

(d) On a strike-slip fault, one block slides laterally past another, so no vertical displacement takes place.

In 1906, an M_w 7.8 earthquake on a transform boundary destroyed San Francisco. This earthquake resulted from sudden slip on the San Andreas fault, along which the Pacific Plate moves north at an average rate of 6 cm (2 inches) per year relative to the North American Plate **(Fig. 8.16a)**. Because of stick-slip behavior along the San Andreas fault, this movement happens in sudden jerks, each of which causes an earthquake. On April 18, 1906, a segment of the fault near San Francisco that was 15 km (9 miles) deep by 477 km (300 miles) long suddenly slipped by as much as 7 m (23 feet; see Fig. 8.5). When, moments later, seismic waves generated by the resulting elastic rebound struck the city, streets undulated like the surface of the sea. Streets cracked, buildings swayed and collapsed, and nearby slopes slumped **(Fig. 8.16b)**. The 1906 San Francisco earthquake has not been the only major event to strike along the San Andreas and nearby related faults, however. An M_w 7.9 event occurred to the south in 1857, and in 1989, an M_w 6.9 earthquake struck Loma Prieta, 100 km (62 miles) south of San Francisco. During the 1989 quake, ground shaking from seismic waves that reached San Francisco collapsed a double-decked freeway **(Fig. 8.16c)**.

CONVERGENT-BOUNDARY SEISMICITY. Convergent boundaries are complicated regions where many earthquakes at a variety of depths take place **(Fig. 8.17a)**. The most dangerous ones occur along relatively shallow thrust faults that delineate the boundary between the base of the overriding plate and the top of the subducting plate. Because the foci of these earthquakes are so shallow, most of the seismic energy produced by elastic rebound reaches the ground surface. Notable examples of devastating convergent-boundary earthquakes include the largest recorded earthquakes: the 1960 M_w 9.5 earthquake in Chile, the 1964 M_w 9.2 Good Friday earthquake in Alaska, the 2004 M_w 9.3 earthquake in Sumatra, and the 2011 M_w 9.0 Tōhoku earthquake in Japan.

Unlike divergent or transform boundaries, convergent boundaries also host intermediate- and deep-focus earthquakes. A plot of the foci of such earthquakes on a cross section through the Earth defines a sloping band of seismicity called a *Wadati-Benioff zone* (named for the seismologists who first recognized it), which extends to a depth of 660 km (410 miles). Earthquakes of this zone occur within subducted lithosphere as it sinks down through the asthenosphere **(Fig. 8.17b)**. The cause of these earthquakes remains a topic of research. They may be due to phase changes in the asthenosphere, which we'll describe in Section 8.9.

Figure 8.15 The distribution of earthquakes at a divergent boundary. Note that normal faults occur along the ridge axis and strike-slip faults occur along active transform faults. Earthquakes do not occur along inactive fracture zones.

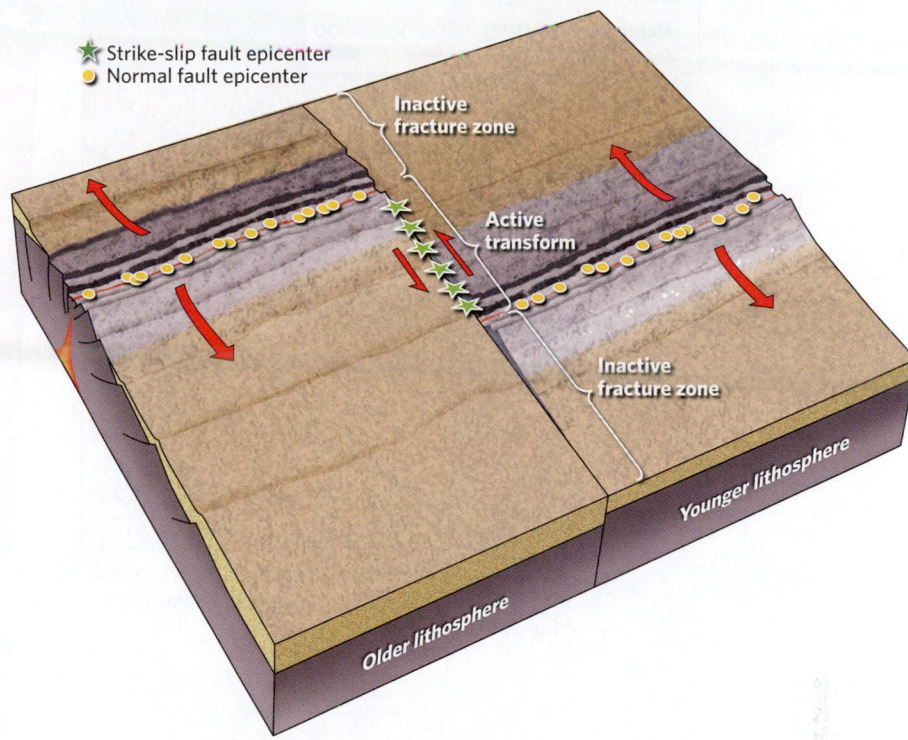

Figure 8.16 Seismicity on a continental transform fault.

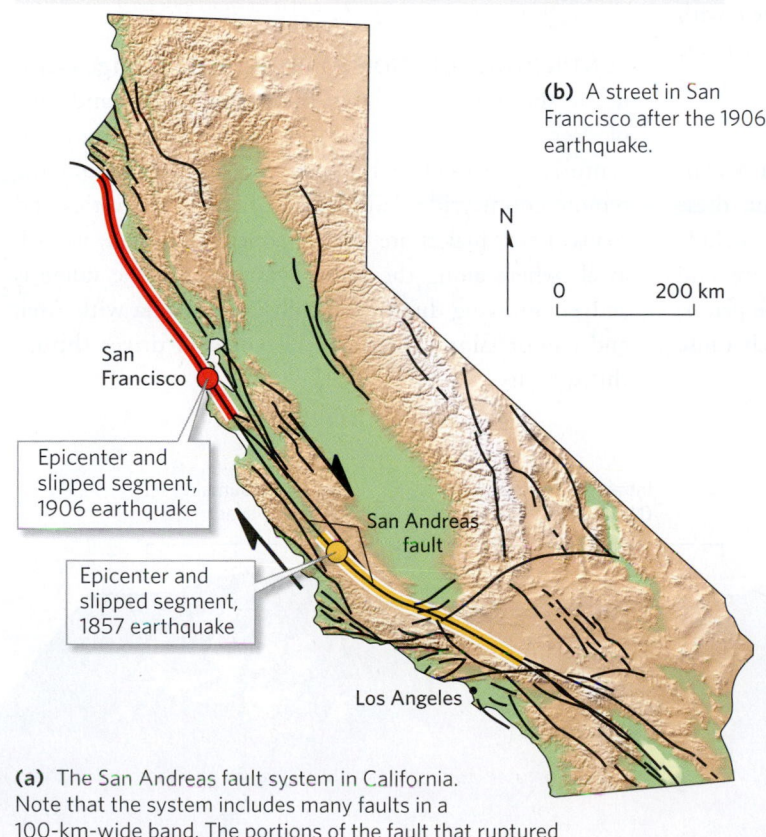

(a) The San Andreas fault system in California. Note that the system includes many faults in a 100-km-wide band. The portions of the fault that ruptured in the 1906 and 1857 earthquakes are indicated.

(b) A street in San Francisco after the 1906 earthquake.

(c) The 1989 Loma Prieta earthquake caused a two-level freeway in San Francisco, 100 km northwest from its epicenter, to collapse as its support columns gave way.

Figure 8.17 Convergent-boundary seismicity.

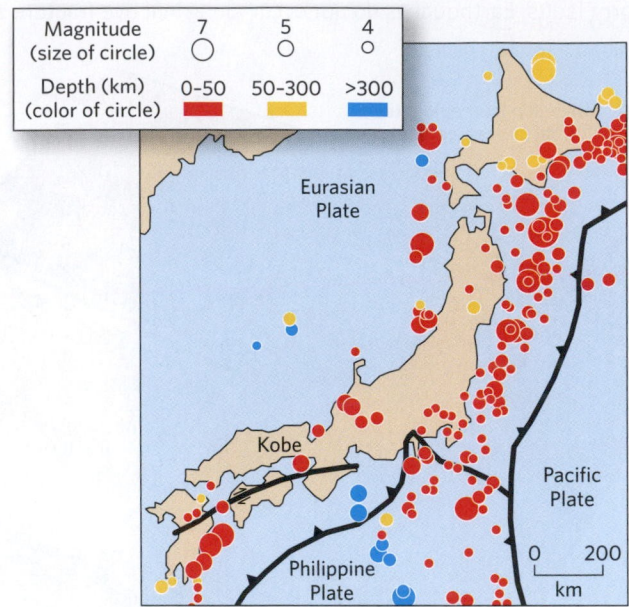

(a) This map of earthquake epicenters and depths in and near Japan shows how complex convergent boundaries can be.

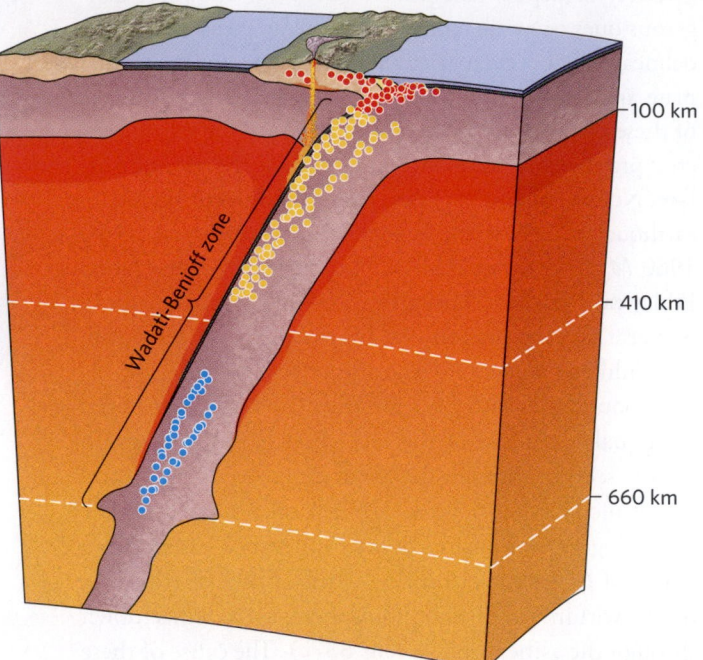

(b) At a convergent boundary, shallow earthquakes (red dots) occur along the contact between the downgoing plate and the overriding plate, as well as in the two plates themselves. Intermediate-focus (yellow dots) and deep-focus earthquakes (blue dots) define the Wadati-Benioff zone. Few earthquakes happen between 400 and 500 km.

Earthquakes within Continents

Not all earthquakes on continents are associated with plate boundaries. Here, we consider a few other geologic settings within continents in which earthquakes occur (Fig. 8.18).

CONTINENTAL RIFTS. The stretching of crust at a continental rift generates normal faults, and slip on these faults produces earthquakes. Active rifts today include the East African Rift, the Basin and Range Province, and the Rio Grande Rift of New Mexico. In all these places, shallow-focus earthquakes occur, some of which cause major damage.

CONTINENTAL COLLISION ZONES. In 2015, a large earthquake shook the ground in the mountainous landscape of Nepal (Fig. 8.19). Monuments that had stood for centuries collapsed in Kathmandu, and throughout the remote countryside, landslides buried communities and roads. Earthquakes are fairly frequent in Nepal, as well as elsewhere along the Himalayas, because the range is actively growing due to the collision of India with Asia, and compression caused by this collision drives slip on thrust faults.

Figure 8.18 The geologic settings in which earthquakes occur in continental lithosphere. (Subduction-related earthquakes in continental crust are not shown.)

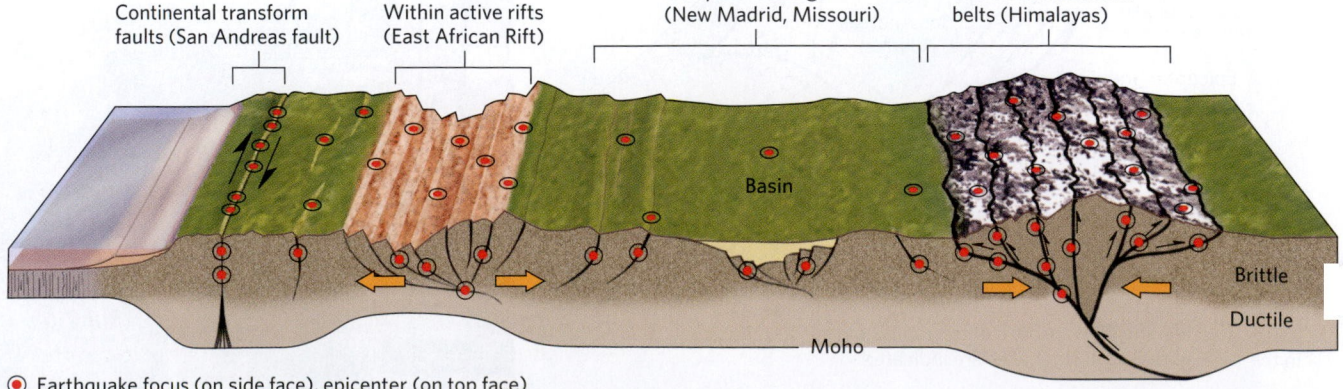

⊙ Earthquake focus (on side face), epicenter (on top face)

Figure 8.19 The 2015 Nepal earthquake destroyed many buildings. For example, this elaborate brick tower, built in 1832, crumbled to rubble.

Before

After

INTRAPLATE EARTHQUAKES. In a given year, 95% of the seismic energy produced on Earth comes from plate-boundary, collision, or rift-related earthquakes. The remaining types, which occur in the interiors of plates, are known as **intraplate earthquakes**. The foci of almost all intraplate earthquakes lie at depths of less than 25 km (15 miles). These earthquakes may happen when weak pre-existing fault zones, some of which initially formed over 1 billion years ago, undergo slip in response to stresses applied to continents.

Intraplate earthquakes are not uniformly distributed. In North America, most take place in southeastern Missouri, eastern Tennessee, eastern South Carolina, and southern Quebec. The largest historical intraplate earthquakes to affect the continental United States struck near New Madrid, in southernmost Missouri. During the winter of 1811–1812, three M_w 7.0 to 7.4 earthquakes, resulting from slip on faults beneath the Mississippi Valley, shook the region. Displacement of the ground surface temporarily reversed the flow of the Mississippi River, and ground movement toppled

cabins (**Fig. 8.20a**). The region remains seismically active (**Fig. 8.20b**).

Take-home message . . .

Most, but not all, earthquakes happen along plate boundaries. Particularly catastrophic earthquakes occur at faults associated with convergent boundaries and transform faults because these earthquakes are relatively shallow. Large earthquakes also occur in rifts and collision zones, and a few take place along weak zones within plate interiors.

Quick Question -
Do people feel most of the earthquakes that take place along divergent boundaries? Why or why not?

8.6 How Do Earthquakes Cause Damage?

Ground Shaking and Displacement

An earthquake starts suddenly and may last from a few seconds to a few minutes. The duration of ground shaking at a given locality depends on how long it took for sliding to occur and on the distance between the locality and the earthquake's focus. The second factor reflects the fact that not all earthquake waves travel at the same velocity, so they don't all arrive at the same time.

Different kinds of seismic waves cause different kinds of ground motion (**Fig. 8.21**). The severity of the shaking at a given location depends on (1) the magnitude of the earthquake, because larger-magnitude events release more energy; (2) the distance from the focus, because seismic energy decreases as seismic waves pass through the Earth; and (3) the strength of the materials just beneath the ground surface, because seismic waves tend to cause more motion in weaker materials.

If you're out in an open field during an earthquake, ground motion alone won't kill you. You may be knocked off your feet and bounced around a bit, but your body is too flexible to break. Human-made structures aren't so lucky, however (**Fig. 8.22**). When seismic waves pass, roads, rail lines, and pipelines buckle or rupture. Buildings sway, twist back and forth, or lurch up and down, so windows may shatter, roofs fail, and building facades crash to the ground. If the floors of a multi-story building or the deck of a bridge bounce up, they slam back down on their support columns, generating enough downward force to crush the columns and cause the structures to collapse. Falling debris from building failure causes the majority of earthquake-related deaths and injuries.

Did you ever wonder . . .
how long an earthquake lasts?

Figure 8.20 Intraplate seismic activity near New Madrid, Missouri.

(a) The earthquakes of 1811–1812 destroyed cabins and disrupted the flow of the Mississippi River.

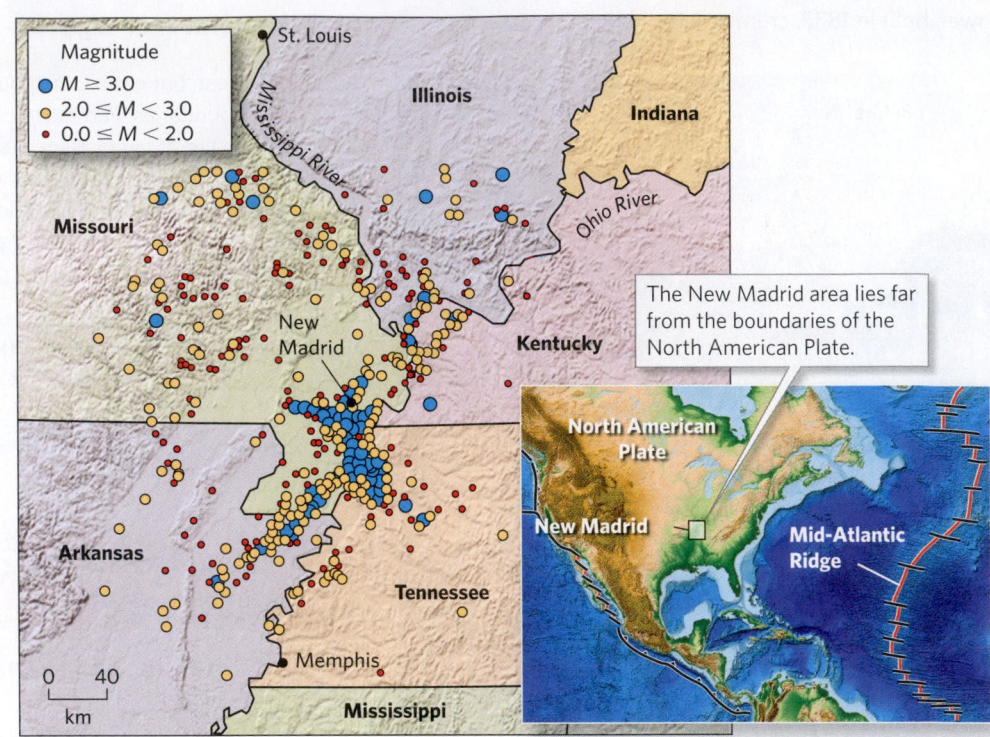

The New Madrid area lies far from the boundaries of the North American Plate.

(b) The epicenters of recent small earthquakes in the New Madrid area, as recorded by modern seismometers. The region remains seismically active. Note that New Madrid is far from a plate boundary.

Figure 8.21 During an earthquake, the ground can shake in many ways at once, causing surface structures to move.

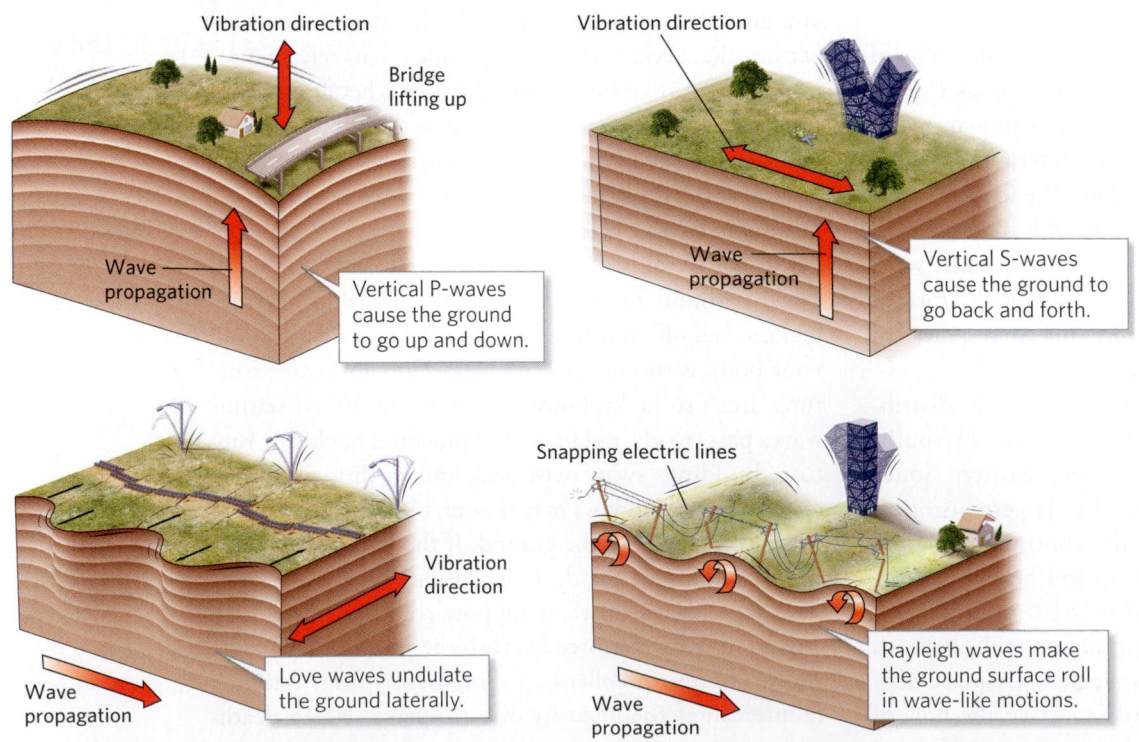

Vibration direction

Bridge lifting up

Wave propagation

Vertical P-waves cause the ground to go up and down.

Vibration direction

Wave propagation

Vertical S-waves cause the ground to go back and forth.

Vibration direction

Wave propagation

Love waves undulate the ground laterally.

Snapping electric lines

Wave propagation

Rayleigh waves make the ground surface roll in wave-like motions.

Landslides

Seismic shaking can cause ground on steep slopes or ground underlain by weak sediment to give way. This movement results in a *landslide*, the tumbling or flow of soil and rock downslope (see Chapter 12). Landslides are often a major cause of earthquake damage. Earthquake-triggered landslides along the coast of California may make headlines, for when steep cliffs facing the Pacific collapse, expensive homes tumble to the beach below **(Fig. 8.23)**. Such events lead to the misperception that "California will fall into the sea" during an earthquake. Although small portions of the coastline do indeed tumble down to the beach, the state as a whole remains firmly attached to the continent, despite what Hollywood scriptwriters say.

Sediment Liquefaction

In 1964, an M_w 7.5 earthquake struck Niigata, Japan, a city that had been partly built on land underlain by wet sand. During the ground shaking, the foundations of over 15,000 buildings sank into the ground, causing walls and roofs to crack. In fact, several buildings simply tipped over (Fig. 8.24a)! In 2011, an earthquake in Christchurch, New Zealand, caused sand to erupt and produce small, cone-shaped mounds on the ground surface (Fig. 8.24b). When the sand moved from underground onto the surface, pit-like depressions developed nearby. Some of these pits became large enough to swallow cars (Fig. 8.24c).

The above examples illustrate the process of **sediment liquefaction**. Liquefaction in wet sand happens when shaking causes the sand grains to try and settle together more tightly. This movement causes the pressure in the water to increase enough to push the grains apart. As a result, what had been stable, load-bearing sand turns into a sand-water slurry, incapable of supporting weight. If cracks open up between the liquefied sand layer and the ground surface, pressure caused by the weight of overlying sediment squeezes the wet sand upward and out onto the ground surface, building cone-shaped mounds called *sand volcanoes* (see Fig. 8.24b). The settling of sediment layers into a liquefied layer can also disrupt bedding and can lead to formation of open fissures and pits on the land surface.

A similar phenomenon happens in a special type of clay called *quick clay*: ground shaking transforms what had been a gel-like solid mass into a liquid, slippery mud by destroying cohesion among the grains. This phenomenon happened during the 1964 Good Friday earthquake in southern Alaska, beneath the Turnagain Heights neighborhood of Anchorage. The neighborhood was built on a small terrace of uplifted sediment. The edge of the terrace was a 20-m (6.5 foot)-high escarpment that dropped down to a bay of the Pacific Ocean. When ground shaking began, a layer of quick clay beneath the development liquefied, and the overlying layers of sediment, along with the houses built on top

Figure 8.22 Examples of earthquake damage due to ground shaking.

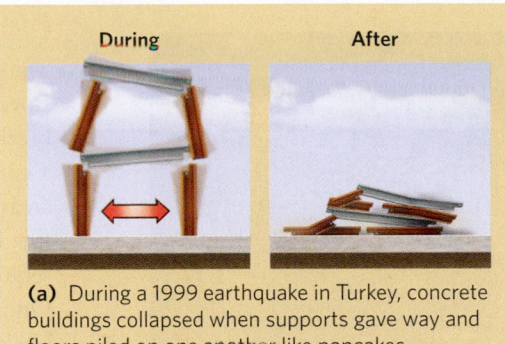

(a) During a 1999 earthquake in Turkey, concrete buildings collapsed when supports gave way and floors piled on one another like pancakes.

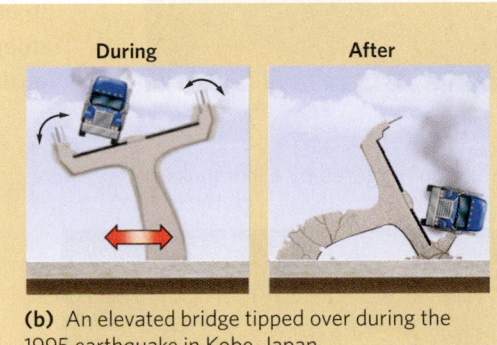

(b) An elevated bridge tipped over during the 1995 earthquake in Kobe, Japan.

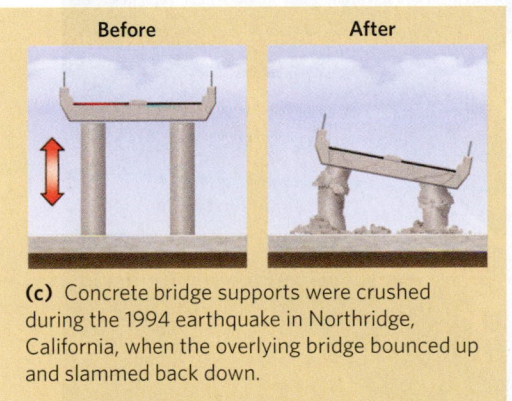

(c) Concrete bridge supports were crushed during the 1994 earthquake in Northridge, California, when the overlying bridge bounced up and slammed back down.

(d) A neighborhood of masonry buildings in Armenia collapsed during a 1999 earthquake because the walls broke apart.

Figure 8.23 During the 1994 Northridge earthquake, a steep slope along the coast of California collapsed, carrying part of a home with it.

Fire

The shaking during an earthquake can make lamps, stoves, or candles with open flames tip over, and it may break wires or topple power lines, generating sparks. As a consequence, areas already turned to rubble, and even areas that are not so badly damaged, may be consumed by fire. Ruptured gas pipelines and oil tanks feed the flames, sending columns of fire erupting skyward **(Fig. 8.26a)**. Once a fire starts to spread, it may become an unstoppable inferno, especially since debris-filled streets block fire-fighting equipment and broken pipes make fire hydrants inoperable. For example, in 1906, most of the damage to San Francisco was caused by fire after the ground had stopped shaking **(Fig. 8.26b)**. And when a large earthquake hit Tokyo in 1923, coals from cooking stoves set

of them, slumped seaward, turning the landscape into a chaotic jumble **(Fig. 8.25)**.

Figure 8.24 Examples of sediment liquefaction triggered by earthquakes.

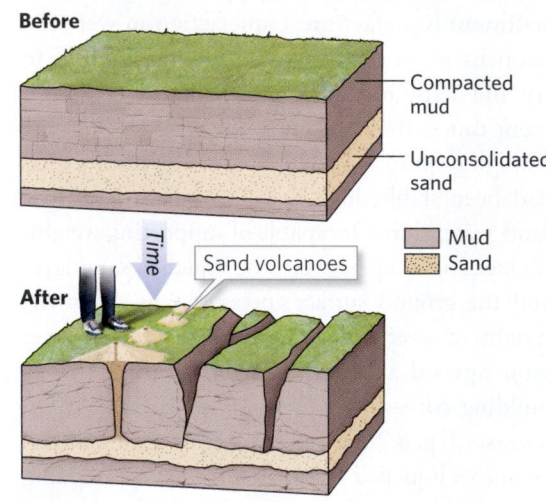

(b) Liquefaction of a sand layer causes the ground to crack and sand volcanoes to erupt.

(a) Liquefaction under their foundations caused these apartment buildings in Niigata, Japan, to tip over during a 1964 earthquake.

(c) During the 2011 Christchurch earthquake, liquefied sand spurted out and spread over the pavement. The process produced open space underground, so the pavement collapsed to form a sinkhole.

Figure 8.25 The 1964 Turnagain Heights disaster.

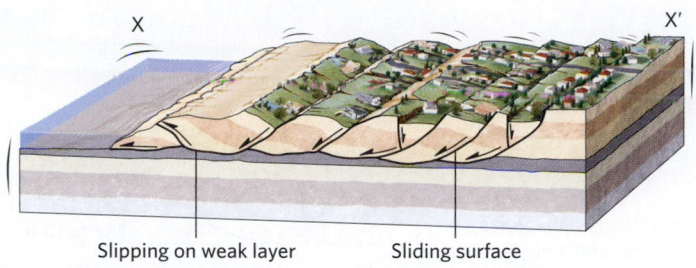

(a) Slip occurred on a weak layer of quick clay.

The dark area in this aerial photograph is the slump that was Turnagain Heights.

(b) The neighborhood slid into the sea. The line (XX') shows the location of the block diagram.

buildings made of wood and paper alight. The fire spread rapidly, and warmed the air above the city. As hot air rose, cool air rushed in from outside the burning area, generating wind gusts of over 160 km per hour (100 mph). The winds stoked the blaze, producing a firestorm that consumed the city **(Fig. 8.26c)**.

Tsunamis

The azure waters and palm-fringed islands of the Indian Ocean's eastern coast hide the Sunda Trench, one of the

Figure 8.26 Fire sometimes follows an earthquake.

(a) Broken gas tanks erupt in fountains of flame after the 2011 Tōhoku, Japan, earthquake.

(b) A street in San Francisco after the 1906 earthquake. Huge fires swept through the city after the quake.

most seismically active plate boundaries on Earth. Just before 8:00 A.M. on December 26, 2004, a 1,300-km (800-mile)-long by 100-km (62-mile)-wide area of a thrust fault along this convergent boundary broke, and the overriding plate lurched westward by as much as 15 m (50 feet), pushing the seafloor up by tens of centimeters. Elastic rebound from this slip triggered the great M_w 9.3 Sumatra earthquake, and the rise of the seafloor shoved the overlying ocean surface upward. Because the area of seafloor that rose was so broad, the volume of water it displaced was immense. Gravity caused the uplifted water to collapse downward and spread out. The uplift and collapse of the broad mound of water generated several broad waves that traveled outward at a velocity of about 800 km per hour (500 mph)—almost the speed of a jet plane.

Geologists use the term **tsunami** for a wave produced by displacement of the seafloor **(Fig. 8.27a)**. The displacement can be due to an earthquake, an explosive volcanic eruption, or, as we'll see in Chapter 12, a submarine landslide. *Tsunami* is a Japanese word that translates literally

Did you ever wonder . . .
whether California will fall into the sea during an earthquake?

See for yourself

Banda Aceh, before Tsunami

Latitude: 5°33′31.82″ N
Longitude: 95°17′13.01″ E

Zoom to 1.5 km (1 mile) and look down.

With the historical imagery tool, you can set the clock back to June 22, 2004, and see dense housing built along the shore of the seaside town of Banda Aceh, on the island of Sumatra. Now set the clock to 2006 and compare the views.

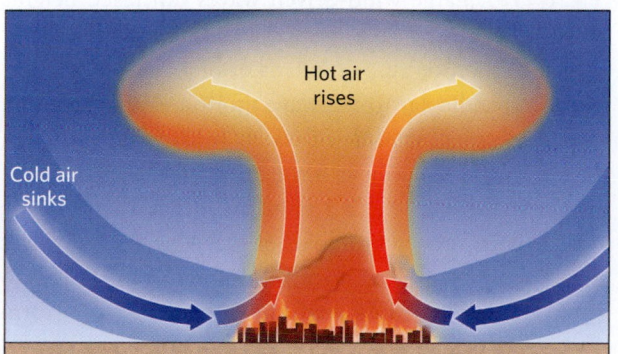

(c) A firestorm develops when cool air rushes in to replace rising hot air above a huge fire. The cool air stokes the blaze, making it larger and hotter.

Figure 8.27 Formation of a tsunami.

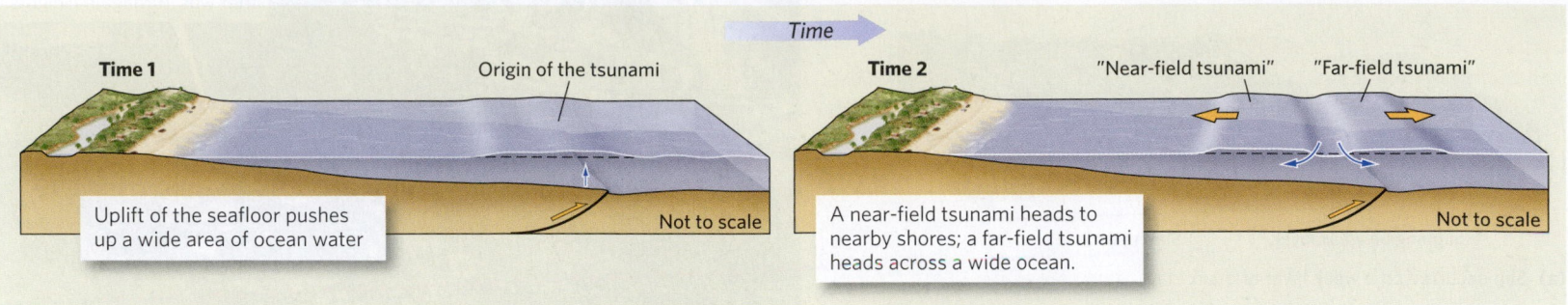

Time 1 — Origin of the tsunami — **Uplift of the seafloor pushes up a wide area of ocean water** — Not to scale

Time 2 — "Near-field tsunami" — "Far-field tsunami" — **A near-field tsunami heads to nearby shores; a far-field tsunami heads across a wide ocean.** — Not to scale

Time

(a) After initial uplift of a mound of water, gravity causes the wave to spread out. A tsunami travels almost as fast as a jet plane.

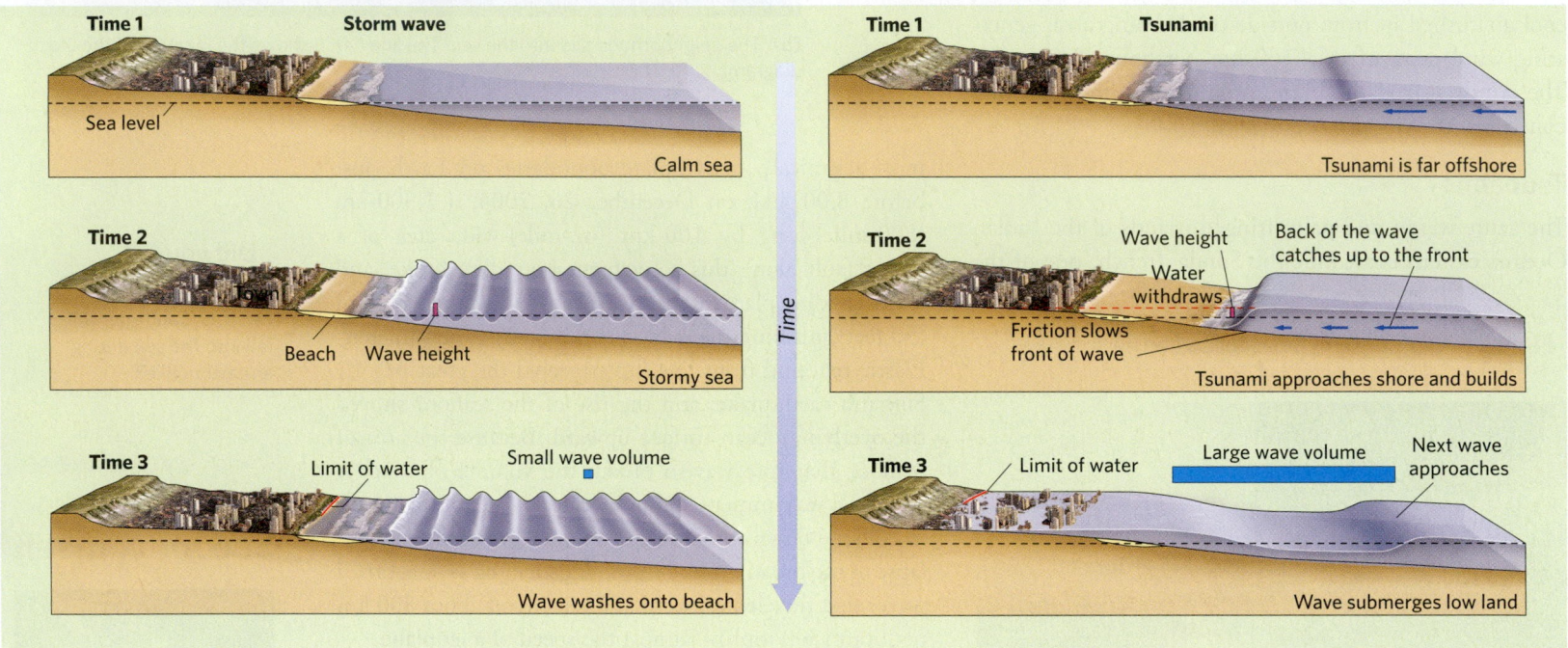

Time 1 — Storm wave — Sea level — Calm sea

Time 2 — Beach — Wave height — Stormy sea

Time 3 — Limit of water — Small wave volume — Wave washes onto beach

Time 1 — Tsunami — Tsunami is far offshore

Time 2 — Wave height — Back of the wave catches up to the front — Water withdraws — Friction slows front of wave — Tsunami approaches shore and builds

Time 3 — Limit of water — Large wave volume — Next wave approaches — Wave submerges low land

Time

(b) The difference between a storm wave and a tsunami. Storm waves can be high, but because they are not very wide, they contain relatively little water. A tsunami is not very high out in the open ocean, but as it approaches the land, the rear catches up and the wave grows higher. Each wave is very wide.

as harbor wave, an apt name because tsunamis can be particularly damaging to harbor towns. Tsunamis differ significantly from familiar, wind-driven storm waves (Fig. 8.27b). Large wind-driven waves can reach heights of 10 to 30 m (32 to 100 feet) in the open ocean. But even such monsters are only tens of meters wide, as measured perpendicular to the wave motion, so they contain a relatively small volume of water. In contrast, although the sea surface rises by at most only a few tens of centimeters at the site where a tsunami forms, the resulting wave may be tens to hundreds of km wide, so it involves an immense volume of water. A tsunami can be 100 to 1,000 times wider than a wind-driven wave, so a storm wave and a tsunami have very different consequences.

When any wave approaches the shore, friction between the base of the wave and the seafloor slows the bottom of the wave, so the back of the wave catches up to the front, and the added water builds the wave higher (see Fig. 8.27b). Near the beach, the top of the wave may fall over the front of the wave, forming a breaker (see Chapter 15). In the case of a wind-driven wave, the breaker may be tall when it washes onto the beach, but because the wave doesn't contain much water, the flow of water stops at the landward edge of the beach. In the case of a tsunami, however, when friction slows the wave and increases its height near the shore, the wave doesn't run out of water after it has crossed the beach. So if the coastal land is low-lying, the tsunami keeps moving inland, eventually submerging a huge area. The largest tsunamis can grow to heights of 30 m (100 feet) and can submerge land several kilometers inland from the shore. Keep in mind that we define a tsunami by its cause and its width, not by its height. Some tsunamis are only tens of centimeters to a meter high and may not even be noticed.

Figure 8.28 The great Indian Ocean tsunami of 2004.

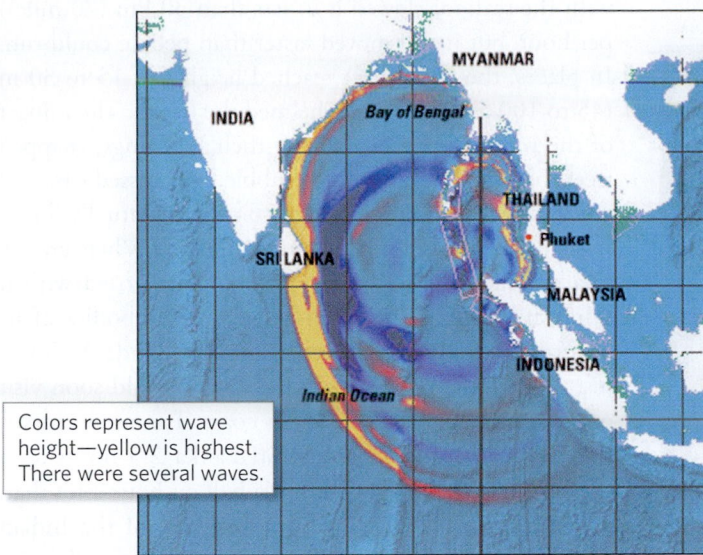

Colors represent wave height—yellow is highest. There were several waves.

At its highest, the tsunami's front was over 15 m high.

Wave height at Banda Aceh

(a) A devastating tsunami was triggered by an earthquake off Sumatra. Three hours later, the leading wave struck the coasts of Sri Lanka and India. A computer model shows the wave. At Banda Aceh, the wave front reached heights of 15 to 30 m.

(b) This snapshot shows the wave rushing toward the coast of Sumatra. Recession of water in advance of the wave exposed a reef.

(c) Here, the wave blasts through a grove of palm trees as it strikes the coast of Thailand.

(d) Satellite photos of the Indonesian province of Aceh before and after the tsunami struck. Note that the city has been washed away and the beach has vanished.

The 2004 Indian Ocean tsunami, the first to be recorded on video, alerted the world to how catastrophic tsunami damage can be **(Fig. 8.28)**. When the tsunami approached Banda Aceh, a city at the northern end of the island of Sumatra, the sea receded much farther than anyone had ever seen, exposing large areas of reefs that normally remained submerged even at low tide. People walked out onto the exposed reefs in wonder. Then, a

Figure 8.29 Damage caused by the tsunami that followed the March 2011 Tōhoku earthquake in Japan.

(a) The tsunami completely destroyed this Japanese coastal town.

Before

Reactor building 4

(b) The Fukushima Daiichi nuclear power plant before the tsunami. Each of these cubic buildings houses a reactor.

After

Reactor building 4

(c) Electrical failures led cooling water to overheat, causing hydrogen explosions to destroy the reactor buildings.

wall of frothing water appeared offshore. With a rumble that grew to a roar, the tsunami approached. Friction with the seafloor slowed it to less than 30 km (20 miles) per hour, but it still moved faster than people could run. In places, the wave front reached heights of 15 to 30 m (45 to 100 feet) as it overwhelmed the beach. The impact of the water ripped boats from their moorings, snapped trees, battered buildings into rubble, and tossed cars and trucks like toys. Water just kept coming, eventually flooding land as far as 7 km (4 miles) inland. When gravity finally pulled the water back seaward, it carried with it a massive jumble of flotsam as well as the bodies of its unfortunate victims. Sadly, the horror of Banda Aceh was just a preamble to the devastation that would soon visit other stretches of Indian Ocean coast. Tsunamis crossed the Indian Ocean and struck Sri Lanka 2.5 hours after the earthquake, the coast of India half an hour after that, and the coast of Africa, on the west side of the Indian Ocean, 5.5 hours after the earthquake. In the end, more than 230,000 people died that day.

The tsunami that struck Japan soon after the 2011 Tōhoku earthquake was also vividly captured in high-definition video that was seen throughout the world, generating a new level of international awareness. The 10-m (30-foot)-high seawalls built to protect Japan's coast were no match for the advance of the wave, which rose to heights of 10 to 30 m (30 to 90 feet) when it reached shore. Racing inland, the wave picked up dirt and debris and evolved into a viscous slurry with the consistency of wet mud, so nothing could withstand its impact. It devastated coastal towns so completely that they looked as though they had been flattened by nuclear bombs (Fig. 8.29a).

But even when the tsunami receded, the catastrophe was not over. The tsunami inundated the Fukushima Daiichi nuclear power plant, where it not only destroyed power lines, cutting the plant off from the electrical grid, but also drowned the backup diesel generators (Fig. 8.29b). As a result, the water pumps driving the plant's cooling system stopped functioning, and the water surrounding the plant's hot reactor cores boiled away. Some of the water became so hot that H_2O molecules separated into H_2 and O_2 gas which, when ignited by a spark, exploded. Explosions blew the tops off three of the plant's four containment buildings, contaminating the surroundings with radioactivity (Fig. 8.29c).

Disease

Once the ground shaking and fires have ceased, disease may still threaten lives in an earthquake-damaged region. Ground movement breaks water and sewer lines, thereby destroying clean-water supplies and exposing the public to pathogens (Box 8.2). Ground rupture and landslides may cut transportation lines, preventing food and medicine from reaching the stricken area. The severity of such

Box 8.2   Consider this . . .

The 2010 Haiti catastrophe

Haiti sits astride the transform boundary along which the North American Plate moves westward at about 2 cm (0.8 inch) per year relative to the Caribbean Plate **(Fig. Bx8.2a)**. Therefore, earthquakes in Haiti are inevitable. But the last major earthquakes on this plate boundary had happened over 200 years ago, so by 2010, elastic deformation and associated stress had been building for quite some time. On the sunny afternoon of January 12, at 4:53 P.M., a 70-km (45-mile)-long segment of a large strike-slip fault suddenly slipped as much as 4 m (12 feet). The motion began at 25 km (15 miles) west-southwest of Port-au-Prince, the capital of Haiti, and 13 km (8 miles) beneath the ground surface **(Fig. Bx8.2b)**. The seismic waves of the resulting M_w 7 earthquake caused the ground to lurch violently for about 35 seconds.

The impact of an earthquake on society depends not only on its size, but also on the nature of the ground materials, on the steepness of the slopes, on construction practices, and on the quality of emergency services in the affected area. Much of Port-au-Prince sits on a basin of weak sediment, which amplified ground movements—in fact, beneath the harbor, sediment liquefied, causing wharfs to sink into the sea. Sadly, the city's buildings were not designed to withstand ground vibration, so many collapsed into rubble **(Fig. Bx8.2c)**. In addition, some of the city's neighborhoods were perched on steep slopes, which slid downhill during the quake, carrying the neighborhoods with them **(Fig. Bx8.2d)**. When the shaking finally stopped, most of Port-au-Prince had collapsed. As a dense cloud of white dust slowly rose over the rubble, survivors began the frantic scramble to dig out victims, a task made more hazardous by aftershocks, of which over 50 had magnitudes between 4.5 and 6.1. The renewed shaking caused still-standing but weakened structures to collapse on rescuers. Some estimates place the death toll at 230,000. In the days that followed, local emergency services were overwhelmed, and access to the victims was nearly impossible. An air caravan of aid arrived to help out, but even so, in the months that followed, cholera spread. Years after the event, recovery from the earthquake remains incomplete.

Figure Bx8.2   The January 2010 earthquake in Haiti and its geologic setting.

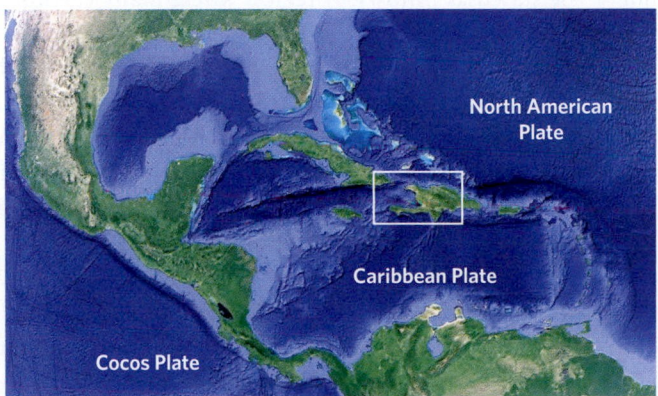

(a) The Caribbean Plate has complex boundaries, delineated by bathymetric features. The white rectangle shows the location of Haiti

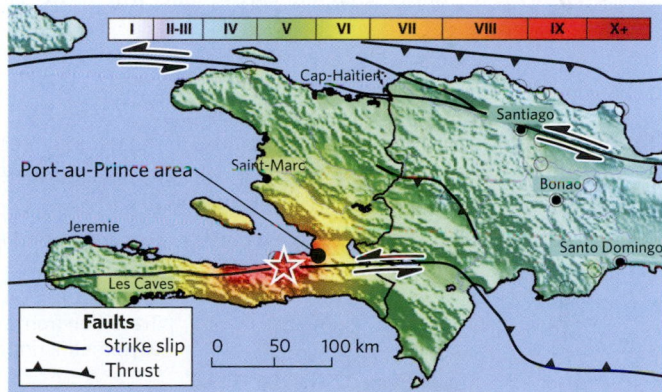

(b) A map showing the intensity of shaking in Haiti. The star marks the epicenter of the quake.

(c) Survivors salvage what they can in Port-au-Prince, the capital of Haiti, after the devastating earthquake.

(d) Ground shaking during the earthquake caused most of the houses in this residential neighborhood to collapse.

problems may exceed the ability of emergency services to cope, so it may take months to years for daily life in a damaged area to return to normal.

Take-home message . . .

🏠 Earthquakes cause devastation in many ways. Ground shaking, landslides, and sediment liquefaction disrupt landscapes and cause buildings to crumble. Though ground shaking itself cannot kill people, falling debris can. In coastal areas, tsunamis may wash over broad areas of low land. Fire and disease may follow earthquakes, causing even more loss of life.

Quick Question -
How does a tsunami differ from a storm wave?

8.7 Can We Predict the "Big One?"

> We learn geology the morning after the earthquake.
>
> —RALPH WALDO EMERSON
> (AMERICAN POET, 1803–1882)

Can seismologists predict earthquakes? The answer depends on the time frame of the prediction. With our present understanding of the distribution of seismic belts and the frequency at which earthquakes occur, we can make *long-term predictions* (on the time scale of decades to centuries). For example, we can say with some certainty that an earthquake will rattle Istanbul, but not north-central Canada, during the next century. Seismologists cannot, however, make accurate *short-term predictions* (on the time scale of hours to years). We cannot say, for example, that an earthquake will happen in San Francisco 40 days from now. New technologies, however, allow seismologists to give people a warning seconds to minutes before seismic waves strike. In this section, we look at the scientific basis of long-term predictions and of earthquake early-warning systems.

Long-Term Predictions

A long-term prediction estimates the probability, or likelihood, that an earthquake will happen during a specified time period. For example, a seismologist may say, "The probability of a major earthquake occurring in the next 50 years in this state is 20%." This sentence implies that there's a 1-in-5 chance that an earthquake will happen before 50 years have passed. Seismologists refer to studies leading to long-term predictions as *seismic-risk assessment*. Urban planners use such assessments to design building codes for a region: codes requiring stronger buildings make sense for regions with greater seismic risk, for the chance that the building will be shaken during its lifetime is greater.

The basic premise of seismic-risk assessment can be stated as follows: A region where many earthquakes have occurred in the past is likely to experience earthquakes in the future. Seismic belts—regions where many earthquakes happen—are, therefore, regions of high seismic risk (see Figure 8.13). This doesn't mean that disastrous earthquakes can't happen far from a seismic belt—they can and do—but the probability that an earthquake will happen in such places in any given time window is less.

To provide a more specific sense of earthquake likelihood, seismologists try to specify the **recurrence interval**, the average time between successive events, for earthquakes of a given size in a region. Since earthquakes do not happen periodically, meaning at predictably spaced time intervals, an earthquake recurrence interval does not give the exact time between successive events. Because the idea of recurrence intervals can be confusing, instead, seismologists sometimes specify the *annual probability*, meaning the likelihood that an earthquake will happen in a given year, where

$$annual\ probability = 1 \div recurrence\ interval$$

For example, if the recurrence interval for an M_w 7 earthquake in a region is 100 years, then the annual probability of such an earthquake is 1/100, or 1%. This means that there is a 1 in 100 chance of an M_w 7 earthquake happening in a given year. Note, however, that because stress builds up over time on a fault, the elastic-rebound

Figure 8.30 Evidence of past earthquakes in the geologic record is used to determine recurrence intervals. The block shows subsurface sediment layers near an active fault. Buried sand volcanoes and disrupted beds (indicated by numbers) indicate when earthquakes happened in the past.

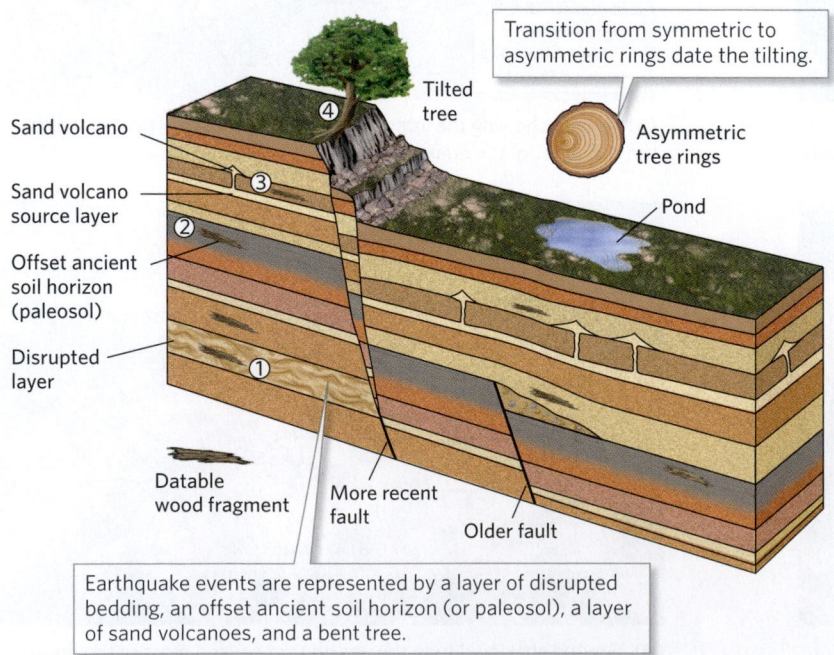

Transition from symmetric to asymmetric rings date the tilting.

Tilted tree

Asymmetric tree rings

Pond

Sand volcano

Sand volcano source layer

Offset ancient soil horizon (paleosol)

Disrupted layer

Datable wood fragment

More recent fault

Older fault

Earthquake events are represented by a layer of disrupted bedding, an offset ancient soil horizon (or paleosol), a layer of sand volcanoes, and a bent tree.

Figure 8.31 Examples of seismic-hazard maps.

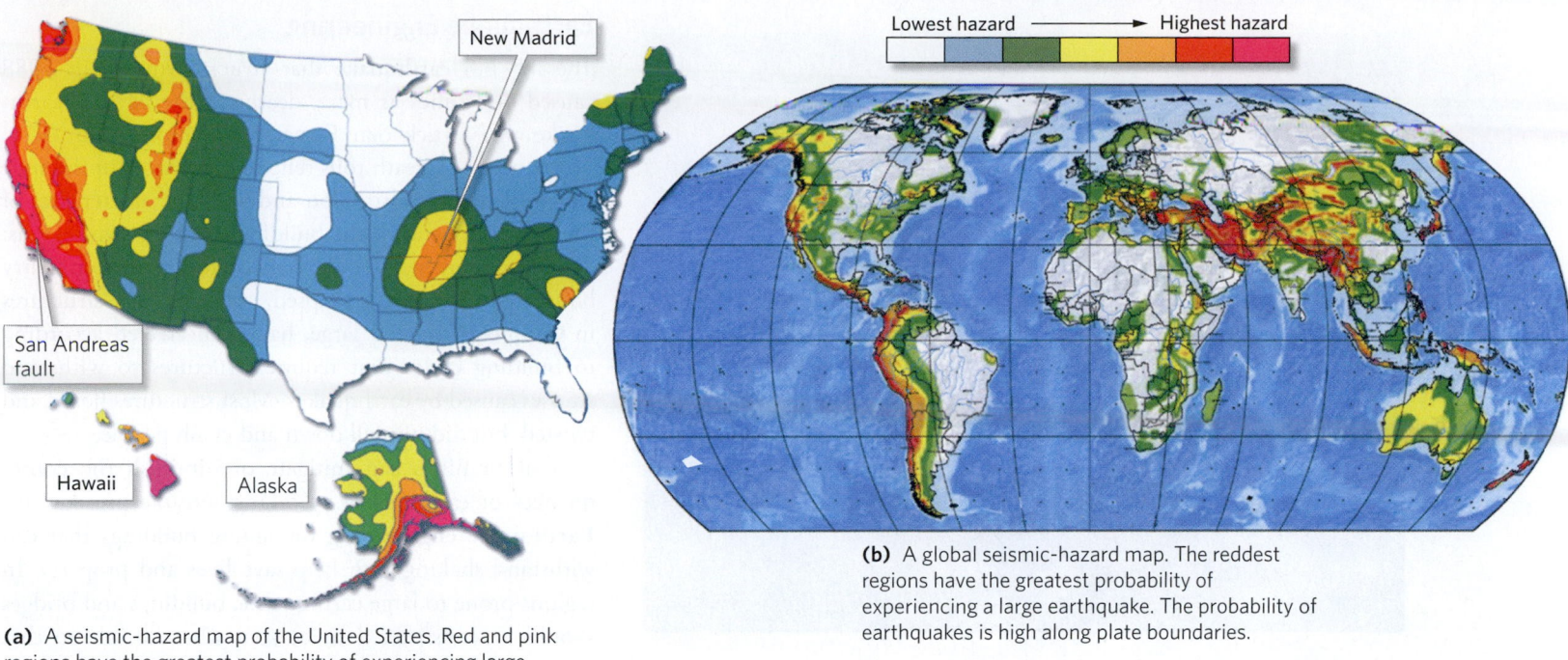

Lowest hazard ➝ Highest hazard

New Madrid

San Andreas fault

Hawaii   Alaska

(a) A seismic-hazard map of the United States. Red and pink regions have the greatest probability of experiencing large earthquakes. The area around the San Andreas fault is particularly prone to large earthquakes. Hazard in the New Madrid area remains subject to debate.

(b) A global seismic-hazard map. The reddest regions have the greatest probability of experiencing a large earthquake. The probability of earthquakes is high along plate boundaries.

theory hints that the annual probability of an earthquake may progressively increase as time passes.

To determine the recurrence interval for large earthquakes within a given seismic belt, seismologists must determine when previous earthquakes happened within that belt. For places where the historical record does not extend far enough back in time to reveal multiple large events, researchers look for evidence of large earthquakes preserved in the geologic record. For example, in places where sedimentary strata accumulated in a basin over a fault, researchers may dig a trench and look for buried layers of sand volcanoes or disrupted bedding in the stratigraphic record. Each such layer, whose age can be determined by using radiocarbon dating of plant fragments (see Chapter 9), records the time of an earthquake **(Fig. 8.30)**. Seismologists calculate the number of years between successive events and then calculate the average to obtain the recurrence interval. Information on recurrence intervals allows seismologists to refine regional maps illustrating seismic hazard **(Fig. 8.31)**.

Earthquake Early-Warning Systems

Short-term predictions, specifying that an earthquake will happen on a given date or within a time window of days to years, are bogus. In fact, such predictions will probably never be reliable. The concept of short-term

prediction should not be confused, however, with the concept of an *earthquake early-warning system*, which is based on a real signal and can potentially save lives. An early-warning system works as follows: When an earthquake happens, the seismic waves it produces start traveling through the Earth. Seismometers positioned between the epicenter and a city will detect seismic waves before they have had time to reach the city. The instant that these seismometers detect the earthquake, a transmitter sends a signal to a control center, which automatically broadcasts emergency signals to the city. These signals, which travel at the speed of light, reach the city several seconds to a minute before the seismic waves. The arrival of the signals can trigger automatic shutdowns, of gas pipelines, trains, nuclear reactors, and power lines. The signals can also set off sirens and trigger broadcasts on radio, TV, and cellular networks, alerting people to take precautions.

Because tsunamis are so dangerous, predicting their arrival can save thousands of lives. At the tsunami early-warning center in Hawaii, observers keep track of earthquakes around the Pacific and use data relayed from tide gauges, buoys, and seafloor pressure gauges to determine whether a particular earthquake has generated a tsunami **(Fig. 8.32)**. If the observers detect a tsunami, they flash warnings to authorities around the Pacific.

Figure 8.32 Buoys can detect a tsunami in the open ocean so that people on land can be warned.

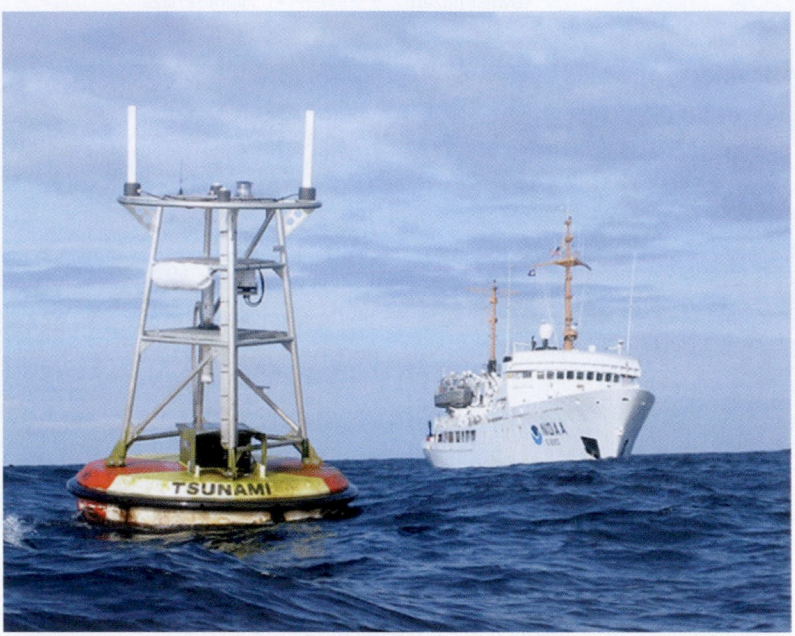

Take-home message . . .

🏠 Seismologists can determine, in general, where earthquakes are most likely to take place, but cannot predict exactly when and where an earthquake will occur. Seismic risk is greater where seismicity has happened more frequently in the past and, therefore, where the recurrence interval for large earthquakes is shorter. Early-warning systems can provide seconds of warning by sending out signals that travel faster than seismic waves.

Quick Question -
What is the relationship between recurrence interval and annual probability?

8.8 Prevention of Earthquake Damage and Casualties

The destruction and death caused by an earthquake of a given size depend on a number of factors, including: the proximity of the epicenter to a population center; the depth of the focus; the style of building construction in the epicentral region; the steepness of slopes, whether faulting displaced the seafloor; the proximity of the affected region to the sea; whether building foundations are on solid bedrock or on weak materials; whether the earthquake happened when people were outside or inside; and whether the government was able to provide emergency services promptly. To minimize the calamity of an earthquake, people can strive to build stronger structures and to choose safer sites to build on.

Earthquake Engineering

The M_w 6.8 earthquake that struck Armenia in 1988 caused 400 times as many deaths as the M_w 6.7 earthquake that struck San Fernando, California, in 1971. The contrast in death tolls reflects differences in the style and quality of construction and in the characteristics of the substrate beneath the buildings in these two regions. The unreinforced concrete-slab buildings and masonry houses in Armenia collapsed, whereas the structures in California, by and large, had been erected according to building codes that require structures to withstand stresses caused by earthquakes. Most structures flexed and twisted, but did not fall down and crush people.

Communities can mitigate or diminish the consequences of earthquakes by taking sensible precautions. **Earthquake engineering** (designing buildings that can withstand shaking) can help save lives and property. In regions prone to large earthquakes, buildings and bridges should be constructed to be somewhat flexible so that ground shaking cannot crack them, but they should have sufficient bracing to dampen their movements **(Fig. 8.33a, b)**. In addition, supports should be strong enough to stand up to the weight of floors that are not static, but may drop down after having bounced upward. In some cases, simple changes in construction practices can make a building stronger. For example, wrapping steel cables around bridge support columns makes them many times stronger, bolting bridge spans to the tops of support columns prevents the spans from bouncing off, bolting buildings to foundations keeps them in place, and adding diagonal braces to frames keeps them from twisting and shearing too much.

In regions with significant seismic risk, certain kinds of construction should be avoided. For example, concrete-block, unreinforced concrete, and unreinforced brick buildings crack and tumble under conditions in which wood-frame, steel-girder, or reinforced concrete buildings remain standing. Traditional heavy, brittle tile roofs can shatter and bury the inhabitants inside, whereas lightweight sheet-metal roofs do not. Inadequate structures can be made safer by *seismic retrofitting*, the process of strengthening existing structures.

Earthquake Zoning

Urban planners in seismically active areas can decrease hazard by **earthquake zoning**. Zoning relies on an assessment of where land is stable and where it is not. Zoning should lead to regulations that discourage building on land underlain by weak mud or wet sand that could liquefy. Similarly, zoning should discourage building on top

Figure 8.33 Preventing earthquake damage and casualties.

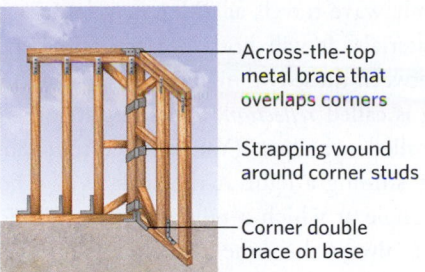

Across-the-top metal brace that overlaps corners

Strapping wound around corner studs

Corner double brace on base

Adding corner struts, braces, and connectors can substantially strengthen a wood-frame house.

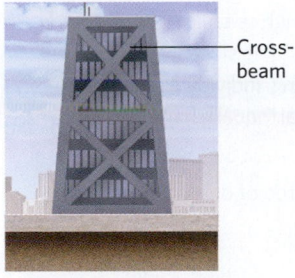

Cross-beam

Buildings are less likely to collapse if they are wider at the base and if crossbeams are added for strength.

Unreinforced building: insufficient shear strength

Reinforced building: Sufficient shear strength

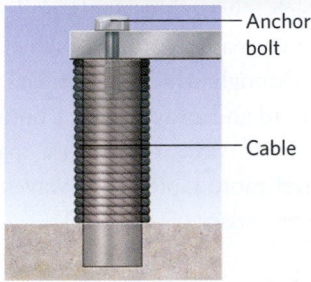

Anchor bolt

Cable

Wrapping a bridge's support columns in cable and bolting the span to the columns will prevent the bridge from collapsing so easily.

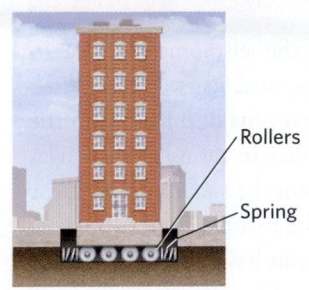

Rollers

Spring

Placing buildings on rollers or shock absorbers lessens the severity of the vibrations.

(a) Damage to structures can be prevented if those structures are designed to withstand vibration.

(b) An unreinforced building will shear side to side in a way that causes floors to shift out of alignment.

(c) If an earthquake strikes, take cover under a sturdy table near a wall.

Did you ever wonder . . .

Can buildings be designed to survive earthquakes?

of, on, or at the base of steep escarpments where landslides might take place, or where the rupture of a dam could cause a flood. Finally, zoning should prevent construction of critical buildings (schools, hospitals, fire stations, communications centers, power plants) over active faults, for fault slip could crack and destroy the buildings.

Prevention of Casualties

Even careful construction and planning can't prevent all earthquake casualties. Therefore, communities in seismic belts should draw up emergency plans to deal with disaster. They should put in place strategies to provide personnel, equipment, and supplies for rescue and recovery.

Individuals living in regions of high seismic risk should also take personal responsibility to protect themselves and their homes from earthquake damage. Simple precautions include bolting or strapping bookshelves and water heaters to walls, adding locking latches to cabinets, and knowing how to shut off the gas and electricity and where to find family members. Schools, factories, and offices should hold earthquake-preparedness drills, and individuals should know where to go to seek protection from falling objects (Fig. 8.33c). As long as lithosphere plates continue to move, earthquakes will continue to shake us. But we can reduce the chances of damage or injury by being prepared.

8.9 Seismic Study of the Earth's Interior

Decades before the invention of the seismometer, researchers had concluded, based on measurements of the Earth's mass and shape, that the Earth consists of three concentric layers that differ from one another in terms of their relative density (see Chapter 1). From the surface down, these layers are (1) the crust, (2) the mantle, and (3) the core **(Fig. 8.34)**. To go beyond this basic understanding to define the specific depths of the boundaries between layers and to characterize the properties of the layers, researchers needed a tool that would allow them to "see" inside the Earth. The study of seismic waves provides such a tool. By measuring how fast seismic waves travel through the Earth, and how they reflect or bend as they travel, seismologists have been able to provide a much more refined image of the Earth's interior.

Controls on the Velocity and Bending of Seismic Waves

The ability of a seismic wave to travel through a material, as well as the velocity at which the wave travels, depends on several characteristics of the material. Factors such as *density* (mass per unit volume), *rigidity* (how stiff, or resistant to bending, a material is), and *compressibility* (how easily a material's volume changes in response to squashing) all affect seismic-wave velocity and therefore the **travel time** of a set of waves, meaning the time it takes them to move from the epicenter to a specific seismometer. Studies of seismic waves reveal the following:

- Seismic waves move at different velocities in different rock types **(Fig. 8.35a)**. For example, P-waves travel 8 km per second in peridotite, but only 3.5 km per second in sandstone. Therefore, waves speed up or slow down as they pass from one rock type into another.
- P-waves travel more slowly in a liquid than in a solid of the same composition. Therefore, P-waves travel more slowly in magma than in solid rock, and more slowly in molten iron alloy than in solid iron alloy **(Fig. 8.35b)**.

- Both P-waves and S-waves can travel through a solid, but only P-waves can travel through a liquid **(Fig. 8.35c)**.

If a seismic wave travels at different velocities in two different materials, it will both bounce and bend at a boundary between those two materials. The phenomenon of bouncing is called *reflection*, and the phenomenon of bending is called *refraction*. (You can see reflection and refraction by shining a light at the surface of a body of water.) The angle at which a reflected wave bounces off a boundary is always the same as the angle at which the incoming, or *incident*, wave strikes that boundary. But the angle by which a refracted wave bends at a boundary depends both on the contrast between the wave velocities in the materials above and below the boundary and on the angle at which a wave hits the boundary. As a rule, as waves pass from a material through which they travel rapidly into one through which they travel more slowly, they bend downward and away from the boundary **(Fig. 8.36a)**. Alternatively, if waves pass into a material through which they travel more rapidly, the waves bend upward and toward the boundary **(Fig. 8.36b)**.

Identifying the Earth's Layers, Seismically

Studies of seismic-wave velocity, refraction, and reflection have been used to locate the major boundaries between layers inside the Earth. Let's see how.

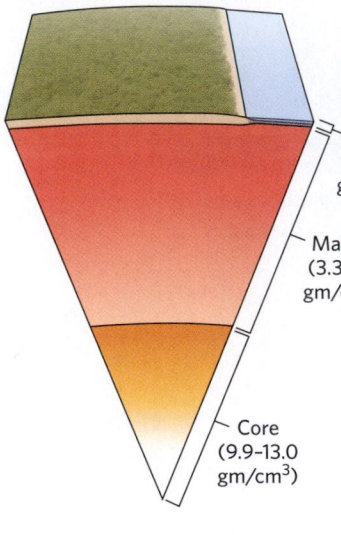

Figure 8.34 Simplified image of the Earth's interior, first proposed in the 19th century. The Earth has a crust, mantle, and core.

Crust
(2.7–3.3 gm/cm³)

Mantle
(3.3–5.7 gm/cm³)

Core
(9.9–13.0 gm/cm³)

Figure 8.35 The propagation of earthquake waves.

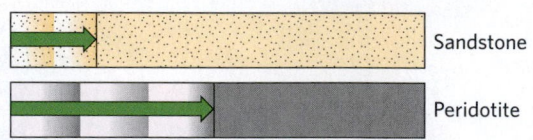

Sandstone

Peridotite

(a) Seismic waves travel at different velocities in different rock types. After a given time, a wave will have traveled farther in peridotite than in sandstone.

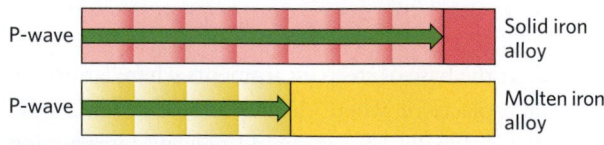

P-wave — Solid iron alloy

P-wave — Molten iron alloy

(b) P-waves travel faster in solid iron alloy than in liquid of the same composition, such as molten iron alloy.

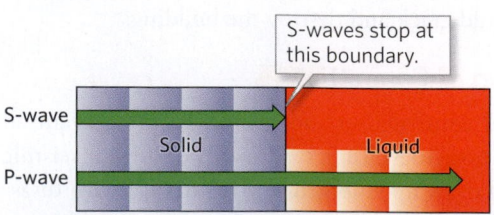

S-waves stop at this boundary.

S-wave
Solid   Liquid
P-wave

(c) Both P-waves and S-waves can travel through a solid, but only P-waves can travel through a liquid.

Figure 8.36 Refraction and reflection of seismic waves. The angle at which a wave is reflected by a boundary between two materials is always the same as the angle at which the incident wave strikes that boundary.

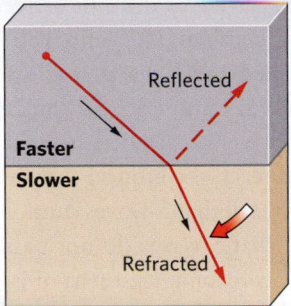

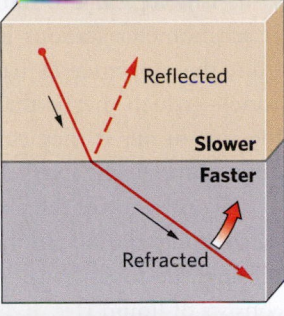

(a) A wave that enters a material through which it travels more slowly bends away from the boundary.

(b) A wave that enters a material through which it travels more rapidly bends toward the boundary.

DISCOVERING THE CRUST-MANTLE BOUNDARY. In 1909, Andrija Mohorovičić, a Croatian seismologist, noted that P-waves arriving at seismometer stations less than 200 km (125 miles) from the epicenter of an earthquake traveled at an average speed of 6 km (4 miles) per second, whereas P-waves arriving at seismometers more than 200 km from the epicenter traveled at an average speed of 8 km (5 miles) per second. To explain this observation, he suggested that P-waves reaching nearby seismometers followed a shallow path that kept them entirely within the crust, in which they traveled relatively slowly, whereas P-waves reaching distant seismometers traveled through the mantle for part of their route, and that their velocity was higher in the mantle. He realized that the waves refracted as they crossed the crust-mantle boundary **(Fig. 8.37a, b)**. Mohorovičić was able to calculate the depth of the crust-mantle boundary from this observation, and he proposed that beneath continents, it occurred at a depth of 35 to 40 km (22 to 25 miles). Later studies showed that the depth of the crust-mantle boundary beneath continents varies from 25 to 70 km (15 to 44 miles), and beneath oceans from 7 to 10 km (4 to 6 miles). The crust-mantle boundary is now called the **Moho**, in honor of Mohorovičić.

DISCOVERING THE STRUCTURE OF THE MANTLE. Seismologists have determined that seismic waves travel at different speeds at different depths in the mantle **(Fig. 8.38)**. Specifically, between depths of about 100 and 200 km (60 and 125 miles) beneath the seafloor, seismic velocities are lower than in the overlying portion of the mantle. This 100- to 200-km-deep layer is now known as the **low-velocity zone** (**LVZ**). Researchers suggest that the LVZ corresponds to a layer in which mantle rock has undergone slight partial melting. Because seismic waves travel more slowly through liquids than through solids, even a tiny bit of melt slows them down. Simplistically, the top of the LVZ delineates the base of the lithosphere and the top of the asthenosphere beneath oceanic plates. Seismologists do not find a well-developed LVZ beneath continents.

Below about 200 km, seismic-wave velocities in the mantle, beneath both continents and oceans, increase with depth (see Fig. 8.38). Seismologists interpret this increase in velocity with depth to mean that mantle peridotite becomes progressively less compressible, more rigid, and denser with depth. This interpretation makes sense, considering that the weight of overlying rock increases with depth, and that as pressure increases, the atoms making up minerals squeeze together more tightly and are less free to move.

At depths between 410 km and 660 km (255 and 410 miles), seismic velocity in the mantle increases in a series of abrupt steps (see Fig. 8.38). A major step occurs at a depth of 660 km. Experiments suggest that such **seismic-velocity discontinuities** occur at depths where pressure causes atoms in minerals to rearrange into more compact minerals of the same composition, a phenomenon called a *phase change* (see Chapter 6). Seismic-velocity discontinuities serve as the basis for subdividing the mantle into the **upper mantle** (above 660 km) and the **lower mantle** (below 660 km). The lowest portion of the upper mantle, between 410 and 660 km, in which several small seismic discontinuities occur, is called the **transition zone**.

DISCOVERING THE STRUCTURE OF THE CORE. In the early 20th century, researchers installed seismometers at many stations around the world, expecting to be able to record

Figure 8.37 Discovery of the Moho.

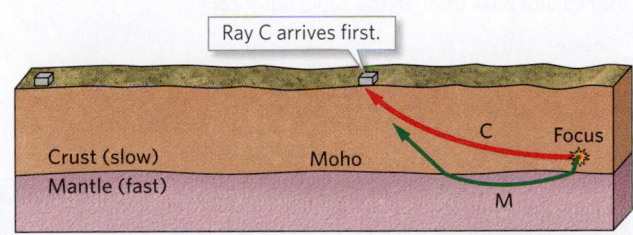

(a) Seismic waves traveling only through the crust reach a nearby seismometer first.

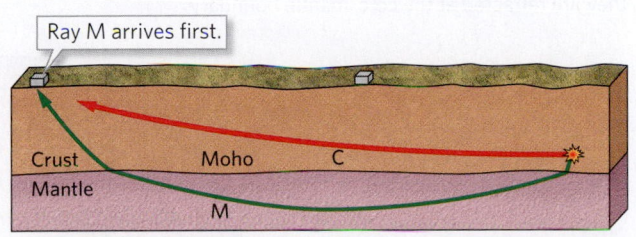

(b) Seismic waves traveling for most of their path through the mantle reach a distant seismometer first.

Figure 8.38 The velocity of P-waves changes with depth in the mantle because the physical properties of the mantle change with depth.

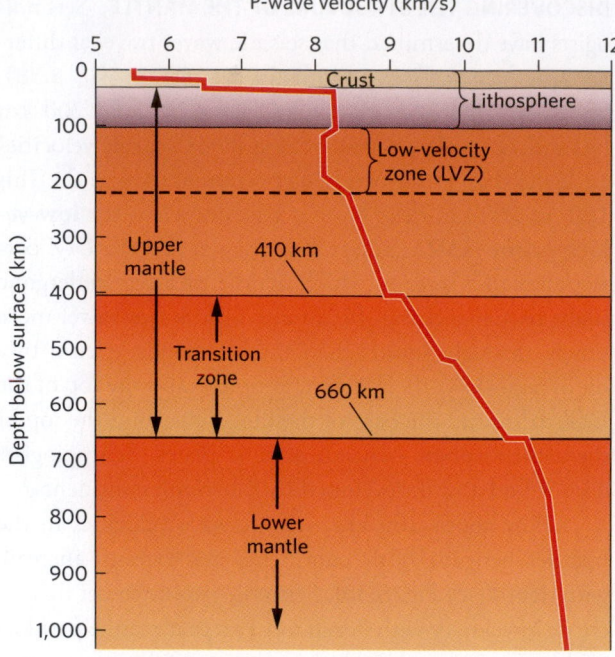

where seismic waves abruptly refract downward because their velocity suddenly decreases. The dimensions of the shadow zone allowed seismologists to calculate that this boundary lies at a depth of 2,900 km (1,800 miles), and they consider it to be the **core-mantle boundary**.

Seismologists also found that S-waves did not arrive at seismometer stations located between 103° and 180° from the epicenter (a band called the **S-wave shadow zone**)—which means that S-waves cannot pass through the core at all. If they could, an S-wave headed straight down through the Earth should reach the ground surface on the other side of the planet. Recall that S-waves cannot pass through liquid, so the fact that S-waves do not pass through the core means that the core, or at least part of it, consists of liquid **(Fig. 8.39b)**.

At first, seismologists thought that the entire core might be liquid iron alloy. But in 1936, a Danish seismologist, Inge Lehmann, discovered that P-waves passing through the core reflected off a boundary within the core. She proposed that the core includes two parts: an **outer core** consisting of liquid iron alloy, and an **inner core** consisting of solid iron alloy. The depth of the boundary between the inner and outer core was eventually located by measuring the time it takes for seismic waves to penetrate the Earth, bounce off the inner core–outer core boundary, and return to the ground surface **(Fig. 8.39c)**. These measurements place the boundary at a depth of about 5,155 km (3,200 miles).

waves produced by a large earthquake anywhere on Earth. In 1914, one of these researchers discovered that P-waves from a given earthquake did not arrive at seismometers within a band between 103° and 143°, as measured along the curve of Earth's surface from the earthquake epicenter. This band is now called the **P-wave shadow zone (Fig. 8.39a)**. The presence of the P-wave shadow zone means that deep in the Earth, a major boundary exists

Figure 8.39 Shadow zones and the discovery of the Earth's core. (Note that the circumference of a circle is 360°, so it is 180° from a given point to a point on the other side of the planet.)

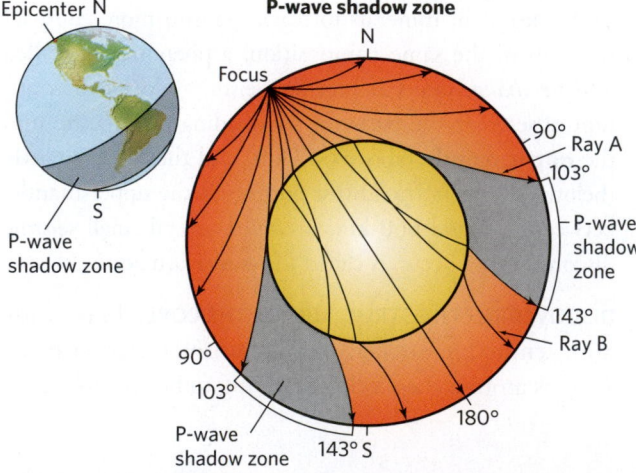

(a) P-waves do not arrive in the P-wave shadow zone because they are refracted at the core-mantle boundary.

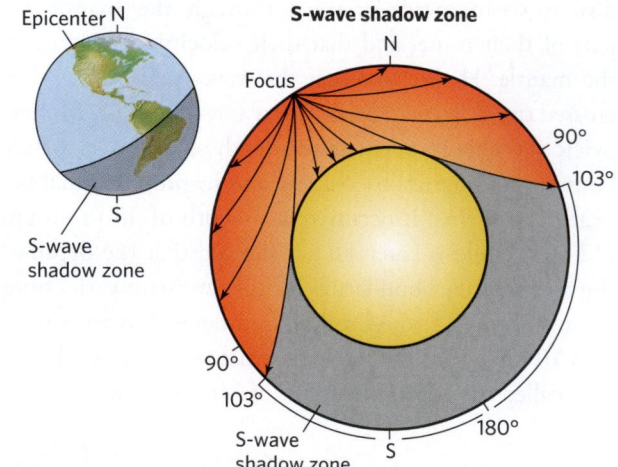

(b) S-waves do not arrive in the S-wave shadow zone because they cannot pass through the liquid outer core.

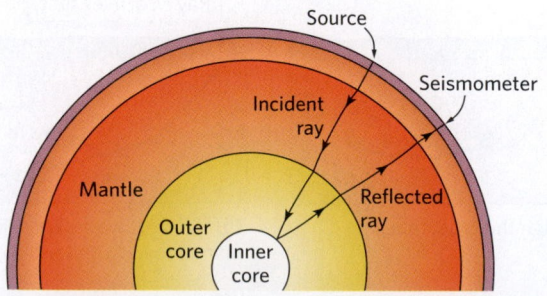

(c) Seismic waves reflect off the inner core–outer core boundary.

A MODERN IMAGE OF EARTH'S LAYERS. Through painstaking effort, seismologists have compiled data on seismic-wave travel times to develop a graph, known as a *velocity-versus-depth curve*, that shows the average depths at which seismic-wave velocity suddenly changes and the average amount of change. Depths at which major changes take place define the principal layers and sublayers of the Earth down to its center **(Fig. 8.40)**.

More detailed studies in recent years have shown that the onion-like layered model of the Earth we've described so far is an oversimplification. Using a technique called *seismic tomography*, seismologists can produce three-dimensional images of seismic-velocity variation in the Earth's interior, just as doctors produce three-dimensional CT (computerized tomography) scans of the human body **(Fig. 8.41a)**. Tomographic studies allow seismologists to identify regions in the mantle where seismic waves travel faster or slower than expected, and these studies have led to the realization that the velocities of seismic waves vary significantly with location at a given depth. The regions of lower velocity probably represent warmer mantle material, and the regions of higher velocity probably represent cooler mantle material, for as rock gets hotter and softer,

Figure 8.40 A graph showing how the velocities of P-waves and S-waves vary with increasing depth in the Earth. Note that the graph does not show a velocity for S-waves in the outer core because S-waves cannot travel through molten iron (a liquid).

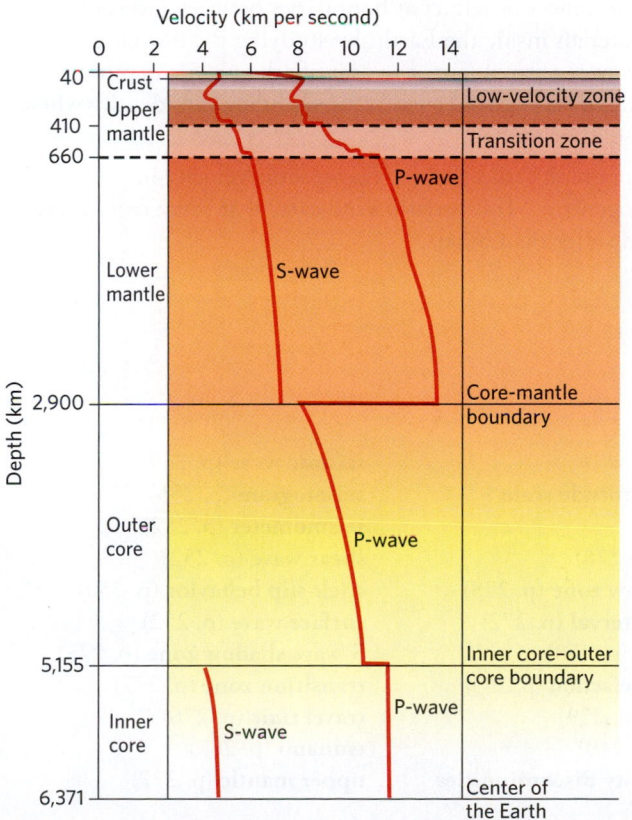

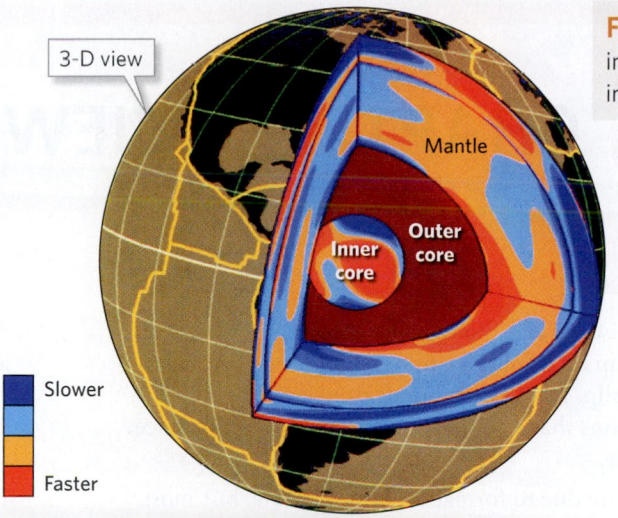

Figure 8.41 Modern images of the Earth's interior.

(a) Three-dimensional tomographic image of the Earth. Areas of lower seismic velocity may be relatively warmer than their surroundings, and areas of higher velocity may be relatively cooler.

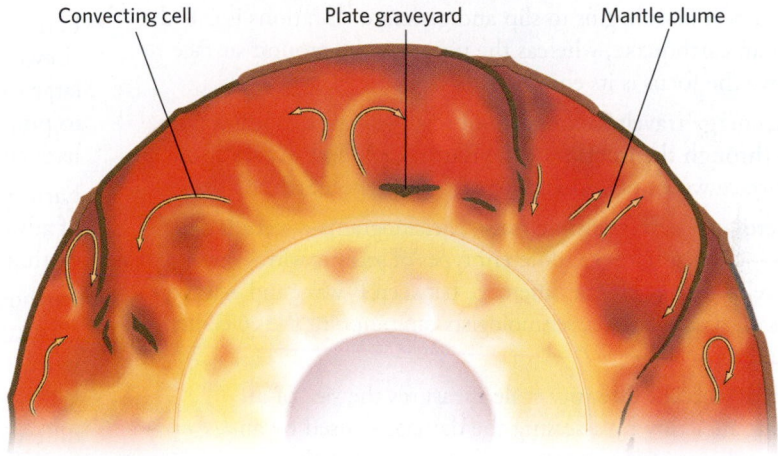

(b) A conceptual image of what the Earth's interior looks like. Hotter mantle material undergoes upwelling, and cooler mantle material sinks. Plates form at the surface and sink back down into the mantle.

it transmits seismic energy less rapidly. The occurrence of warmer and cooler regions is a consequence of convection in the mantle **(Fig. 8.41b)**. The Earth's interior is indeed a dynamic place!

Take-home message . . .

By studying the velocity at which seismic waves travel through the Earth's interior, and by determining the depths at which seismic waves reflect or refract, seismologists have been able to locate the boundaries between Earth's layers, such as the Moho between the crust and mantle. Such work has also demonstrated that the outer core is molten. Using modern techniques, researchers can even identify patterns of convection in the mantle.

Quick Question -
What is the S-wave shadow zone, and why does it exist?

8 CHAPTER REVIEW

- Earthquakes are episodes of ground shaking. They are generally a consequence of slip on a fault, for prior to slip, rock deforms elastically. During slip, the rock rebounds to its original shape, sending out vibrations that, when they reach the ground surface, cause an earthquake.

- Some earthquakes are due to formation of new faults, but most happen when stress overcomes friction on a pre-existing fault and the fault slips. Faults typically exhibit stick-slip behavior as stress builds up until they move in sudden increments.

- The place where a fault begins to slip and generate vibrations is called the focus of an earthquake, whereas the point on the ground surface directly above the focus is its epicenter.

- Earthquake energy travels in the form of seismic waves. Body waves, which pass through the interior of the Earth, include P-waves and S-waves. Surface waves pass along the surface of the Earth.

- A seismometer can detect seismic waves. Seismograms—records of earthquakes—demonstrate that different types of seismic waves travel at different velocities. Using the difference between P-wave and S-wave arrival times at three locations, seismologists can pinpoint the epicenter of an earthquake.

- The Modified Mercalli Intensity scale measures the size of an earthquake at a locality by assessing the damage caused by an earthquake and people's perception of the ground shaking. Earthquake intensity decreases with distance from the epicenter.

- Magnitude scales, which characterize the amount of energy released at the source of an earthquake, are based on the amount of ground motion, at a reference distance from the epicenter, as indicated on a seismogram. There is one magnitude number for any given earthquake. The Richter scale is an early version of a magnitude scale. These days, seismologists use the moment magnitude (M_w) scale.

- An M_w 8 earthquake yields about 10 times as much ground motion as an M_w 7 earthquake, and it releases about 32 times as much energy.

- Seismicity, meaning earthquake activity, takes place mainly in seismic belts, the majority of which lie along plate boundaries. Intraplate earthquakes happen in the interior of plates.

- Earthquake damage results from ground shaking, landslides, sediment liquefaction, fire, and tsunamis. Disease can afflict people afterward.

- Earthquakes are more likely to take place in seismic belts than elsewhere. Seismologists can determine the recurrence interval for large earthquakes in a particular belt, but it may never be possible to pinpoint the exact time and place at which an earthquake will happen.

- Earthquake early-warning systems can provide seconds to minutes of advance notice that an earthquake is coming by detecting an earthquake before the vibrations reach a city.

- Earthquake damage and loss of life can be reduced with better construction practices and zoning and by educating people about what to do during an earthquake.

- Seismic waves reflect or refract at boundaries between layers of different materials inside the Earth. By studying the movements of seismic waves, seismologists can identify the precise depths at which boundaries between layers occur, and they can identify where material is solid or molten.

- Seismic tomography studies show that seismic velocity in the mantle is not homogeneous. This variation indicates that some regions are warmer and softer than others.

aftershock (p. 250)
amplitude (p. 255)
arrival time (p. 255)
body wave (p. 251)
compressional wave (p. 252)
core-mantle boundary (p. 278)
earthquake (p. 247)
earthquake engineering (p. 274)
earthquake zoning (p. 274)
elastic deformation (p. 249)
elastic-rebound theory (p. 250)
epicenter (p. 251)

fault (p. 249)
focus (p. 251)
foreshock (p. 250)
inner core (p. 278)
intensity (p. 257)
intraplate earthquake (p. 263)
lower mantle (p. 277)
low-velocity zone (LVZ) (p. 277)
magnitude (p. 257)
mainshock (p. 250)
Modified Mercalli Intensity scale (p. 257)

Moho (p. 277)
moment magnitude scale (p. 258)
outer core (p. 278)
P-wave shadow zone (p. 278)
recurrence interval (p. 272)
Richter scale (p. 258)
sediment liquefaction (p. 265)
seismic belt (p. 259)
seismicity (p. 249)
seismic-velocity discontinuities (p. 277)

seismic wave (p. 249)
seismogram (p. 255)
seismometer (p. 252)
shear wave (p. 252)
stick-slip behavior (p. 250)
surface wave (p. 252)
S-wave shadow zone (p. 278)
transition zone (p. 277)
travel time (p. 276)
tsunami (p. 267)
upper mantle (p. 277)

Review Questions

The letters following each Review Question refer to the corresponding Learning Objective from the Chapter Opener.

1. Do all earthquakes require that a new fault be formed? Describe elastic-rebound theory and the concept of stick-slip behavior. **(A)**

2. Label the focus and epicenter of the earthquakes shown in the diagram. **(A)**

3. What is the difference between a body wave and a surface wave? How do P-waves and S-waves differ from each other? How do R-waves and L-waves differ from each other? **(A)**

4. Explain how the ground movements produced by an earthquake are detected by a seismometer. **(C)**

5. Explain the differences among the scales used to describe the size of an earthquake. **(B)**

6. How does seismicity on mid-ocean ridges compare with seismicity at convergent or transform boundaries? Do all earthquakes occur at plate boundaries? **(D)**

7. What is a Wadati-Benioff zone? Why do intermediate-focus and deep-focus earthquakes occur there? **(D)**

8. Describe the types of damage caused by earthquakes. Is all damage due to ground shaking? **(E)**

9. What is a tsunami, and why do tsunamis form? How do they differ from storm waves? **(E)**

10. What is sediment liquefaction, and how can it cause damage during an earthquake? **(E)**

11. How are long-term earthquake predictions made? What is the basis for determining a recurrence interval? **(F)**

12. What is an earthquake early-warning system? **(F)**

13. What types of structures are most prone to collapse in an earthquake? What types are most resistant to collapse? What causes most loss of life during an earthquake? **(G)**

14. Why do seismic waves refract at specific depths within the Earth? What causes P-wave and S-wave shadow zones, and what does their presence imply? Which shadow zone does the figure show? **(H)**

15. What clue led to the definition of the Moho? **(I)**

16. What does a tomographic image of the Earth's interior show? **(I)**

On Further Thought

17. Is seismic risk greater in a town on the western coast of South America or in one on the eastern coast? Explain your answer. **(D)**

18. On the seismogram of an earthquake recorded at a seismometer station in Paris, France, the S-wave arrives 6 minutes after the P-wave. On the seismogram obtained by a station in Mumbai, India, for the same earthquake, the difference between the P-wave and S-wave arrival times is 4 minutes. Which station is closer to the epicenter? From the information provided, can you pinpoint the location of the epicenter? Explain. **(C)**

Online Resources

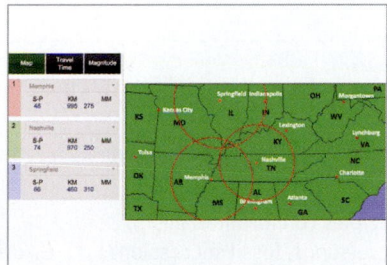

Animations
This chapter features animations that simulate and show the effects of tsunamis and an interactive activity on locating an earthquake's epicenter.

Smartwork5
This chapter features analysis questions on earthquake magnitude, faults, and tsunamis.

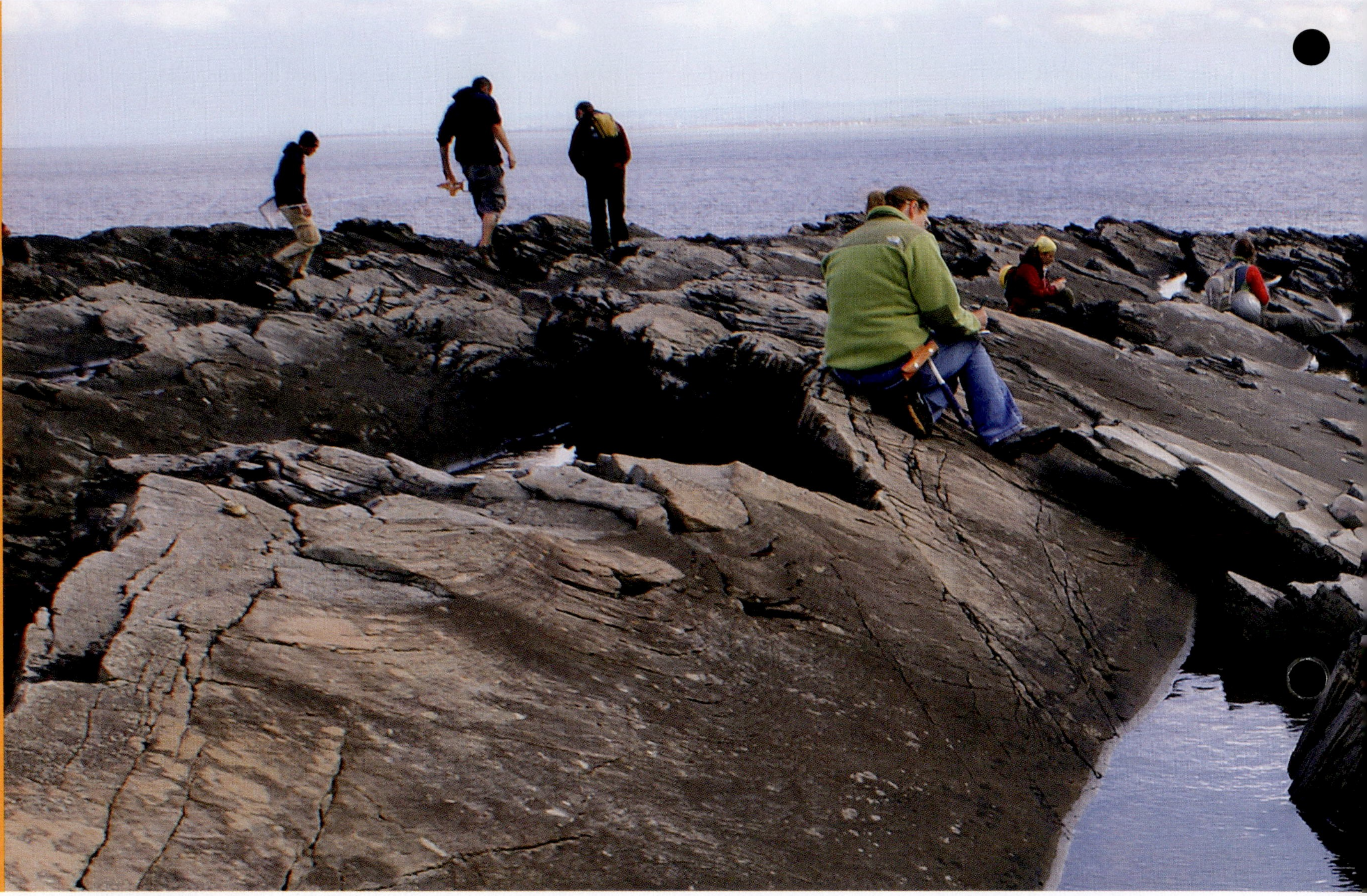

9 DEEP TIME

How Old Is Old?

These dipping layers of sandstone, now exposed along the shore of western Ireland, were deposited almost 100 million years before the first dinosaur took its first step. The length of geologic time is almost unfathomable.

9.1 Introduction

In 1869, a one-armed Civil War veteran named John Wesley Powell, along with nine companions, set out to explore the Grand Canyon, the greatest gorge on Earth. For three months, they drifted down the Colorado River, which flows down the floor of the canyon (Fig. 9.1). During their voyage, seemingly insurmountable walls of rock both imprisoned and amazed the explorers and led them to pose important questions about the Earth and its history, the same questions that tourists visiting the canyon may ponder today: How long did it take to carve the canyon? When did the rocks making up the walls of the canyon form? Was there a time before these rocks accumulated? Thinking about such questions opens the door to thinking about **geologic time**, the span of time since the Earth's formation.

Our modern understanding of geologic time stems from 19th-century work that established principles for describing the relative age of geologic features. By *relative age*, we mean whether one feature is older or younger than another. With this concept in mind, we begin this chapter by discussing relative-age determination. We next introduce fossils, which can help to sort out age relationships, and describe what they tell us about the evolution of life on Earth. Understanding relative ages set the stage for developing the *geologic column*, a chart that divides geologic time into intervals. When geologists developed methods for *isotopic dating* (radiometric dating) in the mid-20th century, it became possible to define the *numerical age*—the age in years—of rocks. Using isotopic dating, geologists determined numerical ages of intervals on the geologic column.

Figure 9.1 John Wesley Powell and his companions floated down the Colorado River, sometimes encountering dangerous rapids, as they explored the Grand Canyon.

This tool led to the production of the *geologic time scale* and, ultimately, to an estimate of when the Earth itself formed.

With the concept of geologic time in hand, a hike down a trail into the Grand Canyon becomes a trip through what popular authors call *deep time*. The

geologic discovery that our planet's history extends billions of years before the start of human history changed humanity's perception of time and the Universe as profoundly as the astronomical discovery that the limit of deep space extends billions of light-years beyond the edge of our Solar System.

9.2 Geologic Principles and Relative Age

The discovery of physical laws by Sir Isaac Newton helped spark the Age of Enlightenment in Europe, a time when scientists began to seek natural, rather than supernatural, explanations for the features and phenomena in the world around them. By the 1850s, geologists had established several principles that ultimately provided a foundation for developing the concept of geologic time.

Uniformitarianism

While wandering in the highlands of Scotland, James Hutton (1726–1797) noticed that many of the features he found in sedimentary rocks resembled features that he could see forming in modern depositional environments. For example, the surfaces of some sandstone beds displayed ripple marks identical to those he saw on a modern beach. These observations led Hutton to speculate that ancient rocks and landscapes were a long-ago product of the same natural processes operating in modern times.

Hutton's idea has come to be known as the principle of **uniformitarianism**. This principle means that physical processes operating in the modern world also operated in the past, at roughly the same rates **(Fig. 9.2)**. More concisely,

Figure 9.2 The principle of uniformitarianism.

(a) Mudcracks formed when a modern-day mud dried up. Note the pen for scale.

(b) A 300-million-year-old rock containing preserved mudcracks. According to the principle of uniformitarianism, they formed the same way as modern mudcracks.

uniformitarianism means that "the present is the key to the past." Hutton further deduced that not all the rocks he observed, or the structures that affected them, could have formed at the same time. Because no one can see the entire process of sediment first turning into rock and then later rising into mountains, Hutton concluded that the production of rocks is very slow, and that the Earth's history must go back a long time before human history began. This proposal, along with several others, became the basis for so much geologic thinking that modern geologists consider Hutton to be "the father of geology."

Determining Relative Ages and Geologic History

Like historians, geologists strive to establish the sequence of events that produced an assemblage of geologic features. The age of one feature with respect to another in a sequence is its **relative age**. The Danish scientist, Nicolas Steno (1636–1686), laid out a set of formal geologic principles, discussed below, which served as a foundation for Hutton's principle of uniformitarianism. Charles Lyell (1797–1875), a British geologist, popularized Steno's ideas along with Hutton's in his book *Principles of Geology*, which was the first comprehensive textbook of geology. This book clarifies how geologic principles, including uniformitarianism, provide a basis for determining relative ages.

- The principle of **original horizontality** states that layers of sediment, when first deposited, are roughly horizontal (**Fig. 9.3a**). Why? Because of gravity, sediments accumulate on relatively flat surfaces, such as floodplains or the seafloor. If they collect on a steep slope, they slide downslope before they can be buried and lithified. With this principle in mind, geologists conclude that folds and tilted beds represent deformation events that happened after deposition.

- The principle of **superposition** states that each layer of sedimentary rock must be younger than the one below it, for a layer of sediment cannot accumulate unless there is already a surface on which it can collect. Therefore, in a sedimentary sequence, the oldest layer lies at the bottom and the youngest at the top (**Fig. 9.3b**).

- The principle of **cross-cutting relations** states that if one geologic feature cuts across another, the feature that has been cut is older. For example, if an igneous dike cuts across a sequence of sedimentary beds, the beds must be older than the dike (**Fig. 9.3c**). If a layer of sediment buries the dike, the sediment must be younger than the dike and if a fault cuts across and

displaces layers of sedimentary rock, then the fault must be younger than the layers.

- The principle of **baked contacts** states that an igneous intrusion "bakes" (metamorphoses) wall rock, so the rock that has been baked (contains the metamorphic aureole) must be older than the intrusion (**Fig. 9.3d**).

- The principle of **inclusions** states that a rock containing an *inclusion* (fragment of another rock) must be younger than the inclusion. For example, a conglomerate containing pebbles of basalt is younger than the basalt, whereas a basalt containing fragments of sandstone must be younger than the sandstone (**Fig. 9.3e**).

Geologists apply the above principles to determine the relative ages of geologic features (rocks, structures, erosional features), each of which is the consequence of a specific *geologic event*. Examples of geologic events include deposition, erosion, intrusion or extrusion of igneous rocks, and deformation (folding or faulting). The sequence of events, in terms of relative age, defines the **geologic history** of the region.

To visualize how to unravel the geologic history of a region, let's decipher the relative ages of the geologic events depicted in **Figure 9.4**. Based on the principle of superposition, deposition of Bed 1 happened first, followed by deposition of Beds 2 through 8. We know that the sill intruded after the deposition of Bed 5 because it contains inclusions of sandstone. Then all the beds, together with the sill, underwent folding. The granite intruded after the folding because we see that the pluton cuts the fold and that the folded rocks have been baked. The fault then slipped because it cuts and offsets the pluton. The dike intruded after the fault because it cuts the fault. Finally, erosion formed the present ground surface, which cuts across all other features. Note that bed 8 has been completely eroded away.

Take-home message . . .

The principle of uniformitarianism—the present is the key to the past—provides a basis for interpreting geologic features and implies that the Earth must be old. This principle, together with others (original horizontality, superposition, and cross-cutting relations) allow geologists to determine the relative ages of features and development of a geologic history of a region.

Quick Question -
What observations led Hutton to propose uniformitarianism?

Figure 9.3 Geologic principles used for determining relative ages.

Modern sediment, exposed at low tide, on the coast of France.

Horizontal sandstone beds in Wisconsin

Bedding plane

Cross beds

What an Earth Scientist Sees

(a) Original horizontality: Gravity causes sediment to accumulate in horizontal sheets. Mud layers settling along the coast of France today (left) are horizontal, so when we see horizontal beds of rock in Wisconsin (right), we assume that the beds are in their original orientation.

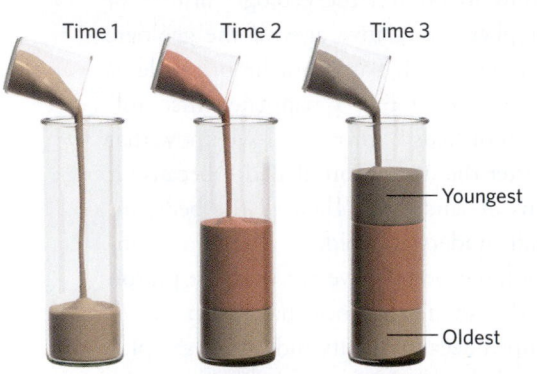

Time 1   Time 2   Time 3

Youngest

Oldest

(b) Superposition: In a sequence of strata, the oldest bed is on the bottom and the youngest on the top. Pouring sand into a glass cylinder illustrates this point. In the photo on the right, the beds get younger going up.

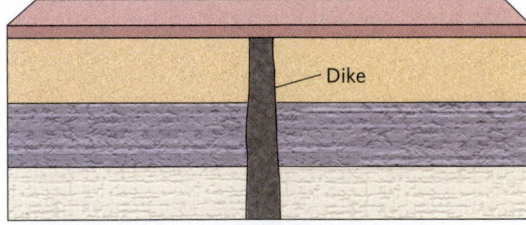

Dike

(c) Cross-cutting relations: The dike is younger than the beds it cuts across. The sediment layer that buries the dike is younger than the dike.

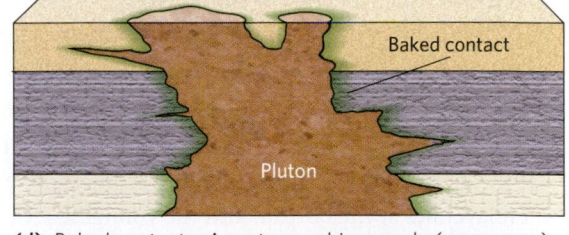

Baked contact

Pluton

(d) Baked contacts: A metamorphic aureole (green area) surrounds a pluton. The pluton is younger than the rock it has baked.

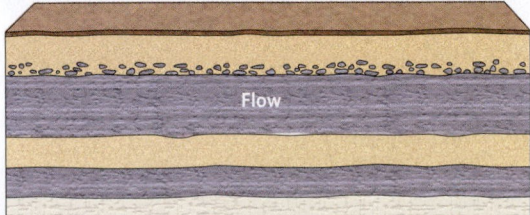

Flow

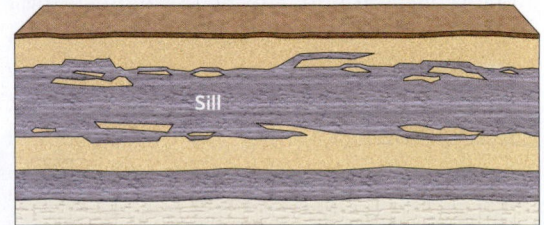

Sill

(e) Inclusions: The pebbles of basalt in a conglomerate (bottom) must be older than the conglomerate. The fragments of sandstone in a basalt sill (top) must be older than the basalt containing them.

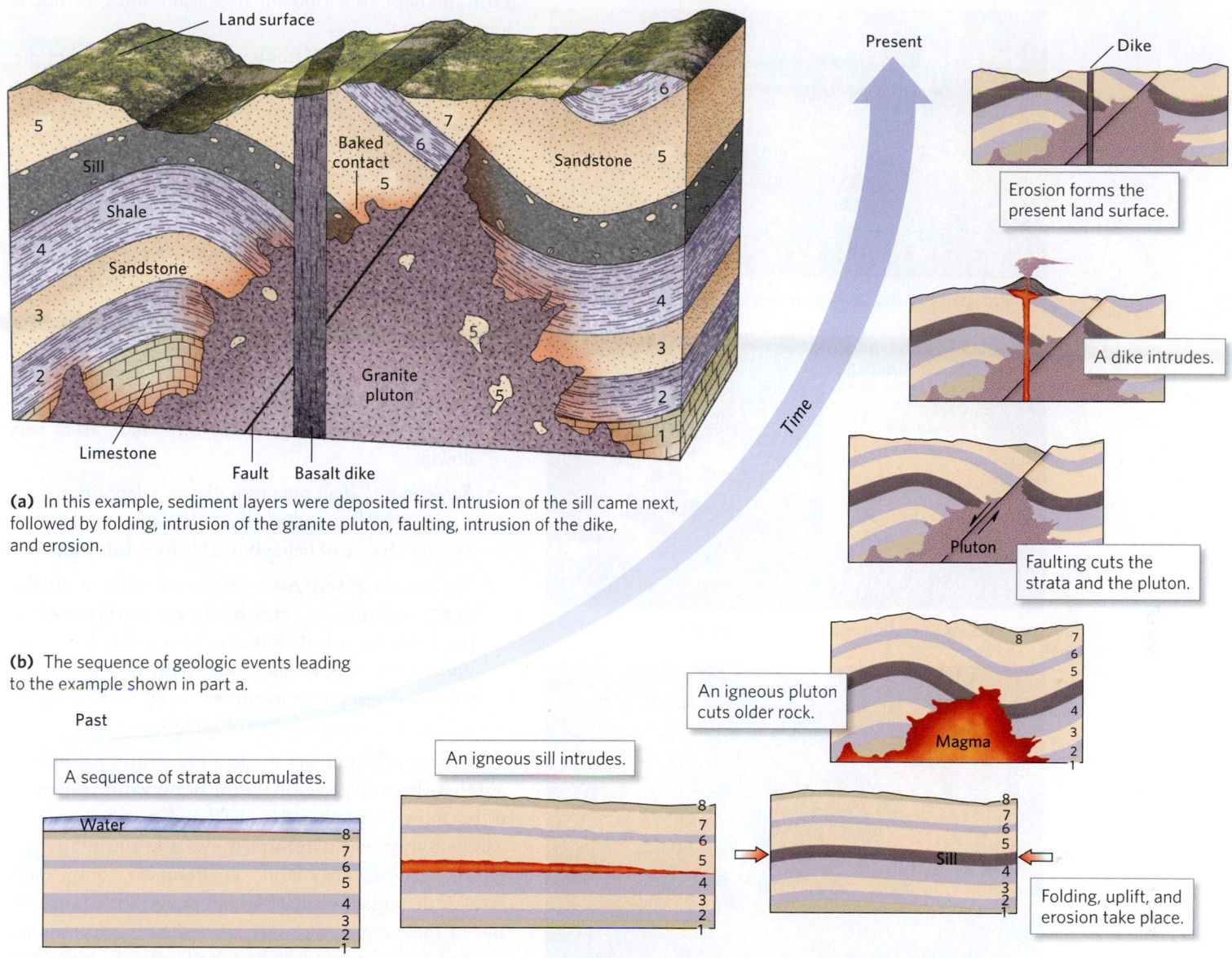

(a) In this example, sediment layers were deposited first. Intrusion of the sill came next, followed by folding, intrusion of the granite pluton, faulting, intrusion of the dike, and erosion.

(b) The sequence of geologic events leading to the example shown in part a.

9.3 Memories of Past Life: Fossils and Evolution

If you look at bedding surfaces of sedimentary rock, you may find shapes that resemble shells, bones, leaves, or footprints **(Fig. 9.5a)**. Researchers consider such **fossils** (from the Latin word *fossilis*, which means dug up) to be remnants or traces of ancient living organisms now preserved in rock.

The 19th century saw **paleontology**, the study of fossils, ripen into a science as museum drawers filled with specimens **(Fig. 9.5b)**. As we'll see, *paleontologists* (biologists or geologists who specialize in studying the fossil record) eventually learned how to use fossils

as a basis for determining the age of one sedimentary rock layer relative to another, so fossils have become an indispensable tool for studying geologic history and the evolution of life **(Fig. 9.5c)**.

Formation and Preservation of Fossils

Fossils form when organisms die and become buried by sediment or ash, or when organisms travel over or through sediment and leave imprints or debris. Paleontologists refer to the process of fossil formation as **fossilization**. To see how a typical fossil develops in sedimentary rock, let's follow the fate of an old dinosaur as it searches for food along a muddy riverbank **(Fig. 9.6)**. On a scorching summer day, the dinosaur succumbs to the heat and collapses dead in the mud. Soon after, scavengers strip the skeleton

Figure 9.5 Examples of fossils and fossil collections.

(a) Fossil skeleton in 200-million-year-old sandstone.

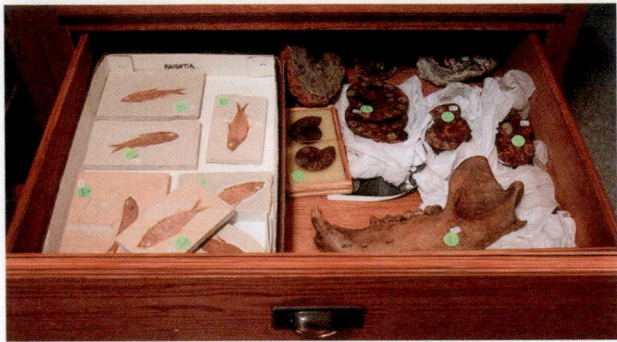

(b) A drawer of fossil specimens in a museum.

(c) A paleontologist collecting specimens.

of meat and may scatter the bones. But before the bones have had time to weather away, the river floods and buries the bones, along with the dinosaur's footprints, under a layer of silt. More sediment buries the bones and prints still deeper, until eventually, the sediment containing the bones and footprints turns to rock (siltstone and shale). The footprints remain outlined by the boundary between the siltstone and the shale, while the bones reside within the siltstone. Minerals from groundwater passing through the siltstone gradually replace some of the chemicals constituting the bones, until the bones themselves become rock-like. The buried bones and footprints are now fos-

sils. One hundred million years later, uplift and erosion expose the dinosaur's grave, so a lucky paleontologist can excavate them. The dinosaur rises again, but this time in a museum.

Not all living organisms become fossils when they die. In fact, only an exceedingly small number do, for it takes special circumstances—one or more of the following three—to produce a fossil and allow it to survive:

- *Death in an anoxic environment:* A dead squirrel by the side of the road won't become a fossil. As time passes, scavengers come along and eat the carcass, and if that doesn't happen, microbes infest the carcass and gradually digest it, or oxidation (chemical reaction with oxygen) breaks it down into gases. A carcass has a better chance of being preserved if it settles in an anoxic (oxygen-poor) environment, where oxidation happens slowly, scavenging organisms aren't abundant, and microbial metabolism takes place very slowly.

- *Rapid burial:* If an organism dies in a depositional environment where sediment accumulates rapidly, it has a better chance of being buried before disintegrating.

- *The presence of hard parts:* Organisms without durable shells, skeletons, or other hard parts usually won't be fossilized, for soft flesh decays long before hard parts under most depositional conditions. For this reason, paleontologists have identified many more fossils of oysters, for example, than of spiders.

By carefully studying modern organisms, paleontologists have been able to estimate the **preservation potential** of organisms, meaning the likelihood that an organism will be buried and transformed into a fossil. In a typical modern-day shallow-marine environment, only about 30% of the organisms have a high preservation potential. But of these organisms, only a few die in a depositional setting where they actually become fossilized, so fossilization is the exception rather than the rule.

The Many Different Kinds of Fossils

Perhaps when you think of a fossil, you picture either a dinosaur bone or the imprint of a seashell in rock. In fact, paleontologists distinguish many different kinds of fossils according to the specific way in which the organisms were fossilized. Let's look at examples of these categories.

- *Frozen or dried body fossils:* In a few environments, whole bodies of organisms may be preserved. Most of these fossils are fairly young by geologic standards— their ages can be measured in thousands of years. Examples include woolly mammoths that became incorporated in the permafrost (permanently frozen ground) of Siberia **(Fig. 9.7a)** or "mummified" fossils preserved in desert caves.

Figure 9.6 The stages in fossilization of a dinosaur.

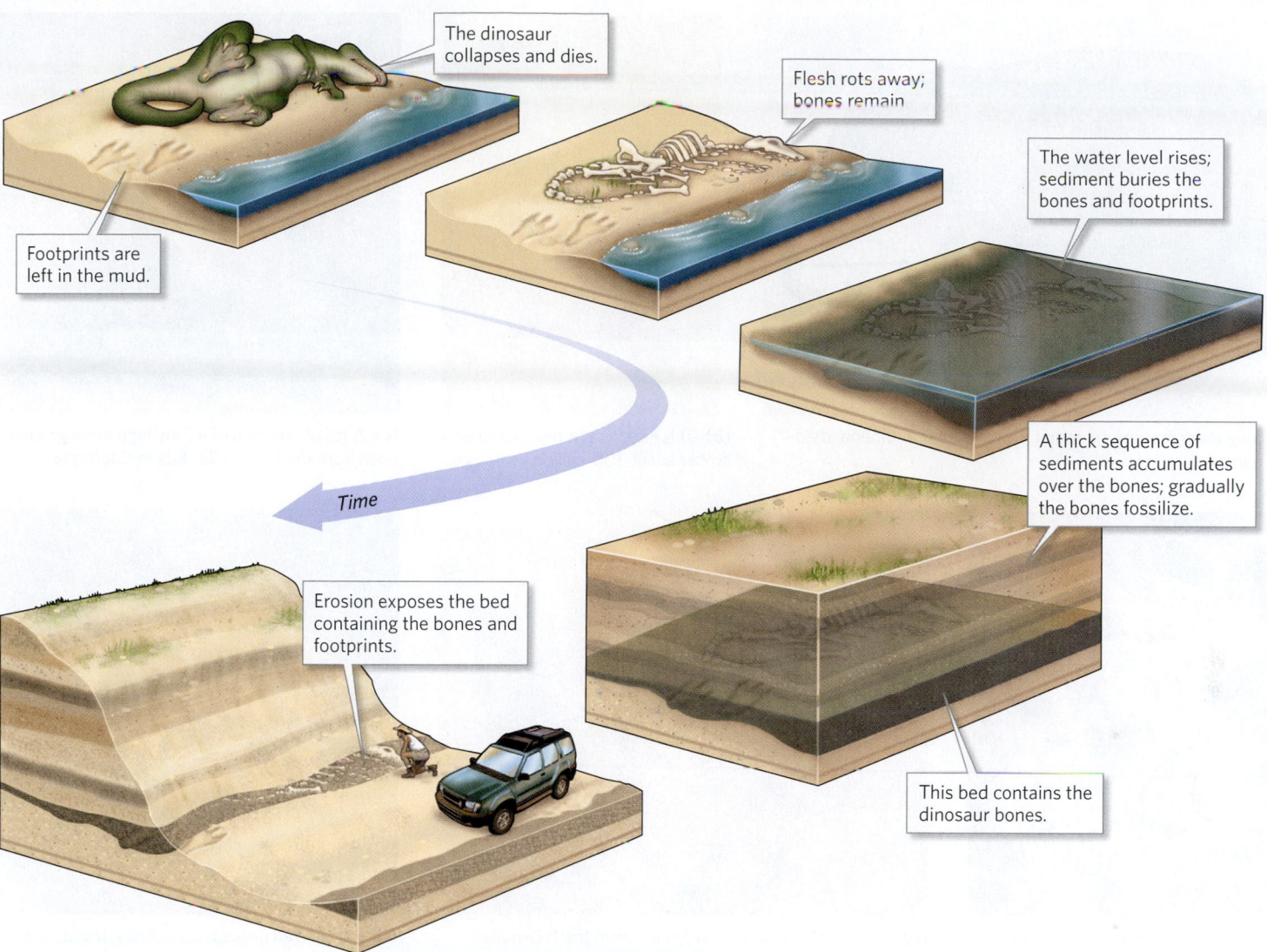

The dinosaur collapses and dies.

Footprints are left in the mud.

Flesh rots away; bones remain

The water level rises; sediment buries the bones and footprints.

A thick sequence of sediments accumulates over the bones; gradually the bones fossilize.

Time

Erosion exposes the bed containing the bones and footprints.

This bed contains the dinosaur bones.

- *Body fossils preserved in amber or tar:* Insects landing on the bark of trees may become trapped in the sticky sap or resin the trees produce. This golden syrup envelops the insects and over time hardens into amber. Amber can preserve insects for 40 million years or more **(Fig. 9.7b)**. Tar similarly acts as a preservative.

- *Preserved or replaced bones, teeth, and shells:* Bones, teeth, and shells consist of durable minerals that may survive in tar or rock **(Fig. 9.7c)**. Some bone, tooth, or shell minerals are not stable and recrystallize over time. But even when this happens, the shape of the original item may be preserved in the rock.

- *Molds and casts:* As sediment compacts around a shell or body, it conforms to the shape of that item **(Fig. 9.7d)**. If the shell or body later disappears because of weath-

ering and dissolution, a cavity called a *mold* remains. If sediment later fills the mold, it, too, preserves the organism's shape. The resulting *cast* protrudes from the surface of the adjacent bed. Usually only hard parts turn into molds or casts. Rarely, the shapes of soft parts may be preserved, forming *extraordinary fossils* **(Fig. 9.7e)**.

- *Carbonized impressions of bodies:* Impressions are flattened molds created when soft or semisoft organisms or their parts (leaves, insects, invertebrates, sponges, feathers, jellyfish) get pressed between layers of sediment. Chemical reactions eventually remove most organic material, leaving only a thin film of carbon on the surface of the impression **(Fig. 9.7f)**.

- *Permineralized organisms: Permineralization* refers to the process by which minerals precipitate from

Figure 9.7 Examples of different kinds of fossils.

(a) This 1-m-long baby mammoth, found in permafrost in Siberia, died 37,000 years ago.

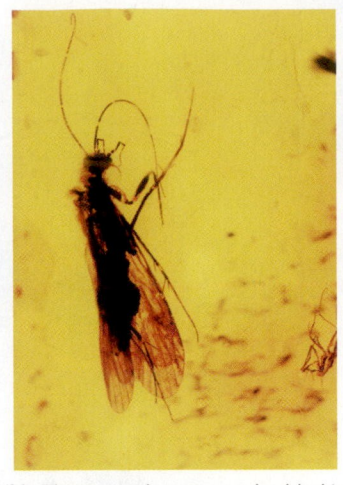

(b) This insect became embedded in amber about 200 million years ago.

(c) A fossil skeleton of a 2-m-high giant ground sloth from the La Brea Tar Pits in California.

(d) The hard parts of invertebrates, such as these fossil shells, are the most likely to be preserved.

(e) *Archaeopteryx*, a very early bird, from the 150-million-year-old Solnhofen Limestone of Germany. The imprints of feathers are clearly visible.

(f) Carbonized impressions of fern fronds in shale.

(g) Petrified wood from Arizona. The chert that has replaced the original wood is so hard that it remains after the rock that surrounded it has eroded away.

(h) These dinosaur footprints from Connecticut are one form of trace fossil.

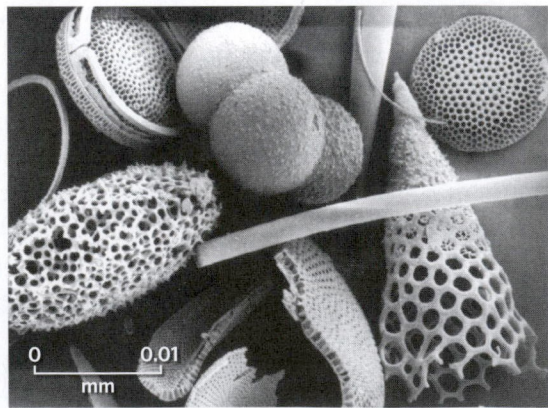

(i) Very tiny fossils, formed from plankton shells, are called microfossils.

groundwater that has seeped into the pores of porous material, such as wood or bone. Petrified wood, for example, forms by permineralization of wood, which transforms the wood into chert (Fig. 9.7g).

- *Trace fossils:* Trace fossils include footprints, feeding traces, burrows, and dung (coprolites) that organisms leave behind in sediment (Fig. 9.7h).
- *Chemical fossils:* Living things consist of complex organic chemicals. Over geologic time, most of these chemicals break down to form different, but still distinctive, chemicals. A distinctive chemical derived from an organism and preserved in rock is called a *chemical fossil* or *biomarker*.

Paleontologists also find it useful to distinguish among different fossils on the basis of their size. *Macrofossils* (like those in Fig. 9.7a–h) are fossils large enough to be seen with the naked eye. But some rocks and sediments also contain abundant *microfossils*, which can be seen only with a microscope (Fig. 9.7i). Microfossils include remnants of plankton, bacteria, and pollen.

Classifying Fossils

Paleontologists classify fossils using the same principles that biologists use to classify modern organisms (Box 9.1). There's nothing magical about classifying fossils. You can recognize common fossils in the field by examining their *morphology*, meaning their form or shape (Fig. 9.8a). If the fossil is complete and has distinctive features, the process can be straightforward, but if it's broken into fragments or if parts are missing, identification can be a challenge (Fig. 9.8b). Many fossil organisms resemble modern ones, so it is relatively easy to figure out how to classify them. For example, a fossil clam looks like a clam and not like, say, a snail. To classify a fossil more specifically, down to genus or species level, identification may involve recognizing such details as the number of ridges on the surface of its shell. Not all fossils resemble known living organisms, and determining their taxonomic relationships may be a challenge. Therefore, building a museum display that depicts fossil organisms as they appeared when alive requires imagination (Fig. 9.9).

The Concept of Extinction

In the 18th century, paleontologists recognized that not all fossils represented the remains of observed living species. But they tacitly assumed that since the world had not been explored completely, all fossils represented species living somewhere on the planet. By the 19th century, it became clear that this interpretation could not be true, for explorers had not found giant animals, such as mastodons or dinosaurs, anywhere. Based on this realization, the French paleontologist Georges Cuvier (1769–1832)

Figure 9.8 Identifying fossils.

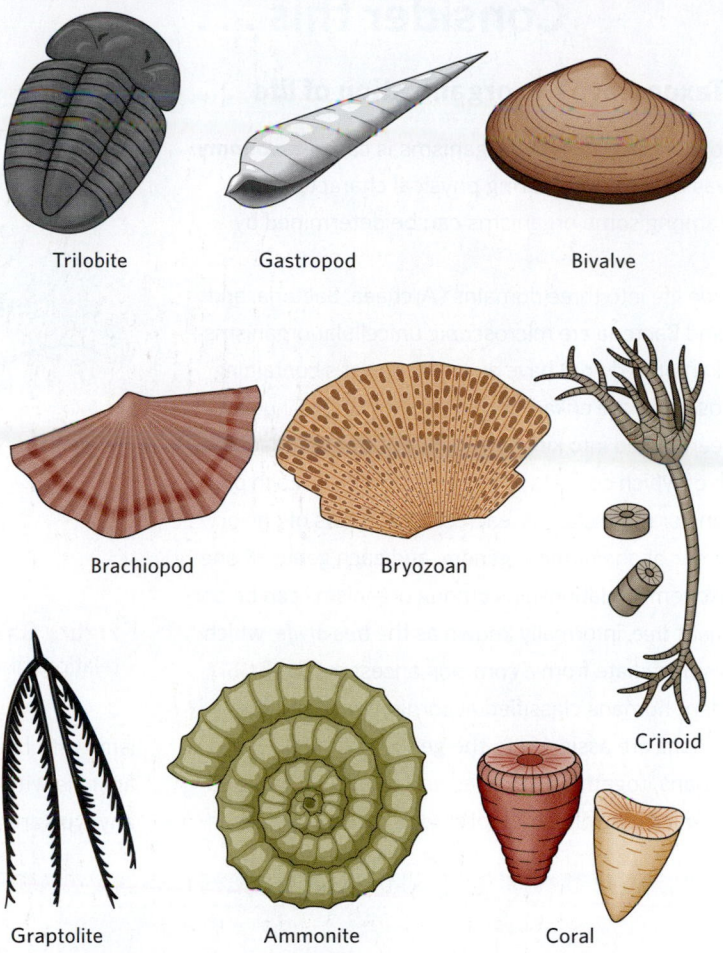

Trilobite   Gastropod   Bivalve

Brachiopod   Bryozoan

Crinoid

Graptolite   Ammonite   Coral

(a) Common types of invertebrate fossils.

(b) The fossils that look like white circles on the bed surface of this 400-million-year-old limestone are crinoids fragments.

argued that some fossil species had gone **extinct**, meaning that all individuals of those species had died. We take the phenomenon of extinction for granted today, since we have seen numerous animals become extinct in historic time, but Cuvier's proposal was revolutionary in his day.

Consider this . . .

Taxonomy: The organization of life

The study of how to identify and name organisms is called **taxonomy**. Traditionally, this was done by comparing physical characteristics. Now, relationships among some organisms can be determined by comparing DNA.

Biologists divide life into three domains (Archaea, Bacteria, and Eukarya). Archaea and Bacteria are microscopic unicellular organisms whose cells are prokaryotic (do not have a central nucleus containing DNA). Eukarya, whose cells are eukaryotic (do have a central nucleus), have traditionally been sorted into kingdoms (Protista, Fungi, Plantae, and Animalia), each of which consists of one or more phyla. Each phylum is divided into one or more classes, each class consists of one or more orders, each order of one or more genera, and each genus of one or more species. Taxonomic relationships among organisms can be portrayed on a *phylogenetic tree*, informally known as the *tree of life*, which shows which organisms radiate from a common ancestor **(Fig. Bx9.1)**.

How are modern humans classified according to the phylogenetic tree? All humans are assigned to the genus *Homo* and the species *sapiens*. Humans, together with apes and monkeys, make up the order Primates, which, together with other warm-blooded organ-

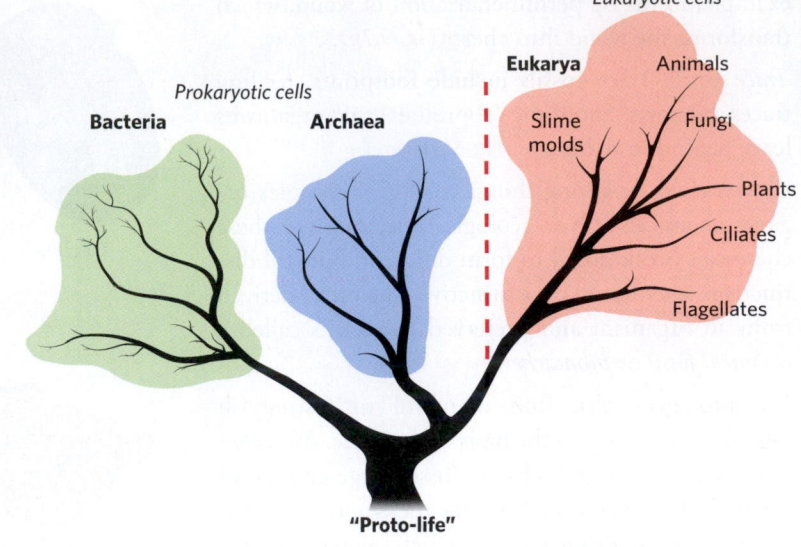

Figure Bx9.1 A simplified phylogenetic tree depicting taxonomic relationships among organisms.

isms with hair, constitute the class Mammalia. Mammals, along with all animals with backbones, are in the phylum Chordata, and all animals of any type are in the kingdom Animalia.

Figure 9.9 An artist's portrayal of what fossil organisms looked like when alive. The paintings are based on fossils found in the Burgess Shale of Canada.

Using Fossils to Determine Relative Ages: Fossil Succession

As Britain entered the industrial revolution in the late 18th century, factories demanded coal to fire their steam engines and needed an inexpensive means to transport raw materials and manufactured goods. Investors decided to construct a network of canals, and they hired an engineer named William Smith (1769–1839) to survey some of the excavations. Canal digging provided fresh exposures of bedrock that had previously been covered by vegetation. Smith learned to recognize distinct layers of sedimentary rock and to identify the **fossil assemblage** (the group of fossil species) that each layer contained. He also realized that a particular assemblage could be found in only a limited sequence of strata and not above or below. In other words, once a fossil species disappears at a horizon in a sequence of strata, it never reappears higher in the sequence. Extinction is forever.

Smith's observation, which has been repeated at millions of locations around the world, has been codified as the principle of **fossil succession**. To see how this principle works, examine **Figure 9.10**, which depicts a sequence of strata. Bed 1 at the base contains Species A, Bed 2 contains Species A and B, Bed 3 contains B and C, Bed 4 contains C, and so on. From these data, we can define the *range* for each species, meaning the interval in the sequence in which fossils of that species occur. In Figure 9.10, the succession of fossils, from oldest to youngest, is A, B, C, D, E, F. Note that the range of one species may overlap with that of others, and that when a species goes extinct, it does not reappear.

Once the relative ages of several fossil species have been determined, the fossils can be used to determine the relative ages of the beds containing them. For example, if a bed contains Fossil A (from Fig. 9.10), geologists can say that the bed is older than a bed containing Fossil F, even if the two beds do not crop out in the same area.

Tracing the History of Life

By some estimates, paleontologists have collected more than 250,000 different species of fossils during the past two centuries. This *fossil record* has allowed paleontologists to reconstruct the history of life over time. Paleontologists realized that, to explain fossil succession, something must have led to the extinction of some species and the appearance of new species. George Cuvier suggested that the fossil record recorded a succession of "catastrophes" during which some species went extinct and new ones suddenly appeared. This interpretation, known as *catastrophism*, seemed at odds with Hutton's concept of uniformitarianism, leading to many lively debates at the time.

Ideas would change after the voyage of HMS *Beagle*, between 1831 and 1836, which carried Charles Darwin (1809–1882) as the onboard naturalist. The observations that Darwin made during his voyage led him to propose the **theory of evolution** by natural selection. In essence, this theory states that over time, traits that make a species more successful in competing with other organisms for survival are passed on to offspring, while those that make it less successful are not. By this process, *survival of the fittest*, existing species may evolve into new species or may go extinct.

Though Darwin's theory of evolution provides a basis for explaining fossil succession, mysteries still remain surrounding the path of life's evolution on the Earth. Known fossils cannot account for every intermediate step in the evolution of every species. Over the billions of years that life has existed, 5 billion to 50 billion species may have lived. Paleontologists have discovered only a tiny percentage of these species. Why is the record so incomplete? First, despite all the fossil-collecting efforts of the past two centuries, paleontologists have not even come close to sampling every cubic centimeter of sedimentary rock exposed on Earth. Just as biologists have not yet identified

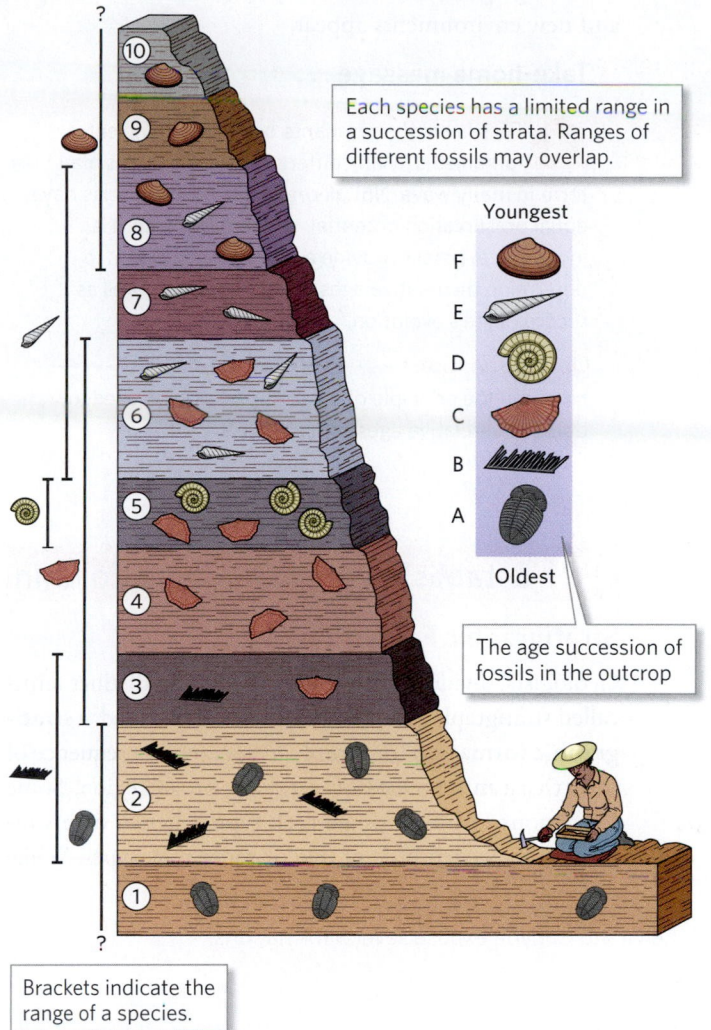

Figure 9.10 The principle of fossil succession.

Each species has a limited range in a succession of strata. Ranges of different fossils may overlap.

Youngest

F
E
D
C
B
A

Oldest

The age succession of fossils in the outcrop

Brackets indicate the range of a species.

every living species, paleontologists have not yet identified every fossil species. Second, not all species are represented in the rock record because not all species have a high preservation potential—only a minuscule fraction of the species that have lived on Earth have left a fossil record. Finally, as you will learn in the next section of this chapter, the sequence of sedimentary strata that exists on Earth does not account for every minute of geologic time at every location. Sediments accumulate only in environments whose conditions are appropriate for deposition, so strata accumulate only episodically.

Because of the incompleteness of the fossil record, the rate of evolution remains a subject of research. Some paleontologists picture evolution as proceeding in fits and starts, rather than at a steady rate. According to this concept, known as **punctuated equilibrium**, periods of relative stability in the number and identity of species alternate with times of rapid change, during which many species go extinct and many new species appear.

Geologists suggest that the times of change could coincide with times when geologic characteristics of the Earth are changing, such as when supercontinents break apart and new environments appear.

Take-home message . . .

Fossils are the remnants or traces of ancient organisms. Many different kinds of fossils may form in many ways. Not all organisms or their parts have equal preservation potential, so the fossil record is incomplete. Fossils provide a basis for geologists to determine the relative ages of rock layers as well as record of life's evolution.

Quick Question -
How can the principle of fossil succession be used to determine relative ages of rock layers?

9.4 Establishing the Geologic Column

Stratigraphic Formations

Geologists divide strata of a region into distinct units called stratigraphic formations. Formally defined, a **stratigraphic formation** (or, simply *formation*) is a sequence of beds that can be traced over a fairly broad region. Some formations include a single rock type, whereas others include interlayered beds of two or more rock types. While most formations consist of sedimentary strata, some also include the products of extrusive volcanism. Typically, a formation represents the products of deposition during a definable interval of time. Therefore, a given formation can be assigned a specific geologic age or age range, and characteristics of rocks in a formation provide clues about the environment at the time of deposition. The formations exposed on the walls of the Grand Canyon stand out as distinct stripes (Fig. 9.11).

Not all formations have the same thickness, and the thickness of a single formation can vary with location. Commonly, geologists name a formation after a locality where it was first identified or first studied. If a formation consists of only one rock type, that rock type may appear as part of the name (for example, the Kaibab Limestone), but if a formation contains more than one rock type, the word *formation* serves as part of the name (such as the Toroweap Formation). Note that all words in the name are capitalized. Several formations in a succession may be lumped together as a *stratigraphic group*, and several adjacent groups, in turn, may be lumped together as a *supergroup*. The boundary surface between two formations is a *depositional contact*, or simply a **contact**. (A fault surface, or a boundary between an igneous intrusion and its wall rock, is also a type of contact.) We can summarize information about the sequence of formations at a location by drawing a **stratigraphic column**, a chart representing the order of formations and their relative thicknesses.

Figure 9.11 The walls of the Grand Canyon expose several formations.

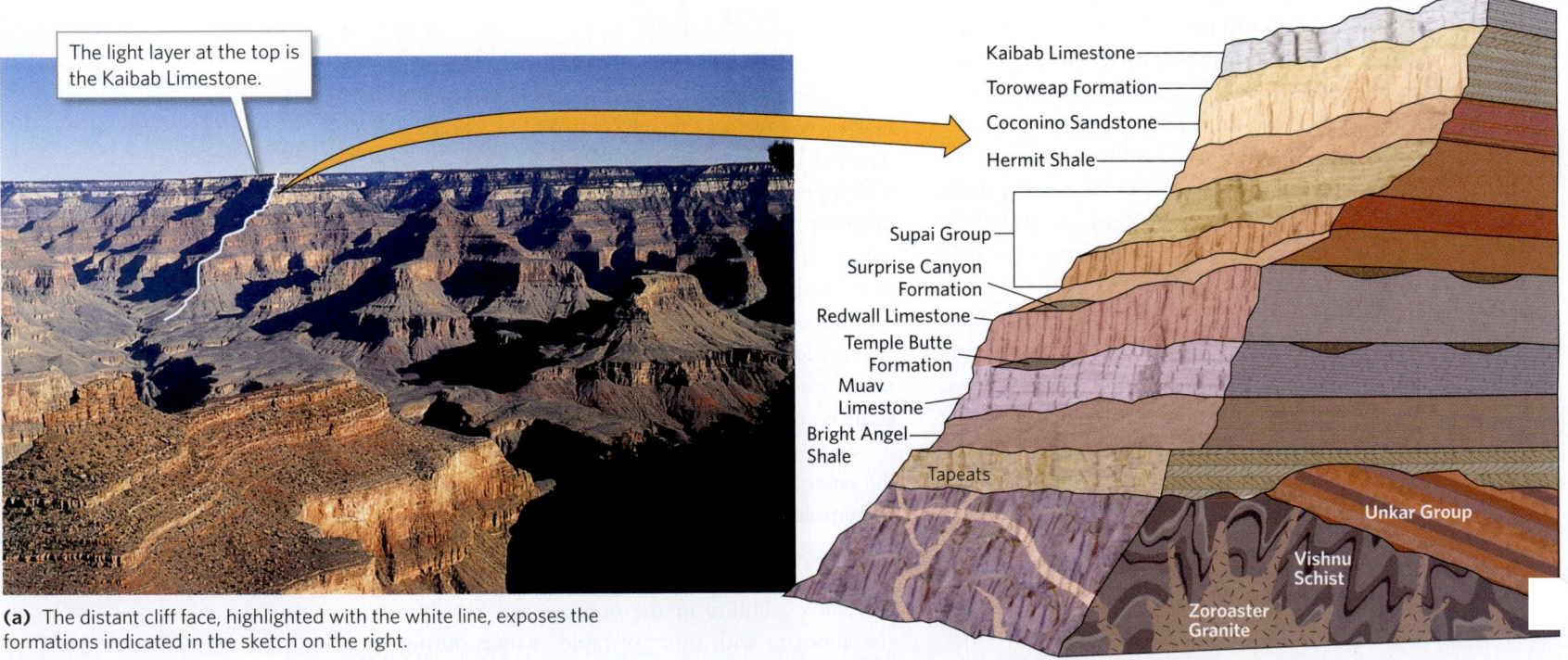

The light layer at the top is the Kaibab Limestone.

Kaibab Limestone
Toroweap Formation
Coconino Sandstone
Hermit Shale
Supai Group
Surprise Canyon Formation
Redwall Limestone
Temple Butte Formation
Muav Limestone
Bright Angel Shale
Tapeats
Unkar Group
Vishnu Schist
Zoroaster Granite

(a) The distant cliff face, highlighted with the white line, exposes the formations indicated in the sketch on the right.

(b) A block diagram representing the stratigraphic formations of the Grand Canyon.

Figure 9.12 James Hutton deduced that the red sandstone beds above the unconformity at Siccar Point were deposited after the gray siltstone and shale beds below had been lithified, tilted, and eroded.

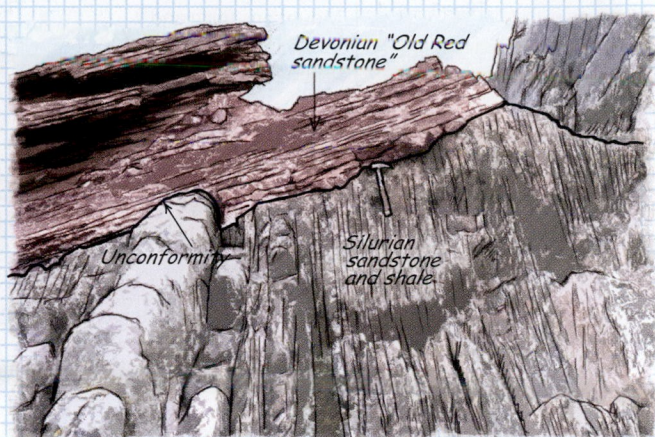

Devonian "Old Red sandstone"

Unconformity

Silurian sandstone and shale

What an Earth Scientist Sees

(a) Siccar Point juts out into the North Sea.

(b) The beds of reddish sandstone above the unconformity dip (are tilted) gently. The beds of gray sandstone and shale below are vertical.

Unconformities: Gaps in the Record

To find good exposures of rock, James Hutton sometimes boated along the eastern coast of Scotland, where waves of the stormy North Sea strip away soil and shrubbery. He was particularly puzzled by an outcrop at Siccar Point, which exposes a contact between two distinct sequences of sedimentary rock (Fig. 9.12). In the lower portion of the outcrop, beds of gray sandstone and shale are nearly vertical, whereas in the upper portion, beds of red sandstone display a dip of less than 20°. Further, the gently dipping layers seem to lie across the truncated ends of the vertical layers, like a handkerchief lying across the top of a row of books. We can imagine that, as Hutton was examining this odd relationship, the tide came in and deposited a new layer of sand on top of the rocky shore. With the principle of uniformitarianism in mind, Hutton suddenly realized the significance of what he saw.

Hutton deduced that the sediments that now make up the gray sandstone–shale sequence had been deposited, lithified, tilted, and truncated by erosion before the sediments that now form the red sandstone had accumulated. Therefore, the contact between the gray and red stratigraphic formations represented a time interval during which new strata had not been deposited at Siccar Point and older strata had been eroded away. Geologists now refer to such a contact, representing a time period of nondeposition, and possibly erosion, as an **unconformity**. The gap in the geologic record represented by an unconformity, meaning the period of time not represented by strata, is called a *hiatus*. We can distinguish among three common types of unconformities:

- *Angular unconformity:* Rocks below an angular unconformity were tilted or folded before the unconformity developed (Fig. 9.13a). Thus, an **angular unconformity** cuts across the underlying layers, and the orientation of the layers below the unconformity differs from that of the layers above. Siccar Point serves as an example.

- *Nonconformity:* A **nonconformity** is a type of unconformity at which sedimentary rocks overlie a *basement* of older intrusive igneous rocks or metamorphic rocks (Fig. 9.13b). These older rocks underwent cooling, uplift, and erosion before becoming the substrate on which sediment accumulated.

- *Disconformity:* Imagine that a sequence of sedimentary beds has been deposited beneath a shallow sea. Then sea level drops, exposing the beds, so no new sediment accumulates for a time, and some of the pre-existing sediment erodes away. Later, sea level rises again, and a new sequence of sediment accumulates over the old. The boundary between the two sequences is a **disconformity** (Fig. 9.13c, d). The beds above and below the disconformity are parallel, but the contact between them represents a hiatus.

The succession of strata at any location provides a record of Earth history there. But because of unconformities, the geologic record preserved in the rock layers at any particular location is incomplete. For example, the stratigraphic column of the Grand Canyon represents only part of geologic time (Fig. 9.14). It's as if geologic history is being chronicled by a data recorder that turns on only intermittently—when it's on (times of deposition), a geologic record accumulates, but when it's off

Did you ever wonder . . .

whether the strata exposed in the Grand Canyon represent all of Earth's history?

Figure 9.13 The three kinds of unconformities and their formation.

Time →

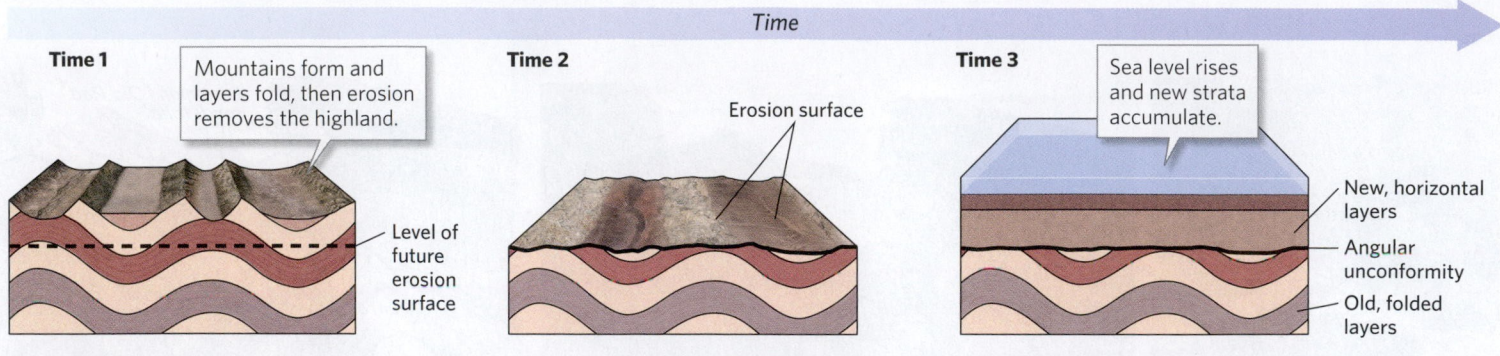

Time 1 — Mountains form and layers fold, then erosion removes the highland. — Level of future erosion surface

Time 2 — Erosion surface

Time 3 — Sea level rises and new strata accumulate. — New, horizontal layers — Angular unconformity — Old, folded layers

(a) Angular unconformity: (1) layers undergo folding; (2) erosion produces a flat surface; (3) sea level rises, and new layers of sediment accumulate.

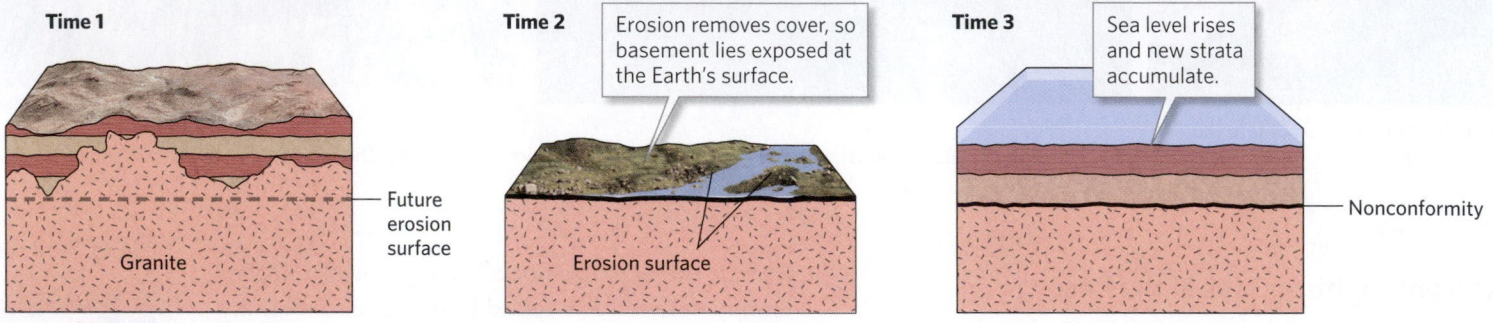

Time 1 — Granite — Future erosion surface

Time 2 — Erosion removes cover, so basement lies exposed at the Earth's surface. — Erosion surface

Time 3 — Sea level rises and new strata accumulate. — Nonconformity

(b) Nonconformity: (1) a pluton intrudes; (2) erosion cuts down into the crystalline rock; (3) new sediment layers accumulate above the erosion surface.

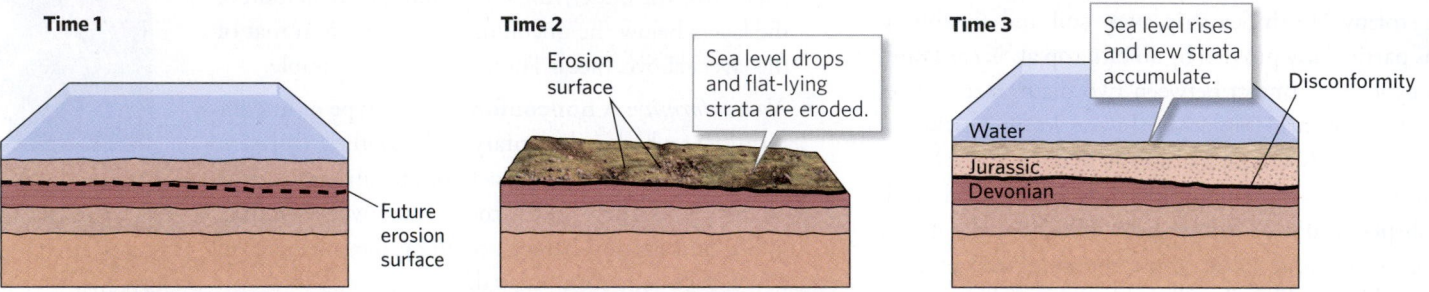

Time 1 — Future erosion surface

Time 2 — Erosion surface — Sea level drops and flat-lying strata are eroded.

Time 3 — Sea level rises and new strata accumulate. — Disconformity — Water — Jurassic — Devonian

(c) Disconformity: (1) layers of sediment accumulate; (2) sea level drops and an erosion surface forms; (3) sea level rises and new sediment layers accumulate.

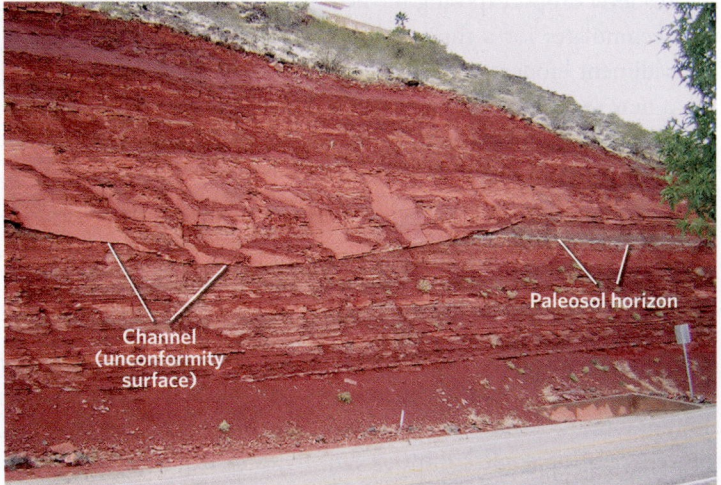

Channel (unconformity surface) — Paleosol horizon

(d) This roadcut in Utah shows a lithified sand-filled channel cut down into floodplain mud. Note that the channel cuts across a paleosol, a preserved, and in this example, lithified soil horizon, which also represents a disconformity.

(times of nondeposition and possibly erosion), an unconformity develops.

Correlation of Stratigraphic Formations

With the concept of stratigraphic formations in mind, we can explore how a succession of beds exposed in one location relates, in terms of relative age, to a succession exposed in another location. This process of determining the age relationship between successions of strata at different locations is called **correlation**.

How does correlation work? Typically, geologists correlate formations between nearby regions based on similarities in rock type, a method known as *lithologic correlation*. For example, the sequence of strata on the southern rim of the Grand Canyon correlates with the sequence on the northern rim because they contain the same rock types in the same order. You can be more confident of a

Figure 9.14 Interpreting the stratigraphic column of the Grand Canyon.

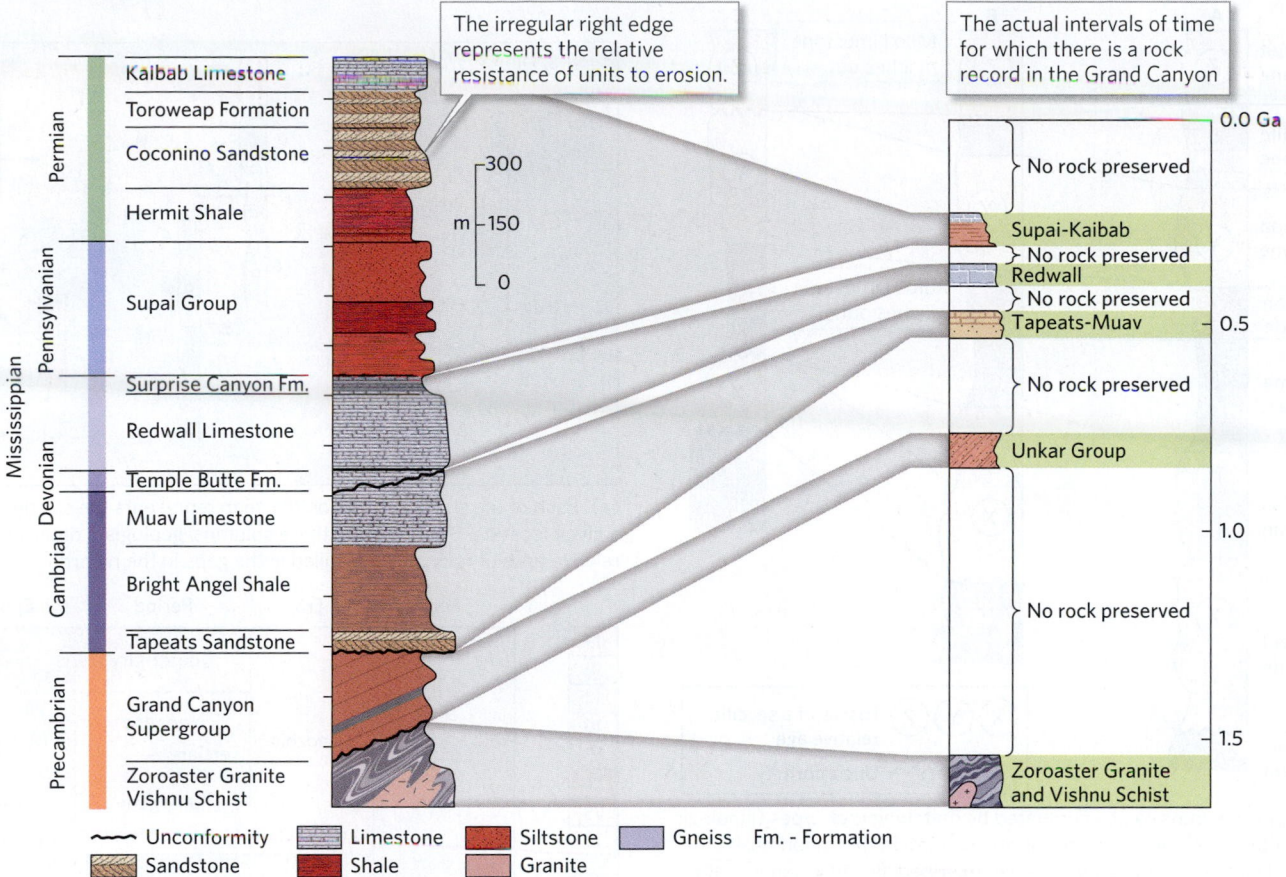

The irregular right edge represents the relative resistance of units to erosion.

The actual intervals of time for which there is a rock record in the Grand Canyon

| Unconformity | Limestone | Siltstone | Gneiss | Fm. - Formation |
| Sandstone | Shale | Granite | | |

(a) On a stratigraphic column, the vertical scale represents thickness and patterns represent different rock types. The Grand Canyon Supergroup includes the Unkar Group.

(b) Because of unconformities, Grand Canyon strata provide only a partial record of geologic time. This chart, with a time scale as the vertical axis, indicates gaps in the record.

proposed correlation if you can trace a distinctive layer, called a *marker bed*, between localities.

To correlate strata over broader regions, however, we may not be able to rely on lithologic correlation. Depositional settings at a given time may be different at various locations in a sedimentary basin, so widely separated localities may contain different successions of strata. To overcome this problem, geologists use fossils to define the relative ages of sedimentary strata at different locations, a method called *fossil correlation*. If fossils of the same age appear at two locations, we can say that the strata at the two locations correlate. Fossil species that are widespread but survived for a relatively short interval of geologic time, called *index fossils*, are particularly useful for fossil correlation because their presence in a bed characterizes its age fairly precisely.

To illustrate correlation, imagine that you measure stratigraphic columns at three locations (**Fig. 9.15**). You divide the succession of strata in column A into several distinct formations to which you assign names. The successions in columns B and C contain some formations with

similar rock types as those in column A, but the formations are not as thick, and not all the formations in column A can occur in the other columns. Lithologic correlation allows you to match the formations of column A with those of column B because you see the same rock types in the same succession. To confirm your correlation, you can compare fossil assemblages—you'll find that the formations in both columns contain fossils of the same ages. Correlation with strata in column C presents more of a challenge because some formations found in the other columns are not present. Fossil correlation, however, allows you to match the sandstone units and locate unconformities.

Dividing Time: From Eons to Epochs

No outcrop on Earth provides a complete record of our planet's history, for two reasons: first, any outcrop exposes only part of the stratigraphic succession in a given region (older strata remain hidden underground, and younger strata have been eroded away), and second, stratigraphic successions contain unconformities. But by correlating rocks from locality to locality at millions of

See for **yourself**

Vermilion Cliffs, Arizona

Latitude: 36°49′4.81″ N
Longitude: 111°37′56.59″ W

Zoom to an elevation of 2 km (~1.2 miles) and look obliquely north.

Marble Canyon (foreground) is the entry to the Grand Canyon. Outcrops in the distance are the Vermilion Cliffs, exposing reddish-brown sandstone and shale of the Moenkopi Formation. The canyon walls consist of underlying Kaibab Limestone.

Figure 9.15 The principles of correlation.

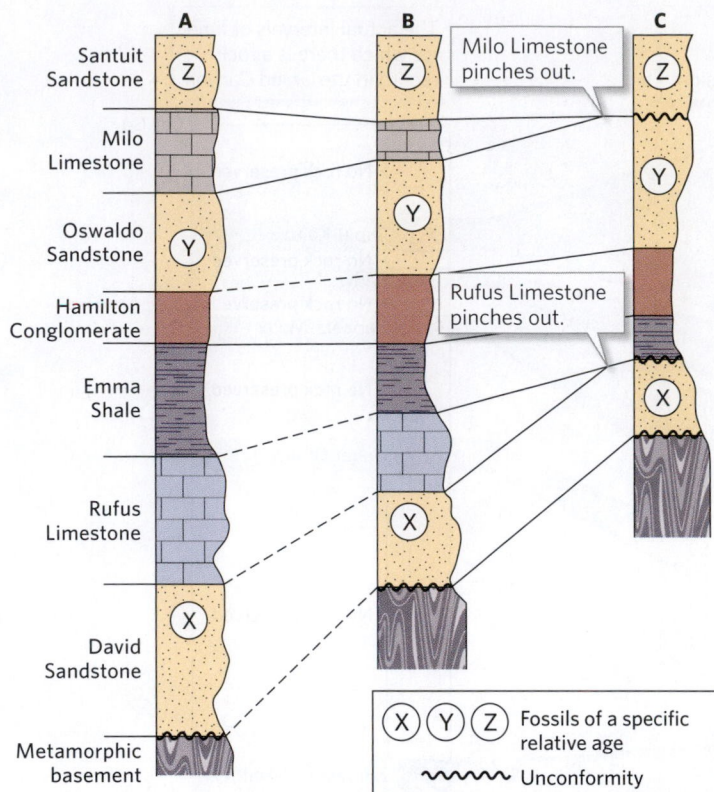

Milo Limestone pinches out.

Rufus Limestone pinches out.

| X | Y | Z | Fossils of a specific relative age |
| ∿∿∿ | | | Unconformity |

(a) Stratigraphic columns can be correlated by matching rock types (lithologic correlation) or by matching fossils (fossil correlation). In this theoretical sequence, the Hamilton Conglomerate is a marker bed. Because some strata pinch out, Column C contains unconformities. Fossil correlation indicates that the youngest formation in C is the Santuit Sandstone.

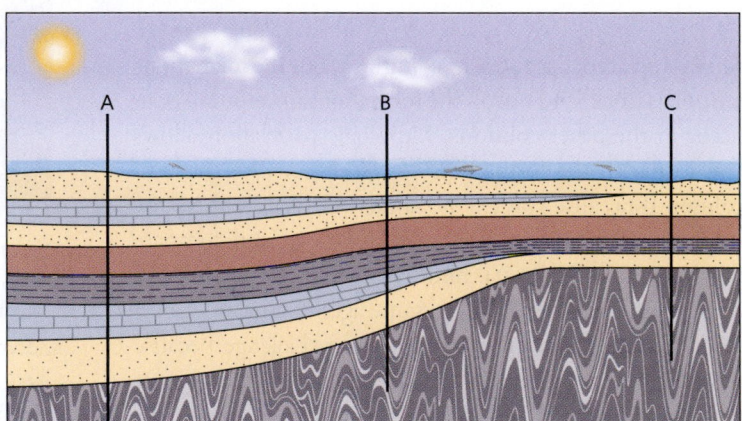

(b) At the time of deposition, locations A, B, and C (which correlate with the columns in part a) were in different parts of a basin. The basin floor was subsiding fastest at A.

places around the world, geologists have pieced together a composite stratigraphic column, called the **geologic column**, that symbolizes the entirety of Earth history **(Fig. 9.16)**.

The geologic column can be divided into segments, each of which represents a specific interval of time. The

Figure 9.16 Global correlation of rock units led to the development of the geologic column.

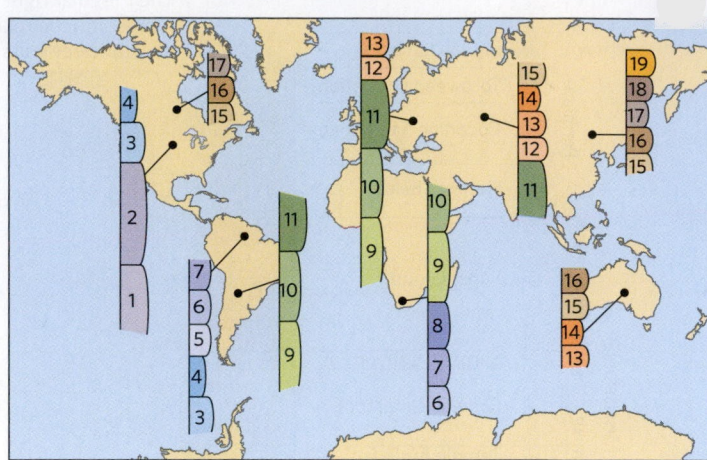

(a) Each of the small columns on this map represents the stratigraphy at a given location. By correlating these columns, geologists determined the relative ages of rock units and filled in the gaps in the record.

	Eon	Era	Period	Epoch
19			Quaternary	Holocene
18				Pleistocene
17		Cenozoic	Neogene	Pliocene
16			Tertiary	Miocene
15				Oligocene
14			Paleogene	Eocene
13				Paleocene
12				
11	Phanerozoic	Mesozoic	Cretaceous	
10			Jurassic	
9			Triassic	
8			Permian	Pennsylvanian
7			Carboniferous	Mississippian
6		Paleozoic	Devonian	
5			Silurian	
4			Ordovician	
3			Cambrian	
2	Precambrian	Proterozoic		
1		Archean		
		Hadean		

(b) Rock units from localities around the world were correlated and stacked in a single chart representing geologic time to create the geologic column. Geologists divided the column into intervals and assigned names to them, but since the column was built without knowledge of numerical ages, it does not show the actual durations of these intervals.

largest subdivisions of Earth history are **eons**. These eons are named, from oldest to youngest, the Hadean, Archean, Proterozoic, and Phanerozoic. Geologists commonly refer to the Hadean, Archean, and Proterozoic together as the **Precambrian**. The suffix –*zoic* means life, so *Phanerozoic* means visible life, and *Proterozoic* means first life. (This 19th-century terminology can be somewhat confusing because in recent decades, geologists have shown that the earliest life—Bacteria and Archaea—appeared in the Archean.) The Phanerozoic Eon is subdivided into **eras**, named in order from oldest to youngest, the Paleozoic (ancient life), Mesozoic (middle life), and Cenozoic (recent life). We further divide each era into **periods** and each period into **epochs**.

Where do the names of the periods come from? They refer either to localities where a fairly complete succession of strata representing that interval was first identified (for example, rocks representing the Devonian Period crop out near Devon, England), or to a characteristic of the time (rocks from the Carboniferous Period contain a lot of coal). The terminology was not organized in a planned fashion that would make it easy to learn. Instead, it grew haphazardly in the years between 1760 and 1845 as geologists began to refine their understanding of geologic history and fossil succession.

Notably, it now appears that the dates that early geologists identified as the boundaries between eras, and between some periods, coincide with **mass-extinction events**, times of short duration when **biodiversity**—the number of different types of organisms living at the same time—decreased substantially **(Fig. 9.17)**. The causes of these mass-extinction events remains a subject of research. They may be the consequence of asteroid impacts, intense volcanic activity, severe climate change, or a combination of these factors. Life diversified after mass extinctions as new organisms appeared, so fossil assemblages in strata deposited before a mass extinction event differ significantly from those in strata younger than the event.

With the concept of the geologic column in mind, let's see how geologists have correlated strata across the Colorado Plateau of the southwestern United States, a desert region that contains the Grand Canyon and other spectacular national parks **(Fig. 9.18)**. The oldest Paleozoic sedimentary rocks of the region crop out near the base of the Grand Canyon, while the youngest rocks form the cliffs of Cedar Breaks and Bryce Canyon. Walking through these parks is like walking through Earth's history: each rock layer gives an indication of the climate and topography of the region at a time in the past **(Earth Science at a Glance**, 304–305). For example, when the Precambrian metamorphic and igneous rocks exposed in the inner gorge of the Grand Canyon first formed, the region was a high mountain range, perhaps as dramatic

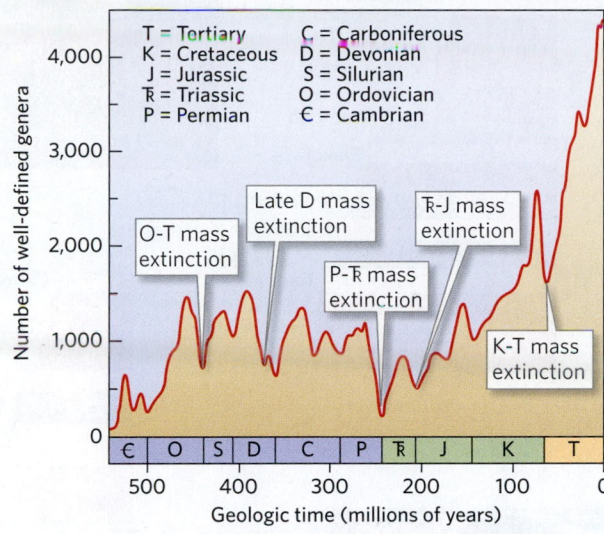

Figure 9.17 Changes in the diversity of life over time. Sudden drops in the number of genera indicate mass-extinction events.

as today's Himalayas. When the fossiliferous beds of the Kaibab Limestone accumulated where we now see the rim of the canyon, the region was a warm, shallow sea. And when the rocks making up the towering red cliffs of sandstone in Zion Canyon were deposited, the region was a Sahara-like desert, blanketed with huge sand dunes. The spatial relationships among such stratigraphic formations can be portrayed by geologic maps **(Box 9.2)**.

Tying Evolution to the Geologic Column

The succession of fossils preserved in the strata of the geologic column defines the course of life's evolution throughout Earth history **(Fig. 9.19)**. Simple Bacteria and Archaea appeared during the Archean Eon, but complex shell-less invertebrates did not evolve until the late Proterozoic. The appearance of invertebrates with shells defines the beginning of the Cambrian Period, the oldest period of the Paleozoic. At this time, a sudden diversification of life occurred, with many new types of organisms appearing over a relatively short interval—this event is called the *Cambrian explosion*. In the 19th century, geologists highlighted the change from the Precambrian (time before the Cambrian) to the Cambrian as an eon boundary because the appearance of shells in the fossil record marked such a radical change.

Progressively more complex organisms populated the Earth during the Paleozoic. For example, the first fish appeared in Ordovician seas, land plants started to spread over the continents during the Silurian, and amphibians climbed out of the sea during the Devonian. Though

Figure 9.18
Correlation of strata among the national parks of Arizona and Utah.

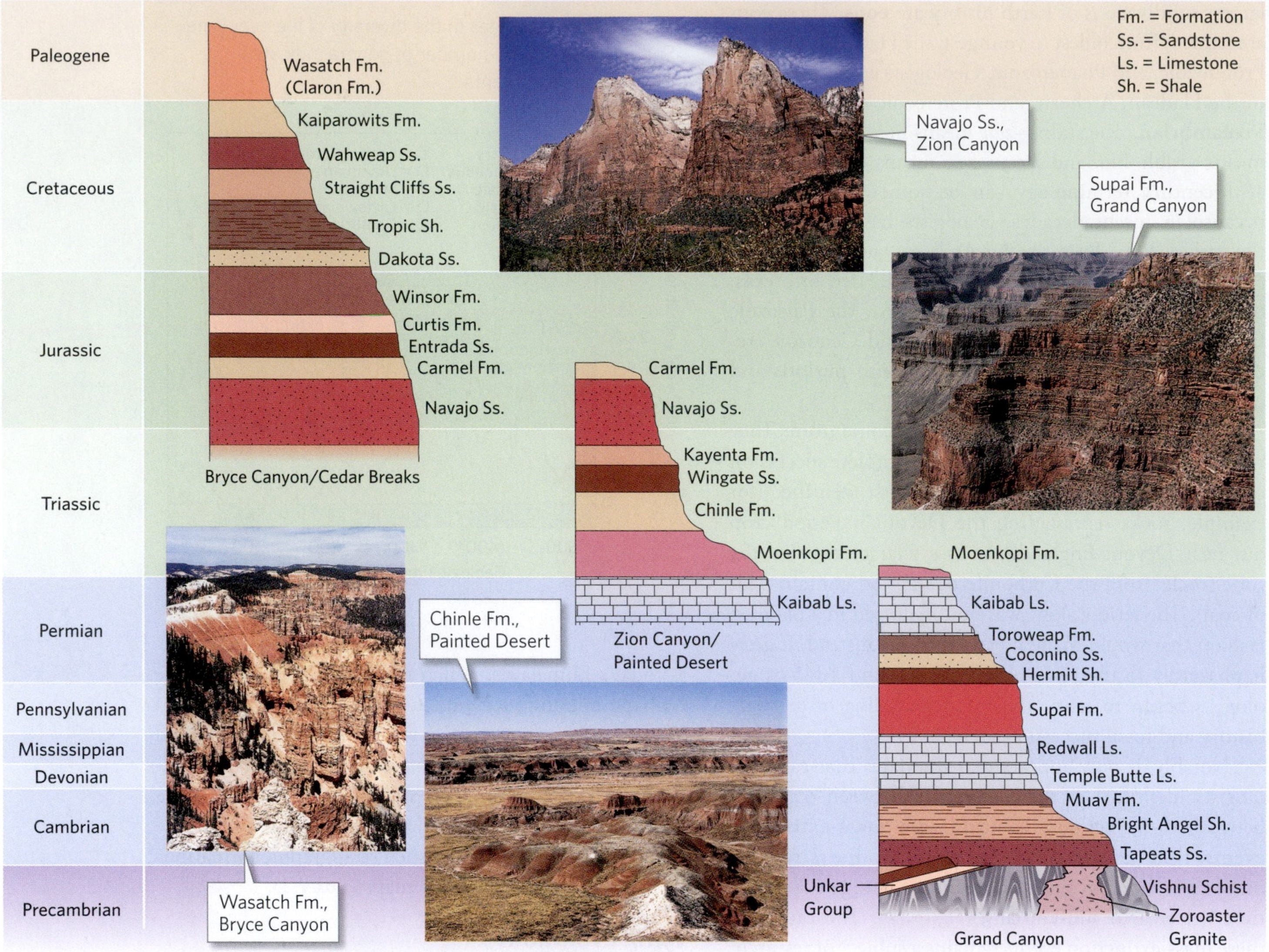

Fm. = Formation
Ss. = Sandstone
Ls. = Limestone
Sh. = Shale

Paleogene

Wasatch Fm.
(Claron Fm.)

Navajo Ss.,
Zion Canyon

Supai Fm.,
Grand Canyon

Cretaceous

Kaiparowits Fm.
Wahweap Ss.
Straight Cliffs Ss.
Tropic Sh.
Dakota Ss.

Jurassic

Winsor Fm.
Curtis Fm.
Entrada Ss.
Carmel Fm.
Navajo Ss.

Carmel Fm.
Navajo Ss.
Kayenta Fm.
Wingate Ss.
Chinle Fm.
Moenkopi Fm.

Moenkopi Fm.

Bryce Canyon/Cedar Breaks

Triassic

Chinle Fm.,
Painted Desert

Kaibab Ls.

Kaibab Ls.

Zion Canyon/
Painted Desert

Permian

Toroweap Fm.
Coconino Ss.
Hermit Sh.

Pennsylvanian

Supai Fm.

Mississippian

Redwall Ls.

Devonian

Temple Butte Ls.
Muav Fm.
Bright Angel Sh.

Cambrian

Tapeats Ss.

Precambrian

Wasatch Fm.,
Bryce Canyon

Unkar
Group

Vishnu Schist
Zoroaster
Granite

Grand Canyon

(a) Different intervals of geologic time are represented by the strata exposed in different parks.

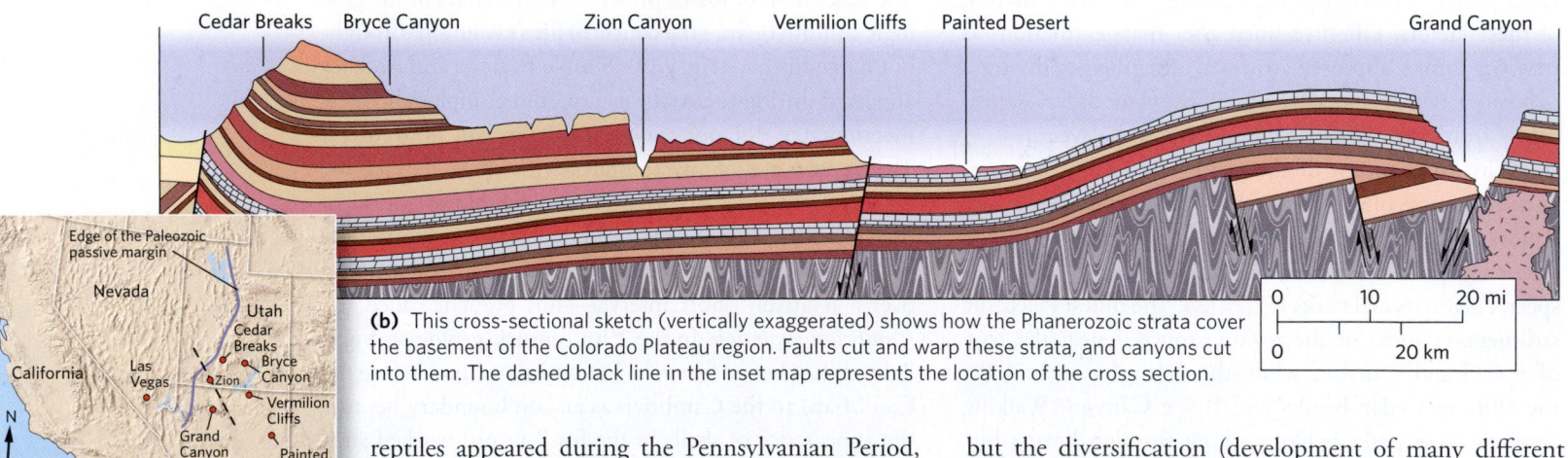

Cedar Breaks   Bryce Canyon   Zion Canyon   Vermilion Cliffs   Painted Desert   Grand Canyon

Edge of the Paleozoic passive margin
Nevada
Utah
Cedar Breaks
Bryce Canyon
Zion
Las Vegas
California
Grand Canyon
Vermilion Cliffs
Painted Desert
Arizona
N
0   300
km

0   10   20 mi
0   20 km

(b) This cross-sectional sketch (vertically exaggerated) shows how the Phanerozoic strata cover the basement of the Colorado Plateau region. Faults cut and warp these strata, and canyons cut into them. The dashed black line in the inset map represents the location of the cross section.

reptiles appeared during the Pennsylvanian Period, the first dinosaurs did not stomp across the land until the Triassic. Dinosaurs continued to inhabit the Earth until the sudden extinction of all but one lineage—birds—at the end of the Cretaceous Period. For this reason, geologists refer to the Mesozoic Era as the *Age of Dinosaurs*. Small mammals appeared during the Triassic, but the diversification (development of many different species) of mammals to fill a wide range of environments did not happen until the beginning of the Cenozoic Era, so geologists call the Cenozoic the *Age of Mammals*. Birds appeared during the Mesozoic (specifically, at the end of the Jurassic Period), but they underwent great diversification in the Cenozoic.

Figure 9.19 Evolution of life in the context of the geologic column. The Earth formed at the beginning of the Hadean Eon. (Not to scale.)

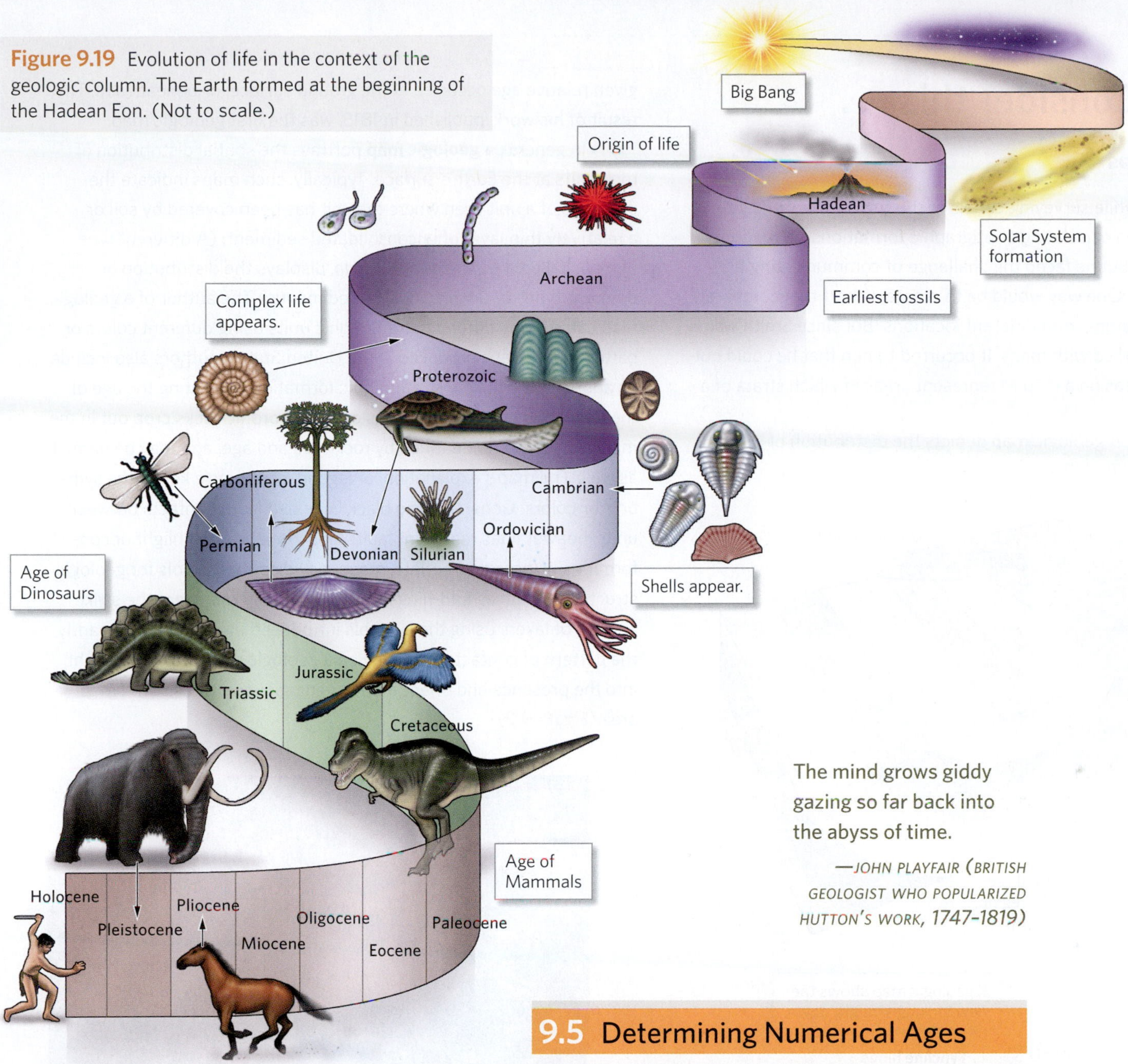

Big Bang

Origin of life

Solar System formation

Hadean

Earliest fossils

Archean

Complex life appears.

Proterozoic

Carboniferous

Cambrian

Permian

Devonian Silurian

Ordovician

Age of Dinosaurs

Shells appear.

Jurassic

Triassic

Cretaceous

The mind grows giddy gazing so far back into the abyss of time.

—JOHN PLAYFAIR (*BRITISH GEOLOGIST WHO POPULARIZED HUTTON'S WORK, 1747–1819*)

Holocene

Pliocene

Pleistocene

Oligocene

Miocene

Paleocene

Eocene

Age of Mammals

Take-home message . . .

A stratigraphic formation is a sequence of beds that can be mapped. Geologists correlate formations based on rock type and fossil content, and can portray the configuration on a geologic map. Due to unconformities, a stratigraphic sequence at any location is incomplete. Correlation led to the creation of the geologic column that shows the entirety of Earth history. It is subdivided into eons, eras, periods, and epochs. We can track life's evolution in the context of the column.

Quick Question -
Is there any place on Earth where an exposed stratigraphic sequence represents all of geologic time? Explain.

9.5 Determining Numerical Ages

Geologists since the time of Hutton could determine the relative ages of rocks, but they had no way to specify rocks' **numerical age** (called *absolute age* in older literature), meaning age specified in years. Therefore, they could not define a timeline for Earth history, nor could they determine the duration of events in years. This situation changed with the discovery of **radioactive elements**. Atoms of radioactive elements spontaneously decay to form different elements. The rate of decay is a constant that can be measured in the laboratory and can be specified in years. In the 1950s, geologists developed **isotopic dating** (or *radiometric dating*) techniques that use the ratios of radioactive elements to their decay products as a basis for calculating numerical ages of rocks. The overall study of numerical ages of rocks is now called **geochronology**. Since the 1950s, isotopic dating techniques have

Box 9.2

Consider this . . .

Geologic maps

William Smith, while surveying canals at the turn of the 19th century, succeeded in correlating stratigraphic formations throughout central England. But he faced the challenge of communicating his results to others. One way would be to create a chart that compared stratigraphic columns from different locations. But since Smith was a surveyor and worked with maps, it occurred to him that he could outline and color areas on a map to represent areas in which strata of a

given relative age occurred at the ground surface, or beneath soil. The result of his work, published in 1815, was the first geologic map.

In general, a **geologic map** portrays the spatial distribution of rock units at the Earth's surface. Typically, such maps indicate the presence of a unit even where the unit has been covered by soil or a relatively thin layer of unconsolidated sediment. (A different type of map, called a *surface-deposits map*, displays the distribution of different types of unconsolidated sediments.) The author of a geologic map can portray different stratigraphic units using different colors or patterns. On maps produced in the United States, authors also include an abbreviation for the stratigraphic formation, indicating the age of the formation. If igneous plutons or metamorphic rocks crop out in the map area, they are identified by rock type and age, and may be named as well. The map's explanation, or legend, serves as a key to the symbols or colors. Generally, thin black lines delineate contacts between units, heavier lines represent faults, and wavy lines highlight unconformities. Where appropriate, maps may display symbols for geologic structures (such as fold-hinge traces), and may represent the strike and dip of layers using the symbols introduced in Box 7.1. Significantly, the pattern of contacts displayed on a geologic map provides insight into the presence and orientation of geologic structures in the map area **(Fig. Bx9.2)**.

Figure Bx9.2 A geologic map depicts the distribution of stratigraphic formations.

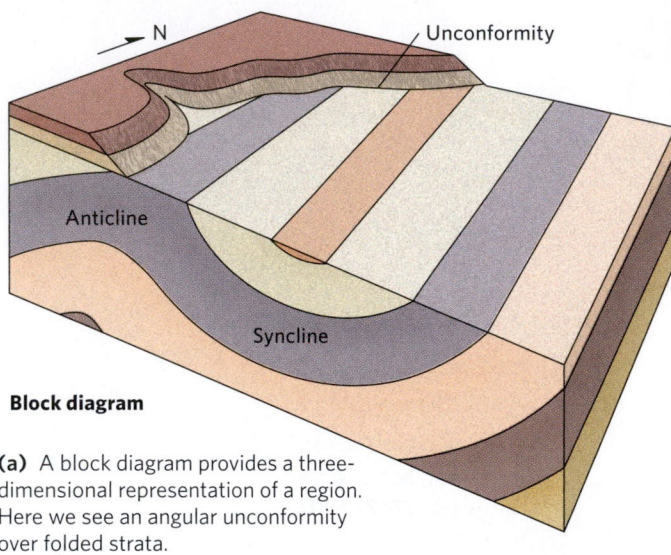

Block diagram

(a) A block diagram provides a three-dimensional representation of a region. Here we see an angular unconformity over folded strata.

A geologic map shows the view looking straight down.

(b) A geologic map portrays the top surface of the block diagram. It shows the distribution of rock units. The black lines represent contacts between units.

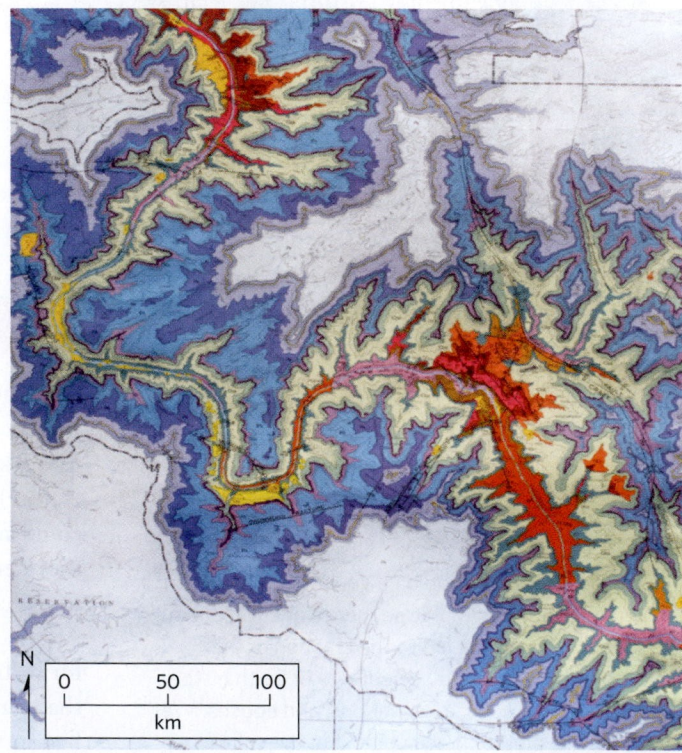

(c) This geologic map portrays the distribution of stratigraphic formations in a portion of the Grand Canyon. Each color represents a different formation.

steadily improved, and geologists can now make very accurate measurements from very small samples. But the basis of the technique remains the same, and to explain it, we must first review the process of radioactive decay.

Radioactive Decay

All atoms of a given element have the same number of protons in their nucleus—we call this number the *atomic number* (see Chapter 1). However, not all atoms of a given element have the same number of neutrons in their nucleus, which means that not all atoms of an element have the same *atomic weight* (roughly, the number of protons plus neutrons). Different versions, or **isotopes**, of an element have the same atomic number but different atomic weights. For example, all uranium atoms have 92 protons, but the uranium-238 isotope (abbreviated ^{238}U) has an atomic weight of 238, whereas the ^{235}U isotope has an atomic weight of 235; the first isotope has 3 more neutrons than the second.

Some isotopes of some elements are *stable*, meaning that they last essentially forever. Radioactive isotopes, however, are *unstable*. Eventually, they undergo a process called **radioactive decay**, which changes the atomic number, and thus changes an atom of one element into an atom of a different element. We refer to the atom that undergoes decay as the **parent isotope** and the decay product as the **daughter isotope**. For example, a parent atom of ^{238}U ultimately decays to become a daughter atom of lead-206 (^{206}Pb). In some cases, the decay that takes place to form the final, stable daughter has many steps, and the process involves the formation and decay of other atoms.

Physicists cannot specify how long an individual atom of a radioactive isotope survives before it decays, but they can measure the **half-life**, the time it takes for half of a group of atoms to decay. Because the half-life for a given decay reaction is constant, the ratio of parent to daughter atoms can serve as a natural clock **(Fig. 9.20)**. (Note that in a real crystal, the number of atoms would be immensely larger.) **Box 9.3** helps you visualize the concept of a half-life.

Isotopic Dating Techniques

Since radioactive decay proceeds at a known rate, it provides a basis for telling time. In other words, because an element's half-life is a constant, we can calculate the age of a mineral by measuring the ratio of parent isotope to daughter isotope in the mineral. To put this concept into practice, geologists have identified isotopes that have long half-lives and occur abundantly within common minerals. **Table 9.1** lists some particularly useful elements, and the minerals that contain them. Note that each radioactive element has a unique half-life. We do not list

Figure 9.20 The concept of a half-life in the context of radioactive decay.

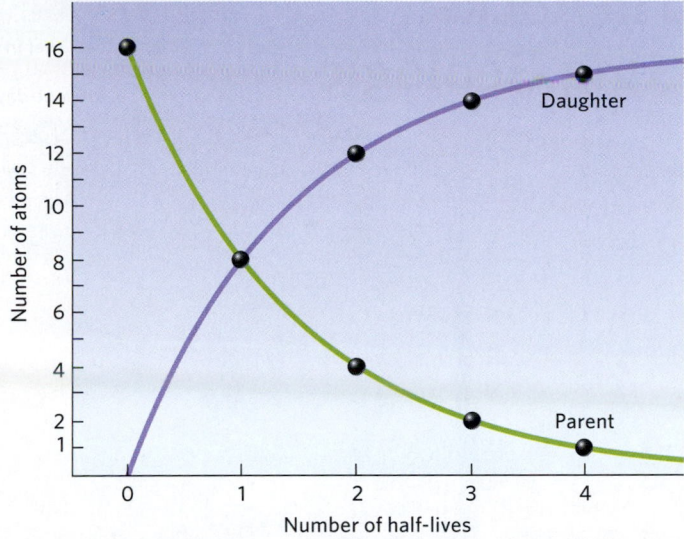

(a) This graph shows how the number of atoms of the parent isotope decreases and the number of atoms of the daughter isotope increases as time passes. The rate of change decreases with time.

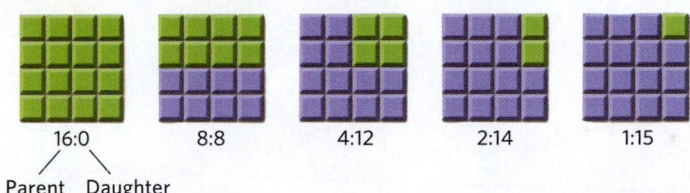

(b) The ratio of parent isotope to daughter isotope changes with the passage of each successive half-life.

(c) In a cluster of atoms undergoing decay, there is no way to predict which atom will decay next.

carbon in Table 9.1 because it is not useful for dating rocks. Radioactive carbon (^{14}C) has a very short half-life (about 5,000 years) and only accumulates in living material, so we can use *radiocarbon dating* only to date organic material that is less than about 70,000 years old.

If you find a rock containing dateable minerals, you can obtain its numerical age by following these steps:

- *Collecting appropriate samples:* For isotopic dating methods to work, you need to obtain unweathered rock samples. Chemical weathering reactions may cause the loss of parent or daughter isotopes, so working with weathered rocks might produce incorrect results.

- *Separating dateable minerals:* Once collected, crush the rocks so that you can pick out and isolate the dateable minerals.

Did you ever wonder . . .

how geologists can specify the ages of rocks in years?

The Record in Rocks: Reconstructing Geologic History

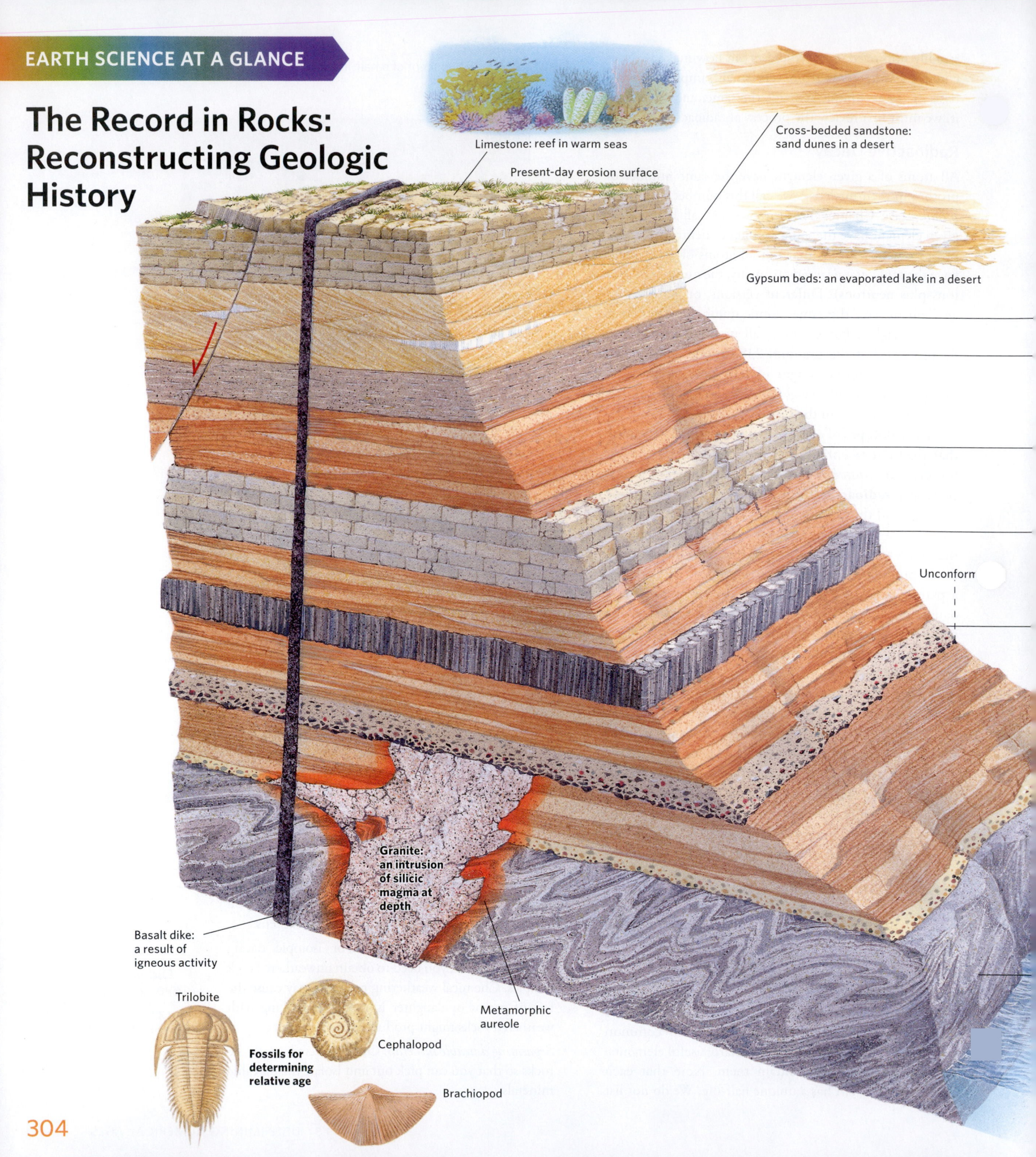

Limestone: reef in warm seas

Present-day erosion surface

Cross-bedded sandstone: sand dunes in a desert

Gypsum beds: an evaporated lake in a desert

Unconform

Granite: an intrusion of silicic magma at depth

Basalt dike: a result of igneous activity

Metamorphic aureole

Trilobite

Cephalopod

Fossils for determining relative age

Brachiopod

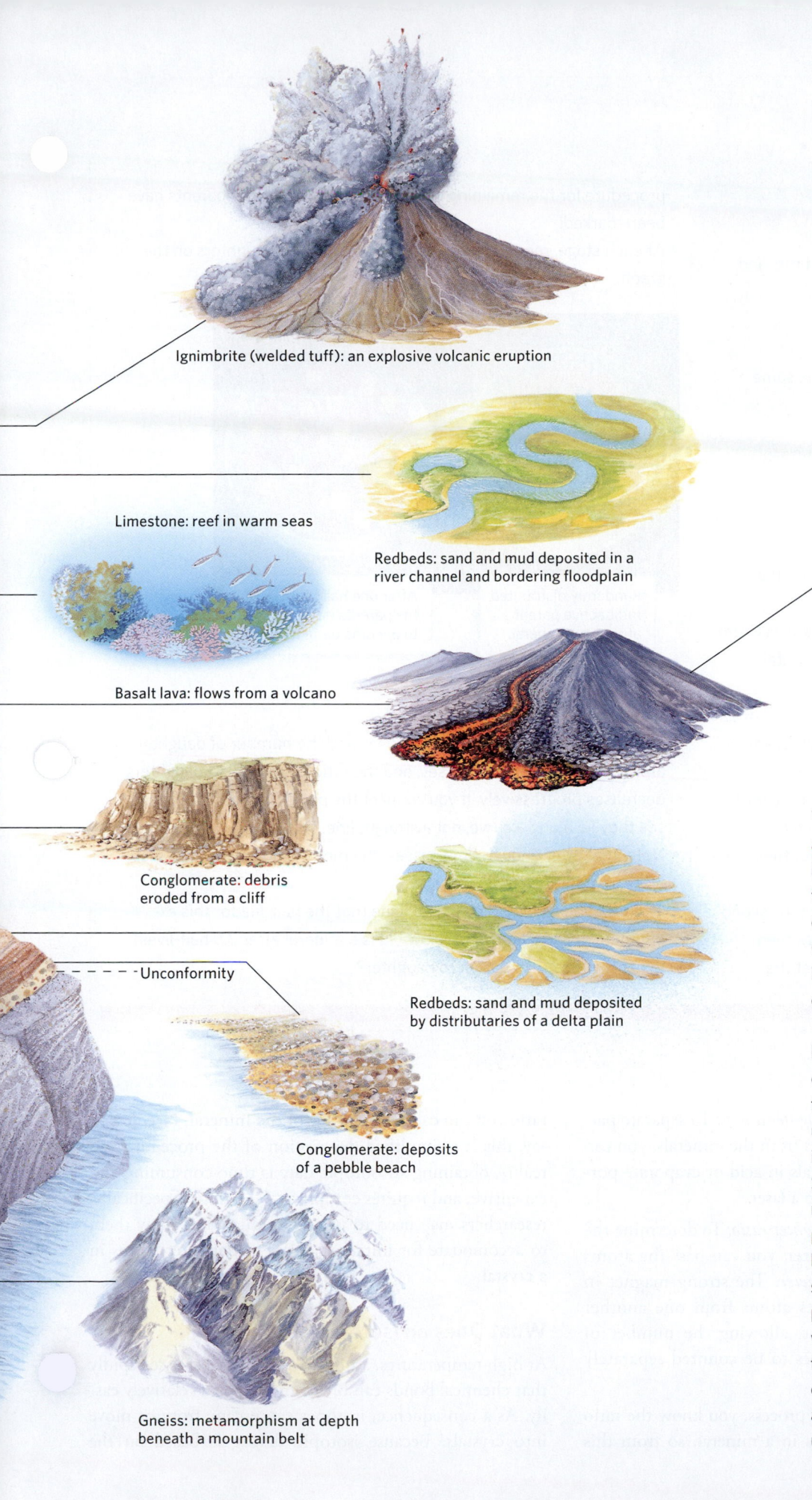

Ignimbrite (welded tuff): an explosive volcanic eruption

Limestone: reef in warm seas

Redbeds: sand and mud deposited in a river channel and bordering floodplain

Basalt lava: flows from a volcano

Conglomerate: debris eroded from a cliff

Unconformity

Redbeds: sand and mud deposited by distributaries of a delta plain

Conglomerate: deposits of a pebble beach

Gneiss: metamorphism at depth beneath a mountain belt

Isotopic dating

Decay

Mineral crystal

Decay

Parent → Daughter

When geologists examine a sequence of rocks exposed on a cliff, they see a record of Earth history that they interpret by applying the basic principles of geology, by searching for fossils, and by using isotopic dating. On this cliff, we see evidence for many geologic events. The layers of sediment (and the sedimentary structures they contain), the igneous intrusions, and the geologic structures tell us about past climates and past tectonic activity.

The insets show the way the region looked in the past, based on the record in the rocks. For example, the presence of gneiss at the base of the canyon indicates that at one time the region was a mountain belt where deeply buried crust underwent metamorphism and deformation. Unconformities indicate that the region later underwent uplift and erosion. Sedimentary successions record transgressions and regressions of the sea. The land surface portrayed in this painting was sometimes a river floodplain or a delta (indicated by redbeds), sometimes a shallow sea (limestone), and sometimes a desert dune field (cross-bedded sandstone). The presence of igneous rocks indicates that, at times, the region was volcanically active, and the presence of faults indicates that it was seismically active. We can gain insight into the age of the sedimentary rocks by studying the fossils they contain, and into the age of the igneous and metamorphic rocks by using isotopic dating methods.

Box 9.3 ## How can I explain . . .

The concept of half-life

What are we learning?
How the ratio of parent to daughter isotopes changes over time, and how we can represent this ratio as a half-life.

What you need:
- 32 small, disk-shaped objects (such as checkers), all of the same color.
- Colored sticky-note paper, cut into squares.
- A cake pan (representing a mineral crystal).
- Graph paper and a pencil.

Instructions:
- Spread the disks around the cake pan. These disks represent the distribution of parent atoms.
- Create a graph that plots number of disks on the vertical y-axis (from 0 to 32) and number of half-lives, from 0 to x, on the horizontal x-axis.
- Plot a point on the graph representing the number of unmarked disks at time 0. Since there are 32 disks, y = 32 and x = 0. These unmarked disks represent parent atoms.
- Place sticky notes randomly on half the disks. Count the number of unmarked disks that remain. The answer, 16, represents the number of parents remaining after the first half-life has passed. Plot the numbers of parents and daughters at 1 half-life on the graph.
- To represent the number of parent atoms remaining after the second half-life, mark half the remaining unmarked disks (8 disks), then record the numbers of marked and unmarked disks. Repeat the

procedure for the remaining half-lives. Continue until all parents have been marked.
- At each stage, record the numbers of parents and daughters on the graph.

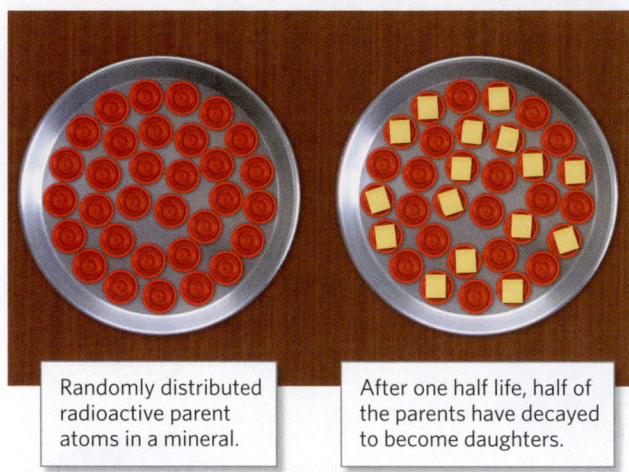

Randomly distributed radioactive parent atoms in a mineral.

After one half life, half of the parents have decayed to become daughters.

What did we see?
- The number of parent atoms decreases and the number of daughter atoms increases as time passes, and the ratio of parents to daughters decreases progressively. If you connect the points on your graph, you see they lie along a curve, not a straight line.
- Note, that after enough time passes, the method no longer works, as no more parents remain to decay.
- To check your understanding, imagine that the half-life for this experiment is 10 million years. How old is a mineral after 2.5 half-lives? What is the ratio of parent to daughter?

- *Extracting parent and daughter atoms:* To separate parent and daughter isotopes from the minerals, you can either dissolve the minerals in acid or evaporate portions of the minerals with a laser.
- *Analyzing the parent-daughter ratio:* To determine the ratio of parent to daughter, you can pass the atoms through a *mass spectrometer*. The strong magnet in this instrument separates atoms from one another based on atomic weight, allowing the number of atoms of specific isotopes to be counted separately **(Fig. 9.21)**.

At the end of the laboratory process, you know the ratio of parent to daughter atoms in a mineral, so from this

ratio you can calculate the age of the mineral. Needless to say, this is a simplified description of the procedure—in reality, obtaining an isotopic date is time-consuming and expensive, and requires complex calculations. Specifically, researchers may need to use techniques that allow them to accomodate for initial amounts of daughter atoms in a crystal.

What Does an Isotopic Date Mean?

At high temperatures, atoms in a crystal vibrate so rapidly that chemical bonds can break and re-form relatively easily. As a consequence, isotopes can escape from or move into crystals. Because isotopic dating is based on the parent–daughter ratio, the isotopic clock starts only when crystals become cool enough for both types of isotopes to be locked into the crystal. The temperature below which isotopes can no longer freely move in and out of a crystal is called the **closure temperature** of a mineral. When we specify an isotopic date for a mineral, we are defining the time at which the mineral cooled below its closure temperature.

The concept of closure temperature provides a basis for interpreting the meaning of isotopic dates. In the case of igneous rocks, isotopic dating tells us when a magma or lava cooled to form a solid, cool igneous rock. In the case of metamorphic rocks, isotopic dating tells us when a rock cooled from a metamorphic temperature above the closure temperature to a temperature below it. Not all minerals have the same closure temperature, so in a rock that cools very slowly, different minerals yield different dates.

Can we use isotopes to date a clastic sedimentary rock directly? No. If we date individual minerals in a sedimentary rock, we determine only when these minerals first crystallized as part of an igneous or metamorphic rock, but not when the minerals were deposited as sediment or when the sediment lithified to form a sedimentary rock. So how can we determine the numerical age of a sedimentary rock? We need to answer this question if we want to add numerical ages to the geologic column. Geologists obtain dates for sedimentary rocks by studying cross-cutting relations between sedimentary rocks and dateable igneous or metamorphic rocks. For example, if we find a sequence of sedimentary strata deposited unconformably on dateable granite, the strata must be younger than the granite **(Fig. 9.22)**. If a dateable basalt dike cuts the strata, the strata must be older than the dike. And if dateable volcanic ash buried the strata, then the strata must be older than the ash.

Table 9.1 Isotopes used in the isotopic dating of rock

Parent → Daughter	Half-Life (years)	Minerals Containing the Isotopes
$^{147}Sm \rightarrow {}^{143}Nd$	106.0 billion	Garnets, micas
$^{87}Rb \rightarrow {}^{87}Sr$	48.8 billion	Potassium-bearing minerals (mica, feldspar, hornblende)
$^{238}U \rightarrow {}^{206}Pb$	4.5 billion	Uranium-bearing minerals (zircon, uraninite)
$^{40}K \rightarrow {}^{40}Ar$	1.3 billion	Potassium-bearing minerals (mica, feldspar, hornblende)
$^{235}U \rightarrow {}^{207}Pb$	713.0 million	Uranium-bearing minerals (zircon, uraninite)

Take-home message . . .

Isotopic dating of rocks specifies numerical ages in years. To obtain an isotopic date, we measure the ratio of parent radioactive isotope to stable daughter isotope in a mineral. An isotopic date gives the time at which a mineral cooled below its closure temperature. We can directly date igneous and metamorphic rocks, but not clastic sedimentary rocks.

Quick Question -
How can you obtain a numerical age for a sedimentary rock?

9.6 The Geologic Time Scale

Adding Numerical Ages to the Geologic Column

In Section 9.4, we pointed out that the geologic column was established a century before researchers had developed isotopic dating methods, so numerical ages could

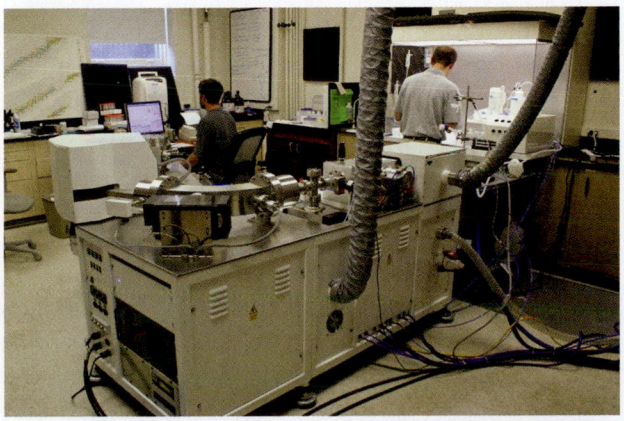

Figure 9.21 In an isotopic dating laboratory, samples are analyzed using a mass spectrometer.

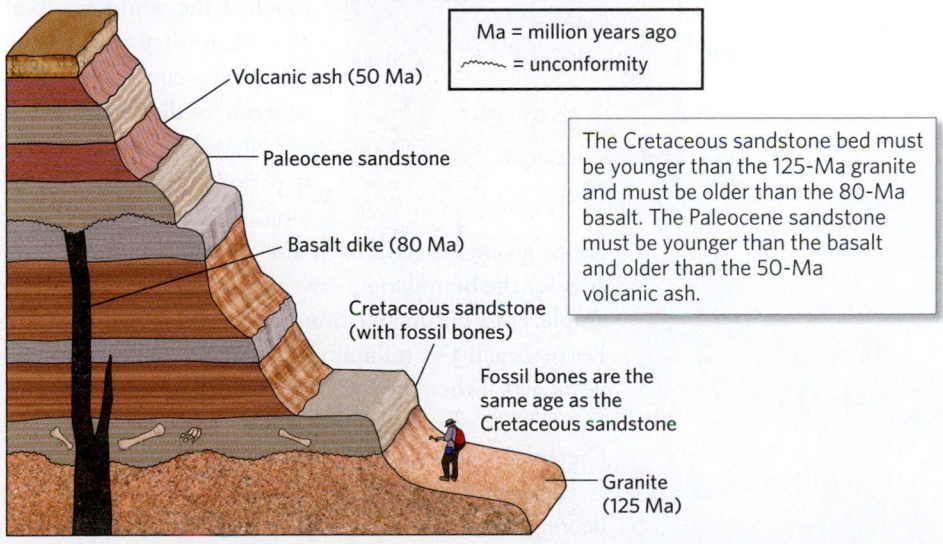

Figure 9.22 Using cross-cutting relations to date sedimentary rocks.

Ma = million years ago
〰〰 = unconformity

Volcanic ash (50 Ma)
Paleocene sandstone
Basalt dike (80 Ma)
Cretaceous sandstone (with fossil bones)
Fossil bones are the same age as the Cretaceous sandstone
Granite (125 Ma)

The Cretaceous sandstone bed must be younger than the 125-Ma granite and must be older than the 80-Ma basalt. The Paleocene sandstone must be younger than the basalt and older than the 50-Ma volcanic ash.

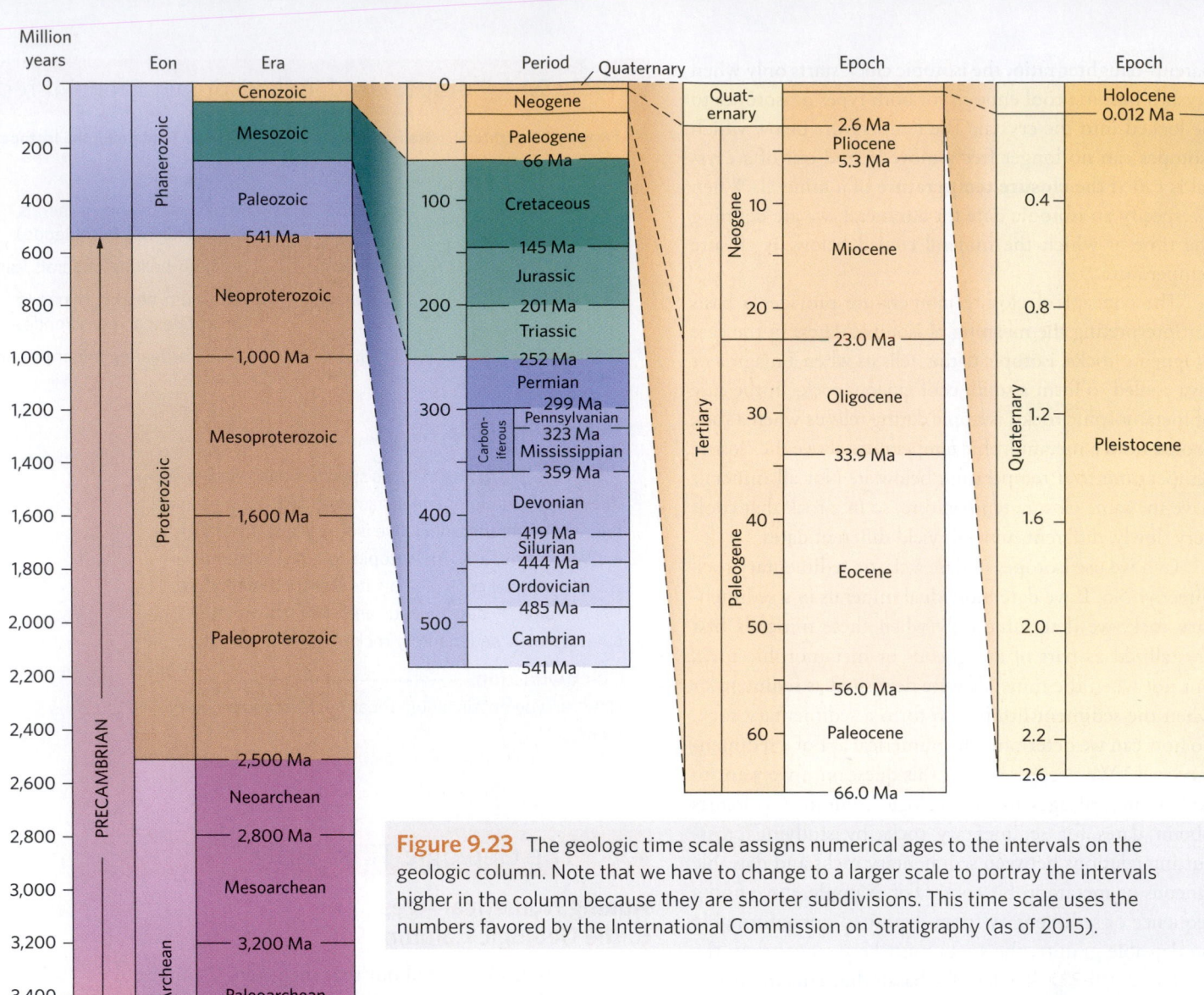

Figure 9.23 The geologic time scale assigns numerical ages to the intervals on the geologic column. Note that we have to change to a larger scale to portray the intervals higher in the column because they are shorter subdivisions. This time scale uses the numbers favored by the International Commission on Stratigraphy (as of 2015).

not initially be assigned to the boundaries between intervals on the geologic column. Once good dating methods became available, geologists searched the world for localities where they could recognize cross-cutting relations between sedimentary rocks and dateable igneous rocks at period boundaries. By isotopically dating the igneous rocks, geologists have been able to determine numerical ages for the boundaries between geologic periods. For example, cross-cutting relations indicate that the Cretaceous Period began 145 million years ago and ended 66 million years ago. (Therefore, the Cretaceous sandstone bed in Figure 9.22 was deposited during the middle part of the Cretaceous, not at the beginning or end of that period.)

As new dates become available, the specific numbers defining the boundaries of geologic periods may shift. For

example, around 1995, new studies on ash layers above and below the Precambrian-Cambrian boundary placed this boundary at 542 Ma, in contrast to previous, less definitive studies that had placed the boundary at 570 Ma. Further studies moved the date to 541 Ma. A chart showing the currently favored numerical ages of eras and periods in the geologic column is called the **geologic time scale (Fig. 9.23)**.

How Old Is the Earth?

During the 18th and 19th centuries, before the discovery of isotopic dating, scientists came up with a great variety of clever answers to the question, "How old is the Earth?"—all of which have since been proved wrong. For example, Lord Kelvin, a renowned 19th-century physicist, estimated that the Earth is 20 million years old by calculating how long it would take a planet the size of the Earth to cool from the temperature of the Sun. Kelvin's estimate contrasted with the longer time estimates promoted by followers of Hutton, who argued that if the concepts of uniformitarianism and evolution were correct, the Earth must be much older than 20 million years.

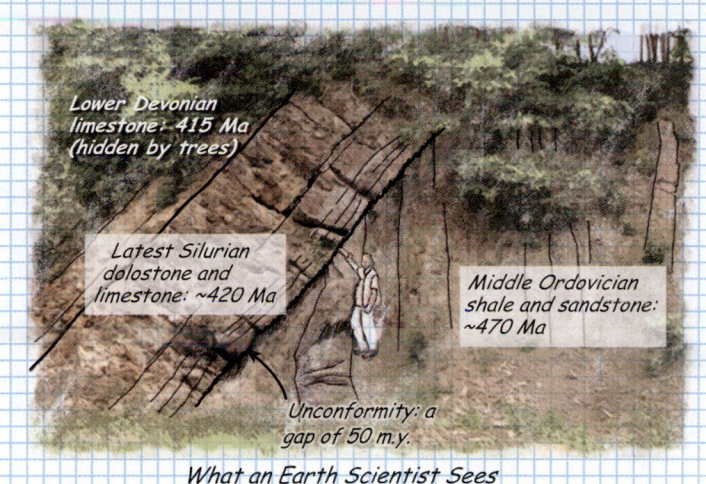

Lower Devonian limestone: 415 Ma (hidden by trees)

Latest Silurian dolostone and limestone: ~420 Ma

Middle Ordovician shale and sandstone: ~470 Ma

Unconformity: a gap of 50 m.y.

What an Earth Scientist Sees

Debate continued until isotopic dating became reliable, at which time geologists determined that the Earth is vastly older than Kelvin predicted. (Kelvin had not known about the radioactive heat generation within the Earth, which causes the planet to cool much more slowly than he realized.) Geologists have scoured the planet to identify its oldest rocks. Rocks that formed more recently than 3.85 Ga are fairly common. (Recall that Ga means billion years ago.) A few rock samples from several localities (Wyoming, Canada, Greenland, and China) have yielded dates as old as 4.03 Ga. Individual grains of the mineral zircon have yielded dates as old as 4.4 Ga, indicating that rock of this age did once exist. To obtain older dates, geologists have analyzed rocks from elsewhere in the Solar System. Meteorites thought to have come from undifferentiated planetesimals have yielded dates as old as 4.57 Ga, so geologists identify 4.57 Ga as the time of planetesimal growth. The oldest meteorites from differentiated planetesimals formed at 4.54 Ga, so if we consider the time of internal differentiation as the time when the Earth formally became a planet, then we can consider geologic time to begin at 4.54 Ga.

Why don't we find rocks with dates between 4.03 and 4.54 Ga in the Earth's crust? Geologists have come up with several hypotheses to explain the lack of extremely old rocks. Some argue that during the first half billion years of its existence, the Earth was so hot that rocks in the crust remained above the closure temperatures of dateable minerals, so isotopic clocks did not start ticking. Others suggest that the inner planets were bombarded so intensely by meteorites at about 4.0 Ga that crust formed earlier than 4.0 Ga was melted or even vaporized (see Chapter 10).

The number 4.54 Ga is so staggeringly large that it's hard to understand. One way to grasp the immensity of geologic time is to equate Earth's history to a single calendar year. On this scale, the oldest rocks preserved on Earth appear in early February, the first bacteria arrive on February 21, the first shelly invertebrates on October 25, the first amphibians crawl out onto land on November 20, and the continents coalesce into Pangaea on December 7. Dinosaurs appear on December 15 and go extinct on December 25. The last week of December represents the Cenozoic, the last 66 million years of Earth history. The first human-like ancestor appears on December 31 at 3 P.M. Our species, *Homo sapiens*, shows up an hour before midnight, and all of recorded human history takes place in the last 30 seconds. To put it another way, human history occupies the last 0.000001% of Earth history. The Earth is so old that it has had more than enough time for its rocks and life forms to have formed and evolved **(Fig. 9.24)**, for mountain ranges to rise and erode away, and for supercontinents to coalesce and disperse.

Take-home message . . .

Isotopic dating techniques can be used to date igneous and metamorphic rocks. Numerical ages for sedimentary rocks can be deduced from isotopic dating of cross-cutting dateable rocks. Such work led to the geologic time scale. The oldest rock of Earth's crust is about 4 billion years old. Dating of meteorites indicates that the Earth became a planet at 4.54 Ga.

Quick Question ---------------------------
Why have the assignments of numerical ages to periods on the geologic time scale changed?

If the Eiffel Tower were now representing the world's age, the skin of paint on the pinnacle-knob at its summit would represent man's share of that age; and anybody would perceive that that skin was what the tower was built for. I reckon they would, I dunno.

—*MARK TWAIN (AMERICAN WRITER, 1835–1910)*

Did you ever wonder . . .
how old the Earth's oldest rock is?

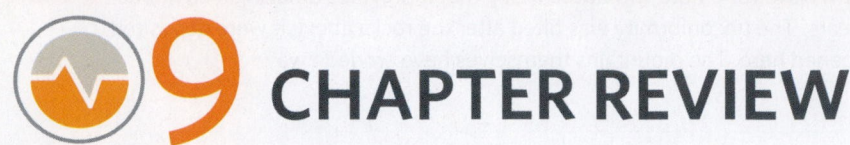

9 CHAPTER REVIEW

Chapter Summary

- Geologic time refers to the time span since the Earth's formation.

- Relative age specifies whether one geologic feature is older or younger than another, whereas numerical age provides the age of a geologic feature in years.

- Using such principles as uniformitarianism, original horizontality, superposition, and cross-cutting relations, we can construct the geologic history of a region.

- Fossils form when organisms or traces of organisms are buried and preserved in rock. Typically, hard parts of organisms are most likely to be preserved.

- Examples of fossils include preserved shells or bones; bodies preserved in amber, tar, or permafrost; molds or casts of shells or bones; permineralized wood or bones; carbonized impressions; and trace fossils. Certain distinctive organic molecules can serve as chemical fossils or biomarkers.

- Fossils provide a record of life's evolution and a way of determining the ages of rock layers relative to one another.

- The principle of fossil succession states that the assemblage of fossils in younger strata differs from that in older strata. Once a species becomes extinct, it never reappears.

- At several times during the Earth's history, a high percentage of species went extinct. Such mass-extinction events may be a consequence of catastrophic events.

- Darwin's theory of evolution by natural selection states that the fittest organisms survive and pass on their traits to their offspring. Over time, organisms become so different from their distant ancestors that they can be considered new species.

- The fossil record is incomplete because preserved strata do not record all of geologic time, not all organisms are preserved, and paleontologists have found only a tiny fraction of fossil species.

- A stratigraphic column shows the succession of strata in a region. A given sequence of beds that can be traced over a fairly broad region is a stratigraphic formation.

- Strata are not necessarily deposited continuously at a location. An interval of nondeposition or erosion is called an unconformity. Geologists distinguish among angular unconformities, nonconformities, and disconformities.

- Correlation defines the age relationship between formations at one location and formations at another.

- The geologic column represents the entirety of geologic time. Its largest subdivisions are eons. Eons are subdivided into eras, eras into periods, and periods into epochs.

- A geologic map shows the distribution of formations and geologic structures.

- The numerical ages of rocks can be determined by isotopic (radiometric) dating. This method is based on the observation that radioactive elements decay at a rate characterized by a known half-life.

- The isotopic age of a mineral specifies the time at which the mineral cooled below a closure temperature. We can use isotopic dating to determine when an igneous rock solidified and when a metamorphic rock cooled.

- To date sedimentary strata, we must examine their cross-cutting relations with dateable igneous or metamorphic rock.

- Isotopic dating indicates that the Earth differentiated and became a planet at 4.54 Ga.

Key Terms

angular unconformity (p. 295)
baked contact (p. 285)
biodiversity (p. 299)
closure temperature (p. 307)
contact (p. 294)
correlation (p. 296)
cross-cutting relations (p. 285)
daughter isotope (p. 303)
disconformity (p. 295)
eon (p. 299)
epoch (p. 299)
era (p. 299)

extinct (p. 291)
fossil (p. 287)
fossil assemblage (p. 292)
fossilization (p. 287)
fossil succession (p. 292)
geochronology (p. 301)
geologic column (p. 298)
geologic history (p. 285)
geologic map (p. 302)
geologic time (p. 283)
geologic time scale (p. 308)
half-life (p. 303)
inclusion (p. 285)

isotope (p. 303)
isotopic dating (p. 301)
mass-extinction event (p. 299)
nonconformity (p. 295)
numerical age (p. 301)
original horizontality (p. 285)
paleontology (p. 287)
parent isotope (p. 303)
period (p. 299)
Precambrian (p. 299)
preservation potential (p. 288)
punctuated equilibrium (p. 293)

radioactive decay (p. 303)
radioactive element (p. 301)
relative age (p. 285)
stratigraphic column (p. 294)
stratigraphic formation (p. 294)
superposition (p. 285)
taxonomy (p. 292)
theory of evolution (p. 293)
unconformity (p. 295)
uniformitarianism (p. 284)

The letters following each Review Question refer to the corresponding Learning Objective from the Chapter Opener.

1. Explain the concept of uniformitarianism. **(B)**

2. Compare the meaning of a relative age and a numerical age. **(A)**

3. Describe the principles that allow us to determine the relative ages of geologic events. **(B)**

4. Which type of unconformity does this drawing show? **(E)**

5. Describe the various processes that can produce fossils. **(C)**

6. Why did geologists determine the change from the Precambrian to the Cambrian to be an eon boundary, and how does this connect to the theory of evolution? **(D)**

7. How does an unconformity develop? Describe the differences among the three kinds of unconformities. **(E)**

8. What is a stratigraphic formation? How are stratigraphic formations portrayed on geologic maps? **(F)**

9. Describe two different methods of correlating rock units. How was correlation used to develop the geologic column? Which of the fossils (X, Y, or Z) in this stratigraphic column is youngest? **(F)**

10. What does the process of radioactive decay entail? **(G)**

11. How do geologists obtain an isotopic date? What are some of the pitfalls in obtaining a reliable one? **(G)**

12. Why can't we date sedimentary rocks directly? How do we obtain numerical ages of periods on the geologic column? **(G)**

13. What is the age of the oldest rock yet found on Earth? What is the age of the oldest meteorite? Why is there a difference? **(H)**

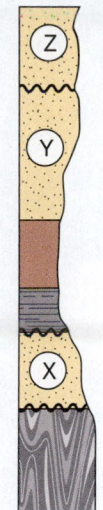

14. Imagine an outcrop exposing a formation consisting of alternating sandstone and conglomerate beds. A geologist studying the outcrop notes the following:

    - The sandstone beds contain land plants, but the fragments are too small to permit identification.
    - A layer of volcanic ash, dated at 300 Ma, overlies the sandstone-conglomerate formation.
    - A paleosol (ancient soil layer) occurs at the base of the ash layer.
    - A 100-Ma basalt dike cuts the ash and the sandstone-conglomerate formation.
    - Pebbles of granite in the conglomerate yield radiometric dates of 400 Ma.

    On the basis of these observations, how old is the sandstone and conglomerate formation? (Specify both the numerical age range and the period or periods of the geologic column during which it formed.) **(F)**

15. Look again at Figure 9.24. The rocks below the unconformity are Middle Ordovician turbidites deposited in deep water, and the rocks above the unconformity are Lower Devonian limestones deposited in shallow water. What type of unconformity does this outcrop expose? Provide a brief geologic history of the outcrop, keeping in mind that the unconformity has been tilted. **(E)**

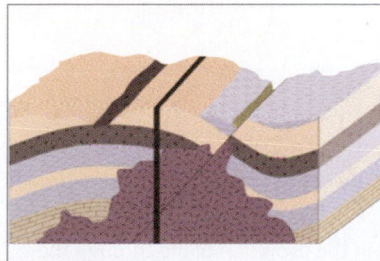

Animations
This chapter features a series of animations demonstrating relative age dating and the different types of geologic unconformities.

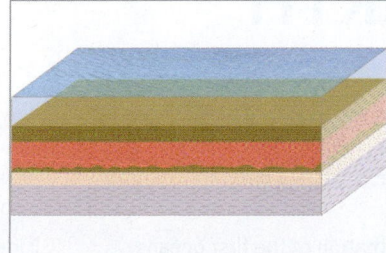

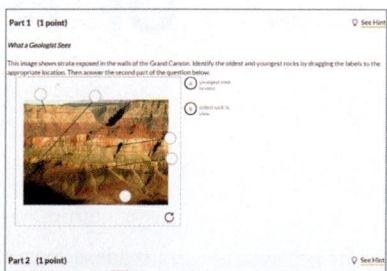

Smartwork5
This chapter covers the principles of defining relative age, unconformities, and identifying relative ages.

10 A BIOGRAPHY OF THE EARTH

By the end of the chapter you should be able to . . .

A. recognize geologic clues that provide a basis for interpreting Earth history.

B. describe a model for the formation of the first oceans and continents.

C. explain the concept and history of supercontinents forming and then drifting apart.

D. describe the earliest life and how life evolution is linked to atmospheric evolution.

E. explain the observation that led paleontologists to define the beginning of the Paleozoic.

F. discuss when and why continental interiors sometimes host shallow seas.

G. identify key stages in the evolution of life through the Phanerozoic.

H. discuss when and why mountain belts formed at various times and places in the past.

I. define global change and provide examples of various types of change.

J. describe the diverse ways in which humans may cause global change.

10.1 Introduction

In 1868, Thomas Henry Huxley, a British biologist, presented a public lecture to an audience in Norwich, England. Seeking a way to convey his fascination with the Earth's history, he focused the audience's attention on the piece of chalk he'd been drawing with. And what a tale the chalk had to tell! Chalk consists of microscopic marine plankton shells. The specific chalk that Huxley held came from 90-Ma beds now exposed in the dramatic white cliffs of England's southeastern coast. Geologists in Huxley's day knew that similar chalk beds cropped out throughout much of Europe and that some of these beds contained fossils of bizarre swimming reptiles that were very different from those of modern times. Clues in his humble piece of chalk allowed Huxley to demonstrate

that the geography and inhabitants of the Earth in the past differed markedly from those of today. In other words . . . the Earth has a history!

In this chapter, we explore Earth's history by offering a concise geologic biography of our home planet, from its birth 4.54 billion years ago to the present. We proceed chronologically, beginning with our planet's formation and finishing with the present day. We see how, over billions of years, the Earth's surface, oceans, and atmosphere changed dramatically as continents drifted, mountain belts grew, climate alternated between warmer and cooler, and sea level rose and fell. In addition, we examine how life evolved and diversified. The chapter ends by introducing the concept of global change and the ways in which humanity may be leaving its mark on Earth history.

These fossil ammonites are marine organisms that lived 200 million years ago. The rocks containing them now lie in the hills of Germany. Clearly, the Earth has a long and complicated history.

Figure 10.1 Scenes from the Hadean Eon.

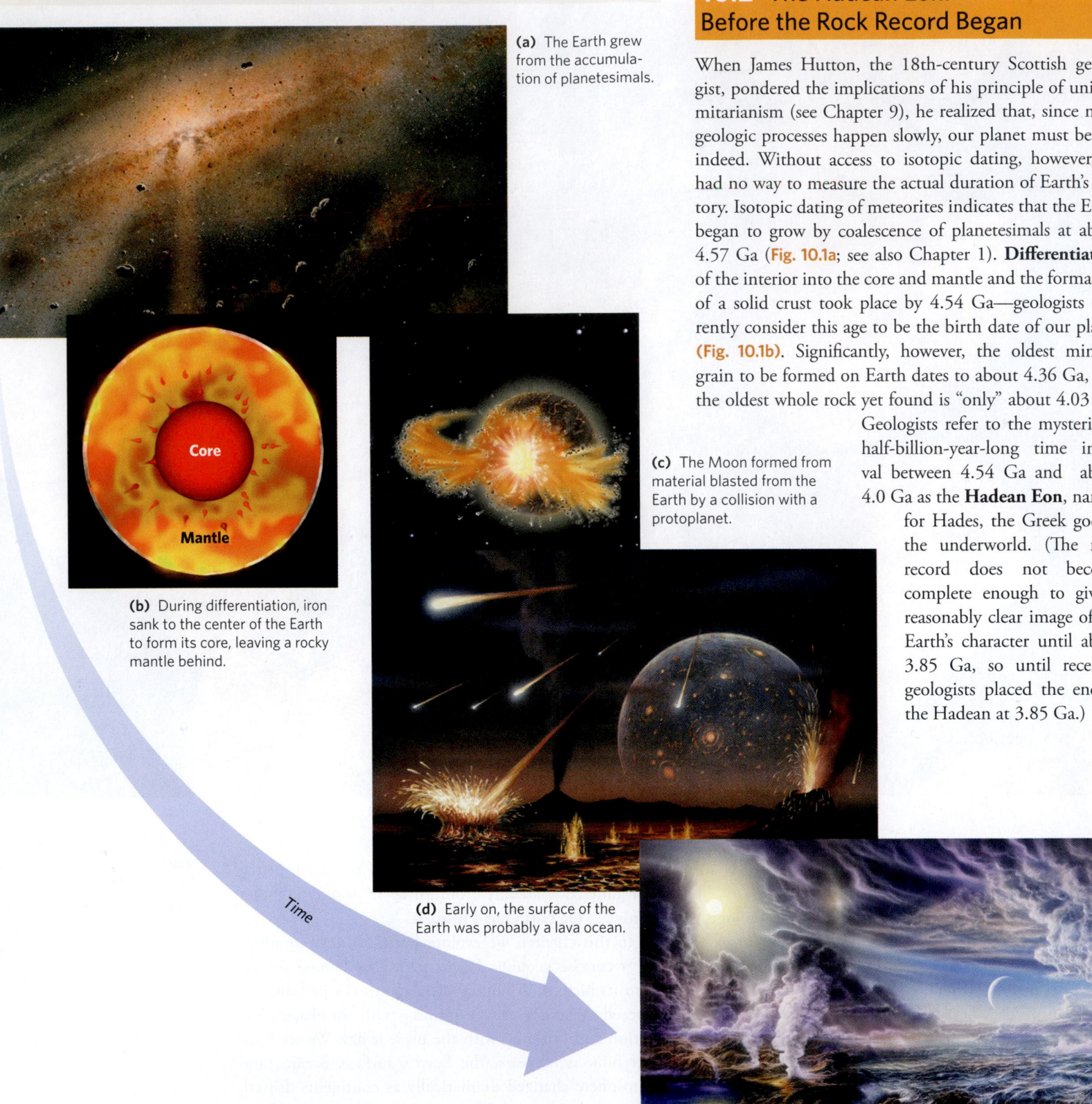

(a) The Earth grew from the accumulation of planetesimals.

(b) During differentiation, iron sank to the center of the Earth to form its core, leaving a rocky mantle behind.

Core

Mantle

(c) The Moon formed from material blasted from the Earth by a collision with a protoplanet.

Time

(d) Early on, the surface of the Earth was probably a lava ocean.

(e) Some crust may have solidified during the Hadean, and an early ocean may have developed.

10.2 The Hadean Eon: Before the Rock Record Began

When James Hutton, the 18th-century Scottish geologist, pondered the implications of his principle of uniformitarianism (see Chapter 9), he realized that, since most geologic processes happen slowly, our planet must be old indeed. Without access to isotopic dating, however, he had no way to measure the actual duration of Earth's history. Isotopic dating of meteorites indicates that the Earth began to grow by coalescence of planetesimals at about 4.57 Ga (Fig. 10.1a; see also Chapter 1). **Differentiation** of the interior into the core and mantle and the formation of a solid crust took place by 4.54 Ga—geologists currently consider this age to be the birth date of our planet (Fig. 10.1b). Significantly, however, the oldest mineral grain to be formed on Earth dates to about 4.36 Ga, and the oldest whole rock yet found is "only" about 4.03 Ga.

Geologists refer to the mysterious, half-billion-year-long time interval between 4.54 Ga and about 4.0 Ga as the **Hadean Eon**, named for Hades, the Greek god of the underworld. (The rock record does not become complete enough to give a reasonably clear image of the Earth's character until about 3.85 Ga, so until recently, geologists placed the end of the Hadean at 3.85 Ga.)

Many major changes happened on Earth during the Hadean. As we saw in Chapter 1, the eon began with differentiation of the Earth's interior, during which gravity pulled molten iron down to the center of the Earth, where it accumulated to form the metallic core, leaving behind a mantle of ultramafic rock. Researchers suggest that about 4.53 Ga, soon after differentiation, a protoplanet collided with the Earth. This impact blasted away a significant fraction of Earth's mantle to yield a ring of silicate debris that coalesced to form our Moon (Fig. 10.1c).

In the wake of differentiation and Moon formation, the Earth may have been so hot that much of its surface was covered with lava. Rafts of solid rock may temporarily have formed on the surface of the lava ocean, but they sank and remelted. It's this image of a hell-like environment that led to the name Hadean (Fig. 10.1d). How long the Earth's surface stayed molten remains a subject of debate. The young planet contained relatively large quantities of radioactive elements. Elements with short half-lives were decaying and releasing heat, so geologists once thought it remained hot through much of the Hadean. But recent calculations hint that it may have taken only a few tens of millions of years after Moon formation for the Earth's surface to have cooled enough to have hosted a crust of solid rock, locally covered by a water ocean (Fig. 10.1e).

During the Hadean Eon, volatile (gassy) elements and compounds that had originally been incorporated into mantle minerals bubbled out of volcanic vents. The products of this *outgassing* accumulated to form the Earth's first atmosphere, a murky mixture consisting mostly of carbon dioxide (CO_2), water vapor (H_2O), nitrogen (N_2), methane (CH_4), ammonia (NH_3), hydrogen (H_2), hydrogen sulfide (H_2S), and sulfur dioxide (SO_2). Some researchers speculate that comets colliding with Earth may have contributed gas molecules, and perhaps even organic matter, to this early atmosphere.

Take-home message . . .

The Hadean Eon began when the Earth differentiated. Shortly after this, a protoplanet collided with the Earth to form the Moon. Initially molten, the Earth's surface eventually cooled and solidified. But no rock record of the Hadean exists because our planet's surface may have been partially molten.

Quick Question -
Where did the gases of the Earth's early atmosphere come from, and what did this atmosphere consist of?

10.3 The Archean Eon: Birth of Continents and Life

Though very few samples of the Earth remain from before 3.85 Ga, geologists now place the start of the **Archean Eon** (from the Greek word *arché*, meaning beginning) at 4.0 Ga, about the age of the oldest known rock. Why are rocks formed before 3.85 Ga so rare? Some researchers argue that rocks older than this date have passed through the rock cycle (see Chapter 6) to become components of younger rocks. Others note that studies of the craters visible on the Moon indicate that meteorites battered the inner planets between 4.0 and 3.9 Ga. This **late heavy bombardment** may have pulverized and melted any crust that existed at the time, so that only after these impacts ceased could long-lasting crust, atmosphere, and oceans survive (Fig. 10.2).

By 3.85 Ga, the near-surface realm of the Earth was cool and stable enough for rocks to form and survive. The cooling Earth System allowed water in the atmosphere to condense, and rains filled permanent oceans. (The evidence for oceans at that time comes from the discovery of 3.85-billion-year-old marine sedimentary rocks.) Most atmospheric CO_2 dissolved into the new oceans. Removal of H_2O and CO_2 from the atmosphere left it with a high concentration of nitrogen (N_2) gas, for nitrogen does not react chemically with or dissolve in other substances. So at the dawn of the Archean, the Earth had a transparent atmosphere composed primarily of N_2 (see Chapter 17).

Continental Crust Appears

The earliest land probably formed at hot-spot volcanoes, which extruded very thick plateaus of mafic lava on the floor of the early oceans. So a visitor to the Earth's surface at the beginning of the Archean would probably

Figure 10.2 The late heavy bombardment may have pulverized and remelted the Earth's crust between 4.0 and 3.9 Ga. Few rocks older than this event remain.

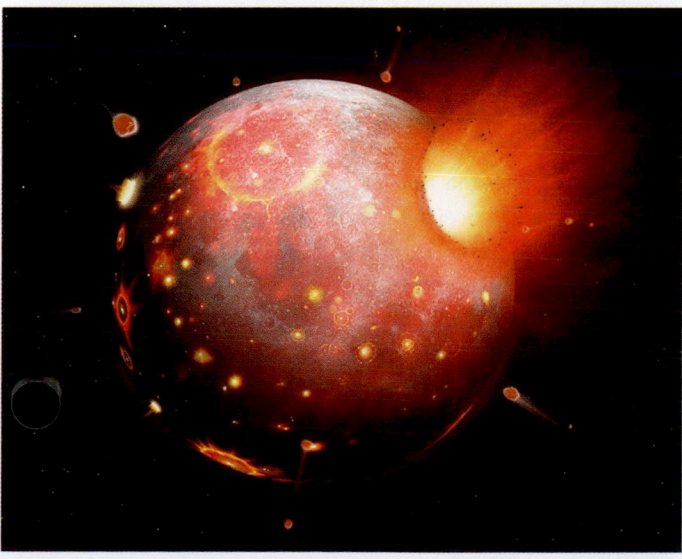

Figure 10.3 A model for crust formation during the Archean Eon.

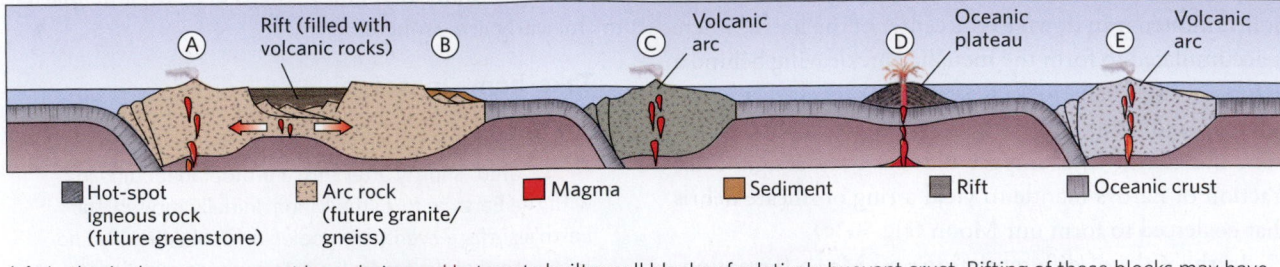

| Hot-spot igneous rock (future greenstone) | Arc rock (future granite/ gneiss) | Magma | Sediment | Rift | Oceanic crust |

(a) In the Archean, convergent boundaries and hot spots built small blocks of relatively buoyant crust. Rifting of these blocks may have produced flood basalts, and erosion of the blocks produced sediment.

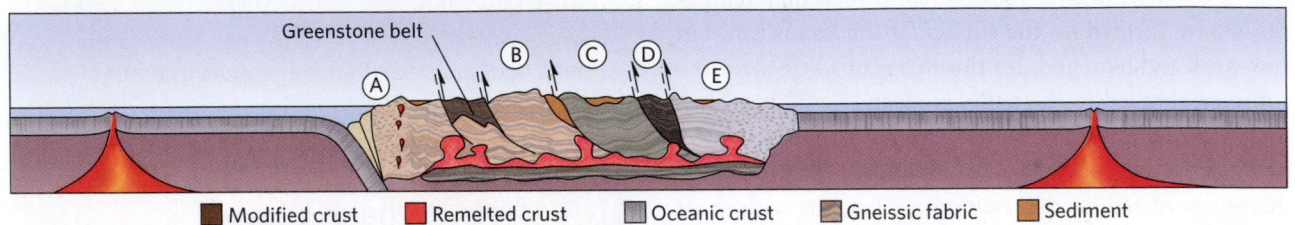

| Modified crust | Remelted crust | Oceanic crust | Gneissic fabric | Sediment |

(b) Buoyant blocks collided and were sutured together, forming protocontinents, which were intruded by granite. Eventually, regions of crust cooled, stabilized, and became continents.

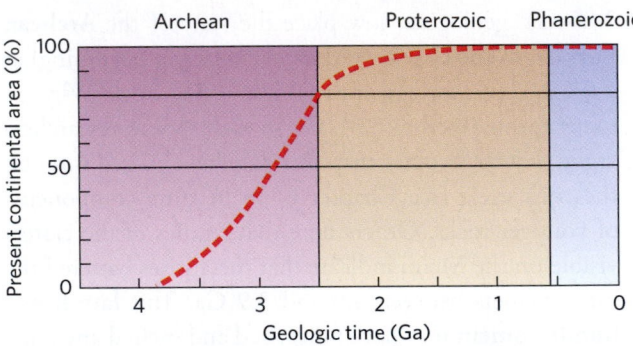

(c) As time progressed, the area of the Earth covered by continental crust increased. Most of the crust that exists today had formed by the end of the Archean.

As collisions continued, the crustal blocks merged to form *protocontinents*. The interiors of these protocontinents cooled over time because they were isolated from the heat of plate-boundary volcanism. When protocontinent interiors cooled, the rock they contained became stronger because it could no longer deform plastically (see Chapter 7), and this strength made the protocontinents durable. By 3.2 Ga, several durable protocontinents had formed, and by 2.7 Ga, some of them had been sutured together to form the first continents. By the end of the Archean, perhaps 80% of the Earth's continental crust existed **(Fig. 10.3c)**.

Early Life Appears

The search for the earliest life on Earth makes headlines in the popular media. While some evidence hints that life appeared before the late heavy bombardment, chemical fossils of life appear only in strata younger than about 3.8 Ga. Shapes resembling cells appear in rocks dated as early as 3.5 Ga, but the oldest undisputed fossil cells of bacteria and archaea, the most primitive living organisms, occur in strata dating from 3.2 Ga **(Fig. 10.4a)**. These rocks contain *stromatolites*, distinct layered mounds interpreted to be made up of fossilized mats of *cyanobacteria* (photosynthetic bacteria). The mounds form because sediment settling out of ocean water sticks to a mucus-like substance secreted by the mats of cyanobacteria. As the mat becomes buried, new cyanobacteria colonize the top of the sediment, building the mound upward **(Fig. 10.4b, c)**.

have seen a water ocean from which only volcanic islands protruded. When plate tectonics began to operate, mafic igneous activity also began to take place at convergent boundaries **(Fig. 10.3a)**. Some of the mafic rock of early Archean island arcs and oceanic plateaus was not subducted, however, because it was less dense than the ultramafic rock of the underlying mantle. When these relatively buoyant blocks collided with one another, they were sutured together to form larger crustal blocks that remained at the Earth's surface **(Fig. 10.3b)**. Metamorphic rocks formed during these collisions. At various locations, rifting took place, and flood basalts extruded. Eventually, melting at depth in crustal blocks produced granite and other lower-density igneous rocks, and erosion of exposed crust yielded sediments. As a result of all these processes, the blocks contained a variety of rock types.

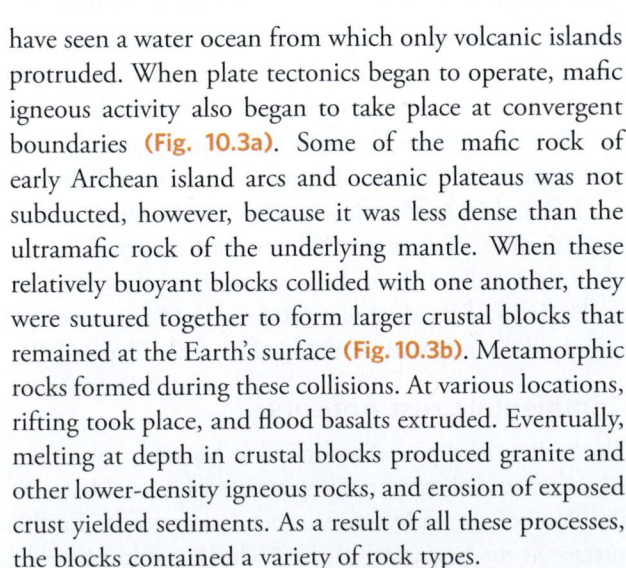

Did you ever wonder . . .

what the oldest relict of life is?

Figure 10.4 Examples of Archean life.

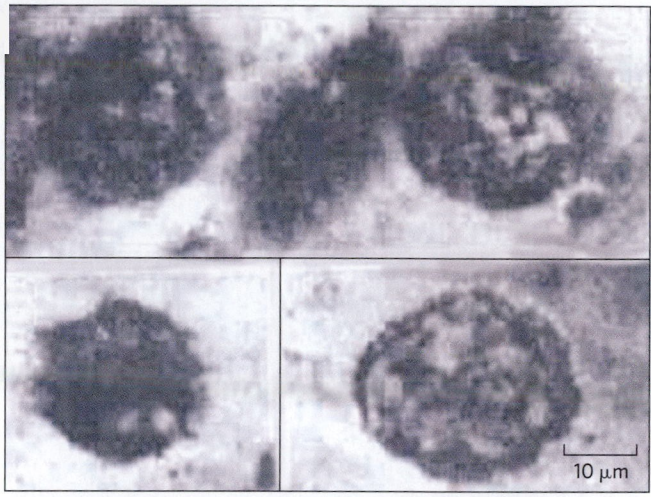

(a) These shapes in 3.2-Ga chert from South America are thought to be fossil bacteria or archaea.

(b) This weathered outcrop of 1.85-Ga dolostone near Marquette, Michigan, reveals the layer-like structure of stromatolites. The delicate ridges represent the fossilized remnants of cyanobacterial mats. Similar stromatolites occur in exposures of Archean rocks.

(c) Modern stromatolites in Shark Bay, Western Australia.

What specific environment on the Archean Earth served as the cradle of life? Laboratory experiments of the 1950s led researchers to speculate that life began in warm pools of surface water beneath a methane- and ammonia-rich atmosphere shocked by bolts of lightning. Studies in the past two decades, however, suggest that deep-sea hydrothermal vents—so-called black smokers (see Fig. 4.28c)—may have hosted the first organisms. These vents emit clouds of ion-charged solutions from which sulfide minerals precipitate. The earliest life in the Archean may well have been heat-loving bacteria or archaea that dined on sulfides emitted from these vents in the dark depths of the ocean. Only later in the Archean, when organisms evolved the ability to carry out photosynthesis (between 3.5 and 3.2 Ga), did life move into shallow, brightly lit water. The oxygen produced by these organisms began to change the composition of the atmosphere.

As the Archean Eon came to a close, the first continents had formed, plate tectonics was under way, continental drift was taking place, collisional mountain belts were forming, and erosion was occurring. Life had colonized not only the depths of the sea, but also the shallow marine realm. The atmosphere had transformed from an H_2O- and CO_2-rich one into an N_2-rich one containing traces of O_2. The stage was set, by about 2.5 Ga, for another major change in the Earth System.

Take-home message . . .

During the Archean (4.0–2.5 Ga), the first continental crust formed from colliding volcanic arcs and oceanic plateaus. As oceans filled with water, the atmosphere lost its H_2O and CO_2 and became nitrogen rich. Early life (archaea and bacteria) appeared in the sea.

Quick Question -
What environment may have served as the cradle of early life?

10.4 The Proterozoic Eon: The Earth in Transition

The **Proterozoic Eon**, the last interval of the Precambrian, spans roughly 2 billion years—almost half of the Earth's history—from 2.5 Ga to the beginning of the Cambrian Period at 541 Ma. During the Proterozoic, Earth's surface environment changed from an unfamiliar world of small continents and an oxygen-poor atmosphere into a world of large continents and an oxygen-rich atmosphere much more like what we see today.

Box 10.1 ▶ Consider this . . .

The nature of the North American craton

To better understand the character of a craton, let's explore the North American craton. As we saw in Chapter 7, cratons can be divided into two areas: a shield and a cratonic platform **(Fig. Bx10.1a)**. Throughout North America's **shield**, outcrops at the ground surface expose Precambrian basement of igneous and metamorphic rocks older than about 1 Ga. The landscape of the shield tends to have fairly low relief—there are small hills and valleys, but no dramatic mountain ranges. Most of North America's shield lies in Canada, so geologists refer to it as the *Canadian Shield*. Throughout North America's **cratonic platform**, which surrounds the shield and also underlies Hudson Bay, a blanket of Paleozoic or Mesozoic strata covers the Precambrian basement. The cratonic platform underlies much of the interior plains of the United States.

Isotopic dating of rock samples has allowed geologists to subdivide North America's craton into distinct crustal provinces, each of which has been assigned a name **(Fig. Bx10.1b)**. The Canadian Shield includes several Archean crustal provinces, which were sutured together during Proterozoic mountain-building events. In contrast, the basement of the cratonic platform in the United States grew when a series of island arcs and continental slivers accreted (attached) to the southern margin of the Canadian Shield between 1.8 and 1.6 Ga. In the Midwest, granite plutons intruded much of this accreted region between 1.5 and 1.3 Ga. At about 1.1 Ga, North America collided with another continent (probably South America) on its eastern and southern margins, adding yet more crust—the Grenville Province—to what is now the craton.

Figure Bx10.1 North America's craton.

(a) The craton occupies much of the continental interior. In the shield, Precambrian rock is exposed at the surface. In the cratonic platform, Phanerozoic strata cover Precambrian rock.

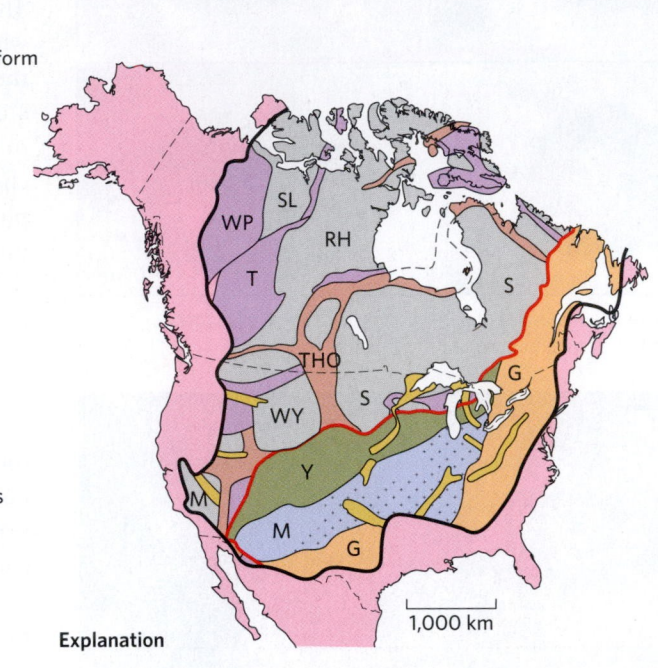

Explanation

G = Grenville; M = Mazatal; Y = Yavapai;

P = Penokean; THO = Trans-Hudson orogen;

WY = Wyoming; WP = Wopmay; T = Thelon;

S = Superior; MO = Mojave; RH = Rae and Hearn;

SL = Slave

— Edge of the craton

— Pre-1.8 and post-1.8 Ga crust boundary

Post-Precambrian crust

Proterozeroic rifts of various ages

Grenville orogen (1.3 – 1.0 Ga)

Granite-rhyolite province (1.5 – 1.3 Ga)

Mazatzal accreted crust (1.7 – 1.6 Ga)

Yavapai accreted crust (1.8 – 1.7 Ga)

Proterozoic collisional orogens (1.9 – 1.8 Ga)

Proterozoic accreted crust (2.0 – 1.8 Ga)

Archean provinces (> 2.5 Ga)

(b) The craton consists of several distinct crustal provinces. Most of what is now Canada and the northwestern United States had assembled by 1.8 Ga. Belts of younger terranes accreted to the craton between 1.8 and 1.1 Ga.

Oceanic Crust
0–20 Ma   20–65 Ma   > 65 Ma

Continental Crustal Province
- Stretched crust
- Large igneous provinces
- Phanerozoic orogens
- Phanerozoic basins
- Phanerozoic platforms
- Precambrian shields
- Archean crustal remnants

U.S. Geological Survey

Large Continents Form

New continental crust continued to form during the Proterozoic, but at progressively slower rates. By the middle of the eon, over 90% of the Earth's present continental crust had formed (see Fig. 10.3c). Some blocks of continental crust became so strong and stable that they have survived to the present time without undergoing significant deformation or metamorphism. During the Proterozoic, several of these blocks collided and were sutured together to form even larger regions of continental crust. A large, relatively stable block of continental crust is called a **craton** (Box 10.1). All the cratons that exist today had formed by about 1 Ga. Each craton consists of many crustal provinces, distinguished from one another primarily by age (Fig. 10.5). A *crustal province* is a region that shares a particular geologic history.

Because of continental drift, the map of the Earth constantly changes. Using a variety of data sources, including correlation of rock units and paleomagnetism, geologists have been able to define the relative positions of continents in the past and to portray their locations on maps. To emphasize that these maps represent times in the past, they are called **paleogeographic maps**. Successive collisions ultimately brought together most continental crust

on Earth into a single supercontinent, named **Rodinia**, by around 1 Ga. On a paleogeographic map of Rodinia (Fig. 10.6a), we can identify the regions that eventually became the familiar continents of today. The last major collision during the formation of Rodinia, an event called the *Grenville orogeny*, affected eastern North America. (Remember that an *orogeny* is a mountain-building event that produces geologic structures and, generally, metamorphic rocks.) Several studies suggest that later, between 800 and 600 Ma, Antarctica, India, and Australia broke away from western North America, swung around, and collided with the eastern margin of North America to form a short-lived supercontinent known as **Pannotia** (Fig. 10.6b).

Life and Climate in the Proterozoic World

Fossil evidence suggests that the Proterozoic saw important steps in the evolution of life. When this eon began, most life was *prokaryotic*, meaning that it consisted of single-celled organisms—Archaea and Bacteria—that do not contain a cell nucleus (an internal, membrane-surrounded region containing the cell's DNA). Though studies of chemical fossils hint that *eukaryotic* life, consisting of cells that do have nuclei, originated as early as 2.7 Ga,

Figure 10.6 Supercontinents in the Proterozoic.

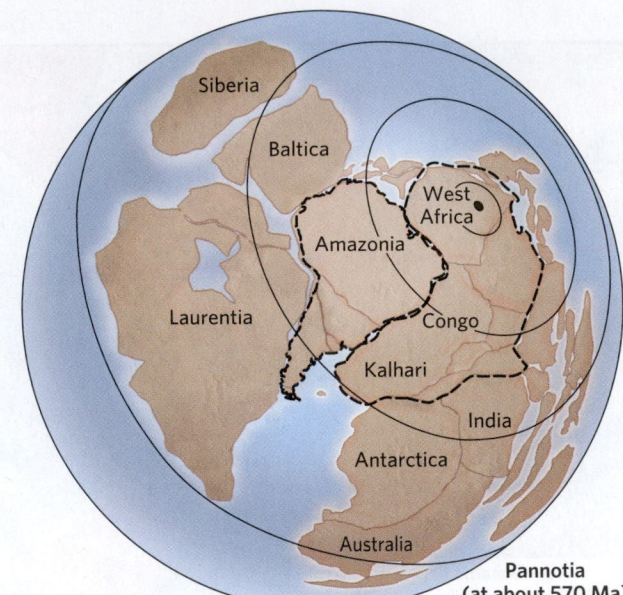

(a) Rodinia formed around 1 Ga and lasted until about 700 Ma. Note that Laurentia consists of the landmasses that are now North America and Greenland.

(b) According to one model, by 570 Ma, Rodinia had broken apart; continents that once lay to the west of Laurentia ended up to the east of Africa. The resulting supercontinent, Pannotia, broke up soon after it formed.

the first possible body fossil of a eukaryotic organism occurs in rocks dated at 2.1 Ga, and abundant body fossils of eukaryotic organisms can be found only in rocks younger than about 1.2 Ga. So the diversification of eukaryotic life, the foundation from which complex organisms, including humans, eventually evolved, took place during the Proterozoic.

The last half billion years of the Proterozoic Eon saw the remarkable transition from simple organisms to complex ones. Sediments deposited perhaps as early as 620 Ma, and certainly by 565 Ma, contain several types of multicellular organisms that together constitute the **Ediacaran fauna**, named for a region in southern Australia. Some of these soft-bodied invertebrate organisms resembled jellyfish, and others resembled worms **(Fig. 10.7a)**.

As the late Proterozoic came to a close, many different types of organisms lived together in the same location, yielding an interacting web of life called an **ecosystem**. In an ecosystem, distinct *food chains*, hierarchies of organisms dependent on eating other organisms, developed. Biologists refer to a particular environment where an ecosystem develops as an *ecological niche*. During the evolution of life from the late Proterozoic onward, very complex ecosystems hosting multicellular organisms of various sizes, including large predators, appeared.

The evolution of life played a key role in changing the composition of Earth's atmosphere. Before life appeared, the atmosphere contained hardly any free oxygen (O_2).

With the appearance of photosynthetic organisms, trace amounts of oxygen accumulated in the atmosphere. But it was not until about 2.4 Ga that the concentration of oxygen reached more than a few percent. This increase, called the **great oxygenation event**, happened when surface rocks and ocean water could no longer absorb or dissolve all the oxygen produced by organisms, so oxygen accumulated as a gas in the atmosphere. When the oceans became saturated with oxygen, the iron that had been dissolved in seawater bonded to oxygen atoms and formed iron oxide minerals that precipitated from the water and settled on the seafloor. In fact, between 2.4 Ga and 1.8 Ga, so much iron oxide settled out of the ocean that thick, colorful sequences of sedimentary rock, called **banded iron formations** (**BIFs**), built up. BIFs, which consists of layers of gray iron oxide minerals (hematite or magnetite) alternating with layers of jasper (red chert), provides most of the iron used by society today **(Fig. 10.7b)**.

The Earth's climate cooled at the end of the Proterozoic Eon, and large ice sheets grew on continents. As a result, glacial till occurs worldwide in late Proterozoic sedimentary sequences. What's strange about the occurrence of this till is that it can be found even in regions that were located at the equator **(Fig. 10.8a)**, implying that the entire planet was cold enough for glaciers to form. Geologists speculate that when continents became entirely glaciated, the entire ocean surface froze as well, and they refer to the resulting ice-encrusted planet as **snowball Earth** **(Fig. 10.8b)**. The

Figure 10.7 Life in the late Proterozoic.

(a) An artist's reconstruction of Ediacaran fauna.

(b) An outcrop of banded iron formation (BIF) from the Upper Peninsula of Michigan. The gray stripes are hematite and the red stripes are jasper.

ice shell may have cut off the oceans from the atmosphere, causing O_2 levels in seawater to drop and many marine life forms to die off. But the ice shell also triggered the end of snowball Earth conditions, for it prevented atmospheric CO_2 from dissolving in seawater, but did not prevent volcanic activity from adding CO_2 to the atmosphere. Carbon dioxide is a greenhouse gas that traps heat in the atmosphere (see Chapter 17), so as the atmospheric CO_2 concentration increased, the Earth warmed up, and eventually the ice shell and continental glaciers melted.

Take-home message . . .

During the Proterozoic (2.5 Ga–541 Ma), the interiors of large continents became cratons. Multicellular organisms appeared, and the atmosphere began to accumulate oxygen. At the end of the Proterozoic, a large supercontinent, Pannotia, existed.

Quick Question -
What was snowball Earth?

Did you ever wonder . . .
whether the oceans have ever frozen over entirely?

Figure 10.8 Snowball Earth.

Strata contain large clasts surrounded by mudstone, for glaciers can carry clasts of all sizes.

Bedding

(a) Layers of Proterozoic glacial till crop out in Africa.

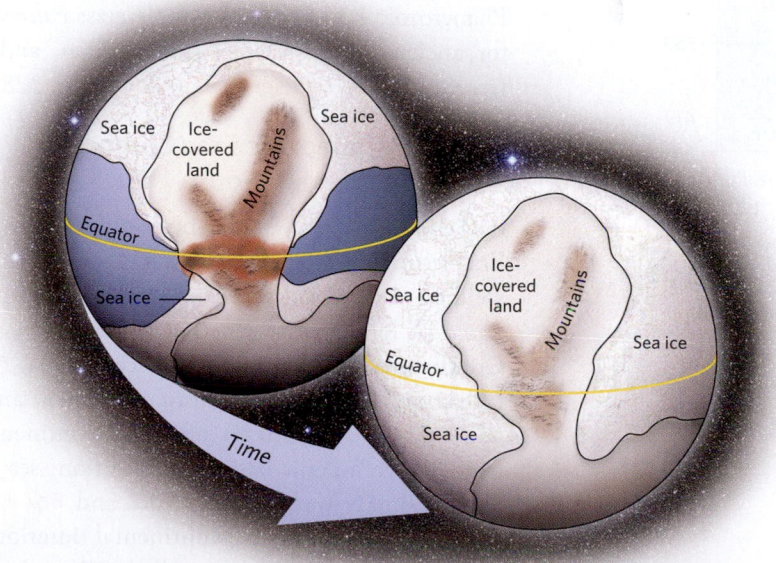

Sea ice — Ice-covered land — Mountains — Sea ice — Equator — Sea ice — Time — Ice-covered land — Mountains — Sea ice — Equator — Sea ice

(b) The planet may have frozen over completely to form "snowball Earth."

Figure 10.9 Land and sea in the early Paleozoic Era.

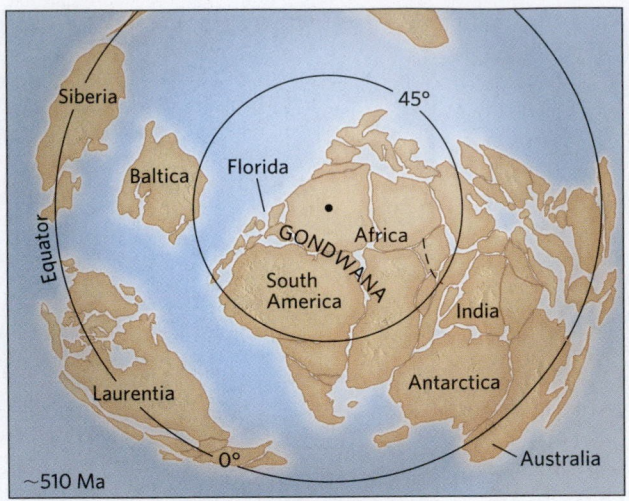

(a) The distribution of continents in the Cambrian Period (510 Ma), as viewed looking down on the South Pole.

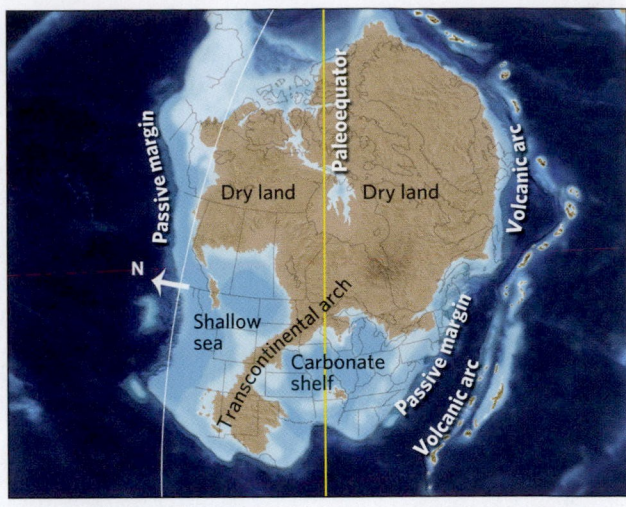

(b) A paleogeographic map of present-day North America shows the regions of dry land and epicontinental seas in the Late Cambrian.

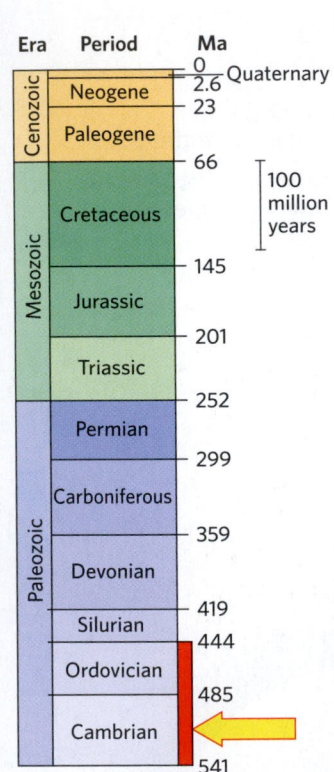

Era	Period	Ma	
Cenozoic	Neogene	0 2.6	Quaternary
		23	
	Paleogene		
		66	}100 million years
Mesozoic	Cretaceous		
		145	
	Jurassic		
		201	
	Triassic		
		252	
Paleozoic	Permian		
		299	
	Carboniferous		
		359	
	Devonian		
		419	
	Silurian		
		444	
	Ordovician		
		485	
	Cambrian		
		541	

10.5 The Paleozoic Era: Continents Reassemble and Life Diversifies

The end of the Proterozoic Eon defines the end of the Precambrian and the start of the **Phanerozoic Eon**. Geologists studying the fossil record recognized the significance of this boundary long before they could assign it a numerical age (currently, 541 Ma) because it marks the appearance of creatures with shells; shells may have evolved as a means of protection against predators. The atmosphere also changed, as photosynthetic organisms prospered. By the beginning of the Paleozoic, O_2 accounted for about 17% of atmospheric gas. The Phanerozoic Eon consists of three eras: *Paleozoic* (Greek for ancient life), *Mesozoic* (middle life), and *Cenozoic* (recent life) (**Earth Science at a Glance**, pp. 328–329).

The Early Paleozoic Era (Cambrian and Ordovician Periods)

PALEOGEOGRAPHY. At the beginning of the Paleozoic Era, rifting broke apart the supercontinent Pannotia, yielding several smaller continents: **Laurentia** (composed of modern-day North America and Greenland), **Gondwana** (South America, Africa, Arabia, Antarctica, India, and Australia), *Baltica* (Europe), and *Siberia* (**Fig. 10.9a**). The coasts of these continents hosted new passive-margin basins. In addition, sea level rose as the climate warmed (**Box 10.2** and **Box 10.3**). As a result, large regions of continental interiors became flooded with shallow seas, called **epicontinental seas** (**Fig. 10.9b**). Sea level did not remain high for the entire

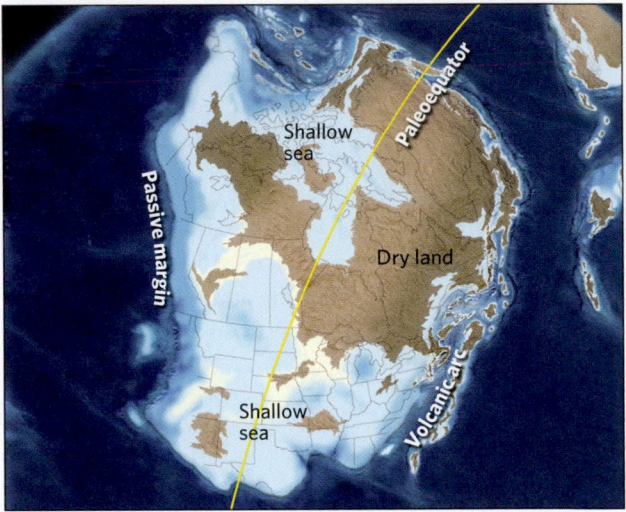

(c) During the Middle Ordovician, epicontinental seas covered much of North America. An island arc off the east coast collided with the continent.

early Paleozoic (see Fig. Bx10.2), however, so continental interiors alternately flooded and dried up. The sediments deposited in epicontinental seas became the strata of cratonic platforms (see Box 10.1).

During the early part of the Paleozoic, passive-margin basins fringed all sides of Laurentia. These basins collected immense amounts of sediment eroded from the continent, and their surfaces became broad continental shelves. In the Middle Ordovician Period, the geologic tranquility of the passive-margin basin on the eastern side of Laurentia came to a close, for the basin rammed into island arcs (**Fig. 10.9c**). These collisions caused the *Taconic orogeny*, which produced a mountain range whose relicts lie within the present-day Appalachians.

Figure 10.10 A museum diorama illustrates what early Paleozoic marine organisms may have looked like.

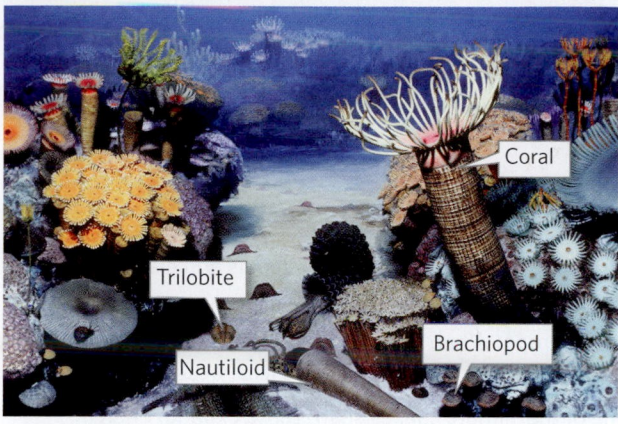

EVOLUTION OF LIFE. The Cambrian began with the appearance of shelled organisms. Soon after, biodiversity increased dramatically. This increase, known as the **Cambrian explosion**, continued over several million years. It may have been triggered by the breakup of Pannotia, for when smaller continents formed and drifted apart, many new ecological niches were formed, and populations of organisms became isolated. By the end of the Cambrian, the seas hosted trilobites, mollusks, brachiopods, nautiloids, gastropods, graptolites, and echinoderms (Fig. 10.10). The Ordovician Period saw the first crinoids as well as the first vertebrate animals (animals with backbones) in the form of jawless fish. In the Middle Ordovician, the first land plants, tiny moss-like liverworts, appeared. At the end of the Ordovician, a mass-extinction event took place, and animal life on Earth) changed significantly.

The Middle Paleozoic Era (Silurian and Devonian Periods)

PALEOGEOGRAPHY. When the Silurian began, land was divided among two very large continents, Laurentia and Gondwana, and several small ones, including Baltica (Scandinavia and western Russia) and Siberia. During the Silurian, these continents began to be sutured together, and each collision produced an orogeny. Specifically, Baltica collided with eastern Laurentia, causing the *Caledonian orogeny*. Today, you can find geologic structures and metamorphic rocks that formed during this event in western Scandinavia, eastern Greenland, and Scotland (Fig. 10.11a). Soon after this, one or more **microcontinents** (continental blocks less than 1,000 km, 600 miles, across), including one called Avalonia, slammed into what is now the eastern United States, causing the *Acadian orogeny*. The mountains that rose during the Caledonian and Acadian orogenies shed massive amounts of sediment onto bordering continental areas (Fig. 10.11b), forming huge deltas. Sediments of these deltas became thick successions of redbeds (red-colored

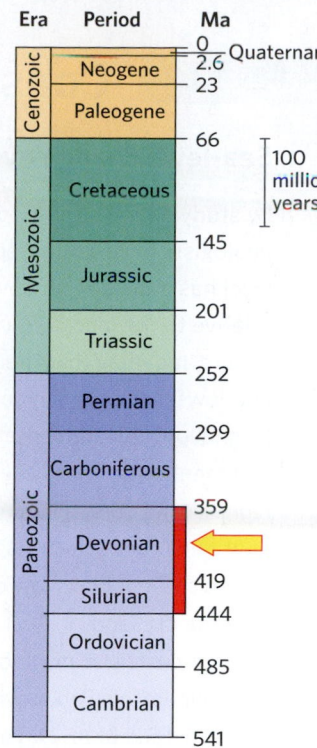

Figure 10.11 Paleogeography of middle Paleozoic time.

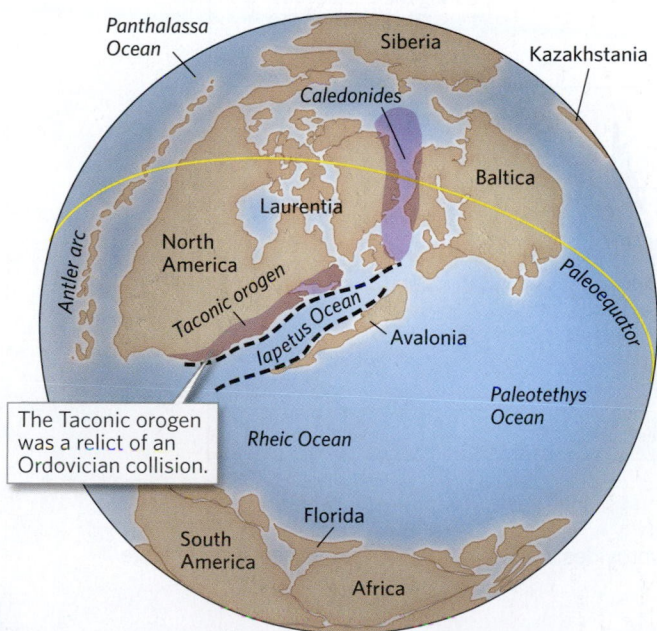

(a) During the Caledonian and Acadian orogenies, Laurentia collided with Baltica and Avalonia. Meanwhile, the Antler island arc formed off the west coast.

The Taconic orogen was a relict of an Ordovician collision.

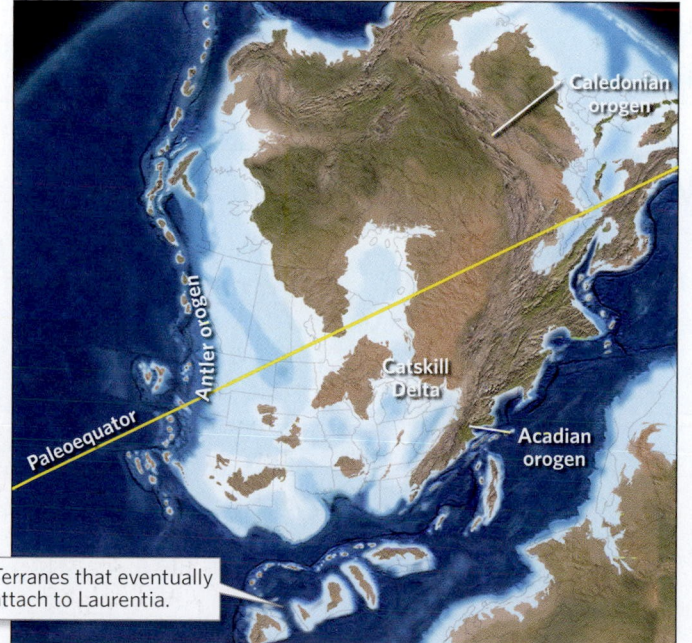

(b) During the Devonian, shallow seas covered parts of North America's interior. The Acadian orogen shed sediment westward to form the Catskill Delta.

Terranes that eventually attach to Laurentia.

Box 10.2 ▸ **Consider this . . .**

Sea-level change over geologic time

By studying the stratigraphic record of continental interiors, geologists have estimated that during the Phanerozoic, global sea level has gone up and down by as much as 300 m (1,000 feet) relative to the land surface. When sea level rises, a *transgression* occurs, meaning that the shoreline migrates inland (see Chapter 5), and low-lying plains in continental interiors become submerged by shallow epicontinental seas. At such times, shallow-marine sediment buries continental regions. When sea level falls, a *regression* occurs, meaning that the shoreline migrates seaward, the continents become dry again, and regional unconformities develop. When sea level falls low enough, large areas of continental shelves become dry land.

A chart tracing global transgressions and regressions during the Phanerozoic Eon provides insight into the rise and fall of global sea level **(Fig. Bx10.2)**. This chart probably does not give us an exact image of sea-level change, however, because the sedimentary record reflects other factors as well, such as changes in sediment supply.

Geologists assume that the total amount of surface water on our planet has stayed fairly constant since the early or middle Archean. If this assumption is correct, then sea-level change does not represent addition or removal of surface water, but rather one of two other processes:

1. *A change in the distribution of surface water:* Most of the Earth's surface water resides in the oceans, but a significant amount can also reside in continental glaciers. When the climate cools and glaciers grow, water that was previously stored in ocean basins gets stored on land, so sea level drops. When the climate warms and glaciers shrink, water returns to the ocean basins, so sea level rises. In addition, the warming of the climate causes seawater to expand. Therefore, changes in sea level reflect climate change (see Chapter 20).

2. *A change in the capacity of the ocean basins:* The capacity (volume) of the ocean basins depends in part on seafloor spreading rates, for as we saw in Chapter 2, the depth of the ocean floor depends on its age. When spreading rates increase, more of the ocean floor is young, so mid-ocean ridges grow wider. When more of the ocean basin becomes shallower, the capacity of the ocean basins decreases, and sea level rises. When spreading rates decrease, mid-ocean ridges become narrower, and the capacity of the ocean basins increases, so sea level falls.

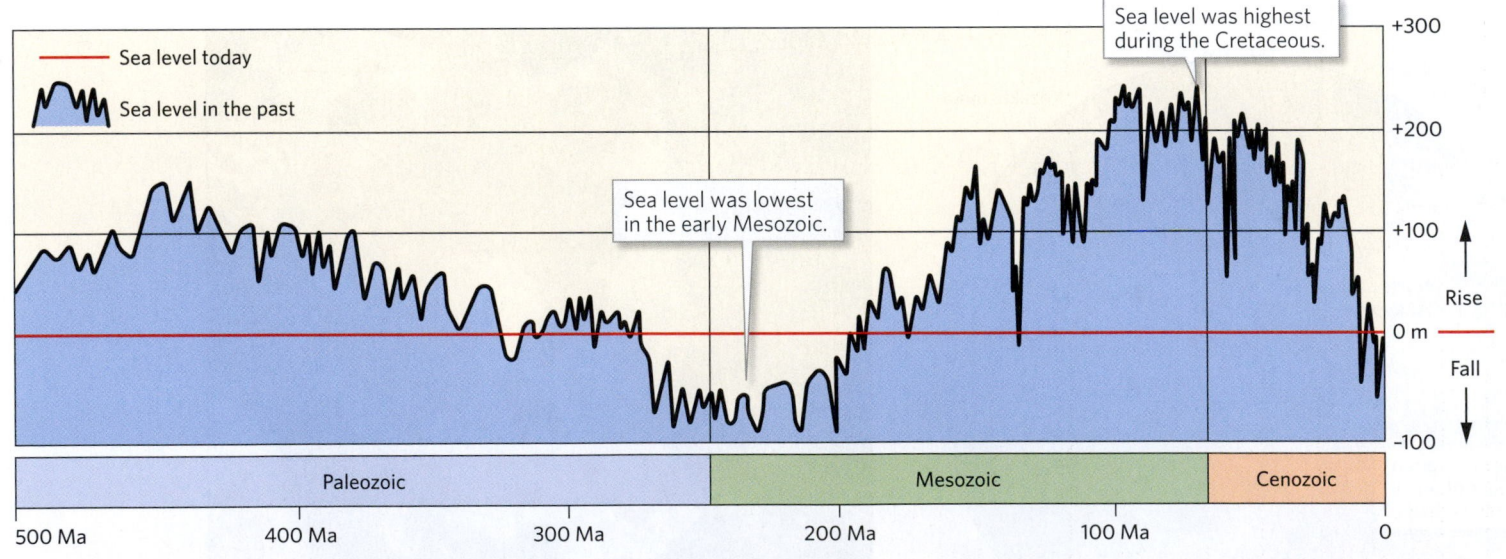

Figure Bx10.2 This chart of transgressions and regressions in the stratigraphic record provides one indication of relative sea-level change during the past half billion years.

Brick = continent   Water = oceans

Brick becomes submerged   Rocks = formation of the mid-ocean ridge

Cup represents the glacial reservoir

When water is stored in glaciers as land, sea level drops.

Box 10.3 # How can I explain . . .

Sea-level change

What are we learning?

The concept that sea level can rise and fall relative to the land surface of continents. over time, and some of the reasons for this change (see Box 10.2).

What you need:

- A plastic basin (about 30 × 40 cm, or one square foot).
- A standard brick.
- A small measuring cup and a ladle.
- Five stones (each about 6 to 10 cm across).

Instructions:

- Place the brick in the middle of the basin.
- Fill the basin with water to just slightly below the top surface of the brick.
- To simulate a rise in sea level due to growth of the volume of a mid-ocean ridge, line up the stones in the basin. As the "ridge" grows, the water surface rises. Eventually, the water submerges the brick.
- To simulate sea-level fall, remove the stones. Water takes the place of the stones, and the top of the brick re-emerges.
- To simulate growth of continental glaciers and the resulting sea-level fall, place the cup on the brick, then ladle some water out of the basin and place it in the cup. Glaciers store water on land, so the amount of water in the sea decreases, and the water surface drops.
- To simulate melting of continental glaciers and the resulting sea-level rise, pour the water from the cup back into the basin.
- To simulate a tectonic rise of the continent relative to sea level, place the pencils underneath the brick so that its surface rises.

What did we see?

Sea-level changes may be due to changes in the capacity of the ocean basin to hold water, removal of water from the oceans and its storage on land, or local tectonic events that cause the land to rise or fall.

Figure 10.12 Examples of middle Paleozoic life.

(a) Vascular plants appeared in the Devonian, allowing the first forests to take root.

~ 20 cm

(b) A Late Devonian fossil of *Tiktaalik*. This lobe-finned fish was one of the first vertebrates to walk on land.

Figure 10.13 Paleogeography of the late Paleozoic.

Approximate area of part (b)

Siberia

North America

Baltica

Paleoequator

South America

North China

Africa

Paleotethys Ocean

South China

India

Tibet

SE Asia

Antarctica

Tethys Ocean

South Pole   Australia

(a) At the end of the Paleozoic, almost all land had combined into a single supercontinent, called Pangaea.

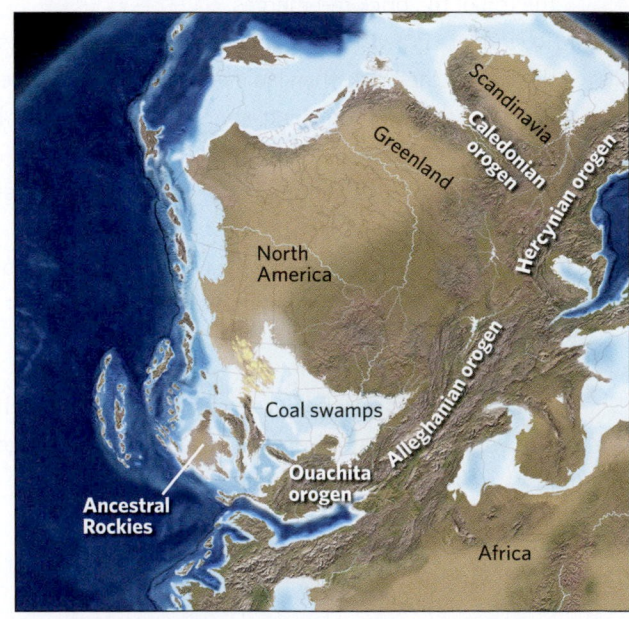

Scandinavia

Caledonian orogen

Greenland

Hercynian orogen

North America

Coal swamps

Alleghanian orogen

Ouachita orogen

Ancestral Rockies

Africa

(b) During the Alleghanian and Hercynian orogenies, a huge mountain belt formed. Coal swamps bordered epicontinental seas. The Ancestral Rockies rose in the western part of the continent.

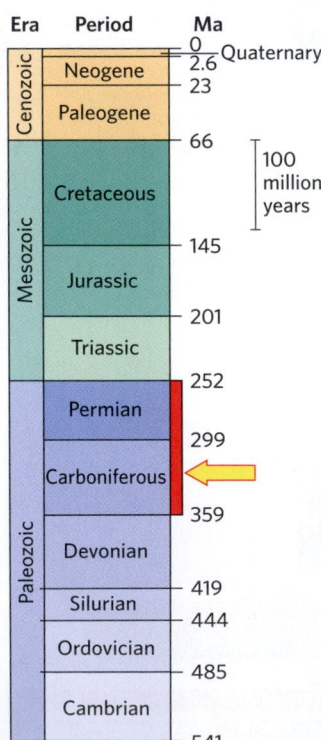

Era	Period	Ma
Cenozoic	Quaternary	0 — 2.6
Cenozoic	Neogene	23
Cenozoic	Paleogene	66
Mesozoic	Cretaceous	145
Mesozoic	Jurassic	201
Mesozoic	Triassic	252
Paleozoic	Permian	299
Paleozoic	Carboniferous	359
Paleozoic	Devonian	419
Paleozoic	Silurian	444
Paleozoic	Ordovician	485
Paleozoic	Cambrian	541

100 million years

clastic sedimentary rocks). Examples of these rocks form cliffs and roadcuts in the Catskill Mountains in New York State. The climate warmed during the Middle Paleozoic, causing sea level to rise. As a result, broad regions of continental interiors that were dry land at the end of the Ordovician flooded once again with epicontinental seas.

Through most of the middle Paleozoic, the western margin of Laurentia remained a quiet passive-margin basin, even as mountains grew on the eastern side of the continent. But in the Late Devonian, the west-coast passive-margin basin collided with an island arc. This event, the *Antler orogeny*, began the long history of mountain building that has dominated the geologic history of western North America ever since (see Fig. 10.11b).

EVOLUTION OF LIFE. Middle Paleozoic seas not only welcomed new species of marine invertebrates, which replaced species that disappeared during the mass-extinction event at the end of the Ordovician Period, but also hosted jawed fishes such as sharks. Even more dramatic changes took place on land. Vascular plants, which have woody tissues, seeds, and veins (for transporting water

and food), rooted on land for the first time in the Silurian, and by the Late Devonian, swampy forests made up of tree-sized relatives of club mosses and ferns covered large areas (**Fig. 10.12a**). The first land animal, a millipede-like organism, appeared in the Silurian. By the Late Devonian, spiders, insects, and crustaceans had exploited both dry-land and freshwater habitats, and the first four-legged vertebrate had crawled out of the sea and inhaled air (Fig. 10.12b).

The Late Paleozoic Era (Carboniferous and Permian Periods)

PALEOGEOGRAPHY. The late Paleozoic Era saw another succession of continental collisions, culminating in the formation of Alfred Wegener's supercontinent, **Pangaea (Fig. 10.13a)**. The largest of these collisions occurred when Gondwana rammed into Laurentia and Baltica. This event caused the *Alleghanian orogeny* and built a vast mountain belt whose eroded remnants crop out in the Appalachian and Ouachita Mountains of the eastern United States (Fig. 10.13b). On the continental side of this mountain belt, compression generated the *Appalachian fold-thrust belt*, in which the sedimentary layer of crust shortened horizontally by as much as 50% (Fig. 10.14). Compression generated during the Alleghanian orogeny was so strong that it caused old, pre-existing faults all across North America to become active again. Movement on these faults produced small mountain ranges and adjacent sediment-filled basins. The largest of these ranges rose, in the late Paleozoic, in the region presently occupied by the Rocky Mountains. Because of their location, these late Paleozoic ranges together make up the *Ancestral Rockies* (see Fig. 10.13b).

The assembly of Pangaea involved other collisions around the world as well. Notably, Africa collided with southern Europe, resulting in the *Hercynian orogeny*. Once formed, Pangaea was so large that it spanned the distance from the South Pole to a latitude of about 70° N (see Fig. 10.13a). Therefore, the southern end of Pangaea had a polar climate and was covered by glaciers, while central Gondwana spanned the equator and enjoyed tropical and semi-tropical conditions that favored lush growth in huge swamps.

EVOLUTION OF LIFE. The fossil record indicates that during the late Paleozoic Era, plants and animals continued to evolve toward forms that are familiar to us today. Huge Carboniferous swamps produced so much O_2 that, for a while, this gas accounted for about a third of the

Figure 10.14 Features of the Appalachian Mountains in the eastern United States.

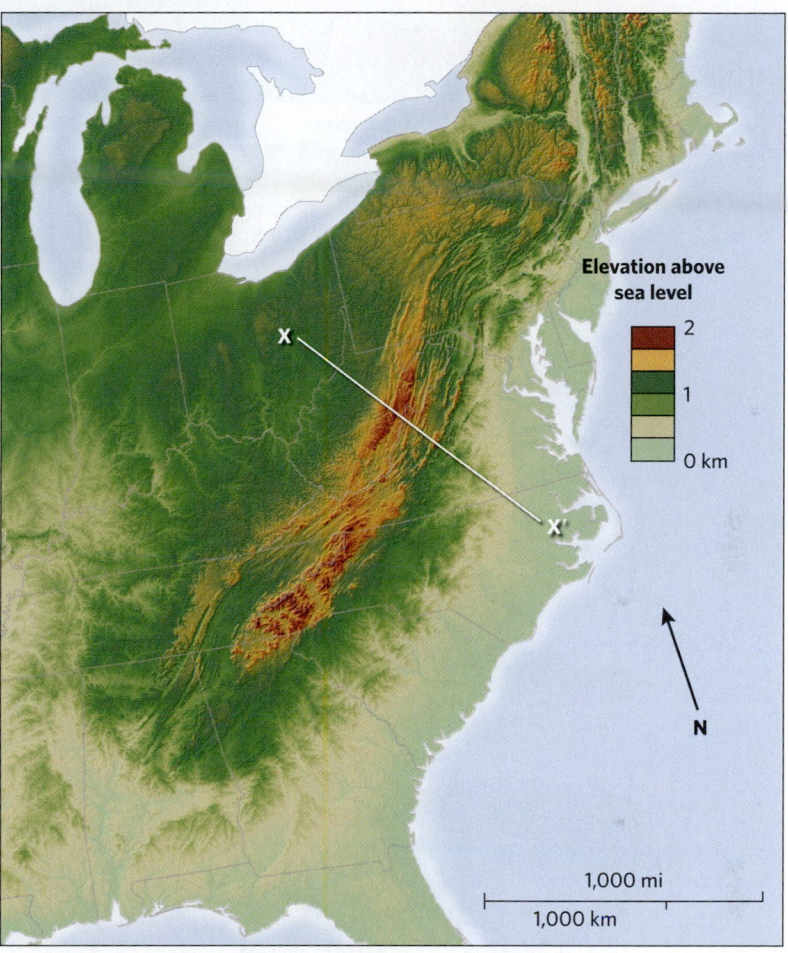

Elevation above sea level

2

1

0 km

N

1,000 mi

1,000 km

(a) The eroded remnants of the Appalachian orogen stand out in this representation of the topography of the eastern United States. Line X–X′ shows the position of the cross section.

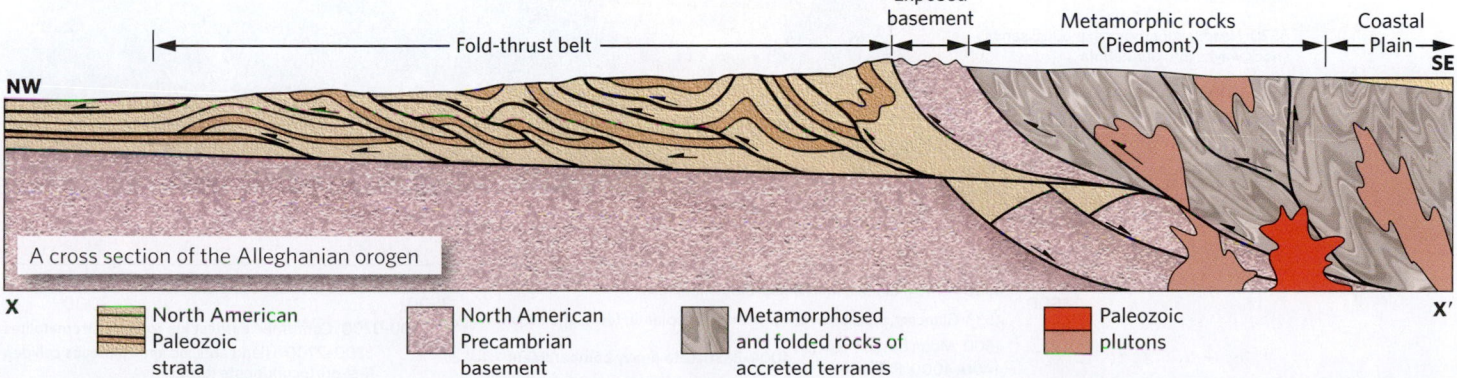

Exposed basement

Fold-thrust belt

Metamorphic rocks (Piedmont)

Coastal Plain

NW

SE

A cross section of the Alleghanian orogen

X

X′

| | North American Paleozoic strata | | North American Precambrian basement | | Metamorphosed and folded rocks of accreted terranes | | Paleozoic plutons |

(b) A cross section of the crust in the Appalachians shows deformation due to Paleozoic orogenies. In the fold-thrust belt, strata have been pushed westward in overlapping thrust sheets.

The Earth Has a History

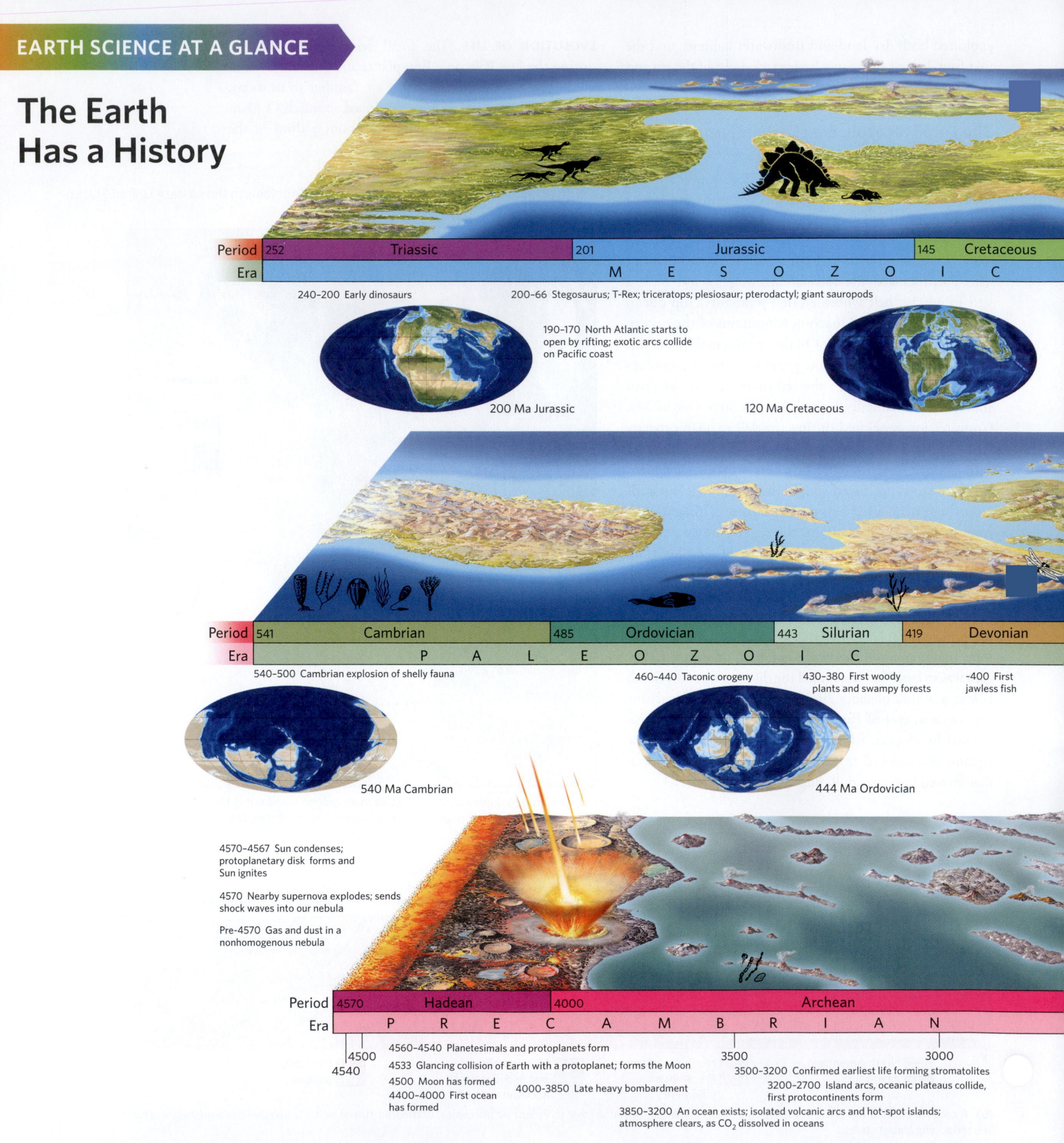

Period	252	Triassic	201	Jurassic	145	Cretaceous
Era				M E S O Z O I C		

240–200 Early dinosaurs

200–66 Stegosaurus; T-Rex; triceratops; plesiosaur; pterodactyl; giant sauropods

190–170 North Atlantic starts to open by rifting; exotic arcs collide on Pacific coast

200 Ma Jurassic

120 Ma Cretaceous

Period	541	Cambrian	485	Ordovician	443	Silurian	419	Devonian
Era				P A L E O Z O I C				

540–500 Cambrian explosion of shelly fauna

460–440 Taconic orogeny

430–380 First woody plants and swampy forests

~400 First jawless fish

540 Ma Cambrian

444 Ma Ordovician

4570–4567 Sun condenses; protoplanetary disk forms and Sun ignites

4570 Nearby supernova explodes; sends shock waves into our nebula

Pre-4570 Gas and dust in a nonhomogenous nebula

Period	4570	Hadean	4000	Archean
Era		P R E C A M B R I A N		

4500
4540

4560–4540 Planetesimals and protoplanets form

4533 Glancing collision of Earth with a protoplanet; forms the Moon

4500 Moon has formed

4000–3850 Late heavy bombardment

4400–4000 First ocean has formed

3500

3000

3500–3200 Confirmed earliest life forming stromatolites

3200–2700 Island arcs, oceanic plateaus collide, first protocontinents form

3850–3200 An ocean exists; isolated volcanic arcs and hot-spot islands; atmosphere clears, as CO_2 dissolved in oceans

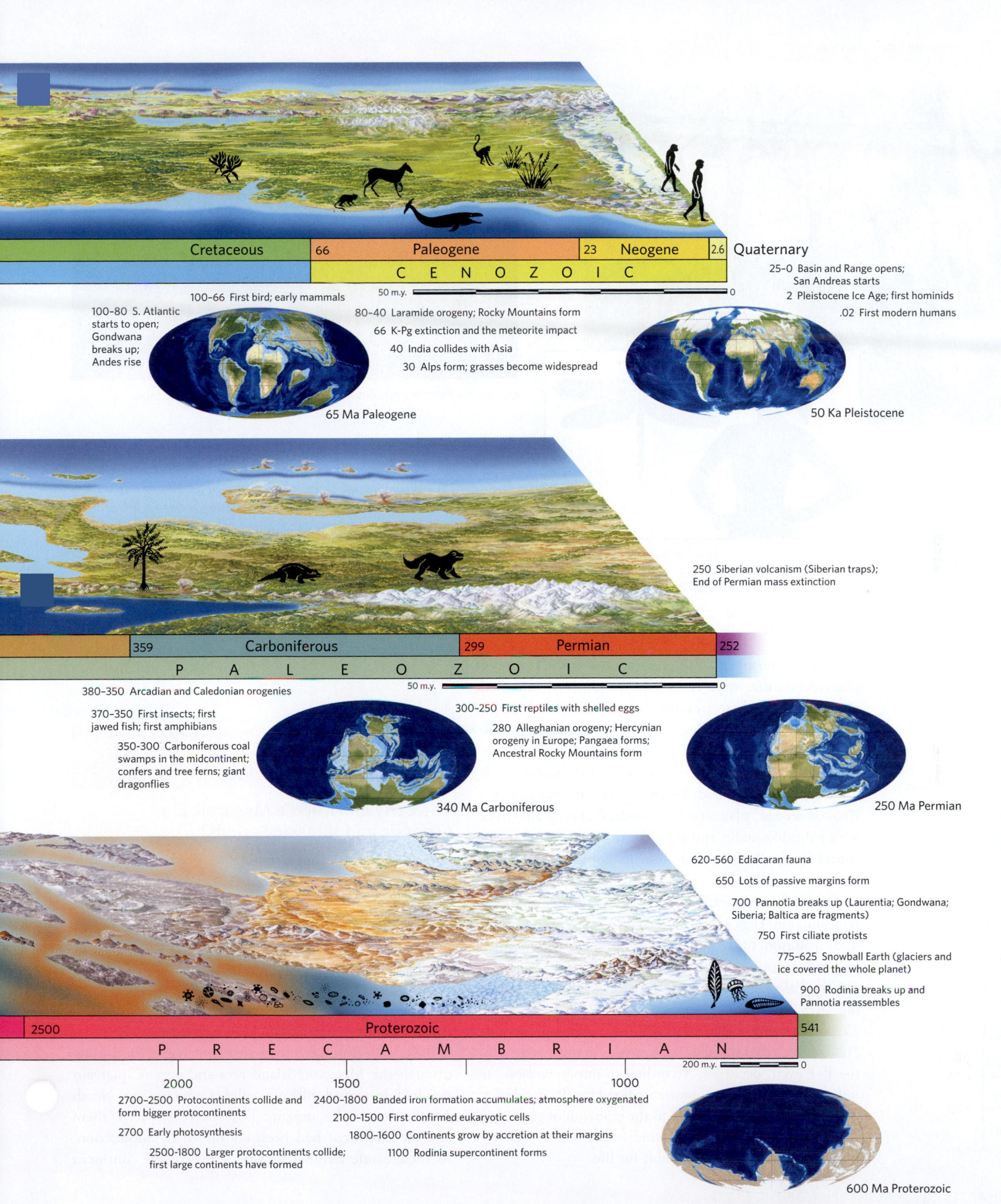

Cretaceous | 66 | Paleogene | 23 | Neogene | 2.6 | Quaternary

C E N O Z O I C

50 m.y. [scale] 0

100–66 First bird; early mammals

100–80 S. Atlantic starts to open; Gondwana breaks up; Andes rise

80–40 Laramide orogeny; Rocky Mountains form

66 K-Pg extinction and the meteorite impact

40 India collides with Asia

30 Alps form; grasses become widespread

25–0 Basin and Range opens; San Andreas starts

2 Pleistocene Ice Age; first hominids

.02 First modern humans

65 Ma Paleogene

50 Ka Pleistocene

250 Siberian volcanism (Siberian traps); End of Permian mass extinction

359 | Carboniferous | 299 | Permian | 252

P A L E O Z O I C

50 m.y. [scale] 0

380–350 Arcadian and Caledonian orogenies

370–350 First insects; first jawed fish; first amphibians

350–300 Carboniferous coal swamps in the midcontinent; confers and tree ferns; giant dragonflies

300–250 First reptiles with shelled eggs

280 Alleghanian orogeny; Hercynian orogeny in Europe; Pangaea forms; Ancestral Rocky Mountains form

340 Ma Carboniferous

250 Ma Permian

620–560 Ediacaran fauna

650 Lots of passive margins form

700 Pannotia breaks up (Laurentia; Gondwana; Siberia; Baltica are fragments)

750 First ciliate protists

775–625 Snowball Earth (glaciers and ice covered the whole planet)

900 Rodinia breaks up and Pannotia reassembles

2500 | Proterozoic | 541

P R E C A M B R I A N

200 m.y. [scale] 0

2000 | 1500 | 1000

2700–2500 Protocontinents collide and form bigger protocontinents

2700 Early photosynthesis

2500–1800 Larger protocontinents collide; first large continents have formed

2400–1800 Banded iron formation accumulates; atmosphere oxygenated

2100–1500 First confirmed eukaryotic cells

1800–1600 Continents grow by accretion at their margins

1100 Rodinia supercontinent forms

600 Ma Proterozoic

Figure 10.15 Life in the late Paleozoic.

(b) By the Permian, reptiles such as *Dimetrodon* inhabited the land.

(a) A museum diorama of a Carboniferous coal swamp. The inset shows the size of a giant dragonfly from the Paleozoic (wingspan of about 1 m) relative to the size of a human.

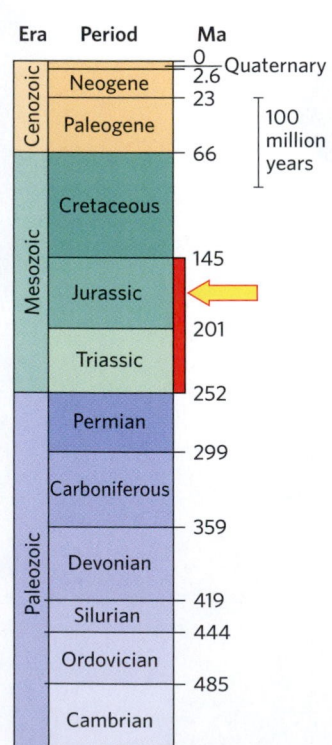

Era	Period	Ma	
Cenozoic	Neogene	0 / 2.6	Quaternary
		23	
	Paleogene		100 million years
		66	
Mesozoic	Cretaceous		
		145	
	Jurassic		
		201	
	Triassic		
		252	
Paleozoic	Permian		
		299	
	Carboniferous		
		359	
	Devonian		
		419	
	Silurian	444	
	Ordovician		
		485	
	Cambrian		
		541	

gas in the atmosphere. This growth also left thick piles of plant debris that were eventually transformed into coal after burial. In these swamps, insects with fixed wings, including huge dragonflies, flew through a tangle of ferns, club mosses, and scouring rushes **(Fig. 10.15a)**, and by the end of the Carboniferous Period, cockroaches, with foldable wings, appeared. Forests containing *gymnosperms* ("naked-seeded" plants such as conifers) and cycads (trees with palm-like stalks and fern-like fronds) became widespread in the Permian. Amphibians populated the land, and reptiles followed **(Fig. 10.15b)**. The success of reptiles on land reflected a radically new component in animal reproduction: eggs with a watertight protective covering. The evolution of eggs permitted reptiles to reproduce without returning to the water.

The Paleozoic Era came to a close with the devastating *Permian-Triassic* (P-Ƭ) *mass-extinction event* about 252 Ma, during which over 96% of marine species and 70% of terrestrial species disappeared. No one is sure why the P-Ƭ event occurred. According to one hypothesis, it followed the eruption of immense flood basalts in Siberia. This eruption could have led to the emission of gases that clouded the atmosphere and acidified the oceans, making many environments inhospitable for life.

Take-home message . . .

Breakup of the late Precambrian supercontinent Pannotia ushered in the Paleozoic Era. During the Paleozoic, shallow seas covered continents at times, and life diversified and moved onto land. As the era ended, continental collisions led to the assembly of Pangaea, and the largest mass-extinction event in geologic history drove most species to extinction.

Quick Question -
How did life on land change during the Paleozoic?

10.6 The Mesozoic Era: When Dinosaurs Ruled

The Early and Middle Mesozoic Era (Triassic and Jurassic Periods)

PALEOGEOGRAPHY. Supercontinents don't last forever. Pangaea, which had assembled at the end of the Paleozoic, existed for about 100 million years until, in the Late Triassic and Early Jurassic, rifting began in the region that is now the east coast of North America. By the end of the Jurassic, the North Atlantic Ocean had started to grow, separating North America from Europe and Africa **(Fig. 10.16a)**.

Meanwhile, along the western margin of North America, convergent-margin tectonics became the order of the day. Beginning in the Late Permian and continuing through the Mesozoic, island arcs and oceanic plateaus that had developed offshore collided with western North America once the oceanic lithosphere between them and the continent had been consumed by subduction. As these exotic terranes became attached, the continent grew westward. Then, starting at the end of the Jurassic, a large continental volcanic arc, now known as the *Sierran arc*, began to form along the western margin of North America itself as the Farallon Plate, part of the Pacific Ocean floor, subducted beneath the continent.

During the Triassic and Early Jurassic, the Earth had a relatively warm climate, so glaciers melted away entirely in polar regions. Sandy deserts covered portions of western North America. The large dunes of these deserts lithified to become the colorful red sandstones exposed in Zion National Park (Fig. 10.16b). In the Middle Jurassic, sea level began to rise, and epicontinental seas once again flooded portions of the continent.

EVOLUTION OF LIFE. During the early Mesozoic Era, a variety of new plant and animal species appeared. Reptiles swam in the oceans, and colonial corals built extensive reefs. On land, gymnosperms and reptiles diversified, and the Earth saw its first turtles and flying reptiles. At the end of the Triassic, the first true dinosaurs appeared (Fig. 10.17). *Dinosaurs* differed from other reptiles in that their legs extended under their bodies rather than off to the sides. By the end of the Jurassic, 100-ton sauropod dinosaurs, with long necks and tails, along with other familiar monsters such as *Stegosaurus*, thundered across the landscape, and the first feathered birds, such as *Archaeopteryx*, took to the skies (see Fig. 9.7e). Small rat-like creatures, the earliest ancestors of mammals, scurried through Late Triassic underbrush.

The Late Mesozoic Era (Cretaceous Period)

PALEOGEOGRAPHY. During the Cretaceous Period, the Earth's climate continued to warm, and sea level rose significantly, reaching heights that had not been attained for the previous 200 million years. In fact, during the Cretaceous, a shark could have swum across North America from the Gulf of Mexico to the Arctic Ocean. Geologists attribute

the high sea level not only to the warming climate, but also to a period of rapid seafloor spreading and to abundant hot-spot volcanic activity (see Box 10.2).

At the dawn of the Cretaceous, the North Atlantic Ocean had grown to a width of 1,500 km (930 miles). Pangaea continued its breakup as the Cretaceous progressed. South America finally broke away from Africa, and the South Atlantic started to grow, by the Middle Cretaceous. By the Late Cretaceous, Antarctica and Australia had

Figure 10.16 Paleogeography of the early Mesozoic.

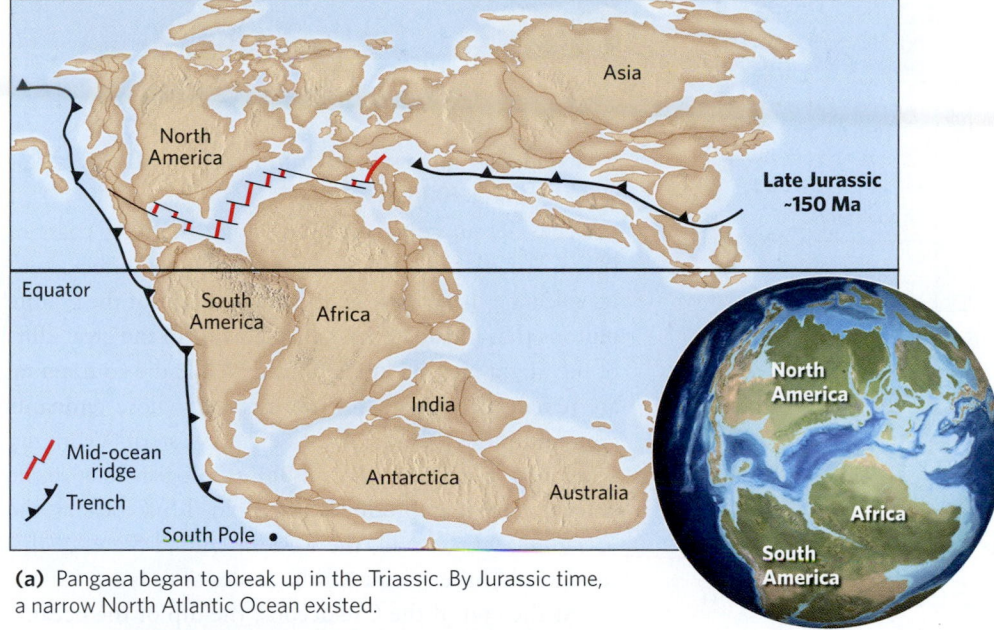

(a) Pangaea began to break up in the Triassic. By Jurassic time, a narrow North Atlantic Ocean existed.

(b) During the Jurassic, immense sand dunes blanketed the southwestern United States. These sandstone beds in Zion National Park are the relics of those dunes.

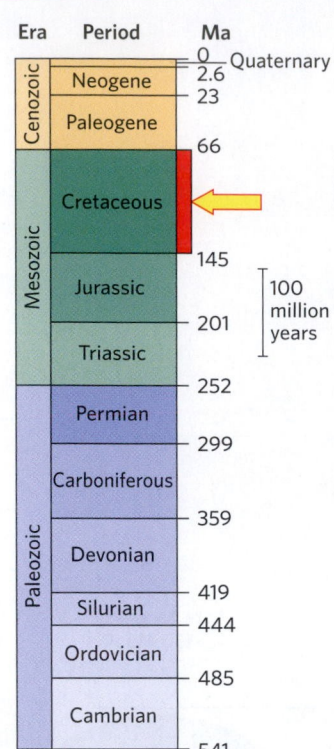

Era	Period	Ma
Cenozoic	Quaternary	0
	Neogene	2.6
		23
	Paleogene	66
Mesozoic	Cretaceous	
		145
	Jurassic	201
	Triassic	252
Paleozoic	Permian	299
	Carboniferous	359
	Devonian	419
	Silurian	444
	Ordovician	485
	Cambrian	541

100 million years

Figure 10.17 During the Jurassic, giant dinosaurs roamed the land. This painting shows several species.

separated, and India started drifting rapidly northward toward Asia **(Fig. 10.18a)**.

The Sierran continental arc along the west coast of North America remained active during the Cretaceous. Its volcanoes have long since eroded away, but the granite plutons that intruded beneath it now form the great cliffs of the Sierra Nevada in California. While the continental arc remained active, a fold-thrust belt, whose remnants crop out in the Canadian Rockies and western Wyoming, developed to the east **(Fig. 10.18b)**. Geologists refer to the deformation that produced this fold-thrust belt as the *Sevier orogeny*. Overall, the western United States of this time would have resembled the Andes of today.

At the end of the Cretaceous, the dip of the oceanic plate subducting beneath the western United States decreased, and the plate started shearing against the base of the continent. As a result, mountain-building activity moved eastward, and large reverse faults became active in Wyoming, Colorado, eastern Utah, and northern Arizona. These faults penetrated deep into the Precambrian basement, so movement along the faults uplifted basement rocks. This movement, in turn, caused layers of overlying Paleozoic strata to warp into large monoclines (folds whose shape resembles the drape of a carpet over a step; see Chapter 7). This series of events, which geologists call the *Laramide orogeny*, resulted in the growth of the present-day Rocky Mountains in the United States **(Fig. 10.18c)**.

EVOLUTION OF LIFE. In the seas of the late Mesozoic world, modern fish appeared and became dominant. In contrast with earlier fish, the new species had short jaws, rounded scales, symmetrical tails, and specialized fins. They served as prey for huge swimming reptiles and gigantic turtles. On land, angiosperms (flowering plants), including hardwood trees, populated the forest **(Fig. 10.19a)**, and dinosaurs inhabited almost all

See for yourself

Rocky Mountain Front, Colorado

Latitude: 39°46′2.32″ N
Longitude: 105°13′45.35″ W

Zoom to 8 km (~5 miles) and look obliquely.

We see the steep face of the Rocky Mountains in Colorado. These mountains were uplifted during the Laramide orogeny. During the event, reactivation of large faults thrust Precambrian rocks up and caused overlying Paleozoic strata to fold.

environments. Mammals also diversified and developed larger brains and more specialized teeth **(Fig. 10.19b)**. Then suddenly, at 66 Ma, the dinosaurs disappeared entirely, along with most other species, during a mass-extinction event now known as the **K-Pg extinction**. This event, which likely resulted from a catastrophic meteorite impact **(Box 10.4)**, brought the era to a close.

Take-home message . . .

During the Mesozoic Era, rifting broke Pangaea apart and produced the Atlantic Ocean. In North America, convergent-boundary tectonics along the west coast produced the granites of the Sierra Nevada and eventually led to the uplift of the Rocky Mountains. Dinosaurs roamed all continents, but they died off, along with many other species, at the end of the era, possibly when a huge meteorite hit the Earth.

Quick Question -----------------------------
Did all of today's continents break off of Pangaea at the same time?

10.7 The Cenozoic Era: The Modern World Comes to Be

PALEOGEOGRAPHY. During the last 66 million years, the map of Earth's surface has continued to change, gradually producing the configuration of continents and plate boundaries that we see today. The continents that once constituted Gondwana drifted northward as subduction consumed the *Tethys Ocean*, which separated Gondwana from Europe and Asia **(Fig. 10.20a)**. Eventually, the southern margins of Europe and Asia collided with various small crustal fragments and then finally with India and Africa, resulting in the growth of the largest orogen on Earth today, the **Alpine-Himalayan chain**. Continued northward movement of India led to uplift of the Tibetan Plateau.

As the Atlantic Ocean grew and the Americas moved westward, convergent-boundary activity occurred along their western margins. In South America, this activity produced the *Andes*, still an active mountain-building region. In North America, the Laramide orogeny continued until about 40 Ma (the Eocene, in the Middle Paleogene). At this time, the mid-ocean ridge separating the Farallon Plate from the Pacific Plate, which lay farther to the west, began to be subducted beneath the west coast of North America. The Farallon Plate had been moving eastward, toward North America, but the Pacific Plate moves northwestward, parallel to the west coast of North America. So, as the ridge was subducted, the convergent boundary between the North American and Farallon

Figure 10.18 Paleogeography of the late Mesozoic.

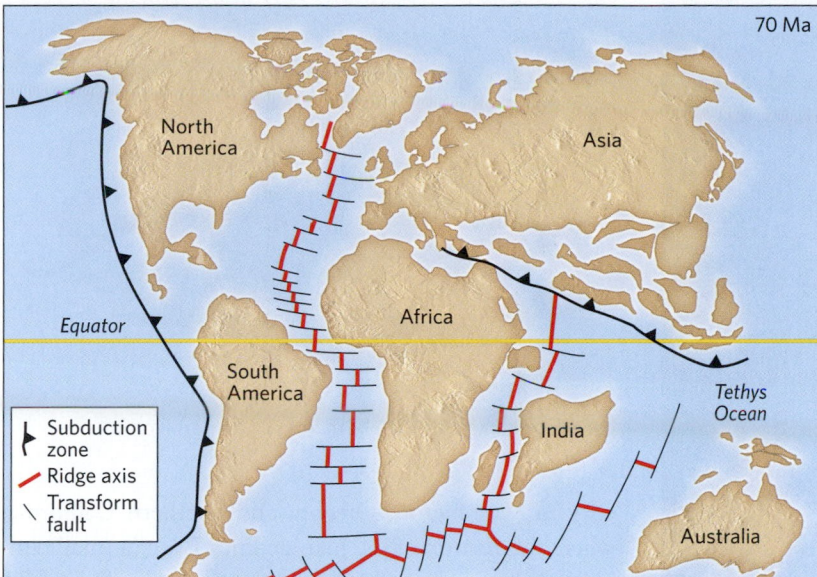

(a) By the Late Cretaceous Period (70 Ma), the Atlantic Ocean had formed, and India was moving rapidly northward toward Asia.

This cross section (XX') shows the relation of the Sierran arc to the Sevier fold-thrust belt.

🦴 Convergent boundary   ⟋ Thrust fault   ▬ Volcanic arc

(b) In the Cretaceous, a long seaway flooded the western interior of North America, a large continental volcanic arc (the Sierran arc) grew on its west coast, and the Sevier fold-thrust belt formed to the east of the arc.

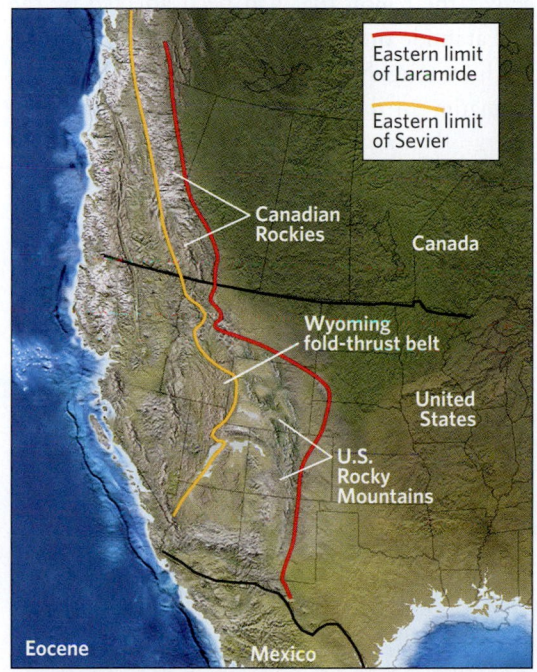

(c) At the end of the Cretaceous, deformation shifted eastward in the United States during the Laramide orogeny, moving from the Sevier fold-thrust belt to the Rocky Mountains. This map shows the end result of the Laramide orogeny in the Eocene, about 40 Ma.

This cross section shows nature of a Laramide basement uplift.

Tilted strata at the east edge of the Rocky Mountains, near Denver, Colorado.

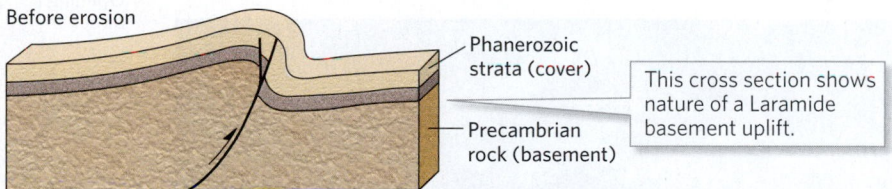

Figure 10.19 Life in the Cretaceous.

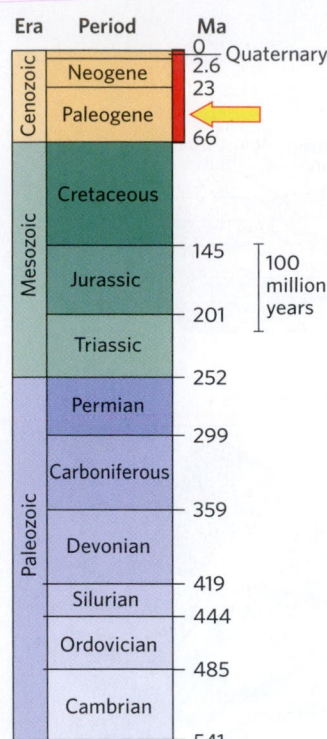

Era	Period	Ma
Cenozoic	Quaternary	0
	Neogene	2.6
		23
	Paleogene	66
Mesozoic	Cretaceous	145
	Jurassic	201
	Triassic	252
Paleozoic	Permian	299
	Carboniferous	359
	Devonian	419
	Silurian	444
	Ordovician	485
	Cambrian	541

100 million years

(a) Flowering plants first appeared in the Cretaceous.

(b) This Cretaceous, rat-like mammal was carnivorous.

only in Washington, Oregon, and northern California, where subduction of the Juan de Fuca Plate (a small remnant of the Farallon Plate) generates the volcanism of the Cascade Range.

At about the same time that the San Andreas fault became active, the region that had been affected by the Sevier and Laramide orogenies began to undergo rifting (extension) in a roughly east-west direction. The result was the formation of the **Basin and Range Province**, a

Plates was gradually replaced by a transform boundary between the North American and Pacific Plates. By 25 Ma, the San Andreas fault system, the trace of this transform boundary, had begun to form along the west coast of the United States. At this strike-slip fault today, the Pacific Plate moves northward with respect to North America at a rate of about 6 cm (3 inches) per year. In the western United States, convergent-boundary activity continues

Figure 10.20 Paleogeography of the Cenozoic.

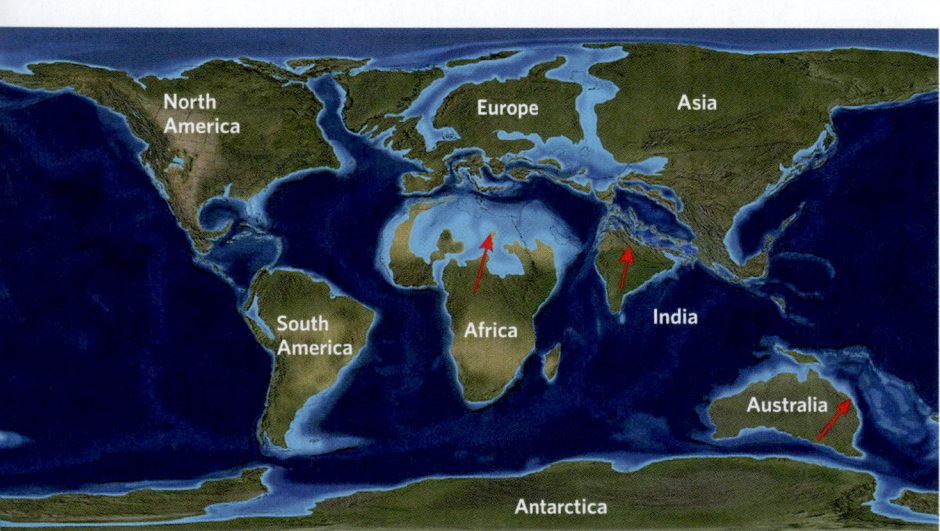

(a) In this Eocene (50 Ma) paleogeographic reconstruction, India has not yet collided with Asia, and Europe and Asia have not yet coalesced into a single landform.

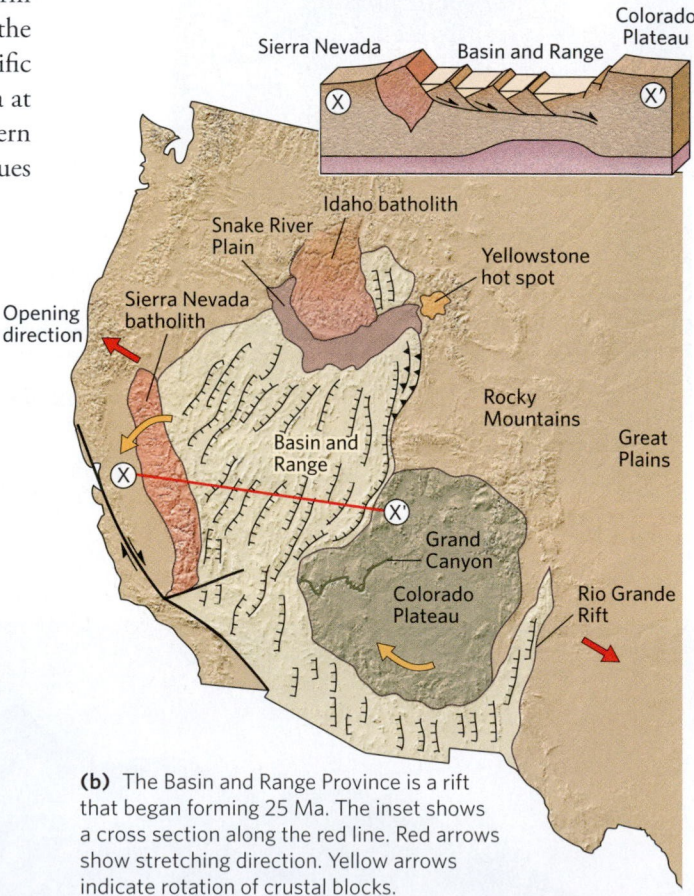

(b) The Basin and Range Province is a rift that began forming 25 Ma. The inset shows a cross section along the red line. Red arrows show stretching direction. Yellow arrows indicate rotation of crustal blocks.

Box 10.4 Consider this . . .

The K-Pg mass-extinction event

In the 18th century, paleontologists defined the end of the Cretaceous by a marked change in fossil species. Until the 1980s, most researchers assumed that the change took place over millions of years. Modern dating techniques, however, indicated that this change actually happened almost instantaneously. Dinosaurs, which had dominated the planet for more than 150 million years, vanished, along with 90% of plankton species in the ocean and up to 75% of plant species. This abrupt mass-extinction event is now known as the *K-Pg extinction*; K stands for Cretaceous and Pg stands for Paleogene. (Prior to recent changes in the geologic time scale, the event was called the *K-T extinction*; T stands for Tertiary.) What kind of catastrophe could cause such a sudden and extensive mass extinction?

The cause of the K-Pg extinction remained a mystery until the late 1970s, when Walter Alvarez, an American geologist, and his colleagues examined a shale layer deposited exactly at the K-Pg boundary. They found that this shale contained relatively high concentrations of iridium, an element that comes primarily from meteorites. Further study showed that the clays of this age contained other unusual materials, such as tiny glass spheres that form when a spray of molten rock freezes, grains of coesite (a mineral that forms when intense shock waves pass through quartz), and even carbon from burned vegetation. All these features pointed to the occurrence of a huge meteorite impact at the time of the K-Pg event **(Fig. Bx10.4a)**. Subsequently, geologists found a meteorite crater, 100 km (62 miles) in diameter and 16 km (10 miles) deep, buried beneath younger strata near the Yucatán Peninsula in Mexico **(Fig. Bx10.4b)**. Isotopic dating indicated that the crater was formed 66 ± 0.4 Ma, the time of the K-Pg event. Because of its age and size, this crater, known as the Chicxulub crater, may be the imprint of the deadly object whose impact with the Earth eliminated so much life.

The K-Pg impact was catastrophic because it not only formed a crater, blasting huge quantities of debris into the sky, but probably also generated 2-km (1.2-mile)-high tsunamis that inundated the shores of continents, and generated a blast of hot air that set forests on fire worldwide. The blast and the blaze together could have ejected enough debris into the atmosphere to cause months of perpetual night and winter-like cold. In addition, chemicals ejected into the air could have combined with water to produce acid rain. These conditions could have broken the food chain and triggered extinctions.

Earth at 66 Ma

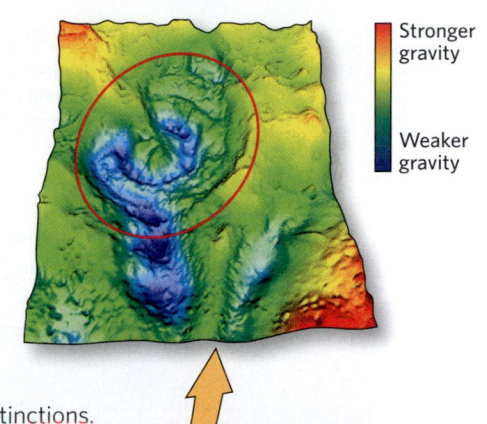

Stronger gravity

Weaker gravity

United States

Mexico

Yucatán Peninsula

(a) An artist's image of the 13-km-wide object as it hit. The inset shows how the Earth looked at the time of impact.

(b) The location of the Chicxulub crater today. Gravity anomalies, as shown on the inset, reveal the shape of the crater.

Figure Bx10.4 The K-Pg meteorite impact.

Figure 10.21 The Alpine-Himalayan orogen formed when Africa, India, and Australia collided with Asia. The Cordilleran and Andean orogens reflect the consequences of continuing convergent-margin tectonics along the eastern coast of the Pacific Ocean.

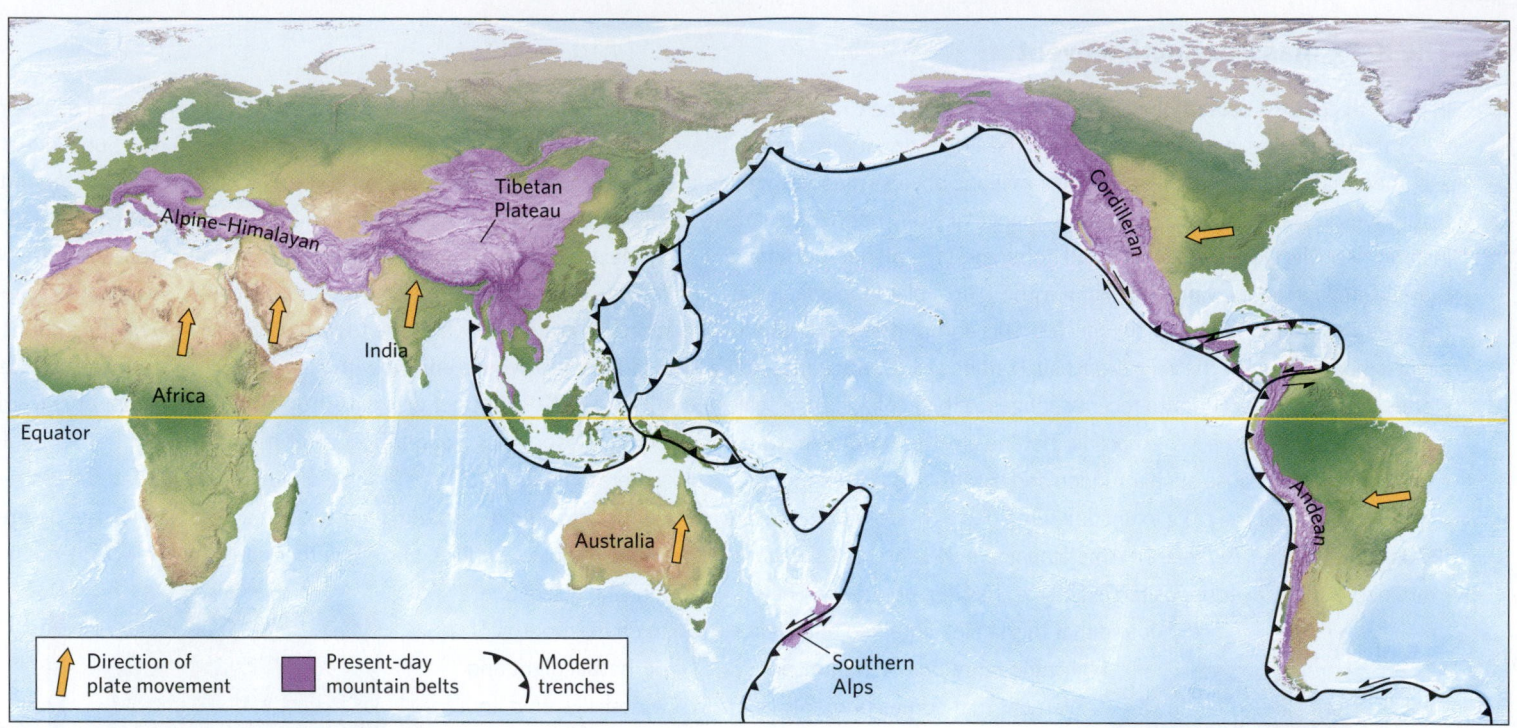

Figure 10.22 The maximum advance of continental glaciers during the Pleistocene Ice Age in North America.

broad continental rift **(Fig. 10.20b)** whose name reflects its topography. The region contains long, narrow mountain ranges separated from each other by flat, sediment-filled basins. This topography formed because the crust of the region broke up when normal faults caused blocks of crust above the faults to slip downward and tilt (see Chapter 7). The crests of the tilted blocks formed the ranges, and the depressions between them, which rapidly filled with sediment eroded from the ranges, became basins.

The Basin and Range Province terminates just north of the Snake River Plain, a feature that marks the track of the hot spot that now lies beneath Yellowstone National Park. As the North American Plate drifted westward, volcanic calderas formed along this track. Yellowstone National Park straddles the most recent caldera (see Chapter 4). Around the world, plate interactions continue to drive seismicity, volcanism, and mountain building **(Fig. 10.21)**.

Recall that in the Cretaceous Period, the world was relatively warm, sea level was high, and extensive areas of continents were submerged. During the Neogene, however, the global climate rapidly became cooler, and in the early Oligocene (about 34 Ma), Antarctic glaciers reappeared for the first time since the Triassic. The climate continued to grow colder through the Late Miocene. In the generally cold climate of the past 2.6 million years, the Quaternary, huge glaciers expanded and retreated across northern continents at least 20 times (see Chapter 14).

Figure 10.23 Life in the Cenozoic.

(a) Grasslands first appeared in the Cenozoic. Giant mammals ruled the land for dinosaurs had gone extinct tens of millions of years earlier.

(b) Our own human lineage evolved in the Cenozoic. Human-like primates such as *Australopithecus*, shown here, first appeared about 4 Ma.

Geologists refer to this time period as the **Pleistocene Ice Age (Fig. 10.22)**. (The Pleistocene is the portion of the Quaternary before the last glacial retreat.) Erosion and deposition by the glaciers created much of the landscape we see today in northern temperate regions. About 11,000 years ago, the climate warmed, the glaciers retreated, and we entered the interglacial time interval we are still experiencing today (see Chapter 14).

EVOLUTION OF LIFE. When the skies finally cleared in the wake of the K-Pg meteorite impact, plant life recovered, and soon forests of both angiosperms and gymnosperms proliferated. By the middle of the Cenozoic Era, grasses had spread across the plains at mid-latitudes, transforming these regions into vast steppes. With the dinosaurs gone, birds and mammals diversified. Most of the groups of mammals that live today originated at the beginning of the Cenozoic, giving this era the nickname *Age of Mammals*. During the latter part of the era (Late Neogene and Early Quaternary), huge mammals, such as mammoths and saber-toothed lions, appeared **(Fig. 10.23a)**. But these animals became extinct during the past 10,000 years, perhaps due to hunting by humans.

According to the fossil record, ape-like primates diversified in the Miocene Epoch (about 20 Ma), and the first human-like primates, *australopithecines*, appeared about 4 Ma **(Fig. 10.23b)**, followed by the first members of the human genus, *Homo*, at 2.4 Ma. *Homo erectus*, a species capable of making stone axes, appeared in Africa about 1.6 Ma, and our species, *Homo sapiens*, diverged from *Homo neanderthalensis* (the Neanderthals) about 500,000 years ago. When modern people first walked the Earth about 200,000 years ago, they shared the planet with two

Did you ever wonder . . .

whether dinosaurs and humans lived at the same time?

See for yourself

Basin and Range Rift, Utah
Latitude: 39°15′1.83″ N
Longitude: 114°38′32.10″ W

Zoom to 250 km (~155 miles) and look down.

In this region of the Cenozoic Basin and Range rift, darker bands are fault-block mountains, whereas lighter areas are sediment-filled basins. White areas are evaporites, from dried-up lakes.

Figure 10.24 The map of the Earth's surface changes over time because of plate motions. Here we see the change that has taken place during the past 200 million years.

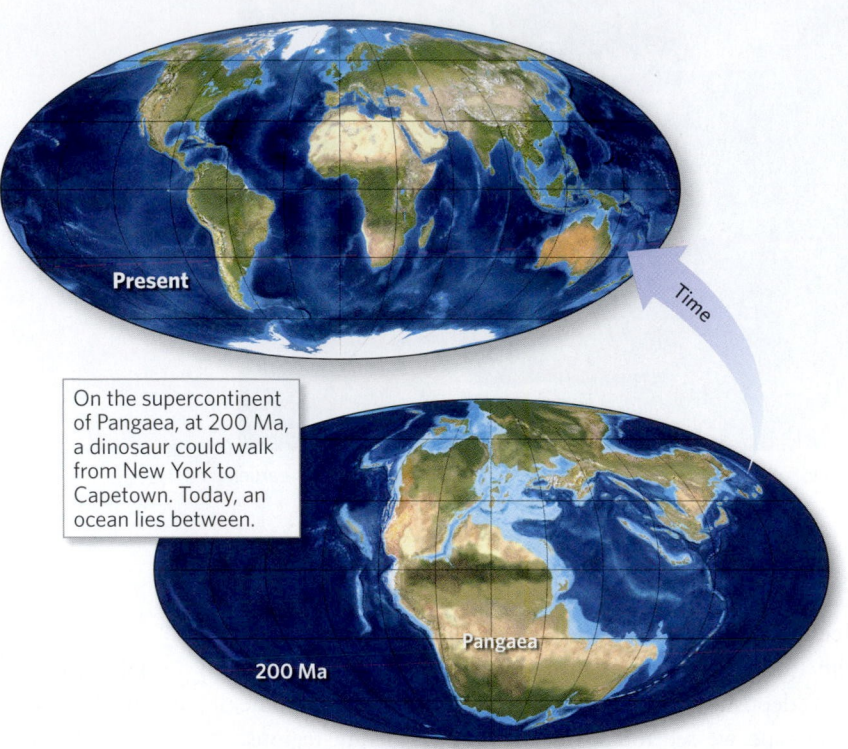

Present

On the supercontinent of Pangaea, at 200 Ma, a dinosaur could walk from New York to Capetown. Today, an ocean lies between.

Time

Pangaea

200 Ma

Take-home message . . .

During the Cenozoic, the mountain belts of today rose, and modern plate boundaries became established. In North America, the San Andreas fault and the Basin and Range Province formed. With the dinosaurs gone, mammals diversified. During the Pleistocene, glaciers covered large areas, and humans appeared.

Quick Question -
Did *Homo sapiens* live during the Pleistocene Ice Age?

10.8 The Concept of Global Change

Did the Earth's surface look the same in the Jurassic as it does today? Definitely not! As we've seen, in Jurassic time, Africa and South America were part of the same continent, and the call of the wild rumbled from the throats of dinosaurs, whereas today the broad South Atlantic Ocean separates the two landmasses (**Fig. 10.24**), and the largest animals are mammals (**Fig. 10.25**). What we see of the Earth today is just a snapshot, an instant in the life story of a constantly changing planet with a long and complex history—an idea that arguably stands as geology's greatest philosophical contribution to humanity's understanding of our Universe (Earth Science at a Glance, pp. 328–329)

To finish our consideration of the Earth's biography, let's focus more closely on the concept of change. In the Prelude, we introduced the concept of the *Earth System*: all the physical and biological realms of the Earth along with

other species of the genus *Homo*—the Neanderthals and the Denisovans. The last Neanderthals died off about 40,000 years ago, and the last Denisovans went extinct at 25,000 years ago, leaving *Homo sapiens* as the only remaining human species on the Earth.

Figure 10.25 Evolution is unidirectional change, in that the transformations never repeat. Because of evolution, the species inhabiting the Earth today are not the same as those that inhabited the planet in the past. For example, the largest animal in the Mesozoic was a reptile, whereas the largest animal today is a mammal.

The largest Mesozoic land animal was a dinosaur. The inset shows an elephant at the same scale.

The largest land animal today is the elephant.

the complex ways in which they interact with one another. In this context, we can define a **global change** as any modification of the Earth System over time. Researchers distinguish among different types of global change, based first on the rate of change: *gradual change*, such as the growth of a mountain belt, takes place slowly, over long intervals of geologic time (millions to billions of years), whereas *catastrophic change*, such as the impact of a meteorite, takes place relatively rapidly (seconds to millennia). But we can also distinguish among types of change based on the way in which the change progresses: *unidirectional change* involves transformations that never repeat; *cyclic change* repeats the same steps over and over, though not necessarily with the same results or at the same rate; and *periodic change* repeats steps with a definable frequency.

Why has the Earth changed so much over geologic time, and how can it continue to change? Ultimately, change can happen both because the Earth's internal heat makes the asthenosphere weak enough to flow and because the Sun's radiation can keep most of the Earth's surface at temperatures above the freezing point of water. Flow in the asthenosphere permits plate tectonics, which in turn leads to continental drift, volcanism, and mountain building. Solar radiation keeps streams, glaciers, waves, and wind in motion, thereby causing erosion and deposition, and it also fuels photosynthesis. If the Earth did not have just the right mix of plate tectonics and solar heat, it would be a frozen dust bowl like Mars, a crater-pocked wasteland like the Moon, or a cloud-choked oven like Venus, and could not host life as we know it.

Let's now briefly re-examine some aspects of Earth history in the context of global change. We'll see that some aspects of this history illustrate the different types of change. And we'll see that humans have become an important agent of change that may be recorded in the stratigraphic record.

Unidirectional Changes during Earth History

The Earth System has changed in many irreversible ways over the course of geologic time. For example, differentiation of the Earth's interior into a core and mantle, which occurred very early in our planet's history, will never happen in the future, because once a core has formed, it can't form again. The filling of the oceans, and the effect that this process has had on atmospheric composition, also represents a unidirectional change in that once it took place, over 3.5 Ga, liquid water persisted on our planet's surface (with the exception of brief, snowball Earth conditions).

The existence of liquid water set the stage for the appearance of life. The fossil record, as we have seen, indicates that life has evolved over the course of geologic time (see Fig. 10.25). Early on, life consisted only of simple archaea and bacteria living in the sea, but during the past 600 million years, multicellular plants and animals living together in complex ecosystems appeared, and some of these populated the land. If another mass-extinction event were to happen, biodiversity would diminish. But evolution, which is unidirectional change in the assemblage of organisms on Earth, would continue. Extinction is forever, so when life rebounded after the next mass-extinction event, it would not be the same as the life before.

Cyclic Changes in the Earth System

During cyclic change, a sequence of stages repeats over time. Some cyclic changes are periodic, in that cycles happen with a definable frequency, but many are not. Some cycles involve only movements of physical components of the Earth System, while others involve transfer of materials among both living and nonliving components of the Earth System. We have seen examples of all of these types of cyclic changes in this chapter:

- *The rock cycle:* As we learned in Chapter 6, atoms making up the minerals of a rock of one rock type may later become part of another rock of the same rock type or of a different rock type. In effect, rocks serve as reservoirs of atoms, and movement of atoms from reservoir to reservoir over time constitutes the rock cycle.

- *The supercontinent cycle:* On a time scale of hundreds of millions of years, continental blocks collide and collect into large supercontinents. A supercontinent survives for tens to hundreds of millions of years until it undergoes rifting to form smaller continents, which drift apart due to seafloor spreading. Eventually, the continents reassemble into a new supercontinent, only to break apart again into different blocks **(Fig. 10.26)**.

- *The sea-level cycle:* Sea level has gone up and down by as much as 300 m (600 feet) during the Phanerozoic, and it probably did the same in the Precambrian. When sea level rises, continental interiors may become shallow seas in which sediment accumulates, and when sea level falls, the land dries up (see Box 10.2).

- *Biogeochemical cycles:* A **biogeochemical cycle** involves the passage of chemicals among nonliving and living reservoirs within the Earth System. Nonliving reservoirs include the atmosphere, the crust, and the oceans, whereas living reservoirs include plants, animals, and microbes. Some stages in a biogeochemical cycle may take only hours, while others may take millions of years. For intervals of time, biogeochemical cycles can attain a **steady-state condition**, meaning that the proportions

All we in one long caravan are journeying since the world began, we know not whither, but we know . . . all must go.

—BHARTRIHARI (INDIAN POET, CA. 500 C.E.)

Did you ever wonder . . .

will the Earth ever look the same in the future as it has in the past?

Figure 10.26 The stages of the supercontinent cycle.

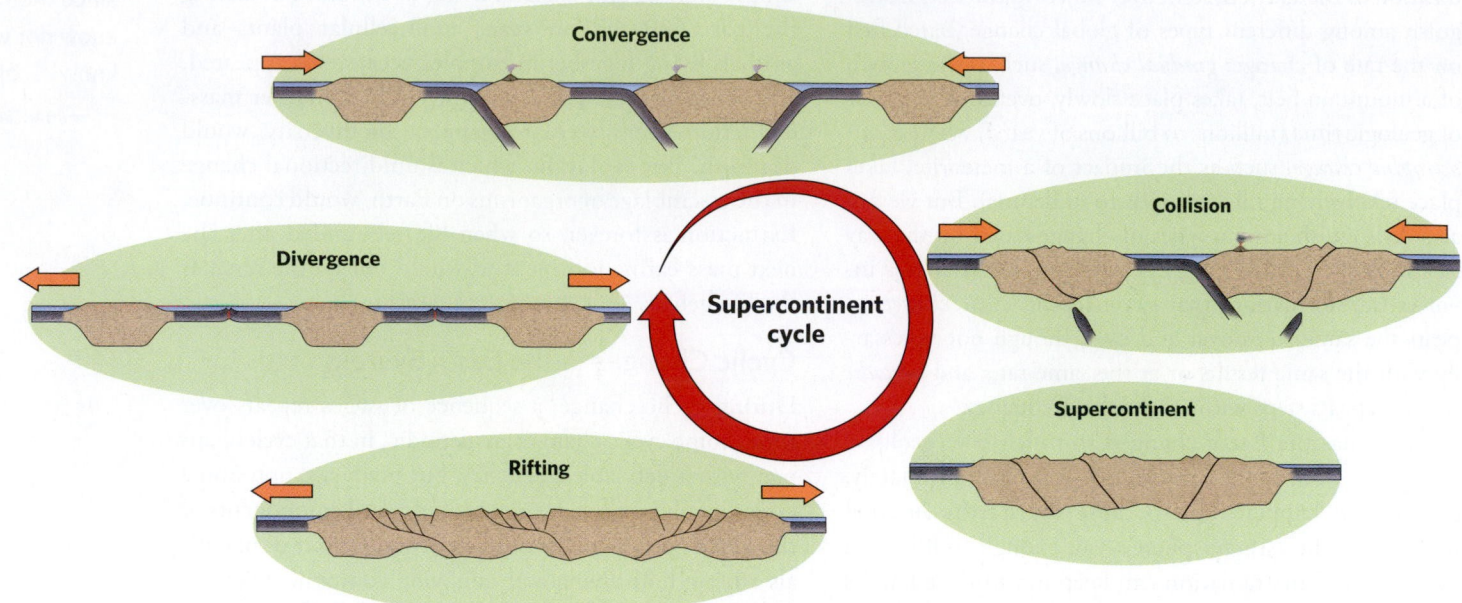

of a chemical in different reservoirs remain fairly constant even though flow among reservoirs continues. A global change in a biogeochemical cycle modifies the proportions of chemicals in different reservoirs, causing a change in the steady-state condition. We'll be discussing examples of biogeochemical cycles, such as the hydrologic cycle and the carbon cycle, later in the book (see Chapters 12 and 20).

Figure 10.27 The human population now doubles about every 44 years. The Black Death pandemic caused an abrupt drop that lasted for a few decades.

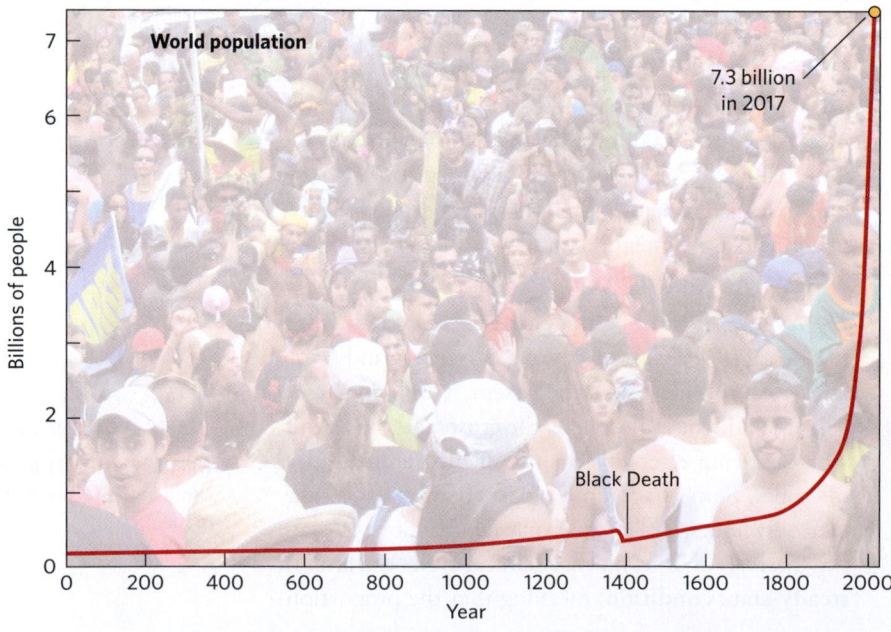

The Geologic Record of Our Time: The Anthropocene?

By the dawn of civilization, at 4000 B.C.E., the human population was at most a few tens of millions. But by the beginning of the 19th century, revolutions in industry and hygiene had substantially lowered death rates and raised living standards, so the population grew at accelerating rates and reached 1 billion in 1850. It then took only 80 years for the population to double, reaching 2 billion in 1930. Today, the doubling time is only 44 years, so the population passed the 6 billion mark just before the year 2000 and surpassed 7.4 billion in 2016 (Fig. 10.27).

As the human population grows and our standard of living continues to improve, our use of the Earth's resources increases. We use land for agriculture and grazing; wood, rock and dirt for construction; oil and coal for energy and plastics; and ores for metals (see Chapter 11). Every time we move a pile of rock, plow a field, drain a wetland, or pave a road, we change a portion of the Earth's crust. Clearly, our use of resources affects the Earth System profoundly, so humanity has become a significant agent of global change. Furthermore, deforestation, overgrazing, agriculture, and urbanization have led to a marked decrease in biodiversity. This decrease may prove to be so severe that it will appear in the geologic record as a mass-extinction event.

Because of all the human-caused changes that have taken place in the last few centuries, some researchers have suggested that this time period marks the start of a new geologic interval, which they refer to as the **Anthropocene**. The geologic record of this time is likely to look distinctly different from the earlier part of the

Holocene. (The Holocene is the portion of time from the last glacial retreat to the present.) Human-caused changes in the atmosphere and the landscape, the production of durable trace fossils (such as concrete and glass), the decrease in biodiversity that is now under way, and the accumulation of long-lived *pollutants* (contaminants that cannot be absorbed or destroyed by natural Earth System processes) may be preserved as a distinct marker bed, discernible by geologists living 100 million years in the future. Use of this term, of course, remains informal and rather controversial. But it does convey the impact that our species will have had on the biography of the Earth as read by observers of the future.

Take-home message . . .

Since our planet first formed, the Earth System has been undergoing major changes. Some of these changes are unidirectional (will never repeat), whereas others are cyclic. Cyclic change can be periodic, repeating at predictable intervals, but often is not. In the present day, human activities are a powerful cause of global change.

Quick Question -
What evidence indicates that sea level rises and falls over time?

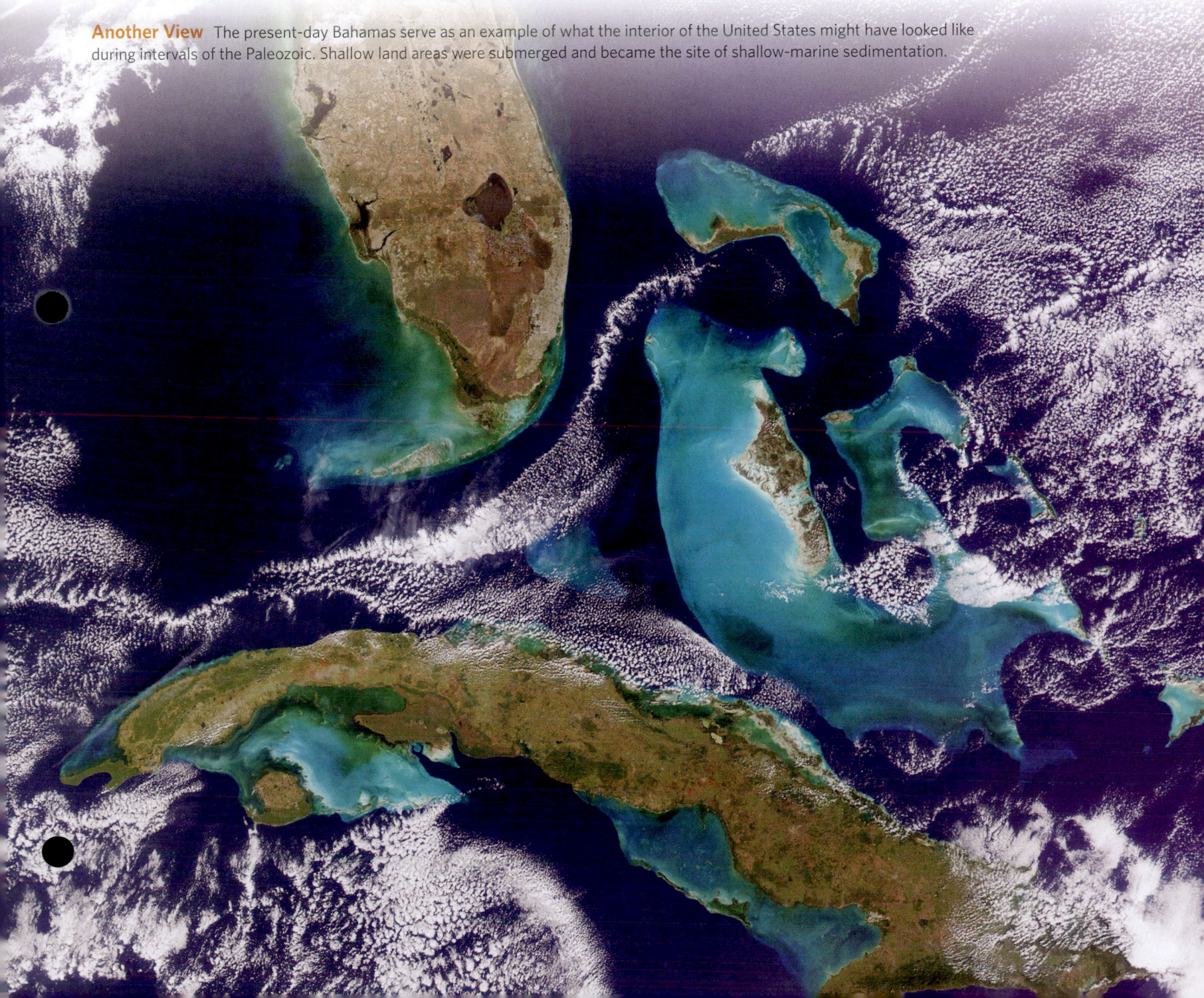

Another View The present-day Bahamas serve as an example of what the interior of the United States might have looked like during intervals of the Paleozoic. Shallow land areas were submerged and became the site of shallow-marine sedimentation.

10 CHAPTER REVIEW

- Isotopic dating of meteorites indicates that Earth differentiated and became a planet about 4.54 billion years ago. For part of its first 600 million years, known as the Hadean Eon, the planet was so hot that its surface was a lava ocean.

- The Archean Eon began about 4.0 Ga. The rock record begins to be readable by about 3.85 Ga, and a permanent liquid-water ocean existed by about that time. Early continental crust was assembled out of volcanic arcs and oceanic plateaus. The first life forms—bacteria and archaea—appeared in the Archean. The atmosphere of that time contained very little oxygen.

- In the Proterozoic Eon, which began at 2.5 Ga, Archean crustal blocks were sutured together to form long-lived cratons. Photosynthesis added oxygen to the atmosphere. By the end of the Proterozoic, soft-bodied marine invertebrates populated the planet, and continental crust had accumulated to form a supercontinent.

- As the Paleozoic Era began, rifting yielded several separate continents. Sea level rose and fell, depositing sequences of strata in continental interiors. Various orogenies took place during the Paleozoic, and by the end of the era, continents combined to form Pangaea. Early Paleozoic life included many invertebrates with shells and jawless fish. Land plants and insects appeared in the middle Paleozoic, and by the end of the era, there were land reptiles and gymnosperm trees.

- In the Mesozoic Era, Pangaea broke apart and the Atlantic Ocean formed. Convergent-boundary tectonics dominated along the western margin of North America. Dinosaurs populated the planet throughout the Mesozoic. During the Cretaceous, the continents flooded, and angiosperms appeared, along with modern fish. A huge mass-extinction event, likely due to a meteorite impact, wiped out the dinosaurs at the end of the Cretaceous Period.

- In the Cenozoic Era, the collision of Africa and India with Europe and Asia formed the Alpine-Himalayan chain. Convergent-boundary tectonics persisted along the margin of South America, creating the Andes, but ceased in part of North America when the San Andreas fault formed. Rifting in the western United States produced the Basin and Range Province. Various kinds of mammals filled ecological niches left vacant by the dinosaurs, and the human genus, *Homo*, appeared. During the Pleistocene, glaciers covered large areas.

- Earth history reflects the process of global change, transformations or modifications of physical and biological components of the Earth System through time. Unidirectional change results in transformations that never repeat, whereas cyclic change involves repetition of the same steps over and over.

- Some types of global change involve biogeochemical cycles, the passage of a chemical among nonliving and living reservoirs within the Earth System.

- Humans have changed landscapes, caused a decrease in biodiversity, produced durable trace fossils, and added pollutants to the environment. The record of human activity during the present time interval, which has been called the Anthropocene, may be preserved in the stratigraphic record.

Alpine-Himalayan chain (p. 332)
Anthropocene (p. 340)
Archean Eon (p. 315)
banded iron formation (BIF) (p. 320)
Basin and Range Province (p. 334)
biogeochemical cycle (p. 339)
Cambrian explosion (p. 323)
craton (p. 319)
cratonic platform (p. 318)
differentiation (p. 314)
ecosystem (p. 320)
Ediacaran fauna (p. 320)
epicontinental sea (p. 322)
global change (p. 339)
Gondwana (p. 322)
great oxygenation event (p. 320)
Hadean Eon (p. 314)
K-Pg extinction (p. 332)
late heavy bombardment (p. 315)
Laurentia (p. 322)
microcontinent (p. 323)
paleogeographic map (p. 319)
Pangaea (p. 327)
Pannotia (p. 319)
Phanerozoic Eon (p. 322)
Pleistocene Ice Age (p. 337)
Proterozoic Eon (p. 317)
Rodinia (p. 319)
shield (p. 318)
snowball Earth (p. 320)
steady-state condition (p. 339)

Review Questions

The letters following each Review Question refer to the corresponding Learning Objective from the Chapter Opener.

1. Why are there no whole rocks on Earth that yield isotopic dates older than 4 billion years? **(A)**

2. Describe the condition of Earth's crust, atmosphere, and oceans during the Hadean Eon. **(B)**

3. How might the first continental crust have formed? When did cratons appear? **(B)**

4. When did life, as recorded by the first fossils, appear? What environment might have served as the cradle of the earliest life? **(D)**

5. How did the atmosphere and plate tectonics change during the Proterozoic Eon? **(C, D)**

6. What evidence suggests that the Earth nearly froze over during the Proterozoic Eon? **(A)**

7. According to the graph, when did most continental crust form? Did supercontinents exist in the Proterozoic? If so, when did they break apart? **(B, C)**

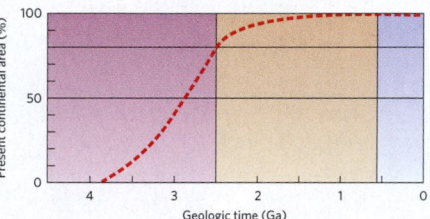

8. Were there multicellular organisms before the Cambrian? How did the Cambrian explosion change the nature of the living world? **(G)**

9. When did continents collide during the Paleozoic? What orogenies mark these collisional events? **(H)**

10. What are the major types of organisms that appeared during the Paleozoic, and in which sequence? Would the land surface at the beginning of the Ordovician have looked different from the land surface at the end of the period? If so, how? **(E, H)**

11. What supercontinent existed at the end of the Paleozoic Eon, and what ocean formed when it broke apart? What event defines the end of the Paleozoic? **(C, G)**

12. Why did broad areas of continental interiors become covered with layers of marine strata during the Paleozoic? **(F)**

13. Describe the plate interactions that led to the formation of the Sierran arc, the Sevier orogeny, and the Laramide orogeny? **(H)**

14. What life forms appeared during the Mesozoic, and what caused a mass-extinction event at the end of the Mesozoic? **(G)**

15. What might have caused the mass extinction at the end of the Paleozoic? **(D, G)**

16. What continents formed as a result of the breakup of Pangaea, and what events eventually produced the Himalayas and the Alps? Which ocean is growing on the attached paleogeographic map? **(C, H)**

17. What major geologic features formed in the western United States during the Cenozoic? **(H)**

18. What major climate changes and biological events happened during the Pleistocene? **(I)**

19. What is global change, and how do unidirectional changes differ from cyclic changes? Do all changes happen at the same rate? **(I)**

20. How have humans become agents of global change, and what might be the record of this change in the stratigraphic record? **(J)**

On Further Thought

21. Geologists have concluded that 80% to 90% of Earth's continental crust had formed by 2.5 Ga. But if you look at a geologic map of the world, you find that only about 10% of the Earth's continental crustal surface is labeled "Precambrian." Why? **(B)**

22. Explain how changes in atmospheric composition may relate to the evolution of life. Keep in mind that organisms that can use oxygen for metabolism can produce much more energy than can organisms that use sulfide minerals for metabolism. **(D)**

Online Resources

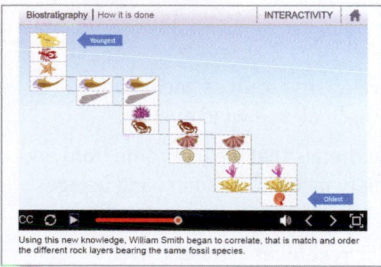

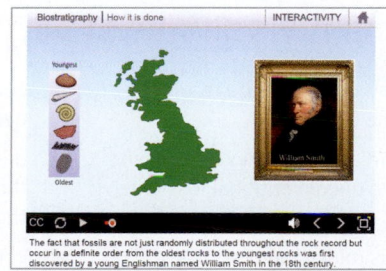

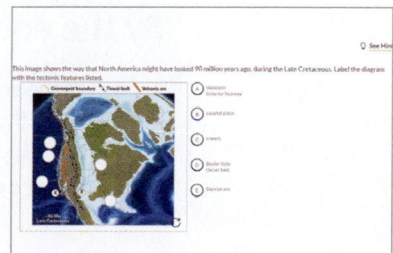

Animations
This chapter features interactive animations testing your knowledge of biostratigraphy.

Smartwork5
This chapter features understanding and ranking exercises about the Earth throughout history, including how the planet's tectonic features have evolved over time.

11 RICHES IN ROCK
Energy and Mineral Resources

By the end of the chapter you should be able to . . .

A. analyze the concept of a resource, and relate resources you use to their sources in the Earth.

B. describe how oil and gas form and can be extracted, and distinguish conventional from unconventional fossil fuel reserves.

C. explain how coal forms and is mined, and discuss challenges associated with coal usage.

D. discuss where nuclear fuel comes from, how a nuclear power plant operates, and what happens to spent nuclear fuel.

E. compare alternative energy sources, including hydroelectric, geothermal, and wind energy.

F. describe where the metals that we use come from and the challenges society faces in obtaining and using them.

G. connect nonmetallic resources (stone, gravel, evaporites) with their sources and uses.

H. recognize the challenge of sustainability as it applies to use of energy and mineral resources.

11.1 Introduction

To survive, our prehistoric ancestors needed little beyond the food and water they could find in their natural surroundings. Humanity's needs changed with the advent of civilization, as even simple agrarian life requires **energy resources** for heat and power and **mineral resources** for construction materials, weapons, and tools (**Fig. 11.1**). Industrialization has greatly increased demands for energy and mineral resources. In fact, to sustain our modern lifestyle, each person today uses about 100 times more resources than one of our prehistoric ancestors did. The rate at which people use resources varies with degree of industrialization. An average resident of an industrialized country uses 30 times more resources than does someone in a developing country. In the United States, for example, one person uses about 16,300 kg (18 tons) of energy and mineral resources each year (**Table 11.1**).

Why do we discuss energy and mineral resources in an Earth Science book? Most of these resources come from geologic materials or are the consequence of geologic, atmospheric, or oceanographic processes. In other words, they are **Earth resources**. In this chapter, we begin our discussion of Earth resources with a focus on hydrocarbons (oil and natural gas) and coal, today's principal energy resources. We then consider other types of resources (nuclear, geothermal, hydroelectric, wind, and solar) that may help satisfy our energy needs in the near future. Next, we turn our attention to mineral resources and describe the origin, distribution, and production of both metallic and nonmetallic materials. For each type of Earth resource, we briefly consider environmental and

In the 1870's, miners excavated this tunnel into the side of a mountain to gain access to ores containing valuable metals. Using small wagons, they wheeled ore out of the mine along the narrow tracks. It was difficult and dangerous work!

Figure 11.1 The discovery of tools and fire led humans to use more Earth resources.

sustainability issues associated with its production and use. Throughout the chapter, we'll learn that supplies of Earth resources occur only where conditions in the Earth System have favored the generation of an **economic reserve** of a resource. This refers to an accessible supply at a concentration that is high enough so that the resource can be extracted or used at a profit.

Table 11.1 Yearly per capita use of earth resources in the united states	
Resource	**Amount (kg)***
Stone	4,100
Sand and gravel	3,800
Oil	3,600
Coal	3,300
Iron and steel	550
Cement	360
Clay	220
Salt	200
Phosphate	140
Aluminum	25
Copper	10
Lead	6
Zinc	5

*1 kg = 2.2 pounds.

11.2 Sources of Energy in the Earth System

What comes to mind when someone asks you to name an energy resource? Perhaps you think about a **fuel**, a substance that burns or undergoes a reaction to produce heat. Or maybe you picture a wind turbine, a hydroelectric dam, or a solar panel. Ultimately, all energy that we use comes from the following sources:

- *Energy directly from the Sun:* Nuclear fusion in the Sun produces radiation that crosses space in the form of electromagnetic waves. This energy can be converted directly to electricity or can be used to heat water.

- *Energy directly from gravity:* The gravitational attraction of the Moon and Sun causes tides. The flow of ocean water due to tides can drive turbines.

- *Energy from both solar radiation and gravity:* Solar radiation and gravity act together to drive atmospheric circulation. The resulting flow of air produces winds that can power turbines. Water, evaporated by solar radiation, blows over land and falls as rain. This rainwater fills streams that flow in response to gravity and drive water wheels or turbines.

- *Energy from inorganic chemical reactions:* Certain inorganic chemicals react to produce energy. For example, combining hydrogen and oxygen can cause an explosion.

- *Energy from nuclear fission:* Splitting radioactive atoms into smaller atoms, a process called *nuclear fission*, yields energy in nuclear power plants and nuclear submarines.

- *Energy from Earth's internal heat:* Earth's interior stays hot because it retains much of the heat produced when the planet originally formed and because radioactive atoms inside the Earth decay to produce additional heat. Extracting heat from underground provides *geothermal energy*.

- *Energy from photosynthesis:* Algae, cyanobacteria, and green plants use energy from sunlight to build organic molecules, a process called **photosynthesis (Fig. 11.2)**. These organisms absorb energy and store it as chemical bonds in the organic molecules that make up their biomass. We can represent the reactions of photosynthesis by a chemical formula:

$$6\,CO_2 + 12\,H_2O + light \rightarrow 6\,O_2 + C_6H_{12}O_6 + 6\,H_2O$$

carbon dioxide   water   energy   oxygen   sugar   water

Over the centuries, the sources of energy on which people rely have changed **(Fig. 11.3)**. For example, before the industrial revolution, most of humanity's energy

Figure 11.2 Plants, by means of photosynthesis, store energy from the Sun in the chemicals that make up their leaves and stems.

needs were satisfied by burning biomass (usually wood or dung). After the industrial revolution, energy needs increased dramatically, and deforestation eliminated wood supplies, so society turned to other resources. **Fossil fuel**, biomass that has been buried and preserved (fossilized) underground in the form of oil, natural gas, or coal, provides most of the energy we use today. This type of fuel dominates because of its accessibility, transportability, and *energy density* (the amount of accessible energy stored per kilogram of a material) (Table 11.2). We can obtain energy from fossil fuels by burning them. Fossil fuel combustion is a chemical reaction in which an oxygen molecule combines with a hydrocarbon molecule. For example, we can represent the burning of methane, a type of natural gas, by the formula

$$CH_4 + 2\,O_2 \rightarrow CO_2 + 2\,H_2O + \text{heat and light}$$

methane oxygen carbon   water   energy
                  dioxide

In power plants, heat produced by burning fossil fuel evaporates water in nearby pipes, thereby producing steam. The pipes carry the steam to a turbine, where the steam's pressure turns the turbine blades. The spinning turbine, in turn, drives an electrical generator.

Figure 11.3 The proportions of different sources of energy that people use have changed over time, as the amount of energy used has increased almost continuously.

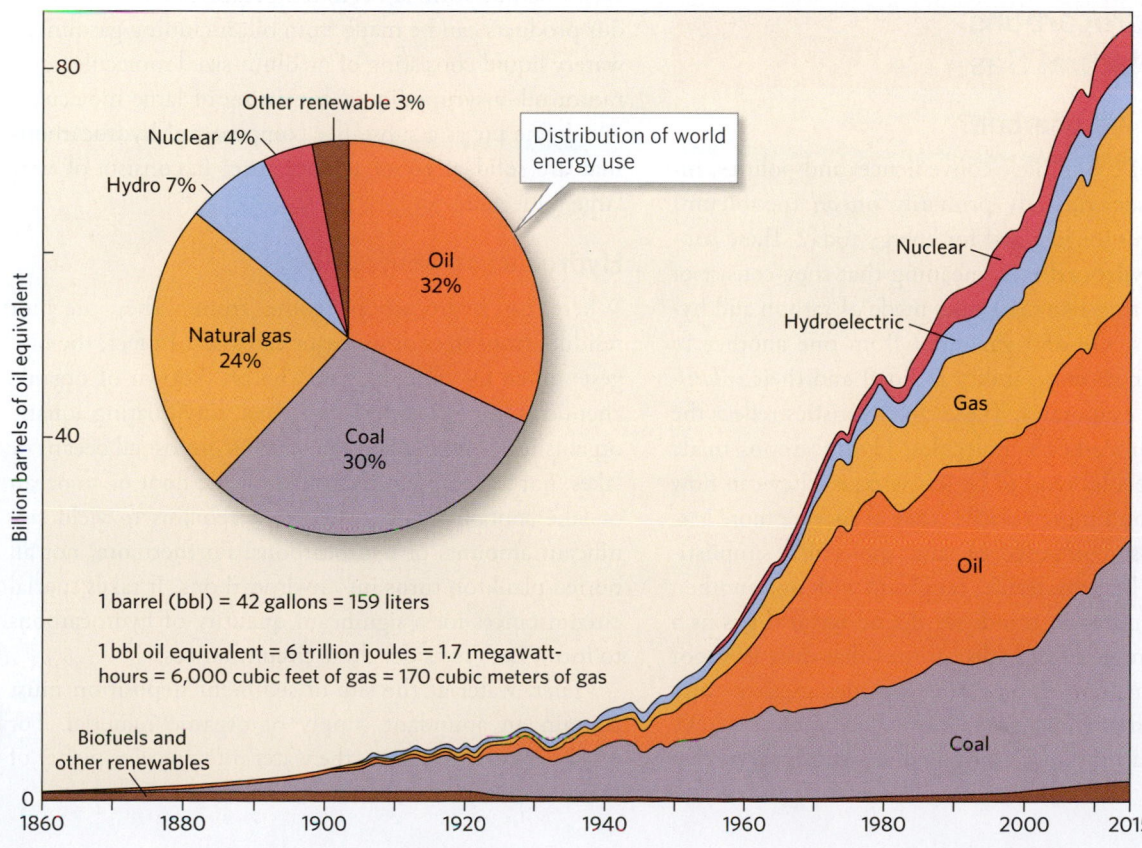

Table 11.2 Energy density of common energy sources

Source	Energy Density (MJ/kg)*
Uranium-235	79,500,000
Gasoline	46
Propane	46
Coal	24
Wood	16
Gunpowder	3
Lithium battery	1.8
Lead-acid battery	0.2

*3.6 MJ (megajoules) = 1 kilowatt-hour (the amount of energy produced by 10 100-watt light bulbs burning for an hour).

Take-home message . . .

The energy we use comes ultimately from several sources: solar radiation, gravity, chemical reactions, nuclear fission, and Earth's internal heat. Photosynthesis stores solar energy in biomass, which can be preserved underground as fossil fuel.

Quick Question -
What sources of energy drive wind and water flow?

Figure 11.4 Oil is a syrupy liquid composed of hydrocarbon molecules.

11.3 Hydrocarbons: Oil and Natural Gas

What Is a Hydrocarbon?

For reasons of economics, convenience, and politics, industrialized societies rely primarily on *oil* (petroleum) and *natural gas* (or just *gas*) for energy today. These substances are **hydrocarbons**, meaning that they consist of chain-like or ring-like molecules made of carbon and hydrogen atoms. Oil and gas differ from one another in terms of their *viscosity* (ability to flow) and their *volatility* (ability to evaporate). These characteristics reflect the size of a hydrocarbon's molecules. Hydrocarbons made of small molecules tend to be less viscous (they can flow more easily) and more volatile (they evaporate more easily) than those composed of large molecules, simplistically because large molecules tangle up with one another. With this concept in mind, we define **natural gas** as a substance composed of hydrocarbons that exist in vapor form at room temperature; this includes methane and propane, both of which are made of small molecules. We define **oil** as a substance composed of hydrocarbons that

exist in liquid form at room temperature **(Fig. 11.4)**. Various products can be made from oil, including gasoline, a watery liquid consisting of medium-sized molecules, and motor oil, a syrupy liquid consisting of large molecules. We define **tar** as a substance composed of hydrocarbons that are solid at room temperature; it consists of very large molecules.

Hydrocarbon Source Rocks

Where do hydrocarbons come from? They are not residues from trees or dinosaur carcasses. In fact, the biggest source for oil and gas is the breakdown of organic chemicals from the cells of *plankton*, tiny floating aquatic organisms. Though plankton grow in almost all oceans or lakes, not all sediment deposited on the floor of an ocean or lake contains enough plankton remains to yield significant amounts of hydrocarbons. Furthermore, not all buried plankton turns into hydrocarbons. It takes special circumstances for a significant quantity of hydrocarbons to form. Let's consider these circumstances.

First, water at the site of sediment deposition must contain an abundant supply of organic material. For organic matter to thrive, the water must receive plenty of

Figure 11.5 The process of hydrocarbon formation begins when organic debris settles with clay. As these materials are buried by additional sediment, heat and pressure transform them into organic shale, in which the organic matter becomes kerogen. At still higher temperatures, kerogen transforms into oil and gas, which can seep upward.

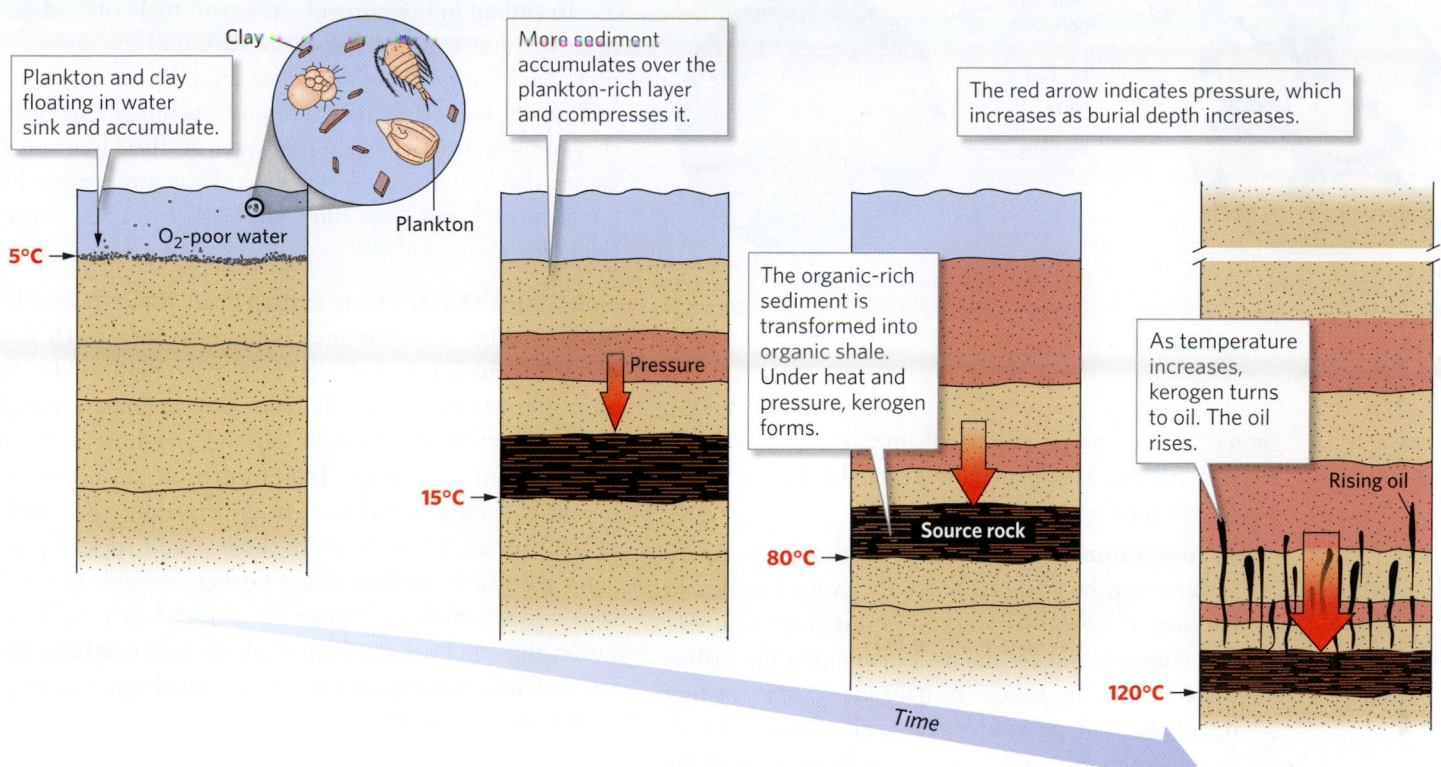

sunlight and must contain abundant nutrients. Second, when plankton die, they must settle in calm water to accumulate along with clay before being washed away by waves or currents. Third, in order for organic molecules not to decay, the water where organic material settles must not contain dissolved oxygen. (Where oxygen is present, microbes or other organisms can consume the organic matter, or the organic matter can degrade by reacting with oxygen.) Organic material that survives mixes with clay to yield a layer of black *organic ooze*. Fourth, the layer of organic ooze must be buried by a thick succession of overlying sediment so that the ooze undergoes lithification into *organic shale*. Organic shale, which is black due to its organic content, contains the raw materials from which hydrocarbons can form, so geologists refer to it as **source rock**.

In order for organic material to transform into hydrocarbons, source rock must undergo burial to a depth of 2 to 4 km (approximately 1.25 to 2.5 miles). At such depths, strata become warm enough (50°C–90°C; 122°F-194°F) for chemical reactions to transform the organic chemicals into a mass of waxy molecules called **kerogen (Fig. 11.5)**. If the rock gets buried deeper and

warms to between 90°C and 160°C (194°F-320°F), kerogen molecules break down into oil and gas molecules, a process known as *hydrocarbon generation*.

Oil molecules can survive to a temperature of about 160°C (320°F). If temperatures increase further, oil molecules break down into the smaller molecules of natural gas, until at temperatures over 250°C (482°F), hydrocarbons lose all hydrogen atoms, and the remaining pure carbon crystallizes into graphite. The limited range of temperatures in which oil forms and survives is called the **oil window**. Gas can form under a broader range of temperatures, known as the *gas window*. Because the energy contained in oil and gas molecules was trapped in the chemical bonds of organisms that lived millions of years ago, oil and gas are fossil fuels.

Hydrocarbon Reserves

Geologists refer to a significant quantity of potentially extractable oil and gas underground as a **hydrocarbon reserve**. If most hydrocarbons are present as oil molecules, it's an *oil reserve*, whereas if most are gas molecules, it's a

Figure 11.6 Porosity and permeability in sedimentary rocks, as seen through a microscope.

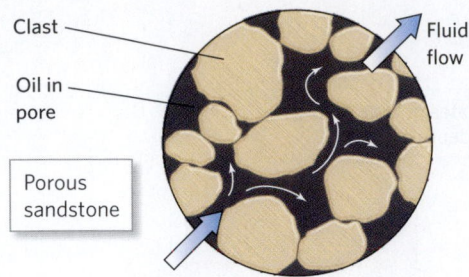
Clast
Oil in pore
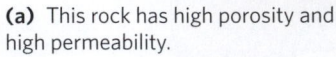
Porous sandstone
Fluid flow

(a) This rock has high porosity and high permeability.

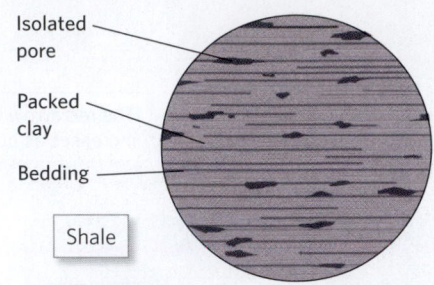
Isolated pore
Packed clay
Bedding
Shale

(b) This rock has low porosity and low permeability.

gas reserve. The energy industry distinguishes hydrocarbon reserves further, based on how easily the hydrocarbons can be pumped up from underground:

- A **conventional reserve** is one from which hydrocarbons can be pumped relatively easily. For this to happen, (1) the hydrocarbons must have relatively low viscosity, and (2) the rock containing the hydrocarbons must be *porous* (contain many *pores*, or tiny openings) and *permeable* (the pores must connect to one another), so that a significant amount of the hydrocarbons can move through the rock.

- An **unconventional reserve**, in contrast, is one in which hydrocarbons cannot flow because (1) the hydrocarbons are too viscous to flow, or (2) the hydrocarbons are locked in *impermeable* rocks (without connecting pores). If either of these conditions exist, the hydrocarbons can be extracted only by using expensive technologies.

Formation of Conventional Hydrocarbon Reserves

Development of conventional hydrocarbon reserves requires two additional steps after a source rock has formed and has passed through the oil window.

MIGRATION INTO RESERVOIR ROCKS. You can't simply drill a well into source rock and pump out oil, for the oil won't flow into the well. That's because a source rock has low porosity and low permeability. (**Porosity** refers to the percentage of open space in a rock, and **permeability** refers to the degree to which pores are interconnected; **Fig. 11.6**). So, a layer of source rock can't be a conventional reserve. Rather, in a conventional reserve, oil or gas is located in a **reservoir rock**, a rock with high porosity and permeability. Poorly cemented sandstones tend to be good reservoir rocks because they have high porosity

Did you ever wonder . . .
whether there are actually pools of oil underground?

(about 35%) and their pores tend to be interconnected. Note that oil and gas underground reside in pores within solid rock; they do not accumulate in underground lakes or pools.

To end up in the pores of a reservoir rock, oil and gas must first move out of the source rock into the reservoir rock. Such *hydrocarbon migration* can take thousands to millions of years **(Fig. 11.7)**. Oil and gas migrate because they are less dense than groundwater, so their buoyancy causes them to rise. If both oil and gas are present in reservoir rock, gas rises above oil because it's less dense than oil.

TRAPS AND SEALS. If oil and gas can move relatively easily through reservoir rock, why don't they reach the surface of the Earth and escape? They sometimes do, at an outlet known as an **oil seep**. For an underground hydrocarbon reserve to exist, oil and gas must be held underground in a **trap**. Trap formation requires two conditions. First, a *seal rock*, an impermeable rock such as shale or salt, must lie above the reservoir rock to prevent hydrocarbons from rising farther. Second, the seal rock and reservoir rock must be arranged in a configuration that encloses the hydrocarbons in a relatively restricted area. Geologists recognize several types of trap geometries **(Fig. 11.8)**.

Formation of Unconventional Hydrocarbon Reserves

Formation of an unconventional hydrocarbon reserve does not require migration into a reservoir rock or confinement within a trap. Geologists distinguish among several types of unconventional reserves:

- An **oil shale** reserve consists of a layer of organic-rich shale in which chemical reactions produced kerogen, but in which temperatures never became hot enough for the kerogen to be transformed into oil or gas.

- A **tar sand** reserve is a sandstone whose pores contain very viscous oil (known as *bitumen*) that cannot flow. It forms when an oil reserve loses small hydrocarbon molecules over time as they evaporate or are consumed by microbes, so that only a residue of large molecules remains.

- A **shale oil** reserve is a layer of source rock in which kerogen transformed into oil, but the oil did not migrate. (Note that shale oil is different from oil shale—unfortunately, the terminology is confusing!)

- A **shale gas** reserve is a layer of source rock that passed into the gas window, so most hydrocarbons it contains occur as small molecules of natural gas.

Take-home message . . .

🏠 Oil and gas form from the remains of plankton. Burial of organic ooze, consisting of these remains mixed with clay, produces organic shale, a source rock. Heat converts organic molecules in source rock into oil or gas. In a conventional reserve, the hydrocarbons have migrated into a porous and permeable reservoir rock and are held underground in a trap. In an unconventional reserve, either the hydrocarbons are too viscous to flow or the rock containing them is impermeable.

Quick Question -
What are the differences between source rock and reservoir rock?

11.4 Oil and Gas Exploration and Production

Birth of the Oil Industry

Prior to the mid-19th century, people used "rock oil," later known as *petroleum* (from the Latin *petra*, meaning rock, and *oleum*, meaning oil) only for greasing wagon axles. Such oil was rare, expensive, and found only at oil seeps. Then, in 1854, a group of investors speculated that petroleum could replace increasingly scarce whale oil as a lamp fuel, if it could be found in reasonable quantities. They hired Edwin Drake to drill for oil at a seep near Titusville, Pennsylvania. On August 27, 1859, Drake and his crew succeeded in finding oil underground. Within a few years, thousands of oil wells had been drilled, and by the turn of the 20th century, civilization was addicted to oil. Initially, most oil was used to produce kerosene for lamps. When electricity replaced kerosene as the favored energy source for illumination, people began using oil as a fuel for generating that electricity, as well as for cars and trucks.

The Modern Search for Oil

The earliest *oil fields* and *gas fields*—areas where numerous wells can be drilled—were found either by searching for seeps or by blind luck. Eventually, energy companies realized that they could hire geologists, who could identify source rocks, reservoir rocks, and traps, to find new fields by means of systematic exploration. Exploration geologists depict the underground configuration of strata and structures in a given area on a cross section, a portrayal of a vertical slice through the Earth. Modern geologists can add detail to these images of the subsurface by producing a **seismic-reflection profile (Fig. 11.9).** Production of such a profile involves the following steps. First, a special vibrating truck, or a dynamite explosion, sends artificial

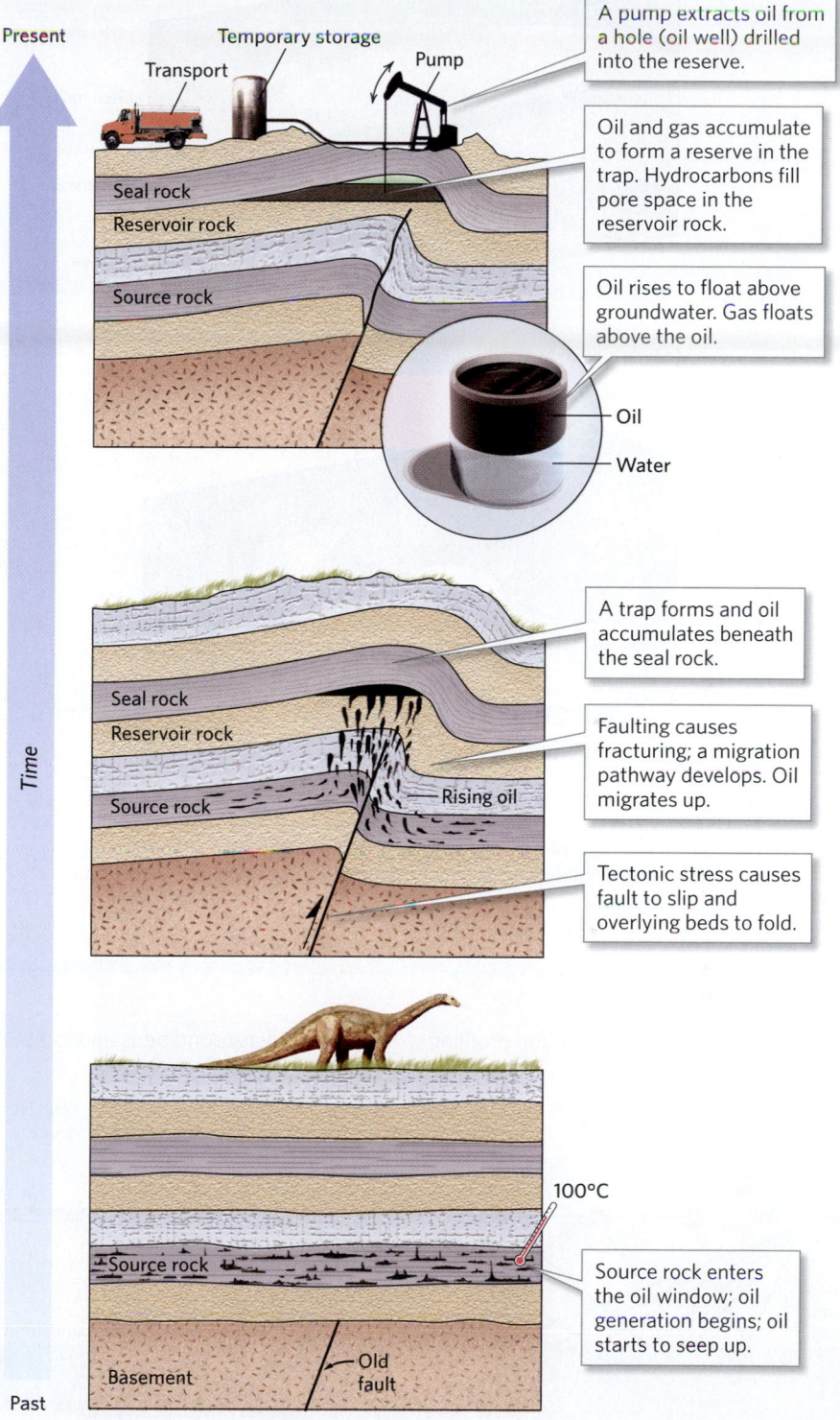

Figure 11.7 Initially, oil resides in source rock. Because it is buoyant relative to groundwater, the oil migrates into the overlying reservoir rock. The oil accumulates beneath seal rock in a trap. In this example, the oil migrates along a fault into a fold.

A pump extracts oil from a hole (oil well) drilled into the reserve.

Oil and gas accumulate to form a reserve in the trap. Hydrocarbons fill pore space in the reservoir rock.

Oil rises to float above groundwater. Gas floats above the oil.

A trap forms and oil accumulates beneath the seal rock.

Faulting causes fracturing; a migration pathway develops. Oil migrates up.

Tectonic stress causes fault to slip and overlying beds to fold.

Source rock enters the oil window; oil generation begins; oil starts to seep up.

seismic waves into the ground. These waves are reflected by contacts between rock layers and return to the ground surface, where portable seismometers record their arrival. Then, by measuring the time between the generation of a

Figure 11.8 A trap is a configuration of seal rock over reservoir rock in a geometry that keeps the oil underground.

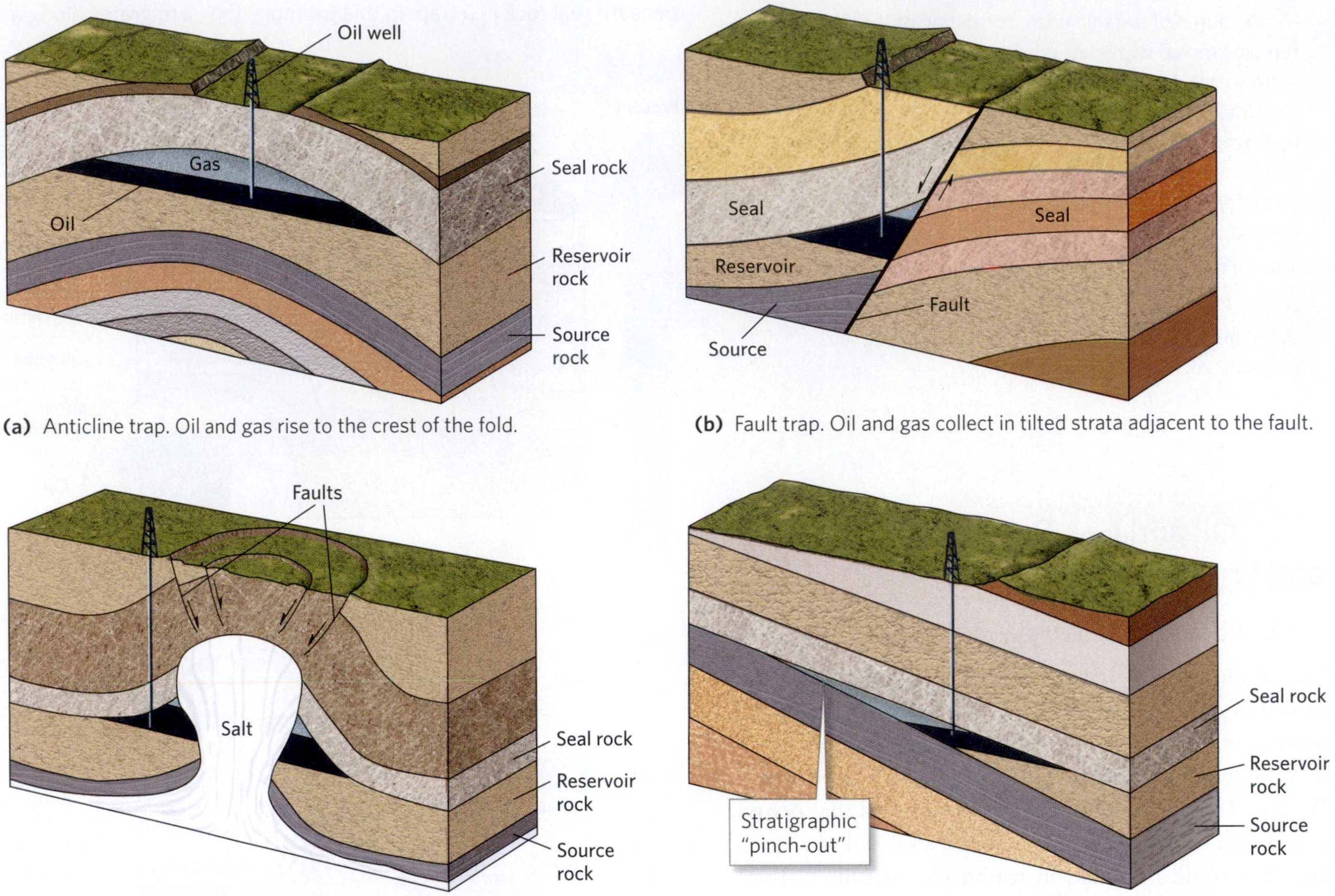

(a) Anticline trap. Oil and gas rise to the crest of the fold.

(b) Fault trap. Oil and gas collect in tilted strata adjacent to the fault.

(c) Salt-dome trap. Oil and gas collect in strata on the flanks of a salt dome, a bulbous intrusion of salt that has risen into overlying strata.

(d) Stratigraphic trap. Oil and gas collect where the reservoir-rock layer pinches out.

Figure 11.9 Using seismic-reflection profiling to describe underground beds and locate hydrocarbon reserves.

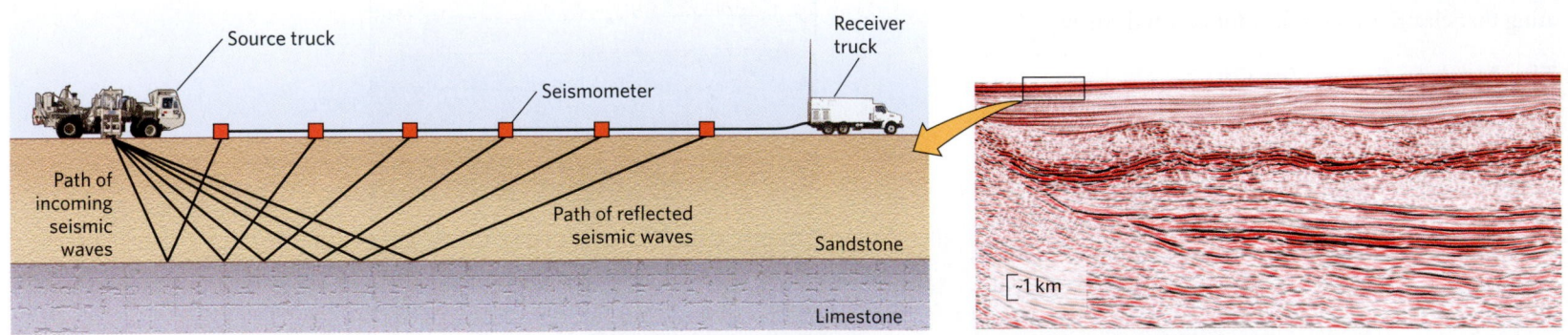

(a) A source truck sends seismic waves into the Earth. The waves are reflected by contacts between rock layers and return to the ground surface, where they are picked up by seismometers. The waves' travel times indicate the depths of the contacts.

(b) A seismic profile can reveal the presence of geologic structures underground.

Figure 11.10 Drilling for oil.

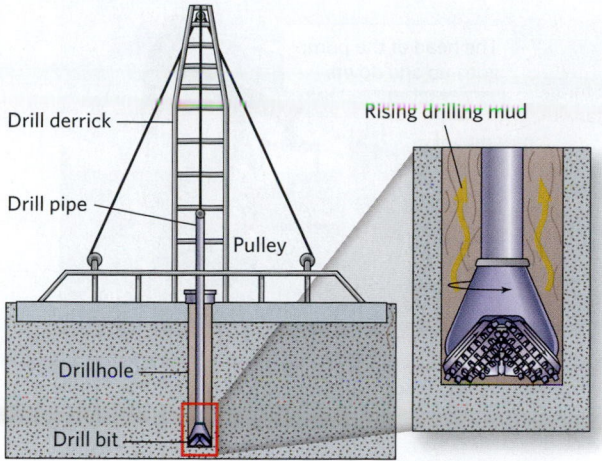

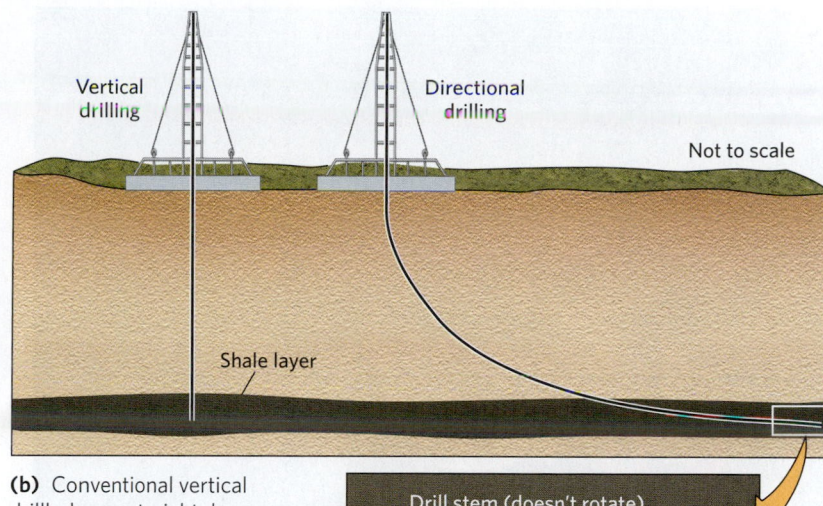

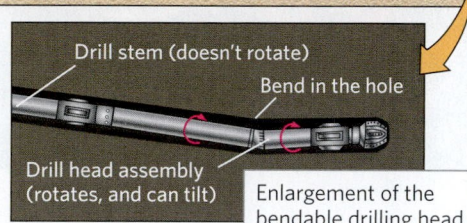

(a) A rotary drill—a rotating pipe tipped by a drill bit—grinds a hole into the ground. Drilling mud, pumped down through the pipe, comes out through holes in the bit, flushing cuttings out of the hole as well as keeping the hydrocarbons underground.

(b) Conventional vertical drillholes go straight down, but directional drilling allows drillers to bend drillholes and hit specific targets. As the inset shows, the head of the drill can bend at an elbow.

seismic wave and its return, a computer can calculate the depths of contacts.

Drilling for Hydrocarbons

When Edwin Drake drilled the first oil well, he used a heavy chisel suspended from a rope. As a steam engine lifted the chisel and dropped it, the chisel broke its way down into rock and made a hole—the process took a long time. Today, drillers use a **rotary drill**, consisting of a metal pipe tipped by a *drill bit*, a bulb of metal studded with hard prongs **(Fig. 11.10a)**. As the bit rotates, it grinds into rock and transforms the rock into *cuttings* consisting of powder and chips. Modern drilling can penetrate rock at a rate of 50 to 100 m (164 to 328 feet) per day. Early drilling methods could produce only vertical drillholes, but today, technology exists to control the path of the drill bit so that the hole can be angled or even horizontal. Such **directional drilling** has become so precise that a driller, using a joystick, can steer the bit to reach an underground target several kilometers away **(Fig. 11.10b)**.

During drilling, workers pump *drilling mud*, a slurry of water and clay, down the center of the drill pipe. The mud comes out of holes at the end of the bit, then flows back up to the surface in the space between the pipe and the hole. This emerging mud cools the bit (which heats up because of friction) and flushes the cuttings out of the hole. The mud's weight also acts like a cork, preventing subsurface hydrocarbons, which tend to be under great pressure, from rushing out of the rock and up the hole. Drilling mud can, therefore, lessen the chances of a **blowout**, during which hydrocarbons burst from a well in a gusher. Blowouts are disastrous because they spill hydrocarbons onto the land or vent them into the air, where they can ignite into an inferno.

Drillers use derricks (towers) to hoist heavy drill pipe into place, segment by segment, during the drilling process. On land, drilling derricks can be installed on a cleared patch of land. But not all oil or gas fields occur on land. To reach oil fields on the continental shelf, drillers build *offshore drilling platforms*, either on stilts anchored to the seafloor or on giant submerged pontoons that allow the platform to float **(Fig. 11.11)**. The drill pipe must be lowered from the platform through water before it penetrates the seafloor.

When they finish drilling, workers "complete" a well. First, they replace the drill pipe with another pipe, called a *casing*, and then they fill the space between the casing and the wall of the hole with concrete in order to seal the wall. Next, the workers use a small explosive charge to puncture the casing and concrete at the level of the hydrocarbon-bearing beds. Finally, they set up a pump to suck oil or gas out of the beds and up to the surface **(Fig. 11.12a)**.

Oil extracted directly from the ground is known as *crude oil*. Once it has been extracted, it is stored in tanks

Figure 11.11 An example of an offshore drilling platform.

Figure 11.12 Pumping, transporting, and refining oil.

Storage tanks

The head of the pump goes up and down.

Hole

(a) When drilling is complete, the derrick is replaced by a pump, which sucks oil out of the ground. This design is called a pumpjack.

(b) The Trans Alaska Pipeline transports oil from fields on the Arctic coast to a tanker port on the southern coast of Alaska.

until it can be transported, by tanker ship or pipeline, to a *refinery* **(Fig. 11.12b, c)**. At the refinery, workers separate crude oil into several different hydrocarbon products by heating it in a vertical pipe called a *distillation column* **(Fig. 11.12d)**. Smaller (lighter) molecules rise to the top of the column, while larger (heavier) molecules stay at the bottom. Natural gas flows from outlets at the top of the column, gasoline and motor oil flow from outlets in the middle, and tar collects at the bottom. Much of the tar is used in the production of plastics. If needed, workers can "crack" larger molecules to break them into smaller ones by heating the molecules to a high temperature.

Surprisingly, simple pumping brings only about 30% of the oil in a conventional oil reserve out of the ground. Energy companies, therefore, commonly use **secondary recovery techniques** to coax out as much as 20% more of the oil. One such technique, **hydrofracturing** (commonly shortened to *fracking*), produces new fractures and opens up pre-existing fractures, thereby increasing the permeability of rock. To frack an oil or gas well, drillers pump *fracking fluid*, a liquid containing 90% water, 9.5% sand, and 0.5% other materials (such as detergents, bactericides, and rust inhibitors), into a sealed-off portion of the drillhole and increase the pressure on the fluid until the rock around the hole breaks **(Fig. 11.13a, b)**. Then they pump out the fluid. Some sand remains, which prevents the cracks from closing tightly **(Fig. 11.13c)**. Oil or gas can then flow from rock into the cracks and up the drillhole.

(c) This oil tanker is capable of carrying a million barrels (enough to supply the entire United States for a few hours).

(d) Distilling columns of an oil refinery transform crude oil into gasoline and other hydrocarbon products.

Figure 11.13 Hydrofracturing technology is able to increase the permeability of rock by producing new fractures and opening existing fractures.

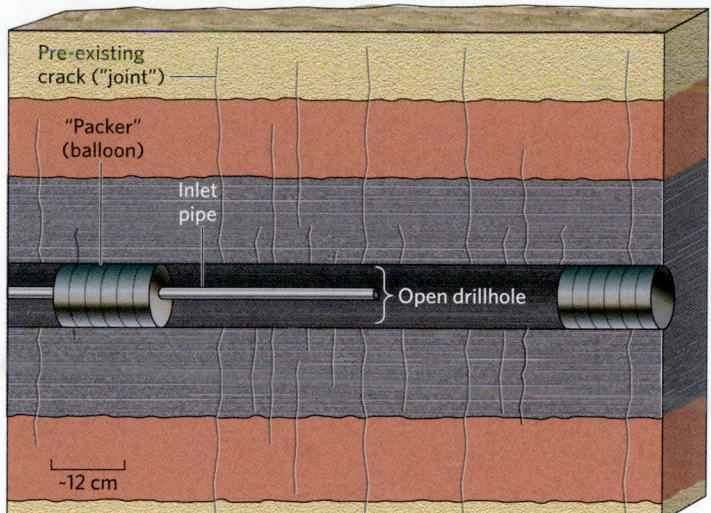

(a) After a hole has been drilled, a portion of the hole is sealed off with expandable cylinders called packers. A pipe is inserted through one of the packers.

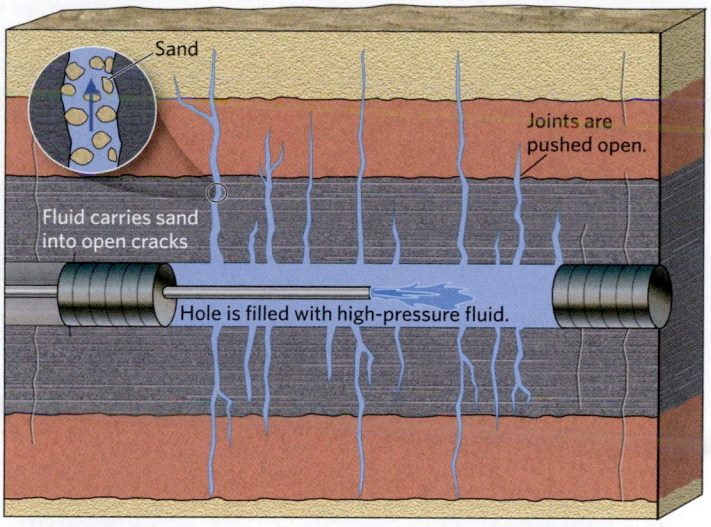

(b) Fluid is pumped under high pressure into the isolated segment of the hole. The pressure pushes open existing cracks and forms new cracks.

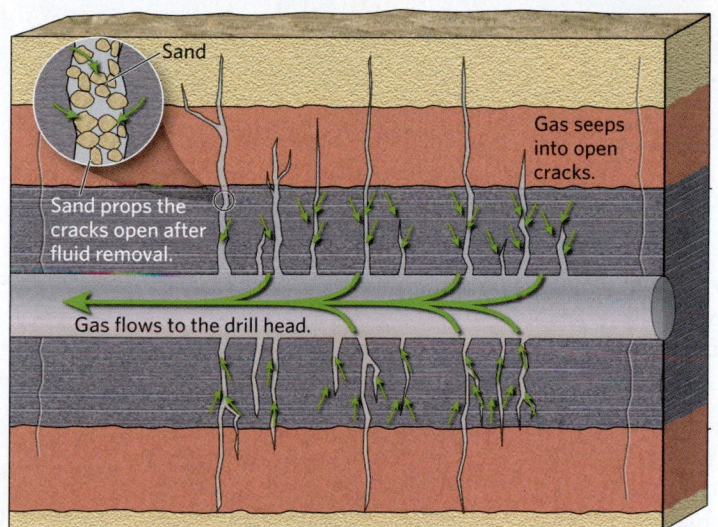

(c) After the packers and the fluid are removed, sand remains behind to prop open the cracks. Oil or gas can seep into the pipe and flow up to the ground, as indicated by the green arrows.

Take-home message . . .

The first oil well was drilled in the mid-19th century. Geologists now use seismic-reflection profiles to locate hydrocarbon traps. Drilling provides access to a reserve, and pumping brings crude oil to the surface, where it can be processed at refineries. New technologies, such as hydrofracturing, can increase the permeability of rock and enhance hydrocarbon recovery.

Quick Question -
What is the purpose of drilling mud?

11.5 Where Do Oil and Gas Occur?

Conventional Hydrocarbon Reserves

Conventional oil and gas reserves, those from which the hydrocarbons can be pumped relatively easily, are not randomly distributed around the planet. Some of these reserves occur along passive margins (onshore and offshore), some in rift basins, some in intracontinental basins, and some in fold-thrust belts. We can specify the volume of oil held in a reserve in *barrels* (1 bbl = 42 gallons = 159 liters) and the volume of gas in cubic feet (as measured at 15°C, 59°F, and 1 atm) **(Fig. 11.14)**.

See for yourself

Oil Field near Lamesa, Texas

Latitude: 32°33′18.42″ N
Longitude: 101°46′55.39″ W

Zoom to 8 km (~5 miles) and look down.

Within a grid of farm fields, 1.6 km or 1 mile wide, small roads lead to patches of dirt. Each patch hosts or hosted a pump for extracting oil. These wells are tapping an oil reserve in Permian sandstone reservoirs.

Figure 11.14 Oil reserves.

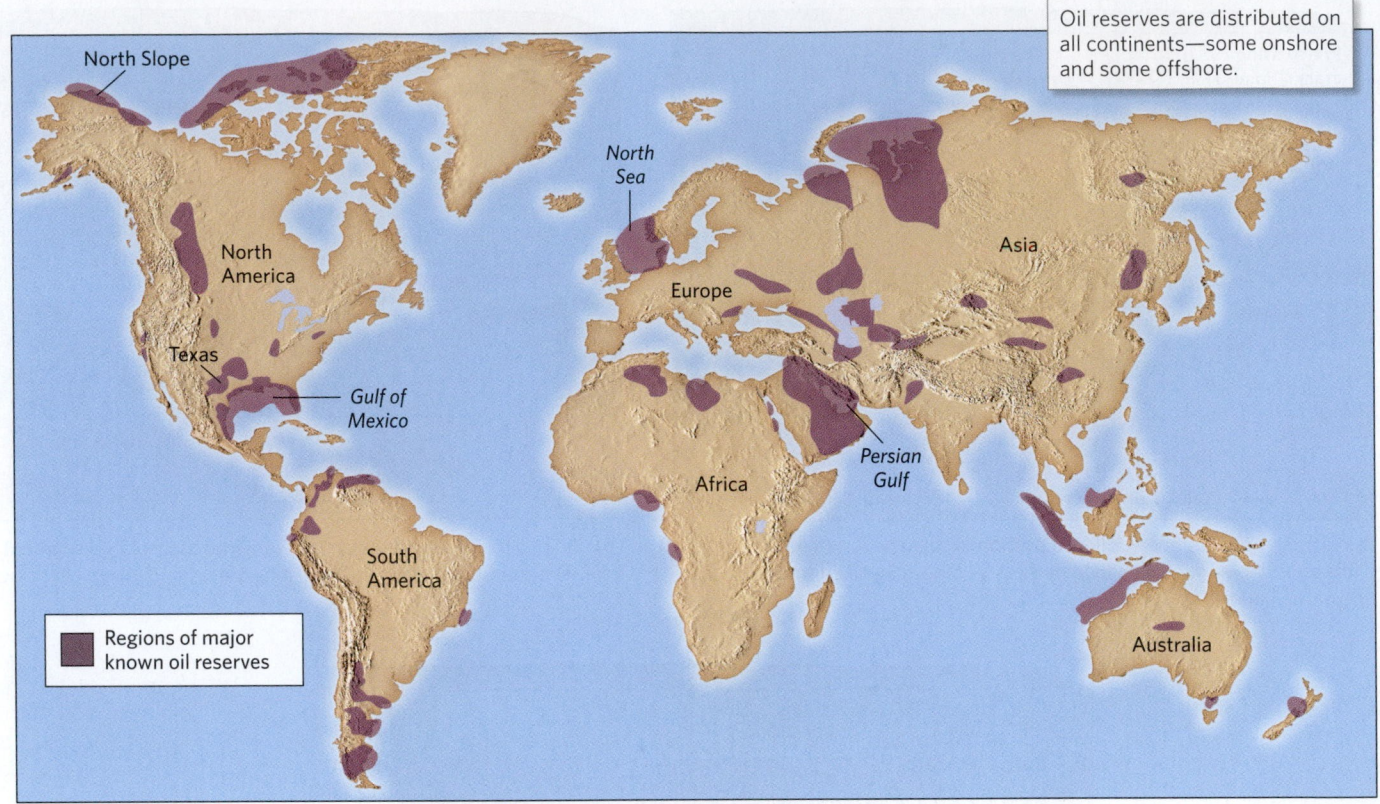

Oil reserves are distributed on all continents—some onshore and some offshore.

North Slope

North
Sea

North
America

Europe

Asia

Texas

Gulf of
Mexico

Persian
Gulf

Africa

South
America

Australia

Regions of major
known oil reserves

(a) The distribution of conventional oil reserves around the world.

Currently, the countries bordering the Persian Gulf contain the world's largest reserves. In fact, this region has almost 60% of the world's conventional reserves. Why? Much of the Earth's crust that now lies in the Middle East was situated in tropical areas during the Cretaceous Period (145 to 66 Ma). Biological productivity in the shallow, brightly lit waters of these regions was high, so the mud that accumulated in these waters was rich in organic material, which lithified to become excellent source rock. Thick layers of sand buried the source rock and eventually became the porous sandstone that serves as reservoir rock. Later, mountain-building processes uplifted and folded these layers, thereby producing traps.

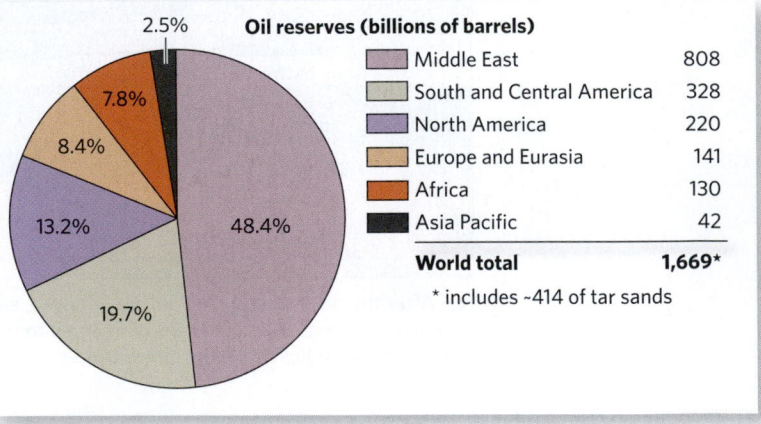

Oil reserves (billions of barrels)	
Middle East	808
South and Central America	328
North America	220
Europe and Eurasia	141
Africa	130
Asia Pacific	42
World total	**1,669***

* includes ~414 of tar sands

(b) Distribution of oil reserves among regions.

Unconventional Hydrocarbon Reserves

In the past 20 years, energy companies have begun to increase their focus on extracting hydrocarbons from unconventional reserves. Because of the high cost of obtaining hydrocarbons from these reserves, their use yields a profit only when the price of oil is high. Let's look at a few examples of unconventional reserves and see how the hydrocarbons they contain can be extracted.

SHALE OIL AND SHALE GAS. Shale oil and shale gas—hydrocarbons that still reside in impermeable source rock (organic shale)—can be extracted only by using directional drilling and hydrofracturing. Directional drilling

Box 11.1

Consider this . . .

Concerns about hydrofracturing

The development of directional drilling and hydrofracturing have led to a boom of drilling into horizontal shale beds to extract shale oil and shale gas. Energy companies have spent billions of dollars to obtain drilling rights in large areas of Pennsylvania, Texas, Oklahoma, and North Dakota and have installed many new wells. Because of the low permeability of the shale, almost all of these wells need to undergo hydrofracturing in order to produce hydrocarbons.

The fracking boom became controversial due to environmental concerns. In many places, a local population boom follows the discovery of a new unconventional field, which strains local services, and once-quiet rural areas become crowded with trucks carrying water, chemicals, and sand (**Fig. Bx11.1a**). Also, residents worry that surface-water supplies may be overused, or that the chemicals in fracking fluids may spill into surface water or leak into groundwater.

Which aspects of fracking pose the greatest risk? Typically, the long horizontal segments of drillholes in source beds lie at depths of over 1.5 km (1 mile), where groundwater tends to be saline and not good for drinking or irrigation, so leakage from the fracked part of the hole at depth might not have immediate effects on water supplies. Fresh, usable groundwater lies at shallower depths, so this region has greater vulnerability (**Fig. Bx11.1b**). Therefore, it is critical that the vertical part of each drillhole is sealed to prevent leakage and that operators take great care not to spill fracking fluid on the ground surface or into surface water.

In some cases, oil and gas extraction brings up large volumes of saline groundwater. Saline water can't be dumped into rivers or used for irrigation, so to dispose of it, drillers pump it underground into deep wells called injection wells. Researchers have shown that pumping at very high pressure and at great depth may trigger small earthquakes. Therefore, injection pressures need to be closely regulated.

Figure Bx11.1 Environmental challenges of hydrofracturing.

(a) A drilling site. The trucks and the holding pond are used during hydrofracturing.

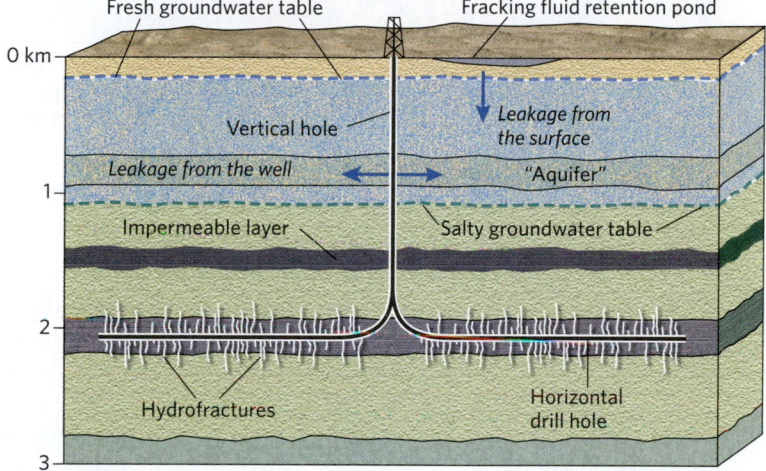

(b) The potential for groundwater contamination by hydrofracturing.

---

allows a driller to bore an angled drillhole into a horizontal shale bed, and then to extend the hole for kilometers within the bed in order to access large volumes of the shale. Once the hole has been drilled, workers hydrofracture the entire length of the horizontal hole. The artificial permeability resulting from hydrofracturing provides a pathway for the hydrocarbons to move into the well. Major reserves of shale oil and shale gas occur in the United States (**Fig. 11.15**), and many of these reserves became a target for drilling starting about 2007. The boom in drilling, however, has led to significant environmental concerns (**Box 11.1**).

TAR SAND. In several locations around the world, vast reserves of tar sand exist. Because of its high viscosity, the bitumen in tar sand can't be pumped directly from the ground. To obtain hydrocarbons from near-surface tar sand deposits, producers dig open-pit mines (**Fig. 11.16**), then heat the extracted sandstone in furnaces until the hydrocarbons become less viscous and can be separated from the sand and refined. About 900 kg (2 tons) of tar sand produce 1 barrel of oil. To extract oil from deep deposits of tar sand, producers drill two parallel wells into the sand: they inject steam or solvent down one well to liquefy the oil, and then pump the oil up from the other well.

Figure 11.15 A map of the major unconventional shale gas and oil fields in North America. The three largest plays are labeled (a *play* is a region being drilled).

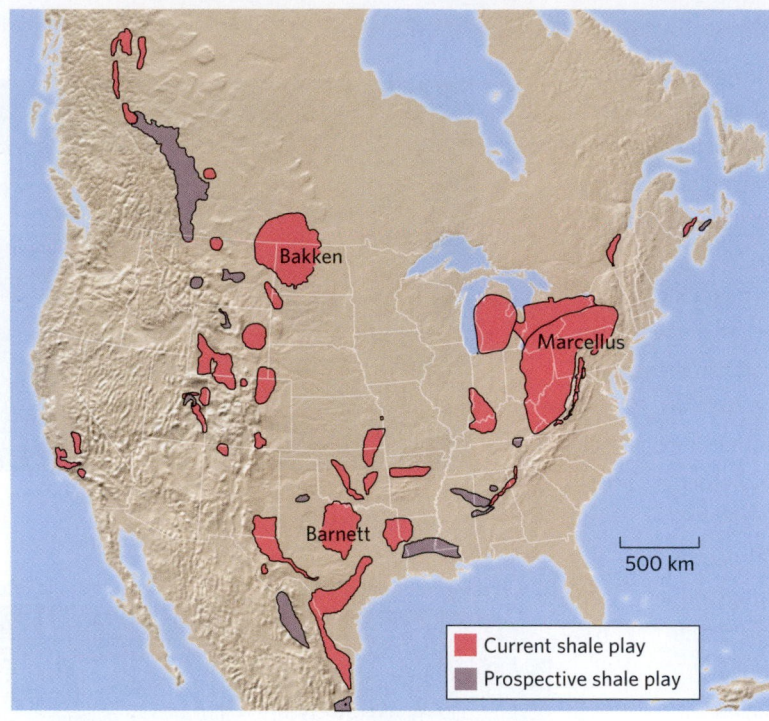

Current shale play
Prospective shale play

OIL SHALE. To obtain usable fuel from oil shale, producers must either mine it and then heat it, so that the kerogen it contains transforms into oil and gas that can be separated from the shale, or they must heat it underground so that it can flow into drillholes and be extracted.

Figure 11.16 An open-pit mine for tar sand in Alberta, Canada.

Take-home message . . .

Conventional hydrocarbon reserves occur in many locations, both onshore and offshore. The largest reserves lie in the Persian Gulf region. Unconventional reserves are more widespread, but extracting hydrocarbons from these reserves is difficult and expensive.

Quick Question ----------------------------
Why is directional drilling used to extract hydrocarbons from shale?

11.6 Coal: Energy from the Swamps of the Past

Coal Formation

Imagine a swamp in a tropical climate. Trees, ferns, leafy plants, vines, and other green plants grow in abundance, producing an immense amount of biomass. This biomass stores energy from the Sun, for photosynthesis, as we have seen, locks energy in the chemical bonds of the molecules that make up the living plants. Now imagine that, over time, plants die and fall into the stagnant water of the swamp. Because this water lacks oxygen, the dead biomass does not rot away entirely and new plants grow on top of it. Eventually, a layer of compacted organic material, known as **peat**, accumulates. Dried peat can be burned directly for heating.

If a peat layer becomes deeply buried by other layers of sediment, it can be preserved over geologic time. When the overlying sediment becomes thick enough, the organic matter in peat undergoes further compaction and heating. Slowly, reactions take place that remove water and other volatile compounds, such as methane and ammonia. What remains is **coal**, a black, brittle sedimentary rock that burns. Coal consists primarily of large carbon molecules, mixed with organic chemicals, quartz, and clay. We consider coal to be a type of fossil fuel because it stores solar energy that reached Earth long ago.

Note that oil and coal have different origins. As we have seen, oil consists of hydrocarbons derived from the remains of buried plankton, whereas coal consists of carbon derived from woody plants. These plants grew in *coal swamps*, regions that resemble the rainforests and wetlands of modern tropical and subtropical coastal areas **(Fig. 11.17a)**. Because coal forms in layers within a sedimentary succession, a coal bed, or **coal seam**, appears in an outcrop as a black band between other kinds of sedimentary rocks **(Fig. 11.17b, c)**. An individual seam may be centimeters to meters thick and may underlie a broad region.

Figure 11.17 Coal forms when plant debris becomes deeply buried.

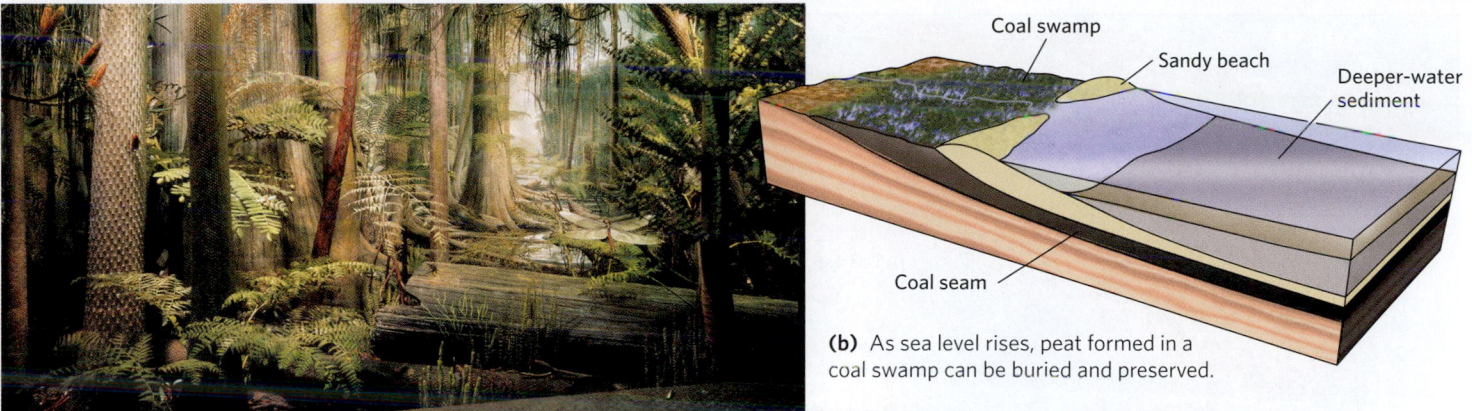

(b) As sea level rises, peat formed in a coal swamp can be buried and preserved.

Coal swamp
Sandy beach
Deeper-water sediment
Coal seam

(a) A museum diorama depicting a Carboniferous coal swamp.

Upper Carboniferous (~ 320 Ma)

■ Coal swamps

Siberia
N. America
Baltica
Equator
S. America
Africa
India
Australia
Antarctica

(c) Coal beds interlayered with beds of sandstone and shale.

(d) When Pangaea existed, during the Late Carboniferous, North America and Europe straddled the equator, so they enjoyed tropical climates in which large coal swamps grew.

The most extensive deposits of coal in the world occur in Carboniferous strata. In fact, the abundance of coal of this age led geologists to give the name *Carboniferous* to the period between 359 and 299 Ma. Why did so much coal form at this time? During the Carboniferous, the continents had assembled to form the supercontinent Pangaea, and large regions of land straddled warm tropical regions with high rainfall **(Fig. 11.17d)**. In these regions, recently evolved large land plants flourished and swampy, oxygen-poor burial conditions existed. Not all coal reserves, however, are Carboniferous. During the Cretaceous (145 to 66 Ma), large coal swamps developed in what is now Wyoming and adjacent states.

Classifying Coal

Peat changes progressively over time as it gets buried more deeply and undergoes more heating **(Fig. 11.18)**. Dried peat contains about 50% to 60% carbon. As we've mentioned, when a layer of peat becomes buried, it loses water, and chemical reactions take place that destroy the

Figure 11.18 Burial of peat leads to the formation of coal of progressively higher carbon content.

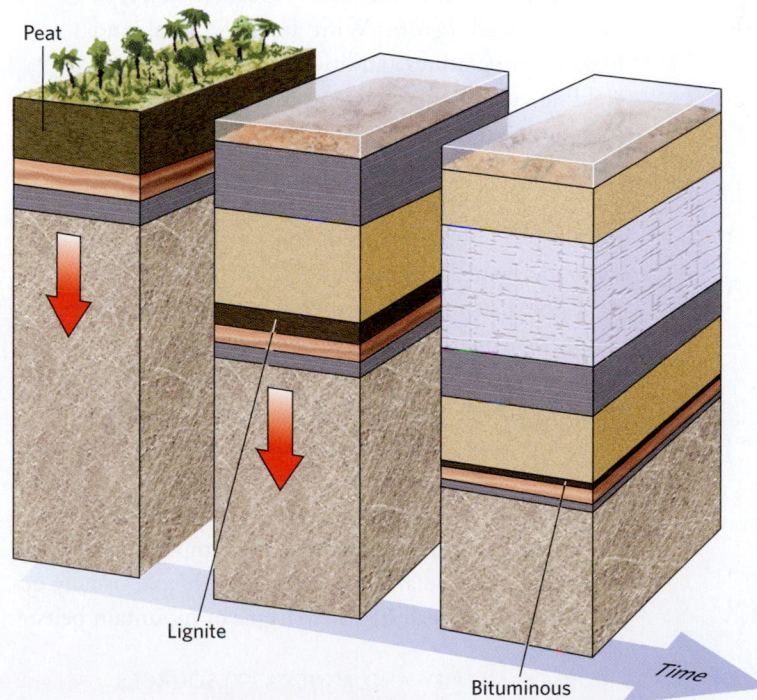

Peat
Lignite
Bituminous
Time

359

Figure 11.19 Coal reserves worldwide.

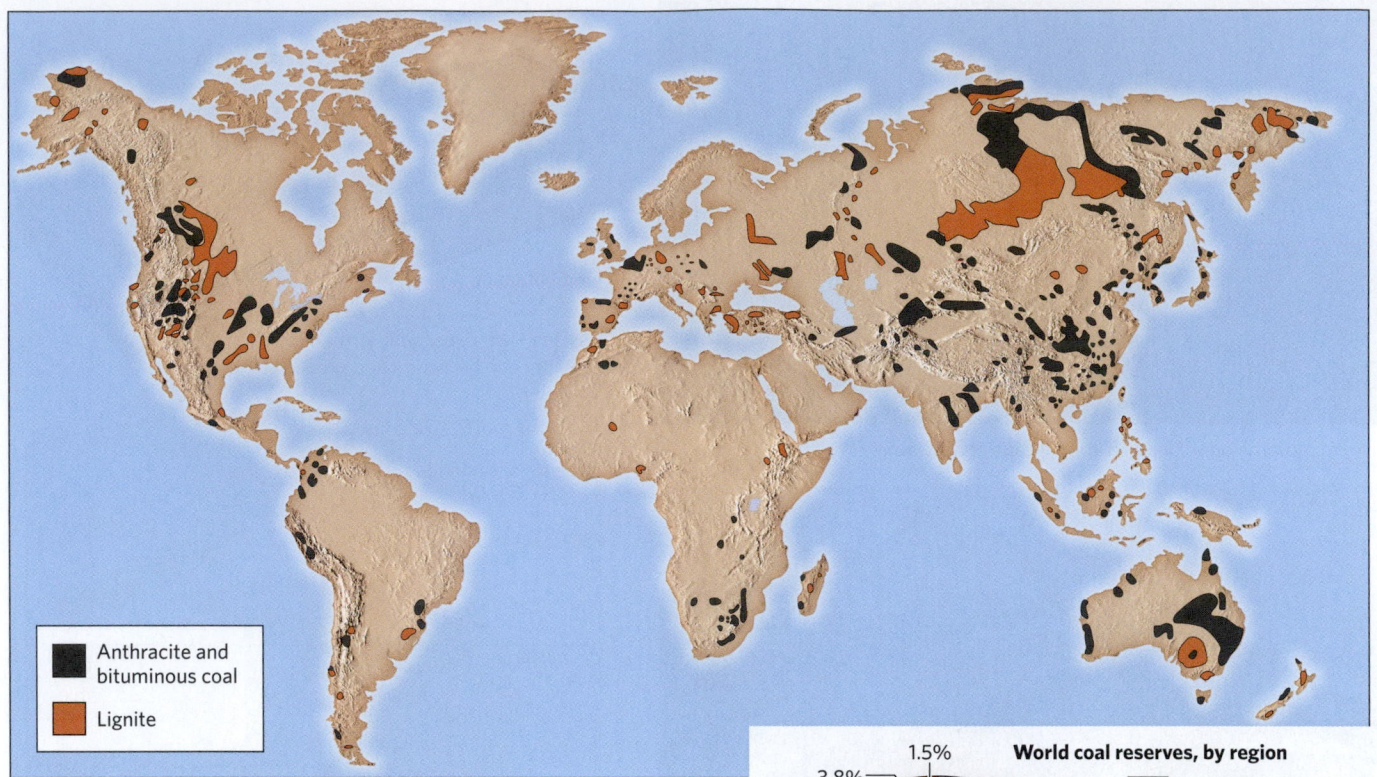

(a) A map showing the global distribution of coal reserves. Most coal has accumulated in mid-continental basins.

World coal reserves, by region

- Europe and Eurasia
- Asia Pacific
- North America
- Middle East and Africa
- South and Central America

1.5%
3.8%
28.5%
35.4%
30.9%

World total = 949,021 million tons

(b) Global coal reserves by region.

original organic chemicals and release volatile compounds. As the resulting gas seeps out of the organic layer, the residue contains a progressively higher percentage of carbon. Once the proportion of carbon in the residue is between 60% and 70%, the deposit becomes true coal—specifically, a type of coal called *lignite*. With further burial and higher temperatures, additional volatiles escape, yielding even higher concentrations of carbon. If the temperature reaches 100°C to 200°C, lignite transforms into dull black *bituminous coal*, which contains 70% to 87% carbon. At still higher temperatures (200°C to 300°C), bituminous coal transforms into shiny black *anthracite*, which contains more than 87% carbon.

Geologists use the term **coal rank** to characterize the extent to which coal has been altered by losing volatiles and becoming enriched in carbon. Lignite is low-rank, bituminous coal is intermediate-rank, and anthracite is high-rank coal. In effect, coal rank depends on the maximum temperature to which coal has been subjected. Lignite and bituminous coal occur in extensive horizontal beds within sedimentary basins. The temperatures needed to form anthracite are so high that such coal usually occurs only in rocks along the margins of mountain belts.

Mining Coal

Sedimentary strata of continents contain huge quantities of discovered coal, called *coal reserves* **(Fig. 11.19a, b)**. The methods by which energy companies extract coal from these reserves depend on the depth of the coal seam. If the coal seam lies within about 100 m (330 feet) of the ground surface, *strip mining* proves to be the most economical method. In a strip mine, miners use a giant shovel called a *dragline* to scrape off the soil and the layers of sedimentary rock that lie above the coal seam **(Fig. 11.20a)**. (Overlying layers of sedimentary rock must be broken up using explosives before they can be removed.) Once the dragline has exposed the seam, miners use other equipment to dig

Figure 11.20 Coal mining.

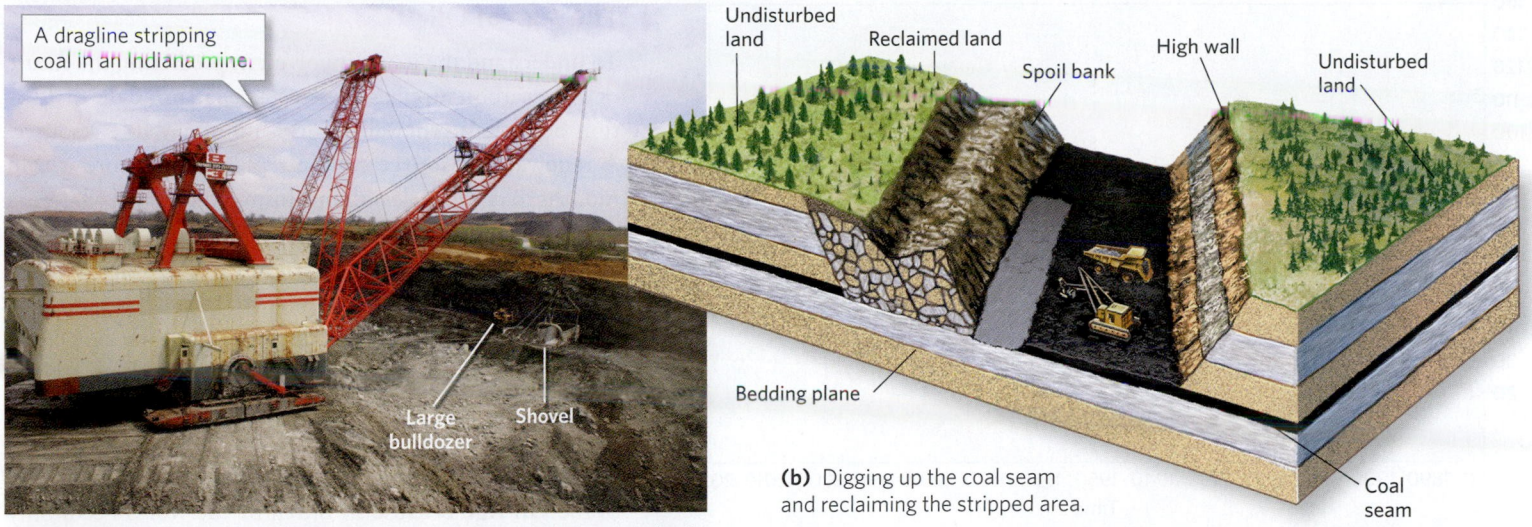

(a) A dragline stripping away the soil and sedimentary rock that overlie the coal seam.

(b) Digging up the coal seam and reclaiming the stripped area.

(c) Piles of recently mined coal.

(d) Underground coal mining.

out the coal and dump it into trucks or onto a conveyor belt **(Fig. 11.20b, c)**. In places where coal underlies plains, miners typically dig a long trench to extract the coal. In hilly areas, they may use a method called *mountaintop removal mining*, meaning that they remove the land above the coal and dump it into an adjacent valley in order to expose the coal.

Before modern environmental awareness took hold, all strip mining left huge scars on the landscape. Without topsoil, the rubble and exposed rock left by mining operations remained barren of vegetation. In many contemporary mines, however, the dragline operator separates out and preserves soil. Then, when the coal has been scraped out, the operator refills the hole with the rock that was stripped to expose the coal and covers the rock with the stockpiled soil, so that grass or trees can grow. In areas

where mining involves mountaintop removal, restoration of the original landscape is impossible, for while the tops of hills may be covered with soil and replanted, they can never be returned to their original shapes.

Deeply buried coal can be obtained only by *underground mining*. To develop an underground mine, miners dig a shaft down to the depth of the coal seam and then excavate a maze of tunnels, using huge grinding machines that chew their way into the coal **(Fig. 11.20d)**. Underground coal mining can be very dangerous, not only because the sedimentary rocks forming the roof of the mine are weak and can collapse, but also because explosive methane gas released by chemical reactions in coal can accumulate in the mine. All coal mining produces dust, so unless they breathe through filters, miners risk contracting black lung disease by inhaling coal dust.

Figure 11.21 The future of hydrocarbon resources.

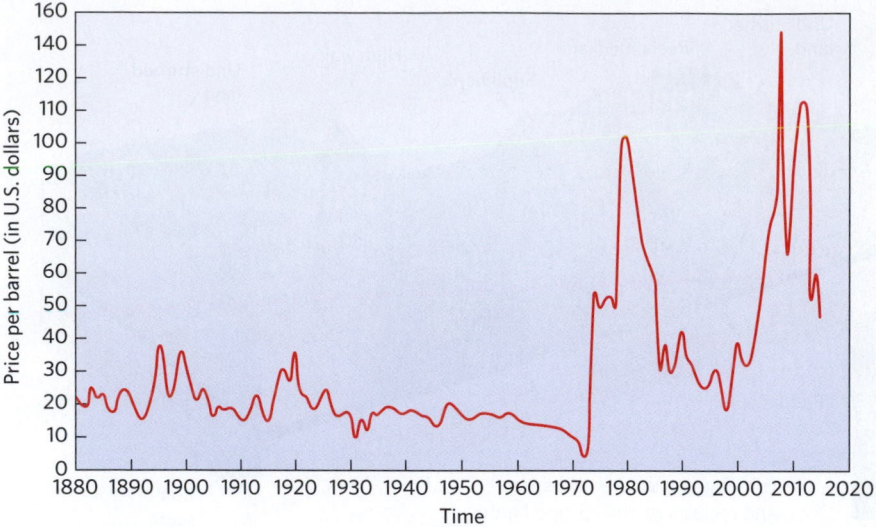

(a) The price of oil held fairly steady for almost 100 years. Starting in 1970, it has risen and fallen dramatically.

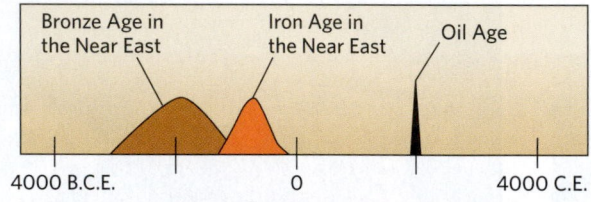

(b) The Oil Age is a relatively short period on the timeline of human history.

M. King Hubbert correctly predicted the peak of production in the United States.

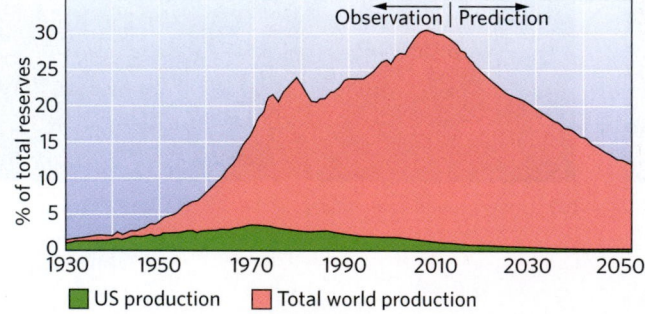

(c) A plot of reserves versus time suggests that in the early 21st century, the global rate of oil production will start to decrease. The time of maximum production is called Hubbert's peak. Production has already declined in the United States.

Take-home message . . .

Coal forms from the remains of plant material. When buried deeply, this organic material undergoes reactions that concentrate carbon. Geologists define different ranks of coal, based on the coal's carbon content. Coal occurs in sedimentary successions and is extracted by strip mining or by underground mining, depending on the depth of the coal bed.

Quick Question -
Why did so much coal form during the Carboniferous?

Did you ever wonder . . .
- - - - - - - - - - - - - - -
how much longer the world's oil supply will last?

11.7 The Future of Fossil Fuel

The Age of Oil

To understand the issues involved in predicting the future of hydrocarbon supplies, we must keep in mind the distinction between a renewable and a nonrenewable resource. A **renewable resource** can be replaced by nature within months to decades (less than a human lifespan), whereas a **nonrenewable resource** can take centuries to millions of years to replenish. Oil and natural gas are nonrenewable, for a reserve takes millions of years to form. Today, despite dramatically fluctuating prices **(Fig. 11.21a)**, the global rate of fuel consumption continues to increase, both due to overall demand and because densely populated countries such as China and India have undergone rapid industrialization. Future historians may refer to the present time as the **Oil Age** because so much of the world's economy depends on oil. How long will the Oil Age last? It's hard to answer this question because of disagreement regarding which numbers should go into the calculation. Opinion varies on whether or not unconventional reserves and natural gas should be included in estimates, and whether or not we can assume that the rate of consumption in the future will stay the same.

Geologists estimate that up to 1,300 billion barrels of oil reside in proven conventional reserves, meaning reserves that have been documented and remain in the ground. Optimistically, an additional 2,000 billion barrels may lie in unproven conventional reserves, meaning areas whose oil has not yet been found but might exist. So the Earth's sedimentary strata may hold up to 3,300 billion barrels of conventional oil reserves. Presently, humanity guzzles this oil at a rate of about 33 billion barrels per year. At this rate, conventional oil supplies will last for about another 100 years.

Of course, the picture of oil's future changes significantly if estimates include unconventional reserves. All told, perhaps 500 billion barrels of shale oil exist, along with 1.5 trillion barrels of tar-sand oil, and 4.5 trillion barrels of oil in oil shale. But wide disagreement remains concerning whether to include all of these reserves in our estimates because a significant proportion would be so difficult and expensive to access that they may never be economical as an energy source.

For the sake of discussion, if we guess that the Earth holds about 3,000 billion barrels of conventional oil reserves, and perhaps another 4,000 billion barrels of relatively accessible unconventional oil reserves, then at current rates of consumption, supplies might last for only another 200 to 250 years. If so, the Oil Age will have lasted a total of about 350 to 400 years. On a timeline representing the 4,000 years since the construction of the Egyptian pyramids, the Oil Age looks like a very short blip, so we may indeed be living during a unique interval of human history (Fig. 11.21b). Some researchers argue that the beginning of the end of the Oil Age has begun because the rate of consumption now exceeds the rate of discovery. If we plot a graph showing how the rate of oil production changes over the years, we see that a peak in production may be reached in the early 21st century. The peak of production for a given reserve is called *Hubbert's peak*, named for the geologist who first discussed it (Fig. 11.21c).

The Future of Coal Reserves

Society still burns a lot of coal. Worldwide coal reserves are estimated at about 850 trillion tons, which could supply approximately the same amount of energy as 11 trillion barrels of oil. But such an estimate of coal reserves does not distinguish clearly between accessible (minable) coal and inaccessible coal, which is too deep to mine. So estimating how long coal reserves can last is a challenge. At current rates of use, many centuries of accessible supplies remain.

Environmental Consequences of Fossil Fuel Use

The extraction, processing, and use of fossil fuel have consequences for the environment and for the climate. Oil drilling requires large drilling rigs and drilling pads to support the rigs, all of which can damage the land. And, as demonstrated by the 2010 *Deepwater Horizon* blowout in the Gulf of Mexico, offshore oil drilling can lead to tragic losses of life and disastrous marine oil spills (Box 11.2). Oil spills from ships and tankers can create oil slicks that spread over the sea surface and foul the shore-

line (Fig. 11.22a, b). On land, oil spills from pipelines or trucks may sink into the subsurface and contaminate groundwater. And as we've seen, if not handled properly, fracking fluids may contaminate water supplies.

Coal mining, in addition to changing the landscape, also has the potential to produce **acid mine runoff**, a dilute solution of sulfuric acid (H_2SO_4). Acid mine runoff develops when sulfide-bearing minerals, such as pyrite (FeS_2) in coal, react with rainwater when exposed by mining. If the runoff enters streams, it can kill fish and plants. Another serious problem arises when, due to arson, accident, or spontaneous processes, unmined coal beds ignite. Once started, a *coal-bed fire* may burn for years and may be difficult or impossible to extinguish. Coal-bed fires produce toxic fumes that rise through joints to the surface to make the overlying land uninhabitable. In addition, when the coal in a seam burns away, the land surface above may sink.

Numerous air pollution issues arise from the burning of fossil fuels. If exhaust from smokestacks is not cleaned before being emitted, it may introduce soot, carbon monoxide, sulfur dioxide, nitrous oxide, and unburned hydrocarbons into the air. These pollutants can cause deadly smog whose presence forces urban dwellers to remain inside. Some fossil fuels, particularly coal, contain sulfur, and burning such fuels produces sulfur-rich gas. When this gas combines with moisture in the air to form dilute sulfuric acid, that moisture falls as **acid rain**. Acid rain can increase the acidity of soils and lakes, causing vegetation and fish to die off in regions downwind of smokestacks. For this reason, many countries now regulate the amount of sulfur that coal can contain when it is burned, so some coal consumers have switched the types of coal they use.

Figure 11.22 Marine oil spills can come from drilling rigs or from tankers.

(a) An oil tanker leaking oil on the sea surface.

(b) Oil spills can contaminate the shore and can be very difficult to clean up.

Box 11.2 ## Consider this . . .

The *Deepwater Horizon* disaster

A substantial proportion of the world's conventional oil reserves reside in the sedimentary basins that underlie the continental shelves of passive continental margins, accessed by offshore drilling platforms. During both onshore and offshore exploration, drillers worry about the possibility of a blowout. A catastrophic blowout occurred on April 20, 2010, when drillers on the *Deepwater Horizon*, a huge pontoon platform, were completing a 5.5-km (18,000-foot)-long well in 1.5-km (5,000-foot)-deep water southeast of the Mississippi

Figure Bx11.2 Offshore drilling and the *Deepwater Horizon* disaster.

Delta. Due to a series of errors, the casing was not sufficiently strong when workers began to replace the drilling mud with clear water. As a result, gassy oil under high pressure in reservoir rock punctured by the well rushed up the drillhole. A backup safety device, called a blowout preventer, was supposed to clamp the wellhead (the outlet of the well shut), but it failed, so the gassy oil reached the platform and sprayed 100 m (330 feet) into the sky. Sparks from electronic gear triggered an explosion, and the platform became a fountain of flame and smoke that killed 11 workers. An armada of fireboats could not douse the conflagration **(Fig. Bx11.2a)**, and after 36 hours, the still-burning platform tipped over and sank.

Robot submersibles sent to the seafloor to investigate found oil and gas billowing from the twisted mess of bent and ruptured pipes at the well's outlet **(Fig. Bx11.2b)**. On the order of 50,000 to 62,000 barrels of oil entered the Gulf's water from the well each day. Stopping this underwater gusher proved to be an immense challenge, and initial efforts to block the well or to cover the wellhead with a containment dome failed. Not until July 15 was the flow finally stopped. Meanwhile, workers drilled another well from a platform a few kilometers away. Using directional drilling, they managed to intersect the 15-cm (6-inch)-diameter *Deepwater Horizon* drillhole, and by September 19, the concrete pumped into the failed well sealed it permanently. All told, about 4.2 million barrels of hydrocarbons from the *Deepwater Horizon* blowout contaminated the Gulf **(Fig. Bx11.2c)**. The spill devastated wetlands, wildlife, and the fishing and tourism industries. Gradually, bacteria have been digesting the oil, and the region's environment has begun to recover.

(a) Fireboats doused the burning platform in vain before it sank.

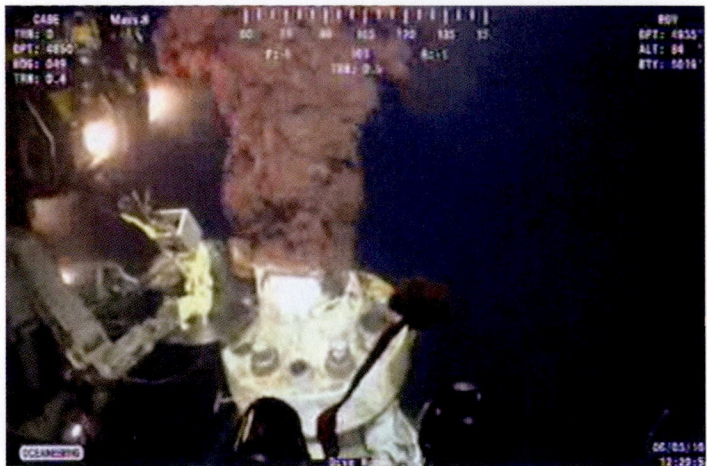

(b) A plume of oil billowed into the water from the wellhead, as viewed by underwater cameras.

(c) A satellite image showing the oil slick that formed in the Gulf of Mexico southeast of the Mississippi Delta.

Some researchers have suggested that pollution problems associated with coal could be overcome by employing a process called **coal gasification**. This process, which involves heating coal in the presence of water, transforms solid coal into a mixture of burnable gases. Sulfur, mercury, and other pollutants that had been within the coal remain as a solid residue that never enters the smokestack.

Even if new technologies can reduce the pollutants emitted by burning fossil fuels, the process still releases carbon dioxide (CO_2) into the atmosphere, as indicated by the chemical formula for methane combustion given in Section 11.2. As we will see in Chapter 20, CO_2 is a greenhouse gas, so an increase in the amount of CO_2 in the atmosphere can make the global climate warmer. Because of concern about the **carbon footprint** of fossil fuels, meaning the amount of CO_2 and other greenhouse gases emitted by their production and burning, governments are encouraging opportunities to decrease CO_2 production. The primary approach is to decrease the amount of fossil fuel burned, both by improving fuel conservation practices (such as increasing gas mileage in cars and improving building insulation) and by switching to other energy sources that we describe next. Another approach is known as **carbon capture and sequestration** (CCS). This process traps CO_2 produced by large sources such as power plants, liquefies the CO_2, and then pumps it into reservoir rocks deep underground.

Clearly, society faces difficult choices about where to obtain energy, and it will need to invest in the research required to discover new alternatives. Some researchers estimate that by 2050, most energy will be coming from alternatives to fossil fuels. In the next two sections of this chapter, we consider some of these alternatives.

Take-home message . . .

Fossil fuel is a non-renewable resource. In fact, at current rates of consumption, oil supplies could run out within a century or two. Production and use of fossil fuel have environmental consequences of concern, so societies are exploring ways to find alternative energy sources.

Quick Question -
Why does acid rain develop?

11.8 Nuclear Power

When you watch a fossil fuel burn, you are seeing a *chemical reaction*. The energy released comes from the breaking of chemical bonds—the bonds that hold atoms together in a molecule. The energy that drives a nuclear power plant, in contrast, comes from a type of *nuclear reaction*, in this case the breaking, or *fission*, of the nuclear bonds that hold protons and neutrons together in an atom's nucleus. Fission occurs when a neutron strikes a radioactive parent atom, causing it to split into smaller daughter atoms. For example, a uranium-235 atom, when struck by a neutron, splits into a barium-141 atom, a krypton-92 atom, and three neutrons. (The numbers in the previous sentence indicate the atomic weights of the atoms; see Chapters 1 and 9.) Neutrons released during the fission of one atom can then strike other atoms, triggering fission of those atoms, in a self-perpetuating process called a **chain reaction**. Uncontrolled, an extremely rapid chain reaction yields the blast of an atomic bomb—the type of bomb used at the end of World War II. Controlled fission can be used to produce energy in a nuclear power plant.

How Does a Nuclear Power Plant Work?

Nuclear power plants were first built during the 1950s to produce electricity. The heart of the plant is a **nuclear reactor**, a container holding *fuel rods*, metal tubes filled with concentrated uranium oxide or plutonium. Fission of these radioactive materials produces energy **(Fig. 11.23a)**. Nuclear reactors also contain *control rods*, which are composed of substances that absorb neutrons and can therefore moderate the rate of overall energy production in the nuclear fuel. Because radioactive materials are dangerous to living organisms, engineers place reactors within a containment building of reinforced concrete **(Fig. 11.23b)**.

A nuclear power plant produces electricity in much the same way a fossil fuel plant does. An array of pipes carries water close to the reactor. Heat produced by fission in the fuel transforms the water into steam. Pipes carry this steam to a turbine, where it pushes blades that, in turn, rotate the shaft of a generator. The steam coming out of the turbine runs through cooling towers, where it condenses back into water that can be reused in the plant.

The Geology of Uranium

Where does the uranium used in nuclear power plants come from? According to astrophysicists, uranium atoms formed during the explosion of a supernova before the formation of the Solar System (see Chapter 23). These atoms became part of the nebula out of which the Earth grew and were therefore incorporated into the planet. They gradually rose into the upper crust, carried in magma that formed granite when it solidified.

Even though granite contains uranium, it does not contain very much, so to be mined economically, the uranium must be concentrated by geologic processes after the granite solidifies. Concentration can take place, for example, when groundwater dissolves the uranium in the granite and then precipitates it elsewhere in a more

Did you ever wonder . . .
- - - - - - - - - - - - - - - -
whether a nuclear power plant could explode like an atomic bomb?

Figure 11.23 Producing electricity at a nuclear power plant.

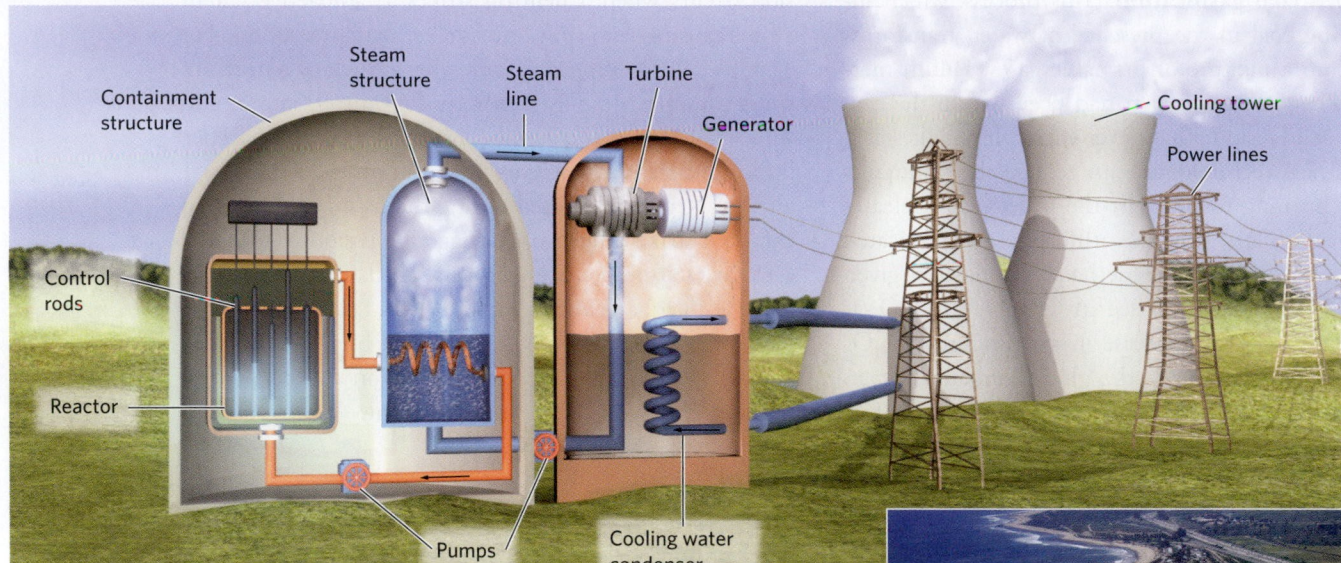

Containment structure

Steam structure

Steam line

Turbine

Generator

Cooling tower

Power lines

Control rods

Reactor

Pumps

Cooling water condenser

(a) A reactor heats water, which produces high-pressure steam. The steam drives a turbine that, in turn, drives a generator to produce electricity. A condenser transforms the steam back into water.

(b) This nuclear power plant in California has two reactors, each in its own containment building.

concentrated form. Uranium extracted from rock can't be used in a reactor directly, however. That's because ^{235}U, the isotope of uranium that serves as the most common fuel for nuclear power plants, accounts for only about 0.7% of naturally occurring uranium—the rest consists of stabler ^{238}U. To make a fuel suitable for use in a reactor, the ^{235}U concentration in natural uranium must be increased by a factor of 2 or 3, an expensive process called **enrichment**. In order to be used as fuel for atomic bombs, uranium must be enriched significantly more than this, so the type of uranium used in a nuclear power plant cannot explode like an atomic bomb.

Challenges of Using Nuclear Power

Maintaining safety at nuclear power plants requires diligence. Not only do control rods need to be in place, but the entire reactor must be cooled by circulating water. Without cooling, the fuel can become so hot that it melts, a severe accident known as a **meltdown**.

Globally, about 435 nuclear power plants currently operate. To date, two major disasters have occurred in which containment buildings were damaged and significant quantities of radiation were released. The first occurred at the power plant in Chernobyl, Ukraine, in April 1986. At Chernobyl, a meltdown began producing so much heat that water molecules in the cooling water separated to form hydrogen and oxygen gases. These gases recombined, producing an explosion that scattered

radioactive debris into the surrounding environment. In addition, the graphite control rods began to burn, sending radioactive smoke skyward, where the wind dispersed it over a broad area. The second disaster took place in 2011 at the coastal Fukushima Daiichi power plant in Okuma, Japan. A tsunami, a water wave generated by the magnitude 9.0 Tōhoku earthquake, knocked out the supply of electricity to the pumps that circulated cooling water (see Chapter 8). The reactors overheated, suffered partial meltdowns, and caused hydrogen gas explosions that breached the containment buildings and dispersed radioactive debris into the environment.

Operation of a reactor produces **nuclear waste**, including used fuel rods as well as water, piping, and concrete contaminated with radioactive materials. Some of the radioactivity in the waste decays relatively quickly (in decades to centuries), but some will remain dangerous for thousands of years. Nuclear waste cannot just be buried in a town landfill, where it could leak into nearby

water supplies. Spent fuel rods need to be kept cool by being submerged in water (Fig. 11.24). Other waste should be isolated from the environment, sealed in containers that will last long enough for the longer-lived radioactive atoms to undergo decay. Finding appropriate places for long-term nuclear waste storage is not easy. To date, most waste remains on the property of the power plants.

Take-home message . . .

Controlled fission in reactors produces nuclear power. The fuel consists of uranium or other radioactive elements obtained by mining. Reactors run the risk of meltdown, but cannot explode like an atomic bomb. They also yield radioactive waste that can be a challenge to store.

Quick Question -
What is the difference between a meltdown in a reactor and the explosion of an atomic bomb?

11.9 Other Energy Sources

Because of concerns about the carbon footprint of fossil fuels, and because of concerns about the safety of nuclear power, society has been working to develop other energy sources. Let's look at a few examples.

Biofuels

Can we transform modern-day biomass into fuels that work like fossil fuels through the use of laboratory chemistry? The answer is yes, and the resulting materials are called **biofuels**. The most commonly used biofuel is *ethanol* (CH_3CH_2OH), a type of alcohol that can substitute for gasoline in car engines. It can be produced commercially either from corn or from sugarcane. More recently, researchers have been developing processes that yield ethanol from cellulose (the fibrous material in plants), permitting perennial grasses to become a source of biofuel. Another promising method uses algae, which naturally synthesizes fatty organic chemicals from which hydrocarbons can be produced. Recent technologies also include the commercial production of **biodiesel**, a fuel that comes from chemical modification of vegetable oils.

Geothermal Energy

As the name suggests, **geothermal energy** comes from the Earth's internal heat and takes advantage of the fact that the crust becomes progressively hotter with increasing depth. Significant sources of geothermal energy can be obtained only in areas of igneous activity, where high temperatures exist at relatively shallow depths. In such settings, groundwater absorbs heat from hot rock and becomes so hot that, when pumped from the subsurface and run through pipes, it can heat houses and buildings directly. Particularly hot groundwater turns to steam when it rises and undergoes decompression, and this steam can drive turbines and generate electricity (Fig. 11.25a). In volcanic areas such as Iceland and New Zealand, geothermal power provides a major portion of energy needs. But on a global scale, access to geothermal energy will always be limited.

The use of the term *geothermal energy* gets a bit confusing, for in recent years, home heating- and cooling-system installers have been using the same term to refer to *geothermal heat pumps*, which consist of a set of pipes installed underground next to a home (Fig. 11.25b). This configuration of pipes takes advantage of the insulating properties of soil and underlying sediment to decrease home energy use. Geothermal heat pumps work because soil and sediment, even when wet, transfer heat to the atmosphere above much more slowly than free air does, so the ground temperature below a depth of a few meters remains nearly constant all year. As an example, consider a location in the north-central United States that has hot summers and cold winters. When the average monthly temperature at this location reaches its summer high of 28°C (83°F), the temperature at a depth of 4 m (12 feet) below the ground surface is only 17°C (63°F). When the average monthly temperature reaches its midwinter low of 3°C (38°F), the temperature at that depth is 14°C (58°F). By running the pipes carrying a home's water

Figure 11.25 Geothermal energy.

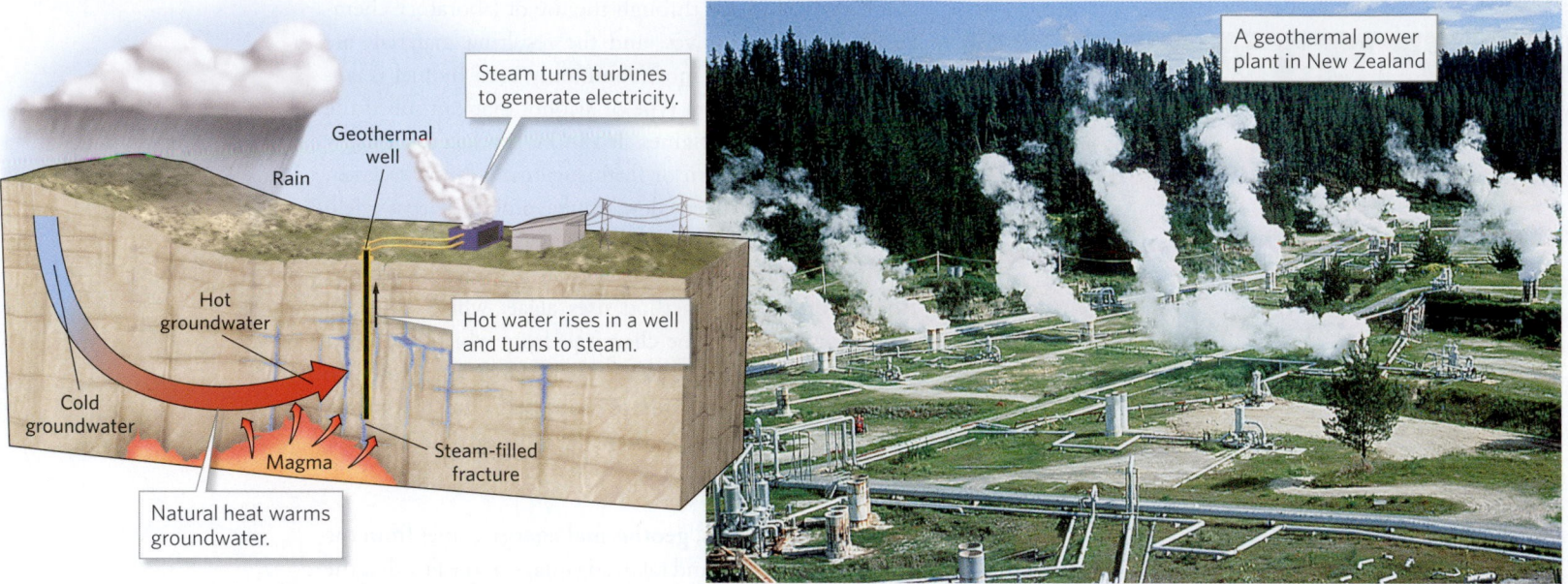

A geothermal power plant in New Zealand

Steam turns turbines to generate electricity.

Geothermal well

Rain

Hot groundwater

Hot water rises in a well and turns to steam.

Cold groundwater

Magma

Steam-filled fracture

Natural heat warms groundwater.

(a) Geothermal power plants use groundwater heated by igneous intrusions. Very hot groundwater turns to steam when it reaches the surface and undergoes decompression.

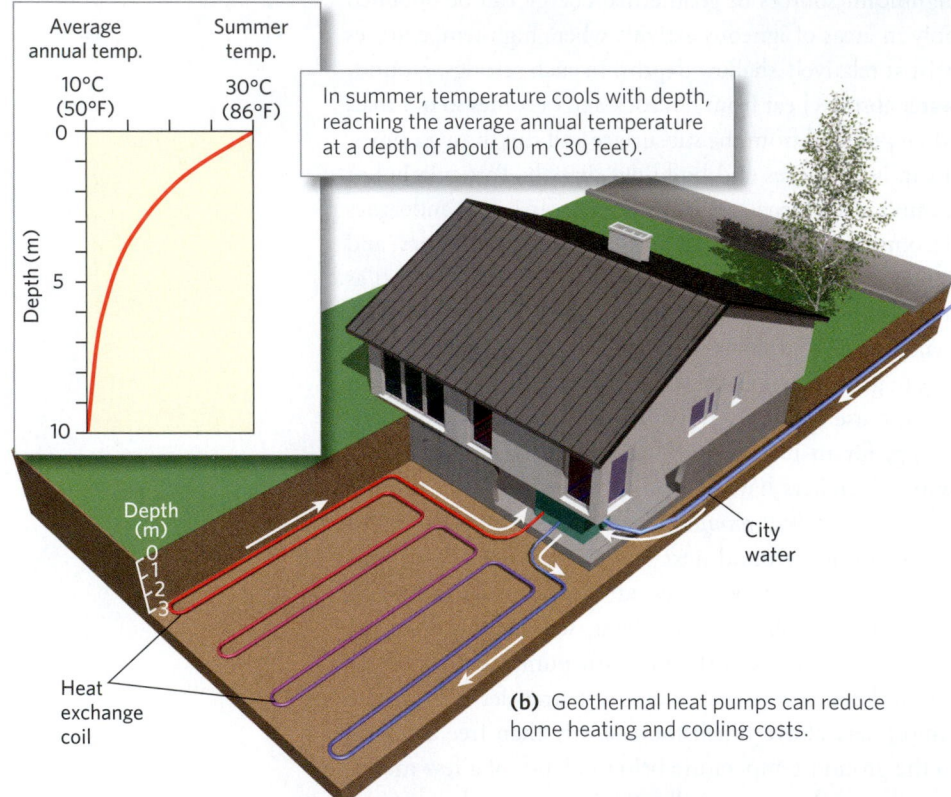

Average annual temp.   Summer temp.

10°C (50°F)   30°C (86°F)

Depth (m)

In summer, temperature cools with depth, reaching the average annual temperature at a depth of about 10 m (30 feet).

Depth (m)

City water

Heat exchange coil

(b) Geothermal heat pumps can reduce home heating and cooling costs.

supply through the ground, homeowners can precool the water used in the home during the summer and preheat it during the winter. This process decreases the amount of energy needed to run furnaces or air conditioners.

Hydroelectric and Wind Power

For millennia, people have used flowing water for energy. Waterwheels powered mills and factories in the years

before electricity, so cities grew up along rivers. Power companies use more complex technology in modern hydroelectric power plants to generate electricity. Flowing water turns the blades of a turbine directly, and the turbine drives an electrical generator. Most hydroelectric plants rely on water from a reservoir held back by a dam **(Fig. 11.26a)**. In effect, the energy produced by a hydroelectric plant comes from the potential energy in the reservoir's elevated water—this potential energy converts into kinetic energy when the water flows to a lower elevation beneath the dam.

Hydroelectric power is clean, in the sense that its production does not release chemical or radioactive pollutants, and does not consume nonrenewable resources. In addition, the large reservoirs behind the dams of hydroelectric power plants may provide irrigation water, flood control, and recreational opportunities. But the construction of dams and reservoirs may also bring unwanted consequences to a region. Damming a river may submerge spectacular scenery and thrilling whitewater rapids, and it may displace riverbank towns or destroy endangered ecosystems. Furthermore, reservoirs trap sediment and nutrients, preventing these materials from reaching floodplains or deltas downstream, where the loss may adversely affect agriculture and delta growth. Sediment trapping also decreases the capacity of a reservoir over time, for the sediment occupies space.

Not all hydroelectric power generation uses flowing river water. Engineers have been developing new means to tap **tidal energy**, the energy associated with the daily

(a) The water held back by the Three Gorges Dam flows through turbines to generate electricity.

(b) A wind farm in southwestern England. The towers are about 50 m high.

rise and fall of tides. One approach involves building a dam, called a *tidal barrage*, across the entrance to a bay or estuary (the flooded mouth of a river) in which there are large tidal ranges. When the tide rises, water flows into the enclosed area through openings in the barrage. When the tide drops, water is trapped behind the barrage, then flows back to the sea via a pipe that carries it through a power-generating turbine.

Modern efforts to harness the wind are rapidly developing on a large scale. To produce wind power, meteorologists identify regions that have steady breezes. In these regions, engineers build *wind farms* that consist of numerous towers, each of which holds a turbine powered by a set of giant fan blades **(Fig. 11.26b)**. Like a hydroelectric turbine, a wind turbine drives an electrical generator as it turns. Some towers are as tall as 100 m (300 feet) with fan blades that are over 40 m (120 feet) long.

Wind power is clean, but it has some drawbacks. Cluttering the horizon with towers may spoil a beautiful view, the loud hum of the turbines can disturb nearby residents, and the fan blades may be a hazard to migrating birds. Also, since the amount of electricity produced by wind turbines depends on wind speed, the amount supplied is not constant.

Solar Energy

The Sun drenches the Earth with energy in quantities that dwarf the amounts stored in fossil fuels. Were it possible to harness all this radiation direct from the Sun, humanity would have a reliable and totally clean solution for powering modern technology. But using this *solar energy* can be challenging because converting light into heat is quite inefficient.

Let's consider two options for producing solar power. The first, a *household solar collector*, consists of a black surface placed beneath a glass plate. The black surface absorbs sunlight that has passed through the glass plate, and heats up. The glass keeps the heat from escaping, so when water passes through pipes placed between the glass and the black surface, the water heats up; it can then be used as is or to heat a home. The second option consists of an array of **photovoltaic cells** (solar cells), shiny panels anchored to the ground or to a roof. Most photovoltaic cells contain two wafers of silicon pressed together. When light strikes the cell, its atoms release electrons that flow from one wafer to the other. If a wire loop connects the back side of one wafer to the back side of the other, this phenomenon produces an electric current **(Fig. 11.27)**.

Figure 11.27 Arrays of photovoltaic cells produce electricity directly from solar radiation.

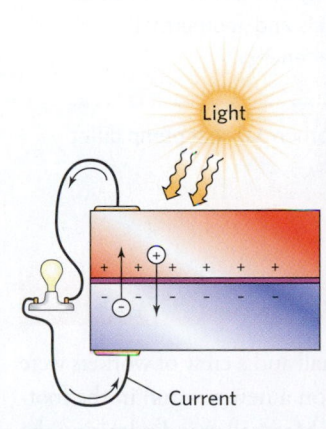

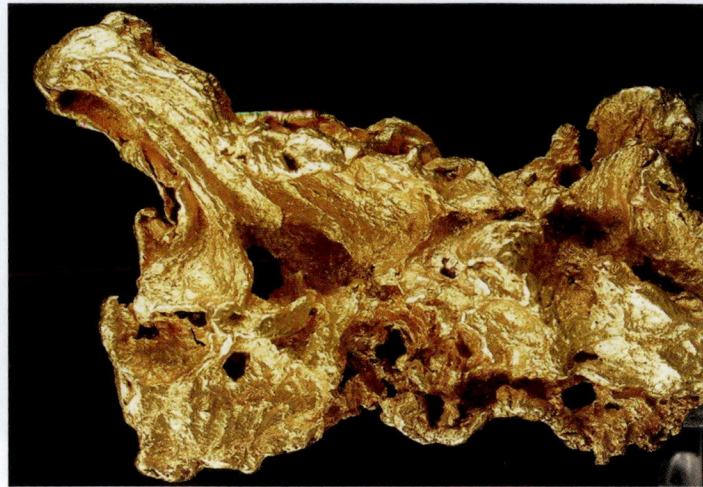

(a) A gold nugget, found near Ballarat, Australia, came out of the ground looking like metallic gold.

(b) Gold bracelets on display in a jewelry market in Kuwait.

Take-home message . . .

Society increasingly uses alternative energy sources that do not rely on fossil fuel or nuclear fuel. Examples include biofuels and geothermal, hydroelectric, wind, and solar energy.

Quick Question -----------------------------
How does energy from a geothermal heat pump differ from geothermal energy?

> All the gold which is under or upon the Earth is not enough to give in exchange for virtue.
>
> —PLATO (GREEK PHILOSOPHER, CA. 428–347 B.C.E.)

11.10 Metallic Mineral Resources

In January 1848, James Marshall and a crew of workers were putting the finishing touches on a new sawmill in the foothills of the Sierra Nevada. As Marshall stood admiring the new building, he noticed a glimmer of metal in the gravel that littered the bed of the adjacent stream. He picked up the metal, banged it between two rocks to test its hardness, and shouted, "Boys, by God, I believe I have found a gold mine!" Word of the gold soon spread, and within weeks, all the workers at Marshall's mill had disappeared into the mountains to seek their own fortunes. Through that year, gold fever spread throughout the country. As a result, 1849 brought 40,000 prospectors to California.

Gold is but one of many *metallic mineral resources*, meaning Earth resources containing metals such as gold, copper, aluminum, or iron. Let's first explore what makes metals special, then examine the various materials from which they can be obtained.

Metal and Its Discovery

A **metal** is an opaque, shiny, smooth solid that can conduct electricity and can be bent, drawn into wire, or hammered into a thin sheet. The first metals that people learned to use—copper, silver, and gold—are those that occur in rock as native metals. A *native metal* consists only of metal atoms, so it looks and behaves like metal. For example, the native gold in a nugget looks just like the processed gold in a bracelet (Fig. 11.28).

While prehistoric people could find sufficient native metals to make weapons, coins, and jewelry, modern societies would quickly run out of metal if our only supply came from native metals. Fortunately, there are many other sources of metal. Most metal atoms we use today originated as ions bonded to nonmetallic elements in minerals that look nothing like metal. Only because some prehistoric genius discovered the process of **smelting** (heating certain rocks to a temperature high enough that the rock decomposes to yield metal plus a nonmetallic residue called *slag*) do we have the ability to fill the metal appetite of industrialized society.

What Is an Ore?

You can't smelt just any rock to obtain a usable amount of metal. That's because most rocks contain too little metal to be worth extracting, or because the metals occur in minerals that don't decompose easily when smelted. Geologists use the term **ore** for a rock that contains sufficient metal to be worth mining. The concentration of a useful metal in an ore determines the **grade** of the ore—the higher the concentration, the higher the grade. Whether or not a company will choose to mine an ore of a given grade, at a given time, depends on the metal's price in the market (Box 11.3).

In an ore, metal may occur either as native metal or in specific minerals, known as **ore minerals**, that contain a high proportion of extractable metal relative to nonmetallic elements. For example, galena (PbS) contains 50% lead, so we consider it to be an ore mineral for lead

Box 11.3 ▸ How can I explain . . .

The economics of mining

What are we learning?

That a lot of waste rock must be produced to obtain a relatively small amount of ore, and that mining a higher-grade ore deposit can be more profitable.

What you need:

- 4 cups of dry rice.
- A tablespoon full of M&M candies.
- 2 large bowls.
- 1 small bowl.
- A 1-cup measuring cup.

Instructions:

- Pour the rice into one of the large bowls. Add the candy and mix thoroughly.
- Scoop out the rice-plus-candy mixture, one cup at a time.
- Sort through each cup and find the candies. Place the candies in the small bowl and the candy-free rice in the other large bowl. Time how long it takes to recover the candy.

- Repeat the experiment, but this time, place the candy into the center of the rice-filled bowl and mix just slightly.
- Then, knowing where the candy is, take a scoop from that part of the bowl, and time how long it takes to recover the candy.

What did we see?

- When the candy was dispersed throughout the rice, you could recover very little in a given time. Therefore, you would have to sift through lots of rice to get the candy. If the time you take scooping and sorting represents the time it takes to mine valuable ore minerals, you are spending a lot of time (which equals money) to get the ore minerals.
- Of course, it's much easier to extract candy if it's concentrated, and if you know where it is. This situation represents the discovery of a high-grade ore deposit. Mining such a deposit is much more efficient.

(Fig. 11.29a). Hematite (Fe_2O_3) and magnetite (Fe_3O_4) are ore minerals for iron, and copper comes from a variety of ore minerals, including chalcopyrite ($CuFeS_2$), bornite (Cu_5FeS_4), and malachite [$Cu_2CO_3(OH)_2$] (Fig. 11.29b).

You can see from the chemical formulas of these examples that many ore minerals are *sulfides*, in which the metal occurs in combination with sulfur (S), or *oxides*, in which the metal occurs in combination with oxygen (O).

Figure 11.29 Examples of ore minerals.

(a) This lead ore, from Missouri, consists of galena (PbS) crystals that grew in dolostone.

(b) Most copper comes from ore minerals that look nothing like metallic copper. This ore consists of azurite (blue) and malachite (green).

Figure 11.30 Some processes that form ore deposits.

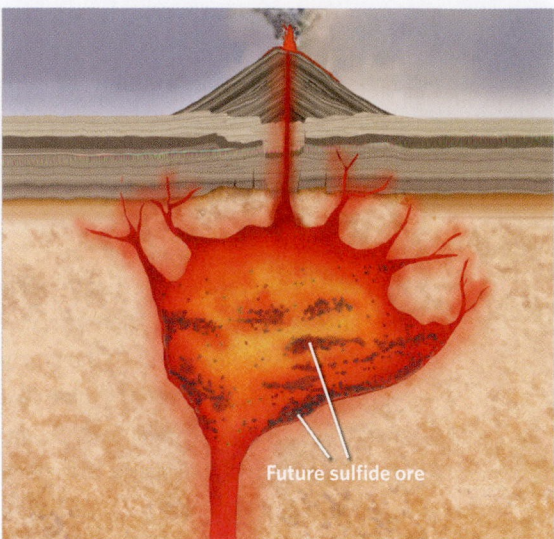

(a) Magmatic ore deposits can form in a magma chamber when ore minerals crystallize.

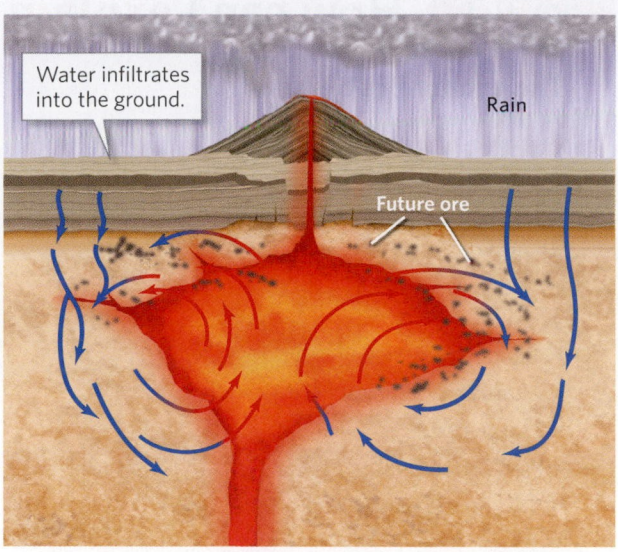

(b) Hydrothermal deposits form when water circulating around and through magma dissolves metals, which then precipitate out of solution elsewhere. (Arrows indicate flowing water.)

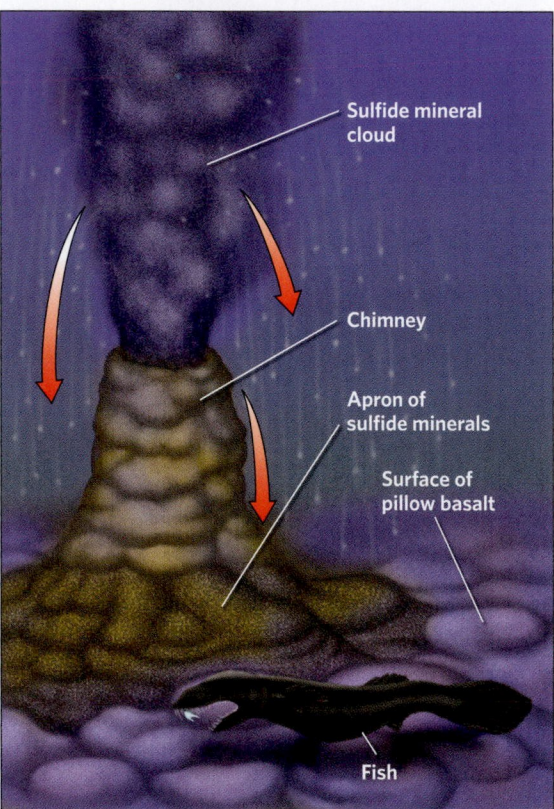

(c) Seafloor massive sulfide deposits form when sulfides precipitate around hydrothermal vents (black smokers) along a mid-ocean ridge.

Formation of Ore Deposits

Ore minerals do not occur uniformly through rocks of the crust. If they did, we would have to process vast quantities of rock in order to extract enough ore minerals to meet demand. Fortunately for humanity, geologic processes con-

centrate ore minerals in regions, known as **ore deposits**, that contain significant amounts of ore. Geologists distinguish among different types of ore deposits based on the processes by which the deposits form. Here are a few examples:

- *Magmatic deposits:* In some magmas, ore minerals crystallize early and accumulate to form lenses of ore, called magmatic deposits, as the magma solidifies and becomes an igneous rock (Fig. 11.30a).

- *Hydrothermal deposits:* Hot groundwater, when circulating through an igneous intrusion and the rocks around it, dissolves metal ions. When the resulting hydrothermal solution enters a different environment (with lower pressure, lower temperature, different acidity, or different oxygen availability), the metals precipitate as ore minerals, yielding a *hydrothermal deposit* (Fig. 11.30b). When the ore minerals precipitate in cracks, they form vein ore, and when they precipitate in pores, dispersed throughout the rock, they form disseminated ore.

- *Seafloor massive sulfide deposits:* Along mid-ocean ridges, hydrothermal vents known as black smokers erupt hydrothermal solutions. The solutions cool when they mix with seawater, and the dissolved components precipitate as tiny crystals of sulfide minerals (Fig. 11.30c). These minerals accumulate in lenses of ore around the vent. Such ore is a *seafloor massive sulfide deposit*.

- *Secondary-enrichment deposits:* In the upper crust, groundwater passing through ore dissolves ore minerals and carries the ions away. When the water flows into a different environment, it precipitates new ore minerals, forming a *secondary-enrichment deposit*.

- *Sedimentary deposits:* A *sedimentary deposit* is one formed from ore minerals that settled or precipitated out of water as sediment. Examples include banded iron formation (Fig. 11.31).

- *Placer deposits:* When rocks containing native metals erode, the resulting sediment contains rock fragments and metal flakes or nuggets (pebble-sized fragments). Moving water carries away lighter grains but can't move the heavier metal grains as easily, so the metal grains concentrate in gravel. Such concentrations constitute *placer deposits* (Fig. 11.32).

- *Residual deposits:* Heavy rains in tropical climates leach all soluble minerals from the soils (see Chapter 5), so the zone of leaching in such soils tends to contain a concentration of insoluble minerals. If the soil formed from a rock that originally contained aluminum, for example, insoluble aluminum oxide minerals that remain in the zone of leaching form ore deposits called *residual deposits*. The aluminum-bearing ore is called *bauxite*.

Where Are Ore Deposits Found?

The Inca Empire of 15th-century Peru boasted elaborate cities and temples, decorated with fantastic statues and masks made of gold. Then, around 1532, ships bearing Spanish conquistadors arrived. The Incas were no match for the armor-clad Spaniards with their guns, horses, and fatal European diseases, and Spanish ships soon began transporting golden treasure back to Spain. Why did the Incas possess so much gold? Or, to ask the broader question, what geologic factors control the distribution of ore?

Several of the ore-deposit types mentioned above occur in association with igneous rocks. As we learned earlier, igneous activity does not happen randomly around the globe; instead, it takes place primarily in the volcanic arcs of convergent-plate boundaries, along mid-ocean ridges, in continental rifts, and at hot spots. Therefore, magmatic and hydrothermal deposits, as well as secondary-enrichment deposits and placer deposits derived from them, develop in these geologic settings. Inca gold, for example, came from placer deposits eroded from hydrothermal ores formed in the Andean continental arc.

Ore Exploration and Production

Imagine prospectors of days past clanking through the wilderness with worn-out donkeys, searching for ore. What, specifically, were those prospectors hoping to see? In some cases, they searched hillsides for outcrops of milky-white quartz veins that might contain native metals, or for brightly colored stains caused by oxidation (rusting) of ore minerals. Sometimes they panned streambed gravels, hoping to find placer gold. (To pan for gold, prospectors would swirl gravel in a pan full of water. Lighter mineral grains wash out with the water, leaving behind a concentration of

Figure 11.31 This banded iron formation from northern Michigan is an example of a sedimentary ore deposit.

Hematite layer

Jasper layer

heavier grains, which may include gold.) On finding a possible ore deposit, a prospector would take a sample back to town for an *assay*, a test to determine how much extractable metal the deposit contained. If the assay indicated a significant concentration of metal, the prospector might "stake a claim" by literally marking off an area of land with stakes.

These days, commercial mining companies employ geologists to survey potential ore-bearing regions. Geologists measure the strength of the local gravitational or magnetic field to help pinpoint promising localities because ore minerals tend to be denser and more magnetic

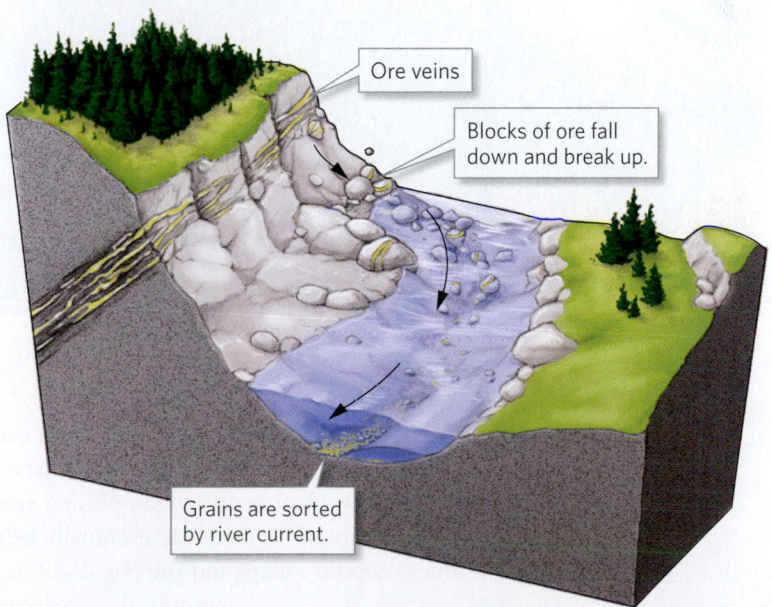

Figure 11.32 Placer deposits form where erosion produces clasts of native metals. Sorting by flowing water concentrates the metals.

Ore veins

Blocks of ore fall down and break up.

Grains are sorted by river current.

Figure 11.33 Mining ore deposits.

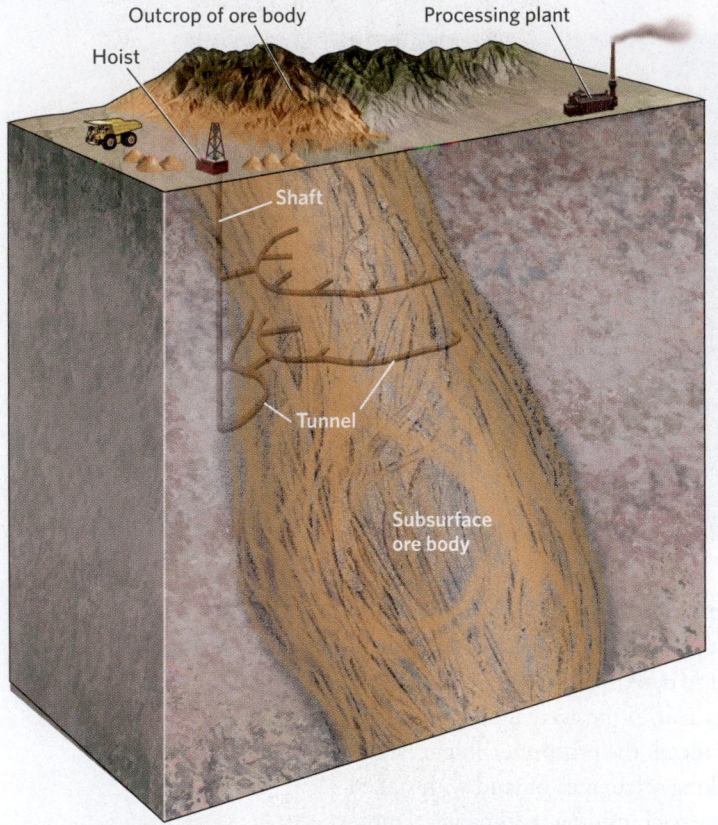

Hoist   Outcrop of ore body   Processing plant

Shaft

Tunnel

Subsurface ore body

(a) The three-dimensional shape of an ore deposit underground. Shafts and tunnels give miners access to the deposit.

(b) Open-pit mining extracts ore that lies fairly close to the ground surface. The steps cut into the wall help ensure its stability.

(c) An underground mine chiseled into a mountainside in Colorado.

than average rocks. They may also sample rocks, soils, and plants to test for metal concentrations. If surface observations hint at ore below, geologists drill to sample and analyze subsurface rock. Subsurface data may eventually help to determine an ore deposit's shape and size **(Fig. 11.33a)**.

If calculations indicate that mining an ore deposit can yield a profit, while accommodating environmental concerns, a company builds a mine. To develop an **open-pit mine (Fig. 11.33b)**, workers drill a series of holes into the bedrock and then fill the holes with explosives. They space the holes carefully, and set off the charges in a precise sequence, so that the rock shatters into appropriate-sized blocks for handling. When the dust settles, front-end loaders dump the rock into giant ore trucks, which can carry as much as 27,000 kg (300 tons) of rock in a single load. The trucks dump waste rock into a *tailings pile*, an artificial hill or mound composed of waste rock, and they load the ore into a crusher, a giant set of moving steel jaws that smashes the ore into small fragments. The fragments go to a processing plant, where workers use various methods to separate ore minerals from other minerals. Finally, the ore-mineral concentrates are smelted to separate their metal atoms from those of other elements.

To reach ore deposits that lie more than about 100 meters (328 feet) below the Earth's surface, miners construct an **underground mine (Fig. 11.33c)**. To do so, they either bore a tunnel into the side of a mountain or sink a vertical shaft. At the level in the crust where the ore deposit appears, they build a maze of tunnels into the ore by drilling holes into the rock and then blasting. The rock removed must be carried back to the surface. Rock columns between the tunnels hold up the ceiling of the mine.

Mining and the Environment

Mining can leave a big environmental footprint (**Earth Science at a Glance**, pp. 376–377). Some of the gaping basins that result from open-pit mining are so big that astronauts can see them from space. And both open-pit and underground mining yield immense tailings piles. Lacking soil, tailings piles may remain unvegetated for decades. In some places, mining companies douse tailings with acidic solutions to leach out metals, and if these acids escape into the environment, they can damage vegetation.

Mining of metal ores, like the mining of coal, can expose sulfide minerals to air and water. These minerals may dissolve to produce acid mine runoff, which can kill vegetation downstream (**Fig. 11.34a**). Similarly, the smoke from ore smelting may contain harmful chemicals, including sulfur, which dissolves in water to produce acid rain that can damage the surrounding countryside (**Fig. 11.34b**). New technologies and regulations, as well as efforts to encourage recycling, have decreased the amount of environmental damage caused by metal mining and processing, but our need for metals continues to grow, so environmental issues associated with their extraction and use will likely continue to challenge future generations.

> **Figure 11.34** Environmental consequences of producing metal resources.

(a) The orange color in this acid mine runoff is due to iron and sulfides in the water, and bacteria that consume them.

How Long Will Metal Resources Last?

Most metallic mineral resources are nonrenewable—once mined, an ore deposit disappears forever. Geologists have calculated reserves for various minerals, just as they have for fossil fuels. Based on current definitions of reserves, which depend on today's prices and rates of consumption, supplies of some metals may run out in only decades to centuries. Further, metallic mineral reserves are not distributed uniformly around the planet, so not all countries have equal access to mineral supplies. This issue is of great importance from the standpoint

Nickel smelter

The "superstack" is 380 m (1,270 ft) high.

Tailings pile

(b) Acidic smelter smoke killed off vegetation near Sudbury, Ontario, in the 1970s. A large tailings pile can be seen in the distance.

Forming and Processing Earth's Mineral Resources

Ore deposits can be obtained either in strip mines or in underground mines.

Mining and processing ore has environmental consequences, including acid runoff, acid rain, and groundwater contamination.

Circulating groundwater may extract and concentrate metals to form ore deposits.

Clay, when formed into blocks and baked, becomes brick.

Gravel itself may be quarried for construction purposes.

Mud, a mixture of clay minerals and water, accumulates in beds.

Ore minerals may collect on the bottom of a magma chamber.

Miners pan for gold in placer deposits where metal flakes and nuggets occur in sand and gravel.

From Mud to Brick

Hydrothermal vents (black smokers) produce accumulations of massive sulfides.

From Magma to Metal

Erosion tears down mountains and produces gravel and sand.

From Stream Channel to Roadbed

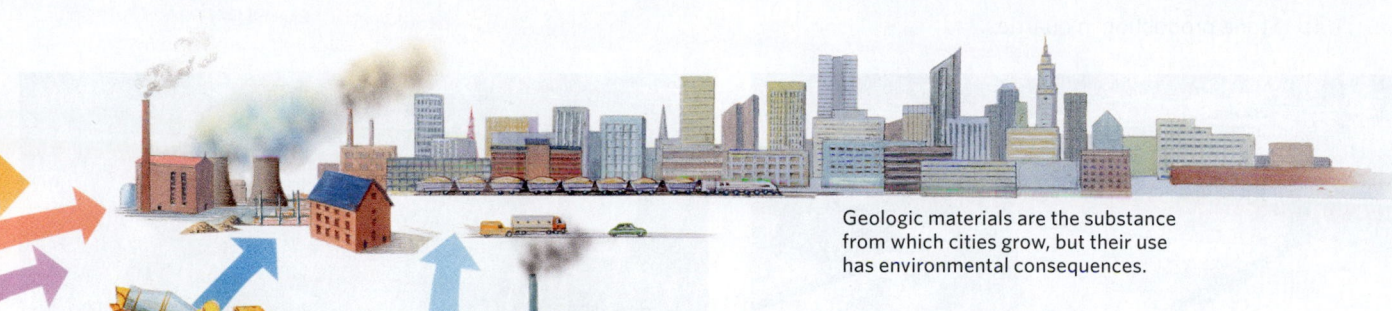

Geologic materials are the substance from which cities grow, but their use has environmental consequences.

A mixture of lime, other elements, sand, and water, when allowed to harden, becomes concrete.

Mixed with water, spread into sheets, and wrapped in paper, gypsum makes drywall.

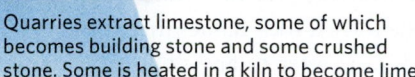

In quarries, operators dig up gypsum, crush it to powder, and ship it to factories.

Quarries extract limestone, some of which becomes building stone and some crushed stone. Some is heated in a kiln to become lime.

Gypsum is a salt that precipitates when saline lakes evaporate. It grows as white or clear crystals.

Over millions of years, shells and shell fragments collect and eventually form beds of limestone.

From Lake Bed to Drywall

The raw materials from which we manufacture the buildings, roads, wires, and coins of modern society were produced by geologic processes. For example, ore deposits—the concentrations of minerals that are a source of metal—formed during a variety of magmatic or sedimentary processes. Limestone, a rock used for buildings and for making concrete, began as an accumulation of seashells. Brick began as clay, a byproduct of chemical weathering. And the gypsum of drywall began as an accumulation of salt along a desert lake. Metal, gravel, lime, and gypsum are all examples of the Earth's mineral resources. We can use some mineral resources right from the Earth, simply by digging them out. But most become usable only after expensive processing.

Organisms extract ions from water and construct shells.

From Seafloor to Sidewalk

Figure 11.35 Stone production in quarries.

(a) An active quarrying operation in Missouri that produces large blocks of cut dimension stone.

(b) Sheets of cut dimension stone being measured for further cutting to become a kitchen countertop.

of national security because some minerals—known as *strategic minerals*—are essential for high-tech equipment such as computers and batteries. Increased efforts at conservation and recycling could dramatically decrease rates of consumption and thereby stretch the lifetime of existing reserves. Recycling rates are likely to grow as the cost of recycling becomes less than the cost of extracting new minerals.

Take-home message . . .

Ores—rocks that can be processed economically to produce metals—contain a variety of ore minerals, many of which are sulfides or oxides. A variety of geologic processes produce ores. Geologists can find ores by measuring gravitational and magnetic fields, studying outcrops, and drilling for samples. Mining takes place either in open-pit mines or underground. Most mineral resources are nonrenewable and exist in limited reserves that are not distributed uniformly around the planet.

Quick Question -----------------------------
Why did many ore deposits form along convergent-plate boundaries?

11.11 Nonmetallic Mineral Resources

Consider the materials you can see in a typical house or apartment—concrete, brick, glass, wallboard. Society uses many such **nonmetallic mineral resources**, meaning Earth resources other than metals. Where do they come from?

Dimension Stone

The Parthenon, a colossal stone temple, has stood atop a hill in Athens, Greece, for almost 2,500 years. No wonder: **dimension stone** (or just *stone*)—an architect's

word for rock—outlasts nearly all other construction materials. We use stone to make facades, roofs, curbs, steps, countertops, and floors. The names that architects, designers, or contractors give to various types of stone may differ from the formal rock names that geologists use. For example, architects refer to any polished carbonate rock as "marble," whether or not it has been metamorphosed, and to any crystalline rock containing feldspar or quartz as "granite," regardless of whether the rock has an igneous or metamorphic texture, or a felsic or mafic composition.

To obtain intact slabs or blocks of stone from a quarry, workers can split the stone from bedrock by hammering a series of wedges into it, thereby causing a crack to propagate, or they can slice it off the bedrock wall by using various power tools **(Fig. 11.35a)**. Examples of such tools include a *wireline saw*, which consists of a braided wire moving around two pulleys (the wire carries abrasives, so as it moves, it grinds down into the rock); a *thermal lance* (a narrow jet of very hot flame whose heat causes minerals on the rock to expand and break off); and a *water jet* (a nozzle that aims a very high-pressure stream, containing abrasive mixed with water, at the rock). Once workers remove blocks of dimension stone from a quarry, the blocks can be cut into smaller pieces **(Fig. 11.35b)**. Rubbing the blocks with abrasive and water creates a shiny polish.

Crushed Stone and Concrete

Crushed stone forms the foundation of highways and railroads, and it serves as the raw material for manufacturing cement, concrete, and asphalt. In crushed-stone quarries **(Fig. 11.36a)**, operators use explosives to break up bedrock into rubble, which they then transport by truck to a crusher. The jaws of the crusher reduce the rubble to chunks of a usable size.

Many buildings constructed in the past century have used concrete for floors, columns, or walls. To make concrete, workers mix cement with *aggregate* (sand and/or gravel) and water to produce a slurry. The term **cement**, as used when discussing concrete, refers to a powder containing various chemicals that dissolve in water. When workers pour wet concrete into molds and let it *set*, it hardens into a solid—basically, human-made rock—because the chemicals in the cement react and precipitate to produce a complex assemblage of new mineral crystals. These crystals bind the grains of the aggregate together, much as the precipitation of crystals from groundwater binds grains of sand together in a sandstone.

What chemicals make up cement? The cement in concrete consists mostly of lime (CaO), with lesser amounts of silica (SiO_2), aluminum oxide (Al_2O_3), and iron oxide (Fe_2O_3). In the 18th and early 19th centuries, workers produced cement simply by placing chunks of a special type of limestone, one that contained a little quartz and clay in addition to calcite, in a kiln. When cooked to a temperature of 1,450°C (2,640°F), the calcite ($CaCO_3$) in the limestone breaks down to form lime (CaO) and carbon dioxide gas (CO_2), and the clay and quartz provide the other oxides in cement. The specific type of limestone needed to make such "natural cement" is fairly rare, so today most concrete production uses *Portland cement*. Workers produce this material by heating a combination of limestone, sandstone, and shale in just the right proportions to provide the proper mix of chemicals to make cement **(Fig. 11.36b)**.

Nonmetallic Minerals in Your Home

What nonmetallic mineral resources make up a modern home? As we've seen, the foundation's concrete comes from crushed and baked limestone, mixed with sand or gravel and water. The *bricks* used to build the walls are made from clay, a mineral produced by the chemical weathering of silicate minerals. To make bricks, workers mold wet clay into blocks and then bake them in order to drive out water and cause metamorphic reactions that lead to the growth of stronger minerals. The *glass* in windows consists of silica (SiO_2). To make glass, workers melt quartz sand and then freeze the resulting molten silica quickly so that it solidifies without crystallizing. To make *drywall* or wallboard, the solid sheets used for interior walls, workers take gypsum, an evaporite mineral precipitated from saline water, crush it into a powder, and mix it with water to form a slurry. They spread the slurry between two sheets of paper and let it dry, so that new crystals of gypsum grow and the board becomes solid. (Evaporite deposits serve as the source of many chemicals used in daily life, including the lithium used in rechargeable batteries.)

The Earth hosts abundant supplies of nonmetallic minerals, but their use comes with a price. For example,

Figure 11.36 Producing the ingredients of cement.

(a) A large crushed-stone quarry in Silurian limestone of Illinois. Drillers are working on the shelf in the distance.

(b) The long tube is a rotating kiln. Rock fed in at one end is heated to a high temperature so that cement comes out the other end.

landscapes must be disrupted to dig quarries, and once dug, a quarry can never be refilled. At present, transportation of nonmetallic resources represents a major part of their cost, so supply companies try to locate quarries as close to consumers as possible. In the future, other land uses may compete with quarrying and restrict supplies of these resources, at least in some locations.

Take-home message . . .

Society uses a great variety of nonmetallic mineral materials. These materials include dimension stone, crushed stone, cement (made from baked limestone), evaporites (including gypsum), and clay (to make bricks).

Quick Question -
What's the difference between natural cement and Portland cement?

Did you ever wonder . . .
how concrete differs from rock?

11 CHAPTER REVIEW

Chapter Summary

- Energy resources come in a variety of forms: energy directly from the Sun; energy from tides, flowing water, or wind; energy from chemical reactions; energy from nuclear fission; and geothermal energy from Earth's internal heat.

- Fossil fuels, such as oil, natural gas, and coal, store energy that came to Earth from the Sun, was locked into the bonds in organic chemicals by photosynthesis, and was then buried and preserved.

- Oil and gas are hydrocarbons formed from the organic remains of plankton, which settle out and become incorporated into organic shale. Chemical reactions at elevated temperatures convert the organic matter to kerogen and then to oil or gas.

- To form a conventional oil reserve, oil must migrate from a source rock into a reservoir rock. A subsurface structure that holds oil or gas underground is called a trap.

- Substantial volumes of hydrocarbons also exist in unconventional reserves such as shale and tar sand. Shale oil refers to oil trapped in shale. Oil shale refers to shale that contains kerogen.

- Obtaining oil from shale typically involves directional drilling and hydrofracturing.

- For coal to form, abundant plant debris must be deposited in an oxygen-poor environment. Compaction of plant debris produces peat, which, when buried deeply and heated, transforms into coal.

- Geologists rank coal based on the percentage of carbon it contains. Coal occurs in beds and can be mined by either strip mining or underground mining.

- We now live in the Oil Age, but conventional oil supplies may last for only another century. Use of fossil fuel resources has many environmental consequences.

- Nuclear power plants generate electricity by using the energy released from the fission of uranium.

- Nuclear reactors must be carefully controlled to avoid overheating or meltdown. Radioactive nuclear waste is difficult to store.

- Geothermal power plants extract the Earth's internal heat from groundwater heated by igneous activity. Hydroelectric and wind power use the energy of flowing water and air, respectively. Photovoltaic cells convert sunlight to electricity.

- Industrialized societies use many types of minerals, all of which come from the upper crust.

- Metals come from ore minerals. An ore is a rock containing native metals or ore minerals in sufficient quantities to be worth mining.

- Ore deposits form in a variety of ways. Magmatic deposits form when ore minerals form in a cooling magma. In hydrothermal deposits, ore minerals precipitate from hot-water solutions. Secondary-enrichment deposits form when groundwater carries metals away from a pre-existing deposit. Sedimentary deposits settle out of water. Placer deposits develop when metal grains formed from erosion of rock accumulate in sediment.

- Mineral resources are nonrenewable. Many are now or may soon be in short supply.

- Nonmetallic mineral resources include dimension stone, crushed stone, clay, sand, and many other materials. A large proportion of materials in your home have a geologic ancestry.

- To produce concrete, workers mix aggregate (sand and gravel) with cement. Cement is made from limestone, mixed with lesser amounts of shale and sandstone, heated in a kiln. When dissolved in water and left to set, it precipitates new minerals.

Key Terms

acid mine runoff (p. 363)
acid rain (p. 363)
biodiesel (p. 367)
biofuel (p. 367)
blowout (p. 353)
carbon capture and sequestration (p. 365)
carbon footprint (p. 365)
cement (p. 379)
chain reaction (p. 365)
coal (p. 358)
coal gasification (p. 365)
coal rank (p. 360)
coal seam (p. 358)

conventional reserve (p. 350)
dimension stone (p. 378)
directional drilling (p. 353)
Earth resource (p. 345)
economic reserve (p. 346)
energy resource (p. 345)
enrichment (p. 366)
fossil fuel (p. 347)
fuel (p. 346)
geothermal energy (p. 367)
grade (p. 370)
hydrocarbon (p. 348)
hydrocarbon reserve (p. 349)
hydrofracturing (p. 354)

kerogen (p. 349)
meltdown (p. 366)
metal (p. 370)
mineral resource (p. 345)
natural gas (p. 348)
nonmetallic mineral resource (p. 378)
nonrenewable resource (p. 362)
nuclear reactor (p. 365)
nuclear waste (p. 366)
oil (p. 348)
Oil Age (p. 362)
oil seep (p. 350)

oil shale (pp. 350)
oil window (p. 349)
open-pit mine (p. 374)
ore (p. 370)
ore deposit (p. 372)
ore mineral (p. 370)
peat (p. 358)
permeability (p. 350)
photosynthesis (p. 346)
photovoltaic cell (p. 369)
porosity (p. 350)
renewable resource (p. 362)
reservoir rock (p. 350)
rotary drill (p. 353)

secondary recovery technique (p. 354)

seismic-reflection profile (p. 351)

shale gas (p. 350)
shale oil (p. 350)
smelting (p. 370)
source rock (p. 349)

tar (p. 348)
tar sand (p. 350)
tidal energy (p. 368)
trap (p. 350)

unconventional reserve (p. 350)
underground mine (p. 374)

Review Questions

The letters following each Review Question refer to the corresponding Learning Objective from the Chapter Opener.

1. What are the fundamental sources of energy? **(A)**

2. What is the source of the organic material in oil, and how is it transformed into oil? **(A)**

3. What is the oil window, and what happens to oil at temperatures higher than the oil window? **(B)**

4. How are hydrocarbons trapped to yield an oil reserve? Explain the difference between a conventional and an unconventional oil reserve. Which type of trap does the drawing show? **(B)**

5. Where is most of the world's oil found? **(B)**

6. What are tar sand and oil shale, and how can oil be extracted from them? **(B)**

7. How is coal formed, and in what class of rocks is coal considered to be? **(C)**

8. What is the difference between a high-rank and a low-rank coal? What are the names of the ranks? **(C)**

9. What is the difference between renewable and nonrenewable resources? **(A)**

10. What is the likely future of fossil fuel production and use in the 21st century? **(C)**

11. Describe how a nuclear power plant produces electricity. **(D)**

12. Where in the Earth's crust do uranium deposits form, and why? **(D)**

13. What are some of the drawbacks of nuclear power? **(D)**

14. What is geothermal energy? What geologic factors limit its use? **(F)**

15. Why don't we use ordinary granite as a source for metals? **(F)**

16. Describe various kinds of ore deposits and how they form. Which type of ore deposit does the drawing show? **(F)**

17. What procedures are used to locate and mine metal resources today? **(F)**

18. What are some environmental hazards of mining? **(C, F)**

19. Will the supply of mineral resources run out? **(H)**

20. How is stone cut from a quarry? **(G)**

21. What are the ingredients of concrete, and how are these substances produced? **(G)**

22. Name materials in your home that come from nonmetallic mineral resources. **(G)**

On Further Thought

23. Do you think it would make sense for an energy company to drill for oil in a locality where beds of anthracite occur in the stratigraphic sequence? Explain your answer. **(B)**

24. An ore deposit in Arizona has the following characteristics: One portion of the ore deposit is an intrusive igneous rock in which tiny grains of copper sulfide minerals are dispersed among the other minerals of the rock. Another nearby portion of the ore deposit consists of limestone in which malachite fills cavities and pores in the rock. What types of ores are these? Describe the geologic history that led to the formation of these deposits. **(F)**

Online Resources

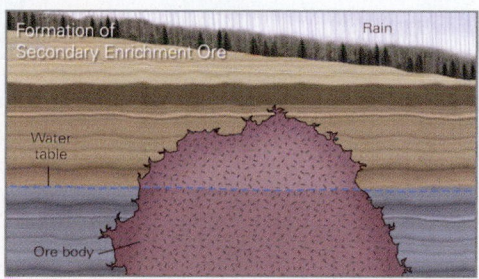

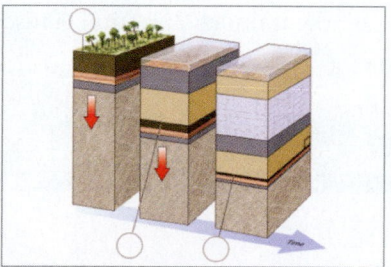

Videos
This chapter features videos on the basic process of hydraulic fracturing and on the different ways ore can form.

Smartwork5
Questions include checks on a basic understanding of energy technologies and the formation of coal and oil.

PART II

Ever-Changing Landscapes

Picture the jagged peaks of a mountain range, the grassy plains of a continental interior, the sand of a desert, or the glassy surface of a pond. Clearly, our home planet hosts an amazing variety of landscapes. In Part II of this book, we explore the characteristics of these landscapes, and the processes that produce them. We see that as soon as tectonic forces cause a region of land to rise or sink, erosion and/or deposition take place to reduce elevation differences. In Chapter 12, we begin our exploration of landscapes by introducing the hydrologic cycle, during which water continuously moves from sea to land and back again—this movement plays a major role in many types of landscape modification. We also discuss downslope movement, the process by which gravity pulls rock and debris from higher to lower elevation. Chapter 13 explores the role of freshwater in landscape evolution—we see how flowing water in rivers, standing water in lakes, and subsurface groundwater all affect our planet's surface. We also consider the impact of human society on water supplies. Finally, in Chapter 14, we focus on the uniquely beautiful landscapes of the Earth's harshest environments by describing characteristic of deserts and consequences of glaciation.

Dark clouds cover the distant sky near Zabriskie Point, overlooking Death Valley, California. Rains that fall from such clouds have carved an intricate network of stream channels into the tan sedimentary beds. We are seeing the never-ending battle between tectonic uplift and surface erosion right before our eyes.

12 SHAPING THE EARTH'S SURFACE

Landscapes, the Hydrologic Cycle, and Mass Wasting

By the end of this chapter you should be able to . . .

A. explain why the Earth's land surface is not static, like that of the Moon, but rather changes constantly over time.

B. differentiate between uplift and subsidence, internal and external energy, and erosion and deposition, and describe their effects on landscapes.

C. sketch a diagram showing processes involved in the hydrologic cycle of the Earth System.

D. describe the characteristics and consequences of different types of mass wasting.

E. understand the concept of a failure surface and recognize factors affecting slope stability.

F. use your knowledge of conditions that lead to mass wasting to assess an area's susceptibility to mass-wasting hazards and to evaluate preventative measures.

12.1 Introduction

A dinosaur, gazing at the Moon, would see the same view that we see today, for with the exception of a few new craters here and there, much of the Moon's solid surface has remained nearly unchanged for over 3 billion years. An observer looking at the Earth, however, would see features of the land surface change radically over time. Some of these changes happen very slowly (over thousands to millions of years), whereas others happen quickly (in seconds to days).

Why do landscapes on the Moon persist while those on the Earth evolve? The Moon's surface remains static because the Moon's outer layer does not consist of moving plates, and because the Moon has no atmosphere, hydrosphere, or biosphere. Therefore, only meteorite impacts or *space weathering* (the breakdown of minerals due to the impact of cosmic rays; see Chapter 22) alter the Moon's surface. The Earth's surface, in contrast, remains dynamic because the Earth's outer layer does consist of moving plates, and because our planet hosts an atmosphere, hydrosphere, and biosphere. Interactions of lithosphere plates with one another and with the asthenosphere below cause earthquakes, volcanism, mountain building, and basin formation (see Chapter 2). Movements associated with these phenomena generate slopes down which rocks and sediments can tumble or slide. Meanwhile, chemical and physical weathering, the flow of streams and glaciers, waves, wind, and the activities of life in the Earth System constantly break down and redistribute surface materials.

Because our planet's land surface continues to be dynamic, it hosts a great variety of landscapes. By **landscape**,

Layers of sedimentary rock, once buried deeply underground, now lie exposed on a cliff in Utah due to erosion. We can see evidence of erosion continuing today. Blocks of tan sandstone tumbled downslope onto gray shale during rockfalls. Water, dropped during heavy rains—a manifestation of the hydrologic cycle— has carved a network of channels into the shale.

Figure 12.1 Examples of the great variety of landscapes on Earth.

(a) Rounded "sugarloaf" mountains surround Rio de Janeiro, Brazil.

(b) Glaciers carved these peaks of the Alps in France.

(c) Buttes of sandstone tower above Monument Valley, Arizona.

(d) Cliffs rise from the forest in the Blue Mountains, Australia.

we mean the character and shape of the land surface in a region. Artists and writers across the centuries have gazed at landscapes for inspiration, for landscapes spark the full range of human emotion **(Fig. 12.1a–d)**. Earth scientists similarly feel inspired when they see a landscape, but they can't help wondering, "How did this landscape come to be? How will it change in the future? Do features of this landscape pose a threat to life and property?" This chapter begins to address these questions by describing: the nature of changes in land-surface elevation and the energy sources that drive these changes; the *hydrologic cycle*, during which water molecules move from ocean to air to land and back to ocean; and *mass wasting*, during which gravity transports rock and sediment down slopes.

12.2 Earth's Ever-Changing Surface

Generation of Relief

If the Earth's surface were perfectly flat, the great diversity of landscapes that we can see would not exist. But the land surface isn't flat because moving lithosphere plates interact with one another and with the underlying asthenosphere.

Figure 12.2 Uplift raises hills, and subsidence forms basins. These processes generate slopes.

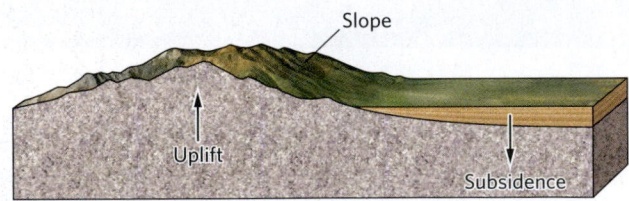

These interactions lead to subduction, continental collisions, rifting, and volcanism, and these phenomena cause portions of the land surface to move up or down relative to adjacent regions. We refer to the upward movement of the land surface as **uplift** and to the sinking or downward movement of the land surface as **subsidence (Fig. 12.2)**. Uplift or subsidence of one location relative to a neighboring location generates **relief**—a difference in elevation—and where relief exists, slopes form. Variations in elevation within a region define the shape of the region's landscape. Geologists use the term **topography** to refer to such variations **(Box 12.1)**.

(a) Uplifted beach terraces form where the coast is rising relative to sea level. Here, present-day wave erosion is forming a new terrace and cutting a cliff on the edge of the old one.

(b) So much erosion can take place during a single hurricane that houses built along the beach become undermined.

When uplift or subsidence generates relief, other components of the Earth System kick into action. Over time, bedrock undergoes physical and chemical weathering, which causes it to disintegrate into pieces that undergo *mass wasting* (tumbling, sliding, or flowing of material downslope from a higher to a lower elevation). During and after weathering and mass wasting, moving water, ice, and air can cause **erosion**, the grinding away and removal of material at the Earth's surface. As we'll see in succeeding chapters, erosion by rivers and glaciers can carve into the land surface and generate steep local relief. Materials or processes that cause erosion are known as *agents of erosion*. Flowing water, ice, or air can transport eroded materials to locations where **deposition**—the settling of sediment—takes place. Overall, mass wasting, erosion, and deposition, acting together, redistribute rock and sediment, ultimately stripping it from higher areas and collecting it in lower areas. Mass wasting and erosion can lower a landscape's elevation, while deposition can cause a landscape's surface to rise.

The energy that drives such landscape evolution comes from three sources:

1. **Internal energy**, the heat from within the Earth, keeps the mantle soft enough to flow, which allows plate movements to take place. These movements, along with the activity of mantle plumes, drive uplift and subsidence (as manifested by volcanism, mountain building, and sedimentary basin formation).

2. **External energy**, the radiation that comes to the Earth from the Sun, causes air and water near the Earth's surface to become warmer. In the Earth's gravitational field, warm air rises and cool air sinks, ultimately resulting in wind. Wind can erode and transport sediment, and it can generate water waves.

External energy also causes water on the Earth's surface to evaporate. This water, when carried over the land by wind, precipitates as rain or snow, filling rivers and feeding glaciers.

3. **Gravitational energy**, an object's potential energy due to the downward pull of gravity, drives mass wasting and, as we have seen, works in concert with other energy sources to drive convective movement in the mantle, oceans, and atmosphere.

Overall, we can think of landscape evolution as a "battle" between what geologists call *tectonic processes* (collision, convergence, rifting, and basin formation) driven by internal energy, which build relief by moving the land surface up or down, and *surface processes* (mass wasting, erosion, and deposition), driven by external energy and gravity, which destroy relief by removing material from high areas and depositing it in low areas. If, in a particular region, the rate of uplift exceeds the rate of erosion, the land surface rises, but if the rate of erosion exceeds the rate of uplift, the land surface becomes lower with time. Similarly, if the rate of subsidence exceeds the rate of deposition, the land surface becomes lower, but if the rate of deposition exceeds the rate of subsidence, the land surface rises.

How rapidly do vertical movements of the Earth's surface take place? Though the Earth's surface can rise or sink by as much as 3 m (10 feet) during a single major earthquake, rates of vertical surface movement, when averaged over time, vary between 0.1 and 10 mm (0.004 and 0.4 inches) per year **(Fig. 12.3a)**. Similarly, erosion during a single storm or mass-wasting event can carve tens of meters from the land **(Fig. 12.3b)**, and deposition by a single mass-wasting event can produce a layer of

> Water flows humbly to the lowest level. In the world, nothing is more submissive or weak than water, yet for attacking what is hard and strong, nothing can surpass it.
>
> —*LAO-TZU*
> (*CHINESE PHILOSOPHER,*
> *604–531 B.C.E.*)

Did you ever wonder . . .

how fast the land surface rises or sinks, on average?

Box 12.1 Science toolbox . . .

Characterizing topography on a map

You can convey information about the character of topography by taking a photograph. But for many applications, it's useful to portray the shape of the land surface on a **topographic map**, which uses contour lines to represent variations in elevation **(Fig. Bx12.1a)**. A **contour line** is an imaginary line on the land surface along which all points have the same elevation. You can picture a contour line as the intersection between the land surface and an imaginary horizontal plane **(Fig. Bx12.1b)**. For example, all points along the 100-m contour line lie at an elevation of 100 m above sea level. The elevation difference between two adjacent contour lines on a topographic map is called the *contour interval*. On a given topographic map, the contour interval is constant. So, for example, if the contour interval is 50 m, the next contour line above the 100-m contour is the 150-m contour, and the one above that is the 200-m contour, and so on. If you walk parallel to a contour line, you stay at the same elevation, but if you walk perpendicular to a contour line, you go upslope (if the next contour you cross is a higher number) or downslope (if the next contour you cross is a lower number). You can picture a slope's angle from the spacing between contour lines. Specifically, closely spaced contour lines represent a steep slope (where you would cross several contour lines while moving a short horizontal distance), whereas widely spaced contour lines represent a gentle slope.

To represent variations in elevation in a given direction, you can sketch a **topographic profile**, the trace of the ground surface as it would appear on a vertical plane that slices into the ground **(Fig. Bx12.1c)**. Put another way, a topographic profile represents the shape of the ground surface as viewed from the side. A topographic profile between two points on a topographic map can help you visualize how the land goes up and/or down between the points. By combining a topographic profile with a representation of geologic features under the ground, geologists produce a **geologic cross section**. In some cases, geologists can gain insight into subsurface geology simply by looking at the shape of a landform **(Fig. Bx12.1d, e)**. For example, a steep cliff in a region of sedimentary strata may indicate the presence of beds that are resistant to weathering, whereas low areas or gently sloping areas may be underlain by nonresistant beds.

In recent years, Earth scientists have developed methods for representing a landscape by using a **digital elevation model** (DEM). Computers construct a DEM from a set of data in which each location on a map has three coordinates: latitude, longitude, and elevation. The shape of the land on a DEM, or on a topographic map, can be

Figure Bx12.1 Topographic maps and profiles.

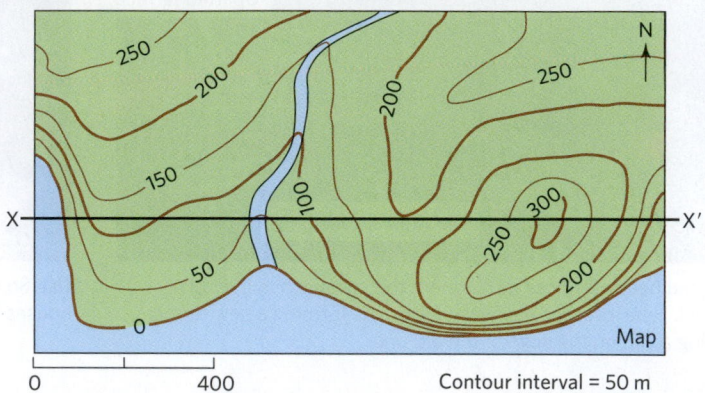

(a) A topographic map depicts the shape of the land surface through the use of contour lines. The difference in elevation between two adjacent lines is the contour interval.

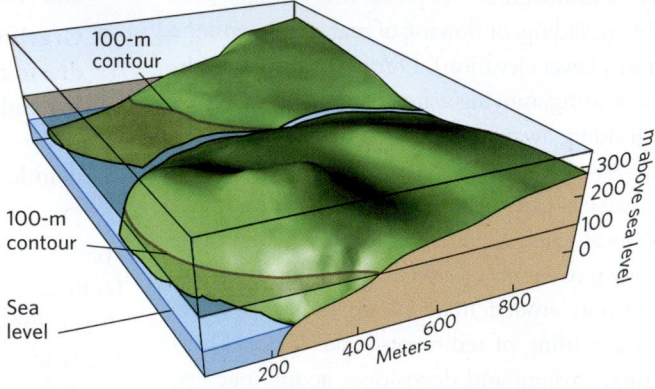

(b) A contour line represents the intersection of a horizontal plane with the land surface. This block diagram shows the area mapped in part a.

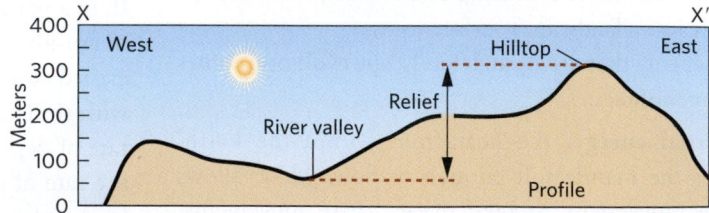

(c) A topographic profile (along section line X–X') shows the shape of the land surface as seen in a vertical slice.

highlighted by adding shadows to simulate the appearance of the land if it were lit by the Sun when it's low in the sky **(Fig. Bx12.1f)**. The resulting image is called a **shaded-relief map**.

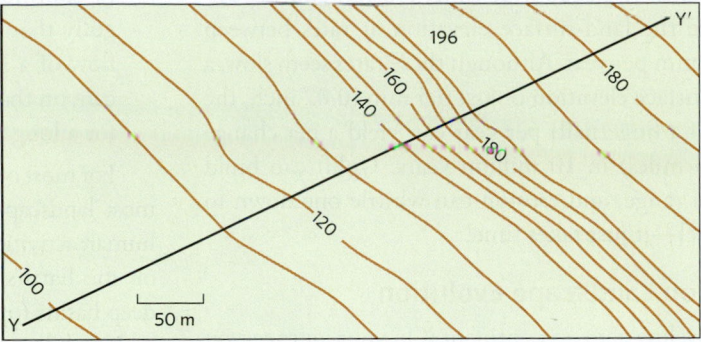

(d) This topographic map shows a distinct cliff.

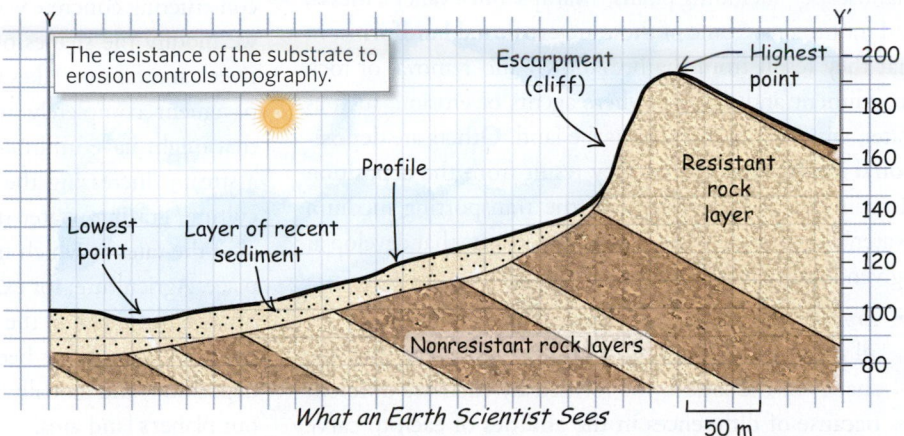

The resistance of the substrate to erosion controls topography.

What an Earth Scientist Sees

(e) A geologic cross-section depicts a geologist's interpretation of the subsurface along the section line Y–Y′ from part (d). The cliff is the edge of a resistant rock layer.

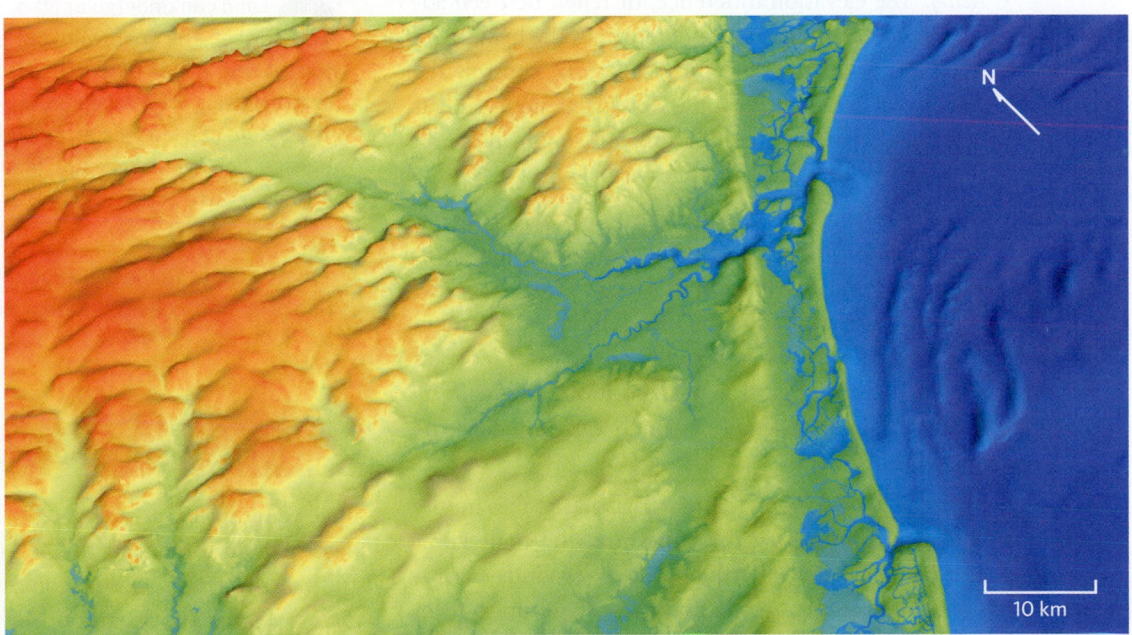

(f) A digital elevation model of the region southwest of Atlantic City, N.J., reveals details of the landscape. Blue represents water; green, lower elevations; orange, higher elevations. All the land shown lies within 20 m of sea level, so the topography is subtle.

debris tens of meters thick in a matter of minutes to days. However, when averaged over time, erosion and deposition change the land-surface elevation at rates between 0.1 and 10 mm per year. Although these rates seem slow, a change in surface elevation of just 0.5 mm (0.02 inch, the thickness of a fingernail) per year can yield a net change of 5 km (3 miles) in 10 million years. Uplift can build a mountain range, and erosion can whittle one down to near sea level—it just takes time!

Controls on Landscape Evolution

Imagine traveling across a continent. On your journey, you see a variety of **landforms** (natural, recognizable features of a landscape), including plains, swamps, hills, valleys, mesas, and mountains. Some of these are **erosional landforms**, in that they result from the breakdown and removal of rock or sediment and develop where agents of erosion, such as water, wind, or ice, carve into the land. Others are **depositional landforms**, in that they result from the deposition of sediment as it drops out of the transporting medium (water, wind, or ice). The specific landforms that develop at a given locality reflect several factors:

- *Eroding or transporting agent:* Water, wind, and ice all cause erosion and all transport sediment. But the shapes of the landforms formed by each are different because of differences in the abilities of each to carve into the land and to carry away debris. Of the three agents, water has the greatest effect on a global basis.
- *Relief:* The elevation difference, or relief, between adjacent places in a landscape determines the height and steepness of a slope. Steepness, in turn, controls the velocity of ice or water flow down the slope, and it determines whether rock or soil can stay in place on the slope despite the pull of gravity.
- *Climate:* The average temperature, seasonal temperature variation, the overall volume of precipitation in a year, and the distribution of precipitation over time defines the *climate* of a region. Climate determines whether running water, flowing ice, or wind serves as the main agent of erosion or deposition, and it affects the lushness of vegetation, which can influence the behavior of slopes.
- *Substrate composition:* The material that makes up the land determines how the **substrate**, the material at and just below the Earth's surface, responds to erosion. For example, strong rocks can stand up to form steep cliffs, whereas soft sediment collapses to generate gentle slopes.
- *Living organisms:* Plants and animals can weaken the substrate (by burrowing, wedging, or digesting) or hold it together (by binding it with roots).

- *Time:* Landscapes evolve through time in response to continued erosion and deposition. For instance, a gully that has just started to form in response to the flow of a stream does not look the same as the deep canyon that develops after the same stream has existed for a long time.

For most of geologic time, water, wind, and ice generated most landscapes. During the past few centuries, however, human activities have had an increasingly important impact on the Earth's surface. We have replaced mountains with deep basins (open-pit mines), have built hills (tailings piles and landfills) where once there were valleys, and have made steep slopes gentle and gentle slopes steep **(Fig. 12.4a–c)**. By constructing concrete walls and by dumping piles of debris, we modify the shapes of coastlines, change the courses of rivers, and fill new lakes (reservoirs). In cities, buildings and pavement completely seal the ground, preventing water that might have infiltrated the ground to spill instead into a stream, increasing the stream's flow. In rural areas, agriculture, grazing, water use, and deforestation substantially alter the rates at which natural erosion and deposition take place. Agriculture, for example, increases the rate of erosion because, for much of the year, farm fields have no vegetation cover. Humanity has become a major agent of erosion and deposition, and globally, humans have modified over half of our planet's land area.

Take-home message . . .

Land can undergo uplift or subsidence to yield relief. The energy driving landscape evolution comes from three sources: Earth's internal energy, external energy from the Sun, and gravitational energy. The nature of a landscape depends on eroding or transporting agents, climate, time, relief, the activity of organisms, and substrate composition.

Quick Question -
What happens to the elevation of the land surface if the rate of uplift exceeds the rate of erosion?

12.3 The Hydrologic Cycle

Since water in its various states—liquid, gas, and solid—plays such an important role in landscape evolution, any study of surface processes requires that we consider how and why water moves around in the Earth System. Our planet's surface and near-surface water occupies several distinct *reservoirs*, a word used here to mean any realm or part of a realm in the Earth System **(Table 12.1)**.

Atmospheric water occurs as vapor, as tiny droplets or ice crystals in clouds, and as rain, snow, or hail that

Figure 12.4 Human influences on a geologic scale.

(a) The pyramids of Egypt are human-made hills that rise above the desert sands. They have lasted for thousands of years.

(b) In the process of highway construction, deep valleys are cut through high ridges. This example borders a highway near Denver.

(c) This stone dam holds back a reservoir in Colorado.

Table 12.1 Major water reservoirs of the earth

Reservoir	Volume (km³)	Percentage of Total Water	Percentage of Freshwater
Oceans and seas	1,338,000,000	96.5	—
Glaciers, ice caps, snowfields	24,064,000	2.05	68.7
Saline groundwater	12,870,000	0.76	—
Fresh groundwater	10,500,000	0.94	30.1
Permafrost	300,000	0.022	0.86
Freshwater lakes	91,000	0.007	0.26
Salt lakes	85,400	0.006	—
Soil moisture	16,500	0.001	0.05
Atmosphere	12,900	0.001	0.04
Swamps	11,470	0.0008	0.03
Rivers and streams	2,120	0.0002	0.006
Living organisms	1,120	0.0001	0.003

Source: Data from P. H. Gleick, *Encyclopedia of Climate and Weather* (New York: Oxford University Press, 1996).

falls to Earth's surface (see Chapter 17). Surface water collects mostly in oceans, but also in lakes, streams, puddles, and swamps on land. Frozen water, sometimes known as the cryosphere (see the Prelude), collects in snowfields and glaciers. Below the ground surface, some water dampens soil and rock near the surface, and some sinks deeper and fills underground pores and cracks as **groundwater**. *Permafrost* exists where soil water and groundwater, extending down to depths of tens to hundreds of meters underground, remains frozen all year. A significant amount of water also resides in living organisms of the biosphere. In fact, about 50% to 65% of your body consists of water. Water constantly flows from reservoir to reservoir, driven by gravity and solar radiation. This never-ending migration is called the **hydrologic cycle** (**Earth Science at a Glance**, pp. 394–395).

To get a clearer sense of how the hydrologic cycle operates, let's follow the fate of a water molecule that has just reached the surface of the ocean. Solar radiation heats the seawater, and the increased thermal energy of the vibrating water molecules allows them to evaporate and drift upward in a gaseous state to become part of the atmosphere. About 0.03% of the total ocean volume evaporates every year. Atmospheric water vapor moves with the wind to higher altitudes, where it cools, undergoes condensation, and *precipitates* (falls out of the air) as rain or snow. About 76% of this water falls directly into the ocean. Of the remainder, which falls on the land surface, most becomes trapped temporarily in the soil or in living organisms and soon returns directly to the atmosphere by *evapotranspiration* (the sum of evaporation from bodies of surface water, evaporation from the ground surface, and release of water by plants and animals). Rainwater that does not become trapped in the soil or in living organisms enters lakes or rivers and ultimately flows back to the sea as surface water, or becomes trapped in glaciers, or sinks deeper into the ground to become groundwater. Groundwater also flows and ultimately returns to the Earth's surface reservoirs.

The average length of time that water stays in a particular reservoir during the hydrologic cycle is called the **residence time**. Water in different reservoirs has different residence times. For example, a typical molecule of water remains in the oceans for 4,000 years or less, in lakes and ponds for 10 years or less, in rivers for 2 weeks or less, and in the atmosphere for 10 days or less. Groundwater residence times are highly variable and depend in part on the path that the groundwater follows between the point where it enters the ground and the point where it exits (see Chapter 13). Water can stay underground for anywhere from 2 weeks to 10,000 years before it inevitably moves on to another reservoir.

Take-home message . . .

Water, which plays a major role in landscape evolution on the Earth, moves among various reservoirs (the ocean, the atmosphere, the land surface, the subsurface, and living organisms) during the hydrologic cycle.

Quick Question -
What energy sources drive water through the hydrologic cycle?

12.4 Introducing Mass Wasting

What Is Mass Wasting?

It was Sunday, May 31, 1970, a market day, and thousands of people had crammed into the Andean town of Yungay, Peru, to shop. Suddenly they felt the jolt of an earthquake, strong enough to topple some masonry houses. Unfortunately, worse was yet to come. The ground shaking had also broken an 800-m (2,600-foot)-wide ice slab off the end of a glacier at the top of Nevado Huascarán, a nearby 6.6-km (4.1-mile) mountain peak. As it tumbled 3.7 km (2.3 miles) down the mountainside, the ice disintegrated into a turbulent mass of chunks. Near the base of the mountain, most of the debris channeled into a valley and thickened into a churning cloud as high as a 10-story building, ripping up rocks and soil along the way. Frictional heating transformed the ice into water, which mixed with loose rock and dust to produce 50 million m³ (1.8 billion cubic feet) of a muddy slurry capable of lifting boulders larger than houses. This slurry, sometimes floating on a cushion of compressed air, traveled over 14.5 km (9.0 miles) in less than 4 minutes.

Near the mouth of the valley, some of the debris came to rest, but part of it rocketed up the sides of the valley and became airborne above a ridge bordering Yungay. As the town's 18,000 inhabitants stumbled out of earthquake-damaged buildings, they heard a deafening roar and looked up to see a wall of debris descending from above. The debris buried the town, along with everyone in it, moments later. Today, only a grassy meadow overlies the site where Yungay once bustled (**Fig. 12.5a, b**).

We commonly assume that the ground beneath us is terra firma, a solid foundation on which we can build our lives. But a catastrophe like the one at Yungay shouts otherwise. Much of the Earth's surface hosts *unstable slopes*, meaning that the materials of the slope, including rock, **regolith** (a general term for soil, unconsolidated sediment, and very weathered rock that lies above coherent bedrock), snow, or ice, might start moving downslope if disturbed. Geologists refer to the gravity-driven tumbling, flowing, or sliding of these materials downslope from higher elevations to lower ones as **mass wasting**, or *mass movement*. Mass wasting plays a critical role in the rock cycle as the first step in the transportation of sediment, and it serves as the most rapid means to modify slope shapes. But, like earthquakes, volcanic eruptions, storms, and floods, mass wasting is a type of *natural hazard*, a feature of the environment that can cause damage to landscapes and to human society. Unfortunately, mass wasting becomes more of a threat to society every year because, as the world's population grows, cities have expanded to include areas of unstable slopes.

Most people refer to any mass-wasting event as a **landslide**. Geologists and engineers, however, find it useful to distinguish among types of mass-wasting events based on four features: (1) the type of material involved (rock or regolith); (2) the velocity of movement (slow, intermediate, or fast); (3) the character of the moving mass (coherent, chaotic, or slurry); and (4) the environment in which the movement takes place (subaerial or submarine). Let's examine the different types of mass-wasting events that occur on land, roughly in order from slow to very fast, then turn to those that occur underwater. The distinction among types tends to be somewhat fuzzy because a given event may start out as one type and evolve into another, or it may display characteristics of two types.

Creep and Solifluction

Have you ever noticed that trees, walls, and gravestones on the side of a hill may be tilted (**Fig. 12.6a**)? If so, you're seeing the visible consequences of **creep**, the gradual movement of regolith down a slope. Creep happens when regolith on a slope alternately expands and contracts in response to freezing and thawing, wetting and drying, or warming and cooling. To see how the process of creep works, let's focus on the consequences of seasonal freezing and thawing. In the winter, when water in regolith freezes, the regolith expands, and particles move outward in a direction perpendicular to the slope. During the spring thaw, when water becomes liquid again, gravity makes the particles sink vertically, so they effectively migrate downslope slightly (**Fig. 12.6b, c**). You can't see creep by staring at a hillside because it occurs too slowly. In fact, trees can continue to grow while creep takes place, as indicated by the curvature of some trees' trunks (see Fig. 12.6a, inset).

In Arctic or high-elevation regions, a warm spell during the summer may cause the uppermost 1 to 3 m (3 to 10-feet) of permafrost to thaw. Since meltwater cannot sink into the still solid permafrost below, the melted layer becomes soggy and weak and flows slowly downslope in overlapping sheets. Geologists refer to this kind of creep as **solifluction** (**Fig. 12.6d**).

Figure 12.5 The May 1970 landslide disaster in Yungay, Peru.

(a) Before the landslide, the town of Yungay perched on a hill near the ice-covered mountain Nevado Huascarán.

(b) The landslide completely buried the town beneath debris. A landslide scar is visible on the mountain in the distance.

Slumping

The majestic Holbeck Hall Hotel had perched on a 60-m (200-foot)-high cliff along the eastern coast of England since 1879. Its guests could enjoy a spectacular view out across the North Sea—until June 5, 1993. On that day, a block of land extending inland 70 m (230 feet) from the cliff face slipped down and rotated seaward, taking half the hotel with it. Fortunately, telltale cracks in the ground weeks before, and a smaller collapse the day before, had led officials to evacuate the hotel, so no one was hurt. But the landmark, along with

The Hydrologic Cycle

Water circulates through a number of reservoirs in the Earth System. The largest reservoir by far is the ocean, which covers 71% of the Earth's surface. Water evaporates from the ocean and enters the atmosphere, where it may remain for quite a while. Thus, the atmosphere serves as another reservoir. Atmospheric water gradually condenses and forms clouds that drop rain or snow onto the oceans or land.

Wind transportation of moisture

Cloud condensation

The Atmospheric Reservoir

Evapotranspiration (from vegetation, trees, etc.)

The Organic Reservoir

Evaporation of surface ocean water

Precipitation over oceans

Surface runoff (returns to sea)

The Ocean Reservoir

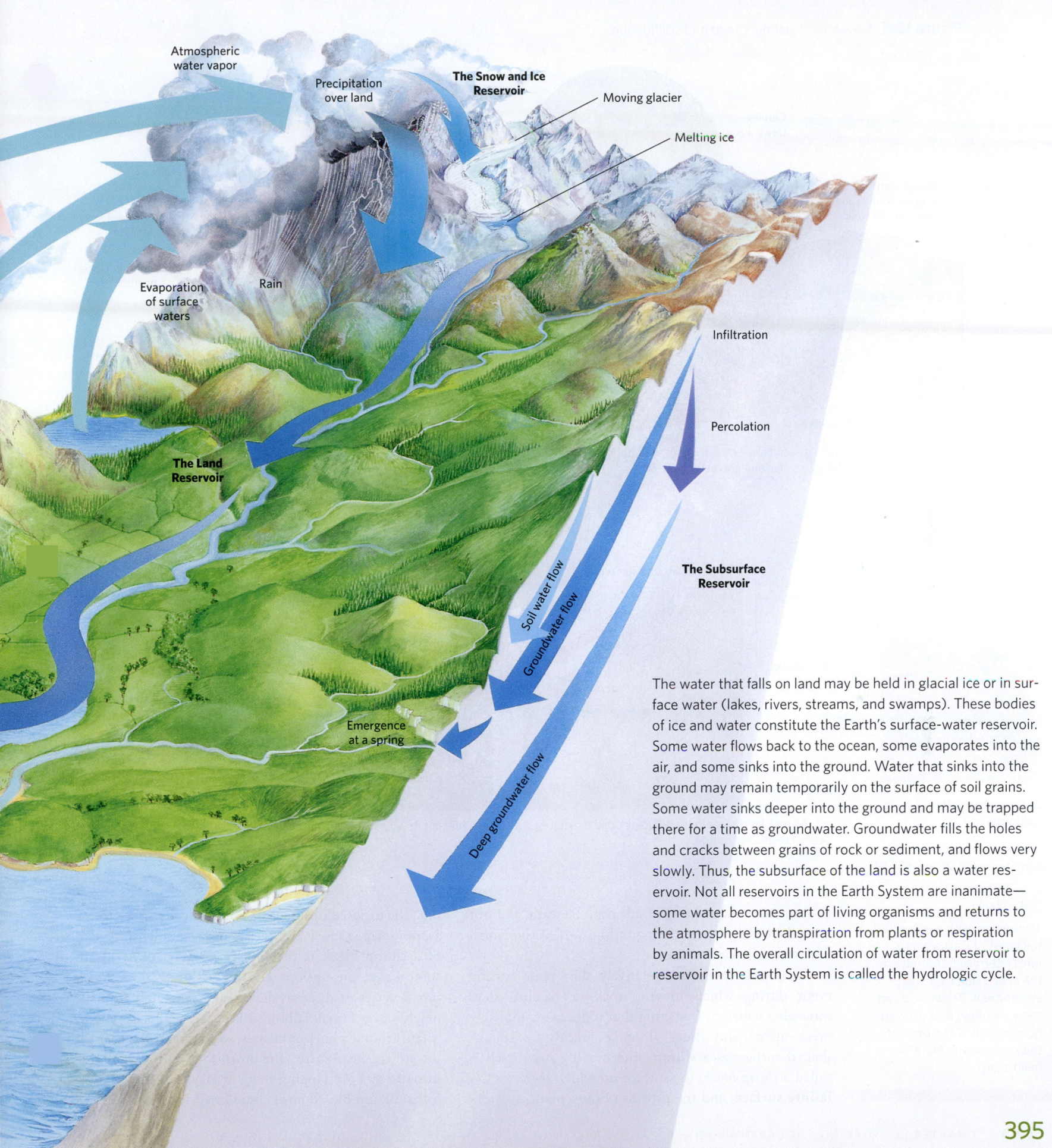

Atmospheric
water vapor

Precipitation
over land

**The Snow and Ice
Reservoir**

Moving glacier

Melting ice

Evaporation
of surface
waters

Rain

**The Land
Reservoir**

Infiltration

Percolation

**The Subsurface
Reservoir**

Soil water flow

Groundwater flow

Emergence
at a spring

Deep groundwater flow

The water that falls on land may be held in glacial ice or in surface water (lakes, rivers, streams, and swamps). These bodies of ice and water constitute the Earth's surface-water reservoir. Some water flows back to the ocean, some evaporates into the air, and some sinks into the ground. Water that sinks into the ground may remain temporarily on the surface of soil grains. Some water sinks deeper into the ground and may be trapped there for a time as groundwater. Groundwater fills the holes and cracks between grains of rock or sediment, and flows very slowly. Thus, the subsurface of the land is also a water reservoir. Not all reservoirs in the Earth System are inanimate—some water becomes part of living organisms and returns to the atmosphere by transpiration from plants or respiration by animals. The overall circulation of water from reservoir to reservoir in the Earth System is called the hydrologic cycle.

Figure 12.6 Slow mass wasting: Creep and solifluction.

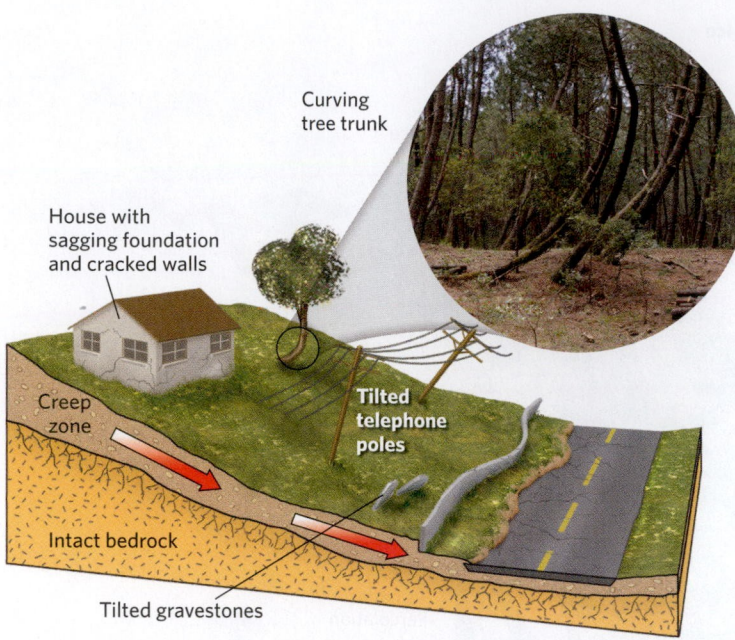

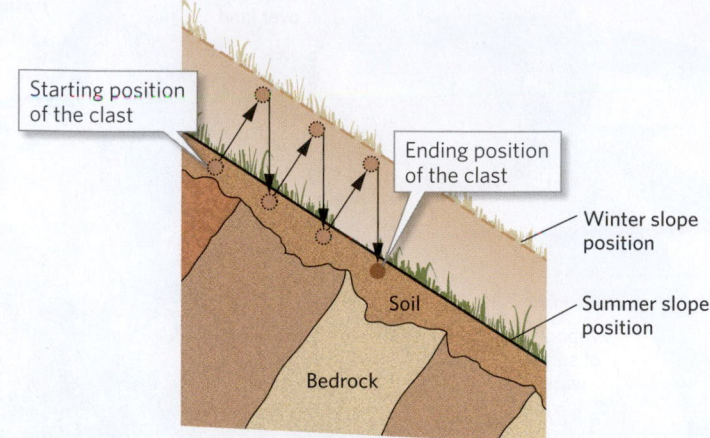

(b) Creep due to seasonal freezing and thawing: A clast rises perpendicular to the ground during freezing, but sinks vertically during thawing. After three years, it migrates to the position shown.

(a) Soil creep causes walls to bend and crack, building foundations to sink, trees to bend, and power poles and gravestones to tilt.

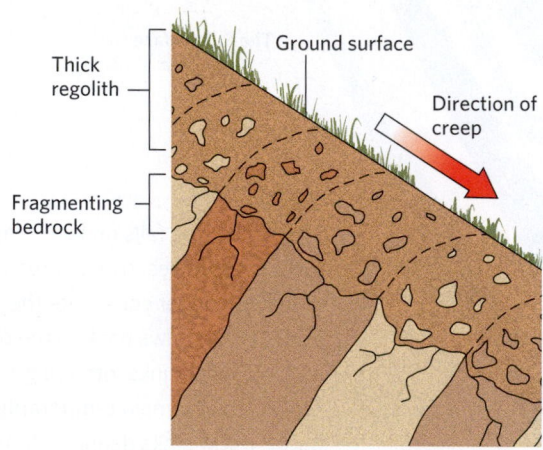

(c) As rock layers weather and break up, the resulting debris creeps downslope.

(d) Solifluction on a hillslope in the tundra.

an immense amount of the substrate beneath, slid out onto the beach, where the pounding surf of the North Sea eventually carried it away.

Geologists refer to a relatively slow mass-wasting event during which moving rock or regolith stays somewhat coherent (meaning that it doesn't completely break apart) and slides along a concave-up spoon-shaped surface as a **slump**. The moving mass itself is called a *slump block*, the surface on which it moves is a **failure surface**, and the process of movement is *slump-*

ing. The exposed upslope edge of a failure surface forms a *head scarp*—a new cliff face—and the downslope end of a slump block is the block's *toe* **(Fig. 12.7a–d)**. In some cases, the upslope ends of a slump block break into a series of discrete slices, each separated from its neighbor by a small failure surface. These blocks tend to rotate around a horizontal axis, so that their surfaces tilt toward the head scarp. The downslope ends may divide into discrete overlapping slices or may break up to form a chaotic jumble. Slumps come in all sizes, from only a

Figure 12.7 The process of slumping on a hillslope.

(a) A head scarp on a hillslope.

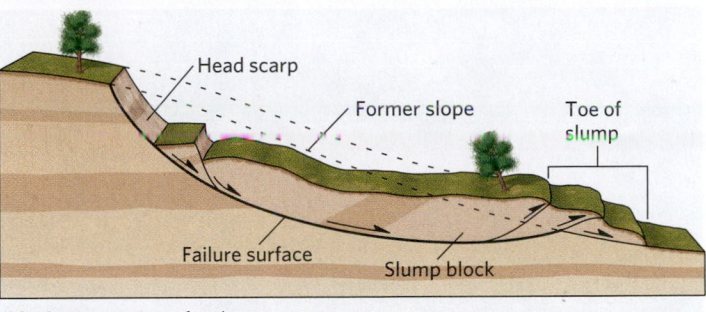

(b) Cross section of a slump.

(c) Slumping has dumped sediment into this river in Costa Rica.

(d) A slump beginning to form along a highway in Utah.

few meters to tens of kilometers across, and they move at speeds from millimeters per day to tens of meters per minute.

Mudflows, Debris Flows, and Lahars

In the tropical climate of Brazil, bedrock undergoes intense chemical weathering and slowly transforms into a thick, clay-rich regolith that becomes saturated with water when it rains heavily. Where slopes are steep, the saturated regolith, containing up to 30% water, can transform into a slurry of mud, with the consistency of wet concrete, that flows downslope. Geologists refer to such a moving slurry of mud as a **mudflow**, or *mudslide*, if it consists mostly of clay and water. If it also carries cobbles and boulders suspended in the mud, geologists refer to it as a **debris flow**.

Mudflows or debris flows may start out as slumps, but because of their high water content, they cannot stay coherent. Their movement strips away forests and can knock down buildings and overwhelm towns **(Fig. 12.8a–d)**.

Mudflows or debris flows travel at speeds of 20 to 80 km per hour (10 to 50 mph). The speed at which a given flow moves depends on both the water content and the slope angle: wetter mudflows or debris flows move faster than dryer ones, and mudflows or debris flows move faster down steep slopes than they do down gentle slopes. Mudflows and debris flows have much greater viscosity than clear water, so they can carry large rock chunks and huge trees as well as houses and cars. They typically follow channels downslope, and at the base of the slope they may spread out into a broad fan **(Box 12.2)**.

Figure 12.8 Examples of mudflows and debris flows.

(a) A 2011 mudslide destroyed houses along this hillside in Brazil.

(b) Mudslides of 2011 stripped away forests on hillslopes in Brazil.

(c) A recent debris flow in Utah. Note the chaotic mixture of rock chunks and mud.

(d) A mudflow down a rain-soaked hillslope buried a highway in Taiwan in 2010.

When mudflows or debris flows spill down the river valleys bordering volcanoes, they consist of a mixture of volcanic ash, coarser pyroclastic debris, and water. The water may come from the snow or ice melted by the volcano's heat or from heavy rains. As we saw in Chapter 4, such volcanic mudflows are known as *lahars* **(Fig. 12.9a)**. Lahars generally follow river valleys and typically travel for distances of tens of kilometers away from the volcano at speeds greater than 35 km per hour (20 mph). In a few cases, lahars have traveled over 100 km (60 miles) from their source. In fact, following the 1877 eruption of a volcano in the Andes of Ecuador, a lahar reached the Pacific Ocean 320 km (200 miles) from the volcano.

When lahars happen in the wilderness, they can tear away forests and bury the landscape with muck. When they happen near populated areas, they spell disaster. And because lahars can be activated by rainwater, not just meltwater, they can be hazards for years after a large eruption of pyroclastic debris. For example, devastating lahars carrying ash from the 1991 eruption of Mt. Pinatubo spilled down valleys into the surrounding lowlands in the Philippines for each of the next four rainy seasons after the eruption **(Fig. 12.9b)**.

Rockslides and Rockfalls

In the early 1960s, engineers built a huge new dam across a river on the northern side of Monte Toc, in the

Box 12.2

Consider this . . .

What goes up must come down: the La Conchita mudflow

Along the coast of California, waves slowly erode the land and produce low, flat areas called wave-cut benches. While this happens, plate movements are slowly pushing the land surface upward. When uplifted, the wave-cut benches form small plateaus, or terraces. One such terrace lies at an elevation of 180 m (600 feet) above sea level, about 500 m (1,600 feet) east of the present-day beach at La Conchita. The west face of this terrace is a cliff-like bluff **(Fig. Bx12.2a)**. Movement on faults associated with the transform boundary between the North American and Pacific Plates has broken up the bedrock of the area, and fragments have weathered substantially, so the substrate of the terrace and bluff consist of a weak, clay-rich regolith.

Relatively little vegetation covers the bluff or the terrace above, so when rain falls on the face of the bluff, it drains away via a network of small, temporary streams. But the water falling on the terrace sinks down into the ground and saturates regolith many meters below the surface of the bluff. Adding water to clay makes mud. Occasionally, the bluff gives way, and the regolith starts to move downslope as a debris flow at rates of up to 35 km per hour (20 mph).

If the region of La Conchita were uninhabited, such mass wasting would simply be part of the natural process of landscape evolution—gravity bringing down land that had been raised by tectonic processes. But when downslope movements take place in La Conchita, they make headlines, because dozens of houses have been built on the present-day wave-cut bench between the shore and the base of the bluff. In 1995, a mud and debris flow overwhelmed 9 of these houses An even more devastating flow happened in 2005, burying 13 houses, damaging 23, and killing 10 people **(Fig. Bx12.2b, c)**. These events have sent a clear message about the importance of reading the landscape when planning construction.

Figure Bx12.2 The 2005 La Conchita mudflow.

(a) A housing development was built on a wave-cut bench between the beach and the bluff.

(c) Rescuers at the toe of the mudflow.

(b) During heavy rains, the bluff gave way, and heavy mud flowed down, burying houses and taking several lives.

Figure 12.9 Examples of lahars.

(a) A lahar that rushed down the side of Mt. St. Helens, Washington.

(b) Damage to a town in the Philippines by a lahar.

See for yourself

Debris Slide, Yungay, Peru

Latitude: 9°7′21.42″ S
Longitude: 77°39′44.86″ W

Looking NE from 6.4 km (~4.3 miles).

Here you can see the steep, glaciated face of Nevado Huascarán. In 1970, a debris slide from the mountain rushed down the valley in the foreground and buried the landscape. More recent, smaller debris flows have accumulated on top of the larger one.

Italian Alps, to create a reservoir for generating electricity. The Vaiont Dam was an engineering marvel, a concrete wall rising 260 m (850 feet)—as high as an 85-story skyscraper—above the valley floor. Unfortunately, the dam's builders did not recognize the hazard posed by Monte Toc. The side of Monte Toc facing the reservoir is underlain by limestone beds interlayered with weak shale beds, all of which dip parallel to the surface of the mountain **(Fig. 12.10a)**. The shale beds serve as failure surfaces. As the reservoir filled, the mountain flank began to crack and rumble, and chunks slid down into the reservoir. Local residents began to refer to Monte Toc as *la montagna che cammina* (the mountain that walks).

On October 9, 1963, after several days of rain, more cracks formed, and Monte Toc started to rumble so much that engineers lowered the water level in the reservoir. They thought the wet ground might slump a little into the reservoir, with minor consequences, so authorities did not order an evacuation of the town of Longarone, a few kilometers down the valley from the dam. Unfortunately, the engineers underestimated the problem. At 10:30 that evening, a huge chunk of Monte Toc—540 billion kg (600 million tons) of rock—detached from the mountain and slid at speeds of up to 110 km per hour (70 mph) down the side of the mountain along a weak shale bed. Some of the debris rocketed up the opposite wall of the reservoir to a height of 260 m (850 feet) above the original reservoir level, and some remained in the reservoir, displacing 115 million m³ (4 billion cubic feet) of water. This water splashed over the top of the dam, then rushed down the valley below **(Fig. 12.10b)**. When the wall of water had passed, nothing remained of Longarone and its 1,500 inhabitants. Though the dam still stands

today, it holds back only debris, and it has never provided electricity.

Geologists refer to such a sudden movement of rock down a non-vertical slope as a **rockslide**, and if the mass consists mostly of regolith, then it's a *debris slide*. If the rock or debris free-falls vertically during part of its journey, the movement can also be referred to as a **rockfall** or *debris fall*. Once one of these mass-wasting events has taken place, it leaves a scar on the slope and forms an accumulation of rock fragments at the base of the slope **(Fig. 12.11a, b)**.

Rockslides happen when bedrock or regolith detaches from a slope, then slips rapidly downslope on a failure surface. Rockfalls happen when rocks on a cliff break off along a joint and topple out away from the cliff. When they strike other rocks downslope, they shatter into smaller fragments. If the fragments accumulate over time at the base of the cliff, they can build an apron or pile of **talus (Fig. 12.12a)**. Talus, like all granular debris, tends to pile up, producing the steepest slope it can without collapsing. The angle of this slope, called the **angle of repose**, has a value between 30° and 37° for most dry, unconsolidated (loose) materials, such as dry sand. Steeper angles of repose, up to 45°, can form where debris consists of irregularly shaped grains that can interlock **(Fig. 12.12b)**.

Rockslides and rockfalls may move at speeds of up to 300 km per hour (185 mph), depending on the steepness of the slope and the distance that they move. They become particularly fast when a cushion of air gets trapped beneath them, for in such cases, there is virtually no friction between the slide and its substrate, so the mass moves like a hovercraft. Large, fast rockslides

Figure 12.10 The Vaiont Dam disaster.

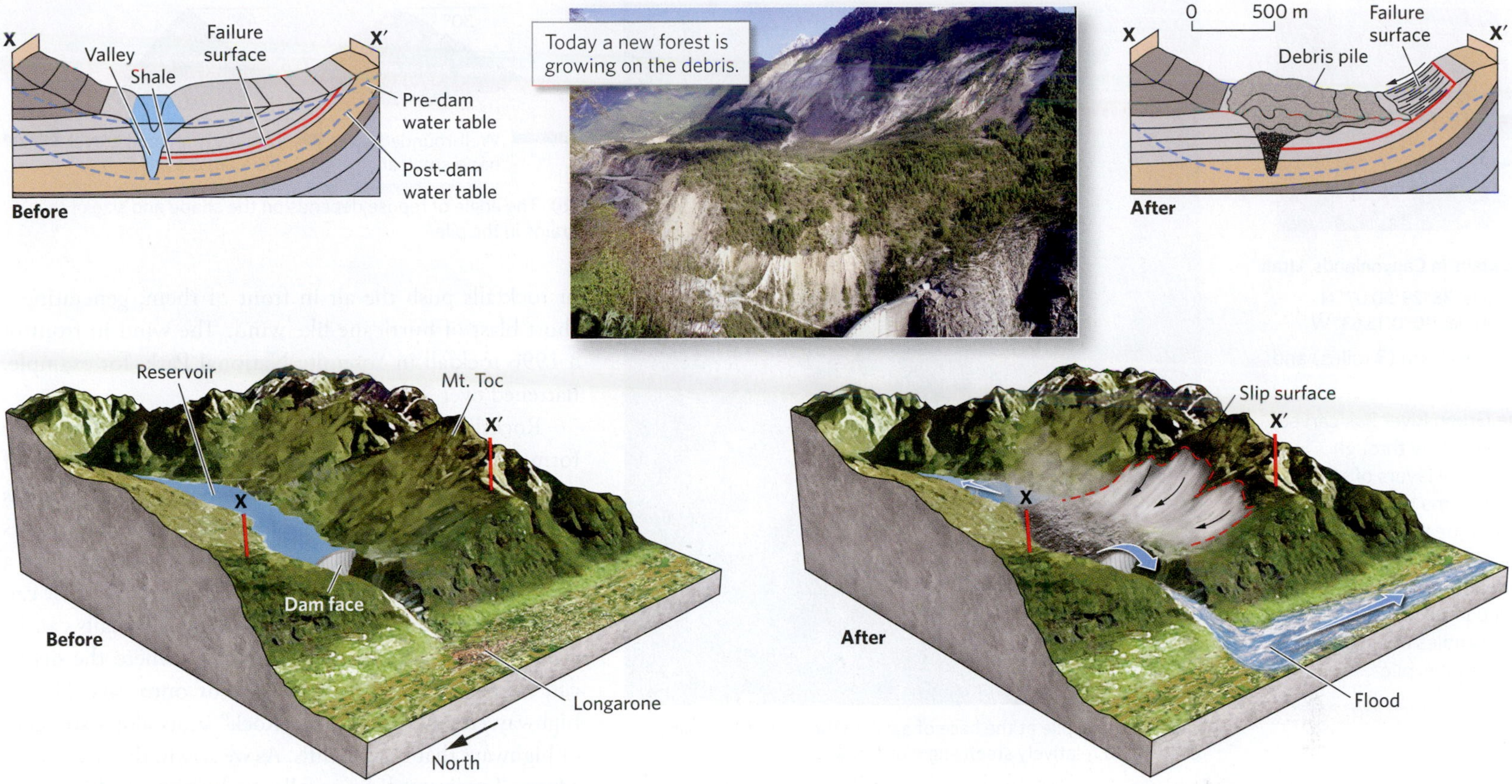

Today a new forest is growing on the debris.

Before diagram labels: X, X', Valley, Shale, Failure surface, Pre-dam water table, Post-dam water table, Before

After diagram labels: 0, 500 m, Failure surface, Debris pile, X, X', After

Before block diagram labels: Reservoir, Mt. Toc, X', X, Dam face, Longarone, Before, North

After block diagram labels: Slip surface, X', X, Flood, After

(a) Before the landslide, the north flank of Monte Toc was forested. When the reservoir was filled with water, the slope became unstable. The cross section shows the shale bed that became the failure surface.

(b) Thirty-three million cubic meters of rock and debris slid down the mountainside and displaced water in the reservoir. The water surged over the dam and swept away a town in the valley below.

Figure 12.11 Examples of rockfalls.

(a) Successive rockfalls have littered the base of this sandstone cliff with boulders. Note the talus at the base of the cliff.

(b) Debris from a rockfall in the Andes of Peru has accumulated downslope. The fresh rock exposed by the rockfall has a lighter color.

Rockfalls in Canyonlands, Utah

Latitude: 38°29'50.07" N
Longitude: 110°0'15.63" W

Zoom to 5 km (3 miles) and look down.

The Green River has carved a deep valley through horizontal layers of strata. Shale forms the slopes, and sandstone forms the top ledge. Erosion of the shale undercuts the sandstone, which breaks away at joints and tumbles downslope to build a talus pile.

Figure 12.12 The angle of repose is the steepest slope at which a pile of unconsolidated sediment can remain stable.

(a) This talus pile at the base of a cliff in the Uinta Mountains, Utah, has a relatively steep angle of repose.

30° | 45°

Well-rounded sand has a small angle. | Irregularly shaped gravel has a large angle.

(b) The angle of repose depends on the shape and size of the grains in the pile.

or rockfalls push the air in front of them, generating a short blast of hurricane-like wind. The wind in front of a 1996 rockfall in Yosemite National Park, for example, flattened over 2,000 trees.

Rockslides and rockfalls, like slumps and any other form of mass wasting **(Fig. 12.13)**, happen at a variety of scales, depending on the amount of material that detaches and on whether the energy of tumbling material moves debris in its path. The smallest events may involve just a few pebbles or fist-sized fragments, while the largest can involve a whole mountainside. Even smaller falls can be a hazard, particularly along highways, where the debris can tumble onto the road surface or onto cars. Hence, highway crews erect "Falling Rock" signs along stretches of highways that border cliffs. As we saw in the example of Monte Toc, large slides or falls can be catastrophic.

Figure 12.13 As we have seen, different kinds of mass wasting can modify the landscape.

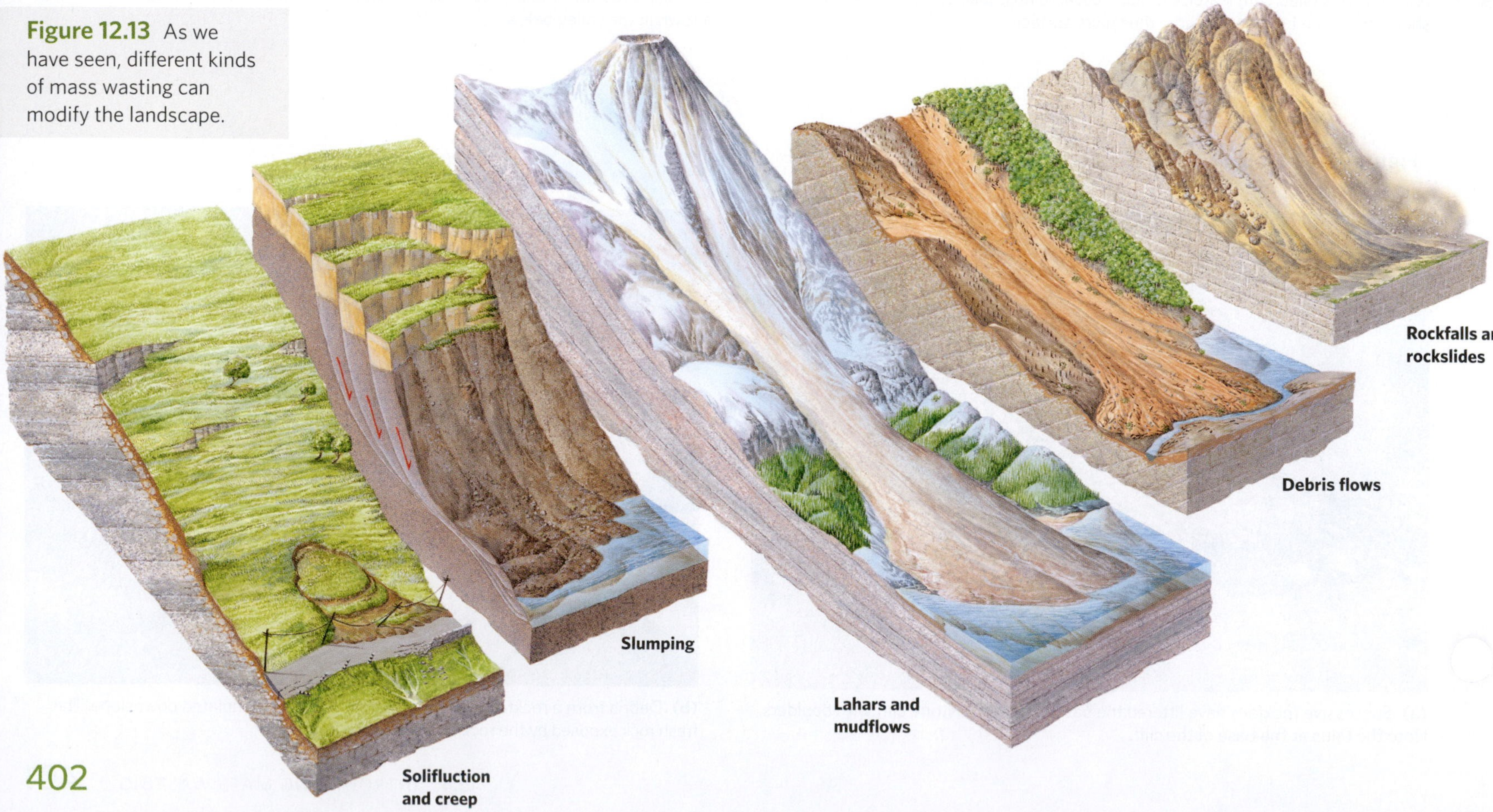

Rockfalls and rockslides

Debris flows

Slumping

Lahars and mudflows

Solifluction and creep

Figure 12.14 Avalanches.

(a) The aftermath of a 1999 avalanche in the Austrian Alps. Masses of snow buried homes, crushed and ripped through cars, and killed 31 people.

Origin of the avalanche

Toe of the avalanche

(b) This dry snow avalanche in Alaska is a turbulent cloud.

Avalanches

In the winter of 1999, an unusual weather system passed over the Austrian Alps. First it snowed. Then the temperature warmed and the snow began to melt. But then the weather turned cold again, and the melted snow froze into a hard, icy crust. This cold snap ushered in a blizzard that blanketed the icy crust with tens of centimeters (1 to 2 feet) of new snow. When the weight of new snow became heavy enough, the icy crust acted as a failure surface, and 180 million kg (200,000 tons) of snow began to slip down the mountain. As it accelerated, the mass transformed into a **snow avalanche**, a turbulent cloud of snow mixed with air, that surged downslope at 290 km per hour (180 mph). At the bottom of the slope, the avalanche overran a ski resort, crushing and carrying away buildings, cars, and trees and killing over 30 people. It took searchers and their specially trained dogs many days to find buried survivors and victims under the 5- to 20-m (16- to 66-foot)-thick pile of snow that the avalanche deposited (Fig. 12.14a).

What triggers snow avalanches? Some happen when a *cornice*, a large drift of snow that builds up on the side of a windy mountain summit, suddenly gives way and falls onto the slopes below, where it knocks additional snow free. Others happen when a broad slab of snow on a moderate slope detaches from its substrate along an icy failure surface. Avalanches behave differently depending on the temperature. *Wet avalanches*, which occur at warmer temperatures, move relatively slowly as a slurry of solid and liquid water, whereas *dry avalanches*, which occur at colder temperatures, tumble rapidly downslope as a cloud of powder (Fig. 12.14b). The Austrian avalanche we described above was a dry avalanche.

Geologists sometimes use the word *avalanche* in a broad sense to refer to any mass-wasting event during which solid fragments become suspended in a fluid, so that the flowing mixture behaves like a turbulent cloud. In this context, a *debris avalanche* is a turbulent cloud containing rock and/or sediment, suspended in air, that surges down a slope. Below, we see that avalanche-like flows can also occur under water.

Submarine Mass Wasting

So far, we've focused on mass wasting that occurs on the land surface—we refer to such events as *subaerial*, to emphasize that they occur beneath the air. But mass wasting also happens underwater. Geologists distinguish among three types of such submarine mass wasting based on whether the mass remains coherent or disintegrates as it moves. In a **submarine slump**, semicoherent blocks slip downslope along a failure surface. In a **submarine debris flow**, the moving mass breaks apart to form a slurry

Figure 12.15 Submarine mass wasting.

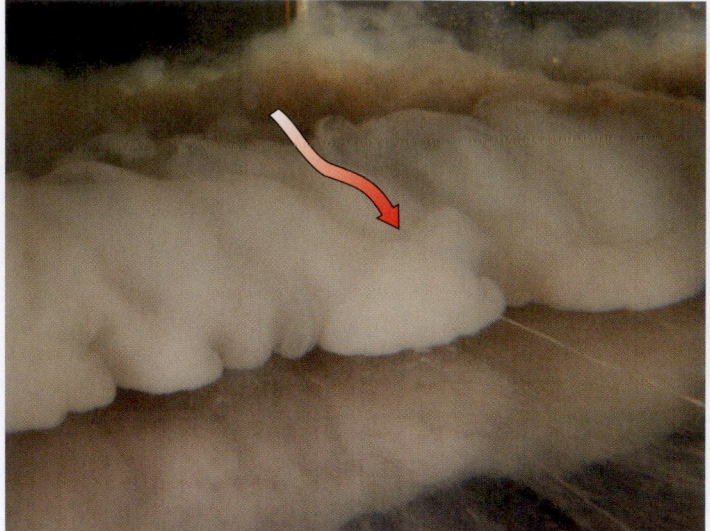

(a) A laboratory model of a turbidity current. The cloud is entirely underwater and consists of fine clay suspended in water.

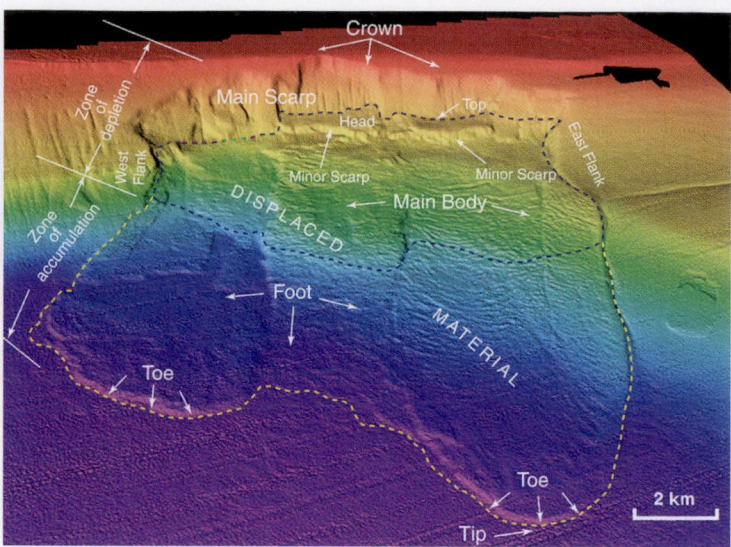

(b) A digital bathymetric model of a submarine slump along the coast of California. The parts of the slump are labeled. Blue is deeper, orange is shallower.

(c) A digital bathymetric model showing the slumps bordering the island of Oahu, Hawaii.

submarine slump can quickly displace a large area of the seafloor, it can trigger a tsunami as devastating as a tsunami caused by an earthquake (see Chapter 8).

Debris from countless slumps over millions of years has substantially modified the flanks of the Hawaiian Islands **(Fig. 12.15c)**. In fact, the cliff that follows the axis of the island of Oahu represents the head scarp of a giant slump. Debris from this slump spread eastward for a distance of 180 km (110 miles) on the floor of the Pacific Ocean. Huge slumps have also been mapped along the coasts bordering the Atlantic Ocean, indicating that passive-margin coasts are not immune to slumping. About 8,000 years ago, for example, the 100-km (60-mile)-wide Storegga Slide slipped down the face of the Atlantic passive margin, west of Norway, to produce a tsunami that washed away Stone Age villages around the coast of the North Sea. Debris from this slump traveled underwater for over 600 km (370 miles) from its source.

containing pebbles, cobbles, and boulders suspended in a mud matrix. And in a **turbidity current**, the moving sediment disperses in water to yield a turbulent cloud of suspended sediment that rushes downslope like an underwater avalanche **(Fig. 12.15a)**.

As revealed by shaded-relief maps of the seafloor, submarine slopes bordering both hot-spot volcanoes and active plate boundaries are scalloped by many immense slumps because earthquakes frequently jar these areas and set masses of material in motion **(Fig. 12.15b)**. Since a

Take-home message . . .

Mass-wasting events differ from one another based on the speed, content, and character of the moving material. Creep, slumping, and solifluction are slow. Mudflows and debris flows move faster, and avalanches and rockfalls move the fastest. Mass wasting occurs both on land and underwater.

Quick Question -
In what way is a snow avalanche like a turbidity current?

12.5 Why Does Mass Wasting Occur?

Mass wasting takes place when the rock or regolith underlying a slope has become weak enough, or the strength of attachment across a failure surface beneath the rock or regolith has diminished enough, that the pull of gravity can cause the material to start moving. If a slope's material becomes susceptible to movement, we say that the slope has become *unstable*. Let's look at various conditions that can make a slope unstable.

Weakening of the Substrate

If the Earth's surface exposed only fresh, intact bedrock, mass wasting would be of little concern, for rock has great strength and could easily hold up stalwart mountain cliffs. Fresh bedrock doesn't occur everywhere, however, for in many places, the ground surface overlies a substrate of regolith. Furthermore, intact rock is rare because in most places, joints or faults break rock into pieces **(Fig. 12.16)**. Regolith and fractured rock are much weaker than intact rock and can collapse in response to gravity. In other words, weathering, jointing, and faulting ultimately make mass wasting possible.

Why are regolith and fractured rock weaker than intact bedrock? The answer comes from looking at the strength of the attachments holding materials together. Intact bedrock is relatively strong because the chemical bonds within its interlocking grains, or within the cements between grains, can't be broken easily. A mass of regolith, in contrast, is relatively weak because the grains are held together by friction, by weak bonds between electrically charged surfaces of grains, or by bonds between water molecules, and none of these attachments can be as strong as the bonds holding atoms together in minerals. To picture this contrast, think about how much easier it is to destroy a sand castle than it is to destroy a sandstone sculpture.

Slope Failure

When material starts moving on an unstable slope, we say that **slope failure** has taken place. Whether or not a slope fails depends on the balance between two forces: the *downslope force*, caused by gravity, and the *resistance force*, which inhibits sliding. The resistance force comes from bonds holding the grains of the slope material together, or from friction across the failure surface. If the downslope force exceeds the resistance force, the slope fails, and mass wasting results **(Fig. 12.17a)**. Note that for a given mass, downslope force is greater on steeper slopes **(Fig. 12.17b)**.

In some locations, downslope movement begins on a weak failure surface beneath the ground surface. Geologists recognize several different kinds of weak surfaces that are likely to become failure surfaces, including: wet clay layers;

wet, unconsolidated sand layers; exfoliation joints; weak beds of shale; and metamorphic foliation planes **(Fig.12.18)**. Weak surfaces that are parallel to the land surface of a slope are particularly likely to fail because the downslope force is parallel to the weak surface. What triggers an individual mass-wasting event? In other words, what causes the balance of forces to change so that the downslope force exceeds the resistance force and causes a slope to suddenly fail? Here, we look at various phenomena—natural and human-caused—that can trigger slope failure.

GROUND SHAKING. An earthquake, a storm, the passage of a large truck, or a blast at a construction site may cause a mass on the verge of moving to begin moving. Ground

Figure 12.16 Jointing has broken up this thick sandstone bed along a cliff in Utah. Blocks of sandstone break free along joints and tumble downslope.

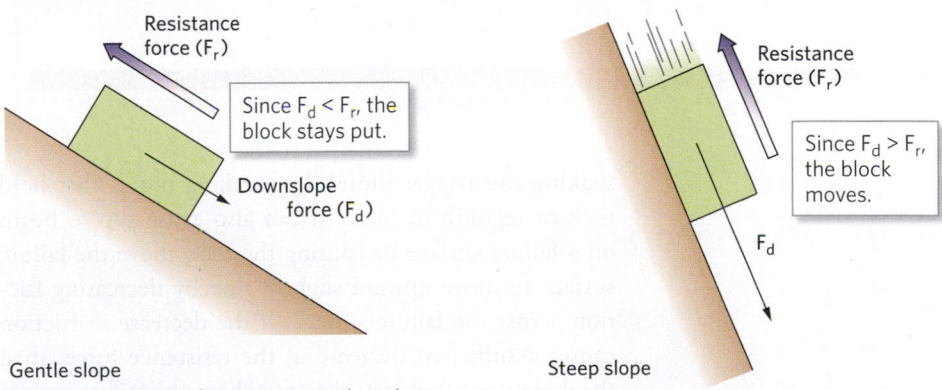

Figure 12.17 Downslope movement is determined by the balance between two forces.

Resistance force (F_r)

Since $F_d < F_r$, the block stays put.

Downslope force (F_d)

Gentle slope
(a) If the resistance force is greater than the downslope force, the block does not move.

Resistance force (F_r)

Since $F_d > F_r$, the block moves.

F_d

Steep slope
(b) If the slope angle increases, the downslope force increases. If the downslope force becomes greater than the resistance force, the block starts to move.

How can I explain . . .

The concept of slope stability

What are we learning?

How slope angle, and friction can affect slope stability.

What you need:

- A brick
- A rigid board (such as a cutting board)
- A protractor
- Medium-coarse sandpaper and tape

Sandpaper

10°

Instructions:

- Place the brick on the surface of the cutting board.
- Slowly lift one end of the board until its angle of slope is about 10°. Does the brick move?
- Continue raising the board, slowly, until the brick becomes unstable and slips. Measure the angle of slope.
- Repeat the tilting experiment, but this time, tilt the board to an angle that is less than the previous angle at which the brick slipped. At this angle, shake the board slightly. Can you get the brick to move?
- Tape the sandpaper to the board, place the brick on it, and press down.
- Repeat the tilting experiment. Measure the angle at which the brick becomes unstable.

What did we see?

- The surface of the board represents a failure surface. Tilting the board represents steepening of the slope. The brick may be stable on gentle slopes, but becomes unstable on steeper slopes.
- Shaking the board (which represents ground shaking) decreases the slope angle at which the brick can remain stable.
- Increasing friction across the surface by adding the sandpaper increases the slope angle at which the brick remains stable.

Figure 12.18 Several kinds of weak surfaces can become failure surfaces.

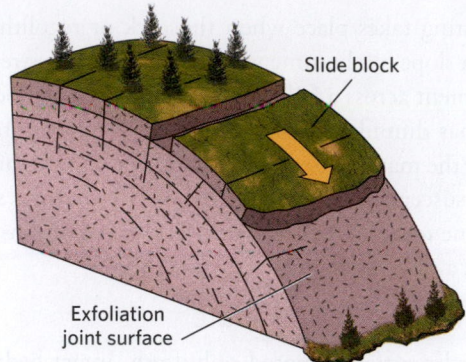

(a) Exfoliation joints, which form parallel to slope surfaces in granite, may become failure surfaces.

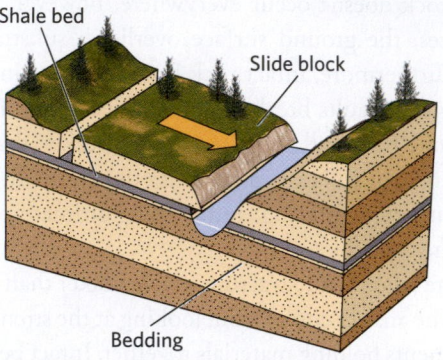

(b) In sedimentary rock, bedding planes (particularly in weak shale) may become failure surfaces.

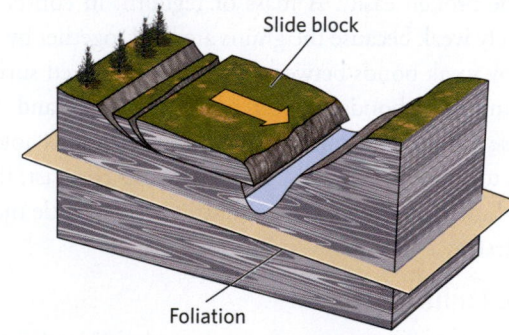

(c) In metamorphic rock, foliation planes (particularly in mica-rich schist) may become failure surfaces.

shaking can trigger sliding by breaking bonds that hold rock or regolith in place. It can also cause slip to begin on a failure surface by causing the rock above the failure surface to move upward slightly, thereby decreasing friction across the failure surface. If the decrease in friction causes a sufficient decrease in the resistance force, then the downslope force sets the mass above the failure surface in motion. Finally, ground shaking can cause liquefaction of wet sediment, either by increasing water pressure in the pores between grains so that the grains are pushed apart or by breaking the cohesion between the grains (see Chapter 8). When liquefaction takes place, the resistance force effectively disappears, so that material above the liquefied layer begins to move even on the slightest slope.

Because earthquakes occur most commonly along plate boundaries where mountain belts grow, steep slopes may exist near the epicenters of earthquakes. Therefore, unfortunately, large mass-wasting events frequently accompany major earthquakes, and these events can have dramatic consequences. For example, during a 1958 earthquake in

Figure 12.19 Events leading to the 1925 Gros Ventre Slide in Wyoming.

Rain

Trace of future scarp
Tensleep Formation
Amsden Shale
Gros Ventre River

Ventre Valley

At depth, the weak Amsden Shale was a potential slip surface because it is parallel to the slope.

Rain weakened the Amsden and made the Tensleep heavier. Downslope force caused a mass of rock to start moving.

Time

Slide debris
Lake

Scarp

The debris filled the valley, blocking a stream and forming Slide Lake. The scarp remained on the hillslope.

Slide scar

Slide debris

Photo of the slide and the lake it trapped

southeastern Alaska, an immense rockslide dumped debris into Lituya Bay and displaced water in the bay. As a result, an immense wave washed down the bay and out to sea. This wave washed forests off the slopes bordering the bay up to an elevation of 300 m (1,200 feet) and carried fishing boats anchored in the bay out to sea.

CHANGES IN SLOPE LOAD AND SUPPORT. Slope loads change when the weight of the material above a potential failure surface changes. If the load increases—for example, due to construction of buildings on top of a slope or due to saturation of regolith by heavy rains—the downslope force increases and may exceed the resistance force. Seepage of water into the ground may also weaken underground failure surfaces, further decreasing resistance force. Water seepage triggered the largest observed landslide in US history, the Gros Ventre Slide, which took place in 1925 on the flank of Sheep Mountain, near

Jackson Hole, Wyoming **(Fig. 12.19)**. Almost 40 million m³ (1.4 billion cubic feet) of rock, as well as the overlying soil and forest, detached from the side of the mountain and slid 600 m (2,000 feet) downslope, filling a valley and forming a 75-m (250 foot)-high natural dam across the Gros Ventre River.

Removal of support at the base of a slope by erosion or by human excavation plays a major role in triggering many slope failures. In effect, the material at the base of a slope acts like a retaining wall. Removing this "wall" allows the material farther up the slope to start sliding down. A disastrous slump triggered by river undercutting took place near Oso, Washington, in 2014 **(Fig. 12.20a, b)**. Over time, a small river ate away at the base of a forested slope until the slope gave way as a 400-m (1,300-foot)-wide slump that moved downslope, burying a housing development and highway. In some cases, erosion by a river or by waves eats into the base of a cliff and produces

Figure 12.20 The Oso, Washington, slump of 2014.

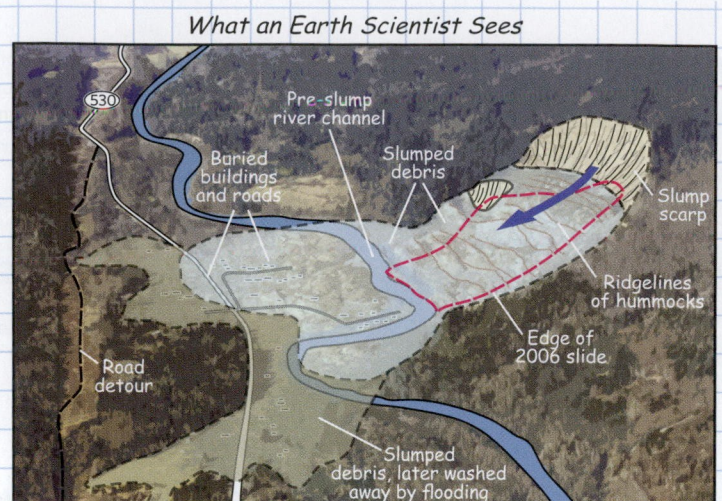

What an Earth Scientist Sees

(a) The slump transformed into a mudflow at the base of the hill, surged over the river, and tragically, buried a small community on the opposite bank.

(b) The mudslide happened when the mountain slope was undercut by a river.

an overhang. When such undercutting has occurred, the rock making up the overhang eventually breaks away from the slope and falls **(Fig. 12.21a, b)**.

CHANGES IN SLOPE STRENGTH. The stability of a slope depends on the strength of the material constituting it. If the material weakens with time, the slope becomes weaker and eventually collapses. Four factors influence the strength of slopes:

- *Weathering:* With time, chemical weathering produces weaker minerals, and physical weathering breaks rocks apart. When this happens, a formerly intact rock composed of strong minerals transforms into regolith. As we have seen, regolith is weaker than intact rock.

- *Vegetation cover:* In the case of slopes underlain by regolith, vegetation tends to strengthen the slope because the roots hold otherwise unconsolidated grains together. Furthermore, plants absorb water from the ground, keeping it from turning into slippery mud. The removal of vegetation therefore has the net result of making slopes more susceptible to downslope movement. Deforestation, either due to logging or fire, can lead to catastrophic mass wasting of the forest's substrate.

- *Water content:* Water affects slope stability in many ways. The attractive forces between molecules in the film of water on grain surfaces may help hold regolith together, but if the water content of the regolith increases too much, the regolith may turn into a muddy slurry that can be susceptible to flow. Also, if the regolith contains *expanding clay*—certain types of clay whose grains absorb water and swell up—the regolith can expand and, as a consequence, break up. Also, water infiltration may make failure surfaces underground more slippery by forming weak films on mineral surfaces, and the weight of added water can increase the downslope force.

- *Water pressure:* Addition of water at the ground surface can change the pressure of water in the rock of the failure surface. The increase in pressure effectively pushes grains apart along the failure surface, decreasing friction and, therefore, the resistance force along the failure surface. This phenomenon probably played a major role in triggering the Vaiont Dam disaster—filling the reservoir at the base of Monte Toc caused the water pressure in pores within the failure surface at the base of the mountain to increase.

Figure 12.21 Undercutting and collapse of a sea cliff.

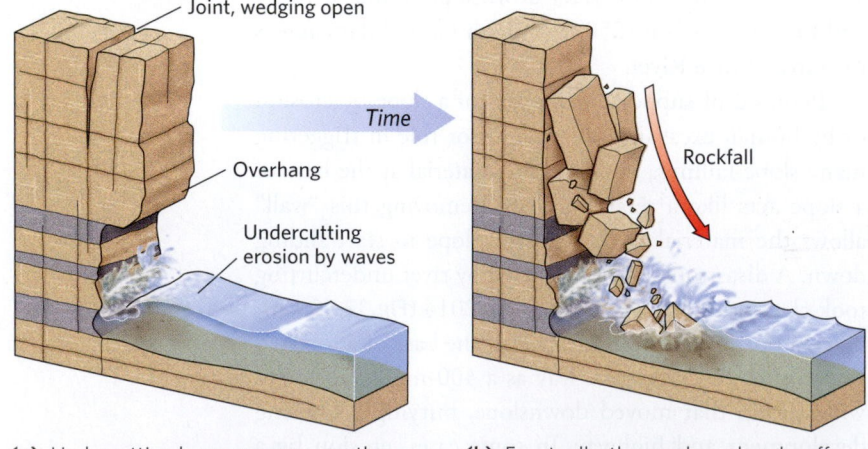

(a) Undercutting by waves removes the support beneath an overhang.

(b) Eventually, the overhang breaks off along joints, and a rockfall takes place.

Figure 12.22 Surface features warn that a large slump is beginning to develop. Cracks that appear at the head scarp may drain water and kill trees. Power poles tilt, and the lines become tight. Fences, roads, and houses on the slump begin to crack.

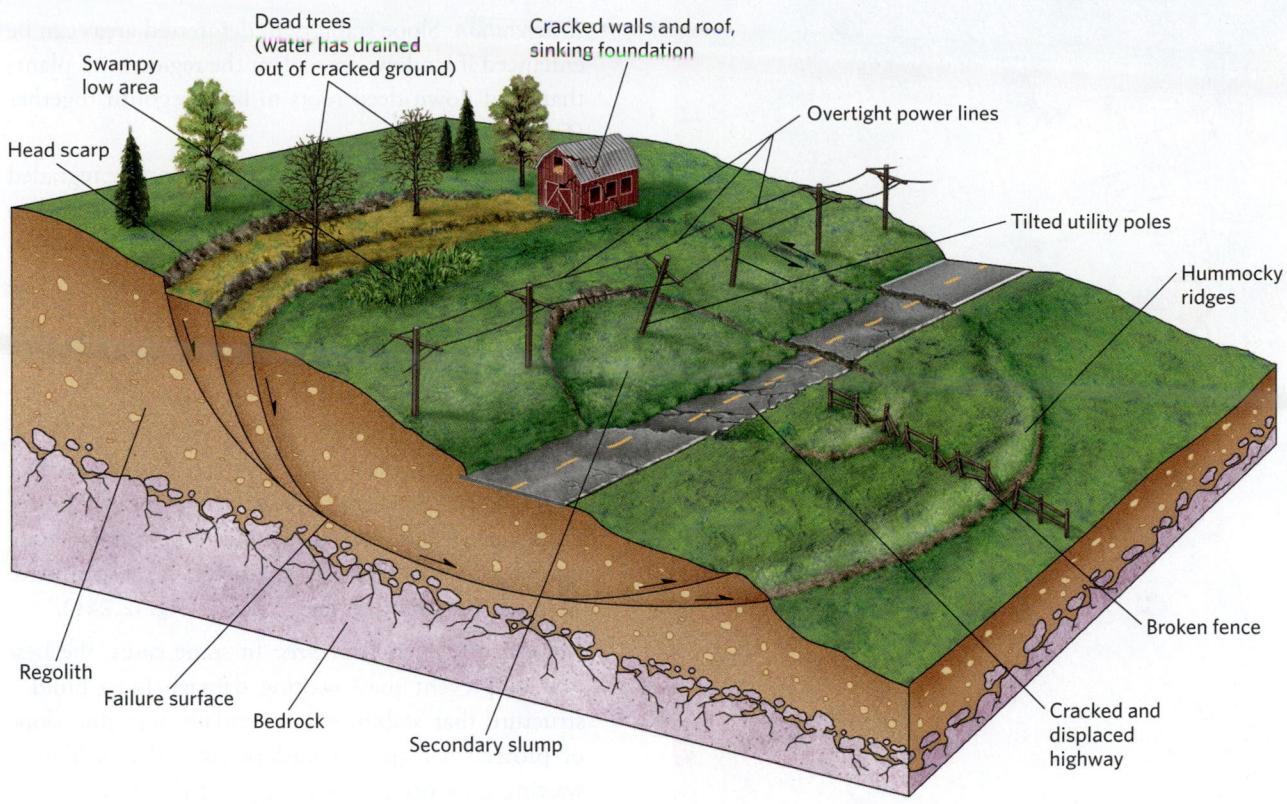

Dead trees
(water has drained
out of cracked ground)

Cracked walls and roof,
sinking foundation

Swampy
low area

Overtight power lines

Head scarp

Tilted utility poles

Hummocky
ridges

Broken fence

Regolith

Failure surface

Bedrock

Secondary slump

Cracked and
displaced
highway

Take-home message . . .

Weathering and fragmentation weaken slope materials and make them more susceptible to mass wasting. Slope failure occurs when downslope force exceeds resistance force due to ground shaking, changing slope angles and strength, changing water content, and changing slope support and loads.

Quick Question -
How does ground shaking trigger mass wasting?

12.6 Can We Prevent Mass-Wasting Disasters?

Clearly, landslides, mudflows, and slumps are natural hazards we cannot ignore. Too many of us live in regions where mass wasting has the potential to kill people and destroy property. In many cases, the best solution is avoidance—don't build, live, or work in an area where mass wasting can take place. But avoidance is possible only if we know where the hazards are.

Identifying Regions at Risk

To pinpoint dangerous regions, geologists look for landforms known to result from mass wasting, for where downslope movement has happened in the past, it might happen again in the future. Features such as head scarps, swaths of forest in which trees have been tilted, piles of loose debris at the bases of hills, and hummocky land surfaces all indicate recent mass wasting.

In some cases, geologists may also be able to detect slopes that are beginning to move **(Fig. 12.22)**. For example, roads, buildings, and pipes begin to crack over unstable ground. Power lines may be too tight or too loose because their support poles have moved together or apart, and trees may tilt. Visible cracks may form on the ground at the potential head of a slump, and the ground may bulge up at the toe of the slump. Locally, subsurface cracks may drain the water from an area and kill off vegetation, or subsidence may cause an area to become swampy.

In recent years, new, extremely precise laser-based surveying methods have permitted geologists to detect the beginnings of mass-wasting events that have not yet

Figure 12.23 A landslide-potential map of the western United States.

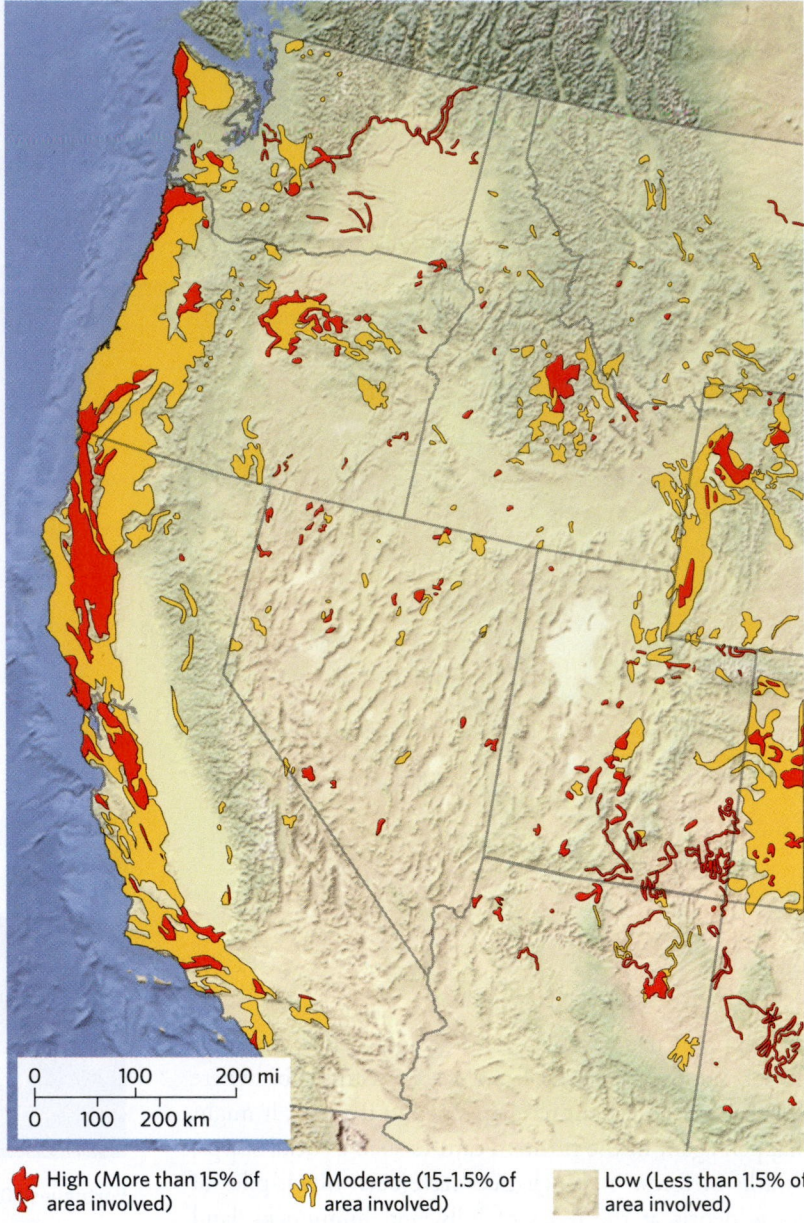

0   100   200 mi
0   100   200 km

High (More than 15% of area involved)
Moderate (15–1.5% of area involved)
Low (Less than 1.5% of area involved)

Preventing Mass Wasting

In areas where landslide potential exists, people can take certain steps to stabilize the slope or mitigate the hazard.

- *Revegetation:* Slope stability in deforested areas can be enhanced if landowners replant the region with plants that send down deep roots to bind regolith together **(Fig. 12.24a)**.

- *Regrading:* A dangerously steep slope can be regraded or terraced so that it does not exceed the angle of repose **(Fig. 12.24b)**.

- *Reducing subsurface water:* Slope instability may be decreased either by improving drainage, so that water does not enter the subsurface in the first place, or by extracting water from the ground **(Fig. 12.24c)**.

- *Preventing undercutting:* In places where a river undercuts a cliff face, engineers can divert the river **(Fig. 12.24d)**. Similarly, in coastal regions, they may build an offshore breakwater or pile riprap (loose boulders or concrete) along a beach to absorb wave energy before it strikes the cliff face **(Fig. 12.24e)**.

- *Constructing safety structures:* In some cases, the best way to prevent mass-wasting damage is to build a structure that stabilizes a potentially unstable slope or protects a region downslope from debris if mass wasting does occur. For example, civil engineers can build retaining walls or bolt loose slabs of rock to more coherent masses in the substrate in order to stabilize highway embankments **(Fig. 12.24f, g)**. The danger from rockfalls can be decreased by covering a roadcut with chain link fencing or by spraying it with concrete. Highways at the base of an avalanche-prone slope can be covered by an avalanche shed whose roof keeps debris off the road **(Fig. 12.24h)**.

- *Controlled blasting of unstable slopes:* When it is clear that unstable ground or snow threatens a particular region, the best solution may be to blast the unstable ground or snow loose at a time when its movement can do no harm.

Take-home message . . .

Various features of the landscape may help geologists to identify unstable slopes and estimate the hazard they pose. Systematic study allows production of landslide-potential maps. Engineers use a variety of techniques to stabilize slopes.

Quick Question --------------------------------
What clues indicate that a slump may be starting to form?

visibly affected the land surface. Geologists have also begun to identify hazardous areas by using computer programs to evaluate factors that trigger mass wasting, including slope steepness, substrate strength, degree of water saturation, orientation of potential failure surfaces relative to the slope, nature of vegetation cover, potential for heavy rains, potential for undercutting, and likelihood of earthquakes. From such hazard assessment studies, geologists compile **landslide-potential maps**, which rank regions according to the likelihood that mass wasting will occur **(Fig. 12.23)**.

Figure 12.24 A variety of remedial steps can stabilize unstable slopes or mitigate landslide hazards.

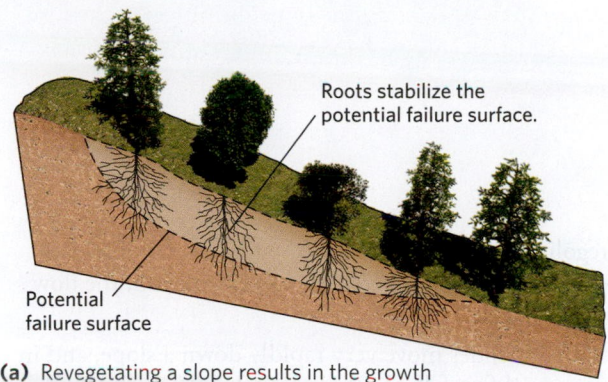

(a) Revegetating a slope results in the growth of roots that can hold the slope together.

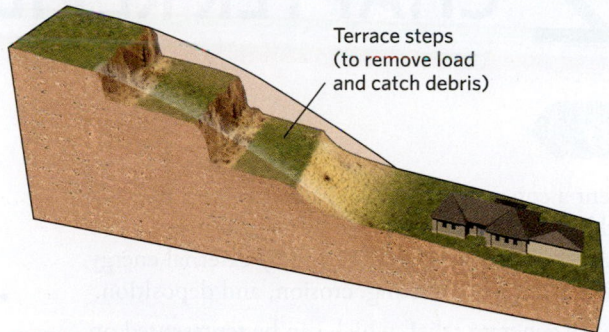

(b) Redistributing the mass on a slope can stabilize it. Terracing can help catch debris.

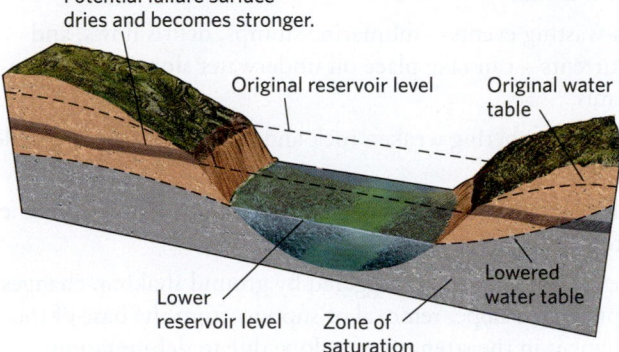

(c) Lowering the water table can strengthen a potential failure surface.

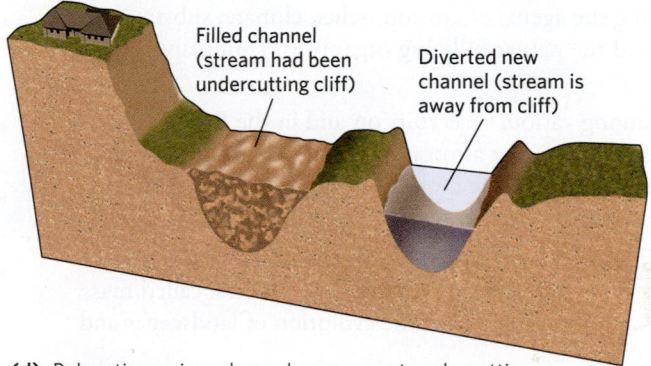

(d) Relocating a river channel can prevent undercutting.

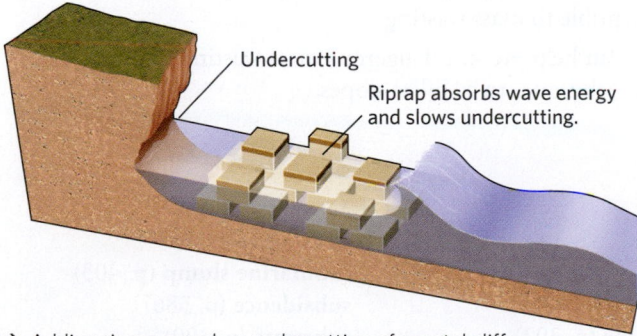

(e) Adding riprap can slow undercutting of coastal cliffs.

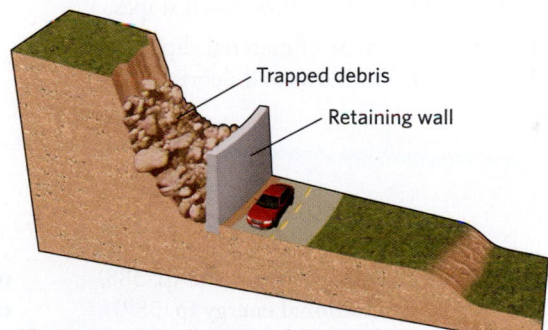

(f) A retaining wall can trap falling rock.

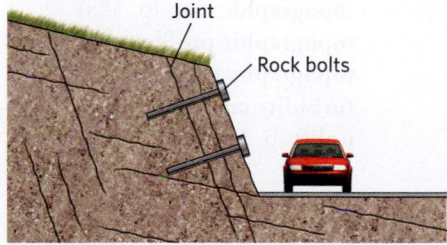

(g) Bolting or screening a cliff face can hold loose rocks in place.

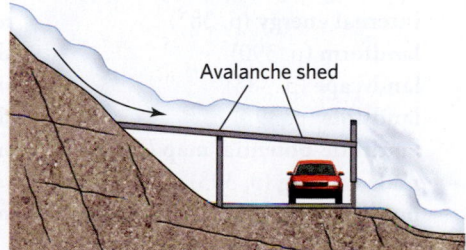

(h) An avalanche shed diverts debris or snow over a roadway.

- Landscapes represent a consequence of the battle between processes driven by the Earth's internal energy, which cause uplift or subsidence, and processes driven by gravity and by external energy from the Sun, which cause mass wasting, erosion, and deposition.

- Uplift and subsidence generate relief, which can be represented on topographic maps or digital elevation models.

- The types of landforms that develop at a location depend on many factors, including the agents of erosion, relief, climate, substrate composition, and the nature of living organisms. Landscapes change over time.

- Water moves among various reservoirs on and in the Earth (including oceans, rivers, the atmosphere, glaciers, groundwater, and life) during the hydrologic cycle. Without this movement, there would be no erosion or deposition by water, ice, or wind.

- Rock or regolith on unstable slopes has the potential to move downslope under the influence of gravity. This process, called mass wasting, plays an important role in the evolution of landscapes and can be a natural hazard.

- Slow mass wasting caused by the freezing and thawing of regolith is called creep. In places where slopes are underlain with permafrost, solifluction causes a melted layer of regolith to flow down slopes.

- During slumping, a relatively coherent mass of material slips down a spoon-shaped failure surface. Mudflows and debris flows occur where regolith has become saturated with water and moves downslope as a slurry. If the mud consists of volcanic ash, the flow is called a lahar.

- Rockslides and debris slides move very rapidly down a slope, and in a rockfall, the material free-falls down a vertical cliff.

- During snow avalanches, snow mixes with air and moves downslope as a turbulent cloud.

- Large mass-wasting events—submarine slumps, debris flows, and turbidity currents—can take place on underwater slopes. Some generate tsunamis.

- Fracturing and weathering weaken rock and make it more susceptible to mass wasting.

- Unstable slopes start to move when the downslope force exceeds the resistance force that holds material in place.

- Downslope movement can be triggered by ground shaking, changes in the steepness of a slope, removal of support from the base of the slope, or changes in the strength of a slope due to deforestation, weathering, or heavy rain.

- Geologists can sometimes detect unstable ground before it begins to move, and they can produce landslide-potential maps to identify areas susceptible to mass wasting.

- Engineers can help prevent dangerous mass wasting by using a variety of techniques to stabilize slopes.

angle of repose (p. 400)
contour line (p. 388)
creep (p. 393)
debris flow (p. 397)
deposition (p. 387)
depositional landform (p. 390)
digital elevation model (p. 388)
erosion (p. 387)
erosional landform (p. 390)
external energy (p. 387)
failure surface (p. 396)

geologic cross section (p. 388)
gravitational energy (p. 387)
groundwater (p. 392)
hydrologic cycle (p. 392)
internal energy (p. 387)
landform (p. 390)
landscape (p. 385)
landslide (p. 393)
landslide-potential map (p. 410)
mass wasting (p. 393)
mudflow (p. 397)

regolith (p. 392)
relief (p. 386)
residence time (p. 392)
rockfall (p. 400)
rockslide (p. 400)
shaded-relief map (p. 388)
slope failure (p. 405)
slump (p. 396)
snow avalanche (p. 403)
solifluction (p. 393)
submarine debris flow (p. 403)

submarine slump (p. 403)
subsidence (p. 386)
substrate (p. 390)
talus (p. 400)
topographic map (p. 388)
topographic profile (p. 388)
topography (p. 386)
turbidity current (p. 404)
uplift (p. 386)

Review Questions

The letters following each Review Question refer to the corresponding Learning Objective from the Chapter Opener.

1. Why does the Earth's surface constantly undergo change? **(A)**
2. Distinguish between internal and external sources of energy in the Earth System. **(B)**
3. What is the hydrologic cycle and what drives it? **(C)**
4. What factors do geologists use to distinguish among various types of mass-wasting events? **(D)**
5. Explain how soil creep operates. Explain the meaning of the arrows in the sketch. **(D)**

6. Identify the key differences between a slump, a debris flow, a lahar, an avalanche, a rockslide, and a rockfall. Which type of mass wasting does the sketch show? Identify parts of the sketch. **(D)**
7. Why is intact bedrock stronger than fractured bedrock? Why is it stronger than regolith? **(E)**

8. Explain the difference between a stable and an unstable slope. What factors determine the angle of repose of a material? What features are likely to serve as failure surfaces? **(E)**
9. Discuss the variety of phenomena that can cause a stable slope to become so unstable that it fails. Identify the forces in the sketch. **(E)**

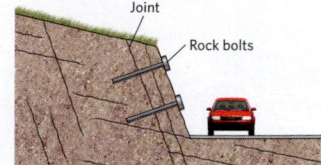

10. How can ground shaking cause fairly solid layers of sand or mud to become weak slurries capable of flowing? **(E)**
11. Discuss the role of vegetation and water in slope stability. Why can fires and deforestation lead to slope failure? **(E)**
12. What factors do geologists take into account when producing a landslide-potential map, and how can they detect the beginning of mass wasting in an area? **(F)**
13. What steps can people take to avoid landslide disasters? What is the purpose of the rock bolts shown in the sketch? **(F)**

On Further Thought

14. Imagine that you have been asked by a large bank to determine whether it makes sense to build a dam in a steep-sided, east-west-trending valley in a small central Asian nation. The local government has lobbied for the dam because the country's climate has gradually been getting drier, and the area's farms are running out of water. The bank is considering making a loan to finance dam construction, a process that would employ thousands of now-jobless people. Initial investigation shows that the rock of the valley wall consists of schist containing strong foliation that dips toward the valley and is parallel to the slope of the valley wall. Outcrop studies reveal abundant fractures in the schist along the valley floor. Moderate earthquakes have rattled the region. What would you advise the bank to do? Explain the hazards and what might happen if the reservoir were filled. **(F)**

Online Resources

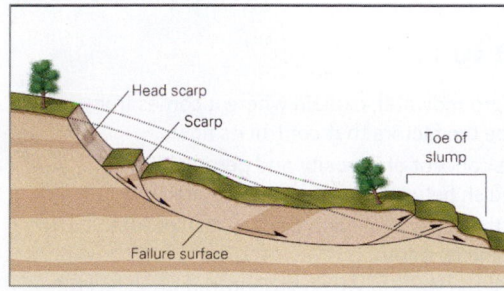

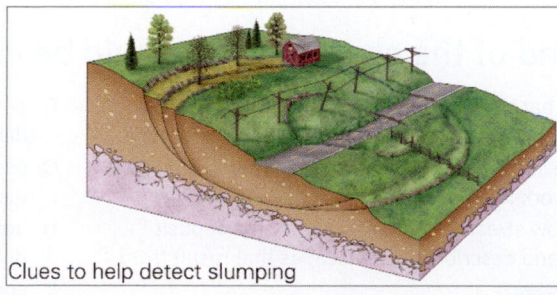

Clues to help detect slumping

Videos
This chapter features a video on developing and detecting slumps.

Smartwork5
This chapter includes questions on features and processes that change the landscape, on mass wasting, and on the water cycle.

⊘13 FRESHWATER
Streams, Lakes, and Groundwater

By the end of the chapter you should be able to . . .

A. recognize that the world's liquid freshwater resources are limited; they occur on the surface and underground.

B. explain the configuration of a drainage network and how these networks form and evolve.

C. describe how streams erode, transport, and deposit sediment, and describe the landscapes that result from these processes.

D. discuss the nature and causes of flooding and how society can protect against it.

E. describe the origin and fate of lakes and why most lakes are fresh while some are salty.

F. define groundwater, explain where it comes from, and describe the factors that control its flow.

G. use the concept of porosity and permeability to distinguish between aquifers and aquitards.

H. define the water table and discuss factors that determine its depth.

I. contrast the ways in which groundwater reaches the surface via wells and via springs.

J. discuss the origin of caves and karst terrains and their relationship to groundwater.

K. recognize sustainability and environmental issues that pertain to freshwater supplies.

13.1 Introduction

In the 1880s, developers built a mud-and-gravel dam across the Conemaugh River, in Pennsylvania, to trap a reservoir of cool water and provide a pleasant setting for summer homes for Pittsburgh's industrialists. Unfortunately, the dam had been poorly designed, and when torrential rain drenched the state on May 31, 1889, water in the reservoir rose so high that it began to flow over the dam. Eventually, the soggy structure collapsed, and a 20-m (65-foot)-high wall of water roared downstream and slammed into the town of Johnstown, transforming bridges and buildings into twisted wreckage and killing 2,300 residents **(Fig. 13.1)**.

Freshwater . . . it's a resource essential to life, a sculptor of the land surface, a means of transportation, a source of power, and, as the inhabitants of Johnstown sadly learned, a cause of destruction. By definition, **freshwater** contains less than about 500 ppm (0.05%) dissolved salts. It accounts for just 2.5% of the total volume of water in the Earth's surface and near-surface realms, but without it, life as we know it could not exist on land **(Fig. 13.2a)**. About 68.7% of the Earth's freshwater resides in glaciers as solid ice, and about 30.1% lies hidden beneath the surface as groundwater. So the water that we actually see—in lakes, streams, swamps, soil, snow, permafrost, the atmosphere, and organisms—surprisingly accounts for only 1.2% of the total **(Fig. 13.2b)**.

In this chapter, we discuss the Earth's liquid freshwater. We begin by focusing on streams to see how they drain the land, how they modify the surface landscape, and why they sometimes cause devastating floods. Next, we examine lakes and learn why most are fresh, but some

The waters of Catskill Creek, in eastern New York State, spill over a small waterfall. This water will eventually drain into the Hudson River, which will carry the water into the Atlantic Ocean. Rivers serve an important role in the Earth's hydrologic cycle.

Figure 13.1 In 1889, raging floodwaters destroyed Johnstown, Pennsylvania.

are not. We then head below the Earth's surface to understand where groundwater comes from, how it flows, and why it sometimes carves caves. This chapter concludes with a look at challenges pertaining to the sustainability of freshwater resources.

13.2 Draining the Land

The Formation of Streams

During the hydrologic cycle, water that enters the atmosphere by evaporation eventually condenses and

Figure 13.2 Amount and distribution of the Earth's freshwater.

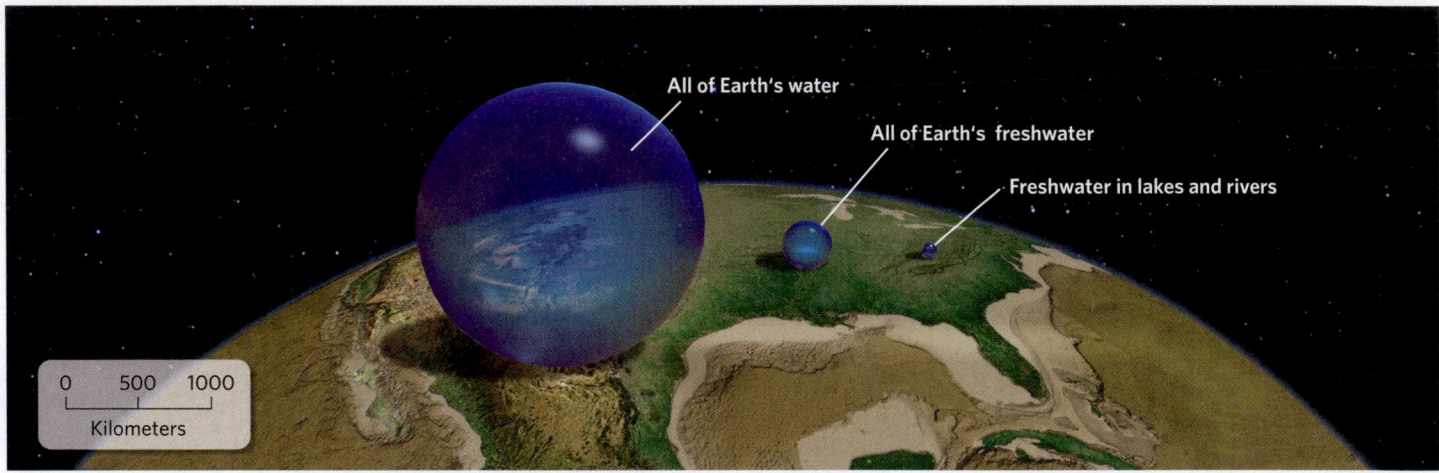

(a) All the water in the world could be collected in a sphere with a diameter of 1,300 km (860 miles). The freshwater portion could be collected in a sphere with a diameter of only 275 km (170 miles).

(b) Most water on the Earth is saltwater. Of the freshwater on the Earth, most is frozen, or underground.

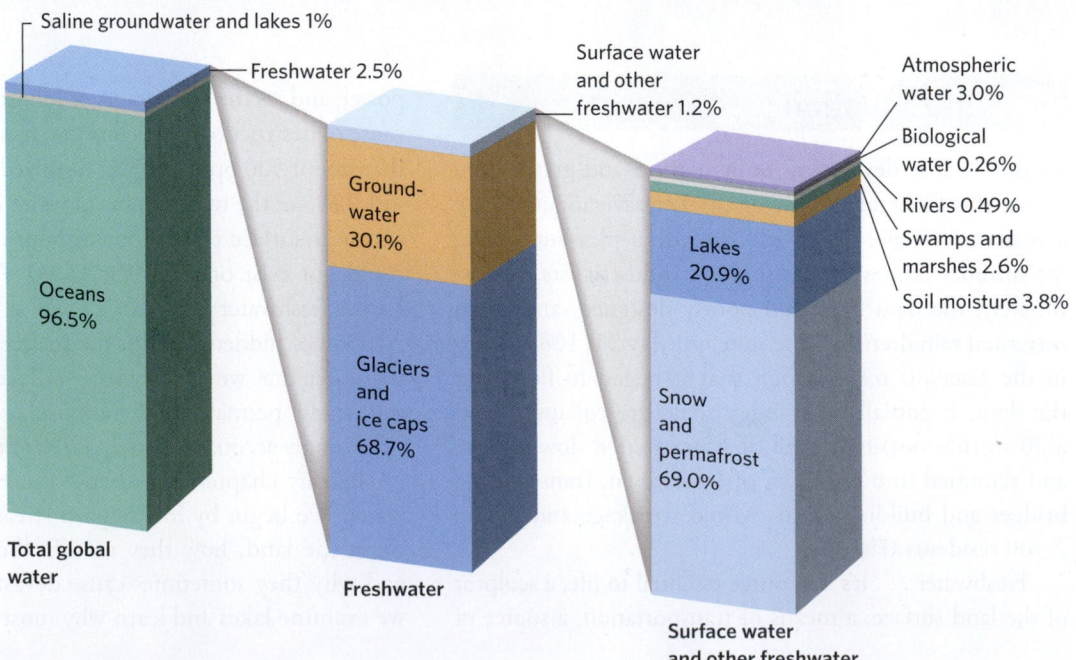

Figure 13.3 The formation of stream channels.

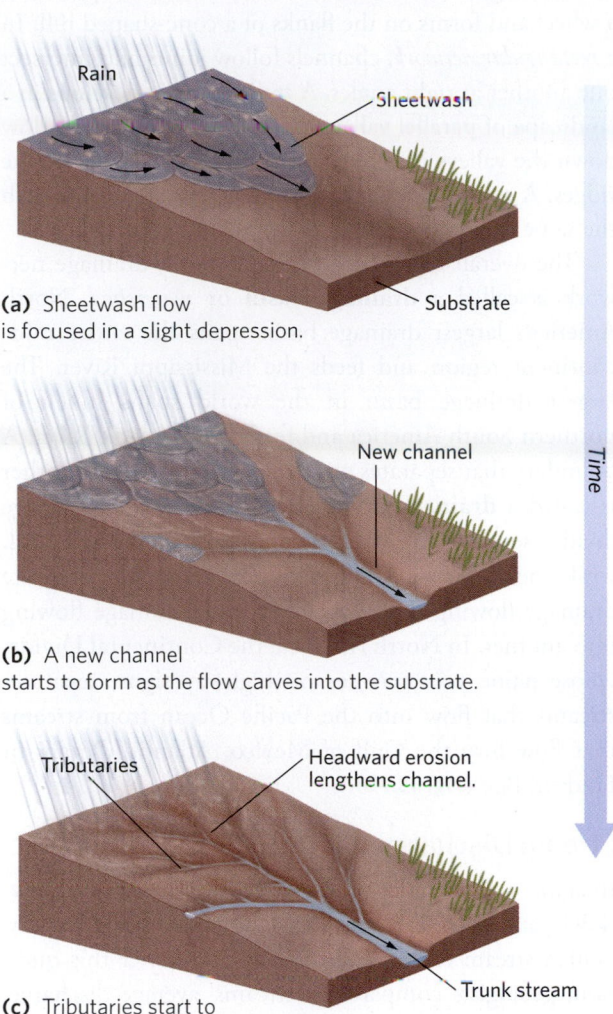

(a) Sheetwash flow is focused in a slight depression.

(b) A new channel starts to form as the flow carves into the substrate.

(c) Tributaries start to develop and feed into the trunk stream.

falls back to the Earth's surface as rain or snow. Some of this *precipitation* accumulates on the land surface as puddles, swamps, lakes, or snowfields, and some sinks into the ground to become soil moisture or groundwater. Gravity causes water that hasn't sunk into the ground or accumulated in a standing body to move downslope as **runoff (Fig. 13.3a)**. When runoff starts flowing, it does so as a thin film called *sheetwash*. The movement of water in sheetwash starts to erode its substrate. In nature, the ground is not perfectly flat, and not all substrate has the same resistance to erosion. Where the substrate is a little weaker, or the flow happens to be a bit faster, the moving water *scours* (digs down into) the substrate to create a trough-like depression called a **channel (Fig. 13.3b)**. Because the channel is lower than the surrounding ground, sheetwash in adjacent areas curves toward the channel, in which the flow becomes stronger. As a result, scouring happens faster in the channel, and the channel gets deeper relative to its surroundings, a process called **downcutting**.

Geologists commonly use the term **stream** for any water flowing down a channel, though in everyday English, we refer to large streams as *rivers* and to medium-sized ones as creeks or brooks. As downcutting deepens a channel, the surrounding land surfaces start to slope toward the channel so that new side channels, or **tributaries**, begin to form and flow into a *trunk stream* in the original channel **(Fig. 13.3c)**.

Streams receive water from many sources in addition to sheetwash **(Fig. 13.4)**. These sources include melting ice and snow, rain, soil moisture, and springs (outlets by which groundwater returns to the surface). Over time, a stream channel may begin to lengthen at its origin, or *headwaters*, by a process called **headward erosion**

Figure 13.4 Excess surface water (runoff) comes from rain, from melting ice and snow, and from springs. On flat ground, water accumulates in puddles or swamps, but on slopes it flows downslope in streams.

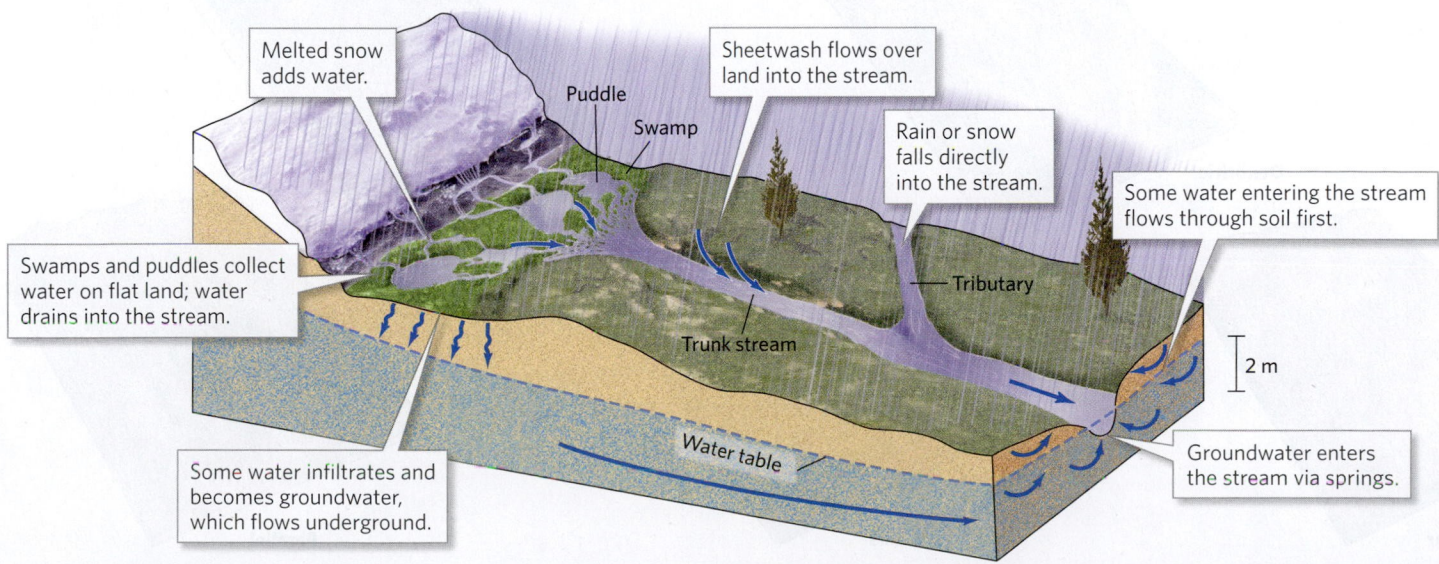

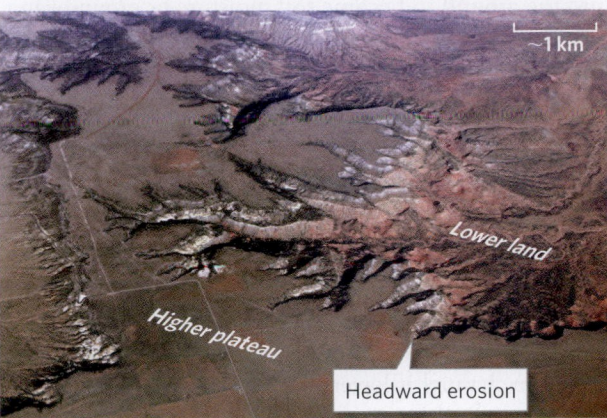

Figure 13.5 An example of headward erosion, as seen from an airplane.

~1 km

Lower land

Higher plateau

Headward erosion

(Fig. 13.5). This process can happen both because sheetwash converging at the entrance to a channel can scour the channel efficiently, and because groundwater seeping out of the ground at the headwaters of the stream can weaken the substrate and cause it to collapse into the channel, where it eventually washes away.

Developing Drainage Networks

Eventually, a **drainage network**, consisting of tributaries draining into a trunk stream, evolves. Such networks reach into all corners of a region, providing conduits for the removal of runoff. Geologists recognize several types of drainage networks on the basis of the network's map pattern **(Fig. 13.6)**. A *dendritic network*, which looks like the pattern of branches on a tree, forms where the sub-

strate consists of horizontal layers of material with uniform strength. A *radial network* looks like the spokes on a wheel and forms on the flanks of a cone-shaped hill. In a *rectangular network*, channels follow joints and intersect one another at right angles. A *trellis network* develops in a landscape of parallel valleys and ridges as tributaries flow down the valleys into a trunk stream that cuts across the ridges. A *parallel network* consists of several streams with the same orientation flowing down a uniform slope.

The overall region drained by a given drainage network is called a **drainage basin** or *watershed*. North America's largest drainage basin spans the entire mid-continent region and feeds the Mississippi River. The largest drainage basin in the world spans much of northern South America and feeds the Amazon River. A boundary that separates one drainage basin from another is called a **drainage divide (Fig. 13.7a)**. Small drainage divides separate the streams on opposite sides of a hill, while the largest ones, called *continental divides*, separate drainage flowing into one ocean from drainage flowing into another. In North America, the Continental Divide, whose name you may see on highway signs, separates streams that flow into the Pacific Ocean from streams that flow into the Gulf of Mexico, Atlantic Ocean, or Hudson Bay **(Fig. 13.7b)**.

Stream Discharge and Turbulence

Imagine two streams, a larger one in which water flows slowly and a smaller one in which water flows rapidly. Which stream carries more water? To answer this question, geologists compare the streams' average discharge. Technically, the **discharge** of a stream is the volume of water that passes through a cross section of the stream in

Figure 13.6 Five types of drainage networks.

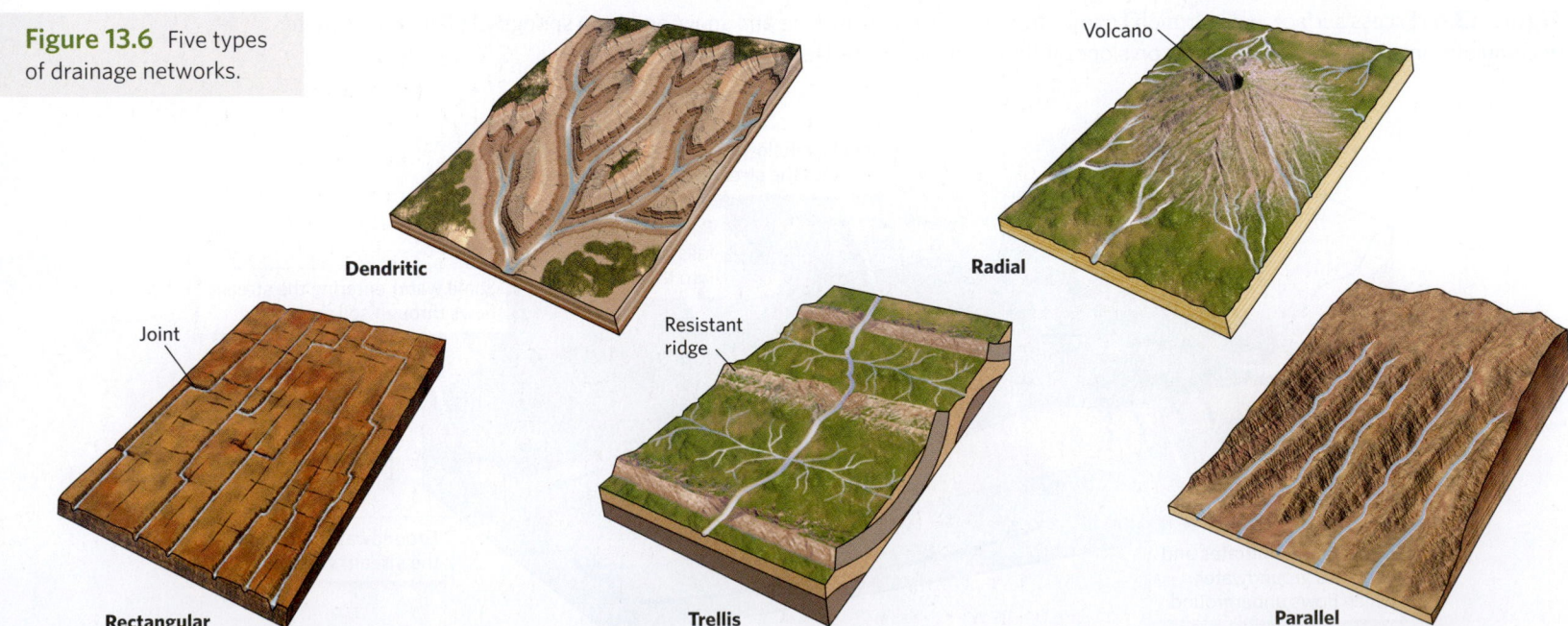

Dendritic

Volcano

Radial

Joint

Rectangular

Resistant ridge

Trellis

Parallel

Figure 13.7 Drainage divides and basins.

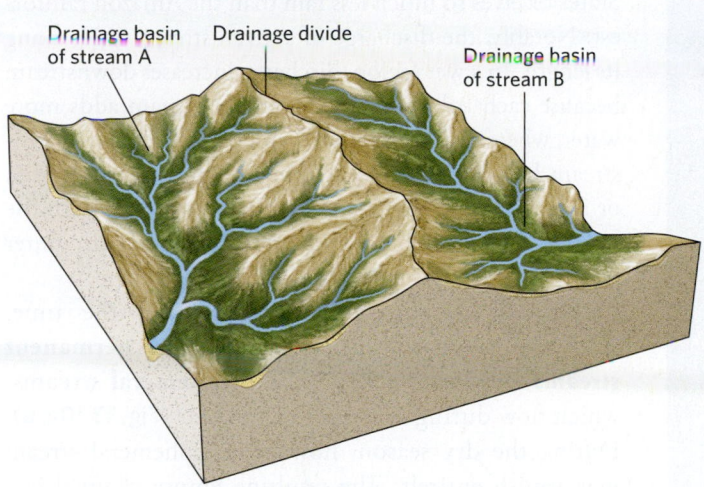

(a) A drainage divide is a ridge that separates two drainage basins.

(b) The major drainage basis of North America. The Great Basin is a region of interior drainage with no outlet to the sea.

a given time. We can calculate discharge by using a simple formula: $D = A \times v$. In this formula, A is the cross-sectional area of the stream, and v is the average velocity at which water moves in the downstream direction. Note that we specify discharge in units of volume (cubic meters or cubic feet) per unit of time.

The average velocity of stream water (v) can be difficult to calculate because water in a stream doesn't all travel at the same velocity. Why? First, friction slows water flow, so water near the channel walls or the *streambed* (the floor of the channel) moves more slowly than water in the middle of the flow **(Fig. 13.8a)**. In fact, the fastest-moving water of the stream lies near the surface in the center of the channel. Second, **turbulence**—the twisting,

swirling motion of a fluid—produces *eddies* in which the water flow direction curves. In an eddy, water may flow upstream or even follow a circular path **(Fig. 13.8b)**.

A stream's average discharge reflects the size of its drainage basin and the climate where the stream flows. For example, the Amazon River, which drains a huge rainforest, has the largest average discharge in the world—about 210,000 m³ (7.4 million cubic feet) per second

Figure 13.8 Flow velocity and its measurement in streams.

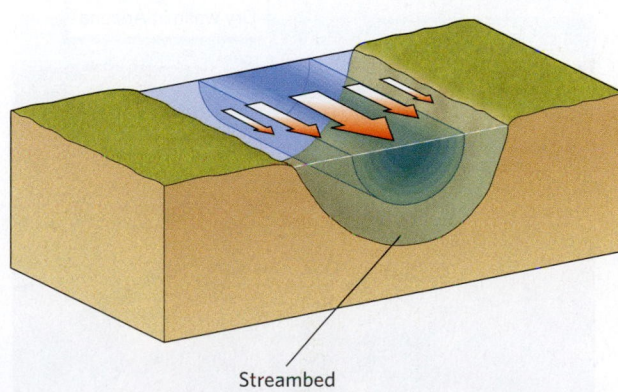

(a) At a stream-gaging station, geologists measure the cross-sectional area (*A*), the depth, and the average velocity of the stream. Velocity is slower near the banks and the streambed.

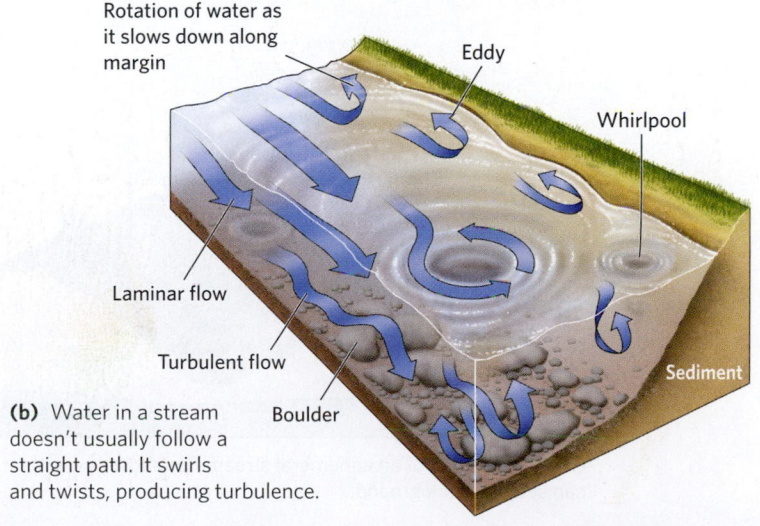

(b) Water in a stream doesn't usually follow a straight path. It swirls and twists, producing turbulence.

Figure 13.9 The Amazon drainage basin of South America stretches from the crest of the Andes to the Atlantic Ocean. It contains a dendritic drainage network and flows east. The Amazon River has the largest discharge of any river in the world.

(Fig. 13.9). The Mississippi River's drainage basin is half the area of the Amazon's, but its discharge is only about 8% of the Amazon's discharge because the central United States receives so much less rain than the Amazon rainforest. Notably, the discharge of a given stream varies along its length. In a wet region, discharge increases downstream because each tributary that enters the stream adds more water, whereas in a dry region, discharge decreases downstream because water seeps into the ground, evaporates, or is removed by humans. Discharge also varies with the seasons—it tends to be greatest in spring, when winter snows are melting, or during the rainy season.

Not all streams discharge water all the time. Therefore, geologists distinguish between **permanent streams**, which flow all year, and **ephemeral streams**, which flow during only part of the year **(Fig. 13.10a, b)**. During the dry season, flow in an ephemeral stream may vanish entirely. The resulting empty channel is a *dry wash*.

Figure 13.10 The contrast between permanent and ephemeral streams.

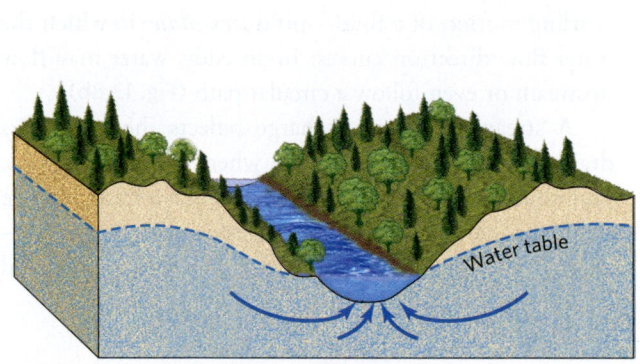

Permanent stream in Wyoming

(a) The bed of a permanent stream in a temperate climate lies below the water table. Springs add water from below, so the stream contains water even between rains.

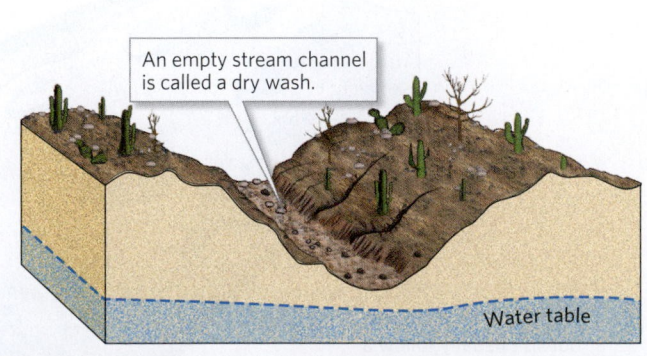

An empty stream channel is called a dry wash.

Dry wash in Arizona

(b) The channel of an ephemeral stream lies above the water table, so the stream flows only when water enters the stream faster than it can seep into the ground.

Figure 13.11 Erosion by streams.

(a) This slot canyon in Arizona was formed by abrasion.

Pothole

(b) A whirlpool formed this pothole in the bed of a stream near Ithaca, New York.

Take-home message . . .

Stream channels form by downcutting and lengthen by headward erosion. Drainage networks, composed of tributaries linked to a trunk stream, carry water from a drainage basin and have a variety of different geometries. Stream discharge, the amount of water passing through a cross section of the stream in a given time, depends on factors such as drainage basin area and climate.

Quick Question -
Is the velocity of stream flow the same everywhere in a stream?

13.3 The Work of Running Water

How Do Streams Erode?

If you spray the ground with a hose and watch the water dig into the soil and carry it away, you are seeing erosion by running water. In the case of a natural stream, gravity causes water to flow downslope and erode the Earth's surface. Erosion by streams takes place in four ways:

- *Scouring:* Running water picks up and carries away loose clasts, such as sand, pebbles, or cobbles, a process called *scouring.*

- *Breaking and lifting:* The force applied by running water can break off or lift chunks of solid rock from the channel floor or walls.

- *Abrasion:* Sediment-laden water acts like sandpaper and rasps away at the channel floor and walls, a

process called **abrasion** (Fig. 13.11a). In places where turbulence produces long-lived circular eddies, or *whirlpools*, abrasion by sand or gravel carves a bowl-shaped depression, called a **pothole**, into the stream-bed (Fig. 13.11b).

- *Dissolution:* Running water dissolves soluble minerals from the channel walls and floor and carries the dissolved ions away in solution.

The efficiency of erosion depends on the velocity and volume of the stream and on its sediment content. A large volume of fast-moving, turbulent, sandy water causes more erosion than a trickle of quiet, clear water. As a result, erosion happens much more rapidly during floods.

How Do Streams Transport Sediment?

The Mississippi River received the nickname "Big Muddy" for a reason: its water tends to be chocolate brown because of all the clay and silt it carries. A stream's **sediment load**, meaning the total volume of sediment carried by a stream, includes three components (Fig. 13.12): (1) *dissolved load*, consisting of ions in solution; (2) *suspended load*, consisting of tiny (silt- or clay-sized) grains that swirl along with the water without settling to the streambed; and (3) *bed load*, consisting of relatively large clasts that bounce or roll along the streambed.

When describing a stream's ability to carry sediment, geologists specify its competence and capacity. The **competence** of a stream refers to the maximum particle size it carries—a stream with more competence can carry large particles, whereas one with less competence can carry only small particles. Competence depends on water velocity, so a fast-moving, turbulent stream has

Figure 13.12 Streams transport sediment in many forms: dissolved ions in solution; tiny suspended grains distributed throughout the water; and larger clasts that slide or roll along the streambed.

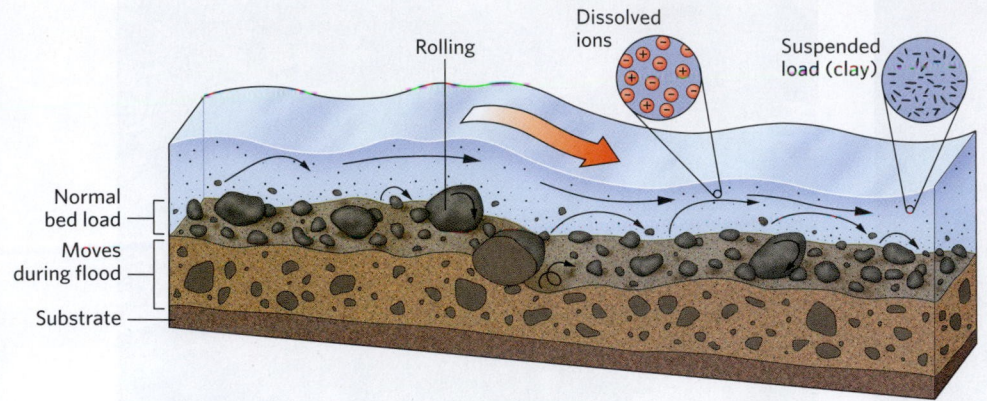

stream varies with location, streams tend to sort clasts by size (see Chapter 5).

Sediments deposited by a stream are called *fluvial deposits* (from the Latin *fluvius*, meaning river) or **alluvium (Fig. 13.13a)**. Coarser alluvium may accumulate along the streambed in elongate mounds, known as **bars (Fig. 13.13b)**. During floods, a stream may overtop the sides, or *banks*, of its channel and spread out over its **floodplain**, a broad, flat area bordering the stream. Friction slows the water on the floodplain, so a sheet of finer-grained alluvium settles out, forming *floodplain deposits*. Where a stream empties into a standing body of water, the water slows, and a wedge of sediment, called a *delta*, may accumulate. We'll discuss floodplains and deltas in more detail in Section 13.4.

How Do Streams Change along Their Length?

more competence than does a slow-moving stream. The **capacity** of a stream refers to the total quantity of sediment it can carry. A stream's capacity depends on both competence and discharge, so a large, fast-flowing river has more capacity than a small, slowly flowing creek.

Depositional Processes

If the flow velocity of a stream decreases, then the competence of the stream also decreases, and its sediment load settles out. The sizes of the clasts that settle at a particular location depend on the flow velocity at that location. Thus, if the stream slows by a small amount, only large clasts settle, but if the stream slows by a lot, medium-sized clasts settle, and if the stream slows almost to a standstill, fine clasts settle. Since the velocity of a

In 1803, the United States bought the Louisiana Territory, a vast tract of land encompassing the western half of the Mississippi drainage basin. President Jefferson asked Meriwether Lewis and William Clark to lead a voyage of exploration from St. Louis to the Pacific and to make a map of what they saw along the way. Lewis and Clark, together with about 40 men, began their expedition at the outlet, or *mouth*, of the Missouri River, where it joins the Mississippi. At this juncture, the Missouri is a wide, languid stream of muddy water. The group found that near its mouth, the river's channel was deep and the water easily navigable. But the farther upstream they went, the more difficult their journey became, for the **stream gradient**, meaning the slope of the stream's surface, became progressively steeper, and the stream's discharge diminished. When Lewis and

Figure 13.13 Examples of alluvium deposited by streams.

(a) Gravel in the bed of a mountain stream in Denali National Park, Alaska. The large clasts were carried during floods.

(b) Bars of sediment deposited by a stream in the Canadian Rockies.

Figure 13.14 The character of a stream changes along its longitudinal profile. A drainage network collects water from a broad drainage basin via numerous tributaries, which carry water to a trunk stream. Points 1 to 5 refer to locations along the longitudinal profile (inset). In general, the longitudinal profile of a stream is concave up.

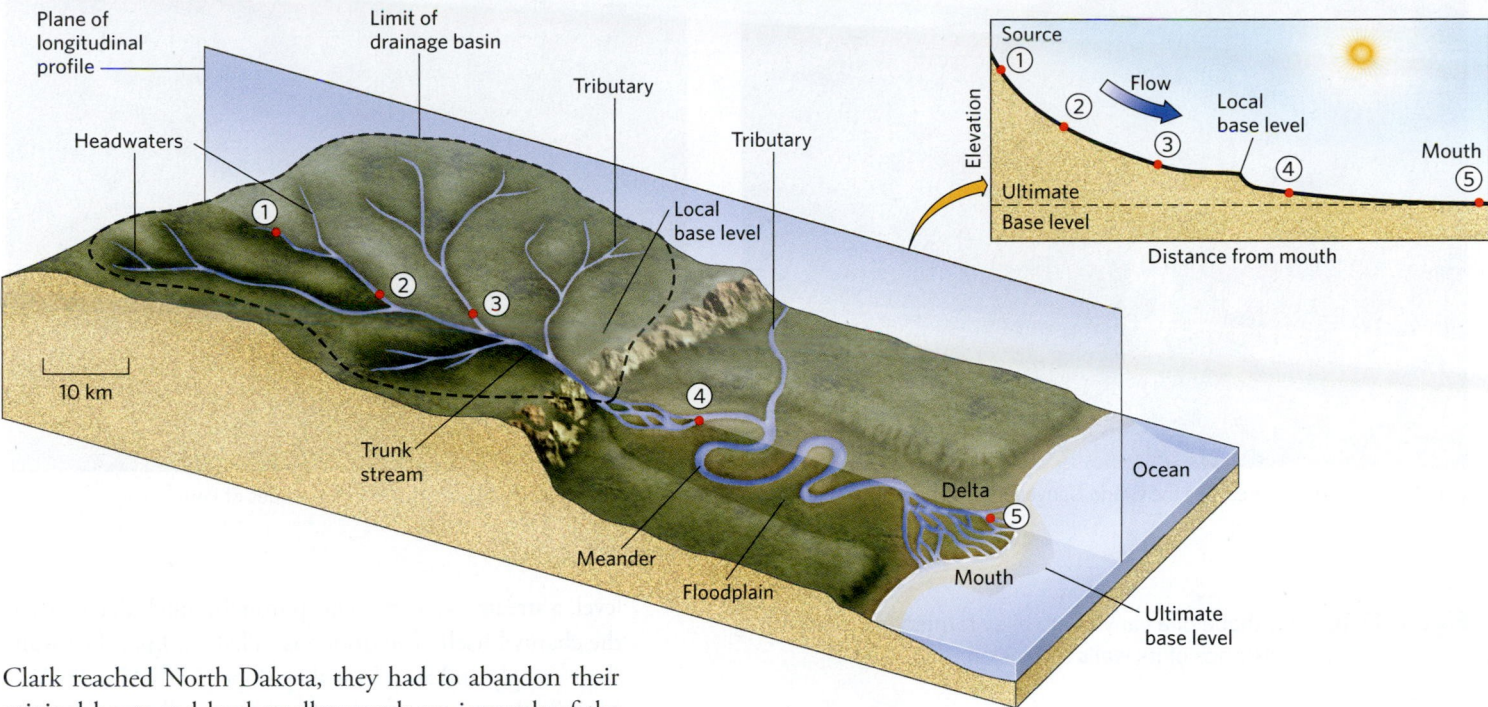

Clark reached North Dakota, they had to abandon their original boats and haul smaller vessels up intervals of the stream where fast, turbulent water plunged and swirled over a rocky bed. When they reached Montana, they abandoned these boats as well and trudged along the stream's banks on foot or on horseback, struggling up steep gradients until they reached the Continental Divide.

Modern geologists represent the change in gradient that Lewis and Clark experienced on a graph called a *longitudinal profile*, which plots elevation on the vertical axis and distance from the mouth on the horizontal axis **(Fig. 13.14)**. An idealized longitudinal profile of a major river is a concave-up curve, emphasizing that streams tend to have steep gradients near their headwaters and gentle gradients near their mouths. Longitudinal profiles of real rivers are not perfectly smooth curves, but rather display little plateaus and steps, representing interruptions by lakes or waterfalls.

The lowest elevation to which a stream can downcut is the **base level** of the stream. A *local base level* is one that occurs at a location upstream of a drainage network's mouth. Lakes or reservoirs can act as local base levels along a stream, and where a tributary joins a larger stream, the surface of the larger stream acts as the base level for the tributary. The *ultimate base level* of a drainage network is the standing body of water at the mouth of the trunk stream. For streams that flow into the ocean, sea level defines the ultimate base level. The surface of such a stream cannot be lower than sea level, for if it were, the stream would have to flow upslope to enter the sea.

Take-home message . . .

Streams erode their substrate and transport and deposit sediment. Competence, a measure of the maximum clast size that a stream can carry, depends on flow velocity, so where the velocity decreases, sediment settles out. The slope of a stream tends to decrease downstream. A stream can cut down to its base level at a locality.

Quick Question -
What's the difference between the competence and the capacity of a stream?

13.4 Streams and Their Deposits in the Landscape

Valleys and Canyons

About 17 million years ago, a large block of crust in the region of the southwestern United States began to rise, eventually becoming the Colorado Plateau. As the land rose, the Colorado River cut down into the landscape and produced the Grand Canyon, whose floor lies 1.6 km (1 mile) below the surface of the plateau **(Fig. 13.15a)**. The formation of the Grand Canyon emphasizes that in regions where the land surface lies well above the base

See for yourself

River-cut gorge in the Himalayas

Latitude: 28°9′41.82″ N
Longitude: 85°23′1.04″ E

Zoom to 6 km (~4 miles) and look obliquely to the NE.

A deep gorge, about 50 km (31 miles) north of Katmandu, was cut by a river flowing out of the Himalayas. This upper reach of the river has a steep gradient. If you look straight upstream, you'll note the valley's V-shaped profile.

Figure 13.15 Examples of landscapes cut by streams.

(a) The Colorado River carved the Grand Canyon, here seen at its western end.

(b) A deep valley cut by a stream in the Andes of Peru.

Figure 13.16 The shape of a canyon or valley formed by downcutting depends on the resistance of its walls to erosion.

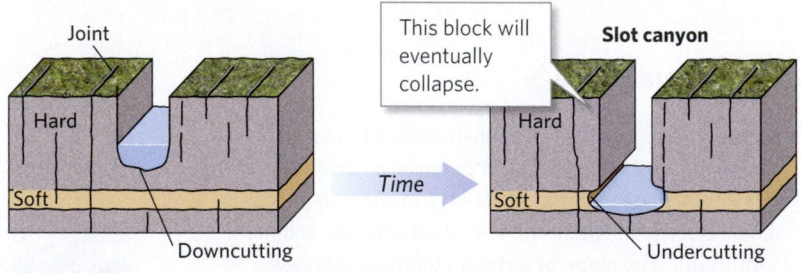

(a) If downcutting by the stream happens faster than mass wasting on the walls, a slot canyon forms. The canyon widens as the stream undercuts the walls.

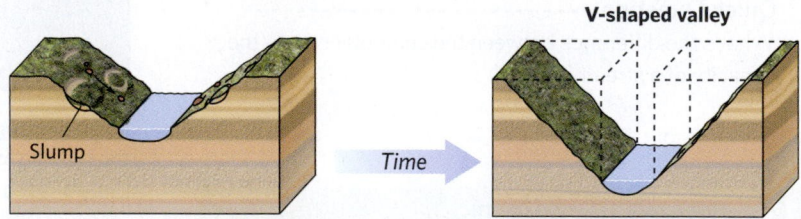

(b) If mass wasting takes place as fast as downcutting occurs, a V-shaped valley develops.

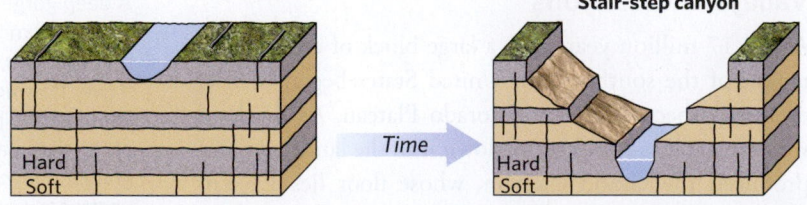

(c) Downcutting through alternating hard and soft layers produces a stair-step canyon.

level, a stream can carve a deep trough, much deeper than the channel itself. The trough is called a *canyon* if its walls slope steeply and a *valley* if they slope gently (**Fig. 13.15b**).

Whether stream erosion produces a valley or a canyon depends on the rate at which downcutting takes place relative to the rate at which mass wasting causes the walls on either side of the stream to collapse. In places where a stream downcuts its substrate faster than the walls collapse, erosion yields a canyon (**Fig. 13.16a**), whereas in places where the walls collapse by mass wasting as fast as the stream downcuts, the slope of the walls tends to be gentler, resulting in a valley (**Fig. 13.16b**). Where a stream cuts through alternating layers of resistant and nonresistant rock, the walls of the resulting canyon have a stair-step shape (**Fig. 13.16c**).

Rapids and Waterfalls

Rafters and kayakers seek out **rapids**, intervals of a stream where the water surface has become particularly turbulent and rough (**Fig. 13.17a**). Rapids can form where water flows over steps or boulders in the streambed, where the channel abruptly narrows, or where its gradient abruptly changes. Turbulence in rapids produces eddies and waves that roil and churn the water surface, in the process creating *whitewater*, a mixture of bubbles and water.

A **waterfall** forms where the gradient of a stream becomes so steep that some or all of the water free-falls above the streambed (**Fig. 13.17b**). The energy of the falling water may excavate a depression, called a *plunge pool*, at the base of the waterfall. Though a waterfall may appear

Figure 13.17 Rapids and waterfalls.

(a) These rapids in the Grand Canyon formed when a flood from a side canyon dumped debris into the channel of the Colorado River.

(b) Iguaçu Falls, at the Brazil-Argentina border, spills across layers of basalt. The basalt acts as a resistant ledge.

Figure 13.18 The geology of Niagara Falls.

(a) The Horseshoe Falls, a part of Niagara Falls.

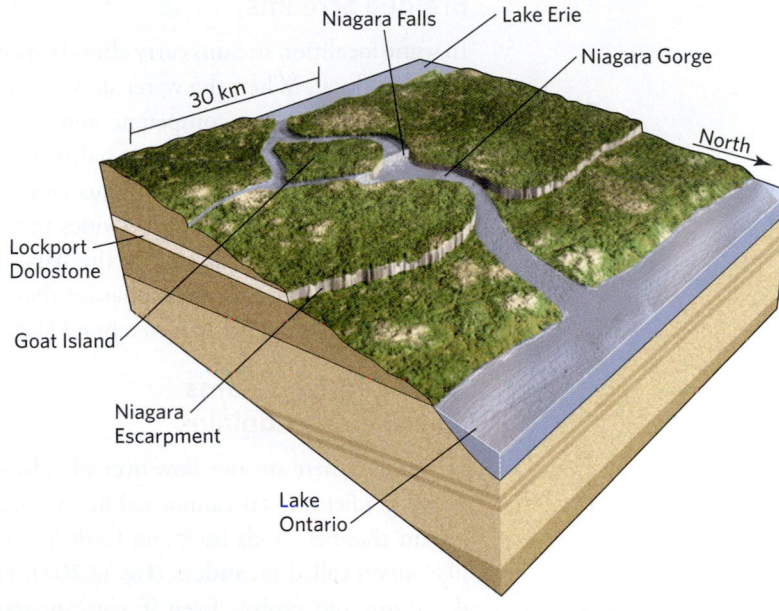

(b) The falls developed where water spills over the Lockport Dolostone, a resistant rock layer that forms the Niagara escarpment.

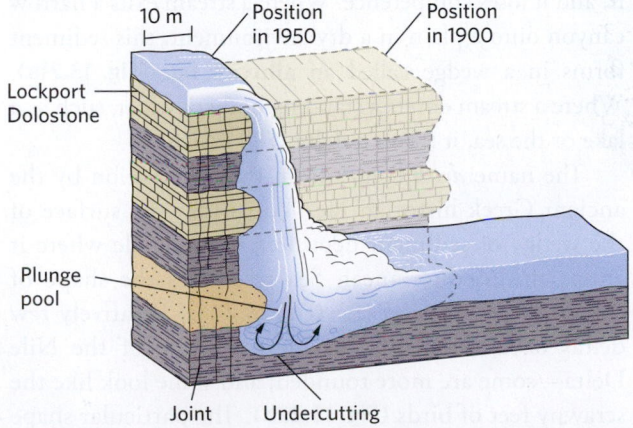

(c) Over time, as erosion of the shale undercuts the dolostone, the position of the falls migrates upstream.

to be a permanent feature of the landscape, all waterfalls eventually disappear because headward erosion slowly eats away the resistant ledge that underlies them. We can see a classic example of headward erosion at Niagara Falls, where water flowing from Lake Erie to Lake Ontario drops over a 55-m (180-foot)-high escarpment of resistant dolostone overlying weak shale **(Fig. 13.18a, b)**. Over time, erosion of the shale undercuts the dolostone, leaving an overhang of dolostone that eventually collapses. Each time this happens, the position of the waterfall migrates upstream **(Fig. 13.18c)**.

Did you ever wonder . . .

whether the position of a waterfall can change over time?

Figure 13.19 Strands of this braided stream carry meltwater from a glacier near Denali, Alaska. The sediment bars were deposited at times when the stream was in flood and had greater competence.

Braided Streams

In some localities, streams carry abundant coarse sediment during a flood. When the water slows after the flood, the stream becomes less competent and cannot carry this sediment. As a result, the coarse sediment settles out and chokes the channel with numerous elongate gravel bars and sandbars, and the stream divides into many strands weaving back and forth among the bars. The result is a *braided stream*—the name emphasizes that the streams intertwine like strands of hair in a braid **(Fig. 13.19)**.

Meandering Streams and Their Floodplains

In regions where streams flow over a landscape with a very gentle gradient, boats cannot sail in a straight line, for the stream channel winds back and forth in a series of snakelike curves called **meanders (Fig. 13.20a)**. How do meanders form and evolve? Even if a stream starts out with a straight channel, the location of the strongest current in the stream tends to wander, so that it sometimes lies nearer the middle of the channel and sometimes nearer the banks **(Fig. 13.20b)**. Water erodes the side of the stream more rapidly where it flows faster, so when the fastest current runs along the bank, it begins to cut into the bank. As a consequence, the channel starts to curve. Once a curve has initiated, the fastest current preferentially flows along the outer arc of the curve, just as a fast race car shifts to the outer arc of a track, so over time, the curve migrates sideways and grows more pronounced until it becomes a meander.

The shape of a meander evolves over time. On the outside edge of a meander, erosion continues to eat away at the channel wall, forming a *cut bank*. On the inside edge, water slows down, so its competence decreases and sediment accumulates, forming a wedge-shaped deposit called a **point bar (Fig. 13.20c)**. With continued erosion, a meander may curve through more than 180 degrees, so that the cut bank at the upstream end of the meander approaches the cut bank at its downstream end, leaving a *meander neck*, a narrow isthmus of land separating the parts of the meander. When erosion eats through a meander neck, a *cutoff* develops. The meander that has been cut off becomes an **oxbow lake** if it remains filled with water, or an *abandoned meander* if it dries out **(Fig. 13.20d)**. Streams that have developed many meanders are known as *meandering streams*. The course of a meandering stream naturally changes over time, on a time scale of years to centuries, as new meanders grow and old ones are cut off and abandoned (see Fig. 13.20a).

Most meandering stream channels cover only a relatively small portion of a broad, nearly flat floodplain **(Fig. 13.20e)**. As we noted earlier, during a flood, water overtops the edge of the stream channel and spreads out over the floodplain. In many cases, a floodplain terminates at its sides along a *bluff*, or escarpment. As the water rises above the channel walls and starts to spread out over the floodplain, friction slows down the flow. This slowdown decreases the competence of the running water, so sediment settles out along the edge of the channel. Over time, the accumulation of this sediment produces a pair of low ridges, called **natural levees**, on either side of the stream. Natural levees, which remain after floodwaters subside, may grow so tall that the floor of the stream channel between them becomes higher than the surface of the floodplain.

The End of the Line for Sediment: Alluvial Fans and Deltas

We've seen that some fluvial sediment gets deposited along streams, in bars between channels, or in point bars along curves. Some of the sediment, however, makes it all the way to the mouth of the stream and accumulates there. As the water spreads out over a broader area, friction slows it, and it loses competence. Where a stream exits a narrow canyon onto a plain in a dry environment, this sediment forms in a wedge called an **alluvial fan (Fig. 13.21a)**. Where a stream enters a standing body of water, such as a lake or the sea, it forms a **delta (Fig. 13.21b)**.

The name *delta* comes from the observation by the ancient Greek historian Herodotus that the surface of the wedge of sediment deposited by the Nile where it enters the Mediterranean Sea resembles the shape of the Greek letter delta (Δ) **(Fig. 13.22a)**. Relatively few deltas have the distinct triangular shape of the Nile Delta—some are more rounded, and some look like the scrawny feet of birds **(Fig. 13.22b)**. The particular shape

Figure 13.20 Development of meandering streams.

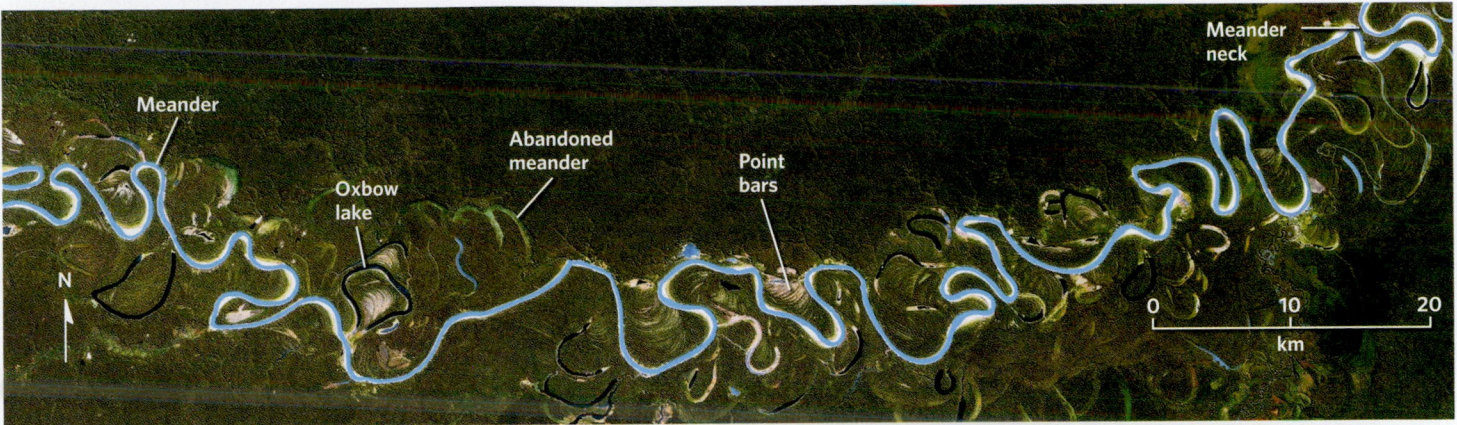

(a) A meandering stream in Brazil, as viewed from space. Note the various landscape features.

Time

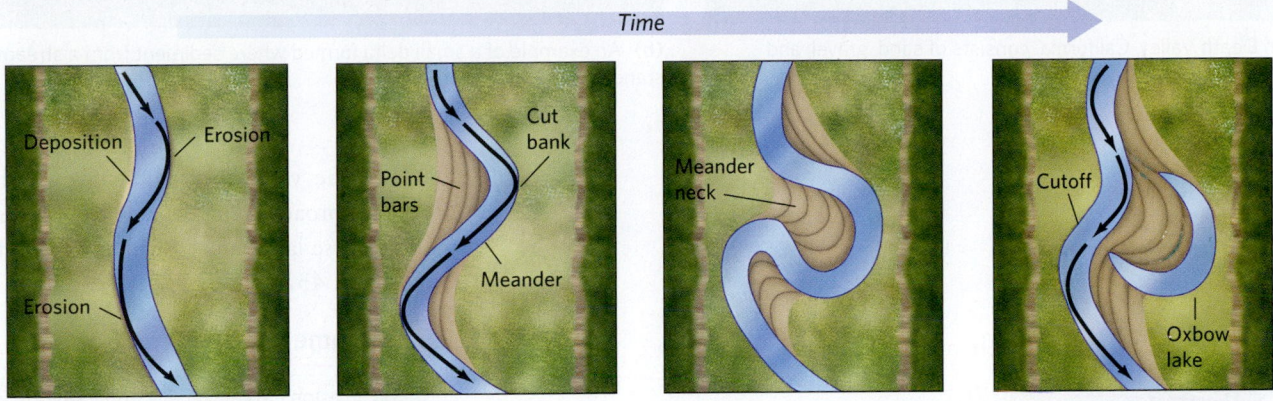

(b) Meanders evolve because erosion occurs faster on the outer bank of a curve, while deposition takes place on the inner bank. Eventually, a cutoff isolates an oxbow lake.

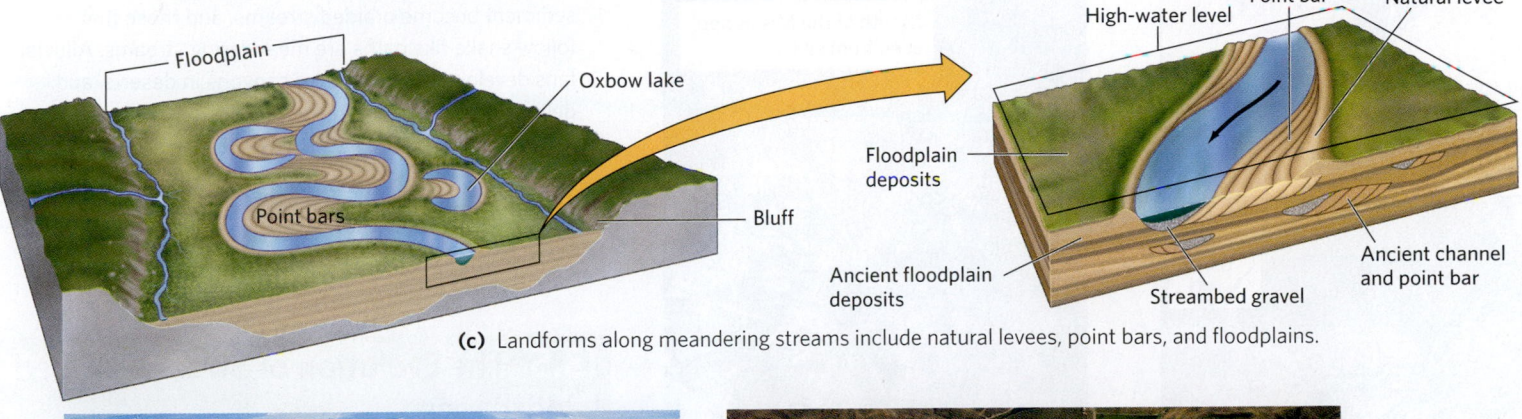

(c) Landforms along meandering streams include natural levees, point bars, and floodplains.

(d) Two oxbow lakes are visible in this aerial photo of a meandering stream.

(e) The flat land of this floodplain hosts farm fields. Trees delineate the stream channel.

Figure 13.21 Examples of depositional landforms formed at the mouths of streams.

(a) This alluvial fan in Death Valley, California, consists of sand, gravel, and debris flows. The curving black line is a road.

(b) An example of a small delta formed where sediment from a stream flows into standing water.

Figure 13.22 Not all deltas look the same.

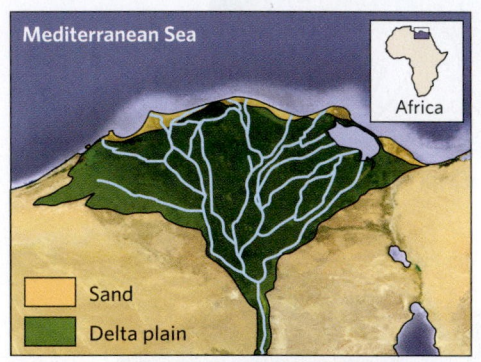

(a) The Nile Delta is shaped like the Greek letter Δ.

(b) The Mississippi Delta is a bird's-foot delta. The inset shows the present-day end of the delta. Note the natural levees.

Mouth of the Mississippi seen from space.

Natural levee

may become very thick and wide, so its surface can become a broad *delta plain*. Clearly, streams can form many diverse landscape features (see **Earth Science at a Glance**, pp. 432–433).

Take-home message . . .

Erosion carves valleys and canyons whose shapes depend on the balance between rates of mass wasting and rates of downcutting. Streams choked with sediment become braided streams, and those that follow snake-like paths are meandering streams. Alluvial fans develop at the mouth of canyons in deserts, and deltas develop in standing water.

Quick Question -
What factors cause the formation of rapids and waterfalls?

13.5 The Evolution of Stream-Eroded Landscapes

Imagine a place where geologic processes cause the land surface of a region to rise. When this happens, the landscape begins to evolve **(Fig. 13.23)**. Streams downcut and produce deep, narrow valleys with steep gradients. Over time, mass wasting and erosion transform the rugged mountains into low, rounded hills, and the valleys broaden into wide floodplains with gentle gradients. Eventually, even the low hills are beveled down, leaving a nearly flat land surface at an elevation near that of the drainage network's ultimate base level.

of a delta depends in part on the interplay between the rate at which the river supplies new sediment and the rate at which waves or currents remove sediment. A delta deposited by a major river along the seacoast

Although the model we have just described makes intuitive sense, research emphasizes that it's an oversimplification. On our dynamic planet, changes in land elevation and sea level prevent the landscape from ever getting shaved down to the elevation of the original base level. When the land surface rises or the base level falls, **stream rejuvenation** takes place, meaning that a stream flowing over the land surface starts to downcut once again. If a rejuvenated stream has a meandering course, *incised meanders*, the type that flow at the base of a canyon, develop **(Fig. 13.24)**.

As stream-eroded landscapes evolve, puzzling landscapes develop locally. For example, some streams flow through a canyon that cuts straight across a mountain ridge. Such a landscape can develop when a stream that had been flowing over horizontal strata downcuts through an unconformity and into underlying folded strata. Geologists call such streams *superposed streams* because their pre-existing geometry has been laid down over underlying geologic structures **(Fig. 13.25)**. When tectonic processes (such as subduction or continental collision) cause a mountain range to rise up beneath an already established stream, the landscape that develops depends on how fast the stream downcuts relative to the rate of uplift. Specifically, if the stream downcuts as fast as the range rises, it can maintain its course and cut right across the range—geologists call such streams *antecedent streams* (from the Greek *ante*, meaning before) to emphasize that they existed before the range was uplifted **(Fig. 13.26a, b)**. If, however, the range rises faster than the stream downcuts, the new highlands divert (change) the stream's course so that it flows parallel to the range **(Fig. 13.26c)**.

In some cases, regional uplift can even reverse the flow direction of a river. For example, the Amazon River, before the late Mesozoic, flowed westward into the Pacific, draining mountains that lay along the boundary between present-day Africa and South America, in the interior of the supercontinent Gondwana. When the South Atlantic

Figure 13.24 The goosenecks of the San Juan River, Utah, are incised meanders.

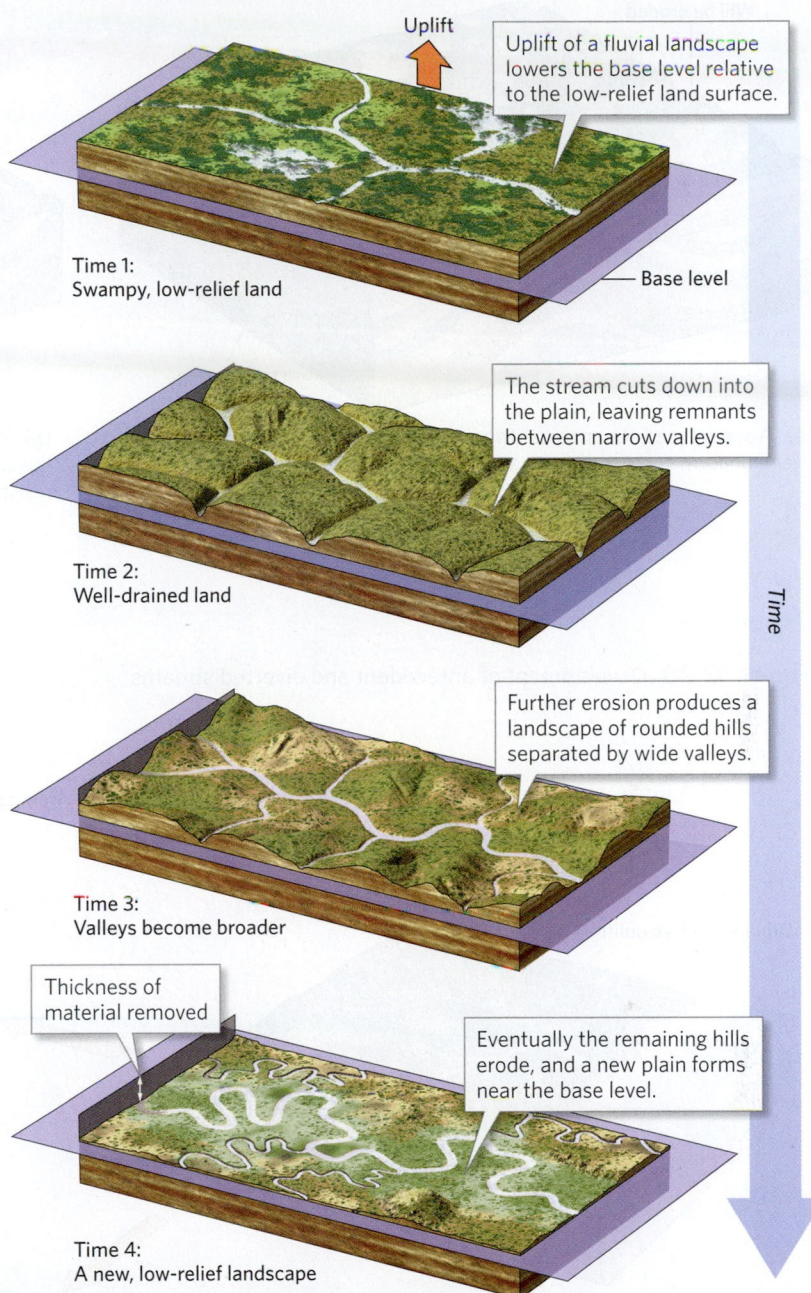

Figure 13.23 Stream-eroded landscapes evolve over time. When the base level drops, streams downcut, and a hilly landscape develops. The resulting relief, however, eventually erodes away.

Uplift

Uplift of a fluvial landscape lowers the base level relative to the low-relief land surface.

Time 1:
Swampy, low-relief land

Base level

The stream cuts down into the plain, leaving remnants between narrow valleys.

Time 2:
Well-drained land

Time

Further erosion produces a landscape of rounded hills separated by wide valleys.

Time 3:
Valleys become broader

Thickness of material removed

Eventually the remaining hills erode, and a new plain forms near the base level.

Time 4:
A new, low-relief landscape

www.gernot-keller.com

Figure 13.25 Development of superposed streams.

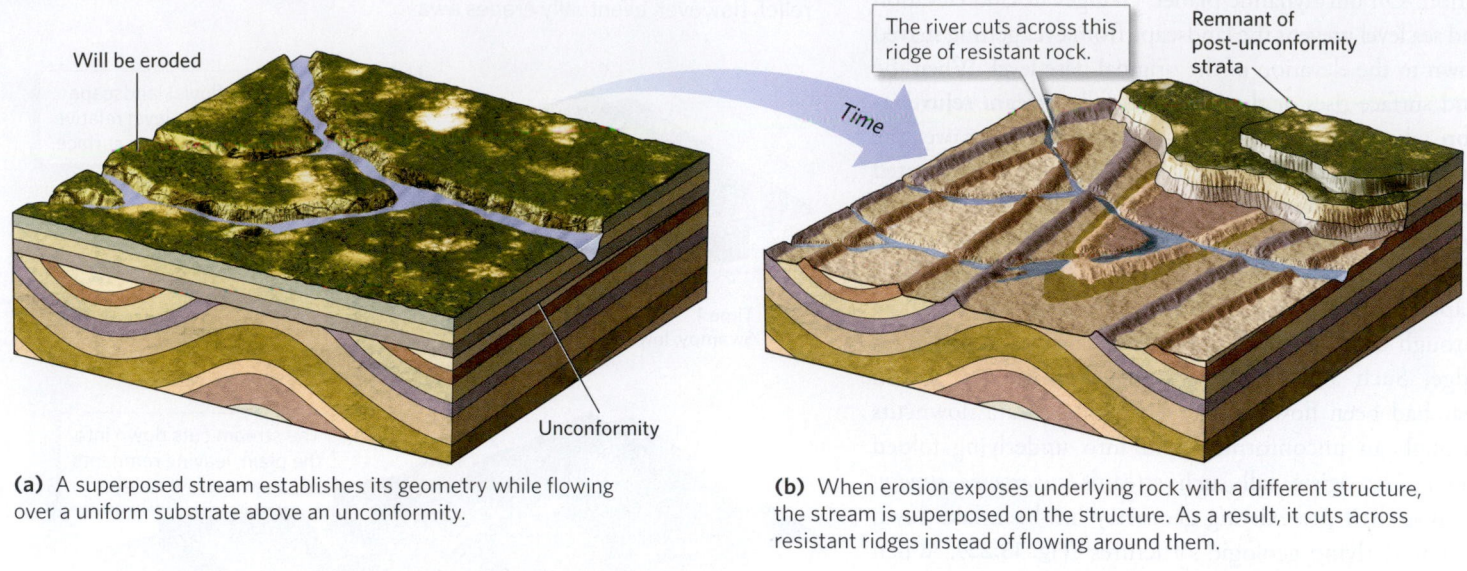

Will be eroded

The river cuts across this ridge of resistant rock.

Remnant of post-unconformity strata

Time

Unconformity

(a) A superposed stream establishes its geometry while flowing over a uniform substrate above an unconformity.

(b) When erosion exposes underlying rock with a different structure, the stream is superposed on the structure. As a result, it cuts across resistant ridges instead of flowing around them.

Figure 13.26 Development of antecedent and diverted streams.

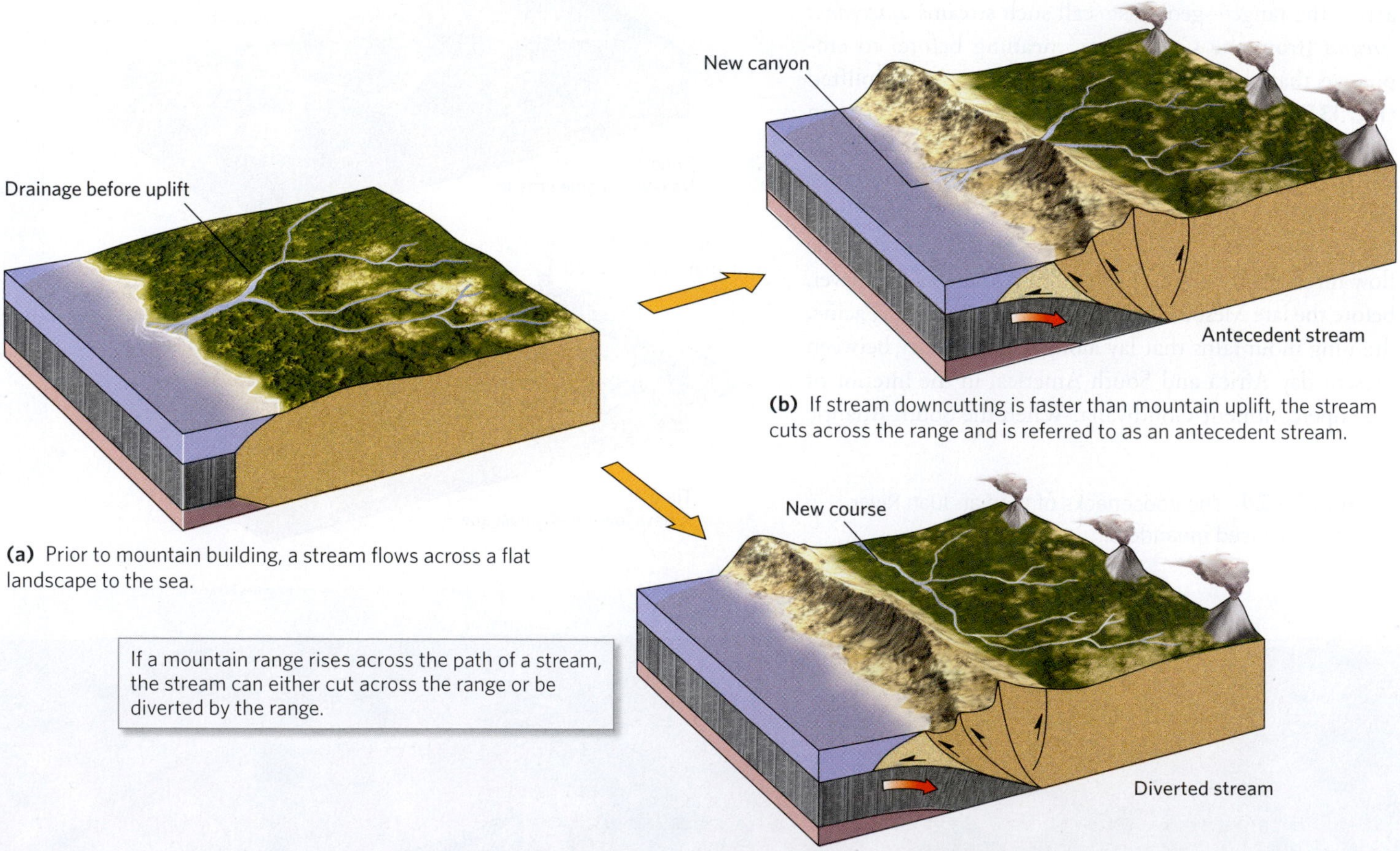

Drainage before uplift

New canyon

Antecedent stream

(a) Prior to mountain building, a stream flows across a flat landscape to the sea.

(b) If stream downcutting is faster than mountain uplift, the stream cuts across the range and is referred to as an antecedent stream.

If a mountain range rises across the path of a stream, the stream can either cut across the range or be diverted by the range.

New course

Diverted stream

(c) If uplift happens faster than downcutting, the stream is diverted and flows along the edge of the range.

opened and subduction led to the uplift of the Andes, the drainage reversed and flowed eastward into the South Atlantic.

Headward erosion by streams may cause one stream to cut through a drainage divide and intersect the course of another stream. When this happens, the stream that cut through the divide "captures" the water of the other. This process, known as **stream piracy**, causes the water of the captured stream to start flowing down the pirate stream, so the channel of the captured stream downstream of the point of capture dries up (Fig. 13.27).

Take-home message . . .

Stream-carved landscapes evolve over time as gradients diminish and ridges between valleys erode away. Superposed streams attain their shape before downcutting into geologic structures, whereas antecedent streams downcut while the land beneath them uplifts. As the landscape evolves, one stream may capture the flow of another.

Quick Question -
What can cause a drainage reversal to take place?

13.6 Raging Waters: River Flooding and Flood Control

Streams can cause havoc during a **flood**, an event in which the volume of water flowing down a stream exceeds the volume of the stream channel. Water overtops the banks and spreads out over a floodplain or delta plain, or fills a canyon to a greater depth than normal. Floods can be classified in different ways. For example, disaster agencies classify them based on location, distinguishing among river floods, coastal floods, and urban floods. Alternatively, floods can be classified based on how fast they develop. We'll take the latter approach here and distinguish between slow-onset floods (which develop over several days and take days to weeks to subside) and flash floods (which develop in minutes to hours and subside in hours).

Slow-Onset Floods

When you read a news story about flooding that takes days to develop, lasts for weeks, and involves the trunk stream of a large drainage network, you're reading about a **slow-onset flood** (Fig. 13.28). Slow-onset floods happen (1) during the spring after a winter of heavy snows, when temperatures rise and widespread melting of the snowpack begins; (2) during the sustained rains of a distinct wet season, such as the monsoon season of Asia

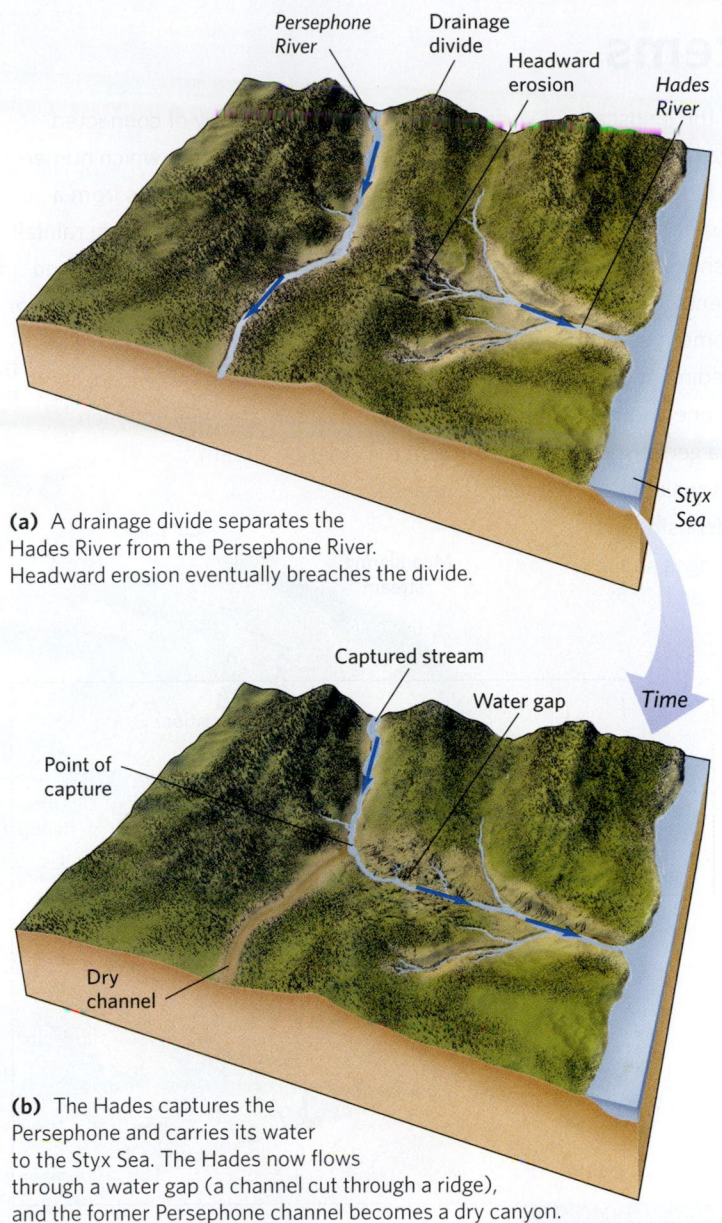

Figure 13.27 Stream piracy.

Persephone River   *Drainage divide*   Headward erosion   *Hades River*

Styx Sea

(a) A drainage divide separates the Hades River from the Persephone River. Headward erosion eventually breaches the divide.

Captured stream   Water gap   *Time*

Point of capture

Dry channel

(b) The Hades captures the Persephone and carries its water to the Styx Sea. The Hades now flows through a water gap (a channel cut through a ridge), and the former Persephone channel becomes a dry canyon.

(see Chapter 18), when winds shift and blow moist air from the oceans over the land for weeks on end and a large region gets drenched; or (3) when a system of storms remains stationary over a broad region for a long time. In all these circumstances, the ground becomes saturated and cannot absorb any more water, so runoff from a broad drainage area flows into tributaries. These streams, in turn, supply water to the trunk stream, whose discharge becomes greater than can be accommodated in its channel. If slow-onset floods are tied to an annual weather pattern, Earth scientists may also refer to them as *seasonal floods* (see Fig. 13.28).

River Systems

Rivers, or streams, drain the landscape of surface runoff. Typically, an array of connected streams called a drainage network develops, consisting of a trunk stream into which numerous tributaries flow. The land drained is the network's watershed. A stream starts from a source, or headwaters. Some headwaters are in the mountains, collecting water from rainfall or from melting ice and snow. In the mountains, streams carve deep, V-shaped valleys and tend to have steep gradients. Locally, a river may flow over a bouldery bed, forming rapids, or it may drop off an escarpment as a waterfall.

Rivers choked with sediment become braided by dividing into numerous entwined channels separated from one another by gravel bars. Where a stream that has not been choked by sediment has a gentle gradient, it becomes a meandering stream, winding back and forth in snake-like curves called meanders. Because of erosion and deposition, a meandering stream changes shape over time.

Developing drainage networks

Transportation along the channel

Rapids

Braided channel

Meandering stream

Cut bank

Deposition

Bank erosion

Terraced floodplain

(present floodplain)

(oldest floodplain)

Deposition of point bar

Back swamps

Yazoo stream

Wide meanders

Neck

Oxbow lake

Cutoff

Natural levees

Wide floodplain

Point bars forming on inner curves

Meanders, abandoned meanders, and cutoffs

A small delta in a mountain lake

Headward erosion

Glaciers

Valleys with
high relief

Melting
ice

Lake

Dendritic
drainage

Rapids

Waterfall

Collection of water
in watershed

Streams contribute to carving mountains

Waterfall in Hawaii spilling over a basalt ledge

Occasionally, a meander may be cut off, leaving a curving lake called an oxbow lake. A broad floodplain, covered with water only during floods, may develop on either side of the stream. Natural levees build up between the channel and the floodplain. They consist of sediment dropped as a flooding river starts to spill out of its channel. Eventually, a river reaches a standing body of water and slows down, and the sediment it carries gets deposited to form a delta.

Deposition
at mouth

Delta

Distributaries

Natural levees

Swamps and
marsh

Tidal flats

Bar

Banks

433

Figure 13.28 Examples of slow-onset flooding.

(a) Flooding along the Missouri River in 1993.

(b) This flooding in Bangladesh submerged whole villages.

Examples of slow-onset floods make headlines every year **(Box 13.1)**. During 2010, particularly intense monsoonal rains in Pakistan caused the Indus River and its tributaries to rise slowly until water covered an area of 800,000 km² (310,000 square miles), or about one-fifth of the country's land area. Flooding affected about 20 million people and led to about 2,000 fatalities. Even more severe slow-onset floods have happened in the past. The 1931 flood of the Yellow and Yangtze Rivers in China may have led to the deaths of over 3 million people. This terrible event happened when a winter with heavy snowfall was followed by heavy monsoon rains and a succession of typhoons (hurricanes). At its peak, water in the Yangtze River was 16 m (53 feet) above normal.

Flash Floods

An event during which the discharge of a stream increases so fast that it may be impossible to escape from the path of the water is a **flash flood (Fig. 13.29)**. Flash floods generally affect a relatively small area. Such floods happen during particularly intense rainfall, when so much water drenches an area that there isn't time for

Figure 13.29 Examples of flash flooding.

(a) A flash flood in a desert region of Israel has washed over a highway, forcing the evacuation of truckers.

(b) During the 1976 Big Thompson River flash flood in Colorado, this house was carried off its foundation and dropped on a bridge.

Consider this . . .

The 1993 Mississippi River flood

The Mississippi River drainage network has had its share of slow-onset floods. The worst in recent decades took place in the summer of 1993, when unusual weather patterns, after the rainy season was supposed to be over, caused a whole year's supply of rain to fall across the upper Midwest of the United States during a period of a few weeks—some regions received 400% more rain than usual. Immense volumes of runoff entered the region's streams and flowed into the Mississippi drainage network. Eventually, the water of the Mississippi and Missouri Rivers spread out over the rivers' floodplains. By July, parts of nine states were underwater **(Fig. Bx13.1)**. Bridges and roads were undermined and washed away, and towns along the river were submerged. Rowboats replaced cars as the favored mode of transportation in towns where only the rooftops remained visible. In St. Louis, Missouri, the river reached a maximum level, or *crested*, at 14 m (47 feet) above normal. For 79 days, the flooding continued. When the water finally subsided, it left behind a thick layer of sediment, filling living rooms and kitchens in floodplain towns and burying crops in floodplain fields. In the end, more than 40,000 km² (15,400 square miles) of the floodplain had been submerged, 50 people died, at least 55,000 homes were destroyed, and countless acres of crops were buried. Comparable flooding struck the Mississippi Valley again in 2011, with water locally rising 5 m (17 feet) above **flood stage** (the level at which the river spreads beyond the channel).

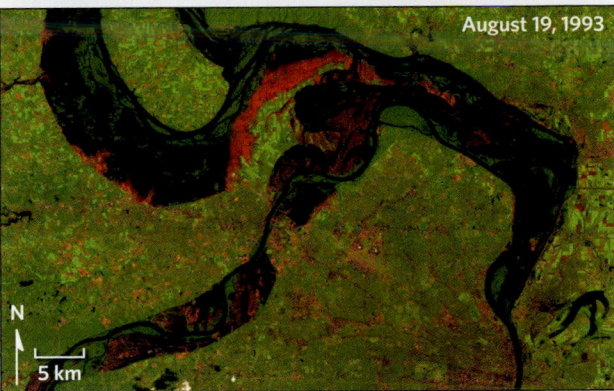

Figure Bx13.1 Satellite photos (before and during) show the extent of inundation during the 1993 Mississippi River flood.

it to sink into the ground, so it becomes runoff instead. Flash floods also happen when dams collapse (as in the 1889 Johnstown flood), or when artificial levees fail. During a flash flood, a narrow canyon or valley may fill to a level many meters above normal in a matter of minutes, and the leading edge of the flood may rush downstream as a wall of water. Some flash floods are so turbulent and fast, and carry so much sediment, that they knock down and carry away everything in their path—boulders, trees, bridges, cars, and houses—and leave devastation in their wake. The floodwaters may subside after a fairly short time, sometimes in minutes to hours. Flash floods can take place in any climate, but are especially dramatic in arid climates. In a desert, runoff from an isolated thunderstorm may suddenly fill the channel of a dry wash, for unvegetated ground allows runoff to reach the channel rapidly.

Living with Floods

FLOOD CONTROL. Mark Twain once wrote of the Mississippi that we "cannot tame that lawless stream, cannot curb it or confine it, cannot say to it, 'go here or go there,' and make it obey." Was Twain right? Since ancient times, people have attempted to control the courses of rivers to prevent undesired flooding. In the 20th century, flood-control efforts intensified as the population living along rivers increased. For example, following a disastrous flood in 1927, the US Army Corps of Engineers began a program to control Mississippi flooding. First, the Corps built about 300 dams along the river's tributaries so that excess runoff could be stored in reservoirs and later released slowly. Second, they built *artificial levees* of sand and mud, as well as concrete *floodwalls* **(Fig. 13.30a, b)**. These structures are designed to increase the channel's volume and to isolate portions of the floodplain from flooding.

Figure 13.30 Flood control strategies.

(a) Artificial levees have been built to protect the town of Galena, Illinois.

High-water marks of past floods.

(b) A concrete floodwall at Cape Girardeau, Missouri. When floods threaten, a crane drops a gate into the slot to keep out the Mississippi River.

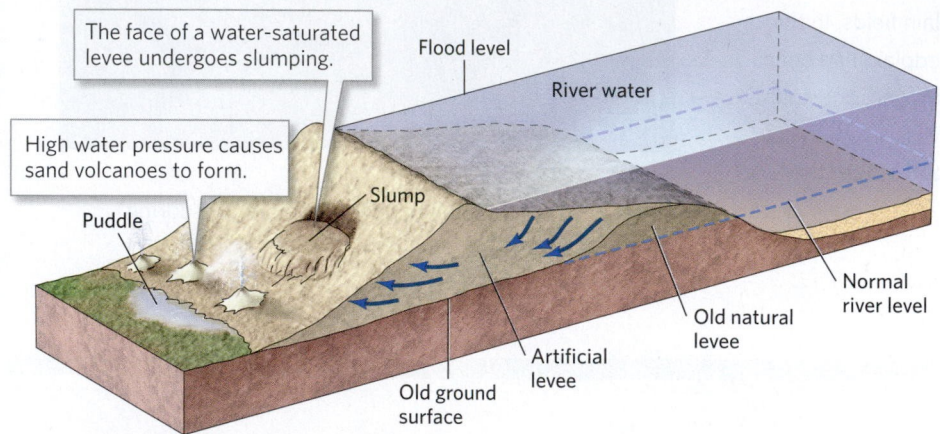

The face of a water-saturated levee undergoes slumping.

High water pressure causes sand volcanoes to form.

Flood level

River water

Puddle

Slump

Normal river level

Old natural levee

Artificial levee

Old ground surface

(c) Artificial levees may be undermined when infiltrated by water.

(d) At a levee breach, water flows from the river to the floodplain.

Although the Corps' strategy has worked for floods up to a certain size, it was insufficient to handle the 1993 and 2011 Mississippi floods, when reservoirs filled to capacity and additional runoff headed downstream. The river rose until it spilled over the tops of some levees and undermined others. *Undermining* occurs when rising water levels increase the water pressure on the river side of the levee, forcing water through sand under the levee **(Fig. 13.30c)**. In susceptible areas, water spurts out of the ground on the dry side of the levee, thereby washing away the levee's support until, eventually, it becomes so weak that it collapses, and water fills the area behind it **(Fig. 13.30d)**.

Flood control in some localities may involve restoration of wetland areas along rivers, for wetlands can absorb significant quantities of floodwater. In addition, planners may prohibit construction within designated land areas adjacent to the channel, creating a *floodway* that can fill without causing expensive damage.

EVALUATING FLOODING HAZARD. When making decisions about investing in flood-control measures, mortgages, or insurance, planners need a basis for defining the hazard or risk posed by flooding. Geologists characterize the risk of flooding in two ways. The **annual probability** of flooding indicates the likelihood that a flood of a given size will happen along a specified portion of a stream during any given year. For example, if we say that a flood with a given discharge has an annual probability of 1%, then we mean there is a 1 in 100 chance that a flood of this size will happen in any given year. The **recurrence interval** of a flood of a given size is the average number of years between successive floods with a given discharge.

Figure 13.31 The relationship between flood size and probability.

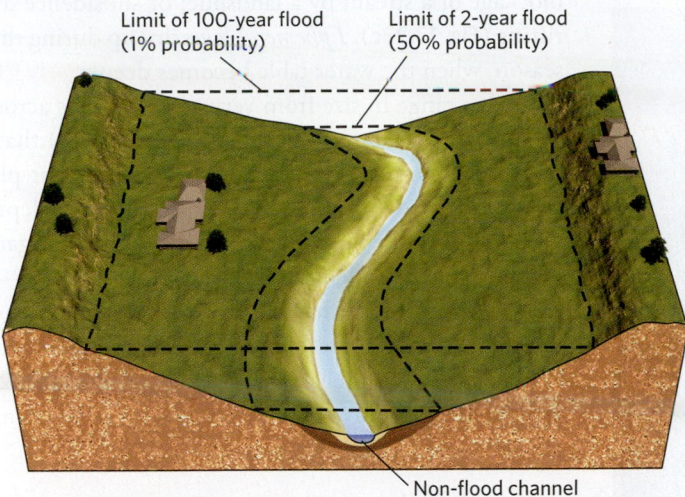

Limit of 100-year flood (1% probability)

Limit of 2-year flood (50% probability)

Non-flood channel

(a) A 100-year flood covers a larger area than a 2-year flood and occurs less frequently.

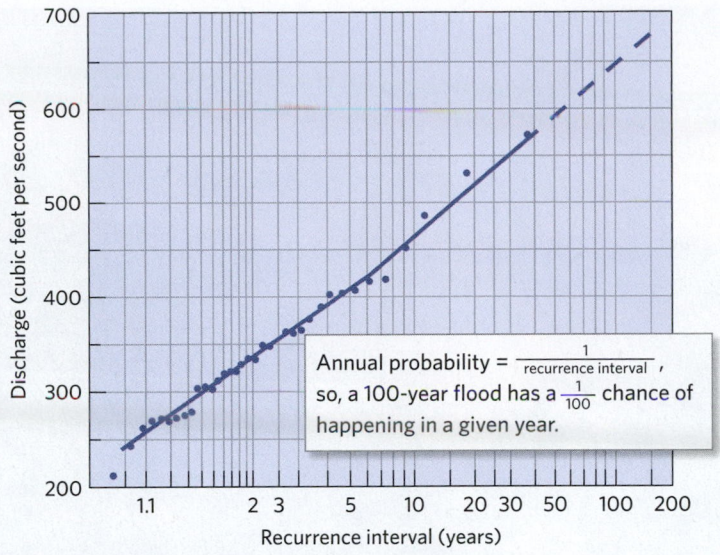

Annual probability = $\frac{1}{\text{recurrence interval}}$, so, a 100-year flood has a $\frac{1}{100}$ chance of happening in a given year.

(b) A flood-frequency graph shows the relationship between the recurrence interval (R) and the discharge for an idealized river.

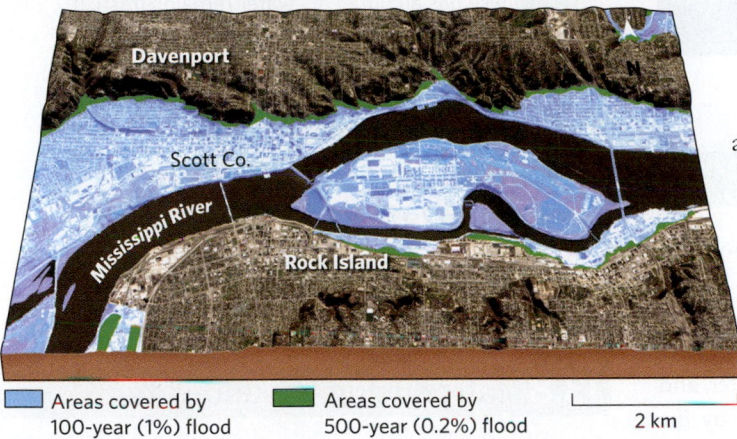

Davenport

Scott Co.

Mississippi River

Rock Island

■ Areas covered by 100-year (1%) flood   ■ Areas covered by 500-year (0.2%) flood   2 km

(c) A flood-hazard map shows areas likely to be flooded. Here, near Rock Island, Illinois, even large floods are confined to the floodplain.

For example, if a flood with a given discharge happens once in 100 years on average, then it has a recurrence interval of 100 years and is called a "100-year-flood." Annual probability and recurrence interval are related by the equation

$$\text{Annual probability} = \frac{1}{\text{recurrence interval}}$$

For example, the annual probability of a 50-year flood is ¹⁄₅₀, which can also be written as 0.02 or 2%. The recurrence interval for a flood with a larger discharge is longer than that for a smaller flood, meaning that large floods happen less frequently than small ones **(Fig. 13.31a, b)**.

Unfortunately, people commonly misunderstand the meaning of a recurrence interval, believing that they do not face future flooding hazard if they buy a home within an area where a 100-year flood has just occurred. Their confidence comes from making the incorrect assumption that because such flooding just happened, it can't happen again until another 100 years have passed. It's important to keep in mind that flooding does not take place periodically. Two 100-year floods can occur in consecutive years, or even in the same year. Alternatively, the interval between such floods could be, say, 210 years.

Knowing the discharge for a flood of a specified annual probability, and knowing the shape of the stream channel and the elevation of the land bordering the stream, allows geologists to predict the extent of land that will be submerged by such a flood. Such data, in turn, permit geologists to produce *flood-hazard maps* **(Fig. 13.31c)**.

Take-home message . . .

Slow-onset floods submerge broad areas, including the floodplains of trunk streams, for days or weeks at a time. Some of these floods may be seasonal. Flash floods are sudden and short-lived. We can specify the probability that a certain size of flood will happen in a given year, but flood-control efforts meet with mixed success.

Quick Question
Does a 100-year flood only happen every 100 years? Why or why not?

Did you ever wonder . . .

what newscasters mean by a "100-year flood"?

stream cuts through a meander neck; glacial erosion of a deep valley **(Fig. 13.33b)**; collapse of a volcanic caldera; blockage of a stream by a landslide; or subsidence due to rifting **(Fig. 13.33c)**. *Ephemeral lakes* dry up during the dry season, when the water table becomes deeper.

Lakes range in size from very small (meters across) to so large (tens to hundreds of kilometers across) that you can't see the other side. The deepest lake on our planet, Lake Baikal in Siberia, reaches a maximum depth of 1.6 km (1 mile), while the longest lake, Lake Tanganyika in Africa reaches a length of 660 km (410 miles)—both are consequences of rifting. Lake Superior, one of the five Great Lakes along the border between Canada and the United States, has the largest area (84,000 km^2, or 32,000 square miles), followed by Lake Victoria (69,500 km^2, or 27,000 square miles) in Africa.

Take-home message . . .

Lakes are bodies of standing water on land. If a lake has an outlet, its water remains fresh. If water leaves only by evaporation, the lake becomes salty. Most lakes are simply places where the land surface lies beneath the water table. Some fill glacially carved valleys and others fill rift valleys.

Quick Question --------------------------
How do tectonic processes contribute to the formation of lakes?

13.7 Lakes

Lakes are bodies of standing water that occur on land and are not directly connected to the sea **(Fig. 13.32)**. (In everyday English, we refer to small lakes as *ponds*.) Geologists distinguish between *natural lakes*, which form where the land surface is relatively low and can collect water, and *reservoirs*, which are lakes that have been trapped by the construction of a dam or a retaining wall.

Most lakes contain freshwater because water can leave them via an outlet stream. As a result, the water in a freshwater lake remains there for only a relatively short time, and most of the dissolved salts it contains leave via the outlet stream, so the concentration of salt remains very low. Some lakes, however, such as the Great Salt Lake of Utah and the Dead Sea along the border of Jordan, contain saltwater. A **salt lake** becomes salty because it has inlets, but no outlets, so water can leave the lake only by evaporation, which removes water molecules but leaves behind salts. Even the freshest water brought into the lake by streams contains a little dissolved salt, and over time, this salt gets concentrated in the lake as the water of the lake evaporates.

Permanent lakes are depressions that fill with water because the ground surface lies beneath the water table, the level below which the ground is saturated with water **(Fig. 13.33a)**. They can form for a variety of reasons, such as: local subsidence to form a low area; the collapse of ground over a cave; the isolation of a meander when a

13.8 Introducing Groundwater

When it rains, some of the water that falls on the land sinks or percolates into the ground, a process called **infiltration**. In effect, the upper part of the Earth's crust behaves like a giant sponge that can soak up water. Of the water that infiltrates downward, some descends only into the soil, where it adheres to the surfaces of mineral grains and organic fragments. This water, called *soil moisture*, may later evaporate, be taken up by the roots of plants, or be pushed back to the ground surface by new soil water. The rest of the infiltrating water sinks deeper into sediment or rock, where, along with water that was trapped in rock at the time the rock formed, it makes up **groundwater**. Groundwater flows slowly, and it may remain underground for anywhere from a few months to tens of thousands or even millions of years before returning to the surface to pass once again into other reservoirs of the hydrologic cycle. Groundwater exists primarily in the Earth's upper crust, for below depths of about 20 to 30 km (12 to 19 miles), metamorphic reactions incorporate water, rocks flow plastically, and water no longer stays in liquid form.

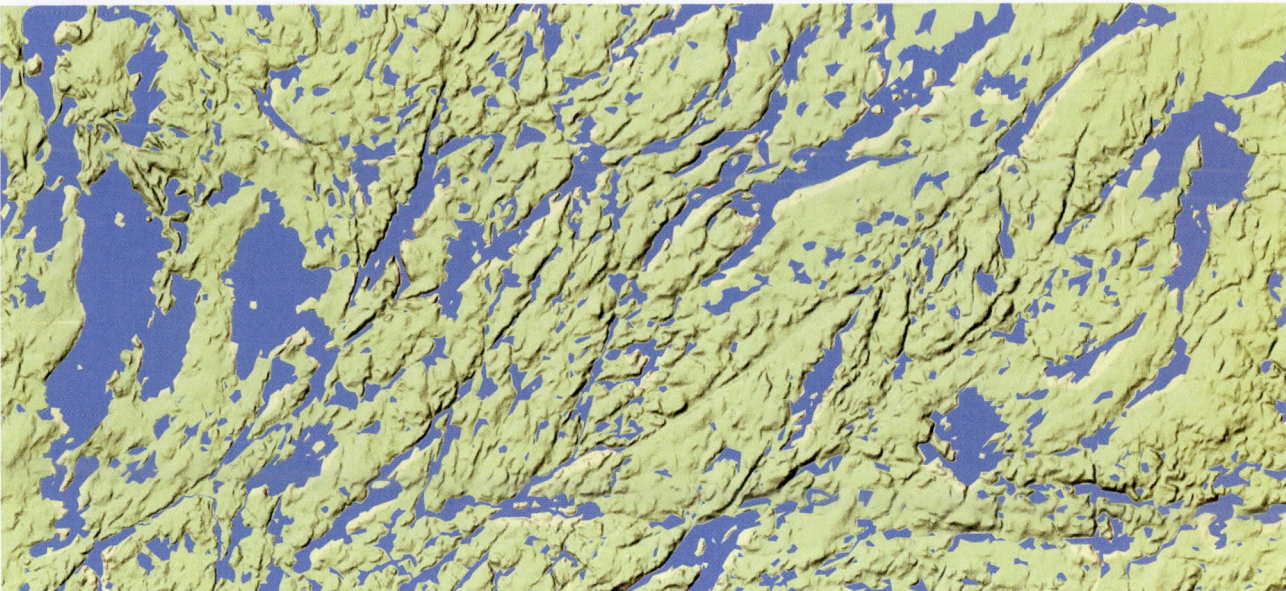

Figure 13.33 Lakes form in a number of ways.

(a) Countless lakes cover the land about 80 km (50 miles) north of the boundary between western Ontario (Canada) and Minnesota. The water table in this cool, rainy region is high, so depressions in the land fill with water.

(b) The Finger Lakes of Central New York State, south of Lake Ontario, fill glacially-carved valleys.

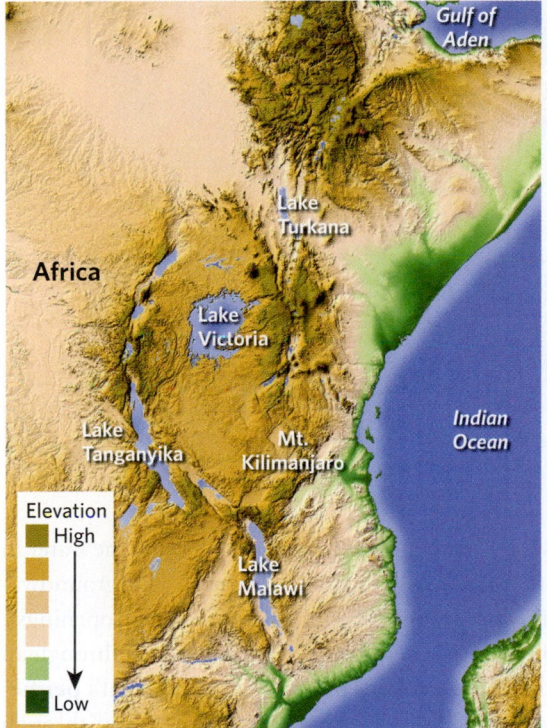

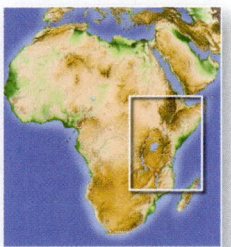

(c) Elongate lakes, such as Lake Tanganyika, fill deep rift valleys in eastern Africa.

Porosity and Permeability

To understand where groundwater resides, we must first review the nature of porosity and permeability. Only a tiny proportion of underground water actually occurs in open caves or underground streams. Most groundwater resides in relatively small open spaces between solid mineral grains in sediment or rock, or within cracks of various sizes. Recall from Chapter 11, in our discussion of how oil occurs underground, that a space or open crack within a volume of sediment or a body of rock is a **pore**, and that **porosity** refers to the total volume of open space within a material, specified as a percentage. For example, if we say that a block of rock has 30% porosity, then 30% of the volume of the block consists of pores.

Hydrogeologists (researchers who specialize in the study of groundwater) distinguish between two basic kinds of porosity. *Primary porosity* develops during sediment deposition or rock formation **(Fig. 13.34a)**. It includes the empty space that exists where the grains don't fit together perfectly. *Secondary porosity* refers to new pore space produced after lithification. Such porosity can form when groundwater passes through rock and dissolves minerals, or it can form due to the development of cracks in association with jointing and faulting **(Fig. 13.34b)**.

Figure 13.34 Porosity and permeability.

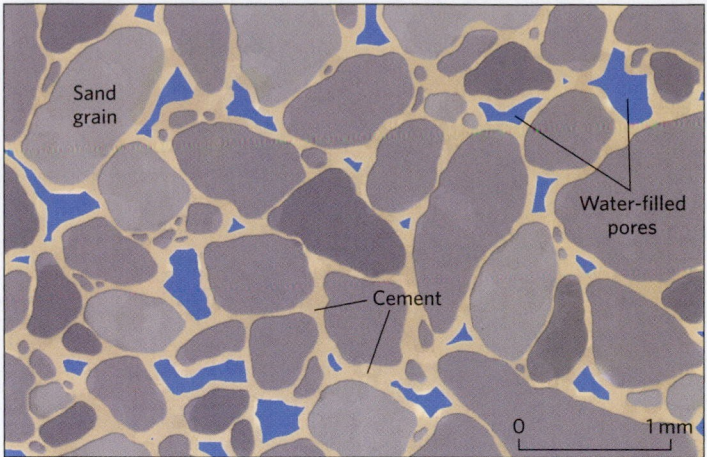

(a) Isolated pores occur in the spaces between grains in a sandstone. Water or air can fill these pores.

These fractures have been enlarged by dissolution.

(b) This limestone outcrop on the coast of Ireland contains abundant fractures that provide secondary porosity.

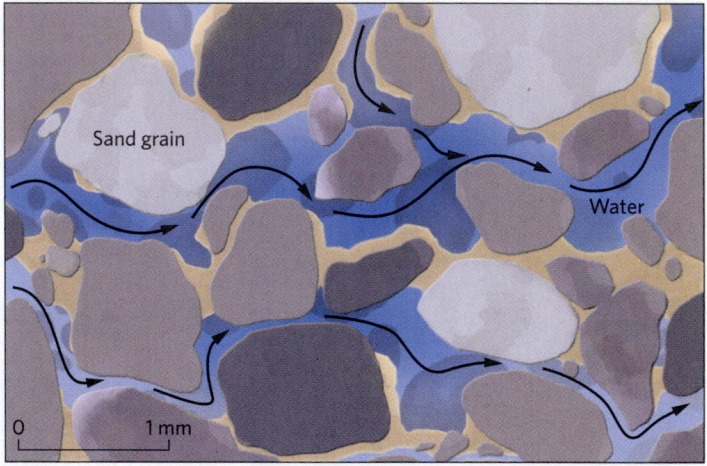

(c) Permeability is the degree to which pores are linked and water can move from pore to pore in rock or sediment.

If solid rock completely surrounds a pore, the water in the pore cannot flow to another location. For groundwater to flow, pores must be linked by conduits (openings). The ability of a material to allow fluids to pass through an interconnected network of pores is the material's **permeability (Fig. 13.34c)**. Groundwater flows easily through a *permeable material* such as loose gravel. Water flows slowly through *low-permeability material* as it follows a tortuous path through tiny conduits. Water cannot flow at all through an *impermeable material*. The permeability of a material depends on several factors:

- *Number of available conduits:* As the number of conduits increases, permeability increases.

- *Size of the conduits:* Fluid travels faster through wider conduits than through narrower ones.

- *Straightness of the conduits:* Water flows faster through straight conduits than it does through crooked ones.

Note that the factors that control permeability in rock or sediment resemble those that control the ease with which traffic moves through a city. Traffic can flow quickly through cities with many straight multilane boulevards, whereas it flows slowly through cities with only a few narrow, crooked streets. Note, too, that porosity and permeability are not the same—a material whose pores are isolated from one another can have high porosity, but low permeability.

Aquifers, Aquitards, and the Water Table

With the concepts of porosity and permeability in mind, hydrogeologists distinguish between an **aquifer**, sediment or rock with high permeability and porosity, and an **aquitard**, sediment or rock with relatively low permeability regardless of porosity. Materials in which water completely fills pore space are said to be *saturated*. We further distinguish between an *unconfined aquifer*, which starts at the ground surface and extends downward, and a *confined aquifer*, which is separated from the ground surface by an aquitard (**Fig. 13.35a**).

Hydrogeologists define the *saturated zone* as the subsurface region in which water completely fills pore spaces. Typically, an *unsaturated zone*, in which water only partially fills pores so that some air-filled pore space remains, extends from the ground surface down for some depth (**Fig. 13.35b**). The underground boundary between the unsaturated zone (above) and the saturated zone (below) is called the **water table**. In effect, the water table forms the top boundary of groundwater. Some groundwater seeps up along grain boundaries due to surface tension and the attraction of water molecules to one another and to surfaces. This rise, known as capillary action, yields a *capillary fringe* of wetness just above the water table.

Box 13.2

How can I explain . . .

Porosity and permeability

What are we learning?
- Where groundwater resides in materials underground.
- The difference between porosity and permeability.
- Factors that affect porosity and permeability.

What you need:
- Two large glass beakers.
- A supply of pebbles, or gravel.
- A supply of sand.
- A measuring cup.
- A screen made of soft plastic mesh.
- Two strong rubber bands.
- A stopwatch.

Instructions:
- Fill an empty beaker to the top in order to determine its volume, then empty the beaker so it can be used for the next part of the experiment.
- Next, fill one beaker with pebbles or gravel, and the other with sand.
- Using the measuring cup, add water to each beaker. Record the volume of water needed to fill each beaker so that the water table lies just at the surface of the sediment.
- Calculate the porosity of each material. Determine which material has higher porosity, and suggest a reason for the result.
- Place a screen over the top of each beaker, and secure it with a rubber band.

- Tilt the beaker so it's horizontal, and time how long it takes for the water to spill out. Determine which material has greater permeability, and suggest a reason for the result.

What did we learn:
- In the subsurface, water fills spaces (pores) between solid grains.
- Not all materials have the same porosity. You can calculate porosity by comparing the total volume of a water that a container holds to the volume of water that fits between solid grains.
- Not all materials have the same permeability. Permeability can be affected by the size of conduits between grains.

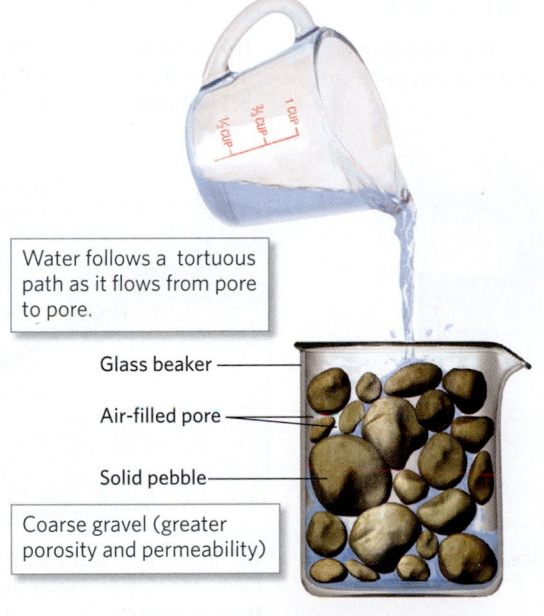

Water follows a tortuous path as it flows from pore to pore.

Glass beaker

Air-filled pore

Solid pebble

Coarse gravel (greater porosity and permeability)

Glass beaker

Sand

Sand (lesser porosity and permeability)

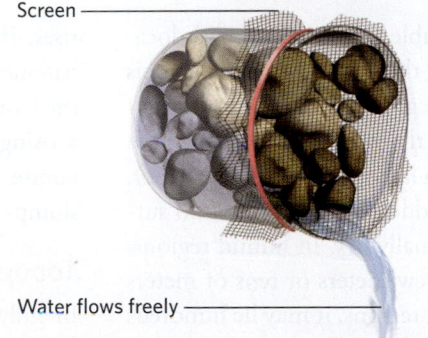

Screen

Water flows freely

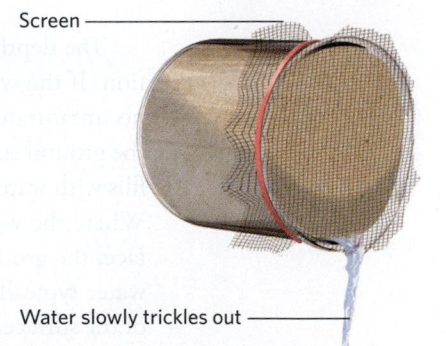

Screen

Water slowly trickles out

Figure 13.35 Water underground: aquifers, aquitards, and the water table.

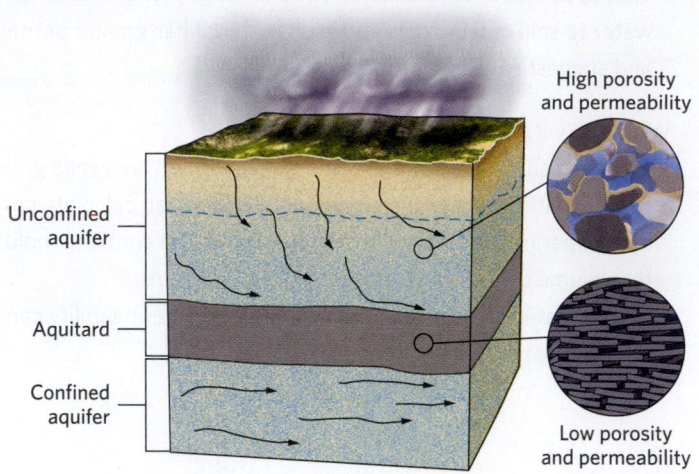

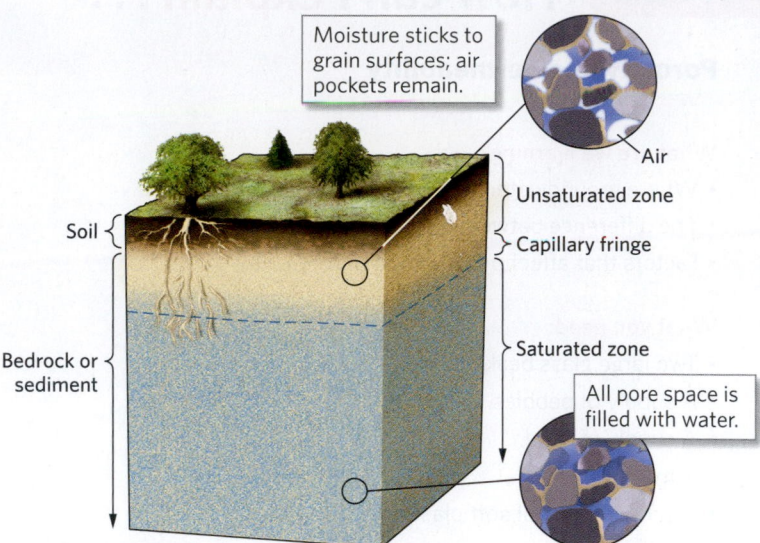

(a) An aquifer consists of high-porosity, high-permeability rock. An aquitard consists of low-permeability rock. Some aquifers are unconfined, and some are confined.

(b) The water table is the top of the groundwater reservoir. It separates the unsaturated zone above from the saturated zone below. The capillary fringe lies just above the water table.

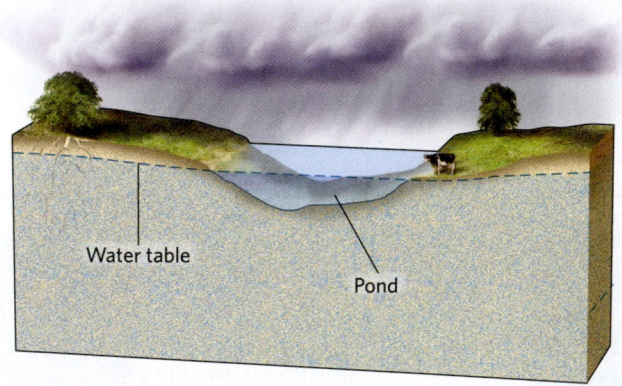

(c) Where the water table lies close to the ground surface, ponds remain filled—the water table is the surface of the pond.

(d) In dry regions and during dry seasons, the water table sinks deep below the surface. Water that collects temporarily in low areas infiltrates the subsurface.

The depth of the water table varies greatly with location. If the water table lies at the ground surface, there is no unsaturated zone, and the ground is wet and soggy. If the ground surface lies below the water table, the low area fills with water and becomes a lake or stream **(Fig. 13.35c)**. Where the water table lies hidden below the ground surface, the ground surface is usually dry. In humid regions, water typically lies within a few meters or tens of meters of the surface, whereas in arid regions, it may lie hundreds to thousands of meters below the surface.

Rainfall can change the water table depth in a given locality, in that the water table drops during the dry season and rises during the wet season. Streams or ponds that hold water during the wet season may dry up during the dry season because their water infiltrates the ground below **(Fig. 13.35d)**. In a particularly wet season, the water table rises. If the water table rises above the level of a house's basement, water seeps through the foundation and floods the basement floor. Catastrophic damage can occur when a rising water table weakens the base of a hillslope or a failure surface underground, triggering landslides and slumps, as we saw in Chapter 12.

Topography of the Water Table

In hilly regions, the shape of the water table tends to mimic, in a subdued way, the shape of overlying topography **(Fig. 13.36a)**. This means that the water table lies at a higher elevation beneath hills than it does beneath valleys. It may seem surprising that the elevation of the water table varies as a consequence of ground-surface topography. After all, when you pour a bucket of water into a pond, the surface of the pond immediately adjusts to remain horizontal. The elevation of the water table varies because groundwater moves so slowly through rock and sediment that it cannot quickly assume a horizontal surface. When rain falls on a hill and water infiltrates to the water table, the water table rises slightly. During a period without rain, the water table sinks slowly, but so slowly that by the time it rains again and causes the water table to rise again, the water table hasn't had time to sink very far.

In some locations, lens-shaped layers of impermeable rock (such as shale) lie within a thick aquifer. A mound of groundwater may accumulate above such an *aquitard lens* at an elevation above the regional water table. Hydrogeologists refer to the top surface of such a mound of groundwater as a **perched water table (Fig. 13.36b)**.

Take-home message . . .

Most underground water fills pores and cracks in rock or sediment. Porosity is the total volume of pore space within a material, whereas permeability is the degree to which that material's pores connect. Aquifers have high porosity and permeability, whereas aquitards do not. The boundary between the unsaturated zone and the saturated zone is called the water table.

Quick Question -
Do we find groundwater in the Earth's core?

Figure 13.36 Factors that influence the position of the water table.

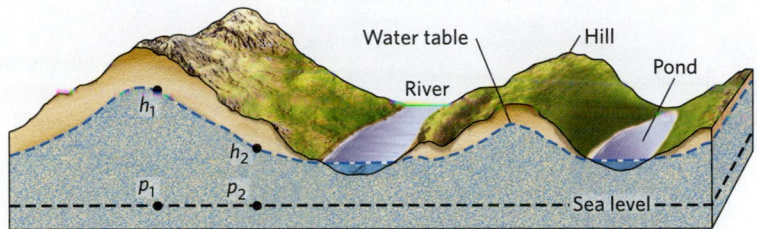

(a) The shape of a water table beneath hilly topography. Point h_1 on the water table is higher than Point h_2 relative to a reference elevation (sea level). The pressure at p_1 is, therefore, higher than the pressure at p_2.

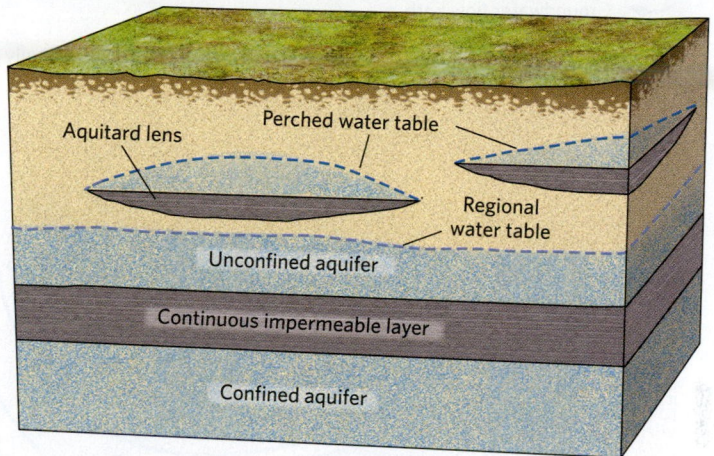

(b) A perched water table occurs where a lens of groundwater becomes trapped above an aquitard lens that lies above the regional water table.

13.9 Groundwater Flow

In the unsaturated zone, water simply percolates downward, like the water passing through a drip coffee maker, for this water moves only in response to the downward pull of gravity. But in the zone of saturation below the water table, groundwater flow is more complex, for this water moves not only in response to gravity, but also in response to differences in pressure. In fact, pressure can cause groundwater to flow sideways, or even upward—if you've ever watched water spray from a fountain, you've seen pressure pushing water upward. Therefore, to understand the nature of groundwater flow, we must first understand the origin of pressure in groundwater. For simplicity, we'll consider only the case of groundwater in an unconfined aquifer.

The pressure on groundwater at any specific point underground comes from the weight of all the overlying water from that point up to the water table. (The weight of overlying rock does not contribute to the pressure exerted on groundwater because contact points between mineral grains bear the rock's weight.) As a result, a point at a greater depth below the water table feels more pressure than a point at a lesser depth. If the water table is horizontal, an imaginary horizontal plane at a particular depth below the water table feels the same pressure everywhere. But if the water table is not horizontal, the pressure exerted at points on a horizontal plane at a particular depth changes from point to point (see Fig. 13.36a).

Both the elevation of a volume of groundwater and the pressure within the water provide energy that, if given the chance, drives groundwater flow. Hydrogeologists have determined that groundwater flow typically follows along concave-up curved paths, as viewed in cross section **(Fig. 13.37a)**. These curved paths eventually take groundwater from regions where the water table is high (such as under a hill) to regions where the water table is low (such as below a valley). A location where water enters the ground—meaning a place where the flow has a downward trajectory—is a **recharge area**, whereas a location where groundwater flows back up to the ground surface is a **discharge area**. Groundwater that follows paths taking it deep into the subsurface generally must travel further from recharge to discharge areas, and therefore stays underground longer, than does groundwater that follows a shallow path **(Fig. 13.37b)**.

Thousands have lived without love, not one without water.

—W. H. AUDEN (1907-1973)

Figure 13.37 The flow of groundwater.

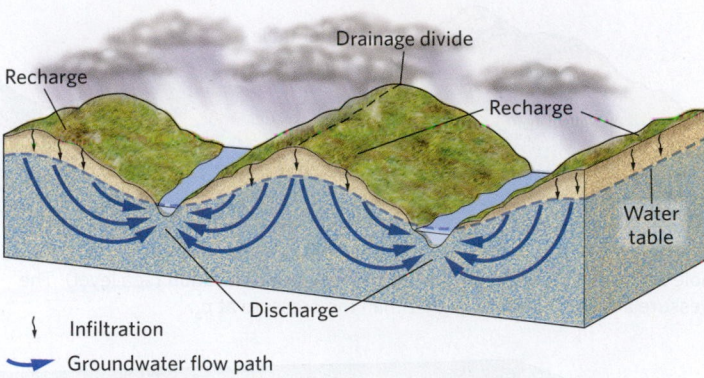

Infiltration

Groundwater flow path

(a) Groundwater flows from recharge areas to discharge areas. Typically, the flow follows curving concave-up paths.

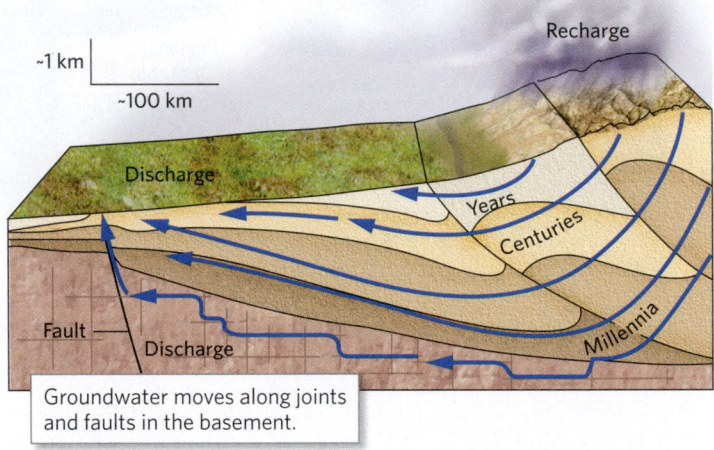

(b) A large elevation difference in the water table may drive groundwater hundreds of kilometers, across regional sedimentary basins. Groundwater that follows deeper flow paths stays underground longer.

Figure 13.38 Spring water nurtures this small oasis in the Sahara.

Flowing water in a steep river channel can reach speeds of up to 30 km (20 miles) per hour. In contrast, groundwater moves at rates between 5 and 500 m (16 to 1,600 feet) per year. Groundwater moves much more slowly than does surface water for two reasons. First, groundwater must follow a crooked network of tiny conduits, so it must travel a much greater distance than it would if it could follow a straight path. Second, friction between groundwater and conduit walls slows down the water flow. Simplistically, the actual velocity of groundwater flow at a location depends on the slope of the water table and the permeability of the material through which the groundwater is flowing. Groundwater flows faster through high-permeability rocks than it does through low-permeability rocks, and it flows faster in regions where the water table has a steep slope than it does in regions where the water table has a gentle slope. A French researcher, Henry Darcy, developed an equation, now known as *Darcy's law*, that describes the controls on groundwater flow more completely.

Take-home message . . .

Gravity and pressure cause groundwater to flow slowly from recharge areas to discharge areas. In essence, the rate of flow depends on the water table's slope and on permeability. Groundwater flow can follow curving paths that take it deep into the crust.

Quick Question -
Does steepening the water table increase or decrease the rate of groundwater flow?

13.10 Tapping Groundwater Supplies

If you have an opportunity to fly over an arid region in an airplane, take a look out the window and watch for circular fields of green crops. In the center of each such field a pump sucks up groundwater to supply a sprinkler that pivots around the field to keep the crops irrigated. Groundwater may be the only source of clean freshwater in arid regions where surface water doesn't exist, or in temperate or humid regions where surface water has been polluted. We can obtain groundwater either from **wells**, holes that people dig or drill to obtain water, or from **springs**, natural outlets from which groundwater flows.

Springs

Imagine an oasis of green palm trees surrounded by the dry sands of a desert **(Fig. 13.38)**. Where does the water that these plants need to survive come from? The water for a desert oasis flows out of a spring, a natural outlet from which groundwater naturally spills or seeps onto the

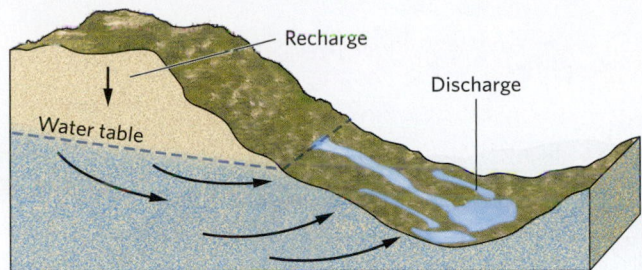

(a) Groundwater reaches the ground surface in a discharge area.

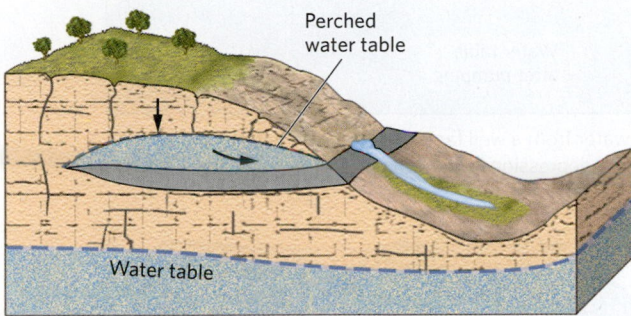

(c) Groundwater seeps from the ground where a perched water table intersects a slope.

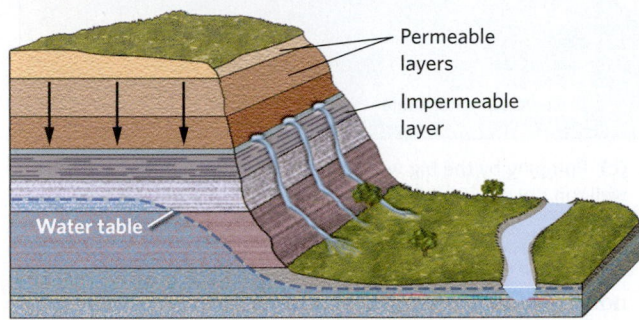

(e) Groundwater seeps out of a cliff face at the top of an impermeable bed.

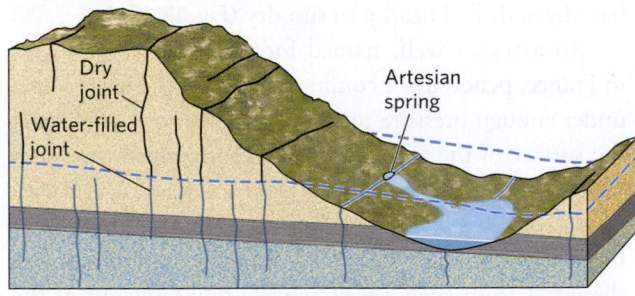

(f) Where water under pressure lies below an aquitard, a crack may provide a pathway for an artesian spring to form.

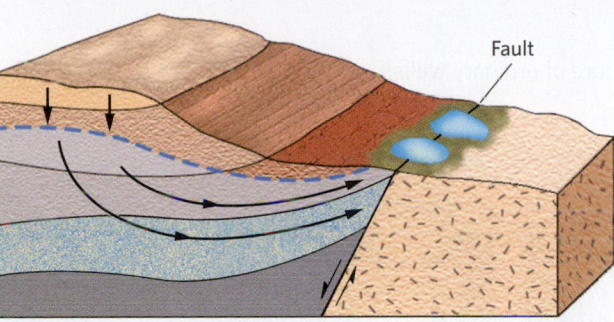

(b) Where groundwater reaches an impermeable barrier, it rises.

Figure 13.39 Springs may form in a variety of geologic settings.

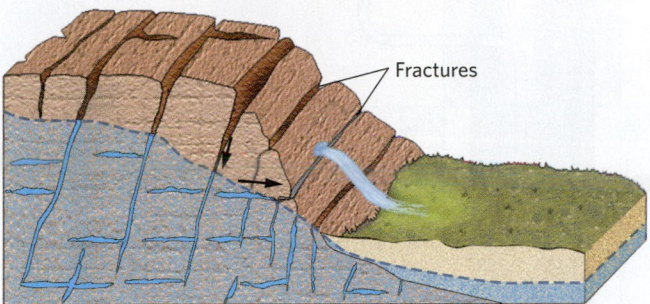

(d) A network of interconnected fractures channels water to the surface of a hill.

(g) A spring on the wall of the Grand Canyon.

ground surface. Springs can provide fresh, clear water for drinking or irrigation without the expense of drilling or digging, so many villages and towns worldwide have grown up adjacent to springs. While we can see springs that spill water onto dry land, many springs are submerged beneath streams or lakes. These springs provide a steady source of the water that helps to keep the stream or lake filled. Springs form in a variety of geologic settings **(Fig. 13.39a–g)**.

Figure 13.40 The nature of ordinary wells.

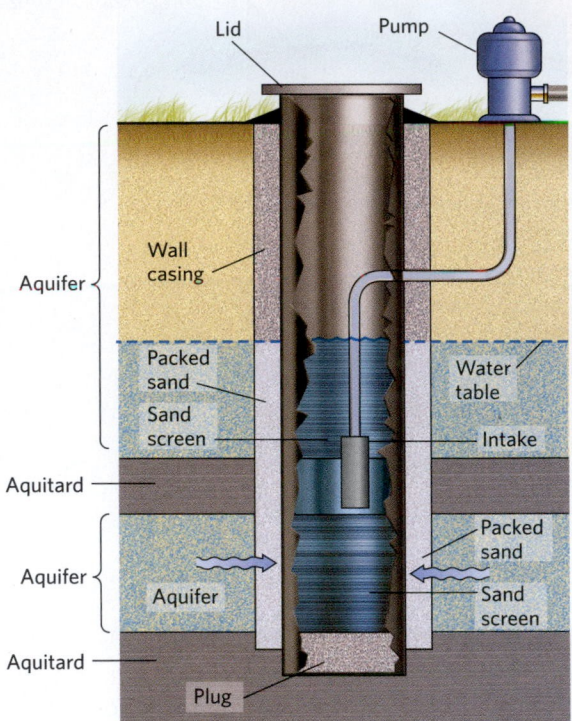

(a) A modern ordinary well sucks up water with an electric pump. The packed sand filters the water.

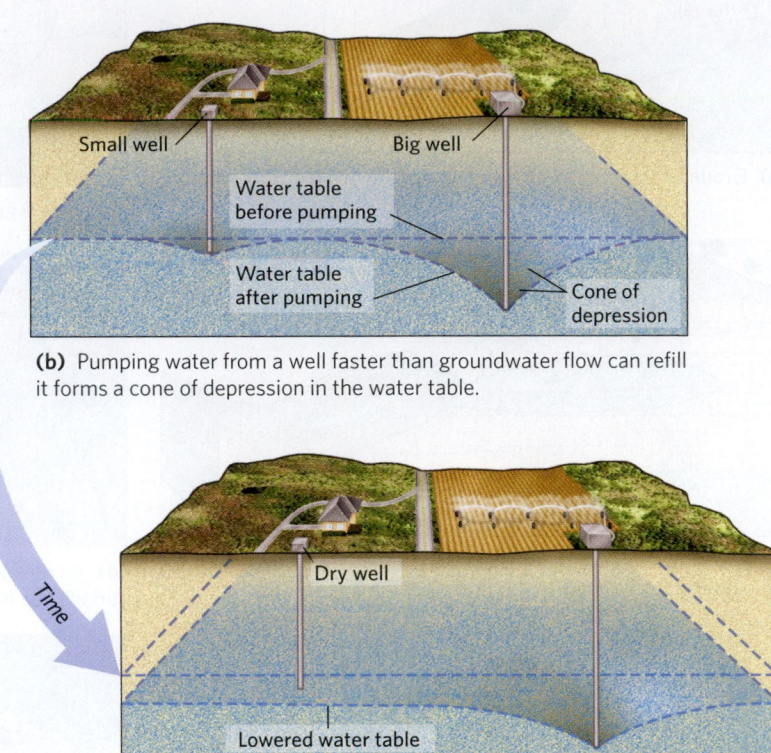

(b) Pumping water from a well faster than groundwater flow can refill it forms a cone of depression in the water table.

(c) Pumping by the big well may be enough to make the small well run dry.

Wells

Millennia ago, farmers and villagers learned that they could access groundwater, even where no springs exist, by digging a hole. Today, people still construct wells by digging holes, but it's usually easier to bore a well by drilling **(Fig. 13.40a)**. Hydrogeologists distinguish between two types of wells based on the pressure of water in the well.

In an *ordinary well*, the base of the well lies below the water table, so water simply seeps from the aquifer into the well and fills it to the level of the water table. Drilling into an aquitard, or into rock or sediment that lies above the water table, will not provide water, and therefore yields a *dry well*. Some ordinary wells are seasonal, meaning that they function only during the rainy season, when the water table rises above the base of the well. During the dry season, the water table lies below the base of such wells, so they becomes dry.

To extract water from an ordinary well, you can either pull the water up in a bucket or you can pump the water out. As long as the rate at which groundwater seeps into the well equals or exceeds the rate at which you remove the water, the level of the water table near the well remains about the same. But if you pump water out of the well too fast, the water table sinks around the well—a phenome-

non called *drawdown*. When this happens, the water table forms a downward-pointing, cone-shaped surface, called a **cone of depression**, surrounding the well **(Fig. 13.40b)**. Drawdown by a deep well may cause shallower wells that have been drilled nearby to run dry **(Fig. 13.40c)**.

An **artesian well**, named for the province of Artois in France, penetrates a confined aquifer in which water is under enough pressure to rise on its own to a level above the surface of the aquifer. If the level to which the water table would rise lies below the ground surface, the well is a *nonflowing artesian well*. But if the level lies above the ground surface, the well is a *flowing artesian well*—the water from such a well actively fountains out of the ground. An artesian well exists when the recharge area for the confined aquifer lies at a higher elevation than the site of the well **(Fig. 13.41a, b)**.

Hot Springs

Around 60 C.E., the Romans occupying Britain build a public bathhouse in what is now the city of Bath, England. Soldiers would spend hours relaxing in pools filled with hot water. Why build such a spa in Bath? The Romans didn't have to build stoves to heat the water, for the town of Bath surrounds a hot spring.

Hot springs, defined as natural springs whose water has a temperature of between 30°C (86°F) and 104°C (219°F), form in two types of geologic settings. Because of the geothermal gradient, temperatures rise above 30°C at a depth of several kilometers underground in most places. Where groundwater can rise from these depths at a rate that is fast enough so that the water doesn't lose its heat as it rises, it will remain hot when it bubbles out of a spring. Such conditions develop in places where faults or fractures provide a high-permeability conduit for rising deep water. Hot springs also develop in regions of igneous activity, where groundwater within magma-heated rock lies relatively close to the Earth's surface (Fig. 13.42a).

Numerous distinctive features form in association with hot springs in regions of igneous activity. For example, natural pools fed by hot springs may become brightly colored by thermophilic (heat-loving) bacteria and archaea that thrive in the hot water (Fig. 13.42b). Many such natural pools have become popular as spas (Fig. 13.42c). In places where the hot water rises into soils rich in volcanic ash and clay, a viscous slurry forms and fills bubbling *mud pots*. When hot water that contains a high concentration of dissolved minerals spills out of hot springs and then cools, the minerals precipitate out of the water, forming colorful mounds or terraces of travertine and other chemical sedimentary rocks (Fig. 13.42d). In some regions of hot springs, scalding water fountains out of the springs periodically. Such fountains are called *geysers* (Box 13.3).

Take-home message . . .

Groundwater can be obtained from springs (natural outlets) and wells (built by people). In ordinary wells, water must be lifted to the surface, but in artesian wells and springs, it rises due to pressure within the aquifer. Pumping of groundwater lowers the water table. In regions of igneous activity, or where groundwater rises quickly from deep in the crust, hot springs develop.

Quick Question -
Why might pumping water from a deep well cause nearby wells to run dry?

13.11 Caves and Karst

The Development of Caves

In 1799, as legend has it, a hunter by the name of Houchins was tracking a bear through the woods of Kentucky when the bear suddenly disappeared on a hillslope. Baffled, Houchins plunged through the brambles trying to sight his prey. Suddenly he felt a draft of surprisingly cool air flowing down the slope from uphill. Curious,

Figure 13.41 The nature of artesian wells.

(a) A flowing artesian well in Wisconsin. Groundwater rises from a confined aquifer into the corrugated pipe without the need of pumping and continuously flows out of the pipe.

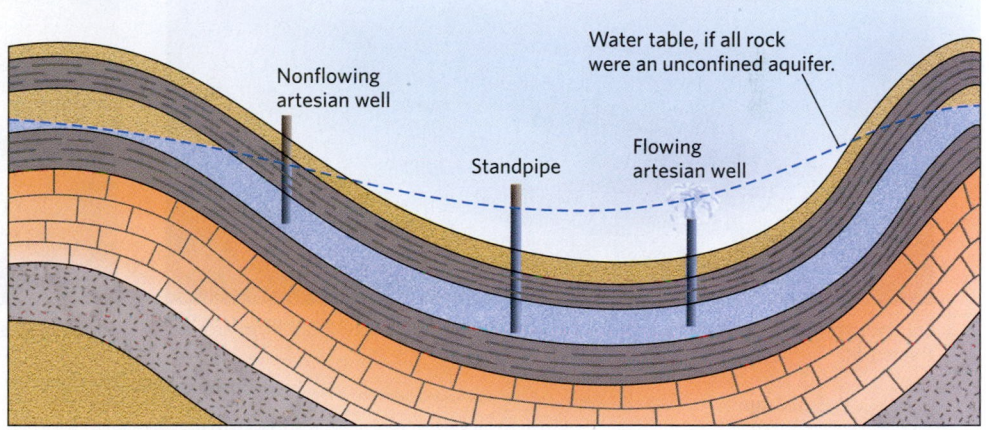

(b) The configuration of a regional artesian system. Water rises above the confined aquifer because the water is under pressure.

Houchins climbed up the hill and found a dark portal into the hillslope beneath a ledge of rocks. With a lantern, he cautiously stepped into the passageway, and after walking a short distance, found himself in a large underground room. Houchins had discovered an entrance to Mammoth Cave, an immense network of natural tunnels and subterranean chambers—a walk through the entire network would extend for 630 km (390 miles)!

Most large cave networks develop in limestone bedrock because limestone dissolves relatively easily in slightly acidic groundwater. Cave formation takes place just below the water table, for in this region groundwater acidity remains high, the mixture of groundwater and newly added rainwater has not become saturated with dissolved ions, and groundwater flows relatively fast.

Did you ever wonder . . .
why huge underground caves form?

Figure 13.42 Hot springs in areas of igneous activity.

(a) Hot springs occur where groundwater, heated at depth, rises to the ground surface.

(b) Colorful bacteria- and archaea-laden pools at Yellowstone National Park, Wyoming.

(c) Hot springs in Iceland, warmed by magma below, attract tourists from around the world.

(d) Terraces of minerals precipitated at Pamukkale, Turkey.

Figure 13.43 The locations of caves are controlled by jointing and bedding in limestone. Passageways follow joints, and chambers preferentially form in more soluble beds.

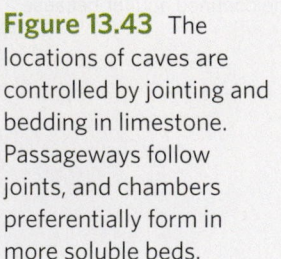

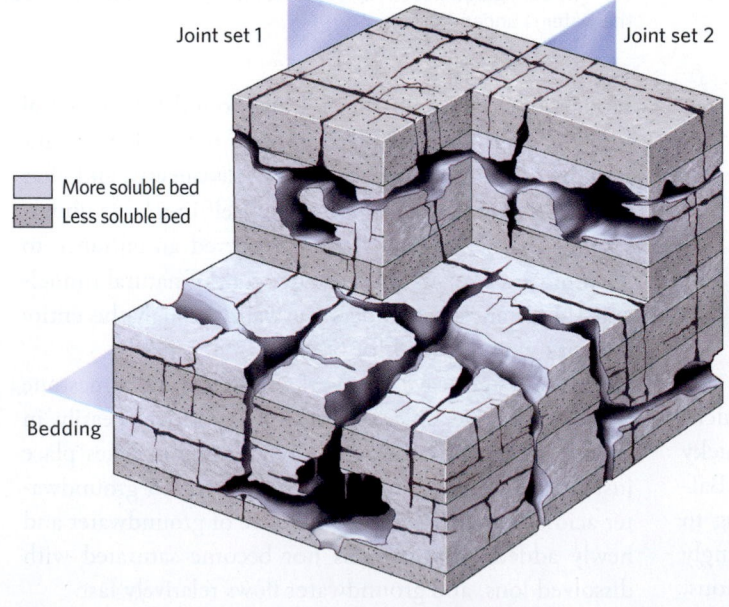

The Character of Cave Networks

Cave networks include rooms, or *chambers*—large, open spaces, sometimes with cathedral-like ceilings—and tunnel-shaped or slot-shaped *passages* (**Earth Science at a Glance**, pp. 450–451). Some chambers may host underground lakes, and some passages may serve as conduits for underground streams. The shape of a cave network reflects variations in the composition and in the permeability of the rock from which the caves formed. Larger open spaces develop where the limestone is most soluble and where groundwater flow is fastest. Passages in cave networks typically follow pre-existing joints, for the joints provide conduits along which groundwater can flow faster (**Fig. 13.43**).

Geysers

Under special circumstances, water emerges from a vent in the ground in a **geyser**, a fountain of steam and hot water that erupts episodically **(Fig. Bx13.3)**. (The word geyser comes from the name of the Icelandic spring *Geysir* and the word for gush.) Why do geysers form? Beneath a geyser lies a network of irregular fractures in very hot rock. Heat transferred from the rock to the groundwater in these fractures raises the water's temperature. Because the boiling point of water (the temperature at which water vaporizes) increases with increasing pressure, hot groundwater at depth can remain in liquid form even if its temperature rises above the boiling point of water at the Earth's surface. When such superheated groundwater begins to rise through a conduit toward the ground surface, the pressure on it decreases until, eventually, some of the water transforms into steam. The resulting expansion causes water higher up in the conduit to burst out at the ground surface. When this burst happens, pressure in the conduit from the weight of overlying water suddenly decreases. This sudden drop in pressure causes the superheated water at depth to turn into steam instantly, and this steam quickly rises, ejecting all the water and steam above it from the conduit in a geyser eruption. Once the conduit empties, the eruption ceases, and the conduit fills once again with water that gradually heats up, starting the eruptive cycle over again.

Figure Bx13.3   The Old Faithful geyser in Yellowstone National Park erupts predictably.

Precipitation and the Formation of Cave Deposits

When the water table drops below the level of a cave, the cave becomes an open space filled with air. In places where downward-percolating groundwater containing dissolved calcite emerges from the rock above the cave and drips from the ceiling, the surface of the cave gradually changes. As soon as groundwater enters the air, it releases some of its dissolved carbon dioxide and evaporates a little. As a result, calcite precipitates out of the water, producing travertine. The travertine forms various intricately shaped cave deposits, or **speleothems**.

Spelunkers (cave explorers) and geologists have developed a detailed nomenclature for different kinds of speleothems **(Fig. 13.44a)**. Where water drips from the ceiling of the cave, the precipitated limestone builds an icicle-like cone called a **stalactite**. Where the drips hit the floor, the resulting precipitate builds an upward-pointing cone called a **stalagmite**. If this process continues long enough, stalagmites merge with overlying stalactites to create massive *columns*. In some cases, groundwater flows along the surface of a wall and precipitates to produce drape-like sheets of travertine called *flowstone* **(Fig. 13.44b)**.

The Formation of Karst Landscapes

When Mae Owens looked out her window on May 8, 1981, she discovered that the large tree in the backyard of her Winter Park, Florida, home had suddenly disappeared. When Owens went outside to investigate, she found that more than the tree had vanished. Her whole backyard had become a deep, gaping pit. The pit continued to grow for a few days until it finally swallowed Owens's house and six other buildings, as well as a municipal swimming pool, part of a road, and several Porsches in a car dealer's lot **(Fig. 13.45a)**.

What had happened? The bedrock under Ms. Owens's town consists of limestone. Water had gradually dissolved

Caves and Karst Landscapes

Limestone pavement, Ireland

Disappearing stream

Sinkhole

Collapsed breccia

Dissolved joint

Stalagmite

Stalactite

Soda straw

Flowstone

Cavern

Stalactite

Limestone column

Underground stream

Underground pool

Corridor

Sinkholes

Underground pool, Mexico

Emerging spring

Natural Bridge, Virginia

Spelunker crawling in a cave

Limestone is soluble in acidic water. Underground openings that develop by dissolution are called caves or caverns. Some of these may be large, open rooms, whereas others are long, narrow passages. Underground lakes and streams may cover the floor. A cave's location depends on the orientation of bedding and joints, for these features localize the flow of groundwater.

Caves originally form at or near the water table. As the water table drops, caves empty of most water and become filled with air. In many locations, groundwater drips from the ceiling of a cave or flows along its walls. As the water evaporates and loses its acidity, new calcite precipitates as speleothems (stalactites, stalagmites, columns, and flowstone).

Distinctive landscapes, called karst landscapes, develop at the Earth's surface over limestone bedrock. In such regions, the ground may be rough where rock has dissolved along joints. Where the roofs of caves collapse, sinkholes develop. If a surface stream flows into a cave network, we say that the stream is disappearing. In some places, the collapse of subsurface openings leaves behind natural bridges.

Figure 13.44 Speleothems in caves.

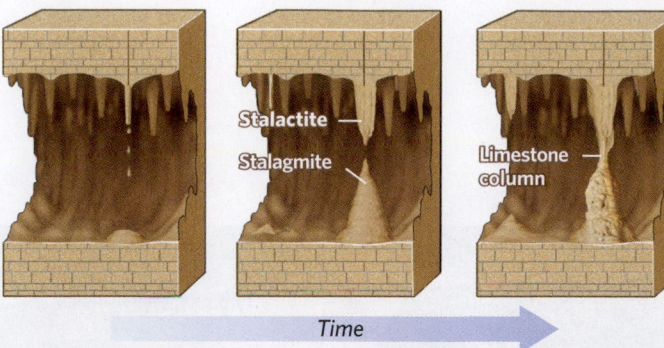

Stalactite
Stalagmite
Limestone column

Time

(a) A drip from the ceiling evolves into a stalactite. Precipitation from water that reaches the floor can form a stalagmite.

1 m

(b) Flowstone on the wall of a cave in Vietnam.

See for yourself

Sinkholes in Central Florida

Latitude: 28°37′50.59″ N
Longitude: 81°23′13.60″ W

Zoom to 20 km (~12.5 miles) and look down.

You can see several sinkholes, ranging from about 100 to 800 m (330 to 2,600 feet) across, that lie within suburban developments. The sinkholes have filled with water and are now lakes.

the limestone to produce caves underground. On May 8, the roof of the cave underneath Owens's backyard began to collapse, forming a circular depression called a **sinkhole** (Fig. 13.45b). Eventually, the sinkhole filled with water, and now it's a circular lake. Similar sinkhole lakes have formed throughout central Florida (Fig. 13.45c). Sinkhole formation can take place without warning and, sadly, has caused fatalities.

Geologists refer to regions such as central Florida, where surface landforms develop as underlying cave networks collapse, as **karst landscapes** or *karst terrains*, named for the Kras Plateau on the eastern side of the Adriatic Sea (Fig. 13.46a). Karst landscapes are characterized not only by sinkholes, but also by several other distinctive landforms. For example, surface streams in karst terrains may flow into cracks or holes that link to caves below (Fig. 13.46b). Such *disappearing streams* re-emerge from a cave entrance downstream. In places where most of the ground has collapsed into sinkholes, the landscape consists of narrow ridges with bowl-like depressions in

between. Over time, the lower parts of a ridge may erode away, leaving a natural bridge. When these bridges collapse, all that remains are tall spires or pinnacles of limestone. A karst landscape dominated by such pinnacles is called a *tower karst* terrain. One of the world's most spectacular examples, in the Guilin region of China, has inspired generations of artists (Fig. 13.47). The development of karst landscapes is summarized in Figure 13.48.

Take-home message . . .

Reaction with acidic groundwater dissolves limestone underground to form caves. Most of this dissolution takes place just below the water table. If the water table sinks, water dripping in caves can produce speleothems. Collapse of a cave network produces a karst landscape.

Quick Question -
How do sinkholes form?

13.12 Freshwater Challenges

Since prehistoric times, humans have used increasing quantities of freshwater for drinking, irrigation, and industry. We conclude this chapter with a discussion of challenges to the sustainability of this essential resource.

Vanishing Rivers

As humans evolved from hunter-gatherers into farmers, areas along rivers became attractive places to settle. Rivers serve as avenues for transportation and sources for food, irrigation water, drinking water, power, recreation, and (unfortunately) waste disposal. Further, agriculture utilizes floodplains, which have particularly fertile soils that are replenished annually by seasonal floods. Sadly, in recent centuries, rivers have come under threat from a host of environmental problems.

POLLUTION. Society dumps sewage, garbage, runoff from pavements, spilled oil and gasoline, toxic chemicals from industrial sites, excess fertilizer, and animal waste into rivers (Fig. 13.49a). Some of these pollutants poison aquatic life, some feed algal blooms that strip water of its oxygen, and some become buried on the streambed along with sediments.

DAM CONSTRUCTION. Today, there are over 38,000 large dams (over 15 m, or 50 feet, high) worldwide. Dam construction impounds reservoirs and, by doing so, floods towns, forests, and rapids upstream. Because a reservoir is a standing body of water, it traps sediments and nutrients carried by the stream, preventing them from reaching floodplains and deltas downstream.

Figure 13.45 Development of sinkholes in central Florida.

(a) The Winter Park sinkhole, as seen from a helicopter.

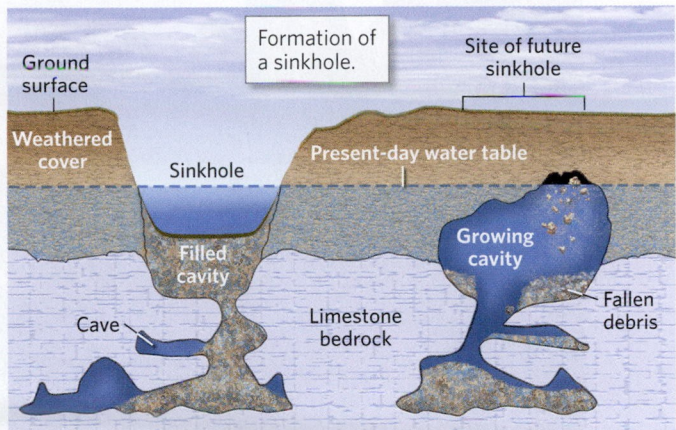

(b) The weathered overburden that forms the roof of a cave slowly washes into the underlying cave, forming a cavity that grows ever closer to the ground surface. When the roof of this cavity collapses, a sinkhole forms (not to scale).

(c) An aerial view of Florida sinkholes that have become lakes.

Figure 13.46 Features of karst landscapes.

(a) The rough, rocky surface of the Kras Plateau, for which karst landscapes were named, is formed by numerous sinkholes.

(b) A small disappearing stream in the Hudson Valley of New York; the water is spilling into a subsurface cave.

See for yourself

Karst Landscape in Puerto Rico

Latitude: 18°23'53.04" N
Longitude: 66°25'49.43" W

Zoom to 7 km (~4.3 miles) and look down.

We see a karst terrain in central Puerto Rico. Each of the rounded depressions is a sinkhole. Ridges of limestone separate adjacent sinkholes.

Figure 13.47 Tower karst in the Guilin region of China.

The landscape is treeless today, a consequence of industrialization policies in the 1950s.

Chinese artists painted scrolls depicting forested towers of karst.

Figure 13.48 The progressive formation of karst landscapes.

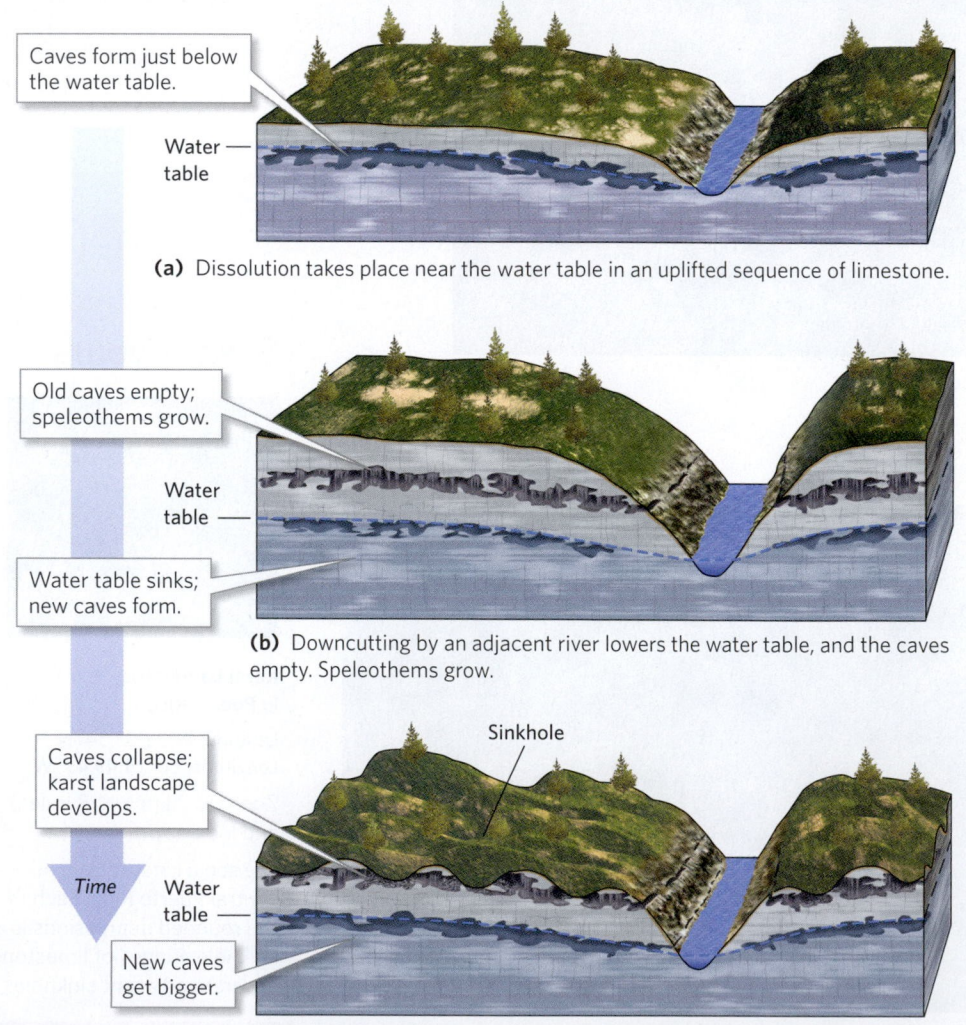

Caves form just below the water table.

Water table

(a) Dissolution takes place near the water table in an uplifted sequence of limestone.

Old caves empty; speleothems grow.

Water table

Water table sinks; new caves form.

(b) Downcutting by an adjacent river lowers the water table, and the caves empty. Speleothems grow.

Sinkhole

Caves collapse; karst landscape develops.

Time   Water table

New caves get bigger.

(c) After roof collapse, the landscape becomes pockmarked with sinkholes.

OVERUSE OF RIVER WATER. As human populations grow, our thirst for river water continues to increase, but the supply of water does not. As a result, in some places, human activity consumes the entire volume of a river's water, so that the channel contains little more than a saline trickle, if that, at its mouth **(Fig. 13.49b)**.

EFFECTS OF URBANIZATION AND AGRICULTURE. Urbanization transforms fields and forests into parking lots, roads, and buildings. The resulting layer of impermeable concrete and asphalt prevents rainfall from infiltrating the ground or being absorbed by plants, so the volume of runoff entering streams increases. This excess can contribute to flooding. Agriculture affects the amount of sediment that enters streams, for although we tend to think of farmland as vegetated land, it hosts much less plant cover than does natural grassland or forest. Sheetwash flowing across the unprotected land surface erodes and carries with it large volumes of sediment, so a river's sediment load increases significantly when farms replace forests.

Threats to Lakes

Lakes are delicate ecosystems. If they are clear and contain lots of dissolved oxygen, they can host a huge variety of life. But the introduction of nutrients (due to an influx of sewage or to fertilizer-rich runoff from fields or lawns) can upset this balance. This process of **eutrophication** can trigger an algal bloom that not only turns a lake bright green, but also diminishes the amount of dissolved oxygen in the water **(Fig. 13.50)**. Respiration by the algal cells at night absorbs oxygen from the water, and when the algae die, bacteria decompose them, a process that uses up even more oxygen. If the dissolved oxygen concentration

Figure 13.49 Sustainability challenges for rivers.

(a) Large quantities of garbage and various pollutants end up in rivers.

(b) The Central Arizona Project canal shunts water from the Colorado River to Phoenix.

decreases too much, fish and other organisms living in the water die off.

Lakes are also under threat in places where river diversion has cut off their supplies of water. Some lakes have dried up as a consequence.

Figure 13.50 An algal bloom produces green slime on the surface of a lake that has undergone eutrophication.

Challenges to Groundwater Resources

Unfortunately, groundwater supplies, though hidden from our view, face several of the same challenges that surface-water supplies face. Some of the damage already done to groundwater supplies cannot be remedied on a human time scale.

Did you ever wonder . . .

where bottled "spring water" comes from?

DEPLETION OF GROUNDWATER SUPPLIES. Is groundwater a renewable resource? In a time frame of 10,000 years, the answer is yes, for the hydrologic cycle will eventually resupply depleted groundwater reserves. But in a time frame of 100 to 1,000 years—the span of a human lifetime or a civilization—groundwater in many regions must be viewed as a nonrenewable resource. By pumping water out of the ground at a rate faster than nature replaces it, people are effectively "mining" groundwater reserves. In some places, reserves of fresh groundwater are already gone, and the only groundwater that remains is too mineralized or too salty to be usable. A number of other problems accompany the depletion of groundwater:

- *Lowering of the water table:* When we extract groundwater from wells at a rate faster than it can be resupplied by nature, the water table drops. As a consequence, existing wells, springs, streams, and swamps dry up **(Fig. 13.51a)**. To continue tapping into the groundwater supply, we must drill progressively deeper.

- *Reversal of groundwater flow direction:* The cone of depression that develops around a well creates a local slope in the water table that may be large enough to reverse the flow direction of nearby groundwater **(Fig. 13.51b)**. Such reversals can allow contaminants to reach wells.

- *Saltwater intrusion:* In coastal areas, fresh groundwater lies in a layer above saltwater that has entered the aquifer from the adjacent ocean **(Fig. 13.51c)**. Pumping water from a well too quickly can suck saltwater up into the well, a phenomenon called *saltwater intrusion.*

- *Pore collapse and land subsidence:* The extraction of water from a pore causes the grains around the pore to pack more closely together. Such *pore collapse* decreases the porosity and permeability of a rock or sediment and lessens its value as an aquifer. Pore collapse also decreases the volume of the aquifer, causing the ground above the aquifer to sink and leading to fissures at the surface **(Fig. 13.51d)**.

NATURAL CONTAMINATION OF GROUNDWATER. In most places, you can drink groundwater right out of the ground, for rocks and sediments serve as natural filters capable of trapping suspended solids. As a result, commercial distribution of bottled groundwater ("spring water") has become a major business worldwide. However, dissolved chemicals and, in some cases, methane make some natural groundwater unusable. For example, deep groundwater may be mineralized and salty and therefore unsuitable for irrigation or drinking. Groundwater that has passed through limestone or dolomite may contain dissolved calcium and magnesium ions; this *hard water* causes problems for consumers because carbonate minerals precipitate from it to form *scale* that clogs pipes, and because soap won't develop a lather in it. Groundwater that has passed through iron-bearing rocks may contain dissolved iron oxide that precipitates to form rusty stains. In recent years, concern has grown about arsenic, a highly toxic chemical that can enter groundwater by the dissolution of arsenic-bearing minerals. In Southeast Asia, a significant percentage of wells, drilled so that people could avoid drinking polluted surface water, unfortunately contain dangerous levels of arsenic.

CONTAMINATION OF GROUNDWATER BY HUMAN ACTIVITIES. In recent decades, increasing amounts of contaminants have been introduced into aquifers by human activities **(Fig. 13.52a)**. The cloud of contaminated groundwater that flows away from a contaminant source is called a **contaminant plume (Fig. 13.52b)**. Fortunately, in some cases, natural processes can clean up groundwater contamination, for clay binds to some contaminants, oxygen in the water reacts with and destroys others, and natural bacteria in the water metabolize still others. But where natural processes can't remove all the contamination, hydrogeologists try to remediate the situation. In some cases, they track the direction of plume flow and close wells in the path of the plume to prevent consumption of contaminated water. They may then attempt to remove and clean the groundwater by drilling a series of steam-injection wells. Steam injected into the ground pushes the contaminants to extraction wells, where they are pumped out of the ground and sent to a treatment plant **(Fig. 13.52c)**. More recently, hydrogeologists have begun exploring novel techniques for cleaning contaminated groundwater underground. For example, they may build a *reactive barrier*, an underground wall of materials (such as iron filings) that react with

Figure 13.51 Effects of groundwater depletion.

Before

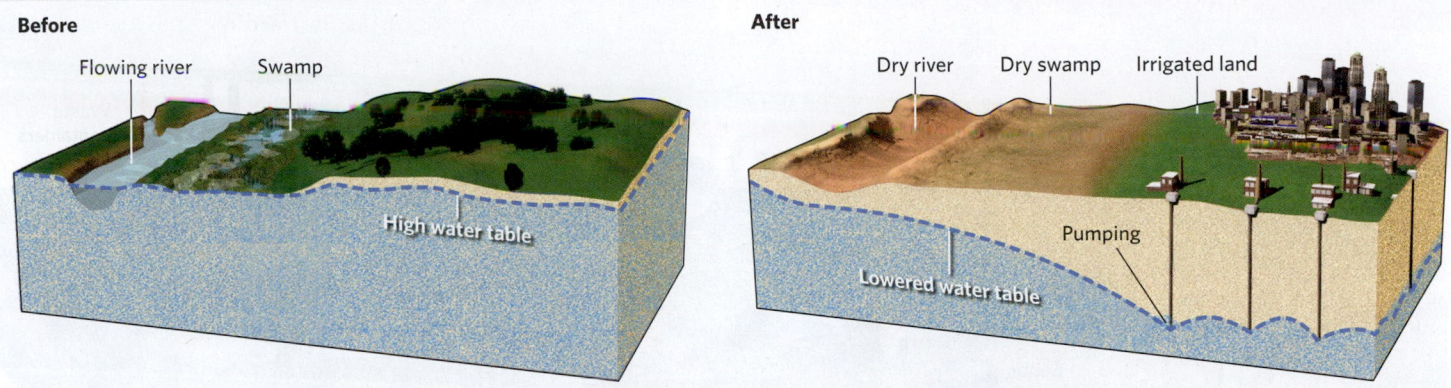

After

(a) Lowering of the water table. (Left) Before humans start pumping groundwater, the water table is high. A swamp and permanent stream exist. (Right) Pumping for consumers in a nearby city causes the water table to sink, so the swamp and stream dry up.

Before

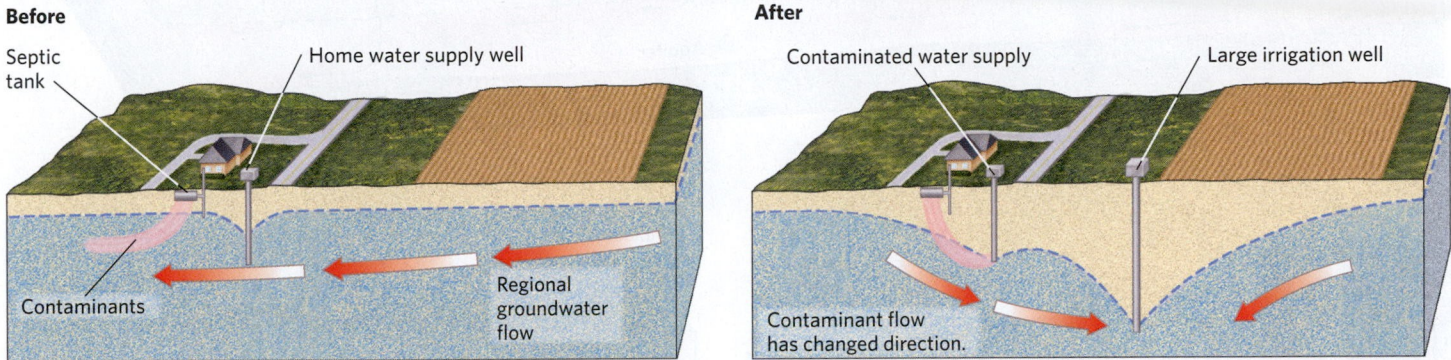

After

(b) Contamination of wells. (Left) Before pumping, effluent from a septic tank drifts with the regional groundwater flow, and the home well pumps clean water. (Right) After pumping by a nearby irrigation well forms a cone of depression, effluent flows into the home well in response to the new local slope of the water table.

Before

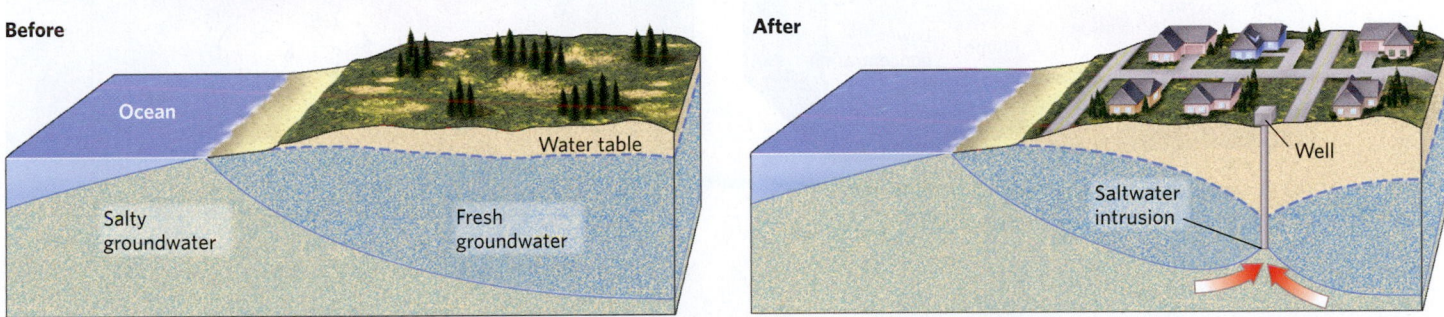

After

(c) Saltwater intrusion. (Left) Before pumping, fresh groundwater forms a lens below the ground. (Right) If the freshwater is pumped too fast, saltwater from below is sucked up into the well.

Before

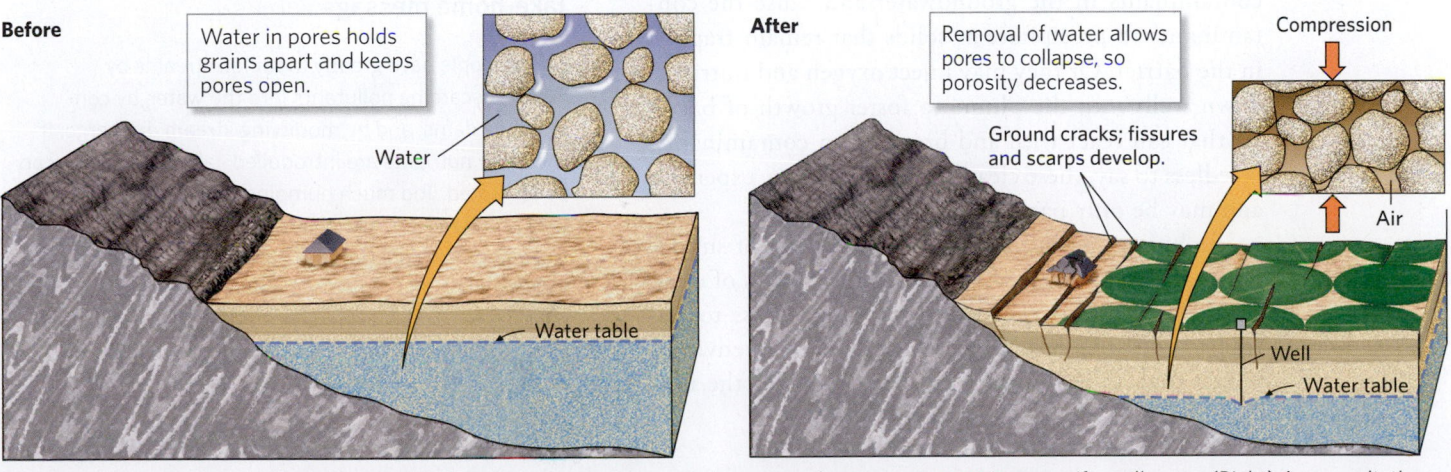

After

(d) Pore collapse and land subsidence. (Left) When intensive irrigation removes groundwater, pore space in an aquifer collapses. (Right) As a result, the land surface sinks, leading to the formation of ground fissures and causing houses to crack.

Figure 13.52 Contamination of groundwater and its remediation.

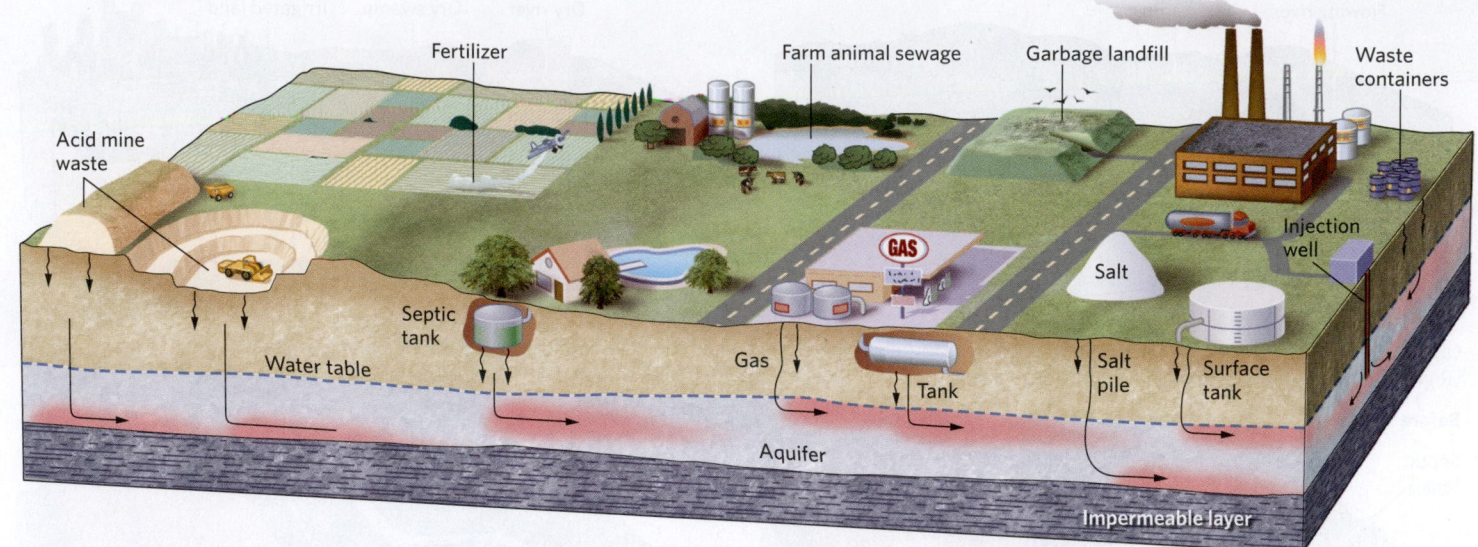

(a) Various sources of groundwater contamination.

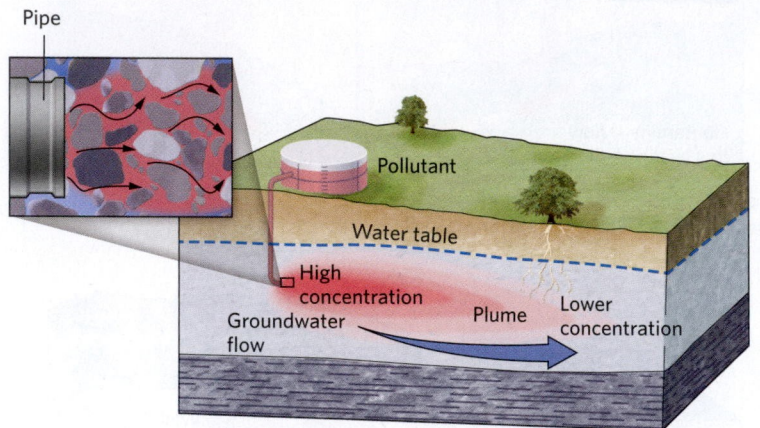

(b) A contaminant plume as seen in cross section. The darker the color, the greater the concentration of the contaminant.

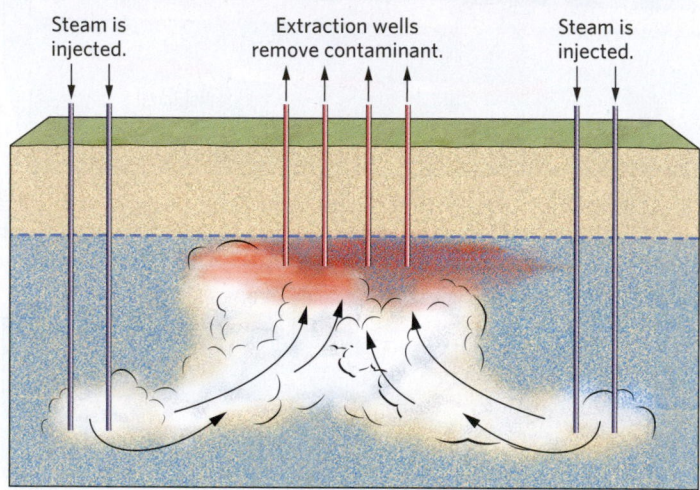

(c) Steam injected beneath the contamination drives the contaminated water upward in the aquifer, where extraction wells remove it.

contaminants in the groundwater and cause the contaminants to precipitate as solids that remain trapped in the barrier. Or they may inject oxygen and nutrients down wells into the plume to foster growth of bacteria that can react with and break down contaminants. Needless to say, these cleaning techniques are expensive and may be only partially effective.

As human populations increase, our use of surface water and groundwater increases, and supplies of clean freshwater become ever more precious. Access to water becomes a particular challenge in extreme environments such as the deserts that we discuss in the next chapter.

Take-home message . . .

People have greatly modified streams by discarding pollutants into the water, by constructing dams, and by modifying stream discharge. If too many nutrients are introduced, lake ecosystems can be disrupted. Too much pumping of groundwater lowers the water table and may cause land subsidence or saltwater intrusion. Contamination can ruin a groundwater supply, and remediation of groundwater contamination is extremely expensive.

Quick Question -
Can contaminants be removed from groundwater?

Another View Swamps are swamps because the land surface lies just below the water table. The bald cypress trees of this swamp in southern Illinois have adapted to life in soggy ground. During dry seasons, the water table sinks below the ground surface, and the land dries out.

13 CHAPTER REVIEW

Chapter Summary

- Streams grow by downcutting and headward erosion. A drainage network consists of many tributaries that flow into a trunk stream. Drainage divides separate adjacent watersheds.

- The discharge of a stream varies with location along the stream and with climate.

- Streams erode the landscape and carry sediment as dissolved, suspended, and bed loads. When stream water slows, its competence decreases, and it deposits alluvium.

- Typically, a stream has a steeper gradient at its headwaters than near its mouth. The base level limits the depth of downcutting.

- Streams cut valleys or canyons, depending on the rate of downcutting relative to the rate of mass wasting on bordering slopes. Locally, they flow over rapids and waterfalls.

- A meandering stream winds back and forth across a floodplain. Eventually, a meander may be cut off. Braided streams consist of many strands separated by bars.

- Streams deposit alluvial fans or deltas at their mouths.

- If more water enters a stream than the channel can hold, flooding takes place. Some floods develop slowly and submerge large regions. Flash floods happen very rapidly and affect a smaller region. Officials try to prevent floods by building reservoirs and levees.

- Lakes are bodies of standing water on land. Most lakes have both an inlet and an outlet, so they contain freshwater. If a lake has inlets but no outlets, the lake becomes salty.

- Groundwater fills the pores and cracks in rock and sediment. It flows more easily through aquifers than through aquitards.

- The water table separates the unsaturated zone from the saturated zone, in which groundwater fills pores. The shape of the water table reflects the shape of the overlying topography.

- Groundwater flows from recharge areas to discharge areas. The velocity of flow depends on permeability and the slope of the water table.

- Groundwater can be extracted in wells. An ordinary well penetrates below the water table, but in an artesian well, water rises on its own. Pumping water out of a well too fast causes drawdown.

- Where limestone dissolves just below the water table, caves develop. If the water table drops, the caves empty out, and travertine precipitates out of water dripping from cave roofs to produce speleothems.

- Regions where caves have collapsed to form sinkholes and natural bridges are called karst landscapes.

- Rivers are being damaged by pollution, damming, and overuse. Some regions have lost their groundwater supply because of overuse or contamination.

Key Terms

abrasion (p. 421)
alluvial fan (p. 426)
alluvium (p. 422)
annual probability (p. 436)
aquifer (p. 440)
aquitard (p. 440)
artesian well (p. 446)
bar (p. 422)
base level (p. 423)
capacity (p. 422)
channel (p. 417)
competence (p. 421)
cone of depression (p. 446)
contaminant plume (p. 456)
delta (p. 426)
discharge (p. 418)
discharge area (p. 443)

downcutting (p. 417)
drainage basin (p. 418)
drainage divide (p. 418)
drainage network (p. 418)
ephemeral stream (p. 420)
eutrophication (p. 454)
flash flood (p. 434)
flood (p. 431)
floodplain (p. 422)
flood stage (p. 435)
freshwater (p. 415)
geyser (p. 449)
groundwater (p. 438)
headward erosion (p. 417)
hot spring (p. 447)
infiltration (p. 438)
karst landscape (p. 452)

meander (p. 426)
natural levee (p. 426)
oxbow lake (p. 426)
perched water table (p. 443)
permanent stream (p. 420)
permeability (p. 440)
point bar (p. 426)
pore (p. 439)
porosity (p. 439)
pothole (p. 421)
rapid (p. 424)
recharge area (p. 443)
recurrence interval (p. 436)
runoff (p. 417)
salt lake (p. 438)
sediment load (p. 421)
sinkhole (p. 452)

slow-onset flood (p. 431)
speleothem (p. 449)
spring (p. 444)
stalactite (p. 449)
stalagmite (p. 449)
stream (p. 417)
stream gradient (p. 422)
stream piracy (p. 431)
stream rejuvenation (p. 429)
tributary (p. 417)
turbulence (p. 419)
waterfall (p. 424)
water table (p. 440)
well (p. 444)

Review Questions

The letters following each Review Question refer to the corresponding Learning Objective from the Chapter Opener.

1. What is a drainage network? Describe the different patterns of drainage networks recognized by geologists. Which does the figure show? **(B)**

2. Distinguish between permanent and ephemeral streams, and explain why they exist. **(B)**

3. How does discharge vary along a stream's length, and how is it affected by climate? **(C)**

4. Describe how streams carry their sediment load. Distinguish between a stream's competence and its capacity. **(C)**

5. What is the difference between a local and an ultimate base level? **(B)**

6. How does a braided stream differ from a meandering stream? **(C)**

7. Describe how meanders form and evolve. **(C)**

8. How do deltas grow and how do they differ from alluvial fans? **(C)**

9. What is stream piracy? What causes a drainage reversal? How does a drainage network change when stream rejuvenation takes place? **(B)**

10. What is the difference between a slow-onset flood and a flash flood? What phenomena can cause flooding, and how do people try to prevent flooding? **(D)**

11. What is the recurrence interval of a flood, and how is it related to the annual probability of flooding? **(D)**

12. What determines whether the water in a lake is fresh or salty? **(E)**

13. How do porosity and permeability differ? Using these terms, contrast an aquifer with an aquitard. **(G)**

14. What is the water table, and what factors affect its level? What factors affect the flow direction of the groundwater? Why did the stream in the diagram dry up? **(H)**

15. What factors control the rate of groundwater flow, and how does the rate of flow compare with that of a surface stream? **(F)**

16. How is an artesian well different from an ordinary well? **(I)**

17. Why do natural springs form? Explain why hot springs form and why geysers erupt. **(I)**

18. Describe the process leading to the formation of caves, speleothems, and various features of a karst landscape. **(J)**

19. How have humans overused and abused streams and lakes? **(K)**

20. Is groundwater a renewable or a nonrenewable resource? Describe ways in which human activities affect groundwater. **(K)**

On Further Thought

21. Records indicate that the height of the Mississippi River above flood stage, for a given amount of discharge, has been getting higher since 1927, when a system of levees began to block off portions of the floodplain. Why? **(D)**

22. The population of a town in the desert Southwest of the United States has been doubling every 10 years. The town has been growing on a flat, gravel-filled basin between two small mountain ranges. Where does the water supply for the town come from? What do you predict will happen to the water table of the area in coming years? **(K)**

Online Resources

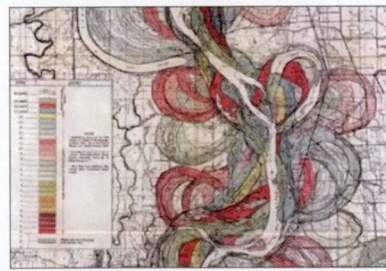

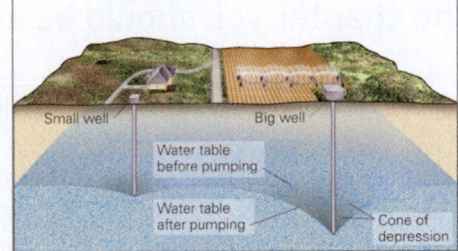

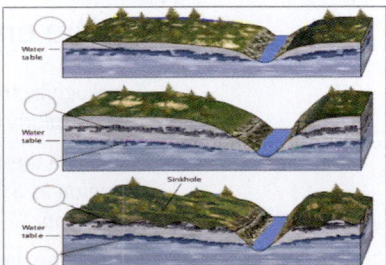

Videos
This chapter features videos about rivers that show how they change over time and the consequences of removing groundwater too quickly.

Smartwork5
This chapter features visual identification and labeling exercises on the effect of water systems on the landscape over time.

14 EXTREME REALMS
Desert and Glacial Landscapes

By the end of the chapter you should be able to . . .

A. recognize the regions of the Earth's surface that are particularly dry or cold.

B. explain the conditions that lead to the designation of a region as a desert and the reasons such conditions can develop.

C. describe the unique processes of weathering, erosion, and deposition that lead to the formation of desert landscapes.

D. analyze the potential of human activities to change regions into deserts.

E. explain how glacial ice forms and flows, and distinguish among various kinds of glaciers.

F. discuss how glaciers advance and retreat, and how they erode the land surface.

G. distinguish between sediments and depositional landforms left by glaciers and those formed in other ways.

H. explain what an ice age is and the evidence used to determine when ice ages have happened.

I. evaluate the ideas proposed to explain why ice ages happen and why glaciations during an ice age are periodic.

14.1 Introduction

By the end of the 18th century, sailors seeking trade routes had compiled maps of coastlines worldwide. But large inland regions remained *terra incognita* (unknown land). Many of the blank spaces on the maps were realms of extremes—our planet's rainforests, deserts, frozen terrains, and high mountains. Europeans didn't begin mapping these realms in earnest until the 19th century **(Fig. 14.1a,b)**. It proved to be a dangerous task, and some expeditions never returned. Gradually, however, the blanks on the maps filled in. Since the 20th century, with the aid of airplanes and, eventually, satellites, researchers have enriched our scientific understanding of these realms, and in the 21st century, you can take a virtual tour of the Earth's harshest, most isolated localities on your own computer screen.

In this chapter, we introduce our planet's realms of extreme dryness, extreme cold, or both: deserts and glaciated landscapes. First, we learn why deserts develop where they do, how erosion and deposition shape their surfaces, and how semiarid landscapes may become deserts. Then, we turn our attention to landscapes that are, or were, covered by glaciers. We see how glaciers form and move, and how they sculpt the ground and deposit sediment. This chapter concludes by describing the consequences and causes of *ice ages*, times when glaciers grew to cover large areas of the continents.

Because of climate extremes, we see vast differences in landscapes on our planet. You can see these differences by comparing a vista of sand dunes and rocky ridges in the hot, parched Mojave Desert of California, with one of snow-dusted, glaciated peaks of the chilly Alps in Switzerland.

(a) Heinrich Barth traversing the sand seas of the Sahara in 1853.

(b) Sledging across polar ice during Robert Scott's 1911 Antarctic expedition.

14.2 The Nature and Locations of Deserts

What Is a Desert?

In northern Africa, nomadic traders cross deserts accompanied by camels, animals that can walk for up to three weeks without drinking or eating (Fig. 14.1a). How do these animals survive such challenges? Camels barely sweat, use their own body fat as a water source, and can withstand severe dehydration. The survival challenges faced by camels emphasize the dryness of deserts. Formally defined, a **desert** is a region that is so *arid* (dry) that it supports vegetation on no more than 15% of its surface. In general, deserts exist where average rainfall (or snowfall

equivalent) totals less than 25 cm per year. Such regions cover about 25% of our planet's land surface **(Fig. 14.2)**. Because of their lack of water, deserts host no permanent streams, except for those that bring water from elsewhere. Note that the definition of a desert depends on a region's aridity, not on its temperature. Geologists distinguish between *cold deserts*, where temperatures generally stay below about 20°C, and *hot deserts*, where temperatures often exceed 35°C (95°F). In some hot deserts, temperatures have reached 58°C (136°F).

Types of Deserts

Each desert on Earth hosts a unique flora that distinguishes it from others. So geologists classify deserts not on the basis of native plant types, as they do for other kinds

Did you ever wonder . . .

how hot it can get in a desert?

Figure 14.2 The global distribution of deserts.

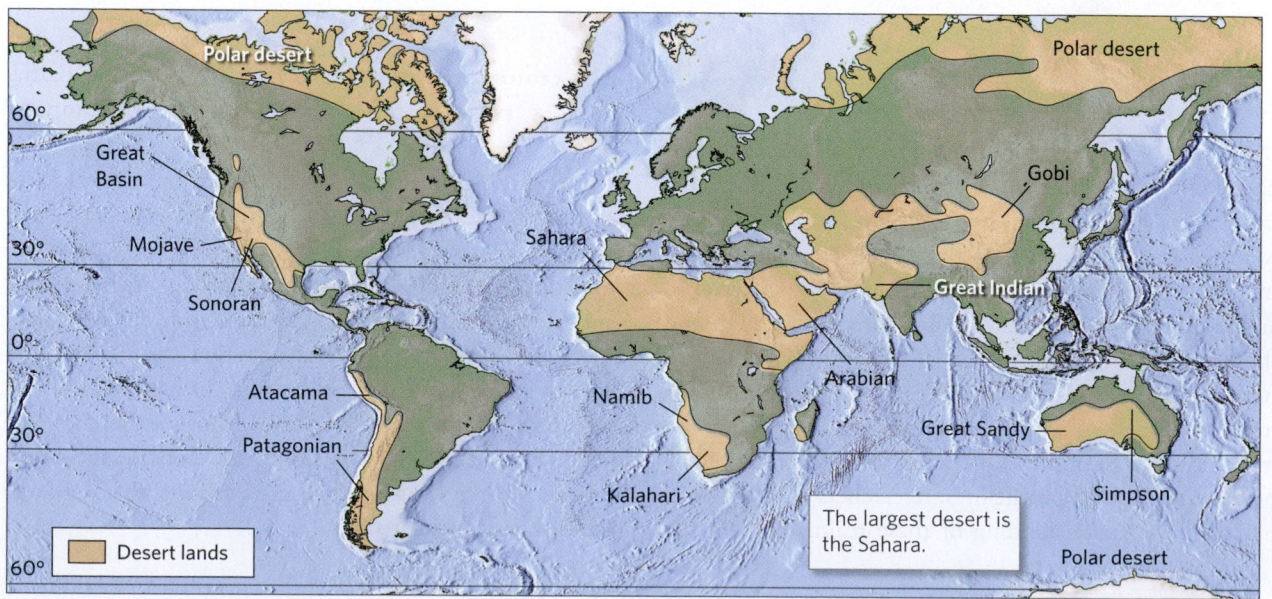

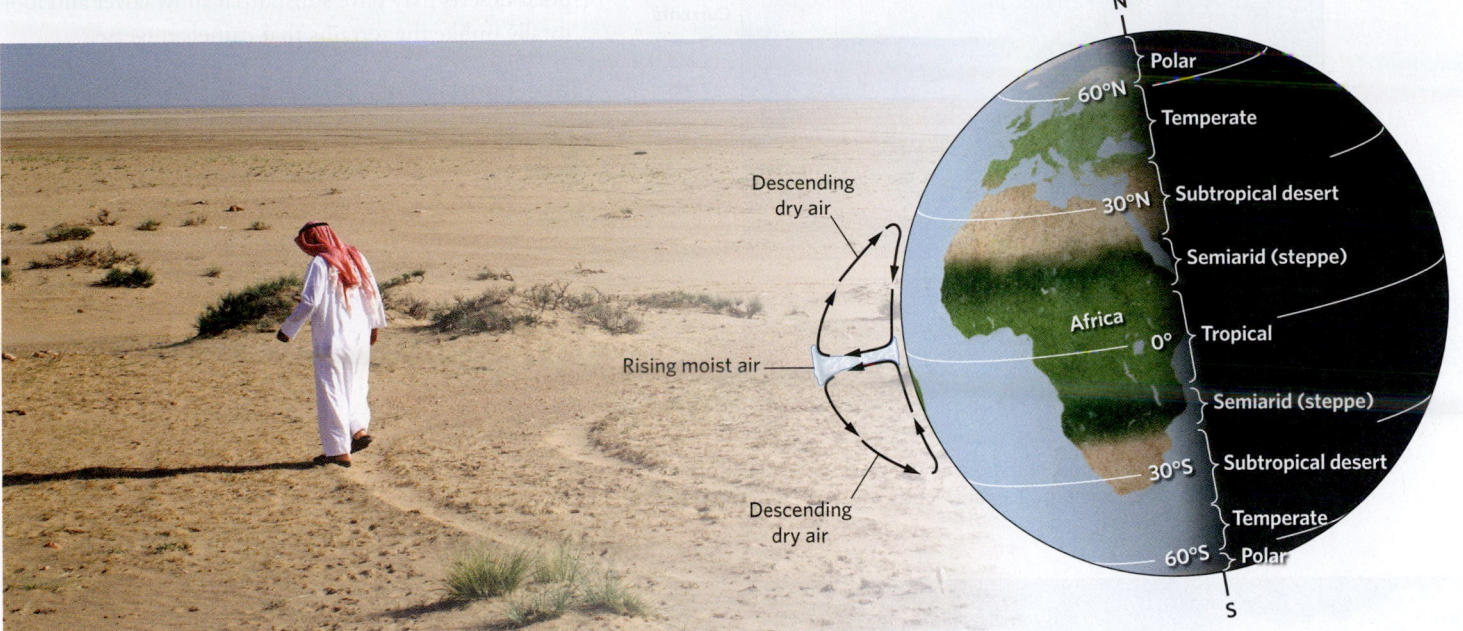

of landscapes, but rather on the reason that the desert formed. Types of deserts include the following:

- *Subtropical deserts:* As we will see in Chapter 18, the air of the lower atmosphere flows convectively between the equator and a latitude of about 30°. This flow results in the formation of two convective cells, known as *Hadley cells*—one in the northern hemisphere and one in the southern hemisphere **(Fig. 14.3)**. Hadley cells exist because intense solar energy warms air at the equator, and because warm air is less dense than cool air, it rises to high elevation, where it expands and cools. The rising warm air contains lots of moisture because the solar energy that warms it also evaporates water from the Earth's surface. Cooling of the rising air causes its moisture to condense and fall as rain, which drenches tropical rainforests. At high elevation, the now-dry air flows away from the equator until, at subtropical latitudes, it sinks. As the air sinks, it undergoes compression, which causes it to warm up. Rain-producing clouds can't form in this dry descending air, so the land below becomes a *subtropical desert*. Most of the Earth's largest deserts, including the Sahara, the Arabian, the Great Indian, the Australian, and the Kalahari are subtropical deserts (see Fig. 14.2).

- *Rain-shadow deserts:* Air flowing toward a mountain range must rise when it encounters the mountains **(Fig. 14.4)**. As it does so, it expands and cools, so the water vapor it

contains condenses and falls as rain, drenching the windward flank of the mountains. When the flowing air descends on the leeward side of the mountains, it has lost its moisture and can no longer provide rain. As a consequence, the land below becomes a desert. Such *rain-shadow deserts* lie east of the Cascade and Sierra Nevada Ranges in the western United States.

- *Coastal deserts:* Where a cold ocean current flows along a coast, it cools the overlying air while adding moisture. The cool air is denser than the overlying air, so it cannot rise and produce rain clouds. Therefore, coastal lands bordering regions of cold currents receive little precipitation, and instead of hosting coastal rainforests, they become *coastal deserts*. One example, the Atacama

Figure 14.4 Formation of rain-shadow deserts. Moist air rises and drops rain on the coastal side of a mountain range. On the leeward side of the mountains, the air is dry.

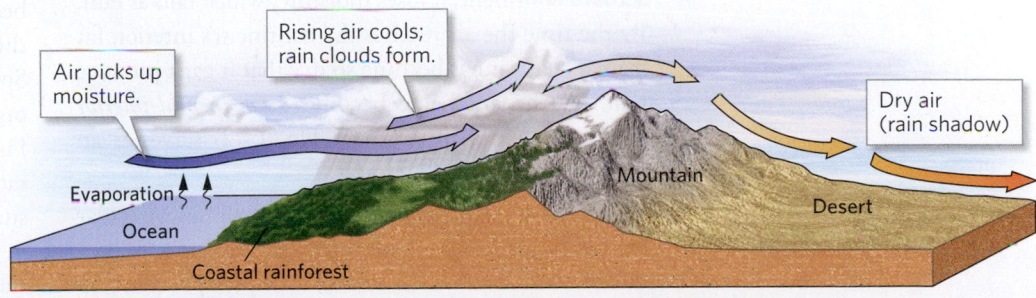

Figure 14.5 Formation of coastal deserts.

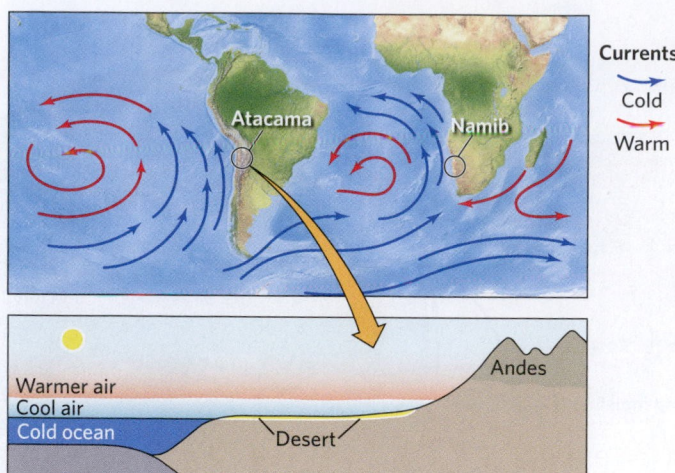

Currents
Cold
Warm

(a) Cold ocean currents cool and moisten the air along the coast. This air is too dense to rise and produce rain clouds.

(b) The Atacama Desert is the driest place on Earth. While clouds form there, they don't produce rain.

Desert along the western coast of South America, has become famous as the driest place on Earth **(Fig. 14.5)**.

- *Continental-interior deserts:* As air flows from the ocean across a continent, it loses moisture, which falls as rain. By the time the air reaches the continent's interior, far from a coast, it has become so dry that it can't produce rain, and the land below becomes a *continental-interior desert.* The Gobi of Asia (see Fig. 14.2) serves as an example of this type of desert.

- *Polar deserts:* The very cold, dense air above polar regions holds very little moisture, and can't rise to produce clouds or snow. Therefore, little precipitation falls

in the Earth's polar regions that these regions are, in fact, very dry (see Fig. 14.2). What snow does fall, however, remains on the ground for a long time. As a result, polar deserts may have substantial snow cover and look totally unlike the terrains that camels traverse.

Take-home message . . .

Deserts receive an average of less than 25 cm of rain a year, making them so arid that they host only sparse vegetation. Deserts develop in several settings: subtropical dry climates, rain shadows, coasts bordered by cold currents, continental interiors, and polar regions.

Quick Question -
Why does the world's largest desert, the Sahara, exist?

14.3 Weathering, Erosion, and Deposition in Deserts

Desert Weathering and Soil Formation

If you imagine a hot desert, you probably picture a landscape of sand and stones. Where does this sediment come from? Physical weathering in the desert begins, as it does in wetter climates, when joints split bedrock into pieces (see Chapter 5). Joint-bounded blocks eventually break free of bedrock, tumble down slopes, and break into smaller pieces. Chemical weathering happens more slowly in deserts than in moister climates because less water is available to react with rock. Nevertheless, rain or dew provides enough moisture for some chemical weathering to occur. When moisture seeps into rock, it leaches (dissolves and carries away) calcite, quartz, and various salts and breaks down silicate minerals. In effect, chemical weathering rots the rock by transforming it into a weak aggregate, which, over time, crumbles into a pile of unconsolidated sediment. Chemical weathering continues to affect the sediment, converting many of the minerals it contains into clay, leaving behind only quartz sand grains.

Although enough rain falls in deserts to leach chemicals out of sediment and rock, there's not enough rain to carry those chemicals away entirely. For this reason, and because of the lack of vegetation, desert soils differ from those formed in wetter climates, as we saw in Chapter 5. Specifically, a desert soil does not have an O-horizon (the organic-rich layer found in topsoils of wetter climates). Furthermore, calcite doesn't wash away entirely, but rather accumulates and precipitates beneath the ground surface to form *caliche.* Notably, the lack of plant cover in deserts allows variations in bedrock and soil color to stand out—slight variations in the concentration of iron,

Figure 14.6 Desert colors.

(a) The red hues of the Painted Desert in Arizona are due to the oxidation (rusting) of iron in the rock.

(b) By chipping away desert varnish to reveal the lighter rock beneath, Native Americans created art and symbols.

or in the degree of iron oxidation, for example, can produce spectacular color banding on the land surface. The Painted Desert of northern Arizona earned its name from the brilliant and varied hues of oxidized iron in the region's shale bedrock **(Fig. 14.6a)**.

Because desert soil develops so slowly, and will not become stabilized by plant roots, it doesn't accumulate on steep slopes. Any soil that might form gets washed away by infrequent, intense rains or by strong winds. Therefore, deserts commonly contain rocky escarpments as well as regions where pebbles, cobbles, or boulders litter the ground surface. Typically, exposed rocks display **desert varnish**, a dark, shiny, rusty-brown coating of iron oxide, manganese oxide, and clay. Recent studies suggest that desert varnish forms when windblown clay settles on the surfaces of rocks in the presence of a small amount of moisture. Chemical reactions or the activity of microbes extract iron and manganese from the dust, incorporating them into shiny oxide minerals that bind the clay to the rock as a resistant coating. Desert varnish takes centuries to millennia to form. Traditional cultures create artistic figures or symbols called *petroglyphs* by chipping away the varnish to reveal the underlying rock **(Fig. 14.6b)**.

Erosion by Water

Although heavy rains rarely fall in deserts, when they do, they can radically alter a landscape in a matter of minutes, for the land surface does not have a protective covering of vegetation. In fact, in most deserts, water causes more erosion than wind does **(Fig. 14.7a)**. Erosion by water begins with the impacts of raindrops, which knock sediment from the ground into the air. On a hill, this sediment lands downslope. When water starts flowing across the ground surface, it carries away loose sediment. Minutes after a heavy downpour begins, an array of small streams develops, and on hillslopes composed of soft substrate, these streams carve closely spaced, nearly parallel channels that merge downslope, producing *badlands topography* **(Fig. 14.7b)**.

Figure 14.7 Evidence of erosion by water in deserts.

(a) These hills in the desert near Las Vegas, Nevada, are bone dry, but their shape indicates erosion by water. Note the numerous stream channels.

(b) Badlands topography, developed by erosion by water.

(c) Gravel and sand have been left behind on the floor of a dry wash after a flash flood in Death Valley. Erosion by water is cutting a channel.

During intense rainstorms, channels in valley floors fill with a turbulent mixture of water and sediment, which rushes downstream as a flash flood, picking up sediment and carrying it along. When the rain stops, the water sinks into the streambed and disappears, so such streams are ephemeral (see Section 13.2). The dry channels that exist between floods are called **dry washes**, *arroyos*, or *wadis* (Fig. 14.7c).

Figure 14.8 Windblown sediment in deserts.

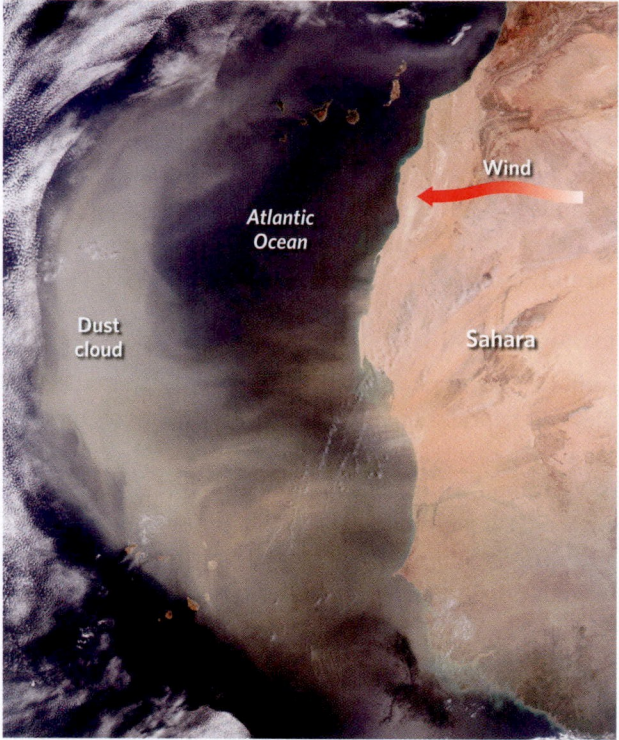

(a) In this satellite image, a huge dust cloud that originated in the Sahara blows across the Atlantic Ocean.

Erosion by Wind

Because vegetation doesn't provide protection, winds have direct access to the ground and can pick up and transport large quantities of sediment. Geologists use the same terminology for this sediment that they use for water-transported sediment (see Section 13.3). **Suspended loads** consist of fine-grained sediment—clay and silt—that can move with the wind and can stay aloft for a long time. In fact, this sediment can be carried so high (up to several kilometers above the Earth's surface) and so far downwind (tens to thousands of kilometers) that it may move completely out of its source region. Dust blown from the Sahara, for example, can cross the Atlantic Ocean (Fig. 14.8a). Particularly strong winds can generate dramatic **dust storms** up to 100 km (60 miles) long and 1.5 km (1 mile) high. An approaching dust storm resembles a roiling, opaque wave. Dust storms are fairly common in the Sahara and the Arabian Desert, where they are known as *haboobs*. A particularly large dust storm engulfed Phoenix, Arizona, in 2011 (Fig. 14.8b).

Surface loads, or bed loads, start moving only in moderate to strong winds, for such winds have the power to roll and bounce sand grains along the ground, a process called **saltation**. Saltation begins when turbulence caused by wind lifts sand grains (Fig. 14.8c). The grains move downwind, following an asymmetrical, arc-like trajectory. When they return to the ground, they strike other sand grains, causing those grains to bounce up and drift or roll downwind. Saltating grains generally rise no more than 0.5 m (1.5 feet), but where sand bounces on bedrock, they may rise 2 m (6 feet).

Just as sandblasting cleans the grime off the surface of a building, windblown sand and dust grind away at surfaces in the desert. Over long periods, such wind abrasion creates smooth faces, or *facets*, on pebbles, cobbles, and boulders. Rocks whose surfaces have been faceted by wind are known as *ventifacts* (Fig. 14.9a, b). The sizes of clasts that wind can carry depend on wind velocity, so wind

(b) A huge dust storm approaches Phoenix, Arizona.

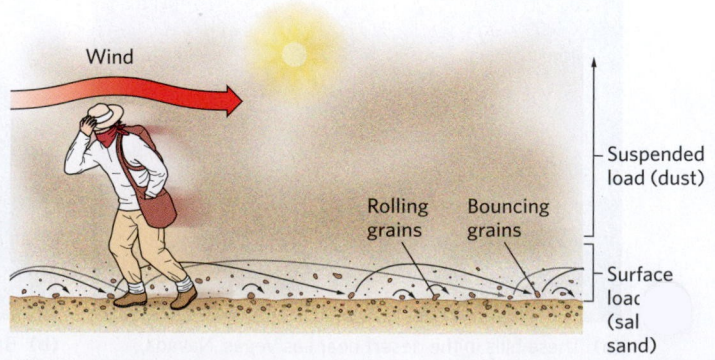

(c) Wind transports desert sediment both in suspension and in a saltating layer.

Figure 14.9 Examples of wind erosion in deserts.

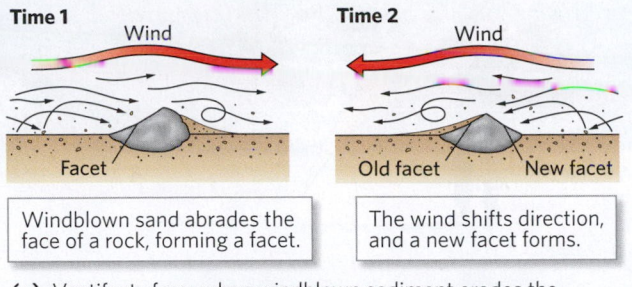

Time 1
Wind

Facet

Windblown sand abrades the face of a rock, forming a facet.

Time 2
Wind

Old facet New facet

The wind shifts direction, and a new facet forms.

(a) Ventifacts form when windblown sediment erodes the surface of a rock.

2 cm

(b) An example of a multifaceted ventifact.

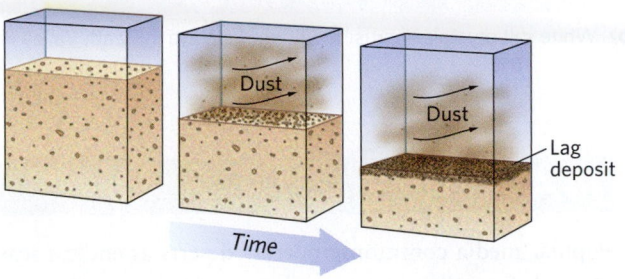

Dust

Dust

Lag deposit

Time

(c) A lag deposit develops when wind blows away finer sediment, leaving behind a layer of coarser grains.

sorts sediment. In some cases, wind carries away so much fine sediment that pebbles and cobbles become concentrated at the ground surface as a **lag deposit (Fig. 14.9c)**.

Deposition in Deserts: Talus Piles and Alluvial Fans

Where does the sediment moved during rockfalls and flash floods in the desert end up? Recall from Chapter 12 that rocky debris falling from cliffs tumbles downslope and may accumulate in a **talus pile**, a steep, sloping apron of debris at the cliff base **(Fig. 14.10a)**. In deserts, talus piles remain unvegetated for a long time, and their rocks may become coated with desert varnish.

Debris transported downstream by flash floods may be left as bars in a dry wash when the water stops moving. While bars deposited by permanent streams may become vegetated if not washed away, bars deposited by ephemeral desert streams remain dry, stony mounds.

If a sediment-laden ephemeral stream flows out onto a plain at the mouth of a desert canyon, it doesn't continue to flow as a tributary to a trunk river, but rather spreads out over a broader surface and slows, subdividing into a number of smaller channels that diverge outward. The flow in these channels disappears as the water seeps into underlying gravel. The network of these distributaries spreads the sediment, or alluvium, out into a broad **alluvial fan**, a broad wedge- or apron-shaped pile of sediment **(Fig. 14.10b)**. But not all distributaries of an alluvial fan flow at the same time. Rather, flows carry sediment along one part of the fan for a while until the

Figure 14.10 Accumulation of debris in deserts.

Cliff

Talus

(a) This talus pile along the base of a desert cliff formed from rocks that broke off and tumbled down the cliff.

Canyon mouth

Road

Fan

(b) This alluvial fan accumulated at the mouth of a small canyon in Death Valley, California.

Figure 14.11 The formation of playas.

(a) This playa in California formed at the base of a bajada, as seen in this aerial, oblique view.

A close-up of salt crystals.

(b) White salt crystals encrust the floor of a playa in Death Valley.

> None of Nature's landscapes are ugly so long as they are wild.
>
> —JOHN MUIR (AMERICAN NATURALIST, 1838–1914)

See for yourself

Death Valley, California

Latitude: 36°12′42.34″ N
Longitude: 116°48′25.43″ W

Zoom to 35 km (~20 miles) and look down.

Death Valley, a narrow basin whose floor lies below sea level, hosts many desert landforms. The white patch is a playa. A bajada forms the slope between the playa and the mountains to the west. Small alluvial fans spill into the basin on the east.

sediment piles up. The next flow finds an easier path, to the side of the previously deposited mound of sediment. Repetition of this process over time causes deposition to move from one part of the fan to another, so overall, the fan maintains its symmetrical shape. Alluvial fans emerging from adjacent valleys may eventually merge and overlap along the front of a mountain range, producing an elongate wedge of sediment bordering the range, called a *bajada*.

Playas and Salt Lakes

During particularly wet seasons, lowlands or basins between mountain ranges in a desert may fill with water from flash floods or springs to form shallow temporary lakes. The water in these lakes, like all freshwater, contains trace amounts of dissolved salt. During drier times, the water evaporates entirely, leaving behind a dry, flat, exposed lake bed known as a **playa (Fig. 14.11a)**. As the water evaporates, the small amount of salt that had been dissolved in the water precipitates. Over multiple cycles of flooding and drying, a crust composed of various salts, such as halite and gypsum, mixed with clay accumulates on the surface of a playa **(Fig. 14.11b)**.

Take-home message . . .

Weathering breaks down rocks in deserts, as in other places, but because of the scarcity of water, soils are thin, and desert varnish may coat rock surfaces. Heavy rains cause significant erosion and sediment transport in deserts, but since streams rarely flow, their channels are usually dry washes. Wind transports sediment and carves ventifacts. Debris formed by erosion accumulates in talus piles and alluvial fans.

Quick Question -
How do playas form, and why do they contain salt?

14.4 Desert Landscapes

Popular media commonly portray deserts as endless seas of sand. In reality, desert landscapes display much more variation. In addition to sand, deserts can host stony plains, a stubble of cacti and other hardy desert plants, or intricate rock formations. In this section, we examine these contrasting landscapes.

Rocky Cliffs, Mesas, and Arches

Because thick soil cover doesn't accumulate on steep slopes in deserts, these slopes become rocky ridges and cliffs. In places underlain by horizontal sedimentary strata, or by horizontal layers of volcanic rock, cliff faces tend to form when rock separates along joints and topples. By breaking off at successive joints, a cliff face steps back into the land while retaining roughly the same form, a phenomenon known as **cliff retreat**. Where beds have different resistance to erosion, cliffs develop a step-like shape—the resistant layers become vertical cliffs, whereas the nonresistant layers become rubble-covered slopes **(Fig. 14.12a)**. Due to cliff retreat, erosion of a flat-topped plateau underlain by horizontal layering produces flat-topped hills, which go by different names depending on their size: a large one is a **mesa** (from the Spanish word for table), a medium one is a **butte**, and a small one is a **chimney (Fig. 14.12b–d)**. In places where bedrock consists of thick sandstone layers cut by widely spaced vertical joints, erosion can transform a layer into a set of sandstone walls known as *fins*. Collapse at the base of the fins yields a **natural arch (Fig. 14.13)**.

Of course, horizontally layered bedrock doesn't underlie all deserts. Where the sedimentary or volcanic layering dips at an angle, an asymmetrical ridge called a *cuesta*

Figure 14.12 Consequences of cliff retreat.

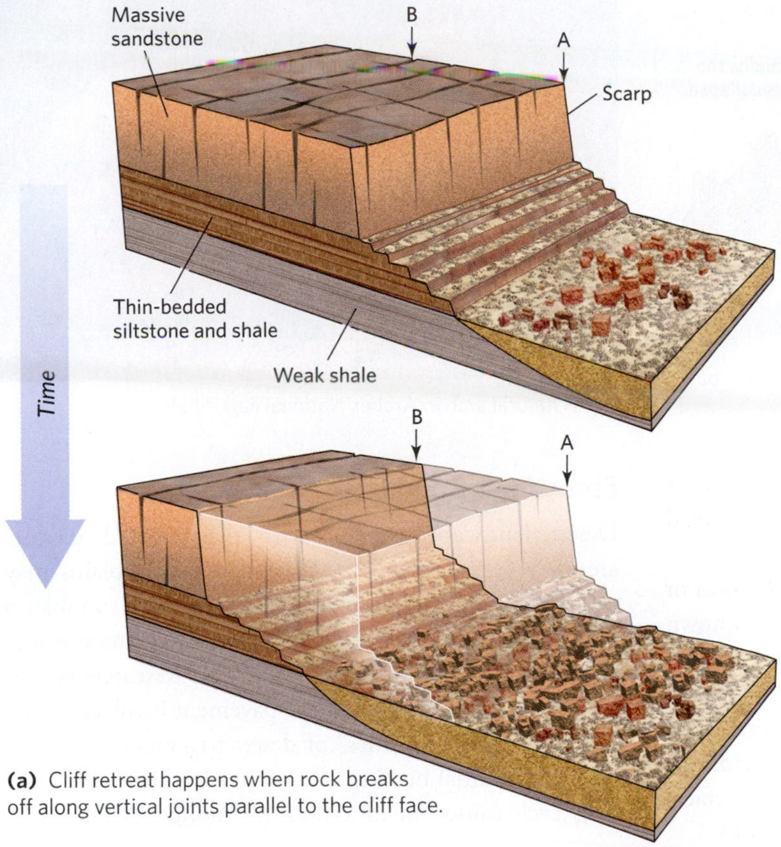

Massive sandstone

B

A

Scarp

Thin-bedded siltstone and shale

Weak shale

Time

B

A

(a) Cliff retreat happens when rock breaks off along vertical joints parallel to the cliff face.

(d) Examples of chimneys in Bryce Canyon, Utah. Chimneys like these are sometimes referred to as hoodoos.

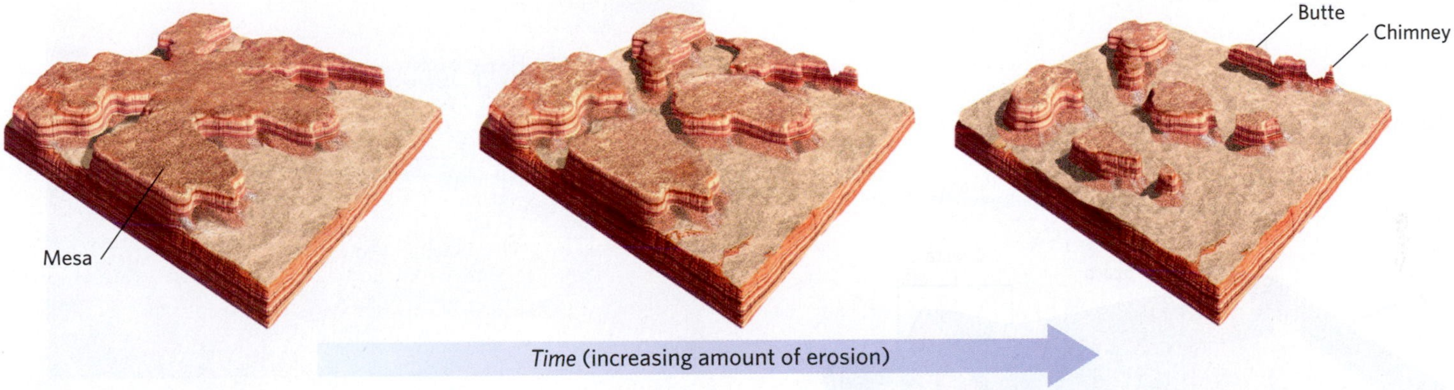

Butte

Chimney

Mesa

Time (increasing amount of erosion)

(b) A once-continuous layer of rock evolves into a series of isolated mesas, buttes, and chimneys. If the bedding is horizontal, the resulting landforms have flat tops.

(c) Buttes and mesas tower above the floor of Monument Valley, Arizona.

Figure 14.13 Formation of natural arches.

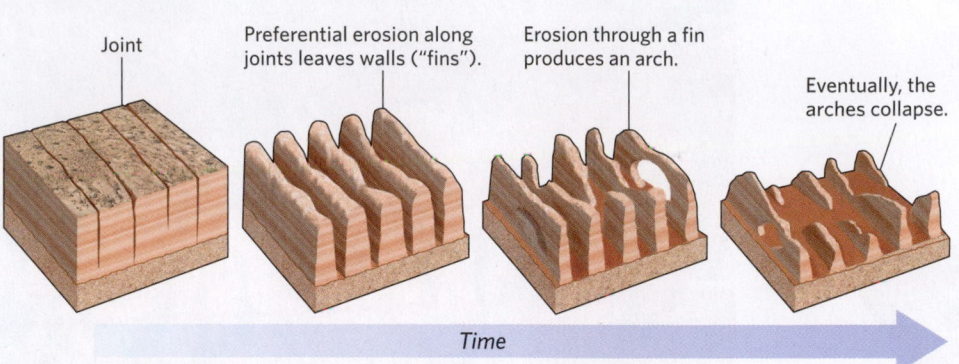

Joint

Preferential erosion along joints leaves walls ("fins").

Erosion through a fin produces an arch.

Eventually, the arches collapse.

Time

(a) Erosion that occurs preferentially along joints produces wall-like fins of rock. When the lower part of a fin erodes, a natural arch remains.

(b) A natural arch in Arches National Park, Utah.

Did you ever wonder . . .

whether all deserts are completely covered by sand?

develops **(Fig. 14.14a)**. Erosion of vertical beds produces a *hogback*. If bedrock consists of complexly deformed strata, intrusive igneous rocks, or metamorphic rocks, erosion yields jagged rocky ridges. Eventually, erosion of a desert landscape leaves isolated ridges of rock, known as *inselbergs* (German for island mountain; **Fig. 14.14b**). In some cases, basins filled with sediment, eroded from the ridges, separate the inselbergs. If water and wind carry away the sediment that accumulates at the base of an inselberg, then a broad, nearly horizontal bedrock surface, called a *pediment*, surrounds the inselberg **(Fig. 14.14c)**.

Stony Plains and Desert Pavement

Desert landscapes strewn with pebbles and cobbles are known as *stony plains*. Portions of these plains may evolve into **desert pavement**, a surface that resembles a tile mosaic in that it consists of countless separate stones that fit together tightly **(Fig. 14.15a, b)**. Researchers suggest that development of desert pavement involves several processes. The rock "tiles" of desert pavement may form when differential heating of rock blocks during the Sun's daily cycle causes the blocks to split along vertical cracks.

Figure 14.14 Formation of cuestas and inselbergs.

Inselberg

Alluvium

(b) Aerial view of an inselberg surrounded by alluvium in the southern California desert.

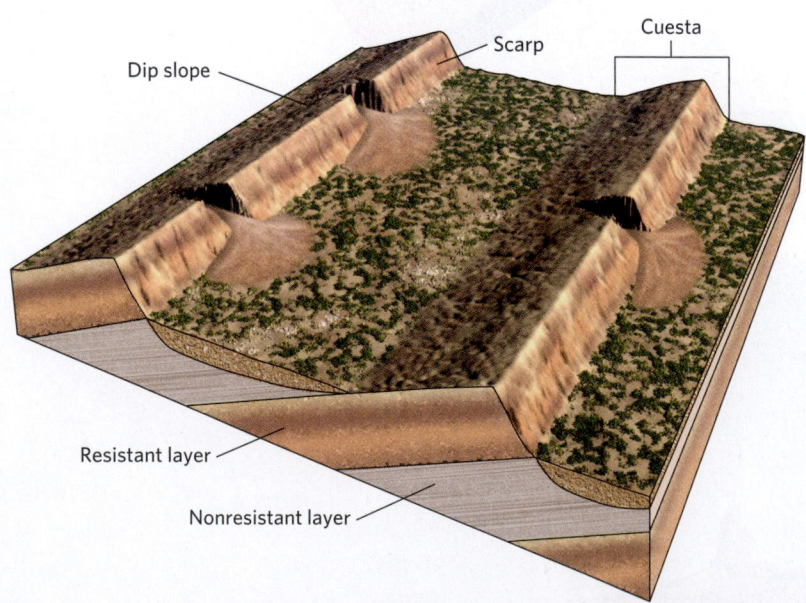

Dip slope

Scarp

Cuesta

Resistant layer

Nonresistant layer

(a) Asymmetrical ridges called cuestas develop where strata are not horizontal.

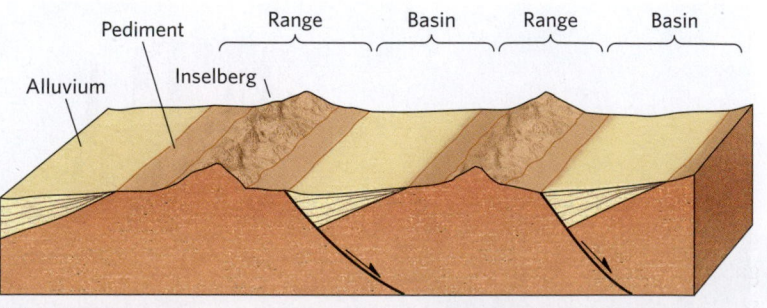

Pediment

Range

Basin

Range

Basin

Alluvium

Inselberg

(c) In the Basin and Range Province of the southwestern United States, tilted fault-block ranges evolve into inselbergs, bordered by pediments and sediment-filled basins.

Figure 14.15 Desert pavement and a hypothesis for its formation.

(a) A well-developed desert pavement in the Sonoran Desert, Arizona. The inset shows a close-up of the pavement.

(b) Students standing at the edge of a trench cut into desert pavement. Note the soil between the pavement and the underlying alluvium.

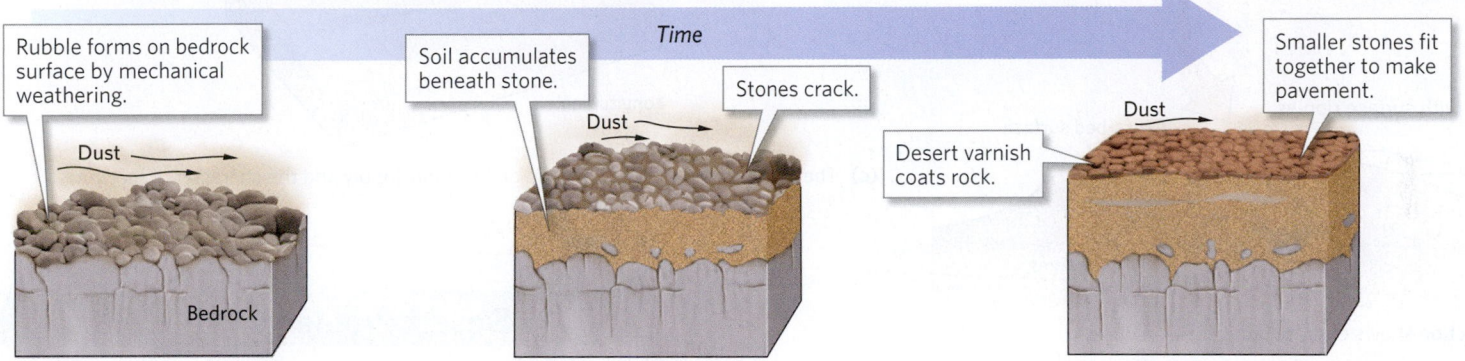

Time

Rubble forms on bedrock surface by mechanical weathering.

Dust

Bedrock

Soil accumulates beneath stone.

Stones crack.

Dust

Desert varnish coats rock.

Smaller stones fit together to make pavement.

Dust

(c) Desert pavement forms in stages. First, loose pebbles and cobbles collect at the surface. Dust settles among the stones and builds up a soil layer below. The stones eventually crack into smaller pieces and settle to form a mosaic-like pavement.

The resulting tiles eventually fall over. Over time, wind-blown dust settles between them. During rains, clay in the dust swells, and during dry times, the clay shrinks. This repeated swelling and shrinking allows the rocks to act like plates, settling into an overall flat surface. As this process continues, additional clay settles between the rocks, and a fine soil builds up beneath them (Fig. 14.15c). Gradually, desert varnish coats the surface of the pavement, making it look shiny from a distance.

Seas of Sand: The Nature of Dunes

In deserts, local wind patterns can build up accumulations of sand that range from just a few kilometers to a few hundred kilometers across. In vast sand seas, known as *ergs* in Arabic, blowing wind builds **sand dunes**, elongate piles of sand (Fig. 14.16a). The process begins when sand becomes trapped on the windward side of an obstacle. Gradually, the sand builds downwind and buries the obstacle. Sand saltates up the windward side of the dune,

blows over the crest of the dune, and then settles on the dune's steeper, leeward face. This face attains the angle of repose, the maximum slope of a freestanding pile of sand. As sand collects on this surface, it eventually becomes unstable and slides down the slope, so geologists refer to the lee side of a dune as the *slip face*. As more and more sand accumulates on the slip face, the crest of the dune migrates downwind, and former slip faces are preserved inside the dune. In cross section, these slip faces appear as cross beds (Fig. 14.16b). Desert sand dunes resemble streambed dunes, but they tend to be larger.

Dunes display a variety of shapes and sizes, depending on the character of the wind and the sand supply (Fig. 14.16c). In regions of scarce sand and steady wind, beautiful crescents called *barchan dunes* develop, with the tips of the crescents pointing downwind. If the wind direction shifts frequently, a group of crescents pointing in different directions overlap one another, resulting in the formation of *star dunes*. Where enough sand accumulates

See for yourself

Uluru, Australia

Latitude: 25°20′47.64″ S
Longitude: 131°2′28.96″ E

Zoom to 5 km (~3 miles) and look down.

The red sandstone of Uluru (Ayers Rock), a inselberg, rises above the stony plains of the central Australian desert. The NW-trending diagonal lines are traces of vertical bedding in the sandstone.

Figure 14.16 Sand dunes in deserts.

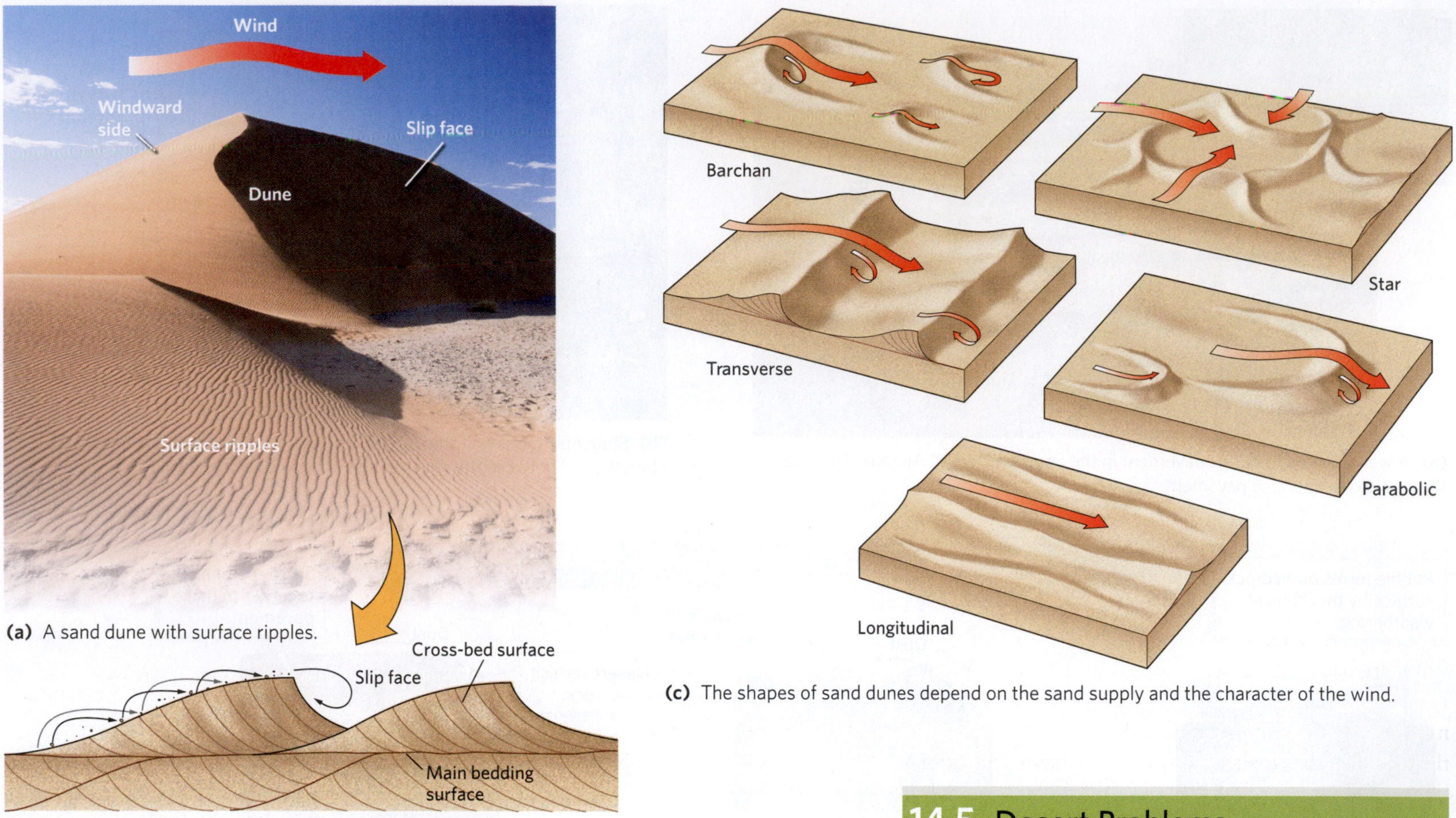

Wind

Windward side

Slip face

Dune

Surface ripples

(a) A sand dune with surface ripples.

Cross-bed surface

Slip face

Main bedding surface

(b) The cross-section shows cross bedding inside a dune.

Barchan

Star

Transverse

Parabolic

Longitudinal

(c) The shapes of sand dunes depend on the sand supply and the character of the wind.

See for yourself

Namib Desert, Namibia

Latitude: 24°44′27.80″ S
Longitude: 15°27′58.92″ E

From 20 km (~12.5 miles), look down and zoom in.

Huge dunes of orange sand border a dry wash in western Africa. The slip faces of dunes are recognizable. A road in the wash provides scale.

to bury the ground surface completely, and only moderate winds blow, sand piles into simple, wave-like shapes called *transverse dunes*. The crests of transverse dunes trend perpendicular to the wind direction. Strong winds may break through transverse dunes and change them into *parabolic dunes* whose ends point in the upwind direction. Finally, if there is abundant sand and a strong, steady wind, the sand streams into *longitudinal dunes*, whose axes lie parallel to the wind direction. Active sand dunes constantly move and change as the wind blows.

Take-home message . . .

In deserts, erosion of horizontal layers by cliff retreat yields buttes and mesas, whereas erosion of tilted strata forms cuestas. Stony plains, some of which develop desert pavements, are common in deserts. Where winds build accumulations of sand, deserts contain sand dunes.

Quick Question ----------------------------
What factors control the shape, dimensions, and orientation of sand dunes?

14.5 Desert Problems

We've seen that deserts contain a great variety of landscapes that differ from those of wetter regions (**Earth Science at a Glance**, pp. 476–477). Can regions that host nondesert landscapes become deserts in a human time frame? Yes. Natural *droughts* (periods of unusually low rainfall), overpopulation, overgrazing, intensive farming, and diversion of water supplies can transform a semiarid grassland into a desert in a matter of years to decades. Geologists refer to this transformation as **desertification**.

The Sahel, the belt of semiarid grassland that fringes the southern margin of the Sahara, displays the consequences of desertification. In the past, the Sahel provided sufficient vegetation to support a small population of nomadic people and animals. But during the second half of the 20th century, many people migrated into the Sahel. As new residents struggled to live off the land, some replaced soil-preserving perennial grasses with seasonal crops, or allowed grazing animals to eat the grass to the ground. As a consequence, the soil dried out. In the 1960s, and again in the 1980s, drought hit the region, bringing catastrophe (**Fig. 14.17**). Without vegetation, the air grew drier, the semiarid grassland of the Sahel became desert, and its inhabitants endured famine.

Figure 14.17 Desertification in the Sahel.

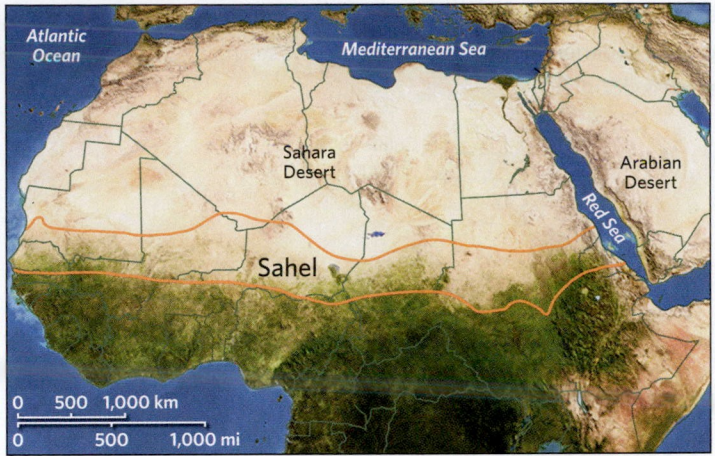

(a) The Sahel is the semiarid land along the southern edge of the Sahara. Large parts have undergone desertification.

(b) In parts of the Sahel, the soil has dried up.

The southern Great Plains of the United States underwent temporary desertification beginning in 1933, the fourth year of the Great Depression. Banks had failed, many workers had lost their jobs, and the stock market had crashed. No one needed yet another disaster—but that year, even nature turned hostile, and seasonal rains did not arrive. Farming had broken up soil that had been held together for millennia by the roots of prairie grasses, and without water, the topsoil of the croplands turned to powdery dust. Strong storms blew across the plains and sent the soil skyward to form dust storms that blotted out the Sun. When the dust settled, it buried houses and roads and dirtied every nook and cranny. What had once been rich farmland turned into

a wasteland that soon acquired a nickname, the Dust Bowl. It stayed that way for almost a decade.

More recently, desertification has devastated the Aral Sea of Central Asia. Once the fourth largest lake in the world, it was over 250 km across in 1990. Now, because the rivers that once fed it have been diverted for irrigation, it has dwindled to only 10 km wide, and fishing ships that once plied its waters lie rusting amid sand dunes built from what was once lake-bed sediment **(Fig. 14.18)**. Unfortunately, exposure of sediment that was once underwater makes it susceptible to wind erosion. Dust carried aloft from the former floor of the Aral Sea, and other places like it, can carry toxic materials and viruses.

Figure 14.18 When the Soviet Union diverted the rivers that flowed into the Aral Sea for irrigation projects, the sea began to shrink. Once home to a fishing fleet, it has been reduced to a few ultra-salty ponds.

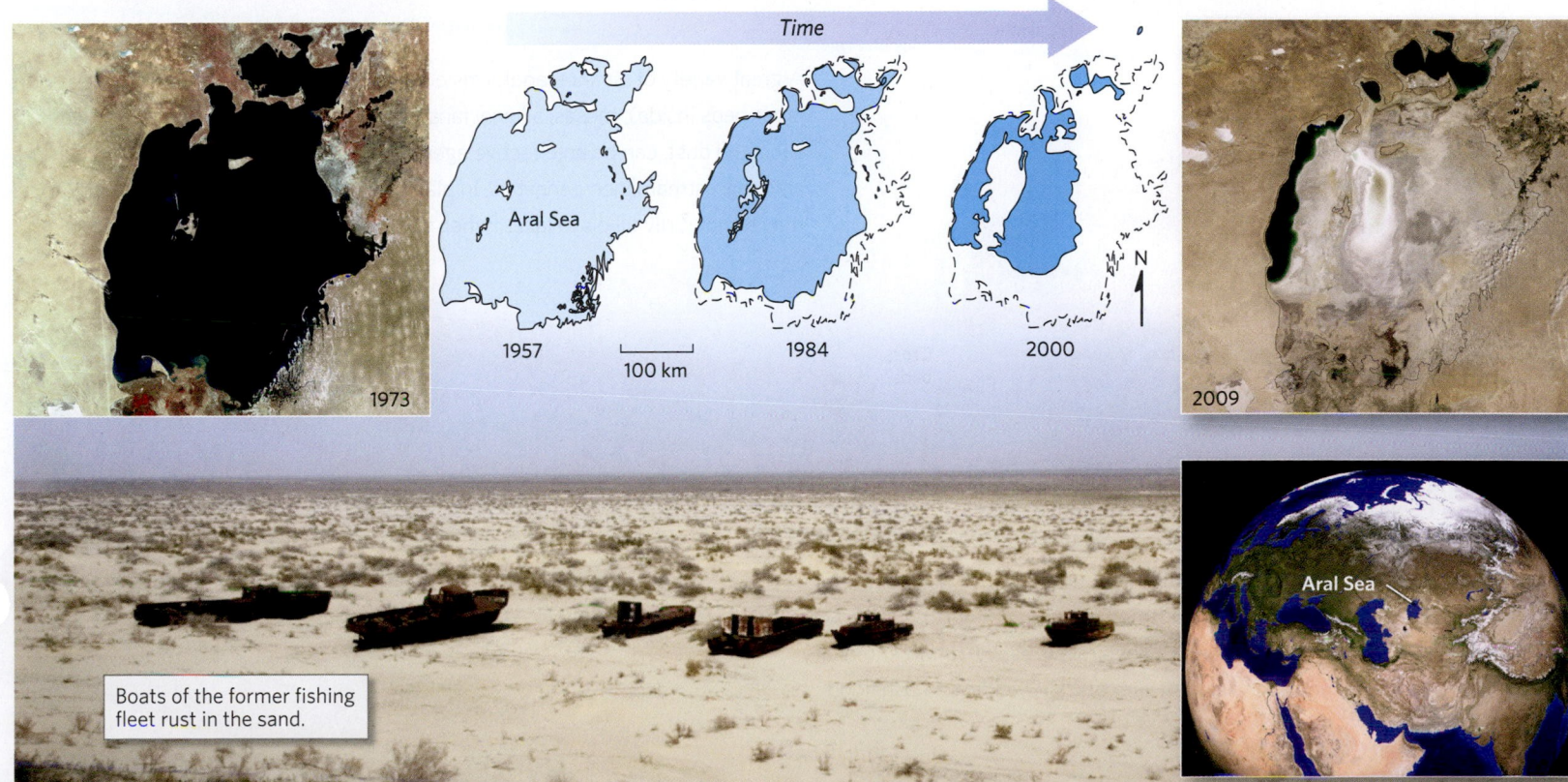

1973

Time

Aral Sea

1957   100 km

1984

2000

N

2009

Aral Sea

Boats of the former fishing fleet rust in the sand.

The Desert Realm

The desert of the Basin and Range Province in Utah, Nevada, and Arizona consists of alternating basins (grabens or half-grabens) separated by narrow ranges (tilted fault blocks), depicted schematically here. The Sierra Nevada, underlain largely by granite, borders the western edge of the province, while the Colorado Plateau, underlain by flat-lying sedimentary strata, borders the eastern edge. The overall climate of the region is dry.

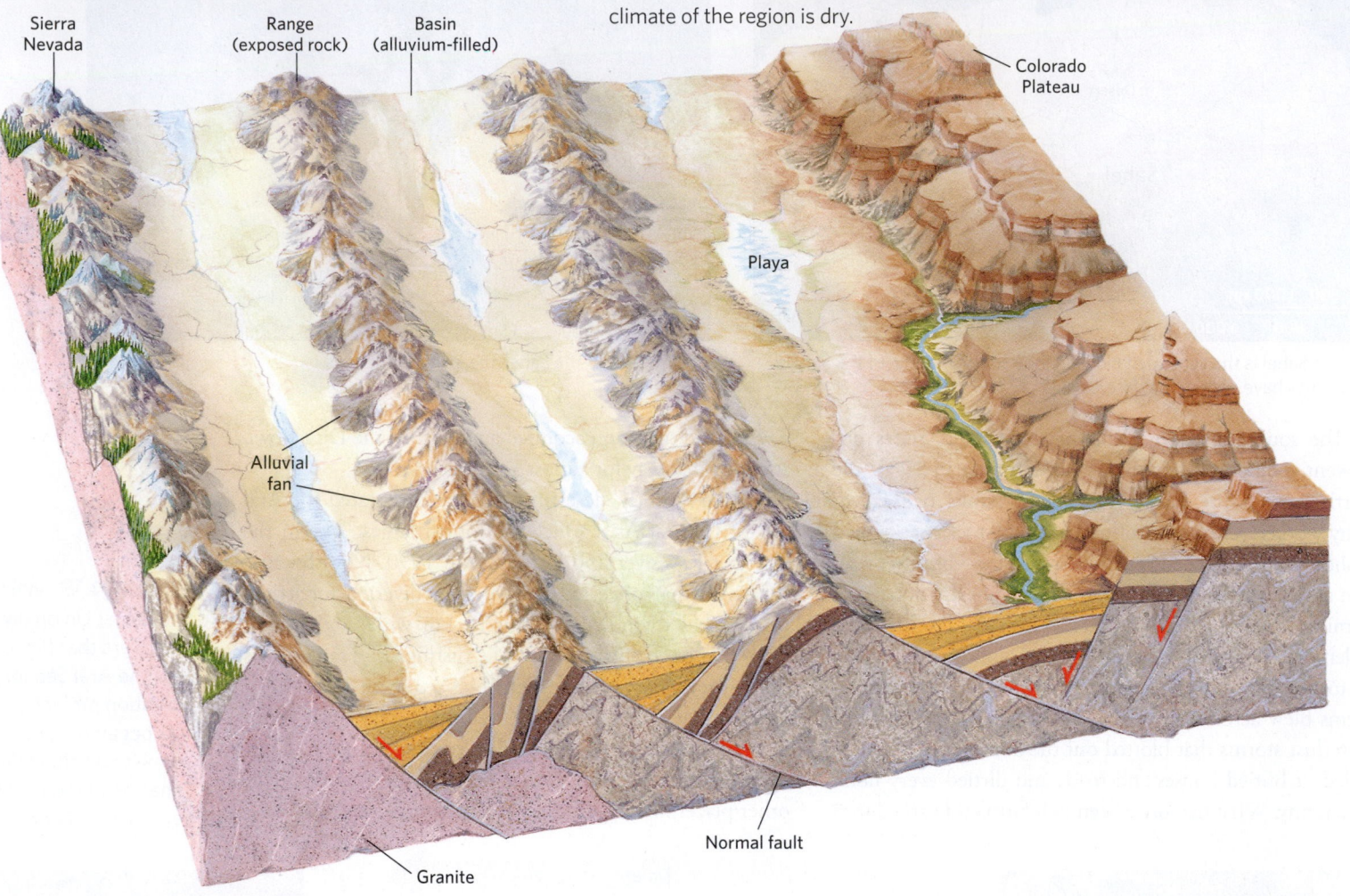

A great variety of distinct landforms occur in deserts, including sand dunes (with cross beds inside), mesas, alluvial fans, arches, and chimneys. Wind, carrying sand and dust, can be an effective agent of erosion in the desert, and can carve yardangs, streamlined pedestals. In places, desert pavements develop. Water may temporarily fill playa lakes; when this water evaporates, it leaves salt.

Most stream channels in deserts fill with water only after heavy rains. At other times, they are dry washes (also known as arroyos or wadis). When rain is heavy, runoff enters dry washes and fills them very quickly, yielding a flash flood.

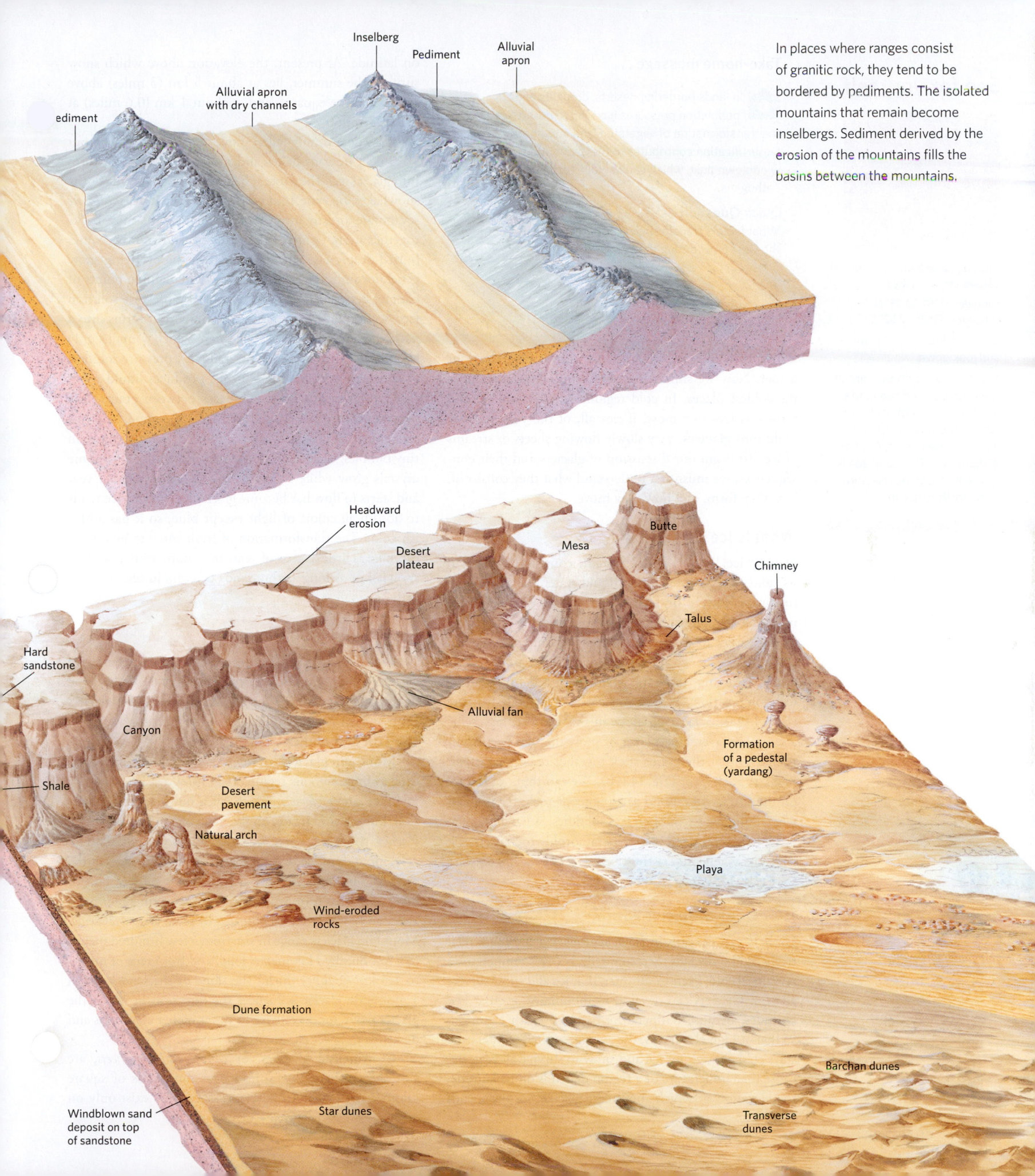

ediment

Alluvial apron
with dry channels

Inselberg

Pediment

Alluvial
apron

In places where ranges consist
of granitic rock, they tend to be
bordered by pediments. The isolated
mountains that remain become
inselbergs. Sediment derived by the
erosion of the mountains fills the
basins between the mountains.

Headward
erosion

Desert
plateau

Mesa

Butte

Chimney

Talus

Hard
sandstone

Shale

Canyon

Desert
pavement

Natural arch

Alluvial fan

Formation
of a pedestal
(yardang)

Playa

Wind-eroded
rocks

Dune formation

Barchan dunes

Windblown sand
deposit on top
of sandstone

Star dunes

Transverse
dunes

Take-home message . . .

In lands bordering deserts, droughts and population pressures have led to desertification, the transformation of vegetated land into desert. Desertification contributes to the distribution of windblown dust, which can contain toxins and pathogens.

Quick Question -
What factors transformed the southern Great Plains into the Dust Bowl during the 1930s?

14.6 Ice and the Nature of Glaciers

We've just discussed some of the hottest places on our planet. Now we shift our attention to consider some of the coldest places. In cold regions, water exists in solid form—as ice—for most, if not all, of the year and can build into **glaciers**, very slowly flowing sheets or streams of ice. To begin our discussion of glaciers and their consequences, we must first understand what they consist of, how they form, and how they move.

What Is Ice?

Natural ice, in the Earth System, consists of water in its solid state, formed when liquid water cools below its freezing point. To understand natural occurrences of ice in a geologic context, let's apply the concepts introduced in our earlier discussions of rocks and minerals. We can think of a single ice crystal as a mineral specimen, a naturally occurring, inorganic solid, with a definite chemical composition (H_2O) and a regular crystal structure **(Fig. 14.19a)**. A layer of snow, in this context, represents a sediment layer, and a layer of snow that has been compacted so that the grains stick together resembles a sedimentary rock layer **(Fig. 14.19b)**. We can think of the ice on the surface of a pond as an igneous rock, for it forms when liquid water (molten ice) solidifies. What about glacial ice? It's comparable to a metamorphic rock, in that it develops when pre-existing ice recrystallizes in the solid state **(Fig. 14.19c)**.

How Does a Glacier Form?

Glaciers can develop in polar regions, even though relatively little snow falls there annually, because temperatures remain so cold, even in summer, that ice and snow survive all year. Glaciers can develop in mountains, even at low latitudes, because temperature decreases with elevation, so that at high elevations, the average annual temperature stays cold enough for ice and snow to survive all year. Because the temperature of a region depends on its latitude, the elevation at which mountain glaciers form depends

on latitude. At present, the elevation above which snow survives the summer, lies at about 5 km (3 miles) above sea level at the equator, and at about 1 km (0.6 miles) at a latitude of 60°.

How does freshly fallen snow, which consists of about 10% delicate snowflakes and about 90% air, transform into glacial ice? The process begins when the points of the snowflakes become blunt because they either *sublimate* (evaporate directly into vapor) or melt, so the snow packs together more tightly. When this packed snow becomes buried deeply enough, the weight of the overlying snow generates sufficient pressure to cause remaining points of contact between snowflakes to melt by pressure solution (see Section 6.2). As a result, the snow transforms into a packed granular material called **firn**, which contains only about 25% air **(Fig. 14.19d)**. Gradually, pressure solution at contacts between grains in firn produces liquid water that crystallizes in the pore spaces between grains. The firn transforms into a solid mass of ice composed of interlocking ice crystals—any remaining air exists only in tiny bubbles. The ice becomes coarser over time, for some crystals grow while others shrink. Ice that lasts all year and starts to flow has become glacial ice. Glacial ice tends to absorb all colors of light except blue, so it has a bluish color. The transformation of fresh snow to glacial ice can take as little as tens of years in regions with abundant snowfall or as long as thousands of years in regions with little snowfall.

Categories of Glaciers

Geologists distinguish between two main categories of glaciers: mountain glaciers and continental glaciers. The distinction depends on the setting in which the glaciers form and on whether the slope of the land surface influences the flow of the glacier.

Mountain glaciers, or *alpine glaciers*, grow in or adjacent to mountainous regions **(Fig. 14.20a)**. This category includes *cirque glaciers*, which fill a bowl-shaped depression, known as a **cirque**, on the flank of a mountain; *valley glaciers*, rivers of ice that flow down valleys; mountain *ice caps*, mounds of ice that submerge peaks and ridges at the crest of a mountain range; and *piedmont glaciers*, fans or lobes of ice that form where a valley glacier emerges from a valley and spreads out over the nearby plain **(Fig.14.20b-d)**. Mountain glaciers range in size from a few hundred meters to a few hundred kilometers long, and from tens of meters to 1.5 km (1 mile) thick. The slope of the land beneath a mountain glacier influences the direction in which the glacier flows, for overall, ice forms at higher elevations and descends to lower elevations.

Continental glaciers, or *continental ice sheets*, are vast sheets of ice that spread over thousands of square kilometers of continental crust. They now exist only on

Figure 14.19 The nature of ice and the formation of glaciers.

A boundary between layers

The layers in the photo at left are part of this glacier in the Alps.

(a) The shapes of snowflakes reflect the hexagonal crystal structure of ice.

(b) Accumulating layers of snow are essentially layers of sediment.

The wall of a tunnel bored into a glacier

Loose snow
(90% air)

Granular snow
(50% air)

Firn
(25% air)

Fine-grained ice
(< 20% air, in bubbles)

Coarse-grained ice
(< 20% air, in bubbles)

10,000 years
(250 m)

130,000 years
(2,000 m)

(c) Glacial ice is blue. As revealed by a microscope, it has coarse grains and contains air bubbles.

(d) Snow compacts and melts to form firn, which recrystallizes into glacial ice in a process comparable to metamorphism. Crystal size increases with depth.

Figure 14.20 A variety of glacier types form in mountainous areas.

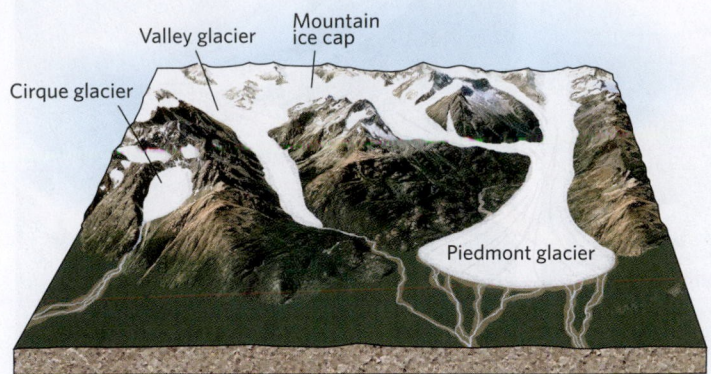

(a) Mountain glaciers are classified by their shape and position.

(b) A valley glacier and cirque glaciers in Switzerland.

(c) Valley glaciers draining a mountain ice cap in Alaska.

(d) A piedmont glacier near the coast of Alaska.

Antarctica and Greenland **(Fig. 14.21a–c)**, but during ice ages, they covered large portions of other continents. Continental glaciers flow outward from their thickest point (which may be up to 4 km, 2.5 miles, thick) and thin toward their margins (where they may be only tens to hundreds of meters thick). The slope of the land surface does not significantly influence the overall movement of a continental glacier.

Geologists also distinguish between types of glaciers on the basis of the temperature of the ice within them. **Temperate glaciers** exist in regions where atmospheric temperatures can become warm enough for glacial ice to be at or near its melting temperature during part or all of the year, whereas **polar glaciers** exist in regions where atmospheric temperatures stay so cold throughout the year that the glacial ice remains below its melting temperature year-round.

The Movement of Glacial Ice

The manner in which ice moves within a glacier typically varies with depth. Under the pressures found at depths of greater than about 60 m (200 feet) below the glacier's surface, ice can undergo *plastic deformation*, meaning that the grains within it change shape very slowly without breaking, and new grains grow while old ones disappear (see Chapter 7). These processes involve rearrangement of water molecules within ice grains. In temperate glaciers, slippery water films form along grain boundaries, so plastic deformation may also involve the microscopic slip of ice crystals past their neighbors.

At depths of less than about 60 m within a glacier, above a boundary known as the *brittle-plastic transition*, ice can't flow. This brittle ice gets carried along as the ice beneath it flows, and it adjusts to changes in the shape of the glacier, as may happen when the glacier flows over a ledge or around a curve, by cracking. A crack that develops by brittle deformation of a glacier is called a **crevasse** **(Fig. 14.22)**. In large glaciers, crevasses can be hundreds of meters long, and they can open up to form gashes up to 15 m (50 feet) across.

Why do glaciers move? Ultimately, they move because the pull of gravity can make ice flow. A glacier flows in

Figure 14.21 The Antarctic ice sheet is one of the two continental glaciers that exist today.

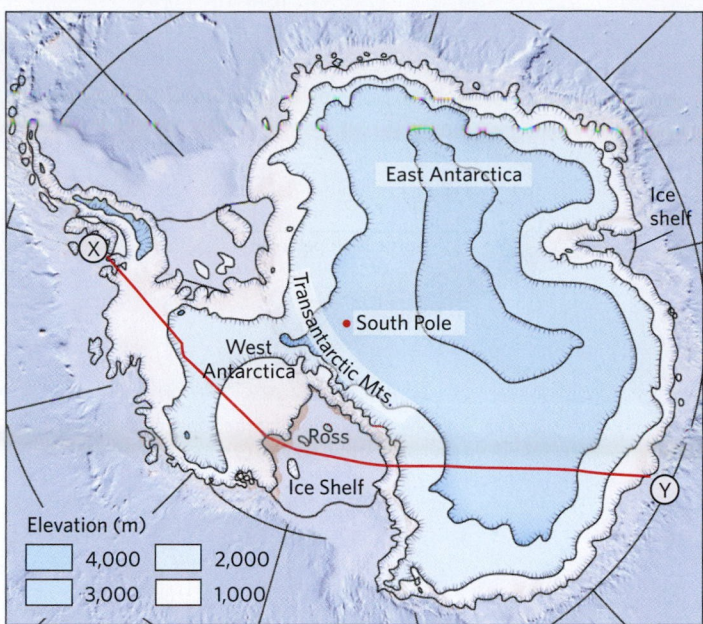

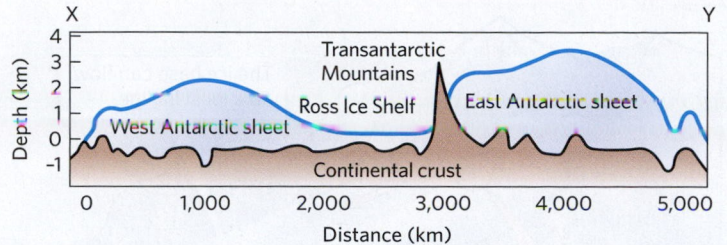

(b) A cross section of the Antarctic ice sheet. The Transantarctic Mountains separate East Antarctica from West Antarctica.

(a) A contour map of the Antarctic ice sheet. Valley glaciers carry ice from the East Antarctic portion of the ice sheet down to the Ross Ice Shelf.

(c) A view across the Antarctic ice sheet. Wind has blown away most loose snow, and the ice surface itself has become rippled.

Figure 14.22 Crevasses form in the upper layer of a glacier, in which the ice is brittle. Commonly, cracking takes place where the glacier bends while flowing over steps or ridges in its substrate.

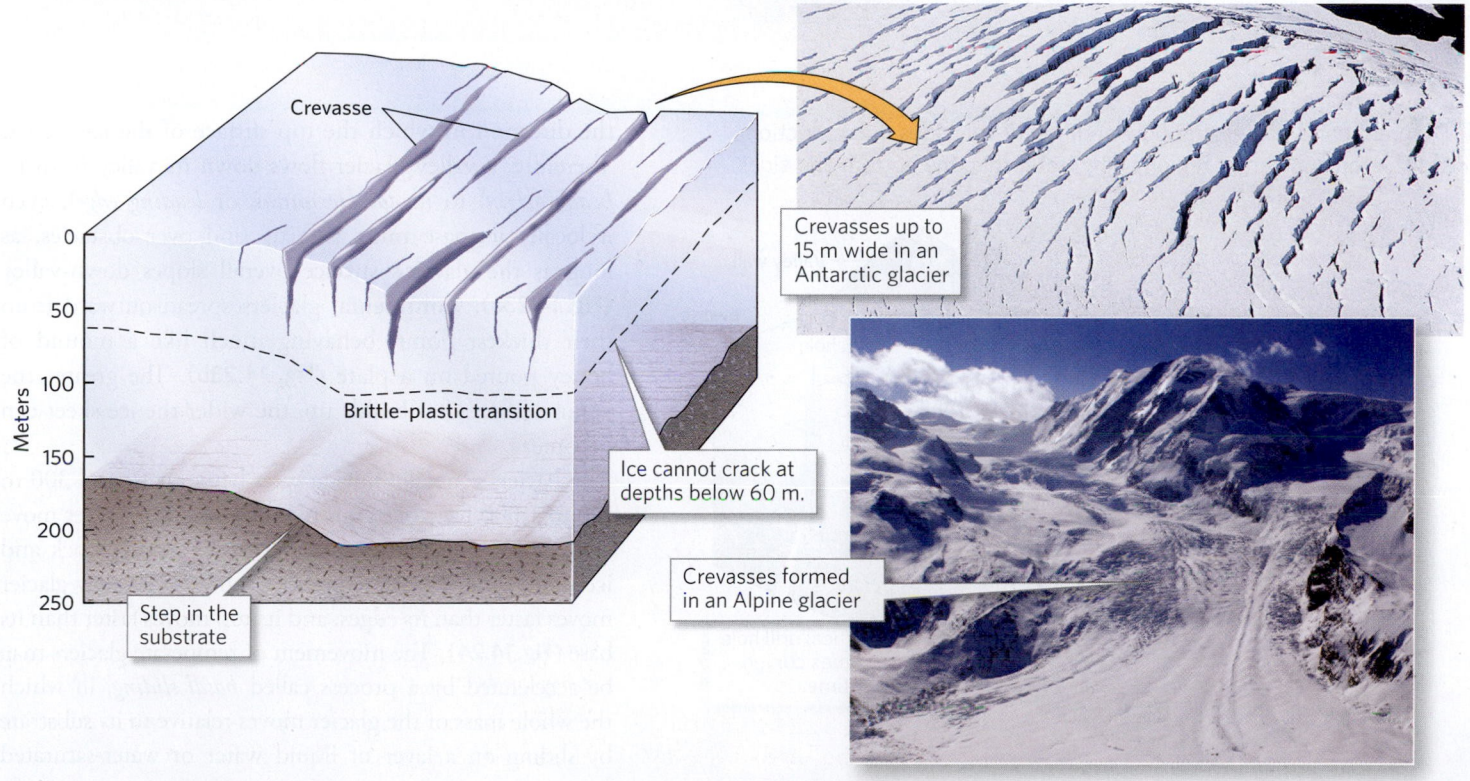

Figure 14.23 Forces that drive the movement of glaciers.

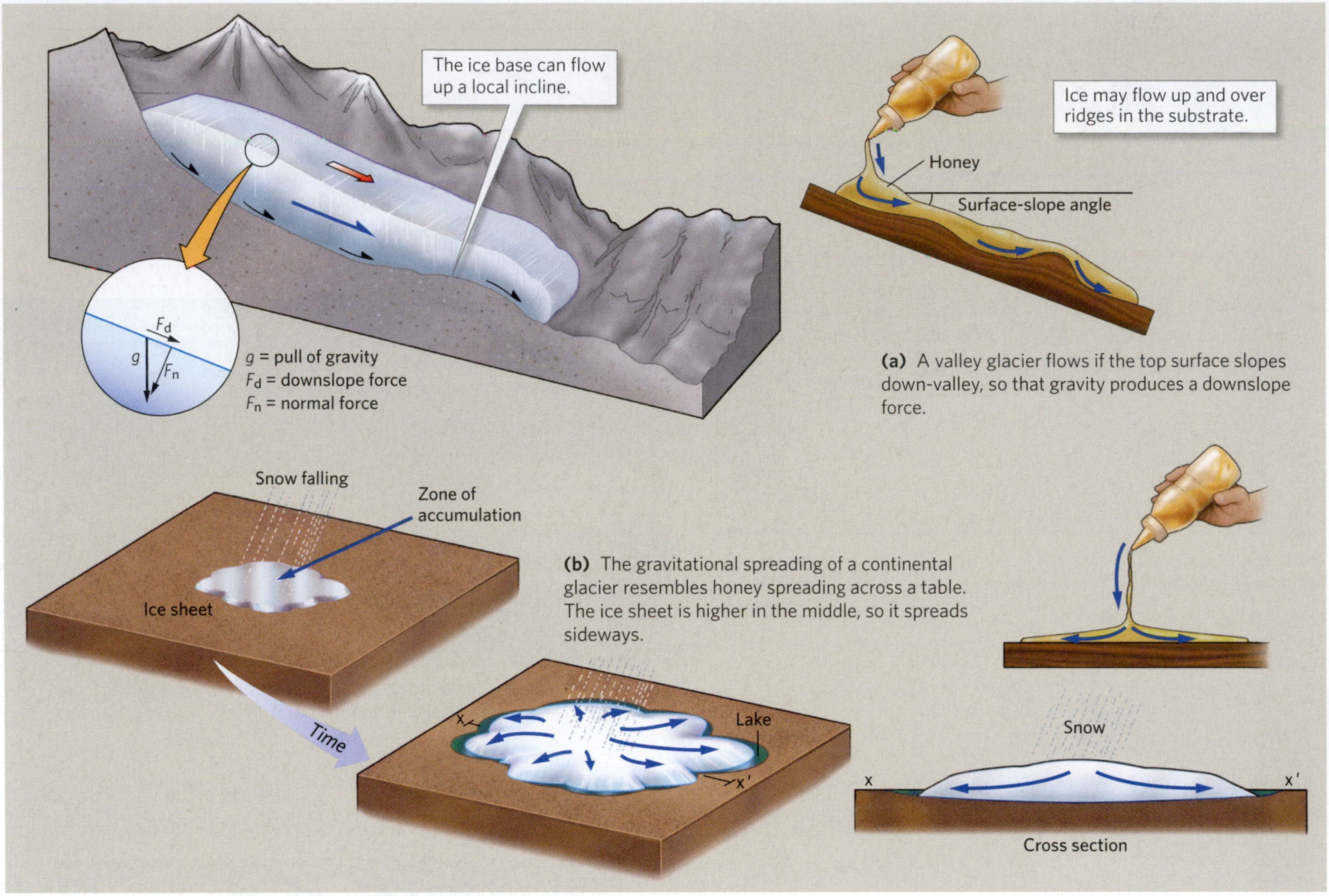

The ice base can flow up a local incline.

g = pull of gravity
F_d = downslope force
F_n = normal force

Ice may flow up and over ridges in the substrate.

Honey

Surface-slope angle

(a) A valley glacier flows if the top surface slopes down-valley, so that gravity produces a downslope force.

Snow falling

Zone of accumulation

Ice sheet

(b) The gravitational spreading of a continental glacier resembles honey spreading across a table. The ice sheet is higher in the middle, so it spreads sideways.

Time

x

x'

Lake

Snow

x

x'

Cross section

Figure 14.24 Different parts of a glacier flow at different velocities due to friction with the substrate. The top and center regions flow faster than the bottom and sides.

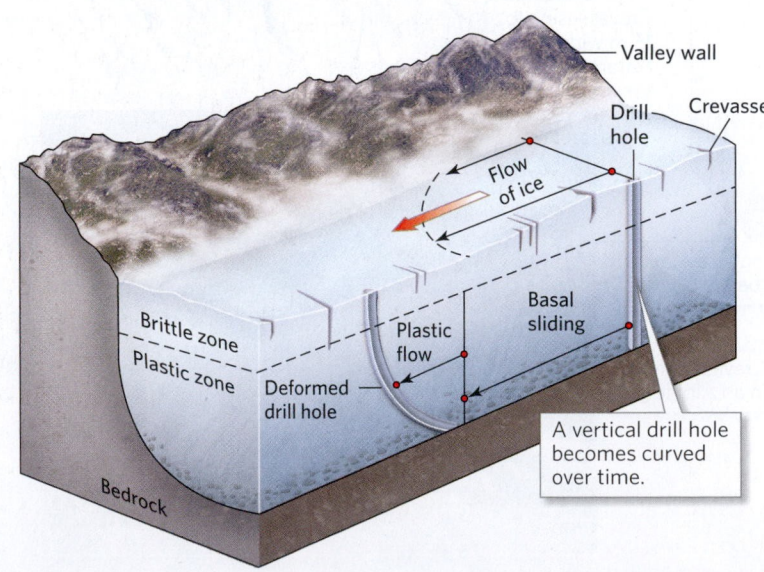

Valley wall

Drill hole

Crevasse

Flow of ice

Brittle zone

Plastic zone

Plastic flow

Deformed drill hole

Basal sliding

A vertical drill hole becomes curved over time.

Bedrock

the direction in which the top surface of the ice slopes. Therefore, a valley glacier flows down its valley from its *head* (*origin*) to its *toe* (*terminus*, or *leading edge*), even if locally its base must ride up and over obstacles, as long as the glacier's surface overall slopes down-valley **(Fig. 14.23a)**. Continental glaciers spread outward from their thickest point, behaving much like a mound of honey poured on a plate **(Fig. 14.23b)**. The greater the volume of ice that builds up, the wider the ice sheet can become.

Glaciers generally flow at rates between 10 and 300 m (30 to 1,000 feet) per year. Not all parts of a glacier move at the same rate, however, for friction between rock and ice slows a glacier. Therefore, the center of a valley glacier moves faster than its edges, and its top moves faster than its base **(Fig. 14.24)**. The movement of temperate glaciers may be accelerated by a process called *basal sliding*, in which the whole mass of the glacier moves relative to its substrate by sliding on a layer of liquid water or water-saturated

sediment. The water or watery sediment holds the glacier above bedrock and thereby decreases friction. Locally, basal sliding may trigger a *glacial surge*, during which ice accelerates to speeds of 10 to 110 m (30 to 360 feet) per day!

Glacial Advance and Retreat

Glaciers resemble bank accounts in that snowfall adds ice to the account, while ablation subtracts ice from the account. **Ablation**, the removal of ice from a glacier, includes several processes: *sublimation* (transformation of ice into water vapor), *melting* (transformation of ice into liquid water, which flows away), and *calving* (breaking off of chunks of ice). Snowfall adds ice to the glacier in the **zone of accumulation**, whereas glacial ice volume decreases in the **zone of ablation**. The boundary between these two zones is the **equilibrium line (Fig. 14.25)**.

If the rate of ice loss in the zone of ablation equals the rate of accumulation above the equilibrium line, then the position of the toe remains fixed, even though ice within the glacier continues to flow toward the toe

Figure 14.25 The equilibrium line separates the zone of accumulation from the zone of ablation. As indicated by arrows, ice sinks while flowing downhill in the zone of accumulation and rises while flowing downhill in the zone of ablation. Overall, it follows a curving path.

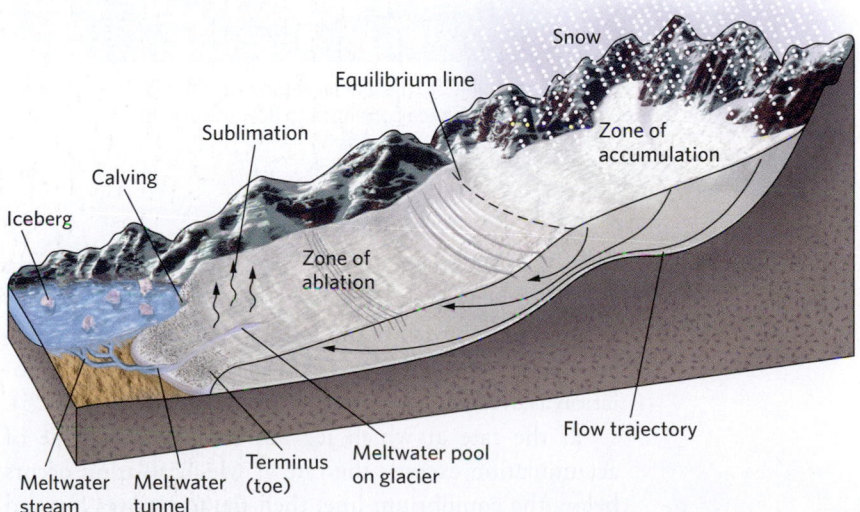

Figure 14.26 Glacial advance and retreat.

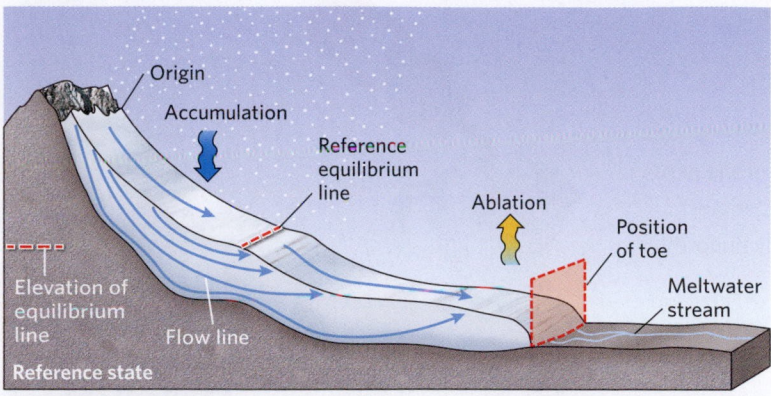

Origin
Accumulation
Reference equilibrium line
Ablation
Position of toe
Meltwater stream
Elevation of equilibrium line
Flow line
Reference state

(a) The position of the toe represents a balance between accumulation and ablation.

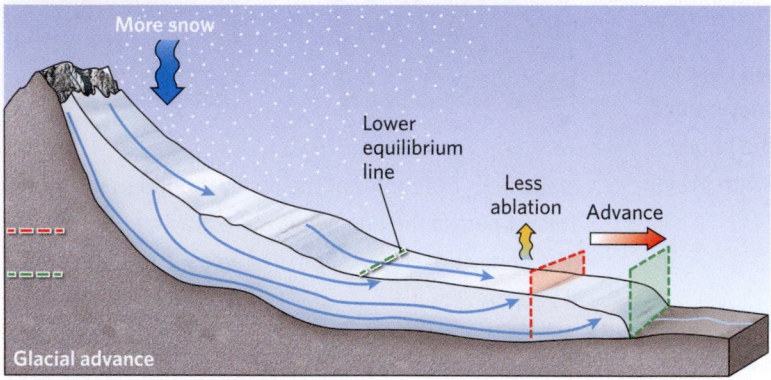

More snow
Lower equilibrium line
Less ablation
Advance
Glacial advance

(b) If accumulation exceeds ablation, the glacier advances, the toe moves farther from the origin, and the ice thickens.

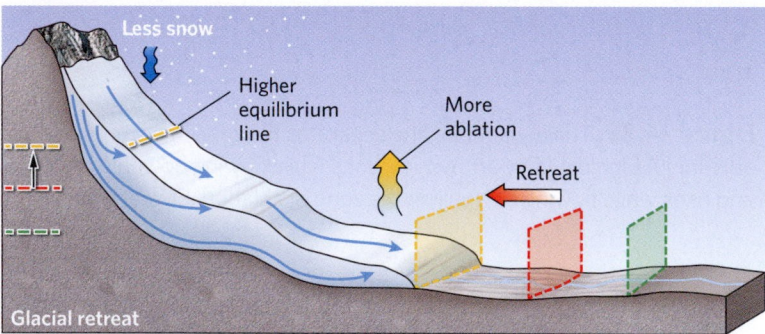

Less snow
Higher equilibrium line
More ablation
Retreat
Glacial retreat

(c) If ablation exceeds accumulation, the glacier retreats and thins. The position of the toe moves back, even though ice continues to flow toward the toe.

(Fig. 14.26a). During this flow, ice follows a curved path, sinking down toward the base of the glacier as new ice accumulates above it in the zone of accumulation, and rising up toward the surface of the glacier in the zone of ablation as overlying ice melts or sublimates (see Fig. 14.25).

If the rate at which ice builds up in the zone of accumulation exceeds the rate at which ablation occurs below the equilibrium line, then the toe moves forward into previously unglaciated regions, a change known as a **glacial advance (Fig. 14.26b)**. In mountain glaciers, the position of the toe moves downslope during an advance, and in continental glaciers, the toe moves outward, away from the glacier's origin, during an advance. If the rate of ablation exceeds the rate of accumulation, then the position of the toe moves back toward the glacier's origin, a change known as a **glacial retreat (Fig. 14.26c)**. During a mountain glacier's retreat, the position of the toe moves upslope, and during a continental glacier's retreat, the toe moves back toward the glacier's origin. Note that when a glacier retreats, it's only the position of the toe that moves back toward the origin; ice itself continues to flow toward the toe, for it cannot move against gravity.

Ice in the Sea

On the moonless night of April 14, 1912, the ocean liner *Titanic* struck an **iceberg**, a large block of floating ice, in the frigid North Atlantic. Lookouts had seen the ghostly mass only minutes earlier and had alerted the ship's pilot, but the ship had been unable to turn fast enough to avoid disaster. The force of the blow split its hull, allowing water to gush in, and less than 3 hours later, the *Titanic* disappeared beneath the surface.

Where do icebergs, such as the one responsible for the *Titanic*'s demise, originate? At high latitudes, mountain glaciers and continental glaciers flow down to, and in some cases, out into coastal waters to become **tidewater glaciers**. A tidewater glacier that forms where a mountain glacier enters the sea may protrude a few kilometers out as an *ice tongue*, whereas a continental glacier entering the sea becomes a broad, flat **ice shelf**. In shallow water, glacial ice remains grounded, but where the water becomes deep enough, the ice floats **(Fig. 14.27a)**. At the toe of a tidewater glacier, blocks of ice calve off and tumble into the water with an impressive splash. The largest calving event ever recorded by video involved a city-sized volume of ice, about 10 km² (4 square miles) and 600 m (2,000 feet) thick, that broke off a tidewater glacier in Greenland over the course of 75 minutes. During the process, the ice broke into countless blocks in various sizes. A free-floating chunk of ice that rises 6 m (20 feet) above the water and is at least 15 m (50 feet) long is formally called an *iceberg*. Since four-fifths of a floating body of ice lies below the surface of the sea, the base of a large iceberg may actually be a few hundred meters below sea level **(Fig. 14.27b)**.

Not all ice floating in the sea originates as glaciers on land. In polar climates, the surface of the ocean itself freezes, forming **sea ice (Fig. 14.27c, d)**. The north polar ice cap of the Earth consists of a 2–6-m (6–20-foot)-thick layer of sea ice formed on the surface of the Arctic Ocean. This sea ice slowly moves with ocean currents at a rate of about 2 km (1.2 miles) per day.

Figure 14.27 Ice in the sea.

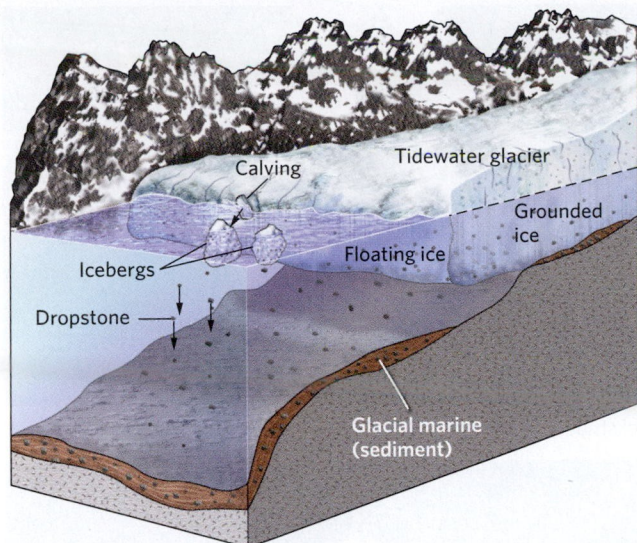

(a) Ice is grounded in shallow water, but floats in deep water.

(b) This artist's rendition of an iceberg emphasizes that most of the ice is underwater.

(c) In summer, some of the sea ice of Antarctica breaks up to form icebergs.

Take-home message . . .

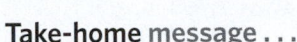 Glaciers form when snow persists all year, gets buried deeply, and turns to ice. Mountain glaciers form at high elevations, whereas continental glaciers form at high latitudes and spread over continents. Ice flows by plastic deformation at depth. The upper, brittle portion of a glacier fractures to form crevasses. The balance between accumulation and ablation controls glacial advance or retreat. In polar regions, sea ice covers large areas.

Quick Question -
Does ice actually flow uphill during a retreat of a valley glacier?

(d) Sea ice covers most of the Arctic Ocean (left) and surrounds Antarctica (right).

14.7 Carving and Carrying: Erosion by Ice

Glaciers serve as very powerful agents of erosion, capable of carving deep valleys and knife-edged ridges in mountains and of beveling and polishing broad areas of continents. In this section, we look at how glacial erosion takes place and at specific landforms it produces.

Glacial Erosion and Its Products

Glacial ice plucks up fragments of its substrate as it moves. Pure ice is too soft to erode rock, but hard clasts embedded in moving ice act like the teeth of a giant rasp and grind away the substrate. Abrasion by incorporated sand-sized grains can polish bedrock at the base or sides of the glacier (**Fig. 14.28a**). Larger clasts, or trains of clasts, embedded in ice gouge out grooves or scratches, called **glacial striations**, that range from 1 mm to 1 m (.04 to 40 inches) across and may be tens of centimeters to tens of meters long (**Fig. 14.28b**). As you might expect, glacial striations trend parallel to the flow direction of the ice.

Mountain glaciers sculpt a variety of erosional landforms. Frost wedging due to freezing and thawing breaks up bedrock bordering the head of the glacier high in the mountains. This fractured rock falls on the ice, or gets picked up at the base of the ice, and moves downslope with the glacier. As a consequence, a bowl-shaped depression, or **cirque**, develops on the side of the mountain (**Fig. 14.28c**).

Figure 14.28 Glaciers are powerful agents of erosion.

(a) A glacially polished outcrop in Central Park, New York City.

(b) Glacial striations in Victoria, British Columbia.

Striation

Cirque

Arête

(c) Examples of a cirque and an arête in the Swiss Alps.

(d) The Matterhorn in Switzerland is bordered by cirques.

See for yourself

Glaciated Peaks, Montana

Latitude: 48°56′33.66″ N
Longitude: 113°49′54.59″ W

Zoom to 8 km (~5 miles) and look down.

Three cirques bound a horn in the Rocky Mountains, north of Glacier National Park. Note the knife-edge arêtes between the cirques. The glaciers that carved the cirques have melted away. The thin stripes on the mountain face are traces of sedimentary beds that make up the bedrock.

An **arête** (French for ridge) separates adjacent cirques. If the glacier later melts, a lake known as a *tarn* may fill the base of the cirque. When cirques form on more than one side of a mountain, the mountain becomes a pointed peak called a **horn (Fig. 14.28d)**.

Glacial erosion dramatically modifies the shape of a valley **(Fig. 14.29)**. To see how, compare a river-eroded valley with a glacially eroded valley. If you look along the length of a river in unglaciated mountains, you'll see that it typically flows down a *V-shaped valley*, with the river channel forming the point of the V. The V shape develops because river erosion occurs only in the channel, and mass wasting causes the valley slopes to approach the angle of repose (see Section 13.4). But if you look down the length of a glacially eroded valley, you'll see that it resembles a U, with steep walls. Such a **U-shaped valley (Fig. 14.30a, b)** forms because glacial erosion not only lowers the floor of the valley but also bevels its sides.

Glacial erosion in mountains also modifies the intersections between tributary valleys and the trunk valley. Recall from Section 13.3 that a trunk stream, of a drainage network transporting water, serves as the base level for its tributaries, so at their point of intersection, a tributary stream and a trunk stream lie at the same elevation. Tributary glaciers, however, cannot erode their valleys to the same depth as the valley carved by a trunk glacier at the same elevation. As a consequence, when the glaciers melt away, the mouths of the tributary glacial valleys perch well above the floor of the trunk valley. Such tributary valleys are known as **hanging valleys**. Often, a present-day tributary stream cascades as a spectacular waterfall where it emerges from a hanging valley **(Fig. 14.30c)**.

The erosional features produced by continental glaciers depend, to a large extent, on the nature of the pre-glacial landscape. Where an ice sheet spreads over a region of low relief, glacial erosion produces a vast region of polished, striated surfaces. Where an ice sheet spreads over a hilly area, it deepens valleys and smooths hills. Glacially eroded hills may be elongate in the direction of flow and asymmetrical because glacial abrasion smooths the upstream part of the hill, yielding a gentle slope, whereas glacial plucking eats away at the downstream part, producing

Figure 14.29 Stages in the development of a glacially carved mountainous landscape. The inset photo shows a tarn.

V-shaped valley

Tributary valley

Before glaciation, valleys are V-shaped, and tributary mouths are the same elevation as the trunk stream.

Trunk valley

Trunk valley

Tributary valley

During glaciation, the valleys fill with ice.

Cirque

Tarn

Mt. Snowdon, Wales

Time

After glaciation, the region contains U-shaped valleys, hanging valleys, truncated spurs, and horns.

Horn

Cirque

Arête

Hanging valley

U-shaped valley

Figure 14.30 Examples of glacially eroded valleys.

(a) A U-shaped valley bordered by Half Dome, in Yosemite National Park, California.

(b) A U-shaped glacial valley in the Tongass National Forest, Alaska.

(c) A waterfall spilling out of a U-shaped hanging valley in the Sierra Nevada.

Figure 14.31 The sculpting of a roche moutonnée.

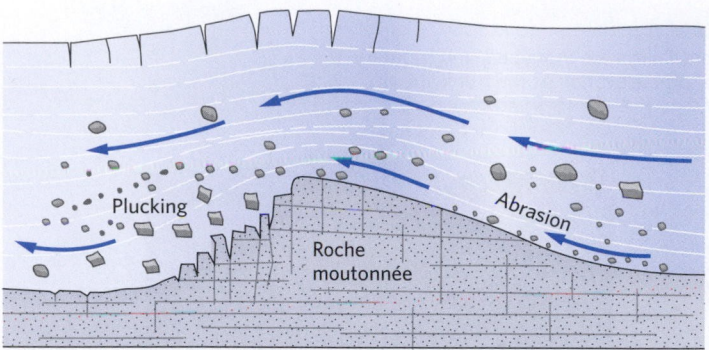

(a) Glacial abrasion rasps the upstream side, and glacial plucking carries away fracture-bounded blocks on the downstream side.

(b) An example of a roche moutonnée in the Sierra Nevada. The glacier that shaped this hill flowed from right to left.

a steep slope. Such a hill is called a **roche moutonnée** (French for sheep rock) because its profile resembles that of a sheep resting in a meadow **(Fig. 14.31)**.

Fjords: Submerged Glacial Valleys

Where a valley glacier meets the sea, the glacier's base remains in contact with the substrate until the water depth exceeds about four-fifths of the glacier's thickness. Therefore, glaciers can carve U-shaped valleys even below sea level. In addition, as we'll see later, sea level rises significantly after ice ages. As a result, the floors of valleys cut by coastal glaciers during ice ages lie far below present-day sea level, so the sea has flooded these deep valleys. A glacially carved valley filled with water is called a **fjord**. Spectacular marine fjords occur along the coasts of Norway, New Zealand, Chile, Maine, and Alaska. Fjords tend to be much longer than they are wide, and their walls—the sides of a U-shaped valley—are very steep cliffs, some of which rise 1,000 m (3,300 feet) **(Fig. 14.32)**. Not all fjords are marine; fjords also develop inland where a glacially carved valley fills with freshwater and becomes a lake. The Finger Lakes of central New York State (see Fig. 13.33b) serve as examples.

Take-home message . . .

A glacier plucks and scrapes up rock from its substrate and carries it, along with debris that falls on its surface, in the direction of flow. Abrasion by these incorporated clasts grinds away at a glacier's substrate. Glacial erosion polishes rock, produces striations, and carves distinctive landforms, such as U-shaped valleys and cirques. U-shaped valleys that fill with water become fjords.

Quick Question -
Why do we find hanging valleys in mountains that have been eroded by glaciers?

See for **yourself**

Baffin Island, Canada

Latitude: 67°8′27.56″ N
Longitude: 64°49′49.31″ W

Zoom to 40 km (~25 miles) and look down.

Two valley glaciers draining the Baffin Island ice cap merge into a trunk glacier that flows NE and then into a fjord, partly filling a U-shaped valley. Note the lateral and medial moraines. In some nearby fjords, meltwater has deposited a delta of outwash at the end of the glacier.

14.8 Deposition Associated with Glaciation

The Glacial Conveyor

Glaciers can carry sediment of any size and, like a conveyor belt, transport it in the direction of flow **(Fig. 14.33)**. Some of the sediment moving with a glacier is plucked from the substrate under the ice. The rest tumbles onto the glacier during rockfalls or slides down slopes bordering the glacier.

Figure 14.32 One of the many spectacular fjords of Norway. The water is an arm of the sea that fills a U-shaped valley. Tourists are standing on Pulpit Rock (Preikestolen).

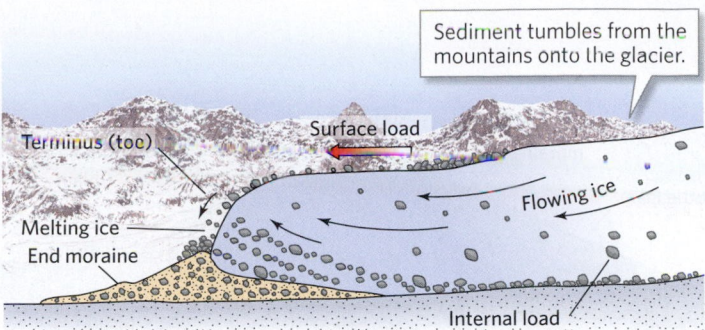

Sediment tumbles from the mountains onto the glacier.

Surface load

Terminus (toe)

Flowing ice

Melting ice

End moraine

Internal load

(a) Sediment falls on a glacier from bordering slopes and gets plucked up from below.

Figure 14.33 The concept of the glacial conveyor.

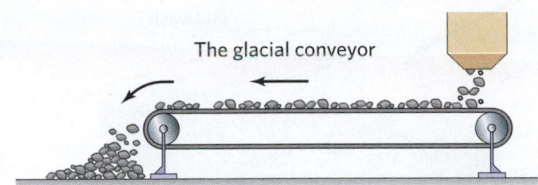

The glacial conveyor

(b) In effect, a glacier acts like a conveyor belt, moving sediment in the direction of flow.

Geologists refer to a pile of sediment carried or left by a glacier as a **moraine**. Sediment tumbling onto the edge of a glacier forms a stripe known as a *lateral moraine* (**Fig. 14.34a**). When a glacier melts, lateral moraines remain stranded along the sides of the glacially carved val-ley, like bathtub rings (**Fig. 14.34b**). Where two valley glaciers merge, two lateral moraines merge to become a *medial moraine,* a stripe of debris that trends parallel to the

Figure 14.34 Medial moraines.

(a) This glacier in the French Alps has a medial moraine and two lateral moraines.

Lateral moraine

Lateral moraine

(b) The large ridge of sediment is a lateral moraine left when the glacier that once filled the valley in the foreground melted away.

Rockfall

Medial moraine

Lateral moraine

Origin of medial moraine 1

Lateral moraine

Moraine 2, also a medial moraine, formed to the right of this image.

(c) These medial moraines formed where lateral moraines of two valley glaciers merged in Alaska.

Figure 14.35 Formation of end moraines.

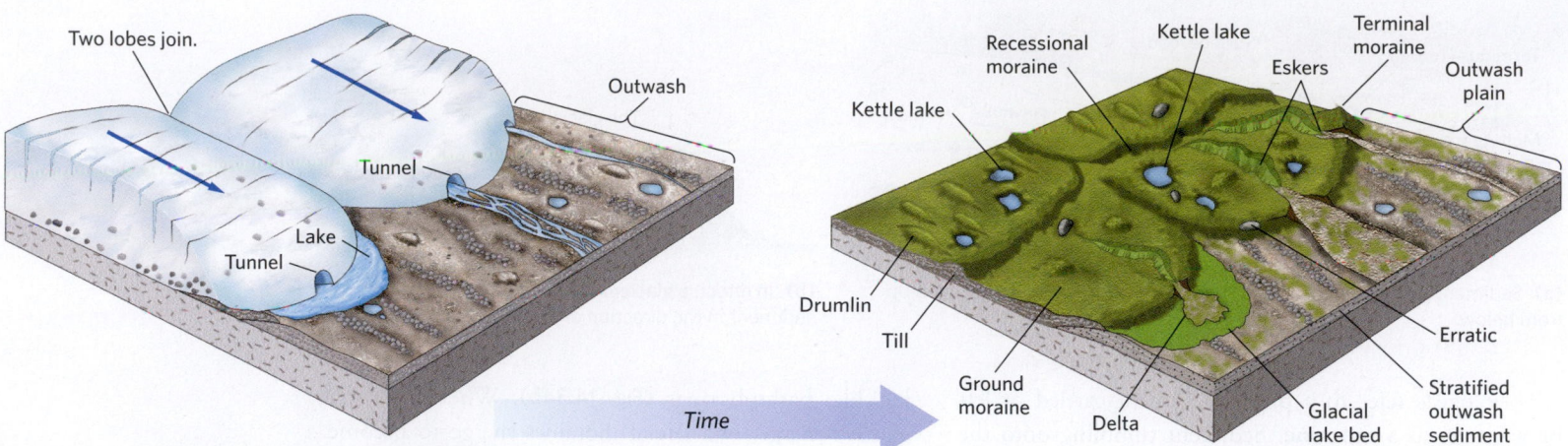

(a) When a glacier begins to recede, it drops sediment and releases meltwater.

(b) A number of distinct depositional landforms, including several types of moraines, form as a consequence of glaciation.

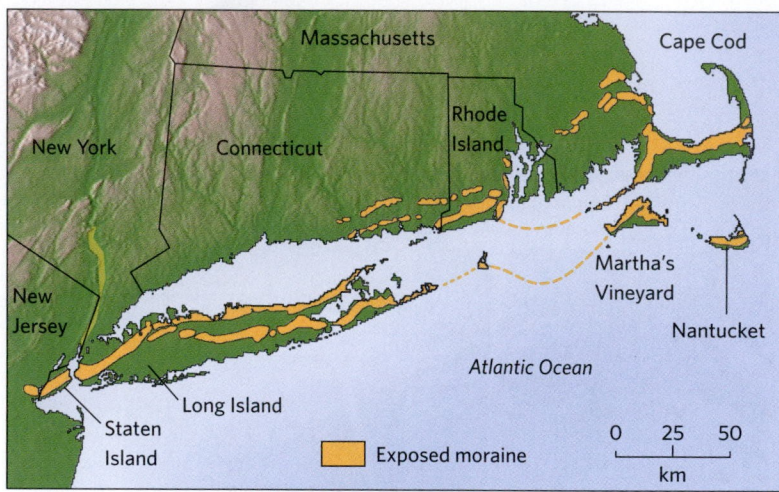

(c) The terminal moraine of a continental glacier underlies Cape Cod and Long Island in the northeastern United States.

flow direction, down the interior of the composite glacier **(Fig. 14.34c)**. Trunk glaciers created by the merging of multiple tributary glaciers contain several medial moraines.

Sediment transported to a glacier's toe by the glacial conveyor accumulates in a pile at the toe and builds up to form an *end moraine* **(Fig. 14.35a, b)**. Geologists refer to the end moraine at the farthest limit of ice flow as the glacier's **terminal moraine**. A large terminal moraine, formed during the last ice age, built the ridge of sediment that now forms Long Island and Cape Cod in the eastern United States **(Fig. 14.35c)**. As a glacier recedes, it may pause temporarily and build up a *recessional moraine*. Sediment left when a glacier recedes without pausing, so a distinct ridge doesn't form, makes up *ground moraine*.

Types of Glacial Sedimentary Deposits

Geologists distinguish among several different types of sediment that accumulate in glacial environments:

- *Till:* Sediment transported by ice and deposited beneath, at the side of, or at the toe of a glacier makes up **glacial till**. Glacial till tends to be unsorted because the solid ice of glaciers carries clasts of all sizes. So till typically consists of cobbles or boulders suspended in clay and silt **(Fig. 14.36a)**.

- *Erratics:* A glacial **erratic (Fig. 14.36b)** is a cobble or boulder, dropped by a glacier. The word comes from the Latin *errare*, which means to wander.

- *Glacial marine sediment:* Where a sediment-laden glacier flows into the sea, icebergs calve off the toe and raft clasts out to sea. As the icebergs melt, the clasts they carry sink to the seafloor. These *dropstones* mix with marine sediment to form *glacial marine sediment*.

- *Glacial outwash:* Till deposited by a glacier at its toe may be picked up and transported by meltwater streams. Such water-transported sediment, known as **glacial outwash**, tends to be sorted by water to form bars of gravel and sand **(Fig. 14.36c)**.

- *Loess:* Strong winds in glacial environments pick up fine clay and silt and transport it away from the glacier. This wind-transported glacial sediment eventually settles out of the air to form deposits of **loess (Fig. 14.36d)**.

- *Glacial lake-bed sediment:* Some of the sediment yielded by glaciers accumulates on the floors of meltwater lakes. Notably, such lake-bed sediments tend to consist of alternate layers of silt (brought into the lake during spring floods) and clay (which settles from still water when the lake freezes in the winter). A pair of layers, representing deposition during an entire year, is called a *varve* **(Fig. 14.36e)**.

Figure 14.36 Sediment types associated with glaciation.

(a) This glacial till in Ireland is unsorted because ice can carry sediment of all sizes.

(b) Glacial erratics resting on a glacially polished surface in Wyoming.

(c) Braided streams choked with glacial outwash in Alaska. The streams carry away finer sediment and leave the gravel behind.

(d) Thick loess deposits underlie parts of the prairie in Illinois.

(e) (Left) In the quiet water of an Alaskan glacial lake, fine-grained sediments accumulate. (Right) Alternating layers (varves) in these lake-bed sediments, now exposed in an outcrop near Puget Sound, Washington, reflect seasonal changes.

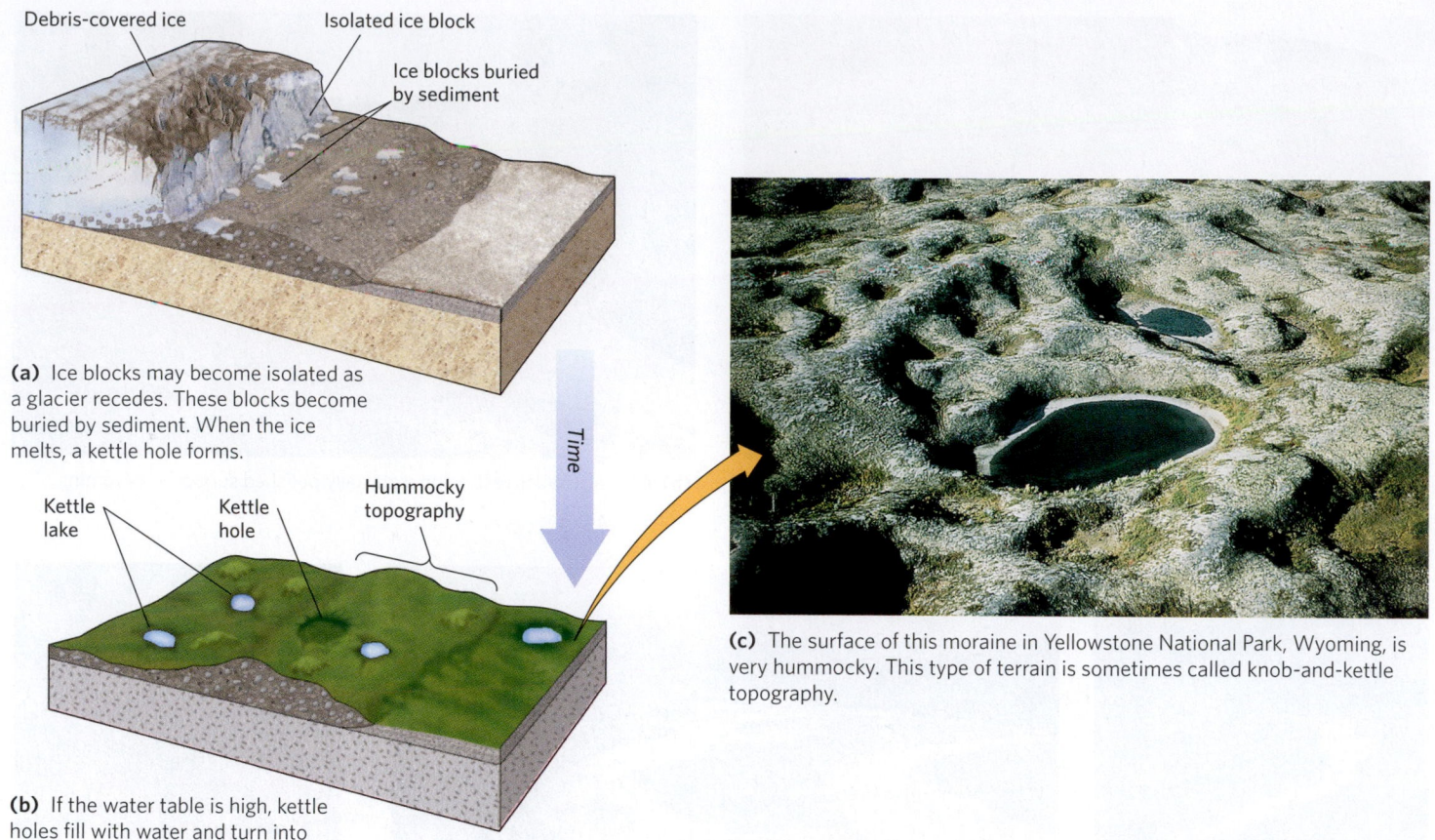

(a) Ice blocks may become isolated as a glacier recedes. These blocks become buried by sediment. When the ice melts, a kettle hole forms.

(b) If the water table is high, kettle holes fill with water and turn into roughly circular lakes.

(c) The surface of this moraine in Yellowstone National Park, Wyoming, is very hummocky. This type of terrain is sometimes called knob-and-kettle topography.

Depositional Landforms of Glacial Environments

Imagine what you would see if you stood on the toe of a retreating glacier. Looking in one direction, you would see nothing but ice. Looking in the other direction, you would see a variety of landscape features produced by glacial erosion and deposition (**Earth Science at a Glance, pp. 494–495**). We've already described landforms produced by glacial erosion—let's shift our focus to landforms built by glacial deposition.

From your perch, you might see a few curving end moraines. Moraines have a *hummocky* texture, meaning that they are bumpy topographic surfaces with countless small hills and depressions. The hummocky surface of a moraine reflects both variations in the amount of sediment supplied by the ice and the development of kettle holes. When a glacier's sediment-covered toe starts to undergo ablation, large ice blocks that calved off the toe, or were left behind as surrounding ice melted away, remain in the wake of the retreating glacier (**Fig. 14.37a, b**). The till covering these blocks may consist of sediment

that moved up to the glacier's surface from the substrate below as the ice flowed, or sediment that fell onto the ice. When each block eventually melts, it leaves behind a roughly circular depression known as a **kettle hole** (**Fig. 14.37c**).

Beyond an end moraine, the land surface may host braided streams of sediment-choked meltwater. These streams deposit bars of gravel and sand which build a broad *glacial outwash plain*. Meltwater may collect as a lake between end moraines or in an *ice-margin lake* adjacent to the glacier's toe. When these lakes later dry up, the land surface left behind tends to be very flat.

In some locations, glacial flow shapes the till beneath the glacier into an elongate hill, generally known as a **drumlin** (from the Gaelic word for small hill or ridge). Drumlins, which commonly occur in swarms and tend to be 10 to 60 m (33 to 200 feet) high and 300 m to 2 km (1,000 feet to 1.2 miles) long, align parallel to the glacier's flow direction. Notably, drumlins taper in the direction of flow—a drumlin's upstream end is steeper and wider than its downstream end (**Fig. 14.38a, b**).

As a glacier recedes, meltwater carves tunnels in the base of the ice. Sediment deposited in these tunnels may remain as a sinuous ridge, known as an **esker**, when the glacier melts away (**Fig. 14.38c, d**).

Figure 14.38 Development of drumlins and eskers.

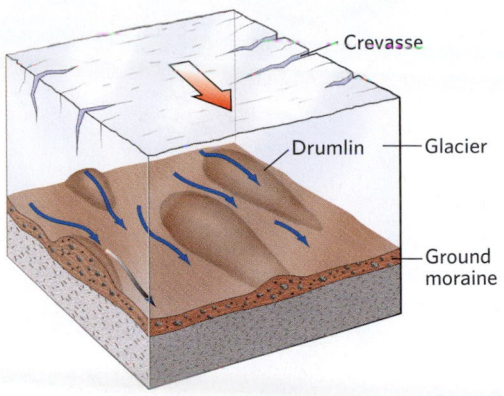

(a) Drumlins are sculpted beneath the ice of a glacier and are aligned with its flow direction.

Shaded relief map of the drumlins in central New York. Their SSE angle gives the direction of glacial flow.

View looking southeast

(b) Drumlins form a set of hills south of Lake Ontario in central New York State.

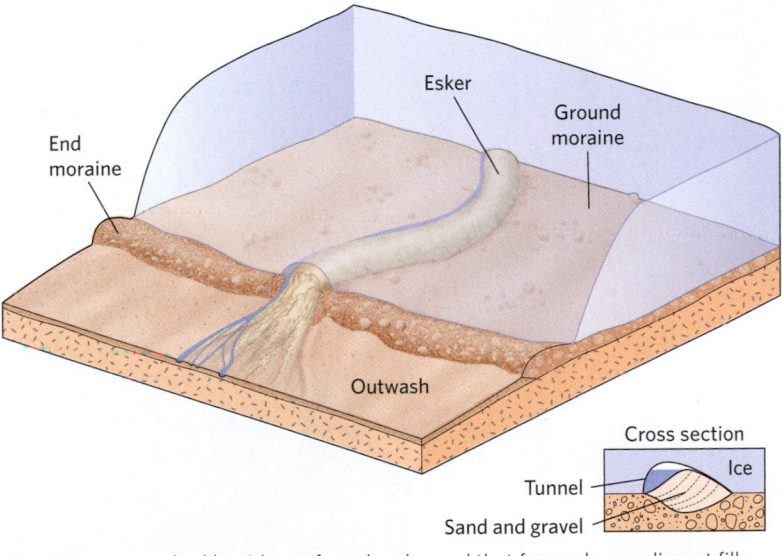

(c) Eskers are snake-like ridges of sand and gravel that form when sediment fills meltwater tunnels at the base of a glacier. The cross section [inset] shows how wedges of sand accumulate in the tunnel.

(d) An example of an esker in an area that was once glaciated.

Take-home message . . .

Moving ice does not sort sediment, so when it melts, a glacier deposits unsorted till. Meltwater streams and wind sort and transport glacial sediment to form glacial outwash plains and loess deposits, respectively. Deposition by glaciers produces distinctive landforms such as moraines, eskers, and kettle holes.

Quick Question -----------------------------
What do the sedimentary deposits that accumulate in glacial lakes look like?

14.9 Additional Consequences of Continental Glaciation

Glacial Subsidence and Glacial Rebound

When a large ice sheet (more than 50 km, or 30 miles, in diameter) grows on a continent, its weight causes the surface of the lithosphere to sink. In other words, an ice load causes **glacial subsidence**. The lithosphere, the Earth's relatively rigid outer shell, can sink because the underlying asthenosphere is soft enough to flow slowly out of

Glaciers and Glacial Landforms

Continental ice sheet

Crevasses

Higher sea level

Ice shelf

Lower sea level

Dropstones

494

Iceberg

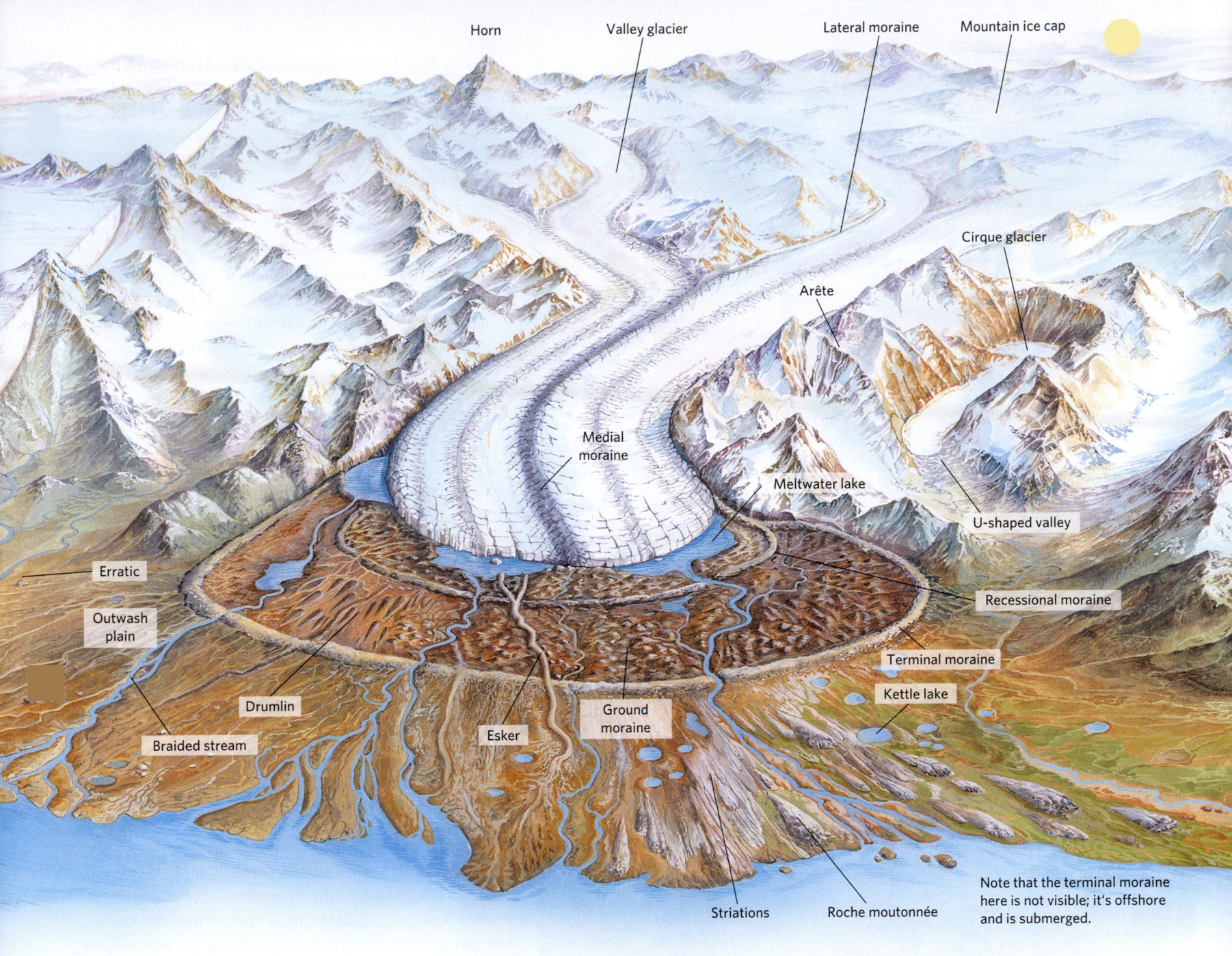

Horn Valley glacier Lateral moraine Mountain ice cap

Cirque glacier

Arête

Medial moraine

Meltwater lake

U-shaped valley

Erratic

Recessional moraine

Outwash plain

Terminal moraine

Kettle lake

Drumlin

Esker

Ground moraine

Braided stream

Striations Roche moutonnée

Note that the terminal moraine here is not visible; it's offshore and is submerged.

Continental glaciers, vast sheets of ice up to a few kilometers thick, covered extensive areas of land during times when Earth had a colder climate. They form from snow that accumulates at high latitudes. When buried deeply enough, the snow packs together and recrystallizes into glacial ice. Although it is solid, ice is weak, so ice sheets spread over the landscape like syrup over a pancake, though much more slowly. When a continental glacier reaches the sea, it becomes an ice shelf. At the edge of the shelf, icebergs calve off and float away, and sediment carried by the ice falls to the seafloor. A second category of glaciers, called mountain or alpine glaciers, grow in mountainous areas because snow can last all year at high elevations. Valley glaciers are confined to valleys, and ice caps cover the peaks of mountains. These glaciers carve a landscape containing U-shaped valleys, cirques, horns, and hanging valleys. Lateral moraines form along their sides.

During an ice age, a mountain glacier advances and may eventually flow out onto the land surface beyond the mountain front. When climate warms, the glacier starts to recede. The landscape in front of the glacier retains erosional features, such as striations and roches moutonée—formed when the area was ice covered by flowing ice.

When the glacier pauses, till (unsorted glacial sediment) accumulates to form an end moraine. Meltwater lakes gather at the toe, and streams pick up till, sort it, and redistribute it as glacial outwash. Sediment that accumulates in ice tunnels, exposed when the glacier melts, make up sinuous ridges called eskers. Flowing ice can mold underlying sediment into drumlins. Ice blocks buried in till melt away and leave behind kettle holes.

Figure 14.39 The concept of glacial subsidence and rebound.

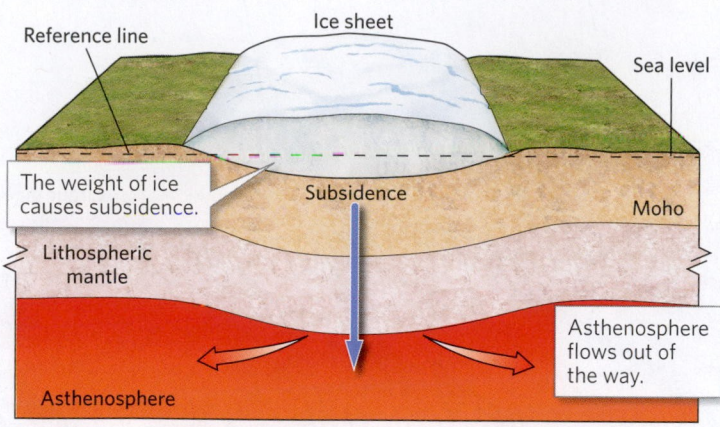

(a) The weight of an ice sheet causes the surface of the lithosphere to sink (subside).

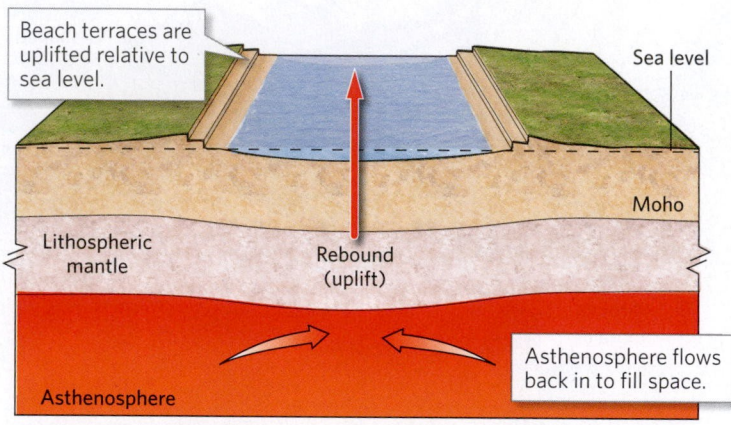

(b) When the glacier melts, the land surface rises (rebounds).

(c) Uplifted beach terraces along the coast of Arctic Canada form as the land undergoes glacial rebound.

the way **(Fig. 14.39a)**. Because of glacial subsidence, large areas of Antarctica and Greenland now lie below sea level.

What happens when continental glaciers melt away? Gradually, the surface of the underlying continent rises back up by a process called **glacial rebound**, and the asthenosphere flows back underneath to fill the space **(Fig. 14.39b)**. This process doesn't take place instantly. Asthenosphere flows so slowly (at rates of a few millimeters per year) that it takes thousands of years for ice-depressed continents to rebound. In fact, glacial rebound still happens in north-central Canada, a region that became ice-free about 6,500 years ago. Because of this rebound, beaches that formed in the past around Hudson's Bay now lie as much as 100 m (330 feet) above sea level **(Fig. 14.39c)**.

Effects of Glaciers on Drainage

A continental glacier disrupts drainage networks, for the glacier may completely cover a drainage network, causing pre-existing streams to find different routes. By the time the glacier melts away, these new streams may have become so well established in valleys that old river courses may remain abandoned.

Glaciation can also lead to the development of large lakes. Some of these lakes form in the low areas between end moraines, some in mountain valleys whose outlets have been blocked by an advancing glacier, and some are ice-margin lakes that occur where land has been pushed down by the weight of the glacier. The largest known ice-margin lake, known as Glacial Lake Agassiz, covered portions of south-central Canada and the north-central United States between 11,700 and 9,000 years ago **(Fig. 14.40)**. At its largest, the lake submerged over 250,000 km² (100,000 square miles), an area greater than that of all the present Great Lakes combined. When Glacial Lake Agassiz burst through the ice dam that retained it, so much cold freshwater surged into the Atlantic Ocean that researchers speculate that it caused sea level to rise by 1 to 3 m (3 to 10 feet) in less than a year. This release may also have disrupted ocean currents bringing warm equatorial water northward, which in turn caused global climate to cool. The short-lived floods caused by the sudden release of glacial lakes can modify landscapes significantly **(Box 14.2)**.

Pluvial and Periglacial Features

During the most recent ice age, the climate in regions to the south of continental glaciers was wetter than it is today. Fed by enhanced rainfall, lakes accumulated in low-lying areas at a great distance from the ice front. Many such **pluvial lakes** (from the Latin *pluvia*, meaning rain) flooded the interior basins of the Basin and Range Province in Utah and Nevada **(Fig. 14.41a)**. The largest

Box 14.2 ▶ Consider this . . .

Glacial torrents

Inevitably, as the most recent continental glacier retreated, the ice dams or moraines that held back glacial lakes melted or broke. In some cases, the contents of the lakes drained in a matter of days, resulting in immense floods known as *glacial torrents*. These floods widened valleys, stripped the land of soil, and left behind mounds of sediment. For example, when Glacial Lake Kankakee, trapped between recessional moraines left by a glacier that once covered northern Illinois, broke through a moraine dam, it scoured the large valley through which the present-day Illinois River flows. This valley is much larger than the valley that the river would have cut by its own flow.

When the ice dam holding back Glacial Lake Missoula in Montana failed, the *Great Missoula Flood* scoured eastern Washington State, creating a barren, soil-free landscape now known as the *channeled scablands* **(Fig. Bx14.2a, b)**. Giant ripple marks formed beneath the floodwaters from the immense amount of sediment carried in the water **(Fig. Bx14.2c)**. The shorelines of the long-gone glacial lake form benches on hills next to Missoula **(Fig. Bx14.2d)**.

Figure Bx14.2 Consequences of the Great Missoula Flood.

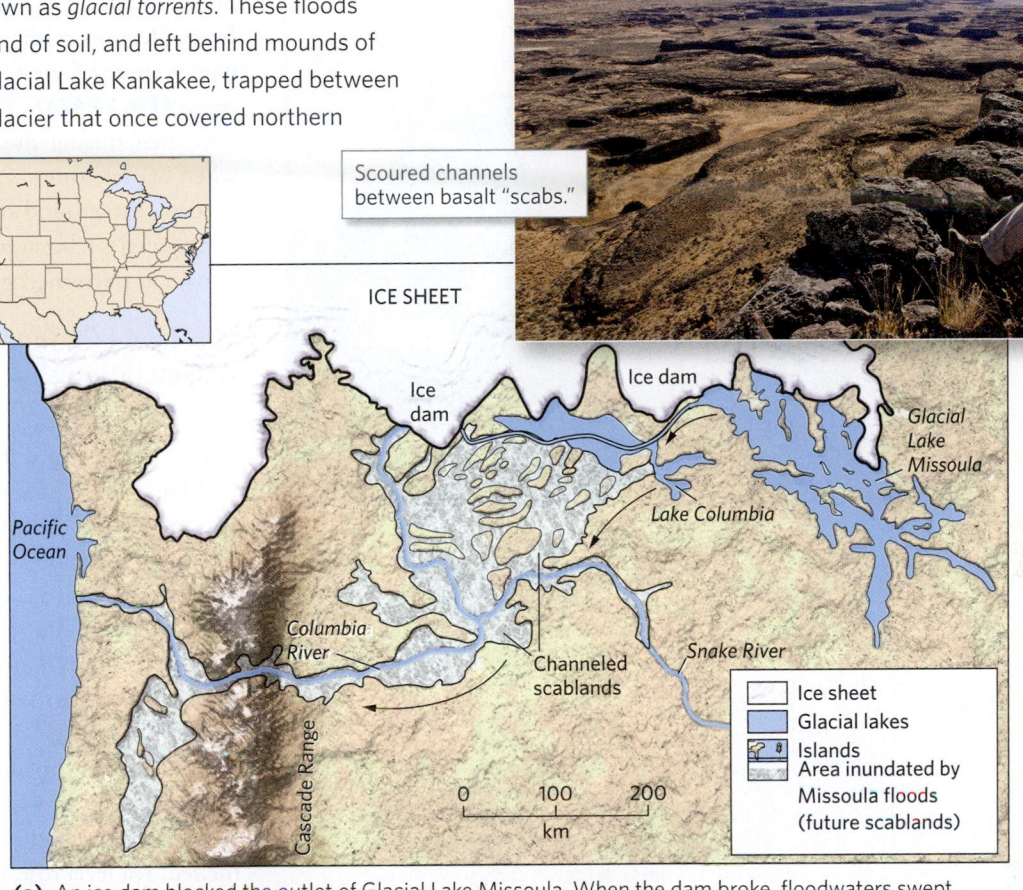

Scoured channels between basalt "scabs."

(b) The soil was stripped off of a region underlain by basalt bedrock. This area is now called the channeled scablands.

(a) An ice dam blocked the outlet of Glacial Lake Missoula. When the dam broke, floodwaters swept westward across central Washington State and down the Columbia River valley.

(c) These giant ripple marks formed during the Great Missoula Flood.

(d) Shorelines of Glacial Lake Missoula.

Figure 14.40 Glacial Lake Agassiz was an ice-margin lake that formed near the end of the most recent glaciation of the Pleistocene Ice Age.

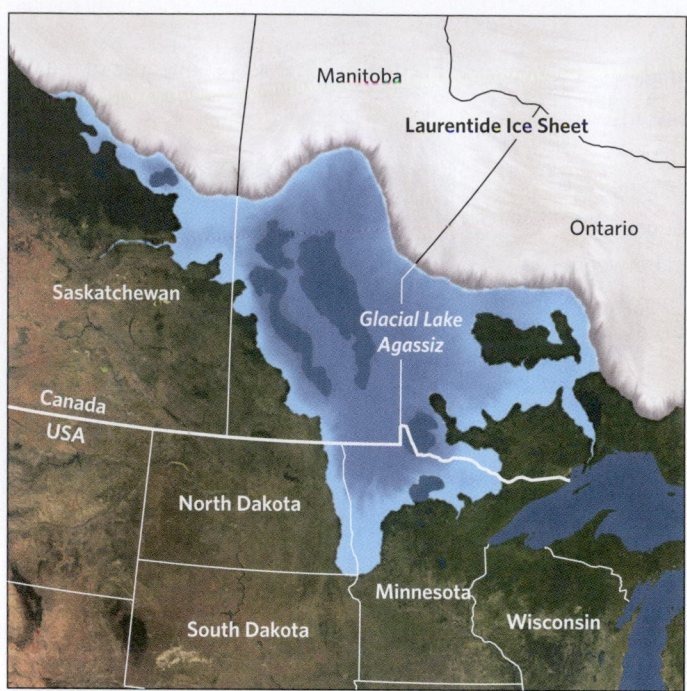

Figure 14.41 Pluvial lakes in the western United States.

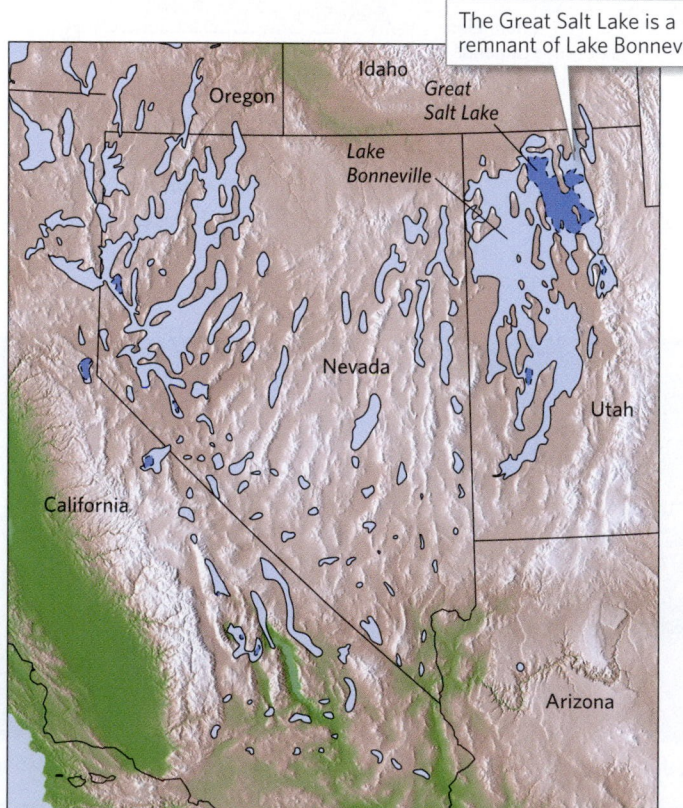

The Great Salt Lake is a remnant of Lake Bonneville.

(a) During the last ice age, a wetter climate produced pluvial lakes throughout the Basin and Range Province of Nevada and Utah.

pluvial lake, Lake Bonneville, covered almost a third of western Utah. When this lake suddenly drained after a natural dam holding it back broke, it left a bathtub ring of shoreline sediments rimming the mountains near Salt Lake City. Today's Great Salt Lake is but a small remnant of Lake Bonneville **(Fig. 14.41b)**.

In polar latitudes today, and in regions adjacent to the fronts of continental glaciers during ice ages, the average annual temperature stays low enough (below −5°C, or 23°F) that the ground freezes and becomes **permafrost**. Regions with widespread permafrost that do not have a cover of snow or ice are called *periglacial environments* **(Fig. 14.42a)**. The upper few meters of permafrost may melt during the summer months, only to refreeze when winter comes. As a consequence of the freeze-thaw process, the ground of some permafrost areas splits into pentagonal or hexagonal shapes, producing a landscape known as **patterned ground (Fig. 14.42b)**.

Sea-Level Changes Related to Glaciation

More of the Earth's surface and near-surface freshwater resides in glacial ice than in any other reservoir on Earth (see Fig. 13.2). At times during the last ice age, glaciers covered almost three times as much land area as they do today, so they also held significantly more water. In effect, water from the ocean reservoir was transferred to the glacial reservoir and remained trapped on land. As a consequence, sea level dropped by as much as 100 m, and extensive areas of continental shelves became dry land **(Fig. 14.43a, b)**. During these times of low sea level, people and animals migrated across a *land bridge* that existed at the location of what is now the Bering Strait between North America and northeastern Asia. When the glaciers melted, sea level rose again **(Fig. 14.43c)**.

Lake Bonneville shoreline

Great Salt Lake

(b) The Great Salt Lake of Utah is a small remnant of Lake Bonneville. Subtle horizontal terraces, the remnants of beaches, now lie about 100 m above the present lake level.

Figure 14.42 Periglacial environments are not ice covered but do include substantial areas of permafrost.

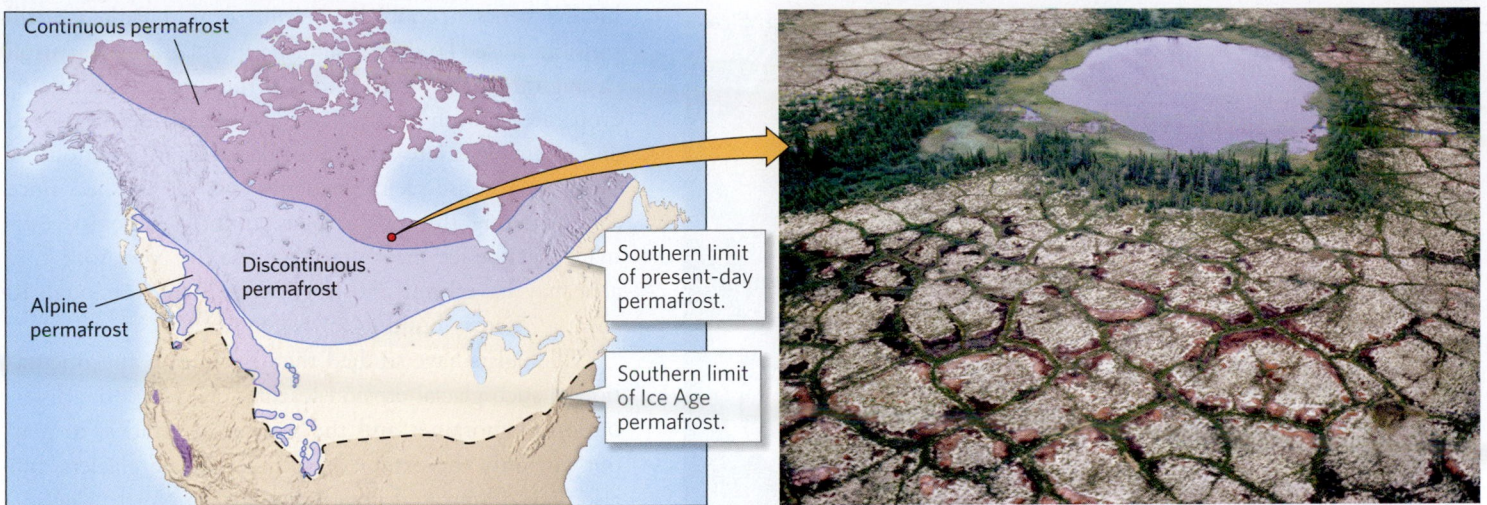

(a) The distribution of periglacial environments in North America.

(b) An example of patterned ground near a pond in Manitoba, Canada.

Take-home message . . .

The weight of a continental glacier can cause the lithosphere to subside, and melting of the ice sheet can cause it to rebound. The land beyond an ice sheet may be covered with permafrost or pluvial lakes. Continental glaciers store large amounts of water, so glacial advance or retreat affects sea level.

Quick Question ------------------------
What features characterize periglacial environments?

The Discovery of the Pleistocene Ice Age

When farmers in temperate regions of northern Europe prepared their land for spring planting, they occasionally broke their plows on large boulders scattered through

Figure 14.43 The link between sea level and global glaciation.

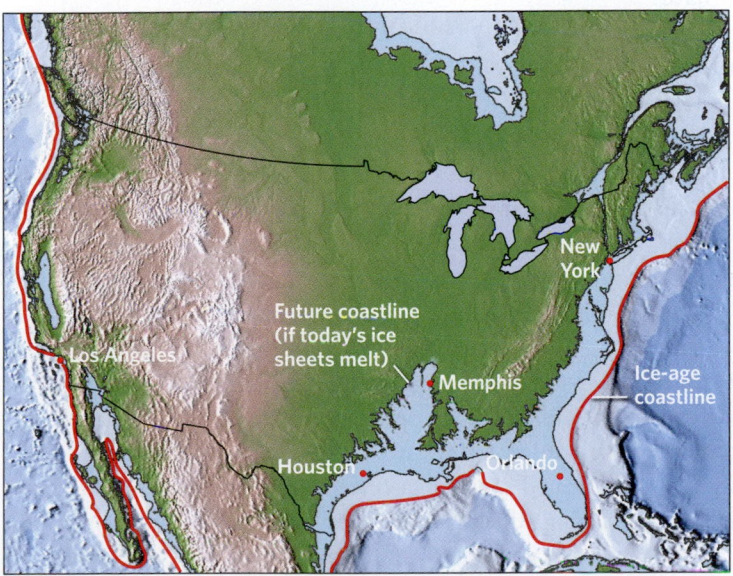

(a) The red line shows the coastline of North America during the most recent continental glaciation, when much of the continental shelf was dry. If the present-day ice sheets melt, areas that are coastal lands today will be flooded.

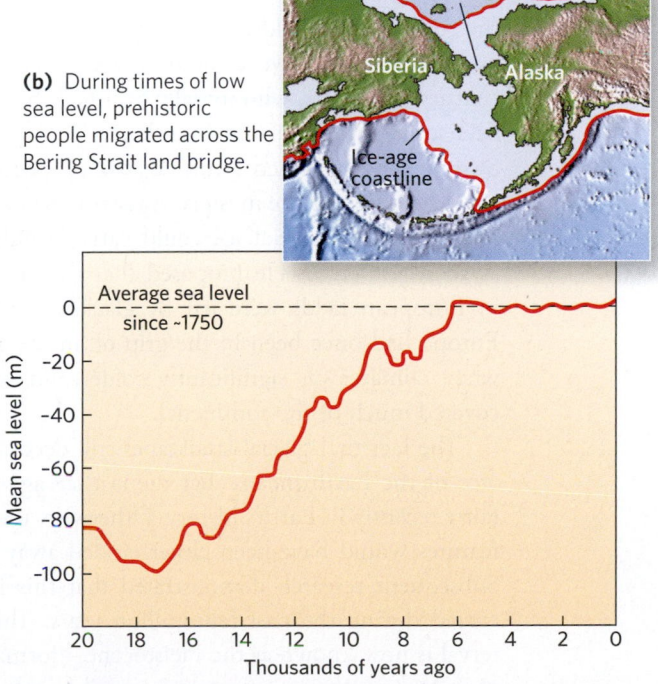

(b) During times of low sea level, prehistoric people migrated across the Bering Strait land bridge.

(c) Sea level rose between 17,000 and 7,000 B.C.E. due to the melting of continental glaciers.

Figure 14.44 Pleistocene ice sheets of the northern hemisphere.

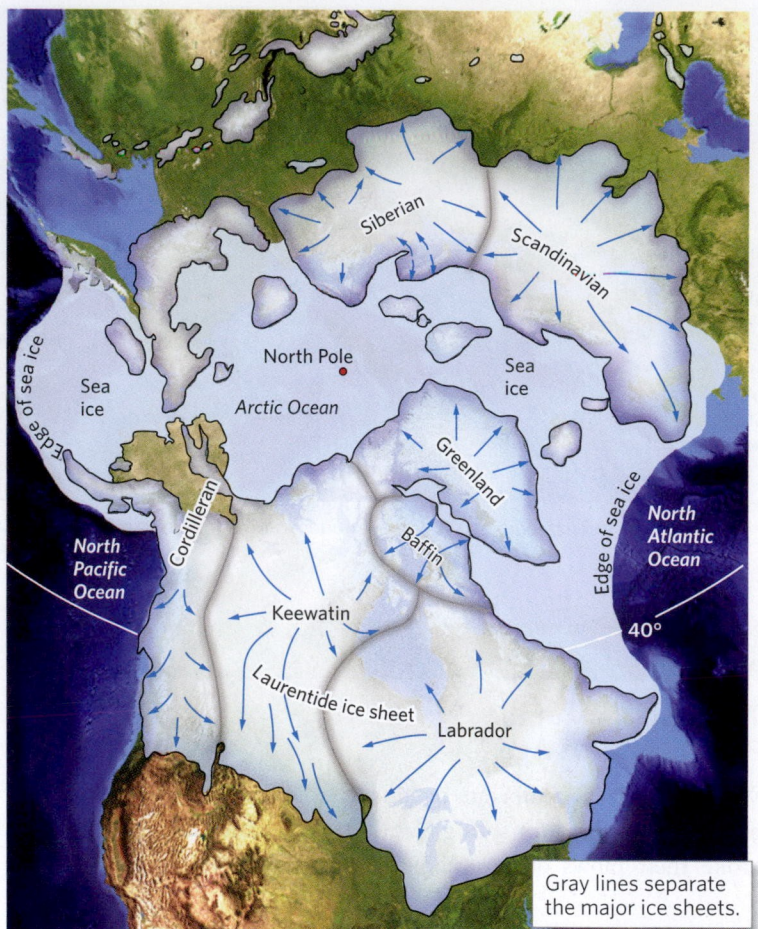

Gray lines separate the major ice sheets.

The Extent of Pleistocene Glaciers

Today, most of the land surface in New York City lies hidden beneath concrete, but in Central Park, it's still possible to see land in a semi-natural state. If you stroll through the park, you'll find that the top surfaces of outcrops are smooth and polished and have been grooved and scratched. Here and there, glacial erratics rest on the bedrock. You are seeing evidence that an ice sheet once scraped along this now-urban ground. Geologists estimate that the ice sheet that overrode the New York City area may have been 250 m (820 feet) thick, enough to cover a 75-story building.

Geologists have studied the distribution and orientation of such glacial erosion features, as well as the distribution of moraines and the sources of erratics, to map out not only the extent of the Pleistocene glaciers, but also their origins and flow directions. In North America, major ice sheets originated in at least three locations **(Fig. 14.44)**. These sheets, together with one or more smaller ones, merged to form the giant *Laurentide ice sheet*, which at times covered all of Canada east of the Rocky Mountains and spread southward over the northern portion of the United States. A separate ice sheet, the *Cordilleran ice sheet*, originated in the mountains of western Canada, then spread westward to the Pacific coast and eastward until it merged with the Laurentide ice sheet. Other ice sheets formed in Greenland, Scandinavia, northern Russia, and Siberia. Because of the extent of the ice sheets, climate belts shifted southward **(Fig. 14.45)**. While Pleistocene glaciers covered the land, sea ice expanded to cover all of the Arctic Ocean and part of the North Atlantic (to a latitude south of Iceland).

Glacial Advances and Retreats during the Pleistocene Ice Age

Louis Agassiz assumed that only one ice age had affected the planet. But close examination of glacial deposits on land revealed that *paleosols* (ancient soils preserved in the stratigraphic record), as well as beds containing fossils of warmer-weather animals and plants, occurred in sediment deposits between layers of glacial sediment. This observation suggested that between episodes of glaciation, glaciers receded and warmer climates prevailed. In the second half of the 20th century, when modern methods for dating geologic materials became available, the age differences between layers of glacial sediment could be confirmed. It's now clear that glaciers advanced and retreated more than once during the Pleistocene. Times during which glaciers grew and covered substantial areas of the continents are called **glaciations**, and times between glaciations are called **interglacials**.

Using the terrestrial sedimentary record, geologists initially recognized five Pleistocene glaciations in Europe

the soils of their fields. The mystery of these wandering boulders, now known as erratics, fascinated early-19th-century geologists, who initially assumed they must have been transported by immense floods. In 1837, a young Swiss geologist named Louis Agassiz proposed an alternative solution to the mystery. Agassiz lived near the Alps and knew that glacial ice could carry boulders as well as sand and mud. He proposed that the erratics found in European fields were left by glaciers, implying that Europe had once been in the grip of an **ice age**, a time when climate was significantly colder and vast glaciers covered much of the continent.

The fact that glacial landscapes still decorate the surface of the Earth means that the last ice age took place fairly recently in Earth's history. Otherwise, the landscape features would have been either eroded away or buried. Subsequent research demonstrated that this ice age occurred during the past few million years. This time interval is now known as the Pleistocene—formally, it's the geologic epoch that spans the interval between 2.6 million and 11,700 years ago. The ice age of this time interval is now called the **Pleistocene Ice Age**.

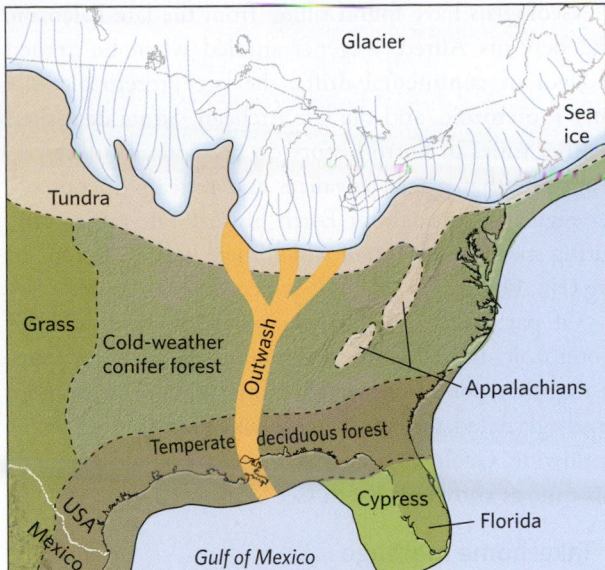

(a) Tundra covered parts of the United States, and southern states had forests like those in New England today.

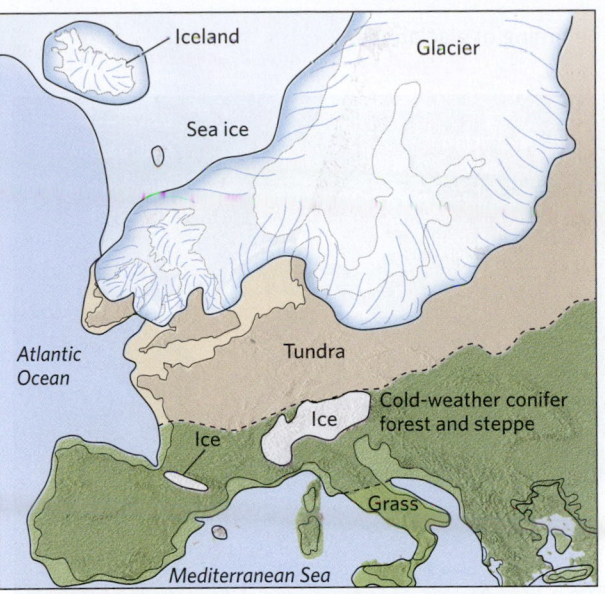

(b) Regions of Europe that support large populations today would have been barren tundra during the Pleistocene.

Figure 14.45 Glacial climates in the northern hemisphere.

(c) Cold-adapted large mammals, now extinct, roamed regions that are now temperate.

and four in the north-central United States (Fig. 14.46a). This interpretation was rejected in the 1960s, when geologists began to study marine sediment accumulations. They found that some layers at a given location contained glacially transported grains, but others did not. Similarly, they found that some layers contained fossils of cold-water plankton, but others contained fossils of warm-water plankton. In marine sediments of the Pleistocene, they found evidence for 20 to 30 distinct glaciations (Fig. 14.46b). The traditionally recognized glaciations of Europe and the United States may represent only the largest of these.

Other Ice Ages during Earth History

So far, we've focused on the Pleistocene Ice Age because of its importance in developing our planet's present landscape in the northern hemisphere. Was this the only ice age during Earth history, or do ice ages happen frequently? To answer such questions, geologists study the stratigraphic record and search for till that has lithified to become rock. Such **tillites** consist of cobbles and boulders distributed in a matrix of sandstone and mudstone, and in some cases, they lie on glacially polished surfaces.

Figure 14.46 The timing of glaciations.

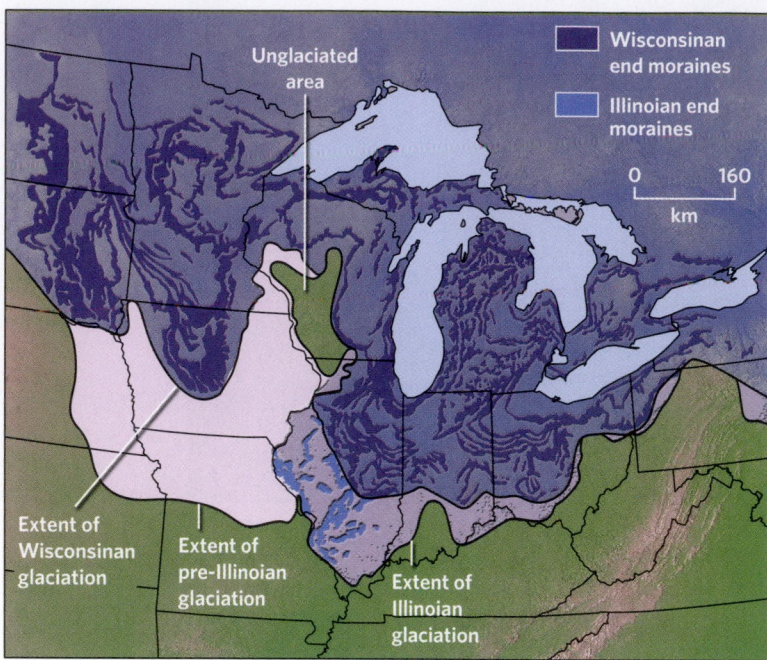

(a) Pleistocene glacial deposits in the north-central United States. The record shows that glaciers advanced more than once. Curving moraines reflect the shape of a glacier's toe.

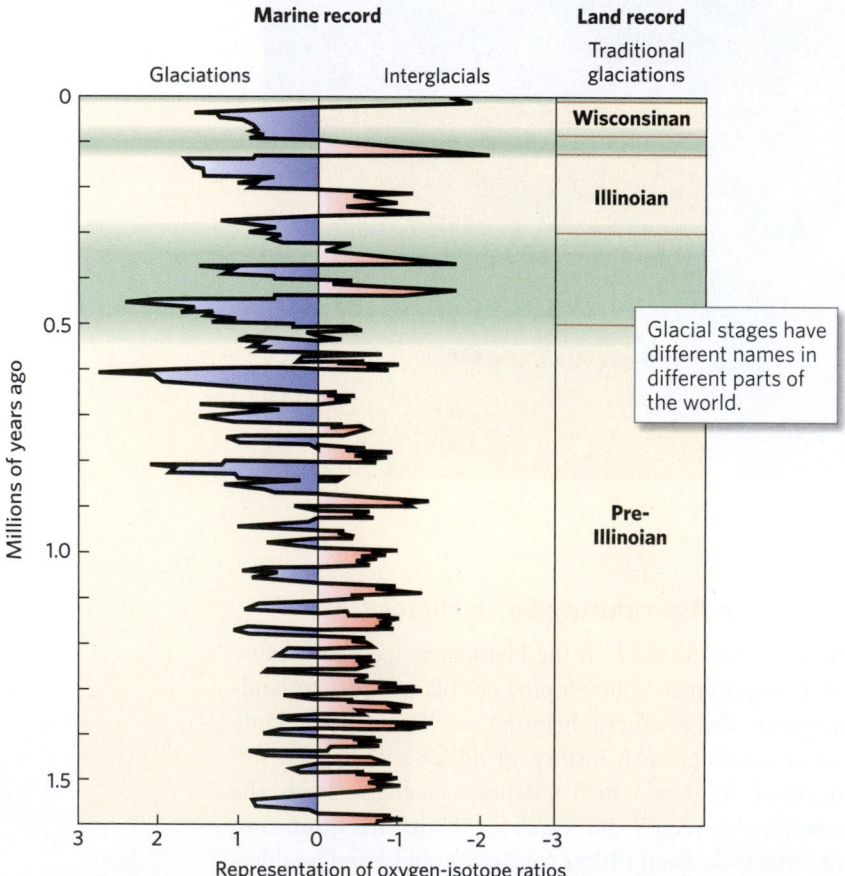

(b) Oxygen-isotope ratios from marine sediment define 20 to 30 glaciations in the Pleistocene. The tan bands represent the traditionally recognized glaciations of the north-central United States.

Geologists have found tillites from the late Paleozoic (the deposits Alfred Wegener studied when he argued in favor of continental drift), the late Proterozoic, the early Proterozoic, and the late Archean. Strata deposited at other times in Earth history do not contain tillites, so it appears that glacial advances and retreats do not occur regularly throughout Earth history, but rather only during specific ice ages, of which there have been four or five **(Fig. 14.47)**.

Of particular note, the sediments from which late Proterozoic tillites formed occur at all latitudes, suggesting that, for at least a short time, all continents were largely glaciated, and the sea may have been covered by ice worldwide. Geologists refer to the ice-encrusted planet of this time as **snowball Earth** (see Section 10.4).

Take-home message . . .

During the Pleistocene Ice Age, ice sheets advanced and retreated many times, causing many glaciations and interglacials. The record on land is less complete than that in marine strata. Ice ages also happened earlier in Earth history. During the Proterozoic, ice may have covered all of "snowball Earth."

Quick Question -
What is the evidence for multiple Pleistocene glaciations?

14.11 The Causes of Ice Ages

As we've just seen, ice ages occur only during restricted intervals of Earth history, hundreds of millions of years apart. But within an ice age, glaciers advance and retreat with a frequency measured in tens of thousands to hundreds of thousands of years. Geologists have searched for both long-term and short-term controls on glaciation to explain this timing.

Long-Term Causes

Tectonic processes play a role in the long-term control of glaciation. For example, continental drift determines the distribution of continents relative to the equator and, therefore, the amount of solar heat that the land receives. Only when large areas of land lie at high latitudes can ice ages begin. The movement of continents and the growth of island arcs may also trigger ice ages, for they influence the configuration of ocean currents, which, as we'll see in Chapter 15, carry heat from equatorial areas to high latitudes.

The concentration of carbon dioxide (CO_2) in the atmosphere probably plays a major role in setting the stage for ice ages. Carbon dioxide is a greenhouse gas,

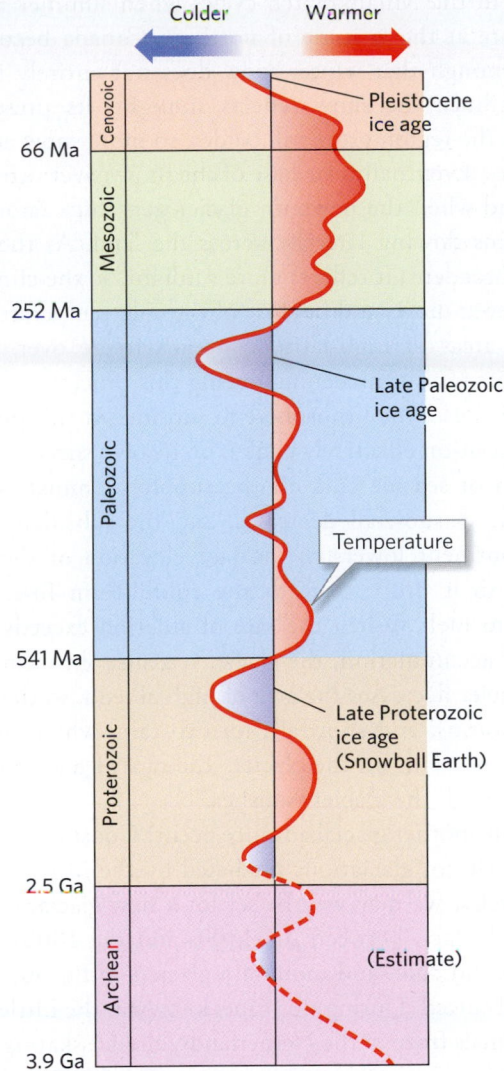

Colder　　　Warmer

Pleistocene
ice age

66 Ma

Cenozoic

Mesozoic

252 Ma

Late Paleozoic
ice age

Paleozoic

Temperature

541 Ma

Late Proterozoic
ice age
(Snowball Earth)

Proterozoic

2.5 Ga

(Estimate)

Archean

3.9 Ga

the atmosphere, thereby helping to trigger glaciations of Pangaea.

Short-Term Causes

Now we've seen how the stage can be set for an ice age to occur, but why do glaciers advance and retreat periodically during an ice age? In 1920, Milutin Milanković, a Serbian astronomer and geophysicist, came up with an explanation. He studied how the Earth's orbit changes shape and how its axis of rotation changes orientation through time, and he calculated the frequency of these changes. In particular, he evaluated three aspects of Earth's movement around the Sun:

- *Eccentricity (orbital shape):* Milanković showed that the Earth's orbit gradually changes from a more circular shape to a more elliptical shape and back again over a period of around 100,000 years **(Fig. 14.48a)**.

- *Tilt of Earth's axis:* We have seasons because the Earth's axis is not perpendicular to the plane of its orbit. Milanković calculated that the Earth's tilt angle has varied between 22.5° and 24.5° over a period of 41,000 years **(Fig. 14.48b)**.

- *Precession:* If you've ever set a top spinning, you've probably noticed that its axis of rotation wobbles; this movement is called *precession* **(Fig. 14.48c)**. Milanković determined that the Earth's axis wobbles over the course of about 23,000 years. Precession determines the relationship between the timing of the seasons and the position of Earth along its orbit around the Sun.

These cycles are now called **Milankovitch cycles**, using an English spelling of his name.

Milanković next examined how cyclic variations in orbital shape, axial tilt, and precession affect the total annual *insolation* (the amount of solar energy arriving at the top of Earth's atmosphere) at mid- to high latitudes and the seasonal distribution of insolation. He found that the variations could cause insolation to change by as much as 25% **(Fig. 14.48d)**. Glaciers advance during times when high latitudes receive a minimum amount of insolation during the summer, for during cool summers, ice from the previous winter can't melt completely. Geologists found that the times of Pleistocene glaciations correlate with the times of minimum insolation predicted by Milanković.

The cooling predicted by Milankovitch cycles, however, might not be enough to trigger a glaciation. Geologists suggest that other factors may come into play to cool the Earth even more and to trigger a glacial advance during times of low insolation. For example, when snow remains on the ground throughout the year, or clouds form in the sky, the *albedo* (reflectivity) of the Earth increases, so Earth's surface reflects incoming sunlight and becomes even cooler.

meaning that it traps infrared radiation rising from the Earth (see Chapter 20), so when CO_2 concentrations are high, the atmosphere becomes warmer. Ice sheets cannot form during such periods, even if other factors favor glaciation. What factors cause long-term changes in CO_2 concentration? Possibilities include changes in the population sizes of marine organisms that extract the dissolved gas from seawater to make shells; changes in the amount of chemical weathering on land (a process that absorbs the gas); and changes in the amount of volcanic activity (which emits the gas). Major stages in evolution may also affect CO_2 concentration. For example, plants incorporate the gas into their bodies during photosynthesis, so the appearance of lush coal swamps at the end of the Paleozoic may have removed CO_2 from

Figure 14.48 Milankovitch cycles cause the amount of insolation received at high latitudes to vary over time.

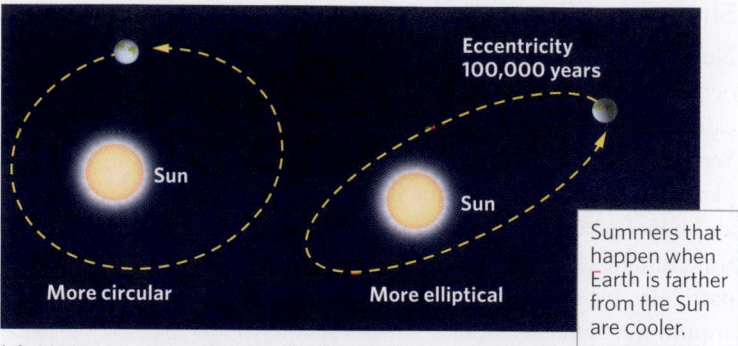

Eccentricity
100,000 years

Sun

Sun

More circular

More elliptical

Summers that happen when Earth is farther from the Sun are cooler.

(a) Variations caused by changes in orbital shape. The shape of the orbit is exaggerated.

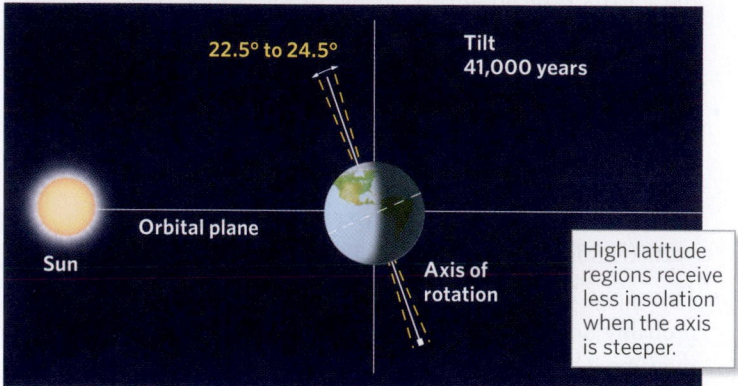

22.5° to 24.5°

Tilt
41,000 years

Orbital plane

Sun

Axis of rotation

High-latitude regions receive less insolation when the axis is steeper.

(b) Variations caused by changes in the tilt of Earth's axis.

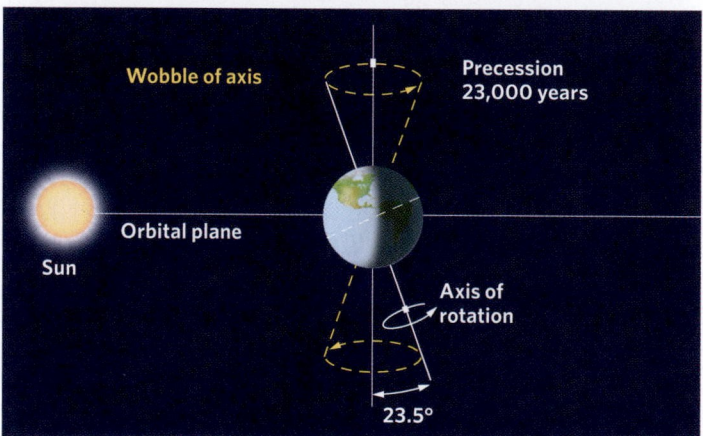

Wobble of axis

Precession
23,000 years

Orbital plane

Sun

Axis of rotation

23.5°

(c) Variations caused by the precession of Earth's axis.

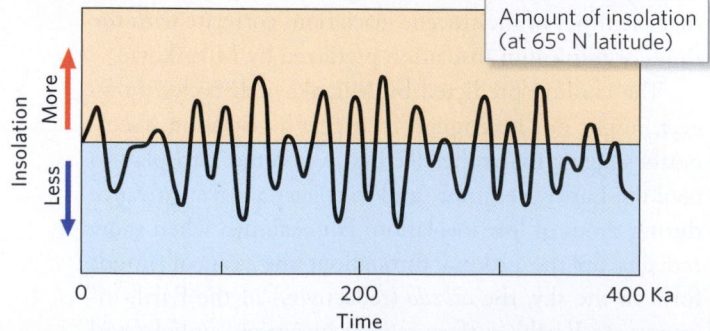

Amount of insolation (at 65° N latitude)

Insolation — More / Less

0   200   400 Ka

Time

(d) Combining the effects of eccentricity, tilt, and precession produces varying amounts of insolation during the summer at mid- to high latitudes.

Taking into account the various inputs that could trigger cooling of the Earth, let's look at a possible case history of a single advance and retreat of the Laurentide ice sheet. The process starts when the Earth reaches a point in the Milankovitch cycles when summer temperature at the latitude of northern Canada becomes cold enough that winter snow does not entirely melt away. Since the snow reflects sunlight, its presence makes the region grow still colder, so more snow accumulates. Eventually, the base of the snow cover turns to ice, and when the newborn glacier gets thick enough, it begins flowing laterally across the land. As the ice sheet broadens, it reflects more sunlight, so the climate cools even more, and because of the cold temperatures, broad areas of high-latitude oceans freeze over. The ice sheet is also thickening during this process, and its weight causes the lithosphere to subside. At this point, the glaciation effectively chokes on its own success. The growth of sea ice cuts off the supply of moisture to the air, so snowfall decreases, and the subsidence of the continent lowers the surface elevation of the ice sheet, so it drops beneath the equilibrium line and starts to melt. When the rate of ablation exceeds the rate of accumulation, the glacier retreats. Shrinking of the glacier decreases the area of high albedo, so the climate warms, and snowfalls turn to rain, which melts ice when it falls on the glacier. The rate of melting increases until the glacier vanishes.

Will another glacial advance occur? Considering the periodicity of glaciations predicted by the Milankovitch cycles, we may well be set for a new glaciation to begin. In fact, between the 1300s and the 1800s, the climate did cool, and mountain glaciers in Europe and Asia advanced. During this time, known as the Little Ice Age, canals froze in the Netherlands, and ice skating became the country's favorite pastime **(Fig. 14.49a)**. But, as we discuss in Chapter 20, global temperatures have been warming in the past two centuries, and glaciers worldwide have been retreating **(Fig. 14.49b)**. Whether ice can return in the next few thousand years remains unknown.

Take-home message . . .

Ice ages occur when the distribution of continents, ocean currents, and the atmospheric concentration of CO_2 are conducive to ice-sheet formation. Glacial advances and retreats during an ice age are triggered by Milankovitch cycles, defined by variations in Earth's orbit and axis of rotation.

Quick Question -
What factors, other than Milankovitch cycles, may cause a glaciation to occur?

Figure 14.49 Global temperature changes since the Ice Age.

(a) Skaters, ca. 1600, on the frozen canals of the Netherlands during the Little Ice Age.

(b) Every summer, the Greenland ice sheet loses ice. Meltwater lakes drain into crevasses.

Another View This mountainous landscape of southern Alaska, as viewed from an airplane, was carved by glaciers. We can see several cirques and arêtes. The view includes several valley glaciers, each cut by crevices, which look like dark gashes cut into the ice.

⬤14 CHAPTER REVIEW

- Deserts receive an average of 25 cm of rain per year or less, so vegetation covers less than 15% of their surfaces.

- Subtropical deserts form between latitudes of 20° and 30° because air at these latitudes is hot and dry. Rain-shadow deserts are found on the leeward side of mountain ranges, coastal deserts develop on land adjacent to cold ocean currents, continental-interior deserts form far from the ocean, and polar deserts form at high latitudes.

- Physical weathering in deserts produces rocky debris. Without constant availability of water, chemical weathering happens slowly.

- Water causes erosion in deserts, during downpours. Flash floods carry large quantities of sediments down ephemeral streams.

- Wind picks up dust as suspended load and causes sand to saltate. Windblown sediment abrades desert surfaces and sculpts ventifacts.

- Talus piles form where rock fragments accumulate at the base of a cliff. Alluvial fans accumulate at a mountain front. When temporary desert lakes dry up, a playa remains. Desert pavements are mosaics of varnished stones.

- In some desert landscapes, erosion causes cliff retreat, eventually resulting in the formation of mesas, buttes, and inselbergs.

- Where sand is abundant, desert winds build it into sand dunes, including barchan, star, transverse, parabolic, and longitudinal.

- Drought and careless agricultural practices may cause desertification.

- Glaciers are streams or sheets of ice that flow slowly in response to gravity. They form where snow persists year-round.

- Mountain glaciers exist in high regions and fill cirques and valleys. Continental glaciers spread over substantial areas of continents.

- Glaciers move by plastic deformation of ice grains or by basal sliding at rates of tens of meters per year. Gravity drives glacial flow.

- Glacial advances or retreats depend on the balance between the rate of accumulation and the rate of ablation.

- Icebergs break off glaciers that flow into the sea. Sea ice forms where the ocean surface freezes.

- Glaciers erode by abrasion and plucking.

- Mountain glaciers carve numerous landforms, including cirques and U-shaped valleys. Fjords are water-filled valleys carved by glaciers.

- Glaciers transport sediment of all sizes. Lateral moraines accumulate till along the sides of valley glaciers, medial moraines form down the middle, and end moraines form at a glacier's toe.

- Glacial depositional landforms include moraines, kettle holes, drumlins, eskers, ice-margin lakes, and glacial outwash plains.

- The weight of a large ice sheet causes the continental lithosphere to subside. When the glacier melts away, the land surface rebounds.

- Permafrost develops when unglaciated land remains frozen all year.

- When water is stored in continental glaciers, sea level drops. When those glaciers melt, sea level rises.

- During the Pleistocene Ice Age, large continental glaciers periodically covered much of North America, Europe, and Asia.

- The stratigraphy of Pleistocene glacial deposits indicates that glaciers advanced and retreated many times during the Pleistocene Ice Age.

- Long-term causes of ice ages include continental drift and changes in the concentration of CO_2 in the atmosphere. Short-term causes include the Milankovitch cycles.

Key Terms

ablation (p. 483)
alluvial fan (p. 469)
arête (p. 486)
butte (p. 470)
chimney (p. 470)
cirque (pp. 478, 485)
cliff retreat (p. 470)
continental glacier (ice sheet) (p. 478)
crevasse (p. 480)
desert (p. 464)
desertification (p. 474)
desert pavement (p. 472)

desert varnish (p. 467)
drumlin (p. 492)
dry wash (p. 468)
dust storm (p. 468)
equilibrium line (p. 483)
erratic (p. 490)
esker (p. 492)
firn (p. 478)
fjord (p. 488)
glacial advance (p. 484)
glacial outwash (p. 490)
glacial rebound (p. 496)
glacial retreat (p. 484)

glacial striation (p. 485)
glacial subsidence (p. 493)
glacial till (p. 490)
glaciation (p. 500)
glacier (p. 478)
hanging valley (p. 486)
horn (p. 486)
ice age (p. 500)
iceberg (p. 484)
ice shelf (p. 484)
interglacial (p. 500)
kettle hole (p. 492)
lag deposit (p. 469)

loess (p. 490)
mesa (p. 470)
Milankovitch cycle (p. 503)
moraine (p. 489)
mountain (alpine) glacier (p. 478)
natural arch (p. 470)
patterned ground (p. 498)
permafrost (p. 498)
playa (p. 470)
Pleistocene Ice Age (p. 500)
pluvial lake (p. 496)
polar glacier (p. 480)

Review Questions

The letters following each Review Question refer to the corresponding Learning Objective from the Chapter Opener.

1. What factors determine whether a region can be classified as a desert? Why do deserts form? **(B)**

2. How do weathering and soil-forming processes in deserts differ from those in temperate or humid climates? **(C)**

3. Describe how water modifies the landscape of a desert. Be sure to discuss both erosional and depositional landforms. **(C)**

4. Explain the ways in which desert winds transport sediment. **(C)**

5. Explain how the following features form: desert varnish; desert pavement; ventifacts. **(C)**

6. Describe the process of cliff retreat and the resulting landforms. **(C)**

7. What type of sand dune is pictured in the figure? **(C)**

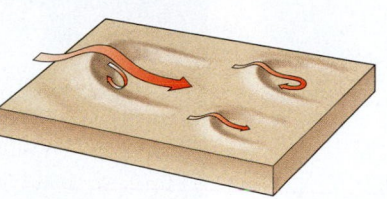

8. What is the process of desertification, and what causes it? **(B, D)**

9. How do mountain glaciers and continental glaciers differ? **(E)**

10. Describe the mechanisms that enable glaciers to move, and explain why glaciers move. **(E)**

11. Identify the zones of ablation and accumulation in the figure. **(F)**

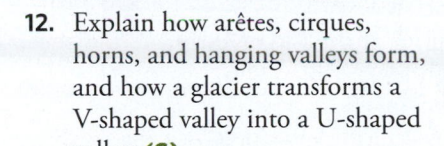

12. Explain how arêtes, cirques, horns, and hanging valleys form, and how a glacier transforms a V-shaped valley into a U-shaped valley. **(G)**

13. Describe the various kinds of glacial deposits. Be sure to note the materials from which the deposits are made and the landforms that result from their deposition. **(G)**

14. How many Pleistocene glaciations were identified using the on-land record? How was this number modified with the study of marine sediment? Were there ice ages before the Pleistocene? **(H)**

15. What are some of the long-term causes that lead to ice ages? What are the short-term causes? **(I)**

On Further Thought

16. The Namib Desert lies to the north and west of the Kalahari Desert, in Africa. The reason that the former region is a desert differs from the reason that the latter is a desert. Explain. **(B)**

17. Have today's deserts always been deserts? (Hint: Keep in mind the consequences of tectonic processes.) **(B)**

18. If you fly over the barren fields of Illinois in early spring, slight differences in soil color outline polygonal shapes tens of meters across. How and when did these patterns form? **(F, G)**

Online Resources

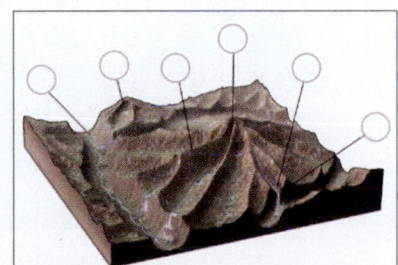

Animations
This chapter features interactive animations that simulate glacial bodies under different conditions.

Smartwork5
This chapter features visual identification and labeling exercises on the effects of desert and glacial systems on the landscape.

PART III

Restless Seas

Most of our planet's solid surface lies hidden beneath a vast global ocean of saltwater, in places deeper than mountains are high. Seawater serves as an incubator of life and as a conveyor of heat. Part III of this book focuses on this watery realm. In Chapter 15, we address the essence of oceanography, the study of seawater. We consider its composition, its movement in currents (both shallow and deep), its rise and fall during daily tides, and its undulations in waves. This chapter also peeks at the life of the sea, from tiny but intricately beautiful plankton shells, to colossal whales. In Chapter 16, we explore marine geology, starting with studies of the seafloor. We discuss why ocean basins exist, and why they display the distinctive bathymetric features that they do. The chapter concludes with a focus on coasts, the boundary between land and sea. You'll learn how the landforms of the coast take shape and evolve, and how the coastal environment in some locations has been changing in recent time.

Ocean waves crash on the beach bordering San Juan, Puerto Rico. We are seeing a dynamic coastline. If you sailed due east from this location, you would have to cross 5,300 km (3,300 miles) of blue saltwater before again touching land. The ocean covers most of the globe.

15 OCEAN WATERS
The Blue of the Blue Marble

By the end of the chapter you should be able to . . .

A. place oceanographic discovery in the context of human history.

B. explain where the ocean's water and salt come from, and how sea level varies.

C. discuss how the salinity, temperature, and density of ocean waters change with location and depth and how these variations cause distinct layers to form.

D. explain what surface currents are and how their configuration relates to wind, the Coriolis force, and pressure variations.

E. relate variations in ocean-water density to the origins of deep currents and consider the importance of thermohaline circulation.

F. discuss why waves form, how they behave, and how big they can get.

G. distinguish among oceanic zones based on water depth and on proximity to land and explain how these zones affect life in the ocean.

15.1 Introduction

During a storm in 1992, a container holding 29,000 plastic bath toys washed off a cargo ship and into the middle of the Pacific Ocean. Ten months later, little yellow duckies started appearing on beaches all around the Pacific **(Fig. 15.1a)**. A small number floated through the Bering Strait, moved with sea ice around the Arctic Ocean to the Atlantic, and years later, ended up on the shores of Great Britain **(Fig. 15.1b)**. The amazing voyage of these toys emphasizes that, though we divide it into several geographic regions with different names, there is really only one global **ocean** on our planet's surface, a layer of saltwater with an average depth of 4 to 5 km (2.5 to 3 miles). The Earth's ocean is a dynamic environment in constant

motion. Its waters flow over vast distances in *currents* (stream-like flows), and it surface elevation changes due to both global-scale *tides* (the daily rise and fall caused by the gravitational attraction of the Moon and other forces) and the development of *waves* (local surface undulations). Despite all this motion, ocean water remains inhomogeneous, for characteristics such as salt content and temperature vary regionally and with depth.

In this chapter, we explore the waters of the world's ocean—the part of our planet's surface that looks blue from space **(Fig. 15.2)**. We begin with a quick survey of ocean exploration. Next, we describe the physical and chemical characteristics of seawater and show how they vary. We then turn our attention to the movement of seawater by describing currents, upwelling and

The ocean covers more than two-thirds of our planet's surface. The texture and color of its surface are forever changing as it interacts with wind and light.

> *The Sea, once it casts its spell, holds one in its net of wonder forever.*
>
> — JACQUES-YVES COUSTEAU
> (FRENCH EXPLORER, 1910–1997)

downwelling, and waves. (Our discussion of tides appears in Chapter 16, for the consequences of tides primarily affect coasts.) We conclude this chapter by focusing on how zones of the ocean, distinguished from one another based on depth and proximity to shore, provide varying environments for different assemblages of living organisms.

Figure 15.2 A view of the Earth from space emphasizes that blue ocean water covers most of its surface.

Figure 15.1 Voyage of the yellow duckies.

(a) A plastic toy, swept off a ship in 1992, ended up on a beach thousands of kilometers away.

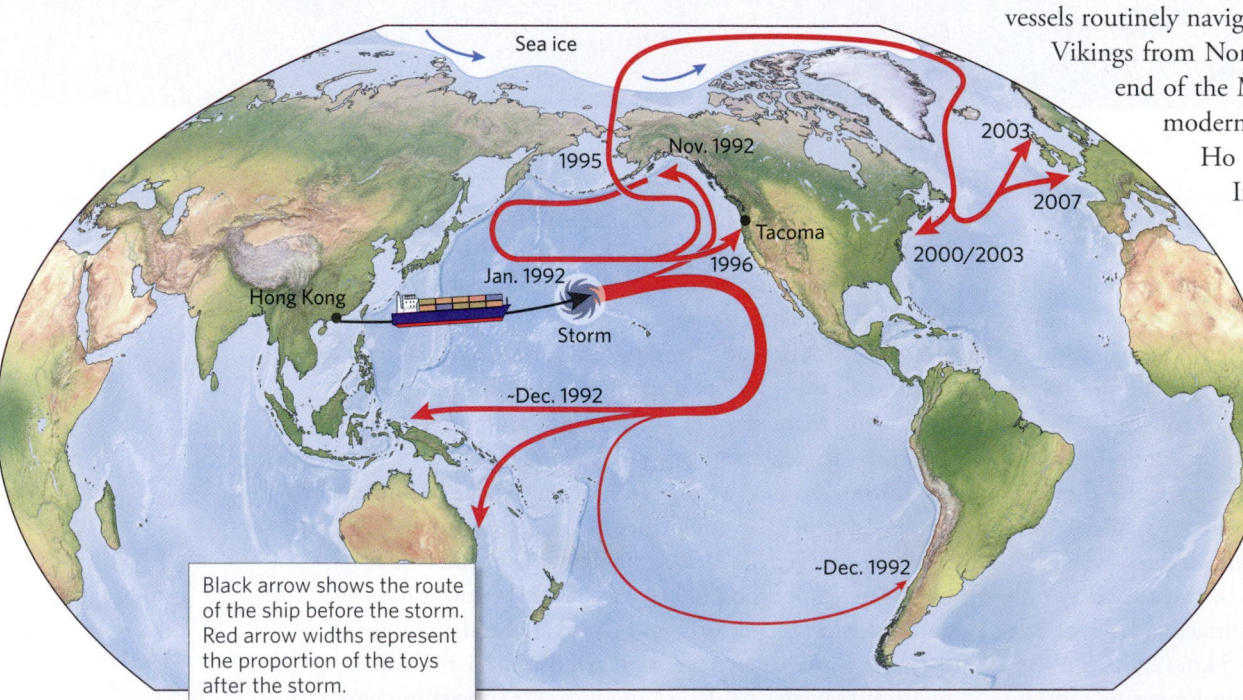

Black arrow shows the route of the ship before the storm. Red arrow widths represent the proportion of the toys after the storm.

(b) The paths that the toys traveled indicates that water flows from ocean to ocean over time. All the world's oceans are interconnected.

15.2 The Blue of the Blue Planet

Discovering the Sea

No one knows when the first person sailed over the horizon, but archeological evidence suggests that as early as 3000 B.C.E., Polynesians built large canoes to travel among isolated islands in the western Pacific. By the time Egyptians constructed the Great Pyramid (2570 B.C.E.), vessels routinely navigated inland seas, and by 1000 C.E., Vikings from Norway had crossed the Atlantic. The end of the Middle Ages saw the beginnings of modern ocean exploration. Admiral Cheng Ho of China explored the Pacific and Indian Oceans between 1405 and 1433 C.E., in ships up to 120 m (400 feet) long (Fig. 15.3a), until the coronation of a new emperor with isolationist views brought a halt to such voyages. The focus of global oceanic exploration then shifted to Europe.

Christopher Columbus's journey across the Atlantic, in 1492, followed by the discovery of sea routes to China in 1498, opened the European *Age of Discovery*. A quarter century later, the sole surviving ship of Ferdinand Magellan's fleet became the first known vessel to journey entirely around the globe. Each voyage of this era brought back a little more information about what the edges, or **shorelines**, of continents looked like, so by the beginning of the 17th century, the basic layout of continents and the oceans between them had been established **(Fig. 15.3b)**. These early maps, however, had major inaccuracies, for although explorers had long known how to measure the angle of the Sun above the horizon at noon to calculate *latitude* (representing the distance north or south of the equator), it was not until the mid-18th century, when the invention of a mechanical clock permitted navigators to keep accurate time on a moving ship **(Fig. 15.3c)**, that it became possible to measure *longitude* (the distance east or west). Measurement of both latitude and longitude then became routine, so explorers could add detail to their maps and could measure the true dimensions of the seas. Improvements in navigation technology, along with aerial photos, satellite images, and GPS satellites, have now characterized every detail of every shoreline.

The early explorers who ventured across the sea undertook their voyages primarily in search of gold and spices, or to find lands to conquer in the name of their monarchs. Starting in the latter half of the 18th century, however, some European explorers also collected data about the sea. Between 1768 and 1780, for example, Captain James Cook of England documented currents, collected biological specimens, and measured water temperatures in addition to mapping coastlines. But the modern science of **oceanography**, the study of the sea, did not began until 1872, when Britain's HMS *Challenger* began a four-year cruise that collected enough new information about ocean water and the seafloor to fill 50 books **(Fig. 15.4a)**. By using research ships, submersibles, and satellites **(Fig. 15.4b–e)**, modern *oceanographers* (who study the physical and chemical

Figure 15.3 Early exploration of the ocean.

(a) A Chinese fleet of large ships set sail in 1405, almost 90 years before Columbus crossed the Atlantic.

(b) A 1598 map by Abraham Ortelius, a Dutch cartographer, showing the coastlines as they were known at the time.

(c) A 19th-century chronometer.

Figure 15.4 Research vessels study the oceans.

(a) HMS *Challenger*, the first dedicated research vessel, set sail in 1872.

(b) A modern oceanographic research vessel.

(c) Submersibles use spotlights to see at depth.

(d) Robotic submersibles provide information without risking lives.

(e) SEASAT is a NASA satellite designed to monitor the world's oceans.

Figure 15.5 Geographers distinguish many oceans and seas. Some are separated by land and some by latitude.

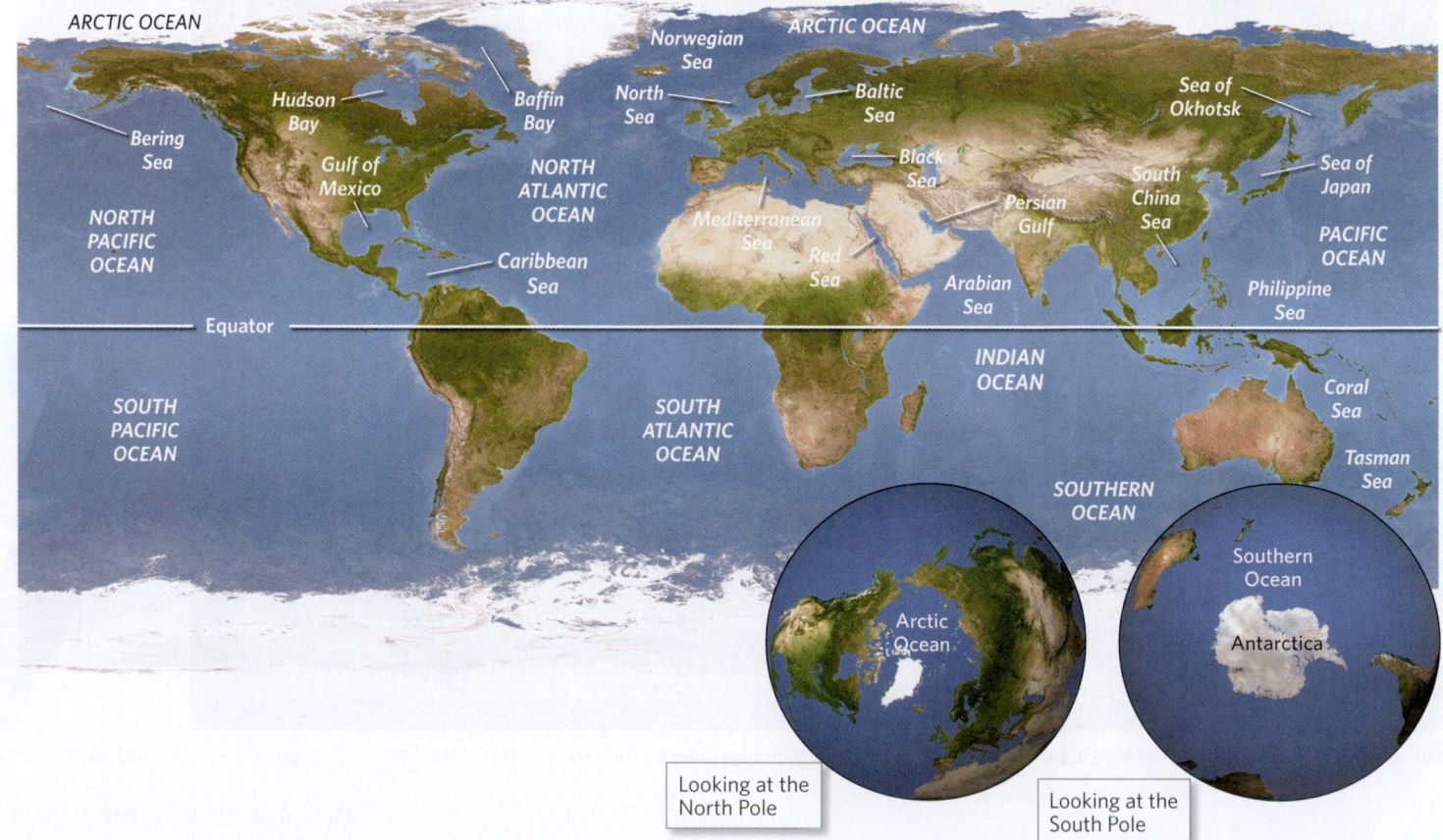

ARCTIC OCEAN

ARCTIC OCEAN

Norwegian Sea

Baffin Bay

North Sea

Baltic Sea

Sea of Okhotsk

Hudson Bay

Bering Sea

Gulf of Mexico

NORTH ATLANTIC OCEAN

Black Sea

South China Sea

Sea of Japan

NORTH PACIFIC OCEAN

Mediterranean Sea

Persian Gulf

PACIFIC OCEAN

Caribbean Sea

Red Sea

Arabian Sea

Philippine Sea

Equator

SOUTH PACIFIC OCEAN

SOUTH ATLANTIC OCEAN

INDIAN OCEAN

Coral Sea

SOUTHERN OCEAN

Tasman Sea

Arctic Ocean

Southern Ocean

Antarctica

Looking at the North Pole

Looking at the South Pole

Figure 15.6 The concept of sea level and its variation over time.

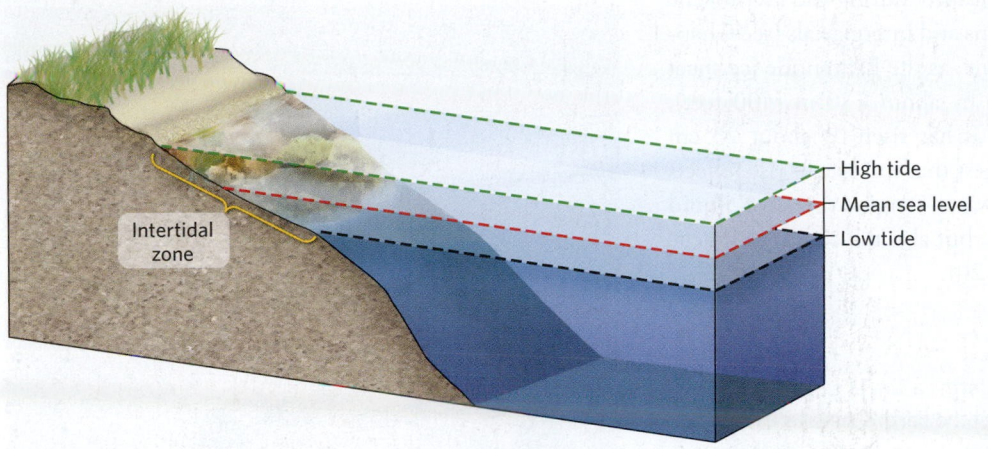

High tide
Mean sea level
Low tide

Intertidal zone

(a) Mean sea level is the average elevation of the water surface over the course of a year. It is halfway between average high and average low tide at a particular location.

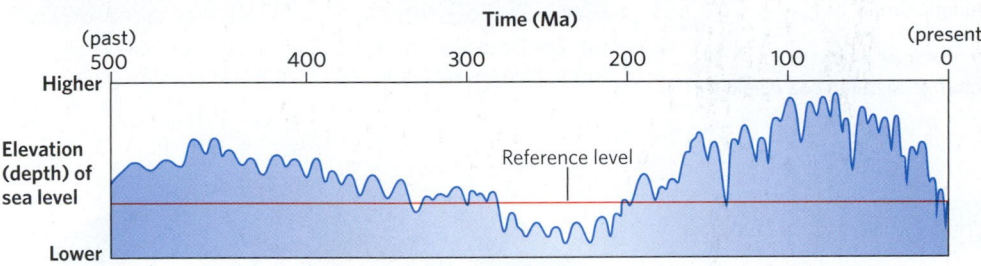

Time (Ma)

(past)
500

400

300

200

100

(present)
0

Higher

Elevation (depth) of sea level

Reference level

Lower

(b) A simplified estimate of sea-level change over the last half billion years.

(c) At times, sea level rises so high, relative to land, that water floods large areas of continental interiors. During the Mississippian, for example, much of the interior of the United States lay beneath an epicontinental sea.

characteristics and movements of ocean water), along with *marine geologists* (who study the seafloor and coasts) and *marine biologists* (who study marine life), continue to enhance our knowledge of our planet's amazing seas.

The Map of the Modern Ocean

Oceans cover 70.8% of the Earth's surface and contain about 1.3 billion cubic kilometers (300 million cubic miles) of water. (Despite these huge numbers, however, ocean water represents only about 0.12% of the Earth's entire volume.) Geographers divide the global ocean into five separate bodies—the Atlantic, Pacific, Indian, Arctic, and Southern Oceans—outlined by continents. The equator divides the Atlantic and Pacific into northern and southern portions. In addition, geographers recognize several *seas* (such as the Mediterranean, Black, Red, and Caribbean Seas), *bays* (such as Hudson Bay), and *gulfs* (such as the Gulf of Mexico), which are smaller regions of seawater that are partially surrounded by land **(Fig. 15.5)**. (A few "seas," such as the Dead Sea and the Caspian Sea, are actually lakes, in that they are completely surrounded by land.) Of course, the shapes and positions of today's oceans and seas differs markedly from the way they appeared in the past, and the way they will be in the future, because of plate motions.

Sea Level and Sea-Level Change

Sea level refers to the elevation of the boundary between ocean water and the air above. When we use the term *sea level*, we are actually talking about *mean sea level*, the average height of this boundary over the course of a year **(Fig. 15.6a)**. At any given moment, the sea surface at a given locality may be higher or lower than mean sea level because of passing waves or the rise or fall of tides.

The geologic record indicates that liquid water has filled the oceans at least since the beginning of the Archean—except, perhaps, for "snowball Earth" episodes in the Proterozoic, when the sea surface was frozen—and that the overall volume of surface water on the Earth has stayed roughly constant (see Chapter 10). But because of ice ages, which transfer large volumes of water into glaciers on land, and because of changes in the capacity of ocean basins resulting from changes in the volume of mid-ocean ridges (see Box 10.2), sea level has gone up and down on the order of tens to a few hundred meters over the course of geologic time relative to today's coastline **(Fig. 15.6b)**. When sea level rises significantly, ocean water may submerge large areas of continents to form *epicontinental seas* **(Fig. 15.6c)**, and when sea level falls, portions of *continental shelves*, the relatively shallow part of the ocean adjacent to continents, may be exposed.

Sea level changed significantly during the Pleistocene in association with glaciations and interglacials (see Chapter 14). During the Holocene, as the Laurentide ice sheet melted away, sea level rose by about 120 m (400 feet), and in the past 150 years, it has risen by about 23 cm (9 inches). Researchers suggest that this recent rise reflects mainly an increase in seawater temperature, for liquid water expands when heated, but also the continued melting of glaciers (see Chapter 20).

Take-home message . . .

A layer of saltwater forms a single global ocean that covers 70.8% of the Earth's surface. The distribution of continents and the equator divide the ocean into distinct geographic regions, and partial enclosure of oceanic regions by land defines seas and bays. Mean sea level, the average elevation of the sea surface, has varied over geologic time.

Quick Question -
What factors can cause a change in mean sea level?

15.3 Characteristics of Ocean Water

Where Does the Water in the Oceans Come From?

When the Earth first formed, most of the hydrogen and oxygen atoms that would later compose molecular water (H_2O) existed instead in hydroxide ions (OH^-) bonded to solid minerals in rock of the crust and mantle. Melting of this rock released hydroxide, which then bonded to hydrogen to yield water molecules that bubbled out of volcanoes as vapor. During the Earth's early history, the planet remained so hot that this water vapor stayed in the atmosphere. Oceans formed when the planet cooled enough for water vapor to condense and fall as liquid water that could collect in ocean basins.

While the volume of ocean water has stayed fairly constant for most of geologic time, individual molecules of water do not remain in the ocean for very long, for the ocean serves as but one of several reservoirs in the hydrologic cycle (see Chapter 12). In fact, on average, a given molecule of water remains in liquid form in the ocean for less than 4,000 years before it evaporates to become vapor in the air, Then, within hours to days, the vapor condenses or crystallizes and falls back to the Earth's surface.

The Salt of the Sea

If you swim in the ocean, you'll immediately notice its salty taste and observe that you float more easily in ocean water

Figure 15.7 Salinity varies within the ocean.

(a) The Dead Sea (which is actually a salt lake, completely surrounded by land) is so salty that people float like corks.

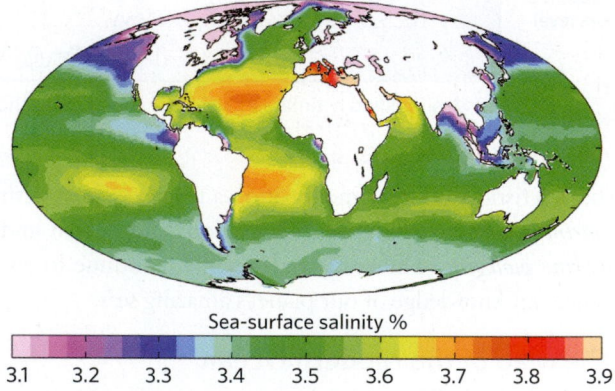

Sea-surface salinity %

| 3.1 | 3.2 | 3.3 | 3.4 | 3.5 | 3.6 | 3.7 | 3.8 | 3.9 |

(b) The salinity of seawater varies with location. These regional variations reflect differences in rates of freshwater addition and evaporation.

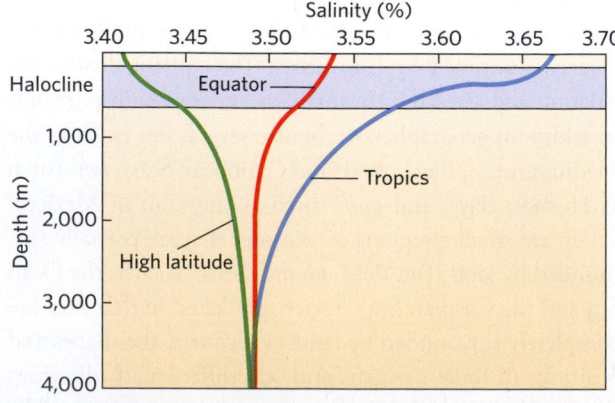

(c) Salinity varies with depth as well as with climate. Variation in salinity is greatest in the uppermost 1 km of ocean water.

than you do in freshwater (Fig. 15.7a). That's because ocean water is a solution containing dissolved salts. In a solution, dissolved ions fit between water molecules without changing the overall volume of the water, so adding salt to water increases the water's density. You float higher in a denser liquid (seawater) than in a less dense liquid (freshwater) because, according to *Archimedes' principle*, an object sinks in water only to a depth at which the weight of the water displaced equals the weight of the whole object.

How salty does the sea get? It has about the same saltiness as a solution you can prepare by stirring a teaspoon of salt into a cup of warm water. More precisely, typical seawater contains about 96.5% water (H_2O) and 3.5% salt by weight. In contrast, typical freshwater contains less than 0.02% salt. Put another way, seawater contains about 220 times the amount of salt in freshwater. There's so much salt in the sea that if all of its water molecules evaporated, the layer of salt left behind would be about 67 m (220 feet) thick, the height of a 25-story building. Oceanographers refer to the concentration of salt in water as **salinity**.

Ocean salinity depends, in part, on water temperature because warm water can hold more salt in solution than can cold water. It also depends on the balance between inputs of freshwater and removal of freshwater by evaporation. For example, in **estuaries**, places where seawater floods the mouth of a river, salty seawater mixes with fresh river water to produce *brackish water* that has less salinity than seawater. Brackish water also develops where glacial meltwater enters the sea or where heavy rain falls on the sea surface. In contrast, along the margins of restricted seas in hot, arid climates, so much water evaporates that water becomes *brine*, meaning that its salinity becomes greater than that of seawater. Therefore, while ocean salinity averages 3.5%, it varies with location, ranging from about 1.0% to about 4.1% (Fig. 15.7b). Ocean salinity also varies with depth (Fig. 15.7c). The largest variations in salinity occur in the upper 1 km of the sea, the portion most affected by evaporation or by freshwater inputs. Salinity in deeper water tends to be more homogeneous. Oceanographers refer to the gradational boundary between surface-water salinities and deep-water salinities as the **halocline**.

What does the salt in the ocean consist of? Chemical analyses indicate that the positive ions in seawater include sodium (Na^+), potassium (K^+), calcium (Ca^{2+}), and magnesium (Mg^{2+}), and that the negative ions include chloride (Cl^-) and sulfate (SO_4^{2-}). The proportions of different ions stay the same throughout the ocean. Most of the ions are sodium and chloride (Fig. 15.8a).

When seawater evaporates, water molecules separate from the liquid and become vapor, but the ions stay behind. If the water becomes supersaturated, ions come out of solution and bond together to form crystals (see Chapter 3). When seawater evaporates entirely, about 85% of the

Figure 15.8 Seawater contains a variety of salts.

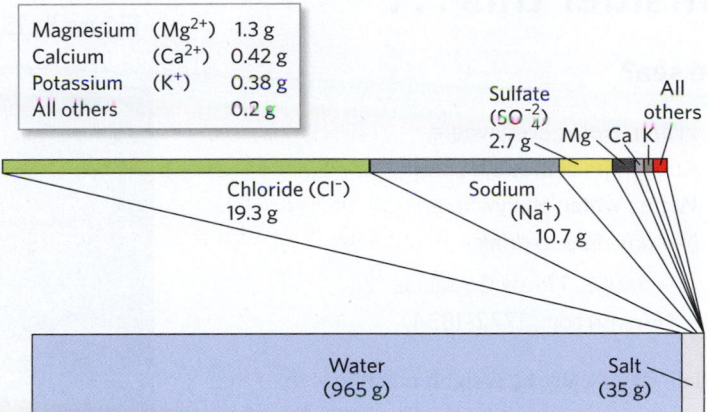

Magnesium	(Mg^{2+})	1.3 g
Calcium	(Ca^{2+})	0.42 g
Potassium	(K^+)	0.38 g
All others		0.2 g

Sulfate (SO_4^{-2}) 2.7 g

All others

Mg Ca K

Chloride (Cl^-) 19.3 g

Sodium (Na^+) 10.7 g

Water (965 g)

Salt (35 g)

(a) The average chemical composition of seawater, in grams per liter.

(b) Harvesting sea salt by evaporating seawater from shallow pools.

salt that precipitates consists of the mineral halite (NaCl). The remainder consists of other mineral salts, including gypsum ($CaSO_4 \cdot 2H_2O$), anhydrite ($CaSO_4$), magnesium sulfate ($MgSO_4$), magnesium chloride ($MgCl_2$), and potassium chloride (KCl). Seawater doesn't need to evaporate entirely to precipitate salt—salt can precipitate directly from brines. Different minerals precipitate from a brine depending on how concentrated the brine has become. As seawater starts to evaporate, NaCl precipitates first, as it can precipitate from less concentrated brines, so geologic salt deposits contain mostly NaCl. Geologic deposits that contain other salts formed where seawater completely, or almost completely, evaporated. Most common table salt, the kind in your salt shaker, comes from purification of geologic salt deposits and contains almost entirely NaCl. The "sea salt" that has become popular in gourmet stores contains 95% to 98% NaCl; the remainder consists of traces of other salts (Fig. 15.8b). Workers produce sea salt by evaporating modern ocean water under controlled conditions. A similar process can be used to extract freshwater from saltwater (Box 15.1).

Box 15.1 **Consider this . . .**

Can we drink the sea?

> *Water, water, everywhere,*
> *And all the boards did shrink;*
> *Water, water, everywhere,*
> *Nor any drop to drink.*
>
> —SAMUEL TAYLOR COLERIDGE
> (ENGLISH POET, 1772–1834)

Human blood has a salinity of 0.9% by weight, much less than that of seawater. Our kidneys regulate the salinity of our blood, so if we drink seawater, our bodies excrete the excess salt in urine. To keep your blood salinity normal, you would have to excrete more water than you drink, so drinking seawater eventually leads to death by dehydration. That is, of course, if you don't die first from seizures or heart failure, for excess salt in your blood can disrupt the conduction of electrical signals by nerves. Clearly, we need freshwater to survive.

As society's need for freshwater increases, we are using up supplies of natural fresh surface water and groundwater. Can we take the salt out of seawater to produce freshwater? The answer is yes. Salt can be removed from water by **desalinization.** How? Desalinization involves either distillation or reverse osmosis.

During *distillation,* workers pump seawater into a tank and heat the water to accelerate evaporation. When seawater evaporates, it leaves its salt behind, so the resulting vapor consists of freshwater— and condensation of the vapor produces liquid freshwater. Most commercial distillation operations use *multi-stage flash distillation* **(Fig. Bx15.1a)**. This process takes advantage of the fact that water's boiling temperature depends on atmospheric pressure—under lower pressures, water boils at lower temperatures. During multi-stage flash distillation, seawater that has been warmed in a furnace flows into a chamber from which air has been extracted. At the lower pressure in this chamber, some of the water boils, producing vapor that can be collected and removed. Evaporation also causes the remaining water to cool (see Chapter 18). This slightly cooler water then flows into another chamber with still lower pressure, where it boils and evaporates. By repeating the process in a series of chambers, each with a pressure lower than the previous one, workers can remove about 85% of the water molecules from seawater.

Reverse osmosis purifies seawater without heating it. During this process, workers apply pressure to push water through a semipermeable membrane that allows water molecules, but not salt molecules, to pass through it **(Fig. Bx15.1b)**. Why is this process called "reverse"

Figure Bx15.1 Desalinization.

(a) This multi-stage flash distillation plant on the coast of a desert produces 470,000,000 liters (125,000,000 gallons) of freshwater a day, enough to supply a small city.

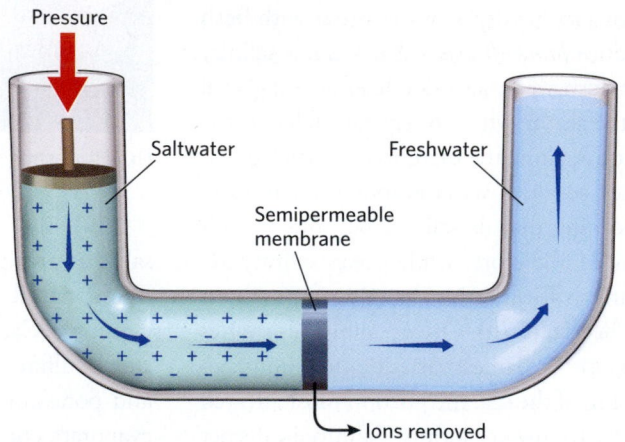

(b) During reverse osmosis, pressure is applied to saltwater to push it through a semipermeable membrane, which filters out salt.

osmosis? During normal osmosis, water from a solution with a lower concentration of salt flows through a membrane into a region with a higher concentration of salt. Reverse osmosis requires the application of pressure to push water from the higher-salt-concentration side to the lower-salt-concentration side of the membrane.

Both distillation and reverse osmosis take energy. By one estimate, the United States would have to increase its energy consumption by 10% if it were to obtain all its water by desalinization. Both processes also produce brine, which must be disposed of.

Where does the salt in the sea come from? In 1715, Sir Edmond Halley suggested that the rivers flowing on the Earth's surface transported salt to the sea, for he realized that even though river water is fresh enough to drink, it contains tiny amounts of dissolved salts. Halley proved to be correct: most dissolved ions in the sea are carried there by flowing groundwater and river water. In fact, rivers deliver over 2.5 billion tons of salt to the sea every year; lesser amounts come from submarine hydrothermal vents (see Chapter 4). The ions in salts originally came from volcanic gases and from chemical weathering of rocks.

Notably, the proportions of ions that enter the ocean from rivers differs from those dissolved in the sea. In fact, more than half of the negative ions in river water consist of HCO_3^-, traditionally known as bicarbonate. Ocean water loses its HCO_3^- because organisms extract the ion to build calcite shells, and when the organisms die, the calcite sinks to the seafloor, where it eventually becomes buried. Na^+ and Cl^- do not get extracted to make shells, so these ions stay behind in seawater and become relatively concentrated.

Temperature Variations in the Ocean

When the *Titanic* sank after striking an iceberg in the North Atlantic, passengers and crew who jumped or fell into the sea died of cold within minutes, for the seawater temperature at the site of the tragedy hovered just below the freezing temperature of freshwater, and such cold water removes heat from a body very rapidly. Yet swimmers can play for hours in the Caribbean, where sea-surface temperatures routinely reach 28°C (83°F). Though global annual sea-surface temperatures average around 17°C, it ranges from almost 35°C (95°F) in restricted tropical seas to –2°C (28°F), the freezing temperature of saltwater, near the poles **(Fig. 15.9a)**.

The correlation of average annual sea-surface temperature with latitude exists because *insolation* (the intensity of solar radiation) varies with latitude. Insolation also varies with the seasons, so sea-surface temperatures vary with the seasons as well, but not by as much as air temperature does, because water has a high *heat capacity*, meaning that it can absorb or release large amounts of heat without changing its temperature very much. For example, in mid-latitudes, the difference between winter and summer temperatures in the ocean averages about 8°C (14°F). In contrast, average

Figure 15.9 Temperature varies within the ocean.

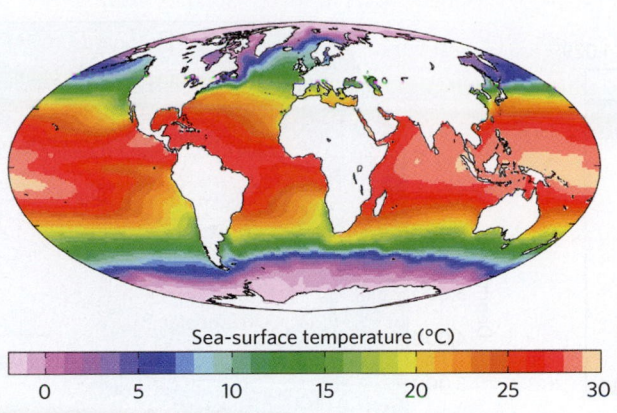

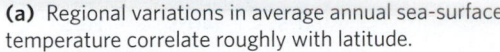
Sea-surface temperature (°C)

(a) Regional variations in average annual sea-surface temperature correlate roughly with latitude.

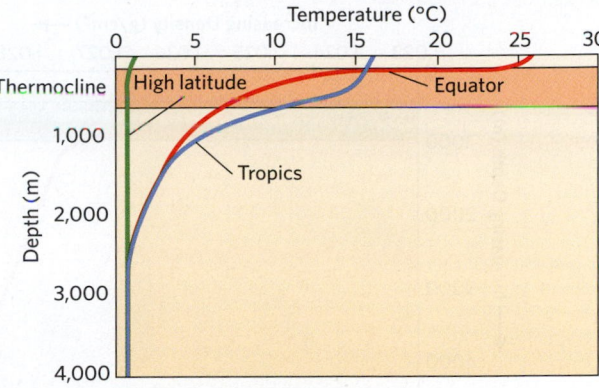

(b) The temperature of seawater changes with depth. Oceanic bottom waters are close to the freezing point of freshwater.

air temperature in the interior of North America changes by 43°C (75°F) over the course of a year.

Water temperature in the ocean varies markedly with depth in tropical and temperate latitudes, for water heated by the Sun expands and becomes less dense than colder water. Therefore, warmer water "floats" above denser, colder water; the gradual boundary between warmer water above and colder water below defines the **thermocline (Fig. 15.9b)**. In the tropics, the top of the thermocline lies at about 200 m (660 feet) below the ocean surface and the bottom at about 900 m (3,000 feet); below the thermocline, seawater temperatures remain fairly constant. A pronounced thermocline doesn't develop in polar seas because surface waters are already so cold that there's hardly any temperature contrast between sea surface and seafloor.

Ocean-Water Density and Water Masses

Both temperature and salinity control the density of seawater, so ocean-water density varies laterally and vertically. Water density at the ocean surface ranges between 1,020 and 1,030 g/L, so it's about 2% to 3% denser than freshwater. Water at the seafloor, due to a combination of the pressure caused by the weight of overlying water, low temperature, and high salinity, can reach a density of 1,050 g/L. In some parts of the world, density changes relatively rapidly across a boundary, called the **pycnocline,** at a depth of about 1 km **(Fig. 15.10a)**. A less dense *shallow layer* extends down from the surface to the pycnocline, and a denser *deep layer* continues from the pycnocline down to the seafloor.

Over time, prevailing conditions at a location may cause a large volume of ocean water to attain a characteristic temperature and salinity and, therefore, a characteristic density; oceanographers refer to such a volume as a **water mass**. Oceanic water masses remain fairly distinct for long

Figure 15.10 Water density varies within the ocean.

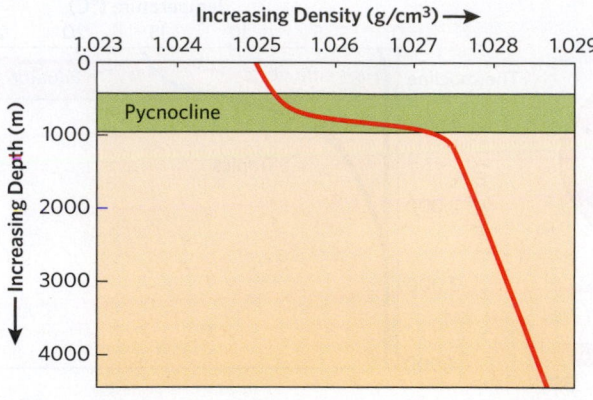
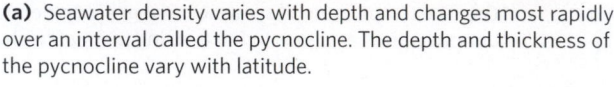

(a) Seawater density varies with depth and changes most rapidly over an interval called the pycnocline. The depth and thickness of the pycnocline vary with latitude.

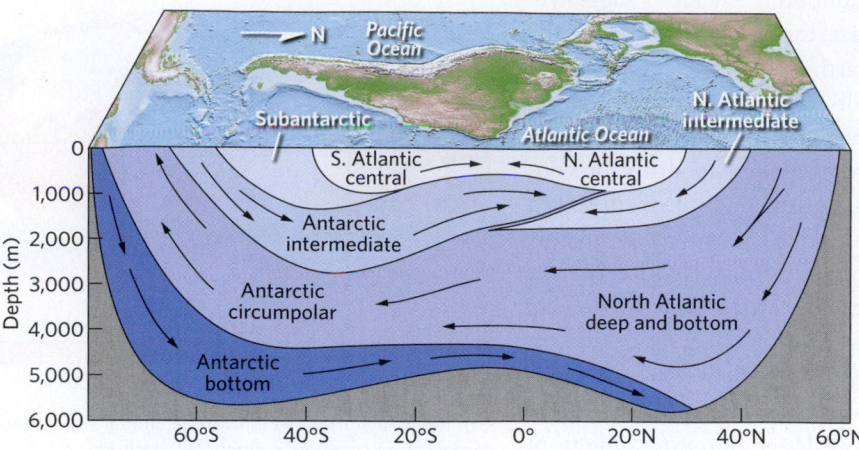

(b) Due to variations in density, the oceans are stratified into distinct water masses.

periods as they move relative to their neighbors because they do not mix easily at their margins. For example, the very salty, relatively dense water of the Mediterranean Sea spills out through the Straits of Gibraltar into the Atlantic, where it sinks to a depth of 1,000 m (3,300 feet) and forms a tongue-shaped layer, a few hundred meters thick, that spreads westward for about 3,000 km (1,900 miles).

At an even larger scale, cold water from the coast of Antarctica sinks to the floor of the ocean, forming a layer on the Atlantic Ocean seafloor known as *Antarctic bottom water*. Similarly, cold water from the Arctic region sinks to form the *North Atlantic bottom water*. The former overlaps the latter, and in equatorial regions, other water masses form above these bottom waters. Interleaving of distinct water masses causes an overall layering, or stratification, of the ocean **(Fig. 15.10b)**.

Take-home message . . .

The ocean's water originated as vapor released during volcanic activity; salts are carried in by rivers. Ocean salinity, temperature, and density vary with depth and location. Differences in these characteristics cause ocean water to form water masses that do not mix readily with others, forming distinct layers.

Quick Question -
What factors affect the salinity of ocean water?

15.4 Currents: Rivers in the Sea

The ocean's waters do not sit still, but rather flow or circulate at speeds between about 1 and 10 km per hour (0.6 and 6 mph). Oceanographers refer to a band of flowing water that moves distinctly faster than adjacent water as a **current.** Unlike a stream on land, an ocean current does not move within a distinct physical channel. But because currents transport water from one environment to another, temperature and salinity may change across the boundary of a current, making it visible in satellite images **(Fig. 15.11)**. Currents flow at two levels in the ocean: **surface currents** affect only the upper 100 to 400 m (330 to 1,300 feet) of water, whereas **deep-sea currents** affect water at great depth and may even move water along the seafloor. Let's now consider the factors that drive currents, starting with surface currents.

Surface Currents

THE PATTERN OF SURFACE CURRENTS. When skippers in the days of sailing ships planned routes between Europe and North America, they paid close attention to the directions of surface currents, for sailing with a current sped up a voyage significantly. For example, when they wanted

Figure 15.11 The Gulf Stream is a current that carries warm water northward along the coast of the eastern United States and then across the North Atlantic. On this map, the colors represent surface water temperature.

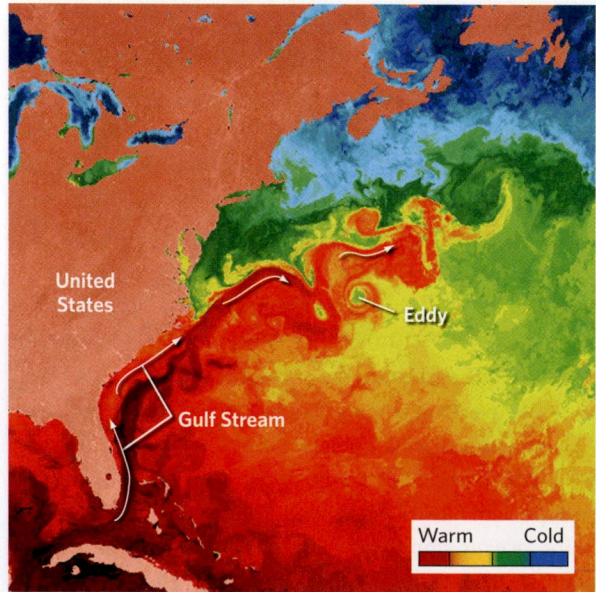

Figure 15.12 Surface currents flow through the upper 400 m (1,300 feet) of ocean water.

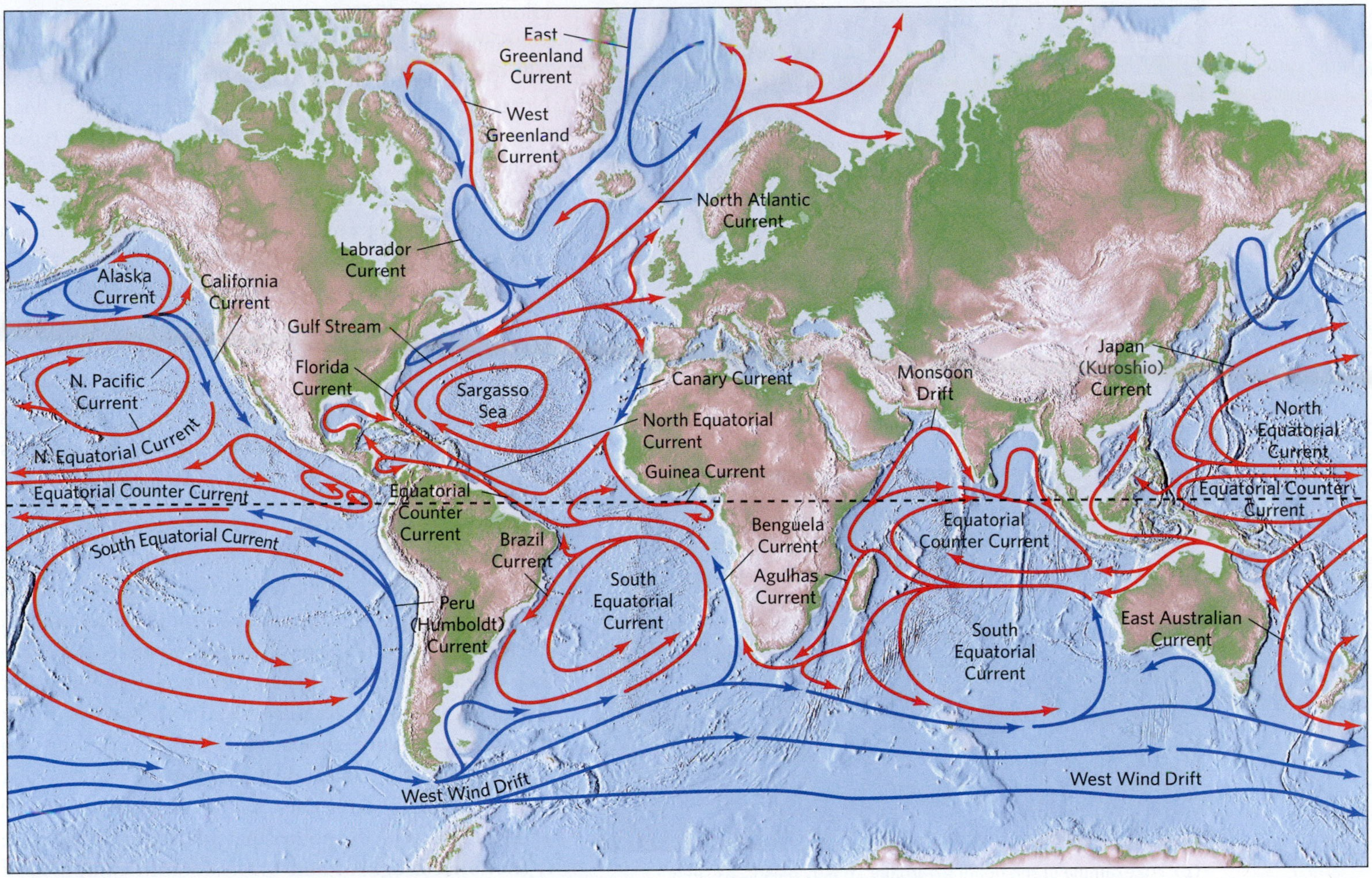

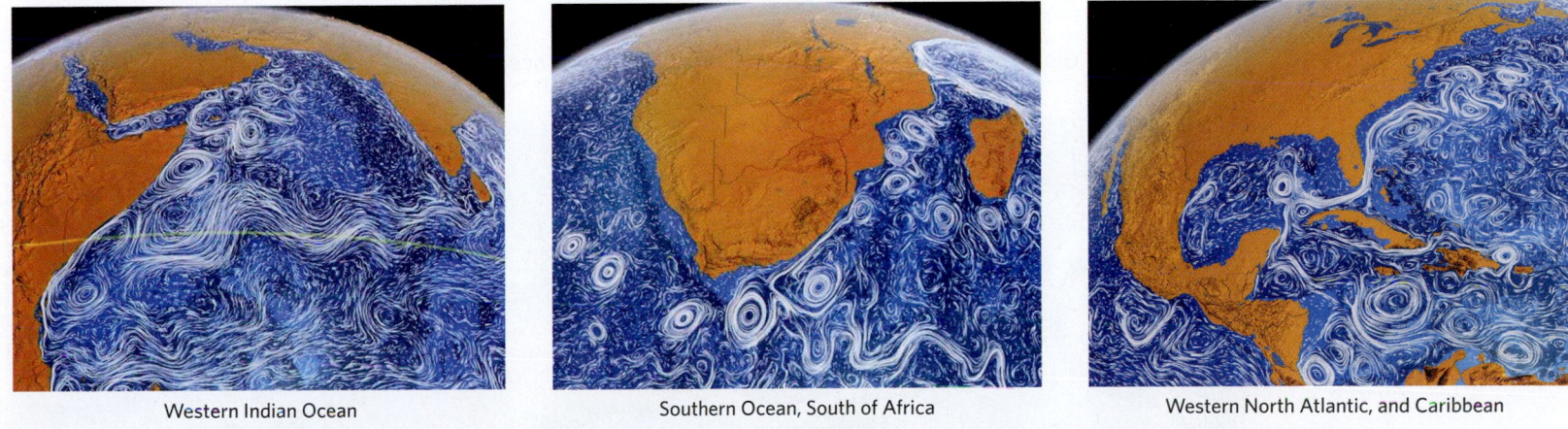

Cold → Warm →

East Greenland Current
West Greenland Current
North Atlantic Current
Labrador Current
Alaska Current
California Current
Gulf Stream
N. Pacific Current
Florida Current
Sargasso Sea
Canary Current
Monsoon Drift
Japan (Kuroshio) Current
North Equatorial Current
N. Equatorial Current
North Equatorial Current
Guinea Current
Equatorial Counter Current
Equatorial Counter Current
Equatorial Counter Current
South Equatorial Current
Brazil Current
Benguela Current
Equatorial Counter Current
Peru (Humboldt) Current
South Equatorial Current
Agulhas Current
South Equatorial Current
East Australian Current
West Wind Drift
West Wind Drift

(a) In each major ocean basin, these currents carry water around large circular paths known as gyres.

Western Indian Ocean   Southern Ocean, South of Africa   Western North Atlantic, and Caribbean

(b) These images from NASA show the presence of many eddies, swirling currents that are smaller than gyres. The white lines represent flow directions.

to head from North America toward Europe, they would steer north so as to stay in an eastward-flowing surface current, the Gulf Stream. The major surface currents on Earth display circle-like paths, called **gyres**, each of which follows the margins of the basin in which it has formed **(Fig. 15.12a)**. Gyres flow clockwise in the northern hemisphere and counterclockwise in the southern hemisphere. If water makes a complete loop within a gyre, it will have

Figure 15.13 The Great Pacific Garbage Patch.

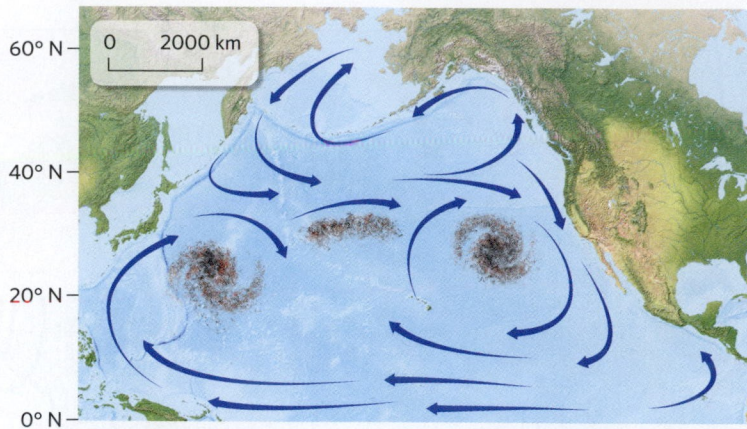

(a) Floating garbage collects within the North Pacific Gyre.

(b) An example of the floating garbage, mostly plastic, as seen from below.

traveled well over 10,000 km (6,200 miles). Locally, currents define smaller swirls, or **eddies**, whose circumference can range from a few hundred to over a thousand kilometers **(Fig. 15.12b)**.

Notably, floating debris gets trapped in the relatively still water that lies in the center of a gyre. The center of North Atlantic Gyre, for example, collects a type of floating seaweed called sargassum, so sailors refer to the region as the *Sargasso Sea*. The interior of the North Pacific Gyre has collected vast quantities of floating plastic garbage and chemical sludge, so the region has come to be known as the *Great Pacific Garbage Patch* **(Fig. 15.13a, b)**.

Why do surface currents form? Why do they flow in the directions that they do? Surface currents begin because of the frictional interaction between moving air and the surface of the sea. But the specific path that a current follows depends on the influence of two other forces—the Coriolis force and the pressure-gradient force—and on the shapes of ocean basins. Let's examine each of these influences.

WIND PUSHING WATER: FRICTIONAL DRAG. You're already familiar with *friction* as resistance to sliding. It's the force that slows down a book when you push it across a table. This force exists because protruding molecules on the table surface push against protruding molecules at the base of the book. In the familiar experiment we've just described, the table stays still while the book slows. If you repeat the experiment by placing a sheet of paper between the book and the table, you'll see that the sliding book "drags" the paper along **(Box 15.2)**. The force applied by a moving material to another material across an interface is **frictional drag.** Air and water are both composed of molecules, so the movement of air against the surface of

Figure 15.14 Global prevailing winds control the overall pattern of ocean currents.

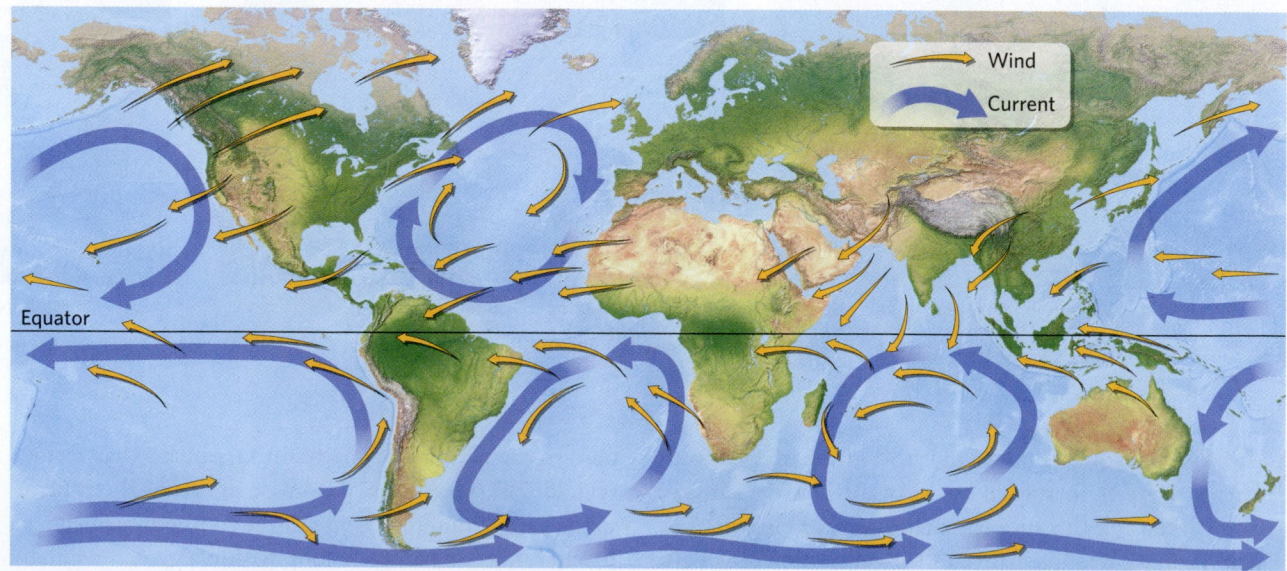

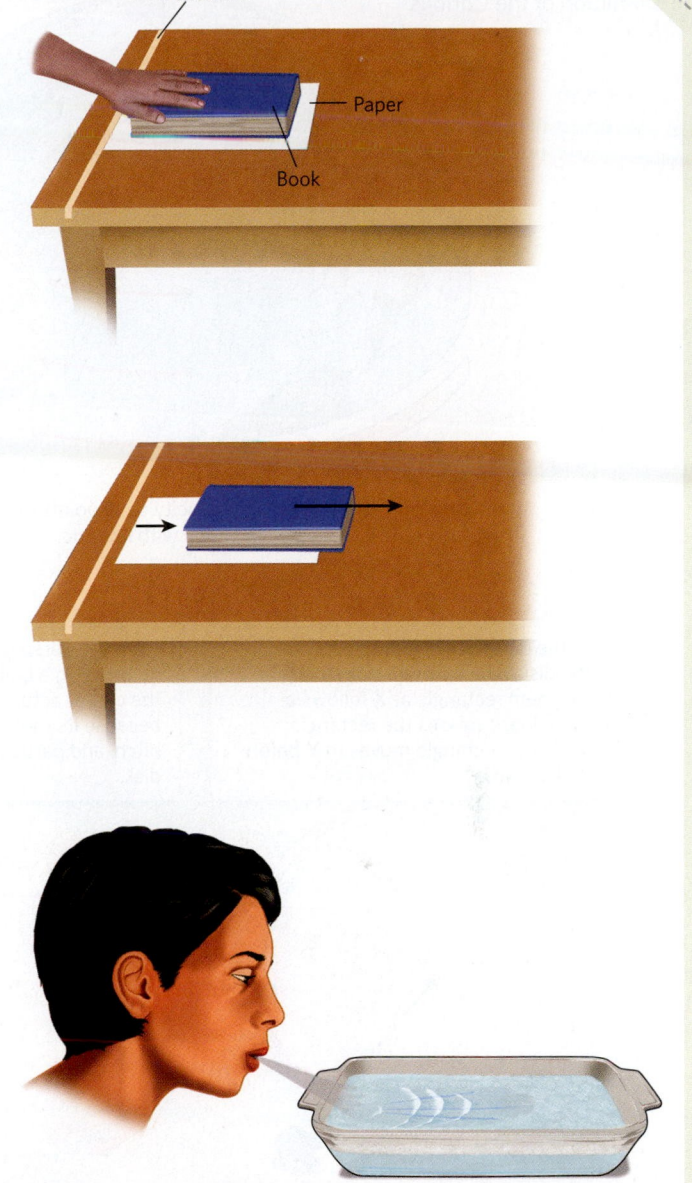

How can I explain . . .

Frictional drag

What are we learning?
Friction caused by a moving object or fluid can carry underlying material along with it.

What you need:
- A textbook, masking tape, a sheet of paper, and a slippery tabletop.
- A shallow pan filled with soapy water, with a few small scraps of paper floating in it.

Instructions:
- Mark out a reference line on the table with masking tape.
- Align the sheet of paper so that its edge is flush with the reference line.
- Place the book on top of the paper, then push it forward by about 15 cm (6 inches).
- Measure the distance that the paper moves.
- Place the pan of soapy water on the tabletop.
- Blow hard at a shallow angle to the water surface from the edge of the basin.

What did we see?
- Friction slows the movement of the book. But frictional drag at the base of the book can move a layer of material beneath it. Materials moved by frictional drag do not move as fast as the materials that drive the motion.
- The same process happens when wind blows on water. You can see this by watching the paper scraps and bubbles move as you blow on the water in the basin. If you add a drop of food coloring, you'll see that the velocity of the water decreases with depth.

water applies frictional drag to the water and causes the water to move. The *wind*—horizontally flowing air—applies frictional drag to water at the sea surface, causing the water to start moving. This movement initiates a current in the same direction as the wind.

In a given region, winds tends to blow in about the same direction, the *prevailing wind direction*, during most of a season (**Fig. 15.14**; see Chapter 18). If prevailing winds didn't exist, well-defined surface currents would not develop, for opposing movements of water would cancel one another out.

DEFLECTIONS ON A ROTATING SPHERE: THE CORIOLIS FORCE. The Earth revolves on its axis once a day, and because of this rotation, a point on the equator moves at a speed of 1,670 km per hour (1,037 mph) relative to an observer outside of the Earth. This speed decreases toward the poles, for a point at a higher latitude doesn't have as far to go in a day; it reaches zero at the poles (**Fig. 15.15a**). Because of the Earth's rotation, the instant that seawater starts to move, it deflects away from its initial straight-line path and starts to travel in a different direction along a curved path. The deflection of a moving object relative to the surface of a spinning object beneath is called the **Coriolis effect**, after the French physicist Gaspard-Gustave de Coriolis (1792–1843), who first explained the phenomenon.

Figure 15.15 A simplified explanation of the Coriolis effect.

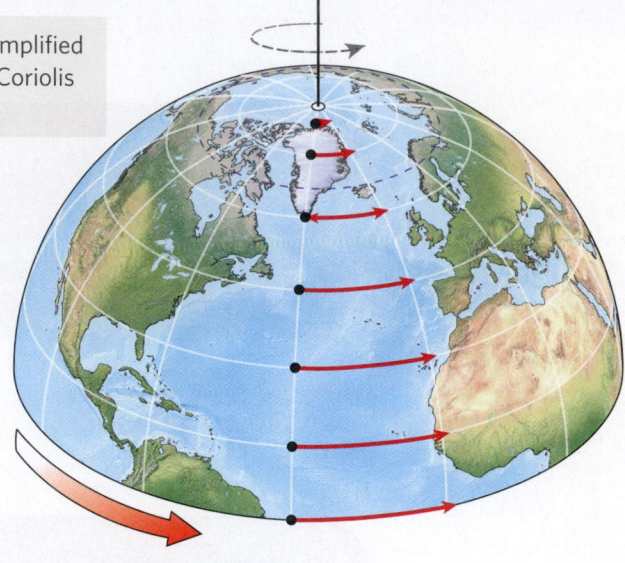

(a) Although the angular velocity of all points on the Earth's surface is the same, the linear velocity varies with latitude.

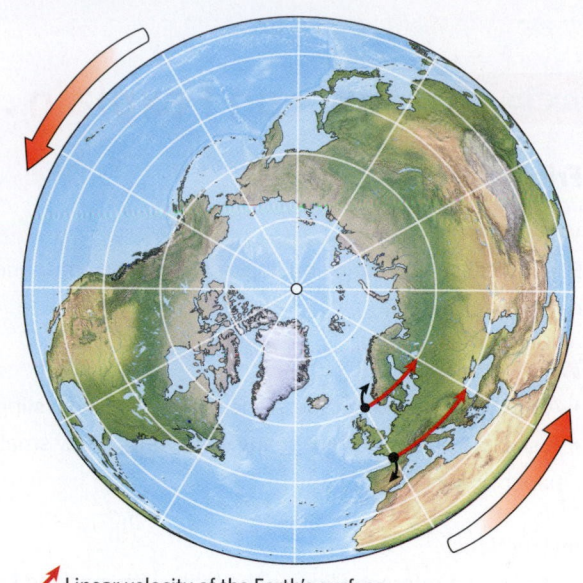

↗ Linear velocity of the Earth's surface
↘ Path of an object that starts heading south
↗ Path of an object that starts heading north

(c) Looking down on the north pole, lines of longitude look like spokes of a wheel, and lines of latitude look like circles.

From the perspective of an observer *not* on the disk, a ball thrown from the center to the green rectangle at **X** follows a straight line. It misses the rectangle because the rectangle moves to **Y** before the ball arrives.

From the perspective of an observer *not* on the disk, a ball that someone aims at the center actually follows the green line because its motion is partly due to the pitch, and partly due to the spin on the disk.

Y

Y

From the perspective of an observer on the disk, a ball thrown from the center to the green rectangle at **X** follows a curved path to the right (west).

From the perspective of an observer on the disk, a ball that someone aims at the center follows a curve to the right (east).

Viewed from the disk

Viewed from the outside

X

(b) If we imagine that, instead of looking down on the Earth, we are looking down on a spinning disk (such as a merry-go-round), we can picture the geometry of the Coriolis effect by seeing how the motion of a ball thrown along a radius of the disk appears to observers on the disk and off the disk.

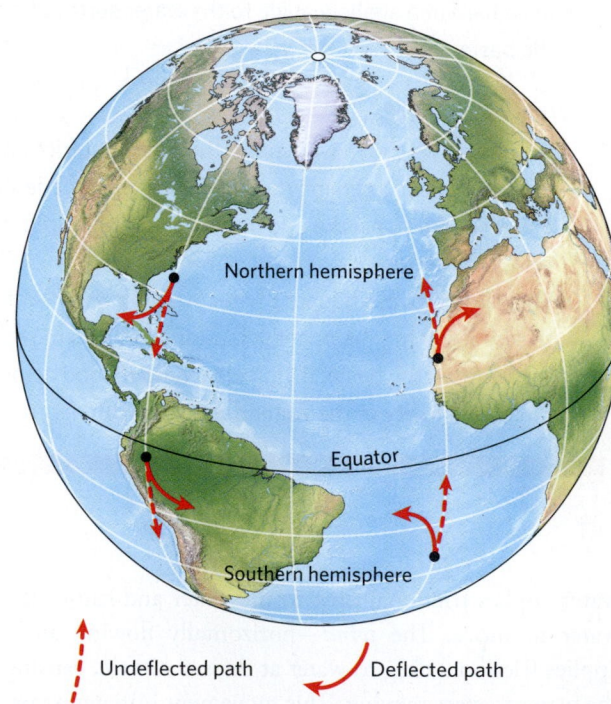

Northern hemisphere

Equator

Southern hemisphere

- - → Undeflected path   ↷ Deflected path

(d) In the northern hemisphere, the Coriolis effect causes objects initially moving north, from a lower to a higher latitude, to deflect to the right (east). Objects initially moving to the south, from a higher to a lower latitude, also deflect to the right (this time, west). In the southern hemisphere, objects moving north deflect to the left (west). Objects moving south also deflect to the left (this time, east).

To picture simplistically what causes the Coriolis effect, start by imagining that you're looking straight down on the top of a two-dimensional disk, like a playground merry-go-round **(Fig. 15.15b)**. What happens to a ball thrown from the rim to the center of the disk, or vice versa? Let's do a few thought experiments to find out:

- *Experiment 1:* First imagine that the disk is standing still, and that you throw a ball from the center to a target on the rim. The ball follows a straight path parallel to a radius and hits the target.

- *Experiment 2:* Now imagine that the disk is spinning counterclockwise and that you again throw the ball from the center toward the target (a green rectangle) on the rim. From the perspective of a stationary observer outside the disk, the ball still follows a straight path—it must, due to *inertia*, the tendency of an object to move in a straight line unless acted on by an outside force (see Fig. 15.15b, top left). But as the ball heads toward the rim, the target moves, so the ball doesn't hit the target. Therefore, from the perspective of an observer sitting on the target and moving with the spinning disk, the ball appears to deflect to its right and follow a curved path (see Fig. 15.15b, lower left).

- *Experiment 3:* Now imagine instead that you throw the ball from the rim to the center of the spinning disk. When launched, the ball not only has motion in the direction in which you aimed it, but also has motion due to the spin of the disk. So, from the perspective of a stationary observer outside the disk, it follows a straight path toward a point to the right of the disk's center (see Fig. 15.15b, upper right). But from the perspective of an observer moving with the target, it deflects to the right and follows a curved path (see Fig. 15.15b, lower right).

We can apply these thought experiments to the spinning Earth. If you imagine that you are looking down on the North Pole, so that you can see the northern hemisphere, the Earth looks like a spinning disk—lines of longitude look like spokes on a wheel, and lines of latitude look like circles **(Fig. 15.15c)**. Just like the disk in our previous thought experiments, the Earth spins counterclockwise, and as it does so, a point on the Earth at a lower latitude (a larger-diameter circle) moves faster than one at a higher latitude (a smaller-diameter circle) because a point at lower latitude has farther to go in a single day.

Now let's rotate the Earth so we're looking at it sideways **(Fig. 15.15d)**. If you launch a projectile (such as a cannonball) due south from a higher latitude toward a lower latitude within the northern hemisphere, the instant the projectile starts on its journey, it not only has southward motion, but is also is moving around the Earth's axis at the same speed as the Earth at its launch point. As the projectile heads south, it moves eastward more slowly than the Earth below, so from the perspective of an observer on the Earth below, it appears to deflect to the right (west) (see Fig. 15.15d). If instead you launch the projectile northward, from a lower latitude to a higher latitude, the projectile not only has northward motion, but also eastward motion. Therefore, as soon as it has moved a distance to the north, it's moving eastward at a rate faster than the Earth below, so, relative to a point on the Earth below, it deflects to the right (east). If you were to repeat the experiments in the southern hemisphere, an object initially heading south would deflect to the left (east), and an object initially heading north would deflect to the left (west).

All moving objects have inertia, so from the perspective of the projectile we described in the previous paragraph, the Coriolis effect must be due to a push by some invisible force, because only an external force can cause a moving object to deflect from its initial straight path. Physicists refer to the apparent force causing the deflection as the **Coriolis force**. It has the following properties: (1) it causes a deflection to the right in the northern hemisphere and a deflection to the left in the southern hemisphere; (2) it affects the direction in which an object moves across the Earth's surface, but not its speed; (3) the magnitude of the force increases as the speed of the object increases; and (4) it has a value of zero on the equator and a maximum at the poles.

While our description above provides a visual image of the Coriolis effect, it doesn't provide a technically correct explanation of the Coriolis force on Earth. The analogy of a spinning disk doesn't explain, for example, why objects undergo a deflection even if their initial motion is due east or due west. On the Earth, objects experience a Coriolis force regardless of the direction in which they move, unless they start by moving exactly parallel to the equator. Therefore, an object that keeps moving long enough with no other forces acting on it follows a circular path, called an *inertial circle*, and ends up back where it started **(Fig. 15.16)**. The diameter of this inertial circle depends on the initial speed of the object and on the latitude: inertial circles get smaller closer to the poles. A more complete explanation of the Coriolis force that accounts for inertial circles reveals that it reflects a balance between gravitational and centrifugal forces acting on an object **(Box 15.3)**.

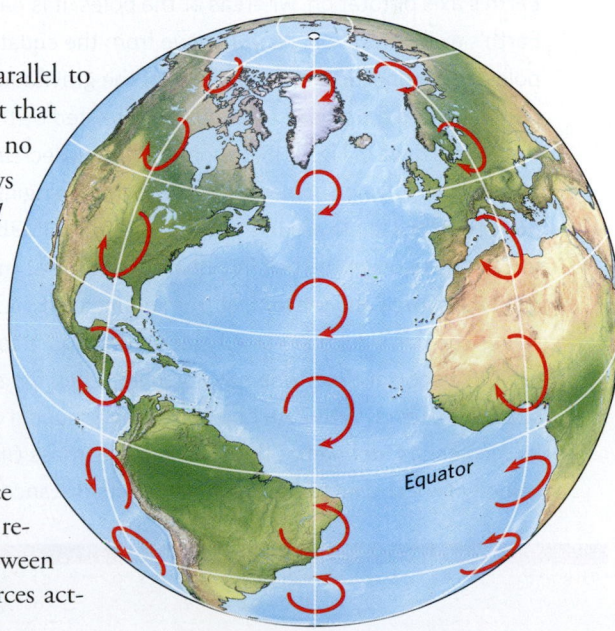

Figure 15.16 Due to the Coriolis effect, an object that keeps moving across the Earth's surface follows a circular path called an inertial circle, whose size varies with latitude and the object's speed. The circles in this diagram represent an object moving at 140 km per hour (87 mph).

Equator

Box 15.3

Science toolbox . . .

What causes the Coriolis force?

To streamline our discussion of the Coriolis force, we must first review a few key terms from physics:

1. *Angular velocity:* The speed of an object around an axis of rotation, given in degrees per unit time (keeping in mind that a circle is divided into 360°). The Earth makes a complete circle around our planet's axis of rotation once a day, so the angular velocity of all points on the Earth is 360° per day.

2. *Linear velocity:* The speed at which an object moves along a line, relative to a fixed point on the line. We specify linear velocity in units of distance per unit time, such as kilometers per hour or miles per hour. An object sitting on the Earth's equator has farther to go than an object at a high latitude, so the object at the equator has a greater linear velocity, even though both objects have the same angular velocity.

3. *Centrifugal force:* The apparent outward-directed force that an object resting on the surface of a spinning disk or sphere experiences, which is a manifestation of inertia. The magnitude of centrifugal force at a point on the spinning Earth depends on the linear velocity at that point, and the distance of the point from the axis, so centrifugal force is greater at the equator than at a high latitude.

4. *Components of gravity:* Gravity is the attractive force between two objects. Because of gravity, centrifugal force doesn't cause objects to fly off the spinning surface of the Earth. The force of gravity acting on an object at the surface of the Earth points toward the center of the Earth. The red arrows in **Figure Bx15.3a** show the direction of gravitational force at different points on the Earth's surface. Note that at the equator, the direction of this force is perpendicular to Earth's axis of rotation, whereas at the poles, it is parallel to the Earth's axis of rotation. As you move from the equator toward the poles, the proportion, or *component*, of the gravitational force that points toward the axis of rotation (represented by the lengths of the blue arrows in Fig. Bx15.3a) progressively decreases.

5. *Conservation of angular momentum:* Imagine swinging a ball attached to a string in a circle around your head **(Fig. Bx15.3b)**. Centrifugal force tries to make the ball fly outward, but the string provides an inward force and prevents it from doing so. If you shorten the string, the ball's angular velocity increases, and it completes an orbit in a shorter time, whereas if you lengthen the string, the opposite occurs. Why? An object moving around a circle has *angular momentum*, defined as the product of its mass (m), its angular velocity (w), and its radius (r), meaning its distance from the axis of

rotation. Written as an equation: angular momentum = $m \times w \times r$. Newton showed that if no external force acts on an object in rotation, there must be *conservation of angular momentum*. In other words, since the mass of the ball doesn't change, a decrease in its the distance from the axis (r) means that its angular velocity (w) must increase.

Using these terms, we can reexamine the origin of the Coriolis force **(Fig. Bx15.3c)**. Let's start by considering the forces acting on a stationary object at the Earth's surface at a latitude in the northern hemisphere (Time 1). Gravity applies a downward force, pulling the object toward the center of the Earth; the Earth's surface provides an upward force to prevent the object from sinking into the Earth's interior. Meanwhile, centrifugal force also acts on the object because the Earth is rotating.

Now, imagine that the object starts moving to the north (Time 2). (For simplicity, we'll ignore friction.) In profile, we see that as the object moves north, it gets closer to the Earth's axis of rotation, so, like the ball when the string is shortened, the object's angular velocity increases to conserve angular momentum. When the object has a greater angular velocity than the Earth below, it deflects to the right (east). Note that as the object turns eastward, it is still moving at the same linear velocity.

As the object moves north, the forces acting on it change. Specifically, the outward centrifugal force increases because the object has a greater angular velocity and is closer to the Earth's axis. At the same time, the component of the gravitational force pointing toward Earth's axis of rotation decreases because the total pull of gravity acts at a smaller angle to the Earth's axis. The changes in these two forces result in a net force outward, away from the Earth's axis (Time 3). The net force pulls the object outward, perpendicular to the Earth's surface, and southward, parallel to the Earth's surface. The outward pull is balanced by the stronger pull of gravity, so the object doesn't fly into space. The southward component of the net force deflects the object southward, back toward the equator. Eventually, the object crosses the latitude where it first started to move, this time heading south (Time 4). It continues moving because of its inertia, much as a pendulum moves past its equilibrium point as it swings.

What happens next? As the object continues southward, it moves farther away from the Earth's axis, so its angular velocity decreases (Time 5). Now, the object has less angular velocity than does the Earth's surface below, so it turns right (west). How did the change in position affect the balance of forces? With the object farther south, the centrifugal force decreases because the object is farther from the axis of rotation and it has a slower angular velocity. Meanwhile, the component of the gravitational force pointing toward the Earth's axis

Figure Bx15.3 The physical origin of the Coriolis force.

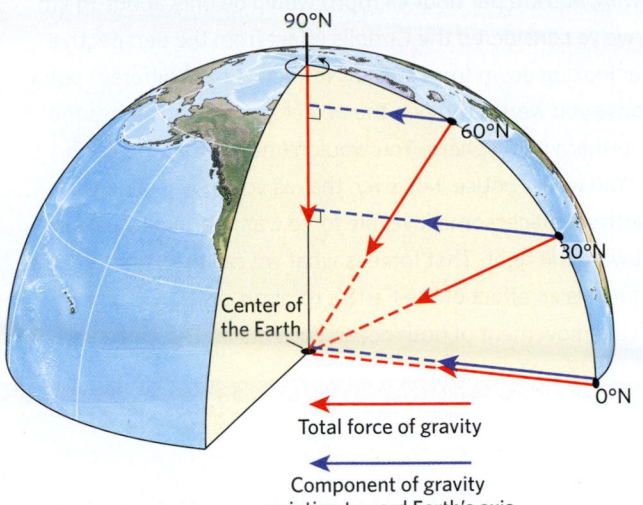

Total force of gravity

Component of gravity
pointing toward Earth's axis

(a) Gravity (red arrows) acts toward the center of the Earth at every point on the Earth. The proportion of gravitational force pointing toward the Earth's rotational axis varies with latitude (as shown by the lengths of the blue arrows). It is strongest at the equator and weakest at the poles.

(b) When swinging a ball overhead, shortening the string increases the ball's angular velocity in order to conserve its angular momentum.

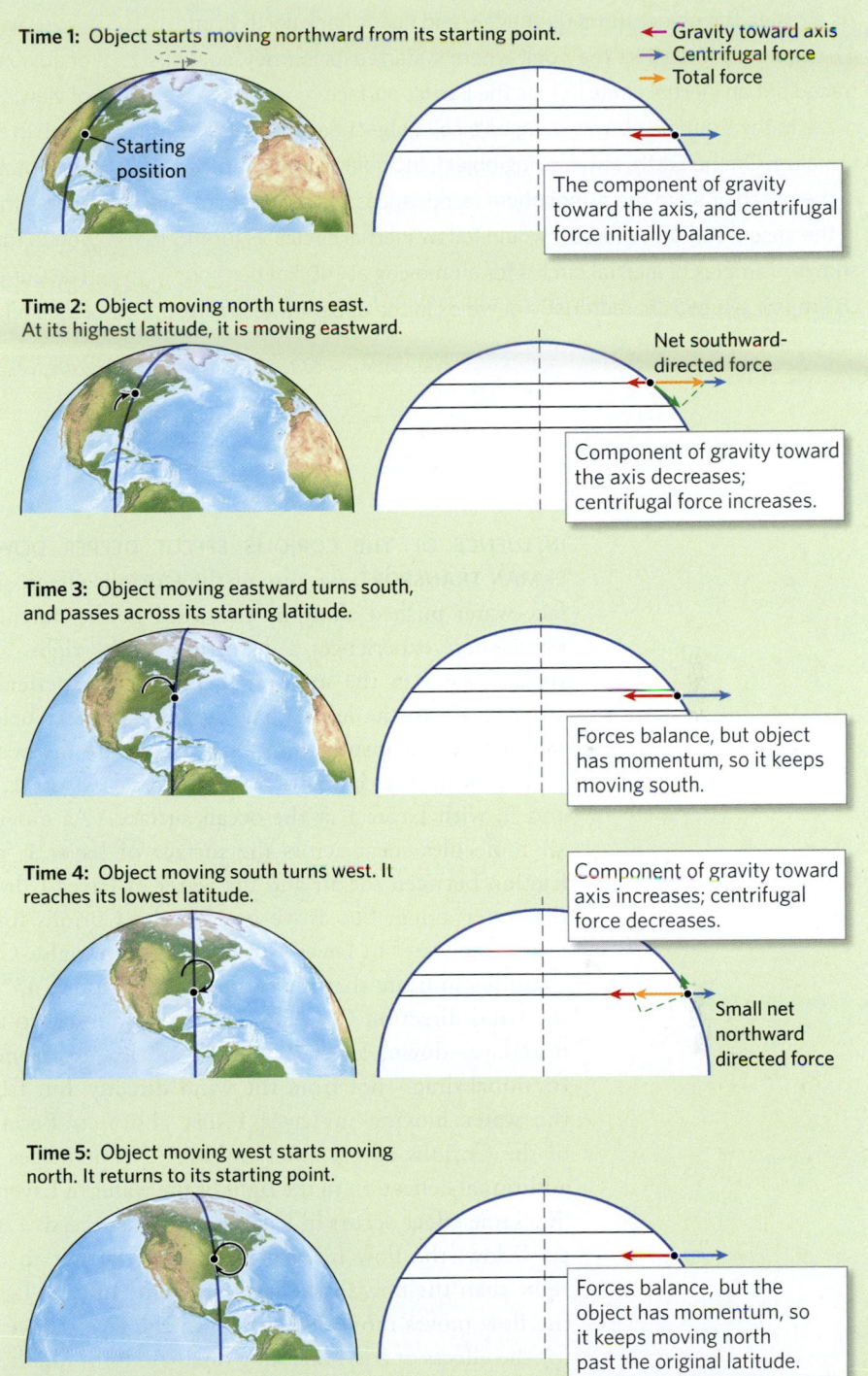

Time 1: Object starts moving northward from its starting point.

← Gravity toward axis
→ Centrifugal force
→ Total force

Starting position

The component of gravity toward the axis, and centrifugal force initially balance.

Time 2: Object moving north turns east. At its highest latitude, it is moving eastward.

Net southward-directed force

Component of gravity toward the axis decreases; centrifugal force increases.

Time 3: Object moving eastward turns south, and passes across its starting latitude.

Forces balance, but object has momentum, so it keeps moving south.

Time 4: Object moving south turns west. It reaches its lowest latitude.

Component of gravity toward axis increases; centrifugal force decreases.

Small net northward directed force

Time 5: Object moving west starts moving north. It returns to its starting point.

Forces balance, but the object has momentum, so it keeps moving north past the original latitude.

(c) As an object moves across the surface of the rotating Earth, the relative magnitudes of centrifugal and gravitational forces acting on it change, so in order to conserve angular momentum, the object changes direction.

(continued)

increases because the total gravitational force is now at a greater angle to the Earth's axis. The net force pulls the object inward, perpendicular to the Earth's surface, and northward, parallel to the Earth's surface. As a result, the object turns right again and heads back north (Time 6). Eventually it reaches the point where it started its journey, so it has traced out an inertial circle across the Earth's surface.

In the example above, we used a solid object because it is easy to visualize. On the Earth, any moving object, including a volume of water in the ocean or air in the atmosphere, experiences the same effect and, in the absence of other forces, would follow inertial circles. Figure 15.16 shows examples of inertial circles for air moving at 140 km per hour (87 mph), a speed characteristic of winds in the upper atmosphere.

Inertial circles are much smaller for ocean currents because they flow much more slowly—for example, at 40° N, the inertial circle traced by a current flowing at 5 km per hour (3 mph) would be only about 14 km.

So far, we've considered the Coriolis effect from the perspective of an observer looking down from space. Now let's take a different point of view. Suppose you were sitting on the object as it moved across the Earth in the northern hemisphere. You wouldn't notice that the Earth was rotating. You would notice, however, that as you moved along across the Earth's surface, some invisible force was continually causing you to turn toward the right. That force is what we call the Coriolis force. The Coriolis force is an effect of the Earth's rotation, and it has a very real effect on the movement of both ocean currents and air currents.

INFLUENCE OF THE CORIOLIS EFFECT DEEPER DOWN: EKMAN TRANSPORT. Because of the Coriolis effect, surface water pushed along by the wind in the northern hemisphere experiences a deflection to the right, and surface water in the southern hemisphere experiences a deflection to the left. What happens to water below the surface? To answer this question, picture the ocean as a stack of thin layers of water (labeled Layers 1, 2, and 3, with Layer 1 at the ocean surface). As moving air molecules shear across the surface of Layer 1, the friction between the air and the water in Layer 1 drags the water along, but as we've seen, the Coriolis force causes the water in Layer 1 to deflect to the right. Calculations indicate that the angle of deflection is 45° to the wind direction **(Fig. 15.17a)**. Now let's move to the next layer down, Layer 2. This layer also experiences frictional drag—not from the wind directly, but from the water moving in Layer 1 just above it. Because of the Coriolis effect, water in Layer 2 experiences an additional deflection to the right of the water in Layer 1. The same effect occurs in Layer 3, and in successive layers below: the flow in each layer turns farther to the right than the flow in the layer above it. In each layer, the flow moves more slowly than in the layer above it, for the effects of frictional drag dissipate until finally, at some critical depth, the wind no longer has an effect, and water no longer moves.

Did you ever wonder . . .

why currents circle the oceans?

The progressive turning of water flow with depth defines the *Ekman spiral* **(Fig. 15.17b)**, named for its discoverer. The existence of an Ekman spiral in a column of water causes the water's overall flow to deflect at an angle of about 90° to the wind shearing the surface of the water. Oceanographers refer to this overall flow as **Ekman**

transport, and as we see next, it plays a major role in governing the geometry of gyres.

A BALANCING ACT: GEOSTROPHIC FLOW AND THE DEVELOPMENT OF GYRES. By this point, you might still be puzzled by the overall flow pattern of surface currents in the ocean. It might seem at first that ocean currents, due to the Coriolis effect, should follow relatively small inertial circles. If, however, you simply compare the paths of surface currents with the paths of prevailing winds, you see that winds and currents actually move in similar directions (see Fig. 15.14). What's going on?

The answer comes from considering a third force acting on ocean water: the **pressure-gradient force**. This force develops because Ekman transport causes a slow drift of water toward the center of an ocean. As a result of this converging flow, sea level in the center of an ocean becomes slightly higher (about 1 m, or 3 feet, higher) than at the ocean's margins—it's as if there is a flat-topped mound of water in the middle of the ocean. (The peak of the mound actually lies to the west of the ocean's center, for reasons beyond the scope of our discussion.) The mound is so broad, and its height so slight, that you don't notice it when you sail across the ocean **(Fig. 15.18a)**, but it is real. If you picture a horizontal surface at a depth beneath this mound, the pressure acting on that surface is greater under the peak of the mound than at its edges. In other words, the slight variation in sea-level height caused by Ekman transport produces a pressure difference, or *horizontal pressure gradient*, in the ocean between the center and periphery of an ocean basin. The resulting outward-directed pressure-gradient force pushes back against the inward-directed Coriolis

Figure 15.17 The concept of Ekman transport.

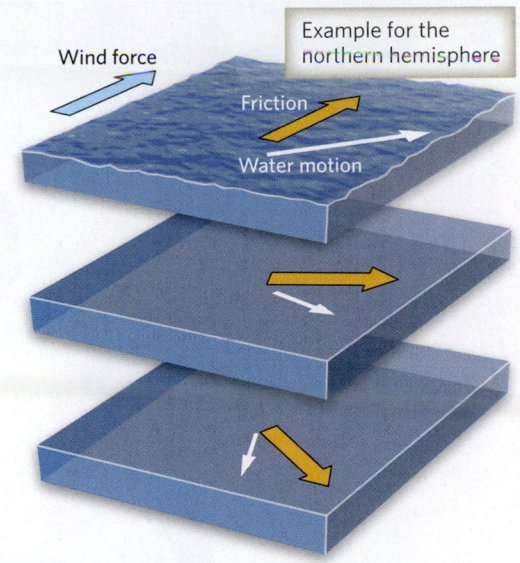

Wind force
Friction
Water motion

Example for the northern hemisphere

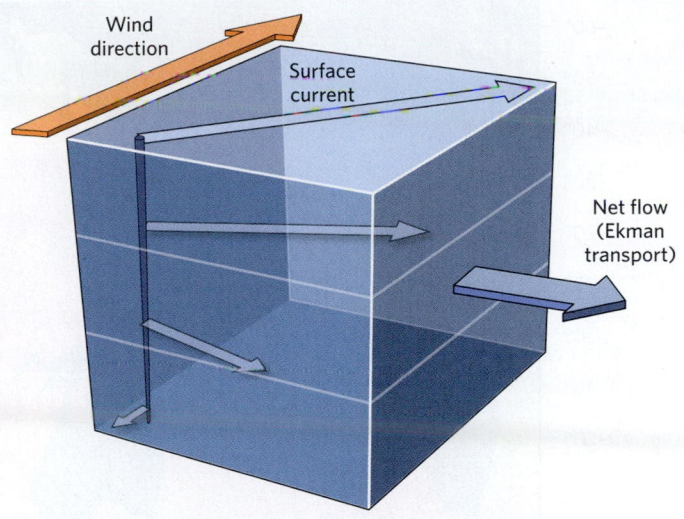

Wind direction
Surface current
Net flow (Ekman transport)

(a) The near-surface realm of the ocean can be pictured as a series of layers. When wind shears the top layer, this layer shears the layer below, and so on. Due to the Coriolis effect, the motion of each layer is deflected relative to the layer above.

(b) The resulting pattern of flow is called an Ekman spiral. As a consequence, the overall flow, known as Ekman transport, trends at 90° to the wind direction.

force responsible for Ekman transport. When the inward and outward forces are equal and opposite, we can say that a *geostrophic balance* exists. When geostrophic balance exists, water must circulate around the circumference of the mound **(Fig. 15.18b)**. The roughly circular currents that result are the gyres, which, because they develop when geostrophic balance has been established, are also known as **geostrophic currents**. Because in reality, the mound rises higher on the western side of an ocean, the current on the western side of an ocean is stronger than on the eastern side **(Fig. 15.18c)**.

THE EFFECT OF OCEAN BASIN SHAPE. As we have seen, continents and the equator separate the global saltwater layer into separate oceans. Because of these divisions, a distinct gyre forms in each of the oceans. Different parts of gyres have different names (see Fig. 15.12a). Gyres don't cross the equator because the direction of the Coriolis force changes there. The southern edges of gyres in the northern hemisphere and the northern edges of gyres in the southern hemisphere produce westward-flowing equatorial currents. (A weaker, eastward-flowing counter current generally exists between the North Equatorial Current and the South Equatorial Current.)

Land acts as an obstacle to ocean currents and can block their flow, causing them to bend and change direction or even to split into two branches flowing in directions opposite to each other. South of 55° S to the shores of Antarctica, however, no land blocks ocean flow, so the *Antarctic Circumpolar Current* carries water

in a circle completely around Antarctica **(Fig. 15.19)**. This current, which ranges between 800 and 2,000 km (500 to 1,200 miles) wide, extends deeper, and carries more water, than any other current. On average, it transports about 140 times the amount of water in all the world's rivers combined. The existence of the Antarctic Circumpolar Current prevents warm waters from getting close to Antarctica, thereby keeping the continent cold enough to remain covered by ice and fringed by sea ice year-round.

In places where currents flow through relatively narrow spaces between landmasses, such as between Africa and Madagascar, frictional interaction with shallow water bordering the land causes larger currents to break up into smaller eddies (see Fig. 15.12b; Fig. 15.18c). Eddies also form at the boundary between a major gyre and the slower water outside the gyre due to the shear between the moving water masses. We can see this phenomenon clearly in the satellite image of the Gulf Stream in Figure 15.11.

Upwelling, Downwelling, and Deep-Sea Currents

Surface currents are not the only movements of water in the ocean. Water also circulates in the vertical direction: oceanographers identify **downwelling** zones as places where near-surface water sinks and **upwelling** zones as places where subsurface water rises.

What causes downwelling and upwelling? Several factors come into play. Coastal downwelling occurs where the wind blows toward the shoreline or Ekman transport causes water to move toward it **(Fig. 15.20a)**. Since the

Figure 15.18 Understanding geostrophic currents.

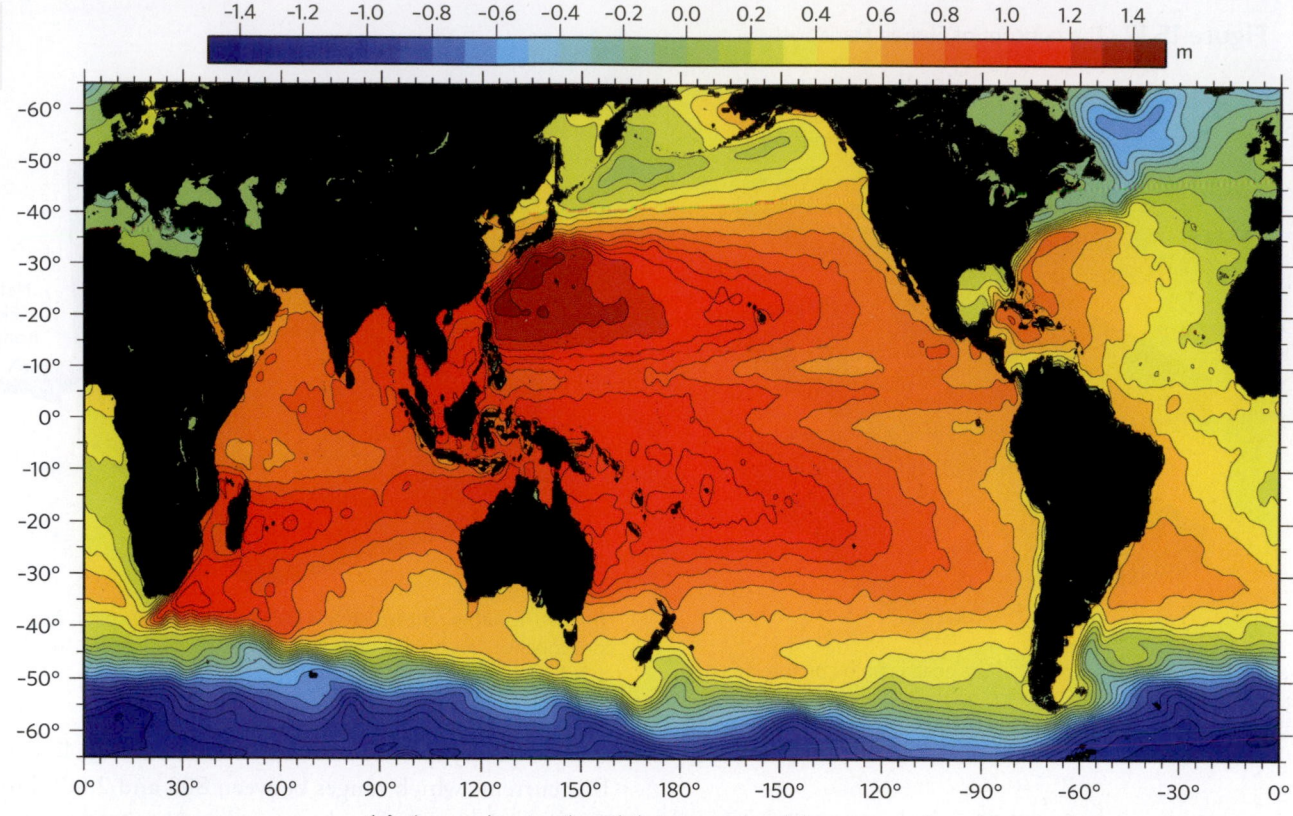

(a) A map showing the subtle topography of the ocean surfaces. The elevated areas rise because of Ekman transport. These elevations do not include the effects of tides and waves.

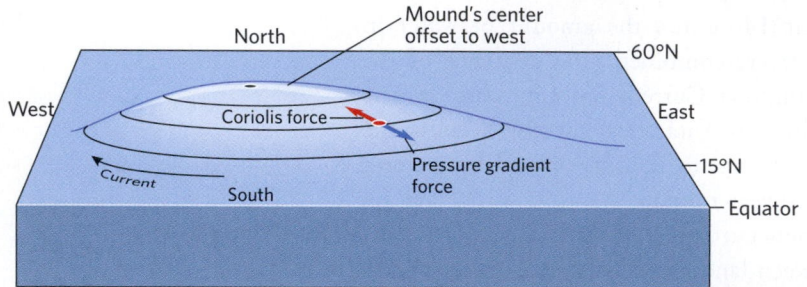

(b) Because of the elevated surface of the sea, the pressure-gradient force can balance the Coriolis force. When this happens, a geostrophic current forms.

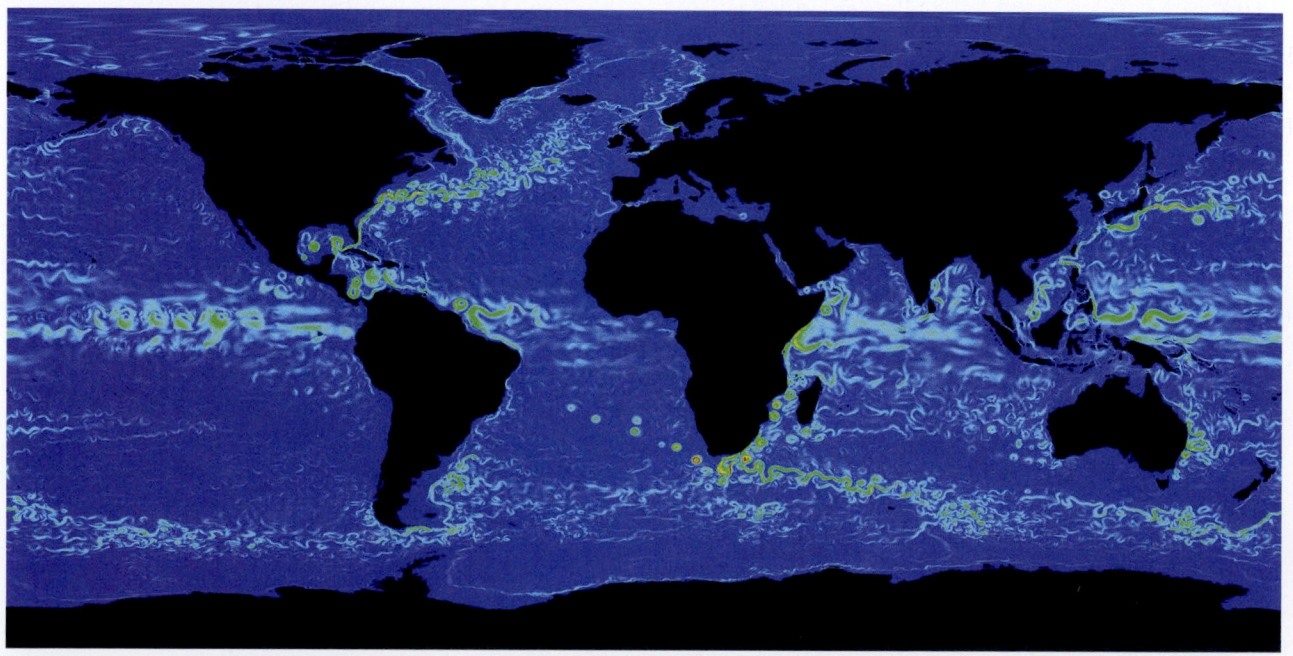

(c) A map showing variations in current velocity reveals that geostrophic currents are fastest on the western sides of oceans, where the sea surface is highest.

Figure 15.19 The Antarctic Circumpolar Current. Note the many eddies along the overall current.

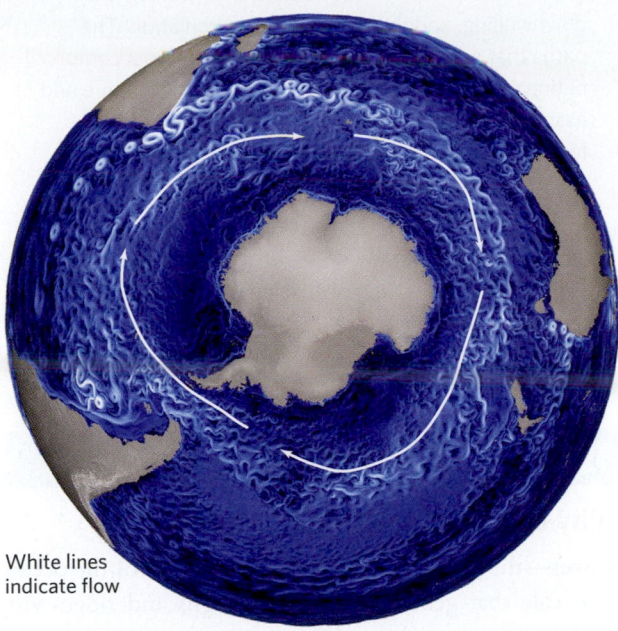

White lines indicate flow

Figure 15.20 Coastal upwelling and downwelling.

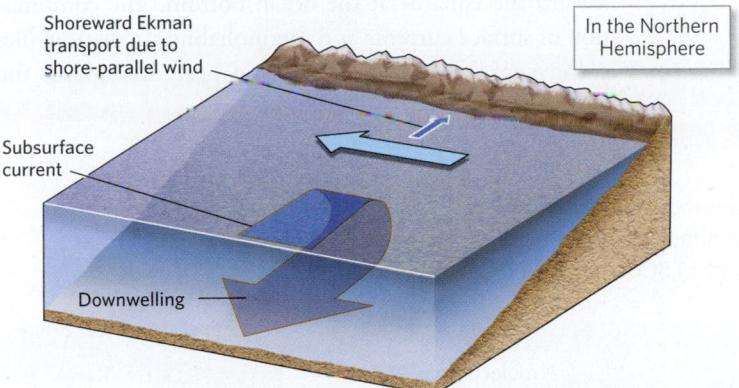

Shoreward Ekman transport due to shore-parallel wind

In the Northern Hemisphere

Subsurface current

Downwelling

(a) A surface flow that moves toward the shore causes downwelling.

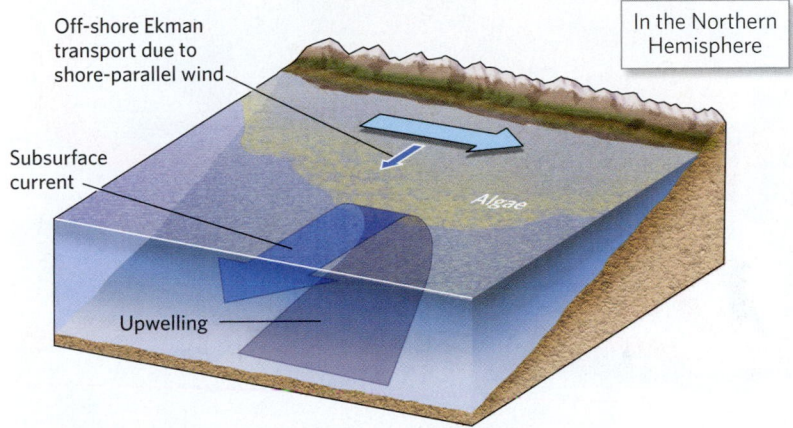

Off-shore Ekman transport due to shore-parallel wind

In the Northern Hemisphere

Subsurface current

Algae

Upwelling

(b) A surface flow that moves water away from the shore causes upwelling. Upwelling brings up nutrients and fosters the growth of algae near the coast.

excess water can't move onto the land, it piles up and sinks. Coastal upwelling occurs near coasts where the wind blows from land toward the ocean or Ekman transport causes surface water to flow away from the shore. As the surface water moves oceanward, deep water rises, or upwells, to replace it **(Fig. 15.20b)**. Indeed, along the west coast of North America, upwelling continually brings cold, nutrient-rich water from the depths to the surface **(Fig. 15.20c)**.

Upwelling of subsurface water also occurs along the equator because the winds there blow steadily from east to west. Due to the Coriolis force, water just to the north of the equator deflects to the right, and water just to the south of the equator deflects to the left. These deflections remove surface water, so upwelling along the equator must take place to replace this water from below.

So far, we've discussed upwelling and downwelling driven ultimately by winds. These processes can also be driven by differences in seawater temperature and salinity because these differences produce contrasts in water density, and in a fluid, density contrasts cause buoyancy differences: denser water sinks, and less dense water buoyantly rises. We refer to the rising and sinking of water driven by density contrasts as **thermohaline circulation (Fig. 15.21)**. As a result of thermohaline circulation, colder or saltier water sinks because it's denser, whereas warmer or less salty water rises because it's less dense. Therefore,

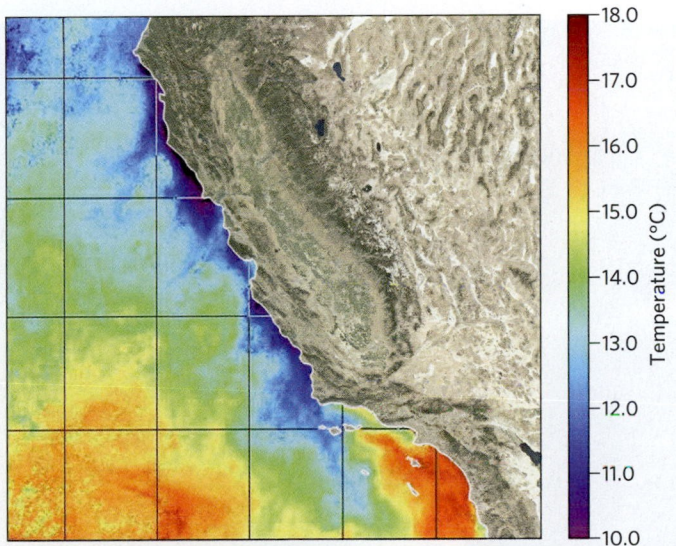

Temperature (°C)

18.0
17.0
16.0
15.0
14.0
13.0
12.0
11.0
10.0

(c) The variations in surface temperature of the eastern Pacific Ocean show upwelling of cold water along the west coast of the United States. The colors indicate water temperature.

cold water in polar regions sinks and flows along the bottom of the ocean as a deep-sea current, flowing back toward the equator at the ocean bottom. The combination of surface currents and thermohaline circulation, like a conveyor belt, moves water and heat throughout the various ocean basins.

Figure 15.21 The thermohaline circulation results in a global-scale conveyor belt that circulates water throughout the entire ocean system. Because of this circulation, the ocean mixes entirely in a 1,500-year period.

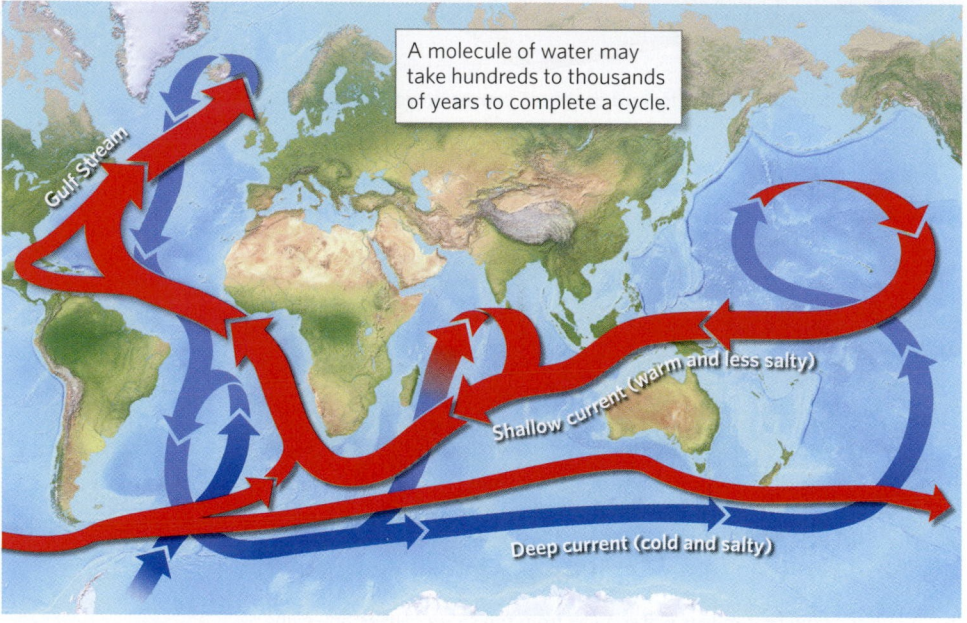

A molecule of water may take hundreds to thousands of years to complete a cycle.

Gulf Stream

Shallow current (warm and less salty)

Deep current (cold and salty)

Figure 15.22 Wind-driven waves provide a playground for surfers off Hawaii.

Take-home message . . .

Ocean water is in constant motion due to wind-driven surface currents, upwelling and downwelling, and the thermohaline circulation. The paths that surface currents follow depend on a combination of factors: wind directions, the Coriolis force and associated Ekman transport, the pressure-gradient force, and the shapes of ocean basins. Overall, the largest surface currents are gyres, which carry water around the periphery of the ocean basins.

Quick Question -
Why does downwelling of ocean water occur at polar latitudes and upwelling at the equator?

15.5 Wave Action

Why Do Wind-Driven Waves Form?

Waves—the periodic up-and-down motions of the ocean at a scale that generates distinct troughs and ridges visible to your eye—make the ocean surface a restless, ever-changing vista **(Fig. 15.22)**. The waves you see when you look out at the sea from the shore, a plane window, or a ship's deck are **wind-driven waves**. These waves form due to the interaction between moving air and the surface of the ocean. (*Tsunamis*, a type of wave caused by sudden displacement of water due to slip on a fault or a landslide, are relatively rare.)

How do wind-driven waves begin? To picture the process, imagine still water in a pond on a windless day. The horizontal surface of the water represents an *equilibrium level*—if you pushed the surface up or down with a paddle, it would return to that level because of gravity. At the surface, water molecules attract one another more strongly than they attract air molecules above. This attraction produces *surface tension*, which makes the water surface behave somewhat like an elastic sheet. When a breeze starts to blow, frictional drag causes this elastic surface to stretch. As soon as it stretches, elastic rebound causes it to twang back like a rubber band, and as a consequence, it wrinkles into small waves called **ripples (Fig. 15.23)**.

Once ripples exist, wind can push against their sides, causing them to build still higher. A ripple can evolve into a wave that lifts water well above the equilibrium level. When this happens, gravity acts on the elevated water in a wave, causing it to sink. Inertia causes the sinking water surface to descend below the equilibrium level **(Fig.15.24a)**. Next, the water surface bounces back up and rises above the equilibrium level, and the process repeats, like the up-and-down motion of a weight suspended from a vertical spring. You can see this phenomenon happening when you drop a pebble into a pond **(Fig. 15.24b)**.

A *wave train* propagates outward, across the water surface, and can move a long distance from the location of the disturbance. Wave trains in the ocean can move far from the location where they were generated by the wind.

Describing Waves and Wave Motion

We can use the same basic terminology for water waves that we use for any waves. The top of a wave is its *crest* and the base is its *trough* (Fig. 15.25a). The vertical distance between the crest and the trough is the *wave height*, and half the wave height represents the wave's *amplitude*. The horizontal distance between two successive troughs or two successive crests defines the *wavelength*. The number of wave crests (or troughs) that pass a point in a given time defines the *wave frequency*, and the time between the passage of two successive crests (or troughs) is the *wave period*. The horizontal velocity at which a crest or trough moves gives the *wave speed*.

When you watch a wave travel, you may get the impression that the whole mass of water constituting the wave moves with the wave. But drop a cork into the water, and you'll see that it bobs up and down and back and forth as a wave passes—it does not move along with the wave. That's because a particle of water within an ideal wave moves in a circle (see Fig. 15.25a). The diameter of the circle is greatest at the ocean's surface, where it equals the amplitude of the wave. With increasing depth, the diameter of the circle decreases until, at a depth equal to about half the wavelength—called the **wave base**—no wave movement takes place at all. Submarines traveling below the wave base enjoy smooth water while ships toss about on the sea surface above.

Although the motion of a water molecule in an ideal wave defines a circle, *wave drift*, an overall lateral motion accompanying the passage of a wave, occurs where waves develop due to the frictional drag of the wind (Fig.15.25b). Typically, wave drift equals about a quarter of the wavelength, meaning that water moves forward by a distance of about one wavelength after four waves have passed. It's because of wave drift that surface currents, as we discussed earlier, develop in the ocean. As we'll see in Chapter 16, the path of a water molecule in a wave changes as the wave approaches the shoreline and friction with the seafloor slows the base of the wave.

The Size of Waves

The size of waves generated by wind in the open ocean depends on the strength of the wind (how fast the air moves), on the **fetch** of the wind (the distance over which it blows), and on the length of time during which the wind blows (Fig. 15.26a). During strong winds, not all the energy of the wind goes into building waves; some blows the tops of the waves over, causing water to mix with air

Figure 15.23 Ripples are small waves that develop due to the elastic behavior of the ocean surface.

Figure 15.24 Development of wave trains.

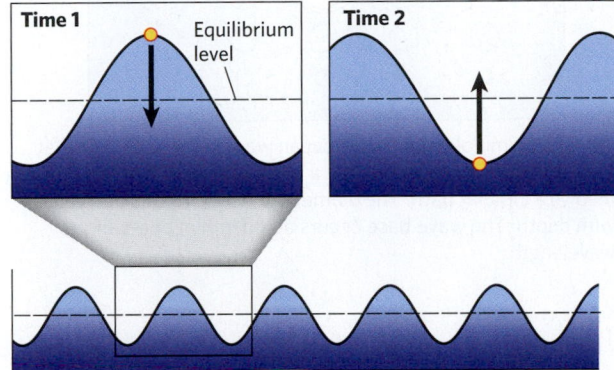

(a) Once a wave builds to a sufficient size, gravity acts as a restoring force that causes its elevated surface to sink below the equilibrium level, as shown here in cross section. Then the wave rebounds like a weight hanging from a string.

(b) An example of a wave train formed when a pebble displaces the surface of a pond.

Figure 15.25 The motion in ocean waves.

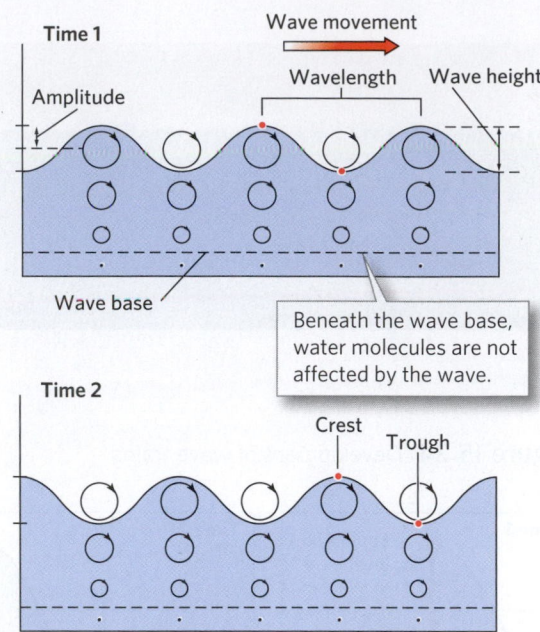

(a) The terminology used for ocean waves is the same as that used for seismic waves. Within a passing wave, a particle of water follows a circular path. The diameter of these circles decreases with depth. The wave base occurs at a depth of one-half the wavelength.

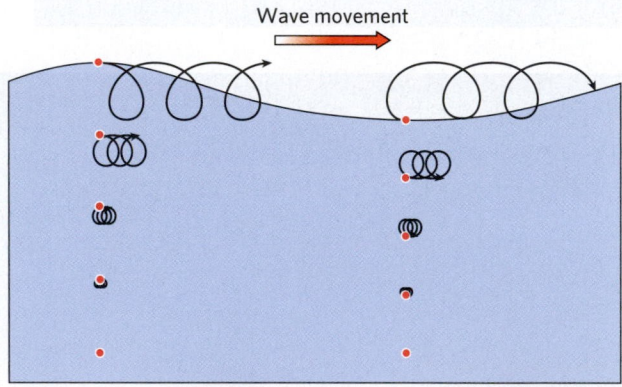

(b) Near the surface of the ocean, wind can cause wave drift. When this happens, particles of water follow a looping path.

Even though waves always move more slowly than the wind that generates them, wave speed depends on wind speed: faster wind causes faster waves. In the open ocean, waves typically travel between 10 km per hour and 50 km per hour (6 and 30 mph). Physicists have shown that the speed of waves depends on the wavelength of the waves: waves with longer wavelengths travel faster than waves with shorter wavelengths. When a storm blows over a region, it generates many different wave trains, each characterized by a different wavelength. Because the wave trains travel outward from their source at different speeds, when they are far from the source, they separate from one another, a phenomenon known as *wave dispersion* (**Fig. 15.26b**). As a result, a ship in a storm gets tossed about on a chaotic sea surface as it rides over wave trains with different wavelengths, whereas a ship that is far from the source may intersect regularly spaced waves. Waves that have traveled away from their source, and are not being driven by the local wind, are called **swells**; they may travel for thousands of kilometers across the ocean (**Fig. 15.26c**).

Particularly large waves may form by *constructive interference*, which takes place when two wind-driven waves moving from different directions come together in such a way that the wave crests overlap to form a single crest that is higher than that of either wave. During the 1979 Fastnet yacht race off Ireland, constructive interference produced waves large enough to capsize 23 out of the 300 participating sailboats. The waves developed when swells from an east-blowing gale collided with those generated by a west-blowing gale. Interactions of wind-driven waves with strong currents, or the focusing of waves into a small area by the shape of the coastline or seafloor, can also form very large waves.

Occasionally, these processes lead to the formation of **rogue waves**, defined as isolated waves that rise more than twice as high as most large waves passing a locality during a specified time interval. Long thought to exist only in the imaginations of sailors, rogue waves have now been documented numerous times. For example, during the 1990s, workers on oil platforms in the North Sea recorded almost 500 encounters with rogue waves—some of which were three to five times higher than other waves present at the time (**Fig. 15.27a**). The decks of large ships have been swamped by rogue waves in the open ocean (**Fig. 15.27b**). The largest wave known to have encountered a ship reached a height of 34 m (112 feet). Mysterious losses of ships in the open ocean may be a consequence of rogue waves. If a rogue wave reaches the shore, it can wash unsuspecting bystanders off piers or beaches.

Did you ever wonder . . .

how big an ocean wave can get?

and resulting in *whitecaps*. How large can wind-driven waves get in the open ocean? With continued blowing over a long fetch, waves with amplitudes of 2 to 10 m (6 to 30 feet) and wavelengths of 40 to 500 m (130 to 1,600 feet) can build. Oceanographers calculate that if hurricane-strength winds were to blow across the width of the Pacific for 24 hours, 15- to 20-m (50- to 70-foot) waves would develop.

Figure 15.26 Waves build progressively as the wind continues to blow.

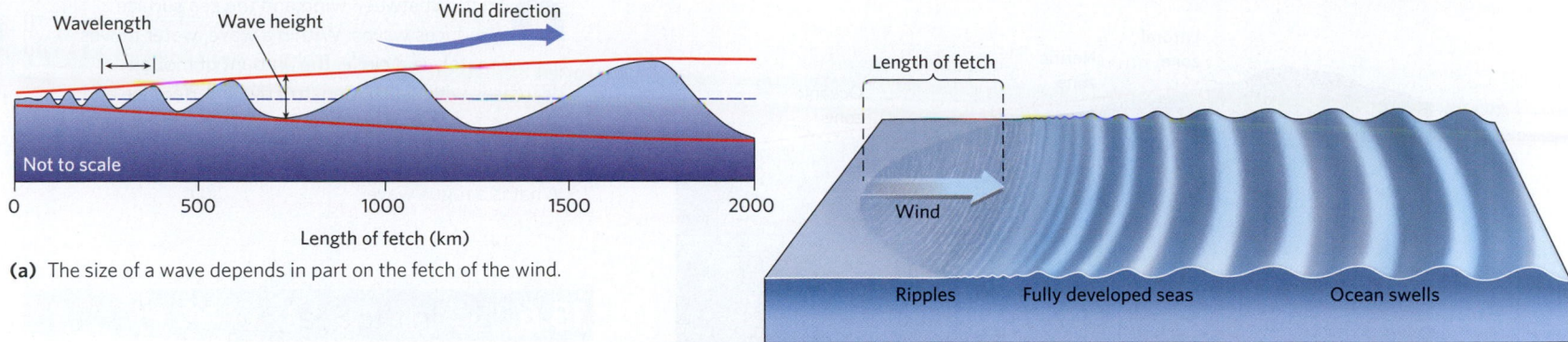

(a) The size of a wave depends in part on the fetch of the wind.

(b) Many waves are generated where the wind blows. They disperse into wave trains of different wavelengths.

(c) An example of swells in the ocean. Long-wavelength waves can travel over long distances as swells.

Figure 15.27 Rogue waves.

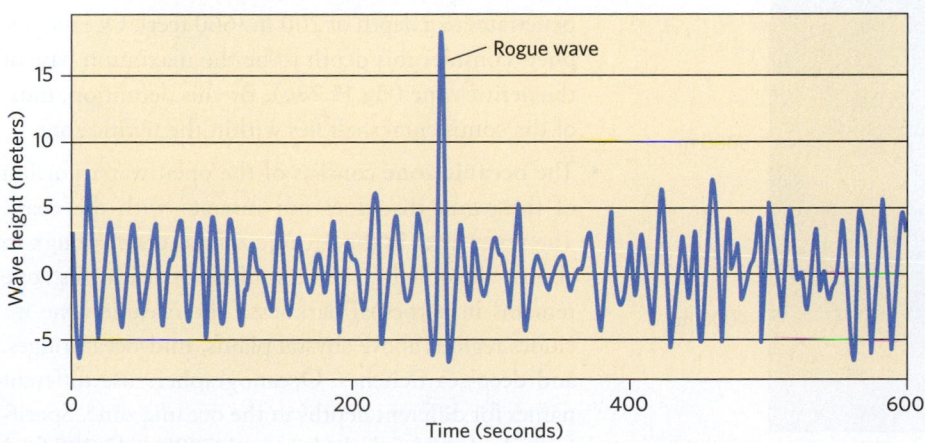

(a) A recording of waves in the North Sea over 10 minutes. Note that most waves are about 5 m (16 feet) high. The rogue wave is 18 m (60 feet) high.

(b) A rogue wave washing over the deck of a large ship.

Figure 15.28 Zones of the ocean.

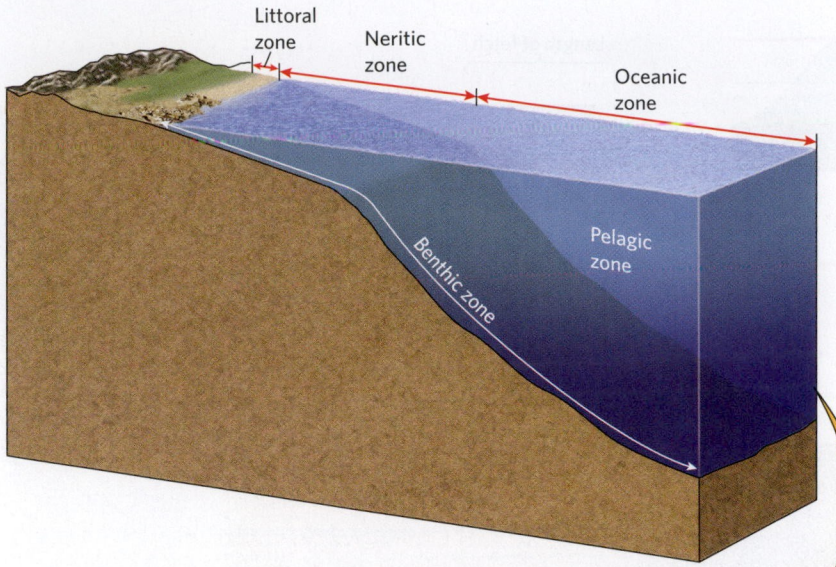

Littoral zone · Neritic zone · Oceanic zone

Benthic zone

Pelagic zone

(a) Oceanographers distinguish among littoral, neritic, and oceanic zones, based on distance from shore, and between pelagic and benthic zones, based on depth.

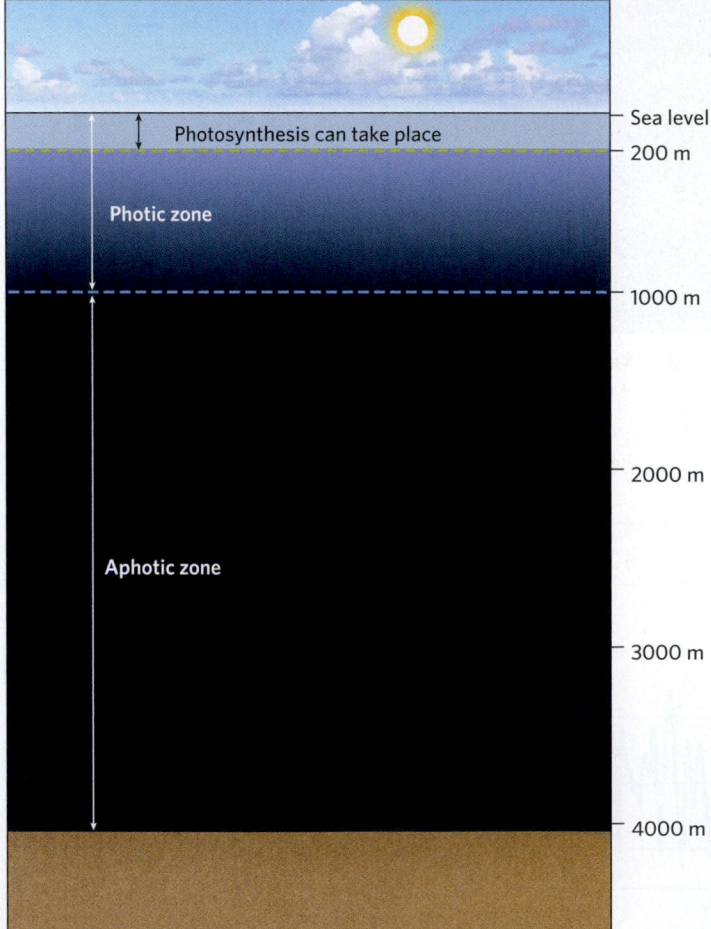

Photosynthesis can take place

Sea level
200 m

Photic zone

1000 m

2000 m

Aphotic zone

3000 m

4000 m

(b) Oceanographers distinguish between photic and aphotic zones based on light penetration.

15.6 Life in the Sea

Oceanic Zones and Their Effect on Life

The character of the ocean changes with distance from the shore. When discussing these differences, oceanographers divide the ocean into three zones (**Fig. 15.28a; Earth Science at a Glance,** pp. 538–539), starting from the shore and moving out:

- The **littoral zone** consists of the intertidal zone (the area that is above water during low tide and under water during high tide) and the adjacent areas of coastal land that become submerged or drenched in spray during unusually high tides or fierce storms. Specialized organisms have evolved to live in these conditions.

- The **neritic zone** is the relatively nearshore, shallow realm of the ocean. Water in the neritic zone receives inputs of chemicals and sediments from the land, and within the neritic zone, sunlight penetrates to the seafloor. Light undergoes *scattering* (absorption by atoms, followed immediately by reemission in a different direction) when it passes through water. Most of the scattered light heads up or sideways, so the intensity of light decreases progressively with depth (**Box 15.4**). So even in clear water, only 1% of sunlight penetrates to a depth of 200 m (660 feet). Oceanographers consider this depth to be the maximum base of the neritic zone (**Fig. 15.28b**). By this definition, most of the continental shelf lies within the neritic zone.

- The **oceanic zone** consists of the open-water portion of the ocean that does not interact with the coast. The water layer in the oceanic zone is deep enough to absorb all sunlight, so the seafloor beneath this zone remains in perpetual darkness. The oceanic zone includes regions above abyssal plains, mid-ocean ridges, and deep-sea trenches. Oceanographers use different names for different depths in the oceanic zone. Specifically, bathyal depths lie between 1,000 m (3,300 feet) and 4,000 m (13,000 feet), whereas abyssal depths lie between 4,000 m and 6,000 m (20,000 feet).

Box 15.4 ▶ Consider this . . .

The color of ocean water

If you've looked at photos of the ocean, had the opportunity to walk along its shores, or sailed across it, you may have noticed that the ocean's color can vary greatly. It can be deep blue, turquoise, dark green, or even gray **(Fig. Bx15.4a–c)**. And when churned to a froth by wind or waves, it becomes white. What gives water its color, and why does its color vary so much?

Clean water in a clear glass looks colorless—you can see through it almost as easily as you can see through the glass itself. That's because you're looking through only about 8 cm (3 inches) of water. When you look at the ocean, you are looking at water that can be meters to kilometers deep. Under a sunny sky, this water looks blue, and the deeper the water, the bluer it looks. You see this color for two reasons. First, water molecules absorb red, green, and yellow light but scatter blue light. Second, when you look at the sea beneath a blue sky, you are seeing the reflection of the blue sky by the water surface. On an overcast day, the water might look greenish gray.

Where water is shallow enough for substantial light to penetrate to the seafloor, the color of the water depends in part on the color of the seafloor: water over white sand will appear lighter than water over dark seaweed. The presence of tiny suspended particles in water also affects its color. For example, the presence of abundant phytoplankton, which contain green chlorophyll, can make the water greenish; the presence of abundant plankton with white calcite shells tends to make water turquoise because the shells reflect light and brighten the water; and the presence of suspended clay can make water a brownish color.

Because water scatters different wavelengths of light by different amounts, the colors that you see reflected from objects in the water change with depth. For example, the color of a fish as seen by a scuba diver at a depth of 20 m (60 feet) isn't the same as the color of the same fish at the surface.

Figure Bx15.4 Colors of the ocean.

(a) On a sunny day, water off a Caribbean island looks blue. The darker-colored water is deeper.

(b) On an overcast day, the ocean water off Cuba looks gray.

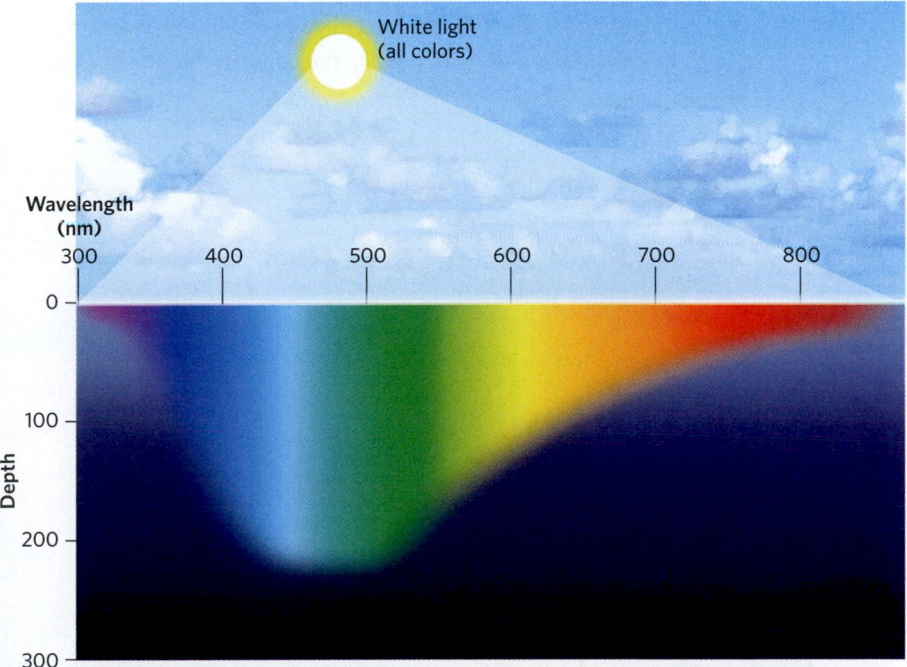

(c) Water absorbs most other colors of light more efficiently than it absorbs blue light. Blue light, therefore, penetrates deeper. In clear water, light can penetrate to a depth of about 200 m. At noon in extremely clear water, it may penetrate deeper.

The Water of the Sea

Our planet's oceans host a variety of realms that differ in temperature, salinity, and brightness. The water stays in motion constantly, at a range of rates.

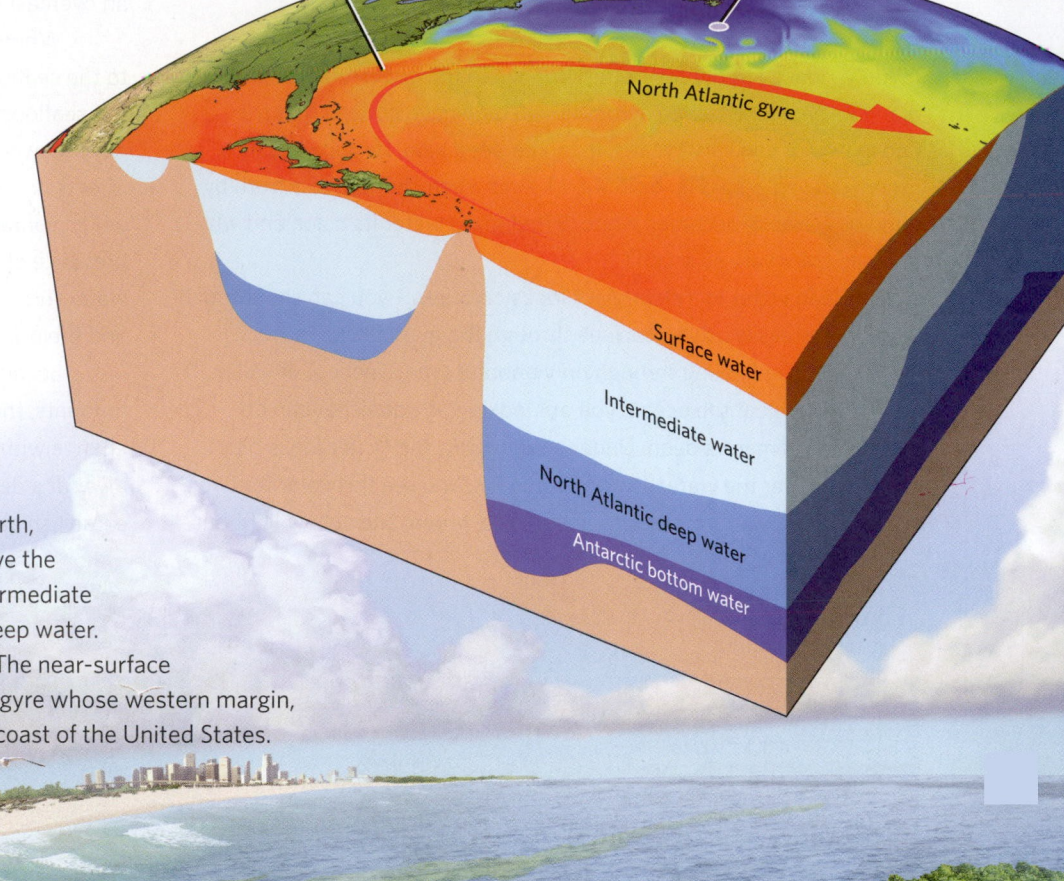

Gulf Stream

Titanic wreck site

North Atlantic gyre

Surface water

Intermediate water

North Atlantic deep water

Antarctic bottom water

Water masses and currents

Oceanographers divide the ocean, at the scale of a whole ocean basin, into different layers, or water masses. In the Atlantic Ocean, for example, the Antarctic bottom water, a cold and saline mass, spreads northward from the southern continent. North Atlantic deep water sinks to depth in the polar regions of the north, and spreads southward to form a layer above the Antarctic bottom water. North Atlantic intermediate water, in turn, overlies the North Atlantic deep water. Relatively warm water forms the top layer. The near-surface portion of the top layer circulates in a huge gyre whose western margin, the Gulf Stream, flows northeast along the coast of the United States.

Seaweed

Coral reef

Benthon

The near-shore neritic zone

The neritic zone of the ocean, the portion where the sea floor receives at least some light, encompasses most of the continental shelf. In its near-shore portion, it hosts a diversity of organisms including coral reefs, seaweed, and a variety of nekton, plankton, and benthon.

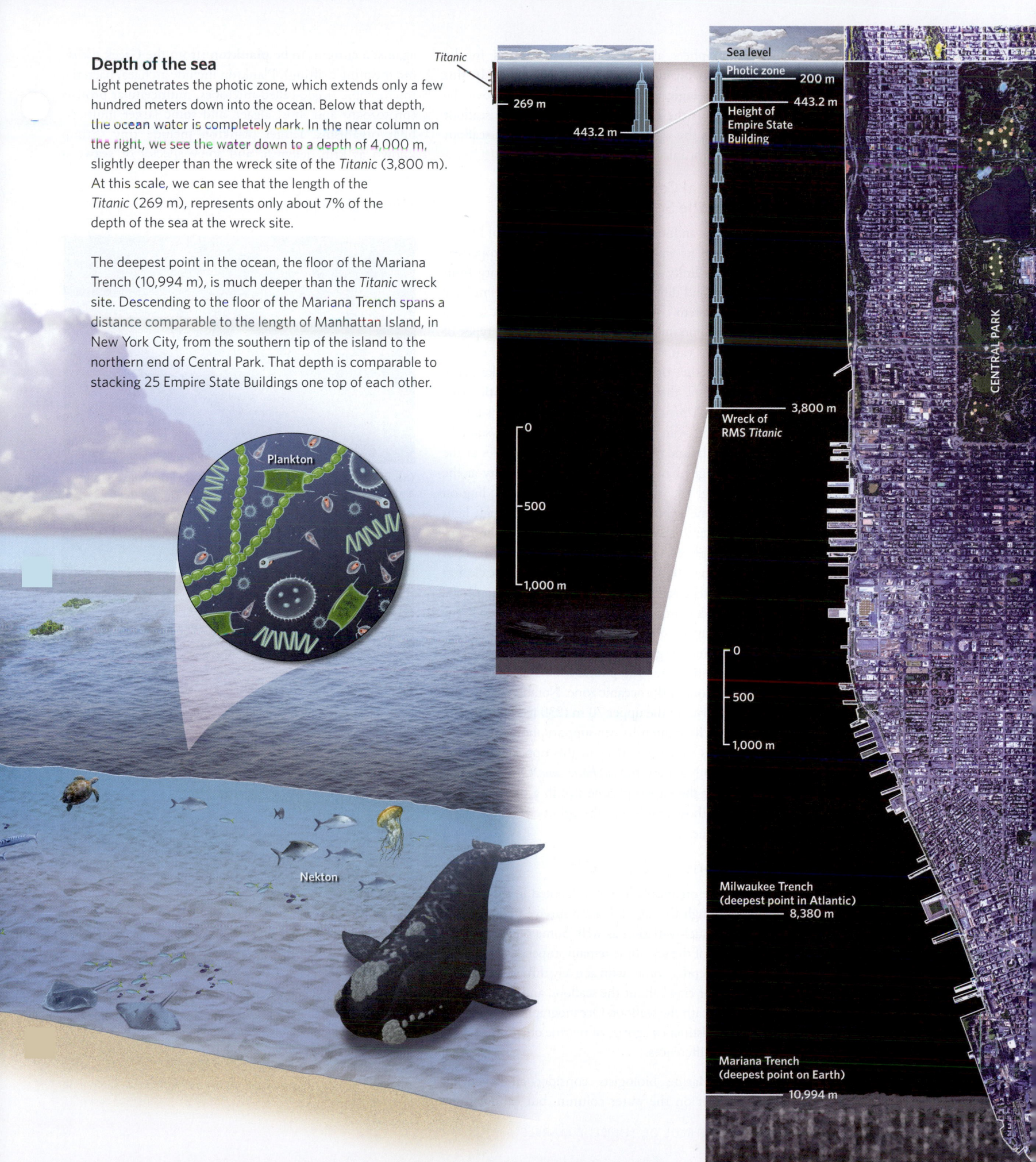

Depth of the sea

Light penetrates the photic zone, which extends only a few hundred meters down into the ocean. Below that depth, the ocean water is completely dark. In the near column on the right, we see the water down to a depth of 4,000 m, slightly deeper than the wreck site of the *Titanic* (3,800 m). At this scale, we can see that the length of the *Titanic* (269 m), represents only about 7% of the depth of the sea at the wreck site.

The deepest point in the ocean, the floor of the Mariana Trench (10,994 m), is much deeper than the *Titanic* wreck site. Descending to the floor of the Mariana Trench spans a distance comparable to the length of Manhattan Island, in New York City, from the southern tip of the island to the northern end of Central Park. That depth is comparable to stacking 25 Empire State Buildings one top of each other.

Titanic

269 m

443.2 m

Plankton

Nekton

0

500

1,000 m

Sea level

Photic zone

200 m

443.2 m

Height of Empire State Building

3,800 m

Wreck of RMS *Titanic*

CENTRAL PARK

0

500

1,000 m

Milwaukee Trench (deepest point in Atlantic)
8,380 m

Mariana Trench (deepest point on Earth)
10,994 m

Oceanographers also describe depth zones in the ocean based on position relative to the seafloor. In this context, the **pelagic zone** includes all water above the seafloor, whereas the **benthic zone** includes the seafloor itself, as well as the water immediately above the seafloor and the sediment immediately below.

Environmental Controls on Living in the Sea

The ocean hosts an incredibly diverse assemblage of life because not all of its water has exactly the same properties. Temperature, salinity, and light penetration all vary from place to place, and different assemblages of organisms have adapted to different conditions. Light penetration serves a particularly important role in determining what types of living organisms can thrive because the availability of light determines whether or not photosynthesis can take place. Marine biologists, therefore, distinguish the **photic zone** (from the Greek *phot*, meaning light), in which sunlight penetrates water, from the **aphotic zone**, in which sunlight does not. The boundary between the two zones lies at the depth where light has less than 1% of its intensity at the surface (see Fig. 15.28b). This depth varies depending on the clarity of the water. In very clear water, enough light penetrates down to about 200 m (660 feet) that photosynthesis can take place. A tiny bit of light may reach a depth of 600 to 1,000 m (2,000 to 3,300 feet) at noon in the clearest water, so this represents the maximum base of the photic zone. If water contains suspended particles, the base of the photic zone may lie at depths much shallower than 200 m.

By the definitions given earlier, all of the neritic zone lies within the photic zone. The photic zone also includes the shallower portions of the oceanic zone. Notably, about 90% of marine life lives in the upper 70 m (230 feet) of the photic zone, where light intensity can support photosynthesis—oceanographers sometimes refer to this upper, brightly lit part of the photic zone as the *euphotic zone*. Organisms that navigate below the euphotic zone live in a dimly lit environment, and those that live in the aphotic zone survive in perpetual darkness.

The Categories of Ocean Life

On land, some organisms live on or rooted in the ground, others fly through the air, and some survive underground. The sea hosts such variation as well. Some organisms float on the surface of the sea, some remain suspended in the water at various depths, some swim actively through the water, some root in or crawl about the seafloor, and some exist in sediments beneath the seafloor. Oceanographers distinguish among three distinct categories of marine organisms to emphasize these differences.

PLANKTON. Marine biologists consider all organisms that float in or on the water column, but cannot move against a current, to be **plankton** (from the Greek *planktos*, meaning drifting). Plankton include archaea, bacteria, microscopic single-celled microbes, multicellular protists (protozoans and most algae), and some larger multicellular organisms. Larger plankton include tiny shrimp known as krill, jellyfish, and some species of **seaweed**

Figure 15.29 Examples of marine plankton.

(a) Phytoplankton, which contain chlorophyll, include diatoms and coccolithophores.

(b) Zooplankton include tiny shrimp, forams, and radiolarians.

(c) The intricate calcareous plates of coccolithophore shells look like wheels.

(non-microscopic multicellular algae). Taken together, plankton account for most of the ocean's living organic matter, or *biomass*.

Most plankton species have no means of moving and consist of immobile living masses that simply float in or on the water. But some can move, though not quickly enough to overcome currents. Examples of moving plankton include *dinoflagellates*, which have whip-like tails; krill, whose tiny legs propel them through water, and jellyfish, which move by making their whole bodies pulsate. Many planktonic organisms are heavier than water, so if the water were perfectly still, they would sink to the bottom. Fortunately for them, currents or local turbulence keep them suspended, just as a breeze keeps dust suspended in air. Some plankton maintain their buoyancy by incorporating gas into their bodies.

Marine biologists distinguish between **phytoplankton**, which carry out photosynthesis to sustain their life processes **(Fig. 15.29a)**, and **zooplankton**, which survive by ingesting other organisms, living or dead **(Fig. 15.29b)**. Many types of phytoplankton have tiny shells. For example, *diatoms* make shells of silica, and *coccolithophores* make shells composed of intricate calcareous (containing calcium carbonate) plates **(Fig. 15.29c)**. Because they are photosynthetic, phytoplankton are considered to be a type of algae.

NEKTON. Organisms that actively swim in the open water of the ocean and can move against a current are known as **nekton** (from the Greek *nektos*, meaning swimming). Some nekton migrate vast distances across the pelagic zone, while others move about a local coastal area of only a few square meters, such as a patch of reef. Most nekton can survive only within limited ranges of such physical conditions as temperature, pressure, and salinity, so they don't stray beyond a relatively small region. Nekton survival also depends on the availability of food, so they often migrate with their food supplies.

Nekton include a great variety of organisms whose sizes range from nearly microscopic to gigantic. Examples of nekton include invertebrates such as squid and octopuses **(Fig. 15.30a)**; swimming reptiles such as sea turtles and sea snakes; all varieties of fish; and marine mammals such as whales, porpoises, and seals **(Fig. 15.30b)**. The largest nekton, the blue whale, reaches a length of 32 m (110 feet) making it the largest animal ever to inhabit our planet. Many kinds of seabirds (such as penguins) can swim as skillfully as fish; marine biologists sometimes consider swimming birds to be a type of nekton.

The largest category of nekton, fish, includes two principal groups. *Bony fish* have a skeleton with a skull, jaws, spine, and many bones protruding from the spine. Examples include familiar species such as tuna, swordfish, cod, and salmon **(Fig. 15.31a)**. Most bony fish have gas-filled swim bladders that allow them to regulate their buoyancy

Figure 15.30 Examples of nekton.

(a) A squid, a type of marine invertebrate, propels itself by ejecting jets of water.

(b) Whales, a type of marine mammal, can grow to become larger than the largest dinosaur.

and remain at a given depth without swimming. Bony fish lay eggs that are fertilized and hatch outside the mother's body. The *non-bony fish*, such as sharks and rays, have cartilaginous skeletons, meaning that their skeletons consist of strong but flexible connective tissue **(Fig. 15.31b)**. Non-bony fish do not have swim bladders, so they must actively swim in order to maintain their depth.

BENTHOS. Organisms that live on, just above, or just below the seafloor are known as **benthos** (from the Greek *benthos*, meaning bottom). The seafloor, as we have seen, ranges from fairly shallow, in the littoral and neritic zones, to moderate depths, above the continental slope and rise, to great depths, on abyssal plains or the floor of deep-sea trenches. The nature of benthos varies radically among these depths. Coastal benthic communities include a variety of animals such as corals, sponges, and bryozoans,

Figure 15.31 Examples of fish.

(a) The bluefin tuna is a type of bony fish.

(b) A shark is an example of a non-bony fish.

which build spectacular reefs **(Fig. 15.32a)**. Even in the great depths of the oceans, where no light penetrates, organisms such as snails, crabs, starfish, and mollusks creep through the mud **(Fig. 15.32b)**. Benthos also include a great variety of seaweed types that anchor to the seafloor. The largest, giant kelp, grows stalks as long as 200 m (600 feet) **(Fig. 15.32c)**. Some seaweed can survive in the intertidal zone **(Fig. 15.32d)**.

Food Webs in the Oceans

Who eats whom in the ocean? The answer to this question defines the ocean's **food web,** the group of organisms associated by feeding relationships. In the food web, organisms that survive by extracting nutrients from nonliving energy sources, and thus bring energy into the web, are known as **primary producers,** or *autotrophs*. In the euphotic zone, phytoplankton serve as the primary producers—these organisms obtain their energy from sunlight. Subsurface exploration in the past few decades has revealed complex benthic food webs surrounding deep-sea hydrothermal vents (black smokers). In these aphotic food webs, the primary producers consist of microbes that use the chemical bonds in sulfide minerals as their energy source. **Consumers,** or *heterotrophs*, in the food web include a vast variety of organisms, ranging from tiny zooplankton to fearsome killer whales.

Nutrients and Dissolved Oxygen

In order to survive, primary producers ingest **nutrients,** elements that an organism uses to construct chemicals and carry out metabolic reactions. Nutrients of particular importance to marine organisms include phosphate (PO_4^{3-}), nitrate (NO_3^-), and iron. The presence or absence of nutrients determines the biomass that a volume of water can support at a given latitude.

Where do nutrients come from? Organisms accumulate nutrients during their lifetimes, and when they die and sink, they carry those nutrients with them to the seafloor, where decomposition releases them back into the water. So, unless new nutrients flow to the sea from rivers or coastal runoff, or rise from the depths back to the ocean surface by upwelling, the concentration of nutrients in the photic zone may become too low for life to thrive. Therefore, if we map the distribution of photosynthetic organisms by means of satellite detection of chlorophyll concentrations at the ocean surface **(Fig. 15.33)**, we see that photosynthetic organisms increase in abundance near the shore, where nutrient sources enter the sea from land, and in offshore localities where upwelling brings nutrients from the aphotic zone back to the photic zone.

Consumers such as zooplankton and marine animals cannot survive without molecular oxygen, which they use for respiration. On land, organisms inhale oxygen in gaseous form from the air. Marine reptiles, birds, and mammals must surface to inhale air, but have adaptations that allow them to hold their breath and stay underwater for relatively long periods—in fact, a sperm whale can stay submerged for an amazing 90 minutes, and during this time can descend to depths of as much as 3 km (9,800 feet). Fish and marine invertebrates never need to surface, although they do require oxygen. They survive by extracting *dissolved oxygen* from water passing through their gills. Typically, dissolved oxygen occurs only in low concentrations, so the gills of marine organisms must be efficient at extracting it from water.

Where does the dissolved oxygen in ocean water come from? Some enters the ocean at the water surface and then diffuses downward, and some comes from algae as a by-product of photosynthesis. Animals remove some dissolved oxygen from the water by breathing, but most removal takes place as a result of decomposition.

Figure 15.32 Examples of benthos.

(a) In the neritic zone, benthos include many varieties of coral and other reef-building organisms.

(b) Organisms, such as this sea cucumber, move through the clay in the aphotic zone of the ocean, leaving tracks across the ocean floor.

(c) Kelp, a type of seaweed, can grow to be hundreds of meters long.

Because decomposition reactions remove oxygen from seawater, an *algal bloom* (a rapid growth of lots of algae) can remove so much oxygen that it makes the water *anoxic* (oxygen free). This may seem counterintuitive, but decomposition of the algae, along with respiration by zooplankton feeding on the algae, removes more oxygen than the algae can produce by photosynthesis. Such anoxic conditions yield a **dead zone** in the water, where fish and shellfish die from lack of oxygen. A large dead zone has developed at the mouth of the Mississippi, where the river dumps vast amounts of nutrients into the Gulf of Mexico. These nutrients, which come from runoff from Midwestern farm fields, feed algal blooms that strip the oxygen from the water.

Clearly, interactions between the sea and the land affects oceans significantly. In the next chapter, we turn our focus to the coast, the boundary where these interactions take place.

Take-home message...

Most marine life lives in the euphotic zone, where sunlight intensity is sufficient to support photosynthesis. Marine organisms can be categorized as plankton, nekton, or benthos, depending on where in the ocean they live and what type of movement they are capable of. Marine life plays an active role in cycling nutrients through the Earth System.

Quick Question --------------
Does classification of an organism as plankton depend on its size?

(d) Intertidal seaweed can survive exposure to air.

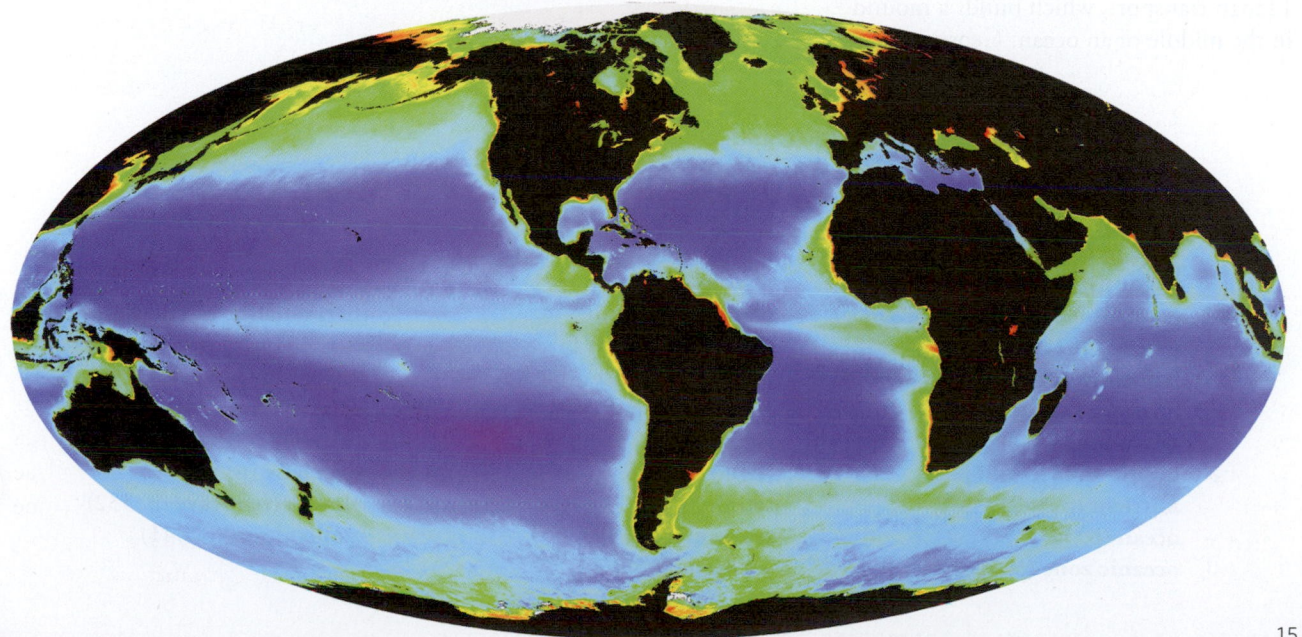

Figure 15.33 The colors on this map represent concentrations of chlorophyll in near-surface seawater. Chlorophyll serves as an indicator of primary production in the ocean's food web.

15 CHAPTER REVIEW

- Mapping of the ocean began in the 15th and 16th centuries, but maps could not be accurate until longitude could be measured. Though they are separated into distinct oceans around the planet, water flows among all the world's oceans.

- Mean sea level represents the elevation of the boundary between ocean water and air averaged over a year. It changes over geologic time by as much as a few hundred meters. The volume of the Earth's surface water stays fairly constant over geologic time, so changes in sea level reflect changes in the volume of continental glaciers and mid-ocean ridges.

- Ocean water originally came from water vapor that bubbled out of volcanoes.

- Most salt in the sea consists of NaCl, but seawater contains other salts as well. The salinity of seawater averages about 3.5%, but varies with location and depth.

- The salt in seawater is carried to the sea by rivers, and to a lesser extent comes from hydrothermal vents. Evaporation of seawater removes water and leaves behind salt.

- Sea-surface temperatures depend on the amount of insolation, so the ocean surface is warmer at the equator than at the poles. Because warm water is less dense than cold water, it floats, so at low latitudes, a thermocline separates warm surface water from cool deep water.

- Variations in salinity and temperature cause variations in water density in the oceans. These variations lead to stratification of the oceans.

- Water in the oceans circulates in currents. Surface currents are driven by frictional drag applied by the wind, but their paths are affected by the Coriolis force and the pressure-gradient force.

- The Coriolis force causes Ekman transport, which builds a mound of slightly elevated water in the middle of an ocean. Geostrophic

balance develops between the resulting outward-directed pressure-gradient force and the inward-directed Coriolis force. As a result, large gyres (geostrophic currents) circle these mounds.

- Smaller currents are, effectively, eddies that form where a larger current shears against nonflowing water or against land.

- Upwelling and downwelling due to density variations leads to thermohaline circulation and the formation of deep currents.

- Waves initiate due to frictional drag where the wind shears across the surface of the ocean. Once waves become large enough, gravity causes the up-and-down movement of the water surface.

- Ideally, water particles move in a circle as a wave passes. The circle gets smaller with depth, so that below the wave base, there is no water movement. In reality, the upper layer of water drifts in the direction of the wind.

- Oceanographers divide the ocean into littoral, neritic, and oceanic zones based on proximity to shore. They also distinguish among pelagic and benthic zones, based on proximity to the seafloor, and photic and aphotic zones, based on light penetration.

- Life in the sea includes floating organisms (plankton), swimming organisms (nekton), and seafloor-dwelling organisms (benthos). Light penetration controls life distribution.

- Phytoplankton serve as the primary producers in the oceanic food web and account for most biomass. The distribution of nutrients affects the abundance of phytoplankton.

- Consumers, organisms that survive by eating other organisms, need dissolved oxygen to survive. Algal blooms can remove dissolved oxygen and produce a dead zone in the ocean.

Key Terms

aphotic zone (p. 540)
benthic zone (p. 540)
benthos (p. 541)
Consumers (p. 542)
Coriolis effect (p. 523)
Coriolis force (p. 525)
current (p. 520)
dead zone (p. 543)
deep-sea currents (p. 520)
desalinization (p. 518)
downwelling (p. 529)
eddy (p. 522)
Ekman transport (p. 528)

estuary (p. 517)
fetch (p. 533)
food web (p. 542)
frictional drag (p. 522)
geostrophic current (p. 529)
gyre (p. 521)
halocline (p. 517)
littoral zone (p. 536)
nekton (p. 541)
neritic zone (p. 536)
nutrient (p. 542)
ocean (p. 511)
oceanic zone (p. 536)

oceanography (p. 513)
pelagic zone (p. 540)
photic zone (p. 540)
phytoplankton (p. 541)
plankton (p. 540)
pressure-gradient force (p. 528)
primary producer (p. 542)
pycnocline (p. 519)
ripple (p. 532)
rogue wave (p. 534)
salinity (p. 517)
sea level (p. 515)
seaweed (p. 540)

shoreline (p. 513)
surface current (p. 520)
swell (p. 534)
thermocline (p. 519)
thermohaline circulation (p. 531)
upwelling (p. 529)
water mass (p. 519)
wave (p. 532)
wave base (p. 533)
wind-driven wave (p. 532)
zooplankton (p. 541)

Review Questions

The letters following each Review Question refer to the corresponding Learning Objective from the Chapter Opener.

1. How much of the Earth's surface do the oceans cover? When were the oceans mapped? **(A)**

2. What do we mean when we refer to sea level, and how has sea level changed over Earth history? **(B)**

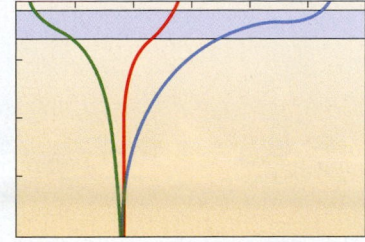

3. Where does the salt in the ocean come from? How does ocean salinity vary with location, and why? Which line in the graph represents the tropics? **(B)**

4. How does the temperature of ocean water vary with latitude and depth, and why? Why does a thermocline develop at the equator, but not at the poles? **(C)**

5. What factors determine ocean-water density? Is the ocean homogeneous or stratified? Explain. **(E)**

6. What forces play a role in establishing surface currents in the ocean? Why do gyres and eddies form? Why does garbage accumulate in the center of the Pacific Ocean? **(D)**

7. Explain the thermohaline circulation and its role in producing deep-sea currents. **(E)**

8. What forces contribute to the formation of ocean waves? Describe the motion of water molecules in a wave. What's the difference between a ripple and a swell? How large can waves become? **(F)**

9. Explain the differences between the littoral, neritic, and oceanic zones. What is the relationship between the neritic and the photic zones? **(G)**

10. Explain the differences between plankton, nekton, and benthon. Which is depicted in the figure? **(G)**

11. Describe the basic components of the oceanic food web and how light penetration affects it. **(G)**

On Further Thought

12. Columbus didn't actually reach continental North America when he sailed west in 1492. Rather, he reached the Bahamas, to the southeast of Florida, about 70 days after leaving Spain, after traveling about 7,000 km (4,300 miles). Why the Bahamas? What was his average speed? How long might the journey have taken if he had just drifted, without sails? **(D)**

13. Compare the water of the Persian Gulf, a confined sea at low altitude, to the water of the northernmost Pacific Ocean, an open ocean at high latitude. **(C)**

Online Resources

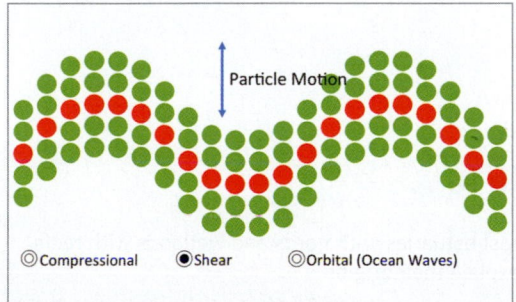

Animations
This chapter features an animation on ocean wave motion.

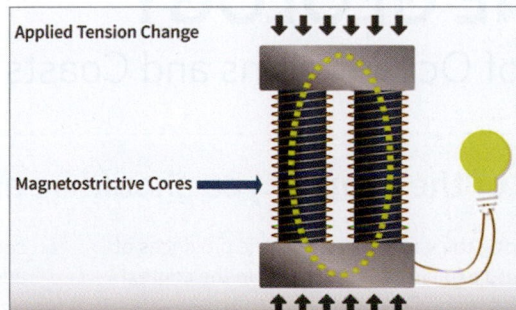

Videos
This chapter features real-world videos on the science of currents, harnessing wave energy to power coastal communities, and other topics.

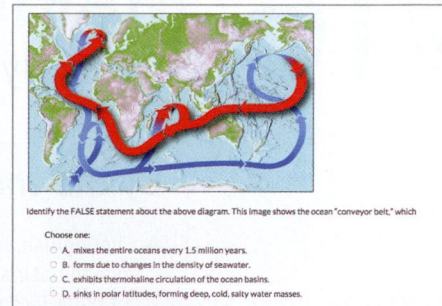

Smartwork5
This chapter features questions on ocean composition and zones, sea level variation, currents and waves, and ocean life.

16 MARINE GEOLOGY
The Study of Ocean Basins and Coasts

By the end of the chapter you should be able to . . .

A. distinguish among the various bathymetric provinces of ocean basins, and explain how they form in the context of plate tectonics theory.

B. discuss ocean tides and their causes, and interpret why some regions have large tidal reaches while others do not.

C. describe the processes that shift sand on beaches and produce observed beach profiles.

D. develop a model explaining the evolution of rocky coasts.

E. contrast estuaries with fjords and wetlands with reefs, and explain their origins.

F. explain the variation in the character of coastlines on the Earth, and the role of relative sea-level change in coastline evolution.

G. describe how wave erosion, and human responses to it, can modify coasts.

16.1 Introduction

A thousand kilometers from the nearest land, two scientists and a pilot wriggle through the entry hatch of the research submersible *Alvin*, ready for a cruise to the floor of the ocean and, hopefully, back. *Alvin* consists of a super-strong metal sphere embedded in a cigar-shaped tube **(Fig. 16.1)**. The sphere protects its crew from the immense water pressures of the deep ocean, and the tube holds motors and oxygen tanks. A cruise begins when, after sealing the hatch, the pilot releases cables tethering *Alvin* to its mother ship, and the submersible sinks like a stone. Most of this journey takes place in utter darkness, for the photic zone extends down only a couple of hundred meters (see Section 15.6). On reaching the seafloor, at a depth of 4.5 km (2.8 miles), the cramped explorers turn on spotlights and gaze at a stark vista of loose sediment, black rock, and an occasional sea creature. For the next 5 hours they take photographs and use a robotic arm to collect samples. When finished, they release some ballast, and *Alvin* rises like a bubble, reaching the surface about 2 hours later.

In the 21st century, human-piloted submersibles like *Alvin* are but one of many tools employed to collect data about the 70.8% of the Earth's surface that lies hidden beneath seawater. Orbiting satellites can characterize regional-scale **bathymetry** (the shape of the seafloor) in general **(Fig. 16.2)**, and shipborne sonar can define local-scale bathymetry in detail. Instruments towed by ships, and robotic submersibles that can cruise at great depths without risking lives, can study physical characteristics of the seafloor, such as variations in its temperature or magnetic field strength. Shipborne tools can

The rocky coast of Brittany, France, highlights the splendor of coasts, where the land meets the sea.

Figure 16.1 *Alvin* is a workhorse submersible for exploring the seafloor.

(a) *Alvin* is lowered into the sea from its mother ship, the *Atlantis*.

(b) The small porthole provides the only view out. The sled in front carries various instruments.

also yield *seismic-reflection profiles*, cross-sectional images that reveal the configuration of layering under the seafloor **(Fig. 16.3)**. By dredging (pulling a chain net along the seafloor) and by coring (plunging hollow tubes into the seafloor), researchers can collect samples of seafloor sediment. And by using *drilling ships* that are capable of

boring holes, researchers have recovered oceanic crustal rocks from depths of up to 4 km (2.5 miles) below the seafloor **(Fig. 16.4)**. Geologists use many of the same tools to study the **coast** (the shore and the adjacent areas of land and sea). But on coasts, they can also make direct field observations. The results of all this research—from land, sea, and sky—provide the knowledge base of **marine geology**, the study of the ocean's floor and margins.

In this chapter, we explore the results of modern marine geology research. We first survey the fundamental features of ocean basins and discuss how these features form. Then we focus on the landforms and habitats that develop along the coast, where over 60% of the global human population lives today. We'll see how coasts interact

Figure 16.2 Bathymetry of the seafloor.

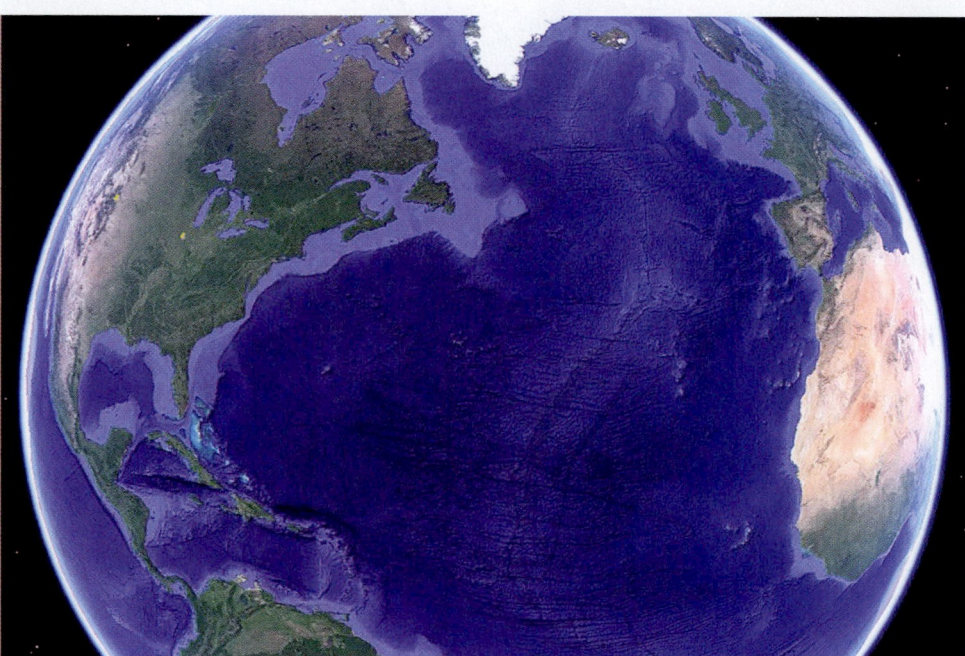

(a) Regional bathymetry of the North Atlantic. The data used to make this *Google Earth*™ image came from satellite measurements of variations in the strength of the Earth's gravitational pull.

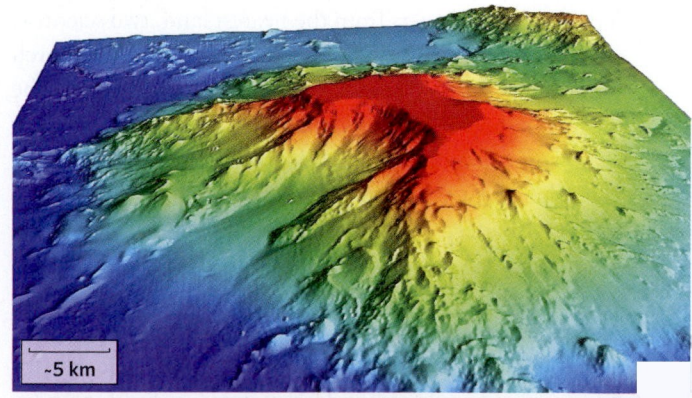

~5 km

(b) Detailed bathymetry can be produced by sonar. This oblique view shows the Turnif Seamount, northwest of Hawaii. The colors indicate depth. Red areas are shallower; blue areas are deeper.

with tides and waves, and how they evolve over time in response to geologic phenomena and human activities.

16.2 What Controls the Depth of the Sea?

If the Earth's crust were at the same elevation everywhere, the water currently held in the oceans would cover the surface of the Earth uniformly to a depth of about 2.5 km (1.5 miles). But the elevation of the Earth's surface is not uniform; rather, it consists of higher areas (the continents) and lower areas (the **ocean basins**) that differ in elevation by an average of 4.5 km (2.8 miles) **(Fig. 16.5a)**. Due to gravity, water flows downslope, so it drains from continents into the ocean basins to fill the oceans, leaving the surface of continents as dry land.

Why do distinct ocean basins exist? Recall from Chapter 2 that oceanic lithosphere and continental lithosphere differ markedly in terms of their composition and thickness **(Fig. 16.5b)**. Oceanic lithosphere has a

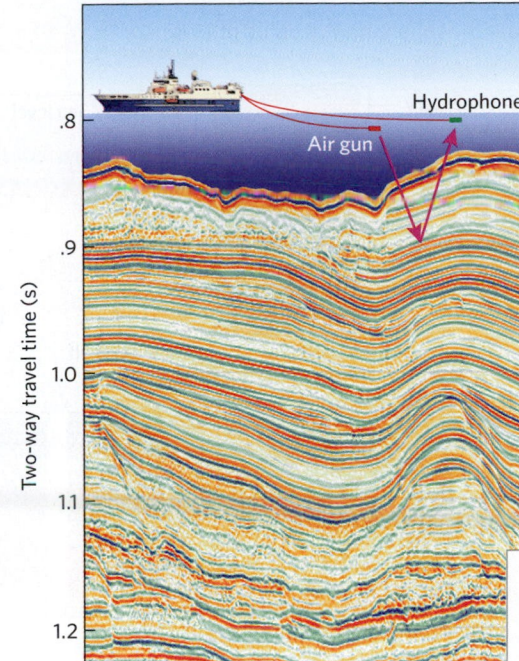

Figure 16.3 Seismic-reflection profiling helps characterize the layering of sediments and rocks beneath the seafloor. Seismic signals are produced by an air gun and are recorded by a string of hydrophones. The resulting profile reveals the layering. In the locality shown, local faulting breaks up the layers.

The two-way travel time of the seismic waves can be translated into the depth of the layers below the surface.

Figure 16.4 Deep-sea drilling and coring bring up sediment and rock from below the seafloor.

(a) The *JOIDES Resolution* drills holes into the seafloor.

(b) Cores can be preserved for future analysis.

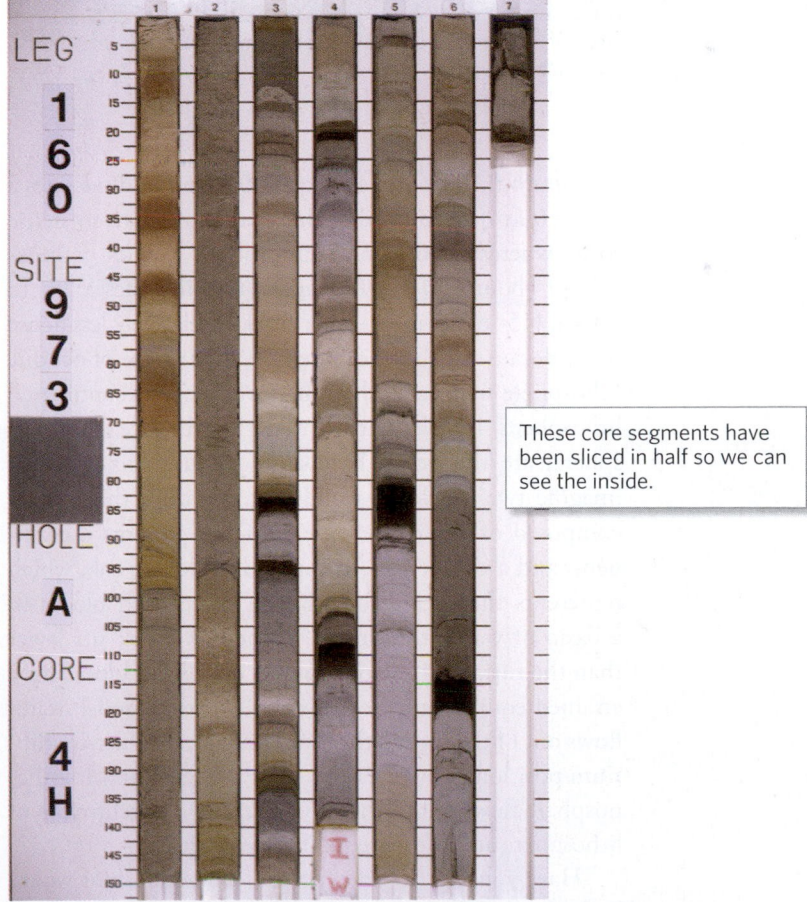

These core segments have been sliced in half so we can see the inside.

(c) Distinct layers of sediment are visible in these core segments. Numbers on the left indicate length in cm.

Figure 16.5 Contrasts between continental lithosphere and oceanic lithosphere.

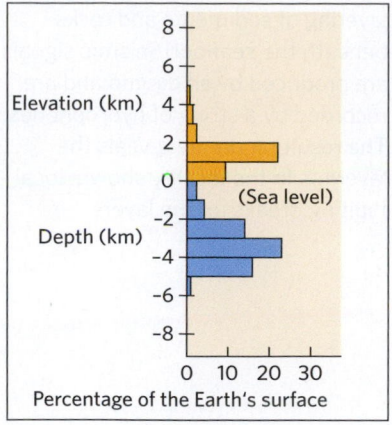

(a) A graph showing the percentage of the Earth's surface at different elevations above sea level, or depth below the sea.

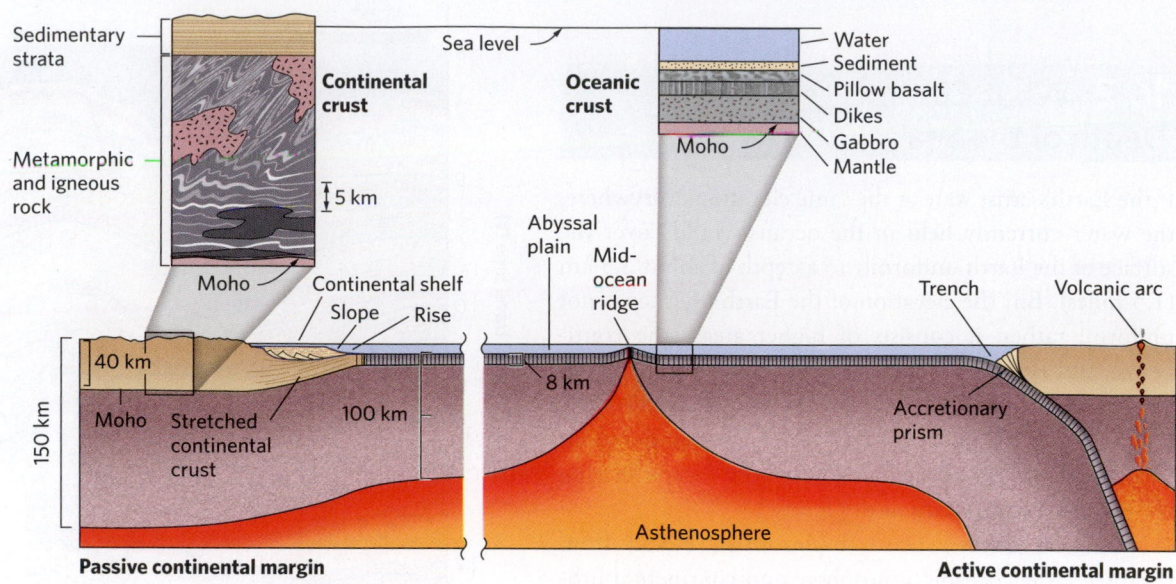

(b) The crustal portion of continental lithosphere differs markedly from that of oceanic lithosphere.

(c) A simple model showing that the surface of a thick pine block sits higher than the surface of a thin, dense oak block when both are placed in water.

maximum thickness of 100 km (60 miles) and includes a 7- to 10-km (4- to 6-mile)-thick crust of relatively dense rock, whereas continental lithosphere reaches a thickness of about 150 km (90 miles) and includes a 25- to 70-km (15- to 45-mile)-thick crust of relatively less dense rock. Because of these differences, the surface of oceanic lithosphere sits deeper than the surface of the continental lithosphere—the low areas, underlain by oceanic lithosphere, are the ocean basins. To picture this contrast, imagine two blocks of wood (Fig. 16.5c): a thicker one composed of less dense pine, which represents a continent, and a thinner one composed of denser oak, which represents the ocean floor. If you place both blocks in a basin of water, the surface of the oak block sits lower than the surface of the pine block, once both blocks have attained equilibrium (see Box 7.3). In this model, water flows out of the way so the blocks can attain their equilibrium positions. In the case of the Earth, the plastic asthenosphere flows out of the way to let the different types of lithosphere attain their equilibrium positions.

Have you ever wondered what the ocean floor would look like if all the water evaporated? Studies of bathymetry indicate that the ocean floor can be divided into *bathymet-*

ric provinces, distinguished from one another by their water depth (Fig. 16.6). We'll now revisit these provinces, first mentioned in Chapter 2, to characterize their surfaces—the landscape of the seafloor—in a bit more detail.

The Shallower Realms: Continental Shelves

Imagine you're cruising in a submersible, heading east from the coast of North America into the North Atlantic. Moving out from the shore for a distance of about 150 to 500 km (100 to 300 miles), you will be above the **continental shelf**, a relatively shallow portion of the ocean that fringes the continent. Over the continental shelf, water depth does not exceed about 200 m (650 feet), and across the width of the shelf, the ocean floor slopes seaward at an angle of only 0.3° (Fig. 16.7a). At its seaward edge, the continental shelf terminates at the *continental slope*, a surface that descends at an angle of about 2° from a depth of 200 m to nearly 4 km (2.5 miles). From about 4 km down to about 4.5 km (2.8 miles), the slope angle decreases, defining a region called the *continental rise*. At a depth of 4.5 km, you find yourself above a vast, nearly horizontal surface known as the **abyssal plain**.

Broad continental shelves, like that of eastern North America, form along passive continental margins. Because these margins are not plate boundaries, they lack seismicity (see Chapter 2). Passive margins originate after rifting succeeds in breaking a continent in two. Rifting ceases when a new mid-ocean ridge forms and seafloor spreading begins. At this time, the stretched continental lithosphere of what had been the rift forms a transitional region between

Figure 16.6 Bathymetric provinces of the South Atlantic Ocean.

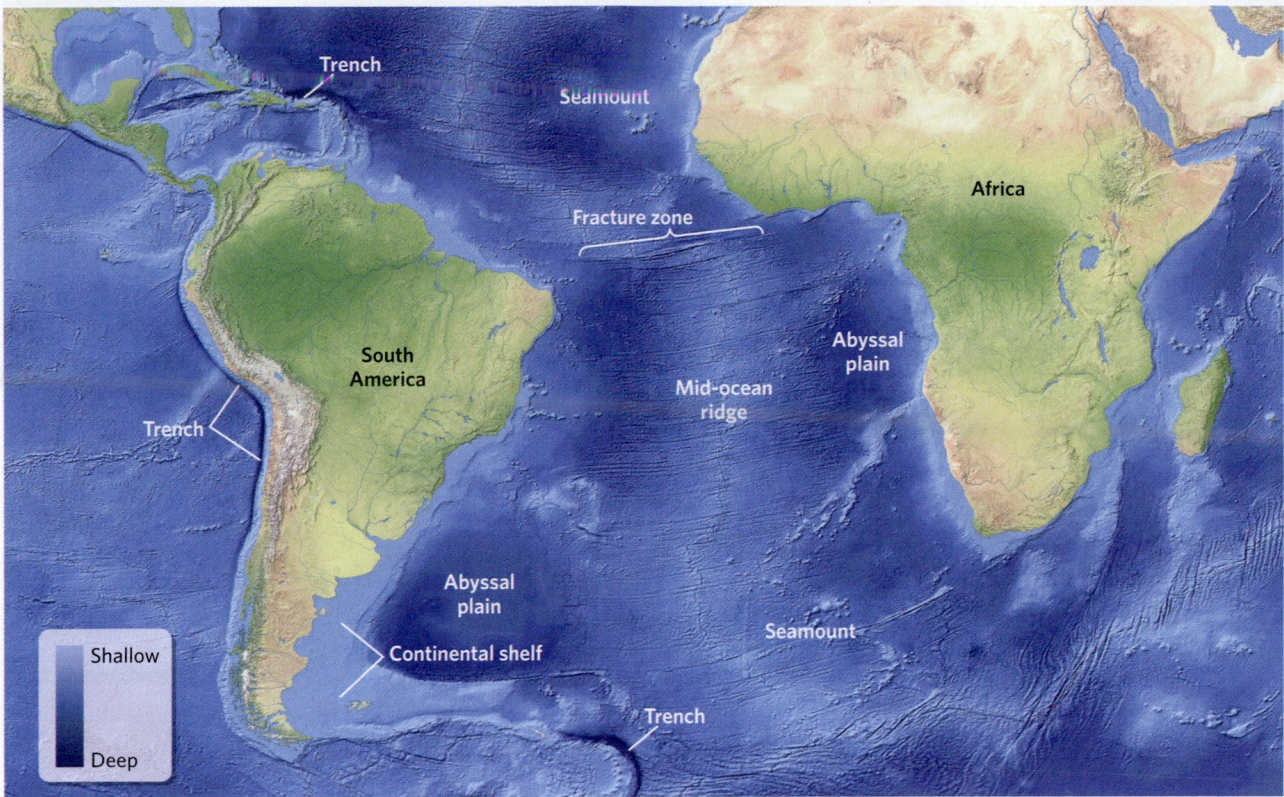

(a) A bathymetric map of the South Atlantic and eastern South Pacific shows the complexity of the seafloor landscape.

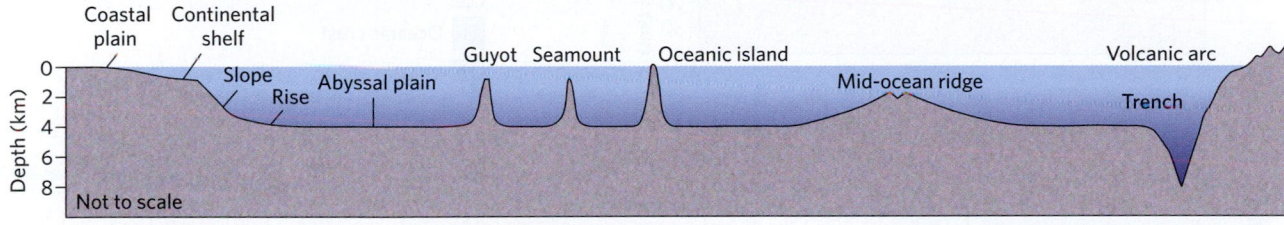

(b) A schematic profile of the ocean shows key geologic features.

newly formed oceanic lithosphere on one side and un-stretched continental lithosphere on the other. Over time, the stretched lithosphere cools and becomes denser and thicker. As a result, its surface slowly sinks, or subsides, just as the deck of a cargo ship moves down, closer to the sea surface, when heavy cargo has been placed in the ship's hold (Fig. 16.7b, c). As it sinks, the surface of the former, now inactive, rift becomes a low area that collects sand and mud washed off the continent as well as shells of plankton and other marine creatures. Deposition of sediment keeps pace with the rate of subsidence, so after tens of millions of years, a very thick layer of sediment covers the region of stretched continental crust. This sediment-filled depres-sion is a **passive-margin basin**, and the flat surface of this sediment layer constitutes the continental shelf. In some cases, the sediment layer in a passive-margin basin reaches a thickness of 15 to 20 km (9 to 12 miles).

If you were to take your submersible to the western coast of South America and cruise out into the Pacific, you would find an active continental margin, where the Pacific Ocean floor subducts beneath the continent (Fig. 16.8a). This margin looks very different from a passive margin. After crossing a relatively narrow continental shelf, you would find yourself over a continental slope that descends at a relatively steep angle of 3.5° into a deep-sea trench whose floor lies at a depth of over 8 km (5 miles). In this location, the continental slope is the face of an accretion-ary prism, a wedge of sediment that has been scraped off the seafloor, like snow in front of a plow, during subduc-tion. The narrow continental shelf consists of a sediment apron that spread out over the top of an accretionary prism (Fig. 16.8b).

Between continental shelves and the abyssal plain or the trench floor are slopes where submarine mass

Figure 16.7 Continental shelf of a passive margin.

1: Continental shelf
2: Continental rise
3: Continental slope
4: Abyssal plain

(a) An oblique view of the continental shelf along the east coast of North America, from South Carolina to Nova Scotia. It is much shallower than the adjacent abyssal plain.

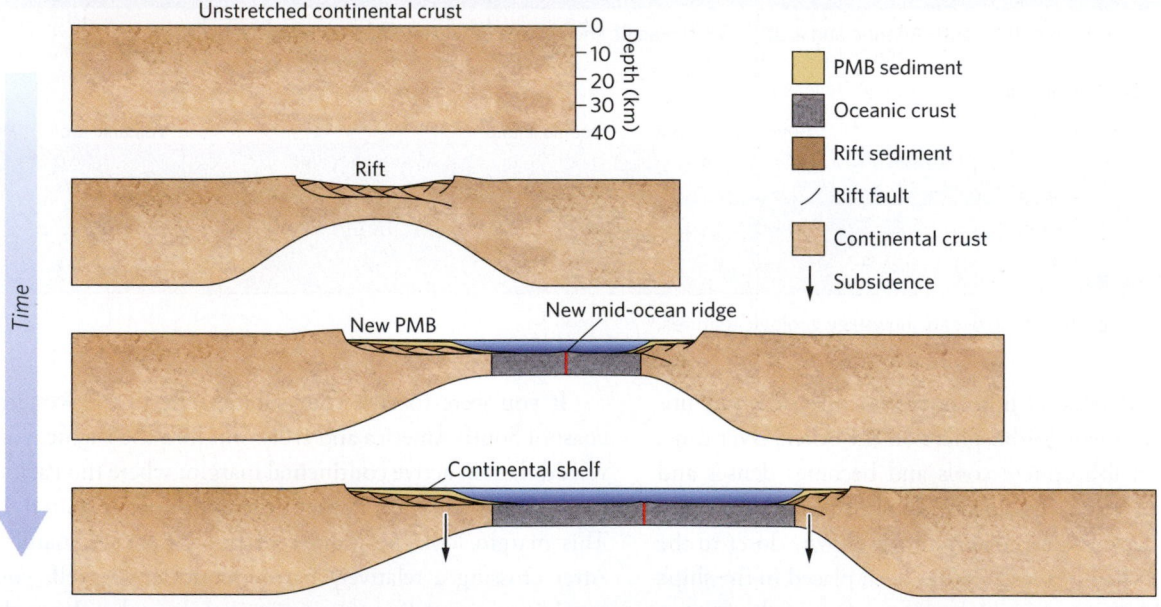

(b) The formation of a passive-margin basin (PMB). The top surface of the PMB is the continental shelf.

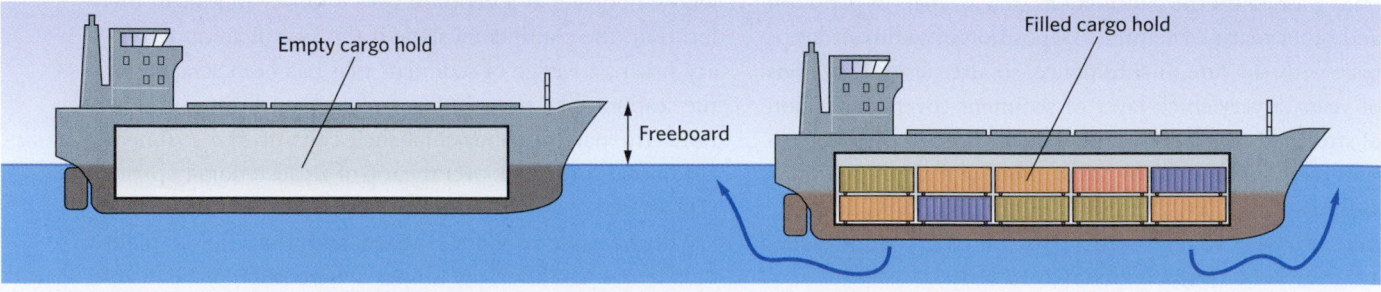

(c) The subsidence of a passive-margin basin happens for the same reason that a heavier ship moves downward relative to a lighter one. As the keel sinks down, water flows out of the way (blue arrows).

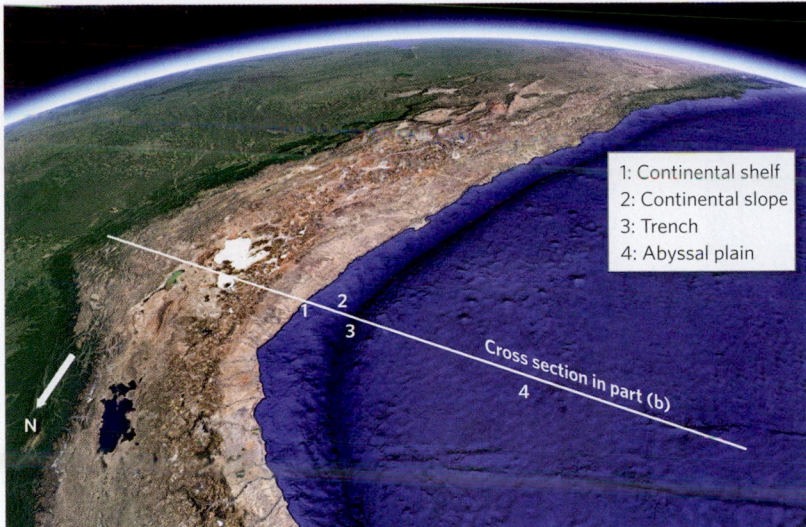

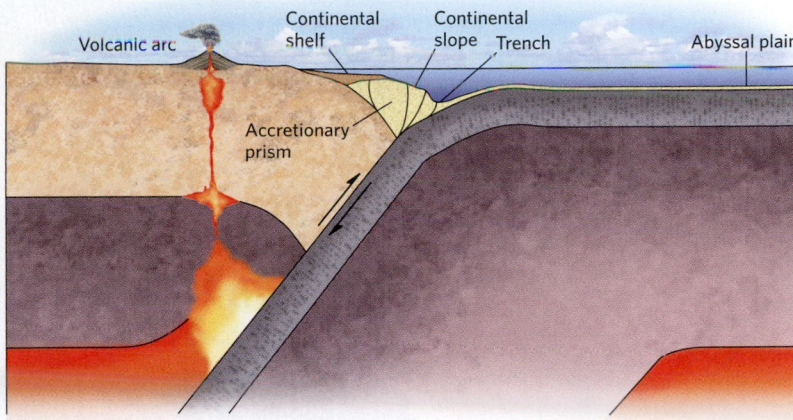

Figure 16.8 Continental shelf of an active margin.

(a) Looking southeast, obliquely, at the western coast of South America, we see a narrow continental shelf, built on top of an accretionary prism. The tan area represents the Andes.

(b) A schematic cross section shows the development of an accretionary prism next to a trench at a convergent boundary.

wasting events occasionally take place. As we mentioned in Chapter 12, large *submarine slumps* may carry immense volumes of debris down continental slopes, and these movements may generate tsunamis. Submarine avalanches, known as *turbidity currents* (see Chapter 5), send abrasive clouds of suspended sediment rushing down continental slopes. In many locations, these turbidity currents focus down a relatively narrow chute and, over time, carve deep underwater valleys, known as **submarine canyons (Fig. 16.9)**. Some submarine canyons start offshore of river mouths, for rivers cut into the continental shelf at times when sea level is low and the shelf is exposed, producing a trough that can focus turbidity currents to the head of a submarine canyon when sea level rises.

When a turbidity current reaches the mouth of a submarine canyon, at the base of the continental slope, its flow velocity decreases, so the sediments it carries settle out to produce *turbidites* made up of graded beds (see Chapter 5). Layer upon layer of turbidites accumulate and build out a **submarine fan**, a wedge of sediment that spreads out over the abyssal plain (along passive margins) or the trench floor (along marine convergent boundaries). These fans underlie the continental rise. The largest submarine fan in the world, the Bengal Fan, has formed at the mouth of the Ganges River in the northeastern Indian Ocean. This fan covers an area of about 3,000 km (1,800 miles) long and 1,000 km (600 miles) wide.

Plate Boundaries on the Seafloor

You can see all three types of plate boundaries by studying the bathymetry of the ocean floor (see Fig. 16.6).

Seafloor spreading at a divergent boundary yields a *mid-ocean ridge*, a submarine mountain belt that rises as much as 2 km (1.2 miles) above the depth of the abyssal plain **(Fig. 16.10)**. Because crust stretches and breaks as seafloor spreading takes place, the ridge axis may be bordered by escarpments (cliffs), a result of normal faulting.

Oceanic transform faults, strike-slip faults along which one plate shears sideways past another, typically link segments of mid-ocean ridges (see Chapter 2). Transform faults are delineated by *fracture zones*, narrow belts of steep escarpments and broken-up rock. These fracture zones can be traced into the oceanic plate away from the ridge axis, where they are no longer seismically active.

Figure 16.9 A submarine canyon off the coast of New Jersey. Turbidity currents flow down the canyon and carry sediment to a submarine fan on the abyssal plain.

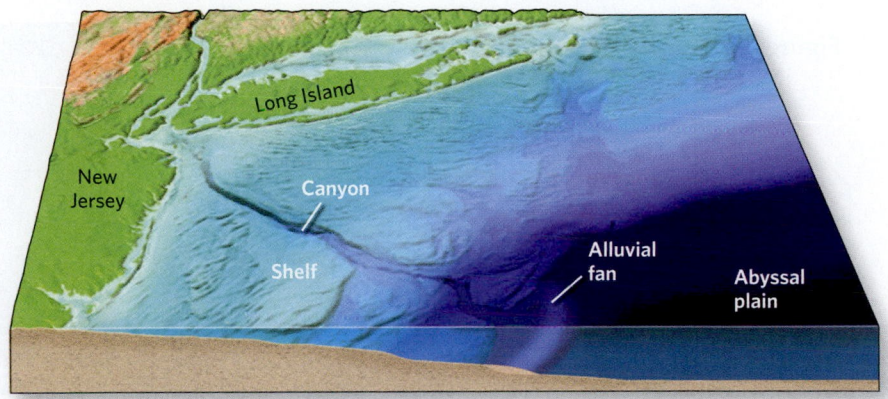

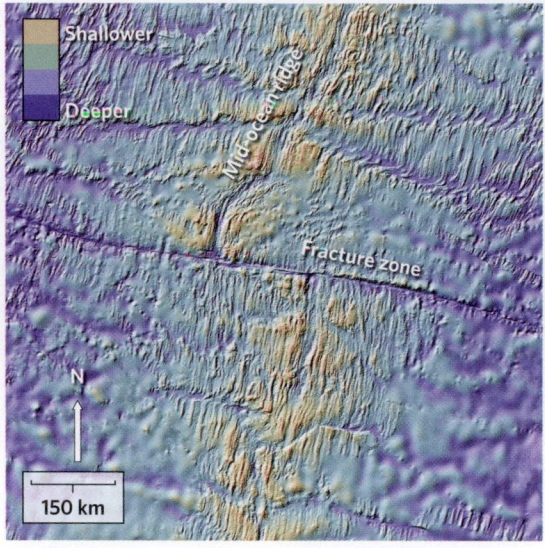

Even though fracture zones away from ridges are not plate boundaries, they delineate the boundary between oceanic lithosphere of different ages (see Fig. 2.22b).

Subduction at convergent boundaries yields a *deep-sea trench*, a deep, elongate trough bordering a volcanic arc (see Fig. 16.8). As we noted, trenches typically reach depths of over 8 km (5 miles). The deepest, the Mariana Trench in the western Pacific, extends to a depth of 11 km (6.8 miles). Some trenches border continents, lying seaward of an active continental volcanic arc, as is the case along the western coast of South America. Others border *island arcs*, curving chains of active volcanic islands.

Abyssal Plains and Seamounts

As oceanic crust moves away from the axis of a mid-ocean ridge and gets progressively older, it cools and thickens, and as it does so, its surface sinks and its slope decreases (see Chapter 2). As a result, the surface of the seafloor becomes a nearly horizontal abyssal plain. A blanket of *pelagic sediment* gradually accumulates and covers the basalt of the oceanic crust (Fig. 16.11a). This blanket consists mostly of microscopic plankton shells and fine flakes of clay (from volcanic ash or windblown dust), which slowly fall like snow from the ocean water and settle on the seafloor. The older the seafloor, the longer the time during which pelagic sediment has been accumulating, so over older oceanic lithosphere of the abyssal plain, it becomes thick enough (up to 0.6 km or 0.4 miles) to bury irregularities in the basaltic crust of the seafloor. The top of this sediment layer forms the nearly featureless surface of the abyssal plain (Fig. 16.11b).

Oceanic islands (whose peaks protrude above sea level) and **seamounts** (whose peaks are submerged) rise above the abyssal plain or mid-ocean ridges (Fig. 16.12a). These features result from hot-spot volcanic activity. Oceanic islands or seamounts that currently lie over hot spots host active volcanoes, whereas those that have moved off the hot spot are extinct (see Chapter 2). Some seamounts are hot-spot volcanoes that did not become tall enough to protrude above sea level (Fig. 16.12b), but many started out as oceanic islands that later sank below sea level. Why do they sink? During and after the end of volcanic activity on an oceanic island, the island erodes and undergoes slumping and, over time, the seafloor beneath it ages and subsides. Oceanic islands and seamounts align in a chain parallel to the motion of the oceanic plate over the hot spot—as we saw in Chapter 4, these chains, called hot-spot tracks, define the direction of plate motion. Where hot-spot igneous activity becomes particularly voluminous, a broad *oceanic plateau*, composed of a layer of basalt up to 3 km (2 miles) thick, develops (Fig. 16.12c). Oceanic plateaus represent submarine large igneous provinces (LIPs; see Chapter 4).

In some cases, the top of a seamount may be beveled flat by erosion as its summit reaches sea level. Or the island may be overgrown by a coral reef as it sinks. When

Figure 16.11 The origin and nature of abyssal plains.

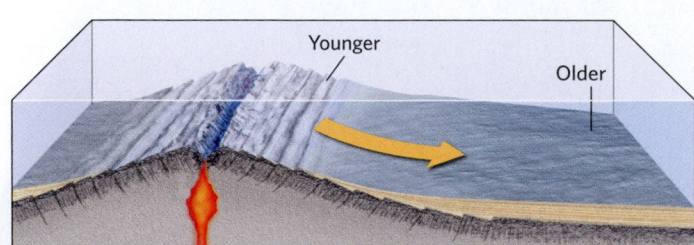

(a) The seafloor sinks progressively as it moves away from a ridge axis. It gets buried by sediment, which thickens away from the ridge axis.

(b) The surface of the abyssal plain has a covering of fine, loose pelagic sediment.

Figure 16.12 Oceanic islands and seamounts.

(a) An oceanic island protruding from the Pacific Ocean.

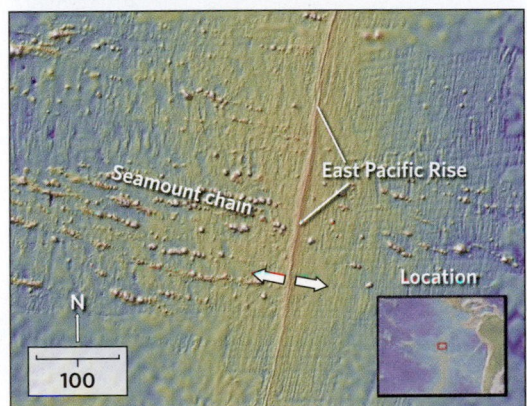

(b) Seamounts adjacent to the East Pacific Rise (112° W, 18° S), a mid-ocean ridge. Note that the seamount chains are perpendicular to the ridge axis and, therefore, are parallel to the direction of plate motion.

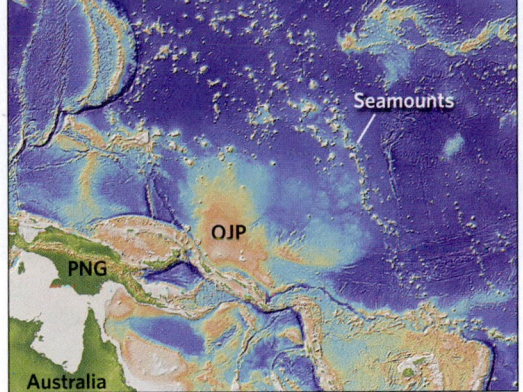

(c) The Ontong-Java Plateau (OJP) lies to the north of Papua New Guinea (PNG). Orange represents shallower water; blue represents deeper water.

such islands sink below sea level, they become flat-topped seamounts, known as **guyots**.

Take-home message . . .

Ocean basins exist because oceanic and continental lithosphere differ in thickness and density. Seafloor bathymetric features (such as ridges, trenches, and fracture zones) reflect plate tectonic processes. Abyssal plains overlie old oceanic lithosphere, and broad continental shelves overlie passive-margin basins.

Quick Question -
How do submarine canyons form?

16.3 The Tides Come In . . . the Tides Go Out . . .

Fishermen hoping to sail from a shallow port must pay attention to the **tide**, the generally twice-daily rise and fall of sea level. At *low tide*, when sea level sinks to its lowest elevation, boats might run aground, whereas at *high tide*, when sea level rises to its highest elevation, they can make it to open water easily **(Fig. 16.13a, b)**. We've already mentioned tides in Chapter 15, for their existence is common knowledge. Here, we look more closely at how they are manifested and why they form. Tides affect

Figure 16.13 Evidence of ocean tides.

(a) Low tide at Perranporth, in Cornwall, England.

(b) High tide from the same viewpoint as in the previous photo.

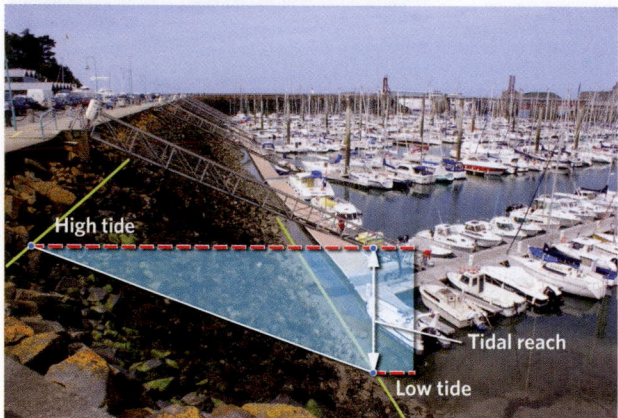

(c) The tidal reach is the vertical distance between low tide and high tide. This photo of a French harbor was taken at low tide.

the entire ocean, so the transition from high tide to low tide, or vice versa, involves the movement of a huge volume of water on a global scale. We discuss tides in this chapter because this water movement primarily affects coastal areas.

Tidal reach, the difference between sea level at high tide and at low tide **(Fig. 16.13c)**, varies greatly with location. Some regions experience a small tidal reach, under 0.5 m (20 inches), whereas others experience a large one. The largest tidal reach on the planet occurs in the Bay of Fundy, on the eastern coast of Canada, where the tidal reach can be as much as 16.3 m (53.5 feet). Large tides also occur along the coast of western Europe. In the open ocean, tidal reach averages about 0.6 m (2.0 feet). At several locations on our planet, the tidal reach is zero.

Tides affect the coast because during a rising tide, or *flood tide*, the **shoreline** (the boundary between water

and land) moves inland, and during a falling tide, or *ebb tide*, the shoreline moves seaward. The surface of the seafloor that lies between the shoreline at high tide and the shoreline at low tide constitutes the **intertidal zone (Fig. 16.14a)**. The horizontal distance over which the shoreline migrates between high and low tides—and, therefore, the width of the intertidal zone—depends on both the tidal reach and the slope of the seafloor surface. Where the tidal reach is large and the slope is gentle, the position of the shoreline can move by kilometers during a tidal cycle, so at low tide, the intertidal zone becomes a broad *tidal flat* exposed to the air **(Fig. 16.14b)**. Large tidal flats can be dangerous places—shellfish hunters digging for clams in the mud at the seaward edge of a tidal flat have drowned when caught by the rising tide because they couldn't make it back to dry land before the water became too deep.

Arrival of a flood tide in the mouth of an estuary, where the tide moves against a river current, can produce a **tidal bore**, a visible wall of water, ranging from a few centimeters to a few meters high, that moves at speeds of up to 35 km per hour (22 mph), faster than a person can run **(Fig. 16.14c)**. In a few places, tidal bores become large enough for surfers to ride. A skilled surfer once rode a tidal bore for 1 hour and 10 minutes, during which he traveled 17 km (11 miles)!

In nontechnical discussions, tides are attributed simply to the gravitational pull of the Moon and Sun. Indeed, gravitational pull does play a key role in driving tides; in fact, the Moon contributes most of the force that causes tides. (The Sun, even though it is vastly larger, lies so far away that its contribution to tides is only 46% that of the Moon.) But the gravitational pull of these objects doesn't

Figure 16.14 The intertidal zone and tidal bores.

(a) The intertidal zone on a beach on Cape Cod. Loose seaweed was deposited at high tide, and a portion of the growing weed is exposed at low tide.

(b) A tidal flat on the coast of France.

(c) The tidal bore that moves up this river in China has become a tourist attraction.

completely explain why tides happen. In detail, tides take place because the water of the ocean moves in response to two forces: (1) the gravitational attraction of the Moon and the Sun, and (2) centrifugal force caused by the revolution of the Earth-Moon system. Oceanographers refer

---

Box 16.1 | How can I **explain** . . .

Variations in tidal reach

What are we learning?
That the slope of the shore determines the tidal reach at a location on the coast.

What you need:
- A small plastic cutting board.
- A rectangular glass pan, large enough to hold the cutting board, deep enough to hold several centimeters (a couple of inches) of water. (An aluminum pan can work, too.)
- Water that has been dyed blue with food coloring.
- A protractor, ruler, and graph paper.
- A ladle, a pitcher, and a measuring cup.

Instructions:
- Fill the pan with blue water.
- Ladle out enough water to lower the water level by half; reserve the water in a measuring cup.
- Repeat the experiment for different slope angles of the cutting board.
- Plot a graph of slope angle (measured with the protractor) against width of the intertidal zone (measured with a ruler). How are the quantities related?

What did we see?
- Simple geometry requires that the gentler the slope, the wider the intertidal zone. The smallest tidal reach and, therefore, the smallest intertidal zones, occur where the shore is a vertical cliff.
- The intertidal zone can be very wide along very gently-sloping coasts.

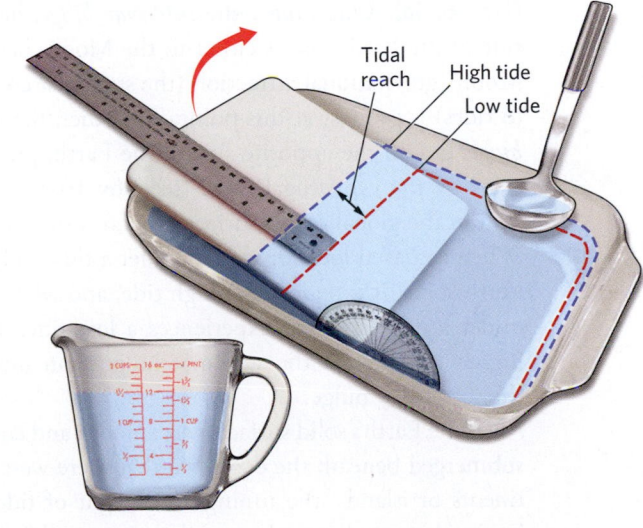

Figure 16.15 Tides.

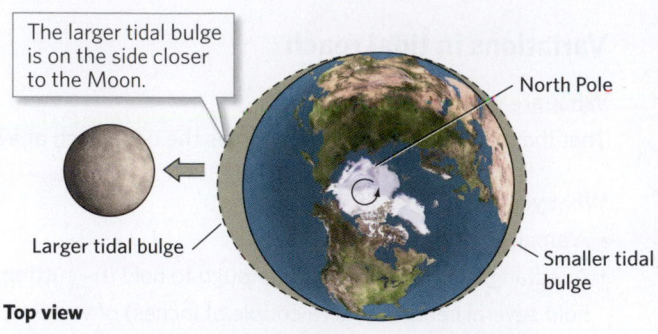

The larger tidal bulge is on the side closer to the Moon.

North Pole

Larger tidal bulge

Smaller tidal bulge

Top view

(a) The larger (sublunar) tidal bulge always faces the Moon, and the smaller (secondary) tidal bulge is always on the opposite side of the Earth.

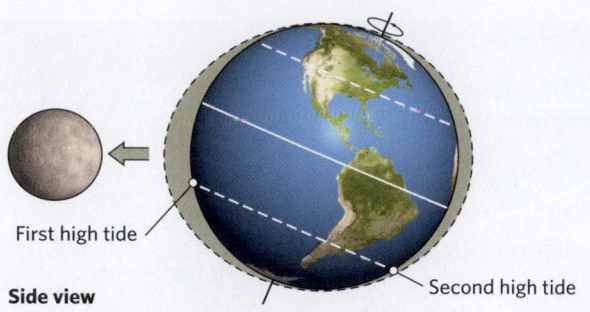

First high tide

Second high tide

Side view

(b) Viewed from the side, the sublunar bulge does not align with the equator.

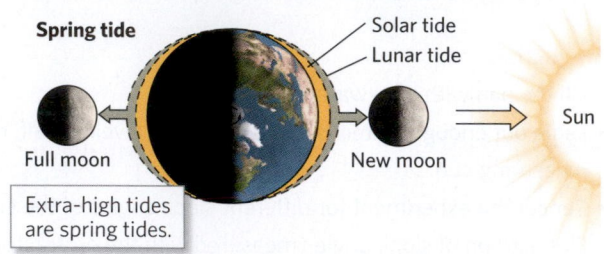

Spring tide

Solar tide

Lunar tide

Sun

Full moon

New moon

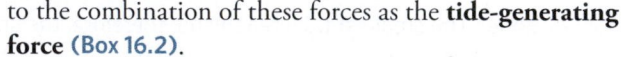

Extra-high tides are spring tides.

(c) When the Sun is aligned with the Moon, stronger, higher tides, called spring tides, result.

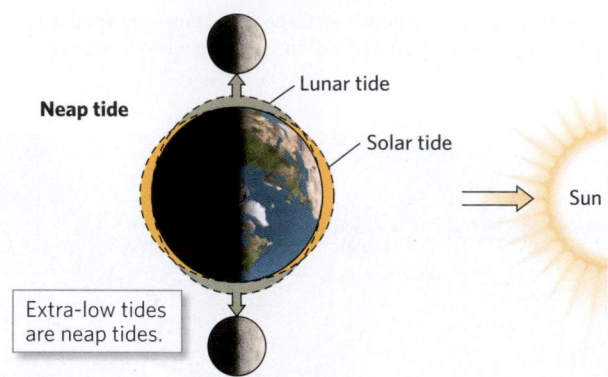

Neap tide

Lunar tide

Solar tide

Sun

Extra-low tides are neap tides.

(d) When the Sun is at right angles to the Moon, weaker, lower tides, called neap tides, result.

to the combination of these forces as the **tide-generating force** (Box 16.2).

The tide-generating force creates two tidal bulges in the global ocean, forming an envelope of water that is more oval shaped than the nearly spherical solid Earth (Fig. 16.15a). One bulge, the *sublunar bulge*, lies on the side of the Earth that is closer to the Moon, because the Moon's gravitational attraction (the strongest contributor to tides) is greatest at this point. The other, the *secondary bulge*, lies on the opposite side of the Earth, pushed outward by the centrifugal force (see Box 16.2). A depression in the global ocean surface separates the two bulges. When a coastal location passes under a tidal bulge as the Earth spins, it experiences a high tide, and when it passes under a depression, it experiences a low tide. Tides are higher underneath the sublunar bulge than underneath the secondary bulge.

If the Earth's solid surface were smooth and completely submerged beneath the ocean, so that there were no continents or islands, the timing and height of tides would be simple to understand—each location would experience two tides a day, and the tidal reach would depend on the position of the location relative to the highest part of the bulge. But the story isn't quite that simple, for many other factors affect the timing and magnitude of tides:

- *Tilt of the Earth's axis:* Because the Earth's axis of rotation is not perpendicular to the plane of the Earth-Moon system, any given point on the Earth passes between a high part of one bulge during one part of the day and a lower part of the other bulge during another part of the day (Fig. 16.15b).

- *The Moon's orbit:* The Moon progresses in its 28-day orbit around the Earth in the same direction as the Earth rotates. High tides arrive 50 minutes later each day because of the difference between the time it takes for the Earth to spin on its axis and the time it takes for the Moon to orbit the Earth.

- *The Sun's gravity:* When the Sun lies on the same side of the Earth as the Moon (a new Moon) or on the side opposite the Moon (a full Moon), we experience particularly high flood tides, called *spring tides*, because the Sun's gravitational attraction adds to that of the Moon (Fig. 16.15c). When the Moon and Sun are 90° apart relative to the Earth (quarter Moon), we experience lower flood tides, called *neap tides*, because the Sun's gravitational attraction counteracts that of the Moon (Fig. 16.15d).

- *Ocean-basin shape:* The shape of an ocean basin influences the sloshing of water back and forth as tides rise

Figure 16.16 A map showing the variation of tidal reach around the world. On a given white line, high tide occurs at the same time.

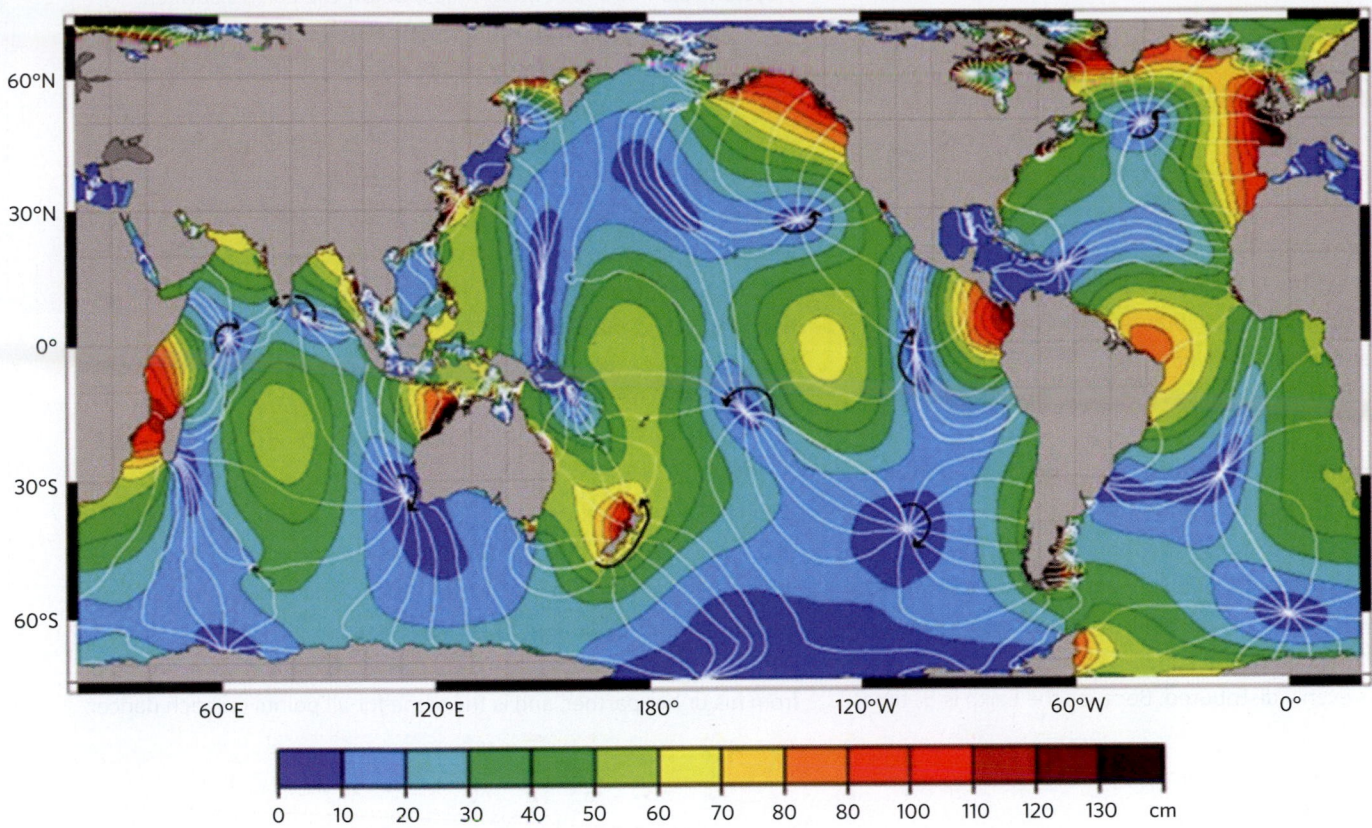

| 0 | 10 | 20 | 30 | 40 | 50 | 60 | 70 | 80 | 80 | 100 | 110 | 120 | 130 | cm |

and fall. Depending on its timing and magnitude, this sloshing can add to or subtract from the tidal bulge on a local scale. In some locations, sloshing cancels one of the daily tides entirely.

- *Focusing effect of bays:* On a still more local scale, the shape of the shoreline influences the tidal reach. In a bay that narrows to a point, the flood tide brings a large volume of water into a small area, so the point at the end of the bay experiences an especially high flood tide.

Because of the complexity of these contributing factors, the timing and magnitude of tides vary significantly around the ocean and along the coast **(Fig. 16.16)**. Nevertheless, at any given location, the tides are periodic and can be predicted. For early civilizations, tides provided a rudimentary way to tell time. In fact, in some languages, the word for tide is the same as the word for time.

Have tides on the Earth been the same through Earth history? Probably not. When the Moon first formed, it was much closer to the Earth than it is now, so the tides in early oceans were larger than they are now. Friction between ocean water and the ocean floor causes the movement of the tidal bulge to lag slightly behind the movement of the Moon around the Earth. The Moon,

therefore, exerts a slight pull on the side of the bulge. This pull acts like a brake and slows the Earth's spin, so that days are growing longer at a rate of about 0.002 seconds per century. Over geologic time, these seconds add up—a day was only 21.9 hours long in the Middle Devonian Period (390 Ma). The process also slows the Moon's orbit, causing the distance to the Moon to increase at a rate of about 4 cm (2 inches) per year. (This change in distance happens to conserve angular momentum, a concept discussed in Box 15.3.)

Take-home message . . .

Tides are the generally twice-daily rise and fall of the sea surface. They are caused by the tide-generating force, a consequence of gravitational pull by the Moon and Sun and of centrifugal force due to the revolution of the Earth-Moon system. Many factors affect the specific tidal reach—the difference between the elevation of high tide and that of low tide—at a given location.

Quick Question -
What factors determine the width of the intertidal zone in a region?

Box 16.2

Science toolbox . . .

The tide-generating force

Tides form in response to two forces acting at the same time: gravitational pull exerted by the Moon and Sun on the Earth, and centrifugal force caused by the revolution of the Earth around the center of mass of the Earth-Moon system. To explain this statement, we must first review a few physics terms:

- *Gravitational pull* is the attractive force that one mass exerts on another. Its magnitude depends on the amount of mass in each object and on the distance between the two masses. (The pull increases as mass increases, and the pull decreases as distance between masses increases.)
- *Centrifugal force*, a manifestation of inertia, is the apparent outward-directed (center-fleeing) force that a body on or in an object feels when the object spins or moves in orbit around a point (see Box 15.3).
- The *Earth-Moon system* refers to our planet and the Moon, viewed as a gravitationally linked pair as they move together through space.
- The *center of mass* is the point within an object, or a group of objects, about which mass is evenly distributed. Because the Earth is 81 times more massive than the Moon, the center of mass of the Earth-Moon system lies 1,700 km (1,060 miles) below the Earth's surface.

With these terms on hand, let's begin by considering the origin and consequence of centrifugal force in the Earth-Moon system. To do this, we must observe the way in which the system moves. Although we usually picture the center of the Earth as following a simple orbit around the Sun, it's actually the center of mass of the Earth-Moon system that follows this orbit **(Fig. Bx16.2a)**. To see why, picture the Earth-Moon system as a pair of dancers, one much heavier than the other. The dancers face each other, hold hands, and whirl as they cross the dance floor **(Fig. Bx16.2b)**. The center of mass of the pair defines their overall path—each dancer crosses back and forth across the overall path as the dance progresses.

With this image in mind, let's turn our attention back to the Earth-Moon system. As the Earth revolves around the system's center of mass, centrifugal forces develop on both the Earth and the Moon that would cause the two bodies to fly away from one another, were it not for the gravitational attraction holding them together. We can understand this statement by thinking again of our dancer analogy— the centrifugal force acting on each dancer points outward, away from his or her partner, and is the same for all points on each dancer

Figure Bx16.2 The concepts of the center of mass of the Earth-Moon system and the tide-generating force.

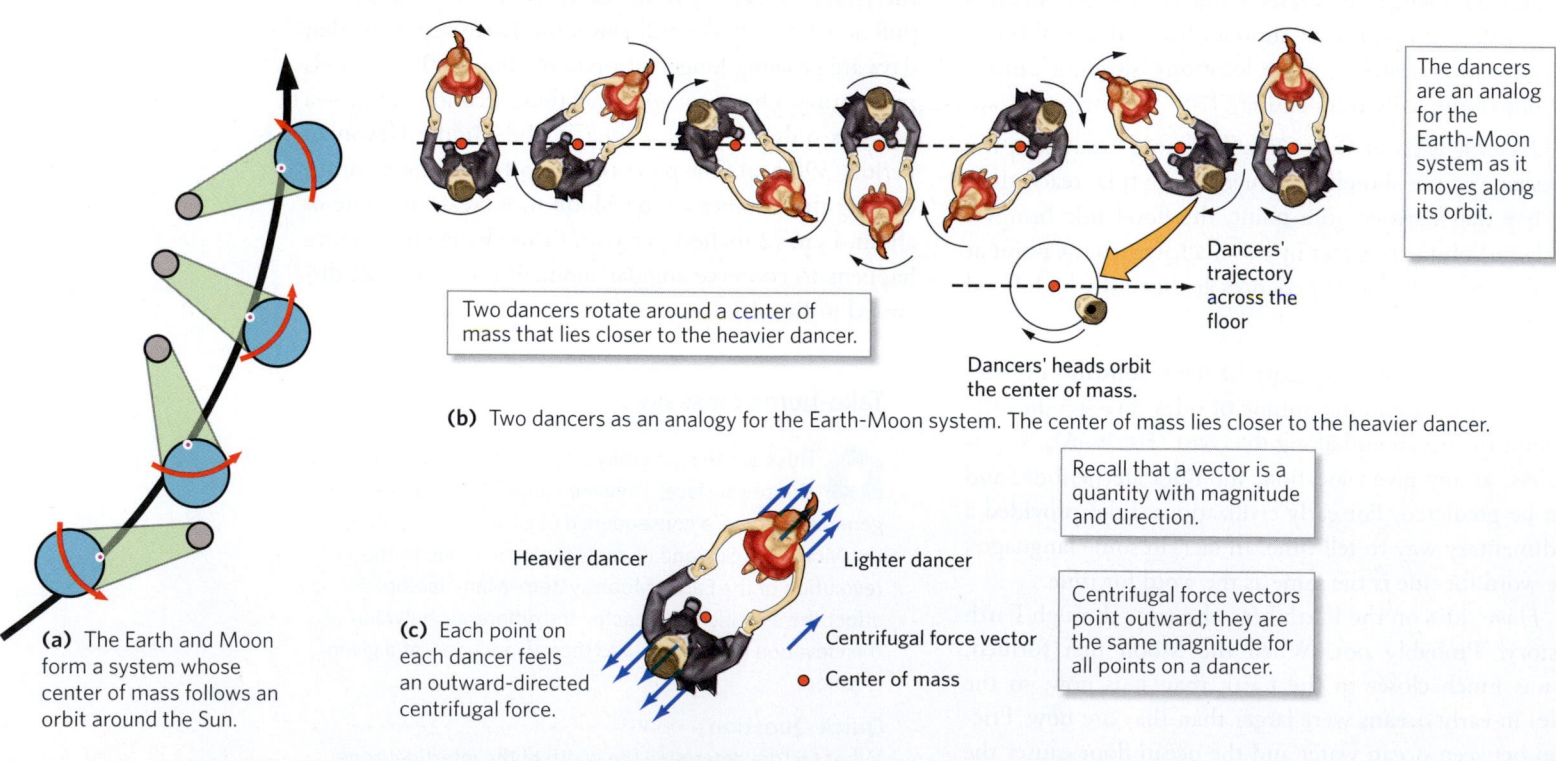

The dancers are an analog for the Earth-Moon system as it moves along its orbit.

Dancers' trajectory across the floor

Dancers' heads orbit the center of mass.

Two dancers rotate around a center of mass that lies closer to the heavier dancer.

(b) Two dancers as an analogy for the Earth-Moon system. The center of mass lies closer to the heavier dancer.

Recall that a vector is a quantity with magnitude and direction.

Heavier dancer   Lighter dancer

Centrifugal force vector

● Center of mass

Centrifugal force vectors point outward; they are the same magnitude for all points on a dancer.

(a) The Earth and Moon form a system whose center of mass follows an orbit around the Sun.

(c) Each point on each dancer feels an outward-directed centrifugal force.

(Fig. Bx16.2c). We can represent the direction and magnitude of centrifugal force by an arrow, or *vector*, whose length represents the size of the force and whose orientation indicates the direction of the force. Centrifugal force vectors at all points all around the surface of the Earth point away from the Moon.

Now, let's consider how the force of gravity comes into play in causing tides. To simplify this discussion, we examine only the effect of the Moon's gravity on Earth. Vectors representing the magnitude and direction of the Moon's gravitational pull, at any point on the surface of the Earth, point toward the center of the Moon. Because the magnitude of gravity depends on distance, the Moon exerts more attraction on the side of the Earth closer to the Moon than on the side of the Earth farther from the Moon.

How do centrifugal force and gravity work together to produce tides? If we draw vectors representing both centrifugal force and gravitational force at various points on or in the Earth, we see that the arrows representing centrifugal force do not have the same length as those representing gravitational attraction, except at the Earth's center **(Fig. Bx16.2d)**. Moreover, the vectors representing centrifugal force do not point in the same direction as the vectors representing gravitational attraction. The force that ocean water feels is the sum of the two forces acting on the water. You can determine the sum of two vectors by drawing the vectors to touch head to tail—the sum is the vector that completes the triangle. This sum of the gravitational vector and the centrifugal force vector at a point on the Earth's surface is the *tide-generating force* **(Fig. Bx16.2e)**.

The magnitude and direction of the tide-generating force vary with location on the Earth. For example, on the side of the Earth closer to the Moon, gravitational force is greater than centrifugal force, so adding the two gives a net tide-generating force that causes the sea surface to bulge toward the Moon. On the side of the Earth farther from the Moon, centrifugal force is the larger vector and causes the surface of the sea to bulge away from the Moon. Therefore, the ocean has two tidal bulges—one on the side of the Earth close to the Moon, and one on the opposite side. The bulge closer to the Moon is always larger.

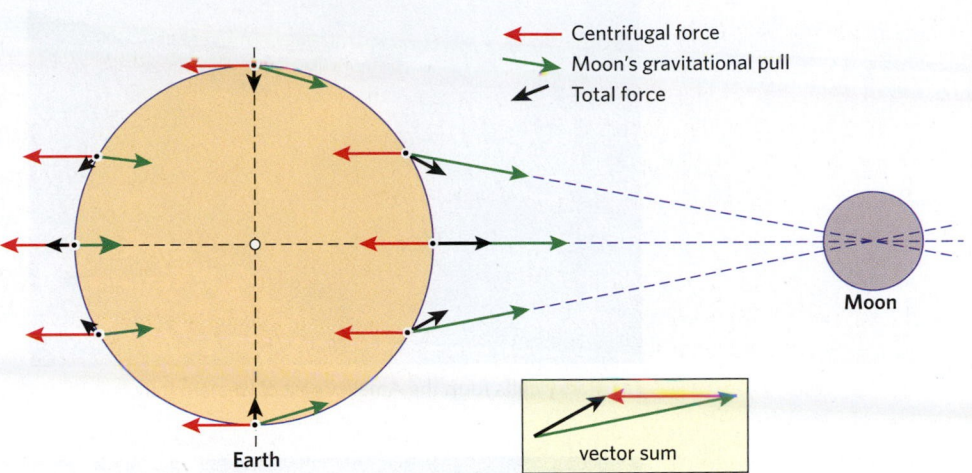

(d) Each point on the Earth's surface feels the same centrifugal force, but also feels a gravitational pull from the Moon. A vector is a quantity with both magnitude and direction.

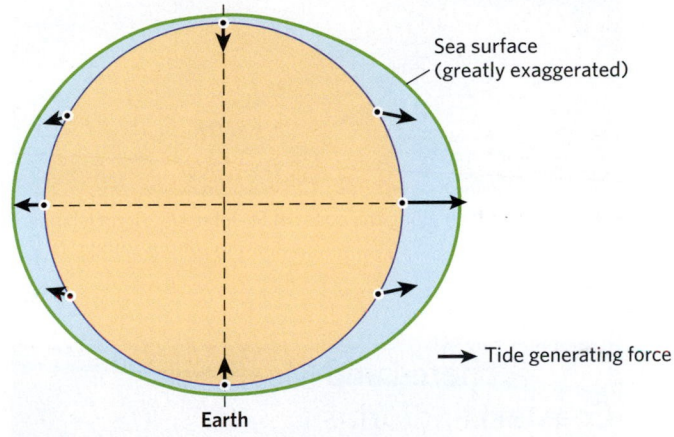

(e) The tide-generating force is the sum of the centrifugal force and the gravitational force vectors. The bulge of the sea surface is exaggerated.

Figure 16.17 Spectacular coastal scenery from localities around the world.

(a) Rocky cliffs form the Amalfi Coast of western Italy.

(b) Steep green slopes form the Na Pali coast of Kauai, Hawaii.

(c) A sandy beach along the coast of St. John, US Virgin Islands.

(d) The abrupt edge of the Nullarbor Plain forms the southern coast of Australia.

The three great elemental sounds in nature are the sound of rain, the sound of wind in a primeval wood, and the sound of the outer ocean on a beach.

—HENRY BESTON (AMERICAN NATURALIST, 1888–1968)

Did you ever wonder . . .

why the breakers that surfers love form near the shore?

16.4 Where Land Meets Sea: Coastal Landforms

Tourists along the Amalfi Coast of Italy thrill to the sound of waves crashing on rocky shores, but in the Virgin Islands, sunbathers can find seemingly endless white sand beaches. A fisherman plying the channels of the Mississippi Delta will find vast swamps along the seacoast, but a sailor approaching Rio de Janeiro, Brazil, will come face-to-face with granite domes rising directly from the sea (see Figure 12.1), and a sailor landing at the southern shore of Australia will have to climb a 100-m (330-foot) vertical cliff **(Fig. 16.17)**. As these examples illustrate, coasts vary dramatically in terms of their topography and associated landforms (see also **Earth Science at a Glance,** pp. 564–565). The shape of a coast depends on many factors, but first and foremost, it reflects deposition or erosion by waves. So we start our discussion of coasts by examining how waves change when they approach the shore.

Waves Approaching the Shore

In Chapter 15, we discussed the motion of waves in the open ocean. In an ocean wave, water molecules follow circular paths. The radius of these paths decreases with depth until, at a distance below the surface equal to half a wavelength, water remains stationary (see Fig. 15.25). In the open ocean, the *wave base*, the lowest level at which water moves in a wave, lies far above the seafloor, so wave motion has no effect on the ocean floor. Nearer shore, however, where the wave base just touches the ocean floor, a passing wave causes a slight back-and-forth motion of sediment. Still closer to shore, as the water gets shallower, friction between the wave and the seafloor slows the deeper part of the wave, and the circular motion in the wave becomes more elliptical **(Fig. 16.18a)**. Eventually, water at the top of the wave curves over its base, and the wave becomes a **breaker**, ready for surfers to ride **(Fig. 16.18b)**. The toppling water at the crest of a breaker mixes with air to form a white froth. Breakers crash onto the shore in the *surf zone*, sending a surge of water up the beach. This upward

Figure 16.18 The interaction of waves with the shore.

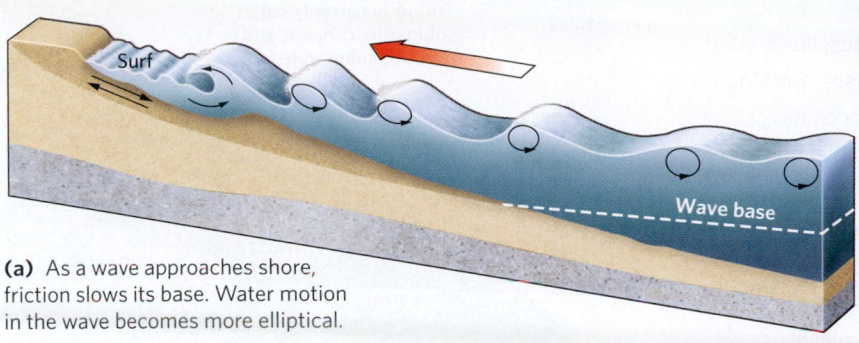

(a) As a wave approaches shore, friction slows its base. Water motion in the wave becomes more elliptical.

(b) Waves becoming breakers as they approach a beach.

surge, or **swash**, continues until friction and gravity bring water motion to a halt. Then gravity draws the water back down the beach as **backwash** (Fig. 16.18c).

The crests of waves that make an angle to the shoreline bend as they approach the shore, a phenomenon called **wave refraction**. Typically, by the time the wave reaches the shore, the angle between the crest and the shoreline decreases to less than 5° (Fig. 16.19). To understand why wave refraction happens, imagine a wave approaching the shore so that its crest initially makes an angle of 45° with the shoreline. The end of the wave closer to the shore touches bottom first and slows down because of friction, whereas the end farther offshore continues to move at its original velocity. This difference swings the whole wave around so that it becomes more parallel with the shoreline.

Although wave refraction decreases the angle at which waves move close to shore, it does not necessarily

(c) Swash and backwash on a beach. The water from the previous wave covered the wet area, but has since flowed down the beach as backwash. The swash of the next wave is moving up the beach.

Figure 16.19 Wave refraction and its consequences along the shore.

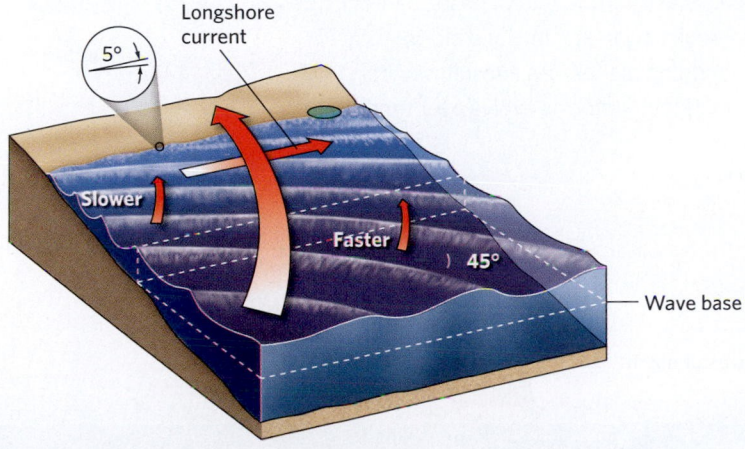

(a) Wave refraction occurs when waves approach the shore at an angle. Waves arriving at the shore obliquely can cause a longshore current.

(b) An example of waves refracting along a beach.

The Seafloor and Coasts

The oceans of the world host a diverse array of environments and landscapes, illustrating the complexity of the Earth System. Tectonic processes and surface processes, working alone or in tandem, generate unique features beneath the sea and along its coasts.

The major structures of the ocean floor reflect plate tectonics. For example, mid-ocean ridges define divergent boundaries, fracture zones initially form along transform faults, and trenches mark subduction zones. Oceanic islands, seamounts, and plateaus build above hot spots. Along passive continental margins, broad continental shelves develop, locally incised by submarine canyons. Trenches delineate convergent boundaries.

Coastal landscapes reflect variations in sediment supply, tides, relative sea-level rise or fall, and climate. Tides cause sea level to rise and fall, and wind builds waves that churn the sea surface, erode shorelines, and transport sediment. Where the supply of sediment is low and the landscape is rising relative to sea level, rocky shores with dramatic cliffs and sea stacks may evolve. Where sediment is abundant, sandy beaches and bars develop. Regions where glaciers carved deep valleys now feature spectacular fjords. Protected coastal areas, especially those in warm climates, host unique coastal ecosystems. Corals may contribute to growth of broad reefs along the shore.

Atoll formation

In warm waters, coral grows to form a fringing reef around an oceanic island. Over time, the island subsides, so the reef becomes a barrier reef. When the island sinks entirely, the reef remains as an atoll. If the atoll eventually sinks below sea level, it becomes the top of a guyot.

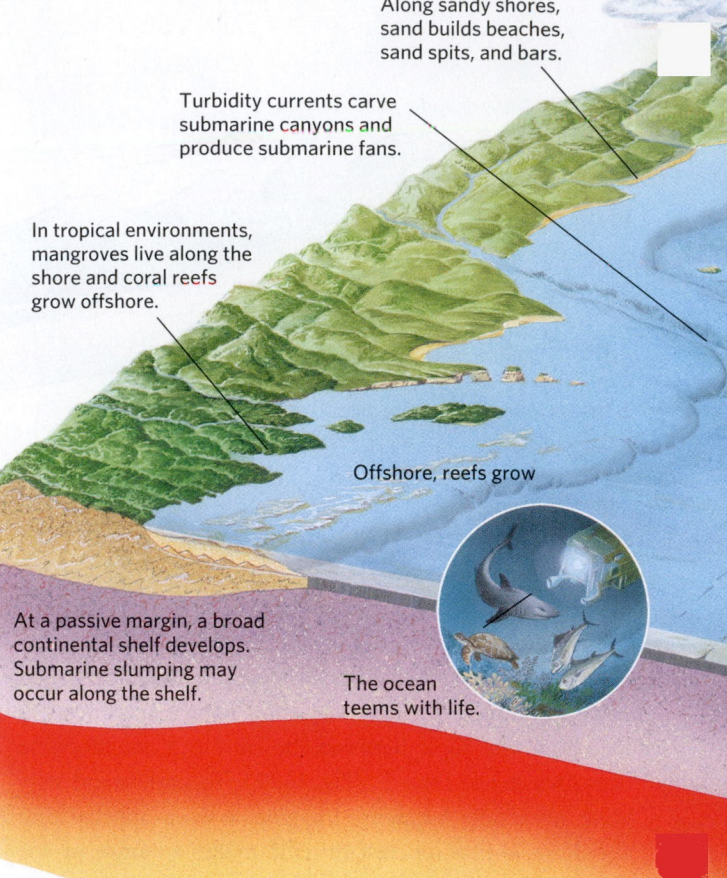

Temperate and tropical coastal landforms

Along sandy shores, sand builds beaches, sand spits, and bars.

Turbidity currents carve submarine canyons and produce submarine fans.

In tropical environments, mangroves live along the shore and coral reefs grow offshore.

Offshore, reefs grow

At a passive margin, a broad continental shelf develops. Submarine slumping may occur along the shelf.

The ocean teems with life.

Fringing Reef

Barrier Reef

Atoll

Time

Rocky coast evolution

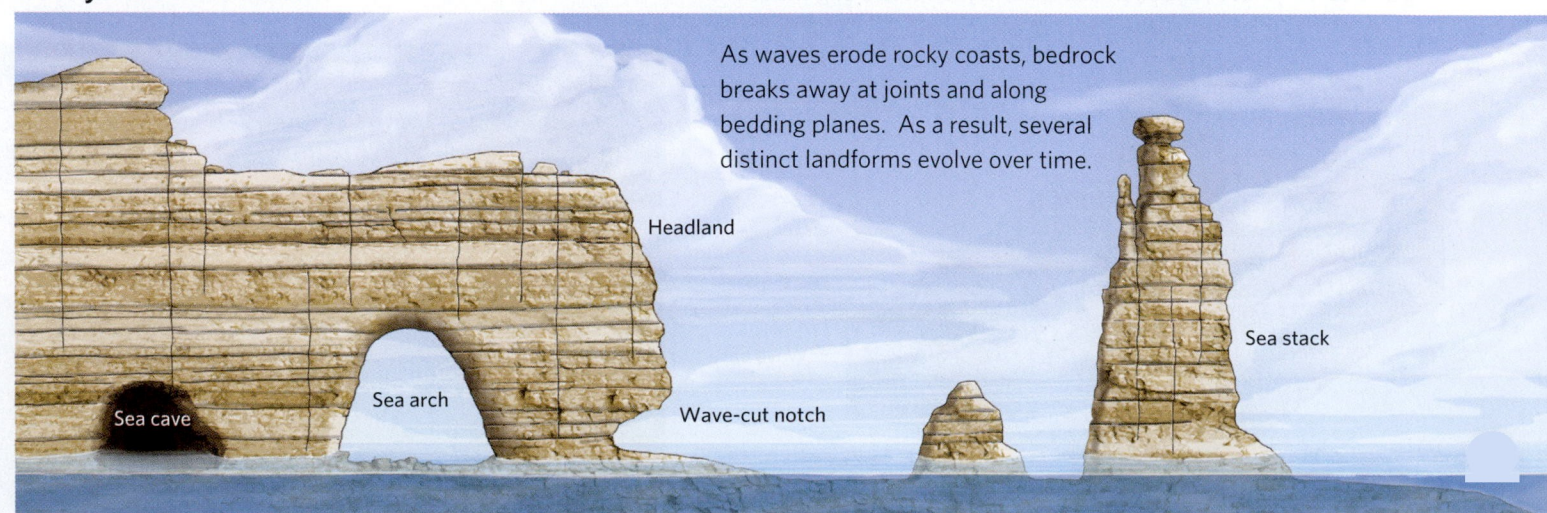

As waves erode rocky coasts, bedrock breaks away at joints and along bedding planes. As a result, several distinct landforms evolve over time.

Headland

Sea stack

Sea cave

Sea arch

Wave-cut notch

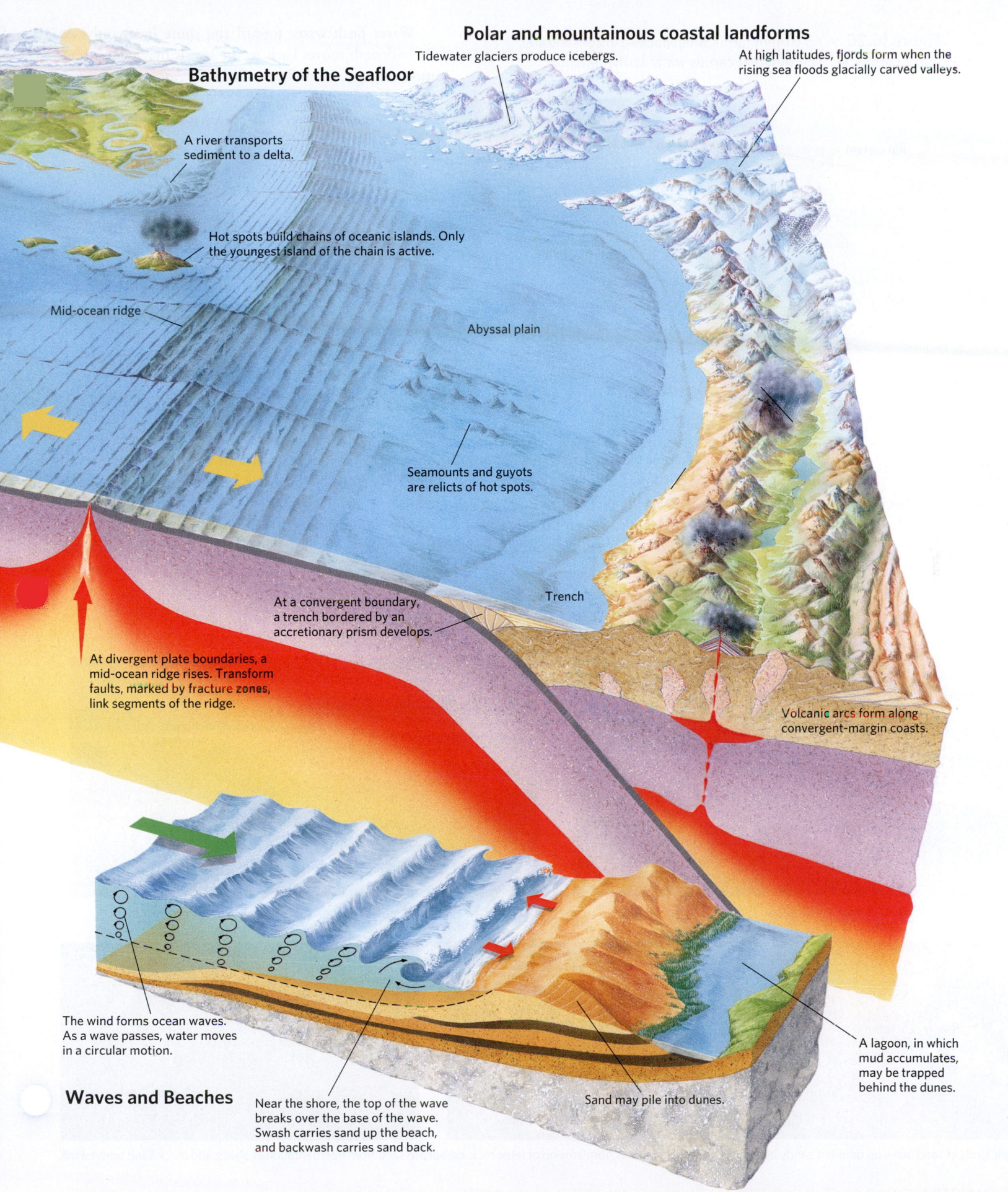

Polar and mountainous coastal landforms

Tidewater glaciers produce icebergs.

At high latitudes, fjords form when the rising sea floods glacially carved valleys.

Bathymetry of the Seafloor

A river transports sediment to a delta.

Hot spots build chains of oceanic islands. Only the youngest island of the chain is active.

Mid-ocean ridge

Abyssal plain

Seamounts and guyots are relics of hot spots.

Trench

At a convergent boundary, a trench bordered by an accretionary prism develops.

At divergent plate boundaries, a mid-ocean ridge rises. Transform faults, marked by fracture zones, link segments of the ridge.

Volcanic arcs form along convergent-margin coasts.

The wind forms ocean waves. As a wave passes, water moves in a circular motion.

Waves and Beaches

Near the shore, the top of the wave breaks over the base of the wave. Swash carries sand up the beach, and backwash carries sand back.

Sand may pile into dunes.

A lagoon, in which mud accumulates, may be trapped behind the dunes.

Figure 16.20 A rip current forms when backwash concentrates into a narrow, local current that carries water farther offshore.

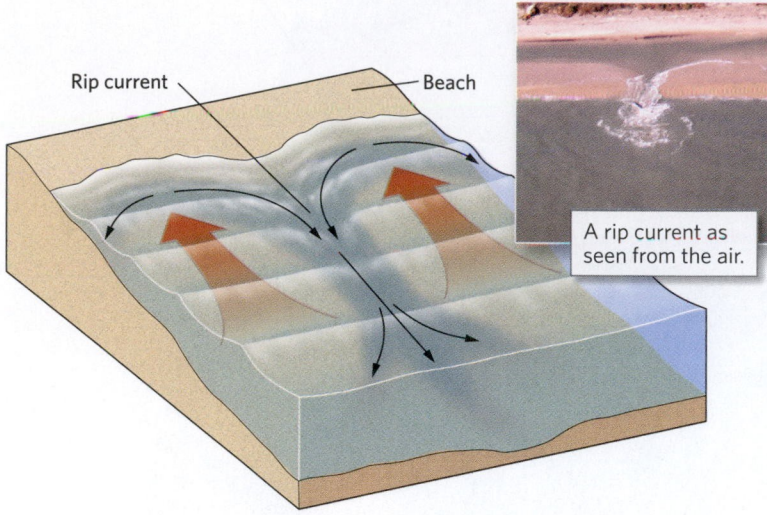

Rip current — Beach

A rip current as seen from the air.

Waves push water toward the shore incessantly. As the backwash moves back to the sea, it may concentrate into a strong seaward flow, called a *rip current*, that moves perpendicular to the shoreline (**Fig. 16.20**). Rip currents cause many drownings along beaches every year because they can carry unsuspecting swimmers out into deeper water.

Beaches and Tidal Flats

For millions of vacationers, the ideal holiday includes a trip to a **beach**, a gently sloping fringe of sediment along the shore. Some beaches consist of pebbles or boulders, whereas others consist of sand grains (**Fig. 16.21a, b**). This is no accident, for waves winnow out finer sediment such as silt and mud and carry it to quieter water offshore, where it settles. Storm waves can smash cobbles against one another with enough force to shatter them, but they have little effect on sand, for sand grains can't collide with enough energy to crack. So cobble beaches persist only where nearby cliffs supply large rock fragments, because without a continuing source of large fragments, cobbles on a beach would eventually be reduced to smaller grains.

The composition of sand itself varies from beach to beach because different sands come from different sources (**Fig. 16.21c**). Sands derived from the weathering

> **Did you ever wonder . . .**
> why beautiful sandy beaches don't form along all coasts?

eliminate that angle. Where waves do arrive at the shore obliquely, water in the nearshore region has a component of motion that trends parallel to the shore. This **longshore current** causes swimmers floating in the water just offshore to drift gradually in a direction parallel to the shoreline (see Fig. 16.19).

Figure 16.21 The type of sediment that makes up a beach varies with location.

(a) Cobble-sized clasts make up a beach in Oregon.

(b) A beach in Puerto Rico consists of quartz sand.

Quartz sand

Carbonate sand

Basalt sand

(c) Different kinds of sand make up different sandy beaches. Quartz sand comes from erosion of felsic rock, carbonate sand from fragmentation of shells, and black sand from basalt.

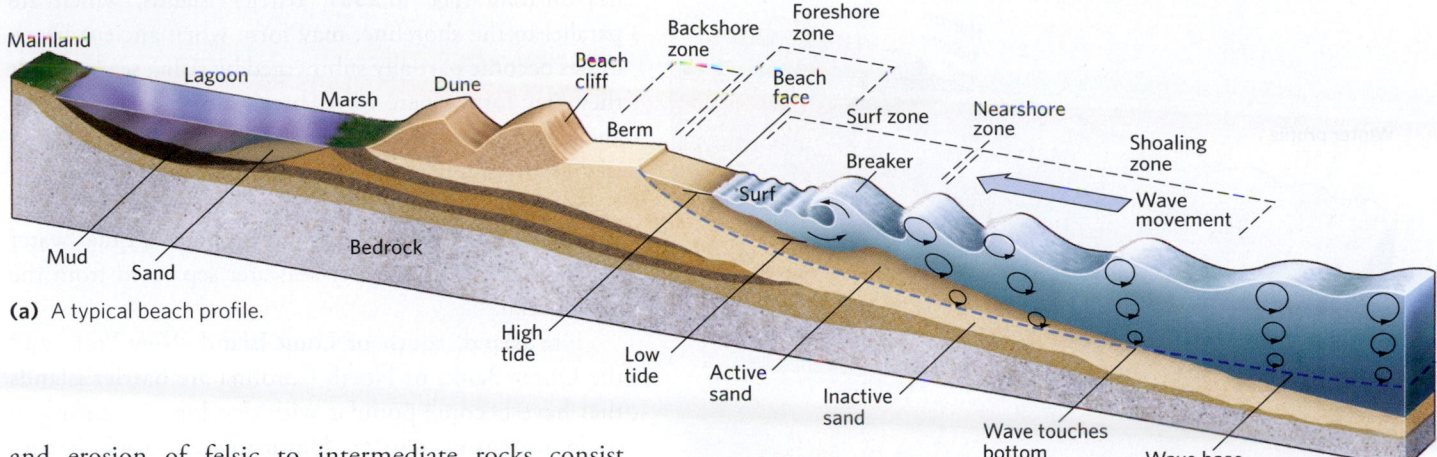

(a) A typical beach profile.

and erosion of felsic to intermediate rocks consist mainly of quartz, a durable mineral. The other minerals in such rocks weather chemically to form clay, which washes away in the waves. Beaches made from the erosion of limestone, coral reefs, or shells consist of carbonate sand, including tiny shell chips. And beaches derived from the erosion of basalt boast black sand, made of tiny basalt grains.

Beaches consist of distinct zones, as illustrated by a **beach profile**, a cross section drawn perpendicular to the shore **(Fig. 16.22a)**. Starting from the sea and moving landward, the *foreshore zone* includes the intertidal zone, across which the tide rises and falls, as well as the *beach face*, a steeper, concave-up area that forms where the swash of the waves actively scours the sand. The *backshore zone* extends from a small step, cut by high-tide swash, to the far edge of the beach farther inshore. The backshore zone includes one or more *berms*, horizontal to slightly landward-sloping terraces made up of sediment deposited during storms **(Fig. 16.22b)**. The sand on the beach, if exposed to persistent winds, may build into sand dunes, similar to those of deserts, along the landward edge of the beach. The inshore edges of some beaches, however, border cliffs or vegetated land not affected by waves.

Beach sand doesn't sit still, for typical wave action moves an *active sand layer* back and forth. (The *inactive sand layer* below moves only during severe storms or not at all.) The width of a beach ultimately reflects the width of the area subjected to frequent wave action. Beach profiles may vary seasonally depending on the types of storms in the area during different seasons. For example, in mid-latitudes, winter storms tend to be stronger and more frequent than summer ones. The larger, shorter-wavelength waves of winter storms wash beach sand into deeper water, making the beach narrower. But smaller summer waves with longer wavelengths bring sand in from offshore and deposit it on the beach **(Fig. 16.23)**.

(b) A berm on a Cape Cod beach.

Where waves roll onto the shore at an angle, grains in the active sand layer follow a sawtooth pattern of movement that results in a gradual net transport of the sediment parallel to the beach; such movement is called **longshore drift** (or *beach drift*). This sawtooth pattern happens because the swash of a wave moves perpendicular to the wave crest, so an oblique wave carries sediment diagonally up the beach. Backwash, however, flows straight down the slope of the beach due to gravity. Over time, longshore drift may move sand tens to hundreds of kilometers along a coast **(Fig. 16.24)**. For this reason, geologists sometimes describe beaches as "rivers of sand."

Where the coastline indents landward, longshore drift can stretch a beach out into open water and produce a **sand spit** **(Fig. 16.25a, b)**. Some sand spits grow across the opening of a bay or estuary to form a *baymouth bar* **(Fig.16.25c)**. In regions where the coast has a very gentle slope and there is an abundant supply of sediment, a

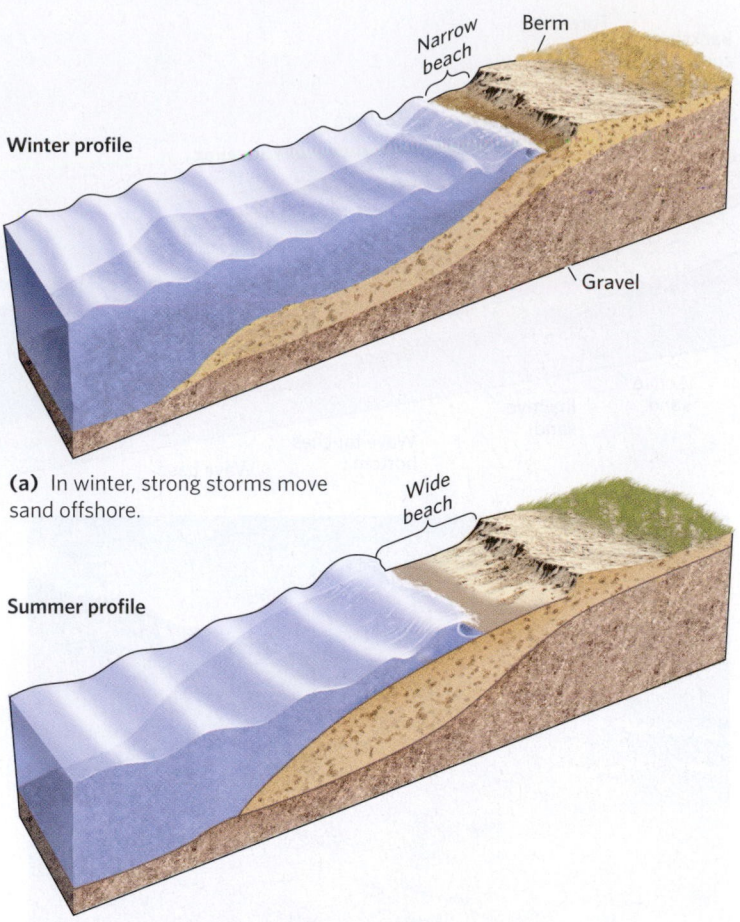

Figure 16.23 The width of a beach can change seasonally.

Narrow beach

Berm

Winter profile

Gravel

(a) In winter, strong storms move sand offshore.

Wide beach

Summer profile

(b) In summer, gentler waves carry sand back to the beach.

narrow ridge of sand, known as an *offshore bar* if submerged, or a **barrier island** if its crest lies above sea level, lies offshore **(Fig. 16.25d)**. Barrier islands, which are parallel to the shoreline, may form when ancient beach dunes become partially submerged by rising sea level. Or they may form where waves break offshore, lose energy, and deposit some of their sediment load. Some barrier islands are simply very long sand spits that built out into a region of shallow sea offshore. The water between a barrier island and the mainland becomes a quiet-water **lagoon**, a body of shallow seawater separated from the open ocean.

Fire Island, south of Long Island, New York, and the Outer Banks of North Carolina are barrier islands that have become popular with developers wanting to build expensive resorts. Unfortunately, such barrier islands, on a time scale of decades to millennia, are temporary geologic features. Wind and waves pick up sand from the ocean side of the island and drop it on the lagoon side, causing the island to migrate landward; longshore drift shifts the sand of the island and modifies its shape, and storms may breach the island to produce an *inlet* (a narrow passage of water between a lagoon and the ocean). So an area might seem like a prime construction site one day, but may be gone in the near future.

Tidal flats, which are broad, nearly horizontal areas of mud and silt that are exposed, or nearly exposed, at low tide but submerged at high tide, develop in regions

Figure 16.24 The concept of longshore drift. Swash carries sand obliquely up the beach, whereas backwash carries it straight downslope. So sand grains follow a sawtooth pattern, yielding longshore drift (gold arrow).

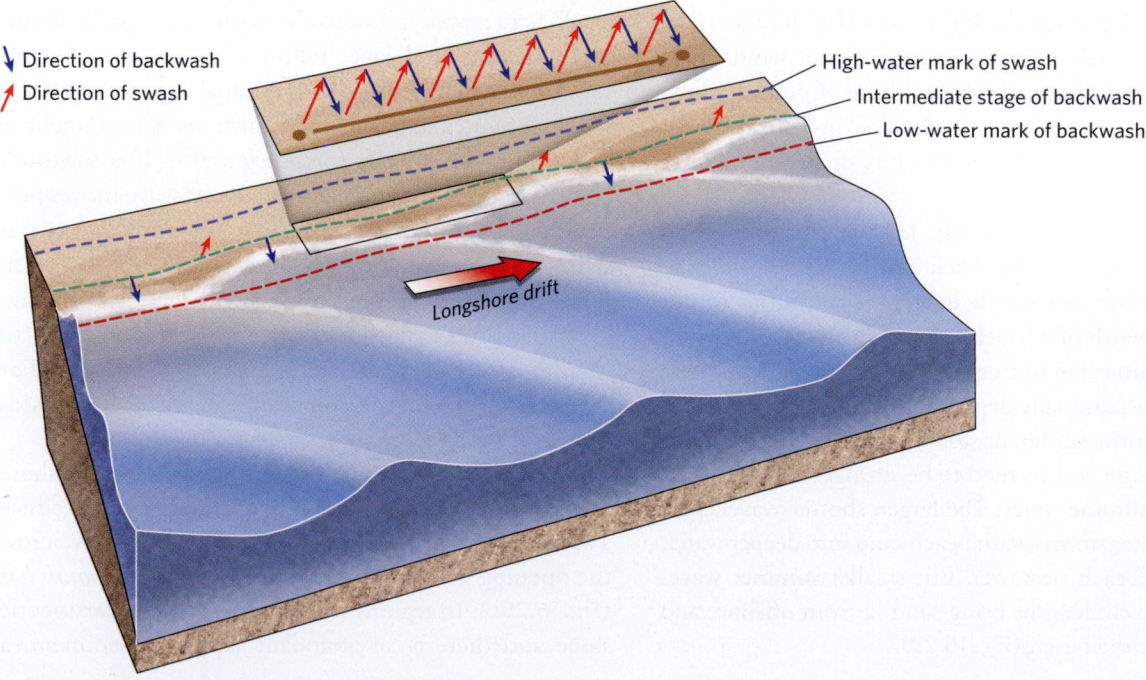

Direction of backwash

Direction of swash

High-water mark of swash

Intermediate stage of backwash

Low-water mark of backwash

Longshore drift

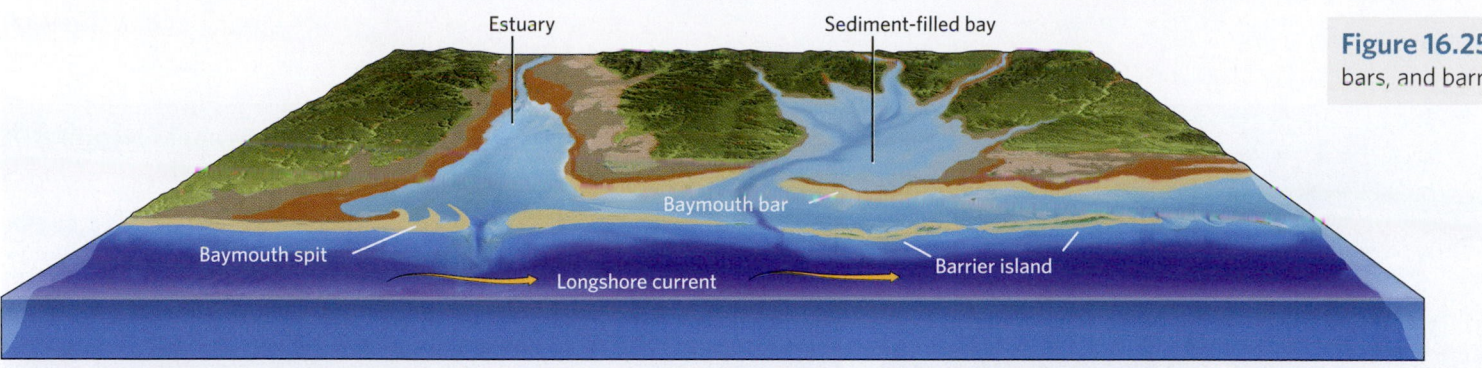

Estuary   Sediment-filled bay

Figure 16.25 Spits, bars, and barrier islands.

Baymouth bar

Baymouth spit

Longshore current   Barrier island

(a) Longshore drift can generate sand spits and baymouth bars. Sedimentation may fill in the region behind the bar.

(b) An example of a sand spit.

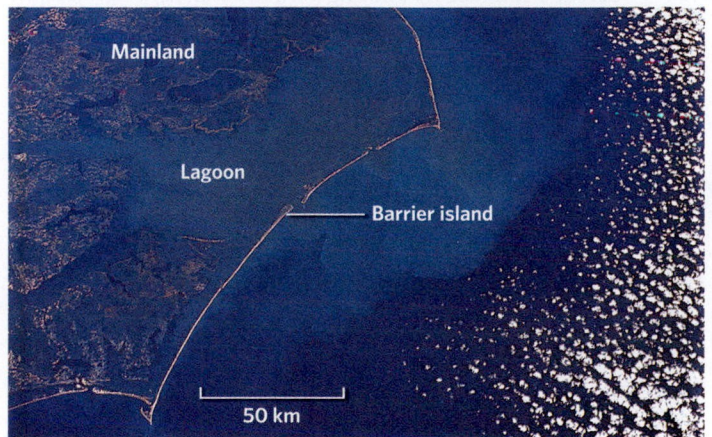

Mainland

Lagoon

Barrier island

50 km

(d) Barrier islands may be separated from the mainland by a lagoon.

(c) An example of a baymouth bar.

protected from strong wave action **(Fig. 16.26)**. They are typically found along the margins of lagoons or on shores protected by barrier islands, where, in relatively quiet water, mud and silt can accumulate in thick, sticky layers. Tidal flats provide a home for burrowing organisms such as clams and worms, so *bioturbation* (which means stirring by life) constantly mixes tidal-flat sediments and disrupts bedding.

Because of the movement of sediment on beaches and in adjacent areas, the **sediment budget**—the

balance between sediment supplied and sediment removed—plays an important role in determining the long-term evolution of a coastal area. Let's look at how the sediment budget works for a small segment of beach **(Fig. 16.27)**. Sand may be added to the segment by local rivers. It may also be brought from just offshore by waves, or from far away by longshore drift. (In fact, the large quantity of sand along the beaches of the southeastern United States may have been carried there by longshore drift from the outwash plains of Pleistocene Ice-Age glaciers in coastal New England.) Conversely, some of the sand may be removed from the segment by longshore drift, be carried offshore by backwash or rip currents, or be blown inland by wind. This sediment either settles in deeper water just offshore or tumbles down a submarine canyon into the deep sea. If new sand does not replace the removed sand, the beach segment grows narrower. If, however, the supply of sand exceeds the amount that washes away, the beach becomes wider.

Rocky Coasts

More than one ship has met its end, smashed and splintered in the spray and thunderous surf, on a **rocky coast**, where

See for **yourself**

Chicago Lakefront Preservation

Latitude: 41°54′59.76″ N
Longitude: 87°37′36.41″ W

Zoom to 3 km (~9,800 feet) and look down.

Sandy beaches fringe the shores of Lake Michigan at Chicago. Waves strike the shore obliquely. As a result, longshore drift carries sand southward. The city constructed a series of groins in an attempt to prevent beach erosion.

Figure 16.26 Tidal flats.

(a) Tidal flats are broad, flat areas, usually floored by mud and silt, that are submerged only during high tide.

(b) This tidal flat exposed at low tide along the coast of Wales is a vast expanse of mud.

Figure 16.27 The sediment budget along a segment of coast. Sediment enters the system from rivers and by erosion of the land along the shore. Some sediment is moved along the coast by longshore drift, and some washes out to sea or avalanches down submarine canyons. Wind may blow beach sand inland.

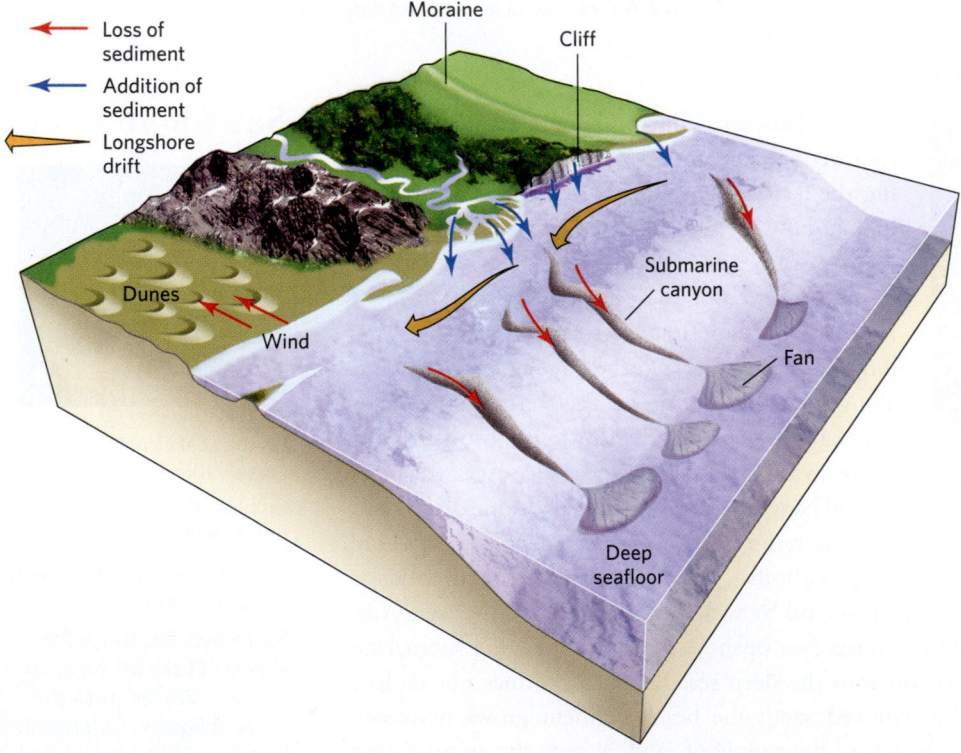

until they shatter, and it can squeeze air into cracks, creating enough force to pluck chunks free from bedrock. Further, because of its turbulence, the water hitting a cliff face carries suspended sand that can abrade the cliff.

The combined effects of shattering, wedging, and abrading, together called **wave erosion**, can gradually undercut a cliff face by making a *wave-cut notch* (**Fig. 16.29a**). Undercutting continues until the overhanging rock becomes unstable and breaks away at a joint, toppling down to form a pile of rubble at the base of the cliff. Then, over time, wave action breaks up the rock until fragments become small enough to be transported offshore by moving water. In this way, wave erosion continually cuts away at a rocky coast, so that the cliff gradually migrates inland. Such *cliff retreat* leaves behind a **wave-cut platform**, or *wave-cut bench*, that becomes visible at low tide (**Fig. 16.29b**).

Other processes besides wave erosion break up rocks along coasts. Salt wedging, for example, occurs as salt spray coats the cliff face above the waves and infiltrates pores in the rock, then dries out so that salt crystals grow. Salt wedging typically produces countless pits, separated by ridges, on the rock surface, a pattern known as *honeycomb weathering*. Biological processes also contribute to erosion as plants and animals in the intertidal zone bore into the rocks and gradually break them up.

Many rocky coasts start out with an irregular coastline, with **headlands** protruding into the sea and **embayments** set back from the sea (**Fig. 16.29c**). Such irregular coastlines tend to be temporary features in the context of geologic time. As a result of wave refraction, wave energy focuses on headlands and disperses in embayments.

bedrock cliffs rise directly from the sea (**Fig. 16.28**). Lacking the protection of a beach, rocky coasts feel the full impact of ocean breakers. Water pressure generated during the impact of a breaker can pick up boulders and smash them together

Figure 16.28 Exposed bedrock along rocky coasts is subjected to powerful, concentrated wave action. Rocky coasts develop unique landforms as a result.

(a) Cliffs rise from the sea along the coast of Wales.

(b) There's no beach to protect these rocky headlands of France from wave erosion.

Figure 16.29 Wave erosion of a rocky coast.

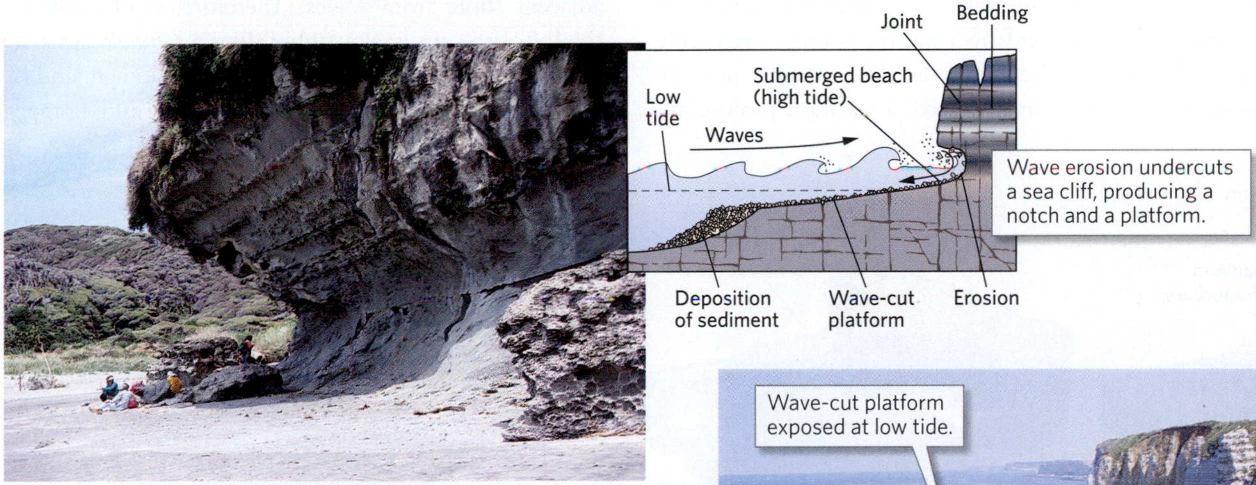

(a) A wave-cut notch.

Joint   Bedding

Submerged beach (high tide)

Low tide   Waves

Wave erosion undercuts a sea cliff, producing a notch and a platform.

Deposition of sediment   Wave-cut platform   Erosion

Wave-cut platform exposed at low tide.

(b) A wave-cut platform at the foot of the cliffs at Êtretat, France.

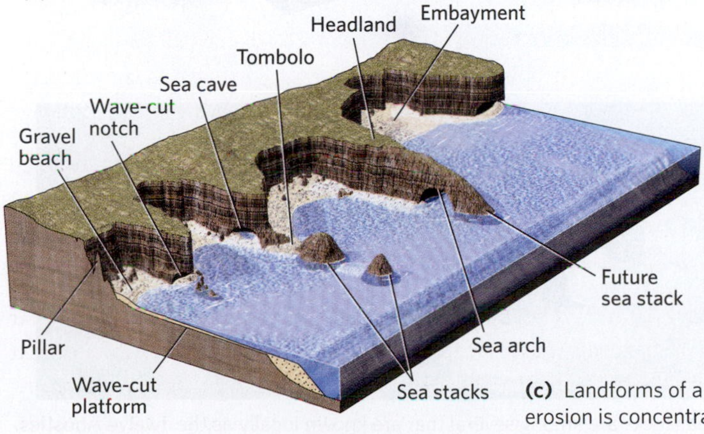

Headland   Embayment

Tombolo

Sea cave

Wave-cut notch

Gravel beach

Pillar

Future sea stack

Sea arch

Sea stacks

Wave-cut platform

(c) Landforms of a rocky shore. Beaches form in embayments, whereas erosion is concentrated at headlands.

Figure 16.30 Wave refraction at headlands.

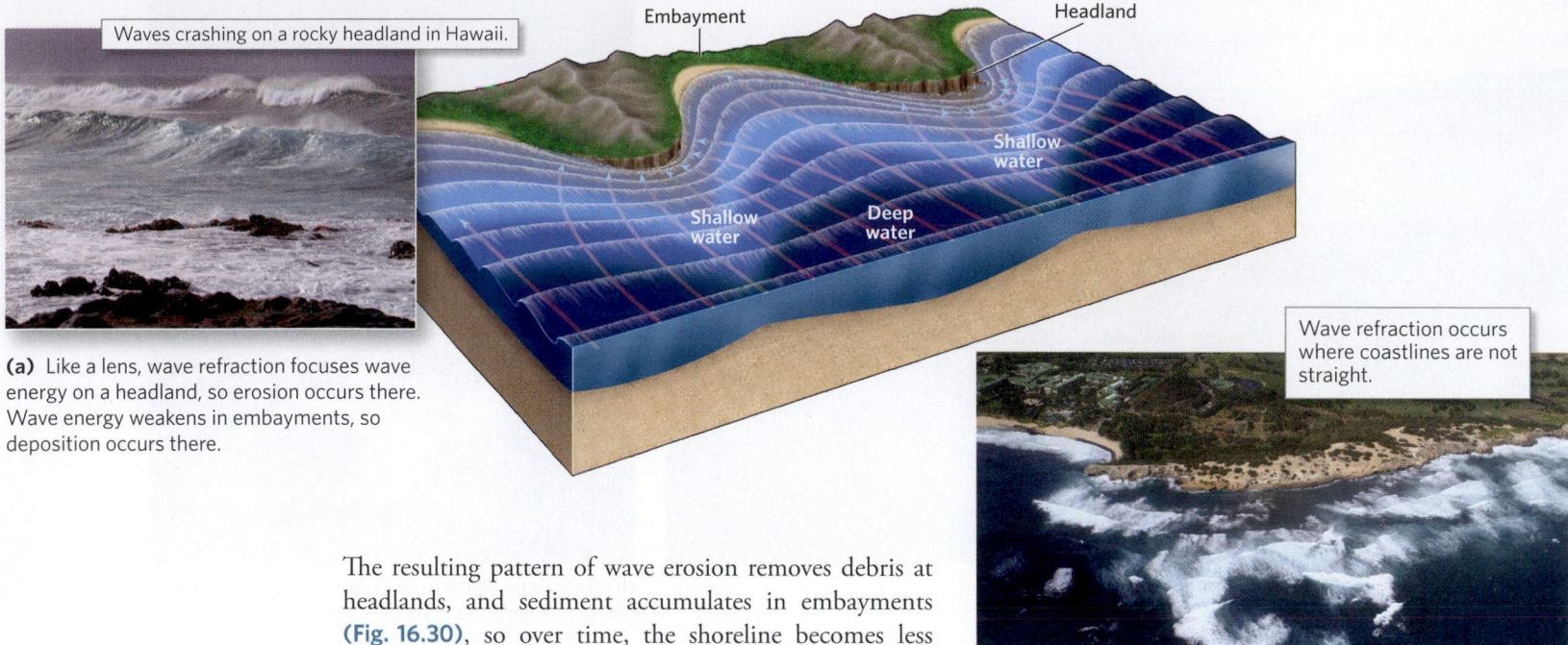

Waves crashing on a rocky headland in Hawaii.

(a) Like a lens, wave refraction focuses wave energy on a headland, so erosion occurs there. Wave energy weakens in embayments, so deposition occurs there.

Embayment

Headland

Shallow water

Shallow water

Deep water

Wave refraction occurs where coastlines are not straight.

The resulting pattern of wave erosion removes debris at headlands, and sediment accumulates in embayments **(Fig. 16.30)**, so over time, the shoreline becomes less irregular.

A headland erodes in stages **(Fig. 16.31a)**. Because of refraction, waves curve and attack the sides of a headland, slowly eating through it to create a *sea arch* connected to the mainland by a narrow bridge. Eventually, the arch collapses, leaving isolated **sea stacks** just offshore **(Fig. 16.31b)**. Once formed, a sea stack protects the

(b) An example of wave erosion of a headland on the coast of Hawaii.

adjacent shore from waves. Therefore, sand collects in the lee of the stack, slowly building a *tombolo*, a narrow ridge of sand that links the sea stack to the mainland (see Fig. 16.29c). Sea stacks eventually crumble and disappear.

Figure 16.31 Stages in the erosion of a headland.

Headland (promontory)

Sea arch

Sea stack

(a) Waves carve out a sea arch, which eventually collapses, leaving a sea stack.

Waves carve sea caves.

A sea arch forms.

The arch collapses.

Time

(b) An example of a sea arch (left) and sea stacks (right) on the southern coast of Australia. The sea stacks are among several that are known locally as the Twelve Apostles.

Estuaries

Along some coasts, a relative rise in sea level causes the sea to flood river valleys that merge with the coast, resulting in **estuaries**, where seawater and river water interact. Estuaries can develop where a river cut a valley below present sea level at a time when sea level was lower, or where the region containing the river valley is slowly subsiding (sinking). You can recognize an estuary on a map by the branching pattern of its river-carved coastline, inherited from the shape of dendritic drainage networks (**Fig. 16.32a**). Estuaries are commonly bordered by marshes that flood at high tide (**Fig. 16.32b**).

Seawater and river water may interact in two different ways within an estuary. In quiet estuaries that are protected from wave action or river turbulence, the water becomes stratified, and the denser seawater flows upstream as a wedge beneath the less dense freshwater during flood tide. Such a *saltwater wedge* migrates about 100 km (62 miles) up the Hudson River in New York, and about 40 km (25 miles) up the Columbia River in Oregon. In turbulent estuaries, such as the Chesapeake Bay, seawater and river water mix to produce nutrient-rich brackish water with a salinity between those of oceans and rivers. Estuaries typically host complex ecosystems inhabited by unique species of shrimp, clams, oysters, worms, and fish that can tolerate large changes in salinity.

Coastal Fjords

During the Pleistocene Ice Age, glaciers carved deep valleys in coastal mountain ranges. When the ice age came to a close, the glaciers melted away, leaving deep, U-shaped valleys (see Section 14.7). When the water stored in the glaciers returned to the sea and caused sea level to rise, it flooded glacial valleys to produce coastal **fjords**, narrow fingers of the sea surrounded by hills or mountains. Because of their deep-blue water and steep walls of polished rock, fjords are distinctively beautiful (**Fig. 16.33**).

Take-home message . . .

A great variety of landscapes form along coasts. In regions with ample sediment supply, beaches build up, each with a distinctive profile. Longshore drift leads to the development of sand spits. In areas protected from wave action, muddy tidal flats develop. Rocky coasts feel the brunt of wave erosion and evolve over time as focused wave energy eats away at headlands. Where the sea floods the mouth of a river valley, an estuary develops, and where the sea floods a glacially carved valley, a fjord results.

Quick Question -
How do lagoons and barrier islands develop?

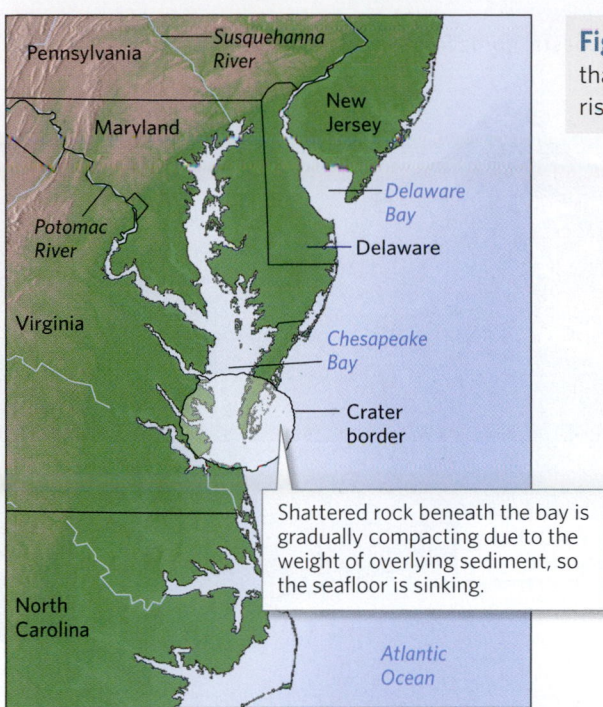

Figure 16.32 River valleys that are flooded by sea-level rise are called estuaries.

(a) Chesapeake Bay is an estuary at the mouth of the Susquehanna and Potomac Rivers. Recent research suggests that the mouth of these rivers subsided because a meteorite crater lies underneath.

> Shattered rock beneath the bay is gradually compacting due to the weight of overlying sediment, so the seafloor is sinking.

(b) The land surrounding the main channel of an estuary floods during high tide along the coast of the Isle of Wight, England.

16.5 Organic Coasts

So far, we've discussed coasts where the nature and distribution of sediments and rocks primarily determine the character of the shore. Along many coasts, however, living organisms control the landforms of shore and nearshore regions. Such **organic coasts** include coastal wetlands, where salt-resistant plants thrive, and coral reefs, where tiny marine animals build mounds of biochemical limestone. The nature of an organic coast depends on the types of organisms that live there, which, in turn, depends on climate.

Coastal Wetlands

It would be hard to find a more dramatic contrast to the crashing waves of rocky coasts than the gentlest type of

See for **yourself**

Fjords of Norway

Latitude: 60°53′56.09″ N
Longitude: 5°12′31.84″ E

Zoom to an elevation of 80 km (50 miles) and look down.

During the last ice age, glaciers carved deep, steep-sided valleys into the mountains of western Norway. When the glaciers melted and sea level rose, the valleys filled with water, forming fjords.

Glacial valleys have a U shape, so fjords have steep sides.

A fjord in Norway

nearshore environment, the coastal wetland. A **coastal wetland** is a vegetated, flat-lying stretch of coast that floods at high tide, becomes partially exposed at low tide, and does not feel the impact of strong waves. Some wetlands grow directly along the shore, whereas others occupy lagoons separated from the sea by a beach. Vast numbers of marine species spawn in wetlands—in fact, wetlands account for 10% to 30% of all marine organic productivity, even though they constitute just a tiny portion of ocean regions.

In mid-latitude climates, coastal wetlands include *swamps* (wetlands dominated by trees), *marshes* (wetlands dominated by grasses; **Fig. 16.34a**), and *bogs* (wetlands dominated by mosses and shrubs). In tropical or subtropical climates (between 30° N and 30° S), *mangrove swamps* thrive along the shore **(Fig. 16.34b)**. Mangroves are trees that have evolved roots that can filter salt out of water, so they can survive in either freshwater or saltwater. Some mangrove species form a broad network of roots above the

water surface, making the tree look like an octopus standing on its tentacles, and some send up small protrusions from the roots that rise above the water and allow the plant to breathe. Dense stands of mangroves absorb the impact of waves and, by doing so, prevent coastal erosion.

Coral Reefs

Along the azure coasts of Hawaii, colorful mounds of living coral form shallow reefs just offshore **(Fig. 16.35a)**. Snorkelers swimming over a reef will see a great variety of different coral species. Some species build mounds that look like brains, others like elk antlers, and still others like delicate fans **(Fig. 16.35b)**. Sea anemones, sponges, clams, and many other organisms grow on and around the coral. Though at first glance, coral looks like a plant, it is actually a colony of tiny invertebrates related to jellyfish. An individual coral animal, or polyp, has a tube-like body with a head of tentacles **(Fig. 16.35c)**. Corals obtain some

Figure 16.34 Coastal wetlands.

(a) A marsh along the coast of Cape Cod.

(b) A mangrove swamp growing along the shore of southern Florida.

Figure 16.35 The character and evolution of coral reefs.

(a) A reef off the shore of Honolulu, Hawaii.

(b) A reef as seen underwater; corals have a great variety of colors and shapes.

(c) A close-up of living coral polyps.

of their nutrition by filtering nutrients from seawater; the remainder comes from algae that live within the corals' tissues. Corals have a *symbiotic* (mutually beneficial) relationship with these algae, in that the photosynthetic algae provide nutrients and oxygen to the corals while the corals provide carbon dioxide and other nutrients to the algae.

Coral polyps secrete calcite shells, which gradually build into a mound of solid limestone. At any given time, only the surface of the mound is alive; the mound's interior consists of shells from previous generations of coral **(Fig. 16.36)**. The realm of shallow water underlain by coral mounds, associated organisms, and debris constitutes a **coral reef**. Living corals must remain submerged,

so the tops of the shallowest reefs lie just below the level of low tide, and they must receive abundant sunlight, so most growing reefs lie at depths of less than 60 m (200 feet). Coral reefs absorb wave energy and thus serve as a living buffer that protects coasts from erosion.

Corals need clear, warm (18°C–30°C, or 64°F–86°F) water with normal ocean salinity, so coral reefs grow only along unpolluted coasts at latitudes below about 30° **(Fig. 16.37a)**. The largest reef in the world, the Great Barrier Reef, extends for a distance of close to 2,000 km (1,200 miles) along the northeastern coast of Australia and reaches a width of 120 km (75 miles) **(Fig. 16.37b)**. Most reefs are much smaller.

Figure 16.36 What's inside a reef?

(a) A quarry in Florida cuts into coral that formed thousands of years ago.

(b) A close-up shows the internal skeleton of a long-dead coral.

Figure 16.37 Reefs of the world.

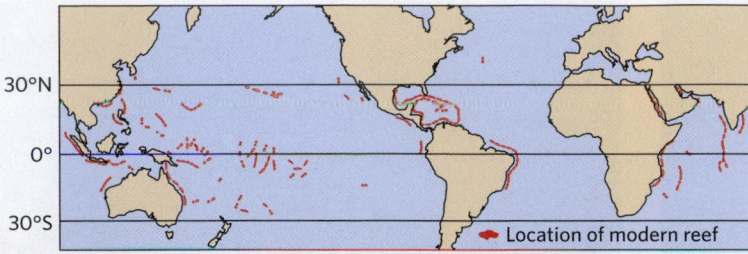

(a) Most coral reefs lie between 30° N and 30° S latitude.

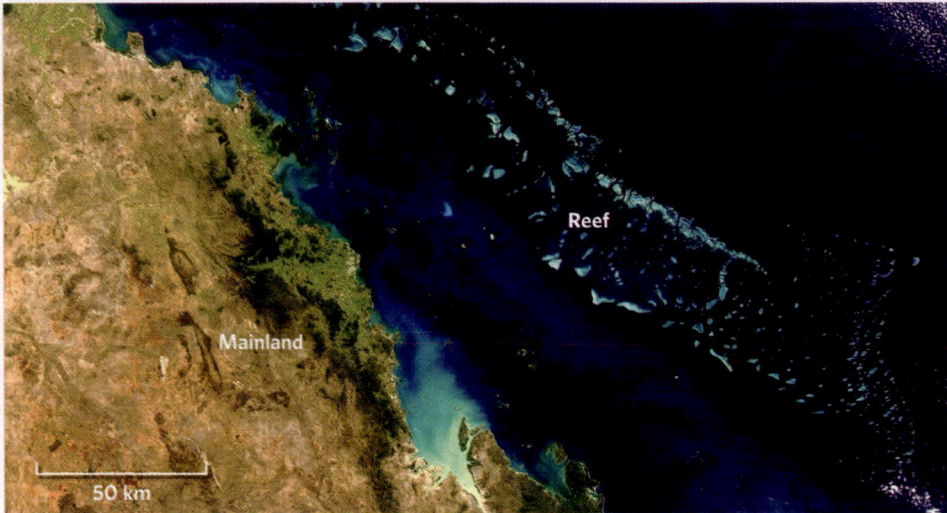

(b) The Great Barrier Reef forms shoals off the coast of northeastern Australia.

Marine geologists distinguish three different shapes of coral reefs **(Fig. 16.38)**. A *fringing reef* forms directly along the coast, a *barrier reef* lies offshore (separated from the coast by a lagoon), and an *atoll* forms a circular ring surrounding a lagoon. As Charles Darwin first recognized back in 1859, coral reefs associated with oceanic islands in the Pacific start out as fringing reefs, then later become barrier reefs and, finally, atolls. Darwin suggested, correctly, that this progression reflects the continued growth of the reef as the island around which it formed gradually subsides. Eventually, the reef itself sinks too far below sea level to remain alive and becomes the cap of a guyot.

Take-home message . . .

> Organic coasts are places where living organisms control landforms along the shore. Some organic coasts consist of wetlands in the intertidal zone, which host grasses, mosses and shrubs, or mangrove trees, depending on climate. Coral reefs, which develop primarily in warm, shallow, clear water, consist of generations of shells produced by corals and other organisms.

Quick Question -
Why do fringing reefs evolve into atolls around oceanic islands of the Pacific?

16.6 Causes of Coastal Variation

Emergent versus Submergent Coasts

Changes in *relative sea level*, meaning the position of sea level relative to the land surface at a given location, can be a consequence either of *global sea-level changes* (the rise or fall of sea level worldwide; see Box 10.2) or of local vertical movement (uplift or subsidence) of coastal land areas, which can take place even when global sea level remains fixed. Vertical movement of the land may be a consequence of plate interactions, as happens where subduction causes compression and thickening in the crust of the overriding plate at a convergent boundary. Vertical movement may also reflect the addition or removal of a load (such as a glacier) on the crust's surface or the cooling and/or heating of lithosphere (which changes lithospheric thickness and density). In some cases, local changes in sea level may result from human activity, as happens when people pump out so much groundwater that pores between grains in the underground sediment collapse and the land surface sinks (see Section 13.12).

Geologists refer to a coastal location where the land is rising relative to sea level as an **emergent coast**. Along emergent coasts, steep-sided cliffs or hills may border the shore. In some cases, several step-like terraces delineate emergent coasts; these terraces form where a wave-cut platform has time to develop before uplift brings the platform above the high-tide level **(Fig. 16.39a)**. Coasts where the land is sinking relative to sea level are known as **submergent coasts (Fig. 16.39b)**. Estuaries and fjords, landforms that develop when the sea floods coastal valleys, characterize some submergent coasts. Submergence of **coastal plains**, nearshore landscapes of low relief, may produce broad wetlands and lagoons.

Changes in Sediment Supply and Climate

The quantity and character of sediment supplied to a shore affect a coast's character. For example, **erosional coasts** form where wave erosion washes sediment away faster than it can be supplied; such coasts recede landward and may become rocky if there's insufficient sand to supply a beach. In contrast, **accretionary coasts**, those that receive more sediment than they lose by erosion, grow seaward and develop broad beaches.

Climate also affects the character of a coast. Shores that enjoy generally calm weather erode less rapidly than those constantly subjected to ravaging storms. A sediment supply may be large enough to generate an accretionary coast in a calm environment, but insufficient to prevent the development of an erosional coast in a stormy environment. The climate also affects biological activity along coasts. For example, in the warm water of tropical climates, mangrove swamps flourish along the shore and

Figure 16.38 Evolution of reefs around oceanic islands in the Pacific.

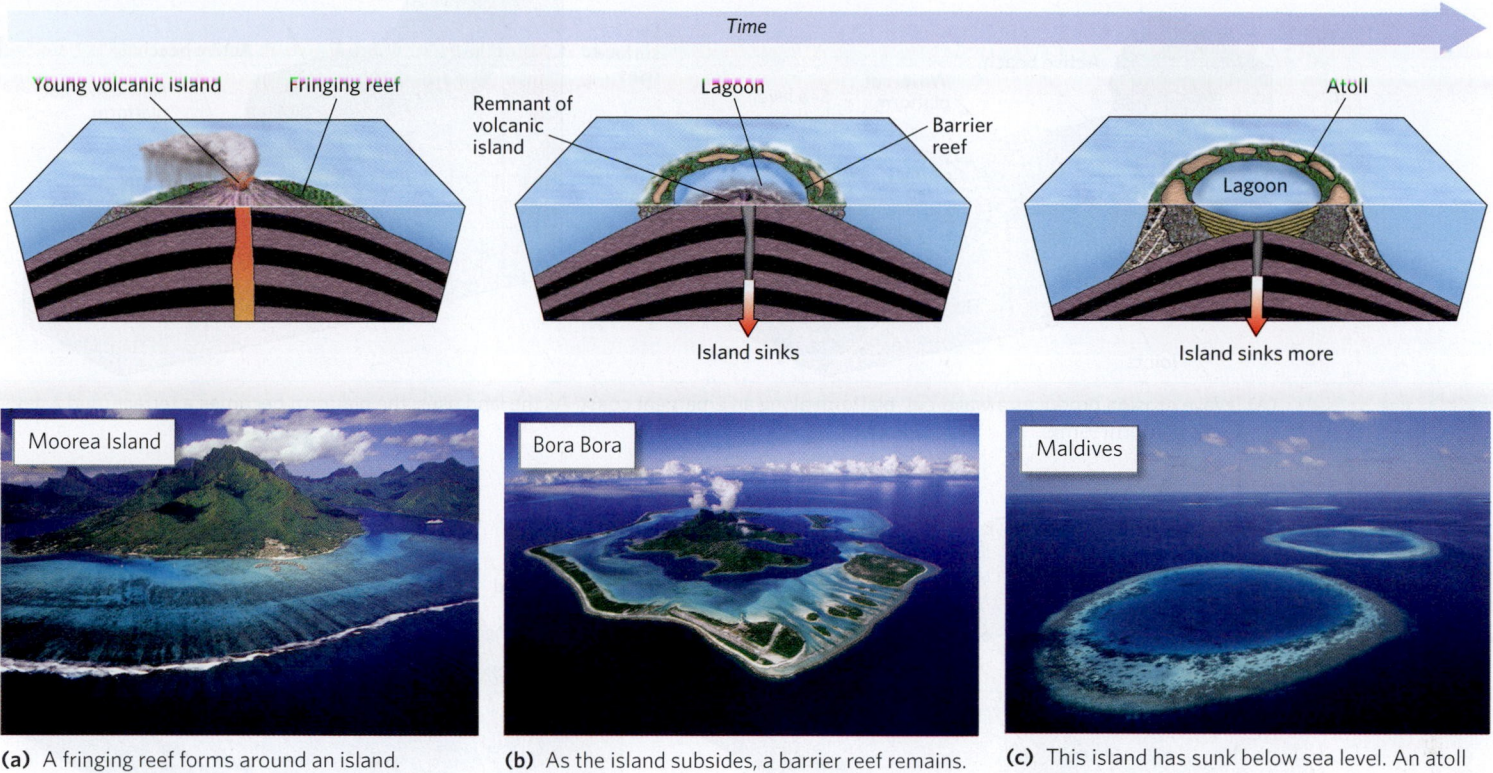

Time

Young volcanic island Fringing reef

Remnant of volcanic island Lagoon Barrier reef

Atoll Lagoon

Island sinks Island sinks more

Moorea Island

Bora Bora

Maldives

(a) A fringing reef forms around an island.

(b) As the island subsides, a barrier reef remains.

(c) This island has sunk below sea level. An atoll remains.

coral reefs form offshore. In cooler climates, marshes develop, and in polar regions, the coast may be a stark environment of lichen-covered rock and barren sediment.

Take-home message . . .

At emergent coasts, the land is rising relative to sea level, and terraces may develop. In contrast, at submergent coasts, sinking land allows the sea to flood river valleys and glacial valleys or to form broad wetlands. Whether coasts are emergent or submergent depends not only on global sea-level change, but also on local uplift or subsidence. The sediment supply relative to the erosion rate determines whether beaches build outward or erode away.

Quick Question -----------------------------
How does climate affect the character of the coast?

16.7 Challenges to Living on the Coast

People tend to view a shoreline as a permanent entity. But, as we have seen in this chapter, shorelines, like many geologic features, are ephemeral, in that they can change on a time scale of hours to millennia. Let's look at some of the changes now taking place along shorelines and how people respond to them.

Coastal Storms

When the winds of large storms, such as hurricanes, blow across the surface of the sea, they can generate immense waves. When these waves reach the shore, they become giant breakers that cause intense erosion and can destroy buildings, ships, and ports. During some types of storms, the wind, as well as changes in atmospheric pressure, causes sea level to rise to form a mound beneath the storm. This **storm surge** is independent of waves and tides (see Section 18.4). When storm surge reaches the coast at the same time as high tide and is accompanied by high waves, sea level temporarily becomes so high that ocean water can inundate land far inland of a beach. In a matter of hours, a storm can radically alter a beach that took centuries or millennia to build. The backwash of storm waves sweeps vast quantities of sand seaward, leaving the beach a skeleton of its former self. Surf during storms can submerge and shift barrier islands and can cut new inlets. Waves and wind together can rip out mangrove swamps and marshes and break up coral reefs, thereby destroying the organic buffer that normally protects the coast and leaving it vulnerable to accelerated erosion for years to come. Surf can also undermine coastal cliffs and trigger landslides.

Of course, major storms also destroy human constructions: erosion undermines seaside buildings, causing them to collapse into the sea; wave impacts smash

See for yourself

Coral Reefs, Pacific

Latitude: 16°47′26.92″ S
Longitude: 150°58′1.27″ W

Zoom to 20 km (~12.5 miles) and look down.

Waves break on the outer edge of the fringing reef of Huahine. Coral buildups lie beneath the surface, and sand partially fills the lagoon. The island itself represents the subsiding remnants of a hot-spot volcano.

Figure 16.39 Features of emergent and submergent coasts.

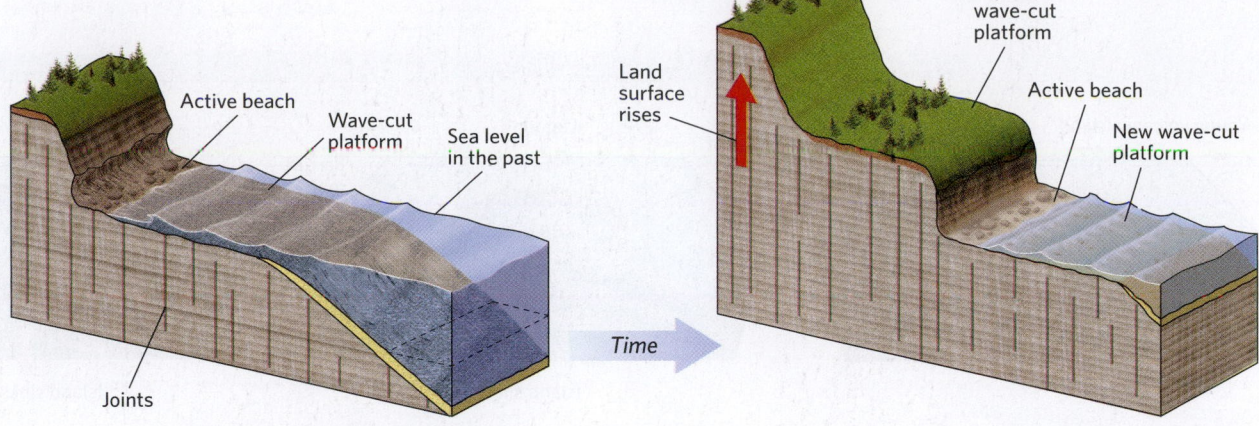

Active beach
Wave-cut platform
Sea level in the past
Joints

Land surface rises
Exposed wave-cut platform
Active beach
New wave-cut platform

Time

(a) Wave erosion produces a wave-cut platform along an emergent coast. As the land rises, the platform becomes a terrace, and a new wave-cut platform forms.

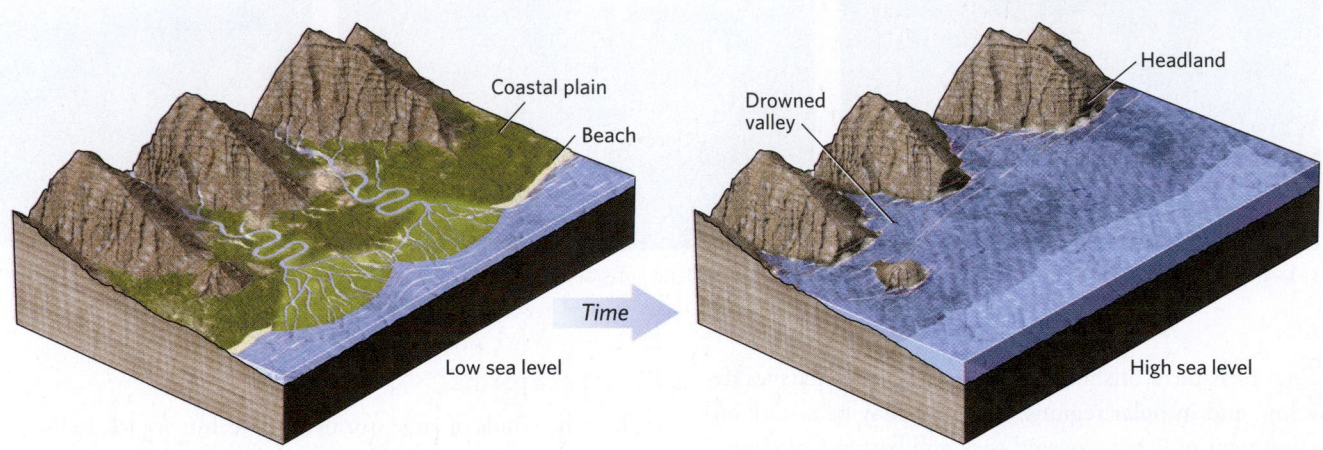

Coastal plain
Beach
Low sea level

Headland
Drowned valley
High sea level

Time

(b) Rivers drain valleys and deposit sediment on a coastal plain. As the land sinks, sea level rises and floods the valleys, and waves erode the headlands.

buildings to bits; and storm surge floats buildings off their foundations **(Fig. 16.40)**. Because storms can be immensely destructive, coastal population centers need safety and evacuation plans, which should ensure that residents know what to do when storms approach.

Ongoing Sea-Level Rise

Global sea level rises today at an estimated rate of about 3.2 mm per year (0.13 inches per year), as we discuss further in Chapter 20. At this rate, water levels will be a meter or two higher on a time frame of centuries. Many of the world's large cities have been built on coastal land less than a few meters above sea level, so sea-level rise will be disruptive. Already, some coastal cities are enduring flooding during times of particularly high tides, or when storms drive storm surge onto the land. By one estimate, 15% of the oceanic islands in the Pacific will become submerged; some low-lying oceanic islands have all but disappeared already. In the United States, flooding will affect a greater area of the east and Gulf coasts than of

the Pacific coast because slopes along the Pacific coast are much steeper **(Fig. 16.41)**. As saltwater encroaches on the land, groundwater may become salty, and as a result, plants without salt tolerance will die off.

How will society deal with coastal flooding due to this ongoing sea-level rise? Solutions remain a work in progress. Some options, all at great cost, include: moving towns inland; adding fill to make the land surface higher; constructing barriers against the sea; or installing pumps and drainage canals to keep rising waters out.

Beach Destruction—Beach Protection?

Storms can cause major changes to beaches overnight. But even less dramatic events, such as the loss of river sediment, a gradual rise in sea level, a change in the shape of a shoreline, or the destruction of coastal vegetation, can alter the sediment budget of a beach. If the sediment supply decreases, erosion eventually removes so much sediment that the beach becomes much narrower; sometimes beach erosion exposes underlying bedrock **(Fig. 16.42)**. Cliff

Figure 16.40 Examples of beach erosion due to a hurricane.

September 9, 2008

September 15, 2008

(a) A community in Galveston, Texas, before Hurricane Ike.

(b) The same location, after destruction by storm surge and storm waves.

retreat at the back edge of a beach may cause the whole beach to migrate landward, a process known as *beach retreat*. Surveys indicate that beach retreat at some locations takes place at rates of 1 to 2 m (3 to 6 feet) per year. Some homeowners react by picking up and moving their houses. Even large lighthouses have been moved to keep them from washing away or tumbling down eroded headlands.

In many parts of the world, beachfront property has great value. Property owners may construct artificial barriers to protect their stretch of coast or to shelter the mouth of a harbor from waves. These barriers alter the natural movement of sand and can change the shape of the beach, sometimes with undesirable results. For example, people may build *groins*—concrete or stone walls perpendicular to the shore—to prevent longshore drift from removing sand from a beach **(Fig. 16.43a, d)**. Sand accumulates on the updrift side of the groin, forming a triangular wedge, but sand erodes away on the downdrift side. Needless to say, the property owner on the downdrift side does not appreciate this process. At the entrance to a harbor, engineers may construct a pair of walls called *jetties* **(Fig. 16.43b)**. Jetties, however, effectively extend the river channel into deeper water, which may lead to the deposition of an offshore sandbar. Engineers may also build an offshore wall called a *breakwater*, parallel or at an angle to the beach, to prevent the full force of waves from reaching the beach **(Fig. 16.43c)**. With time, however, sand builds up in the lee (landward side) of the breakwater, and the beach grows seaward. To protect expensive seaside homes, people build *seawalls* out of riprap (large stone or concrete blocks) or reinforced concrete on the landward side of the backshore zone **(Fig. 16.44)**. But seawalls reflect wave energy, which crosses the beach back to sea, so this process increases the rate of erosion at the foot of the seawall. During a large storm, the seawall may be undermined so much that it collapses.

In some places, people have given up trying to decrease the rate of beach erosion and instead have worked

to increase the rate of sediment supply. To do this, they pump sand from farther offshore or bring in sand from elsewhere by truck or barge to replenish a beach **(Fig. 16.45)**. This procedure, called **beach nourishment** or *beach replenishment*, can be hugely expensive and at best may provide only a temporary fix. The backwash and longshore drift that formerly removed the sand continue unabated as long as the wind blows and the waves break.

Pollution Problems

Bad cases of beach pollution create headlines. Because of longshore drift, garbage dumped in the sea in an urban area may drift along the shore and be deposited on a beach

Figure 16.41 Areas of the United States coast that may be submerged if sea level rises by about 1.0 to 1.5 m.

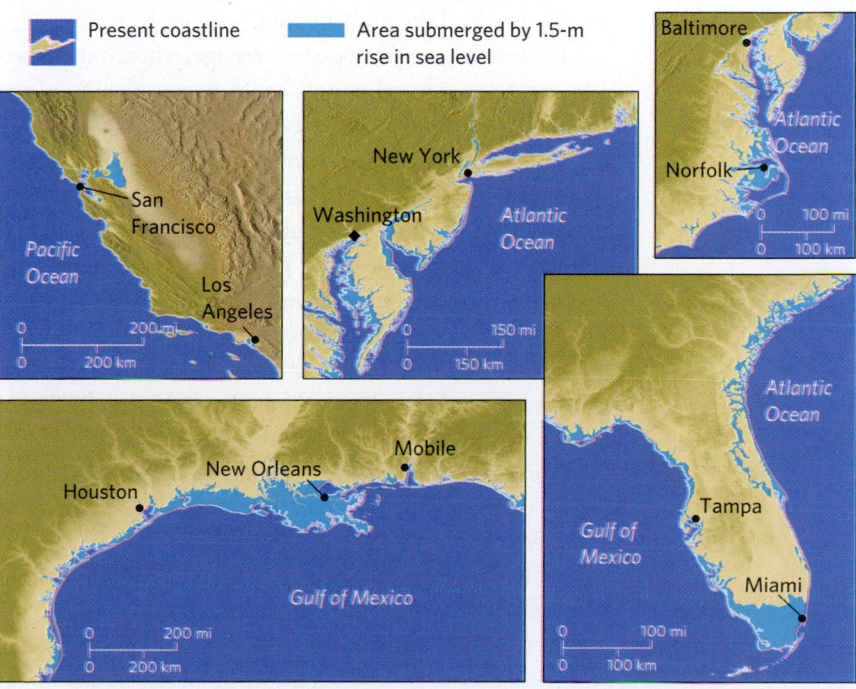

Figure 16.42 Examples of beach erosion.

Figure 16.43 Techniques used to preserve beaches.

(a) On the coast of Cape Cod, wave erosion has removed the entire beach right up to the coastal dune.

(b) Along the shore at Palm Beach, Florida, all the sand has been removed, and a ledge of rock has been exposed.

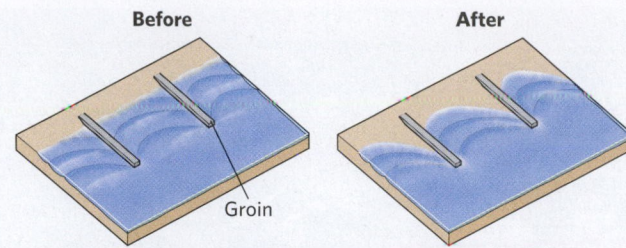

(a) The construction of groins may produce a sawtooth beach.

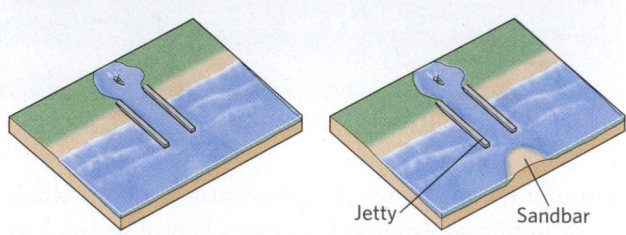

(b) Jetties extend a river farther into the sea, but may cause a sandbar to form at the end of the channel.

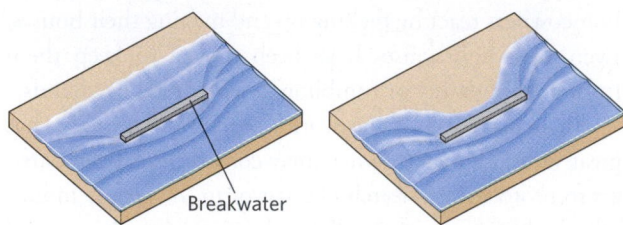

(c) A beach may grow seaward behind a breakwater.

far from its point of introduction. For example, hospital waste from New York City has washed up on beaches tens of kilometers to the south. Oil spills, which may come from ships that flush their bilges with seawater, from tankers that have run aground or foundered in stormy seas, or from blowouts at offshore wells (see Box 11.2), have contaminated shorelines at places around the world.

Threats to Organic Coasts

Organic coasts, a manifestation of interaction between the physical and biological components of the Earth System, are particularly susceptible to changes in the environment. The loss of such landforms can increase a coast's vulnerability to erosion and, because organic coasts provide spawning grounds for marine organisms, can upset the food web of the global ocean.

In wetlands and estuaries, sewage, chemical pollutants, and agricultural runoff cause havoc. Toxins, along with clay, settle and concentrate in sediment, where they contaminate burrowing marine life. Fertilizers and sewage that enter the sea with runoff increase the nutrient content of seawater, stimulating algal blooms that absorb dissolved oxygen from the water and therefore kill animal and plant life. Coastal wetlands face destruction by development: many wetlands have been filled or drained for use as farmland or suburbs, and many have been used as garbage dumps. Throughout the world, between 20% and 70% of coastal wetlands have been destroyed in the last century.

Coral reefs, which depend on the health of delicate coral polyps, can be devastated by even slight changes in the environment. Pollutants, particularly hydrocarbons, will poison the polyps. Sewage fosters algal blooms that rob ocean water of dissolved oxygen and suffocate the corals. And suspended sediment introduced to coastal waters, either by terrestrial runoff or by beach-nourishment

(d) Groins along the shore in southern England.

projects, reduces light levels, killing the algae that live in the coral polyps, and clogs the pores that the polyps use to filter water. Changes in water temperature or salinity, caused by the dumping of wastewater from power plants into the sea or by global warming of the atmosphere, also destroy reefs, for reef-building organisms are very sensitive to temperature changes. People can destroy reefs directly by dragging anchors across reef surfaces, setting off explosives to kill fish, touching reef organisms, or quarrying reefs to obtain construction materials. In the last two decades, marine biologists have noticed that many reefs around the world have lost their color and died **(Fig. 16.46)**. This process, called **reef bleaching**, may be

Figure 16.44 A seawall can protect the sea cliff under most conditions, but during a severe storm, wave energy reflected by the seawall helps scour the beach. As a result, the wall may be undermined and will collapse.

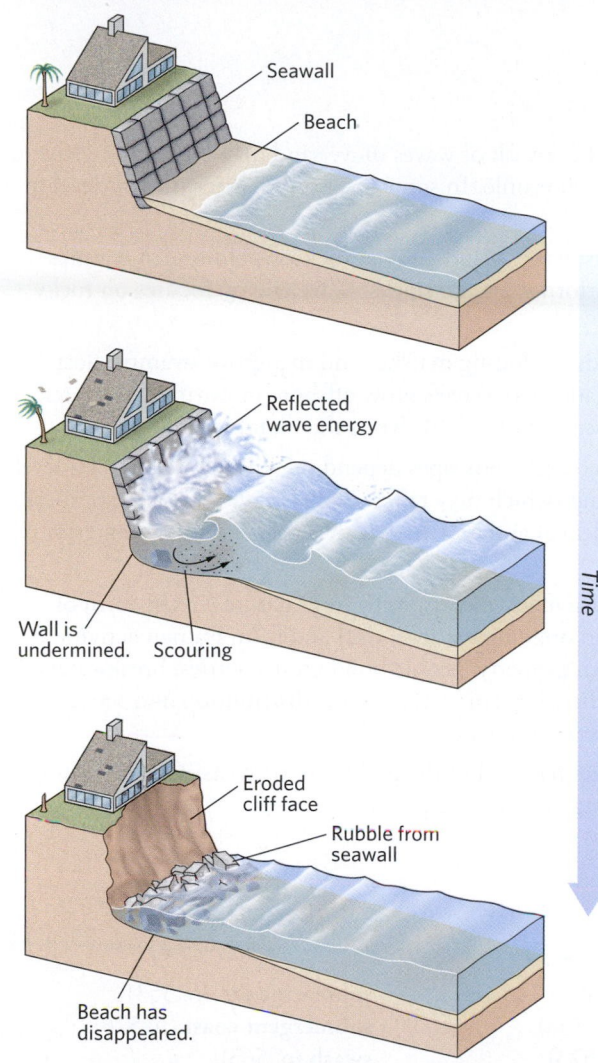

Seawall

Beach

Reflected wave energy

Wall is undermined. Scouring

Time

Eroded cliff face

Rubble from seawall

Beach has disappeared.

A seawall built to protect expensive homes.

Figure 16.45 Beach nourishment involves bringing in vast amounts of sediment by truck, by pipe from offshore, or by barge.

we have discussed in this chapter may help to define a way forward in addressing these problems. We'll return to this topic in Chapter 20, in the context of discussing global climate change.

Take-home message . . .

Coastal landscapes can be affected by storms, human activities, and sea-level change. Beaches naturally erode, and people try to protect them by constructing barriers or replenishing sand, but these actions can lead to other problems. Pollution and warming of seawater are destroying wetlands and coral reefs.

Quick Question -
Why can a beach become polluted even if it is far from a source of pollution?

Figure 16.46 Reef bleaching in the Caribbean. The coral here is now colorless and mostly dead.

due to the removal or death of symbiotic algae in response to the warming of seawater, or it may be a result of the dust carried by winds from desert or agricultural areas.

Clearly, managing the fragile realm of the coast has become a challenge for society. Understanding the processes

⊙16 CHAPTER REVIEW

Chapter Summary

- The landscape of the seafloor depends on the character of the underlying lithosphere. Wide continental shelves form over passive-margin basins. Continental shelves may be cut locally by submarine canyons. Abyssal plains develop on old, cool oceanic lithosphere. Oceanic islands form above hot spots, and trenches, fracture zones, and mid-ocean ridges define plate boundaries.

- The largest tidal reach approaches 16 m (52 feet). Where coastlines have a gentle slope, the position of the shoreline can change by kilometers between high and low tide. Tides form as the Earth spins under two tidal bulges. Because of variations in ocean-basin and coastal shape, tides vary significantly around the globe.

- Tides are caused by the tide-generating force, a combination of forces applied by the gravity of the Moon and Sun and by centrifugal force caused by rotation of the Earth-Moon system around its center of mass.

- The motion of waves changes as they approach the coast, where they can transform into breakers. Where waves intersect the shore at an angle, wave refraction and longshore drift of beach sediment take place.

- The swash and backwash of waves move sand on a beach and yield a distinctive beach profile. In some locations, barrier islands develop offshore.

- On rocky coasts, waves grind away at rocks, yielding such features as wave-cut platforms and sea stacks. Wave energy focuses on rocky headlands.

- Coastal wetlands, including marshes and mangrove swamps, host salt-resistant plants. Coral reefs grow offshore in warm, clear water. Reefs around oceanic islands evolve as the islands slowly sink.

- Differences in coastal landscapes depend on whether relative sea level is rising or falling (which may reflect global sea-level change or local uplift or subsidence) and on whether the coast is undergoing erosion or deposition.

- Coastal areas face many threats, both from nature (in the form of large storms and ongoing sea-level rise) and from human activities. To protect beach property, people build groins, jetties, breakwaters, and seawalls. These structures change the distribution and movement of sediment on beaches.

- Human activities have led to the pollution of coasts. Reef bleaching has become dangerously widespread.

Key Terms

abyssal plain (p. 550)
accretionary coast (p. 576)
backwash (p. 563)
barrier island (p. 568)
bathymetry (p. 547)
beach (p. 566)
beach nourishment (p. 579)
beach profile (p. 567)
breaker (p. 562)
coast (p. 548)
coastal plain (p. 576)
coastal wetland (p. 574)
continental shelf (p. 550)

coral reef (p. 575)
embayment (p. 570)
emergent coast (p. 576)
erosional coast (p. 576)
estuary (p. 573)
fjord (p. 573)
guyot (p. 555)
headland (p. 570)
intertidal zone (p. 556)
lagoon (p. 568)
longshore current (p. 566)
longshore drift (p. 567)
marine geology (p. 548)

ocean basin (p. 549)
oceanic island (p. 554)
organic coast (p. 573)
passive-margin basin (p. 551)
reef bleaching (p. 580)
rocky coast (p. 569)
sand spit (p. 567)
sea stack (p. 572)
seamount (p. 554)
sediment budget (p. 569)
shoreline (p. 556)
storm surge (p. 577)
submarine canyon (p. 553)

submarine fan (p. 553)
submergent coast (p. 576)
swash (p. 563)
tidal bore (p. 556)
tidal flat (p. 568)
tidal reach (p. 556)
tide (p. 555)
tide-generating force (p. 558)
wave-cut platform (p. 570)
wave erosion (p. 570)
wave refraction (p. 563)

Review Questions

The letters following each Review Question refer to the corresponding Learning Objective from the Chapter Opener.

1. How does the lithosphere beneath a continent differ from that beneath an abyssal plain? **(A)**

2. How do the shelf and slope of an active continental margin differ from those of a passive margin? Why do passive-margin basins exist? **(A)**

3. Why is the sediment on the right side of the adjacent figure thicker than near the ridge axis? **(A)**

4. What are tides, and why do they exist? Is the tidal reach the same everywhere? Explain your answer. **(B)**

5. Which tidal bulge on the adjacent diagram is closer to the moon? **(B)**

6. How do waves change as they approach the shore? Why does longshore drift take place along some beaches? What landforms can result from longshore drift? **(B)**

7. Describe the components of a beach profile. What is a barrier island? **(C)**

8. Identify the direction of longshore drift on the adjacent diagram. **(C)**

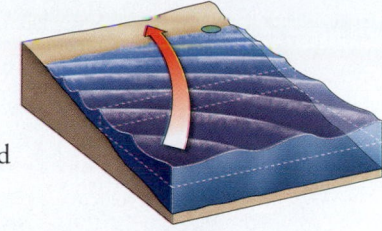

9. Describe how rocky coasts evolve. Why do headlands tend to erode more rapidly than embayments? **(D)**

10. What is the difference between an estuary and a fjord? How do seawater and river water interact in an estuary? **(E)**

11. Discuss the different types of coastal wetlands and the factors that determine which type occurs in a given locality. **(E)**

12. How does a coral reef form? How does a coral reef surrounding an oceanic island change over time? Which type of reef does the diagram show? **(E)**

13. What factors affect the local shape and evolution of a coastline? How does the sediment budget affect a beach? Explain the difference between an emergent and a submergent coast. **(F)**

14. Explain how storms and rising sea level affect coastal areas. **(F)**

15. In what ways do people try to modify or stabilize coasts? How do those efforts threaten coastal areas? **(G)**

On Further Thought

16. A hotel chain would like to build a new beachfront hotel along a north-south-trending stretch of beach where a strong longshore current flows from south to north. The neighbor to the south has constructed an east-west-trending groin on the property line. Will the groin pose a problem for the hotel beach? If so, what solutions could the hotel try? **(B, G)**

17. Much of southern Florida lies at elevations of less than 6 m (20 feet) above sea level. In recent years, as sea level has risen, parts of Miami Beach are flooding. The bedrock beneath the city consists of porous limestone, the remnants of ancient reefs. What protective measures might help mitigate this situation, if any? **(G)**

Online Resources

Animations
This chapter features animations on living with the coasts and ocean wave motion and refraction.

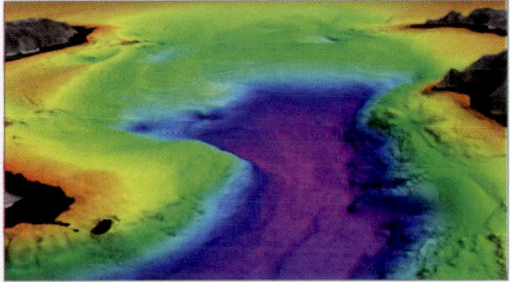

Videos
This chapter features real-world videos on the seabed habitats of oceanic shoals, how marshes respond to sea-level rise, and the seafloor of central California.

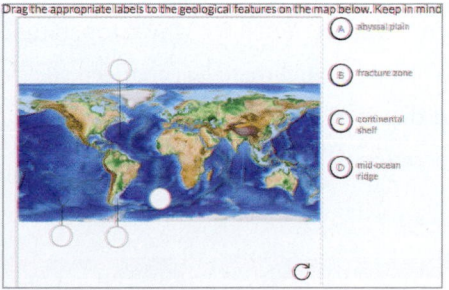

Smartwork5
This chapter features questions on basins, tides, beach profiles, and the role of sea-level change and human impacts on coastline evolution.

583

PART IV

A Blanket of Gas: the Earth's Atmosphere

A thin blanket of gas, the atmosphere, surrounds our planet. Without it, there could be no life on Earth. In Part IV, we explore the properties and behavior of this atmosphere. We begin, in Chapter 17, by describing its composition and character, how these vary with elevation, and how they have changed through Earth history. Next, we turn our attention to clouds, precipitation, and wind, properties that define weather conditions. You'll see that solar radiation drives atmospheric motion, causing weather to vary with location at a given time, and to vary with time at a given location. In Chapter 18, we focus on large weather systems, such as the monsoons and hurricanes of the tropics, and the cyclones that traverse our planet's middle latitudes. Chapter 19 looks at local storms, particularly thunderstorms. You'll see how such storms form, and why they may turn violent and unleash lightning, hail, strong winds, and even tornadoes. The final chapter of Part IV, Chapter 20, examines the Earth's climate. You'll see why temperature and precipitation vary across the planet, that climate has changed significantly over Earth history, and that it continues to change today in ways that may impact future generations.

A view from an airplane window reveals a dreamland of puffy clouds bathed in the red light of the setting Sun. Each cloud consists of countless tiny droplets of water or ice, formed when the moist air in an updraft cools. Since there's vertical air movement associated with these clouds, the plane experiences turbulence (bouncing) as it passes through them, so the seatbelt sign is on.

17 THE AIR WE BREATHE
Introducing the Earth's Atmosphere

By the end of the chapter you should be able to . . .

A. explain how the atmosphere has evolved and why it now contains oxygen.

B. define the quantities that meteorologists use to describe the atmosphere and list the instruments they use to monitor the atmosphere.

C. interpret a drawing that displays atmospheric layers.

D. evaluate the importance of atmospheric aerosols and the role that human activities play in generating them.

E. describe what clouds are, interpret the clouds you see in the sky, and explain how clouds produce precipitation.

F. create a model of the interactions between radiation and the atmosphere that control the temperature of Earth's atmosphere and surface.

17.1 Introduction

Breathe deeply. You've just inhaled **air**, the unique mixture of nitrogen, oxygen, water vapor, carbon dioxide, and other gas molecules that makes up Earth's **atmosphere**, the envelope or blanket of gas that surrounds our planet. Amazingly, some of the nitrogen that just entered your lungs was exhaled by Madame Curie, George Washington, Julius Caesar, Confucius, and Cleopatra; some of the water molecules once flowed down the Congo River, rained from Hurricane Katrina, and watered the gardens of Versailles; and some of the oxygen molecules came from plants on your windowsill, algae in the ocean, trees of the rainforest, and grasses of the prairie. That's because the atmosphere flows and stirs constantly on local to global scales and interacts with many other components of the Earth System.

Without the atmosphere, the Earth would be as barren as the Moon. The atmosphere provides plants with carbon dioxide for photosynthesis and animals with oxygen for metabolism, shields our planet's surface from the dangerous ultraviolet light in solar radiation, and carries water from where it evaporates to where it falls as *precipitation* (rain, snow, and hail). The atmosphere also serves as the medium in which birds, insects, and airliners fly and through which our voices carry. In human culture, the atmosphere's visual splendor—its auroras, rainbows, sunsets, and clouds—have inspired artists and sparked romances. The atmosphere can also unleash violent *storms*, intense episodes of locally strong winds and precipitation **(Fig. 17.1)**.

Though we are merely bottom dwellers in this shared sea of air, we have become its caretakers, for our activities

587

Figure 17.1 Atmospheric phenomena can be both beautiful and dangerous.

(a) A heavy rainstorm in the midwestern United States, pouring from the base of a swirling cloud.

(b) Arcs and haloes at sunrise caused by interaction of sunlight with ice crystals in the air.

(c) A blizzard can bring traffic in a city to a standstill.

(d) Fog at the entrance to San Francisco Bay, California, hides the highway across the Golden Gate Bridge.

have changed, and will continue to change, the composition of the atmosphere. Furthermore, our vulnerability to storms and other atmospheric events increases as cities and farms spread into lands susceptible to flooding and drought. To prevent calamities due to atmospheric phenomena, and to ensure that the atmosphere can continue hosting our planet's rich biosphere into the future, it is essential that we strive to understand our atmosphere's character and behavior, and that we learn how to predict its near-term and long-term future. This chapter provides a foundation for addressing these challenges by discussing several basic questions: What is the atmosphere? How did it form and attain its present composition? What quantities can we use to describe atmospheric conditions? What are clouds, and why are there so many different types? Where

does the energy that drives atmospheric phenomena come from? And how does the atmosphere interact with light?

17.2 Atmospheric Composition

The Recipe for Air

What is the air made of? An average sample of dry air, meaning air from which all water has been removed, consists mainly of two gases: 78% (by volume) molecular nitrogen (N_2) and 21% molecular oxygen (O_2) **(Fig.17.2)**. The remaining 1% includes argon (Ar) and many other gases, such as carbon dioxide (CO_2), ozone (O_3), and sulfur dioxide (SO_2). Although these other gases occur in such minuscule quantities that researchers refer to them as *trace gases*, some, as we'll see, play key roles in controlling atmospheric temperature.

Note that when we describe the composition of the Earth's atmosphere, we consider only dry air. That's because in nature, the amount of water (H_2O) in air varies greatly from place to place at the same time and from time to time in the same place, so we can't specify a single percentage that characterizes the average proportion of H_2O in the whole atmosphere. Water in the atmosphere occurs in three forms: as an invisible gas called **water vapor (Fig.17.3a)**, as a liquid (in tiny *droplets* or larger *drops*) **(Fig. 17.3b)**, and as a solid (in the form of ice, snow, or hail) **(Fig. 17.3c)**. Looking up, we often see **clouds**, collections of countless water droplets or tiny ice crystals suspended in air. Clouds normally appear to be floating above the Earth's surface, but they can also be in contact with the ground, where they are called *fog*.

Atmospheric Aerosols

Press the button on a can of air freshener, and countless extremely tiny particles spray into the air, particles so small that they can be wafted away by even gentle air currents. Researchers refer to such tiny particles—small enough to remain suspended in air—as **aerosols**. Air always contains many types of aerosols **(Fig. 17.4a)**. Inorganic aerosols include specks of mineral dust lofted by winds blowing over soil, salts and sulfates carried into the air by sea spray, fine ash injected into the atmosphere by volcanic eruptions, and soot (carbon) particles billowing from forest

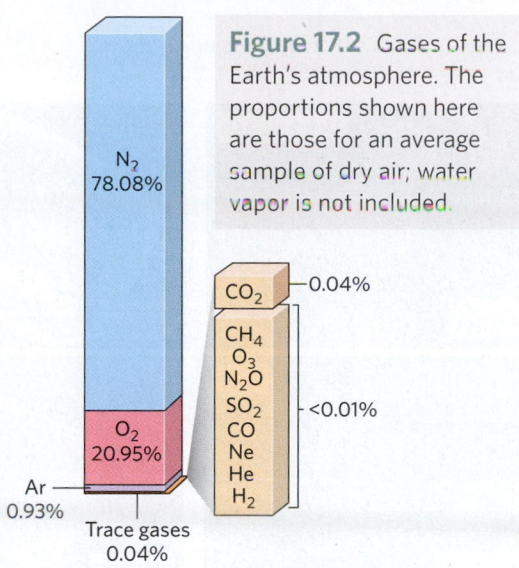

Figure 17.2 Gases of the Earth's atmosphere. The proportions shown here are those for an average sample of dry air; water vapor is not included.

N₂ 78.08%

O₂ 20.95%

Ar 0.93%

Trace gases 0.04%

CO₂ — 0.04%

CH₄
O₃
N₂O
SO₂
CO
Ne
He
H₂

<0.01%

Figure 17.3 Water in the atmosphere occurs in three forms.

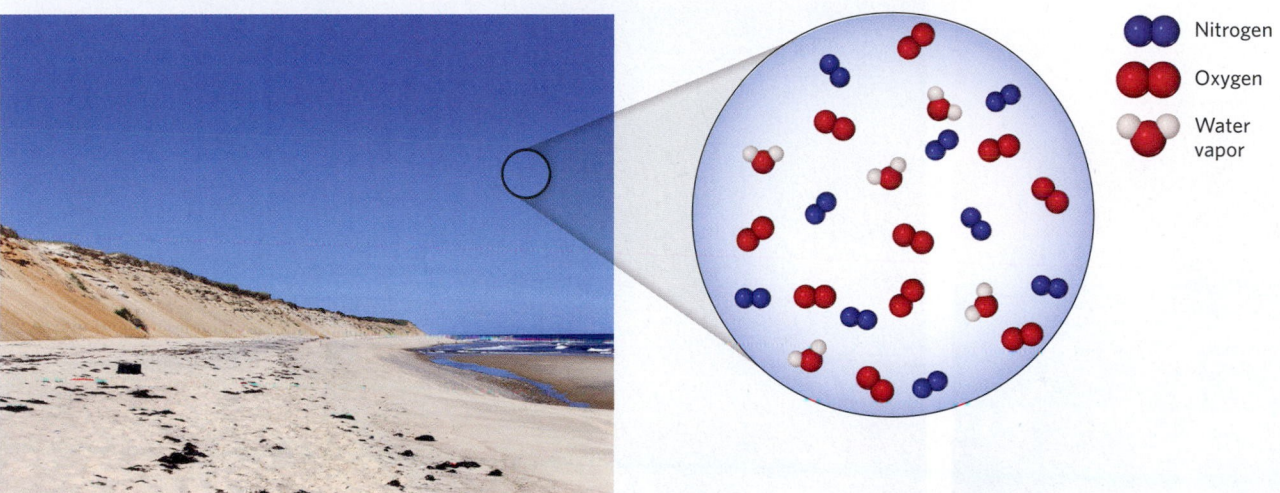

Nitrogen

Oxygen

Water vapor

(a) Air contains invisible water vapor, even on a cloudless day. Water molecules in vapor are mixed in with other gas molecules.

(b) Clouds in the warm sky over Brazil consist of liquid water droplets.

(c) High clouds consist of tiny ice crystals.

Figure 17.4 Atmospheric aerosols.

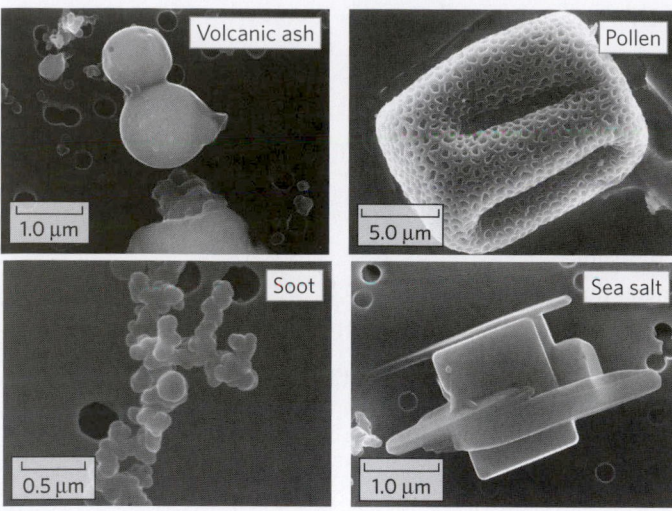

(a) Aerosols viewed through a scanning electron microscope (1 μm = 0.001 mm).

(b) Smoke rising from a forest fire.

(c) Dust blowing from the Sahara out over the Atlantic, as viewed from space.

(d) Haze over the Blue Ridge Mountains of the Appalachians.

(e) Photochemical smog at the Beijing airport includes particles from power plants, factories, and car exhausts.

fires **(Fig. 17.4b)**. Organic aerosols include pollen, bacteria, molds, and viruses as well as detritus from decaying organisms. Many aerosols are produced by human activities such as burning fossil fuels, spraying pesticides on crops, or using household products such as oven cleaner or deodorant.

Atmospheric aerosol concentrations vary with location. They tend to be greater over land surfaces than over the sea because there are more aerosol sources on land. Aerosols from land sources also differ in composition from their oceanic counterparts. Most aerosols found over the oceans are salts and sulfates, whereas over land, most arise from wind-blown dust, fires, and human activities. Aerosols can travel long distances. Saharan dust blown from Africa, for example, reaches the Caribbean **(Fig. 17.4c)**. Aerosol particles absorb light, so an abundance of aerosols in the air limits our ability to see into the distance. Under high-humidity

conditions, some aerosols capture water molecules in the air and dissolve in the water to form microscopic droplets, which produce **haze**. This causes the decrease in visibility that you see on sultry days **(Fig. 17.4d)**.

Aerosols, along with certain gases such as ozone, contribute to air pollution. In discussing pollution, aerosols may be referred to as *fine particulate matter*. The dank, dark *smog* that engulfed industrial cities of the 19th century formed when factories and heating stoves produced coal smoke that mixed with fog. In modern cities, *photochemical smog* develops when exhaust from cars and trucks reacts with air in the presence of sunlight to produce an ozone-rich brown haze **(Fig. 17.4e)**. Government agencies monitor the aerosol concentrations in air to characterize air quality—high concentrations lead agencies to post health-hazard warnings.

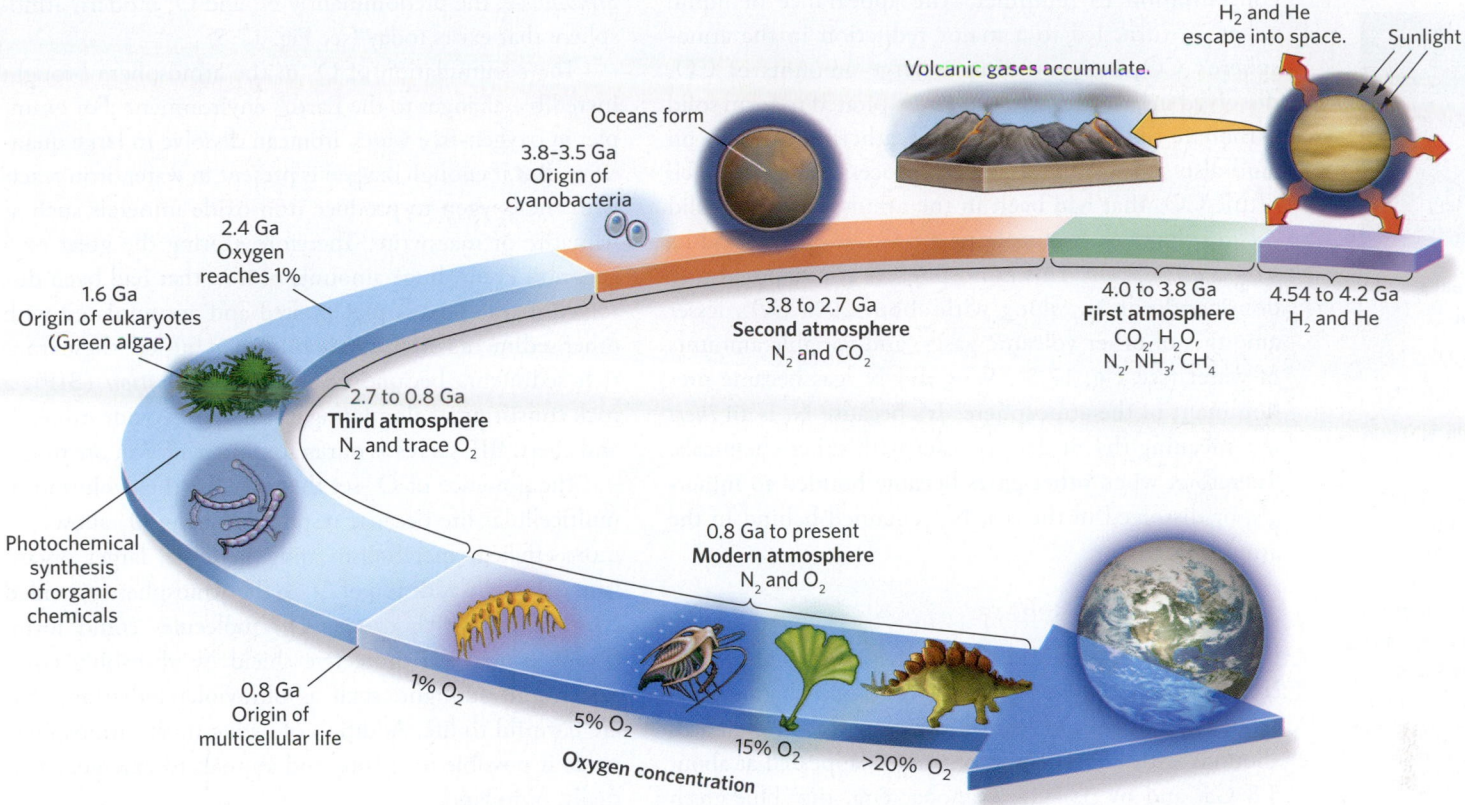

Take-home message . . .

Dry air consists of nitrogen (78%), oxygen (21%), and trace gases (1%). The amount of water in the atmosphere varies considerably over time and space. The atmosphere also contains tiny particles called aerosols, the presence of which can cause haze.

Quick Question -
In what forms does water occur in the atmosphere?

17.3 The Earth's Atmosphere in the Past

If you stepped out of a time machine into the air of the Precambrian, you would suffocate instantly, for the Earth has not always hosted the oxygen-rich atmosphere that we inhale today. Let's look briefly at how our atmosphere has evolved over time, from the Earth's birth to today.

The First Atmosphere

When the Earth first formed, at 4.54 Ga, molecules of hydrogen (H_2) and atoms of helium (He), the most common gases of interstellar nebulae, surrounded our planet as an atmosphere **(Fig. 17.5)**. That atmosphere didn't last long, however, because frequent volcanic eruptions, as well as the arrival of asteroids and comets, added molecules of many other gases (such as H_2O, CO_2, N_2, and SO_2) to the air. In addition, while the Earth's gravity could hold onto heavier gases, lighter gases (H_2 and He) escaped into space. By about 4.0 Ga, the Earth had an atmosphere whose assemblage of gases differed markedly from that of a nebula. This *first atmosphere* consisted of about 55% H_2O, 15% CO_2, 15% N_2, 15% ammonia (NH_3), and traces of methane (CH_4). Most of these gases came from *volcanic outgassing*—the emission, from volcanoes, of gases once bonded to minerals inside the Earth—but some probably came from impacting comets.

The Second Atmosphere

Between about 4.0 and 3.9 Ga, a time called the *late heavy bombardment*, so many meteorites pummeled the Earth that its surface may temporarily have become molten, and water could not remain liquid. Any ocean that had existed before that time evaporated. The geologic record shows that by 3.85 Ga, after heavy bombardment ceased, our planet's surface cooled below the temperature at which water vapor condenses into liquid, and its temperature has remained below the boiling point of water ever since, except, perhaps, for short time intervals after occasional huge impacts. Liquid water collected as surface water (oceans, lakes, and rivers), as groundwater, and in solid form as ice.

Did you ever wonder . . .
- - - - - - - - - - - - - - - - - - -
whether you could breathe the atmosphere of the early Earth?

See for yourself

Evidence of the great oxygenation event

Latitude: 46°26'54.63"N
Longitude: 87°37'31.97"W

Look down from an altitude of 16 km (10 miles) and you can see a large open-pit mine near Ishpeming, Michigan. Miners here are digging up banded-iron formation (BIF), made from sediment that was deposited about 2 billion years ago when ocean water began to contain significant quantities of dissolved oxygen. Iron oxide minerals give the BIF its dark, gray-red color. The rusty, orange color in the settling pool comes from the iron oxide in the water.

Did you ever wonder . . .

Where does the oxygen in the air come from?

Transfer of H_2O from the air into the liquid water reservoirs of the Earth caused the atmosphere's H_2O concentration to plummet. The appearance of liquid water, in turn, led to a major reduction in the atmosphere's CO_2 concentration as large amounts of CO_2 dissolved in seawater and then precipitated to form solid carbonate sediment. Chemical weathering of rocks on land also absorbed CO_2. These processes locked much of the CO_2 that had been in the atmosphere into solid rock of the crust, so by about 3.8 Ga, the Earth had its *second atmosphere*. This atmosphere was composed predominantly of N_2, along with about 20% CO_2, lesser amounts of other volcanic gases, and variable amounts of water (see Fig. 17.5). Why did N_2 gas became predominant in the atmosphere? It's because N_2 is an *inert gas*, meaning that it doesn't react with other chemicals. Therefore, when other gases became bonded to minerals, or dissolved in the sea, N_2 remained behind in the atmosphere.

The Third Atmosphere

If the Earth were devoid of life, the atmosphere would contain hardly any *free oxygen* (O_2 molecules) because volcanoes do not emit O_2 gas. Fortunately for us, the Earth abounds with life. The first organisms appeared at about 3.8 Ga, and by 3.5 Ga, cyanobacteria, tiny blue-green single-celled organisms that carry out photosynthesis, were thriving in the ocean. *Photosynthesis*, a chemical process that takes place within certain types of living cells, extracts CO_2 from the atmosphere or ocean, converts it into organic chemicals, and releases O_2 as a by-product. At first, nearly all of the released O_2 was absorbed by rocks or dissolved in water, so the atmosphere retained only a trace of O_2. Then, at 2.7 Ga, a dramatic transition took place on the Earth: organisms with eukaryotic cells (cells with a nucleus) appeared. Eukaryotic cells are more efficient at carrying out photosynthesis than are bacterial cells, so the global rate of O_2 production started to increase, and between 2.4 Ga and 1.8 Ga, a time known as the **great oxygenation event**, O_2 began to accumulate to yield more than trace quantities in the atmosphere.

Photosynthetic organisms not only release oxygen, but also absorb CO_2. If all *biomass* (the material in organisms) decayed when the organisms died, the atmosphere's CO_2 concentration would have remained at about 20% because decay returns CO_2 to the atmosphere. In the Earth System, however, not all biomass decays. Instead, some gets deposited, along with sediment, on the seafloor or on lake beds. If this organic-rich sediment becomes buried deeply enough to turn into rock, the biomass it contains effectively becomes locked underground. The trapping of organic matter in sedimentary rocks led to a further decrease of the CO_2 concentration in the atmosphere, from about 20% to a trace amount. As a result, the proportion of O_2 in air increased, and the Earth developed its *third atmosphere*, the predominantly N_2 and O_2 modern atmosphere that exists today (see Fig. 17.5).

The accumulation of O_2 in the atmosphere brought incredible changes to the Earth's environment. For example, in oxygen-free water, iron can dissolve in large quantities. But if enough oxygen is present in water, iron reacts with the oxygen to produce iron-oxide minerals such as hematite or magnetite. Therefore, during the great oxygenation event, huge amounts of iron that had been dissolved in the oceans precipitated and accumulated with other sediments on the seafloor. When buried, these iron-rich sediments became *banded iron formation* (BIF), a rock consisting of alternating layers of iron oxide minerals and chert. BIF serves as our main source of iron ore today.

The presence of O_2 set the stage for the evolution of multicellular life because respiration using O_2 allows for more efficient metabolism and, therefore, larger organisms. Also, the addition of O_2 to the atmosphere provided atoms from which ozone (O_3) molecules could form. Ozone provides a protective shield by absorbing components of sunlight, such as ultraviolet radiation, that are harmful to life. Addition of ozone to the atmosphere made it possible for plants and animals to emerge, eventually, onto land.

The atmospheric O_2 concentration remained below 5% until about 600 Ma. At this time, even more efficient oxygen-producing multicellular organisms evolved. During the Phanerozoic, the atmospheric concentration of O_2 has fluctuated. The highest concentrations occurred when vast, O_2-producing forests of the late Paleozoic grew. (The buried debris from these forests became coal.) Notably, these high concentrations of O_2 allowed very large insects, such as 1-m (3-foot)-wide dragonflies, to survive in late Paleozoic forests. The atmosphere's O_2 concentration cannot rise higher than about 35%, however. If it did, immense wildfires would consume O_2-producing land plants, for burning takes place rapidly in O_2-rich air.

Take-home message . . .

The composition of the Earth's atmosphere has changed radically since the Earth first formed. The early atmosphere consisted of volcanic gases. When the oceans formed, H_2O and CO_2 were removed from the atmosphere. Billions of years later, photosynthesis added O_2 to the atmosphere, and today, air consists almost entirely of N_2 and O_2.

Quick Question -
Explain why the atmosphere could never contain 100% O_2.

17.4 Describing Atmospheric Properties

Everywhere in the world, people complain about, marvel at, or just simply button up and prepare for the atmospheric conditions they find outdoors. We refer to the specific atmospheric conditions at a given time and location as the **weather**. Some of the common properties that **meteorologists**—atmospheric scientists who focus on describing and predicting the weather—use to describe atmospheric conditions include the following:

- *Temperature:* a measure of hotness or coldness
- *Atmospheric pressure:* a measure of the weight of a column of air above a location
- *Relative humidity:* a measure of the amount of water vapor in the air
- *Wind speed:* a measure of how fast the air moves horizontally
- *Wind direction:* a measure of the compass direction from which the wind blows
- *Visibility:* a measure of how far one can see through the air
- *Cloud cover:* a measure of the portion of the sky covered by clouds
- *Precipitation:* a measure of the amount of water falling from the air and reaching the ground during a time interval

Let's look at each of these properties in more detail and learn how they can be measured.

Air Temperature

We all pay attention to the air temperature—if it's hot, you might stroll about in shorts, but if it's cold, you need to put on a coat. But what exactly does the *temperature* measure? In the air, gas molecules move rapidly in random directions, vibrate, and occasionally collide with one another. **Air temperature** represents the average speed at which molecules move. The faster the average speed of the molecules, the higher the temperature. Note that *temperature* is not synonymous with *heat*. *Heat* refers to the total thermal energy contained in a material, meaning all the energy due to vibrations and movements of all atoms or molecules in the material. Note that a volume of air near the top of the atmosphere contains few, but very fast-moving gas molecules. It has a very high temperature, but it does not contain as much heat as the same volume of air at the same temperature at sea level, for the sea-level air contains a great many more molecules.

Figure 17.6 Temperature scales and the relationship between centigrade (Celsius) and Fahrenheit values for common temperatures.

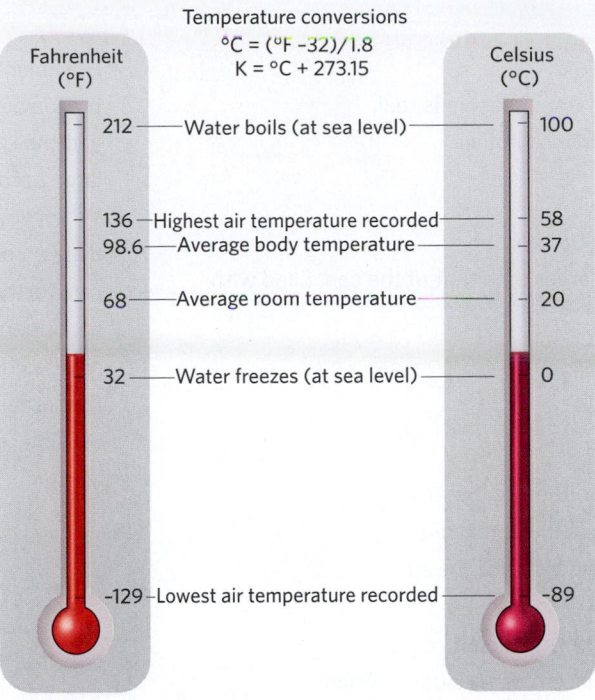

We measure temperature with a **thermometer**, calibrated in degrees according to standard scales **(Fig. 17.6)**. Originally, thermometers consisted of a thin column of mercury in a calibrated glass tube. Today, most thermometers use alcohol because mercury is toxic. When the alcohol warms, it expands, so it rises in the tube. Meteorologists can also measure temperature with a *thermistor*, a sensor containing a conductor in which the resistance to the flow of electrical current depends on temperature. Most thermometers in the United States still use the *Fahrenheit scale*, whereas those in the rest of the world use the *centigrade scale*, also known as the *Celsius scale*. At sea level, water boils at 212°F and freezes at 32°F on the Fahrenheit scale, and boils at 100°C and freezes at 0°C on the centigrade scale. An increment of temperature, a *degree*, represents a greater temperature change on the centigrade scale than on the Fahrenheit scale. Scientists commonly use another scale, the Kelvin scale. *Absolute zero*, the lowest temperature possible, is 0 K (−273.15°C)—at 0 K, molecules stop moving or vibrating. (Note that when we write a temperature in Kelvin, we do not use the word *degree* or the degree symbol.)

At any particular location, air temperature fluctuates daily as the Sun rises and sets, and it also changes seasonally because, depending on the time of year, the number of daylight hours and the Sun's position in the

Box 17.1

How can I explain . . .

Atmospheric pressure

What are we learning?

- The force exerted by atmospheric pressure is real.
- Air pressure is related to the density of air.

What you need:

- An open, empty aluminum soda can.
- A container with a diameter larger than that of the can, filled with water and ice.
- A small hot plate.
- A pair of tongs.

Instructions:

- Put a small amount of water in the can. Set the hot plate on low heat and place the can, open top up, on the heating element. Wait for the air in the can to heat and the water to boil.
- Using the tongs so you don't burn yourself, grasp the can, invert it, and place it open-side-down into the ice water.
- The can will immediately crumple.

What did we see?

- Before being heated, the atmospheric pressure exerted on the inside of the can wall and the outside of the can wall is the same. When you put the can on the heating element, the gas in the can begins to expand, so it escapes from the top opening.
- As air escapes, the amount of air and, therefore, the air pressure inside the can decreases.

- However, because air in the can is warmer than air outside the can, the inside and outside pressure on the walls of the can remain equal, and the can retains its shape. The boiling water inside the can adds a large number of water vapor molecules to the remaining gas.
- When you invert the can in the ice water, the air inside the can immediately cools. In addition, water vapor condenses on the inside of the can, further reducing the interior air pressure. Air pressure in the can no longer equals air pressure outside the can, so the outside pressure crushes the can until the pressure inside and outside the can are the same. The fact that the can crushes emphasizes that air, though invisible, exerts pressure.

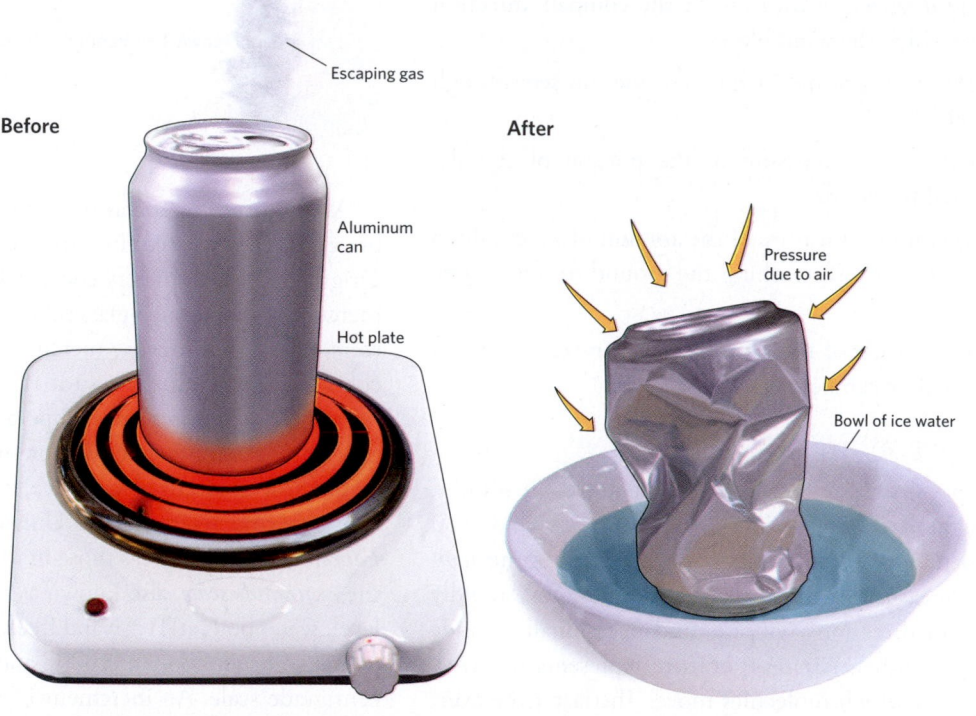

Escaping gas

Before

Aluminum can

Hot plate

After

Pressure due to air

Bowl of ice water

sky vary. Temperature at a location can also change when volumes of cooler or warmer air move in from elsewhere. The highest air temperature recorded on land was measured in Death Valley, California, where the thermometer reached 56.7°C (134°F), and the lowest air temperature was recorded at Vostok Station, Antarctica, where the thermometer dropped to –89.2°C (–128.6°F).

Atmospheric Pressure and the Height of the Atmosphere

Atmospheric pressure refers to the force applied by the overlying air on a surface of a specified area, such as a square meter, a square inch, or a square centimeter. You can picture atmospheric pressure as the weight of a column of air over a unit area at the base of the column.

For example, using English units, imagine a column of air that is 1″ by 1″ at its base and extends from the ground at sea level all the way through the atmosphere to the edge of space. On average, this column weighs 14.7 pounds, so we can say that the average atmospheric pressure at sea level is 14.7 pounds per square inch. Atmospheric pressure exists because air, like any substance, has mass, so in the Earth's gravitational field, it has weight.

How is pressure different from the *directed force* that you feel when your friend pushes you? Pressure, by definition, acts in all directions equally. You don't get knocked over by air pressure in still air because the air pressure pushing on your front is the same as that pushing on your back. Similarly, an empty can doesn't collapse because air pressure pushing on its inside walls equals that pushing on its outside walls **(Box 17.1)**.

Note that pressure affects the density of a gas. As pressure increases, air molecules get pushed closer together, and as pressure decreases, the molecules move farther apart. Density also depends on temperature. If you heat a given volume of air, it expands, because the molecules start moving faster. If the air is confined in a container, the pressure in the container increases. Similarly, if you cool air, it contracts, and if the air is confined in a container, its pressure decreases.

As **Table 17.1** shows, we can use several different units to specify atmospheric pressure. These days, meteorologists prefer either *millibars* (mb) or the numerically equivalent *hectopascals* (hPa). In this book, we'll use the more common unit, millibars. The numbers given for "average atmospheric pressure at sea level" in Table 17.1 are just that—averages. The specific value of atmospheric pressure varies with time and location at a given elevation, but it always decreases as elevation increases because the higher you go, the less gas lies above you.

Meteorologists measure atmospheric pressure with a **barometer**. A traditional mercury barometer consists of a dish of mercury into which a vertical glass tube has been inserted **(Fig. 17.7a)**. The tube is open on the bottom and closed at the top, and the air has been pumped out of the tube, so it contains a vacuum. As the atmosphere pushes down on the surface of the mercury, the mercury rises into the tube. The higher the air pressure, the greater the weight of air pushing on the mercury, so the higher the mercury rises in the tube. When the atmospheric pressure equals the average (mean) pressure at sea level (1,013.25 mb), the column attains a height of 29.92 inches (see Table 17.1). Most television meteorologists and newspaper weather pages still report pressure in "inches of mercury," the height of the mercury column in a mercury barometer. If atmospheric pressure is decreasing, meteorologists say that "the barometer is falling" because the height of the mercury column decreases. Conversely, if air pressure is increasing, they say that

"the barometer is rising." Mercury barometers were actually replaced long ago by *aneroid barometers*, which consist of a flexible metal diaphragm containing a vacuum **(Fig. 17.7b)**. As pressure increases, the walls of the diaphragm move closer together, and if pressure decreases, the walls move farther apart. This movement drives a lever, which in turn moves a dial calibrated in pressure units.

Figure 17.7 How a barometer works.

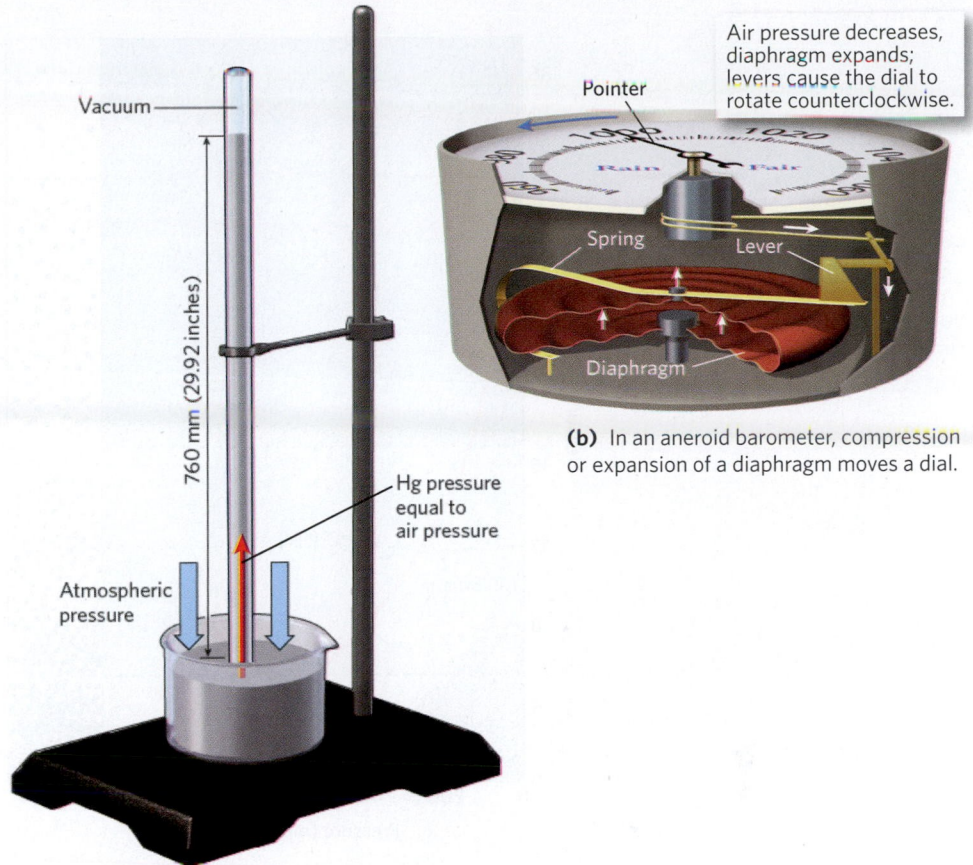

Air pressure decreases, diaphragm expands; levers cause the dial to rotate counterclockwise.

(b) In an aneroid barometer, compression or expansion of a diaphragm moves a dial.

(a) In a traditional mercury barometer, air pressure pushes mercury up a vacuum tube. The greater the pressure, the higher the mercury rises.

Table 17.1   Common units used for measuring atmospheric pressure

Unit	System	Average Atmospheric Pressure at Sea Level
Pounds per square inch	English	14.7 lb/in.^{2}
Pascals (Pa)	metric	101,325 (= 1,013.25 hPa)
Atmospheres (atm)	by convention*	1.000 (= 1.013 bars)
Bars (= 1,000 Pa)	by convention*	1.01325 (= 1 atm)
Millibars (mb) (= 0.001 bar)	by convention*	1,013.25 (= 1,013.25 hPa)
Inches of mercury	traditional	29.92 in.

*"By convention" means that the number was defined by an international agency.

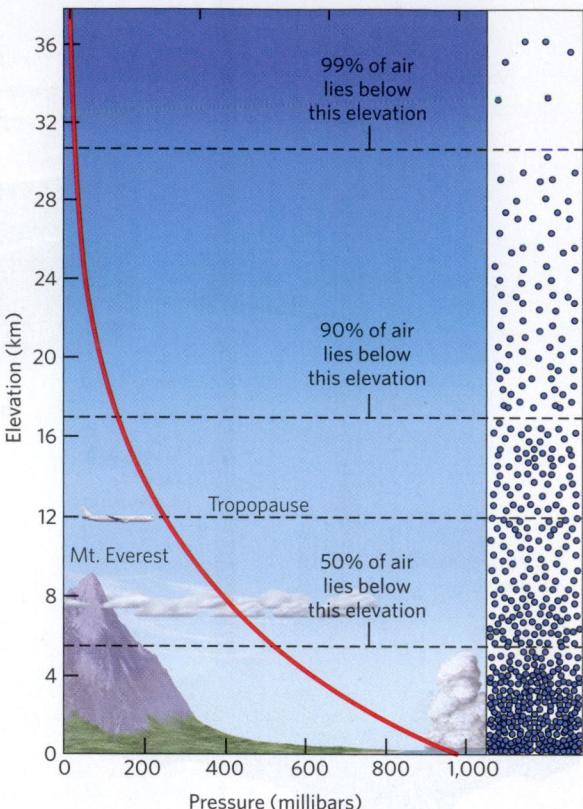

Figure 17.8 Atmospheric pressure and air density decrease with increasing elevation.

99% of air lies below this elevation

90% of air lies below this elevation

Tropopause

Mt. Everest

50% of air lies below this elevation

Pressure (millibars)

Elevation (km)

pressure dropped to 870 mb. Regions of higher pressure develop beneath colder and, therefore, denser air, whereas regions of lower pressure occur beneath warmer, less dense air. As we'll see in Chapter 18, horizontal variations in atmospheric pressure control the speed and direction of the wind.

Moisture and Relative Humidity

As we learned in Section 17.2, water vapor, an invisible gas, consists of freely moving water molecules. These molecules enter the atmosphere by evaporation from the Earth's oceans, lakes, and rivers and from plants by transpiration. Meteorologists use several quantities to characterize the water vapor content of the atmosphere.

The part of total atmospheric pressure exerted by water vapor molecules alone is called the **vapor pressure**. At a given temperature, dry air has a lower vapor pressure than does humid air. The atmosphere has a limited capacity for holding water vapor—otherwise, clouds would never form. When the atmosphere becomes *saturated*, meaning that it's holding as much water as it can, any additional water vapor added to the air will condense and form liquid droplets or solid ice crystals. Meteorologists refer to the vapor pressure at which the atmosphere reaches its holding capacity as the **saturation vapor pressure**.

The atmosphere's saturation vapor pressure strongly depends on temperature (Fig. 17.9). Warm air, in which molecules jostle rapidly, can hold more water vapor than can cold air. In fact, the atmosphere can hold 84 times

Atmospheric pressure, and therefore atmospheric density, decreases by about half for every 5.6 km (3.5 miles) of elevation gain (Fig. 17.8). As a result, about 50% of the atmosphere's gas lies below an elevation of 5.6 km, and 75% lies below an elevation of 11.2 km (7 miles), the elevation at which commercial jets fly. People can't survive long at elevations above about 6 km (3.7 miles) because air doesn't contain enough oxygen, so jet cabins must be pressurized by pumping in air.

Continuing upward, we find that 99.9% of gas in the atmosphere lies below 50 km (31 miles). Nevertheless, there's enough gas above that altitude that meteors begin to heat up and vaporize when they pass below an elevation of 70–120 km (43–75 miles). The "top" of the atmosphere, where the density of the atmosphere equals the density of interplanetary space, varies between 350 and 800 km (217 and 497 miles) in elevation.

Sea-level air pressure varies across the globe at any given time—rarely will a direct measurement of atmospheric pressure at sea level yield a number exactly equal to average sea-level pressure. Corrected for elevation, the highest sea-level pressure ever recorded was 1,085 mb in Siberia, whereas the lowest ever recorded occurred in the center of a typhoon (a Pacific hurricane), where the

Did you ever wonder . . .

why you feel hotter on a steamy, humid day in Florida than on a hot, bone-dry day in the Arizona desert?

Figure 17.9 Saturation vapor pressure (SVP) decreases rapidly as temperature (T) decreases.

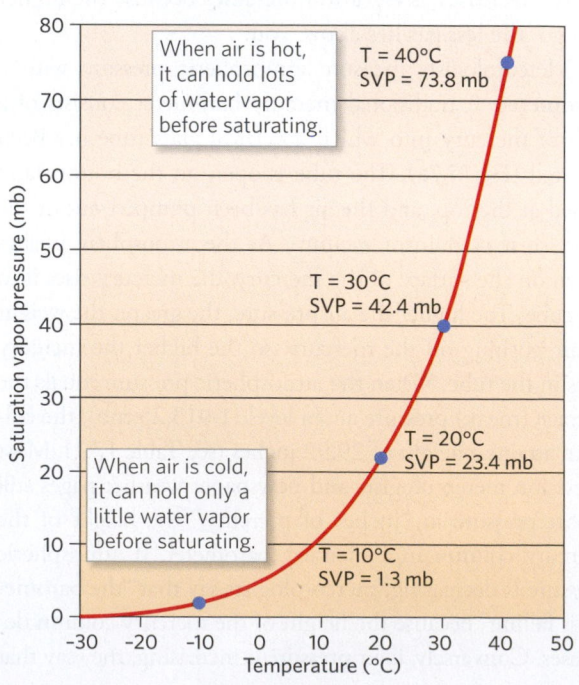

When air is hot, it can hold lots of water vapor before saturating.

T = 40°C
SVP = 73.8 mb

T = 30°C
SVP = 42.4 mb

T = 20°C
SVP = 23.4 mb

When air is cold, it can hold only a little water vapor before saturating.

T = 10°C
SVP = 1.3 mb

Saturation vapor pressure (mb)

Temperature (°C)

Figure 17.10 Relative humidity.

High RH ← Temperature → **Low RH**

(a) Because cold air can hold less water vapor than warm air, the same volume of water vapor in the air makes cold air more humid.

Hot, dry day

Warm, humid day

Cool, dry day

Foggy, chilly day

10°C	20°C	30°C	40°C
RH = 100%	RH = 20%	RH = 80%	RH = 5%

(b) In the real world, both the amount of water vapor in the air and the temperature vary from place to place. The size of each beaker represents the saturation vapor pressure. The ratio of the amount of water in the beaker to the beaker's capacity represents the relative humidity (RH).

more water vapor molecules at 30°C (86°F) than it can at –30°C (–22°F). At 30°C, saturated air at sea level can hold 28 grams of water per cubic meter. By comparison, a cubic meter of dry air at this temperature weighs 1,160 grams (2.6 pounds).

Humans can sense the air's moisture content because our bodies cool by evaporating perspiration. When hot air becomes humid, our perspiration can't evaporate quickly, so we feel hot and sticky. In dry air at the same temperature, our perspiration evaporates quickly, so we feel cool and comfortable. Significantly, it's not the absolute amount of water in the air (as indicated by the vapor pressure) that makes it feel humid, but rather how closely the air approaches saturation. To convey a sense of how damp air feels, meteorologists characterize the air's moisture content by specifying its **relative humidity** (RH), defined as the amount of water vapor in the air (the measured vapor pressure) divided by the air's capacity for holding water vapor (the saturation vapor pressure). We can express this quotient as a percentage:

$$RH = \frac{\text{vapor pressure}}{\text{saturation vapor pressure}} \times 100\%$$

Clearly, relative humidity depends on both the amount of water vapor in the air and the air's holding capacity. For example, if two volumes of air contain the same number of water vapor molecules, the warmer one will have a lower relative humidity **(Fig. 17.10a)**. Therefore, a given location may feel cool and be foggy because the air there has a high relative humidity. Another location may feel hot and dry because it has a low relative humidity, even if it contains the same number of water vapor molecules as the cool, foggy air does **(Fig. 17.10b)**. When the relative humidity is 90%, an actual temperature of 84°F might feel like 98°F to you because your perspiration won't evaporate. The relationship between human comfort, temperature, and relative humidity can be expressed by the *heat index*, a calibration of how hot air feels, given its humidity.

While vapor pressure provides a good way to describe the actual amount of moisture in the air, it's very difficult to measure. Instead, meteorologists rely on another variable, the dew point, to characterize the absolute, as opposed to the relative, amount of water vapor in the air. The **dew point** is the lowest temperature to which air can be cooled, at constant pressure, before becoming

Figure 17.11 The dew point is the temperature at which air becomes saturated with water vapor.

(a) The air where this spider built its web has cooled to the dew point, and water has condensed on the web.

(b) On a cold day, frost forms on leaves.

saturated. To measure the dew point, we simply cool air at constant pressure and measure the temperature at which liquid droplets first appear. When temperatures cool overnight, air may cool to the dew point and become saturated, so that tiny liquid drops, called *dew*, form on exposed surfaces **(Fig. 17.11a)**. If the process happens when the temperature drops below freezing, the temperature where *frost* first forms is called the *frost point* **(Fig. 17.11b)**. We can estimate relative humidity by comparing the dew point with the air temperature. When the dew point approaches the air temperature, the relative humidity is high, as happens on muggy days. In contrast, when the two temperatures are far apart, the relative humidity is low, as happens on crisp, dry days.

Wind Speed and Direction

Everyone knows the feel of flowing air. It can cool your skin, rustle the leaves, or—if stronger—rip off your roof. Atmospheric scientists define **wind** specifically as the horizontal movement of air. When the wind blows against your face, you feel more air pressure on your face than you do on the back of your head. Of course, air in nature doesn't just move horizontally—it can also flow up as an *updraft* or down as a *downdraft*. While up-and-down air movements have real consequences, such as triggering storms or causing discomfort for airline passengers, vertical air movements are not represented by a standard "wind" measurement.

To describe wind, we use two familiar quantities: wind direction and wind speed. By convention, the **wind direction** refers to the direction from which the wind blows. For example, in the northern hemisphere, a north wind brings colder air from the north toward the south. We can measure wind direction with a *wind vane*, mounted so that it swings about a vertical axis; the push of the wind reorients an arrow so the arrowhead points in the direction the wind comes from **(Fig. 17.12a)**. **Wind speed** indicates the horizontal rate of air movement. We can measure wind speed with an *anemometer*, a device

consisting of three cups that catch the wind and, therefore, spin around a vertical axis (see Fig. 17.12a). Meteorologists typically measure wind speed in *knots* (kt), where 1 kt = 1 nautical mile per hour. Since a nautical mile is longer than a statute mile, 1 kt = 1.15 mph. Converting to the metric system, 1 kt = 1.85 km per hour. Meteorologists display winds by placing *wind barbs* on a weather map **(Fig. 17.12b)**. We can also display winds on a map by drawing lines parallel to the wind direction **(Fig. 17.12c)**.

Visibility and Cloud Cover

The old phrase "On a clear day, you can see forever" may be an exaggeration, but it does draw attention to the fact that air doesn't always have the same transparency. Meteorologists describe air transparency by specifying **visibility**, the distance from which a person with normal vision can identify objects, looking through air. In the past, visibility measurements were made by a human observer looking at objects at known distances. Today, meteorologists measure visibility automatically with a sensor that detects air clarity. A fog can reduce visibility to near zero, making driving hazardous, whereas on a clear day, visibility may be "unlimited" in that you can see to the horizon and beyond. What factors affect visibility? Pure air is virtually transparent, so visibility decreases when air contains aerosols, water droplets, or ice particles.

Meteorologists define **cloud cover** as an estimate of how much of the sky is obscured by clouds at a given time. In the past, human observers would determine cloud cover visually by estimating what fraction of the sky contained clouds. Today, laser technology allows us to determine what fraction of the time clouds lie directly overhead. When describing cloud cover, we typically use familiar descriptive terms (clear skies, few clouds, scattered clouds, broken clouds, or overcast skies) to indicate progressively more cloud cover. Cloud cover varies radically with location and time of year. At any given time, clouds hide approximately 70% of the Earth's surface **(Fig. 17.13)**.

Figure 17.12 Measuring wind speed and direction.

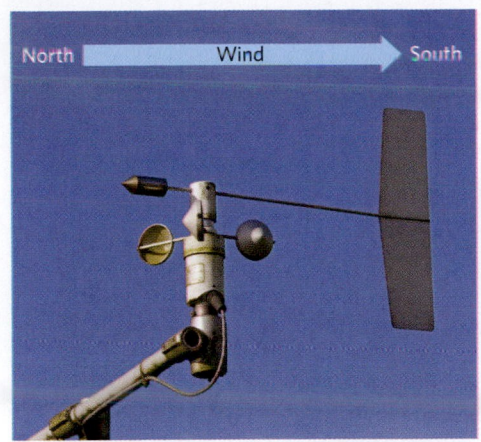

(a) Anemometer and wind vane. In this example, air flows from north to south, so the wind values indicates a "north wind."

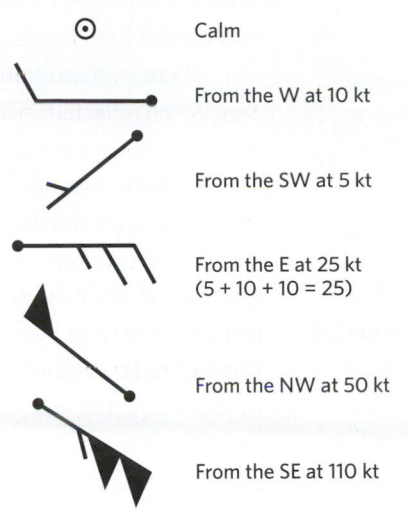

⊙ Calm

⌐ From the W at 10 kt

From the SW at 5 kt

From the E at 25 kt
(5 + 10 + 10 = 25)

From the NW at 50 kt

From the SE at 110 kt

(b) Key to wind barbs used on a weather map.

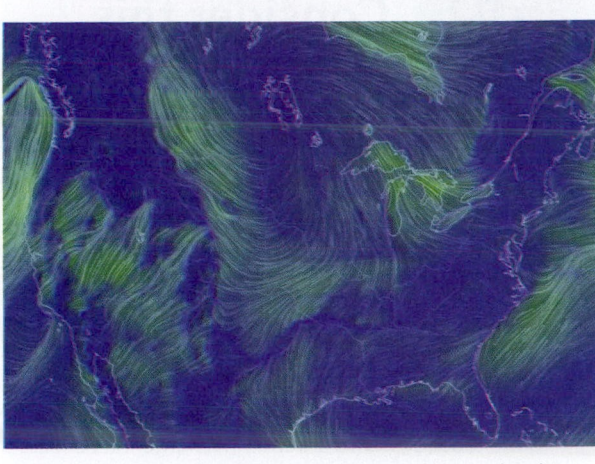

(c) On this wind map, winds are represented by swirling lines; the lines are thicker and more closely spaced where the wind is faster.

Precipitation

In the geology chapters of this book, we generally use the word *precipitation* to refer to the formation of a solid when a liquid solution becomes supersaturated. In atmospheric science, **precipitation** generally refers to water that condenses to a liquid or solid state and falls from the sky. Precipitation includes brief showers, heavy downpours, snow flurries, hailstorms, and blizzards.

To characterize precipitation, meteorologists specify both the *precipitation rate*, a measure of how fast rain or snow is falling in inches per hour or millimeters per hour, and the *total precipitation*, the cumulative amount that fell during a specified time period, such as a single storm, a single month, a season, or a whole year. We can report total rainfall in inches or millimeters. For total snowfall, the measurement indicates either an *accumulated depth*,

Figure 17.13 Composite satellite image showing the Earth's cloud cover.

Box 17.2 ▶ Consider this . . .

Meteorological radars and satellites

Every year, floods, tornadoes, and other types of hazardous weather cause loss of life and property damage. Meteorologists can warn the public of impending hazards by using weather radar **(Fig. Bx17.2a)**. Radar systems transmit microwaves, with a wavelength of about 1 to 10 cm (0.4 to 4 inches), much like the microwaves in a microwave oven. But unlike your oven, a radar transmitter sends out microwaves in pulses that last only about 1 millionth of a second. When the microwaves encounter raindrops and hailstones, some of the microwaves scatter back to the radar system's antenna. The antenna collects this energy, called the *radar* echo, and measures the time it took the microwave pulse to travel out to the precipitation and back. Because microwaves travel at the speed of light, we can calculate the distance to the precipitation from the following equation: distance = velocity × time. By knowing the angle at which the antenna points to the sky, we can precisely determine the elevation of the precipitation within the atmosphere.

The character of precipitation affects the strength of the returned signal, or the *radar reflectivity*. For example, the greater the size and number of particles the pulse intercepts, the greater the radar reflectivity will be. High radar reflectivity values, depicted with red and orange colors on a display, indicate heavy rain or hail. In contrast, low values, depicted with blues and greens, indicate light drizzle or nonprecipitating clouds **(Fig. Bx17.2b)**. By examining radar reflectivity measurements over time, meteorologists can estimate the total amount of rain that fell during the period of observation. These data help predict flash floods.

Radar systems transmit microwave energy at specific frequencies. When reflected energy returns to the antenna, the frequency of the returned energy has typically shifted slightly as a result of the movement of the raindrops or ice particles along the direction of the radar beam. **Doppler radar** systems measure this shift and use it to determine the speed of the wind in the direction parallel to the beam. Doppler radar unfortunately cannot measure air motion across the beam. Nevertheless, by mapping out the wind in the direction of the beam, Doppler radar can identify strong straight-line winds as well as rotation in the air flow, which may be evidence of a tornado. Therefore, meteorologists

Visible channel

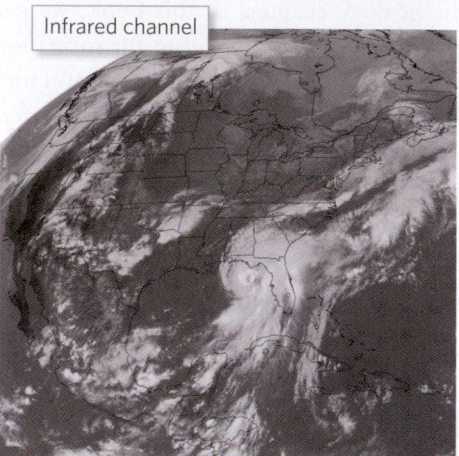
Infrared channel

Figure Bx17.2 Weather radar and satellites.

(a) A radar antenna used in research.

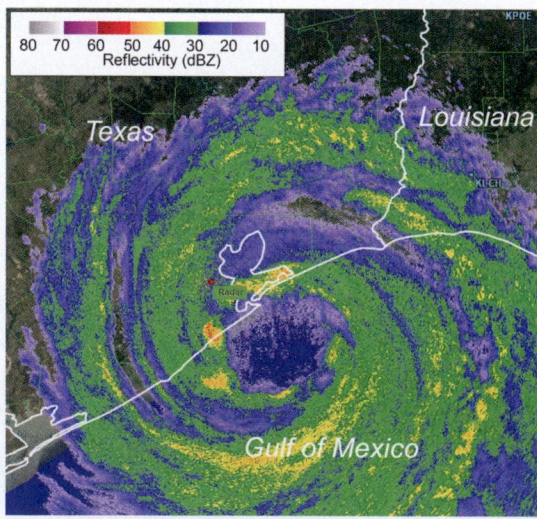

(b) Radar reflectivity from Hurricane Ike as it made landfall near Galveston, Texas in 2008.

Water vapor channel

(c) The same image from a weather satellite, viewed in different channels. The area shown by the map covers the eastern United States.

commonly use Doppler radar to detect severe weather and provide early warning to the public. The development of *polarization Doppler radar*, which employs beams in which the microwaves move only in a given plane, can now help meteorologists identify tornadoes that have lofted airborne debris.

Meteorologists supplement radar measurements from ground-based locations with data collected by orbiting weather satellites, which can acquire images of the atmosphere over a wide region, or even the entire world. Animated satellite images allow meteorologists to watch storms develop and to monitor the progress of a storm as it moves across the Earth. Weather satellites use different wavelengths of radiation to gather different kinds of information **(Fig. Bx17.2c)**. The *visible channel* uses reflected sunlight to show clouds as you would see them from space, in black and white. The *infrared channel* shows the temperatures of cloud tops or the ground, with cold temperatures indicated by bright tones and warm temperatures by dark tones. The *water vapor channel* indicates regions of high moisture content with bright tones and dry areas with dark tones.

the thickness of the snow layer, or its *water equivalent*, the depth of the water that would be formed if the snow melted completely. We measure snowfall in different ways because snow density can vary significantly depending on temperature: warmer snow tends to be dense, whereas colder snow tends to be light and fluffy. Meteorologists traditionally measure precipitation using a *precipitation gauge*, a container that collects either rain or snow and measures its weight or volume **(Fig. 17.14)**. Today, we can also use information from *weather radar* to estimate precipitation over large regions **(Box 17.2)**.

Weather Stations

When you see weather maps in a newspaper or on TV, you're seeing a compilation of data collected at hundreds of *weather stations*, sites at which meteorologists have set up instruments that automatically measure temperature, dew point, wind speed and direction, visibility, cloud type and coverage, and precipitation type and intensity **(Fig. 17.15a)**. In the United States, the National Weather Service operates the *Automated Surface Observing System*, while the Federal Aviation Administration and Department of Defense operate the *Automated Weather Observing System* **(Fig. 17.15b)**. The instruments in these systems update observations 24 hours a day and send electronic records of their measurements to data centers. Weather stations on ships or on buoys provide data for oceanic regions.

Take-home message . . .

To characterize weather, meteorologists measure temperature, atmospheric pressure, dew point, wind speed and direction, cloud cover, visibility, and precipitation. These characteristics vary with location at a given time and with time and elevation at a given location.

Quick Question ----------------------------

Why is total snowfall measured differently than total rainfall?

Figure 17.14 Measuring rainfall using a tipping-bucket rain gauge.

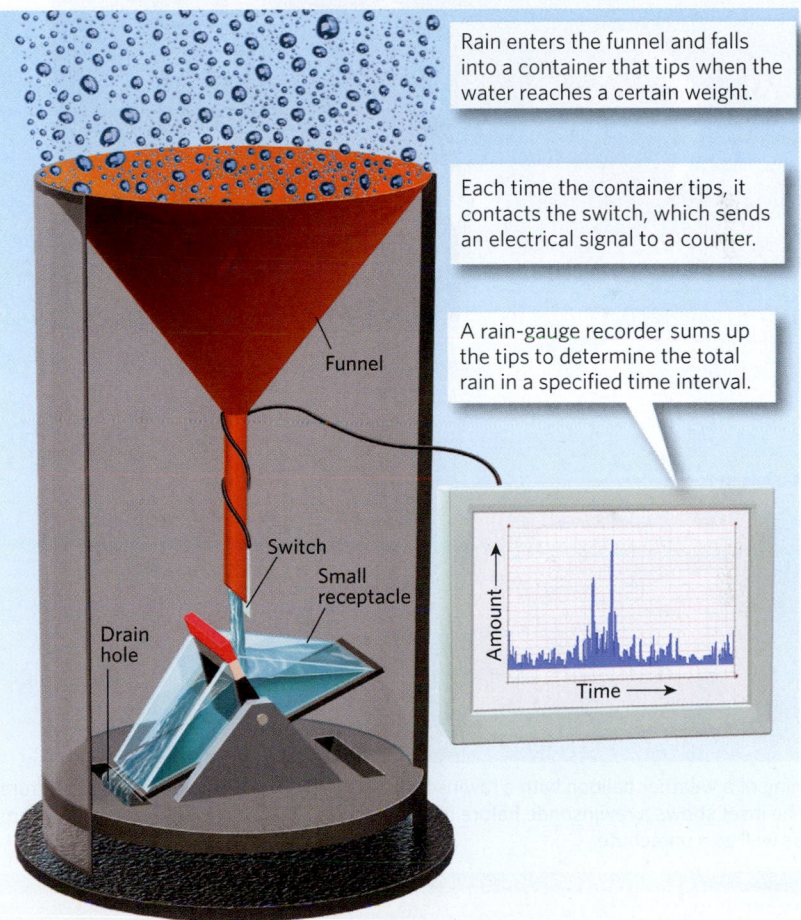

Rain enters the funnel and falls into a container that tips when the water reaches a certain weight.

Each time the container tips, it contacts the switch, which sends an electrical signal to a counter.

A rain-gauge recorder sums up the tips to determine the total rain in a specified time interval.

Funnel

Switch

Small receptacle

Drain hole

Amount

Time

17.5 Vertical Structure of the Atmosphere

Thermally Defined Layers

Imagine that you're sitting outside reading this book on a warm, calm summer day. Right now, only 10 km (6 miles)

Box 17.3 Consider this . . .

Studying the upper atmosphere

To understand and predict the behavior of weather systems, meteorologists need to obtain measurements of atmospheric properties high above the Earth's surface. To do this, they send up *rawinsondes*, compact packages of instruments carried by weather balloons to elevations of about 20 km (12.5 miles) **(Fig. Bx17.3a)**. A rawinsonde radios back information about atmospheric pressure, temperature, dew point, and wind direction and speed continuously as it rises, producing a "sounding" that depicts the vertical structure of the atmosphere **(Fig. Bx17.3b)**. By international agreement, rawinsondes are launched simultaneously twice a day at hundreds of localities worldwide. On average, rawinsonde launch sites on land are located about 500 km (200 miles) apart, so their spacing is much greater than that of ground-based weather stations. As a result, high-elevation data are not as detailed as ground-based data. Because air pressure decreases with elevation, weather balloons expand as they rise. When a balloon reaches about 20 km, it breaks, and the rawinsonde it carried parachutes back to the Earth.

Figure Bx17.3 A rawinsonde unit can be used to obtain information about the atmosphere at high elevations.

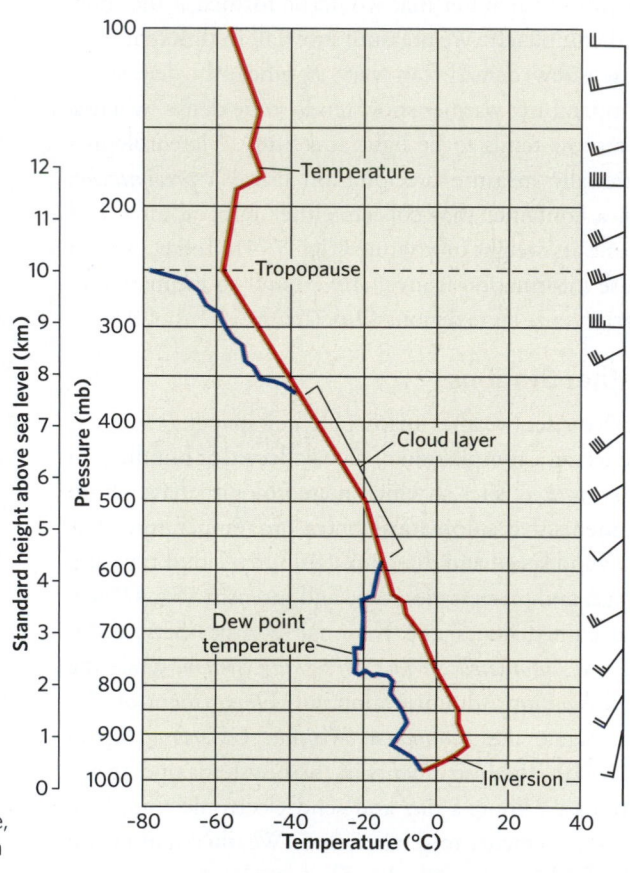

(a) Launching of a weather balloon with a rawinsonde attached. The inset shows a rawinsonde before it was launched, as well as a parachute.

(b) A vertical profile of atmospheric temperature, dew point, and wind from a rawinsonde.

directly above you, the wind may be blowing at over 160 km per hour (100 mph), the temperature may be hovering at a frigid –50°C (–58°F), and the gas in the air has such low density that you couldn't survive by breathing it. If you're a passenger in a jet plane that has climbed to cruising altitude, these conditions exist right outside your window. Clearly, atmospheric conditions change rapidly with elevation. Atmospheric scientists use balloon-lofted instruments called *rawinsondes* to measure temperatures and other properties of the atmosphere at high elevations **(Box 17.3)**.

Figure 17.15 Surface weather stations in North America.

(a) Each white dot on this map represents a weather station.

(b) Equipment set up at a weather station.

In Section 17.4, we saw that air pressure decreases progressively as elevation increases. Temperature also changes with elevation. Research shows that in some layers of the atmosphere, temperature increases as elevation increases, whereas in others, temperature decreases as elevation increases. Elevations at which the direction of change reverses (from decreasing to increasing, or vice versa) delineate boundaries that separate four distinct atmospheric layers **(Fig. 17.16)**. Let's examine these layers in sequence, beginning at the Earth's surface.

Since solar energy reaches the atmosphere from above, it's a common misperception that the air around you becomes warm because it's "broiled from above." In fact, air is transparent to most incoming radiation (visible light)—that's why you can see through air. This energy reaches the Earth's surface, where it gets absorbed. The surface then radiates the energy back into the air as infrared radiation, some of which gets trapped by the air and warms it. (We'll discuss this *greenhouse effect* in more detail later in this chapter.) Put another way, the Earth's surface acts as a heater that warms the air at the base of the atmosphere—in effect, the air is "baked from below."

As we move upward, away from the Earth's surface, the distance from that heat source decreases, and the air gets cooler. So, if you climb a tall mountain, you'll find that as your elevation increases, air temperature decreases. This decrease in temperature with increasing elevation continues up to about the cruising altitude of commercial jets, where air temperature ranges between −40°C and −60°C (−40°F and −76°F). Meteorologists refer to this lowest layer of the atmosphere, in which temperature decreases with elevation, as the **troposphere**. The rate of

Figure 17.16 Major layers in the lower 110 km (70 miles) of the atmosphere.

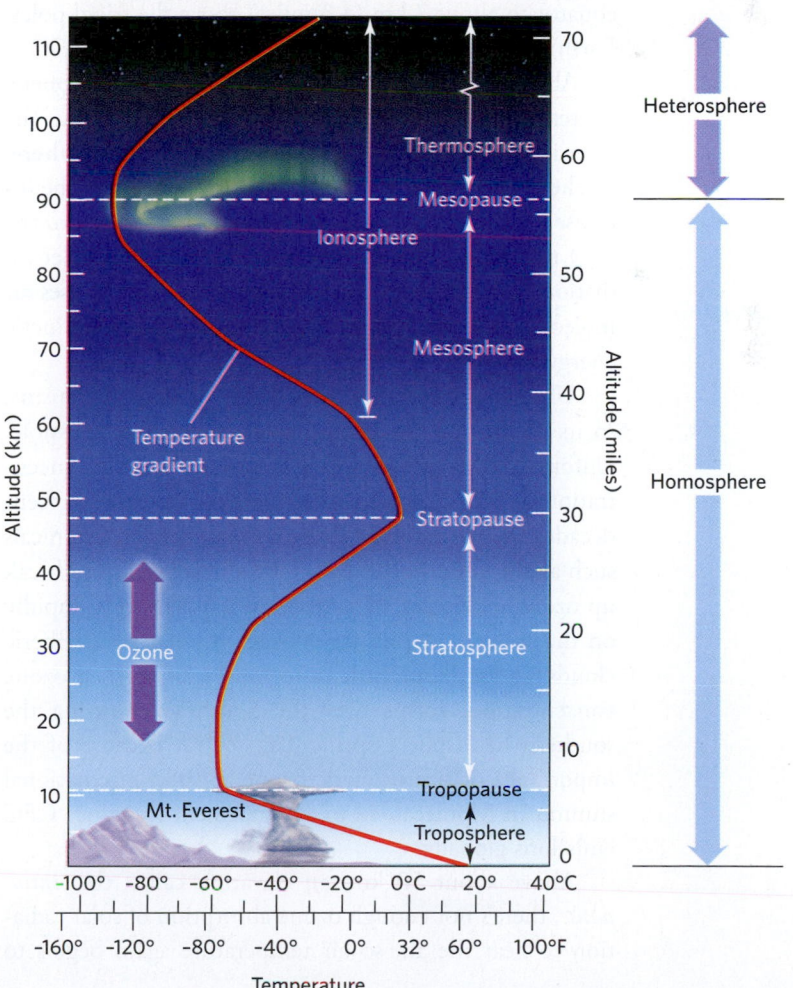

Figure 17.17 High-altitude zones of the atmosphere.

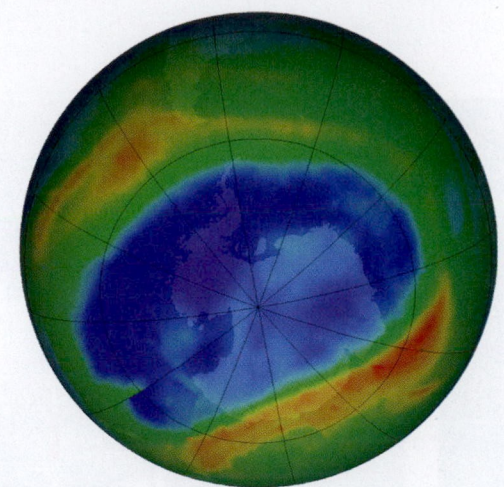

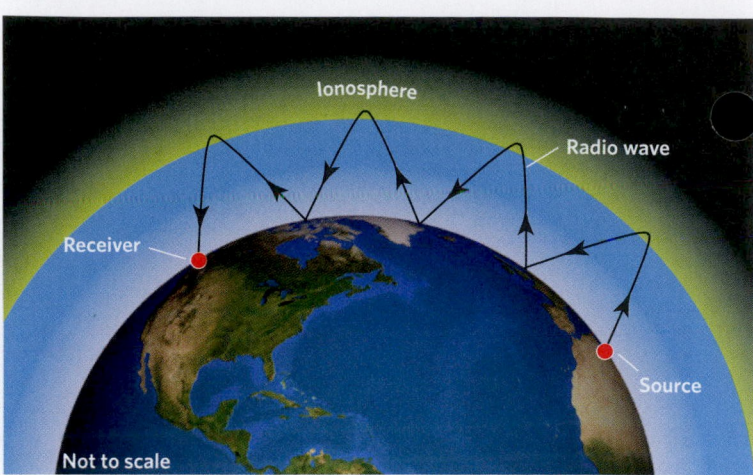

(a) The ozone hole over Antarctica in 2015. A Dobson unit measures the concentration of ozone in the atmosphere; larger numbers mean more ozone.

(b) The ionosphere, a layer containing charged ions, causes certain radio waves to reflect back to Earth, from which they bounce up again. These waves can be detected beyond the horizon.

change of temperature with elevation, a quantity known as the **environmental lapse rate**, varies substantially from place to place and from time to time, but on average has a value of 6.5°C/km (3.5°F/1,000 feet). The thickness of the troposphere varies with latitude and time of year: it decreases from about 20 km (12.5 miles) above the equator to about 7 km (4.3 miles) above the frigid poles. Earth's storms occur entirely within the troposphere.

Above the *tropopause*, the top of the troposphere, air temperature increases with increasing elevation. The layer in which this increase takes place, the **stratosphere**, reaches up to a height of about 50 km (31 miles). The increase in temperature with elevation occurs because ozone (O_3) molecules in the stratosphere absorb ultraviolet radiation from the Sun. This absorbed radiation causes air molecules to move about faster, so their average kinetic energy—their temperature—increases.

Ultraviolet radiation can harm living organisms, so its absorption by ozone makes life on land possible. Unfortunately, human activities have caused the concentration of ozone in the stratosphere to decrease in recent decades. When emitted into the atmosphere, chemicals such as chlorofluorocarbons (CFCs) react with and break up ozone molecules. This reaction happens most rapidly on the surfaces of tiny ice crystals in polar stratospheric clouds, so an **ozone hole**, a region of diminished ozone concentration, forms over the South Pole during the southern hemisphere spring **(Fig. 17.17)**. Because of the importance of stratospheric ozone, a 1987 international summit in Montreal led to an agreement to reduce CFC emissions globally.

Above about 50 km, an elevation called the *strato-pause*, there's not enough ozone absorption of solar radiation to heat the air, so air temperature again begins to

decrease with increasing elevation. This point marks the bottom of the **mesosphere**, which continues upward to the *mesopause*, at an elevation of about 90 km (56 miles). Above the mesopause lies the outermost layer of the atmosphere. In this layer, the **thermosphere**, temperature increases with increasing elevation, reaching a maximum of 1,500°C (2,700°F), as the few remaining air molecules absorb high-energy radiation from the Sun. These high temperatures are deceptive, however, for an astronaut entering the thermosphere without first donning a space suit would freeze instantly. That's because the air density in the thermosphere is so low that a given volume of air holds hardly any heat. All the air molecules colliding with the astronaut, though traveling very fast, could not provide enough heat to counteract the loss of heat by radiation from the astronaut's body. The density of air in the thermosphere decreases upward until, at an elevation that ranges from 350 to 800 km (217 to 500 miles), it equals that of interplanetary space. This elevation varies depending on solar activity (the quantity of incoming atoms brought to the Earth in the solar wind).

Layers Defined by Composition and Electrical Charge

Below an altitude of about 90 km, air circulation mixes the atmosphere so thoroughly that it has a composition that is mostly the same from place to place. Researchers refer to this well-mixed layer as the **homosphere** (see Fig. 17.16). We use the qualifier "mostly" because the concentration of certain gases varies. For example, ozone forms and occurs primarily in the stratosphere. Also, gases that enter the atmosphere from localized sources at the Earth's surface tend to decrease in concentration with distance from the source. And water vapor, which has a short *residence time* in the atmosphere (meaning that it enters and leaves the atmosphere relatively quickly), varies substantially with location and time. Note that the homosphere includes all of the troposphere, stratosphere, and mesosphere. Above the homosphere, gases sort gravitationally according to their molecular weight, with lighter gases rising and heavier ones sinking. This stratified layer, the **heterosphere**, coincides with the thermosphere.

Meteorologists also recognize another layer, not based on temperature or composition, but rather on electrical characteristics. In this layer, known as the **ionosphere**, the atmosphere contains a relatively high concentration of *ions*, molecules that have gained or lost electrons and thus have a positive or negative electric charge. The ionosphere, which occurs above elevations of 60 km (37 miles) during the day and 100 km (62 miles) at night (see Fig. 17.16), serves an important role in human communications because it reflects certain frequencies of radio transmissions. These transmissions, notably AM radio, bounce back and forth between the ionosphere and the Earth's surface, which allows them to travel over long distances (Fig. 17.17b).

Incoming high-energy particles from the Sun are channeled along magnetic field lines to the poles of the Earth. When they interact with certain molecules in the ionosphere, these molecules emit light. We see this light as a shimmering **aurora**, a gauzy curtain of undulating green and red color in the sky (Fig. 17.18). We refer to the aurora above northern polar regions as the *aurora borealis*, or northern lights, and that over southern polar regions as the *aurora australis*, or southern lights.

Take-home message . . .

Atmospheric scientists divide the atmosphere into layers based on changes in its temperature, composition, and electric charge. The layers defined by changes in temperature are the troposphere, stratosphere, mesosphere, and thermosphere. The layers defined by consistency of composition are the homosphere and heterosphere, and the layer defined by electrical charge is the ionosphere.

Quick Question -
What role does the ionosphere play in radio transmission?

17.6 Clouds and Precipitation

From the vantage point of an astronaut in the International Space Station, the pattern of clouds stands out as one of the most striking features of the Earth. Some clouds look like small cotton puffs, others like broad white blankets, and still others like elegant swirls (see Fig. 17.13). What do they consist of? Clouds are not composed of pure water vapor, which, as gaseous water, is transparent. Rather, clouds are collections of countless water droplets and/or tiny ice crystals suspended in the sky. These particles reflect, scatter (emit in random directions), and absorb light, so thin clouds appear whitish and thick clouds look gray and ominous. Let's examine how clouds form, how clouds differ from one another, and why some produce rain or snow while others do not.

Cloud Formation

As we have seen, air can hold only a certain amount of water vapor before it becomes saturated, and the amount it can hold depends on its temperature. If saturated air cools further, it becomes *supersaturated*, and when this happens, water molecules begin to attach to the surfaces of aerosols. In effect, a tiny aerosol can serve as a *condensation nucleus*. If the temperature is above freezing, a droplet of liquid water grows around the nucleus as more and more water molecules attach. If, however, the temperature is below freezing, an aerosol can serve as a *deposition nucleus*, to which water molecules attach to produce a crystal of solid ice. Water vapor is invisible, but water droplets and ice particles are visible—so, simply put, clouds form when water molecules in supersaturated air collect on aerosols to form liquid droplets and ice particles. Every raindrop and snowflake you see initially formed on an aerosol, so if there were no aerosols, there would be no rain, and life on land would never evolve.

Figure 17.18 The aurora borealis, as seen over Alaska.

Figure 17.19 Several mechanisms can cause air to rise.

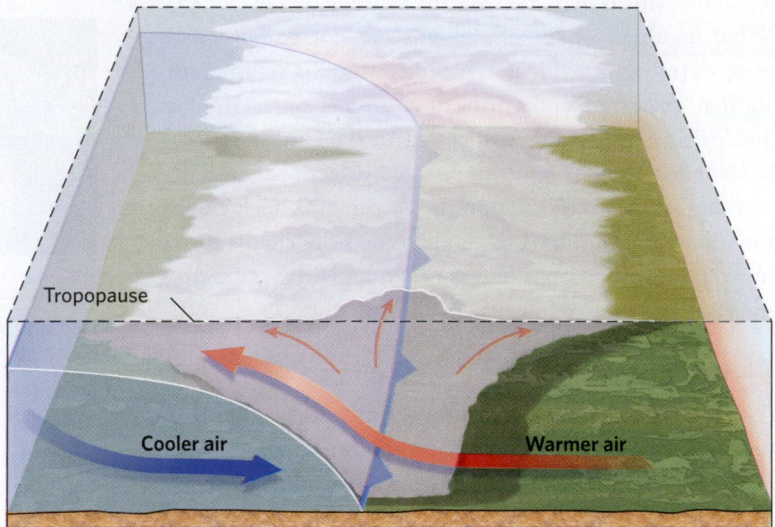

Tropopause

Cooler air   Warmer air

(a) At a front, warm air must rise over an advancing mass of cold air.

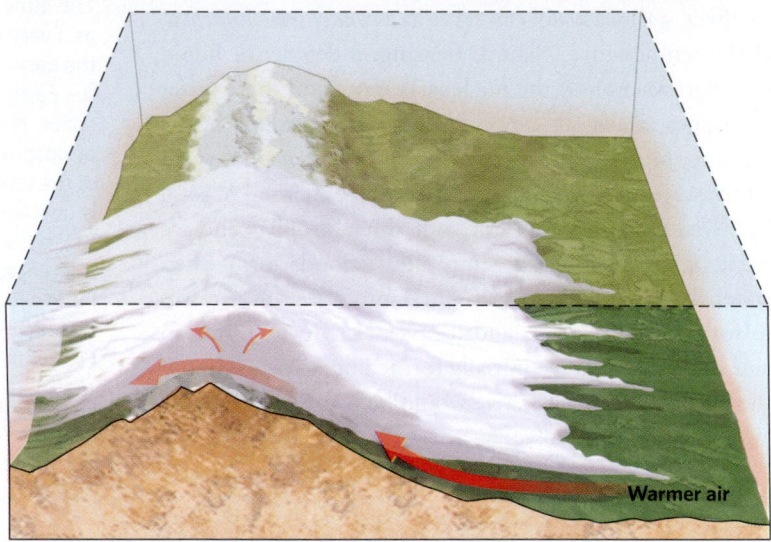

Warmer air

(b) When flowing air reaches a mountain front, it has to rise.

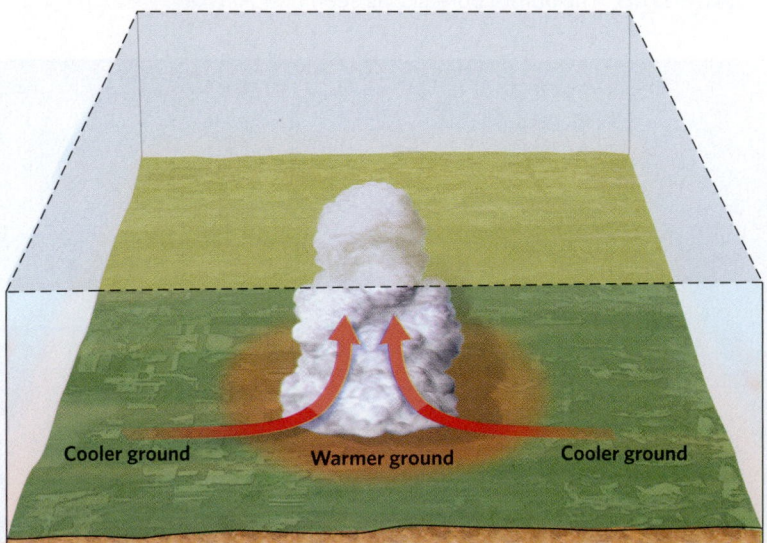

Cooler ground   Warmer ground   Cooler ground

(c) Air becomes buoyant and rises over warmer ground. Cooler air then flows in the front side.

Why does air become locally supersaturated? Generally, supersaturation happens where air rises. Air rises when it undergoes *lifting* by any of several processes that force air upward. Lifting can happen, for example, when less dense, warm air encounters and flows up and over dense, cool air (the boundary where two such contrasting air masses meet is called a *front*) **(Fig. 17.19a)**. A similar phenomenon happens along a shoreline when cool air moves onshore as a sea breeze and lifts warm air. Air can also rise as it encounters and flows over a mountain range **(Fig. 17.19b)**. And air can rise when it becomes *buoyant*, meaning that it becomes less dense than its surroundings **(Fig. 17.19c)**. Buoyancy can develop, for example, when heat rising from the ground below warms air near the ground surface. You've seen this process happening if you've ever seen a hot-air balloon head skyward. We'll discuss additional lifting mechanisms in Chapter 19.

Atmospheric pressure decreases with altitude, so rising air expands as it moves upward. In order to expand, the air must push against surrounding air molecules, so it loses kinetic energy and becomes cooler (see Section 19.3). Cooling, in turn, reduces the air's capacity to hold water vapor, and if its temperature decreases enough, the air reaches its dew point. At this point, any further rising, which leads to more cooling, causes supersaturation and, therefore, cloud production.

Types of Clouds

Clouds fascinate people because of their beauty and the way they constantly change. They come in many shapes and sizes and form at many altitudes within the troposphere. Meteorologists classify clouds by their altitude and shape, and by whether or not they produce precipitation, using terminology first proposed in 1802 by a British pharmacist named Luke Howard **(Fig. 17.20)**. Howard coined the terms *cirrus* for high, wispy clouds, *stratus* for sheets of clouds that cover vast areas, and *cumulus* for tall, puffy clouds with a cauliflower-like appearance. Combining these terms with others provides additional detail about clouds. For example, adding the word *nimbus* to a cloud name indicates that the cloud produces rain, so a *cumulonimbus cloud* is a towering puffy cloud that produces rain. Today, we recognize many more types of clouds, but still use many of the names Howard first proposed in 1802.

Why are there so many types of clouds? The character of a cloud depends on the amount of water vapor available, the nature of the lifting mechanisms that drive cloud formation, the temperature in the troposphere, and the velocity of winds at various altitudes.

HIGH-ALTITUDE CLOUDS. Wispy, elongate clouds that develop very high in the troposphere (above 6 km, or 20,000 feet) are known as **cirrus clouds (Fig. 17.21a)**.

Figure 17.20 The most common cloud types. Clouds come in a variety of sizes and shapes.

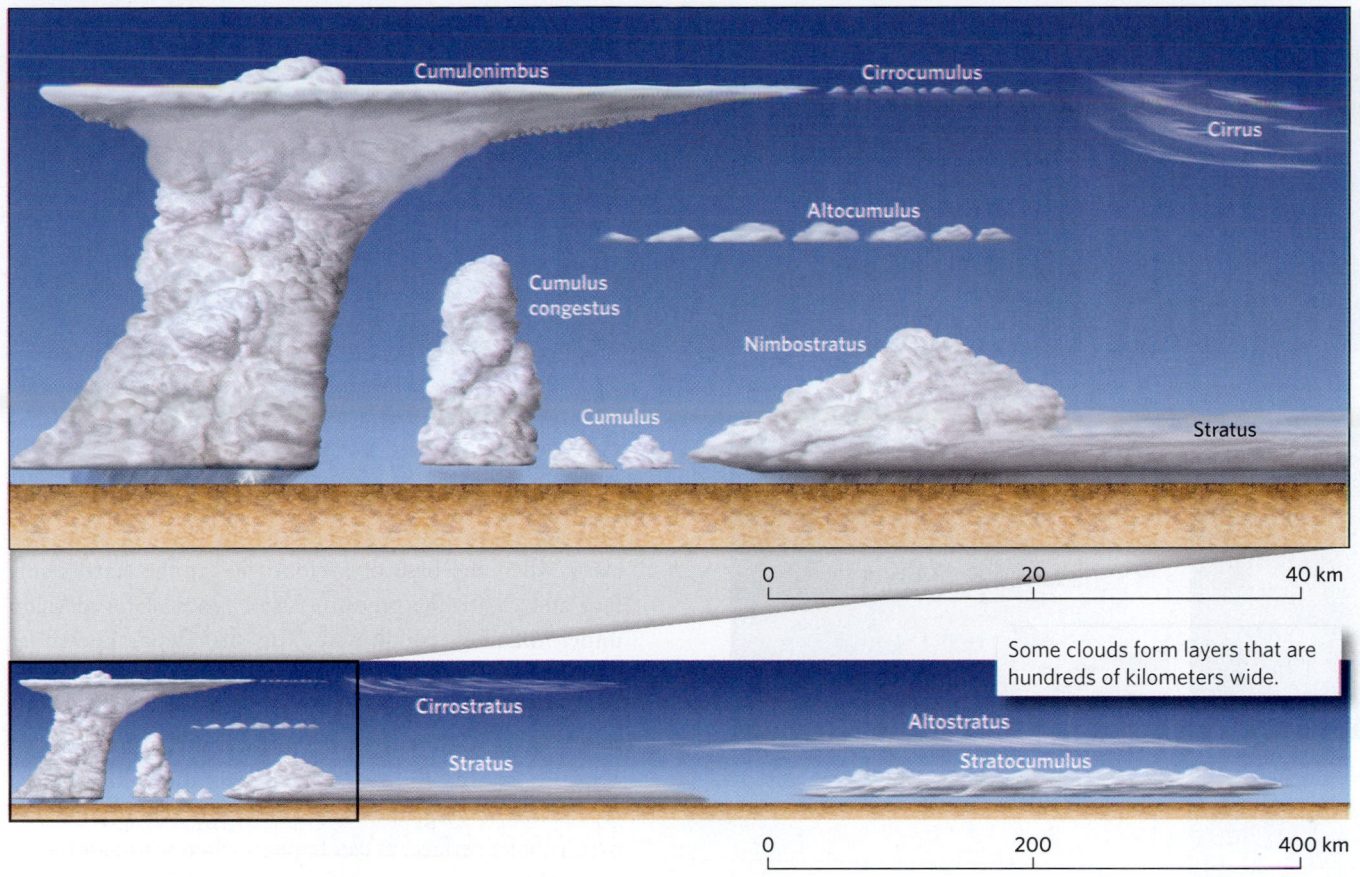

Some clouds form layers that are hundreds of kilometers wide.

The temperature is so cold where cirrus clouds form that these clouds consist entirely of tiny ice crystals. As the ice crystals fall through the atmosphere, winds at different altitudes cause them to line up in wisp-like strands. Cirrus clouds are too high and thin to produce precipitation. If they do accumulate into layers thick enough to dim the Sun, they become *cirrostratus clouds*, and if they grow into puffy billows, they become *cirrocumulus clouds*.

MID-ALTITUDE CLOUDS. Mid-altitude clouds form between 3 and 6 km (10,000 to 20,000 feet) above the Earth's surface **(Fig. 17.21b)**. Meteorologists refer to them as *altostratus clouds* if the clouds are layered, uniformly gray, and thick enough to hide the Sun. If the clouds resemble roll-like patches or puffs, they are *altocumulus clouds*. Altostratus and altocumulus clouds, like cirrus clouds, are too high and too thin to produce precipitation.

LOW-ALTITUDE CLOUDS. Widespread, layered clouds that spread out like great blankets at relatively low altitudes (below 3 km, or 10,000 feet) are called **stratus clouds (Fig. 17.21c)**. The width of stratus clouds greatly exceeds their thickness. If portions of stratus clouds billow into puffy shapes, they become *stratocumulus clouds*, and if stratus clouds thicken sufficiently that

precipitation falls from them, they become *nimbostratus clouds*.

TOWERING CLOUDS. Unlike the altitude-defined clouds described above, *towering clouds* rise across a range of altitudes **(Fig. 17.21d)**. The largest extend from low altitudes to the tropopause. Small towering clouds are called **cumulus clouds** if they are growing vertically and have cauliflower-like lobes. Cumulus clouds form in strong updrafts and, in general, have comparable horizontal and vertical dimensions. If cumulus clouds grow very large, they become *cumulus congestus clouds*, and if they produce precipitation, they become *cumulonimbus clouds*. Some cumulonimbus clouds rise to the base of the stratosphere, where strong winds blow their tops into a broad sheet; these clouds are known as *anvil clouds*.

UNUSUAL CLOUDS. Some clouds take on strange shapes. For example, under special conditions, lens-shaped clouds, known as *altocumulus lenticularis clouds*, may develop; these clouds have been mistaken for flying saucers **(Fig. 17.22)**. Thunderstorms can produce ominous cloud formations with names like wall clouds, shelf clouds, mammatus clouds, and roll clouds, which we'll examine in our discussion of thunderstorms in Chapter 19.

Did you ever wonder...

why clouds have different shapes?

Figure 17.21 Photos of some common cloud types.

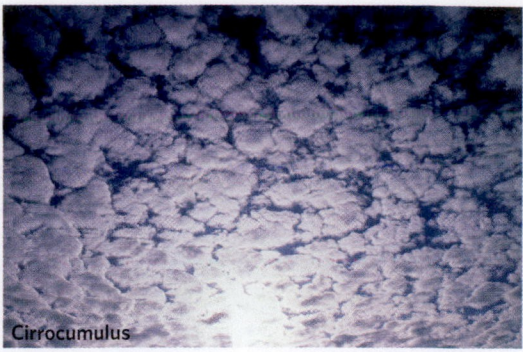

Cirrus

Cirrostratus

Cirrocumulus

(a) High-altitude clouds.

Altostratus

Altocumulus

(b) Mid-altitude clouds.

Fog

On a foggy day, visibility can become so poor that you can't see more than a few meters in front of you. **Fog** exists when the base of a cloud lies at the Earth's surface and cloaks the ground. Three types of fog develop under different conditions. The first type, known as *radiation fog*, tends to form at night when, in the absence of sunlight, the ground cools by radiating heat upward **(Fig. 17.23a)**. The cooled ground absorbs heat from the air above it, causing the air to reach saturation. The second type, *advection fog*, forms when warm, humid air flows over a cooler surface, as can happen when warm air moves over cold ocean water or a field of snow **(Fig. 17.23b)**. The third type, *evaporation fog*, forms where enough surface water evaporates into air that the air becomes saturated **(Fig. 17.23c)**. Such fog might develop over a warm lake or marsh.

Precipitation and Its Causes

Unless you live in a bone-dry desert, you've probably experienced *precipitation* (rain, hail, or snow) many times, and you're familiar with the common terms (drizzle, downpour, flurries, heavy) used to describe precipitation.

Stratocumulus

Stratus

(c) Low-altitude clouds.

Cumulonimbus

Cumulus

Cumulus congestus

(d) Towering clouds.

Figure 17.22 Lenticular clouds (altocumulus lenticularis) may look like flying saucers.

Figure 17.23 Three types of fog.

(a) Radiation fog.

What conditions lead to precipitation, and why do some clouds pass overhead without dropping any?

WARM-CLOUD PRECIPITATION. Meteorologists define a *warm cloud* as one in which the temperature stays above 0°C throughout, so that the cloud consists only of liquid water droplets. In such a cloud, the individual *cloud droplets* are so small that they remain suspended in air. It's only when the droplets grow sufficiently large that they become **raindrops**, spheres of water heavy enough to overcome air resistance and fall toward the ground. *Rain* consists of a vast collection of raindrops—during a downpour, trillions of raindrops fall on a square kilometer of the Earth's surface.

To understand how raindrops form in warm clouds, we must first see how tiny cloud droplets, which range from 0.01 to 0.1 mm (0.0004 to 0.004 inches) in diameter, grow big enough to fall as raindrops (typically 0.5 to 4.0 mm, or 0.02 to 0.16 inches in diameter) **(Fig.17.24a)**. The formation of raindrops begins in a cloud as cloud droplets form, if the air remains supersaturated. In such a cloud, random motions of water molecules cause additional water molecules to collide with and bond to those at the surfaces of the droplets. Such migration of atoms or molecules due to their random motions is called *vapor diffusion*. Vapor diffusion operates very slowly, however, so the process alone can't produce raindrops. For raindrops to form, the cloud droplets themselves must come in contact with one another and merge, a process called **collision-coalescence**. As a somewhat larger droplet falls faster than its smaller neighbors and collides with them, it collects these additional droplets and continues to grow. Small raindrops range from 0.5 to 1.0 mm across, large ones range from 1.0 to 4.0 mm across, and very large ones are 4.0 to 6.0 mm in diameter. For a single large (4-mm) raindrop to form, it must collect 8 million 0.01-mm cloud droplets during its fall!

(b) Advection fog.

(c) Evaporation fog.

Figure 17.24 Raindrop growth and breakup.

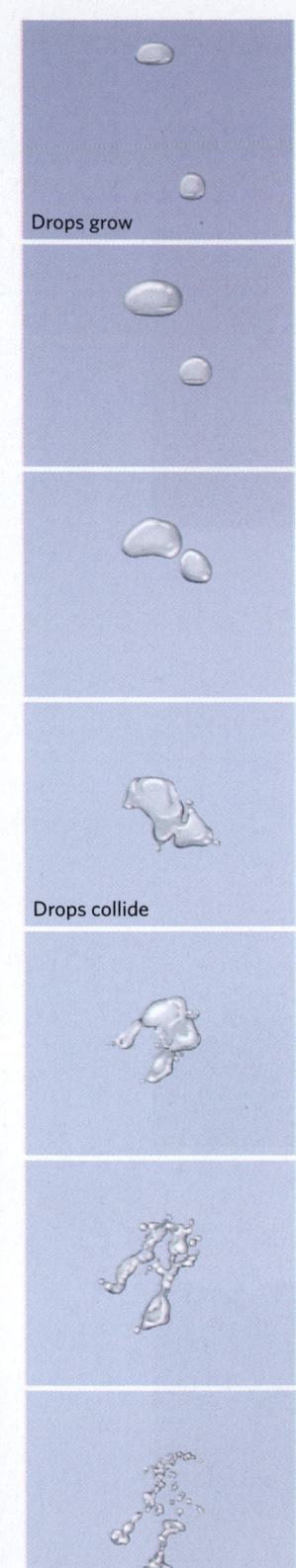

Time

Drops grow

Drops collide

Drops break up

(b) When large raindrops collide, they can break up into many more, smaller raindrops.

Cloud droplet

Condensation nuclei

Drizzle drop

Small rain drop

(a) Cloud droplets form when water vapor condenses around a condensation nucleus. Rain forms as droplets collide with one another and grow.

Popular media typically depict raindrops as having a teardrop shape—round on the bottom, tapering to a point at the top. Real raindrops don't look like teardrops at all. In fact, cloud droplets and small raindrops have a spherical shape because of *surface tension*. Surface tension exists because the water molecules within a drop bond to other water molecules in all directions around them, but can't bond to air molecules around the drop. Because the bonds on the surface of the droplet are all oriented inward, the water drop assumes the shape of a sphere. Due to air resistance, large raindrops flatten into a hamburger shape as they fall. Raindrops larger than about 6.0 mm rarely develop because when large drops collide with their neighbors, they break apart **(Fig. 17.24b)**.

COLD-CLOUD PRECIPITATION. Many clouds extend to altitudes where atmospheric temperatures fall below 0°C. In these *cold clouds*, ice particles grow by one of two paths, depending on whether supercooled droplets are present **(Fig. 17.25)**. *Supercooled droplets* are droplets that remain in a liquid state even though their temperature is well below 0°C (32°F).

First, tiny ice crystals grow on special types of aerosols called *ice nuclei*. Ice nuclei are mineral grains whose crystal structure resembles that of ice. Because of this

similarity, water molecules can find favorable attachment sites. Once these tiny crystals have formed, more water vapor molecules attach to their solid surfaces, eventually building beautiful hexagonal crystals. These crystals grow into a wide variety of shapes, depending on the temperature and humidity. Some crystals have branched arms, so that when they collide with neighboring ice crystals, they latch onto one another and clump together. The resulting composite of numerous ice crystals becomes a **snowflake** (see Fig. 17.25).

Not all ice particles grow entirely by deposition of water vapor directly on ice nuclei, however. If sufficient ice nuclei do not exist in a cloud, water droplets remain in a liquid state, but they become supercooled. When ice crystals falling through clouds encounter these droplets, the droplets instantly freeze onto the ice and form *rime* on the crystals. Ice particles growing by collection of supercooled droplets eventually grow into small, soft spheres of ice called **graupel**. A graupel ball that continues to grow and harden becomes a *hailstone* (see Fig. 17.25). In some cases, raindrops can become supercooled in a layer of cold air near the ground. These drops can reach the ground as *freezing rain*, which freezes to surfaces on contact to form an ice crust on the ground surface, power lines, and trees.

PRECIPITATION INVOLVING BOTH WARM- AND COLD-CLOUD PROCESSES. As we learned earlier in this chapter, temperature decreases with altitude in the troposphere, so even if warm temperatures exist at ground level, the upper parts of towering clouds have temperatures well below freezing. Therefore, water vapor may attach to ice nuclei in the upper part of a tall cumulus congestus cloud. The resulting ice particles collect supercooled water droplets as they fall, forming graupel. But when the graupel particles fall to lower altitudes, where the temperatures rise above freezing, they melt to form raindrops. If strong updrafts exist in the cloud, graupel balls may remain aloft long enough to grow into hailstones. If the hailstones become large enough to fall through the updraft, or if the updraft weakens, they can reach the ground before completely melting and land as hail.

In winter, a warm, above-freezing layer of air may exist aloft while the air beneath it is below freezing. If this happens, raindrops falling from the warm layer aloft sometimes freeze into tiny pellets called **sleet** when they enter the cold air near the ground. At other times, they arrive as freezing rain.

Clouds and Energy

Recall that atmospheric water can exist in three states: gas (vapor), liquid, and solid (ice). A **phase change** happens when water in one state converts to another, such as when ice *melts* (solid → liquid) or *sublimates* (solid → gas), or a water drop *freezes* (liquid → solid) or *evaporates* (liquid → gas). Phase changes can happen when air containing water cools or warms (Fig. 17.26). Since phase changes don't take place instantly, and since temperature varies with altitude, a cloud can contain all three phases at once.

Phase changes absorb or release energy. To understand the relationship between a phase change and energy, imagine a simple experiment. Take a large, full pot of water out of the refrigerator and place it on the red-hot burner of a stove. Heat the water for, say, 10 minutes—time for enough heat to transfer from the burner to the water to bring the water to a boil. Imagine that once the water has come to a boil, you have to continue heating it for another hour for all the liquid water in the pot to become water vapor. If you measure the temperature of the water, you'll see that it remains unchanged during boiling, meaning that all the energy from the burner goes into driving the phase change. This experiment tells us that it took an hour's worth of energy from the red-hot burner to convert the pot's water into vapor—that's a lot of energy! Where did all that energy go? It went into accelerating the movement of the water molecules to a high enough speed that they could break the bonds holding them in the liquid and escape into the air. Therefore, water vapor contains a lot of energy. We call this energy **latent heat** because it's "hidden heat" that a material inherits from a phase change.

We've just seen how water absorbs latent heat when it changes from liquid to gas. What happens when water condenses again? To answer this question, let's continue our experiment. Imagine that you could magically force all the water vapor molecules that had boiled out of the hot pot back into the pot instantly so that they all became part of a liquid again. The latent heat in the water vapor (all the heat added to the water from the burner during an hour's time) would be suddenly released all at once, and the pot (and probably the kitchen) would explode!

The thought experiment we've just described may seem odd, but the processes it describes occur in the atmosphere all the time. Water in the oceans absorbs solar energy, evaporates, and enters the atmosphere as vapor. This vapor stores the solar energy as latent heat. When clouds form, water vapor condenses to liquid, thereby releasing its latent heat. This release, in turn, warms the air. Similarly, since it's necessary to add heat to ice to get it to melt, liquid water stores latent heat that will be released when the liquid water freezes. We'll see in the next two chapters that the energy from phase changes contributes to the development of storms.

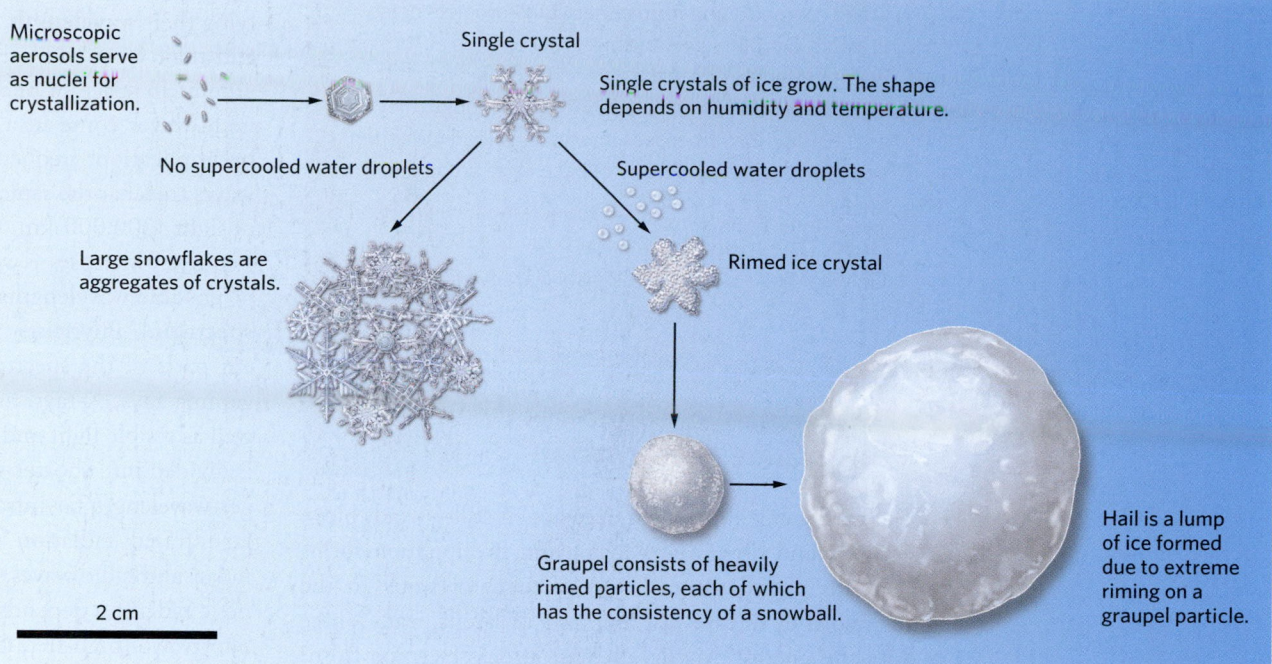

Figure 17.25 Ice particle growth in a cold cloud.

Microscopic aerosols serve as nuclei for crystallization.

Single crystal

Single crystals of ice grow. The shape depends on humidity and temperature.

No supercooled water droplets

Supercooled water droplets

Large snowflakes are aggregates of crystals.

Rimed ice crystal

Graupel consists of heavily rimed particles, each of which has the consistency of a snowball.

Hail is a lump of ice formed due to extreme riming on a graupel particle.

2 cm

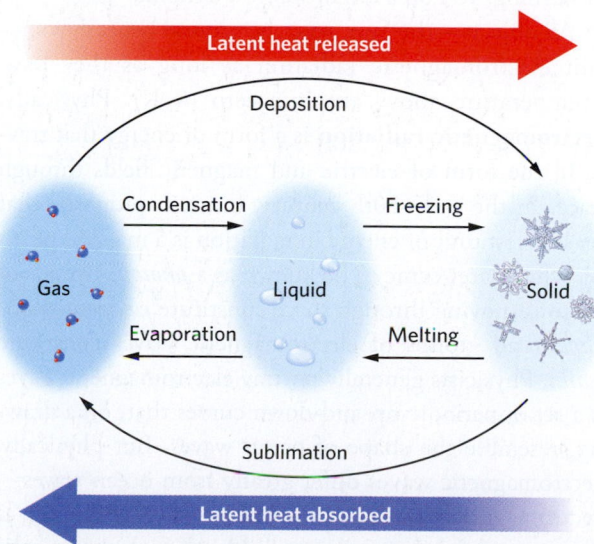

Figure 17.26 Latent heat is either released or absorbed in the atmosphere when water undergoes phase changes.

Latent heat released

Deposition

Condensation

Freezing

Gas

Liquid

Solid

Evaporation

Melting

Sublimation

Latent heat absorbed

Take-home message . . .

> 🏠 Clouds form when water molecules attach to aerosols and form either liquid droplets or ice crystals. Droplets and crystals grow both by absorbing water molecules from the air and by collecting and aggregating. When they become large enough, they fall as precipitation. Processes involved in cloud formation and precipitation absorb or release large amounts of energy.
>
> **Quick Question** --------------------------
> On what basis do meteorologists classify clouds?

17.7 The Atmosphere's Power Source: Radiation

The Earth's envelope of gas does not sit still—winds blow, updrafts and downdrafts form, and precipitation forms and falls. Furthermore, phase changes of water in the air absorb or release vast amounts of latent heat. Where does the energy driving all these processes come from? In a word, the Sun. In this section, we examine how solar energy moves through and interacts with the Earth's atmosphere.

The Nature of Electromagnetic Radiation

Nearly all solar energy arrives at the top of the Earth's atmosphere as radiation. In a general sense, **radiation** is energy that travels away from its source either through a material or through a vacuum. Radiation comes in two general forms: *electromagnetic radiation* (such as visible light) and *particulate radiation* (such as particles emitted from radioactive elements). Here, we focus our attention on electromagnetic radiation—specifically, the solar energy you feel warming (or burning) you on a bright sunny afternoon.

All objects—the Sun, you, the ground beneath you—emit electromagnetic radiation as long as they have a temperature above absolute zero (0 K). Physically, **electromagnetic radiation** is a form of energy that travels in the form of electric and magnetic fields through space. In the early 20th century, Einstein proposed that the smallest unit of energy in radiation is a massless particle, which later came to be known as a *photon*. Groups of photons moving through space constitute *electromagnetic waves*, and groups of electromagnetic waves constitute *beams*. Physicists generally portray electromagnetic waves as a set of periodic up-and-down curves that, on a drawing, resemble the shape of ocean waves. But physically, electromagnetic waves differ greatly from ocean waves—electromagnetic waves can transmit energy through a vacuum and are themselves invisible, though they can be detected by instruments and by our senses. (Ocean waves, in contrast, exist only if water moves; see Section 15.5.)

We can characterize electromagnetic waves by specifying their **wavelength**, the distance between wave crests, and their **frequency**, the number of wave crests passing a point in space in one second (Fig. 17.27). Electromagnetic waves come in a huge range of wavelengths and, by implication, frequencies. Because all electromagnetic waves travel at the same velocity in a vacuum—the speed of light (300,000 km, or 186,000 miles, per second)—frequency increases as wavelength decreases. The full range of possible wavelengths constitutes the **electromagnetic spectrum**. Physicists use different names for different parts of the spectrum. *Shortwave radiation* includes gamma rays, X-rays, and ultraviolet (UV) radiation, as well as visible light and some infrared radiation (infrared is divided into shorter-wavelength near-infrared and longer-wavelength far-infrared). *Longwave radiation* includes far-infrared radiation as well as microwaves (used in radar) and radio waves. The energy carried by electromagnetic radiation depends on its wavelength and frequency. Shortwave (high-frequency) radiation contains more energy than longwave (low-frequency) radiation.

Energy Balance in the Earth-Sun System

THE CONCEPT OF BLACKBODY RADIATION. To understand why the Earth's atmosphere has the temperature that it does, we have to dig into a few background concepts from physics. When discussing the behavior of electromagnetic radiation, physicists picture an imaginary object, called a **blackbody**, that absorbs all radiation that shines on it. Any radiation absorbed by a blackbody must eventually be re-radiated back into space as *blackbody radiation*—otherwise, the object would become infinitely hot. The Earth, the Sun, and for that matter, any object in space acts much like a blackbody. Three laws govern the behavior of blackbody radiation:

- A warmer object emits more radiation, at all wavelengths, than a colder object.
- The total amount of radiation an object emits increases rapidly as its temperature increases.
- As the temperature of an object increases, the wavelength of maximum radiation emission shifts toward shorter wavelengths.

A physics book can explain why these laws operate as they do, but for our purposes, we'll simply keep them in mind as we explore how energy is added to the Earth's atmosphere by sunlight.

THE CONCEPT OF RADIATIVE EQUILIBRIUM. The Sun's surface reaches a temperature of about 5,600°C (10,000°F), hot enough to vaporize the Earth (see Chapter 23). But

Figure 17.27 The electromagnetic spectrum.

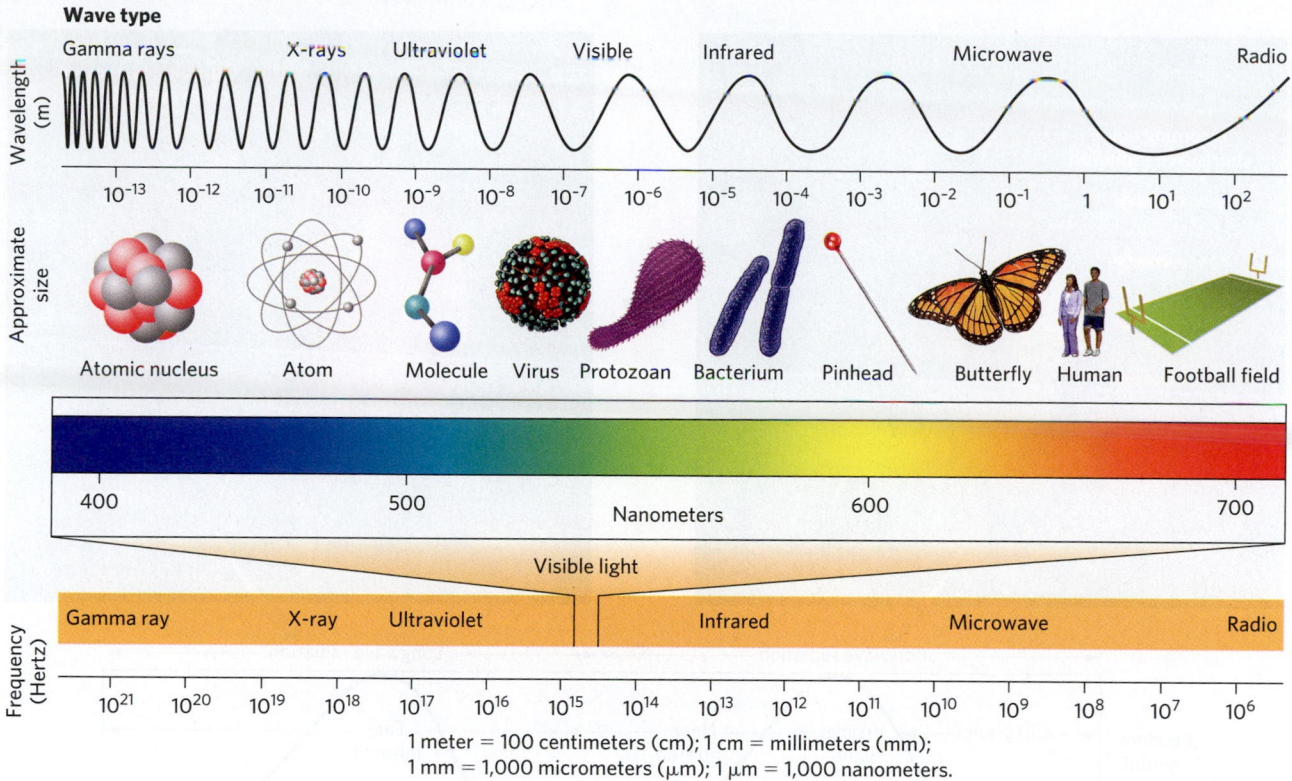

1 meter = 100 centimeters (cm); 1 cm = millimeters (mm);
1 mm = 1,000 micrometers (μm); 1 μm = 1,000 nanometers.

because the Sun sends energy in all directions and lies so far from the Earth, our planet intercepts only a tiny fraction (0.00000015%) of the Sun's total radiation. Nevertheless, if the Earth were to absorb all of the energy that it receives from the Sun, without releasing any back into space, it would eventually vaporize. This, of course, doesn't happen, because the Earth, like any blackbody, constantly re-radiates energy into space. Over time, the total energy arriving at the Earth from the Sun exactly balances the amount of energy re-radiated from the Earth into space. On time scales of a decade, the rate at which solar energy reaches the Earth, a quantity called the *solar energy flux*, remains nearly constant. If the Earth had no atmosphere, the solar energy flux would cause the Earth to stay at a constant −18°C (0°F), a temperature known as the Earth's *radiative equilibrium temperature.*

Because the Sun is so hot and the Earth is so cool, the spectrum of the solar radiation arriving at the Earth differs greatly from the spectrum of the radiation that the Earth re-radiates into space, as the laws of blackbody radiation predict **(Fig. 17.28)**. Specifically, solar energy reaching the Earth arrives predominantly as shortwave radiation— ultraviolet light (the radiation that can give us sunburns), visible light (the radiation our eyes detect), and near-infrared radiation (the radiation we can feel as heat but cannot see). In contrast, the radiation emitted by the Earth consists of longwave radiation (far-infrared radiation).

THE ATMOSPHERIC GREENHOUSE EFFECT. If the Earth's surface were at its radiative equilibrium temperature, you wouldn't be reading this book, for life would not exist. Fortunately, the actual average surface temperature of our planet, 15°C (59°F), is much warmer, so liquid water can exist and life can flourish. This difference is due entirely to the *atmospheric greenhouse effect*—also called, simply, the **greenhouse effect**. We commonly hear about the greenhouse effect in news concerning the Earth's changing climate (see Chapter 20). Here, let's explore how it works and why it is important to our survival.

While the Earth itself acts like a blackbody, its atmosphere does not. Rather, each atmospheric gas absorbs and emits radiation selectively, meaning that it takes in only certain wavelengths and sends out only certain wavelengths. Why does the atmosphere exhibit this behavior? It's because of the nature of the molecules that air contains. Molecules such as O_2 and N_2, which consist of only two atoms apiece, do not absorb or emit very much solar radiation or terrestrial radiation. In contrast, molecules with more than two atoms (such as H_2O, CO_2, O_3, CH_4, and N_2O) are excellent absorbers and emitters of radiation *at certain wavelengths*. This difference in behavior can be traced to how the molecules vibrate when exposed to radiation.

Because of the different absorption and emission characteristics of the various gases in the atmosphere, the

Figure 17.28 In order for the Earth to be in equilibrium, the shortwave radiation that the Earth receives from the Sun must equal the longwave radiation that the Earth emits to space.

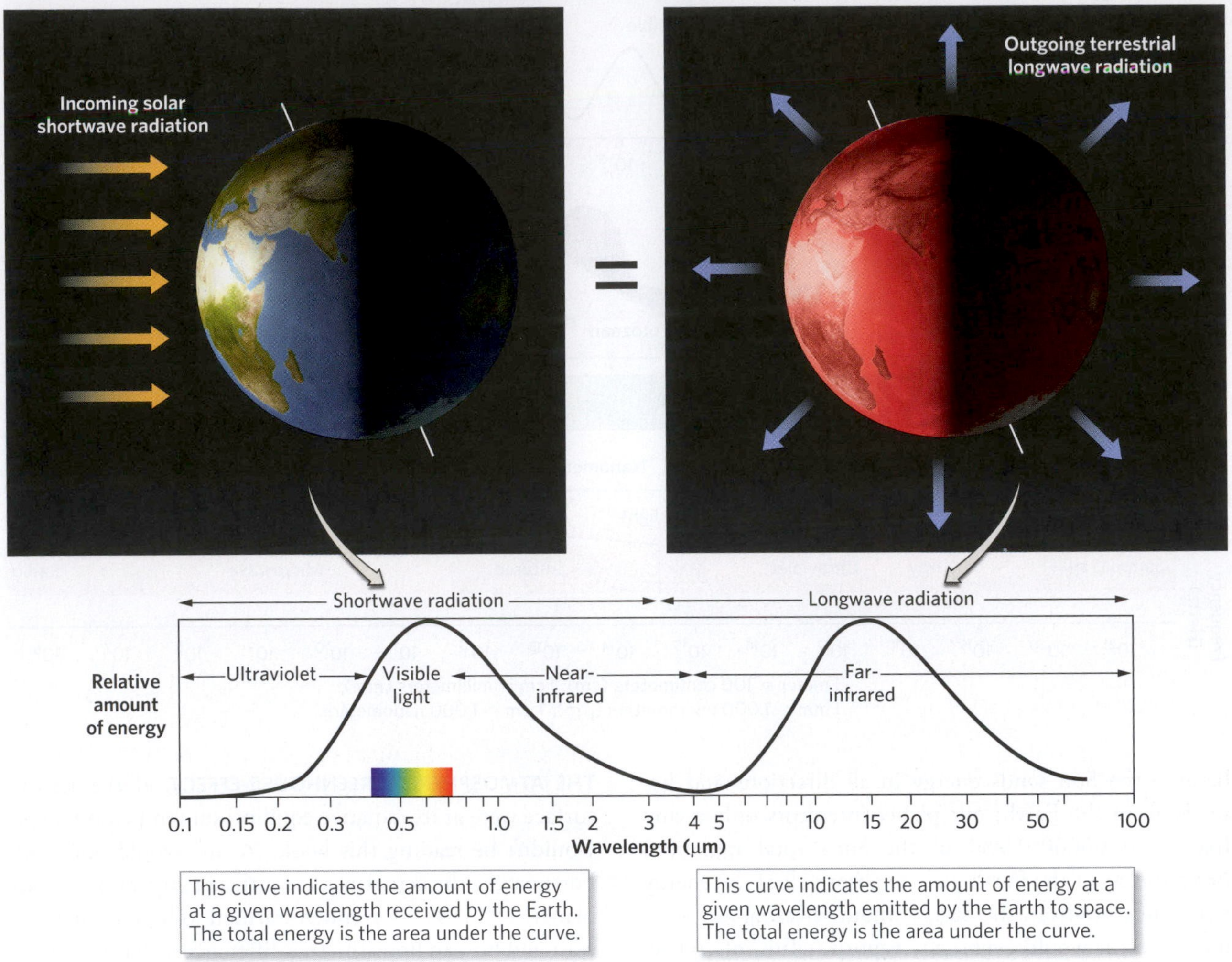

Incoming solar shortwave radiation

Outgoing terrestrial longwave radiation

=

Shortwave radiation — Longwave radiation

Relative amount of energy

Ultraviolet — Visible light — Near-infrared — Far-infrared

0.1 0.15 0.2 0.3 0.5 1.0 1.5 2 3 4 5 10 15 20 30 50 100
Wavelength (µm)

This curve indicates the amount of energy at a given wavelength received by the Earth. The total energy is the area under the curve.

This curve indicates the amount of energy at a given wavelength emitted by the Earth to space. The total energy is the area under the curve.

Figure 17.29 Different atmospheric gases absorb radiation of different wavelengths. Low-absorption regions of the spectrum are called atmospheric windows.

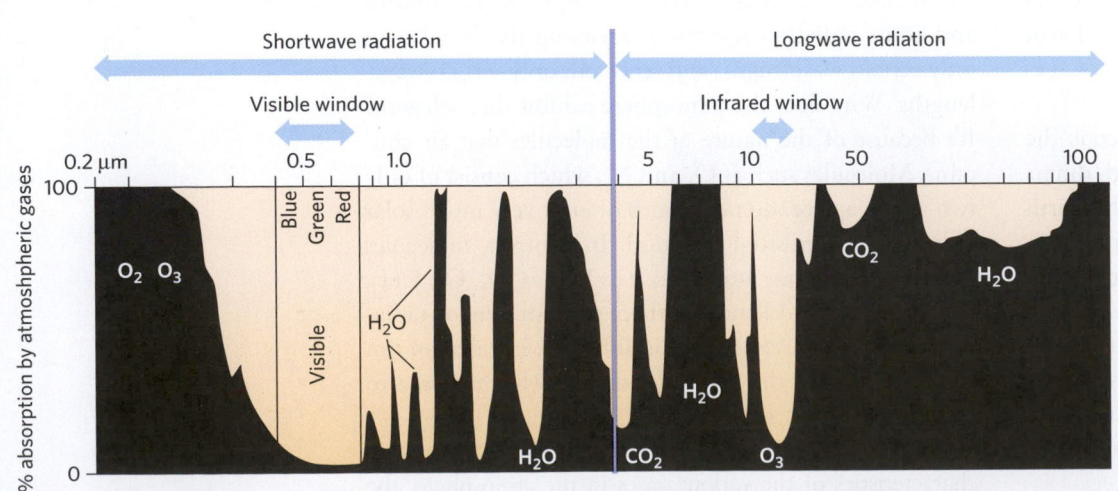

Shortwave radiation — Longwave radiation

Visible window — Infrared window

0.2 µm 0.5 1.0 5 10 50 100

% absorption by atmoshperic gases

O_2 O_3 Blue Green Red Visible H_2O H_2O CO_2 H_2O CO_2 H_2O O_3

atmosphere is transparent to two ranges of solar and terrestrial wavelengths (**Fig. 17.29**). Put another way, the atmosphere has two **radiation windows** through which solar and terrestrial radiation can pass between the Earth's surface and space without interference. The first window includes the visible wavelengths (allowing incoming light to reach the surface, and allowing us to see), and the second window includes certain infrared wavelengths (allowing radiation from the Earth to escape back into space). For all other wavelengths, the atmosphere behaves as a strong absorber. It re-radiates this absorbed energy in random directions. Some of the re-radiated energy heads back to Earth's surface, heating the planet, and some gets absorbed by other atmospheric molecules, increasing the average kinetic energy (that is, the temperature) of the air. Thus, the atmosphere acts like a blanket over the Earth, trapping radiation and preventing it from escaping to space (Fig. 17.30).

The absorption and re-radiation of energy by the atmosphere has come to be known as the atmospheric greenhouse effect because the retention of heat by the atmosphere makes people think of the way in which a glass greenhouse retains heat. The analogy is not perfect, however, because the heat in a real greenhouse gets trapped simply because the warm air inside cannot escape to the outside. In contrast, the atmospheric greenhouse effect works because solar energy can pass through the atmosphere to the ground, but certain atmospheric molecules prevent the Earth's infrared radiation from escaping from the atmosphere back into space. The atmospheric gases that absorb and re-emit infrared radiation are referred to as *greenhouse gases*.

Take-home message . . .

The Sun provides energy to the Earth and its atmosphere in the form of electromagnetic radiation. Visible and some ultraviolet light passes through the atmosphere and is absorbed by the Earth's surface, which in turn re-radiates this energy as infrared radiation. Greenhouse gases trap some of this re-radiated energy, and so keep the atmosphere warm enough for liquid water—and life—to exist on Earth.

Quick Question -

If there were no atmosphere, what would the temperature of the Earth's surface be?

17.8 Optical Effects in the Atmosphere

Why Is the Sky Blue?

When astronauts walked on the Moon and looked upward, they saw a black sky, even though they were bathed in sunlight. Similarly, when they looked into the shadows behind boulders, they saw only black. In contrast, when you look up from the surface of the Earth on a sunny day, you see a blue sky, and if you look into the shadow behind a boulder, you can still see the ground. Why?

Blue skies exist because air molecules cause **light scattering,** a phenomenon that takes place when molecules absorb visible light and then immediately re-emit it in random directions, as we mentioned in Chapter 15. The degree of scattering depends on the size of the molecules relative to the wavelength of light **(Fig. 17.31)**. Because air molecules are small relative to the wavelengths of visible light, short-wavelength light (purple and blue) scatters more effectively than long-wavelength light (orange and red) **(Fig. 17.31a)**. When a beam of sunlight passes through the atmosphere from directly overhead, blue wavelengths

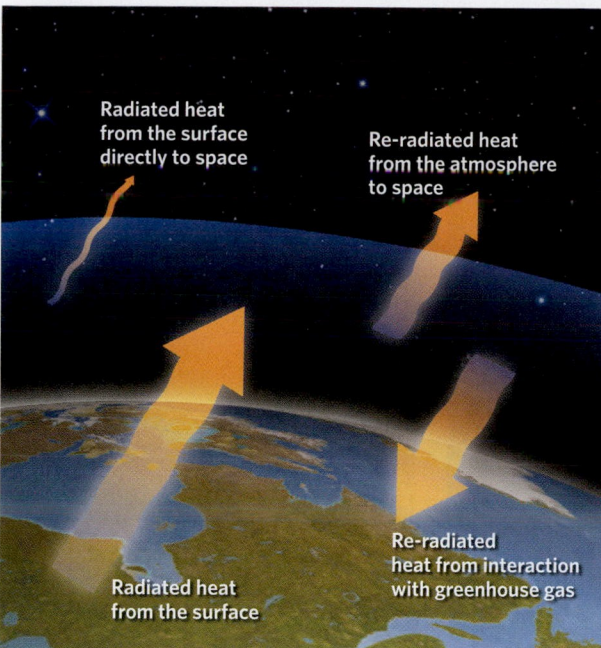

Figure 17.30 A simple depiction of the atmospheric greenhouse effect.

Radiated heat from the surface directly to space

Re-radiated heat from the atmosphere to space

Re-radiated heat from interaction with greenhouse gas

Radiated heat from the surface

within the beam preferentially scatter again and again. After these many scattering events, blue wavelengths arrive at our eyes from somewhere in the sky, making the sky appear blue. Also because of light scattering, some light energy makes it into the shadows behind objects, so shadows on the Earth aren't completely black.

When the Sun lies low in the sky, the distance through the atmosphere that a beam of sunlight must travel to reach our eyes increases. As a result, nearly all blue, green, and yellow wavelengths scatter back into space, leaving only reds and oranges, the color of the rising or setting Sun.

Mirages

Desert travelers sometimes see what looks like a lake of shimmering water on bone-dry land in the distance. Sailors might see ships floating upside down in the sky, while polar explorers look twice when they see ice castles floating above the horizon. Such images are mirages, and they disappear when you approach them. Simply put, a **mirage** is an image formed when the light reflected from an object bends, or *refracts*, though the atmosphere so that the object seems to appear where it doesn't actually exist.

Mirages develop when the air near the ground surface is either exceptionally warm or exceptionally cold relative to the air above. As light passes from less dense warm air into denser cold air (or vice versa), it undergoes refraction. The presence of a hot layer of air beneath a cool layer produces an *inferior mirage* (inferior because the image you see lies below the real object) **(Fig. 17.32a)**, and the presence of a cold layer produces a *superior mirage* (superior because the image you see lies above the real object) **(Fig. 17.32b)**. In deserts, where a layer of very hot air lies

Figure 17.31 The many colors of the sky.

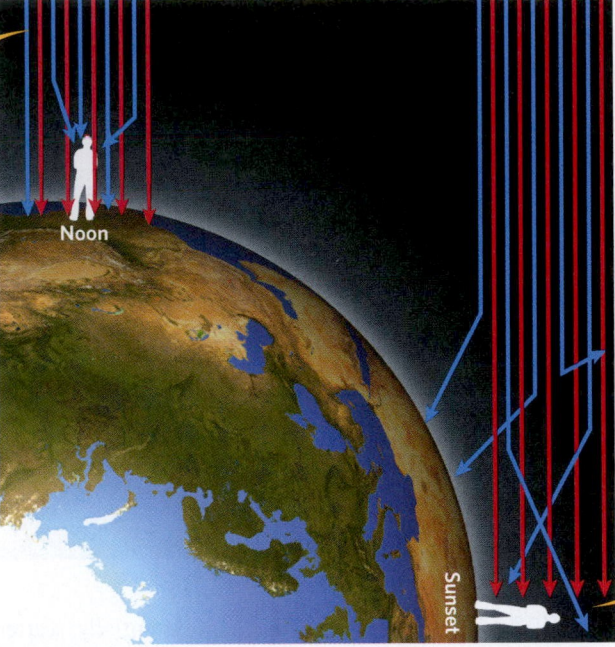

(a) At midday, the sky appears blue.

(b) When sunlight passes through the atmosphere, short wavelengths (primarily blue) are scattered out of the main solar beam.

(c) At sunrise or sunset, the sky appears orange or red.

Figure 17.32 Mirages.

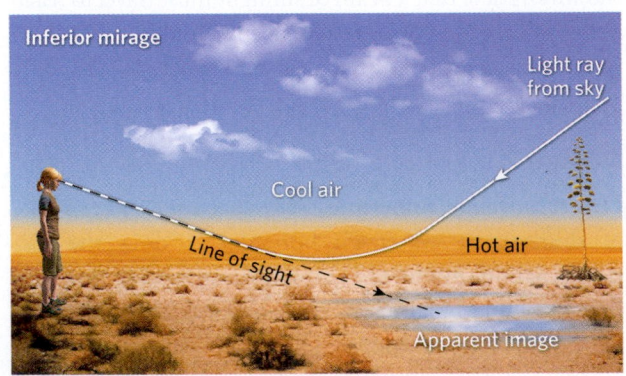

(a) An inferior mirage forms when hot air is present near the ground and bends light upward. The mirage looks like shimmering water because the hot air at ground level is not still.

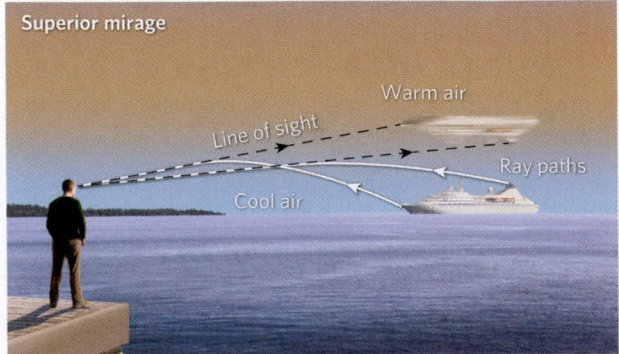

(b) A superior mirage forms when warm air overlies cooler air, and bends light down. Generally, but not always, the mirage is upside down.

at ground level, you'll see inferior mirages of the sky on the ground surface. Because the hot air moves, the light of the mirage shimmers, so it looks like water. Superior mirages develop mostly in the Arctic and Antarctic, where a layer of cold air develops at the Earth's surface, so that light bends downward as it travels from colder air upward into warmer air.

The Atmosphere Showing Off: Colorful Optical Effects

When sunlight passes through air containing water droplets or ice particles, a variety of colorful optical phenomena may develop. For example, when the Sun lies close to the horizon and its rays pass through clear air behind you into rain in front of you, you'll see a **rainbow**, an arc displaying all the colors of visible light in sequence. If you look closely, sometimes you'll see two rainbows, a lower one called the *primary bow* and an upper one called the *secondary bow* (**Fig. 17.33a**). Note that the order of colors in the secondary bow is the opposite of that in the primary bow.

Rainbows develop when light reflects within and refracts through raindrops. Specifically, when an incoming beam of sunlight strikes the spherical surface of a raindrop, the drop acts as a prism and splits the light into colors. The light then reflects off the interior of the drop's back surface and refracts again as it exits the drop (**Fig. 17.33b**). Light in a secondary bow undergoes an additional reflection.

Many other, less frequently observed optical phenomena develop locally in the atmosphere (**Earth Science at a Glance**, pp. 618–619). Examples include *cloud iridescence*, bands of shimmering colors seen when light passes through thin clouds; *glory*, a series of colored rings that surrounds the shadows of aircraft on the tops of clouds when viewed from above; and *halos* or *arcs*, circles and arches of light and color that develop when light passes at a low angle through a thin cloud or fog of ice crystals (see Fig. 17.1b). All of these phenomena are associated with the manner in which light passes around and through water droplets and ice crystals in the air.

Take-home message . . .

Radiation interacting with the atmosphere produces an array of optical phenomena. Mirages develop when the ground-level layer of air has a different temperature than the air above, so that light bends while passing through the air. Rainbows, halos, and other brilliant phenomena form when light interacts with water droplets and ice crystals.

Quick Question -----------------------------
Where is the Sun, relative to you, when you see a rainbow?

Figure 17.33 Rainbows delight the eye.

(a) If you're at the right position, with the Sun behind you and rain in front of you, a rainbow will appear. Sometimes you'll see a double rainbow.

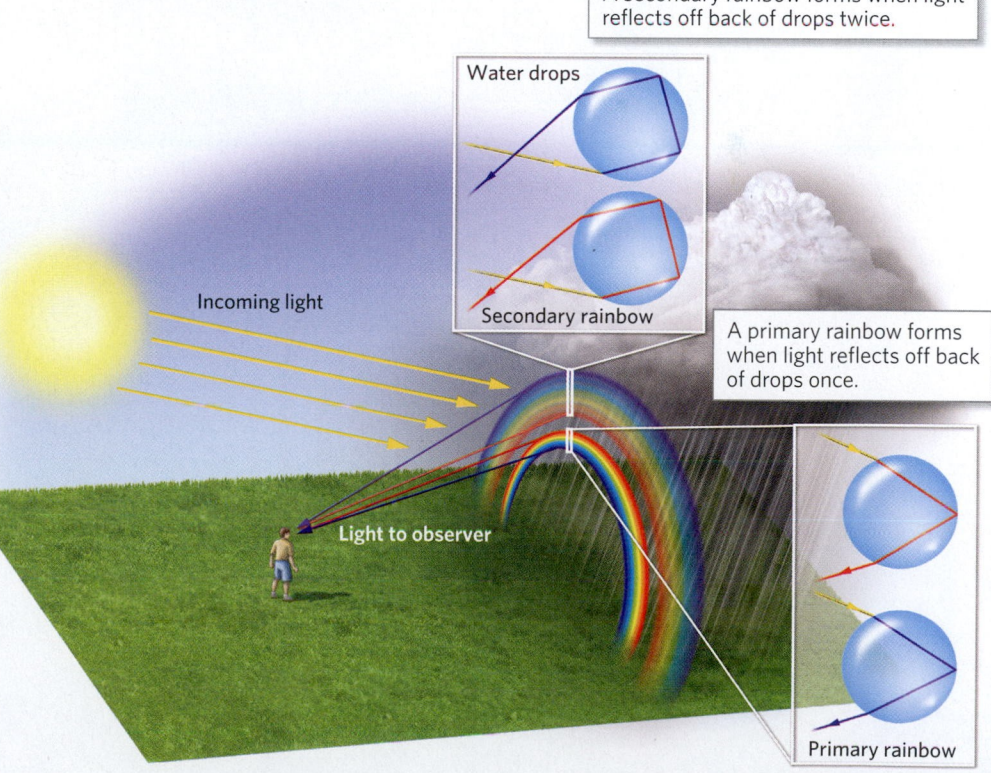

A secondary rainbow forms when light reflects off back of drops twice.

Water drops

Incoming light

Secondary rainbow

A primary rainbow forms when light reflects off back of drops once.

Light to observer

Primary rainbow

(b) A rainbow forms because light refracts and reflects as it interacts with raindrops.

Atmospheric Optics

Light, as it passes through the atmosphere, interacts with aerosols, water droplets, and ice crystals to produce a bewildering array of optical phenomena that delight the eyes, warm the heart, and astonish those are lucky enough to witness them. From a scientific perspective, these effects arise as a result of six different ways light interacts with the material in the atmosphere. These are *absorption*, where light is removed from a beam, *reflection*, where light bounces off a surface like a mirror, *refraction*, where light bends as it passes between denser and less dense substances such as water and air, *diffraction*, where light waves bend around objects such as small raindrops, *scattering*, where light is absorbed and then immediately emitted in a different direction, and *emission*, where light is produced when high-energy particles from the Sun crash into oxygen and nitrogen molecules high in the atmosphere.

Crepuscular rays form when clouds absorb sunlight, shadowing some, but not all solar beams.

Iridescence and glory are diffraction effects caused by light waves bending around tiny water droplets. Glory appears when looking down at the shadow of an aircraft passing over water clouds.

Water

Shadow

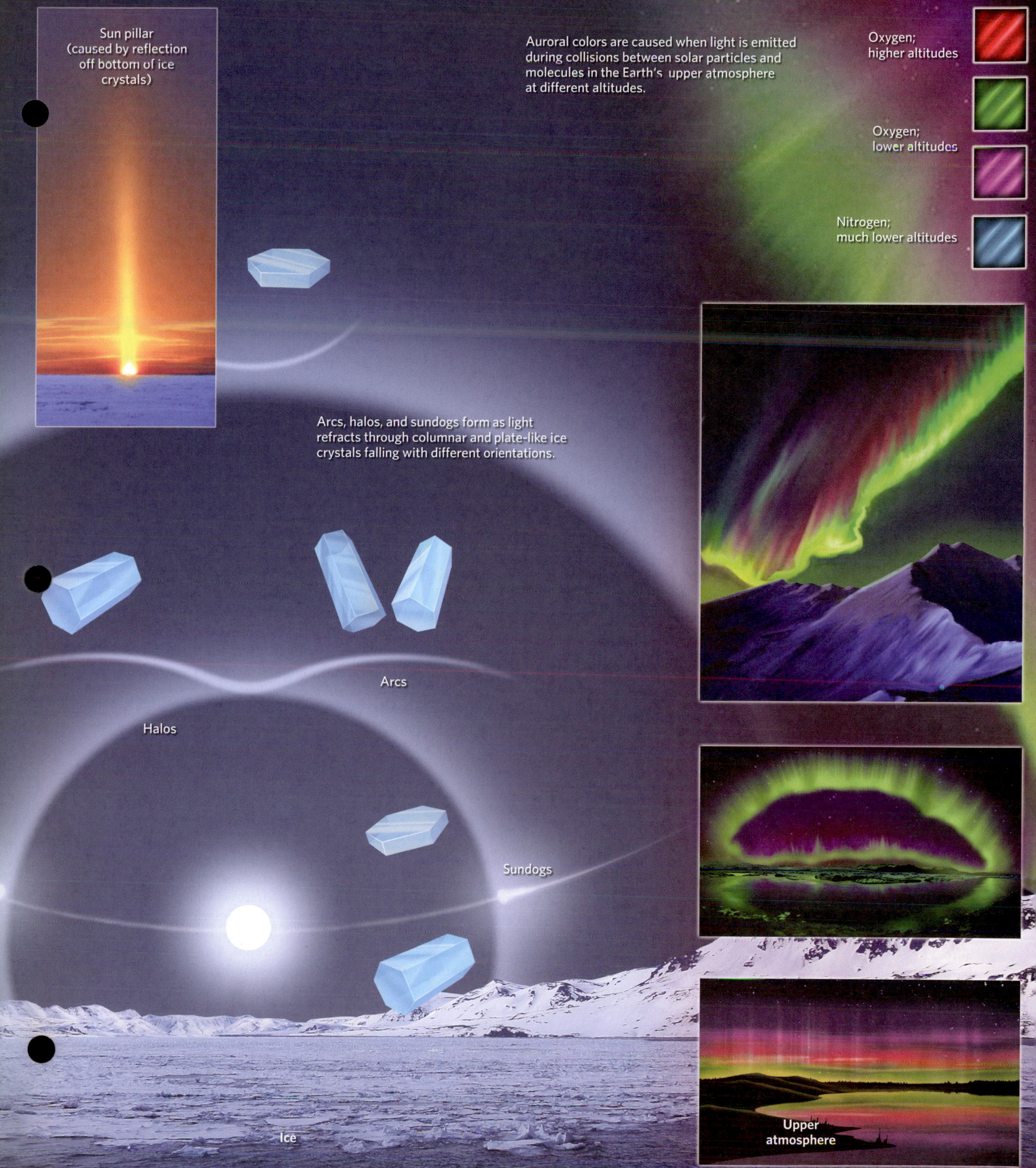

Sun pillar
(caused by reflection
off bottom of ice
crystals)

Auroral colors are caused when light is emitted
during collisions between solar particles and
molecules in the Earth's upper atmosphere
at different altitudes.

Oxygen;
higher altitudes

Oxygen;
lower altitudes

Nitrogen;
much lower altitudes

Arcs, halos, and sundogs form as light
refracts through columnar and plate-like ice
crystals falling with different orientations.

Arcs

Halos

Sundogs

Ice

Upper
atmosphere

⊘17 CHAPTER REVIEW

Chapter Summary

- Dry air consists of nitrogen (78%), oxygen (21%), and trace gases (1%). The amount of water vapor in the atmosphere varies considerably over time and space.

- In addition to gases, the atmosphere contains tiny particles called aerosols.

- Earth's atmosphere evolved over time. Before the appearance of oceans, it consisted of volcanic gases. Formation of the oceans removed most of the atmosphere's water vapor and CO_2, leaving N_2 behind. When photosynthetic organisms evolved, they absorbed CO_2, which became buried, and released O_2.

- Addition of O_2 to the atmosphere also led to the formation of ozone, which absorbs harmful components of sunlight, allowing life on land.

- Atmospheric pressure decreases rapidly with elevation. Half of the atmosphere's mass lies below an elevation of 5.6 km (3.5 miles).

- The atmosphere's capacity to hold water vapor increases as its temperature increases. The relative humidity represents the amount of water vapor in air relative to its holding capacity. Another quantity, the dew point, represents the actual amount of water vapor in air.

- We measure properties of the atmosphere (temperature, pressure, humidity, wind direction and speed, etc.) at ground level using a variety of instruments at a weather station. To measure conditions high in the atmosphere, meteorologists launch rawinsondes on weather balloons.

- Four atmospheric layers are distinguished by patterns of temperature change with elevation: the troposphere, stratosphere, mesosphere, and thermosphere. The atmosphere can also be divided into layers that are well mixed (homosphere) and unmixed (heterosphere).

- Another layer, the ionosphere, is recognized by its high concentration of ions.

- Clouds form in rising air when air is lifted or when it becomes buoyant. We classify clouds based on their altitude, shape, and whether or not they produce precipitation.

- Cloud droplets grow into raindrops by first absorbing water molecules from water vapor, and then by coalescing with neighboring droplets. Ice particles grow to precipitation size by first incorporating water molecules, and then by either linking together to form snowflakes or by collecting supercooled water droplets to form rimed ice particles, graupel, or hail.

- Water in the atmosphere constantly undergoes phase changes, releasing or absorbing latent heat in the process.

- The Sun provides energy to the Earth in the form of electromagnetic radiation, which includes many different wavelengths. An object that absorbs all radiation it receives is called a blackbody. A blackbody re-emits this radiation at wavelengths that depend on its temperature.

- Solar energy arrives at Earth's surface as shortwave radiation, but the much cooler Earth re-radiates this energy as longwave radiation. Certain gases in the atmosphere trap some of this energy, causing the greenhouse effect, without which the Earth's surface would be below freezing.

- Visible light interacts with air molecules and scatters, making the sky appear blue. Mirages form when sunlight bends due to variations in air temperature with elevation. Rainbows, halos, and other brilliant phenomena form when light interacts with water droplets and ice crystals in the atmosphere.

Key Terms

aerosol (p. 589)
air (p. 587)
air temperature (p. 593)
atmosphere (p. 587)
atmospheric pressure (p. 594)
aurora (p. 605)
barometer (p. 595)
blackbody (p. 612)
cirrus cloud (p. 606)
cloud (p. 589)
cloud cover (p. 598)
collision-coalescence (p. 609)
cumulus cloud (p. 607)
dew point (p. 597)

Doppler radar (p. 600)
electromagnetic radiation (p. 612)
electromagnetic spectrum (p. 612)
environmental lapse rate (p. 604)
fog (p. 608)
frequency (p. 612)
graupel (p. 610)
great oxygenation event (p. 592)
greenhouse effect (p. 613)
haze (p. 590)
heterosphere (p. 605)
homosphere (p. 604)
ionosphere (p. 605)
latent heat (p. 611)

light scattering (p. 615)
mesosphere (p. 604)
meteorologist (p. 593)
mirage (p. 615)
ozone hole (p. 604)
phase change (p. 610)
precipitation (p. 599)
radiation (p. 612)
radiation windows (p. 614)
rainbow (p. 617)
raindrop (p. 609)
relative humidity (p. 597)
saturation vapor pressure (p. 596)
sleet (p. 610)

snowflake (p. 610)
stratosphere (p. 604)
stratus cloud (p. 607)
thermometer (p. 593)
thermosphere (p. 604)
troposphere (p. 603)
vapor pressure (p. 596)
visibility (p. 598)
water vapor (p. 589)
wavelength (p. 612)
weather (p. 593)
wind (p. 598)
wind direction (p. 598)
wind speed (p. 598)

The letters following each Review Question refer to the corresponding Learning Objective from the Chapter Opener.

1. What is the present composition of the atmosphere? Which gases have concentrations that vary little over time, and which vary rapidly over time and from location to location? **(A)**

2. How has the atmosphere evolved over time? At what point did it become hospitable to life on land? **(A)**

3. You decide to climb a mountain and take along a thermometer and a barometer. During the climb, you take readings. At the end, you plot graphs of the results. What will the graphs look like? **(B)**

4. Explain why the absolute amount of moisture in the atmosphere over a hot desert can be larger than over the Arctic tundra, even though the relative humidity over the tundra may be 90%, while over the desert it may be 10%. **(B)**

5. A southwest wind blows from where to where? **(B)**

6. Describe the primary layers in the atmosphere as characterized by temperature, atmospheric composition, and electric charge. What are the boundaries between the temperature layers called? **(C)**

7. Explain why aerosols are a critical component of Earth's hydrologic cycle. **(D)**

8. Are you more likely to find clouds on the upwind or downwind side of a mountain range? Why? **(E)**

9. Summarize the primary cloud types. What type of cloud is pictured in the photo at the top of the next column? **(E)**

10. Is latent heat released to the atmosphere or absorbed from the atmosphere when raindrops evaporate? How about when snowflakes melt? How about when a cloud forms? **(E)**

11. Is the surface of the Sun hotter or colder than the electric coils on your kitchen stove when they're fully heated? Explain. (Hint: Recall Figure 17.27 and the laws governing the behavior of blackbody radiation.) **(F)**

12. What is meant by shortwave radiation and longwave radiation? Which is emitted by the Sun? By the Earth? (Hint: Consider the figure on the right) **(F)**

13. Why can we consider radiation arriving at the top of the Earth's atmosphere from the Sun as distinct from radiation emitted by the Earth-atmosphere system? **(F)**

14. Which gases are important selective absorbers of radiation? What relevance does this have to the atmosphere? **(F)**

15. What is the atmospheric greenhouse effect? Which atmospheric gases are most responsible for the greenhouse effect? **(F)**

On Further Thought

16. From weather reports you can obtain on the Internet, determine the current temperature, dew point, and relative humidity in a number of cities. Rank the cities according to their relative humidity, placing the city with the highest relative humidity at the top of the rank. Now subtract the dew point (D) from the temperature (T), and rank the cities by the difference in these two temperatures ($T - D$), but place the city with the smallest difference at the top. Compare the two rankings. Summarize your findings. **(B)**

17. Clouds generally move too slowly for the human eye to appreciate the lifting mechanisms that produce them. High-speed video circumvents this problem. Examine YouTube videos of clouds by searching for "high-speed clouds" or "time-lapse clouds." Summarize what you observe for the different cloud types described in the chapter. **(E)**

Online Resources

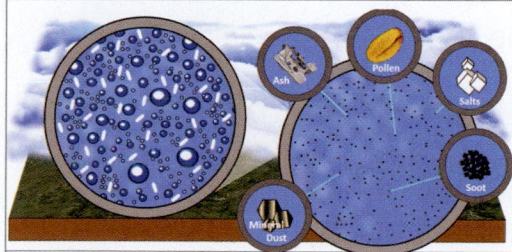

Animations
This chapter features animations on cloud features and how clouds form.

Videos
This chapter features real-world videos on ozone concentrations and air pollution above Asia.

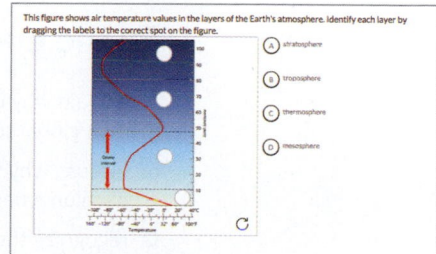

Smartwork5
This chapter features questions on atmospheric composition, layers and measurement tools; cloud formation and types; and the impact of radiation forces and human impacts on the Earth's atmosphere and surface.

18 WINDS OF THE WORLD
The Earth's Major Weather Systems

By the end of the chapter you should be able to . . .

A. explain why our planet's atmosphere constantly flows on a global scale.

B. discuss why tropical rainforests and subtropical deserts develop where they do.

C. recognize the effects of monsoons and ENSO events, and explain why these meteorological phenomena are so important to society.

D. illustrate how forces acting in the atmosphere generate hurricanes.

E. explain how mid-latitude cyclones develop and affect weather.

F. describe the relationship between weather, fronts, and jet streams.

18.1 Introduction

When Sir Francis Drake set sail with the *Golden Hind* and its sister ships in 1577 **(Fig. 18.1)** to begin what would become the second circumnavigation of the globe, he and his crews were literally at the mercy of the wind, and during their three-year voyage, they experienced most of the wild variations in weather that the Earth System can conjure. Sailing south, they blew right into the clutches of a *mid-latitude cyclone*, a swirling storm large enough to cover a wide swath of an ocean or continent. Winds and waves battered the ships so badly that Drake had to return to England for repairs.

Drake's second effort proved more successful, and his ships reached the belt of *trade winds*—so named be-cause of their importance to trans-Atlantic commerce.

These winds filled the explorers' sails and blew them steadily southwestward. When the ships reached the equator, fortunately missing a fierce hurricane to their west, in the Caribbean, they entered the sultry *doldrums*, where the wind goes calm. After bobbing helplessly with limp sails for endless days, they crossed the equator and entered another belt of trade winds. These winds blew to-ward the northwest, so it took all the helmsman's skill to maintain an overall southward route. When the ships reached the southern tip of South America, they encoun-tered perpetually strong winds, the *prevailing westerlies*. Sea spray doused the ships, coated their decks and sails with ice, and thereby disabled part of the fleet. The luckier ships made it into the Pacific and headed north along the west coast of the Americas. Drake and his remaining crew then turned west. As they crossed the Pacific and Indian

Swirling cloud systems that traverse the Earth's middle latitudes bring rain, snow, and occasional hazardous weather.

at successive satellite images that show distributions of clouds and moisture **(Fig. 18.2)**. By compiling measurements from weather stations worldwide, atmospheric scientists can generate maps that display variations in wind direction and speed, around the world (**Earth Science at a Glance**, pp. 634–635). In this chapter and the next, we focus on our atmosphere's rivers of air—both large and small. Our discussion places an emphasis on **storms**, significant atmospheric disturbances involving strong winds and heavy precipitation that can be hazards to landscapes, ecosystems, and society, as well as to ships at sea.

We begin this chapter by investigating the nature of wind itself. We discover why the wind blows and examine factors that control its direction and intensity. With this background, we then show how the characteristics of the general circulation, and associated storms, vary with latitude. We start with the *tropics*, the region that straddles the equator, where intense solar heating drives the planet's largest distinct air circulations: *Hadley cells*, the *Southern Oscillation*, *monsoons*, and finally *tropical cyclones* (called *hurricanes* when they occur in the Atlantic). We then shift to the Earth's mid-latitudes, a region where *jet streams* and *mid-latitude cyclones* play a major role in controlling the weather.

Oceans, they encountered many challenging weather conditions: fierce storms related to *El Niño* as well as *typhoons* and *monsoons*. Finally, after rounding the southern tip of Africa, the *Golden Hind*, now the only remaining ship, headed home, its hold filled with plundered gold and its sailors full of tales about the winds of the world.

Atmospheric scientists refer to the flow pattern of air around the globe as the atmosphere's **general circulation**. Drake had only a vague sense of this circulation. Today, you can see it on a computer screen by looking

18.2 In Constant Motion: Forces Driving the Wind

Why Do the Winds Blow? Forces Acting on Air

Our world would be a very different place without the **wind**, the horizontal component of air flow. You can feel the wind, for when it blows, air molecules bombard your

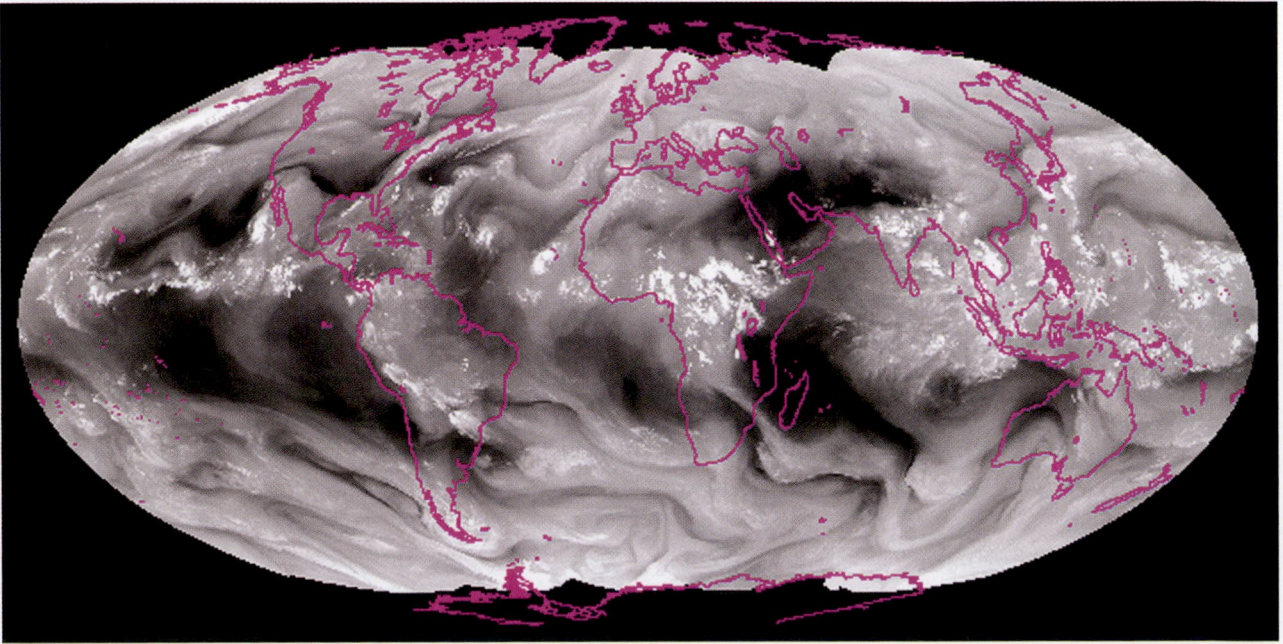

skin, and your nerves sense their combined impact, even though each molecule alone is too tiny to be visible. The faster the molecules move, the more force they exert, so during storms, winds can rip roofs from buildings or blow trains off their tracks. In *steady winds*, air flow maintains the same direction and speed, whereas in *turbulent winds*, the flow of air twists and turns, and the speed at a locality changes constantly. Winds form at all scales, from small gusts that rustle a few leaves on a summer day to vast currents that circle the planet or swirl into gigantic storms.

Have you wondered why the wind blows? Four forces play a key role in determining the direction and speed of the wind: (1) the pressure-gradient force, (2) the gravitational force, (3) the frictional force, and (4) the Coriolis force. Let's look at these forces in turn. You'll recall that we introduced some of these forces in Chapter 15, in our discussion of ocean currents.

PRESSURE-GRADIENT FORCE. In the atmosphere, the air pressure at one location may differ from that at another. We define a **pressure gradient** as a pressure change divided by the distance over which that change occurs. Where a pressure gradient exists, air tries to move from the location of higher pressure to the location of lower pressure. To picture why, think of a bicycle pump—when you push down on the handle, the pressure in the pump's cylinder increases, so the air flows out of the open nozzle. Because the velocity of air changes when this movement takes place, we can say that a pressure gradient causes air to accelerate, and refer to the force causing this acceleration as the *pressure-gradient force*.

To visualize how a pressure-gradient force can exist in the atmosphere, think of a sailing ship on the ocean. If the air pressure behind the ship exceeds the air pressure ahead of it, then the air density behind the ship exceeds the air density in front. Air molecules are in constant motion, so more air molecules strike the back side of the sails than the front, and the ship moves forward **(Fig. 18.3)**. The same concept applies even if the "ship" is the size of a single air molecule—more collisions happen on the side of the air molecule where the pressure is greater, so the molecule, on average, moves in the direction of lower pressure. When this phenomenon affects a vast number of air molecules, the whole air volume moves in the direction of lower pressure, and wind blows.

The wind's speed—measured in meters per second, miles per hour, or *knots* (nautical miles per hour)—depends on the magnitude of the pressure gradient, in that the wind blows faster where the air pressure changes more quickly over a given horizontal distance than where it changes more slowly. To picture why, imagine the difference between stepping on an air-filled plastic bag with all your weight and squeezing the bag gently with your hands. If

Figure 18.3 The relationship between wind and a gradient in air pressure. There are more air molecules to the left of the ship, so the air pressure is greater on the left, and the wind blows from left to right.

there is an opening in the bag, the air escapes faster when you step on the bag than when you squeeze it gently.

We can represent pressure gradients on a map of constant elevation by plotting contour lines, or **isobars**, along which all points have the same air pressure **(Fig. 18.4)**. Where isobars lie closer together, there's a larger (or steeper) pressure gradient, so stronger winds blow, and where isobars lie farther apart, there's a smaller (or gentler) pressure gradient, so weaker winds blow.

GRAVITATIONAL FORCE. Just like everything on the Earth's surface, air molecules are subjected to the downward pull of gravity. Why, then, don't all air molecules fall to the ground? In the atmosphere, an upward-directed pressure-gradient force, which exists because pressure decreases with elevation, balances the downward gravitational force **(Fig. 18.5)**. In fact, were it not for the gravitational force, all the gas molecules of the Earth's atmosphere would soon fly off into space.

FRICTIONAL FORCE. When you picture a *frictional force*, you might think of the force that slows a book that you've slid across a table. Friction occurs in the atmosphere, too, either when a volume of faster-moving air molecules shears against a volume of slower-moving air molecules, or when a volume of air molecules shears against the Earth's surface. The frictional force always acts opposite the direction of motion of an object, so friction reduces the speed of air flow. The frictional force increases near the Earth's surface, both because moving air interacts with the surface and because air density increases toward the base of the atmosphere. The rougher the surface, the greater the frictional force, so more friction exists between moving air and a city of tall buildings than between moving air and a calm sea. For this reason, winds tend to be stronger out over the ocean (see Earth Science at a Glance).

Figure 18.4 A map showing the relationship between pressure gradient and wind speed. Isobars are contour lines that represent air pressure. A large pressure gradient and strong winds occur where the isobars are close together.

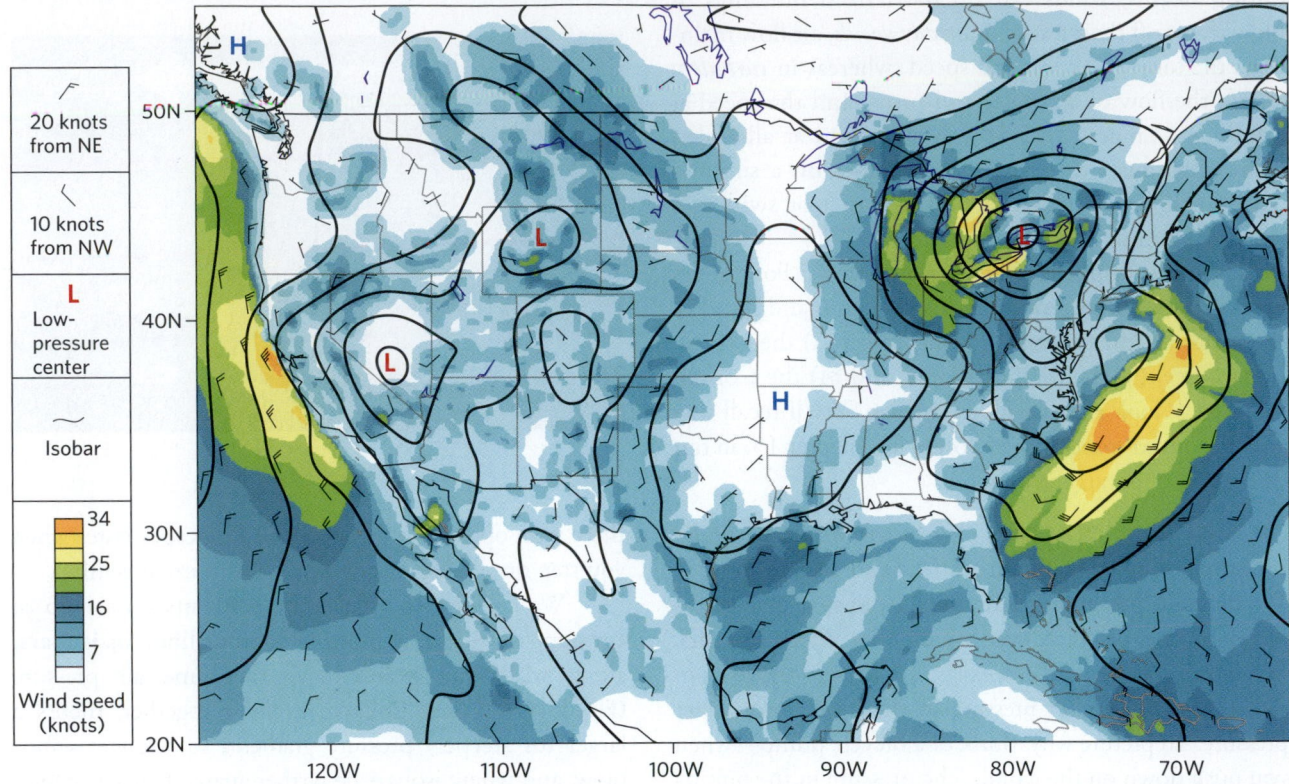

Figure 18.5 The action of the downward-directed force of gravity on air molecules is balanced by the upward-directed pressure-gradient force.

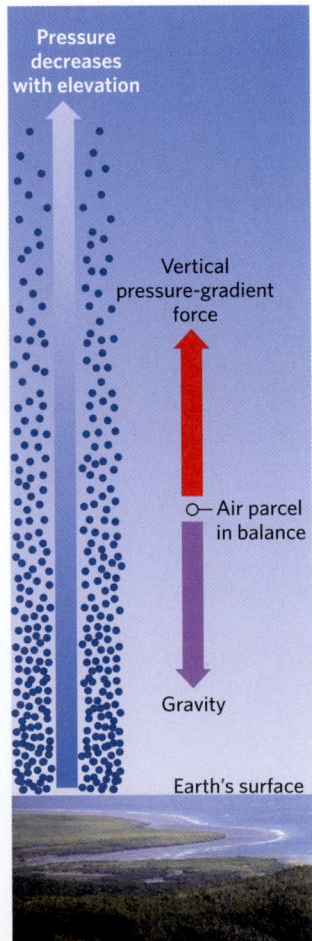

Atmospheric scientists refer to the layer of air adjacent to the Earth's surface in which friction significantly affects air movement as the **boundary layer**. The thickness of the boundary layer depends on several factors: the roughness of the underlying surface, surface heating, the presence of updrafts and downdrafts, and the wind speed. On a cold, calm winter morning at sunrise, the boundary layer over a large ice-covered lake extends upward only a few hundred meters. In contrast, on a hot, windy summer afternoon over a city or over hills, the boundary layer extends upward a few kilometers **(Fig. 18.6a)**. The frictional force can generate turbulence in the boundary layer. *Eddies* (whirlpool-like swirls) tend to develop, for example, when wind speed changes rapidly with distance, or wind shears against the ground, or air passes between buildings or over mountain peaks. The boundary layer is sometimes visible as a layer of pollution or shallow clouds capped by clear air **(Fig. 18.6b)**.

CORIOLIS FORCE. In Chapter 15, we introduced the Coriolis force in the context of ocean currents (see Box 15.3). The Coriolis force also influences wind. To see why, imagine a small volume of air, which we'll call an *air parcel*, that is moving parallel to the Earth's

surface at a constant speed. (To simplify our discussion, let's ignore the frictional force and pressure-gradient force for now.) What path will the air parcel follow? As we learned in Chapter 15, because the Earth rotates on its axis, a moving fluid doesn't move in a straight line relative to the Earth, but rather follows a curved path.

The Coriolis force has four important properties that affect the movement of air parcels: (1) it causes moving parcels to veer to the right of their direction of motion in the northern hemisphere and to the left in the southern hemisphere; (2) it affects the direction in which a parcel moves across the Earth's surface, but has no effect on its speed; (3) it is strongest for parcels that move fast relative to a point on the surface of the Earth and has a value of zero for stationary parcels; and (4) it has a value of zero on the equator and a maximum value at the poles.

Wind Direction: A Consequence of Geostrophic Balance

If the temperature of the atmosphere at each altitude across the Earth were the same everywhere, and if the Earth's surface were uniform, atmospheric pressure at any given altitude wouldn't vary from location to location, and wind would stop blowing. This doesn't happen in reality because various sources add energy to the atmosphere, and the amount of added energy varies with location. For example, solar radiation provides more heat at the equator than at the poles, land and sea absorb and release heat at different rates, the strength of the frictional force depends on the roughness of the Earth's surface below, and the density of air changes with elevation at different rates in different locations. Given all these variables, the atmosphere never achieves a balanced state and remains in constant motion.

While a balanced state can't be achieved permanently everywhere, atmospheric conditions do approach a balanced state locally, particularly at elevations above the boundary layer. This condition, called **geostrophic balance**, develops when the horizontal pressure-gradient force and the Coriolis force are equal and opposite and the frictional force is insignificant.

We first discussed geostrophic balance in the context of ocean currents (see Chapter 15). To understand geostrophic balance in the atmosphere, let's examine how air moves across North America when a pressure gradient exists. Picture an initially stationary air parcel within the region of a pressure gradient (Fig. 18.7a). The pressure-gradient force initially accelerates the air and causes it to move from a region of higher pressure toward one of lower pressure. However, as soon as the parcel of air starts to move, the Coriolis force deflects it, so it starts to turn to the right. As a result, the air doesn't follow a path perpendicular to the isobars, but rather starts flowing at a small angle relative to the isobars. The Coriolis force, and the resulting deflection, increases as the air parcel accelerates to a critical speed, at which point the Coriolis force and the pressure-gradient force become equal and opposite. When this happens, the air has attained geostrophic balance (Fig. 18.7b), and the wind that results, the **geostrophic wind**, flows parallel to isobars (Fig. 18.7c). In nature, air is nearly, but not exactly, in geostrophic balance at elevations above the boundary layer over much of the Earth. (We'll see later in this chapter that deviations from geostrophic balance contribute to the formation of large storms.)

Figure 18.6 The boundary layer occurs at the base of the atmosphere, where friction significantly affects air flow.

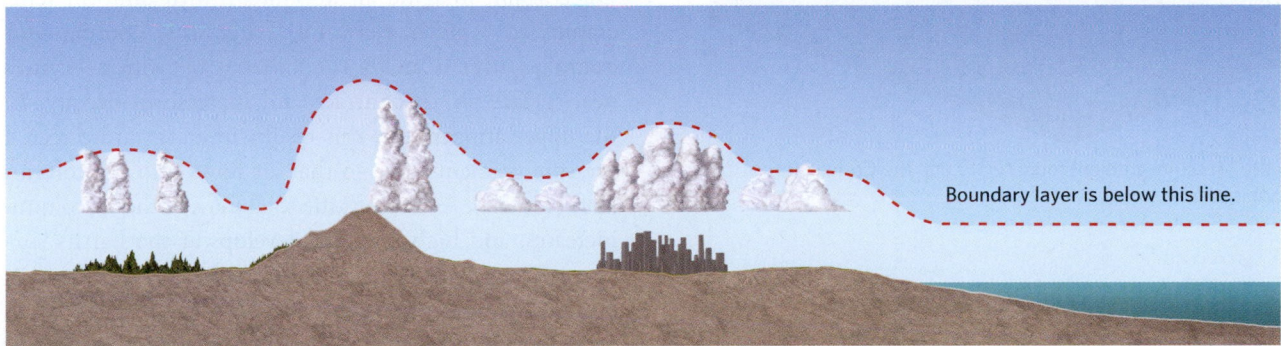

Boundary layer is below this line.

(a) The boundary layer's thickness depends on the roughness and temperature of the surface below; it can vary from a few hundred meters to a few kilometers.

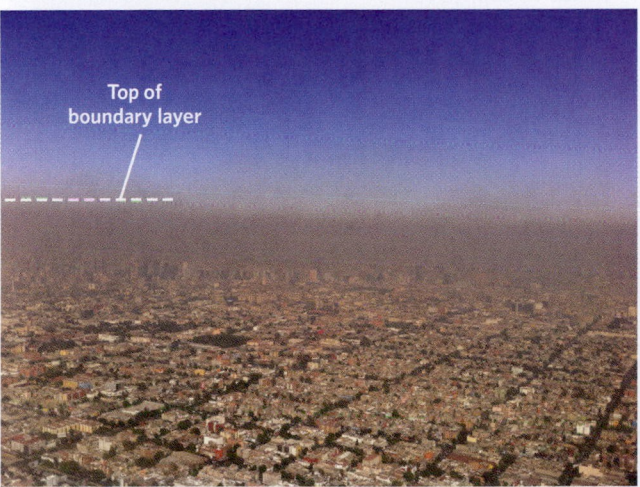

Top of boundary layer

Top of boundary layer

(b) Pollution or low-level clouds can sometimes show where the top of the boundary layer lies.

Figure 18.7 Geostrophic wind.

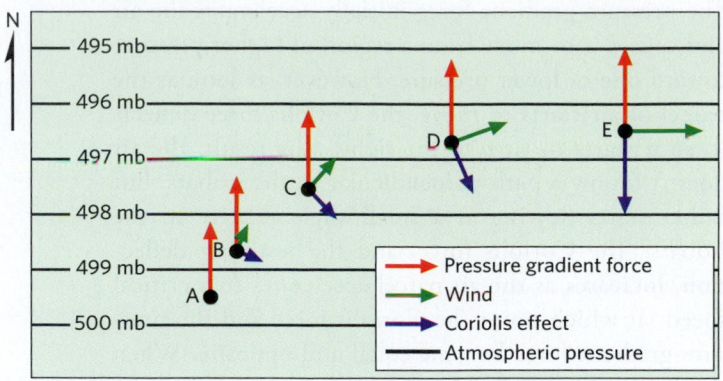

(a) Air at A starts from rest. As it accelerates from A to E, the Coriolis force turns the air's path to the right until it flows parallel to the isobars.

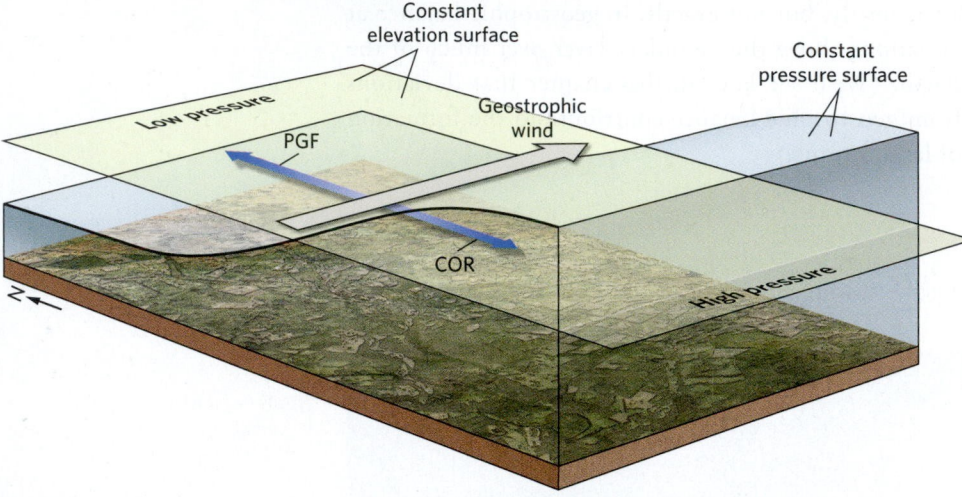

(b) When there is a balance between the pressure-gradient force (PGF) and the Coriolis force (COR), the resulting wind is called the geostrophic wind.

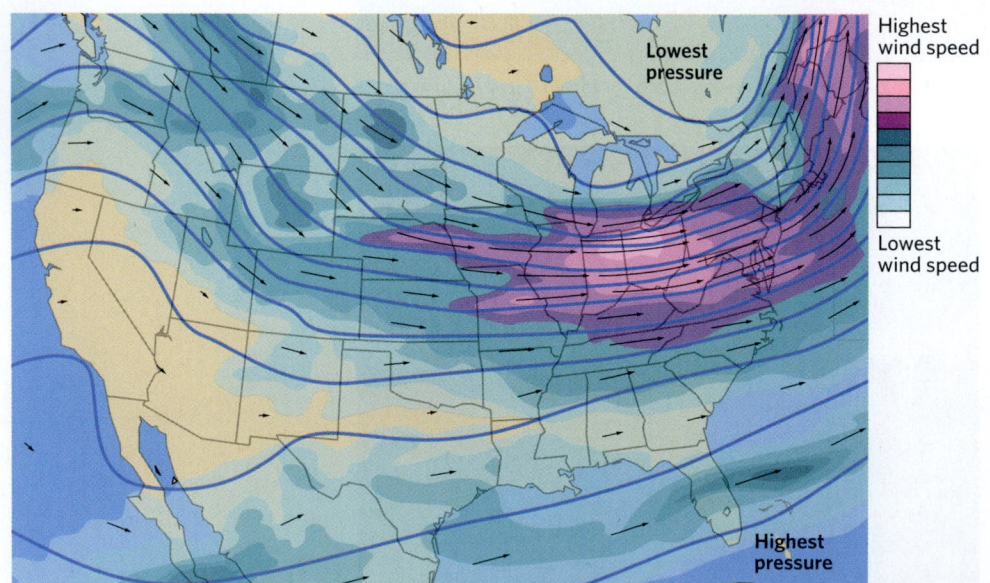

(c) The wind blows roughly parallel to the isobars at 10 km (6 miles) above sea level, the elevation of the jet stream. Winds blow faster where the pressure gradient is greater.

The Effect of Heating and Cooling on Air Pressure

As we saw in Chapter 17, the Earth's surface absorbs incoming solar energy, then re-radiates it as infrared energy, which heats air at the base of the atmosphere. Consequently, the near-surface air expands and decreases in density. This change creates low pressure at the Earth's surface. To see why, let's imagine an atmosphere, at Time 1, where temperature at any elevation is uniform and the pressure surfaces are horizontal (**Fig. 18.8a**). In this situation, pressure decreases uniformly with elevation. Let's now imagine, at Time 2, heating a region of the atmosphere. Expansion of the heated region causes the pressure in a column of air above it to become greater than in surrounding regions. In the upper atmosphere, this situation results in an outward-directed pressure-gradient force, and air starts to flow out of the column. As a result, the total mass of air within the column has decreased by Time 3. Since air pressure at the Earth's surface represents the weight of the overlying column of air, the expansion associated with heating ends up decreasing the air pressure at the surface. This phenomenon is very important to air circulations in the tropics, as we will see.

The opposite effect happens when air cools (**Fig. 18.8b**). Once again, imagine air at Time 1 with uniform temerature at a given elevation. Large-scale cooling—for example, over Canada or Siberia in winter—causes near-surface air to contract and increase in density. In the upper atmosphere, this results in an inward-directed pressure gradient force so that air flows into the column aloft (Time 2). As a result, the total mass of the air column increases, and high pressure develops at the Earth's surface (Time 3). Cooling strongly influences air flows in the Earth's mid- and high latitudes, but it can also affect flows in the tropics. Development of pressure gradients at the Earth's surface, due to heating or cooling, can cause winds to flow.

Take-home message . . .

Four forces (the pressure-gradient, gravitational, frictional, and Coriolis forces) determine the speed and direction of the wind. Above the boundary layer, air accelerates from regions of higher pressure toward regions of lower pressure. As it does so, the Coriolis force increases and eventually balances the pressure-gradient force to produce geostrophic balance. Heating or cooling at the base of the atmosphere lowers or raises air pressure at the Earth's surface.

Quick Question ---------------------------
Why does the thickness of the boundary layer vary with location?

Figure 18.8 The formation of high and low pressure at Earth's surface.

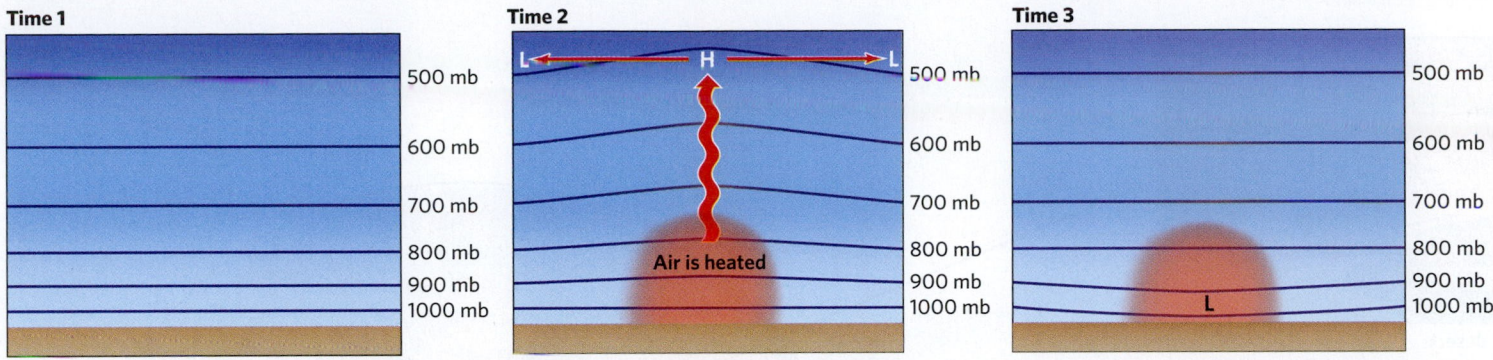

(a) At Time 1, air has a uniform temperature across the region, so isobars are horizontal. At Time 2, heating causes air to expand, which makes the isobars bow upward, and creates an outward-directed pressure-gradient force aloft. By Time 3, air has flowed out of the column aloft, and isobars return to horizontal aloft, but at the ground, a low-pressure center has formed.

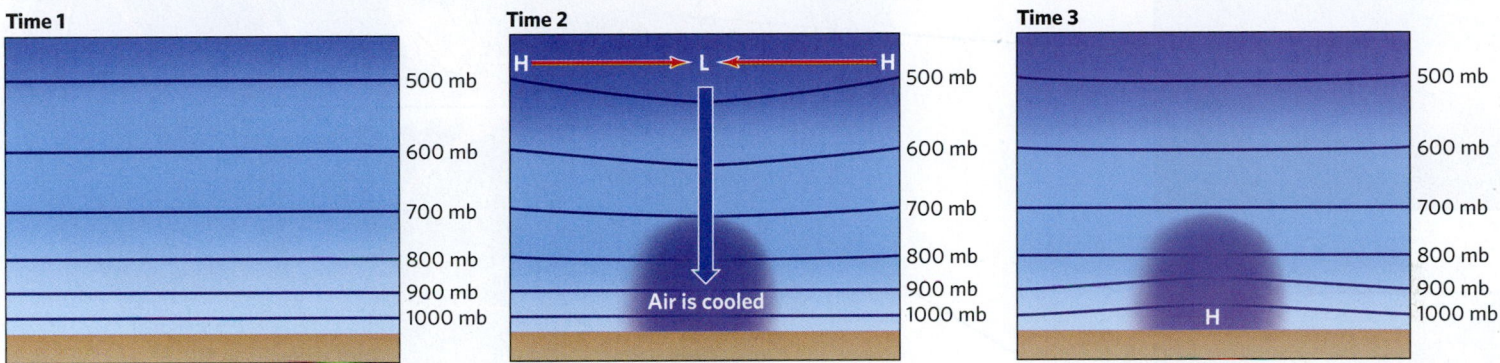

(b) Again, isobars are horizontal when air has a uniform temperature across the region (Time 1). At Time 2, cooling causes air to contract, making the isobars bow down, and creates an inward-directed pressure-gradient force aloft. By Time 3, air has moved into the column aloft, creating high pressure at the ground.

18.3 Air Circulation in the Tropics

What image comes to mind when you hear the word *tropics*—warm sunny breezes and swaying palm trees? In reality, the **tropics**—which lie between the Tropic of Cancer and the Tropic of Capricorn—cover two-thirds of the Earth's surface and host a great variety of landscapes. They include not only rainforests, but also portions of deserts and broad grasslands in which rains falls only seasonally. Weather in the tropics results from a global air circulation system driven by solar heating and by related variations in sea-surface temperatures. Let's explore how this system operates.

Hadley Cells and the Intertropical Convergence Zone

George Hadley, a British lawyer who had a strong interest in meteorology, often wondered why winds blow. In 1735, he published an article in which he suggested that warming of air at the equator—where the Sun heats the Earth most intensely—would cause air to rise from low elevations to the top of the troposphere, where, unable to rise higher, it would start to flow poleward. As the poleward-moving air aloft

cooled and became denser, Hadley proposed that it would sink back toward the surface and, nearer the ground, would flow back toward the equator. Such a circulation caused by convective flow (see Chapter 2) is called a **convective cell**.

Hadley's model was partially correct, but he did not take the Earth's rotation into account. If the Earth did not rotate, the poleward-flowing component of Hadley's proposed convective cell might theoretically extend well into the polar regions. But the Earth does rotate, so the Coriolis force comes into play and deflects the high-elevation poleward flow eastward. As a consequence, air flowing poleward at the top of the troposphere reaches a latitude of only about 25° before it begins flowing roughly parallel to that line of latitude **(Fig. 18.9)**. Furthermore, by 25° latitude, all of the air that rose to the top of the convective cell near the equator has returned toward the Earth's surface. When the sinking air reaches a low elevation, it flows back toward the equator. The Coriolis force deflects this near-surface air westward, so that in the northern hemisphere it flows southwest at a decreasing angle to latitude as it approaches the equator. By the time the air reaches the equator, it's flowing almost parallel to the equator. This tropical surface flow

Figure 18.9 The Hadley cells control the circulation that leads to the distribution of rainfall in low latitudes. They determine where the world's tropical rainforests, grasslands, and deserts occur.

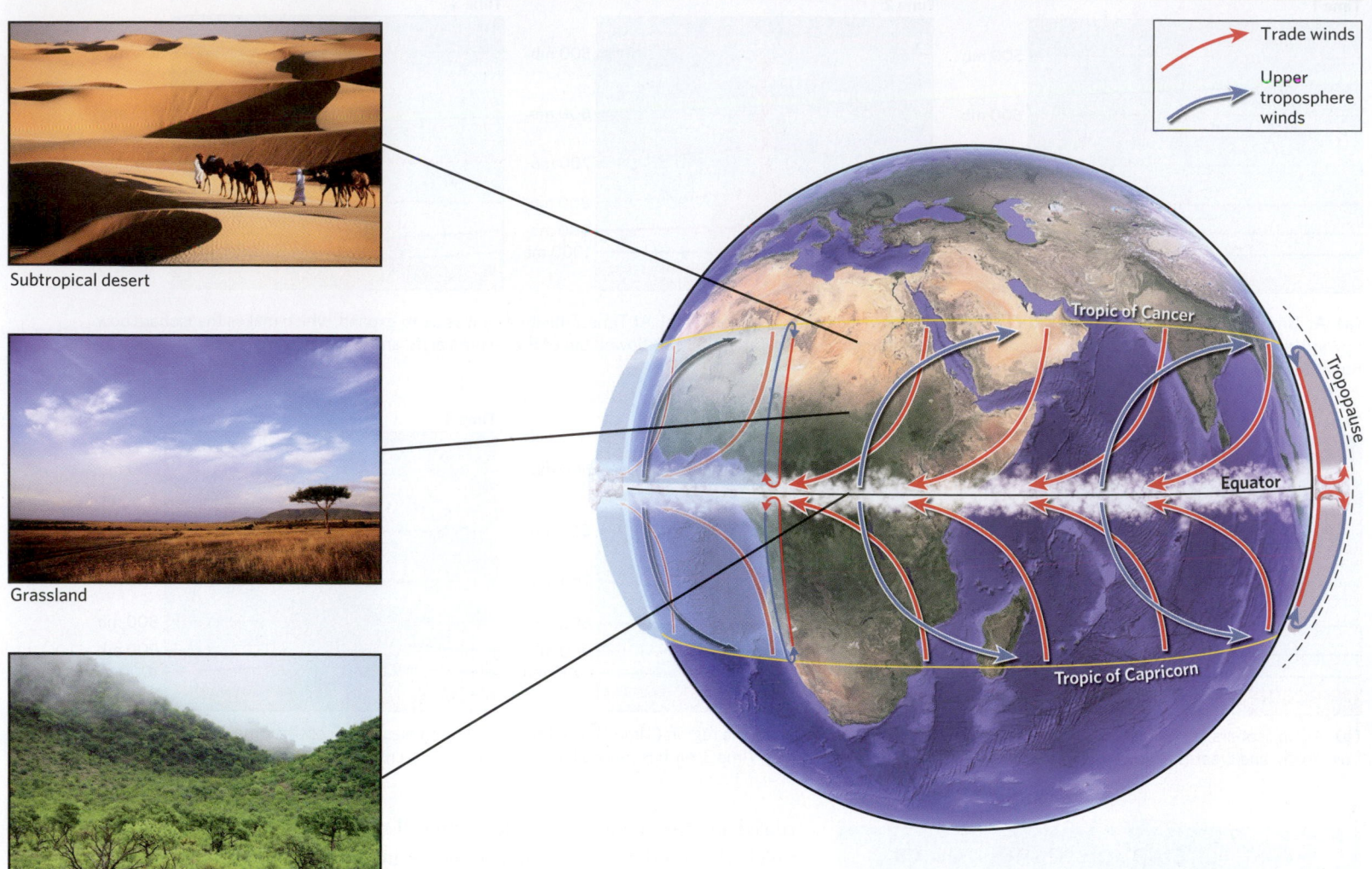

Subtropical desert

Grassland

Tropical rainforest

constitutes the **trade winds**, the same winds that sent Sir Francis Drake westward across the Atlantic and Pacific. Atmospheric scientists refer to this global convective cell, which extends from the equator to a latitude of about 25°, as a **Hadley cell**, in honor of George Hadley. A Hadley cell exists in each hemisphere **(Box 18.1)**.

Near the Earth's surface, the southwest-blowing trade winds of the northern hemisphere and the northwest-blowing trade winds of the southern hemisphere *converge* near the equator, producing the **intertropical convergence zone** or **ITCZ**. The exact position of the ITCZ approximately follows the line along which the Sun's heat is most intense. Because of the Earth's tilt, the position of this line changes over the course of a year. In general, the ITCZ lies north of the equator during the northern hemisphere summer and south of the equator during the northern hemisphere winter. The exact position of the ITCZ also depends on the temperature of the Earth's surface below, so it can be affected by ocean currents, land cover, and topography **(Fig. 18.10a)**.

Air in the ITCZ has nowhere to go but up (since it can't sink into the Earth), so it forms the rising part of the Hadley cell. This air, because it's so warm, absorbs water evaporating from the ocean surface below. For reasons we'll discuss in the next chapter, rising moist air produces **thunderstorms:** localized areas of strong winds and heavy precipitation accompanied by lightning. On a satellite image, clusters of thunderstorms can be seen ringing the globe along the ITCZ **(Fig. 18.10b)**. These storms produce heavy rainfall that provides water for tropical rainforests. Because air in the ITCZ primarily moves upward, surface air below barely moves horizontally. This relatively still air, known to sailors as the **doldrums**, becalms sailing ships.

As we have noted, to the north and south of the ITCZ, in tropical and subtropical latitudes, the air of the Hadley cell descends from the cold upper troposphere to the middle and lower troposphere. This air contains very little water, for most of its moisture rained out in the ITCZ. As this dry air descends, it compresses and warms, so its relative humidity decreases even more. Skies of the subtropics, therefore, tend to host the clear, hot, dry weather that characterizes the

Figure 18.10 The intertropical convergence zone (ITCZ) forms at the junction of the Hadley cells in the two hemispheres.

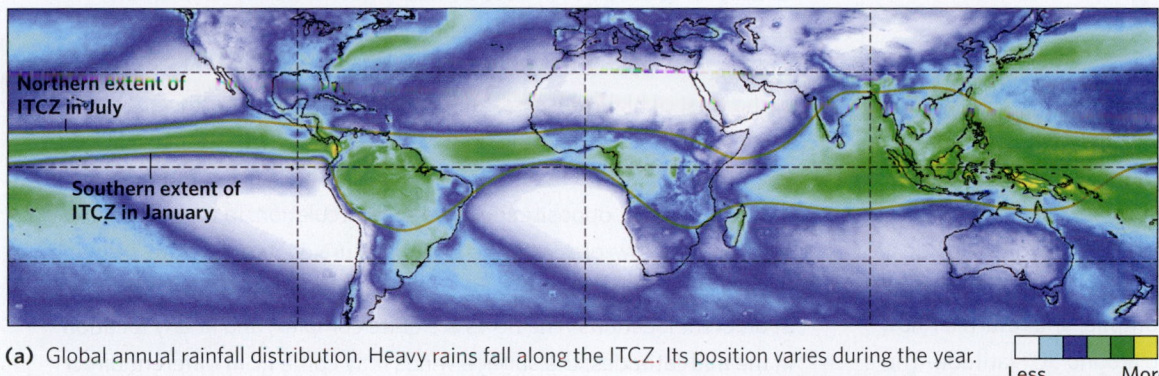

Northern extent of ITCZ in July

Southern extent of ITCZ in January

(a) Global annual rainfall distribution. Heavy rains fall along the ITCZ. Its position varies during the year.

Less rain More rain

(b) The thunderstorms that occur along the ITCZ are visible in this satellite image.

world's largest deserts. The *steppes*, the grasslands between subtropical deserts and tropical rainforests, have a long dry season and a short rainy season. The rainy season occurs during the summer, when the ITCZ moves overhead.

The Subtropical Jet Stream

A strong temperature contrast develops in the upper troposphere above subtropical latitudes, where the poleward-flowing part of the Hadley cell meets the colder air of the mid-latitudes. As we will see in Section 18.5, this temperature change leads to a pressure gradient that accelerates air movement at the top of the troposphere. The Coriolis force deflects the moving air eastward. As a result, a very fast (>160 km per hour, or 100 mph), high-altitude river of air, called the **subtropical jet stream**, encircles the globe. It tends to be relatively narrow, with a width of only 300 to 500 km (180 to 300 miles), measured perpendicular to the air-flow direction. This jet stream can affect weather in the mid-latitudes as well as in the tropics. At any given time, two jet streams exist in each hemisphere; we'll discuss the other one, the polar-front jet stream, later in this chapter.

The Southern Oscillation and El Niño

Often, when a major weather event happens, someone in the media blames El Niño or La Niña. What are these phenomena, and how do they affect the world's weather? To address this question, we first introduce the *Walker circulation*, a cell of flowing air that spans the Pacific Ocean along the equator.

The **Walker circulation**, named after British physicist Gilbert Walker (1868–1958), extends vertically between the Earth's surface and the tropopause and horizontally from South America's western coast to Australia and Indonesia, a distance of more than 16,000 km (9,940 miles). This circulation, which appears as a distinctive feature only within about 5° of the equator, can be thought of as a secondary, east-west-flowing convective cell within the Hadley cells. Due to the Walker circulation, air flows westward in the lower troposphere, rises over the western equatorial Pacific, returns eastward in the upper troposphere, and sinks over the eastern equatorial Pacific **(Fig. 18.11a)**.

Did you ever wonder ...

what an El Niño is and why it may affect your local weather?

Box 18.1 | Consider this . . .

The Ferrel and Polar cells: Do they exist?

Conceptual models of the general circulation between the equator and poles often depict three convective cells in each hemisphere: the Hadley cell and two additional cells, called the *Ferrel cell* and the *Polar cell* (**Fig. Bx18.1a**). The Ferrel cell is supposedly marked by sinking air around 30° latitude, poleward flow at the Earth's surface to about 40° latitude, rising air between 40° and 50° latitude, and return air flow equatorward just below the tropopause to around 30° latitude. The Polar cell supposedly consists of sinking air near the poles, southward flow at the surface, rising air just north of the rising portion of the Ferrel cell, and return flow to the poles in the upper troposphere. Traditionally, the surface winds beneath the Ferrel cells are called the *westerlies,* those beneath the Polar cells are called the *Polar easterlies,* and the boundary between a Polar cell and a Ferrel cell is the *Polar front.* We have seen that the Hadley cell is a dominant feature of the tropics, evident in features such as the trade winds and the ITCZ. Can we say the same about the Ferrel and Polar cells in their respective latitudes?

Unlike those of the tropics, the circulations of the mid-latitudes are dominated by *low-* and *high-pressure centers* rotating in opposite

directions, as we'll see in Section 18.5. **Figure Bx18.1b,** for example, shows a series of low-pressure and high-pressure centers ringing the globe in the mid-latitudes. To the east of low-pressure centers (and to the west of high-pressure centers), warm air flows northward and ascends, consistent with the flow of the Ferrel cell. But to the west of low-pressure centers (and to the east of high-pressure centers) air flows southward, opposite the Ferrel-cell circulation. In the real atmosphere, these low-pressure and high-pressure centers migrate, bringing tropical air northward and polar air southward across the mid-latitudes at different locations. The flows are quite complex, and at any location in the mid-latitudes, the surface winds may come from different directions at different times. In addition, the winds aloft in the mid-latitudes are typically strong, forming a jet stream that flows eastward while undulating in north-south wave-like patterns, as we discuss later.

How do the Ferrel and Polar cells fit into this picture? The Ferrel and Polar cells are in fact statistical averages rather than real circulations. Unlike the flow of the Hadley cells, air flow at any given time rarely conforms to the simple model depicted by the Ferrel and Polar cells, but averaged over months or years, the complex mid-latitude flow patterns transfer air across latitudes in a way that idealized Ferrel and Polar cells would do.

Figure Bx18.1 Are the Ferrel and Polar cells real phenomena?

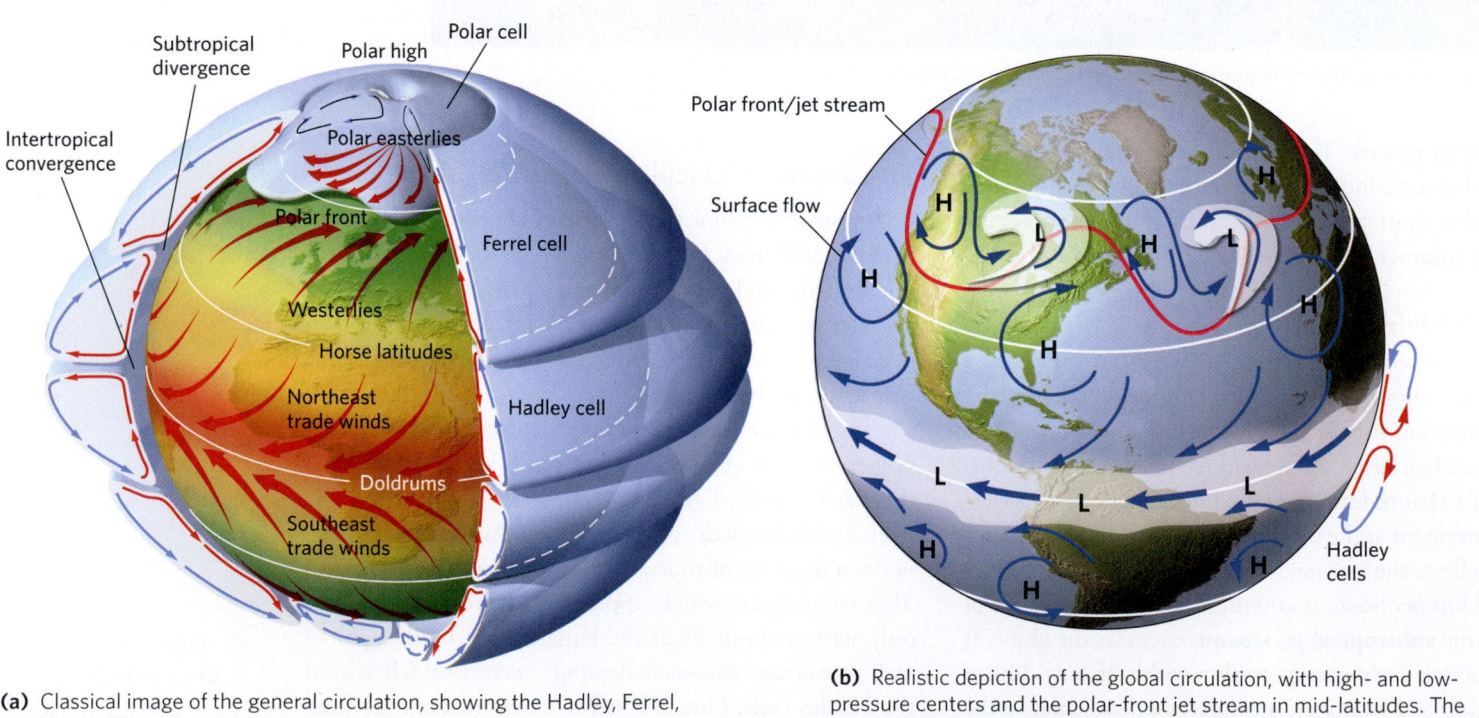

(a) Classical image of the general circulation, showing the Hadley, Ferrel, and Polar cells.

(b) Realistic depiction of the global circulation, with high- and low-pressure centers and the polar-front jet stream in mid-latitudes. The arrows indicate surface winds.

Figure 18.11 The Walker circulation and the Southern Oscillation.

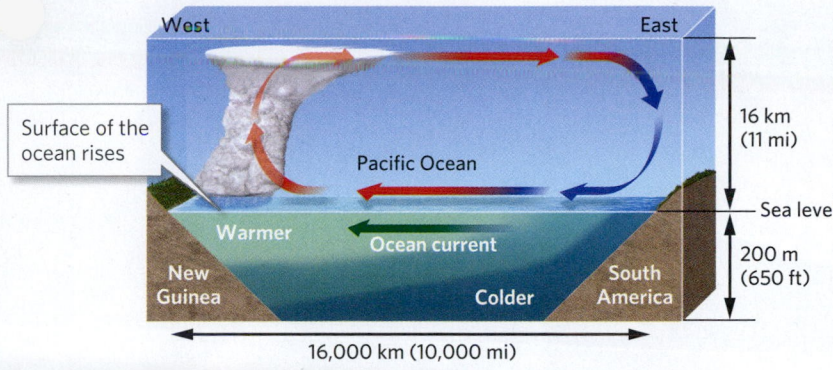

Surface of the ocean rises

Pacific Ocean

16 km (11 mi)

Sea level

200 m (650 ft)

Warmer

Ocean current

New Guinea

Colder

South America

16,000 km (10,000 mi)

(a) Normal Walker circulation. During La Niña, this circulation strengthens.

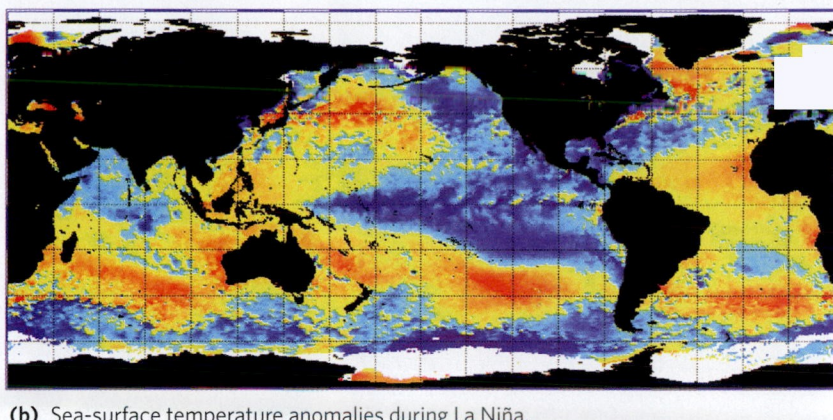

(b) Sea-surface temperature anomalies during La Niña.

−5   −4   −3   −2   −1   0   1   2   3   4   5

Sea-surface temperature anomalies (°C)

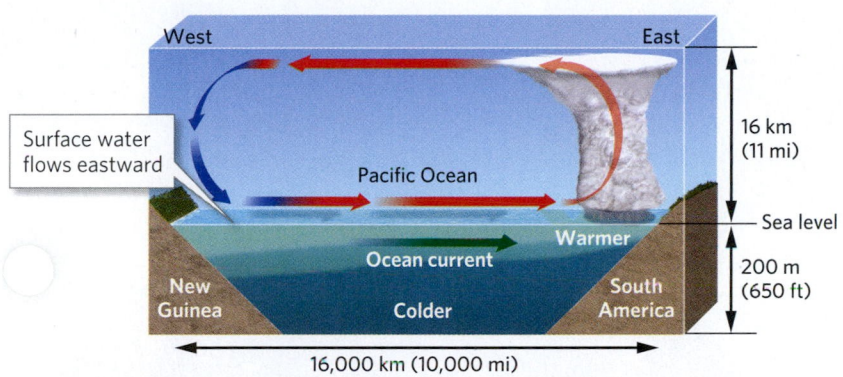

Surface water flows eastward

Pacific Ocean

16 km (11 mi)

Sea level

Warmer

200 m (650 ft)

Ocean current

New Guinea

Colder

South America

16,000 km (10,000 mi)

(c) During El Niño, the Walker circulation weakens.

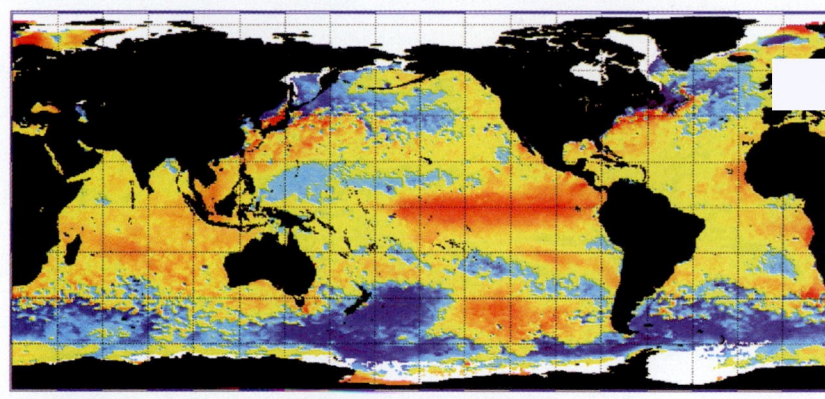

(d) Sea-surface temperature anomalies during El Niño.

Why does the Walker circulation exist? The western Pacific Ocean tends to be warmer than the eastern Pacific because of the configuration of surface currents in the ocean. As a consequence, air over the western Pacific basin becomes warmer than that over the other parts of the Pacific, generating a region of low pressure at the Earth's surface. Meanwhile, a region of high pressure develops at the base of the atmosphere in the eastern Pacific, where the ocean water is cooler, because as air temperature decreases above the cooler water, the air becomes denser. Because the Coriolis force along the equator is very weak (zero at the equator), the pressure-gradient force drives air flow from the high-pressure zone of the eastern Pacific toward the low-pressure zone of the western Pacific.

When the temperature contrast between the eastern and western Pacific becomes large, strong westward-blowing trade winds push surface water westward (see Chapter 15). These winds cause sea level to rise by 10 to 20 cm (4 to 8 inches) in the western Pacific relative to the eastern Pacific. Because the ocean currents carry surface water away from the eastern Pacific, cool water upwells along the coast of South America to replace the surface water (Fig. 18.11b). This water brings up nutrients that nourish the plankton that fish eat, so fish populations thrive.

Every few years, for reasons that atmospheric scientists cannot fully explain, the intensity of the Walker circulation weakens. This weakening reduces the speed of the trade winds blowing west across the equatorial Pacific. When this happens, the tilted sea surface flattens, and warm ocean water from the western Pacific spreads eastward along the equator all the way to the South American coast (Fig. 18.11c, d). As a result, the temperature of the ocean water becomes more uniform across the Pacific, so the pressure-gradient force decreases. In fact, surface water in the eastern Pacific may become warmer than that in the western Pacific, leading to a reversal in the pressure gradient that, in turn, causes a brief reversal in the direction of the equatorial trade winds. Upwelling along the coast of South America shuts off. Over the course of a year, the Walker circulation slowly strengthens, low pressure again

Winds of the World: A Resource for Power

Wind maps

The four global wind maps illustrate the directions and velocities of surface winds at different locations, at the same time, and therefore emphasize that wind direction and velocity vary dramatically around the world. The orientation and taper of the colored lines represent wind direction, and their brightness represents velocity—brighter lines indicate faster velocities. Areas where the lines have a reddish tint are regions where winds are particularly fast (> 70 km/h). Note that surface air flows in a loop around Antarctica, but that locally, it curves into eddies. In the northern hemisphere, several distinct cyclones stand out, as do the trade winds and the doldrums. Overall, surface winds over land are much slower than those over the ocean, due to frictional interaction between air and land.

634

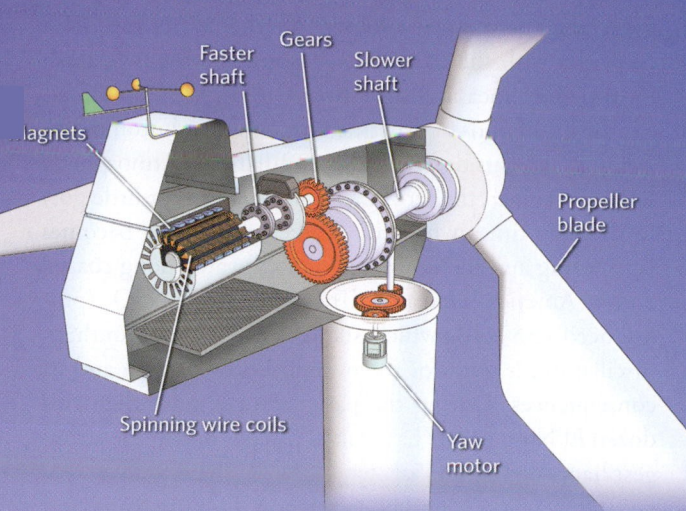

Faster shaft · Gears · Slower shaft

Magnets

Propeller blade

Spinning wire coils

Yaw motor

Map of potential wind power

Because of the variation in the strength and steadiness of winds, some regions have the potential to generate more wind power than do others. On the map below, darker colors represent more potential power. The strongest steady winds are over the oceans, but the technology for building wind farms beyond the continental shelf does not yet exist.

The inside of a wind turbine

The propellers of modern wind turbines translate the flow of air in the wind into the rotation of a shaft. Gears shift the relatively slow rotation of the shaft connected to the spinning blades into the relatively fast rotation of a shaft that enters a generator. In the generator, an iron core rotates within a coil of conducting wire. The movement of the iron core induces an electrical current in the wire.

Figure 18.12 The Southern Oscillation affects global weather.

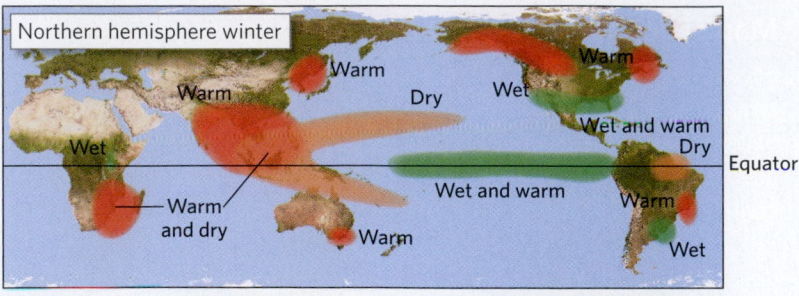

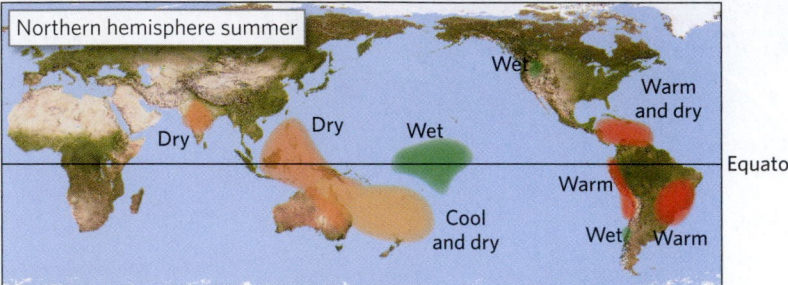

(a) Effects of El Niño.

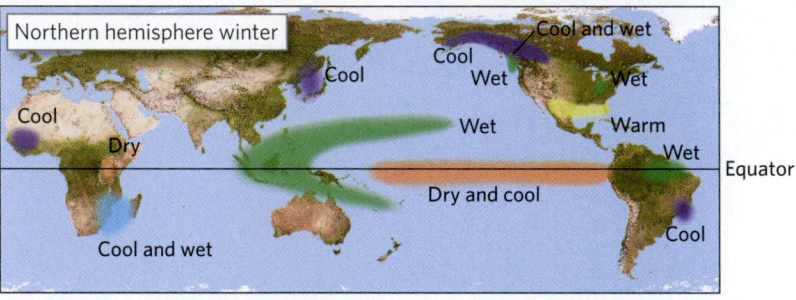

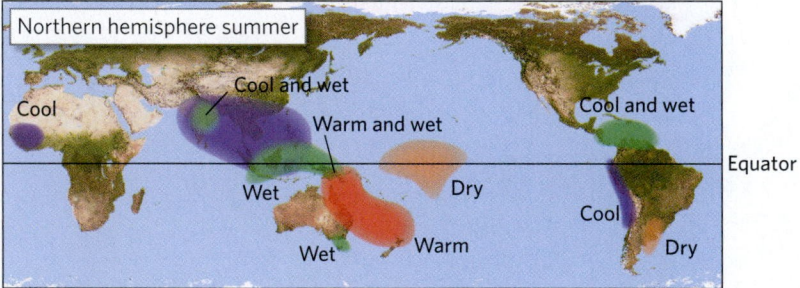

(b) Effects of La Niña.

fisherman because the onset of the warming often coincides with the Christmas season (*El Niño* is Spanish for little boy, meaning the Christ Child). El Niño threatens the livelihoods of fishermen because the shutoff of upwelling nutrients decreases fish populations. Atmospheric scientists coined the term **La Niña** (Spanish for little girl) to describe times when the Walker circulation becomes very strong and the upwelling of cold water along coastal South America increases. The acronym **ENSO** (pronounced enn-soh), which stands for El Niño/Southern Oscillation, is used for the overall seesaw pattern and its consequences. During the past century, more than two dozen El Niño events have taken place—one of the strongest happened in 2015.

If it were just a local phenomenon, El Niño would not have become so famous. But the warming of sea-surface temperatures in the eastern Pacific has far-reaching effects. It shifts the center of tropical thunderstorm activity eastward, and these thunderstorms influence the strength of both the Hadley cells and the subtropical jet stream. Therefore, ENSO modifies atmospheric circulation around the globe, even in the mid-latitudes, influencing weather worldwide. For example, El Niño triggers droughts in Australia and Indonesia, particularly rainy weather in the eastern Pacific, and flooding along South America's northwestern coast **(Fig. 18.12a)**. In North America, El Niño can cause storms that normally move from the Pacific into Washington and Oregon to move northward toward the Gulf of Alaska, or cause storms that normally migrate to the south of the United States to move north instead, triggering floods in California and winter storms farther east. Strong La Niña conditions also affect weather around the globe **(Fig. 18.12b)**.

Monsoons

Think of the word *monsoon* and you probably picture heavy rain and overflowing rivers. In fact, the word has a more general meaning. Technically, a **monsoon** is a seasonally changing wind circulation. In regions that experience monsoons, summer winds blow from the ocean toward the land, and winter winds blow from the land toward the ocean.

Monsoons develop because during the summer months, solar energy heats rock and soil on land more rapidly than it heats water in the ocean **(Fig. 18.13a)**. As a result, air over land becomes warmer than air over the adjacent ocean. The air over land, therefore, expands and rises, so the air pressure over land at Earth's surface becomes less than that over the ocean. The resulting pressure gradient causes moist air to flow from the ocean over the land. Once over the hot surface of the land, the air warms, becomes buoyant, and rises. The rising moist air

develops in the western Pacific and high pressure in the eastern Pacific, and the trade winds again begin to blow strongly westward.

Atmospheric scientists refer to the east-west seesaw in surface air pressure across the equatorial Pacific that accompanies these changes in the Walker circulation as the **Southern Oscillation**. The warming of ocean temperatures in the eastern equatorial Pacific has come to be known as **El Niño**, a term originally coined by Peruvian

produces thunderstorms that drench the land, leading to the common association of the word *monsoon* with flooding. During the winter, the situation reverses **(Fig. 18.13b)**. The land cools faster than the ocean because water cannot lose heat as quickly as rock or soil. Air over the land cools and sinks, forming a region of high pressure. When the air pressure over the land exceeds that over the ocean, the resulting pressure gradient generates a wind that blows seaward; air over the land becomes clear and dry, while rain clouds form over the sea.

The world's largest monsoon occurs over southern Asia, where the seasonal reversals in wind direction cause distinct rainy seasons and dry seasons. Moist air blowing from the Indian Ocean northward over the land rises along the front of the Himalayas and the Tibetan Plateau. This rise spawns strong thunderstorms that produce torrential rainfall. Over a two-month-long period, thunderstorms spread northward across the continent **(Fig. 18.14a)**. The water they supply causes rivers to rise and flood large areas **(Fig. 18.14b)**. Monsoons determine the success or failure of agriculture in southern Asia—abnormally wet or dry monsoon years can destroy a region's crops and livestock and lead to economic disaster **(Fig. 18.14c)**.

Monsoons also occur in other tropical and subtropical regions. For example, the rainy season in South America develops when air warms and rises over the deserts of the Andean region, causing moist winds to blow from the Atlantic westward over the continent. The resulting storms dump rain on the eastern coast of South America and the Amazon rainforest. The summer wet season in the southwestern United States and northwestern Mexico develops when summer sun heats the deserts of this region while the waters of the eastern Pacific remain relatively cool. The rains that fall when moist winds blow eastward over the deserts provide nearly 70% of the region's summer rainfall.

Take-home message . . .

Atmospheric circulations in the tropics are associated with the formation of large convective cells in the atmosphere. Air warmed by strong solar heating near the equator rises to the tropopause along the ITCZ and flows poleward, forming two Hadley cells, one north and one south of the equator. The surface flows of these cells are the trade winds. Other circulations caused by differential temperatures at the Earth's surface drive the Southern Oscillation and monsoons.

Quick Question -
What's the difference between El Niño and La Niña?

Figure 18.13 Formation of monsoons in southern Asia.

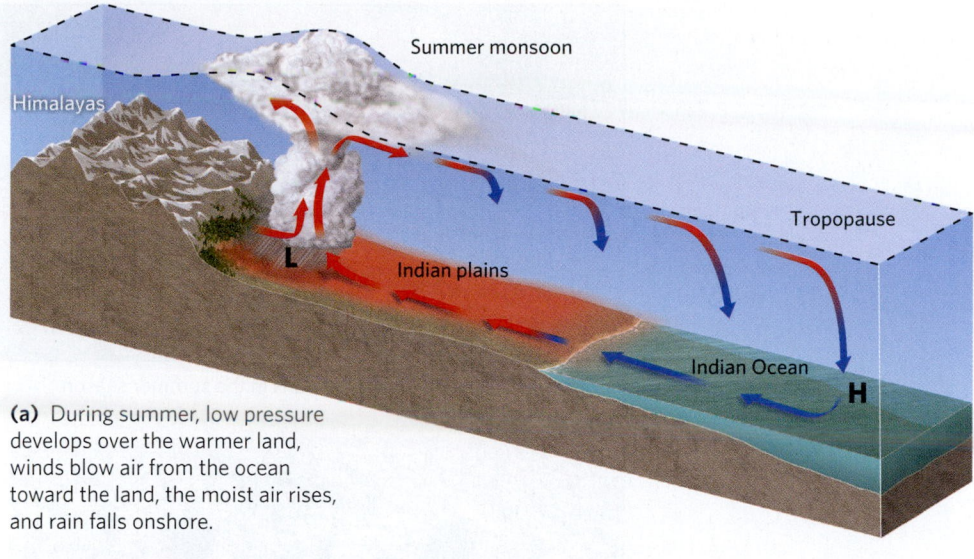

(a) During summer, low pressure develops over the warmer land, winds blow air from the ocean toward the land, the moist air rises, and rain falls onshore.

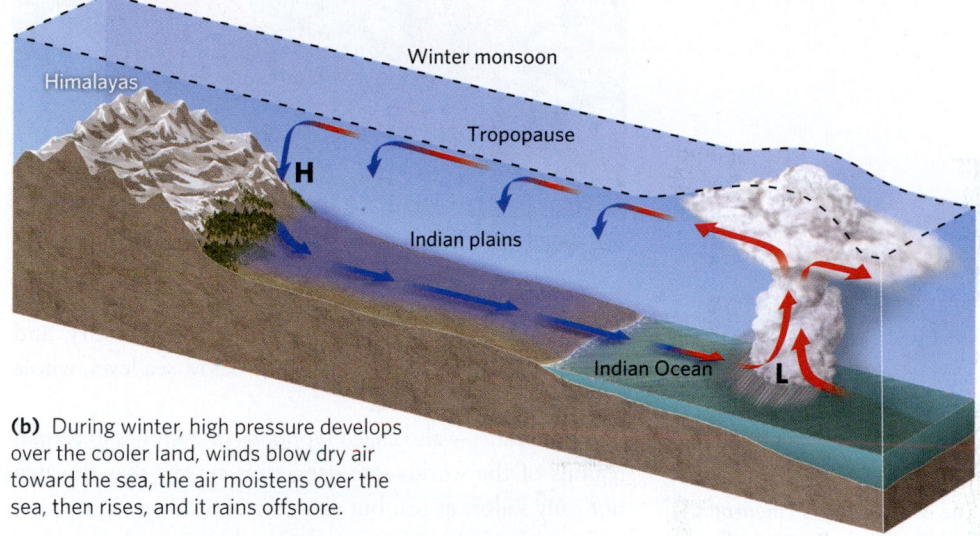

(b) During winter, high pressure develops over the cooler land, winds blow dry air toward the sea, the air moistens over the sea, then rises, and it rains offshore.

18.4 Tropical Cyclones (Hurricanes and Typhoons)

Meteorologists at the National Hurricane Center knew that trouble lay ahead when, in late August 2005, Hurricane Katrina crossed southern Florida into the Gulf of Mexico and ballooned into a monster heading toward New Orleans **(Fig. 18.15)**. The strongest winds and waves of Katrina came ashore to the east of the city and flattened coastal towns in Alabama. But while New Orleans missed the strongest winds, it didn't escape disaster. Water driven by the storm moved into Lake Pontchartrain, a bay north of New Orleans, and along the northern edge of the city, water rose in canals that linked to the lake. Some of the levees bordering the canals couldn't handle the extra

Figure 18.14 The southern Asian monsoon.

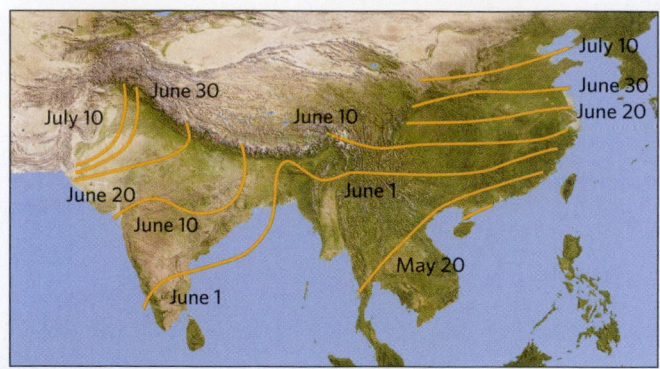

(a) The typical advance of monsoons in the summer season.

(b) Monsoon thunderstorms bring heavy floods.

(c) When the monsoon fails to bring rain, major droughts occur.

See for yourself

The southern Asian monsoon and flooding in Bangladesh

Latitude: 23°57′37.21″ N
Longitude: 89°03′26.15″ E

Look down from an altitude of 1609 km (1,000 miles).

You are looking at the Himalayas to the north, the drainage basins of the Ganges River to the west, and the Brahmaputra River to the east. These rivers join and flow into the Indian Ocean through the Ganges River delta, which is located at the north end of the Bay of Bengal. During the summer, torrential monsoon rains fall on the plains south of the Himalayas, triggering huge floods along these rivers.

pressure and gave way. Water poured into the city, and since large areas of New Orleans lie below sea level, whole neighborhoods flooded.

Hurricanes—also called typhoons or cyclones in certain regions of the world—are dangerous storms that threaten not only sailors at sea, but also coastal residents in tropical and mid-latitudes. Here, we define this type of storm and explain why it forms and why it can cause so much damage.

What Is a Hurricane?

Katrina is an example of a **tropical cyclone**, a rotating spiral-shaped storm that originates over warm tropical ocean waters. Weak tropical cyclones, with sustained winds between 63 and 118 km per hour (39–73 mph), are called **tropical storms**. If a tropical storm in the North Atlantic or eastern Pacific strengthens to the point that it has sustained winds of over 119 km per hour (74 mph), it becomes a **hurricane**. Over the northwestern Pacific, the same type of storm is called a **typhoon**, and over the Indian Ocean and throughout the southern hemisphere, it's simply called a **cyclone**. For simplicity, we'll generally refer to all strong tropical cyclones as hurricanes in this chapter. Due to the Coriolis force, northern hemisphere hurricanes rotate counterclockwise, while those in the southern hemisphere rotate clockwise.

In a hurricane, only the central region, an area 100 to 200 km (60 to 120 miles) in diameter, or less, hosts hurricane-force winds. But the entire storm, in which strong winds spiral around the center and dense clouds hide the sky, may cover an area of 500 to 1,500 km (300 to 900 miles). Because they affect such a broad area, hurricanes are the most destructive storms on the planet. About 100 tropical cyclones originate every year over the world's tropical oceans and seas.

Meteorologists from the US National Weather Service assign human names to Atlantic and eastern Pacific hurricanes. The first hurricane of the season starts with the letter A, and each successive hurricane starts with the next letter of the alphabet, skipping some letters, such as Q, that are uncommon for names. To communicate these storms' severity to the public, meteorologists use the *Saffir-Simpson scale*, which rates hurricanes based on their maximum sustained wind speed (**Table 18.1**).

What's Inside a Hurricane?

Well-developed hurricanes have a distinct internal structure (**Fig. 18.16a**). The top of a hurricane, at the level of the tropopause, looks like a broad sheet of thick clouds. At the center of this cloud sheet is a nearly cloud-free circular area, typically between 10 and 50 km (6 and 30 miles) across at the top, called the **eye**. In three dimensions, a hurricane's eye resembles a vertical downward-tapering funnel, extending from the top of the storm at the tropopause to an elevation of less than 1 km (0.6 miles) above sea level. Its face, the **eye wall**, consists of dense, very rapidly rotating clouds (**Fig. 18.16b**). *Spiral rainbands*, distinct arcs of narrow but tall clouds, some of which contain thunderstorms, extend from the eye wall outward for a distance of several hundred kilometers. As their name suggests, these clouds define the spiral shape of the hurricane. Torrential rain falls from the eye wall clouds, and very heavy rain falls from the spiral rainbands. Tornadoes sometimes form at the base

A radar map shows the precipitation area as Katrina made landfall.

A satellite image shows Hurricane Katrina over the Gulf of Mexico before striking New Orleans.

Figure 18.15 Hurricane Katrina, August 2005.

A wind-swath map of Katrina. Red areas represent hurricane winds, orange areas represent tropical-storm winds.

of the thunderstorms within the spiral rainbands. In the gaps between rainbands, there's less rain or no rain. No rain at all falls right beneath the eye.

The structure of a hurricane reflects the movement of air within the storm. Air near the Earth's surface spirals inward toward the eye of the hurricane. As it does so, it moves progressively faster, like a spinning skater who moves faster as she pulls her arms inward, to conserve angular momentum (see Box 15.3). At the eye wall, the winds reach their greatest velocity. Air then rises along an ascending spiral in the eye wall until it reaches the tropopause. From the tropopause, nearly all rising air then spirals outward and slows, eventually sinking back toward the surface well beyond the margins of the hurricane. A small amount of air from the eye wall flows inward toward the eye. Within the eye, this air slowly descends, and as it does so, it undergoes compression and warms, so its relative humidity decreases, which is why

Anyone who says they're not afraid of a hurricane is either a fool or a liar, or a little of both.

—ANDERSON COOPER (AMERICAN NEWSCASTER, 1967–)

Table 18.1 The Saffir-Simpson scale for hurricane intensity

Category	Sustained Wind Speed	Types of Damage Resulting from Wind
1	74–95 mph 64–82 knots 119–153 km/h	**Dangerous winds produce some damage:** Well-constructed frame homes could have damage to roof shingles, vinyl siding, and gutters. Large branches of trees snap, and shallowly rooted trees may topple. Some power outages occur.
2	96–110 mph 83–95 knots 154–177 km/h	**Extremely dangerous winds cause extensive damage:** Well-constructed frame homes sustain major roof and siding damage. Many shallowly rooted trees are snapped or uprooted, blocking roads. Near-total power outages occur.
3	111–129 mph 96–112 knots 178–208 km/h	**Devastating damage occurs:** Well-built frame homes may incur major damage or removal of roof decking. Large trees snap or are uprooted. Numerous roads become blocked. Electricity and water are unavailable for days to weeks.
4	130–156 mph 113–136 knots 209–251 km/h	**Catastrophic damage occurs:** Well-built homes sustain severe damage. Most trees snap or are uprooted, and most power lines are downed. Debris isolates residential areas, and power outages last weeks to months. The area becomes temporarily uninhabitable.
5	157 mph or higher 137 knots or higher 252 km/h or higher	**Total catastrophic damage occurs:** Most homes are destroyed. Only strongly reinforced buildings remain standing. Debris isolates large areas, and power outages last for weeks to months. Most of the area remains uninhabitable for weeks or months.

Figure 18.16 Internal structure of a hurricane.

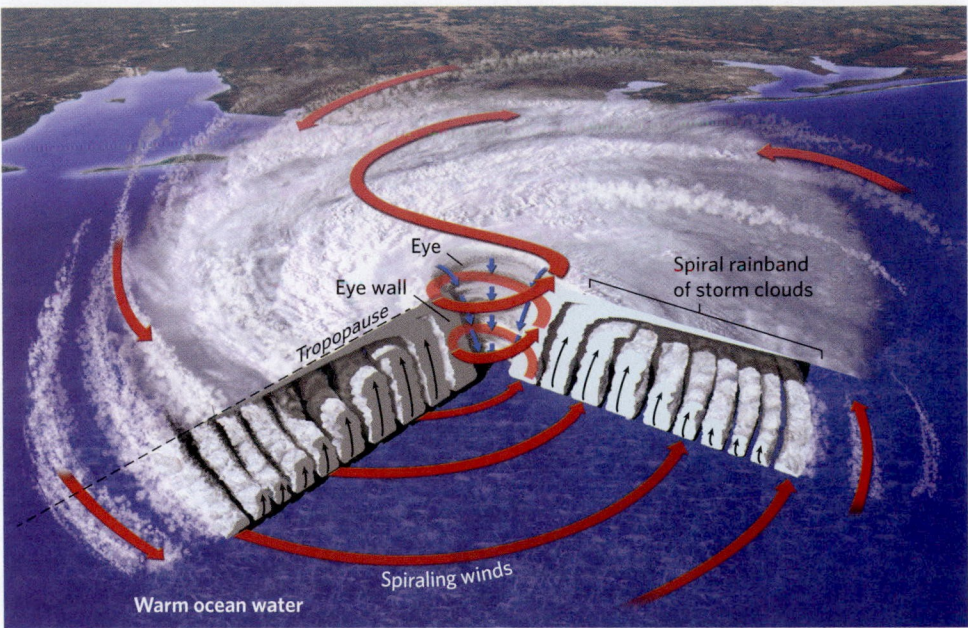

(a) Bands of storm clouds spiral toward a central eye. The strongest winds occur within the eye wall.

(b) The eye and eye wall of a hurricane as viewed from an aircraft in the eye.

energy. The rate at which water evaporates from the ocean surface increases substantially as wind speed increases. Spray generated by strong winds tearing at the ocean surface and whitecaps formed on waves can increase the rate of water vapor transfer to the atmosphere by a factor of 100 to 1,000.

What Is the Process of Hurricane Formation?

A cluster of thunderstorms must form over the warm sea to serve as the "seed" from which a hurricane can grow. The formation of thunderstorms requires the convergence of air currents. When air currents converge at sea level, there's nowhere for the air to go but up. In the rising air, water condenses to form clouds. Its condensation releases latent heat, which warms the air and makes it even less dense, so it rises still higher, where more water condenses. As we'll see in Chapter 19, this process produces thunderstorms.

The convergence of air that forms thunderstorms over the oceans in tropical latitudes typically takes place for two reasons. First, convergence occurs along the ITCZ, where trade winds of the northern hemisphere converge with those of the southern hemisphere; this convergence serves as the primary source of Pacific typhoons and Indian Ocean cyclones. Second, convergence occurs where *easterly waves* develop within trade winds. These air circulations are called *easterly* because their wave-like shape, as seen in map view, moves progressively from east to west, and *waves* because they are north-south oscillations in the direction of the wind. You can see their wave-like shape by looking at a map of variations in wind direction **(Fig. 18.17)**. Easterly waves, which develop when storms originating in Africa move westward into the Atlantic, are the cause of most Atlantic hurricanes. Above the ocean's surface, convergence of air within easterly waves occurs because (in the northern hemisphere) air moves somewhat faster through the southern part of the wave than through the northern part. The result is that air in the south-to-north part of the flow "piles up," converges, and must rise.

What makes a cluster of thunderstorms over the ocean grow into a tropical storm? As the thunderstorms grow, their winds build waves and spray, so more water vapor enters the air. As this water vapor rises and condenses, it releases more latent heat, which in turn warms the air and causes a decrease in air pressure. This decrease makes the pressure gradient steeper, which causes winds to become stronger. Overall, therefore, storm growth represents a *positive feedback* process, meaning that each step enhances the process. For the strong, composite storm to start rotating and become a tropical cyclone, the thunderstorms must be at least 5° north or south of the equator, so that the Coriolis force is strong enough to cause rotation. Hurricanes never form on or cross the equator because, as we noted in Chapter 15, there is no Coriolis force at

Where Does the Energy in a Hurricane Come From?

the eye remains nearly clear. Air also ascends in the rainbands, then generally sinks slowly between them.

Hurricanes don't form over land, and they rarely form outside the tropics. In fact, when a hurricane heads over land, or into higher latitudes, it loses strength and eventually dissipates. Clearly, warm ocean water provides the energy for the hurricane! Hurricanes can form only over ocean water that has a temperature above 26°C (79°F) for at least a depth of 60 m (200 feet).

Energy from the ocean transfers into the base of the atmosphere in two ways. A small portion of the energy in a hurricane comes from the conduction of heat from warm water into the overlying air. The rest of its energy comes from the evaporation of water. The resulting water vapor stores vast amounts of *latent heat* (see Chapter 17). Condensation of the vapor to form the hurricane clouds and rain releases this latent heat, providing the thermal energy that drives the storm. Put another way, evaporated ocean water, incorporated into the storm, supplies the storm's the equator. In addition, for a tropical storm to grow into a hurricane, winds at high altitude must be weak, for if strong winds existed aloft, the spiraling circulation of the storm would be torn apart.

As a tropical storm begins to rotate faster, air pressure at the base of the storm's center decreases. This decrease starts to develop because an outward-directed centrifugal force, due to the rotation of air in the storm, pushes air molecules outward, away from the center of the storm, much as a ball on a string moves outward when you start swinging the ball around your head. Outward air movement, or *divergence*, reduces the weight of the air column at the center of the storm, and therefore reduces the pressure at the base of the storm's center. Once this process has started, the downward pull of gravity overcomes the upward push of the vertical pressure gradient in the storm's center. As a result, air from high elevation begins to sink slowly down in the storm's center. As this air sinks, it undergoes compression and therefore, it warms and dries. Warming causes the air to expand, and the expansion moves more air molecules sideways, out of the storm's center, further reducing the air pressure below the center. Meanwhile, the drying causes clouds to dissipate, so the center becomes a visible eye. When air pressure has become low enough for an eye to form, an extreme pressure gradient exists between the storm's center and the region outside. As we discussed earlier, winds blow because of pressure gradients: the greater the pressure gradient, the faster the wind. The extreme pressure gradient across an eye wall accelerates air sufficiently to develop hurricane-strength winds from the eye wall out for as much as 100 km (60 miles) in the stronger hurricane.

What Path Does a Hurricane Take?

Once a hurricane exists, its "engine" will run as long as it has "fuel" (warm ocean water) and there are no strong winds in its environment in the upper troposphere. The temperature and supply of warm water, the character of the surrounding winds, the depth of the troposphere at the latitude where the hurricane resides, and friction with the Earth's surface all control the strength of a hurricane. Since hurricanes form in the tropics, within the belt of westward-flowing trade winds, they normally drift westward as they form and mature (Fig. 18.18a). In the northern hemisphere, hurricanes may start to drift northward as they move westward, an effect of the Coriolis force. When they reach a latitude of about 30°, they come under the influence of mid-latitude winds, which generally blow from west to east. As a result, hurricanes in the northern hemisphere typically move eastward after passing a latitude of 30° N. Knowledge of these principles allows meteorologists to predict the **hurricane track**, the path that a hurricane will follow, and they can even estimate the width of the belt in which hurricane-force winds will

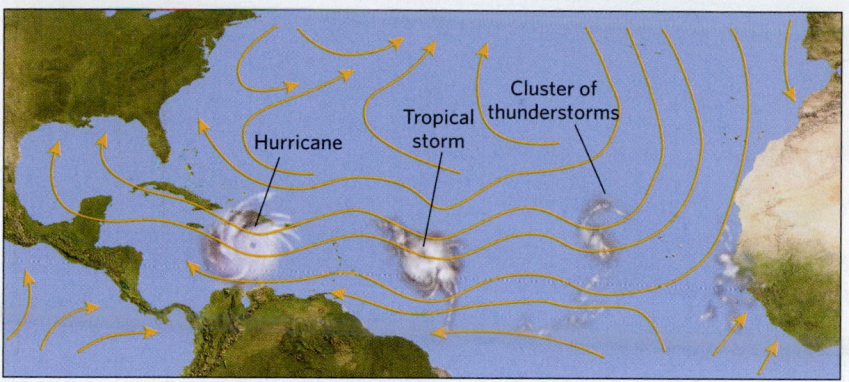

Figure 18.17 A typical Atlantic hurricane evolves from a cluster of thunderstorms off Africa into a tropical depression, then a tropical storm, and finally a hurricane. The thunderstorms are associated with convergence within easterly waves in the trade winds.

occur. Not all hurricanes behave simply, though—in some cases, complicated background winds can lead to irregular tracks (Fig. 18.18b). Hurricanes weaken and eventually dissipate when they move over cold water or over land and lose their energy source. Friction with land hastens their demise. They may also be destroyed if they encounter strong winds in the upper troposphere, for these winds can disrupt the spiraling of hurricane circulation.

How Do Hurricanes Cause Damage?

We've already seen that the winds in the eye wall of a hurricane have immense power (Box 18.2). During Hurricane Patricia in 2015, winds over the eastern Pacific reached a record 345 km per hour (245 mph)—virtually no building can remain standing in such wind (see Table 18.1). Unfortunately, wind is not the only hazard of a hurricane. Because hurricanes come from the sea, they can cause immense damage to the coast, not only because of towering waves, but also because of **storm surge**, the dome of water that forms in front of a hurricane (Fig. 18.19a). Storm surge develops because the wind of the storm drives water in front of it, effectively building a pile of water. Surges develops to a lesser extent because the low atmospheric pressure beneath the storm means that the weight of the atmosphere doesn't push down on the sea surface, allowing it to rise (Fig. 18.19b). When a cyclone moved over the Ganges River delta, at the head of the Bay of Bengal, in 1970, storm surge lifted sea level by about 6 m (20 feet), and 10-m (33-foot)-high waves built on top of the surge. As a result, water inundated the delta, and 300,000 people drowned. If storm surge arrives at high tide, the water level will become even higher. Storm surge was responsible for the failure of New Orleans' levees during Hurricane Katrina and for the destruction along coastal New Jersey during Hurricane Sandy.

Though hurricane winds slow when the storms head over land, their clouds can still dump torrential rains.

Figure 18.18 Hurricane tracks.

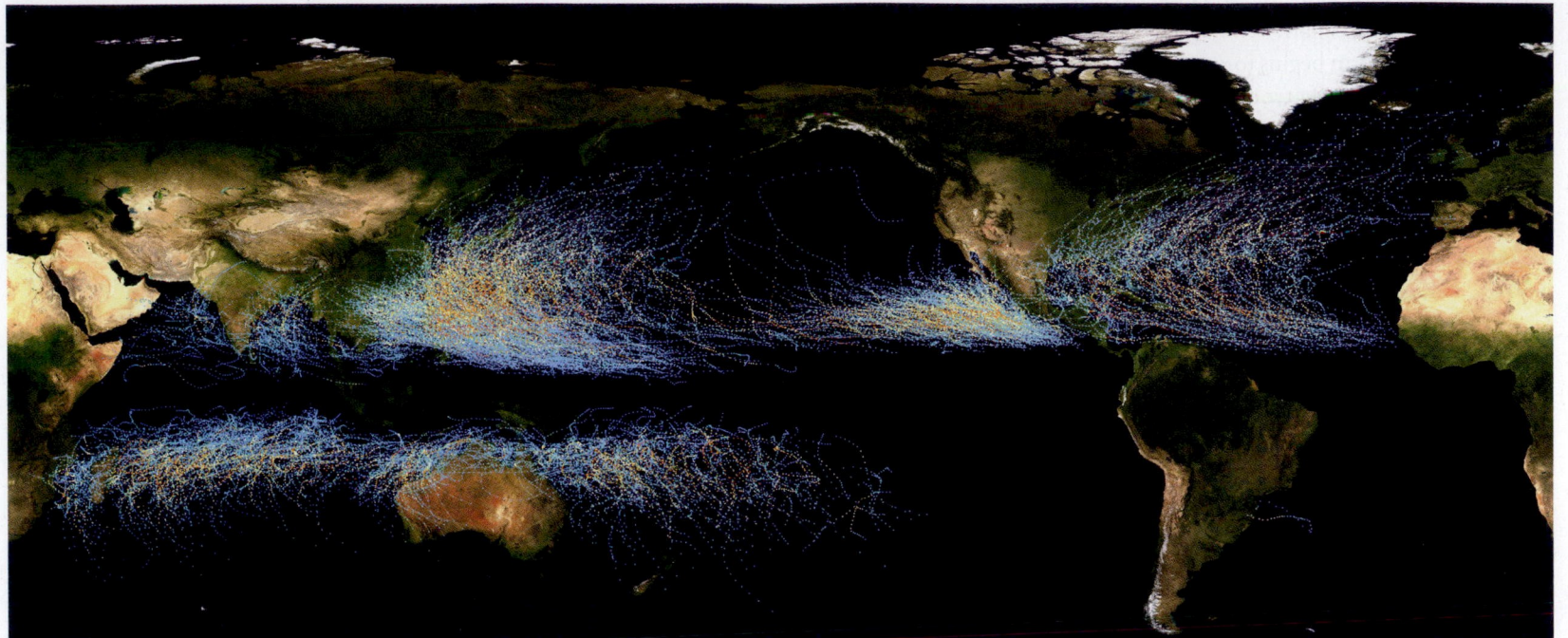

(a) Each line on this map represents a hurricane or tropical storm track. Light blue lines are weaker storms, and yellow lines are stronger storms. Note that none of these tracks crosses the equator.

For example, when Hurricane Mitch stalled over the mountains of Central America in 1998, the downpours it released caused numerous flash floods. Rainwater also soaked into the ground and saturated the soil, causing unstable slopes to give way. The resulting mudslides flowed downhill, and some buried whole towns.

Because hurricanes, typhoons, and cyclones pose such a threat—both to people on land and to ships at sea—many countries have set up tropical cyclone monitoring centers. For example, staff at the US National Hurricane Center watch the world's ocean basins to monitor the development and progress of tropical storms and cyclones. Using data from satellite observations and radar, combined with data from weather stations and aircraft, the center also employs a suite of computer models to predict hurricane tracks and wind speeds in order to provide early warning and evacuation recommendations to governments.

Take-home message . . .

Tropical cyclones—hurricanes and typhoons—are the most destructive storms on the planet. They originate from clusters of thunderstorms and grow into giant spiraling storms, nourished by latent heat in water evaporating from warm seas. The strongest winds occur along the eye wall. Winds, waves, storm surge, flooding, and mudslides caused by these storms cause devastation.

Quick Question -
Why can't a hurricane form on the equator?

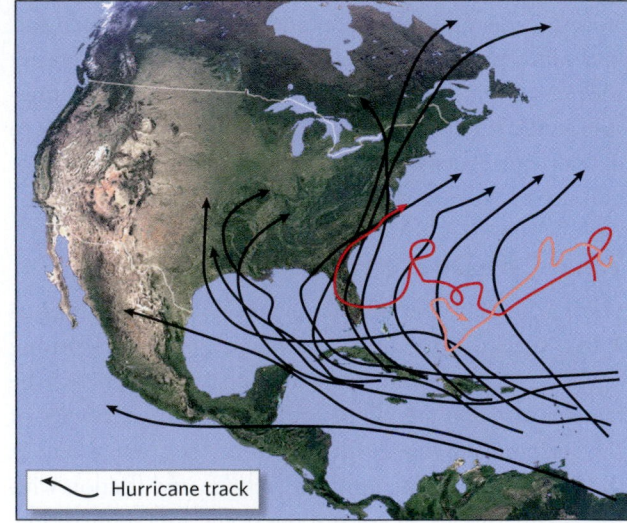

(b) Selected tracks of hurricanes. Most (black lines) first head west and then north. A few (colored lines) follow irregular paths.

18.5 The Atmosphere of the Mid- and High Latitudes

People living at tropical and subtropical latitudes typically experience weeks or months of the same weather—dry during the dry season and wet during the wet season—and do not endure any days of cold and ice. This is not so in the *mid-latitudes*, the regions between latitudes of 30° and 60°. Large portions of these regions have *temperate climates* (see Chapter 20), in which weather changes not only as

Figure 18.19 Storm surge.

(a) Storm surge accompanying a hurricane floods coastal areas and leaves behind devastation.

(b) Wind and low atmospheric pressure combine to produce a bulge of water.

seasons progress, but even during the course of a single day. A typical late summer afternoon in the central United States, for example, may feel hot and humid. In a matter of minutes, skies may grow overcast and towering thunderstorms roll by. An hour later, the Sun shines again, and the air feels cool and dry. In winter, a warm southerly breeze bringing mild air can quickly change to a northern blast of frigid air, followed by heavy snow. Why do such radical changes in weather happen?

To develop an answer, we need to step back and think again about how the Sun heats the Earth's surface. During most of the year, much less solar energy reaches the surface in the polar regions than in the tropics. Because of this imbalance, the polar regions remain cool, and in winter, when they experience near-perpetual night, they become very cold. In the tropics, the Sun rises high in the sky, and incoming solar energy exceeds the energy lost as outgoing infrared radiation.

If there were no atmosphere or oceans, the temperature differences between the poles and the equator would be much greater than they are. But the atmosphere and oceans reduce these differences by transporting heat, in flowing air and water, from the tropics toward the poles, and cold air from the poles toward the tropics. In other words, winds and ocean currents act like giant conveyor belts that redistribute heat between the hot tropics and the cold polar realms in a never-ending attempt to balance their differences in temperature.

Earth's mid-latitudes, which lie between the tropics and the poles, serve as the battleground in the war to balance temperature extremes. When cold polar air wins a battle, it pushes back warm tropical air and envelops a region of the mid-latitudes. When tropical air wins, the reverse happens. Clashes between bodies of air generate the large storms of the mid-latitudes. Let's explore what happens in this complex zone of mid-latitude air circulation.

Air Masses and Fronts

An **air mass** is a broad body of air within which temperature and humidity are relatively uniform. Air masses are typically several kilometers thick and a thousand or more kilometers wide, so a single air mass can cover a

Which side of a hurricane is more destructive?

The wind speed at a location within a hurricane depends on two factors: first, the speed due to the spiraling of the air around the hurricane, which decreases with distance from the eye wall; and second, the speed of the overall movement of the hurricane as the entire storm drifts along, propelled by regional atmospheric circulation. Typical forward speeds of hurricanes range between 15 and 90 km per hour (10–60 mph), though some hurricanes stall so that the eye remains stationary for a while. As a consequence of a hurricane's forward speed, winds on the right side of a counterclockwise-rotating hurricane (as viewed looking in the direction the hurricane is moving) are stronger than those on the left side. For example, if a hurricane rotates at 150 km per hour while moving forward at 40 km per hour, the wind on the right side will be 150 + 40 = 190 km per hour, whereas on the left side the wind will be 150 – 40 = 110 km per hour **(Fig. Bx18.2)**. For this reason, the side of a hurricane rotating in the direction of its forward motion causes more wind damage than the side rotating in the direction opposite its forward motion. In addition, when a hurricane is coming ashore, winds on the right side are generally blowing toward the shore and winds on the left side are blowing offshore. So storm surge builds on the right side, but on the left side, winds blow ocean water seaward, resulting in less destruction from flooding.

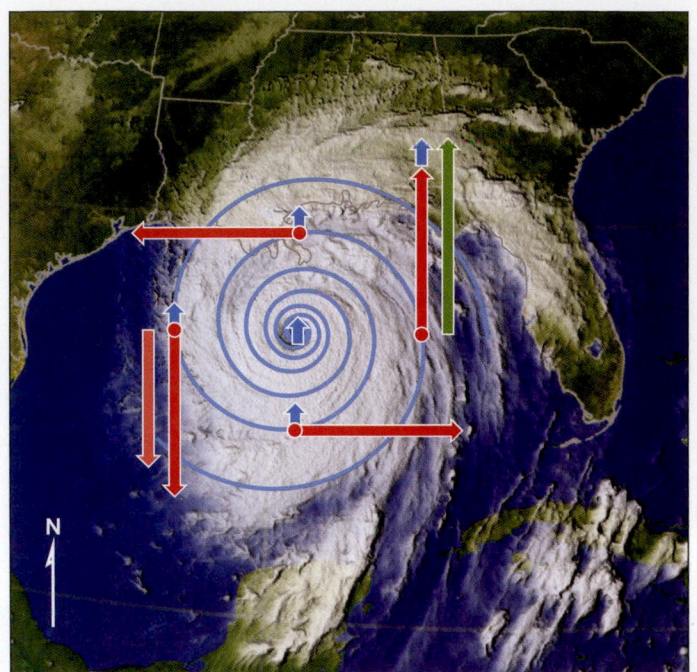

Figure Bx18.2 The wind speed in a hurricane at a specific location (dots) is the sum of the hurricane's rotational wind speed (red arrows) and the overall forward motion of the hurricane (blue arrows). Therefore, as Hurricane Katrina moved north, the wind speed on the east side of the storm (green arrow) exceeded that on the west side (pink arrow).

Figure 18.20 Major air masses affecting North America. Maritime air masses are moist, while continental air masses are dry. Tropical air masses are warm, while arctic and polar air masses are cool.

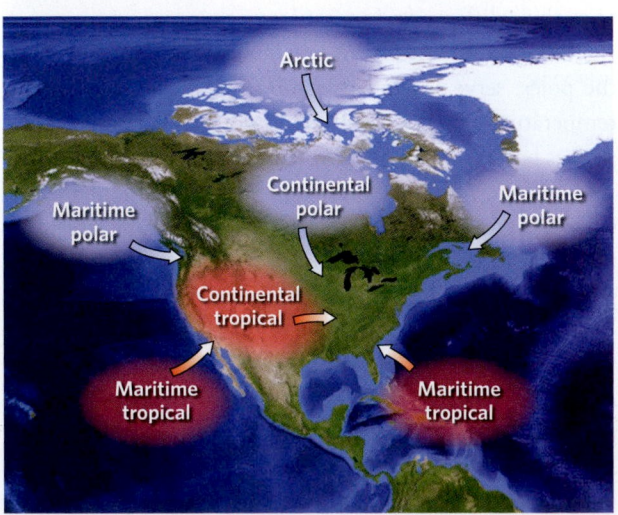

quarter to a half of a continent or ocean. Numerous distinct air masses exist worldwide at any given time. Over a period of days, they may remain nearly stationary, move slowly, or sweep rapidly (at speeds up to 60 km per hour, or 40 mph) across a continent or ocean.

Meteorologists characterize air masses based on their source region (continental or maritime) and their latitude of origin (tropical, polar, or arctic). Seven major air masses typically lie over and around North America and influence its weather **(Fig. 18.20)**. For example, a *continental polar air mass* typically resides over a large part of Canada and the northern United States in winter, and an *arctic air mass* lies over the Arctic Ocean and northern Canada. *Maritime polar air masses* reside over the North Atlantic and North Pacific Oceans. *Maritime tropical air masses*, which are warm and humid, develop over the tropical North Atlantic, the Gulf of Mexico, and the tropical North Pacific. When an air mass moves out of its source location, a new one regenerates due to solar heating (or the lack of it) and, in the case of maritime air masses, evaporation of water from the underlying oceans. In winter, for example, new frigid continental polar air masses form in the Canadian Arctic at the same time as previously formed ones sweep southward from Canada into the United States.

Because cold air is denser than warm air, the area beneath a cold air mass becomes a **high-pressure system** (see Fig. 18.8b). In winter, when large continental land areas become cooler than the surrounding oceans, high-pressure systems develop beneath continental air masses of Canada and Siberia. In summer, when the oceans tend to be cooler than the land, high-pressure systems develop within maritime air masses. Examples of such high-pressure systems include the Bermuda High, over the North Atlantic, and the Pacific High, over the North Pacific (see Chapter 15). As we'll see, air in high-pressure systems tends to descend, which causes the air to warm. Without the addition of more water vapor, the relative humidity in this warming air decreases, so regions beneath high-pressure systems tend to have clear skies.

Because warm air is less dense than cold air, the area beneath a warm air mass becomes a **low-pressure system** (see Fig. 18.8a). Semi-permanent low-pressure systems develop over desert regions, such as the American southwest, in summer, and in winter they form over the far northern Atlantic and Pacific Oceans, which remain warm compared with the adjacent continents. These semi-permanent low-pressure systems are called the Icelandic Low and the Aleutian Low, respectively. As we'll see, air rises within low-pressure systems. As it rises, it cools, so the water it contains condenses, leading to the formation of clouds and storms.

The boundary between two air masses typically occurs along a narrow zone called a **front**. Fronts are delineated by an abrupt contrast in temperature and, commonly, relative humidity as well. Most stormy weather in mid-latitudes, particularly in cooler seasons, develops along fronts because air on the warm side of a front flows up and over air on the cold side. The warm air is often forced upward as the cold air advances. Meteorologists classify fronts based on the direction in which the colder air mass moves.

At a **cold front**, a cold air mass advances under a warm air mass, forcing the warm air to rise (Fig. 18.21a). Commonly, the leading edge of the cold air mass develops a dome-like shape. On a map, meteorologists represent the position of a cold front at the Earth's surface with a line marked by triangles—the triangles lie on the warm side of the front and indicate the direction in which the front is advancing. The type of precipitation that occurs along a cold front depends on the characteristics of the warm air ahead of the front. For example, if the warm air mass lifted by the front is moist and buoyant, thunderstorms develop. If, however, the warm air isn't buoyant, stratus clouds form above the front and produce light rain

or no rain. Where the air ahead of a cold front has very low humidity, no clouds may form at all.

A **warm front** exists where a cold air mass retreats and a warm air mass advances (Fig. 18.21b). Typically, the cold-air dome slopes more gradually ahead of a warm front than it does behind a cold front. On a map, meteorologists represent the position of a warm front at the Earth's surface with a line marked by half circles—the half circles lie on the cold side of the front and indicate the direction in which the cold air is retreating. At a warm front, the warm air flows toward and then over the cold air mass. The type of precipitation that develops along a warm front again depends on the temperature and moisture characteristics of the warm air mass. In general, clouds become thickest and rain falls the hardest near the frontal boundary. Clouds thin and rain diminishes toward the interior of the cold air mass. If the surface temperature within the cold air mass is below freezing, snow, ice pellets, or freezing rain may fall.

In some cases, an advancing cold front overtakes a warm front. When this happens, the cold front lifts the cool air behind the warm front, so the warm front no longer intersects the ground surface. At the ground surface, two cold air masses come into contact. Warmer air forms a layer over the top of both fronts (Fig. 18.21c). Meteorologists indicate the

Did you ever wonder . . .

why weather at a location can change so radically in a matter of minutes?

Figure 18.21 A front marks the boundary separating two air masses of different temperatures.

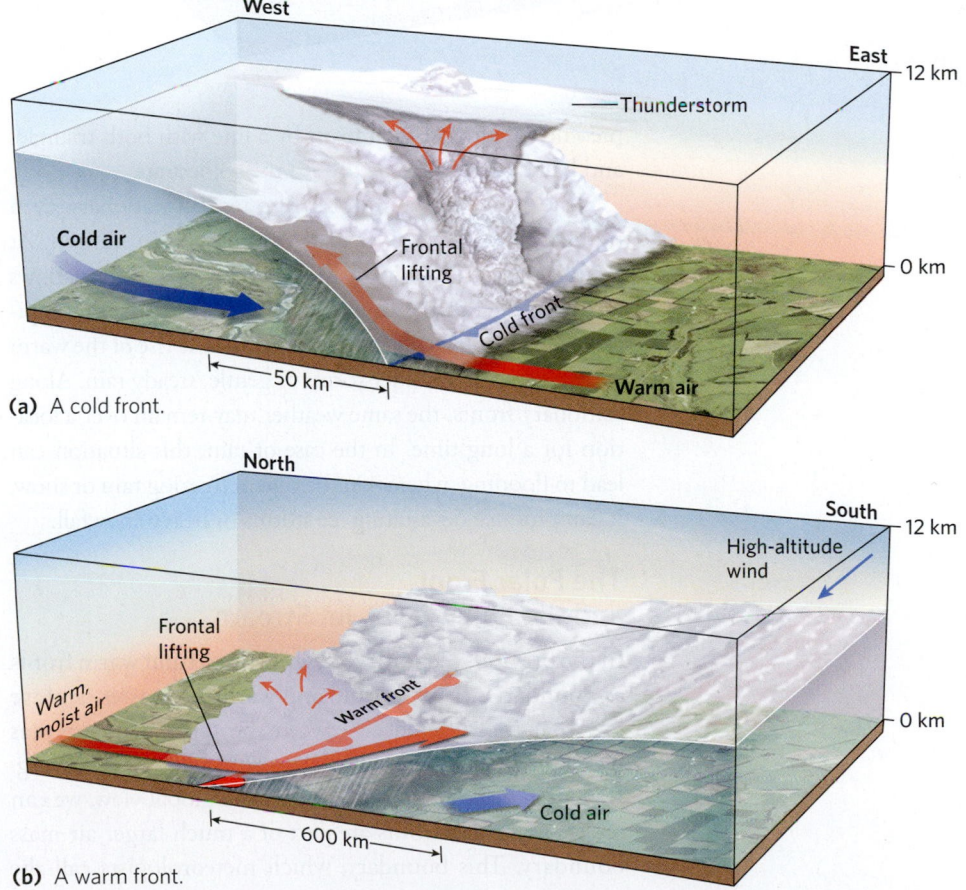

(a) A cold front.

(b) A warm front.

Figure 18.21 Continued

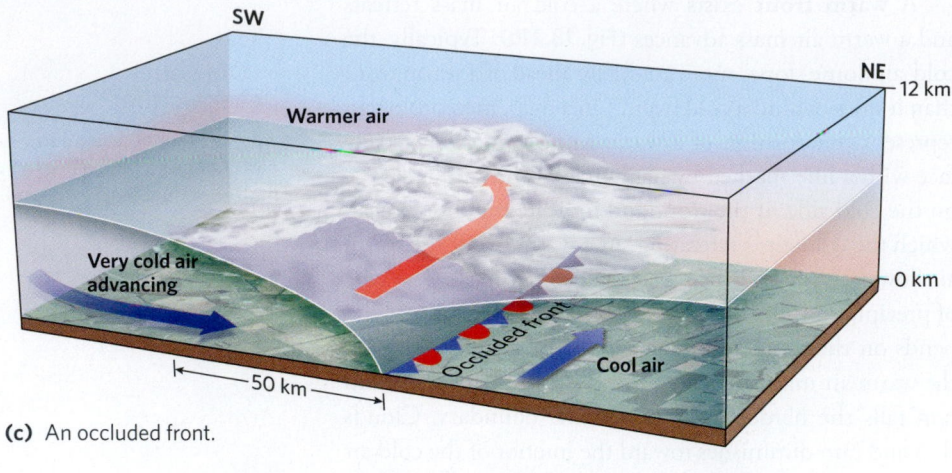

(c) An occluded front.

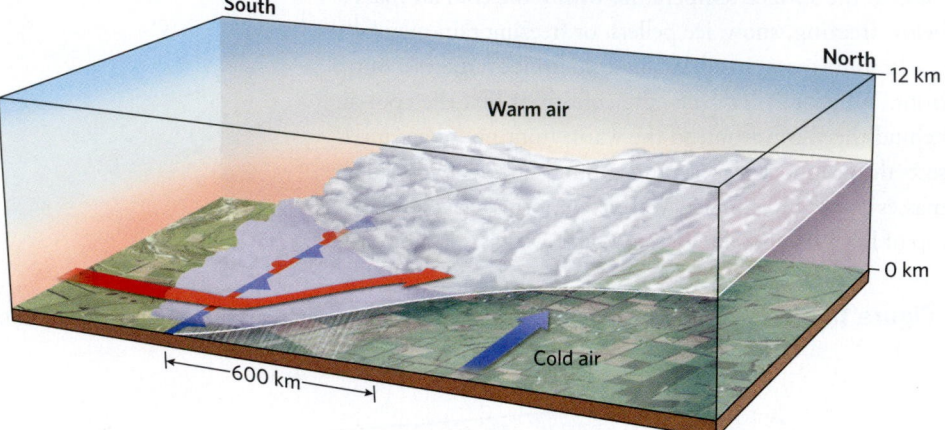

(d) A stationary front.

presence of an **occluded front** by a line with both triangles and half circles, pointing toward the coldest air.

The position of a front can also remain stationary, even though air on both sides of the front stays in motion. At a **stationary front**, air on the cold side of the front flows nearly parallel to the front, but air on the warm side typically rises over the cold air **(Fig. 18.21d)**. The rise of the warm air can generate thunderstorms or gentle, steady rain. Along stationary fronts, the same weather may remain over a location for a long time. In the case of rain, this situation can lead to flooding, whereas in the case of freezing rain or snow, it can produce devastating ice storms or heavy snowfall.

The Polar Front and the Polar-Front Jet Stream

In the previous section, we described cold and warm fronts as boundaries between cold and warm air masses, and we saw that the difference between the types of fronts was based on whether the cold air was advancing or retreating. If we step back for a moment to take a global view, we can see that all these fronts are part of a much larger air mass boundary. This boundary, which meteorologists call the **polar front**, separates warm air originating in the tropics from cold air of the polar and subpolar regions.

The contrast in temperature across the polar front exerts a major influence on wind velocities at the top of the troposphere. Why? Warm air occupies more volume than cold air, so the vertical distance between *isobaric surfaces* (surfaces of constant air pressure) is greater on the warm side of the front than on the cold side. Isobaric surfaces are therefore lower on the cold side of the front than on the warm side, so at the front, isobaric surfaces slope downward from the warm side toward the cold side. Their slopes become progressively steeper at higher altitudes, and they are steepest near the tropopause **(Fig. 18.22)**. As a result, the horizontal pressure-gradient force near the tropopause over the polar front is very strong—much stronger than it is at lower altitudes or far away from the front. Since pressure gradients, as we have seen, drive the wind, this very steep pressure gradient generates very fast winds. The river of fast winds present over the polar front is called the **polar-front jet stream (Fig. 18.23a)**.

Note that we have introduced two jet streams in this chapter: the subtropical jet stream and the polar-front jet stream. Both result from strong pressure gradients that develop across a boundary between different air masses. The subtropical jet stream forms at the boundary between the high-latitude end of a Hadley cell and the cooler air of mid-latitudes. This boundary, called the subtropical front, exists only in the upper troposphere. The polar-front jet stream, in contrast, lies over the polar front, an air mass boundary that extends from the Earth's surface upward to the tropopause. Temperature differences across the polar and subtropical fronts increase in winter, so both jet streams strengthen during winter **(Fig. 18.23b)**. Because of the near balance between the pressure-gradient force and the Coriolis force, each jet stream becomes a nearly geostrophic wind that flows more or less parallel to isobars. The polar-front jet stream tends to occur at an altitude of about 10 km (6 miles), lower than the 10 to 13 km (6 to 8 miles) altitude of the subtropical jet stream. The difference results from the decrease in the altitude of the tropopause with increasing latitude.

As the atmosphere attempts to balance differential heating between the tropics and polar regions, cold air masses north of the polar front frequently advance southward and descend, and warm air masses south of the polar front move northward and ascend. Viewed at a global scale from above one of the poles, the polar front therefore typically follows an undulating wave-like north-south pattern that marks the warm air–cold-air boundary. The polar-front jet stream, which lies over the cold air mass along the polar front, typically also flows in a wave-like pattern, roughly following the air mass boundary **(Fig. 18.24)**. For example, over North America, the polar-front jet stream may flow up over the Pacific Northwest, down along the Rockies into the southeastern United States, turn northeastward, and exit the United States flowing northward toward the North Atlantic (see Fig. 18.23b). Meteorologists describe a region where the jet stream bows toward the poles as a **ridge** and one where it bows toward the equator as a **trough** (see Fig. 18.23b). Warm air lies beneath a ridge, and cold air lies beneath a trough. Why are they called ridges and troughs? Picture an imaginary surface representing uniform air pressure near the top of the troposphere. Because of variations in air temperature below that surface, this isobaric surface has ups and downs, and in three dimensions, it looks wavy. A line along the high points in this pressure surface corresponds to a ridge, and a line along the low points corresponds to a trough **(Fig. 18.25)**.

Air traffic must take into account the position and direction of the jet stream. Specifically, eastbound aircraft can take advantage of the significant tailwind that the jet stream provides to save on time and fuel costs. Planes flying westward, however, face a strong headwind. As a consequence, the difference in travel time for westbound and eastbound flights crossing a continent or ocean can be more than an hour!

The location and intensity of the jet stream changes over time in response to the movements of air masses and the positions of fronts (see Fig. 18.24). As we've noted, the path of the polar-front jet stream, in map view, displays wave-like undulations as it curves around ridges and troughs. Therefore, the latitude at which the polar-front jet stream flows varies greatly around the planet, and at any given time, it may carry air from polar latitudes into mid-latitudes, or vice-versa. As air masses move, the position of

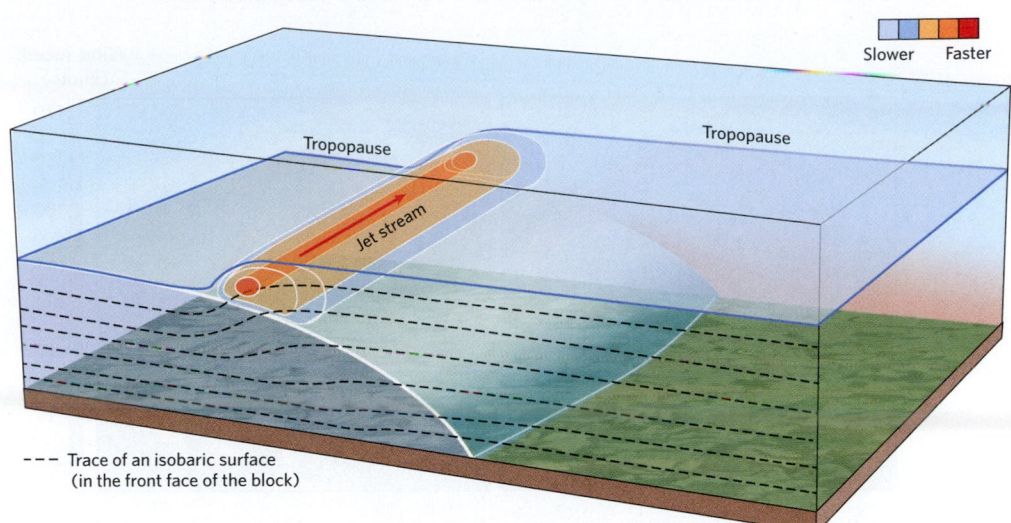

Figure 18.22 The relationship between a front and a jet stream. The strongest winds are at the center of the jet stream.

- - - Trace of an isobaric surface
(in the front face of the block)

troughs and ridges, and therefore the positions and shapes of the waves in the jet stream, change.

Mid-Latitude Cyclones

Huge swirling currents of air occasionally develop at latitudes of between 30° and 60°, north or south of the equator, when tropical air masses and polar air masses (see Fig. 18.20) interact. These currents, known as *cyclones* if they rotate counterclockwise (in the northern hemisphere) and *anticyclones* if they rotate clockwise, play a major role in controlling the weather across the United States, Europe, and other regions occupying the Earth's mid-latitudes. In some cases, cyclones grow into huge storms that span areas as large as half the width of North America.

Figure 18.23 The polar-front jet stream.

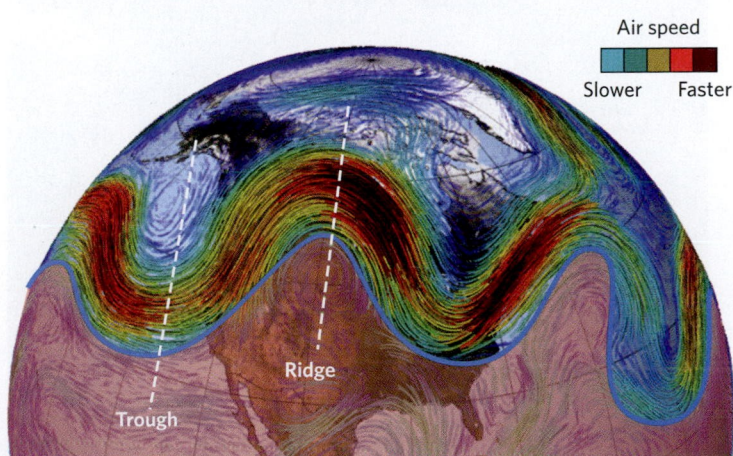

(a) The jet stream takes on a wavy pattern as it circles the globe. It is typically located over fronts with a steep temperature gradient.

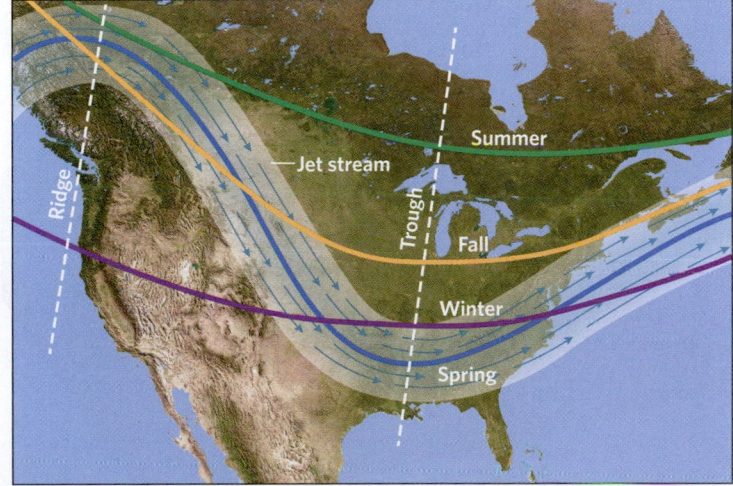

(b) The jet stream's position changes with time. Over North America, it tends to shift north in summer and south in winter.

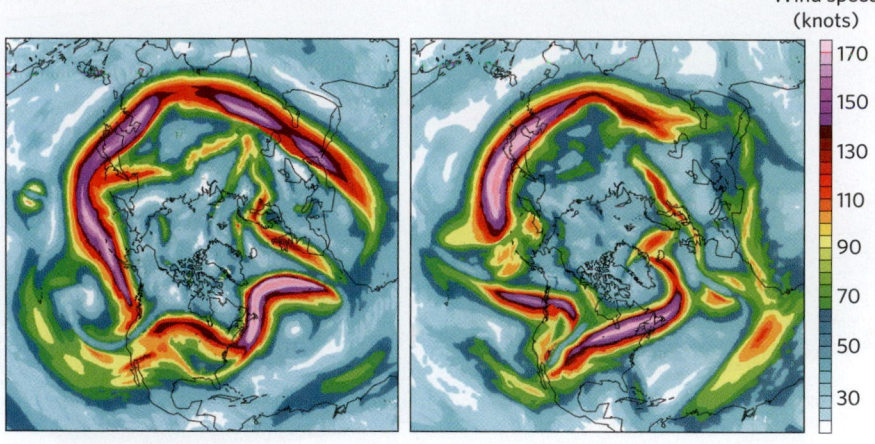

Wind speed (knots)

170
150
130
110
90
70
50
30

These storms, whose cloud cover resembles a huge comma when viewed from a satellite **(Fig. 18.26)**, are called **mid-latitude cyclones**. Because mid-latitude cyclones impact weather across the large population centers and agricultural areas of the world, it's important to understand how they form and how they evolve.

DEVELOPING HIGH- AND LOW-PRESSURE CENTERS. In the previous section, we learned that heating and cooling of the Earth's atmosphere in specific locations leads to the formation of low-pressure and high-pressure zones at the Earth's surface. Cooling plays a particularly important role

in causing high-pressure zones to form in the middle and high latitudes because, during the cold season, these latitudes have long nights when no solar radiation heats the ground. A **high-pressure center** coincides with the region of strongest cooling.

The flow in the jetstream can also contribute to the development of high surface pressure. To see why, recall that the polar-front jet stream flows at the top of the troposphere (see Fig. 18.22), and that the velocity of air in this jet stream varies with location (see Fig. 18.23a). At a locality where air flow in the jet stream slows down, air undergoes **convergence**, meaning that horizontal flow carries more air into the locality (in this case, a segment of the jet stream) than gets carried away. This extra air at the top of the troposphere causes the mass of the underlying column of air to increase, so that a high-pressure center develops at the base of the troposphere **(Fig. 18.27a)**. Air sinks down from the convergence zone aloft toward the high-pressure center at the Earth's surface below.

We also saw in the previous section that heating by the Sun plays an important role in the development of low-pressure zones in the tropics. Solar heating is less significant, however, in the development of low-pressure centers associated with major storms in mid-latitudes. To understand why low-pressure zones form in the mid-latitudes, we must look again at the behavior of the polar-front jet stream. Within the jet stream, where air flow speeds up, **divergence** takes place. This means that horizontal flow carries more air out of a locality (in this case, a segment of the jet stream) than had entered that locality **(Fig. 18.27b)**. The resulting air deficit at the top of

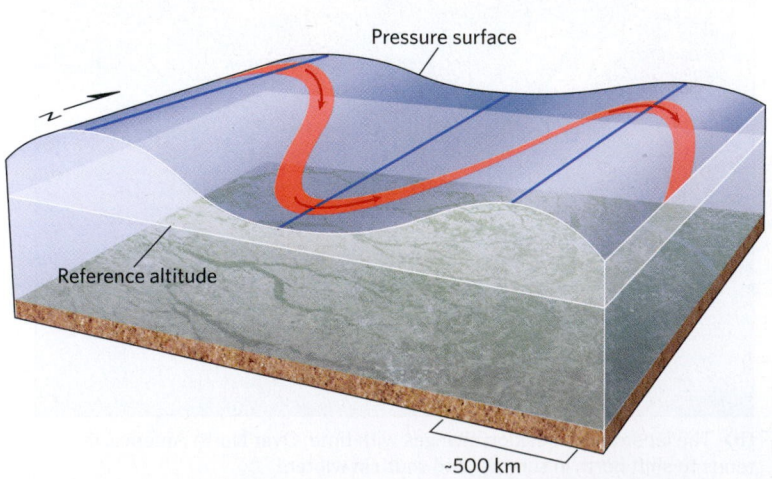

Pressure surface

N

Reference altitude

~500 km

Figure 18.27 The relationship between convergence and divergence in the jet stream, and the development of high- and low-pressure centers at the surface.

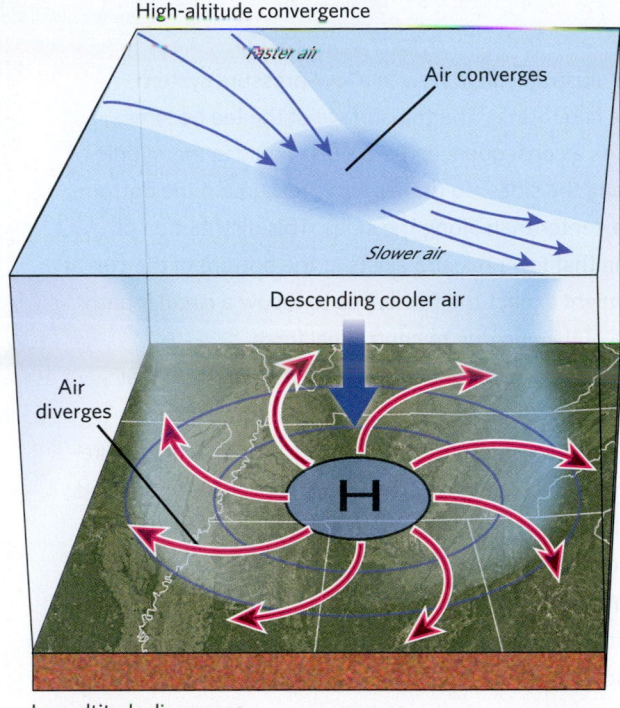

High-altitude convergence

Faster air

Air converges

Slower air

Descending cooler air

Air diverges

H

Low-altitude divergence

(a) Convergence aloft generates a surface high-pressure center.

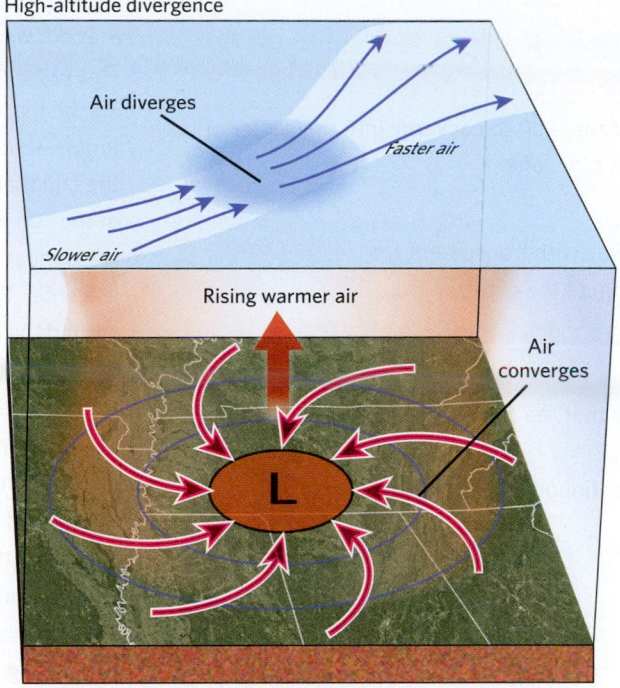

High-altitude divergence

Air diverges

Faster air

Slower air

Rising warmer air

Air converges

L

Low-altitude convergence

(b) Divergence aloft generates a surface low-pressure center.

the troposphere causes the mass of the underlying column of air to decrease, and as a result, air pressure decreases at the base of the troposphere below. The position of the lowest pressure defines a **low-pressure center**. To replace the deficit of air at jet-stream elevation, air from lower in the troposphere must rise in and around the low-pressure center.

At this point, you may be wondering why convergence and divergence take place in the polar jet stream. There are two reasons. First, changes in jet-stream velocity develop in association with curvature that exist where the jet stream flows around ridges and troughs (see Fig. 18.25). When the direction of the jet stream changes, the relative strengths of the Coriolis force and the pressure-gradient force change. This change affects the speed of the flow, to cause convergence or divergence. Atmospheric scientists have determined that the strongest divergence due to curvature develops along the east side of a trough. Second, divergence and convergence develop in association with **jet streaks**, portions of the jet stream where air flows especially fast (see Fig. 18.23a). Jet streaks form because of variations in air temperature that, in turn, cause locally strong pressure gradients to move along the jet stream over time. Divergence develops on the poleward side of the downstream end of a jet streak. When a particularly fast jet

streak is located in the base of a trough, its presence triggers the development of strong divergence on the east side of the trough just north of the wind speed maximum. This jet-streak-related divergence, combined with curvature-related divergence, results in such a deficit of air at the top of the troposphere that a particularly strong low-pressure system occurs at the troposphere's base on the east side of a trough.

Development of high-pressure and low-pressure centers at the base of the troposphere significantly affects the pattern of air flow near the Earth's surface. Specifically, when a high-pressure center is developing at the troposphere's base, it generates a pressure-gradient force that causes air parcels to flow outward from the center into surrounding regions. Recall that when air parcels start to move horizontally on the spinning Earth, the Coriolis force deflects them to the right (in the northern hemisphere; see Chapter 15). If the pressure-gradient force and the Coriolis force were the only forces affecting air, air parcels would circle the high-pressure center in a clockwise direction. But because friction slows the air, and speed determines the magnitude of the Coriolis force, the pressure-gradient force exceeds the Coriolis force, so that air parcels spiral outward from the high-pressure center. This outward-spiraling clockwise flow generates *divergence* at

Flow in a cyclone

What are we learning?

How friction near the ground causes air to converge into the center of a low-pressure system such as a cyclone.

What you need:

- A cup filled about ¾ of the way with transparent tea.
- Loose tea leaves in the bottom of the cup.
- A spoon to stir the tea.

Instructions:

- Let the liquid come to rest so that tea leaves are spread out and resting at the bottom of the cup.
- Stir the tea vigorously in a continuous direction so that the fluid rotates in the cup.
- Observe the tea leaves.

What did we see?

- This experiment illustrates fluid flow in a low-pressure system. As the tea in the cup rotates, the top surface of the tea becomes funnel-shaped. As a consequence, there's less liquid in the middle of the cup than along the sides, so there's less pressure on the bottom of the cup in the center than along the sides. This distribution of mass must mean that low pressure exists at the bottom of the cup at its center. One might expect the tea leaves to follow a circular path around the edge of the cup due to centrifugal force.

- Meanwhile, however, the fluid experiences a pressure gradient that should drive it toward the low pressure at the center of the cup. If a balance were to be established between centrifugal force and pressure-gradient force, the tea leaves would follow a circular path as you stirred. But toward the bottom of the cup, friction between the tea and the cup slows the flow, so the inward-directed pressure-gradient force (toward the center) exceeds the centrifugal acceleration of the spinning fluid, and the leaves spiral toward the center of the cup. The converging liquid then rises in the middle of the cup. A similar process occurs in low-pressure centers in the atmosphere, such as those at the center of both tropical and mid-latitude cyclones. Air near the Earth's surface converges toward the low-pressure center, and as the air approaches the center, it rises, causing clouds and precipitation.

---

the base of the troposphere (see Fig. 18.27a). Meteorologists refer to a high-pressure center, together with the associated three-dimensional pattern of air circulation, as a **high-pressure system**. The downward vertical air movement that takes place in high-pressure systems affects the relative humidity and, therefore, the cloudiness in the air column. When the air within a high-pressure system sinks and spirals downward, it undergoes compression and warms so its relative humidity decreases. As a result, high-pressure systems tend to host relatively dry, cloud-free air—in other words, they produce *fair weather*.

The development of a low-pressure center at the base of the troposphere also leads to the formation of a pressure gradient. Alone, this gradient would cause air parcels from surrounding regions to flow in toward the center. However, the Coriolis force deflects moving parcels to the right (in the northern hemisphere). If only the pressure-gradient and Coriolis forces acted on the moving air, the net result would be that parcels would circle the low-pressure center in a counterclockwise direction. However, because

friction slows the air, the pressure-gradient force exceeds the Coriolis force, so parcels spiral inward toward the low-pressure center (see Fig. 18.27a). This inward-spiraling, counterclockwise flow causes *convergence* at the base of the troposphere (**Box 18.3**; see Fig. 18.27b). The low-pressure center, together with the fronts, clouds, and air circulations within it, is called a **low-pressure system**. The lifting of air along fronts within these low-pressure systems is what produces the huge comma-shaped storms known as *mid-latitude cyclones*. Meteorologists add the prefix "mid-latitude" to these cyclones to distinguish them from tropical cyclones (hurricanes and typhoons) that we discussed in Chapter 17. Mid-latitude cyclones are also known as *extratropical cyclones*—in this context, the prefix "extra" emphasizes that they form outside of the tropics.

LIFE CYCLE OF A MID-LATITUDE CYCLONE. We now focus more closely on mid-latitude cyclones, because they produce much of the severe weather that greatly impacts people who live in their path. As we've seen, a mid-latitude cyclone initiates as a strong low-pressure center develops. During the development of the cyclone, a warm front typically extends eastward of the system's center, while a cold front extends southwestward of the center (Fig. 18.28). Initially, warm air ahead of the cold front undergoes lifting as the cold air advances (Fig. 18.29a). Meanwhile warm air south of the warm front flows up and over the warm front. As low-altitude air starts to spiral counterclockwise around the center of the cyclone, the cold front typically advances rapidly eastward (Fig. 18.29b). Meanwhile, northeast of the cyclone's center, the warm front typically migrates slowly northward. In many cyclones, the cold front eventually catches up with and wraps into the warm front (Fig. 18.29c). When this happens, the cool air north of the warm front undergoes lifting along the face of the cold front, yielding an occluded front (Fig. 18.29d). Once the occluded front forms, cold air from the south side of the cold front comes in contact with cold air from the north side of the warm front, and the warm air that once lay to the south of the warm front becomes a layer that overlies both cold air masses. The clouds within this layer define the comma head of the storm.

A typical mid-latitude cyclone develops and intensifies rapidly, generally reaching its maximum intensity (the lowest air pressure at the low-pressure center) within 36 to 48 hours of initiation. The storm can remain at maximum intensity for another day or two, but then it begins to weaken. Weakening, which may take about a week,

Figure 18.28 The relationship between the jet stream and the formation of a mid-latitude cyclone. Low pressure develops at the Earth's surface below a zone of maximum divergence, which lies to the east of a trough and just north of the strongest winds in the jet stream.

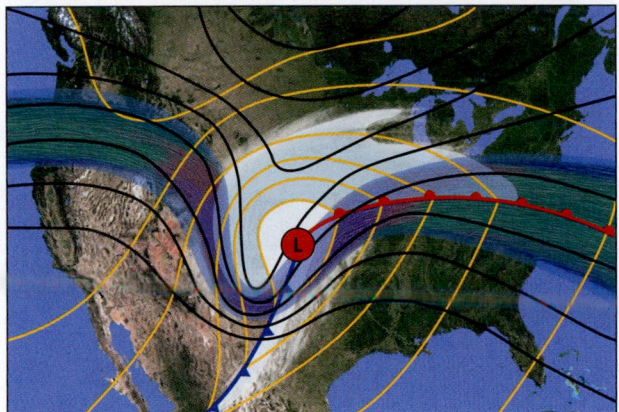

Upper-atmosphere isobars ——————
Air pressure at sea level ——————
Low pressure center   Ⓛ

happens when low-altitude convergence feeds more air into the air column between the ground and the jet stream than gets removed by jet-stream divergence. The addition of air to the column causes the air pressure to rise at the center of the cyclone and, therefore, causes the surrounding

| Box 18.4 | ## Consider this . . . |

How does a mid-latitude cyclone differ from a hurricane?

Like a tropical cyclone (a hurricane), a mid-latitude cyclone rotates about a low-pressure center. But aside from this basic shared characteristic, mid-latitude cyclones and hurricanes have very little in common.

- Mid-latitude cyclones form in association with jet streams, where winds are much stronger aloft than they are at lower elevations. Tropical cyclones form where winds in the atmosphere are uniform with altitude and generally weak.
- Mid-latitude cyclones cover a much larger area than tropical cyclones, and they involve many air masses, whereas tropical cyclones form within a single air mass.

- Tropical cyclones do not include fronts, since they occur entirely within warm regions.
- Severe weather in mid-latitude cyclones is concentrated along fronts, and it can include tornado-producing thunderstorms, blizzards, and ice storms. Hazardous winds in a tropical cyclone occur beneath the eye wall, and these storms bring high winds and torrential rains.
- The clouds of mid-latitude cyclones resemble a comma when viewed from space (see Figs. 18.26 and 18.30b). The clouds of tropical cyclones are more circular in shape, with spiral bands radiating from the center (see Fig. 18.15). Tropical cyclones often have a clear central eye, a feature not found in mid-latitude cyclones.

Figure 18.29 Evolution of a mid-latitude cyclone.

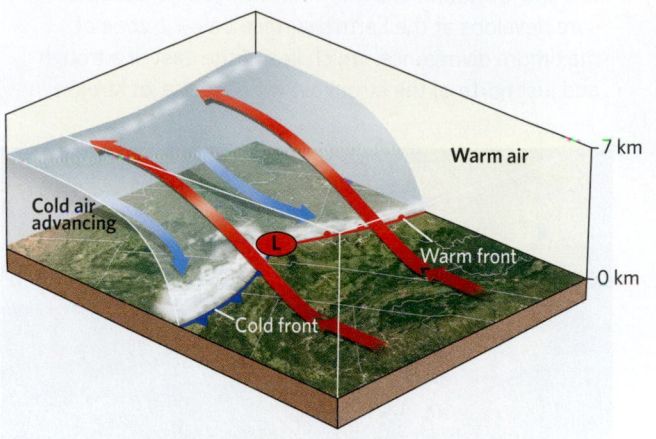

(a) Air initially flows upward over the frontal boundaries.

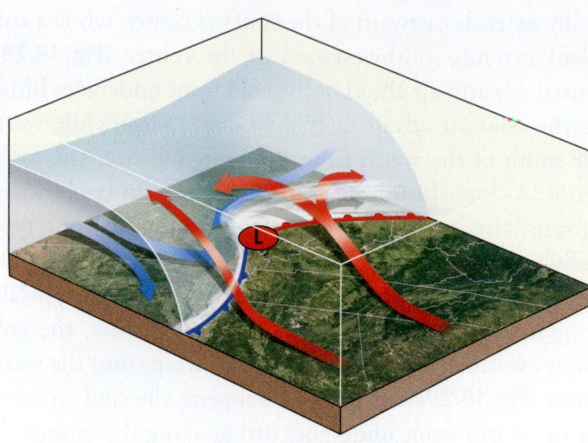

(b) The cold front advances southward and eastward around the center of low pressure.

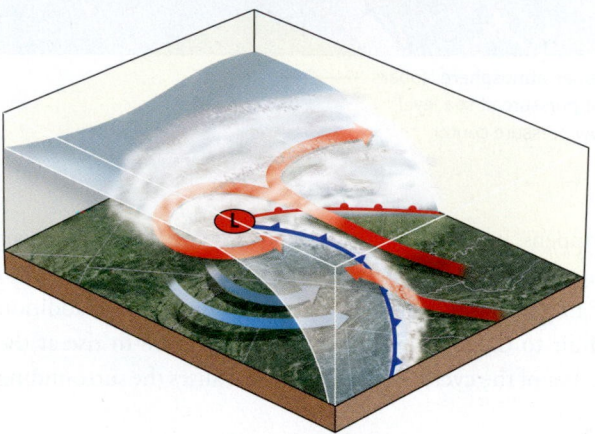

(c) Eventually, the cold front reaches the warm front.

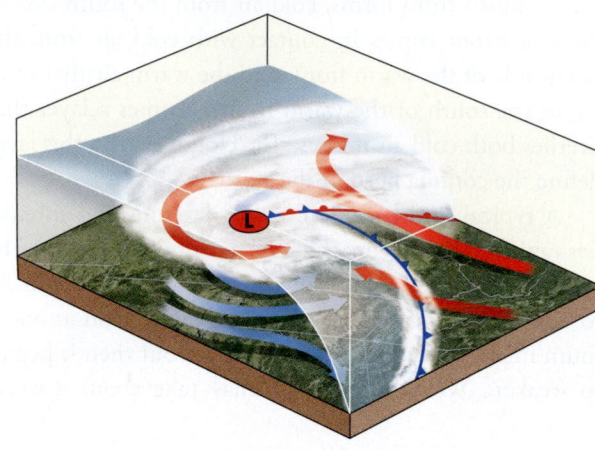

(d) The cold front overtakes the warm front, creating an occluded front.

pressure gradient to diminish. When the low-pressure center weakens sufficiently, the storm dissipates. Overall, a mid-latitude cyclone can survive as a distinct circulation pattern for up to about two weeks, during which it drifts east with the overall movement of the trough with which it is associated. As a consequence, a single cyclone can cross a large part of a continent or ocean. Note that the origin, structure, and evolution of a mid-latitude cyclone differs substantially from that of a hurricane **(Box 18.4)**.

WEATHER IN MID-LATITUDE CYCLONES. The strong pressure gradient that develops in a mid-latitude cyclone generates strong winds which, at ground level (in the northern hemisphere), flow counterclockwise within the cyclone's overall spiral flow **(Fig. 18.30a)**. As we've seen, convergence at the center of the low-pressure system associated with a cyclone, as well as lifting along the system's fronts, causes low-elevation moist air to rise, producing broad areas of cloud cover and precipitation in the shape of a comma **(Fig. 18.30b)**. The "tail" of the comma lies over the cold front and typically consists of a line of precipitating clouds

or, in some cases, a line of thunderstorms **(Fig. 18.30c)**. The "head" of the comma lies over the warm front and the center of the low-pressure zone. It also produces large amounts of precipitation. In warmer weather, precipitation falls as rain, but when it's cold enough, it falls as snow **(Fig. 18.30d)**. Precipitation in mid-latitude cyclones located over the interior plains of the United States becomes especially heavy when warm, moisture-laden air heads north from the Gulf of Mexico and moves along the east side of the cold front (see Fig. 18.29b). During the fall, winter, and spring, when temperature differences between high latitudes and midlatitudes can be large, rain may be falling along the comma's tail, while snow falls around the comma's head.

The strong winds of a mid-latitude cyclone—especially under clouds at the comma head—can produce blizzard conditions if precipitation falls as snow. A **blizzard** is a snowstorm in which winds exceed 56 km per hour (35 mph) for a period of at least three hours. During a blizzard, blowing snow decreases visibility to less than 0.40 km (0.25 miles), and sometimes may

Figure 18.30 Features of a large mid-latitude cyclone.

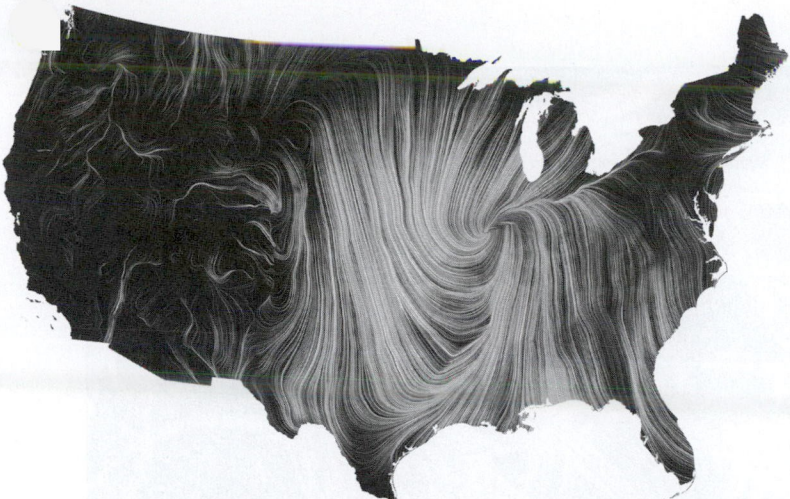

(a) A wind map shows the circulation of air in a mid-latitude cyclone.

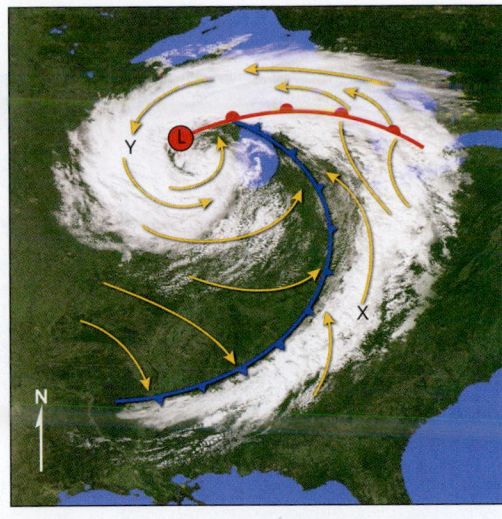

Comma

(b) A mature mid-latitude cyclone can bring different types of hazardous weather over broad geographic regions.

(c) A thunderstorm occurs at point X, shown on the map in part (b).

(d) A blizzard occurs at point Y.

even lead to *whiteout conditions*, when visibility decreases to nearly 0 km **(Fig. 18.31)**. Heavy snowfall in a blizzard can accumulate quickly, and snow blown by the wind—either during or after a snowfall—can build high *snowdrifts* (dunes made of snow). The combination of heavy snowfall, drifting, and decreased visibility can force airports to close and can trap cars on highways. The extreme winds in a blizzard can also cause people to suffer from frostbite or hypothermia, for the heat loss that people feel when it's windy is greater than at the actual air temperature in still air. For example, an air temperature of −7°C (20°F) feels like −16°C (−4°F) when the wind blows at 32 km per hour (20 mph). This *wind chill* happens because flowing cold air removes heat more efficiently from the human body than does calm air. Weather reports in the media often specify the **wind chill temperature**, the temperature that you feel when you go outside, so that you can be prepared.

Ice storms, which result from the accumulation of freezing rain, can also occur in association with a mid-latitude

Figure 18.31 During a blizzard, visibility falls to near zero.

Figure 18.32 Ice storms are another winter hazard associated with mid-latitude cyclones.

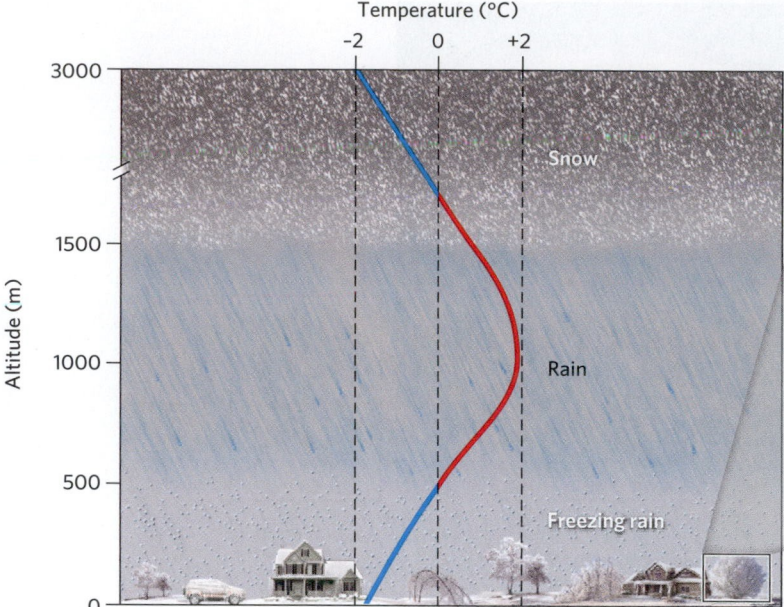

(b) The weight of accumulated ice can cause trees and power lines to collapse.

(a) When surface temperatures are below freezing, but a layer of air aloft is above freezing, snow from high in the clouds can melt into rain and then freeze when it hits the ground.

Figure 18.33 A nor'easter moves up the east coast of North America.

cyclone. *Freezing rain* forms when snow falls from high in the clouds, down into a lower atmospheric layer (typically 0.5 to 1.5 km, or .3 to 1 mile, above the ground) where the temperature exceeds 0°C (32°F) **(Fig. 18.32a)**. Within this layer, the snowflakes melt and continue their fall as raindrops. If the temperature in the layer of air below an elevation of about 0.5 km remains below freezing, the raindrops refreeze when they reach the ground. Freezing rain produces an ice glaze that can become so heavy that it causes trees and power lines to collapse **(Fig. 18.32b)**.

Mid-latitude cyclones formed along coasts have different local names. For example, a mid-latitude cyclone that develops over the northeast Pacific Ocean, off the west coast of North America, is called a *Pacific cyclone*. When such storms, reach the mountain ranges of the west coast, the moisture-laden air that they carry flows up the face of the mountains. This lifting triggers precipitation in the form of torrential rains or heavy snows, depending on the air temperature, over the mountains. A mid-latitude cyclone formed along the east coast of North America is called a *nor'easter*, because the storm's strong winds during the period of heavy snowfall come from the northeast **(Fig. 18.33)**. Some of the worst blizzards in New England have developed in association with nor'easters.

REDISTRIBUTION OF HEAT BY MID-LATITUDE CYCLONES. Let's step back and look at the role that a mid-latitude cyclone plays in redistributing heat in the atmosphere **(Fig. 18.34)**. Prior to the formation of a mid-latitude cyclone, warm, humid air lies near the Earth's surface at subtropical latitudes, while cold air lies near the surface in high latitudes. In addition, cold air occurs at high

Figure 18.34 A mid-latitude cyclone acts like a complicated series of conveyor belts, bringing warm air northward, cold air southward, high-altitude air downward, and low-altitude air upward as the fronts evolve.

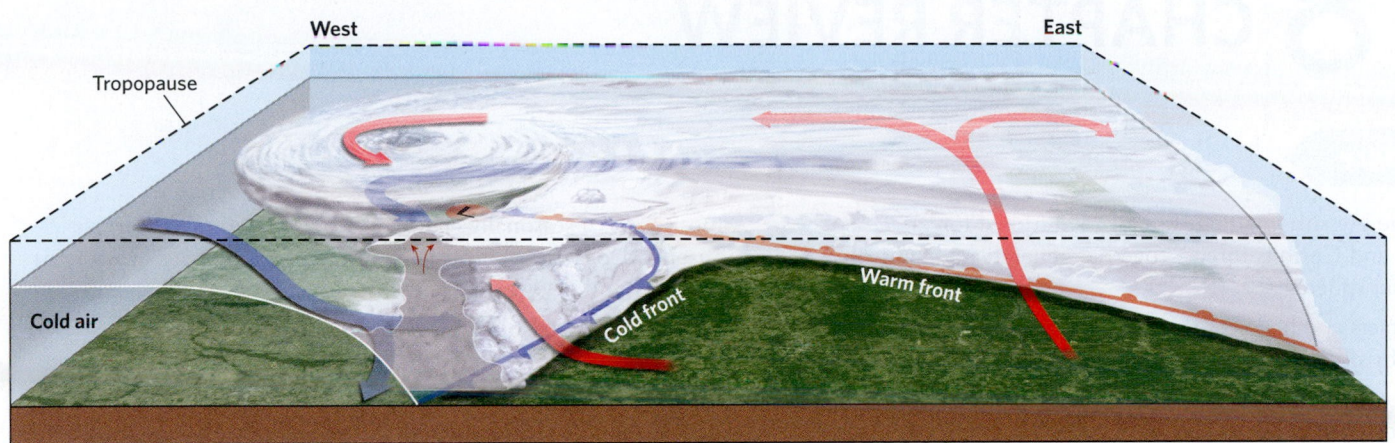

elevations in the troposphere, everywhere. A mid-latitude cyclone, as it forms and evolves, acts like a series of conveyor belts, transporting warm air northward as well as upward to high elevations, while it brings the cold air southward as well as downward to lower elevations. This redistribution reduces the temperature difference between polar and subtropical regions, as well as between the lower and upper troposphere. In this manner, each cyclone contributes to a never-ending process of rebalancing heat distribution in an atmosphere that continually becomes unbalanced due to uneven solar heating. The precipitation associated with mid-latitude cyclones provides water needed for life and benefits agriculture. But the thunderstorms and blizzards associated with these cyclones can be a major hazard.

We've now discussed global-scale and regional-scale air movements, some of which can become quite ferocious. Smaller, local thunderstorms, and associated tornadoes, can form in association with hurricanes and mid-latitude cyclones, as well as for other reasons. In the next chapter, we turn our attention to thunderstorms, tornadoes, and other local weather conditions, and the specific hazards that they produce.

Take-home message . . .

Mid-latitude cyclones are large, counterclockwise-rotating systems that can result in hazardous weather across the mid-latitudes. They form along the polar jet stream in association with low-pressure zones and generally move eastward. The most dangerous weather occurs along fronts, which move as the result of air flow within a cyclone.

Quick Question -
Why does the weather in the mid-latitudes change so frequently?

ANOTHER VIEW This satellite view of the northeastern Pacific, depicting the distribution of moisture in the toposphere, reveals streams of airflow, locally wrapping into swirling cyclones.

Dry   Moist

18 CHAPTER REVIEW

Chapter Summary

- Four forces—the pressure-gradient force, gravitational force, frictional force, and Coriolis force—together generate and control winds in the atmosphere.

- Gravity counteracts and balances the upward pressure-gradient force, so air does not fly off into space or fall to the ground.

- The frictional force, which always acts to slow the winds, is most important in the lower atmosphere, a region called the boundary layer.

- Above the boundary layer, wind speed and direction are largely controlled by the balance between the pressure-gradient and Coriolis forces. Where this balance has been attained, winds are geostrophic and blow parallel to isobars.

- Hadley cells consist of air that rises near the equator, then flows poleward, then descends from the upper troposphere of each hemisphere, and finally returns toward the equator above the Earth's surface. Rain from the Hadley cells strongly influences the distribution of tropical rainforests and deserts, and the air flow near the surface causes the trade winds.

- The Walker circulation is an east-west-flowing convective cell covering the equatorial Pacific. An El Niño event occurs when the Walker circulation weakens, and a La Niña event occurs when it strengthens. The resulting pattern of changes in surface air pressure, called the Southern Oscillation, can affect weather around the world.

- A monsoon is a seasonally changing air circulation. The largest monsoon is located over southern Asia, where it brings heavy rains in summer and dry weather in winter.

- Tropical cyclones, also known as hurricanes or typhoons, form over tropical oceans and are the most destructive storms on the planet. They have distinctive features, including a nearly cloud-free eye, a violently rotating eye wall, and spiral rainbands. Hurricane destruction results from high winds, storm surge, and torrential rains.

- An air mass is a large body of air with relatively uniform temperature and humidity characteristics that covers a large region of the Earth. The boundaries between air masses are called fronts. We distinguish among cold fronts, warm fronts, and stationary fronts by the direction in which the colder air mass moves.

- Jet streams, narrow bands of strong winds that encircle the Earth in the upper troposphere, form at subtropical and mid-latitudes.

- Mid-latitude cyclones develop in the mid-latitudes when a large low-pressure system develops at ground level beneath a zone of divergence in the jet stream.

- From above, the clouds of a mid-latitude cyclone have the shape of a comma. The comma tail follows a cold front, and along it, thunderstorms often form. The comma head overlies a warm front, where rain, snow, or freezing rain may fall. Blizzards and ice storms can occur under the clouds of the comma head.

Key Terms

air mass (p. 643)
blizzard (p. 652)
boundary layer (p. 626)
cold front (p. 645)
convective cell (p. 629)
convergence (p. 648)
cyclone (p. 638)
divergence (p. 648)
doldrums (p. 630)
El Niño (p. 636)
ENSO (p. 636)
eye (p. 638)
eye wall (p. 638)
front (p. 645)

general circulation (p. 624)
geostrophic balance (p. 627)
geostrophic wind (p. 627)
Hadley cell (p. 630)
high-pressure system (p. 645, 650)
high-pressure center (p. 648)
hurricane (p. 638)
hurricane track (p. 641)
ice stork (p. 653)
intertropical convergence zone (ITCZ) (p. 630)
isobar (p. 625)
jet streak (p. 649)

low-pressure center (p. 649)
low-pressure system (p. 645, 650)
mid-latitude cyclone (p. 648)
monsoon (p. 636)
La Niña (p. 636)
occluded front (646)
polar-front (p. 646)
polar-front jet stream (p. 646)
pressure gradient (p. 625)
ridge (p. 647)
Southern Oscillation (p. 636)
stationary front (p. 646)
storm (p. 624)

storm surge (p. 641)
subtropical jet streams (p. 631)
thunderstorms (p. 630)
trade winds (p. 629)
tropical cyclone (p. 638)
tropical storm (p. 638)
tropics (p. 629)
trough (p. 647)
typhoon (p. 638)
Walker circulation (p. 631)
warm front (p. 645)
wind (p. 624)
wind-chill temperature (p. 653)

Review Questions

The letters following each Review Question refer to the corresponding Learning Objective from the Chapter Opener.

1. What forces act on air and cause it to flow? Which of these forces can cause the wind to increase in speed? Which balances gravity and keeps air from falling to the ground? **(A)**

2. Identify the Hadley cells on the diagram. Explain their relationship to the locations of the world's tropical rainforests and subtropical deserts. **(B)**

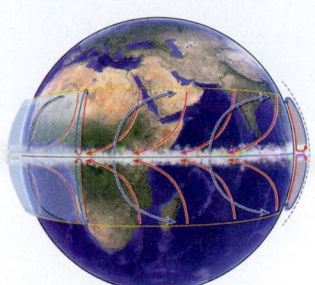

3. The sea surface off the shores of Peru is warmer than normal during an El Niño, and colder than normal during a La Niña. Relate these sea-surface temperature changes to the flow in the Walker circulation and the Southern Oscillation. **(C)**

4. What causes monsoons, and why do they occur only at relatively low latitudes? **(C)**

5. Explain how a hurricane forms and how it gets its energy. **(D)**

6. Identify the key structural features of a hurricane. Where are its strongest winds found? What causes destruction when a hurricane strikes a coastline? **(D)**

7. What is an air mass, and what are the major sources of air masses for North America? What is a boundary between two air masses called, and how can we distinguish among different types of boundaries? **(F)**

8. Explain why jet streams form and describe where they occur. **(F)**

9. What is a mid-latitude cyclone, why does it form, and how is it related to the polar-front jet stream? What kinds of weather form in association with mid-latitude cyclones? **(E)**

10. Aside from where they form, describe three fundamental differences between tropical and mid-latitude cyclones. **(E)**

11. What is the shape of a mid-latitude cyclone? Which types of storms typically form along different parts of the cyclone? **(E)**

On Further Thought

12. The summer/winter transition in weather in the American southwest and northwestern Mexico has been described as the "North American monsoon." Based on your understanding of monsoons, can you propose reasons why meteorologists might use this terminology? **(C)**

13. Airplanes flying from Seattle, on the west coast of the United States, to Boston, on the east coast, take about 5 hours to make the trip in winter. When they return to Seattle, they take about

7 hours. Explain why these times are so different. Will the difference change with the season? **(F)**

14. Outside, it is cool and rainy, and the wind is from the east. Two hours later, the rain has ended, the temperature has increased, and the air now feels humid. The wind is from the south. What type of front passed your location? **(F)**

Online Resources

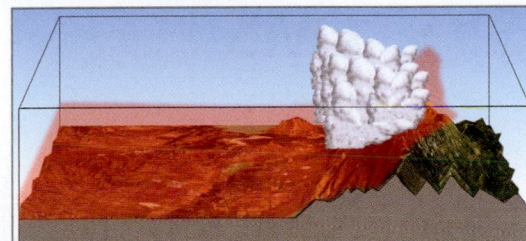

Animations
This chapter features animations on how fronts form and operate.

Videos
This chapter features real-world videos on Hurricanes Katrina and Sandy, El Niño and La Niña, tornado and superstorm tracking, and monsoons.

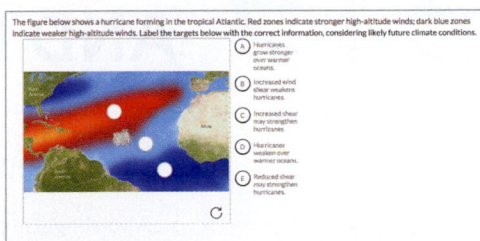

Smartwork5
This chapter features questions on atmospheric forces, climate and weather, ENSO events, hurricanes and cyclones, fronts, and jet streams.

19 THUNDERSTORMS, TORNADOES, AND LOCAL WEATHER SYSTEMS

By the end of the chapter you should be able to . . .

A. distinguish between local-scale and larger weather systems.

B. explain the basic conditions that lead to the formation of a thunderstorm.

C. describe the difference between an ordinary thunderstorm and other types of thunderstorms.

D. interpret the signature of a severe thunderstorm as it appears on radar.

E. differentiate between severe weather watches and warnings.

F. know how to protect yourself against thunderstorm hazards.

G. describe how tornadoes form, how they are classified, and how they cause destruction.

H. discuss how local weather systems develop near land-water boundaries and in association with mountain ranges.

19.1 Introduction

In the late morning of Tuesday, July 13, 2004, a large thunderstorm developed over Illinois. Meteorologists at the National Oceanic and Atmospheric Administration (NOAA) Storm Prediction Center worried that this storm could eventually produce a tornado, so they issued a *tornado watch* for the region, a formal statement that conditions are favorable for tornado development. Over the next few hours, even stronger thunderstorms developed, and at 2:29 P.M., the National Weather Service (NWS) issued a *severe thunderstorm warning*. This warning means that a storm has been observed that is capable of producing dangerous strong winds, as well as lightning, hail, and possibly a tornado.

At the Parsons manufacturing plant in Roanoke, Illinois, the NWS warning interrupted normal broadcasting on the radio in the front office, causing the factory's designated emergency response team leader to look outside. What he saw caused him to gasp. Not only was a towering dark cloud, backlit by lightning, heading straight toward the factory, as the NWS had predicted, but a funnel-shaped cloud extended beneath it. He grabbed a microphone and bellowed the words no one wanted to hear: "This is not a drill. Move immediately to designated tornado shelters . . . now!" Thanks to a training exercise three months earlier, everyone knew exactly where to go and moved quickly.

The plant's owner had witnessed terrible tornado damage years earlier, so when constructing the plant, he had insisted that the restrooms have walls and ceilings of 20-cm (8-inch)-thick reinforced concrete so they could serve as tornado shelters. All 140 employees in the building were

Figure 19.1 The Parsons plant was destroyed by a tornado in July 2004. The inset shows the plant before the storm.

already on their way into these shelters when the NWS broadcast a *tornado warning*, meaning that a tornado had been sighted. This fearsome funnel of destruction, with winds blasting to nearly 290 km per hour (180 mph), was less than a mile away when the employees bolted the shelter doors. Two minutes later, there was a deafening roar, and the tornado struck. Cars in the parking lot tumbled like toys, the roof detached and shredded into pieces, and the walls of the plant collapsed. For a terrifying minute, winds howled and debris clattered and crashed. Then, it stopped—the tornado had passed. The employees emerged from the shelters into a chaotic jumble of twisted beams, smashed machines, and crushed vehicles (Fig. 19.1), yet not one person was seriously injured. Warnings and preparations had saved everyone's lives.

Figure 19.2 Lightning flashes during an intense thunderstorm in Nebraska.

Every year, we hear of devastating thunderstorm and tornado damage, often with results much deadlier than what we've just described. Thunderstorms and tornadoes are examples of **local weather systems**, technically known as *mesoscale weather systems*. These names refer to atmospheric phenomena that affect a region between a few kilometers and several hundred kilometers across—much smaller than the area affected by a hurricane or a mid-latitude cyclone. Local weather systems include tornadoes, thunderstorms, lake-effect storms, coastal sea breezes, and intense mountain windstorms. Some local weather systems take place in isolation, whereas others are part of much larger weather systems; for example, several thunderstorms may align along the front in a mid-latitude cyclone. In this chapter, we focus on local weather systems and discuss how they form and evolve. We also discuss the nature of the hazards—lightning, hail, downpours, and wind—associated with these systems and how to avoid being injured by them. We begin with thunderstorms, the most common type of local weather system.

19.2 Conditions That Produce Thunderstorms

What Is a Thunderstorm?

Imagine sitting outside on a warm, calm summer day. Suddenly, thunder rumbles in the distance. Soon dark, turbulent clouds hide the Sun. Then gusts of cool wind begin to blow, followed by drenching rain, stronger winds, and possibly hail. As the sky crackles with lightning flashes, you hear the booms of thunder. A thunderstorm has arrived (Fig. 19.2).

Formally defined, a **thunderstorm** consists of towering cumulonimbus clouds that produce lightning and thunder. These clouds typically rise from an elevation of less than 2,000 m (6,000 feet) all the way up to the tropopause, at 12 to 18 km (7.5 to 11 miles), depending

Figure 19.3 The global distribution of thunderstorms, as estimated from lightning flashes. Note that thunderstorms concentrate in specific regions.

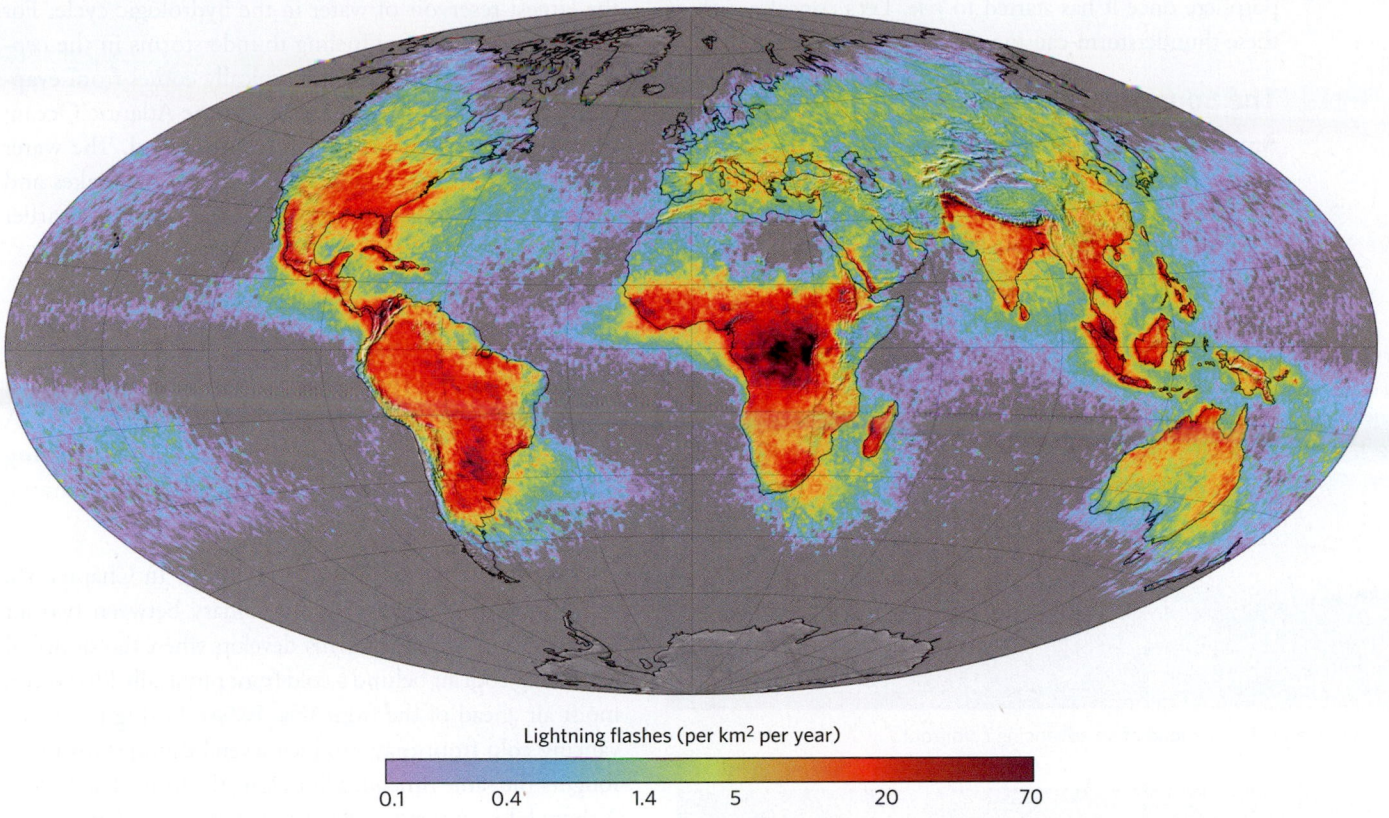

Lightning flashes (per km² per year)

0.1   0.4   1.4   5   20   70

on latitude. In general, thunderstorms are less than 30 km (18 miles) in diameter, and they move across the ground at rates ranging from near stationary to 80 km per hour (0 to 50 mph). A single thunderstorm may pass in less than half an hour, but commonly, a region may experience a succession of thunderstorms that together last for a few hours or more. Thunderstorms can happen during the day or at night. During daytime thunderstorms, thick clouds may block so much light that the sky becomes quite dark. All thunderstorms pose danger because of lightning, but *severe thunderstorms* are particularly hazardous. A **severe thunderstorm** is one that produces hail greater than 2.5 cm (1 inch) in diameter, winds exceeding 50 knots (92 km per hour, or 56 mph), or a tornado.

Thunderstorms take place frequently in some locations on Earth, but rarely or never in other regions **(Fig. 19.3)**. Most develop over the tropics, where they provide the rain that sustains rainforests. During the warm season, they occur fairly frequently in mid-latitudes. They rarely take place in high latitudes, in subtropical deserts, or over large parts of the world's oceans. In the United States, thunderstorms primarily affect regions east of the Rocky Mountains, and they are particularly frequent over the southeastern states. In fact, in the eastern two-thirds of the United States, a given location will experience 30 to 50 *thunderstorm days* (days on which at least one thunderstorm occurs) per year **(Fig. 19.4)**. Note that during a thunderstorm day, only an hour or two

of the day actually hosts a thunderstorm at a particular location. Florida holds the record for thunderstorm days: parts of the state endure 65 to 80 thunderstorm days per year.

Why don't thunderstorms occur everywhere all the time? They form only when these special conditions exist: (1) the lower atmosphere must hold *moist air*; (2) a *lifting mechanism* must be present to initiate an **updraft** (upward-moving air flow, which can transport the moist air

> **Did you ever wonder . . .**
> why thunderstorms don't occur every day?

Figure 19.4 Numbers of thunderstorm days per year in the United States.

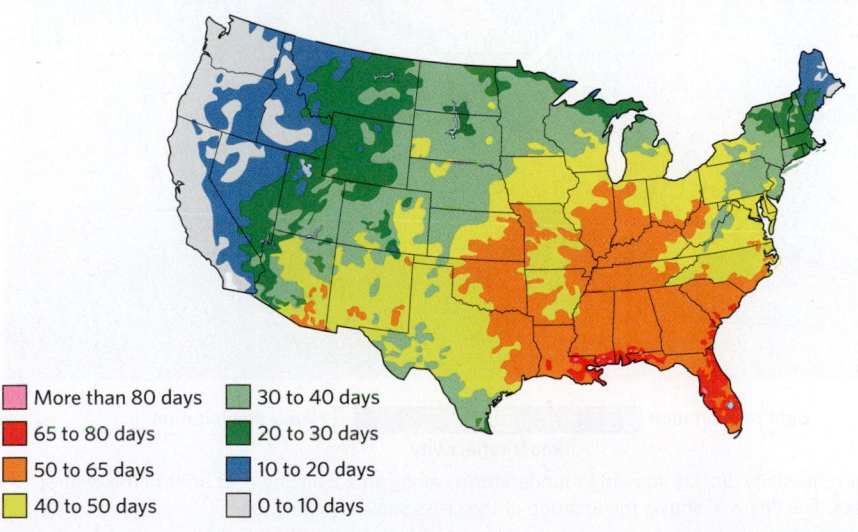

▦ More than 80 days	▦ 30 to 40 days
▦ 65 to 80 days	▦ 20 to 30 days
▦ 50 to 65 days	▦ 10 to 20 days
▦ 40 to 50 days	▦ 0 to 10 days

to a higher elevation); and (3) *atmospheric instability* must allow the air to keep rising buoyantly to the top of the troposphere once it has started to rise. Let's consider each of these thunderstorm-causing conditions in turn.

The Source of Moist Air

Amazingly, a typical thunderstorm contains over 3.7 million tons of water at any given time and, over its lifetime, processes many times that amount. So, air in the lower atmosphere must contain a lot of water vapor (have a high relative humidity) in order for a thunderstorm to form.

Where does all the moisture that feeds a thunderstorm come from? Most evaporates from the world's oceans, the largest reservoir of water in the hydrologic cycle. For example, the moisture fueling thunderstorms in the central and eastern United States typically comes from evaporation from the Gulf of Mexico and the Atlantic Ocean; winds then blow this moist air over the land. The water that enters thunderstorms can also come from lakes and wetlands, from soils that have absorbed water from earlier rainfalls, and from transpiration by plants.

Triggering Thunderstorm Updrafts: Lifting Mechanisms

Development of a thunderstorm begins when moist air starts to rise as an updraft from the lower troposphere. A process that initiates this upward motion is called a **lifting mechanism**. Atmospheric scientists distinguish among several types of lifting mechanisms:

LIFTING DUE TO MOVEMENT OF A FRONT. In Chapter 18, we learned that a *front* is the boundary between two air masses. Many thunderstorms develop when the dome of advancing cool air behind a cold front physically lifts warm, moist air ahead of the front **(Fig. 19.5a)**. Lifting by an advancing cold front may produce several thunderstorms, at roughly the same time, in a line along the front **(Fig. 19.5b)**. Although less common, the lifting of moist air that triggers thunderstorms can also take place along a warm front.

LIFTING DUE TO GUST FRONTS. As we'll see later in this section, when rain starts to fall in a thunderstorm, it produces an intense **downdraft** (a downward flow) of cool air that rushes from the storm to the ground below. When this downdraft reaches the ground, it spreads outward laterally into the regions around the storm. The leading edge of such an outflow of cool, dense air is called a **gust front**. An advancing gust front acts like a local cold front in that it lifts warm air ahead of it. This lifting can trigger new updrafts that can develop into new thunderstorms. At places where gust fronts from different initial storms collide, or where a gust front and a cold front collide, focused lifting and particularly strong updrafts can evolve into strong thunderstorms **(Fig. 19.6)**.

LIFTING ALONG LESS DISTINCT BOUNDARIES. Lifting can take place not only at fronts, but also along less distinct *boundaries* between two local volumes of air that differ in temperature. A difference in temperature between local air masses might arise, for example, where the Earth's surface character within a given region varies, so that different areas heat the overlying air by different amounts. For example, air above the hot concrete of a city may heat up more than air over neighboring forests, and the air above an island may heat up more than the air above the surrounding ocean. At such boundaries, the cooler,

Figure 19.5 Thunderstorms may form as a result of frontal lifting.

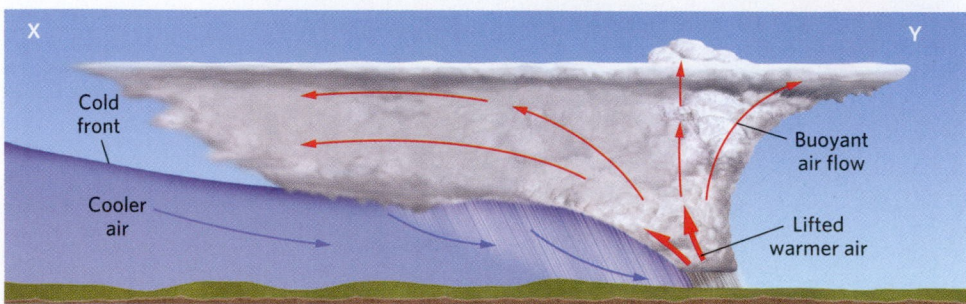

(a) Thunderstorms form as warm air is lifted ahead of an advancing cold front.

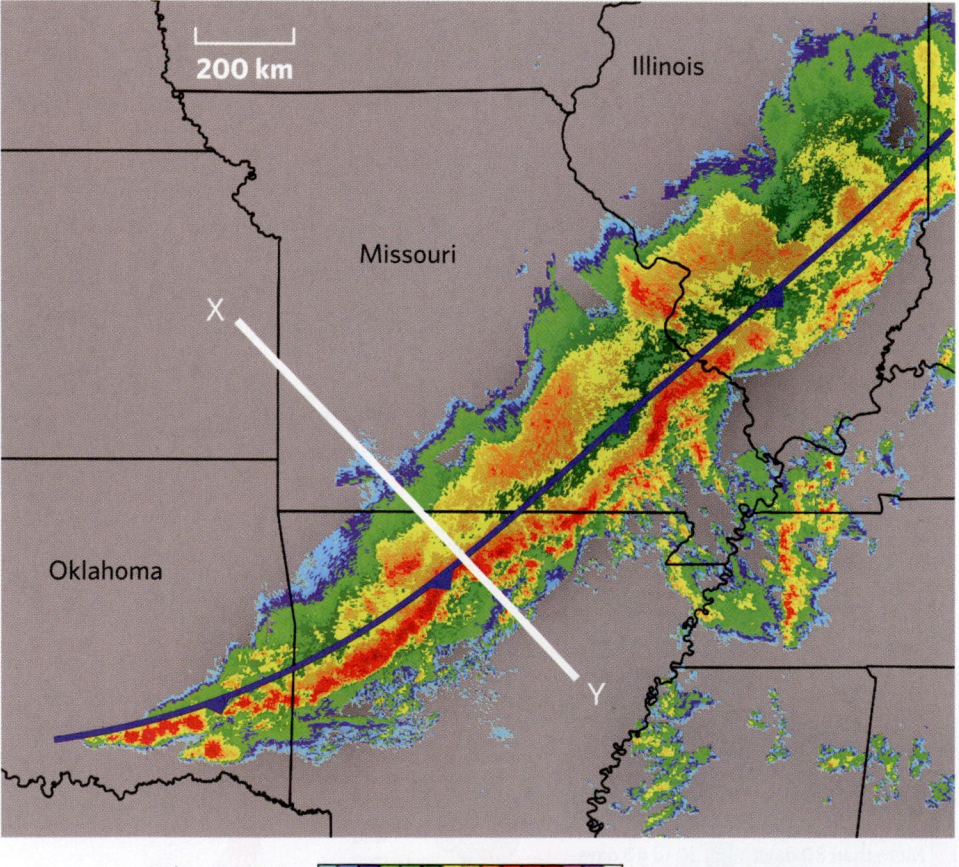

Light precipitation ▮▮▮▮▮▮▮▮▮▮▮ Heavy precipitation
Radar reflectivity

(b) A radar reflectivity display showing thunderstorms along an advancing cold front in the central United States. The line X–Y shows the location of the cross section in part (a).

denser air flows under the warmer air and lifts it.

LIFTING DUE TO HIGH TOPOGRAPHY. When wind reaches hills or mountains, the air can no longer travel horizontally and must rise up the slopes. Such **orographic lifting** can produce updrafts that can trigger thunderstorms.

WARMING FROM BELOW. Air does not always have to be physically pushed upward for an updraft to begin. In some cases, moist air near the ground starts to rise simply because the hot ground beneath warms the air enough to make it buoyant relative to surrounding air. This process typically happens during the summer, when intense sunlight strikes the ground. This type of warming can take place above plains, but it can also occur in mountain ranges when, during a hot afternoon, the slopes of steep mountains warm the adjacent air **(Fig. 19.7a)**. People living near the Front Range of the Rocky Mountains in Colorado witness this process nearly every day during the summer **(Fig. 19.7b, c)**.

Atmospheric Instability

Lifting of air to initiate an updraft, alone, does not produce a thunderstorm. Storm formation also requires **atmospheric instability** (the existence of denser air over less-dense air) in the region where air is rising. Where this condition exists, the rising air, once lifted, doesn't sink back down, but rather continues to rise on its own.

We can simulate the concept of atmospheric instability using familiar kitchen materials. Imagine that you're making salad dressing. Start by filling a jar a third of the way with vinegar, which has a density of 1.05 g/cm³. Then add olive oil, which has a density of 0.85 g/cm³. Because it's less dense, the olive oil forms a horizontal layer floating above the vinegar. Now, plunge a spoon down below the boundary and lift a blob of vinegar up. If you then remove the spoon, the vinegar sinks back down, so the boundary between oil and vinegar returns to horizontal **(Fig. 19.8a)**. We say that the fluid in the jar is stable because if we displace some of the fluid upward, it returns to its original state, even though it has undergone lifting. Now, suppose you invert the layering by pouring the oil in before the vinegar. If you're very careful, when you pour in the vinegar, you can produce a vinegar layer over the oil layer. Left undisturbed, such layering can persist for a while, but it isn't stable, because if you lift a blob of oil, it will keep rising upward. In fact, the rest of the oil will

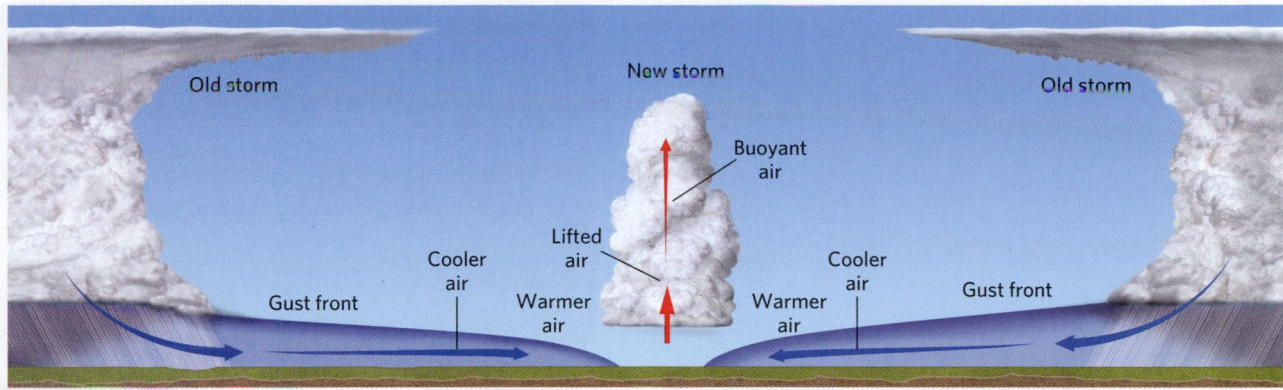

Figure 19.6 Development of thunderstorms along colliding gust fronts.

(a) Each storm sends out a gust front from its base. Where the gust fronts collide, air rises, and a new storm forms.

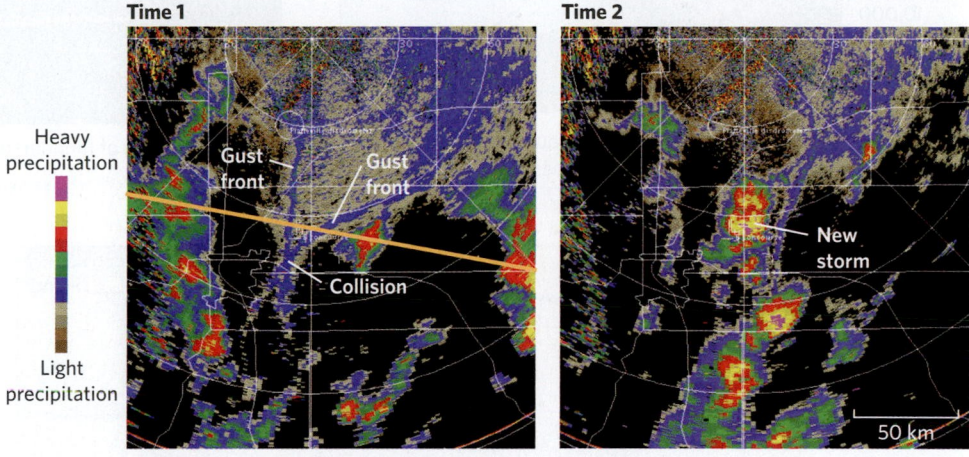

(b) Radar reflectivity images show new thunderstorms forming where two gust fronts collide. The yellow line shows the location of the cross section in part (a).

follow the blob until all the oil ends up on top of the vinegar. You can see this same process in a lava lamp, where buoyant blobs of fluid rise to the top **(Fig. 19.8b)**. In this context, we are using the term *stable* to mean that a denser fluid lies beneath a less dense fluid. The denser fluid, if lifted, has no buoyancy and simply sinks downward, and the system returns to its initial condition. Similarly, we're using the term *unstable* to mean that a less dense fluid underlies a denser fluid. If the less dense fluid receives an initial upward push by a lifting mechanism, its buoyancy continues to carry it upward.

Let's apply the concept of fluid stability to the atmosphere. For convenience, we'll refer to a small amount of air that we watch as it moves through the atmosphere as an *air parcel*. A region contains **stable air** when an air parcel sinks back down after it has undergone initial lifting. In contrast, a region contains **unstable air** when an air parcel continues to rise after it has undergone initial lifting. Thunderstorms develop only where the atmosphere is unstable, so that initial lifting triggers an updraft in which the air remains or becomes buoyant enough to rise toward the tropopause.

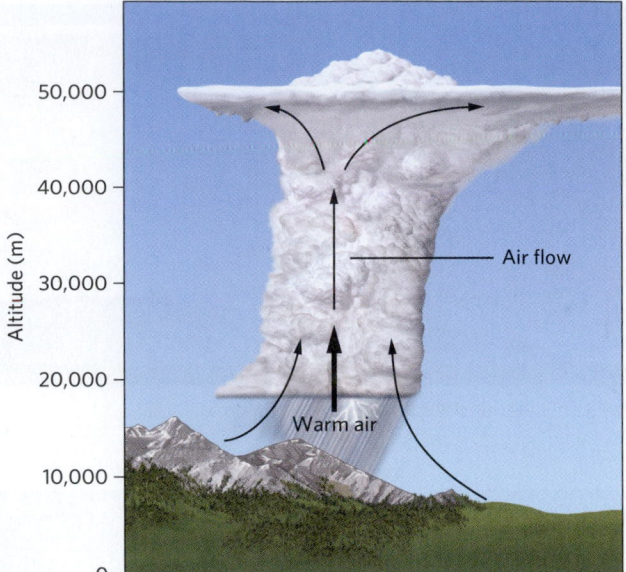

(a) Updrafts can begin over mountain ranges when, during a hot afternoon, the slopes of steep mountains warm the adjacent air.

(b) An example of thunderstorms brewing over mountains.

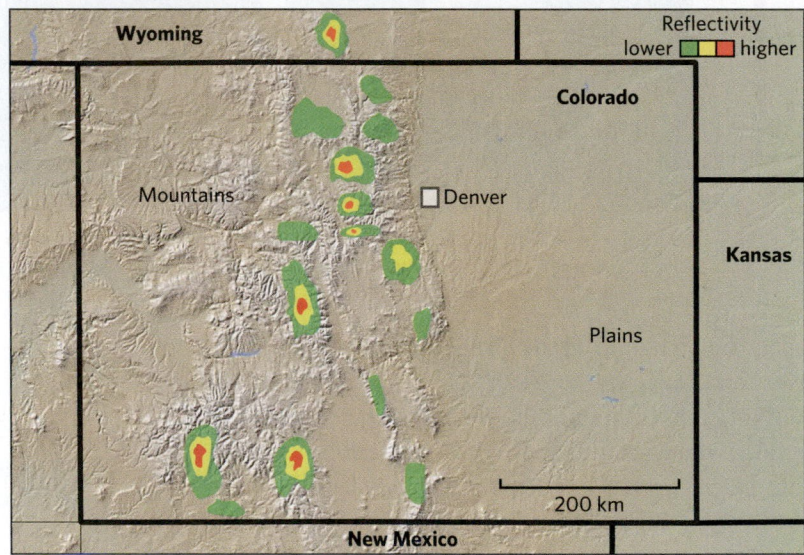

(c) A radar reflectivity image shows intense rain over the Front Range of Colorado.

Instability in the atmosphere is a bit more complicated than instability in salad dressing because the density and temperature of the atmosphere decrease with altitude and because, unlike a blob of oil, a rising air parcel decompresses and expands as it rises. In order to expand, air molecules within the parcel must work to push aside the surrounding air **(Fig. 19.9)**. Since the energy that does this work must come from within the parcel, the air molecules making up the parcel give up some of their heat energy during expansion, so the temperature of the parcel decreases. This process is called **adiabatic expansion** when the air parcel does not exchange heat or mass with its surroundings as it rises.

Due to adiabatic expansion, a rising air parcel may eventually reach an altitude at which it has become cooler and denser than the surrounding air. If this happens, the parcel's rise ceases, and a thunderstorm does not develop. Whether or not an air parcel stops rising depends on two factors. First, stability depends on the **environmental lapse rate**, the rate at which the temperature of the air surrounding the parcel (its environment) decreases with altitude. Typically, environmental lapse rates range between 4°C and 9°C per kilometer (12°F to 26°F per mile), depending on location. Second, it depends on the moisture content of the rising air parcel. *Moist air*—air that is saturated with water vapor—will form clouds as it rises and cools. *Dry air*—which may contain some water vapor, but is unsaturated—will not form clouds when it rises.

The reason that cloud formation within a rising air parcel plays such an important role in determining stability is that the rate at which air cools during adiabatic expansion depends on whether or not the water vapor within it condenses into cloud droplets. Let's see why.

When dry air rises, it undergoes *dry adiabatic expansion*, and even though the air's relative humidity increases as it cools, it doesn't reach saturation, so cloud droplets do not form. Dry adiabatic expansion causes the temperature of the air parcel to decrease by about 10°C for each kilometer (29°F per mile) the parcel rises. This rate, known as the **dry adiabatic lapse rate**, does not depend on altitude: an air parcel undergoes the

Figure 19.8 The concept of fluid stability. Stable layering can remain unchanged for a long time. In unstable layering, the less dense fluid will rise and the denser fluid will sink.

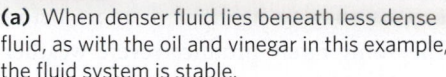
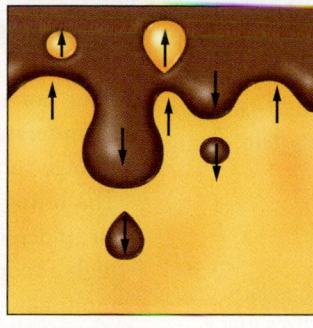

(a) When denser fluid lies beneath less dense fluid, as with the oil and vinegar in this example, the fluid system is stable.

(b) When less dense fluid lies beneath denser fluid, as in this lava lamp, the fluid system is unstable.

same temperature change rising between an altitude of 1 and 2 km (0.6 and 1.2 miles) as it does rising between 9 and 10 km (5.6 and 6.2 miles). If the dry adiabatic lapse rate is greater than the environmental lapse rate, rising dry air eventually becomes colder and denser than the surrounding air and does not continue to rise.

In contrast, rising moist (saturated) air cools at a rate of only 6°C per kilometer (17°F per mile). This rate, known as the **moist adiabatic lapse rate**, is less than the rate for dry air because cloud droplets condense out of saturated air as it rises and cools. As we saw in Section 17.6, conversion of water vapor to liquid water releases latent heat, and adding this heat to the rising air parcel slows the rate of cooling during the parcel's ascent. Since the moist adiabatic lapse rate can be less than the environmental lapse rate, a moist, cloudy air parcel has the potential to remain buoyant as it rises all the way to the tropopause.

Pulling all these ideas together, we see that whether or not a parcel becomes buoyant depends on environmental conditions. When a dry air parcel undergoes lifting, it may rise slightly, but it cools more quickly than the surrounding air, so it becomes denser than the surrounding air and sinks back down—such dry air is stable. When a moist, cloudy air parcel undergoes lifting, it, too, cools as it rises and expands. If it cools so that its temperature becomes colder than the environment, this air, even though it is cloudy, will remain stable **(Fig. 19.10a)**. But if it cools so that its temperature remains warmer than that of the surrounding air, it becomes buoyant; such air is unstable **(Fig. 19.10b)** and will become a buoyant updraft that rises toward the tropopause.

Figure 19.9 A parcel of air expands as it rises. Note that the number of air molecules in the parcel doesn't change as the parcel rises.

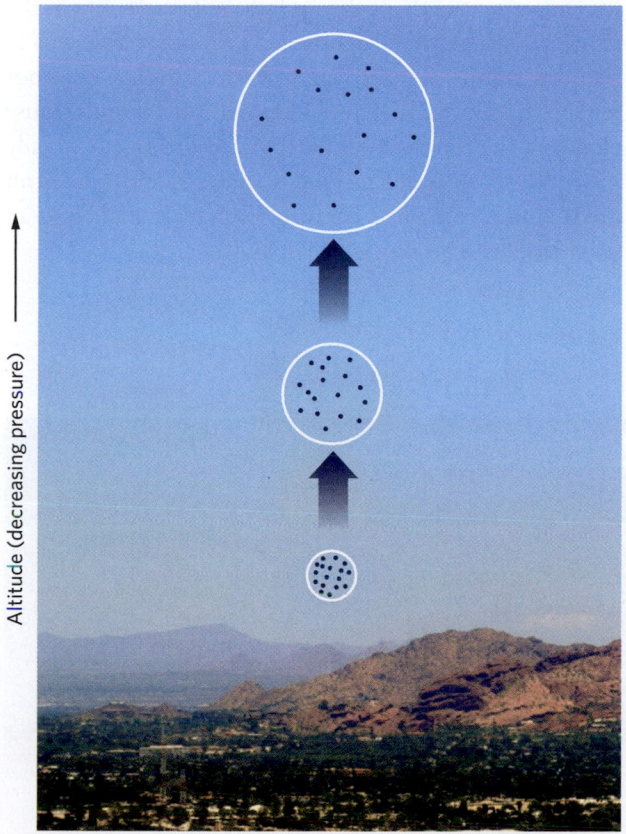

Figure 19.10 Stability under different environmental conditions.

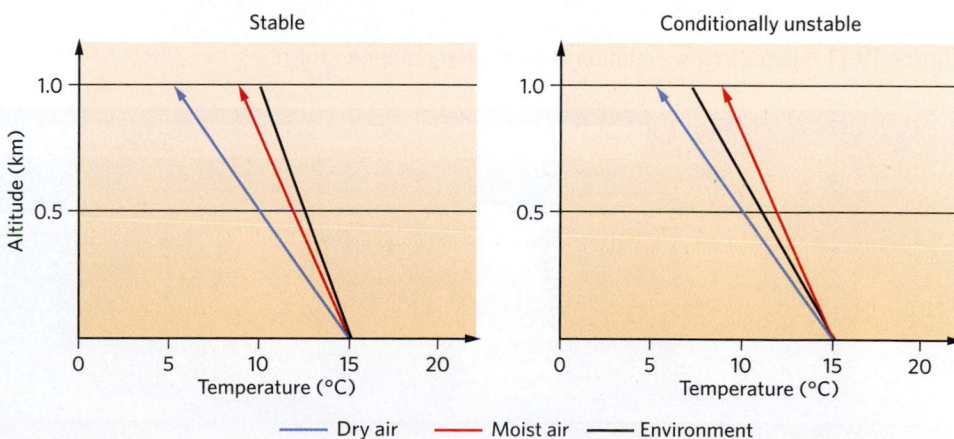

Stable

Conditionally unstable

—— Dry air   —— Moist air   —— Environment

(a) In this environment, both moist and dry air parcels, when lifted, would return to their original altitude.

(b) In this environment, a dry air parcel, when lifted, would return to its original altitude, but a moist air parcel would rise buoyantly.

19.2 CONDITIONS THAT PRODUCE THUNDERSTORMS   **665**

If an updraft is large enough and if it contains enough moisture, billowing clouds form and a thunderstorm develops. In effect, moisture drawn into the base of the storm provides the "fuel" to drive the updraft. Conditions leading to thunderstorm formation exist only when air in the lower troposphere is warm and moist while air in the upper troposphere is cool. These conditions can happen all year in the tropics, but in mid-latitudes they generally become common only during late spring through early fall.

Take-home message . . .

Three conditions are necessary for a thunderstorm to form: moist air near the Earth's surface, a lifting mechanism, and local atmospheric instability. Once lifted, a moist air parcel can become buoyant when its moisture condenses into cloud droplets, because the release of latent heat makes the parcel warmer and less dense than its surroundings. Such instability can produce an updraft that carries rising air up to the tropopause, producing large billowing clouds that evolve into a thunderstorm.

Quick Question -
What mechanisms can lift air to trigger thunderstorm development?

19.3 Different Types of Thunderstorms

Towering dark clouds, flashes of lightning, claps and rumbles of thunder, torrential rains, and sometimes the clatter of hail—these are the signatures of thunderstorms. Where updrafts are strong enough, they billow upward to form huge clouds that produce heavy precipitation and lightning. But even though all thunderstorms share these basic characteristics, not all are alike. Here, we examine the different ways that thunderstorms organize.

Ordinary Thunderstorms

Meteorologists refer to an isolated thunderstorm that does not rotate—meaning that the air in the storm does not spin around a vertical axis—as an **ordinary thunderstorm**. Such storms form where winds do not change substantially in direction or speed with altitude. The lifting that initiates such storms may develop when air undergoes lifting by a front or gust front, encounters high topography, or gets warmed by the ground or a mountain slope. In general, ordinary thunderstorms form during hot afternoons and last an hour or more before they dissipate. They only rarely produce severe weather. Worldwide, at any given time, as many as 2,000 ordinary thunderstorms are active.

Meteorologists divide the life cycle of an ordinary thunderstorm into three stages **(Fig. 19.11)**. During the *development stage*, warm, moist air is lifted to an altitude at which it becomes buoyant. As the air expands and cools, the water vapor within condenses into liquid droplets or, at higher altitudes, solidifies into tiny ice crystals, forming cumulus clouds. If condensation releases enough latent heat to maintain the air's buoyancy, the air becomes unstable and surges upward. As a result of its buoyant rise, the cumulus clouds develop a cauliflower-like appearance. Further rise and condensation produce towering cumulus congestus clouds. If the updraft is strong enough, the rising air reaches the tropopause. The tropopause acts as a lid on storms, for above this level, rising air does not remain buoyant. Therefore, high-altitude winds cause clouds at the top of the thunderstorm to spread laterally into a flat sheet. A storm cloud with this shape is called an **anvil cloud**, so named because of its resemblance to an anvil used by a blacksmith **(Fig. 19.12)**.

Figure 19.11 Stages in the evolution of an ordinary thunderstorm.

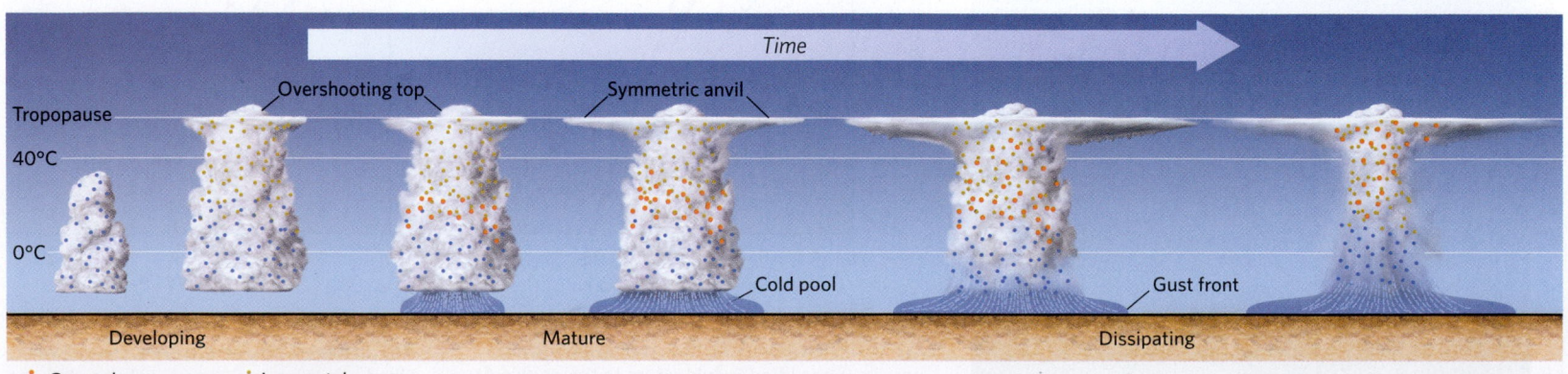

With the formation of the anvil cloud, the thunderstorm enters the *mature stage* of its life cycle. During this stage, ice particles aloft (typically in the form of small hail and graupel) cascade downward through the cloud, falling into the warmer lower atmosphere, where they melt into raindrops. As these particles fall, they push air in front of them, producing a downdraft. The falling rain also encounters drier air drawn into the storm from the sides and top of the cloud. As a result, some of the water evaporates, which removes latent heat from the air and lowers its temperature. (You feel similar cooling when water evaporates from your skin.) Cooling causes the air to become denser and therefore enhances the downdraft. The net effect is that the once-rising air in the storm now descends. When the rain reaches the ground, the cloud, by definition, becomes a *cumulonimbus* cloud.

Because air movement in an ordinary thunderstorm is nearly vertical, the downdraft that forms during the mature stage acts to suppress the updraft that formed during the development stage. In fact, as the downdraft intensifies, it eventually shuts off the source of moisture to the storm, and therefore shuts off the storm's fuel supply. Without a supply of new moisture, the storm dies, and the cloud begins to dissipate. When this happens,

the storm enters its *dissipation stage*. Notably, when the cooled air of the downdraft reaches the ground, it spreads out. The resulting **cold pool** of air near the ground rushes out over the Earth's surface. The leading edge of the cold pool is sometimes marked by the strong winds that form a gust front. As we noted in Section 19.2, the gust front may lift warm air ahead of it, triggering the formation of new thunderstorms.

Squall-Line Thunderstorms

Squall-line thunderstorms form along a distinct line that marks the edge of a front or gust front. The rain that accompanies these thunderstorms may cover a broad area as the line of storms moves over it. Indeed, squall-line thunderstorms produce much of the summer rainfall over the interior plains of the United States.

If you look toward squall-line thunderstorms approaching from a distance, you'll probably see a dark and ominous planar cloud called a *shelf cloud*, which forms over the gust front (**Fig. 19.13**). Typically, only a small amount of precipitation falls from the shelf cloud as it passes, but the wind that accompanies it can be hazardous. Behind the gust front come **straight-line winds**—winds that blow in a single direction, rather than in a spiral as in a tornado.

Figure 19.12 When a thunderstorm rises to the tropopause, the top of the cloud spreads out into a flat layer, making the cloud overall resemble a classic blacksmith's anvil.

Shelf cloud

Heavy rain

A research radar truck, called the "Doppler on Wheels."

Strong straight-line winds can damage buildings, topple trees, and pose a hazard for low-flying airplanes.

Once the shelf cloud has passed overhead, you will see a line of thunderstorms approaching, extending to your right and left. As the storms themselves pass overhead, lightning streaks the sky, heavy rains fall, and the winds become *gusty* (variable in velocity) and occasionally very strong. After the heavy rain passes, lighter rain might continue for hours, with an occasional crack of thunder or a flash of lightning overhead.

How do squall-line thunderstorms form, and why do they last so long? Using weather radar, meteorologists have tracked the life cycle of such thunderstorms. In temperate climates, they typically develop during the late afternoon or early evening during the warm season (late spring through early fall), when lifting of warm, moist air takes place along a boundary and a disorganized cluster of several ordinary thunderstorms grows **(Fig. 19.14a, b)**. As time progresses, downdrafts from the ordinary thunderstorms form a large cold pool at the ground. This cold pool spreads outward, and its leading edge forms a strong gust front. As the initial ordinary thunderstorms go through their life cycle, the cold pool continues to expand outward, and lifting along the gust front triggers the formation of new thunderstorms. These new thunderstorms are aligned with the gust front, forming an organized line of storms—a squall line **(Fig. 19.14c, d)**. As the cold pool advances forward into the warmer air, the squall line widens, updrafts carry moisture over the top of the cold pool, and rainfall becomes more widespread. Eventually, the cold pool spreads out so much that the squall line weakens. Lighter rain may continue for several more hours as the storms eventually dissipate.

In cooler seasons (late fall through early spring), when mid-latitude cyclones develop, lifting becomes focused along strong cold fronts. Very long lines of thunderstorms can develop along such fronts. From a regional perspective, such **frontal squall lines** lie along the "tail" of the comma cloud associated with a mid-latitude cyclone **(Fig. 19.15)**. Typically, frontal squall lines move with the front as it crosses a continent, have relatively long lives (many hours to more than a day), and can produce some severe weather.

Supercell Thunderstorms

On June 22, 2003, the polar-front jet stream flowed southward from the Arctic to Nevada and then curved, so that high-altitude winds sped northeastward across Nebraska. The large curve in this jet stream outlined a relatively cold air mass at ground level. In the high plains, to the east of the front defining the boundary of this air mass, warm, moist air from the Gulf of Mexico flowed northwestward. Frontal lifting caused this air to rise, and it became unstable. Towering cumulus congestus clouds billowed into the sky over the front and developed into thunderstorms. But unlike the ordinary thunderstorms we've already discussed, one of the storms that developed over Nebraska that day became a supercell thunderstorm. This storm produced the largest hailstone ever seen in the world (17.8 cm, or 7.0 inches, across—twice the size of a softball). The storm also spawned a tornado and dumped huge volumes of rain.

How does a supercell thunderstorm differ from an ordinary thunderstorm? Formally defined, a **supercell thunderstorm** is a thunderstorm containing a particularly

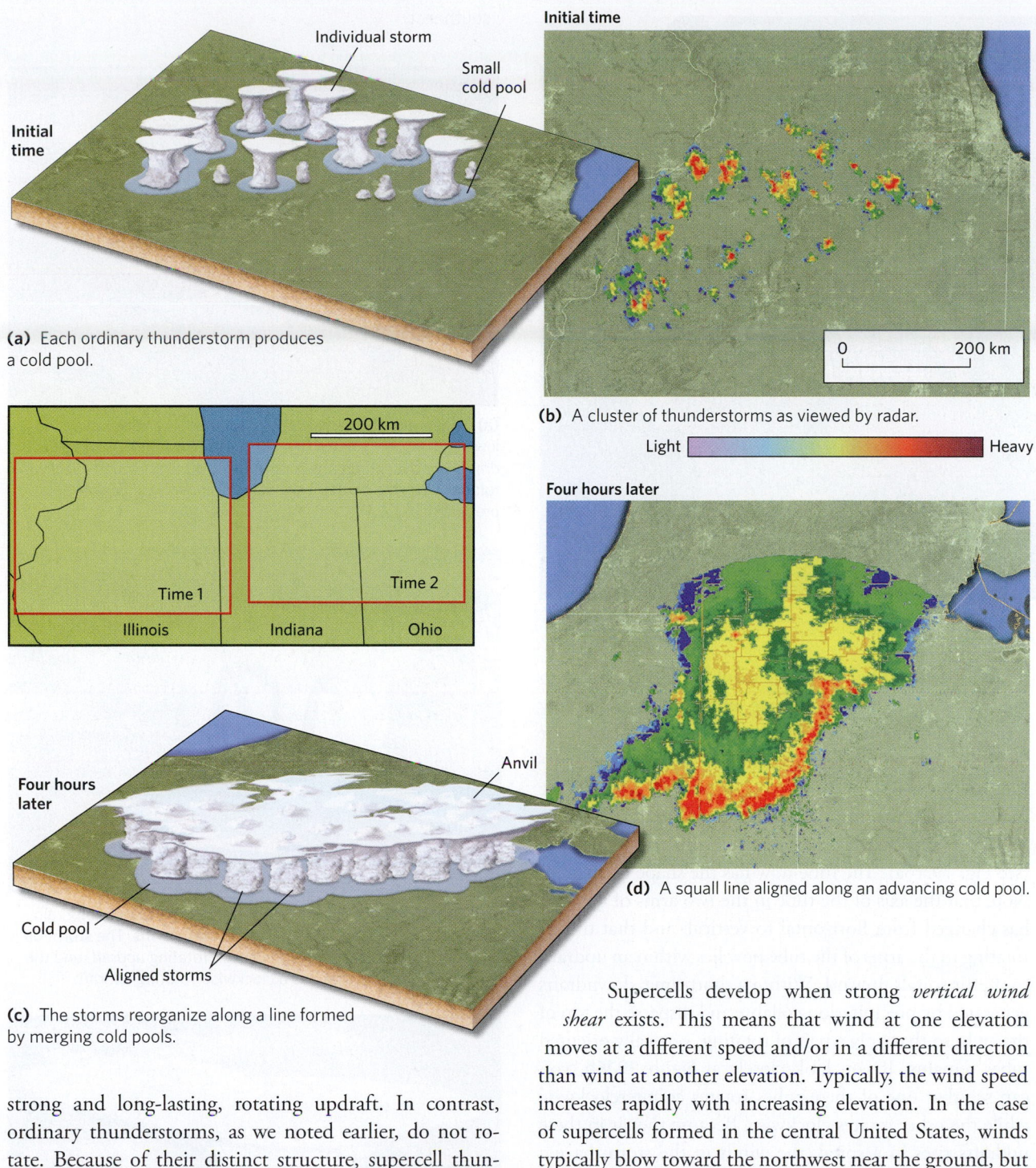

(a) Each ordinary thunderstorm produces a cold pool.

(b) A cluster of thunderstorms as viewed by radar.

(c) The storms reorganize along a line formed by merging cold pools.

(d) A squall line aligned along an advancing cold pool.

strong and long-lasting, rotating updraft. In contrast, ordinary thunderstorms, as we noted earlier, do not rotate. Because of their distinct structure, supercell thunderstorms often become severe and can produce strong tornadoes. Supercell thunderstorms (or *supercells*) occur relatively rarely, and they are typically isolated from neighboring thunderstorms. In some cases, though, several supercells can form in advance of a squall line. They typically have an overall diameter, at the base, of less than 50 km (30 miles); the most dangerous region, beneath the storm's updraft, has a diameter of about 5 to 10 km (3 to 6 miles) or less.

Supercells develop when strong *vertical wind shear* exists. This means that wind at one elevation moves at a different speed and/or in a different direction than wind at another elevation. Typically, the wind speed increases rapidly with increasing elevation. In the case of supercells formed in the central United States, winds typically blow toward the northwest near the ground, but blow toward the northeast at high velocity at the elevation of the jet stream.

To understand why vertical wind shear leads to rotation in a supercell, picture an imaginary horizontal cylindrical tube of air positioned so that its bottom lies near the ground, its top lies a kilometer above the ground, and its axis trends perpendicular to the high-level wind at the top of the tube, so that the wind blows across the tube (**Fig. 19.16a**). Air blowing faster across the tube's top

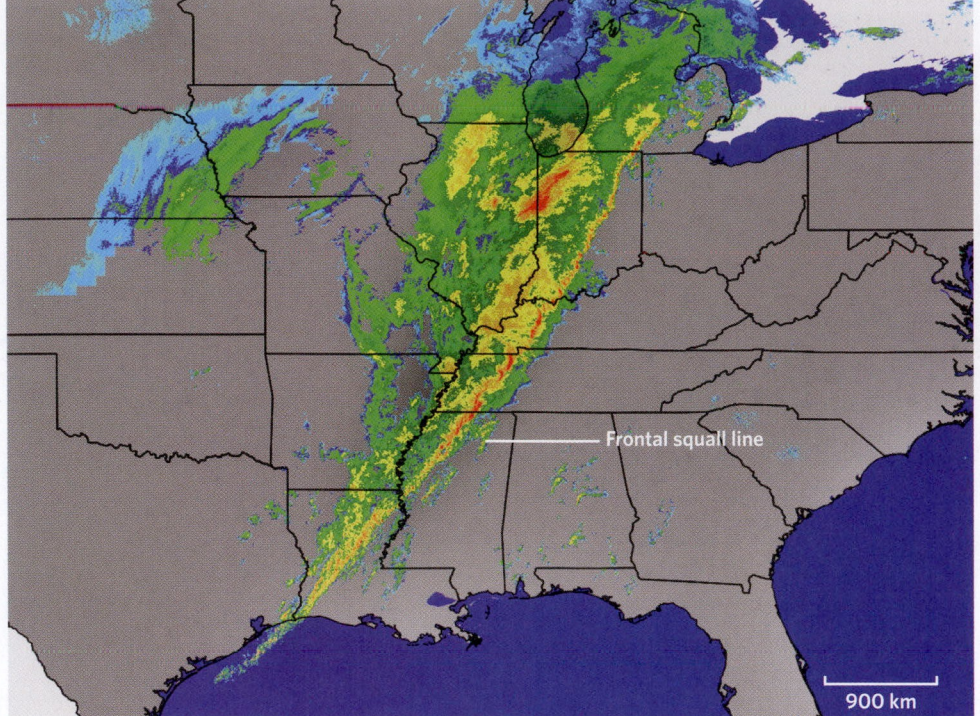

Frontal squall line

900 km

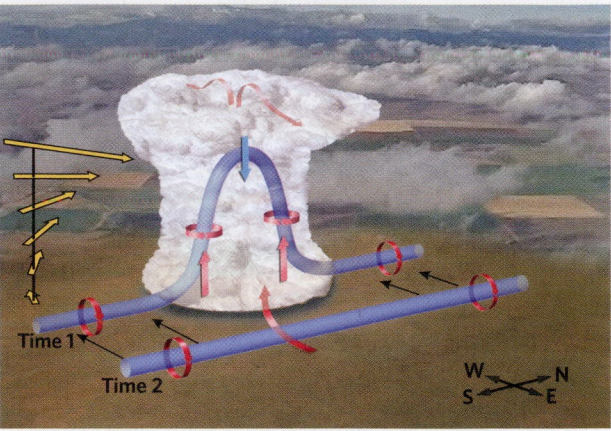

Time 1

Time 2

W N S E

(a) When wind increases with altitude (yellow arrows), air in the lower atmosphere rotates (the horizontal blue tube). As this air is drawn into the updraft, the southern side develops counterclockwise rotation while the northern side rotates clockwise. Blue dots are precipitation.

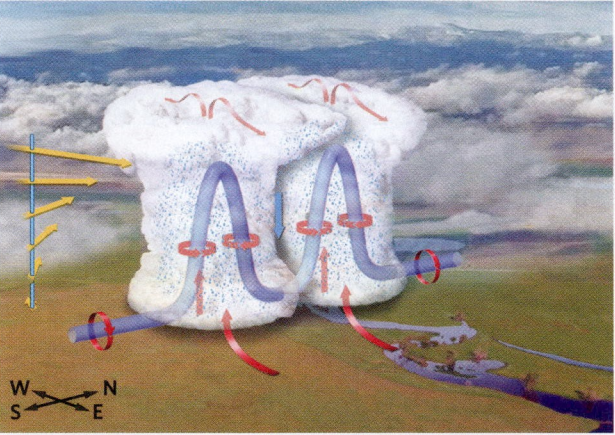

W N S E

(b) As downdrafts develop due to falling and evaporating rain and hail, the original storm splits into two storms. The storm on the south side has a counterclockwise-rotating updraft, and the storm on the north side has a clockwise-rotating updraft.

W N S E

(c) Typically, as the counterclockwise-rotating storm intensifies, the clockwise-rotating storm to the north dissipates.

relative to the tube's bottom, would cause the tube to rotate around its axis. (You can simulate this process by laying a cardboard tube on a table; if you place your hand on the top of the tube and move your hand parallel to the table, the tube starts rotating.) If instability leads to development of an updraft along the rotating air tube, the updraft draws the rotating tube of air upward (see Fig. 19.16a). The tube now has the shape of an arch. Note that the axis of the tube in the two arms of the arch has changed from horizontal to vertical, and that the air rotating in the arms of the tube now lies within an updraft, so the air spirals upward. Within a short time, downdrafts generated by precipitation collapse and disperse the top of the arch, so that only a pair of rotating, vertically oriented tubes remains. If you look closely at **Figure 19.16b**, you can see that one of these tubes rotates counterclockwise while the other rotates clockwise. Because the air in these updrafts rises in a spiral, the winds in the storm have a component of rotation around a steep vertical axis. This movement causes the storm to rotate.

As time progresses, the initial updraft splits into two counter-rotating storms, one counterclockwise and the other clockwise. Typically, the clockwise-rotating updraft dissipates, leaving only the counterclockwise-rotating storm **(Figure 19.16c)**.

To see why, let's look at supercell formation in the United States, where conditions leading to supercells commonly occur in a southwest-to-northeast corridor extending from western Texas and Oklahoma across the Midwest into Ohio. In this region, low-elevation moist air blows northwest from the Gulf of Mexico while jet-stream air aloft flows northeast. Supercells typically migrate from southwest to northeast over time, following the general flow direction of the jet stream. For a typical developing supercell moving from southwest to northeast, the southeastern updraft rotates counterclockwise and the northwestern updraft rotates clockwise (see Figure 19.16b). Moisture—the fuel of thunderstorms—gets carried into the counterclockwise updraft, which lies closer to the moisture source. In fact, this updraft blocks the flow to the northwestern updraft, so the moisture doesn't feed into that clockwise-rotating updraft. As a result, the southeastern counterclockwise-rotating storm grows while the northwestern clockwise-rotating storm dissipates. Therefore, supercells that survive and grow in the central United States contain a counterclockwise-rotating updraft, known as a **mesocyclone**. The updraft within the mesocyclone may have vertical speeds approaching 160 km per hour (100 mph).

Supercell thunderstorms have a distinctive structure because of the direction of flow in the jet stream at their tops and because of the strong mesocyclone within **(Fig. 19.17a, b)**. The jet stream blows the top of a supercell thunderstorm downwind to produce a large anvil, and it causes the whole storm to tilt, so the axis of the updraft is not vertical. Upward flow in the mesocyclone can be so strong that it forms a bubble of billowing clouds, called an *overshooting top*, above the anvil cloud (Fig. 19.17a, b). At the base of the updraft, a lowering of cloud base—called a **wall cloud**—is typically present **(Fig. 19.17c)**. It is from this feature that a tornado, if it occurs, will form. Supercell thunderstorms usually move from southwest to northeast, they tilt toward the northeast. Meteorologists refer to the northeastern side of the storm as the *forward flank* and to the southwestern side as the *rear flank*. Heavy rain on the forward flank produces intense downward-directed air flow, the *forward-flank downdraft* (FFD). A strong downdraft also develops on the rear flank of the storm, producing a *rear-flank downdraft* (RFD). Significantly, because of the storm's northeastward tilt, the FFD lies to the northeast of the mesocyclone, so that precipitation falls outside (to the northeast) of the updraft. Therefore, unlike the updraft of an ordinary thunderstorm, which tends to be destroyed by the downdraft, the mesocyclone of a supercell can survive for a relatively long time (a few hours). A rotating *wall cloud*, which develops beneath the base of the storm's updraft when rain-saturated air is drawn into the mesocyclone from the FFD, outlines the base of the mesocyclone.

Figure 19.17 Supercell thunderstorms have a dramatic appearance that includes several distinct features.

(a) A composite image showing a supercell thunderstorm, as viewed from the southwest.

(b) This view of a supercell thunderstorm from a high elevation shows the anvil and the overshooting top.

(c) A close-up view of the base of a supercell, showing rotating clouds and the wall cloud.

Figure 19.18 A diagram of a radar image of a supercell. Superimposed are the wind flow and other key features.

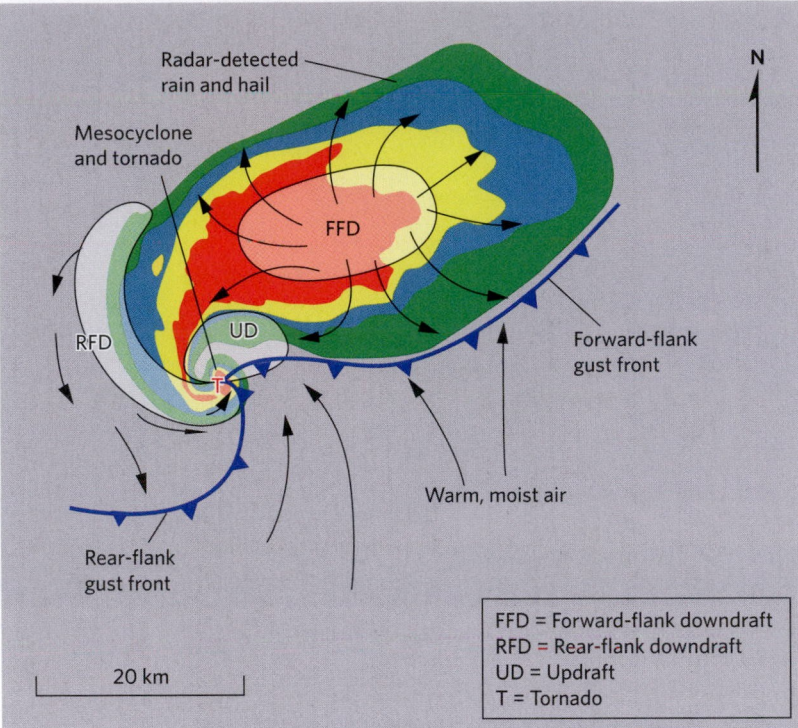

FFD = Forward-flank downdraft
RFD = Rear-flank downdraft
UD = Updraft
T = Tornado

20 km

A radar image of a supercell illustrates the storm's key features (Fig. 19.18). Rain and hail are swept northeastward from the updraft. As the rain falls, some evaporates and cools the air, creating a downdraft. This downdraft, the FFD, descends to the ground and spreads out. A second downdraft, the RFD, develops on the southwest side of the updraft as rain in that region evaporates. As we will see in Section 19.5, the RFD is an important element for tornado formation. A *hook echo* appears on the southwestern side of the supercell, surrounding the updraft region. (Recall that a *radar echo* is the radar energy reflected back from a storm.) The hook is associated with rain wrapping around the southwestern side of the updraft. The updraft itself appears as an echo-free region in the hook because no rain falls out of the updraft. The large echo to the northeast of the updraft indicates the presence of heavy rain. Hail, if present, falls close to the thunderstorm's updraft, primarily on its northeastern side. A tornado, if it occurs, develops at the tip of the hook echo, as we'll see in Section 19.5. Because of its distinct structure, a supercell looks different when viewed from different directions (Box 19.1).

Take-home message . . .

An ordinary thunderstorm has a vertical, non-rotating updraft and tends to dissipate fairly quickly. Squall-line thunderstorms form in a row along a strong front or along storm-generated cold pools. The most violent thunderstorms, supercells, form where vertical wind shear exists in addition to strong instability. A supercell has a rotating updraft, a particularly broad anvil, and an overshooting top.

Quick Question -
Why do supercell thunderstorms tend to survive for a relatively long time?

19.4 Hazards of Thunderstorms

Lightning

Thunderstorms are called thunderstorms because of the clap, crash, and rumble of thunder that accompanies them. Thunder is a consequence of lightning. It's the

Figure 19.19 Lightning is a discharge of static electricity.

(a) A discharge of static electricity between a finger and a doorknob resembles a tiny lightning bolt.

(b) Some lightning bolts go from cloud to cloud, or stay within a cloud. Some go from cloud to ground.

Consider this . . .

Clouds in a supercell thunderstorm

What does a supercell look like? It depends on where you're standing relative to the storm (Fig. Bx19.1). As seen from the northwest (Point 1), the storm's back-sheared (into the wind) anvil and overshooting top are evident, as is the back side of the updraft. The view from Point 2 shows the rear-flank downdraft region. The cloud marking the gust front lies over the rain-cooled air. The wall cloud beneath the updraft is difficult to see with heavy rain behind it. From Point 3, we see the shelf cloud forming over the forward-flank gust front. Point 4, farther to the southeast, gives us a full view of the storm so that the updraft is visible. The final view, Point 5 in the northeast, shows the forward anvil and *mammatus clouds* drooping beneath the anvil. These clouds, which resemble cow udders, form as ice crystals descend from the cloud and sublimate into water vapor, cooling the surrounding air. This cooling produces small downdrafts below the anvil.

Figure Bx19.1 Views of a supercell thunderstorm from different directions. The photos were taken from the approximate locations indicated by the red dots, looking in the directions of the arrows.

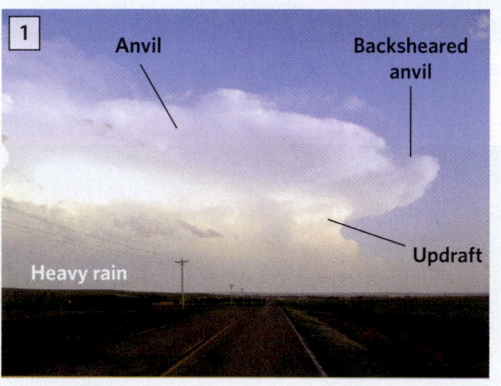

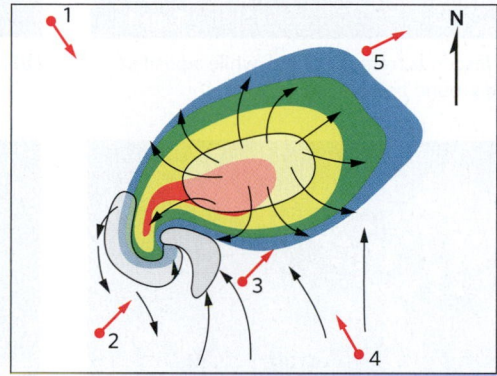

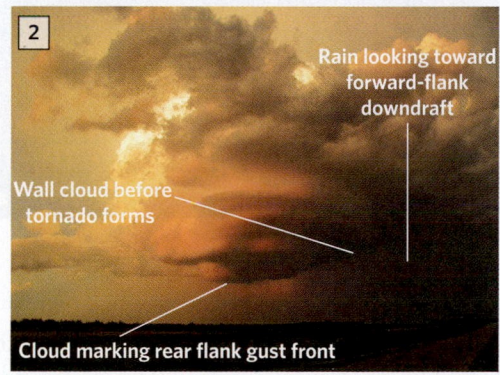

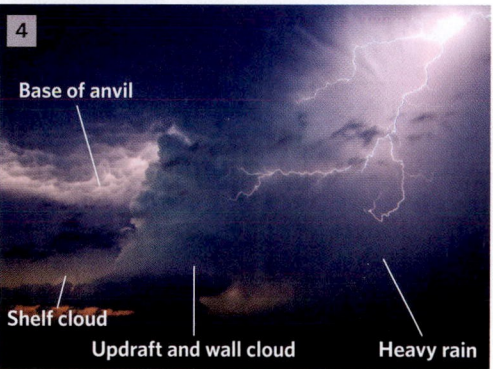

lightning in a storm that can be hazardous—thunder, while frightening, doesn't cause damage. Here we examine why lightning happens.

FORMATION OF LIGHTNING. If you rub your shoes on a carpet and then touch a metal doorknob, you'll see— and feel—an electrical spark (Fig. 19.19a). This spark appears because electrons from the carpet's atoms flow into your body, making your skin negatively charged relative to the doorknob. When your finger gets close enough to the doorknob, electrons suddenly flow from your skin to the doorknob, and this short-lived current generates a spark and tingles your nerves. Physicists refer to such sparks as an *electrostatic discharge*. Simply put, **lightning** is a huge electrostatic discharge, caused when negatively and positively charged particles become separated in the clouds

Figure 19.20 The steps leading to the production of a cloud-to-ground lightning bolt.

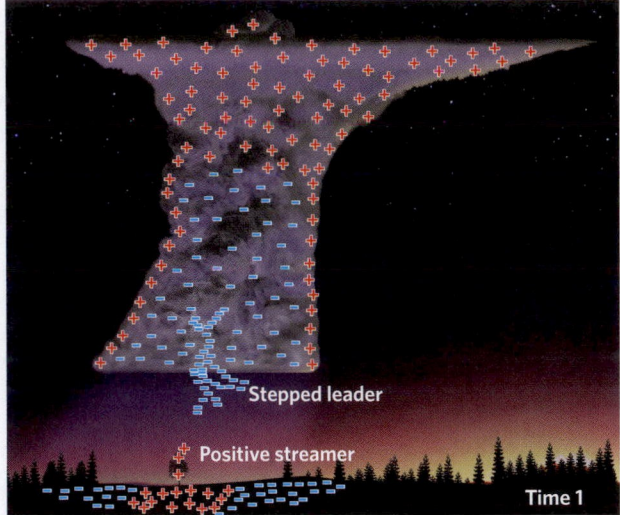

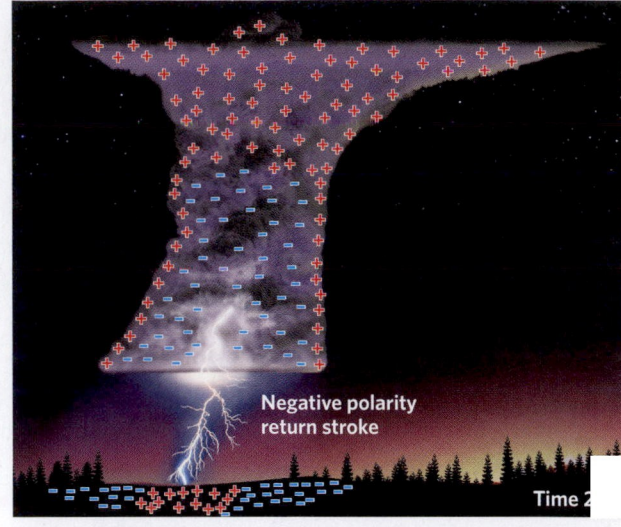

(a) A stepped leader descends from the cloud, while a positive streamer rises from the ground below.

(b) A return stroke makes a brilliant flash.

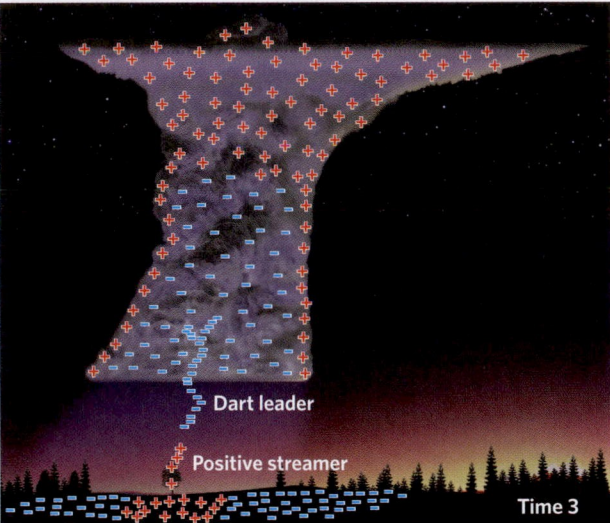

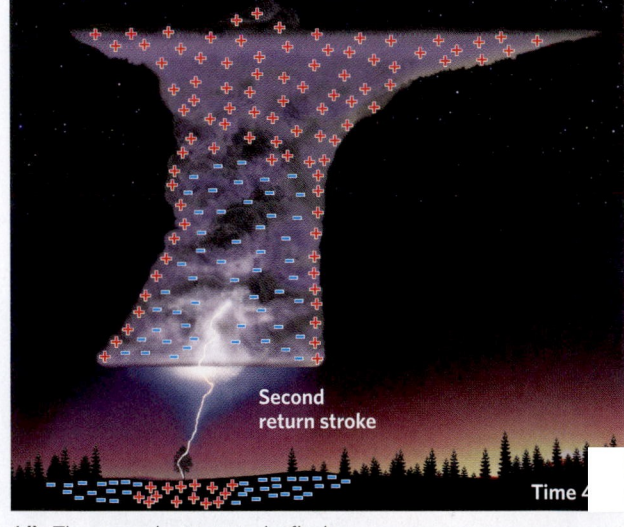

(c) A dart leader descends from the cloud, while a positive streamer flows upward.

(d) The second return stroke flashes.

of a thunderstorm (Fig. 19.19b). The spark that jumps from your finger to a doorknob has a length of about 1 mm and a diameter less than 0.1 mm, and it transmits a current of about 0.0001 amps (an *amp*—short for *ampere*—is a measure of the quantity of electrons flowing in a current during a 1-second interval). In contrast, a **lightning stroke** (or *lightning bolt*), the jagged spark produced by a thunderstorm, can be up to 5 km (3 miles) long and 2 to 3 cm (about 1 inch) in diameter, and it can carry a current of 15,000 to 30,000 amps. In terms of energy, a single lightning stroke produces about the same amount of energy as is required to power a typical house for 6 months. However, almost all of that energy is emitted as heat rather than as electricity. The electrical component of a lightning stroke could power a house for a few hours.

Why does lightning develop? To answer this question, we need to understand how separation of positive and negative charges develops in a cloud. This charge separation happens when electrons jump between tiny ice crystals and larger ice particles, such as hailstones, in the storm. As a consequence, the tiny ice crystals become positively charged, while the larger particles become negatively charged. Updrafts sweep the tiny, positively charged ice crystals upward to high elevations in the cloud, while the larger, negatively charged particles, due to their weight, fall toward the lower part of the cloud.

A charge separation can develop in a cloud only because air is an *electrical insulator*, meaning that it prevents electrons from flowing easily between a negatively charged and a positively charged region. The charge separation builds an *electrostatic potential*, meaning that as long as the charge separation is maintained, potential energy exists that could be available to drive electrons from the region of negative charge to the region of positive charge. A *discharge* (spark) of electrons flows from your finger to a doorknob when the electrostatic potential becomes great enough for the insulating properties of air to be overcome, allowing electrons to jump across the millimeter of insulating air to the doorknob. A lightning bolt forms when the electrostatic potential in a thunderstorm becomes great enough to drive electrons across a large expanse of insulating air.

TYPICAL CLOUD-TO-GROUND LIGHTNING. Since we live on the ground, **cloud-to-ground lightning**, which strikes buildings, trees, people, or other objects, poses the greatest hazard to us. Conditions leading to cloud-to-ground lightning start to develop when negative charges accumulate toward the base of the cloud. Just as the negative end of a magnet repels the negative ends of other magnets, the negatively charged cloud base drives away negative charges on the ground, leaving a region of the ground with a net positive charge.

A cloud-to-ground lightning stroke begins when electrons surge first toward the cloud base and then toward the ground in a series of steps, producing a *stepped leader* (Fig. 19.20a). Each step is 50 to 100 meters (150 to 300 feet) long and takes only a few millionths of a second to form. Typically, electrons take many different paths downward, so the stepped leaders have several branches. One branch happens to approach the ground surface first, and as it does, it attracts positively charged particles. These particles jump upward from an object on the ground, producing a feature called a *positive streamer*. Such streamers typically rise from a narrow high point, such as a tower, tree, chimney, or an unfortunate person. When the positive streamer connects with the stepped leader, a channel for electron flow becomes established. (The channel provides a pathway for current flow because it contains charged particles and, unlike the air around it, which acts as an insulator, behaves like an electrical conductor.) The instant that the positive streamer connects with a branch of the stepped leader, a powerful *return stroke*, the main discharge of lightning, flashes (Fig. 19.20b). The stroke runs from the cloud to the point where the positive streamer started.

In the return stroke, vast numbers of electrons stream from the cloud to the ground. (In this regard, the name *return stroke* may seem puzzling because nothing is actually returning to the cloud. This name remains from a time before researchers understood lightning.) The process of forming a leader, positive streamer, and return stroke typically repeats multiple times, generally within a second or two (Fig. 19.20c, d). Subsequent leaders, after the first stepped leader, are known as *dart leaders*. Those follow the previously established channel, which has become an electrical conductor. The whole process happens so fast that you may perceive it as a single bolt that flashes like a strobe light.

OTHER TYPES OF LIGHTNING. Lightning strokes can occur in several ways (Fig. 19.21). What we've just described is called, technically, a *negative polarity cloud-to-ground stroke*. It occurs when the negative charge in the bottom of the cloud jumps to the positive charge on the ground. A rarer, *positive polarity cloud-to-ground stroke* occurs when a discharge occurs between the storm's anvil and negatively charged ground far beneath it. Since there's a greater distance between the anvil and the ground than between the base of the storm and the ground, a greater electrostatic potential must develop for such a stroke to take place, so these strokes can release an immense amount of energy.

Most lightning, however, does not reach the ground, but rather occurs as **cloud-to-cloud lightning**, formed when a discharge jumps from a negatively charged part of a cloud to a positively charged part. Cloud-to-cloud strokes may hit an airplane flying through a storm. Fortunately, electricity can be conducted along the metal skin of a plane without penetrating its interior, so lightning typically causes minimal damage, if any, to airplanes.

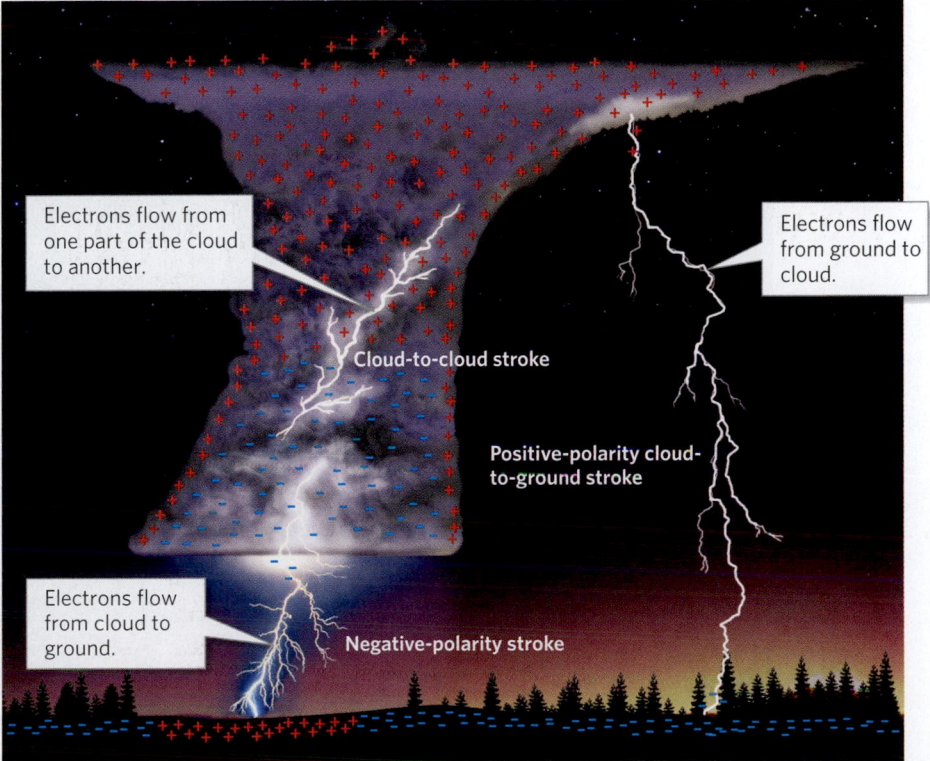

Figure 19.21 Lightning discharges can occur in three ways: as negative polarity cloud-to-ground strokes, positive polarity cloud-to-ground strokes, and cloud-to-cloud strokes.

Electrons flow from one part of the cloud to another.

Electrons flow from ground to cloud.

Cloud-to-cloud stroke

Positive-polarity cloud-to-ground stroke

Electrons flow from cloud to ground.

Negative-polarity stroke

Figure 19.22 When your hair stands on end near a thunderstorm, you are in danger: lightning may strike at any second.

Regardless of whether lightning runs between the base of the storm and the ground, the anvil and the ground, between parts of a cloud, or between different clouds, it generates thunder **(Box 19.2)**.

THE HAZARD OF LIGHTNING. Needless to say, lightning is dangerous. If lightning strikes a tree, it can cause the sap in the tree to vaporize instantly so the tree explodes. If it strikes a house, it can pass into the frame of the house and ignite flammable materials. If it strikes power lines, it can overload the system, blow out transformers, and cause a blackout. While lightning strikes cars, it rarely does them harm because their metal skins, like those of airplanes, conduct the electricity and keep it from entering the vehicle.

When lightning strikes a person, it can cause permanent harm or even instant death. Although people struck by lightning can survive, many suffer burns, neurological damage, and other symptoms. Injury can happen even if a lightning stroke hits the ground or a tree nearby, for the electricity can flow through the ground to the person. In the United States, lightning strokes kill, on average, about 50 people per year.

How can you avoid being a victim of lightning? First and foremost, go inside when a thunderstorm approaches. Don't stand in the middle of an open area or near isolated tall objects, or be on open water where you would be the highest object—lightning tends to strike tall objects because they can serve as the source of a positive streamer. When inside, stay away from appliances, plumbing, and corded telephones that could conduct electricity. If you are outside and feel your hair standing on end **(Fig. 19.22)**, take cover immediately or hop in a car. If there is no cover, assume a stooping position, crouching down as close to the ground as possible with your feet together, and minimize ground contact by staying on your toes or heels. Do not lie down, because lightning travels along the ground and will flow through your body.

Lightning often strikes houses and tall buildings **(Fig. 19.23a)**. It's a good idea to protect your home with

Figure 19.23 Lightning is attracted to tall objects.

(a) Lightning strikes skyscrapers regularly.

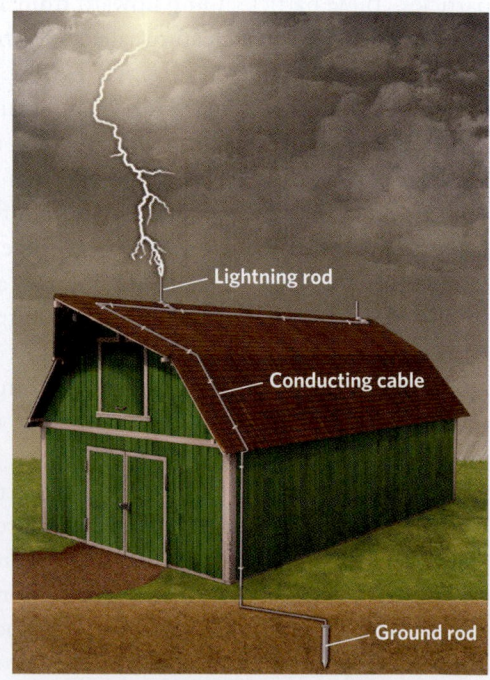

(b) To protect your home (or barn), install lightning rods on its rooftop.

Lightning rod

Conducting cable

Ground rod

Box 19.2

Consider this . . .

The boom of thunder!

The temperature in a lightning stroke channel can reach 30,000°C (54,000°F), five times hotter than the surface of the Sun. The heating causes the air in the channel to expand explosively **(Fig. Bx19.2a)**. When the stroke ceases, the air contracts violently. The sudden expansion and contraction of air generates shock waves that we hear as **thunder**.

If you've experienced a thunderstorm, you know that the character of thunder varies with your distance from the lightning **(Fig. Bx19.2b)**. A nearby lightning stroke sounds like a sudden, deafeningly loud, clap.

Figure Bx19.2 Lightning produces thunder.

Before lightning, the air temperature is uniform.

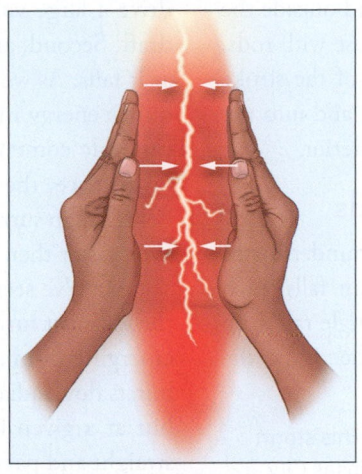

The energy of a lightning bolt causes air to expand almost instantly.

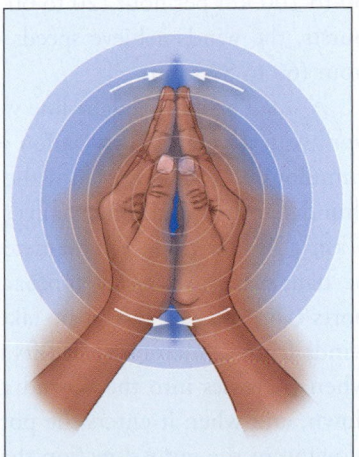

After the lightning bolt disappears, the air slams together, sending out shock waves.

(a) Thunder occurs when air rapidly expands and then contracts.

If you're farther from the stroke, you'll hear a deep rumble that can last for a few seconds. And if you're far enough away, you may see lightning flashes but hear no thunder at all. What causes these differences?

The sound waves generated by a lightning stroke spread outward into the atmosphere. The speed at which the sound travels depends on air density, and therefore on air temperature: sound travels more quickly in warm air than it does in cool air. So, just like seismic waves inside the Earth (see Chapter 8), sound waves of thunder bend (refract) as they cross from warmer to colder air, or vice versa. Because air density and temperature decrease with height in the troposphere, the troposphere bends sound waves upward. The amount of bending depends on the frequency of the sound, so as the sound of thunder propagates, the higher frequencies (higher notes on the musical scale) bend more than do the lower frequencies (lower notes). As a consequence, an observer near the stroke hears both high and low frequencies, an observer standing at a moderate distance hears only the lower frequencies, and an observer far away hears nothing because all the sound travels overhead. Out on a flat plain, thunder may sound like a single boom. In more rugged, forested, or urban landscapes, you'll usually hear it as a long rumble because the sound echoes from hills, trees, and buildings.

Sound travels at approximately 340 m per second (1,236 km per hour, or 768 mph), so it moves about 1 km in 3 seconds (1 mile in 5 seconds). Light travels so fast (300,000,000 meters per second) that it arrives at our eyes in an instant. You can estimate the distance to a stroke of lightning by counting the seconds that pass between the flash and the thunder, then dividing by three (to get the distance in kilometers) or five (to get the distance in miles).

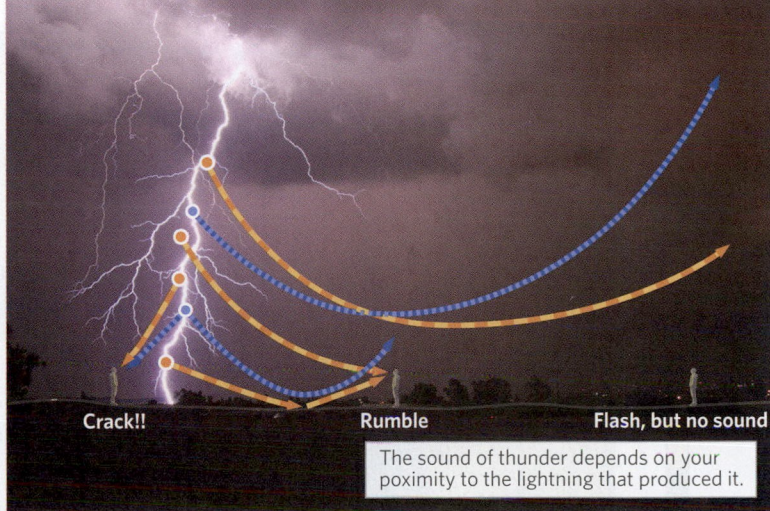

Crack!!   Rumble   Flash, but no sound

The sound of thunder depends on your poximity to the lightning that produced it.

·········· High-frequency sound   **——— Low-frequency sound**

(b) The refraction and reflection of sound waves causes thunder to have a different character at different locations.

Figure 19.24 Microbursts can be a serious hazard in the vicinity of an airport.

(a) A microburst over an airport in Denver, Colorado.

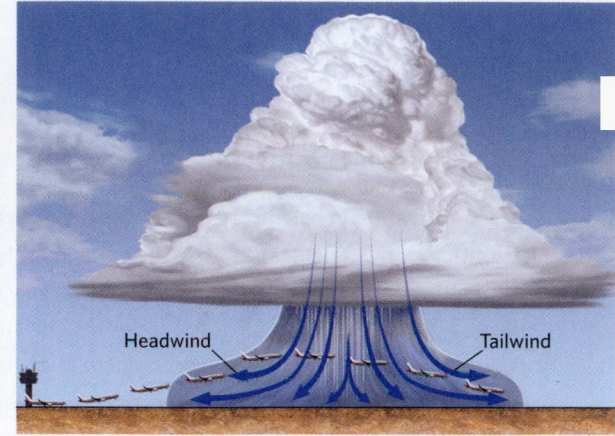

Headwind   Tailwind

(b) As an airplane approaches a microburst, it encounters strong headwinds, then strong tailwinds after it passes the microburst center.

Did you ever wonder . . .

why thunderstorm winds can be so destructive?

lightning rods, pointed metal rods anchored to rooftops and connected by wires directly to the ground outside the house **(Fig. 19.23b)**. If lightning strikes a house with rods, it will typically hit a rod, so the electricity of the stroke travels down the wire, outside of the house, and into the ground, thereby staying out of the house's interior.

Downbursts and Straight-Line Winds

Very strong downdrafts develop within thunderstorms when it rains, for two reasons. First, as rain falls, each drop pushes air ahead of it downward. A single raindrop obviously cannot drive much air, but collectively, the hundreds of billions of raindrops in a downpour can drive a large volume of air downward, causing a downdraft. Second, and more important, some rain evaporates as it falls. As we discussed in Chapter 18, evaporation requires energy in the form of latent heat, and this energy can only come from the surrounding air. As evaporation takes place, the air in the downdraft cools and becomes denser than surrounding air, so it sinks toward the ground even faster then it would otherwise.

As we've seen, when a downdraft reaches the ground, the air must turn and flow outward. The outrushing wind emerging from a downdraft is called a *downburst*. Downbursts flow radially outward from the base of a downdraft, but at a given location, the wind of a downburst flows straight and parallel to the ground, so meteorologists refer to it as a *straight-line wind*. These winds may travel a few kilometers in front of the storm, generally at velocities of 30 to 100 km per hour (20 to 60 mph). In severe downbursts, the winds achieve speeds of 100 to 130 km per hour (60 to 80 mph).

A small area of straight-line wind, known as a *microburst*, can develop in association with rain falling from a cumulonimbus cloud, even if the cloud hasn't become a thunderstorm. Microbursts affect an area of only about 5 km² (2 square miles) in diameter, but even so, they can be dangerous to aircraft approaching or departing airports **(Fig. 19.24a)**. If a plane takes off into approaching winds from a microburst, it lifts into the sky easily. But when it crosses into the downdraft itself, it gets pushed down, and when it enters the portion of the microburst blowing in the same direction that it's flying, it loses lift **(Fig. 19.24b)**.

When straight-line winds from several thunderstorms along a squall line merge, they can affect a broad area. In fact, squall lines occasionally produce a swath of destructive winds extending for hundreds of kilometers along the length of the squall line. Meteorologists refer

Figure 19.25 A squall line produces a bow echo on a radar screen. This storm produced straight-line winds approaching hurricane intensity near Kansas City.

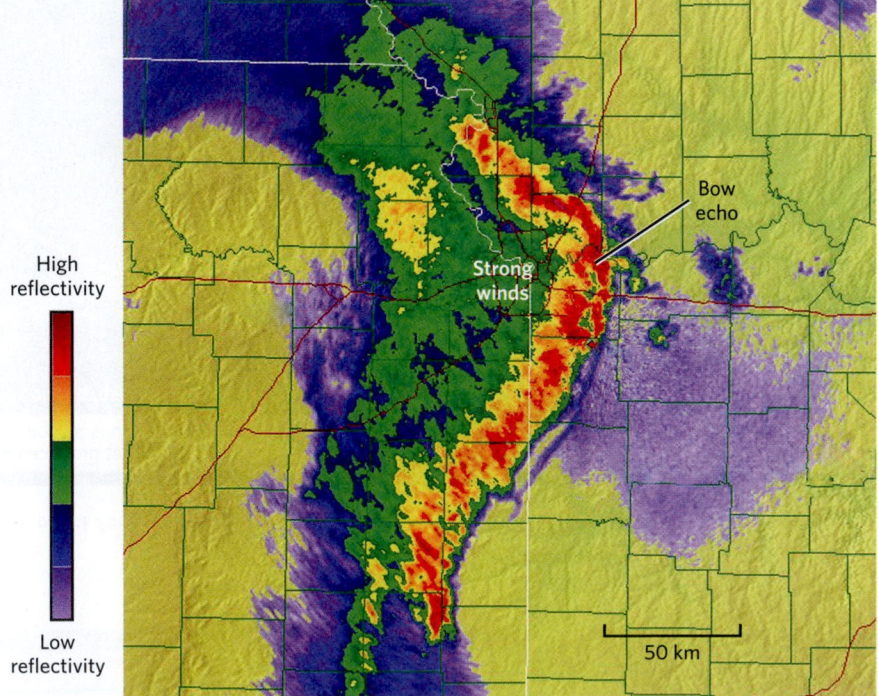

High reflectivity

Bow echo

Strong winds

Low reflectivity

50 km

to these large-scale, destructive windstorms as *derechos*. Particularly strong straight-line winds distort the shape of a squall line into an arc, which on a radar screen appears as a *bow echo* (**Fig. 19.25**). A particularly damaging derecho affected the central and eastern United States over a period of 18 hours during June, 2012. The winds started in Indiana, reaching a velocity of over 145 km per hour (90 mph), and blew all the way to the east coast, resulting in over $3 billion in damage.

Hail

In nearly all thunderstorms, **hail**, which consists of spherical or irregularly shaped lumps of solid ice (each of which is a *hailstone*), forms at high altitudes. Most hail is small (pea-sized to marble-sized), but some hailstones can grow to the size of a softball. The heaviest hailstone ever measured in the United States, which weighed 0.8 kg (1.94 pounds), fell during a 2010 storm in South Dakota (**Fig. 19.26a**). Hail can cause major damage by flattening crops, denting cars, and breaking glass (**Fig. 19.26b**).

Why does hail form? The process begins when updrafts carry liquid cloud droplets up to altitudes where temperatures are well below freezing. These droplets become *supercooled*, meaning that they remain liquid even though their temperature lies below 0°C. When ice particles falling from farther aloft collide with these supercooled droplets, the droplets bond to the ice particles and immediately freeze, so the ice particle quickly grows into a hailstone (see Section 17.6). In an updraft, upward air flow keeps hailstones "floating" in the cloud, where they can keep growing. Eventually, however, a hailstone drifts out of the updraft, or its weight overcomes the force of the updraft, and it starts to fall. It collects more supercooled droplets on its way down, so it continues to grow until it encounters above-freezing temperatures in the lower part of the cloud and starts to melt.

Hailstones reach the ground only if they grow large enough to avoid melting away as they fall. This happens when a thunderstorm contains particularly strong updrafts, so the most damaging hail tends to develop in supercell thunderstorms. Conditions favoring hail reaching the ground may last for minutes to over an hour. During this time, the storm moves, so hail falls on the ground along a path called a *hail swath* or a *hail streak* (**Fig. 19.26c**).

Flash Floods

In Chapter 13, we learned how a *flash flood*, a localized, short-duration flood that happens rapidly with barely any warning, can cause devastation (**Fig 19.27a**). Flash flooding can happen when thunderstorms produce sustained high rainfall rates over a drainage basin. In such cases, more rain falls than can be absorbed by the ground, so the remaining rainfall becomes runoff. Such conditions can

(a) A hailstone that fell in Vivian, South Dakota, on July 23, 2010, holds the US record for diameter (20.3 cm, or 8.0 inches) and weight (0.879 kg, or 1.9375 pounds).

(b) Hail damaged this car's back window.

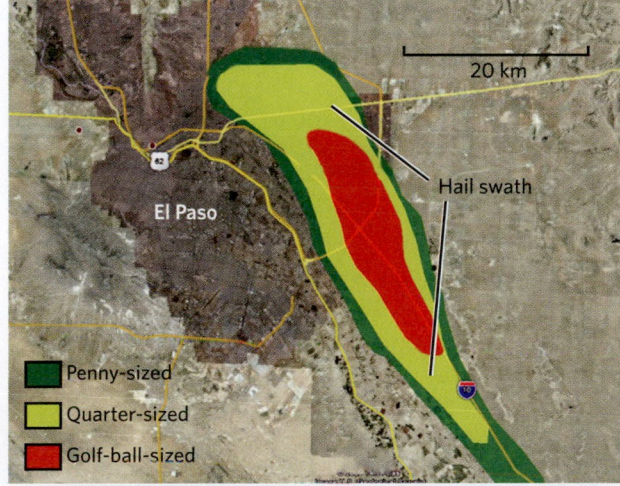

Penny-sized
Quarter-sized
Golf-ball-sized

20 km

El Paso

Hail swath

(c) A hail swath near El Paso, Texas.

occur when a thunderstorm, or a line of thunderstorms, remains fixed in a location for a relatively long time, or when several thunderstorms pass over the same drainage basin in rapid succession.

Did you ever wonder . . .

how big a hailstone can grow?

Slow-moving isolated thunderstorms are most common near mountains, where warm, moist winds approaching a mountain slope undergo orographic lifting. Rain pouring onto mountainous topography, as occurs sometimes along the west coast of the United States, funnels into narrow valleys and canyons, so discharge in tributary streams increases rapidly. The water then flows into a trunk stream that carries floodwaters downstream and, commonly, beyond the mountain front.

Squall lines also can trigger flash floods when winds, both at the surface and aloft, flow almost parallel to the front, so that the position of the front remains stationary (Fig. 19.27b, c). In such situations, individual thunderstorms move along the frontal boundary, one after the other, over the same location. Meteorologists refer to this process as *training* because the succession of storms resembles a train of boxcars passing over the same location on a track.

Take-home message . . .

Thunderstorms produce several dangerous phenomena. When the charge separation in a thunderstorm becomes great enough, a lightning stroke flashes from one part of a cloud to another or from the cloud to the ground. Strong updrafts in thunderstorms can produce large hail. Strong downdrafts that accompany heavy rains can cause damaging straight-line winds, and the heavy rains themselves may trigger flash floods.

Quick Question -
How does a cloud-to-ground lightning stroke develop?

19.5 Tornadoes

Perhaps no other weather phenomenon can be as terrifying as a **tornado**, a violently rotating funnel, or *vortex*, of air that extends from the ground up to the base of a severe thunderstorm (Fig. 19.28a, b). The strongest tornadoes can damage or destroy steel-reinforced concrete structures, can throw trucks long distances, and can sweep houses completely away. At its base (the end touching the ground), a tornado typically ranges in width from about 50 m to 1 km (150 feet to 0.6 miles), but some are much larger. The largest tornado ever recorded, the El Reno, Oklahoma, tornado on May 31, 2013, reached a width of 4.2 km (2.6 miles) at its base. Tornado winds generally blow between 100 and 320 km per hour (65 and 200 mph), but

Figure 19.27 Flash floods.

(a) Flash floods occur rapidly, typically without warning.

(b) Clouds and thunderstorms line up along a stationary front across the southeastern United States.

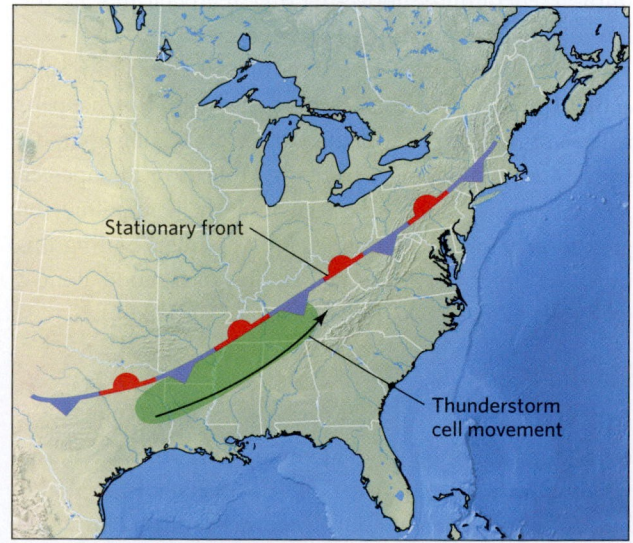

(c) Flash floods can develop when a front is stationary and thunderstorms form and move parallel to the front, so that heavy rain continually falls over a particular location.

Figure 19.28 Examples of tornadoes of different sizes.

(a) Small tornado in Kansas.

(b) Giant tornado (1 km wide) in Oklahoma.

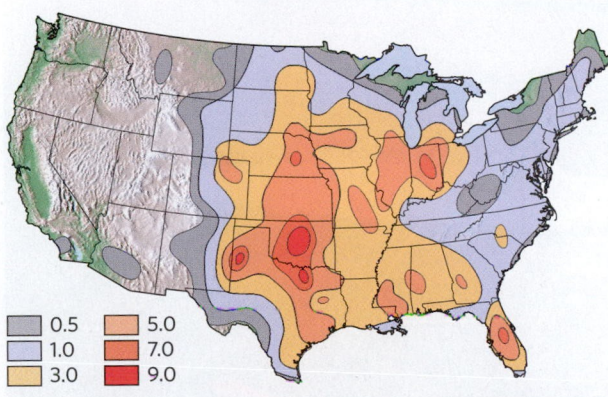

0.5	5.0
1.0	7.0
3.0	9.0

(c) Number of tornadoes per year (per 26,000 square km, for a 27-year period).

the fastest tornado wind speed ever recorded clocked in at 484 km per hour (301 mph). A tornado moves across the countryside along with its host storm, so movement at its base can vary from near-stationary to a forward speed of close to 110 km per hour (70 mph). The central and midwestern United States, where warm, moist air from the Gulf of Mexico borders cold air from Canada along strong fronts in spring and early summer, hosts the world's largest number of tornadoes, giving this region the nickname, Tornado Alley **(Fig. 19.28c)**. Tornadoes are also common across the southeast United States and in Florida.

Most tornados form over land. (When tornadoes develop over water, they are called *waterspouts*; these tornadoes tend to be fairly small.) As a tornado moves, it leaves a swath of destruction, called a **tornado track**, on the ground surface **(Fig.19.29a)**. Most tornadoes are short-lived, traveling only a few kilometers from where they touch down to where they lift and dissipate. But occasionally, a tornado can survive for over an hour and produce a very long track. Perhaps the longest tornado track on record was left by the great Tri-State Tornado of 1925, which tore a 352-km (219-mile) swath of devastation across Missouri, Illinois, and Indiana and

Figure 19.29 Tornado tracks.

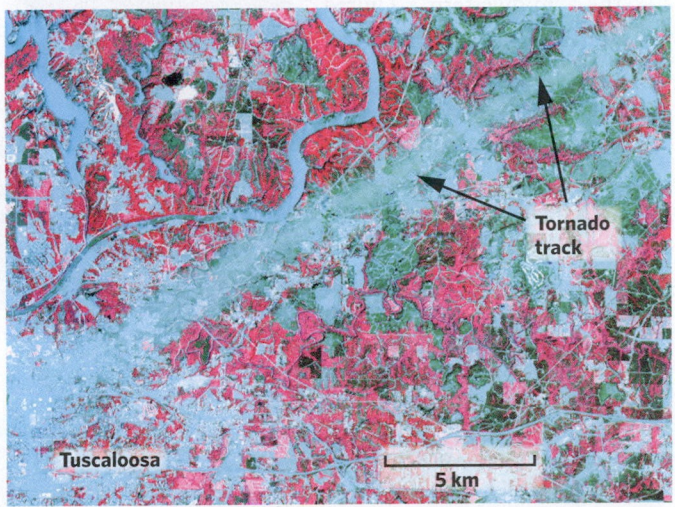

(a) This false-color satellite image shows the track of a tornado near Tuscaloosa, Alabama. Red areas are forested—the winds ripped out the trees.

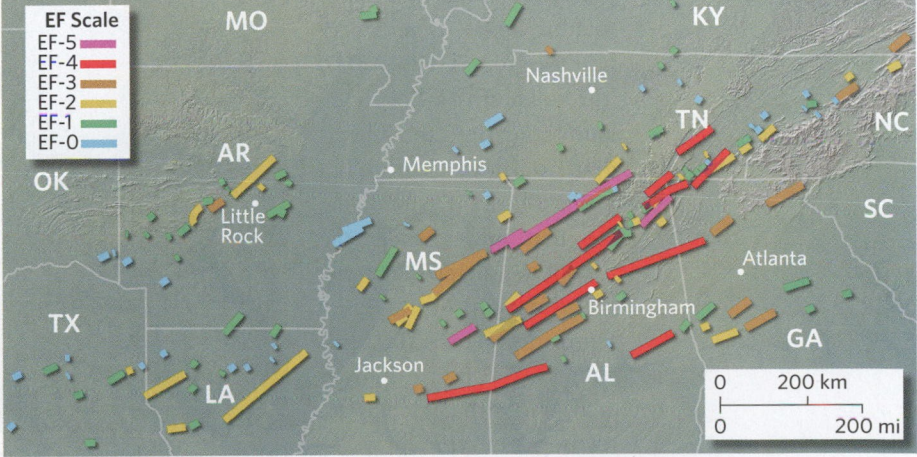

(b) A series of tornado tracks from an outbreak of tornadoes in April 2011.

killed 747 people. On occasion, numerous tornadoes develop during a set of related storms within a relatively short time. The world's largest tornado outbreak took place in late April 2011 (Fig. 19.29b). It produced 363 tornadoes over the course of three days, caused the deaths of 348 people, and destroyed $11 billion worth of property.

Tornadoes develop most commonly within supercell thunderstorms, but they occasionally form along squall lines or within the outer rainbands of hurricanes.

Supercell Tornadoes

To understand tornado formation, let's look at how a tornado evolves in a supercell thunderstorm moving from southwest to northeast across the midwestern United States (Fig. 19.30a). Recall from Section 19.3 that a supercell thunderstorm contains a strong rotating updraft, called a mesocyclone, and that strong downdrafts form on both the forward (northeastern) flank and rear (southwestern) flank of the mesocyclone in association with heavy rainfall (Fig. 19.30b), a necessary condition for tornado formation (Fig. 19.30c). As the rear-flank downdraft approaches the ground, it spreads outward laterally and produces a horizontal roll of rotating air at its periphery near the ground (Fig. 19.31a). The roll forms a horizontal loop bordering the downdraft outflow. A tornado forms when the part of the roll that moves under the base of the mesocyclone gets pulled upward into the mesocyclone (Fig. 19.31b). Initially, the roll warps into an arch, but the clockwise-rotating arm typically weakens and dissipates, while the

counterclockwise-rotating arm, rotating in the same direction as the mesocyclone, is carried directly into the updraft, where it stretches and narrows to becomes the tornado, a spiral or vortex of spinning air (Fig. 19.31c). As the vortex stretches vertically and narrows, the air within it rotates faster and faster in order to conserve angular momentum (see Chapter 15), giving the tornado its immense wind

Figure 19.30 Setting the stage for tornado formation in a supercell thunderstorm.

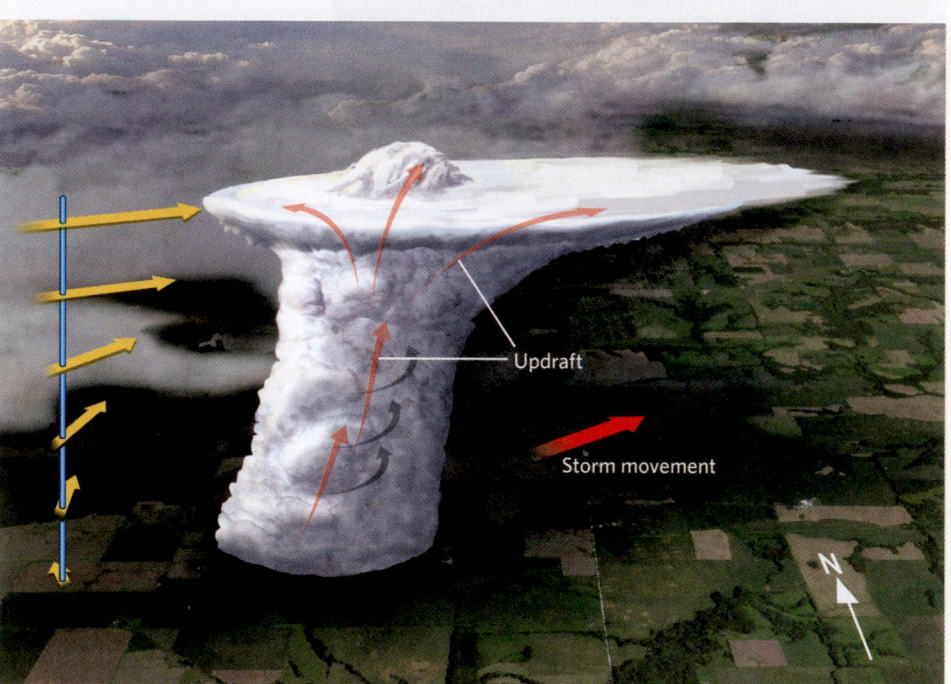

(a) A supercell storm migrates northeastward across the plains.

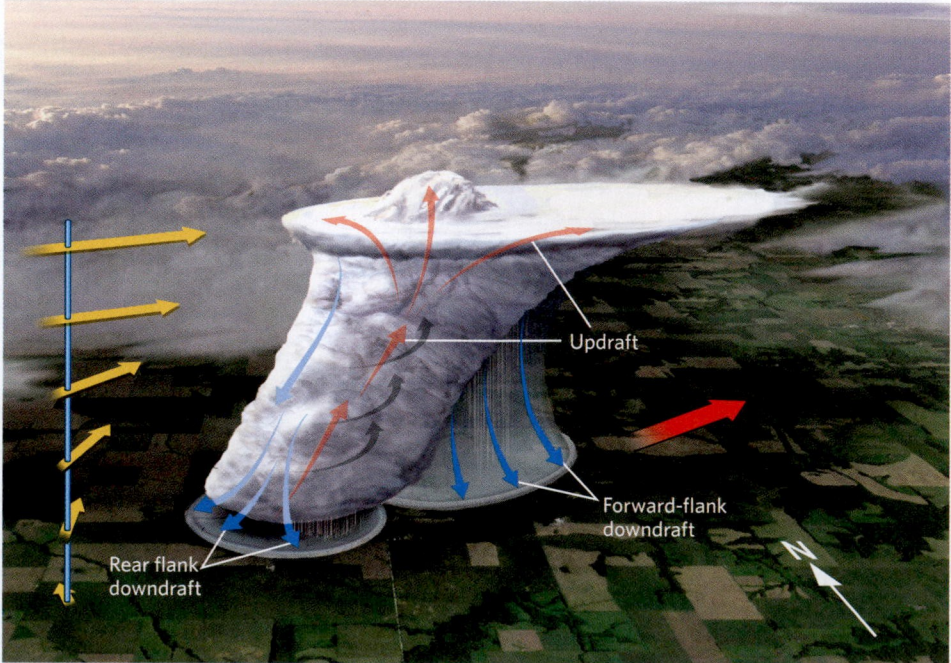

(b) Strong winds aloft cause the updraft (red arrows) to tilt. Both forward-flank and rear-flank downdrafts develop.

speed. In some cases, a tornado can evolve to include several vortices. Such *multiple-vortex tornadoes* can become particularly wide and destructive (Box 19.3). Note that although the process by which a tornado forms resembles the way a mesocyclone forms, a tornado and a mesocyclone are distinctly different phenomena.

The visible tornado extends beneath the wall cloud of the supercell and becomes a funnel-shaped cloud. Within the tornado, the rising air creates very low air pressure, so as moist air gets drawn into the tornado, its water vapor condenses and forms a spinning cloud—the visible funnel that we see. At the ground, the tornado generates debris by tearing apart buildings and trees and even by ripping into the ground. Some of this dust and debris is incorporated into the tornado and contributes to its dark appearance.

Non-Supercell Tornadoes

The largest tornadoes occur in association with supercells, but not all tornadoes form in such storms. Tornadoes can also develop within squall lines and in hurricanes. In such cases, the tornadoes develop where there is *horizontal wind shear*; specifically, where air on one side of a front is moving horizontally in a direction opposite to the air on the other side. Along a northeast-southwest-aligned cold front, for example, air on the warm side may be blowing toward the northwest, whereas air on the cold side may be blowing toward the southeast (Fig. 19.32a). The shear between two such oppositely flowing airstreams, if it is

strong enough, causes air to start rotating around a vertical axis, producing small vertical vortices (Fig. 19.32b). The updraft of an overlying thunderstorm can pull one of these vortices upward, causing it to stretch vertically and spin faster to become a short-lived, weak tornado. This process can take place simultaneously at many locations along a front—in rare cases, producing a line of tornadoes (Fig. 19.32c). Similar types of tornadoes form in the outer rainbands of hurricanes, but these are hard to observe because they are cloaked by heavy rain.

Defining the Intensity of a Tornado

Not all tornadoes cause the same amount of damage. Weak ones may strip the shingles from a roof, while moderate ones may tumble a mobile home, tear off an entire roof, and leave swirling patterns in the grass. Strong tornadoes can sweep the strongest house completely away and can even dig up the ground like a monstrous bulldozer.

The damage caused by a tornado depends, of course, on wind velocity. It also depends on whether the tornado encounters buildings or crosses open fields. It's not possible to measure the wind velocity in every tornado because appropriate instruments are not always nearby. To solve this problem, Ted Fujita, a professor at the University of Chicago, proposed classifying tornado intensities based on the damage caused, and he provided estimates of how damage correlates with wind velocity based on laboratory tests. His classification system, published in 1971, came to be known as the Fujita scale. In 2007, a more accurate version of the Fujita scale, called the **Enhanced Fujita (EF) scale**, became available. The EF scale takes into account newer studies that better constrain the relationship between wind speed and observed damage, and this is the scale that we use today (Table 19.1).

The EF scale divides tornadoes in six categories, labeled EF0 through EF5 (Box 19.4). An EF0 tornado is very weak, whereas an EF5 tornado is catastrophic. During EF5 tornadoes, wind speeds exceed 322 km per hour (200 mph) at the time that damage occurs. The EF rating does not depend on the width of the

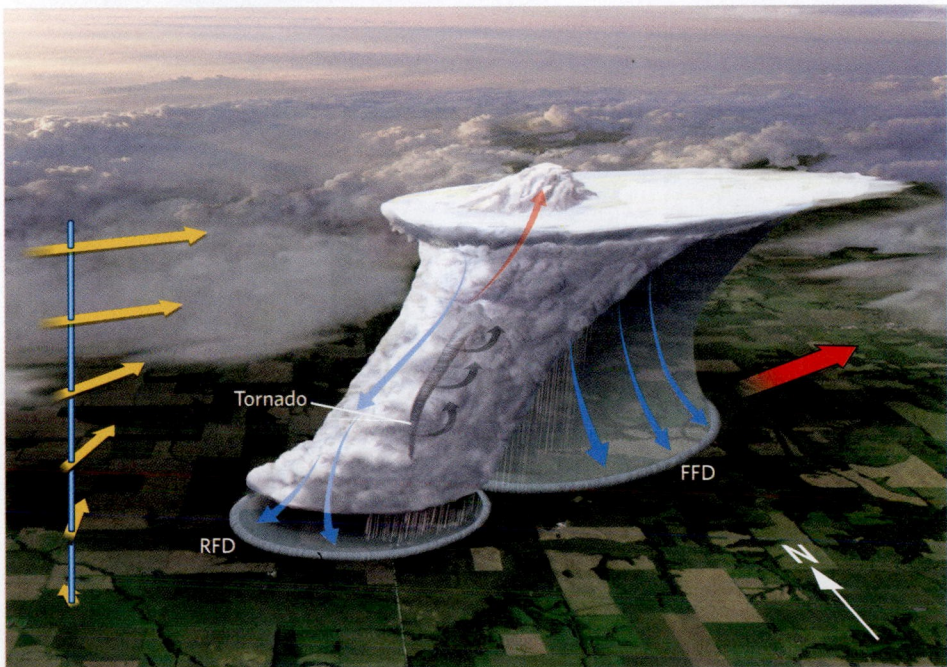

Figure 19.30 continued.

Tornado

RFD

FFD

N

(c) When the rear-flank downdraft reaches the ground, it spreads outward, with some of the air flowing beneath the mesocyclone.

Figure 19.31 A close-up of the rear-flank downdraft of a supercell, showing the formation of a tornado.

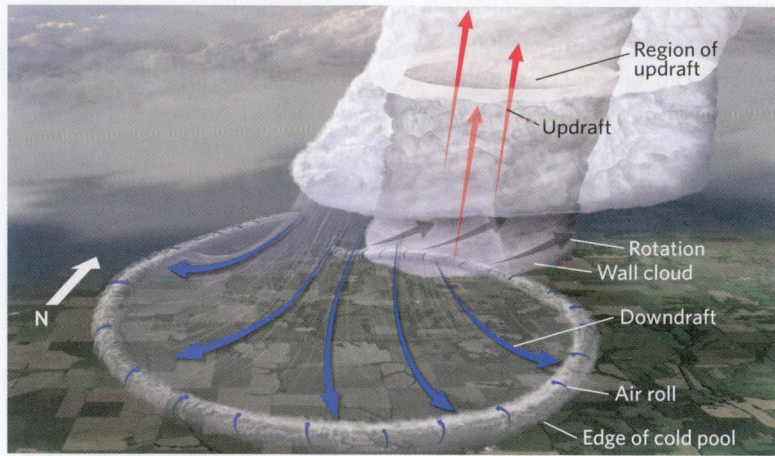

(a) The boundary of the rear-flank downdraft forms a horizontal roll of rotating air close to the ground.

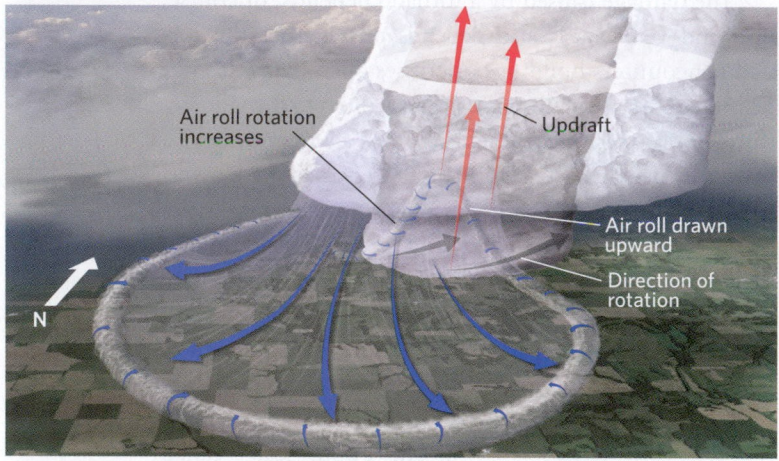

(b) The part of the roll under the updraft gets drawn upward and tilted into the updraft. As it stretches and thins, it spins faster.

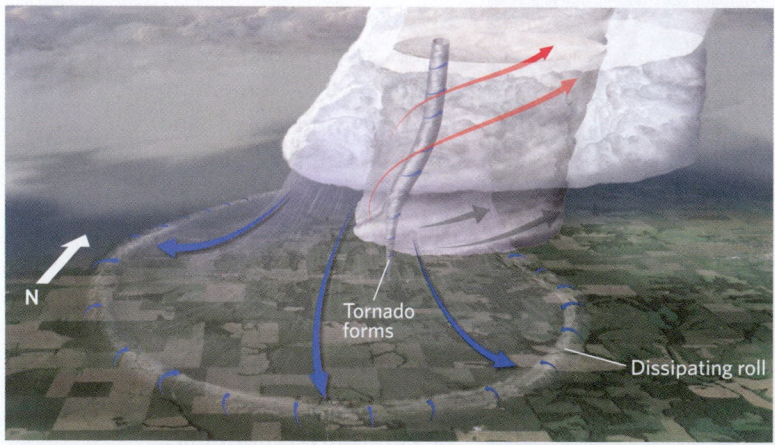

(c) The vortex tightens into a tornado and protrudes from the base of the wall cloud.

tornado, but in general, more intense tornadoes are wider. As a tornado evolves, its EF rating may change. The final rating is based on the worst damage the tornado causes along its track (**Earth Science at a Glance**, pp. 688–689).

Tornado Detection and Safety

In 1925, people in the path of the deadly Tri-State Tornado had virtually no warning that it was coming, unless they happened to be outside scanning the horizon. Fortunately, the skill of tornado detection has become much more advanced. Now, with polarization Doppler radar (see Box 17.2), tornado signatures may appear clearly on a radar screen **(Fig. 19.33)**. For example, flying debris

Figure 19.32 Formation of non-supercell tornadoes.

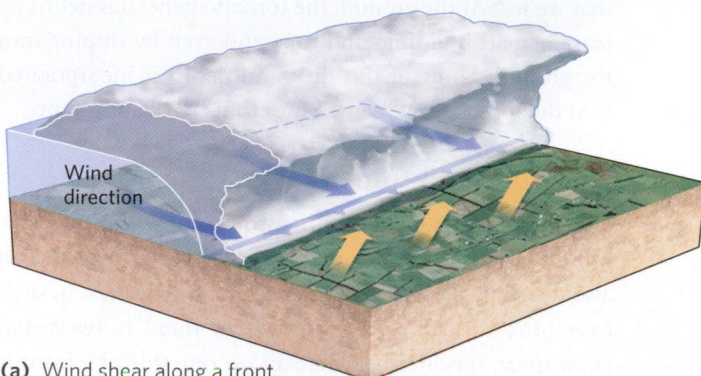

(a) Wind shear along a front.

(b) Horizontal wind shear forms small vortices that may be pulled upward by thunderstorms.

(c) This process can form vortices at several locations, as in this row of tornadoes.

EF rating	Wind (km per hour)	Wind (mph)	Typical damage
EF0	105–137	65–85	*Minor damage:* Peels off some shingles; breaks branches; topples weak trees
EF1	138–177	86–110	*Moderate damage:* Severely strips roofs; overturns mobile homes; breaks windows
EF2	178–217	111–135	*Considerable damage:* Tears roofs off; shifts houses off foundations; destroys mobile homes; uproots large trees; flings debris and lifts cars
EF3	218–266	136–165	*Severe damage:* Destroys upper stories of well-built houses; severely damages large buildings; overturns trains; throws cars; debarks trees
EF4	267–322	166–200	*Extreme damage:* Completely levels well-built houses; cars and trucks thrown about
EF5	> 322	> 200	*Catastrophic damage:* Tall buildings collapse; structures made of reinforced concrete severely damaged; cars and trucks carried for more than a kilometer

scatters radar energy, so the *debris signature* generated by a tornado appears as a bright spot on radar. Doppler radar can detect rotating winds by distinguishing the part of the vortex moving toward the observer from the part moving away. Therefore, when a meteorologist sees a supercell thunderstorm with a hook echo on the southwestern side, rotation within the hook, and a debris signature at the tip of the hook, it's a sure sign that a tornado is on the ground and that a warning should go out.

Tornadoes destroy property and landscapes and often kill (Fig. 19.34). Averaged over many years, about 60 people per year die in tornadoes in the United States. Some years include particularly high death tolls due to tornadoes; in 2011, for example, 553 people lost their lives. Most injuries and fatalities happen when people are struck by flying debris. Even a splinter of wood, when carried by a wind blowing 200 km per hour (125 mph), becomes a deadly projectile.

The NOAA Storm Prediction Center issues a *tornado watch* for specific areas when weather conditions favor tornado formation. A local National Weather Service office issues a *tornado warning* either when observers have spotted a tornado visually or when one has been detected on radar. Warnings trigger weather alerts on radio, TV, and cell phones; in some communities, emergency managers activate sirens. The NWS recommends that families, schools, and businesses have safety plans and hold frequent drills for tornadoes. In general, when a tornado warning goes out, you should move to a predetermined shelter, such as a basement or a storm cellar. If an underground location is not available, the safest location is an interior room or a hallway on the lowest floor, especially an interior bathroom, where plumbing provides additional wall support. Put as many walls as possible between you and the tornado, and avoid windows. Abandon mobile homes, which may well be crushed. If caught outdoors, move away from potential sources of airborne debris and lie in the lowest spot available. In an urban or congested area, abandon your car

for a sturdy shelter. In all cases, when severe weather threatens early action can be the key to your safety and survival.

Take-home message . . .

Tornadoes are rapidly rotating funnels of air that extend from the ground to the base of a severe thunderstorm. They form when a rotating cylinder of air (a vortex) gets drawn into a storm's updraft, which causes it to stretch, narrow, and spin ever faster. The largest tornadoes form within supercell thunderstorms, but tornadoes can also develop within squall lines and in hurricanes. Tornado intensity is classified using the Enhanced Fujita scale, which is based on the damage caused.

Quick Question -
What creates the visible funnel of a tornado?

Figure 19.33 Radar signature of a tornado that struck Alabama in 2013. The bright spot at the tip of the hook echo comes from debris thrown upward by the tornado.

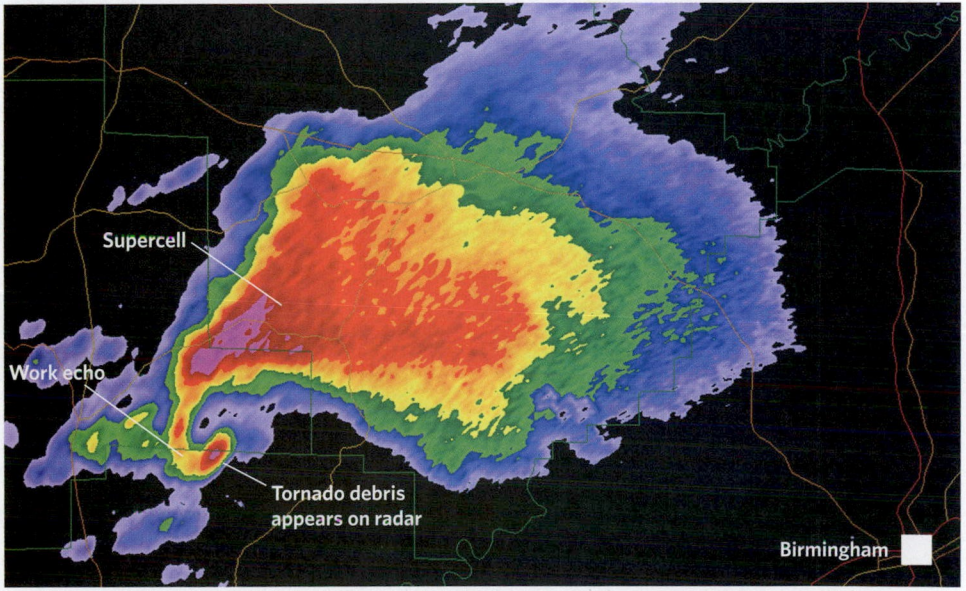

Box 19.3

Consider this . . .

Multiple-vortex tornadoes

Most tornadoes are short-lived and have widths of a few hundred meters at their base. Some tornadoes, however, grow to over 1 km wide and develop multiple vortices **(Fig. Bx19.3)**. Why does this occur? Tornadoes normally consist of rapidly rotating, rising air. In some tornadoes, the extremely low pressure that develops at the base of the vortex can draw air downward within the tornado core. This process generates a downdraft that forms aloft and descends, inside the tornado, toward the ground. The downdraft pushes the sides of the tornado outward, so it becomes wider. When the downdraft reaches the ground, the difference in wind direction between the downdraft outflow and the rotating winds of the tornado causes the main tornado to break down into several smaller tornadoes, called *suction vortices*, moving around the main tornado. Meteorologists refer to the product of this evolution as a **multiple-vortex tornado.** The most violent winds in the tornado occur in the suction vortices because the winds in each one have the combined speeds of the smaller tornado and the main tornado. The most intense tornadoes are nearly always multiple-vortex tornadoes. Such tornadoes have long lifetimes and produce particularly wide swaths of devastation.

Figure Bx19.3 Formation of a multiple-vortex tornado.

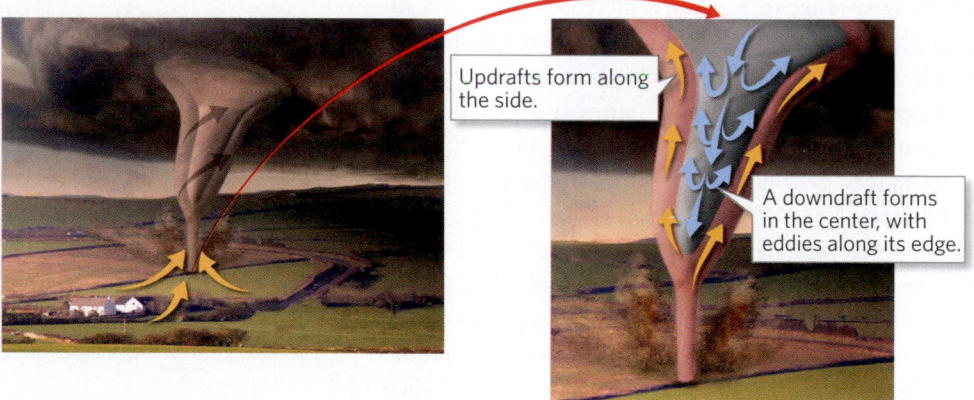

(a) A small tornado, consisting of a single vortex, forms beneath the wall cloud at the base of a supercell.

(b) A downdraft forms in the middle of the tornado and pushes the sides of the tornado outward.

(c) The central downdraft reaches the ground, so the whole tornado is now almost as wide as the wall cloud.

(d) Within the overall huge tornado, a number of short-lived but very intense suction vortices form. These vortices contain the most intense winds.

Figure 19.34 Tornado devastation.

(a) A tornado leaves a swath of destruction.

(b) During a strong tornado, trees snap off and even brick walls fall.

Box 19.4 How can I **explain** . . .

How meteorologists assign an EF rating to a tornado

What are we learning?
• How meteorologists estimate the wind speeds in a tornado from damage indicators.

What you need:
• A picture of a tornado-damaged suburban home taken from the Internet.
• The table to the right.

Instructions:
• Examine the picture carefully and match the damage as well as possible to the damage descriptions in the table.
• Assign an EF rating using the table. This table, one of many used by meteorologists as they look at the damage to structures, is specifically for houses.

What did we see?
• Meteorologists examine damage to structures and assign the EF rating based on the worst damage observed. The tables are based on studies at the Wind Science and Engineering Research Center at Texas Tech University.

Damage description	Expected wind speed	EF rating
Threshold of visible damage	65	0
Loss of roof covering material <20%, gutters, and/or awning; loss of vinyl or metal siding	79	0
Broken glass in doors and windows	96	1
Uplift of roof deck and loss of roof covering material >20%; collapse of chimney; garage doors collapsed inward; failure of porch or carport	97	1
Entire house shifts off foundation	121	2
Large sections of roof structure removed, most walls remain standing	122	2
Exterior walls collapsed	132	2
Most walls collapsed except small interior rooms	152	3
All walls collapsed	170	4
Destruction of residence; slab swept clean	200	5

Life Cycle of a Large Tornado

EF-2 and EF-3

EF-0

EF-1

Examples of damage to houses at different locations along the tornado's track

EF-0

EF-1

EF-2

EF-4 and EF-5

A tornado evolves over time. It starts small, grows larger, and then eventually dissipates. In the record books, the overall rating of a tornado using the Enhanced Fujita (EF) scale reflects the maximum damage that the tornado causes somewhere along a portion of its path. Here, we see the swath of destruction that a tornado produced as it passed through a town, and a representation of the shape and size of the tornado at various times during its life cycle. This example was an EF-5 tornado along only a small part of its track.

EF-2 and EF-3

EF-1

EF-3

EF-4

EF-5

19.6 Region-Specific Weather Systems

Many regions of the world have unique local weather that reflects interactions between the atmosphere and specific surface features, as well as specific seasonal conditions. Particularly notable examples occur along land-water boundaries and along and within mountain belts.

Sea Breezes

Land-sea boundaries generate several types of weather. Oceans have a huge *heat capacity*, which means that they can hold a lot of heat relative to the land. Furthermore, land surfaces warm up rapidly during the day, while the ocean does not—you'll burn your feet walking on the beach in the tropics, but will be quite comfortable immersed in the adjacent sea. Why? The same solar rays warm both, but on opaque, immobile land, the warmth remains at the surface, whereas warming of the ocean affects deeper layers because water is transparent and because it circulates. The contrast between warmer land surfaces and cooler water drives local air circulations along coastlines. Specifically, during the afternoon, air rises over the hot land, flows seaward aloft, descends over the cooler ocean, and flows landward just above the ocean surface **(Fig. 19.35a)**. This circulation, called a **sea breeze**, occurs along many coastlines of the world, and even along the shores of large lakes **(Fig. 19.35b)**. At night, some coastlines experience a *land breeze*, a mirror image of a sea breeze, because the land cools off faster than the sea does, so the direction of the breeze reverses.

Weather associated with a sea breeze depends on the characteristics of the warm and cool air on either side of the *sea breeze front*, the boundary between oceanic cool air and warm air over the land. If the warm air over the land contains sufficient moisture and warms up enough, it becomes unstable, rises, and triggers clouds and thunderstorms. This process takes place daily along many water bodies, including those as small as the Great Lakes (Fig. 19.35b).

Along the Pacific coast of the United States, ocean water remains cold in summer, and the air onshore is dry. When the land heats up in the early part of the day, the sea breeze causes cool air to blow from the ocean over the land. In the Los Angeles Basin, which is surrounded by mountains, the cool, dense marine air collects at low elevations over the city for most of the day, particularly during peak traffic hours. This marine air is stable, so urban air pollutants accumulate within it throughout the day, producing smog **(Fig. 19.36)**.

Lake-Effect Storms

Because water has a high heat capacity, the water of a large lake may remain liquid in early winter even if the air above it drops to temperatures well below freezing. As a consequence, the lake can provide moisture and heat to the atmosphere, and it can therefore produce local weather systems called **lake-effect storms (Fig. 19.37a)**. Lake-effect storms commonly occur over and adjacent to the Great Lakes when cold air from northern Canada flows southward over the lakes. As this air passes, heat and moisture move from the warm lake surface into the boundary layer, while the air above remains cold. The warmed air becomes unstable and starts to rise to form cumulus clouds. These clouds become taller as the layer of warm air increases as it moves across the lake. Eventually, the clouds start to precipitate snow.

When the relatively, warm, moist air arrives at the downwind shore, friction with trees and buildings reduces the wind speed near the ground. Effectively, the near-surface wind piles up as it reaches the shore and is forced to rise. This additional lifting triggers production of even more

Figure 19.35 The sea breeze.

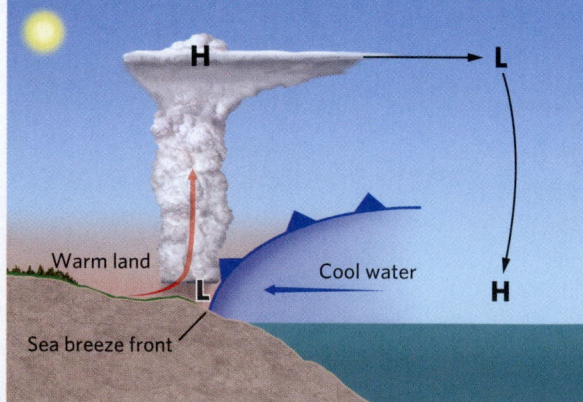

(a) A sea breeze circulation during the daytime along a coastline can trigger thunderstorms.

(b) Clouds along the sea breeze fronts along Lake Erie.

Figure 19.36 Smog trapped in the marine air over the Los Angeles Basin.

clouds and yields heavy snowfalls downwind for several kilometers inland from the shore. As a consequence, regions downwind of the Great Lakes sometimes receive very heavy snow—up to 150 cm (60 inches) may fall during a single lake-effect storm (Fig. 19.37b). Such lake-effect snowfall can continue for days, sometimes at a rate of more than 2.5 cm (1 inch) per hour, resulting in enormous accumulations in areas within a few tens of kilometers of the shore.

Mountain Rain and Snow

Winds blowing toward a mountain range obviously cannot pass through rock. The air has nowhere to go but up, so on the upwind, or *windward*, side of a mountain range, as we have seen, air undergoes orographic lifting. Such lifting produces clouds and precipitation and can trigger thunderstorms in summer and heavy snowfalls in winter. In fact, the heaviest and most persistent precipitation in the world occurs on the windward sides of mountains. When large-scale storm systems, such as tropical or mid-latitude cyclones, bring huge amounts of moist air into mountainous regions, orographic lifting triggers heavy rains that can cause devastating floods and massive mudslides. During the Southeast Asian monsoon rains, for example, local regions in the foothills of the

Did you ever wonder . . .

where on Earth the most rain and snow falls?

Figure 19.37 Lake-effect storms.

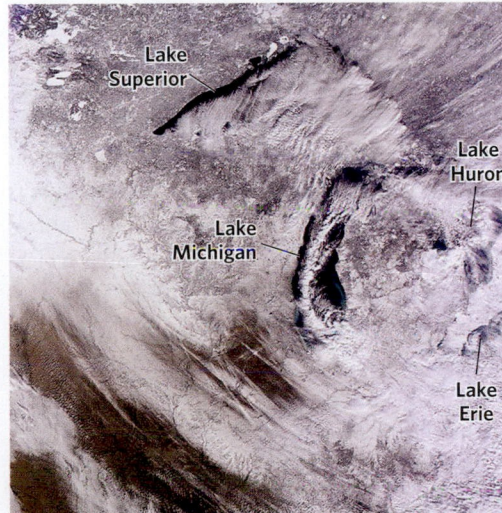

(a) Lake-effect clouds develop when cold air flows over the Great Lakes.

(b) Lake-effect storms produce heavy snowfall.

Himalayas can receive meters of rain. The western slopes of Mt. Waialeale, in Hawaii, receive the greatest average annual rainfall on Earth—nearly 12 m (38 feet)—as winds blowing eastward across the Pacific bring a never-ending supply of moisture.

In winter, the precipitation triggered by orographic lifting can cause *blizzards*, blinding snowstorms that can persist for days and deposit meters of snow on mountain slopes, setting the stage for avalanches. Mountain snow-storms, though dangerous, are extremely important to world water supplies. Snow accumulated in winter represents one of the world's large reservoirs of clean fresh-water **(Fig. 19.38)**, and many of the world's major rivers have their origins in the snowfields and glaciers of high mountain ranges.

Downslope Windstorms in Mountains

Because air moving over a mountain range loses most of its moisture as rain or snow on the windward side of the range, it is relatively dry as it descends on the downwind, or *leeward*, side of the range. In addition, as it descends, the air compresses, becomes denser, and warms up. As a consequence, the leeward side of the range may be a desert, even where the windward side is wet and vege-tated (see Chapter 14). In addition to dry weather, the leeward side of a mountain range occasionally experiences a **downslope windstorm**. This type of windstorm has different names in different locations: along the Rocky Mountains, it's a *chinook wind*; in the Alps, it's a *foehn wind*; and in California, it's a *Santa Ana wind*.

Why do downslope windstorms develop? Wind will blow across a mountain range if higher pressure is present on the windward side than on the leeward side. This sit-uation might occur along the Rockies, for example, if a mid-latitude cyclone lies over the Great Plains at the same time a high-pressure system develops over the Great Basin. As we saw in Chapter 18, stronger horizontal pres-sure gradients cause faster winds. When a particularly strong pressure gradient exists across a mountain range, the stage is set for a downslope windstorm.

In the case of chinook and foehn winds, air flow be-comes more rapid as air rises over the crest of the moun-tains in the shape of a wave, much like the wave that forms when a stream of water flows up and over a large boulder. Clouds form on the upwind side of the moun-tains as air rises. Viewed from the downwind side, the clouds appear likes a wall, called a *chinook cloud wall*. The background pressure gradient above the mountains accelerates the air passing across the range. This wave takes on a special shape when an **inversion**—a layer in the atmosphere in which the temperature increases with altitude—forms upwind of the mountain crest at an elevation just above it **(Fig. 19.39)**. The inversion acts like a lid, confining the wind to a narrow layer above the mountaintop. As a consequence, the wind acceler-ates even more. You may have seen a similar phenom-enon if you've blocked the end of a water hose with your thumb so that water spurts out of the hose more quickly. When this strong wind comes down the lee-ward side of the mountain range in a shooting flow and descends to lower elevations, it accelerates further—the strongest downslope winds can reach 160 km per hour (100 mph). Because the air warms as it descends, it can melt snow rapidly.

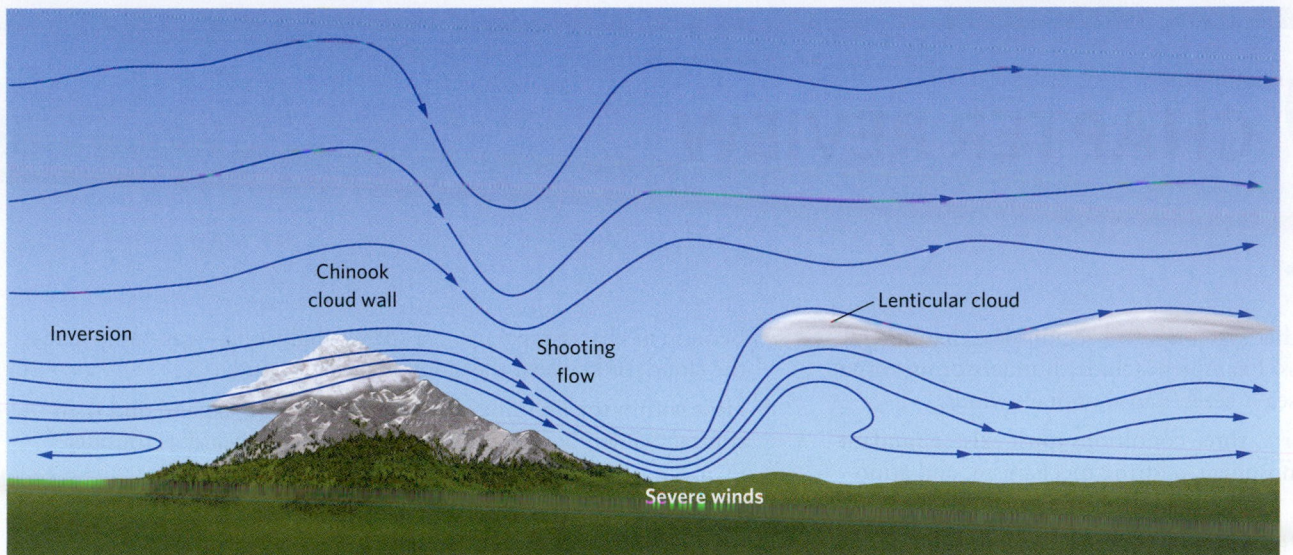

Chinook
cloud wall

Inversion

Shooting
flow

Lenticular cloud

Severe winds

Santa Ana winds blow when high pressure develops over the desert regions to the east of California's mountain ranges. If a low-pressure zone exists to the west, on the coastal side of California, the pressure gradient causes winds to blow from the desert westward over the mountains. These winds rush down the valleys of Southern California, and as with chinook winds, the air warms as it descends. During a dry season, these winds often fan disastrous wildfires.

The downslope windstorms we have just described are driven by strong pressure gradients. A second type of strong windstorm occurs in cold mountainous regions, or in polar regions, where the top surface of a large ice sheet has a higher elevation than its surroundings. These **katabatic winds** develop because air above the ice sheet becomes very cold and therefore relatively dense. As a result, it flows downslope off the ice sheet, sometimes reaching speeds of up to 160 km per hour (100 mph). Katabatic winds can begin abruptly when a dome of extremely cold, dense air has built up over the ice sheet. The winds begin

when gravity finally causes the cold air to spill off the ice sheet and rush outward onto the land or sea below **(Fig. 19.40)**. Katabatic winds are common in Antarctica, along the coast of Greenland, and in mountainous regions of Canada and Alaska that have permanent ice sheets.

Take-home message . . .

Some local weather systems form only at specific locations. For example, the contrast in heat capacity between land and water can drive sea breezes or lake-effect storms. Mountain topography can lead to the development of strong downslope winds, such as chinook and Santa Ana winds. Cold air pools over ice sheets and flows to lower elevations to produce katabatic winds.

Quick Question -
Why does a sea breeze in Florida flow from the ocean toward land?

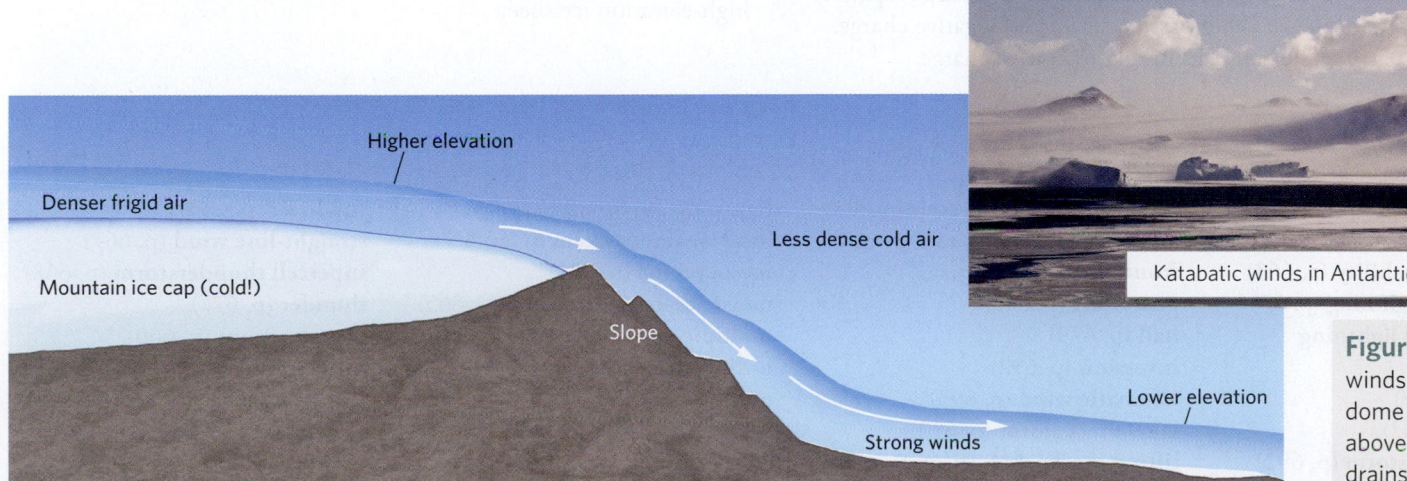

Higher elevation

Denser frigid air

Mountain ice cap (cold!)

Less dense cold air

Slope

Strong winds

Lower elevation

Katabatic winds in Antarctica

Figure 19.40 Katabatic winds develop when a dome of cold air forms above an ice sheet and then drains to lower elevations.

Chapter Summary

- Local, or mesoscale, weather systems affect regions less than several hundred kilometers across. Examples include thunderstorms, tornadoes, lake-effect storms, sea breezes, and mountain winds.

- For a thunderstorm to form, three conditions must exist: abundant moisture in the lower atmosphere, a lifting mechanism, and atmospheric instability.

- Lifting occurs when air is forced upward by a front, gust front, or high topography, or when it is warmed from below by the Earth's surface.

- Rising moist air, because of the release of latent heat due to condensation, becomes buoyant relative to its surroundings when atmospheric instability is present. Updrafts in unstable air can sometimes extend all the way to the tropopause. In stable air, lifted air sinks back down after it is lifted.

- Ordinary thunderstorms form where winds do not vary substantially with altitude, so these storms do not rotate. They evolve through three stages: the development stage, when warm, moist air is lifted and becomes buoyant; the mature stage, when precipitation produces a downdraft; and the dissipation stage, when the updraft ceases and the clouds dissipate.

- In squall-line thunderstorms, several thunderstorms develop along a front or gust front. The combined outflow of cool air from such storms can trigger formation of new thunderstorms.

- Supercell thunderstorms develop if wind changes rapidly in speed and direction with elevation, causing the updraft of the storm to rotate. The updraft in these storms typically tilts, so that rain falls downstream from the updraft. As a result, downdrafts don't counteract updrafts, and the storm can survive for several hours.

- Supercell updrafts are so strong that an overshooting top rises above the anvil cloud of the storm.

- Lightning is a form of static electricity. It is caused by charge separation in a thunderstorm: the cloud base accumulates negative charge, and the upper parts of the cloud accumulate positive charge.

- Because of charge separation, a spark—lightning—can jump across the cloud, or between the cloud and the ground.

- Heat within the lightning channel causes air to expand and then contract violently. The resulting shock wave is heard as thunder.

- Hailstones form when supercooled water droplets freeze onto ice particles in clouds. The largest hailstones form in supercell thunderstorms.

- Flash floods occur when slow-moving thunderstorms, or a succession of thunderstorms, dump a large amount of rain over a drainage basin in a short time.

- Tornadoes typically occur on the rear flank of a supercell thunderstorm. They form when part of the rear-flank downdraft gets drawn into the storm's updraft.

- Tornadoes can cause immense damage along a tornado track. Their intensity is rated using the Enhanced Fujita (EF) scale, which is based on the damage the storms cause.

- A sea breeze is a circulation driven by the difference between daytime heating of the shore and of the ocean. It consists of air flowing shoreward at low altitudes, rising over land, flowing oceanward aloft, and sinking over the ocean. It may trigger thunderstorms.

- Lake-effect storms develop over large lakes during winter as moisture and heat rise from the lake and moisture condenses to form clouds above the lake and its downwind shores.

- On the windward side of a mountain range, heavy rain or snow is common. On the leeward side, dry conditions are common.

- Chinook, foehn, and Santa Ana winds are strong winds that develop occasionally on the leeward side of a mountain range when a strong pressure gradient exists across the range.

- Katabatic winds develop when exceptionally cold air cascades off a high-elevation ice sheet.

Key Terms

adiabatic expansion (p. 664)
anvil cloud (p. 666)
atmospheric instability (p. 663)
cloud-to-cloud lightning (p. 675)
cloud-to-ground lightning (p. 675)
cold pool (p. 667)
downdraft (p. 662)
downslope windstorm (p. 692)
dry adiabatic lapse rate (p. 664)

Enhanced Fujita (EF) scale (p. 683)
environmental lapse rate (p. 664)
frontal squall line (p. 668)
gust front (p. 662)
hail (p. 679)
inversion (p. 692)
katabatic wind (p. 693)
lake-effect storm (p. 690)
lifting mechanism (p. 662)
lightning (p. 673)

lightning stroke (p. 674)
local weather system (p. 660)
mesocyclone (p. 671)
moist adiabatic lapse rate (p. 665)
multiple-vortex tornado (p. 686)
ordinary thunderstorm (p. 666)
orographic lifting (p. 663)
sea breeze (p. 690)
severe thunderstorm (p. 661)
squall-line thunderstorm (p. 667)

stable air (p. 663)
straight-line wind (p. 667)
supercell thunderstorm (p. 668)
thunder (p. 677)
thunderstorm (p. 660)
tornado (p. 680)
tornado track (p. 681)
unstable air (p. 663)
updraft (p. 661)
wall cloud (p. 671)

The letters following each Review Question refer to the corresponding Learning Objective from the Chapter Opener.

1. How does the size of a local-scale weather system compare to that of a hurricane? **(A)**

2. What distinguishes a thunderstorm from other kinds of rainstorms? What characteristics does a thunderstorm have if it is classified as "severe"? **(D)**

3. On a day when thunderstorms occur, do the thunderstorms last all day? **(C)**

4. List the three conditions that must exist for a thunderstorm to develop. **(B)**

5. Where does the moisture that feeds thunderstorms come from? **(B)**

6. Describe the various lifting mechanisms that can trigger initiation of a thunderstorm. **(B)**

7. Explain the difference between stable and unstable atmospheric conditions. **(E)**

8. How does the rate at which temperature changes in an updraft influence whether lifted air will become unstable and trigger a thunderstorm? **(B)**

9. What are squall-line thunderstorms? How can the gust fronts that they create produce new thunderstorms? **(C)**

10. Explain how a supercell thunderstorm forms and how it is different from an ordinary thunderstorm. Why do these storms typically have overshooting tops and produce hail? What do they look like on a radar screen? **(C, D)**

11. Identify the positive polarity and negative polarity cloud-to-ground strokes on the figure. **(F)**

12. What is a downburst, and how can it pose a hazard to aircraft? **(F)**

13. Why is hail more common in supercell thunderstorms than in other thunderstorms? **(F)**

14. What special conditions can cause a thunderstorm or line of thunderstorms to produce particularly devastating flash floods? **(F)**

15. Explain the basic process of tornado formation. Why do tornadoes typically develop on the rear flank of a supercell thunderstorm? **(G)**

16. On what basis do meteorologists classify tornadoes? **(G)**

17. How is a tornado watch different from a tornado warning? What should you do if a tornado is coming your way? **(E)**

18. Identify the flow of warm and cold air in the sea breeze illustration. **(H)**

19. What conditions lead to the development of strong downslope winds on the leeward side of a mountain range? **(H)**

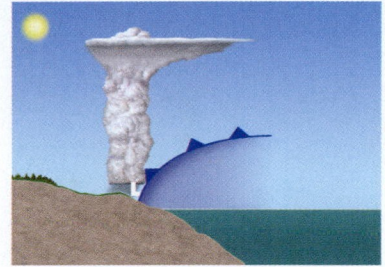

On Further Thought

20. Lake-effect storms typically produce greater snowfall in November and December than in January and February. Why might this be so? (Hint: What changes are likely to occur to the lake as winter progresses?) **(H)**

21. Describe the sequence of weather conditions that you would experience as a supercell thunderstorm, moving southwest to northeast, approached and then progressed over your house. **(F)**

Online Resources

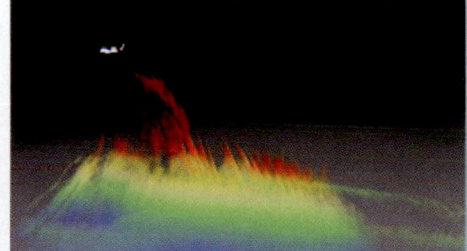

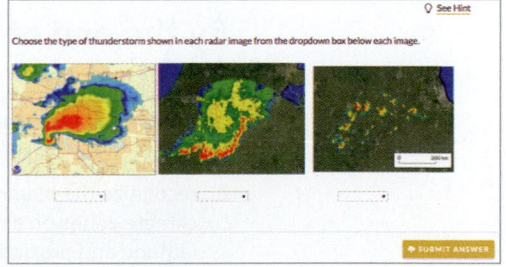

Videos
This chapter features real-world videos on stormy coasts and storm watchers.

Smartwork5
This chapter features questions on weather systems, thunderstorm formation and radar measuring, tornadoes, and human actions to protect from storm hazards.

20 CLIMATE AND CLIMATE CHANGE

By the end of the chapter you should be able to . . .

A. explain the difference between climate and weather, and describe the variables that determine the climate of a region.

B. recognize the boundaries among the Earth's major climate zones on a map and relate their locations to latitude and proximity to the ocean.

C. interpret the factors that control the climate conditions that prevail in a region.

D. describe the techniques that researchers use to reconstruct climates of the past.

E. explain causes of both long-term and short-term climate change.

F. discuss how climate has changed in the recent past, and describe the consequences of this change.

G. evaluate evidence concerning modern-day global warming and its causes.

20.1 Introduction

In 1955, Charles Keeling constructed an instrument that could measure atmospheric carbon dioxide (CO_2) very accurately. This instrument had to be sensitive indeed, for CO_2 is a *trace gas*, one that accounts for far less than 1% of the atmosphere. Keeling made a set of measurements in California and learned that atmospheric CO_2 concentrations can change measurably over the course of a single day. Significantly, he also found that the average of his measurements was higher than the average of measurements made in the mid-19th century. This finding supported a hypothesis, put forward years earlier, that CO_2 concentrations had increased over the previous century.

In 1958, to better understand variations in global CO_2 concentrations, Keeling installed a monitoring station on Mauna Loa, a Hawaiian volcanic peak that rises 3.4 km (11,141 feet) above the Pacific Ocean **(Fig. 20.1a)**. He chose this locality—far from cities, factories, forest fires, and other sources of the gas—so that his measurements would represent only CO_2 that had mixed thoroughly within the atmosphere. When he looked over the measurements he had collected during the first year of his study, Keeling found that the CO_2 concentration ranged between 312 and 318 ppm (parts per million by volume), with an average value of 315 ppm (= 0.0315%) **(Fig. 20.1b)**. The CO_2 concentration dropped in summer, when plants of the northern hemisphere absorbed the gas during photosynthesis, and rose in winter, when those plants shed their leaves, which then decayed and released CO_2. In effect, Keeling had observed the Earth System inhaling and exhaling CO_2!

Lake Mead, along the Colorado River, a critical water supply for California, reaches a new low during a severe drought.

Figure 20.1 The Keeling curve provides a record of atmospheric carbon dioxide concentrations measured at the Mauna Loa Observatory.

(a) The observatory lies at a high elevation in the middle of the Pacific Ocean.

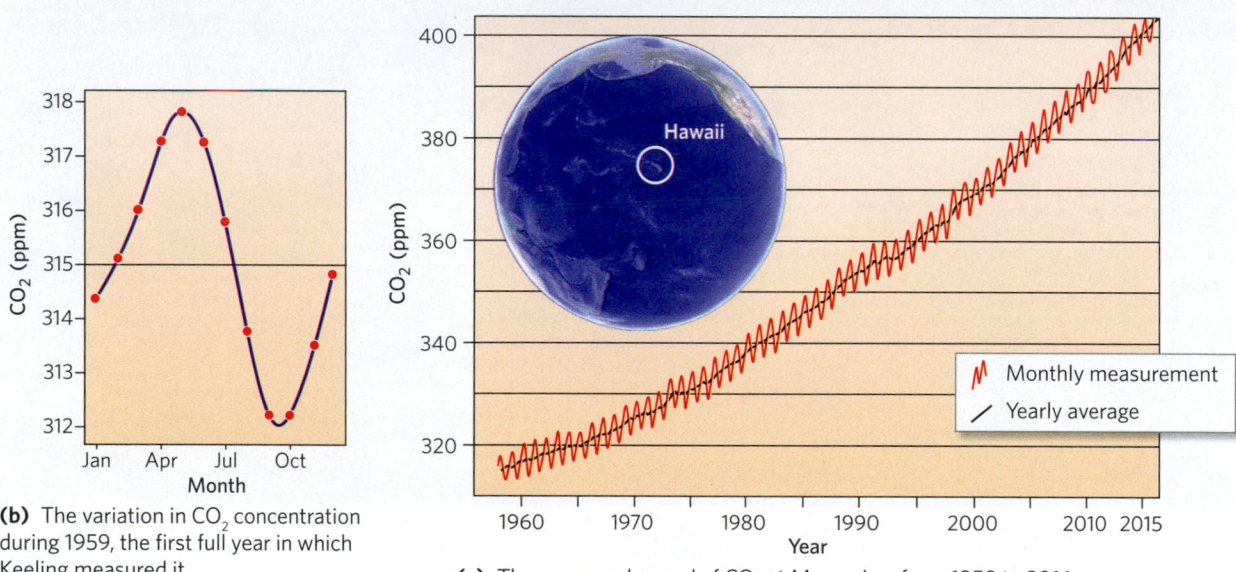

(b) The variation in CO_2 concentration during 1959, the first full year in which Keeling measured it.

(c) The measured record of CO_2 at Mauna Loa from 1958 to 2016.

Keeling continued making measurements on Mauna Loa several times a year during the remainder of his career, and others have continuously maintained the Mauna Loa CO_2-monitoring station to this day. As a result, we now have a nearly 60-year-long record, known as the **Keeling curve**, proving that the atmospheric concentration of CO_2 steadily increases **(Fig. 20.1c)**. In February 2016, air at the station contained 404 ppm of CO_2, an increase of 27% since Keeling's first measurements and an increase of about 40% since the mid-19th century.

Why should society care about the atmospheric concentration of CO_2? Earlier in this book, we pointed out that CO_2 is a greenhouse gas that traps infrared radiation in the atmosphere. By doing so, it plays a crucial role in keeping the Earth System's **critical zone** (the surface and

near-surface realm that provides the resources that sustain life) warm enough to support life. But can there be too much of a good thing? The efficiency of the greenhouse effect depends on the concentration of greenhouse gases, so when CO_2 concentrations increase, the air and oceans warm up. Measurements made by many researchers indicate that this warming is taking place and that too much warming could have a significant impact on society.

Because temperature plays a major role in determining climate, researchers now view changes in average temperature as an aspect of a broader issue, *climate change*. To understand this statement, it's necessary to keep in mind the distinction between climate and weather. **Weather** refers to the conditions of air temperature, relative humidity, wind speed, cloudiness, and precipitation at a

Box 20.1

How can I explain . . .

The difference between weather and climate

What are we learning?
• How meteorologists average data to obtain climate statistics.

What you need:
• A calculator.
• A computer.

Instructions:
• Using the Internet, do a search for "daily average temperature." Choose one of several available websites. Do the same for precipitation.
• Record the daily observations for a nearby location during a particular month.
• Average the data to obtain the average monthly temperature and the average monthly precipitation.

What did we see?
• Temperature and precipitation on any given day depend on the weather occurring on that day. The averages we calculated, in contrast, represent the typical temperature and precipitation for the month, even if that particular temperature and that amount of rainfall did not occur on any day during the month. The average is an estimate of the expected temperature and precipitation for that month. Climatologists average such data to represent a region's climate.

given location at a given time. A description of **climate**, in contrast, characterizes the average weather conditions and the range of weather conditions for a region over the course of decades to millennia (Box 20.1). More specifically, a region's climate depends on its average temperature and typical temperature range, the amount and distribution of precipitation, the nature of storms, typical seasonal variation in weather, and typical extremes of weather. **Climate change**, therefore, refers to a shift in the average of one or more of the above characteristics.

Recognizing climate change in the past, and particularly predicting climate change in the future, has been a major focus of earth science research since the 1970s. Because climate change within our lifetimes could affect water supplies, food production, and sea level, it has also become a subject of political discussion. Addressing climate change concerns many people because steps taken to slow climate change might affect the viability of some industries and the sustainability of some lifestyles.

To help you develop an understanding of climate and climate change, this chapter begins by discussing the Earth's present climate regimes and the factors that control them. Then we look back in time to examine changes in climate during Earth history and phenomena that may have caused these changes. We conclude by examining how human society is modifying Earth's climate today and what the impacts of such climate change might be.

20.2 The Earth's Major Climate Zones

Picture a day in the Amazon rainforest. The air feels sultry, nearly everything in sight has a lush covering of green plants, downpours drench the land, and vibrant sounds rise from countless species of insects, monkeys, and birds. Now, picture the same day in the Sahara. The Sun beats down relentlessly, and the air is so dry that your lips are chapped. If you look to the horizon, you see mostly rocks and sand in hues of brown and tan. Scrubby plants poke up here and there, but you hear hardly any animals. Finally, picture a day in the northern Canadian tundra. It's so cold that your breath freezes, the ground remains frozen solid almost all year, and you see few living creatures, for most animals have migrated south or are hibernating.

By comparing your perception of a rainforest, a desert, and a tundra, you realize that climate affects many of our senses. From a meteorological perspective, however, the major factors that contribute to our experience of climate reflect two variables—temperature and precipitation—and their variation over the course of a year. To a large extent, these variables correlate with latitude, but many other factors, such as elevation and proximity to oceans, influence them, and therefore climate, as well.

Figure 20.2 Average global surface temperatures.

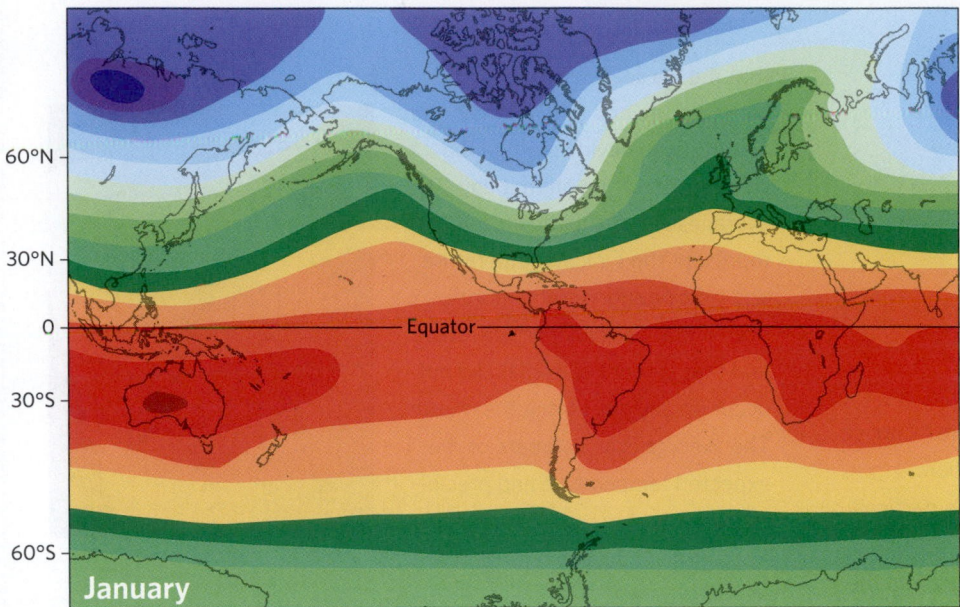

(a) In January, northern-middle latitudes are cool and southern middle-latitudes are warm.

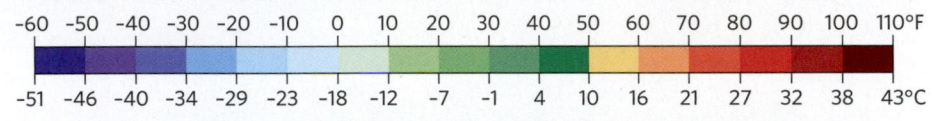

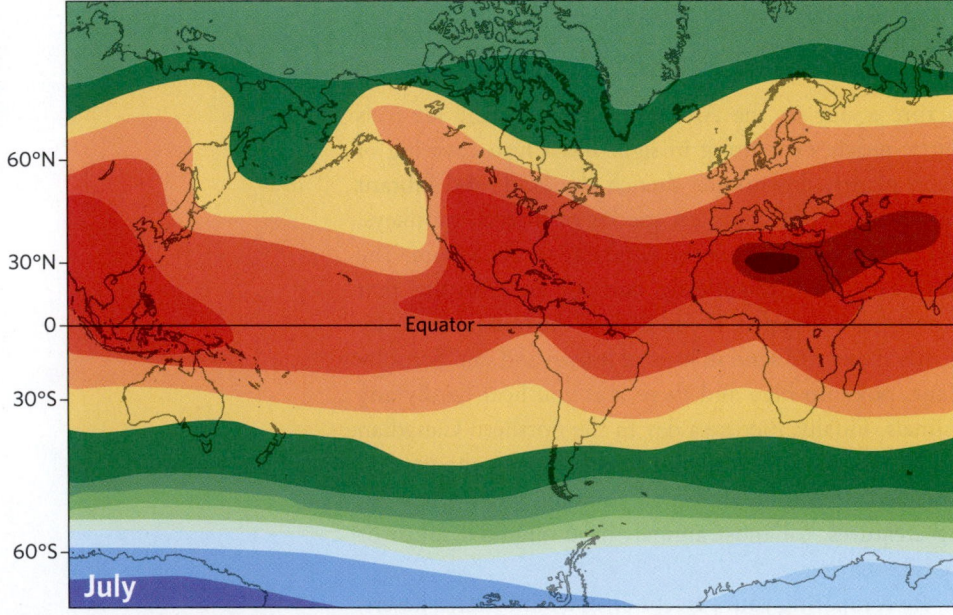

(b) In July, northern-middle latitudes are warm and southern-middle latitudes are cool.

Global Temperature and Rainfall Patterns

On maps depicting average monthly temperatures, each color band represents a region in which the average temperature falls within a given range **(Fig. 20.2)**. By looking at such a map for a given month, we see the overall global pattern of temperature variation, with warmer temperatures dominating at lower latitudes and cooler ones at higher latitudes. We can also see that at a given latitude, warmer temperatures exist at lower elevations and cooler temperatures at higher ones, and that temperatures are influenced by proximity to the ocean. Comparing maps for January and July emphasizes that seasonal variation in the temperature of a region also depends on latitude.

Why does proximity to the ocean affect seasonal temperature variation? As we've seen, water has a high *heat capacity*, meaning that it can absorb a large amount of heat without changing much in temperature, and because water is transparent and circulates, it becomes warm down to the thermocline, 100 to 200 m (330 to 660 feet) below the surface. As a result, oceans slowly absorb a large amount of heat during the summer and release it slowly during the winter, so the water moderates near-shore temperatures. For example, temperatures on the Pacific coast of California display seasonal variation of only 7°C (13°F), whereas an inland region, such as Missouri, at the same latitude may experience seasonal variation of 28°C (50°F). The proximity of a location to a specific ocean current can also influence its climate. For example, Ireland and the United Kingdom have relatively mild winters, despite their high-latitude locations, because the Gulf Stream carries warm water to high latitudes in the Atlantic; at comparable latitudes in the interior of North America, winters are much colder.

The global distribution of precipitation also varies significantly with latitude and season **(Fig. 20.3)**. Large quantities of rain fall along the intertropical convergence zone (ITCZ) at the equatorial edges of the Hadley cells (see Chapter 18), where moist air rises and provides fuel for large thunderstorms. In contrast, lands beneath the descending parts of the Hadley cells host subtropical deserts. Other seasonally varying circulations also affect rainfall. For example, in tropical latitudes subject to monsoons, most rain falls during the summer season.

The Köppen-Geiger Climate Classification

Wladimir Köppen spent most of his adult life in Germany, where he became one of the first *climate scientists* (researchers who focus on understanding climate). In 1884, Köppen published a classification of Earth's climate zones and portrayed the different zones on a map. Later in life, Köppen worked with Rudolf Geiger to modify the original scheme. The resulting **Köppen-Geiger climate classification** (KGCC) remains in use today **(Table 20.1)**. Köppen and Geiger used natural vegetation as the primary clue to characterize the climate of a given area because the assemblages of plant species that thrive in an area reflect the average, range, and seasonal variation of weather conditions over a period of many years.

Figure 20.3 Average global precipitation.

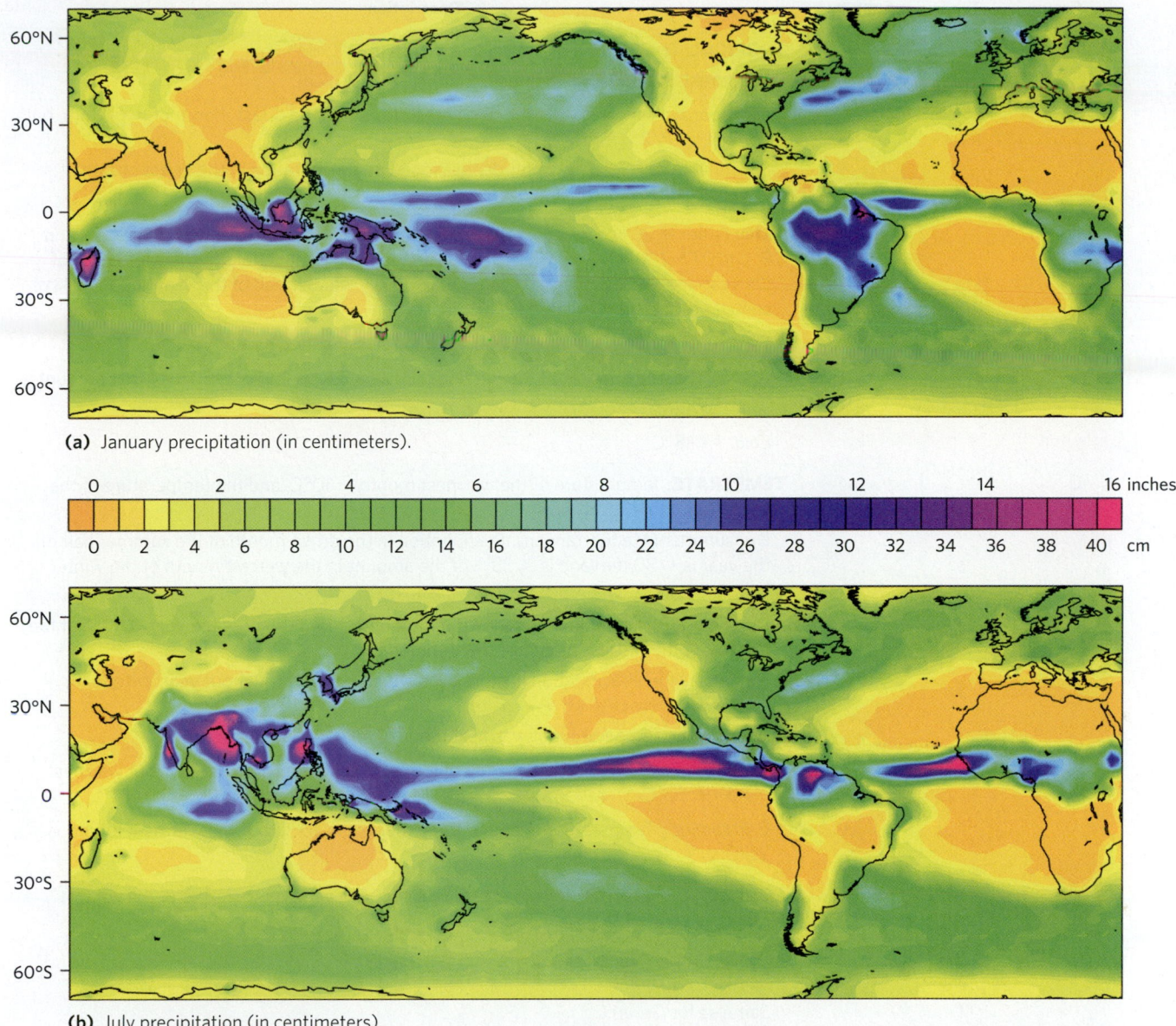

(a) January precipitation (in centimeters).

| 0 | 2 | 4 | 6 | 8 | 10 | 12 | 14 | 16 inches |

| 0 | 2 | 4 | 6 | 8 | 10 | 12 | 14 | 16 | 18 | 20 | 22 | 24 | 26 | 28 | 30 | 32 | 34 | 36 | 38 | 40 | cm |

(b) July precipitation (in centimeters).

The KGCC map shows how variables in the Earth System control climate by distinguishing five broad *climate groups*: tropical, arid, temperate, cold, and polar (Fig. 20.4). These groups differ from one another in their average temperatures, the temperature range between their hottest and coldest months, and their average annual precipitation. Note that the classification scheme uses uppercase letters, derived from German words, for each group. Each climate group, in turn, includes a few *climate types* based on other characteristics. For example, the three types of tropical climates (rainforest, monsoonal, and savannah), can be distinguished from one another based on whether or not rain falls throughout the year. A third level of subdivision (*climate subtypes*)

distiguishes among climates, such as deserts, that can be either hot or cold.

Take-home message . . .

The Earth hosts many different climates. A region's climate primarily reflects temperature and precipitation and their variation during the year. Local vegetation serves as the primary basis for defining climate groups and types in the Köppen-Geiger climate classification.

Quick Question -
How does climate vary with latitude and with proximity to the coast?

Did you ever wonder . . .

why tropical rainforests lie near the equator?

Table 20.1 The Köppen-Geiger climate classification (KGCC)

Group	Type	Subtype	Description
A			**TROPICAL:** Temperature of the coldest month is 18°C or higher.
	f		Rainforest: Precipitation in driest month is at least 6 cm.
	m		Monsoonal: A short dry season, with precipitation in the driest month of < 60 mm, but ≥ [100 − (R/25) mm].[1]
	w		Savannah: Well-defined winter dry season, with precipitation in the driest month of < 60 mm or < [100 − (R/25) mm].
B[2]			**ARID:** Either ≥ 70% of the annual precipitation falls in the summer half of the year and $R <$ [20T + 280 mm], or ≥ 70% of the annual precipitation falls in the winter half of the year and R is less than 20T mm. *Alternatively,* neither half of the year has ≥ 70% of annual precipitation and $R <$ [20T + 140].
	W		Desert: R is < one-half of the upper limit for classification as a B type.
	S		Steppe: R < the upper limit for classification as a B type, but is > 50% of that amount.
		h	*Hot:* T ≥ 18°C
		k	*Cold:* T < 18°C
C			**TEMPERATE:** Temperature of the warmest month ≥ 10°C, and the temperature of the coldest month < 18°C, but > −3°C.
	s		Dry summer (Mediterranean): Precipitation in the driest month of the summer half of the year is < 30 mm and is < 33% of the amount in the wettest month of the winter half.
	w		Dry winter: Precipitation in the driest month of the winter half of the year < 10% of the amount in the wettest month of the summer half.
	f		No dry season: Precipitation is more evenly distributed throughout year; criteria for neither s nor w satisfied.
		a	*Hot summer:* Temperature of warmest month 22°C or above.
		b	*Warm summer:* Temperature of each of four warmest months is ≥ 10°C, but warmest month < 22°C.
		c	*Cold summer:* Temperature of one to three months ≥ 10°C, but the warmest month is < 22°C.
D			**COLD:** Temperature of the warmest month is ≥ 10°C, and temperature of the coldest month is ≤ −3°C.
	s		Same as for Group C
	w		Same as for Group C
	f		Same as for Group C
		a	Same as for Group C
		b	Same as for Group C
		c	Same as for Group C
		d	*Very cold winter*: temperature of coldest month < −38°C (d designation then used instead of a, b, or c).
E			**POLAR:** Temperature of the warmest month is < 10°C.
	T		Tundra: Temperature of warmest month is > 0°C but < 10°C.
	F		Frost (Ice cap): Temperature of warmest month is ≤ 0°C, and ice covers most land.

[1]In the formulas given, R refers to the annual rainfall in millimeters; T refers to the average annual temperature in degrees centigrade. The summer half of the year is defined as the months April–September for the northern hemisphere and October–March for the southern hemisphere.
[2]Any climate that satisfies criteria for designation as a B type is classified as such, regardless of other characteristics.

Figure 20.4 The Köppen-Geiger climate classification (KGCC).

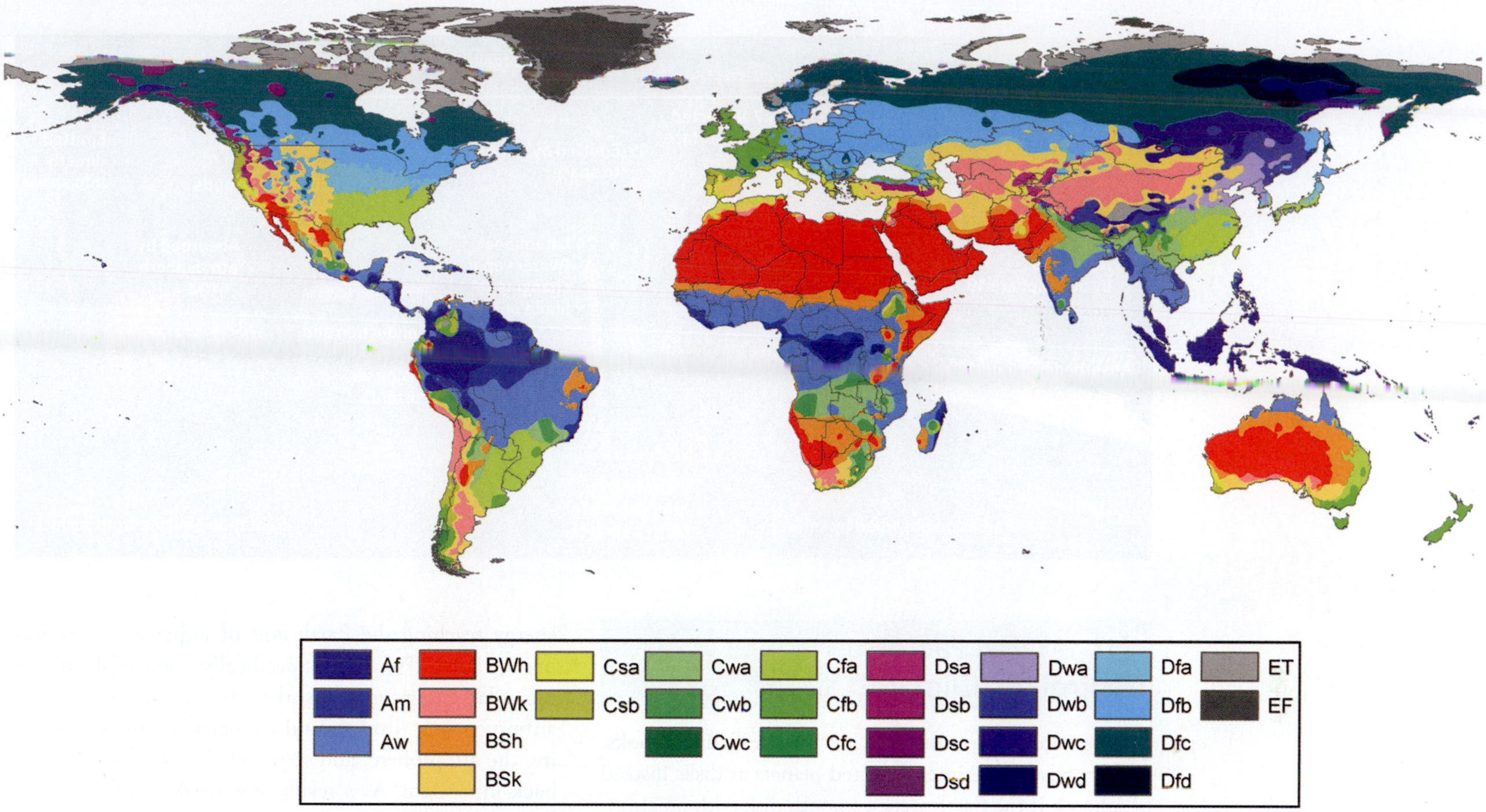

■ Af	■ BWh	■ Csa	■ Cwa	■ Cfa	■ Dsa	■ Dwa	■ Dfa	■ ET
■ Am	■ BWk	■ Csb	■ Cwb	■ Cfb	■ Dsb	■ Dwb	■ Dfb	■ EF
■ Aw	■ BSh		■ Cwc	■ Cfc	■ Dsc	■ Dwc	■ Dfc	
	■ BSk				■ Dsd	■ Dwd	■ Dfd	

(a) The distribution of KGCC climate zones. The labels in the key correspond to the labels in Table 20.1.

(b) Examples of climate zones. The inset in each photo indicates the KGCC climate zone it represents.

Figure 20.5 The energy balance in the Earth's atmosphere. Each arrow represents the transfer of energy. The numbers represent arbitrary units of energy and show relative amounts of energy in each transfer.

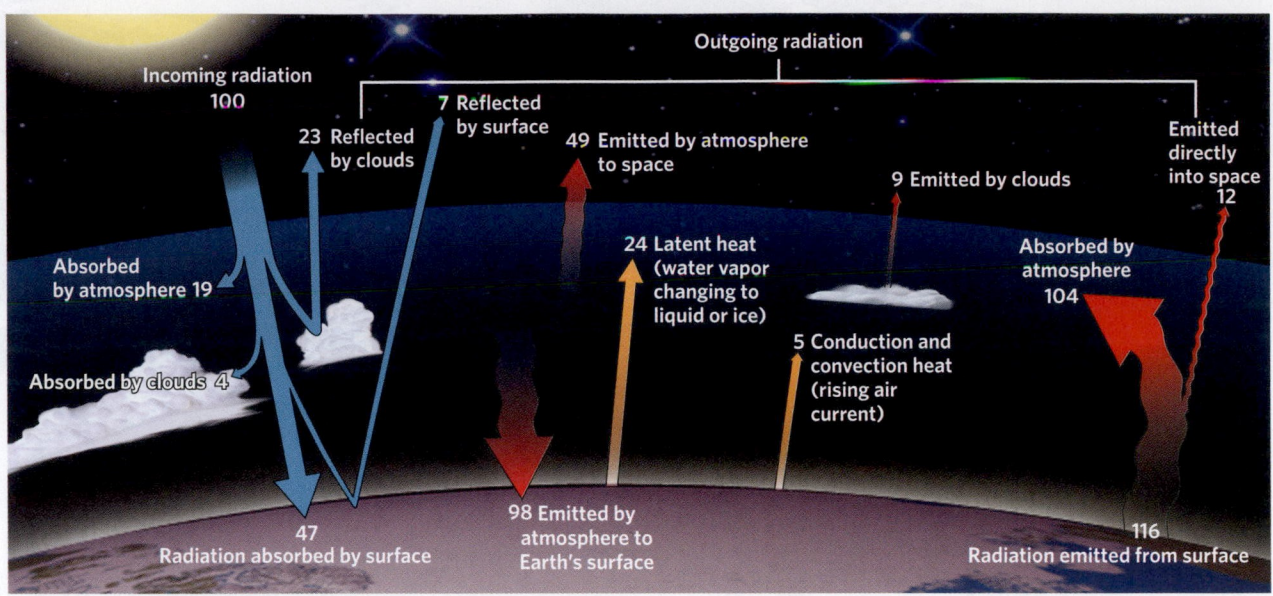

20.3 What Factors Control the Climate?

In the alternative universe of science-fiction books, authors have sometimes depicted planets as disks instead of spheres. If the Earth were a smooth disk with no atmosphere, its climate might be much the same everywhere: cold and dry. But of course the Earth isn't flat—it's a rotating sphere with a tilted axis. Furthermore, it has an atmosphere, its surface consists of both land and ocean, water in the ocean circulates in currents, its land has topography, and large areas of both land and sea sometimes host ice cover. All these characteristics influence overall global temperature and the variation of climate around the planet. In this section, we explore why.

The Effect of the Atmosphere on the Earth's Energy Balance

The Sun bathes the Earth with an immense amount of energy every second. What happens to this energy? If the Earth had no atmosphere, all this energy would reach its surface, and of this energy, a portion would instantly reflect back into space. Most would be absorbed by the Earth's surface, warming it to a certain temperature. Eventually, however, this solar energy, after having been absorbed, would re-radiate back into space as infrared energy. All radiation reaching the Earth must ultimately return to space—otherwise, the Earth would keep warming until it evaporated.

As we discussed in Chapter 17, the existence of our atmosphere significantly changes the fate of both solar

energy reaching the Earth and of radiated energy leaving the Earth (**Fig. 20.5**). Specifically, some of the incoming solar energy gets absorbed by air molecules or by airborne water droplets and ice particles, thereby warming the atmosphere, and some reflects off clouds directly back into space. As a result, the Earth's surface absorbs only about half of the Sun's arriving energy. When this absorbed energy re-radiates, a small amount of it travels directly back into space, but most gets absorbed by clouds or by H_2O, CO_2, and CH_4 molecules in air, causing the atmosphere to warm further. The atmosphere doesn't hold onto this energy permanently, though—some returns to the Earth's surface and contributes to surface warming, while some escapes to space. A small amount of energy from the Earth's surface also rises into the atmosphere due to conduction and convection or due to the evaporation of water and its re-condensation into clouds. When averaged over time, of course, the amount of energy arriving at the Earth from the Sun must equal the amount of energy emitted to space by the Earth. This relationship is called the Earth's **energy balance**. If an energy balance didn't exist, as we noted before, the Earth would keep getting hotter and hotter.

The trapping of re-radiated energy by the atmosphere is a process that atmospheric scientists refer to as the *atmospheric greenhouse effect* (see Chapter 17), and the specific gases that absorb re-radiated energy from the Earth's surface are called *greenhouse gases*. Because of the greenhouse effect, the Earth's atmosphere acts like a blanket, keeping the Earth's surface warmer than it would be if the Earth had no atmosphere. In fact, researchers estimate that were

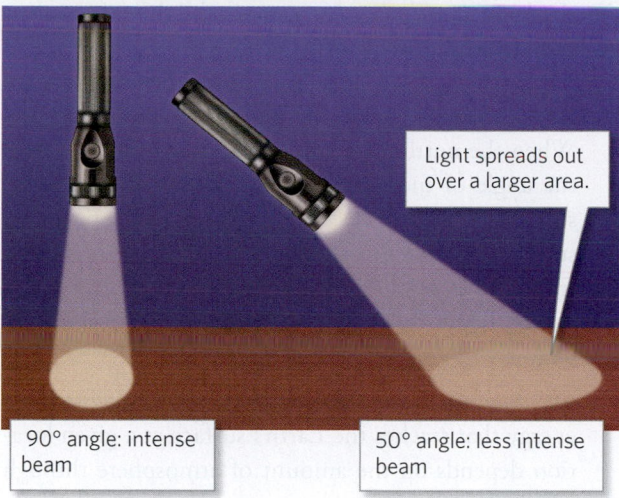

Figure 20.6 The flashlight analogy for explaining why the intensity of solar heating varies with the angle of incidence.

Light spreads out over a larger area.

90° angle: intense beam

50° angle: less intense beam

it not for the greenhouse effect, the Earth's average global surface temperature would be about 33°C (59°F) lower than it is today, and our planet's surface would be a frozen wasteland. Greenhouse gases make the Earth's climate habitable!

The Relationship of Energy Input to Incidence Angle

If all of the Earth's surface received the same amount of solar energy, the temperature of the lower atmosphere would be the same everywhere, and the great variety of climates that Köppen and Geiger identified couldn't exist. The amount of solar energy absorbed by the surface varies with location because the *angle of incidence*, the angle between incoming beams of sunlight and the Earth's surface, varies with location.

To see why the angle of incidence matters, let's do a simple experiment (Fig. 20.6). Aim a flashlight straight down on a tabletop from a distance of about 10 cm (4 inches). The beam intersects the tabletop at an angle of 90° and forms a small circle of bright light. If you then tilt the flashlight and

aim it so that it intersects the tabletop at a shallower angle, the beam spreads out over an ellipse with a larger area. Even though the amount of light energy emitted by the flashlight remains the same, the light in the ellipse looks dimmer because the light has spread out over a larger area. With this concept in mind, note that because the Earth is a sphere, sunlight strikes different latitudes at different angles, at a given time. For example, at noon, a square meter at the equator, where the angle of sunlight is steep like that of the straight-down flashlight beam, receives more solar energy than a square meter at a latitude of 40°, where the angle of sunlight is shallow and light spreads over a broader area.

Seasons and their Consequences

If the Earth's axis of rotation were exactly perpendicular to the plane of the Earth's orbit around the Sun (its *orbital plane*), the variation in temperature from equator to pole that we've just described would be the same all year everywhere on the Earth's surface. It isn't. Most of the Earth experiences distinct **seasons** during the course of the year, each characterized by different overall weather conditions: days are longer and warmer during the summer and shorter and colder during the winter. Fall and spring represent transitions between summer and winter.

The Earth has seasons because its axis of rotation tilts at an angle of 23.5° to a line drawn perpendicular to the orbital plane (Fig. 20.7). The Earth's axis stays in the same orientation, relative to an observer outside of the Solar System, as the Earth goes around the Sun. So during one half of the year, the southern hemisphere

Figure 20.7 The reason for the seasons. The Earth's axis of rotation tilts at an angle of 23.5° to a line perpendicular to the Earth's orbital plane and stays in that orientation as the Earth orbits the Sun. The seasons are labeled for the northern hemisphere.

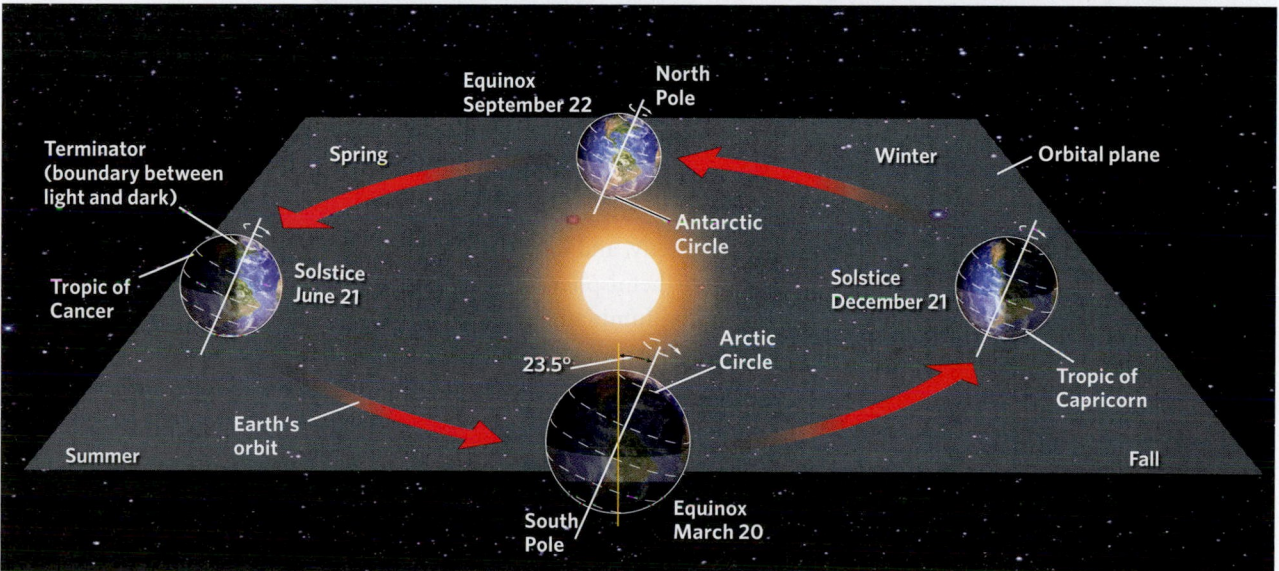

Equinox September 22

North Pole

Spring

Winter

Orbital plane

Terminator (boundary between light and dark)

Antarctic Circle

Solstice June 21

Solstice December 21

Tropic of Cancer

Arctic Circle

23.5°

Tropic of Capricorn

Earth's orbit

Summer

Fall

South Pole

Equinox March 20

Figure 20.8 Solar heating of the Earth's surface varies with season because of the angle of the Earth's axis relative to incoming beams of sunlight.

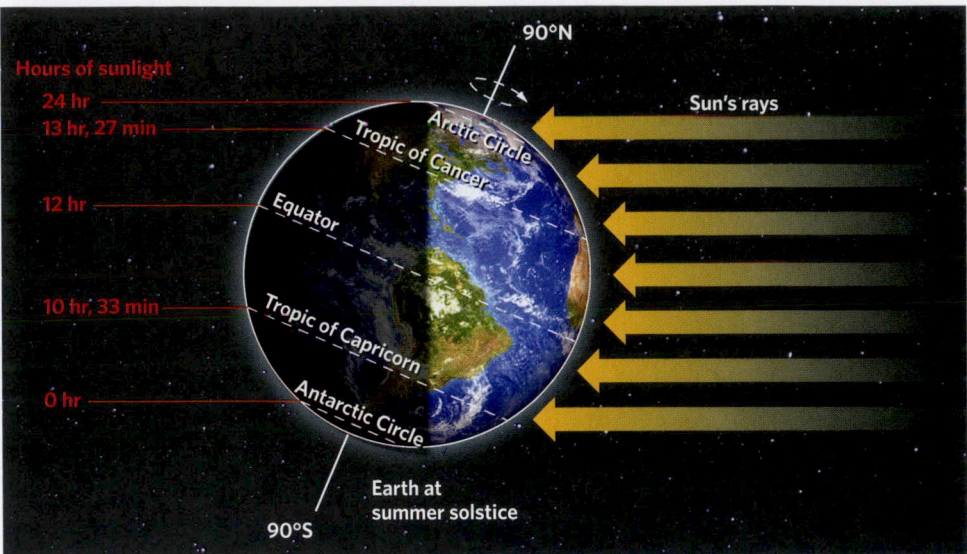

(a) At the northern hemisphere summer solstice, the North Pole receives 24 hours of sunlight, and northern mid-latitudes have long, warm days.

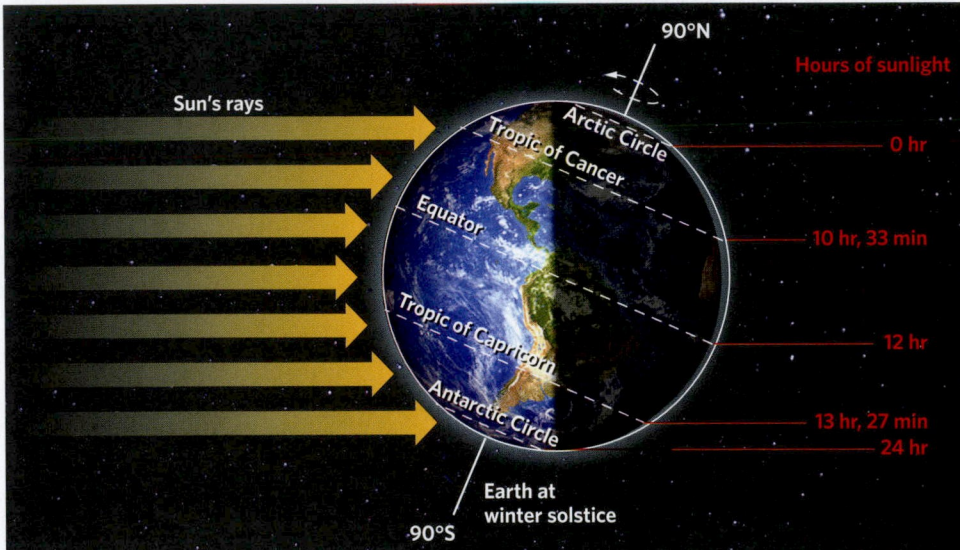

(b) At the northern hemisphere winter solstice, the North Pole is dark all day, and mid-latitudes have short, cool days.

faces the Sun more directly, and during the other half of the year, the northern hemisphere faces the Sun more directly. This change in illumination affects average temperature for three reasons, all of which determine the amount of solar energy that reaches a location on the Earth's surface:

- *Number of daylight hours:* The tilt of the Earth's axis affects the number of daylight hours that a region receives at a given time of year **(Fig. 20.8)**. Since the number of daylight hours is greater in summer than in winter, a region receives more energy during a summer day than during a winter day. In polar regions, no solar radiation at all arrives during winter, but during summer, the Sun shines 24 hours a day.

- *Angle of incidence:* As our flashlight experiment showed us, the angle of incoming sunlight relative to the Earth's surface determines how much solar energy reaches a square meter of the surface at a given time. When the northern hemisphere tilts toward the Sun, the angle of incidence of incoming light is steeper, so the hemisphere becomes warmer and experiences summer. When the northern hemisphere tilts away from the Sun, the angle of incidence is less steep, so the hemisphere becomes colder and experiences winter.

- *The amount of air through which sunlight passes:* The energy that reaches the Earth's surface at a given location depends on the amount of atmosphere through which the Sun's rays must pass. Air molecules scatter some radiation back to space before the radiation can reach the Earth's surface. When the Sun's rays intersect the atmosphere at a steep angle, as happens in summer when the Sun rises high above the horizon, they encounter fewer air molecules before reaching the ground than they do during winter, when the Sun stays closer to the horizon and the rays arrive at a shallow angle.

As a result of these three factors, the Earth's surface at mid- to high latitudes receives less solar energy overall in the winter than it does in the summer. In the tropics, where the Sun rises almost as high in the winter as it does in the summer, hardly any seasonal change takes place.

Some people mistakenly believe that seasons reflect changes in the distance between the Earth and the Sun over the course of a year. This is not the case. The Earth's orbit is elliptical, but only very slightly so (see Chapter 21). In fact, the distance between the Earth and the Sun varies by only 3% during the course of a year, and this variation has almost no effect on climate. In fact, the Earth is closest to the Sun during the northern hemisphere's winter!

The Effect of Atmospheric and Oceanic Circulation on Climate

Regional- to global-scale tropical atmospheric circulations—specifically, the Hadley cells, the Walker circulation, and the monsoons (see Chapter 18)—all redistribute heat and moisture in the atmosphere, so they affect the amount of rainfall and the timing of wet and dry seasons in tropical and subtropical latitudes. For example, because of the Hadley cells, lots of precipitation falls in the tropics whereas very little falls in the subtropics, so we see a transition from rainforest to steppe to desert across the width of each Hadley cell. Because of the Walker circulation, patterns of weather vary in response to ENSO (El Niño and La Niña). And because of monsoons, most rain falls only during certain times of the year across broad regions of some continents.

Atmospheric circulations at mid-latitudes are complicated because, as we saw in Chapter 18, large *air masses*—bodies of air over which temperature and humidity are relatively uniform—form in different locations. Semi-permanent (long-lasting) regions of high atmospheric pressure develop at the Earth's surface beneath cold air masses, whereas semi-permanent regions of low atmospheric pressure develop beneath warm air masses (**Fig. 20.9**). These systems have a dramatic effect on local climates at mid- and high latitudes because the pressure gradients between them drive the winds. At the Earth's surface, winds blow counterclockwise around low-pressure systems and clockwise around high-pressure systems in the northern hemisphere. (The opposite is true in the southern hemisphere.) Therefore, regional winds associated with the semi-permanent high-pressure systems over the North Atlantic and North Pacific in the northern hemisphere summer transport warm air toward the north along the east sides of continents and cold air toward the equator along the west sides of continents. These winds, in turn, drive ocean currents that accelerate heat exchange between the tropics and polar regions (see Chapter 15). Together, the regional winds and associated ocean currents cool the west coasts of continents and warm the east coasts of continents in both hemispheres. In addition, the thermohaline circulation, which acts like a global conveyor belt to transport heat throughout the global ocean, serves to moderate global climates overall.

The Effect of Land Distribution and Elevation on Climate

If the surface of the Earth were all ocean, or all flat land, seasonal variations at a given latitude would be uniform.

Figure 20.9 The northern hemisphere's average sea-level pressure patterns. Winds flow counterclockwise around low-pressure systems and clockwise around high-pressure systems in the northern hemisphere.

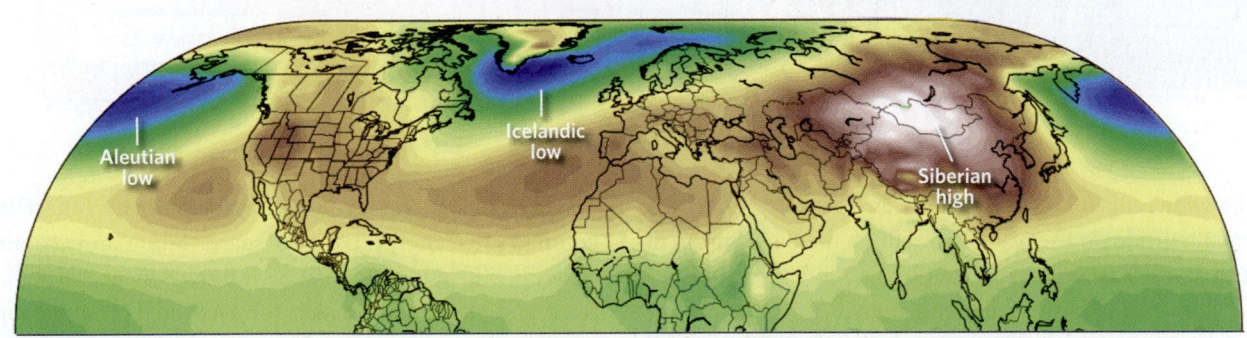

(a) Northern hemisphere winter (December through February). This map indicates names of semi-permanent highs and lows.

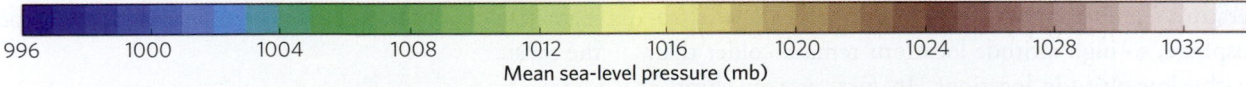

996   1000   1004   1008   1012   1016   1020   1024   1028   1032
Mean sea-level pressure (mb)

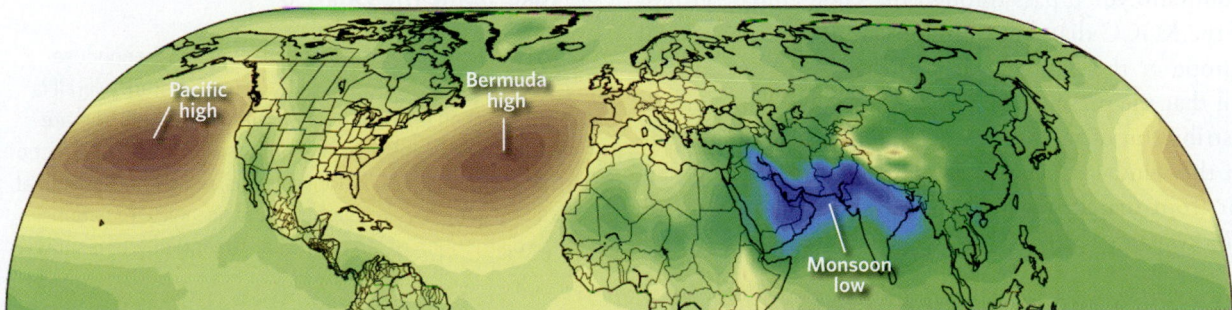

(b) Northern hemisphere summer (June through August). The Bermuda high and the Pacific high become pronounced during the summer.

The wide range of different climates that we can find at a given latitude on the real Earth reflects the distribution of land and sea as well as topography.

We've already seen that continental interiors experience larger seasonal changes in temperature than coastal areas do. The Earth's mountain ranges also affect temperature. Temperature decreases with height in the troposphere, so high-altitude locations remain colder than nearby low-altitude locations. In fact, as you climb a mountain, you'll pass through the same climate groups of the KGCC that you would if you hiked from central Europe or the United States to the Arctic Ocean. Recall that the presence of a mountain range also controls distribution of rainfall (see Chapter 14), so the climate on the windward side of the range may be very different from that in the rain shadow on the leeward side (Fig. 20.10).

How Snow and Ice Cover Affect Climate

If you try to traverse a snowfield or ice field on a sunny day, you have to wear very dark glasses to protect your eyes from glare, for snow and ice efficiently reflect incoming solar energy back to space. Snow and ice also serve as efficient radiators of infrared energy. For this reason, the temperature over snow-covered ground drops very rapidly at night, compared with that over bare or vegetated ground, so snow-covered land has a cooler climate than dry land at the same latitude and distance from the coast.

Take-home message . . .

Because the atmosphere contains greenhouse gases, the Earth's surface temperature overall is warmer than it would be without those gases. Climate variation depends on many factors, including the effect of latitude on solar radiation, seasons, redistribution of heat and moisture by atmospheric and oceanic circulations, the distribution of land and sea, topography, and snow cover.

Quick Question -
How does the atmosphere affect the Earth's energy balance?

20.4 Climate Change

How often have you seen a news report proclaiming, "Record High Temperature Expected Tomorrow"? Does such a headline mean that the climate is changing? Not necessarily—a single hot spell or cold snap may simply represent **natural variation** (random changes that can't be predicted and probably represent the response of a system to input from several subtle causes). But if a new set of conditions—an overall increase in average temperature or an overall change in precipitation—becomes the norm for a region, then climate change has occurred. An increase in average global atmospheric and sea-surface temperatures represents **global warming**, and a decrease represents **global cooling**.

Needless to say, climate change has become a high-profile political issue because of its potential impact on economies and on lifestyles. In this section, we provide a context for understanding *contemporary climate change* (climate change of the past few and next few centuries) by first examining the kinds of data used to characterize climate in the past, and then by considering the record of climate change through the Earth's history.

Characterizing Paleoclimate

What we know about the Earth's past climate, or **paleoclimate**, comes from studying a variety of *paleoclimate indicators*, meaning clues to paleoclimate preserved in wood, shells, ice, sediment, or rock. By determining the numerical age of these indicators, researchers can compile a record of how climate has changed over time. *Paleoclimatologists*, researchers who study the prehistoric record of climate change, investigate climate change at two different scales: *long-term climate change*, which takes place over millions to hundreds of millions of years, and *short-term climate change*, which takes place over decades to hundreds of thousands of years. Notably, the *resolution* (detail) of the paleoclimate record generally permits us to characterize short-term climate changes for only the past few million years or so.

Long-Term Climate Change

The stratigraphic record (see Chapter 5) provides researchers with a basic history of global climate over millions to billions of years. That's because the depositional setting in which sediment accumulates, and the assemblages of organisms that live in or on sediment, depend on the climate. For example, beds of organic debris typically accumulate only in warm climates, whereas beds of till accumulate only in glacial climates. Therefore, a succession of strata that includes coal overlain by tillite (rock made from till) records a change from a warm to a cold climate.

Taken as a whole, the stratigraphic record shows that the Earth's surface temperature has stayed between the freezing point of water and the boiling point of water for almost all of geologic time since the beginning of the Archean. Temperature has not remained uniform, however. Specifically, sometimes the Earth's atmosphere has been significantly warmer than average, and sometimes it has been significantly cooler. Geologists refer to the warmer periods as **hothouse periods** (or *greenhouse periods*) and to the colder ones as **icehouse periods** (Fig. 20.11a). During hothouse periods, even lands at polar latitudes were largely ice free, whereas during icehouse periods, polar regions were ice covered. The more familiar term *ice age* refers to the portions of some icehouse periods when ice sheets advanced and covered substantial areas of continents. During most of the approximately half-dozen ice ages that have occurred during Earth history, ice sheets covered land at mid-latitudes. In the late Proterozoic ice age, however, glaciers appear to have covered all land—even at the equator—and the oceans appear to have frozen over. Researchers refer to this condition as *snowball Earth* (see Chapter 10).

If we focus on the climate record of the past 100 million years, we can get a sense of the range of typical long-term climate changes that happen in the Earth System. The climate of the late Mesozoic was much warmer than that of today. In fact, temperatures may have been 2°C to 6°C (36°F to 43°F) warmer at the equator and 20°C (68°F) warmer at the poles. As a consequence, dinosaurs lived within 1,500 km (1,000 miles) of the poles. By comparison, land that could host most large animals today lies at least 4,500 km (2,500 miles) from the poles. It was so warm that there were no polar ice caps, and sea-surface temperatures in some parts of the ocean may have been as much as 17°C (21°F) hotter than today. Starting about 80 Ma, however, the Earth's surface and atmosphere began to cool. The cooling trend has continued overall, except for one 10 million-year-long interval—known as the *Eocene climatic optimum*—when global warming took place (Fig. 20.11b). Our planet entered an icehouse period at about 34 Ma, the time when the Antarctic ice sheet started to form. Beginning about 2.6 Ma, the Pleistocene Ice Age began, and glaciers advanced repeatedly over large areas of the northern hemisphere continents.

Short-Term Climate Change

About 18,000 years ago, the site of what is now Chicago lay beneath a 1.5-km (0.9-mile)-thick glacier, and sea level was 100 m (300 feet) lower than it is today, so that people could walk across the Bering Strait. Today, in contrast, Chicago has snow cover only during the winter

Figure 20.11 Over geologic time, global climate has changed significantly.

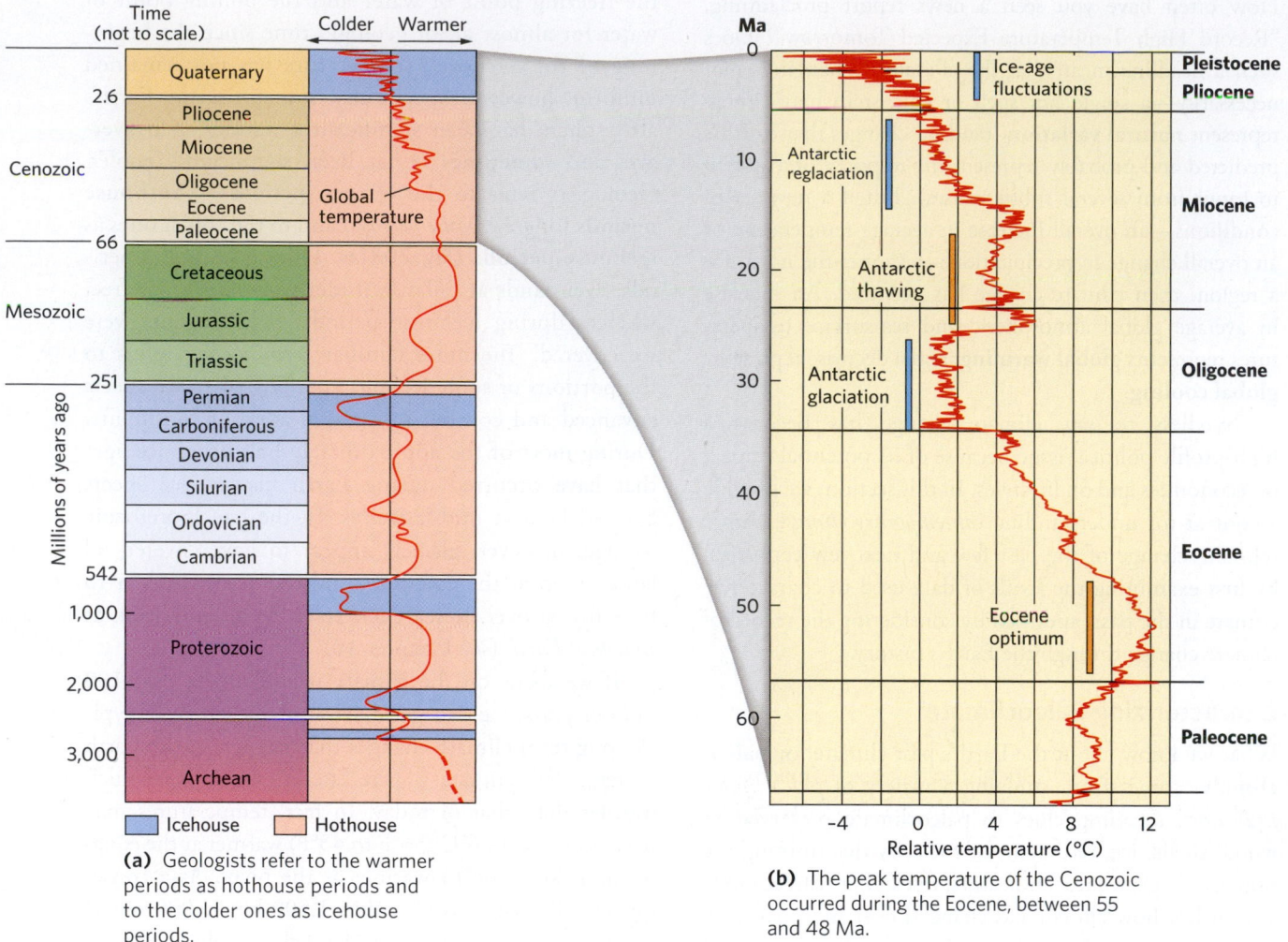

(a) Geologists refer to the warmer periods as hothouse periods and to the colder ones as icehouse periods.

(b) The peak temperature of the Cenozoic occurred during the Eocene, between 55 and 48 Ma.

season, and stormy seas occupy the Bering Strait. When the first farmers cultivated wheat 4,500 years ago in what is now Iraq, the region enjoyed frequent rains and hosted lush vegetation. Today, it hosts an arid climate with desert landscapes. And when Queen Elizabeth I reigned over England, 450 years ago, glaciers were advancing down valleys in the mountains of Europe, and the canals of the Netherlands froze over every winter. Today, the glaciers have retreated and the canals rarely freeze. Clearly, climate can change noticeably on a scale of centuries to millennia—in other words, on a human time scale.

To characterize short-term climate change over the past thousands to millions of years, paleoclimatologists examine several types of paleoclimate indicators:

- *Historical records:* Because human life depends directly on climate conditions, people tend to remember unusual conditions such as droughts, floods, long cold snaps, or severe storms. Prior to the development

of writing, these memories were oral traditions, but with the advent of writing, they appear in diaries and histories.

- *Microfossils and pollen:* Fossilized marine plankton in Pleistocene and Holocene deposits provide a record of climate change because the assemblage of species living at a locality reflects water temperature, and because these plankton have distinctive shells that allow identification of species. Similarly, different plant species produce pollen grains with distinctive shapes, so the study of pollen preserved in lake deposits allows researchers to determine how the ecosystem on land, and thus the climate at a given latitude, has changed **(Fig. 20.12)**.

- *Oxygen-isotope ratios in ice:* The ratio of ^{18}O to ^{16}O isotopes in glacial ice indicates the atmospheric temperature in which the snow that makes up the ice formed. Therefore, the ratio of these isotopes measured in a succession of ice layers in a glacier can indicate temperature change over time.

Paleoclimatologists have obtained *ice cores* in Antarctica that provide a temperature record spanning over 800,000 years **(Fig. 20.13a–c)**. Ice records preserved in mountain glaciers can characterize the climate at nonpolar latitudes, but unfortunately, these glaciers are rapidly disappearing, so researchers are rushing to collect cores from them. Not surprisingly, the logistics of transporting ice from a remote mountaintop to a laboratory halfway around the world can be extremely difficult.

- *Air bubbles in glacial ice:* When glacial ice forms, the ice crystals within surround tiny pockets of air that become preserved as bubbles in the ice. By sampling these bubbles, researchers can measure the composition of air at the time the ice formed.

- *Oxygen-isotope ratios in plankton shells:* The $^{18}O/^{16}O$ ratio in the $CaCO_3$ making up plankton shells in marine sediments provides an indication of past temperatures. Therefore, study of marine sediment cores can extend the record of temperature change back over millions of years **(Fig. 20.13d)**.

- *Growth rings:* Trees, corals, clams, and many other organisms develop distinct rings as they grow because their rate of growth varies with the season **(Fig. 20.14)**. These **growth rings** can be used as markers for a given year and, in some cases, to characterize climate conditions during that year. Growth rings in trees, for example, depend on precipitation and temperature, so variations in rings from year to year serve as an indicator of annual variations of temperature and precipitation. Distinct patterns of growth rings serve as "bar codes" that allow researchers to correlate rings in living trees with rings in dead trees, and even with rings in buried logs. Such information helps paleoclimatologists to extend the record of climate variation far back in time.

By compiling data from the study of paleoclimate indicators, researchers have identified several alternating warming and cooling intervals during the past 800,000 years **(Fig. 20.15)**. And since the retreat of the last Pleistocene ice sheet, there have been trends of cooling or warming that lasted decades to millennia. For example, in the northern hemisphere, the time between about 15,000 and 10,500 B.C.E. was a warming period during which the last continental glaciers of the Ice Age melted away entirely **(Fig. 20.16a)**. This warming trend was followed by an interval of cooler temperatures during which much of Europe was treeless tundra. Researchers have named this interval the *Younger Dryas* after an Arctic flower that became widespread during that time. Temperatures then warmed again, so that between 9,000 and 5,000 years ago—an interval called the *Holocene*

Figure 20.12 Lake sediments contain a record of climate change extending back thousands of years or more.

(a) Example of a lake sediment core, sliced in half. Different colors are different layers of sediment.

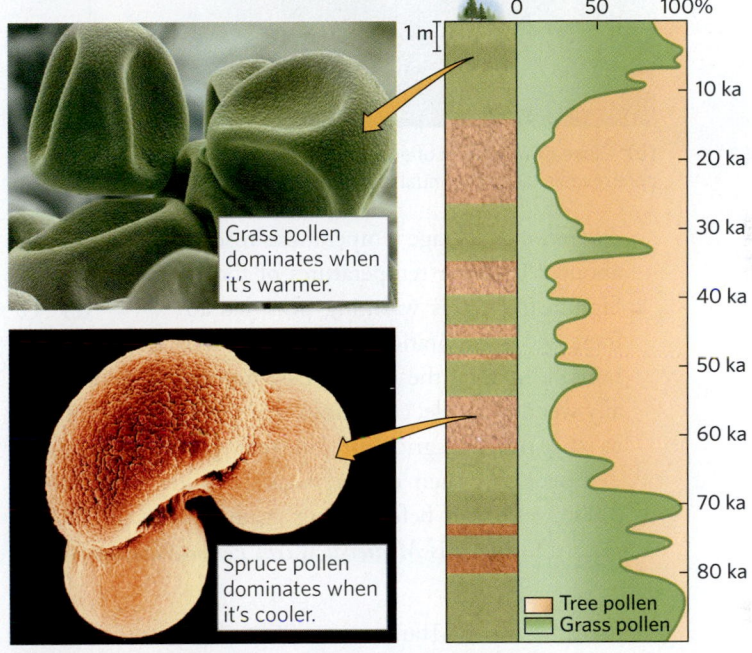

Grass pollen dominates when it's warmer.

Spruce pollen dominates when it's cooler.

Tree pollen
Grass pollen

(b) The pollen preserved in a lake sediment core can be used to determine the atmospheric temperatures at the time the sediment was deposited. ka = Thousand years ago.

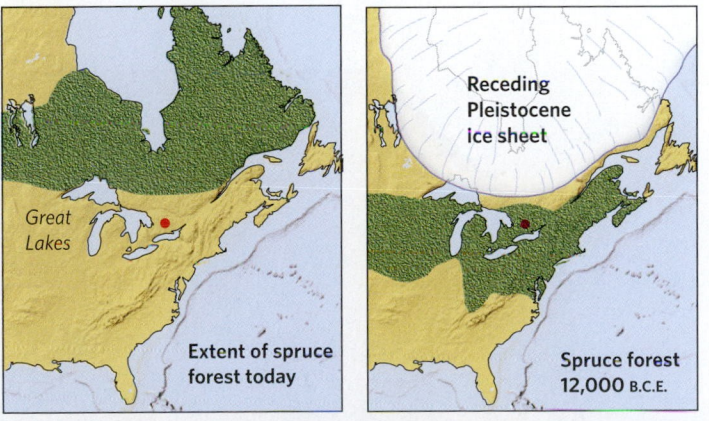

Great Lakes

Extent of spruce forest today

Receding Pleistocene ice sheet

Spruce forest 12,000 B.C.E.

(c) As we see from the pollen analysis, spruce forests (green) grew farther south 12,000 years ago than they do today. The red dots point out the location where the core sample was taken.

Figure 20.13 Paleoclimate data can be obtained by studying oxygen-isotope ratios in ice and sediment cores.

(a) A researcher drilling an ice core.

(b) Close examination of an ice core reveals distinct annual layers.

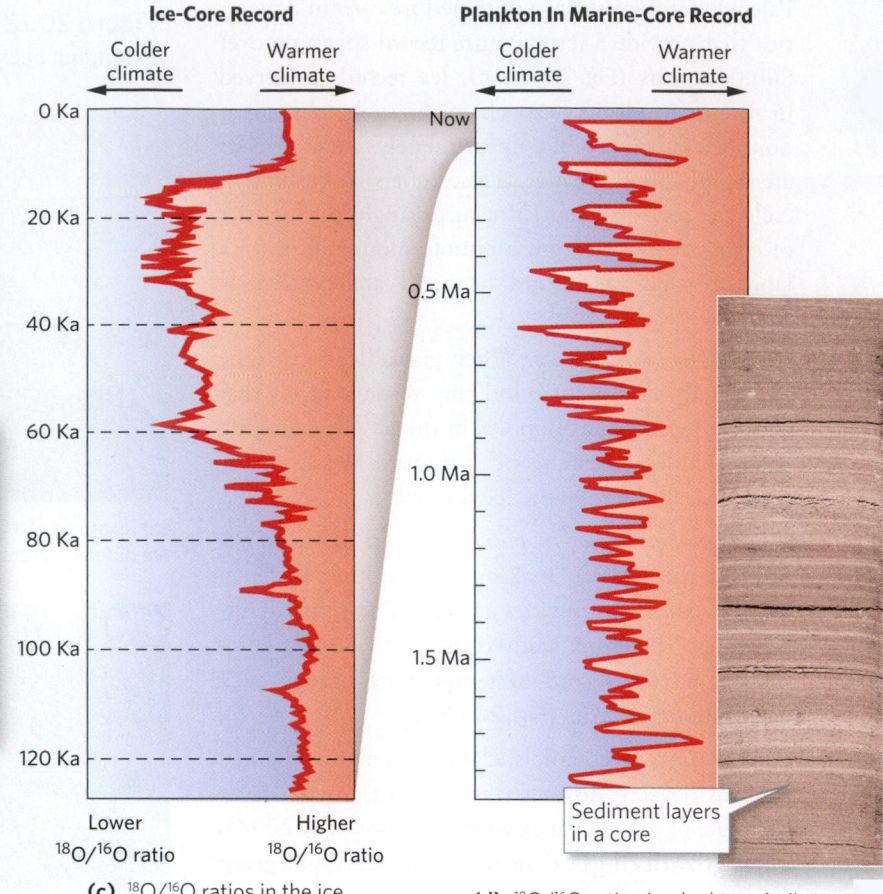

Ice-Core Record

Colder climate ← → Warmer climate

0 Ka
20 Ka
40 Ka
60 Ka
80 Ka
100 Ka
120 Ka

Lower $^{18}O/^{16}O$ ratio   Higher $^{18}O/^{16}O$ ratio

(c) $^{18}O/^{16}O$ ratios in the ice indicate temperatures over the past 120,000 years.

Plankton In Marine-Core Record

Colder climate ← → Warmer climate

Now
0.5 Ma
1.0 Ma
1.5 Ma

Sediment layers in a core

(d) $^{18}O/^{16}O$ ratios in plankton shells in a core of marine sediment record past temperatures over longer time scales.

maximum—average temperatures rose to about 2°C above temperatures of today. Significantly, this warming peak led to increased evaporation and therefore, rainfall, making the Middle East unusually wet and fertile, conditions that may have contributed to the rise of agriculture in that part of the world. Temperatures then dipped back to a low point about 3,000 years ago before rising again during the Middle Ages. During this *Medieval Warm Period*, Vikings settled along the coast of Greenland **(Fig. 20.16b)**. Temperatures dropped again between 1500 C.E. and 1800 C.E., a period known as the *Little Ice Age*, the time when Alpine glaciers advanced **(Fig. 20.16c)** and the canals of the Netherlands

Figure 20.14 The spacing of tree rings records a history of temperature and precipitation during the tree's lifetime. More growth happens in warm, wet years than in cool, dry ones.

Researchers extract cores to see rings in living trees without damaging the tree.

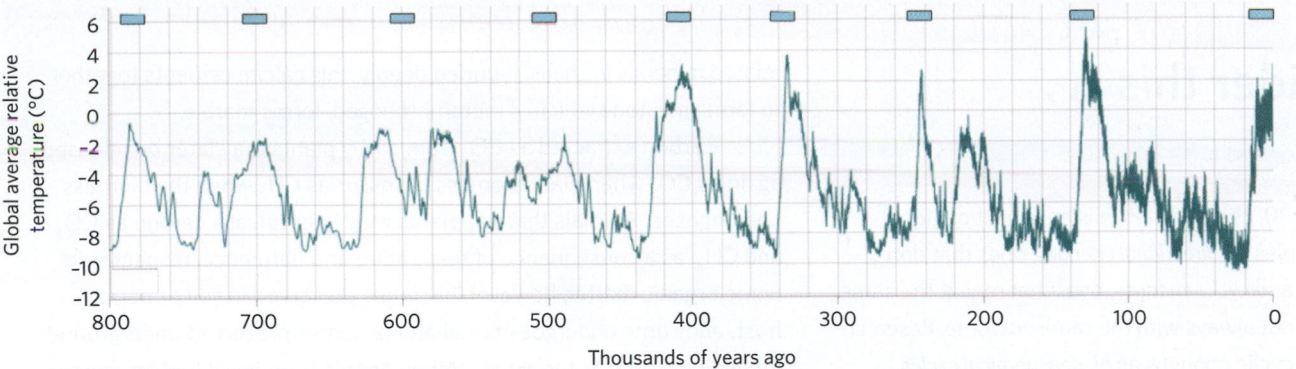

Figure 20.15 Temperature record for the past 800,000 years from an ice core, extracted from the Antarctic ice sheet. Blue bars indicate the duration of interglacials.

Figure 20.16 During the Holocene, temperatures have varied significantly.

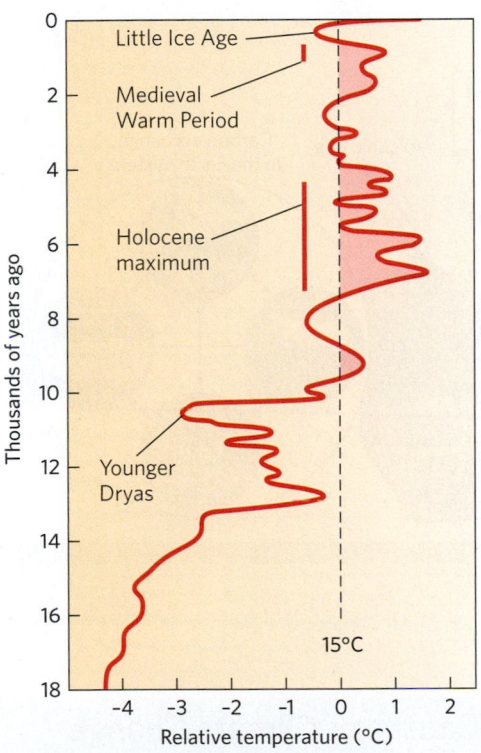

(a) There have been several temperature highs and lows since the most recent glaciation.

(b) The Vikings settled Greenland during the Medieval Warm Period.

At the end of the Little Ice Age, this tributary glacier in France reached the main valley floor.

Today, glaciers extend only part-way down the side valleys.

(c) During the Little Ice Age, glaciers in the Alps advanced. They have since melted away.

Box 20.2 ▸ Consider this . . .

The carbon cycle

As we discussed in Chapter 10, climate change is an aspect of global change. Global change includes *unidirectional changes*, ones that don't repeat during Earth history, and *cyclic changes*, ones that repeat the same steps over and over, though not always with the same outcome. Researchers refer to certain types of cyclic changes as *biogeochemical cycles* because they involve exchanges of chemicals among a variety of living and nonliving reservoirs in the Earth System.

In Chapter 12, we discussed an important example of a biogeochemical cycle, the hydrologic cycle (cycling of H_2O). Another example, the **carbon cycle**, involves the transfer of carbon among various reservoirs **(Fig. Bx20.2)**. This cycle plays a key role in climate change because some carbon occurs in greenhouse gases (CO_2 and methane, CH_4). The Earth System contains many reservoirs of carbon. Carbon initially enters the carbon cycle when it bubbles out of volcanoes and into the atmosphere as CO_2. Some of the CO_2 in air dissolves in the oceans to form HCO_3^- (bicarbonate) ions. This carbon later becomes incorporated into calcite ($CaCO_3$) or

related minerals in shells. If buried deeply, this calcite cements together to form limestone which, if metamorphosed, forms marble.

Carbon extracted as CO_2 from air by plants may later be released again as CO_2 when plant biomass burns, or as CH_4 when the biomass decomposes. Animals that eat plants may later release carbon as CO_2 and CH_4, as a consequence of respiration and flatulence, respectively. Some organic matter, however, becomes preserved in soil or permafrost, and some undergoes burial and becomes preserved underground for geologic time as fossil fuel. When people burn fossil fuel for energy, carbon returns to the atmosphere in CO_2.

Figure Bx20.2 The Earth System contains many reservoirs among which carbon is exchanged during the carbon cycle.

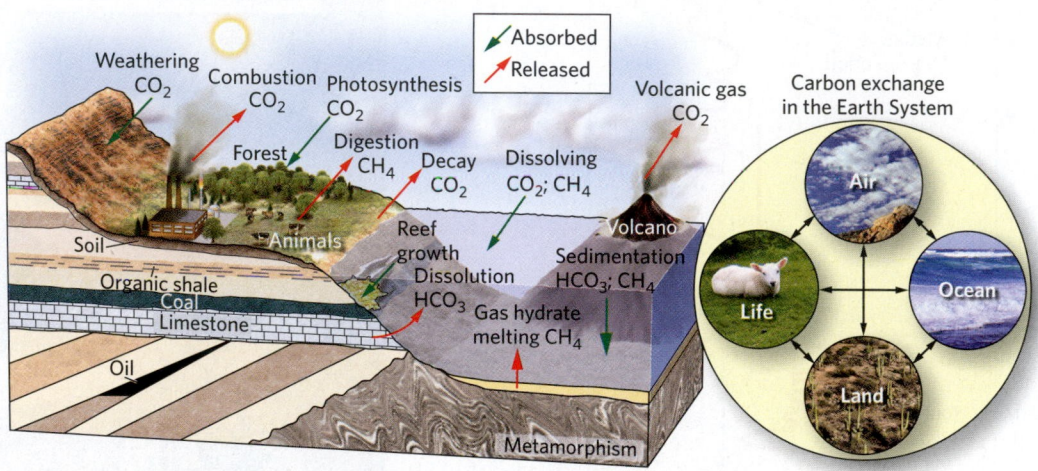

---

froze over in winter. Overall, climate has warmed since the end of the Little Ice Age, becoming as warm, or warmer, than it was during the Medieval Warm Period (see Fig. 20.15c).

Take-home message . . .

Climate change over geologic time includes episodes of both global warming and global cooling. Researchers use the stratigraphic record to characterize long-term changes (taking place over millions to hundreds of millions of years). They use various indicators, such as plankton species, tree rings, pollen, or variations in isotope ratios, to document short-term changes that have happened over centuries to millennia.

Quick Question -
How has climate changed in the northern hemisphere since the end of the last glaciation?

20.5 Causes of Climate Change

What causes climate change in the Earth System? The answer depends, in part, on the duration of the change under consideration, for not all factors responsible for long-term change can cause short-term change, and vice versa. We'll see that the *carbon cycle*, the movement of carbon among components of the Earth System, plays an important role in climate changes of both long and short durations **(Box 20.2)**.

Natural Long-Term Climate Controls

What factors drive long-term climate change on scales of millions of years? There is no one cause. Paleoclimatologists suggest that long-term changes in atmospheric composition, volcanic activity, land and ocean distribution, and mountain building all play a role. The energy output of the Sun, while it has changed over Earth history **(Box 20.3)**, does not correlate directly with long-term climate change.

Consider this . . .

Goldilocks and the faint young Sun paradox

Like Baby Bear's porridge in the tale *Goldilocks and the Three Bears*, the Earth is not too hot, and it's not too cold . . . it's just right for liquid water and, therefore, for life **(Fig. Bx20.3)**. Researchers informally refer to the two related factors that make the Earth "just right" as the *Goldilocks effect*. What are these factors? First, the distance of the Earth from the Sun means that the solar radiation reaching the surface doesn't warm our planet too much or too little. Astronomers define the distance from the Sun where liquid water can exist on the surface of a planet as the *habitable zone* of the Solar System (see Chapter 22). The habitable zone currently extends roughly from 0.8 AU to 1.3 AU. (An AU, or astronomical unit, represents the average distance between the Earth and the Sun.) Second, the atmosphere contains an appropriate concentration of greenhouse gases to keep enough, but not too much, heat trapped in the atmosphere.

Notably, the intensity of the radiation produced by the Sun has increased over the history of the Solar System. This change has taken place because the Sun's energy comes from the fusion of four hydrogen atoms to form one helium atom (see Chapter 23), and one helium atom has less mass than four hydrogen atoms. (The lost mass transforms into energy, which powers the Sun.) Production of helium, therefore, causes the Sun's volume to contract slowly over time, resulting in an increase in the internal pressure and temperature within the Sun, and these changes have caused the rate of fusion reactions to increase.

In fact, the Sun may be over 30% brighter today than it was in early Archean time.

If the Sun's intensity were the only factor controlling Earth's temperature, all water should have been frozen during the Archean. But stratigraphic and fossil records indicate that water has existed in liquid form on our planet's surface at least since the beginning of the Archean (about 4 Ga). Researchers refer to this apparent contradiction between the temperature that should have existed under a weaker Sun and the actual temperature of the early Earth as the *faint young Sun paradox*. Most researchers agree that the paradox can be resolved by keeping in mind that earlier in Earth history, the atmosphere contained more CO_2 than it does today. The greenhouse effect caused by the additional CO_2 would have increased the temperature of the Earth's atmosphere enough to counteract the lower intensity of the faint young Sun, and this would have kept surface temperatures above freezing.

Figure Bx20.3 The Goldilocks effect, as applied to the Earth. Our planet is just right for liquid water, and therefore for life, to exist.

Venus is too hot.

The Earth is just right.

Mars is too cold (and too small).

LONG-TERM CHANGES IN ATMOSPHERIC COMPOSITION. The Earth's early atmosphere consisted mostly of water vapor and carbon dioxide (see Chapter 17). An atmosphere so rich in greenhouse gases would make the Earth's surface realm into an oven, like that of Venus today. Fortunately, the Earth cooled sufficiently for water to precipitate, and once the oceans had formed, CO_2 dissolved in the water and then became incorporated into rock. As a result, the atmospheric concentration of CO_2 decreased substantially, temperatures dropped, and life could exist. Enough CO_2 has remained in the atmosphere, however, to keep the oceans above the freezing temperature of water (see Box 20.3). When photosynthetic organisms appeared, they extracted additional CO_2, and when those organisms died, the extracted carbon was mixed into sediment on the seafloor. Burial of this organic-rich sediment locked even more carbon into strata underground. Overall, the transfer of greenhouse gases from the air into the solid Earth led to a decrease in global temperature. Only after this had happened, by 2.4 Ga, could icehouse conditions even become a possibility. Not surprisingly, the earliest ice age began after 2.4 Ga.

Later changes in atmospheric greenhouse gas concentrations during Earth history caused subsequent climate changes recorded in the stratigraphic record as well. For example, some researchers speculate that the growth of immense coal swamps during the Carboniferous decreased atmospheric CO_2 concentrations enough to cause global cooling and the initiation of the Late Carboniferous ice age.

DISTRIBUTION OF CONTINENTS AND VOLCANIC ARCS. Late Proterozoic, late Paleozoic, and Pleistocene ice ages all occurred when substantial areas of land lay at high latitudes. Why? When large continents drift to high latitudes, less ocean water can reside in their vicinity. As we have discussed, water has a much higher heat capacity than land. This means that, without oceans at high latitudes, these regions of land become colder. If broad areas of high-latitude land lie at a great distance from the warmth of ocean waters, conditions on land can remain cold enough all year for continental ice sheets to grow.

The distribution of continents also affects the amount of rainfall on land in the tropics. For example, tropical rainforests can grow only when large areas of land lie at low latitudes. Also, the distribution of land affects ocean currents, which control heat distribution around the planet. In some cases, even a small amount of land, when strategically located, can affect currents. For example, when the Central American volcanic arc grew in the Cenozoic, it blocked currents from transporting warm Pacific water into the Atlantic, and this change affected temperatures around the Atlantic.

SEA-LEVEL CHANGE. At times during Earth history, the interiors of continents have been flooded by shallow seas (see Chapter 10). When water covers significant portions of the land, the distribution of heat around the Earth changes, and therefore, the climate changes. See **Earth Science at a Glance**, pp. 722–723, for more on sea-level change.

VOLCANIC ACTIVITY. Generally, volcanic activity adds only relatively small amounts of new CO_2 to the air, and the Earth System quickly redistributes this gas into other reservoirs. But at certain times in geologic history, the amount of volcanic activity increased dramatically and may have been sufficient to cause a substantial increase in the atmospheric concentration of CO_2, which in turn would cause global warming. For example, widespread rift-related volcanism associated with the breakup of Pangaea, along with the eruption of large igneous provinces, may have contributed to Cretaceous warming.

UPLIFT OF LAND SURFACES. Tectonic events that lead to the long-term uplift of large areas of land in mountains or plateaus may affect atmospheric CO_2 concentrations because these events expose rock to chemical weathering reactions that absorb CO_2. Uplift of the Himalayas and the Tibetan Plateau, for example, may have contributed to the global cooling that occurred after the warm period in the Eocene. Mountain building also affects global atmospheric circulation patterns and, therefore, the distribution of temperature and rainfall.

Natural Short-Term Climate Controls

During the Pleistocene Ice Age, continental glaciers advanced and retreated perhaps 20 to 30 times. The most recent advance, or *glaciation*, ended only about 11,000 years ago, and since then the planet has been in an *interglacial*. What factors can cause such changes, or comparable ones, that have relatively short durations? Some changes of very short duration (a year or decade) may simply reflect natural variation. But to explain events that last centuries to millennia, researchers suggest the following mechanisms:

MILANKOVITCH CYCLES. As Milutin Milanković recognized in 1920, the shape of the Earth's orbit (eccentricity) changes over a period of 110,000 years, the tilt of its axis changes over a period of 41,000 years, and the axis undergoes precession (wobble) over a period of about 19,000 to 23,000 years. Such periodic changes are called Milankovitch cycles (see Chapter 14), and each affects the amount of solar energy received at high latitudes during the summer **(Fig. 20.17)**. The temperature difference between times when the cycles reinforce one another and times when they cancel one another can be as much as 25°C.

CHANGES IN OCEAN CURRENTS. Recent studies suggest that the configuration of currents can change quickly, and that these changes could affect the climate. The Younger Dryas, for example, may have resulted when a sudden release of freshwater from melting glaciers spread a layer of freshwater over the North Atlantic and temporarily stopped thermohaline circulations throughout the oceans (see Chapter 16). Rapid growth or loss of sea ice may similarly affect currents.

LARGE ERUPTIONS OF VOLCANIC AEROSOLS. Not all of the sunlight that reaches the Earth penetrates its atmosphere and warms its surface. Some gets reflected by the atmosphere. The degree of reflectivity, or **albedo**, of the atmosphere increases if the concentration of volcanic aerosols in the atmosphere increases. Particularly large eruptions can emit enough aerosols (particularly SO_2) to affect global temperature for months to years **(Box 20.4)**.

CHANGES IN SURFACE ALBEDO. Regional-scale changes in the nature of vegetation cover, in the proportion of snow and ice cover, or in the distribution of light-colored volcanic ash can affect the albedo of Earth's surface. Increased albedo causes cooling, whereas decreased albedo causes warming.

Figure 20.17 The Milankovitch cycles.

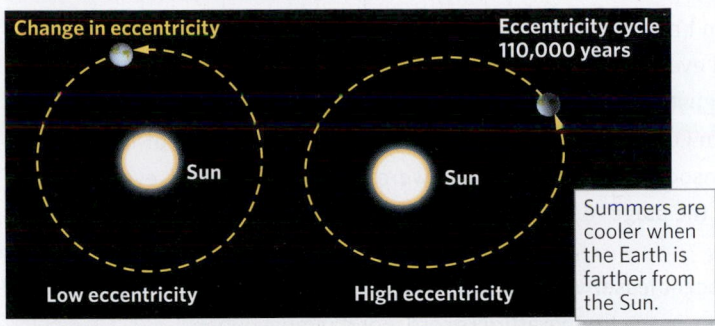

Change in eccentricity

Eccentricity cycle
110,000 years

Sun

Sun

Low eccentricity

High eccentricity

Summers are cooler when the Earth is farther from the Sun.

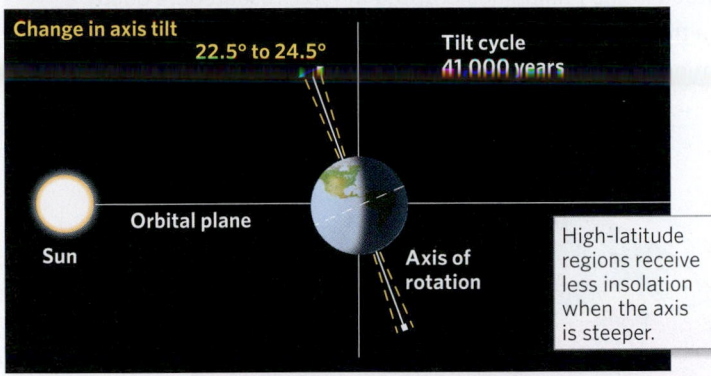

Change in axis tilt

22.5° to 24.5°

Tilt cycle
41,000 years

Sun

Orbital plane

Axis of rotation

High-latitude regions receive less insolation when the axis is steeper.

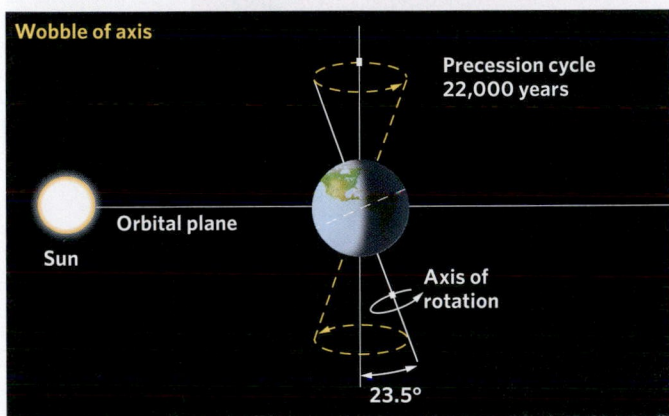

Wobble of axis

Precession cycle
22,000 years

Sun

Orbital plane

Axis of rotation

23.5°

(a) Variations to the amount of incoming solar radiation are caused by changes in the eccentricity of Earth's orbit (its orbital shape), the tilt angle of Earth's axis, and the precession of Earth's axis.

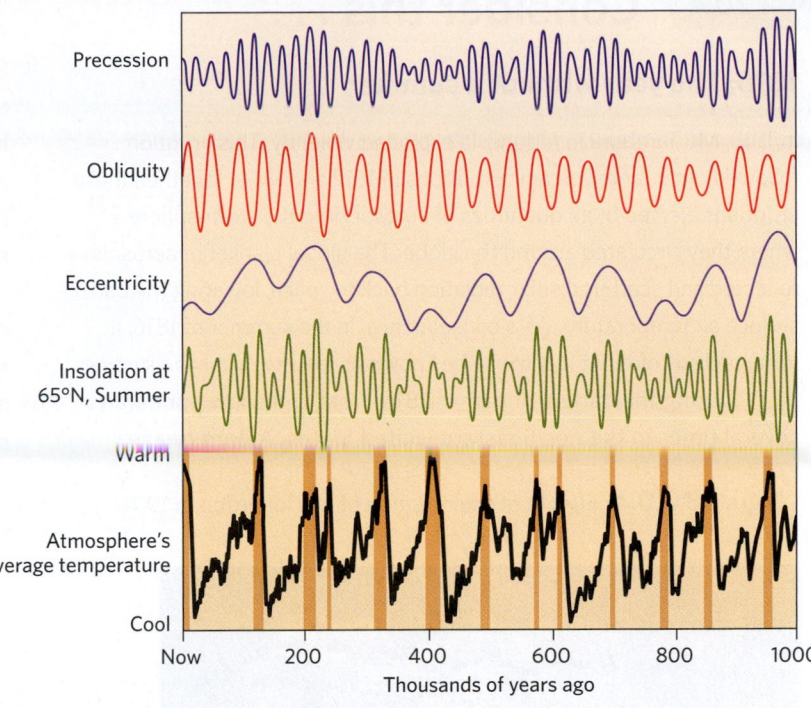

Precession

Obliquity

Eccentricity

Insolation at 65°N, Summer

Warm

Atmosphere's average temperature

Cool

Now   200   400   600   800   1000

Thousands of years ago

(b) When the effects of precession, obliquity change, and eccentricity are combined, they cause changes in the daily insolation (the amount of heat added to the atmosphere) at 65°N. The brown stripes are the warmest times of interglacials.

Take-home message . . .

Researchers are still working to identify causes of climate change. Phenomena causing long-term change include changes in atmospheric CO_2 concentrations, changes in the distribution of continents, and the growth of mountain belts. Those causing short-term climate change include the Milankovitch cycles and large volcanic eruptions.

Quick Question ---------------------------

What factors lead to changes in atmospheric CO_2 concentrations over geologic time?

ABRUPT CHANGES IN CONCENTRATIONS OF GREENHOUSE GASES. A relatively sudden change in greenhouse gas concentrations in the atmosphere can affect climate. Such a change might happen in several ways. If ocean temperatures warmed or sea level dropped, some of the methane-containing ice in seafloor sediments would melt suddenly. Such melting would release CH_4 to the atmosphere. A similar release of CH_4 might happen if permafrost started to melt and the organic matter within it began to rot. Finally, algal blooms and changes in forest cover could also conceivably change atmospheric CO_2 concentrations, for growth of photosynthetic organisms removes CO_2 from the atmosphere.

20.6 Evidence for and Interpretation of Recent Climate Change

Let's now turn our focus to the climate of the past few centuries, the time during which human society has become an active component of the Earth System. Researchers constantly publish new data and interpretations pertaining to this topic. In fact, the amount of information relevant to documenting and understanding climate change has become so overwhelming that in 1988 the World Meteorological Organization, in collaboration with the United Nations Environment Program, founded the

Box 20.4 ▶ **Consider this . . .**

1816: The year without a summer

In 1815, Mt. Tambora in Indonesia exploded violently. This eruption followed other large volcanic eruptions in 1812 and 1814. Together, these eruptions ejected huge quantities of aerosols into the stratosphere, where they circulated around the globe. The global blanket of aerosols reflected and scattered solar radiation back to space, lowering the Earth's surface air temperature. As a consequence, in the summer of 1816, it remained so cold that it came to be known as the *year without a summer*.

During the early summer of 1816, frost and snow appeared several times in the northeastern United States. A hard freeze in June killed animals and destroyed crops, and temperatures in July didn't rise above 7°C (45°F) on many days. Northern New England even experienced crop-destroying frosts in mid-July and mid-August, when the region should have had its hottest days. Northern Europe suffered similar cold during the summer of 1816, and monsoons in India and China were severely disrupted, events that set the stage for famine and outbreaks of disease that killed millions.

Recent lesser volcanic eruptions have caused similar responses in the temperature record. For example, when Mt. Pinatubo in the Philippines erupted in 1991, researchers documented a three-year-long global decrease in temperature and precipitation **(Fig. Bx20.4)**.

Figure Bx20.4 Effects of the eruption of Mt. Pinatubo in 1991.

(a) During the eruption, immense volumes of visible ash and nearly invisible aerosols were blown into the stratosphere.

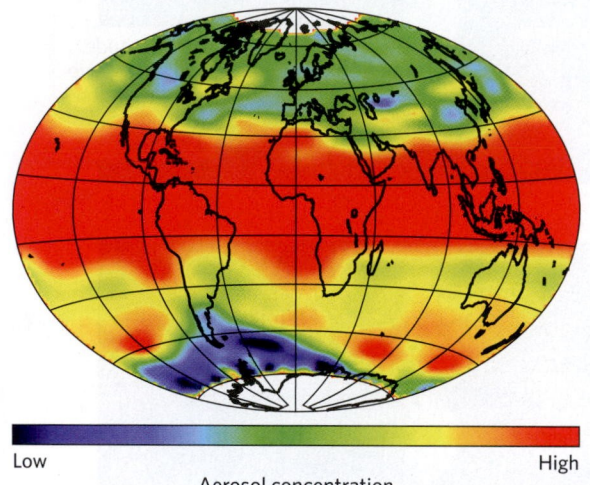

Low — Aerosol concentration — High

(b) Two months after the eruption, volcanic aerosols surrounded the Earth.

(c) A view from the Space Shuttle orbiting the Earth shows the haze caused by Pinatubo's aerosols in the stratosphere.

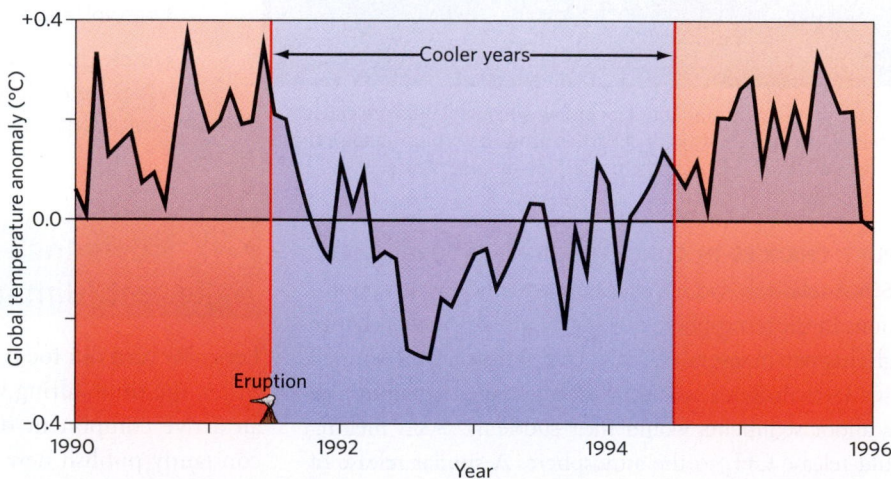

(d) The large injection of dust and aerosols into the atmosphere caused the average global temperature to drop for a few years after the eruption.

Figure 20.18 Measurements of global warming relative to the average for 1951–1980.

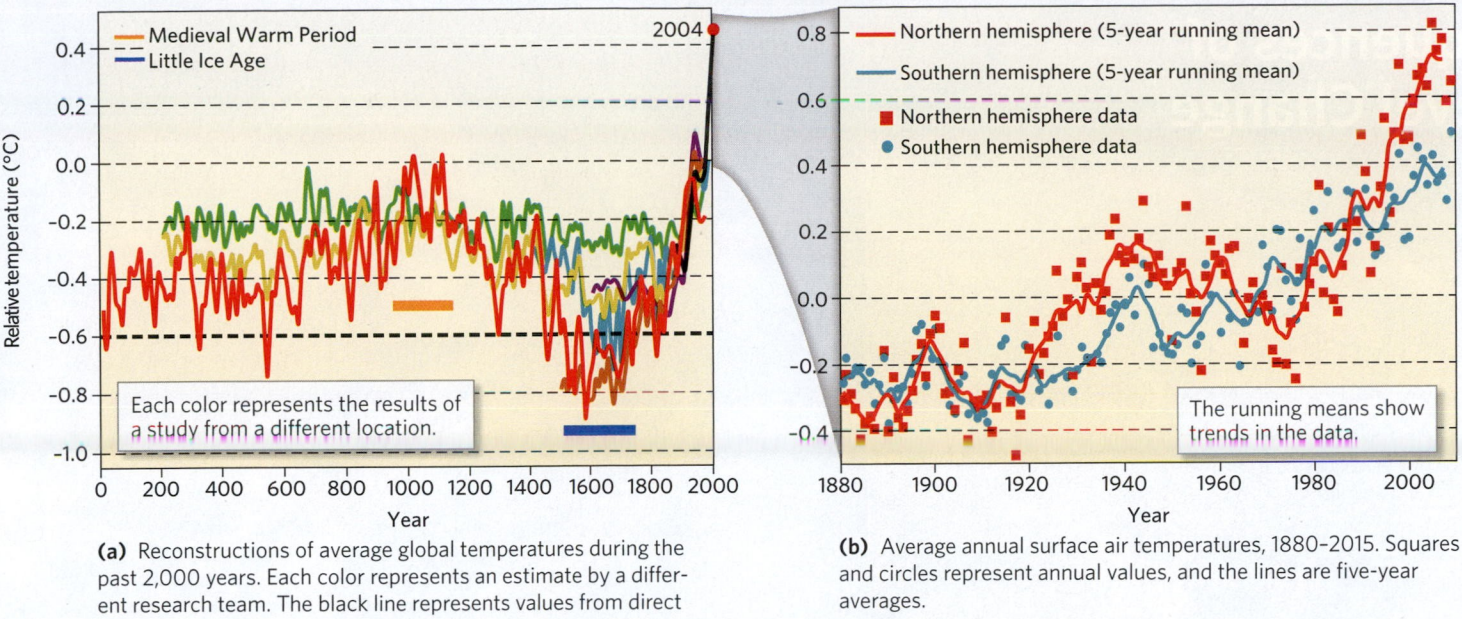

(a) Reconstructions of average global temperatures during the past 2,000 years. Each color represents an estimate by a different research team. The black line represents values from direct measurements.

(b) Average annual surface air temperatures, 1880–2015. Squares and circles represent annual values, and the lines are five-year averages.

Intergovernmental Panel on Climate Change (IPCC) to evaluate published climate-related studies and summarize their conclusions in a report for a general audience. This report, a new edition of which comes out every five years, has had an important influence on political decisions made to address climate change. In this section, we examine evidence for global temperature change and its consequences during the Earth's recent past, as summarized by the IPCC.

Observed Temperature Change in the Recent Past

By combining direct measurements made by thermometers with measurements of paleoclimate indicators, researchers have pieced together a graph showing how *average annual temperature* has changed over the past 2,000 years (Fig. 20.18a). Prior to about 1400, average global temperature rose and fell within a limited range. Over the next four centuries, the Earth experienced a cool period, in which the average temperature was 0.2°C to 0.4°C (0.4°F to 0.7°F) colder than in the previous centuries. Since 1800, the temperature has warmed, with the largest increase after the beginning of the industrial revolution. Because of the shape of the curve depicting this change, the graph has come to be known, informally, as the *hockey-stick diagram*. It remains the subject of research because not all researchers interpret the paleoclimate data it portrays in the same way.

If we zoom in on the recent past (1880–2015), we see that warming hasn't been continuous. We also see that

overall, the northern hemisphere has warmed more than the southern hemisphere (Fig. 20.18b)—the former has warmed by about 1.2°C (2.1°F) and the latter by about 0.9°C (1.6°F). Land areas have been affected most, with the greatest warming taking place in the Arctic, northern North America, and northern Asia (Fig. 20.19). Air over the oceans has undergone less warming because the oceans' high heat capacity moderates change.

Figure 20.19 Pattern of temperature change based on measurements over the past half-century, 1961–2014.

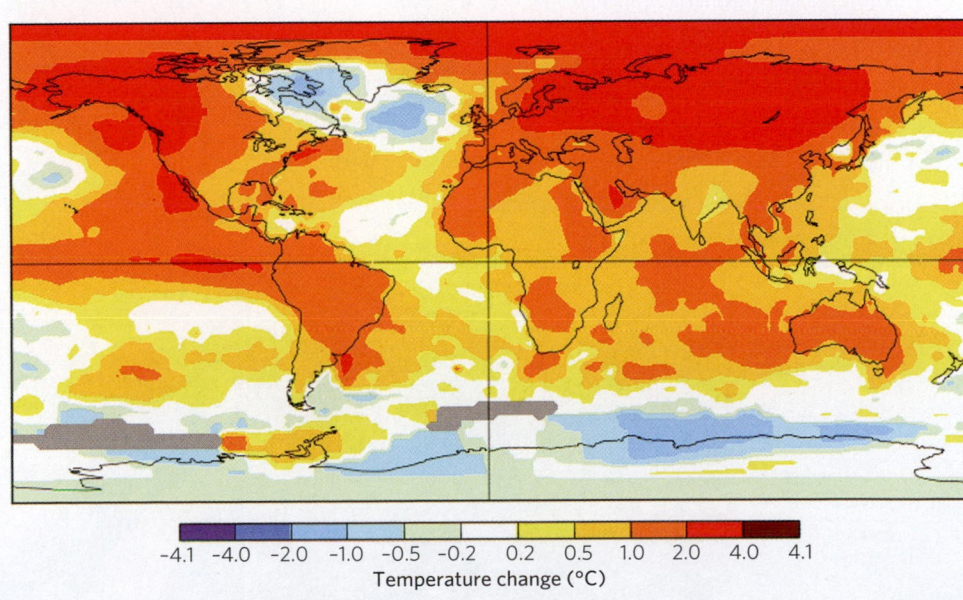

Consequences of Sea-Level Change

Antarctica

If Antarctica suddenly became ice free, most of West Antarctica would be underwater.

A large volume of water resides in the glaciers of Antarctica and Greenland. If all this ice were to melt suddenly, these two continents would become mostly dry land. (Parts would remain submerged until glacial rebound takes place, because subsidence due to the weight of ice has pushed the crust's surface below sea level.) Transfer of water from the glacial reservoir back to the oceanic reservoir would cause a sea-level rise of about 65–70 m (215–230 feet). Even if only 10% of this rise were to happen, coastal cities would flood.

San Francisco (after 7-m rise)

Greenland

A suddenly ice-free Greenland would host a central lake.

New Orleans (after 7-m rise)

Measured sea-level rise varies with location. The maximum current observed rate is about 1 m/century. At this rate, the 7-m rise depicted here in the city images will not happen for several centuries. But even a 1-m rise can lead to a costly increase in nuisance flooding and in storm-surge damage.

A sea-level rise of 70 m would flood large areas of coastal land worldwide (areas shown in purple). In fact, most of Florida would be underwater, and New York City buildings would be submerged up to about the 20th floor. Even a 7-m (20-foot) rise would turn streets in many major cities into canals, as shown in the insets.

New York (after 7-m rise)

Figure 20.20 Changes in sea level since the end of the last glaciation.

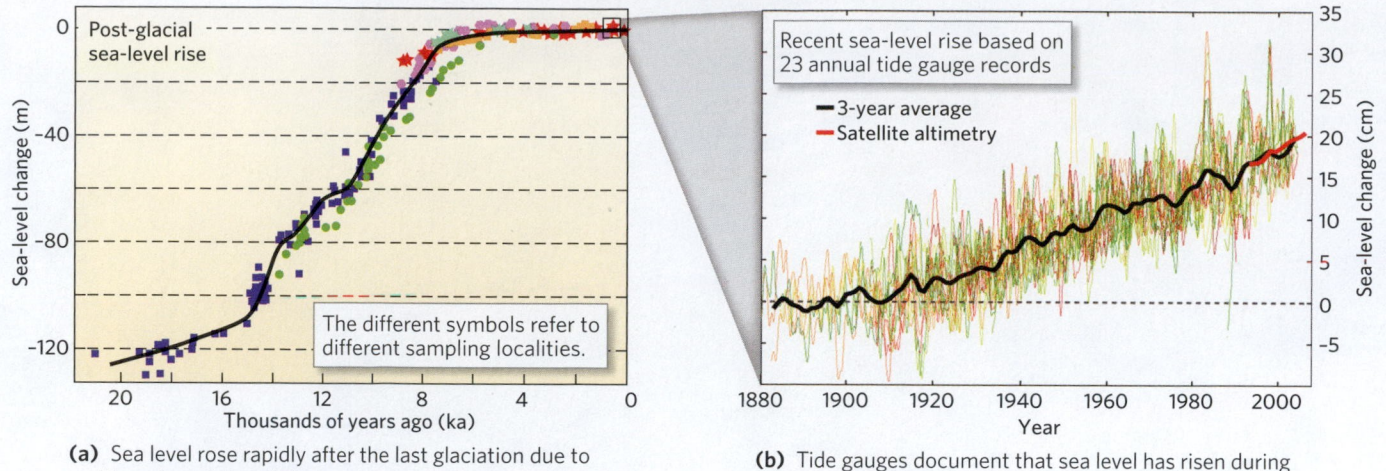

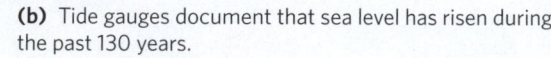

(a) Sea level rose rapidly after the last glaciation due to melting of continental glaciers.

(b) Tide gauges document that sea level has risen during the past 130 years.

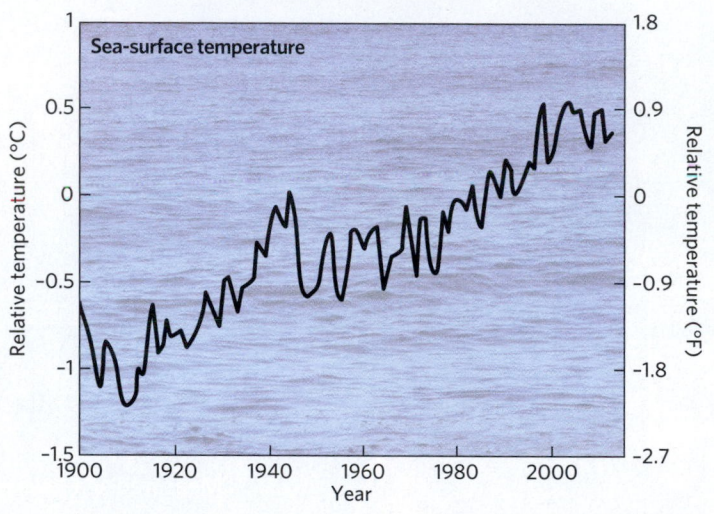

(c) Average global sea-surface temperatures have increased since 1900. This graph shows values relative to the average for 1971–2000.

Observed Recent Global Sea-Level Change

During the last glaciation of the Pleistocene Ice Age, a huge volume of water moved from the ocean reservoir into the glacial reservoir of the hydrologic cycle (see Chapter 12), and sea level dropped. When the glaciers melted, that water returned to the ocean, and sea level rose by about 120 m (about 400 feet) (Fig. 20.20a). This rise tapered off about 8,000 years ago, when the last of the North American and Asian continental glaciers disappeared.

In the past few centuries, sea level has started to rise again. In fact, sea level today lies about 23 cm (9 inches) higher than it was in 1885 (Fig. 20.20b). Researchers link this rise to global warming, for rising temperature affects sea level in two ways. First, liquid water in the ocean expands as it warms (Fig. 20.20c). Because oceans are confined on their bottoms and sides by rock and sediment of the ocean floor, this expansion causes the sea surface

Figure 20.21 The Muir Glacier in Alaska retreated 12 km (7 miles) between 1941 and 2004.

Figure 20.22 Changes in glacial volumes.

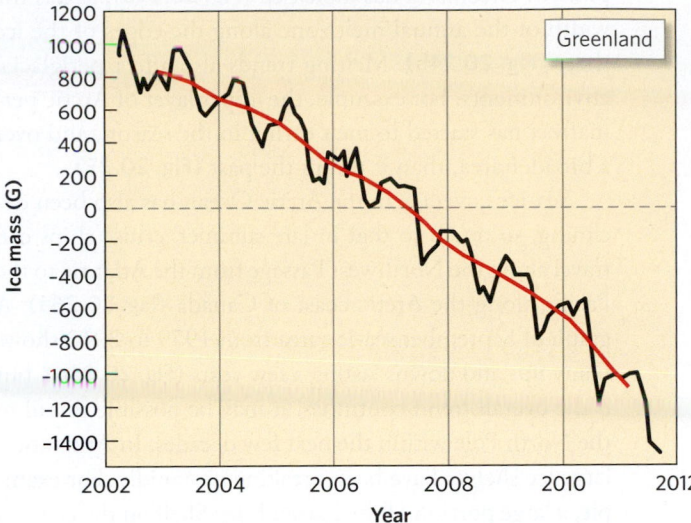

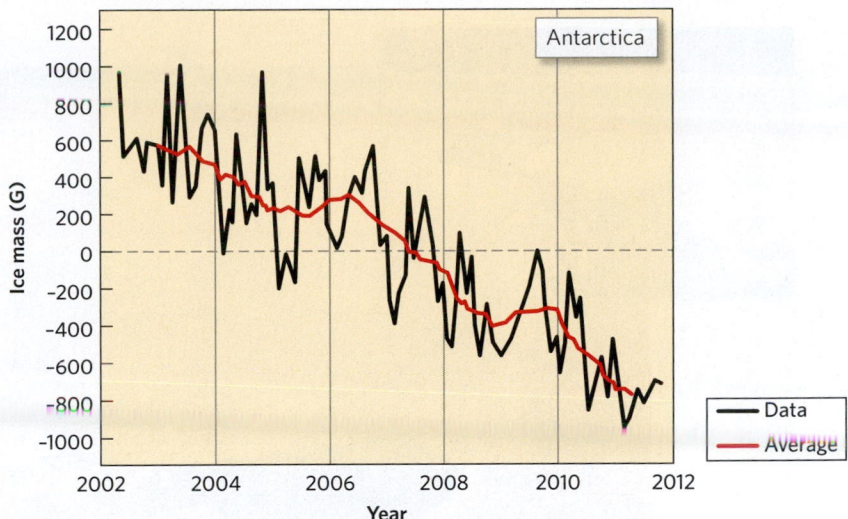

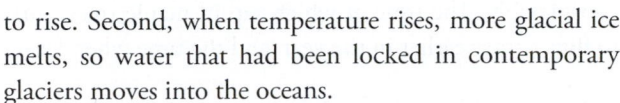

(a) Ice-mass loss in gigatons (G) as measured by the GRACE satellites in Greenland and in Antarctica over the period 2002–2012.

to rise. Second, when temperature rises, more glacial ice melts, so water that had been locked in contemporary glaciers moves into the oceans.

Other Indicators of Recent Climate Change

In order to further test the idea that the Earth's climate has been warming in recent centuries, researchers have examined several other features of the Earth System. They have focused on features whose characteristics reflect temperature trends averaged over a time frame of decades, rather than on indicators of annual or seasonal natural variation.

GLOBAL GLACIAL RETREAT. Compare an old photograph of a valley glacier with a present-day photograph of the same glacier (Fig. 20.21). The difference is obvious. Most valley glaciers worldwide have retreated and thinned dramatically, leaving their lateral and terminal moraines stranded (see Chapter 14). In many cases, glaciers that once filled valleys have disappeared entirely. For example, Glacier National Park in Montana had 150 glaciers in 1850, but now has only 25. The remaining glaciers in the park continue to lose volume by 2% to 6% per year, so all of the glaciers in the park may be gone in just a few decades (Fig. 20.21b).

Are the ice sheets covering Antarctica and Greenland also shrinking? To answer this question, researchers have used the GRACE (Gravity Recovery and Climate Experiment) satellites to measure the gravitational pull of the Earth very precisely. The resulting measurements

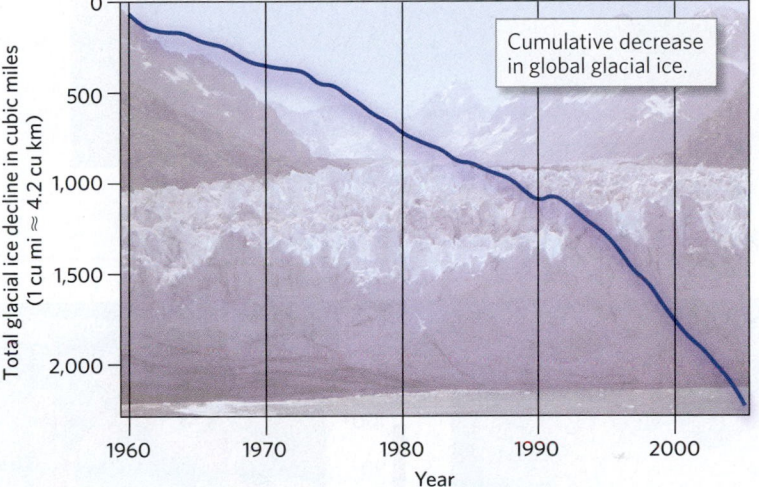

(b) Global glacial ice volume has been decreasing by 400 km³ (about 100 cubic miles) per year.

indicate that the gravitational pull exerted by the mass of ice in the Antarctic and Greenland glaciers has decreased noticeably (Fig. 20.22a). Furthermore, the rate of ice loss has accelerated. For example, in 2003, Greenland lost about 90 km³ (22 cubic miles) of ice per year, whereas in 2010, it lost 220 km³ (53 cubic miles) per year. As a result, the Greenland ice sheet now thins by about 1 m (3 feet) per year. Addition of glacial melting measurements indicates that globally, ice volume has decreased significantly in the last half century (Fig. 20.22b).

These losses of ice have been confirmed by direct observations. Researchers have found that large lakes of meltwater accumulate on the surface of the Greenland ice sheet in the summer; some of these lakes suddenly drain through crevasses down to the base of the glacier, then flow out to sea through tunnels melted into the base of

Figure 20.23 Lakes on the Greenland ice sheet often disappear suddenly as they drain through crevasses to the base of the ice sheet.

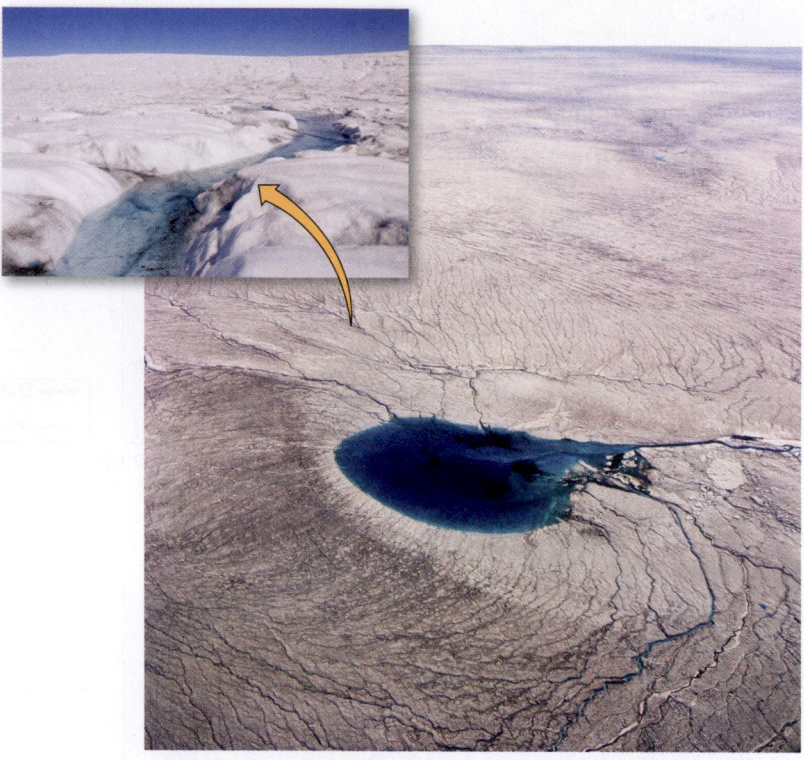

the glacier **(Fig. 20.23)**. Researchers have also found that the number of days during the year when melting takes place in Greenland has increased **(Fig. 20.24a)**, as has the width of the annual melt zone along the edges of the ice sheet **(Fig. 20.24b)**. Melting trends also affect periglacial environments. For example, the upper layer of Arctic permafrost has started to melt earlier in the season, and over a broader area, than it did in the past **(Fig. 20.25)**.

Sea-ice coverage in the Arctic Ocean has also been declining, so much so that in late summer, cruise ships can travel along the Northwest Passage from the Atlantic to the Pacific along the Arctic coast of Canada **(Fig. 20.26a)**. A graph of September sea-ice area from 1979 to 2013 shows many ups and downs lasting a few years **(Fig. 20.26b)**, but if the overall trend continues, it may be possible to sail to the North Pole within the next few decades. In Antarctica, large ice shelves have been breaking up rapidly. For example, a large portion of the Larsen B Ice Shelf on the coast of the Antarctic Peninsula disintegrated in 2002 **(Fig. 20.27)**.

DISTRIBUTION OF ORGANISMS. Biological indicators also point toward global warming. For example, in the northern hemisphere, the growing season for plants has lengthened considerably. The time at which sap in maple trees starts to flow comes earlier in the year, and the time when leaves

Figure 20.24 Melting of the Greenland ice sheet.

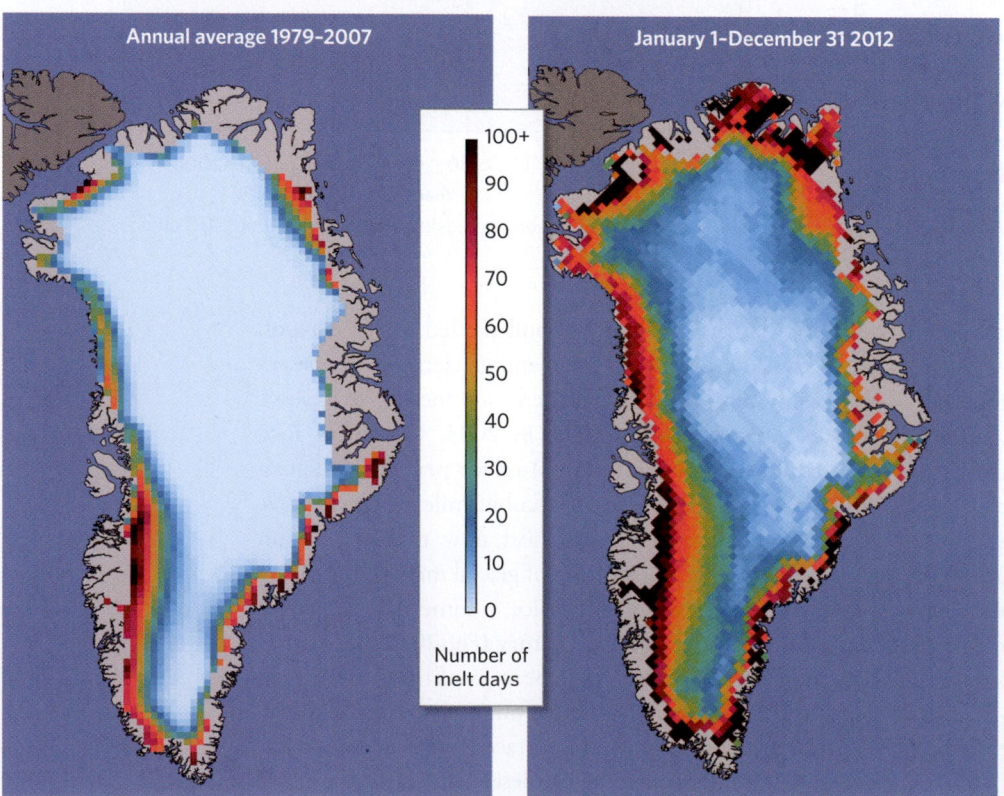

(a) The number of cumulative melt days has increased.

The gray area, spotted with puddles, is the melt zone.

(b) The width of the annual melt zone along the margins of the ice sheet has increased.

Figure 20.26 Sea-ice coverage of the Arctic Ocean has decreased.

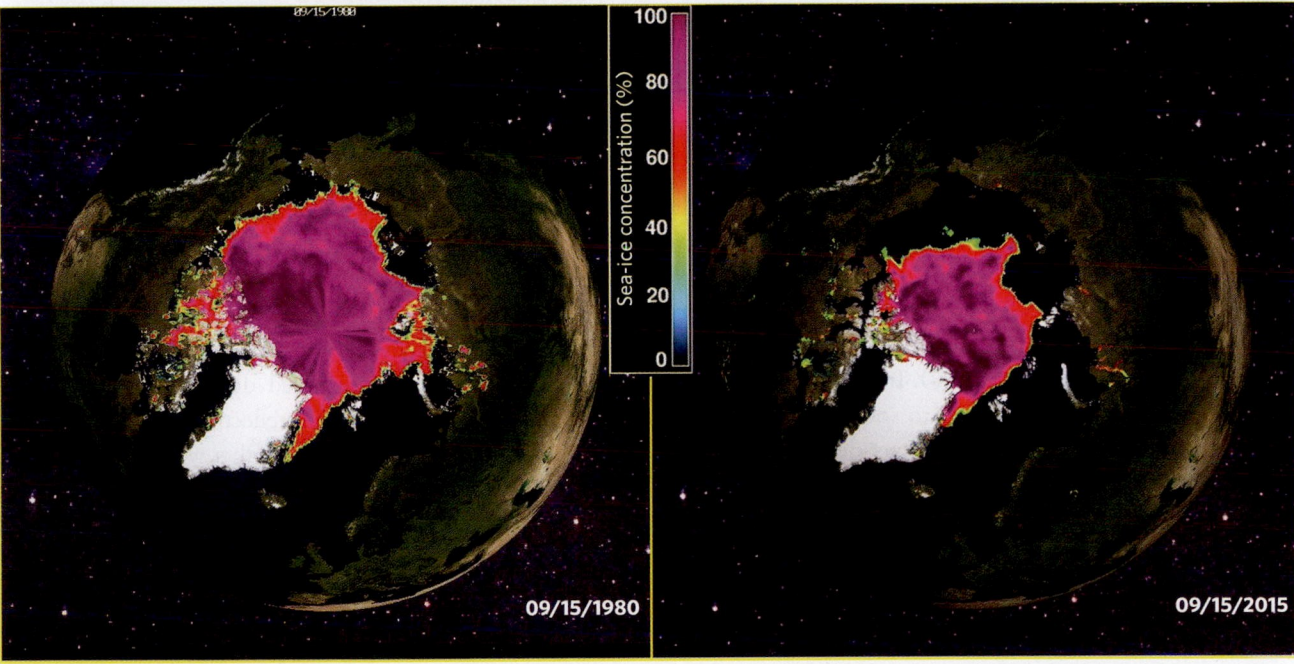

(a) Satellite images of Arctic Ocean sea ice in 1980 and in 2015.

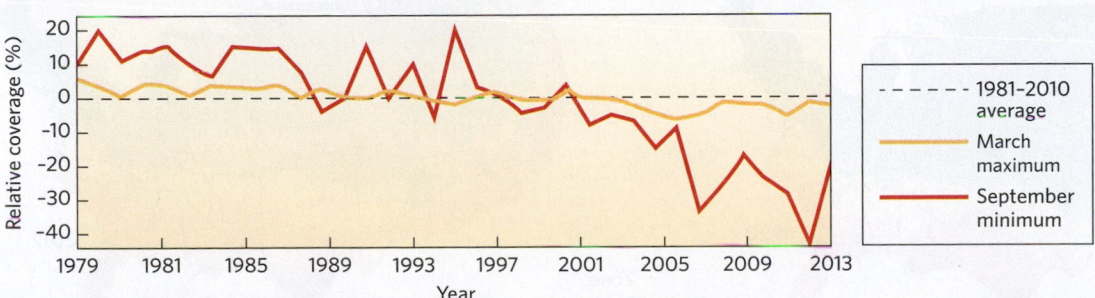

(b) A graph of sea-ice coverage relative to the average for 1979–2013 shows fluctuations, but also shows an overall downward trend.

Figure 20.27 A large portion of the Larsen B Ice Shelf on the coast of the Antarctic Peninsula disintegrated in 2002.

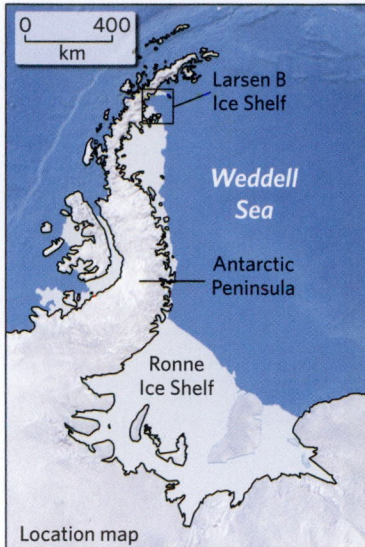

Location map

January 31, 2002

March 7, 2002

See for **yourself**

Northwest Passage, Arctic Ocean

Latitude: 74°22′33.45″ N
Longitude: 101°51′31.42″ W

Look down on this scene from an altitude of 5,246 km (3,260 miles) and you will see the Northwest Passage, the famed path between the Atlantic and the Pacific sought by ships over the centuries to avoid sailing all the way around the tip of South America. The passage has been blocked by sea ice for centuries. That is changing as the Arctic has warmed and Arctic sea ice coverage has diminished. In some recent years, open water has been found in late summer stretching along the coast from Greenland to Alaska's northern coastline.

turn color comes later in the year. Comparing a plant hardiness zone map from 2012 with one from 20 years earlier emphasizes the change in the growing season **(Fig. 20.28)**. The ranges of birds and insects on continents have also changed noticeably in recent decades.

CHANGES IN GLOBAL WEATHER. Researchers have begun to detect changes in weather patterns over time. For example, during the past century, overall precipitation over land areas has increased by about 2% worldwide, and the distribution and character of this precipitation has changed. Some parts of the world (such as central Africa, western South America, and the southwestern United States) have become drier, while other areas have become wetter **(Fig. 20.29)**. In the northern hemisphere, the ratio of rainy days to snowy days in winter has increased while snow cover has decreased. Recent studies hint that the severity of storms has been increasing overall, perhaps because warmer temperatures lead to increased evaporation, which, as we have seen, fuels storms.

Interpreting the Causes of Recent Climate Change

Which of the many causes of climate change that we discussed earlier in this chapter might be responsible for the observed global warming and associated sea-level rise of the past few centuries? The Keeling curve, as well as studies that track CO_2 and CH_4 concentrations over longer time frames **(Fig. 20.30)**, have led nearly all climate scientists to conclude that an increase in greenhouse gas (CO_2 and CH_4) concentrations in the atmosphere serves as the main driver of global warming since the start of the industrial age. Does the change in CO_2 concentrations observed during the past century reflect natural causes, or does it reflect inputs into the Earth System from human activities? In other words, is the change due to **natural forcing** or to **anthropogenic forcing**? Insight into this question comes from several sources.

Figure 20.28 The boundaries between plant hardiness zones, as defined by the US Department of Agriculture, shifted northward between 1990 and 2012. Blues denote shorter growing seasons, and reds indicate longer growing seasons.

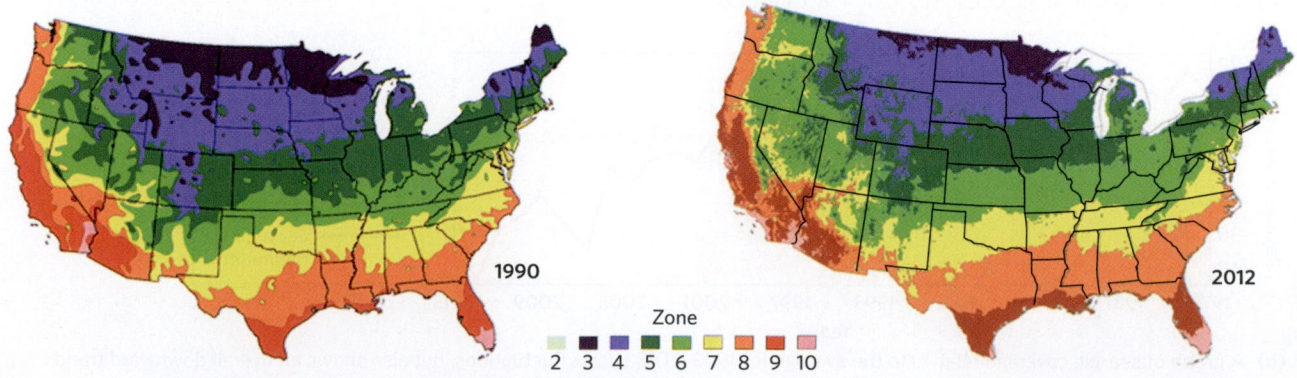

1990

2012

Zone
2 3 4 5 6 7 8 9 10

726   CHAPTER 20   CLIMATE AND CLIMATE CHANGE

Figure 20.29 Trends in measured annual precipitation (% change) between 1900 and 2000. Yellow, orange, and red shading represent increases; green and blue shading represent decreases.

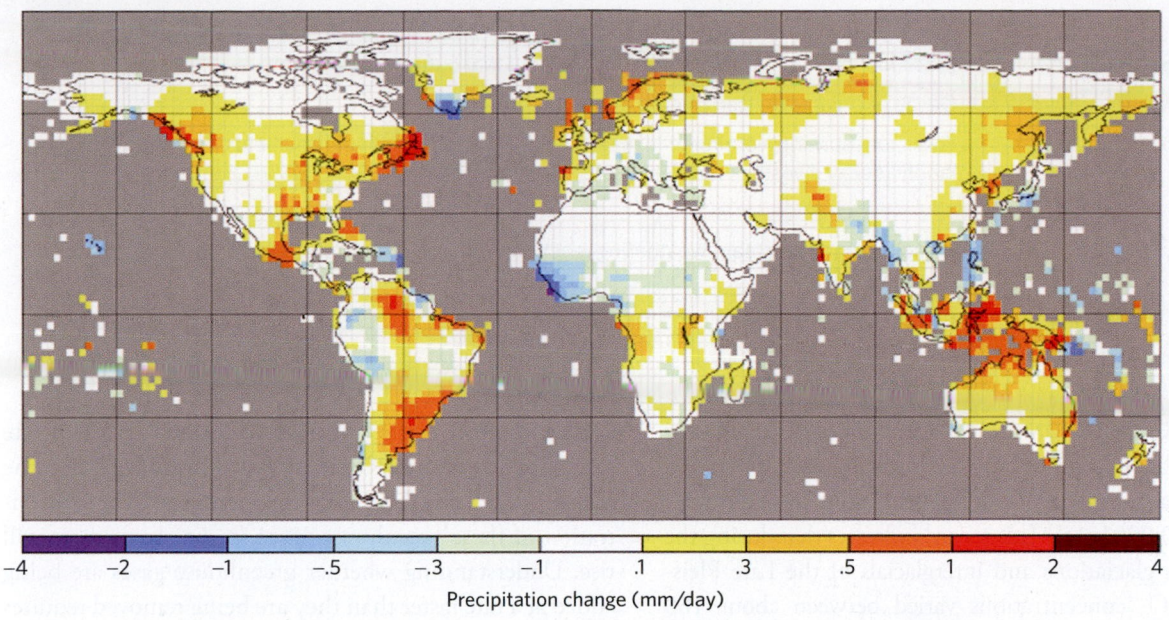

Precipitation change (mm/day)

Figure 20.30 Changes in atmospheric carbon-dioxide and methane concentrations over time.

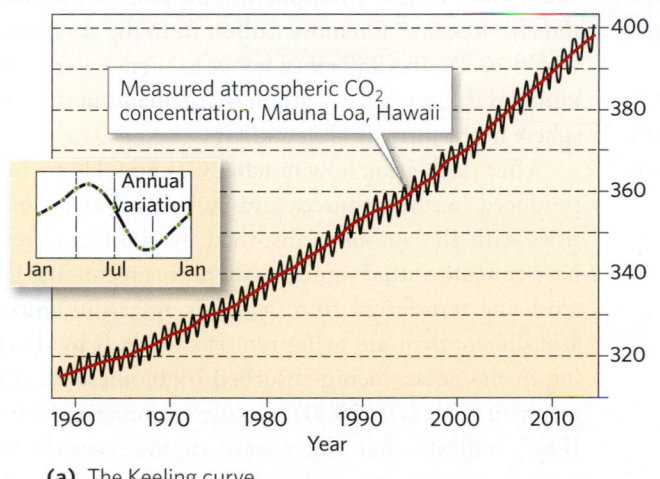

(a) The Keeling curve.

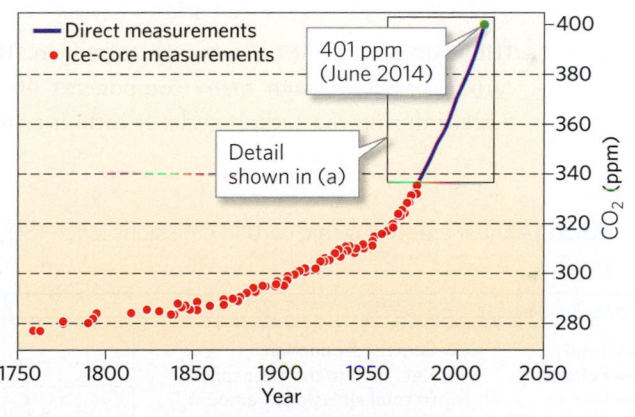

(b) Since the industrial revolution, the atmospheric CO_2 concentration has steadily increased.

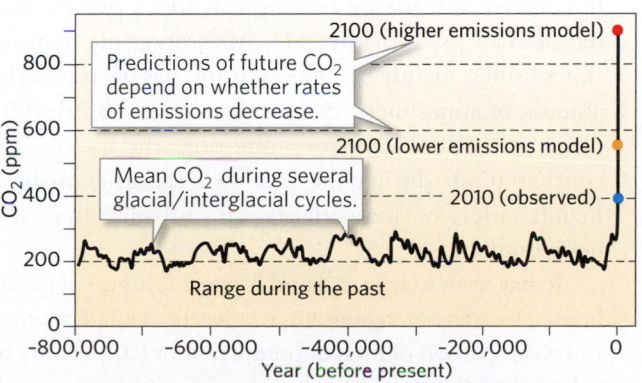

(c) Studies of ice cores from glaciers show that the CO_2 concentration varied between 180 and 300 ppm over the past 800,000 years. Now it is over 400 ppm.

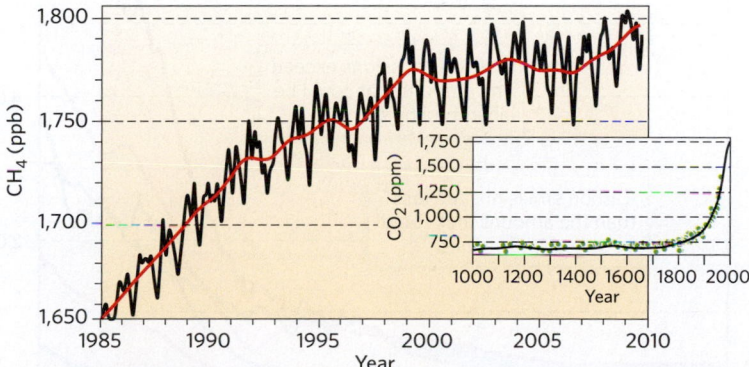

(d) The atmospheric concentration of CH_4, another greenhouse gas, has also been increasing.

Table 20.2 Sources and sinks of carbon-containing greenhouse gases

Sources	Causes
Natural CO_2	Volcanoes; decay of organic matter; animal respiration
Anthropogenic CO_2	Fossil fuel burning; concrete production; industrial output
Natural CH_4	Decay of organic matter; microbial respiration
Anthropogenic CH_4	Fossil fuel burning; venting or leakage of natural gas from oil and gas fields; rice paddy decay; landfill decay; livestock flatulence
Sinks	
Natural CO_2	Dissolution in the ocean; weathering reactions with rocks and soils; incorporation by photosynthetic organisms
Natural CH_4	Chemical reactions in soil and the atmosphere

Earth's climate system is an ornery beast which overreacts even to small nudges.

—WALLACE BROECKER
(AMERICAN GEOCHEMIST, 1931–)

BUBBLE COMPOSITION IN GLACIAL ICE. Measurements of air-bubble composition in Antarctic glaciers provide a record of atmospheric CO_2 concentrations back through almost 800,000 years. This record indicates that during the alternating glaciations and interglacials of the Late Pleistocene, CO_2 concentrations varied between about 180 and 300 ppm (see Fig. 20.30c). Therefore, the observed increases since the mid-19th century are beyond the range of natural variations that have occurred during the last 800,000 years, suggesting that the change is anthropogenic.

THE CARBON BUDGET. As we discussed in Box 20.2, carbon passes through many components of the Earth System during the carbon cycle. If carbon transfers into the atmosphere in the form of CO_2 and CH_4 at a rate faster than it transfers out of the atmosphere into the hydrosphere, biosphere, or geosphere, then the concentration of these greenhouse gases in the atmosphere will rise. Understanding whether greenhouse gases are being added at a rate faster than they are being removed requires that researchers characterize the Earth's **carbon budget** by quantifying *atmospheric carbon sources* (processes that add carbon to the atmosphere) and *atmospheric carbon sinks* (processes that remove carbon from the atmosphere) **(Table 20.2)**. The difference between these two numbers indicates the amount of carbon that remains in the atmosphere in the form of CO_2 and CH_4.

After calculating how much CO_2 and CH_4 are being produced by their sources and comparing those quantities with the amounts absorbed by sinks, researchers have concluded that significantly more greenhouse gases are being transferred from geologic reservoirs into the atmosphere than are being removed from it by dissolving in the ocean, being absorbed by biomass, or reacting with rock **(Fig. 20.31)**. Results summarized by the IPCC indicate that the release of CO_2 comes from burning fossil fuels (combustion produces CO_2) and, to a lesser extent, from concrete production (the cement in concrete is made by heating $CaCO_3$, a process that releases CO_2; see Chapter 11). Anthropogenic destruction of sinks, mainly by deforestation, has decreased the amount of atmospheric carbon that can be absorbed by plants. The IPCC, therefore, interprets the rise of CO_2 concentrations during the past two centuries to be a manifestation of industrialization, urbanization, and deforestation.

It may seem strange that anthropogenic inputs of greenhouse gases are so significant relative to natural sources. But a comparison of human production of CO_2 relative to volcanic production shows that it is. In an average year, all volcanic eruptions together, including both submarine and subaerial eruptions, emit about 0.15 to 0.26 gigatons of

Figure 20.31 Annual anthropogenic carbon emissions since 1800.

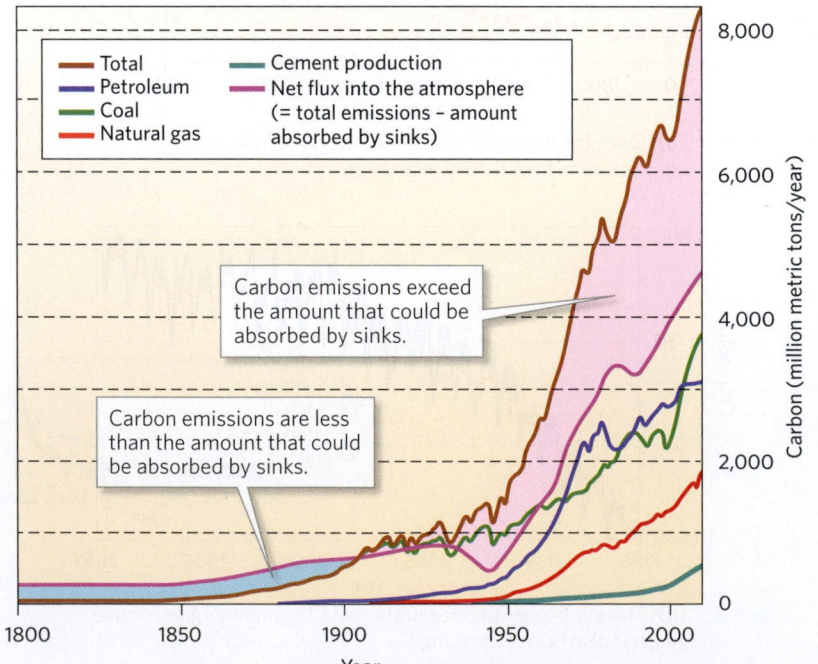

CO_2. By comparison, human society emits about 35 gigatons of CO_2 each year (about 135 times as much).

ISOTOPIC COMPOSITION OF ATMOSPHERIC CO_2. The element carbon exists as three different isotopes (^{12}C; ^{13}C; and ^{14}C). The ratio of these isotopes in a volume of CO_2 gas defines the *isotopic composition* of the gas. Careful analyses demonstrate that the isotopic composition of CO_2 produced by volcanoes and of CO_2 produced by burning organic matter are different: the proportion of ^{13}C tends to be higher in organic matter than in volcanic material because photosynthesis preferentially incorporates ^{13}C. Studies of the isotopic composition of CO_2 in glacial air bubbles show that the proportion of ^{13}C in atmospheric CO_2 has increased since the industrial revolution. This change suggests that much of the new CO_2 entering the atmosphere must come from the burning of organic matter (biomass and fossil fuels), not from volcanoes.

Take-home message . . .

Global temperature and sea level have been rising since the mid-19th century. Many observed changes in the Earth System, such as changes in the volume of glacial ice and in the distribution of vegetation belts, support the conclusion that global warming is taking place. This warming corresponds to the largest increase in atmospheric CO_2 that has happened during the last 800,000 years. Researchers credit this temperature increase to anthropogenic CO_2 production beginning with the industrial revolution.

Quick Question -
How can global warming cause sea level to rise?

20.7 Climate in the Future

Global Climate Models

The development of high-speed supercomputers has allowed climate scientists to write programs, called **global climate models (GCMs)**, that simulate the behavior of climate over time scales of a century or more. These models, which use equations describing the physical and chemical interactions among the atmosphere, ocean, and land, are similar to models used for weather forecasting, but cover much longer times. GCMs allow researchers, for example, to compare climate behavior in an imagined situation in which greenhouse gas concentrations remain at pre-industrial levels with the way the climate has actually behaved to date. These simulations show that if greenhouse gases had remained constant, the Earth would have experienced a slight cooling trend over the last century. The cooling trend would have been caused by decreases in incoming solar energy at high latitudes associated with the Milankovitch cycle. When observed greenhouse gas concentrations are added, the models predict the warming trend that we have actually experienced.

If global climate models have helped us interpret the past, can they help us predict the future? By adjusting the variables programmed into a GCM, researchers can provide a high emissions model (in which greenhouse gas emissions increase above current rates), a medium emissions model (in which the rate of greenhouse gas emissions stays about the same as today), and a low emissions model (in which the rate of greenhouse gas emissions decreases below current rates) **(Fig. 20.32a)**. The medium

Figure 20.32 Model predictions of future global warming.

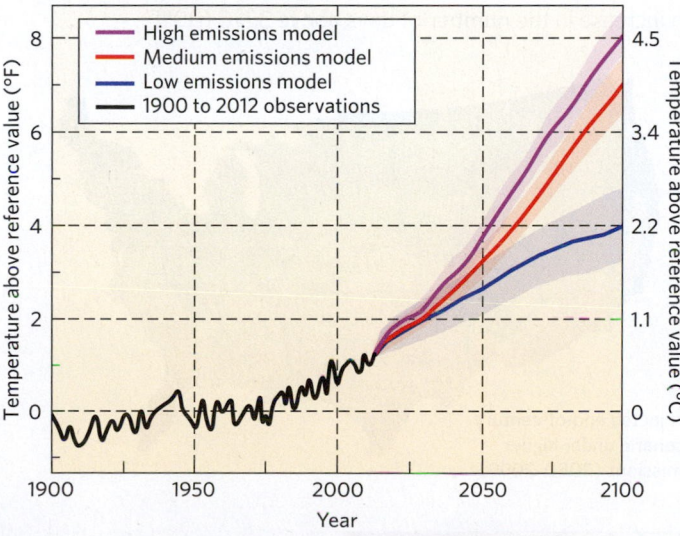

(a) Model calculations for different rates of anthropogenic CO_2 emission. All suggest a significant global temperature increase by 2100.

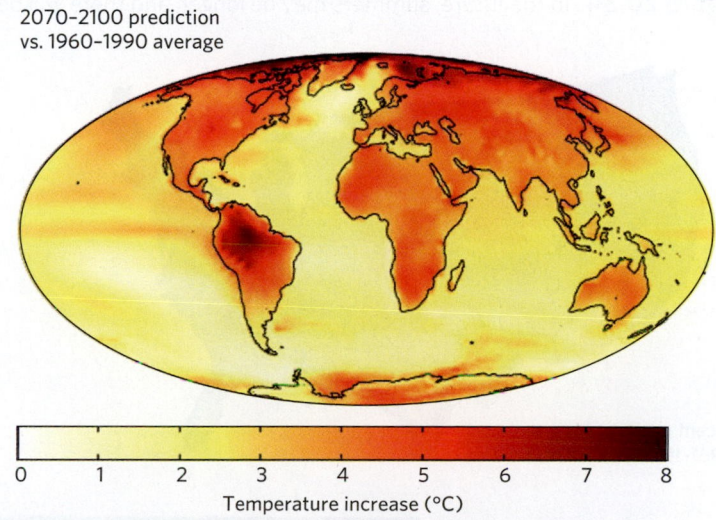

(b) Global warming does not mean that all locations will warm by the same amount. This map shoes a model of possible variations for 2070–2100. Temperatures shown are relative to 1960–1990 averages.

Figure 20.33 In the future, temperate and desert climate belts will probably move to higher latitudes.

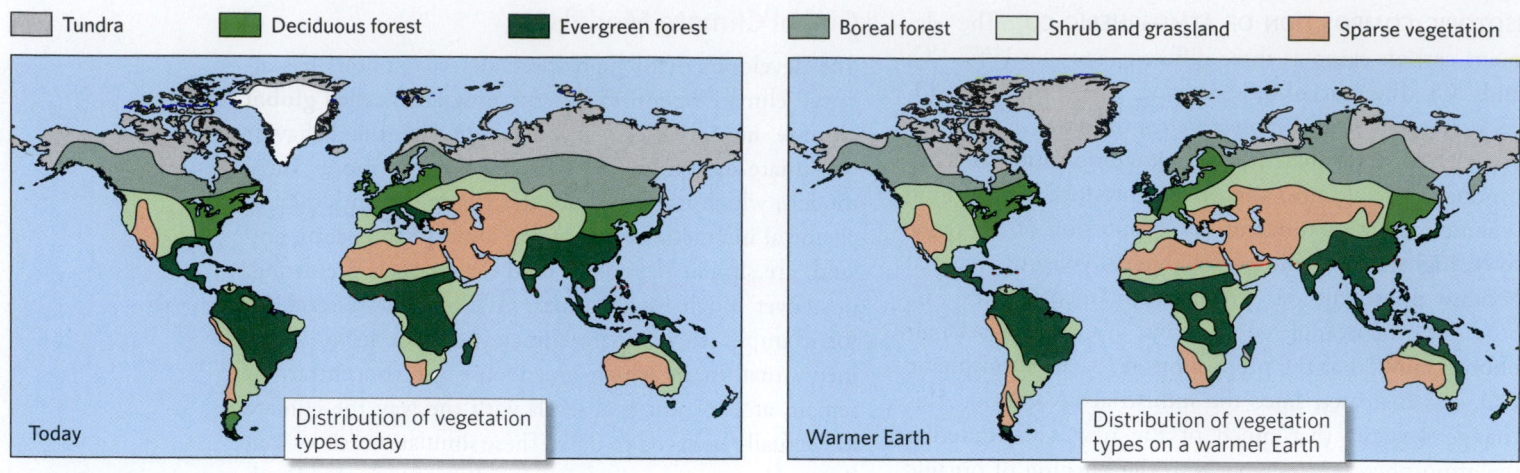

| ⬜ Tundra | 🟩 Deciduous forest | 🟩 Evergreen forest | 🟩 Boreal forest | 🟩 Shrub and grassland | 🟧 Sparse vegetation |

Today — Distribution of vegetation types today

Warmer Earth — Distribution of vegetation types on a warmer Earth

emissions model predicts that atmospheric temperatures will increase by 2°C to 4°C (3°F to 7°F) over the next century, with the greatest warming taking place at high latitudes during the winter because the loss of sea ice will allow ocean water to moderate temperatures **(Fig. 20.32b)**. The exact consequences of global warming remain uncertain, but according to researchers, the following events are possible:

- *A shift in climate belts and season length:* Temperate and desert climate belts will probably move to higher latitudes, so some present-day agricultural lands may become too hot to support the crops that they currently host, and climates appropriate for those crops

will lie farther north **(Fig. 20.33)**. This change will negatively impact global agricultural productivity, for fields farther to the north have thinner soils and receive less sunlight than do those farther south.

- *More heat waves:* Because of warming, summer may lengthen by four to six weeks **(Fig. 20.34)**, and deadly heat waves may occur more often.

- *A change in precipitation patterns:* During the winter, more precipitation will take place in mid-latitudes, especially in western North America, where storms encounter mountain ranges **(Fig. 20.35a)**. During the summer, precipitation will probably increase at higher latitudes and over the ocean, but will decrease over

Figure 20.34 In the future, summers may be longer, and there will be an increase in the number of days above 32°C (90°F).

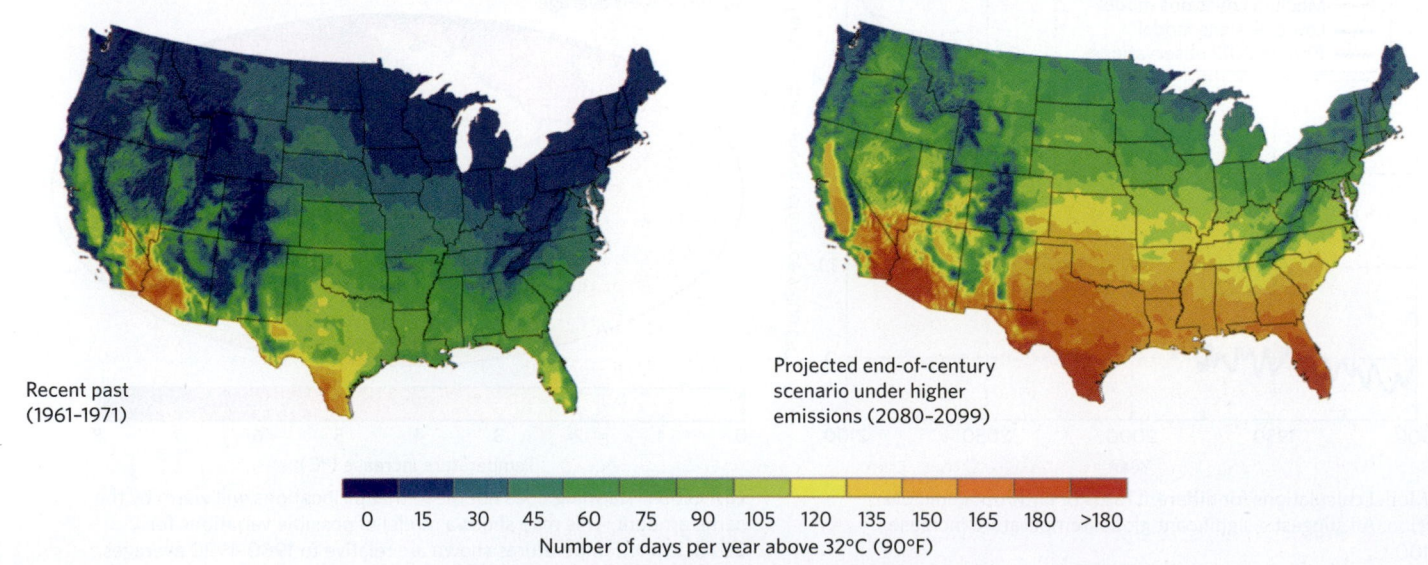

Recent past (1961–1971)

Projected end-of-century scenario under higher emissions (2080–2099)

0 15 30 45 60 75 90 105 120 135 150 165 180 >180
Number of days per year above 32°C (90°F)

Figure 20.35 Projected amounts of precipitation by 2070–2090, relative to those in 1980–2000, in the northern hemisphere, under a medium emissions scenario.

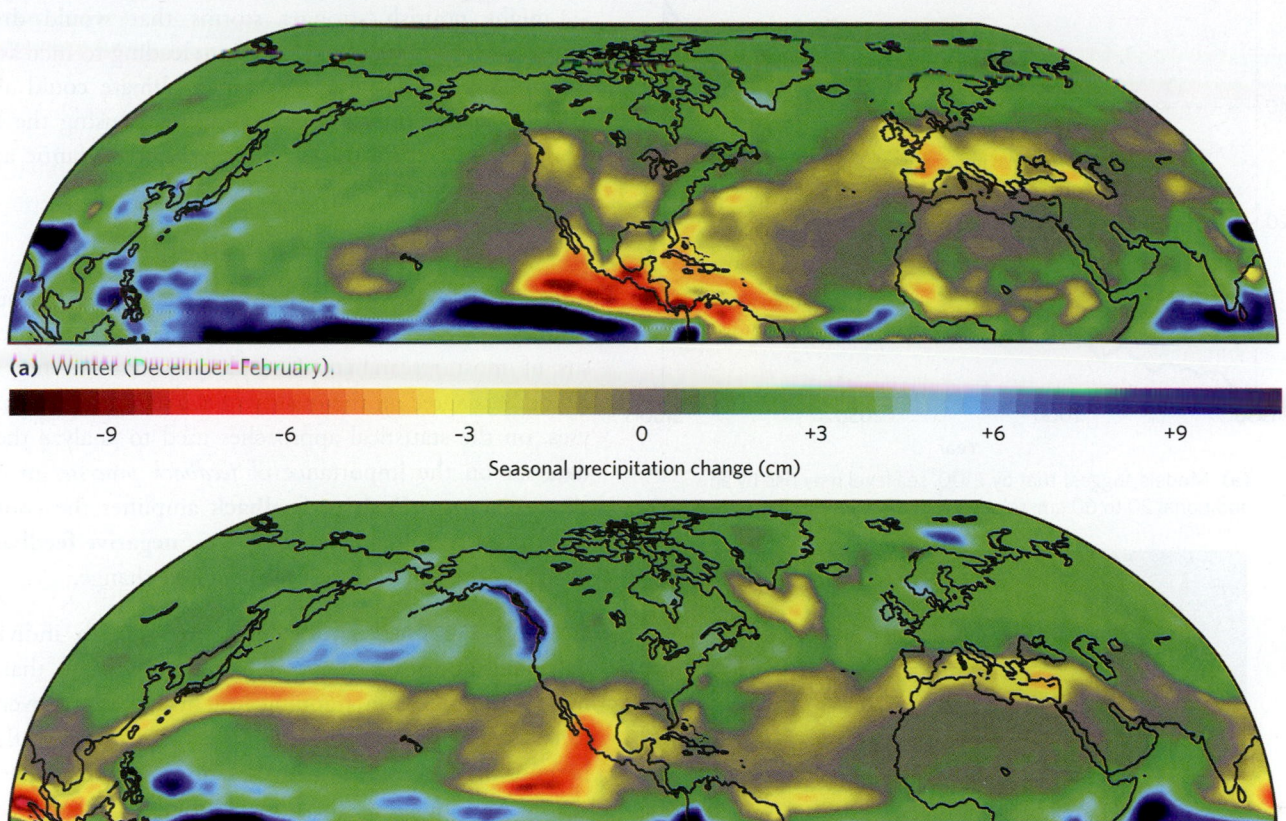

(a) Winter (December–February).

-9   -6   -3   0   +3   +6   +9

Seasonal precipitation change (cm)

(b) Summer (June–August).

the southern United States and western Eurasia, so these regions will experience more frequent droughts **(Fig. 20.35b)**.

- *Ice retreat and snow-line rise:* Warming will drive further retreats of glaciers, and some will disappear entirely. Researchers are concerned that the ice sheet of West Antarctica may collapse (begin to melt and flow rapidly) if ice shelves surrounding the ice sheet disintegrate. In nonpolar latitudes, the snow line in mountains will rise to higher elevations. A decrease in the volume of winter snowpack will cause a decrease in the volume of meltwater, an important component of the water supply in some regions.

- *Melting permafrost and methane ice:* Warming climates at high latitudes will cause areas of permafrost to thaw during the summer. When it defrosts, the organic matter in the former permafrost will decompose and release methane. Methane ice in seafloor sediments may also start to melt and release methane.

- *Additional rise in sea level:* Measurements suggest that sea level is currently rising by about 1.8 mm per year (about an inch per decade). This rise has already caused flooding of coastal areas and has submerged

some oceanic islands (see Earth Science at a Glance, pp. 722–723). Continued rise will only make such problems more widespread and may require that people abandon some communities. Models suggest that by 2100, sea level may rise by an additional 20 to 60 cm (8 to 24 inches) **(Fig. 20.36a)**. A sea-level rise of a meter or two could inundate regions of the world where 20% of the human population lives **(Fig. 20.36b)**. In some neighborhoods of Miami, Florida, particularly high tides already cause flooding **(Fig. 20.36c)**.

- *An increase in wildfires:* Warmer temperatures and drier conditions cause plants to dry out and therefore burn more easily. So, global warming could lead to an increase in the frequency of wildfires.

- *An interruption of thermohaline circulation:* Ocean currents play a major role in transferring heat across latitudes. According to some models, if global warming melts enough glacial ice, the resulting freshwater would dilute ocean surface water at high latitudes. This lower-density water would not sink, so the thermohaline circulation could slow or cease, preventing the currents from conveying heat to high latitudes.

Figure 20.36 Predictions of sea-level change.

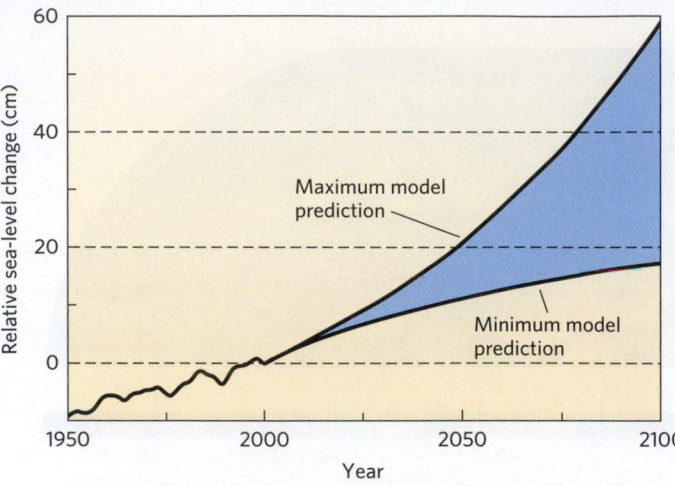

(a) Models suggest that by 2100, sea level may rise by an additional 20 to 60 cm, relative to today's level.

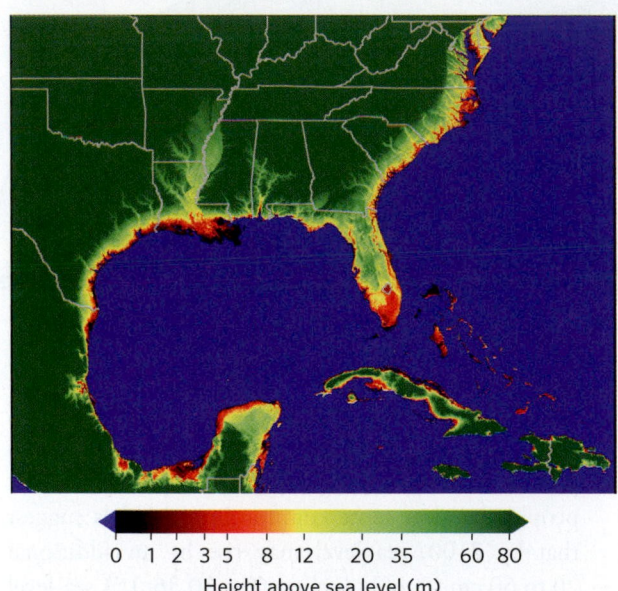

Height above sea level (m)
0 1 2 3 5   8 12   20 35   60 80

(b) If sea level rises by more than a meter, large areas of coastal lowlands, with large populations, may be flooded.

(c) Due to ongoing sea-level rise, when a particularly high tide takes place, parts of Miami Beach, Florida, get flooded with saltwater.

- *Stronger storms:* An increase in average ocean and atmospheric temperatures would lead to increased evaporation from the sea. The additional moisture might nourish stronger storms that would drop greater amounts of precipitation, leading to increased amounts of flooding. A warmer climate could also shift jet streams to higher latitudes, causing the locations of storm tracks over the North Atlantic and North Pacific Oceans to shift.

Complexities in Climate Predictions

The possible consequences of climate change over the next century as we describe in this book represent the consensus of most researchers. Complexities remain; not all researchers agree on the quality of the data used in analyses, on the statistical approaches used to analyze these data, or on the importance of *feedback* processes in the climate system. **Positive feedback** amplifies the consequences of an initial change, whereas **negative feedback** decreases the consequences of the initial change.

POSITIVE FEEDBACK PROCESSES. Even though individual molecules of CO_2 or CH_4 are more efficient than a molecule of H_2O in trapping heat, water vapor overall causes most of the Earth's atmospheric greenhouse effect because its concentration in the atmosphere far exceeds that of CO_2 or CH_4. On a hot, humid day, water vapor accounts for about 2.3% of air, whereas CO_2 accounts 0.04% and CH_4 accounts for 0.0002%. Warming of the atmosphere leads to an increase in its water vapor concentration, which leads to more warming and a further increase in atmospheric water vapor—this process goes on and on. This repeating cycle is a positive feedback process that probably accelerates global warming.

As climate warms, permafrost and methane ice melt. This melting releases CH_4 into the atmosphere, which causes more warming, so this process also acts as a positive feedback. Another positive feedback process results from the decrease in ice and snow cover, which in turn decreases the albedo of Earth's surface, leading to less solar radiation being reflected and less infrared radiation being re-radiated. The increased warming that results leads to further decreases in ice and snow cover.

NEGATIVE FEEDBACK PROCESSES. Release of CO_2 into the atmosphere will warm the atmosphere in such a way that plants and trees may be able to grow at high latitudes where currently only tundra exists. These new plants, in turn, would absorb CO_2, reducing the amount of CO_2 in the atmosphere, and therefore would slow the rate of warming.

UNCERTAIN FEEDBACK PROCESSES. The effect of clouds on global temperature remains uncertain. Global warming increases the amount of water vapor in the air and thus leads to more clouds. Clouds reflect solar radiation, reducing the amount reaching the Earth's surface and cooling the Earth, so an increase in clouds could be a negative feedback. But clouds also trap infrared radiation from the Earth's surface and warm the Earth further, so they could also act as a positive feedback. It is unclear which of these effects will dominate in the future, particularly since so many types of clouds occur at different levels within the atmosphere (see Chapter 17).

Climate Policy

The predictions described above, along with estimates of the large economic cost of global warming, explain why the issue of climate change should be a subject of international discussion. But what can be done? The nations that signed the *Kyoto Protocol* at a 1997 summit meeting held in Japan proposed that the first step would be to slow the input of greenhouse gases into the atmosphere by decreasing the burning of fossil fuels. This approach was reinforced by the *Paris Agreement*, signed in April 2016, which commits 195 countries to efforts that will keep warming from exceeding 2°C above pre-industrial levels.

Worldwide, individuals, private companies, nongovernmental organizations, and some governments are working to find ways to decrease greenhouse gas emissions. Approaches to this problem include decreasing the waste of energy, developing more efficient vehicles and appliances, and switching to alternative energy sources with a smaller **carbon footprint** (the total amount of greenhouse gases

produced by a process or phenomenon). Another, more controversial, approach involves collecting CO_2 produced at large sources, such as power plants, so that it can be condensed and then injected into deep wells to be stored in rock underground—this overall process is called *carbon capture and sequestration*.

Regardless of controversies concerning sources of greenhouse gases or policies aimed at decreasing our carbon footprint, the warming trends already happening will require society to adapt. Change has taken place throughout Earth history, and in fact, the Earth has seen higher atmospheric CO_2 concentrations in its past: 50 million years ago, during the Eocene, they reached 1,000 ppm. But the rate of change now under way is unprecedented. The coming century will pose unique challenges that will demand innovative solutions to problems of energy production, water and food security, and coastal preservation.

Take-home message . . .

Predictions of future climate change based on global climate models remain imperfect because of the complexity of possible feedback processes. Nevertheless, the consensus of researchers is that significant changes will occur in the distribution of climate belts and in sea level in the coming decades, suggesting that society needs to decrease its carbon footprint. The politics of how to do so remains contentious.

Quick Question -
How might global warming affect global agricultural productivity?

ANOTHER VIEW Polar bears rely on sea ice to gain access to food sources. Sea-ice cover has diminished greatly in the past few decades, so polar-bear populations have decreased.

20 CHAPTER REVIEW

- *Weather* refers to atmospheric conditions at a given location and time. Climate refers to the range and variation of conditions in a region over the course of decades or more. Climate change is a shift in climate conditions.

- Two variables, temperature and precipitation, determine the climate a region will experience.

- Daily and seasonal temperature changes depend on latitude and proximity to the ocean because of the ocean's heat capacity.

- Researchers classify the Earth's climate zones using the Köppen-Geiger climate classification, with five climate groups—tropical, arid, temperate, cold, and polar—and several subcategories.

- Over time, the Earth System's energy must balance, with the amount of solar energy arriving from the Sun balancing the amount of infrared energy radiated away from the Earth. The atmosphere, due to the greenhouse effect, traps energy and keeps the atmosphere and the Earth's surface warmer than it might otherwise be.

- Because of the tilt of the Earth's axis, the amount of solar radiation arriving at a particular latitude changes over the course of the year and causes seasons.

- Global atmospheric circulations, ocean currents, distance from an ocean, elevation, and position relative to a mountain range also affect climate in a region.

- Various paleoclimate indicators allow researchers to characterize paleoclimate, the climate of the past. These studies indicate that there has been both long-term and short-term climate change over the course of Earth history.

- The stratigraphic record, which permits reconstruction of long-term climate change, indicates that icehouse and hothouse periods alternate over time, but that the durations of these periods vary.

- The record of short-term climate change comes from studies of isotope ratios and pollen in sediment cores. Evidence also comes from studies of isotope ratios and air bubbles in ice cores from glaciers, as well as from the study of growth rings in trees and other organisms.

- Since the most recent glaciation, there have been significant episodes of global warming and global cooling, including the Medieval Warm Period and the Little Ice Age.

- Long-term climate change may be a consequence of changes in atmospheric concentrations of greenhouse gases, continental positions, ocean currents, and the distribution of mountain ranges.

- The primary controls on short-term climate change appear to be the Milankovitch cycles, volcanic aerosols in the atmosphere, modifications of surface albedo, and abrupt changes in greenhouse gas emission or absorption.

- Since the industrial revolution, average global temperature appears to have increased at rates much greater than in the previous millennia. Observations of glacial and sea-ice retreat, sea-level rise, and shifting of plant hardiness zones support this observation.

- The atmospheric CO_2 concentrations observed during the past two centuries are much greater than at any time during the past 800,000 years. Global climate models suggest that, in the absence of anthropogenic CO_2 emissions, the climate might be experiencing a weak cooling trend rather than warming.

- Global warming could significantly affect habitability of coastal areas, agricultural productivity, and even the frequency of wildfires.

- Positive feedback processes (increasing evaporation, releases of CH_4, decreasing albedo) may be accelerating warming. Negative feedbacks, such as expansion of regions where plants can grow, may be slowing warming.

- Because of the likely role that anthropogenic CO_2 plays in driving global warming, world governments negotiate polices to decrease emissions of greenhouse gases worldwide.

albedo (p. 716)
anthropogenic forcing (p. 726)
carbon budget (p. 728)
carbon cycle (p. 714)
carbon footprint (p. 733)
climate (p. 699)
climate change (p. 699)

critical zone (p. 698)
energy balance (p. 704)
global climate model (GCM) (p. 729)
global cooling (p. 709)
global warming (p. 709)
growth ring (p. 711)

hothouse period (p. 709)
icehouse period (p. 709)
Intergovernmental Panel on Climate Change (IPCC) (p. 719)
Keeling curve (p. 698)
Köppen-Geiger climate classification (KGCC) (p. 700)

natural forcing (p. 726)
natural variation (p. 709)
negative feedback (p. 732)
paleoclimate (p. 709)
positive feedback (p. 732)
season (p. 705)
weather (p. 698)

Review Questions

The letters following each Review Question refer to the corresponding Learning Objective from the Chapter Opener.

1. Distinguish between *weather* and *climate*. What factors control a region's climate? **(A)**

2. What characteristic did Köppen and Geiger use to map out climate zones? **(B)**

3. What factors control a region's rainfall? What factors control a region's temperature and temperature variation? **(C)**

4. Explain the cause of the atmospheric greenhouse effect and how it affects the Earth's energy balance. **(E)**

5. What is the cause of the Earth's seasons? **(A)**

6. Why does the distribution of landmasses and mountain ranges affect climate in a region? **(C)**

7. Explain the difference between long-term and short-term climate change. **(E)**

8. How does a large volcanic eruption affect global temperature in the short term? **(E)**

9. What are the Milankovitch cycles, and what role do they play in controlling climate? **(E)**

10. What techniques can researchers use to determine how climate varied in the past? **(D)**

11. Describe the key factors that play a role in causing long-term climate change and those involved in short-term climate change. **(E)**

12. What is a global climate model? **(E)**

13. Based on paleoclimate records, how has climate changed over the course of the last few millennia? **(D)**

14. What evidence led researchers to the conclusion that the Earth's climate has warmed during the past few centuries? **(F)**

15. What areas in the diagram have warmed most over the past fifty years? (See the temperature change scale of Fig. 20.19 on p. 719.) Why have these regions warmed more dramatically than others? **(A, C)**

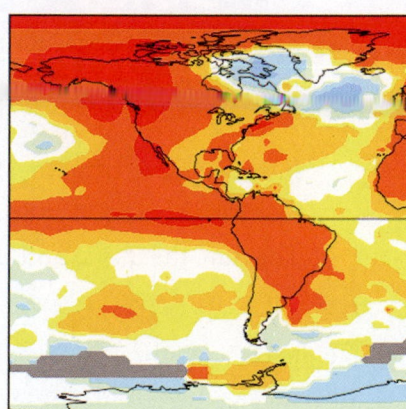

16. What kinds of feedbacks may amplify the process of global warming, and why? **(G)**

17. What efforts may help to diminish human society's carbon footprint? **(G)**

On Further Thought

18. If the CO_2 concentration in the atmosphere doubles during the next century, which of the values for Earth's energy in the diagram (see Fig. 20.5) would increase, and which would decrease? **(E)**

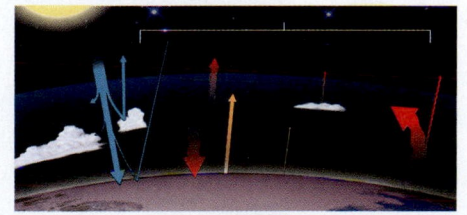

19. Mosquitoes carry diseases (malaria, dengue fever, Zika) that are dangerous to humanity. The species of mosquitoes that carry such diseases prefer warmer climates. If current observed trends in global temperature change continue, will human populations susceptible to these diseases increase or decrease during the next century? Explain your answer. **(G)**

Online Resources

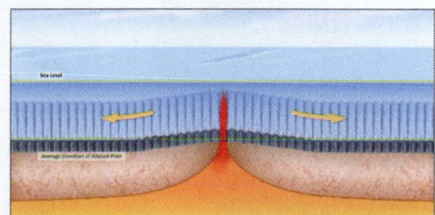

Animations
This chapter features animations on global sea-level change, including glacial ice volume, thermal expansion, seafloor-spreading rate, and oceanic volcanic plateaus.

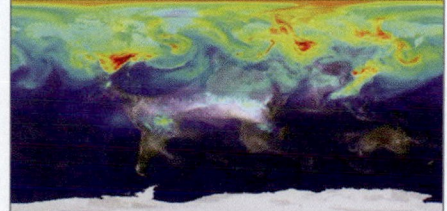

Videos
This chapter features real-world videos on rising CO_2 levels, deforestation, ice sheet stratigraphy, and more.

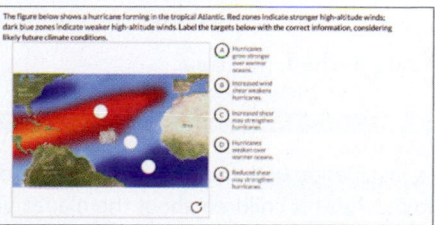

Smartwork5
This chapter features questions on the causes and impacts of long- and short-term climate change, and the factors that control and impact regional climate conditions.

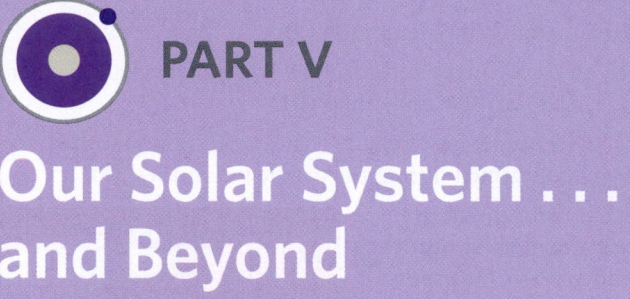

PART V

Our Solar System . . . and Beyond

To understand how our planet formed, and why it's so unique, we need to look into the darkness of deep space. In Chapter 21, we start this exploration by reviewing the history of astronomy and the tools that astronomers use to see what's out there. We'll learn the reason why planets orbit the Sun, why the Moon has phases, and why eclipses occur. Chapter 22 then takes us on a journey around our neighborhood, the Solar System, during which we'll visit each of the planets, their moons, and the other objects that orbit our Sun. We will be astounded by the strange moons that populate our Solar System and learn how we have used space probes to image them. You'll appreciate that there's no place like Earth. Chapter 23, the final chapter of Part V, dives into the Sun's interior to see what it's made of and how it generates so much energy. We learn how stars in general evolve over time, and eventually die. The chapter then describes galaxies, associations of stars bound together by the invisible tendrils of gravity, and other strange-but-true objects (such as supermassive black holes, quasars, and more) that inhabit the Universe. We conclude this chapter, and the book, with a look at models addressing the formation and fate of the Universe.

Humanity's fascination with the heavens carries from generation to generation. Here, a dad teaches his children about the planet Venus.

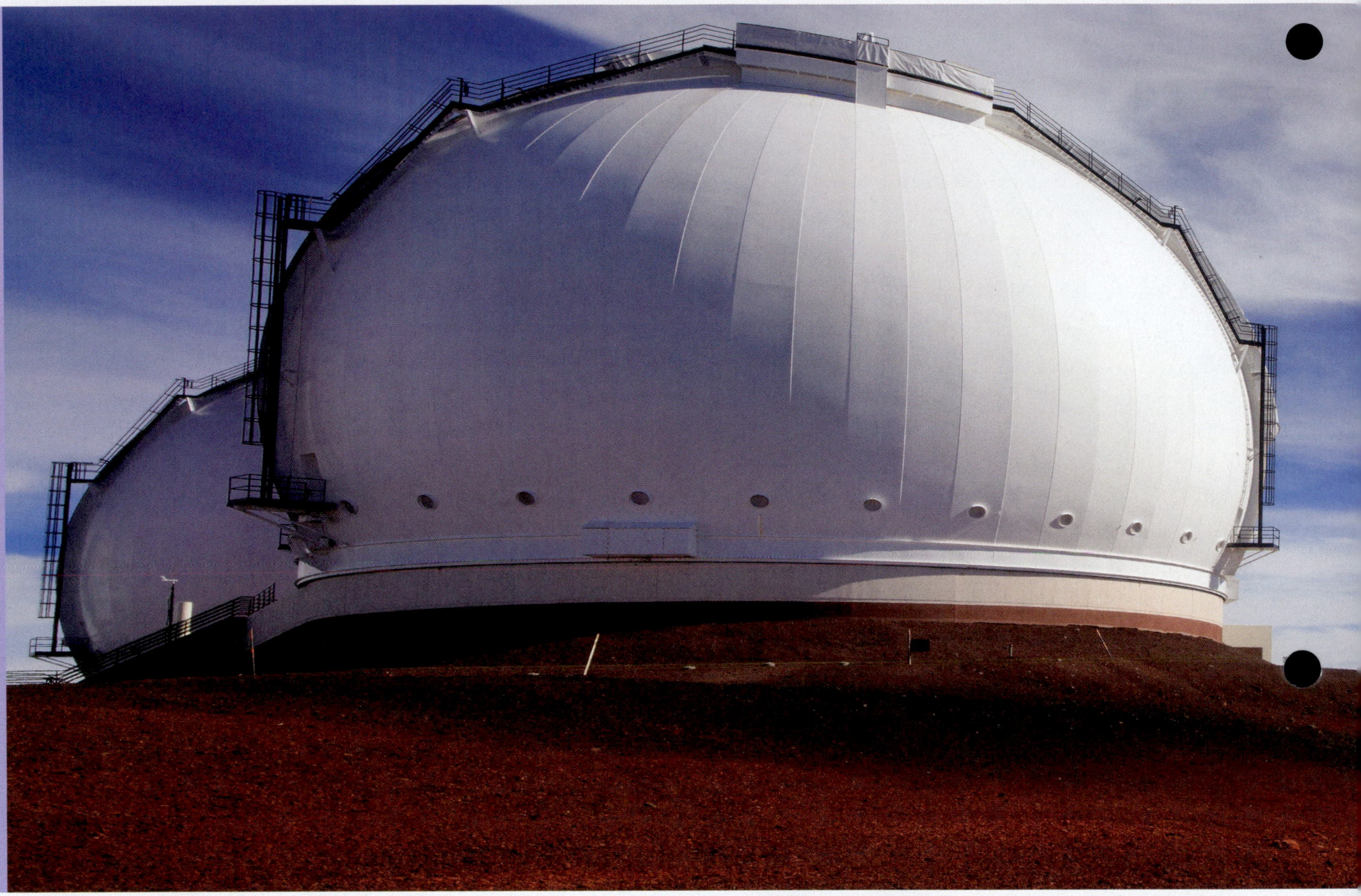

 **21** **INTRODUCING ASTRONOMY**
Looking Beyond the Earth

21.1 Introduction

Centuries before Columbus, Polynesian sailors guided sailing canoes across vast reaches of the Pacific Ocean and colonized over 100 island groups **(Fig. 21.1)**. When a crew journeyed 4,200 km (2,600 miles) from Tahiti to Hawaii, they had to find a target only 120 km (75 miles) across—a feat equivalent to hitting a small coin from a distance of 35 km (22 miles)—without using GPS satellites, compasses, clocks, or charts. How did they do it? To read their surroundings as we read a map today, Polynesian navigators merged knowledge of where distinct groups of stars rose above the horizon with knowledge of how ocean swells move, birds migrate, and cloud characteristics vary with latitude. They then embedded navigational information in songs and myths that they could pass on to the next generation.

You can see the same lights that led Polynesian sailors across the ocean by gazing upward on a dark night. Most of the lights—the *stars*—look like twinkling points that stay fixed in position relative to their neighbors even as they make their nightly journey across the sky **(Fig. 21.2a)**. A few of the lights—the *planets*—have a steadier glow and move relative to the backdrop of stars and relative to one another, so their name, from the Greek word for wanderers, seems apt **(Fig. 21.2b)**. If it's a particularly clear and dark night, you'll also see a glowing haze—the *Milky Way*—forming a band that spans the entire arc of the heavens **(Fig. 21.2c)**. On many nights, however, you'll be able to see only the brighter stars and planets, for the

Mauna Kea Observatory located atop Mauna Kea Volcano on the Island of Hawaii.

Figure 21.1 Polynesian sailing canoes, similar to these modern replicas, were used for trans-Pacific voyages.

what you see when you look up and beyond the Earth. This chapter provides an introduction to the study of astronomy by tracing the path of ideas that early Western philosophers and, later, scientists took as they tried to fathom humanity's place in the cosmos. We then introduce the modern concept of how celestial objects move in the sky, knowledge that reflects scientific discoveries over several centuries. This concept relies on the study of electromagnetic radiation that comes from space, so we next explore the basic properties of this radiation and describe the tools astronomers use to measure this radiation in ways that allow us to peer at unfathomably distant places and see events that happened in the distant past.

21.2 Building a Foundation for the Study of Space

Ancient Attempts to Interpret a Sky of Mystery

The oldest hint of humanity's interest in the heavens dates to 35,000 B.C.E., for at that time, cave dwellers chiseled petroglyphs depicting celestial objects into rock surfaces **(Fig. 21.3a)**. By 14,000 B.C.E., cave paintings found at many localities portrayed distinctive clusters of dots that resemble star groups **(Fig. 21.3b)**. As civilization dawned, observers knew that the Sun, Moon, and stars rose at specific points on the horizon at times of seasonal change, so the study of the sky became a basis for calendars that people used to plan the sowing or reaping of crops. As a result, celestial objects influenced the layout of *megalith*

glow of the Moon masks the fainter lights. During the day, of course, the Sun's intense light completely obscures all planets and stars.

Humans have wondered about the nature of all these **celestial objects** since conscious thought began. In historic time, this wonder evolved into the science of **astronomy**, the systematic study of the **Universe** (or *cosmos*), meaning all of space and all the celestial objects within it. In Part V of this book, we offer a brief survey of astronomy to provide a context within which you can understand the origin and evolution of our home, the Earth. Our study of astronomy will help you complete your personal image of your natural surroundings by explaining

Figure 21.2 Stars, planets, and galaxies.

(a) Look into the sky on a dark night, and you'll see a myriad of stars.

(b) Four planets rise over the Andes Mountains just before dawn: from top to bottom, Venus, Mercury, Jupiter, and Mars.

(c) On a particularly clear, moonless night, in a dark location, you can see the Milky Way.

structures, geometric placements of large rocks. Stonehenge, for example, built between 3,000 and 2,000 B.C.E. in England, points to the location where the Sun rises at the beginning of summer **(Fig. 21.3c)**. The importance of calendars led some societies to appoint **astronomers**, people who systematically study the character and movements of celestial objects, to help priests refine the timing of festivals. Following the invention of writing, early astronomers began to track movements of specific stars, to record them in *star catalogs*, and to assign names to the planets and to bright stars. A star catalog compiled in Mesopotamia in 1,000 B.C.E. proves that people have carefully documented stellar movements for at least three millennia.

The Geocentric Model: Picturing the Earth as a Special Place

Can the changing positions of celestial objects be explained? Early astronomers asking this question were, in effect, exploring **cosmology**: the study of the overall structure and evolution of the Universe. The obvious motion of the Sun, Moon, and stars across the sky, relative to an observer on Earth, coupled with humanity's perception of its self-importance, led many societies to favor a **geocentric model** as an explanation of how celestial objects moved **(Fig. 21.4a)**. In this model, the Earth lay at the center of the Universe, other celestial objects followed orbits around the Earth, and all stars lay on the surface of a distant sphere outside. (Note that an **orbit**, in a general sense, is a closed path in space that one object takes around another; the complete orbit of one object around another is called a *revolution*.) Early astronomers assumed that the Universe must have a "perfect" structure, and because Aristotle (384-322 B.C.E.), the Greek philosopher, had stated that the circle was the most perfect geometric form, astronomers believed that the orbits of celestial objects were perfect circles. How big were these orbits, and how big were celestial objects? No one knew. In fact, it wasn't until a century after Aristotle's death that the Earth's size became known **(Box 21.1)**.

Ptolemy (90–186 C.E.), an Egyptian mathematician, realized that for the geocentric model to be correct, it would need to explain the wanderings of planets. In contrast to the Sun and stars, which always display **prograde motion**, meaning that they always move from east to west across the sky, planets occasionally follow loops that take them a short distance backward (from west to east) across the sky relative to the stars **(Fig. 21.4b)**. To explain this **retrograde motion**, Ptolemy suggested that each planet moved around a small circle called an *epicycle*, whose center, in turn, followed a larger circle, known as *deferent*, around the Earth **(Fig. 21.4c)**. For the next 15 centuries, Ptolemy's calculations were taken as proof of the geocentric model. During much of this time, questioning the

Figure 21.3 Early observations of celestial objects.

(a) Petroglyphs of the Sun and stars.

Painted star cluster

Star cluster in the night sky

(b) Cave paintings thought to depict groups of stars.

(c) On the summer solstice, the Sun rises between two of the megaliths at Stonehenge.

Figure 21.4 The geocentric model.

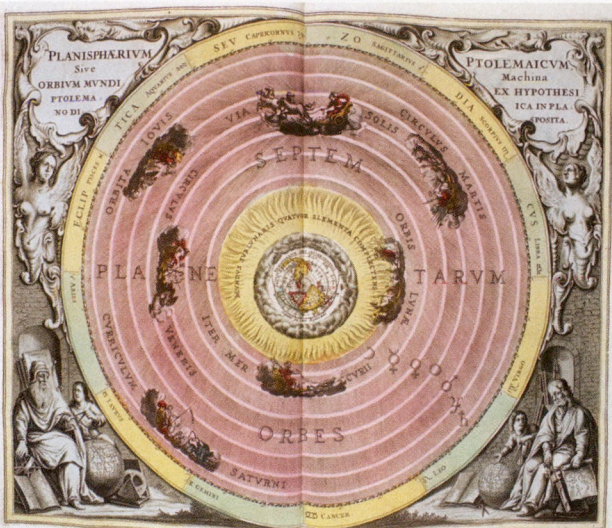

(a) An artist's depiction of Ptolemy's geocentric model.

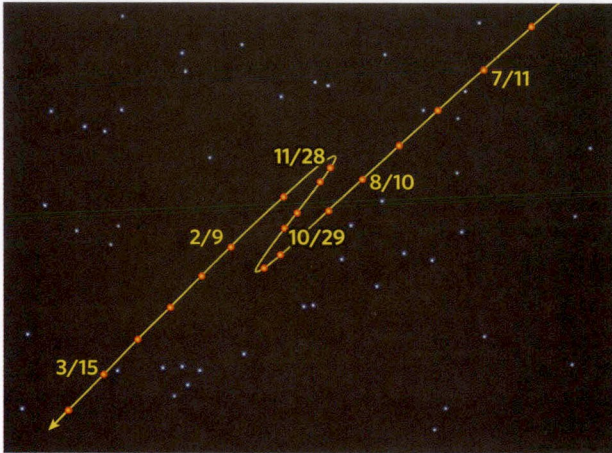

7/11

11/28

8/10

2/9

10/29

3/15

(b) The observed looping path of a planet over part of a year.

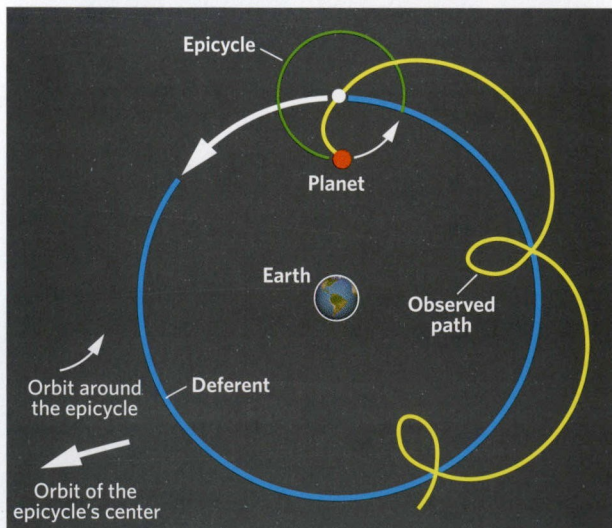

Epicycle

Planet

Earth

Observed path

Orbit around the epicycle

Deferent

Orbit of the epicycle's center

Did you ever wonder . . .
what the shape of the Earth's orbit looks like?

(c) Ptolemy's concept of epicycles was an attempt to explain the wanderings of planets.

geocentric model was deemed heresy because the model fit so well with the philosophical view that humanity had a central role in the Universe. Those who argued against it faced prison or death.

The Copernican Revolution

A few astronomers who lived centuries before Ptolemy had argued that the Sun, not the Earth, occupied the center of the Universe and that the Earth revolved around the Sun. Most people ignored this **heliocentric model**, however, until the Renaissance, a time when philosophers began to question many long-standing dogmas. In this new climate of discovery, Nicolaus Copernicus (1473–1543) reexamined planetary orbits and concluded that the geocentric model, and Ptolemy's epicycles, could not be real. He wrote a book, *De Revolutionibus Orbium Coelestium* (*On the Revolutions of the Heavenly Spheres*), in which he demonstrated that a heliocentric model provided a better explanation of celestial motions. Copernicus delayed publication of his book until he was on his deathbed, perhaps to avoid facing charges of heresy. Subsequently, the heliocentric model came to be known as the **Copernican model (Fig. 21.5)**.

The Copernican model did not gain widespread acceptance for many decades. During this time, Tycho Brahe (1546–1601), who was born in Denmark a few years after Copernicus died, spent his career measuring the precise movements of stars, planets, and comets by using carefully constructed instruments **(Fig. 21.6)**. He merged Copernicus's ideas with those of Ptolemy by envisioning a Universe in which the Sun and Moon orbit the Earth, while all planets except the Earth orbit the Sun.

It took the work of Galileo Galilei (1564–1642), the great Italian scientist, to provide the observations that ultimately led to acceptance of the Copernican model. Galileo, the first astronomer to examine the skies with a telescope, saw that Jupiter has moons, and that the moons cross the face of Jupiter and disappear behind the planet. He reasoned, correctly, that if moons orbit Jupiter, then not all celestial objects orbit the Earth. His book, *Starry Messenger*, which laid out these observations and several others, directly challenged geocentric dogma. Church leaders publicly banned the work of both Galileo and Copernicus in 1616, but Galileo persisted in his thinking, and in 1632 he published a second book, *Dialogue Concerning the Two Chief World Systems*, in which he presented a more complete case for the Copernican model. Because of this book, Galileo was convicted of heresy and spent the rest of his life under house arrest.

Kepler and His Laws of Planetary Motion

Johannes Kepler (1571–1630), a German astronomer, made the next great astronomical discovery. Using Brahe's detailed observations as a starting point, Kepler developed

Box 21.1

Consider this . . .

Discovering the Earth's circumference

When the earliest geographers in Western culture drew maps of the world, they pictured it as a flat disk that didn't extend much beyond the lands surrounding the Mediterranean Sea. This picture changed when a Greek astronomer, Eratosthenes (ca. 276–194 B.C.E.), came across a report noting that in the southern Egyptian city of Syene, the Sun lit the base of a deep vertical well precisely at noon on the first day of summer. Eratosthenes believed that the Earth was a sphere, but he didn't know the size of the sphere. The report from Syene gave him a tool for calculating our planet's circumference.

Eratosthenes deduced that if the Earth were a sphere, then the Sun's rays could not simultaneously be perpendicular to the Earth's surface at Syene and at Alexandria, 5,000 stadia (800 km, or 500 miles) to the north. So, on the first day of summer, Eratosthenes measured the shadow cast by a vertical tower in Alexandria at noon and determined that the tower made an angle of 7.2° with respect to the Sun's rays **(Fig. Bx21.1)**. Knowing that a circle contains 360°, Eratosthenes calculated the Earth's circumference using a simple geometric equation, and came up with a number within 2% of the modern value of 40,008 km (24,865 miles).

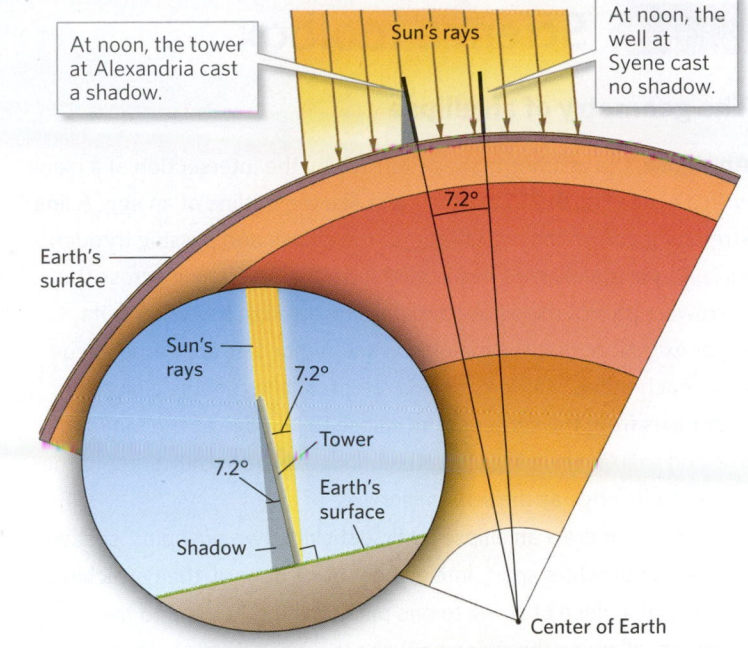

Figure Bx21.1 The geometric calculation that Eratosthenes used to determine the circumference of the Earth.

Eratosthene's calculation:

$$\frac{360°}{x} = \frac{7.2°}{5,000 \text{ stadia}}$$

$$x = \frac{360° \times 5,000 \text{ stadia}}{7.2°}$$

$$x = 250,000 \text{ stadia}$$

250,000 stadia × 0.1572 km/stadium = 39,300 km (or, 24,421 miles)

---

a set of simple mathematical relationships that completely explained the motions of planets in the context of the Copernican model. In these relationships, now known as **Kepler's laws** of planetary motion, Kepler for the first time portrayed planetary orbits as ellipses instead of as circles **(Box 21.2)**. Kepler's laws can be stated as follows:

1. The orbit of every planet is an ellipse, not a circle, with the Sun at one of the ellipse's two foci **(Fig. 21.7a)**.

2. An imaginary line extending from the Sun to a planet sweeps out equal areas during equal intervals of time as a planet moves along its orbit. Therefore, a planet's speed along its orbit changes over the course of a year **(Fig. 21.7b)**.

3. Put simply, this law states that planets farther from the Sun have longer orbital periods, and that we can calculate a planet's orbital period if we know the dimensions of its orbit **(Fig. 21.7c)**. In detail, the square of a planet's *orbital period*, the time it takes for the

Figure 21.5 An artist's depiction of Copernicus's heliocentric model.

Box 21.2

Science toolbox . . .

The geometry of an ellipse

An **ellipse**, the geometric shape formed by the intersection of a plane with a cone (**Fig. Bx21.2a**, left), looks like the outline of an egg. A line stretching across the widest part of the ellipse and passing through its center is the ellipse's *major axis*, and a line stretching across the narrowest part of the ellipse and passing through its center is its *minor axis*. Note that the major axis and minor axis are perpendicular to each other. Mathematicians refer to a line running along the major axis from the center of the ellipse to its edge as the *semi-major axis* and to a line running along the minor axis from the center of the ellipse to its edge as the *semi-minor axis*.

You can draw an ellipse by first sticking two pushpins, spaced several centimeters apart, into a sheet of cardboard, then attaching one end of a piece of string to one pin and the other end to the other pin, allowing the string between the pins to remain loose (**Fig. Bx21.2b**). Hold a pencil perpendicular to the cardboard, with the point touching the cardboard, and press against the string to make the string taut. If you now move the pencil in an orbit around the pins, you'll trace out an ellipse. Each pin is a *focus* of the ellipse.

All ellipses have two foci, positioned on the major axis. The farther apart the foci, the more *eccentric* (elongate) the ellipse. We can say that an ellipse with *high eccentricity* looks more flattened and that an ellipse with *low eccentricity* looks more circular. A circle, in

effect, is a special ellipse whose major and minor axes have the same length—namely, the diameter of the circle. In a circle, both foci lie at the same point, the circle's center. You can produce a circle from the intersection of a plane with a cone if the plane is perpendicular to the cone's axis (see Fig. Bx21.2a, right). **Figure Bx21.2c** shows the mathematical definition of eccentricity. Note that it is a ratio whose value is between 0 and 1. A circle has an eccentricity of 0.

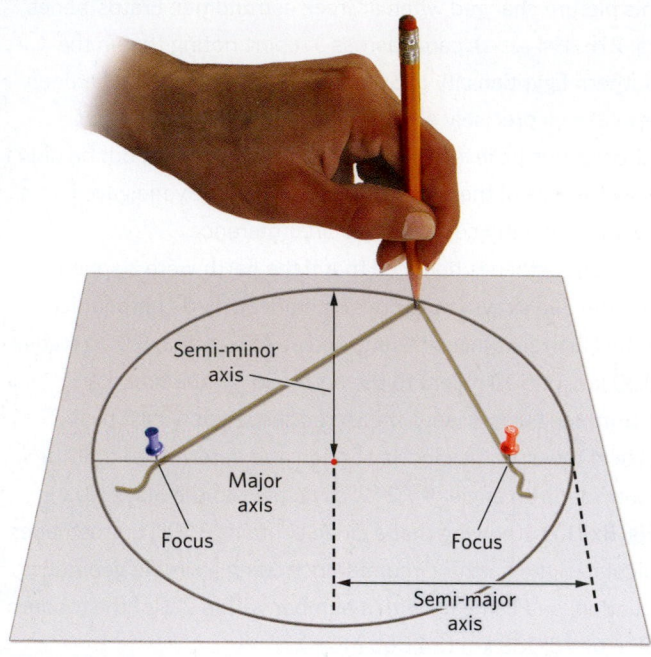

(b) An ellipse has two foci, positioned along the major axis. You can draw an ellipse by tying a loose piece of string to two pushpins.

Figure Bx21.2 Geometry of an ellipse.

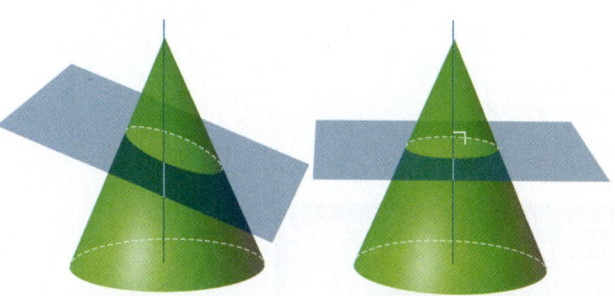

(a) (Left) Intersection of a plane with a cone yields an ellipse if the plane tilts relative to the cone's axis. (Right) Intersection of a plane perpendicular to the cone's axis yields a circle.

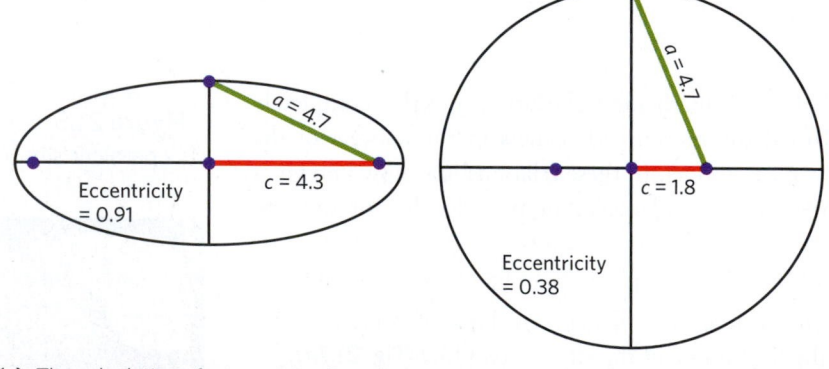

(c) The calculation of eccentricity: E = c/a.

planet to go around the Sun, is directly proportional to the cube of the semi-major axis of the orbit.

Newton and His Laws of Physics

Isaac Newton (1642–1726) was born in England in the year of Galileo's death. To say that Newton was a genius would

be an understatement, for Newton contributed more to the establishment of modern science than any other person. Not only was he one of the inventors of calculus, but he also discovered that white light can be broken into a spectrum of colors, and he invented the type of telescope favored by research astronomers today. Newton's most important

Figure 21.7 Kepler's laws of planetary motion. (The eccentricity of the ellipses shown are exaggerated.)

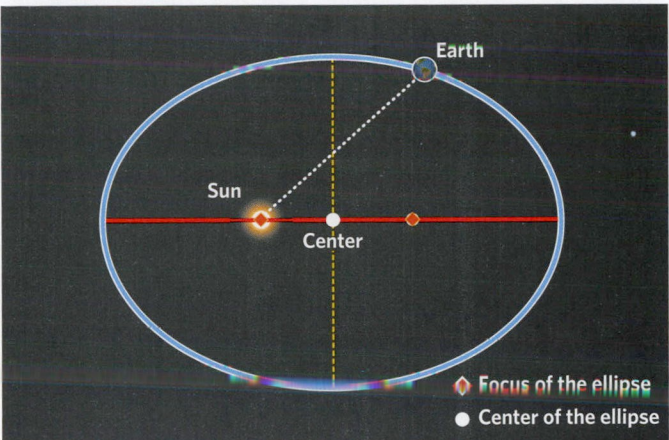

(a) Kepler's first law: The orbit of every planet is an ellipse, with the Sun at one of the ellipse's two foci. The ellipse in this diagram is exaggerated.

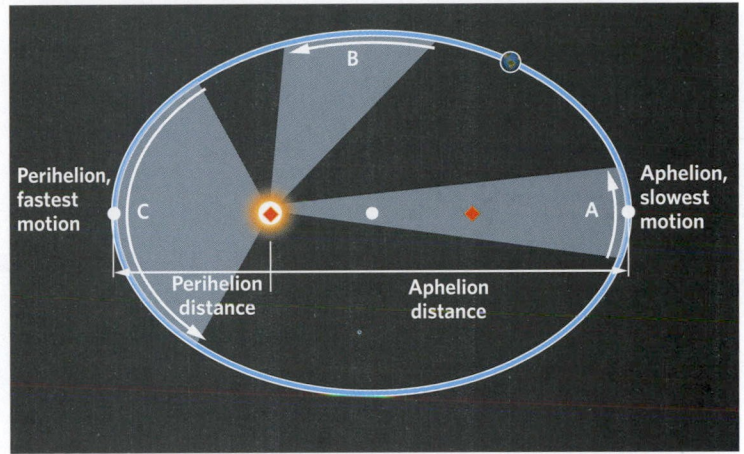

(b) Kepler's second law: As a planet orbits the Sun, an imaginary line joining that planet and the Sun sweeps out equal areas (labeled A, B, and C) during equal intervals of time. As a result, planets move fastest at perihelion, when they are closest to the Sun, and slowest at aphelion, when they are farthest away.

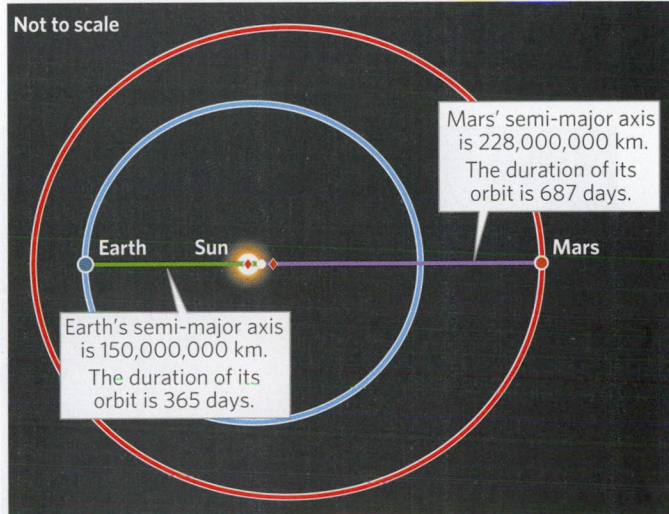

(c) Kepler's third law: The larger the semi-major axis of a planet's orbit (*a*, a measure of the planet's distance from the Sun), the longer the planet's orbital period, the time it takes the planet to orbit the Sun.

contributions appeared in *Philosophiae Naturalis Principia Mathematica*. In this book, generally known simply as *Principia*, Newton proposed the three *laws of motion*, as well as the *law of gravity*. Using these laws, he mathematically derived Kepler's laws, and by doing so he showed why planets have the motions and orbits that they do **(Box 21.3)**.

- *First law of motion:* Every object has *inertia*, meaning that an object will remain in a state of rest, or in a state of uniform motion in a straight line, unless compelled to change by the application of a force.

- *Second law of motion:* When a force (*F*) acts on a mass (*m*), it causes the mass to change its speed and/or direction, meaning that it causes an acceleration (*a*). This law can be written as $F = ma$.

- *Third law of motion:* For every action, there is an equal and opposite reaction. So when we apply a force to a ball by hitting it with a bat, the ball applies the same force to the bat in the opposite direction.

- *Law of gravity:* Every object in the Universe attracts every other object. Objects with larger masses produce greater gravitational attraction than objects with smaller masses, and objects closer to each other attract each other more strongly than do those that are farther apart.

Science toolbox . . .

What is an orbit?

What does it take to put an object into orbit **(Fig. Bx21.3)**? If you throw a baseball, it follows an arc-like path. The change in its direction of movement means that, as Newton's second law requires, a force has acted on the ball. Of course, that force comes from the gravitational pull of the Earth—the ball falls toward the Earth. The horizontal distance that the ball travels before it lands on the ground depends on its velocity and on the angle from horizontal at which the ball started on its path. If you shoot a cannon at a given angle, the cannonball goes much farther than a baseball launched at the same angle because the cannonball travels faster. But the cannonball, like the baseball, eventually falls back to Earth. If a rocket propels a satellite into space fast enough, the arc of the satellite's fall due to the pull of gravity doesn't bring it back to the Earth's surface, but rather takes it entirely around the planet—in other words, the satellite goes into *orbit*. Gravity still pulls the satellite downward, but the Earth curves away from the satellite just as quickly, so the satellite stays at the same altitude above the Earth's surface.

Because orbits can be explained by Newton's laws, they can be calculated. If a friction force doesn't slow down the satellite, it remains in orbit indefinitely. For a space probe to leave orbit and head into space, it must achieve *escape velocity*, meaning that it's going too fast to remain in orbit.

Figure Bx21.3 Neither a baseball nor a cannonball moves fast enough to attain an orbit around the Earth. To attain an orbit, a satellite must be propelled by a rocket at very high speeds.

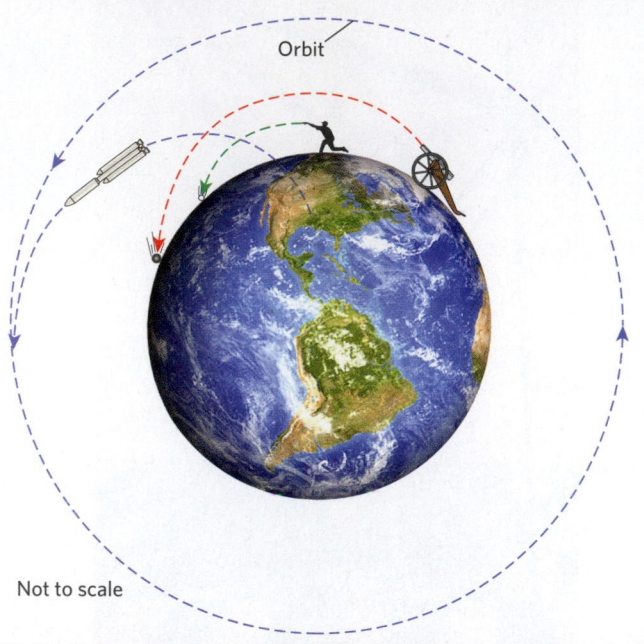

Orbit

Not to scale

Astronomy Enters the Modern Era

In the 18th and 19th centuries, astronomers developed new and more powerful telescopes that allowed them to see celestial objects more clearly and to see deeper into space. Such work allowed astronomers to understand details of the structure of our **Solar System** (the Sun and all the objects that orbit around it) and provided new information about the physical nature of celestial objects. As the 20th century dawned, astronomers learned how to measure distances to celestial objects, and with this knowledge, they came to realize that our Sun lies in the Milky Way, and that the Milky Way is a **galaxy**, a collection of hundreds of billions of stars distributed in a spiral or disk that slowly spins around a *galactic center* **(Fig. 21.8)**. Astronomers then discovered that some of the features that earlier observers knew only as puzzling hazy patches of light were actually **nebulae**, vast clouds of dust and gas, and that others were independent galaxies located far beyond the edge of the Milky Way. Eventually, astronomers learned that the Universe contains over a trillion galaxies, and the distant galaxies are all moving away from one another, meaning that the Universe is expanding. This realization set the stage for development of the *Big Bang theory*, a scientific model for the origin of the Universe (see Chapter 1). Meanwhile, physicists developed new

Figure 21.8 The M-81 Galaxy looks much like our galaxy, the Milky Way. Our Sun resides on one of our galaxy's spiral arms.

theories to explain the behavior of matter and energy, which provided astronomers with ways of using measurements of electromagnetic radiation emitted by celestial objects to characterize those objects.

As our modern image of the cosmos took shape, humanity had to conclude that our home planet's location isn't all that special after all. In fact, our Solar System does not lie at center of the Universe, but rather on an outer arm of the Milky Way. Together with neighboring stars and their planets, the Earth orbits the galactic center about once every 230 million years. This means that the Sun zips through space at about 250 km (155 miles) per second relative to an observer outside our galaxy. Even the Milky Way doesn't sit still—like the other galaxies, it has been hurtling through space for billions of years **(Fig. 21.9a)**. A photo of the Earth and Moon as viewed from Mercury emphasizes that we are all riding on a small speck as it journeys through the vast reaches of space **(Fig. 21.9b)**. But because we live here, it is a very important speck indeed.

Take-home message . . .

Astronomy began when society realized that the movements of celestial objects provided a basis for a calendar. Early astronomers, therefore, focused on cataloging the motions of stars and planets. At first, most observers thought the Earth lay at the center of the Solar System, with planets and the Sun following circular orbits around it. This idea had to be discarded in the wake of discoveries by Copernicus, Galileo, Kepler, and Newton. Astronomers can now demonstrate that planets follow elliptical orbits around the Sun, and those orbits can be explained by Newton's laws. Modern discoveries reveal that the Sun is one of hundreds of billions of stars within one of more than a trillion galaxies.

Quick Question -
Does the velocity of a planet along its orbit remain the same all year?

21.3 Motions in the Heavens

The Celestial Sphere: A Frame of Reference for Describing the Sky

To portray the locations of celestial objects relative to one another, as we see them in the night sky, astronomers mark points representing these objects on the **celestial sphere**, an imaginary surface in space that surrounds the Earth. We can picture the celestial sphere as a giant glass shell **(Fig. 21.10)**. To determine the position of an object on the celestial sphere, astronomers draw a line from the object to the center of the Earth. The point

Figure 21.9 The Earth is but a speck in the vastness of space.

(a) A collection of galaxies in a piece of sky smaller than your thumbnail when you view it at arm's length.

(b) The two tiny specks close to each other in this image are the Earth and the Moon, as viewed by the *Messenger* space probe orbiting Mercury.

Figure 21.10 The concept of the celestial sphere.

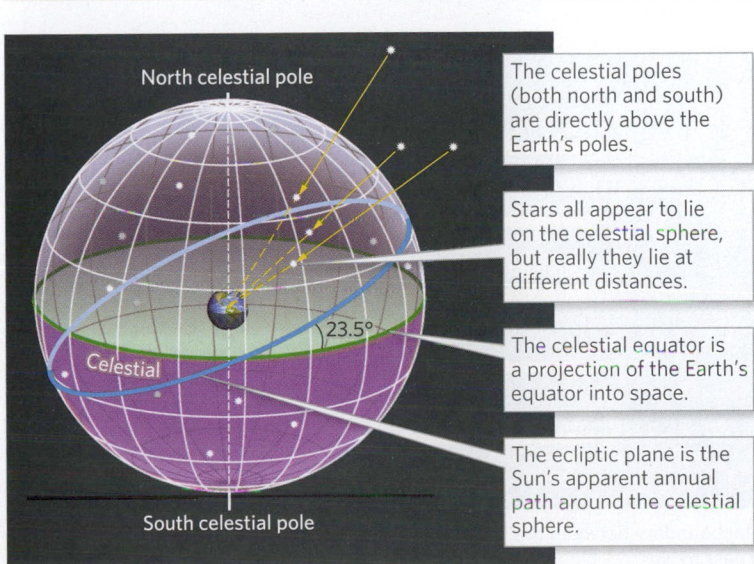

North celestial pole

The celestial poles (both north and south) are directly above the Earth's poles.

Stars all appear to lie on the celestial sphere, but really they lie at different distances.

23.5°

Celestial

The celestial equator is a projection of the Earth's equator into space.

The ecliptic plane is the Sun's apparent annual path around the celestial sphere.

South celestial pole

Figure 21.11 With increasing height above the Earth's surface, the horizon appears farther away, and the visible area of the Earth increases.

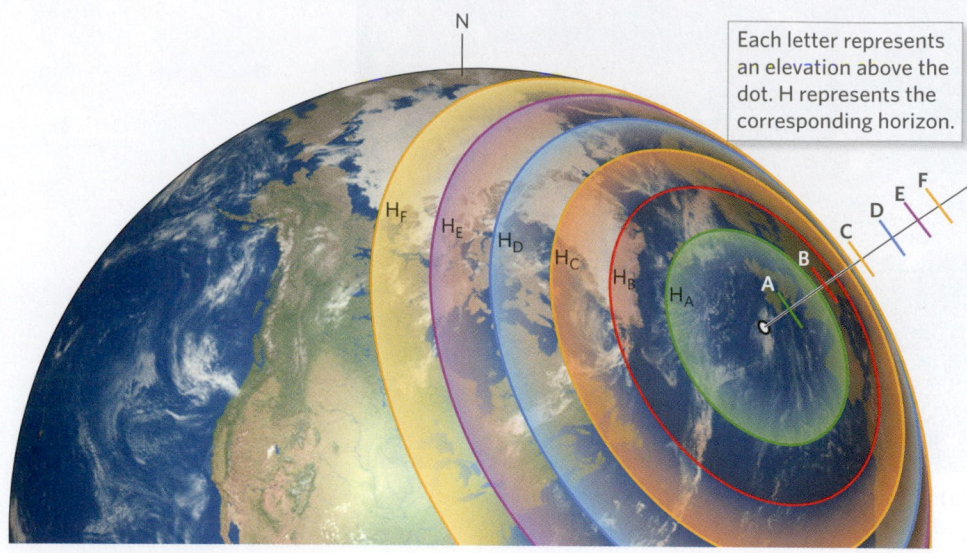

Each letter represents an elevation above the dot. H represents the corresponding horizon.

Figure 21.12 The path of the Sun across the sky.

The path of the Sun, as seen from northern latitudes, during the winter solstice.

(a) If you watch the Sun during the course of the day, you see that it rises at dawn, arcs across the sky, reaching its highest point at noon, and sets at dusk.

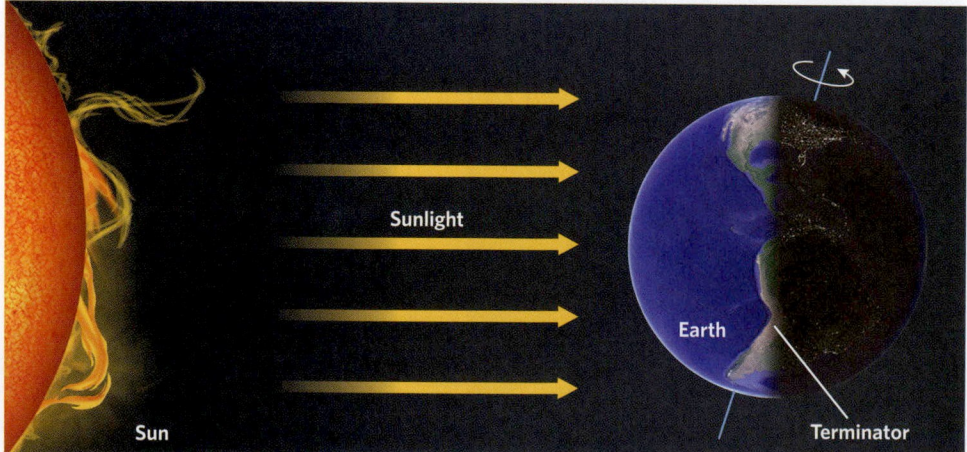

Sunlight

Earth

Sun

Terminator

(b) The terminator is the boundary between the lit and dark sides of the Earth. As the Earth rotates, the terminator moves from east to west.

where the line pierces the sphere represents the object's position. Note that specifying this position does not provide any information about the distance of the object from the Earth.

Astronomers orient the celestial sphere so that its axis aligns with the Earth's **axis of rotation**, the imaginary line passing through the center of the Earth around which our planet spins. The axis of rotation intersects the surface of the Earth at our planet's **geographic poles**. On the celestial sphere, we position the *north celestial pole* as the point directly above the north geographic pole of the Earth, and the *south celestial pole* as the point directly above the Earth's south geographic pole. Similarly, the *celestial equator* is a circle that lies on the celestial sphere halfway between the poles, and therefore in the same plane as the Earth's equator. The celestial equator divides the sky into northern and southern hemispheres.

Imagine that you're standing on the surface of the Earth at night, looking straight up. The point directly above your head is your **zenith**. If you walk to another location, the position of your zenith changes. Now, turn your head and look horizontally. You will see the **horizon**, the line separating the surface of the Earth from the sky as viewed along your line of sight. Because of the Earth's nearly spherical shape, the horizon that you see as you turn completely around while looking horizontally can be plotted as a circle on the surface of the Earth **(Fig. 21.11)**. For a person who stands about 2 m (6 feet) tall, the horizon lies at a distance of about 5 km (3 miles), as measured along the Earth's surface. The horizon moves farther from you as you climb higher, so if you were standing on top of Mount Everest, the horizon would lie about 336 km (209 miles) away from you.

The Journey of Helios: A Consequence of the Earth's Rotation

In ancient times, people thought that the Sun moved along the celestial sphere while the Earth stood still, and they concocted various myths to explain the Sun's path. The Greeks, for example, attributed the movement to the god *Helios*, who drove the chariot of the Sun across the sky once a day. Now, of course, we understand that the Sun's apparent movement actually occurs because the Earth rotates on its axis **(Fig. 21.12a)**. Because the Earth spins counterclockwise, as viewed looking down on the North Pole, the *terminator*, meaning the boundary between night and day, moves from east to west. When the terminator moves over a point on the Earth's surface, the Sun rises or sets at that point **(Fig. 21.12b)**. Sunset and sunrise do not take place instantaneously because the atmosphere scatters sunlight.

How long does the Earth take to make one **rotation**, one complete turn on its axis? If we use the Sun as a reference and measure the amount of time it takes for the Sun to appear again in the same point in the sky, the Earth rotates once every 24 hours. Astronomers call this measure of time a **solar day**. A **sidereal day** (pronounced side-e-re-al), in contrast, represents the time it takes for a distant star to appear at the same point in the sky on successive days. A sidereal day has four fewer minutes than a solar day because the Earth changes its position relative to the Sun over the course of a day as it orbits the Sun. A sidereal day provides the true measure of one full rotation of the Earth.

Surprisingly, no one figured out a way to demonstrate directly that the Earth rotates until the middle of the 19th century, when Léon Foucault (1819–1868), a French physicist, suspended a heavy pendulum from a cable attached to the high roof of the Panthéon, a building in Paris **(Fig. 21.13)**. He set the pendulum in motion and sat back to watch. A pendulum must swing back and forth in the same plane because of inertia. If the Earth did not rotate, that plane would stay in the same orientation relative to the room all day long. But if the Earth did rotate, then the room would slowly spin beneath the pendulum, and the plane of the pendulum's swing would change its orientation relative to the room over time. As the hours passed, the plane of Foucault's pendulum did indeed change, thereby demonstrating that the Earth rotates.

The Earth's Orbit and the Ecliptic Plane

Our planet orbits the Sun once each year, traveling a distance of 940 million kilometers (584 million miles) over a period of 365.256 solar days. So, when you think you're sitting still, you—and the ground beneath you—actually zip along an orbit at an average speed of about 107,000 km per hour (66,000 mph) relative to the Sun. To accommodate the extra quarter day, the *Gregorian calendar*—the standard calendar of the Western world since its introduction by Pope Gregory in 1582—includes a *leap year* every four years, when we add an extra day to February. At the turn of each century, except those centuries divisible by 400, the calendar skips adding the leap day to take into account the extra 0.006 day.

As Kepler recognized, the Earth follows an elliptical orbit, and the Sun lies at one focus—not the center—of the ellipse (see Fig. 21.7a). The distance between the Earth and the Sun, therefore, changes over the course of a year, ranging from 147.1 million km (91.4 million miles) at the *perihelion*, when the Earth is closest to the Sun, to 152.1 million km (94.5 million miles) at the *aphelion*, when the Earth is most distant from the Sun (see Fig. 21.7a). Note that these distances differ by only about 4%, so the Earth's orbit is actually fairly close to being circular **(Fig. 21.14)**.

Figure 21.13 Proving the Earth rotates on its axis.

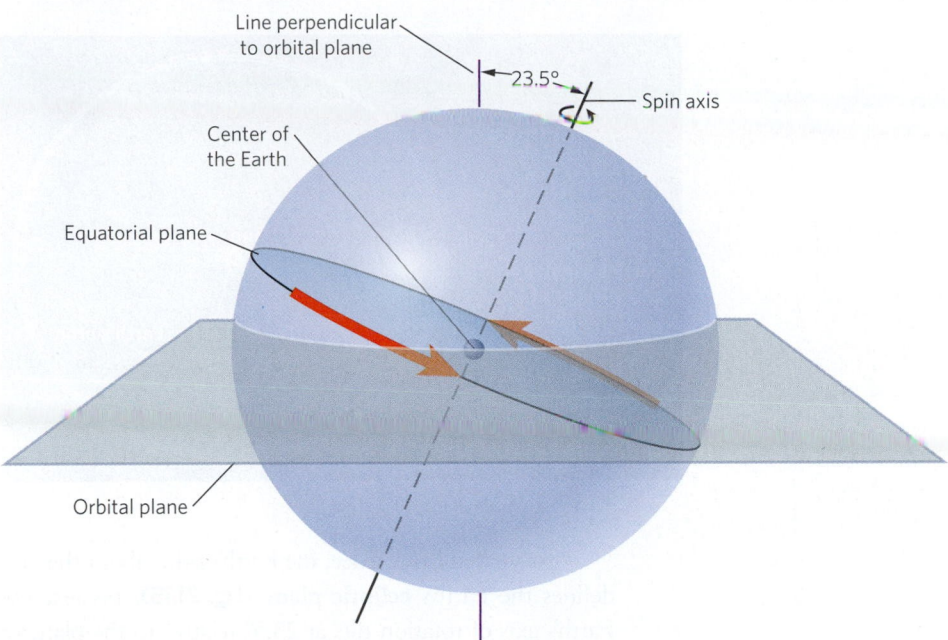

(a) The Earth is a sphere, rotating around its axis. In this representation of a transparent Earth, we see that the axis pierces the Earth at the poles, goes through the center of the planet, and is tilted relative to the ecliptic plane.

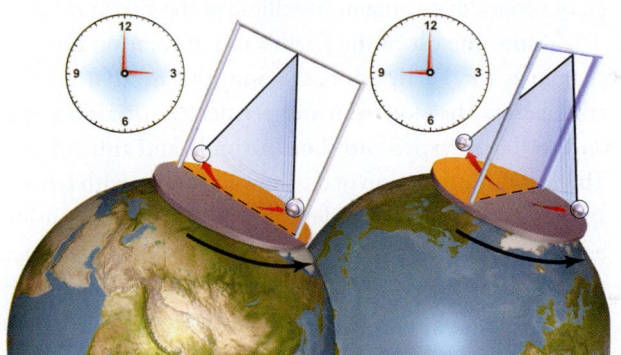

(b) Foucault's experiment as viewed with the pendulum placed over the North Pole. At Time 1, the plane in which the pendulum swings is the same as the plane of its frame. At Time 2, six hours later, the plane is perpendicular to the frame.

(c) A replica of Foucault's pendulum on display in the Panthéon, Paris.

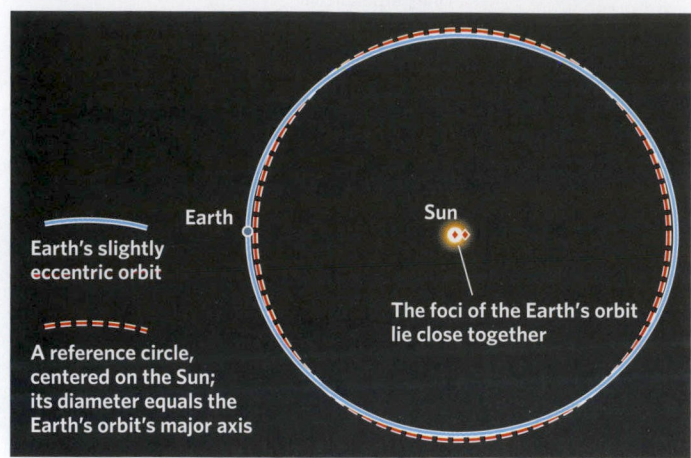

Figure 21.14 The eccentricity of the Earth's real orbit is so small (0.017) that it is actually almost circular.

Earth

Sun

Earth's slightly
eccentric orbit

The foci of the Earth's orbit
lie close together

A reference circle,
centered on the Sun;
its diameter equals the
Earth's orbit's major axis

Did you
ever wonder . . .

why the North Star is called
the North Star?

As viewed from space, the Earth's orbit about the Sun defines the Earth's **ecliptic plane (Fig. 21.15)**. Because the Earth's axis of rotation tilts at 23.5° relative to the plane of the Earth's orbit around the Sun (see Fig. 20.7), the ecliptic plane tilts at an angle of 23.5° relative to the celestial equator (see Fig. 21.10). We can picture the ecliptic plane as an imaginary surface that contains the ellipse of the Earth's orbit.

Since the tilt of the Earth's axis of rotation does not change as the Earth orbits the Sun, the arc that the Sun takes across the sky, from the perspective of an observer on the Earth, varies with both latitude and time of year. Therefore, the duration of daylight also varies with latitude and time of year. To understand this concept, consider how the Sun progresses across the sky from the perspectives of three observers at different latitudes—one standing on the equator, one at mid-latitude in the northern hemisphere, and one at the Arctic Circle—at three different times of the year **(Fig. 21.16a)**.

Note that during half of the year, the northern hemisphere inclines toward the Sun, and during the other half of the year, it inclines away from the Sun **(Fig. 21.16b)**. This tilt, as we saw in Chapter 20, gives the Earth its seasons (see Fig. 20.7). In the northern hemisphere, the time when the Earth's axis transitions from tilting toward the Sun to tilting away defines the *autumnal equinox* (September 22), and the time when it transitions from tilting away from the Sun to tilting toward it defines the *vernal equinox* (March 20).

On the scale of a human lifetime, the Earth's orbit doesn't change noticeably over time. But on the scale of millennia, it does, due to gravitational interactions with other planets. Specifically, the orientation of the orbit's major axis changes relative to distant stars over the course of several millennia, and the eccentricity of the orbit changes over the course of millennia. These changes cause the Milankovitch cycles that we described in Chapter 20.

Navigating Darkness: The Night Sky and Its Constellations

Imagine yourself on a boat with no lights, floating in the middle of the ocean on a moonless, cloudless night. How many stars can you see? Astronomers estimate that during the course of a night, a person with good vision could identify about 3,000 individual stars. If observers from around the world were to get together and compile a list of all the stars visible to the naked eye, taking into account that they would see different stars from different locations, the list would include about 6,000 entries.

Now imagine that you're standing at the North Pole, so that your zenith aligns with the north celestial pole. If you were to aim a camera straight up and set it to record

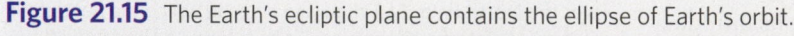

Figure 21.15 The Earth's ecliptic plane contains the ellipse of Earth's orbit.

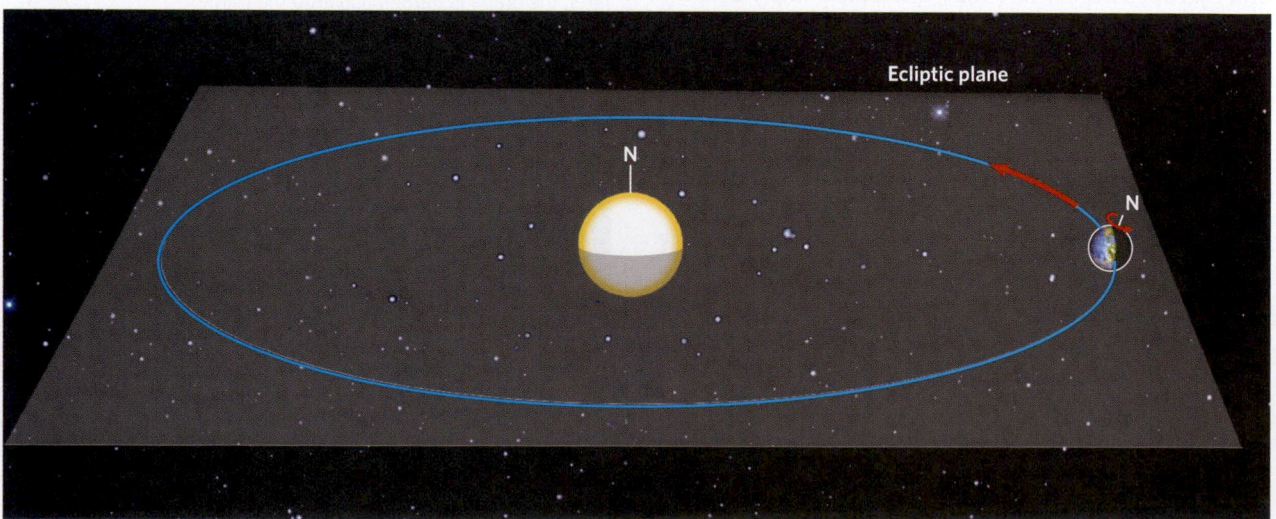

Ecliptic plane

starlight on a single image all night long as the Earth spins, the image would show circular *star streaks*. In the northern hemisphere, the star Polaris, positioned directly above the North Pole, would lie closest to the center of the star streak circles, so we refer to Polaris as the *North Star* **(Fig. 21.17a)**. If you were to move to a lower latitude, you would find that

the North Star does not lie directly overhead, but rather sits lower in the sky. Furthermore, you would record star streaks that have larger diameters and are partial circles that run into the horizon **(Fig. 21.17b)**. If you were to go farther south, you would see the position of the North Star progressively lower in the sky **(Fig. 21.17c)**, and when you reached the equator, you'd see the North Star right at the horizon. You wouldn't see the North Star at all from the southern hemisphere.

The identity of the star that serves as the North Star changes over millennia because the Earth's axis undergoes **precession**. This means that it wobbles like a top, so that the axis of rotation points to different stars on the celestial sphere **(Fig. 21.18a)**. It takes about 26,000 years for the Earth's axis to make a complete circle. Because of precession, about 12,000 years from now, the North Star will be Vega, not Polaris **(Fig. 21.18b)**. This wobble is one of the Milankovitch cycles discussed in Chapter 20.

Ancient observers in many cultures found that if they wished to identify specific locations in the night sky, it helped to group stars into recognizable patterns or arrangements, which came to be known as **constellations** **(Fig. 21.19a)**. These patterns were typically named after animals, warriors, or mythical beings. Each culture came up with its own set of constellations—Greek constellations are not the same as Chinese constellations—emphasizing that constellations are entirely a human invention.

The times and locations at which specific constellations appear at the horizon after sunset provide a useful basis for navigation and calendars. Western culture recognizes 88 constellations, which stargazers use like ZIP codes to direct attention not just to the stars in the constellation, but to a given sector of the celestial sphere. On successive months throughout the year, different constellations make their appearance in the night sky because our planet has moved to a different position along its orbit.

Figure 21.16 The path of the Sun across the sky varies with latitude and season.

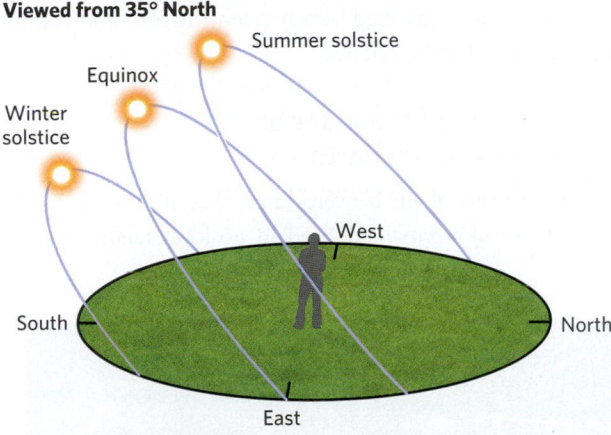

Viewed from Arctic Circle, 66.5° North

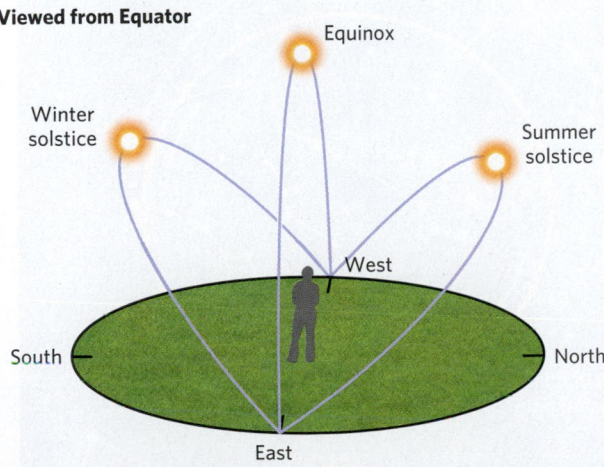

Viewed from 35° North

Viewed from Equator

(a) The paths the Sun takes across the daytime sky at three times of year as viewed from a point on the equator, at about 35° N, and at the Arctic Circle. At the Arctic Circle, the Sun is up for 24 hours at the summer solstice, and it only briefly peeks above the south horizon at noon at the winter solstice before disappearing for the remainder of the day.

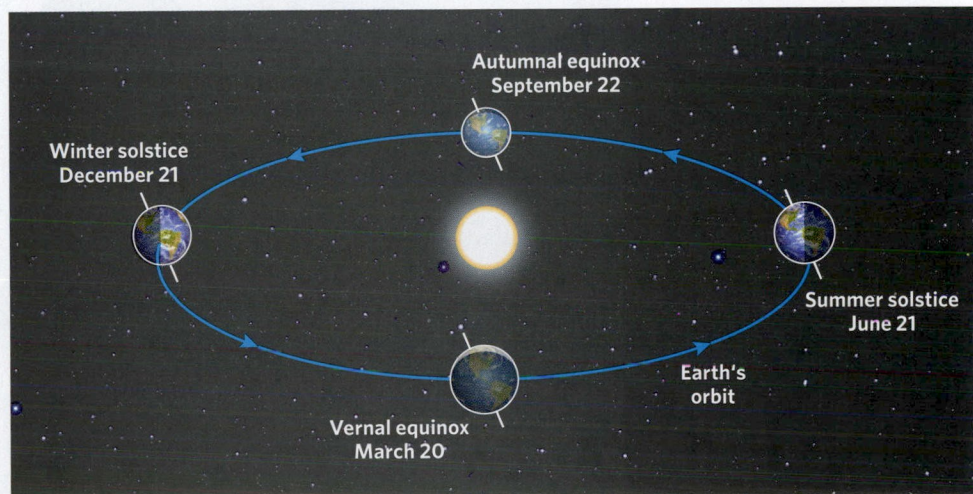

(b) The Earth revolves around the Sun once a year. The Earth's axis tilts 23.5° with respect to Earth's orbital plane, so that Earth's polar regions are in darkness at the winter solstice and in perpetual light at the summer solstice.

Figure 21.17 Star streaks.

(a) Viewed from the North Pole at midwinter, star streaks form circles with Polaris, the North Star, at the center.

(b) Viewed from a lower latitude, star streaks form partial circles that intersect the horizon.

(c) At still lower latitudes, the North Star appears close to the horizon and star streaks form half circles.

The 12 constellations that lie along the ecliptic plane constitute the **zodiac (Fig. 21.19b)**. Each constellation lies within a 30°-wide arc of the ecliptic plane. Star charts and, more recently, apps on smartphones, can be used to locate constellations. Some people imagine that the positions of constellations and planets influence human behavior and fate and determine good or bad luck. This premise under-

lies *astrology*, not astronomy. The premises of astrology have never been justified by any scientific test, and cannot be considered to be science.

Movements of the Planets as Seen from the Earth

Early observers of the heavens knew that all but five of the stars in the sky remained fixed in position relative to one

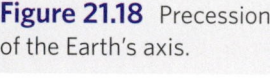

Figure 21.18 Precession of the Earth's axis.

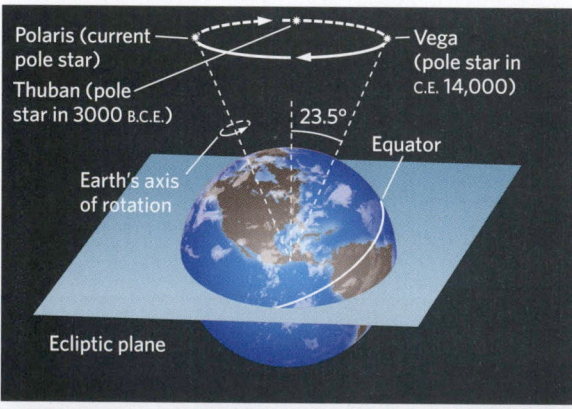

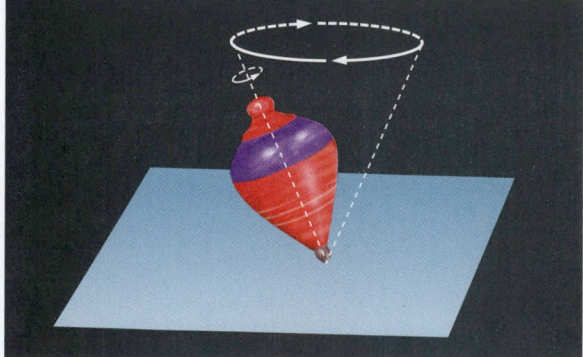

(a) Earth's axis of rotation wobbles, much like a top. This movement is called precession.

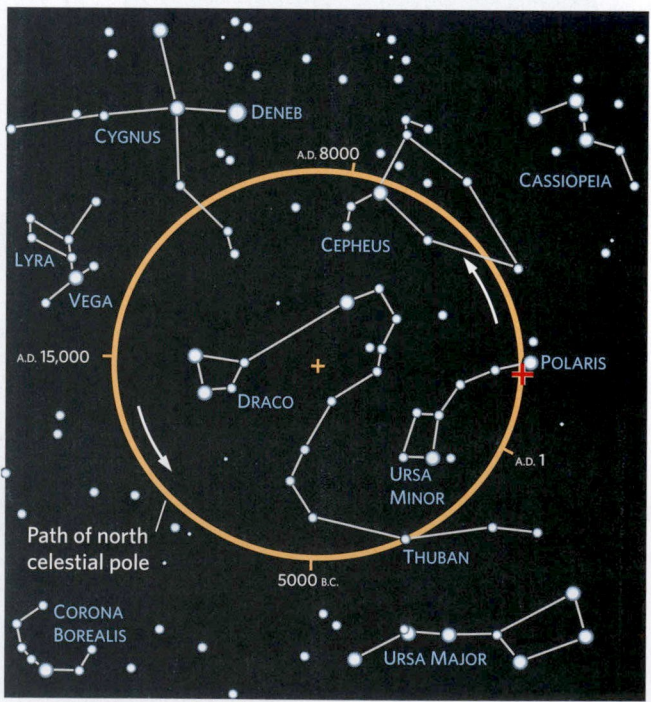

(b) The yellow circle traces the position of the north celestial pole, relative to the positions of distant stars, over 26,000 years. Currently, the pole lies at the red +, near Polaris.

Figure 21.19 Constellations.

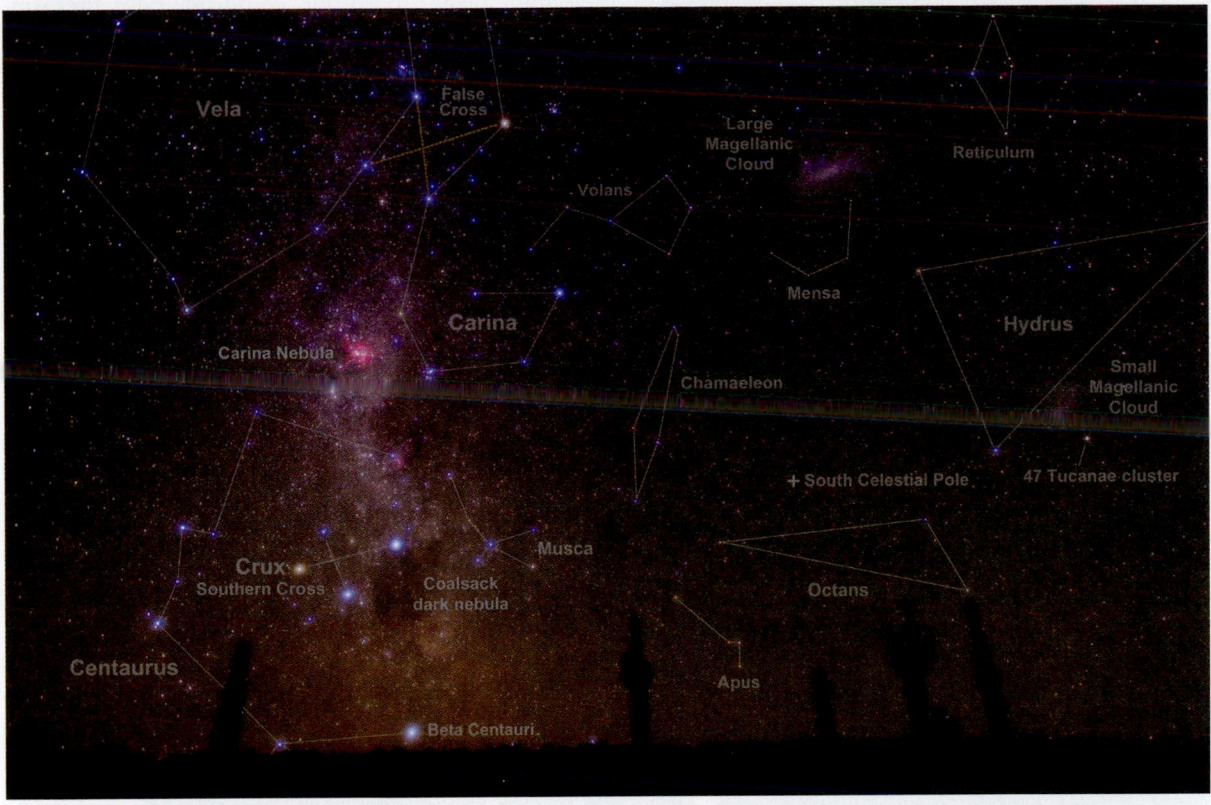

(a) An example of the constellations seen at one time and location in the southern hemisphere. Note that most of these are not visible in the northern hemisphere.

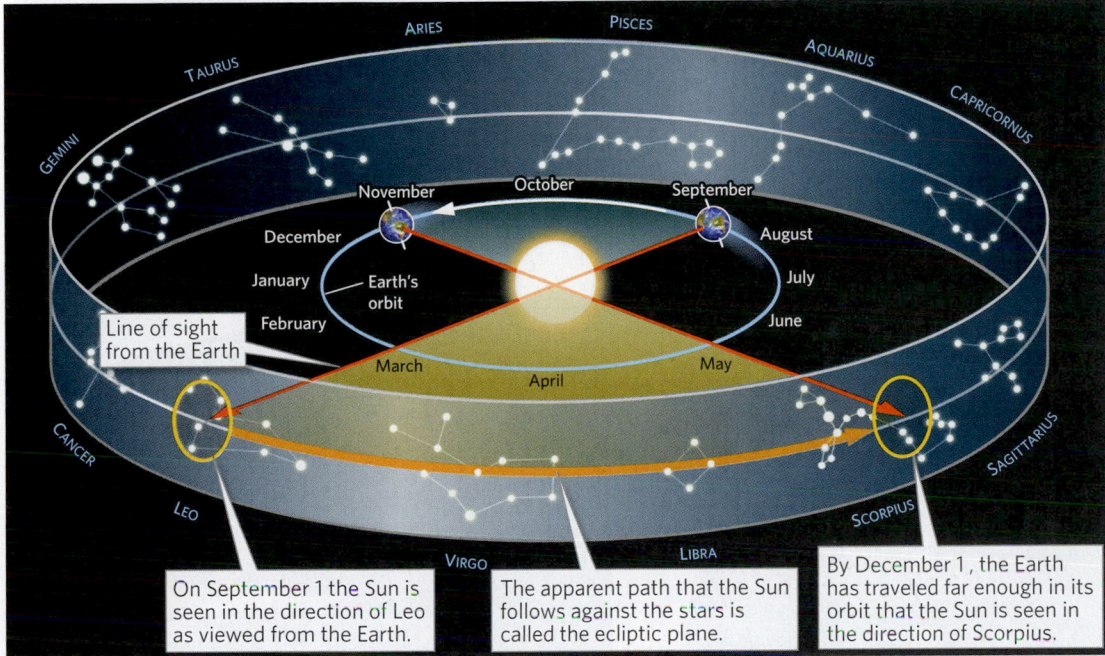

(b) The zodiac is the set of 12 constellations that lie along the ecliptic plane.

another as they moved smoothly across the night sky, and that night after night, each of these stars retained the same brightness. These five celestial objects, however, are not so well behaved. Although their overall path remains close to the ecliptic plane, these objects seem to vary in brightness and to "wander," changing position relative to the stars and to one another as the months pass. As we've noted, these objects are not stars at all. They are the nearer planets of our

Figure 21.20 You can find the ecliptic plane in the night sky by drawing an imaginary line connecting visible planets with the setting or rising Sun. Spica is a bright star.

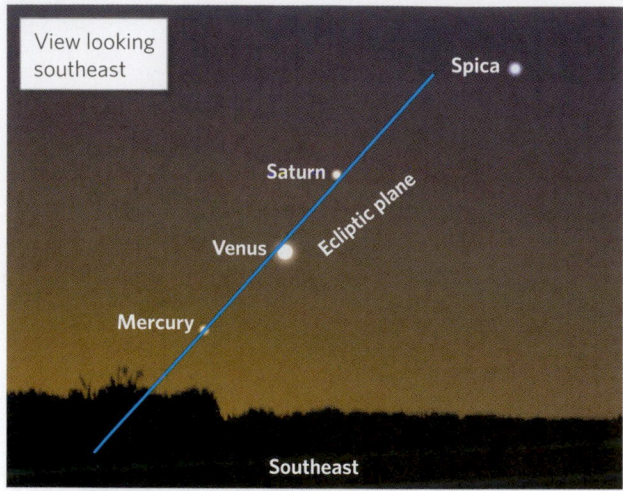

View looking southeast

Spica

Saturn

Venus

Ecliptic plane

Mercury

Southeast

Because their orbital planes roughly parallel the Earth's orbital plane, you can use the positions of planets to find the ecliptic plane. To do this, go outside right after sunset or right before sunrise and find one or more planets. Because they are brighter than most stars, they are among the first celestial objects (not counting the Moon) to be visible at night and the last to be visible in the morning, so you can often identify them at dusk or dawn. Once you've picked out a planet, draw an imaginary line connecting it with the position of the setting or rising Sun. This line shows the trace of the ecliptic plane across the celestial sphere **(Fig. 21.20)**.

Solar System: Mercury, Venus, Mars, Jupiter, and Saturn. Two more distant planets, Uranus and Neptune, which can be seen only with telescopes, follow the same pattern of motion as the other planets.

All planets orbit the Sun in a counterclockwise direction, as viewed looking down on the Sun's north pole. From an earthbound observer's viewpoint, therefore, the planets usually display prograde motion, meaning that they traverse the celestial sphere from west to east. Occasionally, as we've noted, a planet displays retrograde motion (east to west) on the celestial sphere **(Fig. 21.21a)**.

Figure 21.21 Why do planets display retrograde motion?

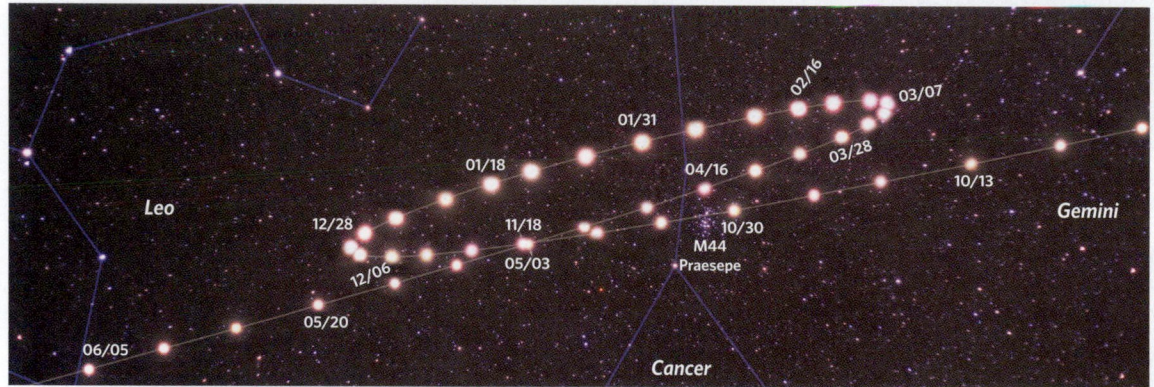

Leo

02/16
01/31
03/07
01/18
03/28
04/16
10/13
12/28
11/18
Gemini
10/30
M44
12/06
05/03
•Praesepe
05/20
06/05
Cancer

(a) Observed apparent retrograde motion of Mars.

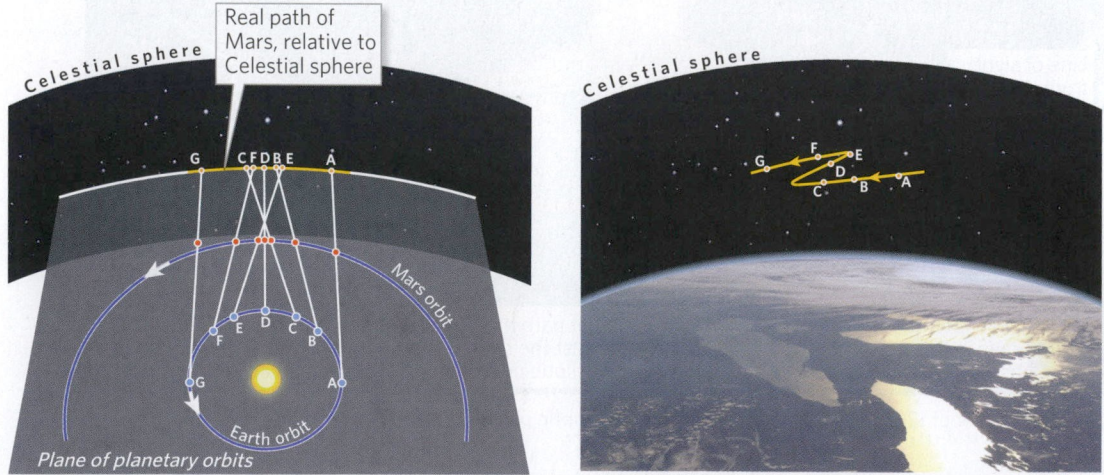

Celestial sphere

Real path of Mars, relative to Celestial sphere

G   CFDBE   A

F E D C B

G   A

Mars orbit

Earth orbit

Plane of planetary orbits

Celestial sphere

G   F   E
D
C   B
A

(b) The retrograde motion of Mars in the night sky occurs because the Earth and Mars orbit the Sun at different speeds, so the view of Mars from the Earth changes. Apparent path of Mars across the sky, relative to the Earth's horizon. The path appears to go up-and-down because, due to the Earth's tilt, the trace of the orbital plane above the horizon moves up-and-down over the course of a year.

754   CHAPTER 21   INTRODUCING ASTRONOMY: LOOKING BEYOND THE EARTH

Its movement appears to be retrograde because planets orbit the Sun at different distances and, therefore, at different velocities. As a result, the line of sight from the Earth to another planet continually changes **(Fig. 21.21b)**.

Why does the brightness of planets, as seen from the Earth, vary over time? Unlike stars, planets do not produce their own light. The light that we see when looking at them comes from reflected sunlight. The brightness of a planet as viewed from Earth depends, therefore, on several factors: its distance from the Sun, which determines how much sunlight it receives; its size, which determines the dimensions of its reflecting surface; its *albedo*, meaning the reflectivity of its surface; and its position relative to the Earth at a particular time, which determines its distance from the Earth and how much of its lit side faces the Earth. As the distance to the planet from the Earth increases, or when the planet attains a position that allows us to see light reflected from only part of its Sun-facing surface, the planet's light appears fainter to an Earth-bound observer.

The Earth's Traveling Companion: Movements of the Moon

The Moon, our planet's closest neighbor, has a fascinating surface displaying both smooth, dark *mare* (Latin for sea), and rugged, light-colored highlands **(Fig. 21.22a)**. Ancient Greeks used geometric methods to determine that the Moon lies about 30 times the Earth's diameter away from the Earth. Modern measurements indicate that, on average, it's 384,400 km (238,900 miles) away. The Moon is the only celestial object outside of the Earth that humans have visited. Because of its orbit around the Earth, the Moon's position in the sky arcs from east to west along a path oriented at about 5.2° to the ecliptic plane **(Fig. 21.22b)**. Relative to a fixed point on the Earth, the Moon speeds along its orbit at 1 km per second (3,600 km per hour, or 2,200 mph).

PHASES OF THE MOON. Even casual watchers of the sky know that if you watch the Moon night after night, you'll see the shape of the Moon's illuminated face change in a predictable cycle, with each successive stage known as a **phase** of the Moon. A single cycle between identical phases takes 29.5 days, a period called a **lunar month**. Lunar months were once used as the basis for calendars, but they don't work well over multiple years because they get out of synchrony with the seasons.

When we see a *new Moon* from the Earth, the Moon rises at approximately the same time as the Sun, so the Sun is behind the Moon, and the face we see is in the shade. Every night after the new Moon, the visible part

Figure 21.22 The Moon.

(a) Distinct features of the Moon's surface are visible from Earth with a telescope.

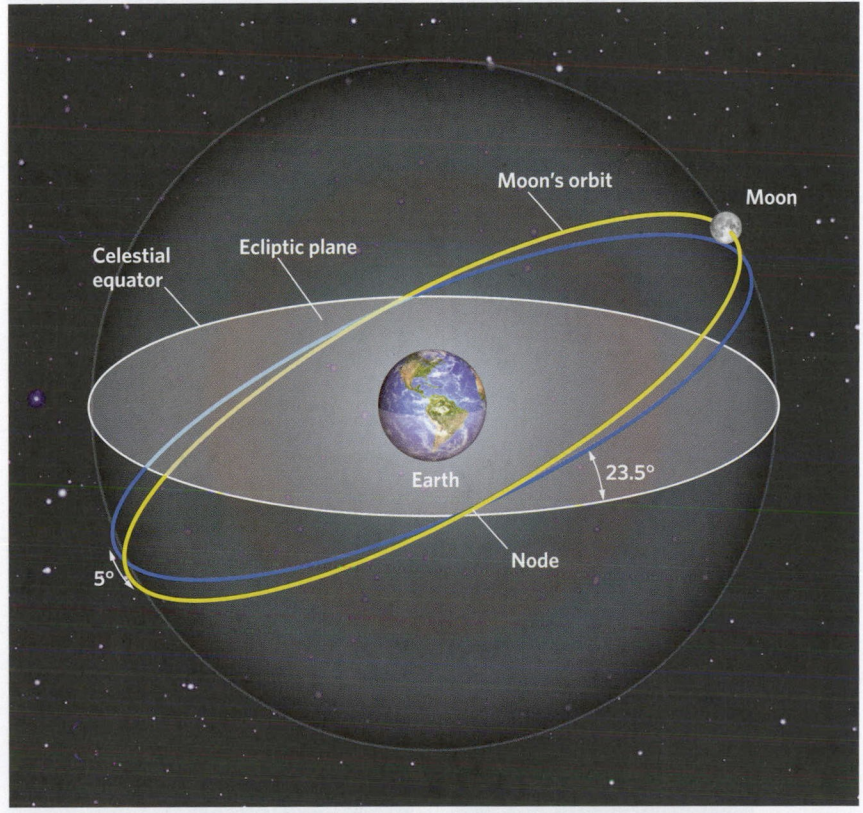

(b) The Moon's orbit is 5.2° off the ecliptic plane as seen on the celestial sphere.

New
Moon

Waning
Crescent

Last
Quarter

Waning
Gibbous

Full
Moon

Waxing
Gibbous

First
Quarter

Waxing
Crescent

New
Moon

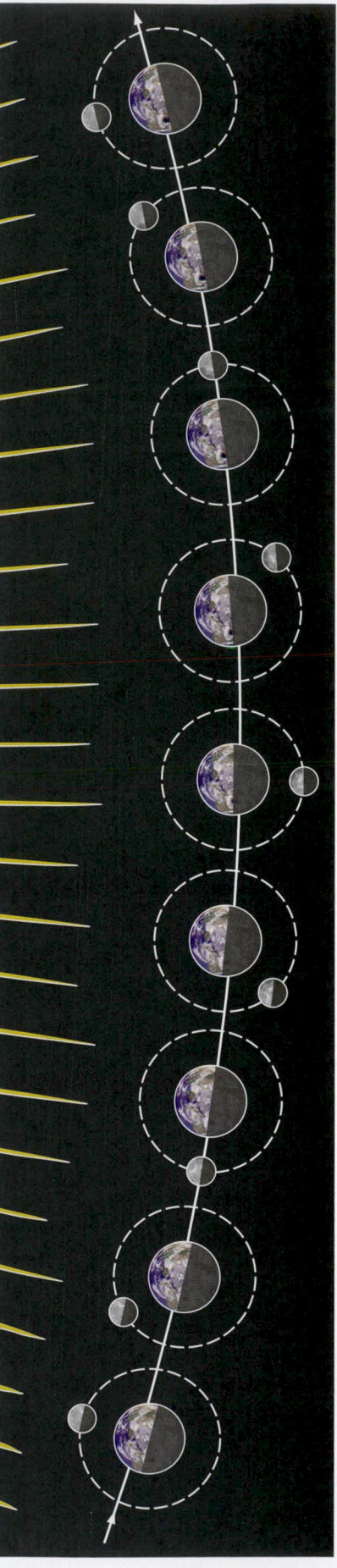

(a) The Moon completes its phases every 29.5 days.

(b) Its position relative to the Sun and the Earth determines how much of the illuminated part of the Moon can be seen from the Earth.

Figure 21.23 The phases of the Moon.

of the Moon *waxes*, or grows, so that sunlight illuminates a progressively larger part of the Moon's surface as we see it from the Earth **(Fig. 21.23a)**. The Moon passes sequentially through phases known as the *crescent Moon*, *quarter Moon*, and *gibbous Moon*. Halfway through a lunar month, the Moon rises as the Sun sets, and the entire face of the Moon that we see from the Earth is lit, producing a *full Moon*. On subsequent days, the Moon rises progressively later than the Sun sets, so the lit portion seen from the Earth *wanes*, or shrinks, through gibbous, quarter, and crescent phases. The cycle comes to completion when we once again see a new Moon.

Why do phases of the Moon happen? Except during times when the Moon passes into the Earth's shadow—an event called a lunar eclipse, which we will discuss shortly—sunlight always shines on the Moon, lighting up half of its surface. This light, which reflects from the Moon's surface, is what we see when looking at the Moon **(Fig. 21.23b)**. When the Moon and the Sun lie on opposite sides of the sky, so that the Sun sets as the Moon rises, we see the Moon's entire lit face, and the Moon appears full. As the Moon orbits the Earth, we see a progressively smaller portion of the lit face until the Moon lies on the same side of the Earth as the Sun, and the side of the Moon facing us receives no direct sunlight.

THE NATURE OF ECLIPSES. The Moon and the Earth occasionally pass through each other's shadows. When they do, an *eclipse* takes place. If the Moon blocks the light of the Sun from reaching the Earth, we see a **solar eclipse**, whereas if the Earth blocks the light of the Sun from reaching the Moon, we see a **lunar eclipse**. Because such shadows can be cast only when the Earth, Sun, and Moon line up, and because the Moon's orbit inclines at 5.2° to the ecliptic plane, eclipses do not happen frequently. They happen only when a line drawn from the Sun to the Earth also crosses the orbital plane of the Moon and when the Moon is either full or new **(Fig. 21.24a)**.

Our Moon, with a radius of 1,737 km (1,079 miles) is clearly much, much smaller than the Sun, which has a radius of 695,800 km (432,450 miles), but the distance to the Moon from Earth is much less than the distance to the Sun, so when viewed from Earth, both objects seem almost identical in size **(Box 21.4)**. As a result, the Moon can occasionally block all the light of the Sun. In a general sense, the shadow that forms when an object obscures light coming from a non-point source (meaning a source that has a significant width) has two parts: an *umbra* of complete darkness and a *penumbra* of partial

Figure 21.24 Eclipses.

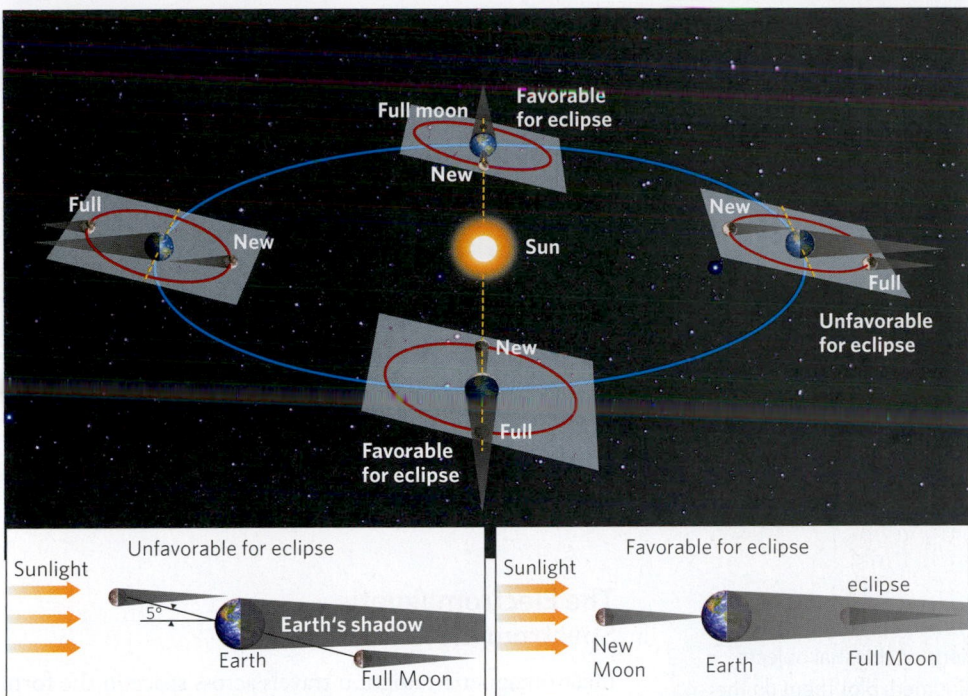

(a) Because the Moon's orbital plane is oriented at an angle to the ecliptic plane, there are only two times in a year when the Moon's orbit and the ecliptic plane intersect and an eclipse is possible. The Moon must be new (for a solar eclipse) or full (for a lunar eclipse) during these brief times.

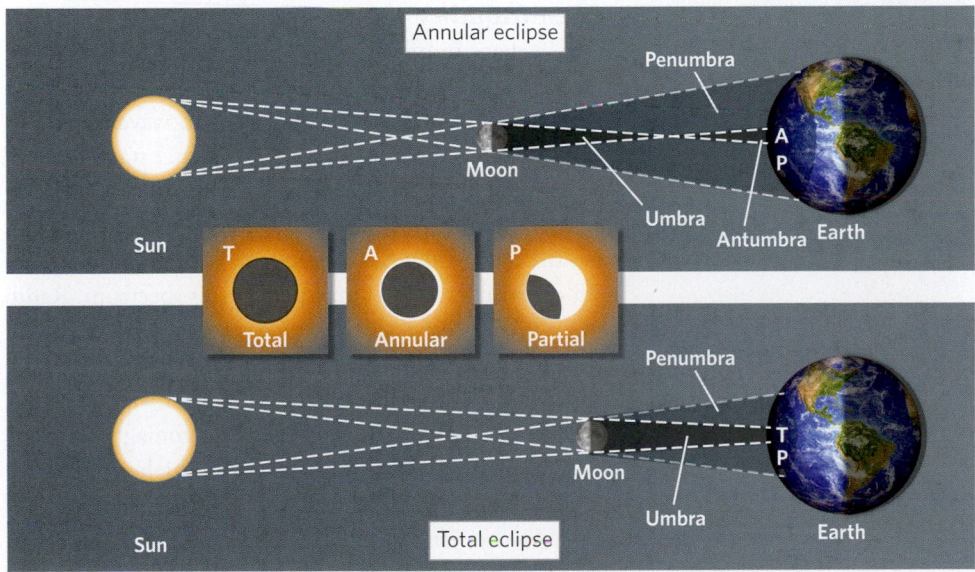

(b) The distance between the Earth and Moon determines the sizes of the umbra and penumbra (or antumbra, in the case of an annular eclipse), and thus whether a solar eclipse will be viewed from Earth as partial, total, or annular. T is the area of total eclipse, and P is the area of partial eclipse.

darkness. We see a *total solar eclipse* when the Earth passes through the Moon's umbra and the Sun becomes completely obscured **(Fig. 21.24b)**. The umbra track on the Earth's surface is quite small, so relatively few people on Earth see a particular total solar eclipse, and umbra tracks differ for different eclipses. The movement of

the Moon and the Earth along their orbits prevents a total solar eclipse from lasting more than 7.5 minutes. For observers in the penumbra, the Moon only partially obscures the Sun, producing a *partial solar eclipse*. When the Moon lies in a position far from the Earth, its disk appears to fit completely within the disk of the Sun, so that an outer ring of the Sun is still visible and the shadowed region is within the antumbra **(Fig. 21.24b)**. This phenomenon is called an *annular eclipse*. Enough radiation reaches the Earth during a solar eclipse that viewing even a total eclipse directly can severely damage your eyes. Never look at a solar eclipse without taking precautions!

A lunar eclipse takes place when the Moon passes through the Earth's shadow. The Earth has a large shadow relative to the Moon, so total lunar eclipses happen more frequently than do total solar eclipses, and they can usually be seen from anywhere on Earth **(Fig. 21.25a)**. During a lunar eclipse, the Moon doesn't go black, but rather turns a ruddy red **(Fig. 21.25b)**. Why? Sunlight that passes through the Earth's atmosphere

undergoes scattering and bending, and some of this light, after leaving the atmosphere, follows a path to the Moon. Because the atmosphere preferentially scatters blue light (see Chapter 17), the light that remains after passing through the atmosphere and reaching the Moon is red.

Figure 21.25 The nature of lunar eclipses.

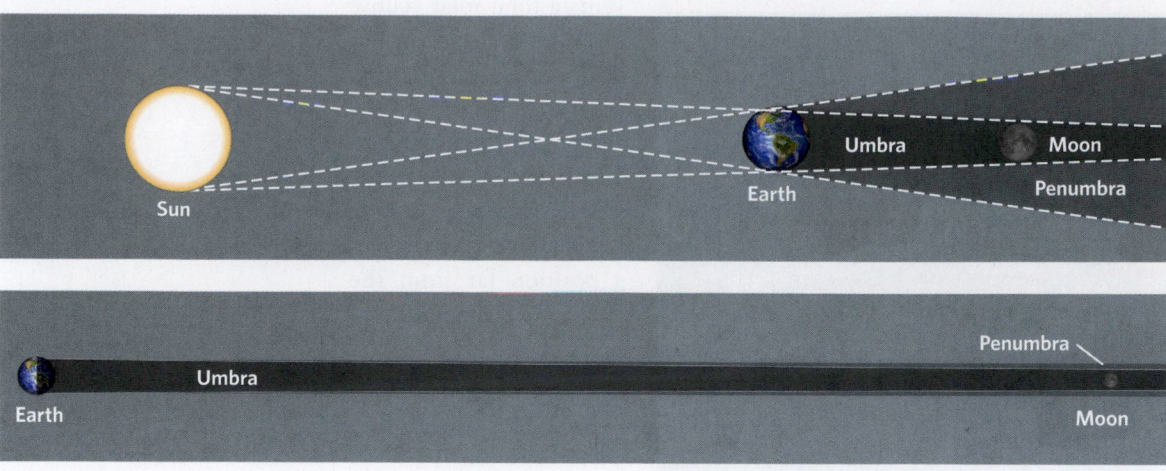

(a) We see a total lunar eclipse when the Moon passes through the Earth's umbra and a partial lunar eclipse when it passes through the penumbra.

(b) The Moon during a total lunar eclipse appears red because light from the Sun passes through the Earth's atmosphere and is bent toward the Moon.

Take-home message . . .

To represent the positions of celestial objects seen from Earth, astronomers plot them on the celestial sphere. We see the movement of celestial objects in the sky because the Earth rotates on its axis, which tilts with respect to its orbital plane. The Earth revolves around the Sun once a year, causing an observer on Earth to see the Sun, Moon, constellations, and planets move across the sky in predictable paths. Because the Moon orbits the Earth, we see phases of the Moon, and because its orbital plane lies close to that of the Earth, we occasionally see eclipses.

Quick Question -
Why do planets sometimes exhibit retrograde motion?

21.4 Light from the Cosmos

Our ancestors knew that celestial objects existed because they could see the visible light that travels from these objects to the Earth. **Visible light** is a type of electromagnetic radiation that can be detected by the nerves in our eyes. In Chapter 17, we introduced electromagnetic radiation to explain how energy from the Sun heats the Earth's atmosphere. Modern astronomers study electromagnetic radiation—not just visible light, but other types as well—because examining this radiation helps astronomers to understand the composition, size, temperature, and movement of celestial objects. Therefore, to provide a foundation for discussing these objects, let's review the characteristics of electromagnetic radiation and consider how radiation interacts with the Earth's atmosphere.

The Electromagnetic Spectrum, Revisited

Electromagnetic radiation travels across space in the form of **electromagnetic waves**, which, unlike sound or water waves, can move through a vacuum. We distinguish among different types of electromagnetic waves by specifying their **wavelength** (the distance between wave crests) or their **frequency** (the number of wave crests passing a point in space in one second). All wavelengths of electromagnetic radiation travel at the same speed, the speed of light, so frequency (f) can be related to wavelength (w) by the equation $f = c \div w$. (In this equation, c stands for the speed of light.) Electromagnetic radiation occurs in a vast range of wavelengths, the entire range of which constitutes the **electromagnetic spectrum (Fig. 21.26a)**. Visible light constitutes only a small part of this spectrum. Radio waves, microwaves, and infrared radiation have longer wavelengths than visible light, whereas ultraviolet (UV) light, X-rays, and gamma rays have shorter wavelengths.

The amount of energy carried by electromagnetic waves depends on the frequency of the waves: higher-frequency (shorter) waves carry more energy than do lower-frequency (longer) waves. In the early 20th century, researchers realized that to explain the way electromagnetic radiation interacts with atoms, radiation must be pictured not only as a succession of waves, but also as a stream of particles. These particles, known as **photons**, have no mass. Each photon contains a very specific amount of energy determined by the wavelength of the radiation.

Looking through Atmospheric Windows

Astronomers can determine many characteristics of a radiation source—be it a star, planet, or nebula—by

How can I explain . . .

The relative sizes of the Sun and the Moon

What are we learning?

How the apparent sizes of celestial objects in the sky depend on their distance from the Earth.

What you need:

- A basketball.
- A measuring device that uses millimeters.
- A piece of paper containing a typewritten period.
- A calculator.

Instructions:

- Use the basketball as your scale model for the Sun. Measure the diameter of the basketball and convert it to millimeters (if you don't have a basketball, look up its diameter on the Internet).
- To calculate the scale-model size of the Moon, we need to calculate proportions. For example,

$$\frac{true\ diameter\ of\ Sun\ (km)}{true\ diameter\ of\ Moon\ (km)} = \frac{measured\ diameter\ of\ basketball\ (mm)}{scaled\ diameter\ of\ Moon\ (mm)}$$

The unknown in this equation is the scaled diameter of the Moon. Use the data for the radius of the Sun and Moon in the discussion of eclipses to obtain the diameter of the Sun and the Moon (twice the radius). Now calculate what the diameter of the Moon would be if the Sun were the size of a basketball.

What did we see?

- If you did the calculation correctly, you arrived at a very small number (0.5 mm). That is the diameter of a period at the end of a sentence on the sheet of paper with typewritten text on it. If the Sun were the size of a basketball, the Moon would be the size of a period. Yet they appear the same size in the sky. The Moon is clearly much closer to the Earth than the Sun.

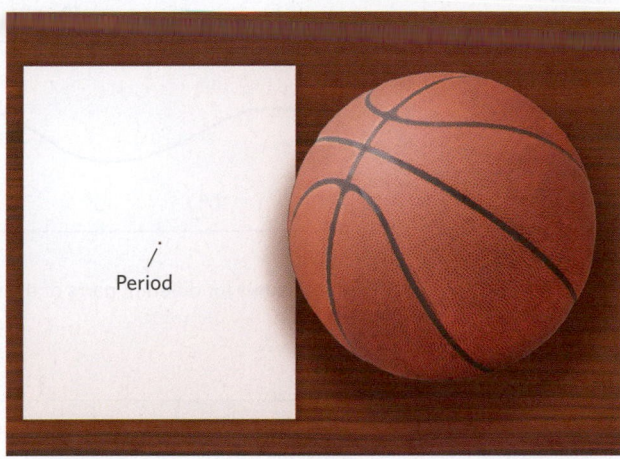

---

analyzing the spectrum that the source emits. If all wavelengths of electromagnetic radiation could pass through the atmosphere, we would not need to send up satellites to study the heavens. Unfortunately for ground-based astronomers, gas molecules in the atmosphere absorb most wavelengths of electromagnetic radiation. In fact, there are only two *atmospheric windows*, meaning ranges of wavelengths that can pass through air without being absorbed **(Fig. 21.26b)**. The **optical window** is *transparent* to visible light and *translucent* to infrared energy. This means that nearly all visible light waves penetrate the atmosphere, but only some infrared radiation does. The **radio window** is transparent to radio waves with wavelengths between 1 cm (0.4 inches) and 10 m (33 feet). We refer to the study of visible wavelengths as *optical astronomy* and the study of radio wavelengths as *radio astronomy*, and these different parts of the spectrum can be observed with *optical telescopes* and *radio*

telescopes, respectively. As we'll see, *space telescopes* can detect other wavelengths of radiation—such as X-rays and gamma rays—that can't pass through the atmospheric windows.

Using Spectroscopy to Answer Questions about Celestial Objects

WHAT'S IT MADE OF? If you've ever played with a prism, you've seen how it spreads a beam of white light into a spectrum of different colors **(Fig. 21.27a)**. That's because different wavelengths of light *refract* (bend) by slightly different amounts as they pass through the prism. Astronomers use **spectroscopes**, sophisticated prisms that spread incoming electromagnetic radiation into a very broad spectrum, to analyze the radiation from stars.

If you look closely at an image produced by a spectroscope, you'll see that the light from celestial objects does

Figure 21.26 The electromagnetic spectrum.

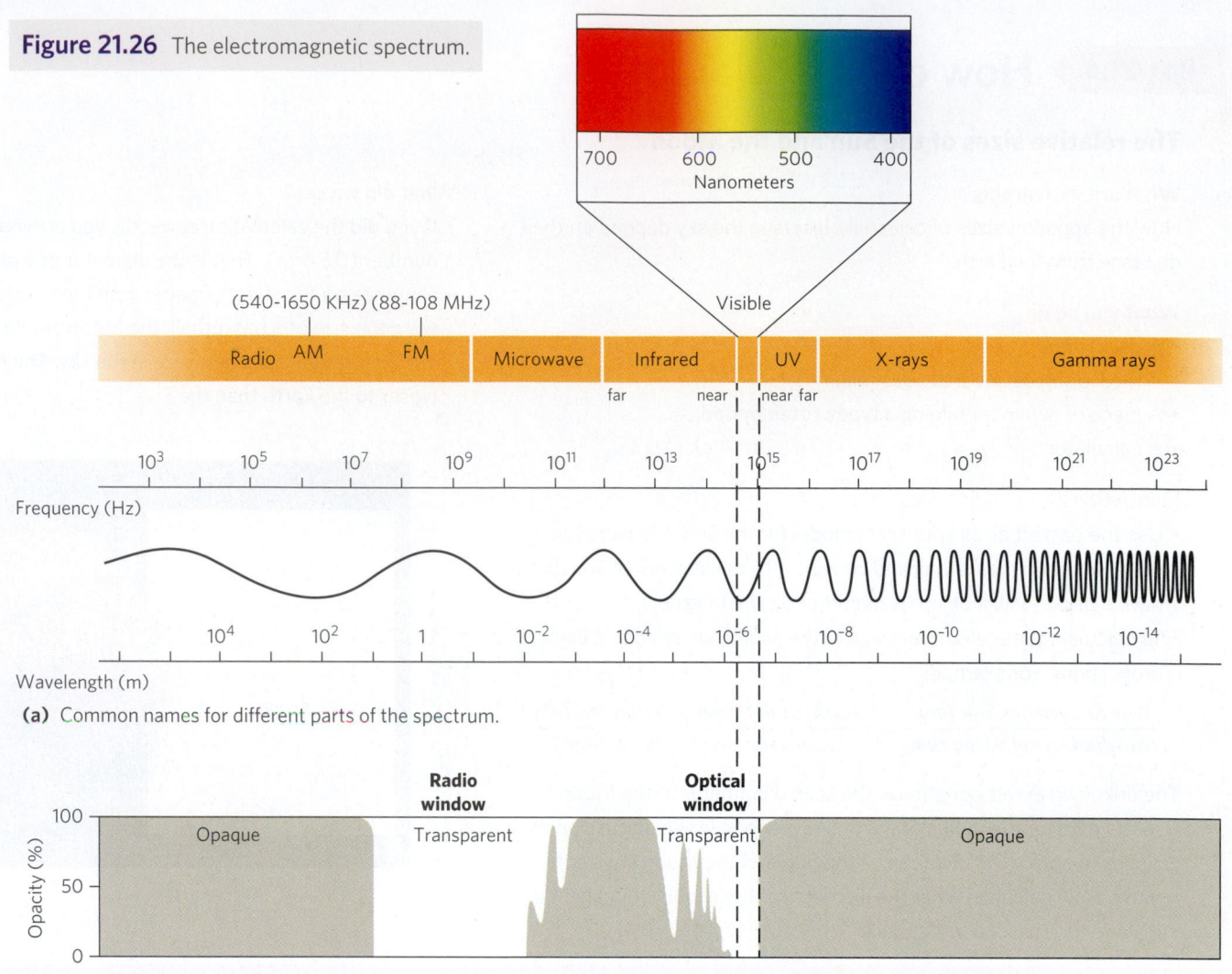

(a) Common names for different parts of the spectrum.

(b) Absorption of specific wavelengths by the atmosphere. Electromagnetic radiation from space can reach the Earth only through the radio and optical windows.

not contain all possible wavelengths, but rather displays distinct **spectral lines**, bright or dark bands at specific wavelengths **(Fig. 21.27b)**. Spectral lines can serve as a fingerprint to characterize the chemical composition of the radiation source. *Emission lines* show up as narrow bright bands in the spectrum. Each emission line corresponds to the wavelength of the energy emitted by a specific element within the radiation source. By examining emission lines, for example, astronomers can determine the specific elements glowing in a star. In fact, by comparing the brightness of the lines, they can estimate the relative abundance of the elements within the star. If the radiation emitted by a source gets absorbed by a material between the source and the Earth, distinct wavelengths of the radiation do not reach the Earth. When this happens, the spectrum displays dark bands, called *absorption lines*, at specific wavelengths. Astronomers can use absorption lines, for example, to determine the chemical composition of nebulae (gas and dust clouds) lying between a star and the Earth.

HOW FAST IS IT MOVING? Listen to the sound of a whistling train as it approaches you, passes you, and then moves away **(Fig. 21.28a)**. As it approaches, the sound gets louder, but its *pitch*, the note on the musical scale, remains constant. The instant that the train passes, however, the pitch of the sound changes to a lower note and stays at this note even as the train rumbles away and the sound gets softer. What's happening?

The pitch of sound depends on the frequency of the sound: higher-pitched sound has a higher frequency, and therefore a shorter wavelength, than does lower-pitched sound. As the train approaches, the sound waves "compact" into shorter wavelengths because both the source and the waves are moving toward the observer. Each successive wave has a shorter distance to travel, so the interval between the arrival time of successive waves decreases. An observer hears the higher-frequency waves as a higher pitch. When the train passes and starts moving away, sound waves from its whistle

Figure 21.27 Spectral lines.

(a) A prism can be used to break white light into a spectrum of colors.

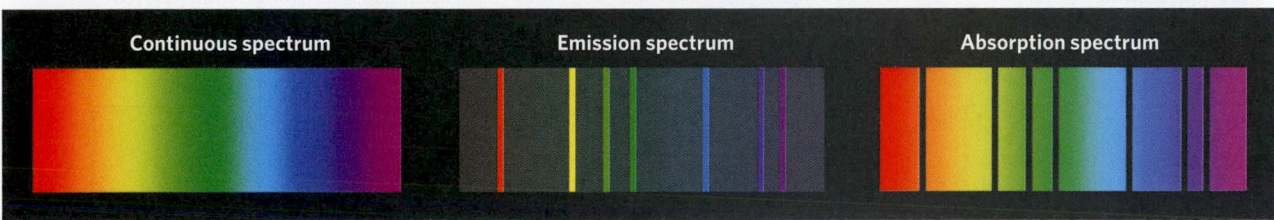

(b) The spectral lines of an element compared with the continuous visible spectrum. Bright lines are emission lines; dark lines are absorption lines. Note that emission and absorption lines occur at the same wavelength for a given element.

still move toward the observer, but they "stretch" because the whistle and the waves are moving in opposite directions. Each successive wave has a longer distance to travel, so arrival times spread farther apart. Therefore, the frequency of the arriving waves decreases, and the observer hears a lower pitch. The faster the train, the greater the frequency shift that the observer hears. Scientists refer to a shift in frequency caused by the movement of a source as the **Doppler effect**, named for the Austrian physicist Christian Doppler (1803–1853), who first explained it.

Just as the pitch of sound depends on the frequency of sound waves, the color of light depends on the frequency of electromagnetic waves. Electromagnetic waves can undergo a Doppler shift when emitted by objects moving toward or away from an observer. The faster the object travels, the greater the Doppler shift. A shift toward a higher frequency (shorter wavelength) happens when the object moves toward the observer, whereas a shift toward a lower frequency (longer wavelength) happens when the object moves away from the observer **(Fig. 21.28b)**. The wavelength of blue light is shorter than that of red light. Therefore, a light source moving toward an observer exhibits a **blue shift**, whereas a light source moving away from an observer exhibits a

red shift. In Box 17.2, we learned that meteorologists use the Doppler shift of radar beams to identify possible tornadoes. Astronomers use the Doppler shift to determine the movements of celestial objects.

To detect blue or red shifts, astronomers need to compare the spectrum received from a celestial object with the spectrum produced by a stationary source of light. They typically use the emission spectrum of our own Sun as this frame of reference **(Fig. 21.28c)**. For example, when we say that a distant galaxy exhibits a red shift, we mean that spectral lines have shifted toward the red end of the spectrum, relative to equivalent spectral lines displayed on the spectrum of light from our Sun.

How do astronomers use Doppler shifts to determine the velocities of celestial objects? Analysis of the Doppler effect for an individual star provides a basis for characterizing the star's rotation speed because the side of the star moving toward us exhibits a blue shift, while the side moving away from us exhibits a red shift, as the star spins on its axis. Doppler shifts can also indicate how fast a star is moving toward or away from the Earth. The same type of analysis can allow astronomers to determine how fast a galaxy is rotating and how fast it is moving toward or away from the Earth. When astronomers studied spectra from

Figure 21.28 The Doppler effect.

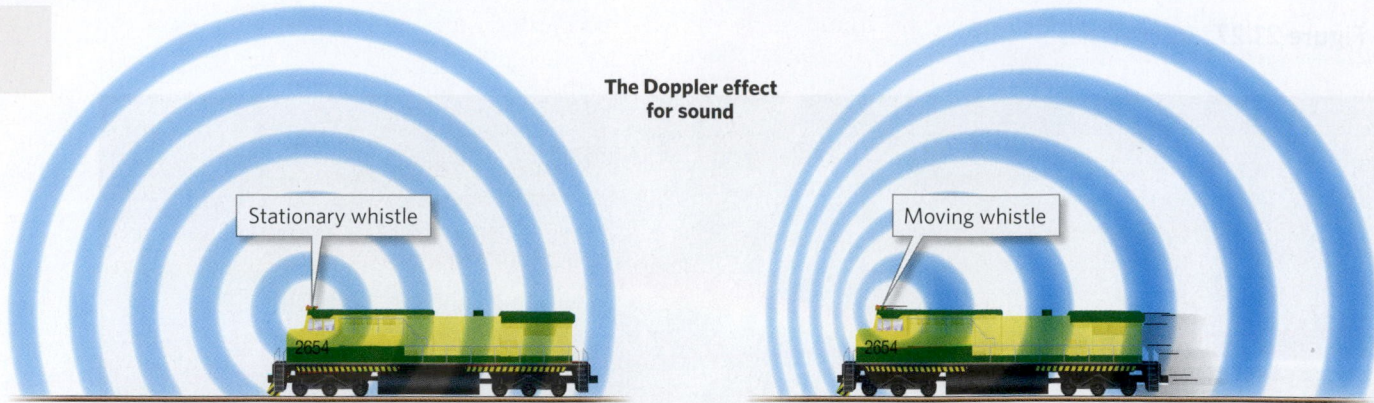

The Doppler effect for sound

Stationary whistle

Moving whistle

(a) The wavelength of the sound waves emitted by a stationary train whistle is the same in all directions. The sound waves behind a moving train whistle have longer wavelengths than those in front of it.

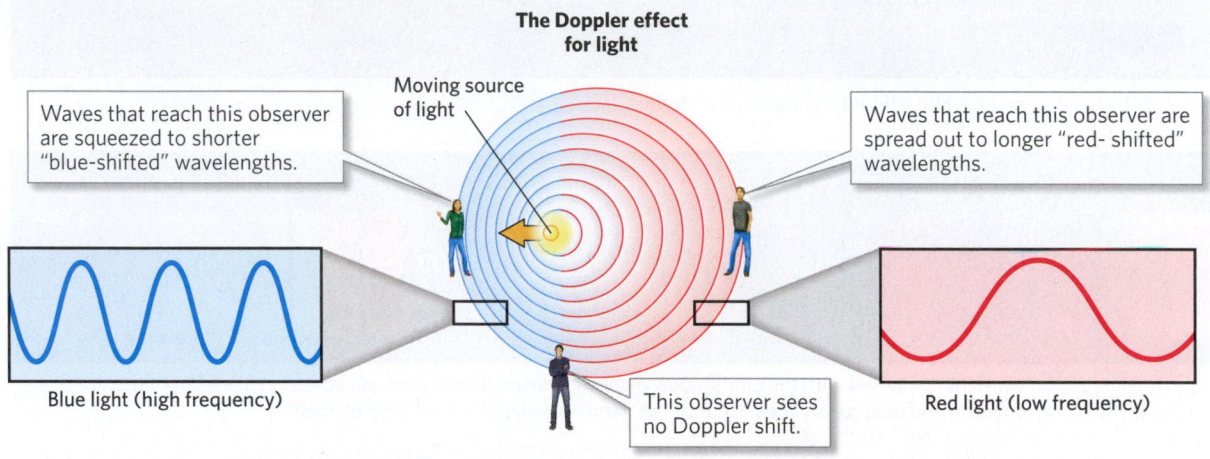

The Doppler effect for light

Moving source of light

Waves that reach this observer are squeezed to shorter "blue-shifted" wavelengths.

Waves that reach this observer are spread out to longer "red-shifted" wavelengths.

Blue light (high frequency)

This observer sees no Doppler shift.

Red light (low frequency)

(b) If a light source moves toward an observer, the radiation detected by the observer will be shifted toward shorter wavelengths (blue-shifted). If the light source moves away, the radiation will be shifted to longer wavelengths (red-shifted).

Sun

Distant galaxy

(c) The shift in wavelengths between radiation from a distant source, such as a galaxy or a star, and from the Sun can be used to determine the velocity of the source's movement toward or away from the Earth.

distant galaxies, they discovered that all display a red shift, indicating that all distant galaxies are moving away from the Earth. They also learned that galaxies farther from Earth are moving away faster. This realization led Edwin Hubble to propose the theory of the expanding Universe (see Chapter 1).

WHAT'S ITS TEMPERATURE? According to the laws of *blackbody radiation* that we discussed in Chapter 17, the wavelength at which a glowing object emits the greatest amount of energy depends on the surface temperature of the object. Hotter objects emit shorter wavelengths of light (toward the blue end of the spectrum), while cooler objects emit longer wavelengths of light (toward the red end of the spectrum). By using this concept,

astronomers have determined that blue stars have a surface temperature of 20,000°C (36,000°F), yellow stars have a surface temperature of 6,000°C (10,800°F), and red stars have a surface temperature of 3,000°C (5,400°F). (We'll take a closer look at these spectral classes of stars in Chapter 23.)

WHAT'S ITS SIZE? Astronomers define the size of a star by giving its radius as a multiple, or as a fraction, of our Sun's radius. Modern telescopes are sensitive enough to allow direct measurement of the sizes of a few nearby stars, using special techniques discussed in more advanced books. Astronomers using these telescopes have found a huge range of star sizes. For distant stars, direct measurement of a radius isn't possible because the resolution of currently available instruments is not sufficient. Instead, astronomers calculate the dimensions of these stars indirectly using equations that relate the temperature and brightness of a star to its surface area. From the surface area, simple geometric formulas yield the radius.

Take-home message . . .

Energy from stars passes through space in the form of electromagnetic radiation, but only part of this radiation makes it through the Earth's atmosphere. Astronomers use the wavelengths of radiation emitted by stars to determine their compositions, how fast they are moving toward or away from Earth, how fast they are spinning, and what their surface temperatures are.

Quick Question --------------------------------
What does the Doppler shift tell us about motion of a star or galaxy relative to the Earth?

21.5 A Sense of Scale: The Vast Distances of Space

The Units That Astronomers Use

Though the ancient Greeks used basic geometry to figure out the radius of the Earth and the distance to the Moon, they knew of no way to determine distances to the Sun, planets, or stars. As we'll see in this section, modern astronomers now have a set of tools to measure such distances, and they have shown that we live in a Universe of vast dimensions. Even nearby celestial objects lie far away. For example, the Sun lies nearly 150 million km (93 million miles) from the Earth. If Pheidippides, the first person to run a marathon, had not died in 490 B.C.E., but was immortal and ran a marathon (42 km, or 26 miles) every day for the past 2,500 years, he would be only about a quarter of the way to the Sun by now. The nearest star to our Sun, Proxima Centauri, glows from a distance of 40 trillion km (25 trillion miles) away. A person aboard the *Voyager 1* spacecraft, now speeding through interstellar space at about 62,000 km per hour (38,500 mph), would take about 73,000 years to reach this star.

To describe the incredible distances between objects in space, astronomers use three units that are much bigger than the familiar ones we use in daily life. The first, called an **astronomical unit** (**AU**), represents the average distance from the center of the Earth to the center of the Sun and equals about 150 million km (93 million miles). The second, called a **light-year**, became available when researchers determined that light travels at a constant, definable speed of about 300,000 km (186,000 miles) per second in a vacuum. Because light moves at a constant speed, it travels a specific distance in a given time. A light-year, the distance light travels through space in one year, equals about 9.5 trillion km (5.9 trillion miles). Nontechnical discussions of astronomy generally describe distances using light-years because it's fairly easy to grasp this concept. Professional astronomers, however, use an even larger unit, the *parsec*—roughly 3.3 light-years—for measuring distances. We'll discuss the basis for defining a parsec later in this section.

Because light takes time to travel, we're looking into the past when we look into space (Fig. 21.29). The light we see when looking at the Moon left the Moon 1.3 seconds ago, light arriving from the Sun left the Sun 8 minutes ago, and light arriving from Alpha and Beta Centauri left the stars 4.2 years ago. The farther we peer into deep space, the further we peer back into deep time. Light arriving from the most distant visible object in the Universe left that object billions of years before the Earth even existed.

How Can We Measure Distances in Space?

Modern astronomers can measure distances from the Earth to the planets, to the stars, and even to galaxies that are millions of light-years from the Earth. How do they do it? Astronomers today rely on five basic tools for measuring the great distances of space.

RADAR. Astronomers can use radar to determine the distances to objects in the Solar System. To make a radar measurement, they use a transmitter to send a pulse of radio waves out from an antenna. When the pulse bounces off an object, it returns to the antenna and can be detected by a receiver. Since a radar pulse travels at the speed of light, the distance to the object can be calculated by precisely measuring the time between transmission of the pulse and reception of the echo.

GEOMETRIC PARALLAX. If you hold your thumb out at arm's length, and first close one eye and then the other, you'll see an example of **geometric parallax**, the apparent displacement of a foreground object relative to background objects when an observer's position, and therefore their line of sight, changes (Fig. 21.30a). The movement of the Earth from one side of its orbit around the Sun to the other side provides enough distance so that astronomers can detect the slight geometric parallax that closer stars display relative to more distant stars (Fig. 21.30b). By precisely measuring the angles at which these changes occur, astronomers can use trigonometry to calculate the distance to relatively nearby stars, meaning those less than 500 light-years away.

The concept of geometric parallax provides the basis for defining a parsec. A **parsec** represents the distance that an imaginary star would have to be from Earth for its parallax to be exactly one arc-second (1/360 of a degree) if an observer

Did you ever wonder . . .
how astronomers can tell how far a star is from Earth?

Figure 21.29 Looking into deep space and back in time.

(a) The Sun, our nearest star, is about 150 million km away. Light from the Sun takes about 8 minutes to arrive at the Earth.

(b) Jupiter, the largest planet in the Solar System, is 778 million km from the Sun. Light from Jupiter takes about 42 minutes to reach the Earth.

(c) By contrast, Alpha and Beta Centauri, the nearest stars to our Sun, are 39,740,000,000,000 km (4.2 light-years) from the Earth. Light from Alpha and Beta Centauri takes 4.2 years to reach the Earth.

(d) Our home galaxy, the Milky Way, is 100,000 light-years wide. Light arriving at the Earth from the edge of the Milky Way left that location 100,000 years ago.

(e) Magnification of a tiny area in a Hubble Space Telescope image of distant galaxies reveals the furthest object yet found in the known Universe. The light in this image traveled through space for 13.3 billion years before it reached the lens of the telescope.

were to move from a point at the Sun's center to a point on the Earth's surface, one AU away.

STELLAR BRIGHTNESS. The geometric parallax method can't be used to determine how far away a distant star lies because the displacement of the star against its background is too small to detect. For distant stars, astronomers rely on the simple relationship between apparent brightness and distance. The *apparent brightness* of a light source (the brightness you see) decreases by the square of its distance from an observer because its light energy spreads over an increasing area, even if its true brightness (the amount of energy emitted) remains unchanged (we'll discuss these

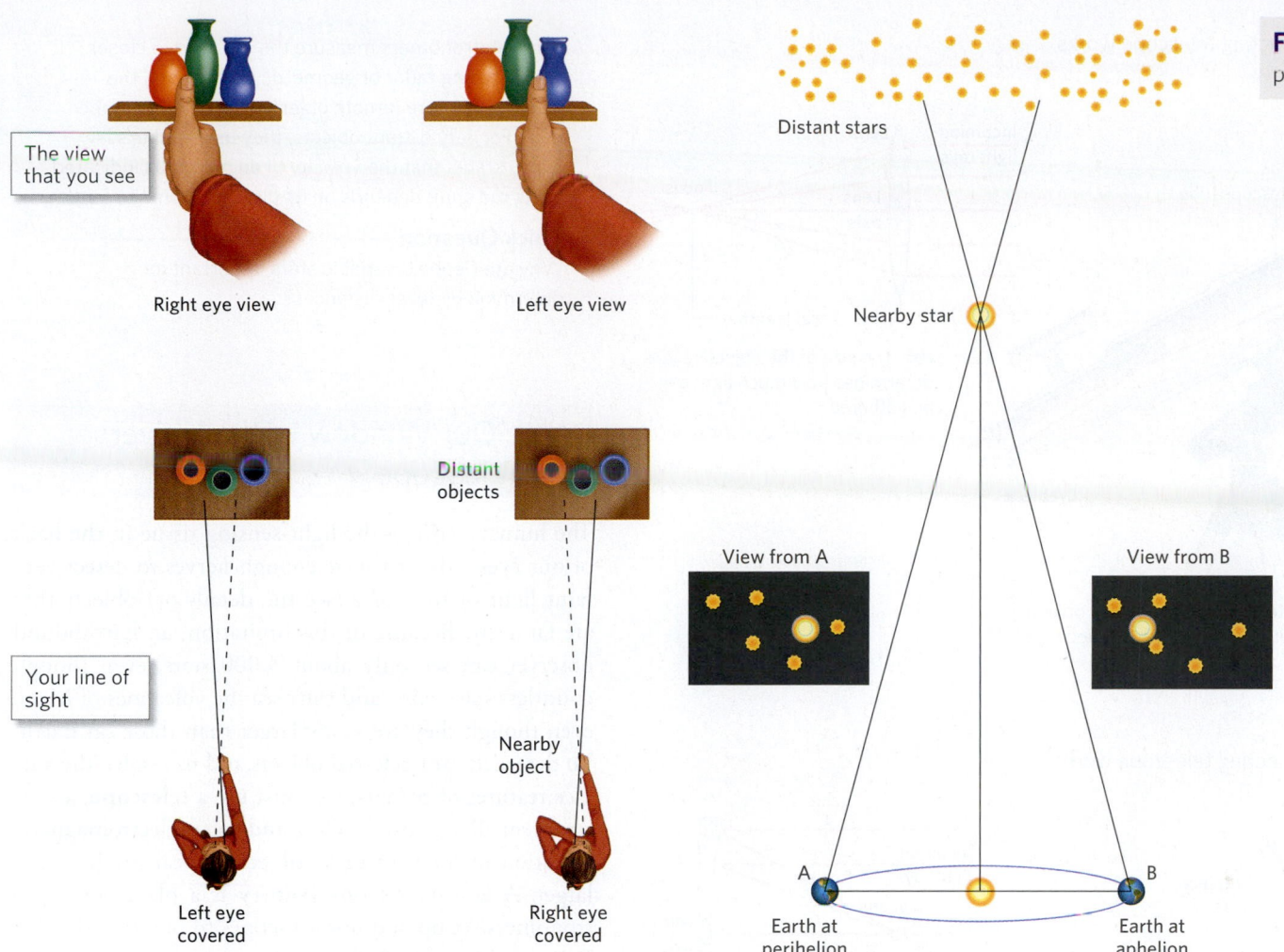

The view that you see

Right eye view

Left eye view

Distant objects

Distant stars

Your line of sight

Left eye covered

Right eye covered

Nearby object

Nearby star

View from A

View from B

Earth at perihelion

A

B

Earth at aphelion

Figure 21.30 Geometric parallax.

(a) You can observe geometric parallax by holding your thumb out at arm's length. When you close one eye and then the other, your thumb covers different distant objects.

(b) A nearby star shifts its position relative to the background stars when viewed from opposite sides of the Earth's orbit.

discuss these measurements of brightness in more detail in Chapter 23). Astronomers can estimate the true brightness of a star from measurements of the star's spectrum and radius, so they can calculate the star's distance from the Earth by measuring how bright it appears from the Earth.

CEPHEID VARIABLES. Measurements of brightness can determine stellar distances up to about 10,000 parsecs. Measurements of even greater distances come from the study of special kinds of stars that pulsate, meaning that they expand and contract periodically. When a pulsating star expands, it has a larger surface area, but its temperature doesn't change, so its apparent brightness increases. Henrietta Leavitt (1868–1921) discovered that one class of pulsating stars, known as *Cepheid variables*, display a predictable relationship between true brightness variation and the periodicity of their pulsations. So, if astronomers know the periodicity of a Cepheid variable star's pulsations (usually between 1 and 100 days), they also know its true brightness. Then, by comparing this true brightness with the apparent brightness of the star, astronomers can calculate the star's distance.

If a Cepheid variable star is within a distant galaxy, the distance from the Earth to the galaxy can be measured by examining the brightness of that star.

DOPPLER SHIFTS. For very distant galaxies, none of the techniques we have described so far will work. A breakthrough came when astronomers measured red shifts associated with closer galaxies using Cepheid variable stars. They determined that with greater distance between a galaxy and the Earth, the galaxy displays a greater red shift, and therefore, the velocity of its movement away from the Earth is greater. This relationship can be expressed in the form of an equation: $v = Hd$. (In this equation, v is velocity, H is a constant, and d is distance.) Edwin Hubble studied these phenomena and determined the value of H. To recognize the importance of Hubble's works, astronomers now refer to H as *Hubble's constant*, and to the equation relating velocity to distance for distant galaxies as **Hubble's law**. Once Hubble's law had been calibrated using Cepheid variables, it could be used to measure the distance to the most remote objects in space.

Figure 21.31 How a refracting telescope works.

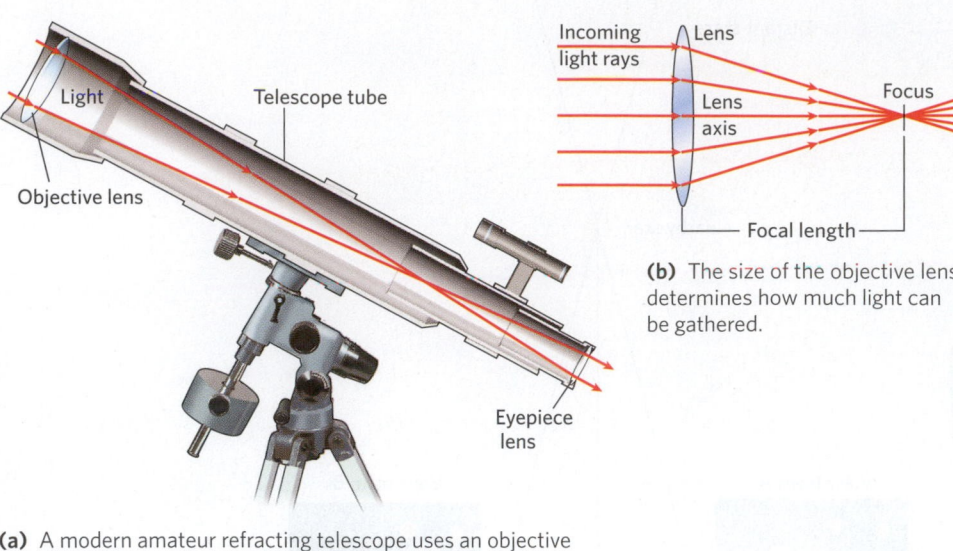

(a) A modern amateur refracting telescope uses an objective lens to gather light and direct it to a focus. The eyepiece magnifies the light.

(b) The size of the objective lens determines how much light can be gathered.

Figure 21.32 How a reflecting telescope works.

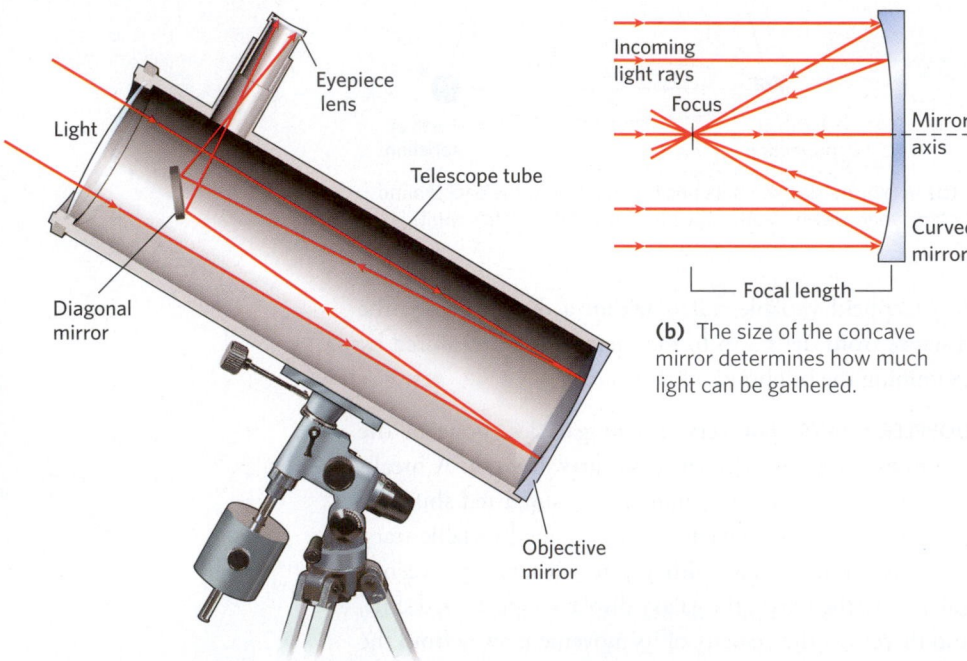

(a) A reflecting telescope uses a concave mirror to gather light and direct it to a focus.

(b) The size of the concave mirror determines how much light can be gathered.

Take-home message . . .

Astronomers use three units of measure to describe the vast distances in the Universe: astronomical units, light-years, and parsecs. The method used to determine distances to objects in space depends on the distances of those objects from the

Earth. Astronomers measure the distance to closer objects using radar or geometric parallax and the distance to more remote objects using stellar brightness. For very distant objects, they use Hubble's law, which states that the velocity of an object, as indicated by its red shift, depends on its distance from the Earth.

Quick Question --------------------------------
Why are Cepheid variable stars important for determining stellar distances?

21.6 Our Window to the Universe: The Telescope

The human retina—the light-sensing tissue in the back of our eyes—doesn't have enough nerves to detect very faint light or to *resolve* (see the details of) objects that are far away. Because of this limitation, an Earthbound observer can see only about 3,000 stars, even though countless stars exist, and can't see the volcanoes of Mars, even though they are vastly larger than those on Earth. To detect distant celestial objects and to resolve the surface features of planets, we must use a **telescope**, an instrument designed to collect and focus electromagnetic radiation in order to make objects appear brighter and larger. A land-based **observatory** is a place where astronomers set up and use telescopes to observer the sky. Because the atmosphere permits transmission of visible light and radio waves, astronomers can use both **optical telescopes**, which collect and magnify visible light, and **radio telescopes**, which collect and amplify radio waves. **Space telescopes**, which observe space from satellites orbiting above the atmosphere, can detect other wavelengths of the electromagnetic spectrum, such as X-rays and gamma rays (**Earth Science at a Glance**, pp. 770–771). To understand how telescopes work, let's look at the designs of a few different types.

Optical Telescopes

TYPES OF OPTICAL TELESCOPES. Optical telescopes work by gathering all the light striking a broad area and concentrating it into a small area. Telescopes can be configured to collect light in two different ways. *Refracting* telescopes were developed first (in 1608), followed by *reflecting* telescopes (in 1668).

When Galileo spotted the moons of Jupiter for the first time, he used a **refracting telescope**. In this format, an *objective lens*—a disk of glass that has been ground and polished so that it has curved surfaces on its top and/or bottom—refracts incoming light and concentrates it at a *focus* behind the lens (**Fig. 21.31**). A second lens, the *eyepiece*, magnifies the concentrated light so that objects appear larger. Most hand-held telescopes are refracting telescopes. The quality of a refracting telescope depends on how clear and how perfectly curved its lenses are, and its light-gathering ability depends on the size of the objective lens.

When Newton began studying optics, he developed the **reflecting telescope**, in which a curved mirror reflects light to a focus in front of the mirror **(Fig. 21.32)**. The larger the area of the mirror, the more light the telescope can gather. A smaller mirror at the focus sends the concentrated light to an eyepiece at the back or at the side of the telescope.

Astronomers using a modern research telescope don't stare through the eyepiece directly, but rather examine photographs of the light coming through the eyepiece. Large research telescopes **(Fig. 21.33)** can be moved by motors so that they stay aimed at exactly the same spot in space for a long time, even as the Earth spins. Therefore, the camera can accumulate all the light collected for many minutes or even hours. Prior to the electronic age, cameras used film to record light coming through the eyepiece. Today, film has been replaced by an electronic screen containing millions of light-sensitive pixels. Such a screen can be configured to produce a digital visual image or to serve as a photon counter that can accurately determine the brightness of an object.

CONSIDERATIONS IN TELESCOPE DESIGN AND LOCATION. The ability of a ground-based optical telescope to collect light depends on the objective lens size in a refracting telescope, and on the mirror size in a reflecting telescope. Because it's difficult and expensive to grind lenses, and because lenses are very heavy and can warp due to their own weight, the largest objective lens ever made for a refracting telescope has a diameter of only 100 cm (40 inches). It's easier to design and manufacture large curved mirrors, and such mirrors can be supported from behind to prevent warping. As a result, reflecting telescopes can be much bigger than refracting telescopes, and today's observatories rely on reflecting telescopes for their research. The largest reflecting telescope mirror has a diameter of 10.4 m (409 inches).

Even if a telescope has great light-gathering power, the telescope's **resolving power**—its capacity to distinguish between two faraway objects that lie close together—limits an observer's ability to discern details. The resolving power of a telescope can be affected by several phenomena, including *light diffraction*, which happens when straight waves bend into curves due to interaction with an obstacle. The best resolution possible in a modern telescope permits an observer to distinguish two coins placed side by side at a distance of 60 km (37 miles).

Figure 21.33 Large research telescopes.

(a) The Hale Telescope at the Palomar Observatory has a mirror that is 5.1 m across.

(b) At night, the dome opens at the McDonald Observatory in Texas.

Even with the best telescopes, ground-based astronomers must contend with **atmospheric seeing**, the blurring or fuzzing of an object's image that happens because light from the object had to pass through the atmosphere **(Fig. 21.34)**. Random variations in air density associated with turbulence cause incoming light to refract in an irregular way. Atmospheric seeing can be reduced by using *adaptive optics*, a technology that uses computers to warp mirrors slightly as needed to compensate for distorted light. This technology has vastly improved the quality of images obtained by ground-based telescopes.

Any atmospheric phenomenon that decreases atmospheric clarity, telescope stability, or the darkness of the sky limits the image quality provided by ground-based telescopes. Snow, rain, clouds, and wind, for example,

Did you ever wonder . . .

why astronomers like to locate observatories on isolated mountaintops?

Figure 21.34 Atmospheric seeing. The path of light passing between space and an observer at the ground is altered by atmospheric turbulence, causing images to go out of focus.

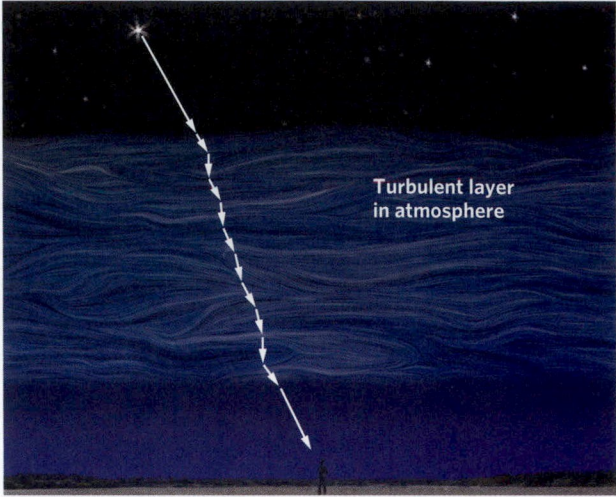

Turbulent layer
in atmosphere

Figure 21.35 Light pollution.

(a) A photo of the United States from space at night shows the extent of urban areas with bright lights.

(b) From the ground, in a well-lit city, only the brightest celestial objects are visible, even on a clear night.

obscure views and blur images. In recent decades, **light pollution** (the glow of the sky due to reflection and scattering of light from sources such as streetlights, windows, and billboards) has become a significant problem for observatories. This glow can wash out faint starlight coming from space **(Fig. 21.35)**.

For all these reasons, when locating observatories, astronomers must take into account elevation, climate, and proximity to urban areas. The largest observatories built in the last century perch on isolated mountaintops in arid climates, where there is less air between the telescope and space, so there's less atmospheric seeing, less bad weather, and less light pollution **(Fig. 21.36)**. Examples include the Paranal Observatory of Chile, the Mauna Kea Observatory of Hawaii, the Kitt Peak National Observatory of Arizona, and the Roque de los Muchachos Observatory on the Canary Islands. All of these observatories lie at elevations between 2 and 5 km (1.2 and 3.1 miles) above sea level.

Radio Telescopes

In the late 1920s, Karl Jansky, a scientist at Bell Labs, began to investigate sources of *static* (unwanted electronic noise) that interfered with Earth-based radio transmissions. Jansky built a dish-shaped antenna—shaped much like the mirror of a reflecting telescope—capable of collecting very weak radio signals, and pointed it skyward. He discovered that some of the static came from thunderstorms. But he also recorded a faint, mysterious background hiss. After measuring this signal over a period of several days, he realized that a particularly strong hiss repeated every 24 hours, and that this hiss appeared when the telescope faced a particular location in space. Jansky concluded that the hiss must be coming from space to the Earth, and he eventually determined that the loudest hiss originated in the galactic center of the Milky Way. In effect, Jansky's dish-like antenna was detecting electromagnetic radiation that had come from space through the radio window of the atmosphere, so it can be considered to be the first radio telescope.

Modern radio telescopes work much like reflecting telescopes. They consist of a dish that reflects incident radio-wave energy to a focus, where instruments collect and analyze the energy. Because radio-wave energy coming from space is weak and the wavelengths of this energy are long, radio telescopes need to be very large **(Fig. 21.37a,b)**. In recent decades, astronomers have built **radio interferometers** that superimpose the waves collected by numerous radio telescopes arranged in an array **(Fig. 21.37c)**. Radio interferometers effectively act like a single huge radio telescope whose diameter equals the distance from one side of the array to the other.

Even with huge antennas **(Fig. 21.37d)**, the resolution of radio telescopes can't come close to that of optical telescopes. Nevertheless, radio telescopes have advantages. The Sun does not emit significant radio-wave energy, so radio telescopes can be used both day and night. Also, since radio waves pass through clouds and precipitation, radio telescopes work in all types of weather. Finally, radio telescopes can detect some objects that cannot be seen with optical telescopes. For example, nebular dust absorbs visible light coming from the center of the Milky Way, so we can't see it optically. Radio waves pass through this dust, so radio telescopes can detect energy sources at the galactic center.

Space Telescopes

A *space telescope*, one placed on a satellite or spacecraft that orbits above the Earth's energy-absorbing atmosphere, has the distinct advantage of being able to detect any part of the entire electromagnetic spectrum **(Box 21.5)**. Astronomers use space telescopes, therefore, to detect particularly short wavelengths (X-rays and gamma rays) and particularly long wavelengths (infrared radiation) that can't pass through the optical or radio windows to the ground. As a result, space telescopes can detect particularly hot objects (which emit short-wavelength energy), as well as particularly cold ones (which emit long-wavelength energy). Furthermore, because space telescopes do not have to contend with atmospheric seeing, they can

Figure 21.36 Examples of large observatories that employ optical telescopes.

European Southern Observatory, Chile

Mauna Kea Observatory, Hawaii

Roque de los Muchachos Observatory, Spain

Kitt Peak Observatory, Arizona

Observatories of the World

Hubble Space Telescope

Visible

James Webb Space Telescope

Infrared

Spitzer Space Telescope

Radio

Ground-based radio telescopes

Astronomical observatories on the Earth and in space observe the Universe using a wide array of telescopes that employ different wavelengths.

For ground-based telescopes, the wavelengths used are those for which the atmosphere is transparent to radiation arriving from space. Two types of ground-based telescopes are radio telescopes and the more traditional, optical telescopes.

Scientists use radio telescopes, which employ large antennas, to study our galactic center, distant objects beyond our galaxy, and other strong radio-wave sources. Radio telescopes can be used in the daytime as well as at night. They study astronomical radio-wave sources, such as stars, nebulas, and galaxies, in the distant reaches of the Universe. Optical telescopes are used to image a wide array of objects that emit light in visible wavelengths that human eyes can detect. Optical telescopes have a long history of opening windows to the Universe, from the first observations by Galileo to the sophisticated mountaintop observatories of today.

Chandra X-ray Observatory

Fermi Gamma-ray Space Telescope

Gamma

X-ray

UV

Space observatories, on the other hand, have no limitations with regard to wavelength. Telescopes designed for these observatories typically sample a small range of wavelengths, with each observatory focusing on a different region of the electromagnetic spectrum.

In the infrared spectrum, the Spitzer Space Telescope has revealed the existence of extrasolar planets and allows scientists to study forming stars, while the James Webb Space Telescope studies the history of the Universe, from the first luminous glow after the Big Bang, to the formation of solar systems capable of supporting life. The Hubble Space Telescope (dealing with the visible spectrum) has astounded the world with fantastic views of space, from star nurseries shrouded with dust, to clusters of galaxies at the far reaches of the Universe. The Chandra X-ray Observatory allows scientists to study very hot regions of the Universe, such as exploded stars, clusters of galaxies, and matter around black holes. The Fermi Gamma-ray Space Telescope targets strange phenomena of the Universe, such as supermassive black holes, neutron stars, and streams of hot gas moving close to the speed of light.

Ground-based optical telescopes

Figure 21.37 Radio telescopes and radio interferometers.

(a) An air view shows the 300-m (900-foot) wide Arecibo Observatory in Puerto Rico. It was made by smoothing a sinkhole in a karst landscape.

(b) A close-up of the Arecibo receiver. This 900-ton instrument dangles 137 m (450 feet) above the dish.

(c) The Very Large Array, a radio interferometer in New Mexico.

(d) The telescope at the National Radio Astronomy Laboratory in West Virginia has a dish 105 m (16 feet) in diameter.

produce amazingly sharp images. Space telescopes are, however, expensive and difficult to operate.

Five satellite-based telescopes have been launched so far, and some remain in operation today. The James Webb Space Telescope will be launched in 2018. Space telescopes have not only provided insight into the architecture of galaxies and other celestial objects, but have also yielded thousands of mind-boggling images of distant galaxies and nebulae. With every new image released, people the world over gasp at the astounding beauty of the Universe.

Take-home message . . .

A telescope can gather much more electromagnetic radiation than a human eye can, allowing us to detect and resolve objects deep in space. Optical telescopes collect visible light and provide visual images. Radio telescopes collect radio waves and can detect some objects that cannot be seen with optical telescopes. Space-based telescopes have allowed astronomers to view all parts of the electromagnetic spectrum and have led to revolutionary discoveries about the nature of the Universe.

Quick Question -
Why can we detect energy from the center of the Milky Way with a radio telescope, but not with an optical telescope?

Box 21.5

Consider this . . .

NASA's space telescopes

Astronomers working with the US National Aeronautics and Space Administration (NASA) used the space shuttle fleet to place five scientific space telescopes in orbit. Each was named in honor of a famous astronomer or physicist. With their capacity to detect all parts of the electromagnetic spectrum, these telescopes have revolutionized our understanding of our Universe.

The first, most famous space-based observatory, the Hubble Space Telescope, was launched in 1990 **(Fig. Bx21.5a)**. It's the only space telescope to be serviced in space by astronauts. The original mirror of Hubble was flawed and could not provide sharp images, so engineers designed a separate, corrective lens (effectively, a pair of glasses) that could focus Hubble's image. Once the new lens was installed, the telescope worked spectacularly. Hubble has given us amazingly sharp visual images of celestial objects **(Fig. Bx21.5b)**, as well as improved measurements of the distances to stars and a better estimate of the rate at which the Universe is expanding. When Hubble turned its lenses toward patches of the night sky once thought to be empty, it found countless new galaxies, billions of light-years away.

Engineers have designed each space telescope for a distinct mission. For example, the Compton Gamma Ray Observatory, launched in 1991, detected gamma rays, the most powerful form of radiation, generated by gases with temperatures exceeding 10 billion degrees centigrade (18 billion degrees Fahrenheit). Its observations yielded insight into new classes of very energetic objects that researchers are still struggling to understand. The observatory operated until 2000, when operators intentionally sent it plummeting into the Pacific Ocean. The Chandra X-ray Observatory, placed in orbit by the space shuttle *Columbia* in 1999, also detects radiation at the high-energy end of the spectrum and has provided new insight into the nature of exploding stars and their products. Launched in 2003, the Spitzer Space Telescope detects infrared radiation, the cooler end of the spectrum, and thus has the capability of studying the nebulae in which new stars form. The Kepler Space Telescope has made headlines since its 2009 launch, for its instruments have detected thousands of *exoplanets*, planets orbiting other stars. Based on observations made by Kepler, astronomers now estimate that as many as 40 billion Earth-sized planets may be in the Milky Way.

Figure Bx21.5 The Hubble Space Telescope.

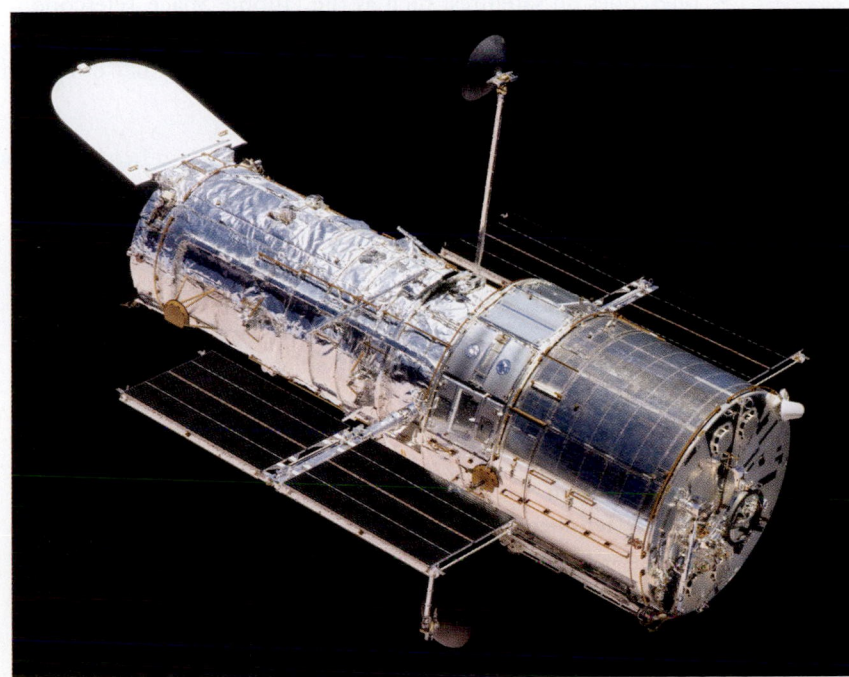

(a) The telescope.

(b) The Pillars of Creation, a spectacular Hubble image of the birth place of stars.

21 CHAPTER REVIEW

Chapter Summary

- Prior to the Renaissance, most people thought that the Earth was the center of the Universe. Giants of science—Copernicus, Galileo, Kepler, and Newton—showed through observation and theory that the Sun lay at the center of the Solar System.

- Discoveries in the modern era have shown that the Sun is a star in the Milky Way Galaxy, one of hundreds of billions of galaxies.

- We can plot the relative positions of objects on the celestial sphere.

- We can divide the night sky into recognizable groups of stars called constellations. The zodiac consists of 12 constellations located along the ecliptic plane.

- Our Moon completes a cycle of phases about every 29 days. When the sunlit side of the Moon faces the Earth, the Moon is full. When it faces away, the Moon is new.

- Solar eclipses occur when the Moon lies between the Earth and the Sun. Lunar eclipses occur when the Earth lies between the Sun and the Moon.

- The electromagnetic spectrum consists of all wavelengths of radiation. Only two portions of the electromagnetic spectrum can pass through the atmosphere, defining the optical window and the radio window.

- Scientists study the spectra of light coming from celestial objects using a spectroscope. This instrument allows detection of spectral lines, which provide information about the composition of energy sources.

- Spectral lines from moving sources shift slightly due to the Doppler effect. Scientists measure the shift to determine how fast stars or galaxies are moving relative to the Earth.

- To describe vast distances in space, we can use one of three units of measure: the astronomical unit or AU (the distance from the Earth to the Sun), the light-year (the distance light travels in a year), or the parsec (3.3 light-years).

- An optical telescope gathers and magnifies visible light. Refracting optical telescopes use an objective lens to focus light. Reflecting telescopes use a mirror.

- Larger telescopes are better light gatherers, but diffraction and atmospheric seeing limit the resolution of optical telescopes. Modern astronomers collect light coming through a telescope over time by using a digital camera.

- Most research telescopes are built on isolated mountaintops in arid climates to minimize the effects of weather, light pollution, and atmospheric seeing.

- Radio telescopes use dish-like antennae to gather weak radio-wave energy from sources throughout and beyond our galaxy.

- Telescopes on satellites, such as the Hubble Space Telescope, allow astronomers to use the full electromagnetic spectrum to study the Universe, and they can provide spectacular images of deep space.

Key Terms

astronomer (p. 741)
astronomical unit (AU) (p. 763)
astronomy (p. 740)
atmospheric seeing (p. 767)
axis of rotation (p. 748)
blue shift (p. 761)
celestial object (p. 740)
celestial sphere (p. 747)
constellation (p. 751)
Copernican model (p. 742)
cosmology (p. 741)
Doppler effect (p. 761)
ecliptic plane (p. 750)
electromagnetic spectrum (p. 758)
electromagnetic wave (p. 758)

ellipse (p. 744)
frequency (p. 758)
galaxy (p. 746)
geocentric model (p. 741)
geographic pole (p. 748)
geometric parallax (p. 763)
heliocentric model (p. 742)
horizon (p. 748)
Hubble's law (p. 765)
Kepler's laws (p. 743)
light pollution (p. 768)
light-year (p. 763)
lunar eclipse (p. 756)
lunar month (p. 755)
nebula (p. 746)
observatory (p. 766)

optical telescope (p. 766)
optical window (p. 759)
orbit (p. 741)
parsec (p. 763)
phase (p. 755)
photon (p. 758)
precession (p. 751)
prograde motion (p. 741)
radio interferometer (p. 768)
radio telescope (p. 766)
radio window (p. 759)
red shift (p. 761)
reflecting telescope (p. 767)
refracting telescope (p. 766)
resolving power (p. 767)
retrograde motion (p. 741)

rotation (p. 748)
sidereal day (p. 749)
solar day (p. 749)
solar eclipse (p. 756)
Solar System (p. 746)
space telescope (p. 766)
spectral line (p. 760)
spectroscope (p. 759)
telescope (p. 766)
Universe (p. 740)
visible light (p. 758)
wavelength (p. 758)
zenith (p. 748)
zodiac (p. 752)

Review Questions

The letters following each Review Question refer to the corresponding Learning Objective from the Chapter Opener.

1. Contrast the geocentric model of the Solar System with the heliocentric model. What problem with the geocentric model led Ptolemy to invent epicycles? **(A)**

2. Describe the contributions of Copernicus, Galileo, and Kepler to modern astronomy. **(A)**

3. What is the celestial sphere? What is the ecliptic plane, and how can you determine its location on the celestial sphere? **(B)**

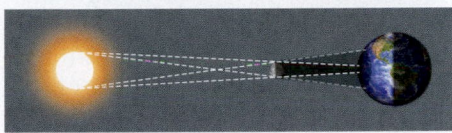

4. Identify the perihelion and aphelion in the diagram. Does the Earth move faster in its orbit when it is at perihelion or aphelion? **(A, B)**

5. How do the orientations of the orbital planes of other planets compare with the Earth's ecliptic plane? **(B)**

6. What are constellations? What determines which constellations constitute the zodiac? **(C)**

7. What is a solar eclipse, and what is the difference between a total eclipse and a partial eclipse? Identify the areas where a total and a partial eclipse would be seen in the diagram. **(D)**

8. What is a lunar eclipse, and why can most people see a total lunar eclipse? **(D)**

9. What is meant by an atmospheric window in the electromagnetic spectrum? What atmospheric window do earthbound astronomers use to peer into space? **(E)**

10. What is a spectroscope? What are the spectral lines visible in the spectrum produced by a spectroscope, and what do those spectral lines tell us about the composition of stars? **(E)**

11. What information can astronomers obtain about stars and galaxies using the Doppler effect? **(E)**

12. How do astronomers determine the surface temperature of a star? **(E)**

13. What is the difference between an astronomical unit and a light-year? Which is best for measuring the distance between objects in our Solar System? **(F)**

14. What is geometric parallax, and how do astronomers use it? **(F)**

15. Why do large astronomical observatories all use reflecting telescopes? **(G)**

16. How does a radio telescope work? Why must radio telescopes be so large? **(G)**

17. Give two reasons why the world's great astronomical observatories are all located on isolated mountaintops. **(G)**

18. What advantages do space telescopes have over earthbound telescopes? **(G)**

On Further Thought

19. Imagine you are an astronomer who has just used a spectroscope to measure the emission lines of radiation from a distant galaxy. When you look at the emission lines of hydrogen, you note that all of the lines you measured are shifted from their expected positions toward the red end of the spectrum. Is the galaxy moving toward you or away from you? **(E)**

20. Why do you think cell phones and weather radar systems transmit signals in the radio and microwave regions of the electromagnetic spectrum, rather than the infrared or ultraviolet regions? **(E)**

Online Resources

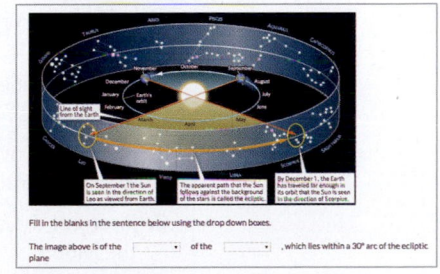

Videos
This chapter features real-world videos on lunar eclipses and the moon's motions over the course of a year.

Smartwork5
This chapter features questions on the paths and features of the Moon, planets, and other celestial objects; the evolution of astronomical thought; and the tools and processes for exploring the cosmos.

⊙22 OUR NEIGHBORHOOD IN SPACE
The Solar System

By the end of the chapter you should be able to . . .

A. define a planet and understand the difference between true planets and dwarf planets.

B. explain the overall structure of the Solar System and why that structure developed.

C. interpret the surface features of our nearest neighbor, the Moon.

D. distinguish the terrestrial planets from one another.

E. identify the key characteristics of the gas-giant and ice-giant planets.

F. recognize that the Solar System hosts a great variety of moons.

G. describe the origin and characteristics of smaller objects in the Solar System.

H. visualize where comets and meteoroids come from and evaluate the threat they pose.

22.1 Introduction

On August 24, 2006, the International Astronomical Union (IAU), the world's principal organization of professional astronomers, held a fateful vote. When the ballots had been counted, Pluto had been impeached. Henceforth, it would no longer be a planet. For three generations, students had memorized the names of the nine planets of the Solar System in order of their distance from the Sun: Mercury, Venus, Earth, Mars, Jupiter, Saturn, Uranus, Neptune . . . and Pluto. The next generation of students would stop at eight **(Fig. 22.1)**.

Clyde Tombaugh, a 23-year-old amateur astronomer, discovered Pluto in 1930. Headlines blared that "Planet X" had been found, and a schoolgirl, thinking about how lonely it must be so far from the Sun, suggested

naming it after the god of the underworld. Pluto might still be the ninth planet were it not for the discovery in the 1990s that Pluto is, in fact, only one object in a broad ring of icy objects known as the *Kuiper Belt*. If Pluto is a planet, then are all other large Kuiper Belt objects planets, too? This was the dilemma facing the IAU, so their vote on Pluto was really a decision on the fundamental question, "What is a planet?"

After much deliberation, the IAU decided that to be a **planet**, an object must pass the following tests: (1) it must orbit a star directly, so **moons**, natural objects that orbit a planet, are not planets; (2) it must be spherical, implying that it's large enough for internal gravitation to smooth out bumps and dimples on its surface; and (3) it must have cleared its orbit of other objects, either by incorporating them through collision or by

Curiosity, a car-sized rover, has explored Gale Crater on the planet Mars since landing on August 6, 2012. It was still roving the planet at the end of 2016.

Figure 22.1 The eight planets of the Solar System.

(b) All eight planets are much smaller than the Sun.

(a) Images of the eight planets emphasize that each is unique.

this background, we analyze distinctive features of the planets, starting with those closest to the Sun. We finish the chapter by discussing the smaller objects of the Solar System, such as the *asteroids* that lie between Mars and Jupiter. As you wander our Solar System, keep in mind the following themes: (1) There's no place like home! Each object in our Solar System is unique, and none shares the features of the Earth that make our planet livable. (2) Without data returned to us from space probes, we would know very little about our neighbors. (3) While many discoveries have been complete surprises, established scientific principles provide a basis for understanding newly discovered features of the Solar System.

Figure 22.2 A composite photo in true color of Pluto, as seen by the *New Horizons* space probe in 2015.

trapping them as moons. Pluto passed the first two tests, but not the third—it has not cleared its neighborhood of other Kuiper Belt objects. Astronomers now refer to objects that pass only the first two tests as **dwarf planets**. The demotion of Pluto didn't dim its mystery, however, so in 2006, NASA launched a space probe, *New Horizons*, to study it. In July 2015, humanity received its first close-up photos of this dwarf planet **(Fig. 22.2)**.

This chapter introduces all components of the Solar System except the Sun, which we'll describe in Chapter 23. We begin by reviewing key events in the history of planetary exploration. We then characterize the overall structure of the Solar System and see how it relates to the process by which the Solar System formed. With

22.2 Discovery and Structure of the Solar System

An Age of Exploration Begins

We can think of Solar System exploration, from its beginning to the present, as having two stages. The early stage began when Galileo, using a small telescope, discovered the four large moons of Jupiter and realized that our Solar System hosts other objects besides the classically known planets (see Chapter 21). Scanning the sky with telescopes led to the discovery of craters on the Moon, rings around Saturn, additional planets (Uranus and Neptune), and the larger asteroids. But, because of atmospheric seeing (see Chapter 21), and the large distances to the planets, we can discover only so much from the Earth using ground-based telescopes. The second, modern stage of research began with the launch of *Sputnik 1*, the first *artificial satellite* (a human-made object that orbits the Earth outside of the atmosphere), by the USSR in 1957. Within a decade, the United States and the USSR had launched *space probes*, vehicles that achieve escape velocity and visit other celestial objects.

As outlined in **Table 22.1**, researchers have benefited from information obtained by using increasingly sophisticated instruments. Recent studies have used space telescopes orbiting the Earth, space probes that have flown by, orbited, crashed onto, or landed on extraterrestrial objects, and spacecraft that have carried astronauts to the Moon **(Fig. 22.3)**.

The Arrangement of Objects in the Solar System

Ongoing exploration continues to add items to the inventory of objects that make up the Solar System. In fact, in 2016, scientists announced, based on their computer simulations of the orbits of small Kuiper Belt objects, that a large (Neptune-sized), yet-unseen planet may be lurking in the far reaches of the Kuiper Belt. Astronomers now recognize many different types of objects based on their size, composition, and orbital characteristics, as summarized in **Table 22.2**.

How are the objects of the Solar System arranged? The planets follow slightly elliptical orbits, and as we saw in Chapter 21, they display *prograde motion*, meaning that they move counterclockwise around the Sun, when viewed from above the Sun's north pole. Nearly all have orbital planes that lie within 3° of the plane of the Earth's orbit (the *ecliptic plane*) **(Fig. 22.4)**. In fact, a side-on view emphasizes that only Mercury's orbit deviates noticeably (7°) from the ecliptic plane. Mercury's orbit is even more tilted than that of the Moon, whose orbit is slightly greater than 5° from the ecliptic plane. When astronomers defined the orbits of asteroids and Kuiper Belt

Table 22.1 Key discoveries in understanding the Solar System

Event	Year
Copernican (heliocentric) model of the Solar System	1543
Kepler's first two laws of planetary motion	1609
Galileo finds the large moons of Jupiter	1610
Observation of Saturn's rings	1659
Newton proposes the laws of motion and gravity	1675
Discovery of Uranus	1781
Discovery of Neptune	1846
Discovery of Pluto	1930
Sputnik 1, first satellite to orbit the Earth	1957
Luna 3, first image of the Moon's far side	1959
Mariner 2, first flyby of Venus	1962
Mariner 4, first flyby of Mars	1964
Luna 9, first successful lunar landing	1966
Apollo 8, first manned orbit of the Moon	1968
Apollo 11, first manned landing on the Moon	1969
Mariner 9, first satellite to orbit Mars	1971
Pioneer 10, first Jupiter flyby	1972
Venera 8, first landing on Venus	1972
Pioneer 11, first Saturn flyby	1973
Mariner 10, first Mercury flyby	1973
Viking 1, first pictures from the surface of Mars	1975
Voyager 2, first flyby of Uranus and Neptune	1986
Galileo flyby of an asteroid; first Jupiter orbiter	1989
Hubble Space Telescope launches	1990
Pathfinder, the first rover on the surface of Mars	1996
Cassini, the first satellite to orbit Saturn	1997
International Space Station launched	1998
Observation of the first (non-Pluto) Kuiper Belt object	1992
Spirit and *Opportunity* rovers start exploring Mars	2003
Hayabusa returns a sample from an asteroid	2003
Rosetta drops the first lander on a comet	2004
Messenger orbits Mercury	2004
Kepler space telescope launches	2009
Curiosity rover starts exploring Mars	2012
Voyager 1 (launched 1977) passes Solar System limit	2012
New Horizons flies by Pluto	2015

The 35-story-high Saturn V rocket that launched the Apollo missions.

The lunar-landing module and the lunar rover on the Moon in 1972.

Figure 22.3 Space probes.

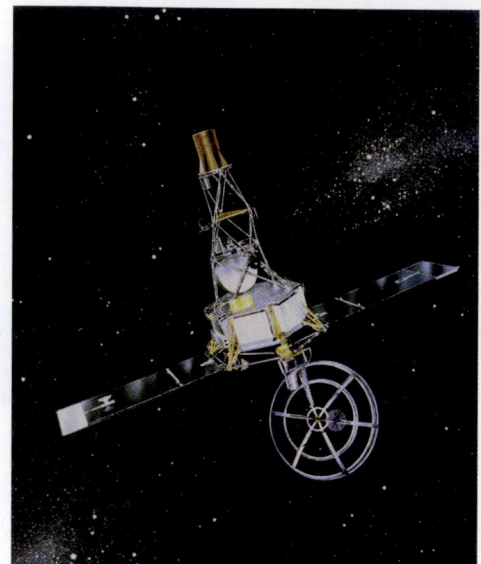

(a) *Mariner 2* became the first spacecraft to reach another planet when it flew by Venus.

(b) Engineers prepare the *New Horizons* space probe, which flew by Pluto in 2015 and is continuing into the Kuiper Belt. The gold foil reflects sunlight to prevent the probe from overheating.

(c) Launch of the *New Horizons* space probe in 2006.

Did you ever wonder . . .

why all planets orbit the Sun in the same direction?

objects, they realized that the majority of these objects also display prograde motion, but some follow highly elliptical paths, called **eccentric orbits**. Also, while some have orbital planes close to the ecliptic plane, others follow paths that incline significantly to the ecliptic plane.

Table 22.3 reveals that the planets differ from one another in many ways. As they orbit the Sun, all the planets spin counterclockwise on their axes—in the prograde direction—except for Venus, which rotates in the *retrograde direction*, meaning clockwise as viewed looking down on the Sun's north pole. Astronomers describe the orientation of a planet's axis of rotation by specifying its *tilt*: an axis perpendicular to the planet's orbital plane has a 0° tilt, and an axis parallel to the orbital plane has

Table 22.2	Objects in the Solar System
Objects	**Description**
• 1 star	Our Sun, a nuclear inferno emitting intense energy
• 8 planets	Relatively large spherical objects that orbit the Sun and have cleared their orbits of other objects
• 5 dwarf planets	Spherical objects that orbit the Sun, but have not cleared their orbits of other matter; there could be as many as 100 more to be discovered
• 176 moons	Objects orbiting planets or dwarf planets; some moons are spherical and some are not
• Asteroids	Rocky and/or metallic objects in the asteroid belt, which lies mostly between Mars and Jupiter; 3 have diameters over 500 km (300 miles) and are spherical; millions more are small and irregularly shaped
• Kuiper Belt	A band of icy objects in orbit around the Sun beyond the orbit of Neptune; about 1,000 have been seen, of which 100 have diameters exceeding 300 km (180 miles); 100,000 more may exist with diameters over 100 km (60 miles), and countless smaller ones may also exist
• Oort Cloud	A hypothetical spherical cloud of icy objects, held by the Sun's gravity but extending a quarter of the distance to the nearest star
• Comets	Icy and dusty objects, with diameters from tens of meters to a few kilometers, that orbit the Sun in highly elliptical orbits; comets emit a tail of glowing gas when they approach the Sun; over 5,000 are known
• Meteoroids	Relatively small objects, with diameters ranging from less than a millimeter up to 100 m (330 feet), that orbit the Sun

a 90° tilt. If we compare the planets, we see that they display a range of tilts (Fig. 22.5). The tilt determines whether or not a planet has seasons, because the axis stays in the same orientation as the planet orbits the Sun. When a planet's axis has a tilt, the *insolation* (the amount of incoming solar radiation) striking a location on the planet varies over a year.

Different planets take different amounts of time to make a complete rotation. The Earth, as we know, takes one day (24 hours) to spin on its axis. In comparison, Venus is a slowpoke, taking 243 Earth days to rotate. Jupiter, Saturn, Uranus, and Neptune whirl completely around in less than 17 hours. Different planets also take different amounts of time to orbit the Sun. As Kepler's third law states, the farther a planet lies from the Sun, the longer the planet takes to orbit the Sun (Box 22.1). For example, Mercury makes its journey in just 88 Earth days, but Neptune takes 165 Earth years.

When we consider planetary size and density, we see that the planets fall into two categories, as Table 22.3 makes clear. The *inner planets*, those closer to the Sun (Mercury, Venus, Earth, and Mars) are relatively small (Earth-sized or smaller) and have relatively high average densities (greater than the density of basalt). High density indicates that the mass of a planet consists mostly of rock or metal. Because of the similarities of the inner planets to the Earth, they are known collectively as the **terrestrial planets**. The *outer planets*, those farther from the Sun (Jupiter, Saturn, Uranus, and Neptune) are all much larger than the Earth. In addition, they all have much lower average densities, indicating that they are composed primarily of gases or ices. In this context, **ice** refers to not only frozen water, but also to the solid version of other compounds (such as frozen carbon dioxide or ammonia).

Figure 22.4 The arrangement of the planets and the orientations of their orbital planes.

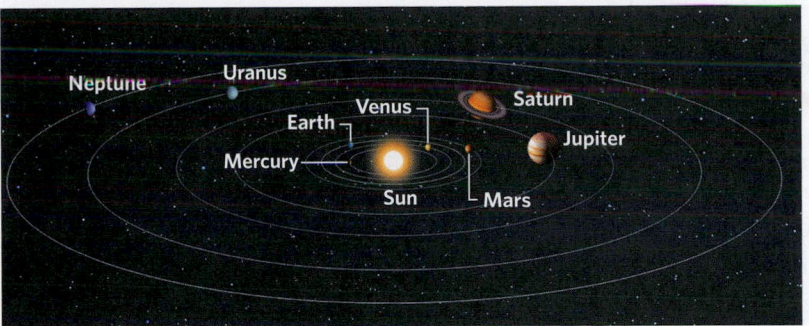

(a) Oblique view of all eight planetary orbits.

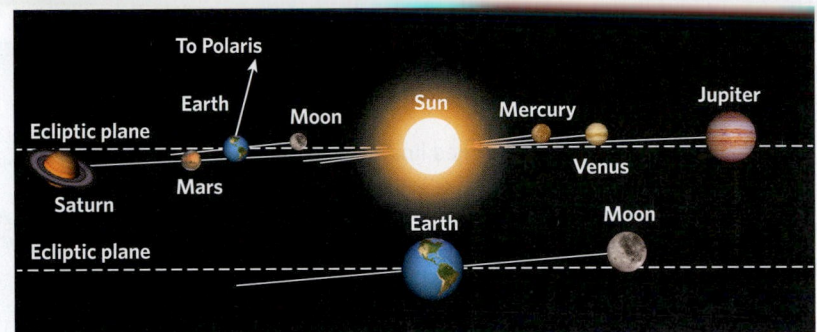

(b) The orientation of the planetary orbits (top) and of the Moon's orbit (bottom) relative to the ecliptic plane.

Because of their similarity to Jupiter, the outer planets have been traditionally known as the **Jovian planets**.

Finally, all planets except Mercury and Venus have moons, but no two planets have the same number of moons. In addition, all Jovian planets have **rings**, thin bands of tiny objects in orbit along the planet's equator.

Table 22.3 Comparing the planets

Name	Duration of Day*	Axial Tilt	Spin Direction†	Duration of Year‡	Known Moons	Mass§	Radius (km)	Radius (× Earth)	Distance to Sun (AU)	Density (g/cm³)
Mercury	59 d	0.1°	CW	88 d	0	0.06	2,440	0.38	0.38	5.4
Venus	243 d	177°	CCW	224 d	0	0.8	6,052	0.95	0.72	5.2
Earth	1 d	23°	CW	365 d	1	1.0	6,371	1.00	1.00	5.5
Mars	1 d	25°	CW	687 d	2	0.1	3,390	0.53	1.52	3.9
Jupiter	10 h	3°	CW	12 y	67	317.8	69,911	10.97	5.20	1.3
Saturn	10 h	27°	CW	29 y	62	95.2	58,232	9.14	9.54	0.7
Uranus	17 h	98°	CW	84 y	27	14.5	25,362	3.98	19.18	1.2
Neptune	16 h	28°	CW	165 y	14	17.2	24,622	3.86	30.06	1.7

*A day is the time it takes for a planet to rotate once on its axis, specified in Earth days or hours (h = hour, d = day).
†CW = clockwise; CCW = counterclockwise.
‡A year is the time it takes a planet to complete one orbit around the Sun, specified in Earth days or years (d = day; y = year).
§The number given for mass is relative to the mass of the Earth.

Box 22.1

How can I explain . . .

Distances in the Solar System

What are we learning?

- How far apart planets are relative to the Sun and to one another.

What you need:

- A 100-yard or -meter-long line on a sidewalk, parking lot, or, ideally, a football field.
- One soccer ball.
- Eight golf balls.

Instructions:

- The goal of this exercise is to visualize relative distances in the Solar System. The scale has been provided by setting 100 yards, or meters, equal to 30.1 AU.
- Place the soccer ball at one end of the marked line.
- Then, use the measurements in the accompanying table to place a golf ball at the location of each planet (numbers have been rounded). To highlight the locations, have a student stand next to each ball.

What did we see?

- This exercise provides a simple but powerful visual illustration that contrasts the proximity of the terrestrial planets to the Sun and to one another with the vast distances of the Jovian planets from the Sun.

Planet	Distance from Sun (AU)	Yards (or meters) from soccer ball
Mercury	0.39	1
Venus	0.72	2
Earth	1.00	3
Mars	1.52	5
Jupiter	5.20	17
Saturn	9.50	32
Uranus	19.20	63
Neptune	30.10	100

The Solar System contains many other objects in addition to the planets (see Table 22.2). All display prograde motion, but some have very eccentric orbits that may be inclined to the ecliptic plane. Despite their great numbers, all the objects of the Solar System, other than the Sun, together account for only a tiny fraction (0.15%) of the Solar System's mass; 99.85% of the mass lies within the Sun itself.

Origin of the Solar System, Revisited

Let's now review the processes by which the Solar System formed. We first introduced these in Chapter 1, and will now describe some of them in more detail.

If you could somehow travel back in time 5 billion years, you would find that our Solar System didn't exist. At its location, you would find a **nebula**, a cloud of gas and

Figure 22.5 Each planet's axis of rotation tilts differently with respect to the ecliptic plane. Uranus's axis is almost parallel to the ecliptic plane, and Venus' axis is nearly upside down.

| Mercury 0.1° | Venus 177° | Earth 23° | Mars 25° | Jupiter 3° | Saturn 27° | Uranus 98° | Neptune 28° |

Red arrow = counterclockwise rotation
Green arrow = clockwise rotaion

Solid particles form in a nebula, and start falling inward.

The nebula evolves into a spinning accretionary disk. Matter falls onto the disk.

The proto-Sun forms. Proto-planets grow from planetesimals.

Planets orbit the new-born Sun.

(a) Stages in the development of the Solar System.

dust in space **(Fig. 22.6)**. Most of this nebula consisted of hydrogen (74%) and helium (24%), gases left over from the Big Bang. The remainder contained all the other 90 naturally occurring elements of the periodic table in varying proportions. Recall that these elements were formed by nuclear fusion reactions in stars supernova explosions. The nebula included **volatile materials** (those that can exist as gases or liquids under conditions found near the Earth's surface today), such as hydrogen (H_2), helium (He), methane (CH_4), ammonia (NH_3), water (H_2O), and carbon monoxide (CO), as well as **refractory materials** (those that melt only at high temperatures), such as silica (SiO_2), iron oxide (Fe_2O_3), and magnesium oxide (MgO).

Initially, the nebula was so hot that it was expanding, as any cloud of gas does as it heats up (see Chapter 18). How did that expanding nebula become our Solar System? According to the **nebular theory** introduced in Chapter 1, gravity pulled the matter in the nebula together, and it collapsed inward **(Fig. 22.7a)**. While this statement describes what happened, it doesn't give the whole picture. Today, astronomers explain the details of the process using a more sophisticated model, called the **condensation theory**, which emphasizes that if the nebula contained only gas, it would have remained hot enough to continue expanding and would never have collapsed. It was the presence of tiny (0.01-mm; 0.0004-inch) specks of ice (solidified volatile materials) and **dust** (solidified refractory materials) that allowed the nebula to cool sufficiently

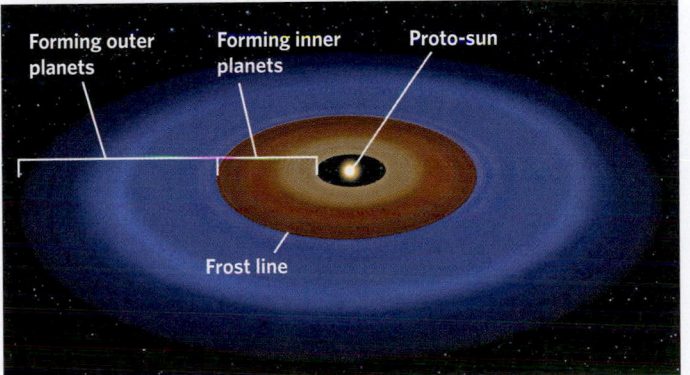

Forming outer planets   Forming inner planets   Proto-sun

Frost line

Inside the frost line, metals and minerals condense into planets.

Outside the frost line, low temperatures allow condensing planets to include primarily gases and volatile molecules.

(b) An oblique view of the developing Solar System, not to scale, showing the frost line.

so it could collapse. Such cooling occurs because solids radiate heat outward more efficiently than gases do. In addition, ice and dust specks acted as "seeds" to which atoms attached, allowing larger solid particles to grow. The process of condensation may have been triggered when the compressional waves emitted by a supernova explosion passed through the nebula and pushed particles closer to one another, as we'll see in Chapter 23.

With time, growing solid particles collided and clumped together, producing larger masses that could exert stronger gravitational attraction and pull in surrounding matter. The part of the nebula containing the most matter exerted the greatest gravitational pull, so more matter fell toward this area, producing a ball of dust and gas material. As this ball grew larger, its gravitational pull became greater, and the remaining nebular material began to collapse inward rapidly. As the originally swirling material of the nebula fell inward, the cloud began to rotate, and as more matter moved inward, the rotation became faster due to *conservation of angular momentum* (a principle of physics discussed in previous chapters). Once the rate of rotation became fast enough, the *centrifugal force*, the apparent outward-directed force that an object experiences when it spins or moves in orbit around a point, became significant, causing the collapsing nebula to flatten. At this stage, it became an **accretionary disk**.

As the central ball of the accretionary disk became larger and denser, it grew warmer, for gas heats up when compressed. Eventually, the ball became hot enough to radiate significant heat, and it became a **protostar**. When the protostar compressed sufficiently under the influence of gravity, it became so hot that, at about 4.57 Ga, nuclear fusion reactions began, and the ball of matter became a true star, our Sun.

Why didn't all the material of the accretionary disk fall inward into the newborn Sun? The material of the flattened part of the accretionary disk was revolving around the proto-Sun fast enough that it stayed in orbit. Once the Sun had formed, that material became the **protoplanetary disk**, which served as the source of planets and other objects. Initially, the protoplanetary disk was a homogeneous mixture of dust, ice, and gas. But when the Sun ignited, the inner part of the disk became hotter, causing its ice specks, composed of volatile materials, to turn into gas. The *solar wind*, a stream of fast-moving particles flowing away from the Sun, blew this gas to the outer part of the protoplanetary disk. When the gas passed a boundary called the *frost line*, the volatile materials refroze into ice **(Fig. 22.7b)**. Eventually, the inner part of the disk ended up with mostly refractory dust, while the outer part ended up with large volumes of volatile gas and ice.

As time passed, solids began to clump together under the influence of gravity in the protoplanetary disk. Soot-sized

particles formed, and these particles merged into pea-sized clots, which collected into boulder-sized blocks. Eventually, these blocks amalgamated into bodies called **planetesimals**, whose diameter exceeded 1 km (0.6 miles). Because of its mass, a planetesimal exerted enough gravitational attraction to pull in nearby objects, so it acted like a vacuum cleaner, collecting small pieces of dust and ice as well as other, smaller planetesimals, and became still larger. Eventually, the victors in this competition to attract mass grew into **protoplanets**, bodies approaching the size of today's planets. Once a protoplanet succeeded in clearing its orbit of debris, it became a full-fledged planet. Astronomers estimate that it took only about 100 million years for the planets to finish forming. The terrestrial planets, consisting almost entirely of rock and metal, formed on the inside of the frost line, closest to the Sun, whereas the Jovian planets, containing large amounts of ice and gas, formed beyond the frost line.

Because the accretionary disk was fairly flat, the orbits of all the planets lie in similar orbital planes, and because the accretionary disk rotated in one direction, all planets orbit the Sun in the prograde direction. Initially, they all had a prograde spin, with an axis of rotation perpendicular to the plane of the accretionary disk. Objects that now display eccentric orbits or tilted axes must have been acted on by an outside force, such as a collision with or the gravitational pull of another object.

When terrestrial protoplanets grow to exceed a diameter of several hundred kilometers, the material inside them became warm and soft enough to flow, so under the influence of their own gravity, they transformed into spheres. As this reshaping took place, molten, dense iron sank inward, causing each protoplanet's interior to differentiate into a metallic core and a rocky mantle. The larger, Jovian planets, consisting of gases and ices, were also modified by gravity to become spherical shapes as they grew.

The process described in this section that led to the formation of our Solar System was probably repeated in most of the stars in our galaxy. Today, with modern space telescopes, we are finding other planetary systems around distant stars **(Box 22.2)**.

Take-home message . . .

The Solar System formed from a nebula. According to the condensation theory, specks of dust and ice played an important role in the cooling of the nebula and provided seeds for the growth of larger particles, ultimately causing matter to collect into an accretionary disk. The center of this disk became the Sun. The outer parts collected into planetesimals, then protoplanets, then planets.

Quick Question -
Why do the Jovian planets lie farther from the Sun?

Did you ever wonder . . .
why planets are round?

Consider this . . .

Planets beyond our Solar System

Do planets form around other stars? Astronomers had long hoped to learn the answer to this question by finding **exoplanets**—planets outside our Solar System—but until the 1990s, they didn't have the technology to detect them. In 1995, new instruments led to the discovery of one exoplanet. During the next five years, astronomers found about twenty more. The rate of discovery accelerated following the launch of the Kepler space telescope in 2009, and by 2017, over 3,550 exoplanets had been found.

From their investigations, astronomers have discovered that some exoplanets resemble planets in our own Solar System, but others seem to be different. Currently, astronomers divide exoplanets into two temperature categories (*hot planets* lie within 0.1 AU of their host star, while *cold planets* lie farther out) and several size categories (based on their sizes relative to planets in our Solar System). The majority of exoplanets found so far are about the size of Jupiter. In 2015, observers found the first rocky planet orbiting in the *habitable zone* of a Sun-like star. In the habitable zone, the surfaces of planets receive enough energy from their star to allow water to exist in its liquid state.

Could any exoplanets host life? All life as we know it must have access to water because cells contain over 50% water. Therefore, the necessary first step for finding possible homes for life outside of the Earth would be to identify other planets with possible reservoirs for liquid water. The search for extraterrestrial life, and efforts to understand how life might appear and survive on other celestial objects, has become a field of study called **exobiology**.

22.3 Our Nearest Neighbor: The Moon

History of Lunar Exploration

Long before the invention of the telescope, people realized that the Moon orbits the Earth, that we see it because it reflects sunlight, and that its surface has dark and light patches. They were also able to measure its dimensions, so they knew that the Moon has a radius of 1,737 km (1,079 miles), about one-fourth of the Earth's radius. But beyond that, our ancestors knew very little about our nearest neighbor.

Scientific understanding of the Moon took a great leap forward in the 1960s, when unmanned space probes and later, Apollo spacecraft orbited the Moon and landed on its surface. Between 1969 and 1972, the United States landed six spacecraft, with two astronauts each, on the Moon. All together, these explorers brought back 381 kg (840 pounds) of *Moon rocks*, providing researchers with the opportunity to analyze their chemical composition and determine the numerical age of the Moon. Modern orbiting space probes have produced high-resolution digital topographic maps of the entire lunar surface, and seismometers placed on the Moon have even detected moonquakes. Let's look at the lessons learned from all this work.

General Characteristics of the Moon

If you watch the Moon night after night, you'll see the same distinct features, more or less, because the Moon spins on its axis at just the right speed to keep the near side facing us as the Moon orbits the Earth (**Fig. 22.8**). Put another way, the time it takes for the Moon to rotate once on its axis is about the same as the time it takes for the Moon to orbit the Earth once. This synchrony developed over geologic time because the tug of the Earth's gravity generates a tidal bulge of the Moon's surface.

That's one small step for a man, one giant leap for mankind.

— NEIL ARMSTRONG (AMERICAN ASTRONAUT, ON TAKING HIS FIRST LUNAR STEP, 1969)

Figure 22.8 The Moon's synchronous orbit around the Earth causes the same side of the Moon to face the Earth continuously.

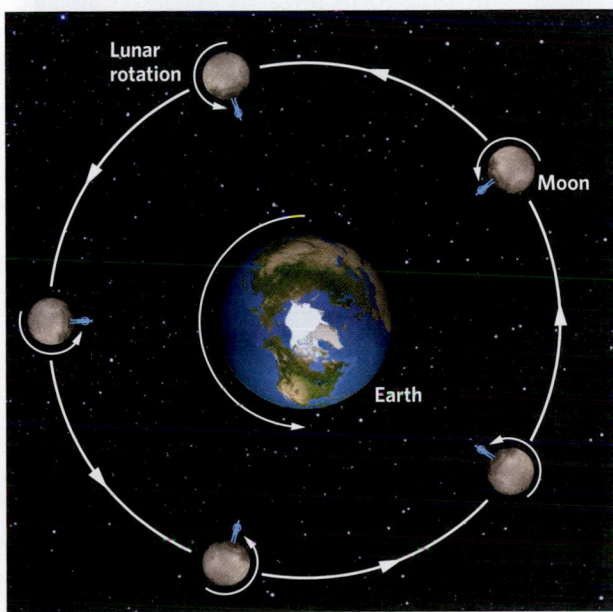

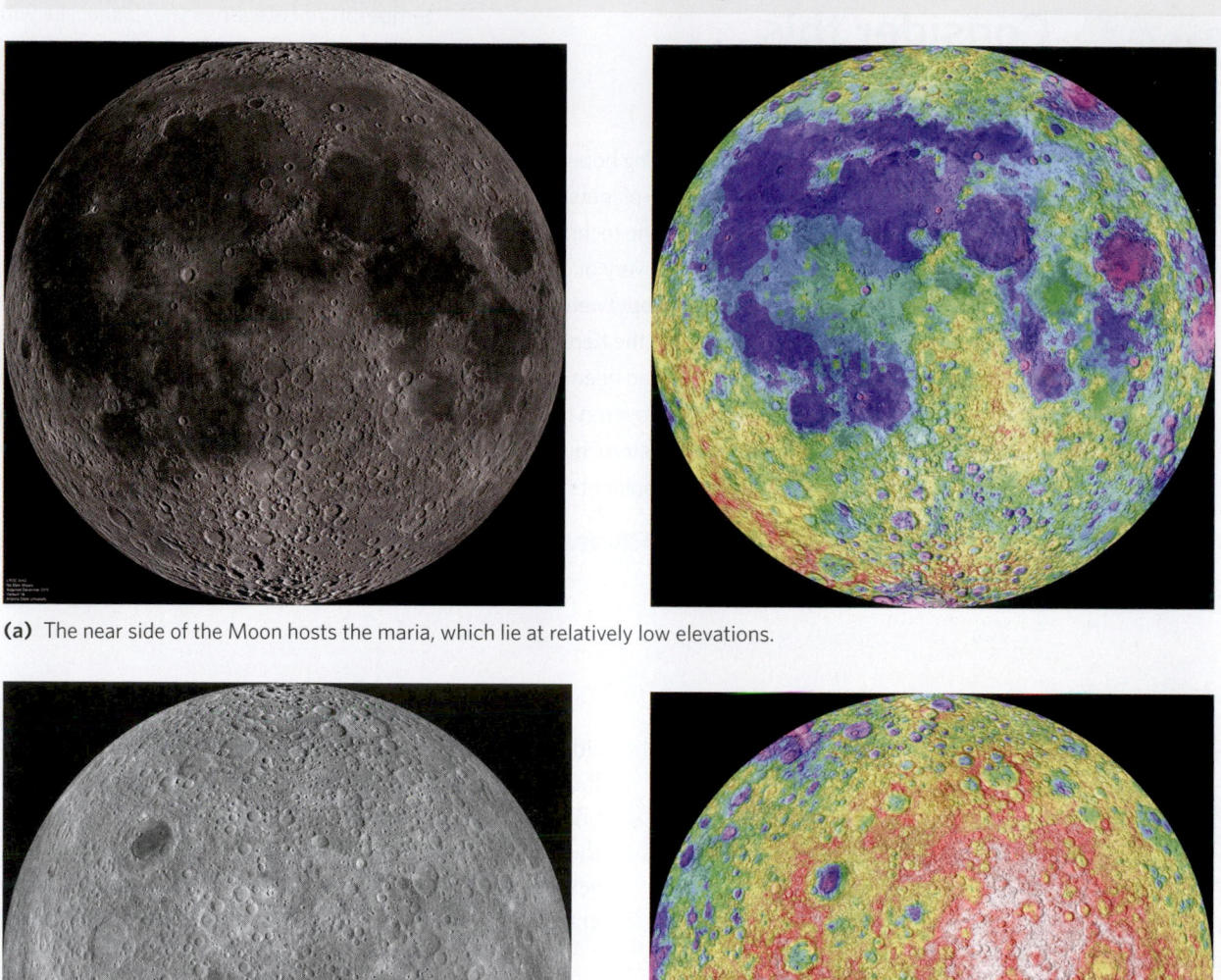

(a) The near side of the Moon hosts the maria, which lie at relatively low elevations.

(b) The far side of the Moon consists entirely of heavily-cratered highlands.

Friction doesn't allow this bulge to move as fast as the Moon spins, so the Earth pulls on one side of it slightly more than on the other. This pull slowed the Moon's spin until it locked into a rate that keeps the bulge facing the Earth. The match in rotation speed isn't perfect, however, so over the course of a month, we actually see 59% of the Moon's surface. In fact, humans didn't see the *far side* (or *dark side*) of the Moon until 1959, when the *Luna 3* space probe sent back pictures from its orbit around the Moon.

The Moon has a lower average density than the Earth (3.4 g/cm³ for the Moon vs. 5.5 g/cm³ for Earth). This characteristic, along with its smaller size, means that the Moon has a mass of only about ⅛₀ of the Earth's mass. Less mass means weaker gravity, so a 68-kg (150-pound) astronaut weighs only 11 kg (25 pounds) on the Moon. Unlike the Earth, the Moon does not have active volcanoes or plate tectonics, for most of its mantle has become too cool to permit convection. Without volcanism to supply gases, the Moon has no atmosphere and, therefore, no oceans, rivers, glaciers, or life. And without an atmosphere or ocean to hold and redistribute heat, temperatures on the Moon undergo violent swings, ranging from over 120°C (248°F) on the sunlit lunar equator to −175°C (−283°F) on the Moon's dark side.

The lack of an atmosphere, and of the light scattering that it causes, also means that we can see the lunar surface clearly. This surface displays varied landscapes with topography ranging from a high of 10.8 km (6.7 miles) above mean (average) elevation to 9.1 km (5.7 miles) below. Lower-elevation regions of the Moon, the **maria** (singular: **mare**, pronounced mar-ay), have fairly smooth, dark surfaces **(Fig. 22.9a)**. Their name comes from the Latin word for sea because they were originally thought to be flooded with water. In fact, the maria consist of plains underlain by basalt lava flows, which have a low *albedo* (reflectivity). Notably, maria lie only on the near side of the Moon, so the near and far sides of the Moon look very different from each other **(Fig. 22.9b)**. The regions at higher elevation, the **lunar highlands**, are relatively rugged and have a high albedo, so they display a whitish glow when viewed from the Earth. They are underlain by a light-colored rock, called *anorthosite*, that consists largely of the mineral plagioclase.

Lunar Craters and Regolith

When Galileo turned his telescope toward the Moon in 1609, he discovered that bowl-shaped indentations, or **craters**, pockmark the surface. For centuries after Galileo, researchers assumed that the craters formed during volcanic eruptions. In 1892, however, an American geologist, G. K. Gilbert, argued that the craters were due to meteorite impacts and, to prove his point, made laboratory models of craters by dropping clay balls onto wet sand. Lunar exploration has confirmed that the lunar craters formed as a result of meteorite impacts. There are over 100,000 craters with a diameter of 1 km (0.6 miles) or more.

Why do we see so many more craters on the Moon than on the Earth? First, fewer objects strike the Earth than the Moon because many of those that reach the Earth burn up or explode in its atmosphere. Second, plate tectonics and erosion destroy craters over time on Earth, but these processes don't happen on the Moon, so a lunar crater, once formed, can survive for billions of years, and we see craters that have accumulated over lunar history. Though cratering has happened throughout lunar history, and continues today, the most intense cratering occurred during the **late heavy bombardment**, a time between 4.0 and 3.9 Ga when the gravitational pull of the outer planets sent huge swarms of objects careering across the orbits of other planets.

Not all lunar craters look the same. Their character depends on the size and speed of the impacting object. Smaller *simple craters* tend to be smooth-floored bowls, whereas larger *complex craters* have a central uplift and may be surrounded by concentric ridges **(Fig. 22.10a)**. The

Figure 22.10 The lunar surface.

(a) Lunar craters, as seen from an orbiting spacecraft. Larger complex craters have a central uplift and several rings. Smaller simple craters are smooth bowls.

(b) Astronaut footprint in lunar regolith.

central uplift forms when the impact disrupts the crust deeply enough to cause it to spring upward. Impacts also send out debris, which forms an *ejecta blanket* around the crater. In some cases, the ejecta forms a pattern of spoke-like streaks.

As the first astronauts to walk on the Moon discovered, a layer of **lunar regolith**, consisting of dust and debris, covers the Moon's surface. This regolith consists of fragmented rock blasted out of craters, glass droplets frozen from impact melt **(Fig. 22.10b)**, and accumulations of *micrometeorites* (meteorites less than 2 mm, or 0.08 inches, in diameter). The top part of the regolith tends to be very fine, probably due to the pulverization of surface material by micrometeorite impacts and by absorption of *cosmic rays* (high-energy atomic nuclei from space). Researchers refer to such breakdown of material as **space weathering**. Because of space weathering, older craters tend to be blunter and somewhat smoothed, while newer ones have sharper features.

Age and Origin of the Moon

Based on analyses of Moon rocks, together with other data, most researchers favor a model in which the Moon formed from debris ejected into orbit around the Earth following a catastrophic collision of the newborn Earth with a protoplanet. This event happened at about 4.53 Ga, shortly after the interior of the Earth had undergone differentiation. This debris, mostly from the mantles of the colliding object and Earth, first formed a ring around the Earth, then accumulated to become a sphere **(Fig. 22.11)**.

When the Moon first formed, its surface was probably a magma sea. As the Moon cooled, denser mafic minerals formed and sank, while less dense felsic minerals, such as plagioclase, rose. The outer layer of less dense rock solidified into an anorthosite crust, exposed in the lunar

highlands today. This crust, like a shell, surrounds a mantle of ultramafic rock, which in turn surrounds a small iron core. The Moon's core makes up only about 20% of the diameter of the Moon, in contrast to the Earth's core, which makes up about 50% of the diameter of the Earth. It's the small size of the Moon's core that led to the hypothesis that the Moon formed primarily from mantle rock ejected from the Earth.

Geologists use standard isotopic dating methods to determine the ages of Moon rocks. Dates obtained so far fall between 4.36 Ga and 3.16 Ga. This range suggests that for a few hundred million years after it formed, the Moon remained too hot for isotopic clocks to start. The younger ages come from mare basalts, which erupted after the crust had completely cooled and solidified. How can we determine ages for still younger lunar features? The **lunar time scale**, which divides the history of the Moon into distinct periods, was developed using three key criteria: (1) whether a feature pre-dates or postdates the distinct ejecta blanket that covered the entire Moon after a huge impact very early in its history; (2) the **cratering density** (the number of craters per unit area) of a feature, for the longer a surface exists, the more impacts it receives; and (3) the sharpness of crater edges or mountain ridges, for as time passes, space weathering and regolith accumulation blunt lunar features. The oldest lunar landscapes are those that host the most craters and have the most weathered features, whereas the youngest landscapes are those that display relatively few craters and less weathered features.

How the maria formed remains a subject of research. According to one model, they began as immense basins carved by meteorite impacts between 4.3 and 3.8 Ga. Melt from the Moon's mantle eventually broke through the thin crust on the floors of these basins in a succession of eruptions. These eruptions began about 100 million years after the basins had formed and continued

Figure 22.11 The formation of the Moon may have begun with an immense impact that blasted debris into orbit.

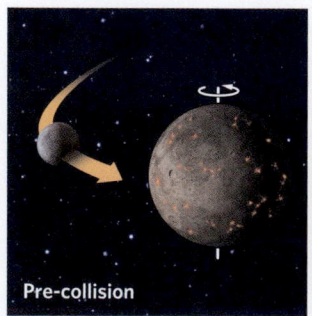    

Pre-collision   Collision   Debris sprays into orbit   Orbiting debris coalesces   The Moon becomes a sphere

Time

on and off for the next several hundred million years. The crust of the Moon's near side is thinner and weaker than that of its far side, so while the crust could crack and permit volcanic eruptions on the near side, it could not on the far side, explaining the difference between the two sides. No one knows for sure why the melt formed. Some melt may have remained from the Moon's formation, but some may have formed due to decompression melting or frictional heating associated with tidal movements.

Take-home message . . .

The Moon has no atmosphere, ocean, or active plate tectonics. Its surface displays dark, low-lying maria and light-colored, elevated lunar highlands. Craters pockmark the Moon's surface, and a layer of regolith covers most regions. The Moon formed around 4.53 Ga, most likely from debris blasted into space when a protoplanet struck the Earth.

Quick Question -
Why are there so many more craters on the Moon than on the Earth?

22.4 The Terrestrial Planets

In 1877, an Italian astronomer named Giovanni Schiaparelli peered through his telescope at Mars. He described features that he called *canali*, the Italian word for channels. The English-speaking press, however, translated the word into "canals," channels constructed by people, and some people immediately concluded that intelligent beings lived on Mars. Schiaparelli's depiction of the canali as straight lines on his drawing reinforced his interpretation **(Fig. 22.12)**. Subsequent observations proved that the canals were an optical illusion, and despite what science fiction says, Mars does not host intelligent beings. Nevertheless, early observations of Mars led to the realization that the three inner planets of the Solar System accompanying Earth are somewhat Earth-like. In this section, we explore these *terrestrial planets*. We'll see that, while many similarities indeed exist between these planets, the Earth, and the Earth's Moon, there are many significant differences **(Fig. 22.13)**.

Mercury: A Moon-Like Landscape

THE FAST PLANET. The English name for Mercury came from the Roman messenger god known for his speed because ancient astronomers saw that the planet moves across the sky faster than any other planet. It takes Mercury only 88 days to go around the Sun—the planet orbits rapidly because it is the planet closest to the Sun. Mercury's orbit

Figure 22.12 Giovanni Schiaparelli's illustration (1877) of Martian *canali*, which he defined based on vague color contrasts. They turned out to be illusions.

has greater eccentricity than that of any other planet. At its closest, it lies 70 million kilometers (46 million miles) from the Sun, and at its farthest, it lies 74 million kilometers (43.5 million miles) away. Though it revolves around the Sun quickly, it rotates on its axis relatively slowly; one day on Mercury equals 59 days on the Earth.

Because Mercury orbits so close to the Sun, and because its surface consists of dark rock that absorbs radiation, surface temperatures on the planet can rise far above the boiling points of volatile materials. These characteristics, along with the lack of volcanic activity to produce new gas, mean that the planet has no ocean and virtually no atmosphere. The trace of gas surrounding Mercury consists of atoms released by space weathering, and the strong solar wind eventually blows these atoms into space. Without an ocean or atmosphere to transport heat around the planet, the temperature at a location depends only on the insolation angle. Mercury's axis lies almost exactly perpendicular to its orbital plane, so at its poles, the temperature remains a constant −93°C (−136°F) all year, whereas at its equator, the temperature ranges from 427°C (800°F) at noon to −173° (−280°F) at midnight; this difference represents the largest temperature range of any planet in the Solar System.

Mercury's mass is about 5% that of the Earth, so a 68-kg (150-pound) astronaut would weigh only 27 kg (60 pounds) on Mercury. Researchers suggest that Mercury's metallic core accounts for a large portion of its interior (see Fig. 22.13), perhaps because the accretionary disk at the orbit of Mercury contained a higher proportion of refractory material than existed farther from the Sun. Mercury is the only terrestrial planet besides Earth to possess a magnetic field, perhaps because of its metallic core.

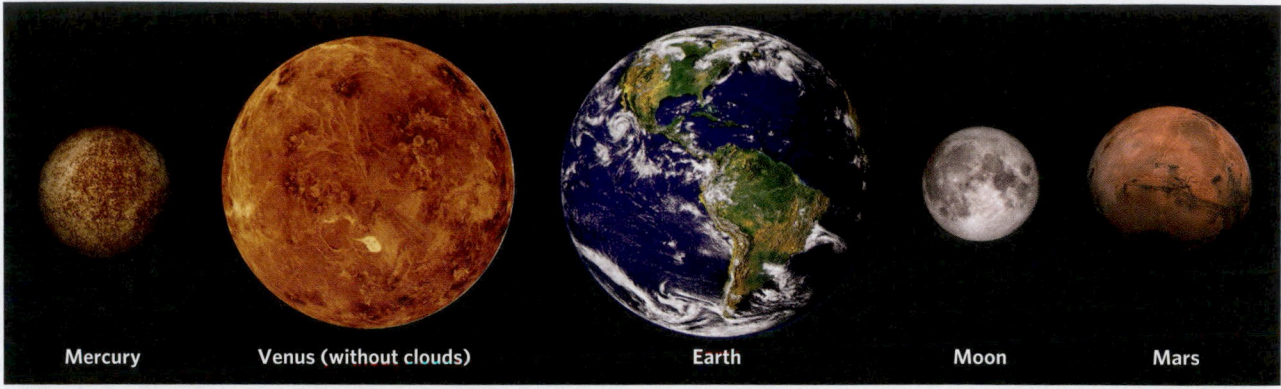

Figure 22.13 Mercury, Venus, Earth, the Earth's Moon, and Mars, with representations of their internal structures.

(a) The surfaces and diameters of the terrestrial planets differ markedly from each other.

Mercury | Venus (without clouds) | Earth | Moon | Mars

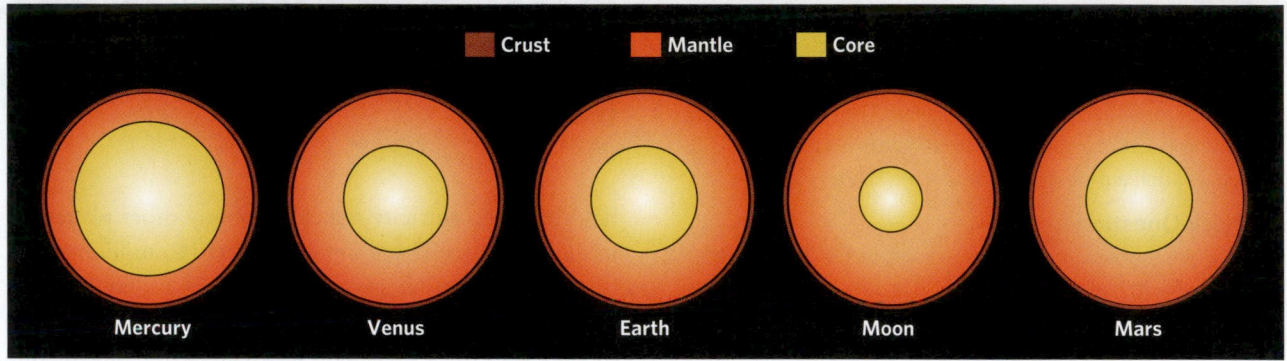

Crust | Mantle | Core

Mercury | Venus | Earth | Moon | Mars

(b) The interiors of the terrestrial planets differ from one another, too. In this figure, all the planets are resized to the same diameter, so as to emphasize differences in the proportions of crust, mantle, and core.

SURFICIAL SIMILARITIES TO THE MOON. Mercury resembles the Earth's Moon in many ways. They are similar in size (Mercury's diameter is 1.4 times that of the Moon's), both have heavily cratered surfaces, both lack active volcanoes and moving plates, both have a surface layer of regolith, and neither has an atmosphere or ocean. The

Messenger space probe, which began to orbit Mercury in 2011, has mapped the topography of Mercury in exquisite detail **(Fig. 22.14)**, revealing that Mercury hosts wide plains that look like lunar maria. [Unlike maria, Mercury's plains may be covered with pyroclastic debris instead of lava flows. Close-up photos indicate that large faults have offset craters on Mercury **(Fig. 22.15)**.]

Venus: Veiled by Clouds

AN UPSIDE-DOWN PLANET? Venus, the second planet from the Sun **(Fig. 22.16a)**, shines brighter than any star in the Earth's night sky. Named for the Roman goddess of beauty, Venus shines brightly because it has a high albedo, and lies fairly close to the Earth. Also, because it is relatively close to the Sun, it receives lots of solar radiation. The planet resembles the Earth in size, but it rotates much more slowly on its axis, taking 243 Earth days to spin once. Since Venus takes 224 days to orbit the Sun, its rotation rate exceeds its orbital period, so a day on Venus takes longer than a year on Venus. Venus and the Earth have about the same average density, suggesting that, like the Earth, Venus has an iron core. In addition, as we noted in Section 22.2, Venus differs from all other planets in that it rotates clockwise around its axis when viewed from the Sun's north pole (see Fig. 22.5). Astronomers speculate that a collision with a protoplanet early in the history of the Solar System flipped

Figure 22.14 The *Messenger* space probe provided the first detailed views of Mercury.

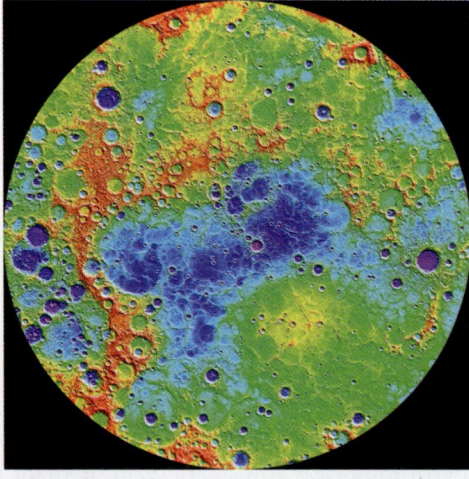

(a) A photograph, in false color, of the cratered surface of Mercury.

(b) A digital topographic map of Mercury's surface. Blue is lowest; red is highest.

Figure 22.15 A large fault dropped part of Mercury's surface to a lower elevation.

the planet over. In other words, Venus might have a retrograde rotation because the planet is upside down.

A CAUSTIC ATMOSPHERE. Venus hosts an atmosphere consisting almost entirely of carbon dioxide (96.5% by volume), with the remainder made up of nitrogen molecules (3.5%) and trace amounts of water vapor, sulfur dioxide, and carbon monoxide. This atmosphere has a much greater density than that of the Earth—it's so dense that if you were to stand on Venus's surface, the air pressure (93 atm) would equal the pressure you would feel at a depth of 800 m (0.5 miles) beneath the sea on Earth. Venus's air pressure exceeds Earth's both because Venus's atmosphere contains more gas and extends farther into space than does the Earth's atmosphere, and because the weight of CO_2

molecules exceeds that of N_2 or O_2 molecules **(Fig. 22.16b)**. The presence of sulfuric acid clouds makes Venus's atmosphere so caustic that it would burn through metal.

The intense greenhouse effect caused by the abundance of CO_2 in Venus's air keeps the planet's surface temperature at about 450°C (840°F), hot enough to melt lead. Venus may once have had milder conditions, but at some point, the greenhouse effect warmed the planet enough for all surface water to evaporate, and the additional water vapor in its atmosphere amplified the greenhouse effect. The result—a *runaway greenhouse effect*—made the planet's atmosphere so hot that water molecules in the air broke apart; hydrogen atoms drifted into space, and oxygen atoms were absorbed by weathering reactions with surface rocks, so Venus's water disappeared for good. Winds blow so strongly on Venus that its atmosphere circulates very rapidly. This atmospheric movement ensures that hardly any temperature difference exists between Venus's poles and equator or between its night and day sides.

A WORLD OF VOLCANOES. We can't see the surface of Venus from Earth, or even from orbit around Venus, because of its dense, cloudy atmosphere. In fact, the surface has been photographed only twice, by Soviet space probes that landed on the Venusian surface **(Fig. 22.17a)**. Each probe, before rapidly succumbing to the extreme conditions on the planet, returned images of flat, broken rocks. To map the surface, researchers use radar, which can penetrate the veil of gas **(Fig. 22.17b)**. Radar mapping reveals lava domes, shield volcanoes, and volcanic craters, but few impact craters **(Fig. 22.17c)**. The lack of craters on Venus suggests that the planet was resurfaced by lava flows since the late heavy bombardment. Like a layer of

Did you ever wonder . . .

if you could breathe the air on Venus?

Figure 22.16 Venus has the densest atmosphere in the Solar System.

(a) A *Mariner 10* image of Venus shows a thick, cloudy atmosphere.

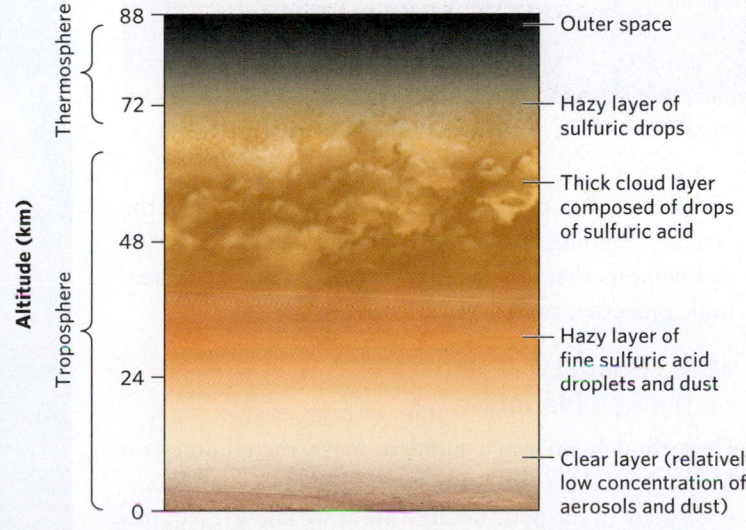

(b) A cross section of Venus's atmosphere. The entire atmosphere consists mostly of carbon dioxide. Its clarity depends on the concentration of sulfuric acid aerosols and dust.

Figure 22.17 Images of the surface of Venus.

(a) The surface of Venus, as seen by the USSR's *Venera 13* space probe.

(b) A radar image of the planet from the *Magellan* space probe.

there. Growing evidence suggests that long ago, Mars indeed had liquid water, as well as a thicker atmosphere. That environment has long since vanished, however, and the Mars of today is a dry, cold, desolate place, covered with rock and dust that contains oxidized iron. This "rust" led to the name Mars, after the Roman god of bloodshed, and to the planet's nickname, *the Red Planet*.

(c) An oblique radar image of Maat Mons, a volcano on Venus.

Figure 22.18 Mars, the most studied planet beyond the Earth.

asphalt filling in potholes on a highway, lava filled in the craters. Mapping has not revealed any plate boundaries, so it appears that hot-spot activity, rather than plate tectonic processes, caused Venus's volcanism.

Mars: Could Life Exist on the Red Planet?

Even though no canal builders make their homes on Mars, researchers still wonder whether microbial life existed there in the past. Because life as we know it requires water, much of the effort to search for life on Mars has focused on finding evidence that liquid water existed

Figure 22.19 The moons of Mars.

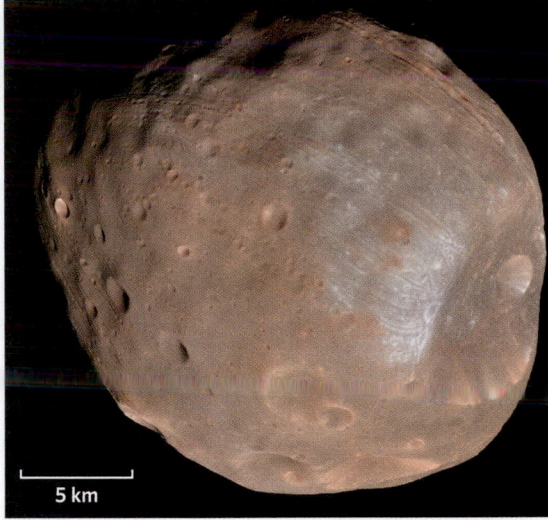

(a) Phobos.

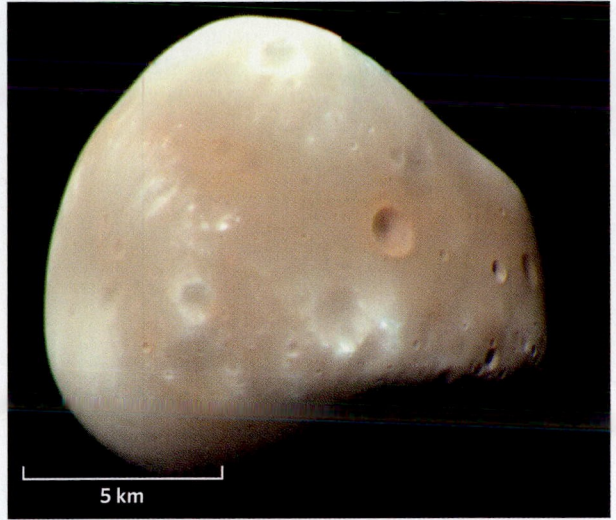

(b) Deimos.

EXPLORING MARS. Because of its proximity to Earth and its mostly transparent atmosphere, we can see fascinating rocks and structures on Mars **(Fig. 22.18)**. Several space probes have landed successfully on the planet, and several have surveyed Mars from orbit. The United States has also deployed three Martian rovers (*Spirit* and *Opportunity* landed in 2004, *Curiosity* in 2012), vehicles that can travel across the Martian surface (see chapter opening photo). These rovers have discovered minerals and sedimentary structures that may have formed underwater in the distant past. Since 2006, the *Mars Reconnaissance Orbiter* has sent back high-resolution images of landscape features on Mars, and in 2008, *Phoenix*, a stationary lander, touched down in Mars's north polar region and drilled into the surface, confirming the presence of subsurface water ice.

A THIN ATMOSPHERE. Air on Mars is much less dense than Earth's air, so the air pressure on Mars's surface reaches only about 0.6% that on Earth. Compositionally, the Martian atmosphere resembles that of Venus— it contains 95.3% carbon dioxide, 2.5% nitrogen, 1.6% argon, and just traces of oxygen, carbon monoxide, and water vapor. Even though CO_2 makes up most of the Martian air, there's not enough to cause a significant greenhouse effect, so the surface of Mars remains frigid, with an average temperature of −55°C (−67°F). The air might have been denser earlier in the history of the planet when volcanoes were active, but most of the gas formed then either froze, reacted with the surface, or escaped into space. The thin air does move, and researchers have clocked winds in the range of 30 to 100 km per hour (20 to 60 mph). These winds can churn up fine dust, producing haze that can last for months, but because the air has such low density, the winds cannot move larger objects.

GENERAL CHARACTERISTICS OF MARS. A day on Mars (24.6 hours) approximately equals a day on Earth, and because it has an axial tilt of 25°, Mars, like Earth, has seasons. But seasons on Mars are more extreme than Earth's, for the orbit of Mars is more eccentric. Summer in the Martian southern hemisphere occurs when the planet lies closest to the Sun, so the southern hemisphere summer is warmer than the northern hemisphere summer.

The radius of Mars is only about half that of the Earth. Like the Earth, Mars has an iron core, and the core's radius is about one-half that of the entire planet. Mars has only about one-tenth of the Earth's mass, so a 68-kg (150-pound) astronaut would weigh just 27 kg (60 pounds) on Mars.

Mars has two moons, Phobos (Fear) and Deimos (Panic), only 22 km (13.7 miles) and 13 km (8 miles) in diameter, respectively **(Fig. 22.19)**. Because of their small size, they have large bumps and dimples that gravity didn't smooth out. Numerous craters speckle their surfaces. The moons, which are probably asteroids that were captured by Mars's gravity, now orbit Mars at such low altitudes (9,400 km and 23,500 km, or 5,841 and 14,600 miles, respectively) that a human standing on the Martian surface could easily see them with the naked eye, even though they're so small.

THE SURFACE OF MARS. The surface of Mars is marked by distinct topographic features **(Fig. 22.20)**. Researchers

See for yourself

Valles Marineris: The Grand Canyon of Mars

Switch *Google Earth*™ to *Google Mars*™ by selecting the "Saturn" button on the button bar and selecting Mars.

Latitude: 14°35′57.61′ S
Longitude: 62°31′ 11.52′ W

From an elevation of 800 km (500 miles), you will see Valles Marineris, the Grand Canyon of Mars, lying below you. This canyon dwarfs the Grand Canyon of Arizona. It is 4,000 km (2,500 miles) long, 600 km (373 miles) wide, and reaches depths of 8 km (5 miles). The canyon covers almost 20% of the circumference of the planet.

Figure 22.20 Digital topographic maps of both sides of Mars.

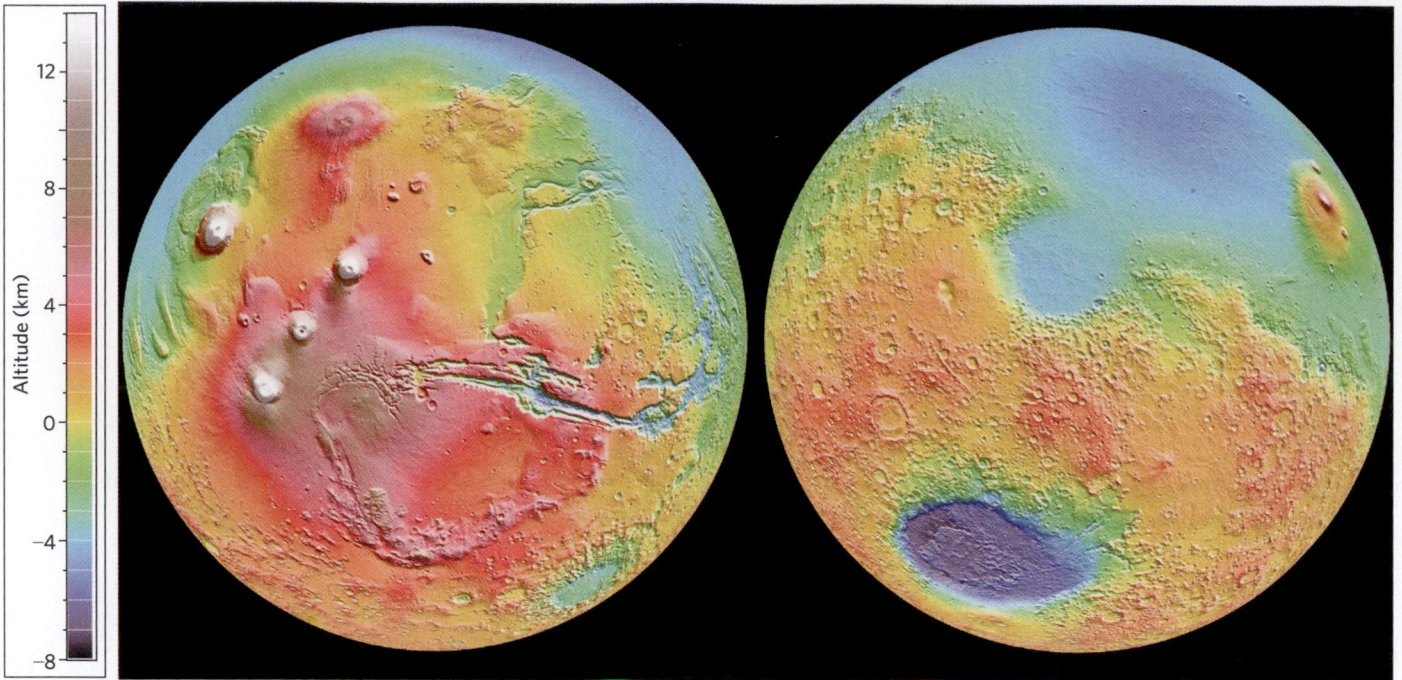

Altitude (km)

12
8
4
0
-4
-8

Figure 22.21 Dramatic landscape features on Mars

(a) Valles Marineris, the largest canyon in the Solar System.

(b) Oblique view of Olympus Mons, with a caldera visible at its summit.

divide the planet into two topographic provinces. The northern province, Vastitas Borealis, a gigantic lava-covered plain encompassing 40% of the planet's surface, lies at a low elevation. It has fewer craters than the rest of the planet, suggesting that basaltic lava covered it after the late heavy bombardment. The rest of the Martian surface consists of cratered highlands, with the exception of two huge impact basins, one of which—Hellas Planitia—includes the lowest point on Mars. Near the equator, the Tharsis Bulge, a plateau as big as North America, rises above the other highlands. Relatively few craters exist on the plateau's surface, suggesting that it's younger than the rest of the highlands. An immense canyon, Valles Marineris, defines the northern edge of the Tharsis Bulge **(Fig. 22.21a)**. This structure, which probably formed as a tectonic rift, is 4,000 km (2,500 miles) long, 600 km (373 miles) across, and up to 8 km (5 miles) deep. So it dwarfs the Earth's Grand Canyon, which is only 250 km (155 miles) long, 10 to 15 km (6.2 to 9.3 miles) wide, and up to 1.6 km (1 mile) deep.

Olympus Mons, the largest volcano in the Solar System, rises from the plains to the west of the Tharsis Bulge **(Fig. 22.21b)**. With a base diameter of 700 km (435 miles) and an elevation of 25 km (15.5 miles), this volcano is 2.5 times the size of the Earth's largest volcano, Hawaii's Mauna Loa. Like the volcanoes of Hawaii, Olympus Mons formed as a shield volcano that erupted relatively low-viscosity basalt. Notably, some of the flows on the flanks of Olympus Mons have so few craters that they may have spilled out of the volcano

Figure 22.22 Martian landscapes and features that could have been formed by flowing water in the past.

(a) Kasei Valles, showing what appears to be an outflow channel that, at times, contained flowing water.

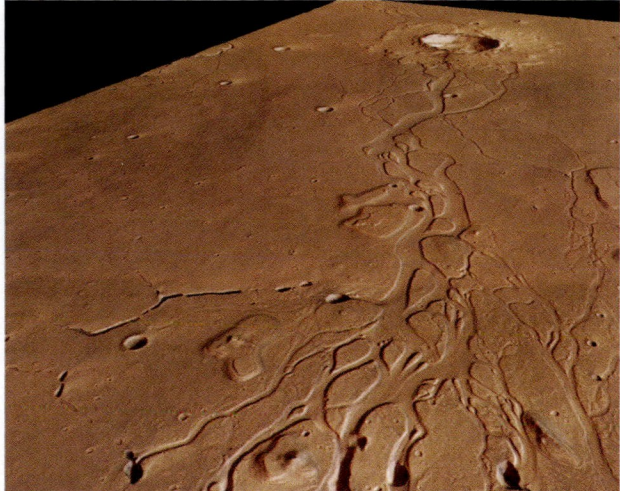

(b) Close-up of stream-like channels, as viewed obliquely from a space probe.

(c) The thin laminations in this rock look like cross beds deposited by currents.

once an ocean. Does any liquid water exist on Mars today? Maybe. In 2015, the *Mars Reconnaissance Orbiter* obtained images of streaks on hillslopes that become darker, as if damper, during part of the year.

Take-home message . . .

The terrestrial planets, although similar in their overall composition, differ in many ways. Mercury resembles the Earth's Moon, but is hotter on its sunlit side. A very dense, CO_2-rich atmosphere cloaks Venus. Mars is now a dry, cold planet with a very thin, CO_2-rich atmosphere. All but Venus contain abundant craters. Venus and Mars also host large volcanoes.

Quick Question -
Do plate tectonic processes happen on all terrestrial planets?

within the last 100 million years. An 80-km (50-mile)-wide caldera collapsed at the crest of the volcano.

Mars has an ice cap at each pole. *Martian ice caps* consist of seasonal ice—frozen CO_2 (dry ice) that accumulates when temperatures at the poles drop in the winter—on top of permanent ice, which lasts all year. The permanent ice of the north pole consists of a 1–3-km (0.6–1.9-mile)-thick sheet of water ice, while that of the south pole consists of frozen CO_2.

Data collected by probes suggest that large amounts of liquid water existed on Mars in the past, but has since vanished. For example, surface soils contain clay, calcite, and gypsum, minerals that form only in the presence of water. Images reveal landscape features that look like river-carved channels, drainage networks, flow-eroded islands, and even deltas **(Fig. 22.22)**. The presence of deltas and of terraces that look like remnants of beaches along the edge of Vastitas Borealis suggest that the plain was

22.5 The Jovian Planets and Their Moons

Each of the Jovian planets beyond Mars is a giant, in comparison to the terrestrial planets, and each holds onto many moons. Based on the state of the material within the Jovian planets, astronomers refer to the two nearer to the Sun (Jupiter and Saturn) as the **gas giants** and the farther two (Uranus and Neptune) as the **ice giants (Fig. 22.23)**. We knew little about these distant orbs until space probes sent back stunning images and other data. Some observations raise the possibility that liquid water may lie beneath the icy surfaces of some of the moons circling the Jovian planets . . . and if there's water, could there be life? This mystery remains to be solved.

Figure 22.23 Character-istics of the giant planets.

(a) The diameters and colors of the giant planets differ markedly from each other. All of the giant planets are much larger than the Earth.

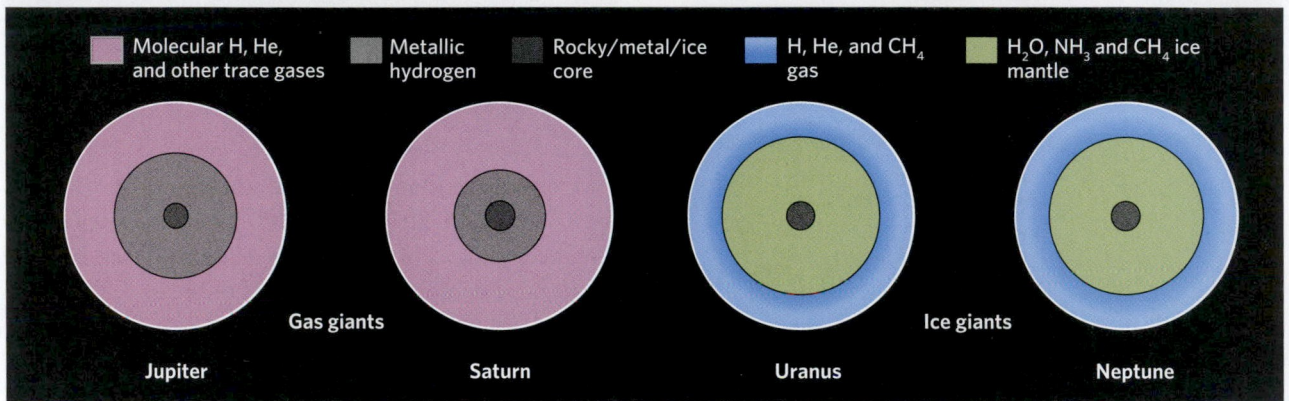

(b) The interiors of the giant planets differ from one another too. In this figure, all of the planets are resized to the same diameter, so as to emphasize the differences of their interior components.

Jupiter: The Giant among Giants

Jupiter, the Roman king of the gods, serves as an apt namesake for the fifth planet from the Sun, for Jupiter contains more matter than all the other planets combined. In fact, 70% of our Solar System's total mass, not including the Sun, lies within Jupiter. This huge planet rotates much faster than any other planet and completes a day in only 10 hours. Centrifugal force associated with this fast rotation makes the planet's equator bulge, so its equatorial radius (11 times the Earth's) exceeds its polar radius by 7%.

Jupiter produces its own internal heat, radiating twice as much energy as it receives from the Sun. This heat does not come from nuclear fusion inside Jupiter, for the planet would have to be a hundred times more massive for fusion to occur. Rather, it's left over from the compression of gas during Jupiter's formation.

THE MOST COLORFUL ATMOSPHERE OF ALL. Even an amateur astronomer looking through a telescope at Jupiter from the Earth can see that the giant planet's atmosphere differs markedly from the Earth's. Orange and tan bands, oriented parallel to the equator, surround the planet **(Fig. 22.24a)**. The composition of these bands remained uncertain until 1994, when *Galileo* entered orbit around Jupiter and dropped a probe into the atmosphere. The probe sent back data for 78 minutes as it descended for a few hundred kilometers, until its instruments succumbed to the intense temperature and pressure of its surroundings.

The *Galileo* orbiter and its probe revealed that Jupiter's outer atmosphere consists primarily of hydrogen (86.1%) and helium (13.8%), but also contains traces of methane (CH_4), ammonia (NH_3), ammonium hydrosulfide (NH_4SH), and water vapor (H_2O). Different components concentrate in distinct layers. Jupiter has no solid surface, but the gases become denser and denser toward the center of the planet. Researchers arbitrarily define the base of the atmosphere—and, therefore, the "surface" of the planet—as the elevation at which the atmosphere has a pressure of 10 atm (10 times that of the Earth). Temperatures at this depth reach about 300°C (570°F).

The colors of the visible atmospheric bands come from trace elements, such as sulfur and phosphorus, in the cloud layers. Researchers suggest that the bands indicate the presence of alternating upwelling (brighter colors) and downwelling (darker colors) regions of the atmosphere. Due to the Coriolis force, these bands align parallel to the equator. They flow at different speeds and even in different directions, so swirling eddies form at the boundaries between bands **(Fig. 22.24b)**.

A giant ellipse of reddish gas, the **Great Red Spot**, dominates images of Jupiter's southern hemisphere **(Fig. 22.24c)**. Detailed cloud patterns in and around the spot indicate that it's a huge, counterclockwise-rotating storm, big enough to contain two to three Earths. The Great Red Spot has existed for centuries and may be permanent. Its origin and longevity remain a mystery. Other huge storms form on Jupiter, but they exist for a shorter time before dissipating or combining with other storms.

WHAT'S INSIDE JUPITER? While the volume of 1,320 Earths could fit inside the volume of Jupiter, the giant planet's mass is only 318 times that of the Earth (see Table 22.3). That's because Jupiter has an average density that's only about one-fourth that of the Earth, for Jupiter contains mostly hydrogen and helium while the Earth consists mostly of rock and metal.

What would we see if we could slice through Jupiter? At the base of the atmosphere, the planet consists of compressed hydrogen and helium gas. Perhaps 100 km (60 miles) farther down, the pressure becomes great enough that the gas turns into liquid. The hydrogen of this liquid still consists of H_2 molecules, but the molecules can get close enough to one another to interact and bond.

The liquid hydrogen layer extends downward for perhaps 20,000 km (12,400 miles). Below this depth, where pressures reach 3 million atmospheres and temperatures reach 11,000°C (20,000°F), hydrogen undergoes a transition into a strange material, called **metallic hydrogen**, that doesn't exist on Earth. Metallic hydrogen flows like a liquid, but the atoms have been squeezed together so tightly that they can't hold their electrons in orbit, so the electrons flow freely, as in a metal. The flow of electrons in Jupiter's metallic hydrogen layer generates a very strong magnetic field around the planet and causes brilliant aurorae near its poles. The magnetic field, in turn, traps charged particles to produce intense radiation belts thousands of times stronger than the Earth's Van Allen belts. Because of these radiation belts, space probes exploring Jupiter must have strong shielding to prevent their electronics from frying. The metallic hydrogen layer continues down to a depth of 60,000 km (37,300 miles). Below that depth, researchers speculate that Jupiter has a rocky and metallic core that may contain about 12 to 45 times the mass of the Earth, though it is only a small part of Jupiter's mass.

AMAZING MOONS. Jupiter has 67 confirmed moons (**Earth Science at a Glance**, pp. 806–807). The four largest, Io, Europa, Ganymede, and Callisto, known as the *Galilean moons* because Galileo observed them in 1610, reach planetary dimensions **(Fig. 22.25)**. Ganymede, the largest moon of Jupiter, has a diameter 8% larger than Mercury's and 1.5 times larger than that of the Earth's Moon, which makes it the largest moon in the Solar System.

Figure 22.24 Color bands, winds, and storms on Jupiter.

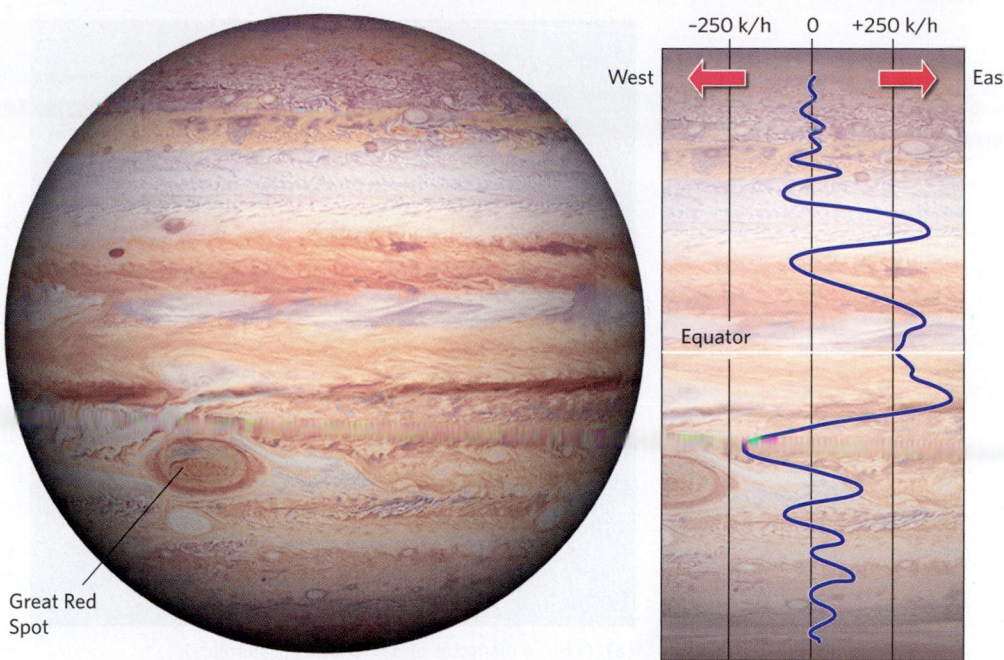

(a) Jupiter hosts distinct color bands.

(b) Wind velocity and direction on Jupiter vary with latitude.

(c) A close-up of the Great Red Spot, a giant storm that rotates counter-clockwise. The spot is bigger than the Earth.

The other Galilean moons are also comparable to the Earth's Moon in size, but all four differ markedly from our Moon and from one another in several ways.

Io, the closest Galilean moon to Jupiter, looks like a weirdly colored pizza, with splotches and spots of rich reds, golds, yellows, browns, and blacks (see Fig. 22.25a). The colors come from sulfur and other elements in the ash and

Figure 22.25 Jupiter's large (Galilean) Moons.

(a) Io has a diameter of 3,636 km (2,259 miles).

(b) Ganymede has a diameter of 5,268 km (3,273 miles).

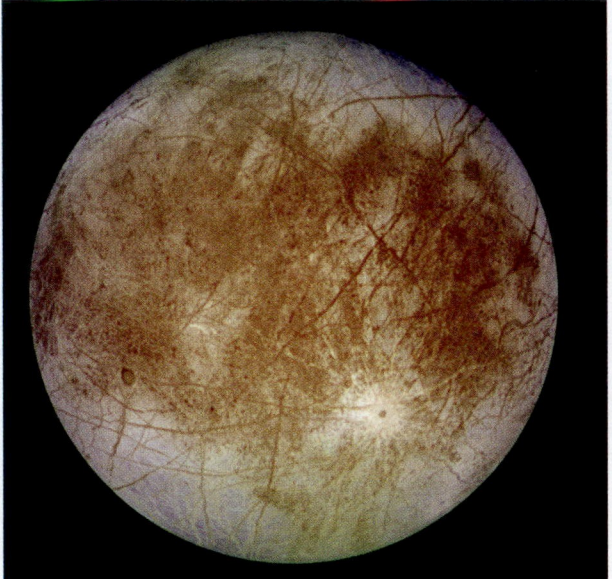

(c) Europa has a diameter of 3,100 km (1,930 miles).

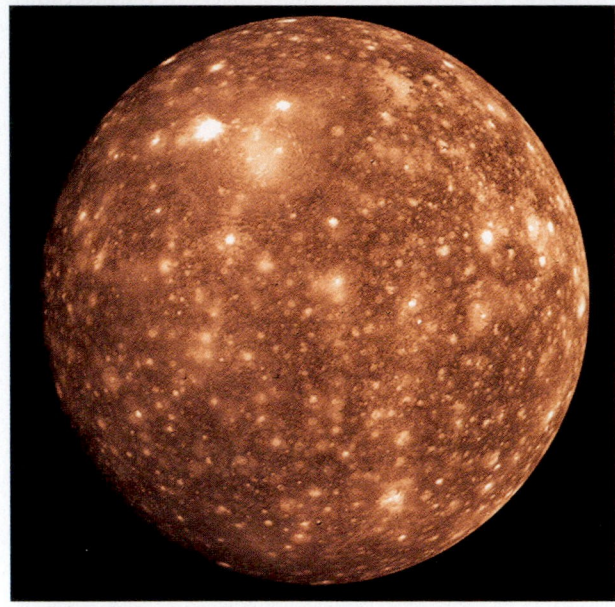

(d) Callisto has a diameter of 4,821 km (2,995 miles).

lava erupted by hundreds of volcanoes. The *Voyager* space probe caught eight of these volcanoes in the act of erupting. What provides the heat that drives this volcanism? Jupiter's immense gravity generates huge tides on Io, and as the tidal bulge moves around the planet, the up-and-down motion of material in its interior generates enough frictional heat to cause melting. The volcanoes erupt explosively, constantly resurfacing the moon, so it retains no impact craters. Beneath the colorful surface, Io probably resembles the terrestrial planets in having a rocky mantle and an iron core.

The remaining three Galilean moons, Europa, Ganymede, and Callisto (see Fig. 22.25b–d), have water-ice surfaces surrounding rocky interiors. Measurements made by the *Galileo* space probe suggest that both Ganymede and Europa have a solid, brittle outer crust of water ice on top of a warmer, softer ice layer **(Fig. 22.26)**. Europa may host a vast subsurface ocean of liquid water, up to 100 km (62 miles) deep, beneath an ice layer. On both moons, the water layers overlie an internal shell of rock that surrounds an iron core. The existence of an ocean beneath Europa's surface has scientists wondering whether

Figure 22.26 Speculative models of the possible watery interiors of Ganymede and Europa.

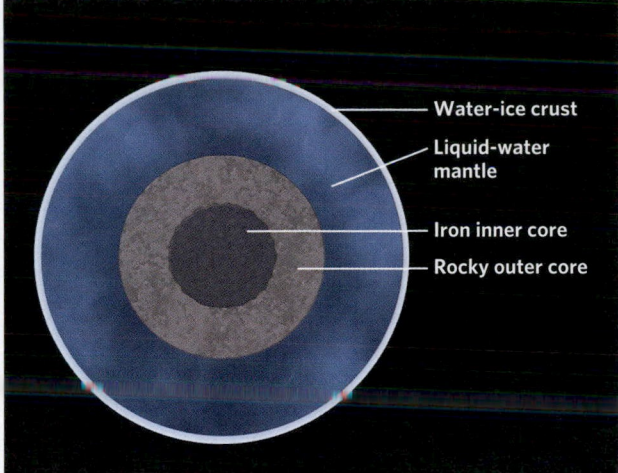

- Water-ice crust
- Liquid-water mantle
- Iron inner core
- Rocky outer core

Ice crust

Liquid ocean under ice

Model of Europa's outer layers

- Metallic core
- Water layer
- Rocky interior

(a) Ganymede may have a liquid-water mantle.

(b) An ocean may exist beneath an ice crust on Europa.

primitive life inhabits this moon. Callisto differs from other large moons, and from any terrestrial planet, in an important way: it has not differentiated into a core and mantle, which suggests that this moon accumulated from debris orbiting Jupiter later in Solar System history, when the accreting material was too cool to melt.

Jupiter has a ring, but it can't be seen from the Earth. It was detected for the first time by the *Voyager* space probe. This ring, which circles the planet in the plane of its equator, lies inside the orbit of Jupiter's innermost moon. It's a few thousand kilometers wide from its inner to its outer edge. Two small moons lie close to the ring and may have been the source of the debris within the ring.

Saturn: The Ringed Planet

Saturn, named for the Roman god of agriculture, has such a high albedo and is so big that you can see it without the aid of a telescope, even though it's so far away. Saturn resembles its bigger brother Jupiter in many ways, but Saturn's rings, in contrast to those of Jupiter, are very large and bright enough to be seen from the Earth.

SATURN'S ATMOSPHERE AND INTERIOR. Like Jupiter, Saturn has a low average density because it consists mostly of hydrogen (96% by volume). In fact, Saturn has the lowest average density of all the planets (0.7 g/cm³, less than that of water). We know its density is less than that of Jupiter because it doesn't exert as much gravitational pull, so the hydrogen within Saturn doesn't undergo as much compression as in Jupiter.

Like Jupiter, Saturn lacks a solid surface, so researchers arbitrarily define the base of its atmosphere as the elevation where pressure is about 10 atm. And like Jupiter, Saturn rotates so fast—a day on Saturn lasts 10.75 hours—that

the planet bulges at its middle, and its equatorial radius exceeds its polar radius by 11%. Saturn's thick atmosphere, composed mostly of hydrogen (92.4%) and helium (7.4%), contains cloud layers that resemble Jupiter's in composition, but Saturn's atmosphere doesn't convect as vigorously as Jupiter's, so Saturn has duller stripes and an overall tannish hue (see Fig. 22.23). Nevertheless, Saturn has intense winds that race along at up to 1,500 km per hour (930 mph).

The core of Saturn, which probably consists of refractory material (rock and metal), has a mass about 10 to 20 times that of the Earth and a temperature of about 11,700°C (21,000°F). Outside this refractory core, Saturn consists of a layer of dense ice (composed mostly of ammonia and water), a layer of metallic hydrogen, a layer of helium (which probably developed when drops of liquid helium formed and fell like rain toward the interior), and a layer of liquid molecular hydrogen. Because Saturn has less internal pressure than Jupiter does, Saturn's metallic hydrogen layer is thinner than Jupiter's, and Saturn has a weaker magnetic field. Saturn generates about 2.5 times as much energy as it receives from the Sun. While some of this energy comes from compression of the planet's gases, astronomers speculate that some comes from the friction generated by the downward rain of helium.

AMAZING RINGS. When Galileo first looked at Saturn, he saw ear-like protrusions to either side of the planet, but had no idea what they could be. A half century later, an astronomer using a more powerful telescope recognized the protrusions as rings of matter that surrounded the planet and reflected light. In 1675, Giovanni Cassini showed that Saturn has several distinct rings extending over 100,000 km (62,000 miles) out into space. No other planet displays rings as bold and beautiful as Saturn's.

Did you ever wonder . . .

why Saturn has spectacular rings?

A Menagerie of Moons

The moons of the solar system vary in size, shape, composition, and surface characteristics. Some, like Io, have active volcanoes or geysers. Others, like Enceladus, are covered with water ice, and may harbor an ocean beneath the icy surface. Some are cratered, while others have smooth surfaces. One, Titan, has an atmosphere. Many are spherical, but the smallest, such as Deimos, are not large enough for gravity to compress them into spherical shapes.

Deimos, Mars's small moon

Mimas, Saturn's cratered moon

Titan

Mercury

Callisto

Io

Moon

Europa

Iapetus, Saturn's icy and cratered moon

Triton

Pluto

Titania

Oberon

Tethys

Enceladus

Mimas

Enceladeus, Saturn's icy moon

Ganymede

Mars

Venus

Earth

Io,
Jupiter's erupting moon

Titan,
Saturn's moon
with an atmosphere

Europa, Jupiter's water- and
ice-covered moon

Figure 22.27 *Cassini* views of the rings of Saturn.

(a) Picture of the rings of Saturn during an eclipse (the Sun is behind Saturn). Note the fainter, more distant rings.

(b) False colors represent different particle sizes. Purple means that particles larger than 5 cm (2 inches) are common; green means that particles smaller than 5 cm are common. White indicates dense rings.

Our understanding of the rings improved greatly when the *Cassini* space probe sent back high-resolution photographs revealing fainter rings, which extend beyond the main rings, and that each main ring consists of hundreds of thin ringlets (10 to 15 m, or 32 to 50 feet, thick) with tiny gaps in between **(Fig. 22.27)**. The rings consist of countless particles of water ice and rocky dust, ranging in size from less than a millimeter to a few meters wide.

Why do Saturn, and the other Jovian planets, have rings? Astronomers suggest that the rings were formed by tidal forces acting on moons orbiting the planets. A planet's gravitational tug stretches the moons toward the planet, and moons closer to the planet experience a stronger pull. When close moons were subjected to strong tidal

forces, their internal gravity couldn't hold them together, and eventually, each moon shattered into fragments, which dispersed into a ring in what had been the orbital plane of the moon. Some gaps between rings exist because they have been cleared by small moonlets, bodies 10 to 20 km (6 to 12 miles) in diameter that gravitationally vacuum in debris. Other gaps have formed because Saturn's moons act as shepherds that gravitationally "herd" particles into rings and out of gaps.

THE MANY MOONS OF SATURN. Saturn hosts 62 known moons, of which only 13 have diameters larger than 50 km (30 miles). Most of the smaller moons don't spin on an axis, but rather tumble through space as they follow their orbit. They probably represent captured planetesimals or fragments from colliding planetesimals.

Each of the seven largest moons of Saturn—those that became big enough to be roughly spherical—has unique characteristics **(Fig. 22.28a)**. Titan, the largest by far, has a diameter larger than Mercury's and an atmosphere 10 times denser than the Earth's. Unlike the atmosphere of Saturn itself, Titan's atmosphere consists of nitrogen (98%), methane (2%), and other trace gases, including a variety of hydrocarbons. Images made by the *Huygens* space probe, which parachuted to the surface of Titan in 2005, suggest that methane clouds hover in Titan's lower atmosphere and that liquid methane rains from these clouds, collecting in hydrocarbon lakes and rivers on the surface. Radar mapping reveals volcanic vents that emit gases, as well as huge dunes, presumably composed of ice and frozen methane grains.

The next six large moons are much smaller than Titan. None have atmospheres, and most appear to be similar in composition, with rocky interiors beneath a water-ice shell. Regions on the surfaces of these moons have high crater densities, suggesting that these areas have been tectonically inactive since the early days of the Solar System. But some of the moons display cracks or ridges indicative of tectonic activity in the past, and one, Enceladus, has been substantially resurfaced by younger material. Photographs of Enceladus reveal geysers of water vapor and ice crystals erupting from a set of beautiful blue fissures, informally called tiger stripes, near the moon's south pole **(Fig. 22.28b)**.

Uranus and Neptune: The Ice Giants

DISCOVERY AND BASIC CHARACTERISTICS. Uranus and Neptune lie so far from the Earth that they cannot be seen with the naked eye, even on the darkest night. Astronomers using telescopes first detected Uranus in 1781 and named it after the Greek god of the sky. (Alas, there is no polite way to pronounce the name of the planet in English—Earth Science teachers wish that we could revert to the original Greek pronunciation, oo-ron-ohs, to avoid snickering in class!) Irregularities in Uranus's orbit, which could only have been caused by the pull of yet another planet farther out, led astronomers to find Neptune in 1846. Uranus and Neptune are both much larger than the Earth—their radii are 4.0 times and 3.9 times the Earth's radius, respectively—but they are only about one-third the size of Jupiter.

Uranus's axis of rotation has a tilt of 98° (see Fig. 22.5), so it lies almost parallel to the planet's orbital plane! In effect, Uranus lies on its side. As a result, Uranus has the most extreme seasons of any planet in the Solar System. During its orbit, which takes 84 Earth years, its polar regions remain in complete darkness for 42 years and then in constant sunlight for 42 years. Neptune's axis also tilts, but at an angle of 28°, just slightly greater than that of the Earth's axis. Both planets rotate faster than the Earth: a day lasts 17.2 hours on Uranus and 16.1 hours on Neptune.

INTERIORS AND ATMOSPHERES OF THE ICE GIANTS. On Uranus and Neptune, interior pressures don't become great enough for metallic hydrogen to form, as on Jupiter and Saturn. The rocky and metallic cores of Uranus and Neptune appear, instead, to be surrounded by a thick layer of slush composed of water, ammonia, and methane ice. A layer of liquid molecular hydrogen, helium, and methane surrounds the slush (see Fig. 22.23). The presence of icy slush inside Uranus and Neptune led to their designation as ice giants.

From space, all we can see when looking at Uranus is the thick, frigid (–215°C) haze that forms the top of its atmosphere, so Uranus appears as a featureless teal-colored globe without banding or storm spots **(Fig. 22.29a)**. Neptune, in contrast, has a deep blue atmosphere with identifiable features, including bands and white streaks caused by clouds **(Fig. 22.29b)**. Uranus's and Neptune's

Figure 22.28 The moons of Saturn.

Mimas (363 km)

Titan (5,152 km)

Enceladus (505 km)

Dione (1,122 km)

(a) Four of Saturn's 62 known moons.

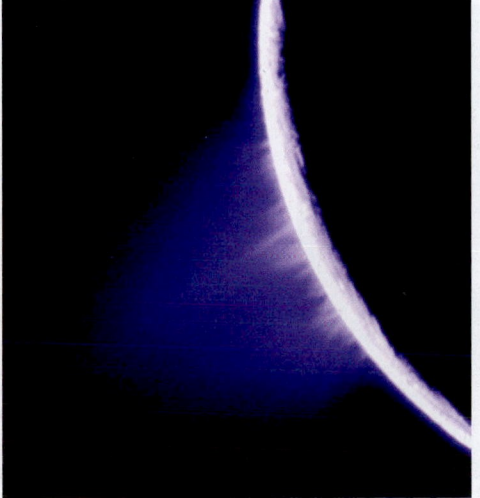

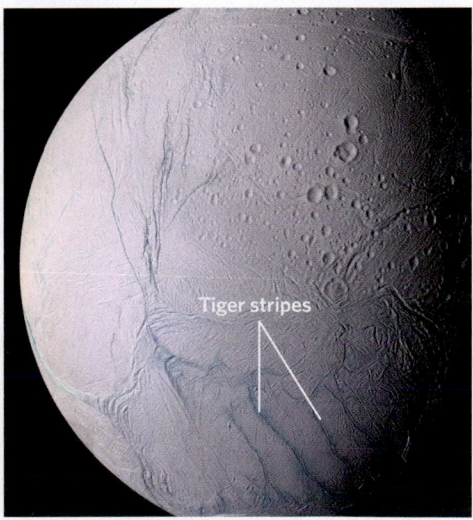
Tiger stripes

(b) Jets of volatiles erupt from the "tiger stripes" of Enceladus's south pole.

Figure 22.29 The ice giants.

(a) The atmosphere of Uranus appears featureless. Although the planet has rings, they are not visible without enhancing the image.

(b) Neptune's atmosphere has bands and storms. Neptune has rings, but they're even fainter than those of Uranus.

atmospheres consist primarily of hydrogen (84%) and helium (14%). The concentration of methane determines the richness of the blue color of these planets, so Neptune (with 3% methane) appears deep blue, while Uranus (with 2% methane) appears paler.

STILL MORE MOONS AND RINGS. At present, astronomers have identified 27 moons orbiting Uranus and 13 orbiting Neptune, as well as thin rings around each planet

(Fig. 22.30). The five largest moons of Uranus are all smaller than the Earth's Moon. They appear to be composed of ice and rock and are heavily cratered **(Fig. 22.31a)**. Neptune has one large moon, Triton, which has a diameter of about 2,700 km (1,680 miles) **(Fig. 22.31b)** and is the only large moon in the Solar System to orbit in a retrograde direction. The frigid (–236°C; –393°F) surface of Triton consists of water ice with polar caps of nitrogen ice.

Figure 22.30 Rings of the ice giants.

(a) The rings of Uranus.

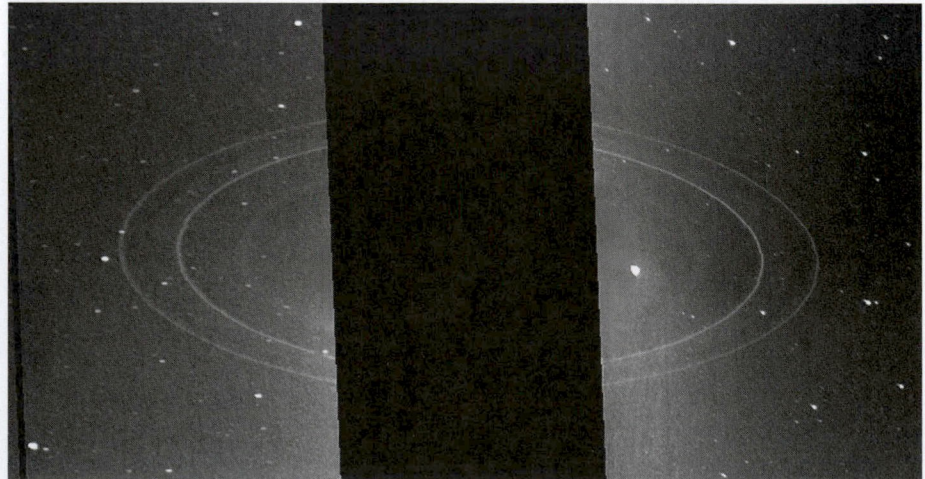

(b) The rings of Neptune. The planet itself has been blacked out in this image, so that its brightness doesn't mask the rings.

Figure 22.31 Moons of the ice giants.

(a) Titania, a moon of Uranus.

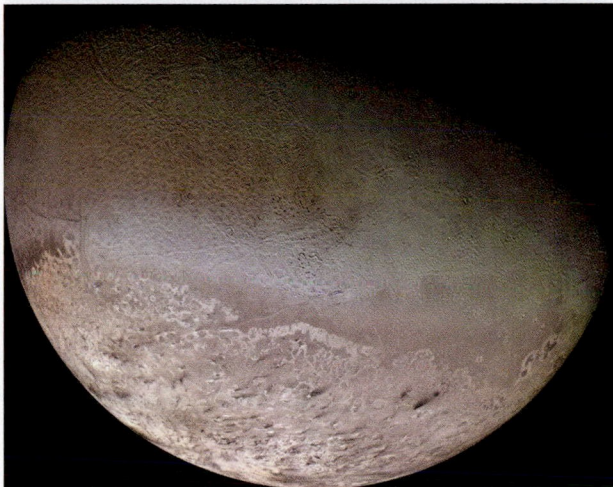

(b) Triton, a moon of Neptune.

Take-home message . . .

The Jovian planets all consist primarily of hydrogen and helium surrounding rocky cores. In Jupiter and Saturn, the larger of the Jovian planets, internal pressures transform hydrogen into a metallic form. The composition of clouds in the atmospheres of the Jovian planets determines their colors as seen from space. All have a large number of moons, and all have rings, with Saturn's the most prominent by far. Some of the moons may contain liquid water. The bluish color of Uranus and Neptune comes from the presence of methane.

Quick Question -
Why do we refer to Jupiter and Saturn as the gas giants and to Uranus and Neptune as the ice giants?

22.6 The Other "Stuff" of the Solar System

The eight planets and their moons represent the vast majority of the non-solar mass in the Solar System, but only a tiny fraction of the total number of objects. Researchers estimate that billions of other objects—nearly all too tiny or too far away to observe—move with the Sun as it orbits the Milky Way's galactic center. We can classify these objects into five primary groups—asteroids, comets, Kuiper Belt objects, Oort Cloud objects, and meteoroids—that differ from one another in terms of size, orbital characteristics, and location relative to the planets. (Astronomers now use the term *dwarf planet* for larger, spherical asteroids or Kuiper Belt objects.) An understanding of these objects not only enhances our understanding of how the Solar System developed, but also allows astronomers to evaluate the remote, but real, hazard that a large object may someday collide with our planet.

Asteroids

By convention, astronomers consider rocky or metallic objects larger than 100 m (330 feet) across to be **asteroids**. Astronomers estimate that there may be about 200 asteroids with diameters larger than 100 km (60 miles), 750,000 with diameters larger than 1 km (0.6 miles), and millions of smaller ones. Nearly all asteroids orbit the Sun within the **asteroid belt**, a band between the orbits of Jupiter and Mars **(Fig. 22.32a)**.

Where did asteroids come from? Researchers suggest that they include primitive planetesimals that never coalesced into a larger body as well as remnants of large planetesimals that had differentiated into a core and mantle before breaking up. The gravitational pull of Jupiter disturbed the orbits of the asteroids so much that they could never amalgamate into a planet.

Typical asteroids have irregular shapes, and may be elongate in one direction, because they are too small to have been remolded into spheres. Three asteroids, including Ceres, with a diameter of 940 km (584 miles) have become spheres, so they qualify as dwarf planets **(Fig. 22.32b)**. Ceres probably has a rocky interior, surrounded a shell of water ice, which in turn is coated with a surface layer of micrometeorite dust. The *Dawn* space probe went into orbit around Ceres in 2015 and sent back images of its intensely cratered surface, on which highly reflective mountains, possibly exposures of its icy crust, have formed. Space probes have also taken close-up images of other asteroids **(Fig. 22.32c)**. In 2001, one landed on Eros, and in 2010, one landed on Itokawa, grabbed a sample, and returned it to Earth.

Figure 22.32 Asteroids.

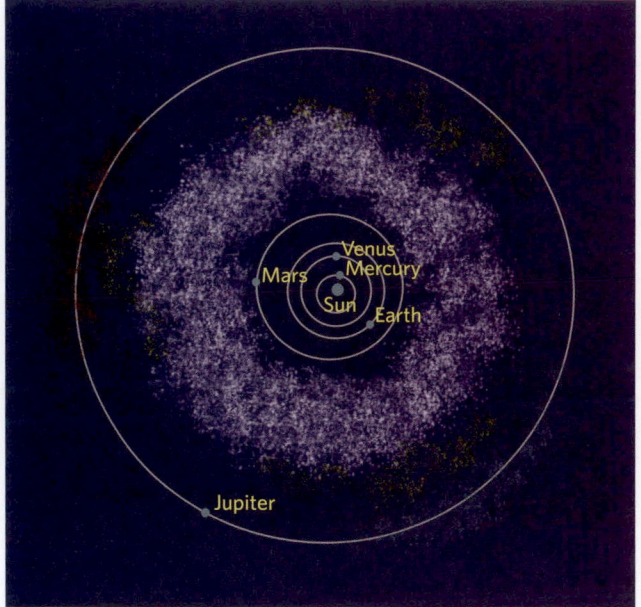

(a) Asteroids lie in a broad ring between the orbits of Mars and Jupiter.

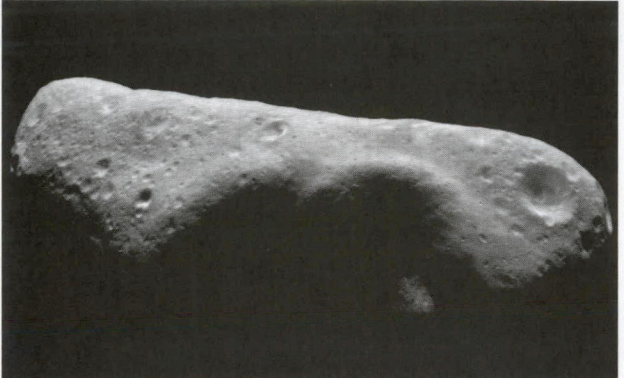

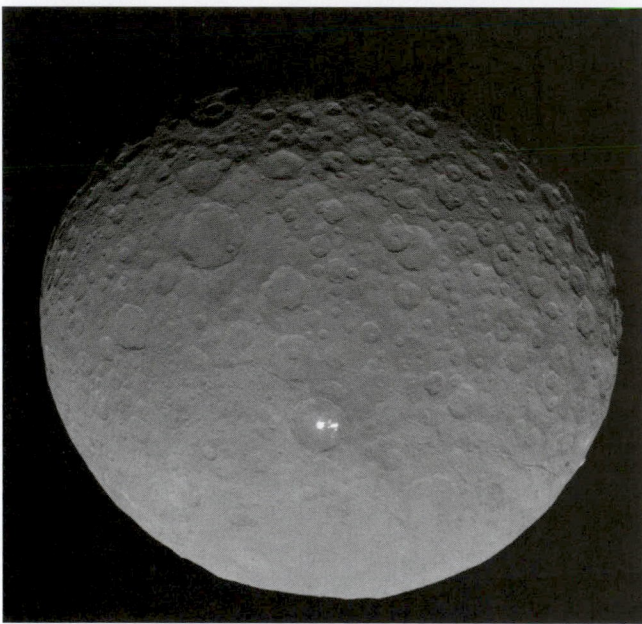

(b) The largest asteroid, Ceres, is now referred to as a dwarf planet. It was photographed by the *Dawn* spacecraft in 2015.

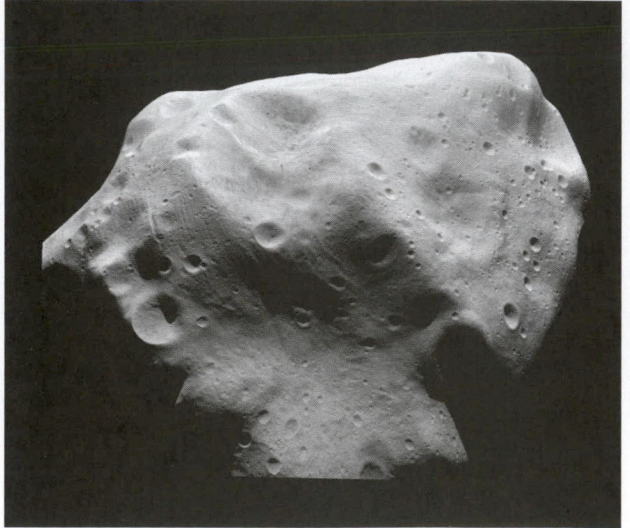

(c) Examples of small asteroids.

Pluto, Other Kuiper Belt Objects, and the Oort Cloud

Even before the dramatic vote demoting Pluto from planet to dwarf planet, astronomers realized that Pluto was an oddball among the other planets. Most notably, Pluto moves differently. Its orbit not only lies at an angle of 17° relative to the ecliptic plane, but is also highly eccentric: at its aphelion, Pluto lies 1.7 times farther from the Sun than it does at its perihelion. In fact, for part of a Pluto year, Pluto lies closer to the Sun than Neptune does.

Pluto's surface characteristics remained unknown until 2015, when the *New Horizons* probe sent back amazingly detailed photos **(Fig. 22.33a)**. The photos reveal that this small world, which has a diameter about a fifth that of Earth's, has a surface with plains that display complex textures resembling snakeskin and mountains that rise to elevations of 3.5 km (2 miles) (see Fig. 22.2). Pluto has a low density (1.9 g/cm³), which gives it a mass that's about 0.2% of Earth's, so it probably consists mostly of water ice and frozen nitrogen. Four moons orbit Pluto, of which the

Figure 22.33 *New Horizons* images of Pluto and Charon.

(a) Close-ups of Pluto's surface reveal icy mountains and fractured plains.

(b) Charon has few impact craters, suggesting that the moon has been resurfaced.

largest, Charon, has a diameter of 1,300 km (800 miles) **(Fig. 22.33b)**. All the moons appear to consist of water ice.

Pluto is one of several objects with diameters in the range of 1,000 to 2,500 km (600 to 1,500 miles) that lie outside the orbit of Neptune. One of these, Eris, may contain more mass than Pluto. Pluto and Eris are the largest known objects of the **Kuiper Belt**, named for the astronomer Gerard Kuiper (1905–1973). These objects, millions of which are tiny, but perhaps over 100,000 of which have a diameter larger than 100 km (60 miles), are icy bodies that circle the Sun in a region that extends from the orbit of Neptune (30 AU) out to a distance of about 50 AU from the Sun. Overall, many objects in the Kuiper Belt orbit near the ecliptic plane, but many other individual objects within it, such as Pluto, have orbits that are inclined to the ecliptic plane.

In 1950, a Dutch astronomer, Jan Oort (1900–1992), proposed that even more icy objects lie beyond the Kuiper Belt but remain gravitationally attached to the Sun. Researchers envision these objects as occupying a somewhat spherical shell, now called the **Oort Cloud**. Its inner edge lies at a distance of about 2,000 AU from the Sun, and its outer edge extends to a distance of 100,000 AU (1.5 light-years), more than a quarter of the distance to the nearest star. The Oort Cloud probably consists of debris that condensed during Solar System formation but never fell into the accretionary disk.

Comets

Imagine how baffled ancient observers must have been when they looked skyward at an area that had previously contained only stars and saw a bright object with a long, glowing tail **(Fig. 22.34a)**. As they watched the object over successive nights, it moved across the backdrop of stars, dimmed, and then disappeared. It's no wonder that, for millennia, such objects, called **comets**, were thought to be omens of historical events. It wasn't until 1705 that the English astronomer Edmond Halley (1656–1742), by applying Newton's laws, calculated the orbit of a comet (now called Halley's comet) and showed that it orbits the Sun. Other astronomers in the 18th and 19th centuries proposed that comets were solid objects that emitted gas when they came close to the Sun, but it wasn't until 1950 that the current image of a comet—as a dirty snowball—came into favor. Modern space probes have provided spectacular images of comets. In 2005, the *Deep Impact* probe blasted debris off a comet that could be analyzed remotely, and in 2014, the *Rosetta* probe went into orbit around a comet and successfully placed a lander on its surface.

We now understand that comets follow highly elliptical orbits with the Sun at one focus **(Fig. 22.34b)**. At the head of a comet lies a solid mass, the comet's *nucleus*. All well-studied comets have an irregularly shaped nucleus with a rubbly looking surface **(Fig. 22.35)**. They range from 0.5 to 50 km (0.3 to 30 miles) long and 0.1 to 10 km (0.06 to 60 miles) wide, and they seem to consist of loosely bound aggregates of rock and ice (water ice, frozen methane, dry ice, and frozen ammonia), along with traces of more complex organic chemicals. As a comet gets to within about 3 to 4 AU of the Sun, its dark surface absorbs enough solar radiation for the comet's interior to start vaporizing. Volatiles vent into space through cracks on the comet's surface, producing geyser-like spouts of ionized (electrically charged) gas that carry refractory dust particles along with them. Some of the gas surrounds the nucleus to form a glowing atmosphere, or *coma*. The rest of the gas, along with all the dust, streams into space as a *comet tail*. The gas forms an *ion tail* whose orientation lies parallel to the direction of the solar wind, and always points away from the Sun. The dust forms a *dust tail* that tends to curve, for the dust particles in it are partly following the comet's orbit, but they are also affected progressively by the solar wind **(Fig. 22.36)**. Some tails can be more than 1 AU long. The closer a comet comes to the Sun, the longer and brighter its tails become.

Did you ever wonder . . .

why a comet has a tail?

Figure 22.34 Comets.

(a) Hale-Bopp, a particularly dramatic comet, crossed the night sky in 1997.

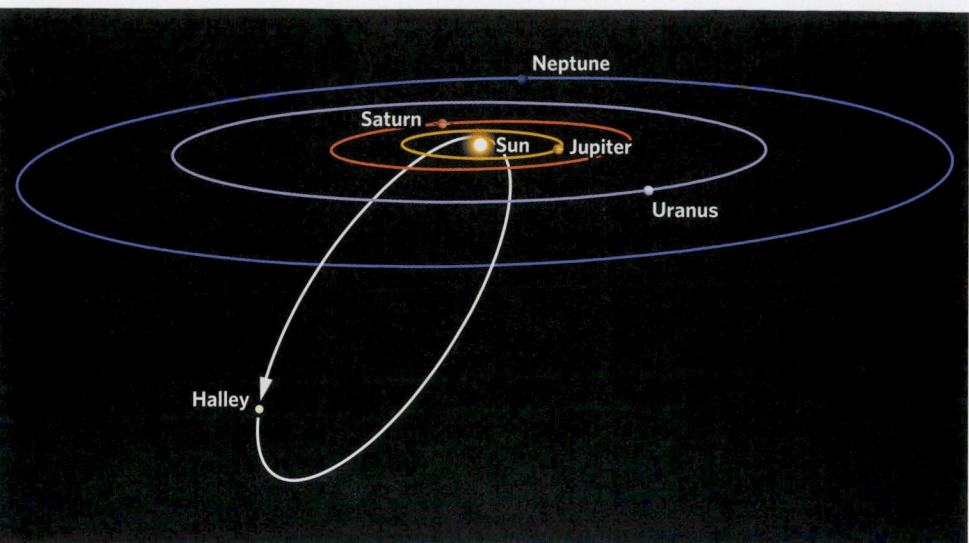

(b) Comets have very eccentric orbits.

The formation of tails causes a comet to lose material each time it approaches the Sun. Small comets probably have little chance of survival beyond a few orbits, and even a large comet probably disintegrates after a few thousand orbits. Therefore, there must be a source that, over time, sends new icy bodies into the inner Solar System. Researchers conclude that comets come from both the Kuiper Belt and the Oort Cloud.

Researchers classify comets based on the duration of their orbit. *Short-period comets* have less eccentric orbits and complete an orbit in less than 200 years, whereas *long-period comets* have more eccentric orbits that take more than 200 years—thousands of years in some cases—to complete. Researchers suggest that short-period

comets originate as Kuiper Belt objects and that long-period comets come from the Oort Cloud. These objects become comets when gravitational tugs or collisions with other bodies send them careering toward the inner Solar System.

Meteoroids, Meteors, and Meteorites

Astronomers refer to small objects (less than 1 m, or 3 feet, across) traveling through space as **meteoroids**. Larger objects are referred to as small asteroids, small comets, or fragments of planets, moons, or asteroids ejected into space by collision with another object.

Meteoroids whose paths cross the Earth's orbit may collide with the Earth. Those that enter the Earth's

Figure 22.35 The nucleus of a comet has an irregular shape.

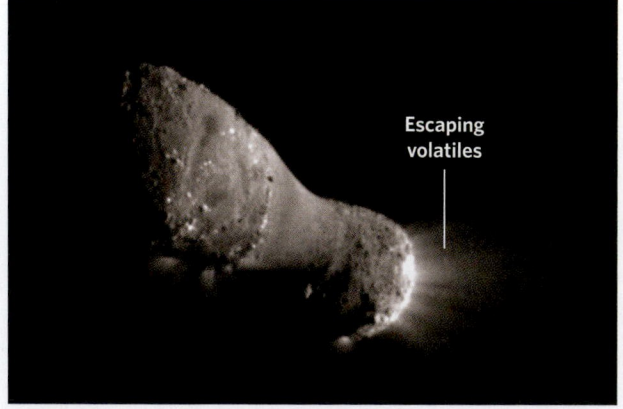

(a) Comet Hartley 2, releasing volatiles. It is 2.0 km (1.2 miles) long.

(b) Comet 67P/Churyumov-Gerasimenko. It is 6.7 km (4.2 miles) long.

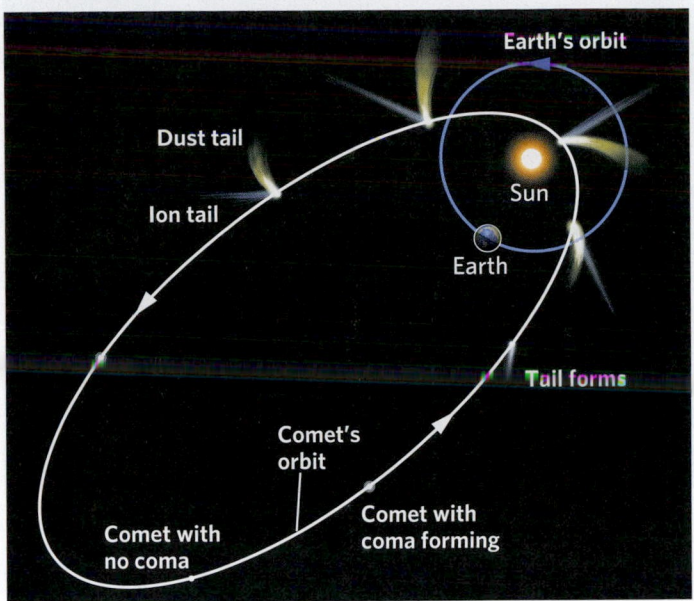

atmosphere arrive at speeds of around 20 km per second (50,000 mph)—100 times the speed of a jet plane. As the speeding object enters the atmosphere, it intensely compresses the air in its path. Compression of a gas heats it up, so a meteoroid entering the air generates immense heat, enough to cause some or all of the object to vaporize. Glowing gases from the meteoroid vapor, along with glowing superheated air molecules, produce a blazing light streak behind the object **(Fig. 22.37a)**. The light streak is called a **meteor** (or, misleadingly, a "shooting star"). Note that by this definition, a meteor is an atmospheric phenomenon. To produce a meteor, an incoming meteoroid must be larger than a sand grain. Note that frictional heating is not what produces the light of a meteor; air isn't dense enough in the upper atmosphere for friction to have much effect.

During a **meteor shower**, an observer can see from ten to over a hundred meteors per hour, all of which appear to emerge from a common point in the sky (see Fig. 22.37a). These showers, which last for one to three days, happen when the Earth passes through the orbit of a comet and intersects its dust tail. The Earth passes specific cometary orbits at the same time every year. Occasionally, meteor showers become *meteor storms*, during which over 1,000 meteors streak across the sky every hour. Not all meteoroids arrive in showers, however; astronomers refer to those that arrive independently as *lone meteoroids*.

When a larger object enters the atmosphere, it becomes a very bright *fireball* that leaves a visible smoke-like trail across the sky. If it explodes, a fireball becomes a *bolide*. During the last century and a half, two bolides have ex-

ploded over Siberia. The first, known as the Tunguska bolide, released about 10 megatons of energy—1,000 times more than the Hiroshima atomic bomb—and flattened about 2,100 km² (830 square miles) of Russian forest. It was probably about 120 m (400 feet) across and blew up at an elevation of 5 to 10 km (3 to 6 miles). A similar but smaller (0.5-megaton) bolide explosion took place over Russia in 2013 **(Fig. 22.37b)**. Dashboard video cameras recorded the streak of the fireball's tail, the blinding light of its explosion, and the chaos caused when its shock wave blew out windows and knocked down walls.

If a meteoroid makes it to the ground without completely burning up, it strikes the Earth. The meteoroid then becomes a **meteorite (Fig. 22.37c)**. Smaller, slower meteoroids can excavate a small indentation when striking the ground, and the rare ones that have landed in populated areas have punched holes through roofs and dented cars. Larger impacts blast out a crater, whose size depends on the size, speed, and density of the meteoroid. The 1.2-km (0.75 mile)-wide Meteor Crater of Arizona was formed about 50,000 years ago from the impact of a dense meteorite 50 m (160 feet) in diameter **(Fig. 22.37d)**.

During the course of Earth history, very large lone objects (1 to 15 km, or 0.6 to 9 miles, across) have collided with the Earth **(Box 22.3)**. The resulting impacts have generated craters up to 180 km (110 miles) across and have disrupted the crust for several kilometers down; the largest impacts have had effects all the way down to the Moho. Such impacts produce fractures and faults, impact melts, ejecta that include ultra-high-pressure minerals, and an unusual pattern of cone-shaped cracks known as

Figure 22.37 Meteors and meteor craters.

(a) The Perseids meteor shower.

(b) The bolide that created a huge shock wave over Russia on February 15, 2013.

2 cm

(c) An example of a meteorite.

(d) Meteor Crater in Arizona.

shatter cones. Geologists refer to the overall assemblage of impact-related deformation as an **impact structure**. As we have seen, giant impacts can cause global calamity and even mass extinction.

Researchers distinguish among three general classes of meteorites: *stony meteorites* (composed of rock), *iron meteorites* (consisting of metallic iron alloy), and *stony-iron meteorites*. Stony meteorites account for 94% of known examples, iron meteorites for about 5%, and stony iron meteorites for only 1%. Most stony meteorites are *chondrites*, so named because they contain small spheres, known as chondrules, formed by the freezing of melt in space. They represent the remains of planetesimals that did not differentiate into a core and mantle. A smaller proportion of stony meteorites are *achondrites*. These do not contain chondrules and, therefore, must have come from the rocky parts of objects that had undergone differentiation. Iron meteorites may represent fragments of cores, and stony-irons may represent fragments of the core-mantle boundary region of differentiated planets or planetesimals.

The Edge of the Solar System

Between the objects of our Solar System, we traverse the vacuum of **interplanetary space**. This vacuum contains between 5,000 and 100,000 atoms or molecules per liter. (By comparison, air at sea level contains about 2.5×10^{22} molecules per liter.) Is there a boundary that defines the edge of the Solar System? Yes, and surprisingly, it lies closer to the Sun than the inner boundary of the Oort Cloud. Astronomers consider a bubble-like invisible surface, called the **heliosphere**, that lies 200 AU from the Sun to be the edge of the Solar System **(Fig. 22.38)**.

Inside the heliosphere, the atoms in space consist primarily of solar wind particles (electrons and protons ejected into space by our Sun). Outside the heliosphere, with the exception of Oort Cloud objects, most of the rare atoms that exist consist of cosmic rays, particles that were ejected from other objects in space and have migrated toward our Solar System. In other words, the heliosphere marks the "front" at which the solar wind butts up against the *cosmic wind*. Astronomers refer to the region outside the heliosphere—the volume between stars, which contains about 1,000 atoms per liter—as **interstellar space**.

Only one human-made space probe, *Voyager 1*, has crossed the heliosphere. Launched in 1977, this tiny metal ball of instruments, attached to a dish-like antenna, was 136 AU (approximately 20 billion kilometers, or 12.4 billion miles) from Earth as of 2016. *Voyager 1* is therefore the farthest-journeying probe ever launched, and it continues to move outward at 17 km (10.6 miles) per second. Astronomers hope it will continue to send back data about

interstellar space until 2025. At that time, *Voyager 1* will see the Sun as just another star, and sunlight will no longer be strong enough to charge the probe's batteries. At its current rate of travel, *Voyager 1* will traverse the distance to the nearest star about 40,000 years from now. In the next chapter, we turn our attention to that star and others (including our Sun), the galaxies they occupy, and the Universe beyond.

Take-home message . . .

The Solar System encompasses an enormous number of small objects in orbit around the Sun. They primarily occur in three locations: the asteroid belt between Jupiter and Mars, the Kuiper Belt beyond Neptune, and the Oort Cloud, which lies beyond the outer fringes of the Solar System but remains tied to the Sun by gravity.

Quick Question ----------------------------
What defines the edge of the Solar System?

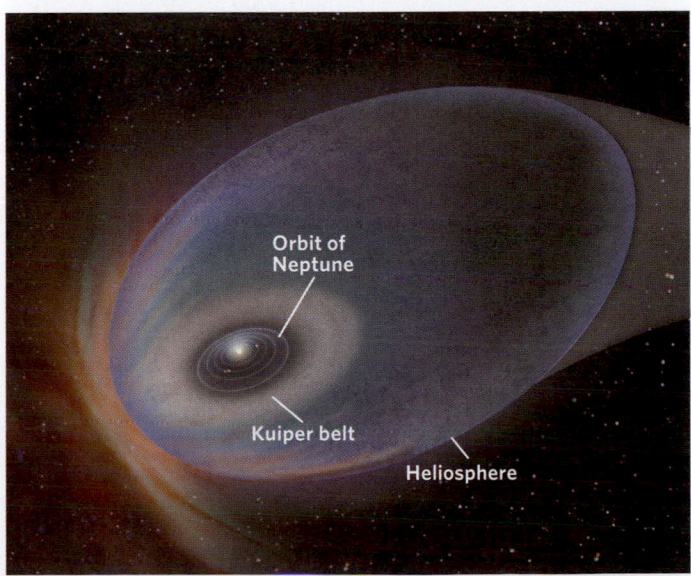

Figure 22.38 The heliosphere encompasses all the planets and the Kuiper belt; it lies at the center of the much larger Oort Cloud.

○22 CHAPTER REVIEW

Chapter Summary

- A planet is a celestial body in orbit around the Sun that has a spherical shape and has cleared the neighborhood around its orbit. A dwarf planet is large enough to be spherical, but has not cleared the neighborhood around its orbit.

- The exploration of the Solar System advanced quickly with the launch of space telescopes and space probes.

- A moon is a large object in orbit around a planet. All planets except Mercury and Venus have moons.

- According to the condensation theory of Solar System formation, tiny specks of solid ice and dust allowed a nebula to cool enough to collapse and acted as "seeds" on which materials could gather, leading to the formation of larger and larger solid bodies in the early Solar System.

- The planets formed from a revolving accretionary disk of dust and ice surrounding the newborn Sun.

- Because the solar wind blew volatile materials to the outer portion of the accretionary disk, terrestrial planets, with rocky mantles surrounding metal cores, formed closer to the Sun, while Jovian planets, consisting of gases and ice, formed farther from the Sun.

- The Earth's Moon rotates once each time it orbits the Earth, so that the near side always faces the Earth. Its surface includes both light-colored lunar highlands and dark-colored plain-like maria underlain by basalt. The Moon's entire surface is covered with regolith.

- Venus has a dense atmosphere, Mars a thin one, and Mercury has hardly any. Volcanoes rise from the surface of Venus and Mars.

- Mars is a cold planet with a very thin atmosphere composed of carbon dioxide. Except for water in its ice cap, and possibly local intermittent flows of liquid water, Mars is now dry.

- Mars has a relatively smooth basin in its northern hemisphere, highlands and volcanoes along its equatorial region, and highlands in the south. It hosts a canyon much bigger than the Earth's Grand Canyon.

- The Jovian planets have rocky and metallic cores, but consist mostly of volatile materials. Jupiter and Saturn, the gas giants, contain gases in various states. Uranus and Neptune, the ice giants, contain icy slush. All have dense, colorful atmospheres.

- All Jovian planets have rings and a variety of moons.

- The asteroid belt, between the orbits of Mars and Jupiter, contains millions of rocky and metallic chunks, fragments of planetesimals that never coalesced into a planet.

- The Kuiper Belt, which includes millions of icy objects, the largest of which are the dwarf planets Pluto and Eris, is a part of the accretionary disk that didn't become incorporated into planets. The Oort Cloud lies farther out.

- A comet consists of dust and ice. It follows a highly eccentric orbit that takes it close enough to the Sun to start vaporizing and develop a coma and tail.

- A meteoroid has a diameter smaller than about 1 m (3 feet). When a meteoroid enters the Earth's atmosphere, it generates a streak of light known as a meteor. A meteorite is a meteoroid that strikes the ground.

- NASA monitors near-Earth objects that have the potential to cross the Earth's orbit and cause a collision within the next 100 years.

- Astronomers define the edge of the Solar System as an invisible surface, called the heliosphere, at which the solar wind butts up against the cosmic wind.

Key Terms

accretionary disk (p. 784)
asteroid (p. 805)
asteroid belt (p. 805)
comet (p. 807)
condensation theory (p. 783)
crater (p. 787)
cratering density (p. 788)
dust (p. 783)
dwarf planet (p. 778)
eccentric orbit (p. 780)
exobiology (p. 785)
exoplanet (p. 785)

gas giant (p. 795)
Great Red Spot (p. 797)
heliosphere (p. 810)
ice giant (p. 795)
impact structure (p. 810)
interplanetary space (p. 810)
interstellar space (p. 811)
Jovian planet (p. 781)
Kuiper Belt (p. 807)
late heavy bombardment (p. 787)
lunar highland (p. 787)
lunar regolith (p. 788)

lunar time scale (p. 788)
mare (p. 787)
metallic hydrogen (p. 797)
meteor (p. 809)
meteorite (p. 809)
meteoroid (p. 808)
meteor shower (p. 809)
moon (p. 777)
near-Earth object (p. 811)
nebula (p. 782)
nebular theory (p. 783)
Oort Cloud (p. 807)

planet (p. 777)
planetesimal (p. 784)
protoplanet (p. 784)
protoplanetary disk (p. 784)
protostar (p. 784)
refractory material (p. 783)
ring (p. 781)
space weathering (p. 788)
terrestrial planet (p. 781)
volatile material (p. 783)

Review Questions

The letters following each Review Question refer to the corresponding Learning Objective from the Chapter Opener.

1. What advances in technology allowed researchers to improve our understanding of the Solar System? **(B)**

2. List the succession of celestial objects in the Solar System, starting from the Sun. **(B)**

3. Are the axes of all the planets parallel to one another? If not, how do they differ? **(B)**

4. What characteristics must an object have to be formally considered a planet? How does a dwarf planet differ from a planet? **(A)**

5. Explain the condensation theory of the formation of the Solar System. Why is the presence of dust and ice in the nebula so important? **(B)**

6. A massive impact strikes our planet, as in the figure. Sketch how this impact could lead to the formation of the Moon. **(C)**

7. What do the different landscapes of the Moon consist of, and how did they form? How can we determine which landscapes are older than others? **(C)**

8. What characteristics do Mercury and the Earth's Moon share, and in what ways are they different? **(F)**

9. Why do the Earth and Venus have vastly different atmospheres and surface temperatures? **(D)**

10. What is the evidence for water on Mars in the present and in the planet's ancient past? **(D)**

11. Why does Jupiter have distinct rings, and what is its Great Red Spot? **(E)**

12. Summarize the distinctive characteristics of the largest moons of the Jovian planets. Which host active volcanoes? **(F)**

13. What does a comet consist of? Where do comets come from, and why do they have two tails, as shown in the illustration? **(H)**

14. Distinguish among a meteoroid, a meteor, and a meteorite. Why do meteor showers occur at predictable times? **(H)**

15. What feature defines the edge of the Solar System? Do Oort Cloud objects lie within or outside this boundary? **(G)**

On Further Thought

16. Name three unique characteristics of each planet that make it different from the other seven. **(D, E)**

17. When the nebula that formed our Solar System condensed, why did it flatten into an accretionary disk? How does the geometry of the disk relate to the orbital planes of the planets? Why is the retrograde rotation of Venus puzzling? **(B)**

18. Of the terrestrial planets, why is abundant water and oxygen found only on the Earth? **(D)**

Online Resources

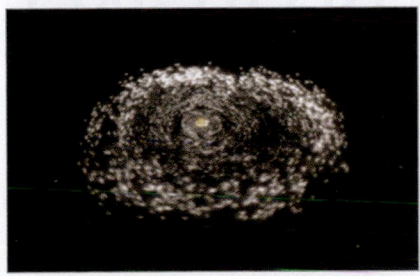

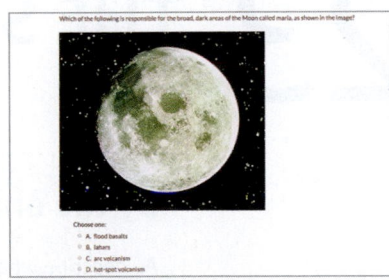

Videos
This chapter features a video on the formation of the Solar System, as well as real-world videos exploring the planets it contains.

Smartwork5
This chapter features questions on the composition, formation, and characteristics of the Sun, the Solar System, the Moon, the planets, and other celestial objects.

⊙23 THE SUN, THE STARS, AND DEEP SPACE

By the end of the chapter you should be able to . . .

A. describe what's inside the Sun and other stars and explain how they produce energy.

B. explain the relationships among a star's luminosity, its apparent brightness or magnitude as seen from the Earth, and its true or absolute magnitude.

C. demonstrate how astronomers classify stars using an H-R diagram.

D. discuss how stars form and evolve, explain what happens when they run out of fuel, and describe the exotic remnants of dead stars such as white dwarfs, neutron stars, and stellar black holes.

E. list the characteristics of nebulae and explain the differences among nebulae.

F. describe the overall structure of a galaxy, the differences among galaxies, and the relationship of a supermassive black hole to a galaxy.

G. visualize current ideas concerning the overall structure of the Universe.

H. compare models for the expansion of the Universe and for the distant future of the Universe.

23.1 Introduction

Gazing at a moonless black sky, awash with bright stars, makes for a truly memorable night. Such a view may be hard to come by for an urban dweller, given the light pollution and haze of cities, but look up from a desert mountain peak and prepare to be amazed. In addition to stars and planets, you'll see the central part of the Milky Way Galaxy slicing across the sky from horizon to horizon. The Sun, and every individual star that you can see from Earth, lies within the Milky Way. But not all the points of light in the night sky are single stars. Some are nebulae, some are clusters of stars, some are galaxies, and some are unusual objects that astronomers still struggle to interpret. Observations made with the Hubble Space Telescope emphasize that even a tiny part of the sky that looks dark to the naked eye contains billions upon billions of celestial objects **(Fig. 23.1)**. Most are so far away that the light you see has traveled for millions or even billions of years since it began its journey across *deep space*—the Universe far beyond the limits of our Solar System—to your eyes.

In this chapter, we complete our introduction to astronomy by exploring the vast variety of objects, in addition to the planets and moons of our Solar System, that constitute the Universe. We begin with our own Sun because it provides a basis for understanding all stars. Next, we turn our attention to other stars, to galaxies, and finally to a host of objects with strange-sounding names—including nebulae, dwarf stars, giant stars, neutron stars, quasars, novae, supernovae, and black holes—that no one had even dreamed of before the 20th century. This chapter, and this book, ends by exploring alternative models for the fate of

The Milky Way, our home galaxy, brilliantly rings the night sky.

Figure 23.1 The immensity of space as revealed by the Hubble Space Telescope.

(a) The Hubble Space Telescope orbits above the atmosphere, so it has an unimpeded view of space.

(b) A speck of the sky, 5% of the area covered by the Moon, looks black when viewed from the Earth.

(c) When Hubble magnifies this speck, it reveals hundreds of galaxies, each containing hundreds of billions of stars. The point of light with spikes is a relatively nearby star in our own galaxy.

Did you ever wonder . . .
why the Sun shines?

the Universe in a time long after the Earth will have ceased to exist. We hope that this exploration will leave you with a new understanding of our beautiful planet and its place in an amazing Universe.

23.2 Lessons from the Sun

While a few ancient philosophers might have wondered whether stars might be distant, light-emitting orbs like the Sun, it wasn't until 1590 that the idea began to gain favor in Western society. At that time, an Italian friar named Giordano Bruno promoted not only the concept that the stars are glowing balls like our Sun, but also the idea that other stars might host planets and even life. In 1838, an astronomer named Friedrich Bessel first measured the distance to a star. By doing so, he proved that stars are much farther away than the Sun **(Box 23.1)**, and therefore, that stars could indeed be huge objects, like the Sun. A few decades later, when other researchers had shown that stars and the Sun have similar compositions, the concept that stars are themselves distant suns became generally accepted.

Studies completed by the early 20th century demonstrated that stars vary greatly in temperature and size, and that our Sun is a fairly ordinary star—not too big or too small, and not too hot or too cool. Because of its proximity **(Fig. 23.2)**, we can observe our Sun in fine detail, and our observations have led to an understanding of its energy generation, its composition, and its internal structure. This knowledge provides a basis for interpreting distant stars.

Figure 23.2 The Sun, as seen by astronauts orbiting the Earth.

How can I explain . . .

The vastness of space

What are we learning?
• How to think about the vast distances of interstellar space.

What you need:
• A small ball to represent the Sun.
• A tiny ball to represent the Earth.
• A small ball to represent the Sun's nearest neighbor, Proxima Centauri.
• A map of your local area that extends at least 25 km (15 miles) beyond your location.

Instructions:
• Place the ball representing the Sun on the floor.
• Now carefully place the ball representing the Earth 30 cm (1 foot) away from the first ball. Note that that this 30-cm distance represents 148 million kilometers (92 million miles).

• So where would the nearest star be located? That star is Proxima Centauri, and it is 4.3 light-years, or 40 trillion kilometers (25 trillion miles) away from Earth. Scaled to the two balls, Proxima Centauri would be 19 km (12 miles) distant. Find a familiar spot on the map that is 19 km (12 miles) away from where you are. That is where the next ball should be placed!
• Now, using the same scale (30 cm = 148 million kilometers), determine where the nearest galaxy, Andromeda, would be located. (Andromeda is 2.5 million light-years from the Sun.)

What did we see?
• The goal of this exercise is to get some sense of the vastness of space. Picturing the distances you would have to travel from your scale-model Earth to your scale-model Proxima Centauri provides a dramatic example of the vastness of our Universe.
• Amazingly, if 30 cm represents 148 million km, the nearest galaxy, Andromeda, would still be about 48 million km (30 million miles) away from the ball representing the Sun!

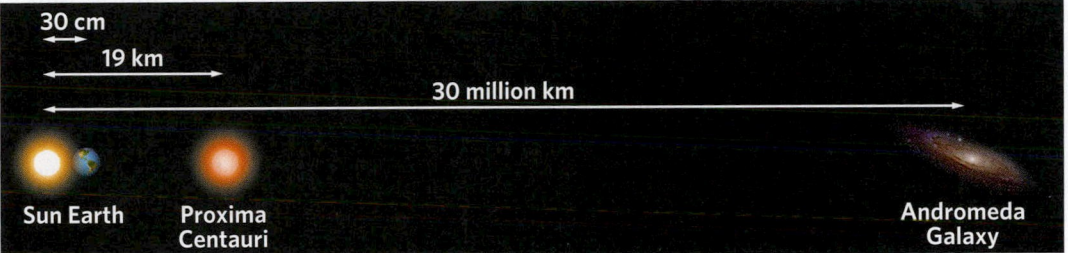

Composition of the Sun

By studying spectra of sunlight and starlight (see Chapter 21), astronomers found that the Sun, and stars like it, consist primarily of hydrogen (71.0% by mass) and helium (27.1%). Because hydrogen atoms weigh less than helium atoms, 91.2% of the individual atoms in the Sun are hydrogen, and only 8.7% are helium. The next eight most common elements, in order of their contribution to the Sun's mass, are oxygen, carbon, nitrogen, silicon, magnesium, neon, iron, and sulfur. The Sun contains traces of all the other naturally occurring elements as well.

Hydrogen and helium in the Sun do not exist in the same form in which we find these elements on Earth, for at the temperatures found in the Sun, atoms lose some or all of their electrons and attain an electrical charge. A material consisting almost entirely of such *ionized atoms*, circulating along with *free electrons* (which move independently of atoms), is called **plasma**. Physicists consider plasma to be the fourth state of matter.

The Source of the Sun's Energy

People in ancient times attributed the Sun's energy to sources familiar to them. For example, in some myths, the Sun is a cauldron of molten metal, and in others, it's a ball of flaming coal. Of course, modern calculations emphasize that such sources could not possibly produce as much energy as the Sun does for as long as it has. Discoveries made during the first half of the 20th century provided an explanation for the Sun's energy generation. The key to solving this mystery came from the work of Albert Einstein (1879–1955). According to his iconic equation, $E = mc^2$, annihilation of a tiny amount of mass produces an enormous amount of energy. In the 1920s, physicists discovered that under special conditions, two atomic nuclei can bind together to form a single atom, and that during this process, called **nuclear fusion**, a tiny amount of matter converts into a large amount of energy. During the 1930s, further work suggested that nuclear fusion takes place in the Sun. By 1939, researchers

Figure 23.3 A nuclear fusion reaction in the Sun produces helium (He) nuclei from hydrogen (H) nuclei. The reaction, which takes place in stages, releases energy in the form of gamma rays, as well as subatomic particles (positrons and neutrinos).

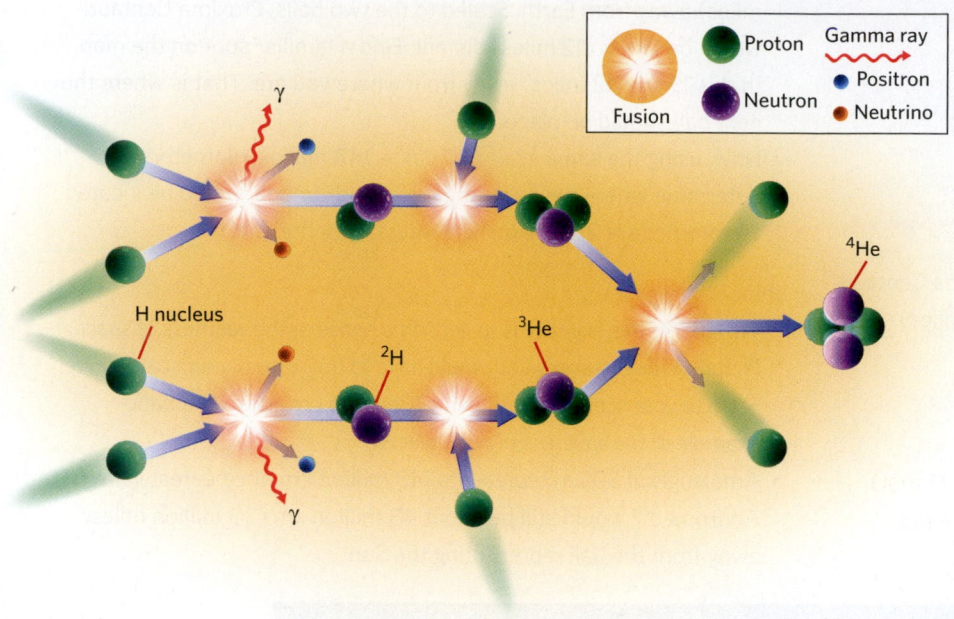

(To picture why, place two ring magnets on a pencil, oriented so that the positive side of one faces the positive side of the other. If you lower one magnet slowly, the electromagnetic force will levitate it, but if you drop it fast, the magnets will touch before the electromagnetic force pushes them apart.) In the case of a high-speed collision between two nuclei, the electromagnetic force doesn't get the opportunity to push the nuclei apart because when nuclei get very close, another, much stronger force starts to operate. This force, called the *strong nuclear force*, exerts influence only over extremely small distances (namely, the dimension of a nucleus), so it plays no role in familiar chemical reactions. During nuclear fusion, however, it acts like superglue, overcoming the electromagnetic force and binding nuclei together.

Why did fusion start in the Sun, but not in the Earth? During the Sun's initial formation, its great mass produced such intense gravitational force that gas atoms deep inside it squeezed tightly together, and this compression raised the temperature of the gas. Eventually, temperatures became so high, and nuclei moved so rapidly, that collisions brought some nuclei close enough for the strong nuclear force to fuse them together and release enormous energy.

The Solar Constant

Because of nuclear fusion, the Sun produces an inconceivable 4×10^{26} watts of power. (Recall that a watt is unit of power; it represents the expenditure of 1 joule of energy in 1 second.) By comparison, burning all the hydrocarbon reserves on the Earth in one second would yield 10^{21} watts, and a typical nuclear power plant produces 5×10^8 watts. If all of the Sun's energy were focused on the Earth, the planet would evaporate. Fortunately for us, the Earth receives only a tiny fraction of the Sun's energy, while most of the rest spreads out into space. A square meter at the top of the Earth's atmosphere, aligned perpendicular to the Sun's rays, receives 1,362 watts. Researchers refer to this number, 1,362 watts/m², as the **solar constant** because it has varied by less than 0.2% over the last 400 years.

At any given time, the Sun illuminates only half of our planet. Because of the Earth's spherical shape, much of its surface lies at a shallow angle to the Sun. As a result, over the course of a year, *insolation*, meaning the radiation arriving at the top of the atmosphere, averages only 350 watts/m². Of this radiation, the atmosphere and its clouds absorb or reflect about half, so only about 175 watts actually reaches each square meter of the Earth's surface (see Chapter 20).

had worked out the series of fusion reactions that take place inside the Sun and could, therefore, explain its energy production.

Almost all of the nuclear fusion in the Sun involves the fusion of protons (the nuclei of hydrogen atoms) to form helium. This reaction, known as the *proton-proton chain reaction*, involves several steps **(Fig. 23.3)**. Simplistically, the reaction begins when two protons fuse to form deuterium (²H), an isotope of hydrogen whose nucleus contains one proton and one neutron; during the reaction, one of the colliding protons transforms into a neutron. Next, a deuterium atom collides with another proton to form ³He, an isotope of helium containing two protons and only one neutron. In the final step, two ³He isotopes collide to form normal helium (⁴He), releasing two excess protons. The complete chain reaction releases energy.

At first glance, it may seem that fusion shouldn't happen, for the process requires two positively charged particles (protons) to stick together, and *like charges* repel one another due to the *electromagnetic force*, the same force that makes it difficult to push together the positive ends of two magnets. In fact, at the relatively low temperatures found on and in the Earth, the electromagnetic force indeed prevents fusion from happening. But at the extremely high temperatures found deep inside the Sun (up to 10 million °C, or 18 million °F), nuclei not only have been stripped of their protective electron shells, but they move so rapidly, and collide at such high speeds, that they can overcome the electromagnetic force and get very close to each other.

Internal Structure of the Sun

The radius of the Sun, defined as the distance from its center to its visible surface, is 696,000 km (432,000 miles),

Consider this . . .

Is the core of the Sun like a hydrogen bomb?

Military arsenals include two types nuclear weapons: *atomic bombs*, whose energy comes from fission reactions of radioactive elements (see Chapter 11), and *hydrogen bombs*, whose energy comes primarily from fusion reactions of hydrogen. The bombs used at the end of World War II were atomic bombs. Hydrogen bombs, also known as thermonuclear bombs, are vastly more powerful and fortunately have never been used in war. The largest hydrogen bomb ever tested released as much energy as 50 million tons of TNT at a remote site in the former Soviet Union. An explosion that size completely flattens an area about 50 km in radius, and causes severe burns out to a distance of 100 km.

While the entire Sun produces about 10 billion times the power of a hydrogen bomb in one second, the amount of energy it produces per kilogram of solar core is only about as much as that produced by a burning log. Clearly, the core of the Sun is not like a nuclear explosion, but rather generates energy at a slow, continuous rate. So how can the Sun produce so much energy? The answer to this question comes from considering the mass of the Sun. A hydrogen bomb contains only a tiny amount of mass compared with the solar core. So, even though each kilogram of a bomb produces vastly more energy in a single second than does a kilogram of the Sun, there's so much more mass in the Sun that in a given second, the Sun produces much more energy than a bomb does. At its current rate of fuel consumption, the Sun has enough fuel to burn for another 5 billion years.

---

110 times that of the Earth. Researchers distinguish three concentric layers within the Sun: the *solar core* at the center, the *radiative zone*, and the *convective zone* **(Fig. 23.4)**. Let's examine the characteristics of these layers, starting from the center.

THE SOLAR CORE. The **solar core** extends from the center outward for 160,000 km (100,000 miles), and therefore accounts for about 23% of the Sun's radius. Temperatures reach 15 million °C (27 million °F) at the center of this layer and diminish to about 7 million °C (12.6 million °F) at its top. Because of the inward gravitational pull of the Sun's immense mass, the pressure at the center of the solar core may be over 10,000 times that found at the center of the Earth. About 99% of the Sun's energy production comes from nuclear fusion reactions in the core. These reactions convert 600 million tons of hydrogen into helium every second **(Box 23.2)**. The remaining 1% of the Sun's energy comes from the base of the radiative zone. Temperatures elsewhere in the Sun are too cool for fusion to be possible.

THE RADIATIVE ZONE. Researchers refer to the thick layer surrounding the solar core as the **radiative zone** because energy passes through this zone in the form of electromagnetic radiation. The radiative zone extends from the core out to about 490,000 km (300,000 miles) from the center of the Sun, a distance of 70% of the radius, so it has a thickness of about 330,000 km (205,000 miles). It accounts for about 43% of the Sun's radius and about 48% of the Sun's mass. In the radiative zone, photons of energy emitted by fusion reactions in the core travel only a short distance before they are absorbed by particles of matter. The particles then re-radiate the energy as new photons in a random direction. These new photons travel a short distance before they, too, are absorbed again and new photons are re-radiated in another direction. Energy, in effect, wanders nearly randomly as it travels through the radiative zone, so a given packet of energy may take about 170,000 years to reach the zone's outer surface.

Figure 23.4 The internal structure of the Sun. The axes show how physical characteristics of the Sun change from the core to the top of the convective zone.

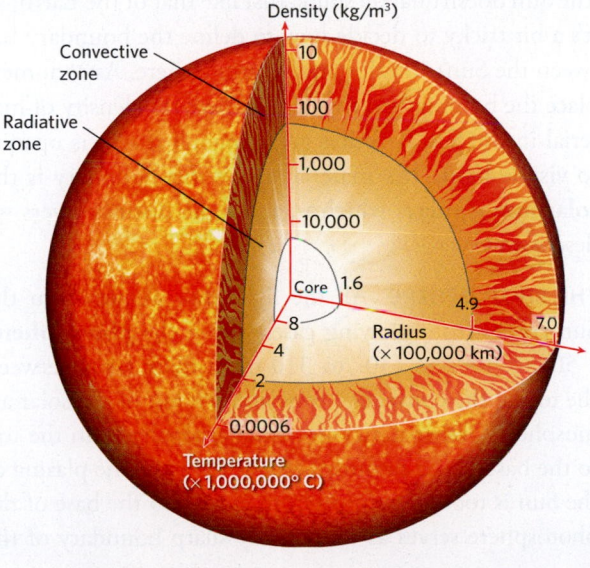

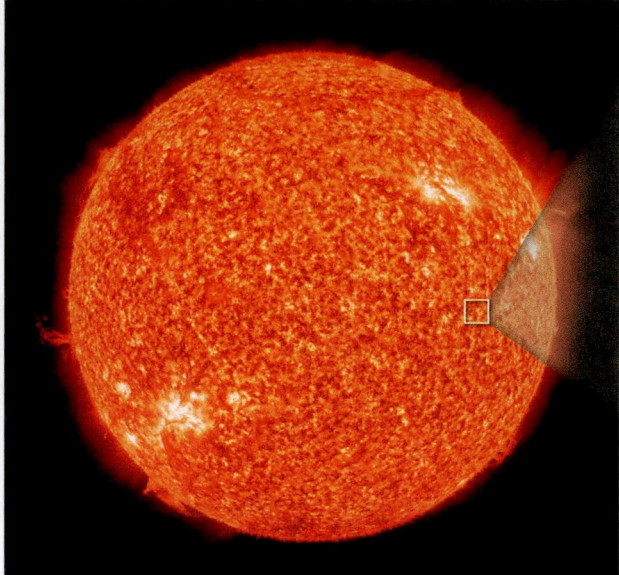

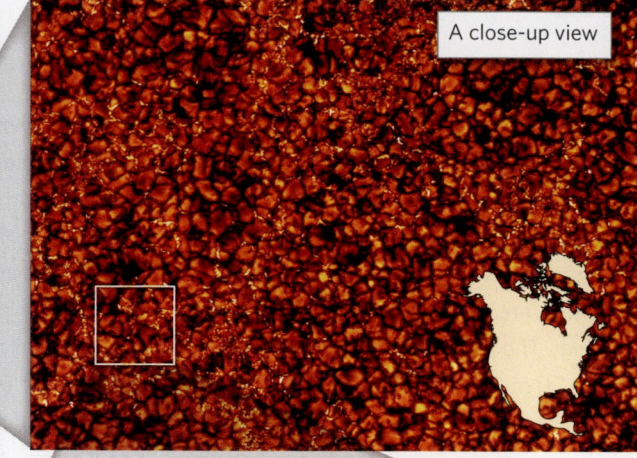

A close-up view

(a) Granulation affects the entire surface of the Sun. Each granule is tiny compared with the Sun, but huge compared with a continent on the Earth. The outline represents North America, for scale.

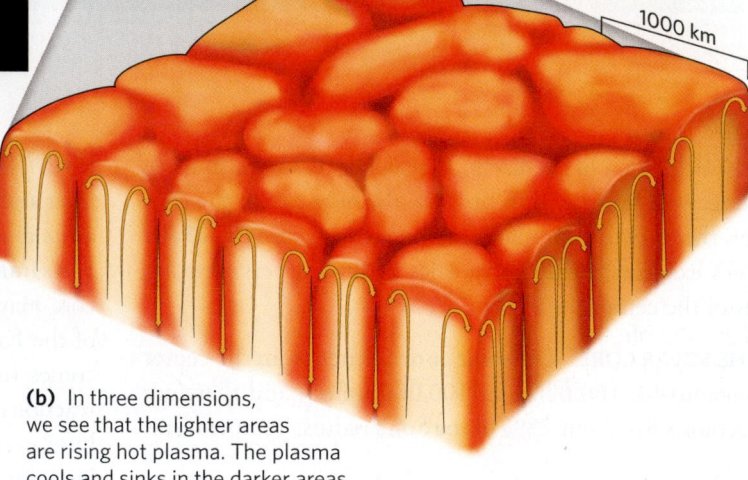

1000 km

(b) In three dimensions, we see that the lighter areas are rising hot plasma. The plasma cools and sinks in the darker areas.

THE CONVECTIVE ZONE. When energy reaches the top of the radiative zone, it heats the base of the overlying layer of plasma to about 2 million °C (3.6 million °F). This heat decreases the density of plasma, making it relatively buoyant. The buoyant plasma rises upward and carries thermal energy with it. As it rises, the plasma cools, and it eventually sinks to replace other rising buoyant plasma at the base of the convective zone. Astronomers refer to the region of the Sun in which this convective transport of energy takes place as the **convective zone**. This zone, which surrounds the radiative zone, has a thickness of about 200,000 km (125,000 miles).

The upwelling and downwelling of plasma organizes into *convective columns*, which are very tall convective cells. Plasma rises at the center of each convective column and sinks back down along the sides. Columns range from 500 to 2,000 km (300 to 1,200 miles) across—about a tenth to a third the width of North America. The column center, where hot plasma rises, is hotter, and therefore brighter, than the sides, where cooler plasma sinks. This pattern of light and dark gives the Sun's surface a grainy appearance, known as **solar granulation (Fig. 23.5)**. At any given time, about 4 million granules (the visible tops of convective columns) exist, but the pattern constantly evolves, with each granule lasting for only a few minutes before dissipating as new granules form.

The Solar Atmosphere

The Sun doesn't have a solid crust like that of the Earth, so it's a bit tricky to decide how to define the boundary between the Sun's interior and its atmosphere. Astronomers place the boundary at the depth where the density of material in the Sun becomes great enough that it is opaque to visible light. The material above this boundary is the **solar atmosphere**, which includes the distinct layers we describe next.

THE PHOTOSPHERE. All the light that we see from the Sun comes from glowing gases within the **photosphere**, a 50- to 500-km (30- to 310-mile)-thick layer between the top of the convective zone and the rest of the solar atmosphere. Transparency decreases gradually from the top to the base of the photosphere. At the base, the plasma of the Sun is too dense for light to escape, so the base of the photosphere serves as the visually sharp boundary of the

Sun that you see at sunset or sunrise (Fig. 23.6). The range of temperatures from the top (4,300°C; 7,800°F) to the base (5,700°C; 10,300°F) of the photosphere resembles the range of temperatures found at the center of the Earth, but is vastly cooler than the inferno at the center of the Sun. Electromagnetic energy emitted by the photosphere heads off into space, reaching the Earth about 8 minutes later.

THE CHROMOSPHERE AND CORONA. If you observe the Sun during a total eclipse (with appropriate eye protection, of course), you'll see a thin bright band surrounding the Sun (Fig. 23.7a). This band, the **chromosphere** (sphere of color) is so named because when viewed during an eclipse, it displays flashing tints of brilliant red. The chromosphere, which is 3,000 to 5,000 km (1,800 to 3,000 miles) thick, has a temperature of about 4,500°C (8,100°F), much less than that of the photosphere—in fact, at the temperature of the chromosphere, some matter exists as gas instead of plasma. The density of the chromosphere is only about 0.00000001 that of the Earth's atmosphere at sea level.

At the top of the chromosphere, across a thin (100-km, or 60-mile) *transition zone*, the temperature increases dramatically. In the overlying **corona**, the outer layer of the solar atmosphere, temperatures rise to 1 million °C (1.8 million °F) at a distance of 10,000 km (6,200 miles) above the photosphere. We can see the corona during an eclipse. The inner part appears as a wispy glowing cloud (Fig. 23.7b). When viewed in ultraviolet light, the outer part of the corona appears as bright streaks radiating far out from the Sun's surface (Fig. 23.7c).

The Solar Wind

At high temperatures, particles move very fast, so it's no surprise that some of the particles in the extremely hot corona achieve escape velocity, break free of the Sun's gravitational pull, and head off into space. The high-speed stream of particles—mostly protons and electrons—flowing permanently away from the Sun makes up the **solar wind**. Note that the solar wind consists of moving matter, not just radiation, so it really is a "wind" like the winds of the Earth's atmosphere, except that solar wind particles move at enormous speeds and are ionized, whereas molecules in the Earth's wind are neutral and travel slowly. Because of the solar wind, the Sun loses about 1.8 billion kg (2 million tons) of matter every second. But even at this seemingly huge rate, only about 0.1% of the Sun has been lost to space since its nuclear furnace first ignited. Solar wind particles travel at about 500 km (300 miles) per second, a speed much less than that of light, so they take a few days to reach the Earth. Because they have an electrical charge, the Earth's magnetosphere deflects most of them.

Figure 23.6 At sunset, the edge of the Sun that you see is the surface of the photosphere.

Figure 23.7 The solar chromosphere and corona.

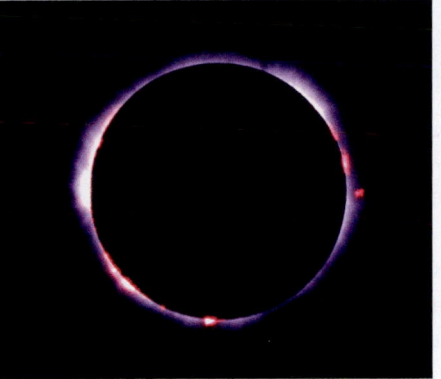 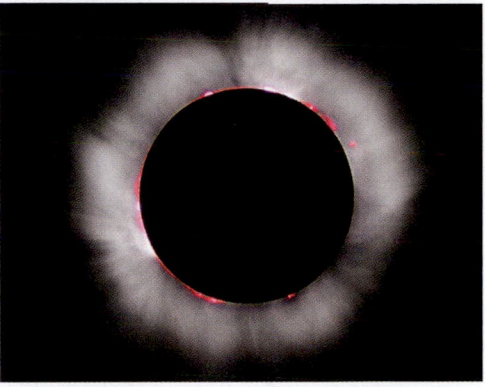

(a) The chromosphere, the thin red and white ring close to the Sun's surface, as seen in visible light during an eclipse. To make this layer visible, the light of the corona has been blocked.

(b) This photo shows the same image as part (a), but here the gases of the inner part of the corona are visible. Photo courtesy of Luc Viatour/www.Lucnix.be.

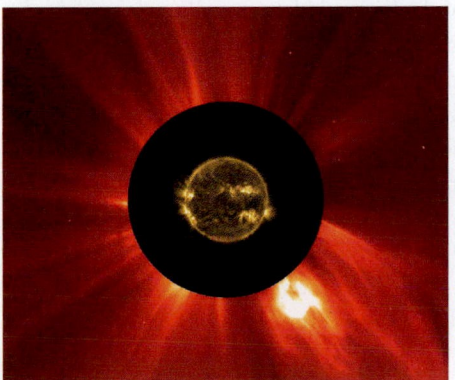

(c) In this photo, the light of the inner corona has been blocked. The outer corona, as viewed in ultraviolet light, streams far into space.

But some reach the atmosphere and stream toward the poles to produce aurorae (see Chapter 17).

Take-home message ...

The Sun is a star, and like other stars, it produces huge amounts of energy by nuclear fusion reactions. These reactions can take place because it's so hot inside the Sun that nuclei can get close enough for the strong nuclear force to bind them together. The interior of the Sun can be divided into layers. Most nuclear fusion takes place in the solar core. The resulting energy passes through the radiative zone as electromagnetic energy and then heats the base of the convective layer, producing convective columns that carry energy to the surface of the Sun. In the photosphere, the energy is emitted into space as electromagnetic radiation. Beyond the photosphere, the solar atmosphere consists of the chromosphere and corona. Particles escaping from the solar atmosphere head into space as solar wind.

Quick Question -
What do the Sun's atmosphere and solar wind consist of?

23.3 The Sun's Magnetic Field and Solar Storms

The Sun possesses an intense magnetic field, much stronger than that of the Earth. This field exhibits complex behavior that yields *sunspots* (dark patches) on the surface. At times, particularly strong emissions of matter blast from the surface. These *solar storms* send out such strong solar winds that they can have adverse effects on the Earth. Let's examine the Sun's magnetic field and sunspots, as well as their relationship to solar storms.

The Sun's strong magnetic field develops in response to the rapid circulation of plasma within the convective zone, because plasma is an electrical conductor. Unlike the magnetic field lines of the Earth, which arc from pole to pole, those of the Sun trend at a small angle to the solar equator, and they aim in one direction in the northern hemisphere and in the opposite direction in the southern hemisphere. This unusual orientation of magnetic field lines reflects the nature of the Sun's rotation. The Sun is a fluid whose rotation rate varies with latitude—plasma at the equator rotates faster around the Sun's axis (once every 25 days) than does plasma near the poles (once every 38 days). This differential motion effectively wraps magnetic field lines around the Sun **(Fig. 23.8a)**.

Recall that on the Earth, the polarity of the magnetic field reverses (flips) at intervals of thousands to millions of years. On the Sun, reversals happen about once every 11 years. If, before a reversal, magnetic field lines in the Sun's northern hemisphere point eastward and those in the southern hemisphere point westward, then after a reversal, those in the northern hemisphere point westward and those in the southern hemisphere point eastward.

Sunspots

Over the course of several years, the wrapping of magnetic field lines around the Sun squeezes the lines together, making the magnetic field locally so intense that the field lines arc into space **(Fig. 23.8b, c)**. At the entrance and exit points of these magnetic fountains, the magnetic field can be strong enough to inhibit the upwelling of hot plasma in the underlying convective zone. Therefore, a patch on the surface of the photosphere becomes about 1,300°C to 2,700°C (800°F to 1,700°F) cooler relative to brighter regions of the photosphere. These patches are known as **sunspots** because their lower temperatures make them look darker than the surrounding regions **(Fig. 23.9a, b)**. Individual sunspots move with the Sun's rotation and can survive for days. But during this time, they constantly change in shape and dimensions as the local magnetic field evolves. Sunspots always occur in pairs, one located at the point where the magnetic field lines arc upward and the other at the point where they arc downward.

The number of sunspots on the Sun's surface varies over time as a consequence of the reversals in the Sun's magnetic field **(Fig. 23.9c)**. Over the course of 11 years, the field strengthens, and the number of sunspots increases to a maximum of about 50 to 120 per month. Then the polarity of the field reverses, after which the number of sunspots decreases to nearly zero. During the next 11 years, the pattern repeats, but the sunspots have the opposite magnetic polarities. The 11-year cycle is known as the *sunspot cycle* or **solar cycle**, and the 22-year cycle is called the *magnetic cycle*.

Solar Storms

Occasionally, local intensification or disruption of the Sun's magnetic field triggers the ejection of particularly large amounts of high-energy particles into space. These events, called **solar storms**, seem to take place in association with sunspot maximums in the solar cycle.

Astronomers distinguish among several different types of solar storms. During a **solar prominence**, outbursts of glowing gas and plasma emerge from the Sun's surface and then follow magnetic field lines back to the surface **(Fig. 23.10a)**. Solar prominences look like particularly huge arcs of plasma—some extend beyond the visible corona—and they can last for hours, days, or weeks.

Figure 23.8 The Sun's magnetic field and the formation of sunspots.

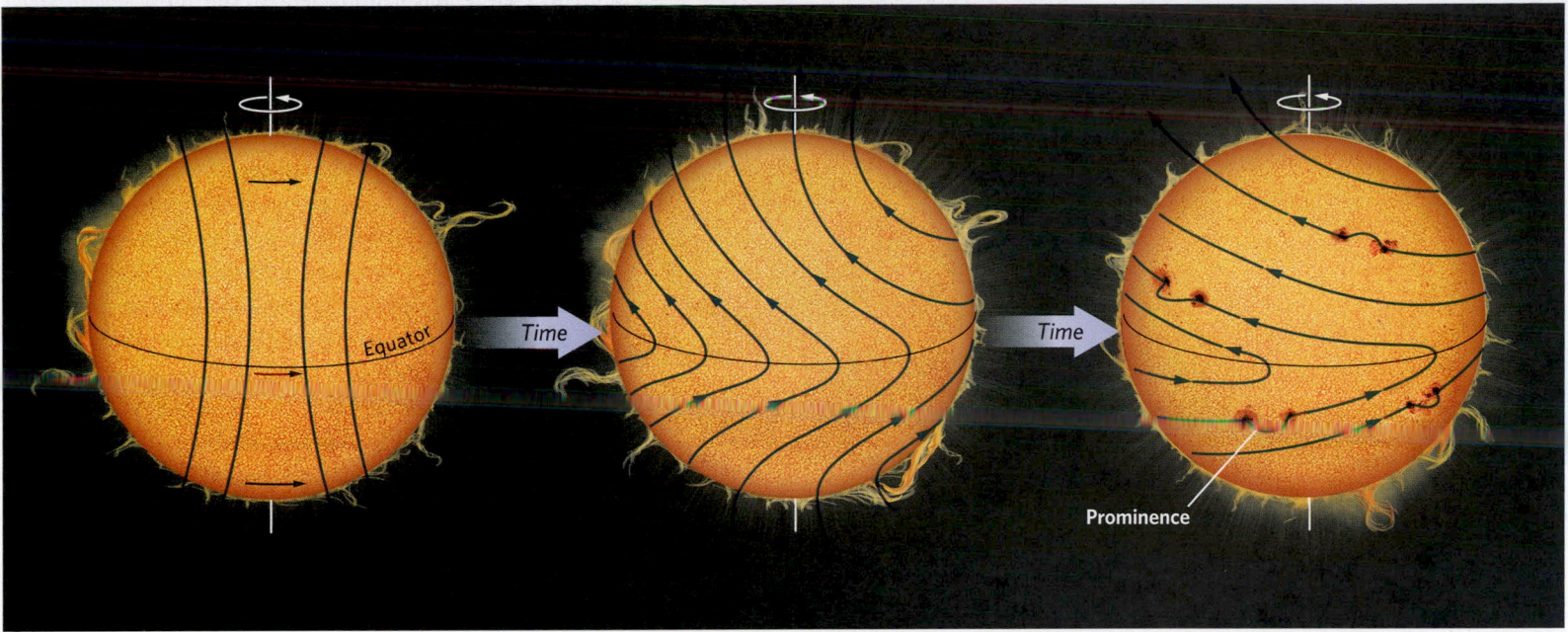

(a) The rotation rate of the Sun is faster at the equator than at the poles, so over time, magnetic field lines bend and become almost parallel to the equator.

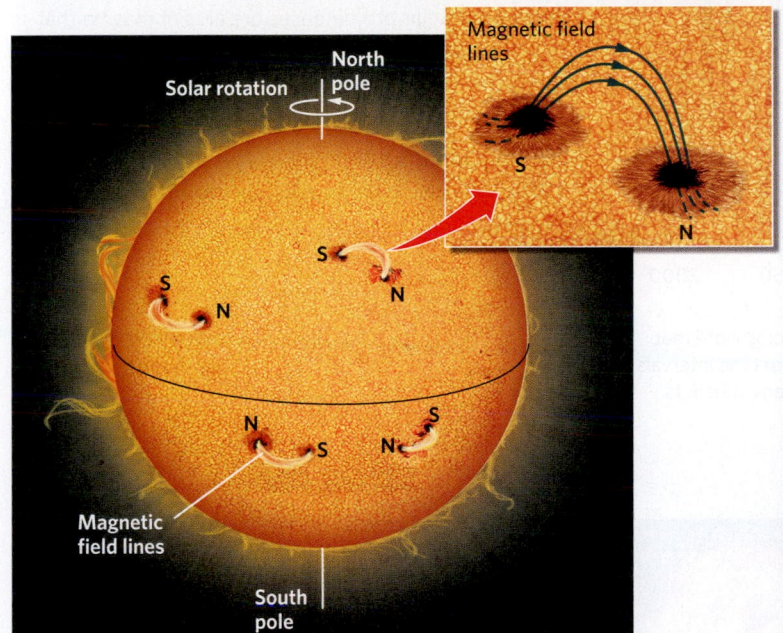

(b) The magnetic field can become so intense that field lines arc out from the solar surface, rising from one sunspot and reentering at another.

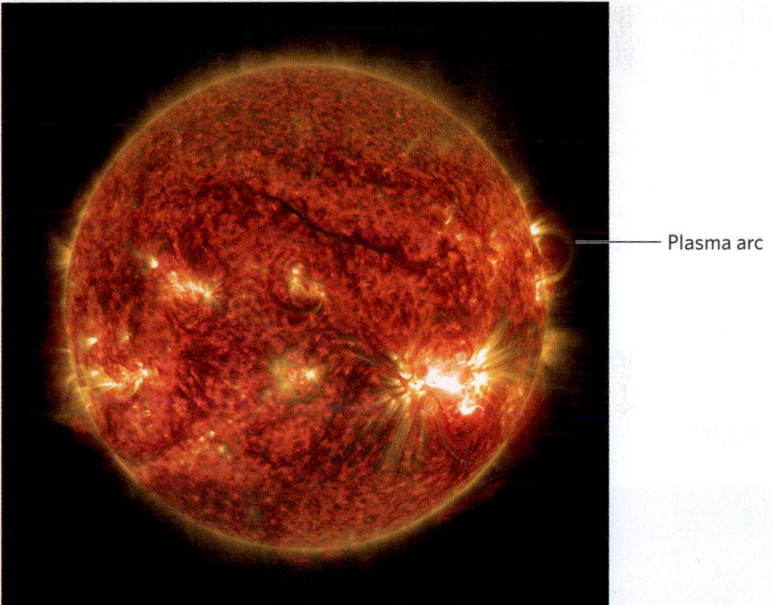

(c) A photo of the Sun showing arcs of plasma associated with sunspots.

Occasionally, a **solar flare**, an even brighter eruption of plasma, shoots out of the Sun's surface **(Fig. 23.10b)**. A solar flare releases a huge number of particles as well as intense energy in the form of X-rays and UV rays, all in a matter of minutes, so a flare effectively represents an explosion at the Sun's surface. In fact, one flare emits 15% of the total average energy output of the Sun over the same period. The plasma in a flare can be extremely hot, reaching 200,000°C to 1,500,000°C (360,000°F to 2,700,000°F). Prominences and flares together produce particularly intense solar winds **(Fig. 23.10c)**.

Astronomers refer to conditions in space surrounding the Earth caused by particles coming from the Sun as **space weather**. Mild space weather happens when typical

Figure 23.9 Sunspots appear as dark patches on the Sun's surface.

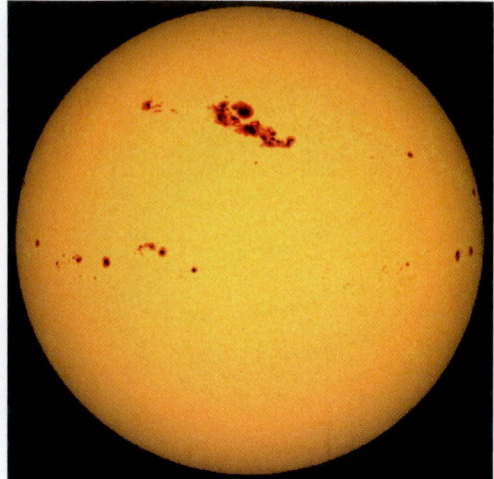

(a) At any given time, sunspots cover 0% to 0.4% of the area of the Sun.

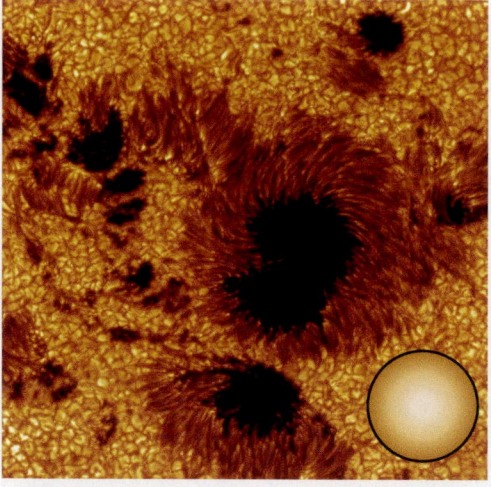

(b) A close-up shows that a sunspot has a cooler, darker central area and a warmer outer area. The circle represents the Earth, for scale.

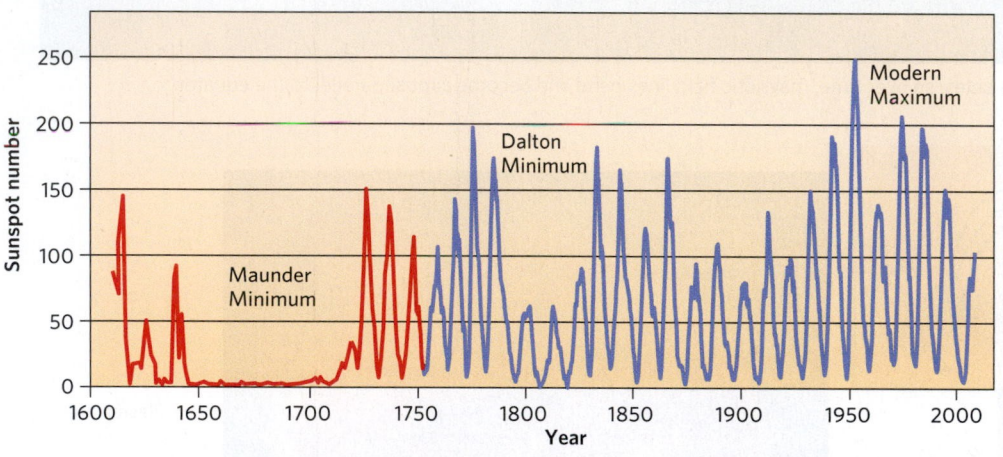

(c) The number of sunspots changes over time, showing that the solar cycle is about 11 years long. Note that the number of sunspots is not the same during each cycle. Astronomers have assigned names to time intervals with an anomalous number of sunspots. Note that from about 1650 to 1710, there were hardly any sunspots.

solar winds sweep toward the Earth. More intense space weather develops a few days after the eruption of a large solar prominence. The most severe space weather is unleashed by a solar flare. Severe space weather can cause real problems for human society because the influx of ionized particles can disrupt electronics, particularly in satellites. In fact, severe space weather can potentially hobble communications networks, electrical grids, computer systems, cell phones, and GPS navigation. Space weather in 1859 generated intensely bright aurorae, visible even at low latitudes, and caused telegraph systems to spark and fail. The radiation due to these particles can also be a threat to the safety of astronauts in space. Because of the threat posed by space weather, the National Weather Service has set up a Space Weather Prediction Center to watch for solar storms and send out warnings so that operators can protect sensitive equipment.

Take-home message . . .

The Sun generates an intense magnetic field. Local intensification of this field causes magnetic field lines to arc into space, forming sunspots. Solar prominences are arcs of plasma that can flow between sunspots. Particularly intense disturbances yield solar flares that send highly energetic plasma into space.

Quick Question -
What is space weather, and how can it affect human society?

Figure 23.10 Examples of solar storms.

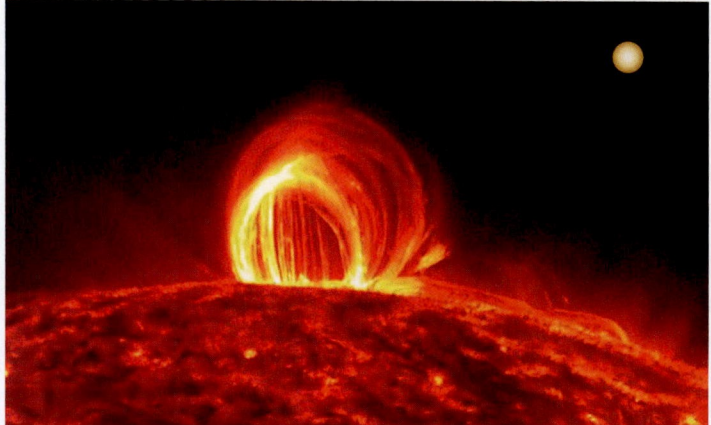

(a) Solar prominences loop into the corona, following curving magnetic field lines. The circle represents the Earth, for scale.

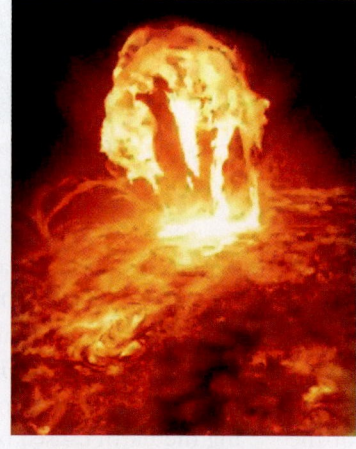

(b) Solar flares are like giant explosions that release energy quickly into space.

(c) Prominences and flares together eject large amounts of plasma into space, yielding a strong pulse of solar wind.

23.4 A Diversity of Stars

From the Earth, all stars look like points of light, but some definitely appear brighter than others. Are the differences in brightness that we see due to variations in the amount of energy that a star emits, in the distance of the star from the Earth, or in the size of the star? Research demonstrates that all three factors play a role. Stars do indeed lie at different distances from the Earth, but they also vary in diameter, mass, temperature, and composition, and all these variables affect the amount of light that a star emits. These variables also affect the wavelengths of light that stars emit, so some stars look yellow, some white, some red, and some blue.

Variations in Stellar Brightness

Astronomers refer to the amount of electromagnetic energy emitted by a star in a specified amount of time as the star's **luminosity**. You can think of luminosity as a measure of power: a more luminous star produces more power than a less luminous star does, just as a more powerful light bulb produces more light than a less powerful light bulb does. Luminosity doesn't depend on the distance between an observer and the star, or on the direction from which an observer views the star.

The brightness that an earthbound observer sees when looking at a star, however, depends not only on the star's luminosity, but also on the star's distance from the Earth. That's because as energy radiates from a source, it spreads out. You see the same phenomenon when you look toward a streetlight, or a light bulb, at night. At a distance, the light looks small and dim, but close up it appears large and bright **(Fig. 23.11a)**. The size and power of the light hasn't changed, but your eye detects a smaller amount of the energy from a distant light than from a nearby light.

The amount of energy from a star that passes through a square meter at the surface of the Earth in one second is the star's **apparent brightness**. Objects we see in space have a huge range of apparent brightness. When astronomers describe the apparent brightness of stars, they use a scheme that dates back to the ancient Greek astronomer Hipparchus (ca. 190–120 B.C.E.). Hipparchus's original scheme classifies stars visible to the naked eye using a scale divided into **apparent magnitudes**, with 1 being the brightest and 6 being the faintest **(Fig. 23.11b)**. Catalogs of stars used by amateur and professional astronomers provide the apparent magnitudes of objects in the sky. Smaller numbers on the apparent magnitude scale imply brighter objects. Therefore, objects such as the Sun that are brighter than stars with an apparent magnitude of 1 have a negative apparent magnitude. Specifically, the Sun's apparent magnitude is −26.7, the full Moon's is −12.5, and Venus's is −4.4. Stars that you can't see with the naked eye, but which

Figure 23.11 The concept of stellar magnitude.

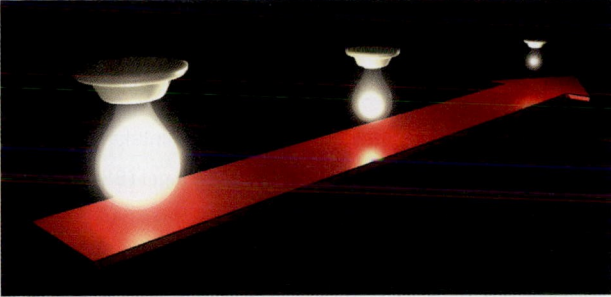

(a) The same light bulb looks dimmer to you as its distance from you increases.

(b) Sirius looks particularly bright relative to the other stars around it in the night sky, so its apparent magnitude is greater than that of its neighbors.

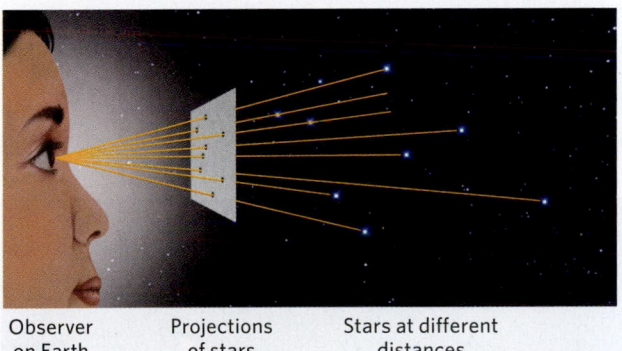

Observer on Earth   Projections of stars   Stars at different distances

(c) The absolute magnitude is the brightness a star would have if positioned at a specific distance (33 light-years) from the Earth.

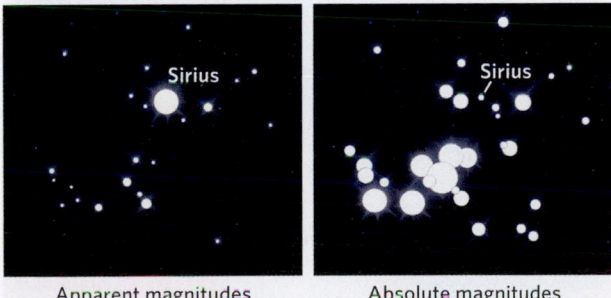

Apparent magnitudes   Absolute magnitudes

(d) A comparison of apparent magnitude and absolute magnitude for the stars around Sirius.

Did you ever wonder . . .

whether all stars are the same?

Table 23.1 Spectral classification of stars

Spectral Class	Approximate Surface Temperature	Color	Example
O	30,000°C (54,000°F)	Electric blue	Mintaka (O9)
B	20,000°C (36,000°F)	Blue	Rigel (B8)
A	10,000°C (18,000°F)	White	Vega (A0)
F	7,000°C (12,600°F)	Yellow-white	Canopus (F0)
G	6,000°C (10,800°F)	Yellow	Sun (G2)
K	4,000°C (7,200°F)	Orange	Arcturus (K2)
M	3,000°C (5,400°F)	Red	Betelgeuse (M2)

can be detected with telescopes, have apparent magnitudes greater than 6. The Hubble Space Telescope can detect objects with an apparent magnitude of 30.

To distinguish among stars based on luminosity, astronomers use two different scales. The first, the **absolute magnitude** scale, represents the value of a star's apparent magnitude if the star were magically placed at a distance of exactly 33 light-years (10 parsecs) from Earth **(Fig. 23.11c, d)**. The second scale, the one preferred by astronomers, expresses a star's luminosity in solar units. A **solar unit** is equal to the luminosity of the Sun. Another star's luminosity, expressed in solar units, is that star's true luminosity, divided by the Sun's luminosity. Rigel, the brightest star in the constellation Orion, has a luminosity of 10,000 solar units, meaning that it generates 10,000 times as much energy per second as the Sun. Our nearest neighbor, Proxima Centauri, has a luminosity of less

than 0.0001 solar units, so it generates only one ten-thousandth as much energy per second as the Sun.

Variations in Stellar Surface Temperature and Color

Different stars emit visible light of different colors. As we discussed in Chapter 17, the wavelength of radiation that an object emits depends on the temperature of the object (according to the rules of blackbody radiation), so the color of starlight tells us the temperature of the star's photosphere. The coolest stars are red, warmer ones (like our Sun) are yellow, hot stars are white, and the hottest stars are blue. There are gradations between these color categories, so each category includes stars in a range of temperatures. Astronomers divide stars into **spectral classes** based on their surface temperatures; these categories, named from hottest to coldest, are O, B, A, F, G, K, and M **(Table 23.1)**. To add greater specificity, each letter category can be divided into ten subcategories, numbered 0 through 9 (with 0 being hotter than 9). According to this classification scheme, our Sun, a G2 star, is cooler than an A0 star and hotter than an M2 star.

Variations in Stellar Dimensions

Stars come in a tremendous range of diameters and masses **(Fig. 23.12)**. When classifying a star, astronomers usually also indicate the star's spectral class because color doesn't necessarily correlate with size; our Sun, for example, can be called a *yellow dwarf*. Astronomers have found numerous examples of very small stars (with diameters only slightly larger than that of the Earth) that glow white because of their high surface temperature; these stars are known as *white dwarfs*. Astronomers also recognize red dwarfs and brown dwarfs, which are in the same size range but are much cooler. **Giant stars** have diameters up to 50 times that of the Sun, and **supergiant stars** have diameters from 50 to over 500 times that of the Sun. If the red supergiant Betelgeuse were placed at the position of our Sun, its surface would extend close to the orbit of Jupiter.

Stars also come in a huge range of masses. Astronomers describe the masses of stars by comparing them with our Sun: one **solar mass** (M_S) represents the mass of our Sun. Low-mass stars contain less than $0.5M_S$, intermediate-mass stars contain 0.5 to $10M_S$, and high-mass stars contain $10M_S$ to $40M_S$. (There are relatively few *supermassive* stars, with masses greater than $40M_S$; the largest of these stars has a mass of $150M_S$.) As we'll see later in this chapter, stellar radius does not necessarily correlate with stellar mass, and stellar mass controls the life history of a star.

The H-R Diagram

In the early 20th century, two astronomers, Ejnar Hertzsprung of Denmark and Henry Norris Russell of the United States, independently examined the relationship between

Figure 23.12 The diameters of stars vary significantly, as we can see by comparing several stars that astronomers have named. Betelgeuse has a diameter 1,200 times that of our Sun.

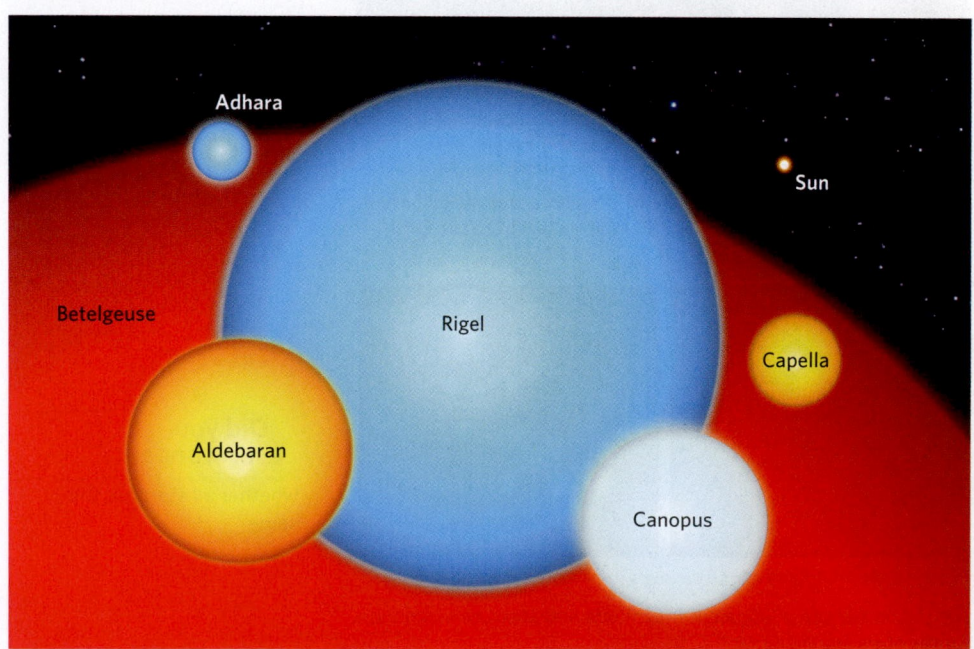

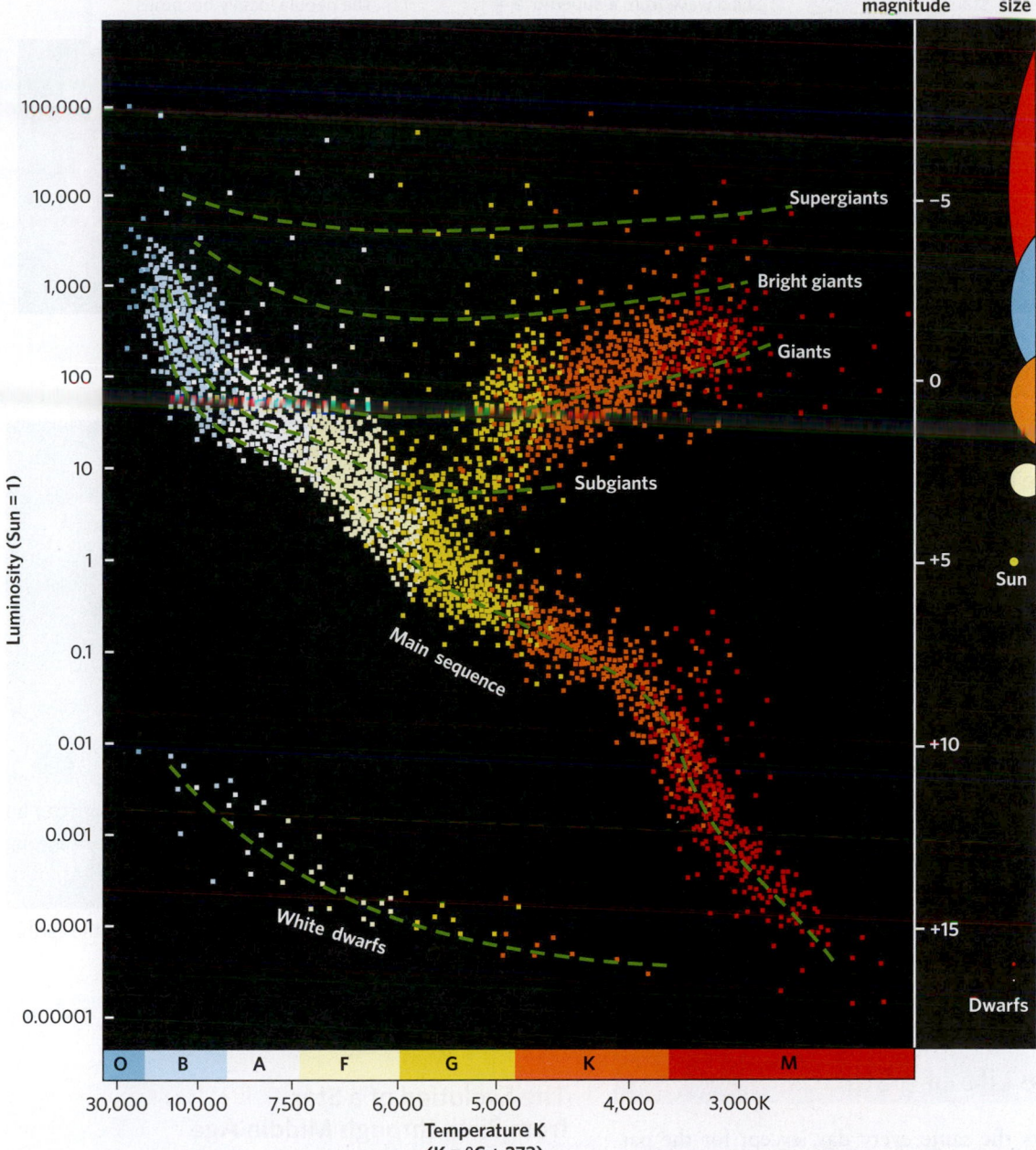

Figure 23.13 The H-R diagram shows the relationship between luminosity and temperature (represented by spectral class). Most stars, including the Sun, fall into the main sequence.

stellar luminosity and surface temperature. The diagram they developed, now known as the *Hertzsprung-Russell diagram* (commonly abbreviated as **H-R diagram**), has become one of the most important tools for understanding stellar life cycles **(Fig. 23.13)**. An H-R diagram is a graph with surface temperature (corresponding directly to color or spectral class) on the horizontal axis and luminosity on the vertical axis. Plotting observations of luminosity and temperature for thousands of real stars on the diagram shows a remarkable order to the properties of stars, as indicated by the relationship between luminosity and color. Most stars, including our Sun, lie in a narrow band, called the **main sequence**, that stretches diagonally across the diagram from the upper left to the lower right. A star that

falls on or below the main sequence, and has a mass less than about 20 solar masses, is classified as a **dwarf star**. This classification includes our Sun which, in its current stage of life, is a yellow dwarf star.

While the main sequence of the H-R diagram includes a great variety of stars that differ from one another in terms of mass, radius, temperature, and luminosity, not all stars fall on the main sequence. Smaller numbers of stars fall into three other areas. *Red giants* populate an area above and to the right of the main sequence. These stars have radii of 5 to over 100 times that of the Sun, but have surface temperatures of only 3,000°C to 5,000°C (5,400°F to 9,000°F). Above the red giants on the diagram lie the even rarer *bright giants* and *supergiants*. White dwarfs fall below the main sequence.

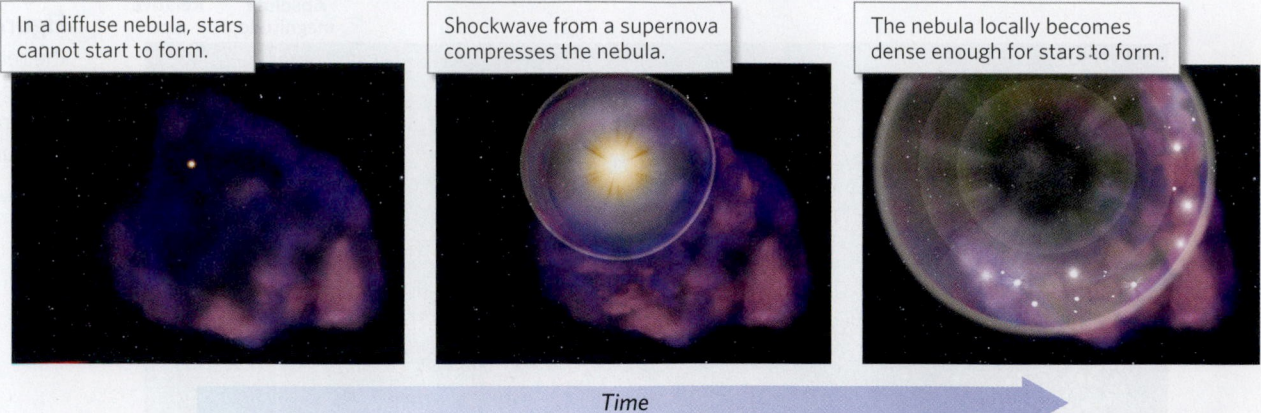

In a diffuse nebula, stars cannot start to form.

Shockwave from a supernova compresses the nebula.

The nebula locally becomes dense enough for stars to form.

Time

(a) Compression due to a shock wave causes pockets of matter to become dense enough for gravitational collapse to begin.

Take-home message . . .

🏠 The apparent brightness (apparent magnitude) of a star that we see from the Earth depends not only on the star's luminosity (the amount of energy it emits), but also on its distance from the Earth. Absolute magnitude—the brightness we would see if the star were exactly 33 light-years away—provides a better characterization of a star's luminosity. Astronomers usually indicate luminosity using solar units, a comparison with the luminosity of the Sun. Stars vary greatly in their luminosity, temperature, size, and mass. We can represent variations in luminosity and temperature on an H-R diagram. Most stars fall on the diagram's main sequence.

Quick Question -
What characteristic defines a star's spectral class?

(b) The Henize 206 Nebula, 163,000 light-years from the Earth, contains stars that are less than 10 million years old.

23.5 The Life of a Star

Our Sun looks the same every day, except for the patterns of sunspots that speckle its surface, and the other stars look the same every night, except for their movement across the sky as Earth rotates. A single glance at the sky provides only an instantaneous snapshot of the Sun and stars. If you could observe stars over the course of billions of years, and in some cases, just millions of years, you would see significant changes in their character, manifested primarily by changes in their luminosity, radius, and mass. In other words, stars evolve over time. The rate of change, however, is not continuous, for stellar evolution happens rapidly during the birth and infancy of a star, slowly during most of the star's life, and then very rapidly at the end of the star's life. In this section, we consider stellar evolution from a star's birth through the beginning of its end. Astronomers deciphered this process, even though its steps take vastly more time than all

of human history, by examining different stars, each at a different stage of life.

The Evolution of a Star from Birth through Middle Age

We discussed the birth of a star in Chapters 1 and 22, which explained that stars are formed by gravitational collapse within nebulae, clouds of gas, dust, and ice floating in space **(Box 23.3)**. Here, we revisit this process, but this time we focus on understanding how the mass involved in the formation of a star affects the star's size, temperature, and luminosity.

The process of star formation probably begins when a disturbance causes the interstellar medium to become relatively denser in a localized region of a nebula. Such disturbances can be caused by the passage of a **shock wave**—a pulse of pressure—sent outward by an immense supernova explosion as the pressure squeezes mass together **(Fig. 23.14a)**. Typically, many distinct pockets of higher density develop within a given nebula, and each pocket becomes a separate star **(Fig. 23.14b)**.

Consider this . . .

Nebulae: A closer look

Between the stars of our galaxy, within the vast reaches of interstellar space, lurk immense quantities of gas and dust. If you flew through most of interstellar space, however, you'd barely detect this, for on average, only about one gas atom occurs in each cubic centimeter and one dust or ice particle occurs in each million cubic meters (a volume that's 100 m on a side). Of the gas atoms in space, 90% are hydrogen and 9% are helium; all the remaining elements constitute just 1% of the atoms. Dust particles consist mostly of silicates, carbon, and iron, whereas ice particles consist mostly of solid water, methane, and ammonia. The tiny quantities of gas, dust, and ice in interstellar space—material that astronomers refer to as the **interstellar medium**—might seem inconsequential, but given the vast volume of space, calculations indicate that their total combined mass significantly exceeds the total combined mass of all stars put together.

In some places within the Universe, "clouds" develop in which the density of the interstellar medium has become greater than elsewhere. In these clouds, known as **nebulae** (from the Latin word for mist), the density of gas and dust reaches 100 to 10,000 particles per cubic centimeter (ten billion times greater than in normal interstellar space). Such clouds affect light passing through them from stars beyond. In fact, nebulae with particularly high concentrations of dust can partially or completely absorb starlight, making their region of sky appear dark and empty of stars from the perspective of Earth.

Astronomers divide nebulae into two types. **Emission nebulae** consist of heated regions of interstellar medium that glow because their gas molecules have been warmed by absorption of UV radiation from very hot stars that recently formed within them **(Fig. Bx23.3a)**. The spectrum of the light produced by these nebulae indicates that the gas in these nebulae consists mostly of hydrogen. **Reflection nebulae** are cooler regions of interstellar medium whose light comes from the scattering of starlight by dust and ice particles **(Fig. Bx23.3b)**. These nebulae look bluish for the same reason that our sky, in which the atmosphere scatters sunlight, looks blue (see Chapter 17). In some nebulae, both emission and reflection take place, leading to a blend of many colors that can make the nebula strikingly beautiful.

Most nebulae are wispy and do not have a distinct geometry; astronomers refer to such nebulae as **diffuse nebulae**. Some emission nebulae, however, have a distinct shape, taking the form of a shell of glowing gas expanding outward from a central point. These nebulae have the somewhat confusing name **planetary nebulae (Fig. Bx23.3c)**. In fact, they have nothing to do with planets. Astronomers use the adjective *planetary* in this context only to indicate that these nebulae surround a central point.

Figure Bx23.3 Examples of nebulae.

(a) The Eagle Nebula, an emission nebula that lies 7,000 light-years from Earth, emits red light after absorbing UV radiation from recently formed stars.

(b) A reflection nebula known as M-78 lies 1,600 light-years from Earth. M stands for Messier (1730–1817), a French astronomer who cataloged comet-like objects, most of which turned out to be nebulae. The light is reflected from nearby stars.

(c) A planetary nebula consists of glowing gases expanding outward from a dying red-giant star. This example, the Cat's Eye Nebula, lies 3,300 light-years from the Earth.

Figure 23.15 Formation of a star from a protostar.

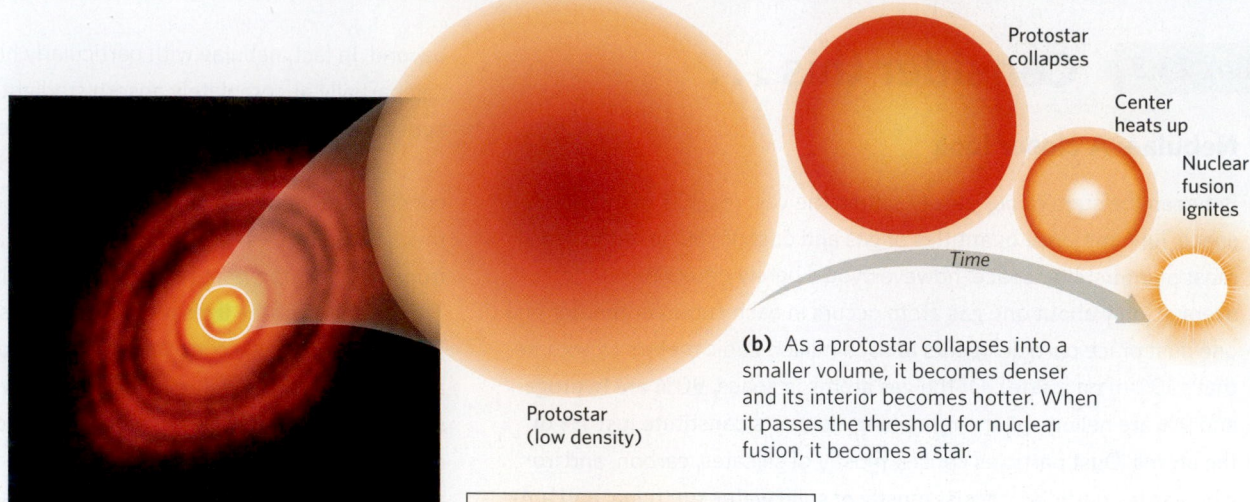

Protostar
collapses

Center
heats up

Nuclear
fusion
ignites

Time

Protostar
(low density)

(b) As a protostar collapses into a smaller volume, it becomes denser and its interior becomes hotter. When it passes the threshold for nuclear fusion, it becomes a star.

(a) A Hubble photo of HL Tauri, an accretionary disk about 450 light-years from the Earth. Astronomers estimate that this disk is only 100,000 years old.

Pressure out

Gravity in

(c) The concept of equilibrium pressure in a star.

Even when a shock wave causes local compression in a nebula, the compressed region won't undergo gravitational collapse if it consists only of gas because gas tends to expand and dissipate. According to condensation theory (see Chapter 22), gravitational collapse requires the presence of dust and ice, for these materials provide condensation nuclei on which other particles collect, building into masses that become large enough that their gravity can pull other matter, including gas, inward. Recall that as gravity pulls matter inward, conservation of angular momentum leads the collapsing pocket of the nebula to start spinning. This spin generates a centrifugal acceleration that flattens the collapsing material into an *accretionary disk* **(Fig. 23.15a)**. At the center of this disk, where rotation is slower and centrifugal force is less, a nearly spherical ball—a *protostar*—grows.

As we have seen, compression of a gas increases its temperature. At the outer edges of the protostar, heat generated through compression escapes to space, so the temperature remains relatively low. But in the opaque interior of the protostar, radiation can't escape, so as gravitational collapse progresses and the protostar shrinks, the temperature in its core progressively rises. A protostar that contains an amount of mass comparable to that of our Sun reaches a milestone about 20 million years after gravitational collapse begins: its core temperature reaches the threshold for nuclear fusion (10 million °C, or 18 million °F). When this happens, the protostar becomes a true star and begins to emit bright light into space **(Fig. 23.15b)**. Over the next 30 million years or so, a Sun-sized star contracts further, its core temperature increases to about 15 million °C (27 million °F), and its photosphere

temperature reaches about 6,000°C (10,800°F). At this point, the star reaches an *equilibrium condition*, in that its temperature and radius stabilize and remain much the same for a long time (billions of years for a star such as our Sun). The radius of the star at this time is called the *equilibrium radius*, and the luminosity represented by its surface temperature is called the *equilibrium luminosity*.

What determines the equilibrium radius and luminosity of a star? At the equilibrium radius, the outward push of internal pressure balances the inward pull of gravity **(Fig. 23.15c)**. The outward-directed internal pressure comes from two sources. First, because pressure depends on the density of a gas or plasma, an outward pressure gradient becomes established, with higher pressure in the core and lower pressure at the surface of the star. Second, when the interior of a star heats up, the plasma in the interior tries to expand outward. Once the star's nuclear furnace starts burning, the temperature rise increases this outward-directed thermal pressure. If it weren't for the outward-directed pressure force in a star, gravity would collapse all of its matter down to a tiny volume at the center of the star. Effectively, the outward-directed pressure force balances the inward-directed gravitational force, and this balance yields an equilibrium condition.

The temperature of a star at equilibrium depends on its mass. A more massive star has a denser core and a hotter core temperature; therefore, fusion happens faster in its core. Mass affects luminosity at equilibrium as well: specifically, a star with more mass produces more energy than one with less mass.

Once a star has achieved an equilibrium condition, it changes only very slowly, but it does change. Over the course of time (a few billion years for a Sun-like main-sequence

Figure 23.16 An example of a binary star. To the naked eye, Albirio, which lies about 430 light-years from the Earth, looks like a single star with an apparent magnitude of 3. A telescope can resolve it into two separate stars that are orbiting each other.

star), the radius of the core slowly decreases. This decrease happens because a helium nucleus, the product of fusion, occupies less volume than did the four hydrogen nuclei from which it formed. Because this decrease in radius results in an increase in density, it causes the temperature of the star, and therefore its luminosity, to increase. Astronomers estimate that due to such contraction, our Sun may be about 30% more luminous now than it was in the Archean.

Significantly, not all accretionary disks produce a single star. In some cases, an object forming from rings of matter in the accretionary disk becomes large enough (over 12 times the size of Jupiter) that nuclear fusion begins within it. This situation yields a **binary star**, a pair of stars orbiting a common center of mass **(Fig. 23.16)**. More than half of all stars are binary stars, but of these, only about 5% to 10% are *visual binaries*, meaning that we can see both members of the system as separate stars from Earth. If the two individual stars of a binary star are farther apart than about 1,000 times the radius of the larger star, then the stars evolve independently. But if the two stars lie relatively close together, they will exchange mass as they evolve, as we'll see in Section 23.6.

Depicting the Evolution of a Star on an H-R Diagram

The position of a given star on the H-R diagram reflects the changes that take place as a star forms and evolves. Here, we summarize these changes for a star the size of our Sun **(Fig. 23.17)**. The numbers on the H-R diagram in Figure 23.17 refer to the numbers indicated in the following steps.

Initially, a volume of mass that will become a star, but has not yet undergone gravitational collapse, has low luminosity and a low surface temperature. As time passes, this volume collapses inward and becomes a protostar. A protostar has a large radius, and because it has so much surface area, it can have a relatively high luminosity (Point 1). As it continues to contract, because it does not contain a nuclear furnace, it has a low surface temperature (about 2,700°C, or 4,900°F), so it appears red (Point 2). As a protostar contracts further, its surface temperature increases and its size decreases. Therefore, even though its temperature is greater, its luminosity decreases (Point 3). Once fusion reactions begin, the star's surface temperature and luminosity both increase, and its position on the H-R diagram begins to move toward the main sequence (Point 4). Eventually, the star achieves an equilibrium size and luminosity. At this time, the star arrives at a position on the main sequence of the H-R diagram (Point 5). Subsequently,

Did you ever wonder . . .
what will happen to the Earth when the Sun's hydrogen fuel is exhausted?

Figure 23.17 The life of a star with the mass of the Sun can be depicted by a succession of points on an H-R diagram. The star moves from the upper right of the diagram, down to the main sequence.

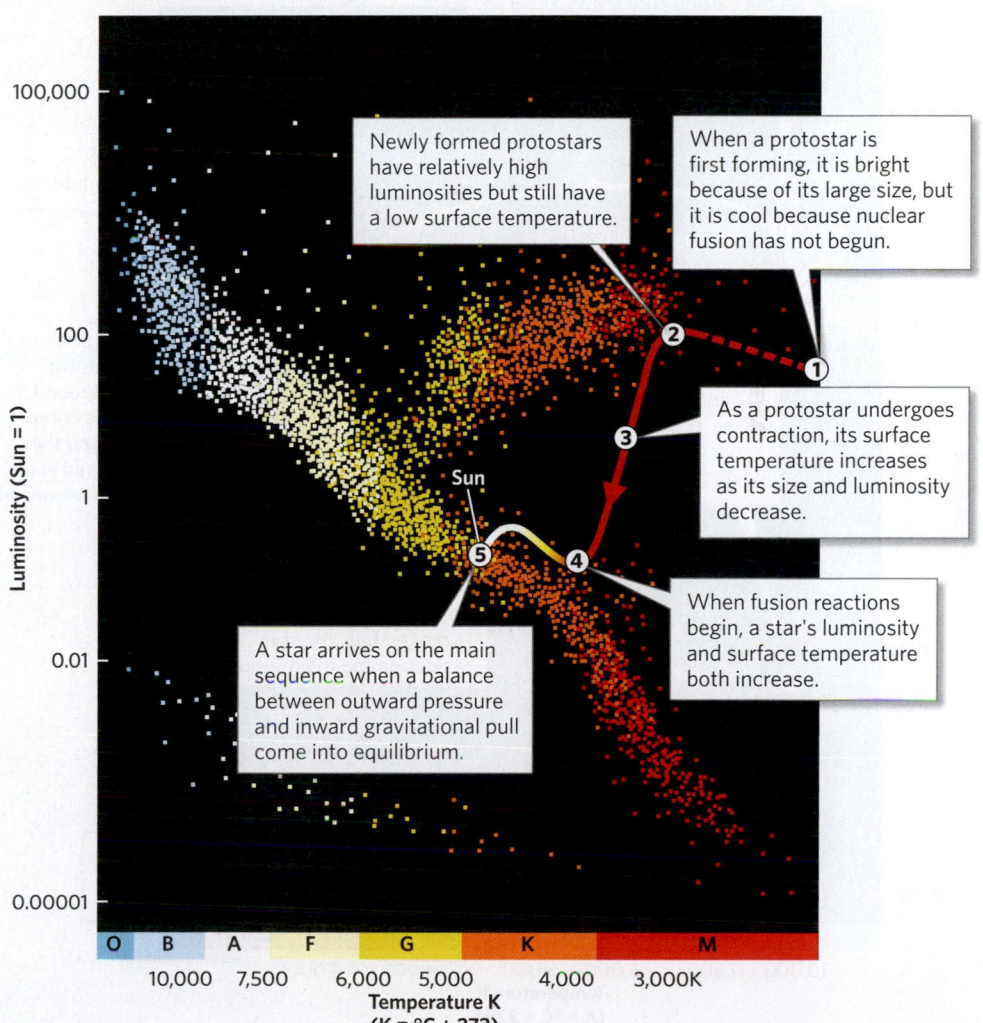

its position doesn't change much until it starts to die. In other words, the main sequence represents stars that have achieved equilibrium. Stars stay on the main sequence for most of their lifetime, as long as they remain in equilibrium.

The specific spot where a given star falls on the main sequence of the H-R diagram depends on the mass of the star. Stars with a mass similar to the Sun's follow tracks similar to that shown in Figure 23.17. Those with a mass higher than the Sun's have a higher equilibrium size and luminosity, and they end up at the upper left of the main sequence. Stars with a mass less than the Sun's have a lower equilibrium size and luminosity and end up at the lower right of the main sequence.

How Long Does a Star Survive?

A star remains in equilibrium until the hydrogen fuel in its core has been used up. How long this equilibrium lasts depends on the star's total mass. The more massive the star, the higher the temperature in the core of the star, and the faster the rate at which hydrogen fuses to form helium. A Type-G star with a mass of $1M_S$ can remain on the main sequence for about 10 billion years. In contrast, a Type-B star with a mass of $5M_S$ resides on the main sequence for only about 100 million years, and a Type-O star with a mass of $10M_S$ exhausts its hydrogen fuel in just 20 million years. Thus, lower-mass stars have longer lives. In fact, stars with masses between $0.08M_S$ and $0.25M_S$ can survive for hundreds of billions of years. Such low-mass stars are called **red dwarfs**. Given the current estimated age of the Universe (13.8 billion years), no star with a mass this low has yet consumed its hydrogen and left the main sequence.

Take-home message . . .

A star forms from the protostar at the center of an accretionary disk when the core of the accretionary disk becomes hot enough for fusion reactions to begin. During its infancy, a star evolves rapidly, but eventually its luminosity and surface temperature reach an equilibrium that depends on its mass. Such stars lie on the main sequence of the H-R diagram. High-mass stars have shorter lives than low-mass stars because they consume their hydrogen fuel at a faster rate.

Quick Question ----------------------------
What forces balance each other inside a star when the star reaches an equilibrium condition?

Figure 23.18 The death of a star with the mass of the Sun can be depicted by a succession of points on an H-R diagram. The star leaves the main sequence and follows a large loop around the diagram.

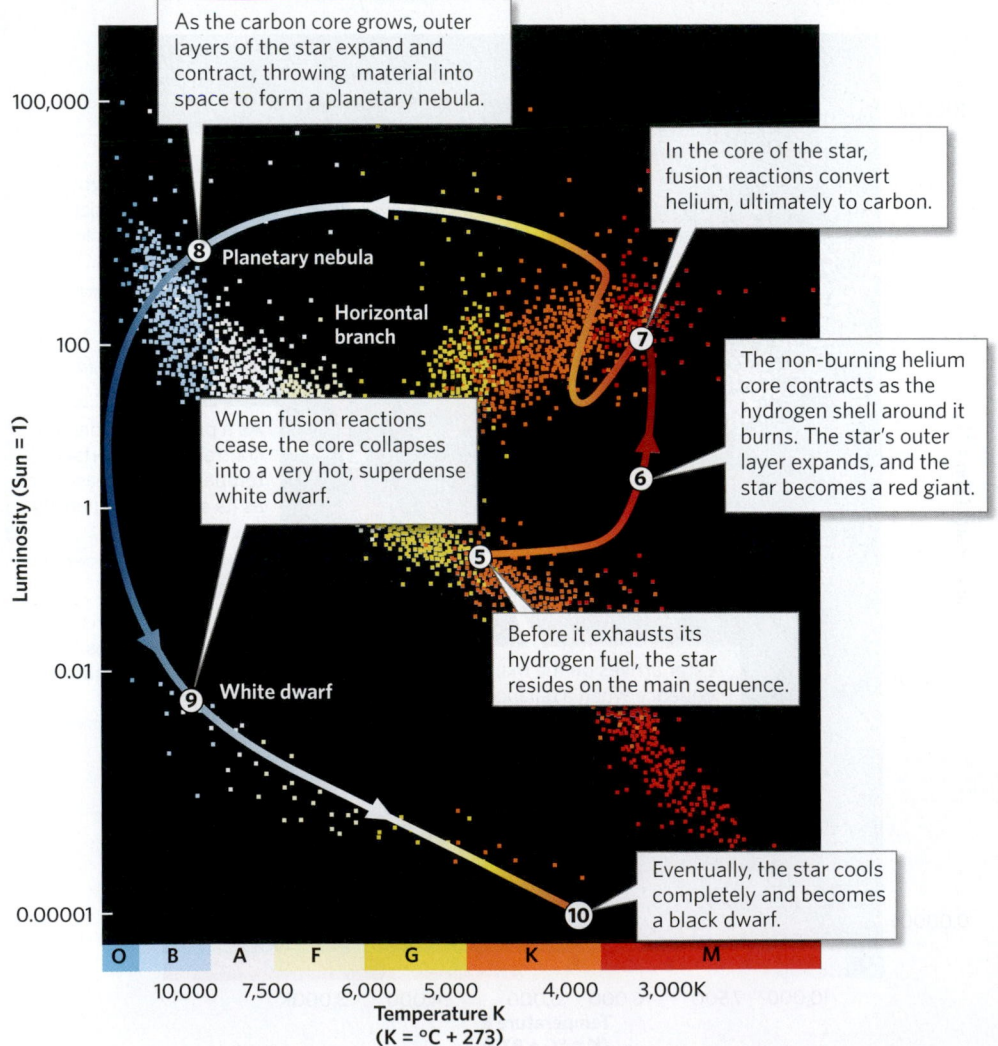

As the carbon core grows, outer layers of the star expand and contract, throwing material into space to form a planetary nebula.

In the core of the star, fusion reactions convert helium, ultimately to carbon.

⑧ Planetary nebula

Horizontal branch

When fusion reactions cease, the core collapses into a very hot, superdense white dwarf.

The non-burning helium core contracts as the hydrogen shell around it burns. The star's outer layer expands, and the star becomes a red giant.

⑨ White dwarf

Before it exhausts its hydrogen fuel, the star resides on the main sequence.

Eventually, the star cools completely and becomes a black dwarf.

Luminosity (Sun = 1)

100,000
100
1
0.01
0.00001

O B A F G K M
10,000 7,500 6,000 5,000 4,000 3,000K

Temperature K
(K = °C + 273)

23.6 The Deaths of Stars and the Unusual Objects Left Behind

In this section, we explore the strange and cataclysmic events that take place when a star finally dies. We'll see that the products of stellar death, which can be quite spectacular, depend on the mass of the star.

Death of an Intermediate-Mass Star and White-Dwarf Formation

Though it seems huge to us, astronomers consider our Sun to be a dwarf star with intermediate mass. As noted in Section 23.4, intermediate-mass stars range from $0.5M_S$ to $10M_S$. Low-mass stars with masses lower than $0.5M_S$ burn so slowly that none have yet exhausted their hydrogen fuel, and they remain red dwarfs. An intermediate-mass star, in contrast, eventually exhausts its hydrogen fuel so that only helium remains in its core. It then begins to undergo a succession of major changes in luminosity and temperature, as indicated by the numbered points on the H-R diagram in **Figure 23.18**.

Just before it begins to die, a Sun-like intermediate-mass star lies on the main sequence of the H-R diagram (Point 5 in Fig. 23.18). Without hydrogen-to-helium fusion occurring in the core, the outward-directed thermal pressure decreases, so the inward gravitational force causes the stellar core to contract. At this time, due to compression, temperatures become high enough for fusion to take place in a hydrogen shell surrounding the helium core (Point 6).

Once ignited, fusion reactions in this shell continue at a furious rate, even as the helium core continues to contract. The outward pressure generated by this burning shell causes the outer layers of the star to expand. Therefore, the diameter of the star increases, and its luminosity increases, but the surface temperature of the expanding star actually decreases. Over the next 100 million years, the hydrogen shell burns outward, like a spreading forest fire, and the outer layers of the star continue to expand. Eventually, the star grows to be 100 times its former size, and its photosphere cools to about 3,000°C. The dying star has become a **red giant** (Point 7).

As time passes, the helium core of a red giant collapses inward and its temperature rises. When the temperature reaches 100 million degrees centigrade, fusion reactions that bind helium nuclei together to produce carbon commence. As a carbon core forms and increases in size, the outer edges of the burning shells of helium and hydrogen progress outward, and the star expands until its photosphere reaches a diameter close to that of the orbit of Mars—when our Sun passes through this stage, some 5 billion years from now, the Earth will have been incinerated.

At the end of the red-giant stage, the burning helium and hydrogen shells become unstable and start to alternately expand and contract (Point 8). This process spews matter from the star's outer layers into space to form a planetary nebula (see Box 23.3). Finally, with its nuclear fuel exhausted, the carbon core cools and thermal pressure decreases, so the star contracts. Over tens of thousands of years, as the nuclear fires of this residual carbon core die, the outer clouds of expelled gas become sufficiently transparent that the white-hot core of the star becomes visible. This Earth-sized remnant is known as a **white dwarf** (Point 9). A white dwarf contains a mass equal to about half that of the Sun packed into a volume only about 1% the volume of the Sun, which gives it an immense density: a cubic centimeter (0.06 cubic inch) of a white dwarf contains about 1,000 kg (1.1 tons) of matter. Ultraviolet radiation emitted by the white dwarf causes the gas of the planetary nebula surrounding it to glow.

As the energy of a white dwarf radiates to space, the star cools and dims, for nuclear reactions are no longer taking place within it. Eventually, the star gets darker until it disappears from the night sky, becoming a cold cinder known as a **black dwarf** (Point 10).

Death of a Binary Star and Nova Formation

The term *nova*, from the Latin word for new, refers to a bright star that suddenly appears where there had been no bright star before. Specifically, a **nova** is a star that was once too dim to see, but for a period ranging from 20 to 80 days, it may attain a luminosity of 50,000 to 100,000 solar units **(Fig. 23.19a)**.

Modern research indicates that novae develop when mass passes from one star to another in a dying binary star in which one of the individual stars has become a white dwarf. The exchange of mass typically begins when the remaining star enters its red-giant stage and swells in size **(Fig. 23.19b)**. When the surface of the red giant gets close enough to the white dwarf, the white dwarf's gravity starts pulling matter away from the red giant. As it undergoes compression within the dense white dwarf, and becomes exposed to the white dwarf's intense radiation, the captured gas heats up to such extreme temperatures that nuclear reactions ignite within it. The sudden onset of fusion reactions in the captured hydrogen results in a violent hydrogen-bomb-like explosion on a stellar scale, causing the star's luminosity to increase quickly and dramatically, and much of the remaining accreted mass gets blasted back into space. It's this explosion that we see from Earth as a nova.

The nuclear fuel that brightens a nova burns up quickly. Once the hydrogen that was captured by the white dwarf has been converted to helium, both the dwarf and its red-giant companion return to their original states. Later on, the white dwarf again draws in mass from its companion red giant until another nova occurs. White dwarfs can undergo several nova events during the lifetime of a binary star.

Death of a High-Mass Star, and Supernova Formation

The evolution of a high-mass star, one with a mass greater than about $10M_S$, differs dramatically from that of an intermediate-mass star because so much gravitational compression takes place in the high-mass star that its core becomes much hotter than that of a lower-mass star. As a high-mass star's core temperature rises, the rate at which hydrogen fusion takes place in its core increases. In fact, fusion takes place so rapidly in a large star that the star can exhaust its hydrogen in just a few million years and suffer an early death.

The early stages in the death of a high-mass star resemble those for an intermediate-mass star: when the core's hydrogen fuel has been used up, gravity compresses the helium that remains until helium fusion ignites and produces carbon. Meanwhile, the outer layers of the star expand.

Figure 23.19 Development of a nova from a binary star.

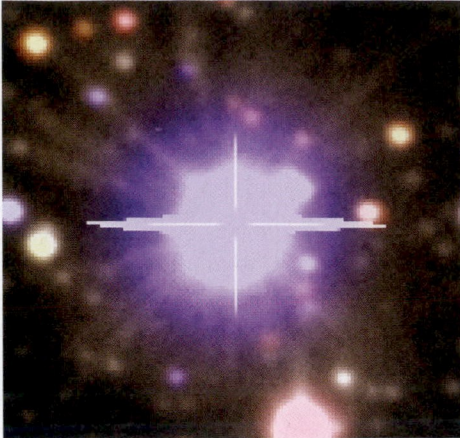 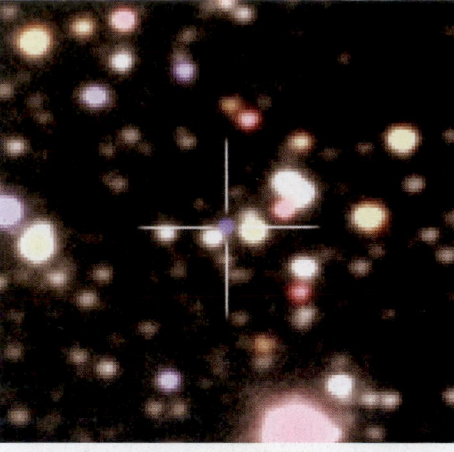

(a) This nova, seen here through a telescope, suddenly appeared in May 2009 and maintained intense brightness for about a week. It continued to fade over the next seven years. The left image shows a sector of space during the nova, whereas the right image shows the same sector after the nova.

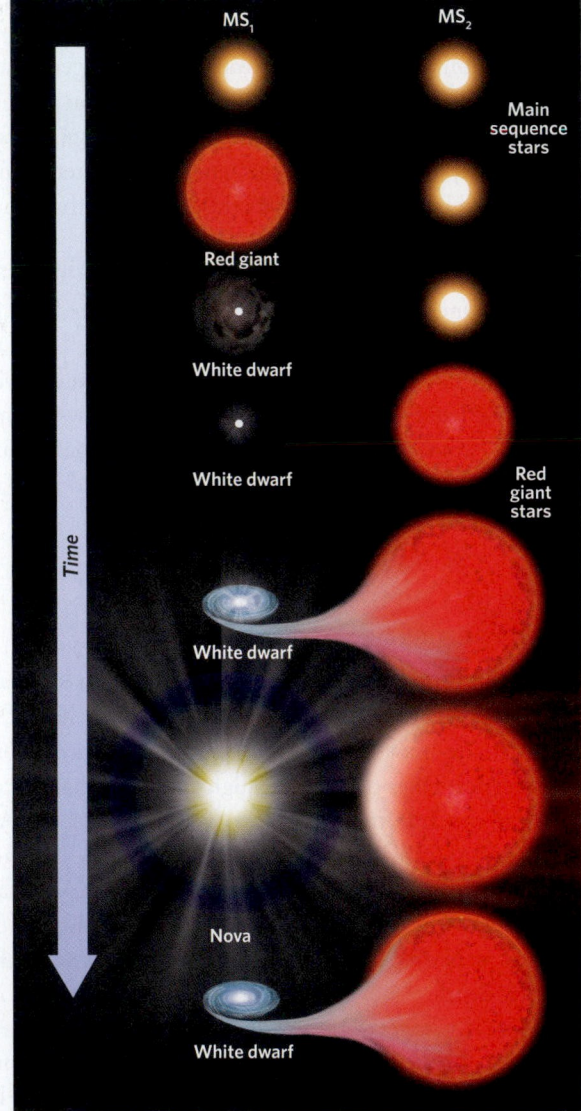

(b) Stages during the development of a nova, which forms when one star in a binary star gravitationally draws in the mass of its partner and eventually explodes.

Because the star contains so much mass, however, a *red supergiant*, instead of a red giant, forms. Because of its immense mass, the late stages of a supergiant's evolution differ from those of a giant. Gravity causes much more inward compression within the core of a supergiant than within the core of a giant. This extra-intense compression, in turn, drives the temperature of the supergiant's core so high that carbon atoms can collide with enough energy to fuse. Such reactions don't take place in the somewhat cooler core of a giant, because carbon atoms don't collide at such high speed. Carbon fusion reactions produce atoms of even heavier elements, such as neon, sodium, magnesium, and oxygen. When carbon fuel runs out, gravity overcomes outward pressure, and the core of the red supergiant collapses even more, and becomes even hotter. As a result, the products of carbon fusion themselves fuse to produce still heavier elements. In fact, a succession of fusion reactions, each producing progressively heavier atoms, continues until iron atoms have been produced. At each stage, fusion reactions involving successively lighter atoms take place in shells around the star's core. So, for example, a shell of oxygen fusion surrounds a shell of neon fusion, and a shell of carbon fusion surrounds the shell of oxygen fusion. In fact, when very heavy atoms are fusing in the core of a red supergiant, hydrogen fusion (to form helium) still takes place in the outer shells of the star **(Fig. 23.20)**.

Once the core of a high-mass star contains only iron, its fusion reactions end. Without the outward thermal pressure produced by the energy of fusion reactions, nothing can counter the relentless inward pull of gravity, and the star suddenly collapses inward. During this *implosion*, matter in the core undergoes such intense compression that its temperature rises to 10 billion °C (18 billion °F). At these incredible temperatures, it takes only about one second for all the core's atomic nuclei to split apart into free protons and neutrons. Further compression of the core causes protons and electrons to combine and form neutrons until the core consists entirely of neutrons packed tightly together in an inconceivably dense mass.

But gravity hasn't finished with the star yet. The neutrons rush inward into an ever-smaller space until they compact into a point with a density approaching 1 trillion kilograms per cubic centimeter (67 million tons per cubic inch). This mass suddenly becomes catastrophically unstable and rebounds outward, generating a violent shock wave that destroys the star and blows all the mass lying outside the core into space. This explosion generates a brilliant but short-lived light in the night sky called a **supernova** **(Fig. 23.21a, b)**. Recorded supernovae have had apparent magnitudes of up to −7.5. Some have been so bright that they can be observed in the daytime sky.

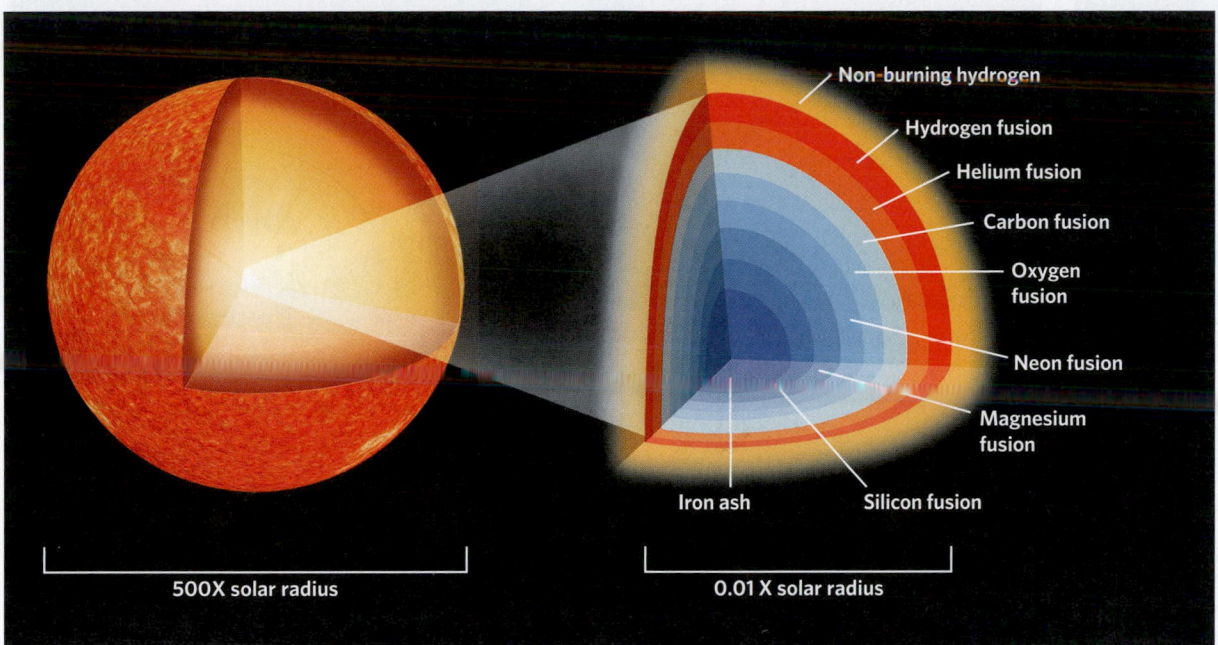

Figure 23.20 Production of elements in shells within the core of a high-mass red-supergiant star just before a supernova.

Non-burning hydrogen

Hydrogen fusion

Helium fusion

Carbon fusion

Oxygen fusion

Neon fusion

Magnesium fusion

Iron ash

Silicon fusion

500X solar radius

0.01 X solar radius

See for yourself

Crab Nebula

Switch *Google Earth*™ to *Google Sky*™ by selecting the "Saturn" button on the button bar and selecting "Sky." Either search for the Crab Nebula in the search box and then zoom in, or use the following sky coordinates:

RA: 5h34m31.97s
Dec: 22°00'29.82"
Arc degrees: 0°10'30.44"

The Crab Nebula is an expanding mass of gas and particles resulting from a giant supernova that was observed from Earth in the year 1054. The Crab Nebula is within an arm of the Milky Way and is located about 6,500 light-years from Earth. It is expanding at a rate of 1,500 km (932 miles) every second. At its center is a neutron star, also called a pulsar. The pulsar is estimated to be about 28–30 km (17–19 miles) in diameter.

The matter blasted into space by a supernova forms an expanding nebula **(Fig. 23.21c)**. For example, the spectacular Crab Nebula, so named because the radiating streaks of glowing gas within it look like the legs of a crab **(Fig. 23.21d)**, came from a supernova observed in 1054 C.E. During supernova explosions, elementary particles such as protons and neutrons move so fast that when they collide, fusion reactions can produce all the natural elements, including heavy elements such as uranium. These newly forming elements blast into the expanding nebula and can then be incorporated into the next generation of stars—stars like our Sun. Indeed, the elements that compose your body today originated in supernovae that occurred early in our Universe's evolution.

Superstitious people thought that supernovae were omens, predictors of major historical events. For example, the 1054 supernova was later taken to be an omen of the great Battle of Hastings in 1066, the start of the Norman conquest of England. Astronomers estimate the recurrence interval for supernovae in the Milky Way at about 100 years. Hundreds of supernovae have been observed in other galaxies. Notably, some of the supernovae in other galaxies have actually outshone their parent galaxies.

Corpses of High-Mass Stars: Neutron Stars and Black Holes

What's left after a supernova blast? We've seen that most of the star's matter spreads out into space as a nebula. The innermost core of the star, however, composed of highly compressed neutrons, survives. This remnant object, known as a **neutron star**, is tiny by stellar standards—only about 10 to 20 km (6 to 12 miles) in diameter—but its mass can be $1.1M_S$ to $3M_S$. A cubic centimeter of neutron star weighs about 100 billion kilograms (100 million tons)!

Astronomers can detect neutron stars, despite their small size, because neutron stars generate very strong magnetic fields and rotate rapidly. As a result, they emit radio waves at regular intervals. The pulses of energy from rotating neutron stars arrive on Earth with clockwork accuracy, so researchers also refer to these objects as **pulsars**. There are well over 1,000 pulsars known in the Milky Way, including one at the center of the Crab Nebula. A pulsar's rotation rate slows and its magnetic field weakens over millions of years, so a pulsar eventually stops sending out its beacon-like radiation.

What happens when a very high-mass star, one with a mass greater than about $25M_S$, enters its death throes? After such a star undergoes a supernova explosion, the remnant neutron star can contain the mass of four Suns or more, crushed into a sphere with a diameter of 15–20 km (9–12 miles), the width of New York City. The overwhelming inward pull of gravity in this mass catastrophically crushes the neutrons into an even smaller volume. Eventually, the gravitational force of this mass becomes so great that within a region whose boundary is called the **event horizon**, nothing can escape, not even electromagnetic radiation or high-speed atomic particles. Astronomers refer to the region within

Did you ever wonder . . .

what's black about a black hole?

Figure 23.21 A supernova produces an expanding nebula.

(a) The M-82 Galaxy (11 million light-years from Earth) before and after a supernova. The supernova was visible in 2014.

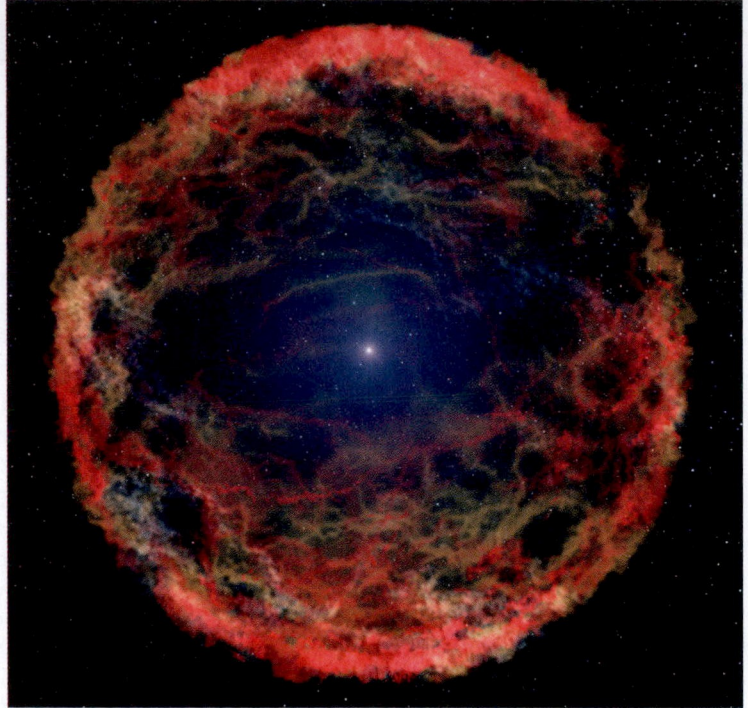

(c) Gas and dust from the explosion expand outward to form a nebula. A tiny, but supermassive bit of the star remains.

(b) A supernova is an immense explosion as portrayed in this artist's rendering

(d) The Crab Nebula, shown here, was formed by a supernova explosion in 1054 C.E.

the event horizon as a **black hole (Fig. 23.22)**. Because no light comes out of a black hole, it would look completely dark to an observer.

Black holes are not all the same size. *Stellar black holes*, formed from individual stars, typically contain $10M_S$ to $20M_S$. Currently, the smallest known black hole contains 3.8 solar masses. The largest black holes, which we examine in Section 23.7, are called *supermassive black holes* and are probably a product of galaxy formation. They can contain the mass of millions of stars within the volume of a single star or less.

Since black holes have mass, they are real physical objects. Formation of a black hole does not destroy matter, so a black hole should not be thought of as a hole in space. Because black holes emit no light or par-

ticles, the only way they can be detected is by identifying their gravitational influence on other bodies in their vicinity. A lone black hole can be completely undetectable, but if a black hole has a binary companion, the behavior of the companion can be used to determine the black hole's mass and location. Science-fiction stories sometimes depict black holes as deadly galactic scavengers, lurking in space and gobbling hapless planets that pass near their location. In reality, black holes act like all other objects in space. Beyond the event horizon, objects orbit black holes just as they do other celestial bodies.

A Summary of Stellar Evolution

The early stages in the life story of a star, involving the formation of a star by collapse of a nebula, are much the same

Figure 23.22 A black hole is a tiny but extremely massive object formed by the collapse of a supermassive star. No radiation or matter can escape from within a boundary called the event horizon.

for all stars, regardless of their mass. But the end-of-life behavior of a star varies radically depending on the star's mass. This relationship emphasizes the importance of mass in determining gravitational force. While a star burns, outward thermal pressure can counter gravity's inward pull. But when a star dies and its fusion furnaces shut down, its mass collapses under the force of gravity. The greater the mass there is to start with, the greater the gravitational force. Smaller stars end up as white dwarfs, larger stars as neutron stars, and the largest stars as black holes **(Fig. 23.23)**.

Take-home message . . .

No low-mass stars (red dwarfs) have yet exhausted their nuclear fuel. Intermediate-mass stars die by expanding to become red giants and, eventually, collapsing to become dense white dwarfs surrounded by a planetary nebula consisting of expelled gas. A high-mass star passes through a red-supergiant stage before eventually dying in a violent explosion called a supernova, which creates a nebula relatively rich in heavy elements. After the supernova, an extremely dense object, known as a neutron star or pulsar, remains. If this object is sufficiently massive, it collapses into a tiny object of incredible density called a black hole, from which not even light can escape.

Quick Question -
What's the difference between a nova and a supernova?

23.7 Galaxies

Our Sun is one of hundreds of billions of stars in the Milky Way Galaxy. Beyond the Milky Way lie hundreds of billions to possibly over a trillion other galaxies. Each **galaxy** contains hundreds of billions of stars, as well as all of their planets and moons, and collections of interstellar gas, dust, and ice. Gravity binds all of this matter together, so it travels through the Universe together. In this section, we look at the character of galaxies and the variation among them.

The Milky Way Galaxy: A Closer Look

The Milky Way is our home galaxy, so when we look at it from the Earth, we can't see its overall form **(Fig. 23.24a)**. But, by analyzing images and by comparing it to other galaxies, astronomers have determined that the Milky Way is a *barred spiral galaxy* **(Fig. 23.24b)**. The term *barred* refers to a concentration of stars that form a straight bar-like shape along the central axis of the galaxy, and the term *spiral* refers to the existence of concentrations of stars in bands, or arms, that curve into the center of the galaxy. The spiral arms of the Milky Way lie within the **galactic disk**, which flattens toward its rim **(Fig. 23.24c)**. Overall, the Milky Way has a diameter of 100,000 light-years and a thickness on average of 1,000 light-years, and it contains between 200 and 400 billion stars. The Sun resides along the outer edge of one of the spiral arms, about 27,000 light-years from the galactic center. The Milky Way slowly rotates, so our Sun makes one orbit around the galactic center about once every 230 million years, a period known as a *galactic year*.

At its center, the galactic disk of the Milky Way thickens into a **galactic bulge** containing a relatively dense concentration of stars. Starlight in the galactic bulge is so bright that night on a planet orbiting a star within it would be as bright as day. The galactic bulge contains an abundance of interstellar gas and dust and serves as a nursery for new stars. It's also a violent place, where stars collide and supernovae explode. At the center of the bulge,

Figure 23.23 Life cycles of stars with different masses.

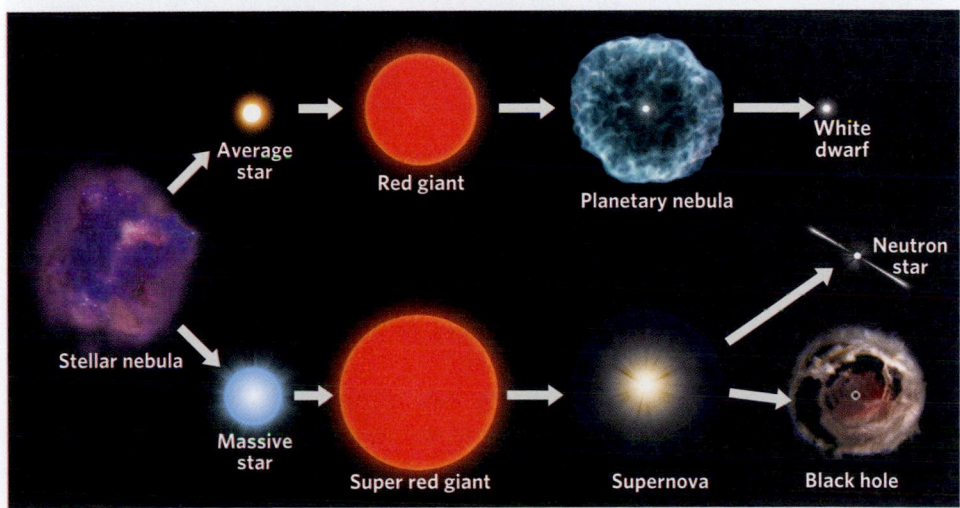

Figure 23.24 The Milky Way Galaxy.

(a) The Milky Way as seen on a dark night from the Earth. The laser beam points directly at the galactic center. The dark patches are regions where interstellar dust blocks our view of stars.

(b) Computer graphic of the Milky Way, as viewed from the top, with components labeled. Note the position of the Sun.

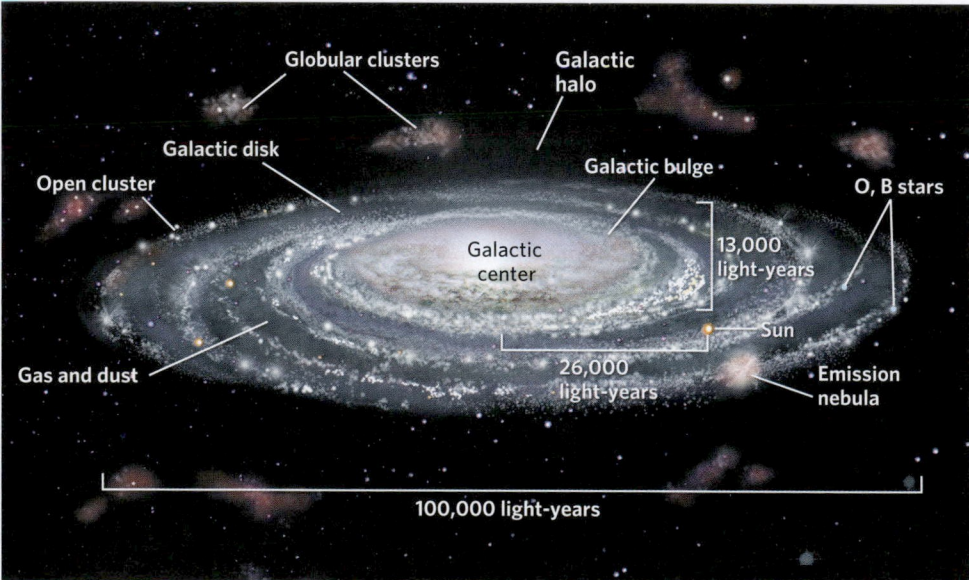

(c) Artist's rendering of the Milky Way as viewed from the side. The galactic bulge lies at the center.

apart **(Fig. 23.25)**. Most stars in globular clusters are small reddish stars that have probably been burning since the formation of the galaxy. New stars do not form in these clusters, and all the larger stars that once existed there have long since died.

Galaxies and Quasars

With the spectacular images of galaxies that we now have, it's hard to imagine that less than a century ago, astronomers didn't even realize that galaxies outside the Milky Way existed. They had seen objects they called "spiral

Figure 23.25 This globular cluster, known as M-13, is 25,000 light-years from the Earth and lies within the Milky Way. It has a diameter of about 145 light-years and contains about 300,000 stars.

at the galactic center, lies a **supermassive black hole**, an object with the mass of 4 million Suns crammed into a volume with a radius of 30 AU (less than that of Neptune's orbit) and an event horizon that extends 3 million kilometers (1.9 million miles) from the object's center. Needless to say, the density of supermassive black holes is beyond comprehension.

A **galactic halo** surrounds the Milky Way. This halo contains hardly any dust or gas, but includes over 150 **globular clusters**, spherical accumulations of hundreds of thousands of stars that may be as close as 1 light-year

(a) Spiral galaxy NGC 1365, known as the Pinwheel Galaxy.

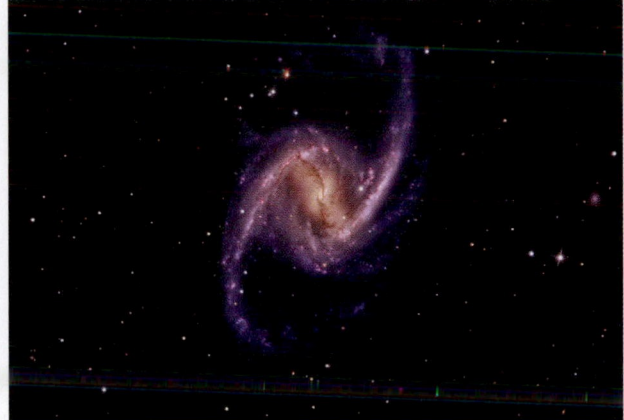

(b) Barred spiral galaxy M-86.

(c) Irregular galaxy NGC 1569.

(d) Elliptical galaxy ESO 325-G004.

nebulae," but thought that these objects lay within the Milky Way. It was only in the 1920s, when astronomers determined that spiral nebulae were actually spiral galaxies outside the Milky Way, that they began to catalog distant galaxies. Edwin Hubble classified galaxies into four basic types based on their shapes: *spiral, barred spiral, elliptical,* and *irregular* **(Fig. 23.26)**.

In addition to distinguishing among galaxies based on shape, astronomers now categorize them based on the nature of the energy they emit. About 60% of galaxies, our own Milky Way included, are **normal galaxies** in that they emit their greatest amounts of radiation at or near visible wavelengths. Normal galaxies range in size from *dwarf galaxies,* with luminosities starting at about 1 million solar units, to *giant galaxies,* with luminosities of up to 1 trillion solar units. **Active galaxies**, unlike normal galaxies, emit most of their energy in the X-ray and radio regions of the spectrum, so their energy cannot simply be the sum of all the light produced by stars in the galaxy. Furthermore, most of the energy of an active galaxy comes from the galactic center.

Astronomers refer to a subset of active galaxies as **quasars** (an abbreviation of *quasi-stellar radio source*). Quasars

differ from other active galaxies in that they display an extreme red shift, meaning that they are the most distant objects detectable within the Universe **(Fig. 23.27)**. In fact, the energy we receive from quasars today started its journey at a very early stage in the Universe's history, long before our Sun formed (see Chapter 21). Quasars emit enormous amounts of radiation, each having a luminosity approaching that of 1,000 Milky Ways.

Figure 23.27 A quasar 10 billion light-years from Earth, as seen in an X-ray photograph by the Chandra Space Telescope.

Figure 23.28 An artist's depiction of the accretion disk around a supermassive black hole in an active galaxy.

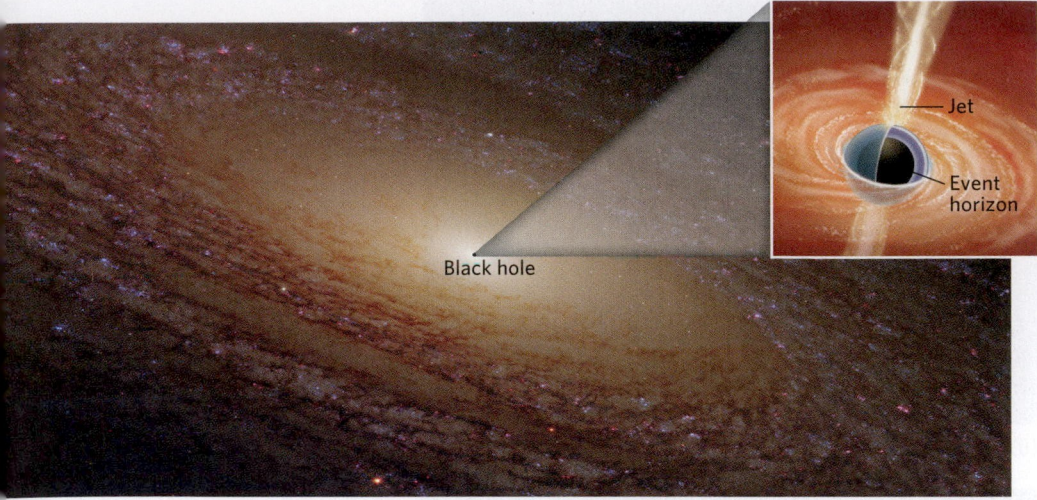

Jet

Event horizon

Black hole

As we've seen, a supermassive black hole occupies the center of the Milky Way. Recent research indicates that similar supermassive black holes occur in most, and perhaps all, galaxies. The discovery of supermassive black holes provided a basis for explaining the difference between normal and active galaxies. In a normal galaxy, not much matter surrounds the central supermassive black hole, whereas in an active galaxy, the space around the central black hole contains abundant matter. Therefore, in a normal galaxy, the region around the black hole is fairly quiet, whereas in an active galaxy, it's a violent place, where extreme tidal forces rip apart stars. The resulting stellar matter collects into an accretion disk that initially surrounds the event horizon of the black hole and eventually spirals into it **(Fig. 23.28)**. Friction between particles in this material, as well as compression of the material, generates the immense amount of energy that active galaxies emit. But active galaxies don't stay active forever. Once all the stellar material in the vicinity of the supermassive black hole has been drawn into it, energy production decreases, and the active galaxy becomes a normal galaxy.

What conditions lead to the formation of an active galaxy? In the case of quasars, because they are so old, we are probably seeing the earliest galaxies in their formation stage, a time during which their central supermassive black holes were still acquiring their mass. In the case of nearby active galaxies, we are probably seeing the consequence of a collision between two galaxies **(Fig. 23.29)**. As galaxies pass through each other's space, gravitational forces can deflect stars dangerously close to the central supermassive black hole, where gravity traps them. Once near the black hole, the material of these stars swirls into an accretion disk and then eventually into the black hole itself.

Take-home message . . .

A galaxy is an assemblage of stars, their planets and moons, and interstellar materials, all bound together by gravity. Evidence suggests that most, if not all, galaxies have a supermassive black hole at the center, whose effects determine whether a galaxy is normal or active.

Quick Question -
What is a quasar?

Figure 23.29 Examples of colliding galaxies.

(a) These colliding galaxies (NGC 4676B and NGC 4676A, 300 light-years from Earth) are called The Mice because of their long tails of stars and gas.

(b) Two spiral galaxies (NGC 2207 and IC 2163, 80 million light-years away) interacting.

23.8 The Structure and Evolution of the Universe

The Large-Scale Structure of the Universe

When astronomers mapped the distribution of galaxies, they discovered that galaxies are not randomly scattered across the Universe and that they do not move independently from one another, for they feel the gravitational pull of their neighbors. For example, the Milky Way has links to 14 smaller *satellite galaxies*, including the Large and Small Magellanic Clouds. Our galaxy and its satellites, together with a few other large galaxies, including the Andromeda Galaxy **(Fig. 23.30)**, and their satellites, constitute the **Local Group**, which is about 10 million light-years across **(Fig. 23.31a)**. All together, the Local Group includes 54 galaxies. The Andromeda Galaxy, our nearest large neighbor in the Local Group, lies about 2.5 million light-years away.

Within a local group, galaxies may be moving toward or away from one another. Andromeda and the Milky Way, for example, are moving toward each other at about 400,000 km per hour. At this rate, the two galaxies will collide in about 4 billion years. Stars in a galaxy are so far apart that during a collision, individual stars probably don't bash into one another. Rather, gravitational forces may merge the two galaxies into a single elliptical galaxy.

Local groups, in general, contain less than about 100 galaxies. Larger associations of galaxies, which may contain hundreds to thousands of galaxies, are known as *clusters*, and even larger associations, containing on the order of a hundred thousand galaxies, are called *superclusters* **(Fig. 23.31b)**. The Milky Way is part of the Virgo Supercluster, which in turn is part of the Laniakea Supercluster, which is over 500 million light-years across **(Earth Science at a Glance**, pp. 850–851).

At an even larger scale, the configuration of galaxies defines a three-dimensional mesh composed of string-like *filaments* and screen-like *walls*, whose long dimensions are on the order of 30 to 300 million light-years **(Fig. 23.31c)**. Walls and filaments surround *voids*, bubble-like regions of space on the order of 300 million light-years across in which galaxies do not occur. Astronomers remain puzzled as to how gravity can extend across such large areas to hold together filaments and walls. Some speculate that **dark matter**, invisible material that has mass, exists within these regions. The composition and character of this undetectable dark matter remains unknown. If it exists, then each galaxy, and the Universe as a whole, contains vastly more mass than telescopes can detect.

Universe Expansion

As we noted in Chapter 21, astronomers have observed that all distant galaxies display a red shift and are therefore

Figure 23.30 Andromeda, the nearest large galaxy to the Milky Way, lies 2.5 million light-years from the Earth.

moving away from the Earth at a high rate of speed. According to *Hubble's law*, as we saw in Chapter 21, the speed at which distant galaxies are moving away from the Earth is directly proportional to their distance from the Earth. At current rates of expansion, defined by Hubble's constant, the distance between any two objects in the Universe will double over the next 9.8 billion years. One can choose any location in the Universe, and from that location, all distant galaxies will appear to be receding.

If all galaxies are moving away from one another, then earlier in the history of the Universe, galaxies must have been closer together. Mathematical simulations led astronomers to conclude that about 13.8 billion years ago, all the matter and energy in the Universe was located in a single point in space with an indescribable temperature and density—a **singularity**. For reasons that we have yet to understand, the Universe came into being when this singularity exploded and began to expand. As we discussed in Chapter 1, this event, the beginning of the Universe, is called the **Big Bang**. In the instants after the Big Bang, the Universe consisted only of energy, but minutes later, as the Universe cooled, matter formed **(Fig. 23.32a)**. Initially, the Universe was dark, as all matter existed in diffuse nebulae. But by the time the cosmic clock passed the 400-million-year mark, the first stars had formed. As stars interacted gravitationally, galaxies formed, and soon linked into groups and clusters. All the while, space enlarged as the Universe expanded, and galaxies outside of groups and clusters moved farther apart.

Has the Universe expanded at the same rate throughout its history? Astronomers now think that the answer is no. During the first few hundred thousand years of the Universe's history, a time that astronomers refer to as the *period of inflation*, the rate of expansion was faster than

Figure 23.31 Galaxies are organized in space at different scales.

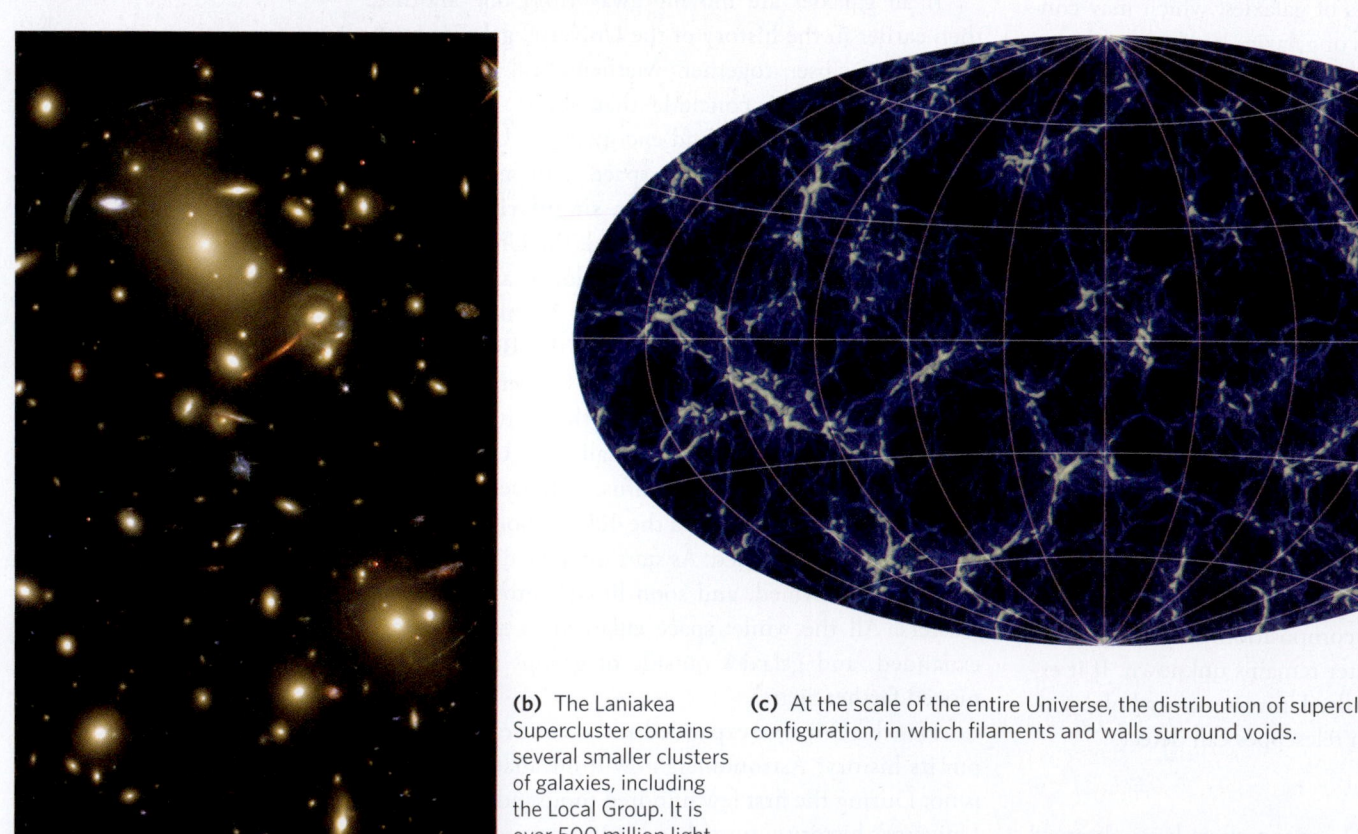

NGC 404

Small Magellanic Cloud

M31 Andromeda

M 110

M 32

M33 Triangulum

NGC 147
NGC 185

Globular clusters

NGC 278

Large Magellanic Cloud

The Sun

Milky Way

(a) The Local Group of galaxies in the vicinity of the Milky Way.

(b) The Laniakea Supercluster contains several smaller clusters of galaxies, including the Local Group. It is over 500 million light-years across.

(c) At the scale of the entire Universe, the distribution of superclusters displays a web-like configuration, in which filaments and walls surround voids.

it is today (see Fig. 23.32a). Then expansion slowed to roughly the rate that we observe today and stayed that way for billions of years. Recent studies suggest that expansion may be accelerating and that billions of years in the future, it will be much faster than it is today. No one knows for sure why expansion should accelerate. Some astronomers have attributed it to the existence of **dark energy**, a yet unknown and unseen force that presumably operates throughout space.

The Future of the Universe

Will the Universe continue to expand forever, or will it, at some point billions of years in the future, reach its maximum extent and start to contract, eventually collapsing back into a singularity? This question lies at the heart of cosmological research today.

The answer to this question depends on whether gravity is strong enough to slow and eventually reverse the expansion of the Universe. If it is, then like a ball that you throw up into the sky that then falls back down, galaxies will eventually stop moving outward and start moving inward, coming together in a *Big Crunch* **(Fig. 23.32b)**. As matter falls inward, galaxies will collide and supermassive black holes will form. They will then combine to form even larger black holes, emitting immense amounts of energy, until, in the end, all matter and energy will converge on a single point, perhaps yielding the precondition for another Big Bang. If the Big Crunch model is correct, then the Universe alternately expands and contracts, like an accordion. Astronomers have calculated that the *critical density* of matter in the Universe necessary to trigger a Big Crunch is only about 5 atoms per cubic meter, and conclude, therefore, that a Big Crunch lies in the realm of possibility.

Recent studies, however, argue that the Universe has less than the critical density necessary for a Big Crunch and will therefore expand forever. As the Universe becomes larger and larger, its matter will become more and more dispersed. Massive stars will continue to explode and their remnants will wink out. But the new nebulae they form, at some point, may be too diffuse to yield another generation of stars, so those nebulae will cool and dissipate further. Eventually, in this scenario, all that will be left are very long-lived red-dwarf stars, which, in hundreds of billions of years, will also dim. So, perhaps a trillion years in the future, our Universe will be an expanding cloud of cold gases and dark cinders, and the atoms that now make up our home planet, the Earth, will be dispersed across vast distances . . . or maybe not. Science being science, such hypotheses will continue to be tested, and models will be challenged. New data, and

Figure 23.32 Alternative futures of the Universe.

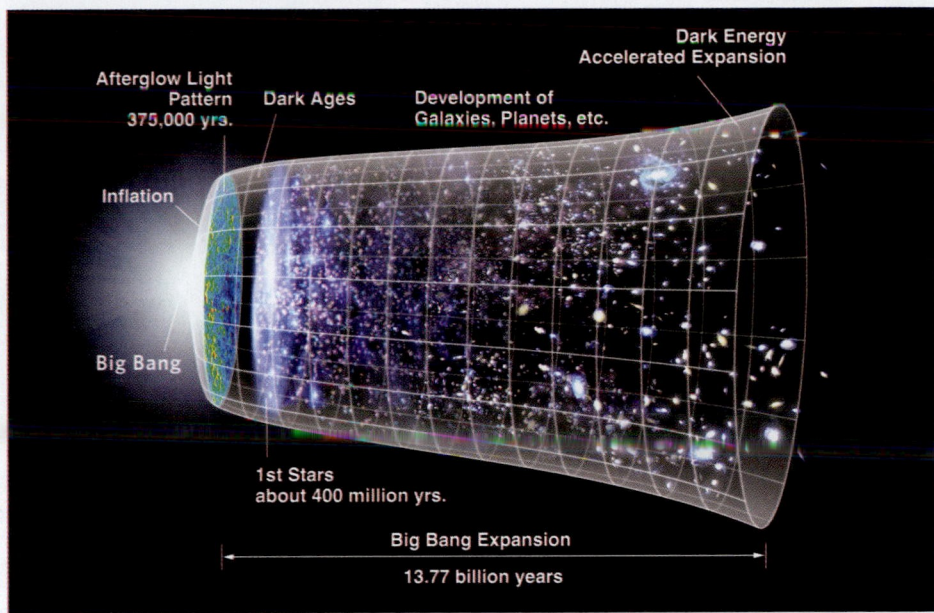

(a) The Universe expanded rapidly after the Big Bang and has continued to expand ever since.

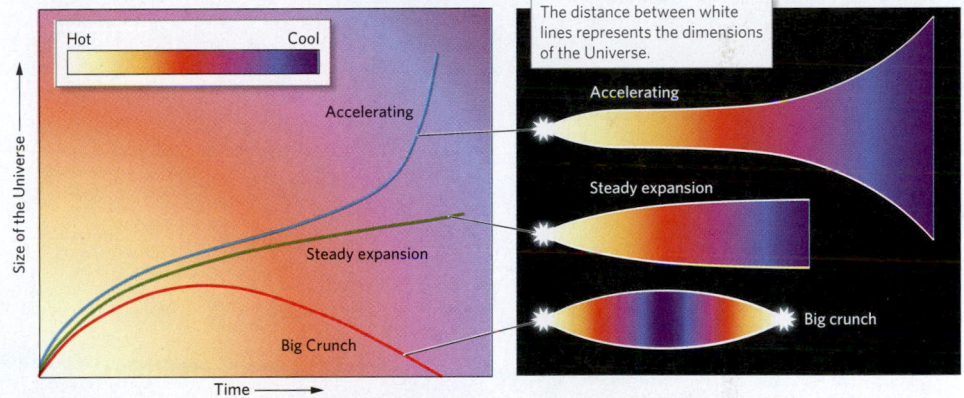

(b) If a Big Crunch happens, the Universe will collapse back to a singularity. Alternatively, the Universe may expand steadily, or at an accelerating rate.

new thinking, may suggest yet another alternative for the future of everything.

Take-home message . . .

Galaxies are arranged in clusters throughout the Universe and are moving away from one another at a constant rate as a consequence of the Big Bang. The Universe may end in a Big Crunch that reverses the Big Bang, or it may expand forever. Only more data will tell us which, if either, is the fate of the Universe.

Quick Question -----------------------------
What is the critical density of the Universe?

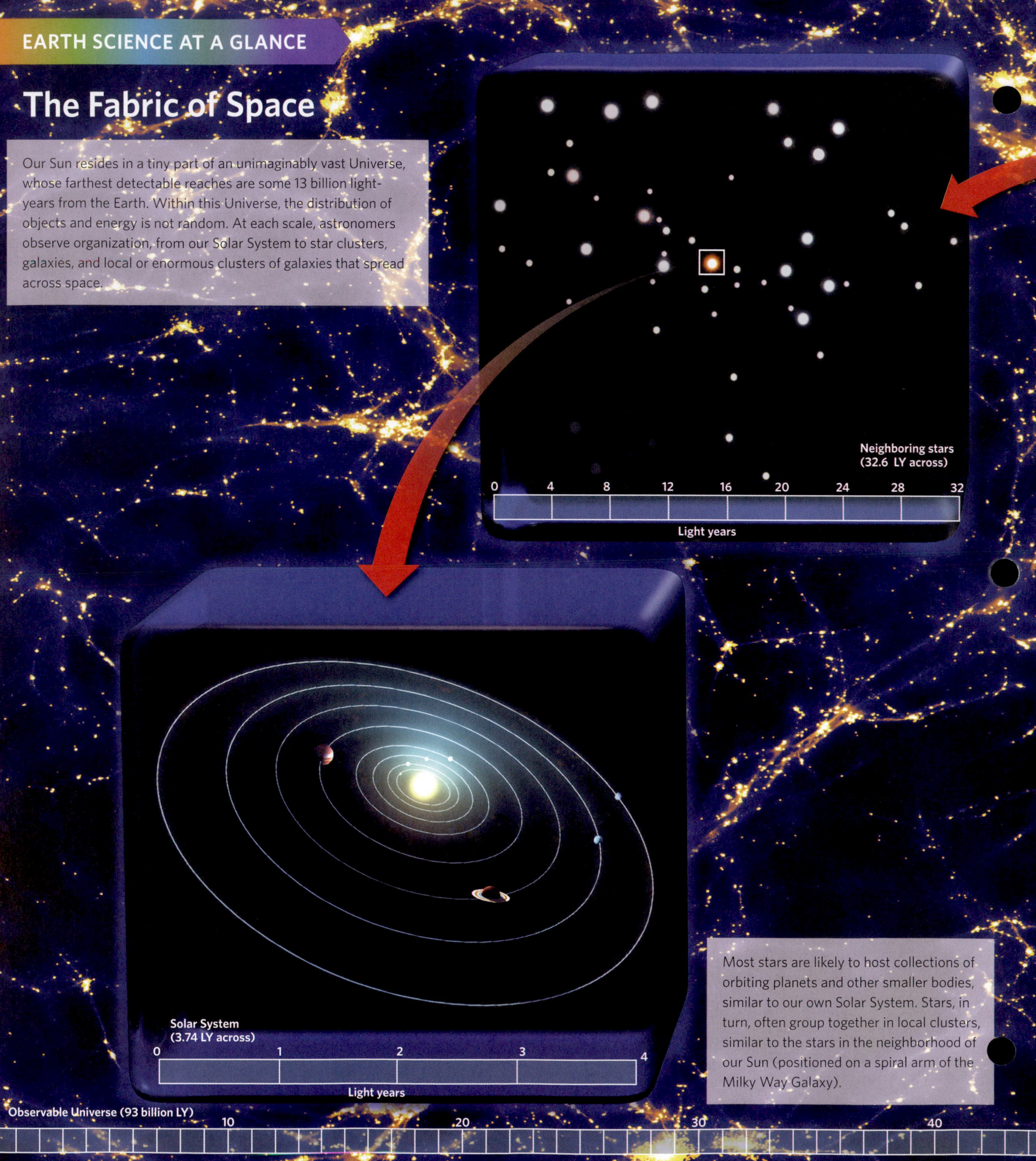

The Fabric of Space

Our Sun resides in a tiny part of an unimaginably vast Universe, whose farthest detectable reaches are some 13 billion light-years from the Earth. Within this Universe, the distribution of objects and energy is not random. At each scale, astronomers observe organization, from our Solar System to star clusters, galaxies, and local or enormous clusters of galaxies that spread across space.

Neighboring stars
(32.6 LY across)

0   4   8   12   16   20   24   28   32

Light years

Solar System
(3.74 LY across)

0   1   2   3   4

Light years

Most stars are likely to host collections of orbiting planets and other smaller bodies, similar to our own Solar System. Stars, in turn, often group together in local clusters, similar to the stars in the neighborhood of our Sun (positioned on a spiral arm of the Milky Way Galaxy).

Observable Universe (93 billion LY)

10   20   30   40

Countless numbers of stars organize in galaxies. In most, if not all galaxies, the stars are believed to be gravitationally bound to, and orbit, a supermassive black hole located at the center of each galaxy.

Milky Way
(100,000 LY across)

| 0 | 20 | 40 | 60 | 80 | 100 |

Thousand light years

Local Galactic Group
(50 galaxies, 10 million LY across)

| 0 | 2 | 4 | 6 | 8 | 10 |

Million light years

Across the Universe, galaxies gather in clusters. Clusters appear as thread-like tendrils across the fabric of space, with vast, apparently empty, regions of space between the tendrils.

Virgo Supercluster
(110 million LY across)

| 0 | 20 | 40 | 60 | 80 | 100 | 110 |

Million light years

| 50 | 60 | 70 | 80 | 90 |

⊙23 CHAPTER REVIEW

- The Sun consists primarily of hydrogen and helium. Other elements make up only about 1% of its mass.

- The Sun has a solar core, in which nuclear fusion of hydrogen to helium generates energy. Surrounding this core is the radiative zone, through which radiation from the core passes. In the outer layer the convective zone, convection transports energy outward.

- The Sun has an atmosphere consisting of the photosphere, the chromosphere, and the corona. The light of the Sun radiates from the photosphere. The corona is an extremely hot, low-density gas layer that can be seen only during an eclipse.

- A large number of high-energy particles flow outward from the Sun at high speeds. This flow is called the solar wind.

- The Sun has a very strong and complicated magnetic field. Magnetic field lines sometimes arc upward from the photosphere. Where they do, sunspots, solar prominences, and solar flares occur. These phenomena affect space weather on the Earth.

- Stars vary in luminosity, a measure of the energy they emit, and in surface temperature. These two quantities are closely related to a star's size and mass.

- Astronomers use several scales to classify a star's brightness. Apparent magnitude indicates how bright a star appears in the sky; absolute magnitude indicates how bright a star would appear to be if it were located 33 light-years from Earth.

- The H-R diagram plots luminosities of stars against their temperatures, provides a tool for understanding the life cycles of stars.

- Stars form from the gravitational collapse of gas and dust within nebulae, which is triggered by disturbances, such as shock waves, that pile up mass locally.

- As gravity compresses a protostar's mass into a smaller and smaller volume, its core temperature increases. A protostar becomes a star when its core reaches a temperature of 10 million degrees centigrade, at which nuclear fusion can occur.

- The lifetime of a star depends on its mass. Less massive stars reside much longer on the main sequence than do more massive stars because fusion reactions happen faster in more massive stars.

- After a star consumes the hydrogen in its core, the core collapses and ignites nuclear reactions of heavier elements, and the outer layers of the star swell. An intermediate-mass star becomes a red giant, whereas a high-mass star becomes a red supergiant.

- When one member of a binary star is a white dwarf, it may capture matter from its companion. When this matter ignites, a nuclear explosion called a nova takes place.

- A star's life ends when it burns through its nuclear fuel. An intermediate-mass star's core collapses into an extremely dense white dwarf, while its outer layers disperse into space, creating a planetary nebula.

- The core of a high-mass star collapses to become a neutron star or black hole, while most of the star's mass is blasted into space in a titanic explosion called a supernova. White dwarfs, neutron stars, and black holes contain superdense matter.

- A black hole's gravity is so strong that no matter or energy can escape from it. The sphere surrounding a black hole within which nothing escapes is called the event horizon.

- A galaxy is a collection of stars, planets and moons, black holes, and interstellar dust and gas bound together by gravity. Hundreds of billions of galaxies exist. Our home galaxy, the Milky Way, consists of a galactic disk with spiral arms surrounding a central star-filled galactic bulge anchored by a supermassive black hole at the galactic center.

- Distant galaxies are moving away from one another at rate defined by Hubble's law, implying that the Universe is expanding.

- The expansion of the Universe implies that about 13.8 billion years ago, all of the Universe's mass and energy emerged from a single point during the Big Bang.

- The future of the Universe remains uncertain. Its expansion may slow down and reverse, yielding a Big Crunch during which matter converges into a single point. Alternatively, its expansion may continue into the future.

Key Terms

absolute magnitude (p. 826)	**chromosphere** (p. 821)	**event horizon** (p. 835)	**interstellar medium** (p. 829)
active galaxy (p. 839)	**convective zone** (p. 820)	**galactic bulge** (p. 837)	**Local Group** (p. 841)
apparent brightness (p. 825)	**corona** (p. 821)	**galactic disk** (p. 837)	**luminosity** (p. 825)
apparent magnitude (p. 825)	**dark energy** (p. 843)	**galactic halo** (p. 838)	**main sequence** (p. 827)
Big Bang (p. 841)	**dark matter** (p. 841)	**galaxy** (p. 837)	**nebula** (p. 829)
binary star (p. 831)	**diffuse nebula** (p. 829)	**giant star** (p. 826)	**neutron star** (p. 835)
black dwarf (p. 833)	**dwarf star** (p. 827)	**globular cluster** (p. 838)	**normal galaxy** (p. 839)
black hole (p. 836)	**emission nebula** (p. 829)	**H-R diagram** (p. 827)	**nova** (p. 833)

nuclear fusion (p. 817)
photosphere (p. 820)
planetary nebula (p. 829)
plasma (p. 817)
pulsar (p. 835)
quasar (p. 839)
radiative zone (p. 819)
red dwarf (p. 832)

red giant (p. 833)
reflection nebula (p. 829)
shock wave (p. 828)
singularity (p. 841)
solar atmosphere (p. 820)
solar constant (p. 818)
solar core (p. 819)
solar cycle (p. 822)

solar flare (p. 823)
solar granulation (p. 820)
solar mass (p. 826)
solar prominence (p. 822)
solar storm (p. 822)
solar unit (p. 826)
solar wind (p. 821)
space weather (p. 823)

spectral class (p. 826)
sunspot (p. 822)
supergiant star (p. 826)
supermassive black hole (p. 838)
supernova (p. 834)
white dwarf (p. 833)

Review Questions

The letters following each Review Question refer to the corresponding Learning Objective from the Chapter Opener.

1. What are the Sun's three internal layers? In which layer is energy produced, and by what mechanism? How does energy move across the other layers? **(A)**

2. What is a solar flare? How are solar flares, solar prominences, and sunspots related? **(A)**

3. Distinguish among luminosity, apparent magnitude, and absolute magnitude. **(B)**

4. What information does the H-R diagram give us about a star? What stars appear at the bottom of the graph? What stars appear at the top right? **(C)**

5. The Sun currently falls on the main sequence of the H-R diagram. Explain how it will evolve as it nears the end of its life as a visible star. **(C, D)**

6. What is a nebula? How are nebulae classified? What role(s) do nebulae play in the life cycle of stars? **(E)**

7. What property of a star is most important in determining whether it will end its life as a white dwarf, a neutron star, or a black hole? **(D)**

8. What are the differences between black-, white-, and red-dwarf stars in terms of their size, mass, temperature, formation mechanism, and life cycle stage? **(D)**

9. What is the difference between a quasar and a pulsar? **(D, F)**

10. What provides the energy of a nova? What provides the energy of a supernova? **(D)**

11. Describe the characteristics of a black hole. Can a planet orbit a black hole? Identify the event horizon in the figure. **(D)**

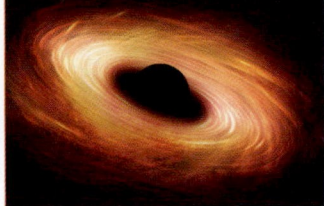

12. What are the key structural features of the Milky Way? How do galaxies differ from one another? **(F)**

13. Describe the current model of the overall structure of the Universe in terms of how galaxies are arranged over vast distances. **(G)**

14. Has expansion of the Universe continued at a constant rate since the Big Bang? How might expansion rates change in the future, and what would such changes mean for the future of the Universe? **(H)**

On Further Thought

15. When astronomers plot the positions of thousands of stars on the H-R diagram, most of those stars fall along the main sequence. Why are so many stars found along this line rather than in other parts of the diagram? Why does the mass of a star determine its residence time on the main sequence? **(C)**

16. What objects in your pocket provide clear evidence that the Sun is not a first-generation star; that is, that it was not one of the first stars that formed after the Universe came into existence? Explain. **(D)**

Online Resources

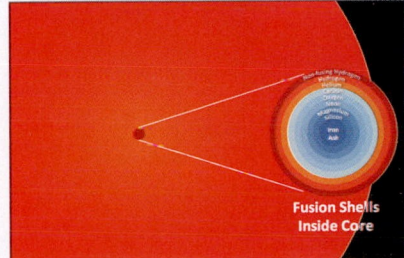

Animations
This chapter features animations examining the life cycle of a star.

Videos
This chapter features real-world videos on black holes, solar wind, sunspots, the magnetic Sun, and space weather.

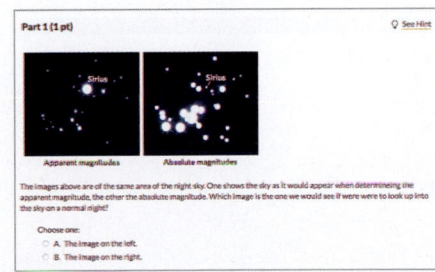

Smartwork5
This chapter features questions on the formation, life cycle, and classification of stars, and the structure and behavior of the Universe.

847

ADDITIONAL CHARTS

Metric Conversion Chart

Length

1 kilometer (km) = 0.6214 mile (mi)

1 meter (m) = 1.094 yards = 3.281 feet

1 centimeter (cm) = 0.3937 inch

1 millimeter (mm) = 0.0394 inch

1 mile (mi) = 1.609 kilometers (km)

1 yard = 0.9144 meter (m)

1 foot = 0.3048 meter (m)

1 inch = 2.54 centimeters (cm)

Area

1 square kilometer (km²) = 0.386 square mile (mi²)

1 square meter (m²) = 1.196 square yards (yd²)

= 10.764 square feet (ft²)

1 square centimeter (cm²) = 0.155 square inch (in²)

1 square mile (mi²) = 2.59 square kilometers (km²)

1 square yard (yd²) = 0.836 square meter (m²)

1 square foot (ft²) = 0.0929 square meter (m²)

1 square inch (in²) = 6.4516 square centimeters (cm²)

Volume

1 cubic kilometer (km³) = 0.24 cubic mile (mi³)

1 cubic meter (m³) = 264.2 gallons

= 35.314 cubic feet (ft³)

1 liter (l) = 1.057 quarts

= 33.815 fluid ounces

1 cubic centimeter (cm³) = 0.0610 cubic inch (in³)

1 cubic mile (mi³) = 4.168 cubic kilometers (km³)

1 cubic yard (yd³) = 0.7646 cubic meter (m³)

1 cubic foot (ft³) = 0.0283 cubic meter (m³)

1 cubic inch (in³) = 16.39 cubic centimeters (cm³)

Mass

1 metric ton = 2,205 pounds

1 kilogram (kg) = 2,205 pounds

1 gram (g) = 0.03527 ounce

1 pound (lb) = 0.4536 kilogram (kg)

1 ounce (oz) = 28.35 grams (g)

Pressure

1 kilogram per
square centimeter (kg/cm²)* = 0.96784 atmosphere (atm)

= 0.98066 bar

= 9.8067×10^4 pascals (Pa)

1 bar = 0.1 megapascals (Mpa)

= 1.0×10^5 pascals (Pa)

= 29.53 inches of mercury (in a barometer)

= 0.98692 atmosphere (atm)

= 1.02 kilograms per square centimeter (kg/cm²)

1 pascal (Pa) = 1 kg/m/s²

1 pound per square inch = 0.06895 bars

= 6.895×10^3 pascals (Pa)

= 0.0703 kilogram per square centimeter

Temperature

To change from Fahrenheit (F) to Celsius (C):

$$°C = \frac{(°F - 32°)}{1.8}$$

To change from Celsius (C) to Fahrenheit (F):

$$°F = (°C \times 1.8) + 32°$$

To change from Celsius (C) to Kelvin (K):

$$K = °C + 273.15$$

To change from Fahrenheit (F) to Kelvin (K):

$$K = \frac{(°F - 32°)}{1.8} + 273.15$$

*Note: Because kilograms are a measure of mass whereas pounds are a unit of weight, pressure units incorporating kilograms assume a given gravitational constant (g) for Earth. In reality, the gravitational for Earth varies slightly with location.

Periodic Table of Elements

Alkali metals

Inert gases

Nonmetals

Transition elements (metals)

Legend:

Symbol	2 (Atomic number)
He	
Helium	(Name)
4.002	(Atomic weight)

H 1 Hydrogen 1.007																	He 2 Helium 4.002
Li 3 Lithium 6.941	Be 4 Beryllium 9.0121											B 5 Boron 10.811	C 6 Carbon 12.011	N 7 Nitrogen 14.006	O 8 Oxygen 15.999	F 9 Fluorine 18.998	Ne 10 Neon 20.179
Na 11 Sodium 22.989	Mg 12 Magnesium 24.305											Al 13 Aluminum 26.981	Si 14 Silicon 28.085	P 15 Phosphorus 30.973	S 16 Sulfur 32.066	Cl 17 Chlorine 35.452	Ar 18 Argon 39.948
K 19 Potassium 39.098	Ca 20 Calcium 40.078	Sc 21 Scandium 44.955	Ti 22 Titanium 47.88	V 23 Vanadium 50.941	Cr 24 Chromium 51.996	Mn 25 Manganese 54.938	Fe 26 Iron 55.847	Co 27 Cobalt 58.933	Ni 28 Nickel 58.693	Cu 29 Copper 63.546	Zn 30 Zinc 65.39	Ga 31 Gallium 69.723	Ge 32 Germanium 72.61	As 33 Arsenic 74.921	Se 34 Selenium 78.96	Br 35 Bromine 79.904	Kr 36 Krypton 83.80
Rb 37 Rubidium 85.467	Sr 38 Strontium 87.62	Y 39 Yttrium 88.905	Zr 40 Zirconium 91.224	Nb 41 Niobium 92.906	Mo 42 Molybdenum 95.94	Tc 43 Technetium 98.907	Ru 44 Ruthenium 101.07	Rh 45 Rhodium 102.905	Pd 46 Palladium 106.42	Ag 47 Silver 107.868	Cd 48 Cadmium 112.411	In 49 Indium 114.82	Sn 50 Tin 118.710	Sb 51 Antimony 121.757	Te 52 Tellurium 127.60	I 53 Iodine 126.904	Xe 54 Xenon 131.29
Cs 55 Cesium 132.905	Ba 56 Barium 137.327	La 57 Lanthanum 138.905	Hf 72 Hafnium 178.49	Ta 73 Tantalum 180.947	W 74 Tungsten 183.85	Re 75 Rhenium 186.207	Os 76 Osmium 190.2	Ir 77 Iridium 192.22	Pt 78 Platinum 195.08	Au 79 Gold 196.966	Hg 80 Mercury 200.59	Tl 81 Thallium 204.383	Pb 82 Lead 207.2	Bi 83 Bismuth 208.980	Po 84 Polonium 208.982	At 85 Astatine 209.987	Rn 86 Radon 222.017
Fr 87 Francium 223.019	Ra 88 Radium 226.025	Ac 89 Actinium 227.027															

Lanthanides

Ce 58 Cerium 140.115	Pr 59 Praseodymium 140.907	Nd 60 Neodymium 144.24	Pm 61 Promethium 144.912	Sm 62 Samarium 150.36	Eu 63 Europium 151.965	Gd 64 Gadolinium 157.25	Tb 65 Terbium 158.925	Dy 66 Dysprosium 162.50	Ho 67 Holmium 164.930	Er 68 Erbium 167.26	Tm 69 Thulium 168.934	Yb 70 Ytterbium 173.04	Lu 71 Lutetium 174.967

Actinides

Th 90 Thorium 232.038	Pa 91 Protactinium 231.035	U 92 Uranium 238.028	Np 93 Neptunium 237.048	Pu 94 Plutonium 244.064	Am 95 Americium 243.061	Cm 96 Curium 247.070	Bk 97 Berkelium 247.070	Cf 98 Californium 251.079	Es 99 Einsteinium 252.083	Fm 100 Fermium 257.095	Md 101 Mendelevium 258.10	No 102 Nobelium 259.100	Lr 103 Lawrencium 262.11

The modern periodic table of the elements. Each column groups elements with related properties. For example, inert gases are listed in the column on the right. Metals are found in the central and left parts of the chart.

CREDITS

Front Matter

Page ii: Stephen Marshak; **p. ix (left):** Stephen Marshak; **(right):** NASA; **p. x (left):** Arctic Images/Getty Images; **(right):** Stephen Marshak; **p. xi (left):** Caters News Agency; **(right):** Stephen Marshak; **p. xii: (left):** Stephen Marshak; **(right):** Prof. Kurt Burmeister, University of the Pacific; **p. xiii (left):** Aurora Photos / Alamy Stock Photo; **(right):** Stephen Marshak; **p. xiv (both):** Stephen Marshak; **p. xv (both):** Stephen Marshak; **p. xvi (both):** Stephen Marshak; **p. xvii (left):** Stephen Marshak; **(right):** Jeffrey Frame, University of Illinois at Urbana-Champaign; **p. xviii (left):** SSEC/University of Wisconsin- Madison; **(right):** Jeffrey Frame, Department of Atmospheric Science, University of Illinois; **p. xix (left):** Ken Dewey; **(right):** Bruce Beck/Alamy Stock Photo; **p. xx (left):** JPL/NASA; **(right):** Alan Dyer /VWPics / agefotostock/Superstock; **p. xxvi (both):** Images provided by Google Earth mapping services/NASA, © DigitalGlobe, © Terra Metrics, © GeoEye, © Europa Technologies, Copyright 2017; **p. xxix (left):** Courtesy Stephen Marshak; **(right):** Karen Eichelberger.

Prelude

Pages 2-3: Stephen Marshak; **p. 4 (all):** Stephen Marshak; **p. 5:** NASA, ESA, and the Hubble Heritage Team (STScI/AURA); **p. 6 (a):** Greenshoots Communications / Alamy Stock Photo; **(b):** Paul A. Souders/Corbis/Getty Image; **(c):** robertharding / Alamy Stock Photo; **(d):** Tony Freeman/Science Source; **p. 7 (a, c):** Stephen Marshak; **(b):** NASA / Lauren Harnett; **(a, right):** Toby Dogwiler, Missouri State University; **(b, right):** ©UCAR. Image courtesy Mel Shapiro, NCAR; **p. 8:** Stephen Marshak; **p. 15 (a):** Stocktrek/Getty Images RF; **(b):** Carsten Peter/National Geographic/Getty Images; **(c):** Stephen Marshak; **(a, bottom):** AP Photo/Pool, Smiley N.; **(b, bottom):** Gary Braasch; **p. 18 (a):** Reuters/Matthias Strauss; **(b):** AP Photo/Hurriyet; **(a, bottom):** John Finney/Getty Images; **(b, bottom):** NOAA; **p. 19:** NASA; **p. 20 (a and bottom):** Stephen Marshak; **(b):** Kevin Frayer/Getty Images; **p. 21:** Stephen Marshak.

Chapter 1

Pages: 22-23: Stephen Marshak; **p. 24:** NASA; **p. 26 (a-b):** Stephen Marshak; **(bottom):** John Shaw/Science Source; **p. 27(a):** SOHO - EIT Consortium, ESA, NASA; **(b):** NASA/JPL-Caltech; **p. 30:** NASA; **p. 31 (top):** Moonrunner Design; **(b):** NASA, ESA; **p. 32:** Reuters; **p. 33 (a):** NASA, esa, AND m. Livio and the Hubble 20th Anniversary Team (STScl); **(b):** NASA, esa, AND m. Livio and the Hubble 20th Anniversary Team (STScl); **p. 34 (a):** NASA; **(b):** Photo by H. Raab, 2005, https://creativecommons.org/licenses/by-sa/3.0/deed.en; **(c):** NASA; **p. 38:** Images provided by Google Earth mapping services/NASA, © DigitalGlobe, © Terra Metrics, © GeoEye, © Europa Technologies, Copyright 2017; **p. 39 (a):** NASA; **(b):** NASA Earth Observatory image by Robert Simmon, using Suomi NPP VIIRS data provided courtesy of Chris Elvidge (NOAA National Geophysical Data Center). Suomi NPP is the result of a partnership between NASA, NOAA, and the Department of Defense; **p. 40:** NASA; **p. 41 (top):** NASA; **(bottom):** Images provided by Google Earth mapping services/NASA, © DigitalGlobe, © Terra Metrics, © GeoEye, © Europa Technologies, Copyright 2017; **p. 42 (melt):** Arctic Images / Corbis/ Getty Images; **(mineral):** DeAgostini/Getty Images; **(glass):** DeAgostini/ Getty Images; **(sediment, metal, rock):** Stephen Marshak; **p. 43:** Images provided by Google Earth mapping services/NASA, ©DigitalGlobe, © Terra Metrics, © GeoEye, © Europa Technologies, Copyright 2017; **p. 44: (a):** NASA; **(b):** NOAA; **(c):** NASA/JPL; **p. 47:** Stephen Marshak; **p. 49:** NASA, esa, AND m. Livio and the Hubble 20th Anniversary Team (STScl).

Chapter 2

Pages 50-51: Arctic Images/Getty Images; **p. 52 (a):** Alfred Wegner Institute for Polar and Marine Research; **(b):** Ronald Blakey, Colorado Plateau Geosystems; **(bottom):** Images provided by Google Earth mapping services/NASA, © DigitalGlobe, © Terra Metrics, © GeoEye, © Europa Technologies, Copyright 2017; **p. 53:** Pete M. Wilson/ Alamy; **p. 54:** Images provided by Google Earth mapping services/NASA, © DigitalGlobe, © Terra Metrics, © GeoEye, © Europa Technologies, Copyright 2017; **p. 55 (b):** Courtesy of Thomas N. Taylor; **p. 56:** NOAA; **p. 57:** Images provided by Google Earth mapping services/NASA, © DigitalGlobe, © Terra Metrics, © GeoEye, © Europa Technologies, Copyright 2017; **p. 60:** Images provided by Google Earth mapping services/NASA, © DigitalGlobe, © Terra Metrics, © GeoEye, © Europa Technologies, Copyright 2017; **p. 63:** Images provided by Google Earth mapping services/NASA, © DigitalGlobe, © Terra Metrics, © GeoEye, © Europa Technologies, Copyright 2017; **p. 67:** J. R. Delaney and D. S. Kelley, University of Washington; **p. 68:** Images provided by Google Earth mapping services/NASA, © DigitalGlobe, © Terra Metrics, © GeoEye, © Europa Technologies, Copyright 2017; **p. 69:** Images provided by Google Earth mapping services/NASA, © DigitalGlobe, © Terra Metrics, © GeoEye, © Europa Technologies, Copyright 2017; **p. 70:** Images provided by Google Earth mapping services/NASA, © DigitalGlobe, © Terra Metrics, © GeoEye, © Europa Technologies, Copyright 2017; **p. 71:** Images provided by Google Earth mapping services/NASA, © DigitalGlobe, © Terra Metrics, © GeoEye, © Europa Technologies, Copyright 2017; **p. 72:** Kevin Schafer/ Alamy; **p. 73:** Images provided by Google Earth mapping services/NASA, © DigitalGlobe, © Terra Metrics, © GeoEye, © Europa Technologies, Copyright 2017; **p. 74:** Cindy Ebinger University of Rochester; **p. 77:** NASA Earth Observatory; **p. 81:** Magnetic Anomaly Map of the World, 2007 Equatorial scale: 1: 50 000 000 © CCGM-CGMW; Authors: J.V. Korhonen,J. Derek Fairhead, M. Hamoudi, K. Hemant, V. Lesur, M. Mandea, S. Maus, M. Purucker, D. Ravat, T. Sazonova & E. Thébault.

Chapter 3

Pages 88-89: Stephen Marshak; **p. 90 (a-b, bottom b):** Stephen Marshak; **(bottom, a):** Yakub88/ Dreamstime.com; **p. 91 (from left to right):** Mark A. Schneider/ Science Source; incamerastock / Alamy; Darryl Brooks/Dreamstime; **p. 94:** © 1996 Jeff Scovil; **p.95 (person volcanic gas):** Maremagnum /Getty Images; **(Close-up salt crystals):** Stephen Marshak; **(person studying salt crystals outdoors):** Stephen Marshak; **(microstructure):** 2005 Olev Vinn; https://creativecommons.org/licenses/by-sa/3.0/deed.en; **p. 97 (a-b, d, f):** Richard P. Jacobs/ JLM Visuals; **(c):** Breck P. Kent/ JLM Visuals; **(e):** Javier Trueba/MSF /Science Source; **(g):** GIPhotoStock/ Science Source; **p. 98 (a):** Farbled / Dreamstime.com; **(b):** Dennis Kunkel Microscopy, Inc.; **(c):** Construction Photography / Alamy; **p. 100 (a, d-e):** Richard P. Jacobs/JLM Visuals; **(b):** 1992 Jeff Scovil; **(c):** Marli Miller/Visuals Unlimited, Inc.; **(g, left):** Arco Images GmbH/ Alamy; **(g, right):** Ann Bryant www.Geology.com; **p. 101:** Images provided by Google Earth mapping services/NASA, © DigitalGlobe, © Terra Metrics, © GeoEye, © Europa Technologies, Copyright 2017; **p. 102:** Images provided by Google Earth mapping services/NASA, © DigitalGlobe, © Terra Metrics, © GeoEye, © Europa Technologies, Copyright 2017; **p. 103 (a):** 1996 Smithsonian Institution; **(b):** Albert Copley/Visuals Unlimited; **(b inset):** Stephen Marshak; **(bottom):** © 1998 Jeff Scovil; **p. 104 (a):** Michael Langford/Gallo Images/Getty Images; **(b):** Ken Lucas/Visuals Unlimited; **(c):** Danita Delimont/ Gallo Images/Getty Images; **p. 105 (a, left):** Stephen Marshak; **(b, left):** sciencephotos/Alamy; **(a, center):** Courtesy David W. Houseknecht, USGS; **(b, center):** Courtesy of Kent Ratajeski, Dept. of Geology and Geophysics, U of Wisconsin,

Madison; **p. 106 (a-b):** Stephen Marshak; **p. 107 (all):** Stephen Marshak; **p. 108 (a):** © Tom Bean; **(b):** Stephen Marshak; **p. 109 (all):** Stephen Marshak; **p. 110: (c):** Stephen Marshak; **(d):** Scenics & Science/ Alamy; **p. 111 (a):** Product photo courtesy of JEOL, USA; **(bottom):** Stephen Marshak; **(b):** Courtesy of Joseph H. Reibenspies, Texas A & M University; **p. 112:** Richard P. Jacobs/ JLM Visuals; **p. 113 (top):** Courtesy David W. Houseknecht, USGS; **(bottom):** Courtesy of Kent Ratajeski, Dept. of Geology and Geophysics, U of Wisconsin, Madison.

Chapter 4

Pages 114-115: Caters News Agency; **p. 116 (a):** Yale Center for British Art, Paul Mellon Collection / Bridgeman Images; **(b-c):** Stephen Marshak; **p. 119:** Stephen Marshak; **p. 121 (both):** Stephen Marshak; **p. 122 (a):** Thomas Hallstein/ Alamy; **(b):** Stephen Marshak; **p. 123 (both):** Stephen Marshak; **p. 124 (both):** Stephen Marshak; **p. 126 (top):** Marli Miller/ Visuals Unlimited; **(a left):** Images provided by Google Earth mapping services/NASA, © DigitalGlobe, © Terra Metrics, © GeoEye, © Europa Technologies, Copyright 2017; **(a right):** Stephen Marshak; **(b):** Robert Francis / Agefotostock; **(c):** Stephen Marshak; **(d):** USGS; **(e):** Stephen Marshak; **p. 127 (a-d):** Stephen Marshak; **p. 128 (a):** AFP/Getty Images; **(b):** Photo by Suzanne MacLachlan, British Ocean Sediment Core Research Facility, National Oceanography Centre, Southampton; **(c-d):** Stephen Marshak; **(bottom left):** Images provided by Google Earth mapping services/NASA, © DigitalGlobe, © Terra Metrics, © GeoEye, © Europa Technologies, Copyright 2017; **p. 129 (top a):** AP Photos; **(top b):** USGS; **(c):** Stephen Weaver; **(d; bottom a-b):** Stephen Marshak; **p. 130(a):** Stephen Marshak; **(inset a):** Dr. Kent Ratajeski; **(b):** Mark A. Schneider/ Science Source; **(inset b):** Omphacite. 2006. Wikimedia; http://en.wikipedia.org/wiki/Public_domain; **(c):** Doug Sokell/Visuals Unlimited; **(inset c):** Dr. Matthew Genge; **p. 131 (down from top left):** geoz/ Alamy; Stephen Marshak; Siim Sepp/ Alamy; **(down from top right):** Stephen Marshak; Joyce Photographics /Science Source; Mark A. Schneider/Science Source; **p. 132 (all):** Stephen Marshak; **p. 133 (top a):** Liysa/Pacific Stock/Agefotostock; **(top b):** National Geographic/Getty Images; **(bottom b):** Stephen Marshak; **p. 136:** USGS; **p. 137 (top center):** Martin Rietze/agefotostock; **(top right):** SUDRES Jean-Daniel / hemispicture.com/Getty Images; **(center):** N. Banks/USGS; **(bottom left):** Marli Miller/Visuals Unlimited; **(bottom b):** © Tom Bean 1985; **p. 139 (top):** Lyn Topinka/ USGS; **(center):** Stephen Marshak; **(bottom):** USGS; **p. 140 (top):** Robert Harding World Imagery/ Alamy; **(bottom):** Marli Miller/Visuals Unlimited; **p. 143 (top b):** Image courtesy of Submarine Ring of Fire 2002 Exploration, NOAA-OE; **(c):** Chris German, ©Woods Hole Oceanographic Institute; **(bottom both):** Images provided by Google Earth mapping services/NASA, © DigitalGlobe, © Terra Metrics, © GeoEye, © Europa Technologies, Copyright 2017; **p. 144 (a):** Jurg Alean; **(top b):** Daryl Balfour /Science Source; **(bottom b):** Images provided by Google Earth mapping services/NASA, © DigitalGlobe, © Terra Metrics, © GeoEye, © Europa Technologies, Copyright 2017; **p. 145:** Stephen Marshak; **p. 146 (top):** Stephen Marshak; **(bottom):** Images provided by Google Earth mapping services/NASA, © DigitalGlobe, © Terra Metrics, © GeoEye, © Europa Technologies, Copyright 2017; **p. 147 (a):** USGS; **(b):** Vittoriano Rastelli/Corbis via Getty Images; **(c):** AP Photo; **(d):** Stephen Marshak; **(e):** Stocktrek Images, Inc. / Alamy Stock Photo; **(f):** Photo by Aldnonymous, 2014: https://creativecommons.org/licenses/by-sa/3.0/deed.en; **(g):** Magnus T. Gudmundsson, University of Iceland; **(h):** USGS; **p. 148:** NASA; **p. 150 (a):** USGS; **(b):** Planetary Visions/NERC-COMET/JAXA/ESA; **p. 151 (a):** Vittoriano Rastelli/Corbis/Getty Images; **(b):** Sigurgeir Jonasson/Frank Lane Picture Agency/Corbis/Getty Images.

Chapter 5

Pages 154-155: Stephen Marshak; **p. 156:** Bettmann/Getty Images; **p. 157 (a):** Cordelia Molloy /Science Source; **(b, all):** Stephen Marshak; **p. 158 (b):** Stephen Marshak; **(c):** Visuals Unlimited; **(bottom, all):** Stephen Marshak; **p. 159:** Stephen Marshak; **p. 160 (all):** Stephen Marshak; **p. 161:** Stephen Marshak; **p. 165 (top):** Stephen Marshak; bJim Richardson; **p. 166 (both):** Stephen Marshak; **p. 167 (both):** Stephen Marshak; **p. 168:** Stephen Marshak; **p. 169:** Images provided by Google Earth mapping services/NASA, © DigitalGlobe, © Terra Metrics, © GeoEye, © Europa Technologies, Copyright 2017; **p. 170 (all but noted)** Stephen Marshak; **(arkrose):** Scottsdale Community College; **p. 171 (all):** Stephen Marshak; **p. 172: (both):** Stephen

Marshak; **p. 173 (top b left):** Visuals Unlimited; **(top b right):** Stephen Marshak; **(bottom a):** Marli Miller, University of Oregon; **(bottom b):** Stephen Marshak; **p. 174 (all top):** Stephen Marshak; **(left):** Images provided by Google Earth mapping services/NASA, © DigitalGlobe, © Terra Metrics, © GeoEye, © Europa Technologies, Copyright 2017; **p. 175 (all):** Stephen Marshak; **p. 176 (left):** All Canada Photos/ Alamy; **(right):** Stephen Marshak; **p. 177 (a, d):** Stephen Marshak; **(b):** Imagina Photography/ Alamy; **p. 178 (all but noted):** Stephen Marshak; **(graded beds):** Marli Miller/Visuals Unlimited; **p. 179:** Images provided by Google Earth mapping services/NASA, © DigitalGlobe, © Terra Metrics, © GeoEye, © Europa Technologies, Copyright 2017; **p. 180 (a):** Emma Marshak; **(b, d, e):** Stephen Marshak; **(c):** Marli Miller/Visuals Unlimited; **(f):** Callan Bentley; **p. 181:** Stephen Marshak; **(d):** David Wall / Alamy Stock Photo; **p. 182:** Stephen Marshak; **p. 184: (a):** The Natural History Museum / Alamy Stock Photo; **(b):** G. R. Roberts ©/ Natural Sciences Image Library; **(left):** Images provided by Google Earth mapping services/NASA, © DigitalGlobe, © Terra Metrics, © GeoEye, © Europa Technologies, Copyright 2017; **p. 187:** Stephen Marshak; **p. 189:** Ronald Blakey Colorado Plateau Geosystems, Inc.

Chapter 6

Pages 190-191: Stephen Marshak; **p. 192 (caterpillar):** Scott Camazine /Science Source; **(butterfly):** John Serrao / Science Source; **(shale):** Science Stock Photography / Science Source; **(schist):** Science Stock Photography / Science Source; **p. 197 (a):** Stephen Marshak; **(b left):** Emma Marshak; **(b right):** Stefano Clemente / Alamy Stock Photo; **p. 198 (all):** Stephen Marshak; **p. 199 (both):** Stephen Marshak; **p. 201 (all):** Stephen Marshak; **p. 204:** Images provided by Google Earth mapping services/NASA, © DigitalGlobe, © Terra Metrics, © GeoEye, © Europa Technologies, Copyright 2017; **p. 206:** Images provided by Google Earth mapping services/ NASA, © DigitalGlobe, © Terra Metrics, © GeoEye, © Europa Technologies, Copyright 2017; **p. 207 (a):** Michael Stewart, University of Illinois; **(b):** Stephen Marshak; **(right):** Images provided by Google Earth mapping services/NASA, © DigitalGlobe, © Terra Metrics, © GeoEye, © Europa Technologies, Copyright 2017; **p. 211 (all):** Stephen Marshak; **(right):** Images provided by Google Earth mapping services/NASA, © DigitalGlobe, © Terra Metrics, © GeoEye, © Europa Technologies, Copyright 2017; **p. 213:** Stephen Marshak.

Chapter 7

Pages 216-217: Prof. Kurt Burmeister, University of the Pacific; **pp. 218-219:** NOAA / ETOPO1382; **p. 220 (all):** Stephen Marshak; **p. 221 (b, d):** Stephen Marshak; **p. 223:** Images provided by Google Earth mapping services/NASA, © DigitalGlobe, © Terra Metrics, © GeoEye, © Europa Technologies, Copyright 2017; **p. 224 (a):** Aurora Photos / Alamy Stock Photo; **(b):** Stephen Marshak; **p. 225:** Stephen Marshak; **p. 227 (a):** Stephen Marshak; **(b):** USGS; **(c):** Lloyd Cluff/Corbis/Getty Images; **p. 228 (a-c):** Stephen Marshak; **p. 229:** Images provided by Google Earth mapping services/NASA, © DigitalGlobe, © Terra Metrics, © GeoEye, © Europa Technologies, Copyright 2017; **p. 230 (a-d):** Stephen Marshak; **(e):** Landsat/USGS; **p. 231:** Images provided by Google Earth mapping services/NASA, © DigitalGlobe, © Terra Metrics, © GeoEye, © Europa Technologies, Copyright 2017; **p. 234:** Stephen Marshak; **p. 235 (both):** © Images provided by Google Earth mapping services /NASA, © DigitalGlobe, © Terra Metrics, © GeoEye, © Europa Technologies, Copyright 2017; **p. 236:** Stephen Marshak; **p. 237 (a-b):** Ronald Blakey, Colorado Plateau Geosystems, Inc.; **p. 238:** Prof. Kurt Burmeister, University of the Pacific; **p. 241:** Stephen Marshak; **p. 242 (a-b):** Stephen Marshak.

Chapter 8

Pages 246-247: Aurora Photos / Alamy Stock Photo; **p. 248 (a):** AFP/Getty Images; **(b):** JIJI Press/ AFP/ Getty Images; **p. 249 (left):** AP Photo/ Kyodo News; **(right):** Stephen Marshak; **p. 251 (a):** NOAA/NGDC, USGS; **(b):** Photo courtesy of Paul "Kip" Otis-Diehl, USMC, 29 Palms CA; **(bottom):** Images provided by Google Earth mapping services/NASA, © DigitalGlobe, © Terra Metrics, © GeoEye, © Europa Technologies, Copyright 2017; **p. 261 (b):** National Archives; **(c):** AP Photo/Paul Sakuma, file; **p. 263 (top):** Anna Kompanek/CIPE; **(bottom):** Omar Havana/Getty Images; **p. 264:** Wikimedia, public domain; **p. 265 (a):** AP Photo; **(b):** Pacific Press Service/ Alamy; **(c):** M. Celebi, USGS; **(d):** Reuters; **p. 266 (top):** Barry Lewis/ Alamy; **(a):** NOAA/ National Geophysical Data Center (NGDC); **(c):**

Photo by Martin Luff; http://www.flickr.com/photos/23934380@N06/5469769673 https://creativecommons.org/licenses/by-sa/2.0; **p. 267 (top, b):** Karl V. Steinbrugge Collection, University of California, Berkeley; **(a):** National Geophysical Data Center (NGDC); **(bottom b):** Library of Congress; **(right):** Images provided by Google Earth mapping services/NASA, © DigitalGlobe, © Terra Metrics, © GeoEye, © Europa Technologies, Copyright 2017; **p. 269 (a):** Vasily V. Titov, Associate Director, Tsunami Inundation Mapping Efforts (TIME), NOAA/PMEL-UW/JISAO, USA; **(b):** AFP/ Getty Images; **(c):** Photo by David Rydevik. 2004. Wikimedia http://en.wikipedia.org/wiki/Public_domain; **(d, both):** IKONOS IMAGES/CRISP; NATL UNIV. SINGAPORE/GEOEYE; **p. 270 (a):** AP Photo/Kyodo News; **(b):** AP Photo/Yomiuri Shimbun, Masamine Kawaguchi; **(c):** Air Photo Service/Reuters; **p. 271 (b):** USGS; **(c-d):** Reuters/Eduardo Munoz; **p. 274:** NOAA/ NOA Center for Tsunami Research.

Chapter 9

Pages: 282-283: Stephen Marshak; **p. 284 (top):** National Parks Service, pd; **(a-b):** Stephen Marshak; **p. 286 (all):** Stephen Marshak; **p. 288 (a):** Stephen Marshak; **(b):** Photo by William L. Jones from the Stones & Bones Collection, http://www.stones-bones.com; **(c):** Reynold Sumayku / Alamy Stock Photo; **p. 290 (a):** Jovfoto/ UIG via Getty Images; **(b):** Dirk Wiersma/ Science Source; **(c-d, g-h):** Stephen Marshak; **(e):** Naturfoto Honal/Getty Images; **(f):** Kevin Schafer/Corbis/Getty Image; **(i):** Biophoto Associates / Science Source; **p. 291:** Stephen Marshak; **p. 292:** Illustration by Karen Carr and Karen Carr Studio, Inc. © Smithsonian Institution; **p. 294:** Stephen Marshak; **p. 295 (both):** Stephen Marshak; **(bottom):** Images provided by Google Earth mapping services/NASA, © DigitalGlobe, © Terra Metrics, © GeoEye, © Europa Technologies, Copyright 2017; **p. 296:** Stephen Marshak; **p. 297:** Images provided by Google Earth mapping services/NASA, © DigitalGlobe, © Terra Metrics, © GeoEye, © Europa Technologies, Copyright 2017; **p. 300 (all):** Stephen Marshak; **p. 302:** 1980 Grand Canyon Natural History Association; **p. 307:** Stephen Marshak; **p. 309:** Stephen Marshak.

Chapter 10

Pages 312-313: Stephen Marshak; **p. 314 (a):** © William K. Hartmann; **(d):** Richard Bizley/Science Source; **(e):** artwork © Don Dixon / www.cosmographica.com; **p. 315:** Science Photo Library/Science Source; **p. 317 (a):** Courtesy of Dr. J. William Schopf/UCLA; **(b):** Stephen Marshak; **(c):** Bill Bachman / Alamy Stock Photo; **p. 319:** USGS; **p. 321 (a):** John Sibbick; **(b):** Stephen Marshak; **(bottom):** Courtesy of Dr. Paul Hoffman, Harvard University; **p. 322 (b-c):** Ronald Blakey; Colorado Plateau Geosystems, Inc.; **p. 323 (top):** Tom McHugh/Science Source; **(b):** Ronald Blakey Colorado Plateau Geosystems, Inc.; **p. 326 (a):** Chase Studio/Science Source; **(b):** Ted Daeschler / Academy of Natural Sciences of Drexel University; **(b. right):** Ronald Blakey; Colorado Plateau Geosystems, Inc.; **p. 327:** Mackenzie, J. 2012. Hillshaded Digital Elevation Model of the Continental US. Http://www.udel.edu/johnmack/data_library/usa_dem.png; **pp. 328-329:** Ronald Blakey, Colorado Plateau Geosystems, Inc.; **p. 330 (a):** Juan Gaertner/Shutterstock; **(b):** Stephen J. Krasemann/Science Source; **p. 331 (inset):** Ronald Blakey, Colorado Plateau Geosystems, Inc.; **(b):** Stephen Marshak; **(b, right):** Ronald Blakey, Colorado Plateau Geosystems, Inc.; **p. 332 (top):** Richard Bizley; **(bottom):** Images provided by Google Earth mapping services/NASA, © DigitalGlobe, © Terra Metrics, © GeoEye, © Europa Technologies, Copyright 2017; **p. 333 (b):** Ronald Blakey, Colorado Plateau Geosystems, Inc.; **(bottom):** Stephen Marshak; **p. 334 (a):** © Karen Carr/Australian Museum; **(b):** De Agostini Picture Library/Getty Images; **(bottom):** Ronald Blakey, Colorado Plateau Geosystems Inc.; **p. 335 (a):** NASA, JPL; **(inset):** Ronald Blakey, Colorado Plateau Geosystems, Inc.; **(b):** Images provided by Google Earth mapping services/NASA, © DigitalGlobe, © Terra Metrics, © GeoEye, © Europa Technologies, Copyright 2017; **p. 337 (a):** Mauricio Anton / Science Source; **(b):** Science Photo Library /Science Source; **(bottom right):** Images provided by Google Earth mapping services/NASA, © DigitalGlobe, © Terra Metrics, © GeoEye, © Europa Technologies, Copyright 2017; **p. 338 (top, both):** Ronald Blakey, Colorado Plateau Geosystems, Inc.; **(bottom right):** DiBgd, 2007, https://creativecommons.org/licenses/by-sa/3.0/deed.en; **p. 340:** Stephen Marshak; **p. 341:** Jacques Descloitres, MODIS Team, NASA Visible Earth; **p. 343:** Ronald Blakey, Colorado Plateau Geosystems, Inc.

Chapter 11

Pages 344-345: Stephen Marshak; **p. 346:** Time Life Pictures/ Mansell / The LIFE Picture Collection /Getty Images; **p. 347 (top):** Stephen Marshak; **(bottom):** Data courtesy of BP; **p. 348:** Reuters/Athit Perawongmetha; **p. 352:** Data courtesy of Fugro. Credit: Virtual Seismic Atlas http://www.seismicatlas.org; **p. 354 (top left):** Omar Torres /AFP/Getty Images; **(a-b, d):** Stephen Marshak; **(c):** Calvin Larsen/ Science Source; **p. 355:** Images provided by Google Earth mapping services/NASA, © DigitalGlobe, © Terra Metrics, © GeoEye, © Europa Technologies, Copyright 2017; **p. 356 (b):** Data from BP Statistical Review of World Energy (2013); **p. 357:** Andrew Harrer / Bloomberg via Getty Images; **p. 358:** Ben Nelms/Bloomberg via Getty Images; **p. 359 (a):** Field Museum Library/Getty Images; **(c):** Department of Natural Resources, Alaska; **p. 360 (b):** Data from BP Statistical Review of World Energy (2013); **p. 361 (a, c):** Stephen Marshak; **(d):** Cultura Creative (RF)/ Alamy; **p. 362 (c):** Data from BP Statistical Review of World Energy (2013); **p. 363: (a):** AP Photo /Stapleton; **(b):** AFP/Getty Images; **p. 364 (a):** Science Source; **(b):** Reuters; **(c):** NASA; **p. 366:** Ron Chapple/Dreamstime; **p. 367:** The Asahi Shimbun via Getty Images; **p. 368:** G. R. Roberts © Natural Sciences Image Library; **p. 369 (all):** Stephen Marshak; **p. 370 (a):** Rob Lavinsky, iRocks.com; https://creativecommons.org/licenses/by-sa/3.0/deed.en; **(b):** Stephen Marshak; **(inset):** Layne Kennedy /Getty Images; **p. 371 (a-b):** Richard P. Jacobs/ JLM Visuals; **p. 373:** Stephen Marshak; **p. 374 (b-c):** Stephen Marshak; **p. 375 (a):** Doug Sokell/Visuals Unlimited; **(b):** A. J. Copley/Visuals Unlimited; **(right):** Images provided by Google Earth mapping services/NASA, © DigitalGlobe, © Terra Metrics, © GeoEye, © Europa Technologies, Copyright 2017; **p. 378 (a):** Richard P. Jacobs/ JLM Visuals; **(b):** Stephen Marshak; **p. 379 (a-b):** Stephen Marshak.

Chapter 12

Pages 382-383: Stephen Marshak; **pp. 384-385:** Stephen Marshak; **p. 386 (a-d):** Stephen Marshak; **p. 387 (a):** G. R. Roberts © Natural Sciences Image Library; **(b):** Julie Dermansky; **p. 389:** NOAA; **p. 391 (all):** Stephen Marshak; **p. 393 (a):** Lloyd Cluff Consulting Earthquake Geologist, San Francisco, California, lloydcluff@gmail.com; **(b):** Lloyd Cluff/Getty Images; **p. 396 (a):** Stephen Marshak; **(d):** Marli Miller/Visuals Unlimited, Inc.; **(left):** Images provided by Google Earth mapping services/NASA, © DigitalGlobe, © Terra Metrics, © GeoEye, © Europa Technologies, Copyright 2017; **p. 397 (all):** Stephen Marshak; **p. 398 (a):** BrazilPhotos.com / Alamy Stock Photo; **(b):** Cascades Volcano Observatory/USGS; **(c):** Stephen Marshak; **(d):** Reuters /National Airborne Service Corps/Handout; **p. 399(a):** National Geographic Image Collection/ Alamy; **(b):** Bob Schuster, USGS; **(c):** Rob Varela/Ventura County Star; **p. 400 (a):** Stephen Marshak; **(b):** Arco Images GmbH / Alamy Stock Photo; **(left):** Images provided by Google Earth mapping services/NASA, © DigitalGlobe, © Terra Metrics, © GeoEye, © Europa Technologies, Copyright 2017; **p. 401 (all):** Stephen Marshak; **p. 402 (a):** Stephen Marshak; **(left):** Images provided by Google Earth mapping services/NASA, © DigitalGlobe, © Terra Metrics, © GeoEye, © Europa Technologies, Copyright 2017; **p. 403 (a):** Jacques Lange /Getty Images; **(b):** Alaska Stock/Alamy; **p. 404 (a):** Jerome Neufield and Stephen Morris, Nonlinear Physics, University of Toronto; **(b):** USGS/ Barry W. Eakins; **(c):** Images provided by Google Earth mapping services/NASA, © DigitalGlobe, © Terra Metrics, © GeoEye, © Europa Technologies, Copyright 2017; **p. 405:** Stephen Marshak; **p. 407:** Breck P.Kent/JLM Visuals; **p. 408:** AP Photo/Ted S. Warren; **p. 409:** Images provided by Google Earth mapping services/NASA, © DigitalGlobe, © Terra Metrics, © GeoEye, © Europa Technologies, Copyright 2017.

Chapter 13

Pages 414-415: Stephen Marshak; **p. 416:** Niday Picture Library / Alamy Stock Photo; **p. 418 (left):** Images provided by Google Earth mapping services/NASA, © DigitalGlobe, © Terra Metrics, © GeoEye, © Europa Technologies, Copyright 2017; **(right):** Stephen Marshak; **p. 420 (a-b):** Stephen Marshak; **p. 421 (a):** © Ron Niebrugge/ www.WildNatureImages.com; **(b):** Stephen Marshak; **p. 422 (a-b):** Stephen Marshak; **p. 423:** Images provided by Google Earth mapping services/NASA, © DigitalGlobe, © Terra Metrics, © GeoEye, © Europa Technologies, Copyright 2017; **p. 424 (a-b):** Stephen Marshak; **p. 425 (a-b):** Stephen Marshak; **p. 426:** Stephen Marshak; **p. 427 (a):** Stephen Marshak; **(d):** 1998 Tom Bean;

(e): Stephen Marshak; **p. 428 (a):** Marti Miller/ Visuals Unlimited; **(b):** Christina Neal/Alaska Volcano Observatory /USGS; **(inset):** NASA/GSFC/meti/ersdac/jaros, and US/Japan ASTER Science Team; **p. 429:** Photography: www.gernot-keller.com; **p. 432 (all):** Stephen Marshak; **p. 433 (both):** Stephen Marshak; **p. 434 (top, a):** Missouri Department of Transportation; **(top, b):** Stringer /India /Reuters; **(bottom, a):** Reuters/Yoray Cohen/Eilat Rescut Unit; **(bottom, b):** USGS; **p. 435 (both):** NASA images created by Jesse Allen, Earth Observatory, using data provided courtesy of the Landsat Project Science Office- copyright 2008; **p. 436 (a-b):** Stephen Marshak; **(d):** AP Photo/The News-Star, Margaret Croft; **p. 438:** Stephen Marshak; **p. 439 (a-b):** Images provided by Google Earth mapping services/NASA, © DigitalGlobe, © Terra Metrics, © GeoEye, © Europa Technologies, Copyright 2017; **p. 440:** Stephen Marshak; **p. 444:** Vladimir Melnik/Dreamstime; **p. 445:** Stephen Marshak; **p. 447:** Stephen Marshak; **p. 448 (b):** Stephen Marshak; **(c):** Allan Tuchman; **(d):** Thom Foley; **p. 449:** Stephen Marshak; **p. 450 (Ireland):** Lois Kent; **(sinkholes):** G. R. 'Dick' Roberts/NSIL/Visuals Unlimited; **(Mexico):** SCPhotos / Alamy Stock Photo; **p. 451 (Virginia):** Lois Kent; **(Spelunker):** Ashley Cooper/Corbis/Getty Images; **p. 452 (b):** Stephen Marshak; **(bottom):** Images provided by Google Earth mapping services/NASA, © DigitalGlobe, © Terra Metrics, © GeoEye, © Europa Technologies, Copyright 2017; **p. 453 (a):** USGS; **(c):** Images provided by Google Earth mapping services/ NASA, © DigitalGlobe, © Terra Metrics, © GeoEye, © Europa Technologies, Copyright 2017; **(bottom a-b):** Stephen Marshak; **(right):** Image provided by Google Earth mapping services/ DigitalGlobe, Terra Metrics, NASA, Europa Technologies, Copyright 2017; **p. 454 (left):** George Steinmetz/ Corbis/Getty Images; **(right):** Stephen Marshak; **p. 455 (a):** Chinch Gryniewicz / UIG/Science Source; **(b):** Photo courtesy of the Bureau of Reclamation; **(bottom):** Stephen Marshak; **p. 459:** Stephen Marshak.

Chapter 14

Pages 462: Stephen Marshak; **p. 464 (a):** Wikimedia, public domain; **(b):** Bettmann/Getty Images; **p. 465:** Stephen Marshak; **p. 466:** Professor Andre Danderfer; **p. 467 (all):** Stephen Marshak; **p. 468 (a):** Image courtesy Jacques Descloitres MODIS Rapid Response Team; **(b):** Scott Wood / Barcroft Media / Getty Images; **p. 469 (all):** Stephen Marshak; **p. 470 (top, all):** Stephen Marshak; **(bottom):** Images provided by Google Earth mapping services/NASA, © DigitalGlobe, © Terra Metrics, © GeoEye, © Europa Technologies, Copyright 2017; **p. 471 (c-d):** Stephen Marshak; **p. 472 (top):** Photo by Flicka. Sept 2007. Wikimedia. Http://creativecommons.org/licenses/by-sa/3.0/deed.en; **(bottom):** Stephen Marshak; **p. 473 (top, all):** Stephen Marshak; **(bottom):** Images provided by Google Earth mapping services/ NASA, © DigitalGlobe, © Terra Metrics, © GeoEye, © Europa Technologies, Copyright 2017; **p. 474 (a):** Stephen Marshak; **(bottom):** Images provided by Google Earth mapping services/NASA, © DigitalGlobe, © Terra Metrics, © GeoEye, © Europa Technologies, Copyright 2017; **p. 475 (top):** Eye Ubiquitous / Newscom; **(center, both):** USGS; **(bottom left):** BDR/ Alamy; **(bottom right):** Stocktrek Images, Inc./ Alamy; **p. 478:** Images provided by Google Earth mapping services/NASA, © DigitalGlobe, © Terra Metrics, © GeoEye, © Europa Technologies, Copyright 2017; **p. 479 (a):** Shutterstock; **(b, both):** Stephen Marshak; **(c, top):** Emma Marshak; **(c, bottom):** Ted Spiegel/National Geographic Creative; **p. 480 (b-c):** Stephen Marshak; **(d):** SRTM Team NASA/JPL/NIMA; **p. 481 (c and center):** Stephen Marshak; **(bottom):** National Geophysical Data Center/NOAA; **p. 485 (b):** Ralph A. Clevenger / Corbis/ Getty Images; **(c):** Stephen Marshak; **(bottom, both):** ESA; **p. 486 (a-d):** Stephen Marshak; **(left):** Images provided by Google Earth mapping services/NASA, © DigitalGlobe, © Terra Metrics, © GeoEye, © Europa Technologies, Copyright 2017; **p. 487 (top and b):** Stephen Marshak; **(a):** Shutterstock; **(c):** 1986 Keith S. Walklet/Quietworks; **p. 488 (b):** Marli Miller/Visuals Unlimited, Inc.; **(bottom):** Wolfgang Meier/Corbis/Getty Images; **(left):** Images provided by Google Earth mapping services/NASA, © DigitalGlobe, © Terra Metrics, © GeoEye, © Europa Technologies, Copyright 2017; **p. 489 (a-c):** Stephen Marshak; **p. 491 (all but noted):** Stephen Marshak; **(e, right):** Kevin Schafer/ Alamy; **p. 492:** Stephen Marshak; **p. 493 (b):** Stephen Marshak; **(d):** Tom Bean; **p. 496:** Economic Development and Transportation, Government of Nunavut; **p. 497 (b-c):** © Tom Foster; **(d):** © Marli Miller; **p. 498:** Stephen Marshak; **p. 499:** Lynda Dredge/ Geological Survey of Canada; **p. 501:** Detail of mural by Charles R. Knight, American Museum of Natural History, #4950(5), Photo by Denis Finnin; **p. 505 (a):** Hendrick Averkamp, *Winter Scene with Ice Skaters*, ca. 1600. Courtesy of Rijksmuseum, Amsterdam; **(b):** Robbie Shone/Science Source; **(bottom):** Stephen Marshak.

Chapter 15

Pages 508-509: Stephen Marshak; **p. 510:** Stephen Marshak; **p. 512 (a):** Photo by Alexander Kaiser; **(right):** NASA's Earth Observatory; **p. 513 (a):** Erich Lessing / Art Resource, NY; **(b):** Library of Congress; **(c):** Stephen Marshak; **p. 514 (a):** TopFoto/ The Image Works; **(b):** Woods Hole Oceanographic Institution; **(c):** © Harbor Branch Oceanographic Institution; **(d):** Photo by Chris Griner, Woods Hole Oceanographic institution; **(e):** NASA; **(background):** Stephen Marshak; **p. 515:** Ronald Blakey, Colorado Plateau Geosystems, Inc.; **p. 516** Stephen Marshak; **p. 517:** Premraj K.P / Alamy Stock Photo; **p. 518:** Juan José Pascual/agefotostock; **p. 520:** NASA; **p. 521:** NASA/SVS; **p. 522:** NOAA; **p. 530 (top):** ESA/European Space Agency; **(bottom):** Los Alamos National Laboratory; **p. 531 (left):** Los Alamos National Laboratory; **(c):** NOAA; **p. 532:** Stephen Marshak; **p. 533 (top):** Stephen Marshak; **(bottom):** Blend Images / Alamy Stock Photo; **p. 535 (top):** Photo by Phillip Capper, 2008; https://creativecommons.org/licenses/by/2.0/deed.en; **(bottom):** NOAA; **p. 537: (a):** Emma Marshak; **(b):** Stephen Marshak; **p. 539:** Images provided by Google Earth mapping services/NASA, © DigitalGlobe, © Terra Metrics, © GeoEye, © Europa Technologies, Copyright 2017; **p. 540 (a):** Andrew Syred / Science Source; **(b):** D.P. Wilson/ FLPA/Science Source; **(c):** Steve Gschmeissner /Science Source; **p. 541: (both):** WaterFrame / Alamy Stock Photo; **p. 542 (a):** Lorna Roberts / Alamy Stock Photo; **(b):** Andrew J. Martinez /Science Source; **(left):** Images provided by Google Earth mapping services/NASA, © DigitalGlobe, © Terra Metrics, © GeoEye, © Europa Technologies, Copyright 2017; **p. 543 (a):** Reinhard Dirscherl/ullstein bild via Getty Images; **(c):** Mark Conlin / Alamy Stock Photo; **(d):** Stephen Marshak; **(bottom):** SeaWiFS Project and NASA/ Goddard Space Flight Center; **p. 545:** WaterFrame / Alamy Stock Photo.

Chapter 16

Pages 546-547: Stephen Marshak; **p. 548 (top, a):** Mountains in the Sea Research Team; the IFE Crew; and NOAA/OAR/OER; **(top, b):** NOAA; **(bottom, a):** Images provided by Google Earth mapping services/NASA, © DigitalGlobe, © Terra Metrics, © GeoEye, © Europa Technologies, Copyright 2017; **(bottom, b):** Chris Kelley/ HURL, data from Schimdt Ocean Institute's ship Falkor; **p. 549 (a, c):** William Crawford, Integrated Ocean Drilling Program/TAMU; **(b):** International Ocean Discovery Program (IODP); **p. 552:** Images provided by Google Earth mapping services/NASA, © DigitalGlobe, © Terra Metrics, © GeoEye, © Europa Technologies, Copyright 2017; **p. 553 (a):** Images provided by Google Earth mapping services /NASA, © DigitalGlobe, © Terra Metrics, © GeoEye, © Europa Technologies, Copyright 2017; **(bottom):** NOAA; **p. 554 (top):** http://www.geomapapp.org, Global Multi-Resolution Topography (GMRT) Synthesis by Ryan et al., 2009; **(bottom):** NOAA Office of Ocean Exploration and Research; **p. 555 (a):** © Jose Fuste Raga/Age Fotostock; **(b):** LDEO:Marine Geoscience Data System. Reference: Ryan, W.B.F., S.M. Carbotte, J.O. Coplan, S. O'Hara, A. Melkonian, R. Arko, R.A. Weissel, V. Ferrini, A. Goodwillie, F. Nitsche, J. Bonczkowski, R. Zemsky (2009), Global Multi-Resolution Topography synthesis, Geochem. Geophys. Geosyst., 10, Q03014, 10.1029/2008GC002332; **(c):** Image created using GeoMapApp and shows elevation data from GMRT, the Global Multi-Resolution Topography synthesis. http://www.geomapapp.org Ryan, W.B.F., S.M. Carbotte, J.O. Coplan, S. O'Hara, A. Melkonian, R.Arko, R.A. Weissel, V. Ferrini, A. Goodwillie, F. Nitsche, J. Bonczkowski, and R. Zemsky (2009), Global Multi-Resolution Topography synthesis, Geochem. Geophys. Geosyst., 10, Q03014, doi:10.1029/2008GC002332; **p. 556 (a-b):** www.michaelmarten.com; **(c):** Stephen Marshak; **p. 557 (a-b):** Stephen Marshak; **(c):** Imaginechina via AP Photos; **p. 559:** R.Ray, TOPEX /Poseidon: Revealing Hidden Tidal Energy, GSFC, NASA; **p. 562 (a-c):** Stephen Marshak; **(d):** Manfred Gottschalk/ Alamy; **p. 563 (b-c):** Stephen Marshak; **(bottom):** Education Images/UIG via Getty Images; **p. 566 (all):** Stephen Marshak; **p. 567:** Stephen Marshak; **p. 569 (b):** Panther Media GmbH / Alamy Stock Photo; **(c):** David Wall / Alamy Stock Photo; **(d):** NASA; **(right):** Images provided by Google Earth mapping services/NASA, © DigitalGlobe, © Terra Metrics, © GeoEye, © Europa

GLOSSARY

a'a' A lava flow with a rubbly surface.

ablation The removal of ice at the toe of a glacier by melting, sublimation (the evaporation of ice into water vapor), and/or calving.

abrasion The process in which one material (such as sand-laden water) grinds away at another (such as a stream channel's floor and walls).

absolute magnitude A measure of the intrinsic brightness, or luminosity, of a celestial object, generally a star. Specifically, the apparent magnitude an object would have if it were located at a standard distance of 10 parsecs.

absolute plate velocity The movement of a plate relative to a fixed point in the mantle.

abyssal plain A broad, relatively flat region of the ocean that lies at least 4.5 km (2.8 miles) below sea level.

accretion (geology) The addition of material. It can pertain to the addition of: crust to the margin of a continent, sediment to the accretionary prism of a subduction zone, or sediment to a coastline.

accretionary coast A coastline that receives more sediment than erodes away.

accretionary disk A flat, rotating disk of gas and dust surrounding an object, such as a young stellar object, a forming planet, a collapsed star in a binary system, or a black hole.

accretionary prism A wedge-shaped mass of sediment and rock scraped off the top of a downgoing plate and accreted onto the overriding plate at a convergent plate margin.

acid mine runoff A dilute solution of sulfuric acid, produced when sulfur-bearing minerals in a mine react with rainwater and flow out of the mine.

acid rain Precipitation in which air pollutants react with water to make a weak acid that then falls from the sky.

active galaxy A galaxy that emits most of its energy in the X-ray and radio regions of the electromagnetic spectrum, with most of the energy coming from the galactic center.

active margin A continental margin that is also a plate boundary.

active volcano A volcano that has erupted within the past few centuries and will likely erupt again.

adiabatic expansion An expansion of a volume of air during which no mass or energy is exchanged with the environmental air surrounding the volume.

aerosol Microscopic solid particles or liquid droplets that remain suspended in the atmosphere.

aftershocks The series of smaller earthquakes that follow a major earthquake.

air The mixture of gases that make up the Earth's atmosphere.

air mass A large body of air (typically about 1,500 km, or 932 miles across) that has relatively uniform physical characteristics.

air pressure The push or force that air exerts on its surroundings.

air temperature The temperature of air as measured by an outdoor thermometer.

albedo The reflectivity of a surface.

alloy A metal containing more than one type of metal atom.

alluvial fan A gently sloping apron of sediment dropped by an ephemeral stream at the base of a mountain in arid or semiarid regions.

alluvium Sorted sediment deposited by a stream.

Alpine-Himalayan chain The collisional orogen that formed where southern continents collided with Eurasia. It includes the Alps and the Himalaya.

amplitude The height of a wave from crest to trough.

angle of repose The angle of the steepest slope that a pile of uncemented material can attain without collapsing from the pull of gravity.

angular unconformity An unconformity in which the strata below were tilted or folded before the unconformity developed; strata below the unconformity therefore have a different tilt than strata above.

annual probability The liklihood of an event, such as a flood or an earthquake, happening in a given year, as specified by a percent. It is equal to 1 ÷ recurrence interval.

Anthropocene The current geologic age, when human activity has been the dominant influence on climate and the environment.

anthropogenic forcing Changes in the climate resulting from human activities.

anticline A fold with an arch-like shape in which the limbs dip away from the hinge.

anvil cloud The upper part of a large cumulonimbus cloud that spreads laterally at the tropopause to form a broad, flat top.

aphotic zone The deeper part of the ocean into which sunlight does not penetrate.

apparent brightness The amount of energy from a star that passes through a square meter at the surface of the Earth in one second.

apparent magnitude A simple scale defining the apparent brightness of a celestial object, generally a star.

apparent polar-wander path A path on the globe along which a magnetic pole appears to have wandered over time; in fact, the continents drift, while the magnetic pole stays fairly fixed.

aquifer Sediment or rock that transmits water easily.

aquitard Sediment or rock that does not transmit water easily and therefore retards the motion of the water.

Archean Eon The middle Precambrian Eon.

arête A residual knife-edge ridge of rock that separates two adjacent cirques.

arrival time (seismology) The time at which a specific seismic wave, such as a P-wave or an S-wave, arrives at a given seismometer.

artesian well A well in which water rises on its own, due to pressure within the groundwater.

ash (volcanic) Very fine particles of glass or pulverized rock erupted by a volcano.

assimilation The process of magma contamination in which blocks of wall rock fall into a magma chamber and dissolve.

asteroid One of the fragments of solid material, left over from planet formation or produced by the collision of planetesimals, that resides between the orbits of Mars and Jupiter.

asteroid belt The zone between Mars and Jupiter where most asteroids reside.

asthenosphere The layer of the mantle that lies between 100 to 150 km (62 to 93 miles) and 350 km (217 miles) deep; the asthenosphere is relatively soft and can flow when acted on by force.

astronomer A scientist who studies space (the Solar System, stars, galaxies, and the Universe).

astronomical unit (AU) The average distance from the Sun to Earth. It is approximately 150 million kilometers (km) (93 million miles).

astronomy The scientific study of planets, stars, galaxies, and the Universe as a whole.

atmosphere A layer of gases that surrounds a planet.

atmospheric instability The condition in the atmosphere where a parcel of air, when displaced vertically, will accelerate away from its original position.

atmospheric pressure The force applied by air on a unit area of surface. Equivalent to the weight of a column of air above a unit area.

atmospheric science The study of the air layer that surrounds the Earth.

atmospheric seeing The blurring or fuzzing of a celestial object's image in a telescope caused by turbulence in the atmosphere.

atom The smallest piece of an element that has the properties of the element; it consists of a nucleus surrounded by an electron cloud.

aurora A curtain of varicolored light that appears across the night sky when charged particles from the Sun interact with the ions in the ionosphere.

axis of rotation An imaginary line around which an object spins.

backwash The gravity-driven flow of water back down the slope of a beach.

baked contact Wall rock that has been altered by heat emitted from an igneous intrusion.

banded iron formation (BIF) Iron-rich sedimentary layers consisting of alternating gray beds of iron oxide and red beds of iron-rich chert.

bar (1) A sheet or elongate lens or mound of alluvium; (2) a unit of air-pressure measurement approximately equal to 1 atm.

barometer An instrument that measures atmospheric pressure.

barrier island An offshore sand bar that rises above the mean high-water level, forming an island.

basalt A fine-grained, mafic, igneous rock.

base level The lowest elevation a stream channel's floor can reach at a given locality.

basin A fold or depression shaped like a right-side-up bowl.

Basin and Range Province A broad, Cenozoic continental rift that has affected a portion of the western United States in Nevada, Utah, and Arizona; in this province, tilted fault blocks form ranges, and alluvium-filled valleys are basins.

batholith A vast composite, intrusive, igneous rock body up to several hundred km long and 100 km wide, formed by the intrusion of numerous plutons in the same region.

bathymetric map A map illustrating the shape of the ocean floor.

bathymetry Variation in depth.

beach A band of sediment, parallel to the shoreline, that undergoes sorting and shifting by the swash and backwash of waves.

beach nourishment The process of adding sediment, generally pumped or dredged from offshore, to replace beach sediment that has been eroded away.

beach profile The variation in elevation of a beach as measured along a line perpendicular to the shoreline.

bearing In the context of measuring the orientation of geologic structures, it is the compass heading of a line as projected onto a horizontal surface.

bed An individual layer of sedimentary rock.

bedding Layering or stratification in sedimentary rocks.

bedrock Rock still attached to the Earth's crust.

benthic zone The region encompassing the seafloor and the region just below the seafloor.

benthos Organisms that live on or just beneath the sea floor (i.e., in the benthic zone).

Big Bang A cataclysmic explosion that scientists suggest represents the formation of the Universe; before this event, all matter and all energy were packed into one volumeless point.

Big Bang nucleosynthesis The formation of low-mass nuclei (H, He, Li, Be) during the first few minutes after the Big Bang.

binary star A system in which two stars are in gravitationally bound orbits about their common center of mass.

biochemical sedimentary rock Sedimentary rock formed from material (such as shells) produced by living organisms.

biodiesel A type of liquid hydrocarbon fuel manufactured from plant or animal oils.

biodiversity The number of different species that exist at a given time.

biofuel Gas or liquid fuel made from plant material (biomass). Examples of biofuel include alcohol (from fermented sugar) and biodiesel from vegetable oil or wood.

biogeochemical cycle The exchange of chemicals between living and nonliving reservoirs in the Earth System.

biosphere The region of the Earth and atmosphere inhabited by life; this region stretches from a few km below the Earth's surface to a few km above.

black dwarf The last stage of life for a low mass star in which its dense matter core has lost sufficient energy that it no longer emits detectable amounts of radiation.

black hole An object so dense that its escape velocity exceeds the speed of light; a *singularity* in spacetime.

black smoker The cloud of suspended minerals formed where hot water spews out of a vent along a mid-ocean ridge; the dissolved sulfide components of the hot water instantly precipitate when the water mixes with seawater and cools.

blackbody An object that absorbs all radiation incident upon it and radiates energy according to the three laws of blackbody radiation.

blizzard A severe weather condition characterized by high winds and reduced visibilities due to falling or blowing snow.

block A general term for a three-dimensional piece of the Earth.

blowout A deep, bowl-like depression scoured out of desert terrain by a turbulent vortex of wind.

blue shift The phenomenon in which a source of light moving toward you appears to have a higher frequency.

body waves Seismic waves that pass through the interior of the Earth.

boundary layer A layer of the atmosphere adjacent to the Earth's surface, where friction is important. Depth can vary from a few hundred to a few thousand meters. Also called "friction layer."

Bowen's reaction series The sequence in which different silicate minerals crystallize during the progressive cooling of a melt.

breaker A water wave in which water at the top of the wave curves over the base of the wave.

brittle deformation The cracking and fracturing of a material subjected to stress.

butte A medium-size, flat-topped hill in an arid region.

caldera A large circular depression with steep walls and a fairly flat floor, formed after an eruption as the center of the volcano collapses into the drained magma chamber below.

Cambrian explosion The remarkable diversification of life, indicated by the fossil record, that occurred at the beginning of the Cambrian Period.

capacity The amount of a substance or energy that a material can contain.

carbon budget The balance between additions and subtractions of carbon within the different reservoirs of the Earth System.

carbon capture and sequestration A process of decreasing the input of carbon dioxide into the atmosphere by first trapping the CO_2 produced by power plants or factories, then putting it through a process of liquefaction, and finally injecting the CO_2 into porous rocks in the subsurface.

carbon cycle The exchange of carbon between the Earth's carbon reservoirs.

carbon footprint The amount of carbon injected into the atmosphere by humans, as a result of either individual or group activities.

carbonate rock Rocks containing calcite and/or dolomite.

celestial object An object or feature visible in the night sky.

celestial sphere An imaginary sphere with celestial objects on its inner surface and the Earth at its center. The celestial sphere has no physical existence but is a convenient tool for picturing the directions in which celestial objects are seen from the surface of the Earth.

cement Mineral material that precipitates from water and fills the spaces between grains, holding the grains together.

cementation The phase of lithification in which cement, consisting of minerals that precipitate from groundwater, partially or completely fills the spaces between clasts and attaches each grain to its neighbor.

centigrade scale A temperature scale in which the melting point of ice is set at zero and the boiling point of water at sea level pressure is set at 100.

chain reaction A succession of fission reactions that takes place when particles released from one reaction trigger further reactions.

channel A trough dug into the ground surface by flowing water.

chemical sedimentary rock Sedimentary rock made up of minerals that precipitate directly from water solution.

chemical weathering The process in which chemical reactions alter or destroy minerals when rock comes into contact with water solutions and/or air.

chimney (1) A conduit in a magma chamber in the shape of a long vertical pipe through which magma rises and erupts at the surface; (2) an isolated column of strata in an arid region.

chromosphere The region of the Sun's atmosphere located between the *photosphere* and the *corona*.

cinder cone A subaerial volcano consisting of a cone-shaped pile of tephra whose slope approaches the angle of repose for tephra.

cirque A bowl-shaped depression carved by a glacier on the side of a mountain.

cirrus cloud A high, wispy cloud that tapers into delicate, feather-like curls.

clast A single particle or grain of sediment.

clastic rock Sedimentary rock consisting of detritus, derived from the weathering of preexisting rock that has been cemented-together.

clastic sedimentary rock Sedimentary rock consisting of detritus, derived from the weathering of preexisting rock that has been cemented-together.

cleavage (1) The tendency of a mineral to break along preferred planes; (2) a type of foliation in low-grade metamorphic rock.

cliff retreat The change in the position of a cliff face due to erosion.

climate The average of weather conditions, along with the range of those conditions in a given region over a long time period, such as a decade.

climate change Long-term change in climate conditions, typically occurring over a period of several decades, centuries, or longer.

climate science The study of a region's overall annual pattern of weather averaged over many years (its climate) and how that pattern changes over time.

closure temperature The temperature at which parent and daughter atoms are locked into a mineral grain, with the ratio of parent to daughter indicating the age of the mineral.

cloud A visible mass of condensed water vapor (consisting of tiny water droplets, ice crystals, or both) floating in the atmosphere; typically high above the ground.

cloud cover A measure of the percentage of the sky covered by cloud.

cloud-to-cloud lightning Lightning occurring between cloud towers that does not come into contact with the ground.

cloud-to-ground lightning Any lightning stroke that extends from a cloud to the ground.

coal A black, organic rock consisting of greater than 50% carbon, formed from the buried and altered remains of plant material.

coal gasification A process of transforming solid coal into various flamable gases.

coal rank A measurement of the carbon content of coal; higher-rank coal forms at higher temperatures.

coal seam A layer or bed of coal.

coast The region of land adjacent to the shoreline.

coastal plain Low-relief regions of land adjacent to the coast.

coastal wetland A shore area that becomes submerged by shallow water for all or part of the day, which hosts salt-resistant vegetation and many other organisms.

cold front The boundary between a cold airmass and a warm airmass where the cold air advances forward, lifting the warmer airmass.

cold pool The cold surface air that develops in association with thunderstorm downdrafts due to the evaporation of rain.

collision The process of two buoyant pieces of lithosphere converging and squashing together.

collision-coalescence The process by which different-sized clouds and raindrops collect each other as they fall at different speeds.

collisional orogen A mountain belt formed when two relatively buoyant blocks of crust push into each other after the ocean floor between them has been entirely subducted.

columnar jointing A type of fracturing that yields roughly hexagonal columns of basalt; columnar joints form when a dike, sill, or lava flow cools.

comet An object composed of ice and dust, probably remaining from the formation of the Solar System, which orbits the Sun in a highly elliptical orbit, producing a tail as it nears the Sun.

compaction The phase of lithification in which the pressure of the overburden on the buried rock squeezes out water and air that was trapped between clasts, pressing the clasts tightly together.

competence A measure of the ability of a stream to carry clasts, as indicated by the size of the clasts that the stream transports. A more competent stream can carry larger clasts than can a less competent stream.

compression A push or squeezing felt by a body.

compressional waves Waves in which particles of material move back and forth parallel to the direction in which the wave itself moves.

conchoidal fractures Smoothly curving, clam-shell-shaped surfaces along which materials with no cleavage planes tend to break.

condensation theory The theory describing how a nebula contracts to form a central star and planetary system, such as our Solar System.

cone of depression The downward-pointing, cone-shaped surface of the water table in a location where the water table is experiencing drawdown because of pumping at a well.

constellation An imaginary image formed by patterns of stars. These 88 defined areas on the celestial sphere are used by astronomers to locate celestial objects.

consumers Organisms in the food chain or food web that eat other organisms.

contact The boundary surface between two rock bodies, such as between two stratigraphic formations, between an igneous intrusion and adjacent rock, between two igneous rock bodies, or between rocks juxtaposed by a fault.

contact force An applied force where you push or pull on one object with another.

contact metamorphism A synonym for thermal metamorphism. It refers to the change in the mineral assemblage in a rock due to heating by an igneous intrusion.

contaminant plume A subsurface cloud of pollution mixed with or dissolved in groundwater.

continental arc An arc that formed where an oceanic plate subducted beneath a continent.

continental drift The idea that continents have moved and are still moving slowly across the Earth's surface.

continental glacier (ice sheet) A vast sheet of ice that spreads over thousands of square km of continental crust.

continental margin A continent's coastline.

continental shelf A broad, shallowly-submerged fringe of a continent. Ocean-water depth over the continental shelf is generally less than 200 meters (656 feet). The widest continental shelves occur over passive margins.

contour line A line on a map along which a parameter has a constant value; for example, all points along a contour line on a topographic map are at the same elevation.

convective cell A distinct flow configuration for a volume of material that is moving during convective heat transport; simplistically, a loop-like path of material that rises when warm and sinks when cool.

convective flow A process of circulation and heat transfer that happens when a deeper layer of material warms up and, therefore, expands and becomes less dense than the overlying layer of cooler material. The deeper material becomes buoyant, and rises, while the cooler material sinks to take its place.

convective zone In reference to thunderstorms, the region of the cumulonimbus cloud or thunderstorm complex with a strong updraft.

conventional reserve A type of hydrocarbon accumulation that can be accessed simply by drilling into it, and then pumping it out of the reservoir rock of a subsurface trap.

convergence (atmospheric science) A net inflow of air molecules into a region of the atmosphere with the result of increasing surface pressure. Convergence can be caused by changing wind speeds and/or the changing direction of air flow; (geology) a situation that exists where the relative motion of two plates is toward each other.

convergent boundary A plate boundary, associated with a trench, where plates don't butt into each other, but rather, one (the downgoing plate) sinks beneath the other (overriding plate).

convergent-boundary orogen A mountain belt, such as the Andes, built on the overriding plate next to a convergent plate boundary.

Copernican model The model for the Solar System, proposed by Copernicus, with the Sun at the center and other objects, including the Earth, orbiting the Sun.

coral reef A mound of coral and coral debris forming a region of shallow water.

core-mantle boundary An interface 2,900 km (1,802 miles) below the Earth's surface separating the mantle and core.

Coriolis effect The deflection of objects, winds, and currents on the surface of the Earth or ocean, or in the atmosphere, owing to the planet's rotation.

corona The hot, outermost part of the Sun's atmosphere. Compare *chromosphere* and *photosphere*.

correlation The process of defining the age relations between the strata at one locality and the strata at another.

cosmologic time Large time intervals related to events in the history of the Universe.

cosmology The study of the overall structure of the Universe.

crater (1) A circular depression at the top of a volcanic mound; (2) a depression formed by the impact of a meteorite.

cratering density The number of craters per unit area of a planet's surface.

craton A long-lived block of durable continental crust commonly found in the stable interior of a continent.

cratonic platform A province in the interior of a continent in which Phanerozoic strata bury most of the underlying Precambrian rock.

creep The gradual downslope movement of regolith.

crevasse A large crack that develops by brittle deformation in the top 60 meters (197 feet) of a glacier.

critical zone The spatial realm of the biosphere that contains the resources and conditions which sustain life.

cross bed A sedimentary structure represented by subtle planes inclined at an angle to the main bedding. They represent preserved slip faces of dunes or ripples.

cross-cutting relations The nature of the intersection between two geologic features that provides information on the relative ages of the features.

crust The rock that makes up the outermost layer of the Earth.

cryosphere The frozen component of the hydrosphere.

crystal A single, continuous piece of a mineral bounded by flat surfaces that formed naturally as the mineral grew.

crystal face Flat surfaces on a crystal that intersect at sharp edges and that developed naturally as the crystal grew.

crystal habit The general shape of a crystal or cluster of crystals that grew unimpeded.

crystal structure The orderly arrangement of atoms inside a crystal.

crystalline igneous rock A rock, formed by solidification of a melt, with a fabric characterized by interlocking crystals that grew together.

crystalline rock Any rock (igneous, sedimentary, or metamorphic) that has a texture characterized by interlocking crystals that grew together.

crystalline solid A material made of atoms or molecules locked into an orderly arrangement.

cumulus cloud A vertically developing cloud with puffy lobes and clear skies surrounding the cloud.

current A relatively narrow fast moving flow. Examples include water within the ocean or a charge in a conductor.

cycle A series of interrelated events or steps that occur in succession and can be repeated, perhaps indefinitely.

cyclone (1) A large, low-pressure system with a surface airflow that moves counterclockwise around its center in the Northern Hemisphere. Large storms forming in the middle latitudes are called "mid-latitude cyclones," while storms forming in the tropics are called "tropical cyclones." Hurricanes over the Indian Ocean are also called cyclones.

dark energy A form of energy that permeates all of space (including the vacuum), producing a repulsive force that accelerates the expansion of the Universe.

dark matter Matter in galaxies that does not emit or absorb electromagnetic radiation. Dark matter is thought to constitute most of the mass of the Universe.

data A set of observations, measurements, or calculations.

daughter isotope The decay product of radioactive decay.

dead zone The region of the sea near the mouth of a river, in which nutrients brought in by the river caused algae blooms. The death of these organisms, in turn, depletes the water of oxygen, so it cannot sustain oxygen-breathing, marine organisms.

debris flow A downslope movement of mud mixed with larger rock fragments.

decompression melting Melting that occurs when hot mantle rock rises to shallower depths in the Earth, so that pressure decreases but the temperature remains unchanged.

deep-sea currents A flow of water that moves at depth in the ocean. Such currents are not driven by the wind; most are due to density differences between volumes of water.

deep-sea trench An elongate trough occurring in or along the edges of an ocean basin, where the ocean floor can reach depths of 7 to 11 km (4 to 7 miles).

deformation A change in the shape, position, or orientation of a material, by bending, breaking, or flowing.

delta A wedge of sediment formed at a river mouth when the running water of the stream enters standing water, the current slows, so the stream loses competence, and sediment settles out.

density Mass per unit volume.

deposition The process by which sediment settles out of a transporting medium. Also, the process by which water vapor deposits on an ice crystal.

depositional environment A setting in which sediments accumulate; its character (fluvial, deltaic, reef, glacial, etc.) reflects local conditions.

depositional landform A landform resulting from the deposition of sediment, where the medium carrying the sediment evaporates, slows down, or melts.

desalinization The process of removing salt from seawater to make freshwater.

desert A region so arid that it contains no permanent streams, except for those that bring water in from elsewhere, and has very sparse vegetation cover.

desert pavement A mosaic-like, stone surface forming the ground in a desert.

desert varnish A dark, rusty-brown coating of iron oxide and magnesium oxide that accumulates on the surface of desert rock.

desertification The conversion of a semi-arid region to desert conditions due to human activities.

dew point The temperature at which air will become saturated if it is cooled at constant pressure.

diagenesis All of the physical, chemical, and biological processes that transform sediment into sedimentary rock, as well as those processes that alter the characteristics of the sedimentary rock after it has been formed.

diamictite A sedimentary rock composed of large clasts surrounded by a muddy matrix, so that the large clasts do not touch each other. Such rocks can form from glacial till or from muddy debris flows.

differentiation The process by which materials of higher density sink toward the center of a molten or fluid planetary interior.

diffuse nebula Luminous or dark formations of irregularly distributed dust and gas seen within spiral galaxies.

diffusion The migration of atoms or molecules through a material.

digital elevation model A computer-produced portrayal of elevation differences commonly using shading to simulate shadows; the data used assigns elevations to each point.

dike A tabular (wall-shaped) intrusion of rock that cuts across a layering of country rock.

dimension stone An intact block of granite or marble to be used for architectural purposes.

dip The angle that a plane makes with respect to the horizontal, as measured in a vertical plane. This angle is also the steepest possible slope on the plane.

dip-slip fault A fault in which sliding occurs up or down the slope (dip) of the fault.

dipole An imaginary arrow that points from the south end to the north end of a magnet.

directional drilling The process of controlling the trajectory of a drill bit to make sure that the drill hole goes exactly where desired.

discharge The volume of water in a conduit or channel passing a point in one second.

discharge area A location where groundwater flows back up to the surface and may emerge at springs.

disconformity An unconformity parallel to the two sedimentary sequences it separates.

displacement (or offset) The amount of movement or slip across a fault plane.

dissolution A process during which materials dissolve in water.

divergence A net outflow of air molecules from a region of the atmosphere with the result of decreasing surface pressure. Divergence can be caused by changes in wind speeds and/or direction of airflow.

divergent boundary (or spreading boundary) A plate boundary where two oceanic plates move apart by the process of seafloor spreading.

doldrums A belt along the equator with very slow winds.

dolostone A carbonate rock that contains a substantial proportion of the mineral dolomite.

dome Folded or arched layers with the shape of an overturned bowl.

Doppler effect The phenomenon in which the frequency of observed wave energy changes as a result of a source of wave energy moving toward or away from an observer.

Doppler radar An instrument that transmits microwave pulses and detects precipitation and wind motions along the direction of the radar beam.

dormant volcano A volcano that has not erupted for hundreds to thousands of years, but does have the potential to erupt again in the future.

downcutting The process in which water flowing through a channel cuts into the substrate and deepens the channel relative to its surroundings.

downdraft Downward-moving air.

downslope windstorm General term given to windstorm in which strong winds flow down the side of a mountain. Common North American examples include the Chinook and Santa Ana.

downwelling The downward flow of water in the ocean due to the action of surface winds.

drainage basin The region between drainage divides. Within a drainage basin, all the tributaries feed into the same trunk stream.

drainage divide A highland or ridge that separates one watershed from another.

drainage network (or basin) An array of interconnecting streams that together drain an area.

drumlin A streamlined, elongate hill formed when a glacier overrides glacial till.

dry adiabatic lapse rate The rate (10 °C/km) at which an unsaturated parcel of air will change temperature if it is displaced vertically in the atmosphere without exchanging mass or heat with its environment.

dry wash The channel of an ephemeral stream when empty of water.

dune A pile of sand generally formed by deposition from the wind.

dust Particles made of rocky or metallic solids, known to chemists as refractory materials because they melt only at high temperatures.

dust storm An event in which strong winds hit unvegetated land, strip off the topsoil, and send it skyward to form rolling dark clouds that block out the Sun.

dwarf planet A body with characteristics similar to those of a planet except that it has not cleared smaller bodies from the neighboring regions around its orbit. Compare *planet* (definition 2).

dwarf star A star of relatively small size and low luminosity. The majority of main sequence stars, including the Sun, are dwarf stars.

dynamic metamorphism The change that takes place in the texture of a rock when it undergoes intense shear under metamorphic conditions.

dynamothermal metamorphism Metamorphism that occurs as a consequence of shearing alone, with no change in temperature or pressure.

Earth resource A material from the Earth that human society uses for construction, to produce energy, or to make tools and fertilizer from.

Earth Science The study of the nature, origin, and evolution of all of our natural surroundings through the disciplines of geology, oceanography, atmospheric sciences, and astronomy.

Earth System The global interconnecting web of physical and biological phenomena involving the solid Earth, the hydrosphere, and the atmosphere.

earthquake A vibration caused by the sudden breaking or frictional sliding of rock in the Earth.

earthquake engineering The design of buildings that can withstand shaking.

earthquake zoning The determination of the susceptibility of a region to earthquake damage, used as a basis for recommending building codes.

eccentric orbit A highly elliptical path taken by one object around another.

ecliptic plane The plane defined by a planet's orbit.

economic reserve The occurrence of an Earth resource in a particular area, of sufficient concentration to be worth mining or extracting.

ecosystem An environment and its inhabitants.

eddy An isolated, curved-shaped current of water or air.

Ediacaran fauna An assemblage of shell-less invertebrates that lived during the last ~100 million years of the Proterozoic.

effusive eruption An eruption that yields mostly lava, not ash.

Ekman transport The overall movement of a mass of water, resulting from the Ekman spiral, in a direction 90° to the wind direction.

El Niño The flow of warm water eastward in the equatorial Pacific Ocean that reverses the upwelling of cold water along the western coast of South America and causes significant global changes in weather patterns.

elastic deformation The change in shape of a solid that can be accommodated by the stretching and bending, but not breaking, of chemical bonds.

elastic-rebound theory An explanation for the generation of seismic waves which states that elastic deformation builds up in rock adjacent to faults prior to an earthquake, and is then released when the fault slips to generate vibrations (seismic waves).

electromagnetic radiation A traveling disturbance in electric and magnetic fields caused by accelerating electric charges. In quantum mechanics, a stream of photons. Light.

electromagnetic spectrum The spectrum made up of all possible frequencies or wavelengths of electromagnetic radiation, ranging from gamma rays through radio waves and including the visible radiation our eyes can detect.

electromagnetic wave A wave consisting of oscillations in the electric- and magnetic-field strength.

electromagnetism A field force produced by magnetic objects or by electric currents.

electron A negatively charged subatomic particle that orbits the nucleus of an atom; electrons are about 0.0005 × the size of a proton.

element A material consisting entirely of one kind of atom; elements cannot be subdivided or changed into other elements by chemical reactions.

ellipse A conic section produced by the intersection of a plane with a cone when the plane is passed through the cone at an angle to the axis other than 0° or 90°.

embayment A low area of coastal land.

emergent coast A coast where the land is rising relative to sea level or sea level is falling relative to the land.

emission nebula A nebula that shines with its own light.

energy The capacity to do work. *Energy* can exist in a variety of forms, such as electrical, mechanical, chemical, thermal, or nuclear, and can be transformed from one form to another.

energy balance For the atmosphere, the balance between solar energy arriving at the Earth, and infra-red energy radiated by the Earth back to space.

energy resource Something that can be used to produce work; in a geologic context, a material (such as oil, coal, wind, flowing water) that can be used to produce energy.

Enhanced Fujita (EF) scale A scale for rating the destruction caused by tornadoes. It is a revised version of the original Fujita scale, which takes into account the structural integrity of buildings using degree of damage indicators.

enrichment The process of increasing the concentration of the more radioactive isotope of uranium in a quantity of uranium.

ENSO El Niño Southern Oscillation. See "El Niño."

environmental lapse rate The rate at which the environmental temperature changes with height in the atmosphere.

eon The largest subdivision of geologic time.

ephemeral stream A stream which flows during only part of a year. Compare with *permanent stream*.

epicenter The point on the surface of the Earth directly above the focus of an earthquake.

epicontinental sea A shallow sea overlying a continent.

epoch An interval of geologic time representing the largest subdivision of a period.

equant A term for a grain that has the same dimensions in all directions.

equilibrium line The boundary between the zone of accumulation and the zone of ablation on a glacier.

era An interval of geologic time representing the largest subdivision of the Phanerozoic Eon.

erode To grind or bevel away.

erosion The grinding away and removal of the Earth's surface materials by moving water, air, or ice.

erosional coast A coastline where sediment is not accumulating and wave action grinds away at the shore.

erosional landform A landform that results from the breakdown and removal of rock or sediment.

erratic A boulder or cobble that was picked up by a glacier and deposited hundreds of kilometers away from the outcrop from which it detached.

eruptive style The character of a particular volcanic eruption; geologists name styles based on typical examples (e.g., Hawaiian, Strombolian).

esker A ridge of sorted sand and gravel that snakes across a ground moraine; the sediment of an esker was deposited in subglacial meltwater tunnels.

estuary An inlet in which seawater and river water mix; created when a coastal valley is flooded because of either rising sea level or land subsidence.

eutrophication A transformation of a body of water from having sufficient oxygen to support life, to being oxygen poor, due to the blooming of algae and cyanobacteria.

evaporite An accumulation of salts, such as halite and gypsum, due to the evaporation of salty water.

event horizon The effective "surface" of a black hole. Nothing inside this surface—not even light—can escape from the black hole.

exhumation The process (involving uplift and erosion) that returns deeply buried rocks to the surface.

exobiology The study of life beyond the Earth.

exoplanet A planet orbiting a star other than the Sun.

expanding Universe theory A theory based on observation of galaxies that the Universe will continue to expand in the future.

explosive eruption A violent volcanic eruption that produces clouds and avalanches of pyroclastic debris.

external energy Energy brought into the Earth System in the form of electromagnetic radiation from the Sun.

extinct volcano A volcano that was active in the past, but has now shut off entirely and will not erupt in the future.

extinction The dying off of the last representative of a species of organism.

extrusive igneous rock Rock that forms by the freezing of lava above ground, after it flows or explodes out (extrudes) onto the surface and comes into contact with the atmosphere or ocean.

eye The relatively calm, circular, clear region in the center of a hurricane.

eye wall A violently-rotating, funnel-shaped cylinder of clouds that surrounds the eye of a hurricane.

facet The smooth, shiny faces on a gem that form sharp angles with their neighbors, created by gem cutters through the grinding and polishing of the stone, rather than having occurred as natural cleavage planes or as natural crystal faces.

Fahrenheit scale An English-system measure of temperature in which the difference between the freezing point (32°F) and the boiling point (212°F) is divided into 180 units.

failure surface A plane on which sliding occurs beneath a slump, rock fall, or other downslope mass movement.

fault A fracture on which one body of rock slides past another.

fault scarp A small step on the ground surface where one side of a fault has moved vertically with respect to the other.

felsic magma A molten rock that has a relatively high proportion of silica.

fetch The distance across a body of water along which a wind blows.

field force A push or pull that applies across a distance (i.e., without contact between objects); examples are gravity and magnetism.

firn Compacted granular ice (derived from snow) that forms where snow is deeply buried; if buried more deeply, firn turns into glacial ice.

fissure eruption The ejection or flow of lava out of an elongate crack, rather than out of a circular vent.

fjord A deep, glacially carved, U-shaped valley flooded by rising sea level.

flash flood A flood that occurs during unusually intense rainfall or as the result of a dam collapse, during which the floodwaters rise very fast.

flood An event during which the volume of water in a stream becomes so great that it covers areas outside the stream's normal channel.

flood basalt Vast sheets of basalt that spread from a volcanic vent over an extensive surface of land; they may form where a rift develops above a continental hot spot, and where lava is particularly hot and has low viscosity.

flood stage The level at which a rising river begins to flood agricultural lands or has the potential to damage property.

floodplain The flat land on either side of a stream that becomes covered with water during a flood.

flux melting The transformation from hot solid to liquid, which occurs when a volatile material injects into the solid.

focus The location where a fault slips during an earthquake (hypocenter).

fog A cloud that forms at ground level.

fold A bend or wrinkle of rock layers or foliation; folds form as a consequence of ductile deformation.

fold-thrust belt An assemblage of folds and related thrust faults that develop above a detachment fault.

food web The interconnection network of organisms defined by which organisms eat other organisms.

force A push or a pull on an object.

foreshock A smaller earthquake that precedes a major earthquake.

fossil The remnant, or trace, of an ancient living organism that has been preserved in rock or sediment.

fossil assemblage A group of fossil species found in a specific sequence of sedimentary rock.

fossil fuel An energy resource such as oil or coal, which comes from organisms that lived long ago and which store solar energy that reached the Earth at the time the organisms lived.

fossil succession The concept that in a sequence of strata, the fossils found higher in the sequence differ from those found lower in the sequence. As a result, the the assemblage of fossil species in strata can be used as a basis for determining the relative age of the strata.

fossilization The process of forming a fossil.

fractional crystallization The process by which a magma becomes progressively more silicic as it cools, because early-formed crystals settle out.

fracture zone A narrow band of vertical fractures in the ocean floor; fracture zones lie roughly at right angles to a mid-ocean ridge. The actively slipping part of a fracture zone is called a transform fault.

fragmental igneous rock An igneous rock formed of debris erupted from a volcano. The debris is either cemented or welded together to form the rock.

frequency The number of waves that pass a point in a given time interval.

freshwater Water that contains very little, if any, salt.

frictional drag The slowing of an object or fluid due to the force of friction.

front A boundary between airmasses of differing density (due to different temperatures and/or moisture). They are classified on the basis of the thermal and moisture characteristics of the airmasses, and the direction of movement of the airmasses.

frontal squall line A line of thunderstorms forming along a frontal boundary such as the cold front of an extratropical cyclone.

frost wedging The process in which water trapped in a joint freezes, forces the joint open, and may cause the joint to grow.

fuel An energy source that can be easily transported; the energy of a fuel is potential energy stored either in chemical or nuclear bonds.

galactic bulge The central region of spiral galaxies where stars cluster in a bulge that extends above and below the galactic disk.

galactic disk The plane in which most stars and spiral arms of a galaxy reside.

galactic halo An extended, roughly spherical component of a galaxy that extends beyond the main, visible features.

galaxy An immense system of hundreds of billions of stars.

gas A state of matter in which atoms and molecules move about freely. A gas can flow, but it can also expand or contract when the size of its container changes.

gas giant A giant planet formed mostly of hydrogen and helium. In the Solar System, Jupiter and Saturn are the gas giants. Compare *ice giant*.

gem A mineral specimen that has been cut and polished and is particularly beautiful or valuable.

general circulation The global-scale pattern of winds, characterized by semi-permanent features such as the subtropical highs, the subpolar lows, the trade winds and the jetstream.

geocentric model An ancient Greek idea suggesting that the Earth sat motionless in the center of the Universe, while the stars, other planets, and Sun orbited around it.

geochronology The science of dating geologic events in years.

geode A cavity in which euhedral crystals precipitate out of water solutions passing through a rock.

geographical pole The locations (north and south) where the Earth's rotational axis intersects the planet's surface.

geologic column A composite stratigraphic chart that represents the entirety of the Earth's history.

geologic cross section A graphical representation of a vertical slice through the Earth, which displays the configuration of rock units and geologic structures underground, as they would appear on a vertical plane.

geologic history The sequence of geologic events that has taken place in a region.

geologic map A map showing the distribution of rock units and structures across a region.

geologic structure A feature that has formed as a consequence of deformation. Examples include faults, folds, joints, and foliation.

geologic time The span of time since the formation of the Earth.

geologic time scale A scale that describes the intervals of geologic time.

geology The study of the Earth, including our planet's composition, behavior, and history.

geometric parallax The apparent displacement or difference in the apparent position of an object relative to a background as viewed along two different lines of sight.

geosphere The composition of the solid Earth.

geostrophic balance The balance that exists when the pressure gradient force and the Coriolis force are equal and opposite.

geostrophic current An ocean current that results from a balance between the pressure gradient and the Coriolis forces.

geostrophic wind The wind that would exist if the pressure gradient and Coriolis forces were in balance.

geotherm The change in temperature with depth in the Earth.

geothermal energy Heat and electricity produced by using the internal heat of the Earth.

geothermal gradient The rate of change in temperature with depth.

geyser A fountain of steam and hot water that erupts periodically from a vent in the ground in a geothermal region.

giant star A star with a substantially larger radius and luminosity than a main-sequence star with the same surface temperature.

glacial advance The forward movement of a glacier's toe when the supply of snow exceeds the rate of ablation.

glacial outwash Coarse sediment deposited on a glacial outwash plain by meltwater streams.

glacial rebound The process by which the surface of a continent rises back up after an overlying continental ice sheet melts away and the weight of the ice is removed.

glacial retreat The movement of a glacier's toe back toward the glacier's origin; glacial retreat occurs if the rate of ablation exceeds the rate of supply.

glacial striation A scratch on rock produced when a glacier, containing clasts embedded in the ice, flowed over the rock.

glacial subsidence Sinking of the surface of the Earth due to the weight of a large glacier or ice sheet.

glacial till Sediment transported by flowing ice and deposited beneath a glacier or at its toe.

glaciation A period of time during which glaciers grew and covered substantial areas of the continents.

glacier A river or sheet of ice that slowly flows across the land surface and lasts all year long.

glass A solid in which atoms are not arranged in an orderly pattern.

glassy igneous rock Igneous rock consisting entirely of glass, or of tiny crystals surrounded by a glass matrix.

global change The transformations or modifications of both physical and biological components of the Earth System through time.

global climate model (GCM) A numerical model consisting of mathematical equations used to simulate changes in global climate conditions.

global cooling A fall in the global average atmospheric temperature.

global positioning system (GPS) A satellite system people can use to measure rates of movement of the Earth's crust relative to one another, or simply to locate their position on the Earth's surface.

global warming A rise in the global average atmospheric temperature.

globular cluster A spherically-symmetric, highly-condensed group of stars, containing tens of thousands to a million members.

gneiss A compositionally banded metamorphic rock typically composed of alternating dark- and light-colored layers.

gneissic banding Compositional layering, in a metamorphic rock, as manifested by alternating lighter (more felsic) and darker (more mafic) intervals.

Gondwana A supercontinent that consisted of today's South America, Africa, Antarctica, India, and Australia. Also called Gondwanaland.

grade (See metamorphic grade)

graded bed A sedimentary layer in which grain size changes progressively from coarser at the base, to finer at the top. Graded bedding forms by deposition from turbidites.

grain A fragment of a mineral crystal or of a rock.

granite A coarse-grained, intrusive, silicic igneous rock.

graupel A small (< 3–4 mm), soft ball of ice that results when an ice crystal collects supercooled water droplets that freeze on its surface.

gravitational energy The potential energy stored in a mass if the mass lies within a gravitational field.

gravity The attractive force that one mass exerts on another; the magnitude depends on the size of the objects and the distance between them.

great oxygenation event The time period between 2.4 and 1.8 Ga when the Earth's atmosphere was enriched with oxygen produced by photosynthetic organisms.

Great Red Spot The giant, oval, brick-red rotating storm seen in Jupiter's southern hemisphere.

greenhouse effect The trapping of heat in the Earth's atmosphere by carbon dioxide and other greenhouse gases, which absorb infrared radiation.

groundwater Water that resides under the surface of the Earth, mostly in pores or cracks of rock or sediment.

growth ring A rhythmic layering that develops in trees, travertine deposits, and shelly organisms as a consequence of seasonal changes.

gust front The leading edge of the outflow of a thunderstorm's rain-cooled downdraft air. Passage is often marked by a sudden increase of wind speed.

guyot A seamount that had a coral reef growing on top of it, so that it is now flat-crested.

gyre A large, circular flow pattern of ocean surface currents.

H-R diagram The Hertzsprung-Russell diagram, which shows the luminosity of stars relative to their temperature. Stars organize into distinct groups on the diagram.

Hadean Eon The oldest of the Precambrian eons; the time between Earth's origin and the formation of the first rocks that have been preserved.

Hadley cells The name given to the low-latitude convection cells in the troposphere.

hail Frozen precipitation particles with diameters ranging from 3 mm to 20 cm resulting from the collection of supercooled liquid droplets by graupel in the strong updrafts of thunderstorms.

half-life The time it takes for half of a group of a radioactive element's isotopes to decay.

halocline The boundary in the ocean between surface-water and deep-water salinities.

hand specimen A fist-sized piece of rock that you can look at more closely with a hand lens (magnifying glass).

hanging valley A glacially carved tributary valley whose floor lies at a higher elevation than the floor of the trunk valley.

hardness A measure of the relative ability of a mineral to resist scratching; it represents the resistance of bonds in the mineral to being broken.

hazard assessment map A portrayal of the distribution of the susceptibility of locations to the consequences of natural hazards; examples include landslide-hazard maps or flooding-risk maps.

haze The reduction in atmospheric visibility due to suspended aerosol and small droplets in the air.

headland A protrusion of land, along the coast, into the sea.

headward erosion The process by which a stream's channel length increases by erosion at the stream's origin.

heat Thermal energy resulting from the movement of molecules.

heat flow The rate at which heat transfers from one location to another, such as rising from the Earth's interior up to the surface.

heat-transfer melting Melting that results from the transfer of heat from a hotter magma to a cooler rock.

heliocentric model A model in which objects in the Solar System, including the Earth, orbit the Sun.

heliosphere A bubble-like region in space in which solar wind has blown away most interstellar atoms.

heterosphere A term for the upper portion of the atmosphere, in which gases separate into distinct layers on the basis of composition.

high-pressure center A location at which the pressure exceeds the pressure at all surrounding points.

high-pressure system The larger region surrounding a high-pressure center, generally associated with subsiding air and clear skies.

hinge The portion of a fold where curvature is greatest.

homosphere The lower part of the atmosphere, in which the gases have stirred into a homogenous mixture.

horizon The distant visual boundary separating the sky from the ground.

horn A mountain peak that has been carved by glaciers so that it has cirques on three sides.

hornfels A non-foliated metamorphic rock that forms due to thermal metamorphism. In such rocks, metamorphic mineral crystals are randomly oriented.

hot spot A location at the base of the lithosphere, at the top of a mantle plume, where temperatures can cause melting.

hot spring A source at which very warm, to even boiling groundwater spills out of the ground.

hot-spot track A chain of now-dead volcanoes transported off a hot spot by the movement of a lithosphere plate.

hothouse period An interval of geologic time during which global climate was substantially warmer and glacial ice caps did not cover polar regions.

Hubble's law The law stating that the speed at which a galaxy is moving away from the Earth is proportional to the distance of that galaxy.

hurricane A strong tropical cyclone over the Atlantic or Eastern Pacific Oceans in which sustained wind speeds reach 119 km/hr (74 mph) or higher.

hurricane track The path a hurricane follows.

hydrocarbon A chain-like or ring-like molecule made of hydrogen and carbon atoms; petroleum and natural gas are hydrocarbons.

hydrocarbon reserve An accumulation of oil or gas underground, in sufficient quantity and concentration to be worth extracting.

hydrofracturing The process of injecting high-pressure water and other chemicals into a drill hole to generate cracks in the surrounding rock.

hydrologic cycle The continual passage of water from reservoir to reservoir in the Earth System.

hydrolysis The process in which water chemically reacts with minerals and breaks them down.

hydrosphere The Earth's water, including surface water (lakes, rivers, and oceans), groundwater, and liquid water in the atmosphere.

hydrothermal fluid Hot water, steam, or supercritical fluid circulating through rock underground; typically, these fluids contain dissolved ions.

hypothesis A possible, natural reason that can explain a set of data.

ice The solid form of a volatile material; sometimes the *volatile material* itself, regardless of its physical form.

ice age An interval of time in which the climate was colder than it is today, glaciers occasionally advanced to cover large areas of the continents, and mountain glaciers grew; an ice age can include many glacial and interglacial periods.

ice giant A giant planet formed mostly of the liquid form of volatile substances (ices). In the Solar System, Uranus and Neptune are the ice giants. Compare *gas giant*.

ice shelf A broad, flat region of ice along the edge of a continent formed where a continental glacier flowed into the sea.

ice storm A winter storm in which there is a substantial accumulation of freezing rain or freezing drizzle at the surface.

iceberg A large block of ice that calves off the front of a glacier and drops into the sea.

icehouse period A period of time when the Earth's temperature was cooler than it is today and ice ages could occur.

igneous activity The overall process during which rock deep underground melts, producing molten rock that rises up into the crust and, in some cases, makes it all the way up to the Earth's surface.

igneous rock Rock that forms when hot molten rock (magma or lava) cools and freezes solid.

impact structure A feature formed by the impact of a meteorite on the Earth. Relatively young impact structures may exhibit a crater, but even if the crater has been eroded away with age, large, older impact structures can still be recognized by the occurrence of distinctive structures and rocks.

inclusion A material—solid, liquid, or gas—that was trapped inside a mineral crystal as the crystal grew.

inequant A term for a mineral grain whose length and width are not the same.

infiltration The migration of a liquid through a porous material.

inner core The inner section of the core, extending from 5,155 km (3,203 miles) deep to the Earth's center at 6,371 km (3,159 miles) and consisting of solid iron alloy.

intensity A measure of the relative size of an earthquake (the severity of ground shaking) at a location, as determined by examining the amount of damage caused.

interglacial A period of time between two glaciations.

Intergovernmental Panel on Climate Change (IPCC) An international assembly of scientists charged with determining how the Earth's climate is changing and how humans are influencing those changes.

intermediate magma A magma whose silica concentration is between that of mafic and felsic magmas.

internal energy Thermal energy that rises from the interior of the Earth.

interplanetary space The space between planets.

interstellar medium The gas and dust that fill the space between the stars within a galaxy.

interstellar space The space between stars.

intertidal zone The area of coastal land across which the tide rises and falls.

intertropical convergence zone (ITCZ) A zone of convergence of winds near the surface in the vicinity of the equator. The converging winds are the low-level branches of the Hadley Cells of the Northern and Southern Hemispheres.

intraplate earthquakes Earthquakes that occur away from plate boundaries.

intrusive igneous rock Rock formed by the freezing of magma underground.

inversion An increase of temperature with altitude. An inversion represents the opposite of the more common tropospheric situation in which temperature decreases with altitude.

ionosphere The region of the upper atmosphere, above 60 to 100 km (37 to 62 miles), that has a high concentration of ions.

island arc An area of volcanic activity that builds up on the seafloor when one oceanic plate subducts beneath another oceanic plate.

isobar A line on a map along which the air has a specified pressure.

isostasy The condition that exists when the buoyancy force pushing lithosphere up equals the gravitational force pulling lithosphere down.

isotope A different version of a given element that has the same atomic number but a different atomic weight.

isotopic dating (or radiometric dating) A tool used by geologists that can provide the age of a rock in years.

jet stream A fast-moving current of air that flows at high elevations.

jet streak A region of exceptionally strong winds within the jetstream.

joint A naturally formed crack in rocks.

Jovian planet One of the largest planets in the Solar System (Jupiter, Saturn, Uranus, or Neptune), typically 10 times the size and many times the mass of any *terrestrial planet* and lacking a solid surface.

K-Pg extinction (formerly known as the K-T extinction) A sudden mass die-off of all species of dinosaurs, as well as of many other species, at the end of the Cretaceous (at 66 Ma). It has been attributed to an immense meteorite impact.

karst landscape A region underlain by caves in limestone bedrock; the collapse of the caves creates a landscape of sinkholes separated by higher topography, or of limestone spires separated by low areas.

katabatic wind Downslope wind caused by gravitational drainage of very cold, dense air. Most common at the edges of ice sheets (Antarctica, Greenland) and large glaciers.

Keeling curve A graph, first developed by Charles Keeling, showing the ongoing change in concentration of carbon dioxide in the Earth's atmosphere since the 1950s.

Kelvin scale A measure of temperature in which 0 K is absolute zero and the freezing point of water is 273.15 K; divisions in the Kelvin scale have the same value as those in the Celsius scale.

Kepler's laws The three rules of planetary motion inferred by Johannes Kepler from the data collected by Tycho Brahe.

kerogen The waxy molecules into which the organic material in shale transforms on reaching about 100°C. At higher temperatures, kerogen transforms into oil.

kettle hole A circular depression in the ground made when a block of ice calves off the toe of a glacier, becomes buried by till, and later melts.

Köppen-Geiger climate classification (KGCC) A characterization of the Earth's climates based on temperature and precipitation, developed by Wladimir Köppen and Rudolf Geiger.

Kuiper Belt A diffuse ring of icy objects, remnants of Solar System formation, that orbit our Sun outside the orbit of Neptune.

La Niña A condition characterized by colder-than-normal surface waters in the eastern equatorial Pacific and by a strengthening of the Walker Cell and the trade winds. The phase of the Southern Oscillation opposite to the El Niño phase.

laccolith A blister-shaped, shallow intrusion formed when magma injects in between layers underground.

lag deposit The coarse sediment left behind in a desert after wind erosion removes the finer sediment.

lagoon A body of shallow seawater separated from the open ocean by a barrier island.

lahar A thick slurry formed when volcanic ash and debris mix with water, either in rivers or from rain or melting snow and ice on the flank of a volcano.

lake-effect storm A snowstorm over and immediately downstream of water bodies such as the Great Lakes, triggered by the flow of very cold air over relatively warm lake water, which supplies moisture and destabilizing heat to the atmosphere.

landform A distinctive shape of the land surface.

landscape A general term for the character and shape of the land surface in a region.

landslide A sudden movement of rock and debris down a nonvertical slope.

landslide-potential map A map on which regions are ranked according to the likelihood that a mass movement will occur.

lapilli Any pyroclastic particle that is 2 to 64 mm in diameter (i.e., marble-sized); the particles can consist of frozen lava clots, pumice fragments, or ash clumps.

large igneous province (LIP) A region in which huge volumes of lava and/or ash erupted over a relatively short interval of geologic time.

late heavy bombardment An event that happened between 4.1 and 3.8 Ga, during which a great many meteorites struck the Earth, perhaps pulverizing the surface and causing any liquid water in existence to evaporate.

latent heat The energy required for, or released by, a transition between two phases (solid, liquid, gaseous) of a substance.

Laurentia A continent in the early Paleozoic Era composed of today's North America and Greenland.

lava Molten rock that has flowed out onto the Earth's surface.

lava dome A dome-like mass of rhyolitic lava that accumulates above an eruption vent.

lava flow A sheet or mound of lava that flows onto the ground surface or seafloor in molten form and then solidifies.

lava fountain A pressurized eruption of lava during which the lava spurts skyward for tens to as much as a few hundred meters.

lava tube The empty space left when a lava tunnel drains; this happens when the surface of a lava flow solidifies while the inner part of the flow continues to stream downslope.

lifting mechanism A process that causes air parcels to rise in the atmosphere. Examples include the ascent of air at a frontal boundary, airflow over mountains, and convergent winds near the surface.

light pollution The reduction in the ability to observe stars in the night sky due to light from human activities.

light scattering The absorbtion of light, and immediate reemission in a different direction, by molecules and particles in the atmosphere or ocean.

light-year The distance that light travels in one Earth year (about 9.5 trillion km or 6 trillion miles).

lightning An electrical discharge in the atmosphere, representing a rapid flow of electrical charge between a cloud and the ground, between two clouds, or between two portions of the same cloud.

lightning stroke A giant spark and pulse of current produced by a thunderstorm.

limb (of fold) The side of a fold, showing less curvature than at the hinge.

limestone Sedimentary rock composed of calcite.

liquid A state of matter in which atoms or molecules can move relative to one another, but remain in contact. A liquid can flow and conform to the shape of its container.

lithification The transformation of loose sediment into solid rock through compaction and cementation.

lithosphere The relatively rigid, nonflowable, outer 100- to 150-km (62- to 93-miles) -thick layer of the Earth, constituting the crust and the top part of the mantle.

lithospheric mantle The part of the Earth's mantle that lies within the lithosphere.

littoral zone The near-shore area of a body of water. In the ocean, it includes the intertidal zone but may also include adjacent shallow waters.

Local Group The nearby group of galaxies that includes the Milky Way and Andromeda galaxies as members.

local weather system A weather system affecting a small area, such as a county. A thunderstorm is a local weather system.

loess Layers of fine-grained sediments deposited from the wind; large deposits of loess formed from fine-grained glacial sediment blown off outwash plains.

longshore current The flow of water parallel to the shore just off a coast, because of the diagonal movement of waves toward the shore.

longshore drift The movement of sediment laterally along a beach; it occurs when waves wash up a beach diagonally.

low-pressure center A location at which the pressure is less than the pressure at all surrounding points.

low-pressure system The region surrounding a low-pressure center, generally associated with rising air, clouds, and precipitation.

low-velocity zone (LVZ) A region of the mantle, just below the oceanic lithosphere, in which seismic waves travel more slowly than they do in the lithospheric mantle above, or in mantle deeper down. It is thought to be due to the existence of a small amount of melt on grain boundaries.

lower mantle The deepest section of the mantle, stretching from 670 km (416 miles) down to the core-mantle boundary.

luminosity The total amount of light emitted by an object. It is measured in units called watts (W).

lunar eclipse An eclipse that occurs when the Moon is partially or entirely in the Earth's shadow. Compare *solar eclipse*.

lunar highland Mountain ranges on the moon that appear light from the Earth relative to the darker Maria plains.

lunar month The time period between successive new moons (roughly 29-1/2 days).

lunar regolith The loose, surface material, including dust, soil, and broken rock found on the moon.

lunar time scale The time scale over which the moon formed and evolved to its present state.

luster The way a mineral surface scatters light.

mafic magma A magma that has a relatively high proportion of iron and magnesium oxides, relative to silica. Solidification of such magmas produces basalt or gabbro.

magma Molten rock beneath the Earth's surface.

magma chamber A space below ground filled with magma.

magnetic anomaly The difference between the expected strength of the Earth's magnetic field at a certain location and the actual measured strength of the field at that location.

magnetic field The region affected by the force emanating from a magnet.

magnetic pole The end of a magnetic dipole; all magnetic dipoles have a north pole and a south pole.

magnetic reversal The change of the Earth's magnetic polarity; when a reversal occurs, the field flips from normal to reversed polarity, or vice versa.

magnetic-reversal chronology The history of magnetic reversals through geologic time.

magnitude A system used by astronomers to describe the brightness or luminosity of stars. The brighter the star, the lower its magnitude.

main sequence The strip on the H-R diagram where most stars are found. Main-sequence stars are fusing hydrogen to helium in their cores.

mainshock The main or largest earthquake of a sequence; it may be preceded by foreshocks and is always followed by aftershocks.

mantle The thick layer of rock below the Earth's crust and above the core.

mantle plume A column of very hot rock rising up through the mantle.

marble A metamorphic rock composed of calcite and transformed from a protolith of limestone.

mare The broad, darker areas on the Moon's surface; they consist of flood basalts that erupted over 3 billion years ago and spread out across the Moon's lowlands.

marine geology The study of the seafloor and of oceanic coasts.

marine magnetic anomaly An occurrence of a magnetic field, over the seafloor, that is either greater than, or lesser than the expected field. The map pattern of positive and negative anomalies defines seafloor stripes.

mass The amount of matter in an object; mass differs from weight in that its value does not depend on the strength of gravity.

mass-extinction event A time when vast numbers of species abruptly vanish.

mass wasting The gravitationally caused downslope transport of rock, regolith, snow, or ice.

matter The material substance of the Universe; it consists of atoms and has mass.

maturity The degree to which a sediment has changed from the place it originated in to the place it was deposited.

meander A snake-like curve along a stream's course.

melt Molten (liquid) rock.

meltdown The melting of the fuel rods in a nuclear reactor that occurs if the rate of fission becomes too fast and the fuel rods become too hot.

mesa A large, flat-topped hill (with a surface area of several square km) in an arid region.

mesocyclone Cyclonic circulation within the updraft region of a supercell thunderstorm. Normally coincides with the region where tornadoes form. Typically several kilometers in diameter.

mesosphere The cooler layer of atmosphere overlying the stratosphere.

metaconglomerate A metamorphic rock, derived from a sedimentary conglomerate protolith. Stretching or flattening clasts in a metaconglomerate defines a foliation.

metal A solid composed almost entirely of atoms of metallic elements; it is generally opaque, shiny, smooth, malleable, and can conduct electricity.

metallic hydrogen A special form of hydrogen that only occurs under the tremendous pressures found near the core of large planets like Jupiter.

metamorphic aureole The region around a pluton, stretching tens to hundreds of meters out, in which heat transferred into the country rock and metamorphosed the country rock.

metamorphic facies A set of metamorphic mineral assemblages indicative of metamorphism under a specific range of pressures and temperatures.

metamorphic foliation A fabric defined by parallel surfaces or layers that develop in a rock as a result of metamorphism; schistocity and gneissic layering are examples.

metamorphic grade An informal indication of the intensity of metamorphism that a rock has been subjected to.

metamorphic mineral A mineral that grows during metamorphism of a rock.

metamorphic rock Rock that forms when preexisting rock changes into new rock as a result of an increase in pressure and temperature and/or of shearing under elevated temperatures; metamorphism occurs without the rock first becoming a melt or a sediment.

metamorphic texture The arrangement of minerals in a rock that develops as a consequence of metamorphism. In a hornfels, metamorphic minerals grow in random orientations. In dynamothermal metamorphic rocks, metamorphic minerals align to define a foliation.

metamorphic zone The region between two metamorphic isograds, typically named after an index mineral found within the region.

metamorphism The process by which one kind of rock transforms into a different kind of rock.

metasomatism The process by which a rock's overall chemical composition changes during metamorphism because of reactions with hot water that bring in or remove elements.

meteor The incandescent trail produced by a small piece of interplanetary debris as it travels through the atmosphere at very high speeds.

meteor shower A larger-than-normal display of meteors, occurring when the Earth passes through the orbit of a disintegrating comet, sweeping up its debris.

meteorite A piece of rock or metal alloy that fell from space and landed on the Earth.

meteoroid A small cometary or asteroidal fragment, ranging in size from 100 microns (μm) to 100 meters (328 feet). When entering a planetary atmosphere, the meteoroid creates a *meteor*.

meteorologist A trained person who studies and forecasts the weather and its consequences.

meteorology The study of the weather, meaning the condition of the atmosphere at a given location and time, as well as the movement of air and its consequences.

microcontinent A small block of continental crust, too small to be considered a continent geographically. Geologists tend to use the term in reference to blocks of continental crust that collide with a larger continent.

mid-latitude cyclone A large, swirling low-pressure system that forms along the jetstream between about 30° and 70° latitude and produces many types of weather, including hazardous weather.

mid-ocean ridge A 2-km (1-1/4-mile) -high submarine mountain belt that forms along a divergent oceanic plate boundary.

Milankovitch cycle Climate cycles that occur over tens to hundreds of thousands of years because of changes in the Earth's orbit and tilt.

Milky Way Galaxy The galaxy in which the Sun and Solar System reside.

mineral A homogenous, naturally-occurring, solid, inorganic substance with a definable chemical composition and an internal structure characterized by an orderly arrangement of atoms, ions, or molecules in a lattice. Most minerals are inorganic.

mineral resource A mineral extracted from the Earth's upper crust for practical purposes.

mirage An optical illusion caused by specific atmospheric conditions that cause light rays to bend.

Modified Mercalli scale An earthquake characterization scale based on the amount of damage that the earthquake causes.

Moho The seismic-velocity discontinuity that defines the boundary between the Earth's crust and mantle.

Mohs hardness scale A list of ten minerals in a sequence of relative hardness, with which other minerals can be compared.

moist adiabatic lapse rate The rate (°C per km) at which a saturated air parcel will cool, as a result of expansion during its ascent, if there is no exchange of heat or mass with the surrounding environment.

molecule The smallest piece of a compound that has the properties of the compound; it consists of two or more atoms attached by chemical bonds.

moment magnitude scale An indication of the relative size of an earthquake based on the energy that the earthquake releases. It has replaced the Richter scale in modern descriptions of earthquake size.

monocline A fold in the land surface whose shape resembles that of a carpet draped over a stair step.

monsoon A seasonal reversal in wind direction that causes a shift from a very dry season to a very rainy season in some regions of the world.

moon A less massive satellite orbiting a more massive object. Moons are found around planets, dwarf planets, asteroids, and Kuiper Belt objects. The term is usually capitalized when referring to the Earth's Moon.

moraine A sediment pile composed of till deposited by a glacier.

mountain (alpine) glacier A glacier that exists in or adjacent to a mountainous region.

mountain belt An elongate region of uplifted crust. Typically, erosion carves such regions to form steep-sided mountains.

mountain building The geologic process of producing a mountain belt.

mudcrack A sedimentary structure formed when mud dries, shrinks, and breaks, producing open cracks.

mudflow A downslope movement of mud at slow to moderate speed.

multiple-vortex tornado A tornado that contains multiple suction vortices within it.

natural arch An arch that forms when erosion along joints leaves narrow walls of rock, and then the lower part of the wall erodes while the upper part remains.

natural forcing Causes of climate change that are not a consequence of human activity.

natural gas Hydrocarbons, such as methane and propane, which consist of short carbon chains and can exist in gaseous form at the Earth's surface.

natural levee A pair of low ridges that appear on either side of a stream and develop as a result of the accumulation of sediment deposited naturally during flooding.

natural variation The range of changes in conditions that happen in a system due the variability natural phenomena affecting the system.

near-Earth object An asteroid, comet, or large meteoroid whose orbit intersects the Earth's orbit.

nebula A cloud of gas or dust in space.

nebular theory The concept that planets grow out of rings of gas, dust, and ice surrounding a new-born star.

negative feedback A process which leads to a change that reduces the rate at which the original process occurs.

nekton Multicellular organisms, such as fish, that actively swim water.

neritic zone The region from the edge of the continental shelf to the shore, in which the sea is relatively shallow, and sunlight can penetrate.

neutron A subatomic particle, in the nucleus of an atom, that has a neutral charge.

neutron star The superdense neutron stellar core left behind by a supernova.

nonconformity A type of unconformity, meaning a surface representing a gap in the geologic record, at which sedimentary strata are in contact with underlying igneous or metamorphic rocks.

nonmetallic mineral resource Mineral resources that do not contain metals; examples include building stone, gravel, sand, gypsum, phosphate, and salt.

nonrenewable resource A resource that nature will take a long time (hundreds to millions of years) to replenish, or may never replenish.

normal fault A fault in which the hanging-wall block moves down the slope of the fault.

normal galaxy A galaxy that emits most of its radiation at or near visible wavelengths of light.

normal polarity Polarity in which the paleomagnetic dipole has the same orientation as it does today.

nova (pl. **novae**) A stellar explosion that results from runaway nuclear fusion in a layer of material on the surface of a white dwarf in a binary system.

nuclear fusion The process by which the nuclei of atoms fuse together, thereby creating new, larger atoms.

nuclear reactor The part of a nuclear power plant where the fission reactions occur.

nuclear waste Radioactive materials left over after they have been used in medical applications or in energy generation.

nucleus The central core of an atom that consists of protons and neutrons (except for hydrogen, whose nuclei contains only a proton).

numerical age The age of a geologic material or feature, as specified in years.

nutrient A chemical essential for the survival of living organisms.

observatory A place where astronomers use telescopes to observe objects and features in space.

obsidian An igneous rock consisting of a solid mass of volcanic glass.

occluded front The airmass boundary that occurs when the air behind a cold front comes in direct contact with the cool air north or east of a warm front. It develops in the latter stages of a mid-latitude cyclone's life cycle when the cold front catches up and wraps into the warm front.

ocean A large body of salt water between continents.

ocean basin Broad, low areas of the Earth's surface, floored by oceanic crust, that filled with water to become the oceans.

oceanic island A peak of igneous rock, produced by volcanic eruptions, that built up above the surrounding seafloor and rises above sea level.

oceanic zone The region of ocean beyond the edge of the continent shelves, where water depth is substantially deeper that it is over the shelves.

oceanography The study of the water and life in the oceans, as well as the way in which ocean water moves and interacts with land and air.

oil A liquid hydrocarbon that can be burned as a fuel.

Oil Age The period of human history, including our own, so named because the economy depends on oil.

oil seep A site where hydrocarbons naturally seep out onto the surface of the Earth from underground.

oil shale Shale containing kerogen.

oil window The narrow range of temperatures under which oil can form in a source rock.

Oort Cloud A cloud of icy objects, left over from Solar System formation, that orbit the Sun in a region outside of the heliosphere.

open-pit mine A large surface excavation to access ore minerals, coal, or diamonds.

optical telescope A type of telescope that employs visible radiation to observe objects in space.

optical window The range of visible and near-infrared wavelengths for which the atmosphere is transparent, so that objects in space can be observed.

orbit The path taken by one object moving around another object under the influence of their mutual gravitational or electric attraction.

ordinary thunderstorm A thunderstorm that does not become severe or become part of an organized thunderstorm complex; often referred to as an airmass thunderstorm.

ore Rock containing native metals or a concentrated accumulation of ore minerals.

ore deposit An economically significant accumulation of ore.

ore mineral A mineral that has metal in high concentrations and in a form that can be easily extracted.

organic chemical A carbon-containing compound that occurs in living organisms, or that resembles such compounds; it consists of carbon atoms bonded to hydrogen atoms along with varying amounts of oxygen, nitrogen, and other chemicals.

organic coast A coast along which living organisms control landforms along the shore.

organic sedimentary rock Sedimentary rock (such as coal) formed from carbon-rich relicts of organisms.

original horizontality The geologic principle stating that sedimentary beds, when originally deposited, are nearly horizontal.

orogen A linear range of mountains.

orogeny A mountain-building event.

orographic lifting The forced ascent of air that occurs on the windward (upslope) side of mountains.

outcrop An exposure of bedrock.

outer core The section of the core, between 2,900 and 5,150 km (1,802 and 3,200 miles) deep, that consists of liquid iron alloy.

oxbow lake A meander that has been cut off yet remains filled with water.

oxidation A reaction in which an element loses electrons; an example is the reaction of iron with air to form rust.

ozone hole An area of the stratosphere over the Antarctic polar region, from which ozone has been depleted.

P-wave shadow zone The region of the Earth that does not receive P-waves from a given earthquake, because of the pattern of seismic-wave refraction inside the Earth.

pahoehoe A lava flow with a surface texture of smooth, glassy, rope-like ridges.

paleoclimate The past climate of the Earth.

paleogeographic map A portrayal of the distribution of land and sea at times in the past.

paleomagnetism The record of ancient magnetism preserved in rock.

paleontology The study of fossils.

paleopole The supposed position of the Earth's magnetic pole in the past, with respect to a particular continent.

Pangaea A supercontinent that assembled at the end of the Paleozoic Era.

Pannotia A supercontinent that may have existed sometime between 800 Ma and 600 Ma.

parent isotope A radioactive isotope that undergoes decay.

parsec Short for *parallax second*. The distance to a star with a parallax of 1 arcsecond (arcsec) using a base of 1 astronomical unit (AU). One parsec is approximately 3.26 light-years.

partial melting The melting in a rock of the minerals with the lowest melting temperatures, while other minerals remain solid.

passive margin A continental margin that is not a plate boundary.

passive-margin basin A thick accumulation of sediment along a tectonically inactive coast, formed over crust that stretched and thinned when the margin first began.

patterned ground A polar landscape in which the ground splits into pentagonal or hexagonal shapes.

peat Compacted and partially decayed vegetation accumulating beneath a swamp.

pelagic zone Microscopic plankton shells and fine flakes of clay that settle out and accumulate on the deep-ocean floor.

perched water table A quantity of groundwater that lies above the regional water table because an underlying lens of impermeable rock or sediment prevents the water from sinking down to the regional water table.

peridotite A coarse-grained ultramafic rock.

period The time it takes for a regularly repetitive process to complete one cycle.

permafrost Permanently frozen ground.

permanent stream A stream that flows year-round because its bed lies below the water table, or because more water is supplied from upstream than can infiltrate the ground.

permeability The degree to which a material allows fluids to pass through it via an interconnected network of pores and cracks.

Phanerozoic Eon The most recent eon, an interval of time from 542 Ma to the present.

phase (1) One of the various appearances of the sunlit surface of the Moon or a planet caused by the change in viewing location of the Earth relative to both the Sun and the object. Examples include crescent phase and gibbous phase; (2) the state of a substance (solid, liquid, gas).

phase change The change of state of a substance, such as water to ice, or water to vapor.

photic zone The upper layer of the sea into which sunlight penetrates.

photomicrograph A photograph taken through the lens of the petrographic microscope, which focuses a beam of electrons onto a small part of a grain to reveal the mineral's chemical composition.

photon Also called *quantum of light*. A discrete unit or particle of electromagnetic radiation.

photosphere The apparent surface of the Sun as seen in visible light. Compare *chromosphere* and *corona*.

photosynthesis The process by which plants and algae convert energy from sunlight into chemical energy.

photovoltaic cell A wafer of materials that are able to transform solar radiation into electricity.

phyllite A fine-grained metamorphic rock with a foliation caused by the preferred orientation of very fine-grained mica.

physical weathering The process in which intact rock breaks into smaller grains or chunks.

phytoplankton Very tiny, floating organisms that are able to carry out photosynthesis.

pillow basalt Glass-encrusted basalt blobs that form when magma extrudes on the seafloor and cools very quickly.

planet (1) A large body that orbits the Sun or other star and shines only by light reflected from the Sun or star; (2) in the Solar System, a spherical body that orbits the Sun and has cleared smaller bodies from the neighborhood around its orbit. Compare *dwarf planet*.

planetary nebula The expanding shell of material ejected by a dying giant star. A planetary nebula glows due to intense ultraviolet light coming from the hot, stellar remnant at its center.

planetesimal Tiny, solid pieces of rock and metal that collect in a planetary nebula and eventually accumulate to form a planet.

plankton Tiny plants and animals that float in sea or lake water.

plasma A gas that is composed largely of charged particles but may also include some neutral atoms.

plastic deformation The deformational process in which mineral grains behave like plastic and, when compressed or sheared, become flattened or elongate without cracking or breaking.

plate One of about 20 distinct pieces of the relatively rigid lithosphere.

plate boundary The border between two adjacent lithosphere plates.

plate interior A region away from the plate boundaries that consequently experiences few earthquakes.

plate tectonics *See* Theory of plate tectonics.

playa The flat, typically salty lake bed that remains when all the water evaporates in drier times. Playas form in desert regions.

Pleistocene Ice Age The period of time from about 2 Ma to 14,000 years ago, during which the Earth experienced an ice age.

plunge The angle that a line makes with respect to a horizontal plane, as measured in a vertical plane. (i.e., the angle at which a linear structure tilts).

pluton An irregular or blob-shaped intrusion; they can range in size from tens of meters across to tens of km across.

pluvial lake A lake formed to the south of a continental glacier as a result of enhanced rainfall during an ice age.

point bar A wedge-shaped deposit of sediment on the inside bank of a meander.

polar front The global boundary between the cold airmass located over the polar and subpolar regions and the semitropical air to the south.

polar glacier A glacier that remains completely frozen inside all year long.

polar-front jet stream The band of strong winds found above the polar front and encircling the middle or high latitudes of each hemisphere in a wavelike pattern; most prominent during the winter months, although it is found in all seasons.

polarity chron The time interval between polarity reversals of the Earth's magnetic field.

pore A small, open space within sediment or rock.

porosity The total volume of empty space (pore space) in a material, usually expressed as a percentage.

positive feedback A consequence of a phenomenon that causes the phenomenon to continue happening at the same or greater rate.

pothole A bowl-shaped depression carved into the floor of a stream by a long-lived whirlpool carrying sand or gravel.

Precambrian The interval of geologic time between the Earth's formation about 4.57 Ga and the beginning of the Phanerozoic Eon 542 Ma.

precession The gradual conical path traced out by the Earth's spinning axis; simply put, it is the "wobble" of the axis.

precipitation (1) The process by which atoms dissolved in a solution come together and form a solid; (2) rain, snow, or hail.

preferred orientation The metamorphic texture that exists where platy grains lie parallel to one another and/or elongate grains align in the same direction.

preservation potential The likelihood that a metamorphic mineral will be preserved in the stratigraphic record.

pressure Force per unit area, or the "push" acting on a material in cases where the push (compressional stretch) is the same in all directions.

pressure gradient The rate of pressure change over a given horizontal distance.

pressure-gradient force The force applied to a small region of air or water due to the variation of pressure over a small distance around the region.

primary producer Organisms in the food web that do not survive by eating other organisms, but rather by photosynthesis or from the chemical energy in minerals.

prograde motion (1) Rotational or orbital motion of a moon that is in the same direction as the planet it orbits; (2) the counterclockwise orbital motion of Solar System objects as seen from above the Earth's orbital plane. Compare *retrograde motion*.

Proterozoic Eon The most recent of the Precambrian eons.

protolith The original rock from which a metamorphic rock formed.

proton A positively charged subatomic particle in the nucleus of an atom.

protoplanet A body that grows by the accumulation of planetesimals but has not yet become big enough to be called a planet.

protoplanetary disk The remains of the accretion disk around a young star from which a planetary system may form.

protostar A dense body of gas that is collapsing inward because of gravitational forces and that may eventually become a star.

pulsar A rapidly-rotating neutron star that beams radiation into space in two searchlight-like beams. To a distant observer, the star appears to flash on and off.

pumice A glassy igneous rock that forms from felsic frothy lava and contains abundant (over 50%) pore space.

punctuated equilibrium The hypothesis that evolution takes place in fits and starts; evolution occurs very slowly for quite a while and then, during a relatively short period, takes place very rapidly.

pycnocline The boundary between layers of water of different densities.

pyroclastic debris Fragmented material that sprayed out of a volcano and landed on the ground or seafloor in solid form.

pyroclastic flow A fast-moving avalanche that occurs when hot volcanic ash and debris mix with air and flow down the side of a volcano.

pyroclastic rock Rock made from fragments that were blown out of a volcano during an explosion and were then packed or welded together.

quartzite A metamorphic rock composed of quartz and transformed from a protolith of quartz sandstone.

quasar Short for *quasi-stellar radio source*. The most luminous of distant forming galaxies, seen only at great distances from the Milky Way.

radiation Electromagnetic energy traveling away from a source through a medium or space.

radiation windows The ranges of wavelengths for which the atmosphere is transparent.

radiative zone A region within a star where energy is transported outward by radiation. Compare *convective zone*.

radio interferometer Instrumentation, used in astronomy, designed to observe interference patterns of electromagnetic radiation at radio wavelengths in order to study characteristics of certain classes of celestial objects.

radio telescope An instrument for detecting and measuring radio frequency emissions from celestial sources.

radio window The range of radio wavelengths for which the atmosphere is transparent, so that objects in space can be observed.

radioactive decay The process by which a radioactive atom undergoes fission or releases particles, thereby being transformed into a new element.

radioactive element A material composed of atoms that will eventually, spontaneously decay by releasing particles from its nucleus. When an atom decays, its atomic number changes so it becomes another element.

rainbow An arch of colored light formed in the sky due to the refraction and reflection of light through water droplets in the atmosphere.

raindrop A drop of water, with a diameter greater than 0.5 mm, that falls through the atmosphere.

rapid A reach of a stream in which water becomes particularly turbulent; as a consequence, waves develop on the surface of the stream.

recharge area A location where water enters the ground and infiltrates down to the water table.

recurrence interval The average time between successive geologic events.

red dwarf A small and relatively cool star that resides on the main sequence on an H-R diagram.

red giant A huge red star that forms when Sun-sized stars start to die and expand.

red shift The phenomenon in which a source of light moving away from you shifts to a lower frequency; that is, toward the red end of the spectrum.

redbed A sedimentary rock that has a reddish color due to the presence of iron oxide, generally in the form of hematite. Redbeds form because they reacted with oxygen during diagenesis. Redbeds occur in non-marine sedimentary sequences.

reef bleaching The death and loss of color of a coral reef.

reflecting telescope A telescope that uses mirrors for collecting and focusing incoming electromagnetic radiation to form an image in their focal planes. The size of a reflecting telescope is defined by the diameter of the primary mirror. Compare *refracting telescope*.

reflection nebula A nebula whose light results from the reflection of light from stars.

refracting telescope A telescope that uses objective lenses for collecting and focusing incoming electromagnetic radiation to form an image. Compare *reflecting telescope*.

refractory material A substance that has a relatively high melting point and tends to exist in solid form.

regional metamorphism Changes in the mineral assemblage and texture of rock over a broad region, generally as a consequence of deep burial and deformation of rock as a consequence of mountain building. (Equivalent to dynamothermal metamorphism.)

regolith A general term for unconsolidated material at the Earth's surface. It includes soil and uncemented sediment.

regression The seaward migration of a shoreline caused by a lowering of sea level.

relative age The age of one geologic feature with respect to another.

relative humidity The ratio of the amount of water vapor in the atmosphere to the atmosphere's capacity for moisture at a given temperature (expressed as the vapor pressure divided by the saturation vapor pressure).

relative plate velocity The movement of one lithosphere plate with respect to another.

relief The difference in elevation between adjacent high and low regions on the land surface.

renewable resource A resource that can be replaced by nature within a short time span relative to a human life.

research The systematic study of materials and sources in order to establish facts and discover new information.

reservoir rock Rock with high porosity and permeability, so it can contain an abundant amount of easily accessible oil.

residence time The average length of time that a substance stays in a particular reservoir.

resolving power The ability of a telescope or film to separate or distinguish small or nearby images.

retrograde motion (1) Rotation or orbital motion of a moon that is in the opposite direction to the rotation of the planet it orbits; (2) the clockwise orbital motion of Solar System objects as seen from above the Earth's orbital plane. Compare *prograde motion*; (3) apparent retrograde motion is a motion of the planets with respect to the "fixed stars", in which the planets appear to move westward for a period of time before resuming their normal eastward motion.

reverse fault A steeply dipping fault on which the hanging-wall block slides up.

reversed polarity Polarity in which the paleomagnetic dipole points north.

revolution The time to complete an orbit.

Richter scale A scale that defines earthquakes on the basis of the amplitude of the largest ground motion recorded on a seismogram.

ridge (1) On the Earth, an elongated hill; (2) in the atmosphere, an elongated area of high atmospheric pressure, normally accompanied by anticyclonic (cyclonic) flow in the Northern Hemisphere.

ridge-push force When the lithosphere beneath a mid-ocean ridge sits higher than the lithosphere beneath an adjacent abyssal plain, causing the lithosphere to spread sideways (due to gravity). This motion, in turn, drives plates in a direction pointing away from the ridge axis.

rift A distinct linear belt where rifting has occurred.

rifting The process of stretching and breaking a continent apart.

ring A thin band of tiny objects in orbit around a planet's equator. All of the Jovian planets have rings.

ripple A very small wave formed in response to the shear of the wind on the surface of water.

ripple mark A small ridge of sediment formed due to the shear of wind or flowing water over unconsolidated sediment.

roche moutonnée A glacially eroded hill that becomes elongate in the direction of flow and is asymmetric; glacial rasping smoothes the upstream part of the hill into a gentle slope, while glacial plucking erodes the downstream edge into a steep slope.

rock A coherent, naturally occurring solid, consisting of an aggregate of minerals or a mass of glass.

rock cycle The succession of events that results in the transformation of Earth materials from one rock type to another, then another, and so on.

rockfall A type of mass wasting that involves the sudden drop of a mass of rock from a vertical cliff or overhang, so that for part of its downward journey, the rock free falls through air.

rockslide A type of mass wasting that occurs when a mass of rock detaches from its substrate on a failure surface and moves downslope.

rocky coast An area of coast where bedrock rises directly from the sea, so beaches are absent.

Rodinia A proposed Precambrian supercontinent that existed around 1 billion years ago.

rogue wave A wave that is much taller (has greater amplitude) than other waves in an area.

root wedging The process by which growing roots of trees push open joints.

rotary drill A machine for producing a hole in bedrock. The hole is created by the rotation of a metal tube with a drill bit (composed of material hard enough to grind into rock) at its end.

rotation The time required to complete one turn around an axis or center.

runoff Water that flows over the land surface, as sheetflow or in stream channels.

S-wave shadow zone A band between 103° and 180°, as measured from either side of an earthquake's epicenter, where S-waves will not arrive at seismometer stations---revealing that S-waves cannot pass through the core of the Earth.

salinity The degree of the concentration of salt in water.

salt lake A body of water, submerging the land, that has no outlet, so water can leave only by evaporation. As a result of this evaporation, salt stays behind, making the lake high in salinity.

salt wedging The process in arid climates by which dissolved salt in groundwater crystallizes and grows in open pore spaces in rocks and pushes apart the surrounding grains.

saltation The movement of a sediment in which grains bounce along their substrate, knocking other grains into the water column (or air) in the process.

sand dune A relatively large ridge of sand built up by a current of wind (or water); cross bedding typically occurs within the dune.

sandspit An area where the beach stretches out into open water across the mouth of a bay or estuary.

saturation vapor pressures The vapor pressure at which air is saturated at a given temperature.

schist A medium-to-coarse-grained metamorphic rock that possesses schistosity.

science The systematic analysis of natural phenomena based on observation, experiment, and calculation.

scientific law A concise statement that completely describes a specific relationship or phenomenon which applies without exception for a defined range of conditions.

scientific method A sequence of steps for systematically analyzing scientific problems in a way that leads to verifiable results.

scientist A person who spends their career in search of ideas and explanations to explain the way our Universe operates.

scoria A glassy, mafic, igneous rock containing abundant air-filled holes.

sea breeze A circulation that develops along shore lines during daytime in which warm air rises over land and moves shoreward aloft, while cool air offshore descends and moves onshore to replace the warm air.

sea ice Ice formed by the freezing of the surface of the sea.

sea level The elevation that the sea surface would have if there were no waves. Sea level on the map usually refers to "mean sea level" which is the average level over time; it lies roughly halfway between high and low tide.

sea stack A chimney-shaped column of rock produced by wave erosion along a rocky coast.

seafloor spreading The gradual widening of an ocean basin as new oceanic crust forms at a mid-ocean ridge axis and then moves away from the axis.

seamount An isolated submarine mountain.

season Each of the four divisions of the year (spring, summer, autumn, and winter) characterized by different weather and climate.

seaweed Multicellular algae that grows in the sea; many varieties of seaweed root to the seafloor.

secondary recovery technique A method for removing additional hydrocarbons from underground, after the easy-to-get hydrocarbons; the method typically involves hydrofracturing or steam injection.

sediment An accumulation of loose mineral grains, such as boulders, pebbles, sand, silt, or mud, that are not cemented together.

sediment budget The relation between inputs of sediment into a region and the removal of sediment from a region.

sediment liquefaction A phenomena that happens when vibrations cause wet sediment which had been consolidated, but not cemented, to disaggregate and form a slurry of sediment and water.

sediment load The total volume of sediment carried by a stream.

sedimentary basin A depression, created as a consequence of subsidence, that fills with sediment.

sedimentary rock Rock that forms either by the cementing together of fragments broken off of preexisting rock or by the precipitation of mineral crystals out of water solutions at or near the Earth's surface.

sedimentary structure A form or surface on or in sedimentary rock that develops as a consequence of depositional processes. Examples include bedding, ripple marks, cross beds, mudcracks, and graded beds.

seismic belt The relatively narrow strips of crust on the Earth under which most earthquakes occur.

seismic wave Vibrations, generated by an earthquake, that pass through the Earth, or along the surface of the Earth.

seismic-reflection profile A cross-sectional view of the crust made by measuring the reflection of artificial seismic waves off boundaries between different layers of rock in the crust.

seismic-velocity discontinuities Boundaries in the Earth at which seismic velocity changes abruptly.

seismicity Earthquake activity.

seismogram The record of an earthquake produced by a seismograph.

seismometer An instrument that can record the ground motion from an earthquake.

severe thunderstorm A storm with the potential to produce hail with diameters one inch or larger, and/or winds which equal or exceed 93 kilometers/hour (58 mph), or a tornado.

shaded-relief map A map that emphasizes the topography or an area by simulating shadows produced by a low-angle Sun.

shale gas Volatile hydrocarbons trapped in an organic shale.

shale oil Liquid hydrocarbons trapped in an organic shale.

shear A stress that moves one part of a material sideways past another part.

shear wave A seismic wave in which particles of material move back and forth perpendicular to the direction in which the wave itself moves.

shield An older, interior region of a continent.

shield volcano A subaerial volcano with a broad, gentle dome, formed either from low-viscosity basaltic lava or from large pyroclastic sheets.

shock wave A vibration caused by a sudden push against an elastic, or somewhat elastic, material. For example, a meteorite impact produces a shock wave.

shoreline The boundary between the water and the land.

sidereal day The Earth's period of rotation with respect to the stars—about 23 hours 56 minutes—which is the time that it takes for the Earth to make one rotation and face the exact same star on the meridian. It differs from the *solar day* because of the Earth's motion around the Sun.

silicate mineral A mineral composed of silicon-oxygen tetrahedra linked in various arrangements; most contain other elements too.

silicon-oxygen tetrahedron A fundamental building block of silicate minerals that consists of one silicon atom surrounded by four oxygen atoms, producing a pyramid-like shape with four triangular faces.

sill A nearly-horizontal, tabletop-shaped tabular intrusion that occurs between the layers of country rock.

singularity A point in space containing a very large mass, such as exists with a *black hole*.

sinkhole A circular depression in the land that forms when an underground cavern collapses.

slab-pull force The force that downgoing plates (or slabs) apply to oceanic lithosphere at a convergent margin.

slate Fine-grained, low-grade metamorphic rock, formed by the metamorphism of shale.

sleet Precipitation consisting of frozen raindrops.

slickenside A fault surface that has been polished by the slip of a fault.

slip The displacement or movement on a fault.

slope failure The downslope movement of material on an unstable slope.

slow-onset flood Flooding of a region that develops by the slow rise of water, as can happen when snows melt, or moderate rainfall occurs for a long time.

slump A type of mass movement that involves the displacement of rock or regolith above a spoon-shaped failure surface. They take place at slow to moderate rates.

smelting The heating of a metal-containing rock to high temperatures in a fire so that the rock will decompose to yield metal plus a nonmetallic residue (slag).

snow avalanche The downslope movement of a mass of snow.

snowball Earth The condition thought to have happened in the late Proterozoic when nearly all of the Earth's land was covered by glaciers, and nearly all of the sea was covered by a thick layer of sea ice.

snowflake A loosely packed collection of individual ice crystals. Also called an aggregate.

soil Sediment that has undergone changes at the surface of the Earth, including reaction with rainwater and the addition of organic material.

soil erosion The removal of soil by wind and runoff.

soil horizon Distinct zones within a soil, distinguished from each other by factors such as chemical composition and organic content.

soil profile A vertical sequence of distinct zones of soil.

solar atmosphere The photosphere, the chromosphere, and the corona of the Sun.

solar constant The energy from the Sun crossing a unit area perpendicular to the Sun's rays in one second, at a distance of one astronomical unit, as measured in Watts per square meter (roughly 1,362 Watts/square meter).

solar core The central part of the Sun where nuclear reactions occur.

solar cycles The 11- and 22-year cycles of sunspot frequency that occur on the Sun's surface.

solar day The day in common use—24 hours—that is the Earth's period of rotation which brings the Sun back to the same local meridian where the rotation started. Compare *sidereal day*.

solar eclipse An eclipse that occurs when the Sun is partially or entirely blocked by the Moon. Compare *lunar eclipse*.

solar flare An explosion on the Sun's surface associated with a complex sunspot group and a strong magnetic field.

solar granulation Small features on the photosphere of the Sun caused by convection currents within the Sun's convective zone.

solar mass The mass of the Sun.

solar prominence A large, bright arc of plasma extending outward from the Sun's surface.

solar storm A disturbance of the Earth's magnetic field caused by radiation and streams of charged particles from the Sun.

Solar System Our Sun and all of the materials that orbit it (including planets, moons, asteroids, Kuiper Belt objects, and Oort Cloud objects).

solar unit A measure of a star's luminosity found by dividing the star's luminosity by the luminosity of the Sun.

solar wind A stream of particles with enough energy to escape from the Sun's gravity and flow outward into space.

solid A state of matter in which atoms or molecules remain fixed in position. A solid retains its shape, regardless of changes in the size of its container.

solid-state diffusion The slow movement of atoms or ions through a solid.

solifluction The type of creep characteristic of tundra regions; during the summer, the uppermost layer of permafrost melts, and the soggy, weak layer of ground then flows slowly down slope in overlapping sheets.

sorting (1) The range of clast sizes in a collection of sediment; (2) the degree to which sediment has been separated by flowing currents into different-size fractions.

source rock A rock (organic-rich shale) containing the raw materials from which hydrocarbons eventually form.

Southern Oscillation The shift in sea-level pressure back and forth across the Pacific Ocean, in association with El Niño.

space telescope A telescope mounted on a satellite in orbit about the Earth or in space.

space weather Disturbances in the Earth's magnetic field resulting from interactions with the solar wind.

specific gravity A number representing the density of a mineral, as specified by the ratio between the weight of a volume of the mineral and the weight of an equal volume of water.

spectral class A classification system for stars based on the presence and relative strength of absorption lines in their spectra. Spectral type is related to the surface temperature of a star.

spectral line A dark or bright line in an otherwise uniform spectra resulting from emission

or absorption of light in a narrow frequency range.

spectroscope An instrument used to identify spectral lines.

speed of light The speed at which light travels through a vacuum, 299,792 kilometers per second (186,282 miles per second).

speleothem A formation that grows in a limestone cave by the accumulation of travertine precipitated from water solutions, which drips in the cave or flows down the wall of the cave.

spring A natural outlet from which groundwater flows up onto the ground surface.

squall-line thunderstorms A line of thunderstorms where individual thunderstorms are sufficiently close together that they appear as one continuous line on a radar screen.

stable air The condition in which a parcel of air, displaced vertically, will return to its original position.

stalactite A type of speleothem that resembles a downward-pointing cone; it forms where water drips from the ceiling of a cave. In caves formed in limestone, stalactites consist of travertine.

stalagmite An upward-pointing cone of limestone that grows when drips of water hit the floor of a cave.

standing wave A wave whose crest and trough remain in place as water moves through the wave.

star An object in the Universe in which fusion reactions occur pervasively, producing vast amounts of energy; our Sun is a star.

states of matter The different ways in which matter can appear—as a solid, liquid, gas, or plasma.

stationary front A boundary between airmasses where the colder airmass is neither advancing nor retreating at the surface.

steady-state condition The situation that occurs in cycles when the relative amounts of a material in different reservoirs stays constant, even though there is flux of material among reservoirs.

stellar nucleosynthesis A process in which smaller nuclei fuse together to form larger ones within a star.

stick-slip behavior Stop-start movement along a fault plane caused by friction, which prevents movement until stress builds up sufficiently.

storm An episode of weather in which winds, precipitation, and in some cases lightning become strong enough to be bothersome or even dangerous.

storm surge A rise of and onshore surge of seawater resulting primarily from the winds of a storm, and secondarily from the surface pressure drop near the storm center.

straight-line wind Severe thunderstorm winds that blow in a single direction, rather than in a rotational pattern as found with tornadoes.

strain The change in shape of an object in response to deformation (i.e., as a result of the application of a stress).

strata A succession of sedimentary beds.

stratigraphic column A cross-section diagram of a sequence of strata summarizing information about the sequence.

stratigraphic formation A recognizable layer of a specific sedimentary rock type or set of rock types, deposited during a certain time interval, which can be traced over a broad region.

stratosphere The stable, stratified layer of atmosphere directly above the troposphere.

stratovolcano A large, cone-shaped, subaerial volcano consisting of alternating layers of lava and tephra.

stratus cloud A sheet-like, widespread stable cloud.

streak The color of the powder produced by pulverizing a mineral on an unglazed ceramic plate.

stream A ribbon of water that flows in a channel.

stream gradient The slope of a stream's channel in the downstream direction.

stream piracy An event that happens when headward erosion of a stream allows the stream to capture the water that had been flowing down another stream.

stream rejuvenation The renewed downcutting of a stream into a floodplain or peneplain, caused by a relative drop of the base level.

stress The push, pull, or shear that a material feels when subjected to a force; formally, the force applied per unit area over which the force acts.

strike The compass orientation of a horizontal line on a plane. Strike and dip together uniquely define the orientation of a plane.

strike-slip fault A fault in which one block slides horizontally past another (and therefore parallel to the strike line), so there is no relative vertical motion.

subduction The process by which one oceanic plate bends and sinks down into the asthenosphere beneath another plate.

subduction zone The region along a convergent boundary where one plate sinks beneath another.

submarine canyon A steep, narrow canyon that dissects a continental shelf and slope.

submarine debris flow A mixture of loose sediment, blocks of rock, and water that flows down a slope, under water, as a slurry.

submarine fan A wedge-shaped accumulation of sediment at the base of a submarine slope; fans usually accumulate at the mouth of a submarine canyon.

submarine slump The underwater, downslope movement of a semicoherent block of sediment along a weak mud detachment.

submergent coast A coast at which the land is sinking relative to sea level.

subsidence (1) The slow, vertical sinking of the Earth's surface in a region, relative to a reference plane; (2) the descent of an airmass.

subsoil A layer of soil (the B-horizon), which lies beneath the topsoil or A-horizon (or, in some cases, the E-horizon), that represents the zone of accumulation.

substrate A general term for material just below the ground surface.

subtropical jet stream A band of strong winds typically found between 20° and 35° latitude. The maximum speeds are typically found at altitudes between 12 and 14 km (7 and 8 miles).

Sun The star at the center of our Solar System.

sunspot A cooler, transitory region on the solar surface produced when loops of the Sun's magnetic field break through the surface of the Sun.

supercell thunderstorm A large, rotating thunderstorm in which the updraft and downdraft circulations maintain the storm's structure for a long time period. Supercells often produce hail, strong straight-line winds, and tornadoes.

supergiant star A star that is thousands of times brighter than the Sun and has a short lifespan—only about 10 to 50 million years.

supermassive black hole A black hole of 1,000 solar masses or more that resides in the center of a galaxy.

supernova A short-lived, very bright object in space that results from the cataclysmic explosion marking the death of a very large star; the explosion ejects large quantities of matter into space to form new nebulae.

superposition The geologic principle which states that, in a succession of sedimentary rock, younger rocks were deposited over older rocks.

supervolcano A volcano that erupts a vast amount (more than 1,000 cubic km) of volcanic material during a single event; none have erupted during recorded human history.

surface current An ocean current in the top 100 m (328 feet) of water.

surface load The sediment transported by rolling or saltation, along a surface, by wind or flowing water.

surface wave A seismic wave that travels along the Earth's surface.

suspended load Tiny solid grains carried along by a stream without settling to the floor of the channel.

sustainability The question of whether we can continue to maintain or improve our standard of living without running out of resources.

suture The surface, formed during the collision of relatively buoyant crustal blocks, that marks the boundary between what had been separate crustal blocks prior to collision.

swash The upward movement of water on the surface of a beach.

swell Relatively-long wavelength, periodic waves that can travel long distances across an ocean, beyond the region where wind shear produced the waves.

symmetry The condition in which the shape of one part of an object is a mirror image of the other part.

syncline A trough-shaped fold whose limbs dip toward the hinge.

talus Fallen rock along the base of a cliff.

talus pile The accumulation of fallen rock along the base of a cliff, typically, in a steep apron whose surface is at the angle of repose.

tar Hydrocarbons that exist in solid form at room temperature.

tar sand Sandstone reservoir rock in which less viscous oil and gas molecules have either escaped or been eaten by microbes, so that only tar remains.

taxonomy The study and classification of the relationships among different forms of life.

telescope The basic tool of astronomers. Working over the entire range from gamma rays to radio waves, astronomical telescopes collect and concentrate electromagnetic radiation from celestial objects.

temperate glacier A glacier in which or on which meltwater forms during part or all of the year.

temperature A measure of the average speed of molecules in a fluid or the vibration of molecules in a solid. Commonly, the hotness or coldness of a substance as measured with a thermometer.

tension A stress that pulls on a material and could lead to stretching.

tephra Unconsolidated accumulations of pyroclastic grains.

terminal moraine The end moraine at the farthest limit of glaciation.

terrestrial planet An Earth-like planet, made of rock and metal and having a solid surface. In the Solar System, the terrestrial planets are Mercury, Venus, Earth, and Mars. Compare *giant planet*.

texture The arrangement of grains in a rock (the way grains connect to one another) and the degree to which inequant grains align parallel to one another.

theory A scientific idea supported by an abundance of evidence that has passed many tests and failed none.

theory of evolution The model theorizing that new species appear on the Earth through the modification of older species, over time, allowing them to compete more successfully for resources and, therefore, to survive; the process is commonly referred to as "natural selection by the survival of the fittest". As a consequence of evolution, the assemblage of species on the Earth changes over time.

theory of plate tectonics The theory that the outer layer of the Earth (the lithosphere) consists of separate plates that move with respect to one another.

thermal energy The total kinetic energy in a material due to the vibration and movement of atoms in that material.

thermal metamorphism Metamorphism caused by heat conducted into country rock from an igneous intrusion.

thermocline A boundary between layers of water with differing temperatures.

thermohaline circulation The rising and sinking of water driven by contrasts in water density, which is due in turn to differences in temperature and salinity; this circulation involves both surface and deep-water currents in the ocean.

thermometer An instrument used to measure temperature, often constructed by partially filling a sealed glass tube with a liquid such as alcohol.

thermosphere The hot, outermost layer of the atmosphere containing very few gas molecules.

thin section A 3/100-mm-thick slice of rock that can be examined with a petrographic microscope.

thrust fault A gently dipping reverse fault; the hanging-wall block moves up the slope of the fault.

thunder The booming sound created by the rapid expansion of air within a lightning channel.

thunderstorm A cumulonimbus cloud that produces lightning and thunder.

tidal bore A visible wall of water that moves toward shore with the rising tide in quiet waters.

tidal energy Electricity produce by using tidal flow to drive generators.

tidal flat A broad, nearly horizontal plain of mud and silt, exposed or nearly exposed at low tide, but totally submerged at high tide.

tidal reach The difference in sea level between high tide and low tide at a given point.

tide The daily rising or falling of sea level at a given point on the Earth.

tide-generating force The force, caused in part by the gravitational attraction of the Sun and Moon, and in part by the centrifugal force created by the Earth's spin, that generates tides.

tidewater glacier A glacier that has entered the sea along a coast.

tillite A rock formed from hardened ancient glacial deposits and consisting of larger clasts distributed through a matrix of sandstone and mudstone.

topographic map A map that uses contour lines to represent variations in elevation.

topographic profile A sketch of the trace of the ground surface as it would appear on a vertical plane that slices into the ground, to represent elevations in a given direction.

topography Variations in elevation.

topsoil The top soil horizons, which are typically dark and nutrient-rich.

tornado A near-vertical, funnel-shaped cloud in which air rotates violently around the axis of the funnel.

tornado track The path a tornado takes across the ground.

trade winds The low-level winds found throughout the tropics that flow from east to west with a component toward the equator.

transform boundary A plate boundary where one plate slips sideways relative to the other, but no new plate forms, and no old plate subducts. It is a vertical fault on which the slip direction parallels the Earth's surface.

transform fault A fault marking a transform plate boundary; along mid-ocean ridges, transform faults are the actively slipping segment of a fracture zone between two ridge segments.

transgression The inland migration of shoreline resulting from a rise in sea level.

transition zone The middle portion of the mantle, from 400 to 670 km (249 to 416 miles) deep, in which there are several jumps in seismic velocity.

trap A configuration of seal rocks and reservoir rocks that confines hydrocarbons underground.

travel time The interval of time that it takes for a seismic wave to travel from the focus of an earthquake to a given seismometer.

tributary A smaller stream that flows into a larger stream.

triple junction A point where three lithosphere plate boundaries intersect.

tropical cyclone A storm that originates in the tropics, and has an organized rotation and low pressure at its center.

tropical storm A cluster of thunderstorms of tropical origin that has an organized circulation and sustained winds between 35 and 64 knots.

tropics Latitudes of the Earth between 30°N and 30°S where cold weather is rare. The higher latitudes between 20° and 30° are sometimes referred to as the subtropics since cold air occasionally reaches these latitudes in winter.

troposphere The layer of the Earth's atmosphere that extends from the Earth's surface to the tropopause and contains all of the Earth's weather.

trough An elongated area of low atmospheric pressure, normally accompanied by a cyclonic (counterclockwise) flow in the Northern Hemisphere.

tsunami A large wave along the sea surface triggered by an earthquake or large submarine slump.

tuff A pyroclastic igneous rock composed of volcanic ash and fragmented pumice, formed when accumulations of the debris cement together.

turbidity current A submarine avalanche of sediment and water that speeds down a submarine slope.

turbulence The chaotic twisting, swirling motion in a flowing fluid.

typhoon The equivalent of a hurricane in the western Pacific Ocean.

U-shaped valley A steep-walled valley shaped by glacial erosion into the form of a U.

ultramafic magma A term used to describe magmas that are rich in iron and magnesium and very poor in silica.

unconformity A boundary between two different rock sequences representing an interval of time during which new strata were not deposited and/or were eroded.

unconventional reserve A hydrocarbon reserve that can only be accessed by using special, and expensive, technology.

underground mine A mine, consisting of tunnels and shafts, to access valuable mineral reserves underground.

uniformitarianism The geologic principle stating that physical processes that we observe today happened in the past at relatively the same rate. Simply put, the present is the key to the past.

Universe All of space, and everything within it.

unstable air The condition in which a parcel of air, displaced vertically, will accelerate away (upward or downward) from its original position.

updraft Upward-moving air.

uplift The upward vertical movement of the surface of the Earth.

upper mantle The uppermost section of the mantle, reaching down to a depth of 400 km (249 miles).

upwelling The rising of deep, cold ocean water to the ocean surface.

vacuum A region of space that contains very little matter.

vapor pressure That part of the total atmospheric pressure due to water vapor molecules. The force exerted by water vapor molecules on a unit area.

vesicles An open hole in igneous rock formed by the preservation of bubbles in magma as the magma cools into solid rock.

viscosity The resistance of material to flow.

visibility The maximum distance that a person with normal vision can distinguish objects when looking through the atmosphere.

visible light Light detectable with the human eye.

volatile material Elements or compounds such as H_2O and CO_2 that evaporate at relatively low temperatures and can exist in gaseous forms at the Earth's surface.

volcanic arc A curving chain of active volcanoes formed adjacent to a convergent plate boundary.

volcanic bomb A large piece of pyroclastic debris thrown into the atmosphere during a volcanic eruption.

volcanic breccia A rock composed of angular fragments of extrusive igneous rock.

volcanic eruption An event during which lava and/or pyroclastic debris comes out of the Earth at a vent.

volcanic gases Volatile substances, such as water vapor, emitted by a volcano.

volcaniclastic deposit A sedimentary accumulation of volcanic fragments that have been transported by streams.

volcano (1) A vent from which melt from inside the Earth spews out onto the planet's surface; (2) a mountain formed by the accumulation of extrusive volcanic rock.

Walker circulation An atmospheric circulation in the equatorial regions of the Pacific Ocean characterized by rising motion over the western Pacific, eastward flow near the tropopause, sinking motion over the eastern Pacific, and westward flow near the surface.

wall cloud A region of rotating clouds that extends below the rain-free base of a supercell thunderstorm. The formation of the wall cloud often precedes tornado formation.

wall rock The rock adjacent to the surface of an igneous intrusion.

warm front A boundary between airmasses where the cold airmass is retreating and the warm air is advancing at the surface.

water mass A large volume of water in the ocean that has density characteristics due to its temperature and/or salinity.

water table The boundary, approximately parallel to the Earth's surface, that separates substrate in which groundwater fills the pores from substrate in which air fills the pores.

water vapor Water in its gaseous form.

waterfall A place where water drops over an escarpment.

wave A periodic motion that carries energy from one location to another.

wave base The depth, approximately equal in distance to half a wavelength in a body of water, beneath which there is no wave movement.

wave erosion The grinding or breaking away of the coastline by the action of water waves.

wave refraction The bending of waves as they approach a shore so that their crests make no more than a 5° angle with the shoreline.

wave-cut platform A shelf of rock, cut by wave erosion, at the low-tide line that was left behind a retreating cliff.

wavelength The horizontal difference between two adjacent wave troughs or two adjacent crests.

weather Local-scale, atmospheric conditions as defined by temperature, air pressure, relative humidity, wind speed, and precipitation.

weathering The processes that break up and corrode solid rock, eventually transforming it into sediment.

weight The mass of an object multiplied by the acceleration due to gravity. Compare *mass*.

well A hole in the ground dug or drilled in order to obtain water.

white dwarf A stellar object formed when a low-mass star has exhausted all its central nuclear fuel, undergone gravitational collapse to develop an ultra-dense core, and lost its outer layers as a planetary nebula.

wind The horizontal movement of air.

wind chill A parameter used in winter to account for the effect of both temperature and wind on the rate at which exposed flesh will cool. Reported numerically as the wind chill temperature.

wind direction The direction from which the wind blows.

wind speed The speed that air moves across the Earth's surface as measured by an anemometer.

wind-chill temperature The temperature that human skin feels due to heat loss caused by the combined effects of both cold and wind.

wind-driven wave A wave that forms due to the interaction between moving air and the surface of the ocean.

zenith The point on the celestial sphere located directly overhead from an observer.

zodiac The array of constellations that lies along the plane of the ecliptic.

zone of ablation The area of a glacier in which ablation (melting, sublimation, calving) subtracts from the glacier.

zone of accumulation (1) The layer of regolith in which new minerals precipitate out of water passing through, thus leaving behind a load of fine clay; (2) the area of a glacier in which snowfall adds to the glacier.

zone of accumulation The region of a glacier in which more snow falls than melts away, so that ice is added to the glacier.

zone of leaching The top interval of a soil in which downward percolating water picks up ions and clay and transports it downwards.

zooplankton Microscopic floating organisms that are animals, not plants.

INDEX

natural arch, **470**, *472*
    rocky cliffs, mesas, and arches, 470, *471*, 472, *472*, 474, *477*, 506
    stony plains, 472–73, **472**, 474
mirages, **615**, *616*, 617, 620
soil formation in, 466–67
types of, 464–66, *465*, *466*
    coastal deserts, 465–66, *466*, 506
    cold deserts, 464
    continental-interior deserts, 466, 506
    hot deserts, 464
    polar deserts, 466, 506
    rain-shadow deserts, 465, *465*, 506
    subtropical deserts, 465, *465*, 506
weathering, 466–67
detachment, 235
Devils Tower (Wyoming), *122*
Devonian Period
    about, *298*, *308*
    Earth history during, 323, **323**, 326–27, **326**, *328–29*
    evolution of life, *301*
    global climate during, *710*
    Grand Canyon (Arizona), *300*
dew, 598
dew point, 597–98, **597**, *598*, 620
diagenesis, 185–86, **185**
Dialogue Concerning the Two Chief World Systems (Galileo), 742
diamictite, *169*, **179**
diamond mines, 101, 102, 104, *104*
diamonds
    composition of, 91
    Cullinan Diamond, 104
    diamond mines, 101, 102, 104, *104*
    faceting, 102, *103*, 105
    formation of, 104, *104*
    gem quality, 101, 102, *103*, 104
    Hope Diamond, 101, *103*
    largest diamond, 104
    synthetic diamonds, 104
diatoms, 541
differential weathering, *160*, 161
differentiation, **35**, 36, **36**, 48, *314*, **314**
diffraction, of light, *618*
diffuse nebulae, **829**
diffusion, **94**, *95*
digital elevation model (DEM), **388**, *389*, 412
dikes, 66, *123*, **123**, 130
dimension stone, **378**
dinoflagellates, 541
dinosaurs, 331, 332, *332*, *335*, 342, 709
Dione (Saturn's moon), 802–3, *803*
diorite, 131, *131*
dip, **223**, *225*, **225**
dip-slip faults, *226*, **226**
dipole, 38, 39, *39*, **76**
directional drilling, **353**, 356–57, 380
disappearing streams, 452, *453*
disaster prevention
    earthquakes, 274–76, *275*, 280
    mass wasting, 409–10, *410*, *411*, 412

discharge, 418–20, **418**, *419*, 421, 460
discharge area, **443**
disconformity, **295**, *296*
disease, following earthquakes, 270, 272, 280
displacement, 219, *220*, **223**
distillation, *518*
distillation column, 354, *354*
distance
    astronomical units of, **763**, 774
    measuring in space, 763–66
distortion, 219, *220*, 221
divergence, 641, 648–49, **648**, *649*
divergent boundaries
    earthquake activity at, 259, *261*
    igneous activity at, *143*, 152
    mid-ocean ridges, *564*, *565*
divergent-boundary seismicity, 259, *261*
divergent-plate boundaries, 62, *62*, **63**, *64–67*, 66–67, 71, 86
diverted streams, *430*
documentation, 9
doldrums, 623, **630**
dolomite, *101*, 172
dolostone, **172**, 173, *371*, 425, *425*
domes, *229*, **229**, 243, *243*, 244
doping stick, 102
Doppler, Christian, 761
Doppler effect, **761**
Doppler radar, **600**, 684
Doppler shift, 761–62, 765
dormant volcanoes, **149**
downbursts, 678
downcutting, **417**
downdraft
    about, 598, 662
    defined, **662**
    forward-flank downdraft (FFD), 671–72
    rear-flank downdraft (RFD), 671, 672, *673*
    in supercell thunderstorms, 682, *684*
    in thunderstorms, 662, 667, 668, 680
    in tornadoes, 686
downgoing plates, 69, *69*, 71
downslope force, 405, *405*, 412
downslope windstorms, 692–93, **692**, *693*, 694
downwelling, **529**, 531, *531*, 544
dragline, 360
drainage
    glaciers and, 496
    rivers, 578
drainage basin, **418**, 419, *419*, 420
drainage divide, **418**, *419*
drainage networks, *418*, **418**, 421, 460
Drake, Edwin, 351, 353
Drake, Sir Francis, 623–24, 630
drawdown, 446
drill bit, 353
drilling
    Deepwater Horizon disaster, 363, *364*
    for hydrocarbons, 353–55, *353–55*

drilling mud, 353
drilling ships, 548
dropstone, 490, *494*
drought, 474, 506, *697*
drumlin, **492**, *493*, *495*
dry adiabatic expansion, 664
dry adiabatic lapse rate, 664–65, **664**
dry air, 664
dry avalanches, 403
dry washes, 420, *420*, **468**
dry well, 446
drywall, manufacture of, 379
dunes, **174**, 176, *177*, *180*, 473–74, **473**, *474*, *477*, 506
Durant, Will, *18*
dust
    defined, **33**, 506, **783**
    formation of solid bodies in Solar System, *34*, 35, *37*, 48
Dust Bowl, 475
dust storms, *165*, 468, **468**
dust tail, 807
dwarf galaxies, 839
dwarf planets, 35, **778**, *780*, 805, 812
dwarf stars
    about, 826, **827**, 832
    black dwarfs, **833**
    brown dwarfs, 826
    red dwarfs, 32, 826, **832**, 837
    white dwarfs, 826, 827, 833, **833**, *834*, 837, 846
    yellow dwarfs, 826
dynamic metamorphism, **204**, 214
dynamothermal metamorphism, 204–5, **204**, *205*, 207, 214

E

E-horizon, 162–63, *162*
Eagle Nebula, *829*
early-warning systems, earthquakes, 273, 280
Earth, 313–42
    age of, 12, 21, 308–9, *309*, 310, 314, 342
    albedo of, 503, **716**, 734
    apparent polar-wander path, 78–79, **78**, *79*, 81, 86
    atmosphere. *See* atmosphere
    average annual temperature, 719, *719*
    axial tilt, *781*, *782*
    axis of rotation, **748**, 750
    change in, 14–15, *15*
    circumference of, *743*
    climate of. *See* climate
    continental collision, 72–73, **72**, *73*, 86
    Coriolis effect, **523**, *524–28*, 525, 528
    Coriolis force, 522, **525**, *526–28*, 528, 531, 544, **626**, 638, 649, 656
    craters in, 43, *43*
    differentiation in, **35**, 36, *36*, 48
    Earth-Moon system, *560*
    eclipses, 756–58, *757*, *758*, 774
    ecliptic plane, **750**, *750*, 753, 754, *754*, 779, *781*
    electronic disruption from space weather, 824, 846

characteristics of, 96, *97*, 99–100, *99, 100*
chemical composition of, 100–101, *101*
cleavage of, 99–100, **99**, *100*
closure temperature, **307**
defined, **42**, 91–93, **91**, *92*, 112
examples of, *42*
formation of, 93–94, *94–95*, 96, 112
gems, 101–2, *101–4*, **101**, 104–5
as health hazards, 98, *98*
identification of, 96–101, 112
    by chemical composition, 100–101, *101*, 112
    by physical properties, 96, *97*, 99–100, *99, 100*, 112
impurities in, 96
internal structure of, 93, *93*
Mohs hardness scale, *99*, **99**
structure of, *102*
synthetic, 91
See also rocks
mining, 376–78
    acid mine runoff, **363**, *375*
    coal mining, 360–61, *361*, 363, 380
    economics of, *371*
    environmental impact of, 375, *375*
    of metallic mineral resources, 374, *374*, 375, *375*
    open-pit mines, *358*, **374**, 375
    underground mines, 361, **374**, 375
minor axis (ellipse), 744
Miocene Epoch, *298, 301*, 336, *710*
mirages, **615**, *616*, 617, 620
mirrors, for telescopes, 767
Mississippi River, 418, 420–22
Mississippi River flood (1911), 436
Mississippi River flood (1993), 435, *435, 436*
Mississippian Period, *298, 300, 308*
Missouri, 243
Missouri River, 422
Missouri River flooding (1993), *434*
mixture, 92
MMI scale. *See* Modified Mercalli Intensity scale
mobilist view, 54
models, 6, *7*
Modified Mercalli Intensity (MMI) scale, *257*, **257**, 280
Moenkopi formation (Arizona), *297*
Moho, *46*, **46**, 60, *277*, **277**, 279
Mohorovičić, Andrija, 277
Mohs hardness scale, *99*, **99**
moist adiabatic lapse rate, **665**
moist air, thunderstorm formation and, 661, *662*, 664, 666, 694
Mojave Desert, *161, 251*
molecule, **29**, 92
molten material, 42
moment magnitude scale, **258**, 280
monoclines, **228**, *229*, 244
monsoons, 636–37, **636**, *637, 638*, 656, 707
Monument Valley (Arizona), *471*
Moon (of Earth), 785–89
    about, 755, 756, 774, *779*, 785
    age of, 788
    apparent magnitude of, 825

cratering density, **788**
craters on, *43*, 787–88, *787*, 789
Earth-Moon system, *560*
Earth tides and, 556–61, *556*, *558–61*, 582
eclipses, 756–58, *757, 758*, 774
far side, 786, *786*
formation of, 25, *36–37*, 315, 788–89, *788*
general characteristics of, 785–87
lack of atmosphere, 786–87
lunar highlands, **787**, 788
lunar time scale, **788**
maria, **787**, 788
near side, *786*, 812
orbit of, 558, *755, 758*, 785, *785*, 812
phases of, 755–56, *755*, **755**, *756*
radius of, 785
regolith, *787*, **788**, 812
relative size of, *759*
secondary bulge, 558
speed of, 755
sublunar bulge, 558, *558*
surface of, 755, *755, 787*, 812
tidal bulge and, 785–86
view of from space, *747, 786, 787*
Moon rocks, 785, 788
moons
    about, *780*, 781, *800–801*, 812
    defined, **27**, *777*, *780*
    Galilean moons, 797
    of Jupiter, 797–99, *798, 799*, 801, 812
    of Mars, 793, *793*
    of Neptune, 804, *805*, 812
    of Pluto, 806–7, *807*
    of Saturn, *800*, 802–3, *803*, 812
    of Uranus, 804, *805*, 812
moraines, *489*, **489**, 492, *495*
motion
    Newton's laws of motion, 745, *779*
    prograde motion, *741*, 754, 779
    retrograde motion, *741*, 754–55, *754*, 780
mountain belts
    continental collision and, 74
    continental drift and, 54, *55*
    defined, **217**
    formation of, 238, 244
    of North America, *237*
    related to subduction, 234, *234*, 244
mountain building, 217–18
    causes of, 231–38
    collision, 234–35
    continental rifting, 235
    cratons, *241, 242*, 243, 244
    deformation, 218–19, **218**, *220, 221, 221*
    folds and foliation, 228–31
    Himalayas, 232–33
    measuring mountain building in progress, 235–36
    North America, *237*
    orogeny, 217–18, **217**, *238*, 323, *323*, 326
        in collisional orogeny, *65*, **234**, 235, 238
        in convergent-boundary orogens, *234*, **234**, 238

        in rift-related orogeny, 235, *236*
    in present time, 236, *238*
    rock types near, 238–39, *238*
    stream diversion, 429, *430*
    stress and, 222–23, *222*, 244
    subduction and, 234, *234*, 244
mountain glaciers, **478**, 480
mountain rain and snow, 691–92, *692*
mountain stream deposits, sedimentary rocks from, 179
mountains, 244
    cirques, **478**, **485**, *486*
    downslope windstorms, 692–93, **692**, 693, *693*, 694
    formation of, 231–38
    isostasy, *239*, **239**, *240*
    major mountain ranges, *218–19*
    rain and snowstorms, 691–92, *692*, 693
    rock types near, 238–39, *238*
    thunderstorms near, 663, *664*
    windward side, 691, 694, 708
mountaintop removal mining, 361
mouth (of a stream), 422
mudcracks, 177, **177**, *178*, 179
mudflows, 397–98, **397**, *398, 399*, 404, 412
mudstone, 169, *169*, 179
Muir, John, *239*, 470, *470*
Muir Glacier (Alaska), *722*
multi-stage flash distillation, *518*
multiple-vortex tornadoes, 683, *686*, **686**
Murray River floodplain (Australia), *169*
Murzug Sand Sea (Libya), *478*
muscovite, 198
mylonite, *205*

N

Namib Desert, *474*
narrative themes, 10
NASA (US National Aeronautics and Space Administration), 773, *773*, 811
National Radio Astronomy Laboratory (West Virginia), *772*
native metal, 370, *370*
natural arch, 470, *472, 477*
natural change, 14, *15*
natural forcing, **726**
natural gas
    defined, **348**
    environmental consequences of fossil fuel use, 363–65, *363, 364*
    exploration and production, 351, *351–55*, 353–54
    formation of, 380
    as nonrenewable resources, 362
    reserves of, 355
natural hazards
    mass wasting, 393, 409
    volcanoes, *134, 146, 147*, 148, 150–51, *151*, 152
    weather, 18–19
    See also disaster prevention
natural levees, **426**
natural selection, 293, 310

Y

Yangtze River flooding (China) (1931), 434
yardangs, *176*
yellow dwarfs (stars), 826
Yellow River flooding (China) (1931), 434
yellow stars, 762
Yellowstone Canyon, *145*
Yellowstone National Park, 75, *122*, *138*, *139*, 144, *145*, *173*, 336, *448*, *492*

Yosemite National Park (California), 125, 402, *487*
Younger Dryas, 711, *713*, 716
Yungay (Peru) landslide (1970), 392, *393*, *400*

Z

zenith, **748**
Zion Canyon (Utah), *176*, 300, *331*
zodiac, **752**, *753*, 774
zone of ablation, **483**, *484*

zone of accumulation, *162*, **162**, **483**, *484*
zone of leaching, **161**, *162*
zoning, earthquake zoning, 274–75, **274**
zooplankton, *540*, **541**